BIOLOGY
LIFE ON EARTH

THIRD EDITION

BIOLOGY
LIFE ON EARTH

THIRD EDITION

GERALD AUDESIRK
&
TERESA AUDESIRK
UNIVERSITY OF COLORADO AT DENVER

Macmillan Publishing Company
NEW YORK
Maxwell Macmillan Canada
TORONTO

To Heather, Joe, Jack, and Lori

Editor: Bob Pirtle
Development Editor: Madalyn Stone
Production Supervisor: York Production Services
Production Manager: Aliza Greenblatt
Text and Cover Designer: Hudson River Studio
Art Direction: Hudson River Studio
Cover Photograph: John Netherton
Photo Researcher: Yvonne Gerin/Chris Migdol
Illustrations: Hudson River Studio: Edmond Alexander, Cynthia Turner, Howard S. Friedman, David Mascaro, Ellen Palm, Michel Craig, Susan Posmentier, Dominique Fitch, Patrick Hannagan, Bob Crimi, Naomi Ganor, Inez Sovjani.

This book was set in Palatino by York Graphic Services and was printed and bound by Von Hoffmann Press. The cover was printed by Lehigh Press.

Acknowledgments and credits for all photographs appear on pages P-1—P-8, which constitute an extension of the copyright page.

Macmillan Publishing Company
866 Third Avenue, New York, New York 10022

Macmillan Publishing Company is part of the Maxwell Communication Group of Companies.

Maxwell Macmillan Canada, Inc.
1200 Eglinton Avenue East
Suite 200
Don Mills, Ontario M3C 3N1

Library of Congress Cataloging in Publication Data

Audesirk, Gerald.
    Biology : life on earth / Gerald Audesirk and Teresa Audesirk;
    3rd ed.
        p. cm.
    Includes bibliographical references and index.
    ISBN 0-02-304811-5
    1. Biology. I. Audesirk, Teresa. II. Title.
QH308.2.A93 1993   92-19232
574—dc20   CIP

Printing: 3 4 5 6 7   Year: 3 4 5 6 7 8 9

Preface

We have designed the third edition of *Biology: Life on Earth* to meet the needs of both major and nonmajor students. We have included sufficient background and detail to give biology majors a broad foundation on which to build in their upper division courses. At the same time, students taking biology to fulfill a science requirement will have a text written in a clear conversant style that effectively communicates the principles of biology. We believe that all students, majors and nonmajors alike, will appreciate our consistent use of examples that are relevant to their own lives. Keeping the various needs of our audience in mind, we have organized the third edition according to a few important principles.

First, a general biology text must capture and hold students' interest. Users and reviewers have responded enthusiastically to our writing style, calling it lively, friendly, and interesting. In this revision, we have tried to retain the comfortable style of the first two editions.

Second, any text, but especially one designed for nonmajors and mixed courses, must begin with basic principles and work up from there. We have tried to make our text sufficiently flexible so that instructors can choose the level of material that they wish their students to master. Some instructors may wish to assign only the essential principles of difficult topics. Others will want to assign both the principles and the anatomical, biophysical, and biochemical details. As often as possible, we have explicitly written difficult material at two levels to serve these two audiences. Subjects such as photosynthesis, respiration, DNA replication, transcription, translation, the Hardy-Weinberg principle, and kidney function are first explained as a "take-home lesson," describing the essential facts and mechanisms that all biology students, at any level, must know. Only then do we fill in the details, either in a subsequent section of the text, in boxed figures, or in essays entitled "A Closer Look."

Third, people learn best when they have mental "hooks" on which to hang new information. Wherever possible, we integrate new concepts in biology with everyday experience. We also offer unusual examples in the hope that they will help to fire students' imagination. We think that this blend of the familiar and the exotic gives a big assist to learning.

New terminology is an important part of any introductory biology course. However, the wealth of unfamiliar terminology can also prove challenging to many students. We have addressed this issue in two ways. First, we included only those terms that we felt were essential to student understanding and that would be used extensively in our discussion. Second, these terms appear in **boldface** and are immediately and explicitly defined in the text. They then appear with their correct pronunciation in the end-of-chapter glossary, a feature that is unique to our text. We hope this approach will make multisyllabic biological terms less intimidating and easier to master for students. Reviewers have praised our sensible use of terminology in all editions of the text. For the third edition, we have also added **boldfaced statements of important concepts** in the text to reinforce student comprehension.

New and Revised Features

This edition of our text contains many new and revised features designed to enhance student interest and learning, and to cover emerging aspects of biology in more depth. These features include:

- A new design with larger photos and a more legible typeface. We have added 250 new photos.

- All new, full-color art. The entire art program has been revised and redrawn to appear vivid and three-dimensional. New labels and cell icons have been added to clarify the presentation. In addition, there are 100 new figures.

- A new chapter on genetic engineering (Chapter 14, "Molecular Genetics and Biotechnology").

- The division of several chapters into more manageable units. Separate chapters are now devoted to:

    —photosynthesis
    —glycolysis and cellular respiration
    —mitosis
    —meiosis
    —mechanisms of evolution
    —results of evolution
    —fungal diversity
    —plant diversity
    —plant structure
    —plant nutrition and transport
    —circulation
    —respiration
    —senses and perception
    —muscles and skeleton

- Boxed essays in easily identifiable categories, including:

 "Planet Watch": discussions of the environmental implications of diverse topics in biology (new to this edition).

 "Health Watch": discussions relating topics in human health (new to this edition).

 "Methods in Biology": brief surveys of important methodologies in biology and medicine (new to this edition).

 "A Closer Look": in-depth discussions of difficult topics.

- Expanded coverage of several topics, including biotechnology, scientific research methods, mechanisms of evolution, immunology, AIDS, sexually transmitted diseases, weather and climate, environmental concerns, freshwater ecosystems, tissue types and homeostasis, and *in vitro* fertilization.

- New, thought-provoking discussion questions at the end of every chapter.

- New, revised study questions.

- Updated suggested readings.

Organizational Themes

We have organized the text by using two main themes: *evolution* and *adaptation to the environment*. As Theodosius Dobzhansky so aptly phrased it, "Noth-

ing in biology makes sense, except in the light of evolution." *Life on Earth* has been forged in the crucible of evolution, and any modern biology text must reflect that fact. Our second organizational theme, adaptation to the environment, flows naturally from the first—organisms are adapted to the environment because that environment places selective pressure upon them that helps to shape their evolution. Nevertheless, organisms are not perfect, and not all biological structures are optimal solutions brought about by natural selection. Genetic drift, chance catastrophes both ancient and modern, and the inevitable requirement that evolutionary change must build upon pre-existing structures have led to many unusual and apparently unlikely structures and behaviors. In the words of Sydney Brenner, "Anything that is produced by evolution is bound to be a bit of a mess," and we have tried to show that, too.

Although not an organizational theme, our own concern for the environment is interwoven into this text. Whenever appropriate, we have tried to present students with the biological rationale for making sound environmental decisions in their own lives.

Chapter Sequence

There is no "correct" sequence of topics in biology. Virtually every subject would be more understandable if everything else preceded it. Some texts claim that their chapters and units are completely self-contained, and can be shuffled into whatever sequence the instructor wishes. Being teachers ourselves, we know that complete interchangeability is impossible. However, we have included elements in our text to maximize flexibility. In Chapter 1, in addition to sections on the nature of life and the scientific method, we have two other important features. First a "Gallery of Life" briefly introduces the five kingdoms of life, with representative organisms, giving students early exposure to diversity. Second, we provide a brief, understandable explanation of the principles of evolution. Thus, instructors can refer to the evolutionary significance of many topics before formally presenting evolution. The comprehensive discussion of evolution comprises Unit III.

We also make use of selective repetition to enhance flexibility. For example, learning how plants adapted for life on land is essential to understanding their initial conquest of dry land, as well as plant diversity, physiology, and ecology. We have repeated key concepts—not in tiresome detail, but with enough thoroughness to be understandable—separately in each discussion. Such selective repetition not only aids sequence flexibility, but also provides added emphasis, strengthening students' comprehension of important points.

Artwork

This edition features all new, full-color art. Each illustration has been completely reworked to appear alive and three-dimensional. "Cell icons" are used extensively, especially in Chapters 5, 6, 7, and 8. These "icons" of a plant or animal cell allow us to locate a particular cell structure (such as mitochondrion, chloroplasts, etc.) within the cell. The structure is then, in effect, "magnified" so that students get a detailed depiction while seeing how it fits into the overall cell (for an example, see our "A Brief Walk Through, *Biology: Life on Earth*" pp. *xxii–xxvii*). When illustrating cycles, systems, and processes we have added labels and numbered steps, making it easier for students to follow the sequence of events. These illustrations, along with their detailed captions, provide a visual summary of the text discussion and form a concise and powerful tool for student study.

This exciting and instructional art program was made possible by the generosity and foresight of Macmillan Publishing Company, and through the efforts of Hudson River Studio and their talented artistic director, Ed Burke, who supervised dozens of outstanding biological illustrators throughout the country. Thanks must also go to Paula Nicholas, who was the biological illustrator for the second edition, and served as a consultant on this edition. Without the input and collaboration of these many talented and committed people, this superb art program could never have been realized. Their contributions have greatly enhanced the pedagogy and beauty of this edition.

Learning Aids

We have incorporated many learning aids into the text. Some are new to this edition, others are popular holdovers from previous editions. These include:

- Chapter Outlines
- Chapter Summaries organized to follow chapter subheadings
- Phonetic end-of-chapter glossaries
- End-of-chapter study questions, discussion questions, and suggested readings
- Boxed essays in specific categories
- Full-color illustrations and photographs that superbly complement the text discussion

- Multilevel explanations of complex materials such as photosynthesis, respiration, and protein synthesis
- Simplified and expanded explanations of basic chemistry, including a new box on interpreting chemical structures and formulas.

Each chapter begins with an *outline* consisting of the first two levels of section headings with page numbers, so students and instructors can easily locate needed information. The *chapter summaries* are organized to follow the first level section headings, as an aid to review.

We have retained our popular *end-of-chapter phonetic glossaries*. In addition, all terms are combined into a single glossary at the end of the text. These elements provide valuable assistance to the student, both in understanding and reviewing the material in each chapter.

Each chapter ends with a set of *study questions, discussion questions*, and *suggested readings*. The *study questions* provide a straightforward review of chapter material, testing students' recall and comprehension; many new questions have been added in this edition. *Discussion questions* ask students to apply their knowledge to more complex and challenging questions; they are new to this edition. The *suggested readings* have been revised and updated to include the latest research findings. We have tried to limit our selections to those a student could understand and enjoy.

We have added a number of new and fascinating essays to this edition. Essays play several important roles in our text. First, many of our essays relate biology to the world of the student in ways that might not be appropriate to the in-text discussion. To neglect this vital link with the real world is to make biology more remote, uninteresting, and forgettable. "Health Watch" essays relate material in the chapter to specific health issues of importance to students. "Planet Watch" essays explain the biological bases of specific environmental issues and concerns—an area of intense student interest. "Methods in Biology" essays describe the scientific methodology used to discover some of the basic principles presented in the chapter, giving students a better understanding of biological inquiry as well as specific methods and techniques. We have continued to use "Reflections" at the end of many chapters. These draw relationships between diverse concepts and try to pique student interest by expressing opinions, ideas, and general food for thought. Finally, essays titled "A Closer Look" provide more in-depth discussion of difficult or complex topics. These allow the instructor more flexibility in choosing the level of presentation. Each type of essay has been given an easily recognizable design, with a distinctive symbol and color scheme.

The Supplements Package

A comprehensive package of supplements is available, coordinating closely with the text. The supplements were prepared by a team of biologists at Virginia Commonwealth University who have taught extensively using earlier editions of *Biology: Life on Earth*. Supplements include:

- A *Study Guide* by Joseph P. Chinnici, contains a review outline of each chapter along with review exercises and activities.

- An *Instructor's Manual* by Margaret L. May, includes an overview of each chapter, teaching suggestions, thought questions for written assignments or discussion, sources of films or computer software, and answers to the study questions in the text.

- A *Test Bank* by Joseph P. Chinnici, Gail C. Turner, J. Lewis Payne, and Margaret L. May contains over 1800 test questions. It is available in computerized form for IBM-PC and Macintosh.

- A *Laboratory Manual* by Donald M. Fritsch has been developed through extensive class-testing and directly references the text.

- A set of *full-color transparencies* draws key illustrations from the text.

- A set of *Transparency Masters* of figures in the text allow instructors to produce readable transparencies as needed.

Acknowledgments

This text, like all textbooks, is not the sole work of its authors. We have benefitted enormously from the opinions and expert advice of numerous colleagues, users, and reviewers. A complete list of reviewers is given below.

Usually unacknowledged, but perhaps most valuable of all, are the thousands of students whom we have taught over the years. Students, by the questions they ask and by their performance on tests, cast the deciding vote on whether an explanation is really helpful and understandable. Student feedback constantly refines our teaching, and therefore, also our writing. Although they often remain unaware of it, our students are actually partners with us in a combined teaching/learning endeavor, and their role is really much more important than ours.

Despite all of this good advice from colleagues and students, we undoubtedly have slipped up now and then, and we remain solely responsible for errors and ineffective explanations. In cases where we have failed to follow the advice of users and reviewers, we ask their indulgence and request their continued gentle remonstrance.

Finally, we wish to thank everyone involved in the development and production of this text. We are particularly indebted to Madalyn Stone, for her cheerful and untiring work in editing our manuscript, constantly seeking for ways to explain difficult topics more clearly; Chris Migdol and Yvonne Gerin, the Sherlock Holmes of photo researchers, who pursued beautiful, fascinating and pedagogically essential photographs across the globe; Marilyn James of York Production Services, who shepherded text and art back and forth among the many contributors, and kept track of everything; and Edward Burke of Hudson River Studio, who supervised dozens of artists and created a beautiful new look for our third edition. Most of all, we thank our editors, Bob Rogers and Kristin Watts Peri, and Executive Editor Bob Pirtle, whose enthusiasm and energy proved invaluable to everyone involved in the third edition.

As authors, we now have an extended classroom that reaches across most of North America. We hope that these colleagues and students will find this edition to be both enjoyable and valuable as a tool for understanding biology. As we wrote in Chapter 1, biology can be much more than just another course to take and another set of facts to memorize. Maybe we're biased, but what can be more fascinating than learning about life on Earth?

Gerry and Terry Audesirk
Golden, Colorado
March, 1992

Reviewers

Brief Contents

Detailed Contents

Essays

xxi

A Brief Walk Through
Biology: Life on Earth

On the following pages you will find an overview of the many new and revised features we have included in *Biology: Life on Earth*, third edition. These carefully developed features will not only make the text more accessible and useful to students, but will also convey some of the excitement and relevance of learning about biology.

Chapter Table of Contents–NEW! Each chapter opens with an outline of the major sections in the chapter, with page references, to help students and instructors locate information quickly and easily.

The History of

16

Principles of Evolution

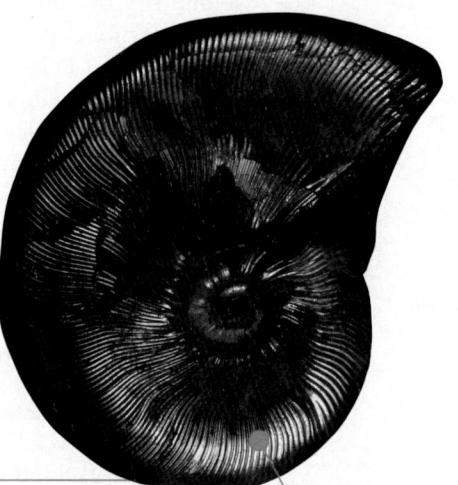

"When on board H.M.S. 'Beagle,' as naturalist, I was much struck with . . . the distribution of the inhabitants of South America, and . . . the geological relations of the present to the past inhabitants of that continent. These facts seemed to me to throw some light on the origin of species—that mystery of mysteries, as it has been called by one of our greatest philosophers."

Charles Darwin in On the Origin of Species by Means of Natural Selection

These words introduce what is perhaps the most important work in biology. In the *Origin of Species*, Charles Darwin proposed that over eons of time, species arise from other, preexisting species through the process of "descent with modification," or evolution.

Before Darwin, how species originated remained the "mystery of mysteries" for a very simple reason. Over the time span of recorded human history, let alone the life of a single human being, no new species had been recognized (although undoubtedly many new species had appeared, especially plant species). It is quite difficult to decide how something happens if there aren't any witnesses.

Nevertheless, one of the most striking features about our world is the remarkable variety of organisms inhabiting it. Why are there dozens of species of pine trees and scores of species of warblers? With no evidence to go on, nearly all peoples of the world historically turned to hypotheses of **creationism**. The most common of these hypotheses is that a supernatural being created each type of organism separately at the beginning of the world, and that all modern organisms are essentially unchanged descendants of these ancestors.

As we pointed out in Chapter 1, one of the fundamental principles of science is that Earthly phenomena are produced by natural, Earthly causes. A nineteenth century English essayist wrote, " . . . with regard to the material world, we can at least go so far as this—we can perceive that events are brought about not by insulated interpositions of Divine power exerted in each particular case, but by the establishment of general laws [of nature]." Science cannot say whether or not divine power originally established those general laws, but science firmly adheres to the principle that natural events have causes that arise from the operation of natural laws.

Therefore, throughout history, scientists have sought natural causes for the origin of species. However, it was only in the nineteenth century that a truly coherent theory—evolution by descent with modification, driven by natural selection—was developed. This theory was published by two British naturalists,

Charles Darwin and Alfr
and today still forms th
standing of evolution. L
work in a vacuum. Centu
tion preceded them and
us begin, then, with a
thought.

The History of
Evolutionary T

What Is a Species

Before we can study the
first decide what a spe
human history, "species
cept. When using the wo
ans meant one of the o
ferred to in the Bible. H
two organisms belonge
Since no one was present
criteria of the Creator, o
species by visible differe
word "species" is Latir
pines and warblers are
blers are different from e
biologists distinguish a
Today, biologists define
tions of organisms that
under natural conditions
isolated from other pop
members of a species c
selves, but usually not
cies. If interbreeding wit
the hybrid offspring are
capped in some way (se
Biologists have found
ance do not always me
long to different species
published in the 1970s I
Audubon's warbler (Fi
more recently the Am

Dynamic Photo Program *Biology: Life on Earth* is lavishly illustrated with beautiful and informative photographs to clarify important concepts and reinforce students' awareness of biological diversity.

Key Terms Key biological terms are set off in **boldface type** and are defined where they first appear.

Key Statements—NEW! To reinforce important biological concepts, key statements appear in **boldface:** an excellent tool for review!

---

588 *Ch 27 Nutrition and Transport in Land Plants*

Water Movement in Xylem

According to the **cohesion-tension theory,** water is pulled up the xylem, powered by the evaporation of water from the leaves (Fig. 27-6). As its name suggests, this theory has two essential parts.

1. *Cohesion:* water within the xylem holds together like a solid rope.
2. *Tension:* this "water rope" is pulled up the xylem, with evaporation providing the necessary energy.

Let's briefly examine both of these propositions.

Cohesion Among Water Molecules

You will recall from Chapter 2 that water is a polar molecule, with the oxygen carrying a slight negative charge while the hydrogens carry a slight positive charge. As a result, nearby water molecules attract one another, forming weak **hydrogen bonds.** However, just as individually weak cotton threads together make a strong seam in your jeans, so too **the network of hydrogen bonds within water is very strong, giving water a high** *cohesion,* **or tendency to resist being separated.** This is essential, because if the water is pulled up the xylem from the top, then the column of water has to be strong enough to bear its own weight without breaking—all the water molecules within the xylem are "hanging on" to the uppermost molecules by the hydrogen bonds connecting this handful of molecules to the rest of the column.

Experiments have found that the column of water within the xylem is at least as strong as a steel wire of the same diameter, and therefore doesn't break (see

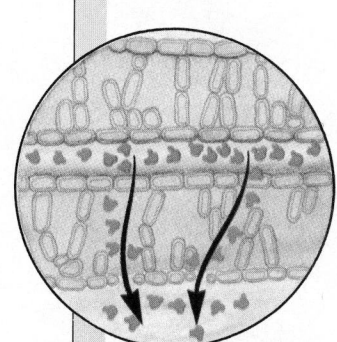

Process Figures Art and captions work together to provide a visual summary of the text.

Figure 27-6 The cohesion–tension theory of water flow from root to leaf in xylem. (1) Water evaporates out of the leaves, and other water molecules replace them from the xylem of the leaf veins. (2) Within the xylem, hydrogen bonding holds nearby water molecules together so firmly that the column of water behaves very much like a rope. The top of the "water rope" is pulled up by evaporation, and the rest of the rope comes along as well, all the way down to the roots. (3) As the molecules of the water rope retreat up the xylem in the roots, the decreased water concentration within the root xylem and the surrounding extracellular space causes water to enter from the soil water by osmosis, thus steadily replenishing the bottom of the rope.

① evaporation from leaves

② cohesion of water molecules to one another and to xylem walls, by hydrogen bonds

③ water enters vascular cylinder of root by osmosis

ments and tracheids. This upward and inward movement of water finally causes water to move into the vascular cylinder by osmosis through the endodermal cells. The force generated by the evaporation of water from the leaves, transmitted down the xylem to the roots, is so strong that water can be absorbed from quite dry soils.

SUMMARY OF WATER TRANSPORT IN XYLEM:

Transpiration from the leaves (and, to a lesser extent, the stem) removes water from the top of a xylem tube. This water is replaced by water further down in the tube, so that water moves by bulk flow up the xylem. Finally, this upward flow removes water from the root xylem and the extracellular space surrounding it, which promotes osmosis of water from the soil water into the vascular cylinder of the root. **The flow of water in the xylem is unidirectional, from root to shoot, because only the shoot can transpire.**

The Control of Transpiration

Transpiration has both positive and negative effects on a plant. On the plus side, transpiration provides the force that transports water and minerals to the leaves at the top of the plant. On the minus side, transpiration is by far the largest source of water loss—a loss that may threaten the very survival of the plant, especially in hot, dry weather.

Most water transpires through the stomata of

In-Text Summaries Extended discussion of complex processes are briefly summarized so students can check their understanding.

The Transport of Water and Minerals **589**

leaves and stem, so you might think that a plant could prevent water loss by simply closing its stomata. However, don't forget that photosynthesis requires carbon dioxide from the air, which diffuses into the leaf mainly through open stomata. **Therefore, a plant, by opening and closing its stomata, must achieve a balance between carbon dioxide uptake and water loss.**

Stoma Structure and Function

A **stoma** consists of a central opening surrounded by two kidney-shaped **guard cells** that regulate the size of the opening (Fig. 27-7). With some exceptions, stomata open during the day and close at night, but they will also close if the leaf begins to dehydrate. The selective value of this arrangement is obvious. First, the stomata open only during the day, when sunlight allows photosynthesis. Even then, the stomata close if water loss becomes too great.

Guard cells change the size of the opening between them by changing their own shape. How does this happen? There are two levels to this question: first, how does changing the shape of the guard cells open and close the opening; and second, what physiological processes cause the change in shape?

The Relationship Between Guard Cell Shape and Opening a Stoma

Stomata open when the guard cells take up water and swell, and close when guard cells lose water and shrink. This might seem paradoxical, since swollen guard cells must take up a larger volume than shrunken ones, and therefore you might expect that the potential central hole would be shut more tightly than ever. The key lies in the construction of the cell wall of guard cells (Fig. 27-8a). Cellulose fibers in the wall encircle the guard cells like a series of inelastic belts. Thus, when water enters the guard cells and their volume increases, they cannot become fatter but must become longer. Each pair of guard cells is attached at both ends, so the only way the cells can become longer is by bowing outward like a cooked sausage, opening a hole between them (Fig. 27-8b).

Osmosis of Water Into and Out of Guard Cells

According to the principles of osmosis, water will enter a guard cell if its cytoplasm has a lower water concentration than the cytoplasm of surrounding cells, and will leave the guard cell if it has a higher concentration of water. Large changes in potassium concentration within the guard cells cause correspondingly large changes in water concentration, driving the osmotic fluxes that open and close a

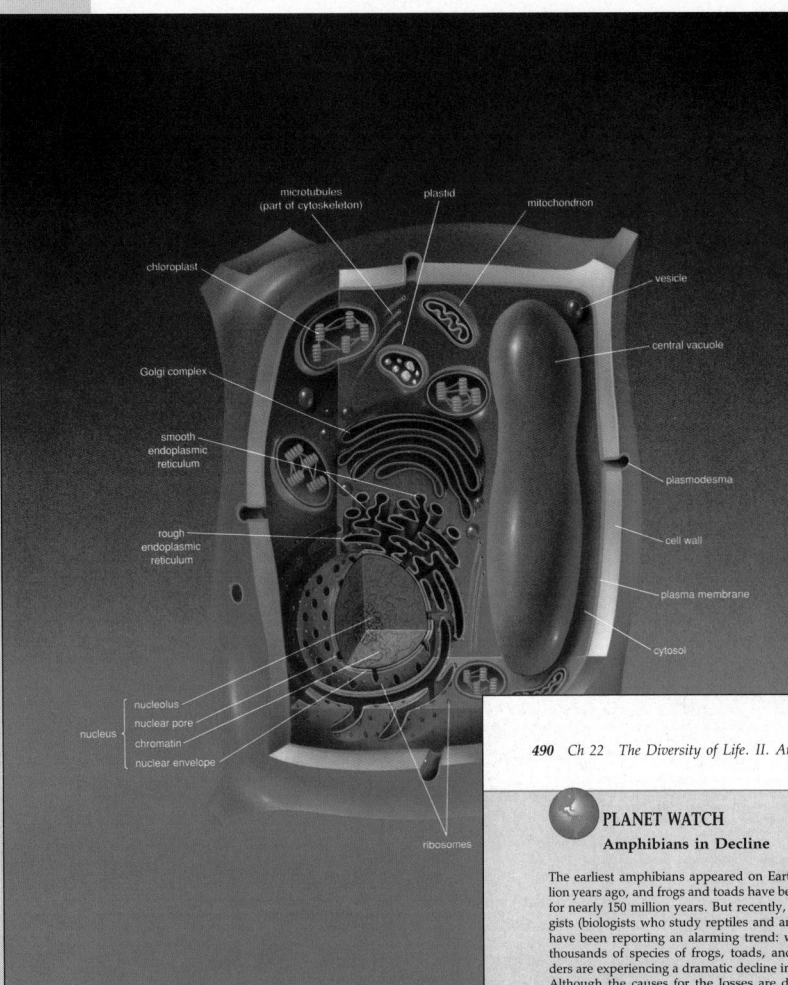

microtubules (part of cytoskeleton) · plastid · mitochondrion · chloroplast · vesicle · central vacuole · Golgi complex · plasmodesma · smooth endoplasmic reticulum · cell wall · rough endoplasmic reticulum · plasma membrane · cytosol · nucleolus · nuclear pore · chromatin · nuclear envelope · nucleus · ribosomes

New Full-Color Illustrations All new, dynamic full-color illustrations will help students to accurately visualize organisms, processes, and concepts, while making biology come to life.

Planet Watch—NEW! These boxed essays present relevant environmental concerns and their relation to biology. (For a complete listing of essay topics, see p. xxi.) In addition, new "Health Watch" essays provide important information concerning human health.

490 Ch 22 *The Diversity of Life. II. Animals*

PLANET WATCH
Amphibians in Decline

The earliest amphibians appeared on Earth 350 million years ago, and frogs and toads have been around for nearly 150 million years. But recently, herpetologists (biologists who study reptiles and amphibians) have been reporting an alarming trend: worldwide, thousands of species of frogs, toads, and salamanders are experiencing a dramatic decline in numbers. Although the causes for the losses are diverse and poorly understood, they can be traced to a single source: human modification of the biosphere—that portion of the Earth that sustains life.

Habitat destruction, particularly in the tropics, is one major cause of the decline. However, the unique biology of amphibians makes them vulnerable even where their habitat is not threatened. The "double life" of amphibians exposes them to toxins in a wider range of habitats, including water, air, and soil. Many amphibians' eggs develop in ponds and streams during the spring. Acid precipitation (see Chapter 45) has made spring a dangerous time for aquatic organisms. The melting of acid snow and ice causes a springtime "pulse" of intense acidity in freshwater ecosystems, just as many amphibian eggs are undergoing critical stages of development. Further, at all stages of life, amphibians are covered by a thin, permeable skin through which toxins carried in air or water can easily penetrate. Some also feed on insects that have accumulated insecticides in their bodies.

Many scientists believe that the decline in the amphibian population signals an overall deterioration of the Earth's ability to support life; and the decline is worldwide. Yosemite toads and yellow-legged frogs are disappearing from the mountains of California,

while tiger salamanders have been nearly wiped out of the Colorado Rockies. Leopard frogs, eagerly chased by rural children throughout the United States, are suddenly becoming rare. While logging destroys the habitats of amphibians from the Pacific Northwest to the tropics (Fig. E22-1), even those in preserves are dying. In the Monteverde Cloud Forest Preserve in Costa Rica, the golden toad (see Fig. 39-7) was common in the early 1980s, but has now almost disappeared. The gastric breeding frog fascinated biologists by swallowing its eggs, brooding them in its stomach, and later regurgitating fully formed offspring. The species was abundant and seemed safe within a national park in Australia. Then suddenly, in 1980, the gastric brooding frog disappeared and has never been seen since. Evidence is mounting that setting aside small islands of nature in preserves amidst surrounding environmental contamination and destruction is not enough to save species from extinction.

Amphibians are not just sensitive indicators of the health of the biosphere, they are themselves a crucial component of many ecosystems. They may keep insect populations in check, while in turn serving as food for larger carnivores. Their decline will further disrupt the balance of these delicate communities. Margaret Stewart, an ecologist at the State University of New York, Albany aptly summarized the problem: "There's a famous saying among ecologists and environmentalists: 'everything is related to everything else.' . . . You can't wipe out one large component of the system and not see dramatic changes in other parts of the system."

Figure E22-1 The corroboree toad, shown here with its eggs, is rapidly declining in its native Australia. Tadpoles can be seen developing within the eggs. The thin water- and gas-permeable skin of the adults, and the jellylike coating surrounding the eggs, provide inadequate protection against pollutants.

Cell Icons These "icons" of plant or animal cells help students locate cell structures within the overall cell. The particular structure appears "magnified" so students can recognize their structure, location, and function.

162 *Ch 8 Glycolysis and Cellular Respiration: Harvesting Energy from Food*

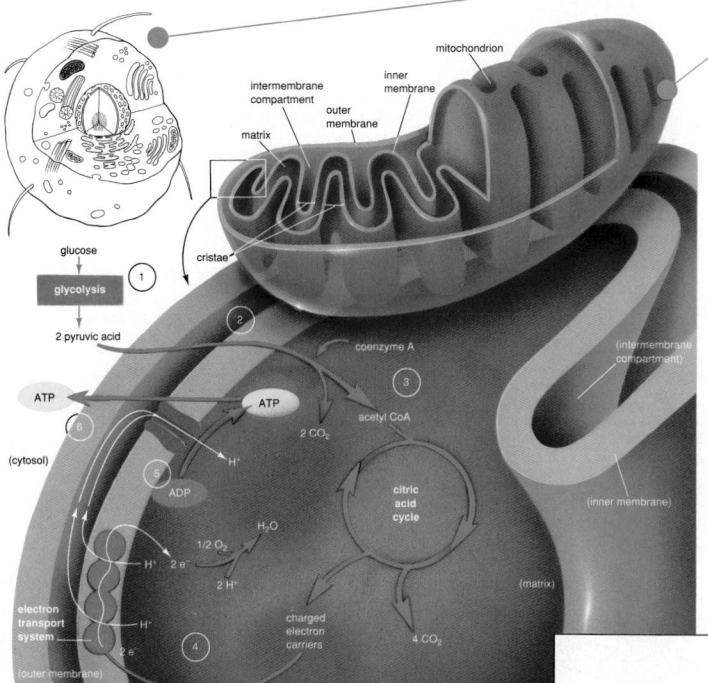

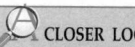

Figure 8-4 Cellular respiration takes place in the mitochondria, whose structure, like that of a chloroplast, reflects the compartmentalized reactions that occur there. The inner membrane separates the inner compartment, containing the soluble enzymes of the matrix, from the intermembrane compartment (between the inner and outer membranes). The lower diagram summarizes the six essential steps in glucose metabolism, from the initial glycolysis in the cytosol to the final transport of ATP out of the mitochondrion and into the cytosol. See the text for details of these steps.

A Closer Look These boxed essays present topics that are discussed in the main text with a greater level of insight and detail, providing a flexible level of presentation for difficult concepts.

506 *Ch 23 The Diversity of Life. III. Fungi*

CLOSER LOOK

At Evolutionary Ingenuity in Fungi

Natural selection, operating over millennia on the diverse forms of fungi, has produced some remarkable adaptations by which fungi solve the problems of dispersing their spores and obtaining nutrients. A few of these are highlighted in this essay.

The Rare, Sexy Truffle

Although many fungi are prized as food, none are as avidly sought as the truffle. A single specimen resembling a small, shriveled, blackened apple (Fig. E23-1) may sell for $100. Truffles are the reproductive structures of an ascomycete that forms a mycorrhizal association with the roots of oak trees. Formed underground, the truffle is faced with the problem of dispersing its spores. Its solution is to entice animals to dig it up. Although human beings cannot smell underground truffles, their odor serves as an irresistible attractant to certain mammals, especially wild pigs. Why? It was recently discovered that the truffle releases a chemical that closely resembles the pig's sex attractant. As aroused pigs dig up and devour the truffle, millions of spores are scattered to the winds.

Human truffle hunters use muzzled pigs to hunt their quarry; a good truffle-pig can smell an underground truffle 50 meters away! Although a truffle growth can be encouraged by sowing spores throughout oak groves, they have eluded commercial cultivation. Thus they remain so rare and costly that few of us will ever taste one.

The Shotgun Approach to Spore Dispersal

The delicate structures in Fig. E23-2 are actually fungal shotguns, the reproductive structures of the zygomycete *Pilobolus*. Only by closely scrutinizing piles of horse manure are you likely to observe this miniature beauty. Hyphae penetrating the dung send up clear bulbs capped with sticky black spore cases. As they mature, the sugar concentration in the bulbs increases, drawing in water by osmosis. Meanwhile, the bulb begins to weaken just below its cap. Suddenly, like an overinflated balloon, it bursts, blowing its spore-carrying top up to a meter away. *Pilobolus* bends toward the light. This may increase the probability that its spores will land on open pasture. Here they adhere to grass blades until consumed by a grazing herbivore, perhaps a horse. Later (some distance away) the spores are deposited unharmed in a pile of their favorite food: manure. Growing hyphae penetrate this rich source of nutrients, sending up new projectiles to continue this ingenious cycle.

Figure E23-1 The truffle. This rare ascomycete is a gastronomic delicacy.

Methods In Biology—NEW! These brief essays describe important methods in biology and medicine to help students appreciate the nature of scientific inquiry as well as specific techniques.

 METHODS IN BIOLOGY
How Do We Know What Goes on Inside Xylem?

As you study biology, we hope you are not only absorbing a lot of facts about living organisms, but also learning how biologists discover those facts in the first place: what tools biologists use in their research, how they design experiments, how they analyze the resulting data, and how they draw conclusions. In this chapter, we presented the cohesion–tension theory for water transport in xylem from roots to leaves. You may have wondered how such a scheme could have been thought up at all, why it has been accepted by botanists, and whether other, simpler, explanations might not suffice.

The problem is how to get water from the soil to the topmost leaves of a plant, which might be as much as 100 meters away. Various hypotheses have been proposed, usually soon to be discarded. For example, consider capillarity. You may have noticed that if one end of a thin tube is immersed in water while the other end sticks out in the air, water rises a short way up the tube (Fig. E27-1). A little experimentation reveals that water ascends farther in thinner tubes. Perhaps in tubes as thin as those of xylem, water simply creeps up by capillarity. Xylem cells, however, with diameters of 30 to 50 microns (about 1/500th of an inch), would only allow capillarity to raise water up about a meter—not even close to the top of a tree.

A second, more serious, proposal is that the roots might pump water up the xylem, a phenomenon called **root pressure.** If you cut the top off a tomato plant and seal a pressure-monitoring device onto the remaining stump, you would find that the roots do indeed push water up: hard enough to reach 30 or 40 meters. Root pressure arises because of the accumulation of minerals in root xylem. As you recall, water moves by osmosis into the extracellular space and xylem tubes in a root vascular cylinder. This water has no place to go but up the plant, and generates root pressure. In some plants—strawberries, for example—root pressure occasionally forces water c...

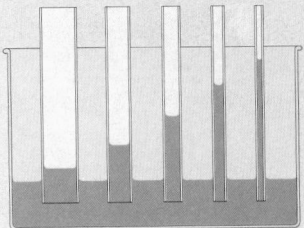

Figure E27-1 If a thin tube is touched to the surface of a dish of water, some water will spontaneously move up the tube. How high the water goes is a function of tube diameter: the thinner the tube, the farther the water rises.

Figure E27-2 Under favorable conditions of high water availability in the soil and high humidity in the air, the root pressure of some plants will force water out the tips of the leaves. Some botanists speculate that the exuded water might carry wastes out of the plant.

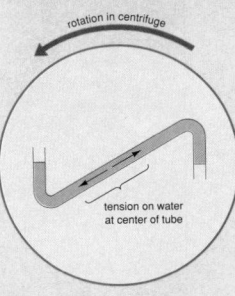

rotation in centrifuge

tension on water
at center of tube

Figure E27-3 An experiment to test whether water cohesion is strong enough to support the weight of a column of water in the xylem of a tall tree. A capillary tube is bent into a Z shape, filled with water, and spun at high speed in a centrifuge. Centrifugal forces push the water toward the ends of the Z. At high enough speeds, the column of water breaks apart in the middle. However, the forces required to break the water column in the Z tube are equivalent to the weight of water in a xylem tube more than 500 meters tall, far taller than any tree that ever lived.

of high flow rates, just as a soda straw collapses if you suck on it too hard. Although it is difficult to imagine a tree trunk shrinking, measurements do show that trunks are thinner during the day (when water is evaporating rapidly from the leaves and therefore is being rapidly pulled up the xylem) than at night (when evaporation and flow rates are low). Finally, botanists can measure the tension within a stem, and have found tensions strong enough to pull water up *200 meters!*

Clearly, the cohesion–tension theory has passed several tests with flying colors, while competing hypotheses have failed. Nevertheless, as with all scientific theories, future experiments may someday come up with data that the cohesion–tension theory cannot explain. Such is the nature of science: a theory stands only as long as experimental data allow it to. So far, the cohesion–tension theory stands as tall as the redwood trees whose water transport it explains.

Summary of Key Concepts As an aid to review, the chapter summaries are organized to correspond to the major sections of the chapter.

---

422 Ch 19 The History of Life on Earth

SUMMARY OF KEY CONCEPTS

Origins
Before life arose, lightning, ultraviolet light, and heat formed organic molecules from water and the components of the primordial Earth's atmosphere. These molecules probably included nucleic acids, amino acids, short proteins, and lipids. By chance, some molecules of RNA may have had enzymatic properties, catalyzing the assembly of copies of themselves from nucleotides in the Earth's waters. These may have been the precursors of life. Protein–lipid microspheres enclosing these RNA molecules may have formed the first cell-like organisms.

The Age of Microbes
The first fossil cells are found in rocks about 3.5 billion years old. These cells were prokaryotes that fed by absorbing organic molecules that had been synthesized abiotically. Since there was no free oxygen in the atmosphere at this time, energy metabolism must have been anaerobic. As the cells multiplied, they depleted the organic molecules that had been formed by prebiotic synthesis. Some cells evolved the ability to synthesize their own food molecules using simple inorganic molecules and the energy of sunlight. These earliest photosynthetic cells were probably ancestors of today's cyanobacteria.

Photosynthesis releases oxygen as a by-product, and by about 2.2 billion years ago significant amounts of free oxygen had accumulated in the atmosphere. Aerobic metabolism, which generates more cellular energy than does anaerobic metabolism, probably arose about this time.

Eukaryotic cells evolved about 1.5 billion years ago. The first eukaryotic cells probably arose as symbiotic associations between predatory prokaryotic cells and bacteria. Mitochondria may have evolved from aerobic bacteria engulfed by predatory cells. Similarly, chloroplasts may have evolved from photosynthetic cyanobacteria.

Multicellularity
Multicellular organisms evolved from eukaryotic cells, first appearing about 1 billion years ago. Multicellularity offers several advantages, including increased speed of locomotion and increased size.

Multicellular Life in the Sea
The first multicellular organisms arose in the sea. In plants, increased size due to multicellularity offered some protection from predation. Specialization of cells allowed plants to anchor themselves in the nutrient-rich, well-lit waters of the shore. For animals, multicellularity allowed more efficient predation and more effective escape from predators. This in turn provided selection pressures for faster locomotion, improved senses, and greater intelligence.

The Invasion of the Land
The first land organisms were probably plants, appearing 500 to 600 million years ago. Although the land required special adaptations for support of the body, reproduction, and the acquisition, distribution, and retention of water, the land also offered abundant sunlight and protection from aquatic herbivores. Around 400 million years ago, arthropods invaded the land. Absence of predators and abundant land plants for food probably facilitated the invasion of the land by animals.

The earliest land vertebrates evolved from lobefinned fishes, which had leg-like fins and a primitive lung. A group of lobefins evolved into the amphibians about 350 million years ago. Reptiles evolved from amphibians, with several further adaptations for land life: internal fertilization, waterproof eggs that coul[...] better lungs. A[...] mammals ev[...] groups of rept[...] constant body [...] body surface.

Human Evolu[...]
One group of [...] ing primates. [...] for human ev[...] lar vision, col[...] large brains. [...] some primates [...] the ancestors [...] pithecines aro[...] These hominid[...] their forebears[...] of Australopith[...]

Phonetic Glossary
Popular with students and instructors alike, the chapter-ending glossaries are a valuable tool for review.

Suggested Readings 423

GLOSSARY

cultural evolution: changes in the behavior of a population of animals, especially humans, by learning behaviors acquired by members of previous generations.
endosymbiotic hypothesis: the hypothesis that certain organelles, especially chloroplasts and mitochondria, evolved from bacteria captured by ancient predatory prokaryotic cells.
microsphere: a small, hollow sphere formed from proteins or proteins complexed with other compounds.

preadaptation: a feature evolved under one set of environmental conditions that, purely by chance, helps an organism adapt to new environmental conditions.
prebiotic evolution: evolution before life existed; especially abiotic synthesis of organic molecules.
ribozyme: an RNA molecule that can catalyze certain chemical reactions, especially those involved in synthesis and processing of RNA itself.

STUDY QUESTIONS

1. What is the evidence that life might have originated from nonliving matter on the primordial Earth? What kind of evidence would you like to see before you would accept this hypothesis?
2. Explain the endosymbiotic hypothesis for the origin of chloroplasts and mitochondria.
3. Name two advantages of multicellularity in plants and animals.
4. What advantages and disadvantages would terrestrial existence have had for the first plants to invade the land? For the first land animals?
5. Outline the general trends in the evolution of vertebrates, from fish to amphibians to reptiles to birds and mammals. Explain how these adaptations increased the fitness of the various groups for life on land.
6. Outline the evolution of humans from early primates. Include in your discussion such features as binocular vision, grasping hands, bipedal locomotion, social living, tool making, and brain expansion.

DISCUSSION QUESTIONS

1. What is cultural evolution? Is cultural evolution more or less rapid than biological evolution? Why?
2. Do you think that studying our ancestors can shed light on the behavior of modern humans? Why or why not?
3. Many paleoanthropologists believe that *Australopithecus afarensis* was the ancestor to all later hominids, including, eventually, *Homo sapiens*. This position is based mostly upon two facts: (1) *A. afensis* fossils appear earlier than those of other Australopithecines or of fossils of the genus *Homo*, and (2) *A. afarensis* fossils show many characteristics that appear to be more primitive than, but possibly ancestral to, those of later hominids. Do these facts prove ancestry? Might other fossils, not yet found, displace *A. afarensis* in the descent of humanity? If so, what characteristics and age would you expect them to have?
4. Is the history of life on Earth a steady progression to more advanced life forms? Are organisms alive today "better" in some absolute sense than organisms alive 100 million years ago? Why or why not?

SUGGESTED READINGS

Diamond, J. "How to Speak Neanderthal." *Discover*, January 1990. Could the Neanderthal's speak? How could we tell? Jared Diamond lucidly explains the arguments and methodology behind a scientific dispute.
Hay, R. L., and Leakey, M. D. "The Fossil Footprints of Laetoli." *Scientific American*, February 1982. The actual footprints of a hominid family were discovered by Hay and Leakey in volcanic ash 3.5 million years old.
Horgan, J. "In the Beginning . . ." *Scientific American*, February 1991. An exploration of the controversies surrounding research into the origin of life.
Morell, V. "Announcing the Birth of a Heresy." *Discover*, March 1987. Paleontologists Robert Bakker and Jack Horner speculate that dinosaurs were not the plodding beasts of monster flicks, but warm-blooded, advanced animals that may have even cared for their young.
Shreeve, J. "Argument Over a Woman." *Discover*, August 1990. An account of the often acrimonious debate between evolutionary geneticists and paleontologists over human evolution, here specifically considering the controversy over "mitochondrial Eve."
Waters, T. "Almost Human." *Discover*, May 1990. Artist John Gurche reconstructs fossil hominids by adding clay "muscle" and "skin" to replicas of hominid skulls.

Study Questions This popular feature has been expanded and provides a straightforward test of students' recall and comprehension.

Discussion Questions—NEW! These ask students to apply their knowledge to more complex and challenging questions, requiring them to synthesize and apply the information in each chapter.

Suggested Readings The suggested readings have been revised and updated to include the latest research findings. Readings have been selected for student understanding and enjoyment.

BIOLOGY
LIFE ON EARTH

THIRD EDITION

1

An Introduction to Life on Earth

Life on Earth is confined to a thin film encompassing the Earth's surface: the biosphere.

"Viewed from the distance of the moon, the astonishing thing about the earth, catching the breath, is that it is alive. The photographs show the dry, pounded surface of the moon in the foreground, dead as an old bone. Aloft, floating free beneath the moist, gleaming surface of bright blue sky, is the rising earth, the only exuberant thing in this part of the cosmos."

Lewis Thomas *in* The Lives of a Cell *(1974)*

On your way to class tomorrow morning, notice the astonishing array of creatures that live even in a place as domesticated as a city or a college campus. Of course there are grass, bushes, trees, dogs, cats, sparrows, and lots of people. But if you look more closely, you may also encounter insects and earthworms, a spider spinning its web, and mushrooms in the shade beneath a bush. And in your mind's eye, picture the creatures too small to see: yeasts making bread rise at the bake shop, protozoa squirming around in a pond, even the bacteria on your teeth that give you "morning mouth" when you wake up.

Why is there such an astounding diversity of living things? How do they interact with one another? In what ways are bacteria, plants, and people alike, and what makes them so different? What processes must occur in the body of each organism for it to live? These are a few of the questions biologists ask in their quest to understand life on Earth, and that we invite you to ask too, as we explore biology together. For although we are part of the web of life, we humans are more than just another strand; only we can contribute that particular brand of exuberance that comes from understanding the nature of the Earth and its inhabitants, and appreciating the beauty of it all. We are life's way of understanding itself.

The Characteristics of Living Things

To study the nature of life on Earth, we should begin at the beginning: *What is life?* If you look up "life" in a dictionary, you will find definitions such as "the quality that distinguishes a vital and functioning being from a dead body," but you won't find out what that "quality" is, and for good reason. We all have an intuitive understanding of what it means to be alive. Nevertheless, defining "life" is difficult, because living things are so diverse, and nonliving matter is sometimes so lifelike. We can, however, describe some of the characteristics of living things that, taken together, are not shared by nonliving objects.

1. **Living things have a complex, organized structure, based on organic molecules.**

2. **Living things actively maintain their complex structure and their internal environment; a process called homeostasis.**
3. **Living things grow.**
4. **Living things acquire materials and energy from the environment and convert them into different forms.**
5. **Living things respond to stimuli from their environment.**
6. **Living things reproduce themselves, using a molecular blueprint called *DNA*.**
7. **Living things, taken as a whole, have the capacity to evolve.**

Let's explore these characteristics in more detail.

Complexity and Organization

Compared with nonliving matter of similar size, living organisms are highly complex and organized. A crystal of table salt (Fig. 1-1a), for example, consists of just two elements, sodium and chlorine, arrayed in a precise cubical arrangement: organized but simple. The oceans (Fig. 1-1b) contain some atoms of all the naturally occurring elements, but these atoms are randomly distributed: complex but not organized. In contrast, even the tiny water flea (Fig. 1-1c) contains dozens of different elements, linked together in thousands of specific combinations of atoms called **molecules.** These molecules are further organized into ever larger and more complex assemblies to form structures such as eyes, legs, a digestive tract, and a brain.

Life on Earth is comprised of a hierarchy of structures, each based on the one below it and providing the foundation for the one above (Fig. 1-2). An **atom** is the smallest particle of an element that retains the properties of that element. If it were possible to cut up a diamond, which is pure carbon, into ever-smaller pieces, each would still be carbon, until we had separated the diamond into individual carbon atoms: any further division would produce isolated **subatomic particles** that would no longer be carbon. Atoms may combine in specific ways to form assemblies called **molecules;** for example, one carbon atom can combine with two oxygen atoms to form a mole-

(a)

(b)

(c)

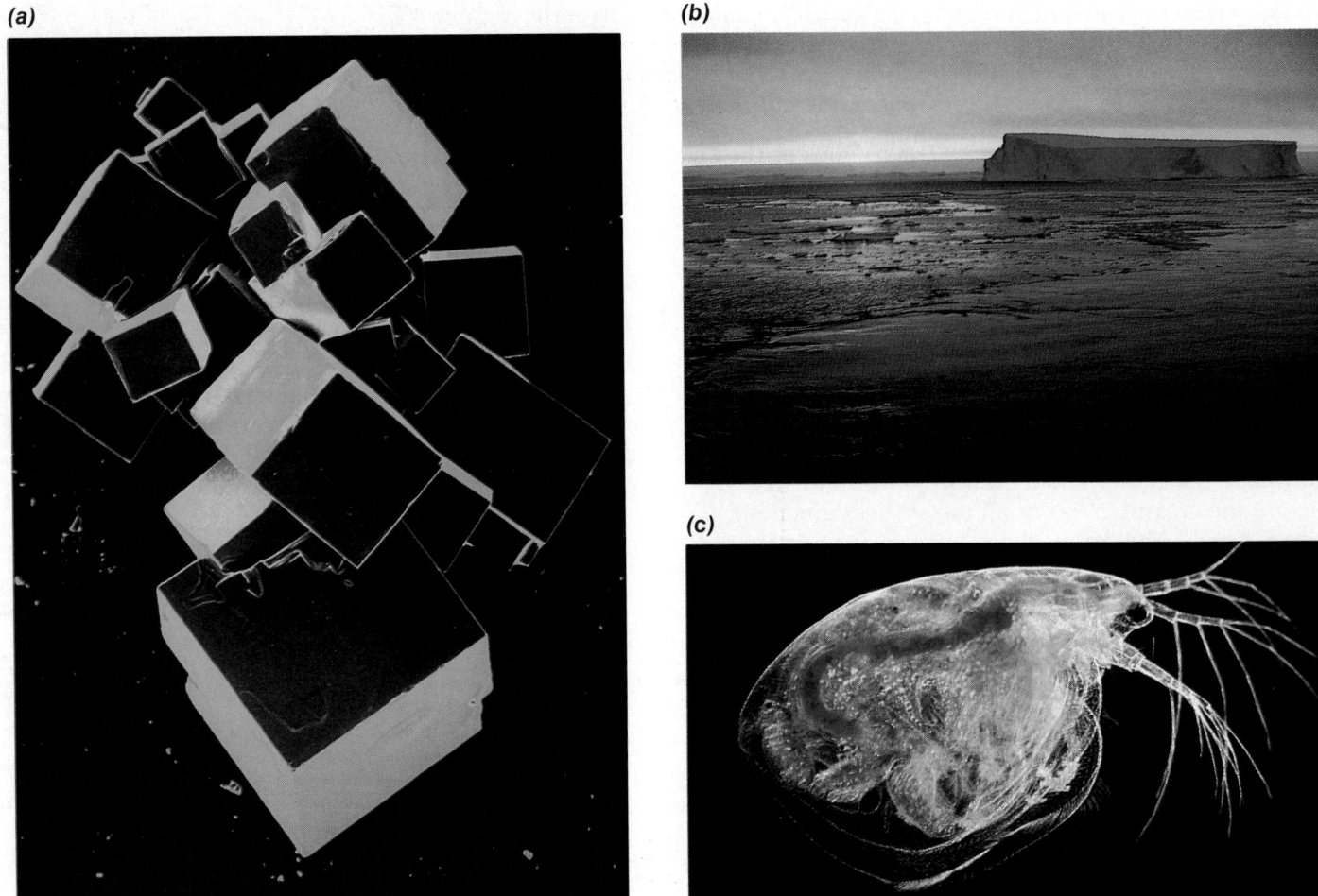

Figure 1-1 Living organisms are complex and organized compared to nonliving matter. **(a)** Each crystal of table salt, sodium chloride, is an exact cube, showing great organization but minimal complexity. **(b)** The water and dissolved materials in the ocean represent nonliving complexity, but there is very little organization. **(c)** Living things have both complexity and organization. The water flea, *Daphnia pulex*, is only 1 millimeter long, yet it has legs, a mouth, a digestive tract, gonads, light-sensing eyes, and even a rather impressive brain considering its size.

cule of carbon dioxide. Although many simple molecules form spontaneously, extremely large and complex molecules are manufactured only by living things.

Just as an atom is the smallest unit of an element, so too the **cell** is the smallest unit of life (Fig. 1-3). Taken at its simplest, a cell consists of genes containing the information needed to control the life of the cell, subcellular structures called **organelles** that act as miniature chemical factories to use the information in the genes and keep the cell alive, and a thin **plasma membrane** that separates the cell from the outside world. Some organisms, mostly microscopic, consist of just one cell, but larger ones are composed of many cells. In these multicellular organisms, cells of similar

Figure 1-2 Levels of organization of matter on Earth. Only the simplest levels of organization are possible without life. At each level, interactions among its components allow the development of the next-higher level of organization.

Term	Definition	Illustration
Biosphere	That part of the Earth inhabited by living organisms; includes both the living and nonliving components	The Earth's surface
Ecosystem	A community together with its nonliving surroundings	snake, antelopes, hawk, bushes, grass, rocks, stream
Community	Two or more populations of different species living and interacting in the same area	snake, antelopes, hawk, bushes, grass
Population	Members of one species inhabiting the same area	herd of pronghorn antelope
Species	Very similar, potentially interbreeding organisms	
Multicellular Organism	An individual living thing comprised of many cells	pronghorn antelope
Organ System	Two or more organs working together in the execution of a specific bodily function	the nervous system
Organ	A structure within an organism usually composed of several tissue types that form a functional unit	the brain
Tissue	A group of similar cells that perform a specific function	nervous tissue
Cell	The smallest unit of life	a nerve cell
Organelle	A structure within a cell that performs a specific funtion	mitochondrion chloroplast nucleus
Molecule	A combination of atoms	water glucose DNA
Atom	The smallest particle of an element that retains the properties of that element	hydrogen carbon nitrogen oxygen
Subatomic Particle	Particles that make up an atom	proton neutron electron

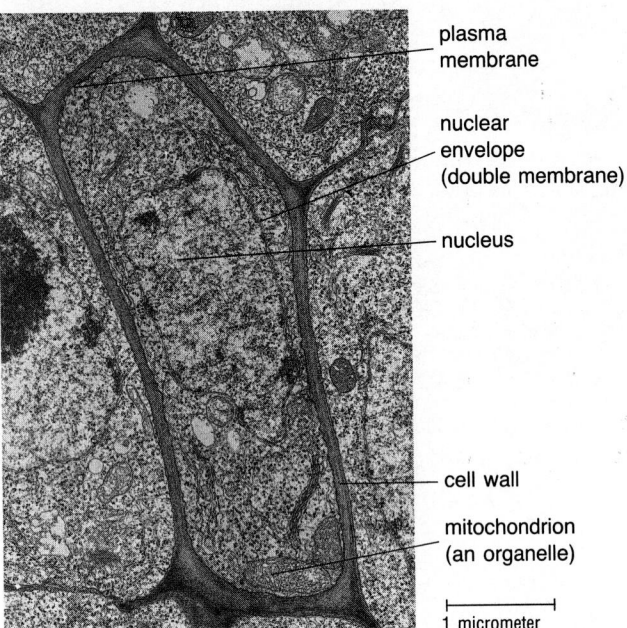

plasma
membrane

nuclear
envelope
(double membrane)

nucleus

cell wall

mitochondrion
(an organelle)

⊢——————⊣
1 micrometer

Figure 1-3 The smallest unit of life is the cell. This electron micrograph clearly shows the plasma membrane that surrounds the cell, separating it from its environment; the nucleus that contains the cell's genes; the cell wall that surrounds and supports the cell while allowing free exchange of materials; and many other specialized structures, called organelles, that perform particular functions in the life of the cell.

type form **tissues,** such as nervous or connective tissue. A variety of tissue types combine to make up a structural unit, called an organ (e.g., the brain). Several organs that collectively perform a single function are called an **organ system;** for example, brain, spinal cord, sense organs, and nerves form the nervous system. All the organ systems functioning cooperatively comprise an individual living thing, the **organism.**

Beyond the individual organisms, a group of very similar, potentially interbreeding organisms comprise a **species.** Members of the same species that live in a given area are considered a **population.** Populations of several different species living and interacting in the same area comprise a **community.** A community plus its nonliving environment constitutes an **ecosystem.** Finally, the entire surface region of the Earth inhabited by living things is called the **biosphere.**

This text will roughly follow the pattern of organization of life on Earth, beginning with atoms and molecules, then moving to cells, organs, and organisms, and concluding with the study of the interactions among organisms that are the focus of ecology.

Homeostasis

Complex, organized structures are not easy to maintain. Whether we consider the molecules of your body or the clothes in your closet, organization tends to disintegrate into chaos unless energy is used to sustain it (we will explore this tendency more fully in Chapter 4). To stay alive, organisms must keep the conditions within their bodies fairly constant, a process called **homeostasis** (derived from Greek words meaning "to stay the same"). Among warm-blooded animals, for example, vital organs such as the brain and heart are kept at a warm, constant temperature despite wide fluctuations in environmental temperature. This is accomplished by a variety of automatic mechanisms, including sweating during hot weather and metabolizing more food when it's cold, and by behaviors such as basking in the sun or even, in the case of people, adjusting the thermostat in the room. In Chapter 25, and in Units IV and V, we will expand on the theme of homeostasis. Of course, not everything stays the same throughout an organism's life. Major changes occur, such as growth and reproduction, but these are not failures of homeostasis. Rather, they are usually specific, genetically programmed parts of the organism's life cycle.

Growth

At some time in its life cycle, every living thing becomes larger—that is, it grows. This is obvious for plants, birds, and mammals, which all start out very small and undergo tremendous growth during their lives. Even single-celled bacteria, however, are small when they are first formed and grow to about double their original size before they divide. In all cases, growth involves the conversion of materials acquired from the environment into the specific molecules of the organism's own body.

Acquisition and Use of Materials and Energy

To maintain homeostasis and to grow, organisms need materials and energy (Fig. 1-4). **Nutrients**—the atoms and molecules of which all organisms are made—may be acquired from the air, water, soil, or other living things. These nutrients must then be incorporated into the molecules of the organism's own body. **Energy**—the ability to do work, including carrying out chemical reactions, growing leaves in the spring, or contracting a muscle—is obtained in one of two basic ways. Plants and some single-celled organisms capture the energy of sunlight and store it

Figure 1-4 Living organisms acquire energy and materials (also called nutrients) from their environment. The plants in this meadow capture energy from the sun and nutrients from the air, water, and soil. In contrast, the grazing cape buffalo extracts both energy and nutrients from the plants. Without a continual supply of solar energy, nearly all life would cease.

in energy-rich sugar molecules, a process called **photosynthesis.** In contrast, neither fungi nor animals can photosynthesize, nor can most bacteria; these must consume the energy-rich molecules contained in the bodies of other organisms. The original sources of these energy-rich molecules are photosynthetic organisms, so nearly all organisms ultimately depend on sunlight for energy.

Responsiveness

Living organisms perceive and respond to stimuli in their internal and external environments. When you feel hungry, you are perceiving internal stimuli including the emptiness of your stomach and low levels of sugars and fats in your blood. You then respond to external stimuli by choosing appropriate objects to eat, such as a piece of pie rather than the plate and fork. Animals, with their elaborate nervous systems and motile bodies, are not the only organisms to perceive and respond to stimuli. The plants on your windowsill grow toward the light, and even the bacteria in your intestine manufacture a different set of digestive enzymes depending on whether you drink milk, eat candy, or both.

Reproduction

The *continuity of life* occurs because organisms reproduce, giving rise to offspring of the same type (Fig. 1-5). The *diversity of life* occurs in part because the offspring are usually somewhat different from their parents, as explained briefly below, and in Chapters 16 and 17. The reason parents produce variable offspring lies in the mechanism by which traits are passed from one generation to the next, through a "genetic blueprint" contained in molecules of DNA.

DNA: The Molecule of Heredity

All known forms of life use a molecule called **deoxyribonucleic acid,** or **DNA,** as the repository of hereditary information (Fig. 1-6). Much of Unit II will be devoted to exploring the structure and function of this remarkable molecule. For now, we should just note that an organism's DNA is its "blueprint" or "molecular instruction manual," a guide both to the construction of its body and, at least in part, to its operation. When an organism reproduces, it passes a copy of its DNA to its offspring. The accuracy of the DNA copy is astonishingly high: only about one mistake occurs for every billion bits of information contained in the DNA molecule. Nevertheless, the occasional errors—called **mutations**—are crucial. Without mutations, all life forms might be identical. Indeed, there is reason to believe that, without mutations, there would be no life at all (see Chapter 19). Mutations in DNA are the ultimate source of genetic variations. These variations, superimposed on a

Figure 1-5 Living organisms reproduce: as it grows, this baby orangutan will resemble, but not be identical to, its parents. The similarity and variability of offspring are crucial to the evolution of life.

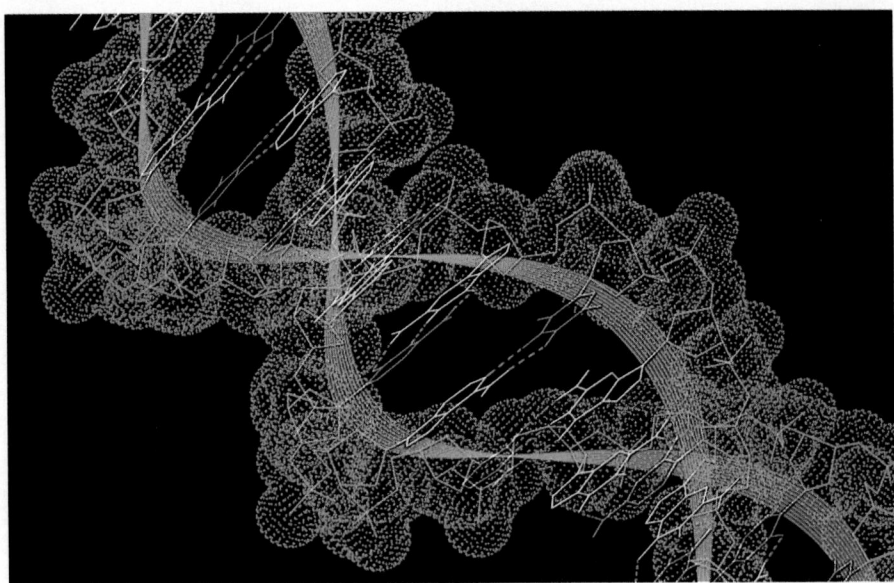

Figure 1-6 A computer-generated model of DNA, the molecule of heredity.

background of overall genetic fidelity, make possible our final property of life, the capacity to evolve.

The Capacity to Evolve

Although the genetic makeup of a single organism remains essentially the same over its lifetime, **the genetic composition of a species as a whole changes over many lifetimes. In other words, the species evolves.** The most important force in evolution is **natural selection,** the process by which organisms with traits that help them cope with the rigors of their environment survive and reproduce more successfully than do others that lack these features. We will return to the concept of evolution later in this chapter.

The Diversity of Life

Although all living things share the general characteristics discussed earlier, evolution has brought forth an amazing variety of life forms. In the following brief description of the features used to classify living organisms and in the essay "A Closer Look at the Five Kingdoms of Life," we introduce you to the diversity of life on Earth. We describe the classification and structures of organisms in detail in Chapters 20 through 24.

Living organisms are grouped into five major categories, called **kingdoms:** Monera, Protista, Fungi, Plantae, and Animalia. Although biologists believe that the groupings reflect evolutionary relationships among organisms, we must classify living things according to the characteristics we can see and measure

today. There are exceptions to any simple set of criteria used to define the kingdoms, but three characteristics are particularly useful: cell type, the number of cells in each organism, and the mode of acquiring energy (Table 1-1).

Cell Type

There are two fundamentally different types of **cells: prokaryotic** and **eukaryotic.** "Karyotic" refers to the **nucleus** of the cell: a membrane-enclosed sac containing the genetic material (see Fig. 1-3). "Eu" means "true" in Greek, and eukaryotic cells are recognized by the presence of a "true," membrane-enclosed nucleus. Eukaryotic cells are larger than prokaryotic cells, and contain a variety of other organelles, many surrounded by **membranes.** Prokaryotic cells do not have a nucleus; their genetic material resides in the cytoplasm of the cell. They are small, only one or two micrometers in length, and lack membrane-bound organelles. "Pro" means "before" in Greek, since many biologists believe that prokaryotic cells evolved before eukaryotic cells (and eukaryotic cells probably evolved from prokaryotic cells, as we will see in Chapter 19). Members of the kingdom Monera consist of prokaryotic cells, while the cells of the other four kingdoms are eukaryotic.

Cell Number

Organisms of the kingdoms Monera and Protista are usually single-celled, or **unicellular,** although a few live in strands or mats of cells with little communica-

Table 1-1 Some Characteristics of the Five Kingdoms

Kingdom	Cell Type	Cell Number	Major Mode of Nutrition
Monera	Prokaryotic	Unicellular	Absorb or photosynthesize
Protista	Eukaryotic	Unicellular	Absorb, ingest, or photosynthesize
Fungi	Eukaryotic	Most multicellular	Absorb
Plantae	Eukaryotic	Multicellular	Photosynthesize
Animalia	Eukaryotic	Multicellular	Ingest

tion, cooperation, or organization among cells. Animals and most plants and fungi are **multicellular;** their lives depend on intimate cooperation among cells.

Energy Acquisition

All organisms need energy to live. Photosynthetic organisms capture energy from sunlight and store it in molecules such as sugars and fats. These organisms, including plants, some monerans, and some protists, are therefore called **autotrophs,** meaning "self-feeders." Organisms that cannot photosynthesize must acquire energy prepackaged in the molecules of the bodies of other organisms; hence these are called **heterotrophs,** meaning "other-feeders." Many monerans and protists, and all animals and fungi, are heterotrophs. Heterotrophs differ in the size of the food they eat. Some, such as bacteria and fungi, absorb individual food molecules; others, including most animals, eat whole chunks of food and break them down to molecules in their digestive tracts.

Biology: The Science of Life

Biology is a science, and its principles and methods are the same as those of any other science. In fact, a basic tenet of modern biology is that living things obey the same laws of physics and chemistry that govern inanimate matter.

Scientific Principles

All scientific inquiry, including biology, is based on a small set of assumptions that we might call scientific principles: natural causality, uniformity in space and time, and common perception.

Natural Causality

Historically, two approaches have been taken to the study of life and other natural phenomena. The first assumes that some events happen through the intervention of supernatural forces beyond our understanding. To the ancient Greeks, Zeus hurled thunderbolts from the sky, while Poseidon made earthquakes and storms at sea. Until relatively recent times, epilepsy was commonly thought to be a visitation from the gods. In contrast, science adheres to the principle of **natural causality: all events can be traced to natural causes.** Today we realize that epilepsy is an organic disease of the brain, in which groups of nerve cells fire uncontrollably.

The principle of causality has an important corollary: the evidence we gather about the causes of natural events is not deliberately distorted to fool us. This corollary may seem foolish, yet not so very long ago people argued that fossils are not evidence of evolution, but were placed in the Earth by God as a test of our faith. If we cannot trust the evidence provided by the universe, then the entire enterprise of science is futile.

Uniformity in Space and Time

A second fundamental principle of science is that **natural laws do not change with time or distance.** The laws of gravity, for example, are the same today as they were a billion years ago, and will hold just as well in Moscow as in New York, or on Mars, for that matter. Uniformity in space and time is especially vital to biology, since many events of great importance to biology happened before humans were around to observe them, such as the evolution of today's diversity of living things. Some people believe that all the different types of living organisms were individually created in the past by the direct intervention of God, a philosophy called **creationism.** As scientists, we freely admit that we cannot disprove this idea. However, creationism is contrary both to natural causality and uniformity in time. The overwhelm-

A CLOSER LOOK
at the Five Kingdoms of Life

Kingdom Monera

Key Characteristics: prokaryotic; unicellular; some autotrophic (photosynthetic), some heterotrophic (absorptive).

The Monera (bacteria) are unicellular, prokaryotic organisms (Fig. E1-1). Some, the cyanobacteria, can photosynthesize, but most cannot. Almost all the nonphotosynthetic bacteria absorb their food molecule by molecule from their surroundings. Most moneran cells are surrounded by a thick, rigid cell wall. In a few, whiplike appendages protrude through the cell wall and enable the cell to move.

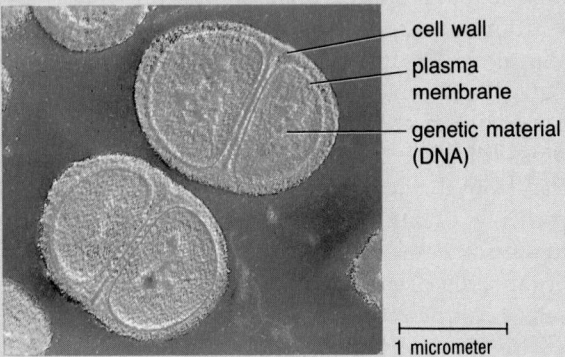

cell wall

plasma membrane

genetic material (DNA)

1 micrometer

Figure E1-1 A color-enhanced electron micrograph of dividing bacteria.

Kingdom Protista

Key Characteristics: eukaryotic; unicellular; some autotrophic (photosynthetic), some heterotrophic (either absorptive or ingestive).

Protists are large, single eukaryotic cells. Some, like *Euglena* and the diatoms, can photosynthesize and are therefore autotrophic, but most are heterotrophic. Some, such as *Amoeba*, can ingest food particles nearly as large as they are, while others, such as the malaria parasite, are limited to absorbing individual food molecules. Most protists, like the *Paramecium* illustrated in Figure E1-2, are mobile, moving with cilia or flagella.

oral groove ("mouth")   food vacuoles   contractile vacuole

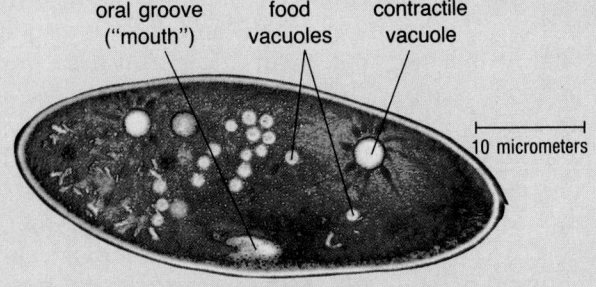

10 micrometers

Figure E1-2 A light micrograph of *Paramecium* shows how complex single eukaryotic cells can be.

Kingdom Fungi

Key Characteristics: eukaryotic, mostly multicellular, heterotrophic (absorptive).

Most fungi are multicellular, often with distinct cell types specialized for feeding or reproduction. The mushroom, for example, consists of a maze of fine underground feeding threads that provide nutrients and energy for the above-ground reproductive stalk and cap (Fig. E1-3). Fungi usually secrete digestive enzymes out of their bodies into their food, which is usually the dead bodies or wastes of plants and animals. These enzymes break down large food particles into separate molecules that can be absorbed by the fungal cells. Most fungi cannot move.

Figure E1-3 An exotic mushroom from Peru.

Kingdom Plantae

Key Characteristics: eukaryotic, multicellular, autotrophic (photosynthetic).

Plants are multicellular, nonmobile eukaryotes with considerable specialization of cell types. The plant kingdom is divided into smaller categories, called divisions, that show different degrees of cell specialization and consequently different habitat requirements.

Algae: The three divisions of algae (Rhodophyta, the red algae; Phaeophyta, the brown algae; and Chlorophyta, the green algae) all live in water, either fresh or marine, and show the least amount of specialization of cells into tissues and organs. Although a few, such as the giant kelps (Fig. E1-4), can grow fairly large, most are quite small.

Figure E1-4 The giant kelp *Macrocystis* forms dense underwater forests along the southern California coast.

Mosses: The mosses and their relatives comprise the division Bryophyta (Fig. E1-5). These are all land plants, but lack true roots for obtaining water and efficient conducting systems for transporting water throughout their bodies. Consequently, bryophytes are quite small and live in moist places.

Figure E1-5 Asexual reproduction in mosses. The capsule is ejecting spores that will drift about on the wind. With luck, a few may land in damp places capable of supporting a new moss plant.

Vascular Plants: This group, also called tracheophytes, includes ferns (Fig. E1-6), conifers, and flowering plants (Fig. E1-7). These are called vascular plants because they contain efficient vessels that transport water and nutrients between roots and above-ground structures. Accordingly, they inhabit drier habitats than do the mosses. Ferns, however, release swimming sperm that thrash their way through a film of water to reach the egg. Conifers and flowering plants enclose their sperm in pollen grains that are carried by wind or animals to their destinations, and consequently they are well adapted for life in drier habitats.

Figure E1-6 (left) Ferns grow in lush profusion on damp forest floors.
Figure E1-7 (right) Flowering plants owe much of their success to a mutually beneficial relationship with insects, in which the flower provides food for the insect, while the insect fertilizes the flower.

Kingdom Animalia

Key Characteristics: eukaryotic, multicellular, heterotrophic.

Animals are multicellular eukaryotes that typically ingest their food in fairly large pieces. Animal bodies contain a wide variety of tissues and organs, composed of specialized cell types. Nearly all animals are motile, at least at some stage in their life cycles. Animals occupy virtually every type of habitat on Earth. The animal kingdom is subdivided into units called phyla (singular, phylum; equivalent to divisions of plants).

Sponges: Sponges (phylum Porifera) are the simplest animals, usually composed of only a few cell types, with minimal coordination among cells. All sponges are aquatic, feeding by pulling water into their bodies through pores (''porifera'' means ''pore bearer'') and filtering out minute plants and animals for food (Fig. E1-8).

Figure E1-8 Sponges draw water into their bodies through tiny pores, filter out their prey, and eject the water out through the large holes visible here.

Cnidarians: The jellyfish (Fig. E1-9), hydras, anemones, and their relatives (phylum Cnidaria) are exclusively aquatic; most are marine, but a few inhabit fresh water. All cnidarians bear stinging tentacles with which they capture prey and bring it to the mouth for ingestion. Since cnidarians have only one opening to their digestive tract, indigestible remnants of a meal have to exit through the mouth.

Figure E1-9 Jellyfish swim constantly by pumping their contractile bells.

Annelids: Segmented worms (phylum Annelida), including earthworms, leeches, and exquisitely colored marine worms (Fig. E1-10), can be found in both aquatic and terrestrial habitats. All, however, require at least damp surroundings, since their moist skin loses water quickly in dry conditions. Annelids, and all the phyla that follow, have digestive tracts with two openings, a mouth and an anus. This allows food to be processed within a series of compartments as it passes through the digestive tract.

Figure E1-10 Annelids are segmented worms, clearly demonstrated by this colorful marine worm that bears a pair of paddles on each segment.

Molluscs: The phylum Mollusca includes animals of very diverse body form, such as snails, clams, scallops, and octopuses (Fig. E1-11). Most molluscs inhabit marine or freshwater environments, although a few snails and slugs live on land. The terrestrial molluscs, like the worms, are not very waterproof, and hence are largely restricted to damp habitats.

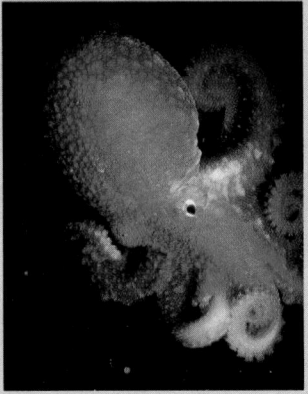

Figure E1-11 The octopus is the brainiest of the invertebrates, with learning abilities almost on a par with the rat.

Arthropods: All the members of the phylum Arthropoda are covered with a stiff suit of armor called an exoskeleton. Like armor, the appendages of the arthropod exoskeleton have joints to allow movement ("arthropoda" means "joint-footed"). In terms of both the number of species and the number of individuals, arthropods are the most successful of the animals, as you will recognize by a short list of representatives: spiders (Fig. E1-12), scorpions, centipedes, millipedes, insects, crabs, shrimp, and lobsters. Arthropods live on land, in the sea, and in fresh water. The weighty exoskeleton, however, limits arthropod size, especially on land; not surprisingly, the largest arthropods inhabit the sea, where the water can support some of the weight of the exoskeleton.

Figure E1-12 A fearsome-looking wolf spider awaits its insect prey, which it will kill with its poisoned fangs. Six of its eight eyes are clearly visible.

Echinoderms: The starfish, sea urchins, sea cucumbers, and sand dollars make up the phylum Echinodermata. The phylum name means "spiny skin," which is especially obvious in sea urchins (Fig. E1-13); however, if you touch a starfish, you will find that the name applies here too. Echinoderms are exclusively marine.

Figure E1-13 The long spines of the red sea urchin deter most potential predators.

Chordates: The familiar vertebrates are members of the phylum Chordata: fish (Fig. E1-14), amphibians (Fig. E1-15), reptiles, birds, and mammals. Chordates are found in fresh water, salt water, and terrestrial habitats. All chordates have an internal skeleton of bone or cartilage, covered with muscles and skin. Internal skeletons are stronger per unit weight than are the external skeletons of arthropods, and not surprisingly all large land animals are chordates.

Figure E1-14 A clownfish cuddles up to its "home" anemone, unharmed by the stinging tentacles that protect the fish from predators.

Figure E1-15 The bright colors of this tropical poison dart frog are a warning: "Don't try to eat me—I'm poisonous." The word "amphibian" means "double life," referring to the fact that amphibians begin life as aquatic larvae, then metamorphose into adults that are usually more terrestrial than the larvae. This adult is carrying its larvae (tadpoles) to water.

ing success of science in explaining natural events through natural causes has led almost all scientists to reject creationism.

Common Perception

A final assumption of science is that we all enjoy common perception: **for practical purposes, all human beings perceive natural events through their senses in fundamentally the same way.** Common perception is, to some extent, a peculiarly scientific principle. Value systems, such as those involved in the appreciation of art, poetry, and music, do not assume common perception. We may perceive the colors on the canvas in a similar way (the scientific aspect of art) but we do not perceive the aesthetic value of the painting identically (the humanistic aspect of art). Moral values also differ radically among people, often owing to their culture or religious beliefs. Since value systems are subjective, not objective, science cannot solve certain types of problems, such as the morality of abortion.

The Scientific Method

Given these assumptions, how do biologists study the workings of life? **Ideally, biology and the other sciences use the scientific method, which consists of four interrelated operations: observation, hypothesis, experiment, and conclusion.** All scientific inquiry begins with an **observation** of the world. Then in a flash of insight, a dream, or more likely after long, hard thought, a **hypothesis** is formulated, that a certain preceding cause produces the observed event. To be useful, a hypothesis must be testable by further observations, or **experiments.** Carrying out these experiments either supports or refutes the hypothesis, and a **conclusion** is drawn about its validity.

Simple experiments test that a single factor, or **variable,** is the cause of a single observation. To be scientifically valid, the experiment must rule out a variety of other variables as the cause of the observation. So scientists design **controls** into their experiments, in which all the variables remain constant. Controls are then compared with the experimental situation, in which only the variable being tested is changed. As an example, let's consider the experiments of the Italian physician Francesco Redi (1621–1697), as he investigated why maggots appear in spoiled meat. Before Redi, this was considered evidence of **spontaneous generation,** the production of living organisms from nonliving matter.

Redi *observed* that flies swarm around fresh meat, and that maggots appear in meat left out for a few days. Redi formed a testable *hypothesis:* the flies pro-

duce the maggots. In his *experiment,* Redi wanted to test just one factor: the access of flies to the meat. Therefore, he took two clean jars and filled them with similar pieces of meat. He left one jar open (the *control* jar) and covered the other with gauze to keep out flies (the *experimental* jar). He did his best to keep all the other variables the same (e.g., type of jar, type of meat, and temperature). After a few days, Redi observed that maggots swarmed over the meat in the open jar, but no maggots appeared in the meat in the covered jar. Redi *concluded* that his hypothesis was correct, and that maggots are produced by flies, and not by the meat itself (Fig. 1-7). Only through controlled experiments could the age-old superstition of spontaneous generation be laid to rest.

It is important to recognize the limitations of the scientific method. In particular, scientists can seldom be sure that they have controlled *all* the variables other than the one they are trying to study. For example, while Redi's gauze kept flies away from the meat, the gauze may also have kept something else away that was the real source of maggots. Or perhaps gauze has some kind of "anti-maggot" effect, like an insecticide. While Redi's experiment supported his hypothesis, it did not preclude all other possible hypotheses. Therefore, **scientific conclusions must always remain tentative, and be subject to revision if new observations or experiments demand it.**

Scientific Theories

In scientific terminology, a **theory** is a hypothesis that has been supported by so many cases that few scientists seriously doubt its validity. To take a trivial example, each time an apple falls *down* to Earth rather than *up* into the sky, the theory of gravitational attraction between masses is supported. Thus, a scientific theory is similar to what we would call a "law" or "fact" in common English usage. However, scientists must be open-minded about even their most cherished theories, and be willing to discard them if new observations render them obsolete.

Probably the foremost theory in biology is evolution. Since its formulation by Charles Darwin and Alfred Wallace in the mid-1800s, the theory of evolution has been supported by fossil finds, geological studies, radioactive dating of rocks, genetics, biochemistry, and breeding experiments. People who refer to evolution as "just a theory" profoundly misunderstand what scientists mean by that term.

Science: A Human Endeavor

Scientists are real people. They are driven by the same ambitions, pride, and fears as other people, and they sometimes make mistakes. As you will read in Chapter 10, ambition played an important role in the discovery by James Watson and Francis Crick of the structure of DNA. Accidents, lucky guesses, controversies with other competing scientists, and, of course, the intellectual powers of individual scientists contribute greatly to scientific advances. To illustrate what we might call "real science," let's consider an actual case.

Chance and the Prepared Mind

When microbiologists study bacteria, they must use pure cultures—that is, plates of bacteria free from contamination by other bacteria, molds, and so on. Only by studying a single type at a time can we learn about the properties of that particular bacterium. Consequently, at the first sign of contamination, a culture is usually thrown out, often with appropriate mutterings about sloppy technique. On one such oc-

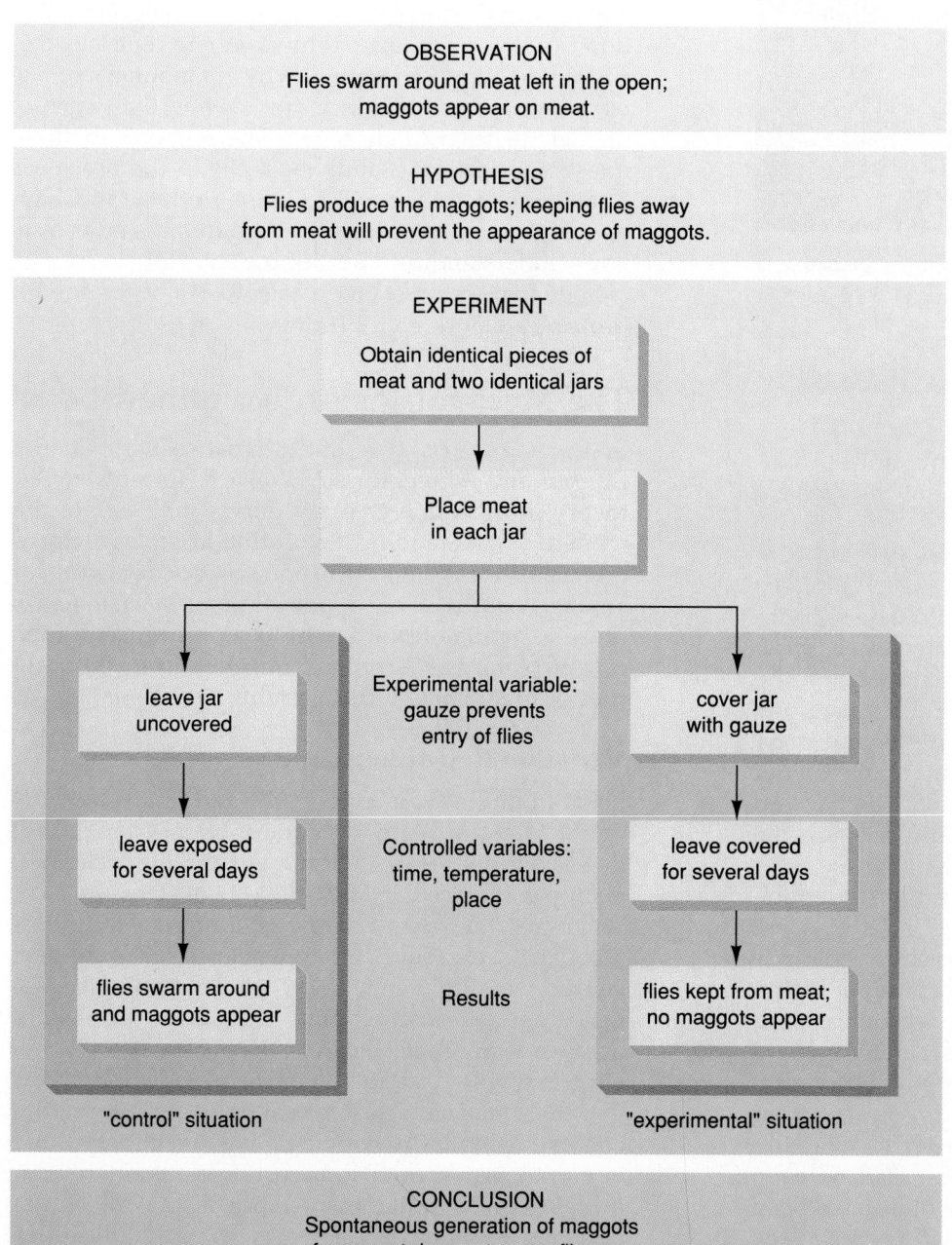

Figure 1-7 The experiments of Francesco Redi.

Figure 1-8 Penicillin diffuses outward from a penicillin-soaked disk of paper, creating a pronounced bald spot in the "lawn" of bacteria on this petri dish.

casion, however, in the late 1920s, Alexander Fleming turned a ruined culture into one of the greatest medical advances in history.

One of Fleming's bacterial cultures became contaminated with a patch of a mold called *Penicillium* (Fig. 1-8). Before throwing out the culture dish, Fleming noticed that *no bacteria were growing near the mold.* Why not? Fleming hypothesized that perhaps the mold releases a substance that kills off bacteria growing nearby. To test this hypothesis, Fleming grew some pure *Penicillium* in a liquid medium, filtered out the mold, and then applied the remaining liquid to an uncontaminated bacterial culture. Sure enough, the bacteria were killed. Further research into these mold extracts resulted in the production of the first antibiotic: penicillin.

Fleming's experiments are a classic in scientific methodology, proceeding from observation to hypothesis to experimental tests of the hypothesis, followed by a conclusion. However, scientific method alone would have been useless without the serendipitous combination of accident and acumen. Had Fleming been a "perfect" microbiologist, he wouldn't have had any contaminated cultures. Had he been less observant, the contamination would have been just another spoiled culture dish. Instead, it was the

beginning of antibiotic therapy for bacterial diseases. As Louis Pasteur said, "Chance favors the prepared mind."

Evolution: The Unifying Concept of Biology

The most important concept in biology is evolution: the theory that modern organisms descended, with modification, from preexisting life forms. In Theodosius Dobzhansky's words, "Nothing in biology makes sense, except in the light of evolution." Why don't snakes have legs? Why are there dinosaur fossils but no living dinosaurs? Why are monkeys so like us, not only in appearance, but even in the structure of their genes and proteins? The answers to these questions, and thousands more, lie in the processes of evolution that we will examine in detail in Chapters 16 through 18. However, evolution is so vital to your understanding and appreciation of the rest of biology that we must take a brief look at its important principles before going further.

The Mechanisms of Evolution

In the mid-1800s two English naturalists, Charles Darwin and Alfred Russell Wallace, formulated the theory of evolution that still forms the basis of our modern understanding. **Evolution arises as a consequence of three natural processes:** *genetic variation* **among members of a population;** *inheritance* **of those variations from parent to offspring; and** *natural selection,* **the survival and enhanced reproduction of organisms with favorable variations.**

Variation and Inheritance

Look around at your classmates and notice how different they are. Although some of this variation is due to differences in environment and life-styles, it is mainly influenced by our genes: most of us could pump iron for the rest of our lives and never develop a body like Arnold Schwartzenegger's. Where does genetic variation come from? As we mentioned earlier, the genetic instructions—the **genes**—of all living organisms are molecules of DNA. Occasionally, the DNA suffers an accident: perhaps radiation strikes the DNA molecule just so, altering its information content. Mistakes in copying DNA during reproduction, although rare, also occur. These accidents and mistakes cause **mutations.** An organism with a mutation may develop characteristics that are a little different from the original. Due to mutations, many of which occurred millions of years ago and have been passed from parent to offspring through countless generations, members of the same species are often slightly different from one another.

Natural Selection

On the average, organisms that best meet the challenges of their environment will leave the most offspring. The offspring will inherit the genes that made their parents successful. Natural selection thus preserves genes that help organisms to flourish in their environment. For example, a mutated gene providing instructions for larger teeth in beavers will be passed from parent to offspring. These offspring will be able to chew down trees more efficiently, build bigger dams and lodges, and eat more bark than "ordinary" beavers. Since these buck-toothed beavers will obtain more food and better shelter than their smaller-toothed relatives, they will probably raise more offspring as a result. The offspring will inherit their parents' genes for larger teeth. Over time, smaller-toothed beavers will become increasingly scarce, and after many generations all beavers will have large teeth. Structures, physiological processes, or behaviors that aid survival and reproduction in a particular environment are called **adaptations.** Most of the features that we admire so much in our fellow life forms, such as the long clean limbs of deer, the wings of eagles, and the mighty columns of redwood trunks, are adaptations molded by millions of years of mutation and natural selection.

In the long run, however, what helps an organism to survive today may become a liability tomorrow. If environments change—for example, as ice ages come and go—then the genetic makeup that best adapts organisms to their environment will also change over time. When new mutations that increase the fitness of an organism in the altered environment happen to appear, these mutations in their turn will spread throughout the population.

Over millenia, the interplay of environment, variation, and selection results in evolution: the modification of the genetic makeup of species. In environments that are reasonably constant, such as the oceans, some well-adapted forms persist relatively unchanged, and are sometimes called "living fossils." For example, sharks (Fig. 1-9) have retained essentially the same body form for tens of millions of years, as their sleek shape, acute sense of smell, and formidable teeth have made them predators *par excellence.* In changing environments, some species do not

experience the genetic changes that allow them to adapt. These species become extinct; the dinosaurs are a classic example. The petrified forests of Arizona (Fig. 1-10) are the remnants of giant trees that could not withstand the conversion of tropical lowland to high desert. Other species experience chance mutations that adapt them to meet new challenges. Fossil horses show gradual changes over time, caused by mutations that happened to produce adaptations to their changing environment (Fig. 1-11).

Figure 1-9 This tiger shark, photographed in the clear waters of the Bahamas, possesses features that have characterized sharks for millions of years: streamlined shape, a long, powerful tail, an acute sense of smell, and rows of sharp teeth.

Figure 1-10 Logs in Petrified Forest National Park. The petrified trunks littering the ground are remnants of vast forests of trees that lived here about 160 million years ago. As their habitat gradually changed from lowland flood plain to high desert, the trees became extinct.

Figure 1-11 An artist's rendering of the evolution of the horse, starting from *Hyracotherium* (also called *Eohippus*) and leading through several intermediate forms to the modern horse, *Equus*. The changing backgrounds through which the horses run depict the changing environments that molded horse evolution, from forests through open woodlands to hard, flat plains. See Chapter 16 for more on horse evolution.

R̄eflections on The Continuing Journey

Some people regard science as a "dehumanizing" activity, feeling that too deep an understanding of the world robs us of vision and awe. Nothing could be further from the truth, as we repeatedly discover anew in our own lives. A few years ago, we watched a bee foraging at a spike of lupines. Lupines have a complicated structure, with two petals on the lower half of the flower enclosing the pollen-laden male reproductive structures (stamens) and sticky female pollen-capturing stigma (Fig. E1-16). In young lupine flowers, the weight of a bee pushing on these petals compresses the stamens, extruding pollen onto the bee's abdomen. In older flowers, the stigma protrudes through the lower petals, and when a pollen-dusted bee visits, it usually leaves behind a few grains of pollen.

Did our new-found insights into the functioning of lupine flowers detract from our appreciation of them? Far from it. Rather, we now looked on lupines with new delight, understanding something of the interplay of form and function, bee and flower, that shaped the evolution of the lupine. A few months later we ventured atop Hurricane Ridge in Olympic National Park, where the alpine meadows fairly burst with color in August (Fig. E1-17). As we crouched beside a wild lupine, an elderly man stopped to ask what we were looking at so intently. He listened with interest as we explained the structure to him, and then went off to another patch of lupines to watch the bees foraging. He too felt the increased sense of wonder that comes with understanding.

We have tried to convey to you that dual sense of understanding and wonder throughout this text. We have also tried to emphasize that biology is not a completed work, but an exploration that we have really just begun. Lewis Thomas, physician and natural philosopher, has stated our situation in his usual elegant way: "The only solid piece of scientific truth about which I feel totally confident is that we are profoundly ignorant about nature. Indeed, I regard this as the major discovery of the past hundred years of biology . . . but we are making a beginning."

We cannot urge you strongly enough, even if you are not contemplating a career in biology, to join in the journey of biological discoveries throughout your life. Don't look upon biology as just another course to take, just another set of facts to memorize. Biology can be much more than that. It can be a pathway to a new understanding of yourself and the living Earth around you.

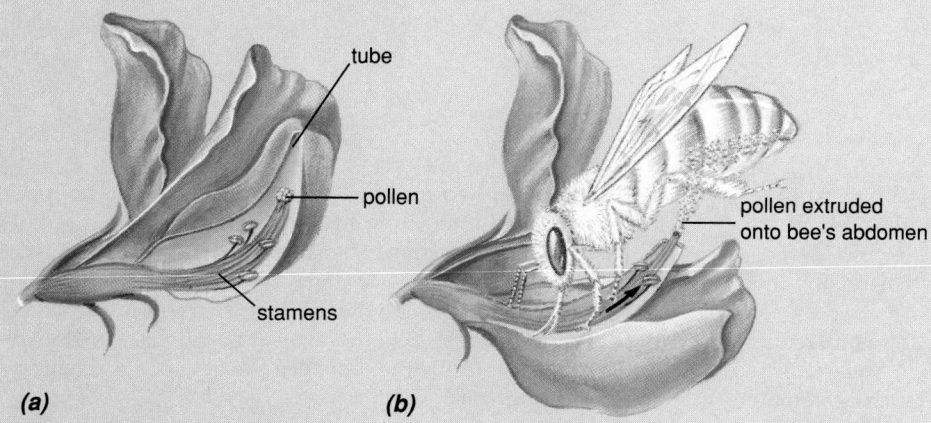

Figure E1-16 Lupines, like many members of the pea family, have complex flowers. **(a)** The reproductive structures are enclosed within the lower petals. In young lupine flowers, the lower petals form a tube within which the stamens fit snugly like the plunger of a bicycle pump. The stamens shed pollen within the tube. **(b)** When the weight of a foraging bee pushes on the lower petals, the stamens are thrust forward, extruding pollen out the end all over her abdomen. Some pollen adheres to the abdomen and may come off on the sticky, pollen-receiving stigma of the next flower visited by the bee.

Figure E1-17 Wild lupines and subalpine fir on Hurricane Ridge in Olympic National Park. Thousands of people visit Hurricane Ridge each summer to gaze in awe at Mt. Olympus, but few bother to investigate the wonders at their feet.

SUMMARY OF KEY CONCEPTS

The Characteristics of Living Things

Living organisms usually possess the following characteristics: their structure is complex and organized; they maintain homeostasis; they grow; they acquire energy and materials from the environment; they respond to stimuli; they reproduce; and they have the capacity to evolve.

The Diversity of Life

Living organisms can be grouped into five major categories, called kingdoms: Monera, Protista, Fungi, Plantae, and Animalia. Some features used to classify organisms into kingdoms are the type of cell the organism possesses, the number of cells per organism, and the mode of acquiring energy.

Cell Type: Eukaryotic cells have their genetic material enclosed within a membrane-bound nucleus. Prokaryotic cells do not have a nucleus.

Cell Number: Organisms may consist of a single cell (unicellular) or of many cells bound together and working cooperatively (multicellular).

Energy Acquisition: Autotrophic organisms obtain energy from inanimate sources, usually converting sunlight energy into energy-rich molecules through photosynthesis. Heterotrophic organisms obtain energy by eating energy-rich molecules (food) synthesized in the bodies of other organisms. The food may be eaten in large chunks (ingestion) or may be absorbed molecule-by-molecule from the environment (absorption). The features of the five kingdoms are summarized in Table 1-1.

Biology: The Science of Life

Biology is based on the scientific principles of natural causality, uniformity in space and time, and common perception. These principles are assumptions that cannot be directly proven, but that are validated by experience.

Knowledge in biology is acquired through the application of the scientific method. First, an observation of the world is made. A hypothesis is formulated, which suggests a natural cause for the observation. The hypothesis is used to predict the outcome of further observations or experiments. A conclusion is then drawn about the validity of the hypothesis.

Evolution: The Unifying Concept of Biology

Evolution is the theory that modern organisms descended, with modification, from preexisting life forms. Evolution occurs as a consequence of (1) genetic variation among members of a population caused by mutation; (2) inheritance of those variations by offspring; and (3) natural selection of the variations that best adapt the organism to its environment.

GLOSSARY

adaptation: a specific structure, physiological mechanism, or behavior that promotes the survival and/or reproduction of an organism.

atom: the smallest particle of an element that retains the properties of the element.

autotroph (aw'-tō-trōf): an organism that can manufacture all its high-energy organic molecules (e.g., sugars, proteins) from simple inorganic molecules (such as carbon dioxide, water, and minerals), using a nonliving energy source (usually sunlight); an organism that does not have to consume organic molecules as food.

biosphere (bī'-ō-sfēr): that part of the Earth inhabited by living organisms; includes both the living and nonliving components.

causality: the scientific principle that natural events occur as a result of preceding natural causes.

cell: the smallest unit of life, consisting, at a minimum, of an outer membrane enclosing a watery medium containing organic molecules, including genetic material composed of DNA.

community: two or more populations of different species living and interacting in the same area.

control: that portion of an experiment in which all possible variables are held constant; in contrast to the "experimental" portion, in which a particular variable is altered.

deoxyribonucleic acid (dē-ox-ē-rī-bō-nū-klā'-ik; DNA): a type of molecule that encodes the genetic information of all living cells.

ecosystem (ēk'-ō-sis-tem): A community together with its nonliving surroundings.

energy: the ability to do work.

eukaryotic (ū-kar-ē-ot'-ik): referring to the type of cells that have their genetic material enclosed within a membrane-bound nucleus, contain other membrane-bound organelles, and are usually larger than prokaryotic cells.

evolution: Any change in the overall genetic composition of a population of organisms from one generation to the next.

gene: the physical unit of inheritance; part of a DNA molecule.

heterotroph (het'-er-ō-trōf): an organism that cannot use inanimate energy sources (such as sunlight) to synthesize all its energy-rich organic molecules; hence heterotrophs must acquire energy-rich organic molecules manufactured by other living organisms as food.

homeostasis (hō-mē-ō-stā'-sis): maintenance of a stable internal environment in the face of changes in the external environment.

hypothesis (hī-poth'-e-sis): in science, a supposition based on previous observations, which is offered as an explanation for an event, and used as the basis for further observations or experiments.

inheritance: the transmission of inborn characteristics from parent to offspring.

kingdom: the most inclusive category in the classification of organisms. The five kingdoms are Monera, Protista, Fungi, Plantae, and Animalia.

membrane: in cells, a thin sheet of lipids and proteins that surrounds the cell or its organelles, separating them from their surroundings.

molecule: a relatively stable combination of atoms.

multicellular: a term describing an organism whose body consists of many cells.

mutation: a change in a gene; usually refers to a genetic change that is significant enough to change the appearance or function of the organism.

natural selection: the differential survival and reproduction of organisms due to environmental forces, resulting in the preservation of favorable adaptations. Usually, natural selection refers specifically to differential survival and reproduction based on genetic differences among individuals.

nucleus (nū-klē-ús): the membrane-bound organelle of eukaryotic cells that contains the cell's genetic material.

nutrients: the atoms and molecules that living organisms need to acquire in their diets.

organ: a structure of an organism (e.g., intestine), usually composed of several tissue types, that is organized into a functional unit.

organelle (ōr-ga-nel'): a structure inside a cell that performs a specific function.

organism (ōr'-ga-niz-em): an individual living thing.

organ system: two or more organs working together in the execution of a specific bodily function (e.g., digestive tract).

photosynthesis: The series of chemical reactions in which the energy of light is used to synthesize high-energy molecules.

plasma membrane: the outer membrane of a cell, enclosing its contents and separating them from its environment.

population: a group of organisms of the same species living in the same area and interbreeding.

prokaryotic (prō-kar-ē-ot'-ic): referring to the type of cell characteristic of members of the kingdom Monera (bacteria). Prokaryotic cells do not have their genetic material enclosed within a membrane-bound nucleus, lack other membrane-bound organelles, and are usually smaller and simpler in structure than eukaryotic cells.

spontaneous generation: the proposal that living organisms can arise from nonliving matter.

subatomic particle: the particles of which atoms are made: electrons, protons, and neutrons.

theory: in science, an explanation for natural events that is based on a large number of observations and is in accord with scientific principles, especially causality.

tissue: a group of (usually similar) cells that together carry out a specific function.

unicellular: a term describing an organism whose body consists of a single cell.

variable: a condition, particularly in a scientific experiment, that is subject to change.

STUDY QUESTIONS

1. What are seven characteristics that can distinguish living from nonliving things? Which would be particularly applicable in distinguishing a living organism from one that has just died? For each characteristic, name a nonliving thing that comes close to possessing it.
2. Name the five kingdoms of living organisms. For each, list the type of cell, whether its members are unicellular or multicellular, and their method of acquiring energy.
3. Define homeostasis. Why must organisms continually acquire energy and materials from the external environment to maintain homeostasis?
4. Distinguish between prokaryotic and eukaryotic cells.
5. Distinguish between autotrophic and heterotrophic organisms, and give an example of each. Which kingdoms are composed exclusively of heterotrophic organisms?

DISCUSSION QUESTIONS

1. Describe the scientific method. In what ways do you use the scientific method in everyday life, for instance in finding out why your car won't start?
2. What is evolution? Briefly describe how evolution occurs.
3. Distinguish between a scientific theory and a hypothesis. Explain how each is used by scientists.

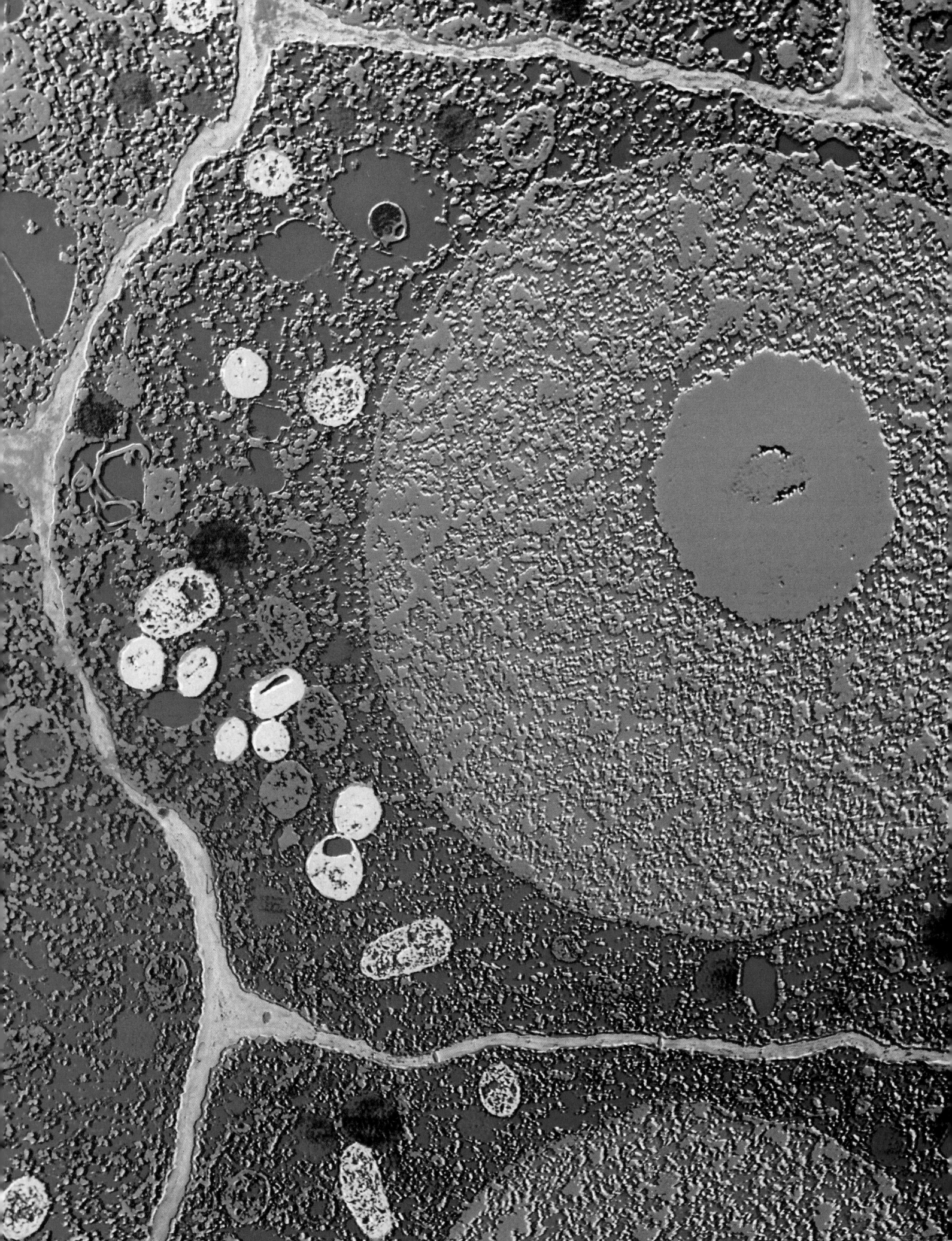

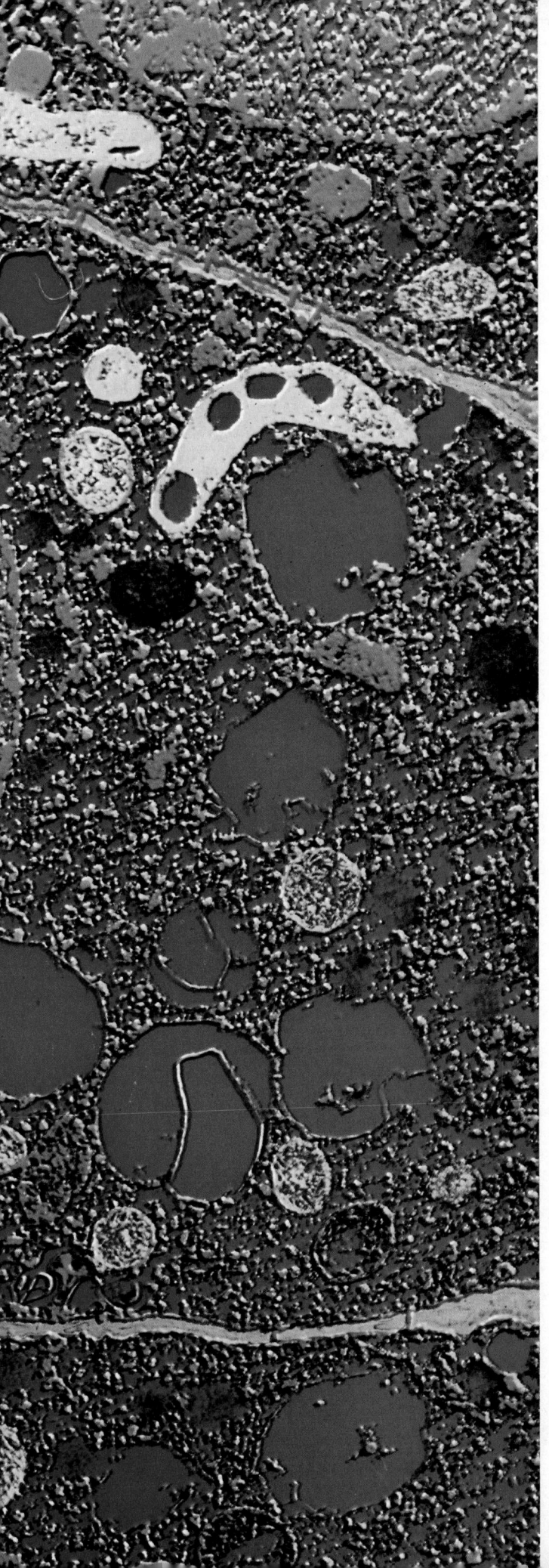

UNIT I

The Life of a Cell

◀ A colorized electron micrograph highlights the complexity of living cells. This corn root cell shows several features typical of plant cells: the nucleus with its darker nucleolus (orange); the cell wall (pale green) surrounds the plasma membrane that separates the cytoplasm from the extracellular environment; and several small vacuoles (yellow).

2

The Chemistry of Life.
I. Atoms and Molecules

A few centuries ago, physics, chemistry, and biology were very discrete sciences. Physicists studied the movements of large objects here on Earth and of the planets about the sun; chemists sought to define the properties of matter, and even tried to transmute base metals into gold; and biologists collected and classified plants and animals. The past 200 years, however, have witnessed the steady merging of the three sciences. Physicists have discovered a strange world of subatomic particles and forces that determine the properties of atoms; chemists study physics so they can understand how atoms interact to form molecules; and biologists learn organic chemistry to help them study the large, complex molecules that form the bodies of living things.

The study of life on Earth often involves the study of the molecules that compose living organisms. How does photosynthesis convert the energy of sunlight into the energy of sugar molecules? What is the structure of the plasma membrane, and how does it control the movement of materials into and out of a cell? How do nerve cells in your brain communicate with one another? What causes cancer? What is acid rain, and why does it endanger the life of many freshwater lakes? To answer these questions, you must first learn a little about energy and matter, the properties of atoms, and how atoms interact with one another to form molecules.

Matter and Energy

Technically speaking, matter and energy are interchangeable, as expressed by Albert Einstein's famous equation $E = mc^2$. However, for the chemical reactions that occur within living organisms, we can treat matter and energy as quite distinct from one another: **matter** is the physical material of the universe, while **energy** is the capacity to do work, usually manifested by moving pieces of matter around.

Forms of Energy

Energy can exist in several forms, and may be converted from one form to another. The two major categories of energy are kinetic energy and potential energy. **Kinetic energy** is the energy of movement. This includes not only movement of large objects, but also other forms of movement such as electrical energy (movement of electrons) and heat (movement of atoms and molecules). **Potential energy** is "stored" energy that can be released as kinetic energy under the right conditions. A ripe coconut high in a palm tree has potential energy owing to its location. Its potential energy is converted into kinetic energy when it falls. The food you eat has chemical potential energy, some of which is converted into kinetic energy of movement and heat when you run (Fig. 2-1).

Transfers of energy regulate all interactions among

Figure 2-1 When 20,000 people run the New York Marathon each year, they do more than just make the bridges quake. They also collectively convert over 50,000,000 kilocalories of food into movement, heat, and sore muscles—that's equivalent to burning off over 14,000 pounds of fat or eating about 30,000 pounds of chocolate cake.

Table 2-1   Common Elements Important in Living Organisms

Element	Symbol	Atomic Number[a]	Percent in Universe[b]	Percent in Earth[b]	Percent in Human Body[b]
Hydrogen	H	1	91	0.14	9.5
Helium	He	2	9	Trace	Trace
Carbon	C	6	0.02	0.03	18.5
Nitrogen	N	7	0.04	Trace	3.3
Oxygen	O	8	0.06	47	65
Sodium	Na	11	Trace	2.8	0.2
Magnesium	Mg	12	Trace	2.1	0.1
Phosphorus	P	15	Trace	0.07	1.0
Sulfur	S	16	Trace	0.03	0.3
Chlorine	Cl	17	Trace	0.01	0.2
Potassium	K	19	Trace	2.6	0.4
Calcium	Ca	20	Trace	3.6	1.5
Iron	FE	26	Trace	5.0	Trace

[a] Atomic number = number of protons in the nucleus.
[b] Approximate percentage of atoms of this element, by weight, in the universe, in the earth's crust, and in the human body.

bits of matter. In this chapter, we examine the structure of matter, from atoms to molecules. Chapter 3 focuses on the molecules that make up the bodies of living organisms, and Chapter 4 examines how the flow of energy among atoms and molecules governs the chemical reactions that occur within cells.

The Structure of Matter

An **element,** such as carbon or oxygen, is a substance with specific properties that can neither be broken down nor converted to different substances by ordinary chemical reactions. Although there are 92 naturally occurring elements in our universe, most are quite rare, and only a relative handful are essential to life on Earth. Table 2-1 lists the most common elements in the universe, the Earth, and the human body.

A **compound** is a substance composed of precise proportions of two or more elements, joined together in a specific geometric pattern. Carbon dioxide, for example, consists of one part carbon and two parts oxygen, arranged like this:

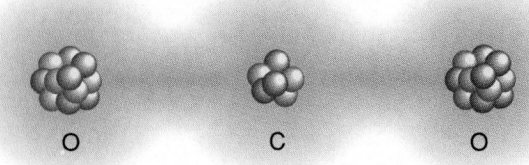

A **mixture** is also composed of two or more elements, but in variable proportions. For example, a can of soda is a mixture that contains, among other things, variable amounts of carbon dioxide (which makes it fizz), depending on how long ago you popped open the top.

Atomic Structure

If you took a diamond (a form of carbon) and cut it into pieces, each piece would still be carbon. If you could continue to make finer and finer divisions, you would eventually produce a pile of submicroscopic carbon **atoms,** the smallest possible particles of the element carbon. Atoms themselves, however, are composed of two parts: a central **nucleus** and one or more outer **orbitals** (Fig. 2-2). Within the nucleus and orbitals reside subatomic particles. The nucleus contains heavy, positively charged **protons** and equally heavy but uncharged **neutrons.** The orbitals contain light, negatively charged electrons. An isolated atom has an equal number of electrons and protons, and is therefore electrically neutral.

Nuclei and orbitals play complementary roles in the structure and function of atoms. Nuclei resist disturbances by outside forces. Ordinary sources of energy, such as heat, electricity, and light, hardly affect them at all. The stability of its nucleus is the reason why a carbon atom remains carbon whether it is part of a diamond, carbon dioxide, or sugar. Electron or-

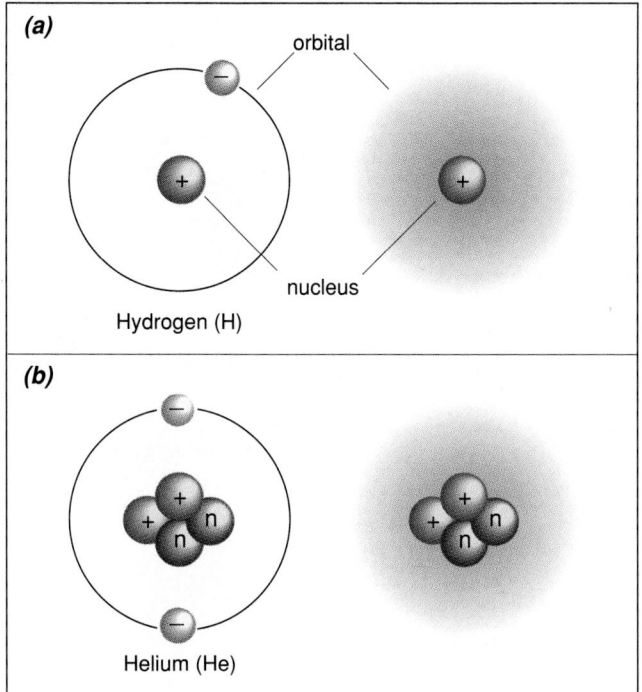

Figure 2-2 Structural representations of the two smallest atoms, hydrogen **(a)** and helium **(b)**. On the left are shown "planetary" models, in which the electrons ($\ominus$) are imagined as miniature planets, circling in precisely defined orbits around a nucleus that contains protons ($\oplus$) and neutrons ($\textcircled{n}$). On the right are more accurate "electron cloud" models. Electrons may be found almost anywhere, but most of the time they stay within the shaded region, which represents the electron orbital. The denser the dots are, the higher the probability that electrons will be found at that place. Although electron cloud models of atoms are more accurate, planetary models are much easier to visualize and understand, and we will use them most of the time.

bitals, however, are dynamic: as you will soon see, atoms interact with one another by gaining, losing, or sharing electrons.

The Nucleus

The number of protons in the nucleus, called the **atomic number,** is characteristic of each element. For example, every hydrogen atom has one proton in its nucleus, every carbon atom has six, and every oxygen atom has eight (see Table 2-1). Different atoms of a given element may, however, have different numbers of neutrons. Atoms with the same number of protons (i.e., atoms of the same element) but with different numbers of neutrons are called **isotopes.** All the isotopes of an element are virtually identical in their chemical reactions, but may vary in their physi-

cal properties. The nuclei of some isotopes spontaneously disintegrate, releasing radioactivity as they do. Radioactive isotopes are extremely useful as "labels" in studying biological processes (see Methods in Biology: The Use of Isotopes in Biology and Medicine).

Electron Orbitals, Shells, and Energy Levels

At one time, it was believed that electrons were like tiny planets, circling the nucleus in well-defined orbits. However, electrons are really much harder than that to pin down. They zip about in regions of space outside the nucleus called orbitals (see Fig. 2-2). At different times a given electron may be found anywhere within its orbital.

Electrons repel one another, owing to their negative electrical charge (as you may know, like charges repel and opposite charges attract). Because of this mutual repulsion, only two electrons can fit into a single orbital, and only a few orbitals can fit at a given distance from the nucleus. Large atoms accommodate many electrons by having layers of orbitals, called **electron shells,** at increasing distances from the nucleus (Fig. 2-3). Electrons in shells closest to the nucleus have the least energy and those farthest away have the most. For this reason, electron shells are often called **energy levels.**

The electron shell closest to the nucleus, with the lowest energy level, has only one orbital and therefore can hold two electrons. The second shell has four orbitals, and can hold up to eight electrons. The electrons in an atom normally occupy the shells closest to the nucleus. Thus a carbon atom, with six electrons, has two electrons in the first shell, closest to the nucleus, and four electrons in its second shell (see Fig. 2-3).

Why do we speak of energy *levels* and *shells*? Because orbitals are found only at specific distances from the nucleus—distances that correspond to discrete energy levels. Most people aren't used to thinking about energy occurring in steps, but perhaps an analogy with gravity can help. Picture yourself ascending a flight of stairs. You must spend a certain amount of kinetic energy to climb each step, and no intermediate amount will suffice. In other words, you can't lift your foot half a step and stand between steps. The energy used in climbing each step is stored as the potential energy of a higher position relative to the bottom of the stairs. This potential energy can be transformed back into kinetic energy by jumping down the steps: the higher up you start, the harder you land.

Energy levels in atoms are analogous to the staircase (Fig. 2-4). It takes energy to force an electron out of a low-energy shell into a higher-energy shell. An

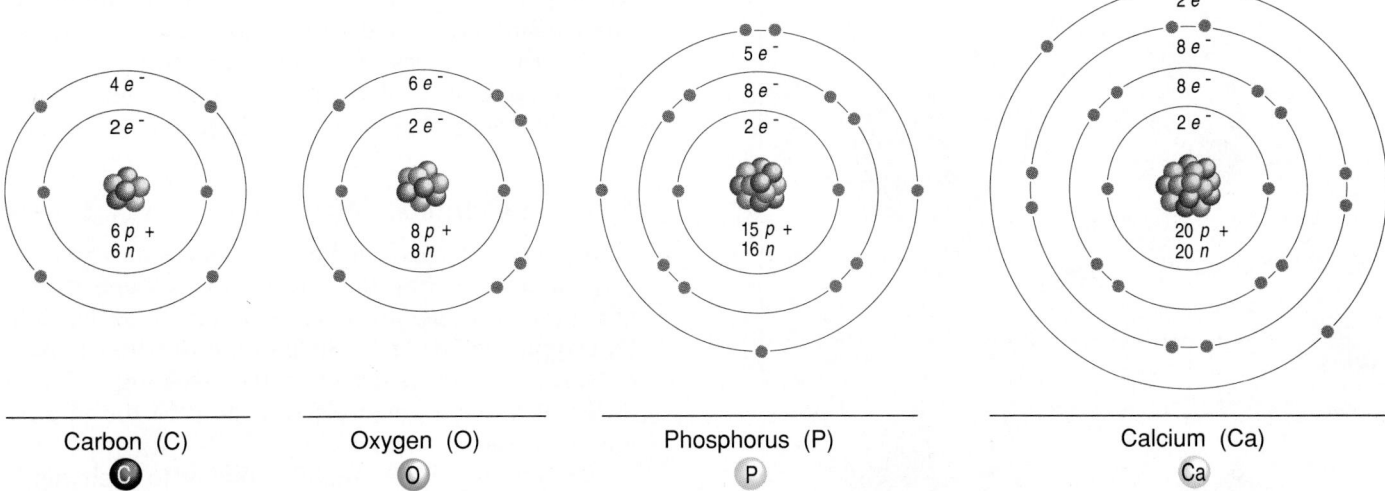

Carbon (C)   Oxygen (O)   Phosphorus (P)   Calcium (Ca)

Figure 2-3 Electron shells in atoms. Most biologically important atoms have at least two shells of electrons. The first shell, closest to the nucleus, can hold two electrons, while the next three shells usually hold a maximum of eight electrons.

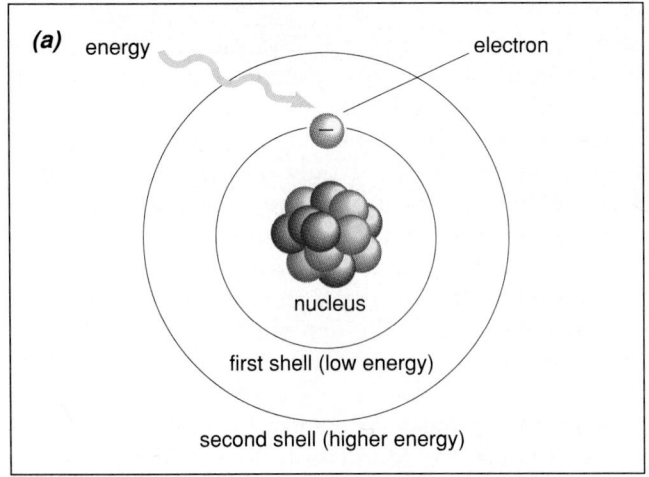

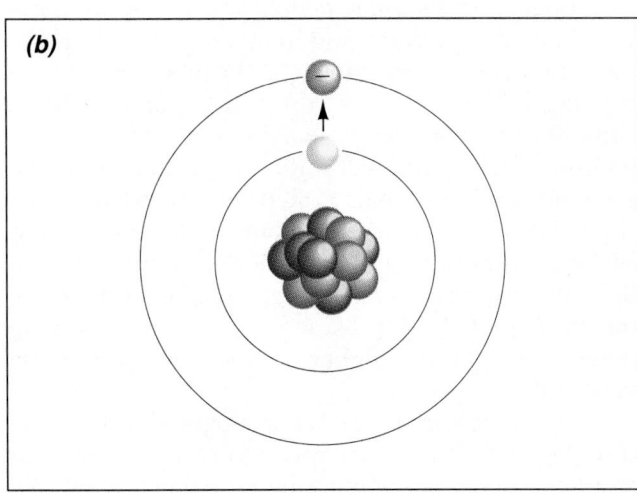

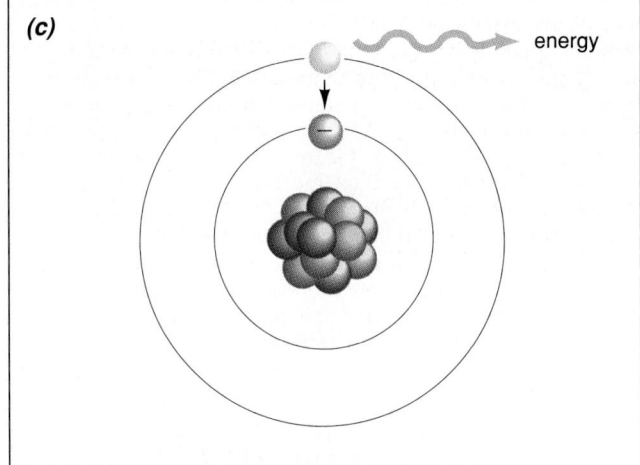

Figure 2-4 Energy absorption and emission by an atom. The electron shells of an atom are found at discrete energy levels; the lowest energy level is the one closest to the nucleus.
(a) Energy (i.e., electrical energy or light) enters an atom and is absorbed by an electron.
(b) Only the specific amounts of energy that boost the electron from a low-energy inner shell to a higher-energy outer shell can be absorbed. Intermediate amounts of energy are not absorbed.
(c) The electron spontaneously returns to a low-energy shell. The "extra" energy is released, and may drive energy-requiring chemical reactions (for example, photosynthesis), or be emitted, often in the form of light.

METHODS IN BIOLOGY

The Use of Isotopes in Biology and Medicine

As you read this text, you will encounter a lot of statements that may cause you to wonder: "How do they know *that*?" How do biologists know that DNA is the genetic material of cells (Chapter 12)? How do paleontologists measure the ages of fossils (Chapter 16)? How do botanists know that sugars made in plant leaves during photosynthesis are transported to other parts of the plant in the sieve tubes of phloem (Chapter 27)? These discoveries, and many more, have been possible only through the use of isotopes.

Atomic nuclei contain protons and neutrons. Every atom of a particular element has the same number of protons, but the number of neutrons may vary. An isotope is named after the element and the total number of protons plus neutrons in its nucleus. A chlorine atom with 17 protons and 18 neutrons is therefore called ^{35}Cl (pronounced "chlorine-35"), while one with 17 protons and 19 neutrons is called ^{36}Cl.

Neutrons don't affect the chemical reactivity of an atom very much, but they do make their presence felt in other ways. First, neutrons add to the mass of an atom. Sophisticated instruments, such as mass spectrometers, can detect the difference in mass between ^{35}Cl and ^{36}Cl. Second, the ratio of neutrons to protons helps to determine how stable a nucleus is. Nuclei with "too many" neutrons break apart spontaneously, often emitting radioactive particles in the process. Instruments such as Geiger counters can detect the radioactivity, and therefore detect the presence of radioactive isotopes.

A particularly fascinating and medically important application of radioactive isotopes is positron emission tomography, more commonly known as PET scans (Fig. E2-1). In a common application of PET scans, a subject is given glucose that has been labeled with a harmless radioactive isotope of fluorine, ^{18}F. When the nucleus of ^{18}F decays, it emits a positron (the antimatter equivalent of an electron). The positron almost immediately collides with a nearby electron, annihilating both particles and converting their mass into two bursts of energy that travel in opposite directions along the same line. Energy detectors arranged in a ring around the subject record the nearly simultaneous arrival of the two energy bursts. Powerful computers register which detectors were struck and calculate the location within the subject at which the decay must have occurred. By recording the arrival of many thousands of energy pairs, the computer generates a map of the frequency of ^{18}F decays at various locations within the subject. Since each labeled

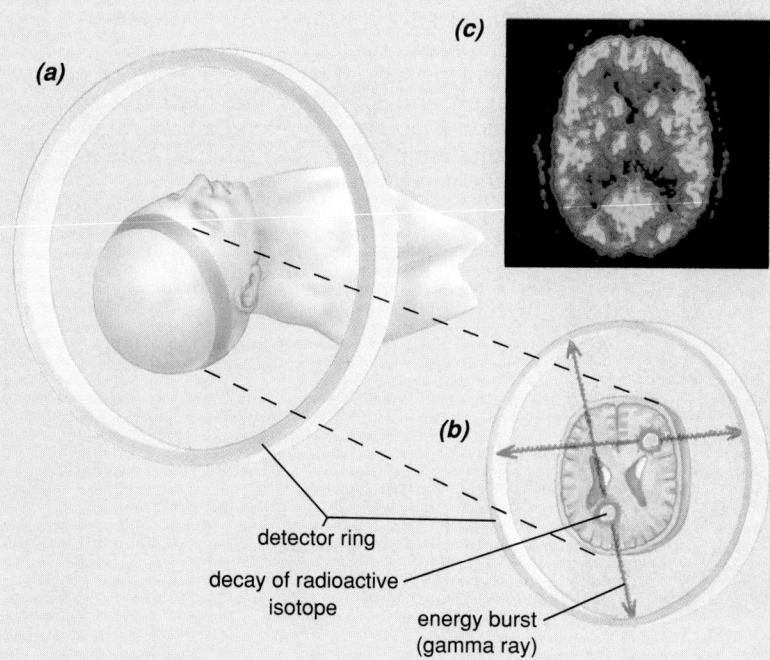

(a)

(b)

(c)

detector ring

decay of radioactive isotope

energy burst (gamma ray)

Figure E2-1 How positron emission tomography (PET scans) work. **(a)** A subject is given a substance, often glucose or water, containing a harmless radioactive isotope that emits positrons. The subject's head is inserted within a ring of detectors and very precisely aligned. The geometry of the radioactive decay and the detector ring, along with sophisticated computerized analysis, combine to provide an image of the "slice" of the head that lies within the plane of the detector ring. **(b)** Inside the "head slice," radioactive decay leads to the production of high-energy bursts of gamma rays that activate the detectors. Computer analysis locates the site of the radioactive decay. **(c)** The numbers of decays at each location are color-coded and projected onto a television screen: usually red and yellow mean many decays, green and blue mean few decays. The resulting image provides a visual representation of the activity and structure of the brain.

glucose molecule contains one ^{18}F atom, the frequency of decay is directly proportional to the glucose concentration, and the map of radioactive decay also provides a map of relative glucose concentrations.

When a subject is injected with the ^{18}F-labeled glucose, the glucose travels throughout the body, but particularly to the brain, which uses prodigious amounts of this sugar for energy. Although all brain cells burn glucose, the more active a brain cell is, the more glucose it uses. How can this be used in biological research?

Let's suppose that a neuroscientist is trying to locate the areas of the brain that are involved in memory. She might give ^{18}F-labeled glucose to a few volunteer subjects, and then ask them to memorize a list of words that she reads aloud. Brain regions active during this process would need more energy than regions that, say, control running or eating. The energy-demanding cells would take up more ^{18}F-glucose molecules, and so would have more ^{18}F decays. PET scans taken during the memorization pinpoint brain regions active in storing memories. However, brain regions involved in simple hearing, breathing, and recording the sensations of being in the PET scan machinery would also be active and light up in the scan. The neuroscientist would therefore do a series of control scans—for example, when the subjects are in the machine but not being spoken to, when the word list is read aloud but the subjects are not asked to memorize the list, or when the subjects are asked to memorize lists that they read silently to themselves from a sheet of paper. By comparing the PET scan images, the researcher could then locate brain structures that are active only during memorization but not during other procedures.

Physicians also use PET scans in the diagnosis of brain disorders. For example, brain regions that are the origin of epileptic seizures generally have excessively high metabolic activity, and show up in PET scans as "hot spots." Many brain tumors also light up in PET scans (Fig. E2-2). Abnormal metabolism of certain brain regions may also appear in patients with some mental disorders, such as schizophrenia.

Biology and medicine have profited immensely from close interactions with the other sciences, especially chemistry and physics. The development of PET scans required close cooperation with chemists (developing and synthesizing the radioactive probes), physicists (understanding of positron/electron interactions and the geometry of the resulting energy emissions), and engineers (designing and building the electronic apparatus). Continued teamwork among scientists promises further advances in both fundamental understanding of biologic processes and applications in medicine and agriculture.

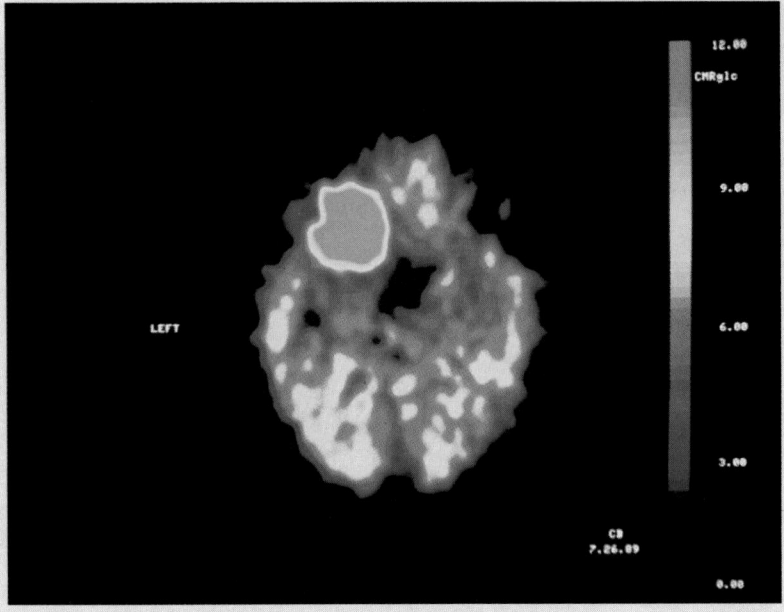

Figure E2-2 A brain tumor is usually composed of actively dividing, metabolically energetic cells. It usually has a rich blood supply and consumes both oxygen and glucose rapidly. Therefore, a brain tumor lights up in a PET scan (orange) with a variety of radioactive probes.

electron cannot occupy an intermediate position between shells, so it takes a certain minimum amount of energy to boost an electron to the next shell. Eventually, the electron returns to the low-energy shell, releasing energy as it does. When plants photosynthesize, the energy of specific wavelengths (colors) of sunlight kicks electrons in chlorophyll molecules out of their low-energy shells to higher-energy shells. As the electrons return to the low-energy shells, some of their "extra" energy is used to power the synthesis of sugar from carbon dioxide and water, as we will describe in Chapter 7.

Atomic Reactivity

Just as an atom is the smallest possible particle of an element, so too a **molecule** is the smallest possible particle of a compound. **A molecule consists of two or more atoms (of the same or different elements) held together by interactions among their outermost electron shells.** The basic principles of atomic reactivity are these:

1. An atom is stable (i.e., will not react with other atoms) when its outermost electron shell is completely full.
2. Conversely, an atom is reactive when its outermost electron shell has one or more vacancies.

To illustrate these principles, consider the two smallest atoms, hydrogen and helium. Hydrogen has one proton in its nucleus and one electron in its outermost electron shell, which can hold two. Helium has two protons in its nucleus and two electrons completely fill its outermost electron shell. Therefore, we predict that helium atoms, with a full shell, should be stable, while hydrogen atoms, with a half-empty shell, should be highly reactive. Our predictions are correct. Except for nuclear fusion reactions, helium is almost completely inert. Hydrogen, however, is highly flammable; that is, it reacts readily with oxygen: the space shuttle and many other rockets use liquid hydrogen as fuel to power lift-off.

Chemical Bonds: Joining Atoms to Make Molecules

An atom with an outermost electron shell that is partially full can gain stability by losing electrons (emptying the shell completely), gaining electrons (filling the shell), or sharing electrons with another atom (with both atoms behaving as though they had full shells). Losing, gaining, and sharing electrons result in forces called chemical bonds that hold atoms together in molecules. Chemical bonds are not *things*, like epoxy glue, but are *energy relationships*, such as the electrical attraction between positive and negative charges.

Ionic Bonds

Some atoms have almost empty outermost electron shells, and some atoms have almost full outermost shells. **These atoms may become stable by losing electrons, completely emptying their outermost shells, or by gaining electrons, completely filling their outermost shells.** The formation of table salt—sodium chloride—illustrates this principle. Sodium has an outermost electron shell containing only one electron, while chlorine has an outer shell that lacks only one electron to be completely full (Fig. 2-5a). Sodium, therefore, can become stable by losing the electron from its outer shell, leaving that shell empty, while chlorine can fill its outer shell by gaining an electron. This is precisely what each of them does to form sodium chloride: sodium loses an electron, becoming positively charged, and chlorine picks up an electron, becoming negatively charged (Fig. 2-5b). Atoms that have lost or gained electrons, and hence are charged, are called ions.

Opposite charges attract; therefore, sodium ions and chloride ions tend to stay near one another, forming crystals (Fig. 2-5c). The electrical attraction between oppositely charged ions that holds them together in crystals is called an **ionic bond.**

Covalent Bonds

An atom with a partially full outermost electron shell can become stable by sharing electrons with another atom in a *covalent bond*. To understand what this means, consider the hydrogen atom, which has one electron in a shell built for two. A hydrogen atom can become reasonably stable if it shares its single electron with another hydrogen atom, forming a molecule of hydrogen gas, H_2 (Fig. 2-6a). Since the two hydrogen atoms are identical, neither can capture the other's electron. Both electrons spend part of the time in the shell of both hydrogen atoms, with the result that each hydrogen behaves almost as if it had two electrons in its shell.

The stability of a covalent bond depends on the atoms involved. Some covalent bonds, such as those in water (H_2O; Fig. 2-6b) and carbon dioxide (CO_2), are extremely stable—that is, it takes a lot of energy to break the bonds. Other bonds, such as those in oxygen gas or gasoline, are less stable, coming apart more easily. When a chemical reaction occurs in which less-stable bonds are broken and more-stable

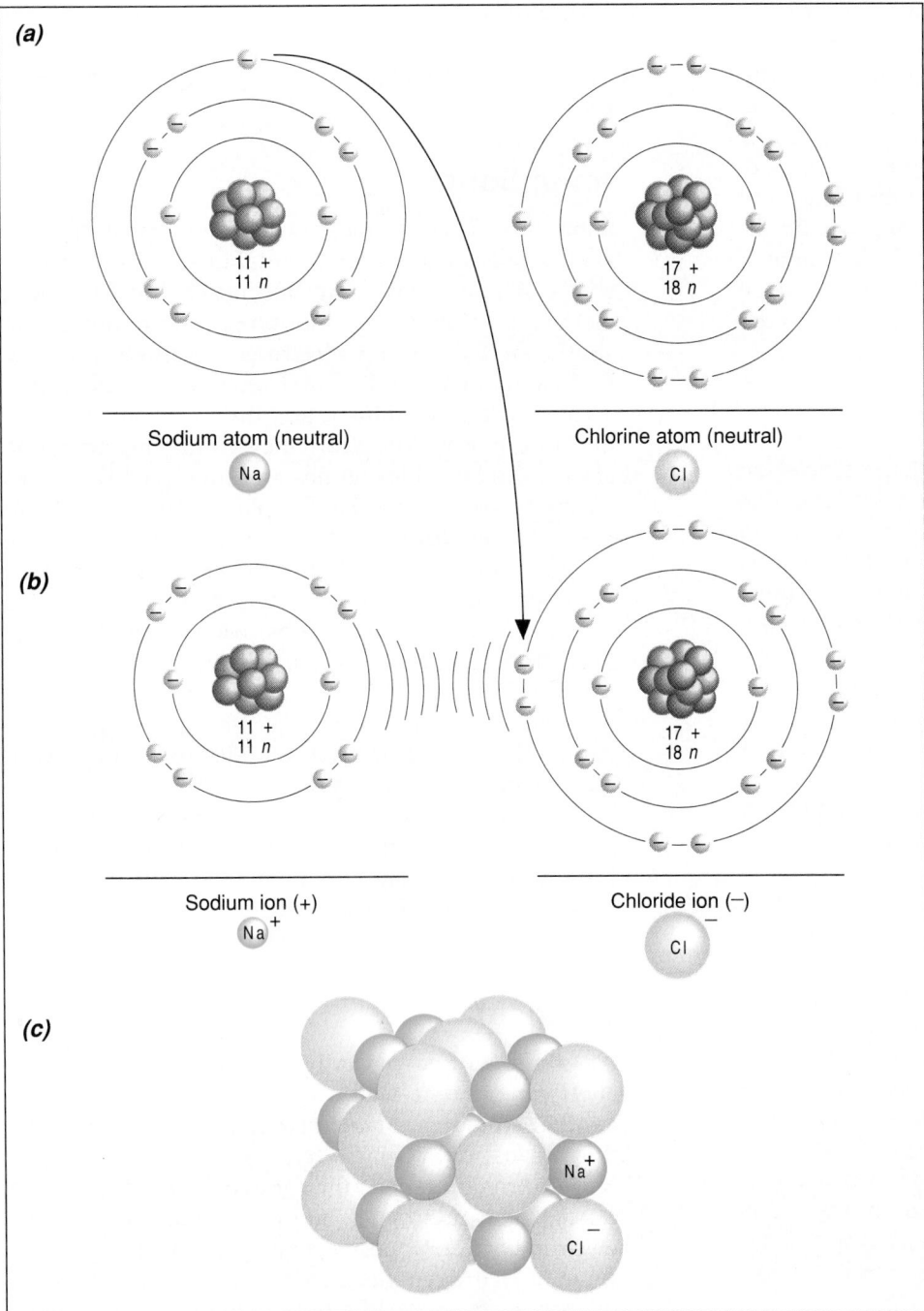

Figure 2-5 The formation of ions and ionic bonds. **(a)** Sodium has only one electron in its outer electron shell, while chlorine has seven (lacking only one electron to fill the outer shell). **(b)** Sodium can become stable by losing an electron, while chlorine can become stable by gaining an electron. When this happens, both have empty or full outer shells. Sodium therefore becomes a positively charged ion, and chlorine a negatively charged ion. **(c)** Since oppositely charged particles attract one another, the resulting sodium and chloride ions nestle closely together in a crystal of salt, NaCl.

bonds are formed (such as burning gasoline with oxygen to form carbon dioxide and water), energy is released (see Chapter 4).

Covalent Bonding in Biological Molecules

The atoms in most biological molecules are joined by covalent bonds. Hydrogen, carbon, oxygen, nitrogen, phosphorus, and sulfur are the most common

atoms found in biological molecules. Except for hydrogen, each of these atoms lacks two or more electrons to fill its outermost electron shell, and can share electrons with two or more other atoms. Hydrogen can form a covalent bond with one other atom, oxygen and sulfur with two, nitrogen with three, and phosphorus and carbon with up to four other atoms (Table 2-2). This diversity of bonding arrangements permits biological molecules to be constructed in almost infinite variety and complexity.

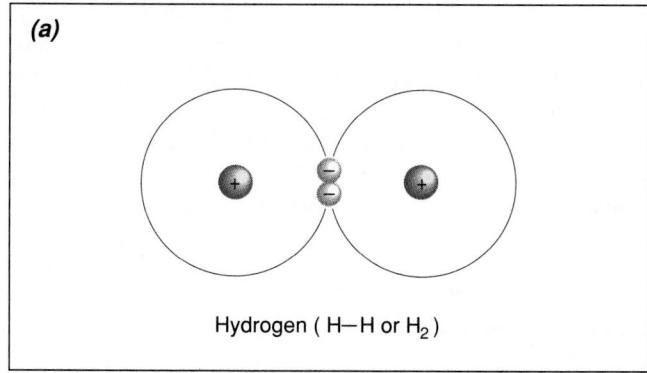

(a)

Hydrogen (H—H or H_2)

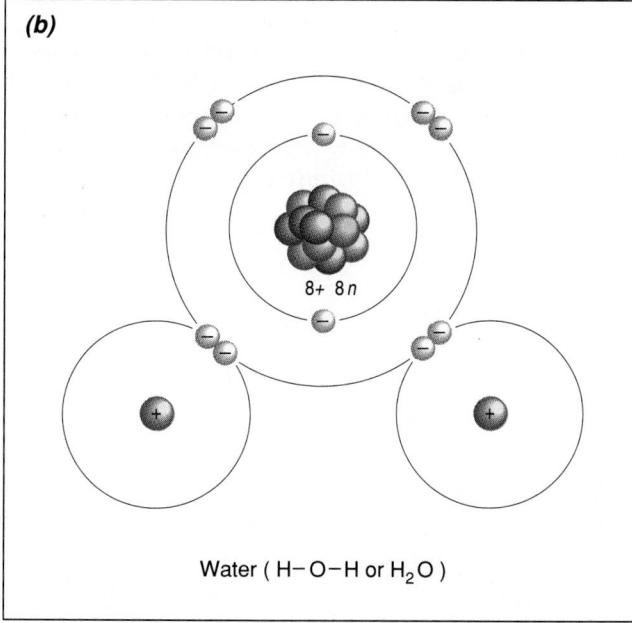

(b)

Water (H—O—H or H_2O)

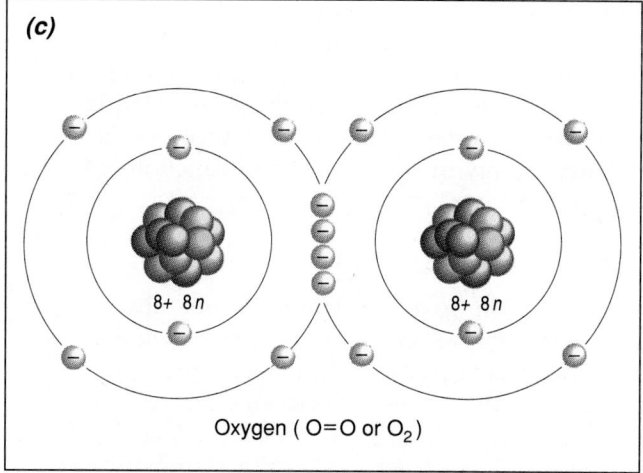

(c)

Oxygen (O=O or O_2)

Figure 2-6 Electrons are shared between atoms to form covalent bonds.
(a) In hydrogen gas, one electron from each hydrogen atom is shared, forming a single covalent bond. The resulting bond is usually represented by a single dash connecting the atomic symbols, and the molecule of hydrogen gas may be represented as H—H or H_2.
(b) Oxygen lacks two electrons to fill its outer shell, and so oxygen can make two single bonds, one with each of two hydrogen atoms to form water (H—O—H or H_2O).
(c) In oxygen gas, two oxygen atoms share four electrons, making a double bond, usually represented by a double dash or equal sign (O=O or O_2).

Table 2-2   Bonding Patterns of Atoms Commonly Found in Biological Molecules

Atom	Capacity of Outer Electron Shell	Electrons in Outer Shell	Number of Covalent Bonds Usually Formed	Common Bonding Patterns
Hydrogen	2	1	1	—H
Carbon	8	4	4	—C— =C= =C= —C≡
Nitrogen	8	5	3	—N— —N= N≡
Oxygen	8	6	2	—O— O=
Phosphorus	8	5	5	—P=
Sulfur	8	6	2	—S—

Sometimes, an atom will share two pairs of electrons with another atom, forming a **double covalent bond** (each atom contributes two electrons; see Fig. 2-6c). If atoms share three pairs of electrons, a **triple covalent bond** is formed. Double and triple bonds create further variety in the shapes and functions of biological molecules.

Polar Covalent Bonds

Electron sharing in covalent bonds is not always equal. In hydrogen gas, the shared electrons spend equal time near each hydrogen nucleus. Therefore, not only is the molecule as a whole electrically neutral, but also each end, or pole, of the molecule is electrically neutral. Such an electrically symmetrical bond is called a **nonpolar covalent bond.** In many other molecules, one nucleus may attract the electrons more strongly than the other nucleus does. This forms a **polar covalent bond.** Although the molecule

as a whole is electrically neutral, it has charged parts: the atom that attracts the electrons more strongly has a slightly negative charge (the negative pole of the molecule), while the other has a slightly positive charge (the positive pole). In water, for example, oxygen attracts electrons more strongly than hydrogen does, so that the oxygen end of a water molecule is negative and each hydrogen is positive (Fig. 2-7).

Hydrogen Bonds

Because of the polar nature of their covalent bonds, nearby water molecules attract one another, the negatively charged oxygens attracting the positively charged hydrogens. The electrical attraction between polar parts of molecules is called a **hydrogen bond** (Fig. 2-8). As we will see shortly, hydrogen bonds between molecules give water several unusual properties that are essential to life on Earth.

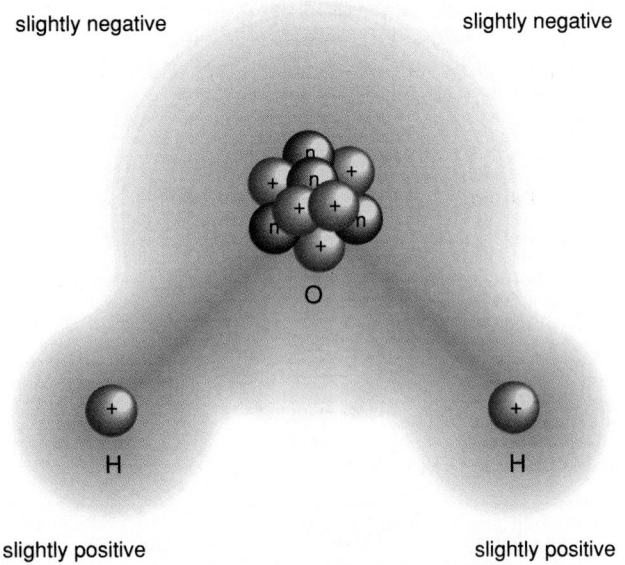

Figure 2-7 The electrons in polar covalent bonds are unequally shared. In water, the oxygen atom attracts the electrons more strongly than do the hydrogen atoms. The oxygen atom therefore has a partial negative charge while the two hydrogen atoms are left somewhat deprived of electrons and are partially positive.

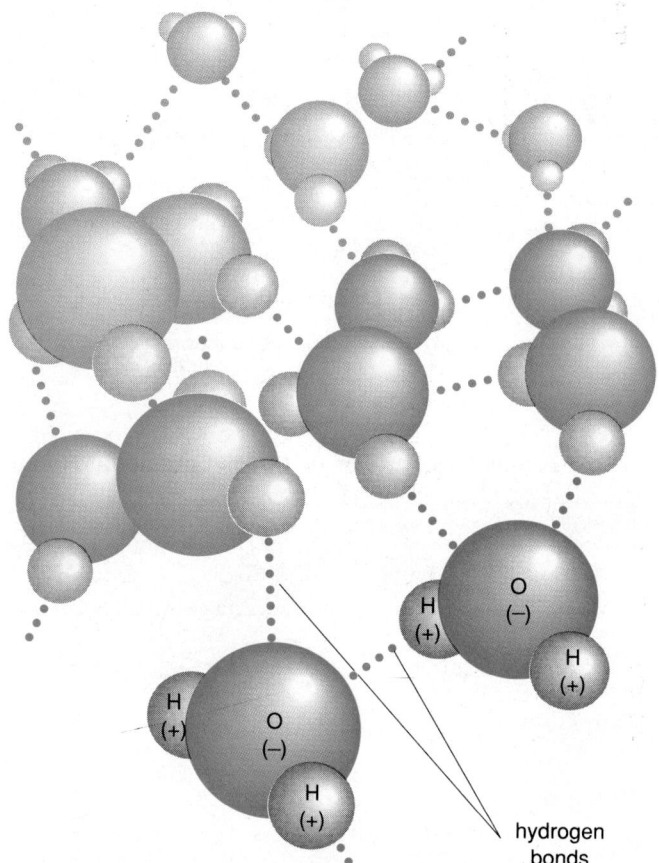

Figure 2-8 The partial charges on different parts of water molecules produce weak attractive forces called hydrogen bonds (dotted lines) between the hydrogens of one water molecule and the oxygens of other molecules.

Interpreting Chemical Structures and Formulas

Many of you have never taken a chemistry course and may not know how to interpret drawings and formulas of molecules. Here, we will introduce you to the ways in which simple molecules are illustrated.

Most people understand and remember something more easily if they can picture it. Consequently, physicists have developed visual models of atoms and molecules. Different types of models serve different intellectual purposes. Using water as an example, let's look at six ways of representing molecules, as shown below.

One of the first, and still very useful, ways to visualize atomic structure is the planetary model (a). The atomic nucleus is pictured as the center of a "solar system" with the electrons orbiting about it, much as the planets orbit around the sun. When you are trying to figure out how two atoms bond together to make a molecule, planetary models are convenient because they allow you to count electrons and see what arrangements result in filled electron shells. For example, oxygen has six electrons in its outer shell; eight electrons would fill the shell. Hydrogen has one electron in its outer shell, with two electrons needed to fill the shell. The planetary model shows that filled orbitals occur when two hydrogens each share one electron with a single oxygen atom.

Planetary models, however, are inaccurate: electrons don't revolve about the nucleus in well-defined orbits. In addition, planetary models cannot show the three-dimensional structure of atoms and molecules. Electron cloud models (b) are the most accurate atomic portraits, because electrons roam about over a relatively large volume of space. The density of dots in an electron cloud model corresponds to the likelihood that an electron will be found at any given place

at any given time. However, except for their accuracy, electron cloud models aren't very useful, partly because they are tedious to draw. It is easier to visualize the three-dimensional structure of a molecule with a space-filling or ball model (c). You can think of the ball as representing the "normal" outer edge of the electron cloud of an atom. Ball models portray atoms within a molecule or even groups of molecules in reasonably accurate spatial relationships.

Ball-and-stick models (d), while not quite as accurate as space-filling models, also represent the three-dimensional structure of molecules. They are also easier to draw, and can depict single bonds (one stick connecting balls), double bonds (two sticks), and triple bonds (three sticks). We will usually use space-filling or ball-and-stick models in this text to portray simple molecules.

Really quick representations of molecules can be drawn using only the atomic symbol for the individual atoms, with lines for the bonds joining them (e). Single bonds are shown as dashes and double bonds as double dashes. Finally, if the structure of the molecule doesn't need to be shown at all, the molecule can be represented by its chemical formula (f). The subscript number after each atomic symbol is the number of atoms of each element in the molecule; for example, "H_2O" means "two hydrogens and one oxygen" make up water.

Large molecules, including virtually all organic molecules, are often drawn using even further shortcuts, such as leaving out carbon and hydrogen atoms. These abbreviated portraits of organic molecules are explained in Figures 3-2 and 3-9 in the next chapter.

(a)   (b)   (c)   (d)   (e)   (f)

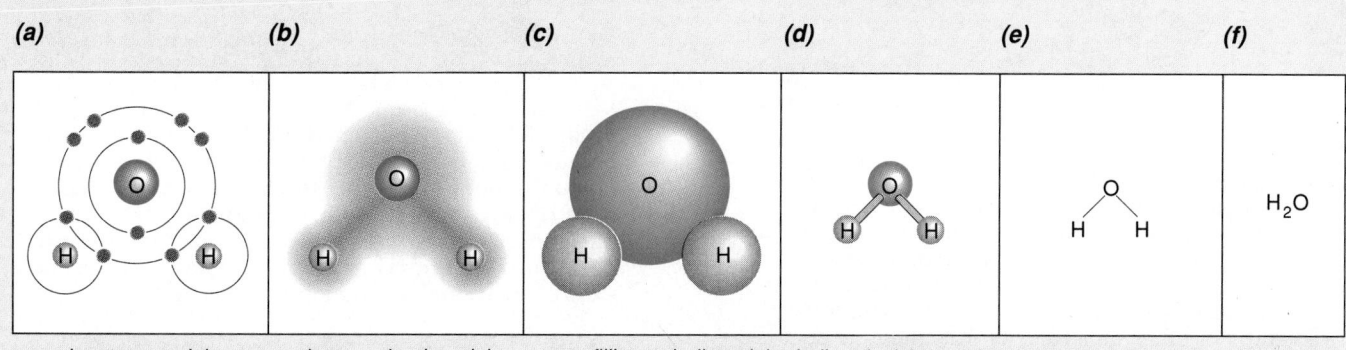

planetary model | electron cloud model | space-filling or ball model | ball-and-stick model | molecular structure | chemical formula

Many biological molecules also form hydrogen bonds. Both nitrogen and oxygen atoms attract electrons more strongly than hydrogen atoms do, becoming slightly negative themselves and leaving the hydrogen slightly positive. The resulting polar parts of the molecules can form hydrogen bonds with water, with other biological molecules, or with distant, polar parts of the same molecule. As we will see in Chapter 3, hydrogen bonds play crucial roles in shaping the three-dimensional structures of proteins and nucleic acids.

Organic and Inorganic Molecules

Chemists classify molecules as organic or inorganic. The word "organic" originally signified that these molecules could only be manufactured within living organisms. Today, however, organic chemists can synthesize many of these molecules in the laboratory. Thus, the modern definition of an **organic molecule** is any molecule containing both carbon and hydrogen. Most organic molecules are large, with complex structures. **Inorganic molecules** include carbon dioxide and all molecules lacking carbon.

Many inorganic molecules are important to living organisms, including, for instance, minerals in the soil and carbon dioxide in the air, which are the raw materials that plants use to construct their bodies. However, one inorganic molecule is extraordinarily abundant on Earth, has unusual properties, and is so essential to life that it merits special consideration. That molecule is water.

Water and Life

Life almost certainly arose in the waters of the primeval Earth. The first life form was probably a small sac enclosing water with an array of dissolved enzymes and simple genetic material. Living organisms are still composed of about 60% to 90% water, and all life on Earth depends intimately on the properties of water. Why is water so crucial to life?

The Effects of Water on Other Molecules

Water Dissolves Many Molecules

Water is an extremely good **solvent**—that is, it is capable of dissolving a wide range of substances, espe-

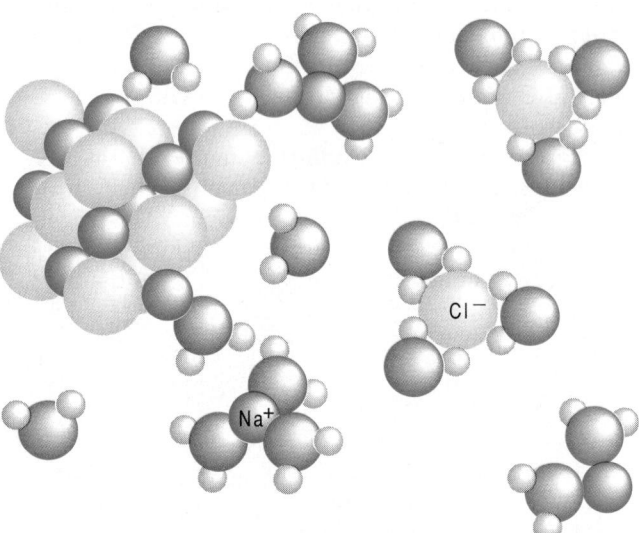

Figure 2-9 The polarity of water molecules allows water to dissolve polar or charged substances. When a salt crystal is dropped into water, the water molecules worm their way around the outsides of the sodium and chloride ions, surrounding them with oppositely charged ends of the water molecules. Thus insulated from the attractiveness of other molecules of salt, the ions float away, and the whole crystal gradually dissolves.

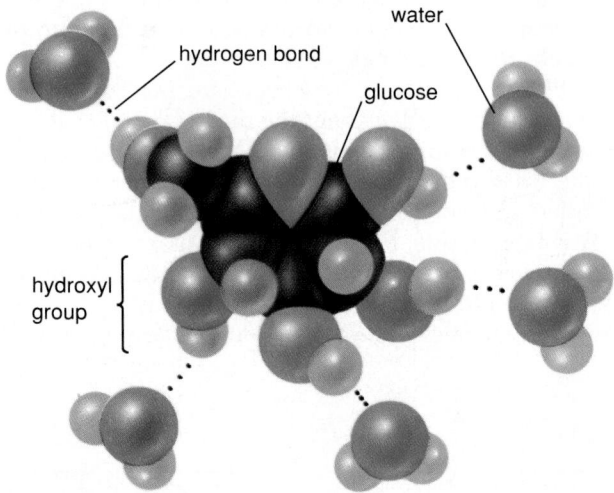

Figure 2-10 Many biological molecules dissolve in water because they have polar parts, particularly $-NH_2$ (amino) groups and $-OH$ (hydroxyl) groups, that can form hydrogen bonds with water molecules. This illustration depicts the hydrogen bonds that can form between the hydroxyl groups on a glucose molecule (a simple sugar) and surrounding water molecules.

cially salts. Recall that a crystal of table salt is held together by the electrical attraction between positively charged sodium ions and negatively charged chloride ions. Since water is a polar molecule, it has positive and negative ends. If a salt crystal is dropped into water, the positively charged hydrogen ends of water molecules will be attracted to and surround the negatively charged chloride ions, while the negatively charged oxygen ends of water molecules will surround the positively charged sodium ions. As water molecules enclose the sodium and chloride ions, the ions separate from the crystal and drift away in the water—the salt dissolves (Fig. 2-9).

Water dissolves polar molecules in a similar fashion, as its positive and negative poles are attracted to oppositely charged regions of dissolving molecules. Charged or polar molecules are termed **hydrophilic** (Greek for "water-loving"), because of their electrical attraction for water molecules. Many biological molecules, such as sugars and amino acids, are hydrophilic and dissolve readily in water (Fig. 2-10). Water also dissolves gases such as oxygen and carbon dioxide.

Other liquids can dissolve some of these substances, but not all of them. Alcohol, for example, dissolves some sugars and proteins but not salts. By dissolving such a wide variety of molecules, the watery cytoplasm of a cell provides a suitable environment for the myriad chemical reactions essential to life on Earth.

Water Influences Nonpolar Molecules

Molecules that are uncharged and nonpolar, including fats, oils, and gasoline, usually do not dissolve in water, and hence are called **hydrophobic** ("water-fearing"). Nevertheless, water has an important organizing effect on such molecules. Gasoline and oil, for example, form globs and slicks when spilled into water.

Consider a container of water before and after the addition of gasoline. In pure water, each water molecule can hydrogen-bond with the maximum possible number of other water molecules. If two molecules of gasoline are added, at distant locations, each disrupts the hydrogen bonding of nearby water molecules (Fig. 2-11). Any water molecule directly adjacent to the gasoline molecules cannot make hydrogen bonds in the direction of the gasoline. Each narrow, elongated gasoline molecule has two surfaces that prevent hydrogen bonding among water molecules, for a total of four "interfering surfaces."

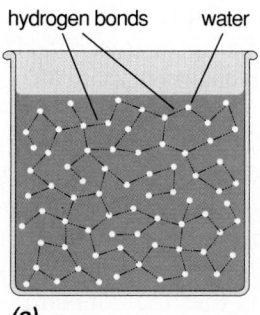 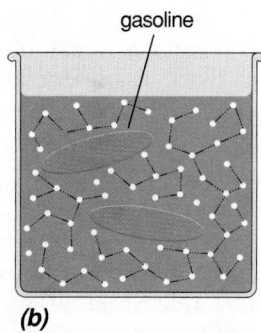 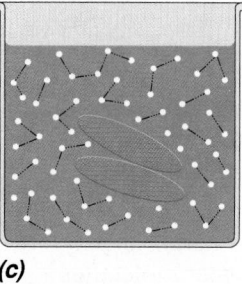

(a)   *(b)*   *(c)*

Figure 2-11 In so-called hydrophobic interactions, hydrophobic molecules such as gasoline are forced together by the power of hydrogen bonding among water molecules.
(a) *Pure water:* maximum number of hydrogen bonds formed among water molecules.
(b) *Two gasoline molecules added to water:* gasoline molecules separate water molecules, reducing hydrogen bonding among water molecules.
(c) *Gasoline molecules coalesce:* fewer water molecules are separated by gasoline, so hydrogen bonding among water molecules increases.

Suppose now that the two gasoline molecules, randomly moving in the water, encounter one another. The instant they make contact, only two surfaces interfere with hydrogen-bonding between water molecules. Since it takes energy to separate water molecules that are attracted to one another by hydrogen bonds, the water prevents the gasoline molecules from moving apart again. Therefore, the gasoline molecules remain together, forming an insoluble glob. (Gasoline is less dense than water, so the gasoline globs float, coalescing to form a thin film on the water surface, which further reduces the "interfering surfaces" to just one.) The tendency for hydrophobic molecules to aggregate is often termed a **hydrophobic interaction,** although in fact it is the hydrogen bonding between water molecules that forces the hydrophobic molecules together, rather than any special affinity among the hydrophobic molecules.

As we will see in Chapter 5, the membranes of living cells owe much of their structure to the organizing effect of water on hydrophobic molecules.

Water Takes Part in Many Chemical Reactions

Water enters into many of the chemical reactions that occur in living cells. The oxygen that green plants release into the air is derived from water during photosynthesis. When your body manufactures a protein, fat, nucleic acid, or sugar, it produces water in the process, and, conversely, when you digest proteins, fats, and sugars in the foods you eat, water is used in the reactions.

Acids, Bases, and Buffers

Although water is generally regarded as a stable compound, individual water molecules constantly gain, lose, and swap hydrogen atoms. As a result, at any given time about two of every billion water molecules are ionized:

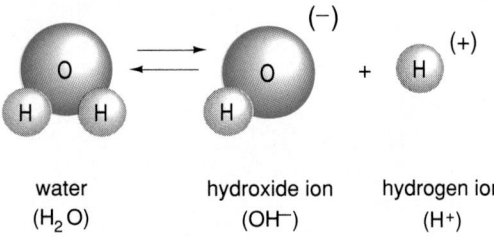

water	hydroxide ion	hydrogen ion
(H_2O)	(OH^-)	(H^+)

Pure water has equal concentrations of hydrogen ions (H^+) and hydroxide ions (OH^-), but in many solutions the concentrations of H^+ and OH^- ions are not the same. If the concentration of H^+ ions exceeds the concentration of OH^- ions, the solution is **acidic;** if the concentration of OH^- ions is greater, the solution is **basic.** The degree of acidity is expressed on the **pH scale** (Fig. 2-12), in which neutrality (equal numbers of H^+ and OH^- ions) is assigned the number 7. Acids have a pH below 7, while bases have a pH above 7. The pH scale is logarithmic: each unit of change in pH represents a tenfold change in the concentration of H^+ ions.

An **acid** is a substance that gives off hydrogen ions. When hydrochloric acid (HCl), for example, is added to pure water, almost all of the HCl molecules separate into H^+ and Cl^- ions. Therefore, the concentration of H^+ ions greatly exceeds the concentration of OH^- ions, and the resulting solution is acidic. A **base** is a substance that combines with H^+ ions. If sodium hydroxide (NaOH) is added to water, the NaOH molecules separate into Na^+ and OH^- ions. The OH^- ions combine with H^+ ions to form water; thus, the resulting solution contains more OH^- ions than H^+ ions, and is basic.

Buffers

In most mammals, including humans, both the cell cytoplasm and the fluids that bathe the cells are nearly neutral (pH about 7.3 to 7.4). Small increases or decreases in pH may cause drastic changes in both structure and function, leading to the death of cells or entire organisms. Nevertheless, living cells seethe with chemical reactions that take up or give off H^+ ions. How, then, does the pH remain constant? The answer lies in the many buffers found in living organisms. A **buffer** is a compound that tends to maintain a solution at a constant pH by accepting or releasing H^+ ions in response to small changes in H^+ ion concentration. If the H^+ ion concentration rises, buffers combine with them; if the H^+ ion concentration falls, buffers release H^+. The result is that the concentration of H^+ ions is restored to its original level. Common buffers in living organisms include bicarbonate (HCO_3^-) and phosphate ($H_2PO_4^-$ and HPO_4^{-2}), both of which can accept or release H^+ ions, depending on pH. If the blood becomes too acidic, for example, bicarbonate can accept H^+ ions to form carbonic acid:

$$HCO_3^- + H^+ \rightarrow H_2CO_3$$

If the blood becomes too basic, bicarbonate can liberate H^+ ions that combine with the excess OH^- ions, forming water:

$$HCO_3^- + OH^- \rightarrow CO_3^{-2} + H_2O$$

In either case, the end result is that the blood pH is maintained at its normal level.

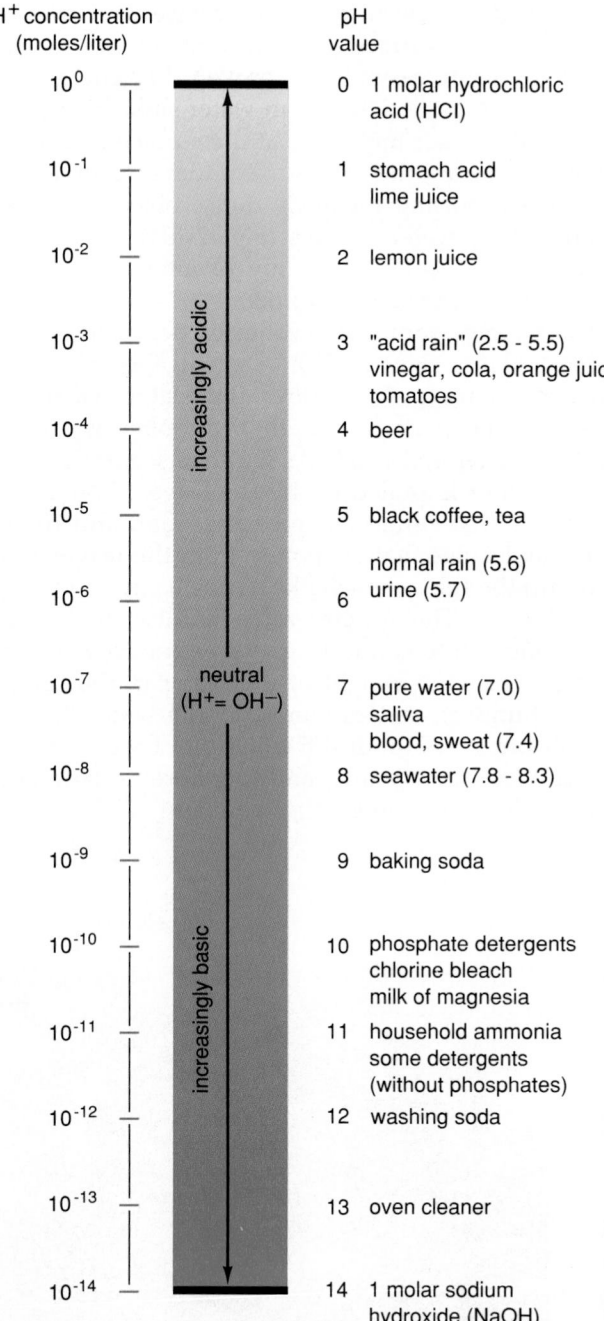

H⁺ concentration (moles/liter)	pH value	

H⁺ concentration (moles/liter) — pH value

10^0 — 0 — 1 molar hydrochloric acid (HCl)

10^{-1} — 1 — stomach acid lime juice

10^{-2} — 2 — lemon juice

10^{-3} — 3 — "acid rain" (2.5 - 5.5) vinegar, cola, orange juice, tomatoes

10^{-4} — 4 — beer

10^{-5} — 5 — black coffee, tea

normal rain (5.6)

10^{-6} — 6 — urine (5.7)

increasingly acidic

10^{-7} — 7 — pure water (7.0) saliva blood, sweat (7.4)

neutral (H⁺ = OH⁻)

10^{-8} — 8 — seawater (7.8 - 8.3)

10^{-9} — 9 — baking soda

10^{-10} — 10 — phosphate detergents chlorine bleach milk of magnesia

10^{-11} — 11 — household ammonia some detergents (without phosphates)

increasingly basic

10^{-12} — 12 — washing soda

10^{-13} — 13 — oven cleaner

10^{-14} — 14 — 1 molar sodium hydroxide (NaOH)

Figure 2-12 The pH scale expresses the concentration of hydrogen ions in a solution on a scale of 0 (very acidic) to 14 (very basic). The pH scale is logarithmic: each unit of change in pH represents a tenfold change in the concentration of acid or base. Lemon juice, for example, is about 10 times more acid than orange juice, and the most severe acid rains in the Northeastern United States are almost 1000 times more acidic than normal rainfall. Except for the insides of your stomach, nearly all the fluids in your body are finely adjusted to a pH of 7.4; a pH of 7.0, although it may not sound like much of a change, is enormously more acidic than normal.

Water as a Temperature Moderator

Organisms can survive only within a limited temperature range: both high and low temperatures can be fatal. High temperatures may damage the protein enzymes that guide the chemical reactions essential to life (see Chapter 4): most enzymes cease to function at temperatures well below boiling. Low temperatures are also dangerous, because enzyme action slows as the temperature drops. Subfreezing temperatures within the body are usually lethal, because cells are ruptured by spear-like ice crystals.

Water has three properties that moderate temperature. First, water has a high **specific heat,** the amount of energy needed to raise the temperature of 1 gram of a substance by 1° C. Temperature measures the velocity of molecules: the higher the temperature, the greater the average velocity. Generally speaking, if heat energy enters a system, the molecules of that system move more rapidly, and the temperature of the system rises. Individual water molecules, however, are weakly linked to one another by hydrogen bonds (see Fig. 2-8). When heat enters a watery system such as a lake or a living cell, much of the heat energy goes into breaking hydrogen bonds rather than speeding up individual molecules. Thus, 1 **calorie** of energy will heat 1 gram of water 1° C, while it takes only 0.6 calories per gram to heat alcohol 1° C, 0.2 calories for table salt, and 0.02 calories for common rocks such as granite or marble. A sunbather on a hot summer day, therefore, can pick up a lot of heat energy without sending his body temperature soaring.

Water moderates high temperatures through its great **heat of vaporization,** the amount of heat required to convert liquid water to water vapor. Water has one of the highest heats of vaporization known, 539 calories per gram. Again, this is due to the hydrogen bonds interconnecting individual water molecules. For a water molecule to evaporate, it must move quickly enough to break all the hydrogen bonds holding it to the other water molecules in the solution. Only the fastest-moving water molecules, carrying the most energy, can break their hydrogen bonds and escape into the air as water vapor. The remaining liquid is cooler for the loss of these high-energy molecules. If our sunbather's body temperature begins to rise, he perspires, covering his body with a film of water. Heat energy is transferred from his skin to the water. Evaporating just 1 gram of water cools 539 grams of his body 1° C, so a great loss of heat can occur without much loss of water.

Finally, water moderates low temperatures through its high **heat of fusion,** the energy that must be removed from the molecules of liquid water before they form the precise crystal arrangement of ice. Very low temperatures are required for a long time before a sizable body of water turns to ice, which helps to protect living organisms, especially aquatic organisms, from the damaging effects of freezing.

Water as a Solid

Water, of course, will become a solid after prolonged exposure to temperatures below its freezing point. But even here the unique properties of water come into play, to the advantage of living things: ice floats. Most liquids become more dense when they solidify, and the solid sinks. Ice, however, is less dense than liquid water, so when a pond or lake starts to freeze in winter, the ice stays on top, forming an insulating layer that delays the freezing of the rest of the water. If ice sank, ponds and lakes in much of North America would freeze solid during the winter, eliminating fish and most other animal life.

Cohesion Among Water Molecules

Because of the hydrogen bonds that interconnect individual molecules, water has high **cohesion**—that is, water molecules tend to stick together. Cohesion among water molecules at the surface of a lake or pond produces **surface tension,** the tendency for the water surface to resist being broken. In general, objects that are more dense than water sink. However, because the water molecules at the surface of a pond cohere to one another, the surface film acts almost as a solid, supporting relatively dense objects such as fallen leaves, water striders (Fig. 2-13a), and, for a painful split second, the body of an inept human diver belly flopping into a pool.

A more important role of cohesion occurs in the life of land plants (Fig. 2-13b). A plant absorbs water through its roots. How does the water reach the above-ground parts, especially if the plant is a hundred-meter-tall redwood? As we shall see in Chapter 27, the water is pulled up by the leaves. Water fills tiny tubes that connect the leaves, stem, and roots. Water molecules that evaporate from the leaves pull water up the tubes, much like a rope being pulled up from the top. The system works because the hydrogen bonds interconnecting water molecules are stronger than the weight of the water in the tubes, even a hundred meters worth, so the water "rope" doesn't break. Without the cohesion of water, there would be no land plants, and therefore no land animals, including people.

(a)

(b)

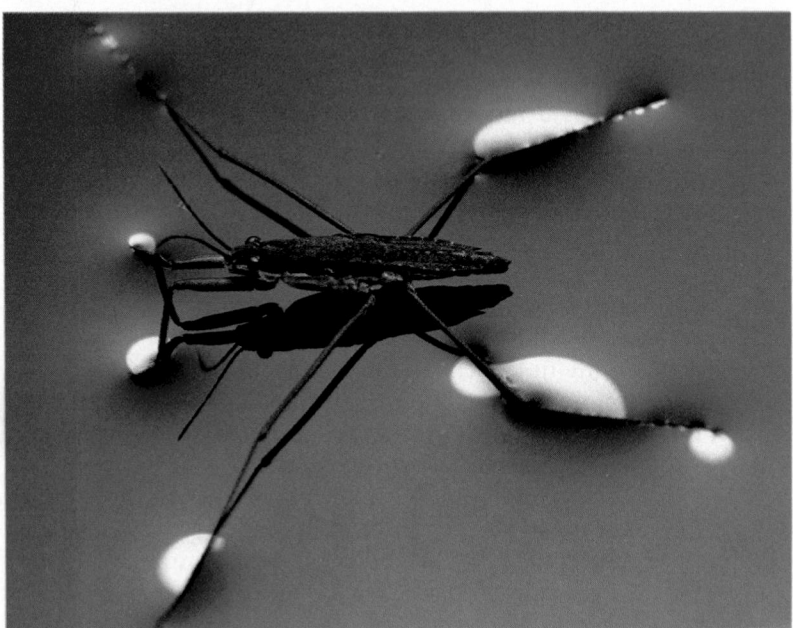

Figure 2-13 (a) Cohesion among water molecules allows water striders to skate across the surface of still waters. **(b)** In giant redwoods, cohesion holds water molecules together in continuous strands from the roots to the topmost leaves hundreds of feet above the ground.

SUMMARY OF KEY CONCEPTS

Matter and Energy

Matter is the physical material of the universe. Energy is the capacity to do work. Energy can exist in several forms that may be converted from one to another. The two major categories are kinetic energy (the energy of motion) and potential energy (stored energy).

The Structure of Matter

An element is a substance that can neither be broken down nor converted to different substances by ordinary chemical means. The smallest possible particle of an element is the atom, which is itself composed of a central nucleus containing protons and neutrons and outer orbitals containing electrons. Every atom of a given element has the same number of protons, which is different from the number found in the atoms of any other element.

Electron orbitals are found at specific distances from the nucleus, called electron shells. Each shell can contain a fixed maximum number of electrons. The chemical reactivity of an atom depends on the number of electrons in its outermost electron shell; an atom is most stable, and therefore least reactive, when its outermost shell is either completely full or completely empty.

Chemical Bonds: Joining Atoms to Make Molecules

Atoms may combine to form molecules. The forces holding atoms together in molecules are called chemical bonds. There are two principal types of chemical bond, called ionic and covalent bonds. If one atom fills its outermost shell by acquiring electrons, while another atom empties its shell by losing electrons, this results in negatively and positively charged particles called ions. Ionic bonds are electrical attractions between charged ions, holding them together in crystals. Covalent bonds involve the sharing of electrons by two atoms, in which neither atom completely gains nor loses an electron. In a nonpolar covalent bond, both atoms share electrons equally. In a polar covalent bond, one atom may attract the electron more strongly than the other atom does; in this case, the strongly attracting atom bears a slightly negative charge, and the weakly attracting atom bears a slightly positive charge. Some polar covalent bonds give rise to hydrogen bonding, which is the attraction between charged regions of individual polar molecules or distant parts of a large polar molecule.

Organic and Inorganic Molecules

An organic molecule is one that contains both carbon and hydrogen; all other molecules are inorganic. The most important inorganic molecule for life on Earth is water.

Water and Life

The water molecule has several properties that facilitate life, including its ability to dissolve many polar and charged substances, to force nonpolar substances to assume certain types of physical organization, to participate in chemical reactions, to maintain a fairly stable temperature in the face of wide temperature fluctuations in the environment, and to cohere to itself.

GLOSSARY

acid: a substance that releases hydrogen ions (H^+) into solution; a solution with a pH of less than 7.

atom: the smallest particle of an element that retains the properties of the element.

atomic number: the number of protons in the nuclei of all atoms of a particular element.

base: a substance that is capable of combining with and neutralizing H^+ ions, producing a solution with a pH greater than 7.

buffer: a compound that minimizes changes in pH by reversibly taking up or releasing H^+ ions.

calorie: the amount of energy required to raise the temperature of 1 gram of water 1° C.

chemical bond: the force of attraction between neighboring atoms that holds them together in a molecule.

cohesion: the tendency of the molecules of a substance to hold together.

compound: a substance composed of two or more elements

that can be broken into its constituent elements by chemical means.

covalent bond (ko-vā'-lent): a chemical bond between atoms in which electrons are shared.

double covalent bond: a covalent bond that occurs when two atoms share two pairs of electrons.

electron: a subatomic particle, found in the orbitals outside the nucleus of an atom, bearing a unit of negative charge and very little mass.

electron shell: all the electron orbitals at a given distance from the nucleus of an atom.

element: a substance that cannot be broken down to a simpler substance by ordinary chemical means.

energy: the capacity to do work.

energy level: the specific amount of energy characteristic of a given electron shell in an atom.

heat of fusion: the energy that must be removed from a compound to transform it from a liquid into a solid at its freezing temperature.

heat of vaporization: the energy that must be supplied to a compound to transform it from a liquid into a gas at its boiling temperature.

hydrogen bond: the weak attraction between a hydrogen atom bearing a partial positive charge (due to polar covalent bonding with another atom) and another atom, usually oxygen or nitrogen, bearing a partial negative charge. Hydrogen bonds may form between atoms of a single molecule or of different molecules.

hydrophobic (hī-drō-fō'-bik): pertaining to a substance that does not dissolve in water.

hydrophobic interaction: the tendency for hydrophobic molecules to cluster together when immersed in water.

hydrophilic (hī-drō-fil'-ik): pertaining to a substance that dissolves readily in water, or to parts of a large molecule that form hydrogen bonds with water.

inorganic molecule: any molecule that does not contain both carbon and hydrogen.

ion (ī'-on): an atom or molecule that has either an excess of electrons (and hence is negatively charged) or has lost electrons (and is positively charged).

ionic bond: a chemical bond formed by the electric attraction between positively and negatively charged ions.

isotope: one of several forms of a single element, the nuclei of which contain the same number of protons but different numbers of neutrons.

kinetic energy (kin-et'-ik): the energy of movement.

matter: the material of which the universe is made.

mixture: a substance composed of two or more elements in variable proportions.

molecule (mol'-e-kūl): a particle composed of one or more atoms held together by chemical bonds. A molecule is the smallest particle of a compound that displays all the properties of that compound.

neutron: a subatomic particle found in the nuclei of atoms, bearing no charge and having mass approximately equal to that of a proton.

nonpolar covalent bond: a covalent bond with equal sharing of electrons.

nucleus: the central region of an atom, composed of protons and neutrons.

orbital: the region of an atom, outside the nucleus, in which an electron is likely to be found.

organic molecule: a molecule that contains both carbon and hydrogen.

pH scale: a scale with values from 0 to 14, used for measuring the relative acidity of a solution. At pH 7 a solution is neutral, pH 0 to 7 is acidic, and pH 7 to 14 is basic. Each unit on the pH scale represents a tenfold change in the concentration of hydrogen ions.

polar covalent bond: a covalent bond with unequal sharing of electrons, so that one atom is relatively negative while the other is relatively positive.

potential energy: "stored" energy, usually chemical energy or energy of position within a gravitational field.

proton: a subatomic particle found in the nuclei of atoms, bearing a unit of positive charge and a relatively large mass roughly equal to the mass of the neutron.

radioactive: pertaining to an atom with an unstable nucleus that spontaneously disintegrates with the emission of radiation.

single covalent bond: a covalent bond that occurs when two atoms share a single pair of electrons.

solvent: a liquid that is capable of dissolving (uniformly dispersing) other substances in itself.

specific heat: the amount of energy required to raise the temperature of 1 gram of a substance 1° C.

surface tension: the property of a liquid to resist penetration by objects at its interface with the air, due to cohesion between molecules of the liquid.

triple covalent bond: a covalent bond that occurs when two atoms share three pairs of electrons.

STUDY QUESTIONS

1. Draw the structure of an atom.
2. Distinguish among atoms and molecules; elements, compounds, and mixtures; protons, neutrons, and electrons.
3. What factors influence the reactivity of an atom?
4. Describe how ionic bonds are formed.
5. Describe how covalent bonds are formed. What determines the number of covalent bonds that an atom can form? What is the difference between a nonpolar and a polar covalent bond?
6. What is a hydrogen bond? How does it influence the structure of water and of biological molecules?
7. List the elements commonly found in biological molecules.
8. Define acid, base, and buffer. How do buffers reduce changes in pH when hydrogen ions or hydroxide ions are added to a solution? Why is this important to living organisms?
9. Describe how water dissolves a salt. How does this compare with the effect of water on a hydrophobic substance such as corn oil?

DISCUSSION QUESTIONS

1. Some isotopes, like ^3H and ^{14}C, are radioactive and decay at predictable rates, called *half-lives* (the time it takes for half the radioactive isotopes to change). The half-life for ^{14}C is 5600 years. If a fossilized tree log initially contained 128 grams of ^{14}C but now contains only 1 gram of ^{14}C, how many years old is the log?
2. Helium (atomic number = 2), neon (atomic number = 10), and argon (atomic number = 18) all have similar chemical properties. Why are their properties similar? Would you expect these elements to easily enter into chemical reactions with other elements, or to be rather inert? Explain.
3. Explain why water (H_2O) is a polar molecule, having slightly positively charged and negatively charged regions, while both hydrogen gas (H_2) and oxygen gas (O_2) are nonpolar molecules.
4. How does panting help a dog regulate its body temperature?

SUGGESTED READINGS

Atkins, P. W., *Molecules*. New York: Scientific American Library, 1987. A layman's introduction to atoms and molecules, with superb illustrations.

Morrison, P., and Morrison, P. *Powers of Ten*. New York: W. H. Freeman and Co., 1982. A fascinating journey from the universe to the nucleus of an atom.

Storey, K. B., and Storey, J. M., "Frozen and Alive." *Scientific American*, December 1990. By triggering ice formation here, suppressing it there, and loading up their cells with antifreeze molecules, some animals, including certain lizards and frogs, can survive with 60% of their body water frozen solid.

3

The Chemistry of Life.
II. Biological Molecules

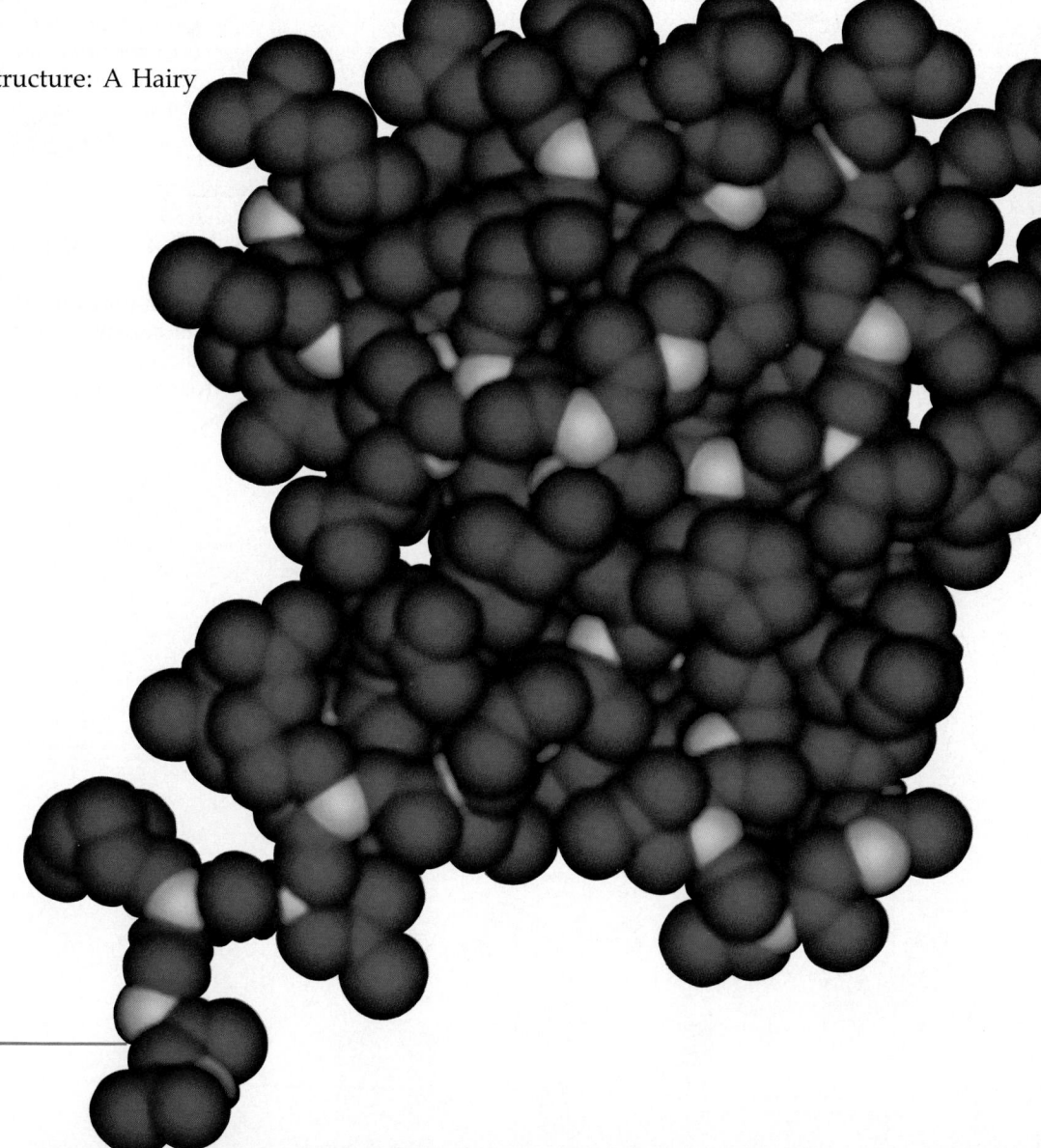

Living things contain an amazing variety of molecules, from the proteins of horn and hoof to the carbohydrates of sugars and tree trunks. All organic molecules owe their versatility to the carbon atom. A carbon atom has four electrons in an outermost shell that has room for eight. Therefore, carbon atoms become stable by sharing four electrons with other atoms, forming up to four single covalent bonds or smaller numbers of double or triple bonds. Molecules with many carbon atoms can assume complex shapes, including chains, branches, and rings.

Organic molecules are much more than just complicated skeletons of carbon atoms, however. Attached to the carbon backbone are groups of atoms, called **functional groups,** that determine the characteristics and chemical reactivity of the molecules. The common functional groups found in organic molecules are diagrammed in Table 3-1.

Synthesizing Organic Molecules: A Modular Approach

In principle, there are two ways to manufacture a large, complex molecule: one could synthesize the molecule atom-by-atom according to an extremely detailed blueprint, or one could take preassembled smaller molecules and hook them together. Just as trains are made by coupling engines, boxcars, coal cars, and cabooses, so too life on Earth takes the modular approach. Small molecules (for example, amino acids) are used as **subunits** with which to synthesize longer molecules (for example, proteins), like cars in a train (Fig. 3-1a). The individual subunits are often called **monomers** (from Greek words meaning "one part"); long chains of monomers are called **polymers** ("many parts").

Table 3-1 Important Functional Groups in Biological Molecules

Group	Structure	Properties	Types of Molecules
Hydrogen (—H)	—H	Polar or nonpolar, depending on what atom hydrogen is bonded to; involved in condensation and hydrolysis	Almost all organic molecules
Hydroxyl (—OH)	—O—H	Polar; involved in condensation and hydrolysis	Carbohydrates, nucleic acids, alcohols, some acids and steroids
Carboxylic acid (carboxyl; —COOH)		Acid; negatively charged when H^+ dissociates, involved in peptide bonds	Amino acids, fatty acids
Amino (—NH$_2$)		Basic; may bond an additional H^+, becoming positively charged; involved in peptide bonds	Amino acids, nucleic acids
Phosphate (—H$_2$PO$_4$)		Acid; up to two negative charges when H^+'s dissociate; links nucleotides in nucleic acids; energy-carrier group in ATP	Nucleic acids, phospholipids
Methyl (—CH$_3$)		Nonpolar; tends to make molecules hydrophobic	Many organic molecules; especially common in lipids

Condensation and Hydrolysis

Biological molecules almost always use the same type of chemical reaction, called a **condensation reaction,** to join subunits to one another. In a condensation reaction, a hydrogen (—H), removed from one subunit, and a hydroxyl (—OH), removed from a second subunit, "condense" to form a molecule of water (H_2O), as the subunits are joined by a covalent bond (Fig. 3-1b). The reverse reaction, called **hydrolysis** (literally, "to break apart with water"), can split the molecule into individual subunits again (Fig. 3-1c). During hydrolysis, water is added back, a hydrogen to one subunit and a hydroxyl to the other.

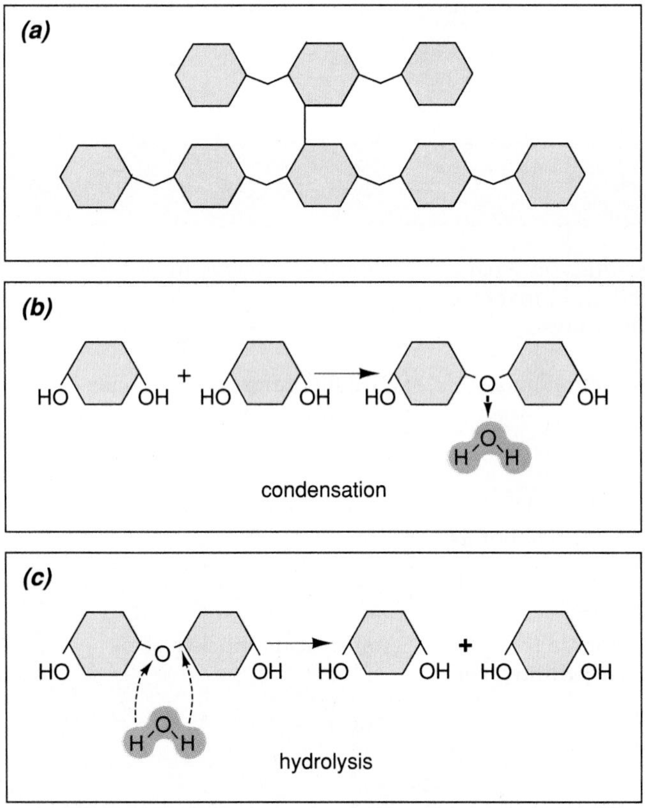

Figure 3-1 Synthesis and breakdown of organic molecules.
(a) A typical organic molecule is composed of similar or identical subunits linked together by covalent bonds.
(b) In a condensation reaction, two subunits are joined by a covalent bond. Simultaneously, a hydroxyl group is removed from one subunit and combines with a hydrogen removed from a second subunit to form water.
(c) Hydrolysis is the reverse of condensation. Hydrogen and hydroxyl from water are added to the subunits as the large organic molecule is broken apart into its subunits.

The Principal Types of Biological Molecules

Although the bodies of living things often include thousands of different organic molecules, nearly all fall into one of four categories: carbohydrates, lipids, proteins, or nucleic acids.

Carbohydrates

All **carbohydrates** are either small, water-soluble **sugars** (glucose, fructose) or chains made by stringing sugar subunits together (starch, cellulose). If a carbohydrate consists of just one sugar molecule, it is called a **monosaccharide** (Greek for "single sugar"). When two or more monosaccharides are linked together, they form a **disaccharide** ("two sugars") or a **polysaccharide** ("many sugars").

Carbohydrates such as sugars and starches are important energy sources for most organisms. Other carbohydrates, such as cellulose and similar molecules, provide structural support for individual cells or even for the entire bodies of organisms as diverse as plants, fungi, bacteria, and insects.

Sugars usually have a backbone of three to seven carbon atoms (Fig. 3-2a, left). Most of the carbon atoms have both a hydrogen (—H) and a hydroxyl group (—OH) attached to them. Therefore, the general formula for a sugar is $(CH_2O)_n$, where n is the number of carbons in the backbone. This formula explains the origin of the name "carbohydrate," which means simply "carbon plus water." When dissolved in water, such as in the cytoplasm of a cell, the carbon backbone of a sugar usually "circles up" into a ring (Fig. 3-2a, right). It is in this ring form that sugars link together to make disaccharides and polysaccharides (see Fig. 3-3).

Most small carbohydrates are water soluble. As in water molecules, the O—H bond in a hydroxyl group is polar, because oxygen attracts electrons more strongly than hydrogen does. Hydrogen bonds between water molecules and the polar hydroxyl groups keep the carbohydrate in solution (see Fig. 2-10 in the previous chapter).

Monosaccharides

Glucose (Fig. 3-2a) is the most common monosaccharide in living organisms, and is the subunit of which most polysaccharides are made. Glucose has six carbons, and hence has the chemical formula $(CH_2O)_6$, or $C_6H_{12}O_6$. Many organisms also synthesize other monosaccharides that have the same chemical formula as glucose, but with slightly different structures (Fig. 3-2b). These include fructose (the "corn sugar"

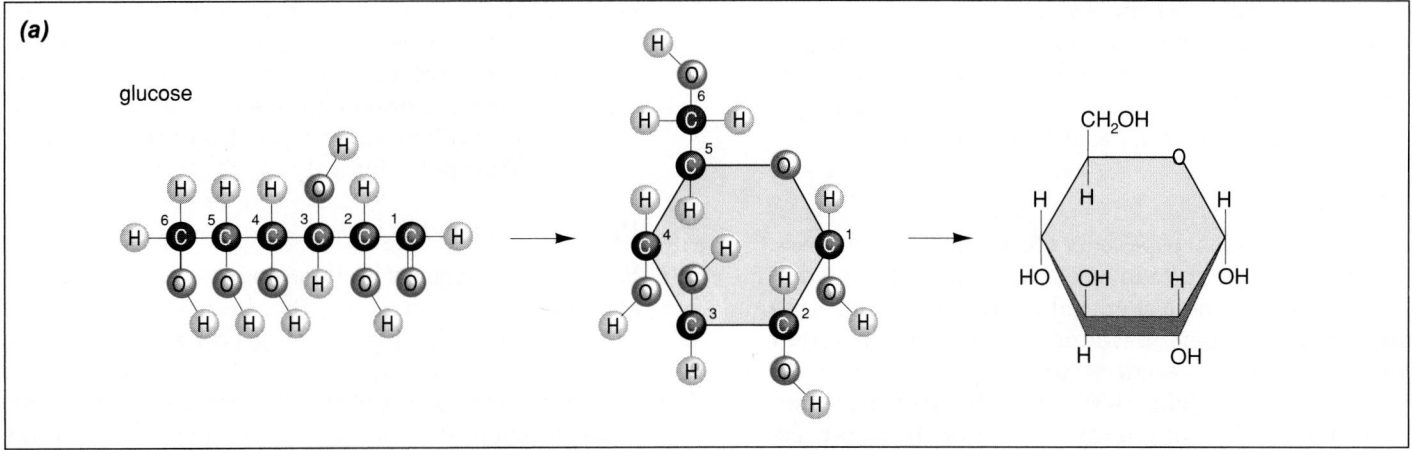

(a) glucose

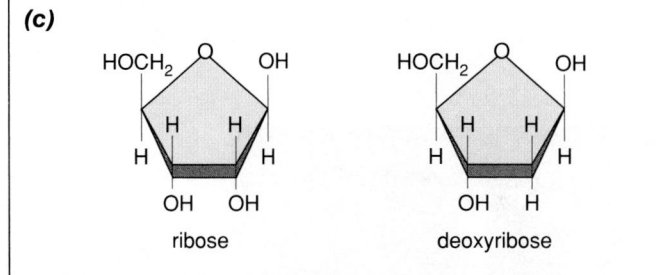

(b) fructose galactose **(c)** ribose deoxyribose

Figure 3-2 Monosaccharide structure.
(a) The most common monosaccharides have "backbones" of either five or six carbon atoms (chain diagram of glucose, left). When dissolved in water, however, the chain bends upon itself to form a ring (center). The ring forms are usually drawn on paper as if you were looking at the edge of the ring (right). The thick edge projects out of the paper toward you, the thin edge recedes behind the paper, and the —H, —OH, and —CH₂OH groups are perpendicular to the ring, in the plane of the paper. For convenience, carbon atoms at the corners of the polygons are omitted, but other atoms (such as oxygen) are usually shown. **(b)** Fructose and galactose have the same atomic composition as glucose, but a different structure. **(c)** Ribose and deoxyribose are five-carbon monosaccharides that form parts of nucleic acids.

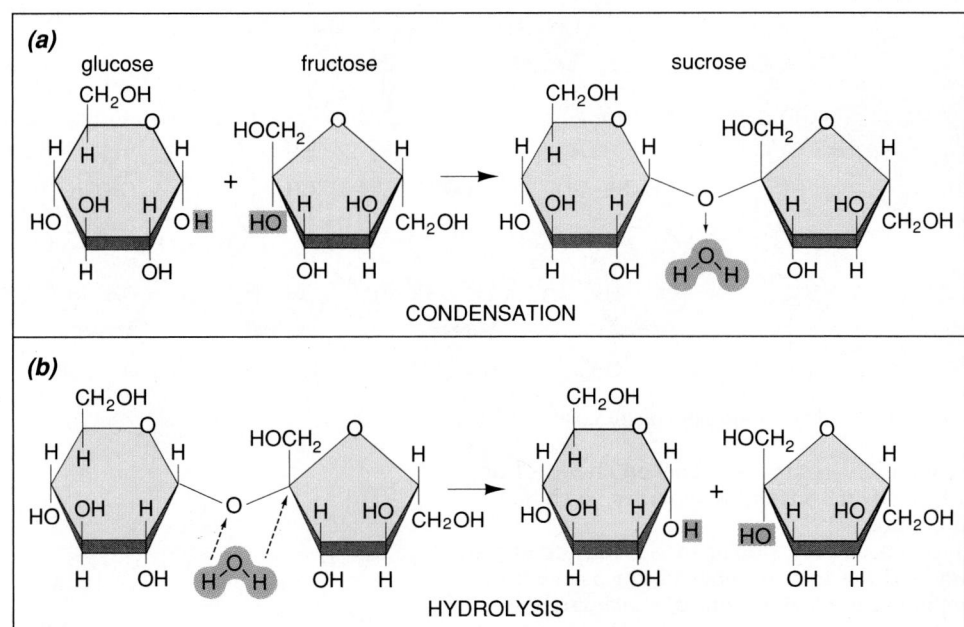

(a) glucose fructose sucrose CONDENSATION

(b) HYDROLYSIS

Figure 3-3 Synthesis and breakdown of a disaccharide.
(a) The disaccharide sucrose is synthesized by a condensation reaction, in which a hydrogen (—H) is removed from glucose and a hydroxyl group (—OH) is removed from fructose, forming a water molecule and leaving the two monosaccharide rings joined by single bonds to the remaining oxygen atom.
(b) Hydrolysis of sucrose is just the reverse of its synthesis, as water is split and added back to the monosaccharides.

of the food industry) and galactose (part of lactose, or "milk sugar"). Other important monosaccharides, such as ribose and deoxyribose, have five carbons (Fig. 3-2c). Ribose and deoxyribose are parts of the genetic molecules RNA and DNA, respectively.

Disaccharides

Monosaccharides, especially glucose and its relatives, have a short life span in a cell. Most are either metabolized, freeing their chemical energy for use in driving needed cellular reactions, or are linked together by condensation reactions to form disaccharides or polysaccharides (Fig. 3-3). Disaccharides are often used for short-term energy storage or transport, especially in plants. Common disaccharides include

sucrose (table sugar: glucose plus fructose), **lactose** (milk sugar: glucose plus galactose), and **maltose** (glucose plus glucose, formed during the digestion of starch). When the organism needs energy, the disaccharides are broken apart again into their monosaccharide subunits by hydrolysis.

Polysaccharides

For long-term energy storage, monosaccharides, usually glucose, are joined together into polysaccharides, forming **starch** (in plants) or **glycogen** (in animals; Fig. 3-4). Starch, commonly found in roots and seeds, may occur as coiled, unbranched chains of up to 1000 glucose subunits, or, more frequently, as huge branched chains of up to half a million glucose mole-

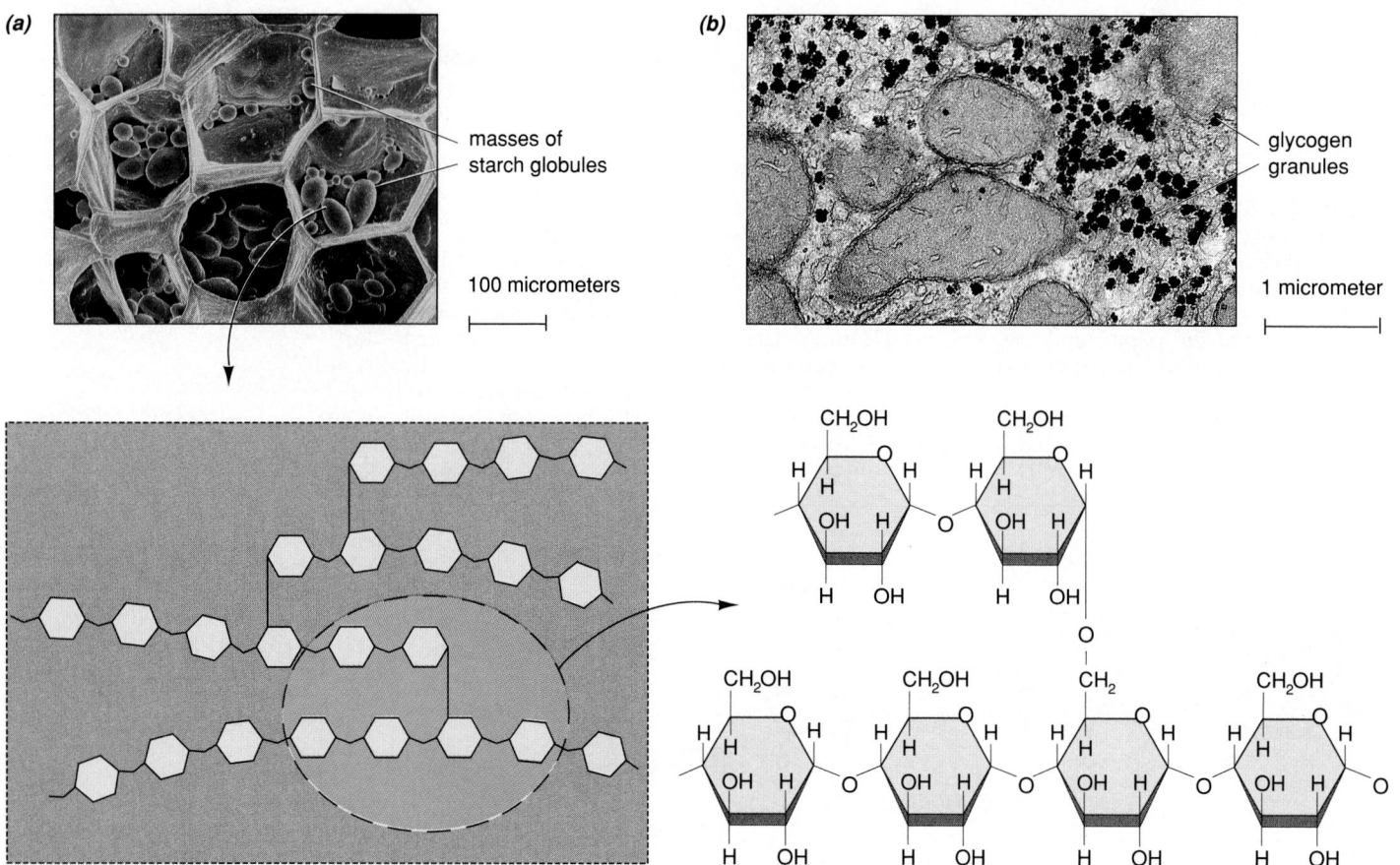

Figure 3-4 Both plants and animals use polysaccharides composed of glucose subunits for energy storage.
(a) Most plants synthesize starch, which commonly occurs as branched chains of up to half a million glucose subunits. Starch forms water-insoluble globules, such as these that make up most of the bulk of a potato.
(b) The animal storage polysaccharide, glycogen, is very similar in structure to starch, except that branches occur more frequently and the total number of glucose subunits is smaller. The electron micrograph shows granules of glycogen in a salamander's liver.

cules. Glycogen, stored in the liver and muscles of vertebrates, is usually much smaller than starch, with branches every 10 to 12 glucose subunits. Having many small branches probably makes it easier to split off the glucose subunits for quick energy release.

Many organisms use polysaccharides as structural materials. The most familiar structural polysaccharide is **cellulose,** which makes up a large part of the cell walls of plants and about half the bulk of a tree trunk (Fig. 3-5). When you picture the vast fields and forests blanketing much of our planet, you will not be surprised to learn that there is probably more cellulose on Earth than all other organic molecules put together. Ecologists estimate that about a *trillion tons of cellulose* are synthesized each year!

Like starch, cellulose consists of glucose subunits bonded together; however, most animals can easily digest starch, while only a few microbes, such as those in the digestive tracts of cows and termites, can digest cellulose. Why? In cellulose the orientation of the bonds between subunits is different, so that every other glucose is "upside down" (compare Fig. 3-5 with Fig. 3-4). This difference in orientation prevents the digestive enzymes of animals from attacking the bonds between glucose subunits. Wholly different enzymes synthesized by certain microbes can break these bonds. As a result, cellulose is food for these microbes, but for most animals cellulose is "roughage," passing unscathed through the digestive tract.

Polysaccharides are also the starting point for the

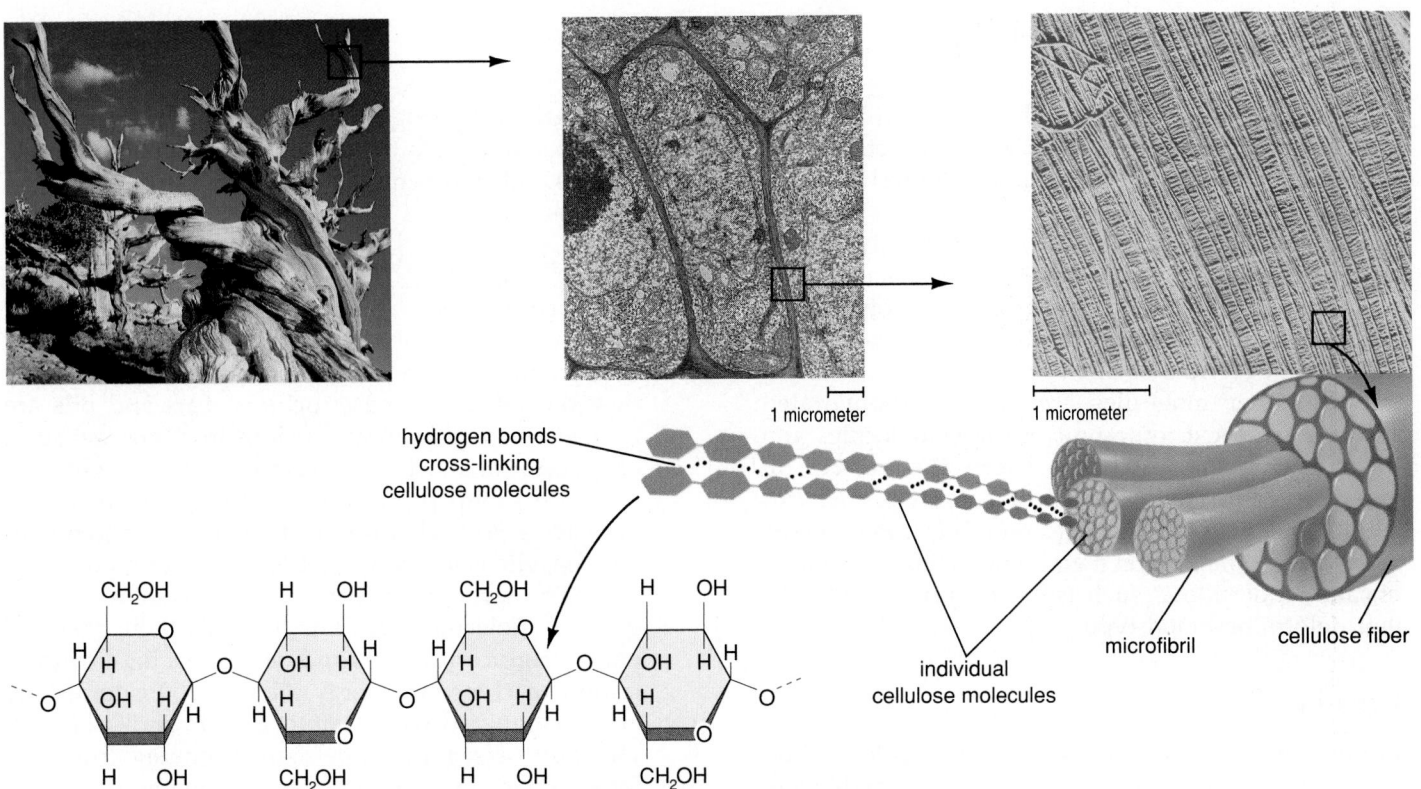

Figure 3-5 Cellulose structure and function. Cellulose, like starch, is composed of glucose subunits, but the orientation of the bond between subunits is different (compare with Fig. 3-4a), so that every other glucose molecule is "upside down." Unlike starch, cellulose has great structural strength, due partly to the difference in bonding and partly to the arrangement of parallel molecules of cellulose into long, cross-linked fibers. Plant cells often lay down cellulose fibers in layers that run at angles to each other, resulting in resistance to tearing in both directions. The final product can be incredibly tough, as this 3,000-year-old bristlecone pine in California's White Mountains testifies.

Figure 3-6 Chitin has the same "alternating upside down" bonding of glucose molecules as cellulose; the difference is that the glucose subunits are modified by replacement of one of the hydroxyl groups with a nitrogen-containing side group (yellow). Tough, slightly flexible chitin supports the otherwise soft bodies of arthropods and fungi.

synthesis of many other important molecules. The hard outer coverings (exoskeletons) of insects, crabs, and spiders are made of **chitin,** a polysaccharide in which the individual glucose subunits have been chemically modified (Fig. 3-6). Interestingly, chitin also stiffens the cell walls of many fungi. Bacterial cell walls contain still other types of modified polysaccharides, as do the lubricating fluids in our joints and the transparent corneas of our eyes.

Many other molecules are partly carbohydrate. Perhaps the most important of these molecules are the nucleic acids (discussed later), the carriers of hereditary information in all organisms. Other examples include mucus, some hormones, and many molecules in the plasma membrane, including "identification molecules," such as those on red blood cells that determine blood type.

Lipids

Lipids are a diverse assortment of molecules, all of which share two important features: (1) Lipids contain large regions composed almost entirely of hydrogen and carbon, with nonpolar carbon–carbon or carbon–hydrogen bonds. (2) These nonpolar regions make lipids insoluble in water. Lipids serve a wide variety of functions. Some lipids are energy-storage molecules; others make up the bulk of all of the membranes of a cell; others are hormones; and still others form waterproof coverings on both plant and animal bodies.

Lipids are classified into three groups: (1) oils, fats, and waxes, which are similar in structure and contain

only carbon, hydrogen, and oxygen; (2) phospholipids, structurally akin to oils but also containing phosphorus and nitrogen; and (3) the fused-ring family of steroids.

Oils, Fats, and Waxes

These compounds are related in three ways: they contain only carbon, hydrogen, and oxygen; they contain one or more fatty acid subunits; and usually they do not have ring structures. **Fats** and **oils** are formed by condensation reactions from one molecule of **glycerol** (a short, three-carbon molecule with one hydroxyl group per carbon) and three molecules of **fatty acids** (long chains of carbon and hydrogen with a carboxylic acid group [—COOH] at one end [Fig. 3-7]). This structure of three fatty acids joined to one glycerol molecule gives fats and oils their chemical name, **triglyceride.** Fats and oils have a high concentration of chemical energy, about 9300 calories per gram (compared with 4100 for sugars and proteins). They are used for semipermanent energy storage—for example, in bears that feast during summer and fall, putting on fat to tide them over during their winter hibernation.

The difference between a fat (solid at room temperature) and an oil (liquid at room temperature) lies in their fatty acids. Fats have fatty acids with all single bonds in their carbon chains. Hydrogens occupy all the other bond positions on the carbons. The resulting fatty acid (e.g., stearic acid, Fig. 3-8a) is called **saturated** because it is "saturated" with hydrogens—that is, it has as many hydrogens as possible. If there are one or more double bonds between carbons, and

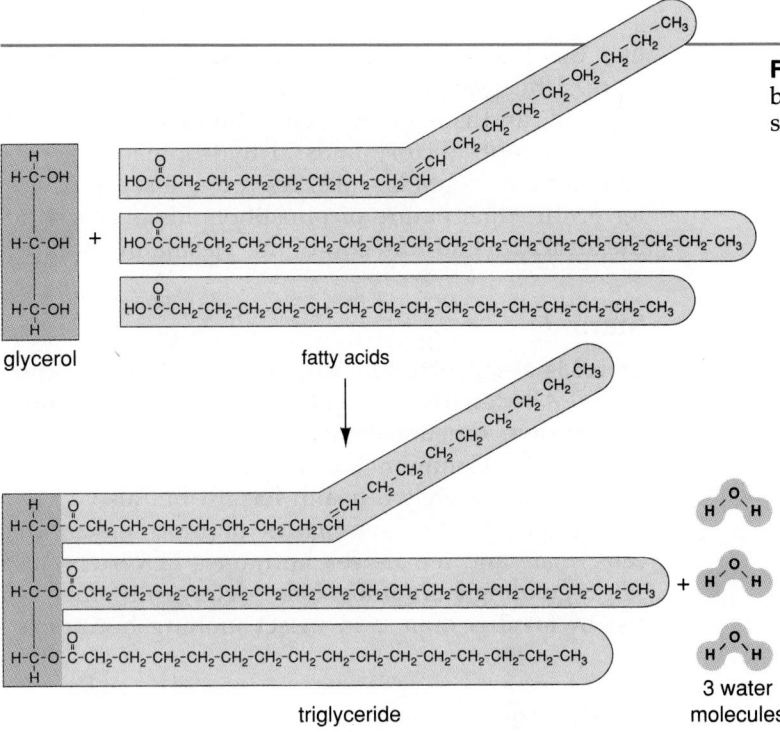

Figure 3-7 Triglycerides (fats and oils) are synthesized by a condensation reaction linking three fatty acids to a single glycerol molecule.

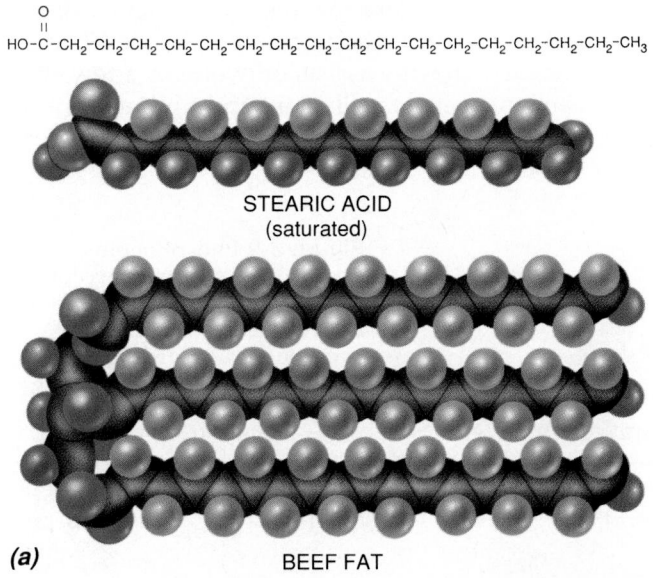

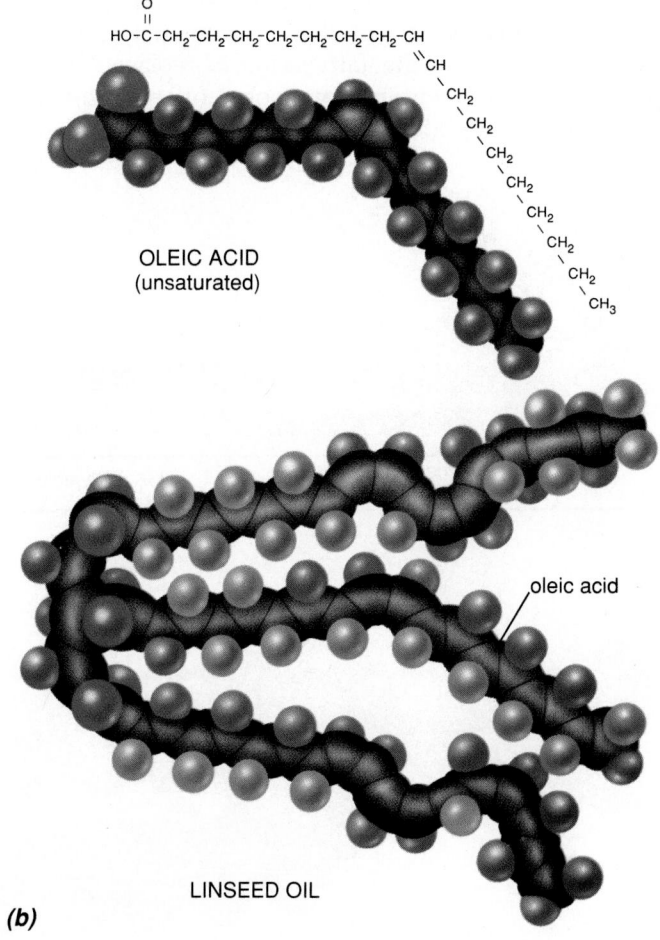

(a) BEEF FAT

Figure 3-8 Fats and oils differ in the degree of saturation of their fatty acids.
(a) A saturated fatty acid, such as stearic acid, has all single bonds between carbons. In beef fat, which is composed of triglycerides with a high proportion of stearic acids, the fatty acid chains pack closely together, forming a solid at room temperature.
(b) An unsaturated fatty acid, such as oleic acid, has one or more double bonds between carbons, which cause kinks in the chain. The kinks in the fatty acids of linseed oil prevent close packing, so that the oil is liquid at room temperature.

STEARIC ACID (saturated)

OLEIC ACID (unsaturated)

oleic acid

LINSEED OIL

(b)

consequently fewer hydrogens, the fatty acid is called **unsaturated** (oleic acid, Fig. 3-8b). Oils have mostly unsaturated fatty acids. The saturated fatty acids of fats can nestle closely together, forming solid lumps at room temperature (beef fat, Fig. 3-8a). The double bonds in the unsaturated fatty acids of oils, on the other hand, form kinks in the fatty acid chains (linseed oil, Fig. 3-8b). The kinks keep oil molecules apart, with the result that an oil is liquid at room temperature. An oil can be converted to a fat by breaking the double bonds between carbons, replacing them with single bonds, and adding hydrogens to the remaining bond positions. This is the "hydrogenated oil" listed in the ingredients on a box of margarine.

Waxes are similar to fats and oils except that the fatty acids are linked to large, long-chained alcohols instead of to glycerol. Waxes form a waterproof coating over the leaves and stems of land plants. Animals also synthesize waxes, as waterproofing for mammalian fur and and insect exoskeletons, and, in a few cases, to build elaborate structures such as beehives.

Phospholipids

The plasma membrane that separates the inside of a cell from the outside world contains several types of **phospholipids.** A phospholipid is similar to an oil, except that one of the fatty acids is replaced by a phosphate group with a short, polar, often nitrogen-containing group attached to the end (Fig. 3-9). Un-

like the fatty acid "tails," which are insoluble in water, the phosphate–nitrogen "head" is polar or charged, and is water soluble. Thus a phospholipid has two contradictory ends: a hydrophilic head attached to hydrophobic tails. As you will see in Chapter 6, this dual nature of phospholipids is crucial to the structure and function of the plasma membrane.

Steroids

Steroids are structurally different from all the other lipids. All steroids are composed of four fused rings with various functional groups protruding from them (Fig. 3-10). Common steroids include cholesterol, which is not merely a health hazard but also a vital component of the membranes of most eukaryotic cells; male and female sex hormones in vertebrates; salt-regulating hormones; bile "detergents" that assist in fat digestion; and insect molting hormones.

Proteins

Proteins are molecules composed of one or more chains of amino acids (see below). Proteins perform many functions. Protein enzymes guide almost all the chemical reactions that occur inside cells (see Chapter 4). Since each enzyme assists only one or a few specific reactions, cells usually contain hundreds of different enzymes. Other proteins are used for struc-

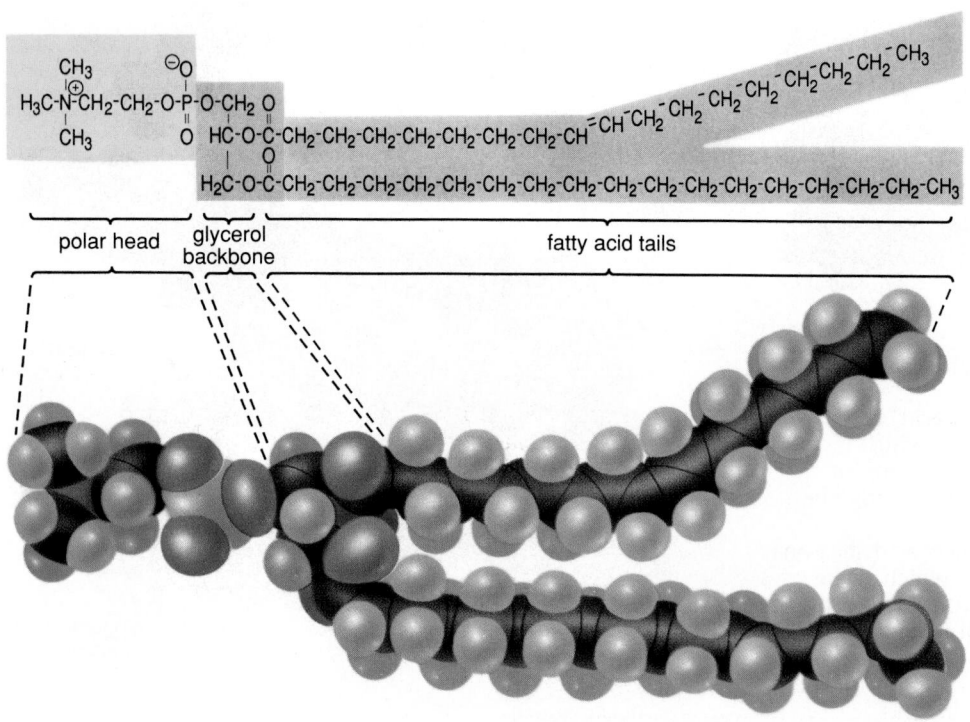

Figure 3-9 Phospholipids are similar to fats or oils, except that only two fatty acid tails are attached to the glycerol backbone. The third position on the glycerol is occupied by a polar head composed of a phosphate group ($-PO_4^-$) to which is attached a second, often nitrogen-containing group.

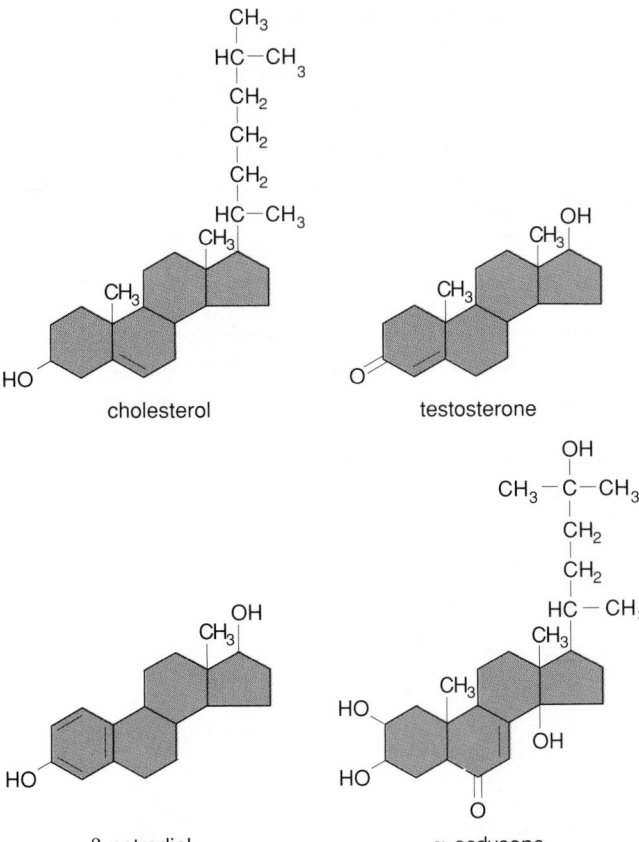

cholesterol

testosterone

β-estradiol

α-ecdysone

Figure 3-10 Steroids are synthesized from cholesterol, and all have almost the same molecular structure (colored rings; note that the carbon atoms at the corners of the rings and the hydrogen atoms attached to the rings have been omitted from these drawings). Differences in functional groups attached to the rings cause great differences in physiologic functioning, including the male sex hormone testosterone, the female sex hormone estradiol (a type of estrogen), and the insect molting hormone ecdysone.

tural purposes, such as elastin, which gives skin its elasticity; keratin, the principal protein of hair, horns, and claws; and the silk of spider webs and silkmoth cocoons (Fig. 3-11). Still other types of proteins are used for energy and material storage (albumin in eggs, casein in milk), transport (hemoglobin to carry oxygen in the blood), and cell movement (contractile proteins in muscle). Hormones (insulin, growth hormone), antibodies, and many poisons (rattlesnake venom) are also proteins.

Amino Acids

Proteins are polymers of **amino acids** (Fig. 3-12). Every amino acid has the same fundamental structure, consisting of a central carbon bonded to four different functional groups: a nitrogen-containing amino group (—NH_2); a carboxylic acid group

Figure 3-11 Common structural proteins include hair **(a)**, horn **(b)**, and spider web silk **(c)**.

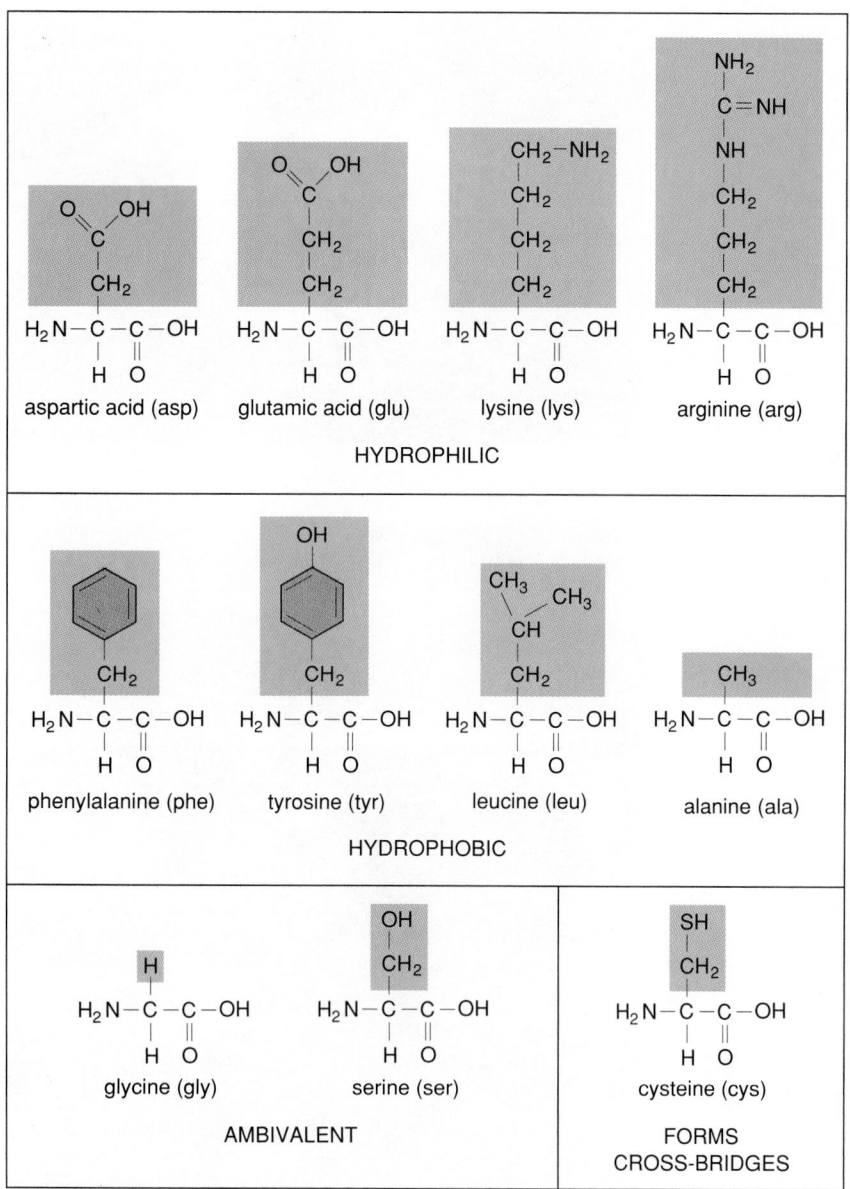

Figure 3-12 A sampling of the diversity of amino acid structure, a consequence of differences in the variable R group (colored blue). Most amino acids are classified according to the variable R group as hydrophilic, hydrophobic, or ambivalent (usually with small R groups that have little effect on the water solubility of the amino acid). Cysteine stands in a class by itself. Two cysteines in distant parts of a protein molecule can form a covalent bond between their sulfur atoms, making a "cross-bridge" that brings the cysteines very close together and bends the protein chain (see Fig. 3-14).

(—COOH); a hydrogen; and a variable group (represented by the letter "R"):

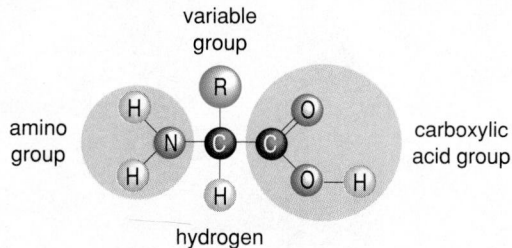

There are 20 amino acids commonly found in the proteins of living organisms. The R group differs among amino acids, and gives each its distinctive properties (see Fig. 3-12). The 20 amino acids can be grouped into a few functional types, based on the nature of the R group: (1) hydrophilic (with acidic, basic, or polar R groups), (2) hydrophobic (large, nonpolar R groups), (3) ambivalent (small, uncharged R groups, not very hydrophilic or hydrophobic), and (4) the sulfur-containing cysteine, two of which can link protein chains together with bonds called **disulfide bridges** (see "A Closer Look at Protein Structure").

Twenty amino acids may seem like a small number of subunits from which to construct thousands of different proteins, but perhaps an analogy with language will help you understand protein structure and function more easily. The English language uses 26 letters to construct thousands of words, each with a distinct meaning based on the exact sequence of letters. However, only a small fraction of all possible letter combinations are used as words; for example, the combination "protein" has meaning, but "nteiopr" does not. Similarly, living organisms construct thousands of different proteins—about 50,000 different ones in your body—from a common alphabet of 20 amino acids. Amino acids differ in their chemical and physical properties—size, water solubility, electric charge—because of their different R groups. **Therefore, the exact sequence of amino acids dictates the function of each protein:** whether it is water-soluble or not, whether it is an enzyme or a hormone or a structural protein. Scrambled sequences of amino acids are useless. In some cases, just one wrong amino acid may cause a protein to function incorrectly.

In addition to their roles as the building blocks of proteins, some water-soluble amino acids are also used by the nervous system as messenger molecules. For example, some nerve cells in your brain release glutamic acid, which binds to specific parts of the plasma membranes of neighboring nerve cells, changing their activity. In fact, the molecular changes that produce some kinds of learning are probably triggered by glutamic acid.

Protein Synthesis

Like lipids and polysaccharides, proteins are synthesized by a condensation reaction. The nitrogen of the amino group of one amino acid is joined to the carbon of the carboxylic acid group of another amino acid by a single covalent bond (Fig. 3-13). This bond is called a **peptide bond,** and the resulting chain of two amino acids is called a **peptide.** More amino acids are added, one by one, until the protein is complete. Protein chains found in living cells vary in length from three to thousands of amino acids. Biochemists have given a variety of names to proteins, depending on their length, including "dipeptide" (two amino acids),

"tripeptide" (three), and "polypeptide" (three or more). Often, the word "protein" is reserved for long peptides, say 50 or more amino acids in length. For simplicity, in this text all amino acid chains, regardless of length, will be called peptides or proteins.

Protein Structure

The phrase "amino acid chains" may evoke images of proteins as floppy, monotonous structures, but this is not correct: proteins are highly organized molecules. Biologists recognize four levels of organization in protein structure (Fig. 3-14). The **primary structure** is the sequence of amino acids that make up the protein (Fig. 3-15). Different types of proteins (e.g., insulin vs. growth hormone) have different sequences of amino acids. However, every molecule of the same type of protein (e.g., every molecule of human insulin) has the same primary structure.

Hydrogen bonds cause many protein chains to form one of two simple, repeating **secondary structures.** Looking back at Figure 3-13, notice that every amino acid subunit retains a —C=O from its carboxylic acid group and an —N—H from its amino group. Since oxygen attracts electrons more strongly than carbon, the oxygen is relatively negative. Similarly, nitrogen attracts electrons more strongly than hydrogen, leaving the hydrogen relatively positive. Therefore, hydrogen bonds can form between the —C=O and —N—H groups of a protein chain. Many proteins, such as the hair protein keratin, have a coiled, Slinky-toy-like shape called a **helix,** in which hydrogen bonds hold together the turns of the coils (Fig. 3-16a). Some proteins, such as silk, are composed of many protein chains lying side by side, with hydrogen bonds holding adjacent chains in a **pleated sheet** arrangement (Fig. 3-16b).

It is rare for an entire protein to be a simple helix or pleated sheet. Most proteins assume complex, three-dimensional **tertiary structures** (see Fig. 3-14). Disulfide bridges may bring otherwise distant parts of a

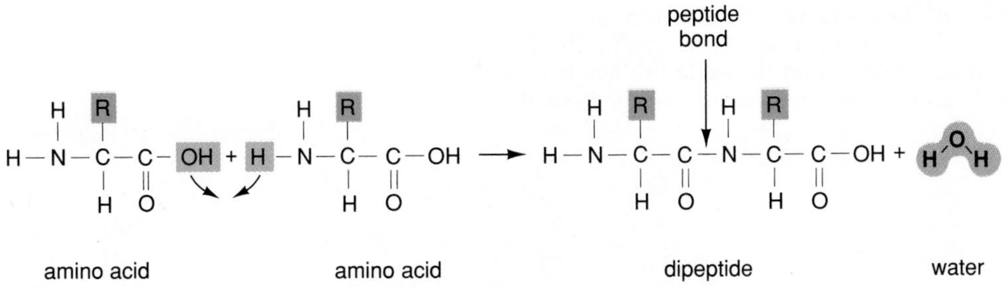

Figure 3-13 In protein synthesis, a condensation reaction joins the carbon of the carboxylic acid group of one amino acid to the nitrogen of the amino group of a second amino acid. The resulting covalent bond is called a peptide bond.

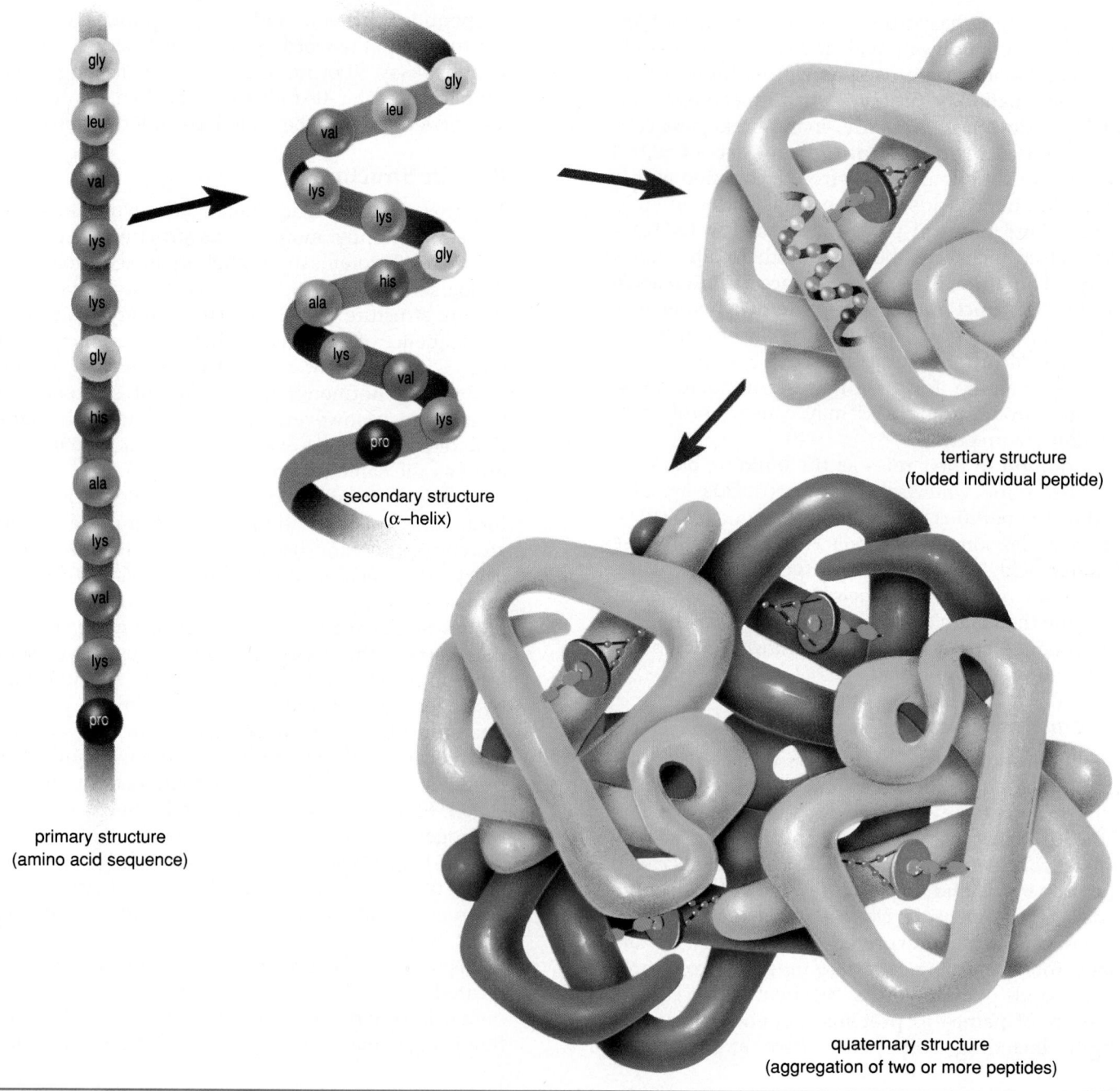

primary structure
(amino acid sequence)

secondary structure
(α–helix)

tertiary structure
(folded individual peptide)

quaternary structure
(aggregation of two or more peptides)

Figure 3-14 The four levels of protein structure, represented here by hemoglobin, the oxygen-carrying protein in red blood cells. All levels of protein structure are determined by the amino acid sequence of the protein, interactions among the R groups of the amino acids (primarily hydrogen bonds and disulfide bridges between cysteines), and interactions between the R groups and their surroundings (usually water or lipids).

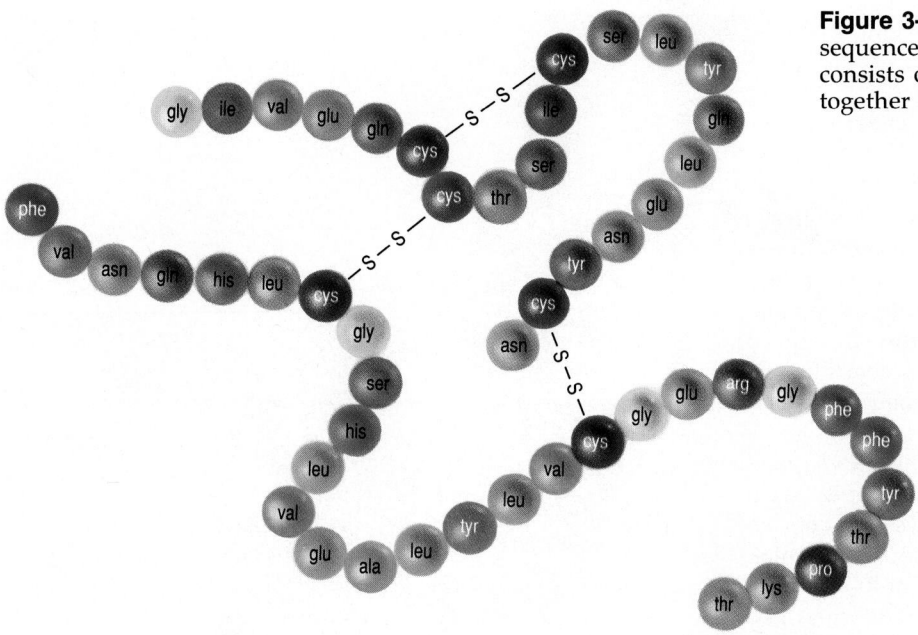

Figure 3-15 The primary structure (amino acid sequence) of insulin. The insulin molecule actually consists of two small amino acid chains, held together by two disulfide bridges between cysteines.

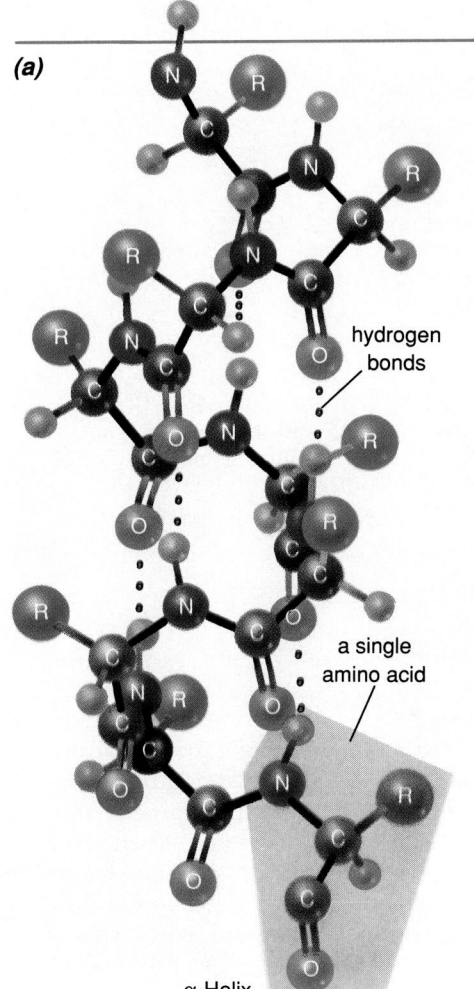

α-Helix

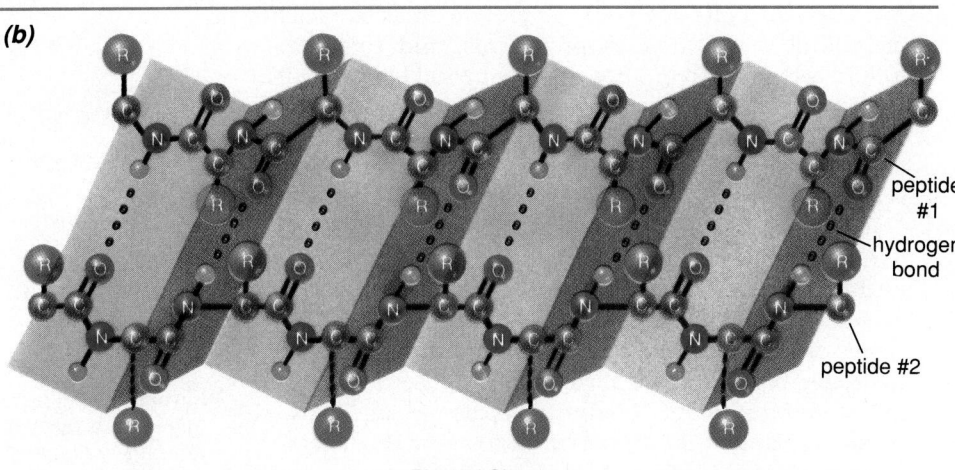

Pleated Sheet

Figure 3-16 The two most common secondary structures of proteins are the helix and the pleated sheet.

(a) In a helix, hydrogen bonds (dots) form between the acid group oxygen of one amino acid and the amino group hydrogen of the third amino acid "up" the helix. These hydrogen bonds hold the protein chain in a spiral, in which about $3\frac{1}{2}$ amino acids make up each turn of the helix. The variable R groups (green) project outward from the helix.

(b) In a pleated sheet, several peptide chains lie side-by-side (running zig-zag *across* the sheets left to right). Hydrogen bonds between peptides (lengthwise *along* the sheets top to bottom) hold each sheet together. The R groups (green) project alternately above and below the sheet. Despite its accordion-like appearance, each peptide chain is in its fully extended state, and cannot be easily stretched. For this reason, pleated sheet proteins such as silk are strong but not elastic.

A CLOSER LOOK

At Protein Structure: A Hairy Subject

A single strand of human hair, about 20 millionths of a meter in diameter and not even alive, is nonetheless a highly organized, complex structure. Hair is composed mostly of a single, helical protein called keratin. (There are also pigments that give hair its color, but these won't concern us here.) If we look closely at the structure of hair, we can learn a great deal about biological molecules, chemical bonds, and why human hair behaves as it does.

In the molecular structure of hair, nature anticipated the technique of sailors who made strong ropes out of weak individual fibers of hemp: a series of fibers are twisted about one another, with bundles of fibers making up the final product. A single strand of hair consists of a hierarchy of structures (Fig. E3-1). The outermost layer is a set of overlapping shingle-like scales that protect the hair and keep it from drying out. Inside the hair lie closely packed, dead cells, each filled with long strands running parallel to the axis of the hair. These strands in turn are composed of thinner strands called microfibrils, embedded in a protein matrix. Each microfibril is a bundle of protofibrils, and each protofibril is composed of three or more helical keratin molecules twisted together. As a hair grows, living cells in the hair follicle embedded in the skin whip out new keratin at the rate of ten turns of the protein helix every second.

If you pluck a hair from your head, you will notice that it is rather strong: it smarts a bit to break off a healthy hair. Hair gets its strength from three types of chemical bonds (see Fig. E3-1). First, the individual molecules of keratin are held in their helical shape by many hydrogen bonds. Before a hair will break, all the hydrogen bonds of all the keratin molecules in one cross-sectional plane of the strand break to allow the helix to be stretched to its maximal extent. Second, each molecule is cross-linked to neighboring keratins by disulfide bridges between cysteines. Some of these bridges have to separate as the hair stretches. Finally, at least one peptide bond in each keratin molecule must break before the strand as a whole breaks.

Hair is also fairly stiff. The stiffness arises from hydrogen bonds within the individual helices of keratin and from disulfide bridges holding neighboring keratin molecules together. When hair gets wet, however, the hydrogen bonds between turns of the helices are replaced by hydrogen bonds between the amino acids and the water molecules surrounding them, so the helices collapse. Wet hair is therefore very limp. If wet hair is rolled onto curlers and allowed to dry, the hydrogen

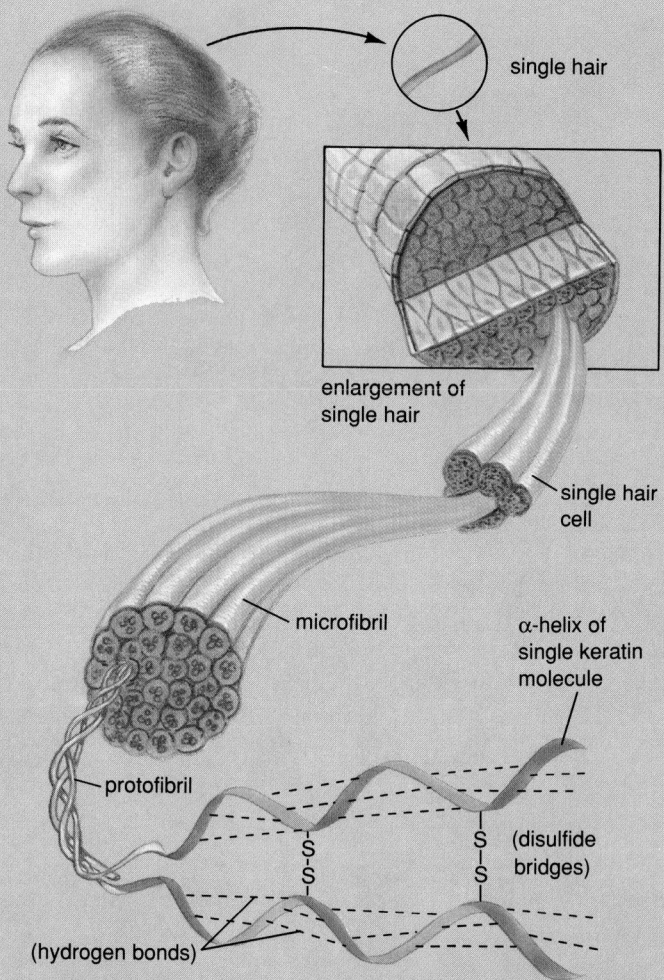

Figure E3-1 The microscopic organization of a single hair is one of bundles of fibers within further bundles of fibers. Hydrogen bonds and disulfide bridges between cysteines impart strength and elasticity to individual hairs.

bonds reform in slightly different places, holding the hair in a curve. The slightest moisture, even humid air, allows these hydrogen bonds to rearrange into their natural configuration, and the hair straightens out again.

Pull gently and you will discover still another property of hair. It stretches and then springs back into shape when you release the tension. When hair stretches, many of the hydrogen bonds within each keratin helix are broken, allowing the helix to be extended. Most of the disulfide bonds between different levels of the neighboring helices, on the other hand, are distorted by stretching but do not break. When tension is released, these disulfide bridges contract, returning the hair to its normal length.

Finally, each hair has a characteristic shape: it may be straight, wavy, or curly. The curliness of hair is genetically specified, and is determined biochemically by the arrangement of disulfide bridges (Fig. E3-2). Curly hair has disulfide bridges crosslinking the various keratin molecules at *different levels*, while straight hair has bridges mostly at the *same level*. When straight hair is given a "permanent wave," two lotions are applied. The first lotion breaks disulfide bonds between neighboring helices. The hair is then rolled tightly onto curlers, and a second solution, which reforms the bridges, is applied. The new disulfide bridges connect helices at different levels, holding the strands of hair in a curl. These new bridges are more or less permanent, and genetically straight hair can be transformed into biochemically curly hair. The new hair that grows in, of course, will have the genetically determined arrangement of bridges, and will not be curly.

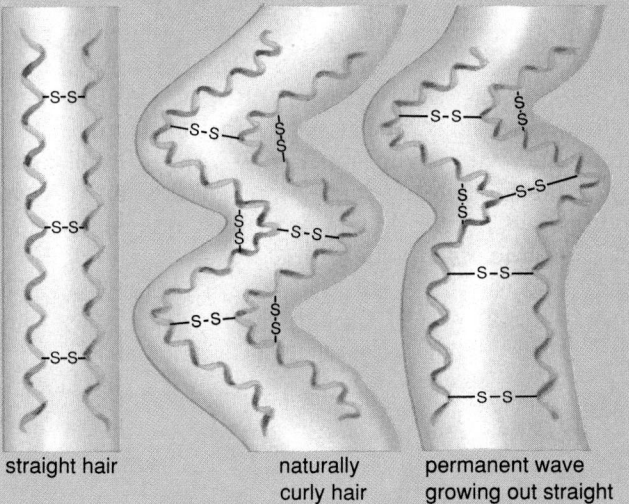

straight hair   naturally   permanent wave
               curly hair   growing out straight

Figure E3-2 If disulfide bridges join individual keratin molecules in a hair at the same level, the hair will be fairly straight. If disulfide bridges connect different levels within or between molecules, the hair will have a natural curl. The same effect can be obtained with a permanent wave that breaks and reforms the disulfide bridges of naturally straight hair. The new hair growing out, however, will be straight.

protein close together. Internal stresses, owing to the particular amino acids present, also may contort a protein. For example, amino acids with very large R groups, such as phenylalanine (see Fig. 3-12), are too bulky to fit side by side in a simple helix. As a result, the helix bends.

Probably the most important influence on the tertiary structure of a protein is its cellular environment, specifically whether the protein is dissolved in the water of the cytoplasm, in the lipids of the membranes, or half in one and half in the other. Hydrophilic amino acids can form hydrogen bonds with nearby water molecules, whereas hydrophobic amino acids cannot. Therefore, a protein dissolved in water folds into an irregular glob, with its hydrophilic amino acids facing the outside watery environment and its hydrophobic ones clustered in the center of the molecule.

Peptides often join with other peptides to form the fourth level of protein organization, the **quaternary structure.** The best-studied such "super-protein" is hemoglobin (see Fig. 3-14), the oxygen-carrying pigment in red blood cells. A single molecule of hemoglobin consists of two pairs of very similar peptides, held together by hydrogen bonds. Although no one knows precisely why hemoglobin contains four peptides, it is known that a single peptide binds oxygen more tightly than the four-peptide molecule does. This would be great for picking up oxygen in the lungs, but, since the single peptides would not release the oxygen again under normal conditions, the rest of the body's cells would die from lack of oxygen. Interactions among the four peptides allow hemoglobin to bind oxygen tightly enough to acquire oxygen in the lungs and still be able to give it up again to the body tissues.

Protein Function

The exact type, position, and number of different R groups of the amino acids of a protein determine both the structure of the protein and its biological function. In any given protein, however, some R groups are more important than others. In hemoglobin, for example, certain amino acids must be present in precisely the right places to hold the iron atoms that bind oxygen. Many of the other amino acids are interchangeable to some extent, if they are functionally equivalent. For example, the amino acids on the outside of a hemoglobin molecule mostly serve to keep it dissolved in the cytoplasm of a red blood cell. Therefore, as long as they are hydrophilic, it doesn't matter too much exactly which amino acids are where.

Nucleic Acids

As you will learn in later chapters, the amino acid sequence of every protein in your body is specified by the genetic instructions residing in the nuclei of your cells. In fact, that's precisely what most genes are: a set of instructions spelling out the amino acid sequences of your body's proteins. Genes are composed of the fourth major category of biological molecule, **nucleic acids.** Nucleic acids are chains of similar but not identical subunits called **nucleotides.** There are two distinct classes of nucleotides, the ribose nucleotides and the deoxyribose nucleotides. All nucleotides have a three-part structure: a five-carbon sugar (ribose or deoxyribose), a phosphate group, and a nitrogen-containing base that differs among nucleotides:

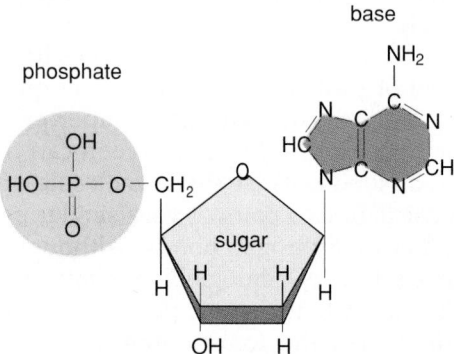

DEOXYRIBOSE NUCLEOTIDE

Four different bases—adenine, guanine, cytosine, and thymine—yield four different deoxyribose nucleotides. Similarly, there are four types of ribose nucleotides, with adenine, guanine, cytosine, or uracil as the base (see Chapters 12 and 13).

Nucleotides may be strung together in long chains, with the phosphate of one nucleotide covalently bonded to the sugar of another:

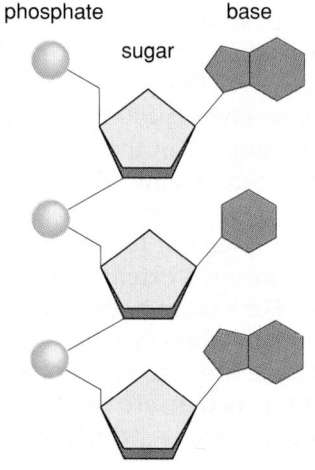

NUCLEOTIDE CHAIN

Deoxyribose nucleotides form chains millions of units long called **deoxyribonucleic acid,** or **DNA.** DNA is found in the chromosomes of all living things, and its sequence of nucleotides, like the dots and dashes of a biological Morse code, spells out the genetic information needed to construct the proteins of each organism (see Chapter 12). Chains of ribose nucleotides, called **ribonucleic acid,** or **RNA,** are copied from the central repository of DNA in the nucleus of each cell. RNA carries DNA's genetic code into the cytoplasm and directs the synthesis of proteins (see Chapter 13).

Other Nucleotides

Not all nucleotides are part of DNA or RNA molecules. Some exist separately in the cell or occur as parts of other molecules. **Cyclic nucleotides,** such as cyclic adenosine monophosphate (cyclic AMP; Fig. 3-17a) are intracellular messengers that carry information from the plasma membrane to other molecules in the cell. Cyclic AMP is synthesized when certain hormones come in contact with the plasma membrane. Cyclic AMP then stimulates essential reactions in the cytoplasm or nucleus (see Chapter 35).

Some nucleotides have extra phosphate groups. These diphosphate and triphosphate nucleotides, such as **adenosine triphosphate** (ATP; Fig. 3-17b), are unstable molecules that carry energy from place to place within a cell. They capture energy where it is produced (during photosynthesis, for example) and give it up to drive energy-demanding reactions elsewhere (say, to synthesize a protein). Finally, certain nucleotides assist enzymes in their action: these are called **coenzymes,** and usually consist of a nucleotide combined with a vitamin (Fig. 3-17c). You will learn more about energy-carrier nucleotides and coenzymes in Chapter 4, where we discuss energy production and use in the cell.

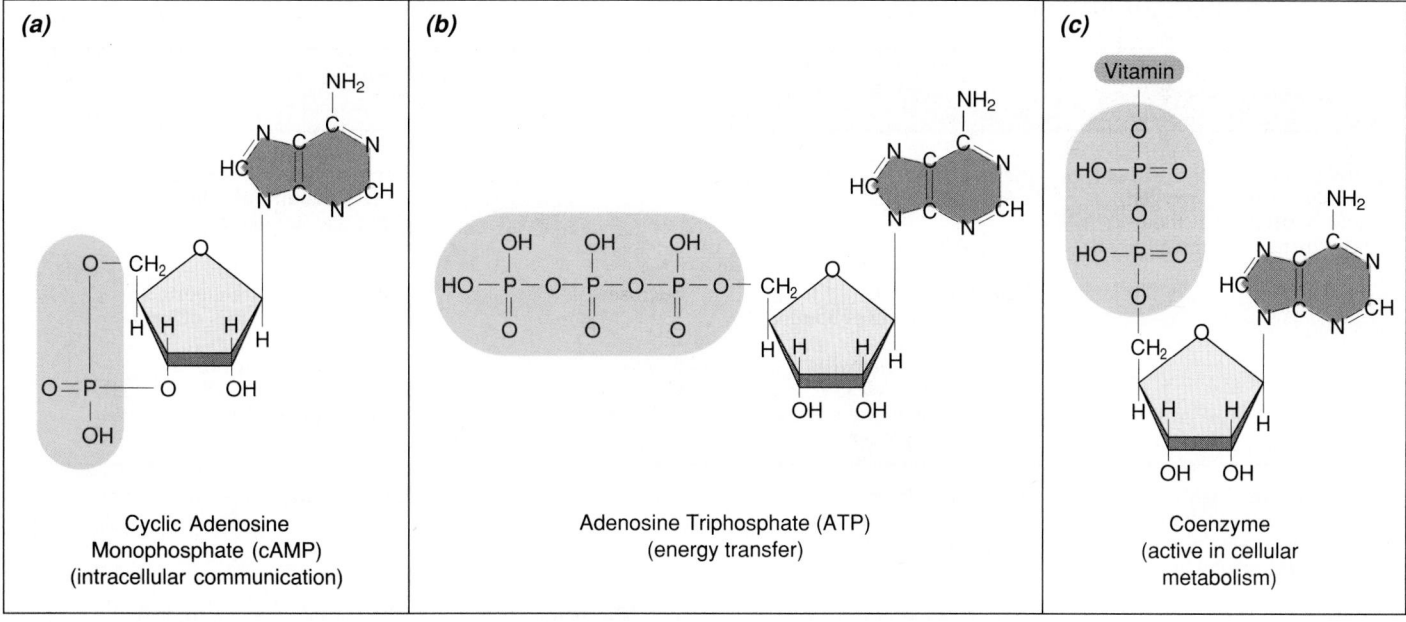

Figure 3-17 A sampling of the diversity of nucleotides. Individual nucleotides, often modified by the addition of functional groups, serve a variety of cellular functions.

SUMMARY OF KEY CONCEPTS

Synthesizing Organic Molecules: A Modular Approach

Most large biological molecules are polymers synthesized by linking together many smaller subunit monomers. Chains of subunits are connected by covalent bonds through condensation reactions; the chains may be broken apart again by hydrolysis reactions.

The Principal Types of Biological Molecules

The most important organic molecules fall into one of four classes: carbohydrates, lipids, proteins, and nucleic acids. Their major characteristics are summarized in Table 3-2.

Carbohydrates include sugars, starches, and cellulose. Sugars (monosaccharides and disaccharides) are used for temporary storage of energy and for the construction of other molecules. Starches and glycogen are polysaccharides that serve for long-term energy storage in plants and animals, respectively. Cellulose and related polysaccharides form cell walls of bacteria, fungi, plants, and some protists.

Lipids are nonpolar, water-insoluble molecules of diverse chemical structure, and include oils, fats, waxes, phospholipids, and steroids. Lipids are used for energy storage (fats and oils), as waterproofing for the outer layers of land organisms (waxes), as the principal component of cellular membranes (phospholipids), and as hormones (steroids).

Proteins are chains of amino acids. The structure and function of a protein is determined by the sequence of amino acids in the chain. Proteins may be enzymes (biological catalysts), structural molecules (hair, horn), hormones (insulin), or transport molecules (hemoglobin).

Nucleic acid molecules are chains of nucleotides. Each nucleotide is composed of a phosphate group, a sugar group, and a nitrogen-containing base. The two types of nucleic acids are deoxyribonucleic acid (DNA) and ribonucleic acid (RNA). Other nucleotides include energy-carrier molecules (ATP), intracellular messengers (cyclic AMP), and coenzymes.

Table 3-2 The Principal Biological Molecules

Class of Molecule	Principal Subtypes	Example	Function
Carbohydrate: Usually contains carbon, oxygen, and hydrogen, in the approximate formula $(CH_2O)_n$	*Monosaccharide:* Simple sugar	Glucose	Important energy source for cells; subunit of which most polysaccharides are made
	Disaccharide: Two monosaccharides bonded together	Sucrose	Principal sugar transported throughout bodies of land plants
	Polysaccharide: Many monosaccharides (usually glucose) bonded together	Starch Glycogen Cellulose	Energy storage in plants Energy storage in animals Structural material in plants
Lipid: Contains high proportion of carbon and hydrogen; usually non-polar and insoluble in water	*Triglyceride:* Three fatty acids bonded to glycerol	Oil, fat	Energy storage in animals, some plants
	Wax: Variable numbers of fatty acids bonded to long-chain alcohol	Waxes in plant cuticle	Waterproof covering on leaves and stems of land plants
	Phospholipid: Polar phosphate group and two fatty acids bonded to glycerol	Phosphatidylcholine	Common component of membranes in cells
	Steroid: Four fused rings of carbon atoms with functional groups attached	Cholesterol	Common component of membranes of eukaryotic cells; precursor for other steroids such as testosterone, bile salts
Protein: Chains of amino acids; contain carbon, hydrogen, oxygen, nitrogen, and sulfur		Keratin	Helical protein, principal component of hair
		Silk	Pleated sheet protein produced by silkmoths and spiders
		Hemoglobin	Globular protein composed of four subunit peptides; transport of oxygen in vertebrate blood
Nucleic acid: Made of nucleotide subunits; may consist of a single nucleotide or long chains of nucleotides	*Long-chain nucleic acids*	Deoxyribonucleic acid (DNA)	Genetic material of all living cells
		Ribonucleic acid (RNA)	Genetic material of some viruses; in living cells, essential in transfer of genetic information from DNA to protein
	Single nucleotides	Adenosine triphosphate (ATP)	Principal short-term energy carrier molecule in cells
		Cyclic adenosine monophosphate (cyclic AMP)	Intracellular messenger

GLOSSARY

adenosine triphosphate (a-den'-ō-sēn trī-fos'-fāt; ATP): a nucleotide with three phosphate groups that serves as an energy-carrier molecule in cells.

amino acid: the individual subunit of which proteins are made, composed of a central carbon atom to which is bonded an amino group (—NH₂), a carboxylic acid group (—COOH), a hydrogen atom, and a variable group of atoms denoted by the letter "R."

carbohydrate: a compound composed of carbon, hydrogen, and oxygen, with the approximate chemical formula $(CH_2O)_n$; includes sugars and starches.

cellulose: an insoluble carbohydrate composed of glucose subunits; forms the cell wall of plants.

chitin (kī'-tin): a compound found in the cell walls of fungi and the exoskeletons of insects and some other arthropods, composed of chains of nitrogen-containing, modified glucose molecules.

coenzyme (kō-en'-zīm): an organic molecule that assists enzymes in their actions.

condensation reaction: a chemical reaction in which two molecules are joined by a covalent bond with the simultaneous removal of a hydrogen from one molecule and a hydroxyl group from the other, forming water.

cyclic nucleotide (sik'-lik nū'-klē-ō-tīd): a nucleotide in which the phosphate group is bonded to the sugar at two points, forming a ring. Cyclic nucleotides serve as intracellular messengers.

deoxyribonucleic acid (dē-ox-ē-rī-bō-nū-klā'-ik; DNA): a molecule composed of deoxyribose nucleotides; the genetic information of all living cells.

disaccharide (dī-sak'-a-rīd): a carbohydrate formed by the covalent bonding of two monosaccharides.

disulfide bridge: the covalent bond formed between the sulfur atoms of two cysteines in a protein; often causes the protein to fold, by bringing otherwise distant parts of the protein close together.

fat: a lipid composed of three saturated fatty acids covalently bonded to glycerol; fats are solid at room temperature.

fatty acid: an organic molecule composed of a long chain of carbon atoms, with a carboxylic acid (—COOH) group at one end. Fatty acids may be saturated (all single bonds between the carbon atoms) or unsaturated (one or more double bonds between the carbon atoms).

functional group: one of several groups of atoms commonly found in organic molecules, including hydrogen, hydroxyl, amino, carboxyl, and phosphate groups.

glucose: the most common monosaccharide, with the molecular formula $C_6H_{12}O_6$. Most polysaccharides, including cellulose, starch, and glycogen, are made of glucose subunits covalently bonded together.

glycerol (glis'-er-ol): a three-carbon alcohol to which fatty acids are covalently bonded to make fats and oils.

glycogen (glī'-kō-gen): a polysaccharide composed of branched chains of glucose subunits, used as a carbohydrate storage molecule in animals.

helix (hē'-licks): a spiral structure similar to a corkscrew or a spiral staircase; a type of secondary structure of a protein.

hydrolysis (hī-drol'-i-sis): the chemical reaction that breaks a covalent bond through the addition of hydrogen to the atom forming one side of the original bond, and a hydroxyl group to the atom on the other side.

lactose (lak'-tōs): a disaccharide composed of glucose and galactose; found in mammalian milk.

lipid (li'-pid): one of a number of water-insoluble organic molecules, containing large regions composed solely of carbon and hydrogen. Lipids include oils, fats, waxes, phospholipids, and steroids.

maltose (mal'-tōs): a disaccharide composed of two glucose molecules.

monomer (mo'-nō-mer): a small organic molecule, several of which may be bonded together to form a chain called a polymer.

monosaccharide (mo-nō-sak'-a-rīd): the basic molecular unit of all carbohydrates, usually composed of a chain of carbon atoms to which are bonded hydrogen and hydroxyl groups.

nucleic acid (nū-klā'-ik): an organic molecule composed of nucleotide subunits. The two common types of nucleic acids are ribonucleic acids (RNA) and deoxyribonucleic acids (DNA).

nucleotide (nū'-klē-ō-tīd): an organic molecule composed of a phosphate group, a five-carbon monosaccharide (ribose or deoxyribose), and a nitrogen-containing base.

oil: a lipid composed of three fatty acids, some of which are unsaturated, covalently bonded to a molecule of glycerol. Oils are liquid at room temperature.

peptide (pep'-tīd): a chain composed of two or more amino acids linked together by peptide bonds.

peptide bond: the covalent bond between the amino group nitrogen of one amino acid and the carboxyl group carbon of a second amino acid, which joins the two amino acids together in a peptide or protein.

phospholipid (fos-fō-li'-pid): a lipid consisting of glycerol to which two fatty acids and one phosphate group are bonded. The phosphate group bears another group of atoms, often containing nitrogen, and often bearing an electric charge.

pleated sheet: a type of secondary structure of a protein. In a pleated sheet, protein chains lie side-by-side, bonded to one another by hydrogen bonds.

polymer (pa'-li-mer): a molecule composed of three or more (may be thousands) smaller subunits called monomers. The monomers that make up a single polymer may be identical (e.g., the glucose monomers of starch) or different (e.g., the amino acids of a protein).

polysaccharide (pol-ē-sak'-a-rīd): a large carbohydrate molecule composed of branched or unbranched chains of repeating monosaccharide subunits, usually glucose or modified glucose molecules. Polysaccharides include starches, cellulose, and glycogen.

primary structure: the amino acid sequence of a protein.

protein: an organic molecule composed of one or more chains of amino acids.

quaternary structure (kwat'-er-na-rē): the complex three-dimensional structure of a protein that is composed of more than one peptide chain.

ribonucleic acid (rī-bō-nū-klā'-ik; RNA): a molecule composed of ribose nucleotides; transfers hereditary instructions from the nucleus to the cytoplasm; also the genetic material of some viruses.

saturated: referring to a fatty acid with as many hydrogen atoms as possible bonded to the carbon backbone; a fatty acid with no double bonds in its carbon backbone.

secondary structure: a repeated, regular structure assumed by protein chains, held together by hydrogen bonds; usually either a helix or a pleated sheet.

starch: a polysaccharide composed of branched or unbranched chains or glucose molecules, used by plants as a carbohydrate-storage molecule.

steroid: a lipid composed of four fused rings of carbon atoms to which functional groups are attached.

subunit: a small organic molecule, several of which may be bonded together to form a larger molecule. See also monomer.

sucrose: a disaccharide composed of glucose and fructose.

sugar: a simple carbohydrate molecule, either a monosaccharide or a disaccharide.

tertiary structure (ter'-shē-ār-ē): the complex three-dimensional structure of a single peptide chain. The tertiary structure is held in place by disulfide bonds between cysteine amino acids, by attraction and repulsion among amino acid side groups, and by interaction between the cellular environment (water or lipids) and the amino acid side groups.

triglyceride: a lipid composed of three fatty acid molecules bonded to a single glycerol molecule.

unsaturated: referring to a fatty acid with fewer than the maximum number of hydrogen atoms bonded to its carbon backbone; a fatty acid with one or more double bonds in its carbon backbone.

wax: a lipid composed of fatty acids covalently bonded to long-chain alcohols.

STUDY QUESTIONS

1. Which elements are commonly found in biological molecules?
2. List the four principal types of biological molecules and give an example of each.
3. Describe condensation and hydrolysis reactions.
4. Distinguish among the following: monosaccharide, disaccharide, and polysaccharide. Give two examples of each, and their functions.
5. Describe the molecular structures of fats, oils, and steroids. Give one example of each, and its function.
6. Draw the structure of an amino acid. What determines the properties of each amino acid? What are the four general types of amino acids.
7. Describe the synthesis of a protein from amino acids.
8. Define the primary, secondary, tertiary, and quarternary structures of a protein.
9. Draw the structure of an individual nucleotide and a nucleic acid, and label the parts (use boxes to represent the various parts).

DISCUSSION QUESTIONS

1. "Preview" question for Chapter 6: In Chapter 2 you learned that hydrophobic molecules tend to cluster when immersed in water. In this chapter, you discovered that a phospholipid has a hydrophilic head and hydrophobic tails. What do you think would be the configuration of phospholipids that are immersed in water?
2. Another "preview" question (see Chapter 13): Hemoglobin has many hydrophilic amino acids on the outside of the molecule; these are important in keeping the molecule soluble in the cytoplasm of red blood cells. What would happen to a mutant form of hemoglobin that replaced one or more of these hydrophilic amino acids with hydrophobic ones?
3. Many birds must store large amounts of energy to power flight during migration. Which type of organic molecule would be the most advantageous for energy storage? Why?

SUGGESTED READINGS

Atkins, P. W. *Molecules*. New York: W. H. Freeman and Co., 1987. A fascinating, nontechnical introduction to the beauties of chemistry.

Doolittle, R. F. "Proteins." *Scientific American*, October 1985. Proteins are the molecular tools that, directly or indirectly, carry out almost all the functions of a cell.

Dushesne, L. C., and Larson, D. W. "Cellulose and the Evolution of Plant Life." *Bioscience*, April 1989. Chemical and natural history aspects of the most plentiful organic molecule on earth.

Hill, J. W. *Chemistry for Changing Times*. 6th ed. New York: Maxwell Macmillan Publishing Co., 1992. A chemistry textbook that is both clearly readable and thoroughly enjoyable, for nonscience majors.

Sharon, N. "Carbohydrates." *Scientific American*, November 1980. Carbohydrates, although all quite similar in structure, perform a variety of roles in the life of a cell.

4

Energy Flow in the Life of a Cell

A brown bear intercepts a salmon attempting to leap up a waterfall on a stream in Alaska. Both predator and prey convert the chemical energy of their food molecules into the kinetic energy of movement.

"Disorder spreads through the universe, and life alone battles against it."

G. Evelyn Hutchinson

Every autumn, salmon struggle up icy Alaskan streams. And every autumn, brown bears gather at the streams, feeding on the salmon before hibernating through the long arctic winter (facing page). Trillions of coordinated chemical reactions in billions of individual cells make such complex behaviors possible. The cells of predator and prey churn with activity, not only during the hunt, but in quiet times too: taking in nutrients and secreting hormones, contracting muscles, and synthesizing proteins and lipids, to name just a few. These activities all require **energy,** which can be simply defined as *the capacity to do work,* including synthesizing molecules, moving things around, and generating heat and light. In this chapter we discuss the physical laws that govern energy flow in the universe, how energy flow in turn governs chemical reactions, and how the chemical reactions within living cells are controlled by the molecules of the cell itself. Chapters 7 and 8 focus on photosynthesis, the chief "port of entry" for energy into the biosphere, and glycolysis and cellular respiration, the most important pathways for energy release.

Energy Flow in the Universe

As you learned in Chapter 2, there are two fundamental types of energy: **kinetic energy** and **potential energy.** Both types of energy may exist in many different forms. Kinetic energy, or *energy of movement,* includes light (movement of photons), heat (movement of molecules), electricity (movement of electrically charged particles), and movement of large objects. Potential energy, or *stored energy,* includes chemical energy stored in the bonds that hold atoms together in molecules, electrical energy stored in a battery or capacitor, and positional energy stored in a boulder perched high atop a cliff (Fig. 4-1). Under the right conditions, kinetic energy can be transformed into potential energy, and vice versa. For example, you can convert kinetic energy of movement into potential energy of position by pushing a boulder up a hill; when the boulder rolls down again, it converts potential energy back into kinetic energy.

All interactions among pieces of matter are governed by two laws of thermodynamics. In their sim-

Figure 4-1 This balanced boulder has potential energy due to its position atop this pedestal of rock supporting it. Sooner or later, the pedestal will erode away, and the potential energy of the boulder will be converted to kinetic energy of movement as it comes crashing down.

plest forms, the laws of thermodynamics are these:
First Law: Energy can be neither created nor destroyed, although it can be converted from one form to another.
Second Law: The quantity of useful energy declines over time.

In the interaction between your body and our hypothetical boulder, the first law states that the sum of the chemical energy stored in your body before you start pushing, the kinetic energy expended during the push, and the potential energy stored in the boulder remains constant. Before you start, all the energy in question is chemical energy. During the push, chemical energy is converted into muscle contraction, heat, sweat, and the new position of the boulder. No net energy is gained, and none is lost.

The second law states, however, that the amount of useful energy declines. The potential energy of the boulder, which could do work by smashing things as it rolls back down the hill, is less than the energy you spent pushing it up. Why? Grunting, panting, and the friction between the boulder and the hillside used up energy, and merely served to warm up the air and the ground a bit.

As this example illustrates, to understand how energy flow governs interactions among pieces of matter, we need to know two things: (1) the quantity of available energy, and (2) the usefulness of the energy. The quantity and usefulness of energy are the subjects of the laws of thermodynamics, which we will now examine more closely.

The Laws of Thermodynamics

The laws of thermodynamics deal with "isolated systems," which are any parts of the universe that cannot exchange either matter or energy with any other parts. The universe as a whole is the largest isolated system. No small part of the universe is truly isolated from all possible exchange with every other part, but the concept of an isolated system is a useful one in thinking about energy flow.

The First Law of Thermodynamics

The first law is often called the law of conservation of energy. **The *First Law of Thermodynamics* states that within any isolated system, energy can be neither created nor destroyed, although it can be changed in form (e.g., from chemical energy to heat energy).**

In other words, within an isolated system *the total quantity of energy remains constant.* To use a familiar example, let's see how the first law applies to driving your car (Fig. 4-2). To a first approximation, we can

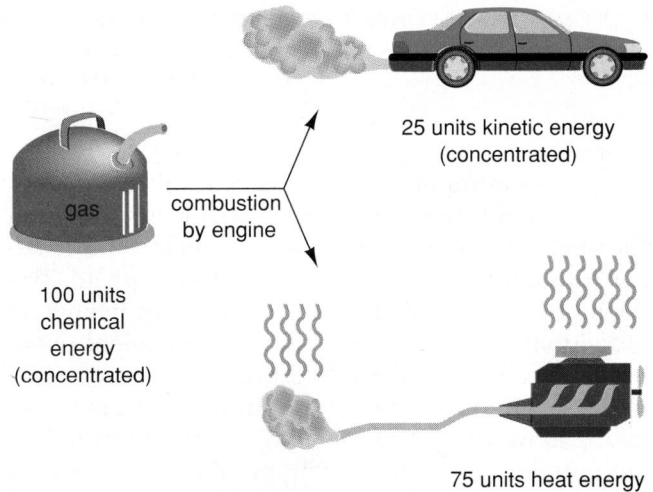

Figure 4-2 Thermodynamics of automobiles. According to the first law, the quantity of energy originally present in the gasoline equals the energy in the resulting products: the molecules in the exhaust, the moving car, and heat. The form of the energy changes from chemical energy into heat or energy of movement. According to the second law, some of the energy in the products is in a less concentrated, less usable form than the chemical energy of the gasoline. In cars, about 75% of the chemical energy of the gas is given off as waste heat to the environment, contributing to random movement of molecules in the air.

consider that your car (with a full tank of gas), the road, and the surrounding air constitute an isolated system. When you drive your car, you convert the chemical energy of gasoline into kinetic energy of movement and heat energy. The total amount of energy in the gasoline is the same as the total amount of this kinetic energy and heat.

An important rule of energy conversions is this: **Energy always flows "downhill," from places with a high concentration of energy to places with a low concentration of energy.** This is the principle behind engines. As you may know, temperature is a measure of the speed of movement of molecules. The burning gasoline in your car's engine consists of molecules moving at extremely high speeds: a high concentration of energy. The cooler air outside the engine consists of molecules moving at much lower speeds: a low concentration of energy. The molecules in the engine hit the piston harder than do the air molecules outside the engine, so the piston moves downward, driving the gears that move the car. Work is done. When the engine is turned off, it cools down. The molecules on both sides of the piston move at the same speed, so the piston stays still. No work is done.

The Second Law of Thermodynamics

The second law relates to the usefulness of energy. **The *Second Law of Thermodynamics* states that any change in an isolated system causes the quantity of concentrated, useful energy to decrease.**

To continue our example of a car engine, the heat of the burning gasoline moves the piston, which moves the car. However, the kinetic energy of the moving car is much less than the chemical energy that was originally contained in the gasoline (see Fig. 4-2). Where is the "missing" energy? Feel the engine and the exhaust: they are both hot. The burning gas not only moved the piston, it also heated up the engine, the exhaust system, and the air around the car. The friction of tires on pavement heated the road up a little, too. So, as the first law dictates, there really isn't any missing energy: the original quantity of energy in the gasoline still exists, only in different forms. However, some of the energy (the heat given off) merely increased the random movement of molecules in the engine block, the road, and the air. Such random movements aren't useful: you can't very well gather up the energy found in the warmed-up road and use it to drive your car a few more blocks.

The second law also tells us something about the organization of matter. Regions of concentrated energy are usually also regions of great orderliness. The eight carbon atoms in a single molecule of gasoline have a much more orderly arrangement than do the eight separate, randomly moving molecules of carbon dioxide that are formed when the gasoline burns. Therefore, we can also phrase the second law in terms of the organization of matter: **All processes in an isolated system result in an increase in randomness and disorder. This increase in randomness, disorder, and low-level energy, and loss of orderliness and high-level energy, is called** *entropy.*

Eventually, several billion years from now, all the energy in the universe will be evenly dispersed as heat, and all the matter will be randomly distributed in small molecules. Without differences in energy, no further work will be possible. Life, therefore, will cease. (This rather gloomy prospect is explored in a marvelous science fiction short story by Isaac Asimov entitled "The Last Question": definitely recommended reading.)

Entropy and Life

If you stop to think about the Second Law of Thermodynamics, you may wonder how life can exist at all. If chemical reactions, including those inside living cells, cause the amount of low-level, unusable energy to increase, and if matter tends toward increasing randomness and disorder, how can organisms accumulate the concentrated energy and precisely ordered molecules that characterize living things? The answer is this: *The second law applies only to isolated systems; living organisms are not isolated systems, because they interact with the world around them.* Nuclear reactions in the sun produce concentrated energy (sunlight) along with vast increases in entropy. Living things on Earth use that concentrated energy to synthesize complex molecules and maintain orderly structures. These small packets of decreasing entropy on Earth do not violate the second law, because the entropy of the solar system as a whole constantly increases.

Energy Flow in Chemical Reactions

A chemical reaction begins with one set of substances, called the **reactants,** and converts them into another set, the **products.** All chemical reactions fall into one of two categories: exergonic and endergonic. In **exergonic reactions** (Greek for "energy out"), the reactants have more energy than the products. Consequently, the reaction releases energy:

EXERGONIC REACTION:

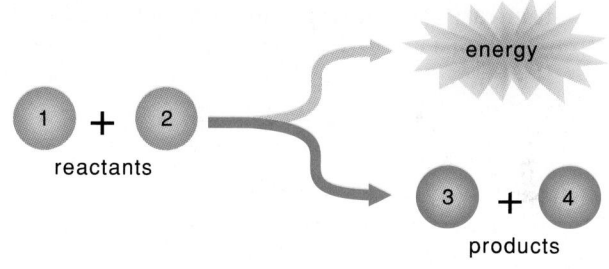

In **endergonic reactions** ("energy in"), the products have more energy than the reactants, and a net input of energy from outside the system is required to drive the reaction:

ENDERGONIC REACTION:

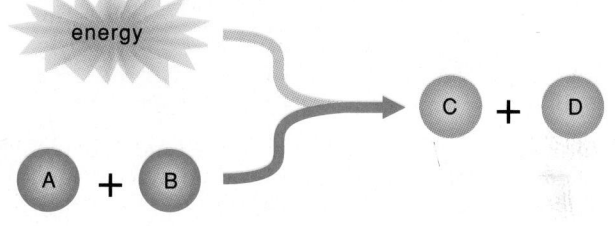

Let's look at two processes that illustrate these types of reactions: burning coal and photosynthesis.

Exergonic Reactions

To simplify things, we will assume that coal is pure carbon, so that burning coal is described by this equation:

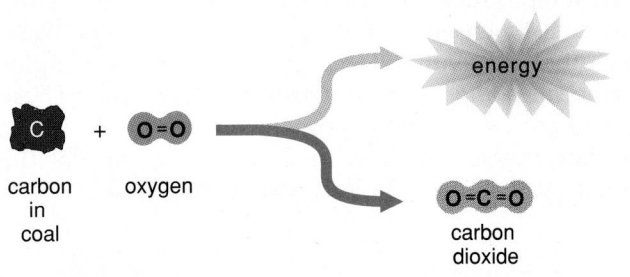

This reaction illustrates two important concepts, diagrammed in Figure 4-3a. First, molecules of carbon and oxygen contain much more energy than molecules of carbon dioxide do, so the reaction releases energy. **Energy release allows exergonic reactions to occur without a net input of energy.** In our example, coal fires, once started, burn by themselves until the coal is consumed: you don't have to keep the fire going by adding outside energy from a flame thrower. (It may be helpful if you think of exergonic reactions as being "downhill," both in the graphical sense shown in Figure 4-3a, and in the everyday sense that cars and bicycles can coast downhill without engines or legs propelling them.)

Although burning coal releases energy, a lump of coal doesn't burst into flames by itself. This leads to the second point: **All chemical reactions require an initial input of energy, called the *activation energy*, to get started.** Chemical reactions require activation energy because atoms and molecules are surrounded by a cloud of negatively charged electrons. For two molecules to react with each other, their electron clouds must be forced together, despite their mutual electrical repulsion. This requires energy. The usual source of activation energy is kinetic energy of movement. Only molecules moving with sufficient speed can collide hard enough to force their electron clouds to mingle and react. Since molecules move faster at higher temperatures, most reactions occur more readily at high temperatures. This is what setting coal on fire does: heat causes the carbon atoms to move faster, some vaporizing completely off the lump of coal. These energetic carbon atoms collide with oxygen molecules and react to form carbon dioxide (CO_2), releasing heat. The heat causes more carbon atoms to vaporize and collide with oxygen molecules, producing more CO_2 and releasing more heat, thus sustaining the reaction until the coal is consumed.

Endergonic Reactions

There are many reactions in living systems in which the products contain more energy than the reactants.

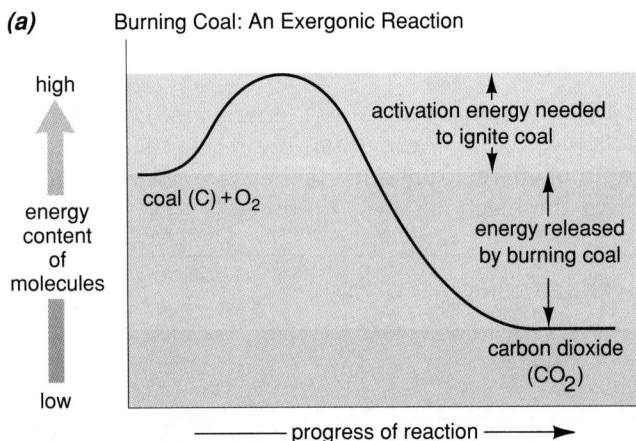

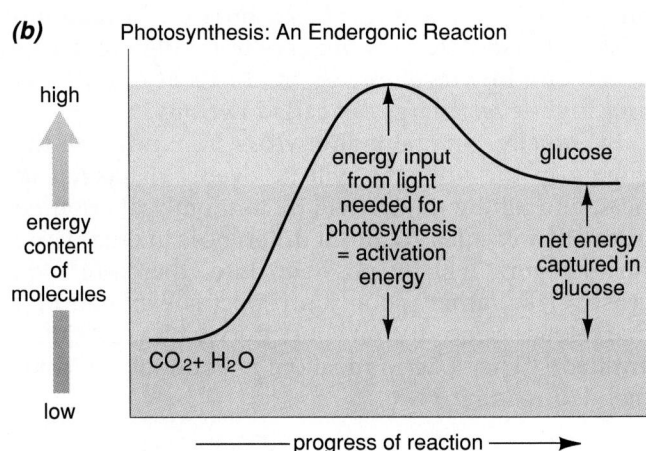

Figure 4-3 Energy relations in exergonic and endergonic reactions. **(a)** An exergonic ("downhill") reaction proceeds from high-energy reactants such as coal to low-energy products such as CO_2. The energy difference between the chemical bonds of the reactants and products is released as heat. To start the reaction, however, an initial input of energy—the activation energy—is required. **(b)** An endergonic ("uphill") reaction, such as photosynthesis, proceeds from low-energy reactants such as CO_2 and H_2O to high-energy products such as glucose, and therefore requires a net input of energy.

These reactions must be driven through an input of energy into the low-energy reactants (Fig. 4-3b). As we shall see in Chapter 7, photosynthesis in green plants takes low-energy water and carbon dioxide and produces high-energy sugar and oxygen:

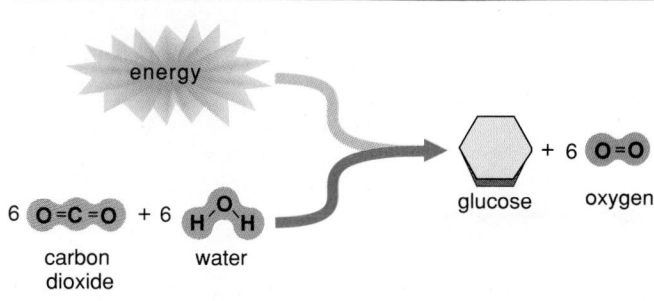

This reaction requires energy, which plants obtain from sunlight. We might call endergonic reactions "uphill," in the same sense that cars and bicycles need energy inputs to go uphill.

Coupled Reactions

If you've been following carefully, about this time you may be saying, "Wait a minute. Photosynthesis is an endergonic reaction in an isolated plant, but what if you include the sun in the system?" You are exactly right. Photosynthesis is no exception to the rule that all reactions are driven by energy flowing downhill. Photosynthesis is an example of a *coupled reaction.* **In a coupled reaction, an exergonic reaction provides the energy needed to drive an endergonic reaction.**

In the case of photosynthesis, the exergonic reaction takes place in the sun, and the endergonic reaction takes place in green plants:

Sun: Nuclear fusion of hydrogen →
helium + large amount of energy released
Plants: $6 CO_2 + 6 H_2O$ + small energy input →
$C_6H_{12}O_6 + 6 O_2$

The reactions occurring in the sun liberate energy as light. The sunlight captured by a plant possesses much more energy than is needed to drive photosynthesis. Therefore, the overall process, if we include the sun, is exergonic.

The fact that sunlight contains more energy than is needed to drive photosynthesis is an example of a general rule of coupled reactions. According to the Second Law of Thermodynamics, the amount of useful energy decreases during every chemical reaction. This means that not all the energy released by an exergonic reaction can be used to drive an endergonic reaction; some energy is lost to the environment as

heat and random molecular movement. Therefore, in coupled reactions, the exergonic reaction must always release more energy than is required to drive the endergonic reaction (Fig. 4-4a).

Living organisms are chemists *par excellence*, constantly using the energy given off by exergonic reactions (metabolism of food, either eaten or manufactured during photosynthesis) to drive essential endergonic reactions (synthesis of complex molecules, brain activity, or movement; Fig. 4-4b). The exergonic and endergonic halves of coupled reactions often occur in different places, so there must be some way to transfer the energy from the exergonic reaction to the endergonic reaction. In photosynthesis, sunlight carries energy from exergonic reactions in the sun to the endergonic reactions in plants. In coupled reactions occurring within cells, energy is usually transferred from place to place by **energy carrier molecules**, of which the most common is **adenosine triphosphate**, or **ATP**. We will examine the synthesis and use of ATP later in this chapter.

Reversibility of Chemical Reactions

So far, we have discussed chemical reactions as if they always proceed in one direction, "forward" from reactants to products, and then are finished. This is usually not the case. **Most chemical reactions are reversible: they may proceed spontaneously in either direction under the appropriate conditions of reactants, products, and energy.** Furthermore, all reactions, if left to themselves, eventually reach a steady state called a **chemical equilibrium**, in which the reaction proceeds at an equal rate in both directions. At equilibrium, "forward" and "backward" reactions occur at exactly equal rates, with the result that, although *individual molecules* are constantly undergoing chemical reactions, the *net concentrations of molecules* do not change.

To understand the reversibility of chemical reactions and the nature of chemical equilibria, let's look at a reaction of vital importance to all of us: the combination of oxygen with hemoglobin in the blood. Hemoglobin, as you will recall from Chapter 3, is the oxygen-carrying protein in your red blood cells. As blood circulates through your lungs, hemoglobin picks up oxygen. When the blood passes through the other organs of the body, the oxygen is released. Why?

The iron atom held by each hemoglobin subunit can reversibly bind one oxygen molecule. During normal breathing, the concentration of oxygen in the air in your lungs is about 15%. When blood enters the lungs, hardly any hemoglobin molecules have oxy-

(a)

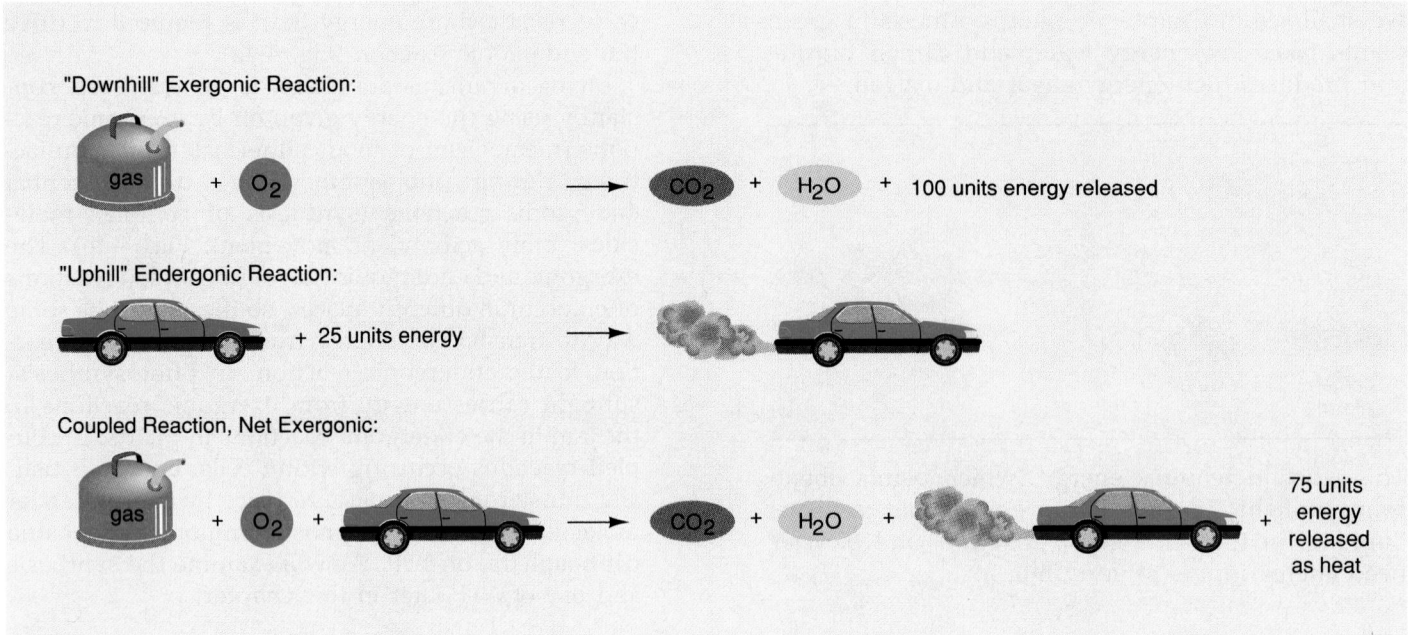

(b)

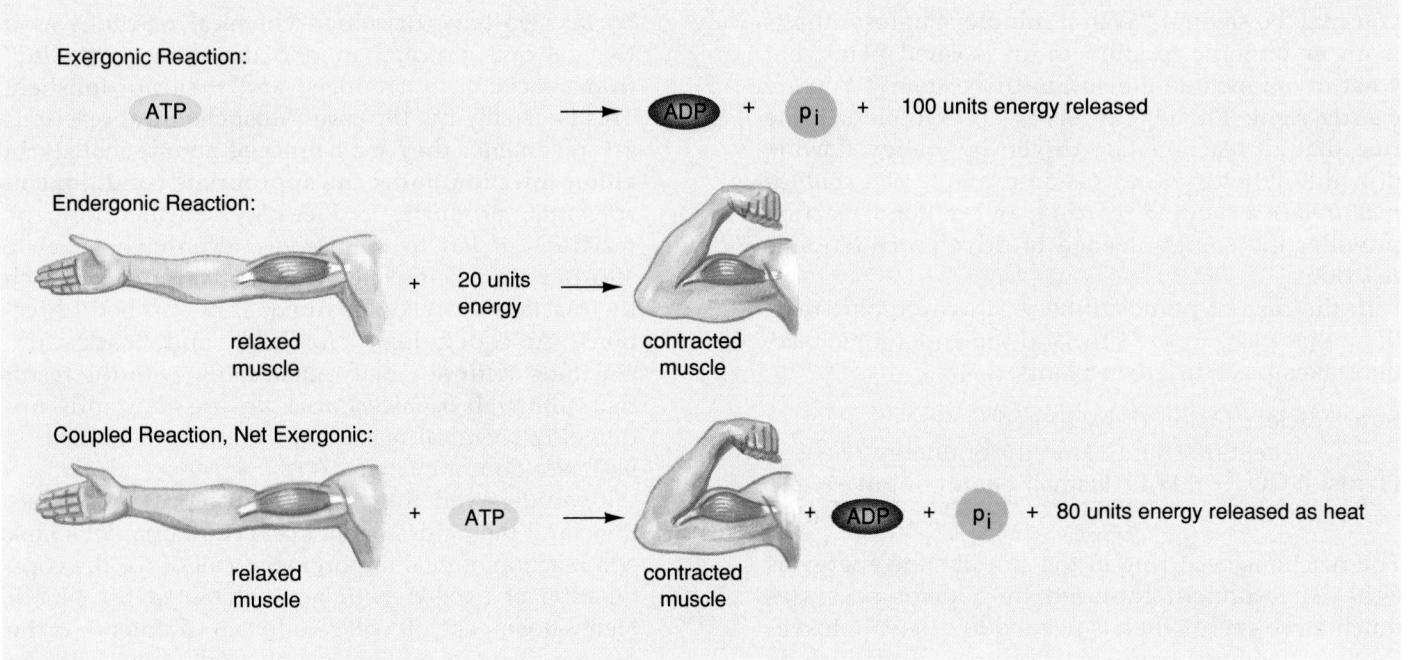

Figure 4-4 Coupled reactions.
(a) Burning gasoline is an exergonic reaction that can be coupled to the endergonic reaction of moving a car. According to the Second Law of Thermodynamics, all reactions are inefficient, so the exergonic reaction must actually release more energy than the endergonic reaction requires. The excess energy is given off to the environment as heat. The overall reaction is exergonic.
(b) Muscle movement is an endergonic reaction coupled to the exergonic reaction of ATP hydrolysis. The energy released by ATP hydrolysis exceeds the energy put into muscle contraction, and the overall reaction is exergonic.

gen bound to them (you'll see why in a minute). Therefore, oxygen binds to hemoglobin, with the initial reaction proceeding from reactants (oxygen and hemoglobin) to products (oxyhemoglobin):

Deoxygenated Blood Entering Lungs

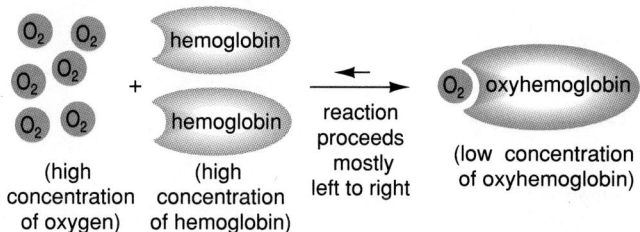

This binding, however, is reversible. Some oxygen molecules leave the hemoglobin again and reenter the air in the lungs. Initially, this happens rarely, because very few of the hemoglobin molecules have oxygen bound to them. As more and more hemoglobin molecules become oxygenated, a few give up their oxygens. Finally, an equilibrium is established: just as many oxygen molecules leave the air and bind to hemoglobin as leave the hemoglobin and reenter the air:

Equilibrium in Lungs

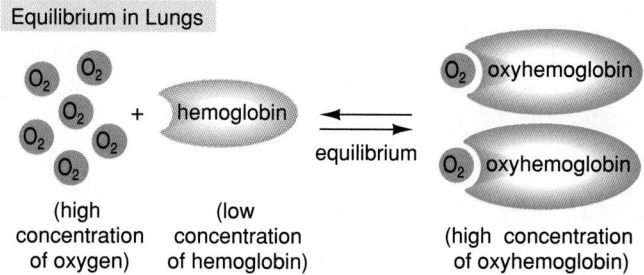

At equilibrium, we can't really speak of products and reactants anymore, since the reaction proceeds in both directions at the same rate.

Chemical equilibria are affected by the concentrations of the molecules involved. At the high concentration of oxygen in newly breathed air, the equilibrium point occurs with most hemoglobin molecules having oxygen bound to them. In body tissues, say in an exercising muscle, chemical reactions use up oxygen, so the tissue concentration of oxygen is very low. When oxygenated blood enters the muscle, oxygen molecules will sporadically leave the hemoglobin molecules, just as they have always done. However, there aren't any free oxygen molecules in the muscle to replace them, so the overall reaction shifts direction, with oxyhemoglobin in the blood losing oxygen to the muscle cells:

Oxygenated Blood Entering Muscle

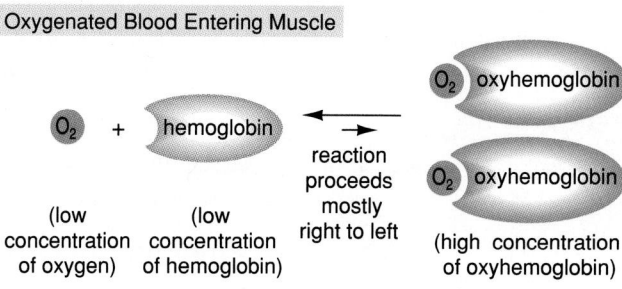

Since the muscle uses up oxygen as fast as the hemoglobin releases it, the oxygen concentration in the muscle remains low. Almost all the oxygen leaves the hemoglobin and enters the muscle cells. A new equilibrium is established, this time with most of the hemoglobin in the deoxygenated state:

Equilibrium in Muscle

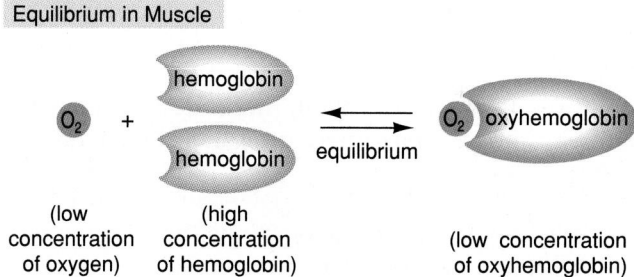

Eventually the hemoglobin returns to the lungs, and the cycle repeats.

The hemoglobin–oxygen reaction illustrates an important principle of reactions occurring in living organisms. **By adding and removing reactants and products, cells drive reversible reactions back and forth as required by the metabolic demands of the organism.**

Controlling the Metabolism of Living Cells

Cells are miniature, incredibly complex chemical factories. The **metabolism** of a cell is the sum of all its chemical reactions. Many of these reactions are linked in sequences called **metabolic pathways** (Fig. 4-5), such as glycolysis, the series of reactions that begin the digestion of glucose (see Chapter 8). Different metabolic pathways may exchange molecules, with the result that all the reactions and all the molecules of a cell are actually interconnected in a single, hugely complicated metabolic pathway.

The chemical reactions within a cell are governed by the same laws of thermodynamics that control any

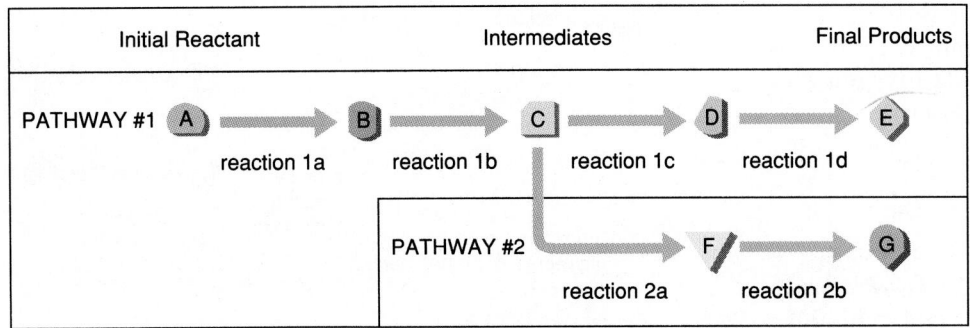

Initial Reactant	Intermediates	Final Products

PATHWAY #1   A ——→ B ——→ C ——→ D ——→ E
   reaction 1a   reaction 1b   reaction 1c   reaction 1d

PATHWAY #2 ——————→ F ——→ G
   reaction 2a   reaction 2b

Figure 4-5 A simplified view of metabolic pathways. The original reactant molecule undergoes a series of reactions, each catalyzed by a specific enzyme. The product of each reaction serves as the reactant for the next reaction in the pathway. Metabolic pathways are often interconnected, so that the product of a step in one pathway may serve as a reactant either for the next reaction in the same pathway or for a reaction in another pathway.

other reactions. How, then, do orderly metabolic pathways arise? Cells finely tune their biochemistry in three ways:

1. Cells regulate chemical reactions through the use of protein catalysts called enzymes.
2. Cells couple reactions together, driving energy-requiring endergonic reactions with the energy released by exergonic reactions.
3. Cells synthesize energy carrier molecules that capture energy from exergonic reactions and transport it to endergonic reactions.

Activation Energy and Reaction Rates

The laws of thermodynamics tell us that energy-releasing reactions can occur spontaneously, but they don't say how fast they will occur. In general, the rate of a reaction is determined by its **activation energy,** which is a measure of the speed with which reactants must collide to force their electron clouds together. Reactions with low activation energies can proceed swiftly, while reactions with high activation energies will be very slow. Therefore, the rate of most reactions is limited by the speed of reactant molecules—the faster they move, the faster the reaction goes. Most reactions can be accelerated by raising the temperature, which increases the speed of molecules.

Without high temperatures, however, many "spontaneous," exergonic reactions proceed extremely slowly. The reaction of sugar with oxygen to yield carbon dioxide and water is spontaneous and exergonic, but has an enormous activation energy. Even quite high temperatures do not provide enough

energy to overcome the activation energy barrier: as any cook knows, you can boil a sugar solution for hours and hardly any breaks down. At the temperatures found in living organisms, sugar and many other energetic molecules would almost never break down and give up their energy. Nevertheless, sugar is an important energy source for life on Earth, because of the catalytic enzymes produced by living cells. Let's see how enzymes and other catalysts influence the progress of chemical reactions.

Catalysts

Catalysts are molecules that speed up the rate of a reaction without themselves being used up or permanently altered. **Catalysts speed up reactions by reducing the activation energy.**

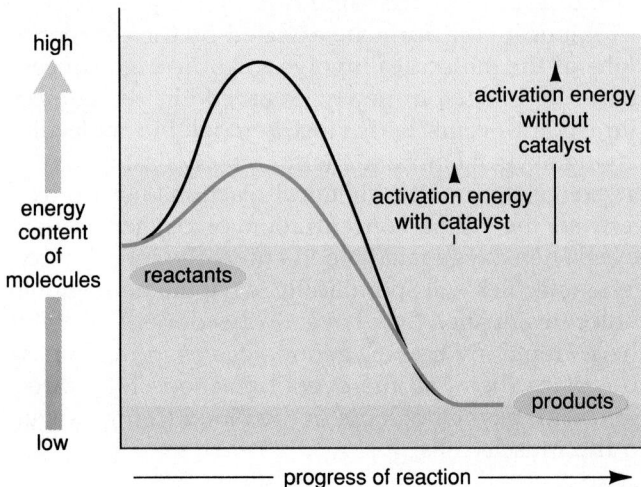

As an example of catalytic action, let's consider the catalytic converters in car engines. When gasoline is completely burned, the final products are carbon dioxide and water:

$$2\ C_8H_{18}\ (octane) + 25\ O_2 \rightarrow$$
$$16\ CO_2 + 18\ H_2O + energy$$

However, flaws in the combustion process generate other substances, including poisonous carbon monoxide (CO). Carbon monoxide reacts spontaneously but slowly with oxygen in the air to form carbon dioxide:

$$2\ CO + O_2 \rightarrow 2\ CO_2 + energy$$

At equilibrium, this reaction results in air with a minuscule concentration of CO. In large cities, however, so many cars emit so much CO that the spontaneous reaction of CO with O_2 can't keep up, and unhealthy levels of carbon monoxide accumulate. Enter the catalytic converter. The platinum catalysts in the converter hasten the conversion of CO to CO_2, thereby reducing air pollution.

Catalytic converters illustrate four important principles about catalysts:

1. **Catalysts speed up reactions.**
2. **Catalysts cannot cause energetically unfavorable reactions to occur.** Catalysts only speed up reactions that would occur spontaneously anyway, although at a much slower rate.
3. **Catalysts do not change the equilibrium point of a reaction.** Catalytic converters cannot reduce the concentration of CO below the naturally occurring minimum level. If fresh mountain air is put through a converter, the concentration of CO in the air coming out is the same as the concentration going in.
4. **Catalysts are not consumed in the reactions they promote.** No matter how many reactions they participate in, the catalysts themselves are not permanently changed. Converters in cars have to be replaced, not because of the numbers of CO molecules they have oxidized to CO_2, but because they become poisoned by trace materials in gasoline, or by a few tankfuls of leaded gas.

Enzymes: Biological Catalysts

Enzymes are catalysts synthesized by living organisms. Almost all enzymes are proteins. Just a few years ago, biology texts stated unequivocally that all enzymes are proteins. However, during the 1980s, Thomas Cech and Sidney Altman discovered that certain molecules of ribonucleic acid also function as enzymes. These molecules, dubbed *ribozymes*, cata-

lyze reactions involved in processing genetic information for use by a cell, and may have been crucial during the early evolution of life (see Chapter 19). Ribozymes notwithstanding, in the rest of this chapter we will discuss only the actions of protein enzymes.

Enzymes possess the four characteristics of catalysts just described. However, enzymes have two additional attributes that set them apart from inorganic catalysts:

1. Although an inorganic catalyst can usually speed up many different reactions, **an enzyme is usually very specific, promoting only a few types of chemical reactions.** In most cases, an enzyme catalyzes a single reaction involving one or two specific molecules, while leaving even quite similar molecules untouched. You may recall from Chapter 3, for example, that animals have enzymes that break apart starch molecules but leave cellulose intact, despite the fact that starch and cellulose are both composed of glucose subunits.
2. Enzyme activity is regulated—that is, enhanced or suppressed—often by the very molecules whose reactions they promote.

Enzyme Structure and Function

Why are enzymes specific, and how are they regulated? Enzyme function is intimately related to enzyme structure. Enzymes are proteins with complex three-dimensional shapes (Fig. 4-6). Each enzyme has a dimple or groove, called the **active site,** into which reactant molecules, called **substrates,** can enter.

The active site of each enzyme has a distinctive shape and distribution of electrical charge. According to the **lock-and-key** model of enzyme action, the active site has a size, shape, and charge distribution that is complementary to its substrate, and allows only a few extremely similar molecules to enter, much as a lock has a size and shape that only allows one or a few keys to fit in. However, enzymes and substrates are much more flexible than locks and keys. For at least some enzymes, a better analogy might be a tight pair of jeans being pulled on over a human body: although the basic fit has to be reasonably close, both garment and wearer change shape. As we shall see, according to this **induced-fit** model (see Fig. 4-7), changes in enzyme and substrate are essential for enzyme action to occur.

In either case, the size, shape, and charge of the active site confer great specificity, allowing only certain molecules to enter and react, while rejecting even fairly similar molecules. For example, there are several protein-digesting enzymes in your intestines that

(a)   **(b)**

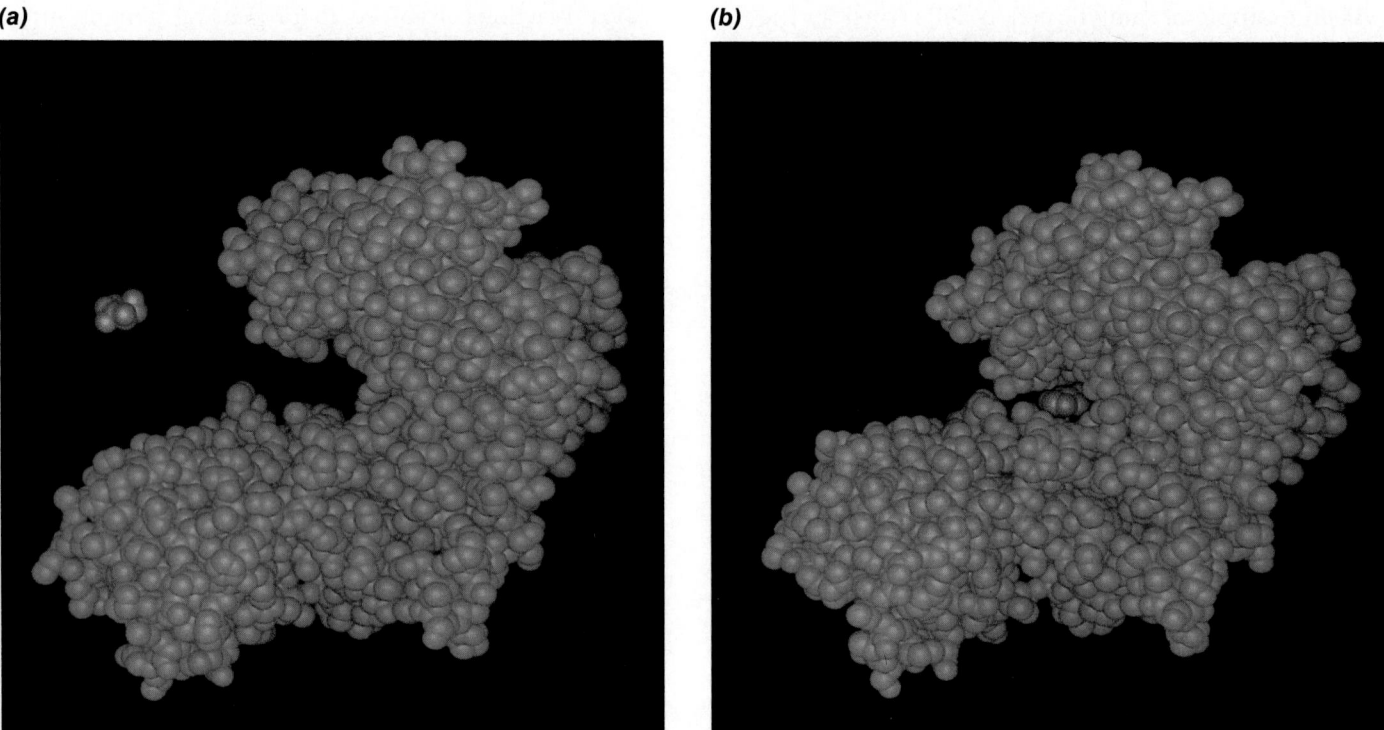

Figure 4-6 Computer-generated models of the enzyme hexokinase and its substrate glucose (hexokinase is the enzyme that catalyzes the first step in glucose metabolism). **(a)** The active site of hexokinase is a groove or dimple into which a glucose molecule may enter. **(b)** As predicted by the induced-fit model of enzyme action, entry of glucose into the active site causes both the glucose and hexokinase molecules to change shape (changes in the glucose molecule cannot be seen because it is too small and is buried).

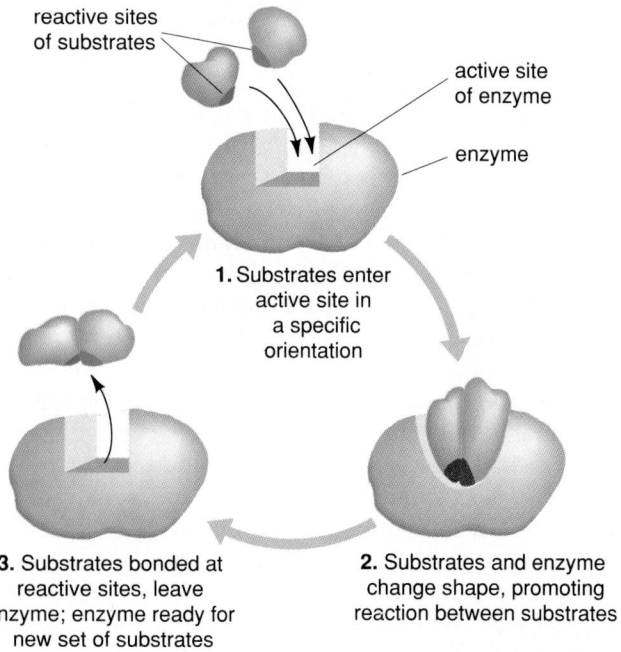

reactive sites
of substrates

active site
of enzyme

enzyme

1. Substrates enter active site in a specific orientation

3. Substrates bonded at reactive sites, leave enzyme; enzyme ready for new set of substrates

2. Substrates and enzyme change shape, promoting reaction between substrates

Figure 4-7 The cycle of enzyme–substrate interactions.

break the peptide bonds that join amino acids. However, each enzyme is quite specific. No single enzyme can digest every type of protein, because only a protein with the right amino acid sequence can fit into the active site of any individual enzyme. Other proteins, with amino acids that are too large or too small, or have the wrong charge, cannot get in and consequently cannot be digested. Complete digestion of all the proteins in the human diet, therefore, requires several enzymes.

Conversely, some molecules may be able to enter the active site of an enzyme but do not have chemical bonds upon which the enzyme can act, and so no reaction occurs. Many poisons, including some insecticides, enter the active site of enzymes essential to brain function and never leave again. The enzymes remain plugged up, and the reactions they normally promote cannot occur. Parts of the brain turn off, or become wildly hyperactive, sometimes with fatal consequences.

How does an enzyme promote a reaction? First, the shape and charge of the active site forces substrates to enter the enzyme in specific orientations (Fig. 4-7, Step 1). Second, when substrates enter the active site, both substrate and active site change shape (Step 2). Certain amino acids that form the active site may make temporary chemical bonds to atoms of the substrates, or electrical interactions between active site and substrates may distort the chemical bonds within the substrates. **The combination of substrate selectivity, substrate orientation, temporary chemical bonds, and bond strain promotes the specific chemical reaction catalyzed by a particular enzyme.** When the final reaction between the substrates is accomplished, the product(s) no longer fit properly into the active site, and are expelled (Step 3). The temporary changes in shape, charge, and bonding patterns within the enzyme revert back to their original configuration, and the enzyme is ready to accept another set of substrates (back to Step 1).

Why does this sequence of events speed up the rate of chemical reactions? The interactions between substrate and enzyme are like mini-reactions with very low activation energies, allowing the overall reaction to bypass its otherwise high activation energy barrier.

Enzyme Regulation

Speeding up reactions is not always desirable. For example, you do not want to metabolize every glucose molecule you eat, immediately after every meal. For one thing, you might starve to death overnight if you couldn't store any energy between supper and breakfast. What's more, glucose is an important ingredient in essential body chemicals, and burning it all up would mean none would be left for synthesizing molecules such as membrane glycoproteins or hormones. Fortunately, cells have several ways of regulating enzyme activity.

1. **A cell regulates the amount of enzyme.** Obviously, if there isn't any enzyme to begin with, the reaction will not occur. As we will see in Chapter 13, cells can regulate the synthesis of enzymes to meet their changing needs.
2. **A cell may synthesize an enzyme in an inactive form, and activate it only when needed.** Certain cells in your stomach, pancreas, and small intestine, for example, produce enzymes that digest food molecules such as proteins and lipids. These enzymes could just as easily digest the proteins and lipids of the cells themselves. This doesn't happen, because the enzymes are synthesized in an inactive form, with the active site blocked off. In your digestive tract, the interfering parts are cut off, thus activating the enzymes (see Chapter 32).
3. **A cell can temporarily activate and inactivate enzymes, depending on the conditions at any given time.** For example, the enzyme threonine deaminase begins the metabolic pathway that converts the amino acid threonine to the amino acid isoleucine. A cell needs suitable concentrations of both amino acids to manufacture proteins. This is accomplished by **end-product inhibition,** in which the activity of an enzyme is inhibited by its product or the final product of the metabolic pathway (Fig. 4-8). If enough isoleucine is present, it inhibits the activity of threonine deaminase, preventing further conversion of threonine.

On the molecular level, there are two common mechanisms of enzyme inhibition (Fig. 4-9). In **competitive inhibition,** two or more molecules compete for entry into the active site (Fig. 4-9b). Obviously, if one molecule occupies the active site, an-

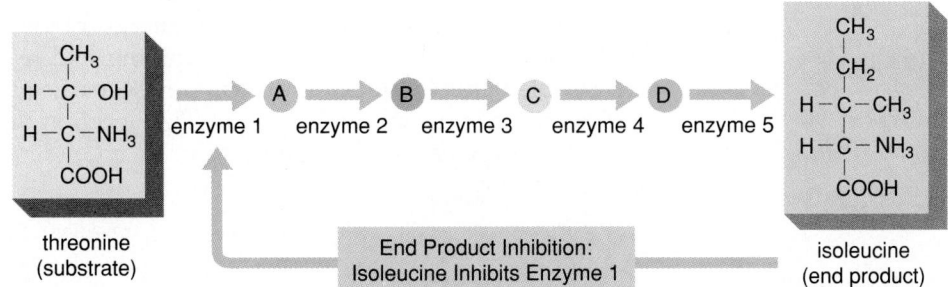

Figure 4-8 Enzyme regulation by end-product inhibition. In this example, the first enzyme in the metabolic pathway that converts threonine to isoleucine is inhibited by high concentrations of isoleucine. Therefore, if a cell lacks isoleucine, the enzyme is not inhibited, and the pathway proceeds rapidly. As isoleucine concentrations build up, the pathway is gradually shut down.

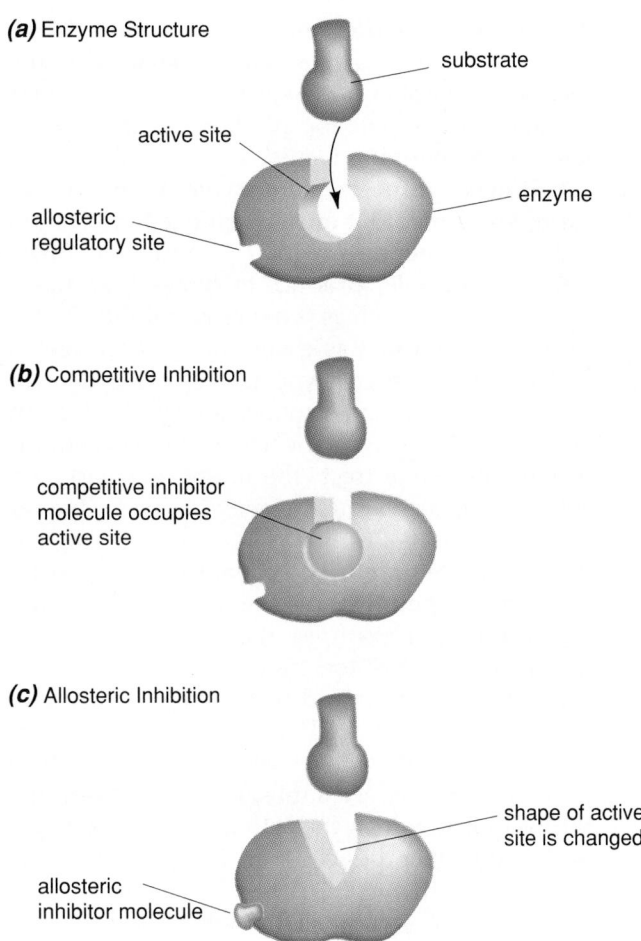

(a) Enzyme Structure

substrate

active site

allosteric regulatory site

enzyme

(b) Competitive Inhibition

competitive inhibitor molecule occupies active site

(c) Allosteric Inhibition

shape of active site is changed

allosteric inhibitor molecule

Figure 4-9 Competitive inhibition and allosteric inhibition are two methods of enzyme regulation. Although they operate by very different mechanisms, they both prevent the substrate from entering the active site of the enzyme.
(a) Many enzymes have an active site into which the substrate enters and an allosteric regulatory site on another part of the enzyme molecule. The enzyme is functional when the substrate can enter the active site.
(b) In competitive inhibition, an inhibitor molecule competes with the substrate for entry into the active site: whenever the inhibitor occupies the active site, the substrate cannot get in.
(c) In allosteric inhibition, an inhibitor molecule binds to the allosteric regulatory site. This causes the shape or charge of the active site to change, thereby preventing the substrate from entering.

other cannot. If the product of a series of reactions can bind to the active site of one of the enzymes in the pathway, the rate of reaction will be slowed down. The second mechanism is **allosteric inhibition,** in which enzyme action is blocked by molecules binding to the enzyme somewhere away from the active site

(Fig. 4-9c). Many enzymes have both an active site that catalyzes the reaction and an inhibitor site on a different part of the enzyme. When the inhibitor site is occupied, the enzyme molecule changes shape (allosteric means "other shape"). This distorts the active site, which keeps the substrate out. The reaction stops.

You might wonder how competitive and allosteric inhibition can really regulate concentrations of molecules in a cell, since an enzyme totally stops working if either the active site or the allosteric site is occupied by inhibitors. However, don't forget what you learned earlier about chemical equilibria: reactions are reversible. This includes the binding of inhibitor molecules to enzymes. Therefore, for any given enzyme, inhibitors are continually binding and releasing. If inhibitors are scarce in the cell, it may be a long while between binding, and the enzyme may be active most of the time. If the concentration of inhibitors is high, then as soon as one inhibitor leaves the enzyme another is likely to take its place, and the enzyme will be inactive most of the time. In end-product inhibition, this means that when there is very little product in the cell, there is very little enzyme inhibition. Therefore, the enzymes rapidly synthesize the product. As the product accumulates, it gradually shuts down the enzymes, preventing further depletion of substrate.

The ability of an enzyme to catalyze reactions is controlled by the concentration of active enzyme within a cell and the concentration of inhibitor molecules. This allows a cell to maintain optimal concentrations of both substrates and products.

Coupled Reactions and Energy Carrier Molecules

As we pointed out earlier, cells control energy flow by coupling reactions, so that the energy released by exergonic reactions is used to drive endergonic reactions. Energy transfer is the role of **energy carrier molecules.** Energy carriers work something like rechargeable batteries, picking up an energy charge at an exergonic reaction, moving to another location in the cell, and releasing the energy to drive an endergonic reaction.

ATP

The most common energy carrier in living cells is **adenosine triphosphate,** or **ATP.** As you learned in Chapter 3, ATP is composed of a nitrogen-containing base, adenine; a sugar, ribose; and three phosphate groups (Fig. 4-10). Energy released in the cell through

ATP: Principal Energy Carrier of the Cell

Molecule	Shorthand Representations	Energy Content
NH₂ (adenine) attached to ribose sugar, linked via CH₂—O—P—O—P structure. **Adenosine Diphosphate (ADP)**	A—P—P or ADP	low
NH₂ (adenine) attached to ribose sugar, linked via CH₂—O—P—O—P—O—P structure with "high-energy" bonds indicated. **Adenosine Triphosphate (ATP)**	A—P—P—P or ATP	high

Figure 4-10 The structures of adenosine diphosphate (ADP) and adenosine triphosphate (ATP). A phosphate group is added to ADP to make ATP. In most cases, only the last phosphate group and its high-energy bond are used to carry energy and transfer it to endergonic reactions within a cell. In the remainder of the text, we will usually employ the shorthand representations of ATP and ADP.

glucose metabolism is used to drive the synthesis of ATP from adenosine diphosphate (ADP) and inorganic phosphate:

ATP synthesis: energy is stored in ATP

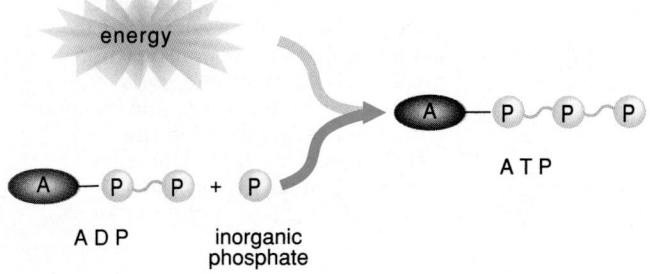

ADP   inorganic phosphate   ATP

Most of the chemical bonds in ADP and ATP are "ordinary" covalent bonds, but the bonds joining the last two phosphate groups of ATP to the rest of the molecule are usually called "high-energy bonds." Under most circumstances, the cell uses only the last high-energy bond of ATP (the one that joins phosphate to ADP to make ATP) to carry energy.

The term "high-energy bond" is somewhat misleading, because in fact the bonds do not require an enormous amount of energy to form, nor do they release an enormous amount of energy when they are broken. Rather, a high-energy bond's relationship to a chemical reaction is a bit like Goldilocks' opinion of the baby bear's porridge: it's just right. When your cells digest sugar, many of the reactions release just enough energy to add a phosphate to ADP to form ATP. Similarly, when ATP is broken down to ADP and inorganic phosphate, it releases an amount of energy that is just right to drive many essential cellular reactions. Nevertheless, the phrase "high-energy bond" is so common that we will continue to use it here.

ATP carries this energy to various sites in the cell that perform energy-requiring reactions, such as muscle contraction. The ATP is then hydrolyzed to ADP and inorganic phosphate:

ATP hydrolysis: energy of ATP is released

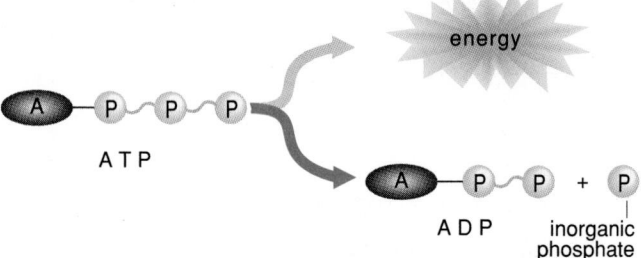

The energy released by ATP hydrolysis may drive the energy-requiring reaction. The ADP and inorganic phosphate are usually recycled back to energy-generating reactions that resynthesize ATP.

Most uses of energy in living cells involve pairs of coupled reactions linked by ATP (Fig. 4-11). In the first coupled reaction, energy release drives ATP synthesis; in the second, ATP hydrolysis drives an ender-

gonic reaction. Each coupled reaction is mediated by a specific enzyme that positions the molecules and ensures that the ATP energy is channeled properly. Enzyme-catalyzed coupled reactions provide the energy and the specificity necessary to construct the different types of molecules needed by the cell.

So much ATP is used by living organisms that the lifespan of any given ATP molecule is very short. For example, a human runner can use up as much as a pound of ATP each minute (obviously, the ADP thus produced must be quickly converted back to ATP, or it will be a very brief run). As a result, ATP is *not* a long-term energy storage molecule. More stable molecules such as sucrose, glycogen, starch, or fat are used to store energy for hours, days, or months.

Electron Carriers

Energy may also be transported around a cell by other carrier molecules. In some exergonic reactions, including both glucose metabolism and the light-capture stage of photosynthesis, energy is transferred to electrons. These energetic electrons (sometimes along with hydrogen atoms) may be captured by **electron carriers** (Fig. 4-12), such as nicotinamide adenine dinucleotide (NAD^+) and its relative, flavin adenine dinucleotide (FAD). Loaded electron carriers then donate the electrons, along with their energy, to other molecules. We will see more about electron carriers and their role in cellular metabolism in Chapters 7 and 8.

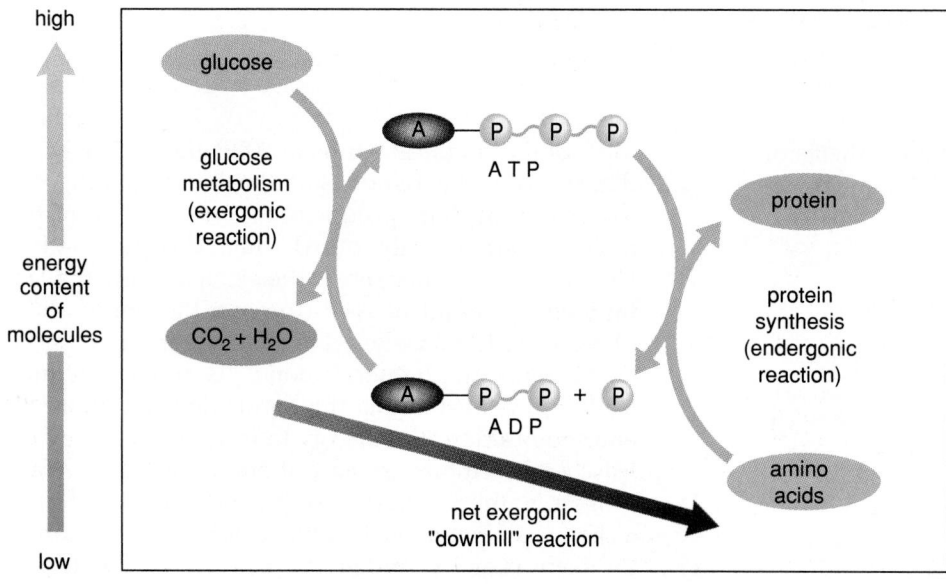

Figure 4-11 Coupled reactions within living cells. Exergonic reactions (e.g., glucose metabolism) drive the endergonic reaction of ATP synthesis from ADP. The ATP molecule moves to another part of the cell, where ATP hydrolysis liberates some of this energy to drive an essential endergonic reaction (e.g., protein synthesis). The ADP and inorganic phosphate are recycled back to the exergonic reactions, where they will be converted to ATP once again. The overall reaction is "downhill": more energy is produced by the exergonic reaction than is needed to drive the endergonic reaction. The extra energy is released as heat.

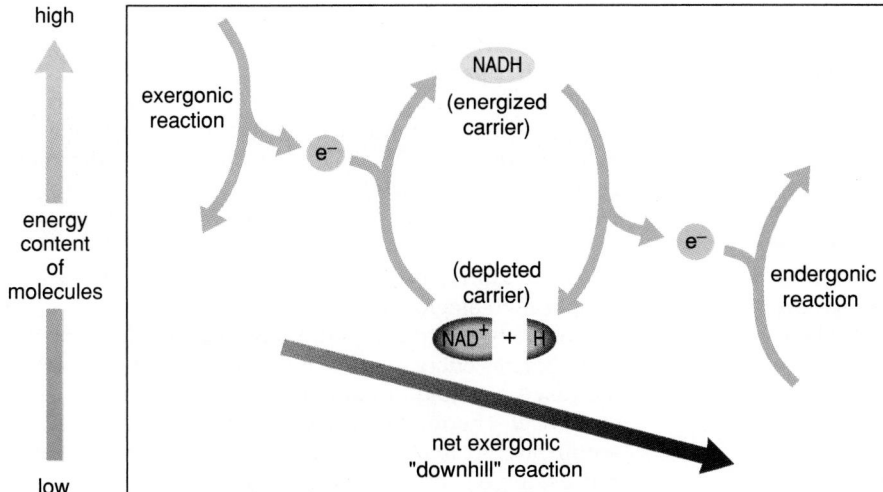

energy
content
of
molecules

low

exergonic
reaction

e^-

NADH
(energized
carrier)

(depleted
carrier)

$NAD^+ + H$

e^-

endergonic
reaction

net exergonic
"downhill" reaction

Figure 4-12 Electron carrier molecules such as nicotinamide-adenine dinucleotide (NAD) pick up electrons generated by exergonic reactions and hold them in high-energy outer electron shells. Hydrogen atoms are often picked up simultaneously. The electron is then deposited, energy and all, with another molecule to drive an endergonic reaction, often the synthesis of ATP.

SUMMARY OF KEY CONCEPTS

Energy Flow in the Universe

Kinetic energy is the energy of movement (light, heat, electricity, movement of large particles). Potential energy is stored energy (chemical energy, energy of position within a gravitational field). The flow of energy among atoms and molecules obeys the laws of thermodynamics. The first law states that within an isolated system the total amount of energy remains constant, although it may change in form. The second law states that any change within an isolated system causes a decrease in the quantity of concentrated, useful energy, and an increase in the randomness and disorder of matter. Entropy is a measure of disorder within a system.

Thermodynamics and Chemical Reactions

Chemical reactions fall into two categories. In exergonic reactions, the product molecules have less energy than the reactant molecules, so the reaction releases energy. In endergonic reactions, the products have more energy than the reactants, so the reaction requires an input of energy. Exergonic reactions can occur spontaneously, but all reactions, including exergonic ones, require an initial input of energy, called the activation energy, to overcome electrical repulsions between reactant molecules. Exergonic and endergonic reactions may be coupled together, so that the energy liberated by an exergonic reaction drives the endergonic reaction. Living organisms couple exergonic reactions such as light-energy capture or sugar metabolism with endergonic reactions such as the synthesis of organic molecules.

Both exergonic and endergonic reactions are reversible and can proceed in either direction, given suitable inputs of products, reactants, and energy. Living organisms provide these inputs in controlled ways to drive reversible reactions in the directions necessary to maintain life.

Controlling Reactions Within the Cell

High activation energies slow many reactions, even exergonic ones, to an imperceptible rate under normal environmental conditions. Catalysts lower the activation energy and thereby speed up chemical reactions. Catalysts are not permanently altered during the reaction. Living organisms synthesize protein catalysts called enzymes that promote one or a few specific reactions. The reactants temporarily bind to the active site of the enzyme, which strains their original chemical bonds and makes it easier to form the new chemical bonds of the products. Enzyme action is regulated in three ways: (1) by altering the rate of enzyme synthesis, (2) by activating previously inactive enzymes, and (3) by inhibiting enzyme activity. One common form of enzyme inhibition is end-product inhibition, in which the end product of a metabolic pathway inhibits the activity of one of the enzymes that mediates an earlier step. Therefore, as products accumulate, they automatically slow down their own rate of synthesis.

Energy released by chemical reactions within a cell is captured and transported about the cell by energy carrier molecules such as ATP and electron carrier molecules. These molecules are the major means whereby cells couple exergonic and endergonic reactions occurring at different places within the cell.

GLOSSARY

activation energy: in a chemical reaction, the energy needed to force the electron clouds of reactants together, prior to the formation of products.

active site: the region of an enzyme molecule that binds substrates and performs the catalytic function of the enzyme.

adenosine triphosphate (a-den'-ō-sen trīfos'-fāt; ATP): a molecule composed of ribose sugar, adenine, and three phosphate groups. The last two phosphate groups are attached by energy-carrier "high-energy bonds." ATP is the major energy carrier in cells.

allosteric inhibition (al-ō-ster'-ik): enzyme regulation in which an inhibitor molecule binds to an enzyme at a site away from the active site, changing the shape or charge of the active site so that it can no longer bind substrate molecules.

catalyst (cat'-a-list): a substance that speeds up a chemical reaction without itself being permanently changed in the process. Catalysts lower the activation energy of a reaction.

chemical equilibrium: the condition in which the "forward" reaction of reactants to products proceeds at the same rate as the "backward" reaction from products to reactants, so that no net change in chemical composition occurs.

competitive inhibition: in enzyme-catalyzed reactions, a condition in which two molecules (at least one a substrate for the enzyme) compete for entry into the active site of the enzyme, thus slowing the rate of reaction.

coupled reaction: a pair of reactions, one exergonic and one endergonic, that are linked together so that the energy produced by the exergonic reaction provides the energy needed to drive the endergonic reaction.

electron carrier: a molecule that can reversibly gain and lose electrons. Electron carriers generally accept high-energy electrons produced during an exergonic reaction and donate the electrons to acceptor molecules that use the energy to drive endergonic reactions.

endergonic (en-der-gon'-ik): pertaining to a chemical reaction that requires an input of energy to proceed; an "uphill" reaction.

end-product inhibition: in enzyme-mediated chemical reactions, the condition in which the product of a reaction inhibits one or more of the enzymes involved in synthesizing the product.

energy: the capacity to do work.

energy carrier: a molecule that stores energy in "high-energy" chemical bonds and releases the energy again to drive coupled endothermic reactions. ATP is the most common energy carrier in cells.

entropy (en'-trō-pē): a measure of the amount of randomness and disorder in a system.

enzyme (en'zīm): a protein catalyst that speeds up the rate of specific biological reactions.

exergonic (ex-er-gon'-ik): pertaining to a chemical reaction that liberates energy (either heat energy or in the form of increased entropy); a "downhill" reaction.

First Law of Thermodynamics: a principle of physics that within any isolated system, energy can be neither created nor destroyed, but can be converted from one form to another.

induced fit: a model of enzyme activity proposing that binding of substrates to an enzyme active site changes the shape or charge of both the substrates and the active site.

kinetic energy: the energy of movement; includes light, heat, mechanical movement, and electricity.

lock-and-key: a model of enzyme activity proposing that substrates fit precisely into the active site of an enzyme.

metabolism: the sum of all chemical reactions occurring within a single cell or within all the cells of a multicellular organism.

metabolic pathway: a sequence of chemical reactions within a cell, in which the products of one reaction are the reactants for the next.

potential energy: stored energy; includes chemical energy and the energy of position within a gravitational field.

product: an atom or molecule resulting from a chemical reaction.

reactant: an atom or molecule that is used up in a chemical reaction to form a product.

Second Law of Thermodynamics: a principle of physics that any change in an isolated system causes the quantity of concentrated, useful energy to decrease, and the amount of randomness and disorder (entropy) to increase.

substrate: the atoms or molecules that are the reactants for an enzyme-catalyzed chemical reaction.

STUDY QUESTIONS

1. State the two laws of thermodynamics. Define an isolated system in which they apply.
2. Why don't living organisms violate the Second Law of Thermodynamics? What is the ultimate energy source for most forms of life on Earth?
3. Define endergonic and exergonic reactions, and give an example of each.
4. What are coupled reactions? How do living organisms use coupled reactions to sustain life?
5. What is activation energy? How do catalysts affect activation energy? How does this change the rate of reactions?
6. What is a chemical equilibrium? Why are reversible reactions important in living organisms?
7. What is an enzyme? In what ways are enzymes similar to, and different from, inorganic catalysts?
8. Describe the structure and function of enzymes. How is enzyme activity regulated?
9. What is an energy carrier? What is the most common energy carrier in living cells? How does it couple exergonic and endergonic reactions?

DISCUSSION QUESTIONS

1. A preview question for ecology: When the brown bear shown in the Chapter opener photo eats the salmon, does it acquire all the energy contained in the body of the fish? Why or why not? What implications do you think this would have for the relative abundance (by weight) of predators and their prey? Does the Second Law of Thermodynamics help to explain the title of the book *Why Big Fierce Animals Are Rare?*
2. The large ears and almost hairless skin of an elephant promote heat loss from its huge body. Where does this heat come from?
3. As you learned in Chapter 3, the subunits of virtually all organic molecules are joined by condensation reactions and can be broken apart by hydrolysis reactions. Why, then, does your digestive tract have separate enzymes to digest proteins, fats, and carbohydrates, and in fact several of each?
4. Suppose someone tried to disprove evolution with the following argument: "According to evolutionary theory, organisms have increased in complexity through time. However, such evolution of increased biological complexity contradicts the second law of thermodynamics. Therefore, evolution is impossible." If you were to support the theory of evolution, what would be your response to this argument?

SUGGESTED READINGS

Baker, J. J. W., and Allen, G. E. *Matter, Energy, and Life.* 4th ed. Reading, Mass.: Addison-Wesley, 1981. An excellent introduction to chemical-energy principles for those interested in biology.

Dickerson, R. E. "Cytochrome c and the Evolution of Energy Metabolism." *Scientific American,* March 1980. Cytochrome c is an electron carrier in mitochondria, and one of the best-studied proteins. Dickerson points out the evolutionary heritage of this molecule in diverse organisms, and also discusses its role in energy acquisition by cells. See also Dickerson's earlier article on cytochrome c, "The Structure and History of an Ancient Protein." *Scientific American,* April 1972.

Fenn, J. *Engines, Energy, and Entropy.* New York: W. H. Freeman and Co., 1982. Elegantly simple introduction to the laws of thermodynamics and its relationship to everyday life.

Koshland, D. E., Jr. "Protein Shape and Biological Control." *Scientific American,* October 1973. Enzyme function is intimately related to structure. Koshland discusses enzyme specificity and regulation in terms of its protein structure.

Sackheim, G. *Introduction to Chemistry for Biology Students.* Redwood City, CA, Benjamin/Cummings Publishing Co., 1991. This book is specifically designed for biology students who don't have much chemistry background. Basic chemical concepts, including energetics, are presented in manageable bites.

5

Cells: The Units of Life.
I. Cell Structure and Function

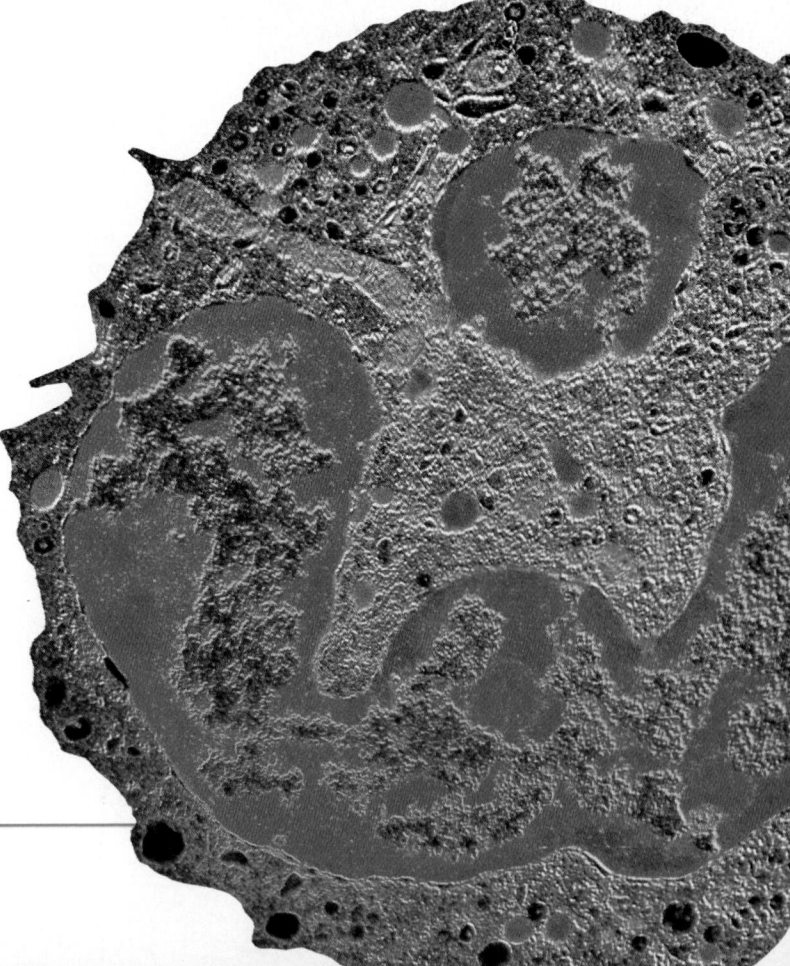

From the bacteria in your large intestine quietly dining on the remnants of your lunch to the cheetah chasing a gazelle on the African plains, the activities of living organisms are possible only because of the activities of the cells that compose them. In this chapter, we examine the overall structure and function of cells and their component parts. In Chapter 6, we focus on the plasma membrane that regulates the interactions of cells with their external environment. But first, what *is* a cell? How do biologists know that living organisms are made of cells, and that the activities of an organism reflect the combined activities of its cells?

The Development of Cell Theory

The Englishman Robert Hooke was a remarkable scientist. He built one of the earliest reflecting telescopes, made detailed observations of Mars, and discovered the law of elasticity. In 1665 he peered through a primitive microscope at an "exceeding thin . . . piece of Cork" and discovered that cork, which is made from the bark of certain trees, is composed of "a great many little Boxes" (Fig. 5-1). Hooke called the boxes "cells," because he fancied that they resembled the tiny rooms, or cells, occupied by monks in monasteries. Later, when he looked at living plants, he found that "these cells [are] fill'd with juices." What these cells were, and how they related to the lives of entire plants, Hooke couldn't say.

In the late 1600s, the Dutch inventor Antonie van Leeuwenhoek observed red blood cells, sperm, and myriads of "animalcules" in pond water, some of them single cells. Nevertheless, more than a century passed before any real progress was made in understanding the role of cells in life on Earth.

Gradually, microscopists noted that many plants are composed of cells, a task made easy by the thick cell wall that outlines every plant cell. Animal cells, however, escaped notice until the 1830s, when Theodor Schwann saw that cartilage contains cells that "exactly resemble . . . [the cells of] plants." In 1839, Schwann was confident enough of the universality and importance of cells to state that cells are the "elementary particles" of both plants and animals. By the mid-1800s, the botanist Mattias Schleiden had grasped the significance of cells: "It is . . . easy to perceive that the vital process of the individual cells must form the first, absolutely indispensable fundamental basis" of life.

Within a few years, several microscopists had

Figure 5-1 Robert Hooke's drawings of the cells of cork, as seen with an early light microscope. The cells of cork are not living, and only the cell walls remain to outline the cell.

noted that living cells could grow larger and divide into two smaller but still living cells. In the late 1850s, the Austrian pathologist Rudolf Virchow wrote, "Every animal appears as a sum of vital units, each of which bears in itself the complete characteristics of life." Furthermore, Virchow predicted that "all cells come from cells." These phrases were the forerunners of the three principles of modern cell theory:

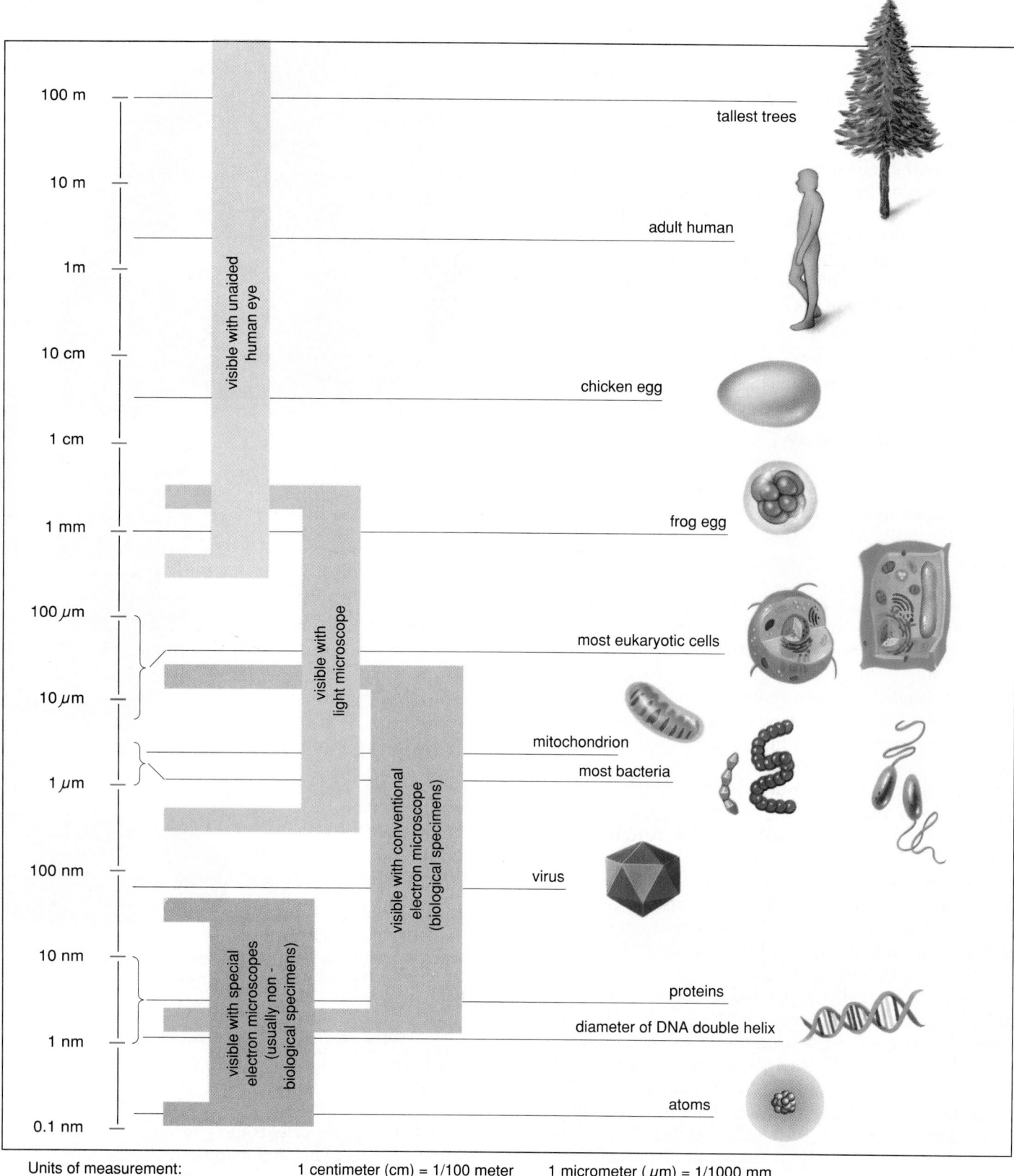

Figure 5-2 Dimensions commonly encountered in biology range from about 100 meters (the height of the tallest redwoods) through a few micrometers (the diameter of most cells) to a few nanometers (the diameter of many large molecules). Note that in the metric system (used almost exclusively in science) new names are given to dimensions that differ by factors of 10 to 1000.

1. Every living organism is made up of one or more cells.
2. The smallest living organisms are single cells, and cells are the functional units of multicellular organisms.
3. All cells arise from preexisting cells.

Generally speaking, cells are small, ranging from about 1 to 100 micrometers (millionths of a meter) in diameter (Fig. 5-2). Living organisms may consist of just one cell (bacteria, protists) or aggregates of interacting, cooperating cells (fungi, plants, and animals). Because of the small size of nearly all cells, biologists use a variety of microscopes to assist their studies of cell structure and function. Some of the diversity of images produced by modern microscopes is shown in "Methods in Biology: Viewing the Cell: A Gallery of Microscopic Images."

An Overview of Cell Structure and Function

All cells have at least three components: plasma membrane, genetic material, and cytoplasm (Fig. 5-3).

Plasma Membrane

The **plasma membrane** (also called the *cell membrane*) is a double layer of phospholipids (see Chapter 3) in which are embedded a wide variety of proteins. The plasma membrane performs three functions: (1) it isolates the cytoplasm from the external environment; (2) it regulates the flow of materials between the cytoplasm and its environment, for example, acquiring nutrients and expelling wastes; and (3) it allows interaction with other cells.

Genetic Material

Each cell has its own hereditary blueprint, in which is stored the instructions for making all the other parts of the cell and for making new daughter cells. In all living cells, the genetic material is **deoxyribonucleic acid (DNA).** In eukaryotic cells (plants, animals, fungi, and protists), the DNA is contained within a separate, membrane-bound structure, the nucleus. In prokaryotic cells (bacteria), the DNA, although localized to a particular place within the cytoplasm called the **nucleoid,** is not separated by membranes from the rest of the cytoplasm.

Cytoplasm

The **cytoplasm** consists of all the material inside the plasma membrane and outside the nucleus (in eukaryotic cells, which contain a membrane-limited nucleus). The cytoplasm includes water, salts, an assortment of organic molecules, and a variety of discrete structures, called **organelles,** each carrying out a distinct function.

Cell Function Limits Cell Size

Why are all living things composed of cells? From the essential components of cells listed above, you can probably guess why microscopic organisms are cellu-

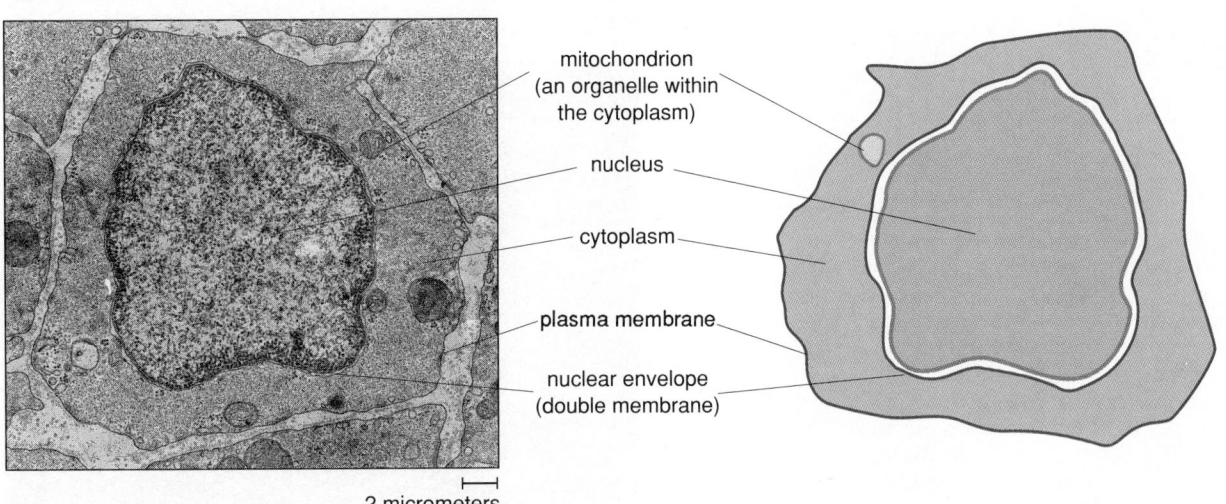

mitochondrion (an organelle within the cytoplasm)

nucleus

cytoplasm

plasma membrane

nuclear envelope (double membrane)

2 micrometers

Figure 5-3 An animal cell as seen with a transmission electron microscope. Note the plasma membrane (thin line surrounding cell), cytoplasm (material enclosed by the plasma membrane), and the nucleus containing the genetic material.

METHODS IN BIOLOGY

Viewing the Cell: A Gallery of Microscopic Images

Collaboration among biologists, physicists, and engineers has resulted in development of a variety of microscopes that allow us to examine structures too small to see with the naked eye. In some cases, biologists use microscopes to look at living cells (Fig. E5-1a through d). In most instances, however, the specimen must be carefully prepared, usually sliced thin, and stained (Fig. E5-1d and e; Figs. E5-2 through E5-5).

Light microscopes use lenses, usually made of glass, to focus and magnify light rays that either pass through or bounce off a specimen. Light microscopes provide a wide range of images, depending on how the specimen is illuminated and whether it has been stained (Fig. E5-1a through e). The wavelength of visible light limits the resolving power of light microscopes to about 1 micrometer (a millionth of a meter).

Electron microscopes use beams of electrons instead of light. The negatively charged electrons are focused by magnetic fields rather than by conventional lenses. Electrons behave as if they have much shorter wavelengths than visible light. Therefore, some types of electron microscopes can resolve structures as small as a few nanometers (billions of a meter). *Transmission electron microscopes* (TEMs) pass electrons through a thin specimen, and can reveal minute subcellular structures (Fig. E5-1f), including organelles and cell membranes (Fig. E5-2). *Scanning electron microscopes* (SEMs) bounce electrons off specimens that have been coated with metals, and provide three-dimensional images. SEMs can be used to view structures ranging in size from entire insects (Fig. E5-3) down to cells (Fig. E5-4) and even organelles (Fig. E5-5).

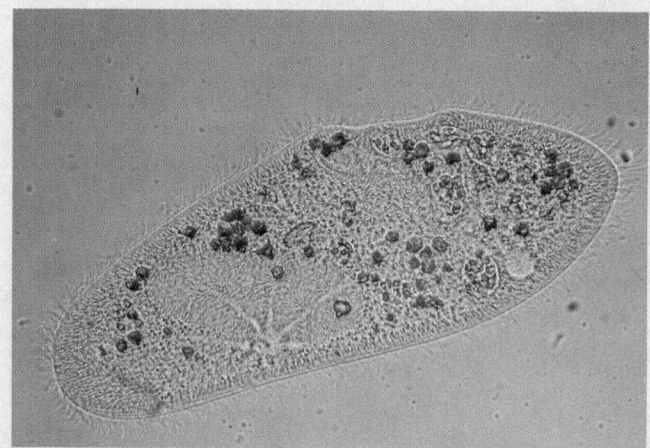

(a)

⊢—————⊣
10 micrometers

(b)

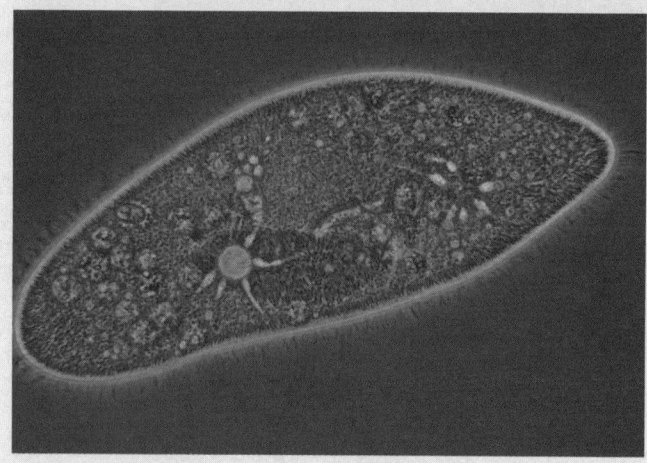

Figure E5-1 Images of the protozoan *Paramecium* seen through different types of microscopes.
(a through d) Living *Paramecia* seen with **(a)** conventional brightfield, **(b)** darkfield, **(c)** phase contrast, and **(d)** differential interference contrast illumination.
(e) This *Paramecium* has been stained to reveal some subcellular structures, and photographed with a brightfield microscope. **(f)** A transmission electron micrograph of part of a *Paramecium*. Note the increased detail compared with the specimens photographed through a light microscope. Because the magnification is greater, the entire *Paramecium* is not shown.

(c)

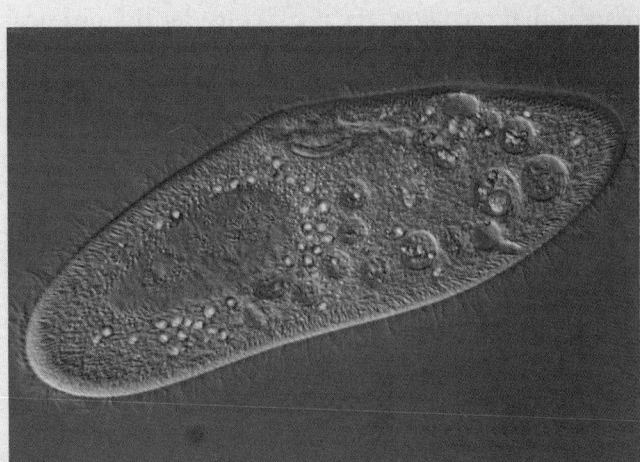

(d)

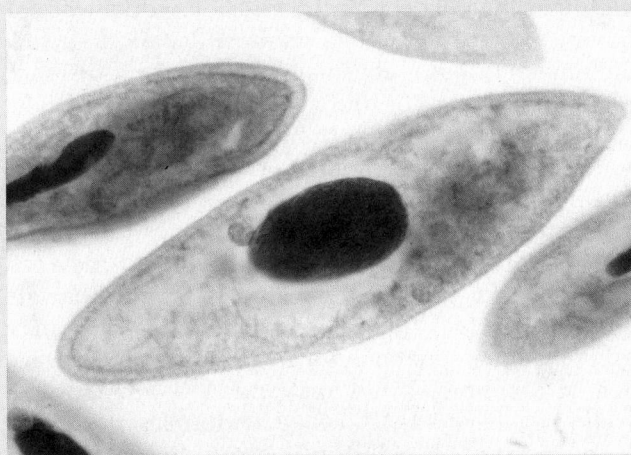

(e)

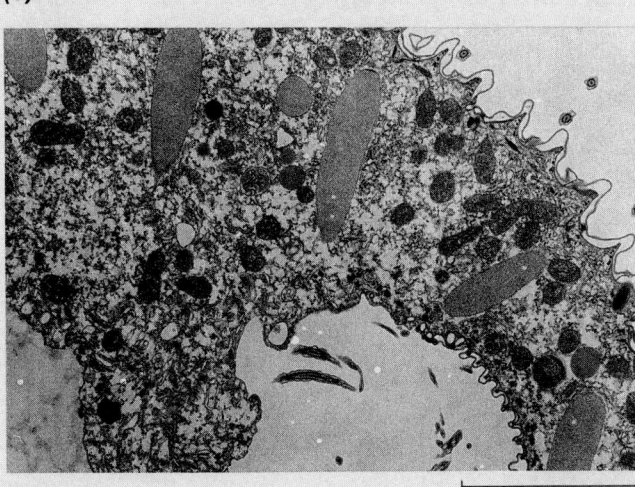

(f)

10 micrometers

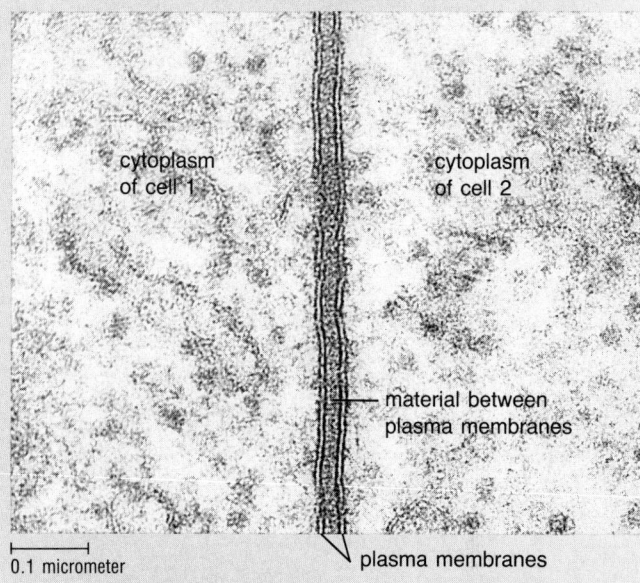

cytoplasm of cell 1

cytoplasm of cell 2

material between plasma membranes

0.1 micrometer

plasma membranes

Figure E5-2 Transmission electron microscopes can magnify much more than light microscopes. These photos show a pair of cell membranes that have been cut perpendicularly (like cutting a sheet of paper with scissors and looking at the paper edge on). In micrographs such as this one, membranes appear as a pair of dark lines (the phospholipid heads) separated by a thin layer of paler material (the phospholipid tails; see Fig. 5-11). A cell membrane is only about 7 billionths of a meter in thickness.

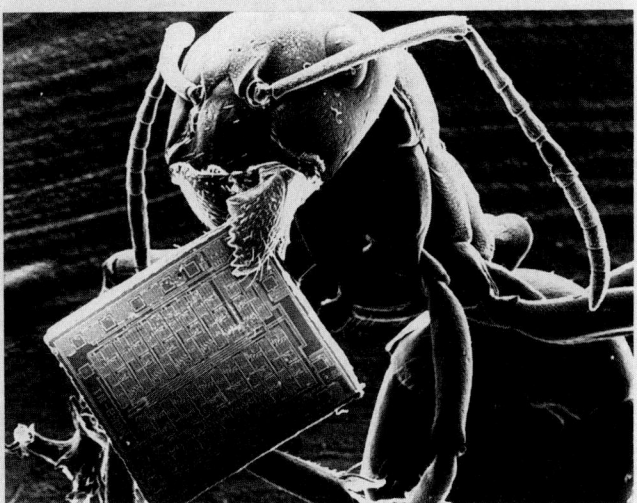

Figure E5-3 Scanning electron microscopes provide images that look three-dimensional. Here a high-tech ant carries off a computer chip.

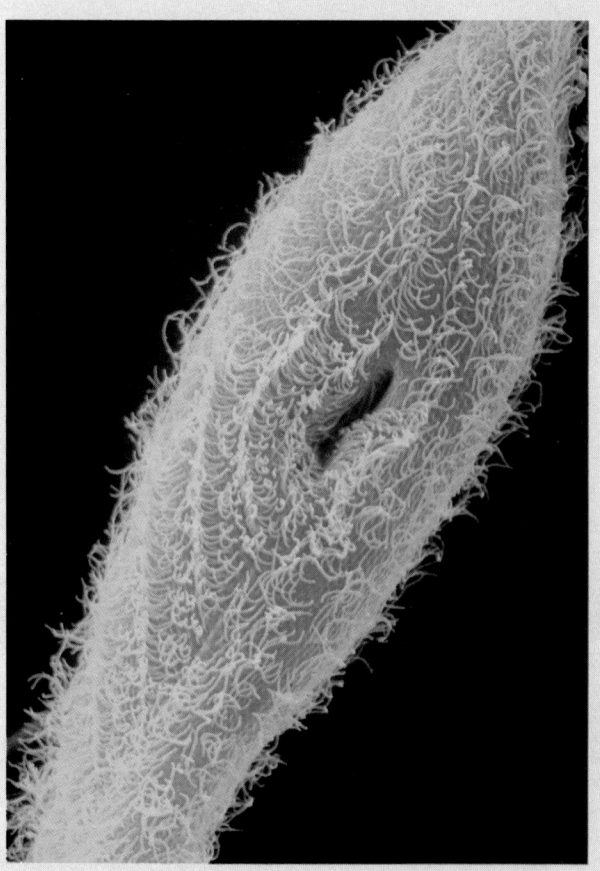

10 micrometers

Figure E5-4 An SEM photograph of *Paramecium* shows the pattern of hairlike cilia that cover the surface of the cell.

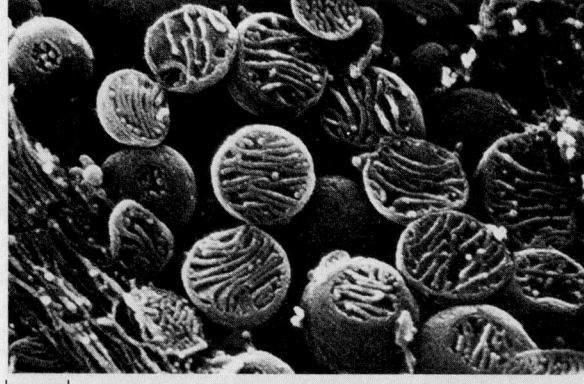

0.5 micrometers

Figure E5-5 An SEM photograph showing part of the inside of a cell. The spherical structures, many partially sliced open, are mitochondria (see Fig. 5-17).

lar: the plasma membrane separates the complex life processes in the cytoplasm from the chaos of the outside world. The genetic material is the data bank that directs those life processes.

What may be less obvious is why large organisms consist of many cells rather than just one large cell. To answer that question, we must consider the physical constraints that limit cell size. All living organisms exchange nutrients and wastes with their external environment. If a large organism were composed of a single cell, this exchange would be limited both by the distance from the center of the cell to its surface and by the surface area of the cell.

Acquiring nutrients and eliminating wastes both occur through the plasma membrane. This would cause two major problems for very large cells (Fig. 5-4). First, as a cell becomes larger, its innermost regions become farther away from the membrane. Most nutrients and wastes move into, through, and out of cells by **diffusion,** the net transport of molecules from places of high concentration to places of low concentration (we will discuss diffusion more thoroughly in the next chapter). Diffusion works well for moving molecules over very short distances (a few micrometers), but is abysmally slow for long-distance transport. Relying on diffusion alone, it would take oxygen molecules more than *200 days* to reach the center of a cell 20 centimeters in diameter (about the thickness of your chest). Clearly, not much life could go on inside a cell that large!

The second difficulty arises from geometry: as a cell enlarges, its volume increases more rapidly than its surface area does (from high school days, you may remember the equations for the volume and surface area of a sphere, $V = 4/3\pi r^3$ and $A = 4\pi r^2$, respectively; see Fig. 5-4). A cell that doubles its radius therefore becomes eight times greater in volume but only four times greater in surface area. In general, as a cell increases in volume, more chemical reactions occur. Thus more nutrients and oxygen are needed, and more waste products must be eliminated, all through the plasma membrane. In a very large cell the surface area of the membrane would be too small to keep up with the cell's metabolic needs. Large organisms, therefore, consist of billions of coordinated cells, with special provisions made for internal cells to exchange materials with the external environment.

Types of Cells:
Prokaryotic Versus Eukaryotic

As we briefly discussed in Chapter 1, there are two fundamentally different types of cells on Earth. The

distance to center (r)	1.0	2.0	3.0	1.0
surface area ($4\pi r^2$)	12.6	50.3	113.1	339.4
volume ($4/3\pi r^3$)	4.2	33.5	113.1	113.1
area/volume	3.0	1.5	1.0	3.0

Figure 5-4 Geometrical considerations limit the size of cells. As a cell enlarges, the distance from the center of the cell to the outside world increases. Further, the volume increases much more rapidly than the surface area. As these spherical cells illustrate, doubling the radius halves the surface area : volume ratio. Thus each unit volume of cytoplasm within the cell has only half the membrane area available to exchange nutrients and wastes with the external environment. If the volume of the largest sphere is divided among cells the size of the smallest sphere, the overall surface area : volume ratio of the resulting "multicellular organism" remains large.

first type, represented by the bacteria, is called **prokaryotic,** Greek for "before the nucleus." The second type of cell, which evolved from the prokaryotic cell and is today found among protists, plants, fungi, and animals, is called **eukaryotic** ("true nucleus"). As their names imply, perhaps the most striking difference between prokaryotic cells and eukaryotic cells is that eukaryotic cells have their genetic material contained within a membrane-limited structure, the nucleus, while the genetic material of prokaryotic cells is not enclosed within a membrane. A variety of other features also differ between the two types of cells, making a closer look worthwhile.

Prokaryotic Cells

Prokaryotic cells are usually very small (less than 5 μm in length), with a relatively simple internal structure (Fig. 5-5; compare with the eukaryotic cell shown in Fig. 5-3). Most prokaryotic cells are surrounded by a relatively stiff cell wall. The materials of the cell wall are secreted by the cell itself. Many bacteria live in fluid environments, from ponds to the human bloodstream, from which water tends to enter the bacterial cell (see Chapter 6). Without a cell wall, water entry would cause the cell to swell and explode; in fact, penicillin combats certain bacterial in-

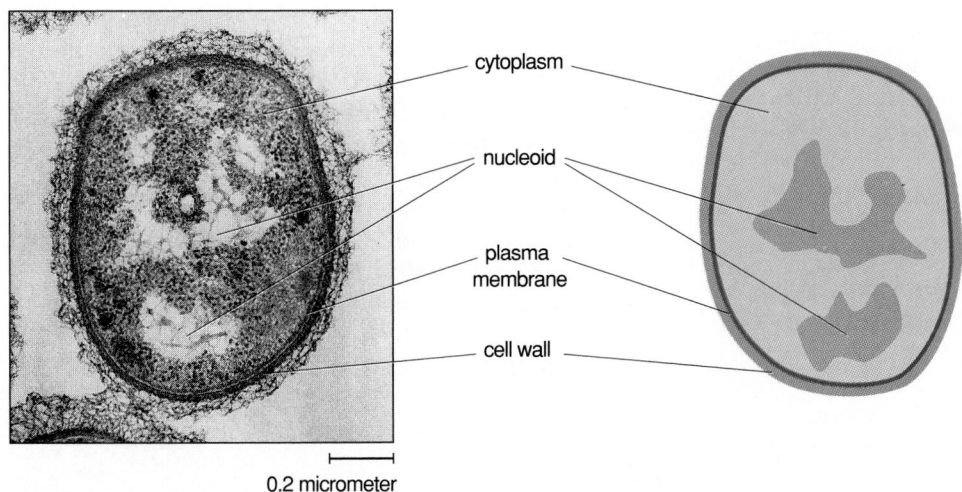

cytoplasm

nucleoid

plasma membrane

cell wall

0.2 micrometer

Figure 5-5 A transmission electron micrograph of a bacterium, a prokaryotic cell. The pale area occupying much of the cell, called the nucleoid, is the location of the cell's DNA; it is not separated from the rest of the cytoplasm by a membrane. Compare the structural simplicity of this prokaryotic cell with the micrograph of a eukaryotic cell in Figure 5-3.

fections by inhibiting cell wall synthesis, causing the bacteria to rupture. In some bacteria, such as the one shown in Figure 5-5, the cell wall has a polysaccharide coating that prevents white blood cells from engulfing the bacterium. Although it plays an important supportive role, the cell wall is quite porous. Movement of materials into and out of a prokaryotic cell is regulated by the plasma membrane that lies just beneath the cell wall.

The cytoplasm of most prokaryotic cells is relatively homogeneous in appearance (although some photosynthetic bacteria have elaborate internal membranes). The DNA is usually coiled, attached to the plasma membrane, and concentrated in a region of the cell called the **nucleoid.** It is not, however, physically separated from the rest of the cytoplasm by a membrane.

Chapter 21 provides a more detailed look at the diversity of prokaryotic cells.

Eukaryotic Cells

Eukaryotic cells differ in many respects from prokaryotic cells. In addition to being larger than prokaryotic cells (usually more than 10 μm in diameter), eukaryotic cells contain a variety of membranous **organelles** that lend structural and functional organization to the cell interior. Technically, the material within the plasma membrane of a eukaryotic cell is divided into the **nucleus,** an organelle consisting of a double layer of membrane that encloses the genetic material, and the **cytoplasm,** which includes everything else. The cytoplasm in turn is composed of several types of organelles, occupying as much as half the volume of the cell, and a fluid matrix, the **cytosol** (literally, "cell solution"), in which the organelles reside. The cytosol is an aqueous solution of salts, sugars, amino acids, proteins, fatty acids, nucleotides, and other materials. Giving shape and organization to the cytoplasm is a network of protein fibers called the **cytoskeleton.** Many of the organelles and even individual molecules of the cytoplasm are thought to be attached to the cytoskeleton.

With a few rare exceptions, all eukaryotic cells contain the following organelles, each with its own structural and functional specialization:

nucleus: contains the cell's genetic information
mitochondrion: captures the chemical energy of food molecules as high-energy bonds of adenosine triphosphate (ATP)
endoplasmic reticulum: a network of membranous channels, often studded with small protein/RNA particles called **ribosomes;** proteins in the endoplasmic reticulum

membrane synthesize lipids, while the ribosomes synthesize many proteins
Golgi complex: stacks of membranous sacs that modify proteins and lipids, synthesize carbohydrates, and package molecules for transport
vesicles: membranous sacs that store and transport molecules around the cell

Eukaryotic cells, however, are not all alike. Figures 5-6 and 5-7 illustrate the structures that are found in animal and plant cells, although few individual cells contain all the features shown in the drawings. You will note that each type of cell possesses a few unique organelles not found in the other. You may want to refer back to these illustrations as we describe the subcellular structures in more detail. As a reference and study guide, Table 5-1 lists the major characteristics of prokaryotic and eukaryotic cells, including a brief description of the functions of the organelles.

In the following pages, we will introduce you to the major components of eukaryotic cells. We will emphasize the union of *structure* and *function* that is crucial to understanding biology. In studying cellular anatomy, remember that each structure originally evolved, and persists today, because it performs a specific function that is essential to the survival and reproduction of the cell.

The Nucleus: Control Center of the Cell

Deoxyribonucleic acid, or **DNA,** is the genetic material of all living cells. A cell's DNA is the repository of information needed to construct the cell and direct its life processes. Just as a blueprint is used selectively, one part at a time, to build a house, so too the hereditary information in DNA is used selectively by the cell, depending on its stage of development and its environmental conditions. In eukaryotic cells, the DNA is housed within the nucleus.

The nucleus consists of three readily distinguishable components. The **nuclear envelope** separates the nuclear material from the cytoplasm. Inside the nu-

Figure 5-6 An illustration of a "typical" animal cell. The entire range of organelles depicted here seldom occurs in a single animal cell. ▶

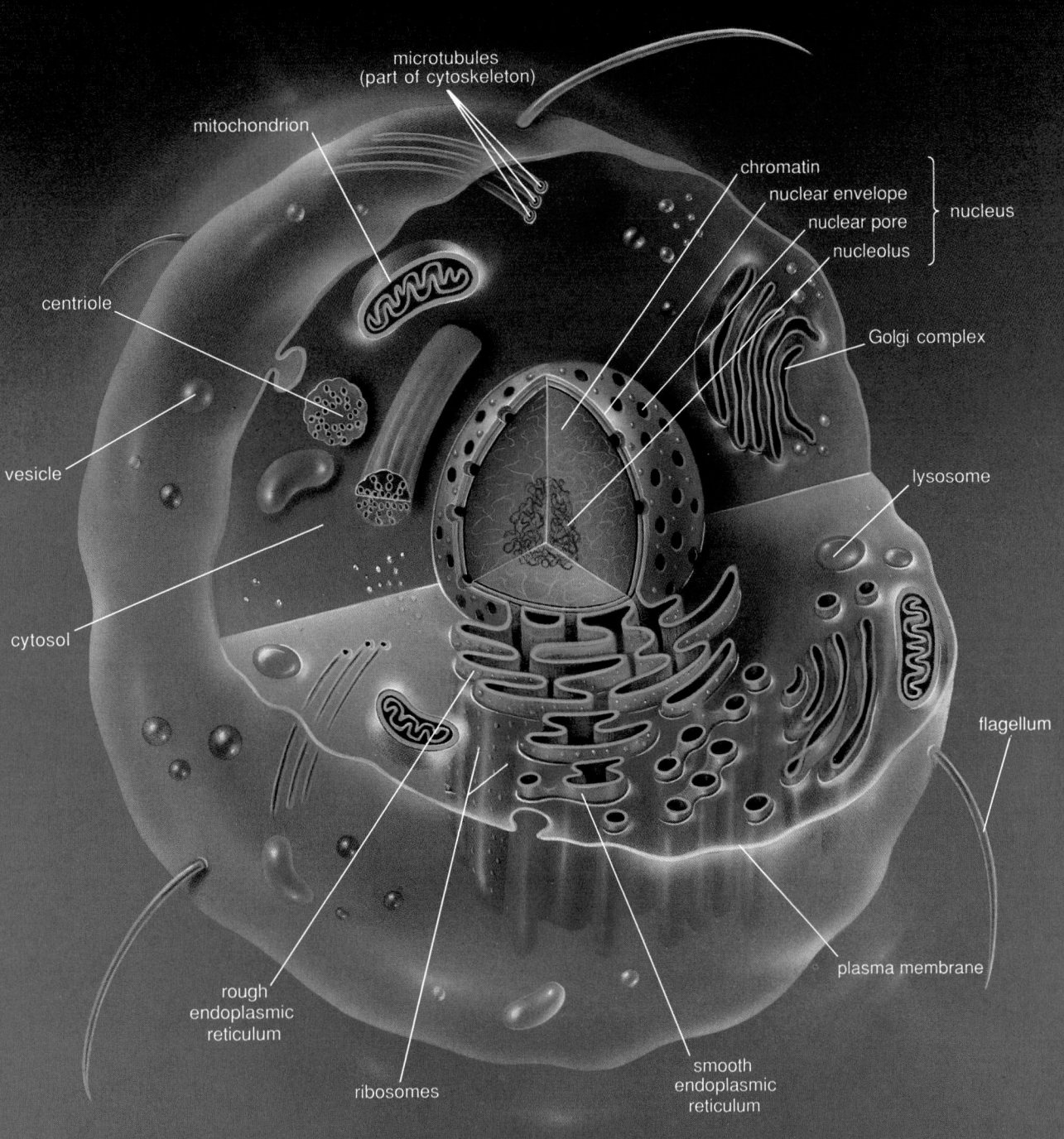

microtubules
(part of cytoskeleton)

mitochondrion

chromatin
nuclear envelope
nuclear pore
nucleolus

} nucleus

centriole

Golgi complex

vesicle

lysosome

cytosol

flagellum

rough
endoplasmic
reticulum

plasma membrane

ribosomes

smooth
endoplasmic
reticulum

Table 5-1   Cell Structures, Their Functions, and Their Distribution in Living Cells

Structure	Function	Prokaryotes[a]	Distribution Eukaryotes	
			Plants	Animals
Cell surface				
Cell wall	Protect, support cell	Present	Present	Absent
Plasma membrane	Isolate cell contents from environment; regulate movement of materials into and out of cell; communicate with other cells	Present	Present	Present
Organization of genetic material				
Genetic material	Encode information needed to construct cell and control cellular activity	DNA	DNA	DNA
Chromosomes	Contain and control use of DNA	Single, circular, no proteins	Many, linear, with proteins	Many, linear, with proteins
Nucleus	Membrane-bound container for chromosomes	Absent	Present	Present
Nuclear envelope	Enclose nucleus; regulate movement of materials into and out of nucleus	Absent	Present	Present
Nucleolus	Synthesize ribosomes	Absent	Present	Present
Cytoplasmic structures				
Mitochondrion	Produce energy by aerobic metabolism	Absent	Present	Present
Chloroplast	Perform photosynthesis	Absent	Present	Absent
Ribosome	Provide site of protein synthesis	Present	Present	Present
Endoplasmic reticulum	Synthesize membrane components and lipids	Absent	Present	Present
Golgi complex	Modify and package proteins and lipids; synthesize carbohydrates	Absent	Present	Present
Lysosome	Contain intracellular digestive enzymes	Absent	Present	Present
Plastid	Store food, pigments	Absent	Present	Absent
Central vacuole	Contain water and wastes; provide turgor pressure to support cell	Absent	Present	Absent
Other vesicles and vacuoles	Contain food obtained through phagocytosis; contain secretory products	Absent	Present (some)	Present
Cytoskeleton	Give shape and support to cell; position and move cell parts	Absent	Present	Present
Centriole	Synthesize microtubules of cilia and flagella; may produce spindle in animal cells	Absent	Absent (in most)	Present
Cilia and flagella	Move cell through fluid or move fluid past cell surface	Absent[b]	Absent (in most)	Present

[a]Many structures are listed as "absent" in prokaryotes; however, their functions are often essential to the life of any cell. In prokaryotes, these functions still occur, but not in discrete structures. For example, prokaryotes synthesize digestive enzymes on ribosomes attached to the cell membrane and immediately secrete the enzymes, where they digest food outside the cell. Therefore, although endoplasmic reticulum, secretory vesicles, and lysosomes are not found in prokaryotes, the functions of these organelles are still carried out.

[b]Many prokaryotes have structures called flagella, but these are not made of microtubules and move in a fundamentally different way than eukaryotic cilia or flagella do.

Figure 5-7 An illustration of the major features of a "typical" plant cell. Not all of these structures occur in every plant cell. Compare this drawing with that of an animal cell in Figure 5-6.

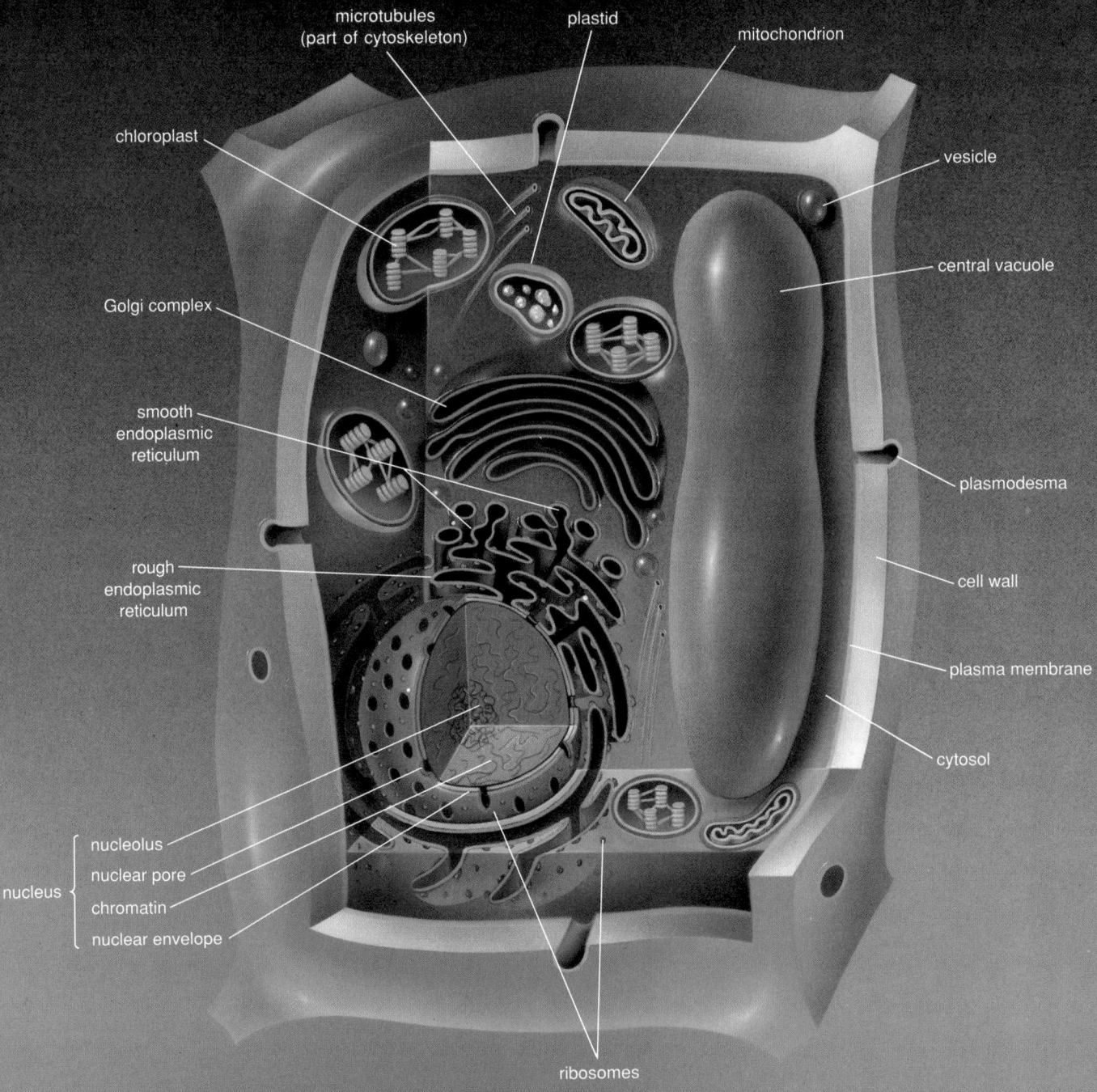

microtubules
(part of cytoskeleton)

plastid

mitochondrion

chloroplast

vesicle

central vacuole

Golgi complex

plasmodesma

smooth
endoplasmic
reticulum

cell wall

rough
endoplasmic
reticulum

plasma membrane

cytosol

nucleolus

nuclear pore

nucleus

chromatin

nuclear envelope

ribosomes

clear envelope, the nucleus contains a granular-looking material called **chromatin** and a darker region called the **nucleolus** (Fig. 5-8).

The Nuclear Envelope

The nucleus is isolated from the rest of the cell by a **nuclear envelope** that consists of two membranes riddled with pores (see Fig. 5-8). Although in some microscopic images the pores appear to be open holes in the envelope, in reality they are complex structures made up of protein and ribonucleic acid (RNA) with a small channel through the middle. Water, ions, and small molecules such as ATP can pass freely through the pores, but the passage of large molecules, partic-

ularly proteins and RNA, is probably regulated. Consequently, the pores may help control the flow of information to and from the DNA.

Chromatin: The Hereditary Material

The bulk of the nucleus appears granular in an electron micrograph (see Figs. 5-3 and 5-8), with darker and lighter regions but no obvious structure. Since the nucleus is often colored very strongly by the common stains used in light microscopy, early histologists named the nuclear material **chromatin,** meaning simply "a colored substance." Biologists have since learned that chromatin consists of DNA complexed with proteins. Although you can't see

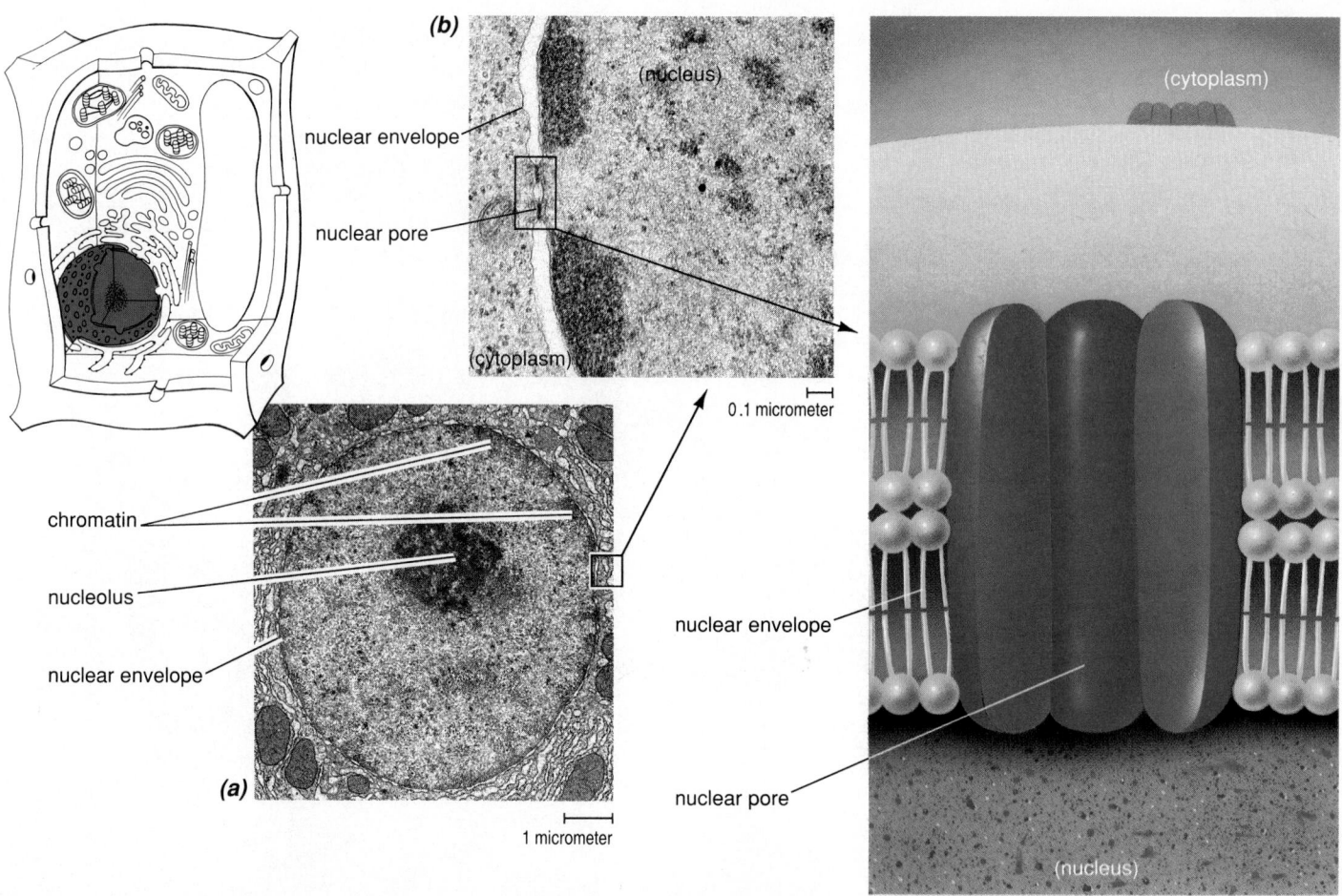

Figure 5-8 The structure of the nucleus and its envelope. **(a)** The eukaryotic nucleus consists of an outer double membrane, termed the nuclear envelope, enclosing the genetic material, DNA. The DNA is complexed with proteins to form chromatin. One region of the chromatin, the nucleolus, stains much more darkly than the rest of the nucleus. **(b)** A transmission electron micrograph of the nuclear envelope clearly shows that it is composed of a double membrane that is fused around the lips of the nuclear pores. Micrographs such as this one can only hint at the structure of the pore, which is shown more clearly in the accompanying drawing.

them in ordinary electron micrographs, eukaryotic DNA and its associated proteins form long strands called **chromosomes** ("colored bodies"). When cells divide, each chromosome coils upon itself, becoming thicker and shorter. The resulting "condensed" chromosomes are easily visible even with light microscopes (Fig. 5-9).

Cellular processes are governed by the information encoded in DNA. Since the DNA stays in the nucleus, while most of the chemical reactions that it controls occur in the cytoplasm, information molecules must be exchanged between the nucleus and the cytoplasm. Genetic information is copied from DNA into molecules of RNA, which move through the pores of the nuclear envelope into the cytoplasm. Ribosomes in the cytoplasm use the information in the RNA molecules to direct protein synthesis (see below). Other molecules, especially proteins, pass from the cytoplasm into the nucleus and regulate the transfer of information from DNA to RNA, depending on what is happening in the cytoplasm and the extracellular environment. We take a closer look at these processes in Chapter 13.

The Nucleolus: Site of Ribosome Assembly

Most eukaryotic nuclei have one or more darkly staining regions called **nucleoli** ("little nuclei"; Fig. 5-8), which are the sites of ribosome synthesis. A **ribosome** is a small particle composed of RNA and protein that serves as a "workbench" for the synthe-

sis of proteins (Chapter 13). Like many human workshops, ribosomes are nonspecific, and any ribosome can be used to synthesize any of the hundreds of proteins made by a cell. In electron micrographs, ribosomes appear as dark granules, either distributed in the cytosol (Fig. 5-10) or clustered along the membranes of the nuclear envelope and the endoplasmic reticulum (see Fig. 5-12).

Ribosomes consist of two subunits, each composed of proteins and special types of RNA molecules called *ribosomal RNA.* Eukaryotic DNA usually contains many genes that encode the information for ribosomal RNA. These genes, often located on several different chromosomes (10 chromosomes in humans), cluster together in the nucleolus, where they direct the synthesis of ribosomal RNA. Ribosomal proteins, like all proteins, are synthesized in the cytoplasm, and they pass from the cytoplasm through the nuclear envelope to the nucleolus. Here, the ribosomal RNA and proteins are assembled into ribosome subunits. Completed subunits then pass through the nuclear pores back into the cytoplasm.

In summary, **the nucleolus consists of DNA (the genes for ribosomal RNA), ribosomal RNA, proteins, and ribosomes in various stages of synthesis.** It stains darkly because of this accumulation of material. During cell division, ribosome synthesis slows. The chromosomes bearing the ribosomal RNA genes move away from one another as they condense. Consequently, the nucleolus disappears until after cell division is complete and the daughter cells begin synthesizing ribosomes.

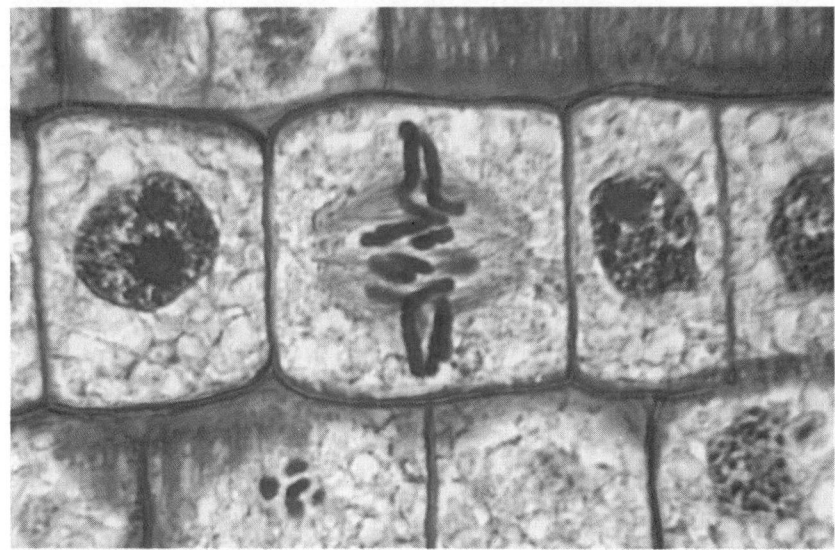

Figure 5-9 Chromosomes, seen here in a light micrograph of a dividing cell in an onion root tip, are the same material as the chromatin seen in nondividing cells (DNA and proteins) but in a more compact state.

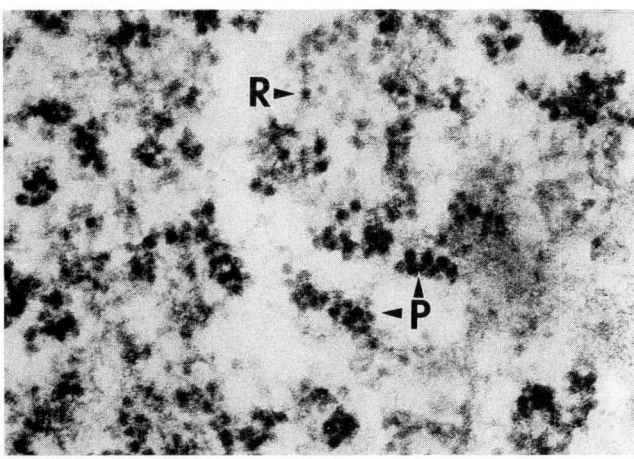

0.1 micrometer

Figure 5-10 Ribosomes may be found free in the cytoplasm either singly (R) or in groups called polyribosomes (P). Free ribosomes synthesize proteins that will be used within the cell. Other ribosomes are attached to the endoplasmic reticulum (see Fig. 5-12).

The Membrane System of the Cell

All eukaryotic cells have an elaborate intercommunicating system of **membranes.** This *membrane system* is composed of the plasma membrane and several organelles within the cytoplasm, including the endoplasmic reticulum, nuclear envelope, Golgi complex, and a variety of membranous sacs such as lysosomes (see Figs. 5-6 and 5-7). All these membranes are composed of lipids and proteins that are synthesized in the endoplasmic reticulum. The membrane components are then sorted and perhaps modified in the Golgi complex and sent off to their appropriate destinations within the cell as small membranous sacs called *vesicles*. We will study this membrane system in several steps. First, we will describe the general structure of membranes, which is quite similar in all these structures. Second, we will briefly examine the structure and function of individual components of the membrane system. Finally, we will trace the flow of membrane throughout the system.

Membrane Structure: An Overview

We will defer a detailed description of membrane structure and function to Chapter 6; however, a brief look at the general structure of membranes will help you to understand the membrane system better.

All the membranes of eukaryotic cells are quite similar in their overall structure, including those of the interconnected membrane system and those of discrete organelles such as mitochondria and chloroplasts. These membranes are all composed of a bilayer of phospholipids in which are embedded cholesterol and several kinds of proteins (Fig. 5-11). The exact types and amounts of phospholipids and proteins vary among organelles, according to the function of the organelle.

Generally speaking, phospholipids and cholesterol form the structural framework of membranes, while proteins carry out specialized functions, such as transporting molecules across the membrane, catalyzing chemical reactions, or attaching one membrane to another. As you will recall from Chapter 3, phospholipids are clothespin-shaped molecules with two very different ends: a polar, hydrophilic "head" and a pair of nonpolar, hydrophobic fatty acid "tails." Membrane phospholipids are always immersed in water: the cytoplasm, the fluid contents of an organelle, or the extracellular environment of a cell. Because water and phospholipid heads are both polar, the phospholipids line up in a double layer called a *bilayer*, with the hydrophilic heads facing the water and the hydrophobic tails hiding inside the bilayer (Fig. 5-11). Cholesterol, an extremely hydrophobic steroid, insinuates itself among the phospholipid tails.

The location of a membrane protein depends on its amino acid sequence. Some proteins are embedded within the phospholipid bilayer, with part of the protein protruding out one or both sides of the membrane. Others are attached to one of the membrane surfaces. In large part, the properties of the proteins determine the functions of the membranes. Since membranes are so important in determining the functions of cells and cell organelles, a considerable part of biology is really the study of membrane proteins.

The Principal Components of the Membrane System

The Plasma Membrane

The plasma membrane forms the outer boundary of the living part of a cell. It is a marvelously complex structure that must perform the seemingly contradic-

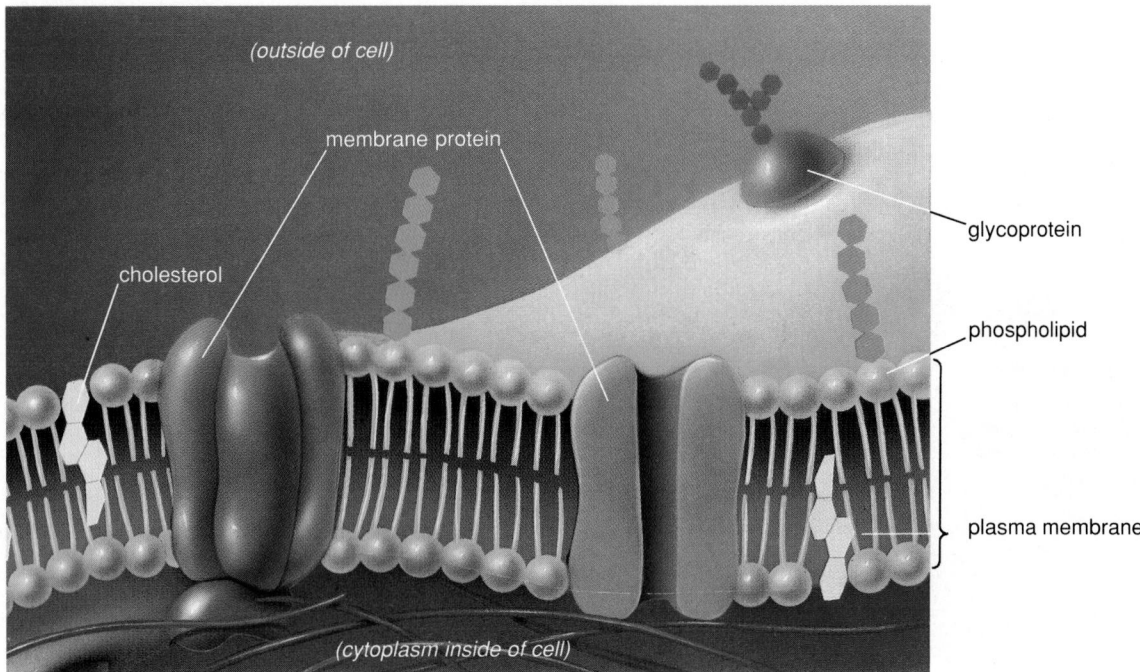

Figure 5-11 The general structure of all biological membranes is similar: a phospholipid bilayer in which are embedded cholesterol and a variety of proteins. Carbohydrates are sometimes attached to certain proteins and phospholipids.

tory functions of separating the cytoplasm of the cell from the outside environment while simultaneously providing for transport of selected substances into or out of the cell. We discuss plasma membrane structure and function in Chapter 6.

Endoplasmic Reticulum

The **endoplasmic reticulum** (ER) is a series of interconnected membranous tubes and channels in the cytoplasm (Fig. 5-12). Most eukaryotic cells have two forms of ER—rough and smooth. Numerous ribosomes stud the outside of **rough endoplasmic reticulum,** while **smooth endoplasmic reticulum** lacks ribosomes.

The different structures of smooth and rough ER reflect different functions. **Enzymes embedded in the membranes of the smooth ER are the major site of lipid synthesis, including the phospholipids of the ER and other membranes.** In some cells the smooth ER synthesizes other types of lipids as well, such as the steroid hormones testosterone and estrogen produced by the gonads of mammals. **The ribosomes on the outside of rough ER synthesize proteins, including membrane proteins.** Therefore, the ER can synthesize itself, both lipid and protein components. Although most of the membrane synthesized in the ER forms new or replacement ER membrane, some of it moves inward to replace nuclear membrane or outward to form the Golgi apparatus, lysosomes, and the plasma membrane (see Fig. 5-15).

The ribosomes on rough ER also manufacture the proteins that secretory cells export into their surroundings, including digestive enzymes and protein hormones. As these proteins are synthesized by the ribosomes on the outside of the ER, they are simultaneously transported into the channels inside. The proteins then move through the ER and accumulate in pockets at the ends, especially the ends near Golgi complexes. These pockets then "bud off," forming membrane-bound sacs called **vesicles** that migrate to the Golgi complex.

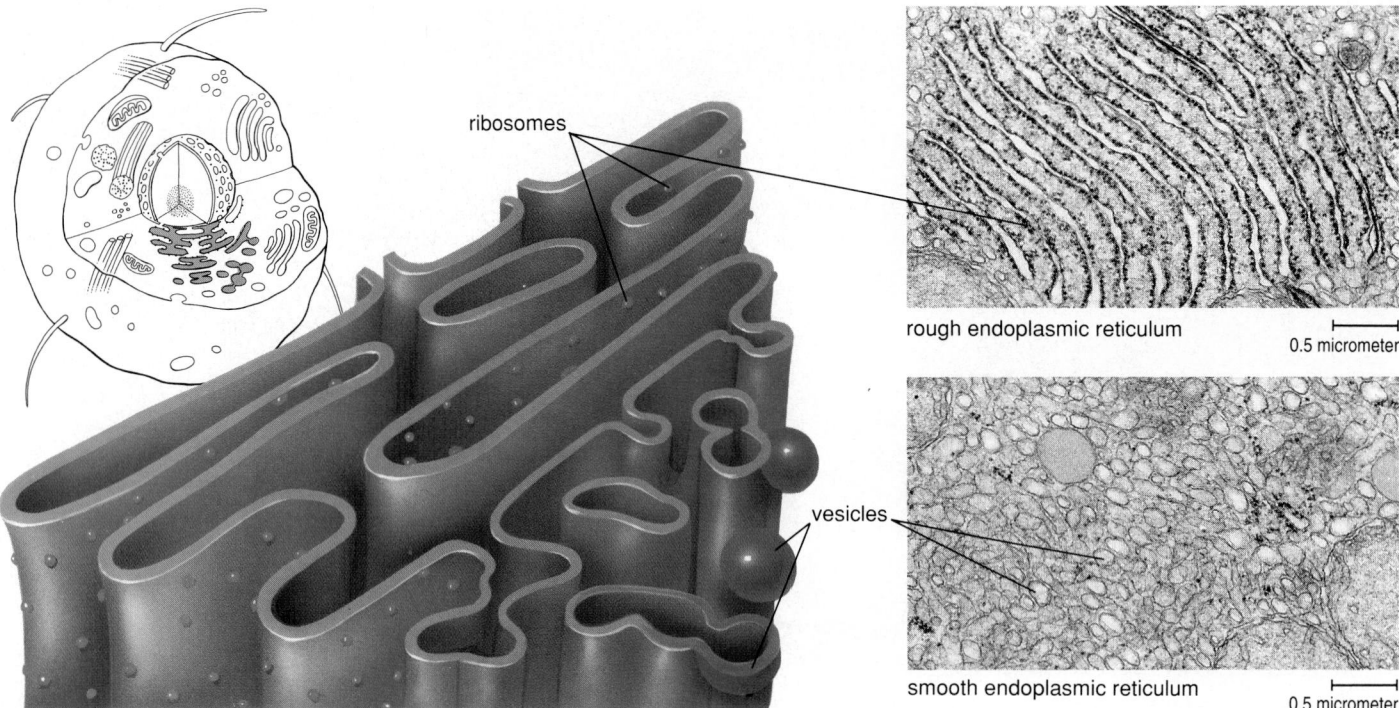

ribosomes

rough endoplasmic reticulum

0.5 micrometer

vesicles

smooth endoplasmic reticulum

0.5 micrometer

Figure 5-12 There are two types of endoplasmic reticulum: rough ER, coated with ribosomes, and smooth ER without ribosomes. Although the ER looks like a series of tubes and sacs in electron micrographs, it is actually a maze of folded sheets and interdigitating channels. In many cells the rough and smooth ER are thought to be continuous, as depicted in the drawing. Ribosomes (red) stud the cytoplasmic face (purple) of the rough ER membrane.

Golgi Complex: Processing and Packaging

The Golgi complex is a specialized set of membranous sacs derived from the ER. In fact, the Golgi looks very much like a stack of smooth ER that has been stepped on, flattening the middle and making the ends bulge out (Fig. 5-13). In reality, vesicles that bud off from the smooth ER fuse with the sacs on one side of the Golgi stack, adding their membrane to the Golgi and emptying their contents into the Golgi sacs. Other vesicles bud off the Golgi on the far side of the stack, carrying away proteins, lipids, and other complex molecules.

The Golgi complex performs three major functions:

1. It separates proteins and lipids received from the ER according to their destinations; for example, the Golgi separates digestive enzymes that are bound for lysosomes from hormones that will be secreted from the cell.

2. It modifies some molecules, for instance adding sugars to proteins to make glycoproteins.

3. It packages these materials into vesicles that are then transported to other parts of the cell or to the plasma membrane for export.

Lysosomes: Intracellular Digestion

Some of the proteins manufactured in the ER and sent to the Golgi are intracellular digestive enzymes that can break down proteins, fats, and carbohydrates into their component subunits. In the Golgi, these enzymes are packaged in membranous vesicles called **lysosomes** (Fig. 5-14). The major function of lysosomes is to digest food particles, which range from individual proteins to complete microorganisms. As we will see in Chapter 6, many cells "eat" by phagocytosis, engulfing extracellular particles with extensions of the plasma membrane. The food particles are then moved into the cytoplasm, enclosed within membranous sacs, now called **food vacuoles.** By an unknown process, lysosomes recognize these food vacuoles and fuse with them. The contents of the two vesicles mix, and the lysosomal enzymes digest the food into amino acids, monosaccharides, fatty acids, and other small molecules. These simple molecules then diffuse out of the lysosome into the cytosol to nourish the cell.

Lysosomes also assist the cell by digesting defective or malfunctioning organelles, such as mitochon-

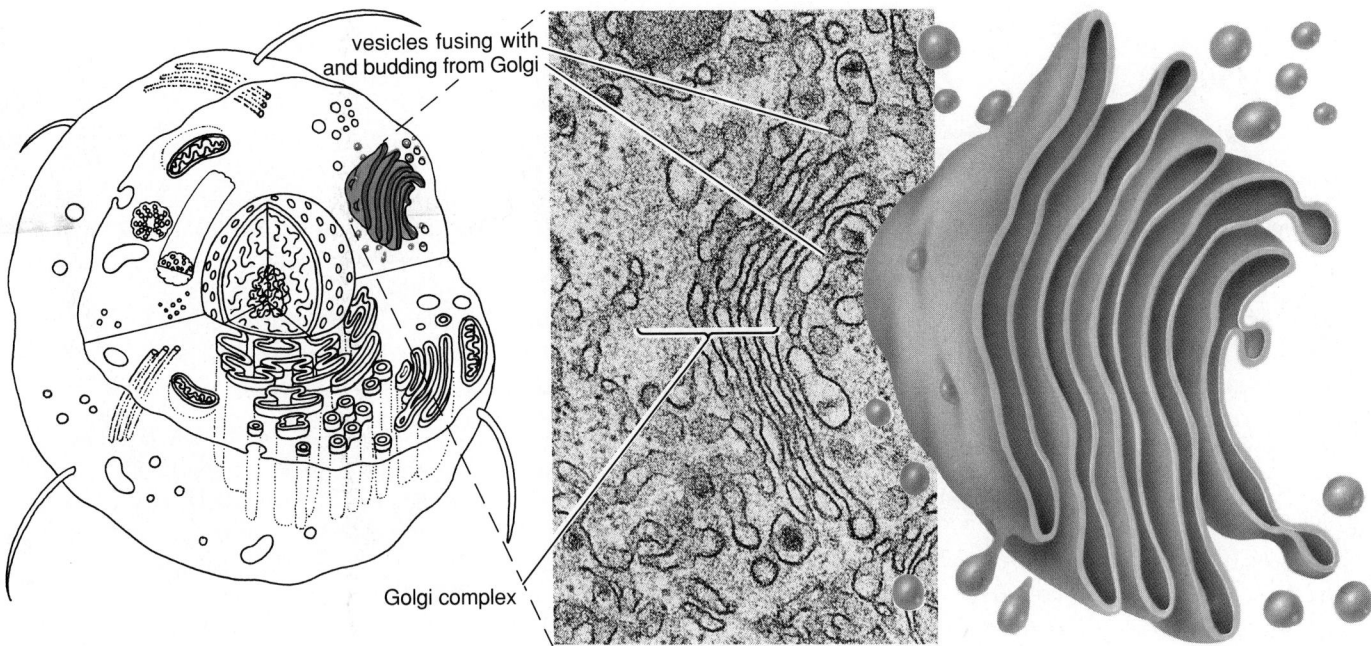

Figure 5-13 The Golgi complex is a stack of flat, membranous sacs derived from the endoplasmic reticulum. Vesicles constantly bud off from and fuse with the Golgi and ER, transporting material in several directions: from the ER to the Golgi and back again, and from the Golgi to plasma membrane, lysosomes, and vesicles.

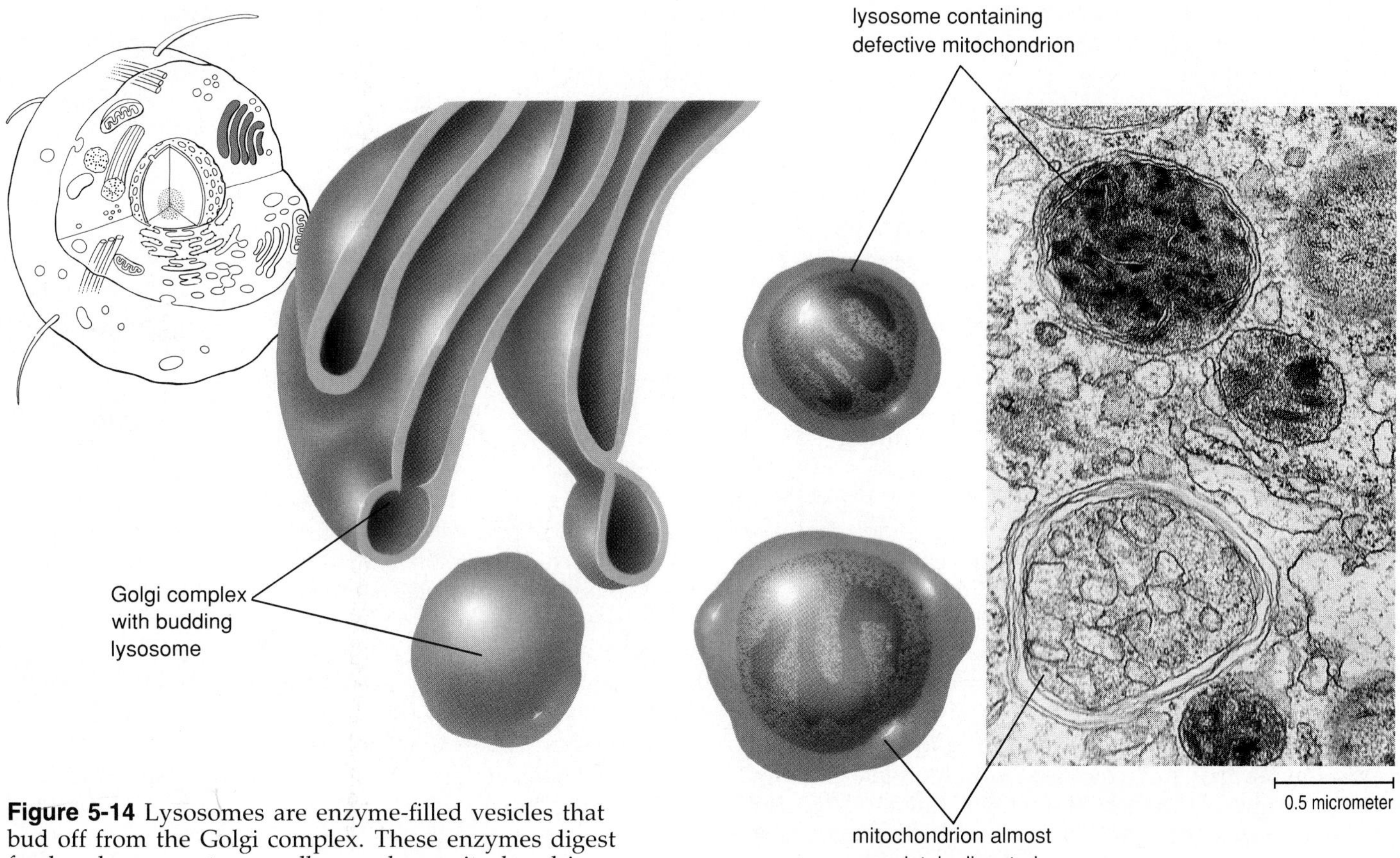

Figure 5-14 Lysosomes are enzyme-filled vesicles that bud off from the Golgi complex. These enzymes digest food and worn-out organelles, such as mitochondria.

lysosome containing defective mitochondrion

Golgi complex with budding lysosome

mitochondrion almost completely digested

0.5 micrometer

A CLOSER LOOK

At the "Essentially Foreign Creatures" Within Us: The Evolution of Chloroplasts and Mitochondria

Many years ago, botanists observed that the chloroplasts of some red algae bear a striking resemblance to certain photosynthetic bacteria. Both have separate, ribbonlike thylakoids, not at all like the stacked, disklike thylakoids found in most plants. Both also have somewhat unusual pigment granules attached to the outside of each thylakoid. Finally, most chloroplasts, even those in land plants, are about the size of an entire photosynthetic bacterium. Could these similarities be coincidence, or are they evidence of an evolutionary relationship?

Further study revealed that chloroplasts and mitochondria are very unusual organelles, in many respects resembling prokaryotic cells more than they resemble the other organelles of eukaryotic cells. First, chloroplasts and mitochondria contain a single circular chromosome composed of DNA without any proteins. Prokaryotic cells also have circular chromosomes of DNA without proteins. Second, chloroplasts and mitochondria contain their own ribosomes and synthesize proteins. The ribosomes are more similar to prokaryotic ribosomes than to the ribosomes found in the rest of a eukaryotic cell. Third, chloroplasts and mitochondria can grow, duplicate their DNA, and reproduce (Fig. E5-6). Fourth, they apparently cannot be manufactured by the cell. During cell division, occasionally one of the daughter cells receives all the chloroplasts or mitochondria, and the other receives none. When this happens, the cell that lacks these organelles never gets them back. Finally, mitochondria and chloroplasts are surrounded by two separate membranes, whereas most other organelles are bounded by one. The inner membrane of a mitochondrion, the thylakoid membranes of a chloroplast (derived from the inner membrane), and the plasma membrane of a prokaryotic cell are all about three quarters protein and can carry out ATP synthesis. Eukaryotic membranes are usually only about half protein and play no role in ATP synthesis.

Why should chloroplasts and mitochondria resemble prokaryotic cells so strongly? Many years ago, a few biologists suggested that they resemble prokaryotes because chloroplasts and mitochondria might be the descendants of prokaryotic cells that took up residence within the cytoplasm of pre-eukaryotic cells. More recently, this **endosymbiotic hypothesis** has been persuasively argued by Lynn Margulis, and today most biologists agree that mitochondria and chloroplasts are, in Lewis Thomas' words, "essentially foreign creatures" within eukaryotic cells.

How could free-living bacteria wind up as organelles of eukaryotic cells? Let's look first at a hypothetical scenario for the origin of mitochondria. Many single-celled organisms feed by phagocytosis (Fig. E5-7; see Chapter 6 for the details of phagocytosis). They engulf entire prey organisms in extensions of the plasma membrane, "swallowing" the prey in a vesicle of membrane. Usually, the prey is doomed, as the vesicle fuses with enzyme-containing lysosomes and the prey inside is digested. According to the endosymbiotic theory, over a billion years ago the ancestors of eukaryotic cells had evolved, and some of them fed by phagocytosis. However, these cells lacked the enzymes needed to use oxygen in the metabolism of food molecules. As you shall see in Chapter 8, such a cell can only extract a small amount of usable energy as ATP from its food. Some bacteria, however, had evolved these enzymes, and these aerobic bacteria could produce much more ATP from the same amount of food.

Bacteria were probably favorite prey of pre-eukaryotic cells, and occasionally some bacteria were captured as prey but not digested. The bacteria took up residence in the cytoplasm of their new host. In most cases this must have harmed the host, for host and bacterium would compete for the same nutrients. The waste products of the bacterium may also have been poisonous. The host cell would likely have died—done in, as it were, by an indigestible bit of food.

However, an aerobic bacterium in this situation would be in prokaryotic paradise. The cytoplasm of an anaerobic host cell would be rich in half-digested food molecules. Awash in food, the bacterium would probably generate large amounts of surplus ATP, some of which might pass into the host cytoplasm and put to use by the host. If the bacterium did *not* kill the host, the relationship would be beneficial for both. The host–bacterium combination would survive better than either type could by itself. The "mitochondrion-bacterium" multiplied inside its host, and when the host cell divided, each daughter cell contained a few of these newfound powerhouses. The aerobic, truly eukaryotic cell—ancestor to all protists, plants, animals, and fungi—had been born.

Later, one of these new mitochondrion-containing cells may have similarly captured a photosynthetic cyanobacterium. It, too, found a congenial home in its host's cytoplasm. It photosynthesized, grew, and multiplied. Surplus sugars synthesized by the prechloroplast may have passed to the host, providing some of the en-

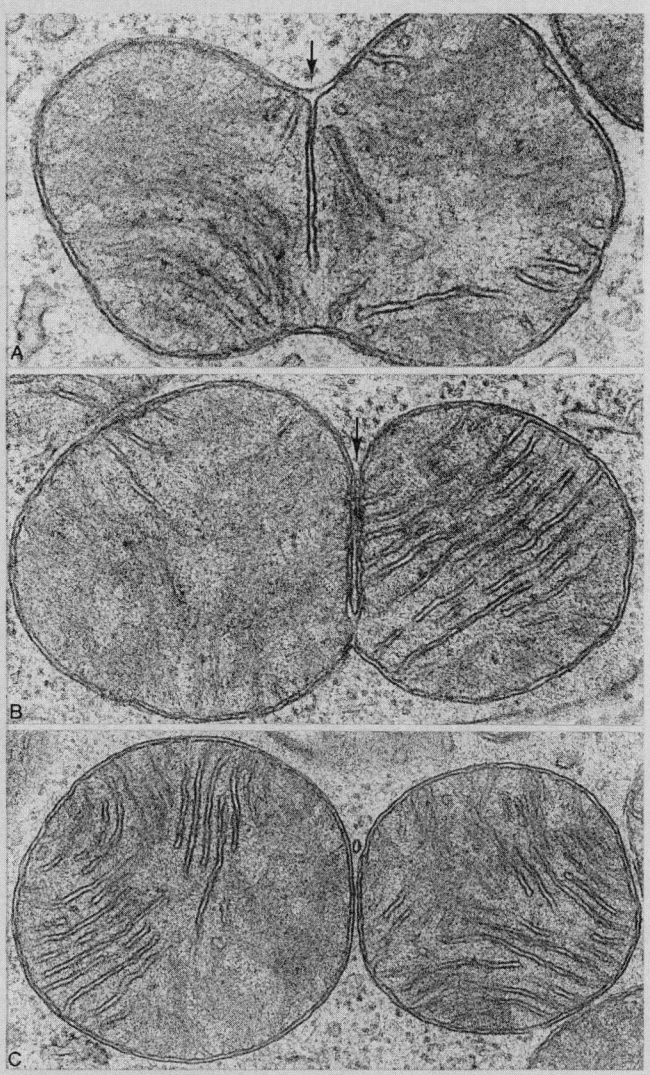

Figure E5-6 This series of micrographs illustrates division of mitochondria. A membranous partition grows inward from one side of the mitochondrion, eventually reaching the other side and producing two daughter mitochondria from the parent. Not visible in these micrographs is the mitochondrial DNA, which duplicates before division, with one copy being incorporated into each daughter organelle.

ergy needed for the host to live and grow. The mitochondria already present in the cell metabolized these sugars, providing plenty of energy. The plantlike protists and all true plants would be the descendants of this cell.

If mitochondria and chloroplasts descended from bacteria, this would explain why they have bacterialike DNA and ribosomes, and can grow and duplicate themselves. The endosymbiotic theory also neatly explains the origin of the two membranes. Look again at Figure E5-7. When feeding by phagocytosis, a cell surrounds its prey with a vesicle made from its plasma membrane (Fig. E5-7, left). A live bacterium has its own plasma membrane. Therefore, the newly engulfed premitochondrion or prechloroplast would have two membranes, an outer one from the eukaryotic predator and an inner one from the bacterium (Fig. E5-7, right).

Needless to say, if chloroplasts and mitochondria arose from such predation/parasitisms, it was a very long time ago, and evolution has not stood still since. Chloroplasts and mitochondria are now fully incorporated into the lives of eukaryotic cells, and cannot survive by themselves. Indeed, many of the proteins found in chloroplasts and mitochondria are not encoded by the organelle DNA, but are specified by the nuclear DNA of the cell. This should not surprise us, because for over a billion years, natural selection has operated not on the "host" alone, nor on the "parasite" alone, but on the functional whole of the entire eukaryotic cell.

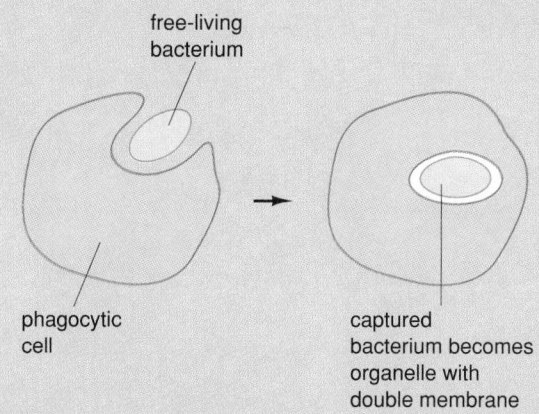

Figure E5-7 A model for the incorporation of bacteria into the cytoplasm of host cells, and the possible origin of the double membrane surrounding mitochondria and chloroplasts.

dria or chloroplasts. It is not known how the cell identifies organelles that have outlived their usefulness, but apparently they are enclosed in vesicles made of membrane from the ER. These vesicles fuse with lysosomes, and digestive enzymes within the lysosome enable the cell to recycle valuable materials from the defunct organelles.

The Flow of Membrane Through the Cell

The nuclear envelope, rough and smooth ER, Golgi complex, lysosomes, food vacuoles, and plasma membrane all form an integrated membrane system.

Membrane is synthesized in the ER and flows back and forth among these structures in an orderly way. As an example, let's look at the movement of materials destined for inclusion in the plasma membrane (Fig. 5-15). The ER synthesizes the phospholipids and proteins that make up the plasma membrane, and simultaneously synthesizes proteins and lipids that belong in the ER itself. How are these sorted out? The ER buds off a vesicle whose membrane is a mixture of ER and plasma membrane components. The vesicle fuses with the Golgi complex, adding its "mixed membrane" to the Golgi membrane. In the Golgi, ER and plasma membrane components are separated. ER material is removed as "recycle" vesicles that

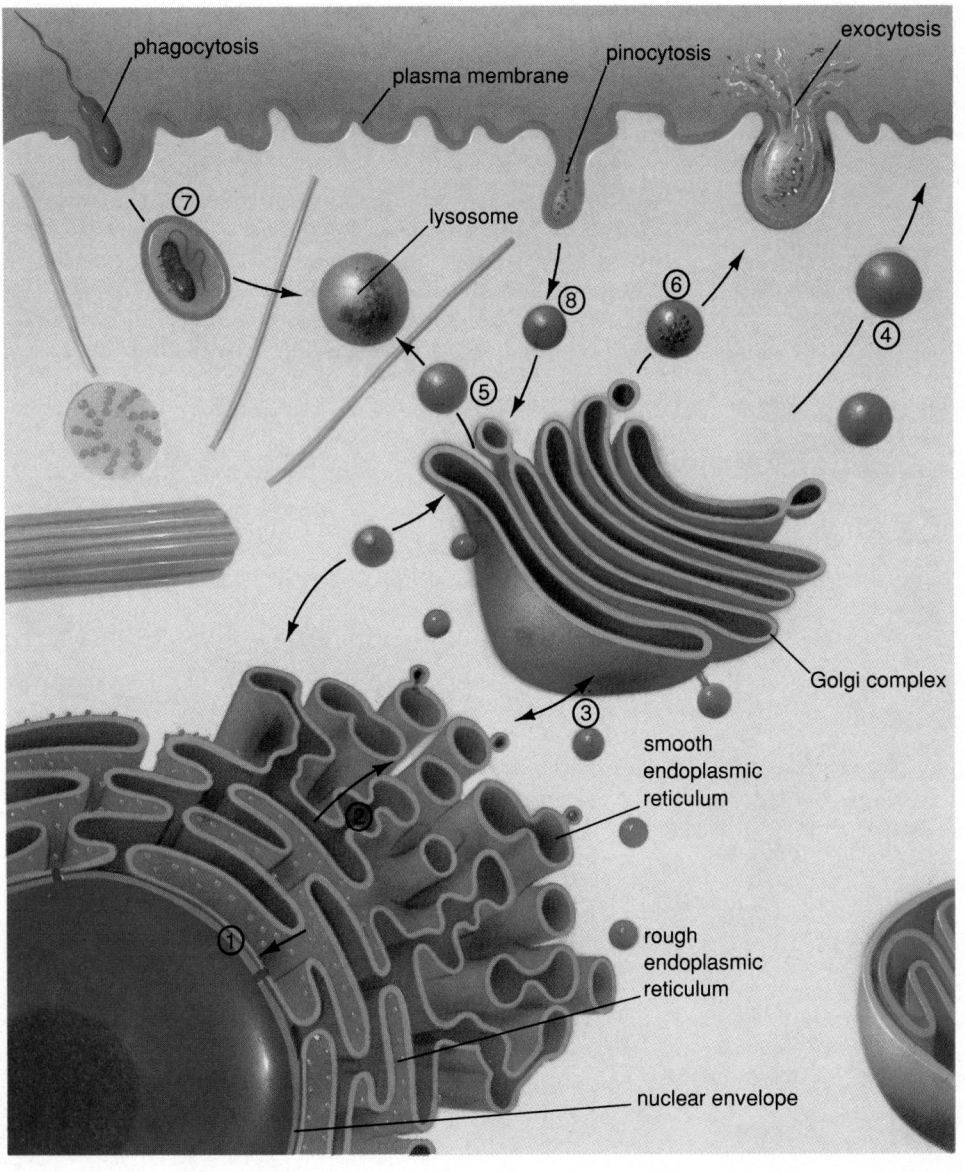

Figure 5-15 The flow of membrane and protein within a cell. Membrane, regardless of its destination, is synthesized by the endoplasmic reticulum. (1) Some of this membrane moves inward to form new nuclear envelope, while most of it moves outward to form (2) smooth ER and (3) Golgi membrane. From the Golgi, membrane moves to form (4) new plasma membrane and (5) the membranes that surround other cell organelles such as lysosomes. Many proteins are synthesized in the rough ER and move through the smooth ER to the Golgi. Here the proteins are sorted according to function; some remain as parts of the Golgi membrane, some are returned to the ER, some are packaged in vesicles bound for the plasma membrane where they will be (6) secreted from the cell, and some are packaged in lysosomes. (7) Lysosomes may fuse with food vacuoles for intracellular digestion of food particles. (8) Plasma membrane brought into the cell is recycled through the Golgi complex. The processes by which plasma membrane moves from the surface to the interior of the cell (called phagocytosis and pinocytosis), and by which the membrane of vesicles fuses with the plasma membrane (called exocytosis) are described in Chapter 6.

pinch off and return to the ER. Plasma membrane material continues on through the Golgi, where it may be modified, for example by adding sugars to make glycoproteins or glycolipids. Eventually the plasma membrane material becomes an "outward bound" vesicle that buds off the far side of the Golgi and moves to the cell surface. The vesicle fuses with the plasma membrane, replenishing and enlarging the membrane.

The Golgi complex processes and packages all membrane-enclosed materials produced by the cell. Many of the vesicles pinched off from the Golgi contain secretory products (e.g., hormones) that are released outside the cell (see Chapter 6). Lysosomes contain digestive enzymes and often fuse with food vacuoles to carry out intracellular digestion. How these diverse materials are recognized, purified, modified properly, separated out, and individually packaged remains a challenge to cell biologists.

Chloroplasts and Mitochondria: Energy Capture and Extraction

Every cell has prodigious energy needs: to manufacture materials, to pick things up from the environment and throw other things out, to move, and to reproduce. As Lewis Thomas put it, "there are structures squirming inside each of our cells that provide all the energy for living." These structures are the **chloroplasts** and **mitochondria,** "essentially foreign creatures" thought to have evolved from bacteria that took up residence long ago within a fortunate eukaryotic cell (see A Closer Look at the "Essentially Foreign Creatures" Within Us).

Chloroplasts and mitochondria are similar to each other in many ways (see Figs. 5-16 and 5-17). Both are usually oblong in shape, about 1 to 5 micrometers long, and are surrounded by a double membrane. Both have enzyme assemblies that synthesize the energy carrier molecule ATP, although the systems are used in a very different manner (see Chapters 7 and 8). Finally, both have many characteristics, including their own DNA, that seem to be remnants of their probable evolution from free-living organisms. However, they also have many differences, corresponding to their vastly different roles in cells: chloroplasts capture the energy of sunlight during photosynthesis and store it in sugar, whereas mitochondria convert the energy of sugar into ATP for use by the cell.

Chloroplasts

Chloroplasts (Fig. 5-16) are found only in plants and certain protists, notably the unicellular algae. Chloroplasts are surrounded by two membranes, although there is very little space between them. The inner membrane encloses a semi-fluid material called the **stroma.** Embedded within the stroma are interconnected stacks of hollow membranous sacs. The individual sacs are called **thylakoids,** while a stack of sacs is a **granum** (plural grana). During chloroplast development, thylakoids probably arise as invaginations of the inner membrane that bud off into the stroma, so that in mature chloroplasts the thylakoids are not connected to the inner membrane.

The thylakoid membranes contain **chlorophyll** and other pigment molecules. During photosynthesis, chlorophyll captures the energy of sunlight and transfers it to other molecules in the thylakoid membranes. These molecules in turn transfer the energy to ATP and other energy carrier molecules. The energy carriers diffuse into the stroma, where their energy is used to drive the synthesis of sugar from carbon dioxide and water. Photosynthesis is described in more detail in Chapter 7.

Mitochondria

Almost all eukaryotic cells have **mitochondria,** often called the "powerhouses of the cell." Well deserving of their nickname, mitochondria extract energy from food molecules and store it in the high-energy bonds of ATP. As you shall see in Chapter 8, various amounts of energy can be released from a food molecule, depending on how it is metabolized. The breakdown of food molecules begins in the cytosol, but the cytosol lacks the enzymes needed to use oxygen in food metabolism. This **anaerobic** (without oxygen) metabolism does not convert very much food energy into ATP energy. Mitochondria are the only places in a cell where oxygen can be used in food metabolism. This **aerobic** metabolism is much more effective than anaerobic metabolism: 18 to 19 times more ATP is generated by aerobic metabolism in the mitochondria than by anaerobic metabolism in the cytosol.

Mitochondria are round, oval, or tubular sacs made of a pair of membranes (Fig. 5-17). Although the outer mitochondrial membrane is smooth, the inner membrane loops back and forth to form deep folds called **cristae** (singular, **crista,** meaning "crest"). As a result, the mitochondrial membranes enclose two fluid-filled spaces, the **intermembrane compartment** between the inner and outer membranes and the **matrix,** or inner compartment, within the inner mem-

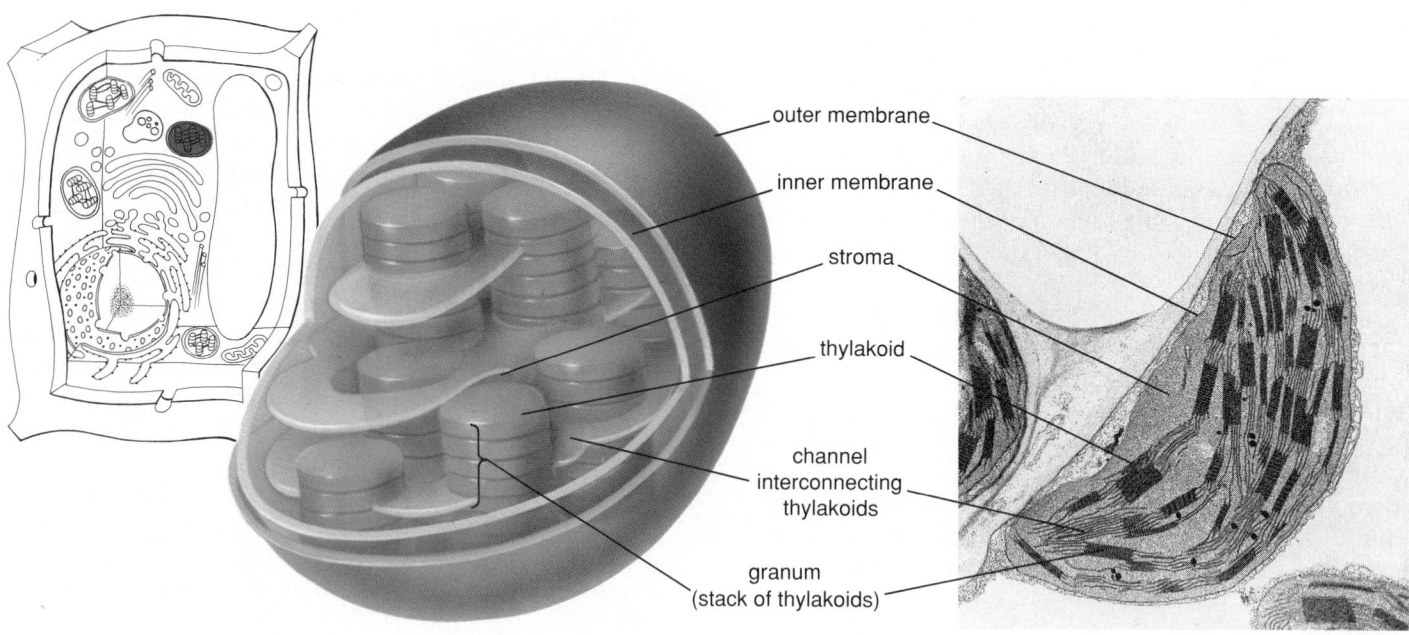

outer membrane

inner membrane

stroma

thylakoid

channel
interconnecting
thylakoids

granum
(stack of thylakoids)

1 micrometer

Figure 5-16 Chloroplasts are surrounded by a double membrane, although the inner membrane is not usually visible in electron micrographs. Enclosed by the inner membrane is the semi-fluid stroma, in which are embedded stacks of sacs collectively referred to as grana. The individual sacs of the grana are called thylakoids. Chlorophyll is embedded in the membranes of the thylakoids.

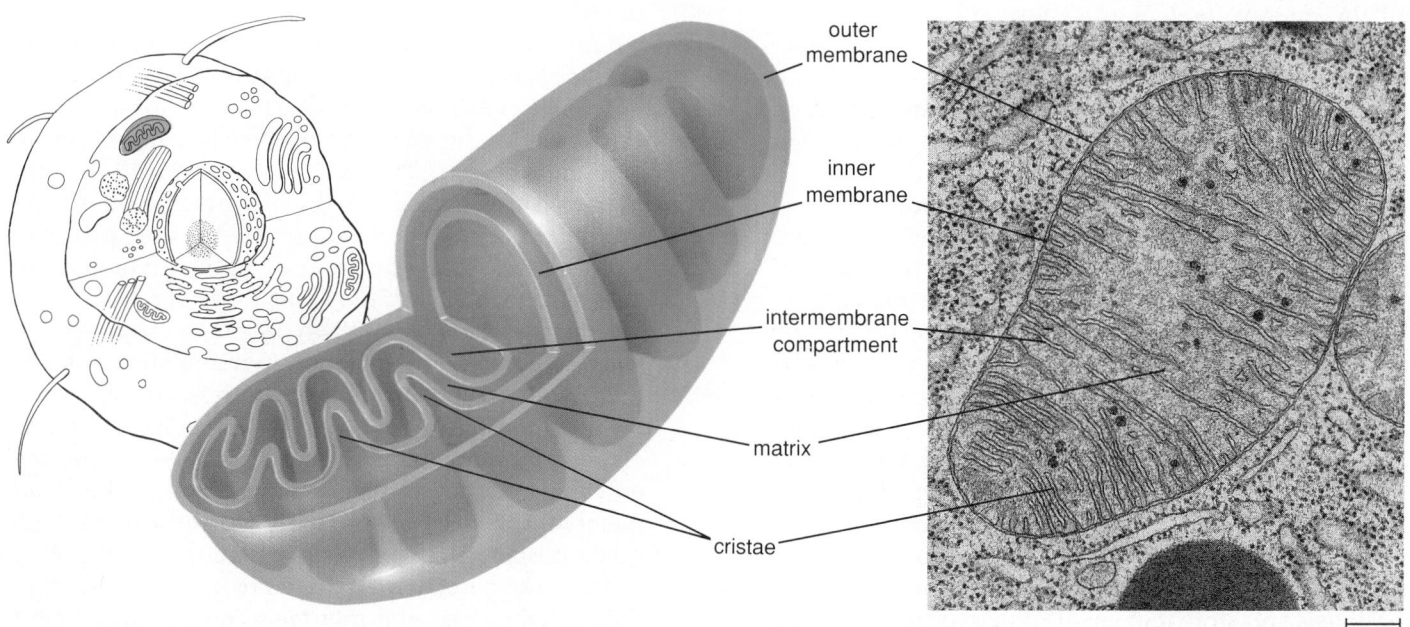

outer
membrane

inner
membrane

intermembrane
compartment

matrix

cristae

0.2 micrometer

Figure 5-17 Mitochondria consist of a pair of membranes enclosing two fluid compartments. Mitochondria are the site of aerobic metabolism.

brane. Some of the reactions of food metabolism occur in the fluid matrix contained within the inner membrane, while the rest are conducted by a series of enzymes attached to the membranes of the cristae.

Plastids and Vacuoles: Storage and Elimination

If a cell finds itself in a favorable environment, some of the molecules that it synthesizes and some of the food that it acquires and digests will be more than it needs at the moment. Cells have evolved organelles in which to store such valuable, if temporarily surplus, molecules. Other materials, such as the waste products of digestion, must be eliminated, and still other organelles serve this function.

Plastids

Besides mitochondria and nuclei, plant cells have other organelles with double outer membranes.

These are the **plastids;** we have already met the most important plastid, the chloroplast. Other types of plastids are used as storage containers for various types of molecules, including pigments such as the carotenoids that give ripe fruits their yellow, orange, or red colors. Especially important, particularly for perennial plants, are plastids that store photosynthetic products from the summer for use during the following winter and spring. Plants usually convert the sugars made during photosynthesis into starch and store it in plastids (Fig. 5-18). Potatoes, for example, are masses of cells stuffed with starch-filled plastids.

Vacuoles

Most cells contain sacs, called **vacuoles,** that are bounded by a single membrane. Eukaryotic cells may contain several types of vacuoles, both temporary and permanent, that serve a variety of functions, including storing food and/or wastes, eliminating water, and supporting the cell. Since many of the

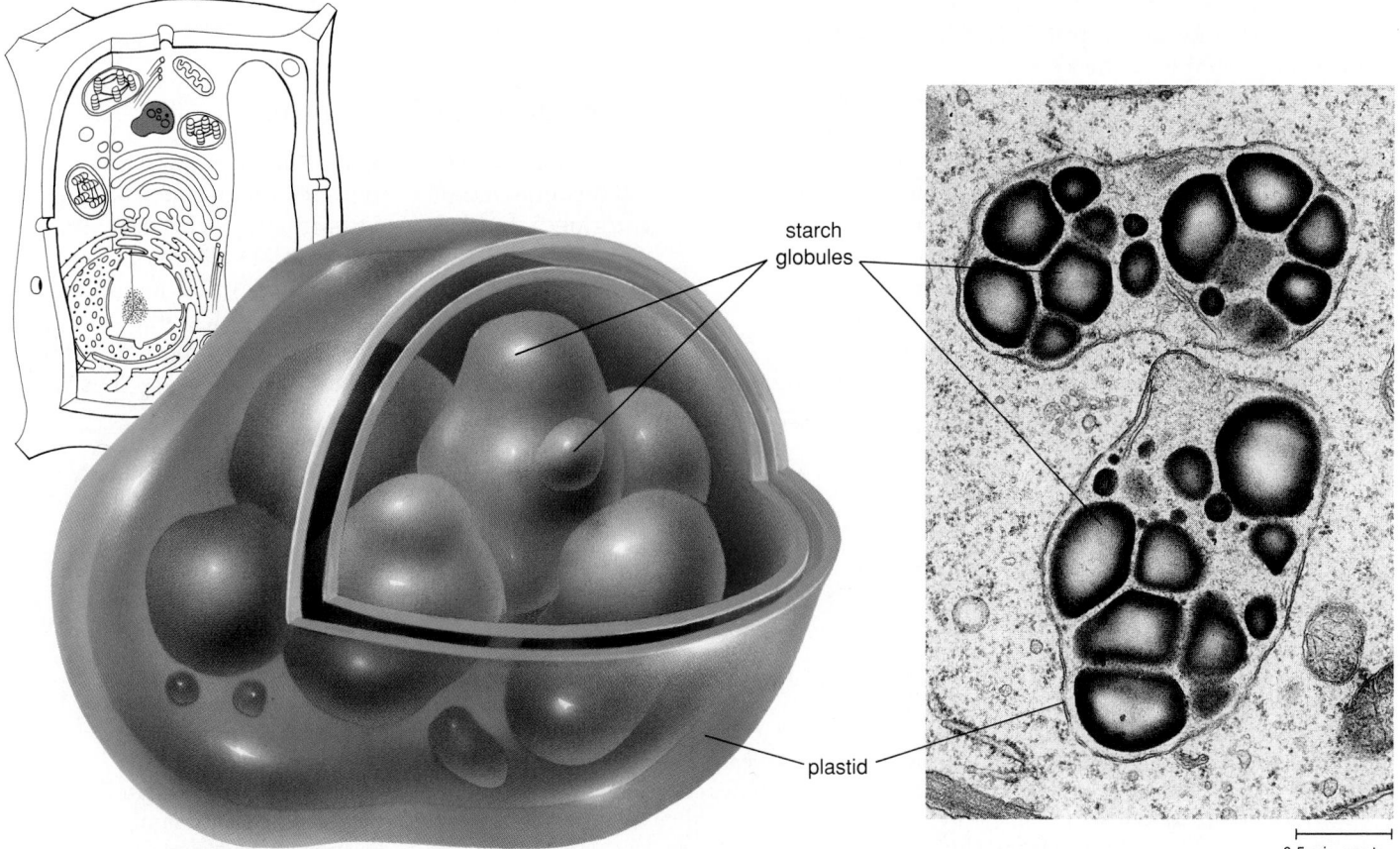

starch globules

plastid

0.5 micrometer

Figure 5-18 Plastids are organelles, found in the cells of plants and plantlike protists, that are surrounded by a double membrane. Chloroplasts are the most familiar plastids; other types store various materials, such as the starch filling these plastids in potato cells.

functions of vacuoles are intimately connected with the movement of water across membranes, we will describe vacuoles in Chapter 6, which discusses membrane structure and function in more detail.

The Cytoskeleton: Shape, Support, and Movement

The organelles we have just described do not float about the cytoplasm haphazardly. Rather, most of the organelles are attached to a network of protein fibers, the **cytoskeleton** (Fig. 5-19). Even individual enzymes, which are often parts of complex metabolic pathways, may be fastened in sequence to the cytoskeleton, so that molecules can be passed from one enzyme to the next. Several types of protein fibers, including thick **microtubules,** medium-sized **intermediate filaments,** and thin **microfilaments,** make up the cytoskeleton (Table 5-2).

The cytoskeleton performs several important functions.

1. **In cells without cell walls, the cytoskeleton, especially networks of intermediate filaments, determines the shape of the cell.**
2. **The assembly, disassembly, and sliding of microfilaments and microtubules causes cell movement.** Cell movement includes both the familiar locomotion of amoebas and white blood cells and the migration and shape changes that occur during the development of multicellular organisms.
3. **Microtubules and microfilaments move organelles from place to place within a cell.** For example, microfilaments attach to vesicles formed during endocytosis (see Chapter 6) and pull the

vesicles into the cell. The vesicles budded off the ER and Golgi complex are probably guided by the cytoskeleton as well.

4. **Microtubules and microfilaments are essential to cell division in eukaryotic cells (see Chapter 9).** First, when eukaryotic nuclei divide, microtubules move the chromosomes into the daughter nuclei. Second, in animal cells, division of the cytoplasm of a single parent cell into two new daughter cells results from the contraction of a ring of microfilaments that pinch the "waist" of the parent cell around the middle.

Microfilaments

Microfilaments are strands that consist mostly of the protein actin, sometimes in association with a second protein, myosin. Myosin has small extensions that can grab onto actin and flex, sliding the actin filament past the myosin, much like a sailor pulling up an anchor line hand over hand. The most familiar function of actin and myosin is muscle contraction (see Chapter 38), but actin and myosin also contribute to shape changes and organelle movement in most, if not all, eukaryotic cells.

Intermediate Filaments

There are at least five types of intermediate filaments, each composed of a different protein. Each variety of intermediate filament is normally found in only one or a few cell types. Intermediate filaments serve a variety of functions, with one common feature: they are usually permanent frameworks that provide shape to cells and anchor various cell parts together. For example, the long axon of a nerve cell is sup-

Table 5-2 Compounds of the Cytoskeleton

	Microfilaments	Intermediate Filaments	Microtubules
Structure	Solid strands about 7 nm in diameter; may be several cm long (muscle cells)	Hollow tubes about 8 to 10 nm in diameter, 10 to 100 μm in length	Hollow tubes about 25 nm in diameter; may be more than 50 μm in length
Protein	Actin (most) and/or myosin	At least five different proteins	Tubulin
Function	Muscle contraction; changes in cell shape, including division of cytoplasm in dividing animal cells; cytoplasmic streaming; movement of pseudopodia	Maintenance of cell shape; attachment of microfilaments in muscle cells; support of nerve cell processes (axons)	Movement of chromosomes during cell division; movement of organelles within cytoplasm; movement of cilia and flagella

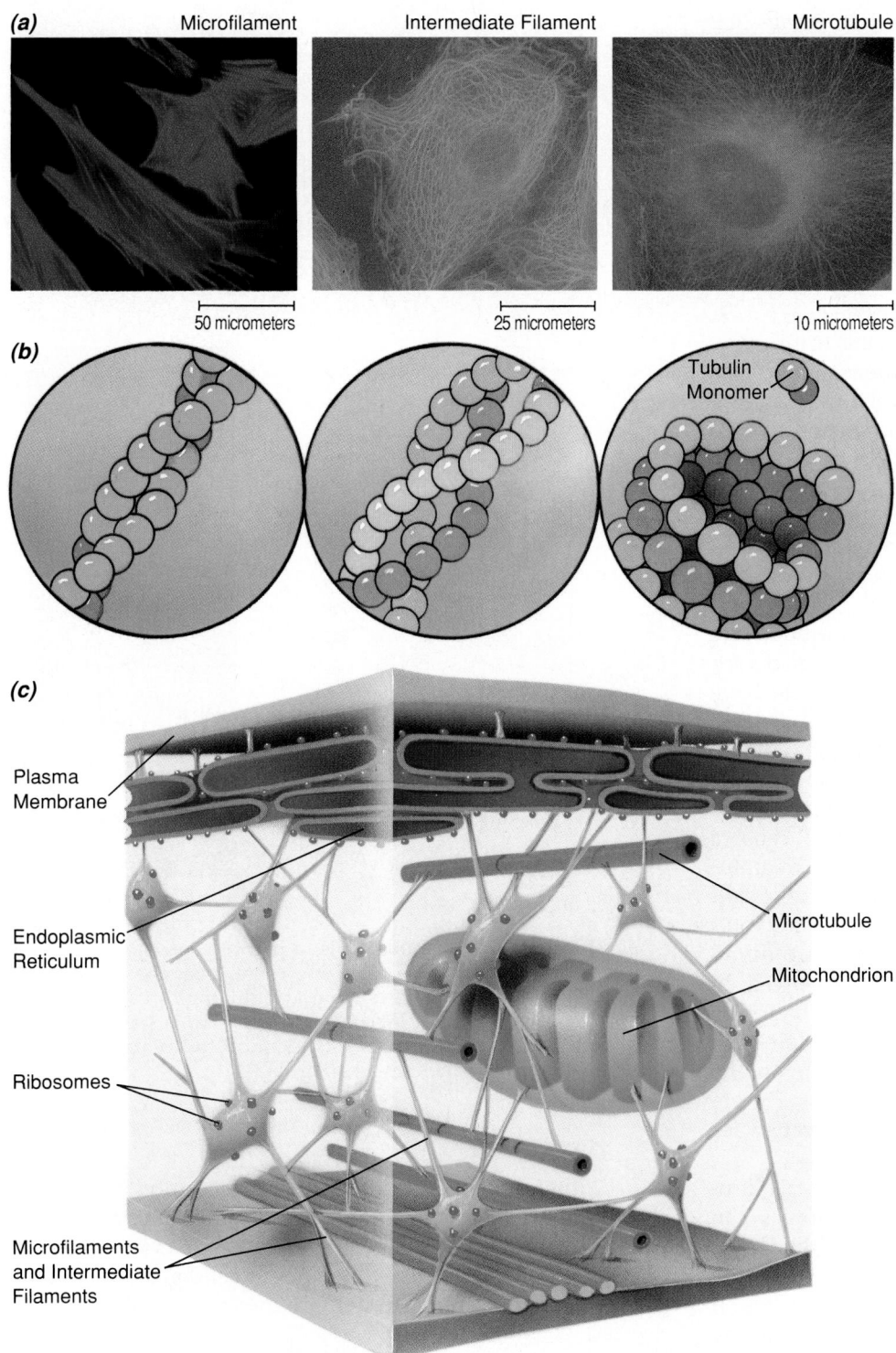

Figure 5-19 Cells are given shape and organization by the cytoskeleton, which consists of three types of proteins: microfilaments, intermediate filaments, and microtubules. **(a)** Fluorescent antibodies that bind specifically to each of the cytoskeletal proteins show their abundance, organization, and orientation within living cells. **(b)** The structures of the cytoskeletal proteins. **(c)** An artist's interpretation of the cytoskeleton. Many organelles are probably attached to the cytoskeleton, which also reinforces the plasma membrane. Chemical reactions, movement of cell parts, and acquisition and release of materials are all coordinated by the cytoskeleton.

ported by an internal skeleton of intermediate filaments in association with microtubules. In many
cells, at least some of the proteins of the plasma
membrane are attached to intermediate filaments.
Finally, intermediate filaments anchor the microfilaments of actin in muscle cells, ensuring that the cells
don't tear themselves apart during strong contractions.

Microtubules

Microtubules are hollow cylinders made from many
dumbbell-shaped subunits of the protein tubulin (see
Fig. 5-19). Individual tubulin subunits are available
dissolved in the cytoplasm, and can be used to construct microtubules when needed by the cell. Some
microtubules are relatively permanent features, such
as those that make up the cilia and flagella by which
many cells move (see below). Many microtubules,
however, are transitory, appearing and disappearing
as required by the cell.

Both temporary and permanent microtubules are
generated by **microtubule organizing centers** in various locations within the cell. All eukaryotic cells have
a microtubule organizing center near the nucleus.
During cell division, this center produces a football-
shaped array of microtubules, called the **spindle
apparatus,** that moves the chromosomes into the two
daughter cells (see Chapter 9). In animal cells, a
prominent pair of **centrioles** is found at the microtubule organizing center near the nucleus (Fig. 5-20).
Centrioles are short, barrel-shaped rings of microtubules. When a cell divides, the centrioles are duplicated, and one pair moves into each daughter cell. In
ciliated cells, the centrioles multiply, and "offspring
centrioles" move to the surface of the cell. Here they
become anchored just beneath the plasma membrane
and give rise to the microtubules of the cilia. Due to
their location at the base of the cilia, these centrioles
are often called **basal bodies.**

Microtubules cause cellular movement in two fundamentally different ways. First, microtubules can
slide past one another, like escalators moving in opposite directions. Any cellular objects attached to the
microtubules move apart in opposite directions too. If
one microtubule is fixed in place, for example to the
plasma membrane, then the second microtubule and
its passenger organelles will move away to another
part of the cell. Microtubule sliding is also responsible
for the whiplike action of cilia and flagella (see
below). The second mechanism of microtubule movement is, at first glance, slightly bizarre. Tubulin subunits can be added to the ends of a microtubule, making it longer, or removed from the ends, making it
shorter, or added at one end and removed from the

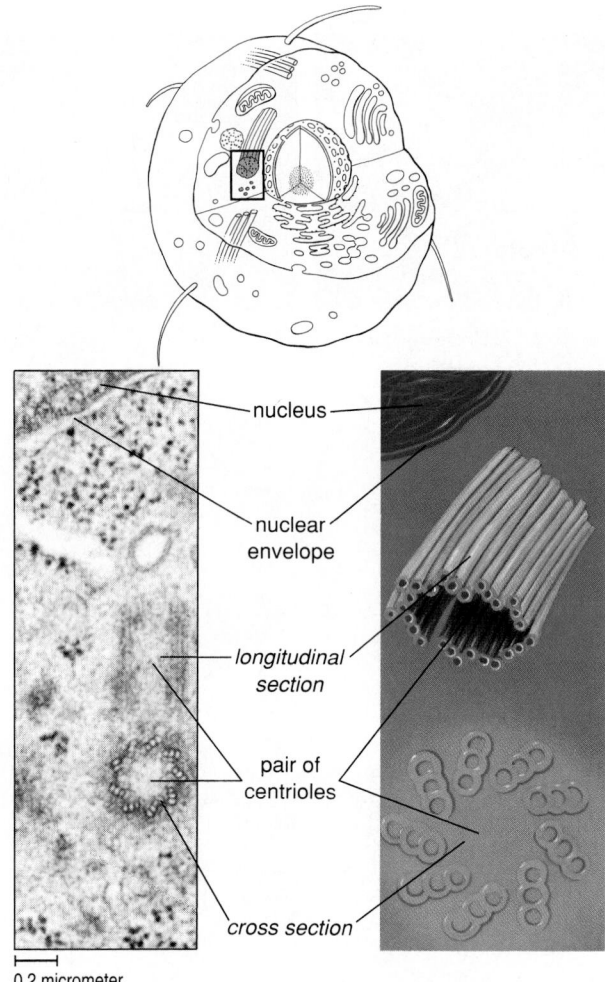

nucleus

nuclear
envelope

longitudinal
section

pair of
centrioles

cross section

0.2 micrometer

Figure 5-20 In animal cells, a pair of centrioles is found
at the microtubule-organizing center near the nucleus.
The centrioles, always lying at right angles to one
another, are composed of nine short triplets of
microtubules. A centriole is identical to the basal body of
a cilium or flagellum (see Fig. 5-21).

other, effectively moving the microtubule along in
one direction. Assembly and disassembly of microtubules contribute to changes in the overall shapes of
cells, for example when nerve cells grow processes
from your spinal cord out to the muscles of your legs.

Cilia and Flagella

Both cilia and flagella are whiplike extensions of the
plasma membrane. Each cilium and flagellum contains a ring of nine fused pairs of microtubules, with
an unfused pair of microtubules in the center of the
ring (the so-called "9 + 2" arrangement; Fig. 5-21).
This pattern of microtubules is produced by a basal
body (centriole) located just beneath the plasma

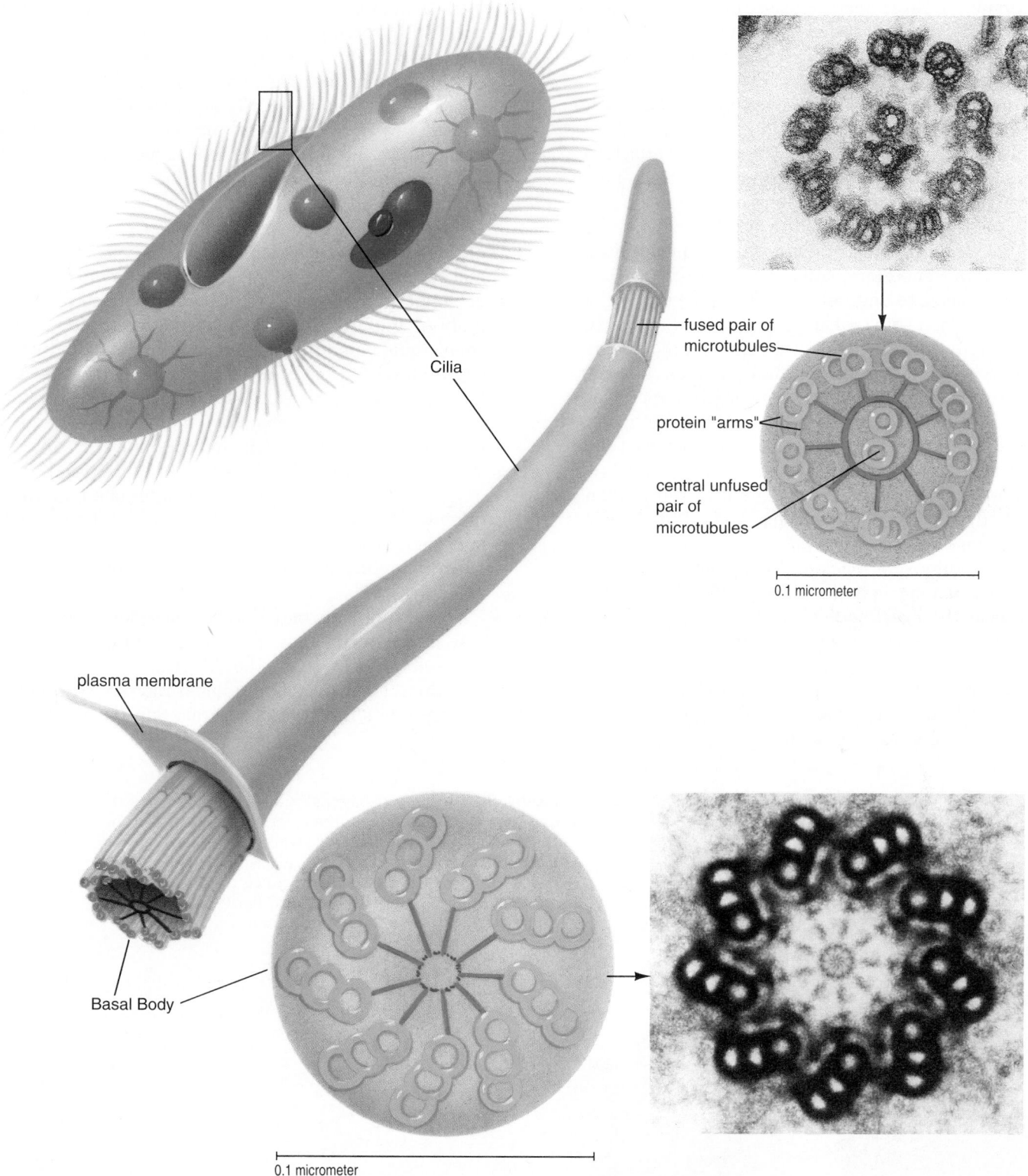

fused pair of
microtubules

protein "arms"

central unfused
pair of
microtubules

0.1 micrometer

Cilia

plasma membrane

Basal Body

0.1 micrometer

Figure 5-21 Both cilia and flagella contain microtubules arranged in an outer ring of nine fused pairs of microtubules surrounding a central unfused pair. The outer pairs have "arms" made of protein that interact with adjacent pairs to provide the force for bending (see Fig. 5-22). Cilia and flagella arise from basal bodies (centrioles) located just beneath the plasma membrane. Basal bodies have nine fused triplets, from which the fused pairs arise, and an indistinct central hub. It is not known how the basal body organizes the somewhat different structure of the cilium or flagellum.

membrane. A basal body consists of a ring of nine triplets of short microtubules. Two of the members of each triplet give rise to the pairs of microtubules in the cilium or flagellum.

Figure 5-22 diagrams how cilia and flagella bend. Tiny protein "arms" project out from each pair of microtubules in the outer ring. These arms attach to the neighboring pair of microtubules and flex, thereby moving the first pair along relative to the second. However, the basal body firmly anchors the "bottom" of all the microtubules in the entire cilium or flagellum. Therefore, adjacent microtubules can only slide past one another if the whole cilium or flagellum bends. (Think of a paint brush: Bristles of equal length are all attached to the handle of the brush. When the bristles are straight, their tips are lined up evenly. When you flex the brush, the bristles slide a bit relative to one another, so that the tips of the bristles on the inside of the bend extend past the tips of the bristles on the outside of the bend.)

ATP energy powers the movement of the protein arms during microtubule sliding. Cilia and flagella often move almost continuously, and consequently require enormous supplies of ATP, which are generated by mitochondria that are usually found in abundance near the basal bodies.

The main differences between cilia and flagella lie in their number, length, and the direction of the force they generate. In general, **cilia** (Latin for "eyelash") are short (about 10 to 25 μm in length) and numerous, and provide force in a direction parallel to the plasma membrane. This is accomplished through a fairly stiff "rowing" motion during the power stroke, with most of the bending occurring at the base of the cilium, and a flexible return stroke (Fig. 5-23a). **Flagella** (Latin for "whip") are long (50 to 75 μm), usually few in number, and provide force perpendicular to the plasma membrane. Flagella undulate with a continuous bending motion, without distinct power and return strokes (Fig. 5-23b).

Unicellular organisms such as *Paramecium* and *Euglena* use cilia or flagella to move about. Most animal sperm also rely on flagella for movement. In multicellular animals, cilia are occasionally used for locomotion, moving the animal through a fluid. Many small aquatic invertebrates, for example, swim by the coordinated beating of rows of cilia, like the oars on a Roman slave galley. More often, however, the organism stays put and its cilia move fluids and suspended particles past a surface. Ciliated cells line such diverse structures as the gills of oysters (moving food- and oxygen-rich water), the oviducts of female mammals (moving the eggs along from the ovary to the uterus), and the respiratory tracts of most land vertebrates (clearing debris and microorganisms from the trachea and lungs).

Prokaryotic cells may also bear slender protrusions that undulate or spin, thereby enabling the cell to move about. However, the "flagella" of prokaryotic cells do not contain microtubules, and have no evolutionary relationship to eukaryotic flagella or cilia.

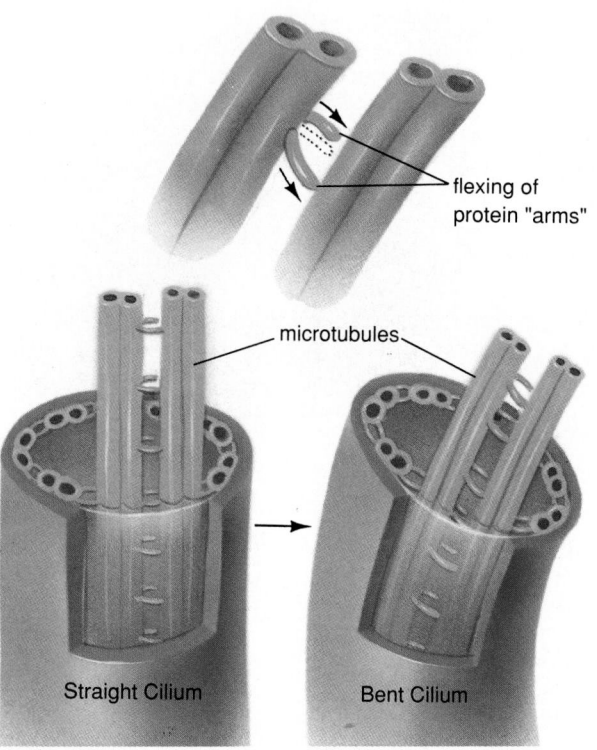

flexing of protein "arms"

microtubules

Straight Cilium

Bent Cilium

Figure 5-22 Cilia and flagella bend when the arms of one pair of microtubules temporarily attach to the shaft of an adjacent pair and flex, "walking" along the shaft. If the microtubules were free, this would result in one pair sliding past the other pair. Since the microtubules are anchored in place by the basal body, the sliding causes the cilium or flagellum to bend.

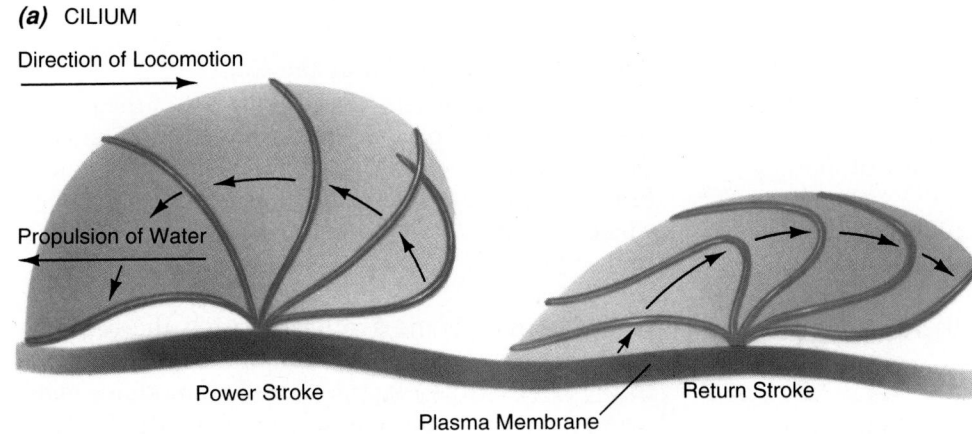

(a) CILIUM

Direction of Locomotion

Propulsion of Water

Power Stroke

Return Stroke

Plasma Membrane

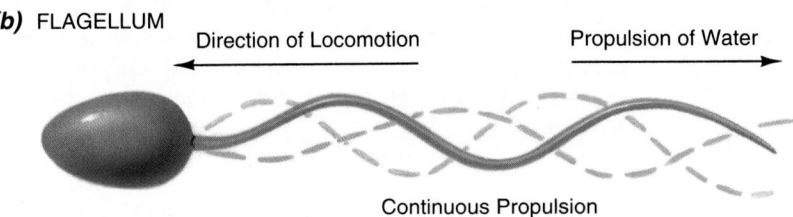

(b) FLAGELLUM

Direction of Locomotion

Propulsion of Water

Continuous Propulsion

Figure 5-23 Characteristic movement patterns of cilia and flagella. **(a)** Cilia usually "row" along, providing a force of movement parallel to the plasma membrane, just as oars provide movement parallel to the sides of a rowboat. **(b)** Flagella often undulate with a continuous bending that starts at the base and moves up to the tip. This provides a force of movement perpendicular to the plasma membrane. In this way a flagellum attached to a sperm can move the sperm straight ahead.

Я⫶R eflections on Cellular Diversity

Every cell in a multicellular organism is highly specialized, down to the level of its organelles, to perform specific tasks. Further on in the text, when we discuss plant and animal physiology, you will see that the functioning of each organ, whether brain or kidney, root or leaf, depends both on the specialization of individual cells and on the organization and coordination of all the cells within the organ. Only by understanding how the subcellular parts work can we hope to understand how each cell contributes to organ function.

Let's look at the endoplasmic reticulum as an example. The ER, along with its associated ribosomes, synthesizes both proteins and lipids. Some of these proteins are enzymes that remain in the ER membrane. Some are pores, carriers, and receptors that will become part of the plasma membrane. Others are export proteins such as hormones, which will be sent outside the cell to other parts of the body, or even, as with milk proteins, outside the body to another organism. Similarly, lipids may be destined for future ER, Golgi, nuclear envelope, or plasma membrane. Which lipids and proteins are synthesized in the ER, and what becomes of them, varies tremendously from cell to cell.

Some nerve cells within your spinal cord send a long fiber all the way from your lower back to your foot, per-

haps a meter away. The ER of these cells manufacture and maintain enormous amounts of plasma membrane. Further, this membrane must contain the specific proteins needed to conduct electrical signals along the fiber. The cells of the pancreas that secrete digestive enzymes, on the other hand, are more or less round cells with comparatively small amounts of plasma membrane. These secretory cells, however, must produce prodigious amounts of enzymes, and are nearly filled with rough ER manufacturing enzyme proteins. Secretory cells of the gonads are specialized to produce lipids, specifically steroid hormones such as testosterone or estrogen. These cells also have massive amounts of ER, but here it is smooth ER lined with lipid-synthesizing enzymes.

Cells differ with respect to other organelles as well. In plants, for example, most leaf cells are packed with chloroplasts, while root cells lack chloroplasts entirely. That is why we began our discussion of cells with the disclaimer that there is no "typical" cell. Although all cells have many features in common, the subcellular specializations of cells yield a variety of functions, allowing the construction of marvelously complex multicellular organisms.

SUMMARY OF KEY CONCEPTS

The Development of Cell Theory

The principles of cell theory are:

1. Every living organism is made up of one or more cells.
2. The smallest living organisms are single cells, and cells are the functional units of multicellular organisms.
3. All cells arise from preexisting cells.

Cell Function Limits Cell Size

Cells are limited in size by two constraints. First, if a cell were too large, the rate of diffusion of essential materials from the outer surface of the cell to the center of the cell would be too slow to sustain life. Second, exchange of nutrients and wastes with the environment occurs through the plasma membrane surrounding the cell. As a cell enlarges, its volume increases more rapidly than its surface area. Therefore, the surface area of a very large cell would be too small to service the metabolic needs of the cell cytoplasm.

Types of Cells: Prokaryotic Versus Eukaryotic

All cells are either prokaryotic or eukaryotic. Prokaryotic cells lack membrane-bound organelles; specifically, the DNA of prokaryotic cells is not enclosed within a membrane-limited nucleus. Eukaryotic cells have several types of membrane-bound organelles, including a nucleus containing DNA. In eukaryotic cells, the cytoplasm includes all the material within the plasma membrane but outside the nucleus. The cytoplasm in turn consists of a fluid cytosol and organelles. With rare exceptions, all eukaryotic cells contain a nucleus, mitochondria, endoplasmic reticulum, Golgi complex, and various types of vesicles (see Table 5-1).

The Nucleus: Control Center of the Cell

Genetic material (DNA) is contained within the nucleus, which is bounded by the double membrane of the nuclear envelope. Pores in the nuclear envelope regulate the movement of molecules between nucleus and cytoplasm. The genetic material of eukaryotic cells is organized into linear strands called chromosomes, which consist of DNA and proteins. The nucleolus consists of the genes that code for ribosome synthesis, together with ribosomal RNA and protein. Ribosomes are complexes of ribosomal RNA and protein that are the sites of protein synthesis.

The Membrane System of the Cell

The membrane system of a cell consists of the plasma membrane, endoplasmic reticulum (ER), Golgi complex, and various vesicles derived from these membranes. Endoplasmic reticulum with ribosomes, called rough ER, manufactures many cellular proteins. Endoplasmic reticulum without ribosomes, called smooth ER, manufactures lipids. The ER is the site of all membrane synthesis within the cell. The Golgi complex is a series of membranous sacs derived from the ER. The Golgi processes and modifies materials synthesized in the rough and smooth ER. Some substances in the Golgi are packaged into vesicles for transport elsewhere in the cell. Lysosomes are vesicles budded off the Golgi that contain digestive enzymes, which digest food particles and defective organelles.

Chloroplasts and Mitochondria: Energy Capture and Extraction

All eukaryotic cells contain mitochondria, organelles that use oxygen to complete the metabolism of food molecules, capturing much of their energy as ATP. Cells of plants and some protists contain chloroplasts, which capture the energy of sunlight during photosynthesis, enabling the cells to manufacture organic molecules, particularly sugars, from simple inorganic molecules. Both mitochondria and chloroplasts probably originated from bacteria that were captured by preeukaryotic cells over a billion years ago, and have become incorporated into the normal functioning of the eukaryotic cell.

Plastids and Vacuoles: Storage and Elimination

Plastids are organelles with double membranes that are found in plant cells, including both chloroplasts and storage plastids that contain pigments or starch. Many cells contain sacs bounded by a single membrane, called vacuoles, that may store food or wastes, excrete water, or support the cell. The size and function of many vacuoles changes as a result of water movement across their membranes.

The Cytoskeleton: Shape, Support, and Movement

The cytoskeleton organizes and gives shape to the cytoplasm of eukaryotic cells. The cytoskeleton is composed of several types of proteins that form strands and filaments within the cytoplasm: microfilaments, intermediate filaments, and microtubules (see Table 5-2). Some proteins of the cytoskeleton, notably the microfilaments and microtubules, also contribute to movement of cell parts. Cilia and flagella are whiplike extensions of the cell membrane that contain microtubules in a characteristic 9 + 2 pattern. Cilia and flagella bend as a result of sliding movements of the microtubules.

Study Note

Figures 5-6 and 5-7 illustrate the overall structure of animal and plant cells, respectively. Table 5-1 lists the principal organelles, their functions, and their occurrence in prokaryotic cells, animal cells, and plant cells.

GLOSSARY

aerobic: using oxygen.

anaerobic: not using oxygen.

basal body: the organelle, structurally identical to a centriole, that gives rise to the microtubules of cilia and flagella.

centriole (sen'-trē-ōl): in animal cells, a microtubule-containing structure found at the microtubule organizing center and the base of each cilium and flagellum. Gives rise to the microtubules of cilia and flagella, and may be involved in spindle formation during cell division.

chlorophyll (klōr'-ō-fil): a pigment found in chloroplasts that captures light energy during photosynthesis.

chloroplast (klōr'-ō-plast): the organelle of plants and plantlike protists that is the site of photosynthesis; surrounded by a double membrane and containing an extensive internal membrane system bearing chlorophyll.

chromatin (krō'-ma-tin): the complex of DNA and proteins that makes up the chromosomes of eukaryotic cells.

chromosome (krō'-mō-sōm): threadlike bodies that contain the genetic material, DNA; eukaryotic chromosomes also contain proteins bound to the DNA.

cilium (sil'-ē-um; pl. cilia): a short, hairlike projection from the surface of certain eukaryotic cells, containing microtubules in a 9 + 2 arrangement. Movement of cilia may propel cells through a fluid medium or move fluids over a stationary surface layer of cells.

crista (kris'-ta; pl. cristae): a fold in the inner membrane of a mitochondrion.

cytoplasm (sī'-tō-plazm): the material contained within the plasma membrane of a cell, exclusive of the nucleus.

cytoskeleton: a network of protein fibers in the cytoplasm that gives shape to a cell, holds and moves organelles, and is often involved in cell movement.

cytosol (sī'-tō-sol): the fluid part of the cytoplasm.

deoxyribonucleic acid (DNA): the molecules that are the genetic material of all living cells.

endoplasmic reticulum (en-dō-plaz'-mik re-tik'-ū-lum; ER): a system of membranous channels within eukaryotic cells; the site of most protein and lipid biosynthesis.

endosymbiotic hypothesis: the hypothesis that chloroplasts and mitochondria (and perhaps some other organelles, such as centrioles) arose as mutually beneficial associations between the ancestors of eukaryotic cells and captured bacteria living within the cytoplasm of the pre-eukaryotic cell.

eukaryotic (ū-kar-ē-ot'-ik): referring to cells of organisms of the kingdoms Protista, Fungi, Plantae, and Animalia. Eukaryotic cells have their genetic material enclosed within a membrane-bound nucleus and contain other membrane-bound organelles.

flagellum (fla-jel'-um; pl. flagella): a long, hairlike extension of the cell membrane. In eukaryotic cells, contains microtubules arranged in a 9 + 2 pattern. Movement of flagella propel some cells through fluid media.

food vacuole: a membranous sac containing food obtained by phagocytosis.

Golgi complex (gōl'-jē): a stack of membranous sacs found in most eukaryotic cells, which is the site of processing and separation of membrane components and secretory materials.

granum (gra'-num; pl. grana): in chloroplasts, a stack of thylakoids.

intermediate filament: part of the cytoskeleton of eukaryotic cells, probably functioning mainly for support.

intermembrane compartment: the fluid-filled space between the inner and outer membranes of a mitochondrion.

lysosome (lī'-sō-sōm): a membrane-bound organelle containing intracellular digestive enzymes.

matrix: the fluid contained within the inner membrane of a mitochondrion.

membrane: in cells, a thin sheet of lipids and proteins that surrounds the cell or its organelles, separating them from their surroundings.

microfilament: part of the cytoskeleton of eukaryotic cells, composed of the proteins actin and (sometimes) myosin; functions in the movement of cell organelles and in locomotion by pseudopodia.

microtubule: a hollow, cylindrical strand found in eukaryotic cells, composed of the protein tubulin; part of the cytoskeleton used in movement of cell organelles, cell growth, and construction of cilia and flagella.

microtubule organizing center: a region of a eukaryotic cell at which tubulin is assembled into microtubules.

mitochondrion (mī-tō-kon'-drē-un): an organelle, bounded by two membranes, that is the site of the reactions of aerobic metabolism.

nuclear envelope: the double membrane system surrounding the nucleus of eukaryotic cells. The outer membrane is often continuous with the endoplasmic reticulum.

nucleoid (nū-klē-oyd): the location of the genetic material in prokaryotic cells; not membrane-enclosed.

nucleolus (nū-klē'-ō-lus): the region of the eukaryotic nucleus engaged in ribosome synthesis, consisting of the genes encoding ribosomal RNA, newly synthesized ribosomal RNA, and ribosomal proteins.

nucleus: the membrane-bound organelle of eukaryotic cells that contains the cell's genetic material.

organelle (or-ga-nel'): a structure found in the cytoplasm of eukaryotic cells that performs a specific function; sometimes used to refer specifically to membrane-bound structures such as the nucleus or endoplasmic reticulum.

plasma membrane: the outer membrane of a cell, composed of a phospholipid bilayer in which proteins are embedded.

plastid (plas'-tid): in plant cells, an organelle bounded by two membranes that may be involved in photosynthesis (chloroplasts), pigment storage, or food storage.

prokaryotic (prō-kar-ē-ot'-ik): referring to cells of the kingdom Monera. Prokaryotic cells do not have their genetic material enclosed within a membrane-bound nucleus and also lack other membrane-bound organelles.

ribosome (rī'-bō-sōm): a particle composed of RNA and protein that is the site of protein synthesis in both eukaryotic and prokaryotic cells.

rough endoplasmic reticulum: endoplasmic reticulum lined on the outside with ribosomes.

smooth endoplasmic reticulum: endoplasmic reticulum without ribosomes.

spindle apparatus: a structure found in dividing eukaryotic cells, composed of microtubules, which appears to guide the movement of chromosomes.

stroma (strō'-ma): the semi-fluid material of chloroplasts in which the grana are embedded.

thylakoid (thī'-la-koyd): a membranous sac within a chloroplast; the thylakoid membranes contain chlorophyll.

vacuole (vak'-ū-ōl): a vesicle, often large, consisting of a single membrane enclosing a fluid-filled space.

vesicle (ves'-i-kul): a small membrane-bound sac within the cytoplasm.

STUDY QUESTIONS

1. Diagram "typical" prokaryotic and eukaryotic cells, and describe their important similarities and differences.
2. Which organelles are common to both plant and animal cells, and which are unique to each?
3. Define nucleus, cytoplasm, and cytosol.
4. Describe the cytoskeleton of a eukaryotic cell.
5. Describe the structure and function of the nuclear envelope.
6. Define chromatin, chromosome, and DNA.
7. What structures make up the nucleolus, and what is its function?
8. What are the functions of mitochondria and chloroplasts? What is the evidence that these organelles originated as captured prokaryotic cells?
9. What is the function of ribosomes? Where in the cell are they typically found?
10. Describe the structure and function of the endoplasmic reticulum and Golgi apparatus.
11. How are lysosomes formed? What is their function?
12. Diagram the structure of cilia and flagella.

DISCUSSION QUESTIONS

1. If muscle biopsies (samples of tissue) were taken from the legs of a world-class marathon runner and a typical couch potato, which would you expect to have a higher density of mitochondria? Why? What about a muscle biopsy from the biceps of a weight lifter?
2. One of the functions of the cytoskeleton in animal cells is to give shape to the cell. Plant cells have a fairly rigid cell wall surrounding the plasma membrane. Does this mean that, for a plant cell, having a cytoskeleton is superfluous? Defend your answer in terms of other functions of the cytoskeleton.

SUGGESTED READINGS

Dustin, P. "Microtubules." *Scientific American,* August 1980. Construction and action of microtubules in cell division, cilia, and flagella.

Lazarides, E., and Revel, J. P. "The Molecular Basis of Cell Movement." *Scientific American,* May 1979. The functioning of the cytoskeletal proteins in diverse types of cell movement.

Margulis, L. "Symbiosis and Evolution." *Scientific American,* August 1971. Evidence for the hypothesis of the evolution of mitochondria and chloroplasts as captured bacteria, by its leading proponent.

Murray, M. "Life on the Move." *Discover,* March 1991. Many cells, including *Amoeba* and white blood cells, crawl about using tiny protein "motors" to manipulate their cytoskeleton.

Rothman, J. "The Compartmental Organization of the Golgi Apparatus." *Scientific American,* September 1985. Although they look homogeneous in electron micrographs, the sacs of the Golgi actually form three biochemically distinct processing units.

Symmons, M., Prescott, A., and Warn, R. "The Shifting Scaffolds of the Cell." *New Scientist,* February 18, 1989. Good overview showing the dynamic nature of the cytoskeleton.

Weber, K., and Osborn, M. "The Molecules of the Cell Matrix." *Scientific American,* October 1985. New techniques in biochemistry and microscopy are used to probe the structure and function of the cytoskeleton.

6

Cells: The Units of Life.
II. Membrane Structure and Function

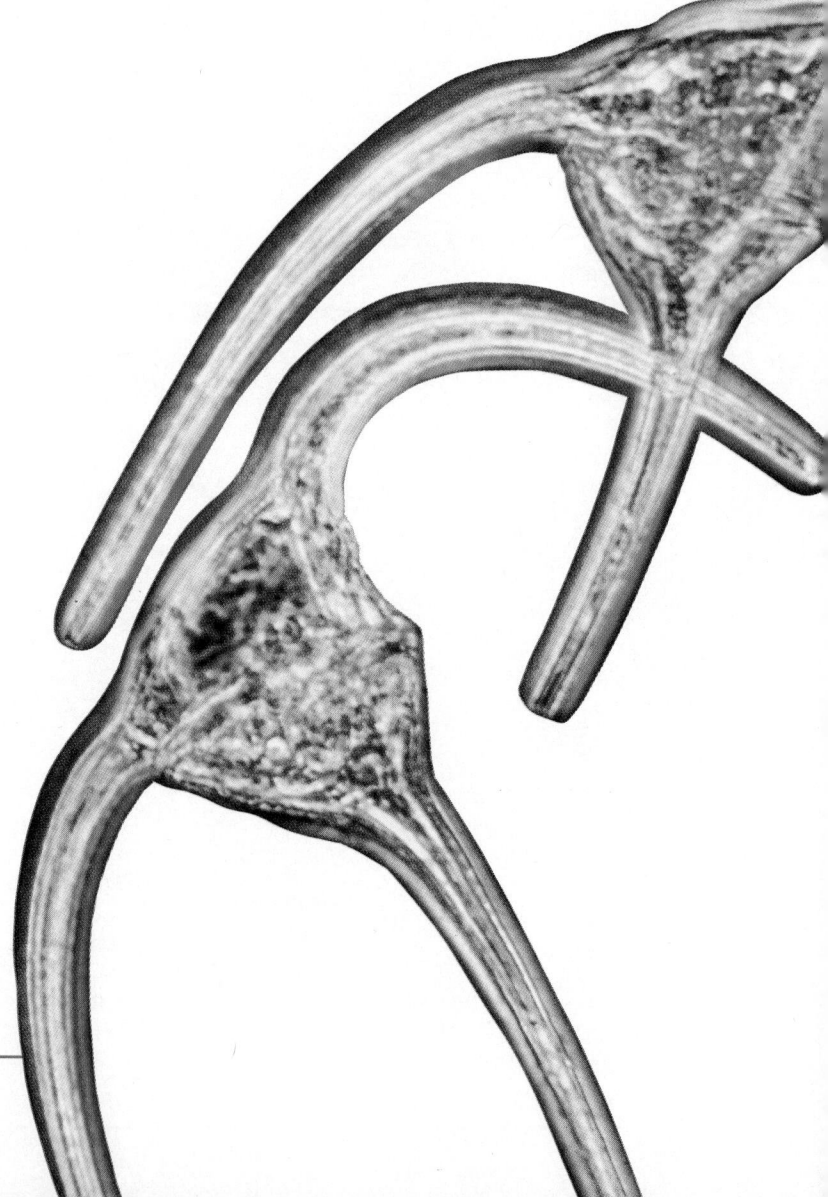

"Inside the cell is alive. Outside the cell is not. The plasma membrane is thus the gate-keeper that separates the quick from the dead."

William S. Stark

To paraphrase the poet John Donne, "No cell is an island, sufficient unto itself." A unicellular organism—for example, a unicellular alga drifting about in a tide pool (Fig. 6-1)—leads a very challenging life. It must defend itself against the vagaries of its environment, including drastic and often rapid changes in temperature and salinity as the sun heats its tiny pool during low tide. However, it cannot simply wall itself off from the world outside. It needs sunlight energy to drive its metabolic processes and nutrients such as nitrogen, carbon, and phosphorus with which to synthesize new cellular components. It also needs to excrete waste products into its environment. Finally, in order to engage in sexual reproduction, during its lifetime it must communicate with at least one another cell of its own species.

Figure 6-1 A tide pool contains a myriad of life forms, including single-celled dinoflagellates (facing page). All cells face formidable challenges in coping with their external environment. The plasma membrane, which both isolates the cell from its environment and regulates interactions with that environment, is the focus of this chapter.

All these interactions with the external environment occur at or across the surface of the cell. In this chapter, we explore the structures that make possible defense, exchange of materials with the environment, and communication.

Cell Walls

The outer surfaces of the cells of bacteria, plants, fungi, and some protists are covered with stiff, non-living coatings called **cell walls.** Plant cell walls are composed of **cellulose** and other polysaccharides (see Fig. 3-5), while fungal cell walls are made of the modified polysaccharide **chitin** (see Fig. 3-6). Bacterial cell walls have a chitinlike framework to which amino acids and other molecules are bound.

Cell walls are produced by the cells they surround.

In plants, membranous vesicles filled with sticky polysaccharides such as pectin (the ingredient that congeals grape juice into jelly) line up across the middle of a dividing cell. As the vesicles fuse, they form the new plasma membranes that separate the two daughter cells (see Chapter 9). The pectins glue the daughter cells together, forming the **middle lamella** (Fig. 6-2). The two cells then secrete cellulose through their plasma membranes, underneath the middle lamella, forming the **primary cell wall.** Many plant cells later secrete additional cellulose molecules and other polysaccharides beneath the primary wall, forming a thick **secondary cell wall.** In some plant cells, the secondary wall may become thicker than the rest of the cell.

Cell walls support and protect otherwise fragile cells. For example, cell walls allow plants and mushrooms to stand erect on land without the skeleton required by animals. Tree trunks are the ultimate cell

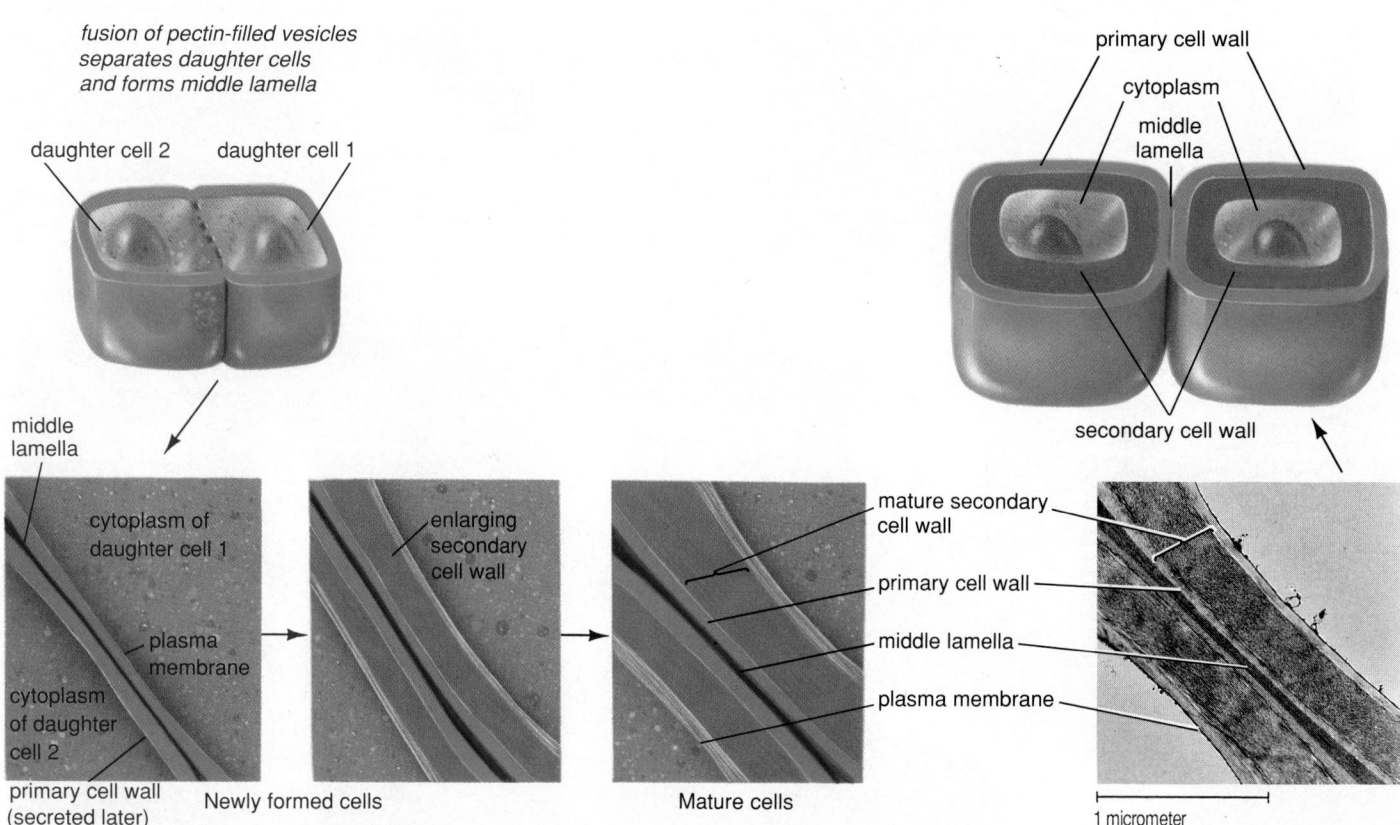

Figure 6-2 A closeup view of plant cell walls. During cell division, the fusion of pectin-filled vesicles forms the middle lamella, which glues together the plasma membranes of the two daughter cells. Each cell then secretes cellulose and other carbohydrates to form its primary cell wall, between the middle lamella and the plasma membrane. Later, many cells may secrete additional cellulose and other carbohydrates beneath the primary wall, forming a secondary cell wall, pushing the primary cell wall and middle lamella further away from the plasma membrane.

wall, so to speak, being almost entirely composed of cellulose and other materials laid down over the years, and capable of supporting impressive loads.

Although strong, cell walls are usually porous, permitting easy passage of small molecules such as minerals, amino acids, and sugars (otherwise, the cell within would soon die). The structure that really governs the interactions between a cell and its external environment is the **plasma membrane.**

The Plasma Membrane

The plasma membrane of a cell must carry out several functions:

1. Isolate the cell cytoplasm from the external environment.
2. Regulate the exchange of essential substances between the cytoplasm and the external environment.
3. Communicate with other cells.
4. Identify the cell as belonging to a particular species and a particular individual member of that species. In multicellular organisms, specific cell types often also bear specific labels on their cell surfaces.

These are formidable tasks for a structure so thin that 10,000 stacked atop one another would scarcely equal the thickness of this page. The key to membrane function lies in membrane structure. Membranes are not simply homogeneous sheets: they are complex, heterogeneous structures, with different parts, indeed even individual molecules, performing very distinct functions.

As we pointed out in Chapter 5, all the membranes of a cell have a similar structure. Therefore, although this chapter focuses on the plasma membrane, the information presented here applies to other cellular membranes as well.

The Fluid Mosaic Model of Membranes

Biologists' current understanding of membranes is embodied in the **fluid mosaic model,** first developed by S. J. Singer and G. L. Nicolson in 1972 (Fig. 6-3). According to this model, when viewed from above, a membrane would look a bit like a lumpy, constantly shifting mosaic of tiles. A bilayer of phospholipids forms a viscous, fluid "grout" for the mosaic, while an assortment of proteins are the "tiles," often sliding sluggishly about within the phospholipid bilayer, so that the overall image slowly changes over time. As

strange as this model may seem, it captures something of the dynamic quality of real membranes. With this model in mind, let's examine the components of the membranes of a living cell.

The Phospholipid Bilayer

As you learned in Chapter 3, a phospholipid consists of two very different parts, a hydrophilic head and a pair of hydrophobic tails:

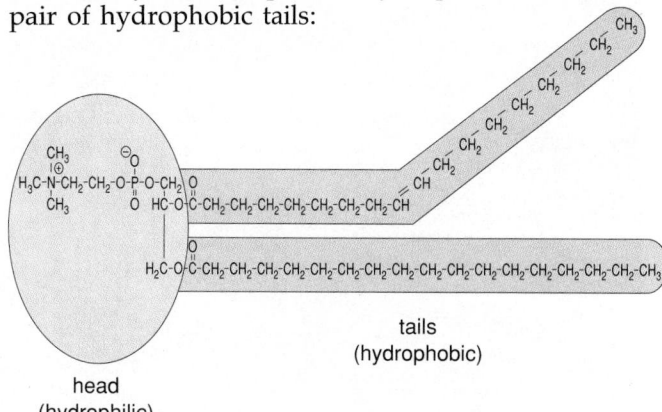

All living cells are surrounded by water, whether it is the pond in which an *Amoeba* spends its life or the extracellular fluid that bathes the cells of animals. The cell cytoplasm is also mostly water. Plasma membranes therefore separate a watery cytoplasm from a watery external environment. Under these conditions, phospholipids spontaneously arrange themselves in a double layer called a **phospholipid bilayer:**

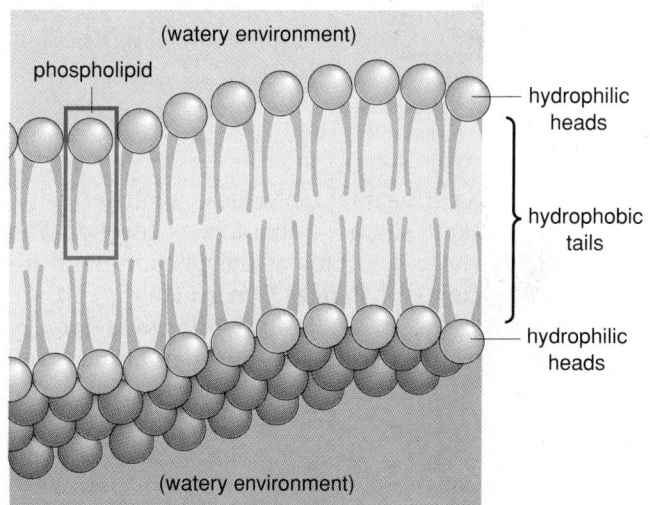

The hydrophilic heads face the cytoplasm or the extracellular fluid, while the hydrophobic tails hide inside the bilayer. Since individual phospholipid molecules are not bonded to one another, this double layer is quite fluid, and individual molecules can move about easily.

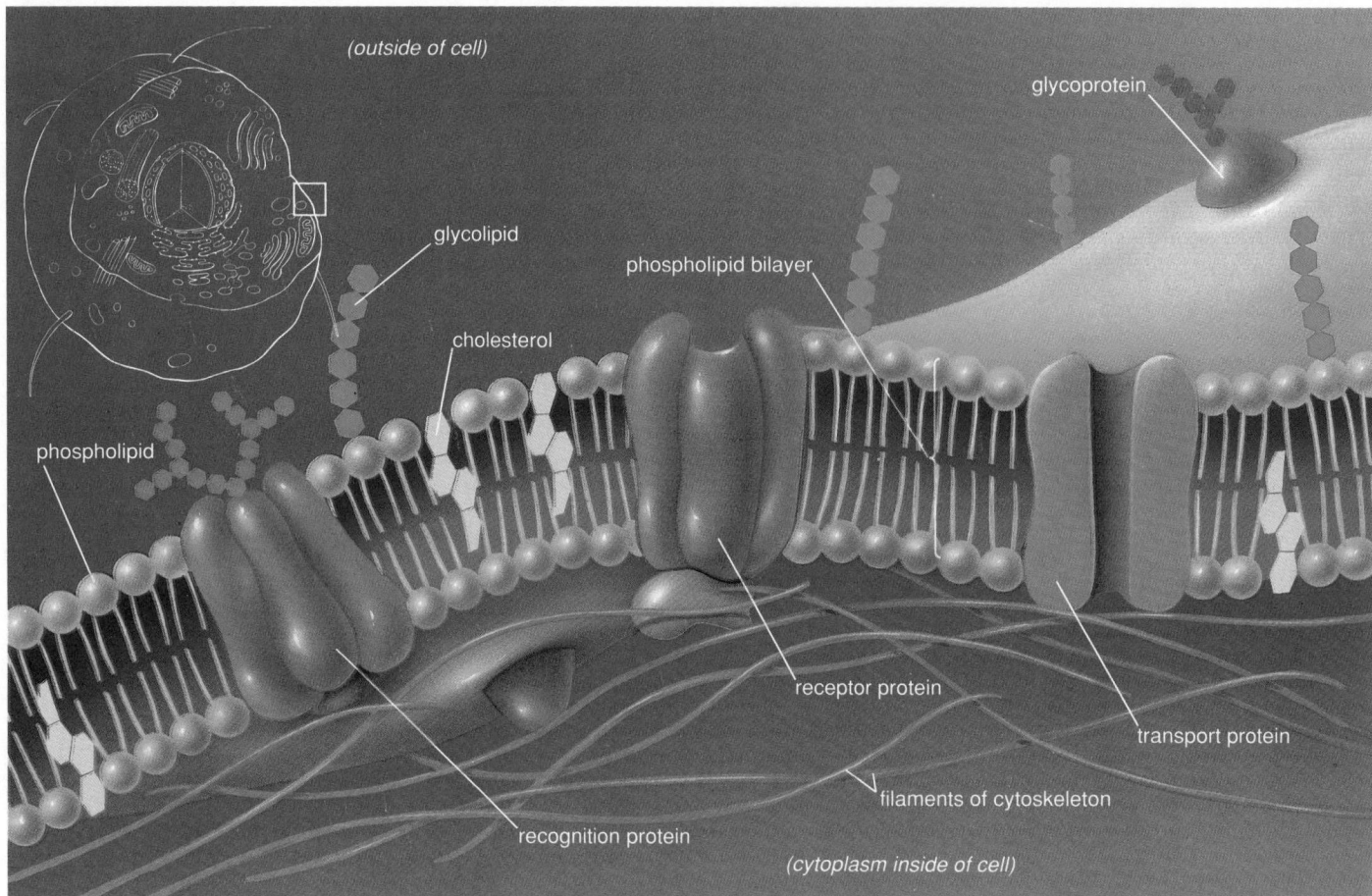

Figure 6-3 The fluid mosaic model of the plasma membrane. According to this model, the plasma membrane is a bilayer of phospholipids in which are embedded various proteins. Many proteins and lipids have carbohydrates attached to them, forming glycoproteins and glycolipids, respectively. The wide variety of proteins fall mostly into three categories: transport proteins, receptor proteins, and recognition proteins.

Most water-permeable molecules, such as salts, amino acids, and sugars, cannot pass through the hydrophobic, fatty acid tails of the phospholipid bilayer. **Since most substances that contact a cell are water soluble, the phospholipid bilayer is largely responsible for the first of the four membrane functions that we listed earlier: it isolates the cell cytoplasm from the external environment.**

The phospholipid bilayer of membranes also contains cholesterol, ranging from mitochondrial membranes that have just a few cholesterol molecules scattered here and there to some plasma membranes that have as many cholesterols as phospholipids. Cholesterol affects membrane structure and function in several ways. It makes the bilayer stronger, more flexible but less fluid, and less permeable to water-soluble substances such as ions or monosaccharides.

The flexible, somewhat fluid nature of the bilayer is very important for membrane function. As you breathe, move your eyes, and turn the pages of this book, cells in your body change shape. If their plasma membranes were stiff instead of flexible, they would break open and die. Further, as we saw in the previous chapter, vesicles of membrane constantly bud off from the endoplasmic reticulum, merge with the Golgi complex, bud off again, and fuse with the plasma membrane. Membranes can bud off vesicles and seal up again, or fuse with vesicles and smooth out again, because of the fluid nature of the bilayer.

The Protein Mosaic

A variety of proteins are embedded within or attached to the surface of a membrane's phospholipid bilayer. Remember that some amino acids are hydrophilic while others are hydrophobic. Membrane proteins are held in the lipid bilayer by interactions between hydrophobic amino acids and the hydrophobic tails of the phospholipids. Hydrophilic parts of the proteins protrude either into the cytoplasm or into the extracellular fluid. Many of the proteins in plasma membranes have carbohydrate groups attached to them, especially to the parts that stick outside the cell. These proteins are called **glycoproteins.**

Owing to the fluid nature of the membrane, many proteins can move about within the phospholipid bilayer. Other membrane proteins, however, are anchored in place by a network of protein strands connected to the cytoskeleton. In animal cells, which lack cell walls, attachments between plasma membrane proteins and the underlying cytoskeleton produce characteristic shapes, from the dimpled discs of red blood cells to the elaborate branching processes of nerve cells.

The Functions of Membrane Proteins

There are three principal categories of membrane proteins: transport proteins, receptor proteins, and recognition proteins (see Fig. 6-3).

Transport proteins **regulate the movement of most water-soluble molecules through the plasma membrane.** Some, called **channel proteins,** form pores or channels that allow small water-soluble molecules to penetrate through the membrane. These proteins form a structure something like a sleeve with a lining: hydrophobic amino acids (the outer material of the sleeve) anchor the protein in the phospholipid bilayer while hydrophilic amino acids form the inside of the pore (the lining of the sleeve). Every plasma membrane bears a large assortment of protein pores, each lined with specific amino acids that allow certain molecules, usually ions, to pass through. Other transport proteins have binding sites, much like the active sites of enzymes, that can grab onto specific molecules on one side of the membrane. The transport protein then changes shape, in some cases through the use of cellular energy, and moves the molecule across the membrane.

Receptor proteins **are molecular triggers that set off cellular responses when specific molecules in the extracellular fluid, such as hormones or nutrients, bind to them.** Most cells bear dozens of receptors on their plasma membranes. When activated by the appropriate molecule, some receptors set off elaborate sequences of cellular changes, such as increased metabolic rate, cell division, movement toward a source of nutrients, secretion of hormones, or even cellular suicide. Other receptors are the gates on protein pores; activation of the receptor opens the gates, allowing ions to flow through the pores. These latter

receptors are activated when nerve cells in your brain communicate with one another (see Chapter 36).

Recognition proteins **and glycoproteins serve as identification tags and cell-surface attachment sites.** The cells of your immune system, for example, recognize a *Salmonella* bacterium as a foreign invader and target it for destruction. These same immune cells ignore the trillions of cells of your own body, because of different identification glycoproteins on their surfaces. During development, the growth of nerve processes from your spinal cord down to the muscles in your feet is guided by attachments between recognition proteins on the nerve cell and the tissues it traverses on the way to the muscle.

As you can see from these brief descriptions, **proteins are largely responsible for the second, third, and fourth membrane functions in our list: moving substances across the membrane, communicating with other cells, and identifying the cell.**

Transport Across Membranes

Nearly all living cells are bathed in liquid, which may be the extracellular fluid of a human body, the pond in which a *Paramecium* swims, or the water-saturated cell walls of a young plant. The plasma membrane separates the fluid cytoplasm of the cell from its fluid environment. Let's begin our study of membrane transport with a brief look at the characteristics of fluids, starting with a few definitions:

1. A **fluid** is a liquid or a gas; that is, any substance that can move or change shape in response to external forces without breaking apart.
2. The **concentration** of molecules in a fluid is the number of molecules per unit volume.
3. A **gradient** is a physical difference between two regions of space, so that molecules tend to move from one region to the other. Cells frequently encounter gradients of concentration, pressure, and electrical charge.

Movement of Molecules in Fluids

The individual molecules in a fluid are in constant random motion, bouncing off one another in every direction. Superimposed atop these random collisions, however, may be a very nonrandom net movement of particular types of molecules in the fluid, or even of the entire fluid, in response to gradients. Molecules move from regions of high concentration to low concentration (e.g., sugar dissolving in tea) or from high pressure to low pressure (e.g., air flowing out of a bicycle pump when you push down the piston). Charged molecules (i.e., ions) move in response

Table 6-1 Transport Across Membranes

Passive transport:	movement of substances across a membrane, going down a gradient of concentration, pressure, or electrical charge. Does not require the expenditure of energy by the cell.
Simple diffusion:	diffusion of water, dissolved gases, or lipid-soluble molecules through the phospholipid bilayer of a membrane.
Osmosis:	diffusion of water across a differentially permeable membrane—that is, a membrane that is more permeable to water than to dissolved solutes.
Facilitated diffusion:	diffusion of (usually water-soluble) molecules through a membrane, assisted by membrane proteins.
Energy-requiring transport:	movement of substances across a membrane, usually against a concentration gradient, using cellular energy.
Active transport:	movement of individual small molecules or ions through membrane-spanning proteins, using cellular energy, usually ATP.
Endocytosis:	movement of large particles, including large molecules or entire microorganisms, into a cell by a process in which the plasma membrane engulfs extracellular material, forming membrane-bound sacs that enter the cytoplasm.
Exocytosis:	movement of materials out of a cell by enclosing the material in a membranous sac that moves to the cell surface, fuses with the membrane, and opens to the outside, allowing its contents to diffuse away.

to electrical gradients, toward unlike charges or away from like charges. The gradient of concentration, pressure, or electrical charge provides the potential energy that can drive the net movement of the molecules. By analogy with gravity, we will refer to such movements as going "down" the gradient.

Movement Across Membranes

Because the cytoplasm of a cell is very different from the extracellular fluid, gradients of concentration, electrical charge, and occasionally pressure span the plasma membrane. In its role as gatekeeper of the cell, the plasma membrane provides for two types of movement (Table 6-1).

1. **Passive transport: substances move down gradients of concentration, electrical charge, or pressure.** The gradients provide the potential energy that drives the movement and controls the direction of movement, into or out of the cell. The plasma membrane acts as a filter; the lipids and protein pores regulate which molecules can

cross, and may influence the rate and timing of movement, but do not influence the direction of movement.

2. **Energy-requiring transport: substances move against gradients of concentration, electrical charge, or pressure.** Transport proteins in the plasma membrane control the direction of movement, using chemical energy from cellular metabolism to drive movement against the gradients. (A helpful analogy is to consider what happens when you ride a bike. If you don't pedal, you can only go downhill; if you expend enough energy pedaling, you can go wherever you wish, even uphill.)

Passive Transport: Movement Down Concentration Gradients

Although gradients of pressure (see Chapter 27) and electrical potential (see Chapter 36) are important in regulating the movement of certain substances in living organisms, most materials that move passively across plasma membranes do so in response to **concentration gradients.**

Principles of Diffusion

Diffusion **is the net movement of molecules in a fluid from regions of high concentration to regions of low concentration, driven by the concentration gradient.** Diffusion can occur from one part of a fluid to another, or across a membrane separating two fluid compartments. We will first examine the simplest case, the diffusion of molecules within a fluid.

To see how concentration gradients can cause molecules to diffuse from one place to another, let's consider what happens when a drop of dye is placed in a glass of water (Fig. 6-4). With time, the drop will seem to become larger and paler, until eventually, even without stirring, the whole glass of water will be uniformly faintly colored.

Why? At the border of the drop, dye molecules are moving in all directions. Those that stay within the drop don't change its composition, which was all dye to begin with, but some, simply due to random motion, go out into the water. Thus, **the net movement of dye molecules is from the region of high dye concentration (in the drop) to the region of low dye concentration (in the water).** The same thing happens with water molecules. Random motion causes some to enter the drop, and the net movement of water is from the high water concentration outside the drop into the low water concentration inside the drop. To the observer, the dye molecules that moved beyond the original borders of the drop have made the drop grow larger; the water molecules that invaded the drop diluted the dye, making the drop paler. At first, when the drop is pure dye and the water is pure water, there is a very steep concentration gradient, and the dye diffuses rapidly. As the concentration differences lessen, the dye diffuses more and more slowly. However, as long as the concentration of dye within the expanding drop is greater than the concentration of dye in the rest of the glass, the net movement of dye will be from drop to water, until the dye becomes uniformly dispersed in the water. With no concentration gradient of either dye or water, diffusion stops. Individual molecules still move about, but there are no changes in concentration of either water or dye.

As you can appreciate from this imaginary experiment, diffusion cannot move molecules rapidly over long distances. Although the drop of dye immediately begins to diffuse into the water, uniform dispersion may take many minutes. As described in the previous chapter, the slow rate of diffusion over long distances is one of the reasons cells are small.

SUMMARY OF THE PRINCIPLES OF DIFFUSION

1. Diffusion is the net movement of molecules down a gradient from high to low concentration.
2. The greater the concentration gradient, the faster the rate of diffusion.
3. If no other processes intervene, diffusion will continue until the concentration gradient is eliminated.
4. Diffusion cannot move molecules rapidly over long distances.

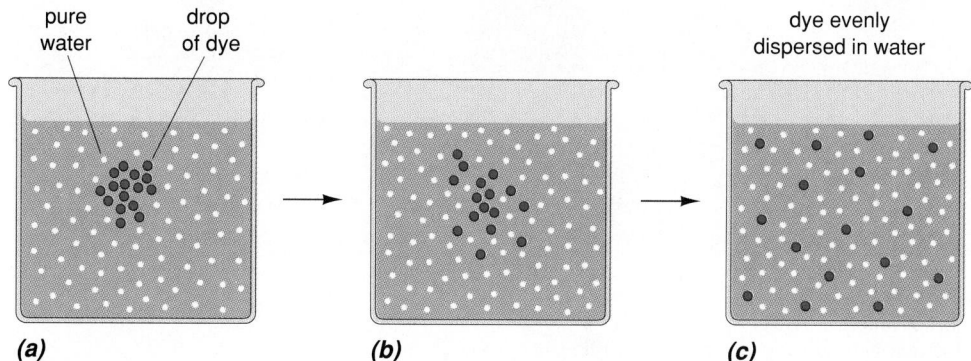

pure water   drop of dye

dye evenly dispersed in water

(a)   (b)   (c)

Figure 6-4 Diffusion of a dye in water. **(a)** A drop of pure dye (red dots) is suspended in a glass of pure water (white dots in blue background). **(b)** Although individual molecules move at random, concentration gradients cause diffusion of dye molecules into the water and water molecules into the drop of dye. **(c)** Eventually, dye and water are both evenly dispersed. Individual molecules still move, but there is no longer any concentration gradient, and therefore no further diffusion.

Diffusion Across Membranes

Many molecules cross plasma membranes by diffusion, driven by differences in the concentration of particular molecules in the cytoplasm versus the external environment. Molecules cross the plasma membrane at different locations and at different rates, depending on the properties of the molecule in question. Therefore, plasma membranes are said to be **differentially permeable:** they allow some molecules to pass through, or **permeate,** more rapidly than others.

Simple Diffusion

Water, dissolved gases such as oxygen and carbon dioxide, and lipid-soluble molecules such as ethyl alcohol and vitamin A easily diffuse across the phospholipid bilayer. This process is called **simple diffusion** (Fig. 6-5a). Generally, the rate of simple dif-

(a)

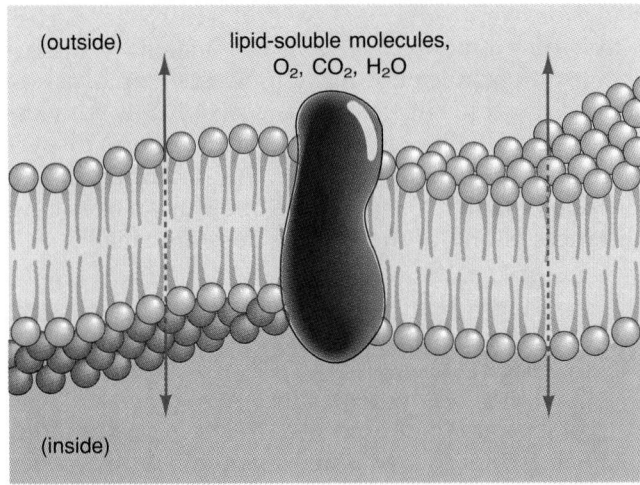

(b)

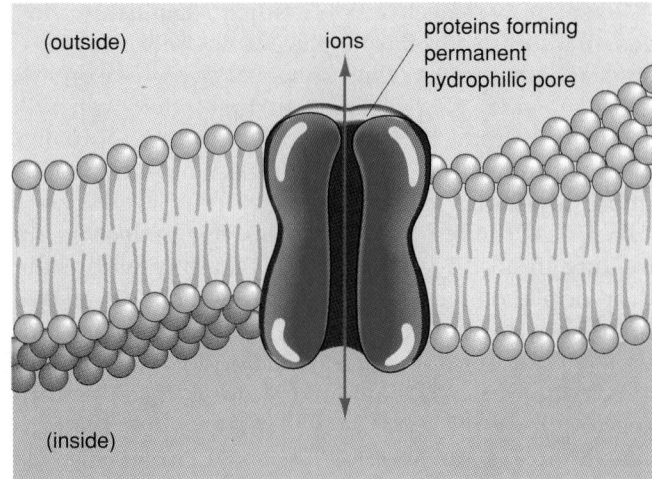

(c)

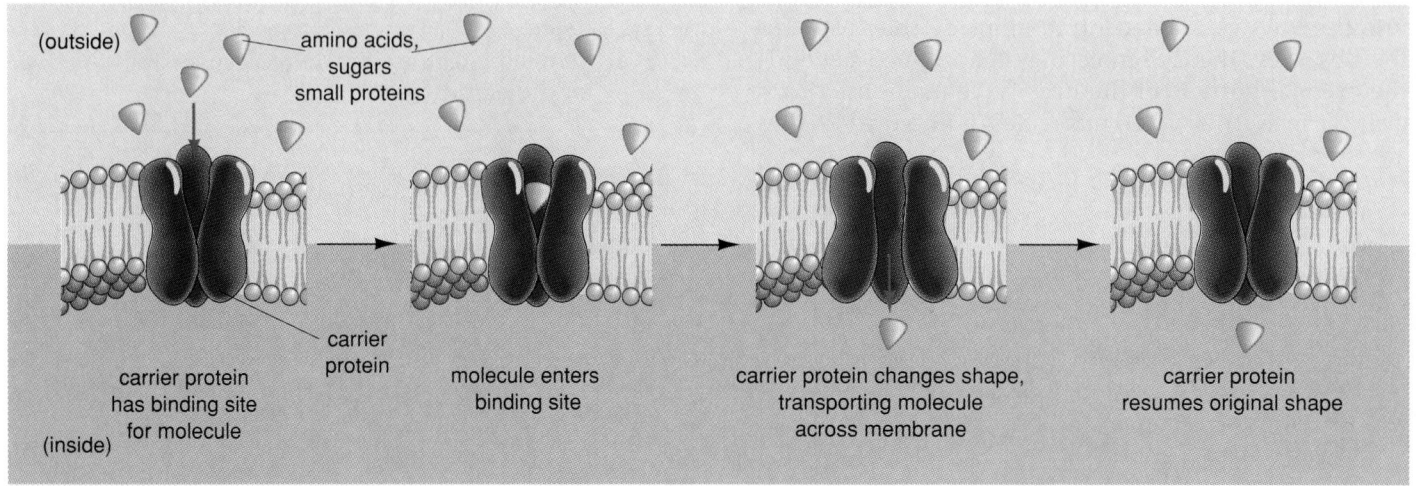

Figure 6-5 Diffusion through the plasma membrane.
(a) Simple diffusion through the phospholipid bilayer: Water, gases such as oxygen and carbon dioxide, and lipid-soluble molecules can diffuse directly through the phospholipids.
(b) Facilitated diffusion through a channel: Water-soluble molecules cannot pass through the phospholipid bilayer. Protein channels pores allow some water-soluble molecules, principally ions, to penetrate the cell.
(c) Carrier-mediated facilitated diffusion: Carrier proteins may bind specific molecules and, as a result, change their shape, passing the molecule through the middle of the protein to the other side of the membrane.

fusion is a function of the concentration gradient across the membrane, the size of the molecule, and its lipid solubility.

Facilitated Diffusion

Most water-soluble molecules, such as ions (K^+, Na^+, Ca^{2+}), amino acids, and monosaccharides, cannot move through the phospholipid bilayer. These molecules can diffuse across only with the aid of two types of transport proteins, called **channel proteins** and **carrier proteins**. This process is called **facilitated diffusion.**

Channel proteins form permanent pores, or channels, in the lipid bilayer through which certain ions can traverse the membrane (Fig. 6-5b). Usually, each channel protein has a specific interior diameter and distribution of electrical charges that allow only particular ions to pass through. Nerve cells, for example, have separate channels for sodium ions, potassium ions, and calcium ions. As we will see in Chapter 36, many of these channels have "gates" on one end, which open and close in response to electrical or chemical signals.

Carrier proteins bear groups of amino acids that bind specific molecules from the cytoplasm or extracellular fluid (Fig. 6-5c). Binding triggers a change in the shape of the carrier that allows the molecule to pass through the protein and thus across the plasma membrane. Carrier proteins that do not use cellular energy move molecules only down their concentration gradients (see Active Transport, below).

The Rate of Facilitated Diffusion

Plasma membranes contain relatively few pores or carriers through which any given molecule can pass. The rate of facilitated diffusion is therefore a function of both the concentration gradient and the density of transport proteins. Molecules that cross the membrane by facilitated diffusion usually do so more slowly than those that cross by simple diffusion through the lipid bilayer.

Osmosis

Water, like any other molecule, moves by diffusion from high water concentrations to low water concentrations. However, the diffusion of water across semipermeable membranes has such dramatic consequences that it has been given a special name: **osmosis.** Let's investigate osmosis in another thought experiment.

A very simple kind of differentially permeable membrane consists of an impervious sheet perforated with tiny pores. The pores allow water molecules to pass through, but not larger molecules such as sugar. Suppose we make a bag out of such a membrane, fill it with a sugar solution, tie off the top, and place the bag in a glass of pure water. The bag will rapidly swell up, and, if it is weak enough, it will burst (Fig. 6-6). Why?

If you could visualize individual molecules, you would see that there are two categories of water molecules in the sugar solution inside the bag (Fig. 6-6): "free" water molecules, well separated from the sugars; and "bound" water molecules, held to the sugars by hydrogen bonds (this is why sugars dissolve in water; see Fig. 2-10 in Chapter 2). In the pure water outside the bag, of course, there are only free water molecules. Now, free water molecules can diffuse through the pores in the membrane, but the bound water molecules cannot, since they are at least temporarily attached to the bulky sugars.

Therefore, the concentration of free water molecules is lower inside the bag than in the pure water outside. This water concentration gradient favors the movement of free water molecules from the pure water outside the bag to the sugar solution inside the bag. The bag swells up as more water molecules enter the bag than leave it. The sugar can't escape at all, so the free water concentration inside the bag is always lower than in the pure water outside. Water continues to enter the bag until it bursts.

Now let's redo our experiment, but with one change. Suppose we tie a piece of the same differentially permeable membrane across the end of a glass tube, suspend the tube in pure water, and put the sugar solution in the tube (Fig. 6-7). Water will diffuse across the membrane from the pure water into the solution in the tube. Since the upper end of the tube is open, as the water enters it will raise the level of the solution in the tube. Eventually, the level will stop rising. Why? As the solution in the tube rises above the level of water in the glass, its weight applies a pressure to the membrane across the bottom of the tube. Therefore, two opposing gradients are set up: a concentration gradient moving water *into* the tube, and a pressure gradient pushing water *out of* the tube. When the two gradients are equal, no further net movement of water will occur across the membrane. The lower the concentration of free water molecules in the tube (i.e., the higher the sugar concentration), the greater will be the tendency for water to move across the membrane and into the tube. The physical pressure that exactly balances the osmosis of water due to the concentration difference between a solution and pure water is defined as the **osmotic pressure** of the solution. A solution with a high osmotic pressure has a low concentration of free water,

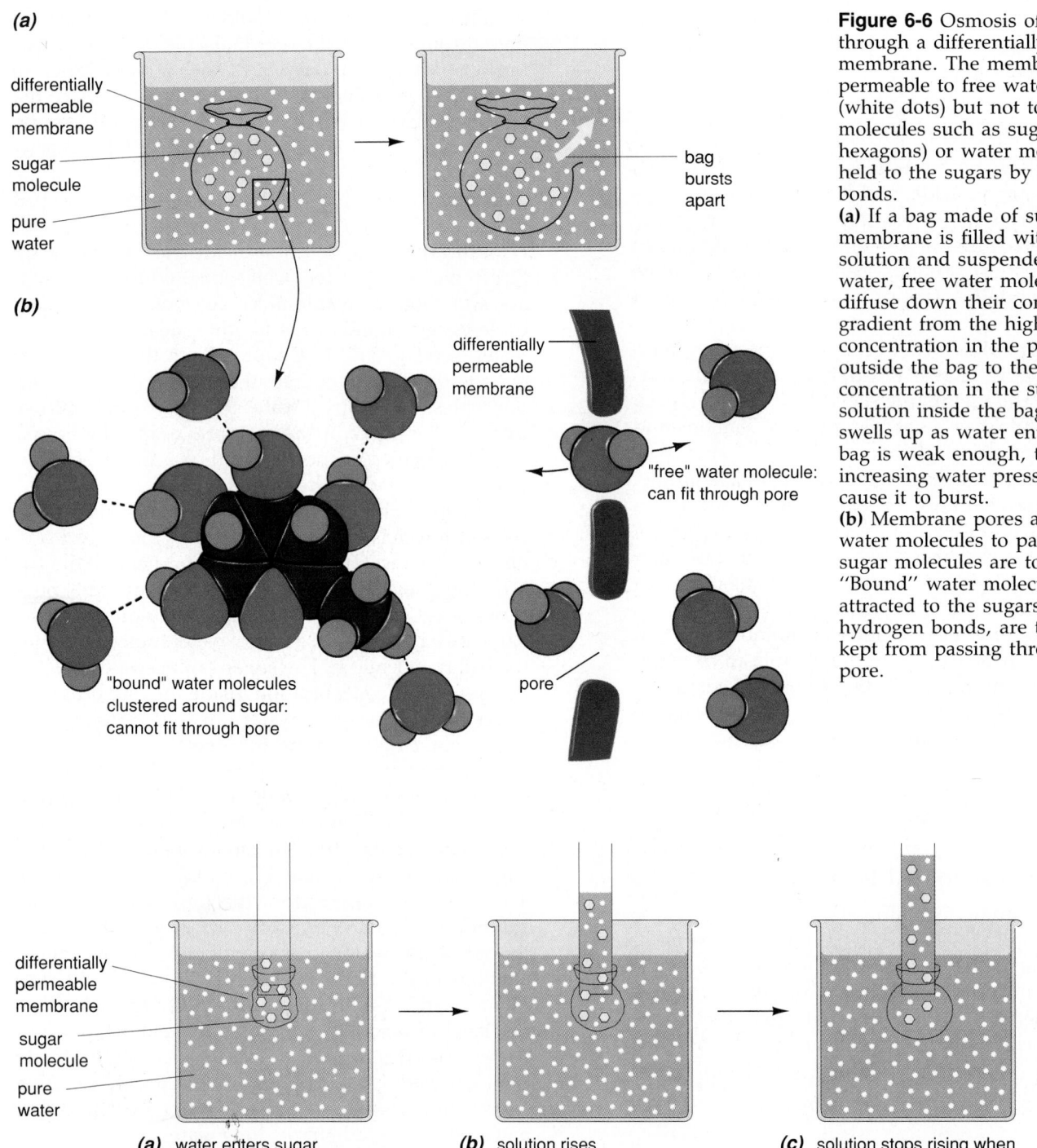

Figure 6-6 Osmosis of water through a differentially permeable membrane. The membrane is permeable to free water molecules (white dots) but not to larger molecules such as sugar (yellow hexagons) or water molecules held to the sugars by hydrogen bonds.
(a) If a bag made of such a membrane is filled with a sugar solution and suspended in pure water, free water molecules will diffuse down their concentration gradient from the high concentration in the pure water outside the bag to the lower concentration in the sugar solution inside the bag. The bag swells up as water enters. If the bag is weak enough, the increasing water pressure will cause it to burst.
(b) Membrane pores allow "free" water molecules to pass through; sugar molecules are too large. "Bound" water molecules, attracted to the sugars by hydrogen bonds, are thus also kept from passing through the pore.

Figure 6-7 Osmotic pressure. **(a)** A membrane permeable to free water molecules (white dots) but not to sugar (yellow hexagons) is fastened across the bottom of a glass tube. A sugar solution is placed in the tube and lowered into a glass of pure water until the level of the sugar solution is even with the surface of the water. **(b)** Water moves by osmosis across the membrane into the tube. As a result, the solution rises within the tube. **(c)** When the pressure of the column of solution in the tube forces water out through the membrane as fast as osmosis causes water to enter, the column stops rising. The resulting hydrostatic pressure equals the osmotic pressure of the solution.

while a solution with a low osmotic pressure has a high concentration of free water.

As we have emphasized, **osmosis is driven by differences in water concentration between solutions**. Therefore, osmosis does not depend on the *type* of dissolved molecules in a solution, but only on their *concentration*. We can produce the same osmotic pressure by adding sucrose, glucose, amino acids, sodium ions, or a mixture of all of these to a solution, as long as the concentration of all types of nonpermeable particles stays the same.

SUMMARY OF THE PRINCIPLES OF OSMOSIS

1. Osmosis is the diffusion of water across a differentially permeable membrane.
2. Water moves across a membrane from a high concentration of free water molecules to a low concentration of free water molecules, or from high pressure to low pressure.
3. Dissolved substances, regardless of what they are, lower the concentration of free water molecules in a solution.

The Effects of Osmosis on Cells

Most plasma membranes are highly permeable to water. Since all cells contain appreciable amounts of dissolved salts, proteins, sugars, and so on, the flow of water across the plasma membrane depends on the concentration of water in the liquid that bathes the cells. The extracellular fluids of animals are usually **isotonic** ("having the same strength") to the insides of the body cells: that is, the concentrations of water and dissolved particles outside the cells are the same as those inside, so there is no net tendency for water either to enter or leave the cells. Note that the *types* of dissolved particles are seldom the same inside and outside the cells, but the *total concentration* of all particles is equal.

If cells, say red blood cells, are taken out of the body and immersed in salt solutions of varying concentrations, the effects of the differential permeability of the plasma membrane to water and dissolved particles become dramatically apparent (Fig. 6-8). If the solution has a higher salt concentration than the cytoplasm (that is, if the solution has a lower water concentration), water will leave the cells by osmosis. The cells will shrivel up until the concentrations of water inside and outside become equal. Solutions that cause water to leave cells by osmosis are called **hypertonic** ("having greater strength"). Conversely, if the solution has little or no salt, water will enter the cells, causing them to swell. If the solution has little

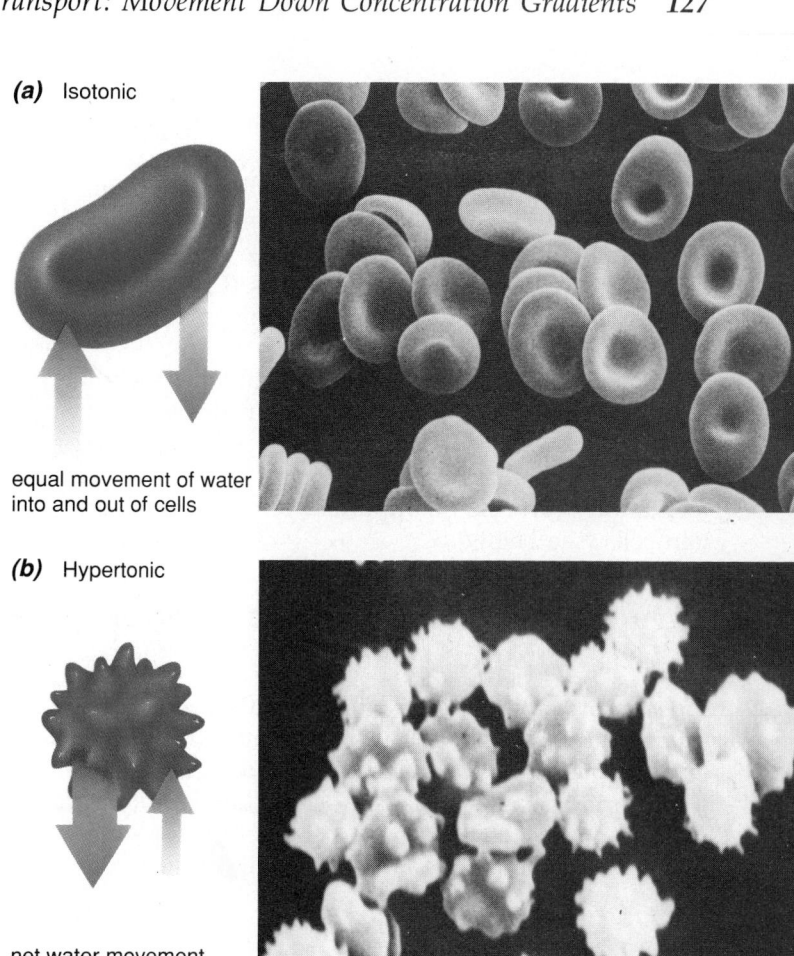

(a) Isotonic

equal movement of water into and out of cells

(b) Hypertonic

net water movement out of cells

(c) Hypotonic

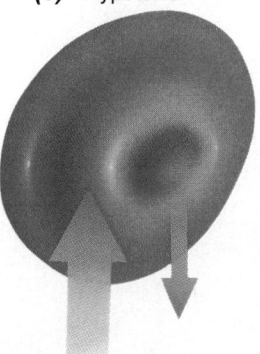
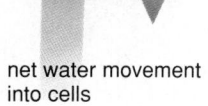

net water movement into cells

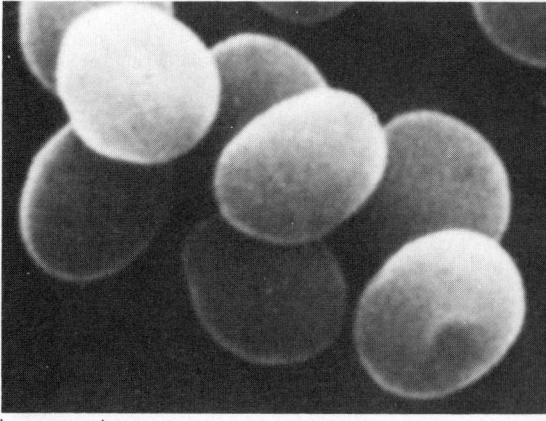

|— 10 micrometers —|

Figure 6-8 The effects of osmosis. Red blood cells are normally suspended in the fluid environment of the blood, and lack the ability to regulate water flow across their plasma membranes. **(a)** If red blood cells are immersed in an isotonic salt solution, which has the same concentration of dissolved substances as the blood cells, there is no net movement of water across the cell membrane, and the red blood cells keep their characteristic dimpled disc shape. **(b)** A hypertonic solution, with too much salt, causes water to leave the cells, shrivelling them up. **(c)** A hypotonic solution, without enough salt, causes water to enter, and the cells swell.

enough salt, the cells will burst. Solutions that cause water to enter cells by osmosis are called **hypotonic** ("having lesser strength").

Osmosis across plasma membranes is crucial to the functioning of many biological systems, including water uptake by plant roots (Chapter 27), absorption of dietary water from your intestines (Chapter 32), and reabsorption of water and minerals in your kidneys (Chapter 33).

Osmosis will occur across any membrane, including the membranes surrounding organelles within a cell, if the water concentration differs on the two sides of the membrane. Osmosis across internal membranes can also be important in the lives of certain cells (see below).

Vacuoles, Osmosis, and the Control of Cell Volume

Most cells contain one or more **vacuoles,** which are fluid-filled sacs surrounded by a single membrane (see Chapter 5). Some, such as the food vacuoles that form during phagocytosis (see below), are temporary features of cells. However, many cells contain permanent vacuoles that have important roles in maintain-

ing the integrity of the cell, especially by regulating the cell's water content.

Contractile Vacuoles

Freshwater protozoans such as *Paramecium* often contain complex **contractile vacuoles** composed of collecting ducts, a central reservoir, and a tube leading to a pore in the plasma membrane (Fig. 6-9). Since fresh water is hypotonic to the cytosol of protozoans, water constantly enters the cell by osmosis. The increasing volume of water might soon burst the fragile protozoan. However, water is pumped into the collecting ducts and drains into the central reservoir (this process requires cellular energy). When the reservoir is full, it contracts, squirting the water up the exit tube and out through the pore in the plasma membrane.

Central Vacuoles

Three quarters or more of the volume of many plant cells is occupied by a large **central vacuole** (see Fig. 5-7). The central vacuole has several functions. It provides a dump site for hazardous wastes, which plant cells often cannot excrete. Some plant cells store extremely poisonous substances in their vacuoles, such

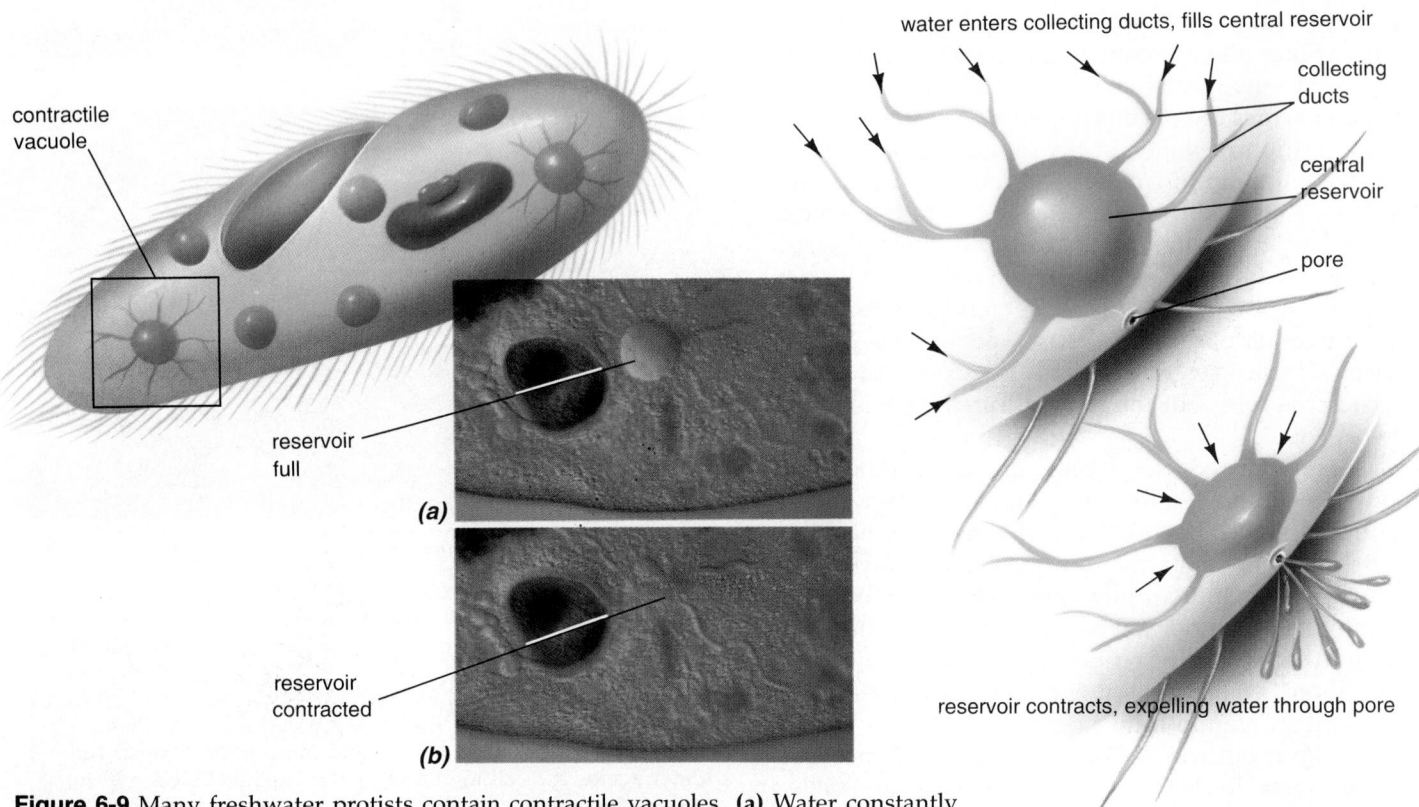

Figure 6-9 Many freshwater protists contain contractile vacuoles. **(a)** Water constantly leaks into the cell, where it is taken up by collecting ducts and drains into the central reservoir of the vacuole. **(b)** When full, the reservoir contracts, expelling the water outside the cell through a pore in the plasma membrane.

cytoplasm   central   cell wall   plasma
              vacuole   membrane

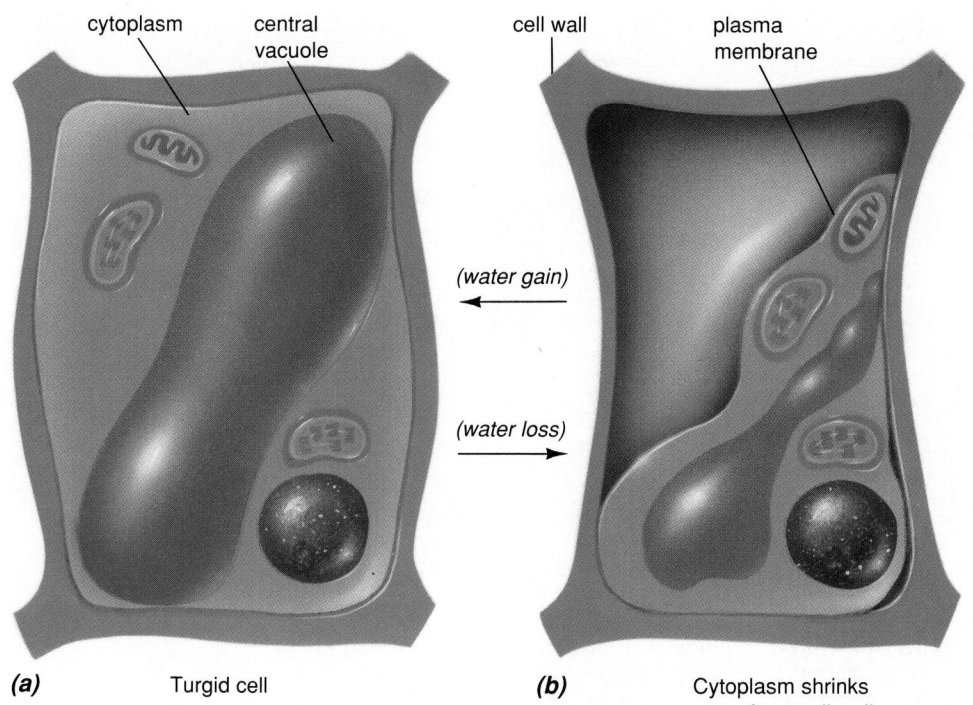

(water gain)

(water loss)

(a)   Turgid cell   *(b)*   Cytoplasm shrinks
                                         away from cell wall

Figure 6-10 The role of turgor pressure in plant cells. **(a)** Water enters the central vacuole by osmosis, so that pressure builds up within the vacuole. This pressure pushes the cytoplasm up against the cell wall, helping the cell to maintain its shape. **(b)** When a plant wilts, the central vacuole loses water and shrinks. The cytoplasm pulls away from the cell wall, and the whole cell may become soft and shrunken.

as sulfuric acid, which deter animals from munching on the otherwise tasty leaves. Vacuoles may also store sugars and amino acids not immediately needed by the cell. The blue or purple pigments stored in the central vacuoles of a flower's cells are responsible for the colors of many flowers.

These dissolved substances make the vacuole contents hypertonic to the cell cytoplasm, which in turn is usually hypertonic to the extracellular fluid bathing the cells. Water therefore enters the vacuole by osmosis, which tends to make it swell. The pressure of the water within the vacuole, called **turgor pressure,** pushes the cytosol up against the cell wall with considerable force (Fig. 6-10). Primary cell walls are usually somewhat flexible, and the overall shape and rigidity of the cell depends on turgor pressure within the cell. Turgor pressure thus provides support for the nonwoody parts of plants. If you fail to water your houseplants, the central vacuoles and cytoplasm lose water and the cell shrinks away from its cell walls. Just as a balloon goes limp when its air leaks out, so too the plant droops as its cells lose turgor pressure.

Energy-Requiring Transport Across Membranes

All cells need to move some materials "uphill" across their plasma membranes, against diffusion gradients.

For example, every cell requires some nutrients that are extremely scarce in its environment. Since these nutrients may be less concentrated in the environment than in the cell cytoplasm, diffusion would cause the cell to lose, not gain, these nutrients. Other substances, such as sodium and calcium ions in your brain cells, must be maintained at much lower concentrations inside the cells than in the extracellular fluid. When these ions diffuse into the cells, they must be pumped out again against their concentration gradients. Finally, many cells acquire or expel some substances, such as whole bacteria or large proteins, that are too large to diffuse across a membrane regardless of concentration gradients. Cells have evolved several processes that use cellular energy to move materials into or out of the cell.

Active Transport

In **active transport,** membrane proteins use cellular energy to move individual molecules across the plasma membrane, usually against their concentration gradients (Fig. 6-11). Active transport proteins span the membrane, and have two active sites. One active site (which may be on the inside or outside face of the plasma membrane, depending on the active transport protein) recognizes a particular molecule, say a sugar, and binds it. The second site (always on the inside of the membrane) binds an energy-carrier

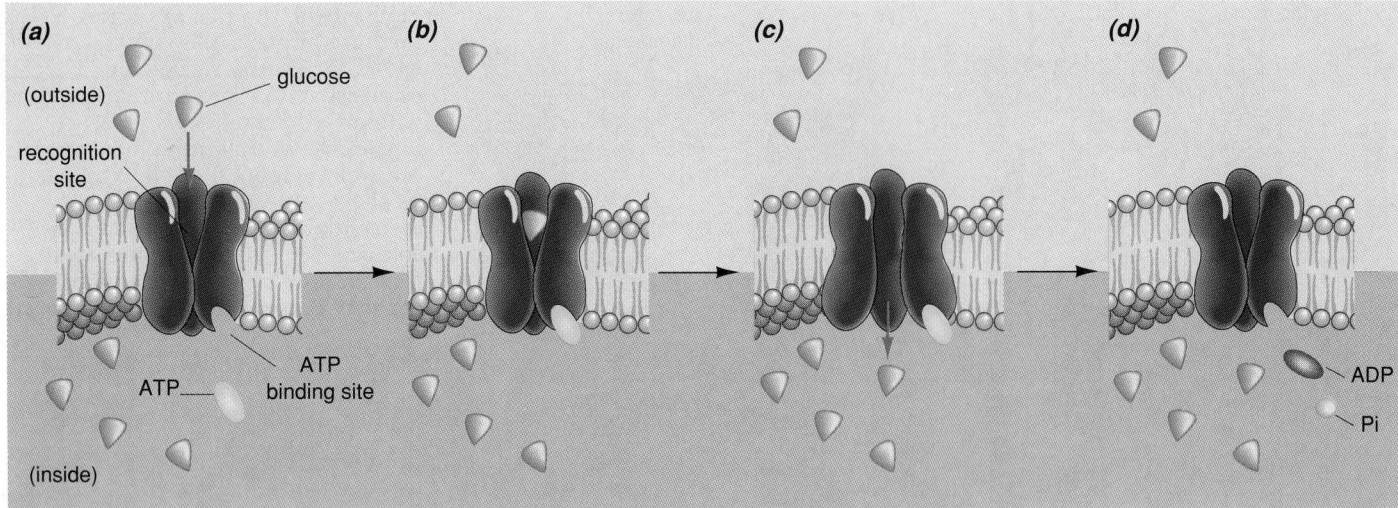

Figure 6-11 Active transport uses cellular energy to move molecules across the plasma membrane, often against a concentration gradient. **(a)** A transport protein (blue) has a recognition site for the molecule to be transported (glucose in this example) and an ATP binding site. **(b)** The transport protein binds ATP and glucose. **(c)** Energy from ATP changes the shape of the transport protein and moves the glucose molecule across the membrane. **(d)** The carrier releases the glucose and the remnants of the ATP and resumes its original configuration.

molecule, usually adenosine triphosphate (ATP). The ATP donates energy to the protein, causing it to change shape and move the sugar molecule across the membrane. Active transport proteins are often called *pumps,* in analogy to water pumps, because they use energy to move molecules "uphill" against a concentration gradient. We will see that plasma membrane pumps are vital in mineral uptake by plants (Chapter 27), mineral absorption in your intestines (Chapter 32), and maintaining concentration gradients essential to nerve cell functioning (Chapter 36).

Endocytosis

Cells can also acquire particles, especially large proteins or entire microorganisms such as bacteria, by a process called **endocytosis** (Greek for "into the cell"). During endocytosis, the plasma membrane engulfs the particle and pinches off a membranous sac called a **vesicle,** with the particle inside, into the cytoplasm (Fig. 6-12). Three types of endocytosis can be distinguished, based on the size of the particle acquired and the method of acquisition.

Pinocytosis

In **pinocytosis** ("cell drinking"), or **fluid-phase endocytosis,** a very small patch of plasma membrane dimples inward and buds off into the cytoplasm as a tiny vesicle (Fig. 6-12a). Pinocytosis moves a droplet of extracellular fluid, contained within the dimpling patch of membrane, into the cell. Therefore, the cell acquires materials in proportion to their concentration in the extracellular fluid.

Receptor-Mediated Endocytosis

Cells can also acquire molecules more efficiently, by a process known as **receptor-mediated endocytosis** (Fig. 6-12b). Most plasma membranes bear many receptor proteins on their outside surfaces, each bearing a binding site for a particular nutrient molecule. In most cases, these receptors move through the lipid bilayer and accumulate in depressions of the plasma membrane called coated pits (Fig. 6-13). If the right molecule contacts a receptor protein in one of these coated pits, it attaches to the binding site. The coated pit deepens into a U-shaped pocket that eventually pinches off into the cytoplasm as a coated vesicle. Both the receptor–nutrient complex and a bit of extracellular fluid move into the cell in the coated vesicle.

Phagocytosis

Phagocytosis ("cell eating") is used to pick up large particles, including whole microorganisms (see Fig.

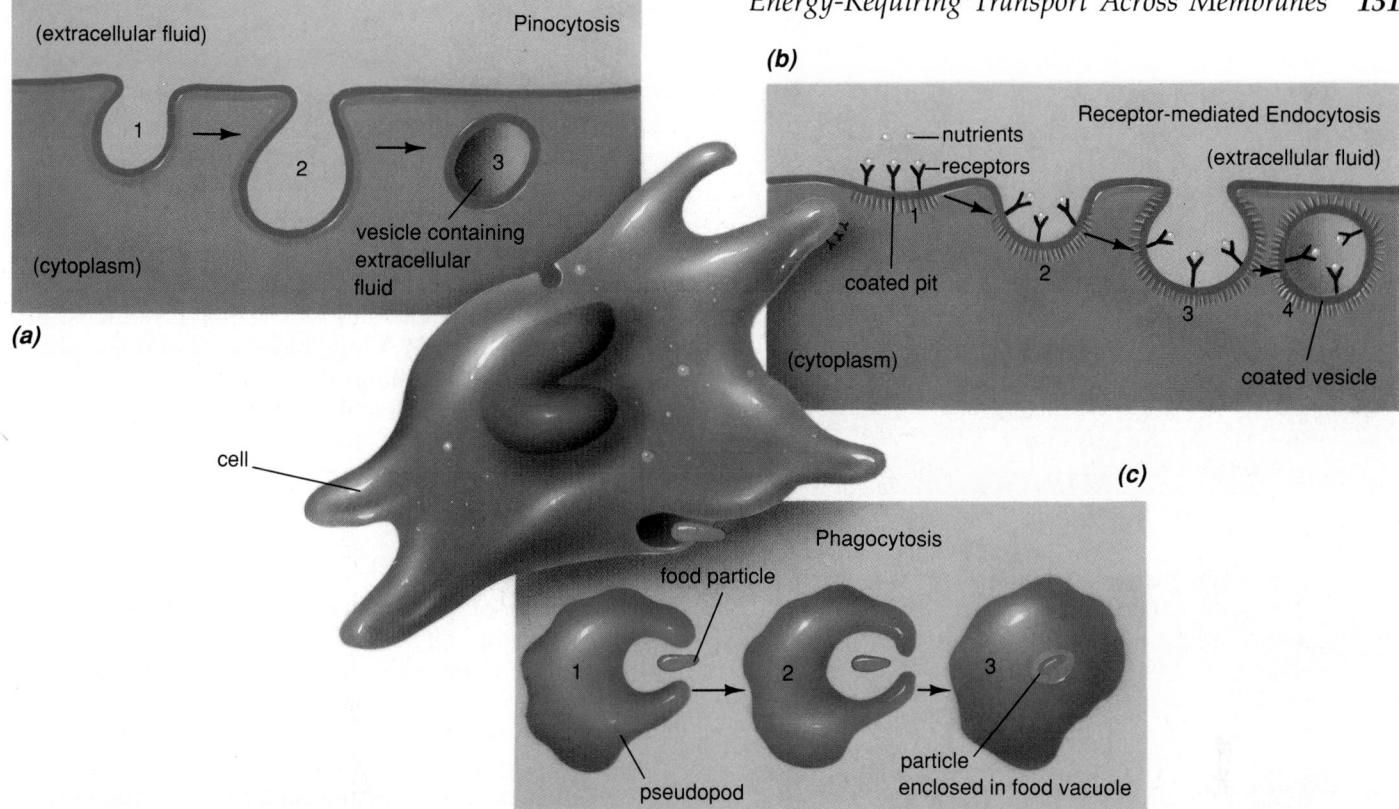

Figure 6-12 Three types of endocytosis.
(a) Pinocytosis: A dimple in the plasma membrane deepens and eventually pinches off as a fluid-filled vesicle. The vesicle contains a random sampling of the extracellular fluid.
(b) Receptor-mediated endocytosis: Receptor proteins selectively bind molecules (e.g., nutrients) in the extracellular fluid. The receptors migrate along the fluid lipid bilayer of the plasma membrane to dimpling sites (coated pit). The membrane dimples inward, carrying the receptor/captured molecule complexes with it. The end of the coated pit buds off a coated vesicle into the cytoplasm of the cell. The vesicle contains both extracellular fluid and a high concentration of the molecules that bind to the receptors.
(c) Phagocytosis: Extensions of the plasma membrane, called pseudopods, encircle an extracellular particle (e.g., food). The ends of the pseudopods fuse, forming a large vesicle (a food vacuole) containing the engulfed particle.

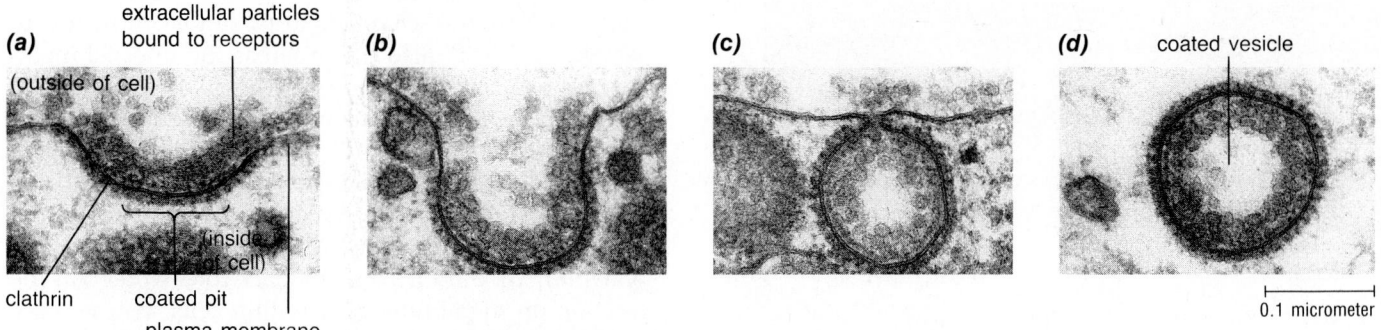

Figure 6-13 These electron micrographs illustrate the sequence of events in receptor-mediated endocytosis via coated pits. **(a)** This type of endocytosis begins with a shallow depression in the plasma membrane, coated on the inside with a protein called clathrin (dark fuzzy substance in the micrographs) and bearing receptor proteins on the outside (not visible in the micrographs). The pit deepens **(b)**, probably pulled in by elements of the cytoskeleton, and eventually pinches off as a coated vesicle **(c, d)**. The clathrin protein is eventually recycled back to the plasma membrane, restarting the cycle.

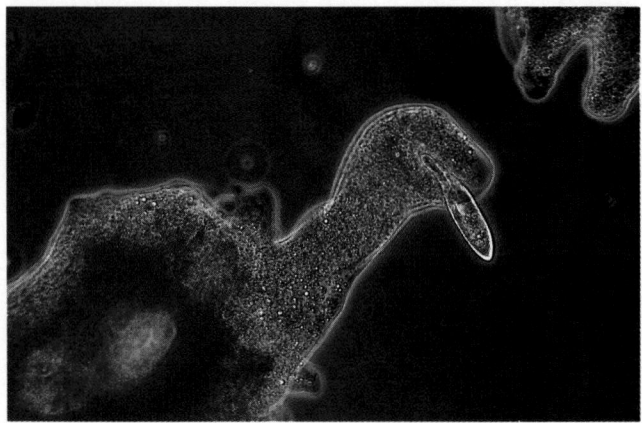

100 micrometers

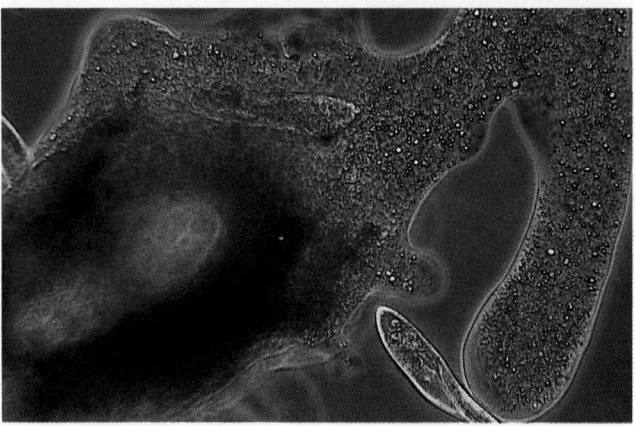

100 micrometers

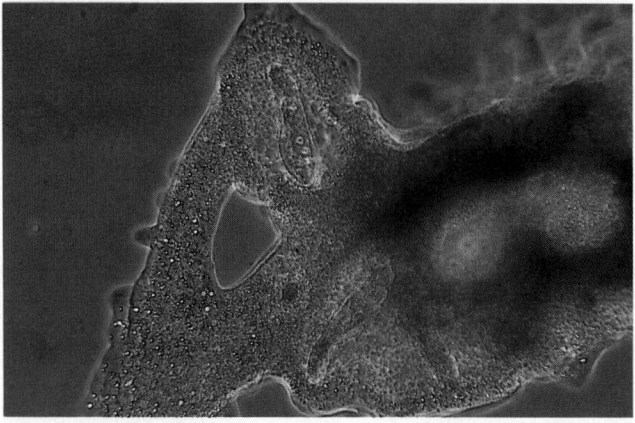

100 micrometers

Figure 6-14 Phagocytosis. An *Amoeba* senses a *Paramecium* nearby, and sends out a pseudopod that encircles the hapless prey and rejoins the main cell body, trapping the *Paramecium*. The prey enters the *Amoeba* enclosed within a food vacuole made of plasma membrane. Fusion of the food vacuole with a lysosome spells the end of the *Paramecium* but continued life for the *Amoeba*.

6-12c). When an *Amoeba*, for example, senses a tasty *Paramecium*, it produces extensions of its surface membrane, called **pseudopodia** (Latin for "false foot"; sing., **pseudopod**). The pseudopodia surround the luckless *Paramecium*, their ends fuse, and the prey is carried into the interior of the *Amoeba* for digestion (Fig. 6-14). The resulting vesicle, called a food vacuole, fuses with lysosomes whose enzymes digest the prey (see Chapter 5). White blood cells also use phagocytosis and intracellular digestion to engulf and destroy bacteria that have invaded your body (Chapter 34).

Exocytosis

The reverse of endocytosis, called **exocytosis** (Greek for "out of the cell"), is often used by cells to dispose of unwanted materials, such as the waste products of digestion, or ι·ν secrete materials, such as hormones, into the extracellular fluid (Fig. 6-15). During exocytosis, a vesicle created by the Golgi apparatus moves to the cell surface, where the membrane of the vesicle fuses with the plasma membrane. The vesicle opens to the extracellular fluid and its contents diffuse out.

Transport Across Intracellular Membranes

As we mentioned earlier, all the membranes of a cell, including those that surround such diverse organelles as chloroplasts, mitochondria, and the nucleus, are similar in structure and function to the plasma membrane. The transport processes that we have described in this chapter, from simple diffusion to active transport, also occur across the internal membranes of a cell. For example, we will see in Chapters 7 and 8 that chloroplasts and mitochondria have specific protein channels that are permeable to hydrogen ions, and that are crucial to ATP production in these organelles. Calcium ions are actively transported across certain internal membranes in muscle cells (see Chapter 38). And if you look back to Chapter 5 at the section on membrane flow within cells, you will see vesicles budding from and fusing with the membranes of the endoplasmic reticulum and Golgi complex, processes virtually identical to endocytosis and exocytosis. Transport across internal membranes is just as vital to the life of a cell as is transport across the plasma membrane, and similar mechanisms are employed.

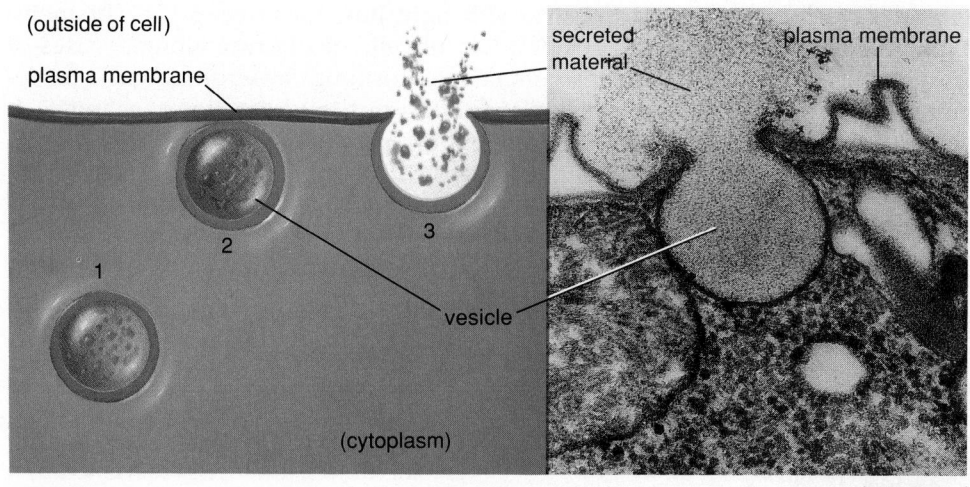

(outside of cell)

plasma membrane

1

2

3

(cytoplasm)

secreted material

plasma membrane

vesicle

0.2 micrometers

Figure 6-15 Exocytosis is functionally the reverse of endocytosis. The material to be ejected from the cell is encapsulated into a membrane-bound vesicle that moves to the cell membrane and fuses with it. As the vesicle opens to the outside, the material within leaves by diffusion.

Cell Connections and Communication

In multicellular organisms, plasma membranes also function in holding together clusters of cells and in providing avenues through which cells can communicate with their neighbors. Depending on the organism and the cell type, one of four types of connection may occur between cells: desmosomes, tight junctions, gap junctions, and plasmodesmata.

Desmosomes: Cell-to-Cell Adhesion

Animals, as you know, tend to be flexible, mobile organisms. Many of an animal's tissues are stretched, compressed, and bent as the animal moves about. If the skin, intestines, stomach, urinary bladder, and other organs are not to tear apart under the stresses of movement, their cells must adhere firmly to one another. Such animal tissues have junctions called **desmosomes** that hold adjacent cells together (Fig. 6-16). In a desmosome, the membranes of adjacent

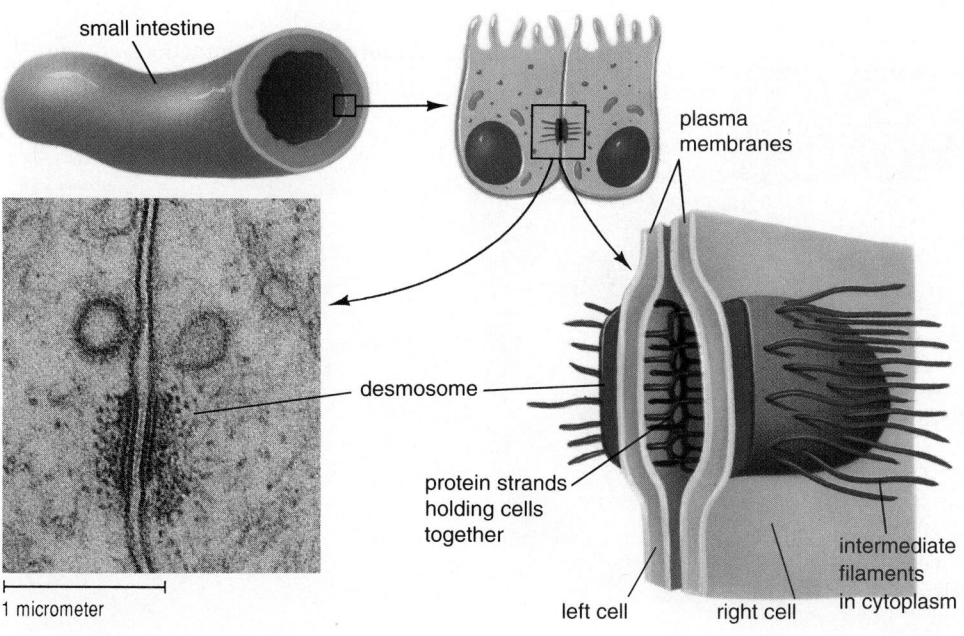

small intestine

plasma membranes

desmosome

protein strands holding cells together

left cell

right cell

intermediate filaments in cytoplasm

1 micrometer

Figure 6-16 Cells lining the small intestine are firmly attached to one another by desmosomes. Intermediate filaments bound to the inside surface of each desmosome extend into the cytoplasm and attach to other elements of the cytoskeleton, strengthening the connection between cells.

cells are glued together by proteins and carbohydrates. Intermediate filaments attached to the insides of the desmosomes extend into the interior of each cell, further strengthening the attachment.

Tight Junctions: Leakproofing

The animal body contains many tubes or sacs that must hold their contents without leaking: a leaky urinary bladder would spell disaster for the rest of the body. The spaces between the cells lining such sacs are sealed with **tight junctions** (Fig. 6-17). The membranes of adjacent cells nearly fuse along a series of ridges, effectively forming waterproof gaskets between cells. Continuous tight junctions sealing each cell to its neighbors prevents molecules from escaping between cells.

Gap Junctions and Plasmodesmata: Cell-to-Cell Communication

Multicellular organisms must coordinate the actions

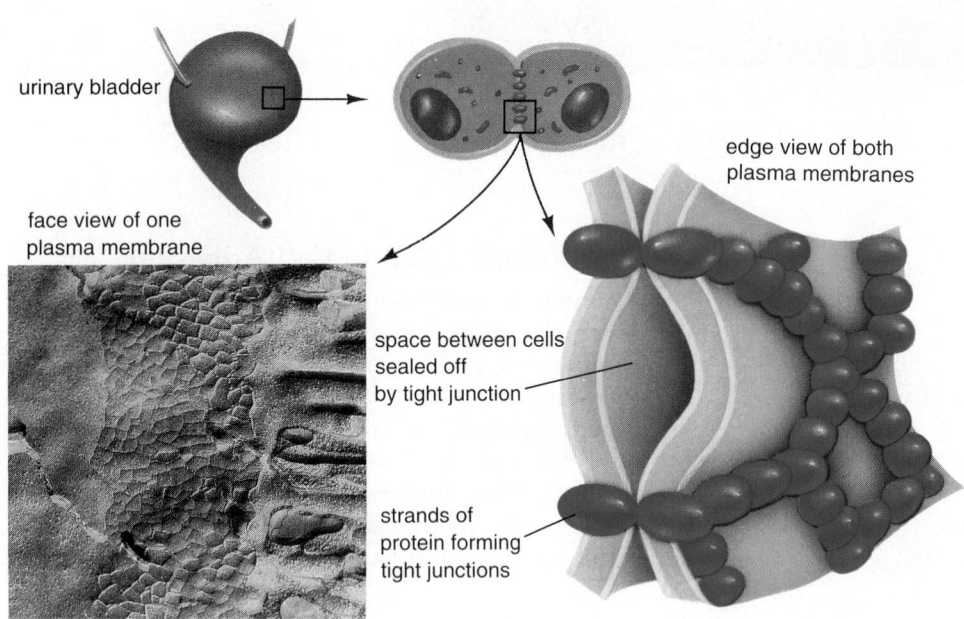

Figure 6-17 Leakage between cells of the urinary bladder is prevented by close-fitting tight junctions.

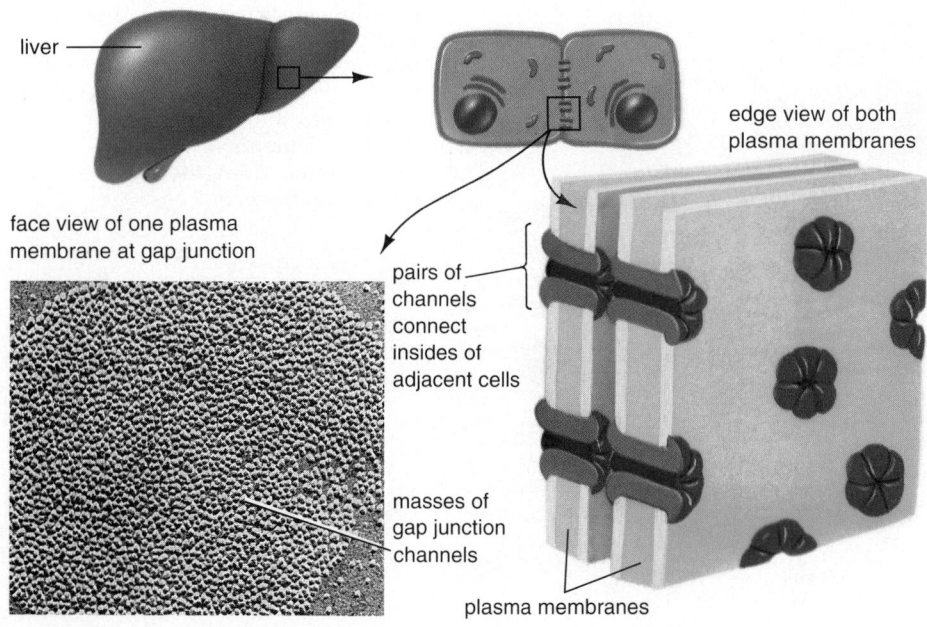

Figure 6-18 Gap junctions contain cell-to-cell channels that interconnect the cytoplasm of adjacent cells.

of their component cells. In animals, many cells, including some brains cells, heart muscle cells, most gland cells, and every cell of very young embryos, communicate through protein channels directly connecting the insides of adjacent cells (Fig. 6-18). These cell-to-cell channels are clustered in specialized regions called **gap junctions.** Hormones, nutrients, ions, and even electrical signals can pass through the channels at gap junctions.

Virtually all the living cells of plants are connected to one another by much larger channels called **plasmodesmata** (Fig. 6-19). Each plasmodesma is a tube, lined with plasma membrane, that penetrates through the cell wall from the cytoplasm of one cell to the cytoplasm of its neighbor. Many plant cells have thousands of plasmodesmata. As a result, water, nutrients, and hormones pass quite freely from one cell to another.

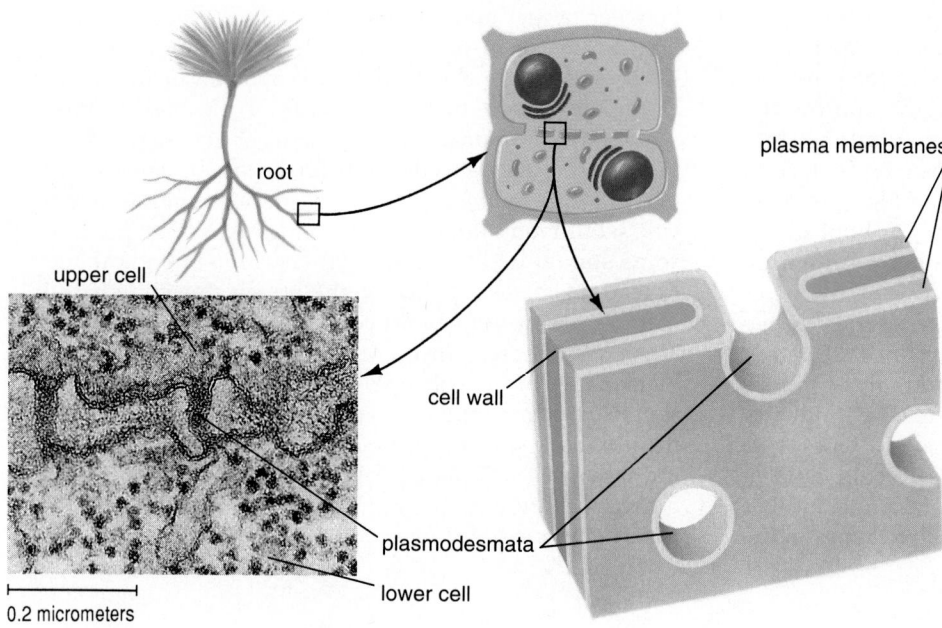

Figure 6-19 Plant cells are widely interconnected by rather large cell-to-cell pores called plasmodesmata.

Reflections on Plasma Membranes

Although the membranes of all cells have a similar structure, membrane function from organism to organism, and from cell to cell within a single organism, varies tremendously. This diversity arises largely from the different proteins and phospholipids in the membrane.

Our discussion of membranes emphasized the unique functions of the membrane proteins. Consequently, you may think that the lipids are just a waterproof matrix that serves as a place for the proteins to reside. This isn't quite true, as we can see by examining the plasma membrane lipids in the legs of caribou (Fig. E6-1). During the long arctic winters, temperatures plummet far below freezing. Keeping their legs and feet really warm would waste precious energy, and caribou have evolved specialized arrangements of arteries and veins that allow the temperature of their lower legs to drop almost to freezing (0° C). The upper legs and main trunk of the body, in contrast, remain at about 37° C. The lipids in the membranes of cells in the upper leg are very different from those near the hooves. Why?

Remember, the membrane of a cell needs to be somewhat fluid. For example, protein receptors that gather materials from the extracellular fluid must move through the lipids to sites of endocytosis. The fluidity of a membrane is a function of the fatty acid tails of its phospholipids: unsaturated fatty acids remain fluid at lower temperatures than saturated fatty acids (see Chapter 3). Caribou have a gradient of lipid types in the plasma membranes of the cells in their legs. The membranes of cells near the chilly hoof have lots of unsaturated fatty acids, while the membranes of cells near the warmer trunk have more saturated ones. This arrangement gives the plasma membranes throughout the leg the proper fluidity despite great differences in temperature.

As important as the phospholipids are, the membrane proteins probably play the major roles in determining cell function and in governing the interactions between a cell and its neighbors. Every nerve cell in your body, for instance, has membrane proteins essential for producing electrical signals and conducting them along the nerves to various parts of the body. Other membrane proteins receive chemical messages from neighboring nerve cells or from hormones and other chemicals in the blood. Each cell in the brain has a specific set of membrane proteins, allowing it to respond to some stimuli while ignoring others. In fact, your ability to read this page depends on the proteins residing in the membranes of your brain cells.

As we progress through this book, we shall return many times to the concepts of membrane structure presented in this chapter. Understanding the diversity of membrane lipids and proteins is the key to understanding not just the isolated cell, but entire organs, which function as they do largely because of the properties of the membranes of their component cells.

Figure E6-1 Caribou browse on the Alaskan tundra. The lipid composition of the membranes in the cells of a caribou's legs varies with distance from the trunk. Unsaturated phospholipids predominate in the lower leg, while more saturated phospholipids are found in the upper leg.

SUMMARY OF KEY CONCEPTS

Cell Walls
Moneran, plant, fungal, and some protist cells are surrounded by a rigid cell wall outside the plasma membrane. The cell wall is produced by the cell that it surrounds. It protects and supports the cell.

The Plasma Membrane
The plasma has four major functions: (1) isolate the cytoplasm from the external environment; (2) regulate the flow of materials into and out of the cell; (3) communicate with other cells; and (4) identify the cell. The membrane consists of a bilayer of phospholipids in which is embedded a variety of proteins. There are three major categories of membrane proteins: (1) transport proteins, which regulate the movement of most water-soluble substances through the membrane; (2) receptor proteins, which bind molecules in the external environment, triggering changes in the metabolism of the cell; and (3) recognition proteins, which identify cells as to species and cell type.

Transport Across Membranes
Unless living things intervene, particles in a fluid move in response to gradients of concentration, pressure, or electrical charge. In cells, passive transport is the movement of substances across the plasma membrane down gradients, usually of concentration. Cellular energy is not used. Energy-requiring transport processes move substances across the membrane, usually against concentration gradients, through the use of cellular energy.

Passive Transport: Movement Down Concentration Gradients
Diffusion is the movement of particles from regions of higher concentration to regions of lower concentration. Many substances diffuse across plasma membranes. In simple diffusion, water, dissolved gases, and lipid-soluble molecules diffuse through the phospholipid bilayer. In facilitated diffusion, water-soluble molecules cross the membrane through protein channels or with the assistance of protein carriers. In both cases, molecules move down their concentration gradients, and cellular energy is not required.

Osmosis is the diffusion of water across a differentially permeable membrane, down its concentration gradient. Dissolved solutes decrease the concentration of free water molecules. Osmosis does not require cellular energy.

Energy-Requiring Transport Across Membranes
In active transport, protein carriers in the membrane use cellular energy to drive the movement of molecules across the plasma membrane, usually against concentration gradients. Large molecules (e.g., proteins), particles of food, microorganisms, and extracellular fluid may be acquired by endocytosis, either by pinocytosis, receptor-mediated endocytosis, or phagocytosis.

The secretion of substances such as hormones from a cell is accomplished by exocytosis. The molecules to be voided are enclosed in a vesicle that moves to the cell surface, fuses with the membrane, and empties its contents outside the cell.

Transport Across Intracellular Membranes
The intracellular membranes that surround organelles such as the endoplasmic reticulum, Golgi complex, mitochondria, and chloroplasts are fundamentally similar in structure and function to the plasma membrane. Substances are transported across these membranes by most of the same mechanisms that are involved in transport across the plasma membrane.

Cell Connections and Communication
Cells may be connected to one another by a variety of junctions. Desmosomes attach cells firmly to one another, preventing tearing of a tissue during movement or stress. Tight junctions seal off the spaces between adjacent cells, leakproofing organs such as the urinary bladder. Gap junctions in animals and plasmodesmata in plants are locations in which the cytoplasm of two adjacent cells is interconnected by pores through adjoining plasma membranes.

GLOSSARY

active transport: the movement of materials across a membrane through the use of cellular energy, usually against a concentration gradient.

carrier protein: a membrane protein that facilitates diffusion of specific substances across the membrane. The molecule to be transported binds to the outer surface of the carrier protein; the protein then changes shape, allowing the molecule to move across the membrane through the protein.

cell wall: a layer of material, usually made up of cellulose or celluloselike materials, found outside the plasma membranes of plants, fungi, bacteria, and plantlike protists.

central vacuole: a large, fluid-filled vacuole that occupies most of the volume of many plant cells. See also *turgor pressure.*

channel protein: a membrane protein that forms a channel or pore completely through the membrane, and that is usually permeable to one or a few a water-soluble molecules, especially ions.

concentration: the number of particles of a dissolved substance per unit volume of fluid.

concentration gradient: the difference in concentration of a substance between two parts of a fluid or across a barrier such as a membrane.

contractile vacuole: a fluid-filled vacuole found in certain protists that takes up water from the cell cytoplasm, contracts, and expels the water outside the cell via a pore in the plasma membrane.

desmosome (dez'-mō-sōm): a strong cell-to-cell junction that functions in attaching cells to one another.

differential permeability: the property of a membrane by which some substances can permeate more readily than other substances.

diffusion: the net movement of particles from a region of high concentration of that particle to a region of low concentration, driven by the concentration gradient. May occur entirely within a fluid or across a barrier such as a membrane.

endocytosis (en-dō-sī-tō'-sis): the movement of material into a cell by a process in which the plasma membrane engulfs extracellular material, forming membrane-bound sacs that enter the cytoplasm.

exocytosis (ex-ō-sī-tō'-sis): the movement of material out of a cell by a process in which intracellular material is enclosed within a membrane-bound sac that moves to the plasma membrane and fuses with it, releasing the material outside the cell.

facilitated diffusion: diffusion of molecules across a membrane, assisted by protein pores or carriers embedded in the membrane.

fluid: a liquid or gas.

fluid mosaic model: a model of membrane structure. According to this model, membranes are composed of a double layer of phospholipids in which is embedded a variety of proteins. The phospholipid bilayer is a somewhat fluid matrix that allows movement of proteins within it.

fluid-phase endocytosis: nonselective movement of extracellular fluid into a cell, enclosed within a vesicle formed from the plasma membrane.

gap junction: a type of cell-to-cell junction in animals in which channels connect the cytoplasm of adjacent cells.

glycoprotein: a protein to which a carbohydrate is attached.

gradient: a difference in concentration, pressure, or electrical charge between two regions of space.

hypertonic (hī-per-ton'-ik): referring to a solution that has a higher concentration of dissolved particles (and therefore a lower free water concentration) than the cytoplasm of a cell.

hypotonic (hī-pō-ton'-ik): referring to a solution that has a lower concentration of dissolved particles (and therefore a higher free water concentration) than the cytoplasm of a cell.

isotonic (ī-sō-ton'-ik): referring to a solution that has the same concentration of dissolved particles (and therefore the same free water concentration) as the cytoplasm of a cell.

middle lamella: a thin layer of pectin and other carbohydrates that separates and sticks together the primary cell walls of adjacent plant cells.

osmosis (oz-mō'-sis): the diffusion of water across a differentially permeable membrane, usually down a concentration gradient of free water molecules. Water moves into the solution that has a lower free water concentration from the solution with the higher free water concentration.

osmotic pressure: a measure of the tendency of water to move from a solution with a lower concentration of free water molecules into one with a higher concentration of free water molecules; the physical pressure which must be applied to a solution to prevent water movement into it from pure water.

passive transport: movement of materials across a membrane down a gradient of concentration, pressure, or electrical charge without using cellular energy.

permeate (per'-mē-āt): to pass through, as through pores in a membrane.

phagocytosis (fa-gō-sī-tō'-sis): a type of endocytosis in which extensions of a plasma membrane engulf extracellular particles and transport them into the interior of the cell.

phospholipid bilayer: a double layer of phospholipids that forms the basis of all cellular membranes; the phospholipid heads face the water of extracellular fluid or the cytoplasm, and the tails are buried in the middle of the bilayer.

pinocytosis (pī-nō-sī-tō'-sis): nonselective movement of extracellular fluid into a cell, enclosed within a vesicle formed from the plasma membrane.

plasma membrane: the outer membrane of a cell, composed of a bilayer of phospholipids in which proteins are embedded.

plasmodesma (plaz-mō-dez'-ma; pl. plasmodesmata): a cell-to-cell junction in plants that connects the cytoplasm of adjacent cells.

primary cell wall: cellulose and other carbohydrates secreted by a young plant cell between the middle lamella and the plasma membrane.

pseudopod (sū'-dō-pod): a temporary extension of the plasma membrane used for locomotion or phagocytosis in certain cells such as the protist *Amoeba* or white blood cells of vertebrates.

receptor-mediated endocytosis: selective uptake of molecules from the extracellular fluid, by binding to a receptor located at a coated pit on the plasma membrane, and pinching off the coated pit into a vesicle that moves into the cytosol.

receptor protein: a protein or glycoprotein, located on a membrane (or in the cytoplasm), that recognizes and binds to specific molecules. Binding by receptors often triggers a response by a cell, such as endocytosis, protein synthesis, or exocytosis.

recognition protein: a protein or glycoprotein protruding from the outside surface of a plasma membrane that identifies a cell as belonging to a particular species, to a specific individual of that species, and often to a specific organ within the individual.

secondary cell wall: a thick layer of cellulose and other polysaccharides secreted by certain plant cells between the primary cell wall and the plasma membrane.

simple diffusion: diffusion of water, dissolved gases, or lipid-soluble molecules through the phospholipid bilayer of a membrane.

tight junction: a type of cell-to-cell junction in animals that prevents the movement of materials through the spaces between cells.

transport protein: a protein in a plasma membrane that binds specific molecules and facilitates their transport across the membrane. See also *channel protein, facilitated diffusion,* and *active transport.*

turgor pressure: pressure developed within a cell (especially the central vacuole of plant cells) as a result of osmotic water entry.

STUDY QUESTIONS

1. Name two chemical compounds in cell walls and give an example of an organism that synthesizes each.
2. Describe the process of cell wall formation in plant cells.
3. Describe and diagram the structure of a plasma membrane. What are the two principal types of molecules in plasma membranes? What are the four principal functions of plasma membranes?
4. What are the three types of proteins commonly found in plasma membranes, and what is the function of each?
5. Define diffusion and osmosis.
6. What types of molecules diffuse through plasma membranes, and through which parts of the membrane do they pass?
7. Define hypotonic, hypertonic, and isotonic. What would be the fate of an animal cell immersed in each of the three types of solution?
8. Describe the following types of transport processes: simple diffusion, facilitated diffusion, active transport, pinocytosis, receptor-mediated endocytosis, phagocytosis, and exocytosis.
9. Name four types of cell-to-cell junctions and the function of each. Which allow communication between the interiors of adjacent cells?

DISCUSSION QUESTIONS

1. Different cells have somewhat different plasma membranes. The plasma membrane of a *Paramecium,* for example, is only about 1% as water permeable as the plasma membrane of a human red blood cell. Referring back to our discussion of the effects of osmosis on red blood cells and the role of contractile vacuoles in *Paramecium,* what do you think is the function of the low water permeability of *Paramecium?* What molecular differences do you think might account for this low water permeability?
2. A preview question for Chapter 34: The integrity of the plasma membrane is essential for cellular survival. Could this fact be utilized by the immune system to destroy foreign cells that have invaded the body? How might cells of the immune system disrupt membranes of foreign cells (two hints: virtually all cells can secrete proteins, and some proteins form pores in membranes)?
3. A preview question for Chapter 27: Plant roots take up minerals (inorganic ions such as potassium) that are dissolved in the water of the soil. The concentration of such ions is usually much lower in the soil water than in the cytoplasm of root cells. Design the plasma membrane of a hypothetical mineral-absorbing cell, with special reference to mineral-permeable channel proteins and mineral-transporting active transport proteins. Justify your choice of channels and active transport proteins.
4. Red blood cells will swell up and burst when placed in a hypotonic solution like pure water. Why don't we swell up and burst when we go for a swim in water that is hypotonic to our cells and body fluids?

SUGGESTED READINGS

Bretscher, M. S. "The Molecules of the Cell Membrane." *Scientific American,* October 1985. Beautifully illustrated, this article explores the structure and function of cell membranes, with special attention to cell junctions.

Dautry-Varsat, A., and Lodish, H. "How Receptors Bring Proteins and Particles Into Cells." *Scientific American,* May 1984. Receptor-mediated endocytosis is an important pathway both for cell nutrition and for cell-to-cell signalling.

Luria, S. E. "Colicins and the Energetics of Cell Membranes." *Scientific American,* December 1975. Active transport is studied by observing the effects of molecules that inhibit various phases of the process.

Sharon, N. "Carbohydrates." *Scientific American,* May 1980. Discusses, along with the functions of carbohydrates in cellular activity, membrane carbohydrates that are involved in cell recognition.

7

Photosynthesis: Tapping Solar Energy

With few exceptions, the flow of energy through life on today's Earth begins with the sun (Fig. 7-1). This was not always so. About 4½ billion years ago, the Earth formed as chunks of matter collided and fused, transforming their energy of movement into heat. Storms and volcanic eruptions released still more energy on the newly formed planet, but no organisms existed to harness the enormous energy fluxes, nor is it likely that any could have withstood the violence of that time. Energy-rich organic molecules formed, their synthesis driven by heat and lightning (see Chapter 19). As the Earth cooled and calmed, living cells arose, feeding on the soup provided by the earlier chemical cauldron. However, the cells gradually consumed the organic molecules, and the soup thinned. Sources of organic energy became scarce.

All the while, another source of energy flowed to the Earth: the light of the sun. Through chance changes in their molecules, some fortunate cells acquired the ability to use the energy of sunlight to synthesize organic molecules, such as glucose, that are rich in chemical energy, from simple inorganic molecules of carbon dioxide and water, that contain little chemical energy. Photosynthesis had evolved (Fig. 7-1, left). These cells prospered, and the seas became filled with their progeny.

Evolution continued, and several types of photosynthesis evolved. The most common type released oxygen as a by-product. Gradually, so much oxygen was released that it began to accumulate in the atmosphere. Free oxygen is an extremely reactive chemical, and it would have been dangerous to early life forms, breaking down their hard-won organic molecules. But this very reactivity also provided an opportunity. When cells metabolize glucose without oxygen, a process called glycolysis, the products are not much lower in energy than the reactants, and therefore not much energy is released (Fig. 7-1, upper right). This is true even for photosynthetic organisms: without oxygen, only a small fraction of the light energy captured by photosynthesis and stored in the chemical energy of glucose molecules can be extracted again to drive essential cellular reactions.

When glucose is broken down in the presence of oxygen, however, the products are extremely low energy molecules of carbon dioxide and water, and therefore much energy is released (Fig. 7-1, lower right). If a cell could control these reactions, it could obtain vast new supplies of useful energy. Eventually, cells evolved the enzymes that enabled them to do just that. The advantages of this new process, called cellular respiration, were enormous. Cellular respiration can extract about 18 or 19 times more energy from each food molecule than can glycolysis (the

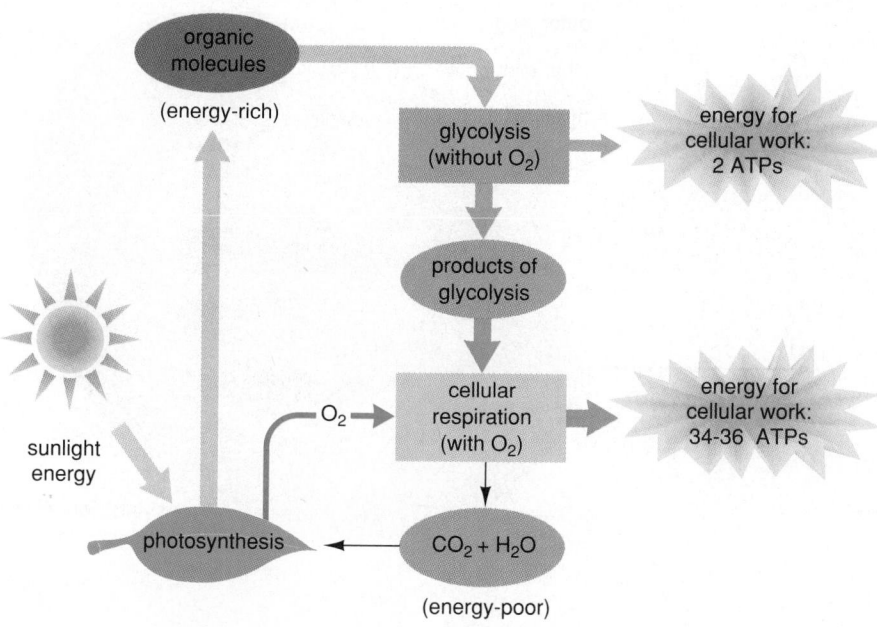

Figure 7-1 Carbon cycles through the environment in the mirror-image reactions of photosynthesis and cellular respiration. Chloroplasts in green plants use the energy of sunlight to synthesize high-energy carbon compounds such as glucose from low-energy molecules of water and carbon dioxide. Plants themselves, and other organisms that eat plants or one another, extract energy from these organic molecules by glycolysis and cellular respiration, yielding water and carbon dioxide once again. This energy in turn drives all the reactions of life. The orange arrows trace the flow of energy through the carbon cycle.

◀ Almost all the energy available to life on Earth comes from the sun. Photosynthesis by plants traps a small fraction of the sunlight energy striking the Earth, and nearly all other life forms obtain their energy, directly or indirectly, from plants.

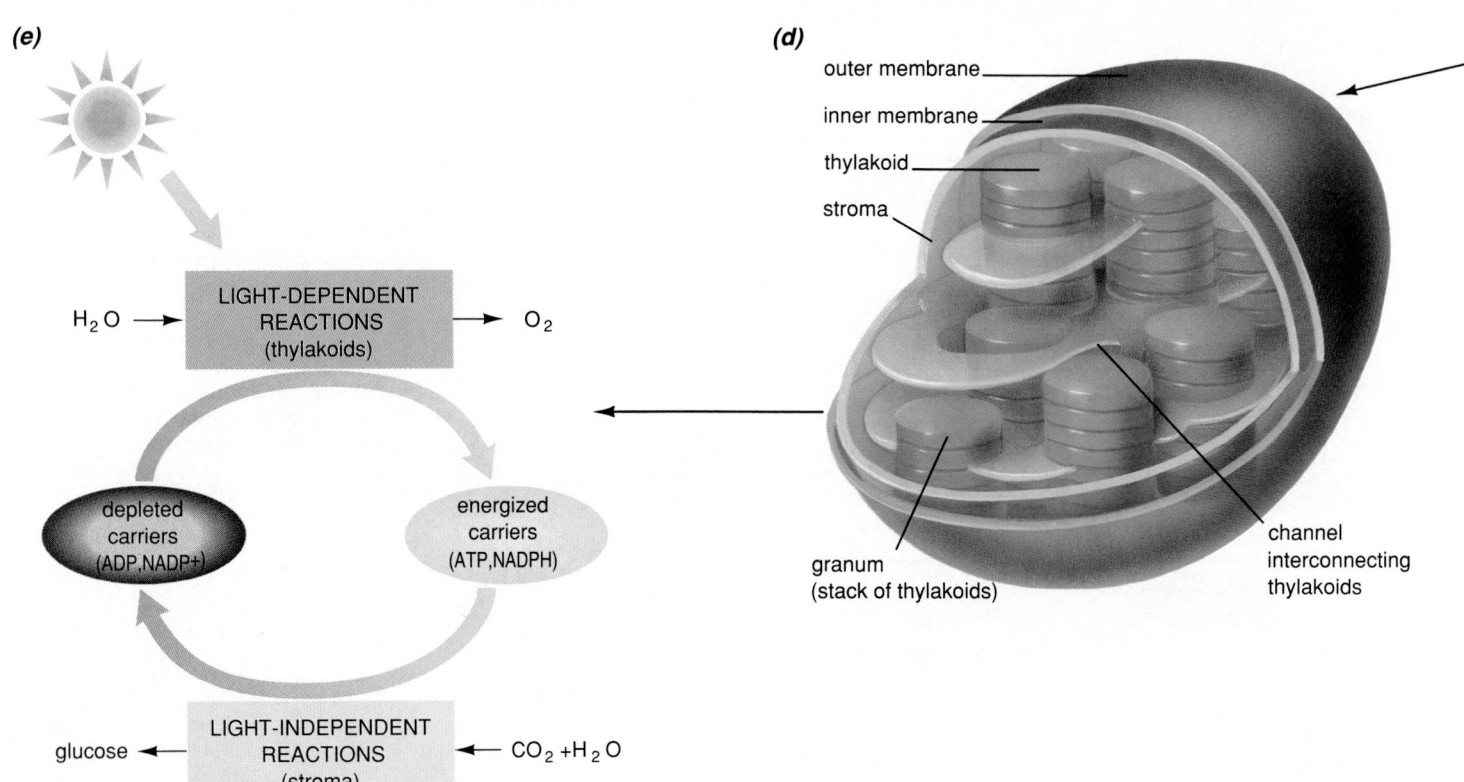

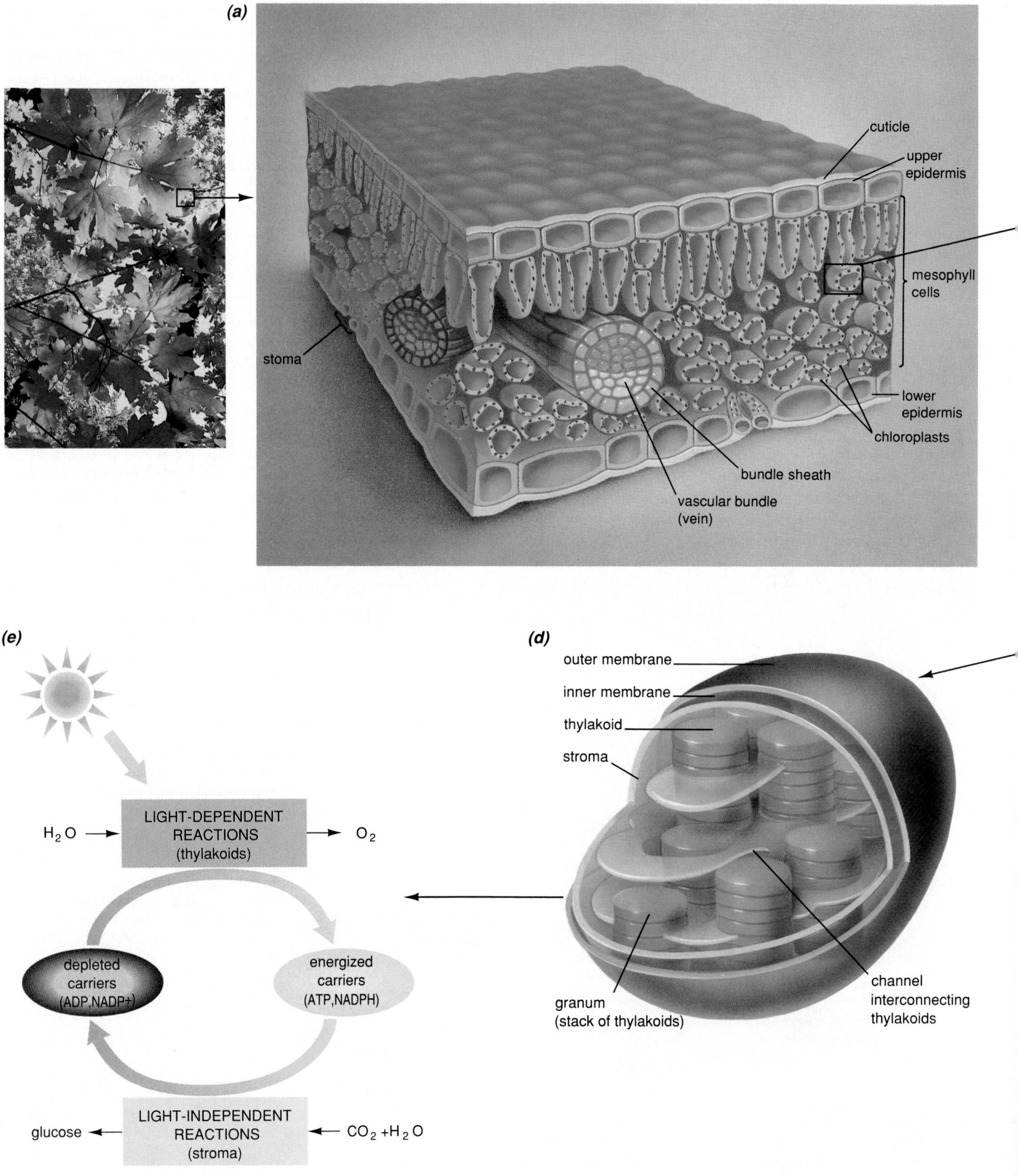

(a)

cuticle

upper epidermis

mesophyll cells

stoma

lower epidermis

chloroplasts

bundle sheath

vascular bundle (vein)

(e)

$H_2O \rightarrow$ **LIGHT-DEPENDENT REACTIONS** (thylakoids) $\rightarrow O_2$

depleted carriers (ADP, NADP+)

energized carriers (ATP, NADPH)

glucose $\leftarrow$ **LIGHT-INDEPENDENT REACTIONS** (stroma) $\leftarrow CO_2 + H_2O$

(d)

outer membrane

inner membrane

thylakoid

stroma

granum (stack of thylakoids)

channel interconnecting thylakoids

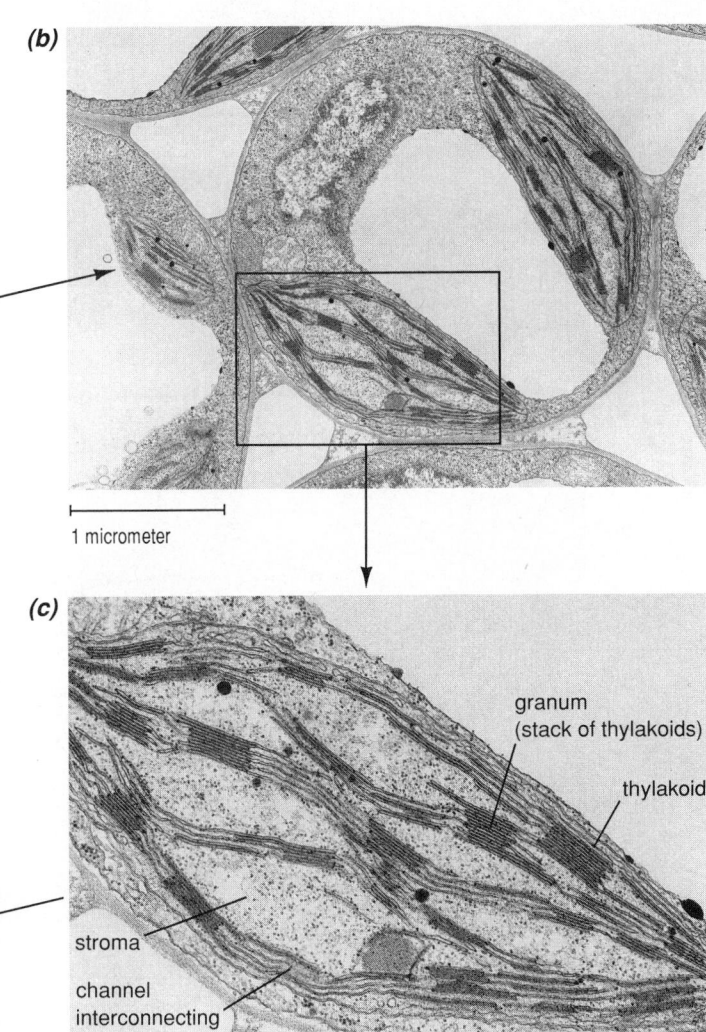

(b)

1 micrometer

(c)

granum
(stack of thylakoids)

thylakoid

stroma

channel
interconnecting
thylakoids

Figure 7-2 (a) In land plants, photosynthesis usually occurs in the leaves. **(b)** Cells in the middle of the leaf, called *mesophyll cells,* contain chloroplasts, the organelles of photosynthesis **(c, d). (e)** Within the chloroplasts, sunlight energy is absorbed by pigments in the membranes of the thylakoids, and this energy is used to drive the synthesis of the energy carrier ATP and electron carrier NADPH (yellow). These are the light-dependent reactions. In the light-independent reactions, enzymes in the stroma of the chloroplast then use ATP and NADPH to synthesize carbohydrates from CO_2 and H_2O. In the process, ADP and $NADP^+$ (gray) are regenerated and pass back to the light-dependent reactions to be recharged.

exact amount differs among cells). Cells that could respire would grow faster and reproduce more rapidly than cells that relied on glycolysis alone.

The complementary reactions of photosynthesis and cellular respiration are perhaps 2 billion years old, and together they drive the flow of energy and the cycling of carbon through individual organisms and ecosystems (see Fig. 7-1). This chapter investigates photosynthesis, the process by which almost all useful energy enters the biosphere. Chapter 8 examines glycolysis and cellular respiration, in which the energy captured by photosynthesis is conveyed to adenosine triphosphate (ATP), the principal energy carrier molecule in living cells.

The molecular events of photosynthesis, glycolysis, and cellular respiration are very complex. Therefore, we will divide our discussion into two parts, both in this chapter and in Chapter 8. In the main body of the text and its accompanying illustrations, we will describe each of these processes in terms that minimize the molecular and biochemical details. Special boxes entitled "A Closer Look . . ." will portray the molecular mechanisms in more depth.

Photosynthesis: An Overview

Photosynthesis **uses the energy of sunlight to synthesize energy-rich products—glucose and oxygen—from energy-poor reactants—carbon dioxide and water.** Therefore, photosynthesis converts the electromagnetic energy of sunlight into chemical energy stored in the bonds of glucose and oxygen. The overall chemical reaction for photosynthesis is:

$$6\ CO_2 + 6\ H_2O + \text{light energy} \rightarrow C_6H_{12}O_6 + 6\ O_2$$

In plants, photosynthesis occurs within the chloroplasts, most of which are contained in leaf cells. Let's begin, then, with a look at the structure of leaves and the chloroplasts they contain.

Leaves, Chloroplasts, and Photosynthesis

The leaves of most land plants are only a few layers of cells thick (Fig. 7-2a). The upper and lower surfaces of a leaf each consist of a layer of transparent cells, the epidermis. The outer surface of both epidermal layers is covered by a waxy, waterproof covering, the cuticle, that reduces the evaporation of water from the leaf. A leaf obtains CO_2 for photosynthesis from the air; adjustable pores in the epidermis, called **stomata,** open and close at appropriate times to admit CO_2. Inside the leaf are a few layers of cells collectively called **mesophyll** (which means simply "middle of the leaf"). The mesophyll cells contain the vast majority of a leaf's chloroplasts (Fig. 7-2b), and consequently the mesophyll cells are the principal sites of photosynthesis. Vascular bundles, or veins, supply water and minerals to the mesophyll cells and carry the sugars produced there to other parts of the plant.

Chloroplasts are organelles that consist of a double outer membrane enclosing a semi-fluid medium, the **stroma** (Fig. 7-2c and d). Embedded in the stroma are disk-shaped, interconnected membranous sacs called **thylakoids**. In most chloroplasts, the thylakoids are piled atop one another in stacks called **grana** (singular, **granum**).

The seemingly simple chemical reaction of photosynthesis actually involves dozens of enzymes catalyzing dozens of individual reactions. Conceptually, however, photosynthesis can be thought of as a pair of reactions coupled together by energy carrier molecules (Fig. 7-2e). Each reaction occurs in a different site in the chloroplast.

1. **In the light-dependent reactions, chlorophyll and other molecules in the membranes of the thylakoids capture sunlight energy and convert some of it into the chemical energy of energy carrier molecules (ATP and NADPH).**
2. **In the light-independent reactions, soluble enzymes in the stroma use the chemical energy of the carrier molecules to drive the synthesis of glucose or other organic molecules.**

The Light-Dependent Reactions: Converting Light to Chemical Energy

The first step in photosynthesis, the light-dependent reactions, converts the energy of sunlight into the chemical energy of two different carrier molecules: the familiar energy carrier ATP and the electron carrier nicotinamide adenine dinucleotide phosphate (NADPH).

Light, Chloroplast Pigments, and Photosynthesis

The electromagnetic radiation emitted by the sun covers a wide spectrum, from short wavelength gamma rays, through ultraviolet, visible, and infrared light, to very long wavelength radio waves (Fig. 7-3a). As you may know, light and the other types of radiation are composed of individual packets of energy, called **photons**. The energy of a photon corre-

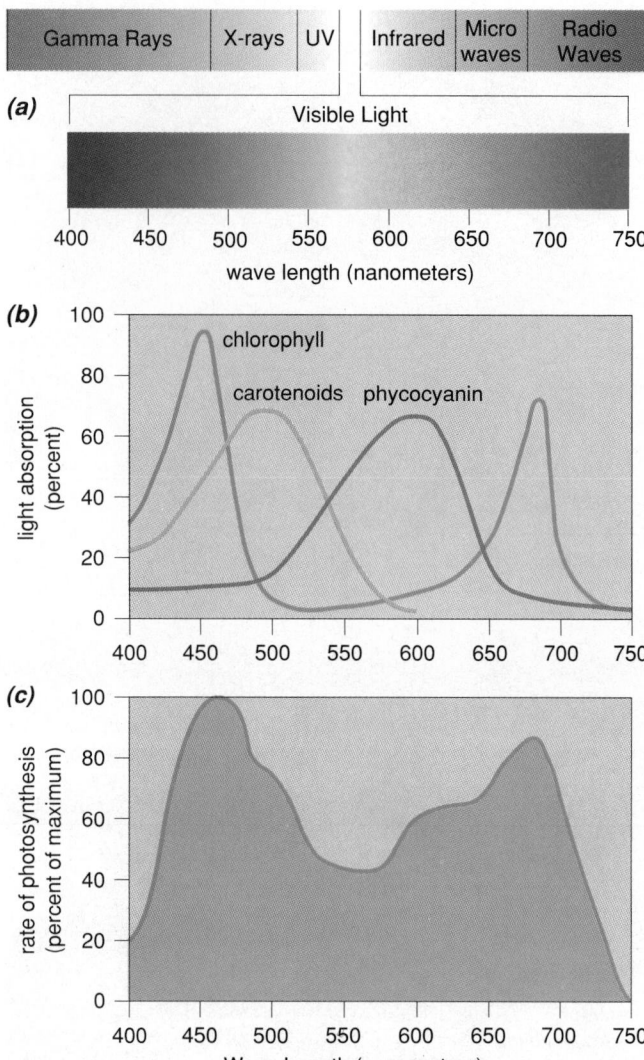

Figure 7-3 Light, chloroplast pigments, and photosynthesis. **(a)** Visible light, a small part of the electromagnetic spectrum, consists of wavelengths that correspond to the colors of the rainbow. **(b)** The first step in photosynthesis is the absorption of light by pigment molecules in the thylakoid membranes of chloroplasts. Different types of pigments selectively absorb certain colors; the height of each line represents the ability of each pigment to absorb light of each color. Chlorophyll (green line) strongly absorbs violet, blue, and red light and therefore looks green. The other pigments in chloroplasts absorb other colors of light. **(c)** Photosynthesis is driven to some extent by all colors of light, owing to the absorption of light by several thylakoid pigments.

sponds to its wavelength: short wavelength photons are very energetic, while longer wavelength photons have lower energies. Visible light, although only a small portion of the spectrum, contains the wavelengths with energies that are most useful to living organisms.

Three processes may occur when light strikes an object such as a leaf: the light may be absorbed, reflected (bounced back again) or transmitted (passed through). Light that is absorbed can drive biological processes, such as photosynthesis. Light that is reflected or transmitted gives an object its color.

Chloroplasts contain several types of molecules that absorb different wavelengths of light. **Chlorophyll,** the key light-capturing molecule in thylakoid membranes, absorbs blue and red light, but reflects green, and therefore appears green to human eyes (Fig. 7-3b). Thylakoids also contain other pigment molecules, called accessory pigments, that can capture light energy and transfer it to chlorophyll. **Carotenoids** absorb blue and green light, and appear yellow or red, while **phycocyanins** absorb green and yellow, and appear blue or purple. Because all wavelengths of light are absorbed to some degree either by chlorophyll, carotenoids, or phycocyanins, all wavelengths can drive photosynthesis to some extent (Fig. 7-3c).

The Light-Dependent Reactions

Photosystems: Site of the Light-Dependent Reactions

In the thylakoid membranes, chlorophyll, accessory pigment molecules, and electron carrier molecules form highly organized assemblies called **photosystems.** Each thylakoid contains thousands of copies of two different kinds of photosystems, named photosystem I and photosystem II (Fig. 7-4a). Each consists of two major parts, a **light-harvesting complex** and an **electron transport system.**

Overview of the Light-Dependent Reactions

The light-harvesting complex is composed of about 300 chlorophyll and accessory pigment molecules. These molecules absorb light and pass the energy to a specific chlorophyll molecule called the **reaction center.** By analogy with TV reception, the light-absorbing pigments are called antenna molecules, because they gather energy and transfer it to the energy-processing reaction center. The reaction center chlorophyll is located adjacent to the electron transport system, which is a series of electron carriers embedded in the thylakoid membrane. When the reaction center chlorophyll receives energy from the antenna molecules, one of its electrons absorbs the energy, leaves the chlorophyll, and jumps over to the electron transport system. This energetic electron passes along from one carrier to another. At some of the transfers, the electron releases energy that drives reactions resulting in the synthesis of ATP or

NADPH. This sequence of events is called **photophosphorylation,** because light energy (the "photo" in the name) is used to phosphorylate (add a phosphate group to) ADP, thus forming ATP. With this overall scheme in mind, let's look a little more closely at the actual sequence of events in the light-dependent reactions (Fig. 7-4b).

Photosystem II Generates ATP

For historical reasons, the photosystems are numbered "backwards," and the usual process of light-energy capture is most easily understood by starting with photosystem II. The light-dependent reactions begin when a photon of light is absorbed by an antenna molecule in photosystem II (Step 1 in Fig. 7-4b). The photon's energy passes from molecule to molecule until it reaches the reaction center, where it boosts an electron completely out of the chlorophyll molecule (Step 2). The first electron carrier of the adjacent electron transport system instantly accepts this energized electron (Step 3). The electron passes from carrier to carrier, releasing energy as it goes. Some of the energy is used to produce a hydrogen ion gradient within the thylakoid. This gradient drives the synthesis of ATP by a process known as chemiosmosis (Step 4). "A Closer Look at Chemiosmosis: The Mechanism of ATP Synthesis in Chloroplasts" describes chemiosmosis in more detail.

Photosystem I Generates NADPH

Meanwhile, light rays have also been striking the light-harvesting complex of photosystem I (Step 5), ejecting an electron from its reaction center chlorophyll (Step 6). This electron jumps to photosystem I's electron transport system (Step 7). Photosystem I's reaction center chlorophyll immediately obtains a replacement for its lost electron from the last electron carrier in photosystem II's electron transport system. Photosystem I's high-energy electron moves through its electron transport system to the "empty" electron carrier $NADP^+$. Each $NADP^+$ molecule picks up two energetic electrons and one hydrogen ion, forming NADPH (Step 8). $NADP^+$ and NADPH are both water-soluble molecules dissolved in the chloroplast stroma.

Splitting Water Maintains the Flow of Electrons Through the Photosystems

Overall, electrons flow from the reaction center of photosystem II, through the photosystem II electron transport system, to the reaction center of photosystem I, through the photosystem I electron transport system, and on to NADPH. To sustain this one-way flow of electrons, photosystem II's reaction center

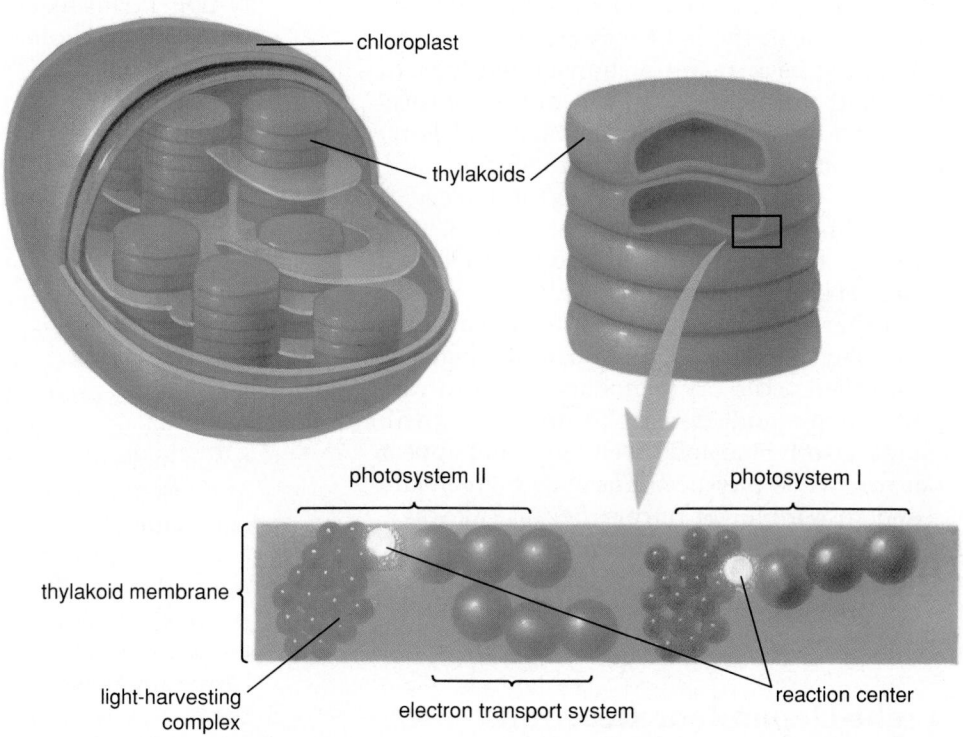

Figure 7-4 Thylakoid structure and the light-dependent reactions of photosynthesis.
(a) The thylakoid membranes contain many copies of photosystems I and II. Each photosystem consists of a light-harvesting complex of pigment molecules and an adjacent electron transport system.
(b) A summary of the light-dependent reactions. (1) Light is absorbed by the light-harvesting complex of photosystem II (light green), and the energy is passed to the reaction center chlorophyll. (2) This energy ejects electrons out of the reaction center. (3) The electrons pass to the adjacent electron transport system. (4) The transport system passes the energetic electrons along and some of their energy is used to

must be continuously supplied with new electrons to replace the ones it gives up. These replacement electrons come from water (Step 9). In a poorly understood series of reactions, photosystem II's reaction center chlorophyll attracts electrons from water molecules within the thylakoid compartment, causing the water molecules to split apart:

$$H_2O \rightarrow O + 2\ H^+ + 2\ e^-$$

For every two photons captured by photosystem II, two electrons are boosted out of the reaction center chlorophyll and are replaced by the two electrons obtained by splitting one water molecule. As water molecules are split, the liberated oxygen atoms combine to form a molecule of oxygen gas, O_2. The oxygen may be used directly by the plant in its own respiration or it may be given off to the atmosphere.

SUMMARY OF THE LIGHT-DEPENDENT REACTIONS:

The light-dependent reactions begin with (1) the absorption of light by the light-harvesting complex of photosystem II. (2) The light energizes electrons that are ejected from the reaction center of the complex. (3) These electrons are transferred to photosystem II's adjacent electron transport system. As the electrons pass through the transport system, they release energy. (4) Some of the energy is used to create a hydrogen ion gradient that drives ATP synthesis. (5) Meanwhile, light is absorbed by the light-harvesting complex of photosystem I. (6) The light energizes electrons that are ejected from the reaction center and (7) are picked up by photosystem I's electron transport system. (8) Some of this energy is captured as

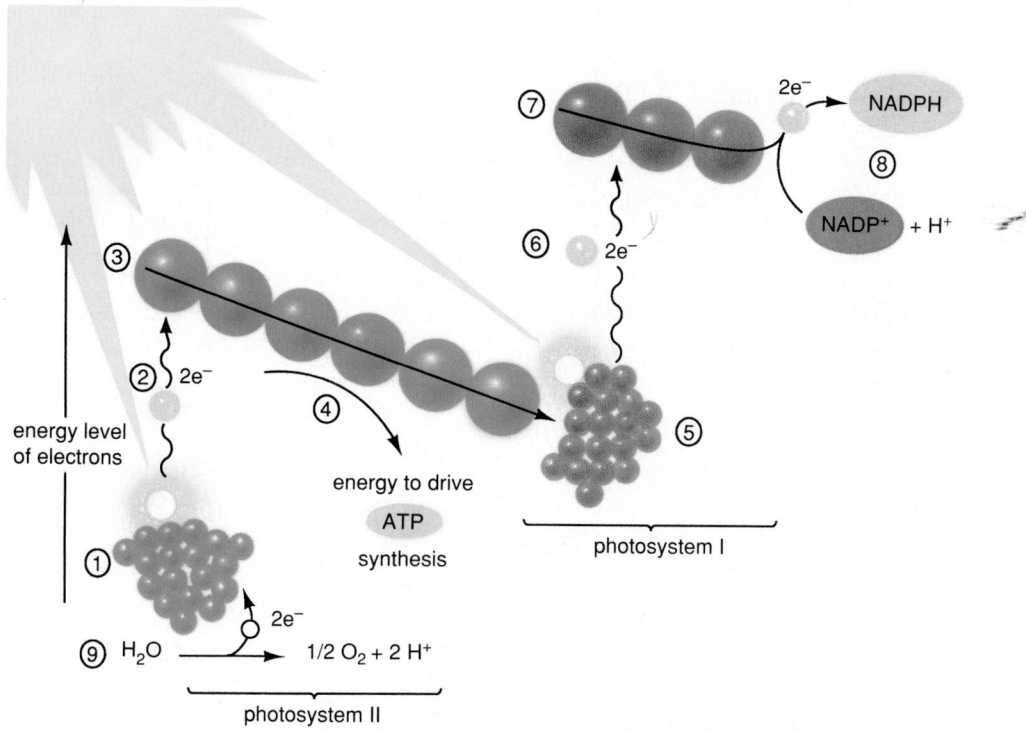

pump hydrogen ions into the thylakoid interior. The hydrogen ion gradient thus generated can drive ATP synthesis (see "A Closer Look at Chemiosmosis" for details). (5) Light strikes photosystem I (dark green), (6) causing it to emit electrons. (7) The electrons are captured by the photosystem I electron transport system. The electrons lost from the reaction center of photosystem I are replaced by those coming from the transport system of photosystem II. (8) The energetic electrons from photosystem I are captured in molecules of NADPH. (9) The electrons lost from the reaction center of photosystem II are replaced by electrons obtained from splitting water. This reaction also releases oxygen.

NADPH. (9) Finally, the "electron-deprived" chlorophyll of photosystem II attracts electrons from nearby water molecules. A water molecule splits apart, donating electrons to the photosystem II chlorophyll and generating oxygen as a by-product.

The Light-Independent Reactions: Securing Chemical Energy in Glucose Molecules

The ATP and NADPH synthesized during the light-dependent reactions are dissolved in the stroma. Here, they provide the energy for the synthesis of glucose from carbon dioxide and water. These reactions are termed the **light-independent reactions,** because they can occur independently of light if ATP and NADPH are available.

Carbon dioxide capture and glucose synthesis occur in a set of reactions known as the **Calvin-Benson cycle** (after its discoverers) or the **C_3 (three-carbon) cycle,** because some of the important molecules in the cycle have three carbon atoms in them (Fig. 7-5). The C_3 cycle requires (1) CO_2 (normally from the air); (2) a CO_2-capturing sugar, ribulose bisphosphate (RuBP); (3) enzymes to catalyze all the reactions; and (4) energy in the form of ATP and NADPH (usually from the light-dependent reactions).

The C_3 cycle can be most easily understood if we

CLOSER LOOK

At Chemiosmosis, the Mechanism of ATP Synthesis in Chloroplasts

In the electron transport system of photosystem II, energetic electrons are passed from carrier to carrier. Until about 30 years ago, it was thought that ATP synthesis was directly coupled to these electron transfers: where the exergonic steps were large enough, the energy given up by the electrons drove ATP synthesis. However, it turns out that ATP synthesis in chloroplasts is *not* a simple coupled reaction. The electron transfers do not directly drive ATP synthesis; rather, the energy released during the transfers is used to generate the concentration gradient of hydrogen ions across the thylakoid membrane. In a completely separate reaction, the energy stored in this gradient then powers ATP synthesis. Let's follow these reactions in some detail.

In the first step of the light-dependent reactions of photosynthesis, a photon strikes the light-harvesting complex of photosystem II, and the energy is absorbed by an antenna pigment molecule. The energy hops around from molecule to molecule within the complex until it reaches the reaction center chlorophyll. Here, an electron absorbs the energy, and is ejected completely out of the chlorophyll molecule. Within a billionth of a second, the electron is captured by the first electron carrier of the adjacent electron transport system.

The electrons pass from carrier to carrier, losing energy as they go. The exergonic reaction of electron movement is coupled to the endergonic reaction of the active transport of hydrogen ions across the thylakoid membrane from the stroma into the thylakoid compartment:

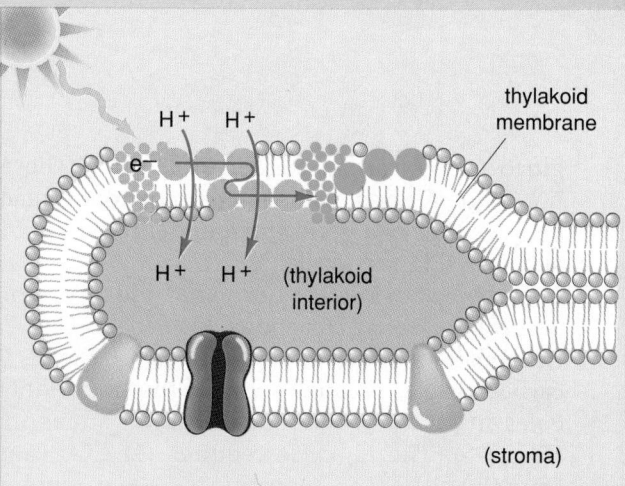

This raises the concentration of hydrogen ions (and therefore the positive charge) inside the thylakoid. The thylakoid membrane is impermeable to hydrogen ions, except at specific protein channels that are coupled to ATP-synthesizing enzymes. When hydrogen ions flow through these channels, down their gradients of charge and concentration, the energy released drives the synthesis of ATP:

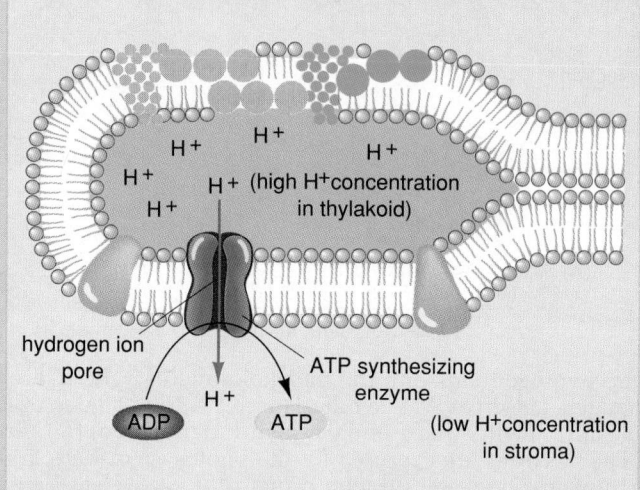

Exactly how this works isn't completely understood, but perhaps we can get a feeling for the process if we compare the hydrogen ion gradient across the thylakoid membrane to a dry-cell battery (the kind used in flashlights). In a battery, the negative pole (the flat end of the battery) tends to release electrons, and the positive pole (knobbed end) tends to absorb electrons and transfer them to chemical reactions going on inside the battery. The positive and negative poles, however, are insulated from each other. Electrons can only flow from negative to positive poles if wires connect the two. The flow of electrons can do work, such as lighting a bulb or running a motor. The chloroplast thylakoid operates in an analogous way. Hydrogen ions in the thylakoid interior can only move down their concentration and charge gradients out into the stroma through the ATP-synthesizing enzyme channels. This flow of hydrogen ions can do work, namely drive ATP synthesis. This mechanism of ATP synthesis was first proposed in 1961 by Peter Mitchell, who called it **chemiosmosis.**

Chemiosmosis has been shown to be the mechanism of ATP generation in chloroplasts, mitochondria (see Chapter 8), and bacteria. For his brilliant hypothesis, Mitchell was awarded the Nobel Prize in 1978.

mentally divide it into three parts: carbon fixation, synthesis of phosphoglyceraldehyde (PGAL), and regeneration of ribulose bisphosphate.

Carbon Fixation

The C_3 cycle begins (and ends) with a 5-carbon sugar, ribulose bisphosphate (RuBP). RuBP combines with CO_2 to form an extremely unstable 6-carbon compound. This compound spontaneously reacts with water to form two 3-carbon molecules of phosphoglyceric acid (PGA), which gives the C_3 cycle its name. Capturing CO_2 is called **carbon fixation,** since it "fixes" gaseous CO_2 into a relatively stable organic molecule (Step 1 in Fig. 7-5).

Synthesis of Phosphoglyceraldehyde

In a series of enzyme-catalyzed reactions, energy donated by ATP and NADPH is used to convert PGA to phosphoglyceraldehyde (PGAL; Step 2). PGAL is a versatile molecule and can take one of several paths. Two PGAL molecules (3 carbons each) may combine, becoming one molecule of glucose (6 carbons; Step 3). PGAL may also be used in the synthesis of lipids or amino acids.

Regeneration of Ribulose Bisphosphate

Alternatively, through complex series of reactions requiring ATP energy, 10 molecules of PGAL (10×3 carbons) can regenerate 6 molecules of RuBP (6×5 carbons; Step 4).

Overall, to fix 6 molecules of CO_2 as one molecule of glucose, 6 molecules of RuBP enter and are regenerated in each "turn" of the C_3 cycle (see Fig. 7-5).

SUMMARY OF THE LIGHT-INDEPENDENT REACTIONS

For the synthesis of one molecule of glucose: 6 molecules of CO_2 are captured by 6 molecules of RuBP. A series of reactions driven by energy from ATP and NADPH produces 12 molecules of PGAL. Two PGAL molecules join to become one molecule of glucose. Further ATP energy is used to regenerate the 6 RuBP molecules from 10 PGAL molecules.

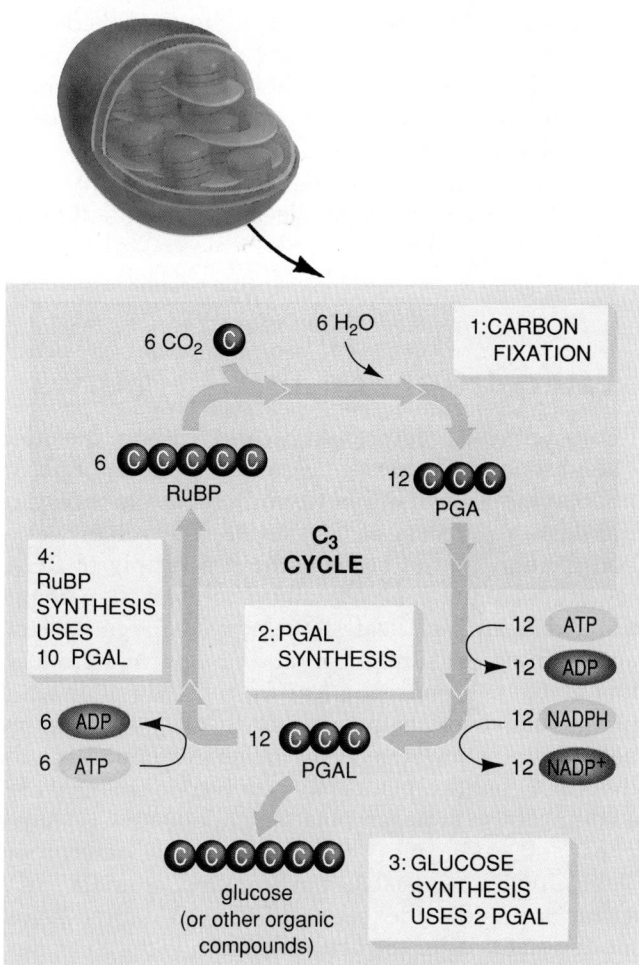

Figure 7-5 The C_3 cycle of carbon fixation. Carbon atoms are gray balls; depleted carrier molecules ADP and $NADP^+$ are gray ovals; and energized carriers ATP and NADPH are yellow ovals. (1) Six molecules of ribulose bisphosphate (RuBP) react with 6 molecules of CO_2 and 6 molecules of H_2O to form 12 molecules of phosphoglyceric acid (PGA). This is **carbon fixation,** the capture of carbon from CO_2 into organic molecules. (2) The energy of 12 ATPs and the electrons and hydrogens of 12 NADPHs are used to convert the 12 PGA molecules to 12 phosphoglyceraldehydes (PGALs). (3) Two PGAL molecules are further processed into glucose or other organic molecules such as glycerol, fatty acids, or the carbon skeleton of amino acids, depending on the needs of the plant. (4) Energy from 6 ATPs is used to rearrange 10 PGALs into 6 ribulose bisphosphates, completing one turn of the C_3 cycle.

The Relation Between the Light-Dependent and the Light-Independent Reactions

As Figure 7-2e illustrates, the light-dependent and light-independent reactions are closely coordinated.

The light-dependent reactions in the thylakoids use light energy to "charge up" the energy carrier molecules ATP and NADPH. These energized carriers move to the stroma, where their energy is used to drive glucose synthesis by means of the light-independent reactions. The depleted carriers, $NADP^+$ and ADP, then return to the light-dependent reactions for recharging.

Water, CO_2, and the C_4 Pathway

Photosynthesis requires light and carbon dioxide. The absence of either of these may limit the rate of photosynthesis. First, photosynthesis cannot occur in the dark, regardless of CO_2 levels. Conversely, carbon fixation cannot occur without a supply of CO_2, regardless of the light intensity. Therefore, you might think that an ideal leaf should have a large surface area to intercept sunlight and be very porous to allow lots of CO_2 to enter the leaf from the air. For land plants, however, being porous to CO_2 also means being porous to water vapor, which evaporates out of the leaf. Broadleaf plants have evolved leaves that are a compromise between obtaining adequate CO_2 supplies and reducing water loss: large, waterproof leaves, with adjustable pores called **stomata** that admit carbon dioxide (see Fig. 7-2; see Chapter 26 for more details of leaf structure). When water supplies are adequate, the stomata open, admitting CO_2. If the plant is in danger of drying out, the stomata close, which reduces evaporation, but at the cost of reducing CO_2 intake as well.

Photorespiration

What happens to carbon fixation when the stomata close? As it happens, the enzyme that catalyzes the reaction of RuBP with CO_2 is not very selective: it can cause *either* CO_2 *or* O_2 to combine with RuBP (Fig. 7-6). The O_2 + RuBP reaction is the first step in a series of reactions, collectively called photorespiration, that generate CO_2 from carbon atoms originally present in the RuBP molecules. Thus photorespiration not only prevents carbon fixation by combining O_2 instead of CO_2 with RuBP; it also squanders some of the plant's hard-won carbon. (Photorespiration, by the way, received its name because it uses up O_2 and produces CO_2. Unlike cellular respiration, it does not produce any useful cellular energy, such as ATP.)

Whether or not this matters very much depends on the relative amounts of carbon fixation versus photorespiration. The atmosphere contains much more O_2 (21%) than CO_2 (0.03%). In addition, the light reac-

tions of photosynthesis release O_2 within the leaf. Therefore, some photorespiration occurs all the time, even under the best of conditions. During hot and/or dry weather, moreover, the stomata seldom open. The CO_2 that was initially present within the leaf is rapidly depleted by photosynthesis, new CO_2 from the air can't get in, and the O_2 generated by photosynthesis can't get out. With CO_2 levels low and O_2 levels very high, photorespiration dominates (Fig. 7-7a). Plants, especially seedlings, may die during hot, dry weather because respiration (both cellular respiration and photorespiration) exceeds photosynthesis, and the plants consume all their carbohydrate supplies by respiration.

Two-Stage Carbon Fixation in C_4 Plants

Some plants have evolved a way to reduce photorespiration and boost photosynthesis during dry weather. In the leaves of "regular" (C_3) plants, almost all the chloroplasts reside in the mesophyll cells, while the bundle-sheath cells surrounding the veins lack chloroplasts. In plants adapted to hot, dry conditions, such as corn and crabgrass, both the mesophyll cells and the bundle-sheath cells contain chloroplasts (Fig. 7-7b). These so-called C_4 plants have a two-stage carbon-fixation pathway.

First Stage: Initial CO_2 Fixation

In C_4 plants, the mesophyll cells contain a 3-carbon molecule called phosphoenolpyruvate (PEP), instead

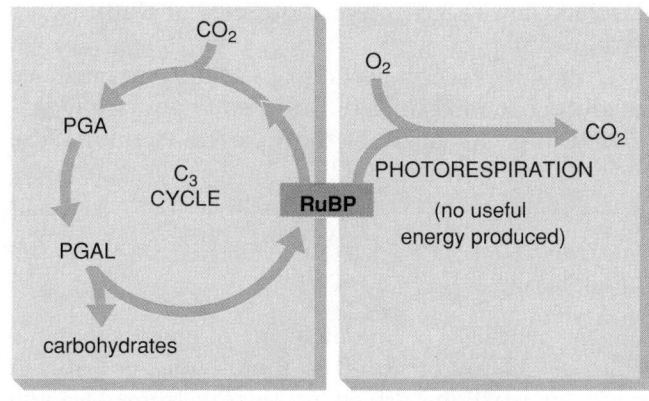

Figure 7-6 Carbon fixation via the C_3 cycle and photorespiration are alternative pathways for ribulose bisphosphate (RuBP). Some photorespiration always occurs. When the stomata of a leaf close during hot, dry weather, CO_2 levels drop while O_2 levels remain high. Under these conditions, photorespiration predominates and the plant loses carbon.

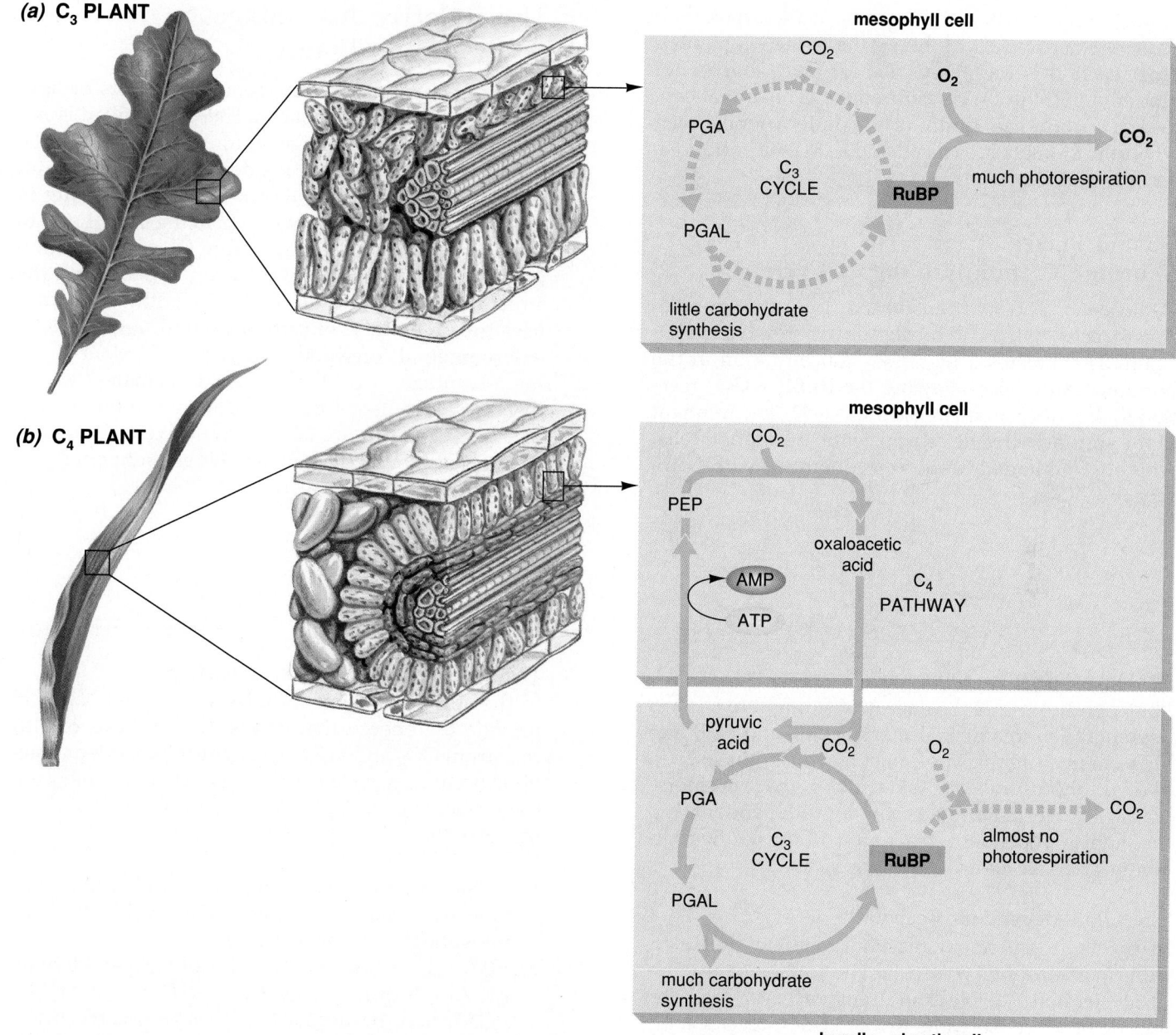

Figure 7-7 Under hot, dry conditions, plants have their stomata closed most of the time, restricting CO_2 entry and O_2 emission. In these circumstances, C₄ plants fix more carbon, with less photorespiration, than C₃ plants do.
(a) In C₃ plants, only the mesophyll cells carry out photosynthesis. All carbon fixation is by the C₃ pathway. With low CO_2 and high O_2 levels, photorespiration dominates in C₃ plants, because the enzyme that should catalyze the RuBP + CO_2 reaction catalyzes the RuBP + O_2 reaction instead.
(b) In C₄ plants, both the mesophyll cells and bundle-sheath cells contain chloroplasts and participate in photosynthesis. The initial carbon fixation step in the mesophyll cells is a reaction between phosphoenolpyruvic acid (PEP) and CO_2, with which O_2 does not compete. A 4-carbon molecule of oxaloacetic acid is produced, giving the C₄ pathway its name. The C₄ shuttle then releases CO_2 in the bundle-sheath cells, thus maintaining a high CO_2 concentration in their chloroplasts. Higher CO_2 levels allow efficient carbon fixation in the C₃ pathway of the bundle-sheath cells with little photorespiration.

of RuBP. CO_2 reacts with PEP to form a 4-carbon molecule of oxaloacetic acid (hence the name, C_4 plants). This reaction is highly specific for CO_2, and is not hindered by high O_2 concentrations. This initial capture of carbon can continue for quite a while even when the stomata are closed and CO_2 concentrations become very low.

Second Stage:
Shuttling Carbon into the C_3 Cycle

Oxaloacetic acid is transported into the bundle-sheath cells, where it breaks down, releasing CO_2 again. This creates a high CO_2 concentration in the bundle-sheath cells, allowing the RuBP + CO_2 reaction to fix carbon in the regular C_3 cycle. The remnant of the shuttle molecule returns to the mesophyll cells, where ATP energy is used to regenerate the PEP capture molecule.

The Relative Advantages of C_3 and C_4 Plants

The C_4 pathway is advantageous when lots of light and little water are available. The C_4 pathway is not *always* better, because regenerating PEP uses up ATP produced by the light-dependent reactions. If water is plentiful (so the stomata can stay open, letting in lots of CO_2) or light is weak (making ATP hard to come by), then the straightforward C_3 carbon fixation pathway is more efficient. However, a plant with the structures and enzymes of the C_4 pathway always uses the C_4 method of carbon fixation, regardless of environmental conditions. Therefore, C_3 plants have the advantage in cool, wet, cloudy climates, where their exclusive use of the C_3 pathway is more energy efficient. C_4 plants are favored in deserts and in midsummer in temperate climates, where light energy is plentiful but water is scarce.

SUMMARY OF KEY CONCEPTS

Photosynthesis: An Overview
Photosynthesis uses the energy of sunlight to convert low-energy inorganic molecules of CO_2 and H_2O into high-energy organic molecules such as glucose. In plants, photosynthesis takes place in the chloroplasts, in two major steps: the light-dependent and the light-independent reactions. Photosynthesis is summarized in Figure 7-8.

The Light-Dependent Reactions
In the light-dependent reactions, light excites electrons in chlorophyll molecules and transfers the energetic electrons to electron transport systems. The energy of these electrons drives three processes.

1. Some of the energy from the electrons is used to pump hydrogen ions into the thylakoid sacs. The hydrogen ion concentration is therefore higher inside the thylakoids than in the stroma outside. Hydrogen ions diffuse down this concentration gradient through ATP-synthesizing enzymes in the thylakoid membranes, providing the energy to drive ATP synthesis.

2. Some of the energy, in the form of energetic electrons, is added to electron carrier molecules of $NADP^+$ to make the highly energetic carrier NADPH.

3. Some of the energy is used to split water, generating electrons, hydrogen ions, and oxygen.

The Light-Independent Reactions
In the stroma of the chloroplasts, ATP and NADPH provide the energy that drives the synthesis of glucose from CO_2 and H_2O. The light-independent reactions occur in a cycle of chemical reactions called the Calvin-Benson or C_3 cycle. The C_3 cycle has three major parts:

1. In the carbon fixation step, CO_2 and water combine with ribulose bisphosphate (RuBP) to form phosphoglyceric acid (PGA).
2. PGA is converted to phosphoglyceraldehyde (PGAL), using energy from ATP and NADPH. PGAL may be used to synthesize organic molecules such as glucose.
3. Alternatively, 10 molecules of PGAL may be used to regenerate 6 molecules of RuBP, again using ATP energy.

The Relation Between the Light-Dependent and the Light-Independent Reactions
The light-dependent reactions produce the energy carrier ATP and the electron carrier NADPH. The energy from these carriers is used in the synthesis of organic molecules during the light-independent reactions. The depleted carriers, ADP and $NADP^+$, return to the light-dependent reactions for recharging.

Water, CO_2, and the C_4 Pathway
The enzyme that catalyzes the reaction between

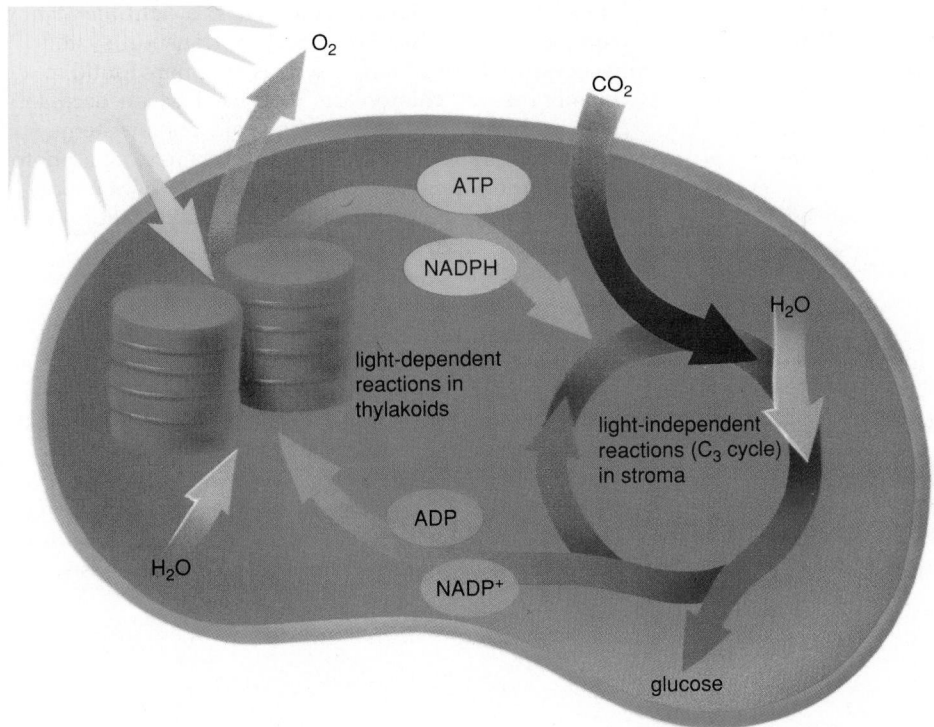

O_2

CO_2

ATP

NADPH

light-dependent
reactions in
thylakoids

H_2O

light-independent
reactions (C_3 cycle)
in stroma

ADP

H_2O

$NADP^+$

glucose

Figure 7-8 A summary diagram for photosynthesis. The light-dependent reactions in the thylakoids convert the energy of sunlight into the chemical energy of ATP and NADPH. Part of the sunlight energy is also used to split H_2O, forming O_2. In the stroma, the light-independent reactions (C_3 cycle) use the energy of ATP and NADPH to convert CO_2 and H_2O to glucose. The depleted carriers, ADP and $NADP^+$, return to the thylakoids to be recharged by the light-dependent reactions.

RuBP and CO_2 is not very selective. It may also catalyze a reaction, called photorespiration, between RuBP and O_2. Photorespiration uses up ATP without producing any sugar. If CO_2 concentrations drop too low, or O_2 concentrations rise too high, photorespiration may exceed carbon fixation. Some plants have evolved an additional step for carbon fixation that minimizes photorespiration. In the mesophyll cells of these C_4 plants, CO_2 combines with phosphoenolpyruvic acid to form the 4-carbon molecule oxaloacetic acid. Oxaloacetic acid is transported into adjacent bundle-sheath cells, where it releases CO_2, thereby maintaining a high CO_2 concentration in the bundle-sheath cells. This CO_2 is then fixed in the C_3 cycle.

GLOSSARY

C_3 **cycle:** the cyclic series of reactions whereby carbon dioxide is fixed into carbohydrates during the light-independent reactions of photosynthesis. Also called Calvin-Benson cycle.

C_4 **pathway:** the series of reactions in certain plants that fixes carbon dioxide into organic acids for later use in the C_3 cycle of photosynthesis.

Calvin-Benson cycle: see C_3 cycle.

carbon fixation: the initial steps in the C_3 cycle, in which carbon dioxide reacts with ribulose bisphosphate to form a stable organic molecule.

carotenoid (ka-rot'-en-oyd): red, orange, or yellow pigments found in chloroplasts that serve as accessory light-gathering molecules in thylakoid photosystems.

chemiosmosis (kē-mē-os-mō'-sis): a process of ATP generation in chloroplasts and mitochondria. The movement of electrons down an electron transport system is used to pump hydrogen ions across a membrane, thereby building up a concentration gradient of hydrogen ions across the membrane. The hydrogen ions diffuse back across the membrane through the pores of ATP-synthesizing enzymes. The energy of their movement down their concentration gradient drives ATP synthesis.

chlorophyll (klōr'-ō-fil): the primary light-trapping molecule in photosynthesis.

electron transport system: molecules found in the thylakoid membranes of chloroplasts and the inner membrane of mitochondria, which extract energy from electrons and generate ATP.

granum (pl. grana): in chloroplasts, a stack of thylakoids.

light-dependent reactions: the first stage of photosynthesis, in which the energy of light is captured as ATP and NADPH; occurs in thylakoids of chloroplasts.

light-harvesting complex: in photosystems, the assembly of pigment molecules (chlorophyll and often carotenoids or phycocyanins) that absorb light energy and transfer the energy to electrons.

light-independent reactions: the second stage of photosynthesis, in which the energy obtained by the light-dependent reactions is used to fix carbon dioxide into carbohydrates; occurs in the stroma of chloroplasts.

photon (fō'-ton): the smallest unit of light.

photophosphorylation (fō-tō-fos-for-i-lā'-shun): ATP synthesis in the light-dependent reactions of photosynthesis; light energy ("photo") is used to add a phosphate group to ("phosphorylate") ADP to yield ATP.

photosynthesis: the complete series of chemical reactions in which the energy of light is used to synthesize high-energy organic molecules, usually carbohydrates, from low-energy inorganic molecules, usually carbon dioxide and water.

photosystem: in thylakoid membranes, a light-harvesting complex and its associated electron transport system.

phycocyanin (fī-kō-sī'-a-nin): a bluish pigment found in the membranes of chloroplasts and used as an accessory light-gathering molecule in thylakoid photosystems.

reaction center: in the light-harvesting complex of a photosystem, the chlorophyll molecule to which light energy is transferred by the antenna pigments. The captured energy ejects an electron from the reaction center chlorophyll, and the electron is transferred to the electron transport system.

stoma (stō'ma; pl. stomata): adjustable openings in plant leaves. Most gas exchange between leaves and the air occurs through the stomata.

stroma (strō'-ma): the semi-fluid medium of chloroplasts, in which the grana are embedded.

thylakoid (thī'-la-koyd): a disk-shaped, membranous sac found in chloroplasts, the membranes of which contain the photosystems and ATP-synthesizing enzymes used in the light-dependent reactions of photosynthesis.

STUDY QUESTIONS

1. Define photosynthesis. What is the overall chemical equation for photosynthesis? Is photosynthesis an exergonic or endergonic process?
2. Diagram and describe the structure of the chloroplast.
3. Briefly describe the light-dependent and light-independent reactions of photosynthesis. In which part of the chloroplast does each occur?
4. What is the difference between carbon fixation in C_3 and C_4 plants? Under what conditions does each method of carbon fixation work most effectively?
5. Describe the process of chemiosmosis in chloroplasts, tracing the flow of energy from sunlight to ATP.

DISCUSSION QUESTIONS

1. If the light-independent reactions were to stop (for example, if they were poisoned by an herbicide), would the light-dependent reactions continue indefinitely? Why or why not? (Hint: What molecules are required for ATP and NADPH synthesis? Where do they come from?)
2. Many lawns and golf courses are planted with bluegrass, a C_3 plant. In the spring, the bluegrass grows luxuriously. In the summer, a weed called crabgrass, a C_4 plant, often appears and spreads rapidly. Explain this sequence of events, given the normal weather conditions of spring and summer, and the characteristics of C_3 versus C_4 plants.
3. Scientists can determine the rate of photosynthesis of aquatic plants by placing them in a test tube of water and measuring the rate at which oxygen gas is released as gas bubbles. If bicarbonate (which contains CO_2) is added to the water, production of oxygen gas increases. Discuss why this is true, considering that oxygen is produced from the light-independent reactions while CO_2 is used in the light-independent reactions of photosynthesis.
4. Suppose an experiment is performed in which "Plant I" is supplied with normal carbon dioxide but with water containing radioactive oxygen atoms. "Plant II," however, is supplied with normal water but with carbon dioxide containing radioactive oxygen atoms. Each plant is allowed to perform photosynthesis and the oxygen gas and sugars produced are tested for radioactivity. Which plant would you expect to produce radioactive sugars and which plant would you expect to produce radioactive oxygen gas? Discuss your answers.

SUGGESTED READINGS

Govindjee and Coleman, W. J. "How Plants Make Oxygen." *Scientific American*, February 1990. The generation of oxygen during photosynthesis is just beginning to be understood.

Govindjee and Govindjee, R. "The Primary Events of Photosynthesis." *Scientific American*, December 1974. A good introduction to photosynthesis.

Hinkle, P. C., and McCarthy, R. E. "How Cells Make ATP." *Scientific American*, March 1978. A good explanation of chemiosmosis, which is a difficult concept for many students.

Miller, K. R. "The Photosynthetic Membrane." *Scientific American*, October 1979. An exploration of the molecules and chemical events that occur in the thylakoid membrane.

Youvan, D. C., and Marrs, B. L. "Molecular Mechanisms of Photosynthesis." *Scientific American*, June 1987. Recent information about how photosynthesis works at the molecular level.

8

Glycolysis and Cellular Respiration: Harvesting Energy from Food

A ruby-throated hummingbird feeding at a flower. The reflection at the end of this chapter examines the energy demands that migrating ruby-throats must meet.

When a ruby-throated hummingbird sips nectar from a flower (facing page), it consumes the products of photosynthesis. As you learned in Chapter 7, photosynthesis converts the energy of sunlight into the chemical energy of organic molecules such as sugars and fats. However, the chemical energy contained in a molecule of sugar or fat is not in a form that the hummingbird's cells can readily use. The hummingbird's wing muscles use adenosine triphosphate (ATP), not sugar or fat, as their immediate energy source when they contract during flight. The plant, too, uses ATP when it synthesizes its flower pigments, grows a root, or sets seed. In this chapter, we explore the metabolic processes by which living organisms convert the energy of organic molecules into the usable energy of ATP.

Most cells can metabolize a variety of organic molecules to produce ATP. In this chapter, however, we focus on the metabolism of a single molecule: glucose. There are three reasons for this focus. First, virtually all organisms, and every cell of multicellular organisms, metabolize glucose for energy at least part of the time. For some, such as the nerve cells in your brain, glucose is the predominant or sole source of energy. Second, glucose metabolism is less complex than the metabolism of most other organic molecules. Finally, when other organic molecules are used as energy sources, they are usually converted into glucose or other compounds that enter into the pathways of glucose metabolism (see "A Closer Look at Metabolic Transformations: Why You Can Get Fat by Eating Sugar," later in this chapter).

We will cover glucose metabolism on two levels. The main text will concentrate on essential concepts and minimize biochemical details. Boxed sections titled "A Closer Look at . . ." will provide more detailed explanations of the chemical pathways and processes involved.

Glucose Metabolism: An Overview

The chemical equations for glucose synthesis by photosynthesis and the complete metabolism of glucose back to CO_2 and H_2O are wonderfully symmetrical:

Photosynthesis:

$$6\ CO_2 + 6\ H_2O + \text{sunlight energy} \rightarrow C_6H_{12}O_6 + 6\ O_2$$

Complete glucose metabolism:

$$C_6H_{12}O_6 + 6\ O_2 \rightarrow 6\ CO_2 + 6\ H_2O + \text{energy}$$

This symmetry might lead you to think that a cell can convert all the chemical energy contained in a glucose molecule to high-energy bonds of ATP. Unfortunately, according to the Second Law of Thermodynamics, "you can't break even": most of the "energy" listed on the right side of the second equation is heat, not ATP. Nevertheless, a cell can extract a great deal of ATP energy from glucose if the glucose molecule is completely broken down to CO_2 and H_2O.

Figure 8-1 summarizes the major steps of glucose metabolism in eukaryotic cells. The first stage, **glycolysis,** does not depend on the availability of oxygen, and proceeds in exactly the same way under both aerobic (with oxygen) and anaerobic (without oxygen) conditions. Glycolysis splits apart a single glucose molecule (a 6-carbon sugar) into two 3-carbon molecules of pyruvic acid. This releases a small fraction of the chemical energy stored in the glucose, some of which is used to generate two ATP molecules. Under anaerobic conditions, the pyruvic acid is usually converted by **fermentation** into lactic acid or ethanol. Fermentation does not produce any more ATP energy. Both glycolysis and fermentation occur in the cytosol. The pyruvic acid produced by glycolysis may also enter the mitochondria. Here, if oxygen is available, **cellular respiration** uses oxygen to break pyruvic acid down completely to carbon dioxide and water, generating an additional 34 to 36 ATP molecules (the exact amount differs among cells).

Glycolysis

Glycolysis (in Greek, "to break apart a sweet") is a complex sequence of reactions in the cytosol of a cell in which a molecule of glucose is broken down into two molecules of pyruvic acid, with a small energy gain of two molecules of ATP and two molecules of the electron carrier NADH. Reduced to its essentials, glycolysis consists of two major processes: glucose activation and energy harvest (Fig. 8-2).

Glucose Activation

In glucose activation, a molecule of glucose undergoes two reactions that each use ATP energy. These reactions convert a relatively stable glucose molecule into a highly reactive molecule of fructose diphosphate, but at the cost of using up two ATP molecules.

Energy Harvest

In the energy harvest steps, fructose diphosphate splits apart into two 3-carbon molecules of phospho-

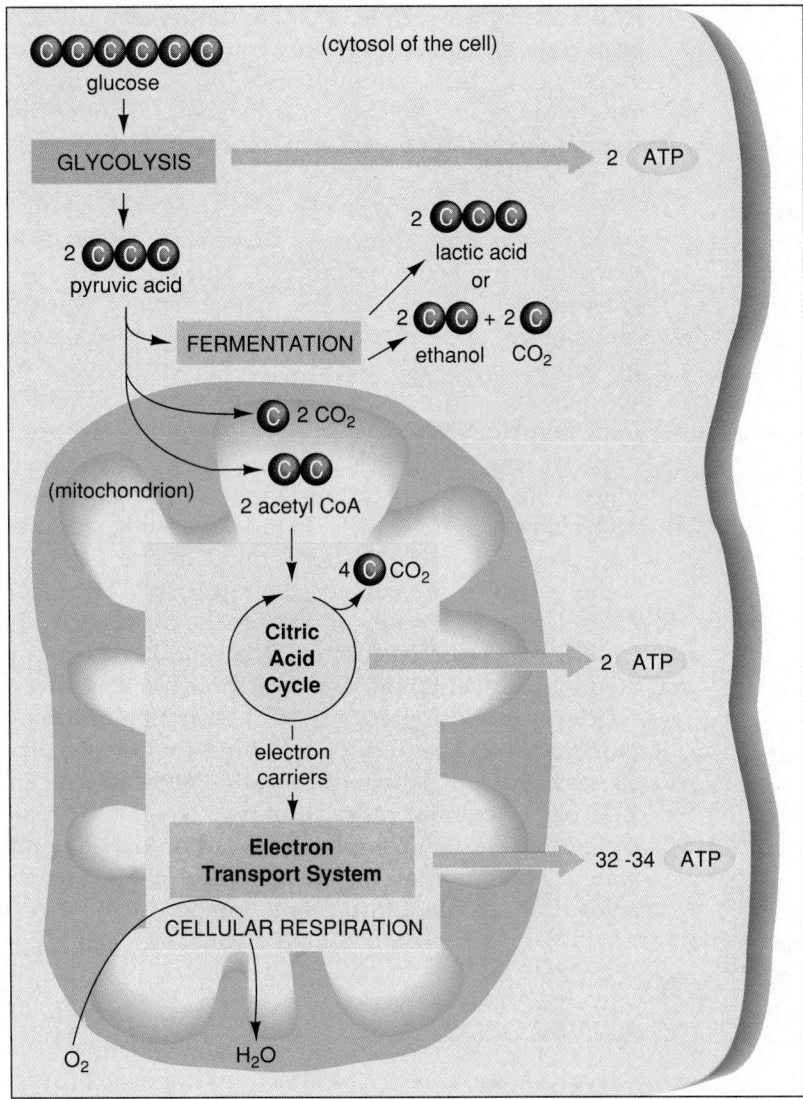

Figure 8-1 A summary of glucose metabolism. Use this diagram as a reference as we progress through the reactions of glycolysis (in the cell cytosol) and cellular respiration (in the mitochondria). Note that the breakdown of glucose occurs in stages, with various amounts of energy harvested as ATP along the way. The vast majority of the ATP is produced in the mitochondria, justifying their nickname of "powerhouse of the cell."

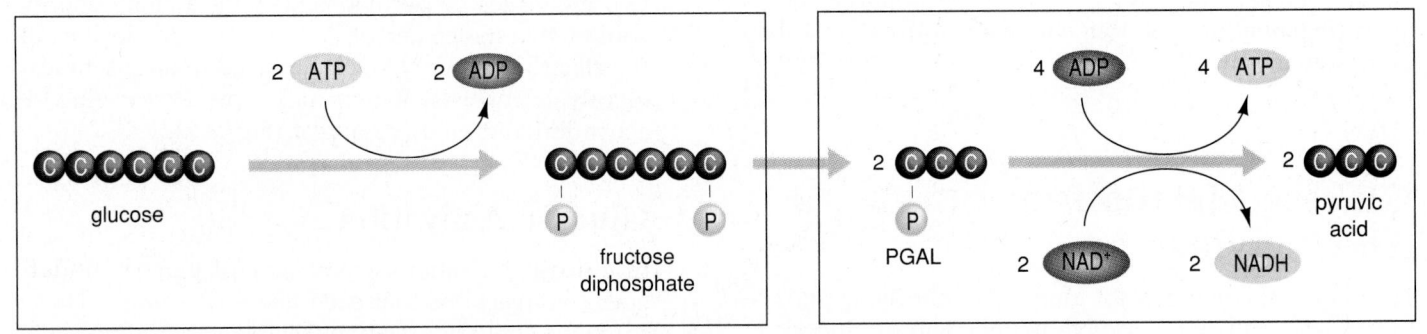

GLUCOSE ACTIVATION

ENERGY HARVEST

Figure 8-2 The essentials of glycolysis. (1) Glucose activation: The energy of two ATP molecules is used to convert glucose to the highly reactive fructose diphosphate. Fructose diphosphate splits into two smaller, but still reactive, molecules of phosphoglyceraldehyde (PGAL). (2) Energy harvest: The PGAL molecules both go through a series of reactions that generate four ATP and two NADH molecules. Therefore, glycolysis results in a net harvest of two ATP molecules per glucose molecule.

glyceraldehyde (we have already encountered PGAL in the C₃ cycle of photosynthesis). The two PGAL molecules then go through a series of reactions that culminate in the production of two molecules of pyruvic acid, one from each PGAL. Two of these reactions are coupled to ATP synthesis, generating two ATP molecules per PGAL, for a total of four ATPs. Since two ATPs were used to activate the glucose molecule in the first place, there is a net gain of two ATPs per glucose molecule. At another step along the way from PGAL to pyruvic acid, an electron and a hydrogen atom are added to the "empty" electron carrier NAD⁺ to make the "energized" carrier NADH (note that these are different molecules than NADP⁺ and NADPH seen in the light-dependent reactions of photosynthesis). Two PGAL molecules are produced per glucose molecule, so two NADH carrier molecules are formed.

For a complete description of glycolysis, see the box "A Closer Look at Glycolysis."

SUMMARY OF GLYCOLYSIS

Each molecule of glucose is broken down to two molecules of pyruvic acid. During these reactions, two ATP molecules and two NADH electron carriers are formed.

The electrons carried in NADH are highly energetic, but their energy can only be used to synthesize ATP when oxygen is available (see "Cellular Respiration," below). Under anaerobic conditions (under which life, and probably glycolysis, evolved), NADH production is not a method of energy capture; it is actually a way of getting rid of hydrogens and electrons during the metabolism of glucose to pyruvic acid. This works fine except for the fact that NAD⁺ is used up as it accepts electrons and hydrogen ions to become NADH. If a cell couldn't dispose of the electrons and hydrogens, and regenerate NAD⁺, glycolysis would have to stop once the supply of NAD⁺ was exhausted.

Fermentation

Fermentation solves this problem by allowing pyruvic acid to accept electrons and hydrogens from NADH, thereby regenerating NAD⁺ for use in further glycolysis. There are two main types of fermentation, one converting pyruvic acid to lactic acid, and the other converting it to carbon dioxide and ethanol.

Lactic Acid Fermentation

When you exercise vigorously, your muscles need lots of ATP as an energy source (Fig. 8-3a). Even though cellular respiration generates much more ATP than glycolysis alone, *cellular respiration is limited by organismal respiration:* you may not be able to get enough oxygen out of the air, into your blood, and delivered to your muscles to keep cellular respiration going at the necessary pace. When this happens, your muscles do not immediately stop working. Instead, glycolysis continues for a while, providing its meager two ATP molecules per glucose, and both pyruvic acid and NADH pile up. To regenerate NAD⁺,

(a)

(b)

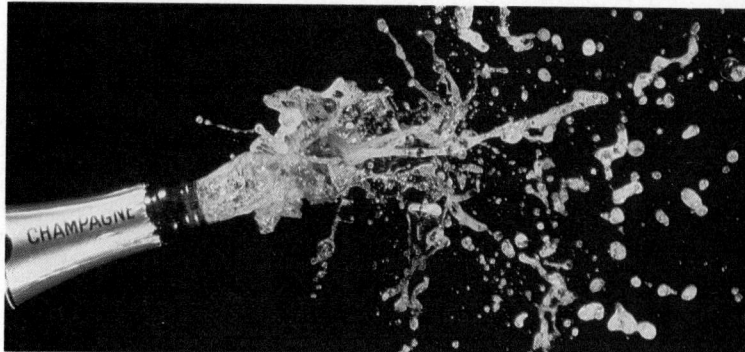

Figure 8-3 Fermentation. **(a)** A sprinter at the end of a race. The runner's respiratory and circulatory systems cannot supply oxygen to his leg muscles fast enough to keep up with the demand for energy, so glycolysis and lactic acid fermentation must provide the energy. Panting after the race brings in the oxygen needed to remove the lactic acid through cellular respiration. **(b)** Champagne spurts out of a newly opened bottle, driven out by CO₂ trapped in the bottle by fermenting yeast.

CLOSER LOOK

At Glycolysis

Glycolysis is a series of enzyme-catalyzed reactions that break down a single molecule of glucose into two molecules of pyruvic acid. To help you follow the reactions, we show only the "carbon skeletons" of glucose and the molecules produced during glycolysis as gray balls. Depleted carrier molecules of ADP and NAD⁺ are gray ovals; energized carriers ATP and NADH are yellow ovals; and phosphate groups attached to the carbons are yellow circles. Each colored arrow represents a reaction catalyzed by at least one enzyme. (A molecular model of hexokinase, the first enzyme in glycolysis, is shown in Fig. 4-6 in Chapter 4.)

1. A glucose molecule is energized by the addition of a high-energy phosphate from ATP.

2. The molecule is slightly rearranged,

3. then a second phosphate is added from another ATP.

4. The resulting molecule of fructose-1,6-diphosphate is split into two three-carbon molecules, one DHAP and one PGAL. Each has one phosphate attached.

5. DHAP rearranges into PGAL, so from now on there are two molecules of PGAL going through the identical reactions.

6. Each PGAL undergoes two almost simultaneous reactions. An electron and a hydrogen atom are donated to NAD⁺ to make the energized carrier NADH, and an inorganic phosphate (Pᵢ) is attached to the carbon skeleton with a high-energy bond. The resulting molecules of 1,3-diphosphoglyceric acid have two high-energy phosphates.

7. One phosphate from each diphosphoglyceric acid is transferred to ADP to form ATP, for a net of two ATPs. This compensates for the initial two ATPs used in glucose activation.

8. After another rearrangement, the second phosphate from each phosphoenolpyruvic acid is transferred to ADP to form ATP, leaving pyruvic acid as the final product of glycolysis. There is a net profit of two ATPs from each glucose molecule.

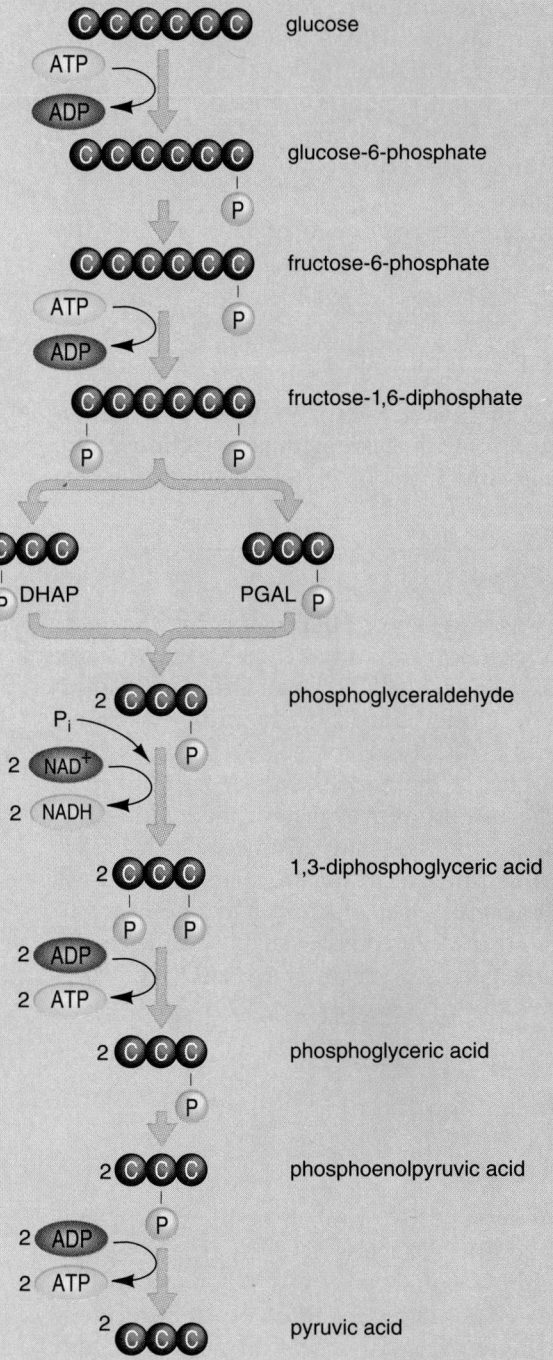

muscle cells convert pyruvic acid molecules to lactic acid, using electrons and hydrogens from NADH:

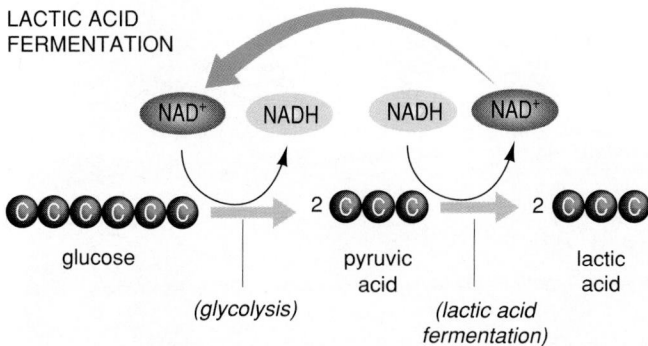

LACTIC ACID
FERMENTATION

glucose

(glycolysis)

pyruvic acid

lactic acid

(lactic acid fermentation)

The regenerated NAD$^+$ can again accept electrons during glycolysis, and energy production can continue. Lactic acid, however, is toxic in high concentrations. Eventually, the muscles suffer from lactic acid poisoning, and fatigue sets in. After exercise ceases, the lactic acid is reconverted to pyruvic acid, which is broken down through cellular respiration to carbon dioxide and water. Interestingly enough, this reconversion occurs not in the muscle cells, which lack the necessary enzymes, but in the liver.

A variety of microorganisms also use lactic acid fermentation, including the bacteria that produce yogurt, sour cream, and cheese.

Alcoholic Fermentation

Many microorganisms use another process to regenerate NAD$^+$ under anaerobic conditions: alcoholic fermentation. This series of reactions produces ethanol and CO$_2$ from pyruvic acid, using hydrogens and electrons from NADH:

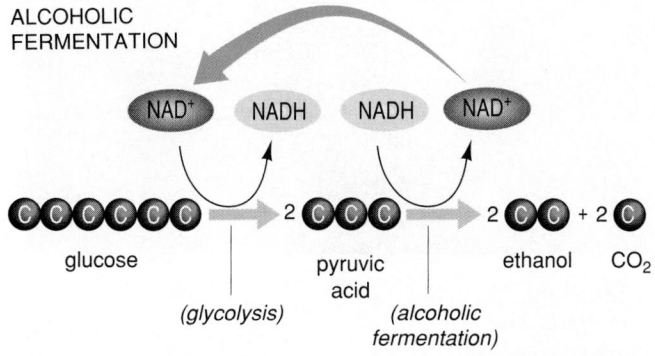

ALCOHOLIC
FERMENTATION

glucose

(glycolysis)

pyruvic acid

ethanol   CO$_2$

(alcoholic fermentation)

Alcoholic fermentation is a somewhat chancy proposition for the microbe, because the alcohol it generates is poisonous. If the alcohol concentration becomes too great, the microbe will die. Most yeasts, for example, die if the alcohol concentration exceeds 12%, although certain strains developed for wine production do not succumb until 18%. (Under natural

conditions, of course, yeasts are unlikely to find themselves confined in a small volume of water, with lots of sugar for energy production and no oxygen. Selective pressure for alcohol tolerance was probably very slight before humans developed a taste for wine.) Sparkling wines, such as champagne, are bottled while the yeasts are still alive and fermenting, trapping both the alcohol and the CO$_2$. When the cork is removed, the pressurized CO$_2$ is released, sometimes rather explosively (Fig. 8-3b).

Cellular Respiration

Cellular respiration is a series of reactions in which the pyruvic acid produced by glycolysis is broken down to carbon dioxide and water. The final reactions of cellular respiration require oxygen.

In eukaryotic cells, cellular respiration occurs in the mitochondria. You may recall from Chapter 5 that a mitochondrion has two membranes, producing two compartments: an inner compartment enclosed by the inner membrane and containing the fluid **matrix,** and an **intermembrane compartment** between the two membranes (Fig. 8-4). Reactions involving enzymes that are dissolved in the matrix, electron transport proteins in the inner membrane, and the movement of hydrogen ions through ATP-synthesizing proteins in the inner membrane yield most of the ATP produced during cellular respiration.

Figure 8-4 summarizes the main events of cellular respiration.

Step 1: Glycolysis in the cell cytosol produces two molecules of pyruvic acid per molecule of glucose.

Step 2: The pyruvic acids are transported across both mitochondrial membranes into the matrix.

Step 3: Each pyruvic acid is split into CO$_2$ and a 2-carbon acetyl group, which enters the citric acid cycle. The citric acid cycle releases the remaining carbons as CO$_2$, produces one ATP, and donates energetic electrons to several electron carrier molecules.

Step 4: The electron carriers donate their energetic electrons to the electron transport system of the inner membrane. The energy of the electrons transports H$^+$ ions from the matrix to the intermembrane compartment. At the end of the system, the electrons combine with O$_2$ and H$^+$ to form H$_2$O.

Step 5: In chemiosmosis, the H$^+$ ion gradient created by the electron transport system discharges through ATP-synthesizing enzymes in the inner membrane, and the energy is used to produce large amounts of ATP.

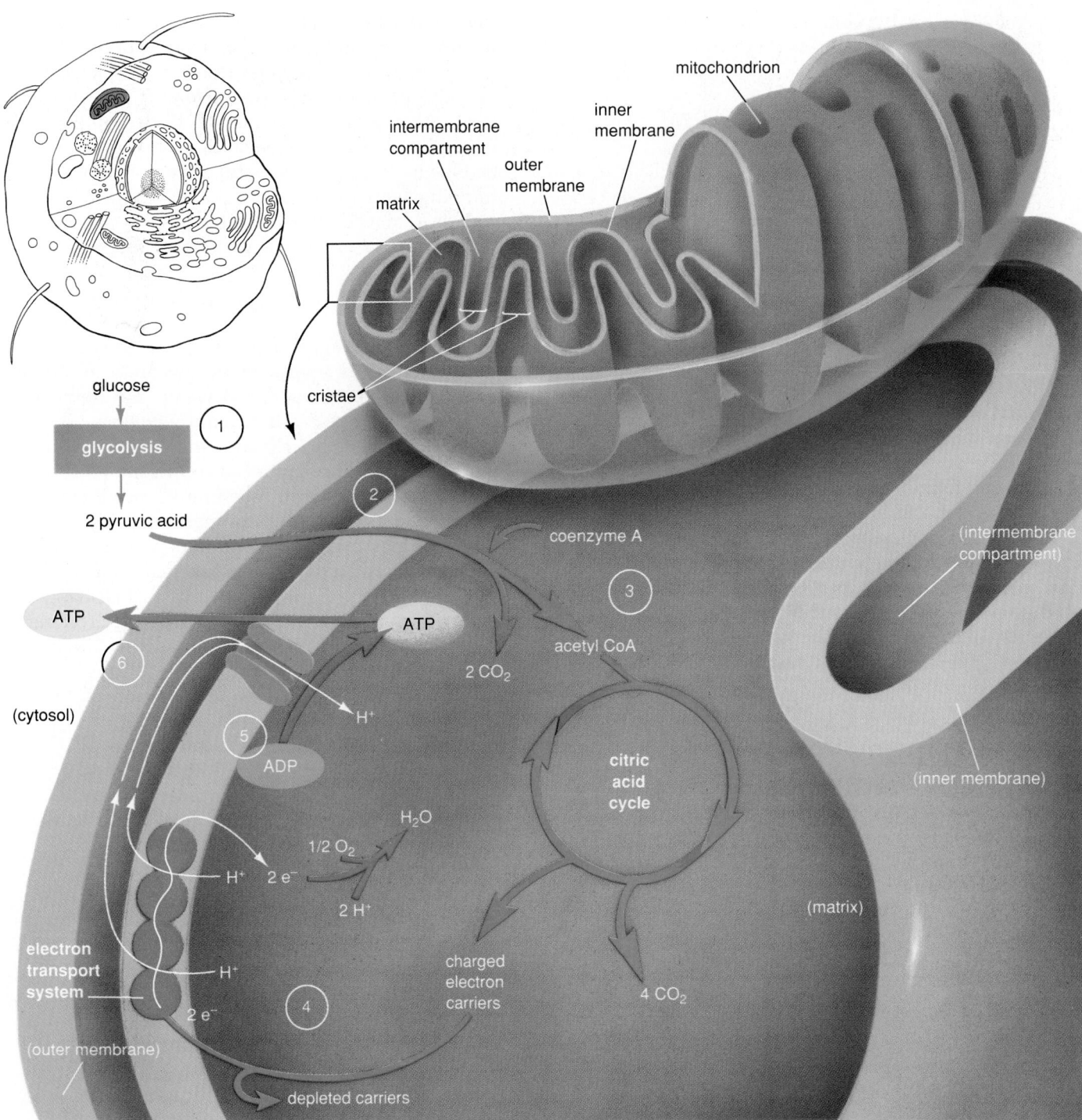

Figure 8-4 Cellular respiration takes place in the mitochondria, whose structure, like that of a chloroplast, reflects the compartmentalized reactions that occur there. The inner membrane separates the inner compartment, containing the soluble enzymes of the matrix, from the intermembrane compartment (between the inner and outer membranes). The lower diagram summarizes the six essential steps in glucose metabolism, from the initial glycolysis in the cytosol to the final transport of ATP out of the mitochondrion and into the cytosol. See the text for details of these steps.

Step 6: The ATP is transported back out of the mitochondrion into the cytosol, available to drive needed reactions in the rest of the cell.

We have already discussed glycolysis; now let's look a little more closely at the processes of cellular respiration in the mitochondria.

Pyruvic Acid Transport

The outer membrane of mitochondria contains many large pores, and is highly permeable to pyruvic acid. Facilitated diffusion through proteins in the inner membrane allows pyruvic acid to pass down its concentration gradient from its site of synthesis in the cytosol to its site of consumption in the mitochondrial matrix.

The Matrix Reactions

In the matrix, pyruvic acid reacts with a molecule called coenzyme A (Fig. 8-5, left). Each pyruvic acid is split into CO_2 and a 2-carbon molecule called an acetyl group, which immediately attaches to coenzyme A to form the acetyl-coenzyme A complex (acetyl CoA for short). During this reaction, an energetic electron and a hydrogen are transferred to NAD^+, forming NADH.

The two acetyl CoA molecules then enter a cyclic pathway known as either the **Krebs cycle,** named after its discoverer Hans Krebs, or the **citric acid cycle,** named after the first product in the reaction sequence, citric acid (Fig. 8-5, right). Each acetyl CoA briefly combines with a molecule of oxaloacetic acid. The 2-carbon acetyl group is donated to the 4-carbon oxaloacetic acid to form the 6-carbon citric acid, and coenzyme A is released once again (coenzyme A, like an enzyme, is not permanently altered during these reactions, and is re-used many times). Mitochondrial enzymes then lead each citric acid through a number of rearrangements that regenerate the oxaloacetic acid, give off two CO_2 molecules, and capture most of the energy of the acetyl group as one ATP and four electron carriers, one $FADH_2$ (flavin adenine dinucleotide), and three NADH.

The box "A Closer Look at the Matrix Reactions" shows the complete set of reactions in the mitochondrial matrix, from acetyl CoA formation through the citric acid cycle.

SUMMARY OF THE MATRIX REACTIONS

Acetyl CoA synthesis produces one CO_2 and one NADH per pyruvic acid molecule. The citric acid cycle produces two CO_2, one ATP, three NADH, and one $FADH_2$ per acetyl CoA. Therefore, at the conclusion of the matrix reactions, the two pyruvic acids produced from a single glucose molecule have been completely oxidized to six CO_2 molecules. Two ATPs, eight NADH, and two $FADH_2$ electron carriers have been formed.

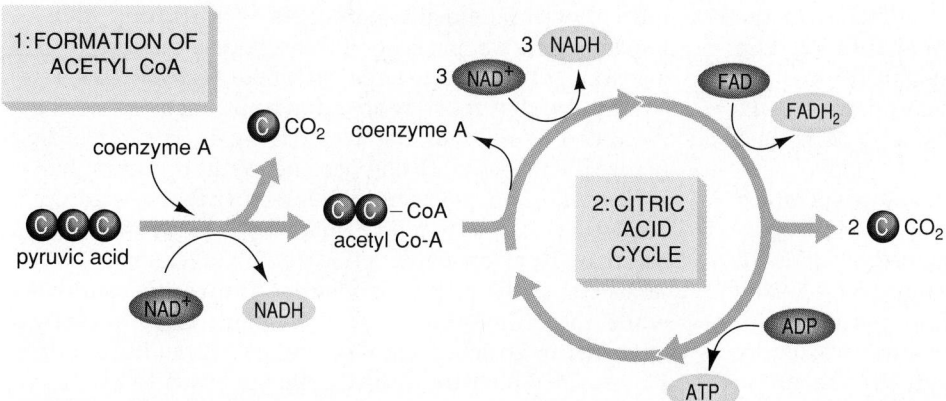

Figure 8-5 The essentials of the metabolic reactions occurring in the mitochondrial matrix. Pyruvic acid reacts with coenzyme A to form CO_2 and acetyl CoA. During this reaction, an energetic electron is added to NAD^+ to form NADH. When acetyl CoA enters the citric acid cycle, the acetyl group combines with oxaloacetic acid to form citric acid, and coenzyme A is released. One turn through the reactions of the cycle produces two molecules of CO_2, one of ATP, three of NADH and one of $FADH_2$ per acetyl CoA. Since each glucose molecule yields two pyruvic acids, the total energy harvest in the matrix is two ATP, eight NADH, and two $FADH_2$.

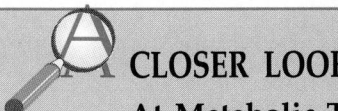

A CLOSER LOOK

At Metabolic Transformations:
Why You Can Get Fat by Eating Sugar

As you know, humans do not live by bread alone, much less by glucose alone. Nor does the typical diet contain exactly the required amounts of each nutrient. Accordingly, the cells of the human body seethe with biochemical reactions, synthesizing one amino acid from another, making fats from carbohydrates, and channeling surplus organic molecules of all types into energy storage or release. Let's look at two examples of these molecular interconversions: producing ATP from fats and proteins during fasting, and synthesizing fats from sugars during surplus food intake.

Metabolizing Fats and Proteins

Even the leanest people have some fat in their bodies. During fasting or starvation, the body mobilizes these energy reserves for ATP synthesis, since even the bare maintenance of life requires a continuous supply of ATP, while seeking out new food sources demands even more energy expenditure. Fat metabolism is very straightforward, flowing directly into the pathways of glucose metabolism.

Chapter 3 illustrated the structure of a typical fat: three fatty acids connected to a glycerol backbone. In fat metabolism, the bonds between the fatty acids and glycerol are hydrolyzed. The glycerol part of a fat, after activation by ATP, feeds directly into the middle of the glycolysis pathway (Fig. E8-1). The fatty acids are transported into the mitochondria, where enzymes in the inner membrane and matrix chop them up into 2-carbon acetyl groups. These attach to coenzyme A to form acetyl coenzyme A (acetyl CoA), which enters the citric acid cycle just as acetyl CoA from pyruvic acid does.

In cases of severe starvation, or of feeding almost exclusively on protein, amino acids too can be used to produce energy. Some amino acids can be readily cut down to pyruvic acid. Other amino acids are more complicated in structure, but can be converted to pyruvic acid, acetyl-CoA, or the compounds of the citric acid cycle. These molecules then proceed through the remaining stages of cellular respiration, yielding various amounts of ATP depending on their point of entry into the pathway.

Converting Carbohydrates to Fat

Just as fats can be funneled into the glucose metabolic pathway for energy production, the sugars and starches in cornflakes or candy bars can be converted into fats for energy storage. Complex sugars, such as starches and sucrose, are first hydrolyzed into their monosaccharide subunits (see Chapter 3). The monosaccharides are broken down to pyruvic acid and thence converted to acetyl-CoA. If the cell needs ATP, the acetyl-CoA will enter the citric acid cycle. If the cell has plenty of ATP, acetyl-CoA will be used to make fatty acids via a series of reactions that are essentially the reverse of fatty acid breakdown. In humans, the liver synthesizes fatty acids, but fat storage is relegated to fat cells, with their all too familiar distribution in the body, particularly around the waist and hips.

Energy use, fat storage, and eating are usually precisely balanced. Where the balance point lies, however, varies considerably from person to person. Some people seem to eat nearly continuously without ever storing much fat, probably because their metabolic rates are high or adjust upward to compensate for excess food intake. Others crave high-calorie foods even when they have a lot of fat already stored. As every chronic dieter knows, many people seem to have bodies that are "biologically set" to have large fat deposits. Constant restraint can slim most people down, but as soon as the discipline is relaxed, the fat once again accumulates. Before the fashionably thin become too smug about their svelte condition, they should remember that overeating during times of easy food availability is highly adaptive behavior from an evolutionary point of view. If hard times come, the plump may easily (if hungrily) survive while the trim succumb to starvation. Only recently have many societies enjoyed regular agricultural surpluses, in which the drive to eat can lead to obesity.

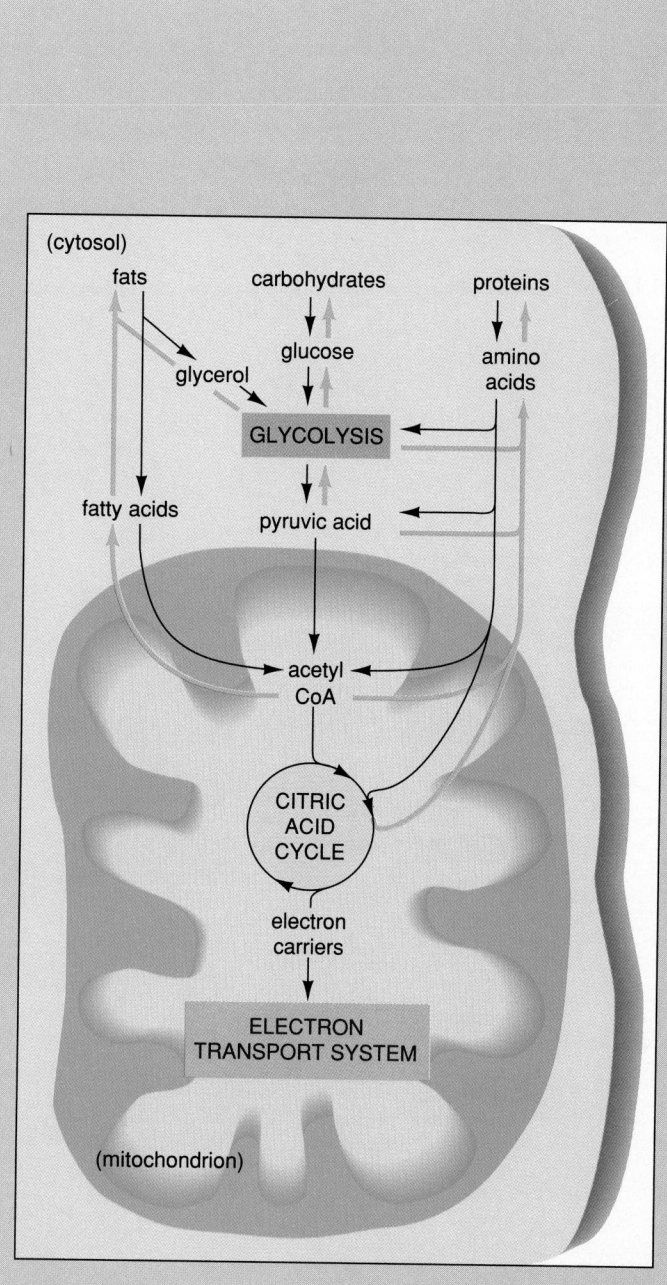

Figure E8-1

Electron Transport

At this point the cell has gained only four ATP molecules from the original glucose molecule: two during glycolysis and two during the citric acid cycle. The cell has, however, captured many energetic electrons in carrier molecules: two NADH during glycolysis, and eight more NADH and two FADH$_2$ from the matrix reactions, for a total of 10 NADH and two FADH$_2$. The carriers deposit their electrons in **electron transport systems** located in the inner mitochondrial membrane (Fig. 8-6). The energetic electrons move from molecule to molecule along the transport systems. Energy released by the electrons during these transfers is used to pump hydrogen ions from the matrix, across the inner membrane, and into the intermembrane compartment (see "Chemiosmosis," below).

Finally, at the end of the electron transport system, oxygen and hydrogen ions accept the energetically depleted electrons: two electrons, one oxygen atom, and two hydrogen ions combine to form water. This clears out the transport system, leaving it ready to run more electrons through. Without oxygen, the electrons "pile up" in the transport system, with the result that the hydrogen ions are no longer pumped across the inner membrane. The hydrogen ion gradient soon dissipates, and ATP synthesis stops.

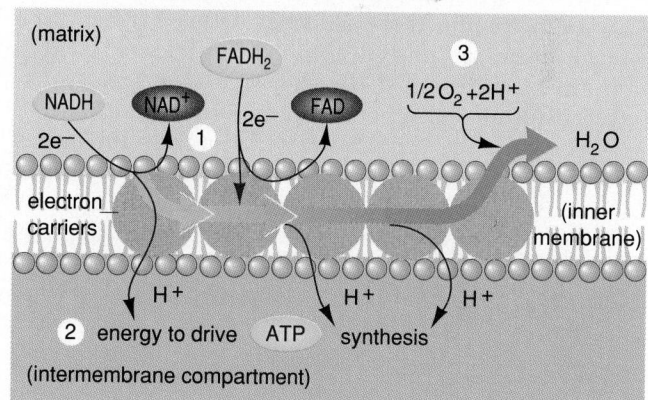

Figure 8-6 The electron transport system of mitochondria. (1) The electron carrier molecules NADH and FADH$_2$ deposit their energetic electrons with the carriers of the transport system located in the inner membrane. (2) The electrons move from carrier to carrier within the transport system. Some of their energy is used to pump hydrogen ions across the inner membrane from the matrix into the intermembrane compartment. This creates a hydrogen ion gradient that can be used to drive ATP synthesis (for details, see "A Closer Look at Chemiosmosis in Mitochondria"). (3) At the end of the electron transport system, the energy-depleted electrons combine with oxygen and hydrogen ions in the matrix to form water.

CLOSER LOOK

At the Matrix Reactions

The reactions that occur in the matrix of a mitochondrion are composed of two stages: the formation of acetyl coenzyme A and the citric acid cycle. The color conventions used in this box are the same as those used in A Closer Look at Glycolysis. Remember that, during glycolysis, there are two pyruvic acids produced from each glucose molecule. Therefore, each set of matrix reactions occurs twice during the metabolism of a single glucose molecule.

First Stage: Formation of Acetyl Coenzyme A

Pyruvic acid is split to CO_2 and an acetyl group. The acetyl group attaches to coenzyme A to form acetyl coenzyme A (acetyl CoA). Simultaneously, NAD^+ receives an electron and a hydrogen atom to make NADH. The acetyl CoA enters the second stage of the matrix reactions, while the NADH will donate its energetic electron to the electron transport system of the inner mitochondrial membrane.

Second Stage: The Citric Acid Cycle

Acetyl CoA donates its acetyl group to oxaloacetic acid to make citric acid. As citric acid proceeds through the remaining reactions, two more molecules of CO_2 are given off. These two CO_2 molecules, along with the one that was released during the formation of acetyl CoA, account for the three carbons of the original pyruvic acid. Three carbons released per pyruvic acid, with two pyruvic acids formed per glucose during glycolysis, account for all six carbons in a molecule of glucose.

The citric acid cycle also produces three molecules of NADH, one $FADH_2$, and one ATP per acetyl CoA. The molecules of NADH and $FADH_2$ donate their electrons to the electron transport system of the inner membrane, where the energy of the electrons will be used to synthesize ATP by chemiosmosis (see A Closer Look at Chemiosmosis).

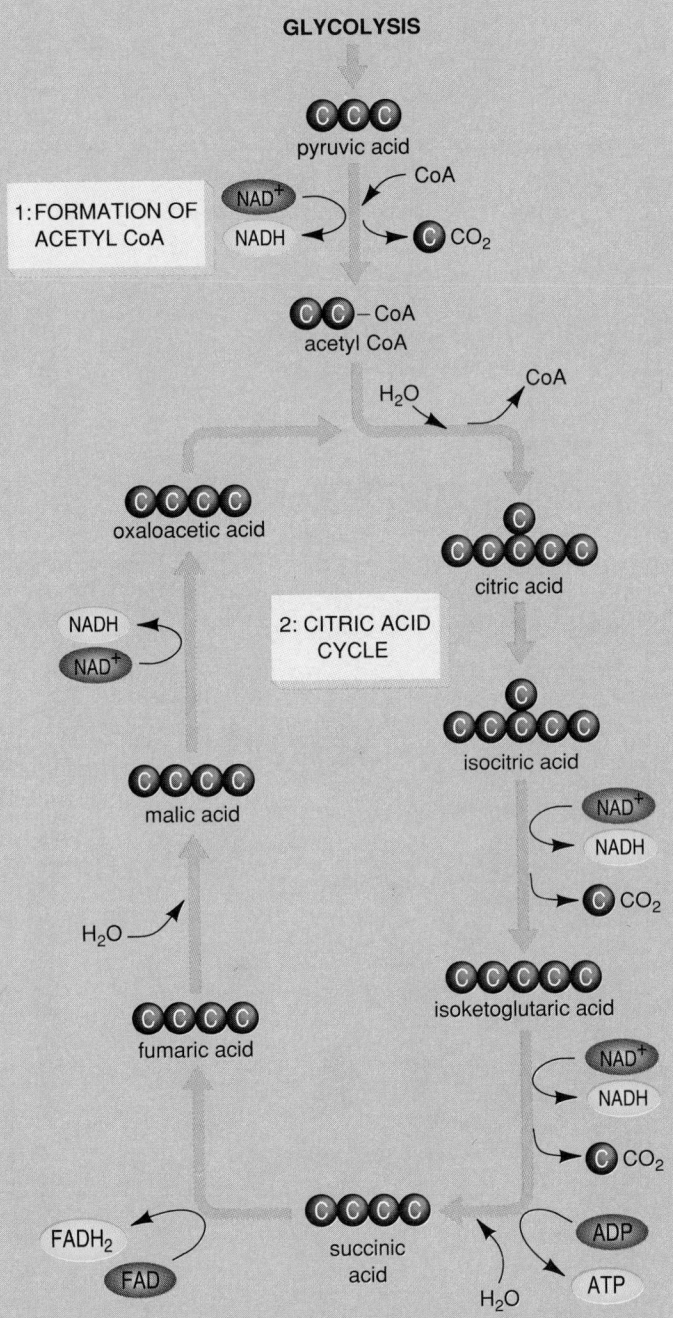

Chemiosmosis

Hydrogen ion pumping across the inner membrane generates a large concentration gradient; that is, a high concentration of hydrogen ions in the intermembrane compartment, and a low concentration in the matrix. The inner membrane is mostly impermeable to hydrogen ions, except at protein pores that are part of ATP-synthesizing enzymes. In the process known as **chemiosmosis,** hydrogen ions move down their concentration gradient from the intermembrane compartment to the matrix through these ATP-synthesizing enzymes. The flow of hydrogen ions provides the energy to synthesize 32 to 34 molecules of ATP from ADP and inorganic phosphate in the matrix. The box "A Closer Look at Chemiosmosis in Mitochondria" examines chemiosmosis in more detail.

ATP Transport

The outer mitochondrial membrane is very permeable to ATP, ADP, and inorganic phosphate. Proteins in the inner mitochondrial membrane exchange ATP from the matrix for ADP and inorganic phosphate, which are diffusing through the outer membrane from the cytosol. The ATP molecules provide most of the energy needed by the cell.

SUMMARY OF ELECTRON TRANSPORT AND CHEMIOSMOSIS

Electrons from the electron carriers NADH and $FADH_2$ enter the electron transport system of the inner mitochondrial membrane. Here their energy is used to generate a hydrogen ion gradient. Movement of hydrogen ions down their gradient through the pores of ATP-synthesizing enzymes drives the synthesis of 32 to 34 molecules of ATP. At the end of the electron transport system, two electrons combine with one oxygen atom and two hydrogen ions to form water.

The Importance of Cellular Respiration

The complete breakdown of pyruvic acid to carbon dioxide and water in the mitochondria results in an impressive gain in ATP molecules for the cell (Table 8-1). Whereas glycolysis only produces two ATP molecules per glucose, the mitochondrial reactions add 34 to 36 more ATPs, for a total of 36 to 38 ATPs per glucose molecule. Most modern organisms are so dependent on the additional ATP produced by cellular respiration that anything that interferes with it, such as lack of oxygen, quickly results in death.

Table 8-1 Summary of Glycolysis and Cellular Respiration

Process	Location	Reactions	Electron Carriers Formed	ATP Yield
Glycolysis	Cytosol	Glucose broken down to 2 pyruvic acids	2 NADH	2 ATP
Acetyl-CoA formation	Matrix of mitochondrion	Each pyruvic acid combined with coenzyme A to form acetyl-CoA and CO_2	2 NADH	
Citric acid cycle	Matrix of mitochondrion	Acetyl group of acetyl-CoA metabolized to two CO_2	6 NADH, 2 $FADH_2$	2 ATP
Electron transport	Inner membrane, intermembrane compartment	Energy of electrons from NADH, $FADH_2$ used to pump H^+ ions into intermembrane compartment, H^+ gradient used to synthesize ATP: three ATP per NADH, two ATP per $FADH_2$		32–34 ATP*

*Glycolysis produces two NADH molecules in the cytosol of the cell. Unlike the NADH and $FADH_2$ molecules generated in the matrix of the mitochondrion, the electrons from these two NADH molecules must be transported into the matrix before they can enter the electron transport system of the inner membrane. In most eukaryotic cells, the energy of one ATP molecule is used to transport the electrons from one NADH molecule into the matrix. Thus, the two "glycolytic NADH" molecules net only two ATPs, not the usual three, during electron transport. The heart and liver cells of mammals, however, use a different transport system, one that does not consume ATP. In these cells, then, the "glycolytic NADH" molecules net three ATPs apiece, just as the "mitochondrial NADH" molecule do.

CLOSER LOOK

At Chemiosmosis in Mitochondria

ATP synthesis in mitochondria is similar to the process of chemiosmosis that we described in chloroplasts in Chapter 7. The inner membrane of a mitochondrion has an electron transport system that functions similarly to the one in photosystem II of a thylakoid. Further, the intermembrane compartment between the outer and inner membranes of a mitochondrion is analogous to the interior of a thylakoid.

Anatomically, the arrangement in mitochondria looks like this:

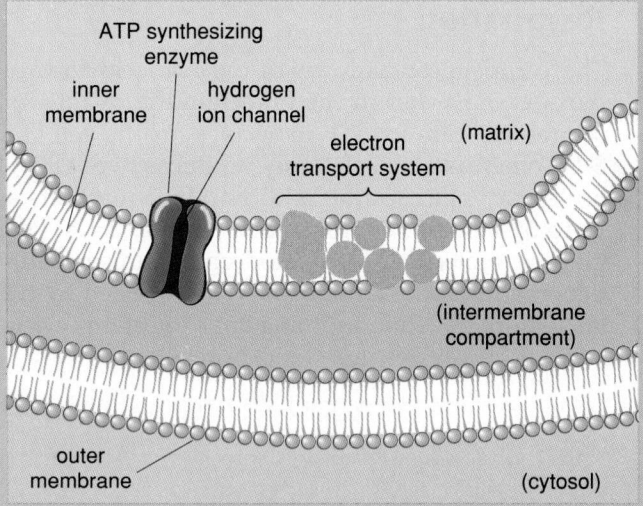

The electron carriers formed during glycolysis and the citric acid cycle—NADH and $FADH_2$— deposit their electrons with the electron transport system of the inner membrane (for clarity, $FADH_2$ is not shown in the illustration). As they pass through the electron transport system, the electrons provide the energy to pump hydrogen ions (H^+) across the inner membrane, from the matrix to the intermembrane compartment:

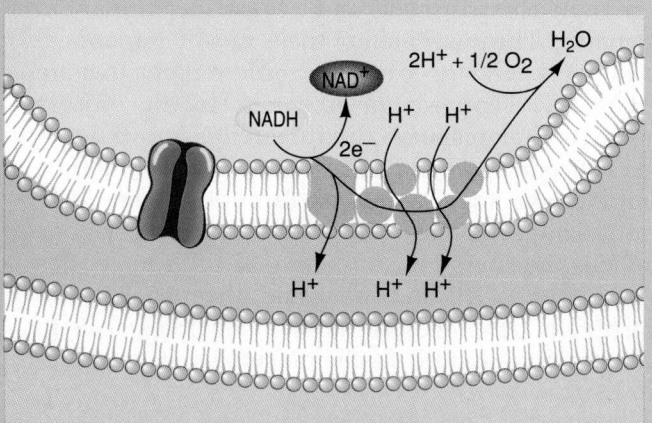

This increases the H^+ concentration in the intermembrane compartment and decreases the H^+ concentration in the matrix; therefore, a H^+ gradient is produced across the inner membrane. As in the thylakoid membrane of a chloroplast, the inner membrane of a mitochondrion is only permeable to H^+ at pores that are coupled with ATP-synthesizing enzymes. The movement of hydrogen ions down their concentration gradient through these pores drives ATP synthesis.

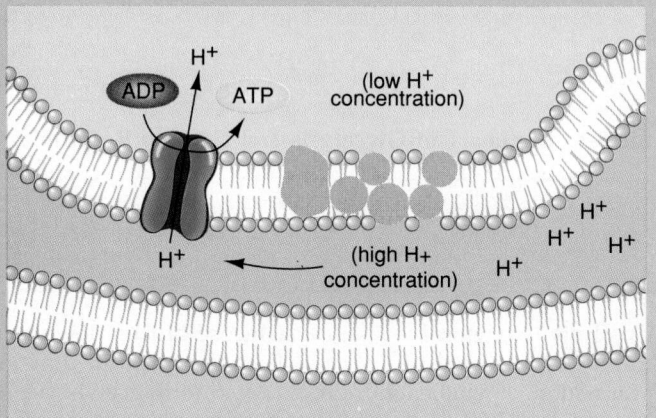

eflections on Cellular Metabolism and Organismal Function

Many students feel that the details of cellular metabolism are hard to learn and don't really help them to understand the living world around them. However, metabolic processes within individual cells have enormous impacts on the functioning of entire organisms. To take just two familiar examples, let's consider the migration of ruby-throated hummingbirds and track events in the Olympics.

Hummingbird Migration

As the days shorten in August, many birds in North America prepare to migrate to Central and South America, thereby escaping the cold weather and food shortages of the northern winter. For some, such as ducks and geese, migration is a relatively leisurely affair, with frequent stops along the way for rest and food. For others, migration is a very dangerous undertaking. Ruby-throated hummingbirds (chapter opening photo), which breed in the eastern United States, fly across the Gulf of Mexico from the southern states to Mexico and Central America. There is nowhere to stop and no food on a journey of over 1000 kilometers of open sea.

This presents real difficulties for a migrating hummingbird. With its short wings, a hummer couldn't fly at all if it weighed too much. Yet flying 1000 kilometers requires a lot of energy that must be stored in the bird's body. Hummers solve this dilemma with a twofold strategy of storing the highest energy molecules possible—fat—and extracting the maximum usable energy during flight.

A ruby-throated hummingbird weighs 2 to 3 grams prior to putting on weight for migration. It adds as much as 2 grams of fat in late summer, nearly doubling its weight. As you recall from Chapter 3, fats contain over twice as much energy per unit weight as do proteins or carbohydrates. If a hummer had to store 4 or 5 grams of glycogen or protein, it would be too heavy to lift off.

Even so, the hummer must still squeeze every ATP molecule possible out of each fat molecule. The hummer that just makes it to Guatemala on 2 grams of fat using cellular respiration would collapse almost within sight of the Gulf Coast if it used lactic acid fermentation instead. A hummingbird's flight muscles are packed with mitochondria, and its respiratory system is exquisitely designed to extract oxygen out of the air (Chapter 31), so that even during strenuous flight cellular respiration never slows down.

The Olympics

Humans, like hummingbirds, must regulate energy reserves and energy use. Why is the average speed of the 5000-meter run in the Olympics slower than that of the 100-meter dash? It's not because the distance runners could not, at any given time, run faster; the finishing kick at the end of the race testifies to that. The reason is that at top speed, their leg muscles consume ATP faster than their lungs can extract oxygen from the air to keep cellular respiration going. Glycolysis and lactic acid fermentation can keep the muscles functional for a short time, but soon the toxic effects of lactic acid build-up cause fatigue and cramps. While runners may be able to do a 100-meter dash anaerobically, 5000 meters is out of the question. Distance runners must therefore pace themselves, using cellular respiration to power their muscles for most of the race and saving the anaerobic sprint for the finish.

Marathon runners face somewhat the same dilemma that hummingbirds do. Nobody ever saw a fat marathoner: it's too much work to lug a lot of weight around for 26 miles. Nevertheless, a marathon may require 3000 kilocalories of energy, and would require even more if cellular respiration did not provide all the ATP. Marathoners train 50 or 100 miles a week, not so much to build up their leg muscles (which are usually pretty stringy) as to build up the capacity of their respiratory and circulatory systems to deliver enough oxygen to their muscles.

As you can see, life on Earth depends on efficiently obtaining, storing, and using energy. Your knowledge of the principles of cellular metabolism should enable you to appreciate the energy-related adaptations of living organisms more fully.

SUMMARY OF KEY CONCEPTS

Glucose Metabolism: An Overview

Cells produce usable energy by breaking down glucose to lower energy compounds and capturing some of the released energy as ATP. In glycolysis, glucose is metabolized in the cytosol to two molecules of pyruvic acid, generating two ATP molecules. In the absence of oxygen, pyruvic acid is converted by fermentation to lactic acid or ethanol and CO_2. If oxygen is available, the pyruvic acids are metabolized to CO_2 and H_2O through cellular respiration in the mitochondria, generating much more ATP.

Figure 8-7 summarizes the locations, major mechanisms, and overall energy harvest for the complete metabolism of glucose from glycolysis through cellular respiration.

Glycolysis

During glycolysis, a molecule of glucose is activated by adding phosphates from two ATP molecules, to form fructose diphosphate. The fructose diphosphate then undergoes a series of reactions that break it down into two molecules of pyruvic acid. These reactions produce four ATP molecules and two NADH electron carriers. Since two ATPs were used in the activation steps, the net yield from glycolysis is two ATPs and two NADHs. Glycolysis, in addition to providing a small yield of ATP, uses up NAD^+ to produce NADH. Once the cell's supply of NAD^+ is consumed, glycolysis must stop. NADH may be regenerated by fermentation, with no additional ATP gain, or by cellular respiration, which also produces additional ATP.

Fermentation

If oxygen is absent, the pyruvic acids undergo either alcoholic fermentation, to produce ethanol and CO_2, or lactic acid fermentation, to produce lactic acid. In both cases, no new ATP is formed. However, both types of fermentation regenerate NAD^+ from NADH, thus replenishing the cell's supply of NAD^+.

Cellular Respiration

If oxygen is available, cellular respiration can occur. The pyruvic acids are transported into the matrix of the mitochondria. In the matrix, each pyruvic acid reacts with coenzyme A to form acetyl-CoA plus CO_2. One NADH is also formed at this step. The 2-carbon acetyl group of acetyl-CoA enters the citric acid cycle, which releases the remaining two carbons as CO_2. One ATP, three NADHs, and one $FADH_2$ are also formed for each acetyl group that goes through the cycle. At this point, each glucose molecule has produced four ATPs (two from glycolysis and one from each acetyl-CoA via the citric acid cycle), 10

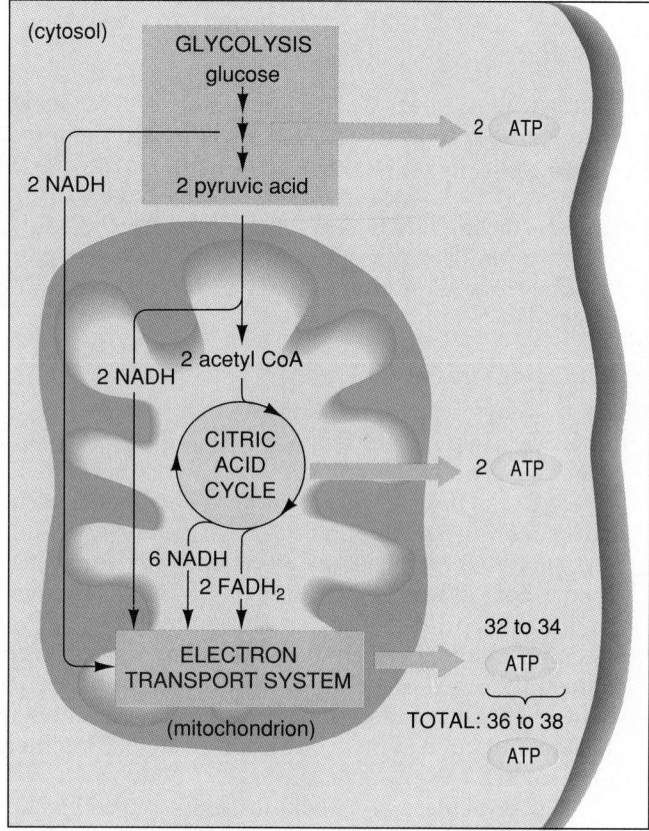

Figure 8-7 A summary of the energy harvest from the complete metabolism of one glucose molecule. Glycolysis and the citric acid cycle each produce two ATP molecules. The reactions within the mitochondrial matrix produce eight NADH molecules and two $FADH_2$ molecules. By donating its electrons to the electron transport system, each NADH molecule yields three ATP molecules, for a total of 24 ATPs. Each $FADH_2$ molecule yields two ATP molecules, for a total of four ATPs. The electrons from the two NADH molecules produced in the cytoplasm during glycolysis must be transported into the mitochondrion to reach the electron transport system. In heart and liver cells, this transport is "free"; in most cells, transport costs one ATP per NADH. The two "glycolytic NADH" molecules therefore yield either four or six ATP molecules, depending on the cell. Therefore, the energy harvest from electron transport is 32 to 34 ATPs. Including two ATPs from glycolysis and two ATPs from the citric acid cycle, the total energy yield from glucose metabolism is 36 to 38 ATPs.

NADHs (two from glycolysis, one from each pyruvic acid during the formation of acetyl-CoA, and three from each acetyl-CoA during the citric acid cycle), and two $FADH_2$s (one from each acetyl-CoA during the citric acid cycle).

The NADHs and $FADH_2$s deposit their energetic electrons in the electron transport system embedded in the inner mitochondrial membrane. The energy of the electrons is used to pump hydrogen ions across the inner membrane from the matrix to the intermembrane compartment. At the end of the electron transport system, the depleted electrons combine with hydrogen ions and oxygen to form water. During chemiosmosis, the hydrogen ion gradient created by the electron transport system is used to produce ATP, as the hydrogen ions diffuse back across the inner membrane through pores in ATP-synthesizing enzymes. Electron transport and chemiosmosis yield 32 to 34 additional ATPs, for a net yield of 36 to 38 ATPs per glucose molecule.

GLOSSARY

cellular respiration: the oxygen-requiring reactions occurring in mitochondria that break down the end products of glycolysis into carbon dioxide and water, while capturing large amounts of energy as ATP.

chemiosmosis (kē-mē-os-mō'-sis): a process of ATP generation in chloroplasts and mitochondria. The movement of electrons down an electron transport system is used to pump hydrogen ions across a membrane, thereby building up a concentration gradient of hydrogen ions across the membrane. The hydrogen ions diffuse back across the membrane through the pores of ATP-synthesizing enzymes. The energy of their movement down their concentration gradient drives ATP synthesis.

citric acid cycle: a cyclic series of reactions in which the acetyl groups from the pyruvic acids produced by glycolysis are broken down to CO_2, accompanied by the formation of ATP and electron carriers. Occurs in the matrix of mitochondria.

electron transport system: a series of molecules found in the inner membrane of mitochondria and the thylakoid membranes of chloroplasts that extract energy from electrons and generate ATP or other energetic molecules.

fermentation: anaerobic reactions that convert the pyruvic acid produced by glycolysis into lactic acid or alcohol and CO_2.

glycolysis (glī-kol'-i-sis): reactions carried out in the cytosol that break down glucose into two molecules of pyruvic acid, producing two ATP molecules. Glycolysis does not require oxygen, but can proceed when oxygen is present.

intermembrane compartment: the space contained between the inner and outer membranes of a mitochondrion.

Krebs cycle: the citric acid cycle (in honor of Hans Krebs, who discovered many of its biochemical details).

matrix: the fluid contained within the inner membrane of mitochondria.

STUDY QUESTIONS

1. Briefly describe glycolysis and cellular respiration. In what part of the cell does each occur? What conditions are necessary if cellular respiration is to occur? What is the overall energy harvest (in terms of ATP molecules generated per glucose molecule) for each?
2. Briefly describe fermentation. Why is fermentation essential if glycolysis is to continue under anaerobic conditions?
3. Diagram and describe the structure of the mitochondrion. In which parts do the various reactions of cellular respiration occur?
4. Describe the citric acid cycle. In what form is most of the energy harvested?
5. Describe the mitochondrial electron transport system and the process of chemiosmosis.
6. Why is oxygen necessary for cellular respiration to occur?

DISCUSSION QUESTIONS

1. The chemical equations for photosynthesis and cellular respiration are symmetrical (see p. 157). Why, then, is photosynthesis *not* the reverse of cellular respiration?
2. In detective novels, "the odor of bitter almonds" is always the telltale clue to murder by cyanide poisoning. Cyanide reacts with one of the proteins of the electron transport system in mitochondria, with the result that the movement of electrons through the transport system stops immediately. Why is this fatal?
3. Over a century ago, the French biochemist Louis Pasteur described a phenomenon, in the wine making process using yeast, that we now call "the Pasteur effect." He observed that in a sealed container of grape juice containing yeast, the yeast will consume the sugar very slowly as long as oxygen remains in the container. As soon as the oxygen is gone, however, the rate of sugar consumption by the yeast increases greatly and the alcohol content in the container rises. Discuss the Pasteur effect based on what you know about aerobic and anaerobic cellular respiration.

SUGGESTED READINGS

Hinkle, P. C., and McCarty, R. E. "How Cells Make ATP." *Scientific American*, March 1978. A summary of cellular respiration, with an emphasis on electron transport in mitochondria.

McCarty, R. E. "H^+-ATPases in Oxidative and Photosynthetic Phosphorylation." *BioScience*, January 1985. A description of the structure and function of the ATP-synthesizing enzymes in mitochondria and chloroplasts.

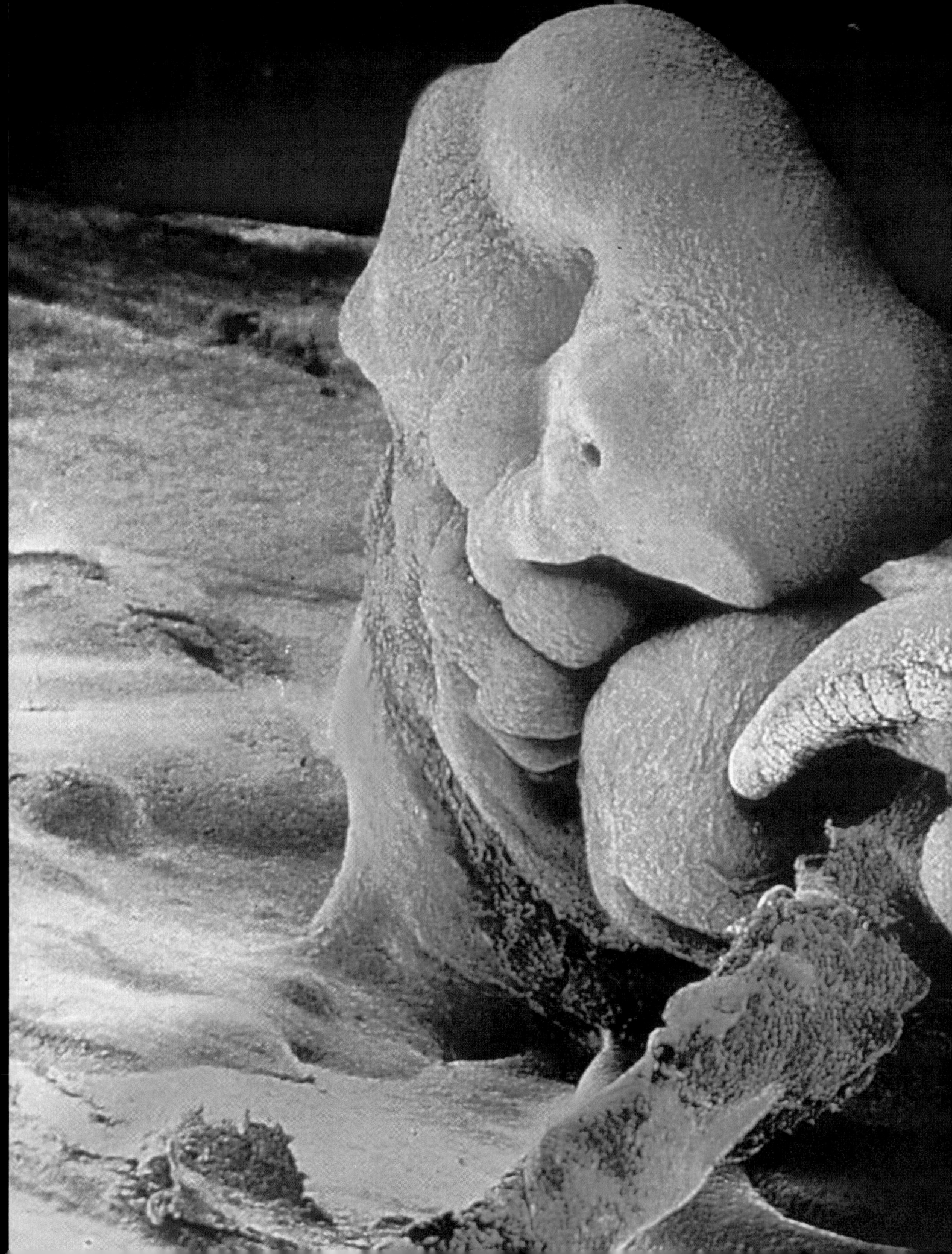

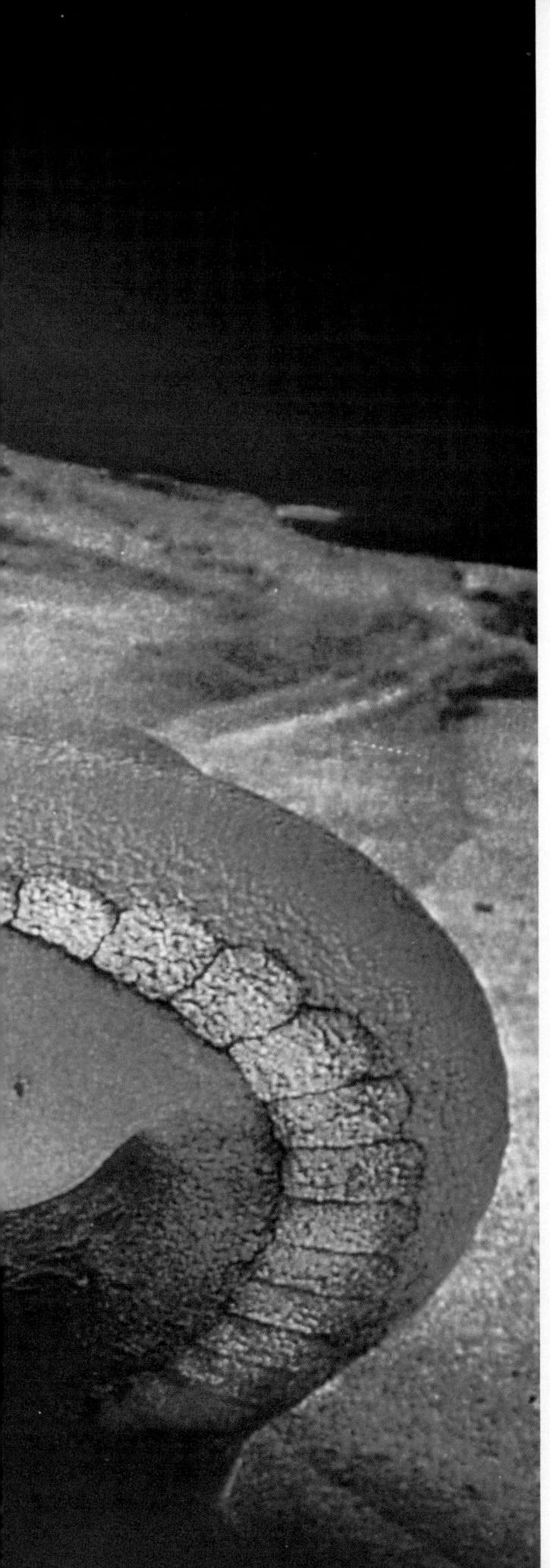

Inheritance

9

Cellular Reproduction

"All cells come from cells." With these words, Rudolf Virchow captured the crucial importance of cellular reproduction for both unicellular and multicellular organisms. Since all living organisms consist of one or more cells, and since all cells are descended from pre-existing cells, then cellular reproduction is absolutely essential for the continued existence of life on Earth.

How do cells reproduce? Although there are many variations, most cells reproduce according to the simple repeating phrases: enlarge, divide in two, enlarge, divide in two. . . . This process, appropriately called **cell division,** produces two daughter cells that are roughly identical copies of the original cell before it started to grow. Each round of growth and cell division is called a **cell cycle.** For unicellular organisms such as bacteria or *Amoeba,* the cell cycle is synonymous with the **life cycle,** which consists of the events in the life of an organism from one generation to the next.

In multicellular organisms, the cell cycle is only part of the life cycle. Multicellular organisms begin life as a fertilized egg. Repeated cell divisions produce the hundreds to trillions of cells that make up the adult organism. Meanwhile, a specialized set of cells in the reproductive organs undergo a different kind of cell division, called **meiosis.** In animals, the daughter cells of meiosis are sex cells—sperm or eggs. A sperm then fertilizes an egg to start the life cycle over again.

In this chapter, we study cell cycles and cell division. We will see that cells have evolved elaborate mechanisms to ensure that each daughter cell receives all the materials it needs to continue the flow of life. In Chapter 10, we examine meiosis and its role in the life cycles of multicellular organisms.

Essentials of Cellular Reproduction

When a cell divides, it must transmit to its offspring cells two essential requirements for life (Fig. 9-1):

1. **hereditary information to direct life processes, and**
2. **materials in the cytoplasm that the offspring need to survive and to use their hereditary information.**

Hereditary Information

The hereditary information of all living cells is **deoxyribonucleic acid,** DNA. Like many large biological molecules, a molecule of DNA consists of a long chain of smaller subunits (see Chapter 3). DNA has four different types of subunits, called nucleotides. Segments of DNA a few hundred to a few thousand nucleotides long are the units of inheritance—the **genes.** The sequence of nucleotides in a gene encodes information for the synthesis of the

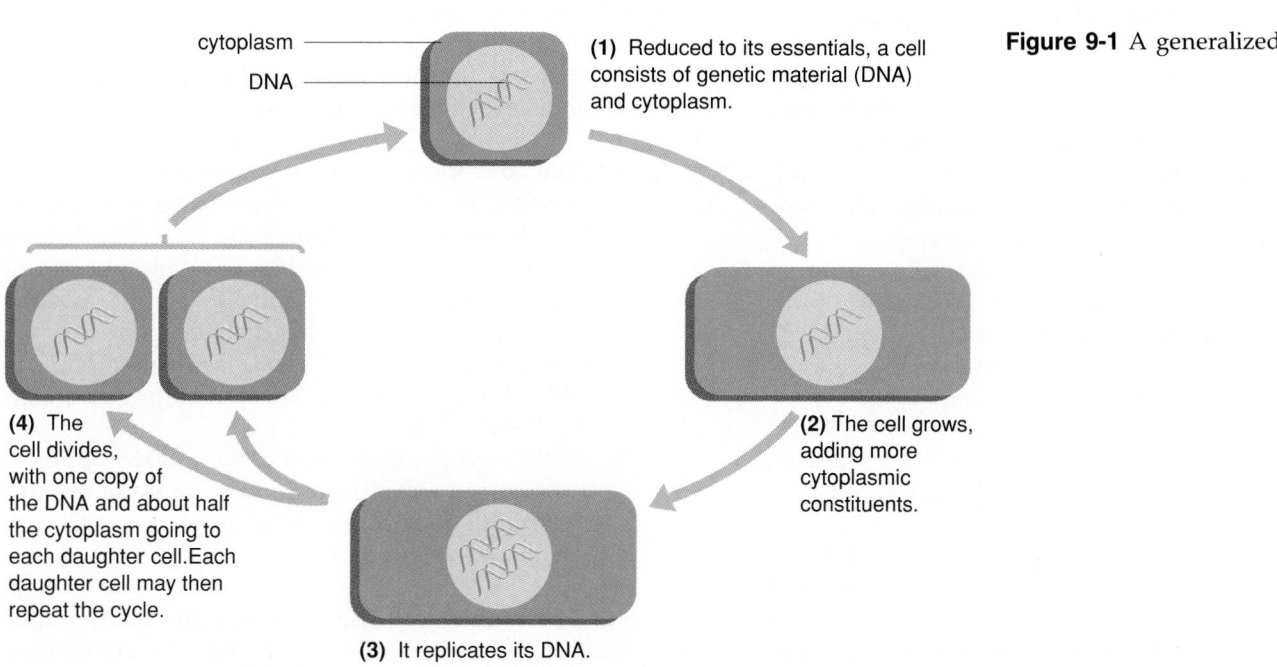

cytoplasm

DNA

(1) Reduced to its essentials, a cell consists of genetic material (DNA) and cytoplasm.

(2) The cell grows, adding more cytoplasmic constituents.

(3) It replicates its DNA.

(4) The cell divides, with one copy of the DNA and about half the cytoplasm going to each daughter cell. Each daughter cell may then repeat the cycle.

Figure 9-1 A generalized cell cycle.

RNA and protein molecules that are needed to build a cell and carry out its metabolic activities. We will see in Chapters 12 through 14 how DNA encodes genetic instructions and how a cell regulates which genes it uses at any given time.

For any cell to survive, it must have a *complete set of genetic instructions.* Therefore, when a cell divides, it cannot simply split its set of genes in half, and give each daughter cell half a set. Rather, the cell must first *duplicate its DNA,* much like making a photocopy of an instruction manual. Each daughter cell then receives a complete "DNA manual" containing all the genes.

Cytoplasmic Materials

Like the blueprints for a house, the instructions encoded by DNA are useless without materials to work with. Each newly formed cell must receive the molecules it needs to read its genetic instructions and to keep it alive long enough to acquire new materials from the environment and to process them into new cellular components. Furthermore, as we described in Chapter 5, a cell cannot synthesize either mitochondria or chloroplasts from scratch: these organelles arise only by the division of previously existing mitochondria and chloroplasts. Usually, when a cell divides, its cytoplasm is divided about equally between the two daughter cells. This simple mechanism normally provides both daughter cells with all the organelles, nutrients, enzymes, and other molecules they need.

Cell Cycles

The complete sequence of activities of a cell from one cell division to the next constitutes the cell cycle. Newly formed cells usually acquire nutrients from their environment, synthesize more of their own materials, and grow larger. After a variable amount of time, depending on the organism, the type of cell, and the nutrients available, the cell divides. This general description applies to both eukaryotic and prokaryotic cells. However, prokaryotic cells are structurally and functionally very different from eukaryotic cells, and accordingly their cell cycle differs in many respects. We will therefore discuss prokaryotic and eukaryotic cycles separately.

The Prokaryotic Cell Cycle

Under favorable conditions, the cell cycle of many prokaryotes proceeds rapidly (Fig. 9-2). The common intestinal bacterium *Escherichia coli,* for example, can complete its cell cycle in 30 minutes or less. During

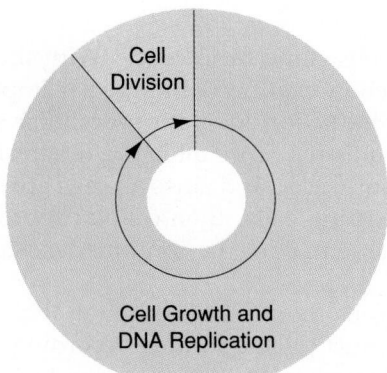

Figure 9-2 The prokaryotic cell cycle. In prokaryotic cells, growth of the cell and DNA replication occur simultaneously. When the DNA has finished replicating, the cell divides. In many bacteria, the whole cycle takes less than half an hour under favorable conditions; cell division may take only a minute or two.

most of this time, the cell absorbs nutrients from the environment, grows, and replicates its DNA. The DNA of a prokaryotic cell is a single, circular **chromosome** (Fig. 9-3), attached at one point to the plasma membrane. As soon as the DNA is replicated, the bacterium divides; in fact, in some bacteria the cell begins dividing even before DNA replication is quite finished. Cell division itself is quite straightforward, and is actually a modification of the normal mechanisms of cell growth, as we will see shortly.

You will recall from Chapter 5 that a prokaryotic cell does not have a membrane-bound nucleus. With only a single chromosome and no separate nucleus, bacterial cell division proceeds by the fairly simple mechanism of **fission** (Fig. 9-4):

1. Prior to DNA replication, one point on the chromosome is attached to the plasma membrane.
2. The chromosome replicates (by a process that we will describe in Chapter 12). The resulting pair of identical chromosomes attach to the plasma membrane at nearby, but distinct, points.
3. The cell elongates, and new plasma membrane is added between the two attachment points, pushing them apart.
4. As the two chromosomes move toward opposite ends of the cell, the plasma membrane around the middle of the cell grows inward.
5. Finally, two new daughter cells are formed.

Each daughter cell thus receives one of the replicated chromosomes and about half the cytoplasm.

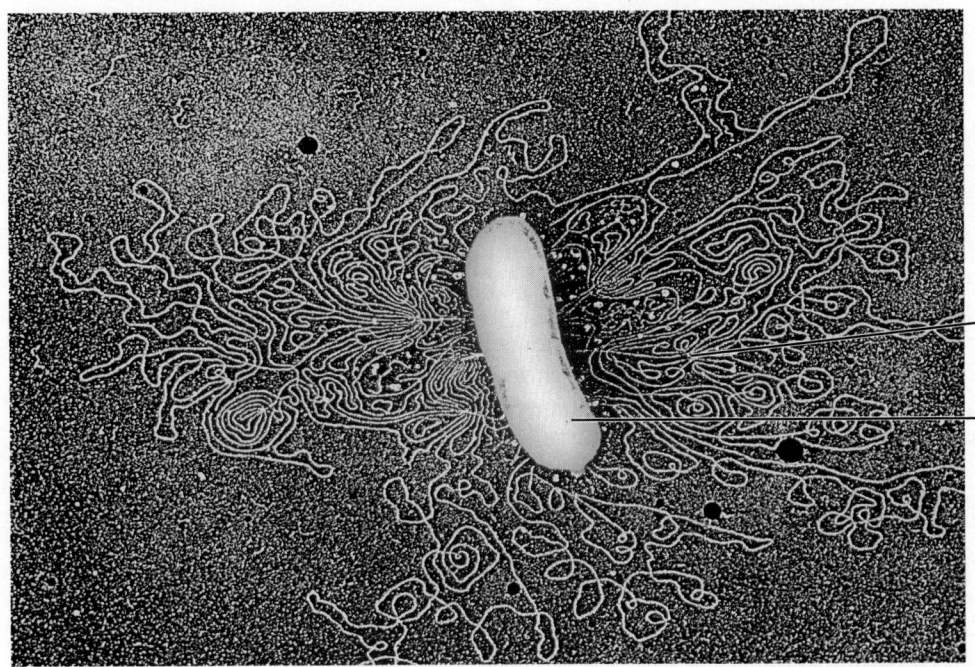

Figure 9-3 This bacterial cell (yellow central oblong) is about 1 × 3 micrometers in size. The cell was ruptured osmotically, freeing its chromosome. If you look carefully, you will see that there are no free ends: if stretched out, the chromosome would be a circular molecule of DNA 1 or 2 mm in circumference.

bacterial chromosome

ruptured bacterium

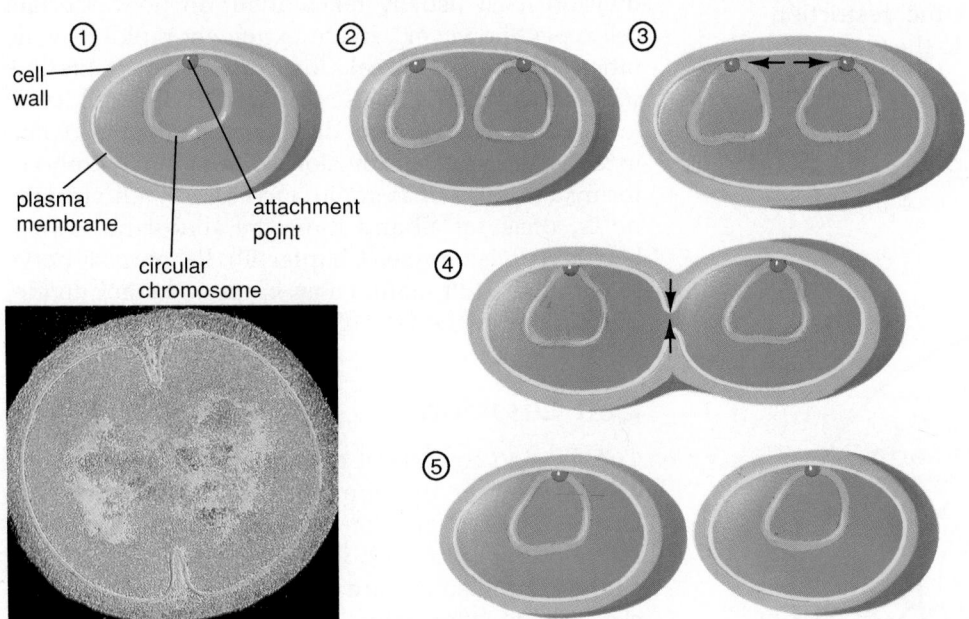

Figure 9-4 Prokaryotic cells divide by fission.
(1) The circular DNA is attached at one point to the cell membrane.
(2) The chromosome is replicated. The two copies are attached to the membrane at nearby points.
(3) The cell elongates; new plasma membrane is added between the attachment points.
(4) The plasma membrane grows inward at the middle of the cell.
(5) The parent cell has divided into two daughter cells.

With a complete set of genes and enough materials to work with, the two daughter cells begin to grow, starting the cycle over again.

The Eukaryotic Cell Cycle

The eukaryotic cell cycle consists of two major phases (Fig. 9-5):

1. **Interphase:** the period between cell divisions, during which the cell acquires nutrients from its environment, grows, and replicates its chromosomes.

2. **Cell division:** one copy of each chromosome, and approximately half the cytoplasm, is parcelled out into each of the two daughter cells.

Interphase

Most of the life of a cell (usually 90% or more) is spent in interphase. Cell biologists divide interphase into three subphases, called G_1, S, and G_2 (see Fig. 9-5).

G_1 Phase

The period that occurs after the most recent cell division and before duplication of the chromosomes is the **G_1 phase** (G_1 stands for "first gap," meaning the first gap or interruption in DNA synthesis). Most of the growth and activity of a cell occurs during the G_1 phase. The cell acquires nutrients from its environment, carries out its specialized functions (e.g., hormone synthesis and secretion), and grows.

Late in the G_1 phase, at a time called the restriction point, the cell undergoes a kind of "internal evaluation" of its ability to complete the cell cycle and produce two viable daughter cells. If the "evaluation" is negative, the cell does not go on to divide. If the "evaluation" is positive, the cell becomes committed to DNA replication and cell division. The commitment to cell division is much like launching the space shuttle: once a cell passes through the restriction point, it cannot turn back and return to the G_1 phase.

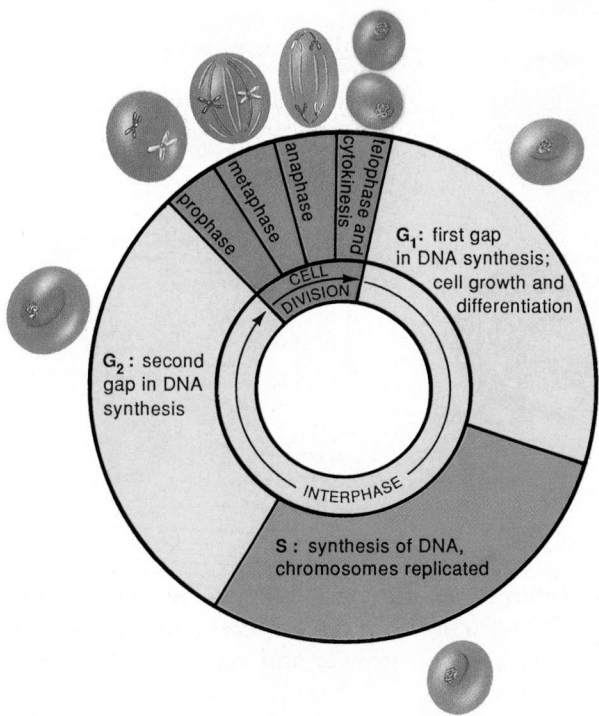

Figure 9-5 The eukaryotic cell cycle consists of two major phases, interphase and cell division. Each is divided into several subphases, as described in the text.

S Phase

Chromosome replication defines the **S,** or **synthesis, phase,** since this is the only time DNA synthesis normally occurs. Each chromosome is duplicated—just once. How a cell keeps from making multiple copies of its chromosomes is unknown. During the S phase, animal cells also duplicate their centrioles.

G_2 Phase

The period after DNA synthesis, but before the next cell division, is called the **G_2 phase** (the "second gap" in DNA synthesis). The cell is already committed to cell division before entering the G_2 phase. Most of G_2 seems to be spent in synthesizing molecules (other than DNA) that are required for cell division.

Progress Through Interphase

Many types of mammalian cells progress steadily through interphase, spending about 5 hours in the G_1 stage, 7 hours replicating their DNA during the S phase, and 3 hours in G_2 preparing for division. Cell division itself usually takes about an hour. Certain cell types, however, divide extremely rapidly, while others may go for weeks or even an entire lifetime without dividing. These differences in cell cycle length usually arise from differences in the time spent in G_1. The early cell divisions of an animal embryo, for instance, occur in rapid succession with virtually no G_1 phase at all and therefore almost no growth between divisions (see Chapter 40). In contrast, nerve cells in the adult mammalian brain no longer divide, but remain in the G_1 phase for life.

Cell Division

Cell division consists of two potentially independent events, nuclear division and cytoplasmic division. During nuclear division (called **mitosis,** or the **M phase** of the cell cycle), identical, complete copies of all the chromosomes are packaged into two new nuclei. During cytoplasmic division (also called **cytokinesis**), the cytoplasm is split into two daughter cells, with each cell receiving one of the newly formed nuclei and (usually) roughly equal amounts of cytoplasm. Therefore, the two daughter cells are essentially identical to each other, both genetically and cytoplasmically. They are also genetically identical to the parent cell.

Not all cells adhere to this scheme. Some cells, including vertebrate skeletal muscle cells and some fungi, undergo mitosis without cytokinesis, a process that produces single cells with many nuclei.

As we mentioned already, a multicellular organism

grows from a single fertilized egg by repeated cell divisions. Therefore, with a few exceptions such as sex cells and the cells of the immune system, **every cell of a multicellular organism is genetically identical.** Nerve cells and liver cells differ greatly in structure and function, but not in the genes that they contain. Rather, they differ in which genes are operating during the life of the cell.

Much of biology, reduced to its essentials, is the study of cellular activities during interphase, which result in such diverse phenomena as photosynthesis, muscle movement, and thought. Most of the rest of this text will be devoted to these topics. Here, we focus on the events of cell division.

The Eukaryotic Chromosome

Eukaryotic cells contain much more DNA than prokaryotic cells do. The circular chromosome of a bacterium is a millimeter or two in circumference, whereas the total length of all the chromosomes in a human cell, for example, is about *2000* millimeters. Eukaryotic cells package these enormous amounts of DNA, along with a roughly equal amount of protein, into many separate chromosomes. Even so, eukaryotic chromosomes are often much longer than prokaryotic chromosomes. Human cells have 46 chromosomes, so the average human chromosome is over 40 millimeters long.

The complicated events of eukaryotic cell division are largely an evolutionary solution for sorting out a large number of long chromosomes. To help you understand the mechanisms of eukaryotic cell division, we will begin by taking a closer look at the structure of the eukaryotic chromosome.

Chromatin and Chromosomes

Under the light microscope, interphase nuclei appear as large, darkly staining bodies without much internal structure. Early microscopists named the material in the nucleus **chromatin,** from the Greek word for color, because chromatin is stained by some commonly used dyes. During cell division, dark thread-like structures form where the nucleus used to be. The microscopists called these threads **chromosomes,** meaning "colored bodies." With the advent of the electron microscope and advances in molecular biology, we now know that chromatin and chromosomes are actually the same thing (DNA complexed with proteins), but in different stages of condensation.

Each eukaryotic chromosome is a single long molecule of DNA, continuous from end to end, complexed with special proteins called histones. During most of

its life (i.e., interphase), a cell must be able to read the information contained in the chromosomes, which can be done only when the chromosomes are extended (see Chapter 13). In this extended state, individual chromosomes are too thin to be visible in light microscopes, or even in ordinary electron microscopes. During cell division, however, the chromosomes must be sorted out and moved into the daughter nuclei. Just as sewing thread is easier to organize when it is wound onto spools, sorting and transporting chromosomes is easier if they are condensed and shortened (Fig. 9-6).

Chromosome Structure During Mitosis

The nomenclature used by biologists to describe chromosomes during mitosis often troubles beginning students. However, you will find it difficult to understand cell division without understanding the terminology, so follow carefully. For simplicity, the drawings below illustrate chromosomes only in the condensed state.

Most chromosomes consist of two arms that extend

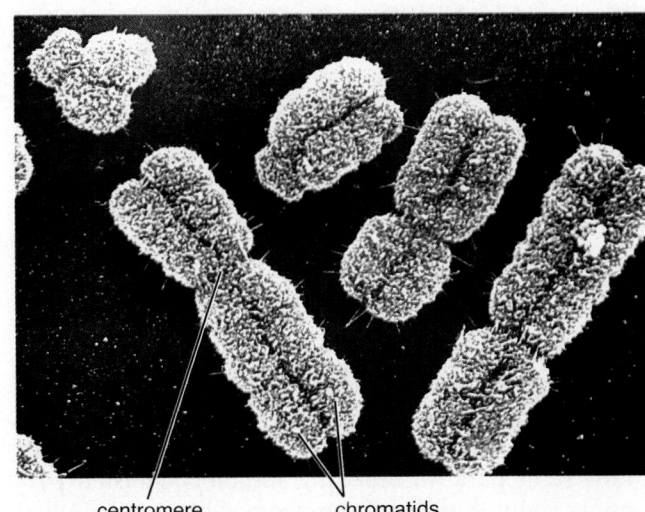

centromere   chromatids

Figure 9-6 A scanning electron micrograph of human chromosomes, as seen during cell division. The DNA and protein of the chromosomes, which are thin and extended during interphase, have coiled up into the thick, short structures seen here. The chromosomes were replicated prior to condensation; the two strands, called chromatids, are attached at the centromere. To get a sense of the degree of compaction, look at the fuzzy texture of the chromosomes: each strand of "texture" is a loop of chromosome, about as thin as the whole chromosome would look during interphase. The condensed chromosomes are about 5 to 20 micrometers long, while the same chromosomes during interphase would relax into thin strands about 10,000 to 40,000 micrometers long.

out from a specialized region called the **centromere** (meaning "middle body"):

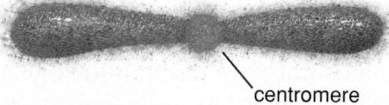

During interphase, chromosomes are duplicated, but the two copies remain attached at their centromeres. **As long as they are attached to one another, the copies are called sister chromatids.** Each chromatid is a single DNA molecule identical to the DNA of the original chromosome before replication. The two DNA molecules are attached to each other at the centromere. The whole structure (two sister chromatids attached at their centromeres) is still considered to be a single chromosome:

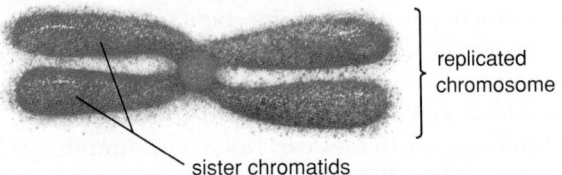

replicated chromosome

sister chromatids

Figure 9-6 therefore shows several single chromosomes each consisting of two chromatids.

During mitosis, the two sister chromatids separate. **When the chromatids separate, each chromatid becomes an independent daughter chromosome:**

independent daughter chromosomes

Numbers of Chromosomes

The chromosomes of each species have characteristic shapes, sizes, and staining patterns (Fig. 9-7). The nonreproductive cells of many organisms contain *pairs of chromosomes* that are the same length and stain in the same pattern. Breeding and biochemical experiments show that the chromosomes of each pair also have similar, although usually not identical, genetic content. Therefore, the members of a pair are called **homologues,** meaning "to say the same thing" in Greek. Cells with pairs of homologous chromosomes are called **diploid.** The 46 chromosomes of the nonreproductive cells of human beings occur as 23 pairs of homologues.

Not all cells have pairs of homologous chromosomes. At some point in the life cycle of all sexually reproducing organisms, a special type of cell division called **meiosis** produces cells that have only one copy of each type of chromosome (see Chapter 10): these cells are called **haploid.** In animals, these haploid

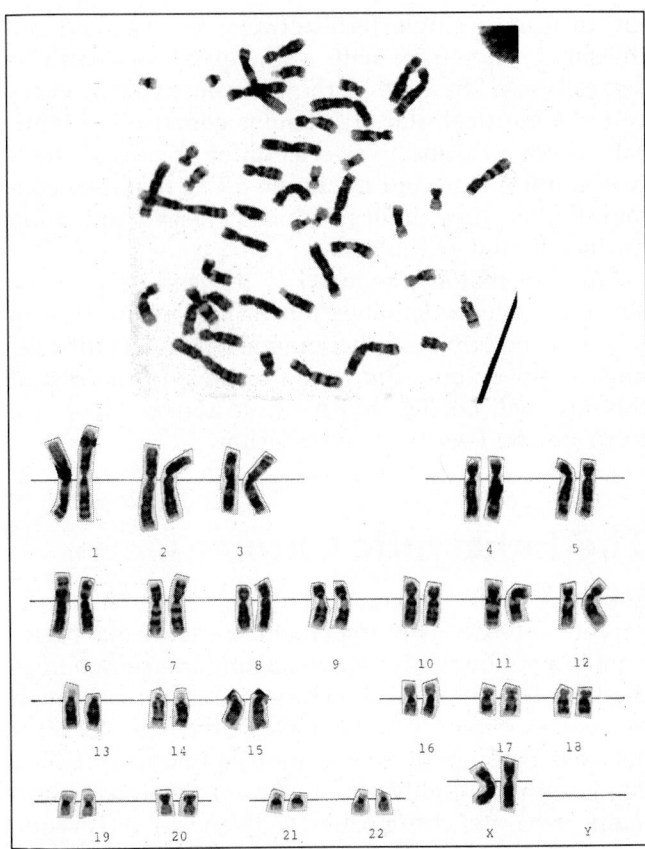

Figure 9-7 The karyotype of a human female, showing the complete complement of chromosomes displayed according to size. First, the chromosomes of a dividing cell are stained and photographed (top). Pictures of the individual chromosomes are cut out and arranged in descending order of size. Note that the chromosomes occur in pairs, which are similar in both size and staining patterns. If these chromosomes were from a male, there would be one X and one Y chromosome. In humans, the Y chromosome is much smaller than the X chromosome.

cells are the sex cells, or **gametes** (sperm and egg). Gametes cannot live independently; male and female gametes fuse to reform a diploid cell, the fertilized egg or **zygote.** In plants, the haploid cells produced by meiosis are not gametes. They undergo ordinary cell division to produce multicellular, haploid organisms. Certain cells of these haploid organisms later become gametes, and fuse to form a zygote. As you may have already guessed, in both plants and animals one of each pair of chromosomes in the diploid zygote is inherited from each parent. We will see how this happens in Chapter 10.

Some organisms have more than two homologues of each type of chromosome in their cells, and are

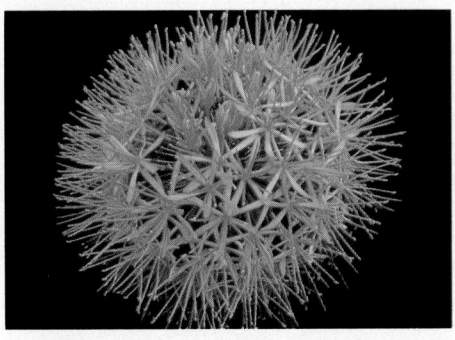

(a) The flower of the African blood lily.

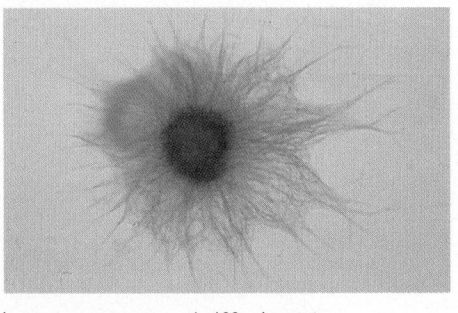

⊢——————————⊣ 100 micrometers

(b) Interphase in a cell of the endosperm (a food storage organ in the seed) before mitosis begins. The chromosomes are in the thin, extended state, and appear as an amorphous mass of stained material in light micrographs such as this one. The microtubules during interphase form a vaguely star-shaped array leading from the nucleus to all parts of the cell.

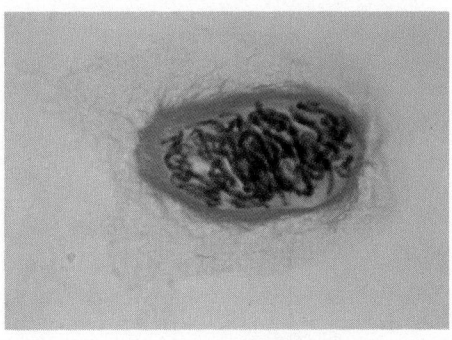

(c) Early prophase: the chromosomes have condensed into thick, separate strands.

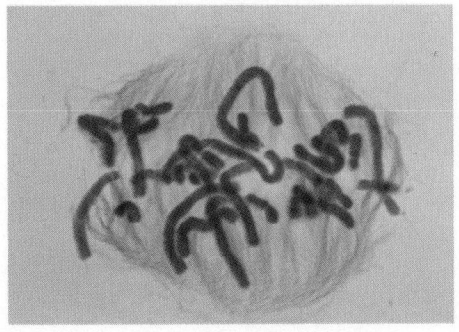

(d) Late prophase: the chromosomes are attached to microtubules of the almost fully developed spindle.

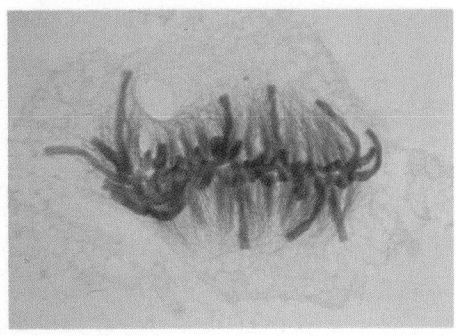

(e) Metaphase: the spindle has moved the chromosomes to the equator of the cell.

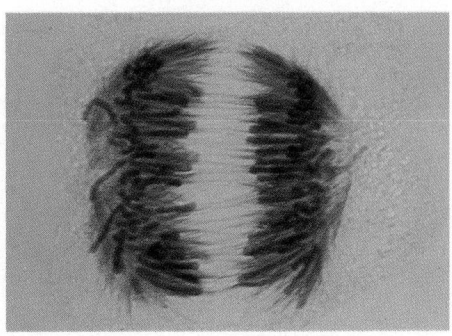

(f) Anaphase: the spindle is moving one set of chromosomes to each pole of the cell.

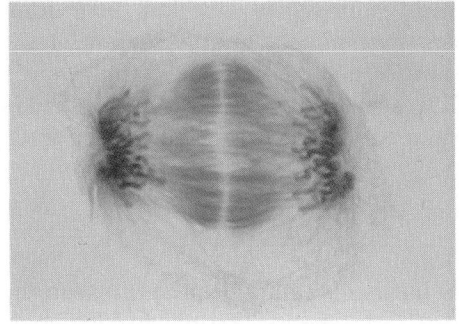

(g) Telophase: the chromosomes have been gathered into two clusters, one at the site of each future nucleus.

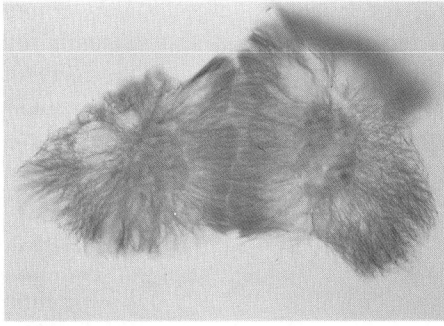

(h) Resumption of interphase: the chromsomes are relaxing again into their extended state. The spindle is disappearing and the microtubules are rearranging into the interphase array.

Figure 9-8 Mitosis as seen in the developing seed of the African blood lily, *Haemanthus katherinae*. In all of the micrographs, the chromosomes are stained bluish-purple, and the microtubules of the spindle are stained pink to red. Compare these micrographs with the drawings of mitosis in an animal cell shown in Figure 9-9. (Note that each micrograph is of a different single cell that has been fixed and stained at a particular stage of mitosis. The cells therefore have different sizes and shapes.)

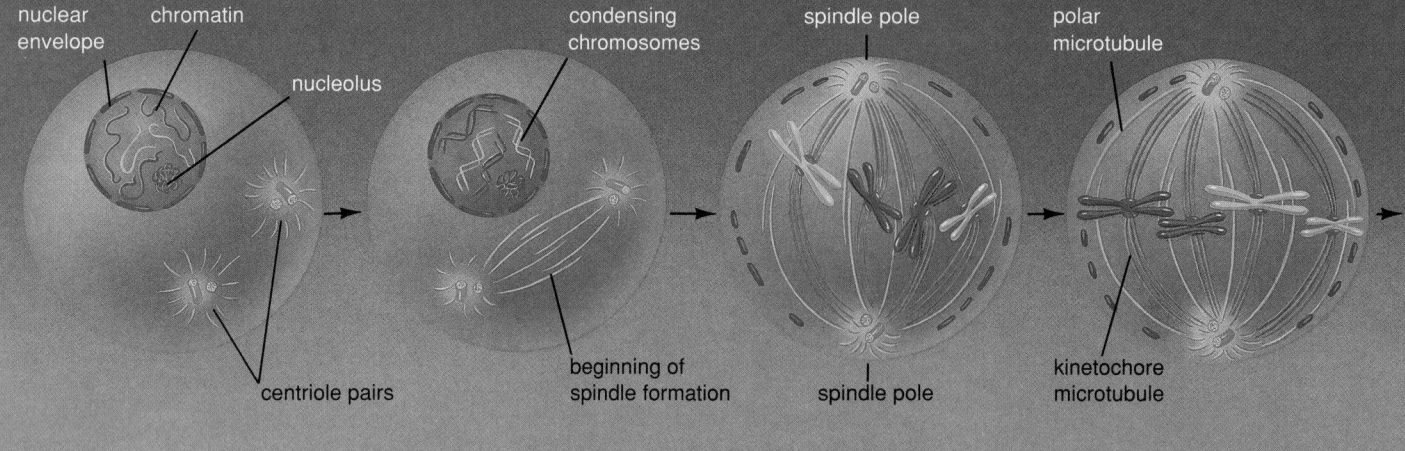

(a) LATE INTERPHASE   **(b) EARLY PROPHASE**   **(c) LATE PROPHASE**   **(d) METAPHASE**

nuclear envelope   chromatin   nucleolus   condensing chromosomes   spindle pole   polar microtubule

centriole pairs   beginning of spindle formation   spindle pole   kinetochore microtubule

(a) Late Interphase: The chromosomes have been replicated but remain elongated and relaxed within the nucleus. The centrioles within the microtubule organizing center have also been replicated.

(b) Early Prophase: The chromosomes condense and shorten, becoming visible with a light microscope. The centrioles begin to move apart, and the spindle apparatus begins to form.

(c) Late Prophase: The nuclear envelope breaks down and the spindle microtubules invade the nuclear region. Some spindle microtubules (the kinetochore microtubules) attach to each chromosome at its kinetochore. Other microtubules (the polar microtubules) extend from each pole to the equator of the cell, where their ends overlap.

(d) Metaphase: The spindle is fully formed. The kinetochore microtubules move the chromosomes to the equator of the cell.

Figure 9-9 Mitosis and cytokinesis in an animal cell.

called **polyploid.** In biological shorthand, **the number of different types of chromosomes is termed the haploid number,** and is designated n. For humans, $n = 23$; for other organisms, n ranges from 1 to several hundred. Diploid cells, with two homologues of each type, are $2n$, while polyploid cells may be $3n$ (triploid), $4n$ (tetraploid), and so on.

Mitosis

As we outlined earlier, eukaryotic cell division consists of mitosis (nuclear division) and cytokinesis (cytoplasmic division). These two processes usually occur together, but, in some cells, mitosis may occur without cytokinesis. Before we begin our discussion of mitosis, remember that the chromosomes are duplicated during the S phase of interphase. Therefore, when mitosis begins, each chromosome consists of two sister chromatids attached to one another at the centromere.

For convenience, mitosis is divided into four phases: prophase, metaphase, anaphase, and telophase (Figs. 9-8 and 9-9). As with most biological processes, these phases are not really discrete events. Rather, they form a continuum, each phase merging into the next.

Prophase

The first phase of mitosis is called **prophase** (meaning "the stage before" in Greek). Prophase develops slowly from the G_2 phase of interphase, as three major events occur: chromosome condensation, assembly of the spindle apparatus, and capture of the chromosomes by the spindle.

Chromosome Condensation

The chromosomes have already been duplicated during interphase, and sister chromatids remain attached to each other at their centromeres. During interphase, the chromosomes are relaxed, allowing the

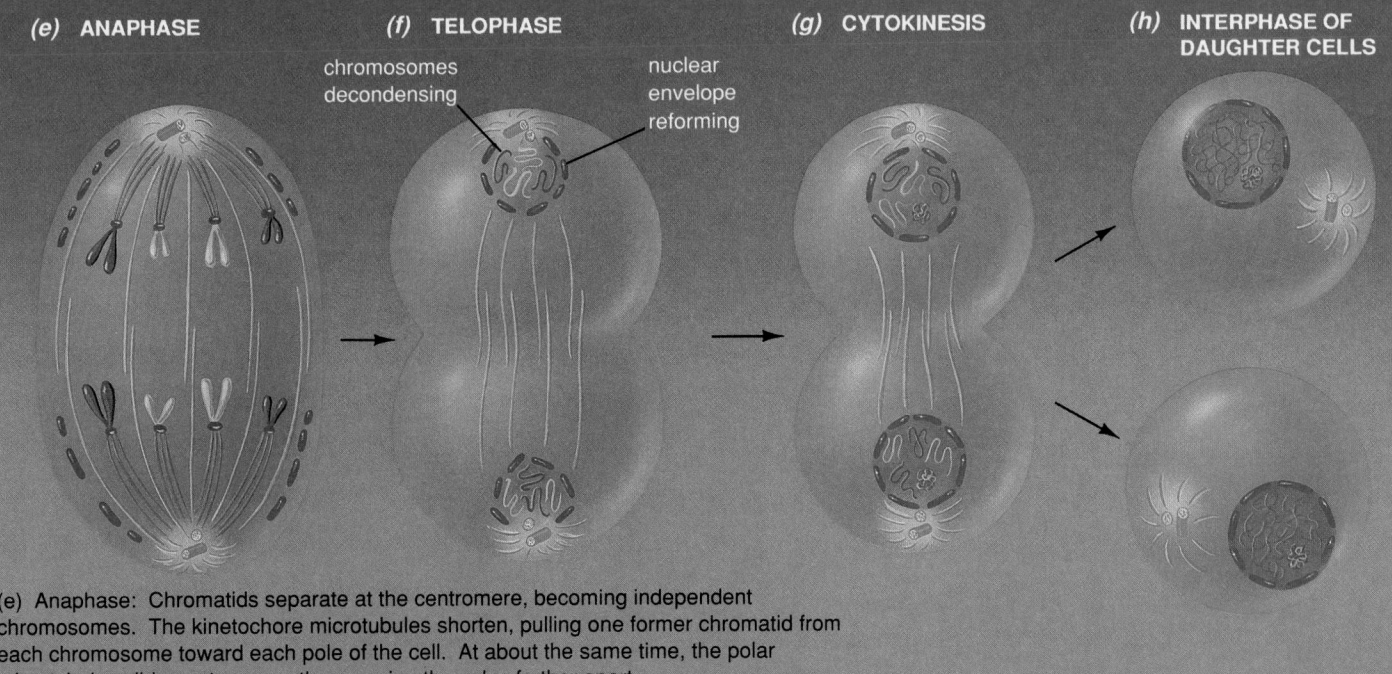

(e) ANAPHASE

(f) TELOPHASE

chromosomes decondensing

nuclear envelope reforming

(g) CYTOKINESIS

(h) INTERPHASE OF DAUGHTER CELLS

(e) Anaphase: Chromatids separate at the centromere, becoming independent chromosomes. The kinetochore microtubules shorten, pulling one former chromatid from each chromosome toward each pole of the cell. At about the same time, the polar microtubules slide past one another, moving the poles farther apart.

(f) Telophase: One complete set of chromosomes has reached each pole of the cell. The chromosomes relax into their extended state, the spindle disappears, and the nuclear envelopes begin to reform.

(g) Cytokinesis: Usually simultaneously with the end of telophase, the cytoplasm is divided along the equator of the parent cell, with each daughter cell receiving one nucleus and about half the original cytoplasm.

(h) Interphase of Daughter Cells: The daughter cells return to interphase. The spindle disappears, the nuclear envelope completely reforms, the chromosomes finish decondensing, and the nucleolus reappears.

information in their DNA to be read by the cell (Fig. 9-9a). During prophase, the chromosomes coil and condense, becoming visible in a light microscope (Fig. 9-9b). As the chromosomes condense, the nucleolus disappears. You may recall from Chapter 5 that the nucleolus contains an aggregation of the parts of several chromosomes that are involved in ribosome synthesis. As the chromosomes condense and separate from one another, this aggregation breaks up.

Spindle Formation

Near the end of prophase, after the chromosomes are well condensed, the **spindle apparatus** begins to form. The spindle apparatus is a football-shaped array of microtubules that will separate the sister chromatids. The spindle arises in the following way.

During interphase, a cell contains an array of microtubules radiating outward from the **microtubule organizing center** near the nucleus (Fig. 9-9a). During prophase, this interphase array disintegrates. The microtubule organizing center splits in two, and each

daughter center begins to form a new array of microtubules (Fig. 9-9b). Gradually the two microtubule organizing centers move to opposite sides of the nucleus (the poles of the cell). The microtubules extending from each pole almost surround the nucleus. This new microtubule array is called the spindle apparatus because of its resemblance to the thread wound around the spindle of a spinning wheel—fat in the middle, and tapering more-or-less to a point on either end.

Although the operation of the spindle during mitosis is similar in all eukaryotic cells, the spindles of animal and plant cells differ in two details (Fig. 9-10). First, in animal cells, each microtubule organizing center contains a pair of **centrioles.** During interphase, the two centrioles separate and a new centriole develops near the base of each "parent" centriole, resulting in a microtubule organizing center containing two pairs of centrioles. When the microtubule organizing center splits, each daughter center contains one pair of centrioles. Second, in animal cells,

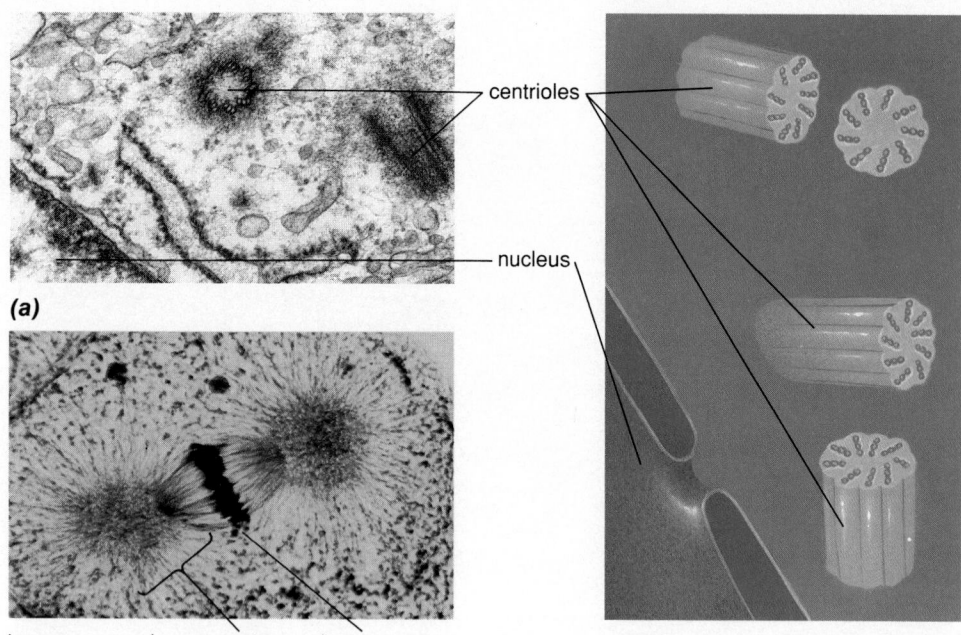

(a)

centrioles

nucleus

aster chromosomes

10 micrometers

(b)

Figure 9-10 (a) An electron micrograph of the microtubule organizing center of an animal cell. The centrioles have been replicated just prior to cell division. The centrioles of each pair lie at right angles to each other. During mitosis, the two pairs migrate to opposite sides of the nuclear region (see Fig. 9-9b).
(b) Asters in embryonic whitefish cells. The asters consist of microtubules that extend both in toward the equator of the cell and out toward the plasma membrane. The centrioles, which are not visible at this magnification, lie in the center of the asters.

spindle microtubules not only extend across the nucleus toward the other pole, but also extend outward in all directions from the microtubule organizing center toward the plasma membrane, forming a star-shaped **aster.** Plant cells, in contrast, do not have centrioles and do not form asters.

Capture of the Chromosomes

When the spindle is fully formed, the nuclear envelope abruptly disintegrates. The double membrane of the envelope breaks up into vesicles that resemble pieces of endoplasmic reticulum. The disruption of the nuclear envelope allows the microtubules of the spindle to invade the nuclear region and capture the chromosomes (see Fig. 9-9c).

Some of the spindle microtubules attach to the chromosomes at sites near the centromeres called **kinetochores** (Fig. 9-11). Each sister chromatid has its own kinetochore. Anywhere from a few to several dozen **kinetochore microtubules** attach to each kinetochore. Somehow, each kinetochore is grabbed by microtubules that run to only one pole of the spindle. Further, the kinetochore of one sister chromatid of each chromosome is captured by microtubules leading to one pole, while the kinetochore of the other sister chromatid is captured by microtubules leading to the other pole. The sister chromatids of each chromosome therefore face opposite poles of the spindle (see Fig. 9-11).

Other microtubules do not attach to chromosomes, but retain free ends that overlap along the cell's equa-

tor. As we will see, these **polar microtubules** push the two spindle poles apart.

Metaphase

During **metaphase** (the "middle stage"), the kinetochore microtubules running to opposite spindle poles engage in a tug of war, each pulling toward its own pole. Apparently, long kinetochore microtubules pull harder than short ones do. Therefore, if a chromosome is farther from one pole than from the other pole, it is dragged toward the more distant pole. The tug of war always ends in a draw: when a chromosome is aligned along the cell's equator, midway between the spindle poles, the kinetochore microtubules leading to each pole pull equally hard. Metaphase ends when all the chromosomes have lined up along the equator (see Fig. 9-9d).

Anaphase

At the beginning of **anaphase** (see Fig. 9-9e), the centromere of each chromosome divides, and the sister chromatids separate into two independent daughter chromosomes. The kinetochore microtubules running to opposite spindle poles no longer fight against each other; they shorten, pulling their attached chromosomes toward their respective poles (Fig. 9-12a). Simultaneously, the polar microtubules grow longer and push the spindle poles apart (Fig. 9-12b).
Note that one sister chromatid of each chromo-

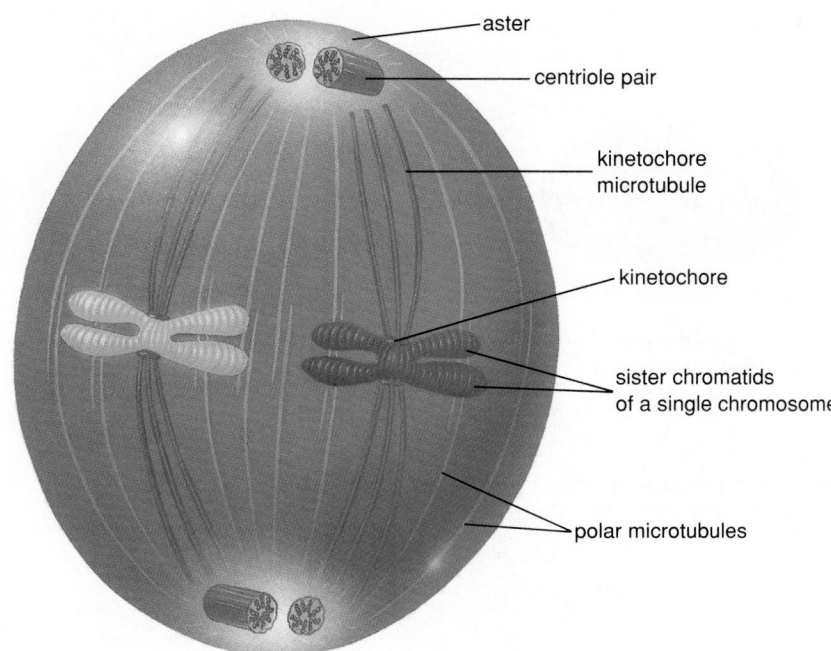

aster

centriole pair

kinetochore microtubule

kinetochore

sister chromatids of a single chromosome

polar microtubules

Figure 9-11 The spindle apparatus in an animal cell. Note that there are three sets of spindle microtubules. The aster consists of microtubules radiating out from the pair of centrioles at the pole toward the plasma membrane (see Fig. 9-10b). Kinetochore microtubules run from one pole to one kinetochore of each chromosome. The two kinetochores of a single chromosome attach to kinetochore microtubules that lead to opposite poles of the cell. Finally, polar microtubules extend from each pole to, and slightly beyond, the equator of the cell. Polar microtubules overlap at the equator.

(a)

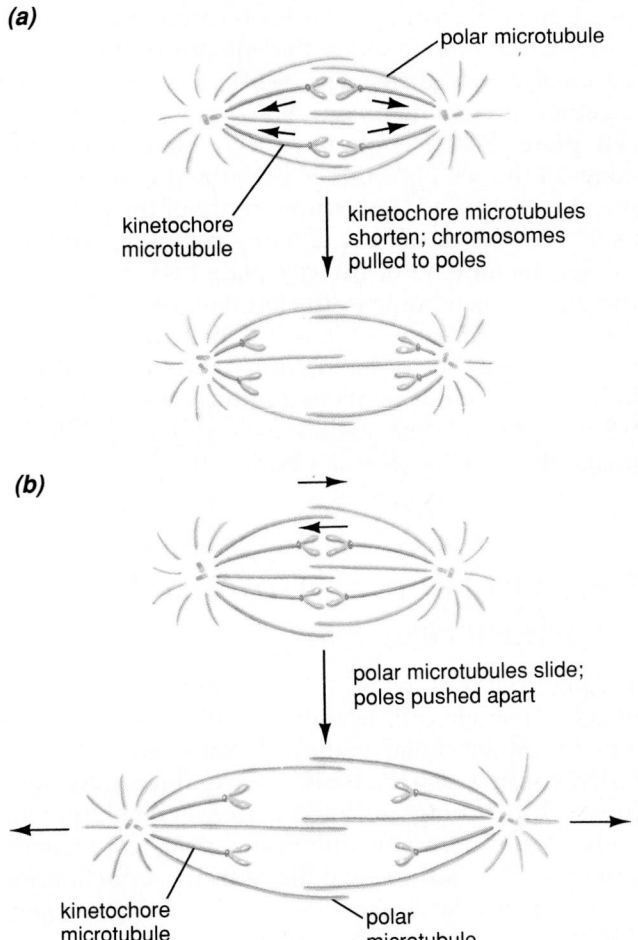

polar microtubule

kinetochore microtubule

kinetochore microtubules shorten; chromosomes pulled to poles

(b)

polar microtubules slide; poles pushed apart

kinetochore microtubule

polar microtubule

Figure 9-12 Spindle movements during anaphase, just after the separation of chromatids to form individual daughter chromosomes.
(a) Movement of chromosomes toward the poles: Kinetochore microtubules shorten, while remaining attached to the kinetochores of the newly formed daughter chromosomes. This pulls the chromosomes toward the poles.
(b) Separation of the poles: At their sites of overlap along the equator of the cell, polar microtubules slide past one another, probably by a mechanism similar to bending of cilia or flagella (see Fig. 5-22). This pushes the poles of the spindle farther apart.

(a)

(b)

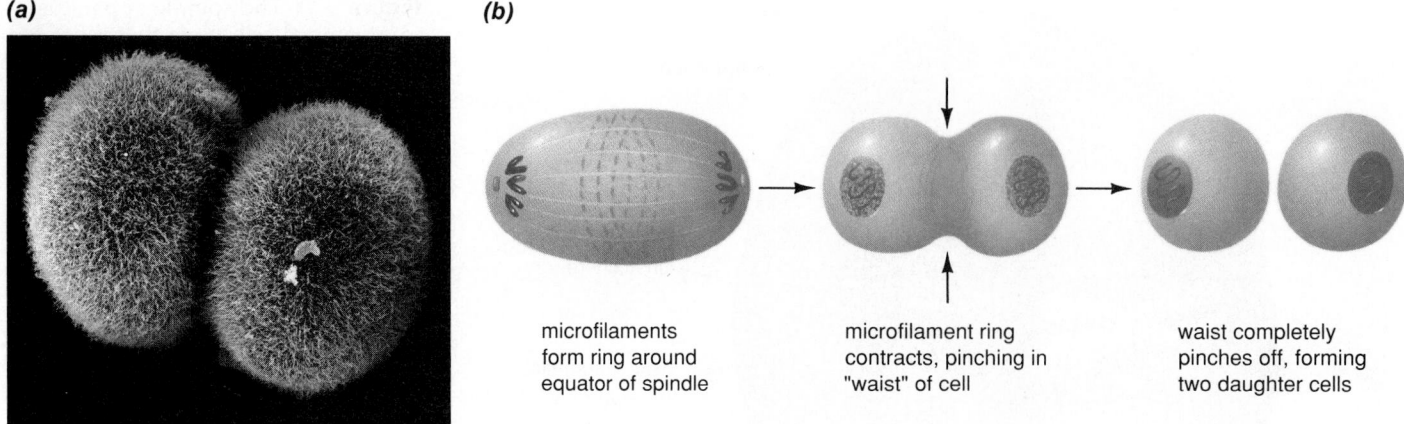

microfilaments
form ring around
equator of spindle

microfilament ring
contracts, pinching in
"waist" of cell

waist completely
pinches off, forming
two daughter cells

Figure 9-13 Cytokinesis in animal cells. **(a)** A ring of microfilaments just beneath the plasma membrane contracts around the equator of the cell, pinching it in two. **(b)** The mechanism of cytokinesis in animal cells.

some moves toward each pole of the cell. Because the sister chromatids are identical copies of the original chromosomes, the two clusters of chromosomes that form each contain one copy of every chromosome. Further, mitosis works equally well in haploid, diploid, or polyploid cells, and whether there are few or many chromosomes.

Telophase

When the chromosomes reach the poles of the spindle, **telophase** (the "end stage") has begun (see Fig. 9-9f). The spindle disintegrates. Vesicles that formed when the old nuclear envelope broke up during late prophase coalesce around each group of chromosomes, forming two new nuclear envelopes. The chromosomes relax into their extended state once again, and the nucleoli reappear. In most cells, cytokinesis occurs during telophase, enclosing each daughter nucleus in its own separate cell.

Cytokinesis

In most cells, the division of the cytoplasm into nearly equal halves begins during telophase. In animal cells, microfilaments composed of actin and myosin form rings around the equator of the cell, surrounding the remnants of the spindle (Fig. 9-13). The microfilaments are attached to the plasma membrane. During cytokinesis, the rings contract and pull in the equator of the cell, much like pulling the drawstring around the waist of a pair of sweatpants. Eventually the "waist" contracts down to nothing, dividing the cytoplasm into two new daughter cells.

Cytokinesis in plant cells is quite different, perhaps because the stiff cell wall makes it impossible to divide one cell into two by pinching at the waist. Instead, the Golgi complex buds off carbohydrate-filled vesicles that line up along the equator of the cell between the two nuclei (Fig. 9-14). The vesicles fuse together, producing a pancake-shaped structure, the **cell plate.** When enough vesicles have fused, the edges of the cell plate merge with the original plasma membrane around the circumference of the cell. Seen as if you were facing the dividing cell in Figure 9-14, the left membrane of the cell plate becomes part of the plasma membrane of the left daughter cell, while the right membrane of the cell plate becomes the plasma membrane of the right daughter cell. The carbohydrate formerly contained in the vesicles remains between the plasma membranes as the middle lamella of the cell wall (see Chapter 6).

Cell Division and Asexual Reproduction

Through cell division and subsequent diversification of cells, a single cell, usually a fertilized egg, generates the multicellular bodies of plants and animals. Cell division is also the basis of **asexual reproduction,** in which offspring are formed from a single parent without the necessity of uniting male and female gametes. This is the usual mode of reproduction for many unicellular organisms, such as *Paramecium, Euglena,* and yeasts (Fig. 9-15a).

Many multicellular organisms can also reproduce asexually. Through cell division, small replicas of the

(a)

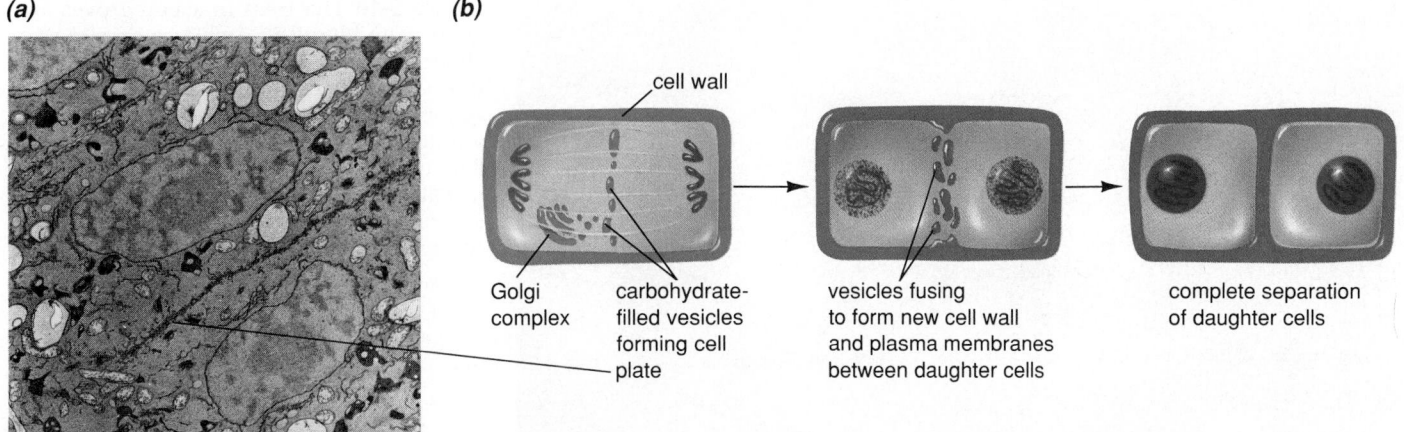

(b)

cell wall

Golgi complex

carbohydrate-filled vesicles forming cell plate

vesicles fusing to form new cell wall and plasma membranes between daughter cells

complete separation of daughter cells

Figure 9-14 Cytokinesis in plant cells. **(a)** Carbohydrate-filled vesicles produced by the Golgi complex congregate at the equator of the cell, forming the cell plate. The vesicles will fuse to form the two plasma membranes separating the daughter cells, while their carbohydrate contents form the middle lamella. **(b)** The mechanism of cell plate formation and cytokinesis in plant cells.

(a)

1 micrometer

(b)

Figure 9-15 Modes of asexual reproduction. **(a)** In unicellular microorganisms, such as the protist *Tetrahymena* shown here, cell division produces two new, independent organisms. **(b)** *Hydra*, a freshwater relative of jellyfish and anemones, grows a miniature replica of itself (a bud) protruding from its side. When fully developed, the bud breaks off and assumes independent life.

Figure 9-16 The trees in aspen groves are genetically identical, each tree growing as shoots from the roots of a single ancestral tree. This photo shows three groves near Aspen, Colorado. The genetic identity within a grove, and genetic difference between groves, is shown by their leaves in fall. One entire grove is still green, the second has turned bright gold, while the third has already lost its leaves.

parent are grown. A *Hydra*, for example, can reproduce by growing a miniature new *Hydra* as a bud (Fig. 9-15b). Eventually the bud separates from its parent, going off to live independently. Because mitosis produces genetically identical cells, these offspring are genetically identical to their parents.

Many plants reproduce both sexually and asexu-

ally. The beautiful aspen groves of Colorado and New Mexico (Fig. 9-16) develop asexually from shoots growing up from the root system of a single parent tree. The entire grove, although seeming to be a population of separate trees to the admiring visitor, is actually a clone of genetically identical individuals, interconnected by their root systems.

SUMMARY OF KEY CONCEPTS

Essentials of Cellular Reproduction

Every cell must begin life with (1) a complete set of genetic information (DNA) and (2) the materials necessary to survive and use the genetic information. Cellular reproduction involves replication of the DNA, with one copy packaged into each daughter cell, and cytoplamic division, with roughly half the parental cytoplasm passing into each daughter cell.

The Prokaryotic Cell Cycle

The DNA of a prokaryotic cell is in the form of a single circular chromosome, attached at one point to the cell membrane. When a prokaryotic cell reproduces, it duplicates the chromosome, moves the two resulting copies to opposite ends of the cell, and divides the cell in two.

The Eukaryotic Cell Cycle

The eukaryotic cell cycle consists of two major phases, interphase and cell division. During interphase, the cell takes in nutrients, grows, and duplicates its chromosomes. Cell division consists of two processes, nuclear division (mitosis) and cytoplasmic division (cytokinesis). Mitosis parcels out one copy of each chromosome into two separate nuclei, and cytokinesis subsequently encloses each nucleus in a separate cell.

The Eukaryotic Chromosome

The chromosomes of eukaryotic cells consist of both DNA and protein. During interphase, the chromosomes are in an extended form, accessible for use in reading their genetic instructions. During cell division, the chromosomes condense to form short, thick structures. At some point in all eukaryotic life cycles, cells contain pairs of chromosomes called homologues. Homologues have virtually identical appearance and similar genetic information. Cells with pairs of homologous chromosomes are diploid. Those with only a single member of each pair are haploid, and those with more than two copies of each homologous chromosome are polyploid.

Mitosis

The chromosomes are duplicated during interphase, prior to mitosis. The two identical replicas, called chromatids, are attached to one another at the centro-mere. During mitosis, the chromatids separate, one going to each daughter nucleus. The two daughter nuclei thus receive identical genetic information. Mitosis consists of four phases:

Prophase: The chromosomes condense and the spindle microtubules begin to form. The nuclear membrane breaks apart into vesicles. The spindle microtubules invade the nuclear region. Some microtubules, now called kinetochore microtubules, attach to the chromosomes at kinetochores located at the centromeres. Other microtubules, called polar microtubules, extend from the spindle poles to the equator, and overlap at the equator with polar microtubules extending from the opposite pole.

Metaphase: The chromosomes move to the equator of the cell.

Anaphase: The two chromatids of each chromosome separate; each former chromatid is now considered to be an individual chromosome. The spindle microtubules move the newly independent chromosomes to opposite poles of the cell.

Telophase: The chromosomes relax into their extended state and nuclear envelopes reform around each new daughter nucleus.

Cytokinesis

Mitosis may occur without cytokinesis, for example in mammalian muscle cells and many fungi. In most cells, however, cytokinesis roughly coincides with the end of telophase. Cytokinesis usually divides the cytoplasm into approximately equal halves. In animal cells, the plasma membrane is pinched in along the equator by a ring of microfilaments, separating the cytoplasm into two daughter cells, each containing one of the daughter nuclei. In plant cells, new plasma membrane forms along the equator by fusion of vesicles produced by the Golgi complex.

Cell Division and Asexual Reproduction

Some organisms reproduce asexually—that is, without the fusion of gametes. Asexual reproduction usually occurs by the growth of a new organism from the body of a single parent organism, by repeated cell divisions and differentiation of daughter cells. Therefore, the offspring are genetically identical to the parent.

GLOSSARY

Note: Roughly equivalent stages of mitosis and meiosis are given the same names (e.g., anaphase). For clarity, this glossary defines the stages only in terms of mitosis. Complete definitions that incorporate both mitosis and meiosis are in the glossary at the end of Chapter 10.

anaphase (an'-a-fāz): the stage of mitosis in which the sister chromatids of each chromosome separate from one another and are moved to opposite poles of the cell.

asexual reproduction: reproduction that does not involve the fusion of haploid gametes.

aster: during cell division in animals and some protists, a star-shaped array of microtubules extending in all directions outward from the centrioles.

cell cycle: the sequence of events in the life of a cell, from one division to the next.

cell division: in eukaryotes, the process of reproduction of single cells, usually into two identical daughter cells, by mitosis accompanied by cytokinesis.

cell plate: in plant cell division, a series of vesicles that fuse to form the new plasma membranes and cell wall separating the daughter cells.

centromere (sen'-trō-mēr): the region of a replicated chromosome at which the sister chromatids are held together.

chromatid (krō'-ma-tid): one of the two identical strands of DNA and protein forming a replicated chromosome. The two sister chromatids are joined at the centromere.

chromatin (krō'-ma-tin): the complex of DNA and proteins that makes up eukaryotic chromosomes.

chromosome (krō'-mō-sōme): in eukaryotes, a linear strand composed of DNA and protein, found in the nucleus of a cell, that contains the genes; in prokaryotes, a circular strand composed solely of DNA.

cytokinesis (sī-tō-ki-nē'-sis): division of the cytoplasm and organelles into two daughter cells during cell division. Usually cytokinesis occurs during telophase of mitosis.

diploid (dip'-loyd): referring to a cell with pairs of homologous chromosomes.

fission: cell division in prokaryotes.

gamete (gam'-ēt): a haploid sex cell formed in sexually reproducing organisms.

gene: a unit of inheritance; the segment of DNA that carries the genetic information required to produce a specific trait.

haploid (hap'-loyd): referring to a cell that has only one member of each pair of homologous chromosomes.

homologue (hō'-mō-log): a chromosome that is similar in appearance and genetic information to another chromosome with which it pairs during meiosis. Also called homologous chromosome.

interphase: the stage of the cell cycle between cell divisions. During interphase, chromosomes are replicated, and other cell functions occur, such as growth, movement, and acquisition of nutrients.

kinetochore (kī-nēt'-ō-kōr): a protein structure that forms at the centromere regions of chromosomes; attaches the chromosomes to the kinetochore microtubules of the spindle.

kinetochore microtubule: a microtubule of the spindle apparatus that connects the kinetochore of a chromosome with a spindle pole; involved in movement of the chromosome to the spindle pole during anaphase.

life cycle: the events in the life of an organism from one generation to the next.

metaphase (met'-a-fāz): the stage of mitosis in which the chromosomes, attached to kinetochore microtubules, are lined up along the equator of the cell.

meiosis (mī-ō'-sis): a type of cell division found in eukaryotic organisms, in which a diploid cell divides twice to produce four haploid cells.

microtubule organizing center: the region(s) of a eukaryotic cell at which tubulin is polymerized into microtubules.

mitosis (mī-tō'-sis): a type of nuclear division found in eukaryotic cells. Chromosomes are duplicated during interphase before mitosis. During mitosis, one copy of each chromosome moves into each of two daughter nuclei. The daughter nuclei are therefore genetically identical to each other.

polar microtubule: a microtubule of the spindle apparatus that extends from one spindle pole to the equator of the cell; involved in moving the spindle poles farther apart during anaphase.

polyploid (pol'-ē-ployd): having more than two homologous chromosomes of each type.

prophase (prō'-fāz): the first stage of mitosis, in which the chromosomes first become visible in the light microscope as thickened, condensed threads and the spindle begins to form. As the spindle is completed, the nuclear envelope breaks apart. The spindle invades the nuclear region and attaches to the kinetochores of the chromosomes.

sexual reproduction: a form of reproduction in which genetic material from two parental organisms is combined. In eukaryotes, two haploid gametes fuse to form a diploid zygote.

spindle apparatus: a football-shaped array of microtubules that moves the chromosomes to opposite poles of a cell during anaphase of meiosis and mitosis.

telophase (tēl'-ō-fāz): the last stage of mitosis, in which a nuclear envelope reforms around each new daughter nucleus, the spindle disappears, and the chromosomes relax from their condensed form.

zygote (zī'-gōt): in sexual reproduction, a diploid cell formed by the fusion of two haploid cells.

STUDY QUESTIONS

1. Define a cell cycle and a life cycle. What is the role of the cell cycle in a bacterium? In a human?
2. Diagram and describe the prokaryotic cell cycle.
3. Diagram and describe the eukaryotic cell cycle. Name the various phases and briefly describe the events that occur during each.
4. Define mitosis and cytokinesis. Do these two processes always occur together?
5. Diagram the stages of mitosis. How does mitosis en-sure that each daughter nucleus receives a complete set of chromosomes?
6. Define the following terms: homologous chromosome; centromere; kinetochore; chromatid; diploid; haploid; polyploid.
7. Diagram the spindle apparatus. How do the two types of spindle microtubules operate during anaphase?
8. Describe the process of cytokinesis in animal and plant cells.

DISCUSSION QUESTIONS

1. Do you think that each type of cell in your body steadily progresses through one cell division after another, or do you think that the rate of division is likely to vary from time to time? Explain and justify your answer in terms of some common events in human life: growing up as a child; healing a wound in your skin; menstruation; or replacing blood donated to the Red Cross, to name a few.
2. A drug called colchicine inhibits the polymerization of microtubules. What specific effects do you think colchicine would have on cell division? What effects would it have on the daughter cell(s)?
3. Suppose a slide containing 100 cells is made from a random sample of cells growing in a petri plate. The average time for a complete cell cycle to occur in the petri plate is 24 hours. Ten cells on the slide are in prophase, 3 cells are in metaphase, 6 cells are in anaphase, 8 cells are in telophase, and 73 cells are in interphase. What percentages of the cell cycle are spent in each of the five phases? How many hours, on the average, does a cell in the petri plate spend in each phase of the cell cycle? Discuss how one could determine how much time an average cell in the petri plate spends in the G_1, S, and G_2 portions of interphase.

SUGGESTED READINGS

Mazia, D. "The Cell Cycle." *Scientific American*, January 1974. A clearly illustrated description of the essentials of cell division.

McIntosh, J. R., and McDonald, K. L. "The Mitotic Spindle." *Scientific American*, October 1989. Beautiful micrographs and lucid drawings describe the structure and function of the spindle apparatus.

Murray, A. W., and Kirschner, M. W. "What Controls the Cell Cycle." *Scientific American*, March 1991. The synthesis, processing, and degradation of a group of proteins seems to control the progression of a cell through the stages of the cell cycle.

Sloboda, R. D., "The role of Microtubules in Cell Structure and Cell Division." *American Scientist*, volume 68, pages 290-298 (May-June) 1980. The various functions of the spindle apparatus are clearly described.

10

Meiosis and the Life Cycles of Eukaryotic Organisms

Whole groves of aspens that are clones of an ancestral parent; runners from strawberry plants, snaking out over the soil and generating new offspring plants; a baby *Hydra* budding off from its parent—asexual reproduction through cell division seems to be so simple and efficient. In contrast, sexual reproduction is such a nuisance. Consider male elephant seals, fighting to obtain a harem. These immense beasts develop long, rubbery snouts as displays to show off in front of rival males. They bash each other with their massive necks, and bite each other until their blood stains the beach. A defeated male may never reproduce at all, but victory is not unalloyed bliss, either. The victor often spends so much energy defending his harem that, by the end of the breeding season, he is drained of all reserves, a mockery of his former magnificence. Many do not survive the winter.

Why not just reproduce asexually, as a strawberry can? Or perhaps we should word the question the other way around—why, if asexual reproduction is so effective, do strawberries nevertheless flower and set seed, reproducing sexually as well as asexually?

Offspring produced through asexual reproduction arise from cell division. As we saw in Chapter 9, this means that the offspring are usually genetically identical to their parents. This may be an advantage if the parent organism is well adapted to its environment, if that environment never changes, and if the organism and its offspring never move to a new location. But suppose the environment changes, the climate grows warmer, or new predators appear. Neither the parent nor its asexually reproduced offspring may be well adapted any longer, and neither may survive. On the other hand, suppose that the offspring vary somewhat, both from the parent and from one another. Perhaps some of them would be better adapted to the changing environment.

Relatively early in the history of life on Earth, organisms evolved a way to produce offspring that are similar but not identical to their parents: **sexual reproduction.** Recall from Chapter 9 that most multicellular organisms are diploid, with *paired homologous chromosomes* in their body cells. As we will see in Chapter 11, segments of the DNA on the homologous chromosomes—their genes—influence the same traits, such as size or tolerance of high temperatures. However, the DNA that makes up the genes of homologous chromosomes is often not identical, and may influence the same trait in different ways. For example, the DNA of the "size gene" on one member of a pair of homologous chromosomes may promote large size, while the DNA of the "size gene" on its homologue may promote small size.

In sexually reproducing, multicellular organisms, parents produce haploid sex cells, or **gametes,** each containing only one of each pair of homologous chromosomes. Two gametes, usually from different parents, fuse to form a diploid cell with paired homologous chromosomes, the **zygote.** Through repeated mitotic cell divisions, the zygote then develops into the diploid offspring organism. Insofar as the homologous chromosomes from the two parents have different genes, their offspring will receive a combination of genes that is unique to itself and different from either parent. As we shall see, sexual reproduction offers virtually unlimited possibilities for producing genetically unique offspring.

Meiosis and Sexual Reproduction

Although the details vary widely from organism to organism, there are three features typical of sexual reproduction in almost all multicellular eukaryotes:

1. **Organisms engaging in sexual reproduction have diploid cells with pairs of homologous chromosomes at some stage in their life cycle.**
2. **At some point, the homologues are separated from one another through a special cell division process called** *meiosis,* **which produces haploid cells.** In animals, these haploid cells are gametes (i.e., sperm or egg). In plants and many fungi, the haploid cells are spores that undergo mitosis to produce a multicellular, haploid body. Certain cells of this haploid form then differentiate into gametes at a later time (see the section on Mitosis, Meiosis, and Eukaryotic Life Cycles later in this chapter).
3. **The haploid gametes fuse to form a diploid cell once again (the zygote), with one copy of each homologous chromosome donated by each parent.**

An Overview of Meiosis

The key to sexual reproduction is **meiosis,** the production of haploid cells with unpaired chromosomes. If organisms produced sex cells with the full diploid complement of chromosomes, then their offspring would be tetraploid ($4n$), having received $2n$ chromosomes from its father and $2n$ from its mother. The next generation would become 8*n*, the next 16*n*, and so on. Only by separating homologous chromosomes in meiosis and fusing the resulting haploid cells can each new generation remain diploid.

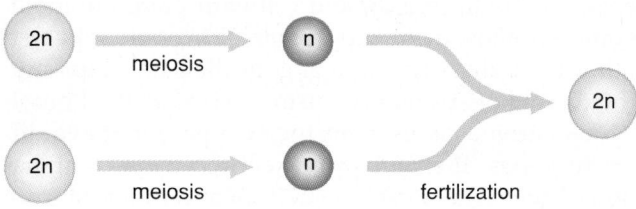

The word "meiosis" comes from a Greek word meaning "to diminish," and that is just what meiosis does: it reduces the number of chromosomes by half. Meiosis is not, however, a random division. **In meio-** **sis, each daughter cell receives one member of each pair of homologous chromosomes.**

The Principal Events and Outcomes of Meiosis

Let's begin with an overview of the primary events of meiosis (Fig. 10-1). Meiosis involves two nuclear divisions, termed **meiosis I** and **meiosis II.** The overall process of meiosis proceeds as follows:

1. All the chromosomes are replicated during interphase prior to meiosis I, with sister chromatids attached at the centromere of each chromosome.
2. During meiosis I, homologous chromosomes are separated; each daughter nucleus I receives one homologue of each pair of chromosomes. There-

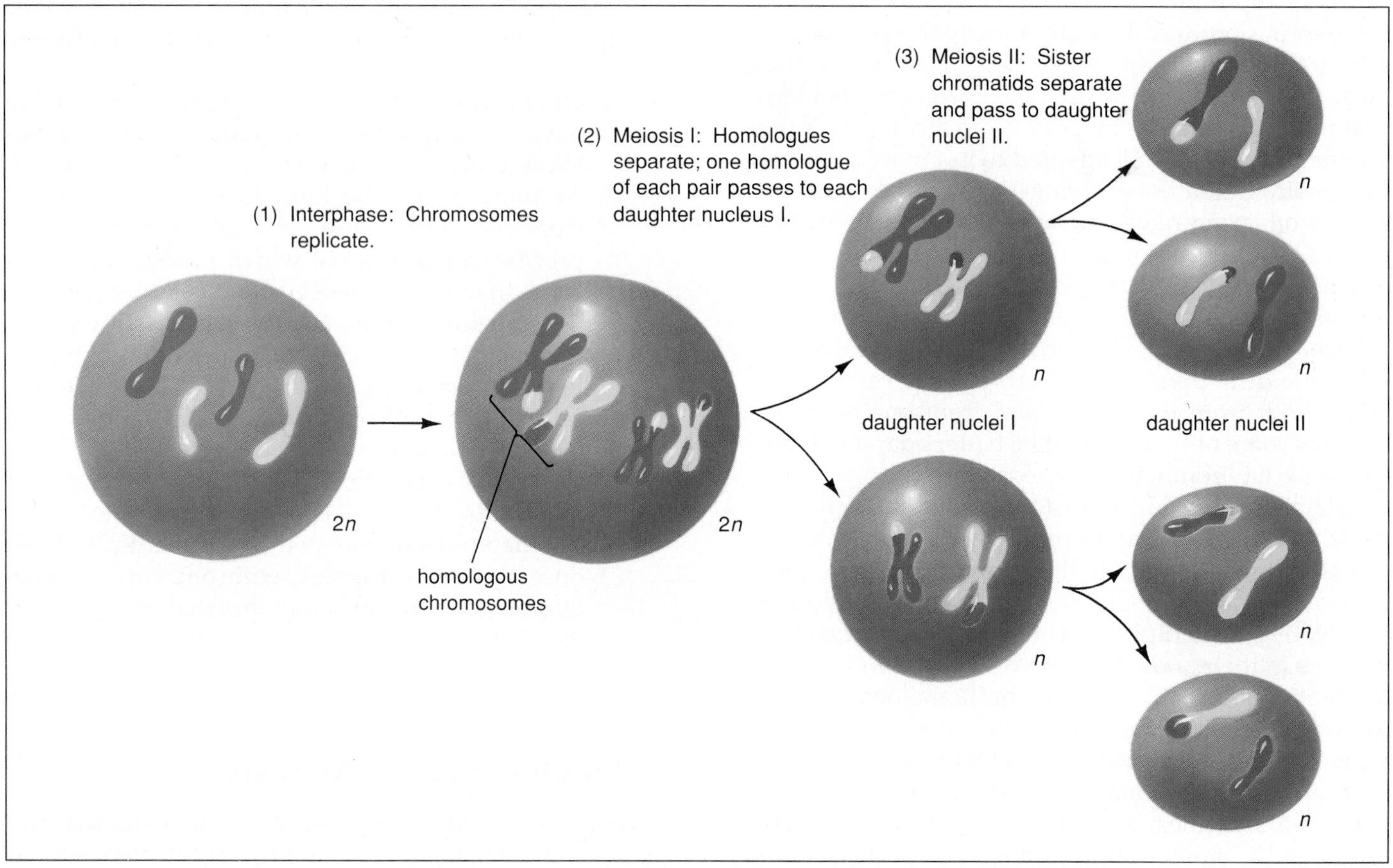

(1) Interphase: Chromosomes replicate.

(2) Meiosis I: Homologues separate; one homologue of each pair passes to each daughter nucleus I.

(3) Meiosis II: Sister chromatids separate and pass to daughter nuclei II.

2*n*

homologous chromosomes

2*n*

n

daughter nuclei I

n

n

n

daughter nuclei II

n

n

n

Figure 10-1 An overview of meiosis. Meiosis consists of two nuclear divisions: meiosis I and meiosis II. Together, the two divisions separate homologous chromosomes and produce haploid nuclei that each contain one member of each pair of homologues.

fore, the daughter nuclei are haploid. The sister chromatids do *not* separate during meiosis I, so at the completion of meiosis I each chromosome still consists of two joined sister chromatids. The chromosomes *do not replicate again* between meiosis I and meiosis II.

3. In meiosis II, the sister chromatids split into two independent daughter chromosomes, with one going into each daughter nucleus II.

Therefore, the reduction in chromosome number from diploid to haploid occurs during meiosis I. Meiosis II is essentially the same as mitosis in a haploid cell, with the number of chromosomes remaining the same. Each nuclear division is usually accompanied by cytokinesis. Therefore, a cell that undergoes both meiosis I and II produces a total of four haploid cells.

The Mechanisms of Meiosis

The phases of meiosis are given the same names as the roughly equivalent phases in mitosis, although, as we shall see, there are important differences, especially in meiosis I (Fig. 10-2).

Meiosis I

In multicellular organisms, meiosis occurs only in a relatively few, specialized cells in the reproductive organs. Meiosis begins in these cells with the replication of the chromosomes. (Since one function of meiosis is to *reduce* the number of chromosomes from diploid to haploid, you may think that to begin by duplicating the chromosomes is an unnecessary step. The cellular mechanisms of meiosis evolved from those of mitosis, and the replication of chromosomes is probably leftover evolutionary baggage. As we will see many times throughout this text, evolution is not an engineer; it's more like a tinker.) As in mitosis, the sister chromatids of each chromosome are attached to one another at the centromere. From this point on, meiosis I differs tremendously from mitosis.

Prophase I

During **prophase I,** homologous chromosomes line up side by side, exchange segments of DNA, and condense (Fig. 10-3). In some way that is not yet understood, homologous chromosomes find one another early in prophase I of meiosis (Step 1 in Fig. 10-3). (To make it easier to keep track of the homologues, we'll call one homologue the "maternal chromosome" and the other the "paternal chromosome," because, as we will see, each was originally inherited from the organism's mother and father, respectively.) The ends of the maternal and paternal chromosomes

attach to the nuclear envelope right next to each other, and the two chromosomes are brought together, side by side, much like zipping the zipper on a jacket (Step 2). Protein strands join the homologous chromosomes so that they match up exactly along their entire length. Then large enzyme complexes assemble at random places along the paired chromosomes (Step 3). The enzymes snip apart the chromosomes and graft the broken ends together again. However, *the enzymes almost always graft a maternal "stump" onto a paternal "stem," and vice versa* (Step 4). Thus, the maternal and paternal chromosomes now intertwine, forming crosses, or **chiasmata** (sing. **chiasma**). Eventually the enzyme complexes leave the chromosomes and the protein zippers that held homologues together disappear. The chromosomes coil and condense, becoming visible with a light microscope, with chiasmata connecting homologues (Step 5 in Fig. 10-3, and Fig. 10-4).

Chiasmata serve two functions. First, the maternal and paternal chromosomes have swapped some segments of their DNA, an event called *crossing over*. If, as is likely, they had slightly different DNA, then neither chromosome is quite the same as it was before the chiasmata formed. We will discuss the genetic consequences of crossing over in the next chapter. The second function of chiasmata is to hold the homologous chromosomes together during their attachment to the spindle microtubules. As we will see, this continued pairing is essential.

As in mitosis, the spindle begins to assemble outside the nucleus during prophase I. Near the end of prophase I, the nuclear envelope breaks up and the spindle microtubules capture the chromosomes by attaching to their kinetochores.

Metaphase I

During **metaphase I,** the spindle microtubules move the chromosomes to the equator of the cell. **Unlike in mitosis, during metaphase I of meiosis the chromosomes are aligned in homologous pairs along the equator. The two homologues of each pair are attached to kinetochore microtubules leading to opposite poles of the spindle.**

Since the attachment of the chromosomes to the spindle is really the key to understanding the movement of chromosomes during meiosis I, let's compare the spindle attachments of mitosis to those of meiosis I (Fig. 10-5). In mitosis, the two sister chromatids of each chromosome attach to kinetochore microtubules leading toward opposite poles of the cell (Fig. 10-5a). Therefore, when the centromeres split during anaphase, the former sister chromatids are pulled to opposite poles. In meiosis I, both chromatids of one

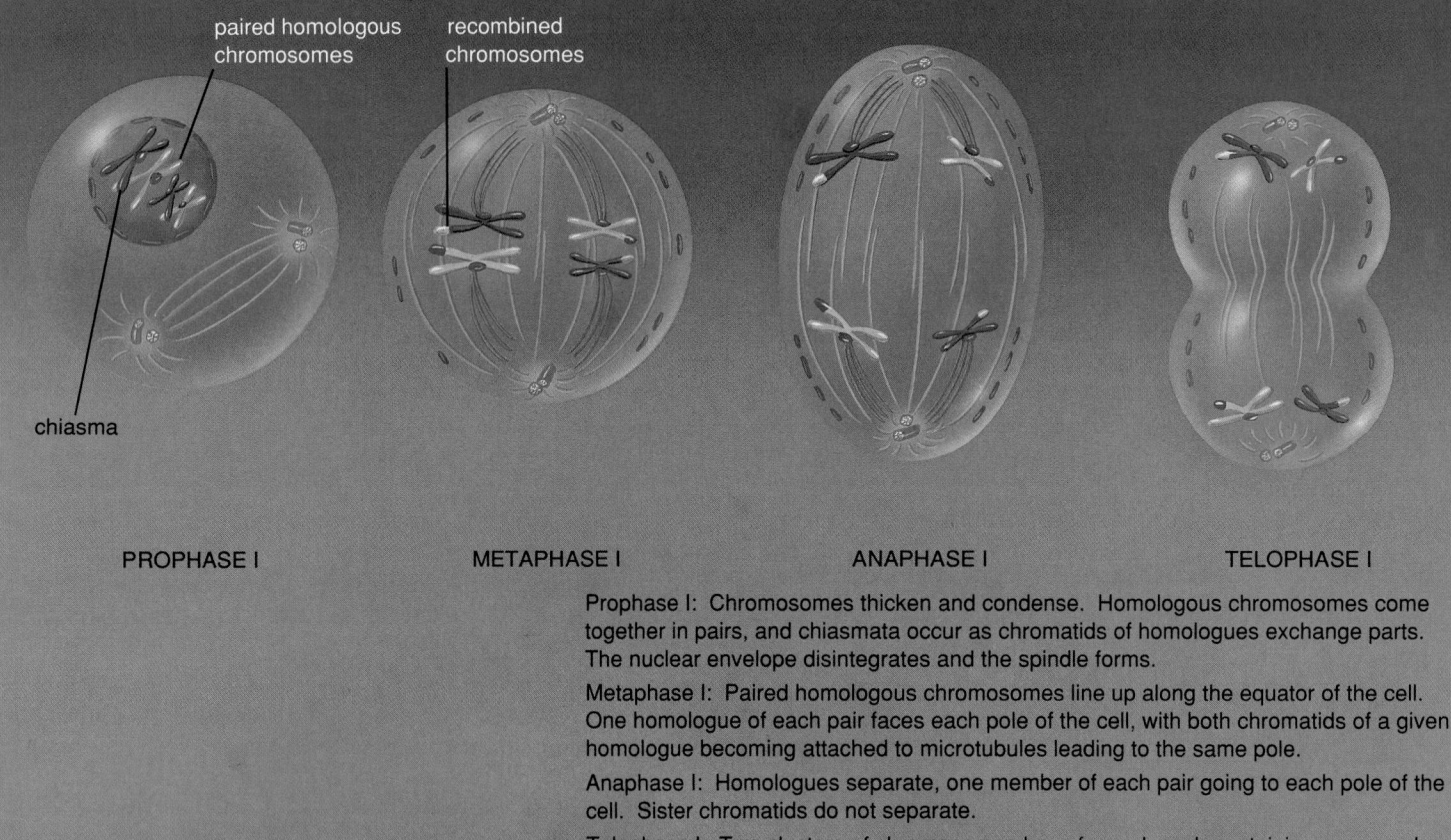

paired homologous chromosomes

recombined chromosomes

chiasma

PROPHASE I METAPHASE I ANAPHASE I TELOPHASE I

Prophase I: Chromosomes thicken and condense. Homologous chromosomes come together in pairs, and chiasmata occur as chromatids of homologues exchange parts. The nuclear envelope disintegrates and the spindle forms.

Metaphase I: Paired homologous chromosomes line up along the equator of the cell. One homologue of each pair faces each pole of the cell, with both chromatids of a given homologue becoming attached to microtubules leading to the same pole.

Anaphase I: Homologues separate, one member of each pair going to each pole of the cell. Sister chromatids do not separate.

Telophase I: Two clusters of chromosomes have formed, each containing one member of each pair of homologues. The daughter nuclei are therefore haploid. Cytokinesis often occurs during telophase I. There is usually little or no interphase between meiosis I and meiosis II.

Figure 10-2 Meiosis is the process by which the homologous chromosomes of a diploid cell are separated. Each of the four resulting haploid daughter cells contains one member of each pair of homologous chromosomes. In these diagrams, two pairs of homologous chromosomes are shown, large and small. The yellow chromosomes are from one parent (e.g., father), and the violet chromosomes are from the other parent.

homologue attach to kinetochore microtubules extending toward one pole of the cell, while both chromatids of its paired homologue attach to kinetochore microtubules leading to the opposite pole (Fig. 10-5b). The kinetochore microtubules will thus separate the paired homologous chromosomes, but not pull apart the sister chromatids of any single chromosome.

Which member of a pair of chromosomes faces which pole of the cell is random. For some pairs, the maternal chromosome may face "north," while for other pairs, the maternal chromosome may face "south." We will look at the genetic consequences of this randomness in a moment.

Anaphase I

During **anaphase I,** homologous chromosomes separate from one another and move to opposite poles of the cell (see Fig. 10-2). Anaphase I begins when the chiasmata holding homologous chromosomes together suddenly loosen up. The kinetochore microtu-

MEIOSIS II

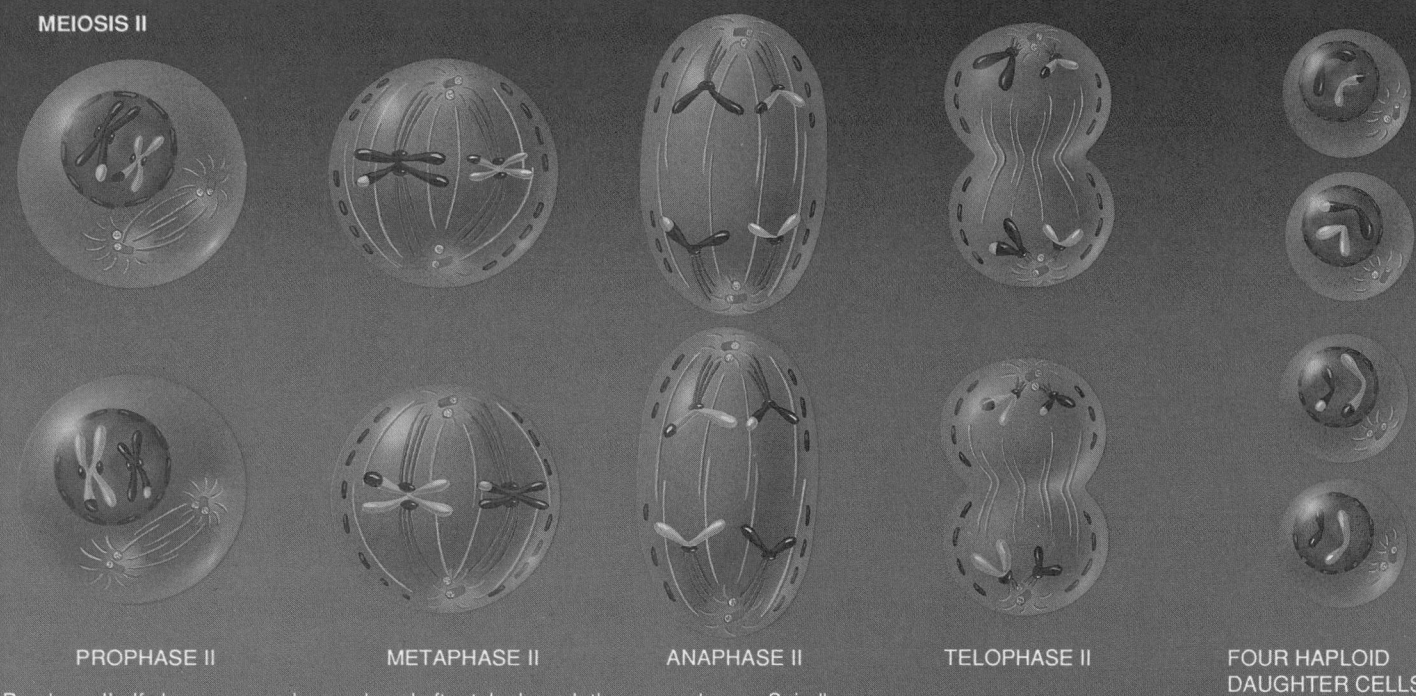

| PROPHASE II | METAPHASE II | ANAPHASE II | TELOPHASE II | FOUR HAPLOID DAUGHTER CELLS |

Prophase II: If chromosomes have relaxed after telophase I, they recondense. Spindles re-form and the spindle microtubules attach to the sister chromatids.

Metaphase II: Chromosomes line up along the equator, with sister chromatids of each chromosome attached to microtubules leading to opposite poles of the cell.

Anaphase II: Chromatids separate into independent daughter chromosomes, one former chromatid moving toward each pole.

Telophase II: Chromosomes finish moving to opposite poles.

Four haploid cells: Cytokinesis results in four haploid cells, each containing one member of each pair of homologous chromosomes.

bules tug the homologues apart, with the chiasmata apparently sliding along to the ends of the chromosomes, eventually letting the two homologues separate. The centromeres holding sister chromatids together, however, do *not* split, so the sister chromatids remain attached to one another. Therefore, one chromosome of each homologous pair (still consisting of two sister chromatids) moves to one pole of the dividing cell while its homologue moves to the other pole.

At the end of anaphase I, there are two clusters of chromosomes. Each cluster contains one member of **each pair of homologous chromosomes and is therefore haploid.**

Telophase I

During **telophase I,** the spindle disappears. In many cases nuclear envelopes do not reform, and, particularly in plants, cytokinesis may not occur either (see Fig. 10-2). Meiosis I is usually followed immediately by meiosis II, with little or no intervening interphase. It is important to remember that *the chromosomes do not*

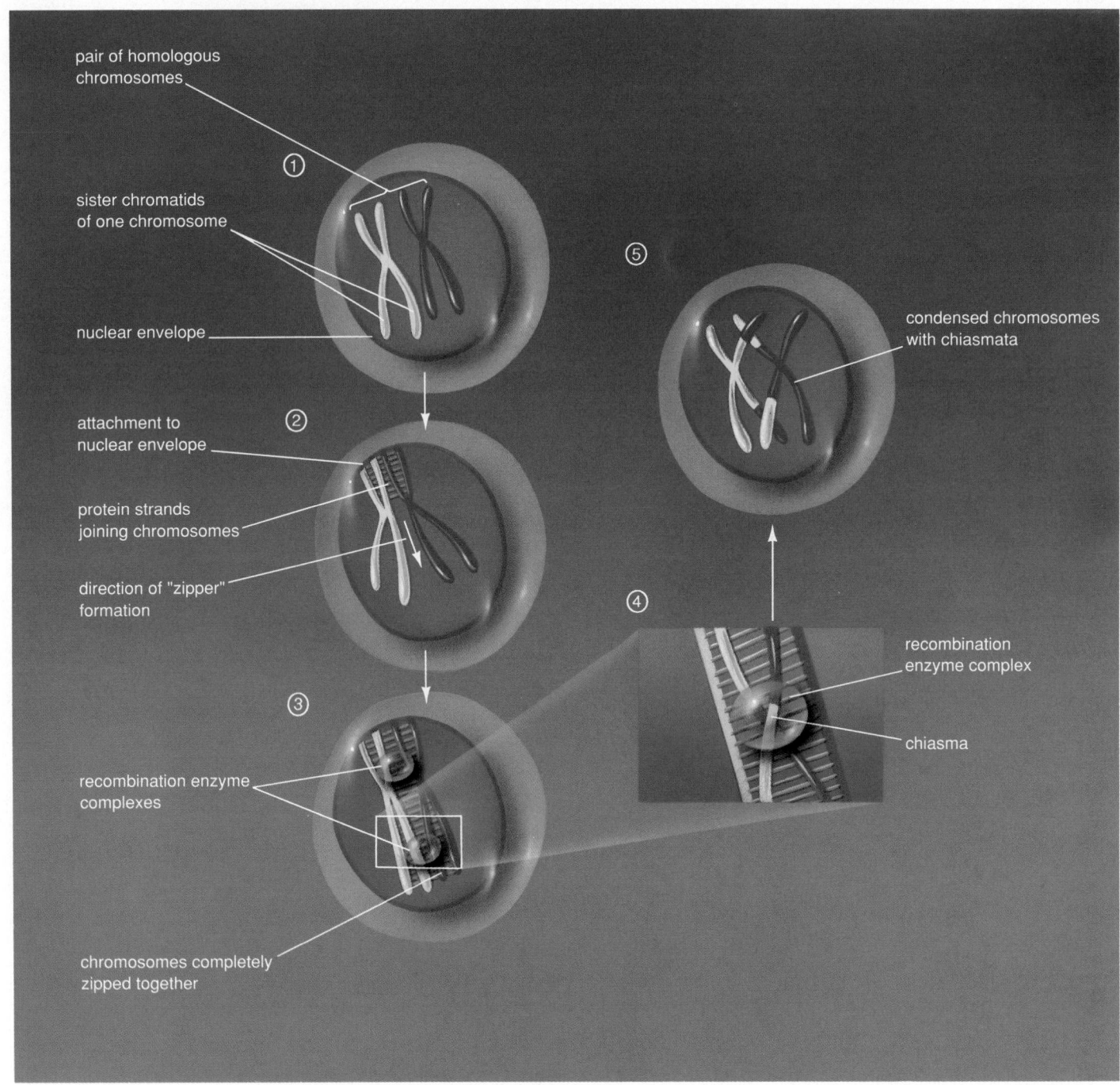

Figure 10-3 The mechanism of crossing over. See text for details.
(1) Homologous chromosomes pair up side by side.
(2) One end of each chromosome binds to the nuclear envelope. Protein strands "zip up" homologous chromosomes together.
(3) Homologous chromosomes are fully joined by protein strands. Recombination enzymes bind to the chromosomes.
(4) Recombination enzymes snip chromatids apart and reattach the chromatids. Chiasmata are formed when one end of a chromatid of a paternal chromosome (yellow) is attached to the other end of a chromatid of a maternal chromosome (violet).
(5) The protein strands and recombination enzymes leave as the chromosomes condense. The chiasmata remain as locations where homologous chromosomes are twisted around each other, helping to hold homologues together.

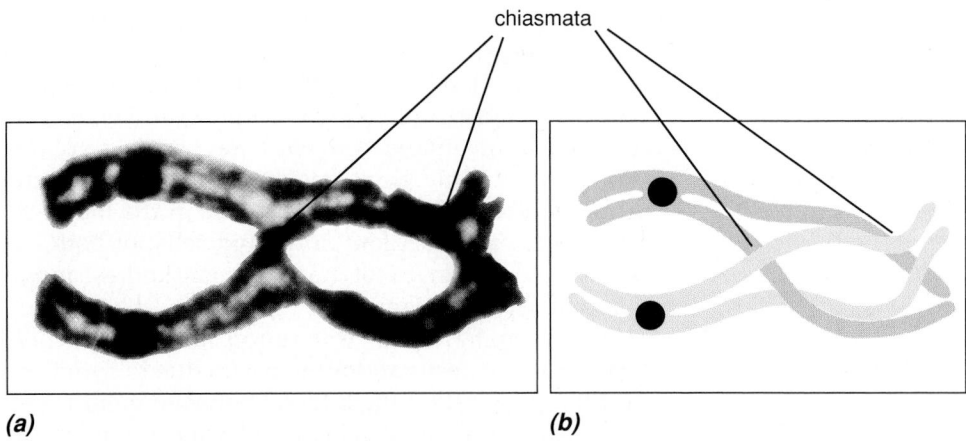

chiasmata

Figure 10-4 Chiasmata in homologous chromosomes during prophase I in a salamander testis. Chiasmata are sites of DNA exchange between homologues.

(a)

(b)

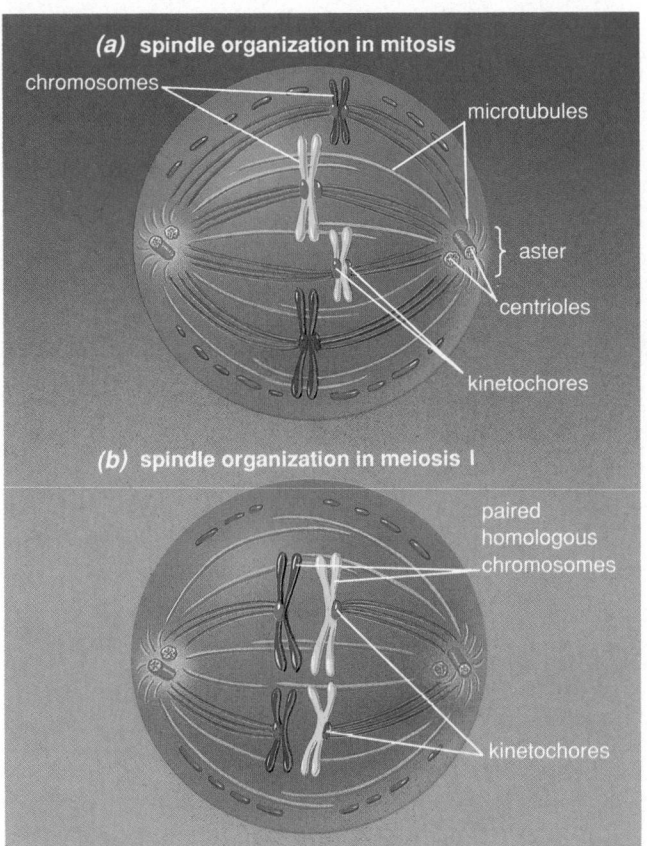

(a) spindle organization in mitosis

chromosomes

microtubules

aster

centrioles

kinetochores

(b) spindle organization in meiosis I

paired homologous chromosomes

kinetochores

Figure 10-5 A comparison of the spindles formed during mitosis and meiosis I.
(a) Homologous chromosomes are not paired. The kinetochores of sister chromatids are attached to kinetochore microtubules that lead to opposite poles. Therefore, when the sister chromatids separate during anaphase, the newly independent daughter chromosomes move to opposite poles of the cell.
(b) Homologous chromosomes are paired. Both kinetochores of the sister chromatids of a single chromosome are attached to kinetochore microtubules that lead to the same pole. Both kinetochores of the sister chromatids of the paired homologous chromosome are attached to kinetochore microtubules that lead to the opposite pole. Therefore, during anaphase I, sister chromatids of each chromosome remain together, moving to the same pole, but homologous chromosomes separate and move to opposite poles.

replicate between meiosis I and meiosis II. Frequently, the chromosomes remain in their condensed, shortened state, and so prophase II can be distinguished from telophase I only by the reappearance of the spindle.

Meiosis II

In meiosis II, the sister chromatids of each chromosome separate. Meiosis II is easy to follow, because it is virtually identical to mitosis in a haploid cell (see

Fig. 10-2). During **prophase II** the spindle reforms. The chromosomes attach to microtubules as they did in mitosis: *the two sister chromatids of each chromosome attach to microtubules that extend to opposite poles of the cell.* **Metaphase II** is marked by completion of the spindle apparatus and migration of the chromosomes to the equator of the cell. During **anaphase II,** the centromeres holding sister chromatids together split, and the chromatids separate from one another. The kinetochore microtubules pull each former chromatid, now an independent chromosome, to opposite poles of the cell. **Telophase II** and cytokinesis conclude meiosis II, as nuclear envelopes reform, the chromosomes relax into their extended state, and the cytoplasm is divided. Since both daughter cells produced in meiosis I undergo meiosis II, a total of four haploid cells are formed by the end of meiosis.

Mitosis, Meiosis, and Eukaryotic Life Cycles

Most eukaryotic organisms have life cycles in which mitosis and meiosis both play crucial roles (Fig. 10-6).

All life cycles have a common overall pattern. (1) A diploid organism produces haploid cells through meiosis. (2) Two haploid cells fuse, bringing together genes from different parent organisms and endowing the resulting diploid cell with new gene combinations. (3) At some point, meiosis occurs again, recreating haploid cells. (4) At some time in the life cycle, mitosis of either haploid or diploid cells, or both, results in the growth of multicellular bodies and/or asexual reproduction.

The seemingly vast differences between the life cycles of, say, ferns and humans are due to variations in three aspects: (1) the interval between meiosis and the fusion of haploid cells; (2) at what points in the life cycle mitosis and meiosis occur; and (3) the rela-

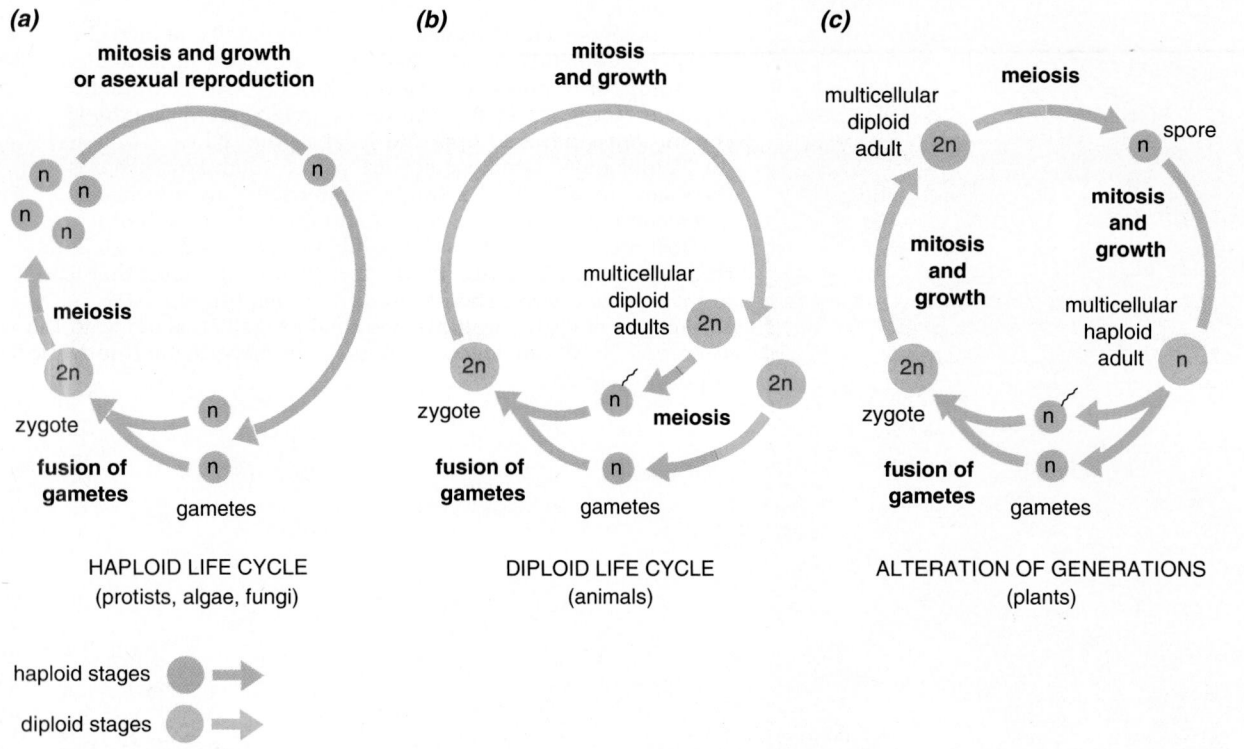

Figure 10-6 Schematic diagrams of the three major types of eukaryotic life cycles. The lengths of the arrows correspond roughly to the proportion of the life cycle spent in each stage.
(a) Haploid life cycle (many protists, algae, and fungi): Most of the life cycle is spent with haploid cells. Mitosis of haploid cells results in either asexual reproduction of unicellular organisms or growth of multicellular organisms. At some point, specialized reproductive haploid cells fuse, and the resulting diploid cell almost immediately undergoes meiosis.
(b) Diploid life cycle (animals): Most of the life cycle is spent with diploid cells. When haploid cells are produced, they fuse to form a diploid zygote. Mitosis of the zygote produces multicellular bodies composed of diploid cells.
(c) Alternation of generations (plants): Diploid and haploid stages are of roughly equal duration. Cells of the diploid stage undergo meiosis to form spores. The spores then undergo mitosis to produce a multicellular haploid body. Later, specialization of haploid cells produces gametes, which fuse to form a zygote. Through mitosis, the zygote develops into a multicellular diploid form.

tive proportions of the life cycle spent in the diploid and haploid states. These aspects of life cycles are interrelated, and we can conveniently label life cycles according to whether diploid or haploid stages predominate.

are produced. Two of these sexual haploid cells fuse, forming a diploid cell. This cell immediately undergoes meiosis, producing haploid cells again. In organisms with a haploid life cycle, mitosis never occurs in diploid cells.

Haploid Life Cycles

Some eukaryotes, such as the unicellular alga *Chlamydomonas*, spend most of their life cycles in the **haploid** state, with unpaired chromosomes (Fig. 10-7). Asexual reproduction by mitosis produces a population of identical, haploid cells. Under certain environmental conditions, specialized "sexual" haploid cells

Diploid Life Cycles

Most animals have life cycles that are just the reverse of the *Chlamydomonas* cycle. Virtually the entire animal life cycle is spent in the **diploid** state (Fig. 10-8). Haploid gametes (sperm in males, eggs in females) are formed through meiosis. These fuse to form a diploid fertilized egg, the zygote. Growth and devel-

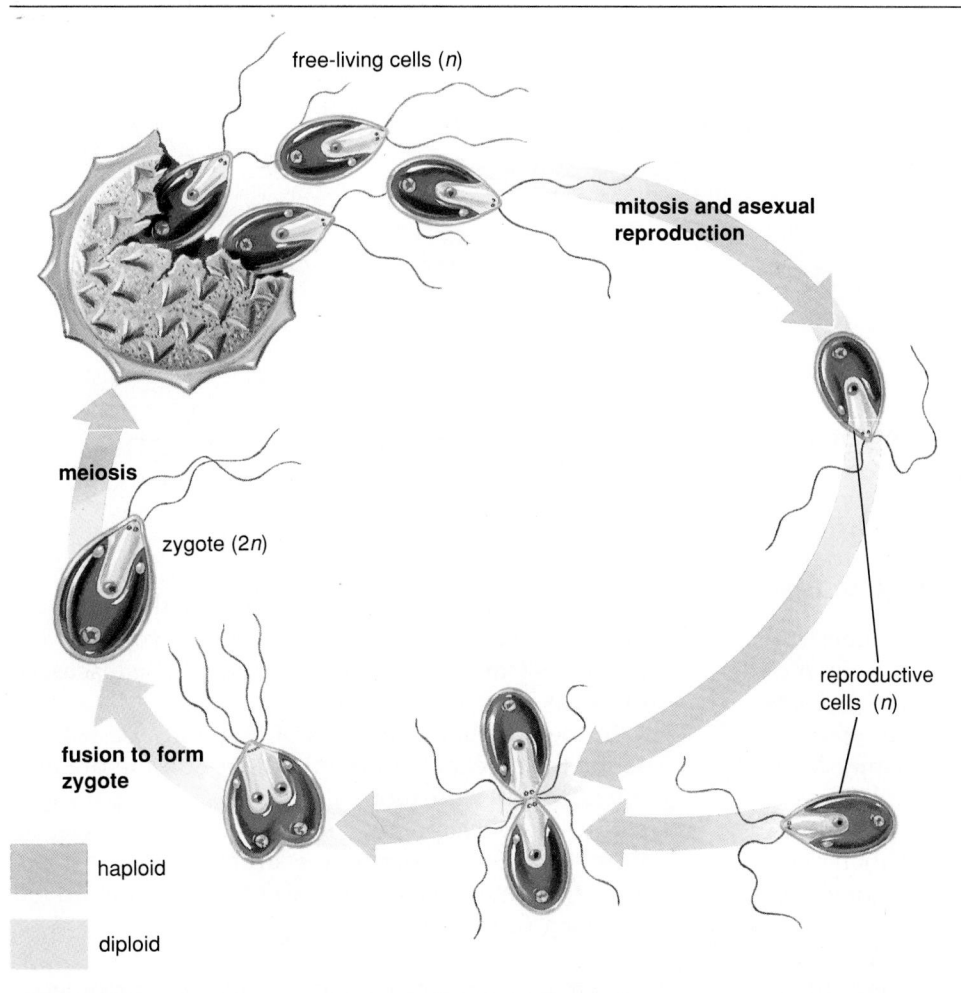

free-living cells (*n*)

mitosis and asexual reproduction

meiosis

zygote (2*n*)

reproductive cells (*n*)

fusion to form zygote

haploid

diploid

Figure 10-7 The life cycle of the unicellular alga *Chlamydomonas*. *Chlamydomonas* multiplies asexually by mitosis of haploid cells. When nutrients are scarce, specialized haploid cells (usually of different strains), fuse to form a diploid cell. Meiosis then immediately produces four haploid cells, usually with different genetic composition than either of the original parental strains.

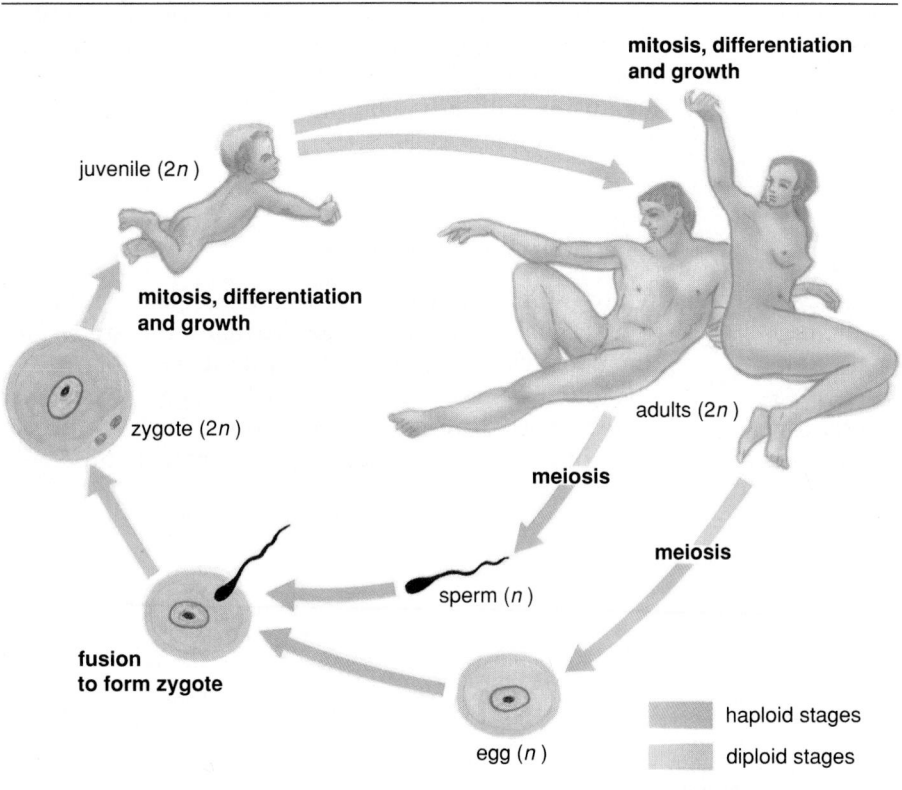

Figure 10-8 The human life cycle. Through meiosis, the two sexes produce different types of gametes— sperm in males and eggs in females— that fuse to form a diploid zygote. Mitosis and specialization of cells produce an embryo, child, and ultimately a sexually mature adult. The haploid state lasts only a few hours to a few days, while the diploid state may survive for a century.

opment of the zygote into the adult organism is a result of mitosis in diploid cells.

Alternation of Generation Life Cycles

The life cycles of most plants include both multicellular diploid and multicellular haploid stages. In the typical pattern (Fig. 10-9), a multicellular diploid body gives rise to haploid cells, called **spores,** through meiosis. The spores then undergo mitosis to create a multicellular haploid stage (the "haploid generation"). At some point, certain haploid cells differentiate into haploid gametes. Two gametes then may fuse to form a diploid zygote. The zygote grows by mitosis into a diploid multicellular body (the "diploid generation"). In "primitive" plants, such as ferns, both the haploid and diploid stages are free-living, independent plants. The flowering plants, however, have reduced the haploid stage to a minimum, represented only by the pollen grain and a small cluster of cells in the ovary of the flower.

Chromosomes, Meiosis, and Inheritance

As you can see from these life cycle summaries, the details of sexual reproduction differ tremendously among the various forms of life on Earth. These are important factors in understanding the evolution and physiology of living organisms, and will be discussed in detail later on in the book.

In the remaining chapters of this unit, we will describe the mechanisms of inheritance on the organismic, cellular, and molecular levels. In preparing to study these chapters, you should concentrate on those aspects of chromosome structure, meiosis, and gamete formation that are essential for your understanding of inheritance:

1. Diploid organisms have pairs of homologous chromosomes.
2. The paired chromosomes are derived from an organism's parents, one homologue coming from the mother and one from the father.

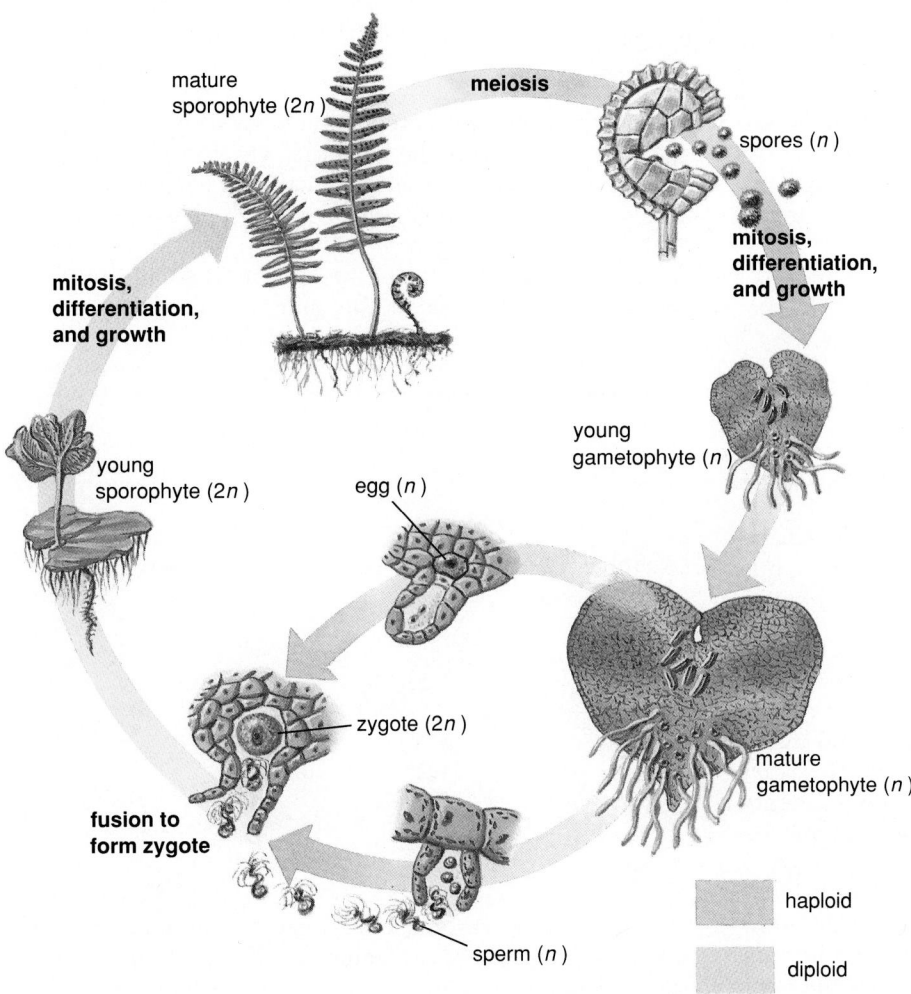

mature
sporophyte (2*n*)

meiosis

spores (*n*)

**mitosis,
differentiation,
and growth**

**mitosis,
differentiation,
and growth**

young
gametophyte (*n*)

young
sporophyte (2*n*)

egg (*n*)

zygote (2*n*)

mature
gametophyte (*n*)

**fusion to
form zygote**

haploid

diploid

sperm (*n*)

Figure 10-9 The life cycle of a fern is representative of alternation of generations. The diploid stage is called the *sporophyte;* this is the fern usually seen in the woods. Specialized cells in the sporophyte undergo meiosis, producing haploid spores. Spores differ from gametes in that spores undergo mitosis and grow into a multicellular haploid body (the *gametophyte*). In ferns, the haploid stage is inconspicuous, a small plant growing in moist places on the forest floor. "Gonads" form on the haploid stage, producing sperm and eggs. Sperm and eggs fuse to make a diploid zygote. Through mitosis, the zygote develops into a new multicellular diploid body, the sporophyte.

3. Homologues contain the same type of genetic information, for example, the genes that control hair color.

4. If the parents differ genetically, homologues will not have identical genetic information (e.g., one homologue may have DNA specifying blond hair color, while the other has DNA specifying brown hair color).

5. Meiosis separates homologous chromosomes, so that gametes have a single copy of each homologue.

6. Which homologue (originally from the mother or from the father) is included in any given gamete is random.

7. Fusion of gametes forms a diploid zygote with paired chromosomes.

Reflections on Meiosis and Paired Chromosomes

All multicellular eukaryotes have paired homologous chromosomes at some stage in their life cycle. Why should a cell have two versions of the same genetic information? And why should meiosis, which separates the pairs, be so complicated? Diploidy and meiosis may provide two crucial advantages.

Genetic Backup Copies

Pairs of chromosomes offer the protection of having two copies of each gene, one on each homologue. If one copy should be damaged—by radiation or toxic chemicals, for example—its counterpart on the homologous chromosome may still provide the genetic information the organism needs to survive. As we will see in Chapter 15, most humans have dozens of defective genes, but still function normally because of intact genes on homologous chromosomes.

Genetic Variability

A second advantage of paired chromosomes and meiosis is the production of variable offspring through sexual reproduction. As we pointed out earlier, environments change and organisms move about, so that what may be a good set of genes for the parent may not be very good for the offspring. Therefore, it is an evolutionary advantage to produce offspring that are somewhat different from the parents and from each other. Meiosis in cells with paired homologous chromosomes allows the generation of genetic variability within a vital background of chromosome constancy.

Chromosome Constancy:
Separation of the Homologues

Meiosis separates homologous chromosomes, so that each haploid daughter cell receives one member of each homologous pair. In humans and other animals, the daughter cells of meiosis are sperm or eggs. Since each gamete contains one member of each homologous pair of chromosomes, fertilization creates a zygote that once again has pairs, and only pairs, of all the chromosomes. Occasionally the homologues do not separate properly, with devastating results. If the zygote is missing a chromosome, or has three homologues instead of two, it almost always aborts early in development, or grows into an organism with severe genetic defects (see Down Syndrome in Chapter 15).

Shuffling Homologues

Maternal and paternal homologues are randomly shuffled out into the daughter cells of meiosis I. Remember, at metaphase I the paired homologues line up at the cell's equator, with the maternal chromosome facing one pole and the paternal chromosome facing the opposite pole. Which homologue faces which pole is random. Let's consider an organism with three pairs of homologous chromosomes: large, medium, and small in size. Further, let's color-code the maternal homologue of each pair yellow, the paternal homologues violet. The alignment of chromosomes at metaphase I can assume four configurations:

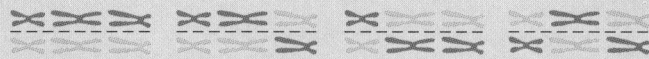

Anaphase I can therefore produce eight possible sets of chromosomes:

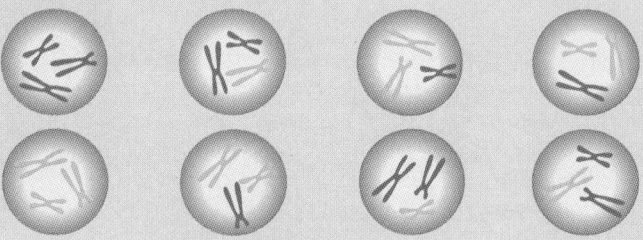

Each of these chromosome clusters will then undergo meiosis II to produce two gametes. Therefore, our hypothetical parent organism could produce gametes with eight different chromosome complements.

Shuffling Genes on Homologues

Finally, crossing over produces chromosomes that differ from those of either parent. Consequently, offspring may receive new combinations of genes that may never before have occurred on single chromosomes. Each time our hypothetical parent produces gametes, its homologous chromosomes cross over in new and different places. Therefore, it can produce many more than eight different types of gametes. *Because of crossing over, it probably never produces any two gametes that are completely identical!*

The potential for genetic variability in human beings is truly staggering. Humans have 23 pairs of homologous chromosomes. A single person, therefore, can theoretically produce about 8 million different gametes, just based on random separation of the homologues. When you add in crossing over, is it any wonder that, with the exception of identical twins, there is truly no one just like you?

SUMMARY OF KEY CONCEPTS

Meiosis and Sexual Reproduction

Sexual reproduction usually combines genetic contributions from two parental organisms. At some point in the life cycle of all sexually reproducing eukaryotes, diploid cells undergo meiosis, which separates homologous chromosomes and produces haploid cells with only one homologue from each pair. These haploid cells or their descendants fuse to form diploid cells that receive one homologue of each pair from each parent, reestablishing pairs of homologous chromosomes.

The Mechanisms of Meiosis

During meiosis, a diploid cell undergoes two specialized cell divisions, meiosis I and meiosis II, to produce four haploid daughter cells. During interphase before meiosis, chromosomes are replicated. The cell then proceeds into meiosis I.

Meiosis I: During prophase I, homologous chromosomes, each consisting of two chromatids, pair up and exchange parts by crossing over. During metaphase I, paired homologues move together to the equator of the cell, one member of each pair facing opposite ends of the cell. Homologous chromosomes separate during anaphase I, and two nuclei form during telophase I. Each daughter nucleus receives only one member of each pair of homologues and is therefore haploid. However, sister chromatids do *not* separate during meiosis I, so each chromosome is still duplicated, consisting of two sister chromatids attached at the centromere. Chromosomes do *not* duplicate again between meiosis I and meiosis II.

Meiosis II: Meiosis II occurs in both of the daughter nuclei, and resembles mitosis in a haploid cell. The chromosomes move to the equator of the cell during metaphase II. The two chromatids of each chromosome separate and are moved to opposite ends of the cell during anaphase II. This second divi-

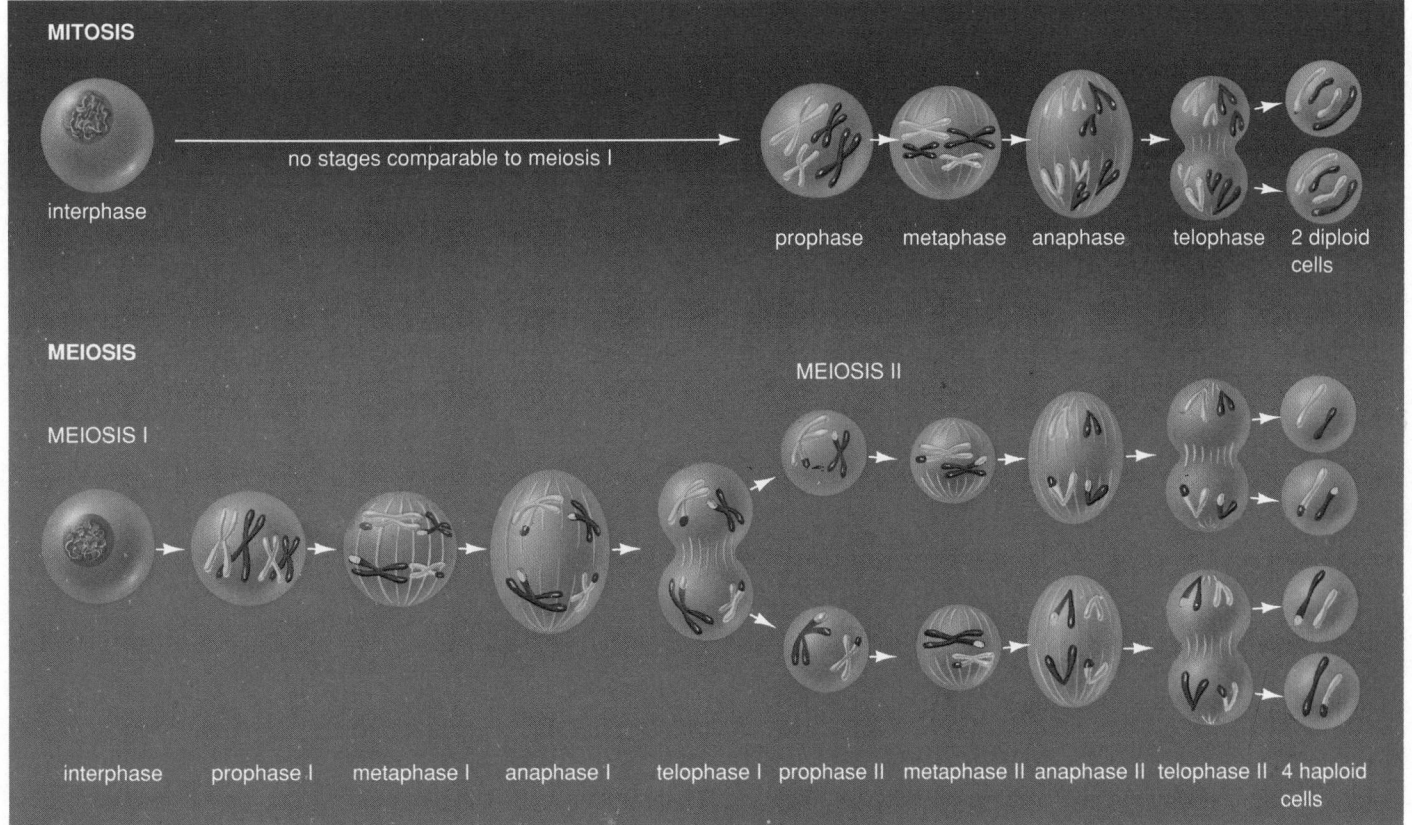

Figure 10-10 A comparison of mitosis and meiosis, with comparable phases aligned. In both, chromosomes are replicated during interphase. Meiosis I, with pairing of homologous chromosomes, formation of chiasmata and exchange of chromosome parts, and separation of homologues to form haploid daughter nuclei, has no counterpart in mitosis. Meiosis II, however, is similar to mitosis occurring in a haploid cell.

sion produces four haploid nuclei. Cytokinesis usually occurs during or shortly after telophase II, producing four haploid cells.

Meiosis separates pairs of homologous chromosomes in diploid cells to form unpaired chromosomes in haploid cells. Homologous chromosomes contain similar, but usually slightly different, genetic information. The separation of homologues during meiosis I is random: either member of each pair of homologues ("maternal" or "paternal") may end up in either daughter nucleus. Therefore, the haploid cells produced by meiosis usually differ in their genetic composition.

Figure 10-10 compares mitosis with meiosis. The illustrations are arranged so that comparable stages are aligned. Note that there are no stages in mitosis that are comparable to meiosis I. Meiosis II is essentially identical to mitosis in a haploid cell.

Mitosis, Meiosis, and Eukaryotic Life Cycles

All eukaryotic life cycles are similar in two respects: (1) sexual reproduction combines haploid gametes to form a diploid cell, and (2) at some point in the life cycle diploid cells undergo meiosis to produce haploid cells once again. Three general types of life cycle can be distinguished, based on the proportion of the life cycle spent in diploid or haploid stages: (1) haploid cycles (mostly microbes and fungi), in which most of the life cycle is spent in the haploid state; (2) diploid cycles (animals), in which most of the life cycle is spent in the diploid state, with meiosis immediately preceding gamete formation; and (3) alternation of generations (plants), in which there are both haploid and diploid multicellular stages.

GLOSSARY

anaphase (an'-a-fāz): the stage of mitosis and meiosis II in which the sister chromatids of each chromosome separate from one another and are moved to opposite poles of the cell. In meiosis I, the stage in which homologous chromosomes are separated.

chiasma (kē-as'-ma; pl. chiasmata): during prophase I of meiosis, a point at which a chromatid of one chromosome crosses with a chromatid of the homologous chromosome. Exchange of chromosomal material between chromosomes takes place at a chiasma.

diploid (dip'-loyd): referring to a cell with pairs of homologous chromosomes.

fertilization: the fusion of male and female haploid gametes to form a zygote.

gamete (gam'-ēt): a haploid sex cell formed in sexually reproducing organisms.

haploid (hap'-loyd): referring to a cell that has only one member of each pair of homologous chromosomes.

homologue (hō'-mō-log): a chromosome that is similar in appearance and genetic information to another chromosome with which it pairs during meiosis. Also called homologous chromosome.

life cycle: the events in the life of an organism from one generation to the next.

meiosis (mī-ō'-sis): a type of cell division found in eukaryotic organisms, in which a diploid cell divides twice to produce four haploid cells.

metaphase (met'-a-fāz): the stage of mitosis or meiosis in which the chromosomes, attached to kinetochore microtubules, line up along the equator of the cell.

prophase (prō-'fāz): the first stage of mitosis or meiosis, in which the chromosomes first become visible with a light microscope as thickened, condensed threads and the spindle begins to form. In meiosis I, the homologous chromosomes pair up and exchange parts at chiasmata.

sexual reproduction: a form of reproduction in which genetic material from two parental organisms is combined. In eukaryotes, two haploid gametes fuse to form a diploid zygote.

spore: in the alternation of generation life cycle of plants, a haploid cell formed by meiosis that undergoes repeated mitotic cell divisions to form a multicellular haploid body.

telophase (tēl'-ō-fāz): the last stage of meiosis and mitosis, in which a nuclear envelope reforms around each new daughter nucleus, the spindle disappears, and the chromosomes relax from their condensed form.

zygote (zī'-gōt): in sexual reproduction, a diploid cell formed by the fusion of two haploid cells.

STUDY QUESTIONS

1. Diagram the events of meiosis. At what stage do haploid cells first appear?
2. Describe the process of homologue pairing and crossing over. At what stage of meiosis does this occur? Name two functions of chiasmata.
3. In what ways are mitosis and meiosis similar? In what ways are they different?
4. Draw a generalized diagram of each of the three types of life cycles of eukaryotic organisms, and give an example of an organism using each type. Label the hap-

loid and diploid stages, and indicate at which point in the cycle meiosis occurs.

5. What are probable advantages and disadvantages of sexual reproduction?

6. Describe how meiosis provides for genetic variability. If an animal had a haploid number of 2 (no sex chromosomes), how many genetically different types of gametes could it produce (assume no crossing over)? If it had a haploid number of 5?

DISCUSSION QUESTIONS

1. A preview question for Chapter 11: Crossing over apparently occurs at random locations along a pair of homologous chromosomes. Would you expect crossing over to separate two genes more often if they are adjacent to one another on the chromosome, or if they are on opposite ends of the chromosome? Why?

2. A preview question for Chapter 15: In humans, meiosis is the first step in the formation of sperm and eggs. Normally, meiosis separates homologous chromosomes during meiosis I, and separates sister chromatids during meiosis II. Occasionally, however, something goes awry. Suppose that during meiosis I, *both members of one pair of homologous chromosomes* moved to the same pole of the cell. If meiosis II occurred normally, what would be the chromosome complement of the resulting gametes? Suppose now that meiosis I occurred normally, but both sister chromatids of one chromosome migrated to the same pole during meiosis II. What would be the chromosome complement of resulting gametes? One relatively common genetic defect is Down syndrome, otherwise known as trisomy 21 (three copies of chromosome 21). Can abnormalities of meiosis explain trisomy 21? How?

3. In the bodies of humans and most other animal species, both mitosis and meiosis occur. Discuss the similar and different functions of mitosis and meiosis in our bodies and why it is important that both processes occur.

4. Among the lower animals, many species can reproduce either asexually or sexually, depending on the state of the environment. Asexual reproduction tends to occur in stable, favorable environments while sexual reproduction increases in unstable and/or unfavorable circumstances. Discuss the possible advantages or disadvantages this behavior might have on survival of the species in an evolutionary sense, or on survival of individuals in the populations.

SUGGESTED READING

Stahl, F. "Genetic Recombination." *Scientific American*, February 1987. A look at the mechanisms of crossing over.

11

Principles of Inheritance

Inheritance provides for both similarity and difference. These animals are both tigers because of the vast majority of identical genes that they inherited from common ancestors; they differ in color because of differences in a single gene.

"In the mingling of the seed, sometimes the woman, with sudden force, overpowers the man, and then the children, born of maternal seed, will resemble more the mother; but if from paternal seed, the father. The children you see resembling both their parents, having the features of both, have been created from father's body and mother's blood, when the seeds course through the bodies excited by Venus, in harmony of mutual passion, breathing as one, with neither conquering and neither being conquered."

Lucretius *in* On the Nature of Things (96–55 BC)

Speculation about reproduction and inheritance goes back thousands of years. As the passage from Lucretius quoted above indicates, fanciful explanations, with no evidence to support them, were once widely accepted. However, the ancients were not totally ignorant about reproduction. People have known for millenia that most animals and plants reproduce sexually, the offspring arising from the mating of male and female parents. Further, organisms mate with other organisms of the same kind, and the resulting offspring are also of the same kind as the parents. Nevertheless, offspring are almost never exactly identical to either parent.

Any hypotheses about how organisms pass their traits on to their offspring must explain at least this short list of observations:

1. All living things arise from preexisting living things.
2. The offspring of two parents usually resemble the parents and each other.
3. Many traits in the offspring are not exactly like those of either parent.

Early Hypotheses about Inheritance

Historically, philosophers and scientists have devised many hypotheses to explain our first two observations. Both Hippocrates (ca. 460–377 BC) and Charles Darwin (1809–1882) believed that each part of the body gives off tiny particles that travel to the gonads. During intercourse, the father's particles supposedly merge with the mother's to form the offspring. Aristotle (384–322 BC) favored a fluid rather than particles, but the general idea was similar. After the invention of the microscope, when sperm and eggs were seen for the first time, two new schools of thought arose. One held that an entire tiny person is contained in each sperm (and that tiny person has even smaller sperm with yet smaller people in them, with still smaller sperm . . .), while the other be-

lieved that the egg contains the tiny person with tiny eggs inside.

In the nineteenth century, many scientists felt that, although the physical mechanisms of inheritance were uncertain, the resulting patterns could be explained by **blending**. This hypothesis held that parents pass on their traits to their offspring, where they blend to produce intermediate traits. At first glance, blending seems quite reasonable. For example, if a tall man marries a short woman, most of the sons are not as tall as the father, and most of daughters are not as short as the mother. Superficially, it appears as if the tallness of the father and the shortness of the mother mix together in the children.

When carried to its logical conclusion, however, the blending hypothesis is unsatisfactory. If large organisms mate with small ones, they would produce medium-sized offspring. The medium-sized offspring, when they mate in their turn, could give rise only to more medium-sized organisms. Over thousands of generations, surely all individuals of any given species would become identical—"medium" everything—as all variation "blended out." However, everyone can see that members of the same species differ in appearance and behavior. Further, if blending *is* the major principle of inheritance, then Darwin's theory of evolution by natural selection cannot be correct. If there is no variation among organisms, then natural selection cannot favor one individual over another. Selection could play no role in shaping a species. When he published the *Origin of Species* in 1859, Darwin recognized that blending could not explain the inheritance of variation demanded by his theory of evolution, but he couldn't offer a better alternative.

Enter Gregor Mendel.

Gregor Mendel and the Origin of Genetics

Gregor Mendel (Fig. 11-1) was a monk in the monastery of St. Thomas in Brünn (now Brno), Czechoslovakia. Before settling down in the monastery, Mendel

tion for the modern science of genetics. At St. Thomas, Mendel carried out both his monastic duties and an elegant series of experiments on inheritance in the common edible pea.

Mendel was not the first person to study inheritance. However, as Mendel himself wrote:

> "Whoever surveys the work in this field will come to the conviction that among the numerous experiments, not one has been carried out to an extent or in a manner that would make it possible to determine the number of different forms in which hybrid progeny appear, permit classification of these forms in each generation with certainty, and ascertain their numerical interrelationships. . . . [H]owever, this seems to be the one correct way of finally reaching the solution to a question whose significance for the evolutionary history of organic forms must not be underestimated."

Doing It Right: The Secrets of Mendel's Success

Why did Mendel succeed where others before him had failed? There are three key steps to any successful experiment in biology: (1) choosing the right organism to work with, (2) designing the experiment correctly, and (3) analyzing the data properly. Mendel was the first geneticist to take all three steps and not stumble along the way.

The Right Organism: The Edible Pea

Mendel's choice of the edible pea as an experimental subject was excellent. The pea flower is built so that pollen normally cannot enter from outside the flower, i.e., from another plant (Fig. 11-2). Instead, pea flowers usually self-pollinate. As a result, the egg cells in

Figure 11-1 A portrait of Gregor Mendel, painted about the time of his pioneering genetics experiments.

tried his hand at several pursuits, including health care and teaching. In an effort to earn his teaching certificate, Mendel attended the University of Vienna for 2 years, where he studied botany and mathematics, among other subjects. This training proved crucial to his later experiments, which were the founda-

Intact Pea Flower

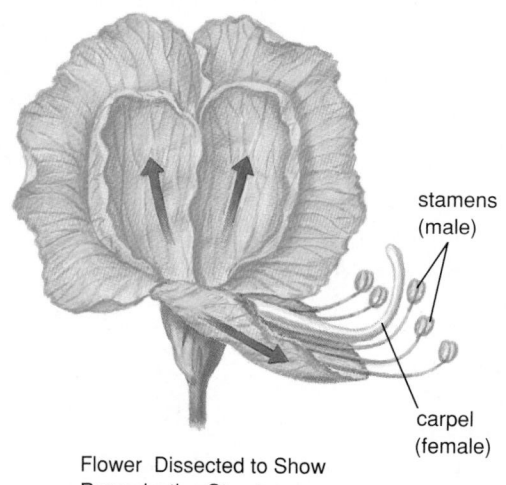

Flower Dissected to Show Reproductive Structures

stamens (male)

carpel (female)

Figure 11-2 The flower of the edible pea. In the intact flower (left), the lower petals form a container enclosing the reproductive structures, the stamens (male) and carpel (female). Pollen normally cannot enter the flower from outside, and consequently peas usually self-fertilize. Plant breeders, however, can pull apart the petals (right) and remove the stamens so that self-fertilization cannot occur. Dusting the carpels with pollen of their choice results in controlled cross-fertilization.

each flower are fertilized by sperm from the pollen of the same flower (this is called **self-fertilization**). Even in Mendel's time commercial seed dealers sold many different types of peas that were **true-breeding**—that is, all the offspring produced through self-fertilization were essentially identical to the parent plant.

Although peas normally self-fertilize, two different plants can be mated by hand. By carefully picking the flowers apart, pollen can be collected from one plant and transferred to a flower on another plant. The sperm from the "foreign" pollen will then fertilize the egg cells of the recipient flower (**cross-fertilization**). In this way, one can mate two different true-breeding plants and see what types of offspring they produce.

Although peas were an excellent subject for studies of inheritance, they were no guarantee of success. Others had experimented with peas long before Mendel, without discovering the principles of genetics. The key reasons for Mendel's success were his methods of experimentation and analysis.

Experimental Design: Keep It Simple

Mendel began by working with only one trait at a time, and he chose traits that had unmistakably different forms of expression, such as white versus purple flowers (Fig. 11-3). Earlier scientists had often crossed plants that differed in several traits, or that

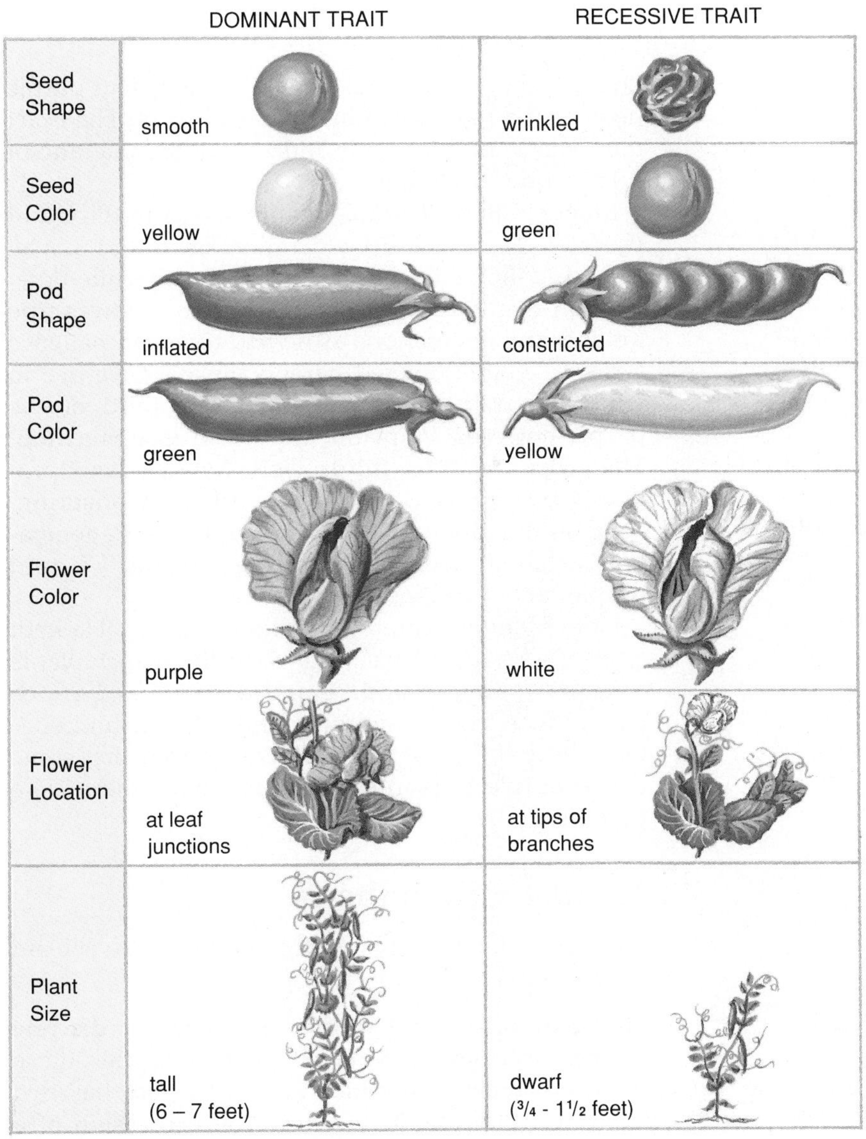

	DOMINANT TRAIT	RECESSIVE TRAIT
Seed Shape	smooth	wrinkled
Seed Color	yellow	green
Pod Shape	inflated	constricted
Pod Color	green	yellow
Flower Color	purple	white
Flower Location	at leaf junctions	at tips of branches
Plant Size	tall (6 – 7 feet)	dwarf ($^3/_4$ - $1^1/_2$ feet)

Figure 11-3 Traits of pea plants used by Mendel in his studies of plant inheritance.

had different forms of a trait that might be confused with one another. Not surprisingly, the results were difficult to interpret. Studying single traits, with clearly different alternate forms, allowed Mendel to see through to the underlying principles of inheritance.

Data Analysis

Perhaps most important, Mendel counted the numbers of offspring bearing the various traits and critically analyzed the numbers. The use of numbers as a tool for finding underlying principles has since become an extremely important practice in biology (see "Methods in Biology: On the Importance of Statistics in Biology").

Single-Trait Experiments: The Law of Segregation

Mendel started as simply as possible. He raised varieties of pea plants that were true-breeding for different forms of a single trait and cross-fertilized them. Such an experiment, involving organisms that differ in only one trait, is called a **monohybrid cross**. (In this usage, **hybrid** refers not to the offspring of parents of two different species, such as a mule, but to offspring of parents of the same species that differ in one or more inherited traits, such as hybrid corn.) Mendel saved the resulting hybrid seeds and grew them the next year.

In one of these experiments, Mendel cross-fertilized a white-flowered pea with a purple-flowered one. (In modern nomenclature, this was the **parental generation,** denoted by the letter P.) When he grew the resulting seeds, he found that all the first-generation offspring ("first filial," or **F₁ generation**) produced purple flowers:

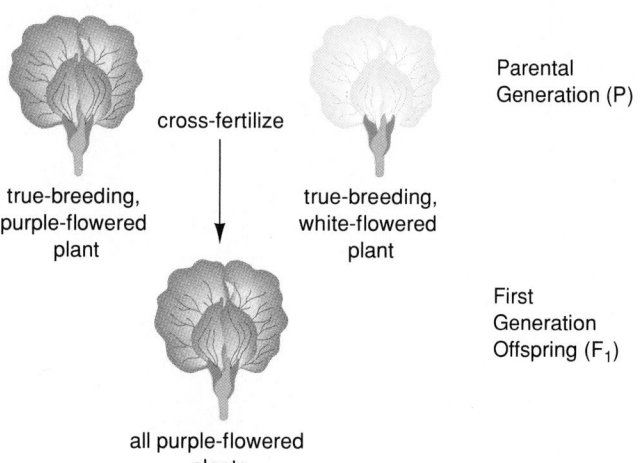

cross-fertilize

true-breeding, purple-flowered plant

true-breeding, white-flowered plant

Parental Generation (P)

First Generation Offspring (F₁)

all purple-flowered plants

What happened to the white color? It hadn't blended in; the flowers of the hybrids were every bit as purple as the flowers of their purple parent. White seemed to have disappeared completely in the F₁ offspring.

Mendel allowed the F₁ flowers to self-fertilize, collected the seeds, and planted them the next spring. The second generation (F₂) had some plants with flowers of each color:

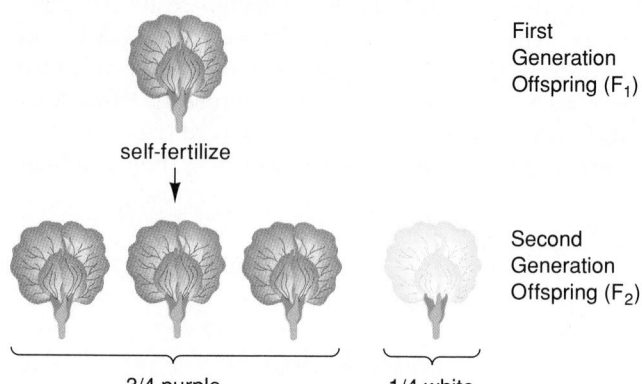

First Generation Offspring (F₁)

self-fertilize

Second Generation Offspring (F₂)

3/4 purple

1/4 white

Overall, about three fourths of the plants had purple flowers and one fourth had white flowers. The exact numbers were 705 purple and 224 white, or a ratio of 3.15 purple to 1 white.

Mendel allowed the F₂ plants to self-fertilize and produce yet a third (F₃) generation. He found that all the white-flowered F₂ plants produced white-flowered offspring; that is, they bred true. This was true for as many generations as he had time and patience to raise: white-flowered parents always gave rise to white-flowered offspring. About one third of the purple-flowered F₂ plants were also true-breeding. The remaining two thirds were hybrids, and produced both purple- and white-flowered offspring, again in the ratio of 3 to 1. Therefore, the F₂ generation included $\frac{1}{4}$ true-breeding purple plants, $\frac{1}{2}$ hybrid purple, and $\frac{1}{4}$ true-breeding white.

Note Mendel's careful attention to detail. He kept track of each individual seed, recording which plants were its "parents" and "grandparents," and which plants were its "children" and "grandchildren." Otherwise, he would never have realized that some of the purple F₂ plants bred true while others were hybrids.

Mendel's Hypothesis

Mendel formed a five-part hypothesis to explain these results.

1. **Each trait is determined by pairs of discrete physical units.** Modern geneticists call these units **genes**. Each individual organism has two genes that together control the expression of a given trait (e.g., two genes for flower color).

2. **Pairs of genes separate from each other during gamete formation.** The is Mendel's **Law of Segregation:** each gamete receives only one of an organism's pair of genes. When a sperm fertilizes an egg, the resulting offspring receives one gene from the father and one from the mother.

3. **Which member of a pair of genes becomes included in a gamete is determined by chance.** If we assign letters to each member of a pair of genes, say G1 and G2, then each gamete is just as likely to receive a G1 as a G2. The gamete will not receive both, and there will not be a preference for one member of the pair.

4. **There may be two or more alternative forms of a gene.** White-flowered peas, for example, have a different form for the gene controlling flower color than true-breeding purple-flowered peas do. Alternative forms of a single gene are called **alleles.** In many instances, one allele, called the **dominant** allele, can completely mask the expression of the other, **recessive,** allele. However, although the dominate allele of a gene masks the *expression* of the recessive allele, it doesn't alter the *physical nature* of the recessive allele, which can be passed unchanged into the individual's gametes. In Mendel's experiments with flower color, the allele for purple flowers is dominant to the allele for white flowers.

5. **True-breeding organisms have two of the same alleles for the trait under study; hybrids have two different alleles.** Since a true-breeding, or **homozygous** ("same pair"), organism has only one type of allele, it can produce only one type of gamete:

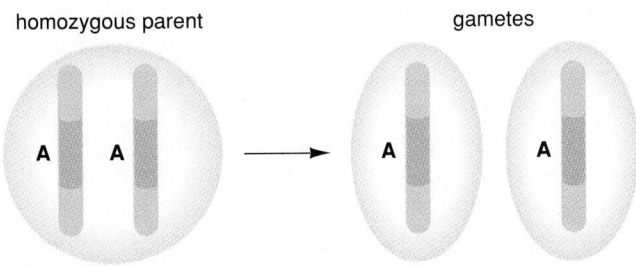

homozygous parent   gametes

A hybrid, or **heterozygous** ("different pair"), individual, with two different alleles, produces equal numbers of gametes with each of the two alleles:

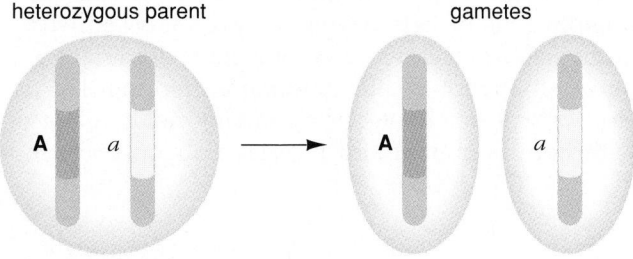

heterozygous parent   gametes

Applying the Hypothesis to a Single-Trait Cross

Let's see how Mendel's hypothesis explains the results of his experiments with flower color. Using letters to represent the different alleles, we will assign the uppercase letter P to the allele for purple (the dominant allele is usually represented by a capital letter), and the lowercase letter p to the allele for white (recessive). A homozygous purple-flowered plant has two alleles for purple (PP), while a white-flowered plant has two alleles for white (pp). A PP plant produces all P sperm and eggs, while a pp plant produces all p sperm and eggs:

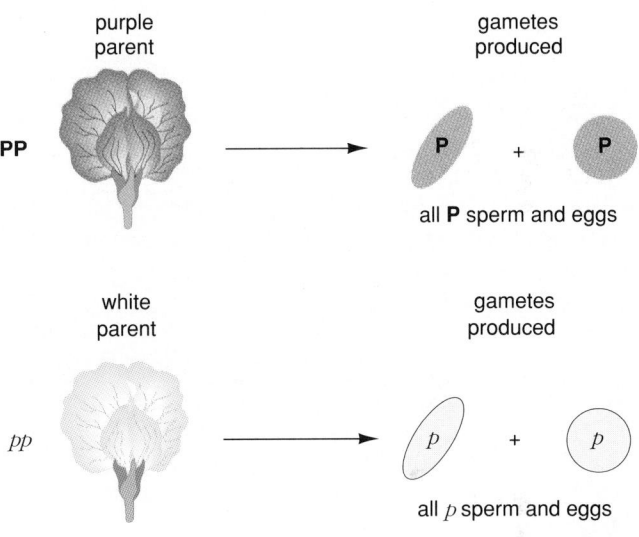

The F_1 hybrid offspring are produced when P sperm fertilize p eggs, or when p sperm fertilize P eggs. In either case, the F_1 offspring are Pp. Because P is dominant to p, all the offspring are purple:

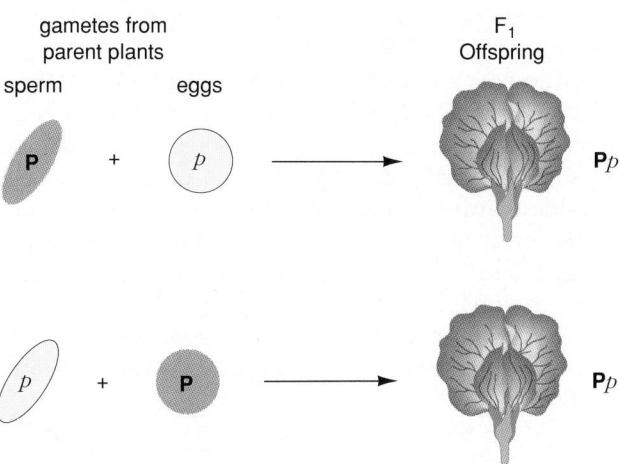

Each gamete produced by a heterozygous *Pp* plant has an equal chance of receiving either the *P* allele or the *p* allele (that is, the plant produces equal numbers of *P* and *p* sperm and equal numbers of *P* and *p* eggs). When a *Pp* plant self-fertilizes, both types of sperm have an equal chance of fertilizing both types of eggs:

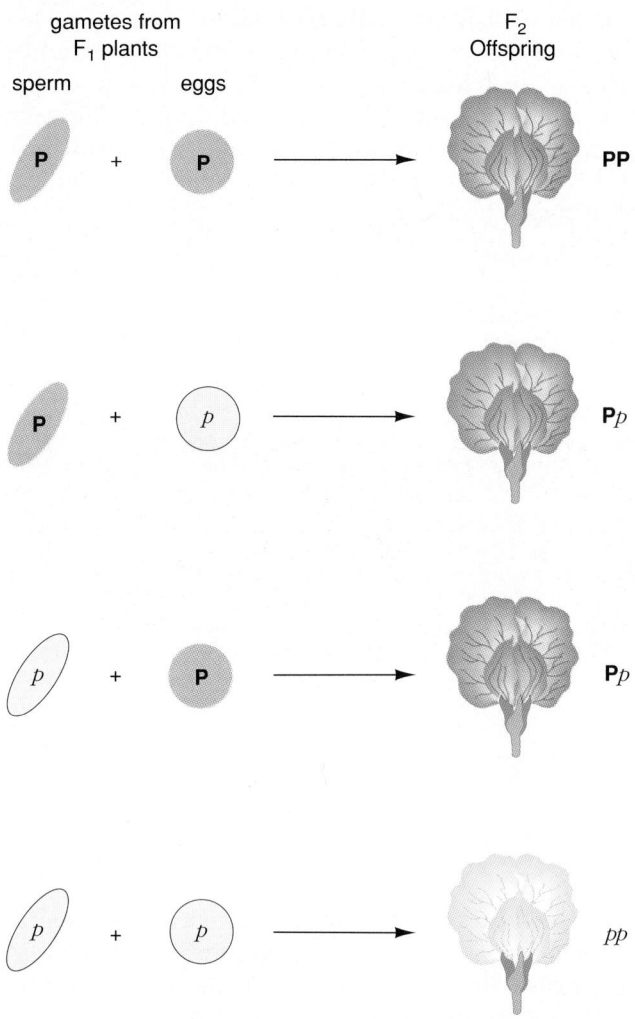

Therefore, three types of offspring can be produced: *PP*, *Pp*, and *pp*. These occur in the approximate proportions of $\frac{1}{4}$ *PP*, $\frac{1}{2}$ *Pp*, and $\frac{1}{4}$ *pp*.

The actual combination of alleles carried by an individual (e.g., *PP* or *Pp*) is its **genotype.** The physical characteristics of an organism (morphology, behavior, digestive enzymes, blood type, or any other observable feature) is its **phenotype.** As we have seen, plants with genotypes *PP* and *Pp* both bear purple flowers. Thus, even though they have different genotypes, they have the same phenotype. Therefore, the F$_2$ generation consists of three genotypes ($\frac{1}{4}$ *PP*, $\frac{1}{2}$ *Pp*, and $\frac{1}{4}$ *pp*) but only two phenotypes ($\frac{3}{4}$ purple and $\frac{1}{4}$ white).

Using Mendel's Hypothesis to Calculate the Offspring Expected from a Single-Trait Cross

Figures 11-4 and 11-5 present two methods of "genetic bookkeeping" for determining the expected proportions of offspring in a monohybrid cross. The first method uses a diagram called a Punnett square (Fig. 11-4), named after a famous geneticist of the early 1900s. The second method relies on probability theory (Fig. 11-5). Both methods yield the same results, so you can use whichever seems easier for you. Whichever method you choose, there is one crucial point to keep in mind: *these calculations give only the most probable proportions of offspring of different genotypes and phenotypes.* In a real experiment, one would only expect that the offspring will occur in *approximately* the predicted proportions. In the F$_2$ generation of Mendel's monohybrid cross, note that he didn't obtain *exactly* $\frac{3}{4}$ purple-flowered and $\frac{1}{4}$ white-flowered plants.

The Test Cross

By now you have probably recognized that Mendel used the scientific method discussed in Chapter 1. He observed the results of his experiments, and then formed a hypothesis. As you know, the scientific method has a third step: to use the hypothesis to predict the results of other experiments, and see if those experiments support or refute the hypothesis. Mendel did just that. If the hybrid F$_1$ flowers have one allele for purple and one for white (*Pp*), then he should be able to predict the outcome of cross-fertilizing these *Pp* plants with homozygous recessive (*pp*) white plants. (Can you?) As Figure 11-6a shows, Mendel's hypothesis predicts that there will be equal numbers of *Pp* (purple) and *pp* (white) offspring. This is precisely what happened.

This type of experiment has practical uses, too. Just by looking at an organism that has the dominant phenotype, one usually cannot tell whether it is homozygous or heterozygous. Cross-fertilization of a phenotypically dominant individual with a homozygous recessive individual is called a **test cross,** because it can be used to test whether the dominant parent is homozygous or heterozygous. When crossed with a homozygous recessive, a homozygous dominant produces all phenotypically dominant offspring (Fig. 11-6b), while a heterozygous dominant yields offspring with both dominant and recessive phenotypes, in a 1:1 ratio (Fig. 11-6a).

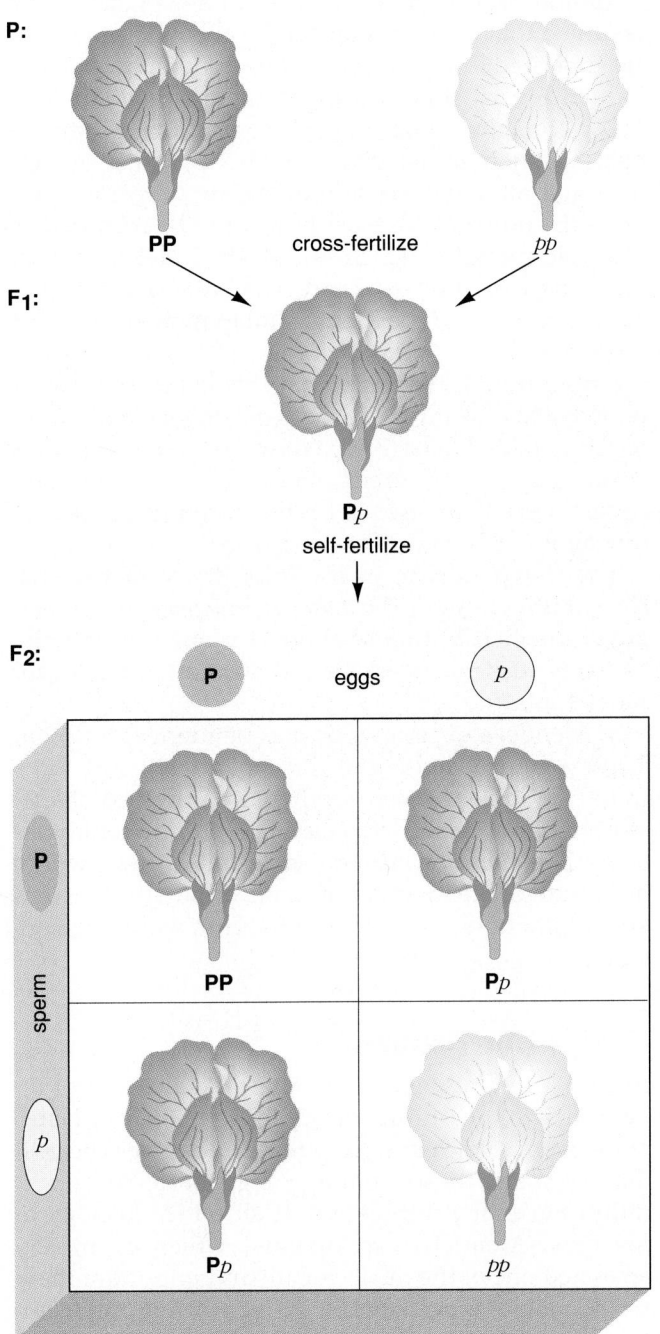

P:

PP   cross-fertilize   pp

F₁:

P*p*

self-fertilize

F₂:

P   eggs   p

sperm

PP   P*p*

P*p*   pp

Figure 11-4 The Punnett square method of determining likely genotypes and phenotypes of offspring. The Punnett square is intuitive and relatively nonmathematical, and is preferred by many students. The example shown here analyzes the inheritance of flower color in peas, based on a cross of F₁ hybrids. Using the Punnett square requires the following six steps.

(1) Assign letters to represent the different alleles; dominant alleles are usually upper case (*P*), recessive alleles lower case (*p*). **Be sure that you can distinguish upper and lower case letters:** for example, *W* and *w* or *S* and *s* may be difficult to tell apart, especially in the heat of a test.

(2) Assuming that the genotypes of the parents are known, determine the types of equally likely, genetically different gametes that can be produced by the male and female parents. In the case of the gene pair shown here, ½ of the gametes should receive the *P* allele, and ½ should receive the *p* allele. (Of course, if a parent is homozygous, then all of its gametes will receive the same allele.)

(3) Draw the Punnett square. The number of rows in the square is the same as the number of different types of sperm, and the number of columns in the square is the same as the number of different types of eggs. Label the rows with the types of sperm, the columns with the types of eggs.

(4) The expected offspring are the "cells" of the square. Each offspring results from the fertilization of the egg in its column by the sperm in its row. Fill out the offspring genotypes systematically, for example, first using row 1 sperm to fertilize all types of eggs, then row 2 sperm to fertilize all types of eggs, etc.

(5) Examine the offspring cells, and add up the number of offspring with the same genotypes.

(6) **Remember that the numbers generated by a Punnett square do not mean that every mating yields exactly four offspring.** Convert the numbers of offspring genotypes to expected fractions of offspring genotypes by dividing the number of each offspring genotype by the total number of offspring cells. If desired, add together those genotypes that produce the same phenotypes (*PP* and *Pp* in this example), to yield the expected fractions of offspring with each phenotype. Remember also that the fractions of different offspring generated by the square are only *expected* fractions. Real matings would only be expected to yield *approximately* these fractions of different offspring.

Multiple-Trait Experiments: Independent Assortment

Having determined the mode of inheritance of single traits, Mendel then turned to the more complex question of multiple traits. He began with a **dihybrid cross**; that is, he cross-bred plants that differed in two traits, for example seed color (yellow or green) and seed shape (smooth or wrinkled). If he crossed a plant that was homozygous for smooth yellow seeds with one that was homozygous for wrinkled green seeds, the F₁ offspring all bore smooth yellow seeds. This was no great surprise, because from separate monohybrid crosses of each of these traits he already knew that smooth (*S*) is dominant to wrinkled (*s*), and that yellow (*Y*) is dominant to green (*y*) (see Fig.

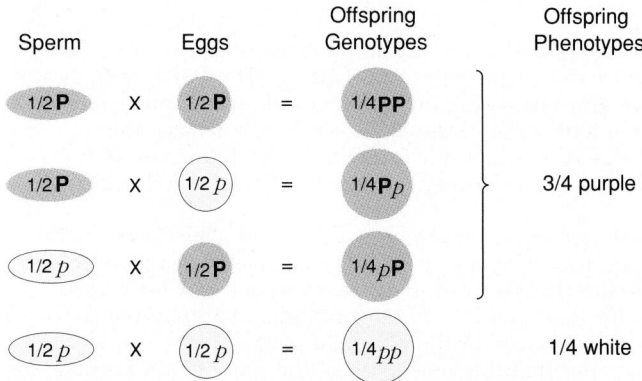

Sperm		Eggs		Offspring Genotypes	Offspring Phenotypes
1/2 **P**	X	1/2 **P**	=	1/4 **PP**	
1/2 **P**	X	1/2 *p*	=	1/4 **P***p*	3/4 purple
1/2 *p*	X	1/2 **P**	=	1/4 *p***P**	
1/2 *p*	X	1/2 *p*	=	1/4 *pp*	1/4 white

Figure 11-5 Using probability theory to determine the distribution of genotypes and phenotypes of offspring. The example is the F_1 cross illustrated in the Punnett square method in Figure 11-4. In probability theory, the probability of two independent events occurring together is the product (multiplication) of their independent probabilities. In this cross, each type of gamete (*P* or *p*) is equally likely to occur; that is, each has a probability of $\frac{1}{2}$. Further, each type of sperm is equally likely to fertilize each type of egg. In terms of probability theory, the first event (which allele is in a particular sperm) does not depend on the second event (which allele is in the egg fertilized by that sperm). We can therefore obtain the probable ratio of offspring by multiplying the probability of obtaining each type of sperm by the probability of obtaining each type of egg.

To use probability theory to analyze genetic crosses, perform the following four steps:
(1) Assign letters to the alleles (the same as Step 1 of the Punnett square method).
(2) Determine the types of equally likely, genetically different gametes that can be produced by the parents (the same as Step 2 of the Punnett square method).
(3) Set up a multiplication table as shown below: $\frac{1}{2}$ of the sperm are *P*, and $\frac{1}{2}$ of the eggs are *P*; multiply these together to obtain $\frac{1}{4}$ *PP* offspring. Continue filling in the table; be sure that all types of sperm have the "opportunity" to fertilize all types of eggs. For a "genotype" table such as this one, the number of rows in the table (4) should equal the number of different types of sperm (2) multiplied by the number of different types of eggs (2).
(4) If desired, add together the genotypes that produce the same phenotypes (the same as Step 6 of the Punnett square method).

11-3). The F_1 offspring, therefore, are all genotypically *SsYy*. Allowing these F_1 plants to self-fertilize, Mendel found that the F_2 generation consisted of 315 smooth yellow seeds, 101 wrinkled yellow seeds, 108 smooth green seeds, and 32 wrinkled green seeds. This works out to roughly $\frac{9}{16}$ smooth yellow seeds, $\frac{3}{16}$ wrinkled yellow seeds, $\frac{3}{16}$ smooth green seeds, and $\frac{1}{16}$ wrinkled green seeds, or a ratio of 9:3:3:1. Other dihybrid crosses produced similar ratios.

Mendel realized that this could be explained if he assumed that the genes for seed color and seed shape are inherited independently of each other and do not influence each other during gamete formation (Fig. 11-7). (Imagine flipping two coins, a dime and a nickel. Whether the dime comes up heads or tails does not affect which side of the nickel comes up.) Thus the outcome for each trait could be regarded as a simple monohybrid cross, in which a 3:1 ratio of offspring would be expected. The laws of probability state that the independent combination of two 3:1 ratios yields a 9:3:3:1 ratio (Fig. 11-8b), and we can see from Mendel's results that this is just what happened. There were 423 smooth seeds (of either color) to 133 wrinkled ones (3.18:1) and 416 yellow seeds (of either shape) to 140 green ones (2.97:1). The Punnett square shows how two 3:1 ratios combine to form an overall 9:3:3:1 ratio (Fig. 11-8a).

The independence in the inheritance of two distinct traits is called the *Law of Independent Assortment*: the distribution of alleles for one trait into the gametes does not affect the distribution of alleles for other traits.

In Mendel's experiments, independent assortment also seemed to hold true for combinations of three traits, say flower color, seed color, and seed shape, although now there were eight different phenotypes in the F_2 generation. Mendel expected that this would hold true for any number of traits, although the numbers of phenotypes of the F_2 offspring would be very large.

Genius Unrecognized

In 1865, Gregor Mendel presented his theories of inheritance to the Brünn Society for the Study of Natural Science, and they were published the next year. It did *not* mark the beginning of genetics. In fact, it didn't make any impression at all on the biology of his time. Mendel's experiments, which eventually spawned one of the most elegant and important theories in all of science, simply vanished from the scientific scene. Apparently, very few biologists read his paper, and those who did probably couldn't understand it. It was not until almost half a century later that biologists rediscovered Gregor Mendel and his principles of genetics.

In 1900, three biologists—Carl Correns, Hugo de Vries, and Eric von Tschermak—working independently and knowing nothing of Mendel's work, rediscovered the principles of inheritance. No doubt to their intense disappointment, when they searched the scientific literature before publishing their results, they found that Mendel had scooped them over 30 years before.

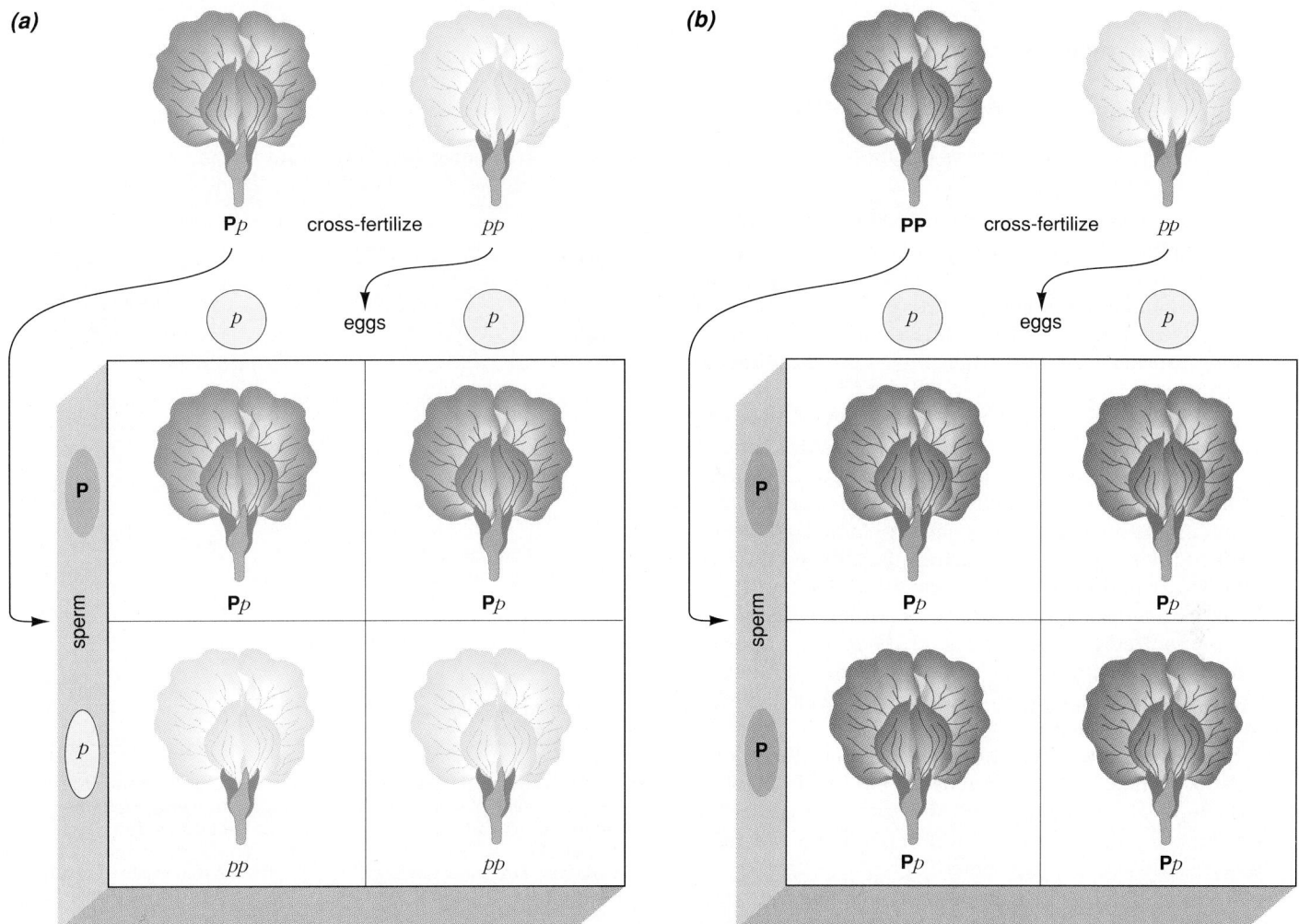

Figure 11-6 Test crosses involve breeding a homozygous recessive organism (in these examples, *pp* plants with white flowers) with a phenotypically dominant organism (here, plants with purple flowers). The resulting offspring differ, depending on whether the phenotypically dominant parent is heterozygous *Pp* **(a)** or homozygous *PP* **(b)**.

Genes and Chromosomes

While Mendel's findings languished in oblivion, other areas of biology, particularly microscopy, flourished. In the 1870s and 1880s, chromosomes were discovered, and their movements during mitosis and meiosis were deduced. Having already studied mitosis and meiosis, you have probably noticed that Mendel's genes behave much like chromosomes during meiosis:

1. There are two alleles of each gene in each organism; likewise, there are two homologous chromosomes in each diploid cell.

2. Only one of each gene is passed on to the offspring in each sperm or egg; during meiosis only one of each pair of homologues ends up in each haploid daughter cell.

3. Different genes sort out independently of each other; the different types of chromosomes also appear to assort independently.

Genes Are Parts of Chromosomes

It wasn't long before someone saw the connection between chromosomes and Mendel's hypotheses. In 1902, William Sutton, a graduate student at Columbia

METHODS IN BIOLOGY

On the Importance of Statistics in Biology

"There are three kinds of lies: lies, damn lies, and statistics." —attributed by Mark Twain to the
British Prime Minister Benjamin Disraeli

The passage we quoted from Gregor Mendel's paper at the Brünn Society for the Study of Natural Sciences provides an insight into one of his most significant contributions. Unlike the biologists who preceded him, Mendel realized that he had to "determine the number of different forms" and "ascertain their numerical interrelationships."

In 1824, over 30 years before Mendel even began his studies, an Englishman named John Goss did some of the same crosses that Mendel later repeated. Goss cross-fertilized true-breeding yellow-seeded peas with true-breeding green-seeded peas, and obtained all yellow seeds in the F_1 generation. Further, he allowed the F_1 to self-fertilize and raised an F_2 generation. He found that the F_2 plants bore both yellow peas and green peas. But if Goss realized the significance of the *numbers* of green and yellow peas, or even counted them at all, he never reported it. Consequently, we have Mendelian genetics, not Gossian genetics.

Since Mendel's time, numbers have become increasingly important, not just in genetics but in all of biology and medicine. Most modern studies are quantitative: the interpretation of the results depends on the numbers obtained in the experiment.

As an example, let's suppose that a biologist suspects that a toxic substance in the waste water from an industrial plant stunts the growth of fish. To test his hypothesis, he might set up two aquaria, each stocked with 10 small fish of the same size and age. All environmental conditions, such as light, temperature, and food, would be kept the same for both tanks, except that the suspected toxin would be added to the water in one tank. At the end of a predetermined time, say 6 months, he would measure the fish in each tank.

What the biologist would probably find is that some fish in the toxic tank would be larger than some fish in the control tank, and vice versa (see the table). How can he reach any conclusions? He would calculate the average size of the fish in each tank, and the standard deviation, which is a measure of the variability of sizes. He would then perform statistical tests to see whether the difference in the average size of the fish in the two tanks reflects chance variation, or might be caused by the toxic substance. In this example, the statistical analysis supports the hypothesis that the toxin inhibits fish growth.

Statistical analyses are extremely important in biology and medicine, especially when a given effect might have several causes. For example, it has been known for many years that people who smoke cigarettes are more

Length of Fish (cm)

Control (without toxin)	Experimental (with toxin)
6.0	5.0
6.0	5.0
7.0	5.0
7.0	6.0
7.0	6.0
8.0	7.0
8.0	7.0
9.0	7.0
10.0	8.0
10.0	8.0
Average 7.8	6.4
Standard Deviation 1.5	1.2

As determined by a statistical procedure called the Student's *t*-test, the probability of obtaining this difference in average lengths of fish in the two tanks purely by chance, and not due to the effect of the pollutant, is less than one in twenty. This is the generally accepted level of "statistical significance," that is, that the toxin really did stunt the growth of the fish.

likely to suffer from lung cancer and heart attacks than nonsmokers. This is what might be called a statistical fact: obviously some nonsmokers die from heart attacks, and some smokers do not contract lung cancer. Nevertheless, on the average, nonsmokers run a much lower risk of these diseases.

Mark Twain expressed a common feeling when he implied that statistics is worse than a "damn lie." Statistics can be confusing, and statistical facts often seem to be contradicted by personal experience, such as knowing someone who smoked three packs of cigarettes a day for 60 years and never developed lung cancer. However, in a world in which effects may have many causes, statistics is an essential tool in our efforts to understand the workings of natural phenomena. It was only through statistical analysis that Mendel could understand pea genetics or the U.S. Surgeon General could draw conclusions between smoking and health. Without the marriage of biology and mathematics begun by Gregor Mendel, we could not evaluate the results of new drug therapies, crop varieties, or thousands of other advances in biology and medicine.

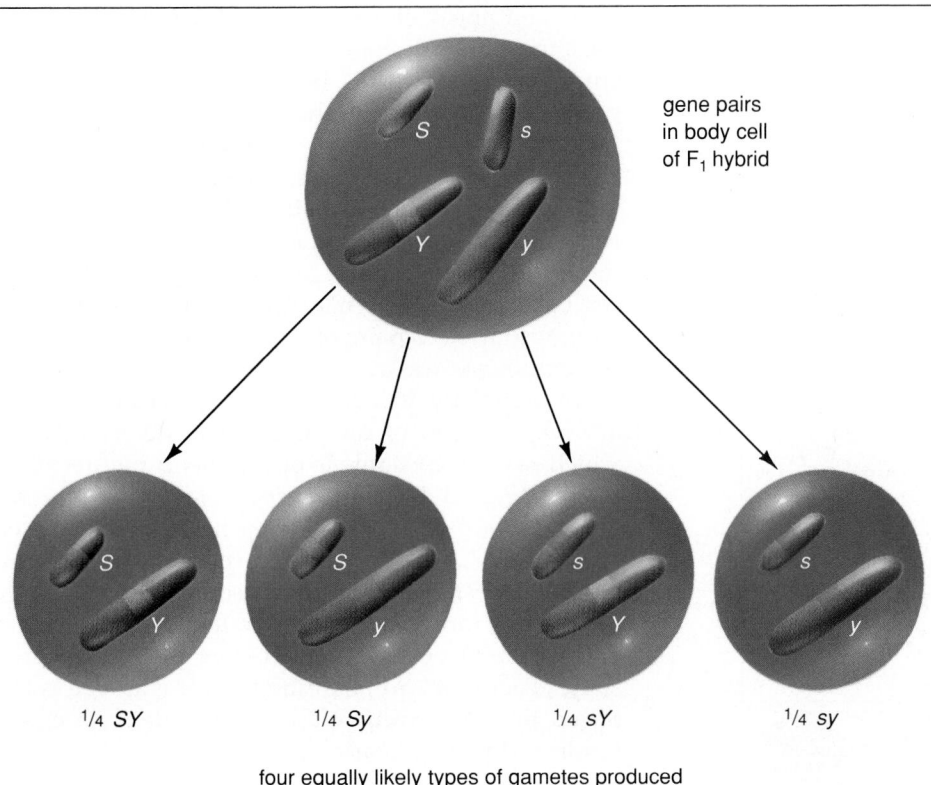

gene pairs
in body cell
of F₁ hybrid

¹/₄ *SY*   ¹/₄ *Sy*   ¹/₄ *sY*   ¹/₄ *sy*

four equally likely types of gametes produced
by independent assortment

University, was studying sperm formation in grass-hoppers. With Mendel's principles of inheritance in mind, he saw the chromosome movements of meiosis in a new light: perhaps Mendel's genes were chromosomes. Of course, there was immediately a great difficulty with this hypothesis—namely, that there have to be many more genes than there are chromosomes. So Sutton's hypothesis became: **genes are parts of chromosomes.**

The proof that genes are parts of chromosomes came from the work of Thomas Hunt Morgan of Columbia University and his students, studying the common fruit fly, *Drosophila melanogaster* (Fig. 11-9). You might be surprised that anyone would study the genetics of such a homely little pest. However, fruit flies are nearly ideal organisms for genetics experiments. To study inheritance in any organism, geneticists must cross-breed the organism, raise the F₁ generation, breed these individuals, raise the F₂ generation, and so on, and end up with enough offspring to draw meaningful conclusions about the ratio of phenotypes. As Mendel found, if only one generation a year is possible, and if the organism needs a lot of space or care, it takes a long time and huge amounts of work to carry out complicated crosses for

several generations. Fruit flies are small, each female lays hundreds of eggs, and a complete generation from egg to egg takes only 2 weeks. Finally, fruit flies have only four pairs of chromosomes. For these reasons, and because Morgan and his colleagues were careful, persistent, and brilliant scientists, it was the fruit fly that provided many of the advances in genetics in the first half of this century.

The Chromosomal Basis of Sex Determination

Sex determination provided the first direct evidence supporting Sutton's hypothesis that genes are parts of chromosomes. In this case, a trait (sex) is almost invariably associated with a particular distribution of chromosomes.

In the 1980's, microscopists observed that not all chromosomes have exact homologues. In mammals and many insects, males have the same number of chromosomes as females do, but one "pair," the **sex chromosomes,** are very different in appearance. Females have two identical sex chromosomes, called **X chromosomes,** while males have one X chromo-

(a)

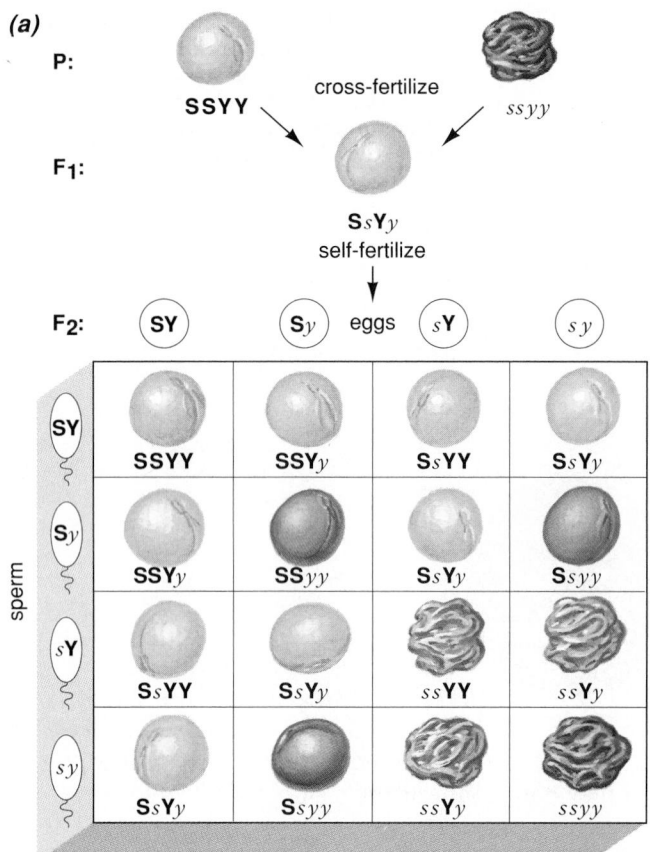

(b)

seed shape		seed color		F₂ offspring
$\frac{3}{4}$ smooth	×	$\frac{3}{4}$ yellow	= $\frac{9}{16}$	smooth yellow
$\frac{3}{4}$ smooth	×	$\frac{1}{4}$ green	= $\frac{3}{16}$	smooth green
$\frac{1}{4}$ wrinkled	×	$\frac{3}{4}$ yellow	= $\frac{3}{16}$	wrinkled yellow
$\frac{1}{4}$ wrinkled	×	$\frac{1}{4}$ green	= $\frac{1}{16}$	wrinkled green

Figure 11-8 Calculating the expected offspring in a dihybrid cross. **(a)** Punnett square analysis. One of the original parents is homozygous for smooth (*SS*) yellow (*YY*) seeds, and the other original parent is homozygous for wrinkled (*ss*) green (*yy*) seeds. The F₁ offspring therefore all have smooth yellow seeds (*SsYy*). If we allow the F₁ to self-fertilize, we obtain the offspring shown within the square. There are now 16 squares in the analysis, but the method is the same as Fig. 11-4. The expected proportions of offspring are $\frac{9}{16}$ smooth yellow, $\frac{3}{16}$ smooth green, $\frac{3}{16}$ wrinkled yellow, and $\frac{1}{16}$ wrinkled green, or 9:3:3:1. Note also that the Punnett square predicts 12 smooth: 4 wrinkled seeds and 12 yellow: 4 green seeds (both 3:1 ratios), just as would be expected from monohybrid crosses if we ignore the second trait. **(b)** Analysis by probability theory. **Remember that independent assortment means that each trait assorts independently of the others.** We know from monohybrid crosses that three quarters of the F₂ offspring will be smooth and one quarter will be wrinkled, and that three quarters will be yellow and one quarter will be green. Multiplying these independent probabilities produces the expected F₂ offspring. These are identical to the offspring ratios generated by the Punnett square.

some and one **Y chromosome.** Although the Y chromosome is morphologically very different from the X chromosome, parts of both sex chromosomes are homologous to one another, so they pair up during prophase of meiosis I and separate during anaphase I. The other chromosomes, which occur in pairs of identical appearance in both males and females, are called **autosomes.** *Drosophila* has four pairs of chromosomes (three pairs of autosomes and one pair of sex chromosomes), while humans have 23 pairs of chromosomes (22 pairs of autosomes and one pair of sex chromosomes).

For organisms in which males are XY and females are XX, the sex of an offspring is determined by which sex chromosome is in the sperm that fertilizes the egg (Fig. 11-10). During sperm formation, the sex chromosomes segregate, and each sperm receives either the X or the Y chromosome (plus one of each pair of autosomes). The sex chromosomes also segregate during egg formation, but since females have two X chromosomes, every egg receives one X chromosome. Therefore, an offspring is male if an egg is fertilized by a Y-bearing sperm, or female if an egg is fertilized by an X-bearing sperm.

Sex Linkage

Proof that chromosomes carry more than one gene came early in the course of Morgan's fruit fly studies.

Figure 11-9 The common fruit fly, *Drosophila melanogaster*, has had an uncommon impact on studies of inheritance.

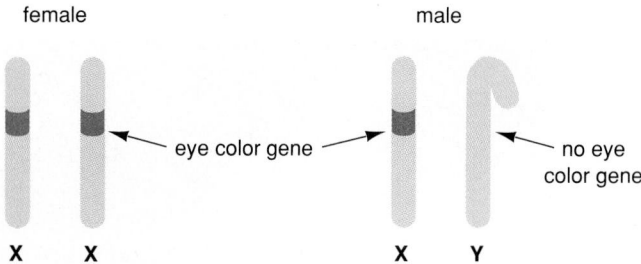

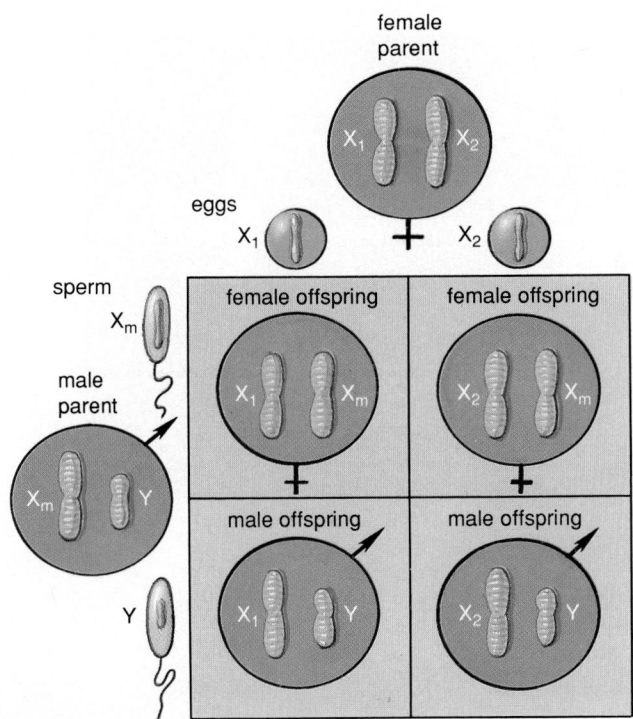

Figure 11-10 Sex determination in species such as humans and fruit flies, in which males carry two dissimilar sex chromosomes (XY) while females carry two similar sex chromosomes (XX). Only the distribution of sex chromosomes is illustrated. Male offspring receive the Y chromosome from the father, while female offspring receive the father's X chromosome (labelled X_m in the drawing). The mother passes one of her X chromosomes (X_1 or X_2 in the drawing) to both male and female offspring.

A male fly was discovered that had white eyes, instead of the normal red eyes (Fig. 11-11). This white-eyed male was mated to a virgin red-eyed female. The resulting offspring were all red-eyed flies (Fig. 11-11a), indicating that white eyes (w) is probably recessive to red eyes (W). The F_2 generation, however, was a surprise. As expected, the ratio of red- to white-eyed flies was about 3:1; however, *there were nearly equal numbers of males with red eyes and with white eyes, and no females with white eyes at all* (Fig. 11-11b)! A test cross of the F_1 red-eyed females and the original white-eyed male yielded roughly equal numbers of red-eyed and white-eyed males and females.

From these data, could you figure out the mode of inheritance of eye color? Morgan made the brilliant hypothesis that *the gene for eye color must be located on the X chromosome, while the Y chromosome has no corresponding gene:*

Let's look first at the F_1 generation. Both male and female F_1's receive an X chromosome, with its W allele for red eyes, from their mother. The F_1 males receive a Y chromosome from their father, with no allele for eye color at all, so the males have a WY genotype. The F_1 females received the father's X chromosome with its w allele, so the females have a Ww genotype. Thus, all male and female F_1 offspring had red eyes.

Crossing two F_1 flies thus results in an F_2 generation with the chromosome distribution shown in Figure 11-11b. All the F_2 females receive one X chromosome from their F_1 male parent with its W (red) allele. They therefore have red eyes. All the F_2 males, on the other hand, inherit their single X chromosome from their F_1 female parent. Since the F_1 females are heterozygous for eye color (Ww), the F_2 males have a 50–50 chance of receiving either an X chromosome with the W allele or one with the w allele. *With no corresponding gene on the Y chromosome, the F_2 males must display the phenotype determined by the allele on the X chromosome.* Therefore, half the F_2 males have red eyes and half have white eyes.

Sex Linkage Is Usually X-linkage

Genes that are found on one sex chromosome but not on the other are called **sex linked**. In principle, the X chromosome might carry genes that are not found on the Y chromosome, and vice versa. However, in many animals, the Y chromosome carries relatively few genes other than those determining maleness, whereas the X chromosome bears many genes, such as the *Drosophila* gene controlling eye color, that have nothing to do with specifically female traits. In most cases, therefore, sex-linked genes are found on the X chromosome, and should probably he called X-linked. There are few Y-linked genes.

Females can be homozygous or heterozygous for X-linked genes, but males cannot. Whatever genes a male carries on his single X chromosome are the only copies of those genes that he has. They cannot be masked by or interact with their corresponding gene on another X chromosome, as they can in females. Genes on the X chromosome are thus fully expressed

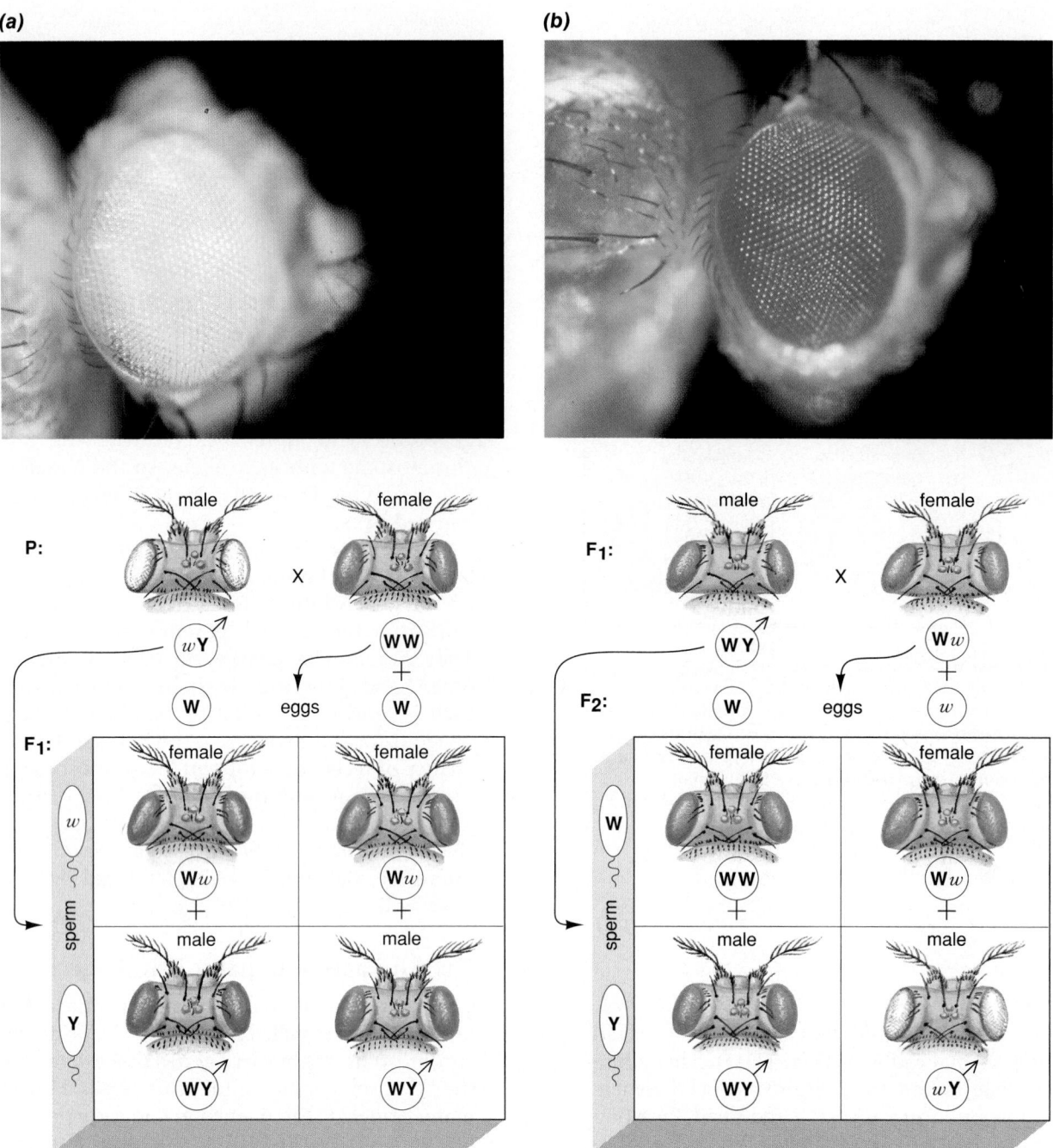

Figure 11-11 Morgan's interpretation of the results of sex-linked inheritance of white eye color in fruit flies. The gene for eye color is carried on the X chromosome, and has no corresponding gene on the Y chromosome. Normal red eyes (*W*) is dominant to the mutant allele for white eyes (*w*).

(a) The F$_1$ Generation: Female offspring receive the *w* allele on their father's X chromosome, but phenotypically this is masked by the dominant *W* allele on their mother's X chromosome. Male offspring receive the *W* allele from their mother, and the Y chromosome with no eye color gene from their father. Therefore all the females are *Ww* and all the males are *WY*. Both males and females have red eyes.

(b) The F$_2$ Generation: All F$_1$ males carry the *W* allele, so all their F$_2$ female offspring receive the *W* allele as well. Therefore, the F$_2$ females are all red eyed. The F$_1$ females are all heterozygous *Ww*. Consequently, half their F$_2$ sons receive the *W* allele and half receive the *w* allele. The F$_2$ males also receive a Y chromosome from their fathers. Therefore half the F$_2$ males are *WY* (red eyes) and half are *wY* (white eyes).

in males, whether they are dominant or recessive. As we shall see in Chapter 15, many human traits, such as red–green color blindness and hemophilia, are X-linked.

Gene Linkage on the Autosomes

Like the X chromosome, the autosomes also bear many genes. **X linkage is merely a special case of the general phenomenon of** *linkage,* **in which two or more genes tend to be inherited together because they lie on the same chromosome.** One of the first pairs of linked autosomal genes to be discovered were those for flower color and pollen grain shape in the sweet pea. Purple flower color (P) is dominant over red (p), and long pollen shape (L) is dominant over round (l). When a homozygous purple-flowered, long-pollen pea is crossed with a homozygous red-flowered, round-pollen pea, all the F_1 offspring have purple flowers and long pollen. However, the genes for flower color and for pollen shape are both carried on the same chromosome. Since chromosomes assort independently during meiosis, both genes tend to assort together, and therefore to be inherited together. Thus, the phenotypes of the F_2 generation do not occur in a 9:3:3:1 ratio, as Mendel's Law of Independent Assortment would predict. Instead, the F_2 generation has about $\frac{3}{4}$ purple-flowered, long-pollen plants and $\frac{1}{4}$ red-flowered, round-pollen plants (Fig. 11-12).

Linkage requires that we modify the Law of Independent Assortment:

1. **Genes located on different chromosomes assort independently during meiosis.** (Mendel was fortunate that the traits he studied were controlled by genes on different chromosomes.)
2. **Genes located on the same chromosome usually will not assort independently.** They tend to be inherited together.

Interactions Between Linkage and Crossing Over

There is just one thing wrong with this tidy scheme: genes on the same chromosome do not *always* stay together. In the sweet pea cross described above, for example, often the F_2 generation will include a few purple-flowered, round-pollen plants and red-flowered, long-pollen plants. How can this be?

As you know, during prophase I of meiosis, homologous chromosomes intertwine, producing X-shaped figures called **chiasmata** (Fig. 11-13). At chiasmata, parts of chromosomes may exchange with each other, a process called crossing over. In fact, there is

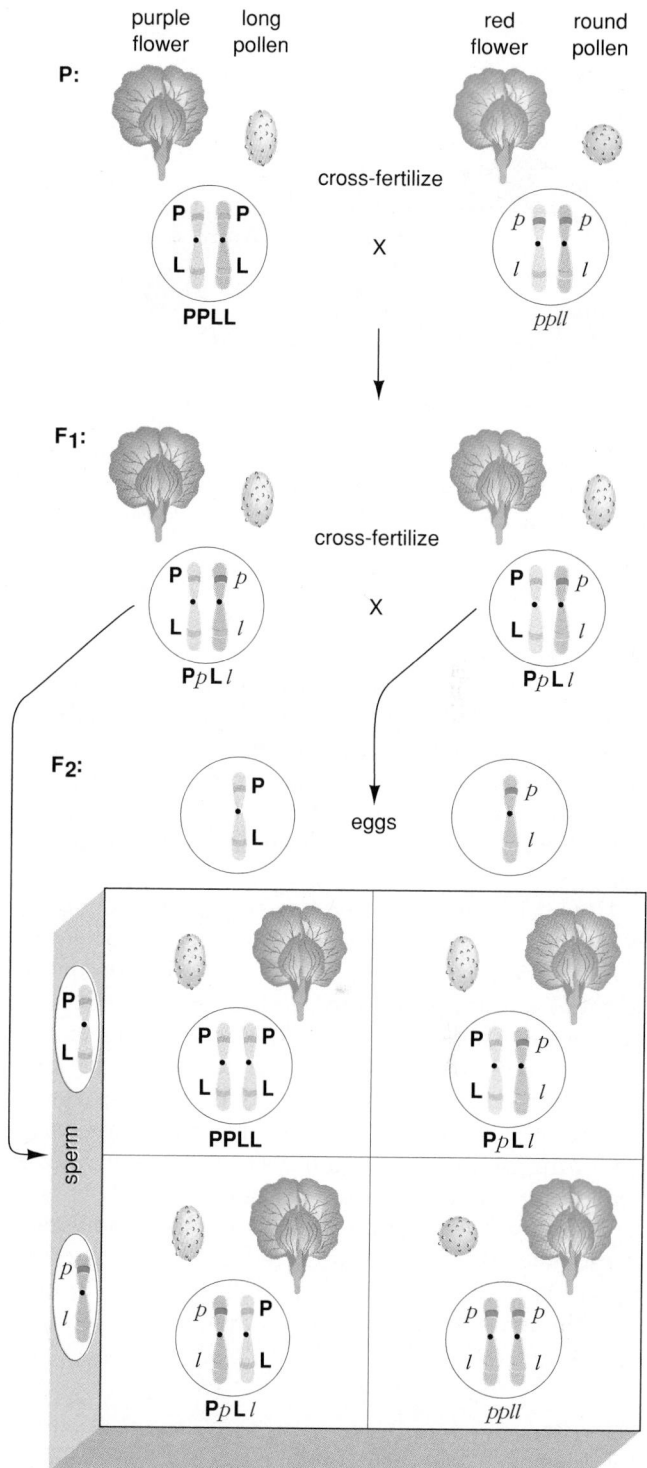

Figure 11-12 Inheritance of linked genes for pollen grain shape and flower color in sweet peas. Since these genes are found on the same chromosome, the configuration *PL* and *pl* in the parental generation tends to be preserved in all subsequent generations. The F_2 offspring show a phenotypic ratio of 3 purple, long: 1 red, round rather than the 9:3:3:1 ratio one would expect from independently assorted genes.

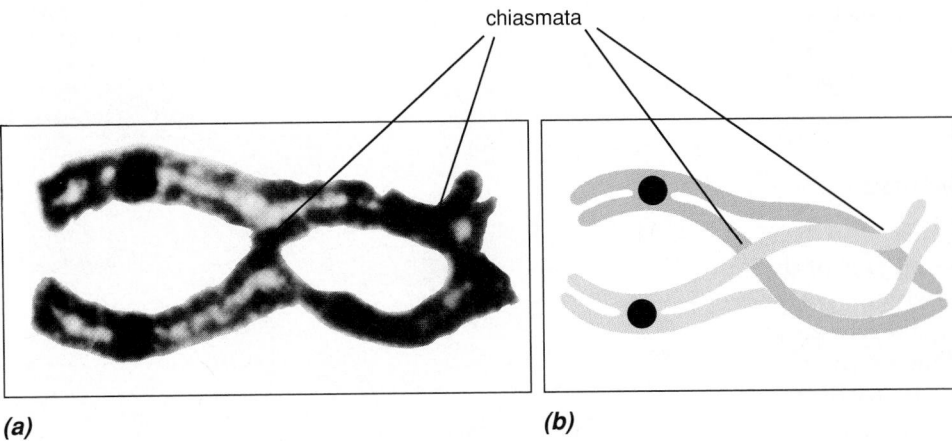

Figure 11-13 A photomicrograph of crossing over during meiosis **(a)** and a drawing **(b)** based on the micrograph. The two pairs of chromatids intertwine, forming chiasmata in several places. Chromosomal material is exchanged at the chiasmata.

(a) *(b)*

almost always at least one exchange between homologues during any meiosis. **Exchange of genetic material during crossing over forms new gene combinations on both homologous chromosomes.** Therefore, when homologous chromosomes separate at anaphase I, the chromosomes that each haploid daughter cell receives are different from those of the parent cell.

How does crossing over exchange genes? Chromosomes are long, continuous strands of DNA. Genes are specific segments of a DNA strand that contain the information needed to produce a particular trait. Genes are arranged linearly along each chromosome, like colored stripes painted on a rope, with each stripe corresponding to a gene. When crossing over occurs, the two chromosome "ropes" swap corresponding segments, so that a series of genes moves over from one chromosome to the other.

Crossing over during meiosis explains why new combinations of traits occurred in the sweet pea cross. At the beginning of prophase I, all the F$_1$ peas had this pair of homologous chromosomes:

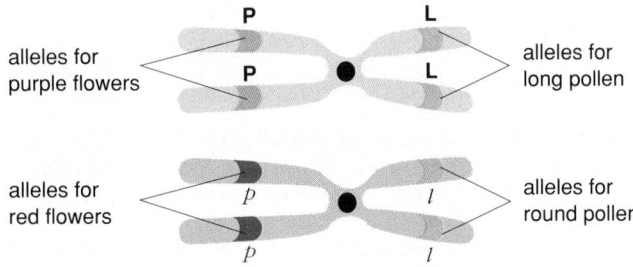

alleles for purple flowers

alleles for red flowers

alleles for long pollen

alleles for round pollen

In a few reproductive cells, crossing over occurred between the locations of the genes for flower color and pollen shape:

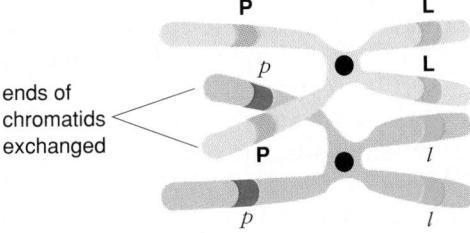

ends of chromatids exchanged

At anaphase I the separated homologous chromosomes had this gene composition:

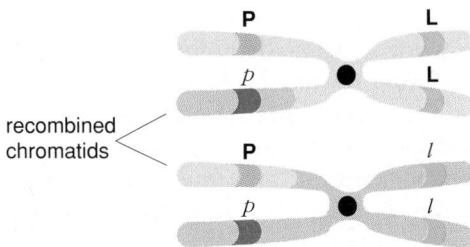

recombined chromatids

Four different types of chromosomes were distributed to the haploid daughter cells during meiosis II:

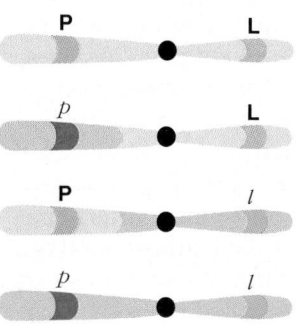

Therefore, some gametes were produced with each of the four chromosome configurations: *PL* and *pl* (the original parental types), and *Pl* and *pL* (recombined chromosomes).

In most of the reproductive cells, crossing over between the flower-color and pollen-shape genes did not occur. Therefore, most of the F₂ offspring received chromosomes with the *PL* and *pl* allele combinations. Fusion of sperm and eggs that formed from cells in which crossing over occurred, however, gave rise to a few plants with purple flowers and round pollen and a few plants with red flowers and long pollen.

This is **genetic recombination:** generating new combinations of genes by the exchange of DNA between homologous chromosomes. In most organisms, **sexual recombination** also occurs, in which offspring receive one homologous chromosome from each of two, genetically different, parents. Together, genetic and sexual recombination provide great genetic variability among organisms.

Chromosome Mapping

Crossing over occurs more often between genes that are far apart on a chromosome than between genes that are close together. (Once again, think of chromosomes as ropes with "gene stripes." If you throw two ropes together onto the ground, one rope might lie on top so that it crosses the other. If you threw the pair of ropes many times and recorded where crossing occurred, you would find that two stripes would be separated by a cross much more often if the stripes lie on opposite ends of the ropes than if they are adjacent to one another.) By swapping alleles between homologous chromosomes, crossing over produces offspring with recombined traits. **Therefore, the proportion of offspring with recombined traits is a reflection of the spacing of genes along a chromosome.**

This idea has been used to map chromosomes. The distance between two genes on a chromosome is defined as the percentage of recombination observed during a cross between one organism homozygous recessive for both genes and another organism heterozygous for both genes. (If no crossing over occurred, what would be the expected ratio of genotypes and phenotypes? Hint: this is a "two-gene" test cross.) If two genes recombine 10% of the time, they are placed 10 "map units" apart on the chromosome. In *Drosophila*, it has been possible to correlate this type of recombination map with the physical appearance of certain chromosomes, and locate the actual sites of

genes (Fig. 11-14). As we will see in Chapter 15, chromosome mapping is important in human genetics, as molecular biologists try to locate the genes that cause severe inherited diseases, such as Huntington's disease.

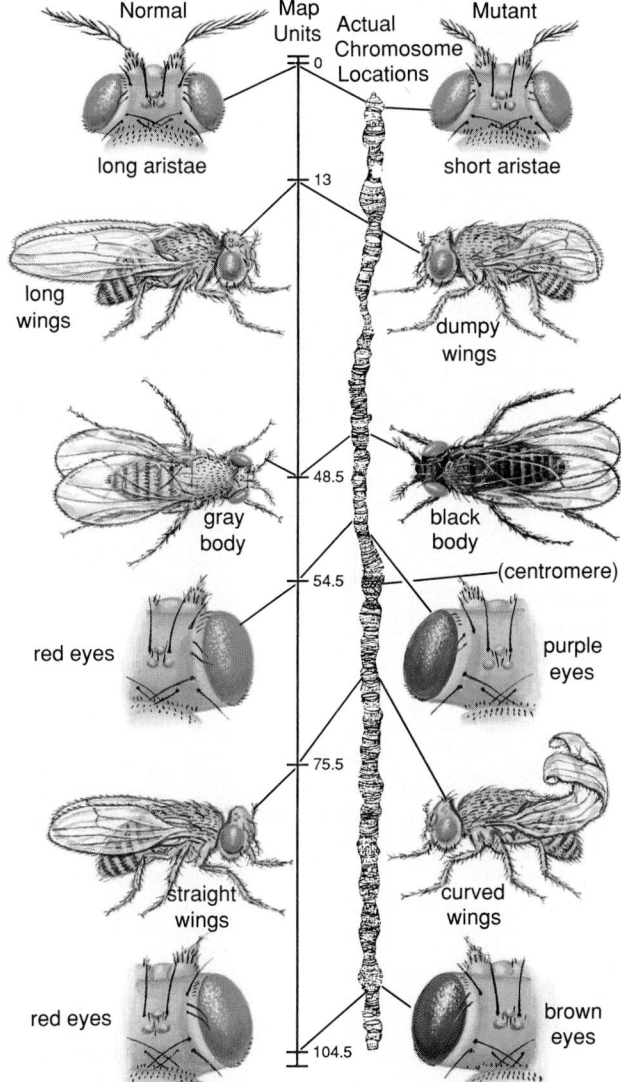

Figure 11-14 A map of chromosome 2 in the fruit fly, comparing the distances between genes obtained by recombination analysis with the physical locations of genes on the chromosomes. The order of the genes is the same in both maps, but the distances between genes are not. Crossing over is apparently hindered in some regions of the chromosome, especially around the centromere, distorting the recombination map.

The Relationships among Genes, Alleles, and Chromosomes

We can now refine the concepts of genes and alleles to reflect the data from mitosis and meiosis, Mendel's and Morgan's experiments, and crossing over. Many of these concepts are illustrated in Figure 11-15.

1. **Chromosomes** are long, continuous strands of DNA that encode genetic instructions.
2. A **gene** is a segment of DNA on a chromosome that encodes the instructions for producing a spe-

cific trait. The location of a specific gene on a chromosome is called its **locus.**

3. The different types of chromosomes in the body cells of most multicellular eukaryotic organisms occur in pairs called **homologues**. Such organisms are called **diploid**. Most diploid eukaryotes have one to several dozen pairs of **autosomes** (the number is characteristic of the species) and one pair of **sex chromosomes**.
4. Homologous chromosomes carry the same genes, located at the same places on the chromosomes. The exception to this rule is the pair of sex chromosomes. Although sex chromosomes have some homologous regions that allow them to pair up during meiosis, they often are very different otherwise. In mammals and many insects, the **X chromosome** contains many genes, including many genes for traits that are unrelated to sex. The **Y chromosome,** in contrast, contains few genes other than those that determine maleness. In these organisms, males are usually XY, and females are XX.
5. Differences in DNA composition at the same gene locus on two homologous chromosomes form different **alleles** of the gene. These differences in DNA may in turn cause variations in the trait controlled by that gene, such as purple versus red flowers.
6. If both homologous chromosomes contain the same allele at a given gene locus, the organism is **homozygous** for that gene locus. If homologous chromosomes contain different alleles at a given gene locus, the organism is **heterozygous** for that locus.
7. Because chromosomes are the structural units that are moved about during meiosis, genes located on a single chromosome are **linked**—that is, they tend to be inherited together. However, **crossing over** during prophase I exchanges segments of DNA of homologous chromosomes. If these two segments differ in their DNA composition (that is, if they are different alleles), crossing over can create new combinations of alleles by "unlinking" old combinations.

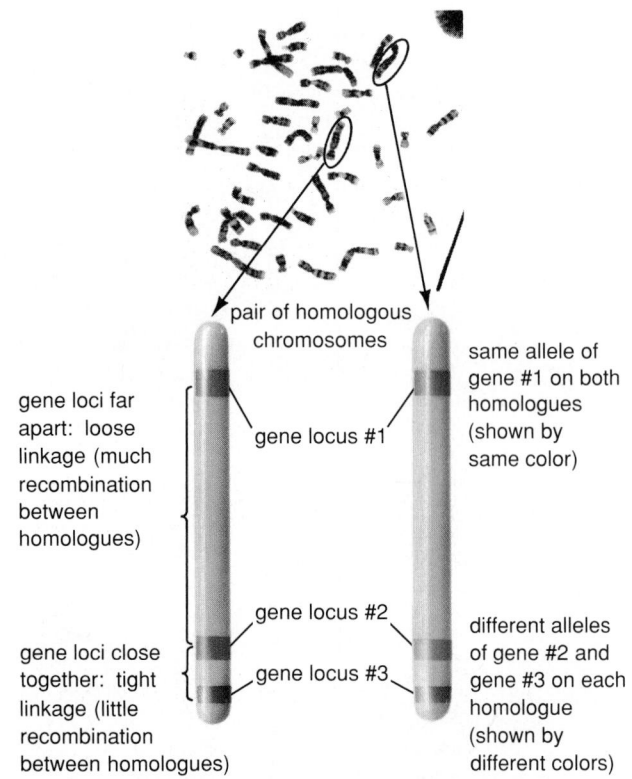

Figure 11-15 The relationships among genes, alleles, and chromosomes. The photograph shows a human karyotype. The drawing illustrates a pair of homologous autosomes (non–sex chromosomes). A given gene is a segment of DNA at the same location (locus) on both homologous chromosomes. Differences in DNA composition at the same gene locus (depicted by different colors) form different alleles of the gene. A chromosome carries many genes; each gene occupies a different locus on the chromosome. If two loci are close together, the genes are tightly linked, and little crossing over will occur between them. If the loci are far apart, the genes are loosely linked, and much crossing over will occur between them, producing a high frequency of recombination of alleles.

Variations on the Mendelian Theme

So far, we have restricted our discussion of inheritance in two major ways. First, we have assumed that genes never change. This is not true: **the molecular composition of genes can change, a process known as *mutation.*** The molecular mechanisms of mutation and allele action will be described in Chapters 12 and 13. For now, we should merely note that alleles originate as mutations. In Mendel's peas, for example, the allele for purple flower color directs the synthesis of purple pigment. The allele for white is probably a mutation of the purple allele that occurred many generations ago, in which the genetic instructions have been scrambled, with the result that no pigment is produced. Second, we have assumed that all traits are inherited in a simple, single-gene, dominant versus recessive manner. Most traits, however, are influenced in more varied and subtle ways than this. In the remainder of this chapter, we will discuss some of the more common variations in inheritance.

Incomplete Dominance

In his pea experiments, Mendel encountered a particularly simple situation: heterozygotes and homozygous dominants had the same phenotype. This is often not the case. In snapdragons, for example, crossing homozygous red-flowered plants (RR) with homozygous white-flowered ones ($R'R'$) does not produce red-flowered F_1 hybrids. Instead, the F_1 flowers are pink. When the heterozygous phenotype is intermediate between the two homozygous phenotypes, the pattern of inheritance is called **incomplete dominance**. At first glance, this looks like blending, as if you had mixed white paint with red paint to make pink. But the F_2 generation shows that the alleles for flower color have not changed (Fig. 11-16). The F_2 offspring include about $\frac{1}{4}$ red, $\frac{1}{2}$ pink, and $\frac{1}{4}$ white flowers. This corresponds to the genotypic ratio of $\frac{1}{4} RR : \frac{1}{2} RR' : \frac{1}{4} R'R'$.

Multiple Alleles and Codominance

A single individual, having only one set of homologous chromosome pairs, can have at most only two different alleles for a given gene. Alleles, however, arise through mutation, and the genes of different organisms may suffer many mutations, each producing a new allele. If we could sample all the individuals of a species, we would often find several, sometimes even dozens, of alleles for every gene. The gene for eye color in fruit flies, for example, has many alleles, each recessive to normal red eyes and producing various shades of yellow or pink when homozygous.

The ABO blood types in humans constitute a familiar system of such **multiple alleles**. The blood types A, B, AB, or O arise as a result of three different alleles of a single gene (usually designated I^A, I^B, and i). This gene directs the synthesis of glycoprotein

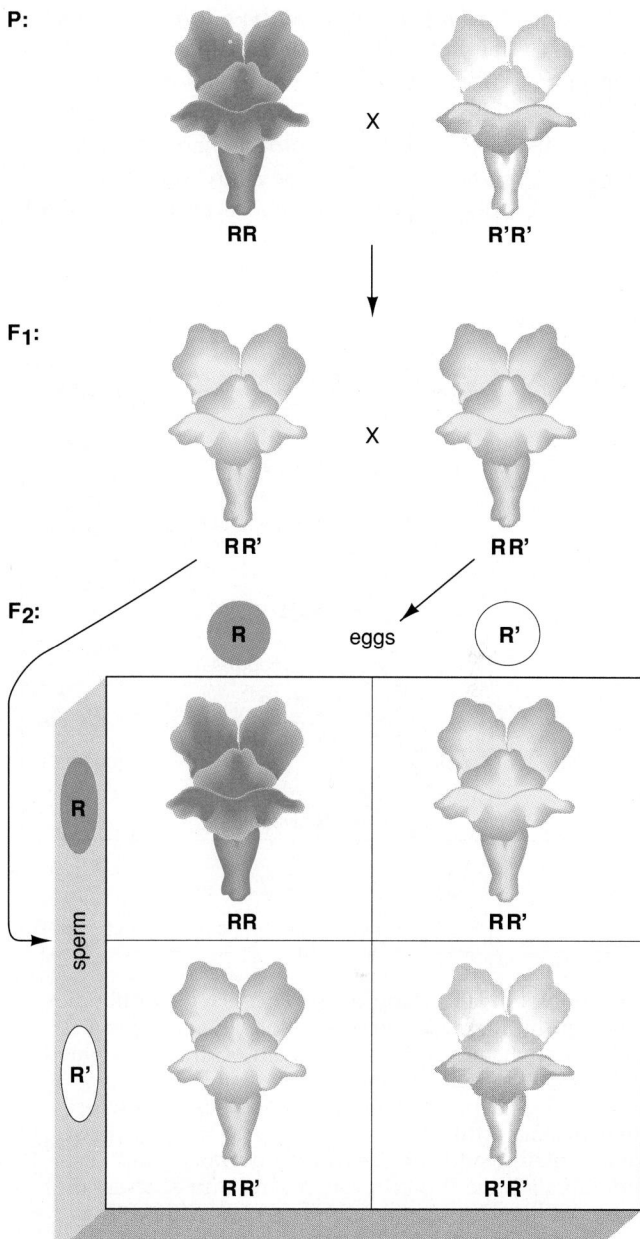

Figure 11-16 Incomplete dominance in the inheritance of flower color in snapdragons. In cases of incomplete dominance, we will use capital letters for both alleles (here R and R') rather than capital and lowercase letters. Hybrids (RR') have pink flowers, while the homozygotes are red (RR) or white ($R'R'$). Since heterozygotes can be distinguished from homozygous dominants, the distribution of phenotypes in the F_2 generation ($\frac{1}{4}$ red: $\frac{1}{2}$ pink: $\frac{1}{4}$ white) is the same as the distribution of genotypes ($\frac{1}{4} RR$: $\frac{1}{2} RR'$: $\frac{1}{4} R'R'$).

''identification markers'' that protrude from the surfaces of red blood cells. Alleles I^A and I^B direct the synthesis of glycoproteins A and B, respectively, while allele i produces no glycoproteins at all (Fig.

11-17). Individual humans may have one of six genotypes: I^AI^A, I^BI^B, I^AI^B, I^Ai, I^Bi, or ii. Alleles I^A and I^B are dominant to i. Therefore, individuals with genotypes I^AI^A or I^Ai have type A glycoproteins on their red blood cells, and have type A blood. Those with genotypes I^BI^B or I^Bi synthesize type B glycoproteins, and have type B blood. Homozygous recessive ii individuals lack these glycoproteins, and have type O blood. However, alleles I^A and I^B are **codominant** to one another—that is, *both are phenotypically detectable in heterozygotes*. I^AI^B individuals have red blood cells with both A and B glycoproteins, and have type AB blood.

These red blood cell glycoproteins may react with

(a)

Genotype	Blood Type	Red Blood Cells	Plasma Antibodies
$I^A\ I^A$ $I^A\ i$	A	A — A glycoprotein	anti-**B**
$I^B\ I^B$ $I^B\ i$	B	B — B glycoprotein	anti-**A**
$I^A\ I^B$	AB	AB A glycoprotein and B glycoprotein	none
$i\ i$	O	O neither glycoprotein	anti-**A** anti-**B**

(a) Red blood cell glycoproteins and plasma antibodies found in individuals with each blood type. Type A individuals have the A glycoprotein on their red blood cell surfaces. Their plasma contains antibodies only against B glycoproteins (anti-B antibodies). Therefore, their plasma antibodies do not bind to their own red cell glycoproteins, and their own cells do not clump. Type B individuals have B glycoproteins and anti-A antibodies. Type AB blood contains red blood cells with both the A and B glycoproteins, but neither antibody. Type O individuals have both anti-A and anti-B antibodies but no reactive glycoproteins on their red blood cells.
(b) The reaction between type A blood cells transfused into type B blood. Type B blood contains many anti-A antibodies. Since each antibody has two sites that can bind the A glycoprotein, the A cells become clumped, held together by the anti-A antibodies. These clumps can become large clots, with serious medical consequences. The permissible donors for each blood type are shown in the table.

(b)

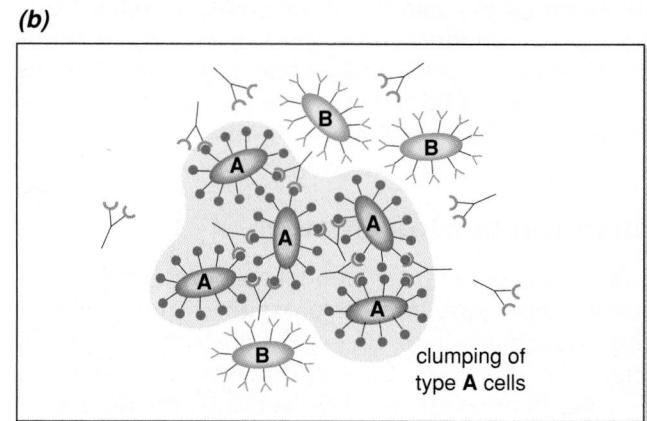

clumping of type **A** cells

Medical Effects of Blood Transfusions			
Donor Type	Recipient Type	Effect on Recipient	Permissible Blood Donation ?
A	A	—	yes
	B	clumping	no
	AB	—	yes
	O	clumping	no
B	A	clumping	no
	B	—	yes
	AB	—	yes
	O	clumping	no
AB (universal recipient)	A	clumping	no
	B	clumping	no
	AB	—	yes
	O	clumping	no
O (universal donor)	A	—	yes
	B	—	yes
	AB	—	yes
	O	—	yes

Figure 11-17 Human ABO blood group reactions. Glycoproteins on the surfaces of the red blood cells determine blood type. If antibodies in the blood plasma bind to red cell glycoproteins, the red blood cells clump together. Each antibody can bind only to one specific glycoprotein; for example, anti-A antibody binds only to glycoprotein A and not to glycoprotein B.

antibodies in the blood plasma. If a patient with type B blood receives a transfusion of type A blood, the anti A antibodies in the patient's serum cause the type A blood cells to clump (Fig. 11-17b).

Polygenic Inheritance

Many traits are not controlled by just one gene, but are influenced by the action of many genes. This is called **polygenic inheritance**. The simplest polygenic inheritance occurs when two genes code for the same trait. In wheat, for example, there are two genes for kernel color, which we might designate 1 and 2. Each gene has two alleles, R and R'. The R allele directs the synthesis of one "unit" of red pigment in the kernel, while the R' allele causes no pigment synthesis at all. If only gene 1 were active, then the inheritance of kernel color would follow simple incomplete dominance: R_1R_1 = red; $R_1R'_1$ = pink; $R'_1R'_1$ = white. Since kernel color is controlled by both genes 1 and 2, the color becomes more finely graded in intensity, depending on the number of R alleles of both genes. Kernels with the genotype $R_1R_1R_2R_2$ synthesize four units of pigment, and therefore are dark red. Kernels with the genotype $R'_1R'_1R'_2R'_2$ produce no pigment, and are white. Intermediate numbers of R alleles yield intermediate intensities of red.

As you can well imagine, the more genes that contribute to a single trait, the greater the number of categories of the trait, with increasingly fine gradation between categories. Continuing our example of color intensity, if three genes are involved, we would have seven phenotypic classes; with four genes, nine classes, and so on. With more than three genes, differences between phenotypes are small, and it is extremely difficult to classify the phenotypes reliably.

Polygenic inheritance leads to small, "quantitative" differences in traits (such as shades of color in wheat kernels) rather than the large "qualitative" differences that are caused by single-gene inheritance (such as the all-or-none, purple-or-white color of pea flowers). Therefore, polygenic inheritance is often referred to as **quantitative inheritance.**

Many common human characteristics, such as height, body build, and the color of skin, hair, and eyes, are influenced by many genes. As you know, the human population shows virtually continuous variation in these traits. Such continuous variation is due partly to polygenic inheritance, and, as we shall describe below, partly to environmental influences.

Gene Interactions

In the inheritance of some traits, there are genes whose actions are required for other genes to be expressed. An example of such gene interaction, called **epistasis**, occurs in the inheritance of hair color in virtually all mammals, including, for example, the mouse (Fig. 11-18a). One gene controls the *synthesis* of melanin, the pigment in hair. The dominant allele of this gene (M) allows melanin to be produced, while the recessive allele (m) does not. A second gene controls the *distribution* of pigment in the hair. The normal fur of the wild house mouse, called agouti, has individual hairs that are black with a yellow tip, giving the mouse an overall brownish-grey appearance. Agouti (A) is dominant to plain black fur (a). These two genes control hair color in the following way. If a mouse has the MM or Mm genotype, melanin will be produced. The fur will then be agouti if the mouse is AA or Aa for the melanin-distribution gene, or black if the mouse is aa. If a mouse is mm, then it cannot produce melanin. All mice with the mm genotype therefore have white hair and are albino. In an albino, the melanin-distribution gene cannot affect the phenotype, since there isn't any melanin to distribute. Therefore, the *melanin-synthesis gene* controls the expression of the *melanin-distribution gene*. A cross of two agouti mice can result in a variety of offspring, as the genes for melanin production and distribution assort independently (Fig. 11-18b).

The expression of almost all genes is influenced to some extent by other genes. To take an obvious, but usually overlooked, example, no genes can be expressed in an adult organism unless the genes that direct development operate properly. An organism is a cohesive, coordinated whole, and all its genes influence its ultimate anatomy, physiology, and behavior.

Multiple Effects of Single Genes

We have just seen that a single phenotype may require the interaction of several genes. The reverse is also true: single genes may have multiple phenotypic effects, a phenomenon called **pleiotropy**. Take the case of albino mice, for example. Albino mice not only lack pigment in their fur; they also lack pigment in their eyes (their eyes are bright pink, because the blood circulating through capillaries inside their eyes shows through the transparent iris). Without any eye pigment, they are extraordinarily sensitive to light: they can't see well in even fairly dim light, and normal daylight rapidly destroys the receptors in their eyes, causing blindness. Consequently, the single

(a)

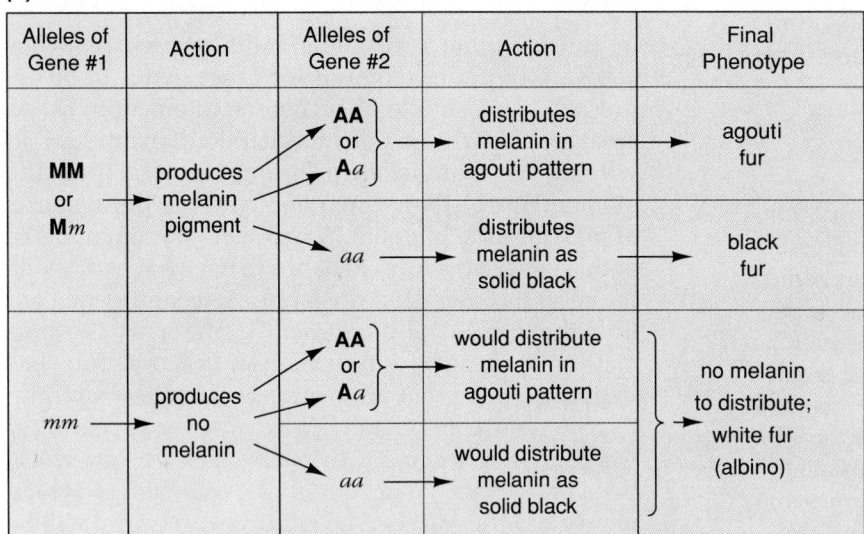

Alleles of Gene #1	Action	Alleles of Gene #2	Action	Final Phenotype
MM or **M***m*	produces melanin pigment	**AA** or **A***a*	distributes melanin in agouti pattern	agouti fur
		aa	distributes melanin as solid black	black fur
mm	produces no melanin	**AA** or **A***a*	would distribute melanin in agouti pattern	no melanin to distribute; white fur (albino)
		aa	would distribute melanin as solid black	

Figure 11-18 Two genes interact in the inheritance of fur color in mice.
(a) To produce any melanin pigment at all, a mouse must have at least one dominant allele *M* for melanin production (*MM* or *Mm* genotypes). If so, the melanin-distributing gene will produce agouti (*AA* or *Aa* genotypes) or black (*aa* genotype) fur. If a mouse is homozygous recessive *mm*, then it produces no melanin, and will have white fur regardless of its genotype for the melanin-distribution gene.
(b) If two mice that are heterozygous for both genes are mated, their offspring do not have a 9:3:3:1 ratio of phenotypes, because of the interaction between the genes. In this cross, the phenotypic ratio of offspring is 9 agouti: 3 black: 4 albino. Other types of crosses (e.g., a heterozygote mated with a homozygous recessive) yield other modifications of the 9:3:3:1 ratio.

(b)

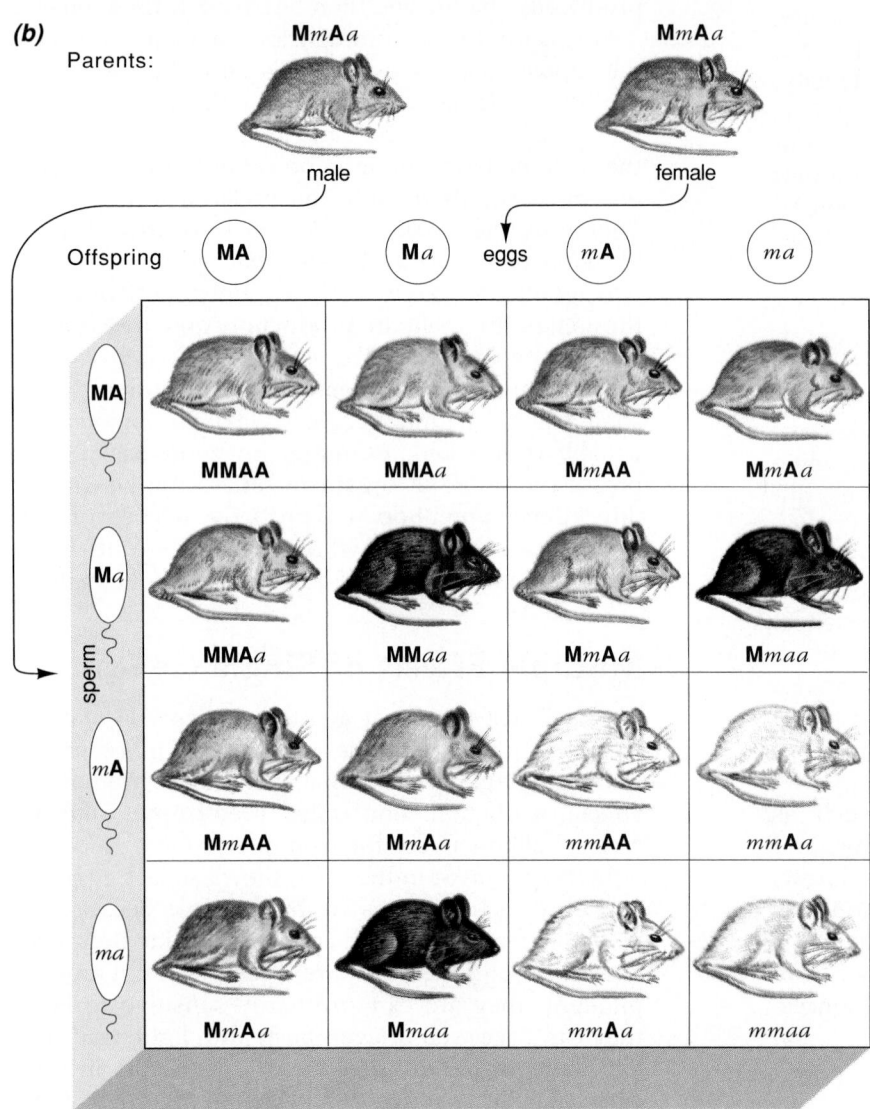

gene for pigment production actually can have multiple phenotypic effects: white fur, pink eyes, and blindness.

Environmental Influences on Gene Actions

An organism is not just the sum of its genes. **Both the genotype and the environment in which an organism lives profoundly affect its phenotype.** This holds true even for simple physical traits.

A striking example of environmental effects on gene action occurs in the Himalayan rabbit, which, like the Siamese cat, has pale body fur but black ears, nose, tail, and feet. The Himalayan rabbit actually has the genotype for black fur all over its body. The enzyme that produces the black pigment, however, is temperature sensitive; above about 34° C, the enzyme is inactive. At the temperatures typical of rabbit hutches, extremities such as the ears and feet are cooler than the rest of the body, and black pigment can be produced there. The main body surface is warmer than 34° C, so this fur is pale. By cooling parts of the rabbit's surface artificially, we can produce bizarre patterns of fur color (Fig. 11-19).

Most environmental influences are more complicated and subtle than this. The interactions between complex genetic systems and varied environmental conditions can create a continuum of phenotypes that defies analysis into genetic and environmental components. This is particularly true of human characteristics. Genetic analysis of all but the simplest human traits is notoriously difficult: the human generation time is long, the number of offspring per couple is small, and in any case one can't very well kidnap a few thousand people and keep them for genetics experiments. Add to these factors the myriad subtle ways in which people respond to their environments, and you can see that a precise determination of the genetic bases of complex traits such as intelligence or musical ability is probably impossible.

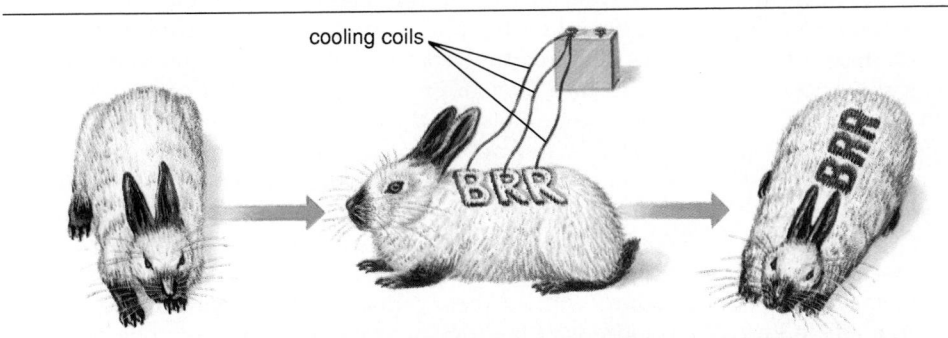

cooling coils

Figure 11-19 A simple case of interaction between genotype and environment in the expression of the gene for black fur in the Himalayan rabbit. Cool areas (nose, ears, feet, or under the cooling coils) allow expression of the gene for black fur.

Reflections on Genetic Diversity and Human Welfare

The science of genetics, which began in a monastery over a century ago, is crucially important to mankind today. Genetics plays a role in many human activities, from possible new sources of energy to inherited human diseases, and from paternity lawsuits to crop improvement. Virtually all the corn grown in the United States today, for example, is hybrid corn, the result of crossing parent strains that are themselves usually not very good as a crop. Faced with high prices, rising populations, and increasing demand for food, agricultural geneticists are constantly trying to develop strains of plants or animals that produce more food per unit input of energy, labor, and money. There are three principal approaches to improving crops and livestock. First, breeders can search for individual plants or animals with superior characteristics, and use only these organisms as breeding stock. The superior individuals could arise from mutations or from chance recombinations of preexisting genes. This approach relies on "good" alleles appearing spontaneously and on our ability to recognize them. Second, molecular geneticists can try to create superior genes in the laboratory or transplant them from one species to another. Gene transplantation is rapidly approaching commercial use, but creating superior genes is not yet feasible. The third approach is to look for desirable traits in wild populations of the same or closely related species. Breeders might then crossbreed or otherwise incorporate the desired genetic information into livestock or crop plants. This last approach shows great promise, and in fact is how many crops and livestock breeds were developed in the first place, millenia ago. Wheat, for example, was a cross between varieties of wild grass, aided by irregularities in meiosis that resulted in a polyploid plant producing large edible kernels.

Today, our reserves of wild genes are diminishing at a frightening rate, largely due to the pressures of human population and development. An estimated 5 to 20 million species of organisms exist on Earth, and only about a million and a half of these have even been identified. What's worse, many, perhaps most, of the unidentified species may become extinct before science ever has a chance to study them. Large tracts of wilderness, both in the United States and in other countries, are diminishing rapidly. When these ecosystems go, myriads of species and varieties go with them.

Rare species and local varieties of otherwise widespread species have been important in crop development in the past, and promise to be even more important in the future. In the 1960s, for example, plant geneticists crossbred commercial strawberries with a wild variety growing in Cottonwood Canyon, Utah. The result: strawberries that set fruit year-round. More recently, Florida breeders have developed a heat-tolerant blueberry bush by crossbreeding domestic blueberries with the rabbit-eye blueberry of the South. Geneticists at Oregon State University are developing a wildflower called meadowfoam into a source of high-quality commercial oil for the pharmaceutical and electronics industries. One of the most useful meadowfoam species is found in a 6-square-mile area near Medford, Oregon, and nowhere else in the world.

This is an important, but little recognized, reason to preserve wilderness: to preserve the genes of the plants and animals that live in them. This is also an important goal of efforts to save endangered species, for once a species becomes extinct, its genes are lost forever. Genes, with all their various alleles, have evolved over hundreds of millions of years, and represent one of our most valuable and irreplaceable natural resources.

SUMMARY OF KEY CONCEPTS

Single-Trait Experiments: The Law of Segregation

Mendel's first experiments dealt with the inheritance of a single trait at a time; these are called *monohybrid crosses*. From these experiments, Mendel hypothesized that the inheritance of each individual trait is determined by physical entities that we now call *genes*. Each organism possesses a pair of similar, but not necessarily identical, genes that influence each trait, but includes only one of each pair of genes in its gametes (the Law of Segregation). An offspring formed by the fusion of two gametes therefore receives pairs of genes, one of each pair inherited from each parent. Each gene may exist in alternative forms, called *alleles*. Each allele causes a different form of the trait (e.g., purple or white flowers). If an individual possesses two different alleles of the same gene, one allele, called dominant, may completely mask the expression of the other, recessive allele (e.g., a pea with alleles both for purple and white flower color will have purple flowers). In individuals with two different alleles of the same gene, which allele is included in any given gamete is determined by chance. Therefore, we can predict the relative proportions of offspring through the laws of probability.

The physical appearance of an organism (its phenotype) may not always be an infallible indicator of its alleles (the genotype), because of the masking of recessive alleles by dominant alleles. Organisms with two dominant alleles (homozygous dominant) have the same appearance as organisms with one dominant and one recessive allele (heterozygous).

Genes and Chromosomes

Genes are parts of chromosomes. The F_2 generation from a dihybrid cross (parental organisms differing in two traits) may have two fundamentally different outcomes:

1. If the genes for the two traits are on different chromosomes, then the F_2 offspring will appear in four different phenotypes, resulting from the independent assortment of the chromosomes (and hence the alleles) during meiosis. This is Mendel's Law of Independent Assortment.

2. If the genes are found on the same chromosome, then (except for crossing over) the F_2 offspring will express only the two parental phenotypes.

Sex linkage is a special and easily observed case of linkage of traits on the same chromosome. In many animals, females have two X chromosomes, while males have one X and one Y chromosome, with many fewer genes. Consequently, males have only one copy of most of the X chromosome genes, and recessive traits are more likely to be phenotypically expressed in males.

In crossing over, corresponding parts of homologous chromosomes are exchanged during meiosis, resulting in different combinations of alleles than existed in the parental chromosomes. The frequency of crossing over can be used to measure the positions of genes on chromosomes.

Variations on the Medelian Theme

Not all inheritance follows the simple dominant-recessive pattern.

1. In incomplete dominance, heterozygotes have a phenotype intermediate between the two homozygous phenotypes.
2. Codominant alleles, such as those determining blood type, are both phenotypically detectable in heterozygotes. The heterozygotes are not intermediate in phenotype between the two parents, but have separately distinguishable features of both parental types.
3. Many traits are determined by several genes, which is called *polygenic inheritance*. Traits that appear to exist in a continuum of finely graded forms are often determined polygenically, with three or more genes controlling the trait.
4. The actions of some genes are required if other genes are to be expressed at all (epistasis).
5. Many genes have multiple phenotypic effects (pleiotropy).
6. The environment plays at least some role in the phenotypic expression of all traits.

GLOSSARY

allele (al-ēl'): one of several alternative forms of a particular gene.

autosome (aw'-tō-sōm): a chromosome found in homologous pairs in both males and females, and which does not bear the genes determining sex.

codominance: the relation between two alleles of a gene, such that both alleles are phenotypically expressed in heterozygous individuals.

cross-fertilization: union of sperm and egg from two different individuals of the same species.

crossing over: the exchange of corresponding segments of the chromatids of two homologous chromosomes during meiosis.

dihybrid cross: a breeding experiment involving parents that differ in two distinct, genetically determined traits.

dominant: an allele that can determine the phenotype of heterozygotes completely, so that they are indistinguishable from individuals homozygous for the allele. In the heterozygotes, the expression of the other (recessive) allele is completely masked.

epistasis (ep-i-stā'-sis): a pattern of inheritance in which the actions of one gene are required for another gene to be expressed.

gene: a unit of heredity containing the information for a particular characteristic. A gene is a segment of DNA located at a particular place on a chromosome.

genetic recombination: the recombining of alleles on homologous chromosomes, due to exchange of DNA during crossing over.

genotype (jēn'-ō-tīp): the genetic composition of an organism; the actual alleles of each gene carried by the organism.

heterozygote (het-er-ō-zī'-gōt): an organism carrying two different alleles of the gene in question; sometimes called a hybrid.

homozygote (hō-mō-zī'-gōt): an organism carrying two copies of the same allele of the gene in question; also called a true-breeding organism.

hybrid: an organism that is the offspring of parents differing in at least one genetically determined characteristic; also used to refer to the offspring of parents of different species.

incomplete dominance: a pattern of inheritance in which heterozygotes have a phenotype intermediate between those of the two homozygotes.

independent assortment: a pattern of inheritance of multiple traits, in which the distribution of alleles for one trait into the gametes does not affect the distribution of alleles for other traits. Occurs with genes that are located on different chromosomes.

linkage: the inheritance of certain genes as a group because they are parts of the same chromosome. Linked genes do not show independent assortment.

locus: the physical location of a gene on a chromosome.

monohybrid cross: a breeding experiment in which the parents differ in only one genetically determined trait.

mutation: a change in the molecular composition of a gene; usually refers to a change that potentially alters phenotype.

phenotype (fēn'-ō-tīp): the physical properties of an organism. Phenotype can be defined as outward appearance (e.g., flower color), as behavior, or in molecular terms (e.g., ABO glycoproteins on red blood cells).

pleiotropy (plē'-ō-trō-pē): a situation in which a single gene influences more than one phenotypic characteristic.

polygenic inheritance: a pattern of inheritance in which the interactions of two or more genes determine phenotype.

recessive: an allele expressed only in homozygotes, and which is completely masked in heterozygotes.

segregation: a principle of inheritance, that the two alleles of a given gene separate from each other during gamete formation, with the result that each gamete receives one allele; occurs because of the separation of homologous chromosomes during meiosis.

self-fertilization: union of sperm and egg from the same individual.

sex chromosome: one of the pair of chromosomes that differ between the sexes and usually determine the sex of an individual; e.g., human females have similar sex chromosomes (XX) while males have dissimilar ones (XY).

sex linkage: a pattern of inheritance characteristic of genes located on one type of sex chromosome (e.g., X) and not found on the other type (e.g., Y) (see also *X linkage*).

sexual recombination: during sexual reproduction, the formation of new combinations of alleles in offspring, due to the inheritance of chromosomes from two different parental organisms.

test cross: a breeding experiment in which an individual showing the dominant phenotype is mated with an individual that is homozygous recessive for the same gene. The ratio of offspring with dominant versus recessive phenotypes can be used to determine the genotype of the phenotypically dominant individual.

true-breeding: pertaining to an individual all of whose offspring produced through self-fertilization are identical to the parental type. True-breeding individuals are homozygous for the trait in question.

X-linkage: referring to a gene carried on the X chromosome, in organisms in which the X chromosome carries many non–sex-related genes, while the Y chromosome carries few if any homologous genes (see also *sex linkage*).

STUDY QUESTIONS

1. Define the following terms: gene, allele, dominant, recessive, monohybrid, dihybrid, true-breeding, homozygous, heterozygous, cross-fertilization, self-fertilization.
2. Explain the meaning of Mendel's Law of Segregation and Law of Independent Assortment. Under what circumstances does the Law of Independent Assortment apply? When is it violated?
3. Explain why genes located on one chromosome are linked during inheritance. Why do linked genes sometimes separate during meiosis?

4. Explain why human skin color does not occur in just two forms, e.g., black and white.
5. What is sex linkage? In mammals, which sex would be most likely to show recessive sex-linked traits?
6. What is the difference between a phenotype and a genotype? Does knowledge of an organism's phenotype always allow you to determine the genotype? What type of experiment would you perform to determine the genotype of a phenotypically dominant individual?
7. Define polygenic inheritance, epistasis, and pleiotropy. Describe an example of each.

DISCUSSION QUESTIONS

1. There are some groups of insects in which females are XX and males are XO (the Y chromosome has not evolved in these species). In some other groups of animals, including moths, butterflies, and birds, females are ZW and males are ZZ for their sex chromosomes. Regarding chromosomal sex determination, discuss several basic ways in which these groups differ from mammals in which females are XX and males are XY.

2. Discuss why it was correct when Mendel predicted that, if he were to test cross all F_2 peas with yellow color (dominant) from a cross beginning with homozygous yellow x homozygous green peas, $\frac{2}{3}$ of the yellow F_2 peas would produce some green offspring and $\frac{1}{3}$ of the F_2 yellow peas would not produce any green offspring.

GENETICS PROBLEMS

(*Note:* An extensive group of genetics problems, with answers, can be found in the Study Guide.)

1. In certain cattle, hair color can be red (homozygous RR), white (homozygous $R'R'$), or roan (a mixture of red and white hairs; heterozygous RR').
   a. When a red bull is mated to a white cow, what genotypes and phenotypes of offspring could be obtained?
   b. If one of these offspring were mated to a white cow, what genotypes and phenotypes of offspring could be produced? In what proportion?
2. The palomino horse is golden in color. Unfortunately for horse fanciers, palominos do not breed true. In a series of matings between palominos, the following offspring were obtained:
   65 palominos, 32 cream-colored, 34 chestnut (reddish brown)
   What is the probable mode of inheritance of palomino coloration?
3. In the edible pea, tall (*T*) is dominant to short (*t*), and green pods (*G*) are dominant to yellow pods (*g*). List the types of gametes and offspring that would be produced in the following crosses:
   a. *TtGg* × *TtGg*
   b. *TtGg* × *TTGG*
   c. *TtGg* × *Ttgg*

4. In tomatoes, round fruit (*R*) is dominant to long fruit (*r*), and smooth skin (*S*) is dominant to fuzzy skin (*s*). A true-breeding round, smooth tomato (*RRSS*) was cross-bred with a true-breeding long, fuzzy tomato (*rrss*). All the F_1 offspring were round and smooth (*RrSs*). When these F_1 plants were bred, the following F_2 generation was obtained:
   Round, smooth: 43
   Long, fuzzy: 13
   Are the genes for skin texture and fruit shape likely to be on the same or on different chromosomes? Explain your answer.
5. In the tomatoes of Question 4, an F_1 offspring (*RrSs*) was mated with a homozygous recessive (*rrss*). The following offspring were obtained:
   Round, smooth: 583
   Long, fuzzy: 602
   Round, fuzzy: 21
   Long, smooth: 16
   What is the most likely explanation for this distribution of phenotypes?
6. In humans, hair color is controlled by two interacting genes. The same pigment, melanin, is present in both brown-haired and blond-haired people, but brown hair has much more of it. Brown hair (*B*) is dominant to blond (*b*). Whether any melanin can be synthesized at

all depends on another gene. The dominant form (*M*) allows melanin synthesis, while the recessive form (*m*) prevents melanin synthesis. Homozygous recessives *mm* are albino. What will be the expected proportions of phenotypes in the children of the following parents:

   a. *BBMM* × *BbMm*
   b. *BbMm* × *BbMm*
   c. *BbMm* × *bbmm*

7. In humans, one of the genes determining color vision is located on the X chromosome. The dominant form (*C*) produces normal color vision, while red–green color-blindness (*c*) is recessive. If a man with normal color vision marries a colorblind woman, what is the probability of their having a colorblind son? A colorblind daughter?

8. In the couple described in Question 7, the woman gives birth to a colorblind daughter. The husband sues for a divorce, on the grounds of adultery. Will his case stand up in court? Explain your answer.

ANSWERS TO GENETICS PROBLEMS

1. a. A red bull (*RR*) is mated to a white cow (*R'R'*). The bull will produce all *R* sperm, while the cow will produce all *R'* eggs. All the offspring will be *RR'*, and have roan hair (codominance).

   b. A roan bull (*RR'*) is mated to a white cow (*R'R'*). The bull produces half *R* and half *R'* sperm, while the cow produces *R'* eggs. Using the Punnett square method:

eggs
R'

		R'
sperm	R	RR'
	R'	R'R'

Using probabilities:

sperm	egg	offspring
$\frac{1}{2}$ R	R'	$\frac{1}{2}$ RR'
$\frac{1}{2}$ R'	R'	$\frac{1}{2}$ R'R'

The predicted offspring will be $\frac{1}{2}$ *RR'* (roan) and $\frac{1}{2}$ *R'R'* (white).

2. The offspring occur in three types, classifiable as dark (chestnut), light (cream), and intermediate (palomino). This suggests incomplete dominance, with the alleles for chestnut (*C*) combining with the allele for cream (*C'*) to produce palomino heterozygotes (*CC'*). We can test this hypothesis by examining the offspring numbers. There are approximately $\frac{1}{4}$ chestnut (*CC*), $\frac{1}{2}$ palomino (*CC'*), and $\frac{1}{4}$ cream (*C'C'*). If palominos are heterozygotes, we would expect the cross *CC'* × *CC'* to yield $\frac{1}{4}$ *CC*, $\frac{1}{2}$ *CC'*, and $\frac{1}{4}$ *C'C'*. Our hypothesis is supported.

3. a. *TtGg* × *TtGg:* This is a "standard" dihybrid cross. Both parents produce *TG*, *Tg*, *tG*, and *tg* gametes. The expected proportions of offspring are $\frac{9}{16}$ tall green, $\frac{3}{16}$ tall yellow, $\frac{3}{16}$ short green, $\frac{1}{16}$ short yellow (see Fig. 11-9).

   b. *TtGg* × *TTGG:* In this cross, the heterozygous parent produces *TG*, *Tg*, *tG*, and *tg* gametes. However, the homozygous dominant parent can only produce *TG* gametes. Therefore, all offspring will receive at least one *T* allele for tallness and one *G* allele for green pods, and thus all the offspring will be tall with green pods.

   c. *TtGg* × *Ttgg:* The second parent will produce two types of gametes, *Tg* and *tg*. Using a Punnett square:

eggs

		T *g*	t *g*
sperm	TG	TTG*g*	T*t*G*g*
	T*g*	TT*gg*	T*tgg*
	*t*G	T*t*G*g*	*tt*G*g*
	T*g*	T*tgg*	*ttgg*

The expected proportions of offspring are $\frac{3}{8}$ tall green, $\frac{3}{8}$ tall yellow, $\frac{1}{8}$ short green, $\frac{1}{8}$ short yellow.

4. If the genes are on separate chromosomes—that is, assort independently—then this would be a typical dihybrid cross with expected offspring of all four types (approximately $\frac{9}{16}$ round smooth, $\frac{3}{16}$ round fuzzy, $\frac{3}{16}$ long smooth, and $\frac{1}{16}$ long fuzzy). However, only the parental combinations show up in the F_2 offspring, indicating that the genes are on the same chromosome.

5. The genes are on the same chromosome, and are quite close together. On rare occasions, crossing over occurs between the two genes, producing recombination of the alleles.

6. a. *BBMM* (brown) × *BbMm* (brown). The first parent can only produce *BM* gametes, so all offspring will receive at least one dominant for each gene. Therefore, all offspring will be brown-haired.

   b. *BbMm* (brown) × *BbMm* (brown). Both parents can produce four types of gametes: *BM*, *Bm*, *bM*, and *bm*. Filling in the Punnett square:

eggs

		BM	B*m*	*b*M	*bm*
sperm	BM	BBMM	BBM*m*	B*b*MM	B*b*M*m*
	B*m*	BBM*m*	BB*mm*	B*b*M*m*	B*bmm*
	*b*M	B*b*MM	B*b*M*m*	B*bmm*	*bb*M*m*
	bm	B*b*M*m*	B*bmm*	*bb*M*m*	*bbmm*

Remembering that all *mm* offspring are albino, the expected proportions are $\frac{9}{16}$ brown-haired, $\frac{3}{16}$ blond-haired, $\frac{4}{16}$ albino.

c. *BbMm* (brown) × *bbmm* (albino):

eggs

bm

	BM	B*b*M*m*
sperm	B*m*	B*bmm*
	*b*M	*bb*M*m*
	bm	*bbmm*

The expected proportions of offspring are: $\frac{1}{4}$ brown-haired, $\frac{1}{4}$ blond-haired, $\frac{1}{2}$ albino.

7. A man with normal color vision is *CY* (remember, the Y chromosome does not have the gene for color vision).

His colorblind wife is *cc*. Their expected offspring will be:

eggs

c

sperm	C	C *c*
	Y	C Y

We therefore expect that all the daughters will have normal color vision, while all the sons will be colorblind.

8. The husband should win his case. All his daughters must receive one X chromosome, with the *C* allele, from him, and therefore should have normal color vision. If his wife gives birth to a colorblind daughter, her husband cannot be the father (unless there was a new mutation for colorblindness in his sperm line, which is very unlikely).

SUGGESTED READINGS

Benzer, S. "Genetic Dissection of Behavior." *Scientific American*, December 1973. Not only physical traits, but behaviors too are under the influence of genes. The fruit fly once again proves a useful model system in which to study inheritance.

Crow, J. F. "Genes That Violate Mendel's Rules." *Scientific American*, February 1979. A close look at genetic recombination and its role in evolution.

Hoyt, E. "Wild Relatives." *Living Wilderness*, Summer 1990. Undisturbed ecosystems are a valuable gene bank with great potential for improving crop plants.

Stern, C., and Sherwood, E. R. *The Origin of Genetics: A Mendel Source Book*. San Francisco, 1966. W. H. Freeman and Company, Publishers. There is no substitute for the real thing, in this case a translation of Mendel's original paper to the Brünn Society.

12

DNA: *The Molecule of Heredity*

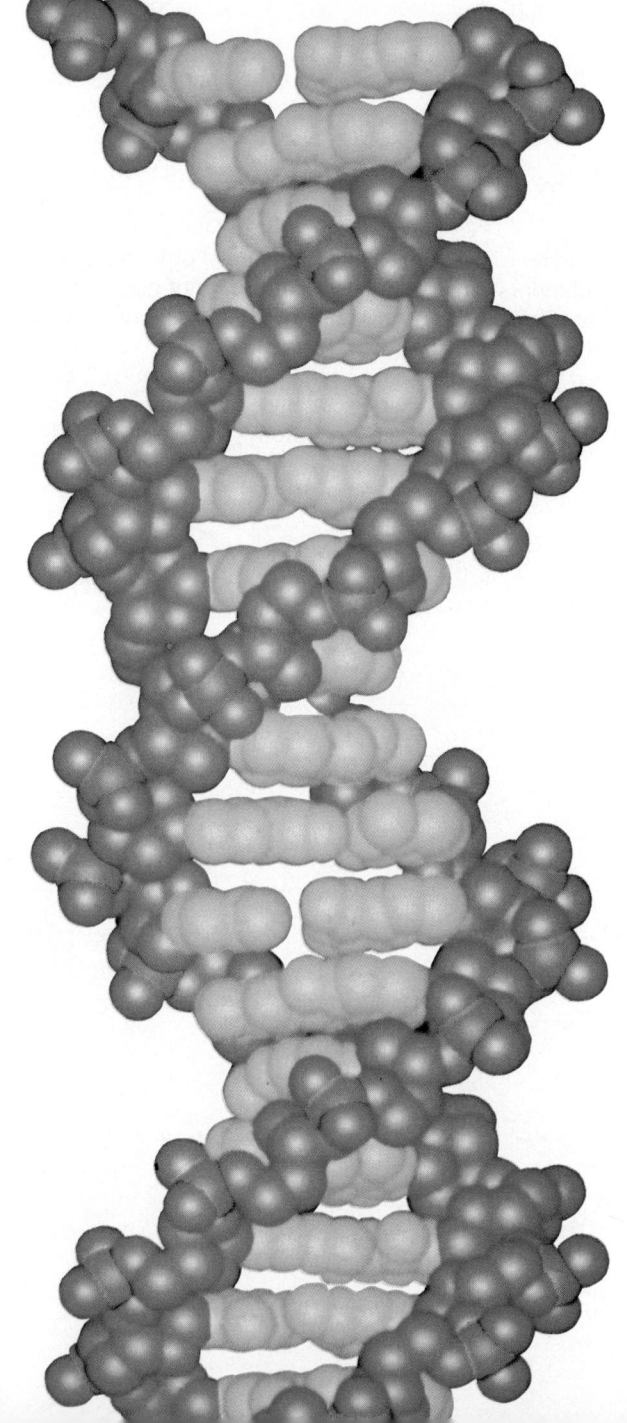

". . . a structure this pretty just had to exist."

James Watson in The Double Helix

The Composition of Chromosomes

Early microscopists discovered many of the organelles of the cell without, of course, knowing the function of most of them. In the early 1870s, the biochemist Friedrich Miescher analyzed one of these organelles, the nucleus. He found that he could extract a previously unknown chemical substance from nuclei, an acidic material with an unusually high phosphorus content. Because of its location in the nucleus and its acidic properties, this material came to be known as nucleic acid.

The discovery that chromosomes are the carriers of genetic information focused attention on their chemical composition. Biochemists discovered that a eukaryotic chromosome is composed of protein and a specific kind of nucleic acid, **deoxyribonucleic acid (DNA).** One of these substances, therefore, must carry the cell's hereditary blueprint.

The Chemical Nature of DNA

DNA seemed to be a particularly simple molecule, composed of just four kinds of subunits, called **nucleotides,** strung together in a long chain. Each nucleotide consists of three parts: a phosphate group; deoxyribose, a 5-carbon sugar; and a nitrogen-containing base:

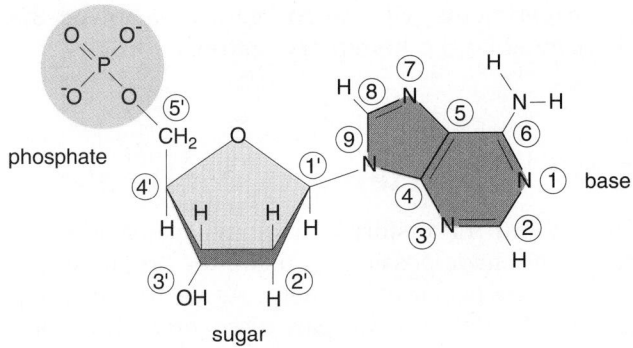

Biochemists have assigned numbers to the carbon and nitrogen atoms that make up the skeletons of the sugar and the base (the numbers for the sugar are given as 1', 2', etc., to distinguish them from the numbers for the base). Later in this chapter we'll see why these numbers are useful.

The four different DNA nucleotides have the same phosphate and sugar but different bases. The bases come in two types, the single-ringed **pyrimidines,** thymine (abbreviated T) and cytosine (C), and the double-ringed **purines,** adenine (A) and guanine (G):

Nucleotides with Pyrmidine Bases:

thymine

cytosine

Nucleotides with Purine Bases:

adenine

guanine

In a single strand of DNA, the phosphate of one nucleotide bonds to the sugar of another. This forms a long strand consisting of a "backbone" of sugars and phosphates, with the bases protruding from the backbone:

Single Strand of DNA:

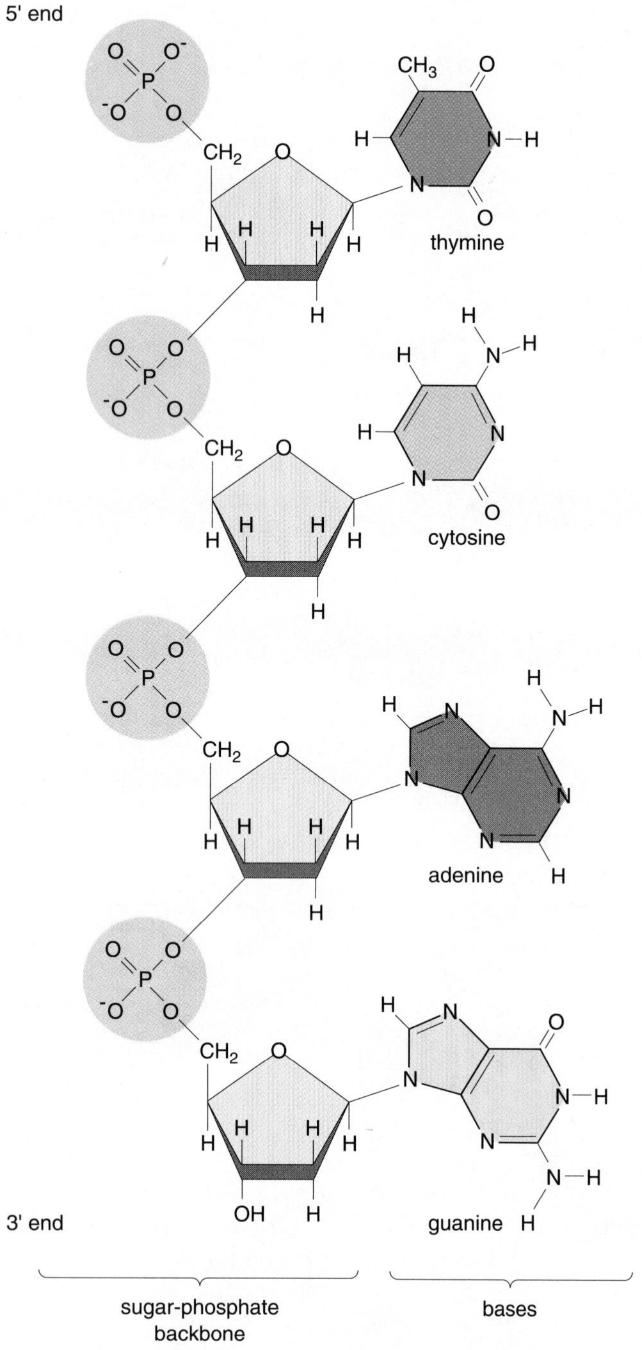

5' end

thymine

cytosine

adenine

guanine

3' end

sugar-phosphate
backbone

bases

Referring back to the numbering system for the carbon atoms of the sugars, the phosphate is attached to the 5' carbon. To form a single strand of DNA, the phosphate of one nucleotide undergoes a condensation reaction with the hydroxyl group attached to the 3' carbon of the next nucleotide. Each nucleotide in

the strand is therefore oriented in the same direction. Notice also that the DNA strand has a "free" sugar on one end (the 3' end of the strand) and a "free" phosphate on the opposite end (the 5' end of the strand).

Is the Genetic Material DNA or Protein?

Since DNA has only four subunits, each quite similar to the others, many biologists were skeptical that DNA could contain the information needed to construct all the molecules of complex organisms. Proteins, on the other hand, are constructed of about 20 different amino acids, and intuitively seemed to offer much more opportunity for carrying the enormous amount of genetic information that each chromosome obviously contains. Besides that, most of the substance of living cells, aside from water, is protein or is synthesized through the action of enzymes, which are also proteins. An attractive hypothesis, therefore, was that the proteins of chromosomes are templates for the proteins of the rest of the cell, a kind of mold from which copies could be made.

Gradually, biologists accumulated information that indicated that proteins might not be the genetic material. For example, chromosomes in the sperm of some fish contain only DNA and one protein, called **protamine.** Like other animals, a fish receives half its genes from its male parent, through the male's sperm. Now, protamine is a very simple protein, mostly composed of a single amino acid. If protamine were the genetic material in fish sperm, could it possibly serve as a template for all the other, more complex, proteins of the whole fish? Despite contradictory evidence such as this, many scientists continued to support the protein-as-gene hypothesis. Finally, two experiments with microorganisms showed that DNA must be the hereditary material.

DNA: The Heredity Molecule

Throughout the history of biology, major advances have been made possible by using the "right" organism for a particular experiment. As we saw in Chapter 11, Mendel and Morgan discovered the fundamentals of classical genetics through experiments with peas and fruit flies because they realized that peas and flies have advantages for genetics studies. Since all life probably evolved from a common, though distant, ancestor, knowing the mechanisms of inheritance in flies may help us to understand inheritance in other organisms, including people. Often, a particular organism is especially suitable for a certain experiment because it is simple in structure or reproduces very rapidly. Such simple, rapidly reproducing organisms were the keys to proving that DNA, not protein, is the chemical of heredity.

Bacterial Transformation

In the mid 1920s, the bacteriologist Frederick Griffith tried to develop a vaccine to protect people against a bacterium that causes pneumonia, named *Streptococcus pneumoniae*. He never produced a successful vaccine, so in that sense his experiments were failures. As it turned out, however, he discovered an intriguing phenomenon, which he called **transformation,** that eventually played a major role in the rise of molecular genetics.

Griffith found two strains of *Streptococcus*. In one strain, a polysaccharide capsule covers each bacterium; in the other form, the bacteria are naked. The presence or absence of the capsule is genetically determined. When *Streptococcus* bacteria invade a mammalian host—for example, a mouse—the host's white blood cells try to ingest the invading microbes. Encapsulated bacteria seem to be "slippery," preventing the white blood cells from getting a grip on the bacterial surface. Consequently, white blood cells cannot ingest encapsulated bacteria very well. Encapsulated bacteria therefore multiply in the host and cause disease. Bacteria without capsules are easily destroyed by white blood cells, so naked bacteria do not cause disease.

Griffith ran four experiments with the two strains of *Streptococcus* (Fig. 12-1).

1. **Infection with live encapsulated bacteria** As expected, when he injected mice with encapsulated bacteria, the mice contracted pneumonia and died (Fig. 12-1a). Autopsy showed that the blood of infected mice contained hordes of encapsulated bacteria.
2. **Infection with live naked bacteria** If mice were injected with the naked strain of *Streptococcus*, they remained healthy (Fig. 12-1b). Presumably, their white blood cells destroyed all the bacteria before any disease symptoms could be produced. No *Streptococcus* could be found in the mice's blood.
3. **Injection with heat-killed encapsulated bacteria** Griffith found that heating encapsulated *Streptococcus* to a high temperature killed the bacteria. Not surprisingly, dead bacteria did not cause pneumonia, and the mice remained free of living bacteria (Fig. 12-1c).
4. **Injection with a mixture of dead encapsulated bacteria and live naked bacteria** Griffith next did a surprising experiment. He injected mice with a mixture of heat-killed encapsulated bacteria and live naked bacteria (Fig. 12-1d). Although neither of these caused pneumonia when injected alone, the mixture did cause pneumonia. What's more, the infected mice teemed with live encapsulated bacteria, and these encapsulated bacteria bred true.

A few years later, in 1933, J. L. Alloway obtained the same results as the fourth experiment without any mice at all. He killed encapsulated bacteria, filtered out the larger bacterial parts, and mixed the resulting solution with living naked bacteria. The mixture produced live, virulent, true-breeding, encapsulated bacteria.

What did these last two experiments mean? A likely hypothesis was that the living bacteria had acquired molecules of genetic information from the dead bacteria. In some cases, the molecules encoded the instructions for making capsules, thereby **transforming** the formerly naked bacteria into the encapsulated form.

The Transforming Molecule Is DNA

In 1944, three researchers at Rockefeller University—Oswald Avery, Colin MacLeod, and Maclyn McCarty—discovered that the transforming molecule is DNA. They isolated DNA from encapsulated bacteria, mixed it with live naked bacteria, and produced live encapsulated bacteria. To prove that transformation was caused by DNA, and not by small quantities of protein contaminating their extracts, they treated different extracts with enzymes. Protein-destroying enzymes did not affect the transforming ability of the extracts, but DNA-destroying enzymes prevented transformation. They concluded that **DNA is the genetic material of bacteria,** and that a live bacterium can take up DNA from its environment and incorporate this DNA into its own chromosome (Fig. 12-2). When DNA carrying the genes for capsules becomes part of the chromosome of another bacterium, the recipient becomes capable of synthesizing capsules.

Despite this evidence, many biologists remained unconvinced that DNA was the universal hereditary molecule. Perhaps, some thought, DNA induces a mutation in naked bacteria, changing a (protein) gene that does not direct capsule synthesis into one that does. Or perhaps DNA is the hereditary molecule of bacteria, but not of other organisms. A second experiment convinced most of the skeptics. This experiment used an even simpler system, a virus that infects bacteria.

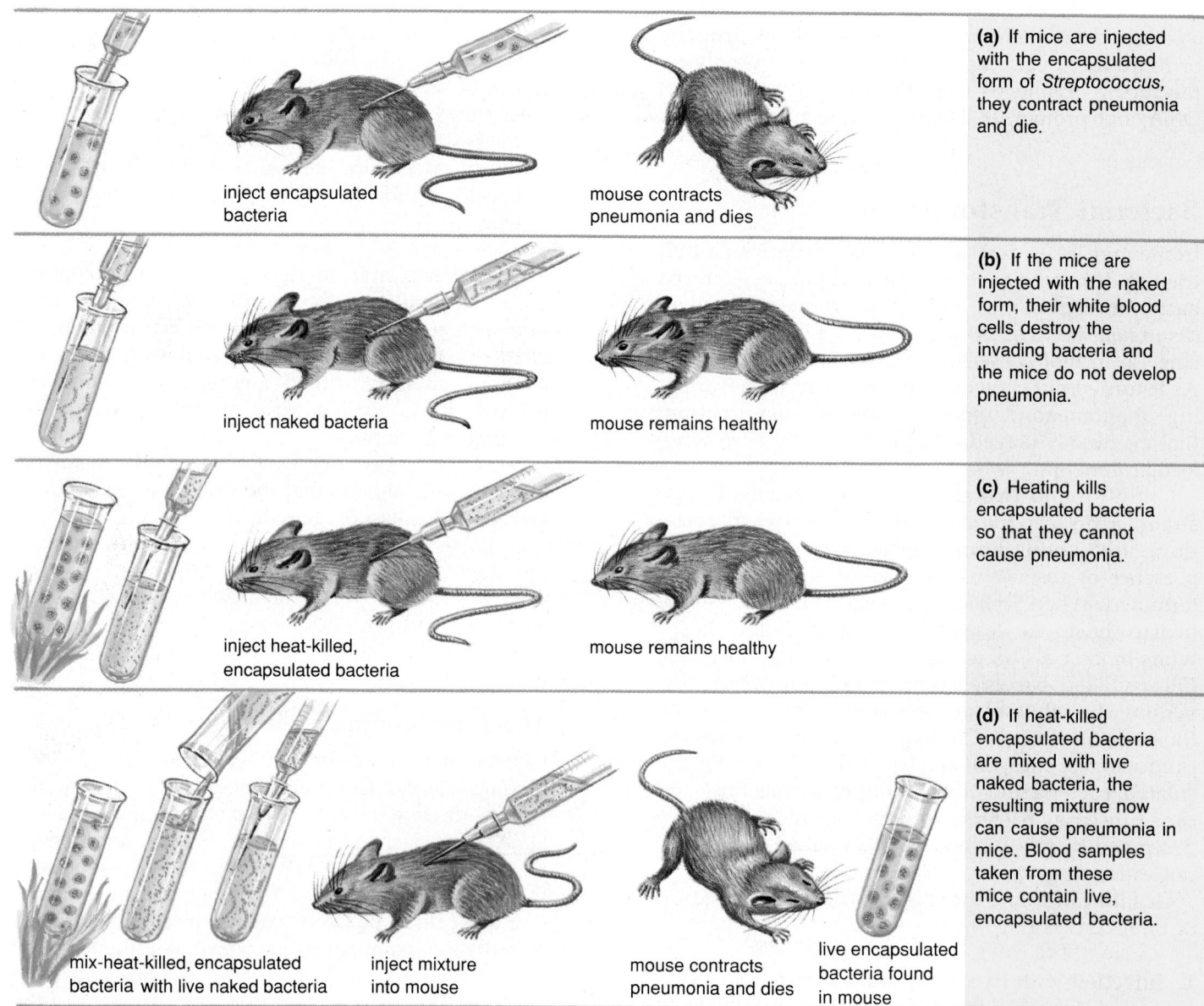

(a) If mice are injected with the encapsulated form of *Streptococcus*, they contract pneumonia and die.

inject encapsulated bacteria

mouse contracts pneumonia and dies

(b) If the mice are injected with the naked form, their white blood cells destroy the invading bacteria and the mice do not develop pneumonia.

inject naked bacteria

mouse remains healthy

(c) Heating kills encapsulated bacteria so that they cannot cause pneumonia.

inject heat-killed, encapsulated bacteria

mouse remains healthy

(d) If heat-killed encapsulated bacteria are mixed with live naked bacteria, the resulting mixture now can cause pneumonia in mice. Blood samples taken from these mice contain live, encapsulated bacteria.

mix-heat-killed, encapsulated bacteria with live naked bacteria

inject mixture into mouse

mouse contracts pneumonia and dies

live encapsulated bacteria found in mouse

Figure 12-1 Griffith's discovery of transformation in pneumonia bacteria.

The Bacteriophage Experiments

Certain viruses infect only bacteria, and are called **bacteriophages** (meaning "bacteria eaters" in Greek). Even though many bacteriophages (phage for short) have elaborate structures (Fig. 12-3), they are chemically very simple, being composed only of DNA and protein. As we will see, this chemical simplicity was the key to showing that DNA is the hereditary molecule of phages.

The Phage Life Cycle

A phage depends on its host bacterium for every aspect of its life cycle (Fig. 12-4). When a phage encoun-

ters a bacterium, it attaches to the bacterial cell wall and injects its genetic material into the bacterium. The rest of the phage (head, tail, tail fibers, etc.) remain outside the bacterium. The phage genes subvert the bacterial metabolism into producing more phages. Finally, the genetic material of the phage directs the synthesis of an enzyme that ruptures the bacterium, liberating the newly manufactured phages.

The Heredity Molecule of Bacteriophages Is DNA

In a brilliant series of experiments published in 1952, Alfred Hershey and Martha Chase used the chemical

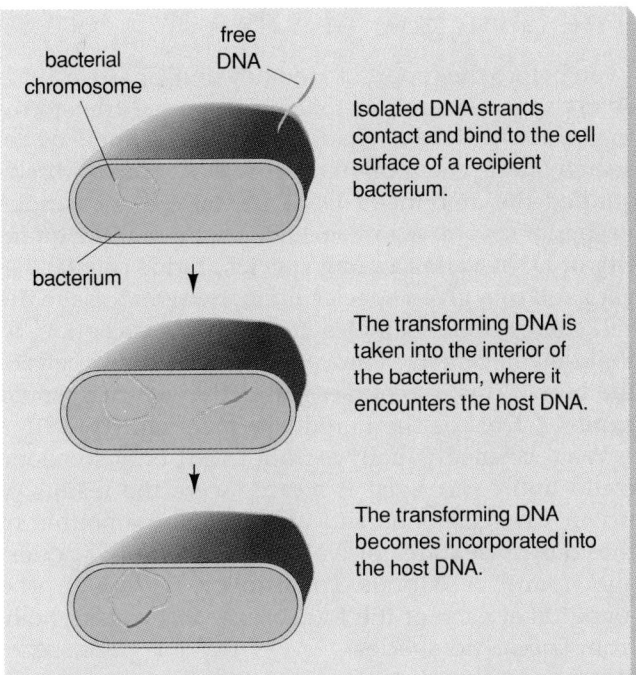

Figure 12-2 Bacterial transformation by free DNA. Transformation is not the "invasion" of a bacterium by free DNA. Rather, the recipient actively acquires the DNA and incorporates it into its chromosome. Transformation allows bacteria to acquire new genes, and confers a selective advantage on the recipient bacteria, which can now colonize new habitats or hosts with the help of their new genes. For example, transformation aids in the spread of antibiotic resistance from one type of bacterium to another.

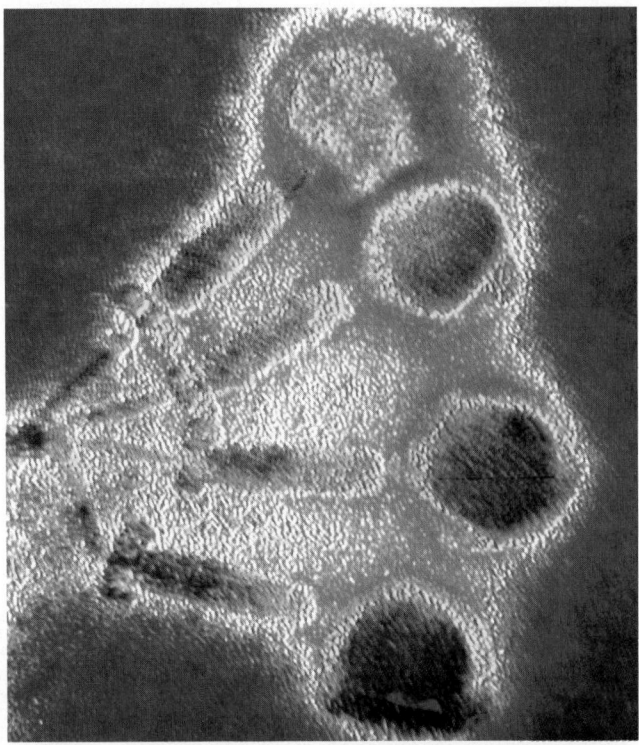

simplicity of bacteriophages to determine whether the genetic material of phages is DNA or protein (Fig. 12-5). Chemically, DNA and protein both contain carbon, oxygen, hydrogen, and nitrogen. However, DNA also contains phosphorus but not sulfur, while protein contains sulfur (in the amino acids methionine and cysteine) but not phosphorus. Hershey and Chase forced one population of phages to synthesize DNA using radioactive phosphorus, and another population to synthesize protein using radioactive sulfur. When bacteria were infected by phages containing radioactively labeled protein, radioactivity did not appear inside the bacteria. Offspring phages were not radioactive either (Fig. 12-5a). When bacteria were infected by phages containing radioactive DNA, the radioactivity was subsequently found inside the bacteria. When the bacteria burst, some of the offspring phages also had radioactive DNA (Fig. 12-5b). These experiments showed that the genetic material that phages inject into their hosts is DNA, not protein.

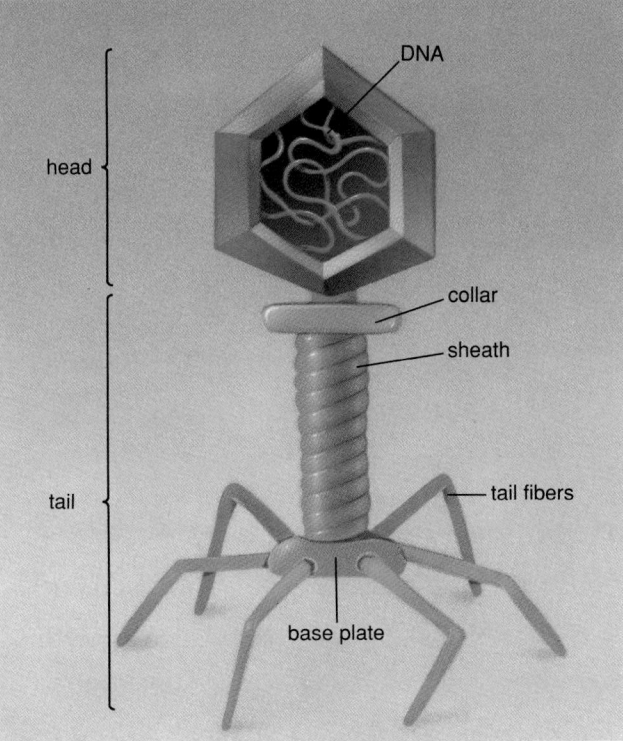

Figure 12-3 The structure of a T2 bacteriophage. The T2 phage consists of a head region composed of a protein coat containing a single DNA molecule, and a protein tail region (the sheath, base plate, and tail fibers) responsible for attachment to its host bacterium and injection of the genetic material.

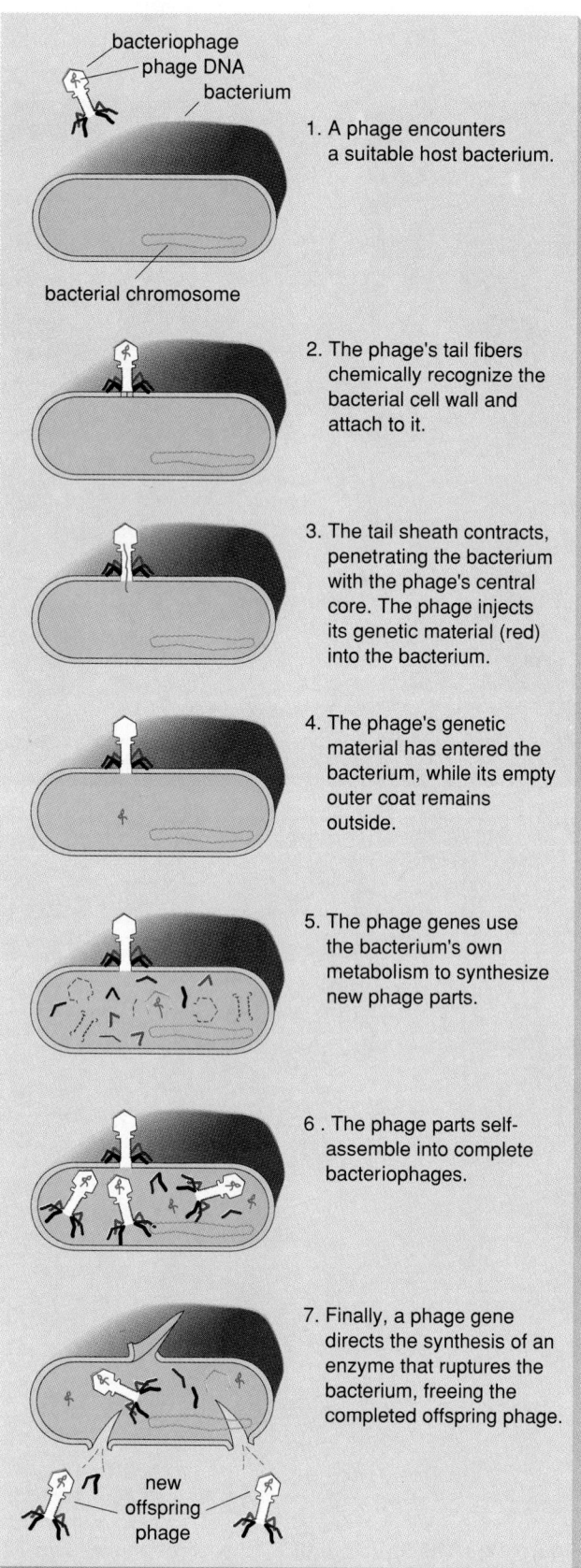

1. A phage encounters a suitable host bacterium.

2. The phage's tail fibers chemically recognize the bacterial cell wall and attach to it.

3. The tail sheath contracts, penetrating the bacterium with the phage's central core. The phage injects its genetic material (red) into the bacterium.

4. The phage's genetic material has entered the bacterium, while its empty outer coat remains outside.

5. The phage genes use the bacterium's own metabolism to synthesize new phage parts.

6 . The phage parts self-assemble into complete bacteriophages.

7. Finally, a phage gene directs the synthesis of an enzyme that ruptures the bacterium, freeing the completed offspring phage.

Figure 12-4 The life cycle of T2 bacteriophages.

The Structure of DNA

Even before the report of Hershey and Chase in 1952, Avery's demonstration that the bacterial transforming factor was DNA had stimulated a burst of research on the chemical nature of DNA. Alfred Mirsky studied the amount of DNA in the cells of various tissues of several organisms. He found that **the quantity of DNA varies among species, but is constant in each cell of a given species no matter what tissue the cell comes from.** Gametes are a notable exception, in that they have half as much DNA as the other cells of the body. This, of course, is exactly what we would expect if DNA is the hereditary material. (Why?)

What seemed equally significant, if only someone could figure out what it meant, were the results of Erwin Chargaff. Chargaff analyzed the amounts of the four nucleotides of DNA in a variety of species, and found a curious consistency. Although the amounts of each of the four bases vary considerably from species to species, *the DNA of any given species contains equal amounts of adenine and thymine, and equal amounts of cytosine and guanine*, within the limits of experimental measurement.

To summarize the data up to 1952:

1. DNA is the molecule of heredity.
2. Every cell in the body, gametes excepted, contains the same amount of DNA. Gametes contain half as much DNA as the other body cells.
3. The amount of cytosine is the same as the amount of guanine; and the amount of thymine is the same as the amount of adenine.

The major questions still remained unanswered: How does DNA encode genetic information? How is it duplicated before mitosis, so that each daughter cell receives exactly the same genetic information? What are mutations and how do they occur? Biologists in the early 1950s agreed that the secrets of DNA function, and therefore of heredity itself, could only be found by understanding the structure of the molecule.

Finding out the structure of any biological molecule is no simple task. Even the most powerful electron microscopes cannot reveal the structure of molecules in atomic detail. To study the structure of DNA, Maurice Wilkins and Rosalind Franklin turned to X-ray diffraction. They bombarded crystals of purified DNA with X-rays and photographed the resulting diffraction patterns (Fig. 12-6). As you can see, the X-ray pattern of DNA does not provide a direct picture of the structure of the molecule. However, the diffraction pattern suggested to Wilkins and Franklin that a DNA molecule is helical (that is, twisted about

(a)

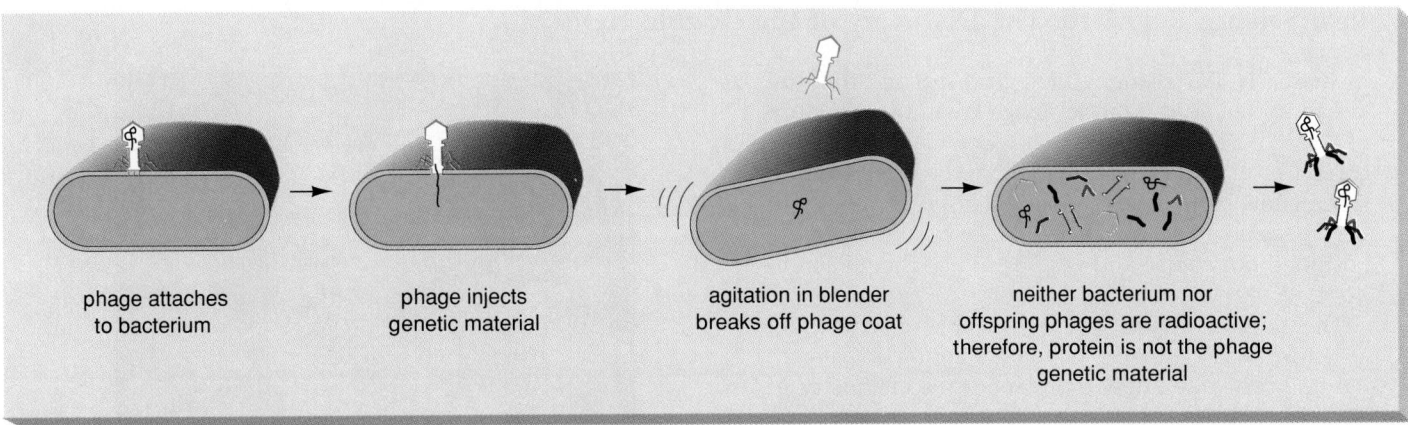

(b)

Figure 12-5 The Hershey–Chase experiment to determine whether DNA or protein is the genetic material of T2 bacteriophages. Hershey and Chase grew one population of phages in a medium containing radioactive sulfur, ^{35}S, which radioactively labels the phage proteins. Another population of phages was grown in radioactive phosphorus, ^{32}P, which labels the phage DNA.
(a) The fate of ^{35}S-labeled protein (red). Phages attach to the cell walls of unlabeled bacteria, and inject their genetic material. Whirling the mixture in a blender breaks off the phage heads and sheaths left on the outside of the bacteria. The ^{35}S radioactivity is found outside the bacteria in the phage coats, while the bacteria, containing the phage genetic material, are not radioactive. New phages resulting from the infection are also not radioactive. The phage genetic material, therefore, is not protein.
(b) The fate of ^{32}P-labeled DNA (red). When the phage-bacteria combinations are broken apart in a blender, the ^{32}P is found inside the bacteria, while the phage pieces are unlabeled. Some of phages synthesized inside the infected bacteria are ^{32}P-labeled. The genetic material must be DNA, carrying the ^{32}P inside the bacteria and to progeny phages.

like a corkscrew), has a uniform diameter of 2 nanometers, and consists of subunits, each of which is separated from neighboring subunits by 0.34 nanometer. One full turn of the DNA helix occurs every 3.4 nanometers. Finally, the X-ray picture suggests that the sugar-phosphate "backbone" of the molecule (see p. 240) is on the outside of the helix, while the bases are on the inside.

The Double Helix

The chemical and X-ray diffraction data were not nearly enough information with which to work out the structure of DNA. Some good guesses were also needed (see "Real Science Revisited: The Discovery of the Double Helix"). Combining a knowledge of how complex organic molecules bond with one an-

Real Science Revisited: The Discovery of the Double Helix

In the early 1950s, many biologists realized that the key to understanding inheritance lay in the structure of DNA. They also knew that whoever deduced the correct structure of DNA would receive recognition from fellow biologists, fame in the popular press, and very possibly the Nobel Prize. Less obvious were the best methods to use and who would be the person to do it.

The betting favorite in the race to discover the structure of DNA had to be Linus Pauling of Caltech. Pauling probably knew more about the chemistry of large organic molecules than any person alive, he was an expert X-ray crystallographer, and he had hit upon the idea that accurate models could aid in deducing molecular structure. Finally, he was almost frighteningly brilliant. In 1950, he demonstrated these traits by showing that many proteins were coiled into single-stranded helices (the α-helix; see Chapter 3). Pauling, however, had two main handicaps. First, for years he had concentrated on protein research, and therefore he had little data about DNA. Second, he was active in the peace movement. During the reign of Senator Joseph McCarthy in the early 1950s, he was considered to be potentially subversive and possibly dangerous to national security. This latter handicap may have proved decisive, as we shall see.

The second most likely competitors were Maurice Wilkins and Rosalind Franklin, English X-ray crystallographers. They set out to determine the structure of DNA by the most direct procedure, the careful study of the X-ray diffraction patterns of DNA. They were the only scientists who had really good data about the general shape of the DNA molecule. Unfortunately for Wilkins and Franklin, their methodical approach was also slow.

This left the door open for the eventual discoverers of the double helix, James Watson and Francis Crick, two young scientists with neither Pauling's tremendous understanding of chemical bonds nor Franklin and Wilkins' expertise in X-ray analysis. They did have three crucial advantages: the lesson of Pauling's work on proteins, that models could be enormously helpful in studying molecular structure; access to the X-ray data; and a driving ambition to be first.

Watson and Crick did no experiments in the ordinary sense of the word; rather, they spent their time thinking about DNA, trying to construct a molecular model that made sense and fit the data. Since they were in England and since Wilkins was very open about his and Franklin's data, they were familiar with all the X-ray information relating to DNA. This was just what Pauling lacked. Because of his presumed subversive tendencies, the U.S. State Department refused to issue Pauling a passport to leave the coun-

James Watson and Francis Crick with a model of the structure of DNA.

try, and consequently he could not attend meetings at which Wilkins presented the X-ray data, nor visit England to talk with Franklin and Wilkins directly.

Watson and Crick knew that Pauling was working on DNA structure, and were terrified that he would beat them to it. In *The Double Helix*, Watson recounts his belief that, if Pauling could have seen the X-ray pictures, "in a week at most, Linus would have the structure."

About this time, you might be thinking, "But wait just a minute! That's not fair. If the goal of science is to advance knowledge, then everybody should have access to all the data. If Pauling was the best, he should have discovered the double helix first." Perhaps so. But science is an activity of scientists, who, after all, are people too. While virtually all scientists want to see the advancement and benefit of humanity, each individual also wants to be the one responsible for the advancement, and to receive the credit and the glory. We should not overlook the fact that the ambition to be first helps to inspire the intense concentration, the sleepless nights, and the long days in the laboratory that ultimately produce results.

At any rate, Pauling remained in the dark about the correct X-ray pictures of DNA, and was beaten to the correct structure. When Watson and Crick discovered the base-pairing rules that were the key to DNA structure, Watson sent off a letter about it to Max Delbruck, a friend and advisor at Caltech. He asked Delbruck not to reveal the contents of the letter to Pauling until their structure was formally published. Delbruck, perhaps more of a model scientist, firmly believed that scientific discoveries belong in the public domain, and promptly told Pauling all about it. With the class of a great scientist and a great person, Pauling graciously congratulated Watson and Crick on their brilliant solution to the structure. The race was over.

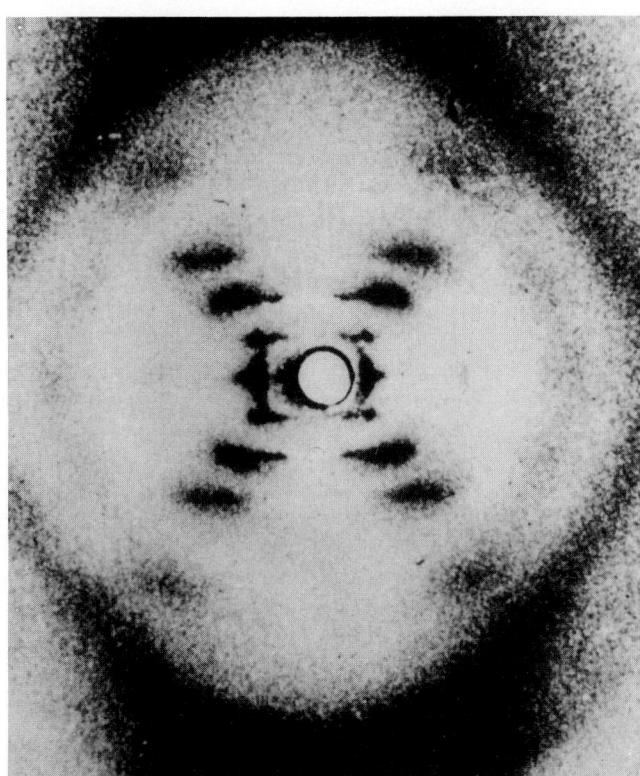

Figure 12-6 The X-ray diffraction pattern of DNA, taken by Rosalind Franklin. The "cross" formed of dark spots is characteristic of helical molecules such as DNA. Measurements of various aspects of the pattern indicate the dimensions of the DNA helix; for example, the distance between spots in the cross corresponds to the distance between turns of the helix.

other with an intuition that "important biological objects come in pairs," James Watson and Francis Crick proposed that the DNA molecule consists of two strands, twisted about each other into a **double helix,** much like a ladder twisted about its long axis into a corkscrew shape (Fig. 12-7). The sugar–phosphate backbones of the two DNA strands are on the outside of the double helix, like the uprights of the ladder. Notice that the sugar–phosphate uprights run in opposite directions, so that one upright has its sugar "foot" on one end of the DNA molecule while the other has its sugar "foot" on the opposite end. The bases are packed into the middle, paired up to form the rungs of the ladder.

Watson and Crick proposed that each rung is composed of a purine and a pyrimidine, held together by hydrogen bonds. If adenine pairs with thymine, and guanine pairs with cytosine, then hydrogen bonds can form between the bases, holding the two halves of the rungs together:

Adenine and thymine are held together by two hydrogen bonds, while cytosine and guanine are held together by three.

Adenine–thymine and guanine–cytosine pairs make sense of Chargaff's data, that the amount of adenine in DNA equals the amount of thymine, and that the amount of guanine equals the amount of cytosine. **In nucleic acids, bases that pair together via hydrogen bonds are called *complementary base pairs.*** In DNA, adenine is complementary to thymine, while guanine is complementary to cytosine. This is called the base-pairing rule.

The structure of DNA was solved. Although further data would be needed to confirm its details, "a structure this pretty just had to exist," as Watson later put it. Watson and Crick published their double helix model for DNA in 1953, a model that revolutionized all of genetics and practically all of biology in just a few years. In 1962, Watson, Crick, and Maurice Wilkins shared the Nobel Prize in recognition of their brilliant work in deciphering the structure of DNA. (Rosalind Franklin, whose work was equally deserving of the Prize, died in 1958. Nobel Prizes are not awarded posthumously, and so her contributions sometimes do not receive the recognition they deserve.)

The Power of the Double Helix

The double helix, of course, is not merely pretty; it also helps to explain two key features of inheritance. First, different types of bases can be arranged in any linear order down one of the DNA strands, matched up with complementary bases on the other strand. Each sequence of bases represents a unique set of genetic instructions, like a biological Morse code. A stretch of DNA just 10 nucleotides long can exist in over a million different possible sequences of the four

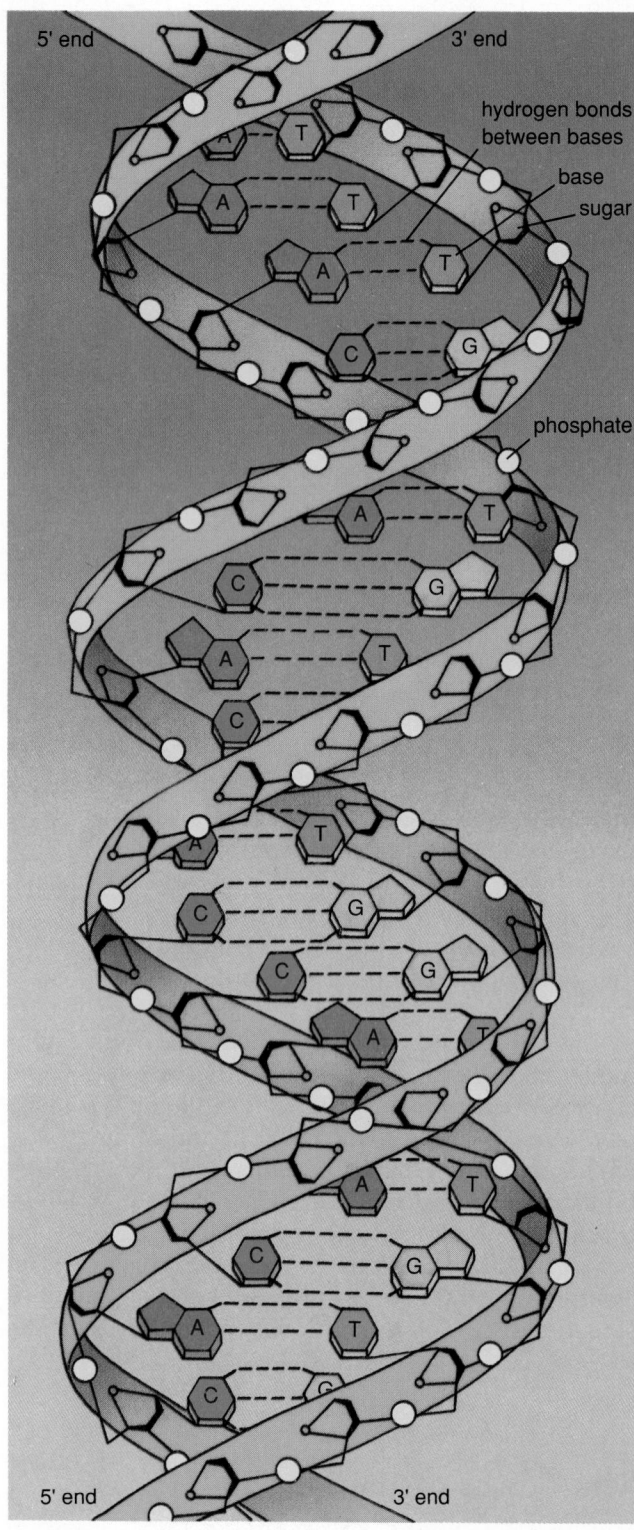

5' end
3' end
hydrogen bonds
between bases
base
sugar
phosphate
5' end
3' end

Figure 12-7 In the Watson–Crick model of DNA, two strands of DNA are wound about each other in a double helix, like a ladder twisted about its long axis. Specific patterns of hydrogen bonding allow complementary bases to pair together in the center of the helix. Three hydrogen bonds hold guanine to cytosine, while two hold adenine to thymine. The two DNA strands run in opposite directions: note that the sugar (the 3' end of the sugar–phosphate backbone) is at the bottom of one strand, while the phosphate (the 5' end of the backbone) is at the bottom of the other strand. (It may be easier for you to follow the "oxygen arrowheads" on the sugars: these run in the 3' to 5' direction.) As Figures 12-8 and 12-9 show, the directionality of the strands is important during DNA replication.

works: how the sequence of nucleotides in DNA directs synthesis of cellular components; how identical copies are formed during DNA replication; how mutations occur and how they change the genetic information. In the remainder of this chapter and the following one, we will explore how the Watson–Crick model of DNA helps provide answers to these questions.

DNA Replication: The Key to Constancy

In classically understated scientific style, Watson and Crick included the following comment in their paper describing the double helix: "It has not escaped our notice that the specific [base] pairing we have postulated immediately suggests a possible copying mechanism for the genetic material." In other words, the base-pairing rule offers a simple hypothesis for the replication of the DNA of chromosomes prior to mitosis and meiosis:

The sequence of bases in one strand of the double helix accurately predicts the sequence of bases in the other, complementary strand. A chromosome could be replicated by separating the two DNA strands, and synthesizing new strands using nucleotides with bases complementary to the parental strands. Each new chromosome would then consist of one parental strand and one complementary daughter strand.

SUMMARY OF DNA REPLICATION

There are three fundamental steps to DNA replication in all living cells (Fig. 12-8).

1. The two DNA strands of the double helix of a parental chromosome unwind and separate.

bases. Second, since an "average" chromosome has millions (in bacteria) to billions (higher plants and animals) of nucleotides, DNA molecules can encode a staggering amount of information.

What remains is to find out how the structure

2. Each parental strand is used as a template for the formation of a complementary daughter strand of DNA. The daughter strand is formed by connecting nucleotides together in an order that is dictated by the nucleotide sequence of the parental strand: the base of each nucleotide added to the daughter strand is complementary to the base of the nucleotide in the corresponding location of the parental DNA.

3. Finally, one parental DNA strand and its newly synthesized, complementary daughter strand wind together into one double helix, while the other parental strand and its complementary daughter strand wind together into a second double helix. Since each new double helix consists of one intact parental DNA strand and one newly synthesized complementary strand, the process of DNA replication is called **semiconservative replication** (see "Methods in Biology: Meselson and Stahl Show That DNA Replication is Semiconservative").

We will now examine these steps in more detail.

Molecular Mechanisms of DNA Replication

Biologists are still uncovering the details of DNA replication. However, in eukaryotic cells, the sequence of events seems to be as follows (Fig. 12-9).

Unwinding the Parental DNA Strands

The first step in DNA synthesis is to separate the two DNA strands, thereby unwinding the double helix (Fig. 12-9a). An enzyme appropriately named DNA helicase ("an enzyme that breaks apart the helix") wiggles its way between the two strands. According to one model, the helicase enzyme then "walks" along one strand, nudging the other strand out of the way as it goes. The result is that the two DNA strands separate, which exposes their bases. The junction between the separated strands and the remaining double helix is called the **replication fork.**

Synthesizing Complementary DNA Strands

Enzymes called **DNA polymerase** then bind to each unwound strand. DNA polymerase performs a dual function. **First, DNA polymerase recognizes bases exposed in a parental strand, and matches them up with free nucleotides that have complementary bases. Second, DNA polymerase bonds together the sugars and phosphates of the free complementary nucleotides to form a chain of DNA** (Fig. 12-9b). In other words, when the DNA polymerase is in contact with an adenine in the parental strand, it adds thymine to the growing daughter DNA strand; when it encounters cytosine in the parental strand, it adds guanine to the daughter strand, and so on.

As far as DNA polymerase molecules are concerned, the parental DNA strands are one-way streets: DNA polymerase can only travel in one direc-

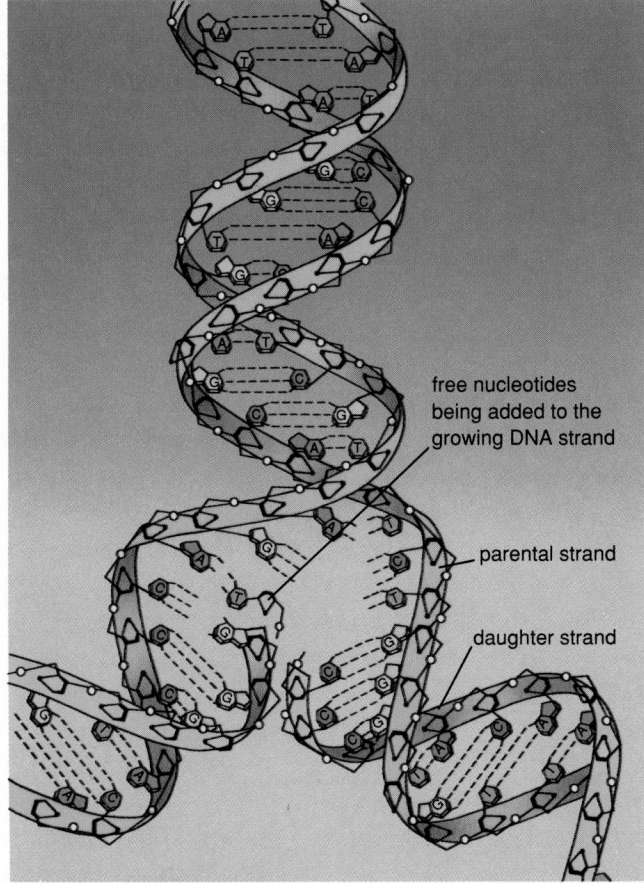

free nucleotides being added to the growing DNA strand

parental strand

daughter strand

Figure 12-8 Replication of DNA. Individual nucleotides of all four types are synthesized by the cell before replication, and are available within the nucleus. The sugar–phosphate backbones of the parental DNA strands are colored light tan and the backbones of the daughter DNA strands are colored reddish tan. The parental DNA strands unwind, and the hydrogen bonds between complementary bases are broken. DNA polymerase enzymes move along the two DNA strands in opposite directions, traveling in the 3' to 5' direction (the direction that the "oxygen arrowheads" are pointing). The polymerase enzymes recognize the nucleotides in the parental DNA strands, pair each one with a complementary nucleotide from the nucleoplasm, and connect the new nucleotides together into daughter strands. This semiconservative replication process produces two new double helices of DNA, each composed of one parental strand (light tan backbone) and one daughter strand (reddish tan backbone) that is an exact copy of the other parental strand. Therefore, each new double helix is an exact copy of the parental DNA.

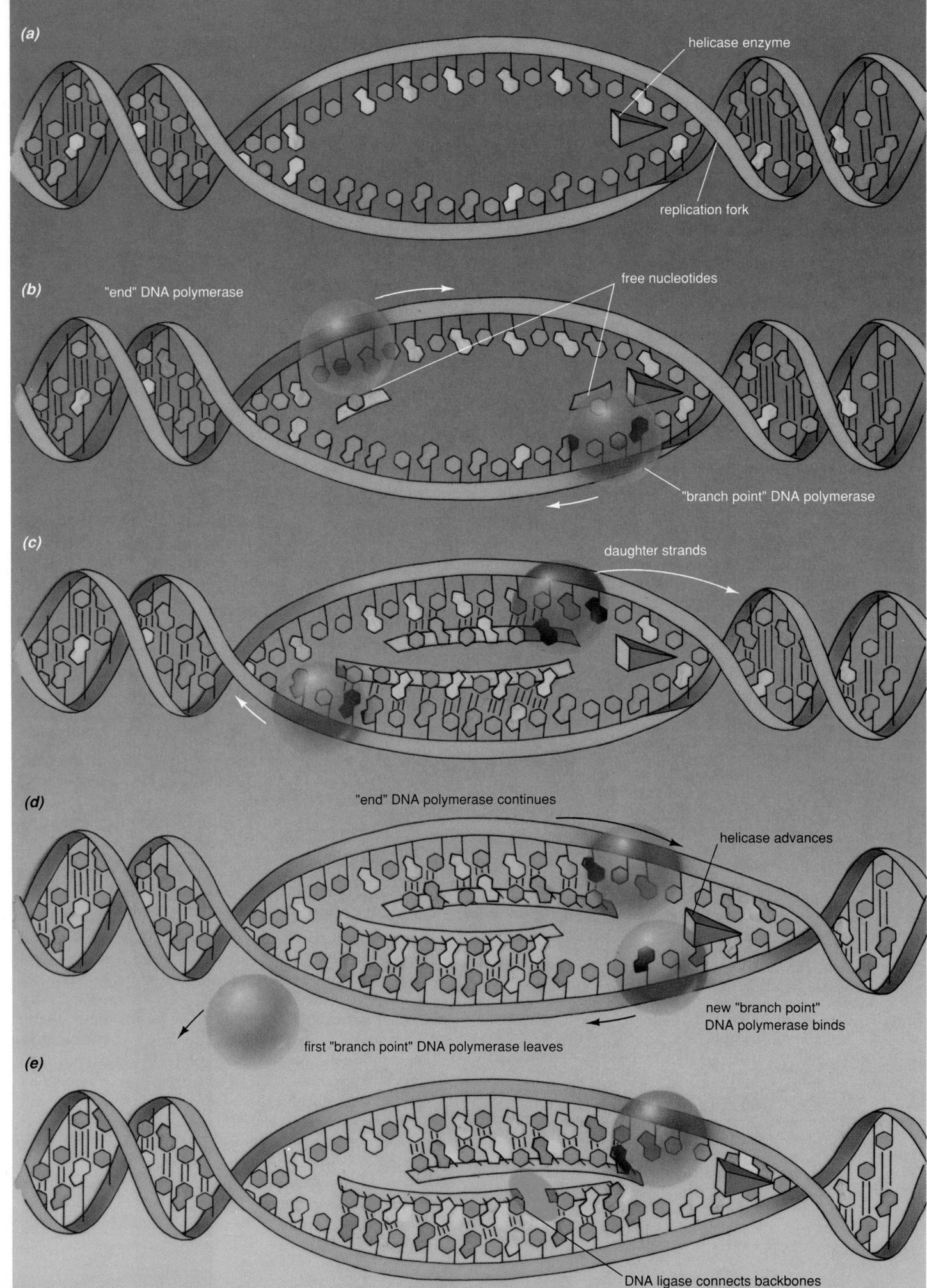

(a)

helicase enzyme

replication fork

(b) "end" DNA polymerase

free nucleotides

"branch point" DNA polymerase

(c)

daughter strands

(d) "end" DNA polymerase continues

helicase advances

new "branch point" DNA polymerase binds

first "branch point" DNA polymerase leaves

(e)

DNA ligase connects backbones

Figure 12-9 Molecular mechanisms of DNA replication.
(a) The enzyme DNA helicase separates the two DNA strands, unwinding a small portion of the double helix.
(b) Two DNA polymerase molecules attach to the separated parental strands. DNA polymerase can only synthesize new DNA strands by proceeding in the 3′ to 5′ direction on the phosphate–sugar backbones. Therefore, the "branch point" polymerase attaches near the helicase and proceeds away from the helicase, while the "end" polymerase attaches at the other end of the unwound DNA region and proceeds toward the helicase (which is also the direction of future unwinding).
(c) The two DNA polymerase molecules match up free nucleotides with the parental DNA strands, using the base-pairing rules, and synthesize daughter DNA strands (see Fig. 12-8).
(d) The helicase advances, unwinding more parental DNA. The "end" DNA polymerase continues to synthesize additional daughter DNA, adding new nucleotides to the end of the strand already synthesized in part (c). The original "branch point" DNA polymerase, however, has run into the end of the unwound DNA and leaves the DNA, having synthesized only a short daughter DNA segment. Meanwhile, a new DNA polymerase molecule attaches at the new "branch point," and proceeds away from the helicase toward the first daughter DNA segment.
(e) When the new "branch point" polymerase has synthesized daughter DNA that reaches to the end of the first daughter DNA segment, it leaves the DNA. Another enzyme, DNA ligase, attaches the second daughter DNA segment to the first daughter DNA segment.

tion on a DNA strand, in the 3′ to 5′ direction (it may be easier for you to visualize this by following the "oxygen arrowheads" that point toward the 5′ end of a DNA strand; see Fig. 12-8). This means that DNA polymerase moves in opposite directions on each strand, which dictates the third step in DNA synthesis.

Stitching Together the Pieces

Let's look a little more closely at the process so far. Suppose that DNA helicase has wiggled its way into the helix and separated about 100 bases, forming a replication fork. Two DNA polymerase molecules then enter the fork. Now, the DNA polymerases will travel in opposite directions on the two strands of DNA, in the direction of the "oxygen arrowheads" on each strand (toward the 5′ end). Let's assume that our two DNA polymerase molecules settle down on opposite ends of the separated DNA, one at the "end" of the fork, one at the "branch point." The "end" DNA polymerase builds a new strand by moving along its parental strand (let's call it the A strand) toward the "branch point." Therefore, as the helicase

continues to separate the parental DNA strands, the "end" polymerase simply follows along on the A strand, synthesizing one long, continuous complementary strand (Fig. 12-9c and d).

The "branch point" polymerase, however, can't do this. It binds to the other parental DNA strand (the B strand) right behind the helicase, and *moves away from the helicase* (Fig. 12-9c). Therefore, as the helicase continues to separate the parental strands, it exposes a new B strand that the "branch point" polymerase can't get to (Fig. 12-9d). Soon, however, a new DNA polymerase attaches to the B strand, becoming a new "branch point" polymerase. Like the first "branch point" polymerase, it moves away from the helicase. What happens when this second "branch point" polymerase reaches the place where the first "branch point" polymerase started? The two short DNA strands synthesized by the two "branch point" DNA polymerases must be joined up somehow, to make a continuous DNA strand. A third enzyme, DNA ligase, bonds the two strands together (Fig. 12-9e). This process is repeated many times (perhaps 10 million or so for a human chromosome), until the entire B strand has been replicated.

Proofreading

DNA polymerase is not infallible. Partly because of the speed of replication (50 to 500 nucleotides per second) and partly because of spontaneous chemical flip-flops in the bases, DNA polymerase occasionally incorporates incorrectly matched bases, perhaps one mistake for every 10,000 base pairs. In mammalian cells, however, the completed DNA strands contain only about one mistake for every billion base pairs. This phenomenal accuracy is ensured by several enzymes, including DNA polymerase itself, that "proofread" each daughter strand as it is synthesized. The directionality of DNA structure allows the proofreading enzymes to recognize the parental strand, running in one direction, as the "right stuff," and to correct any mismatches by changing the daughter strand, which runs in the other direction.

The Final Result: Faithful Inheritance

During prophase of mitosis, each chromatid consists of a double helix of DNA, composed of one of the original strands of the parent DNA plus one new strand that is an exact copy of the other original strand. When sister chromatids separate at anaphase and are parceled out to the daughter cells, each cell receives an exact copy of each of the parental chromosomes. Thus, if there are no mistakes in the whole

METHODS IN BIOLOGY

Meselson and Stahl Show
That DNA Replication is Semiconservative

In the Watson–Crick model of DNA structure, the two strands of the double helix are held together by hydrogen bonds between complementary bases (see Fig. 12-8). Watson and Crick immediately realized that the complementary nature of the two strands of the helix could provide a mechanism for accurate DNA replication. Each strand could serve as a template for the synthesis of a new daughter strand: new nucleotides could be added to the daughter strand only if their bases were complementary to the bases of the parental strand.

Assuming that this hypothesis is correct, there are two likely mechanisms for DNA replication. In the **conservative replication** model, the two parental DNA strands might separate and be used as templates on which to synthesize new daughter strands. Then the two parental strands would rewind together, and the two daughter strands would wind up together. In the **semiconservative replication** model, the parental strands would still separate and be used as templates for the new daughter strands. However, each parental strand would wind up with its complementary daughter strand, so that each double helix would now consist of one parental strand and one daughter strand. How can these two models be distinguished? Caltech researchers Matthew Meselson and Frank Stahl devised an elegant experiment to test the models.

First, a bit of technical background (Fig. E12-1a). The bases of DNA contain a lot of nitrogen. "Ordinary" nitrogen is ^{14}N (that is, its atomic weight is 14). A heavy isotope of nitrogen, ^{15}N, with an extra neutron, can also be obtained. If bacteria are grown in a medium with ^{14}N, they synthesize all ^{14}N-containing DNA, which is relatively light. If they are grown in a medium with ^{15}N, they will synthesize ^{15}N-containing, heavy DNA. Further, DNA with the two nitrogen isotopes can be separated by a method known as density-gradient centrifugation. Centrifuge tubes are filled with a solution of cesium chloride, and DNA is carefully layered on top of the solution. The tubes are then spun at high speed, and

two things happen. First, cesium ions are much denser than water. The spinning centrifuge forces the cesium to sink, so that a gradient of cesium forms in the tube: lots of cesium near the bottom, less and less toward the top. Therefore, the density of the entire solution in the tube also has a gradient, very dense at the bottom, and progressively less dense toward the top. Second, the DNA molecules sink, too. How far? A DNA molecule sinks until the density of the surrounding solution is equal to the density of the DNA. Then it stops sinking.

How can this be used to distinguish conservative from semiconservative replication? Meselson and Stahl grew bacteria for several generations in ^{15}N medium, so that all their DNA would be heavy. Then they transferred the bacteria to ^{14}N medium, so that any newly synthesized DNA would be light. Their predictions were as follows (Fig. E12-1b). Under the conservative replication model, the parental DNA, synthesized in ^{15}N medium, would be composed of two heavy strands of DNA (let's call it H–H DNA). In the F_1 generation, in ^{14}N medium, half the daughter cells would receive H–H DNA, while the other half would receive all-light (L–L) DNA. Centrifuging this DNA would result in two bands of DNA, one near the bottom of the tube (H–H) and one near the top (L–L). Under the semiconservative replication model, the parental generation would also have H–H DNA. The F_1 generation, however, would have DNA composed of one heavy and one light strand (H–L). Centrifuging this DNA, therefore, would produce one band near the middle of the tube.

When they actually performed the experiment, the results were clear (Fig. 12-1c). The F_1 generation produced only one band of DNA, in the middle of the tube. The conservative replication model flunked, while the semiconservative replication model passed with flying colors. Other experiments have also supported the semiconservative model, which appears to be the universal mechanism of DNA replication in every cell of every form of life on Earth.

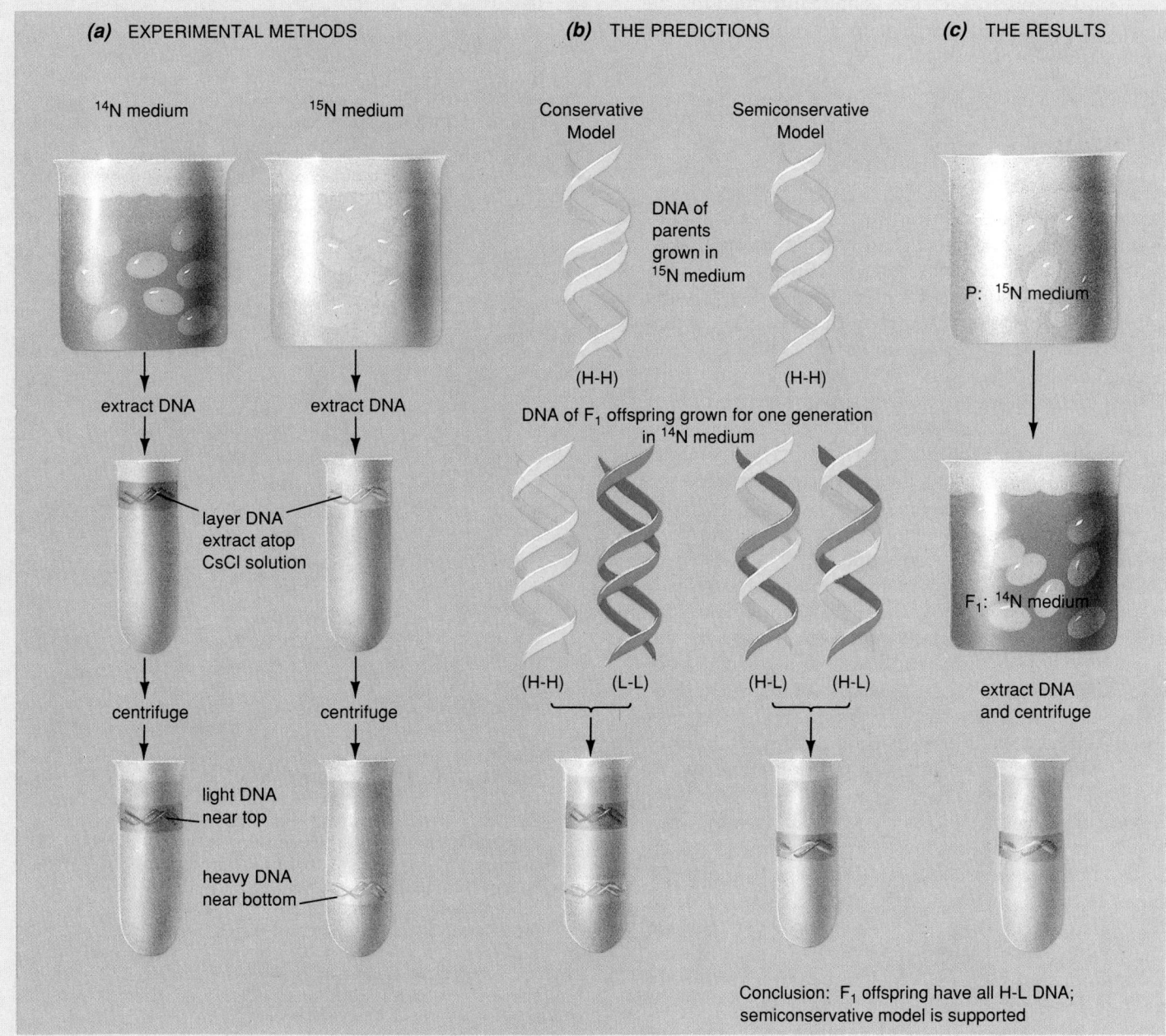

(a) EXPERIMENTAL METHODS

14N medium

15N medium

extract DNA

extract DNA

layer DNA extract atop CsCl solution

centrifuge

centrifuge

light DNA near top

heavy DNA near bottom

(b) THE PREDICTIONS

Conservative Model

Semiconservative Model

DNA of parents grown in 15N medium

(H-H)

(H-H)

DNA of F₁ offspring grown for one generation in 14N medium

(H-H)   (L-L)

(H-L)   (H-L)

(c) THE RESULTS

P: 15N medium

F₁: 14N medium

extract DNA and centrifuge

Conclusion: F₁ offspring have all H-L DNA; semiconservative model is supported

Figure E12-1 The Meselson–Stahl experiment.
(a) The Experimental Methods: Bacteria are grown in either ^{14}N medium (purple) or ^{15}N medium (yellow). Their cells are broken open and the DNA extracted. Centrifuge tubes are filled with a cesium chloride solution, and the DNA is layered atop the solution. The tubes are spun in a centrifuge, and the positions of the DNA are measured. The ^{14}N DNA band (L–L; purple) is near the top of the tube, while the ^{15}N DNA band (H–H; yellow) is near the bottom of the tube.
(b) The Predictions: Bacteria are grown in ^{15}N medium, then transferred to ^{14}N medium for one generation. Both strands of the parental DNA contain ^{15}N, and are heavy (yellow–yellow, or H–H). Under the conservative replication model, the F_1 DNA would be of two types, H–H and L–L (purple–purple). Centrifuged F_1 DNA would show two bands. Under the semiconservative replication model, the F_1 DNA would all be H–L (purple–yellow). Centrifuged F_1 DNA would show one band at an intermediate position.
(c) The Results: Centrifuging the F_1 DNA results in a single band of intermediate density, corresponding to H–L DNA. Therefore, DNA replication is semiconservative.

process, the integrity of the genetic information will be maintained from cell division to cell division and from parent to offspring.

Mutations: The Key to Variability

Complete constancy is not, however, the best evolutionary strategy. When environments change, species must change too, or else face extinction. We have already discussed two ways of generating genetic variability during sexual reproduction: (1) **sexual recombination,** as meiosis randomly shuffles homologous chromosomes into the gametes, and, through fusion of the gametes, the offspring receive one of every pair of homologues from each parent; and (2) **genetic recombination** (crossing over), in which new combinations of alleles arise when parts of the chromosomes are exchanged.

Sexual recombination and crossing over can only generate new combinations of alleles if a number of different alleles already exist. New alleles arise as a result of changes in DNA, called **mutations.** Since the information in a gene is encoded in the specific sequence of bases, **a mutation is a change in the sequence of bases.** How can the base sequence change? As we have seen, one way for a mutation to occur is through a mistake in base pairing during replication. A few base-pairing mistakes occur spontaneously, in the best of circumstances: even with proofreading, replicating several billion bases occasions some mistakes; not many, but a few. Certain chemicals (such as the mustard gas used in World War I) and some types of radiation (such as X-rays) increase errors of base pairing during replication or even induce changes in DNA composition between replications.

Random changes in DNA composition are unlikely to code for improvements in the functioning of the gene products (see Chapter 13), much as typing random words in the midst of a script of *Hamlet* will be unlikely to improve on Shakespeare. Therefore, cells have evolved mechanisms to monitor DNA and repair damaged regions (Fig. 12-10a). Nevertheless, mistakes do occur.

Types of Gene Mutations

Point Mutations

In a **point mutation,** a pair of bases becomes incorrectly matched (Fig. 12-10b). Repair enzymes recognize the mismatch, cut out one of the nucleotides and replace it with a complementary nucleotide. However, the enzymes occasionally replace the wrong nucleotide. Thus, although there is a complementary

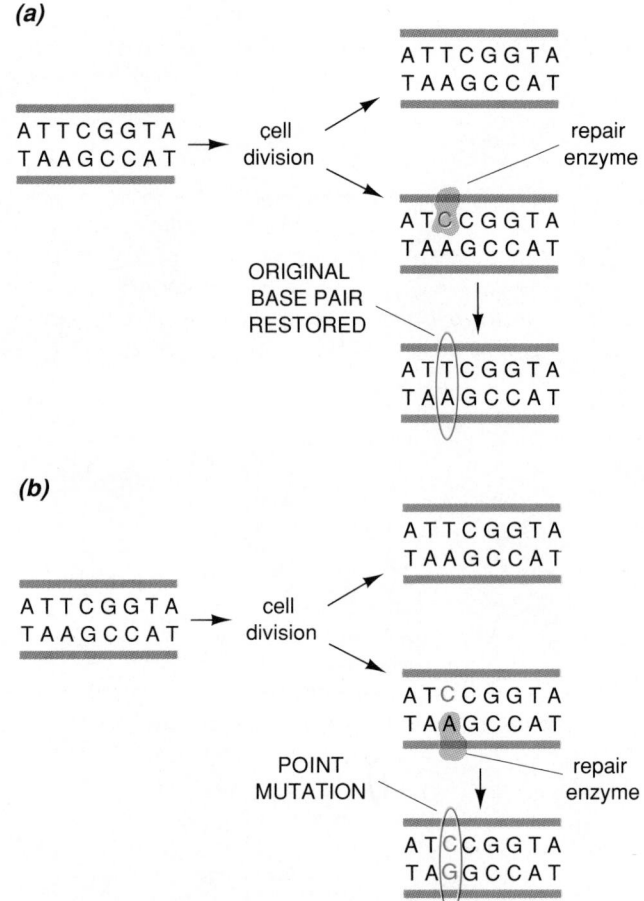

Figure 12-10 The fate of mismatched base pairs. Repair of damaged or mismatched regions of DNA sometimes results in restoration of the original nucleotide sequence and sometimes results in mutations. In this diagram and in Figure 12-11, the sugar–phosphate backbone of DNA is represented by solid lines, and the bases are represented by the letters A, C, G, and T.
(a) Correction of mismatched bases: During DNA replication, a mistake in base pairing occurs, resulting in an incorrect A–C pair instead of the proper A–T pair. Repair enzymes in the daughter cell recognize the mismatch, excise the incorrect base in the daughter DNA, and substitute the correct complementary base (T). No mutation has occurred.
(b) Point mutation: With the same replication error as in part (a), if the repair enzymes excise the mismatched base on the parental strand instead of on the daughter strand, then the new base pair differs from the original. A mutation has occurred.

pair of bases at that location once again, it isn't the same pair that the chromosome originally had.

Insertions and Deletions

As their names suggest, an **insertion** occurs when new nucleotide pairs are inserted in the midst of a gene (Fig. 12-11a), while a **deletion** occurs when nucleotide pairs are removed from a gene (Fig. 12-11b). As we will see in the following chapter, insertion and deletion mutations often have a catastrophic effect on gene function.

Effects of Mutations

Although most mutations are harmful or neutral, mutations are the prerequisite for evolution, because ultimately all genetic variation originates as these random changes in DNA sequence. Natural selection tests new alleles in the crucible of competition for survival and reproduction. Occasionally, a mutation proves beneficial in the organism's interactions with its environment. These alleles may spread throughout the population and become common, as their possessors outcompete rivals bearing the old allele.

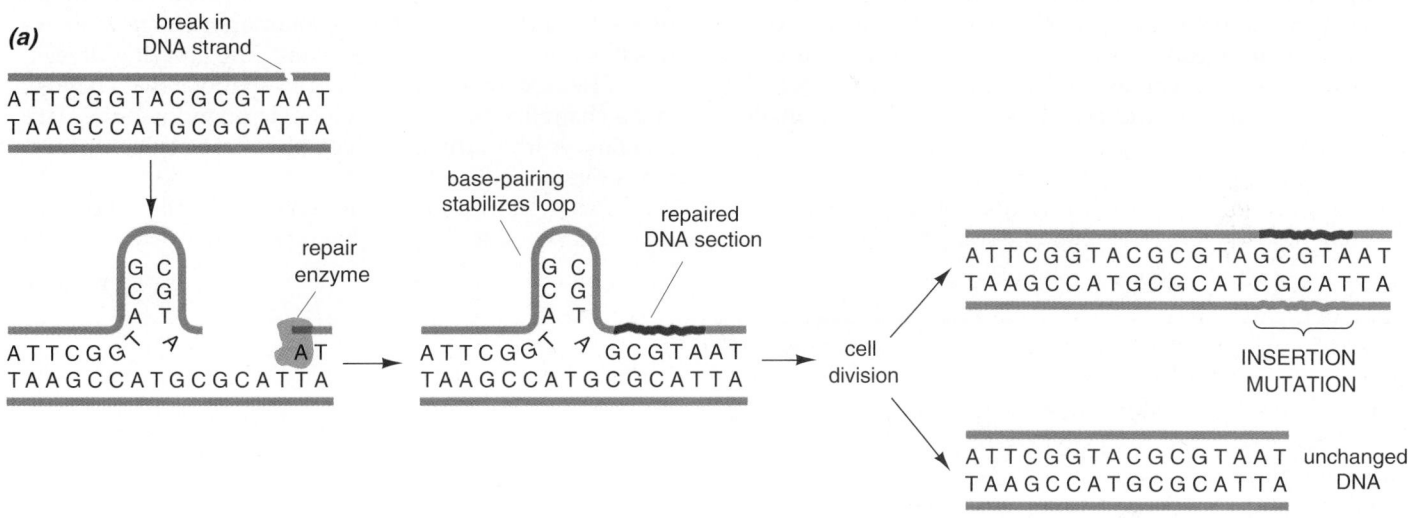

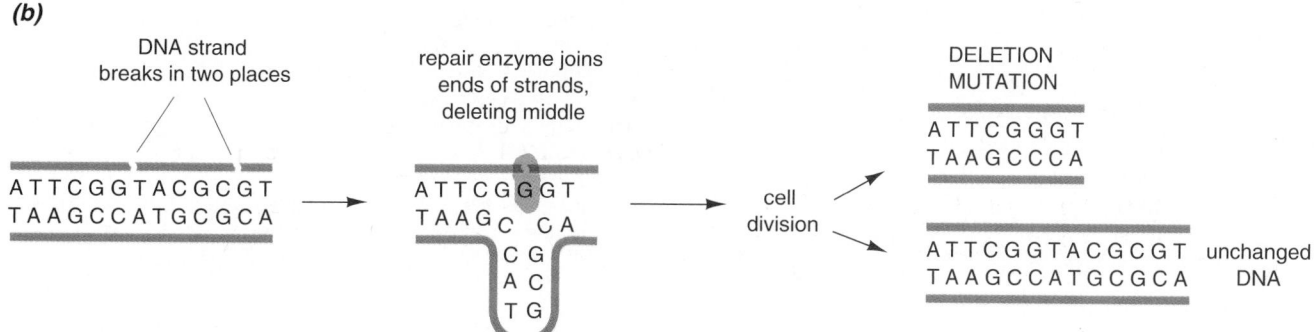

Figure 12-11 Insertions and deletions.
(a) Insertion: Radiation or chemical mutagens cause a break in one of the DNA strands, and the freed DNA end loops out. If the free loop has certain sequences of bases, base pairing may stabilize the loop. Repair enzymes fill in the resulting gap with new nucleotides, making one DNA strand (with the loop) longer than the other (intact) strand. When DNA replication occurs, each strand is used as a template for synthesis of a new complementary strand. The long strand thus carries an insertion mutation to one of the daughter cells.
(b) Deletion: If a DNA strand breaks in two places, a repair enzyme may stitch together the remaining pieces without filling in the missing nucleotides. When the DNA is replicated, one daughter cell receives the deletion mutation.

Reflections on DNA and Evolution

The double helix of DNA is the genetic material of every living cell, whether bacterium or buffalo, hyacinth or human. As we have seen, this unique molecule fulfills the dual requirements of constancy and change, of similarity and variability, that are demanded by the observations of Mendelian genetics and evolution. Constancy of form and function from generation to generation arises because of the almost error-free base pairing of DNA during the replication of chromosomes, and the precise transmission of chromosomes during meiosis and mitosis. In the short term, variability is mostly due to sexual and genetic recombination. In the long term, the ultimate source of variability is mutation, which occurs largely because of the occasional errors that are made during DNA replication.

Given what we know about the role of DNA in inheritance, it is not surprising that modern biologists almost unanimously accept evolution as an inevitable fact of life. New alleles of genes constantly appear, sexual and genetic recombination shuffle alleles into new combinations, and even the numbers of chromosomes occasionally change through errors in meiosis. Through these chance events and the demands of a changing environment over time, new variations appear, are tested, and are retained or discarded.

The centrality of DNA in the life of every organism is both a thread of relationship among us and a wonder of individuality. At the heart of things, a giant redwood, a butterfly, and a mouse share a common heritage written in the DNA of their chromosomes. The amazing diversity of living forms testifies to the variations that chance and a changing environment have wrought over the last 3 billion years in this most flexible of molecules. If humans remain willing to share the world with the myriad creatures around us in their natural habitats, who can say what wonders future generations may see?

SUMMARY OF KEY CONCEPTS

The Composition of Chromosomes

Eukaryotic chromosomes are composed of DNA and protein. DNA is the molecule that carries genetic information. It is composed of subunits called nucleotides, linked together into long strands. Each nucleotide consists of a phosphate group, the 5-carbon sugar deoxyribose, and a nitrogen-containing base. Four different bases occur in DNA: adenine, guanine, thymine, and cytosine.

DNA: The Hereditary Molecule

Transformation in bacteria and infection of bacteria by bacteriophages provided the first evidence that DNA is the hereditary molecule. Griffith discovered that living bacteria can acquire genes from dead bacteria, transforming the genotype of the live bacteria. Avery and his colleagues showed that the genes taken up by living bacteria during transformation are composed of DNA.

A bacteriophage consists solely of DNA and protein. When it infects a bacterium, a phage injects its genetic material into the bacterium, where it directs the synthesis of more phages. Hershey and Chase demonstrated that phages inject DNA, but not protein, into bacteria. Therefore, DNA must be the genetic material of phages.

The Structure of DNA

The DNA of chromosomes is composed of two strands, wound about one another in a double helix. The sugars and phosphates that link one nucleotide to the next form the backbone on each side of the double helix, while the bases from each strand pair up in the middle of the helix. Only specific pairs of bases, called complementary base pairs, can link together in the helix, held by hydrogen bonds: adenine with thymine and guanine with cytosine.

DNA Replication: The Key to Constancy

When a chromosome is replicated prior to mitosis or meiosis, the two DNA strands of each double helix unwind. DNA polymerase enzymes move along each strand, linking up free nucleotides into new DNA strands. The sequence of nucleotides in each newly formed strand is complementary to the sequence on a parental strand. As a result, two double helices are synthesized, each consisting of one parental DNA strand plus one newly synthesized, complementary strand that is an exact copy of the other parental strand. The two daughter DNA molecules are therefore duplicates of the parental DNA molecule.

Mutations: The Key to Variability

A gene is a specific sequence of nucleotides in DNA. A mutation is a change in this sequence. Mutations are caused by mistakes in base pairing during replication, by chemical agents, and by certain kinds of radiation. Common gene mutations include point mutations, insertions, and deletions. Although mutations are usually harmful, occasionally a mutation will promote better adaptation to the environment, and thus will be favored by natural selection.

GLOSSARY

bacteriophage (bak-tēr'-ē-ō-fāj): a virus that infects bacteria.

base: in molecular genetics, one of the nitrogen-containing, single- or double-ringed structures that distinguish one nucleotide from another. In DNA, the bases are adenine, guanine, cytosine, and thymine.

complementary: referring to a nucleotide that can pair with another nucleotide via hydrogen bonding; in DNA, adenine is complementary to thymine, and guanine is complementary to cytosine.

deletion: a mutation in which one or more nucleotides are removed from a gene.

DNA polymerase: an enzyme that covalently bonds DNA nucleotides together into a continuous strand, using a preexisting DNA strand as a template. DNA polymerase catalyzes the replication of the DNA of chromosomes during interphase prior to mitosis and meiosis.

genetic recombination: assembling a new combination of preexisting genes through crossing over.

helix: (hē'-licks): a corkscrew-shaped object, as if a wire were wrapped around a cylinder.

insertion: a mutation in which one or more nucleotides are inserted within a gene.

mutation: a change in the base sequence of DNA.

nucleotide: an individual subunit of which nucleic acids are composed. A single nucleotide consists of a phosphate group covalently bonded to a sugar (deoxyribose in DNA), which is in turn covalently bonded to a nitrogenous base (adenine, guanine, cytosine, or thymine). Nucleotides are linked together by covalent bonds between the phosphate of one and the sugar of the next, to form a strand of nucleic acid.

point mutation: a mutation in which a single base pair in DNA has been changed.

semiconservative replication: the process of replication of the DNA double helix; the two DNA strands separate, and each is used as a template for the synthesis of a complementary DNA strand. Each daughter double helix therefore consists of one parental strand and one new strand.

sexual recombination: the formation of new combinations of alleles owing to inheritance of chromosomes from two different parental organisms during sexual reproduction.

transformation: a method of genetic recombination whereby DNA from one bacterium (usually released after the death of the bacterium) becomes incorporated into the DNA of another, living, bacterium.

STUDY QUESTIONS

1. Draw the structure of a nucleotide. Which parts are identical in all nucleotides, and which can vary?
2. Name the four types of nitrogenous bases found in DNA.
3. What is transformation? In Griffith's experiments, why were the transformed bacteria able to synthesize capsules?
4. Diagram the life cycle of a bacteriophage.
5. Describe the structure of DNA in a chromosome. Where are the bases, sugars, and phosphates in the structure?
6. Which bases are complementary to one another? How are they held together in the double helix of DNA?
7. Describe the process of DNA replication.
8. Define mutation, and give one example of how a mutation might occur. Would you expect most mutations to be beneficial or harmful? Explain your answer.

DISCUSSION QUESTIONS

1. Diagram and describe the Hershey–Chase experiment. Suppose they had found that *both radioactively labeled DNA and radioactively labeled protein* were injected into the bacteria. What would they have been able to conclude from this? (Note that if the bacteriophage chromosome was like the eukaryotic chromosome, i.e., if it contained both DNA and protein, this is exactly what would have happened.)
2. A preview question for Chapter 13: Genetic information is encoded in the sequence of nucleotides in DNA. Let's suppose that the nucleotide sequence on one strand of a double helix encodes the information needed to synthesize a hemoglobin molecule. Do you think that the sequence of nucleotides on the other strand of the double helix would also encode useful information? Why or why not? An analogy to think about: Suppose that English were a "complementary language," with letters at opposite ends of the alphabet complementary to one another (e.g., A is complementary to Z, B to Y, C to X, etc.). Would the complementary sentence to "All the world's a stage" make sense?
3. Suppose mammalian cells are grown and divide many times in a culture medium containing thymine made radioactive with a form of hydrogen (^{3}H). The cells are then removed from the radioactive medium and allowed to replicate several times in normal culture medium. Daughter chromosomes are tested each generation to determine whether they contain radioactive thymine. While in the radioactive medium, every strand of DNA of each chromosome had radioactive thymine. Assuming that each chromosome contains a single long double-stranded molecule of DNA, predict the radioactive status of daughter chromosomes after one, two, and three rounds of cell division in normal medium. Explain how your predictions are consistent with the Watson-Crick explanation of semi-conservative DNA replication.

SUGGESTED READINGS

Felsenfeld, G. "DNA." *Scientific American*, October 1985. The structure of DNA and how its use is regulated in a cell.

Holliday, R. "A Different Kind of Inheritance." *Scientific American*, June 1989. Methyl groups attached to DNA may alter its use in a cell. What's more, patterns of DNA methylation may be passed down from parent to offspring.

Judson, H. F. *The Eighth Day of Creation: Makers of the Revolution in Biology*. New York: Simon and Schuster, 1979. A remarkably readable history of the beginnings of molecular biology.

Kimura, M. "The Neutral Theory of Molecular Evolution." *Scientific American*, November 1979. Kimura persuasively argues that some evolutionary changes can occur without selection.

Mirsky, A. E. "The Discovery of DNA." *Scientific American*, June 1968. The early history of DNA research.

Olby, R. *The Path to the Double Helix*. Seattle, University of Washington Press, 1975. A historical perspective on the development of genetics in this century.

Watson, J. D. *The Double Helix*. New York, Atheneum Publishers, 1968. If you still believe the Hollywood images, that scientists are either maniacs or cold-blooded, logical machines, be sure to read this book. Although scarcely models for the behavior of future scientists, Watson and Crick are certainly human enough!

13

From Genotype to Phenotype: Gene Expression and Regulation

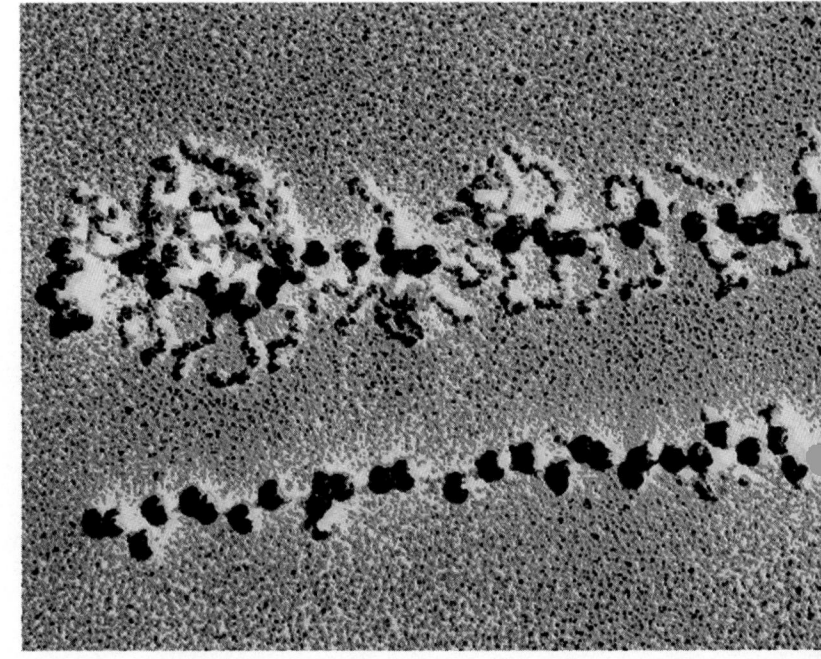

Protein synthesis in action: the large beads are ribosomes, traveling along a central strand of messenger RNA. The smaller beads issuing from the large beads are the beginnings of proteins. The molecules involved in protein synthesis are described in this chapter.

The genotype of an organism, interacting with its environment, produces its actual structure, functioning, and behavior—its phenotype. When Watson and Crick solved the puzzle of the double helix, they provided the framework for understanding the organization of the genotype. But two major questions still remained.

First, how does a cell use its genotype to create a phenotype? You may recall from previous chapters that most of the organic molecules of a cell are proteins or are synthesized through the actions of protein enzymes. Therefore, to go from genotype to phenotype means synthesizing the appropriate proteins. How does DNA direct protein synthesis and function in a cell? We will see that **the sequence of nucleotides in DNA is a code that is translated into the sequence of amino acids in proteins.** Genes are, in fact, a library containing the information needed to construct all the proteins of a cell.

Second, how does an organism use its genotype in the appropriate manner during its development and subsequent interactions with its environment? For example, if you begin life as a fertilized egg, and if all the cells of your body are derived from that egg through mitosis, then it follows that all the cells of your body contain the same genes. How, then, can different cells of the body have different structures and functions? Why do hair follicle cells synthesize hair proteins and not hormones? Further, does the environment influence how the genotype is used? If so, how? The answer is that, within certain limits, **cells can regulate which genes are used and which are not, depending on the function of the cell and the environment in which it lives.**

In this chapter, we examine how genotypes are transformed into phenotypes. With this knowledge, we can return to the observations of Mendelian genetics and understand what mutations really are, why genes have different alleles, and why some alleles are dominant and some are recessive.

The Relationship Between Genes and Proteins

Long before biologists knew anything about the molecular structure of genes, they tried to find out how genes work. In the early 1900s, the English physician Archibald Garrod studied the inheritance of human metabolic disorders. Garrod knew that the human body is a chemical cauldron, churning with reactions that convert molecules of food into other molecules the body needs. These biochemical conversions proceed stepwise, with each step catalyzed by a specific enzyme. The metabolism of the amino acids phenylalanine and tyrosine is an instructive example (Fig. 13-1).

Dietary proteins usually contain both amino acids. Once in the body, the amino acids are used as building blocks for bodily proteins and for a variety of other functions, each catalyzed by a specific set of enzymes. If a single enzyme is defective, it may cause one or both of two types of effects. First, the product of the reaction catalyzed by the enzyme will not be synthesized. For example, defects in the enzymes that convert tyrosine to melanin will result in a lack of melanin, and the affected person will be an albino, with extremely pale skin, white hair, and usually pink eyes. Second, if a substance is not converted to a given product, either the substance itself or products of alternative pathways are likely to accumulate in the body. In phenylketonuria (PKU), the enzyme that converts phenylalanine to tyrosine is defective. Therefore, phenylalanine and metabolites called phenylketones accumulate. ("Phenylketonuria" means that **p**henyl**k**etones are found in the **u**rine: hence, PKU.) In infants, high phenylalanine levels injure developing brain cells, causing severe mental retardation.

Garrod catalogued a number of these "inborn errors of metabolism," and deduced, from their patterns of inheritance, that each defective enzyme was probably due to a rare allele of a single gene. Of course, breeding experiments to confirm this hypothesis cannot be performed with humans.

The One-Gene, One-Protein Hypothesis

The common red bread mold, *Neurospora crassa*, proved to be an ideal organism for studying the relationship between genes and enzymes. Although we commonly see it on stale bread, *Neurospora* is an extremely independent organism that can synthesize almost all the organic compounds it needs. It can grow on a minimal medium containing an energy source such as sucrose, a few minerals to supply essential elements such as nitrogen and phosphorus, and a single vitamin, biotin.

For most of its life cycle, *Neurospora* is haploid, with just one copy of each chromosome. Therefore, mutations in *Neurospora* always show up phenotypically, since they cannot be masked by a normal allele on a homologous chromosome, as they can be in diploid organisms. (A similar phenomenon occurs in male mammals with respect to genes carried on the X chro-

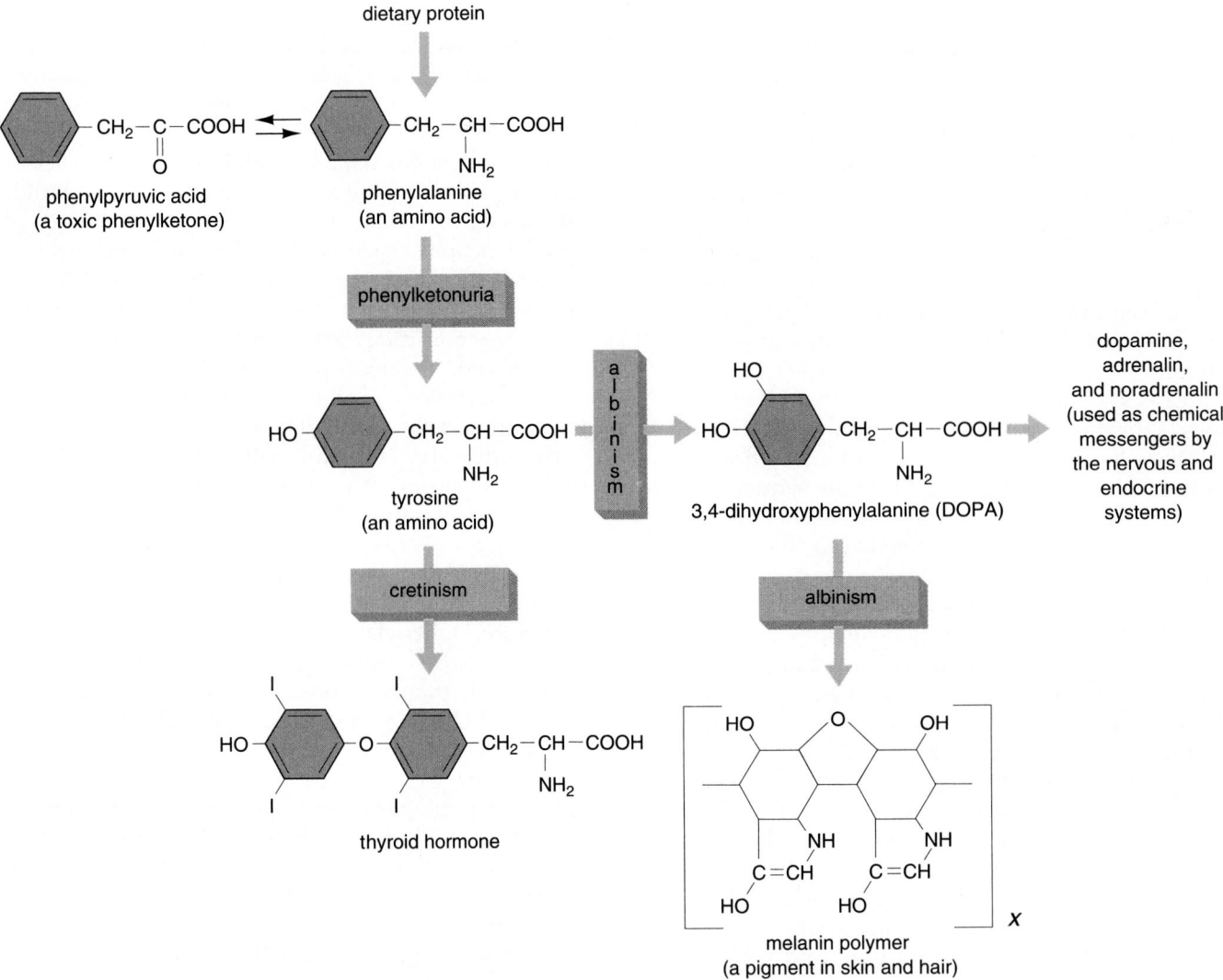

Figure 13-1 Some of the pathways of phenylalanine and tyrosine metabolism in humans. The arrows represent enzyme-catalyzed chemical reaction(s) that produce a product (e.g., tyrosine) from a substrate (e.g., phenylalanine). If the enzyme that converts phenylalanine to tyrosine is defective, phenylalanine and phenylpyruvic acid build up, damaging developing brain cells. Tyrosine is a precursor for several other essential compounds, including thyroid hormone, the nervous system chemicals adrenalin and noradrenalin, and the pigment melanin. Hereditary disorders caused by single-gene mutations that lead to defective enzymes are given in boxes placed across the appropriate arrows.

mosome. With no corresponding gene on the Y chromosome, a male must show the phenotype of whatever genes he carries on his X chromosome.)

Geneticists George Beadle and Edward Tatum bombarded *Neurospora* with X-rays, producing mutant molds that lost the ability to synthesize some substances (Fig. 13-2). One particularly interesting mutant couldn't grow on minimal medium, but did grow if the amino acid arginine was added. Beadle and Tatum therefore concluded that the mutant must have lost the ability to synthesize arginine. By starting with minimal medium and adding one precursor

molecule in the biosynthetic pathway of arginine at a time, they found that the mutant lacked only the enzyme that catalyzed one specific step in arginine synthesis. Genetic analysis showed that the mutant differed from normal molds in a single gene. From this and many other experiments, they concluded that **a gene encodes the information needed for the synthesis of a specific enzyme.** This became known as the "one-gene, one-enzyme" hypothesis. This is the same conclusion that Garrod reached, but couldn't prove, based on his studies of human metabolic disorders.

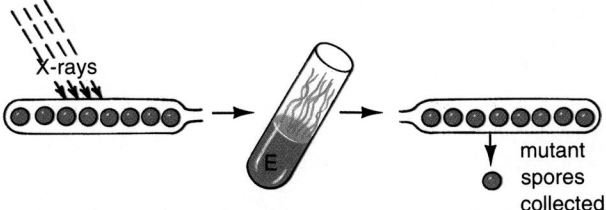

(a) Mold spores are irradiated with X-rays to induce mutations. They are germinated and grown on an enriched medium. When these molds reproduce, their spores are collected.

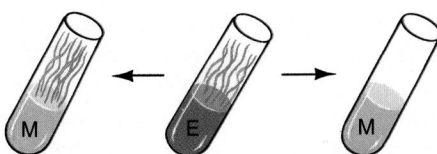

(b) Spores are germinated individually in enriched medium containing all the amino acids. A piece of each resulting mold is placed on minimal medium lacking amino acids. If the mold still grows on minimal medium (left) then no mutation has occurred, and this mold is discarded. If the mold cannot grow on minimal medium (right), a mutation has occurred and the experiments are continued.

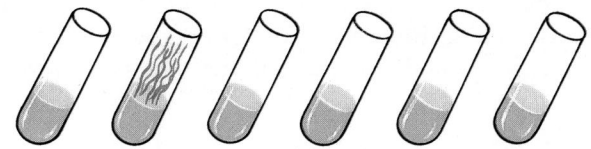

+alanine +arginine +cysteine +leucine +glycine +lysine ... etc.

(c) Pieces of mutated mold are placed in tubes containing minimal medium plus one amino acid. The mutant can grow only if arginine is added. Therefore, the mutant cannot synthesize arginine.

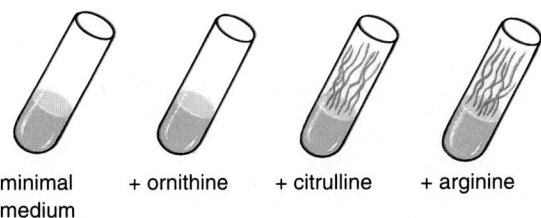

minimal   + ornithine   + citrulline   + arginine
medium

(d) Mutants are tested in media to which precursors in the biosynthetic pathway of arginine have been added. Since the mold can grow if supplied with citrulline or arginine, but not ornithine, the mutation must have ruined a single enzyme, the one that normally catalyzes the conversion of ornithine to citrulline.

◀ Figure 13-2 Beadle and Tatum's experiments with the mold *Neurospora*, showing that a mutation may produce a single defective enzyme.

Geneticists have since learned that not all genes encode the information needed by a cell to produce enzymes. Some genes carry information for the synthesis of structural proteins such as collagen in skin and keratin in hair, or hormones such as insulin. For a few genes, the final product isn't protein at all, but **ribonucleic acid (RNA)**, such as the RNA of ribosomes. Nevertheless, most genes do code for proteins or parts of proteins. As a generalization, then, **each gene encodes the information for a single protein, which might be called the** *one-gene, one-protein hypothesis.*

From DNA to Protein

Since we know that genes are DNA, we can reinterpret the experiments of Beadle and Tatum: **a gene is a segment of DNA that contains the information needed to synthesize a protein.** Different genes have different nucleotide sequences, and different proteins have different amino acid sequences. Therefore, **the sequence of nucleotides in DNA must encode the sequence of amino acids in a protein.** How does a cell translate the message of its DNA into proteins?

As you know, the DNA of eukaryotic cells is located in the nucleus. Protein synthesis, on the other hand, occurs on ribosomes in the cytoplasm. Therefore, DNA cannot directly guide protein synthesis. **There must be an intermediate molecule that carries the information from DNA in the nucleus to the ribosomes in the cytoplasm. This molecule is RNA.**

RNA is similar to DNA, but differs in three respects: (1) RNA is usually single stranded; (2) it has a different type of sugar in its backbone—ribose instead of deoxyribose; and (3) thymine in DNA is replaced by uracil in RNA:

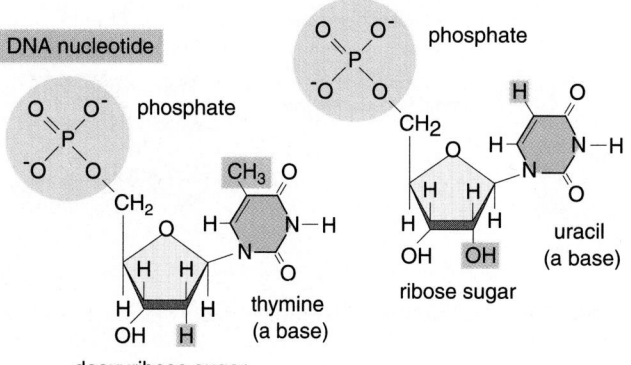

There are three types of RNA in a cell, all of which are involved in translating genetic information in DNA to the amino acid sequence of proteins. These are messenger RNA, transfer RNA, and ribosomal RNA. We will examine their functions in more detail shortly.

An Overview of Information Flow from DNA to Protein

Information flows from DNA to proteins in a two-step process (Fig. 13-3).

1. In *transcription,* **the information contained in the DNA of a specific gene is copied into RNA.** All three types of RNA are transcribed from DNA. One particular type of RNA, appropriately called **messenger RNA,** is transcribed from DNA in the nucleus and travels to the cytoplasm. The sequence of nucleotides in messenger RNA carries information to the ribosomes about the sequence of amino acids in the protein to be manufactured.
2. In *translation,* **two other types of RNA,** *transfer RNA* **and** *ribosomal RNA,* **convert the information of messenger RNA into the correct amino acids and help to synthesize the protein.**

To understand the molecular mechanisms of this information flow from DNA to protein, geneticists first had to break the language barrier: how does the language of nucleotide sequences in DNA and messenger RNA translate into the language of amino acid sequences in proteins? This translation relies on a "dictionary" called the **genetic code.**

The Genetic Code

We have used the word "code" several times to refer to the information stored in DNA and ultimately translated into the amino acid sequence of proteins. This **genetic code** is conceptually similar to Morse code: a set of symbols (bases in nucleic acids or dots and dashes in Morse code) that can be translated into another set of symbols (amino acids in proteins or letters of the alphabet). The question is, what combinations of bases stand for which amino acids?

Since there are four different bases in DNA and RNA, and 20 different amino acids in proteins, the bases cannot serve as a one-to-one code for amino acids: there are simply not enough of them. Perhaps, just as Morse code uses a short sequence of dots and dashes to encode the letters of the alphabet, so too the genetic code might use a short sequence of bases to encode each amino acid. If a sequence of two bases codes for an amino acid, then there would be $4^2 = 16$ possible combinations of bases. This isn't enough either. Three bases per amino acid, however, gives $4^3 = 64$ possible combinations, which is more than enough. Under the assumption that nature operates as economically as possible, biologists hypothesized that **the genetic code must be triplet: three bases specify one amino acid.** In 1961, Francis Crick and three coworkers demonstrated that this hypothesis is

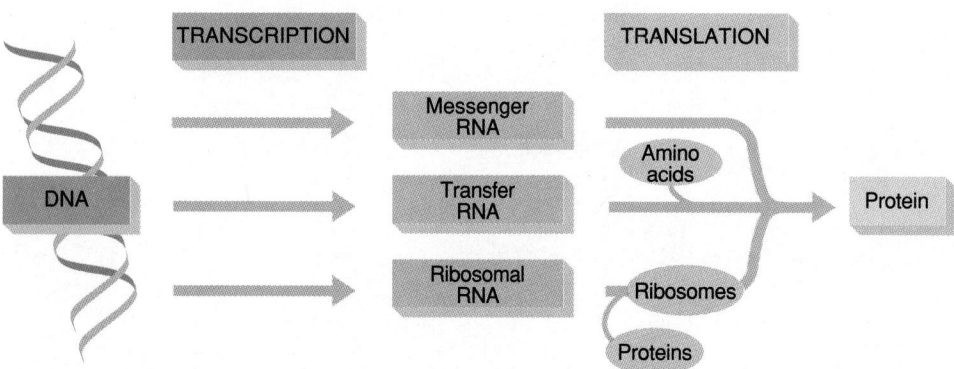

Figure 13-3 A simplified diagram of information flow from DNA to RNA to protein.

correct (see "Methods in Biology: Cracking the Genetic Code.")

The Nature of the Code

A few requirements must be met for any language to be understood. The users must know what the words mean; where words start and stop; and where sentences start and stop. Shortly after the Crick experiments demonstrating that the "words" of the genetic code are all three nucleotides long, researchers began to decipher the code. They ground up bacteria and isolated the components needed to synthesize proteins. Then, by synthesizing artificial messenger RNA and adding it to their protein-synthesizing mixtures, they could see which amino acids were incorporated into the resulting proteins. For example, an RNA composed entirely of uracil (UUUUUUUUUU...) directed the mixture to synthesize a protein composed solely of phenylalanine. Therefore, the triplet specifying phenylalanine must be UUU. Since the genetic code was deciphered using these artificial RNAs, the code is usually written as the nucleotide triplets in messenger RNA that code

for each amino acid (Table 13-1). These messenger RNA triplets are called **codons.**

What about punctuation? How does the cell recognize where codons start and stop and where entire protein codes start and stop? The codon AUG signals "start," that is, the beginning of a protein (**start codon**). Three codons, UAG, UAA, and UGA, signal "stop," that is, the end of a protein (**stop codon**). Now, if all the codons have three nucleotides, and the beginning and end of a protein are specified, then punctuation between codons is unnecessary. To see why this is so, consider what would happen if English used only three-letter words: a sentence such as THEMANSAWTHECAT would be perfectly understandable, even without spaces between the words, as long as the reader knew where the sentence started and stopped. The genetic code doesn't need, and doesn't have, punctuation between codons.

Finally, there are 60 codons in addition to the start and stop codons, and only 20 amino acids to code for. All 60 codons are used in the genetic code. Therefore, the genetic code is highly redundant or **degenerate;** that is, **a single amino acid may be specified by several codons.** For example, six different codons all

Table 13-1 The Genetic Code (Codons of mRNA)

First Base	Second Base								Third Base
	U		C		A		G		
U	UUU	Phenylalanine	UCU	Serine	UAU	Tyrosine	UGU	Cysteine	U
	UUC	Phenylalanine	UCC	Serine	UAC	Tyrosine	UGC	Cysteine	C
	UUA	Leucine	UCA	Serine	UAA	Stop	UGA	Stop	A
	UUG	Leucine	UCG	Serine	UAG	Stop	UGG	Tryptophan	G
C	CUU	Leucine	CCU	Proline	CAU	Histidine	CGU	Arginine	U
	CUC	Leucine	CCC	Proline	CAC	Histidine	CGC	Arginine	C
	CUA	Leucine	CCA	Proline	CAA	Glutamine	CGA	Arginine	A
	CUG	Leucine	CCG	Proline	CAG	Glutamine	CGG	Arginine	G
A	AUU	Isoleucine	ACU	Threonine	AAU	Asparagine	AGU	Serine	U
	AUC	Isoleucine	ACC	Threonine	AAC	Asparagine	AGC	Serine	C
	AUA	Isoleucine	ACA	Threonine	AAA	Lysine	AGA	Arginine	A
	AUG	Start (Methionine)	ACG	Threonine	AAG	Lysine	AGG	Arginine	G
G	GUU	Valine	GCU	Alanine	GAU	Aspartic Acid	GGU	Glycine	U
	GUC	Valine	GCC	Alanine	GAC	Aspartic Acid	GGC	Glycine	C
	GUA	Valine	GCA	Alanine	GAA	Glutamic Acid	GGA	Glycine	A
	GUG	Valine	GCG	Alanine	GAG	Glutamic Acid	GGG	Glycine	G

METHODS IN BIOLOGY
Cracking the Genetic Code

The hypothesis that three nucleotides in DNA code for one amino acid in protein is an attractive and logical possibility: fewer than three nucleotides can't unambiguously code for all 20 amino acids, and more than three are superfluous. But how could you prove that nature really uses a triplet code?

As with the Hershey–Chase experiments (see Chapter 12), the simplicity of the bacteriophage proved invaluable in deciphering a fundamental principle of genetics. Francis Crick and his coworkers exposed bacteriophages to a dye called acridine, which causes insertion mutations: one, two, or three nucleotides were inserted into the DNA molecule at random places within a particular gene. They found that inserting one or two nucleotides into the DNA causes the synthesis of defective enzymes that prevent the phages from reproducing in their host bacteria. Inserting three nucleotides, however, sometimes produces phages that synthesize normal or nearly normal enzymes.

Crick concluded that during RNA synthesis the DNA of a gene is "read" in a linear order, starting at the beginning. Each set of three nucleotides makes up a "word"; that is, three nucleotides in DNA encode a single amino acid. The code must specify the beginning and end of a gene, but within a gene there are no spaces or punctuation between words. How did he arrive at these conclusions?

To understand Crick's reasoning, let's suppose that we have a gene with this repetitive sequence of nucleotides:

| T | A | G | T | A | G | T | A | G | T | A | G | T | A | G | T | A | G | T | A | G | T | A | G |

If DNA always has three letter words, then this nucleotide sequence spells out the English word, "tag," over and over again. You don't need spaces to separate the words if they all have three and only three letters.

Suppose that the acridine dye inserts another thymine somewhere near the beginning of the gene. The gene now reads:

| T | A | G | T | A | T | G | T | A | G | T | A | G | T | A | G | T | A | G | T | A | G | T | A |

first insertion ⌐   incorrect "reading"   ⌐

Protein synthesis is like a computer: garbage in, garbage out. From the point of the insertion to the end of the gene, the gene now reads GTA, GTA, GTA . . . , which is nonsense.

Inserting another T nearby still results in nonsense, as most of the gene now reads AGT, AGT, AGT:

| T | A | G | T | A | T | G | T | T | A | G | T | A | G | T | A | G | T | A | G | T | A | G | T |

second insertion ⌐   incorrect "reading"   ⌐

A third insertion, however, results in most of the gene reading TAG, TAG, TAG again:

| T | A | G | T | A | T | G | T | T | A | G | T | A | G | G | T | A | G | T | A | G | T | A | G |

third insertion ⌐   remainder of gene
reads correctly

How does this explain Crick's results? One insertion near the beginning of the bacteriophage gene causes the rest of the triplets to encode the wrong amino acids, so the enzyme specified by that gene completely malfunctions. The second insertion still leaves most of the triplets calling for the wrong amino acids. If there are three insertions near the start of the gene, however, then the first few amino acids would be wrong, but all the triplets beyond the third insertion would be correct once again. Most of the enzyme would therefore be synthesized correctly. If the incorrect parts of the enzyme were relatively unimportant to its overall function, it may work well enough to allow the phages to reproduce.

Only a triplet code can account for the fact that three insertions, and not one or two or four, restore near-normal enzyme function. Further, only a code without punctuation or spaces between words will be confused by any of the insertions. Finally, a code without punctuation between words can work only if the start and stop points of protein synthesis are clearly marked. All three conclusions have been found to be correct, as we describe in the text.

code for arginine (see Table 13-1). However, although the code is redundant, it is not ambiguous; **each codon specifies only one amino acid.**

RNA: Intermediary in Protein Synthesis

Protein synthesis requires the cooperation of three types of RNA: messenger RNA, transfer RNA, and ribosomal RNA. How are these RNA molecules synthesized? How do they differ? How do they function in protein synthesis?

RNA Synthesis

All three types of RNA are synthesized using molecules of DNA as a template. As we saw earlier in this chapter, RNA nucleotides are chemically very similar to DNA nucleotides. Since the two "languages" are so much alike, RNA synthesis has been named **transcription,** meaning "to copy over."

Transcription of DNA into RNA is restricted in two major ways. First of all, most of the DNA in a cell is never transcribed into RNA. On the other hand, the DNA of certain genes, such as the genes for ribosomal RNA, are copied hundreds or thousands of times during interphase. Therefore, **transcription normally copies only the DNA of selected genes into RNA.**

Second, the useful information of any given gene normally resides on only one strand of the DNA double helix. (A single chromosome contains many genes; some genes may lie on one strand, while other genes lie on the other strand.) Why is this so? Remember, the two strands of DNA are *complementary,* not *identical.* If the sequence of nucleotides on the "gene-containing" strand codes for a sequence of amino acids that forms a functional protein, the complementary strand will consist of a different sequence of nucleotides that codes for different amino acids, which probably would form a completely useless protein. Therefore, **transcription copies only one strand of DNA into RNA.** The DNA strand that actually contains the gene, and is transcribed into RNA, is called the *sense strand* of the DNA (because its sequence of nucleotides "makes sense").

With these constraints in mind, we can visualize transcription as a three-step process: (1) initiation, (2) elongation of the RNA chain, and (3) termination (Fig. 13-4). These three steps correspond to the three major parts of most genes, in both eukaryotes and prokaryotes: a promoter at the beginning of the gene; the "body" of the gene, which for most genes consists of the DNA nucleotides that actually code for amino acids in the protein to be synthesized; and a terminator at the end of the gene.

Initiation

RNA synthesis is carried out by an enzyme called **RNA polymerase.** Now, if a cell needs to transcribe only the DNA of a particular gene, then the obvious first step in transcription is for RNA polymerase to locate the beginning of the gene.

The **promoter** region of a gene is a short sequence of DNA nucleotides located just "upstream" (in the 3' direction) of the body of the gene. RNA polymerase recognizes the promoter nucleotide sequence as marking the beginning of a gene, and binds to the DNA at that site (Fig. 13-4a). RNA polymerase is a fairly large protein that occupies at least 50 nucleotides, including both the promoter and the first dozen or so nucleotides in the body of the gene.

Elongation

Once the RNA polymerase has bound to the promoter site, it changes shape, forcing the DNA double helix to open up at the beginning of the body of the gene. RNA polymerase then moves along the sense strand of the DNA. RNA polymerase travels in the 3'-to-5' direction along the sense strand (in the direction of the "oxygen arrowheads"), just as DNA polymerase does (refer back to Figs. 12-9 and 12-10). Using free RNA nucleotides present in the nucleus, RNA polymerase synthesizes a single strand of RNA that is complementary to the DNA of the sense strand (Fig. 13-4b). The same base-pairing rules are used for RNA synthesis as for DNA replication, except that adenine in DNA is paired with uracil in RNA. The DNA-to-RNA base-pairing rules thus become:

Base in DNA	Complementary Base in RNA
adenine	uracil
cytosine	guanine
guanine	cytosine
thymine	adenine

Although RNA polymerase adds RNA nucleotides to the growing RNA strand according to base-pairing with the DNA nucleotides of the gene, this base-pairing does not persist. After about 10 nucleotides have been added to the growing RNA chain, the beginning of the RNA molecule separates from the DNA (Fig. 13-4c). As the RNA continues to elongate,

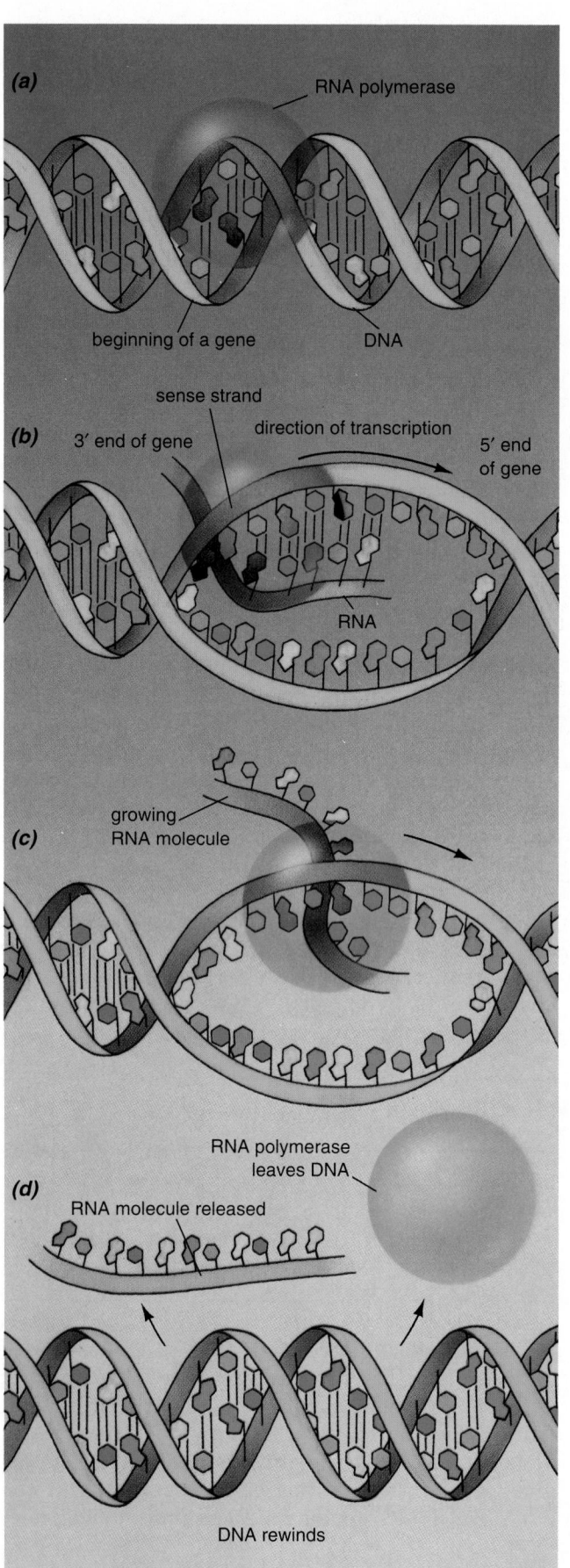

◀ **Figure 13-4** RNA transcription.
(a) The enzyme RNA polymerase binds to the promoter region of DNA near the beginning of a gene.
(b) The DNA double helix unwinds. The RNA polymerase travels along one of the DNA strands (the sense strand), catalyzing the formation of a continuous strand of RNA from free RNA nucleotides. The bases incorporated into the growing RNA strand are complementary to the bases in the sense strand of DNA.
(c) The RNA polymerase continues to the end of the gene.
(d) At the end of the gene, the RNA polymerase leaves the DNA. The DNA rewinds and the RNA molecule is released.

it forms a long "tail" drifting away from the DNA, as can be seen in electron micrographs (Fig. 13-5).

Termination

The end of transcription is not well understood, particularly in eukaryotes. Molecular geneticists have discovered that RNA polymerase continues along the sense strand for some distance past the real body of the gene, in some cases thousands of nucleotides further. Eventually it reaches sequences of DNA nucleotides that trigger two events (see Fig. 13-4d). First, the RNA chain completely separates both from the DNA and from the RNA polymerase. Second, the RNA polymerase leaves the sense strand of the DNA.

Types of RNA

There are three types of RNA: **messenger RNA (mRNA)** carries the code for the amino acid sequences of proteins from the genes in DNA to the ribosomes, the actual sites of protein synthesis. The other two types of RNA, **ribosomal RNA (rRNA)** and **transfer RNA (tRNA),** do not carry information to be translated into protein. Instead, these RNA molecules are the final products of certain genes, and thus are an exception to the generalization that genes code for proteins.

Messenger RNA

Messenger RNA is a long, single-stranded molecule that includes the codons that will be translated into the amino acid sequence of a protein. In prokaryotes, mRNA is directly transcribed from the DNA of a gene, and translation into proteins often begins even before transcription is complete. In eukaryotes, things are a bit more complicated, because the RNA transcribed from DNA contains more nucleotides than will ultimately be translated into protein. We will examine the formation of eukaryotic mRNA in more detail a bit later in this chapter.

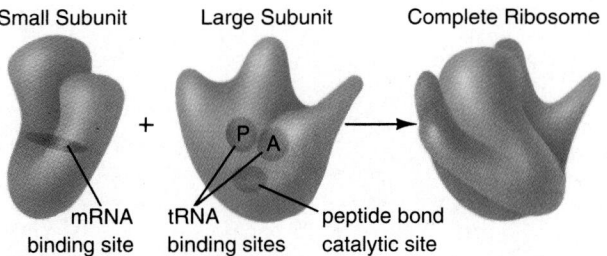

Figure 13-6 A ribosome has two subunits, each composed of protein and rRNA. The small subunit binds messenger RNA. The large subunit has three functional sites. Two, called the P and A sites, bind tRNA, and the third catalyzes the formation of the peptide bond between amino acids of the growing protein.

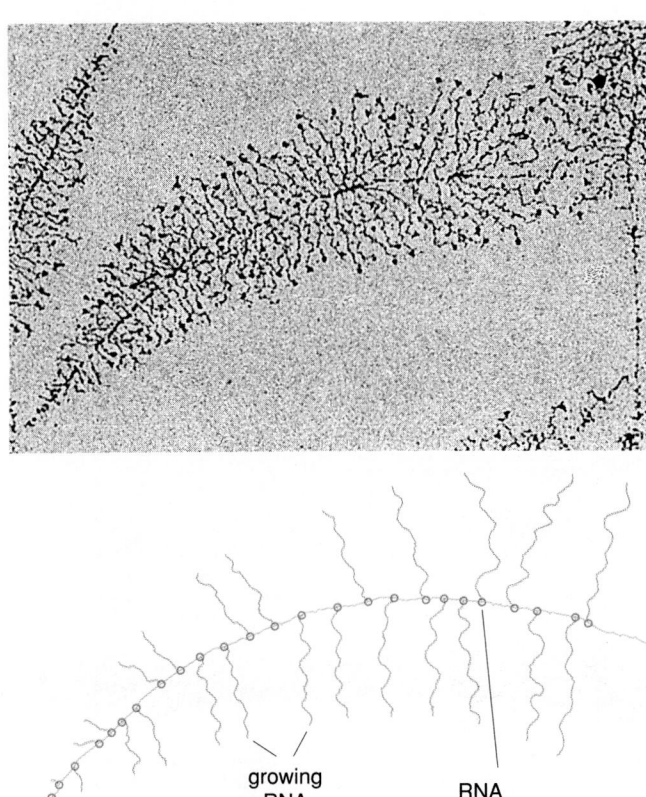

Figure 13-5 This electron micrograph clearly shows the progress of RNA transcription in the egg of the African clawed toad. As shown in the diagram, in each "Christmas tree" structure, the central "trunk" is DNA and the "branches" are RNA molecules. A series of RNA polymerase molecules are traveling down the DNA, synthesizing RNA as they go. The beginning of the gene is on the left. Therefore, the short RNA molecules on the left have just begun to be synthesized, while the long RNA molecules on the right are almost finished.

The bottom line in eukaryotic cells, however, is that mRNA is synthesized in the nucleus and enters the cytoplasm through the pores in the nuclear envelope, carrying its message of nucleotide sequences. In the cytoplasm, mRNA binds to ribosomes, where the codons of mRNA are translated into the language of amino acids in proteins. (You might think of mRNA as a "molecular photocopy" of the gene DNA. The gene itself remains safely stored in the nucleus, like a valuable document in a library, while copies are sent to the cytoplasm to be used in protein synthesis.)

Ribosomal RNA

Ribosomes are composites of rRNA and a variety of proteins. Each ribosome is composed of two subunits (Fig. 13-6). In eukaryotic cells, the small subunit con-

sists of one molecule of rRNA and about 30 proteins. It recognizes and binds mRNA and tRNA. The large ribosomal subunit consists of three rRNA molecules and 45 to 50 proteins. It contains an enzymatic region that catalyzes the addition of amino acids to the growing protein chain, and two sites (usually designated P and A) that bind molecules of tRNA.

Transfer RNA

Transfer RNA molecules bind amino acids and deliver them to the ribosome, where they are incorporated into protein chains. There are many different types of tRNAs, at least one type for each amino acid. Transfer RNAs are like "code books," the only molecules in the cell that can decipher the codons of mRNA and translate them into the amino acids of proteins. Transfer RNAs are complex molecules, twisted about into a shape something like a three-leaf clover with a stem (Fig. 13-7a). For our purposes, the stem and the central leaf are the important parts. Enzymes in the cytoplasm recognize each specific tRNA molecule and attach the correct amino acid to the stem (Fig. 13-7b). The energy of ATP is used to form the tRNA–amino acid bond. Some of the ATP energy is stored in the tRNA–amino acid bond, and this energy will be used to forge the peptide bond when the amino acid is added to a growing protein molecule. The outside bend of the central tRNA leaf bears three exposed bases, called the **anticodon,** that actually decipher the mRNA code: **the anticodon of each tRNA is complementary to the codon of mRNA that specifies the amino acid attached to that tRNA.**

Protein Synthesis

Now that we have introduced all the actors involved in protein synthesis, let's look at the actual events.

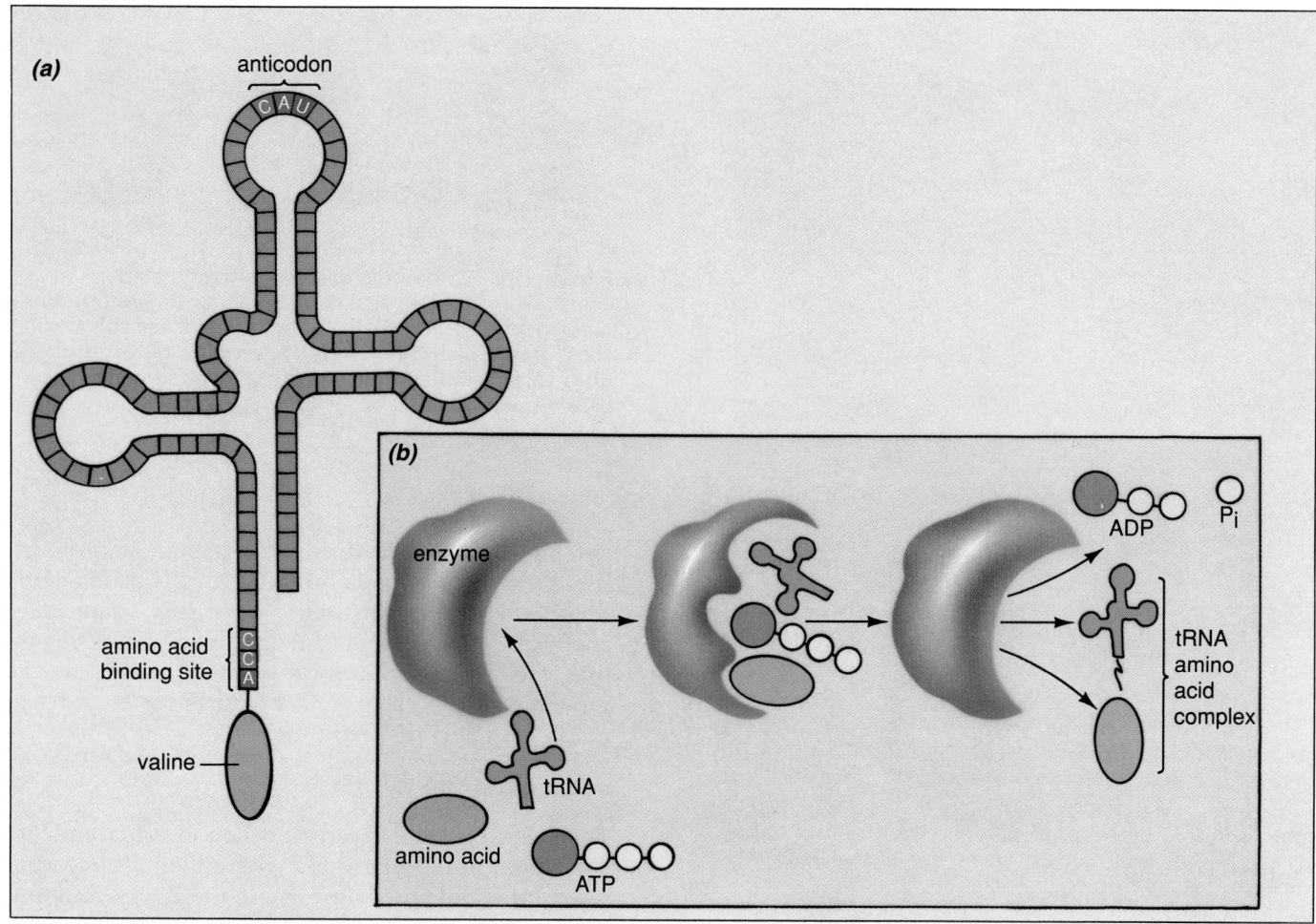

Figure 13-7 Transfer RNA.
(a) Transfer RNA is a single RNA strand, folded back upon itself into loops like a three-leafed clover growing from a single stem. The loops are held together by hydrogen bonds between complementary bases. The central "leaf" of the tRNA clover bears three bases called the anticodon, which base-pair with complementary bases in mRNA during protein synthesis (see Fig. 13-8). The anticodons differ among different tRNA molecules. The "stem" binds the amino acid encoded by the anticodon bases of the leaf. All tRNA stems end with the bases CCA.
(b) Each type of tRNA binds to a specific amino acid. Enzymes in the cytoplasm, one for each amino acid, catalyze the bond between each tRNA and its corresponding amino acid, using the energy of ATP. Some of the ATP energy is stored in the tRNA–amino acid bond, and will be used to drive the formation of the peptide bond when the amino acid is used in protein synthesis.

Assuming that amino acids have already linked up with their appropriate transfer RNAs (Fig. 13-7b), protein synthesis can be considered to occur in two stages:

1. Messenger RNA is transcribed from the DNA template of the genes in the nucleus (see Fig. 13-4). The mRNA travels to a ribosome in the cytoplasm.

2. A ribosome binds mRNA and the appropriate tRNAs. On the ribosome, the codons of mRNA are translated into the amino acid sequence of a protein.

We already discussed the first stage; here we will examine the second stage, **translation** (Fig. 13-8). Translation has three steps: initiation of protein synthesis, elongation of the protein chain, and termination.

Initiation

Two codons of mRNA, along with several protein "initiation factors," bind to the small subunit of a ribosome (Fig. 13-8a). The first codon is always the "start codon" AUG. The second codon (GUU in our example) codes for the next amino acid in the protein. The tRNA bearing the complementary "start anticodon" UAC hydrogen bonds to the start codon (Fig. 13-8b). The large ribosomal subunit then attaches to the small subunit. As it does so, the start tRNA simultaneously binds to the P site on the large subunit (Fig. 13-8c). The ribosome is now fully assembled and ready to begin translation.

Protein Elongation

Proteins are synthesized one amino acid at a time. The anticodon of a tRNA–amino acid complex (valine in our example) recognizes the second mRNA codon, and moves into the A site on the large subunit (Fig. 13-8d). The two amino acids borne by the two tRNAs now lie adjacent to one another. The catalytic site on the large subunit breaks the bond holding the "start" amino acid (methionine) to its tRNA and uses the released energy to form a peptide bond between the methionine and the valine borne by the second tRNA. At the end of this step, the start tRNA is now "empty," while the second tRNA bears a short, two-amino acid, protein chain (Fig. 13-8e).

At this point, the empty start tRNA drops off the ribosome, and the ribosome shifts to the next codon on the mRNA molecule (Fig. 13-8f). The tRNA holding the growing protein chain shifts too, from the A site to the P site on the ribosome. A new tRNA–amino acid complex binds to the emptied A site (Fig. 13-8g). The catalytic site on the large subunit breaks the bond between the dipeptide and its tRNA, and links the dipeptide with the amino acid (histidine) in the A site (Fig. 13-8h). The depleted tRNA in the P site leaves the ribosome, the ribosome shifts over another codon, and the process repeats.

Termination

Near the end of the mRNA, a stop codon is reached. No tRNA recognizes a stop codon. Instead, "termination factors" cut the finished protein chain off the last tRNA, releasing it from the ribosome (Fig. 13-8i).

The Decoding Chain from DNA to Protein

We can now understand how a cell decodes the genetic information stored in its DNA to synthesize a protein (Fig. 13-9).

1. The DNA is organized into genes that are dozens to thousands of nucleotides long. You might think of a gene as a sentence within the genetic information manual of the cell.
2. The "words" that comprise the "gene sentences" are groups of three nucleotides.
3. A codon of mRNA consists of three nucleotides with bases complementary to the three bases of a DNA "word."
4. An anticodon of tRNA, in turn, is complementary to a specific codon of mRNA.
5. The tRNA bears a specific amino acid, which is attached to the tRNA by enzymes that can "read" the anticodon.

This decoding chain, from bases in DNA to codon of mRNA to anticodon of tRNA to amino acid, results in the incorporation of the correct amino acid in the growing protein.

Gene Regulation

Knowing how proteins are synthesized does not provide a complete understanding of how an organism's genotype produces its phenotype. For example, most of the cells of your body have the same DNA, but don't use all the DNA all the time. Individual cells express (produce proteins encoded by) only a small fraction of their genes, those that are appropriate to the function of that particular cell type. Muscle cells, for example, synthesize the contractile proteins actin and myosin, but not insulin or hair proteins. Gene expression also changes over time, depending on the needs of the body from moment to moment. Therefore, understanding gene function requires an understanding of how genes are regulated.

The use of genetic information by a cell is a multistep process, beginning with the transcription of DNA and often ending with an enzyme catalyzing a needed reaction (Fig. 13-10). Regulation can occur at any of these steps.

1. Regulation of Transcription

The rate of transcription of individual genes depends on the type of cell, its stage in the cell cycle, and the metabolic activity of both the cell and the whole organism. Some genes are never transcribed in certain types of cell; for example, the gene for insulin is not transcribed in muscle cells. Transcription of other genes is turned on or off on demand.

2. Processing of Pre-messenger RNA

As we will discuss shortly, the genes of eukaryotic cells are much larger than the final mRNA transcribed

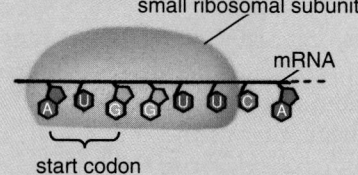

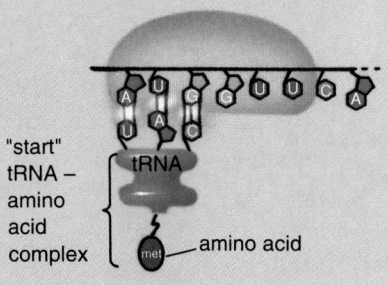

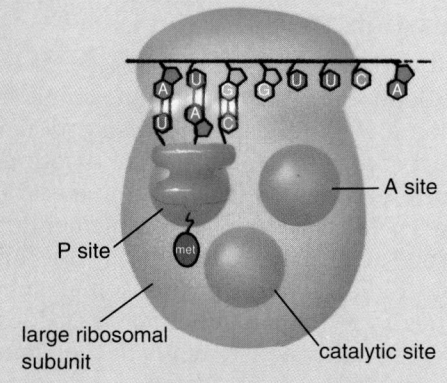

(a) The small subunit of a ribosome binds to two codons of mRNA. The first codon bound is always the start codon.

(b) A tRNA-methionine molecule bearing the anticodon UAC base-pairs with the start codon of the mRNA.

(c) The large ribosomal subunit joins with the small subunit. The start tRNA-methionine complex binds to the P site of the large subunit.

ELONGATION:

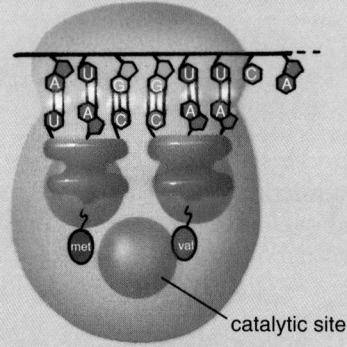

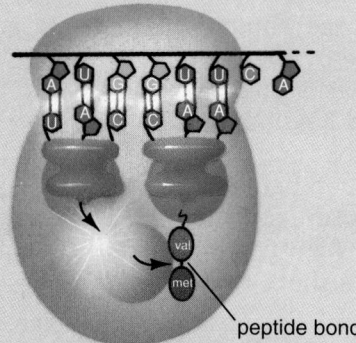

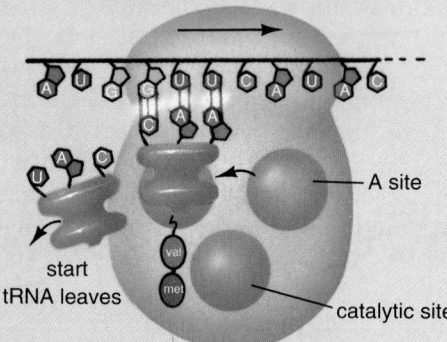

(d) The second mRNA codon (GUU) base-pairs with the CAA anticodon of a tRNA-valine molecule, which enters the A site of the large subunit.

(e) The catalytic site on the large subunit catalyzes the formation of a peptide bond between the amino acids methionine and valine, using the energy stored in the tRNA-methionine bond. The dipeptide remains attached to the second tRNA.

(f) The start tRNA drops off the ribosome, and the ribosome moves one codon to the right on the mRNA. The tRNA bearing the newly formed dipeptide moves to the P site and the A site is emptied.

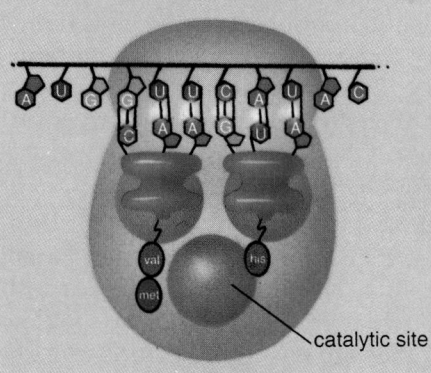

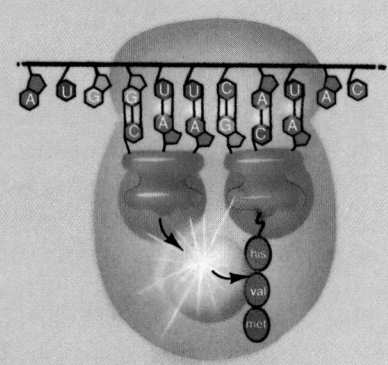

TERMINATION:

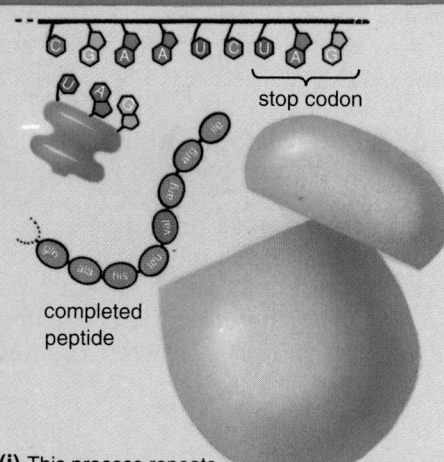

(g) The next tRNA base-pairs with the third mRNA codon and moves into the A site.

(h) A peptide bond is forged between the dipeptide and the new amino acid, forming a tripeptide that remains attached to the third tRNA.

(i) This process repeats until a "stop" codon (UAG, UAA, UGA) is reached. The finished peptide is released from the ribosome. The ribosomal subunits separate.

◀ **Figure 13-8** Protein synthesis is the translation of the sequence of nucleotides in mRNA to the sequence of amino acids in the encoded protein.

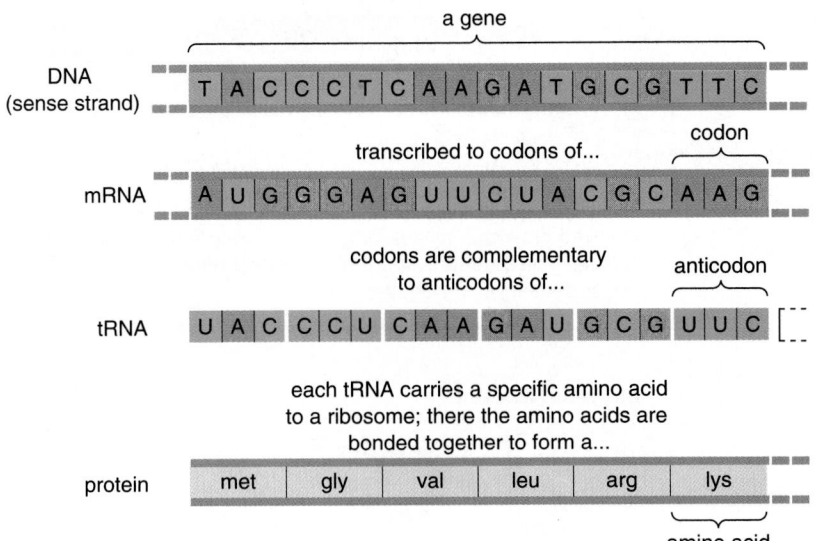

◀ **Figure 13-9** The decoding chain from DNA to protein.

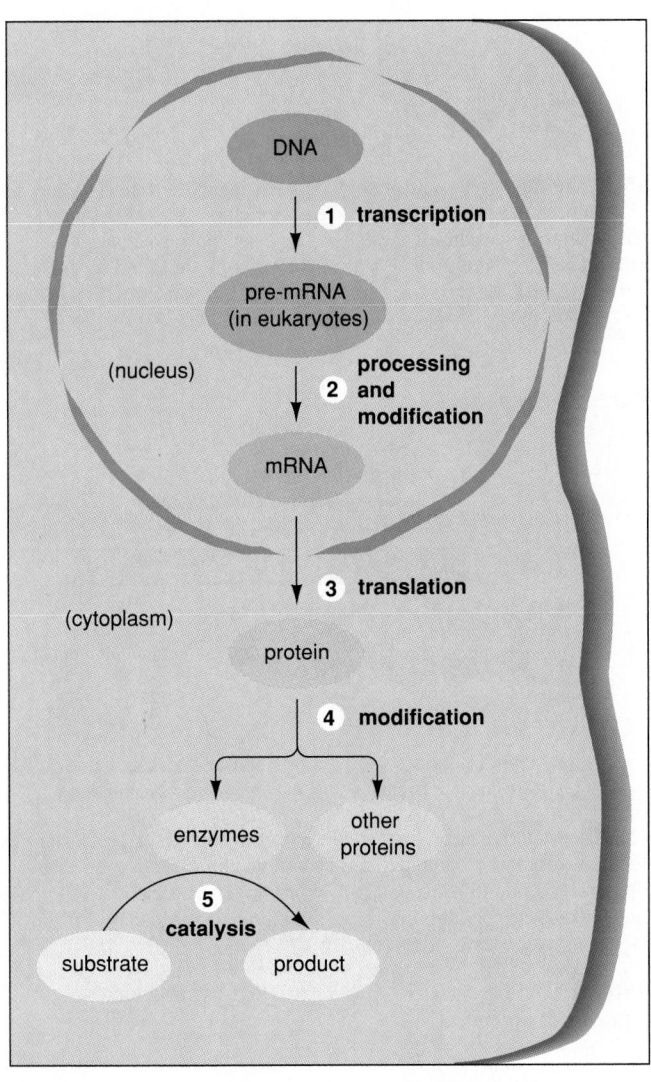

◀ **Figure 13-10** A simplified diagram of information flow in a cell, from DNA to protein to chemical reactions catalyzed by enzymes. Information flow may be regulated at any step.

from them. The genes are first transcribed into very long "pre-messenger RNA" molecules. Differential processing of these "pre-mRNA" molecules can produce different types of final mRNA that are therefore translated into different proteins. Messenger RNA molecules must also be modified, for example by the addition or removal of nucleotides at the ends, before they can be used by the cell.

3. Regulation of Translation

Messenger RNAs vary in their stability and the rate at which they are translated into protein. Some mRNAs are extremely stable, which offers the possibility for repeated translation, while others are rapidly degraded. Further, a cell may block translation of certain mRNAs, depending on its metabolic requirements.

4. Protein Modification

Many proteins must be modified before they become active. For example, the protein-digesting enzymes produced by cells of your stomach wall and pancreas are initially synthesized in an inactive form (this keeps the enzymes from digesting the cells themselves). After these inactive forms are secreted into the digestive tract, portions of the enzymes are then snipped out to unveil the active site.

5. Regulation of Enzyme Activity

Enzyme activity is often controlled by competitive or allosteric inhibition, as discussed in Chapter 4.

These methods of regulating gene activity are all important, and are probably used to some extent by virtually all eukaryotic cells. We will restrict our discussion, however, to the first two steps, the transcription of DNA to RNA in prokaryotic and eukaryotic cells, and the processing of pre-mRNA to true mRNA in eukaryotic cells.

Gene Regulation in Prokaryotes

Prokaryotic DNA is often organized in coherent packages called **operons,** in which the genes for related functions lie next to one another (Fig. 13-11). An operon consists of four parts: (1) a **regulatory gene,** (2) a **promoter** that RNA polymerase recognizes as the place to start transcribing, (3) an **operator** that governs access of RNA polymerase to the promoter, and (4) the **structural genes** that encode the protein sequences of enzymes (Fig. 13-11a). Whole operons are regulated as units, so that related enzymes are synthesized simultaneously when the need arises. Prokaryotic operons are regulated differently, depending on the functions they control. Some operons synthesize enzymes that are needed by the cell just about all the time, such as the enzymes that synthesize amino acids. These operons are usually transcribed continuously, except under unusual circumstances when the bacterium encounters a vast surplus of a particular amino acid. Other operons synthesize enzymes that are needed only occasionally, for example to digest a relatively rare food substance. They are transcribed only when the bacterium encounters the rare food.

As an example of the latter type of operon, consider the common intestinal bacteria *Escherichia coil.* These bacteria have to live on whatever types of nutrients their host eats, and they can synthesize a variety of enzymes to metabolize a potentially wide variety of foods. The genes that code for most of these enzymes are transcribed only when the enzymes are needed. The enzymes that metabolize lactose, the principal sugar in milk, are a case in point. The **lactose operon** contains three structural genes, each coding for an enzyme that aids in lactose metabolism (Fig. 13-11a).

Structure of the Lactose Operon:

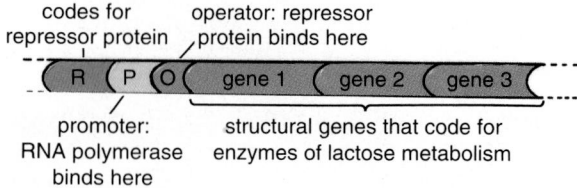

(a) The lactose operon consists of a regulatory gene, a promoter, an operator, and three structural genes that code for enzymes involved in lactose metabolism. The regulatory gene codes for a protein, called a repressor, that can bind to the operator site under certain circumstances.

Lactose Absent:

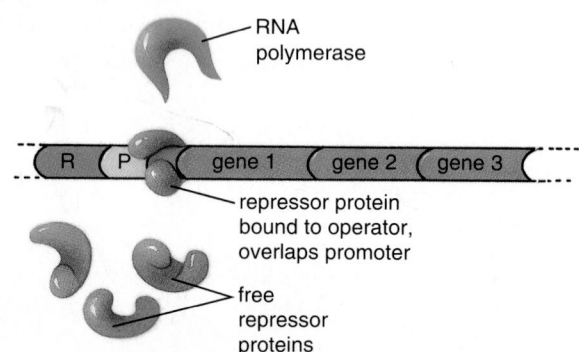

(b) When lactose is not present, repressor proteins bind to the operator of the lactose operon. The repressor protein is much larger than the operator region of the DNA, so it overlaps the promoter site. Therefore, RNA polymerase cannot bind to the promoter and the structural genes cannot be transcribed.

Lactose Present:

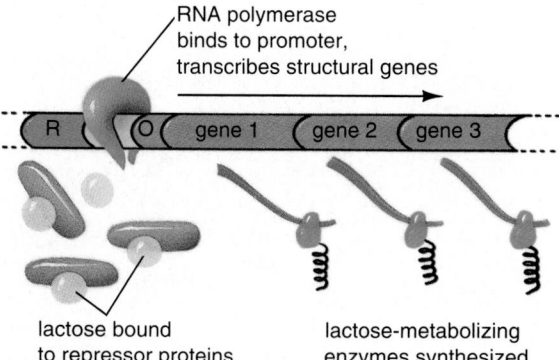

(c) When lactose is present, it binds to the repressor protein. The lactose–repressor complex cannot bind to the operator, so RNA polymerase has free access to the promoter. The RNA polymerase transcribes the three structural genes coding for the lactose-metabolizing enzymes.

Figure 13-11 Structure and regulation of the lactose operon of *E. coli.*

The lactose operon is shut off, or **repressed,** unless specifically activated by the presence of lactose. The regulatory gene of the lactose operon directs synthesis of a protein, called a **repressor protein,** that binds to the operator site. Since the operator site on the DNA is small and the repressor protein is large, the repressor overlaps onto the promoter site (Fig. 13-11b). RNA polymerase is physically prevented from binding to the promoter and starting transcription. Consequently, the lactose-metabolizing enzymes are not synthesized.

When *E. coli* colonize the intestines of a newborn mammal, however, they find themselves bathed in a sea of lactose whenever the host nurses from its mother. Lactose molecules enter the bacteria and bind to the repressor proteins, changing their shape (Fig. 13-11c). The lactose–repressor combination cannot attach to the operator site. Therefore, RNA polymerase can bind to the promoter of the lactose operon and transcribe the structural genes. Lactose-metabolizing enzymes are synthesized, allowing the bacterium to use lactose as an energy source.

When the young mammal is weaned, it usually never consumes milk again. The intestinal bacteria no longer encounter lactose, the repressor proteins are free to bind to the operator, and the genes for lactose metabolism are shut down.

Gene Regulation in Eukaryotes

Gene regulation is quite different in eukaryotes. Not only are genes for related functions sometimes found on entirely different chromosomes, but even the individual genes are split up on the chromosome. As a result, transcription and its regulation are more complex.

Eukaryotic Gene Structure

In the 1970s, molecular geneticists discovered that eukaryotic structural genes have much more DNA than is needed to encode the amino acids of proteins. Each gene consists of two or more DNA segments that encode the protein, interrupted by other DNA segments that apparently code for nothing at all. The coding segments are called **exons** because they are **ex**pressed in protein, while the noncoding segments are called **introns** because they **int**ervene between the exons (Fig. 13-12a).

Each eukaryotic gene has its own promoter. A nearby region of the chromosome, called the **enhancer,** regulates binding of RNA polymerase to the promoter. When specific regulatory proteins bind to the enhancer, they facilitate the binding of RNA

polymerase to the promoter, thus enhancing transcription.

When a eukaryotic gene is transcribed, a very long molecule of RNA is synthesized, starting before the first exon and ending after the last exon (Fig. 13-12b). The resulting RNA contains many more nucleotides than the true codons for the amino acids of the encoded protein. Two major steps convert this RNA molecule into mRNA. First, RNA nucleotides are added at the beginning (the "cap") and the end (the "tail") of the molecule. Second, enzymes in the nucleus precisely cut the molecule apart, splice together the sections that code for the protein, and discard the rest.

Functions of Fragmented Genes

Why are eukaryotic genes split up like this? There appear to be at least two functions served by fragmentation. The first function is to produce multiple proteins from a single gene. In rats, there is a gene that is transcribed in both the thyroid and the brain. In the thyroid, one splicing arrangement results in the synthesis of a hormone called calcitonin. In the brain, another splicing arrangement results in the synthesis of a peptide that is probably used as a messenger molecule for communication among brain cells.

The second function is more speculative, but has some good experimental evidence in its support: fragmented genes may provide a quick and efficient way for eukaryotes to evolve new proteins with new functions. This possibility is explored in "A Closer Look at Rube Goldberg Genetics: Making New Proteins from Old Parts."

Regulation of Transcription

As in prokaryotes, eukaryotes also regulate the rate of transcription of genes. Transcriptional regulation can operate on three levels: the individual gene, large parts of chromosomes, or entire chromosomes.

Single Gene Regulation

Some of the best-known examples of transcriptional regulation at the level of the individual gene are the cellular actions of steroid hormones, for instance the stimulation by estrogen of albumin (egg white) synthesis in female birds (Fig. 13-13). Being lipid-soluble, steroid hormones readily penetrate cell membranes and enter the interiors of cells. During the breeding season, estrogen is secreted into the bloodstream by the birds' ovaries and enters the cells of the oviduct. The estrogen binds to receptor proteins in the cyto-

(a) Eukaryotic Gene Structure

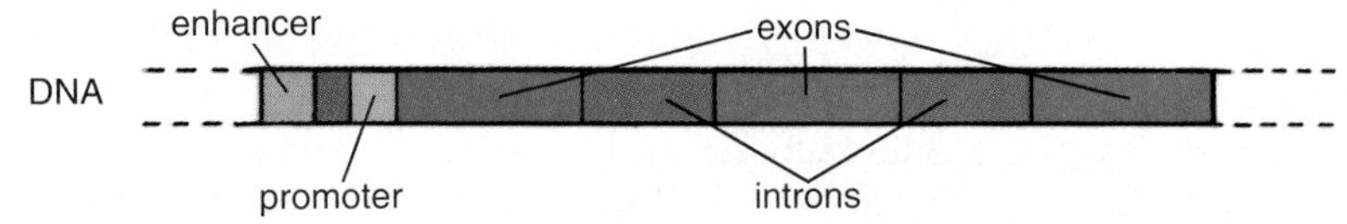

A typical eukaryotic gene consists of sequences of DNA called exons that code for the amino acids of a protein (red), and intervening sequences called introns (purple) that do not. At least two control regions, the promoter (yellow) and the enhancer (pink), regulate the transcription of eukaryotic genes.

(b) RNA Synthesis and Processing in Eukaryotes

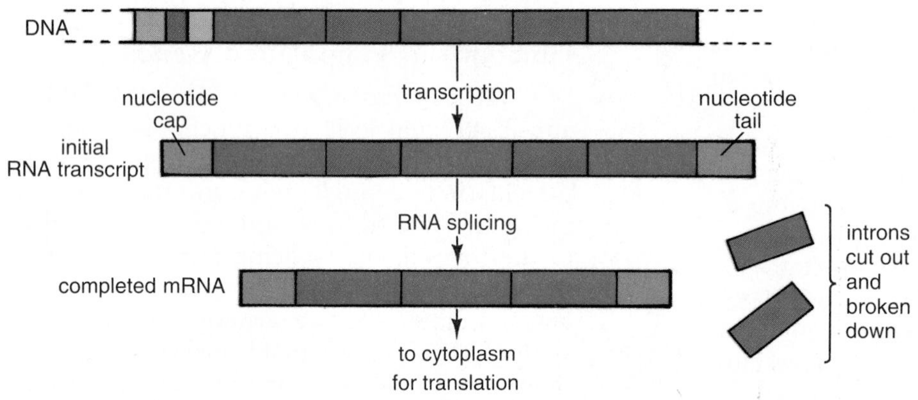

RNA polymerase transcribes both the exons and introns, producing a long RNA molecule. More nucleotides are added at the beginning and end of this initial RNA transcript. Enzymes then cut out the RNA introns and splice together the exons to form the true mRNA, which moves out of the nucleus to be translated on the ribosomes.

Figure 13-12 Eukaryotic gene structure and the processing of RNA.

plasm. The estrogen–protein complex enters the nucleus, where it binds to DNA, probably near the enhancer for the albumin gene. This makes it easier for RNA polymerase to contact the promoter of the albumin gene. Rapid transcription occurs and albumin is synthesized. Similar activation of genes by steroid hormones occurs in other animals, including humans.

Regulation of Parts of Chromosomes

Certain parts of chromosomes are in a highly condensed, compact state, in which the DNA seems to be inaccessible to RNA polymerase. Some of these regions are structural parts of chromosomes that don't contain genes. For example, condensed DNA is usually found at the centromeres that hold sister chromatids together during cell division. In other cases, DNA can change from the condensed state to a looser configuration that allows genes to be tran-

scribed, depending on the stage in the life cycle of the animal or the type of cell (Fig. 13-14).

Regulation of Entire Chromosomes: Inactivation of X Chromosomes

Female mammals have two homologous X chromosomes. However, only one X chromosome is available for transcription in any given cell. The other entire X chromosome is condensed into a tight mass. In the light microscope, the inactivated X chromosome shows up as a dark spot in the nucleus called a **Barr body,** after its discoverer, Murray Barr (Fig. 13-15). Apparently, both X chromosomes are in the "loose" state in fertilized eggs. After a few cell divisions, one or the other condenses and forms a Barr body. (The germ cells in the ovaries are an exception; here, both X chromosomes are in the loose, uncondensed state.) Which X chromosome is inactivated in any given cell is random, but all its daughter cells will then have the

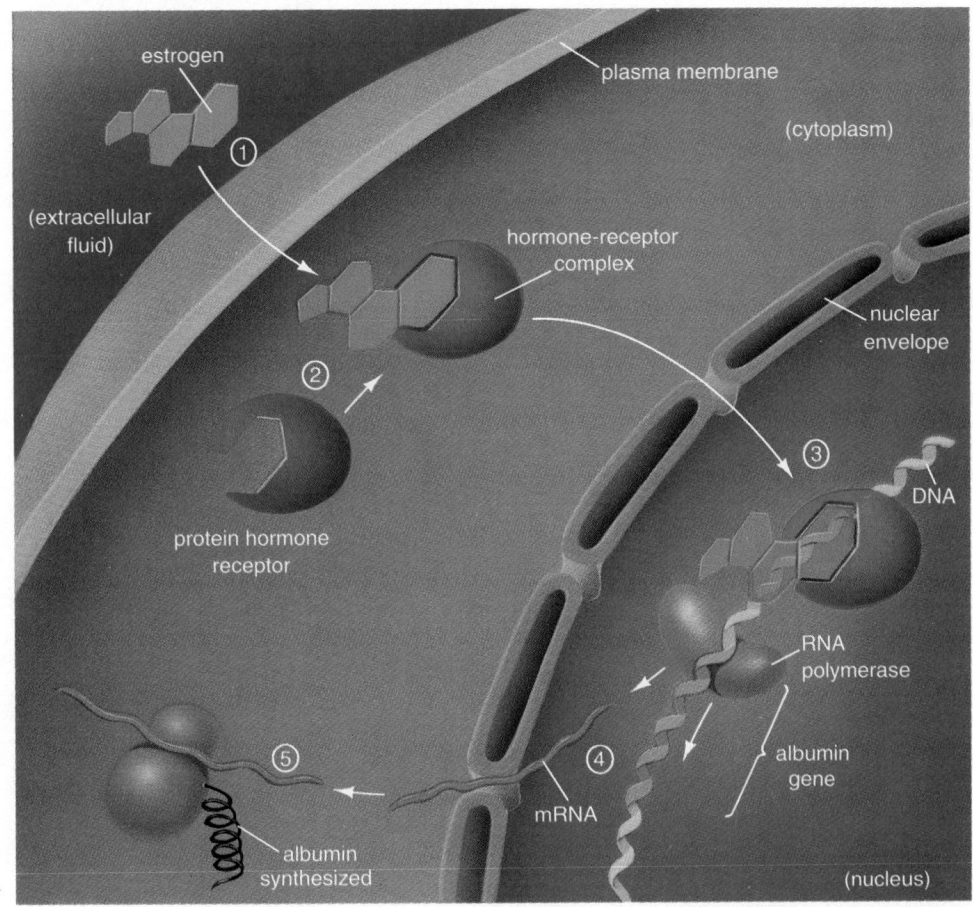

Figure 13-13 Stimulation of transcription by estrogen, a steroid hormone. (1) Estrogen diffuses through the plasma membrane into the cytoplasm. (2) Estrogen combines with a receptor protein. (3) The hormone–receptor complex enters the nucleus and binds to the enhancer of the albumin gene. (4) RNA polymerase transcribes the albumin gene. (5) The mRNA leaves the nucleus and is translated into albumin.

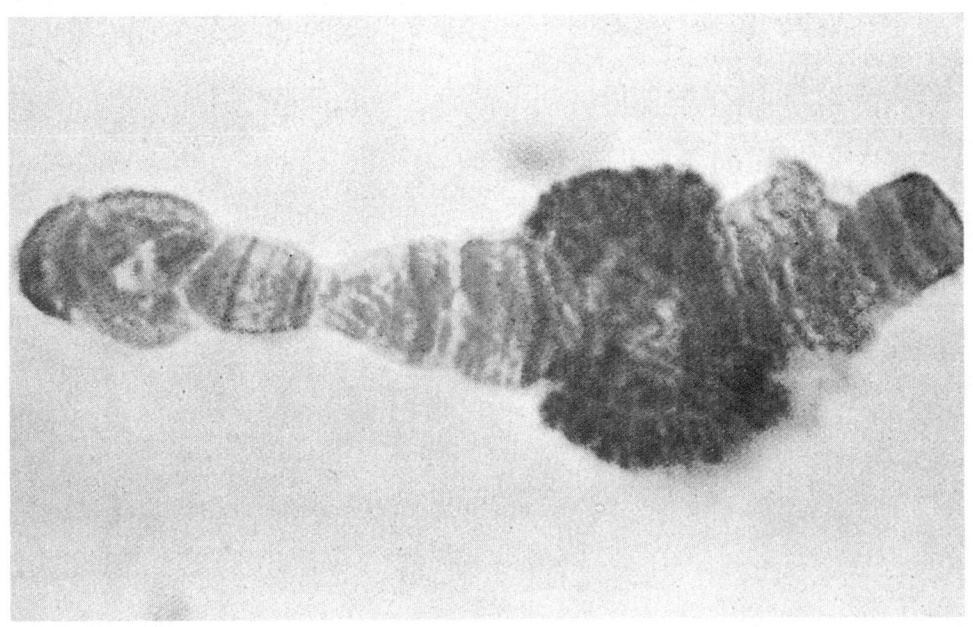

Figure 13-14 A giant chromosome of a midge (a small fly). Most of the chromosome is in a tight, condensed state, while a few regions have puffed out in a much looser configuration. The purple stain binds to RNA, which is mostly transcribed from the loose, puffed regions of DNA. Changing patterns of puffing and condensation of the chromosomes reflect different genes turning on and off during development.

A CLOSER LOOK

At Rube Goldberg Genetics: Making New Proteins from Old Parts

The cartoonist Rube Goldberg created marvelous fictional contraptions to perform simple functions, such as sharpening a pencil (Fig. E13-1). The beauty of a Goldberg "invention" was that you could have really built it, if you wanted to, from ordinary household items and scrap lumber. In many respects, evolution works a lot like Rube Goldberg, modifying ordinary, preexisting structures to perform new functions. Consider, if you will, an elephant's ears: huge flaps of skin laced with blood vessels. The selective advantage of "ordinary" external ears, like those on a wolf or a deer, is that they funnel sound into the ear canal, helping animals locate the source of sounds. From these humble beginnings, the enormous ear flaps on an elephant have been adapted for quite a different role: dissipating heat in the African savanna. They also have the added benefit of looking most impressive when an elephant threatens a rival.

In evolution, it is always difficult to devise anything from scratch. Inventing a new protein, say of 200 amino acids, means putting together a string of 600 nucleotides in DNA in just the right sequence. Mutations cannot assemble a new, useful string of 600 nucleotides in one fell swoop. Instead, evolution has hit upon a very different, much quicker, and quite effective strategy: making new proteins from old parts.

Many proteins consist of several subunits, each with a completely different function. A protein that actively transports potassium across the cell membrane, for example, might have three subunits: one to anchor the protein in the membrane, one to bind potassium ions, and one to bind ATP to power the transport (Fig. E13-2a). The membrane anchor and ATP-binding subunits are modules that might be useful for other transport proteins as well. Exchange the potassium-binding subunit for a calcium-binding subunit, and voila! The cell has a calcium transport protein (Fig. E13-2b).

How can an organism exchange subunits among various proteins? It turns out that chromosomes are not quite the stable, nearly-perfectly-replicating structures that geneticists pictured not very long ago. Chromosomes occasionally break in two, and one end may become attached to a completely different chromosome. Sometimes segments pop out of one chromosome and insert themselves in another. These DNA rearrangements provide a mechanism for rearranging protein subunits as well.

Since genes are segments of DNA on chromosomes, scrambling parts of a chromosome will often scramble parts of genes. You might think that this would completely ruin the genetic instructions in the rearranged genes. However, remember that a eukaryotic gene in-

Figure E13-1

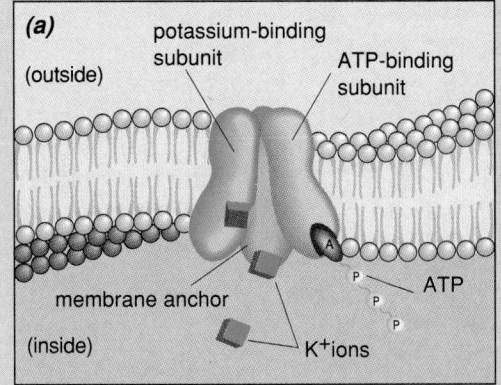

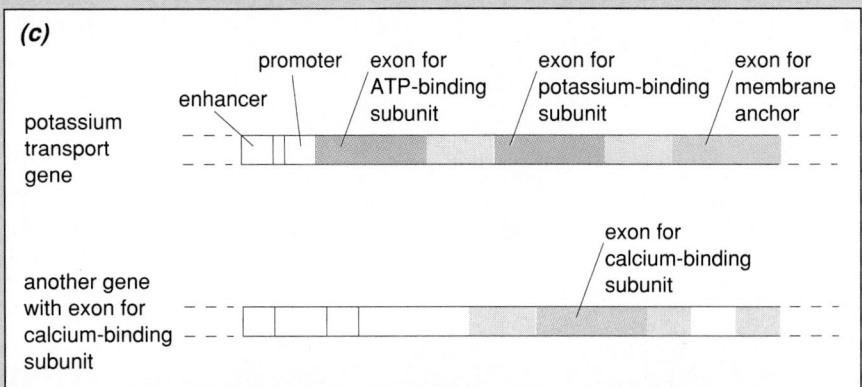

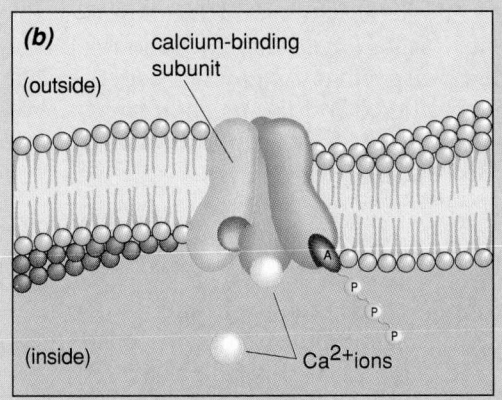

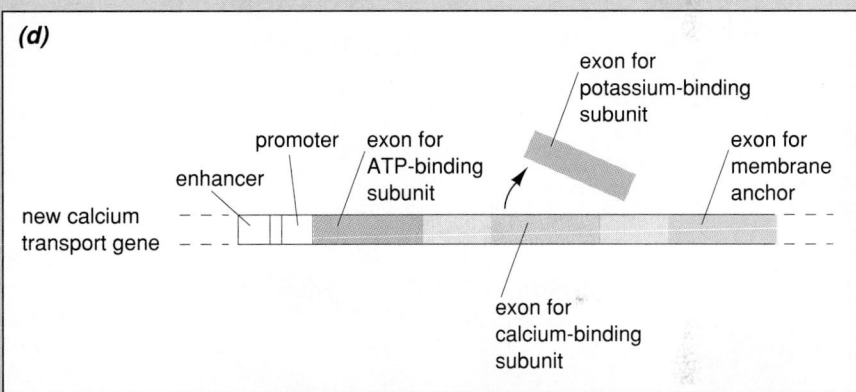

Figure E13-2 Shuffling exons to make new functional proteins.
(a) Hypothetical protein subunits that might make up an active-transport protein for potassium ions.
(b) Substituting a calcium-binding subunit for the potassium-binding subunit creates a calcium-transporting protein.
(c) The potassium-transport gene might have three exons, one coding for each subunit of the transport protein. Another gene, even on another chromosome, has an exon coding for a calcium-binding protein.
(d) If the potassium-binding exon is cut out and replaced with the calcium-binding exon, a new gene is formed, now coding for a calcium-transporting protein.

cludes both expressed regions (exons) and intervening regions (introns). Exons often code for individual protein subunits (Fig. E13-2c). What if chromosomes preferentially break within introns, so that intact, functional exons can be moved from one chromosome to another (Fig. E13-2d)? By interchanging exons among genes, a eukaryotic organism can create new genes and thereby adapt more quickly to changing environmental conditions. Molecular biologists have recently found evidence that this is probably just what happened in the case of ATP-powered transport molecules and enzymes, several of which have essentially the same ATP-binding subunit.

Rube Goldberg would have been proud.

(a)

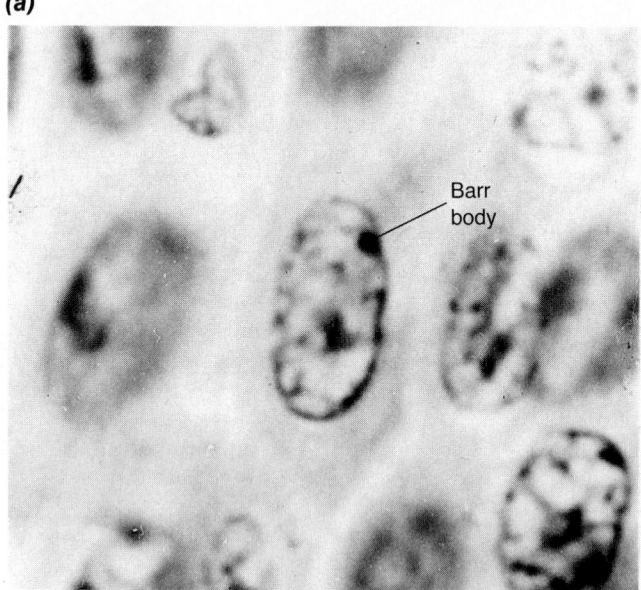

Barr
body

(b)

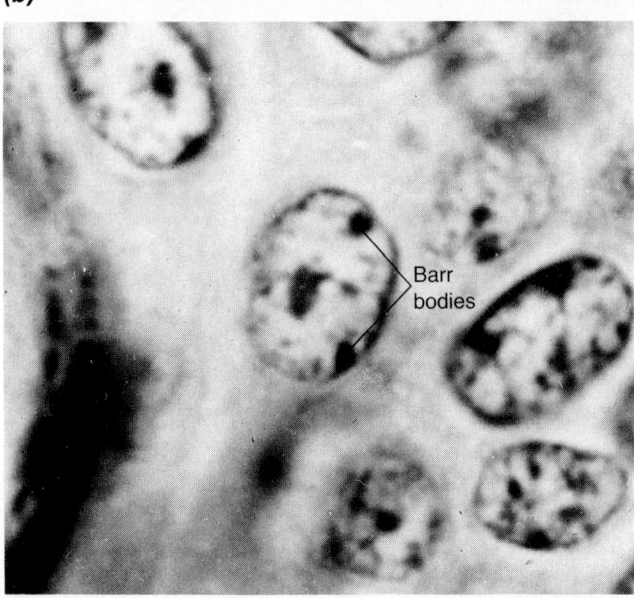

Barr
bodies

Figure 13-15 Nuclei of human cells, stained to show Barr bodies (arrows). Generally, human nuclei have one fewer Barr body than the number of X chromosomes in the nucleus. **(a)** These nuclei, from a normal XX woman, each have one Barr body. Nuclei from men (XY) lack Barr bodies. **(b)** Women with extra X chromosomes have two or more Barr bodies. Female athletes in international sports are sometimes required to submit to examination for Barr bodies, to ensure that they are normal XX women (one Barr body), and not men posing as women (no Barr bodies).

same condensed chromosome. As a result, female mammals (including women) are mosaics: patches of cells with one X chromosome active are interspersed with patches in which the other X chromosome is active. This is strikingly evident in calico and tortoise-shell cats (Fig. 13-16), which are almost invariably female. The X chromosome contains a gene for fur color, with orange and black being the common alleles. Males, with only one X chromosome, are usually either orange *or* black. Females, with two X chromosomes, can be orange (both X chromosomes carry the orange allele), black (both carry the black allele), or orange *and* black in patches (one X chromosome has the orange allele and one has the black allele).

Mendelian Genetics Revisited

You have now learned enough about the mechanisms of heredity to come full circle and reexamine Mendelian genetics, mutations, and evolution from a new, molecular perspective. As you will see, the various types of inheritance and the mechanisms of evolution are outgrowths of the replication and transcription of DNA.

Mutations and the Genetic Code

A **gene mutation** is a change in the sequence of nucleotides in DNA. If a mutation occurs in cells whose progeny become gametes, it may be passed on to future generations. But how does a change in nucleotide sequence affect the organism that inherits the mutated DNA? As the essay on "Cracking the Genetic Code" pointed out, deletions and insertions can have catastrophic effects on a gene, since all the codons following the deletion or insertion will be misread. The enzyme synthesized from such misread directions is almost certain to be nonfunctional.

Point mutations, in which one nucleotide is replaced by another, may lead to more subtle effects. Four different categories of effects may result from point mutations (Table 13-2). As a concrete example, let's consider possible mutations of the DNA sequence CTC, which codes for glutamic acid.

1. **A mutation may not change the amino acid sequence of the encoded protein.** Remember that the genetic code is degenerate, so that one amino acid may be encoded by several different codons. If a mutation changes CTC to CTT, the new triplet still codes for glutamic acid. Therefore, the protein synthesized from the mutated gene remains the same.

Figure 13-16 Male and female cats from the same litter. The female is a calico, with both orange and black patches of fur. (The predominant white color is due to an entirely different gene, which prevents color formation.) The male is simply orange and white, while another male of the same litter was black and white. An occasional calico or tortoiseshell cat appears to be phenotypically male. These cats are almost always sterile, and chromosome analysis usually finds an XXY genotype.

Table 13-2 Examples of Functional Outcomes of Single Substitutions in the Glutamic Acid Codon of DNA

	DNA	mRNA	Amino Acid	Properties	Effect
Original sequence	CTC	GAG	Glutamic acid	Hydrophilic, acidic	—
Mutation 1	CTT	GAA	Glutamic acid	Hydrophilic, acidic	—
Mutation 2	CTA	GAU	Aspartic acid	Hydrophilic, acidic	Neutral
Mutation 3	CAC	GUG	Valine	Hydrophobic, neutral	Lose water solubility; possibly catastrophic
Mutation 4	ATC	UAG	Stop codon	Ends translation	Only synthesize part of protein; catastrophic

2. A mutation may code for an amino acid that is functionally equivalent to the original amino acid. Many proteins have large "background" regions whose exact amino acid sequence is rela-

tively unimportant. For example, in hemoglobin, the amino acids on the outside of the protein must be hydrophilic to keep the protein dissolved in the cytoplasm of red blood cells. Exactly *which* hydrophilic amino acids are on the outside probably doesn't matter too much. A mutation from CTC to CTA, replacing glutamic acid (hydrophilic) with aspartic acid (also hydrophilic), probably wouldn't affect the solubility of hemoglobin. Mutations that do not detectably change the function of the encoded protein are called **neutral mutations.**

3. **A mutation may encode for a functionally different amino acid.** A mutation from CTC to CAC replaces glutamic acid (hydrophilic) with valine (hydrophobic). This substitution, which is the genetic defect in sickle cell anemia (see Chapter 15), causes hemoglobin molecules to stick to each other, clumping up and distorting the shape of the red blood cells. This is a potentially fatal mutation.

4. **A mutation may produce a stop codon.** An inappropriate stop codon will cut short the translation of mRNA before the protein is finished. This is almost certainly catastrophic to protein functioning; if the protein is essential to life, as hemoglobin is, the mutation will be lethal to the organism.

As you might expect, random changes in nucleotides usually result in mutated proteins that function less effectively than the normal protein. On the rare occasion when the altered protein functions better, or functions in a way that is adaptive in a new environment, the mutation will confer a selective advantage on its possessor. In this way new alleles arise, are tested by the environment, and contribute to evolution (see Chapters 17 and 18).

Relationships Among Alleles

The concept that each gene codes for a specific protein is a powerful tool in understanding the multitude of relationships among alleles.

Dominance and Recessiveness

Consider the situation in which a single gene controls the expression of a single trait. There are two different alleles of the gene, each producing a different form of the trait, and one allele is completely dominant to the other allele. To see how such a system might work on the molecular level, let's look at the inheritance of body fat color in rabbits.

In domestic rabbits, body fat is normally white, but rabbits that are homozygous for a recessive allele have yellow fat. Why? As you know, rabbits eat plants. Most plants contain a yellow pigment called

xanthophyll, which is fat-soluble. If not broken down, xanthophyll dissolves in the rabbit's fat, coloring it yellow. Normal rabbits synthesize an enzyme that breaks down xanthophyll to a colorless compound, and so these rabbits have white fat. The yellow-fat allele is a mutation that renders the enzyme nonfunctional. If a rabbit has one normal allele and one mutant allele, the normal allele directs the synthesis of enough normal enzyme to degrade the xanthophyll in the rabbit's diet completely. Therefore, the body fat of heterozygotes will still be white, and the normal allele is dominant to the mutant allele. If a rabbit is homozygous recessive for the mutant allele, it produces no functional xanthophyll-digesting enzymes, so xanthophyll from its diet is not metabolized. The xanthophyll dissolves in the fat, coloring it yellow.

In general, **dominant alleles direct the synthesis of functioning enzymes.** An organism with one dominant allele synthesizes enough enzyme to produce a phenotype that is indistinguishable from the phenotype of organisms with two dominant alleles. In contrast, **recessive alleles usually direct the synthesis of nonfunctional enzymes.**

Incomplete Dominance

The simplest form of incomplete dominance is very similar to complete dominance: one allele directs the synthesis of a functioning enzyme, and the other allele directs the synthesis of a completely nonfunctional enzyme. The difference between complete and incomplete dominance results from the effects of one versus two copies of the functional allele. In incomplete dominance, the amount of enzyme synthesized under the direction of a single functional allele is not enough to catalyze all the reactions that would be needed to produce a dominant phenotype.

For example, the red allele in snapdragon flowers (see Fig. 11-16) probably directs the synthesis of an enzyme that produces a red pigment. If a plant has two red alleles, its flowers produce a lot of red pigment, and are colored red. If a plant has only one red allele, its flowers produce less red pigment, and are colored pink.

Multiple Alleles and Codominance

Mutations may occur anywhere in a gene. Therefore, different organisms may suffer quite different mutations in the same gene. These different mutations may all produce slightly different proteins, giving rise to multiple alleles of the same gene. In some circumstances, each of these multiple alleles may be detectable phenotypically, as is the case with the human ABO blood groups (see Fig. 11-17). The I^A and I^B alleles both code for slightly different, functional enzymes. The I^B enzyme attaches galactose to red blood cell membranes, whereas the I^A enzyme attaches a slightly different compound, galactosamine. The *i* allele is a mutation producing a nonfunctional enzyme that can't attach anything to the red blood cells. People with $I^B I^B$ or $I^B i$ genotypes have red blood cells that bear galactose, and have type B blood. Those with $I^A I^A$ or $I^A i$ genotypes have red blood cells that bear galactosamine, and have type A blood. The $I^A I^B$ genotype results in red blood cells with both galactose *and* galactosamine (type AB blood). Finally, the red blood cells of *ii* individuals bear neither compound (type O blood). Since the phenotypes produced by both the I^A and I^B alleles can be detected in people with type AB blood (by blood clotting reactions), the I^A and I^B alleles are called codominant.

SUMMARY OF KEY CONCEPTS

The Relationship Between Genes and Proteins

Genes are segments of DNA on chromosomes. The ultimate cellular product encoded by a gene is usually a protein. Therefore, with a few exceptions, the specific nucleotide sequence of a gene encodes the amino acid sequence of a protein or a part of a protein.

From DNA to Protein

Information flows from DNA to proteins in a two-step process. (1) *Transcription:* the information contained in the DNA of a gene is copied into messenger RNA (mRNA). (2) *Translation:* the sequence of nucleotides in mRNA provides the information needed to synthesize a protein with the amino acid sequence specified by the nucleotide sequence of the DNA of the gene. The overall sequence of information flow is summarized in Figure 13-17.

The Genetic Code

The sequence of bases in mRNA carries the genetic code for the amino acid sequence in a protein. Sequences of three nucleotides in mRNA, called codons, specify the amino acids of the protein. There are also start and stop codons that signal the beginning and end of protein synthesis.

RNA: Intermediary in Protein Synthesis

Synthesizing proteins from the information in DNA requires RNA molecules as intermediates. RNA is transcribed from one DNA strand by the enzyme RNA polymerase. RNA polymerase recognizes a region of DNA called the promoter as the beginning of a gene. Starting there, RNA polymerase uses free ribose nucleotides to synthesize an RNA strand that is complementary to the DNA of the gene.

There are three types of RNA. The sequence of bases in messenger RNA (mRNA) carries the information needed to determine the amino acid sequence of a protein. Ribosomal RNA (rRNA) and proteins form ribosomes. Ribosomes consist of large and small subunits. The small subunit has binding sites for two codons of mRNA. The large subunit bears a catalytic site that forges the peptide bond between amino acids as a protein is synthesized, and two binding sites for transfer RNA (tRNA). There are at least 20 different tRNAs. Each binds a specific amino acid and transports it to a ribosome. A set of three bases in tRNA, called the anticodon, is complementary to the codon in mRNA that specifies the amino acid borne by that tRNA.

Protein Synthesis

Protein synthesis occurs in the following sequence:

1. Messenger RNA is transcribed from a gene. The mRNA leaves the nucleus and travels to a ribosome.
2. Two codons of mRNA bind to the small subunit of the ribosome. The first codon is the "start" codon, which signals where protein synthesis is to begin.
3. Transfer RNAs, carrying their amino acids, move to the mRNA. The anticodons of two tRNA molecules base pair with the two codons of mRNA, and the tRNAs bind to the large ribosomal subunit.
4. The large subunit of the ribosome catalyzes the formation of a peptide bond between the amino acids carried by the two tRNA molecules. The "first" amino acid detaches from its tRNA. The chain of two amino acids remains attached to the "second" tRNA.
5. The "first" tRNA leaves the ribosome. The ribosome moves one codon over on the mRNA. A "third" tRNA, with its attached amino acid, base pairs with the third codon on mRNA. A new peptide bond is formed between the amino acid of the "third" tRNA and the dipeptide still attached to the "second" tRNA.
6. This process continues until a "stop" codon is reached, whereupon the mRNA and the newly formed protein leave the ribosome.

Gene Regulation

Which genes are transcribed in a cell at any given time is regulated by the function of the cell, the developmental stage of the organism, and the environment. Access of RNA polymerase to the promoter of a gene may be either prevented or enhanced by other molecules in the cell, including nutrients and hormones. Large parts of chromosomes may also be rendered inaccessible to RNA polymerase by changes in DNA structure.

Eukaryotic genes consist of exons (sequences of nucleotides that encode the information for the amino acid sequence of a protein) and introns (nucleotides that separate exons and do not themselves encode amino acid sequences). Therefore RNA transcribed from eukaryotic genes must be processed to form a true mRNA, by cutting out the introns and splicing together the exons.

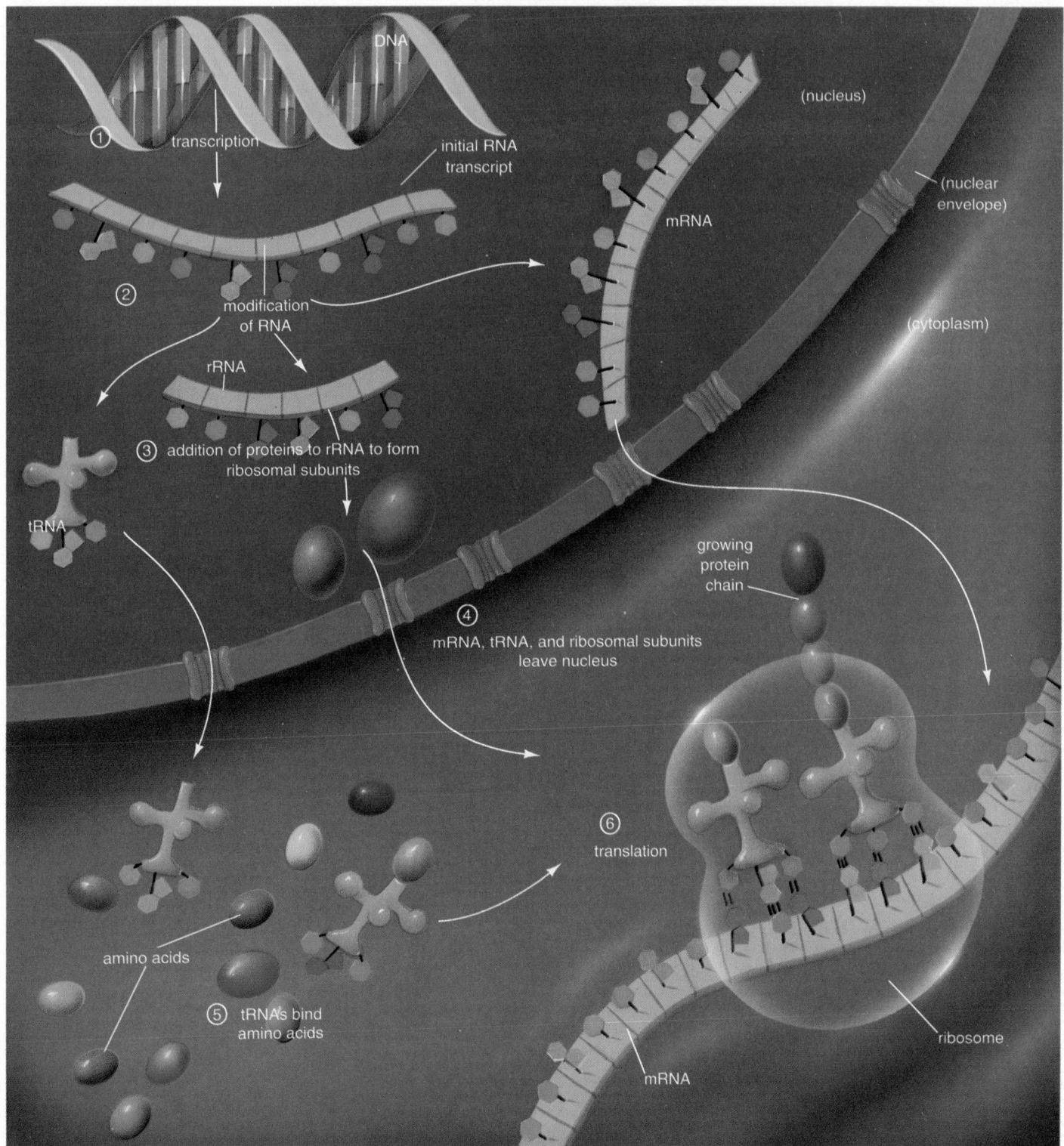

Figure 13-17 An overview of the molecules and pathways involved in information flow from DNA to protein. (1) DNA is transcribed into RNA. (2) In eukaryotic cells, most RNA transcripts are modified by cutting and splicing to produce the final molecules of mRNA, rRNA, and tRNA. Although for convenience the diagram shows only one initial RNA transcript, the three types of RNA are made by processing different initial RNA molecules. (3) Ribosomal proteins and rRNA bind together to form large and small subunits of ribosomes. (4) Ribosome subunits, tRNA, and mRNA leave the nucleus through pores in the nuclear envelope. (5) In the cytoplasm, tRNAs bind their appropriate amino acids. (6) Messenger RNA binds to ribosome subunits; tRNAs match up to the proper codons of mRNA, and proteins are synthesized.

Mendelian Genetics Revisited

Molecular genetics explains many aspects of Mendelian genetics and evolution. A mutation is a change in the sequence of nucleotides in DNA. The different alleles of a gene arise by mutation. Mendelian relationships among alleles are a consequence of the molecular nature of the gene and its resulting protein product. For example, a dominant allele codes for an amino acid sequence that results in a functioning enzyme. A recessive allele is usually a mutation that codes for a nonfunctioning enzyme.

GLOSSARY

anticodon: a sequence of three nucleotides in transfer RNA that is complementary to the three nucleotides of a codon of messenger RNA.

Barr body: an inactive X chromosome found in somatic cells of mammals that have at least two X chromosomes (usually females). The Barr body usually appears as a dark spot in the nucleus.

codon: a sequence of three nucleotides of messenger RNA that specifies a particular amino acid to be incorporated into a protein. Certain codons also signal the beginning and end of protein synthesis.

degeneracy: the property of the genetic code whereby several codons may specify the same amino acid.

enhancer: in eukaryotes, a stretch of DNA that influences the access of RNA polymerase to the promoter region of a structural gene.

exon: a segment of DNA in a eukaryotic gene that codes for amino acids in a protein (see also *intron*).

genetic code: the collection of codons of mRNA, each of which directs the incorporation of a particular amino acid into a protein during protein synthesis.

inactivation: a process whereby certain chromosomes or parts of chromosomes are converted into a dense mass, preventing transcription.

intron: a segment of DNA in a eukaryotic gene that does not code for amino acids in a protein.

messenger RNA (mRNA): a strand of RNA, complementary to the DNA of a gene, that conveys the genetic information in DNA to the ribosomes to be used during protein synthesis. Sequences of three nucleotides (codons) in mRNA specify particular amino acids to be incorporated into a protein.

neutral mutation: a mutation (change in DNA sequence) that has little or no phenotypic effect.

one-gene, one-protein hypothesis: the proposition that each gene encodes the information for the synthesis of a specific protein.

operator: in prokaryotes, a segment of DNA that controls access of RNA polymerase to the promoter.

operon: a unit of organization of prokaryotic chromosomes, in which several genes that specify related functions (e.g., enzymes in the same biosynthetic pathway) are grouped together on the chromosome, are transcribed at the same time, and are regulated together.

promoter: a specific sequence of DNA to which RNA polymerase binds, initiating gene transcription.

regulatory gene: a gene that controls the timing or rate of transcription of other genes.

repression: in the lactose operon, the condition that the genes of the operon are not transcribed unless specifically activated by the presence of lactose.

repressor protein: in the lactose operon, a protein that binds to the operator site, thereby preventing access of RNA polymerase to the promoter.

ribonucleic acid (RNA): a single-stranded nucleic acid molecule composed of nucleotides, each of which consists of a phosphate group, the sugar ribose, and one of the bases adenine, cytosine, guanine, or uracil.

ribosomal RNA (rRNA): a type of RNA that combines with proteins to form ribosomes.

ribosome: an organelle consisting of two subunits, each composed of ribosomal RNA and protein. Ribosomes are the site of protein synthesis, in which the sequence of nucleotides of messenger RNA is translated into the sequence of amino acids in a protein.

RNA polymerase: an enzyme that catalyzes the covalent bonding of free RNA nucleotides into a continuous strand, using RNA nucleotides that are complementary to those of a strand of DNA.

structural gene: a gene that codes for a protein used by the cell for purposes other than gene regulation. Structural genes code for enzymes or for proteins that are structural parts of a cell.

start codon: a codon in messenger RNA that signals the beginning of protein synthesis on a ribosome.

stop codon: a codon in messenger RNA that stops protein synthesis and causes the completed protein chain to be released from the ribosome.

transcription: the synthesis of an RNA molecule from a DNA template.

transfer RNA (tRNA): a type of RNA that (1) binds to a specific amino acid and (2) bears a set of three nucleotides (the anticodon) complementary to the mRNA codon for that amino acid. Transfer RNA carries its amino acid to a ribosome during protein synthesis, recognizes a codon of mRNA, and positions its amino acid for incorporation into the growing protein chain.

translation: the process whereby the sequence of nucleotides of messenger RNA is converted into the sequence of amino acids of a protein.

STUDY QUESTIONS

1. Draw an RNA nucleotide. How does RNA differ from DNA?
2. Describe RNA synthesis. Where does it occur?
3. What are the three types of RNA? What are their functions?
4. Define the following terms: genetic code; codon; anticodon. What is the relationship between the nucleotides in DNA, the codons of mRNA, and the anticodons of tRNA? What does it mean to say that the genetic code is degenerate?
5. Diagram and describe protein synthesis.
6. What is an operon? Are operons found in prokaryotes, eukaryotes, or both?
7. Describe the process of gene regulation in prokaryotes, using the lactose operon as an example.
8. Diagram the structure of a eukaryotic gene, including both the internal structure of the gene and the nearby control regions of the chromosome.
9. How is mRNA formed from a eukaryotic gene?
10. How do steroid hormones regulate eukaryotic genes?
11. Describe the molecular relationships that are involved in genetic dominance and recessiveness. What is the nature of the enzymes synthesized from the directions encoded in many recessive alleles?
12. Describe four functional consequences of substitution mutations.

DISCUSSION QUESTIONS

1. A preview question for Chapter 14: The same genetic code is used in most organisms, including bacteria and people. Suppose that the gene that encodes a human protein, insulin for example, were inserted into a bacterium. If the bacterium transcribes this gene into messenger RNA, would you expect that the amino acid sequence of the protein would be the same as the protein synthesized in a human being?

2. If you look at the genetic code (Table 13-1), you will notice that most of the codons that specify the same amino acid have the same first two bases, and differ only in the third base. Now, most mutations that cause a phenotypic effect are harmful. If you sequenced the DNA for, say, the hemoglobin gene, from 100 people, in which codon position do you think you would find the most differences? Why?

SUGGESTED READINGS

Chambron, P. "Split Genes." *Scientific American*, May 1981. The segmented nature of eukaryotic genes is described.

Crick, F. H. C. "The Genetic Code." *Scientific American*, October 1962. The determination of the triplet nature of the genetic code.

Crick, F. H. C. "The Genetic Code: III." *Scientific American*, October 1966. The genetic code is completely solved.

Darnell, J. E. "RNA." *Scientific American*, October 1985. A description of the types of RNA, their synthesis, and their processing.

Nirenberg, M. W. "The Genetic Code: II." *Scientific American*, March 1963. Nirenberg deciphered much of the genetic code. Here he describes some of those experiments and their reuslts.

Steitz, J. A. "Snurps." *Scientific American*, June 1988. "Snurps" are small nuclear ribonucleoproteins, which snip introns out of eukaryotic pre-messenger RNA. Snurps are probably examples in which the RNA part of the molecule is the real catalyst.

14

Molecular Genetics and Biotechnology

Recombinant DNA technologies can produce genetically identical organisms, such as these calves.

Biotechnology—almost every state and large city in the United States tries frantically to lure "biotech" companies that provide high-paying, low-pollution jobs. Virtually every week newspapers feature articles discussing the potential health benefits of some new medical advance in the works, based on biotechnology. Biotechnology often seems to be portrayed as both the economic and medical miracle of our times. And yet furious debate rages about whether "ice-minus" bacteria should be sprayed onto crops, or whether milk from cows treated with a recombinant hormone should be sold in supermarkets. What exactly *is* biotechnology? Is it boon or bane or both?

Biotechnology is the manipulation of the molecular basis of inheritance by methods collectively called recombinant DNA technology. Biotechnology is usually practiced to achieve one or more of three goals:

1. **to understand more about the processes of inheritance and gene expression;**
2. **to provide better understanding and treatment of various diseases, particularly genetic disorders; and**
3. **to generate economic benefits, including improved plants and animals for agriculture and efficient production of valuable biological molecules.**

The features of biotechnology that provide both vast promise and potential threat are the specificity with which biotechnology can, or will soon be able to, direct genetic changes; the speed with which genetic changes can be made; and the ability to transfer genetic material between species.

This chapter has three major themes. First, we look at a few important recombinant DNA technologies, briefly examining some of the methods that molecular biologists use to manipulate genes. Second, we discuss a few of the applications, both real and potential, of biotechnology. Third, we explore some of the practical and ethical issues that the use of biotechnology may raise.

Before we begin our discussion of human-directed biotechnology, however, it will be useful to provide a bit of background on what we might call "natural biotechnology." Let us begin, then, with a survey of some naturally occurring methods of DNA recombination.

DNA Recombination in Nature

DNA recombination, whether by natural events or human intervention, involves two distinct processes:

(1) changing the nucleotide composition of the DNA of a single cell, a few cells, or an entire organism, and (2) selecting valuable DNA combinations.

Naturally Occurring Methods of Recombining DNA

You are already familiar with two major methods of "recombinant DNA," although we called them "genetic recombination" and "sexual recombination" when we discussed them in Chapters 10 and 11. Genetic recombination occurs through crossing over during meiosis I, when genes from a maternal chromosome and a paternal chromosome produce a new chromosome with a sequence of genes that may never have occurred before. In sexual recombination, chromosomes from two different organisms combine to produce offspring that are often genetically unique. These natural recombinations, of course, usually occur within a single species. Many people have a tendency to consider these single-species recombinations as "natural" and therefore good, while between-species recombinations performed in the lab are "unnatural" and possibly intrinsically bad. However, between-species recombinations occur in nature, too.

Recombination Between Bacterial Species

Bacteria employ several methods of recombination that allow gene transfer between unrelated species. As you know, transformation allows bacteria to pick up free DNA from the environment (Chapter 12). Sometimes, the free DNA acquired during transformation is part of the chromosome of another bacterium, including DNA from a bacterium of another species. Transformation may also occur when bacteria pick up tiny circles of DNA called **plasmids** (Fig. 14-1). Plasmids, which range in size from about 1,000 to 100,000 nucleotides, are self-replicating "parasites" normally found in the cytoplasm of many bacteria. A single bacterium may contain dozens or even hundreds of copies of a plasmid. Although the bacterium's "own" chromosome contains all the genes that it normally needs for survival, plasmid genes may be useful, too. For example, some plasmids contain genes that code for enzymes that digest certain antibiotics, such as penicillin or ampicillin. In hospitals, such plasmids are obviously extremely valuable to their bacterial hosts (see Chapter 21). When a plasmid-containing bacterium dies, its plasmids may be liberated into the environment and transform other living bacteria. In fact, the acquisition of plasmids is probably the most common form of bacterial transformation.

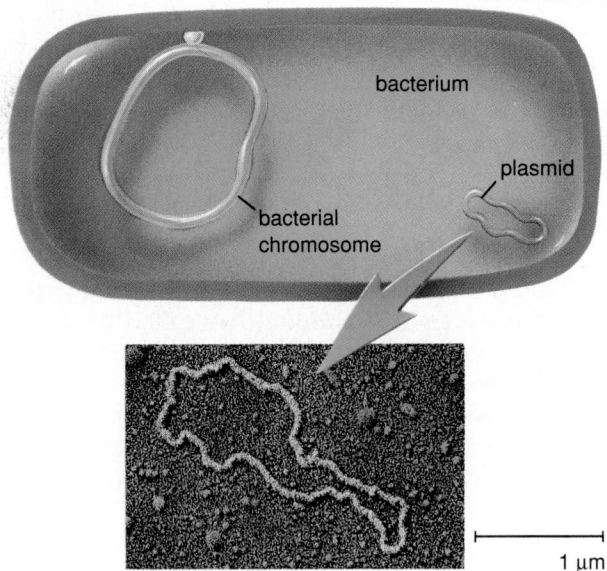

Figure 14-1 An electron micrograph of a single plasmid. These tiny rings of DNA have become important tools in recombinant DNA research.

Recombination Between Eukaryotic Species

Recent evidence suggests that viruses sometimes transfer genes among eukaryotic organisms. The DNA of certain viruses can insert itself into a chromosome of its eukaryotic host, and exist there quietly for days, months, or even years. Then, perhaps in response to a stressful stimulus to the host, the viral DNA leaves the chromosome, occasionally taking a bit of the eukaryotic DNA along with it. The viral DNA then takes over the host cell metabolism, replicates itself, and directs the synthesis of new viruses. The offspring viruses thus may include some host DNA. If one of these offspring infects a new host of a different species, it may insert itself, along with the piece of DNA from the former host, into a chromosome of its new host. In this way, the new host may acquire some genes that originally belonged to an unrelated species.

Recombination and Evolution

Recombination, by whatever mechanism, changes the genetic makeup of organisms. As is the case with mutations, natural recombination is random and undirected—a bacterium, for example, does not deliberately take up a plasmid that contains an antibiotic-resistance gene so that it can flourish in hospitalized patients. Naturally occurring DNA recombinations are tested by natural selection, just as

mutations are tested. Probably most recombinations prove to be harmful or neutral. A few, however, prove helpful in a particular environment. The organisms with the new DNA combinations thrive and pass on the new combinations to their offspring.

Recombination in Nature and the Laboratory

Before we begin our discussion of recombinant DNA technologies, let's compare natural and laboratory DNA recombinations for similarities and differences.

1. Both natural and laboratory DNA recombinations involve exchanges of DNA between organisms, including between species.
2. Naturally occurring DNA recombinations are relatively random and undirected; generally speaking, specific genes are not preferentially moved, and there is no "goal" that drives DNA movements. In laboratory DNA recombinations, specific pieces of DNA are moved between deliberately chosen organisms, to achieve a specific goal.
3. The "usefulness" of naturally occurring DNA recombinations is determined by natural selection. Human interests determine the usefulness of laboratory DNA recombinations.

Recombinant DNA Technology

Within the past few years, the technologies and applications of recombinant DNA have mushroomed. Rather than attempting to summarize all the technologies, we will follow the sequence of procedures that might be followed in actually using recombinant DNA methods to solve a particular problem or produce a specific product. Although many other techniques are also commonly used, one logical sequence of activities would be the following: (1) produce a "DNA library" of an organism; (2) identify individual genes of interest; (3) produce a copy or (preferably) many copies of the gene; and (4) insert the gene into the desired organism and regulate the expression of the gene in a useful way.

Building a DNA Library

The first task in recombinant DNA technology is to produce a **DNA library**—a readily accessible, easily duplicable assemblage of all the DNA of a particular organism. Now, obviously, a human being contains a complete set of human genes; however, they are not very accessible nor can they be easily duplicated the many thousands of times that are required for experiments. This problem has been neatly solved using restriction enzymes, plasmids, and bacteria.

Restriction Enzymes

Many bacteria produce **restriction enzymes** that sever DNA at particular nucleotide sequences. In nature, restriction enzymes defend bacteria against bacteriophage invasion, by cutting apart the phage DNA. The host bacteria protect their own DNA against being cut by the enzymes, probably by attaching methyl groups to some of the DNA bases. Restriction enzymes thus "restrict" phage infections to those phage that are not harmed by the enzymes.

As shown in Figure 14-2, many restriction enzymes sever palindromic DNA, which reads the same in one direction on one strand as it reads in the reverse direction on the other strand (a palindrome in English is a word that reads the same forward and backward, such as "madam"). Furthermore, the DNA is cut between the same two bases on the two strands, in this case between guanine and adenine. For the enzyme illustrated in Fig. 14-2, this results in two pieces of DNA, one with a single-stranded end reading TTAA

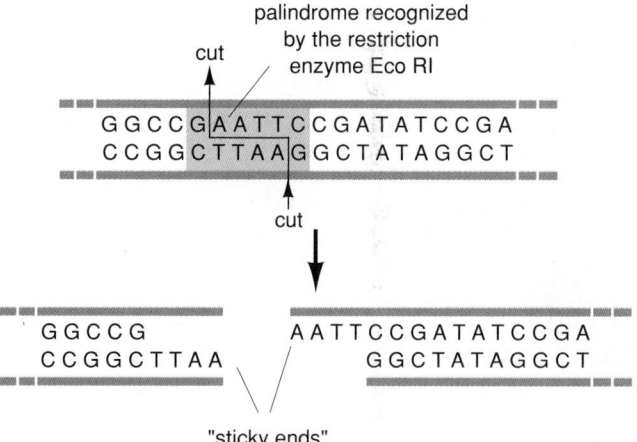

"sticky ends"

Figure 14-2 Restriction enzymes sever double-stranded DNA at specific nucleotide sequences. Many of the most useful restriction enzymes, such as the one called Eco RI illustrated here, sever DNA at palindromes, where the nucleotide sequence reads the same on one DNA strand (GAATTC) as it does in the opposite direction on the other strand. When a double-stranded DNA molecule is cut by Eco RI, two short, single-stranded "sticky ends" remain: AATT on one strand and TTAA on the other. Since these sticky ends are complementary, they can base-pair, temporarily holding these strands, or any strands with the same sticky-end sequences, together again (see Fig. 14-3).

and one with a single-stranded end reading AATT. Complementary DNA regions like these can pair up, held together by hydrogen bonds between the bases. If the appropriate DNA repair enzymes are present, the two pieces can be rejoined.

Many restriction enzymes have been isolated from various species of bacteria. Each cuts DNA apart at different nucleotide sequences. The specificity and variety of restriction enzymes have enabled molecular geneticists to identify and isolate segments of DNA from many organisms, including humans.

Building the DNA Library

How can restriction enzymes be used to build a DNA library? Suppose that DNA is isolated from a human source, say white blood cells, and is cut apart with the restriction enzyme Eco RI (Fig. 14-3a). Everywhere that the GAATTC/CTTAAC pairing occurs, the human DNA will be severed, leaving single-stranded AATT and TTAA "sticky ends" protruding. The next step is to isolate many copies of a bacterial plasmid that includes an easily identifiable "marker gene," such as the gene for ampicillin resistance (we'll see why this is useful in a minute). The plasmids are then exposed to the same restriction enzyme. The plasmids, too, will be cut open, with AATT and TTAA sticky ends protruding.

Now the human DNA fragments and opened plasmids are mixed together. The sticky ends will hydrogen bond, forming human–human, plasmid–plasmid, and, most importantly, human–plasmid DNA combinations (Fig. 14-3b). DNA ligase enzymes are then added. DNA ligase covalently bonds the sugar–phosphate backbones together, inserting human DNA into plasmids (Fig. 14-3c). Ideally, each plasmid receives only a relatively small piece of human DNA. Millions or billions of plasmids collectively would incorporate DNA from the entire human genome.

The new rings of plasmid plus human DNA are mixed with bacteria treated with calcium salts to make them permeable to DNA (Fig. 14-3d). The bacteria take up the plasmid–human DNA. Usually, 100 to 1000 times more bacteria than plasmids are used, so that no individual bacterium ends up with more than one plasmid–human DNA circle. Of course, this procedure also ensures that most of the bacteria don't have any plasmid–human DNA combinations at all. Here's where the ampicillin resistance gene becomes useful (14-3e). The bacteria are plated out on a culture dish containing medium with ampicillin. The antibiotic kills off all the bacteria that have not taken up a plasmid. The result: a population of bacteria, all with a plasmid, and most with the recombined plasmid–human DNA. This constitutes our human DNA li-

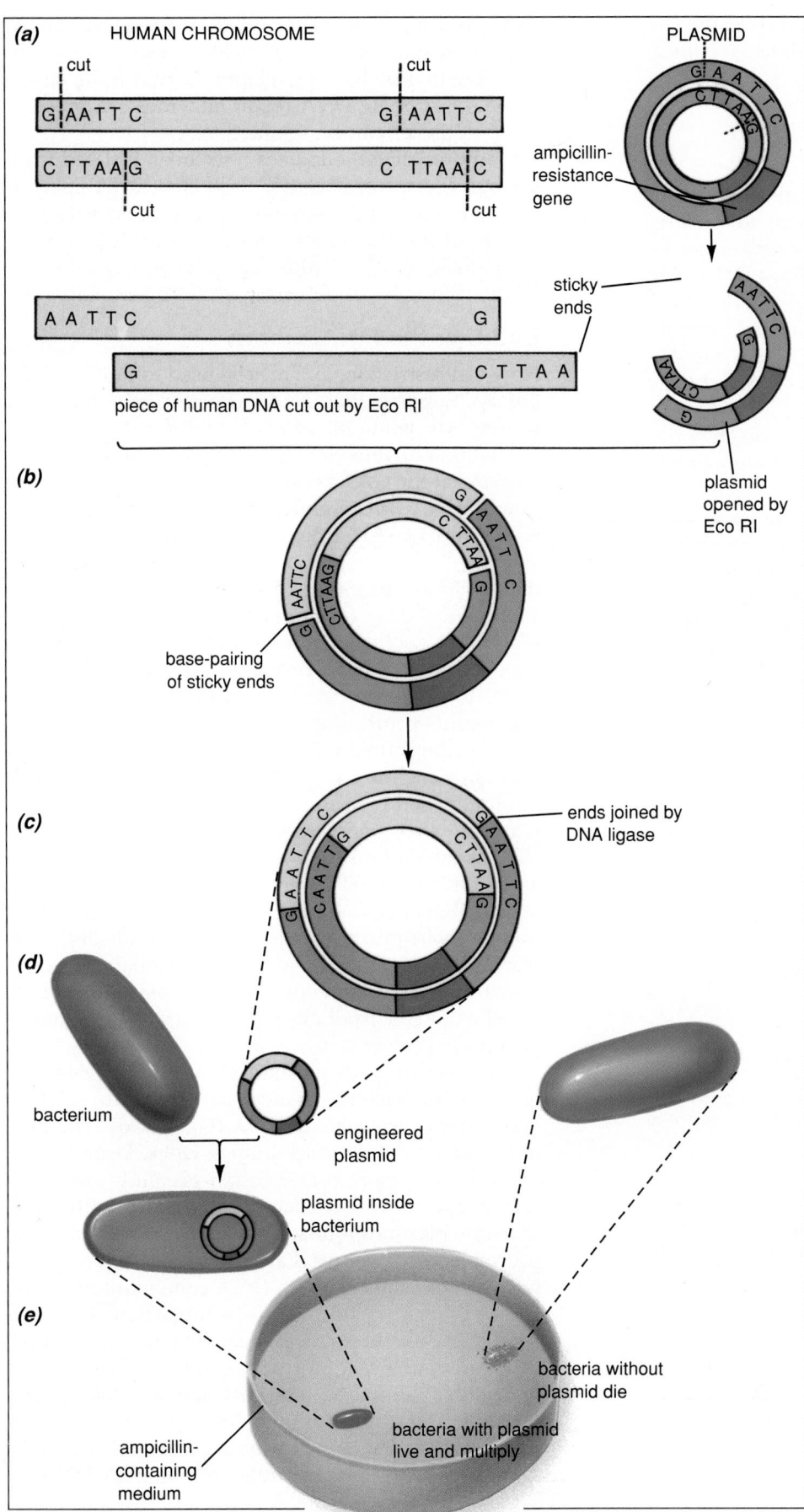

Figure 14-3 The use of restriction enzymes in building a human DNA library.
(a) Human chromosomes and bacterial plasmids (with the ampicillin resistance gene) are both cut with the Eco RI restriction enzyme. Both types of DNA therefore have AATT and TTAA sticky ends.
(b) The severed human and plasmid DNA molecules are mixed. Human–plasmid combinations are temporarily held together by complementary base-pairing of sticky ends.
(c) DNA ligase bonds are backbones of the human–plasmid combinations together.
(d) The new human–plasmid rings are mixed with bacteria that have been treated with calcium salts to make the bacteria permeable to DNA. Some bacteria take up a human–plasmid ring.
(e) Bacteria are plated onto culture dishes filled with medium containing the antibiotic ampicillin. Bacteria without plasmids succumb to the ampicillin; those with plasmids survive. Plasmids within the bacteria constitute a human DNA library.

brary. Before the library can be used, however, the researcher must identify the gene to be manipulated.

Identifying Genes

The human genome contains about 6 billion nucleotides (3 billion pairs). The average protein has 300 to 400 amino acids, so at three nucleotides per amino acid, only about a thousand nucleotides encode the information for a typical protein. Identifying or finding a particular gene within the human genome, then, would appear to be a search for a very small needle within a very large haystack. Fortunately, there are several "tricks" that can be used to identify or locate genes.

It is fairly straightforward to find a gene if you first know the amino acid sequence of its encoded protein (Fig. 14-4). For example, the amino acid sequences of most of the major human hormones are known. From the amino acid sequence, one can then work backward through the genetic code to determine likely DNA sequences (remember, however, that a single amino acid may be encoded by several nucleotide sequences, which complicates things a little). Machines are now commercially available that can synthesize DNA chains from nucleotide sequences typed on a keyboard. The synthesizers cannot make long chains, but they can accurately synthesize a chain a few dozen nucleotides in length. We'll see what to do with this artificial DNA shortly.

Another way of identifying a gene is to find a cell that you know synthesizes vast amounts of a particular protein. Immature red blood cells, for example, synthesize lots of hemoglobin. That means that they have lots of the mRNA for hemoglobin. This mRNA is complementary to the DNA of the hemoglobin gene, and can also serve to locate the gene.

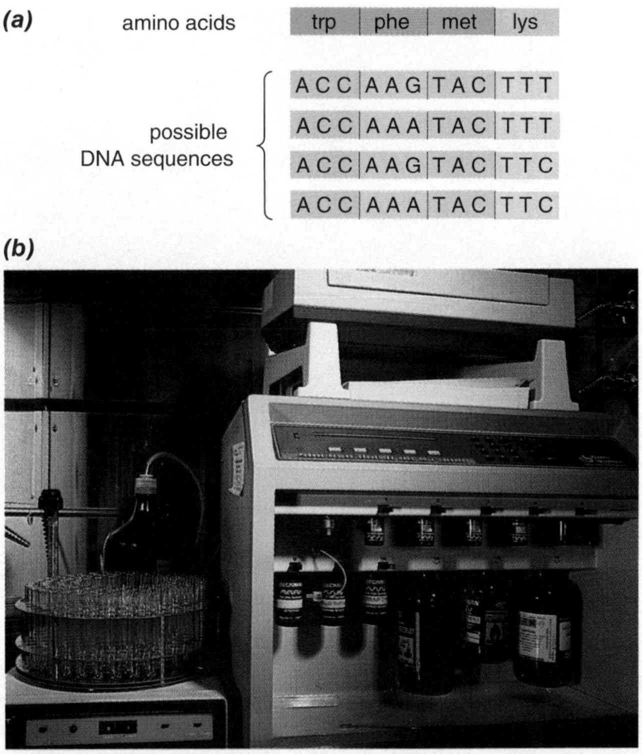

Searching the Library

So far, our DNA library is pretty useless, because it lacks a card file or computer index that would tell us which bacterium contains the plasmid with the gene we want to study. This is where synthetic DNA or isolated mRNA, labeled with radioactive isotopes, comes in. Figure 14-5 explains how these "probes" are used to locate bacteria containing specific genes.

Making Copies

Now that we have located the gene in the library, we can make copies of it for further use. One merely picks bacteria from the appropriate colonies located during the search, and grows them in appropriate culture conditions. Biotechnology firms often use

Figure 14-4 (a) DNA sequences that might code for a peptide with the amino acid sequence tryptophan-phenylalanine-methionine-lysine (trp-phe-met-lys). Note that, because phenylalanine and lysine may each be encoded by two DNA triplets, there are four possible DNA sequences for this peptide. **(b)** A DNA synthesizing machine.

huge vats to produce many pounds of bacteria, all containing a specific human gene (Fig. 14-6).

A recently developed method for making copies of specific stretches of DNA, called the **polymerase chain reaction,** shows enormous promise for both industrial and biomedical applications. The polymerase chain reaction has become so important, in fact, that *Science* magazine named the key enzyme involved in the procedure, a DNA polymerase from a specialized microbe, the "Molecule of the Year" in 1989 (see "A Closer Look at the Polymerase Chain Reaction: Hot Springs and Hot Science").

Using the Gene

At this point, procedures vary greatly, depending on what one wants to do with the gene. Further, the technologies tend to be quite complex, and differ for each application. Therefore, we will discuss only a couple of possibilities, and generally we will describe only the applications, not the methodologies.

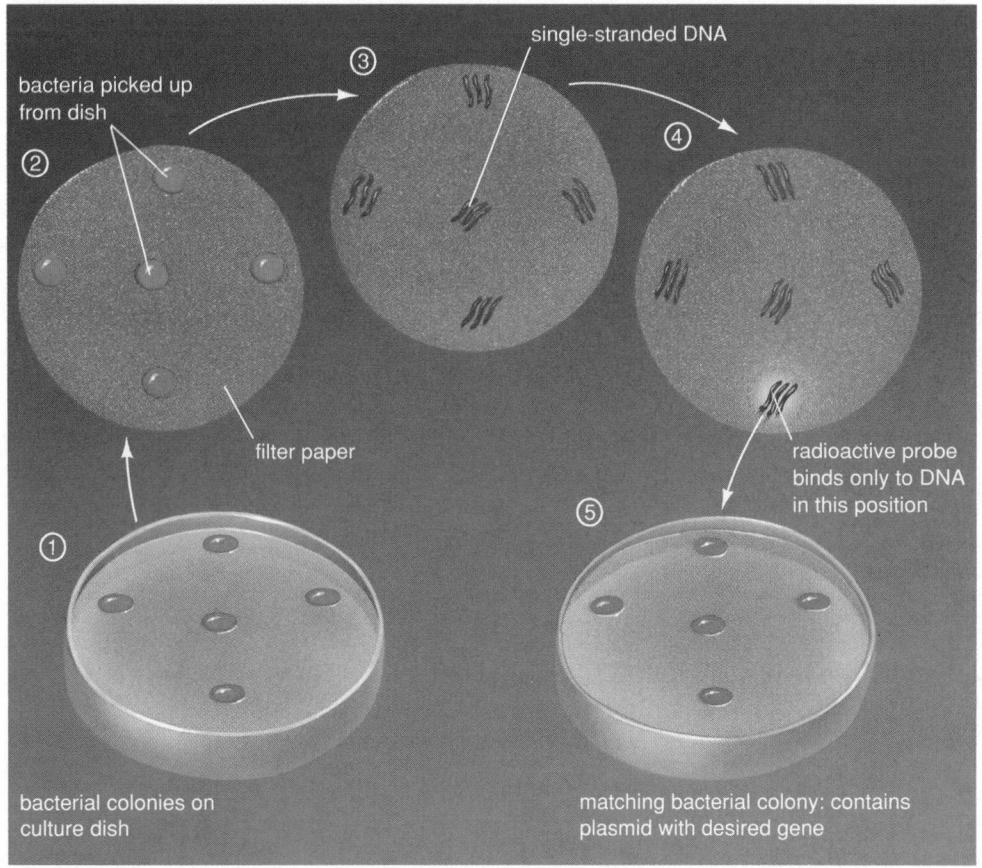

Figure 14-5 Searching a human DNA library for a specific gene.
(1) Bacteria from the DNA library are sparsely spread out onto a culture dish. Each bacterium multiplies into a visible colony. If few enough bacteria were originally plated, then each colony should consist of the descendants of a single bacterium, containing in turn a single type of human–plasmid combination.
(2) A sheet of special filter paper is pressed onto the culture dish. It picks up a few bacteria from each colony, faithfully preserving the colony positions. The original culture dish is saved.
(3) The filter paper is placed in an alkaline solution. This breaks open the bacteria, freeing the plasmids, and separates the double-stranded DNA of the plasmids into single strands.
(4) The paper is now bathed in solution of neutral pH containing a radioactive synthetic mRNA or DNA "probe" (yellow). The probe hydrogen-bonds only to plasmid DNA that is complementary to the nucleotide sequence of the probe; that is, to plasmids that contain the human gene complementary to the probe.
(5) The locations of radioactivity on the paper are matched with bacterial colonies on the culture dish. Colonies in the same position consist of bacteria containing plasmids with the desired gene. Samples of these bacteria are subcultured.

Protein Factories

One prominent application of recombinant DNA is to produce medically useful quantities of human proteins. The genes for human and bovine growth hormone, insulin, blood clotting factors, and enzymes that dissolve blood clots have all been inserted into bacteria or eukaryotic cells for commercial production. The recombinant cells are grown in huge vats.

The desired product is extracted from the culture medium or the cells themselves, and purified.

Why is this useful? Take the case of human growth hormone. Growth hormone, secreted by the pituitary gland, governs bodily growth, protein metabolism, and perhaps aging. A few thousand people don't produce enough growth hormone, remain very short, and age prematurely. Formerly, the only

Figure 14-6 Fermenting apparatus at a biotechnology company raises huge numbers of bacteria, each containing a plasmid that incorporates a valuable human gene.

source of human growth hormone was human pituitaries. Tens of thousands of pituitaries were removed from corpses and processed to purify the growth hormone, which would then be injected into growth hormone–deficient children. This, of course, was very expensive. To make matters worse, in 1985 a few recipients died from a rare viral infection, Creutzfeldt–Jakob disease, apparently caused by viruses that had contaminated the hormone preparation. Meanwhile, a biotechnology company, appropriately called Genentech, inserted the gene for human growth hormone into bacteria and in late 1985 began to sell bacterially produced hormone. Since no human body parts are involved, no human diseases can be transmitted by the Genentech hormone. Similarly, blood

clotting factors produced through recombinant DNA technology are now given to hemophiliacs rather than blood transfusions or blood-clotting factors isolated from donated human blood. (Unfortunately, this development came too late to save hundreds of hemophiliacs from contracting AIDS from contaminated blood products; see Chapter 34.)

Vaccines

Presently, vaccines are effective mostly against viral diseases, such as smallpox or measles. The vaccine itself usually consists of killed or weakened virus, or a strain of virus that is genetically incapable of causing disease. The killed, weakened, or genetically defec-

A CLOSER LOOK
At the Polymerase Chain Reaction: Hot Springs and Hot Science

One major problem with all the recombinant DNA technologies described in this chapter is that they require *lots* of DNA: millions or billions of copies. One DNA molecule, or even a few hundred, is usually not enough for even the most sensitive modern instruments to detect. In principle, lots of DNA can be produced by gene cloning in bacteria, but this takes time and money. For prenatal diagnosis of genetic defects, a lot of DNA is often not available, and a lot of time is not available either. For decades, a dream of molecular geneticists has gone something like this: "If only someone could find a way to get me millions of copies of the gene I'm interested in, really quickly, and cheaply, too, of course." Well, Kary Mullis of the Cetus Corporation did just that: he invented a process, called the **polymerase chain reaction,** or **PCR** for short, that rapidly and selectively replicates specific parts of a DNA molecule.

The key to PCR lies in the nature of DNA polymerase. Remember that DNA polymerase works on single strands of DNA (other enzymes initially unwind the double helix and separate the two strands). If you add DNA polymerase and a batch of free nucleotides to a long single strand of DNA, the polymerase doesn't know where to start. However, if you add a short piece of DNA that is complementary to part of the strand (this piece of DNA is called a **primer**), it will bind to its complementary region of the strand, and DNA polymerase will start adding new nucleotides at the end of the primer. Therefore, by adding a primer that is complementary to a specific place on the original DNA strand, you can effectively "tell" the DNA polymerase where to start copying. The polymerase chain reaction is so sensitive and so fast that it can actually produce billions of copies of a gene in a single afternoon, starting, if necessary, from a single molecule of DNA.

Before beginning the polymerase chain reaction, two primers are made in a DNA synthesizer: one primer complementary to the sense strand at the beginning of the gene (let's call this the "sense primer"), the second primer complementary to the antisense strand at the end of the gene (the "antisense primer").

With a sample of DNA and these two primers in hand, PCR can begin (Fig. E14-1). In these diagrams, DNA is represented as long "railroad tracks" of sugar-phosphates with notched "ties" connecting the two rails. The primers and the gene that they will help to copy are colored blue.

The First Cycle

A DNA double helix is heated to about 98° C (just below boiling), which separates the double helix into two single strands (Steps 1 and 2 in Fig. E14-1). The separated strands are cooled and the primers, DNA polymerase, and free nucleotides are mixed in. The primers bind to their complementary regions of the DNA (Step 3). DNA polymerase begins copying the DNA strand, beginning at a primer and continuing on to the end of the strand. One DNA polymerase molecule binds at the "sense primer" and synthesizes a new strand complementary to the DNA of the sense strand (Step 4, top). A second DNA polymerase molecule binds at the antisense primer and synthesizes a new strand complementary to the antisense strand (Step 4, bottom). One complete copy of the gene, plus a lot of DNA that extends beyond the gene, has been made.

The Second Cycle

The DNA is heated up to 98° C again, separating the DNA into single strands. The DNA is cooled down, and DNA polymerase, nucleotides, and primers are added (Step 5). DNA synthesis proceeds as before, but both the original DNA strands and the first-cycle copies are used as templates. Note that when the first-cycle copies are used as templates, the newly synthesized DNA strands start at one end of the gene and finish at the other end—no extraneous DNA is copied (Step 6).

The Third Cycle

The DNA is heated, cooled, and the ingredients added once again (Step 7). Note that in this cycle, two of the eight DNA strands to be copied include *only* the gene. Therefore, this third cycle of PCR at last synthesizes two DNA double helices that are exact copies of just the gene, with no superfluous DNA at either end (Step 8).

Further Cycles

As further cycles of PCR continue, the proportion of DNA copies that consist only of the gene increases rapidly. In the fourth cycle, 8 out of 16 DNA double helices include only the gene of interest; in the fifth cycle, 22 out of 32, and so on. Since each cycle takes only a few minutes to complete, within a few hours PCR produces many billions of copies of the gene, and most of these copies include just the gene, with no extra baggage.

PCR machines are now standard equipment in virtually all molecular biology labs. PCR has resulted in quicker and more accurate prenatal diagnosis of genetic defects. It provides a more sensitive assay for AIDS than the older antibody tests, by allowing a person's white blood cells to be examined for AIDS virus genes. PCR will, of course, be crucial in the Human Genome Project (see Chapter 15).

PCR promises to revolutionize other areas of biology as well. For example, Svante Paabo of the University of

The First Cycle

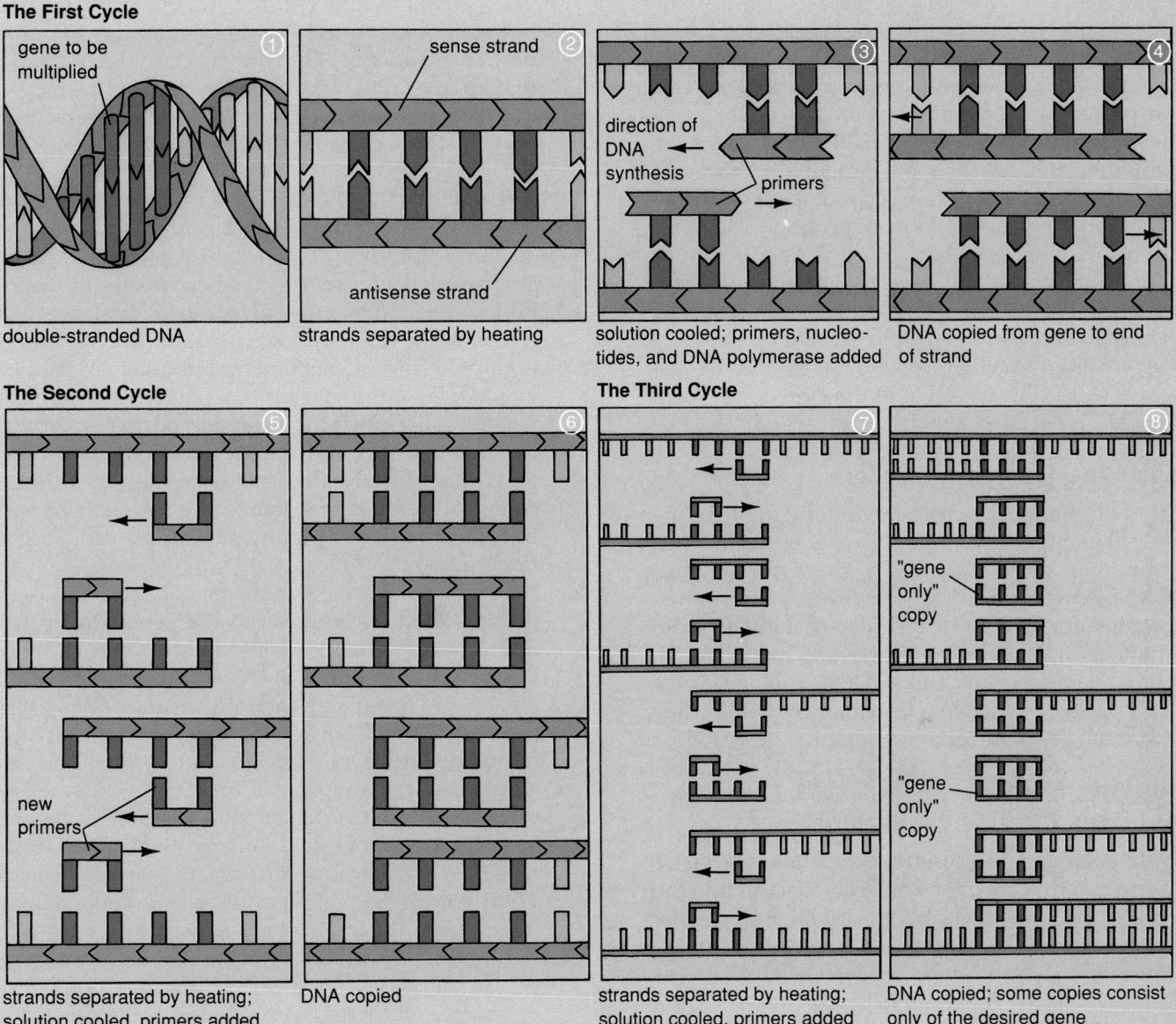

double-stranded DNA | strands separated by heating | solution cooled; primers, nucleotides, and DNA polymerase added | DNA copied from gene to end of strand

Figure E14-1 How the polymerase chain reaction copies genes. The steps are described in the text.

California at Berkeley has used PCR to amplify bits of DNA remaining in Egyptian mummies. Allan Wilson, also at Berkeley, is using PCR to study evolutionary relationships (see Chapter 17).

Oh, yes—what about the hot springs mentioned in the title? Well, as you know, most proteins denature at high temperatures. If you heat DNA up to 98° C to separate the two strands, you irreversibly denature most DNA polymerase molecules as well. Therefore, early versions of PCR required the addition of new DNA polymerase in every cycle, after the DNA sample was cooled down. In 1987, someone at Cetus had a brainstorm: what about the bacteria that live in hot springs,

such as those in Yellowstone National Park? Sure enough, the bacterium *Thermus aquaticus* (literally, "hot water") has evolved a DNA polymerase molecule that survives repeated heating and cooling cycles. Since the free nucleotides and the short single-strand DNA primers are heat-stable as well, a PCR machine is now a rather simple device that mostly just runs heating and cooling cycles over and over again. The investigator loads test tubes up with the ingredients, turns on the machine, and then goes away for a while. With the editor's tongue only partially in his cheek, *Science* magazine named the hot springs' bacterium DNA polymerase "Molecule of the Year" in 1989.

tive virus, although not disease-causing, is recognized by the immune system, triggering the production of defenses that would protect against infection by pathogenic viruses of the same types (see Chapter 34).

A safer and potentially more effective vaccine might be synthesized using recombinant DNA techniques. The immune system usually recognizes not a whole virus, but one or more proteins on the viral surface. Recombinant DNA technologies might be used to clone the genes for the appropriate viral proteins, insert them into bacteria, and produce huge quantities of the protein. A vaccine would then consist of pure protein, and would be completely safe, since no living viruses would ever have been involved in its production. Recombinant protein vaccines are especially attractive in the case of the AIDS virus, because it is so lethal.

Diagnosing Genetic Disorders

A variety of human diseases are inherited, including sickle-cell anemia, Tay–Sachs disease, cystic fibrosis, and muscular dystrophy. So far, there are no cures for these genetic disorders. However, carriers (heterozygotes for recessive disorders who show no symptoms) and/or affected fetuses can sometimes be identified using recombinant DNA technology. Some of the techniques and the ethical concerns that they raise are discussed in Chapter 15.

Modifying DNA in Free-Living Organisms: Bacteria

Genetic engineers are inserting and deleting genes from bacteria that are intended to be released into the outside world. For example, several laboratories are working to produce bacteria that selectively metabolize toxic substances, from crude oil to hazardous industrial wastes. In the 1980s, one of the first intentional releases of bioengineered bacteria—the so-called "ice-minus" bacteria—raised a storm of controversy, even though a gene had been cut out of, rather than engineered into, the bacteria (see the Reflections at the end of the chapter).

Modifying DNA in Free-Living Organisms: Agricultural Applications

Genes can also be inserted into eukaryotic cells. Under the appropriate circumstances, entire multicellular organisms can be genetically engineered to contain foreign genes. A variety of plants, including carrots, tomatoes, and potatoes, can be grown from single cells: cells can be taken from, say, the root of a plant, grown in culture, and then exposed to specific conditions that cause each cell to develop into a complete plant. Genes might be introduced into single plant cells, probably via plasmids or viruses. Complete plants would then be grown from the transformed cells, and used as the progenitors of entirely new crop varieties. One drawback to this scenario is that many valuable traits may be multigenic—the desirable phenotype may be due to the interaction of several, perhaps many, genes, some of which may not yet be known.

A few genetically engineered plants have already been developed. A gene that confers resistance to a common herbicide, glyphosate, has been identified and inserted into several types of plants. If inserted into crop plants, the glyphosate-resistance gene would allow the crops to withstand heavier applications of herbicide. Higher concentrations of herbicide would kill off more weeds that otherwise would compete with the crops for water and nutrients. Whether this is beneficial or not depends on one's views about the safety of herbicides, and how likely it is that the genes for herbicide resistance might be inadvertently transferred to other plants, including weeds (see the Reflections).

One of the most valuable genetic modifications of plants would be the insertion of genes for nitrogen fixation into common crops such as corn or wheat. With a few exceptions such as soybeans and alfalfa, our major crop plants cannot use atmospheric nitrogen as a nitrogen source. Consequently, they require nitrogen fertilizers, which are synthesized and spread using the energy of fossil fuels. Rain washes off some of the fertilizer into nearby streams and lakes, causing pollution. Further, the expense of fertilizers is burdensome to farmers, especially in the Third World. Several types of bacteria can extract nitrogen from the atmosphere and capture it in usable form, a process called nitrogen fixation (see Chapter 45). If the genes for nitrogen fixation could be inserted into wheat and corn, farmers could grow much more food at much lower cost. Especially in poor countries, an improvement in food supplies would have a tremendous impact on human health and well-being.

Modifying DNA in Free-Living Organisms: Human Applications

A few children are born each year with severe combined immunodeficiency disease, in which the affected person has virtually no defense against disease (see Chapter 34). In about 25% of these children, the failure of the immune system can be traced to a single defective gene. In the late 1980s, R. Michael Blaese of the National Cancer Institute and W. French Anderson of the National Institutes of Health inserted nor-

PLANET WATCH

Biotechnology Attacks the Chestnut Blight

Scarcely a century ago, the towering American chestnut tree made up a major portion of the eastern hardwood forest (Fig. E14-2). Today the splendor of the chestnuts is barely a memory; few notice the struggling sprouts emerging from the roots of the former giants. As described in Chapter 45, the American chestnut has been all but driven to extinction by the ravages of a fungus called the chestnut blight, inadvertently imported from Asia around 1900 on Asian chestnuts. However, there is hope that the American chestnut will reappear, thanks to chance and the research techniques of biotechnology.

In 1950, an Italian researcher noted that some European chestnuts, also attacked by chestnut blight, were recovering—had the fungus lost its virulence? In the 1960s, plant pathologists found that some strains of chestnut blight had been weakened, apparently by a new viral infection that left particles containing double-stranded RNA (dsRNA, which encodes genetic information in many fungal viruses) floating throughout the fungal cytoplasm. In an unknown way, the viral dsRNA decreases the virulence of the fungus, allowing the tree to fight it successfully (Fig. E14-3b). By spreading the viral dsRNA infection to virulent chestnut blight fungi, foresters are gradually controlling the blight in Europe. Unfortunately, the European blight's virus is not very useful against the strains of chestnut blight fungus found in the United States.

Although viral-infected strains of chestnut blight also occur in the United States, they do not successfully control the blight. Therefore, the tools of biotechnology are being applied toward weakening the parasitic fungus. Researchers at the University of Kentucky discovered that dsRNA from infected American fungi is genetically distinct from dsRNA from infected European fungi. This suggests that the viral infections arose independently on each continent. Other researchers at Roche Institute in New Jersey are sequencing the dsRNA to determine what proteins it codes for, and how these proteins might weaken the fungus. Researchers are also using an enzyme called reverse transcriptase, which transcribes RNA into DNA, to synthesize DNA from viral dsRNA. These DNA molecules are then inserted into healthy fungi to determine which DNA sequences cause loss of virulence without killing the fungus. A genetically engineered fungus that was less virulent could be introduced to compete with the deadly variety.

Biotechnology may eventually be used, not only to weaken the fungus, but to strengthen the chestnut. There is hope that the gene that confers resistance to the fungus in the Asian chestnut may be identified, sequenced, and introduced into "germlings" of the American chestnut, producing a blight-resistant strain. Although you will never see the chestnut forests that delighted your great-great-grandparents, thanks to biotechnology a chestnut tree may yet shade the homes of your children.

(a)

(b)

Figure E14-2 (a) A healthy American chestnut, which escaped the blight because it had been transplanted far west of the normal range of the chestnut before the invasion of the blight fungus.
(b) Chestnut blight fungus has produced a lesion on this chestnut tree. The lesioned area was treated with infected fungus, and the tree is walling off the infection with a growth of new tissue.

mal alleles of this gene into white blood cells, and obtained reasonably normal gene expression. At 12:52 p.m., September 14, 1990, the first clinical trial of human gene therapy began: a 4-year-old girl received a transfusion of her own white blood cells that were genetically engineered to contain the normal gene. As of December, 1990, she had suffered no detectable side effects, and showed some improvement in her immune response. Whether the treatment will be successful in the long run remains to be seen.

As with immunodeficient children, other likely candidates for genetic engineering in humans are defects involving discrete structures, such as glands or bone marrow, that are normally the only active sites of transcription of certain genes. These therapies would, of course, offer cures only for the affected individual: the patient's germ line cells would not be "fixed," and so he or she could still pass the genetic defect on to future generations. It is likely to be many years before genetic engineers are in a position to modify the genetic composition of human gametes. Whether society will allow it is also an open question, which we discuss in the Reflection at the end of this chapter.

Locating and Sequencing Genes

The applications of biotechnology that we have discussed so far all rely on some prior knowledge: the desired genes are already known either by their protein products, their mRNA, or, in a few instances, at least part of their DNA sequences. Unfortunately, geneticists do not know much, if anything, about the molecular nature of most genes or their protein products. Huntington's disease, for example, is a lethal brain disorder affecting thousands of people worldwide, caused by a single dominant allele (see Chapter 15). However, nothing whatever is known about the gene or the protein that it encodes. How can such information ever be obtained?

Locating Genes: Restriction Fragment Length Polymorphisms

You already know something about methods for locating genes: for example, linkage analysis, which uses the frequency of crossing over to map chromosomes (Chapter 11). To locate a gene by linkage analysis, one calculates the incidence of recombination between two traits during sexual reproduction. If recombination *never* occurs between the two traits, then either the genes controlling those traits lie very close together on a chromosome, or in fact the traits are

controlled by the same gene. If the location of the gene controlling one trait is already known, then it should be easy to find the other gene.

In principle, to locate an unidentified gene in a DNA library, you could make a probe for a known "marker gene" that linkage analysis tells you is close to the unidentified gene. Somewhere nearby should be the gene that you're interested in.

Using ordinary linkage analysis to find a gene in a DNA library, however, has two problems. First, geneticists usually know the location of relatively few genes on any given chromosome, and eukaryotic chromosomes are very large: the average human chromosome contains about 60 million nucleotides. Therefore, if linkage analysis reveals that an unknown gene is "close" to a known gene, say, only 10% of the chromosome's length away, that means that the 1000 nucleotides that make up the unknown gene might lie anywhere within about 6 million nucleotides. That isn't necessarily very useful. A second, related, problem with linkage analysis is the relative scarcity of genes in a eukaryotic genome. Molecular geneticists estimate that nucleotides that code for protein make up less than 5% of the human genome: the rest is introns, regulatory sequences, or vast amounts of other, apparently nonfunctional, DNA. Therefore, classical linkage analysis cannot usually be expected to locate an unknown gene more closely than within a few hundred thousand nucleotides, at best. That still isn't very useful. What geneticists need are a lot more markers, a lot closer together.

Using Restriction Enzymes to Produce Markers on a Chromosome

Restriction enzymes can be used to provide markers on a chromosome, just as known genes can. Remember that restriction enzymes cut DNA only at specific nucleotide sequences. Further, there are a host of known restriction enzymes, and each cuts DNA at a different sequence of nucleotides. Let's suppose that a pair of homologous chromosomes is exposed to a given restriction enzyme (Fig. 14-7). The DNA is cut up into pieces called **restriction fragments** (fragments of DNA produced by restriction enzymes). These fragments can be separated by size, using a method known as gel electrophoresis. Would the pattern of sizes be the same for both homologous chromosomes? Perhaps, or perhaps not, depending on the nucleotide sequences of the chromosomes and the recognition sequence of the enzyme. In the example diagrammed in Figure 14-7, the two homologous chromosomes have different nucleotide sequences near one end. The restriction enzyme Hind III can cut

(a)

chromosome cutting
by restriction enzyme Hind III

(b)

analysis of DNA pieces
by gel electrophoresis

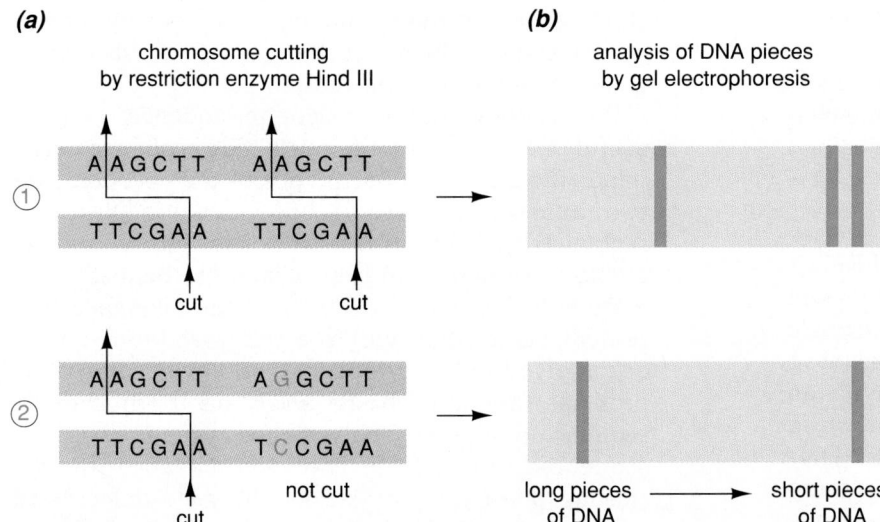

long pieces → short pieces
of DNA   of DNA

Figure 14-7 The nature and detection of restriction fragment length polymorphisms. **(a)** Two homologous chromosomes differ in a single base substitution. A restriction enzyme, Hind III, cleaves DNA at AAGCTT/TTCGAA palindromes. The upper chromosome has two of these sequences, and is cut into three pieces, two short and one medium long. The lower chromosome, because of the base substitution (a G/C pair substituted for an A/T pair), has only one sequence cleaved by Hind III. Therefore, the lower chromosome is cut into only two pieces, one short and one very long.
(b) The DNA pieces produced by Hind III are separated by the technique of gel electrophoresis. DNA migrates through the gel, pushed along by an electric field. Small pieces slip through the gel more easily than large pieces can, and migrate farther to the right. The size and number of DNA pieces differ for the two chromosomes.

apart the upper chromosome in two places, but the lower chromosome in only one place (Fig. 14-7a). Gel electrophoresis therefore shows three pieces of DNA from the upper chromosome, and two pieces of DNA from the lower chromosome (Fig. 14-7b). These differences in the size of the DNA pieces are called **restriction fragment length polymorphisms,** or **RFLPs** for short ("polymorphism" means "many forms" in Greek).

Using RFLPs to Locate a Gene

Gene location by RFLP analysis is conceptually similar to locating a gene by crossing over. In RFLP analysis of a human genetic disorder, researchers look for a restriction enzyme "cut site" that is always, or almost always, inherited along with the disorder. The cut site must therefore be within, or very close to, the gene causing the disorder. In the late 1980s, RFLP analysis was used to locate and sequence the gene for cystic fibrosis, a fairly common, usually fatal, inherited lung disease.

It is easy to understand why a unique restriction enzyme cut site might be located within a defective

allele: if the altered nucleotide sequence that causes the genetic defect is also the one recognized and cleaved by the restriction enzyme, then the enzyme will cut the defective allele but not the normal allele. But why should a unique cut site be located *close to* a defective gene? To understand this, you must consider the inheritance of genetic defects and the nature of linkage. Defective alleles for inherited genetic disorders are usually quite rare. Therefore, all affected individuals may well be descended from a common, though distant, ancestor. Suppose that this person, strictly by chance, had a second altered nucleotide sequence on the same chromosome. How he or she got it, by inheritance or mutation, doesn't matter. If this second altered nucleotide sequence is close enough to the locus of the defective allele, then there will be little or no crossing over between the two sites.

This second nucleotide change can be used to find the locus of the defective allele if one more condition is met: the "normal" and "altered" nucleotide sequences must be affected differently by a restriction enzyme. Either the normal sequence must be cut by the enzyme while the altered sequence is not cut, or vice versa. Given the vast number of restriction enzymes, a "cut–no cut" difference can almost always be found relatively close to any given defective allele, although it may take years of work to find one close enough to be useful for DNA sequencing (see below).

Other Uses of RFLPs

Locating a gene is not the only application of RFLPs. For example, RFLP analysis can also be used for prenatal diagnosis of certain genetic defects, if the nucleotide sequence or at least the location of the gene is known. In Chapter 15, we describe how RFLPs are used to diagnose sickle-cell anemia. A variation on standard RFLP analysis, popularly known as "DNA fingerprinting," has occasionally been used in forensic medicine to identify the perpetrator of a crime, based on matching a class of RFLPs from blood or semen samples left at the scene of a crime with RFLPs from a suspect's tissues. The methodology and some other applications of RFLP analysis are explained in a fascinating *Scientific American* article, "Chromosome Mapping with DNA Markers" (see "Suggested Readings").

Sequencing Genes

Both basic research and medical/industrial applications of biotechnology are greatly enhanced if the sequence of nucleotides in a gene are known. During the 1980s, automated machines were developed that allow fairly large genes (up to a few tens of thousands of nucleotides long) to be sequenced (Fig. 14-8). The process isn't cheap, and even 50,000 nucleotides is short relative to the 60 million in an average human chromosome, which is why precise localization of genes is still very important.

Why is knowing the nucleotide sequence of a gene useful? For one thing, the nucleotide sequence determines the amino acid sequence of the encoded protein. In many instances, it is now easier to locate and sequence a gene than it is to isolate, purify, and sequence a protein. In recent years, biochemists have frequently "forecast" the amino acid sequence of a protein, its location within a cell (membrane-bound or dissolved in the cytoplasm), and even some of its probable functions, based solely on the nucleotide sequence of a gene. The gene for cystic fibrosis was sequenced in late 1989. The gene sequence was the starting point for pinpointing the gene defect, and even, as of early 1991, "curing" the disease in lung cells cultured from patients with cystic fibrosis.

Further, if one knows the nucleotide sequence of a gene, and the nucleotide sequence that is cleaved by a variety of restriction enzymes, one may be able to design a rapid, definitive test for genetic disorders—for example, the DNA of affected people will not be cut by a certain restriction enzyme while the DNA of everyone else will be cut. This type of test is now the basis for prenatal diagnosis of sickle-cell anemia, and is explained in Chapter 15.

Recently, in one of the most remarkable instances of cooperation and collaboration in modern biology, a huge group of scientists, headed by James Watson, have begun the ultimate sequencing task—to sequence the entire human genome. The rationale, approach, and methodology of the Human Genome Project is also explored in Chapter 15.

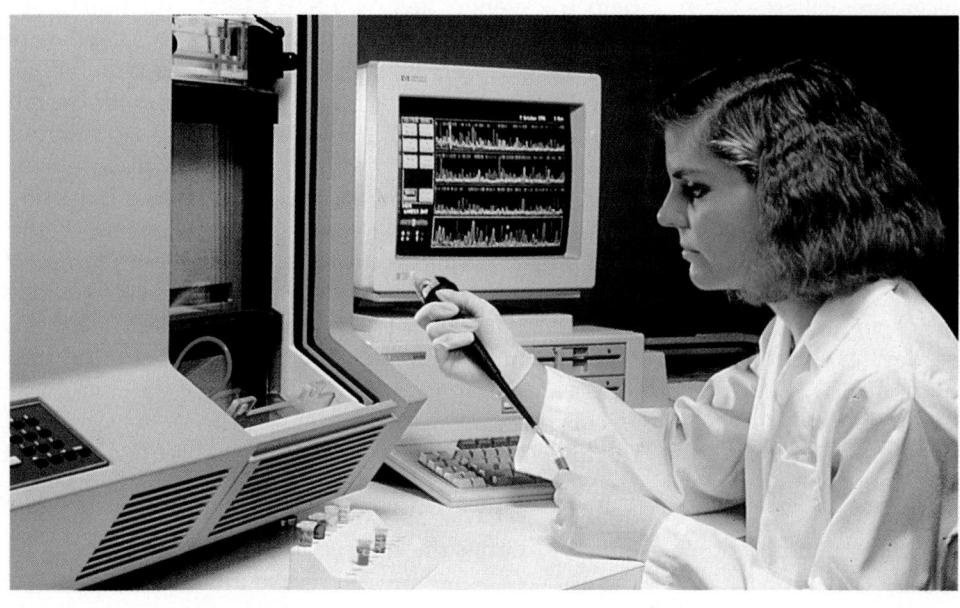

Figure 14-8 A commercially available DNA sequencing machine. Equipment such as this, only a dream a couple of decades ago, are routine equipment in many molecular biology labs.

 # Reflections on the Ethics of Biotechnology

In many respects, biotechnology presents a microcosm of the benefits and dilemmas that science and technology have always offered to humankind. There are few fundamental advances in science that do not have both bright and dark sides—the first spear allowed hunters to nourish their families better, but also provided a more effective means of killing their fellow humans. There are groups in society who would greatly restrict the uses of biotechnology, and who might prefer that the techniques of recombinant DNA had never been developed— Jeremy Rifkin, director of the Foundation on Economic Trends, is perhaps their most forceful spokesperson. Others feel that the benefits greatly outweigh the liabilities, and that any potential for harm can rather easily be overcome by fairly simple legal means. Many others take no stand on biotechnology in general, but have strong opinions about specific applications. In this Reflection, we will explore three aspects of biotechnology that together represent many of the ethical issues that surround this emerging field. In the following chapter, we will examine some ethical concerns in the application of biotechnology to medicine.

"Undesirable" Uses of Biotechnology

Perhaps the easiest ethical issue surrounding biotechnology is the objection that some applications of bioengineering may be socially undesirable. For example, in the late 1980s, a biotech firm began to mass-produce bovine growth hormone by recombinant DNA techniques. Growth hormone, when injected into cows, enhances milk production. Some people are vigorously opposed to this, mostly for two concerns. First, will the recombinant hormone change the composition of the milk? In December 1990, a panel of experts convened by the National Institutes of Health reviewed the data. They concluded that, yes, there are changes, but the changes are minor and pose no threat to human health. Second, there is already a milk glut in both America and Europe. Many small dairy farmers are opposed to the use of recombinant growth hormone because they fear that greater milk production per cow will drive prices down even further and put some small farmers out of business.

Both of these concerns are real. However, neither has any direct bearing on biotechnology *per se.* Millions of cattle and other mammals are slaughtered for meat every year; suppose that a cheap way was found to extract growth hormone from their pituitary glands? The same arguments would arise without biotechnology being part of the issue at all. In instances such as this, society may wish to intervene in *a specific use of biotech-*

nology for reasons of public health, economic issues, or social justice. The outcome, not the methodology, is the crux of the matter.

The Release of Bioengineered Organisms into the Environment

Although bioengineered *products* are relatively simple to deal with, bioengineered *organisms* are more problematic. Two examples illustrate the potential difficulties.

Many crop plants are quite sensitive to frost: oranges and tomatoes, to name just two. Ice formation in the air, in water, or on plant surfaces is greatly facilitated by small particles of dust or other matter that act as "seeds" for ice crystal formation. In the absence of "seed particles," ice crystals do not form even at temperatures somewhat below freezing. One of the primary sources of ice seeding on plants is a common bacterium, *Pseudomonas syringae*. A specific protein on the surface of the bacterium promotes ice formation. Therefore, an obvious use of genetic engineering would be to remove the gene encoding this protein from the bacterium, and try to replace the "wild-type" bacterium on crops with the new "ice-minus" bacterium. The genetic engineering part of this scenario was accomplished in the mid 1980s, but there were several years of controversy and litigation before field trials were permitted (Fig. 14-9).

What was the controversy about? What if the "ice-minus" strain, once released into the environment, drifted beyond the fields onto which it was sprayed? What if it became established in the wild, and became widespread? Susceptibility to frost is a major determining factor in the distribution of plants across the planet— might "ice-minus" bacteria trigger massive changes in vegetation? Like the ice on strawberry plants, raindrops also usually form around "seeds" of dust or debris in the atmosphere. What if wild-type bacteria contribute significantly to normal rain-seeding—might droughts become frequent if "ice-minus" bacteria became the norm? Genetic engineers responded that the altered bacteria were actually less fit than the wild type, and that a minuscule number of bacteria were to be released, compared to the vast hordes of wild-type bacteria in the environment. Therefore, they argued, the chances were infinitesimally small that the "ice-minus" bacteria could actually become established in the wild. At least so far, the genetic engineers appear to be right: a test field was sprayed in 1987, and no ecological effects seemed to occur.

A similar controversy is brewing about herbicide-resistant plants. Some people are opposed to herbicide-resistant plants because they feel that extensive

Figure 14-9 A worker spraying "ice-minus" bacteria onto a strawberry patch. Although the bacteria are completely harmless to humans, the protective clothing was required by law because the bacteria were legally considered to constitute a pesticide that had not yet been safety-tested on people. The "frostbusters" logo, although just a publicity gimmick, was accurate: treated plants showed about 80% less frost damage than nearby, untreated control plants.

herbicide use is hazardous to human health and the environment, and herbicide-resistant crops would almost certainly lead to greatly increased herbicide use. Alternatively, what if herbicide resistance genes were to be transferred from crops to weeds, for example by a plant virus? Such "lateral transfers" of plant genes may be possible. If this happened, herbicide-resistant crops might ultimately lead to herbicides becoming completely useless.

"Ice-minus" bacteria and herbicide-resistant plants are examples of larger issues surrounding releases of

genetically engineered organisms into the environment: People are not omniscient; people make mistakes. Can we be sure that genetically released organisms, or the modified genes that they carry, will not cause ecological harm? If we can determine only that the risk is "small," then how small is small enough?

Genetic engineers correctly assert that humans have practiced crude genetic engineering for millenia, breeding plants and animals with desired properties. Modern biotechnology is merely a faster, more precise version of standard agricultural practice. What's more, various forms of genetic recombination occur all the time in nature. Why should modern genetic engineering be singled out as a special threat?

Some ecologists reply that ecological disasters of varying magnitude do indeed happen: the invasion of pastureland in the western United States by cheatgrass and knapweed, the displacement of bluebirds by sparrows, the evolution of antibiotic-resistant bacteria, and the mutation of a monkey virus into the AIDS virus, to name a few modern examples. (On the other hand, bioengineering may offer the possibility of *correcting* an ecological disaster of major proportions, the destruction of American chestnut trees by chestnut blight; see the Planet Watch essay.) Genetically engineered organisms, such as the "ice-minus" bacteria, are likely to be genetically inferior to their wild-type counterparts, and therefore be outcompeted, but there may remain some risk. Although generally supportive of properly controlled releases of genetically engineered organisms, in 1989 a committee of prominent ecologists wrote the following caveat in the journal *Ecology*: ". . . the direct effects of self-replicating introduced organisms may not necessarily decrease with time or with distance from the point of introduction. The absence of an immediate negative effect does not ensure that no effect will ever occur."

Tinkering with the Human Genome

The greatest concern of many observers is that genetic engineering offers the potential to change the human genome. As we pointed out, it is not yet possible to eliminate, say, insulin-dependent diabetes by cutting the defective gene out of fertilized eggs and inserting a functional one. But suppose that it does become possible (few molecular geneticists would be willing to bet against it). Then what? Could humankind agree on what constitutes a "bad" gene? Even if we could, would we be right? Probably most people could agree on diabetes, cystic fibrosis, and muscular dystrophy. But what about more subtle changes? What about genes that might alter personality? Or length of life? Or predisposition to take risks? Are we wise enough to direct our own evolution—or, if we develop the capability, are we wise enough to refrain?

SUMMARY OF KEY CONCEPTS

DNA Recombination in Nature
There are many naturally occurring forms of DNA recombination, where the genetic makeup of an organism incorporates DNA from two or more other organisms. These include genetic recombination (crossing over), sexual recombination, and transformation (in bacteria). Bacterial transformation often involves exchange of circular pieces of DNA called plasmids, which are found in the cytoplasm of many bacteria. Bacteria may take up plasmids that have been released by the destruction of other bacteria, thereby acquiring plasmid DNA from different species.

Recombinant DNA Technology
Many common uses of recombinant DNA technology include the following steps: (1) production of a DNA library; (2) identification of the desired gene; (3) production of copies of the desired gene; and (4) insertion and regulation of the gene in specific organisms.

A DNA library is a complete set of all the DNA of a particular organism, usually cloned in plasmids. DNA libraries are produced through the use of restriction enzymes that cleave DNA at specific nucleotide sequences, leaving single-stranded "sticky ends" at each side of the cut. A restriction enzyme is used to cleave DNA of both source genome and plasmid. The two sets of DNA fragments are mixed together; the single-stranded sticky ends cause plasmid and source genome DNA to base-pair. DNA repair enzymes are then used to bond the source genome DNA to the plasmid DNA. Bacteria then take up the plasmids, and may be stored indefinitely as a DNA library.

Genes may be identified through a number of techniques. If the amino acid sequence of a protein is known, then artificial DNA may be synthesized that corresponds to the DNA sequence encoding all or part of the amino acid sequence. Alternatively, mRNA may be isolated from a source that is rich in the specific mRNA transcribed from the desired gene. In either case, radioactive mRNA or synthetic DNA can be used to highlight the desired DNA region from a DNA library by base-pairing with the gene.

Copies of a specific gene may be produced in bacteria from the DNA library by culturing those bacteria that contain plasmids with the desired gene. Alternatively, the DNA containing the gene may be amplified using the polymerase chain reaction.

Uses of cloned genes include (1) production of large quantities of specific proteins; (2) synthesis of vaccines; (3) diagnosis of genetic disorders; (4) modifying DNA in intact organisms, which may be bacteria, agricultural plants, or animals, and releasing the organisms into the environment for specific uses; and (5) modifying the DNA of cells of specific organs in human beings to correct genetic defects.

Locating and Sequencing Genes
Restriction fragment length polymorphisms (RFLPs) may be used to locate genes in chromosomes. Restriction enzymes cleave DNA at specific nucleotide sequences. If the same chromosomes from two or more sources are cut with the same restriction enzymes, differences in nucleotide sequences will yield differences in restriction fragment lengths. Inheritance of phenotypic traits, such as genetic diseases, can be compared with the differences in RFLPs to locate genes quite precisely in chromosomes.

DNA nucleotide sequences can be determined by automated processes in the laboratory. The nucleotide sequence of a gene predicts the amino acid sequence of the encoded protein. Knowledge of the amino acid sequence of the encoded protein can reveal the function and location of the protein, and may assist in understanding genetic disorders and devising treatments for them.

Reflections on the Ethics of Biotechnology
Biotechnology has generated ethical controversies regarding its proper uses. These controversies mostly deal with the risks involved in the introduction of genetically engineered organisms into the environment and the potential for making significant changes in the human genome.

GLOSSARY

DNA library: a complete set of all the DNA of a particular organism, usually cloned into bacterial plasmids.

plasmid (plaz'-mid): a small, circular piece of DNA found in the cytoplasm of many bacteria; usually does not carry genes required for the normal functioning of the bacterium, but may carry genes that assist bacterial survival in certain environments, e.g., antibiotic resistance.

polymerase chain reaction: a method of producing virtually unlimited numbers of copies of a specific piece of DNA, starting with as little as one copy of the desired DNA.

restriction enzyme: an enzyme, usually isolated from bacteria, that cleaves double-stranded DNA at a specific nucleotide sequence; the cleavage sequence differs for different restriction enzymes.

restriction fragment: a piece of DNA that has been isolated by cleaving a larger piece of DNA with restriction enzymes.

restriction fragment length polymorphisms (RFLPs): differences in the lengths of restriction fragments, produced by cutting samples of DNA from different individuals of the same species with the same set of restriction enzymes; occurs because of differences in nucleotide sequences among different individuals of the same species.

STUDY QUESTIONS

1. Describe three natural forms of genetic recombination, and discuss the similarities and differences between recombinant DNA technology and these natural forms of recombination.
2. What is a plasmid? How are plasmids involved in transformation in bacteria?
3. What is a restriction enzyme? How can restriction enzymes be used to splice a piece of human DNA within a plasmid?
4. What is a DNA library? Briefly describe the steps involved in creating a DNA library of a mouse.
5. What is a restriction fragment length polymorphism? Describe the use of RFLPs in locating a gene on a human chromosome.
6. Describe the polymerase chain reaction. Why is it so useful in prenatal diagnosis of genetic defects?

DISCUSSION QUESTIONS

1. Let's say that you are a citizen member of a National Institutes of Health advisory panel on the uses of recombinant DNA. Discuss the ethical issues that surround the release of bioengineered organisms into the environment. What levels of benefits and risks do you think would justify such a release?

2. Do you think that using recombinant DNA technologies to change the genetic composition of a human egg cell is ever justified? If so, what restrictions should be placed on such a use?

SUGGESTED READINGS

Baskin, Y. "DNA Unlimited." *Discover*, July 1990. A clear, succinct description of the methods and applications of the polymerase chain reaction.

Mullis, K. B. "The Unusual Origin of the Polymerase Chain Reaction." *Scientific American*, April 1990. Mullis conceived the idea of PCR while on a drive through the mountains of California.

Roberts, L. "Ethical Questions Haunt New Genetic Technologies." *Science* 243:1134–1136, and "Ecologists Wary About Environmental Releases." *Science* 243:1141, both March 1990. These two articles provide a brief primer on the debates concerning the social and ecological issues surrounding the uses of recombinant DNA technology.

White, R., and Lalouel, J.–M. "Chromosome Mapping with DNA Markers." *Scientific American*, February 1988. The authors describe the use of RFLPs to locate genes on human chromosomes.

15

Human Genetics

Nature and nurture. These triplets are genetically identical, but are also growing up in extremely similar environmental situations. The interplay of both heredity and the environment shapes the characteristics of all organisms, but especially of people.

In principle, human genetics is similar to the genetics of peas or fruit flies, except that it is more interesting to most human beings. In practice, the study of human genetics is vastly more complex. First, humans have long life spans and few children per couple. Second, people choose their own mates, so crosses that might be genetically informative may not occur very often. Finally, humans interact extensively with their environment, an environment that is incredibly diverse from culture to culture and even within a single culture. These interactions often obscure underlying genetic patterns. This is particularly true for personality traits, in which learning and experience play such a major role.

Nevertheless, human genetics is a flourishing field of study, and a great deal is known about the inheritance of human traits. From the viewpoint of genetics, two factors partially compensate for the difficulties humans present as experimental subjects: the enormous number of human beings on the Earth, and the extensive documentation of human families, ranging from family Bibles to the pedigrees of royalty. Given enough time, ingenuity, and patience, geneticists can often find records of matings that help them determine how particular human traits are inherited, especially physical features and genetic diseases.

Methods in Human Genetics

Pedigree Analysis

Since experimental crosses are out of the question, human geneticists must search medical, historical, and family records to find crosses that have already been made voluntarily. Records extending across several generations can be arranged in the form of **family pedigrees,** such as those shown in Fig. 15-1. Careful analysis of pedigrees shows that certain traits, such as an unattached earlobe, are inherited as simple dominants, while other traits, such as albinism, are inherited as recessives.

Pedigree analysis requires that two important criteria be met. First, the trait must be clearly defined. This might seem trivial, yet often proves frustratingly difficult. For instance, "everyone knows" that brown eyes are dominant to blue, but eye color is really much more complex than that. Eyes are not simply blue or brown: they may be light blue, violet-blue, green, golden brown, medium brown, dark brown, or almost black. Carefully defining human eye colors presents quite a different picture than a simple dominant versus recessive pattern of inheritance.

The second criterion is closely related to the first: the people who draw the family tree must be certain who possessed which traits. When traits are difficult to define, or are commonplace, accurate pedigrees are rare. Who remembers, or would think to record, whether their great-grandmother's earlobes were attached or unattached? Thus, much of what we know about human genetics relates to striking physical features such as Kirk Douglas' cleft chin, or to diseases such as phenylketonuria (see Chapter 12).

Molecular Genetics

In the past 30 years, great strides have been made in understanding gene function on the molecular level. For instance, geneticists now know the molecular basis of several inherited diseases, such as sickle-cell anemia. Genetic engineering promises to increase our ability to predict genetic diseases and perhaps even to cure them (see Chapter 14 and the essay, "Methods in Biomedicine: Prenatal Diagnosis Through Biotechnology," later in this chapter).

Single Gene Inheritance

Many "normal" human traits, such as freckles, the ability to roll one's tongue, and the length of one's eyelashes, are inherited in a simple Mendelian fashion. Rather than merely cataloging inherited human traits, however, we will concentrate on a few examples of medically important genetic diseases and defects, and explore their consequences for individuals and society.

Recessive Inheritance

The human body depends on the integrated action of hundreds of enzymes and other proteins. A mutation in the gene coding for one of these enzymes almost always impairs or destroys enzyme function. As we described in the case of fat color in rabbits (Chapter 13), however, the presence of one normal allele often generates enough functional enzymes so that heterozygotes are phenotypically indistinguishable from homozygous normals. **Therefore, normal alleles are usually dominant and mutant alleles are recessive. Genetic diseases that are caused by lack of an essential enzyme are mostly inherited as recessives.**

If the metabolic pathway controlled by a particular gene is essential to survival, then homozygous recessives will die. This was especially true before medical care became widely available in recent decades. In evolutionary terms, this means that defective alleles

(a) A Pedigree for a Dominant Trait

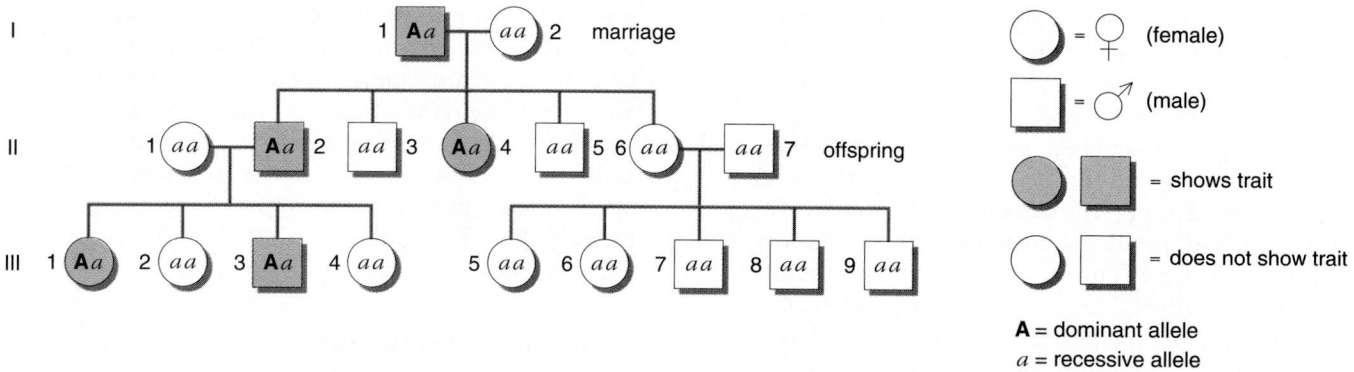

(b) A Pedigree for a Recessive Trait

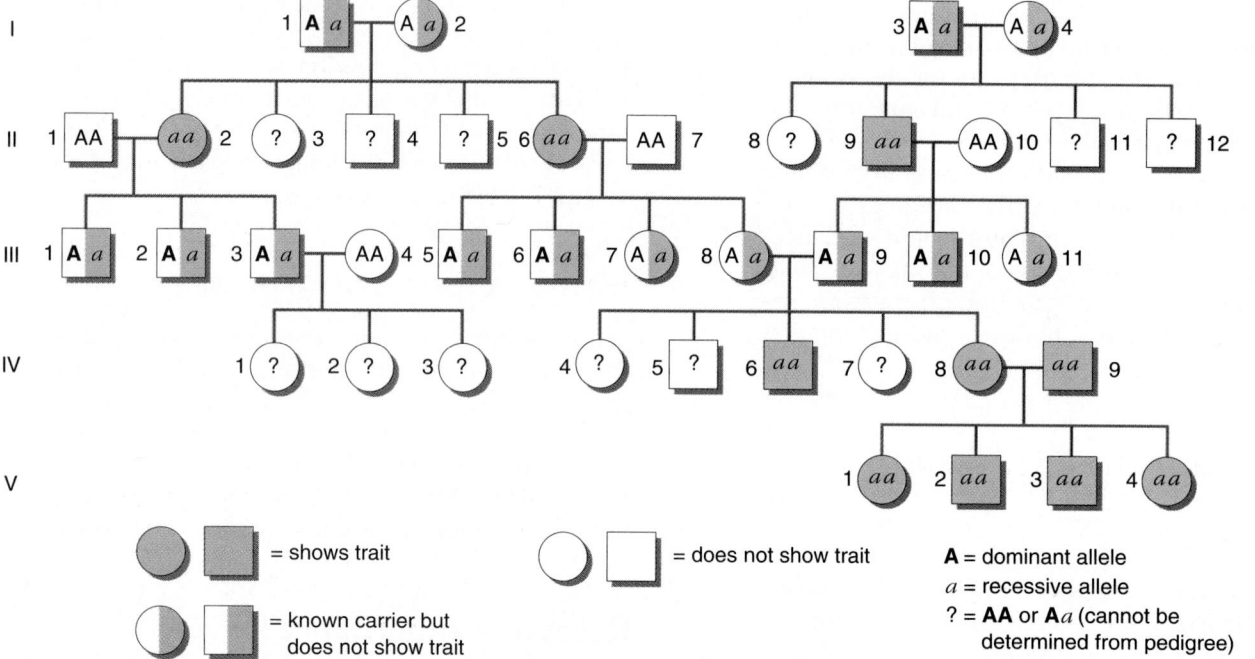

Figure 15-1 Representative family pedigrees.
(a) A pedigree for a dominant trait. Remember that all individuals bearing even one dominant allele will express the trait. Therefore, the trait will appear in every generation, in both males and females, and no child will possess the trait who does not have at least one parent with the trait. On the other hand, since heterozygotes show dominant traits, not all the children of phenotypically dominant parents will necessarily show the trait (e.g., generation III, individuals 2 and 4). In this hypothetical pedigree, the genotypes of the individuals are given; in real pedigrees, genotypes must be deduced from inheritance patterns and cannot always be determined with certainty. **(b)** A pedigree for a recessive trait. Only homozygous recessives will show the trait. People with only one recessive allele will have the dominant phenotype. Parents with the dominant phenotype but who have a child who shows the recessive trait must themselves be heterozygotes. Therefore, the recessive phenotype may not appear in every generation (e.g., generation III, families composed of individuals 5 through 8 and 9 through 11); the recessive allele may remain "hidden" in heterozygotes for several generations until two heterozygotes happen to marry (generation III, individuals 8 and 9) and have a homozygous child (generation IV, individuals 6 and 8). Two parents who both show the recessive phenotype (IV 8 and 9) will have all homozygous recessive children (V 1–4). In a pedigree for a rare trait, it is usually safe to assume that all individuals are homozygous normal (II 1, 7, and 10) unless there is positive evidence of possession of the recessive allele (e.g., an affected child).

are selected against. Consequently, for serious diseases, even heterozygous **carriers,** who are phenotypically normal but can pass on their defective allele to their children, are usually rare. An unrelated man and woman will seldom both possess the same defective allele and produce a homozygous child. Related couples, however, especially first cousins or closer, have inherited some of their genes from recent common ancestors. Therefore, they are much more likely to carry the same allele and have an affected child (Fig. 15-2).

By their very nature, genetic diseases are an integral part of an affected person. They can neither be prevented with vaccines nor cured with antibiotics. Some, such as diabetes, can be treated, with varying degrees of success. As we pointed out in Chapter 14, genetic engineering offers the promise of a permanent genetic cure in certain cases. In instances in which the expression of the defective genes is localized to certain accessible organs such as liver or bone marrow, it may soon be possible to extract some of the patient's cells, insert the proper genes, and reimplant them back into the patient. However, even here the genetic defect would still remain in the reproductive cells that give rise to sperm or eggs, and the affected person might pass on the gene to his or her children.

For the present, then, the only way to prevent genetic diseases is to prevent the birth of affected babies. Reducing the incidence of recessive genetic diseases would require (1) identifying heterozygous carriers and (2) preventing reproduction by couples who are both carriers or (3) screening fetuses and aborting those that are homozygous recessive. Such a course of action is often both scientifically difficult and ethically questionable.

By definition, carriers of recessive genetic diseases cannot be identified by casual observation; if they can be identified at all, it usually requires expensive medical procedures. With a few exceptions, it is impractical to screen the entire population of any country for a genetic disease. Screening tests can be useful, however, when the defective alleles are found almost exclusively within a readily identifiable group. Tay–Sachs disease and sickle-cell anemia are recessive genetic diseases, largely restricted to certain segments of the population, for which both carrier identification and prenatal diagnosis are possible. These two diseases illustrate the procedures used to diagnose recessive diseases and some possible courses of action. The practical and ethical dilemmas implicit in the diagnosis and treatment of genetic diseases are explored in the Reflection on Medical Genetics at the end of the chapter.

Tay–Sachs Disease

Tay–Sachs disease occurs when nerve cells fail to synthesize an enzyme involved in the breakdown of lipids. Lipid accumulates in the brain of a child who is homozygous for the Tay–Sachs allele, causing progressive mental retardation, blindness, and failure of motor control (Fig. 15-3a). There is no cure, and death occurs in early childhood.

The Tay–Sachs allele is very rare in the United States population as a whole; only about one in 400 Americans is a carrier. However, for unknown reasons, about one in 30 American Jews is a carrier, which would result in about one Tay–Sachs child in 3600 Jewish births. Heterozygotes for Tay–Sachs disease can be easily detected by a blood test (Fig. 15-3b), and many prospective Jewish parents avail themselves of this test. If both husband and wife are identified as carriers, they have three choices: (1) forego reproduction entirely; (2) take a 25% chance of having an affected child who will face suffering and a certain early death; or (3) through **amniocentesis** or **chorionic villus sampling** determine whether each fetus is affected and abort homozygous recessives. The techniques of amniocentesis and chorionic villus sampling are described in "Methods in Biomedicine: Collecting Fetal Samples."

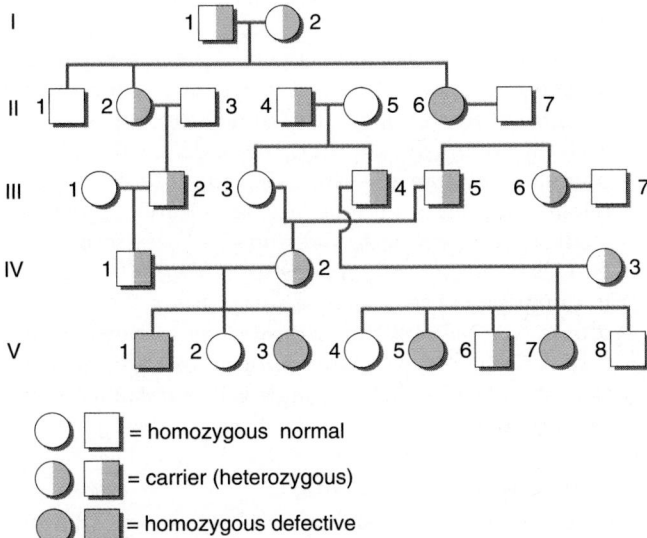

= homozygous normal

= carrier (heterozygous)

= homozygous defective

Figure 15-2 When related people marry, they often carry some of the same alleles, inherited from their common ancestors. This greatly increases the likelihood that they will both carry the same defective recessive allele. As a result, marriages between cousins or even closer relations are the cause of a disproportionate number of recessive diseases. In this family, marriages between cousins were common, including those between III 3 and 5, III 4 and IV 3, and IV 1 and 2.

(a)

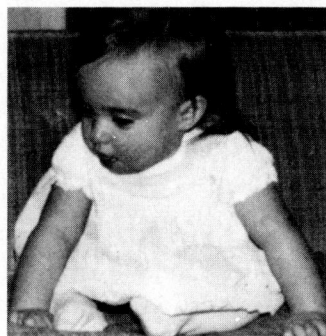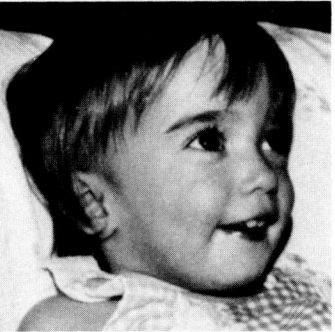

(b)

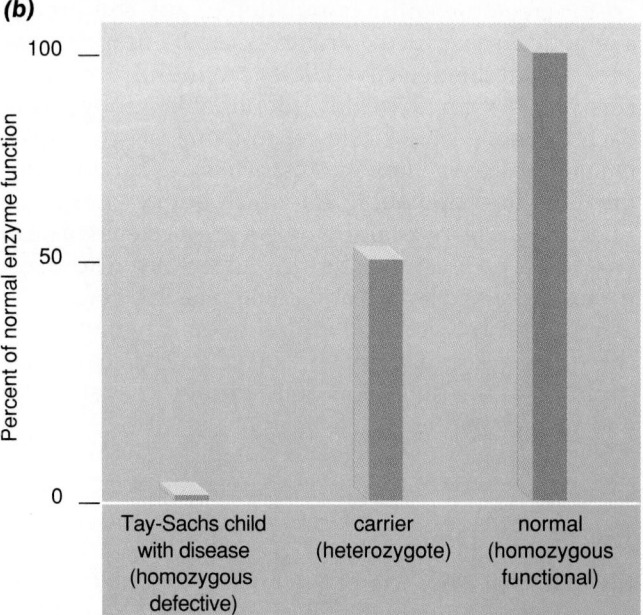

Figure 15-3 Tay–Sachs disease.
(a) The progression of Tay–Sachs disease. Evelyn, beautiful and seemingly bright, happy, and normal for the first few months after birth (left, 9 months), never managed simple tasks such as sitting upright (notice that she is using her hands for balance). As the disease progressed (right, 17 months), she became unable to respond to sounds or other stimuli, and suffered from recurring infectious diseases. She died shortly before her sixth birthday.
(b) The Tay–Sachs allele codes for a nonfunctioning enzyme. Even though heterozygotes may appear phenotypically normal, they often have only about half as much of the normal enzyme as homozygous dominants do, because they have only one normal allele. In Tay–Sachs, this difference in the concentration of normal enzymes in the blood can be used to detect heterozygotes. This is the basis for genetic counseling that can prevent what Evelyn's parents have called "this cruel but preventable disease."

METHODS IN BIOMEDICINE
Collecting Fetal Samples

Prenatal diagnosis of a variety of genetic disorders, including sickle-cell anemia, Tay–Sachs disease, Down syndrome, and thalassemia, requires samples of fetal cells or chemicals produced by the fetus. Presently, there are two main techniques used to obtain these samples: amniocentesis and chorionic villus sampling. Both procedures have their advantages and disadvantages. A few years ago, it was widely predicted that chorionic villus sampling would largely replace amniocentesis by the early 1990s, but this has not yet happened. We will briefly describe both procedures, and some of the tests that can be performed on the resulting fetal samples.

Amniocentesis
The human fetus, like all animal embryos, develops in a watery environment. A waterproof membrane called the amnion (see Chapter 40) surrounds the fetus and holds the amniotic fluid. This fluid contains fatty acids, steroids, free amino acids, enzymes and other proteins, and cells. The chemicals come from both the mother and the fetus, but the cells are all from the fetus, having been shed by the skin, digestive tract, and respiratory tract. When a fetus is 16 weeks or older, enough amniotic fluid has accumulated, and there is enough space between the fetus and the amniotic membrane, so that amniotic fluid can be collected by a procedure called **amniocentesis** (Fig. E15-1). A physician determines the position of the fetus by ultrasound scanning and inserts a sterilized needle through the abdominal wall of the pregnant woman and into the amniotic fluid. Ten to twenty milliliters of fluid are withdrawn. Biochemical analysis may be performed on the fluid immediately. However, there are very few cells in the fluid sample. For many analyses, such as karyotyping for Down syndrome (see below), the cells must first be grown in culture. The cells multiply in culture, and after a week or two there are usually enough cells for karyotyping or other analyses.

Chorionic Villus Sampling
This is a newer and still somewhat controversial procedure. The chorion is a membrane, produced by the fetus, that becomes part of the placenta (see Chapter 40). The chorion produces many small projections, called villi; the loss of a few villi seems to cause no harm. In **chorionic villus sampling,** a physician inserts a small tube into the uterus through the vagina, and suctions off a few villi for analysis (see Fig. E15-1).

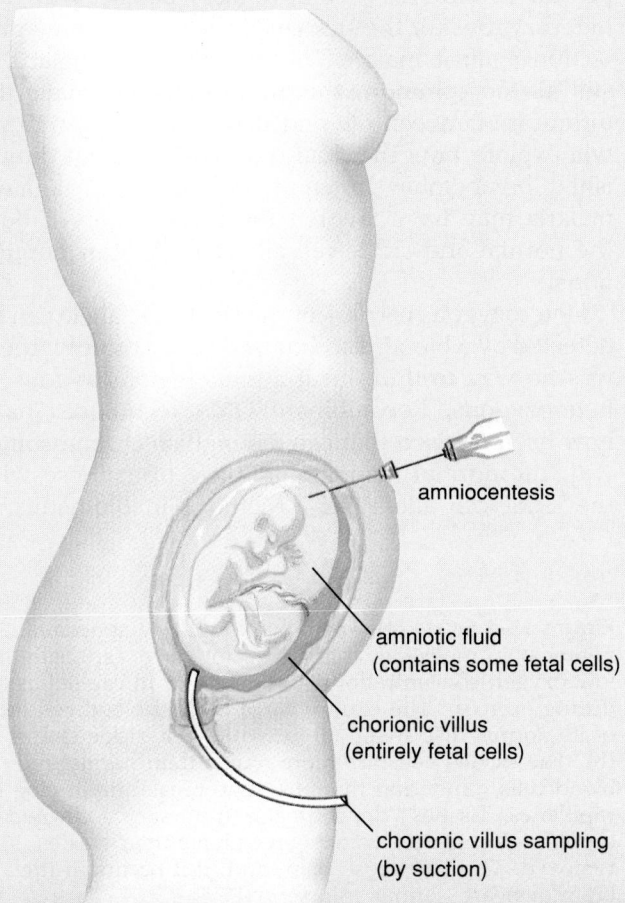

amniocentesis

amniotic fluid
(contains some fetal cells)

chorionic villus
(entirely fetal cells)

chorionic villus sampling
(by suction)

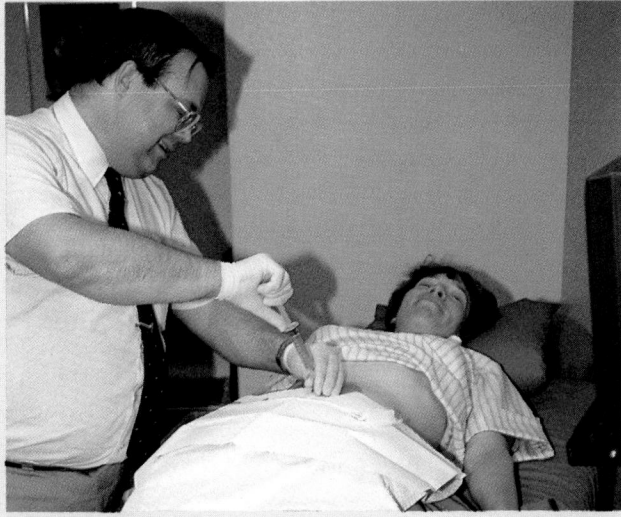

Figure E15-1

Analyzing the Samples

Several analyses can be performed on the fetal cells or on the amniotic fluid. Biochemical analysis is used to determine the concentration of chemicals in the amniotic fluid. For example, Tay–Sachs disease and many other metabolic disorders can be detected by the low concentration of the enzymes that normally catalyze specific metabolic pathways, or by the accumulation of precursors or by-products. Analysis of the DNA of fetal cells with recombinant DNA techniques can detect some defective alleles, such as the Tay–Sachs and sickle-cell anemia alleles (see "Methods in Biomedicine: Prenatal Diagnosis Through Biotechnology"). Analysis of the chromosomes of the fetal cells can show if all the chromosomes are present in their normal number, if there are too many or too few of some, and if there are gross structural abnormalities of any of the chromosomes.

Comparing the Two Techniques

Chorionic villus sampling promises two great advances over amniocentesis. First, it can be done much earlier in pregnancy, perhaps as early as the eighth week. This is especially helpful if the woman is contemplating a therapeutic abortion in case the fetus suffers from a major defect, such as Down syndrome. Second, the sample contains a much higher concentration of fetal cells than amniocentesis can obtain, so analyses can be performed immediately rather than waiting for a week or two. However, thus far chorionic villus sampling has a somewhat higher risk of inducing miscarriages than amniocentesis does, although the risk is still small. Further, chorionic cells seem to be rather likely to have abnormal numbers of chromosomes, even in a normal fetus, which complicates karyotyping. In most large metropolitan areas, both amniocentesis and chorionic villus sampling are available, and both have their proponents and detractors.

Sickle-Cell Anemia

As with Tay–Sachs disease, almost all the carriers of sickle-cell anemia belong to one ethnic group, in this case blacks. In sickle-cell anemia, a single nucleotide substitution in DNA (adenine for thymine) causes valine to be substituted for glutamic acid at one position on the outside of the hemoglobin molecule (see Chapter 13). Glutamic acid is highly charged, while valine is neutral and hydrophobic. Glutamic acid is crucially important in keeping hemoglobin molecules dissolved in the cytoplasm of the red blood cells. Substituting valine at this position causes the hemoglobin molecules to clump together, forcing the red blood cells into the sickle-like shape that gives the disease its name (Fig. 15-4). During exercise or stress, the sickled cells break and clog the capillaries, cutting off circulation. In some instances, this may cause fatal strokes or heart attacks. Heterozygotes have about half normal and half abnormal hemoglobin, which you might expect would lead to partial sickle-cell symptoms. However, heterozygotes usually have few sickled cells, and they show no symptoms whatever (in fact, many world-class black athletes are heterozygotes).

Sickle-cell anemia is surprisingly common among blacks. In the United States, about 8% of the black population is heterozygous; in parts of Africa, two or three times this many may be heterozygous. Given how severe the disease can be to homozygotes, why hasn't natural selection eliminated the sickle-cell allele? The prevalence of the sickle-cell allele is probably an evolutionary compromise: people who are heterozygous for the sickle-cell allele enjoy some protection against malaria. In Africa, where malaria is still all too common, this protection may make the difference between life and death. In Chapter 17 we will explore how the dual selective effects of diminished hemoglobin function and protection against malaria may have favored the preservation of both the normal and sickle-cell alleles in African populations.

Heterozygous carriers of the sickle-cell allele can be detected by a blood test, but until fairly recently there was no way to find out if a fetus is homozygous or heterozygous. Recombinant DNA techniques have now been devised that can distinguish chromosomes with the normal hemoglobin allele from those with the sickle-cell allele (see "Methods in Biomedicine:

(a)

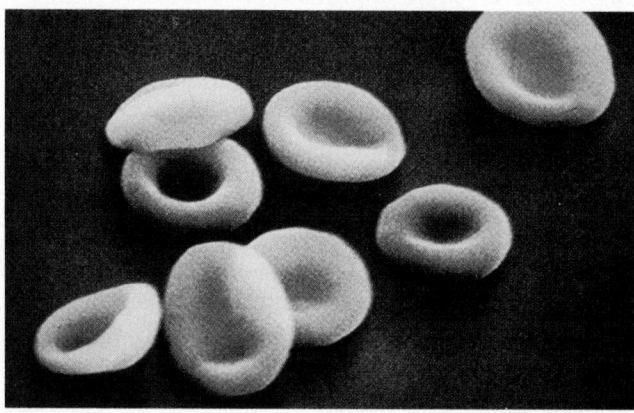

(b)

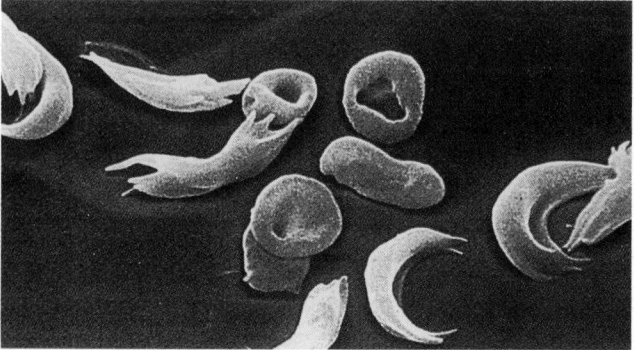

Figure 15-4 Sickle-cell anemia is caused by abnormal hemoglobin molecules that clump together, especially in low oxygen concentrations such as occur in capillaries during exercise. The clumps force the red blood cell out of its normal disc shape **(a)** into a longer, sickle shape **(b).** The sickled cells are more fragile than normal red blood cells, rendering them likely to break and/or clog in capillaries. Tissues "downstream" from such a clogged capillary do not receive oxygen or have their wastes removed. This can cause pain, and, if it occurs in the brain or heart, serious injury.

Prenatal Diagnosis Through Biotechnology"). Analyzing fetal cells collected by amniocentesis or chorionic villus sampling now allows medical geneticists to diagnose sickle-cell anemia in fetuses.

Dominant Inheritance

Many physical traits are inherited as dominants, including cleft chin and freckles. However, for two reasons, few people have serious genetic diseases caused by dominant alleles. First, as we have already explained, when a mutation strikes a normal allele, it usually produces a nonfunctioning, recessive allele. Second, everyone bearing a dominant defective allele will develop the disease: there can be no phenotypically normal carriers. Before the advent of modern medicine, if the disease was serious, then these people died without reproducing, and did not pass on their defective allele to future generations.

An exception to this rule is Huntington's disease (Fig. 15-5). An incurable disease, Huntington's causes a slow, progressive deterioration of parts of the brain, resulting in loss of motor coordination, flailing movements, personality disturbances, and eventual death. Huntington's disease is particularly insidious because symptoms usually do not occur until 30 to 50 years of age. Therefore, a person usually has already had children before he or she ever suffers the first symptoms.

Since everyone who has even one defective dominant allele shows the defective trait (if they live long enough), dominant diseases could be virtually eliminated in a single generation if all the affected people chose not to reproduce. In the case of Huntington's, however, few people know for sure if they have the disease until after they have had children. Therefore, to eliminate Huntington's, all people with a parent suffering from the disease would themselves have to forego having children. Since half of these people would be homozygous normal, and could not pass on the Huntington's gene, the decision not to reproduce would have to be based on statistics. For some, the choice is clear; for others, the dilemma is excruciating.

In 1984, painstaking pedigree analysis, RFLP analysis (see Chapter 14), and a lot of luck combined to localize the Huntington's gene to a relatively small part of chromosome 4. No one knows what the protein encoded by the normal Huntington's gene is, or what it does in the brain, much less what's wrong with the protein encoded by the defective allele. By finding the gene and studying the proteins produced by the normal and defective alleles, perhaps it will be possible to devise therapies that slow or stop the progress of the disease. Unfortunately, although

(a)

(b)

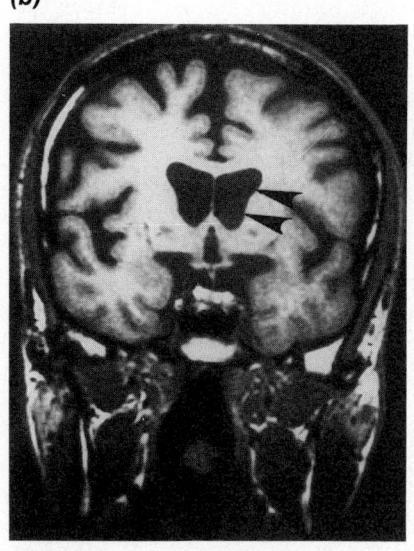 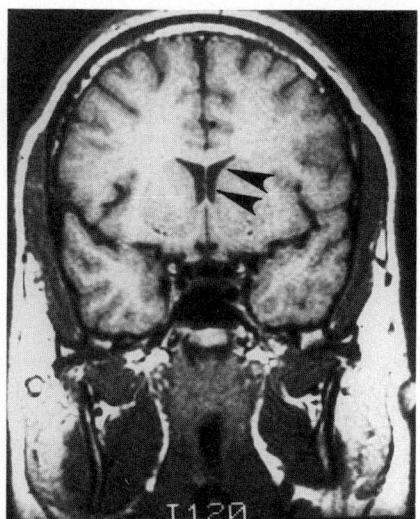

Figure 15-5 The public became aware of Huntington's disease when folksinger Woody Guthrie (a) developed symptoms in the 1950s. The reason for the behavioral and motor deficits that Huntington's patients develop is appallingly clear when we compare magnetic resonance images of the brains of a living patient with Huntington's disease (b, left) and an age-matched normal person (b, right). The brain of the Huntington's victim has greatly enlarged ventricles (dark butterfly-shaped area at the arrows) that are filling up the space formerly occupied by brain tissue.

METHODS IN BIOMEDICINE
Prenatal Diagnosis Through Biotechnology

Recombinant DNA technologies are increasingly valuable tools in medicine. Three applications of genetic engineering in medicine include the synthesis of therapeutic drugs or other agents, the early diagnosis of genetic diseases, and the possible elimination of the effects of genetic diseases by implanting genes.

Synthesizing therapeutic drugs with recombinant DNA techniques has now moved into the mainstream of the pharmaceutical industry. Many formerly independent biotechnology companies have been acquired by the giants of the industry, who see drug manufacturing by recombinant DNA no longer as the wave of the future, but as the technology of the present. Insulin, growth hormone, clotting factors, enzymes to treat heart attacks—these are but a few of the drugs now routinely manufactured this way.

As we described in Chapter 14, gene implantation is just now entering clinical trials in a few special cases. It is not likely to be a major therapeutic technique for some time. Prenatal diagnosis of genetic defects, however, is rapidly becoming the province of biotechnology. In this essay, we describe a simple, elegant application of recombinant DNA techniques to the prenatal diagnosis of sickle-cell anemia.

Sickle-cell anemia is caused by a point mutation in the hemoglobin gene, in which adenine is substituted for thymine. Heterozygotes are usually completely normal, but people homozygous for the sickle-cell allele may suffer painful and potentially fatal blood clots during exercise. Although there are no cures, there are some moderately effective treatments for sickle-cell anemia. Using recombinant DNA techniques, prospective parents can be tested to see if they are carriers, and both fetuses and infants can be tested to see if they are homozygous for the sickle-cell allele. Here, we will describe prenatal diagnosis; similar procedures (without the need for amniocentesis or chorionic villus sampling, of course), can be used for parents or infants.

Prenatal diagnosis of sickle-cell anemia is a three-step process: collecting fetal cells, amplifying the DNA with the polymerase chain reaction, and using a restriction enzyme called Mst II to analyze the DNA for the presence of the defective hemoglobin gene. Mst II cuts apart DNA that has the base sequence of the normal hemoglobin allele, but cannot cut apart DNA that has the mutated base (Fig. E15-2).

First, fetal cells are collected by amniocentesis or chorionic villus sampling (see "Methods in Biomedicine: Obtaining Fetal Samples"). For many applications, amniocentesis has a major drawback: very few fetal cells are collected. Just a few years ago, prenatal diagnosis of sickle-cell anemia or other genetic disorders required thousands or millions of copies of the DNA. Therefore, fetal cells often had to be grown in culture for as long as 2 weeks before they had multiplied enough to be used. For sickle-cell anemia, the development of the polymerase chain reaction (PCR) has eliminated the wait.

As you may recall from Chapter 14, PCR can produce virtually unlimited copies of a gene, starting with as little as one copy. Further, it can pick out and amplify the desired gene from amidst the entire human genome. The second step in prenatal diagnosis of sickle-cell anemia, therefore, is to extract the DNA from a few cells, and to amplify the hemoglobin genes (both normal and sickle-cell versions) with PCR. This takes only a few hours.

The third step is to cut the PCR-amplified DNA with the restriction enzyme Mst II, and determine the lengths of the resulting pieces of the hemoglobin gene. Mst II cuts the normal hemoglobin allele into two pieces, but leaves the sickle-cell allele intact (Fig. E15-2a). Mst II is simply added to the PCR-amplified DNA. The resulting pieces of DNA are sorted according to size by gel electrophoresis (see Chapter 14) and stained to show where the DNA is located (Fig. E15-2b). The exact genotype of the fetus can easily be determined (Fig. E15-2c). If the fetus is homozygous for the normal hemoglobin allele, all the pieces of DNA cut by Mst II will be short. If the fetus is homozygous for the sickle-cell allele, all the pieces will be long. If the fetus is heterozygous, there will be both long and short pieces of DNA on the gel.

White blood cells from parents or infants can be analyzed in the same way. If at least one parent is not a carrier, then the fetus cannot be homozygous for the sickle-cell allele (why?), and prenatal diagnosis is not necessary. If both parents are carriers, and prenatal diagnosis is not carried out, then the infant can be tested (many states now require this). If the infant is homozygous for the sickle-cell allele, some preventive measures can be taken. In particular, regular doses of penicillin greatly reduce bacterial infections that otherwise kill about 15% of homozygous children. Further, knowing that a child has the disorder ensures correct diagnosis and rapid treatment, should "sickling crises" occur.

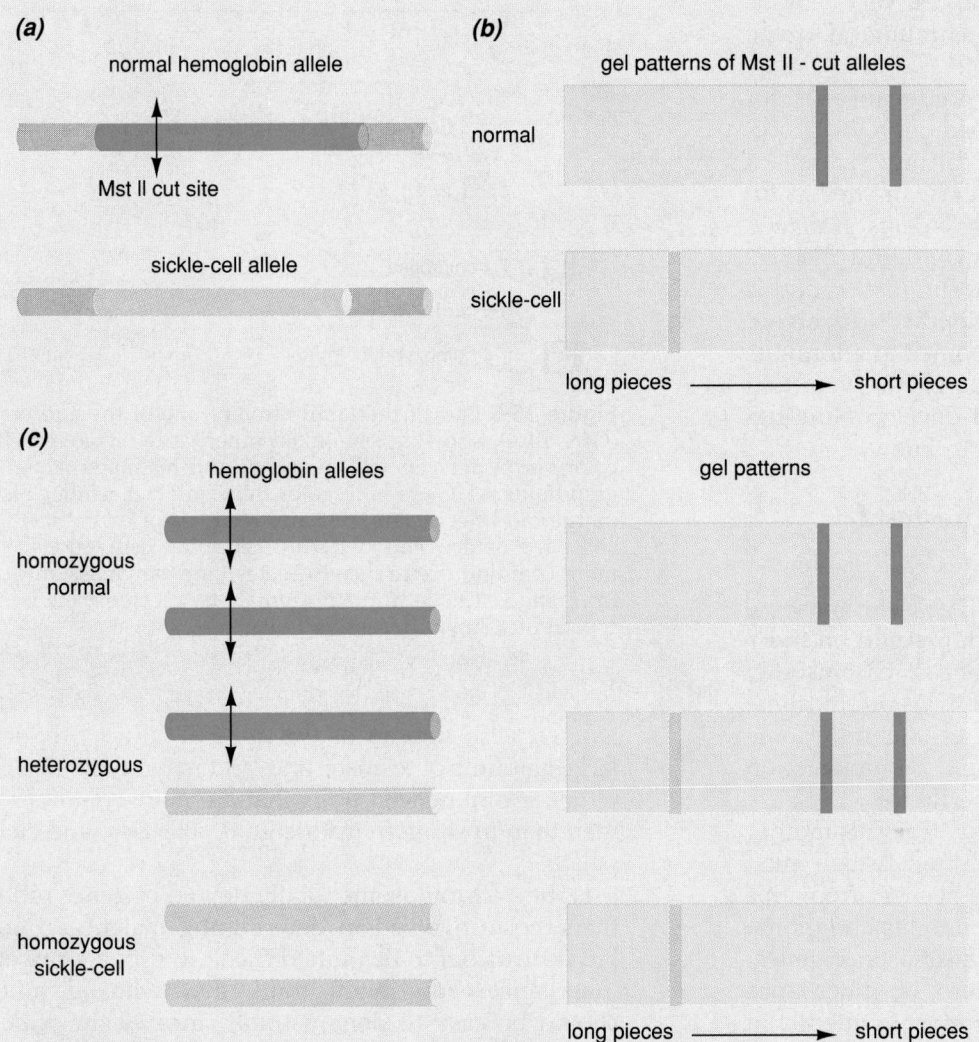

Figure E15-2 Diagnosing sickle-cell anemia.
(a) The restriction enzyme Mst II can cut DNA with the base sequence of the normal hemoglobin allele, but not with the altered sequence of the sickle-cell allele.
(b) PCR-amplified hemoglobin genes are exposed to Mst II, and the resulting pieces are separated by gel electrophoresis. The normal allele is cut into two short pieces, while the sickle-cell allele is left intact in one long piece.
(c) Analysis of fetal genes. If the fetus is homozygous normal, then Mst II will cut both hemoglobin alleles on both homologous chromosomes into two short pieces (red). If the fetus is heterozygous, Mst II will cut the normal hemoglobin allele (on one chromosome) into two short pieces of DNA, but will be unable to cut the sickle-cell allele (on the homologous chromosome). The gel will therefore contain both short pieces of DNA (from the normal allele; red) and long pieces (from the sickle-cell allele; pink). If the fetus is homozygous recessive for the sickle-cell allele, the gel will contain only long DNA pieces (pink).

RFLP analysis will probably work eventually, finding an unknown gene with RFLPs can be very tedious and frustrating: by early 1991, the Huntington's gene itself still had not been found.

RFLP analysis has succeeded in producing a test for Huntington's disease, however. Although the procedure is not infallible, geneticists can now sample a person's DNA and predict whether he or she has inherited the Huntington's allele *before any symptoms appear*. At first glance, this is a wonderful medical advance, which could result in the elimination of this devastating disease. For the individuals involved, however, the knowledge will be gained at enormous personal cost. What would *you* do if you were told that in 20 or 30 years you would develop Huntington's? Many people do not want to know.

Sex-Linked and Sex-Influenced Inheritance

As we described in Chapter 11, the X chromosome bears many genes that have no counterpart on the Y chromosome. With one X and one Y chromsome, males are effectively haploid for X-chromosome genes; that is, recessive alleles of X-chromosome genes are always expressed in men, a phenomenon called **sex-linked** or **X-linked inheritance.**

A son receives his X chromosome from his mother and his Y chromosome from his father. A man must therefore inherit X-chromosome genes from his mother, and can only pass them on to his daughters. For rare, recessive alleles of X-chromosome genes, there is usually a striking pattern of inheritance (Fig. 15-6). Recessive traits appear most frequently in males, and typically skip generations, with an affected male passing the trait on to a phenotypically normal, carrier daughter who in turn bears affected sons. The most familiar genetic defects due to recessive alleles of X chromosome genes are red–green colorblindness (Fig. 15-6) and hemophilia (Fig. 15-7).

Each sex also has its own set of **sex-influenced traits** (in addition to the obvious ones of breast and genitalia development) that occur more commonly or more strongly in that sex but are not coded by genes on the sex chromosomes. The majority of sex-influenced traits affect males more often than females, and seem to be enhanced by male sex hormones. A familiar example is baldness, which appears as if it were dominant in men (heterozygotes become bald) but recessive in women (heterozygotes retain their hair). The hormonal connection in baldness is readily apparent in people with abnormal levels of sex hormones. Castrated men almost never become bald, but if given testosterone treatments, castrated men with the allele for baldness usually lose their hair. Some

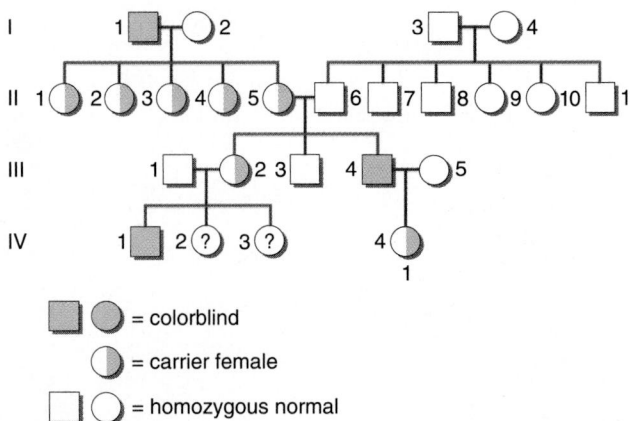

Figure 15-6 Part of the family tree of one of the authors (GJA; III 4), showing sex-linked inheritance of red–green color vision deficiency. The author and his maternal grandfather (I 1) are both color deficient, but neither his mother and her sisters, nor any relatives of his father, show the defect. This pattern of skipping generations, more common occurrence in males, and transmission from affected male to carrier female to affected male is typical of X-linked recessive traits.

women with tumors of the adrenal cortex produce large amounts of testosterone, and may become bald. Other sex-influenced traits that are more common in men than in women include gout, allergies, and cleft palate.

Is the Y chromosome totally devoid of genes other than those involved in determining maleness? No, but it turns out to be quite difficult to prove Y-linked inheritance. You might think that Y-linked traits should be easy to demonstrate, since *all* the males and *only* the males descended from a given affected male ancestor should show the trait. In practice, the situation is complicated by the influence of male sex hormones. Most traits that once were thought to be due to Y-chromosome genes are actually due to autosomal genes whose expression is strongly influenced by testosterone levels.

Complex Inheritance

Many human characteristics are not inherited in a simple either/or fashion. Even traits that are controlled by single genes, such as Huntington's disease, are often influenced by other genes and by environmental factors. For example, although everyone who has at least one allele for Huntington's contracts the disease, the age at which the first symptoms appear is extremely variable: some people are affected in their teens, while others do not develop symptoms until their 50s or even 60s.

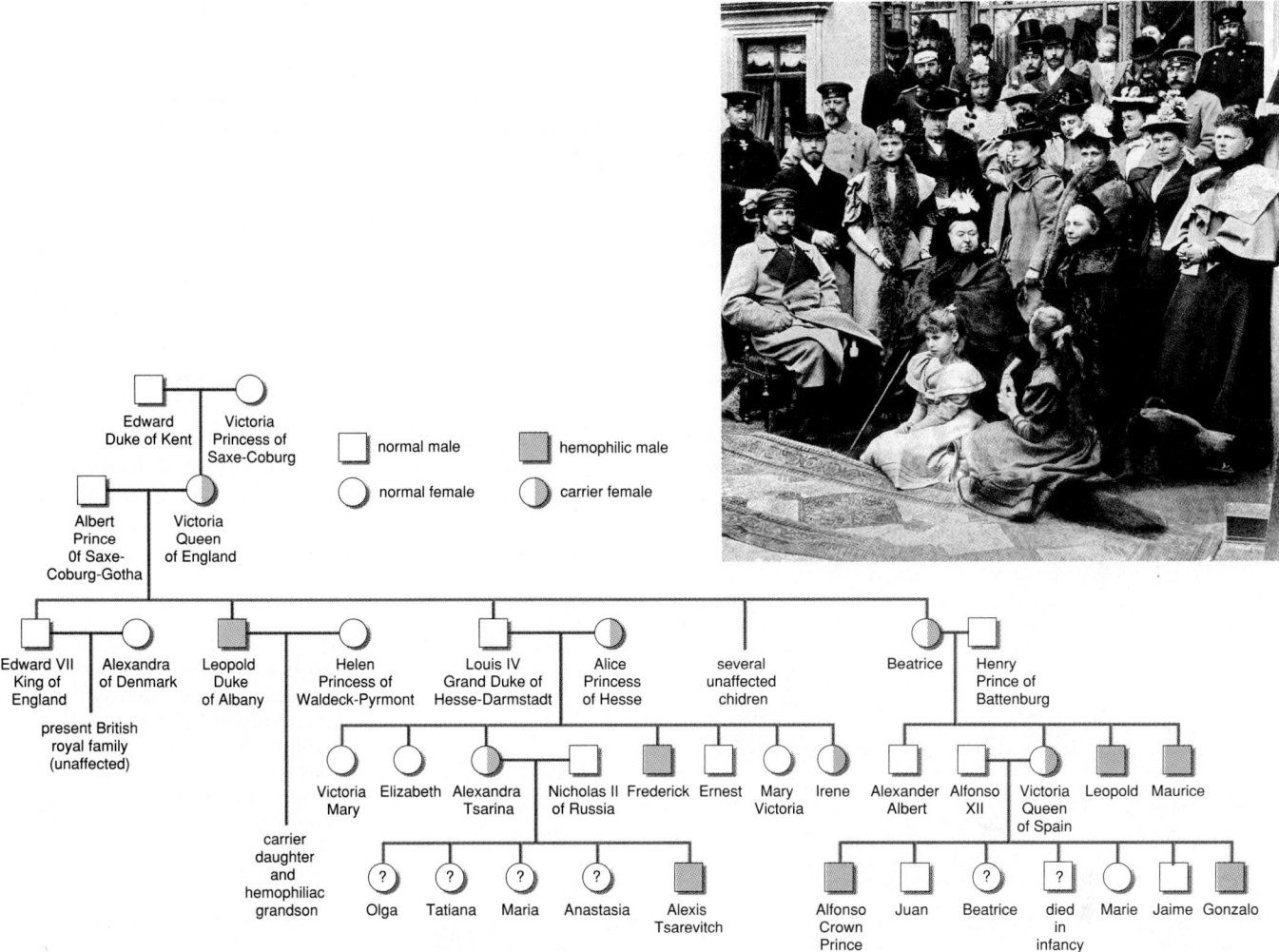

Figure 15-7 Hemophilia among the royal families of Europe. Hemophilia results from a deficiency of one of the factors causing blood clotting, a deficiency inherited as a recessive allele on the X chromosome. Affected males do not bleed to death from the first scratch or bruise; most hemophiliacs can clot off a minor wound. As a result, males often survive to pass on their allele to their daughters. Homozygous hemophiliac women are rare, but some are known. Surprisingly enough, they do not usually die from menstruation, nor even from bleeding following childbirth. In both of these instances, blood flow is stopped not by clotting, but by muscular contraction, shutting off circulation to the uterine wall.

The most famous genetic pedigree in history involves the transmission of sex-linked hemophilia from Queen Victoria of England (seated in right center, with crown) to her offspring, and eventually to virtually every royal house in Europe. Since all of Victoria's ancestors were free of hemophilia, the hemophilia allele must have arisen as a mutation either in Victoria herself when she was an embryo, or in one of her parents.

Extensive intermarriage among royalty, who, after all, are not supposed to marry commoners, spread Victoria's hemophilia allele throughout Europe. Her most famous hemophiliac descendant was great-grandson Alexis, Tsarevitch (crown prince) of Russia. Bleeding episodes in the only son of the Tsar naturally distressed his parents greatly. The Tsarina Alexandra (Victoria's granddaughter) believed that the monk Rasputin, and no one else, could control Alexis' bleeding. Rasputin may actually have been able to do this through hypnosis, perhaps causing Alexis to cut off circulation to bleeding areas by muscular contraction. Although there were many causes underlying the Russian Revolution, the influence that Rasputin had over the imperial family may have contributed to the downfall of the Tsar. In any event, hemophilia was not to be the cause of Alexis' death; along with the rest of his family, he was killed by the Bolsheviks in 1918.

Polygenic Inheritance

Many, perhaps most, human traits are determined by the interaction of two or more genes. Examples of such **polygenic inheritance** in humans are the colors of eyes and skin.

The color of the iris in human eyes varies from very pale blue through green to almost black. Nevertheless, there are no blue, green, or black pigments in the human iris. Eye colors are actually caused by the distribution of a single yellowish-brown pigment, melanin, the same pigment that colors skin and hair. The iris contains two layers of pigment, one at the back and one at the front. If there is little or no pigment at the front, the iris appears blue. The blue color arises from the scattering of light in the front layers, viewed against the dark background of melanin in the rear, just as the sky appears blue because of light scattered by the air, seen against the black background of space. In people with greater amounts of melanin in the front layers, the iris color may be green (blue plus yellowish brown), brown, or almost black. At least two, and probably more, genes direct the synthesis of melanin in the front of the iris, with each gene having two alleles showing incomplete dominance. The simplest scheme of two genes can create five shades of eye colors (Fig. 15-8).

Skin color is another case of different amounts of melanin. People don't really have white, yellow, red, or black skin: all are various shades of brown, with a pinkish tint from surface blood vessels showing through the paler tones. Skin color is inherited through the action of at least three genes with incompletely dominant pairs of alleles. As with eye color, polygenic inheritance explains both a more-or-less continuous gradation of skin colors and the occasional offspring whose skin color differs considerably from that of either parent.

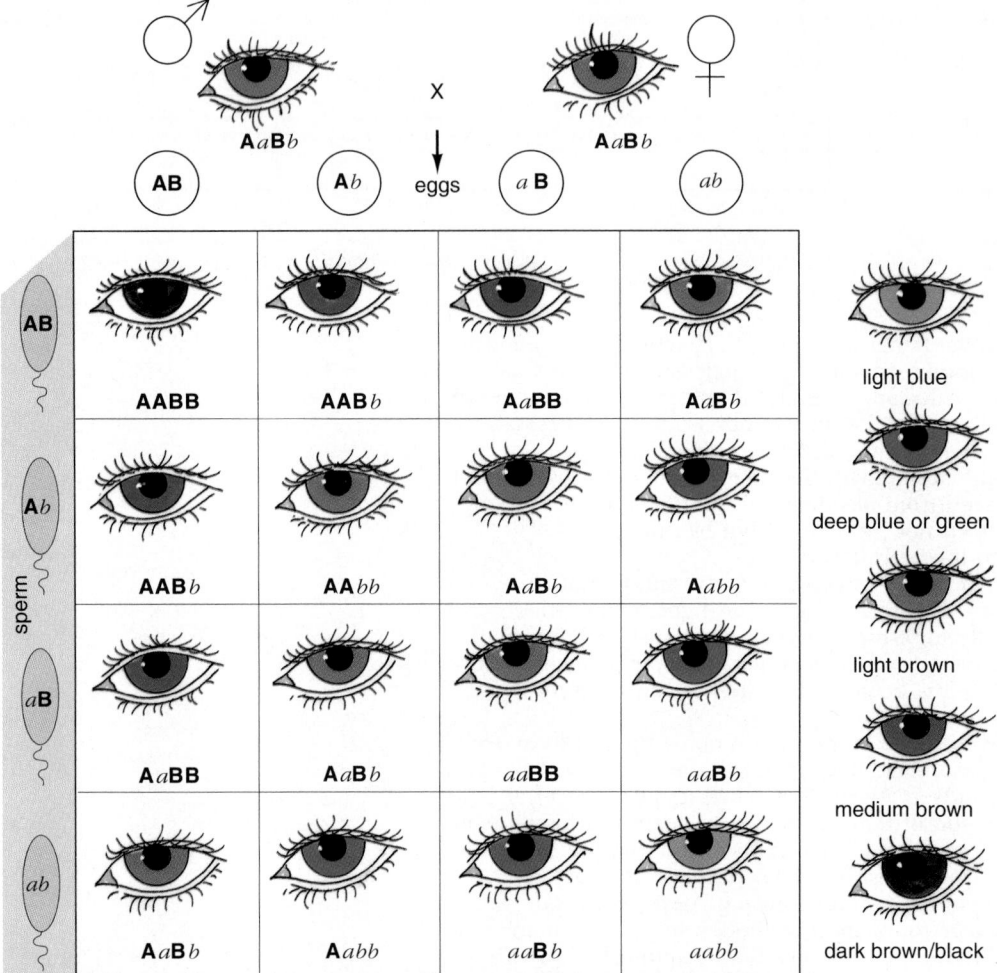

Figure 15-8 At least two separate genes, each with two incompletely dominant alleles, govern human eye color. A man and a woman, each heterozygous for both genes, could have children with five different eye colors, ranging from light blue (no dominant alleles) through light brown (two dominants) to almost black (all four alleles dominant).

light blue

deep blue or green

light brown

medium brown

dark brown/black

drome are sterile, usually short in stature, often have webbed skin around their necks, and, under microscopic examination, lack Barr bodies in their nuclei (see Fig. 13-15). Mentally, they are usually normal, except that they are frequently weak in mathematics and spatial perception. The differences between XO and XX women suggest that the theory of X chromosome inactivation in females, presented in Chapter 13, is oversimplified. Some genes on the "inactivated" X chromosome must be functional in XX females, preventing the Turner's syndrome traits.

Trisomy X (XXX)

About 1 in 1000 women have three X chromosomes. These women usually have no detectable defects at all, except for a higher incidence of subnormal intelligence. Unlike women with Turner's syndrome, XXX women are fertile, and, interestingly enough, almost always bear normal XX and XY children. Some mechanism, presently unknown, must operate during meiosis to prevent the extra X chromosome from being included in the egg.

Klinefelter's Syndrome (XXY)

About one male in 1000 is born with two X and one Y chromosomes. At puberty, these men show mixed secondary sexual characteristics, including partial breast development, broadening of the hips, and small testes. Men with Klinefelter's syndrome are always sterile, but usually not impotent. As is common in people with abnormal chromosome numbers, XXY males have an increased incidence of mental deficiency; in fact, about 1% of all people institutionalized for mental retardation are XXY males.

XYY Males

The last common type of sex chromosome abnormality is XYY, occurring in about one male in 1000. You might expect that having an extra Y chromosome, which presumably has few genes, would not make very much difference, and this seems to be true in most cases. However, XYY males may be affected in two ways: below average intelligence and above average height (about two thirds of XYY males are over 6 feet tall, compared with the average male height of 5'9"). There is some debate about whether XYY males are genetically predisposed to violence. For instance, several studies have shown that a higher than expected percentage of men in prison are XYY. In several countries men accused of murder have attempted to use their XYY constitution as a defense, like the insanity plea. They were not acquitted. The juries were probably right: only a minuscule percentage of XYY males ever commit any sort of crime, so an extra Y chromosome certainly doesn't force anyone into a life of violence.

Sex Chromosomes and Sex Determination

Studies of men and women with abnormal numbers of sex chromosomes lead to the inescapable conclusion that the Y chromosome determines maleness in humans. Having only one X chromosome, as both XY males and XO females do, does not automatically lead to maleness. In most respects, including external genitalia, XO individuals are clearly female. On the other hand, having a Y chromosome produces the male phenotype, no matter how many X chromosomes are present. Even the rare XXXY or XXXXY person is male. In computer terminology, femaleness is the "default" condition for human sex; explicit instructions encoded on the Y chromosome are required to produce a male. After years of determined searching, in 1990 geneticists finally located the gene on the Y chromosome that determines maleness. What instructions this gene encodes will probably be known soon.

Abnormal Numbers of Autosomes

Nondisjunction of the autosomes may also occur, producing eggs or sperm with a missing autosome or two copies of an autosome. Fusion with a normal gamete (one copy of each autosome) leads to an embryo with either one or three copies of the affected autosome. With only one copy of any of the autosomes, the embryo aborts so early in development that the woman never knows she was pregnant at all. Three copies of an autosome (trisomy) usually also causes a spontaneous abortion, but often later in pregnancy. Sometimes, however, trisomic babies are born, especially those with three copies of chromosomes 13, 18, or 21. Of these, trisomy 21 is the most common.

Trisomy 21 (Down Syndrome)

In about one of every 900 births, the child has an extra copy of the twenty-first chromosome. These children have several distinctive physical characteristics, including lack of muscle tone, a small mouth held partially open because it cannot accommodate the tongue, and distinctively shaped eyelids (Fig. 15-10). Much more serious defects include low resistance to infectious diseases, heart malformations, and mental retardation so severe that only about one in 25 ever learns to read and only one in 50 learns to write.

The frequency of nondisjunction is influenced by the age of the parents, especially the mother (Fig. 15-11). As we will see in Chapter 39, meiosis begins in a woman's ovaries while she is still a fetus in her moth-

Figure 15-10 Children with abnormal numbers of chromosomes are almost always both physically and mentally defective, in ways that are characteristic of the particular chromosome abnormality. These girls have the typical relaxed mouth, "Oriental" eyes, and congenital heart defects usually seen in cases of Down syndrome.

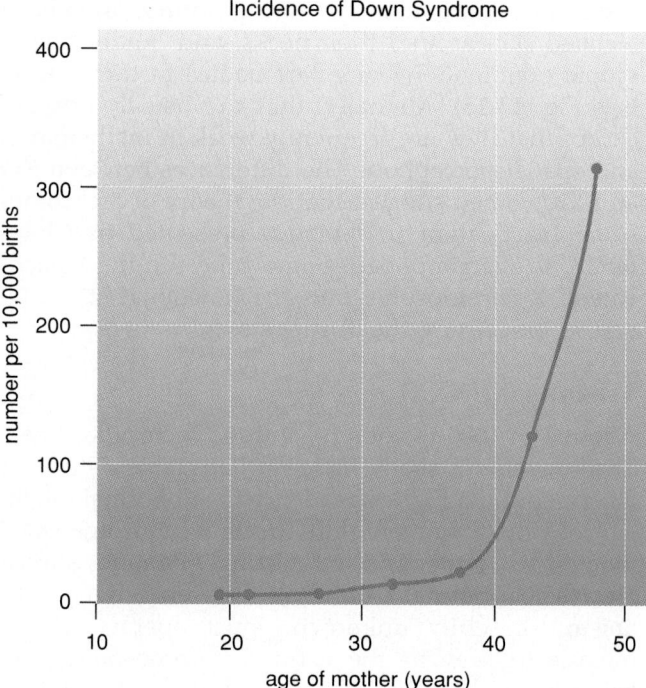

Figure 15-11 The frequency of nondisjunction is influenced by the age of the mother. This graph shows that nondisjunction of chromosome 21, producing children with Down syndrome, greatly increases after the mother reaches age 35. Geneticists have found that this is true even for women whose husbands are much younger.

er's womb, but is suspended during late prophase of meiosis I. When a woman matures, meiosis resumes, a few cells at a time, during the monthly menstrual cycle. Therefore, a 40-year-old woman produces eggs that have been in meiosis for 40 years. Errors in chromosome distribution are more likely in old cells, perhaps just from aging itself, or perhaps from exposure to radiation, toxic chemicals, or viruses. In men, new sperm-producing cells are constantly being produced in the testes, and meiosis takes only a couple of weeks. Therefore, nondisjunction does not increase with age as rapidly in men as in women. Nevertheless, nondisjunction in sperm accounts for about 25% of the cases of Down syndrome, and there is a small age effect.

In the past two decades, it has become increasingly common for couples to delay having children, so that both husband and wife can establish careers. As a result, many women bear children when they are in their late 30s or early 40s, and in general they have husbands of similar age. This inevitably leads to more trisomic fetuses than if the same couples had reproduced earlier. Fetal trisomy can be diagnosed by examining the chromosomes of fetal cells collected through amniocentesis or chorionic villus sampling. A therapeutic abortion may then be used to prevent the birth of a trisomic child.

The Human Genome Project

Beginning with a few small meetings in 1985 and 1986, and now racing along with multimillion dollar budgets, the Human Genome Project is probably science's most ambitious response to the command of the ancient Oracle at Delphi: "Know thyself." The answers to a host of questions about humankind reside in the nucleotide sequences of our genes. Further, humanity is heir to about 4,000 genetic disorders. If we knew the nucleotide sequence of the entire human genome, then, in theory at least, we could devise rapid screening and prenatal diagnosis of many genetic disorders, design better therapies for many diseases, and understand much more about the workings of the human brain.

Ever since DNA sequencing was first developed by Allam Maxam, Walter Gilbert, and Frederick Sanger, molecular geneticists have ruminated about the potential for sequencing the human genome, but the time, effort, and cost seemed daunting—$3 billion was the estimate in 1986. Recent developments in RFLP analysis, automated DNA sequencing, and PCR technologies, however, have combined to make the project now seem quite reasonable. Although many meetings and much research went on before, the Human Genome Project, with James Watson as its head, officially got underway in 1988.

The project has four major themes:

1. **Linkage mapping:** The locations of at least 3,000 genes or other markers (such as restriction enzyme cut sites) will be sought. Remember, linkage maps locate genes by the relative frequency of crossing over (Chapter 11). Linkage maps can be used in pedigree analysis to help locate other genes that are known only from phenotypic effects.

2. **Physical mapping:** Each human chromosome will be cut apart with restriction enzymes. Unique DNA nucleotide sequences will be sought in each piece, with the unique sequences no farther apart than about 100,000 nucleotides. These unique nucleotide sequences can then be used as starting points for walking along the chromosomes, sequencing all the DNA or looking for previously unmapped genes.

3. **Sequencing the human genome:** A complete set of haploid human chromosomes contains about 6 billion nucleotides (3 billion pairs on two complementary strands). Since a total of only 37 million nucleotide pairs, in any organism, had been sequenced up to the end of 1989, this is clearly an ambitious and expensive undertaking. Advocates point out that, if the entire genome sequence is known, and someone later pinpoints the gene locus of a genetic disorder, then one can obtain the gene sequence merely by looking it up in a computer.

4. **Sequencing diversity:** There is, of course, no one human genome—yours is different from everyone else's. By definition, the nucleotide sequence for a defective allele differs from that of the normal allele, and there is considerable variety among normal, functioning alleles, too. Therefore, at least for the DNA sequences that code for genes, it will be important to sequence the DNA from a fairly large sample of people.

The Human Genome Project has its detractors, mostly because of its cost. If the overall science budget remains constant, then the $50 million to $100 million slated to be spent each year on the Human Genome Project must mean decreased funding for other projects. Supporters, however, point to increased understanding of normal functioning, human evolution, and human genetic disorders that the project would produce. Certainly, the Human Genome Project has brought together many of the best molecular geneticists, and is fostering cooperation in a field in which fierce competition is the norm.

Finally, the human genome is not the only one of interest. There are already plans being made to sequence the genomes of the bacterium *Escherichia coli*, the fruit fly *Drosophila melanogaster*, the nematode *Caenorhabditis elegans*, the domestic dog, and at least one plant. These sequences would be used in fields as diverse as microbiology, developmental biology, veterinary medicine, and agriculture.

Reflections on Medical Genetics

Genetics has opened up new vistas of understanding human nature. Analyses of family pedigrees, biochemical tests, amniocentesis, and recombinant DNA technologies have enabled geneticists to identify prospective parents who are carriers for some disorders, such as sickle-cell anemia and Tay–Sachs disease, and to detect fetuses who are homozygous for these and other disorders. These new powers demand a new set of decisions, both ethical and economic, by individuals, physicians, and society. Let us examine these new choices in one specific case, phenylketonuria.

The Biomedical Situation

As you learned in Chapter 13, people with phenylketonuria (PKU) are homozygous for a recessive allele coding for a defective version of the enzyme that converts phenylalanine to tyrosine (see Fig. 13-1). As is usual with recessive diseases, most homozygous children result from the mating of two heterozygous (and therefore phenotypically normal) parents. The level of phenylalanine in the developing fetus is regulated by its pregnant mother, so that homozygous recessive infants are born with normal brain development. Once on their own, however, the infants' phenylalanine levels increase rapidly, damaging the brain. Fortunately, homozygous infants can be detected as early as 4 or 5 days after birth by a simple blood test that costs just a few dollars. Treatment formerly consisted of maintaining these children on a diet low in phenylalanine for at least the first 5 or 6 years of life, when most brain development occurs. Most naturally occurring proteins contain phenylalanine, so the low-phenylalanine diet is quite expensive and not very convenient.

This is not the end of the story, however. Women

with PKU who were treated in infancy have normal mental development, but they are still deficient in phenylalanine metabolism, and therefore have an elevated concentration of phenylalanine in their blood. Children of such women, fathered by homozygous normal men, will be heterozygous, and therefore should be phenotypically normal. The problem is that the phenylalanine levels of an otherwise normal fetus are raised by the high levels of its phenylketonuric mother. High phenylalanine levels during pregnancy, even just the first few months, cause mental retardation. The only protection for the fetus is if the mother adheres to the low-phenylalanine diet.

Thus, treating phenylketonuria in female infants brought with it new responsibilities. Many women now of childbearing age went off their low-phenylalanine diets years ago. Many have since forgotten that they have PKU, or, not understanding genetics, think that they *used* to have PKU, but have now outgrown it. Others planned to resume the low-phenylalanine diet when they wanted to become pregnant, but became pregnant unintentionally. A number of retarded children have already been born to such women, and a nationwide effort is underway to locate all phenylketonuric women of childbearing age (for more information, call the Children's Hospital in Boston, 617-735-7945). Further, the modern medical advice for phenylketonuric women is to stay on the low-phenylalanine diet until menopause.

The Ethical Situation

Insofar as we cannot cure inherited diseases, or foresee a cure in the near future, what are the responsibilities and options open to individuals and societies? A hundred years ago, when no one understood the nature of inherited diseases, the choices were limited and the responsibilities few. A prospective parent could not know whether he or she carried an inherited defect, could not predict the likelihood of having an affected child, and could not tell whether a child was affected until the child was born or perhaps was even several years old.

The situation has changed dramatically. Today, many people *do* know that they carry a seriously defective gene, and genetic counselors can predict their chances of having an affected child. Should such an individual refrain from reproduction? In some cases, such as Tay–Sachs disease and the chromosomal abnormalities, prenatal diagnosis can detect the inherited defect in a fetus. Should the fetus be aborted? Should screening tests, at least for infants, be encouraged or required, as is now the case for PKU and, in some states, sickle-cell anemia? Does this constitute governmental interference in the rights of the individual? What about screening tests for prospective parents, at least in high-risk groups? Should these tests be required?

What, if any, are the rights and responsibilities of society in these decisions? In many Western countries, society pays most or all of the costs of medical care for affected children, often running to hundreds of thousands of dollars each. What, if any, are the obligations of society to children who might be born with genetic defects? Suppose a phenylketonuric woman goes off the low-phenylalanine diet against the advice of her physician, and then conceives a child. Is this child abuse?

Some people argue that the human species is becoming genetically "loaded" with defective alleles: modern medicine allows people to survive and reproduce who have the alleles for diabetes, sickle-cell anemia, and many other genetically influenced diseases, who would have died in earlier times. As a result, some believe that carriers of alleles for serious, incurable diseases should not reproduce, to eliminate not only homozygous recessives who are afflicted with the disease but also future carriers. That is usually the opinion of someone who is not known to be a carrier for anything. Eventually, when we become knowledgeable enough, we may find that almost *all* of us carry recessive alleles for some inherited disease or other, yet few would argue that the majority of the population is morally obligated to remain childless. In a free society, probably the best solution is to give people the best information possible about their genetic constitution and that of their future children. The choices, whatever their consequences for society as a whole, almost certainly must remain with the individual.

SUMMARY OF KEY CONCEPTS

Methods in Human Genetics
The genetics of humans is similar to the genetics of other animals, except that experimental crosses are not feasible. Analysis of family pedigrees and, more recently, molecular genetic techniques, must be used to determine the mode of inheritance of human traits.

Single Gene Inheritance
Many genetic disorders are inherited as recessives; therefore, only homozygous recessive persons show symptoms of the disease, whereas heterozygotes are phenotypically normal and usually cannot be detected by casual observation. Heterozygotes are called carriers, because the offspring of two carriers may be homozygous recessive and show the trait. Recessive genetic disorders include Tay–Sachs disease, sickle-cell anemia, and phenylketonuria.

Many normal traits, and a few diseases, are inherited as simple dominants. Both heterozygous and homozygous dominant individuals show the trait.

The sex chromosomes are paired in women (XX) but unpaired in men (XY). The Y chromosome bears few genes other than those determining maleness. Therefore, women show normal dominant-recessive relationships among alleles of X chromosome genes, while men phenotypically display whatever allele they carry on their single X chromosome, a phenomenon called sex-linked or X-linked inheritance. Sex-linked conditions include red–green color discrimination and hemophilia.

Sex hormones secreted by the gonads help to determine the expression of sex-influenced traits. The majority of sex-influenced defects are more common in males than in females, and include baldness, cleft palate, and gout.

Complex Inheritance
Most human traits are polygenic—that is, they are influenced by the action of many genes. Examples include the color of skin and eyes, and probably many personality traits. The environment affects the phenotypic expression of all genes in all organisms. This is particularly apparent in human beings, especially in intellectual and personality traits.

Chromosomal Inheritance
Sex in humans is determined by the presence of a Y chromosome: individuals with one or more Y chromosomes are phenotypically male; individuals with no Y chromosome are phenotypically female. People with abnormal numbers of sex chromosomes often have mental and physical deficiencies. The most common defect is below-normal intelligence.

Abnormal numbers of autosomes usually lead to spontaneous abortion early in pregnancy. In rare instances, the fetus may survive to birth, but severe mental and physical deficiencies are always found. The likelihood of abnormal numbers of chromosomes increases with increasing age of the mother, and, to a lesser extent, the father.

The Human Genome Project
The largest single project in the history of biology, the Human Genome Project proposes to produce (1) a detailed linkage map; (2) a fine-grained physical map; (3) the complete nucleotide sequence of the "average" human genome; and (4) nucleotide sequences of a sample of the diversity of alleles at selected gene locations.

GLOSSARY

amniocentesis (am-nē-ō-sen-tē'-sis): a procedure for sampling the amniotic fluid surrounding a fetus. A sterile needle is inserted through the abdominal wall, uterus, and amniotic sac of a pregnant woman, into the amniotic fluid. Ten to twenty milliliters of amniotic fluid is withdrawn into a syringe. Various tests may be performed on the fluid and the fetal cells suspended in it to provide information on the developmental and genetic state of the fetus.

carrier: an individual who is heterozygous for a recessive condition. Carriers display the dominant phenotype but can pass on their recessive allele to their offspring.

chorionic villus sampling: a procedure for sampling cells from the chorionic villi produced by a fetus. A tube is inserted into the uterus of a pregnant woman, and a small sample of villi are suctioned off for genetic and biochemical analysis.

Down syndrome: a genetic disorder caused by the presence of three copies of chromosome 21. Common characteristics include mental retardation, abnormally shaped eyelids, a small mouth with protruding tongue, short fingers, heart defects, and unusual susceptibility to infectious diseases.

hemophilia: a recessive, sex-linked disease in which the blood fails to clot normally.

Klinefelter's syndrome: a set of characteristics typically found in individuals who have two X chromosomes and one Y chromosome. These individuals are phenotypically males, but sterile, and have several femalelike traits, including narrow shoulders, broad hips, and partial breast development.

nondisjunction: an error in meiosis in which chromosomes fail to segregate properly into the daughter cells.

pedigree: a diagram showing genetic relationships among a set of individuals, usually with respect to a specific genetic trait.

phenylketonuria (fen-ul-kē-tō-nū′-rē-a): a recessive disease in which the enzyme that catalyzes the conversion of the amino acid phenylalanine to tyrosine is faulty.

sex-influenced inheritance: a mode of inheritance in which traits of a nonsexual nature are more common in one sex than in the other, often due to differing levels of sex hormones.

sex-linked inheritance: inheritance of traits controlled by genes carried on the X chromosome. Females show the dominant trait unless they are homozygous recessive, whereas males will express whatever allele is found on their single X chromosome.

sickle-cell anemia: a recessive disease caused by a single amino acid substitution in the hemoglobin molecule. Sickle-cell hemoglobin molecules tend to cluster together in long chains, distorting the red blood cell shape and causing them to break and clog the capillaries.

Tay–Sachs disease: a recessive disease caused by a deficiency in enzymes regulating lipid metabolism in the brain.

trisomy 21: see *Down syndrome*.

trisomy X: a condition of females who have three X chromosomes instead of the normal two. Most of these women are phenotypically normal, and are fertile.

Turner's syndrome: a set of characteristics typical of a woman with only one X chromosome. These women are sterile, failing to develop normal ovaries. They also tend to be very short and to lack normal female secondary sexual characteristics.

X-linked inheritance: see *sex-linked inheritance*.

STUDY QUESTIONS

(The answers to questions 1 and 2 are given at the end of the question list.)

1. If the frequency of heterozygous carriers for Tay–Sachs disease is one in 30 American Jews, why is the frequency of homozygous recessive babies of Jewish parents one in 3600?

2. If one parent of a couple has Huntington's disease (assume that this parent is heterozygous), calculate the fraction of their children that would be expected to develop the disease. What if both parents were heterozygous?

3. Why are most genetic diseases inherited as recessives rather than dominants?

4. How is sex determined in humans? What is the evidence for this?

5. Define polygenic inheritance. Why could polygenic inheritance allow parents to produce offspring that are notably different in eye or skin color than either parent?

6. What is sex-influenced inheritance? What is the evidence that hormonal levels control the expression of sex-influenced traits?

7. Define nondisjunction, and describe the common syndromes caused by nondisjunction of sex chromosomes and autosomes.

8. Describe amniocentesis and chorionic villus sampling, including the advantages and disadvantages of each. What are their medical uses?

ANSWERS TO QUESTIONS 1 AND 2

1. The probability of two or more independent events occurring simultaneously is the product of their individual probabilities. The probability of the husband being heterozygous is 1/30; the probability that the wife is heterozygous is also 1/30; the probability that two heterozygotes will produce a homozygous recessive child is 1/4 (see Chapter 11). Multiplying $1/30 \times 1/30 \times 1/4$ gives 1/3600.

2. Let H = the Huntington's allele (dominant) and h = the normal allele (recessive). Then the first set of parents would be Hh and hh. Half of their offspring would be expected to inherit Huntington's disease. If both parents are heterozygous, then the cross is Hh × Hh. We would expect the offspring to be 1/4 HH, 1/2 Hh, and 1/4 hh; therefore, 3/4 would develop the disease.

SUGGESTED READINGS

DeLisi, C. "The Human Genome Project." *American Scientist*, May 1988. An overview of the promises and pitfalls of the Human Genome Project.

Friedmann, T. "Prenatal diagnosis of genetic disease." *Scientific American*, November 1971. The techniques of amniocentesis are explained, and some of its biological and social consequences are explored.

Mange, A. P., and Mange, E. J. *Genetics: Human Aspects,* 3rd ed. Philadelphia: W. B. Saunders Company, 1990. Clearly and eloquently written descriptions of the biological and social aspects of human heredity.

McGue, M. "Nature-nurture and intelligence." *Nature* 1989; 340:507–508. Nature works via nurture in producing the outward manifestations of intelligence.

Pines, M. "In the Shadow of Huntington's." *Science 84*, May 1984; Grady, D. "The Ticking of a Time Bomb in the Genes." *Discover*, June 1987; and Roberts, L., "Huntington's Gene: So Near, Yet So Far." *Science* 1990; 247:624–627. The story of the scientific detective work involved in the diagnosis of Huntington's disease using recombinant DNA techniques, the social dilemma it created, and the frustrations of searching for 6 years without yet finding the gene. Particularly poignant, because one of the lead investigators may herself be a victim.

Roberts, L. "To Test or Not to Test." *Science* 1990; 247:17–19. Some of the alleles causing cystic fibrosis cannot be detected—should the entire childbearing population be screened for carriers for this devastating disease?

Watson, J. D. "The Human Genome Project: Past, Present, and Future." *Science* 1990; 248:44–48. Although fairly technical, this article spells out the biological and political goals of the Human Genome Project.

UNIT III

Evolution

16

Principles of Evolution

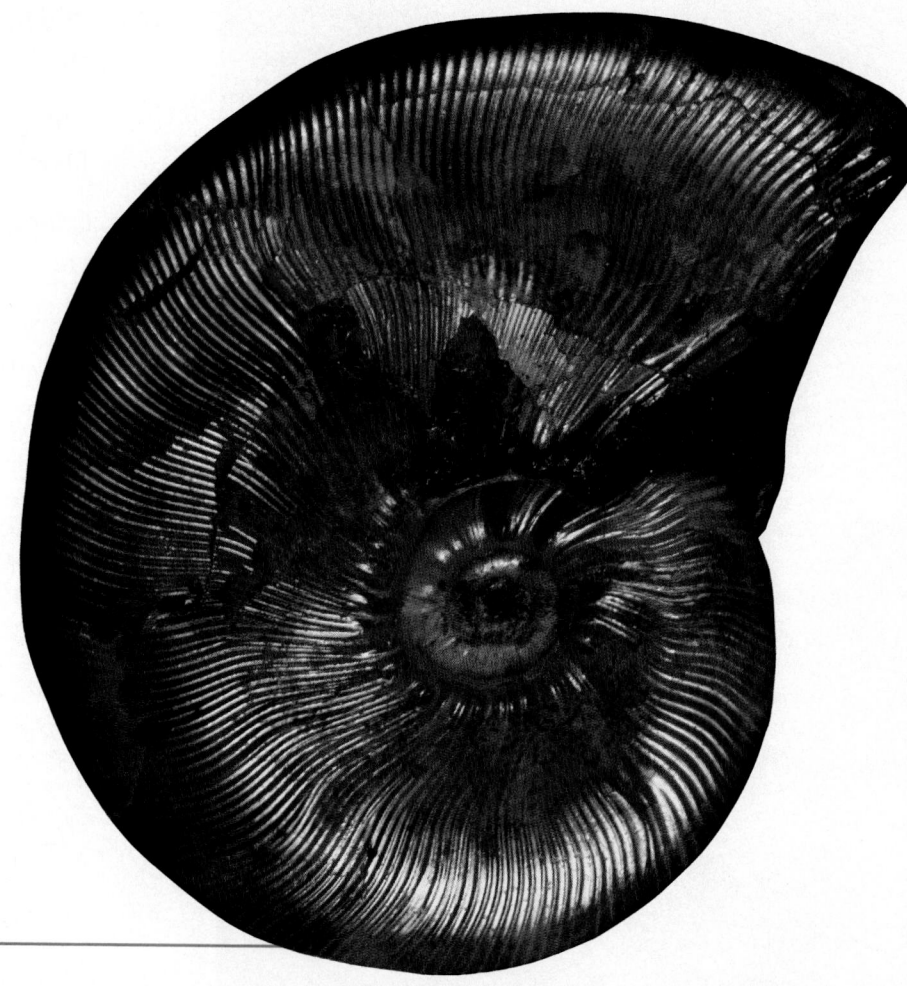

"When on board H.M.S. 'Beagle,' as naturalist, I was much struck with . . . the distribution of the inhabitants of South America, and . . . the geological relations of the present to the past inhabitants of that continent. These facts seemed to me to throw some light on the origin of species—that mystery of mysteries, as it has been called by one of our greatest philosophers."

Charles Darwin in On the Origin of Species by Means of Natural Selection

These words introduce what is perhaps the most important work in biology. In the *Origin of Species*, Charles Darwin proposed that over eons of time, species arise from other, preexisting species through the process of "descent with modification," or evolution.

Before Darwin, how species originated remained the "mystery of mysteries" for a very simple reason. Over the time span of recorded human history, let alone the life of a single human being, no new species had been recognized (although undoubtedly many new species had appeared, especially plant species). It is quite difficult to decide how something happens if there aren't any witnesses.

Nevertheless, one of the most striking features about our world is the remarkable variety of organisms inhabiting it. Why are there dozens of species of pine trees and scores of species of warblers? With no evidence to go on, nearly all peoples of the world historically turned to hypotheses of **creationism.** The most common of these hypotheses is that a supernatural being created each type of organism separately at the beginning of the world, and that all modern organisms are essentially unchanged descendants of these ancestors.

As we pointed out in Chapter 1, one of the fundamental principles of science is that Earthly phenomena are produced by natural, Earthly causes. A nineteenth century English essayist wrote, " . . . with regard to the material world, we can at least go so far as this—we can perceive that events are brought about not by insulated interpositions of Divine power exerted in each particular case, but by the establishment of general laws [of nature]." Science cannot say whether or not divine power originally established those general laws, but science firmly adheres to the principle that natural events have causes that arise from the operation of natural laws.

Therefore, throughout history, scientists have sought natural causes for the origin of species. However, it was only in the nineteenth century that a truly coherent theory—evolution by descent with modification, driven by natural selection—was developed. This theory was published by two British naturalists,

Charles Darwin and Alfred Russell Wallace, in 1858, and today still forms the foundation of our understanding of evolution. Darwin and Wallace did not work in a vacuum. Centuries of thought and observation preceded them and influenced their ideas. Let us begin, then, with a brief survey of evolutionary thought.

The History of Evolutionary Thought

What Is a Species?

Before we can study the origin of species, we must first decide what a species is. Throughout most of human history, "species" was a poorly defined concept. When using the word "species," most Europeans meant one of the originally created "kinds" referred to in the Bible. How could a naturalist tell if two organisms belonged to two different species? Since no one was present at the Creation to record the criteria of the Creator, one had to distinguish among species by visible differences in structure; in fact, the word "species" is Latin for "appearance." Clearly pines and warblers are different species, and warblers are different from eagles and ducks. But how do biologists distinguish among species of warblers? Today, biologists define a **species** as all the populations of organisms that are capable of interbreeding under natural conditions, and that are reproductively isolated from other populations. In other words, the members of a species can interbreed among themselves, but usually not with members of other species. If interbreeding with another species does occur, the hybrid offspring are usually infertile or handicapped in some way (see Chapter 17).

Biologists have found that differences in appearance do not always mean that two populations belong to different species. For example, field guides published in the 1970s listed the myrtle warbler and Audubon's warbler (Fig. 16-1) as distinct species; more recently the American Ornithological Union

(a)

(b)

Figure 16-1 The myrtle warbler **(a)** and Audubon's warbler **(b)** were formerly considered to be two separate species, but are now considered to be merely local varieties of one widespread species.

decided that they are, after all, merely local varieties of the same species. The main reasons for the initial splitting were differences in range and in the color of the throat feathers. Ornithologists now consider them to be a single species because, where their ranges overlap, interbreeding occurs, and the offspring are just as vigorous as the parents.

Early European naturalists didn't have to worry too much about the definition of species. Europe has a rather scanty assortment of flora and fauna, having lost much of its diversity to widespread extinctions during the last Ice Age. Most European species, at

least among the more prominent land plants, mammals, and birds, differ quite a bit from their nearest relatives. Therefore, the permanence of species, persisting unchanged from an original creation, seemed quite reasonable. As we will see, this comfortable situation changed with the exploration of new lands in Southeast Asia, Africa, and the Americas.

The Greek Philosophers

Two principal lines of thought descended from the ancient Greeks to influence later Western ideas about the nature and origin of species. The philosophy of Plato (427–347 BC) rested on the foundation of the "ideal Form": each object on Earth, whether animate or inanimate, is a mere temporary reflection of its nonmaterial Form. Thus every dog and every human is an imperfect version of the ideal Dog or Human that exists somewhere beyond the Earth in the world of Forms. The Forms are perfect and unchangeable, having come into existence in some unknown way but persisting without alteration forever into the future. Plato's concept of unchanging Forms greatly influenced early Christian thought, and came to be embodied in the idea that every species of living thing was created by God at the beginning of time. Although minor variations may occur among individual members of a given species, each species as a whole remains unchanged, very like a Platonic Form.

Plato's student Aristotle (384–322 BC), one of the first great naturalists, categorized all the living things that he encountered. Aristotle thought that all organisms fit into an orderly scheme, later called the *Scala Naturae*, or Ladder of Nature (Fig. 16-2). The ladder stood, so to speak, upon nonliving matter, and ascended rung by rung from fungi and mosses to higher plants, through primitive animals such as molluscs and insects, and finally culminating in human beings. Aristotle's ideas, even more than Plato's, were incorporated into Christian thought. The *Scala Naturae* was considered to be permanent and immutable: each organism has its place on the ladder, ordained by God during creation.

Evolutionary Thought Before Darwin

Creationism, the idea that each species was created individually by God and never changed thereafter, reigned unchallenged for nearly 2000 years. The *Scala Naturae* was completely compatible with medieval thought: the Earth was the center of the Universe, and Man stood atop Creation. With the limited fauna and flora of Europe to consider, a fixed ladder of nature seemed obvious. Naturalists felt it to be their task to catalog the diversity of organisms, describing

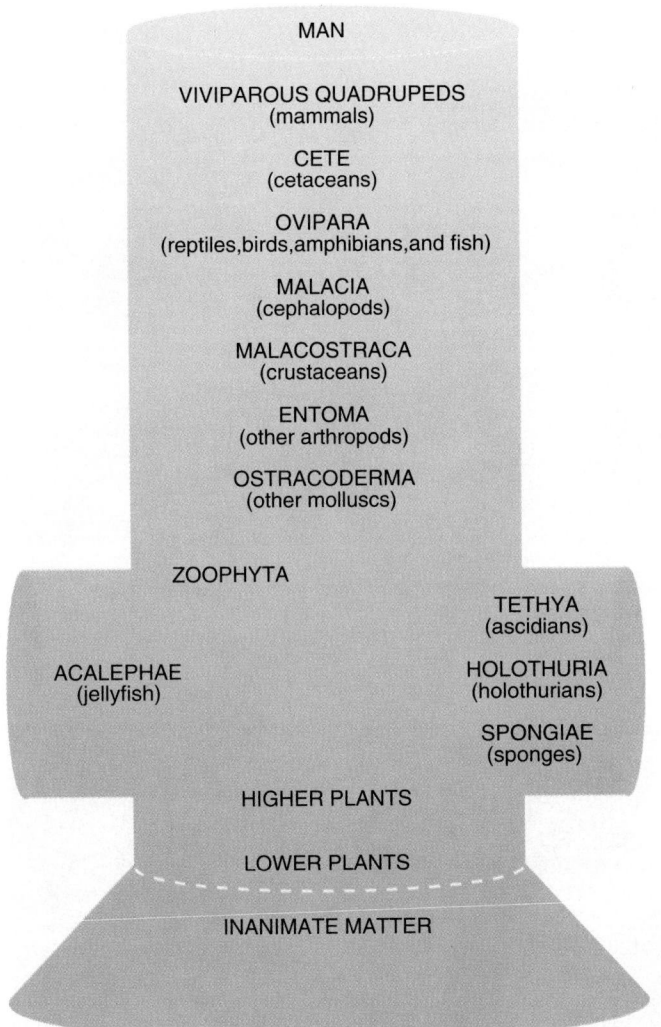

Figure 16-2 The *Scala Naturae*. It is not clear exactly how Aristotle thought the Scala came into being, but he often sounded almost like an evolutionist: "Thus Nature passes from lifeless objects to animals in such unbroken sequence . . . that scarcely any difference seems to exist between two neighboring groups owing to their close proximity." (Drawing and translation from Singer, 1959.)

the glory of God's creation. By the eighteenth century, however, new evidence suggested that this static view of creation might be incorrect.

The Diversity of Living Organisms

As naturalists explored the newly discovered lands of Africa, Asia, and America, they found that the diversity of living things was much greater than anyone had suspected. Further, some of these exotic species closely resembled one another. This embarrassment of riches led some naturalists to consider that perhaps species could change after all, and that some of the

similar species might have developed from a common ancestor.

Fossils

At the same time, excavations for roads, mines, and canals revealed that rocks often occur in layers (Fig. 16-3). Sometimes, a few strangely shaped rocks were found embedded within one of these layers. These rocks, called **fossils** (from the Latin, meaning "dug up"), often resembled parts of living organisms. At first, fossils were thought to be ordinary rocks that wind, water, or people had worked into lifelike forms. As more and more fossils were discovered, however, it became obvious that they were in fact plants and animals that had died long ago and been changed into rock (Fig. 16-4). Upon careful study, William Smith (1769–1839) realized that certain fossils were always found in the same layers of rock. Further, the organization of fossils and rock layers was consistent: fossil type A could always be found in a rock layer resting atop an older layer containing fossil B, which in turn rested atop a still older layer containing fossil C, and so on.

Fossil remains also showed a remarkable progression in form. Fossils found in the lowest (and therefore oldest) rock layers were invariably primitive looking, with a gradual advancement to greater complexity and greater resemblance to modern species in younger rocks, as if there were a *Scala Naturae* stretch-

Figure 16-3 The Grand Canyon of the Colorado River. Layer upon layer of sedimentary rocks form the walls of the Canyon, exposed in cliffs and mesas. The Grand Canyon strata cover over a billion years of evolutionary history.

(a) The bipedal carnivore *Allosaurus* confronts a *Stegosaurus*, but both are buried under volcanic ash.

(b) Their skeletons become impregnated with minerals from the surrounding rocks.

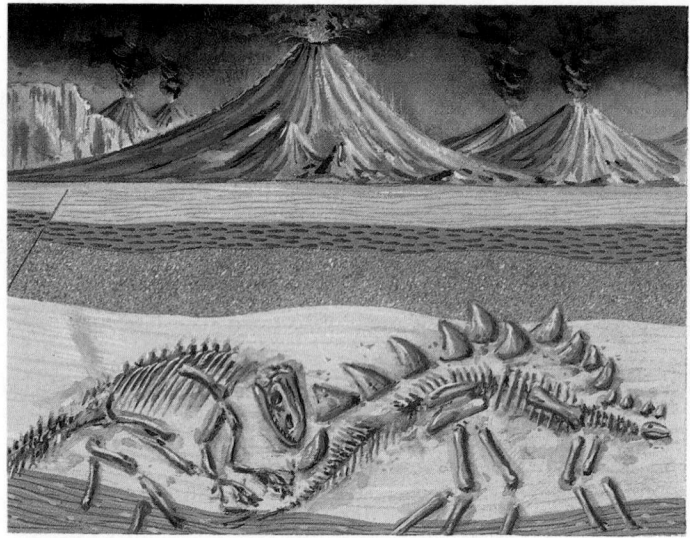

(c) Further eruptions bury the fossil skeletons more deeply, where they remain for tens of millions of years.

(d) Finally, stream erosion cuts through the overlying rock layers and exposes the fossils, where they may be found by human paleontologists.

Figure 16-4 Fossilization occurs when living organisms are buried beneath mud, silt, sand, or volcanic ash. This example might have occurred in Montana or Colorado.

ing back in time. Many of these fossils were the remains of plants and animals that no longer lived on Earth (Fig. 16-5). Putting these facts together, the conclusion became inescapable that different types of organisms had lived at various times in the past.

But what did this newfound richness of organisms, both living and extinct, mean? Was each organism produced by a separate act of creation? If so, why? And why bother to create so many, letting thousands become extinct? George-Louis Buffon (1707–1788)

(a)　　　　　　**(b)**　　　　　　**(c)**

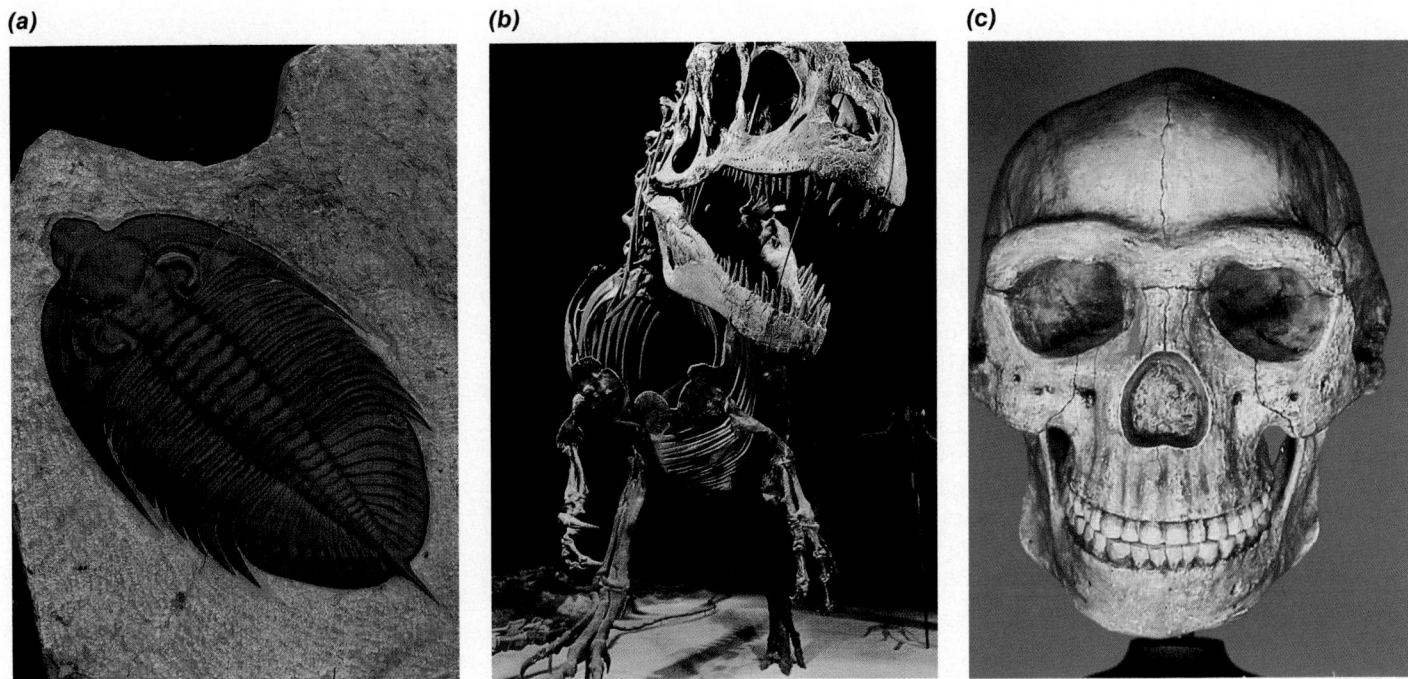

Figure 16-5 Fossils of extinct animals. **(a)** Trilobites were early arthropods (relatives of spiders and crabs) that flourished for scores of millions of years. Dozens of species of fossil trilobites are found in many parts of the world, but there have been no living trilobites for hundreds of millions of years. **(b)** *Allosaurus,* a predatory dinosaur similar to the more familiar *Tyrannosaurus rex.* Allosaurus preyed on large and formidable prey, including *Stegosaurus* (see Figs. 16-4 and 19-14). **(c)** The skull of *Zinjanthropus,* an extinct relative of human beings. Clearly unlike either today's humans or apes, *Zinjanthropus* probably left no modern descendants.

suggested that perhaps the original creation provided a relatively small number of founding species, and that some of the modern species had been "conceived by Nature and produced by Time"—that is, they had evolved through natural processes. Most people were not convinced. First, Buffon could not provide any mechanism whereby Nature could "conceive" new species. Second, no one thought that there was Time enough for their "production."

Geology and the Age of the Earth

In the early 1700s, few scientists suspected that the Earth could be more than a few thousand years old. Counting generations in the Old Testament, for example, yields a maximum age of 4,000 to 6,000 years. By reading the descriptions of plants and animals from ancient writers such as Aristotle, it was clear that wolves, deer, lions, and other European organisms had not changed in over 2,000 years. How, then, could whole new species arise, if the Earth was created only a couple of thousand years before Aristotle?

To account for a multitude of species, both extinct and modern, while preserving creationism, the French paleontologist Georges Cuvier (1769–1832) proposed the theory of **catastrophism.** Cuvier hypothesized that a vast supply of species was created in the beginning. Successive catastrophes (akin to Noah's Flood) produced the layers of rock and destroyed many species, fossilizing some of their remains in the process. The reduced flora and fauna of the modern world, he theorized, are the species that survived the catastrophes. However, if modern *species* have survived from an original creation, then many *individuals* of those species should have died in the ancient catastrophes. Surely some of them would have been fossilized, and even the lowest and oldest rock layers should contain at least a few fossils of present-day species. Unfortunately for Cuvier's hypothesis, they do not. A rescue attempt by Louis Agassiz (1807–1873) proposed that there was a new creation after each catastrophe, and that modern species result from the most recent creation. The fossil record forced Agassiz to postulate at least 50 separate catastrophes and creations!

Alternatively, perhaps the Earth *is* old enough to

METHODS IN BIOLOGY

Radioactive Dating

Early geologists could date rock strata and their accompanying fossils only in a *relative* way: Fossil A would be older than fossil B if the rock layer containing A was beneath that containing B, but neither fossil could be assigned an actual age. With the discovery of radioactivity, it became possible to determine *absolute* dates, within certain limits of uncertainty. The nuclei of radioactive elements spontaneously break down, or decay, into other elements. For example, carbon-14 (usually written ^{14}C) emits an electron to become nitrogen-14 (^{14}N). Each radioactive element decays at a rate that is independent of temperature, pressure, or the chemical compound of which the element is a part. The rate is geometric, so that half the radioactive nuclei decay in a characteristic time, called the half-life. The half-life of ^{14}C, for example, is 5,730 years.

How are radioactive elements used in determining the age of rocks? A particularly straightforward dating technique uses the decay of potassium-40 (^{40}K) into argon-40 (^{40}Ar), which has a half-life of about 1.25 billion years. Potassium is a very reactive element and is a common constituent of igneous rocks such as granite and basalt. Argon, on the other hand, is essentially inert, being unable to enter into chemical bonds with other atoms. What's more, argon is a gas. Let us suppose that a volcano such as Hawaii's Kilauea volcano erupts with a massive lava flow, covering the countryside. All the ^{40}Ar, being a gas, will bubble out of the molten lava, so that when the lava solidifies into rock, it will start out with no ^{40}Ar. Potassium-40 present in the hardened lava will decay to ^{40}Ar, half the ^{40}K decaying every 1.25 billion years. The ^{40}Ar gas will be trapped in the rock. A geologist could take a sample of the rock and determine the proportion of ^{40}K to ^{40}Ar (Fig. E16-1). If the analysis finds equal amounts of the two elements,

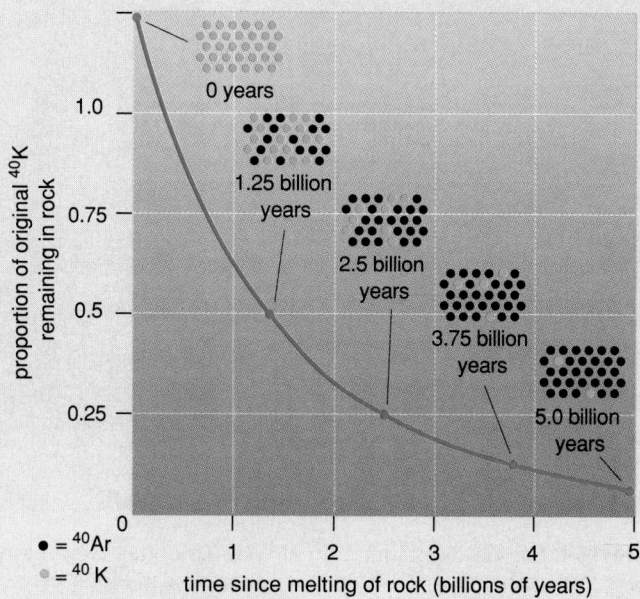

Figure E16-1 The relationship between time and the decay of radioactive ^{40}K to ^{40}Ar.

the geologist would conclude that the lava hardened 1.25 billion years ago.

With appropriate care, such age estimates are quite reliable. In the case of $^{40}K/^{40}Ar$ analysis, the age estimate will be, if anything, too low, because some of the ^{40}Ar gas may have escaped even though the rock was solid, thus increasing the ratio of ^{40}K to ^{40}Ar. If a fossil is found beneath a lava flow dated at, say, 500 million years, then we know that the fossil is at least that old.

Elements Used in Radioactive Dating

Original Element	Final Element	Half Life (years)	Useful Dating Ages (years)
Carbon-14	Nitrogen-14	5730	100–50,000
Uranium-235	Lead-207	0.70 billion	over 500,000
Potassium-40	Argon-40	1.25 billion	over 500,000
Uranium-238	Lead-206	4.47 billion	over 100 million
Rubidium-87	Strontium-87	48.8 billion	over 100 million
Samarium-147	Neodymium-143	106 billion	over 1 billion

Some radioactive decay pairs, especially uranium-238 to lead-206, can even give an estimate of the age of the solar system. Analysis of Earthly rocks, meteorites, and rocks collected from the Moon by the Apollo astronauts all agree that the solar system, and hence the Earth, is about 4.5 billion years old.

Radioactive dating has other uses, too. Probably the most interesting recent application was the use of ^{14}C decay to date the Shroud of Turin. This relic, discovered in the 1300s, bears an image of a man who was scourged and crucified. Many people believed that the cloth was the burial shroud of Christ. Others claimed that the shroud was painted during the middle ages. Now, all natural cloths contain carbon, which was fixed from atmospheric CO_2 by the plants from which the threads were made. Therefore, if the shroud is the burial cloth of Christ, it should be almost 2,000 years old. If it is a medieval reproduction, it should be more like 700 years old (except in the unlikely event that the medieval artist somehow obtained 1,300-year-old cloth). The Vatican and the Archbishop of Turin agreed to radiocarbon dating of the shroud, and small samples were taken on April 21, 1988, by three independent dating laboratories in Arizona, England, and Switzerland. All three laboratories found that the cloth of the shroud dates from about 1300; that is, the cloth is only 700 years old. Whether the shroud was painted as a religious art object or as a fraud, we will probably never know.

allow for the production of new species. Geologists James Hutton (1726–1797) and Charles Lyell (1797–1875) contemplated the forces of wind, water, earthquakes, and volcanism. They concluded that there was no need to invoke catastrophes to explain the findings of geology. Do not rivers in flood lay down layers of sediment? Do not lava flows produce layers of basalt? Why, then, should we assume that layers of rock are evidence of anything but ordinary natural processes, occurring repeatedly over long periods of time? This concept, called **uniformitarianism,** satisfies a scientific axiom often called Occam's Razor: the simplest explanation that fits the facts is probably correct. The implications of uniformitarianism were profound. If slow natural processes suffice to produce layers of rock thousands of feet thick, then the Earth must be old indeed, many millions of years old. Hutton and Lyell, in fact, concluded that the Earth was eternal: "no Vestige of a Beginning, no Prospect of an End," in Hutton's words. (Modern geologists estimate that the Earth is about 4.5 billion years old: see "Methods in Biology: Radioactive Dating"). Thus, Hutton and Lyell provided the time for evolution, but there was still no convincing mechanism.

Proposed Mechanisms for Evolution

One of the first to propose a mechanism for evolution was Jean Baptiste Lamarck (1744–1829). Lamarck was impressed by the progression of forms in the fossil record. Older fossils tend to be simpler, while younger fossils are more complex and more like existing organisms. In 1801, Lamarck hypothesized that organisms evolved through the **inheritance of acquired characteristics:** living organisms can modify their bodies through use or disuse of parts (which is correct to some extent), and these modifications can be inherited by their offspring (which is not correct). Why would organisms modify their bodies? Lamarck proposed that all organisms possess an innate drive for perfection, an urge to climb the Ladder of Nature. In his most celebrated example, Lamarck hypothesized that ancestral giraffes stretched their necks to feed on leaves growing high up in trees, and as a result their necks became slightly longer. Their offspring inherited these longer necks, and in their turn stretched even further, to reach still higher leaves. Eventually, this process might produce modern giraffes with very long necks indeed (Fig. 16-6).

Today, Lamarck's theory seems silly: the fact that a prospective father pumps iron doesn't mean that his children will look like Arnold Schwartzenegger. Remember, though, that in Lamarck's day no one had the foggiest idea how inheritance worked. Gregor Mendel wouldn't even be born for another 20

giraffe ancestor ——————— time ————————→ modern giraffe

Figure 16-6 Lamarck's hypothesis for the evolution of the giraffe. Time flows from left to right. Short-necked proto-giraffes stripped the leaves from the lower branches of trees, and stretched and strained to reach leaves higher up, making their necks slightly longer. Their offspring inherited these longer necks, did more stretching in their turn, and passed on still longer necks to their offspring. The modern giraffe on the right would be the outcome of this continual striving to feed on tree leaves high off the ground.

years, and the incorporation of his principles of inheritance into mainstream biology didn't happen until the early twentieth century.

Although Lamarck's theory fell by the wayside, by the mid-1800s many biologists realized that the fossil record and the similarities between fossil forms and modern species could be best explained if present-day species had evolved from preexisting ones. The question remained: *but how?* In 1858, two English naturalists, Charles Darwin and Alfred Russell Wallace, independently provided convincing evidence that the driving force behind evolutionary change was natural selection.

Evolution by Natural Selection

Although their social and educational backgrounds were very different, Darwin and Wallace were quite similar in some respects. Both had traveled extensively in the tropics (see the essay on the Voyage of the *Beagle*), and had studied the staggering variety of plants and animals living there. Both found that some species differed only in a few fairly subtle, but ecologically important, features (Fig. 16-7). Darwin and Wallace were familiar with the fossil record, which showed a trend of increasing complexity through time. Finally, both were aware of the studies of Hutton and Lyell, who proposed that the Earth is ex-

tremely ancient. These facts suggested to both men that species change over time; that is, they evolve. Both sought a mechanism that might direct change over many generations, causing new species to arise.

Part of the answer came to both men from an unlikely source: the writings of an English clergyman, Thomas Malthus. In his *Essay on Population*, Malthus wrote, "It may safely be pronounced, therefore, that [human] population, when unchecked, goes on doubling itself every 25 years, or increases in a geometrical ratio." Darwin and Wallace realized that a similar principle holds for plant and animal populations. In fact, most organisms can reproduce much more rapidly than humans (consider the dandelion and the housefly), and consequently could produce overwhelming populations in short order. Nonetheless, the world is *not* chest-deep in dandelions or flies: natural populations do not grow "unchecked," but tend to remain approximately constant in size. Clearly, vast numbers of organisms must die in each generation, and most must not reproduce. Population growth is restrained by countless environmental factors, including food supply, predation, diseases, and weather (see Chapter 43).

From their experience as naturalists, Darwin and Wallace realized that the members of a species often differ from one another in form and function. Further, *which organisms die in each generation is not arbitrary, but depends to some extent on the structures and*

(a)

Camarhynchus
pallidus

Camarhynchus
heliobates

Camarhynchus
psittacula

Camarhynchus
parvulus

Geospiza magnirostris

Geospiza fortis

Geospiza
fulginosa

Geospiza scandens

Certhidia olivacea

Geospiza
difficilis

seed
crushers

finchlike
insectivores

warblerlike
insectivores

ancestral South American finch

(b)

(c)

Figure 16-7 Darwin's finches, residents of the Galapagos Islands. The Galapagos are a group of volcanic islands about 600 miles off the coast of Ecuador. On these inhospitable-looking islands, Darwin found many species of finches, all similar to a species found on the South American mainland. Ordinarily, finches are seed-eating birds, with large bills adapted to crushing hard seeds. Apparently, thousands of years ago a finch or small flock of finches became lost during migration or were blown off course by a storm, and arrived at the Galapagos. There they found few other birds. With few competitors, the finches found rich pickings both of seeds and insects. Over time, different groups of finches evolved adaptations to exploit different food sources. Modern biologists agree with Darwin that the variety of finch species on the Galapagos evolved from a common ancestor **(a)**. Particularly significant are the differences in beaks, which vary from a slender, warbler-like bill in the insect-catching *Certhidia olivacea* **(b)** to the massive seed-crushing bill of *Geospiza magnirostris* **(c)**. Beak differences are also important in mate recognition. From behind, many Galapagos finches are very similar, and a male may approach a female of the wrong species. As soon as he catches sight of her beak, however, he immediately recognizes his mistakes and looks elsewhere for a mate.

The Voyage of the *Beagle*

Like many modern students, Charles Darwin excelled only in subjects that intrigued him. Although his father was a physician, Darwin was uninterested in medicine and unable to stand the sight of surgery. He eventually obtained a degree in theology from Cambridge, although this too was of minor interest to him. What he really liked to do was to tramp over the hills, observing plants and animals, collecting new specimens, scrutinizing their structures, and categorizing them. As he himself later put it, "I was a born naturalist." Fortunately for Darwin (and for the development of biology), some of his professors at Cambridge had similar interests, notably the botanist John Henslow. So constant was their companionship in field studies that Darwin was sometimes called "the man who walks with Henslow."

In 1831, when Darwin was only 22 years old (Fig. E16-2), the British government sent Her Majesty's Ship *Beagle* on a surveying expedition along the coast of South America. As was common on such expeditions, the *Beagle* would carry along a naturalist, to observe and collect geological and biological specimens encountered along the route. Thanks to Henslow's recommendation to the captain, Robert FitzRoy, Darwin was offered the position of naturalist aboard the *Beagle*. (When they first met, FitzRoy almost rejected Darwin for the post because of the shape of Darwin's nose! Apparently FitzRoy felt that a person's personality could be predicted by the shape of the facial features, and Darwin's nose failed to measure up, as it were. Later, Darwin expressed the opinion that FitzRoy was "afterwards well satisfied that my nose had spoken falsely.")

The *Beagle* sailed to South America, making many stops along the coast and visiting the now-famous Galapagos Islands (Fig. E16-3). Along the way, Darwin observed the fauna and flora of the tropics, and was stunned by the diversity of species compared to Europe. Although he boarded the *Beagle* convinced of the permanence of species, his experiences soon led him to doubt this. He discovered a snake with rudimentary hind limbs, calling it "the passage by which Nature joins the lizards to the snakes." Another snake vibrates its tail like a rattlesnake, but has no rattles and therefore makes no noise. Penguins have wings that look like paddles, almost flying through the water. What could be the purpose behind these makeshift arrangements, if the Creator had individually created each animal in its present form, to suit its present environment?

Perhaps the most significant stopover of the voyage was the month spent on the Galapagos Islands. Here Darwin found huge tortoises (*galapagos* in Spanish); different islands bore distinctively different

Figure E16-2 A painting of Charles Darwin as a young man.

types. On islands without tortoises, prickly pear cactus grew in the common style, with juicy though spiny pads spread out over the ground. On islands where tortoises lived, the prickly pears grew substantial trunks, bearing the succulent pads high above the reach of the voracious and tough-mouthed tortoises (Fig. E16-4). Several varieties of mockingbirds and finches occurred, and as with the tortoises, different islands had subtly different forms. Unfortunately for Darwin, he "was not aware of these facts [about the finches] till my collection was nearly completed," and so did not bring a systematically labeled collection back to England with him. "It is the fate of every voyager, when he has just discovered what object in any place is worth his attention, to be hurried from it." The diversity of tortoises and birds "haunted" him for years afterward, as he pondered how such diversity might have occurred.

In 1836, when Darwin returned to England after 5 years on the *Beagle*, he was already somewhat famous, based on letters and journals forwarded home on other ships. The specimens and further journals that he brought back with him enhanced his fame, and to all outward appearances he settled down to become one of the foremost naturalists of his day. However, constantly gnawing on his mind was the problem of the origin of species. For over 20 years, Darwin pondered the problem, collecting evidence of any sort that might provide insight into its solution. When he finally published *On the Origin of Species* in 1859, his evidence had become truly overwhelming. Although its full impact would not be realized for decades, Darwin's theory of evolution by natural selection has become a unifying concept for virtually all of biology.

Events that change the world sometimes hinge on minute details, even the shape of a nose!

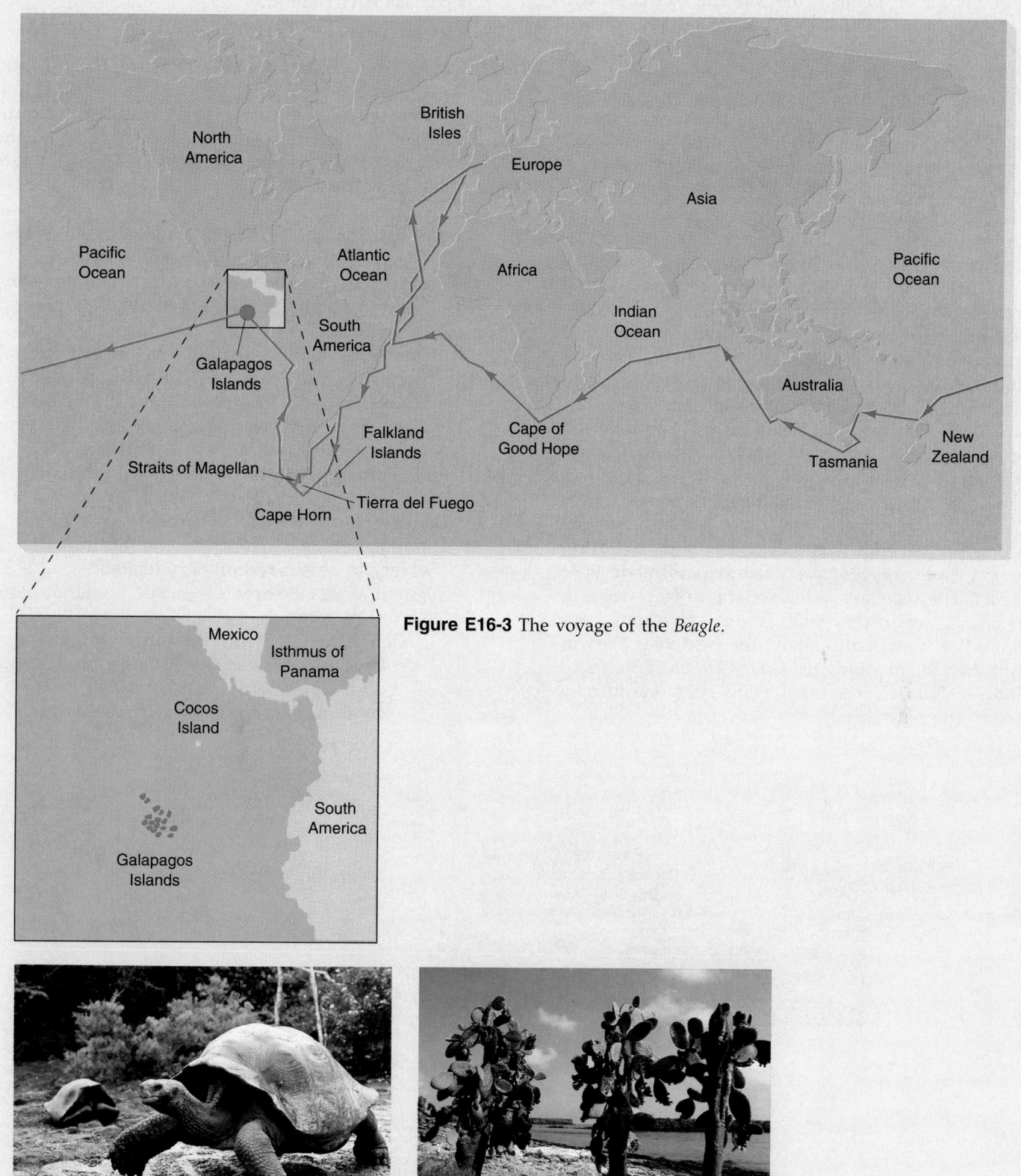

Figure E16-3 The voyage of the *Beagle*.

Figure E16-4 Galapagos tortoises (left) feed on prickly pear cactuses. On islands with tortoises, a young cactus quickly grows a tall trunk (right), which lifts the succulent pads beyond the reach of the tortoises.

abilities of the organisms. This was the source of the theory of evolution by natural selection. As Wallace put it, " . . . those which, year by year, survived this terrible destruction must be, on the whole, those which have some little superiority enabling them to escape each special form of death to which the great majority succumbed . . . " That "little superiority" might be better resistance to cold, more efficient digestion, or any of hundreds of other advantages, some very subtle. Everything now fell into place. Darwin wrote: " . . . it at once struck me that under these circumstances favorable variations would tend to be preserved, and unfavorable ones to be destroyed." If the favorable variations were inheritable, then the entire species would eventually consist of individuals possessing the favorable trait. With the continual appearance of new variations (due, as we now know, to mutations), which in turn are subject to further selection, "the result of this would be the formation of new species. Here, then, I had at last got a theory by which to work."

In 1858, remarkably similar papers from Darwin and Wallace were presented to the Linnaean Society in London. As with Gregor Mendel's manuscript on the principles of genetics, their papers made little impact. The secretary of the society, in fact, wrote in his annual report that nothing very interesting happened that year. Fortunately, the next year Darwin published his monumental *On the Origin of Species by Means of Natural Selection*, forcing everyone to take note of the new theory.

The Essentials of Evolutionary Theory

The essence of the Darwin–Wallace theory is very simple, consisting of three conclusions based on four observations. Rather than restricting ourselves to the terminology and information available to Darwin and Wallace, we will summarize their theory in modern terms (Fig. 16-8).

Observation 1: Natural populations of all organisms have the potential to increase rapidly, since organisms can produce far more offspring than are required merely to replace the parents.

Observation 2: Nevertheless, the sizes of natural populations are relatively constant over time.

Conclusion 1: Therefore, in each generation, many organisms must die young, fail to reproduce, produce few offspring, or produce less fit offspring that fail to survive and reproduce in their turn.

Observation 3: Individual members of a population differ from one another in their ability to obtain resources, withstand environmental extremes, escape predation, and so on.

Conclusion 2: Which organisms produce the largest number of viable offspring depends on their ability to deal with their environment; the most well-adapted

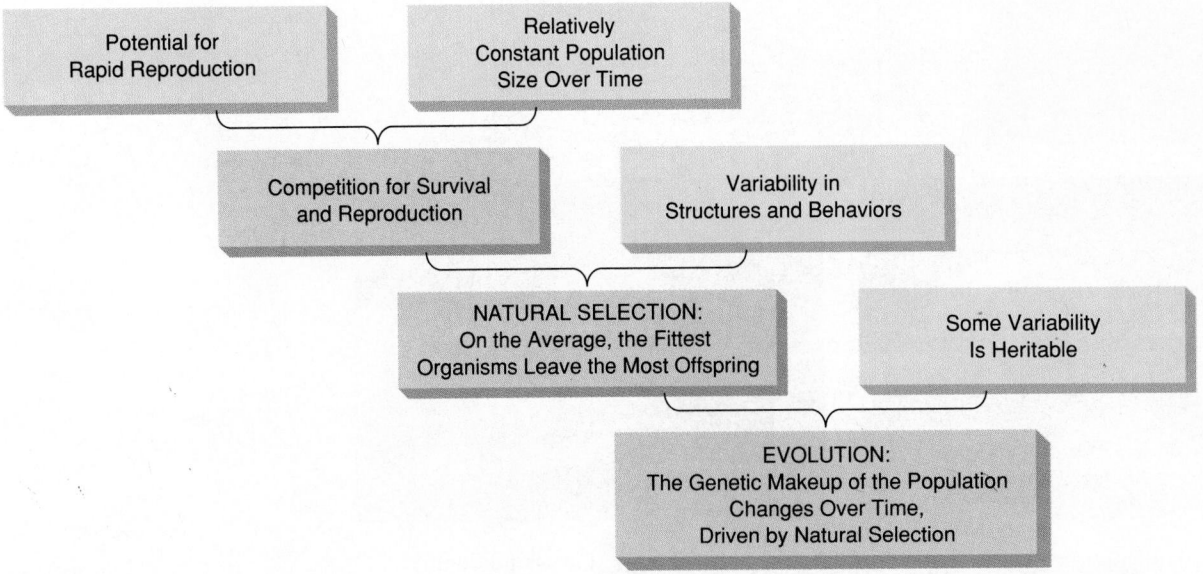

Figure 16-8 A flow chart of evolutionary reasoning, based on the hypotheses of Darwin and Wallace, but incorporating ideas from modern genetics.

organisms in one generation will probably leave the most offspring in the next generation. This is **natural selection.**

Observation 4: At least some of the variation in adaptedness among individuals is due to genetic differences that may be passed on from parent to offspring.

Conclusion 3: Over many generations, differential reproduction among individuals with different genetic makeup changes the overall genetic composition of the population. This is **evolution.**

As you know, the principles of genetics had not yet been discovered when Darwin wrote the *Origin of Species.* Our Observation 4, therefore, was an untested assumption for Darwin and a grave weakness in his theory. Although he could not explain how inheritance operated, Darwin's theory made an important prediction that we now know is correct. According to Darwin, the variations that appear in natural populations arise purely by chance. Unlike Lamarck, he postulated no internal drives for perfection or any other mechanisms that would ensure that variations would be favorable. Molecular genetics has shown that Darwin was correct: variations arise because of chance mutations in DNA (see Chapters 12 and 13).

How might natural selection among chance variations change the makeup of a species? In the *Origin of Species,* Darwin proposed the following example. "Let us take the case of a wolf, which preys on various animals, securing [them] by . . . fleetness. I can under such circumstances see no reason to doubt that the swiftest and slimmest wolves would have the best chance of surviving, and so be preserved or selected. . . . Now if any slight innate change of habit or structure benefited an individual wolf, it would have the best chance of surviving and of leaving offspring. Some of its young would probably inherit the same habits or structure, and by the repetition of this process, a new variety might be formed . . . " The same argument would apply to the wolf's prey, in which the fastest or most alert would be the most likely to avoid predation, and would pass on these traits to its offspring. This interplay between predator and prey is an example of **coevolution,** and is thought to be the cause of many of the exquisite adaptations that we see in the animal kingdom.

Although it is easiest to understand how natural selection would cause *changes within a species,* under the right circumstances, the same principles might produce *entirely new species.* In Chapter 18, we discuss the circumstances that give rise to new species. In the rest of this chapter, we briefly review some of the evidence for evolution.

The Evidence for Evolution

Virtually all biologists consider evolution to be a fact. Although debates still rage over the *mechanisms* of evolutionary change, exceedingly few biologists dispute that evolution occurs. Why? Because an overwhelming body of evidence permits no other conclusion.

Fossils

If fossils are the remains of members of species that are ancestral to modern species, then one would expect to find graded series of fossils leading from an ancient, primitive organism, through several intermediate stages, and culminating in the modern form. (Just how fine the gradations should be is currently a bone of contention among evolutionary biologists; see Chapter 18.) Probably the best-known progressive series are the fossil horses (Fig. 16-9), but giraffes, elephants, and several molluscs all show a gradual evolution of body form over time, suggesting that one species evolved from and replaced previous species. Certain sequences of fossil snails have such slight gradations in form between successive fossils that paleontologists cannot easily decide where one species leaves off and the next one begins.

Comparative Anatomy

Homologous Structures

Modern organisms are adapted for a wide variety of habitats and life-styles. The forelimbs of birds and mammals, for example, are variously used for flying, swimming, running over several types of terrain, and grasping objects such as branches and tools. Despite this enormous diversity of function, the internal anatomy of all bird and mammal forelimbs is remarkably similar (Fig. 16-10). It is inconceivable that these bone arrangements are optimal for such different functions, as would be expected if each animal were created separately. Such similarity is exactly what we would expect, however, if bird and mammal forelimbs evolved from a common reptilian ancestor. Through natural selection, each has been modified to perform a particular function. Such internally similar structures are termed **homologous structures,** meaning that they have a similar evolutionary origin, despite possible differences in function. Studies of comparative anatomy have long been used to determine the relationships among organisms, on the grounds that the more similar the internal structures of two species, the more closely related they must be; that is, the more recently they must have diverged from a common ancestor.

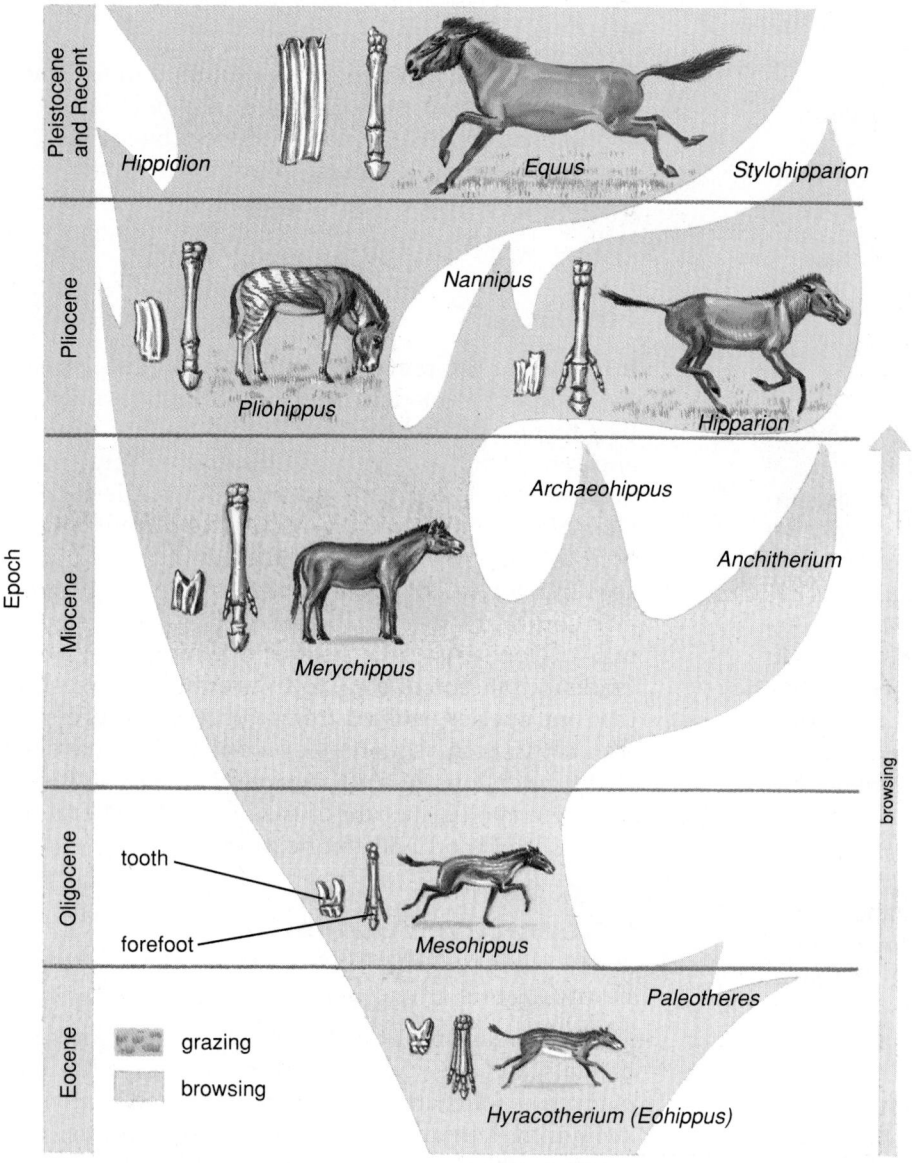

Pleistocene and Recent

Hippidion   *Equus*   *Stylohipparion*

Pliocene

Nannipus
Pliohippus   *Hipparion*

Archaeohippus

Miocene

Anchitherium

Merychippus

Oligocene

tooth
forefoot   *Mesohippus*

Paleotheres

Eocene

grazing
browsing

Hyracotherium (Eohippus)

grazing   browsing

Figure 16-9 Over the past 50 million years, horses evolved from small woodland browsers to large plains-dwelling grazers (see Figure 1-11). Three major changes include size, leg anatomy, and tooth anatomy. No one is certain why horses became larger, but it may have been an antipredator adaptation. In any case, a large body running over hard plains favored the evolution of large, hard hooves, attached by springlike, shock-absorbing joints to legs with stout bones. Finally, the teeth became larger, with more enamel. This reflects a change in diet from the relatively soft leaves and buds of bushes to the silicon-containing, abrasive blades of grasses. If a modern horse had teeth like *Hyracotherium*, they would be ground away while it was still very young, leaving it to starve.

Convergent Evolution

Evolution by natural selection also predicts that, given similar environmental demands, unrelated organisms might independently evolve superficially similar structures, a process called **convergent evolution.** Such outwardly similar body parts in un-related organisms, termed **analogous structures,** are often completely different in internal anatomy, since the parts are not derived from common ancestral structures. The wings of flies and birds, and the fat-insulated, streamlined shapes of seals (which are mammals) and penguins (birds), are two examples of analogous structures that have arisen through convergent evolution (Fig. 16-11).

Vestigial Structures

Evolution also helps to explain the curious circumstance of **vestigial structures.** These are structures that serve no apparent purpose, and include such things as molar teeth in vampire bats (which live on a diet of blood and therefore don't chew their food) and pelvic bones in whales and certain snakes (Fig. 16-12). Both of these vestigial structures are clearly homologous to important structures found in other vertebrates. Their continued existence in animals that have no use for them is best explained as a sort of "evolutionary baggage." For example, the ancestral mammals from which whales evolved had four legs and a well-developed set of pelvic bones. Whales do

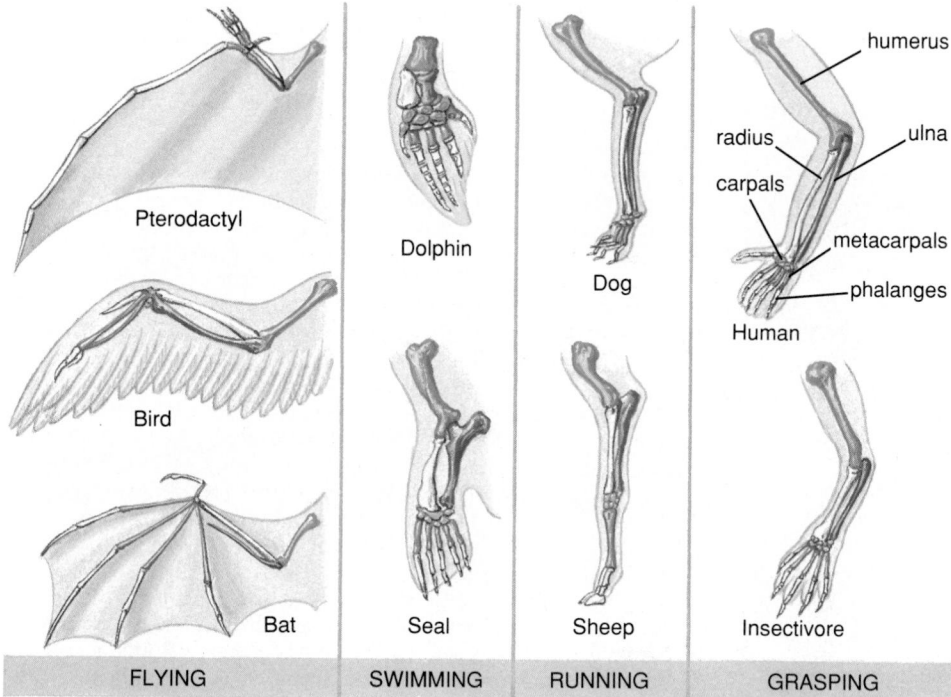

Figure 16-10 Homologous structures. The bones in the forelimbs of amphibians, reptiles, birds, and mammals are all similar to one another, despite wide differences in function. The bones have been tinted different colors to highlight the similarities among the various species.

Figure 16-11 Similar selective pressures acting on unrelated animals may result in the evolution of outwardly similar structures. The wings of insects **(a)** and birds **(b)** and the sleek, streamlined shapes of seals **(c)** and penguins **(d)** are examples of such analogous structures.

CLOSER LOOK
at Molecular Evolutionary Biology

Evolution deals with inherited changes over time. Since DNA is the molecule of heredity, evolutionary changes must be reflected in changes in DNA. Chapter 17 will explore a few of the genetic mechanisms that contribute to evolution. In this essay, we will consider two methods of studying DNA and how these methods are applied to determining the relatedness of organisms.

Cross-species DNA Hybridization

The DNA in the chromosomes of all living organisms is double-stranded, with the two strands held together by hydrogen bonds between complementary bases. As you may recall from our discussion of the polymerase chain reaction in Chapter 14, the two strands of DNA can be separated by heating. Heat energy causes the two strands to jiggle, and, when the temperature becomes high enough, the hydrogen bonds break and the two strands separate (molecular biologists call it "melting"). The separation can be measured with an instrument called a spectrophotometer. Single DNA strands absorb ultraviolet (UV) light more effectively than double helices do, so the higher the proportion of single strands, the more opaque to UV light the solution becomes (Fig. E16-5a).

Suppose now that you have two single strands of DNA in which only half of the bases were complementary. The complementary bases would still hold the two strands together, but with only half the normal number of hydrogen bonds. If you heated this "semi-double helix," you would find that, with fewer hydrogen bonds holding the strands together, the strands would separate at a lower temperature.

In the technique called DNA–DNA hybridization, single strands of DNA from two different species are allowed to form "hybrid" double helices (the procedures are fairly involved, but this is the final result). They are then heated up, and the temperature at which the two strands separate is measured. If we call the separation temperature for two strands from the same species T, then separation temperatures for strands from different species will be less than T. The principle is this: Closely related species will have more similar nucleotide sequences in their DNA than more distantly related species do. The more similar the nucleotide sequences, the higher the temperature required for the hybrid helices to separate. Therefore, one can generate a tree of evolutionary relationships, based on the separation temperature of the hybrid helices (Fig. E16-5b).

Direct Comparison of Nucleotide Sequences

"Melting" DNA double helices is, of course, a somewhat indirect way of measuring similarities in nucleotide sequences. With the advent of DNA sequencing

Figure E16-5 DNA–DNA hybridization as a tool for determining evolutionary relatedness.
(a) When DNA is dissolved in water and heated, the double helix separates into two single strands, or "melts." Single DNA strands absorb more UV light than the double helix does. At low temperatures all DNA is double-stranded; at high temperatures all DNA is single-stranded. The steepest point of the curve of UV absorption, at which roughly half the DNA is separated into single strands, is the "melting temperature." (Modified from Strickberger, M.W., *Genetics*, Macmillan Publishing Co., 1985.)
(b) In this study by Charles Sibley and Jon Ahlquist, double helices of DNA were made by combining single strands from two different species of primate. The melting temperatures were then determined (graphed here as the change in melting temperature compared with double helices made from DNA taken from only one species). When the melting temperature for hybrid, two-species DNA is similar to the melting temperature for single-species DNA, then the nucleotide sequences of the two species must be similar, and the species are probably closely related. When hybrid DNAs (e.g., human–Old World Monkey) melt at much lower temperatures than single-species DNAs, the species must have a common ancestor a very long time ago, and their common branch point is low down on the tree (branch point labeled [a]). Hybrid DNAs for which melting temperatures are very similar (e.g., the two chimpanzee species, branch point [b]) have a recent common ancestor. If two types of hybrid DNA (e.g., human–chimp [c] vs. human–gorilla [d]) have similar melting temperatures, then it is difficult to be certain about the relationships. Biologists still debate whether chimps are more closely related to humans or to gorillas.

machines, it is now possible to compare the nucleotide sequences of single genes from a variety of species. The number of nucleotide substitutions between genes encoding the same or similar proteins is taken to be a reflection of the evolutionary relatedness between species: the fewer the substitutions, the more closely the organisms are related. Of course, comparisons of a single gene may give an incorrect evolutionary tree, because the rate of change in nucleotide composition of genes may differ, depending on, among other things, the strength and direction of natural selection acting upon different organisms. This is, in fact, an argument often advanced on behalf of DNA–DNA hybridization, because hybridization looks at changes over millions of nucleotides, not just the few hundred of any single gene. However, as sequencing becomes more automated and more economical, sequence data promise to offer insight into the molecular nature of evolution.

(a)

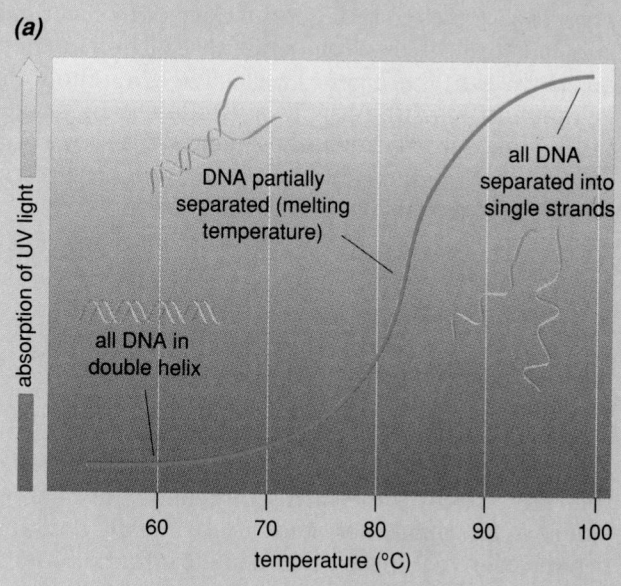

(b)

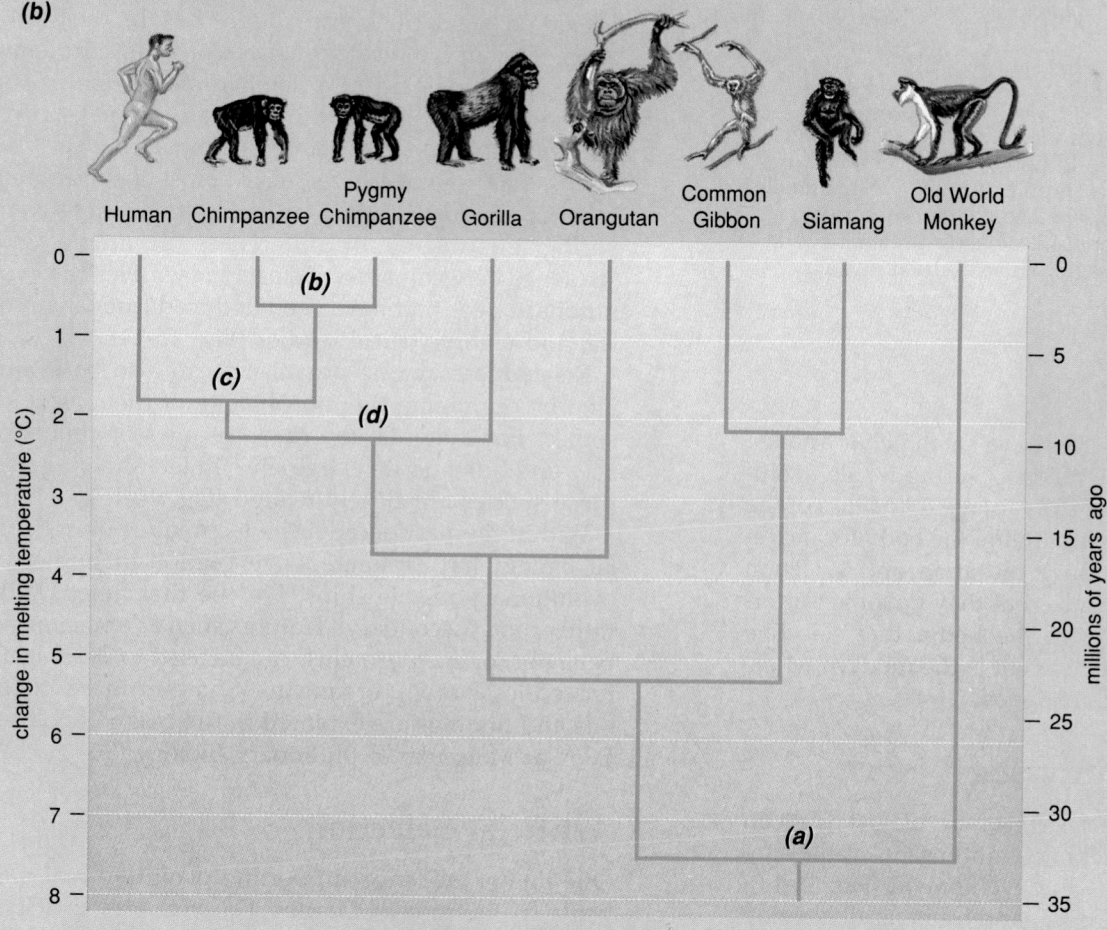

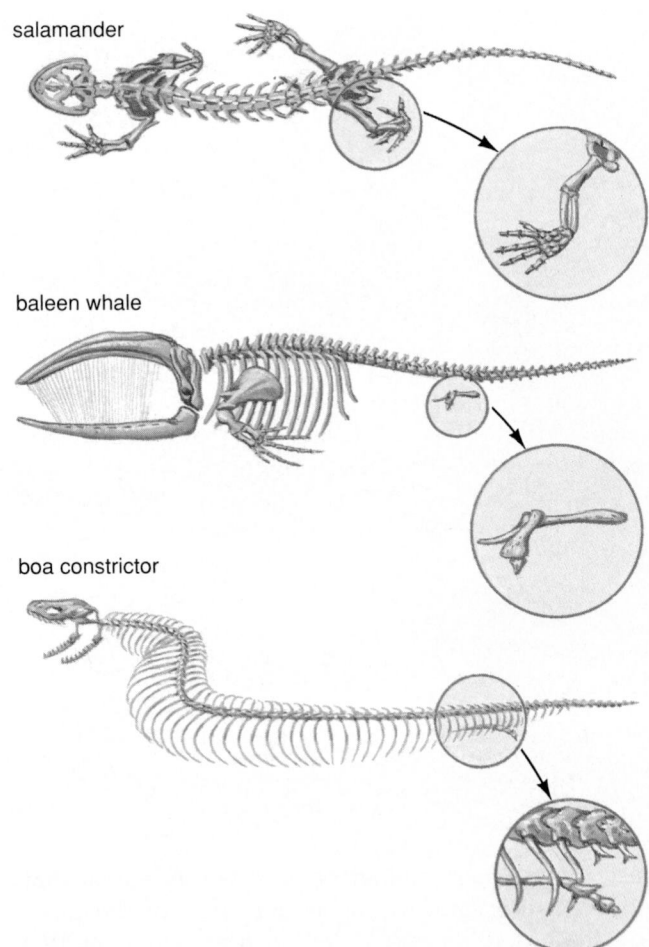

salamander

baleen whale

boa constrictor

Figure 16-12 The functional hindlimb of a salamander (top) is probably similar to the limb of the common ancestor of all amphibians, reptiles, birds, and mammals. Baleen whales (center) and boa constrictors (bottom) have no functional legs but still develop vestigial pelvic girdles and even miniature leg bones buried in their sleek sides.

not have hind legs, and yet have small pelvic and leg bones embedded in their sides. During whale evolution, there was a selective advantage to the loss of the hind legs, the better to streamline the body for movement through water. Once mutation and selection reduced the pelvic bones so that they no longer interrupted the smooth line of the body, that selective pressure diminished. The result is the modern whale with small, useless pelvic bones.

Comparative Embryology

In the early 1800s, the Prussian embryologist Karl von Baer noted that all vertebrate embryos look quite similar to one another early in development (Fig. 16-13). Fish, turtles, chickens, mice, and humans all develop tails and gill arches early in development. Only fish go on to develop gills, and only fish, turtles, and mice retain substantial tails. Why do such diverse vertebrates have similar developmental stages? The only plausible explanation is that ancestral vertebrates possessed genes that direct the development of gills and tails. All of their descendants still retain those genes. In fish, these genes are active throughout development, resulting in gill- and tail-bearing adults. In humans and chickens, these genes are active only during early developmental stages, and the structures are lost or inconspicuous in adults.

Comparative Biochemistry and Molecular Biology

Biochemistry and molecular biology provide striking evidence for the evolutionary relatedness of all living organisms. At the most fundamental biochemical levels, all living cells are very similar. For example, all cells have DNA as the carrier of genetic information; all use RNA, ribosomes, and approximately the same genetic code to translate that genetic information into proteins; all use roughly the same set of 20 amino acids to build proteins; and all use ATP as an intracellular energy carrier.

Evolutionary relationships among species are reflected in similarities and differences in their proteins. The amino acid sequences of a few proteins, such as hemoglobin and cytochrome c (one of the electron carriers in the mitochondrion), have been determined for many different species. The sequences are remarkably similar across a huge spectrum of species. Further, an evolutionary tree comparing the degree of differences in amino acid sequence between species closely resembles the evolutionary trees that have been deduced from anatomical and embryological studies (Fig. 16-14).

Relatedness among organisms can also be evaluated by examining the morphology of their chromosomes. For example, the chromosomes of chimpanzees and humans are extremely similar, showing that these species are closely related (Fig. 16-15).

Within the past decade, the techniques of molecular biology have triggered a revolution in studies of evolutionary relationships. For the first time, DNA, rather than "secondary" features such as appearance, behavior, or even proteins, can be used to investigate relatedness among organisms. Some of these methods and findings are explored in the essay "A Closer Look at Molecular Evolutionary Biology."

Artificial Selection

One line of evidence supporting evolution that particularly impressed Charles Darwin was **artificial**

(a)

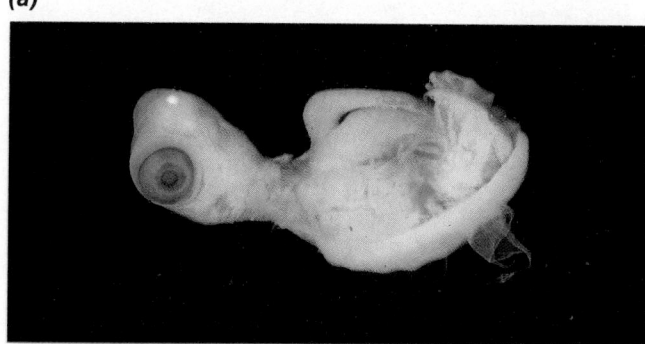

(b)

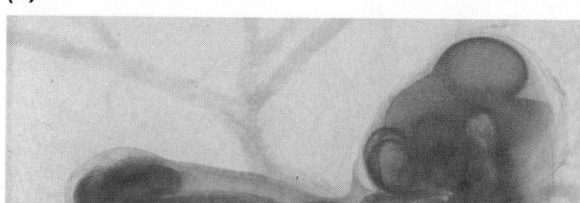

(c)

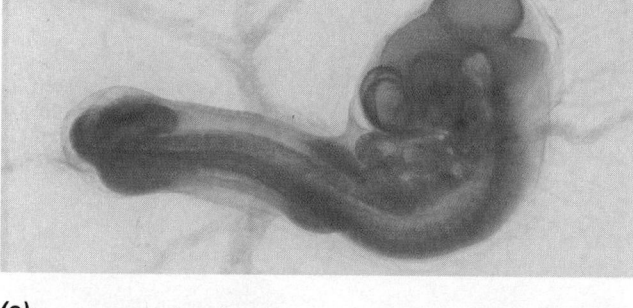

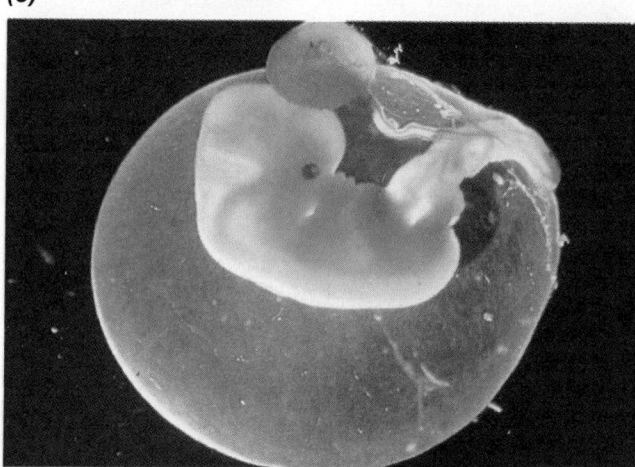

(d)

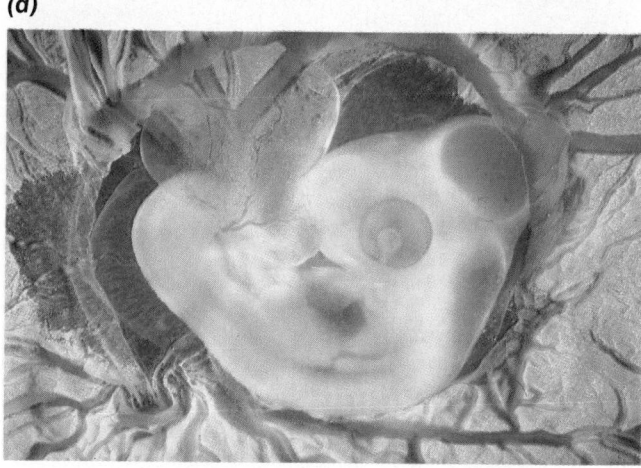

Figure 16-13 Early embryonic stages of a **(a)** turtle, **(b)** mouse, **(c)** human, and **(d)** chicken, showing strikingly similar anatomical features.

selection: breeding domestic plants and animals to produce specific desirable features. The various breeds of dogs provide a striking example of artificial selection (Fig. 16-16). Dogs descended from wolves, and even today the two will readily cross-breed. However, with rare exceptions, few modern dogs resemble wolves. Some breeds, such as the Chihuahua and the Great Dane, are so different from one another that they would be considered separate species if they were found in the wild. Interbreeding would hardly be possible without a lot of human assistance. If humans could breed such radically different dogs in a few hundred to at most a few thousand years, Darwin reasoned, it seemed quite plausible that natural selection could produce the spectrum of living organisms in hundreds of millions of years.

Present-Day Evolution

One problem that many people have with the theory of evolution is that they think it all happened in the past. Evolutionary biologists, however, maintain that evolution is *not* merely a phenomenon of the past, but that it continues today. Strong evidence for present-day evolution has come as a fortuitous side effect of an otherwise undesirable circumstance: the rise of industrial pollution in the nineteenth and twentieth centuries.

Great Britain is famous for its cool, damp climate. As a result, tree trunks in British woodlands usually support a lush growth of mottled grey lichens (a symbiotic association of an alga and a fungus; see Chapter 23). Before the Industrial Revolution, most peppered moths, *Biston betularia*, were white with scattered specks of black pigment. This coloration matched the color and pattern of the lichens growing on the trees. Since the moths sat quietly on the lichens during the day, predatory birds could not easily see them (Fig. 16-17a). Occasionally, mutant black individuals appeared. These black moths were extremely conspicuous against the pale lichens, were easily spotted by birds, and didn't live long.

The Industrial Revolution changed everything. The growing industries of nineteenth-century Britain burned coal for fuel. With no pollution-control technology, soot from the smokestacks soon blanketed the countryside around the mills and factories (Fig. 16-18). The lichens on the trees became covered with pollutants and died off, leaving the trunks sooty black. The pale form of the peppered moth was no

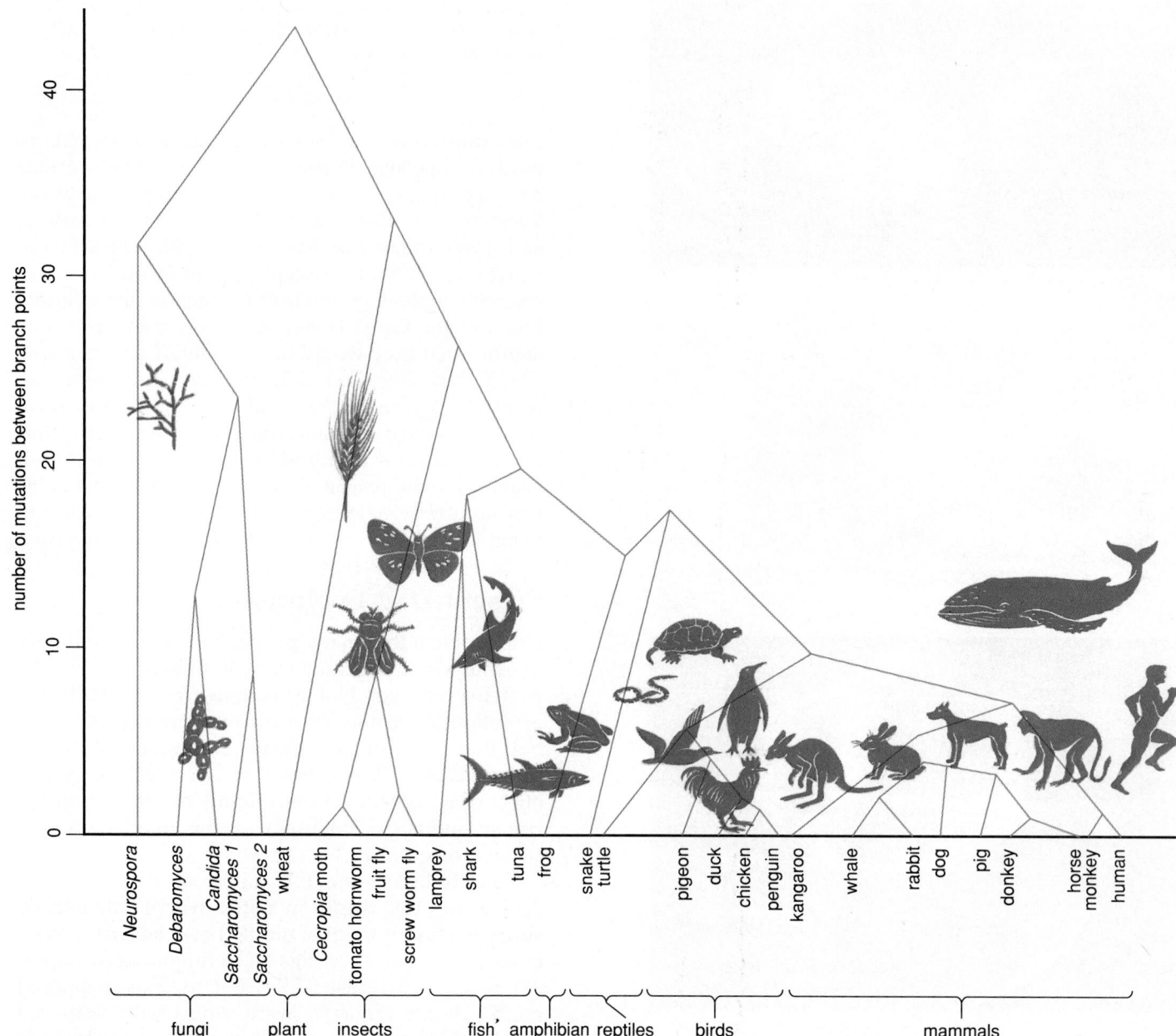

Figure 16-14 Evolutionary trees (see Chapter 20) were originally devised by comparative anatomists, botanists, and microbiologists. In recent years, biochemistry and molecular biology have offered independent evidence for the accuracy of these trees. The amino acid for cytochrome c, for example, has been determined for scores of species. One would expect that there should be fewer genetic differences (mutations) between closely related species than between more distantly related species. Thus, the more closely related two species are, the more similar one would expect the amino acid sequences in cytochrome c to be. This graph depicts the probable number of mutations that would be required to produce the observed differences in amino acid sequences between organisms connected by branch points. As you can see, the tree is very similar to "classical" trees based on anatomy.

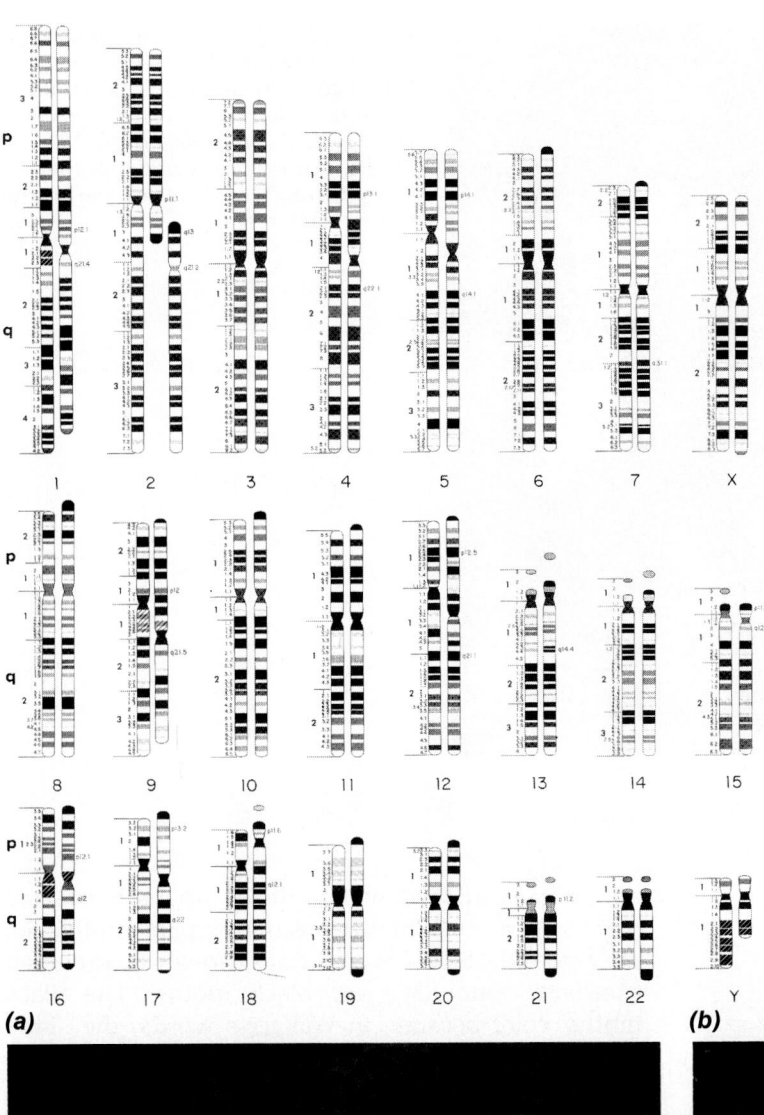

(a)

Figure 16-15 A comparison of the chromosomes of humans (left member of each pair) and chimpanzees (right member of each pair), stained to reveal detailed banding patterns. The numbers are those assigned to human chromosomes; note that chimps have two chromosomes that, combined, are virtually identical to human chromosome number 2. The close evolutionary relationship between humans and chimps is clearly revealed by the striking similarity of the chromosomes.

(b)

Figure 16-16 A comparison of the ancestral dog (the gray wolf, *Canis lupus*, **a**) and various breeds of dog (**b**). Artificial selection by humans has caused a great divergence in size and shape in only a few thousand years. Huge differences in size, particularly of the reproductive organs, would make it extremely difficult for the large dogs in the rear of the photo to interbreed with the small terriers in the foreground.

(a)   (b)

Figure 16-17 The two color forms of the peppered moth, resting on a lichen-covered tree trunk **(a)** and a soot-blackened trunk **(b)**. The pale form is well camouflaged on the lichen, but is conspicuous on the sooty trunk, while the reverse is true of the black form.

Figure 16-18 The Peak District near Manchester, England, is crisscrossed by black stone fences. If you break one of the stones, you can see that they are actually a bright golden sandstone. The black outer crust is the remnant of pollution from the Industrial Revolution of over a half century ago.

longer camouflaged while sitting on the blackened trunks (Fig. 16-17b). Increasingly, pale moths fell prey to birds. Soot-covered trees, however, provided excellent camouflage for black moths. The black moths' color became, in Wallace's words, the "little superiority enabling them to escape" predation by birds (see "Methods in Biology: Demonstrating that Industrial Melanism Is an Example of Evolution by Natural Selection"). Black moths, once rare, survived and reproduced, passing on their genes for dark pigmentation to succeeding generations. As the years passed, ever-increasing numbers of black moths appeared, until by the end of the nineteenth century, about 98% of the moths around the industrial city of Manchester were black. This phenomenon of "industrial melanism" (becoming pigmented as a consequence of industrial activities) was not restricted to *Biston betularia*. Dark varieties of many moth species appeared and spread.

Incidentally, during the past few decades, pollution-control laws have dramatically reduced emissions of soot and other pollutants, and lichens grow once again on British trees. As predicted by the principles of evolution, the pale moths are making a comeback, and the black form is becoming increasingly rare.

METHODS IN BIOLOGY
Demonstrating that Industrial Melanism Is an Example of Evolution by Natural Selection

There is no denying the fact that black forms of the peppered moth became increasingly common in polluted industrial areas of Britain during the Industrial Revolution. It is also obvious that black moths matched the sooty trees better than the pale variety of moth did (see Fig. 16-17). Are these two facts causally related? Was industrial melanism a coincidence, or was predation by birds the selective force that drove industrial melanism? How could one decide?

Ecologist H. B. Kettlewell set out to test the hypothesis that coloration influenced survival in the wild, using the "mark-and-recapture" method. He chose two sites for his studies. In Dorset, on the western coast of Britain, the woods were clean and lichen-covered, and most of the moths were pale. In Birmingham, in the midst of the industrial region, the woods were blackened, and about 85% of the moths were black. Kettlewell captured and marked a large number of moths, and then released moths of both varieties at the two locations. After a few days, he recaptured as many moths as he could. Kettlewell reasoned that the fraction of marked moths that were recaptured would be an indication of the rate of predation.

Obviously, a moth that was eaten would no longer be available to recapture. In this way, he could compare the survival of the two types of moths at the two sites.

Two striking findings emerged (Table E16-1). First, far more moths survived at Birmingham than at Dorset. Although the reason for this is not known with certainty, it may have been because there were fewer birds in polluted industrial areas. Second, and more importantly, at Dorset twice as many pale moths as black moths survived, while at Birmingham twice as many black moths survived.

The conclusion was inescapable: moth survival was directly related to their camouflage coloration. Where the pale coloration provided camouflage, the pale form dominated, but where black was the camouflage color, natural selection favored black moths, and consequently the black form prevailed. In just half a century, the black variety almost completely took over in the industrialized areas. Thus evolution does occur today, and natural selection by environmental pressures is the mechanism driving changes in characteristics over the generations, just as Darwin and Wallace hypothesized.

Table E16-1 Recaptures of Pale and Black Moths

Location	Pollution Level	Trunk Color	Pale Moths Recaptured (%)	Black Moths Recaptured (%)
Dorset	low	mottled gray	13.2	6.3
Birmingham	high	solid dark	13.1	27.5

Although we will come back to these points in the next chapter, there are a few additional comments that we should make here. First, the black coloration was not *produced* by industrial activities. The mutation for black coloration arises spontaneously now and then in both clean and soot-covered areas. **The variations upon which natural selection works are produced by chance mutations.** This is further shown by other moth species in both England and the eastern United States. Some species had occasional dark mutants, and the dark forms spread as the environment became dark. Other species lacked black mutants, and became extinct in industrial areas. A second point to remember, then, is that **selection does not necessarily produce well-adapted species.** If the right raw material isn't there—in this case black mutant moths—then selection may drive a species to extinction. Finally, some people tend to regard evolution as a modern mechanism for producing the *Scala Naturae*, with ever-greater degrees of perfection appearing over time. This is not correct. **The processes of evolution select for organisms that are best adapted to a particular environment.** Neither black nor pale colors are "best" in any objective sense, but only with respect to their environment, which may change at any time in an unpredictable fashion.

Industrial melanism is but one of many instances of modern-day evolution. Other examples include antibiotic resistance in bacteria (Chapters 12, 14, and 21) and the appearance of the AIDS virus (HIV) (Chapter 34). Evolution is easiest to observe in rapidly reproducing, short-lived organisms such as insects or bacteria, but occurs constantly in all the species of life on Earth.

Reflections on Evolution

"It is interesting to contemplate an entangled bank, clothed with many plants of many kinds, with birds singing on the bushes, with various insects flitting about, and with worms crawling through the damp earth, and to reflect that these elaborately constructed forms . . . have all been produced by laws acting around us. These laws, taken in the highest sense, being Growth with Reproduction; Inheritance [and] Variability . . . ; a Ratio of Increase so high as to lead to a Struggle for Life, and as a consequence to Natural Selection, entailing Divergence of Character and Extinction of less-improved forms . . . . There is grandeur in this view of life, with its several powers, having been originally breathed into a few forms or into one; and that, whilst this planet has gone cycling on according to the fixed law of gravity, from so simple a beginning endless forms most beautiful and most wonderful have been, and are being, evolved."

—the concluding sentences of *On the Origin of Species* by Charles Darwin

SUMMARY OF KEY CONCEPTS

The History of Evolutionary Thought
A species consists of all the populations of organisms that can potentially interbreed and that are reproductively isolated from other populations. Historically, the most common explanation for the origin of species has been creationism, that a divine being created each species in its present form, and that species have not significantly changed since the creation. Since the middle of the last century, however, scientists have concluded that species originate by the operation of natural laws, as a result of changes in the genetic makeup of the populations of organisms. This process is called evolution.

Evolution by Natural Selection
Charles Darwin and Alfred Russell Wallace independently proposed the theory of evolution by natural selection. Their theory can be concisely expressed as three conclusions based on four observations. In modern biological terms, these are:

Observation 1: Each organism can produce far more offspring than are required merely to replace it.

Observation 2: Nevertheless, the sizes of natural populations are usually constant.

Conclusion 1: Therefore, in each generation, many organisms must die young, fail to reproduce, produce few offspring, or produce less fit offspring that fail to reproduce in their turn.

Observation 3: The individuals in a population vary in their ability to obtain resources, withstand environmental extremes, escape predation, etc.

Conclusion 2: Which organisms produce the largest number of viable offspring depends on their ability to deal with their environment; the most well-adapted organisms reproduce the most. This is natural selection.

Observation 4: At least some of the variation in adaptedness is due to genetic differences that may be passed on from parent to offspring.

Conclusion 3: Over many generations, differential reproduction among individuals with different genetic makeup changes the overall genetic composition of the population. This is evolution.

The Evidence for Evolution
Many lines of evidence indicate that evolution has occurred, and that natural selection is the chief mechanism driving changes in the characteristics of species over time. The evidence includes:

1. Fossils of ancient organisms are simpler in form than modern organisms. Sequences of fossils have been discovered that show a graded series of changes in form. Both of these facts would be expected if modern forms evolved from older forms.

2. Organisms thought to be related through evolution from a common ancestor show many similar anatomical structures. Examples include the limbs of amphibians, reptiles, birds, and mammals. Similarly, stages in embryological development, similarities in chromosome structure, sequences of amino acids in proteins, and similarities in DNA composition all support the notion of descent of related species through evolution from common ancestors.

3. Rapid, heritable changes have been produced in domestic animals and plants by selectively breeding organisms with desired features (artificial selection). If differences as vast as those between chihuahuas, bulldogs, and Great Danes can be produced in a few thousand years of artificial selection by humans, than it seems likely that much larger changes could be wrought by hundreds of millions of years of natural selection.

4. Both natural and human activities may drastically change the environment over short periods of time. Significant changes in the characteristics of species have been observed in response to these environmental changes. A well-studied example is the evolution of black coloration among moths in response to the darkening of their environment by industrial pollutants.

GLOSSARY

analogous structures: structures that have similar functions and superficially similar appearance but very different anatomy, such as the wings of insects and birds. The similarities are due to similar selective pressures.

catastrophism: the hypothesis that the Earth has experienced a series of geological catastrophes, much like Noah's Flood, probably imposed by a supernatural being.

coevolution: the evolution of adaptations in two species due to their extensive interactions with one another, so that each species acts as a major force of natural selection upon the other.

convergent evolution: the independent evolution of similar structures among unrelated organisms, due to similar selective pressures. See *analogous structures*.

creationism: the hypothesis that all species of organisms on Earth were created in essentially their present form by a supernatural Being, and that significant modification of those species, specifically their transformation into new species, cannot occur through natural processes.

evolution: the descent of modern organisms from preexisting life forms; strictly speaking, any change in the proportions of different genotypes in a population from one generation to the next.

fossil: the remains of an organism, usually preserved in rock. Fossils include petrified bones or wood; shells; impressions of body forms such as feathers, skin, or leaves; and markings made by organisms such as footprints.

homologous structures: structures that may differ in function but that have similar anatomy, presumably because of descent from common ancestors.

inheritance of acquired characteristics: the hypothesis that organisms' bodies change during their lifetimes by use and disuse, and that these changes are inherited by their offspring.

natural selection: the unequal survival and reproduction of organisms due to environmental forces (e.g., physical factors such as climate and living organisms such as predators or prey) that act differently upon genetically different members of a population.

species (spē'-sēs): the sum of all of the populations of organisms that are potentially capable of interbreeding under natural conditions and that are reproductively isolated from other populations.

uniformitarianism: the hypothesis that the Earth developed gradually through natural forces similar to those at work today.

vestigial structures (ves-tij'-ē-ul): structures with no known function, but which are homologous to functional structures in related organisms.

STUDY QUESTIONS

1. Distinguish between catastrophism and uniformitarianism. How did these hypotheses contribute to the development of evolutionary theory?
2. What is a fossil? What sorts of evidence do fossils provide about past life on Earth and how it arose?
3. Describe Lamarck's theory of inheritance of acquired characteristics. Why is it invalid?
4. Define species.
5. What is natural selection? Describe how natural selection might have caused differential reproduction among the ancestors of a fast-swimming predatory fish, such as a barracuda.
6. Describe how evolution occurs through the interactions among the reproductive potential of a species, the normally constant size of natural populations, variation among individuals of a species, natural selection, and inheritance.
7. Distinguish between homologous and analogous structures, and give examples of each.
8. What is convergent evolution? Give an example.
9. What is a vestigial structure? Give two examples of vestigial structures.
10. How do biochemistry and molecular genetics contribute to the evidence that evolution occurred?

DISCUSSION QUESTIONS

1. Does evolution through natural selection produce "better" organisms in an absolute sense? Are we climbing the *Scala Naturae?* Defend your answer.
2. The hypotheses of Cuvier and Agassiz failed to provide a convincing explanation of the enormous numbers of extinctions that have occurred in the history of life on Earth. How are extinctions explained by the theory of evolution by natural selection?

3. Describe and compare the views of Plato and Darwin on the existence of variation in natural populations.
4. The idea of special creation and the study of fossils have each had an impact on evolutionary thought. Discuss why one is considered scientific endeavor and the other not scientific.

SUGGESTED READINGS

Altman, S. A. "The Monkey and the Fig." *American Scientist*, May–June 1989. An amusing, informative discussion of many evolutionary themes, cast as a Socratic dialog.

Bishop, J. A., and Cook, L. M. "Moths, Melanism and Clean Air." *Scientific American*, January 1975. This follow-up to Kettlewell's work shows that cleaning up the air reverses industrial melanism in moths, providing graphic evidence that evolution only adapts organisms to existing environments.

Darwin, C. *On the Origin of Species by Means of Natural Selection*, 1859. Garden City, N.Y.: Doubleday and Company, Inc., 1960. An impressive array of evidence amassed to convince a skeptical world.

Eiseley. L. C. "Charles Darwin." *Scientific American*, February 1956. An essay on the life of Darwin, by one of his foremost American biographers. Even if you need no introduction to Darwin, read this anyway, as an introduction to Eiseley, author of many marvelous essays.

Gould, S. J. *Ever Since Darwin*, 1977; *The Panda's Thumb*, 1980; and *The Flamingo's Smile*, 1985. New York: W.W. Norton & Company, Inc., A series of witty, imaginative, and informative essays, mostly from *Natural History* magazine. Many deal with various aspects of evolution.

Grant, P. R. "Natural Selection and Darwin's Finches." *Scientific American*, October 1991. A drought in the Galapagos Islands provides dramatic evidence for natural selection as an agent of evolutionary change.

Kettlewell, H. B. D. "Darwin's Missing Evidence." *Scientific American*, March 1959. Industrial melanism as an example of modern-day evolution.

Lewontin, R. C. "Adaptation." *Scientific American*, September 1978. Lewontin shows that perfect adaptation is not always achieved.

Noonan, D. "Dr. Doolittle's Question." *Discover*, February 1990. The protein fibrinogen provides evidence on evolutionary relationships.

17

The Mechanisms of Evolution

*"What but the wolf's tooth whittled so fine
  The fleet limbs of the antelope?
What but fear winged the birds, and hunger
  Jewelled with such eyes the great goshawk's head?"*

Robinson Jeffers in The Bloody Sire

In Chapter 16 we discussed the history of the theory of evolution and presented some of the evidence that evolution actually happens. But what processes drive evolutionary change? Is natural selection the only cause of evolution? Does evolution always occur all the time in all populations of organisms? In this chapter, we examine evolutionary processes in more detail. As we do, you will see that *evolution is an inevitable consequence of the nature of living things.* It occurs as a direct result of the chemical structure of genes and the interactions between organisms and their environment.

Evolution and the Genetics of Populations

Individual organisms live, reproduce, and die. Individuals, however, do not evolve: **evolution is the genetic change occurring in a population of organisms over many generations.** Inheritance, therefore, is the link between the lives of individual organisms and the evolution of **populations.** We will begin our discussion of the processes of evolution by reviewing the principles of genetics as they apply to individuals and then extend these principles to the genetics of populations. You may want to refer back to Unit II to refresh your memory on specific points.

Gene Function in Individual Organisms

Each cell of every organism contains a repository of genetic information encoded in the DNA of its chromosomes. A gene is a segment of DNA located at a particular place on a chromosome. Its sequence of nucleotides encodes the sequence of amino acids of a protein, usually an enzyme that catalyzes one particular reaction in the cell. Slightly different sequences of nucleotides at a given gene's location, called alleles, generate different forms of the same enzyme. The specific alleles borne on an organism's chromosomes (its genotype), interacting with the environ-ment, determine its physical and behavioral traits (its phenotype).

Let's illustrate these principles with an example that should be familiar to you from Unit II. A pea flower is colored purple because a chemical reaction in its petals converts a colorless molecule to a purple pigment. When we say that a pea plant has the allele for purple flowers, we mean that a particular stretch of DNA on one of its chromosomes contains a sequence of nucleotides that codes for the enzyme catalyzing this reaction. A pea with the allele for white flowers has a different sequence of nucleotides at the corresponding place on one of its chromosomes. The resulting enzyme cannot produce purple pigment. If a pea is homozygous for the white allele, its flowers produce no pigment, and are white.

Genes in Populations

A branch of genetics, called **population genetics,** deals with the frequency, distribution, and inheritance of alleles in populations. Since evolution is a change in the genetic makeup of populations over generations, you will need to learn the principles of population genetics to understand the mechanisms of evolution.

In population genetics, the **gene pool** for a particular gene is defined as the total of all the alleles of that gene that occur in a population. (The **total gene pool** for the population is the total complement of alleles for all the genes.) For example, in a population of 100 pea plants, the gene pool for flower color would consist of 200 alleles (peas are diploid, so there are two color alleles per plant, times 100 plants). If we could analyze the genetic composition of every plant in the population, we might find that some have alleles for white flowers, some have alleles for purple flowers, and some have both alleles. If we added up the color alleles of each plant in the population, we could determine the relative proportions of the different alleles, a number called the **allele frequency.** Let's say that the gene pool for flower color consisted of 140 alleles for purple and 60 alleles for white. The allele frequencies would then be purple, 0.7 (70%), and white 0.3 (30%).

Population Genetics and Evolution

What does all this have to do with evolution? Quite a bit. Suppose a flower-eating cow comes along and, being enamored of purple flowers, eats all the purple flowers before they set seed. As you know from Chapter 11, the allele for purple flowers is dominant to the allele for white. Therefore, all the purple alleles in the entire population are in the purple-flowered plants. If none of these plants reproduce, while the white-flowered plants do reproduce, then the next generation will consist entirely of white-flowered peas. The allele frequency for purple will drop to 0, while the allele frequency for white will rise to 1.0. Because of the selective eating habits of the cow, *evolution will have occurred in that field*. The gene pool of

the pea population will have changed, and natural selection, in the form of foraging by the cow, will have caused the change.

This simple example illustrates four important points about evolution.

1. **Natural selection does not cause genetic changes in individuals.** The alleles for purple or white flower color arose spontaneously, long before the cow ever found the pea field. The white peas did not possess some sort of "foresight" and acquire the alleles for white flowers in anticipation of the cow. The cow, in turn, did not cause white alleles to appear. It merely favored the differential survival of white alleles compared with purple alleles.

The Hardy–Weinberg equilibrium predicts that, if a large population undergoes no mutation, migration, or natural selection, and if all members of the population mate randomly, then the frequencies of alleles will not change from generation to generation. To see how this can be so, consider our familiar pea plants. Pea seeds can be round (*R*: dominant) or wrinkled (*r*: recessive; see Fig. 11-8). To determine the genotypes of the offspring of two individuals, for example two heterozygotes, we would draw a Punnett square:

	R	*r*
R	**RR**	**R***r*
r	**R***r*	*rr*

Each parent produces both *R* and *r* gametes. The expected offspring are $\frac{1}{4}$ *RR*, $\frac{1}{2}$ *Rr*, $\frac{1}{4}$ *rr*. There are two ways to arrive at these frequencies. The first is to add up the offspring in each box of the square. Another way of doing it is by probabilities. Each gamete has an equal probability of containing either allele. Therefore we can assign probabilities to the gametes in the Punnett square: *R* = 0.5, *r* = 0.5. From the laws of probability, **the**

probability of two independent events occurring simultaneously is the product of their individual probabilities. If you flip a coin, the probability of a head is $\frac{1}{2}$. If you flip two coins simultaneously, the probability of two heads is $\frac{1}{2} \times \frac{1}{2} = \frac{1}{4}$. Similarly, we can obtain the probability of obtaining each type of offspring by multiplying the relative proportions of each allele: We obtain 0.25 *RR*, 0.5 *Rr*, and 0.25 *rr*.

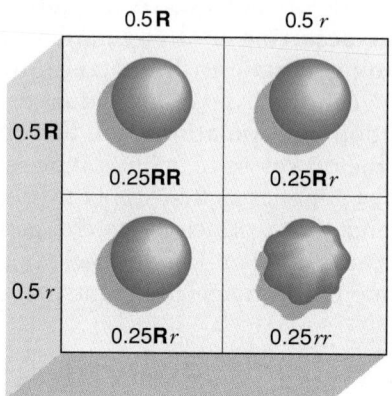

Let us suppose, now, that we have a population of 100 peas, that we collect sperm and egg cells from all of them, and determine their genotypes. We may find, for example, that there are 60% *R* alleles and 40% *r* alleles in the gametes. The proportions of the two alleles *R* and *r* are identical to the probability that any given offspring will receive either *R* or *r*. We can thus draw a "population Punnett square":

Figure 17-1 The genetics of equilibrium populations.

2. **Natural selection befalls individuals, but evolution occurs in populations.** Individual pea plants either reproduced or not, but it was the population as a whole that evolved.

3. **Evolution is a change in the allele frequencies of a population, owing to differential reproduction among organisms bearing different alleles.** In evolutionary terminology, the **fitness** of an organism is a measure of its reproductive success: in our example, the white flowers had greater fitness than the purple flowers, because they produced more viable offspring.

4. **Evolutionary changes are not "good" or "progressive" in any absolute sense.** The white alleles were only favorable because of the dietary preferences of this particular cow; in another environment, with other predators, the white allele may well be selected against.

The Equilibrium Population

It will be easier to understand the forces that cause populations to evolve if we first consider the characteristics of a population that would *not* evolve. In 1908, G. H. Hardy and W. Weinberg defined an **equilibrium population** as one in which the allele frequencies and the distribution of genotypes remains constant with succeeding generations (Fig. 17-1). If allele frequencies do not change, evolution does not occur. A population can remain in equilibrium only if several restrictive conditions are met:

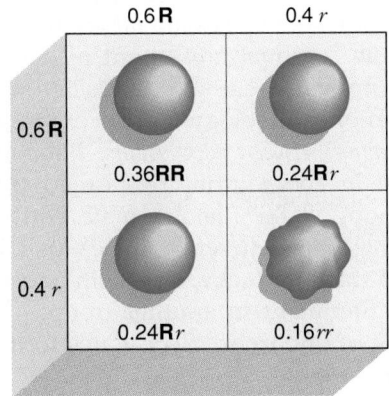

In the "population F_1 generation," we expect the following proportions of genotypes: 0.36 *RR*, 0.48 *Rr*, and 0.16 *rr*. If the population remains the same size, 100 peas, then we would have 36 *RR*, 48 *Rr*, and 16 *rr* peas. What gametes would this F_1 generation in its turn produce? Under the Hardy–Weinberg conditions, each plant produces equal numbers of gametes, and by the principles of Mendelian genetics, each plant produces equal numbers of gametes with each of its two alleles for seed shape. To keep things simple, let's assume that each plant contributes two gametes, one with each of its two alleles. We therefore collect 72 *R* alleles from the homozygous dominants, 48 *R* alleles and 48 *r* alleles from the heterozygotes, and 32 *r* alleles from the homozygous recessives, for a total of 120 *R* and 80 *r* alleles. The allele frequencies of the gametes from the "population F_1 generation," then, are 0.6 *R* and 0.4 *r*,

just as we started out with. Therefore, the F_2 generation has the same distribution of genotypes as the F_1. If there are no disturbances, this process will go on indefinitely: the population remains in equilibrium.

Rather than going through Punnett squares, there is an easier way of calculating allele and genotype frequencies. The sum of all allele frequencies must equal 1. Let the frequency of the *R* allele be represented by *p*, and the frequency of the *r* allele by *q*. Then the sum of the frequencies $p + q = 1$. Just as we generated the genotype frequencies in the "population Punnett square" by multiplying allele frequencies, we can do the same with this equation:

$$(p + q) \times (p + q) = p^2 + pq + qp + q^2$$
$$= p^2 + 2pq + q^2 = 1$$

For our particular example, $p = 0.6$ and $q = 0.4$, so the genotypes of the population F_1 generation will be:

$$(0.6)^2 \ RR + 2 \times (0.6) \times (0.4) \ Rr + (0.4)^2 \ rr$$
$$= 0.36 \ RR, \ 0.48 \ Rr, \ \text{and} \ 0.16 \ rr.$$

This is the same set of frequencies that we calculated with the "population Punnett square."

As these calculations show, in an equilibrium population allele frequencies and the distribution of genotypes remain constant, generation after generation. In actual experiments, if measurements of allele frequencies in a population show significant changes over time, then evolution is occurring in that population.

1. There must be **no mutation.**
2. There must be **no differential migration** of alleles into the population (immigration) or out of the population (emigration).
3. The **population must be large** (theoretically infinite).
4. **All mating must be random,** with no tendency for certain genotypes to mate with specific other genotypes.
5. All genotypes must reproduce equally well; that is, there must be **no natural selection.**

Under these conditions, allele frequencies within a population will remain the same indefinitely. If one or more of these conditions are violated, then allele frequencies will change, i.e., evolution will occur.

As you might expect, few if any natural populations are truly in equilibrium. If so, then what is the importance of the Hardy–Weinberg principle? The Hardy–Weinberg conditions are useful starting points for studying the mechanisms of evolution. In the following sections, we will examine each condition, show why it is often violated by natural populations, and illustrate the consequences of its violation. In this way, you can better understand both the inevitability of evolution and the forces that drive evolutionary change.

The Mechanisms of Evolution

As the Hardy–Weinberg conditions predict, there are five major causes of evolutionary change within a population: mutation, migration, small populations, nonrandom mating, and natural selection.

Mutations: The Raw Material of Evolution

Cells have efficient mechanisms that protect the integrity of their genes. Enzymes constantly scan the DNA, repairing flaws caused by radiation, chemical damage, or mistakes in copying. Nevertheless, changes in nucleotide sequence can happen. These are **mutations,** and they vary tremendously in their impact. As we explained in Chapter 13, some changes in DNA have virtually no effect on the organism; many, perhaps most, are harmful; and a few may be beneficial, or may aid the organism in coping with new or changed environments.

How significant is mutation in altering the gene pool of a population? Mutations are rare, occurring once in 10,000 to 1,000,000 genes per generation per individual. Therefore, mutation is not a major force in evolution by itself. However, *mutations are the source of new alleles,* new heritable variations upon which other evolutionary processes can work. As such, they are the foundation of evolutionary change.

As we mentioned earlier, *mutations are not goal-directed.* A mutation does not arise as a result of, nor in anticipation of, environmental necessities (Fig. 17-2). A mutation simply happens, and may in turn produce a change in the structure or function of the organism. Whether that change is helpful or harmful, now or in the future, depends on environmental conditions over which the organism has little or no control. The mutation provides *potential;* other forces, such as migration and especially natural selection, acting on that potential, may favor the spread of a mutation through the population.

Migration: Redistributing Genes

In biology, the word migration has two distinct meanings. In the most familiar context, migration refers to the seasonal movement of many species between summer breeding grounds and winter refuges. In evolutionary biology, however, **migration** *is the flow of genes between populations.* Baboons, for example, live in social groupings called troops. Within each troop, all the females mate with a handful of dominant males. Juvenile males usually leave the troop. If they are lucky, they join and perhaps even become dominant in another troop. Thus the male offspring of one troop carry genes to the gene pools of other troops.

Migration has two significant effects.

1. **Gene flow spreads advantageous alleles throughout the species.** Suppose that a new allele arises in one population, and that this new allele benefits the organisms that possess it. Migration can carry this new allele to other populations of the species.
2. **Gene flow helps to maintain all the organisms over a large area as one species.** If migrants constantly carry genes back and forth among populations, then the populations can never develop large differences in allele frequencies. Isolation of populations, with no gene flow to or from other populations of the same species, is an important factor in the origin of new species.

Small Populations: Random Changes in Allele Frequencies

To remain in equilibrium, a population must be large. To see why, let's return to the thoughts of Darwin

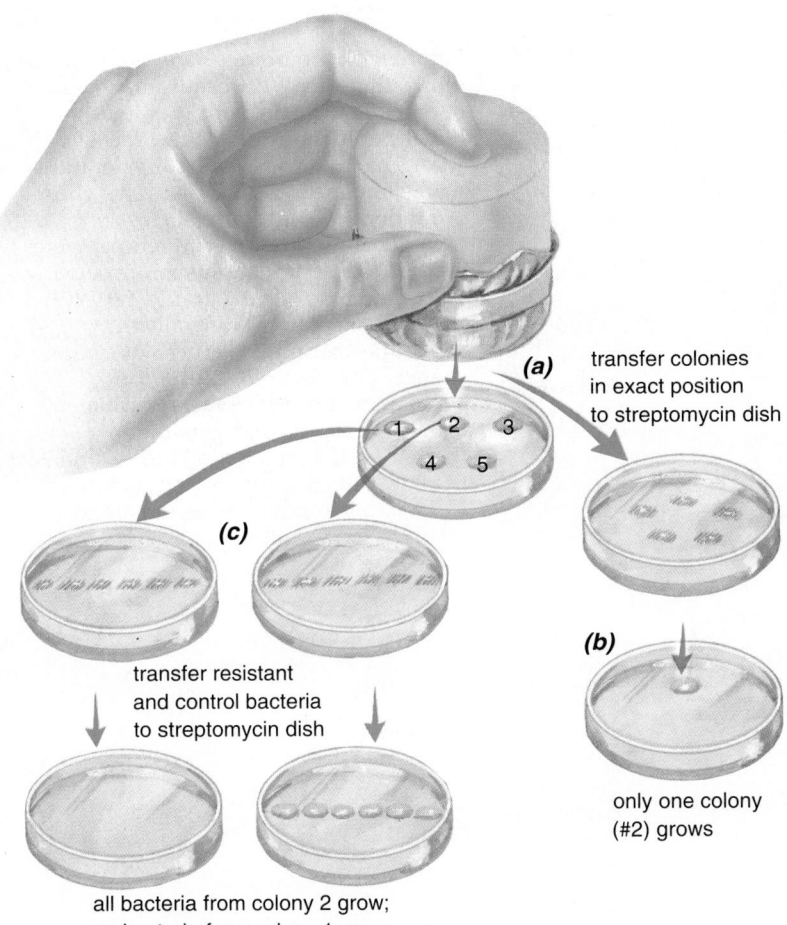

Figure 17-2 Proof that mutations occur spontaneously, and not in response to specific selective pressures. **(a)** Clumps of bacteria (colonies) are grown on a solid nutrient medium in a dish. These bacteria have never been exposed to antibiotics. A piece of velvet the exact size of the dish is lightly pressed into the bacterial colonies, and then touched to the surface of nutrient medium containing the antibiotic streptomycin in a second dish. A few bacteria from each original colony adhere to the velvet and then come off the velvet into the second dish. Thus the exact positions of the "parent" colonies are duplicated in the second dish. **(b)** Only one daughter colony, in position 2, grows on the streptomycin-containing medium in the second dish. There are two possible explanations for this result: First, the bacteria of parent colony 2 may have been already resistant, due to a spontaneous mutation, while the bacteria of the other parent colonies were not. In this case, we would expect that the other bacteria of parent colony 2 will also be resistant to streptomycin. Second, perhaps none of the parent bacteria were resistant, and one mutated in the second dish in response to streptomycin. If this happened, then the other bacteria from parent colony 2 will not, in general, be resistant to streptomycin. **(c)** Samples of the original colonies 1 and 2 are transferred to streptomycin-containing medium. All bacteria from colony 2 grow, while none from colony 1 grow, proving that the bacteria of colony 2 already possessed the mutation for streptomycin resistance prior to exposure, and that the presence of streptomycin in the medium does not induce an adaptive mutation for streptomycin resistance.

and Wallace. In general, all populations have tremendous potential for growth, but are limited to a relatively constant size by the available resources (e.g., food, nesting sites, hiding places). This means that most organisms must die without reproducing, or produce offspring that cannot successfully reproduce in their turn. Put another way, *only a small sample of a population actually serves as parents for succeeding generations.*

Which individuals reproduce depends both on fitness and on chance. Obviously, those individuals that are better adapted to their environment are more likely to survive and leave offspring to carry on their genes. Nevertheless, chance is also important, because disaster may befall even the fittest organism. The maple seed that falls into a pond never sprouts; the deer and elk blasted away by Mount St. Helens left no descendants.

Genetic Drift

It is much more likely that chance events will change allele frequencies in a small population than in a large

population, by a process called **genetic drift.** Consider, for example, two hypothetical populations of ladybugs, in which the carapace is either spotted or solid-colored, controlled by alternate alleles of a single gene. In each population, half the ladybugs are spotted and half are solid-colored (i.e., the frequencies of both alleles are 0.5), but one population has only four bugs while the other has 1000. Let us assume that each individual that survives to maturity produces two offspring identical to itself. Therefore, if population sizes remain constant, exactly half the individuals will reproduce in each generation. Let us further assume that whether an individual ladybug reproduces or not is determined entirely by chance. In the larger population, 500 bugs will be parents to the next generation. It would be astronomically unlikely that all 500 parents would be spotted; in fact, it would be extremely unlikely for even 300 parents to be spotted. In this large population, then, we would not expect a major change in allele frequencies to occur from generation to generation (Fig. 17-3). In the small population, on the other hand, only two individuals will reproduce. There is a 25% chance that

(a)

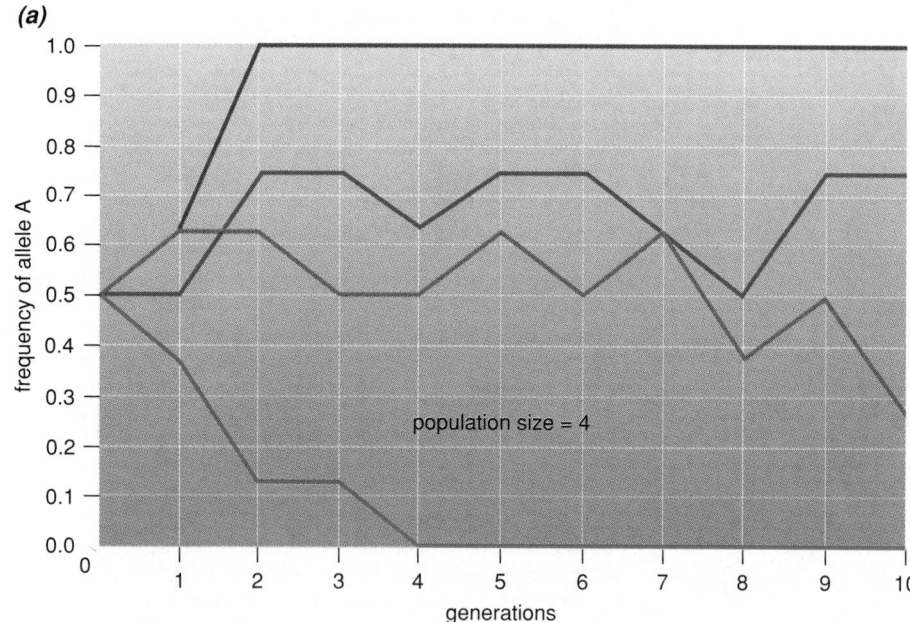

Figure 17-3 Computer-generated graphs illustrating the effect of population size on genetic drift. In both graphs, the initial population was composed of half *A* and half *a* alleles, and 10 generations were simulated, with individuals chosen at random to contribute alleles to the next generation. Four simulations were run for each population size, producing the four lines on each graph. **(a)** With a population size of 4, one allele sometimes became "extinct" due to chance. For example, in the top simulation run, the *a* allele became extinct by the second generation (therefore, the frequency of the *A* allele became 1.0). **(b)** With a population size of 1000, allele frequencies remained relatively constant.

(b)

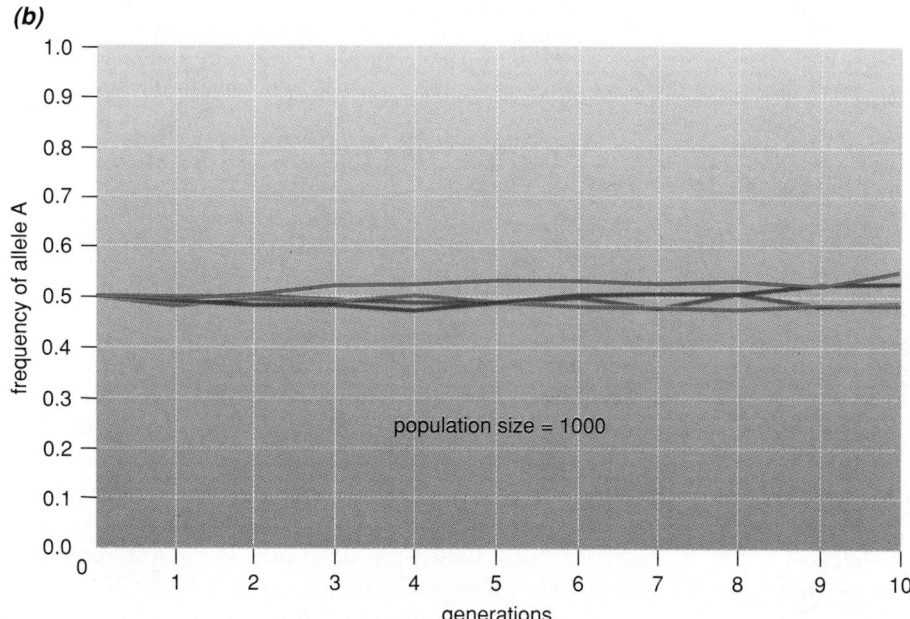

both parents will be spotted (this is the same likelihood as flipping two coins and having both come up heads). If this happens, then the next generation will consist entirely of spotted ladybugs.

Figure 17-3a illustrates two important points about genetic drift. (1) **Genetic drift tends to reduce genetic variability within a small population.** In extreme cases, all members of a population may become genetically identical (Fig. 17-3a, top line). (2) **Genetic drift tends to increase genetic variability between populations.** Purely as a result of chance, separate populations may evolve extremely different allele frequencies (Fig. 17-3a, top versus bottom lines).

Two special cases of genetic drift, called the population bottleneck and founder effect, further illustrate the enormous consequences that small population size may have on the allele frequencies of a species.

Population Bottleneck

In a **population bottleneck,** a species undergoes a drastic reduction in population size, so that only a few individuals contribute genes to the entire future

(a)

(b)

Figure 17-4 Both the northern elephant seal **(a)** and the cheetah **(b)** passed through a population bottleneck in the recent past, resulting in an almost total loss of genetic diversity.

population of the species. As our ladybug example showed, population bottlenecks may cause both *differences in allele frequencies* and *reductions in genetic variability*. Even if the population then rebounds, and the species becomes common, these genetic effects of the bottleneck may remain for hundreds or thousands of generations.

Loss of genetic variability has been documented in the northern elephant seal and the cheetah (Fig. 17-4). The elephant seal was hunted almost to extinction in the 1800s; by the 1890s only about 20 survived. Since elephant seals breed harem-style, a single male may have fathered all the offspring at this extreme bottleneck point. The population today has expanded to about 30,000, but biochemical analysis shows that all northern elephant seals are almost genetically identical. Other species of seals, whose populations have historically always remained large, are much more variable. The rescue of the northern elephant seal from extinction is rightly regarded as a triumph

of conservation; however, with very little genetic variation, the elephant seal cannot evolve in response to environmental changes. No matter how many elephant seals there are, the species must be considered to be threatened with extinction. Cheetahs are also genetically homogeneous, although the reason for the bottleneck is unknown. Consequently, cheetahs too could be gravely threatened by small changes in their environment.

Founder Effect

A special case of a population bottleneck is the **founder effect,** which occurs when isolated colonies are founded by a small number of organisms. A flock of birds, for instance, may become lost during migration, or may be blown off course by a storm (this is thought to have happened in the case of Darwin's finches in the Galapagos Islands). Among humans, small groups may migrate for religious or political reasons (Fig. 17-5). Such a small group may have al-

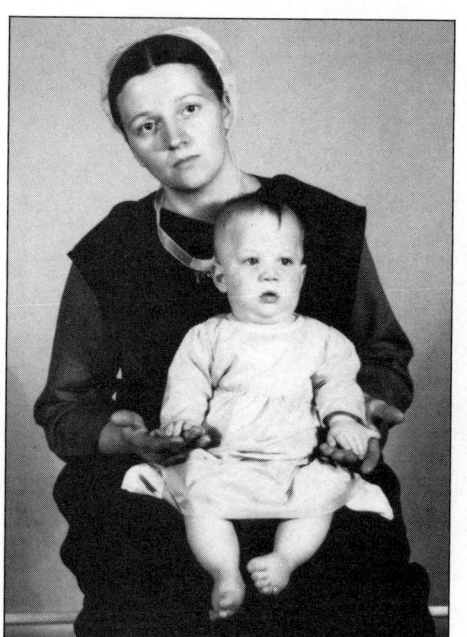

Figure 17-5 An Amish woman with her child, who suffers from a set of genetic defects known as the Ellis–van Creveld syndrome (short arms and legs, extra fingers, occasionally heart defects). Fleeing from religious persecution, about 200 members of the Amish religion migrated from Switzerland to Pennsylvania between 1720 and 1770. Since that time, virtually all the Pennsylvania Amish moved to Lancaster County, and have remained reproductively isolated from non–Amish Americans. The population increased to about 8000 by 1964. In that year, geneticist Victor McKusick surveyed the Lancaster County Amish and discovered that they had an unusually high frequency of Ellis–van Creveld syndrome. McKusick found that the Amish had an allele frequency for Ellis–van Creveld of about 0.07, compared with a frequency of less than 0.001 in the general population. Why? One couple who immigrated in 1744 carried the allele. Inbreeding among the Amish passed the allele along to their descendants, a clear example of a founder effect. In addition, by chance the Ellis–van Creveld carriers had more children than the Amish average, further increasing the allele frequency by genetic drift. The combination of an initially high frequency in the immigrants (1 or 2 out of 200) plus genetic drift resulted in the modern situation, in which more cases of Ellis–van Creveld syndrome are known from Lancaster County than from the rest of the world combined.

lele frequencies that are very different from the frequencies of the parent population. If the isolation of the founders is maintained for a long period of time, a sizable new population may arise that differs greatly from the original population.

The Importance of Genetic Drift in Evolution

How much does genetic drift contribute to evolution? No one really knows. Only rarely are natural populations extremely small or completely cut off from gene flow from other populations. The effective breeding population of mosquitoes in a swamp may be millions, that of wildebeest in the Serengeti plains of Africa may be tens of thousands, and that of even relatively rare species, like lions, is usually in the hundreds. However, populations occasionally do become very small, and it may be precisely these small populations that contribute most to major evolutionary changes. As we will see in the next chapter, biologists believe that new species often arise in small populations.

Random Mating and Sexual Selection

Organisms seldom mate strictly randomly. For example, most animals have limited mobility and are most likely to mate with nearby members of their species. Further, they may choose to mate with certain individuals of their species rather than with others. The white-crowned sparrow is a case in point. Although all white-crowned sparrows sing a fundamentally similar song, each local population has its own song dialect. A female usually chooses a mate that sings the same dialect that her father sang (Fig. 17-6). Among animals, there are three common forms of nonrandom mating: harem breeding, assortative mating, and sexual selection.

In some species, like elephant seals, baboons, and bighorn sheep, only a few males fertilize all the females. Following some sort of contest, which may involve showing off with loud sounds or flashy colors, making threatening gestures, or actual combat, only certain males succeed in gathering a harem and mating (Fig. 17-7).

Many animals mate assortatively—that is, they se-

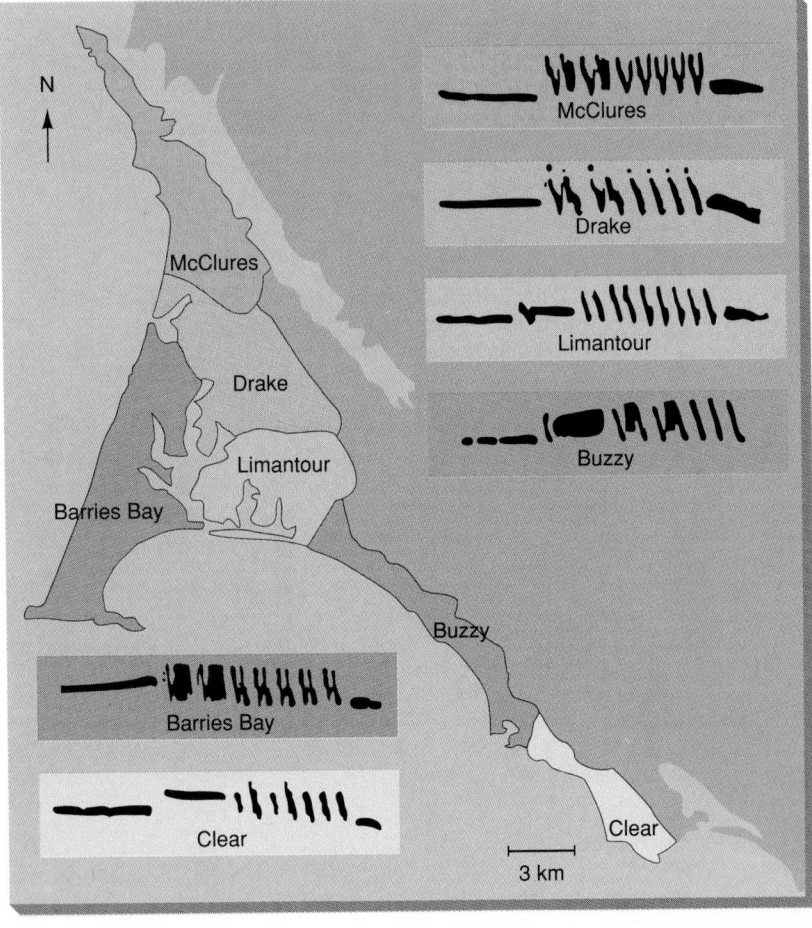

Figure 17-6 Song dialects among populations of white-crowned sparrows at Point Reyes National Seashore north of San Francisco. As the sonograms show, the songs are fairly similar, but both birds and human listeners can recognize the different dialects. Male birds of each population learn their local dialect while in the nest, and sing it when they mature. Females preferentially mate with males that sing the dialect sung by the females' own fathers, i.e., the females' own local dialect.

Figure 17-7 Sparring contests between males result in extremely nonrandom mating among many animals, including deer, elk, seals, and many monkeys. Here, two male bighorn sheep square off against each other during the fall rutting season. Although the horns are potentially lethal weapons, they are used in ritualized ways that minimize the danger of injury to either contestant.

Figure 17-8 Many male birds, including peacocks, attract mates by displaying their wares. The features evolved for female attraction are often irrelevant, or even harmful, to the day-to-day survival of the males. Sometimes, as this photo suggests, the female isn't interested anyway!

lect mates that are similar to themselves. Humans, for example, tend to marry members of the opposite sex that are similar in height, race, I.Q., and social status.

Finally, in many mammals and birds, mate selection is primarily the prerogative of one sex, usually the female. Males display their virtues, which may be bright plumage, as in peacocks (Fig. 17-8), or rich territories, as in many songbirds. A female evaluates the males and chooses her mate. Choosing a male with a good territory is obviously advantageous, but why do females often prefer elaborate "fashions" in their mates? A popular, although hotly debated, hypothesis is that nonadaptive structures and colors provide the females with an outward sign of the males' fitness. Only vigorous, energetic males can survive when burdened with conspicuous coloration or large tails. Similarly, heavily parasitized or sick males may be dull and frumpy compared with healthy males. Whatever the exact selective mechanisms, it is thought that many of the elaborate structures and behaviors found only in males have evolved through the selective pressure of female mate choice: only the flashy males transmitted their genes to the next generation.

Darwin was so impressed with these structures that he coined the term **sexual selection** to designate the process of evolution through mate choice, and considered it a distinct category from natural selection. Since conspicuous structures and bizarre behav-

iors render the males more vulnerable to predators, sexual selection often seems to work in opposition to other forms of natural selection. However, nonsexual selective forces may also oppose one another; the height of a giraffe, for example, is a compromise between the advantage of reaching higher leaves for food and the disadvantage of vulnerability while drinking water (Fig. 17-9). In both sexual and nonsexual selection, then, some aspect of the environment (in sexual selection the "opinion" of the opposite sex, which is part of the social environment) influences reproductive success.

Equivalence of Genotypes

Genetic equilibrium requires that all genotypes must be equally adaptive—that is, none has any selective advantage over the others. It is probably true that some alleles are adaptively neutral, so that organisms possessing any of several alleles will be equally likely to survive and reproduce. However, this is clearly not

(a)

(b)

Figure 17-9 (a) The long neck and legs of a giraffe are a decided advantage in feeding on acacia leaves high up in trees. **(b)** However, a giraffe has to get into an extremely awkward and vulnerable position to drink. Feeding and drinking thus place opposing selective pressures on the length of neck and legs.

true of all alleles in all environments. Any time an allele confers, in Wallace's words, "some little superiority," natural selection will favor the enhanced reproduction of the individuals possessing it.

Natural selection is not the *only* evolutionary force. As we have seen, mutation provides initial variability in heritable traits. The chance effects of genetic drift may change allele frequencies, even spawning new species. Further, evolutionary biologists are just now beginning to appreciate the power of random catastrophe in shaping the history of life on Earth—mass extinctions that may exterminate flourishing and floundering species alike. Nevertheless, it is natural selection that prunes the growth of a species, molding it to fit its environment, and indeed shaping the evolution of adaptations that humankind has admired for millenia. For this reason, we will examine the mechanisms of natural selection in some detail.

Natural Selection

To most people, the words *natural selection* are synonymous with the phrase *survival of the fittest*. Natural selection evokes images of wolves chasing caribou, of lions snarling angrily in competition over a zebra carcass. However, natural selection is not really about *survival*, but about *reproduction*. It is certainly true that an organism must survive at least for a while, to live

long enough to reproduce. In some cases, it may also be true that a longer-lived organism has more chances to reproduce. But no organism lives forever, and the only way that its genes continue into the future is through successful reproduction. When the ancient organism that fails to reproduce eventually dies, its genes die with it. The organism that dies at a much younger age, but that reproduces first, lives on, in a sense, through the genes that it has passed on to its offspring. Therefore, although evolutionary biologists often discuss survival, partly because survival is usually easier to measure than reproduction, natural selection is really an issue of **differential reproduction:** individuals bearing certain alleles leave more offspring (who inherit those alleles) than other individuals with different alleles.

Genotype vs. Phenotype

The agents of natural selection cannot directly detect an organism's genotype. Rather, selection acts on phenotypes: the actual structures and behaviors that the organisms in a population display. Genotype and phenotype, however, are related in the following way. If you were to measure the phenotypes of a specific trait in all the individuals in a population, you would find a range of values (Fig. 17-10). This range of phenotypes would arise from differences both in the genotypes of the organisms and in the environments in which they live. However, just as bad calls

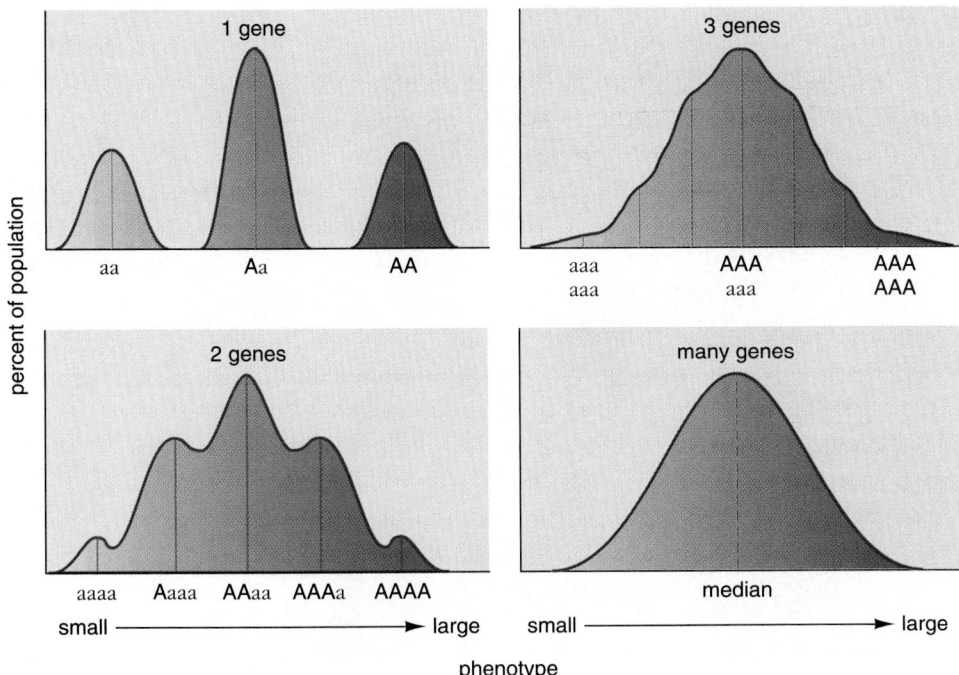

Figure 17-10 Both genes and environment contribute to the phenotype of an organism. This series of graphs illustrates the distribution of phenotypes that would be expected if one, two, three, or many genes, each with two incompletely dominant alleles (see Chapter 11), contributed to a particular body characteristic (e.g., size). The vertical lines represent the precise size expected due to genotype alone. In each case, environmental conditions (e.g., amount of available food) create some variation in size, represented by the colored curves. As the number of genes contributing to the characteristic becomes large, the distribution of phenotypes approximates a smooth curve called a normal distribution (lower right-hand graph). The most common value for the phenotype is the middle value, also called the median.

by umpires tend to average out over a long baseball season, so that the best team wins the most games, so too environmental differences influencing phenotypes average out in a large population. On the average, genotype predicts phenotype: most large plants will have genes promoting large size, while most small plants will have genes promoting small size. In our discussion of selection, therefore, we will ignore environmental causes of variability.

Types of Selection

Biologists recognize three major categories of natural selection (Fig. 17-11):

1. **Directional selection** favors individuals possessing values for a trait at one end of the distribution, and selects against both average individuals and individuals at the opposite extreme of the distribution (e.g., favors small size, selects against both average and large individuals).
2. **Stabilizing selection** favors individuals possessing an "average" value for a trait, and selects against individuals with extreme values.
3. **Disruptive selection** favors individuals possessing relatively extreme values for a trait at the expense of individuals with average values. Disruptive selection favors organisms at both ends of the distribution of the trait (e.g., favors both large and small body size).

Directional Selection

If environmental conditions change in a consistent direction—for example, if the climate becomes colder—then a species may evolve in a consistent direction in response, for example with thicker fur (Fig. 17-11a). The evolution of long necks in giraffes was almost certainly due to directional selection: pre-giraffes with longer necks obtained more food and therefore reproduced more prolifically than their shorter-necked contemporaries did. Antibiotic resistance in bacteria is another example of directional selection (Chapters 12 and 21).

How fast can directional selection change genotypes? That depends on both the genetic nature of the variability in the population and the strength of selection. The increased frequency of the black form of the peppered moth in Britain earlier this century was an extremely rapid case of directional selection (Chapter 16). In this instance, the color did not vary in a finely graded manner, but was either black or pale, controlled by two alleles of a single gene. Predation by birds was also a very strong selective force. Together, these two factors produced a dramatic change in the population in just a few years. If little variability exists in the population, or if the different alleles produce only slightly different phenotypes, then directional selection will drive much slower changes. In some instances, a population may not be able to respond fast enough to the selective forces, and may become extinct.

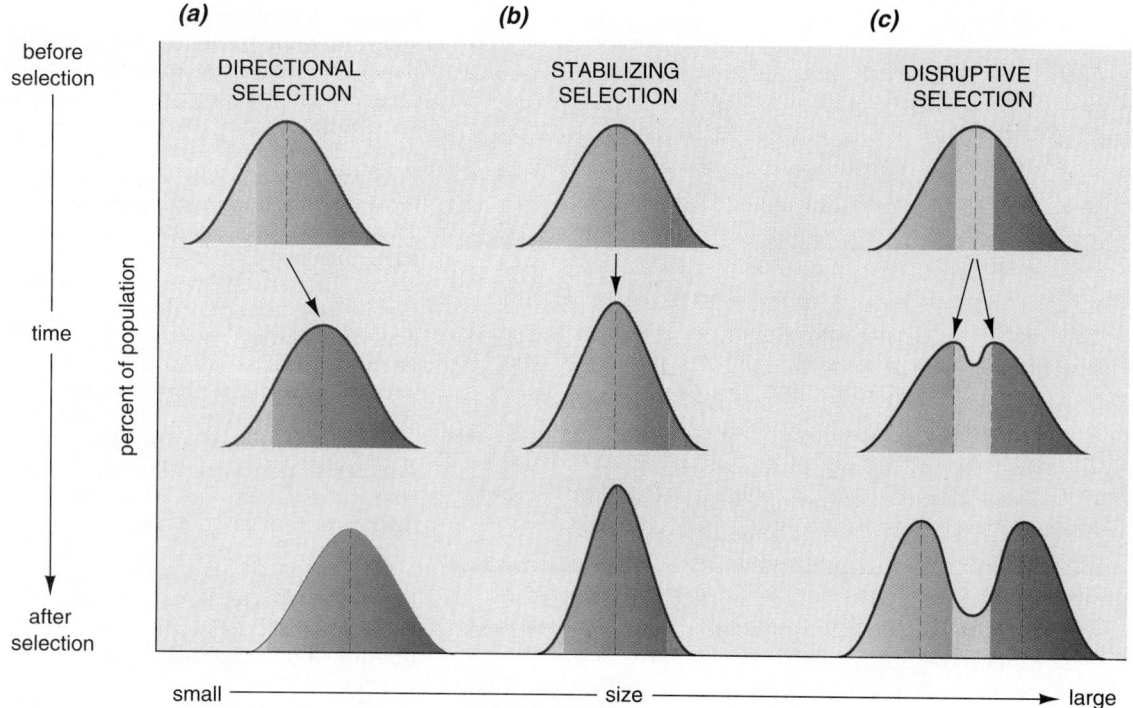

Figure 17-11 Three types of natural selection, acting upon a normal distribution of phenotypes (in these examples, size). In all graphs, the blue area represents individuals that are selected against—that is, do not reproduce as frequently.
(a) In stabilizing selection, the organisms most likely to reproduce are those with phenotypes close to the average for the population. The variability of phenotypes may decline, but the average value remains the same.
(b) In disruptive selection, phenotypes that are both larger and smaller than average are favored. The population splits into two phenotypic groups.
(c) In directional selection, phenotypes that are either larger or smaller than average (larger illustrated here) are favored. The average phenotype shifts position over the generations.

Stabilizing Selection

Directional selection can't go on forever. Once a species is well adapted to a particular environment, and if the environment doesn't change, then most variations that appear through new mutations or recombination of old alleles will be harmful. Therefore, the species will often undergo stabilizing selection, which favors the survival and/or reproduction of "average" individuals (see Fig. 17-11b). Stabilizing selection often occurs when a single trait is under opposing selective pressures from two different sources. M. K. Hecht, for example, studied lizards of the *Aristelliger* genus. He found that small lizards had a hard time defending territories, but large lizards were more likely to be preyed upon by owls. Therefore, *Aristelliger* lizards were under stabilizing selection favoring an "average" body size.

It is widely assumed, although difficult to prove, that many traits are under stabilizing selection. We

have already mentioned several of these. Although the lengths of legs and necks in giraffes probably originated under directional selection for feeding on leaves high up in trees, they are almost certainly now under stabilizing selection, balancing the demands of feeding and drinking. Similarly, female mate choice probably drove the evolution of elaborate sexual displays in many birds, but now increased predation may exert stabilizing selection: if a peacock's tail became so long that he couldn't fly, he would be unlikely to live long enough to woo a female.

Under certain circumstances, stabilizing selection may act not to eliminate variability, but to maintain it. Opposing selective pressures often give rise to **balanced polymorphism,** in which two or more alleles of a gene are maintained in a population because each is favored by a distinct selective force. This seems to have occurred with the hemoglobin alleles in blacks (see Chapter 15). In people who are homozygous for sickle-cell anemia, hemoglobin molecules clump up into long chains, distorting and weakening their red blood cells. This causes severe anemia. Before the advent of modern medicine, people homozygous for sickle-cell anemia were strongly selected against. Heterozygotes have only mild anemia, but some may still suffer ill effects during strenuous exercise. Under these circumstances, you might wonder why natural selection has not eliminated the sickle-cell allele. Far from being eliminated, however, the sickle-cell allele is carried by nearly half the people in some areas of Africa. This seems to result from the counterbalancing effects of anemia and malaria, which was formerly very common in equatorial Africa.

Malaria parasites multiply rapidly within the red blood cells of homozygous normal individuals. Before effective medical treatments were discovered, homozygous normals consequently often died of malaria. Heterozygotes, on the other hand, enjoy some protection against malaria. Malaria parasites inside a heterozygote's red blood cells use up oxygen, which causes the sickle-cell hemoglobin to clump and the cells to sickle. Infected, sickled cells are destroyed by the spleen before the parasites can complete their development. Heterozygotes, therefore, have mild anemia but do not succumb to malaria. During the evolution of African populations, heterozygotes survived better than either type of homozygote, and reproduced the most. As a result, both the normal hemoglobin allele and the sickle-cell allele have been preserved (Fig. 17-12).

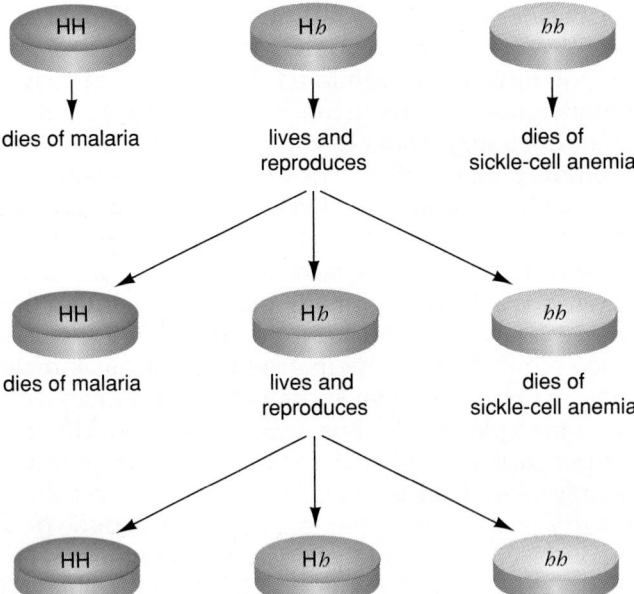

Figure 17-12 Sometimes two or more alleles, each producing a different phenotype, can be maintained in a population by opposing selection pressures. The alleles for normal (*H*) and sickle-cell (*h*) hemoglobin are maintained by selection against both homozygotes. Heterozygotes (*Hh*) reproduce the most, thereby keeping both alleles present in the population.

Disruptive Selection

Disruptive selection (see Fig. 17-11c) may occur when different microhabitats are available to a species, and different characteristics best adapt individuals to each microhabitat. For example, an island, such as one of the Galapagos, may have several species of plants, some producing large, hard seeds and others small, soft seeds. Large seeds provide the most food per seed, but can only be cracked and eaten by birds with large bills. Although large birds can easily eat small seeds, they would probably spend too much energy lugging their large bodies about looking for tiny seeds. If a single species of bird colonizes such an island, what will happen? We would expect that larger-bodied, larger-beaked birds will specialize on large seeds, while small-bodied, small-beaked birds will specialize on small seeds. Medium-sized birds might not be able to crack open the large seeds, and might not get enough energy from small seeds, and so would be selected against. Disruptive selection would favor the survival and reproduction of both large and small, but not medium-sized, birds. Dis-

ruptive selection has not been extensively studied, although it has been documented or at least supported by studies in both butterflies and birds.

The Forces of Natural Selection

Natural selection acts by eliminating individuals that do not have the characteristics needed for survival and reproduction in their environment. It follows, then, that **the end result of natural selection is adaptation to the environment.** An organism's environment can be divided into two components: the nonliving *abiotic* part and the *biotic* part that consists of other organisms. Adaptations to both biotic and abiotic components occur through natural selection.

The abiotic environment includes physical factors such as climate, availability of water, and minerals in the soil. The abiotic environment provides the "bottom line" requirements that an organism must have to survive and reproduce. However, the vast majority of the adaptations that we see in modern organisms have arisen because of interactions with other organisms. As Darwin wrote, ". . . the structure of every organic being is related . . . to that of all other organic beings, with which it comes into competition for food or residence, or from which it has to escape, or on which it preys." A simple example illustrates this.

Consider a patch of soil 1 meter square in the eastern Wyoming plains. The soil contains enough minerals and receives enough rain for plant life. Let's say a buffalo grass sprouts there. Its roots must be able to take up enough water and minerals for growth and reproduction, and to that extent it must be adapted to its abiotic environment. Even in the dry prairies of Wyoming, this is a relatively trivial requirement *provided that the plant is alone and protected in its square meter of soil.* In reality, many plants, including other grasses, sagebrush bushes, and annual wildflowers also sprout in that same patch of soil. If our buffalo grass is to survive, it must compete for resources with the other plants. Its long, deep roots and efficient mineral uptake processes have evolved not so much because the plains are dry, but because it must share the dry prairies with other plants. Further, cattle (formerly bison) graze the prairies. Buffalo grass is extremely tough, with silica compounds reinforcing the blades, an adaptation that discourages grazing. Over millenia, tougher plants were harder to eat, so survived better and reproduced more—another adaptation to the biotic environment.

When two species or two populations of a single species interact extensively, each exerts strong selective pressures on the other. When one evolves a new feature, or modifies an old one, the other often evolves new adaptations in response. As the Red Queen told Alice in *Through the Looking Glass*, "Here, you see, it takes all the running you can do to keep in the same place." This constant, mutual feedback between two species is called **coevolution** (Table 17-1).

Table 17-1 Interactions Among Organisms

Type of Interaction	Effect on Organism A[a]	Effect on Organism B[a]
Competition between A and B	−	−
Predation by A on B	+	−
Symbiosis		
Parasitism by A on B	+	−
Commensalism of A with B	+	0
Mutualism between A and B	+	+
Altruism by A on behalf of B	−	+

[a] +, benefits; −, harms; 0, neutral or no effect.

Competition

One of the major selective forces in the biotic environment is **competition** with other members of the same species. As Darwin wrote in the *Origin of Species:* ". . . the struggle almost invariably will be most severe between the individuals of the same species, for they frequent the same districts, require the same food, and are exposed to the same dangers." In other words, no competing organism has such similar requirements for survival as another member of the same species. For example, both Lazuli buntings and Western bluebirds are brightly colored in blue, red, and white, and both nest and rear their young in the foothills of the Rockies in the summer. However, they do not compete very much with each other, because they eat different foods: bluebirds mostly catch insects, while buntings specialize in seeds. Each mosquito picked off by a bluebird makes little difference to a bunting, but makes it harder for other bluebirds to find enough to eat.

Different species may also compete for the same resources, although generally to a lesser extent. As we will discuss more fully in Chapter 46, whether a particular plot of prairie is covered with grass, sagebrush, or trees is at least partly determined by competition among these plants for scarce soil moisture.

Predation

Although we commonly think of predation as one animal preying upon another animal, predation actually includes any situation in which one organism eats another. In some instances, coevolution between predators and prey is a sort of "biological arms race," with each side evolving new adaptations in response to escalations by the other. Using Darwin's example of wolves and deer, wolf predation selects against slow or incautious deer, thus leaving faster, more alert deer to propagate the species. In their turn, alert, swift deer select against slow, clumsy wolves, since such predators cannot acquire enough food.

Symbiosis

Symbiosis is any relationship in which individuals of different species closely interact with one another for much of their lives. One species usually benefits from the relationship, but the other may suffer injury, enjoy benefits, or not be affected at all (see Table 17-1). The different types of symbiosis are described in Chapter 44. From an evolutionary perspective, symbiosis leads to the most intricate coevolutionary adaptations. Although a given predator usually preys on several species, and may interact with a particular species only sporadically, partners in symbiosis live together virtually their entire lives (Fig. 17-13). At least one of the partners, and usually both, must continually adjust to any evolutionary changes developed by the other.

Altruism

Evolution is often portrayed in the popular press as being "red in tooth and claw." While it is true that competitive and predatory interactions influence the evolution of most species, cooperation and even self-sacrifice can be important selective forces too. **Altruism** is a behavior that endangers an animal or reduces its reproductive success, but that benefits other members of its species. People helping other people immediately springs to mind, but altruistic behaviors are common in the animal kingdom. A mother killdeer flutters just out of reach of a predator, feigning an injured wing and luring the predator away from her nest (Fig. 17-14); female worker bees

Figure 17-14 Altruism between mother and offspring: a female killdeer lures a predator away from its nest by feigning injury. The mother places herself in some small danger (she can always fly away if the predator comes too close), but saves her offspring from much greater danger.

Figure 17-13 Several species of clownfish live in a symbiotic relationship with anemones, each species of fish frequenting its own species of anemone. The fish nestle within the stinging tentacles of the anemone, thus protected from predation by other fish. The clownfish evolved specialized skin secretions and behaviors, protecting it from being eaten by the anemone. The fish may accidentally drop food onto the anemone once in a while, but the benefits are probably fairly one-sided.

forego reproduction and devote their lives to raising the offspring of the hive queen (see Chapter 42); and young male baboons scout around the edges of the troop, even though this increases their danger from leopards.

You might think that altruism runs counter to natural selection: if altruism is encoded in an organism's genes, those genes are placed at risk every time the altruist performs one of its gallant behaviors. However, natural selection can indeed select for altruistic genes, if the altruist benefits relatives who possess the same genes. This special case of natural selection is an example of **kin selection,** and is explored in "A Closer Look at Kin Selection and the Evolution of Altruism."

Extinction

Natural selection produces not only the fleet limbs of the antelope and the exquisite eyes of the goshawk. It may also lead to **extinction.** Trilobites, dinosaurs, saber-tooth cats—all are extinct, known only from fossils. Paleontologists estimate that *at least* 99.9% of all the species that ever existed are now extinct. Why? Two characteristics seem to predispose a species to extinction: localized distribution and overspecialization. However, environmental events are usually the immediate cause of extinction.

CLOSER LOOK

At Kin Selection and the Evolution of Altruism

Altruism is any behavior that is potentially harmful to the future reproduction of an organism, but enhances the reproductive potential of other organisms. Altruism includes birds and mammals defending their offspring, worker bees rearing the offspring of queen bees, and juvenile Florida scrub jays helping their parents to rear the next brood (Fig. E17-1). Note that altruism does not imply conscious, voluntary decisions to engage in selfless behavior. Indeed, no one would argue that worker bees consider the welfare of the queen and her offspring and deliberately sacrifice their own reproduction in favor of the queen's. Rather, most altruistic behaviors have a strong instinctive component; that is, many animals have altruism programmed in their genes.

From an evolutionary viewpoint, how can this be? Surely, if a mutation arose that caused altruistic behavior, and the bearers of that mutation lost their lives or failed to reproduce because of their self-sacrificing behaviors, their "altruistic alleles" would disappear from the population. Maybe, or maybe not. To understand the evolution of altruism, we will need to introduce a new concept: **inclusive fitness.** As formulated by W. D. Hamilton, the inclusive fitness of an allele is the fitness conferred on *all* organisms that have the allele. Therefore, *if an altruist benefits related members of its own species that bear the same altruistic allele, then the altruistic allele may be favored by natural selection.*

To see how altruism might increase the inclusive fitness of an allele, let's consider the Florida scrub jay. Yearling jays usually do not mate and reproduce. Instead, they remain at their parents' nest and help out with next year's brood. Let's assume, for simplicity, that this altruistic behavior is controlled by a single "altruistic" allele, and that in the distant past helper jays had the altruistic allele, while non-helpers had another, "selfish" allele.

At least for 1 year, altruistic yearlings do not reproduce; some probably die from predation or accidents and never reproduce at all. How, then, can this be adaptive behavior? It all has to do with relatedness and the probability of successful reproduction. First, an animal's offspring inherit 50% of its genes (the other 50% of the offspring's genes come from the other parent). On the average, an animal also shares 50% of its genes with its siblings. Therefore, *a scrub jay is just as related to its siblings as it would be to its own offspring.* The second factor influencing scrub jay reproduction is that there is a limited number of good jay habitat. Inexperienced yearling jays would probably have a hard time acquiring a good nest site and would be hard put to feed their offspring. Their best "reproductive bet," then, is to put their energy into helping their parents. Selfish yearling jays that

Figure E17-1 In Florida scrub jays, young birds usually do not go off and mate in their first year, even though they are sexually mature. Instead, they remain at their parents' nest and assist in feeding the next year's brood.

try to nest on their own will probably contribute fewer genes to the next generation than the altruistic yearlings do. This phenomenon, whereby the actions of an individual increase the survival and/or reproductive success of its relatives, is called **kin selection.**

As this example suggests, *kin selection can favor the evolution of altruism if the altruistic behavior benefits relatives that bear the same altruistic allele.* In most cases, an animal will not know if another carries the altruism allele, but the animal must at least be able to distinguish relatives from strangers: relatives stand a good chance of possessing the altruism allele, but you never can tell with strangers. A yearling jay that helped out at the nest of unrelated adult jays would probably waste its time and effort.

Identification of relatives isn't too hard to imagine in the case of jays and their parents. Many biologists objected to other proposed instances of altruistic behaviors, however, arguing that animals cannot evaluate degrees of relatedness. This concern has been alleviated by two findings. First, many social groups, including wolf packs and baboon troops, are actually family groups. Therefore, an animal would not have to identify relatives in order for its altruistic behaviors to benefit them the most. Second, many animals, including birds, monkeys, tadpoles, bees, and even tunicate larvae, can indeed identify relatives. Given the choice between relatives and strangers, these animals preferentially associate with their relatives, *even if they were separated at birth and have never seen those relatives before.* If animals selectively form related groups, then altruistic behaviors will most likely benefit relatives.

Although not the only mechanism, kin selection has been a powerful selective force in the evolution of altruism in many species, probably including humans.

Susceptibility to Extinction

Localized Distribution

Species vary widely in their range, and hence in their susceptibility to extinction. Some species, such as herring gulls, white-tailed deer, and humans, inhabit entire continents or even the whole Earth, while others, such as the Devil's Hole pupfish (Fig. 17-15), have extremely limited ranges. Obviously, if a species occurs only in a very small area, any disturbance of that area could easily result in extinction. If Devil's Hole dries up from climatic change or well-drilling nearby, its pupfish will immediately vanish. Wide-ranging species, on the other hand, usually do not succumb to local environmental catastrophes.

Specialization

Another factor that may make a species vulnerable to extinction is extreme specialization. Each species evolves a set of genetic adaptations in response to pressures from its particular environment. Sometimes these adaptations imprison the organism in a very narrow ecological niche. The Everglades kite, for example, feeds only on a certain freshwater snail (Fig. 17-16). As the swamps of the American Southeast are drained for farms and developments, the snail population shrinks. Should the snail become extinct, the kite will surely go extinct along with it.

In the fossil record, such behavioral specialization is hard to recognize. Structural specializations, however, may be just as restrictive. A case in point is

Figure 17-16 The Everglades kite feeds on the apple snail, found in swamps of the southeastern United States. Such behavioral specialization renders the kite extremely vulnerable to any environmental change that may exterminate its single species of prey.

Figure 17-15 The Devil's Hole pupfish is found in only one spring-fed waterhole in the Nevada desert. During the last glacial period, the southwestern deserts received much more rainfall, forming numerous lakes and rivers. As the rainfall decreased, pupfish populations were isolated in shrinking small springs and streams. Isolated small populations and differing environmental conditions caused the ancestral pupfish species to split up into several very restricted modern species, all of which swim on the brink of extinction.

giantism. For poorly understood reasons, many animals have evolved huge size: certain amphibians, dinosaurs, and giant mammals (including mammoths, ground sloths, and titanotheres). To support their bulk, these animals must have consumed enormous amounts of food. If environmental conditions deteriorated, these giants may have been unable to find enough food, and thus died out. Smaller animals that ate the same food but needed less of it survived.

Extinction and the Environment

The actual cause of extinction is probably always environmental change, either in the living or the nonliving parts of the environment. Three major changes that drive species to extinction are competition among species, novel predators or parasites, and habitat destruction.

Competition

Competition for limited resources occurs in all environments. If a species' competitors evolve superior adaptations, and it doesn't evolve fast enough to keep up, it may become extinct. A particularly striking example of extinction through competition occurred in South America 2 to 3 million years ago. For millions of years North and South America were isolated from one another, and each developed a distinctive fauna. When the Panamanian land bridge arose, connecting the two continents, massive migrations took place. In general, North American animals displaced their South American counterparts, and many South American species became extinct.

Predators or Parasites

When formerly isolated populations encounter one another, not only competitors migrate between the areas: predators and parasites do too. With the exception of humans, who have exterminated hundreds of species, predators probably cause few extinctions. Parasites, on the other hand, can be devastating. In North America, Dutch elm disease and chestnut blight are well known instances of introduced parasites that almost completely destroyed widespread native species. We cannot tell much about prehistoric parasite invasions, but the extinction of South American animals, mentioned above, might have been at least partly due to diseases carried south by resistant North American migrants.

Habitat Destruction

Habitat destruction may be the leading cause of extinction, both contemporary and prehistoric. Presently, habitat destruction due to human activities is proceeding at a frightening pace. Perhaps the most rapid extinction in the history of life will occur over the next 50 years, as tropical forests are cut for timber and to clear land for cattle and crops. As many as half the species presently on Earth may be lost because of tropical deforestation.

Prehistoric habitat destruction usually occurred over a longer time span, but nevertheless had serious consequences. Climate changes, in particular, caused many extinctions. Several times, moist, warm climates gave way to drier, colder climates with more variable temperatures. Many plants and animals failed to adapt to the new rigors and became extinct. One cause of climate change is continental drift (Fig. 17-17). As the continents flow about over the surface of the Earth, they change latitudes. Much of North America was tropical many millions of years ago, but drift carried the continent up into temperate and arctic regions.

An extreme, and very sudden, type of habitat destruction might be caused by catastrophic geological events, such as massive volcanic eruptions. Several prehistoric eruptions, which would make the Mount St. Helens explosion look like a firecracker by comparison, wiped out every living thing for scores of miles around, and probably caused global climatic changes as well.

The fossil record reveals episodes of extensive worldwide extinctions, especially among marine life (Fig. 17-18). Enormous meteorites, several miles in diameter, may have hit the Earth at these times. If a huge meteorite struck land, it would kick up enormous amounts of dust. The dust might be thick enough, and spread widely enough, to block out most of the sun's rays. Fires started by the impact might be widespread, adding soot to the atmosphere. Many plants would die because they couldn't photosynthesize. Many animals, all of which ultimately depend on plants for food, would also die. Smaller amounts of dust might still block out enough sunlight to cause global cooling, perhaps even triggering an

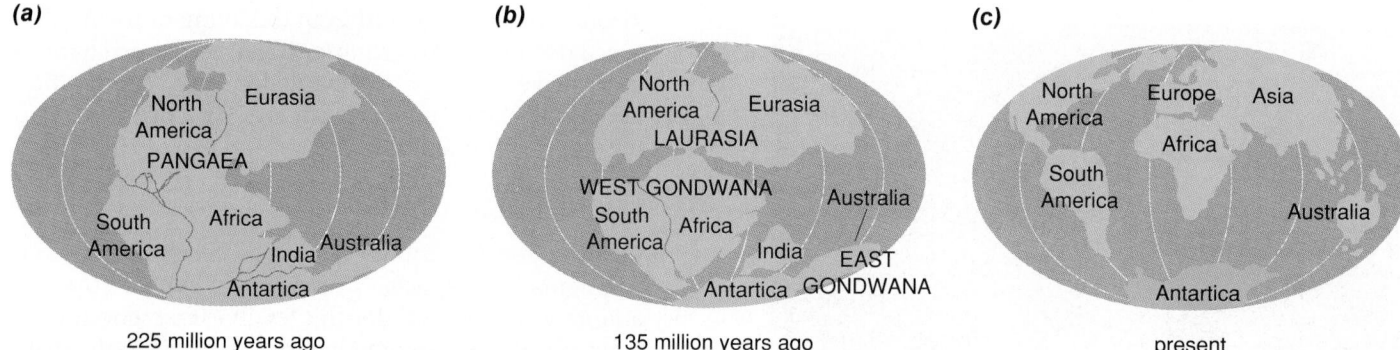

(a) 225 million years ago   **(b)** 135 million years ago   **(c)** present

Figure 17-17 Although slow, continental drift can cause tremendous environmental changes, as land masses are moved about on the surface of the Earth. The solid surfaces of the continents slide about over the viscous, but fluid, subterranean mantle. **(a)** About 225 million years ago, all the continents were fused together into one gigantic land mass which geologists call Pangaea. **(b)** Gradually Pangaea broke up into Laurasia and Gondwana. **(c)** Further drift eventually resulted in the modern positions of the continents. Continental drift continues today: the Atlantic Ocean, for example, widens by a few centimeters each year.

ice age. In either case, widespread extinctions would result.

Did such massive meteorite strikes really occur, and if so, would they cause extinctions? No one knows for sure, but considerable evidence points to meteorites as the causes of at least some major extinctions (Fig. 17-19). Recently, two groups of researchers have suggested that the Chicxulub crater near the Yucatan Peninsula of Mexico was the impact site of the meteorite that might have killed the dinosaurs.

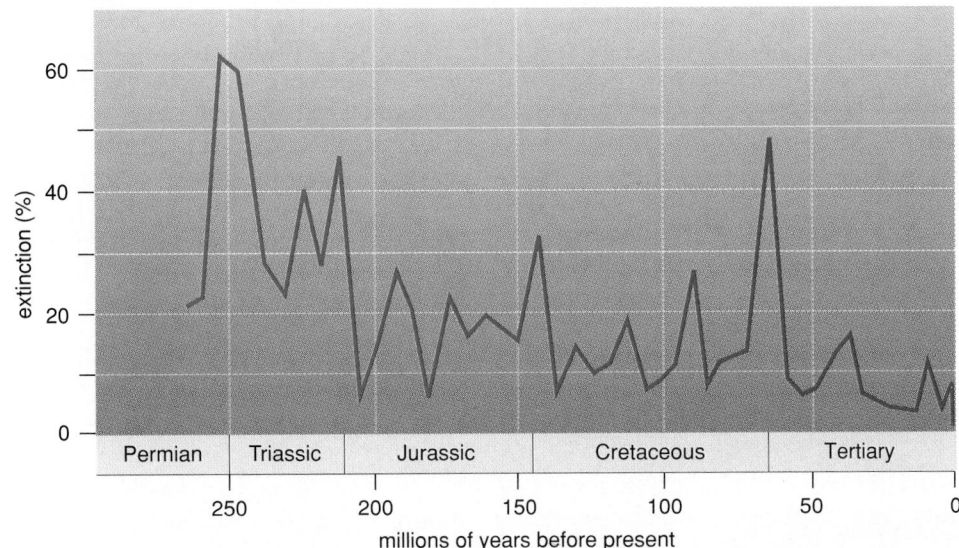

Figure 17-18 Life on Earth has undergone times of mass extinction in the past. This graph plots the percentage of genera of marine animals that have become extinct during geologic time. The higher the peak, the greater the extent of extinctions. Marine animals were chosen for this study because their fossils are abundant, and easily dated, in sedimentary rocks. The large peak of extinctions at the boundary between the Cretaceous and Tertiary periods also approximately coincides with the end of the dinosaurs. Many paleontologists are convinced that the dinosaurs were going downhill for millions of years before this time, so there is hot debate over whether the Cretaceous–Tertiary extinction event (such as an asteroid impact) provided the final blow to the dinosaurs or whether the dinosaurs would have died out at that time anyway.

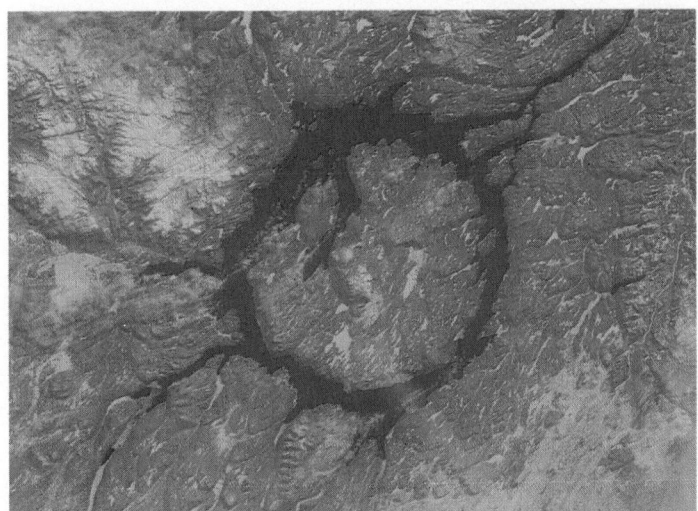

Figure 17-19 The Manicouagan crater in Quebec is about 45 miles in diameter. Giant impact craters such as this one often contain a central dome of rock that splashes up after the meteorite has buried itself in the earth, leaving a ring-shaped depression between the outer crater wall and the inner dome. In this satellite photo, water backed up behind a dam fills the crater ring. The Manicouagan asteroid struck a little over 200 million years ago, leading geologist Paul Olsen to suggest that its impact triggered the mass extinction at the Triassic–Jurassic boundary (see Fig. 17-19).

ℝ ℝeflections on Natural Selection, Genetic Diversity, and Endangered Species

Ever since the Endangered Species Act was passed in 1976, the United States has had an official policy of protecting rare species. In fact, the real goal of the Act is not protection but recovery; as one U.S. Fish and Wildlife official put it, "the goal is to get species *off* the list." Wildlife biologists try to determine how large a population a species needs to have before it is no longer in danger of extinction from unpredictable events, like a couple of years of drought or an epidemic of parasites. If a species reaches this population size, it is no longer legally "endangered" with extinction.

Does a "large enough" population (which usually is still very small by historical standards) really ensure a species' survival? From our discussion of genetic drift and population bottlenecks, you probably realize that the answer is "no." If the population of a species has been reduced to the point where it is placed on the endangered species list, then it has probably lost much of its genetic diversity. As ecologist Thomas Foose aptly put it, loss of habitat and consequent reduction in population size means that "gene pools are being converted into gene puddles." Even if the species recovers in numbers, it cannot recover its gene pool. When the forces of natural selection change at some future time, the species may not have the genetic capability to respond appropriately, and may become extinct.

What can be done about this? The best solution, of course, is to leave enough habitat of diverse types so that species never become endangered in the first place. The human population, however, has grown so large and appropriated so much of the Earth's resources that this is not feasible in many places. If we are to be, as we imagine ourselves, the stewards of the planet and not merely its ultimate consumers, then protection of other life forms will be a continuing responsibility as long as humankind exists.

Figure E17-2 The island of Madagascar, located off the east coast of Africa, is home to many species of plants and animals that are found nowhere else in the world. Unfortunately, many of those species, like the black lemur pictured here, are in danger of extinction from habitat destruction, mostly due to massive deforestation.

SUMMARY OF KEY CONCEPTS

Evolution and the Genetics of Populations
The gene pool of a population is the total of all the different alleles of all the genes carried by the members of a population. In its broadest sense, evolution is a change in the frequencies of alleles in the gene pool of a population, due to enhanced reproduction by individuals bearing certain alleles.

The Equilibrium Population
Allele frequencies in a population will remain constant over generations only if the following conditions are met: (1) There must be no mutation. (2) There must be no differential migration of alleles into or out of the population. (3) The population must be large. (4) All mating must be random. (5) All genotypes must reproduce equally well (i.e., no natural selection).

These conditions are rarely, if ever, met in nature. Understanding why they are not met leads to an understanding of the mechanisms of evolution.

The Mechanisms of Evolution

1. Mutations are random, undirected changes in DNA composition. Although most mutations are neutral or harmful to the organism, some prove advantageous in certain environments. Mutations are usually rare, and do not change allele frequencies very much. However, they provide the raw material for evolution.
2. Migration is the flow of genes between populations. If migrants carry different alleles than the populations from which they come or to which they migrate, then migration will cause changes in allele frequencies.
3. In any population, chance events kill or prevent reproduction by some of the individuals. If the population is small, chance events may eliminate a disproportionate number of individuals bearing a particular allele, thereby greatly changing the allele frequency in the population. This is termed genetic drift.
4. Many organisms do not mate randomly. If only certain members of a population can mate, then the next generation of organisms in the population will all be offspring of this select group, whose allele frequencies may differ from those of the population as a whole.
5. The survival and reproduction of organisms is influenced by their phenotype. Since phenotype depends at least partly on genotype, natural selection will tend to favor the reproduction of certain alleles at the expense of others.

Natural Selection
Three types of natural selection are:

1. *Directional selection:* organisms with characteristics that are different from average in one direction (e.g., smaller) are favored both over average organisms and over those that differ from average in the opposite direction.
2. *Stabilizing selection:* organisms of the "average value" for a characteristic are favored over organisms of extreme values.
3. *Disruptive selection:* organisms of extreme characteristics are favored over organisms with average values.

The Forces of Natural Selection
Natural selection occurs as a result of the interactions of organisms with both the biotic and abiotic parts of their environments. The biotic parts, however, usually exert the stronger selective pressures. When two or more species interact extensively so as to exert mutual selective pressures on each other for long periods of time, they both evolve in response. Such coevolution can occur as a result of any type of relationship between organisms, including competition, predation, and symbiosis.

Extinction
Over time, many species become extinct. Two factors that contribute to the likelihood of extinction of a species are localized distribution and overspecialization. Factors that actually cause extinctions include competition among species, novel predators or parasites, and habitat destruction.

GLOSSARY

adaptation: a characteristic of an organism that helps it to survive and reproduce in a particular environment; also the process of acquiring such characteristics.

allele frequency: for any given gene, the relative proportion of each allele of that gene found in a population.

altruism: a behavior that benefits another organism, usually at some risk to the altruistic organism.

balanced polymorphism: the prolonged maintenance of two or more alleles in a population, usually because each allele is favored by a separate selective force.

coevolution: the evolution of adaptations in two different species due to their extensive interactions with one another, so that each acts as a major force of natural selection upon the other.

competition: a relationship between individuals or species in which both require the same resource, which is not available in sufficient quantity to satisfy the needs of all users.

differential reproduction: differences in reproductive output among individuals of a population, usually as a result of genetic differences.

directional selection: a type of natural selection in which one extreme phenotype is favored over all others.

disruptive selection: a type of natural selection in which both extreme phenotypes are favored over the average phenotype.

equilibrium population: a population in which allele frequencies do not change from generation to generation.

extinction: the death of all members of a species.

fitness: the reproductive success of an organism, usually expressed in relation to the average reproductive success of all individuals in the same population.

founder effect: a type of genetic drift in which an isolated population founded by a small number of individuals may develop allele frequencies that are very different from those of the parent population, because of chance inclusion of disproportionate numbers of certain alleles in the founders.

gene flow: the movement of alleles from one population to another owing to migration of individual organisms.

gene pool: for a single gene, the total of all the alleles of that gene that occur in a population; the total gene pool is the total of all alleles of all genes in the population.

genetic drift: a change in the allele frequencies of a small population purely by chance.

inclusive fitness: the reproductive success of all organisms bearing a given allele, usually expressed in relation to the average reproductive success of all individuals in the same population. Compare with fitness.

kin selection: selection favoring a certain allele because of benefits accruing to relatives bearing the same allele.

migration: in population genetics, the flow of genes between populations.

natural selection: the differential survival or reproduction of organisms due to environmental forces that act differently upon genetically different members of a population.

population: a group of individuals of the same species, found in the same time and place, and actually or potentially interbreeding.

population bottleneck: a form of genetic drift in which a population becomes extremely small, which may lead to differences in allele frequencies as compared with other populations of the species, and to a loss in genetic variability.

population genetics: the study of the frequency, distribution, and inheritance of alleles in a population of organisms.

sexual selection: a type of natural selection in which the choice of mates by one sex is the selective agent.

stabilizing selection: a type of natural selection in which those organisms displaying extreme phenotypes are selected against.

symbiosis: a sustained relationship between two organisms of different species.

STUDY QUESTIONS

1. What is a gene pool? How would you determine the allele frequencies in a gene pool?
2. Define an equilibrium population, and outline the conditions that must be met for a population to remain in equilibrium.
3. How does population size affect the likelihood of changes in allele frequencies by chance alone? Can significant changes in allele frequencies (i.e., evolution) occur due to genetic drift?
4. Describe the three types of natural selection. Which type(s) are most likely to occur in stable environments and which type(s) in rapidly changing environments?
5. What is sexual selection? How is sexual selection similar to and different from other forms of natural selection?
6. Briefly describe competition, predation, symbiosis, and altruism, and give an example of each.
7. Define kin selection and inclusive fitness. Can these concepts help to explain the evolution of altruism?

DISCUSSION QUESTIONS

1. Assume that the environment plays a major role in determining a particular phenotype, for example plant size. Would you expect directional selection for increased size to be rapid or slow? Defend your answer.
2. A preview question for Chapter 18: A species is all the populations of organisms that potentially interbreed with one another but that are reproductively isolated from (cannot interbreed with) other populations. Using the five assumptions of the Hardy–Weinberg equilibrium population as a starting point, what factors do you think would be important in the splitting of a single ancestral species into two modern species?
3. Several species that had been reduced to just a few individuals have been "rescued" from becoming extinct by the work of conservationists, and these species now are growing in size. What special evolutionary problems do such populations have as a result of being rejuvenated from just a few survivors?
4. In the early 1900s, some students of evolution thought that dominant alleles should constantly be increasing in frequency in populations, eventually "driving out" the recessive alleles. In response to this concept, Hardy and Weinberg independently developed their theories about population genetics. Discuss how the Hardy-Weinberg concept disputes the erroneous concept of these students. What false assumption led these students astray?

SUGGESTED READINGS

Allison, A. C. "Sickle Cells and Evolution." *Scientific American*, August 1956. The story of the interaction between sickle-cell anemia and malaria in Africa.

Alvarez, W., and Asaro, F. "An Extraterrestrial Impact." and Courtillot, V. E. "A Volcanic Eruption." *Scientific American*, October 1990. Leading geologists debate the question: What caused the mass extinction of the dinosaurs at the end of the Cretaceous period?

Ayala, F. "The Mechanisms of Evolution." *Scientific American*, September 1979. Our increasing understanding of the molecular mechanisms of heredity and mutation reveals that organisms are extremely variable, providing tremendous amounts of material on which selection can operate.

Diamond, J. "Founding Fathers and Mothers." *Natural History*, June 1988. Interesting article on the importance of genetic drift in the evolution of human populations.

Kimura, M. "The Neutral Theory of Molecular Evolution." *Scientific American*, November 1979. Kimura presents his theory that much evolutionary change is not driven by natural selection, but by genetic drift, even in large populations, as chance increases the frequency of alleles that are no more adaptive than the alleles they replace.

May, R. M. "The Evolution of Ecological Systems." *Scientific American*, September 1979. Coevolution accounts for much of the structure of natural communities of plants and animals.

O'Brien, S. J., Wildt, D. E., and Bush, M. "The Cheetah in Peril." *Scientific American*, May 1986. According to molecular and immunological techniques, a population bottleneck has reduced the genetic variability of the world's cheetahs almost to zero.

Ryan, M. J. "Signals, Species, and Sexual Selection." *American Scientist*, January–February 1990. Ryan explores a variety of experiments on sexual selection, including the genetic basis of male characteristics and female choice.

Smith, J. M. "The Evolution of Behavior." *Scientific American*, September 1979. A leading mathematical evolutionist explains how complex, apparently unlikely behaviors might be selected for.

Stebbins, G. L., and Ayala, F. "The Evolution of Darwinism." *Scientific American*, July 1985. A synthesis of molecular and classical evolutionary methodologies.

18

The Origin of Species

The great bird of paradise (*Paradisea apoda*) is one of almost 140 species of birds living on New Guinea, many of them unique to New Guinea and its neighboring islands. Why there are so many species, how they arose, and how they maintain reproductive isolation from other, often similar, species are major questions in evolutionary biology.

A tribe on New Guinea has 137 separate names for local birds, almost exactly the same as the number of species determined by Western ornithologists. The larvae of each of the 750 species of fig wasps on Earth can grow only in its own particular species of fig (Fig. 18-1). All three—New Guinea islanders, ornithologists, and fig wasps—perceive qualitative differences between members of different species; that is, differences in kind. Mere quantitative differences—this fig tree is a bit larger than that one, or this bird's feathers are a bit brighter blue than that one's—are usually ignored. To biologists, the major qualitative difference between species is the *ability to interbreed.* Thus, **a species consists of all the populations of organisms that are potentially capable of interbreeding under natural conditions and that are reproductively isolated from other populations.**

Our discussion of the mechanisms of evolution in the previous chapter dealt exclusively with quantitative changes within a population: the frequency of the sickle-cell anemia allele, the length of a giraffe's neck, or the size of a peacock's tail. At what point do these quantitative changes add up to a large enough total difference so that the population is a new species, and cannot or will not interbreed with any other populations? Or is there a sudden jump, a quantum leap of speciation distinct from, and perhaps unrelated to, the accumulation of small, quantitative differences?

And what about the grand sweep of evolutionary history? Is all of evolution, from the first prokaryotic cell to the human brain, simply the accumulation of small changes over eons? Or are other processes at work?

These are some of the most contentious questions in biology. Virtually all biologists agree that evolution occurs and that the processes described in Chapter 17 produce quantitative changes in populations. What they don't agree on are the larger questions: How do species arise? What is the relative importance of many small changes vs. a few major changes? Is natural selection within populations a major force in generating the stunning diversity of living organisms? In this chapter, we will explore current hypotheses that try to explain the origin of species.

Speciation

Although we have described several evolutionary forces that lead to changes *within* species, we have not yet outlined a mechanism of **speciation,** whereby *new species* may be formed. To produce a new species, evolution must generate large enough genetic

Figure 18-1 A female fig wasp uses her long ovipositor to inject her eggs into a fig. Each species of fig wasp is restricted to a single species of fig.

changes between populations so that mating cannot occur or hybrid offspring are less fit. Speciation depends on the isolation and genetic divergence of two populations.

1. **Isolation of populations:** If two populations are to become sufficiently distinct, genetically, so that interbreeding is difficult or impossible, then there must be relatively little gene flow between them. If there is a great deal of gene flow, then genetic changes in one population will soon become widespread in the other as well.

2. **Genetic divergence:** What if two formerly isolated populations are reunited? They will have become separate species only if they had evolved sufficiently large genetic differences during the period of isolation so that they cannot interbreed and produce vigorous, fertile offspring when they are reunited. If isolated populations are small, chance events may generate significant genetic differences by genetic drift (see Chapter 17). In both small and large populations, different selective pressures in separate environments may favor the evolution of large genetic differences.

Speciation has seldom been observed in the wild (with the exception of "instant speciation" in plants by polyploidy, described below). However, based partly on theoretical considerations and partly on experiments and observations, evolutionary biologists have devised plausible hypotheses for the origin of new species in two different cases: **allopatric speci-**

ation (two populations are geographically separated from one another) and **sympatric speciation** (two populations share the same geographical area) (Fig. 18-2).

At first glance, you might think that sympatric speciation violates our first principle of speciation, isolation of populations, because the speciating populations live in the same locale. However, it is *isolation from gene flow* that is crucial to speciation. Isolation from gene flow is simplest to envision if two populations are geographically separated from one another by some kind of physical barrier, such as a river. Indeed, most cases of speciation, at least in animals, probably have occurred through such geographical isolation. However, as we will see below, even two populations living in the same area may have very little gene flow if they select different habitats within the area (e.g., marshes vs. forests) or have differences in chromosome numbers so that they cannot form fertile hybrids. Therefore, the principle still holds: *isolation from gene flow is the key to both allopatric and sympatric speciation.*

Allopatric Speciation

In Greek, the word *allopatric* means "having a different fatherland." Allopatric speciation occurs when two populations become **geographically isolated** from one another; that is, they are physically separated either by distance or by an impassible barrier (Fig. 18-3). As we described in Chapter 17, migration allows gene flow, which reduces genetic differences

between populations and probably eliminates any possibility that two populations could ever become reproductively isolated from one another. If two populations become separated by a physical barrier, or even sheer distance, little or no gene flow can occur between them. If the pressures of natural selection differ in the two locations, or if the populations are small enough for genetic drift to occur, then the two populations may accumulate large genetic differences and become separate species. Founder events, in which a few members of a species become isolated from the main body of the species, may also be important in initiating genetic differences between populations. There is some debate as to whether genetic drift or natural selection normally plays a major role in allopatric speciation. Probably the majority opinion is that different selective pressures in the two geographical locales provide the major impetus to speciation. In either case, evolutionary biologists believe that geographical isolation is involved in most cases of speciation, especially in animals.

Sympatric Speciation

Sympatric means "having the same fatherland." As the name implies, sympatric speciation refers to speciation occurring within a single population and in a single geographical area. Sympatric speciation, like allopatric speciation, requires limited gene flow. There are two likely mechanisms whereby gene flow can be reduced between members of a single population: ecological isolation and chromosomal abberations.

Figure 18-2 Models of allopatric and sympatric speciation.
(a) Allopatric speciation: (Top) A single species (white mice) occupies a relatively homogeneous habitat. (Middle) An impassable geographical barrier (here, a river changing course) splits the habitat into two parts, separating the species into two isolated populations. Due to genetic drift or different selective pressures, the two populations diverge genetically (tan vs. white mice). (Bottom) The barrier is removed (the river changes back to its original course), and the members of the two populations can share the same habitat. If the genetic differences between the two populations have become large enough so that interbreeding cannot occur (i.e., they are reproductively isolated from one another), then the two populations constitute separate species (brown vs. white mice).
(b) Sympatric speciation: (Top) A single species occupies a homogeneous habitat. (Middle) Climate change or other factors form two distinctly different habitats that are still physically part of the same general region; i.e., there are no barriers to movement between habitats. Different selective pressures in the two habitats lead to genetic divergence of organisms living in each (tan vs. white mice) (Bottom) Sufficient genetic divergence causes reproductive isolation; former occupants of the two different habitats are now separate species (brown vs. white mice).

(a) Allopatric Speciation

(b) Sympatric Speciation

time

geographic isolation

ecological isolation

genetic divergence

genetic divergence

reproductive isolation

reproductive isolation

(a)   **(b)**

Figure 18-3 Geographic isolation that prevents gene flow is usually necessary for animal populations to split into two species. The tassel-eared squirrels that live on the rims of the Grand Canyon are a classic example. In the distant past, a single population of squirrels became separated as the Canyon was carved into the plateau of northwestern Arizona. The two populations are split by the treeless desert of the canyon depths. Today, the Kaibab squirrel **(a)** is confined to the north rim, while the Abert squirrel **(b)** is found on the south rim and in forests throughout northern Arizona, Utah, New Mexico, and Colorado. It is not known whether the two are genetically distinct enough to constitute separate species.

Ecological Isolation

If the same geographical area contains two distinct types of habitats (e.g., food sources, nesting places, etc.), different members of a single species may begin to specialize in one habitat or the other. If conditions are right, natural selection for habitat specialization may cause the formerly single species to split into two species. This seems to be occurring, right before biologists' eyes, so to speak, in the case of the fruit fly *Rhagoletis pomonella* (Fig. 18-4; see also Chapter 44 for some other fascinating features of *Rhagoletis*).

Rhagoletis is a parasite of the American hawthorn tree, laying its eggs in the hawthorn's fruit. When the maggots hatch, they eat the fruit. About 150 years ago, entomologists noticed that *Rhagoletis* began to infest apple trees, which were introduced into North America from Europe. Now, it appears that *Rhagoletis* is splitting into two species, based on their preference for apples or hawthorns. There are substantial genetic differences between apple-liking flies and hawthorn-liking flies. At least some of these genetic differences, such as the timing of emergence of the adult flies (see below), are important for survival on a particular host plant. Since apples and hawthorns are often found quite close together, and flies, after all, can fly, why don't apple-flies and hawthorn-flies interbreed and cancel out any incipient genetic differences? There appear to be at least two reasons. First, female flies usually lay their eggs in the same type of fruit in which they themselves developed. Males also

tend to rest on the same type of fruit in which they developed. Therefore, apple-liking males are likely to encounter and mate with apple-liking females. Second, apple fruits mature 2 or 3 weeks later than hawthorn fruits, and the two types of flies emerge with a timing appropriate for their chosen host fruits. Thus the two types of flies have very little chance of meeting. There is still some interbreeding between the two types of flies, but it looks as if they are well on their way to speciation. Will they make it? Entomologist

Figure 18-4 *Rhagoletis pomonella,* the apple maggot fly, seems to be undergoing sympatric speciation through ecological isolation on two host trees: apples and hawthorns.

Guy Bush suggests, "Check back with me in a few thousand years."

Chromosomal Aberrations

In some instances, new species can arise nearly instantaneously. This may occur through changes in chromosome configuration or number, due to irregularities during meiosis. A common speciation mechanism in plants is **polyploidy,** the acquisition of multiple copies of each chromosome. As you know, most plants and animals have paired chromosomes, and are called diploid. Occasionally, especially in plants, a fertilized egg duplicates its chromosomes but doesn't divide into two daughter cells. The resulting cell thus becomes tetraploid, with four copies of each chromosome. If all of the subsequent cell divisions are normal, this tetraploid zygote will develop into a plant that consists of tetraploid cells. Tetraploid plants are usually vigorous and healthy, and many can successfully complete meiosis to form viable gametes. The gametes, however, are diploid. Therefore, if sperm from a tetraploid plant fertilizes an egg cell of the diploid "parental" species, the resulting offspring will be triploid. The triploid offspring may develop normally, but when this triploid offspring begins meiosis, it will be unable to pair up all of its chromosomes, because there will be an odd number (three) of each. Meiosis fails, gametes are not formed, and so the triploid hybrids are sterile. The tetraploid plant is therefore reproductively isolated from its diploid parent.

Why is speciation by polyploidy common in plants but not in animals? Many plants can either self-fertilize or reproduce asexually, or both. If a tetraploid plant self-fertilizes, then its offspring will also be tetraploid. Asexual offspring, of course, are genetically identical to the parent, and are also tetraploid. In either case, the new tetraploid plant may perpetuate itself and form a new species. Animals, on the other hand, usually cannot self-fertilize or reproduce asexually. Therefore, if an animal produces a tetraploid offspring, the offspring would have to mate with a member of the diploid parental species, and would produce all triploid offspring. As we pointed out, the triploid offspring would almost certainly be sterile. Speciation by polyploidy is extremely common in plants: in fact, nearly half of all species of flowering plants are polyploid, many of them tetraploid.

Maintaining Reproductive Isolation Between Species

Once a species has formed, it may remain **reproductively isolated** from other species in two ways (Table 18-1). First, members of the species may not mate with members of other species. This usually has a clear adaptive value. Separate species are usually genetically different in ways that adapt them to different environments. Any individual that mates with a member of another species will probably produce unfit or sterile offspring, thereby "wasting" its genes and contributing nothing to future generations. Thus there is strong selective pressure to avoid mating between different species. Incompatibilities between species that prevent mating are called **premating isolating mechanisms.**

Table 18-1   Mechanisms of Reproductive Isolation

Premating isolating mechanisms: any structure, physiological function, or behavior that prevents organisms of two different populations from mating.

1. Geographical isolation: the separation of two populations by a physical barrier.
2. Ecological isolation: lack of mating between organisms belonging to different populations that occupy distinct habitats within the same general area.
3. Temporal isolation: the inability of organisms to mate if they have significantly different breeding seasons.
4. Behavioral isolation: lack of mating between species of animals that differ substantially in courtship and mating rituals.
5. Mechanical incompatibility: the inability of male and female organisms to exchange gametes, usually because of incompatibility of the reproductive structures.

Postmating isolating mechanisms: any structure, physiological function, or developmental abnormality that prevents organisms of two different populations, once mating has occurred, from producing vigorous, fertile offspring.

1. Gametic incompatibility: the inability of sperm from one species to fertilize eggs of another species.
2. Hybrid inviability: the failure of a hybrid offspring of two different species to survive to maturity.
3. Hybrid infertility: reduced fertility (often complete sterility) in hybrid offspring of two different species.

Sometimes premating isolation fails or has not yet evolved, and members of different species do mate. If the resulting hybrid offspring die during development, then naturally the two species are still reproductively isolated from one another. In some cases, however, viable hybrid offspring are produced. Even so, if the hybrids are less fit or infertile, the two species may still remain separate, with little or no gene flow between them. Incompatibilities between species that prevent the formation of vigorous, fertile hybrids are called **postmating isolating mechanisms.**

Premating Isolating Mechanisms

Mechanisms that prevent mating between different species include geographical isolation, ecological isolation, temporal isolation, behavioral isolation, and mechanical incompatibility.

Geographical Isolation

Members of different species obviously cannot mate if they never get near one another. As we have already seen, geographical isolation often provides the conditions for speciation in the first place. However, we cannot tell if geographically separated populations constitute distinct species. Should the barrier separating the two populations disappear (an intervening river changes course, for example), it may well turn out that the reunited populations would interbreed freely and not be separate species at all. If they cannot interbreed, then other mechanisms, such as different courtship rituals, must have developed during their isolation. Geographical isolation, therefore, is usually considered to be a mechanism that *allows new species to form* rather than a mechanism that *maintains reproductive isolation between different species.*

Ecological Isolation

If two populations have different resource requirements, they may use different local habitats within the same general area. White-crowned and white-throated sparrows, for example, have extensively overlapping ranges. The white-throated sparrow, however, frequents dense thickets, whereas the white-crowned sparrow inhabits fields and meadows, seldom penetrating far into dense growth. The two species may coexist within a few hundred yards of one another and yet seldom meet during breeding season. The *Rhagoletis* fruit flies discussed earlier are at least partially isolated by habitat selection for apples vs. hawthorns. Although ecological isolation may slow down interbreeding, it seems unlikely that it could prevent gene flow entirely. Other mecha-

nisms usually also contribute to interspecific isolation.

Temporal Isolation

Even if two species occupy similar habitats, they cannot mate if they have different breeding seasons. Bishop pines and Monterey pines coexist near Monterey on the California coast (Fig. 18-5). Viable hybrids have been produced between these two species in the laboratory. However, in the wild, they release their pollen at different times: the Monterey pine releases pollen in early spring, the bishop pine in summer. Therefore, the two species never cross-breed under natural conditions. Hawthorn-liking and apple-liking *Rhagoletis* fruit flies are also partially isolated from one another because they emerge from their host fruits and breed at somewhat different times of year.

Behavioral Isolation

Among animals, the elaborate courtship colors and behaviors that so enthrall human observers have evolved not only as recognition and evaluation signals between male and female; they may also aid in distinguishing among species. The striking colors and calls of male songbirds, for example, may be siren songs for females of their own species, but are treated with the utmost indifference by other females. Among frogs, males are often impressively indiscriminate, jumping on every female in sight, regardless of the species, when the spirit moves them. Females, however, only approach male frogs croaking the correct "ribbet!" If they do find themselves in an unwanted embrace, they utter the "release call," which causes the male to let go. As a result, few hybrids are produced.

Figure 18-5 Bishop pines produce fertile hybrids in the laboratory. In nature, however, they do not interbreed because they release pollen at different times of the year.

Mechanical Incompatibility

In rare instances, ecological, temporal, and behavioral isolating mechanisms fail, and male and female of different species attempt to mate. Among animals with internal fertilization, in some cases the male and female genitalia simply won't fit together. Among plants, differences in flower size or structure may prevent pollen transfer between species, if, for example, the species attract different pollinators (see Chapter 28 for a description of the interesting deceptions orchids use to lure specific types of pollinators).

Postmating Isolating Mechanisms

Sometimes premating isolation fails, and mating occurs between members of different species. However, if vigorous, fertile hybrids are not produced, there will still be little gene flow between the two species.

Gametic Incompatibility

Even though a male inseminates a female, his sperm may not fertilize her eggs. For example, the fluids of the female reproductive tract may weaken or kill sperm of other species. Among plants, chemical incompatibility may prevent the germination of pollen from one species that lands on the stigma (pollen-catching structure) of the flower of another species.

Hybrid Inviability

If fertilization does occur, the resulting hybrid may be weak or even unable to survive. The genetic programs directing development of the two species may be so different that hybrids abort early in development. Even if the hybrid survives, it may display behaviors that are mixtures of the two parental types. In attempting to do some things the way species A does them, and other things the way species B does, the hybrid may be hopelessly uncoordinated. Hybrids between certain species of lovebirds, for example, have great difficulty learning to carry nest materials during flight, and probably could not reproduce in the wild (see Chapter 42).

Hybrid Infertility

Animal hybrids, such as the mule (a cross between a horse and a donkey), are usually sterile. A common reason is the failure of chromosomes to pair properly during meiosis, so that eggs and sperm never develop. Among plants that have speciated by polyploidy, any offspring produced by mating between the diploid "parent" species and the tetraploid "daughter" species will be triploid and sterile.

Phyletic vs. Divergent Speciation

The mechanisms of speciation and reproductive isolation that we have just described all apply only to "splitting," or **divergent speciation**: isolation from gene flow and genetic divergence between two populations of a single parental species causes them to become reproductively isolated; in other words, to become different species (Fig. 18-6a). Divergent speciation is often called "true" speciation, because it is possible, in principle at least, to determine if the two presumed species are in fact reproductively isolated.

In the fossil record, however, things are not so simple. Fossils, being dead and petrified, cannot breed. Therefore, paleontologists make distinctions among species purely on morphological grounds: if two fossils are structurally as different from one another as two modern species are, then the fossils are assigned to different species. Suppose that a whole species, under massive directional selection, changes over time so that the later specimens are very different from the earlier ones (Fig. 18-6b). If the differences are large enough, it may be assumed that the later organisms could not possibly interbreed with the earlier ones, and thus form two separate species. However, reproductive isolation between early and late forms cannot be directly tested. This process is called **phyletic speciation,** or sometimes pseudospeciation.

Owing to its incompleteness, the fossil record does not reveal whether divergent or phyletic speciation dominates evolutionary history. If divergent speciation is most common, as many evolutionary biologists believe, but few enough fossils are found, then most evolutionary lineages will appear to show phyletic speciation (Fig. 18-6c). Beginning with a modern species and working backward, one may be able to reconstruct a "phyletic evolutionary history" back to an ancient ancestor. This phyletic history may, however, simply be selective sampling of a much larger, divergent history in which most of the branches left no fossils.

The Genetics of Speciation

When plants speciate by polyploidy, it is clear that a sudden, major genetic change has taken place. But what about other speciation events, whether allopatric or sympatric? There are two genetic models for speciation: the gradual accumulation of many small changes or the sudden appearance of a few major changes.

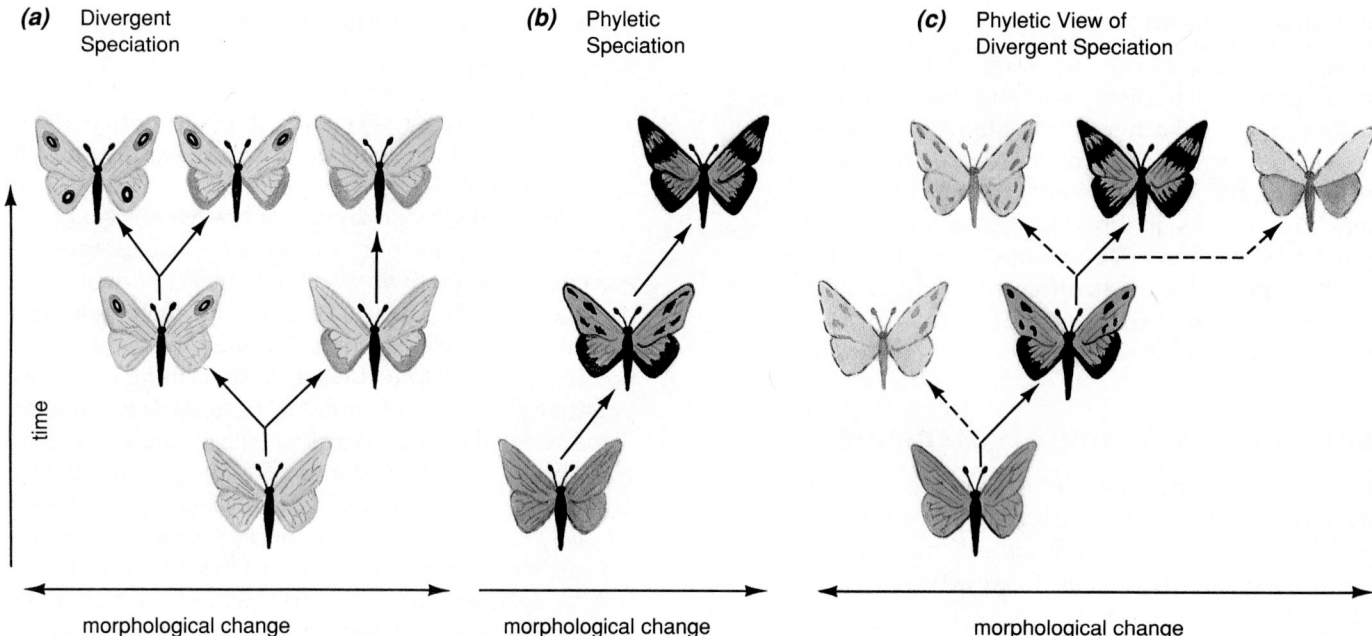

(a) Divergent Speciation

(b) Phyletic Speciation

(c) Phyletic View of Divergent Speciation

time

morphological change · morphological change · morphological change

Figure 18-6 Divergent speciation vs. phyletic speciation.
(a) In divergent speciation, a single species splits up into two or more daughter species that exist simultaneously and are reproductively isolated from one another.
(b) In phyletic speciation, the whole population of a single species evolves over time. Sufficient genetic changes occur so that the later populations are considered to be a new species, different from earlier populations.
(c) The fossil record is incomplete. Most species may become extinct without leaving either fossils or living descendants (dashed butterflies). Therefore, even if divergent speciation is most common, the fossil record may appear to show phyletic speciation (solid butterflies).

Gradual Accumulation of Many Small Changes

According to this model, two populations, whether by genetic drift or through differential natural selection pressures, gradually accumulate many small genetic changes. Each mutation or different allele by itself has only a minor phenotypic effect. Over time, however, the accumulation of many small changes may result in reproductive isolation between the two populations, and they may become separate species.

Sudden Appearance of a Few Major Changes

This model starts out with the idea that single regulatory genes, or small numbers of genes, control major developmental pathways. In certain aspects of the development of animals, this is known to be true. If so, then mutations in just a few regulatory genes might result in such significant changes in develop-

ment that the mutants would immediately be reproductively isolated from their parental populations. A new species would therefore arise almost instantaneously.

Which Model Is Correct?

Since so few speciation events have been observed, and since so little is known about the genetics of development, it is difficult to choose between these models. The extreme genetic similarity between certain species, for example chimps and humans (see Figs. 16-16 and E16-5), suggests to some observers that mutations in a few key regulatory genes may have been the cause of speciation. On the other hand, the "small change" model is supported by the finding that many phenotypic characteristics, including size, coloration, and various behavioral traits, are under the control of several to many genes (see Chapter 11 for a discussion of quantitative inheritance). Thus, one might expect that changes in many genes would be required to generate differences at the species level. Further, both theoretical modeling and common observation suggest that one-shot, massive genetic changes are usually harmful. This debate is not likely to be settled until molecular developmental biology comes to the rescue with a better understanding of the genetic basis of development and of developmental differences between species.

Rates of Speciation

Over the course of evolutionary history, species continually form, exist for a time, and become extinct. Geographical isolation, genetic drift, mutations, the invasion of new habitats, and coevolution all ensure that speciation never stops. The rate of speciation varies considerably over evolutionary time, however, and bursts of speciation can be seen both in the fossil record and in the distribution of modern organisms.

Adaptive Radiation

Sometimes a species gives rise to many new species in a relatively short time. This process, called **adaptive radiation,** occurs when populations of a single species invade different habitats and evolve in response to the differing selective pressures in those habitats. Adaptive radiation has occurred many times and in many groups of organisms. Adaptive radiation usually results from one of two causes. First, a species may encounter a wide variety of unoccupied habitats; for example, when the ancestors of Darwin's finches colonized the Galapagos Islands, or when marsupial mammals first invaded Australia. With no competitors except other members of their own species, all the available ecological roles were rapidly filled with new species that evolved from the original invaders. The ultimate in unoccupied habitats results from mass extinctions, such as those that might follow meteorite impacts (see Chapter 17). With most potential competitors exterminated by the effects of the impact, the survivors speciate prolifically, filling a host of empty habitats.

Adaptive radiation also occurs if a species develops a fundamentally new and superior **adaptation,** enabling it to displace less well adapted species from a variety of habitats. This has apparently happened several times during evolutionary history, for example, when warmblooded mammals diversified extensively at the expense of reptiles.

The Progress of Evolution

Although the processes that drive evolutionary change, such as genetic drift and natural selection, are well understood, there is considerable debate among evolutionary biologists concerning *which processes actually play the major roles in shaping evolutionary history.* Three subjects of particular interest are (1) the rate and timing of evolutionary change over time; (2) the role of speciation in producing morphological change; and (3) the relative importance of natural selection vs. nonselective forces in determining structures and behaviors.

In both the popular press and scientific journals, these topics are often discussed within the framework of two opposing views of evolution, termed the **gradualism** and **punctuated equilibrium** models. These two models are actually at the opposite ends of a spectrum, as we will see. In all likelihood, the evolution of real organisms usually falls somewhere closer to the middle of the spectrum, sometimes evolving mostly according to "gradualistic" rules and sometimes according to "punctuated" rules. We will briefly outline the two extreme positions, and then attempt to synthesize a coherent scheme.

Gradualism

The extreme **gradualism** model of speciation takes the following positions:

1. Speciation occurs from the accumulation of dozens or hundreds of small genetic differences between two populations. The two populations thus become separate species in a gradual fashion, as less and less interbreeding occurs and any hybrids that are formed become less and less fit. Speciation may take hundreds of thousands or even millions of years (Fig. 18-7a).
2. Morphological change within a lineage (e.g., horses; see Fig. 16-10) occurs continuously, although the rate may change somewhat from time to time.
3. Morphological change and speciation are not closely linked. Two populations of a species may undergo considerable morphological change without necessarily becoming reproductively isolated from one another.
4. Both continual morphological changes and speciation are driven largely by natural selection among individuals, with fitter organisms leaving more offspring than less fit organisms do.

Punctuated Equilibrium

The extreme form of the **punctuated equilibrium** model of evolution takes contrasting stands on every issue:

1. Speciation is a geologically rapid event, driven by a relatively small number of genetic changes,

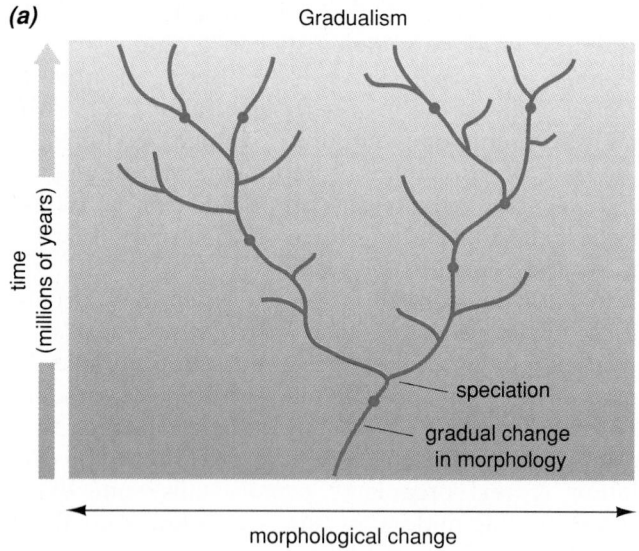

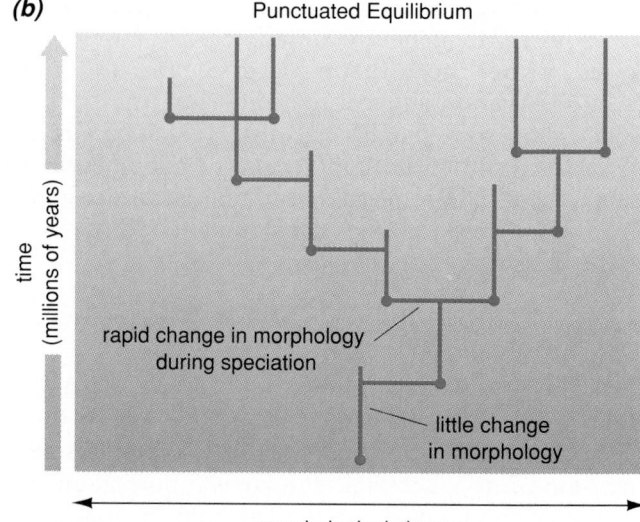

Figure 18-7 Extreme views of gradualism vs. punctuated equilibrium. Solid dots represent fossils considered to be separate species. Lines that do not reach the top of the graph represent extinctions.
(a) Gradualism: Morphological change occurs gradually over time, although the rate of change may vary (more rapid during times of strong directional selection, much more slowly during times of stabilizing selection). Both phyletic and divergent speciation probably occur.
(b) Punctuated equilibrium: Morphological change and speciation are simultaneous and very rapid, interrupted by long periods during which neither morphological change nor speciation occur. All speciation is divergent.

such as mutations in a few developmentally important regulatory genes. The rapidity of speciation is the "punctuated" part of the model (Fig. 18-7b).

2. Species remain morphologically the same for long periods of time; this is the "equilibrium" part of the model.

3. Morphological change and speciation are very tightly linked. Virtually all morphological change occurs during the brief period of speciation, and the newly formed species then remain essentially unchanged until the next speciation event.

4. Large-scale evolutionary changes are driven by selection among species, not individuals. Further, "species selection" often may be driven not by natural selection but by random chance.

Applying Gradualism and Punctuated Equilibrium Models to the Evolution of the Horse

In the evolution of the horse, the fossil record shows a progression from *Hyracotherium*, the "dawn horse,"

through many intermediate steps to modern *Equus* (see Fig. 16-10). Extreme gradualism would propose that fossil horses are representative samples of changes that occurred gradually throughout the evolution of the horse. As environments changed from forest to open woodlands to prairies, natural selection favored larger, faster horses with strong, shock-absorbing legs, hard hooves, and large, grinding teeth. Divergent speciation occurred from time to time as genetic changes accumulated between, for example, woodland populations and prairie populations, but phyletic speciation also occurred, as whole populations changed into new species under selective pressures. During phyletic speciation, selection *between individuals within a population* drove the whole population to change, gradually becoming a new species.

One problem with this view, which provided the original stimulus for the development of the punctuated equilibrium model, is the nature of the fossil record. Every type of fossil horse is quite different from every other type; that is, the fossil record does not show a continuous accumulation of small changes, but rather a series of jumps from one fossil type to the next. Why haven't fossils been found that represent all the presumed intermediate stages? Gradualism replies that fossilization is a rare event; for every organism that becomes fossilized, and is found by modern humans, probably millions do not. Further, many thousands of years may pass during which suitable conditions for fossilization (e.g., massive eruptions of volcanic ash, flash floods, sedimentation in shallow seas) do not occur. Therefore, many intermediate stages, perhaps even whole species, never show up in the fossil record.

The gaps in the fossil record are the starting point for the punctuated equilibrium model, championed by Niles Eldredge, Stephen Jay Gould, and Steven Stanley. Punctuated equilibrium assumes that the gaps are not accidents: they are produced by such rapid evolutionary change that fossil remains of intermediate forms would be extremely rare. Punctuated equilibrium hypothesizes that *Hyracotherium* underwent little change for millions of years (the "equilibrium"). Small fringe populations of *Hyracotherium* split off now and then. Some evolved very rapidly, perhaps mostly by genetic drift, to become entirely new species. At some point, perhaps due to a major change in the environment, one of these new species, *Mesohippus,* quickly replaced *Hyracotherium* as the dominant form of horse. The speciation of *Mesohippus* and its replacement of *Hyracotherium* may have taken place within a few thousand years; a mere instant, geologically speaking (the "punctuation" between equilibria). Punctuated equilibrium assumes that evolution is largely driven by this frequent, rapid divergent speciation. Selection *between species* then results in the extinction of some species and the survival of others.

A Synthesis

As we have already cautioned, the gradualism and punctuated equilibrium models just presented are extreme positions. Although there are certainly major disagreements among evolutionary biologists, something like a majority view could be summarized as follows.

Rates of Evolution

The rate of evolution, as measured by morphological changes in fossils, varies greatly. In some organisms, there are indeed long periods of constancy—sharks, for example, have remained outwardly similar for scores of millions of years. Morphological constancy might arise from several causes. In a constant environment, stabilizing selection may limit the rate of evolutionary change. A rapidly fluctuating environment—for example, one with wide but frequent tem-

perature changes—would promote equally rapid, fluctuating directional selection, with a net effect of preserving the status quo. Or even if the environment changes in a constant direction, species may simply "track" favorable habitats. For example, during the ice ages, caribou and other tundra species simply followed the tundra vegetation: south when the ice sheets advanced, back north when the ice retreated. Great morphological changes were not required, because the organisms kept to the same habitat, even though its geographical location moved.

Rapid morphological changes, which may in some instances correspond with times of speciation, also occur. Genetic drift in small populations, directional selection among populations that do not track suitable habitats, coevolution owing to competition or cooperation with other species, or sexual selection may all produce rapid evolutionary jumps.

Just how fast must evolution occur to be perceived as punctuated equilibrium? Returning to the fossil horses, let's suppose that a 60-centimeter tall *Hyracotherium* evolves into a 100-centimeter tall *Mesohippus* in 10,000 years, and that the time between generations is 5 years. This allows 2000 generations to produce a height increase of 40 centimeters, or 0.2 millimeter per generation. This would be far too small a difference to detect with the naked eye, and is more than a hundred times less than the variability in size among modern horses, even of the same breed. To a geneticist, this rate of change doesn't seem very fast at all. In the fruit fly *Drosophila,* for example, mild selection pressure can cause evolutionary changes that are more than 100 times faster than those required for *Hyracotherium* to evolve into *Mesohippus* in 10,000 years. Geologically speaking, however, 10,000 years is a very short time. No one would be surprised if intermediate fossils could not be found. Therefore, "punctuated" speciation would seem quite slow to a geneticist, but very fast to a geologist.

Speciation and Morphological Change

Morphological change may occur whenever a small population is isolated from gene exchange with other populations and is subjected to new selective pressures. Considerable change may also occur by genetic drift and/or founder events, if the population is very small. Selection or genetic drift may also cause speciation, with the small, isolated population rapidly becoming a separate species from its larger parent population. Therefore, in these cases speciation may be linked with major morphological change.

Morphological change may occur in large populations subjected to widespread directional selection. However, if gene flow among subpopulations is adequate, speciation is unlikely to occur. In this situation, morphological change will not be linked to speciation.

As we discussed previously, speciation could occur through the accumulation of many small genetic changes or through a few key mutations. Genetic data, although scanty, suggest that each mechanism may operate at different times in different populations. Which is more common cannot be decided with the evidence now available.

Species Selection versus Individual Selection

If two species are in direct competition for a particular resource in a particular habitat, one will probably move, evolve rapidly to use a different resource, or become extinct (see Chapters 17 and 44). In this case, species, not individuals, may be considered to be the unit of selection. However, the members of the species that "wins" in such a competition may well have attained their superior adaptations through natural selection acting on individual organisms. Further, the "winning" species may continue to evolve even after the extinction of the "losing" species, because of competition among individual members of the species. In these instances, the unit of selection is the individual. Most evolutionary biologists feel that selection occurs to some extent at both species and individual levels.

The Importance of Natural Selection

Is natural selection the paramount force operating during evolution? Several "punctuationists," notably Stephen Jay Gould, emphasize the importance of nonselective, random events during evolution. They have two major objections to what they call the "adaptationist program," that is, that most features evolved as an adaptation to natural selection. First, it is extremely difficult to prove that a particular feature, say the size of a bird's beak, was shaped by natural selection. If a biologist observes a bird with a large bill cracking a large, tough nut, it is tempting to assume that the large bill evolved because of the advantage it conferred in nutcracking. But maybe the bird acquired the large bill for other reasons, say for courtship or even through genetic drift, and eats nuts because it "accidentally" has a suitable bill. On one of the Galapagos Islands, Peter Grant demonstrated quite conclusively that natural selection strongly influences beak size in a species of Darwin finch. In the (more usual) absence of such evidence, however, Gould and others complain that adaptationist explanations are a biologist's version of Aesop's fables, and should be paid about as much attention.

The second objection is that random forces do indeed operate in nature. In the previous chapter, we discussed the likelihood that gigantic meteorites strike the Earth from time to time, causing extinctions of many species. Species that were superbly adapted to warm, sunny climates might go extinct within a couple of years in the cold and gloom following the impact. The survival or extinction of species might have little to do with adaptations to their habitats, because the habitats would disappear. Smaller-scale catastrophes, such as floods, sea-level changes, or massive forest fires, could similarly cause the extinction of well-adapted but localized species.

The consensus view is that both natural selection and random forces shape the evolutionary history of life on Earth. No one doubts that natural selection is the driving force behind the evolution of the magnificent structures and behaviors that we witness in the living world around us. And no one doubts that a 10-kilometer meteorite would drive many otherwise well-adapted species to extinction. Which processes dominate, over what time scales and over what fraction of species on Earth, is the real debate.

 **eflections on Evolution and Evolutionary Hypotheses**

In the popular press, conflicts among evolutionary biologists are sometimes seen as conflicts about evolution itself. One occasionally reads statements to the effect that new theories are overthrowing Darwin and casting doubt on the reality of evolution. Nothing could be further from the truth. Selectionists and gradualists, and everyone in between, unanimously agree that evolution occurred in the past and is still occurring today. The only argument is over the mechanisms of evolutionary change, their relative importance in the history of life on Earth, and which forces were most important in shaping the evolution of particular species. Meanwhile, wolves still tend to catch the slowest caribou, small populations still undergo genetic drift, habitats still change or disappear, and maybe even Nemesis, dark sister-star to the Sun, is swinging another asteroid our way: evolution continues, still generating "endless forms most beautiful."

SUMMARY OF KEY CONCEPTS

Speciation

A species is defined as all the populations of organisms that can potentially interbreed with one another under natural circumstances and that are reproductively isolated from other populations. Speciation, the development of new species, requires that two populations be isolated from gene flow between them and develop significant genetic divergence. Allopatric speciation occurs by geographical isolation and subsequent divergence of the separated populations through genetic drift or natural selection. Sympatric speciation occurs by ecological isolation and subsequent divergence, or by rapid chromosomal changes, such as polyploidy.

Maintaining Reproductive Isolation Between Species

Reproductive isolation between species may be maintained by one or more of several mechanisms, collectively called premating isolating mechanisms and postmating isolating mechanisms. Premating isolating mechanisms include geographical isolation, ecological isolation, temporal isolation, behavioral isolation, and mechanical incompatibility. Postmating isolating mechanisms include gametic incompatibility, hybrid inviability, and hybrid infertility.

Phyletic Versus Divergent Speciation

In divergent speciation, a single species splits into two or more subpopulations that evolve into separate species. Daughter species therefore exist simultaneously. In phyletic speciation, the entire population of a species evolves in concert over time, so that the later population is recognized as being a new species. No one knows which form of speciation dominates evolutionary history, but it is likely that phyletic speciation is a "tunnel vision" of divergent speciation as seen through a spotty fossil record.

The Genetics of Speciation

Separate species must have large enough genetic differences to prevent interbreeding. There are two models to explain the development of genetic differences between species: the accumulation of many small genetic differences or the acquisition of a few key differences, probably in regulatory genes. It is likely that both mechanisms occur, and each may account for the genetic divergence of some species.

Rates of Speciation

Rates of speciation vary greatly over geologic time. Invasion of new, unoccupied habitats or the development of fundamentally superior adaptations may trigger extremely rapid speciation, a process called adaptive radiation.

The Progress of Evolution

There are two extreme models for the overall pattern of evolution. The gradualism model asserts that genetic and morphological changes accumulate slowly over time, with speciation being a relatively slow process not closely linked to observable morphological change. Natural selection among individuals is the prime driving force behind evolutionary change. The punctuated equilibrium model asserts that morphological change and speciation are very rapid, simultaneous events. Speciation events are separated by long periods during which species undergo little or no morphological change. Evolutionary change is driven by natural selection among simultaneously existing species, and also by random, chance events leading to speciation and extinction. Most evolutionary biologists believe that some aspects of both models probably occur during evolution, and that evolutionary history may more closely resemble either model at different times or in the lineage of different organisms.

GLOSSARY

adaptation: a characteristic of an organism that helps it to survive and reproduce in a particular environment; also, the process of acquiring such characteristics.

adaptive radiation: extensive speciation occurring among related organisms as a result of adaptation to a wide variety of habitats.

allopatric speciation: (al-ō-pat'-rik): speciation that occurs when two populations are separated by a physical barrier that prevents gene flow between them (geographical isolation).

behavioral isolation: lack of mating between species of animals that differ substantially in courtship and mating rituals.

divergent speciation: speciation in which subpopulations of a species split to form two or more new species that exist simultaneously. See phyletic speciation.

ecological isolation: lack of mating between organisms belonging to different populations that occupy distinct habitats within the same general area.

gametic incompatibility: the inability of sperm from one species to fertilize eggs of another species.

geographical isolation: the separation of two populations by a physical barrier.

gradualism: a model of evolution, stating that morphological change and speciation are slow, gradual processes that are not necessarily simultaneous or linked.

hybrid infertility: reduced fertility (often complete sterility) in hybrid offspring of two different species.

hybrid inviability: the failure of a hybrid offspring of two different species to survive to maturity.

mechanical incompatibility: the inability of male and fe-

male organisms to exchange gametes, usually because of incompatibility of the reproductive structures.

phyletic speciation: a form of evolution in which the gene pool of a species changes so greatly over time that a new species is formed. Old and new species do not exist simultaneously. See divergent speciation.

premating isolating mechanism: any structure, physiological function, or behavior that prevents organisms of two different populations from exchanging gametes.

postmating isolating mechanism: any structure, physiological function, or developmental abnormality that prevents organisms of two different populations, once mating has occurred, from producing vigorous, fertile offspring.

punctuated equilibrium: a model of evolution, stating that morphological change and speciation are rapid, simultaneous events (the "punctuation"), separated by long periods during which a species remains unchanged (the "equilibrium").

reproductive isolation: the failure of organisms of one population to breed successfully with members of another population; may be due to premating or postmating isolating mechanisms.

speciation: the process whereby two populations achieve reproductive isolation.

sympatric speciation (sim-pat'-rik): speciation of populations that are not physically divided; usually due to ecological isolation or chromosomal abnormalities (e.g., polyploidy).

temporal isolation: the inability of organisms to mate if they have significantly different breeding seasons.

STUDY QUESTIONS

1. Define the following terms: species, speciation, allopatric speciation, and sympatric speciation. Explain how allopatric and sympatric speciation might work, and give a hypothetical example of each.
2. Describe how polyploidy can give rise to "instant species."
3. List and describe the different types of premating and postmating isolating mechanisms.

4. Define punctuated equilibrium and gradualism. Look up the evolutionary history of a lineage (e.g., dinosaurs, humans) and describe how each mechanism might account for the evolution of that lineage.
5. Describe phyletic and divergent speciation. Do you think that you could distinguish between these two mechanisms in the fossil record?

DISCUSSION QUESTIONS

1. Do phyletic speciation and divergent speciation coincide with the gradualism and punctuated equilibrium models of evolution? Defend your answer.
2. Many species of animals are capable of mating, at least in the laboratory or in zoos, that seldom, if ever, mate

in the wild because of differences in mating behaviors. Explain the development of species-specific mating behaviors in terms of the likely success of hybrid offspring.

SUGGESTED READINGS

Ayala, F. "The Mechanisms of Evolution." *Scientific American*, September 1979. Our increasing understanding of the molecular mechanisms of heredity and mutation reveals that organisms are extremely variable, providing tremendous amounts of material on which selection can operate.

Kimura, M. "The Neutral Theory of Molecular Evolution." *Scientific American*, November 1979. Kimura presents his theory that much evolutionary change is not driven by natural selection, but by genetic drift, even in large populations, as chance increases the frequency of alleles that are no more adaptive than the alleles they replace.

Smith, J. M. *Evolution Now: A Century After Darwin.* San Francisco: W. H. Freeman and Company, 1982. The section on "Evolution—Sudden or Gradual?" contains essays by Russell Lande and Smith himself on the "gradualist" side and by Stephen Jay Gould on the "punctuationist" side.

Stebbins, G. L., and Ayala, F. "The Evolution of Darwinism." *Scientific American*, July 1985. A synthesis of molecular and classical evolutionary methodologies.

Strickberger, M. W. *Evolution.* Boston: Jones and Bartlett Publishers, 1990. Excellent descriptions of speciation.

19

The History of Life on Earth

The Great Nebula in Orion, a cloud of gases in which stars may be forming, as they did in the beginning of the Universe.

" . . . thence life was born,
Its nitrogen from ammonia, carbon from methane,
Water from the cloud and salts from the young seas . . . the cells of life
Bound themselves together into clans, a multitude of cells
To make one being—as the molecules before
Had made of many one cell. Meanwhile they had invented
Chlorophyll and ate sunlight, cradled in peace
On the warm waves; but certain assassins among them
Discovered that it was easier to eat flesh
Than feed on lean air and sunlight: thence the animals,
Greedy mouths and guts, life robbing life,
Grew from the plants; and as the ocean ebbed and flowed many plants and animals
Were stranded in the great marshes along the shore,
Where many died and some lived. From these grew all land-life,
Plants, beasts, and men; the mountain forest and the mind of Aeschylus
And the mouse in the wall."

From The Beginning and the End by *Robinson Jeffers*

Fifteen billion years ago, before the Big Bang, there was no universe as we know it: only a dense, minuscule mass in which energy and matter were melded into a single incomprehensibly ferocious state. Then the mass erupted. Particles of matter and antimatter formed, annihilating each other in great bursts of energy when they collided. So great was the energy released that astrophysicists can still detect its faint radiation.

Gradually, local accumulations of matter formed as gravity drew particle to particle. Many accumulations grew so large that their centers became dense and hot, triggering the thermonuclear reactions that fuel the light of the stars. Small, simple atoms of hydrogen fused to form helium, and these fused to give rise to still larger atoms. The nuclear forces of some stars grew so intense that they exploded, spewing their matter out into space.

About 5 billion years ago, far out in one of the arms of a spiral galaxy, a small cloud of matter began to condense, enriched with heavy elements provided by the self-destruction of ancient stars. The center of the cloud collapsed into the yellow dwarf star we call the sun. Farther out, other local aggregations appeared, forming the planets. The first four planets were small and contained a high proportion of heavy elements. The innermost two, being close to the sun, became very hot, while the fourth cooled into an eternal winter of temperatures more than 100 degrees below zero. The third settled into an orbit that would receive just the right intensity of sunlight to permit water to exist as a liquid. This became the Earth, although perhaps it is more appropriate to name it, in Jacques Cousteau's phrase, the Water Planet.

On this third planet, and so far as we know upon no other in our solar system, life evolved (Table 19-1). Where did life come from? How did it develop into the myriad forms we see today? What is its future?

Origins

How and when did life first appear on Earth? Just a few centuries ago, this question would have been considered trivial. Although no one knew how life *first* arose, people thought that new living things appeared all the time, through **spontaneous generation** from both nonliving matter and other, unrelated forms of life. In 1609, a French botanist wrote, "There is a tree—not, it is true, common in France, but frequently observed in Scotland. From this tree leaves are falling; upon one side they strike the water and slowly turn into fishes, upon the other they strike the land and turn into birds." Medieval writings abound with similar observations and delightful prescriptions for creating life—even human beings. Microorganisms were thought to arise spontaneously from broth, maggots from meat, mice from mixtures of sweaty shirts and wheat, and people from sperm injected into a cucumber!

In 1668 the Italian physician Francesco Redi disproved the maggots-from-meat hypothesis simply by

Table 19-1 The History of Life on Earth

Era	Period	Epoch	Years Ago* (millions)	Major Events
Precambrian			4600–3500 3500–590	Origin of solar system and Earth. Origin of first living cells; dominance of bacteria; origin of photosynthesis and evolution of oxygen atmosphere; origin of algae and soft-bodied marine invertebrates.
Paleozoic	Cambrian		590–505	Primitive marine algae flourish; origin of most marine invertebrate types.
	Ordovician		505–438	Invertebrates, especially arthropods and molluscs, dominant in sea; first fish, fungi.
	Silurian		438–408	Many fish, trilobites, molluscs in sea; first vascular plants; invasion of land by plants; invasion of land by arthropods.
	Devonian		408–360	Fishes and trilobites flourish in sea; origin of amphibians and insects.
	Carboniferous		360–286	Swamp forests of tree ferns and club mosses; dominance of amphibians; numerous insects; origin of reptiles.
	Permian		286–248	Origin of conifers; massive marine extinctions, including last of trilobites; flourishing of reptiles and decline of amphibians; continents aggregated into one land mass, Pangaea.
Mesozoic	Triassic		248–213	Origin of mammals and dinosaurs; forests of gymnosperms and tree ferns; breakup of Pangaea begins.
	Jurassic		213–144	Dominance of dinosaurs and conifers; origin of birds; continents partially separated.
	Cretaceous		144–65	Flowering plants appear and become dominant; mass extinctions of marine life and some terrestrial life, including last dinosaurs; modern continents well separated.
Cenozoic	Tertiary	Paleocene Eocene Oligocene Miocene Pliocene	65–54 54–37 37–24 24–5 5–2	Widespread flourishing of birds, mammals, insects, and flowering plants; drift brings continents into modern positions; mild climate at beginning of period, with extensive mountain building and cooling toward end.
	Quaternary	Pleistocene Recent	2–0.01 0.01–present	Evolution of *Homo*; repeated glaciations in northern hemisphere; extinction of many giant mammals.

*From Harland, W. B., et al. *A Geologic Time Scale.* Cambridge, England: Cambridge University Press, 1982.

keeping flies (whose eggs hatch into maggots) away from uncontaminated meat (see Chapter 1). Then in the mid-1800s Louis Pasteur in France and John Tyndall in England and disproved the broth-to-microbe idea as well (Fig. 19-1), effectively demolishing the notion of spontaneous generation. After this, speculation about life originating from nonliving matter, now or on the primeval Earth, was not likely to further one's scientific career. For almost half a century the subject lay dormant.

Eventually, biologists returned to the question of the origin of life. In the 1920s and 1930s, Alexander Oparin in Russia and J. B. S. Haldane in England speculated that, given the correct conditions, perhaps life could have risen from nonliving matter through ordinary chemical reactions. This process is called chemical evolution or **prebiotic evolution**—that is, evolution before life existed.

Prebiotic Evolution

The primordial Earth differed greatly from the equable planet we now enjoy. As rock after rock smashed into the forming planet, their energies of motion were converted into heat. Radioactive atoms decayed, re-

leasing still more heat. Soon the rock melted, and heavier elements such as iron and nickel sank to the center of the mass, where they remain molten still. Gradually the Earth cooled, and elements combined to form compounds of many sorts. Virtually all the oxygen combined with hydrogen to form water, carbon to form carbon dioxide, or heavier elements to form minerals. After millions of years, the Earth cooled enough to allow water to exist as a liquid, and for millenia it must have rained, as water vapor condensed out of the cooling atmosphere. As the water struck the surface, it dissolved many minerals, forming a weakly salty ocean. Lightning from storms, heat from volcanoes, and intense ultraviolet light from the sun all poured energy into the young seas.

Judging from the chemical composition of the rocks formed at this time, geochemists have deduced that the primitive atmosphere may have contained substances such as carbon dioxide, methane, ammonia, hydrogen, nitrogen, hydrochloric acid, hydrogen sulfide, and water vapor. However, because oxygen atoms were bound up in water, carbon dioxide, and minerals, there was virtually no free oxygen in the early atmosphere. This is an important factor in all hypotheses and experiments dealing with prebiotic evolution.

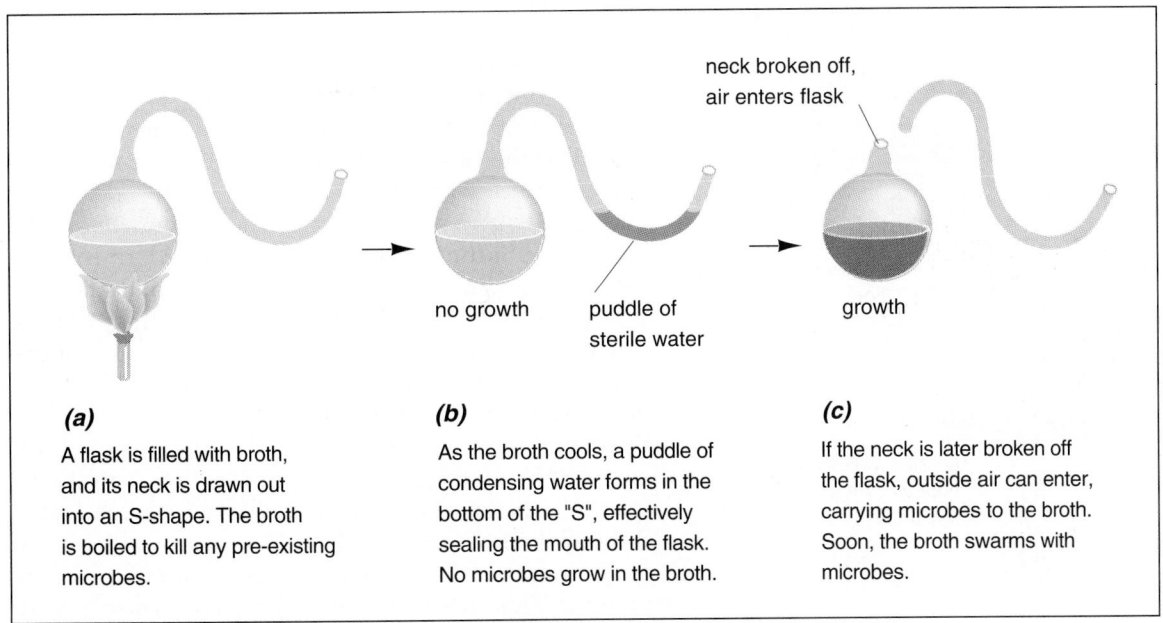

(a) A flask is filled with broth, and its neck is drawn out into an S-shape. The broth is boiled to kill any pre-existing microbes.

(b) As the broth cools, a puddle of condensing water forms in the bottom of the "S", effectively sealing the mouth of the flask. No microbes grow in the broth.

(c) If the neck is later broken off the flask, outside air can enter, carrying microbes to the broth. Soon, the broth swarms with microbes.

Figure 19-1 Louis Pasteur's experiment disproving the spontaneous generation of microbes in broth. Microbes did not appear in sterilized (boiled) broth, so they must not rise through spontaneous generation. They do appear when outside air is let into the flask, showing that contamination by air-borne microbes must be the source of any microbes that grow in the broth.

Prebiotic Synthesis of Organic Molecules

In 1953, Stanley Miller, a graduate student at the University of Chicago, mixed water, ammonia, hydrogen, and methane in a flask, and provided energy with heat and electric discharge (to simulate lightning). He found that simple organic molecules appeared after just a few days (Fig. 19-2). In these and similar experiments, Miller and others have produced amino acids, short proteins, nucleotides, adenosine triphosphate (ATP), and other molecules characteristic of living things. Interestingly, the composition of the "atmosphere" used in these experiments is unimportant, providing that hydrogen, carbon, and nitrogen are available and that free oxygen is excluded. Similarly, a variety of energy sources, including ultraviolet light, electric discharge, and heat, all seem about equally effective. Even though geochemists may never know exactly what the primordial atmosphere was like, it is certain that organic molecules were synthesized on the ancient Earth.

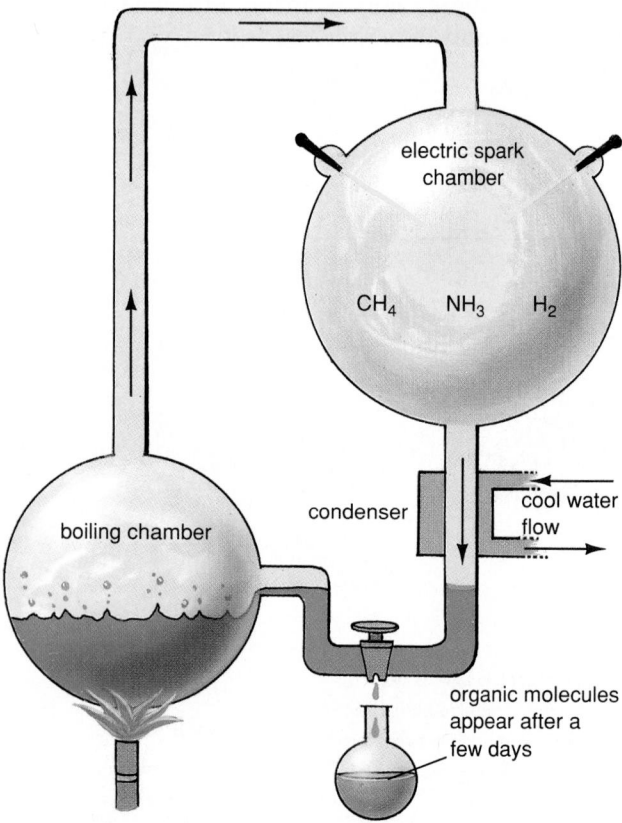

Figure 19-2 The experimental apparatus of Stanley Miller. Energy from heat and electrical discharge causes amino acids and other organic molecules to form from methane, ammonia, hydrogen, and water, all of which are thought to have been present in the atmosphere of the early Earth.

Stability and Accumulation of Organic Molecules

Prebiotic synthesis would not have been very efficient or very fast. Nevertheless, over the course of a few hundred million years, a lot of organic molecules could accumulate, if they didn't break down. On the modern Earth, organic molecules break down, even if not digested by living organisms, by reacting with atmospheric oxygen. Since the primeval Earth lacked both life and free oxygen, these sources of degradation were absent. On the other hand, the primordial atmosphere also lacked an ozone layer, and ultraviolet bombardment, which can also break apart organic molecules, must have been fierce. Some places, however, such as those beneath rock ledges or at the bottoms of even fairly shallow seas, would have been protected from ultraviolet radiation. In these locations, organic molecules may have accumulated to relatively high concentrations.

These concentrations of organic molecules may have been crucial in the evolution of life. First, they may have provided the molecules that would form the first living organisms. Second, the chemical energy stored in these molecules would be food for the first cells.

RNA: The First Living Molecules?

In modern cells, proteins carry out most of the cellular functions, while DNA encodes the information the cell needs to synthesize these proteins. Like the old "chicken-and-egg" riddle, which came first? If useful protein catalysts came first, how was the information needed to synthesize them passed from protocell to protocell? If nucleic acids came first, what would be the function of an information-storage molecule if there were no information to store?

In the 1980s, Thomas Cech of the University of Colorado and Sidney Altman of Yale offered an intriguing solution to this riddle. They found that some types of RNA molecules act as enzymes that, among other things, can cut apart RNA and synthesize more RNA molecules. During hundreds of millions of years of prebiotic chemical synthesis, RNA nucleotides may have occasionally bonded together to form short RNA chains. Let us suppose that, purely by chance, one of these RNA chains was a catalyst—dubbed a **ribozyme**—that could synthesize copies of itself from the free ribonucleotides in the surrounding waters. This first ribozyme probably wasn't very good at its job, and made lots of mistakes. These mistakes, of course, were the first mutations. Like modern mutations, most undoubtedly ruined the catalytic abilities of the "daughter molecules," but a few may have been improvements. Molecular evolution could begin, as ribozymes with increased speed and accuracy of replication reproduced faster, making more and more copies of themselves.

Just as RNA-containing ribosomes are crucial to protein synthesis today, perhaps some early ribozymes began to bind amino acids and catalyze the synthesis of short proteins. Further ribozyme mutations might lead to the formation of the first protein enzymes. At the same time, these protein-synthesizing ribozymes were vulnerable to being cut up by other ribozymes, destroying the information so slowly accumulated over millions of years of random nucleotide substitutions. Further mutations might have allowed certain ribozymes to copy themselves over into DNA molecules that would be safe from the ravages of their fellow ribozymes. In this hypothesis, then, RNA occupies center stage as the first living molecule, with both DNA and proteins evolving later (Fig. 19-3). RNA gradually receded into its present role as an intermediary between DNA and the protein enzymes that carry out most of the work of modern cells.

The First Living Cells

If proteins and lipids are agitated in water, simulating waves beating against ancient shores, hollow structures called **microspheres** are formed (Fig. 19-4).

These hollow balls resemble living cells in several respects. They have a well-defined outer boundary, separating internal contents from the external solution. If the composition of the microsphere is right, a "membrane" is formed that is remarkably similar in appearance to a real cell membrane. Under certain conditions microspheres can absorb more material from the solution ("feed"), grow, and even divide (See Fig. 19-4).

If a microsphere happened to surround the right ribozymes, something very much like a living cell would have been formed. The ribozymes and their protein products would have been protected from free-roaming ribozymes in the primordial soup. Nucleotides and amino acids might have diffused across the membrane and have been used to synthesize new RNA and protein molecules. After sufficient growth, the microsphere may have divided, with a few copies of both ribozymes and proteins becoming incorporated into each daughter microsphere. If so, the first cells would have evolved.

But Did All This Happen?

The above scenario, although plausible and supported by a lot of research findings, is by no means certain. In fact, one of the most striking aspects of origin-of-life research is the great diversity of as-

Figure 19-3 A hypothesis for the origin of life, with RNA as the first "living molecule."

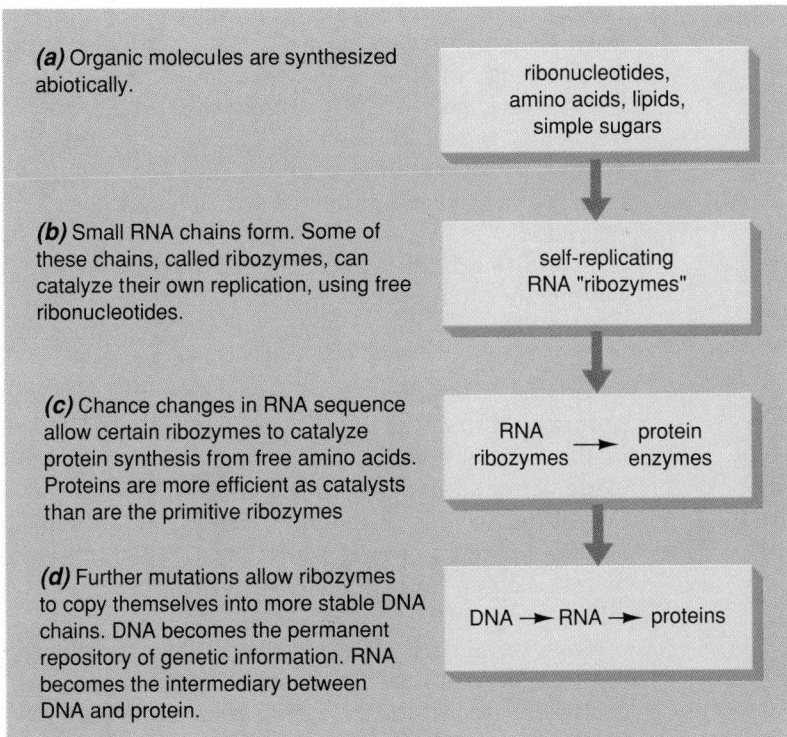

(a) Organic molecules are synthesized abiotically.

ribonucleotides, amino acids, lipids, simple sugars

(b) Small RNA chains form. Some of these chains, called ribozymes, can catalyze their own replication, using free ribonucleotides.

self-replicating RNA "ribozymes"

(c) Chance changes in RNA sequence allow certain ribozymes to catalyze protein synthesis from free amino acids. Proteins are more efficient as catalysts than are the primitive ribozymes

RNA ribozymes → protein enzymes

(d) Further mutations allow ribozymes to copy themselves into more stable DNA chains. DNA becomes the permanent repository of genetic information. RNA becomes the intermediary between DNA and protein.

DNA → RNA → proteins

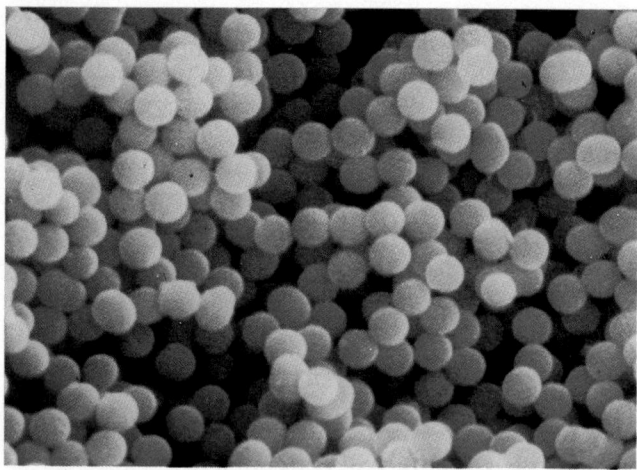

Figure 19-4 Cell-like microspheres can be formed by agitating proteins and lipids in a liquid medium. Such microspheres can take in material from the surrounding solution, grow, and even "reproduce," as these are.

sumptions, experiments, and contradictory hypotheses. In fact, it almost seems as if every researcher has his or her own hypothesis that is at odds with everyone else's (see the Suggested Reading, "In the Beginning . . ." for a taste of the controversies). Some researchers think that life arose in quiet pools; others in the sea; others in moist films on the surfaces of clay or iron pyrite (fool's gold); still others in furiously hot deep-sea vents. A few even argue that life may have arrived on Earth from space. Can any conclusions really be drawn from the research done so far? No one really knows the answer to that question, but we can offer a few observations.

First, the experiments of Miller and others show that amino acids, nucleotides, and other organic molecules would have been formed in abundance on the primordial Earth. Second, no one imagines that prebiotic evolution would have formed any molecules as large or sophisticated as, say, hemoglobin. The first ribozymes, if they existed, may have been only a dozen nucleotides long. Third, ribozymes or other primitive catalysts need not have been very efficient to have accelerated the synthesis of important molecules. The stone tools of early humans weren't as handy as a Swiss army knife, either, but they were better than fingernails. Therefore, to a biologist, the origin of life means the origin of a few small molecules with terribly inefficient catalytic activities, perhaps surrounded by a film vaguely resembling a primitive membrane. Finally, several hundred million years probably elapsed between the appearance of organic molecules and the first primitive cell-like

structures, and spontaneous synthesis must have proceeded over huge areas of the Earth. Given enough time and space, even extremely rare events may in fact happen quite often. Most biologists accept that the origin of life is probably an inevitable consequence of the working of natural laws. We should emphasize, however, that *this proposition is not proven and may never be.* Biologists investigating the origin of life have neither millions of years nor trillions of liters of reaction solutions with which to work!

The Age of Microbes

The fossil record indicates that the earliest living cells arose about 3.5 billion years ago (see Table 19-1). What were these early cells like? The first cells were prokaryotic; that is, their genetic material was not sequestered from the rest of the cell within a membrane-limited nucleus. These cells probably obtained nutrients and energy by absorbing organic molecules from the primordial soup. Since there wasn't any free oxygen in the atmosphere, the cells must have metabolized the organic molecules anaerobically. You will recall from Chapter 8 that anaerobic metabolism yields only small amounts of energy.

As you probably have already recognized, the earliest cells were primitive anaerobic bacteria. As these proto-bacteria multiplied, however, they must have eventually used up the organic molecules produced by prebiotic synthesis. Simpler molecules, such as carbon dioxide and water, were still very abundant, as was energy, in the "dilute" form of sunlight. What was lacking, then, was not *materials* or *energy itself*, but *energetic molecules.* Eventually some cells evolved the ability to use the energy of sunlight to drive the synthesis of their own complex, high-energy molecules from simpler molecules: photosynthesis appeared.

Several kinds of photosynthetic bacteria evolved, but the ones that proved most important in the evolution of life were the cyanobacteria. Their photosynthetic reactions converted water and carbon dioxide to organic compounds, releasing oxygen as a byproduct. At first, the oxygen reacted with iron atoms in the Earth's crust, forming huge deposits of iron oxide. After all the iron turned to rust, the concentration of free oxygen in the atmosphere rose. Chemical analysis of rocks suggests that appreciable amounts of free oxygen appeared in the atmosphere about 2 billion years ago.

Now, oxygen is potentially very dangerous to living things because it reacts with organic molecules, destroying them and releasing their stored energy. The accumulation of oxygen in the atmosphere provided the selective pressure for the next great advance in the Age of Microbes: the ability to use oxygen in metabolism, channeling its destructive power through aerobic respiration to generate useful energy for the cell. Since the amount of energy available to a cell is vastly increased when oxygen is used to metabolize food molecules, aerobic cells had a significant selective advantage.

The Rise of the Eukaryotes

Hordes of bacteria would offer a rich food supply to any organism that could eat them. There are no fossil records of the first predatory cells, but paleobiologists speculate that predation would have evolved quickly, once a suitable prey population appeared. These predators would have been specialized prokaryotic cells, lacking cell walls and consequently able to engulf whole bacteria as prey. According to the most widely accepted hypothesis, these predators were otherwise quite primitive, being capable of neither photosynthesis nor aerobic metabolism. Although they could capture large food particles, namely bacteria, they metabolized them inefficiently. About 1.4 billion years ago, however, one predator probably gave rise to the first eukaryotic cell.

Eukaryotic cells differ from prokaryotic cells in many ways, but perhaps most fundamental are the membrane-bound nucleus containing the genetic material and the inclusion of organelles for energy metabolism, mitochondria and (in plants) chloroplasts. How did these organelles evolve?

The Origin of Mitochondria and Chloroplasts

The **endosymbiotic hypothesis,** championed most forcefully by Lynn Margulis, proposes that cells acquired the precursors of mitochondria and chloroplasts by engulfing certain types of bacteria. Let us suppose that an anaerobic predatory cell captured an aerobic bacterium for food, as it often did, but for some reason failed to digest this particular prey (Fig. 19-5a; see also "A Closer Look at the Evolution of Mitochondria and Chloroplasts" in Chapter 5). The aerobic bacterium remained alive and well. In fact, it was better off than ever, because the cytoplasm of its predator/host was chock full of half-digested food molecules, the remnants of anaerobic metabolism. The aerobe absorbed these molecules and used oxygen to complete their metabolism, gaining enormous amounts of energy as it did so. So abundant were its food resources, and so bountiful its energy production, that the aerobe must have leaked energy, probably as ATP or similar molecules, back into its host's cytoplasm. The mitochondrion had been born. Interestingly, the amoeba *Pelomyxa palustris* is almost a living fossil in this respect (Fig. 19-6). Unlike almost all other eukaryotic cells, it lacks mitochondria; however, a permanent population of aerobic bacteria carries out much the same role.

The predatory cell with its symbiotic bacteria could metabolize food aerobically, gaining a great selective advantage over its anaerobic compatriots. Soon its progeny filled the seas. One of these daughter cells carried out a second coup: it captured a photosynthetic cyanobacterium and similarly failed to digest its prey (see Fig. 19-5a). The cyanobacterium flourished in its new host, and gradually evolved into the first chloroplast. Some modern organisms resemble this hypothetical ancestral condition. A variety of corals, some clams, a few snails, and at least one species of *Paramecium* harbor a permanent collection of algae in their cells (Fig. 19-7). These algae share some of their photosynthetically produced food molecules with the host cells.

Other eukaryotic organelles may have also originated through endosymbiosis. Many paleobiologists believe that cilia, flagella, centrioles, and microtubules may all have evolved from a symbiosis between a spirochete-like bacterium (see Chapter 21) and a primitive eukaryotic cell. Supporting this hypothesis was the discovery in 1990 that centrioles contain their own minute supply of DNA, which some interpret as a remnant of the DNA originally contained within the symbiotic spirochete.

The Origin of the Nucleus

The evolution of the nucleus is more obscure. One possibility is that the plasma membrane folded inward, surrounding the DNA (see Fig. 19-5b). However the nucleus originated, having the DNA sequestered within the nucleus seems to have conferred great advantages in regulating the use of the genetic material.

Multicellularity

Once predation had evolved, increased size became an advantage. A larger cell could more easily engulf a smaller cell, while in turn being more difficult for other predatory cells to ingest. Larger organisms can usually also move faster than small ones, making successful predation and escape more likely. However, enormous single cells have problems. Oxygen and nutrients going in and waste products going out of the cell must diffuse through the plasma membrane. As we pointed out in Chapter 5, the larger a cell becomes, the less surface membrane is available per

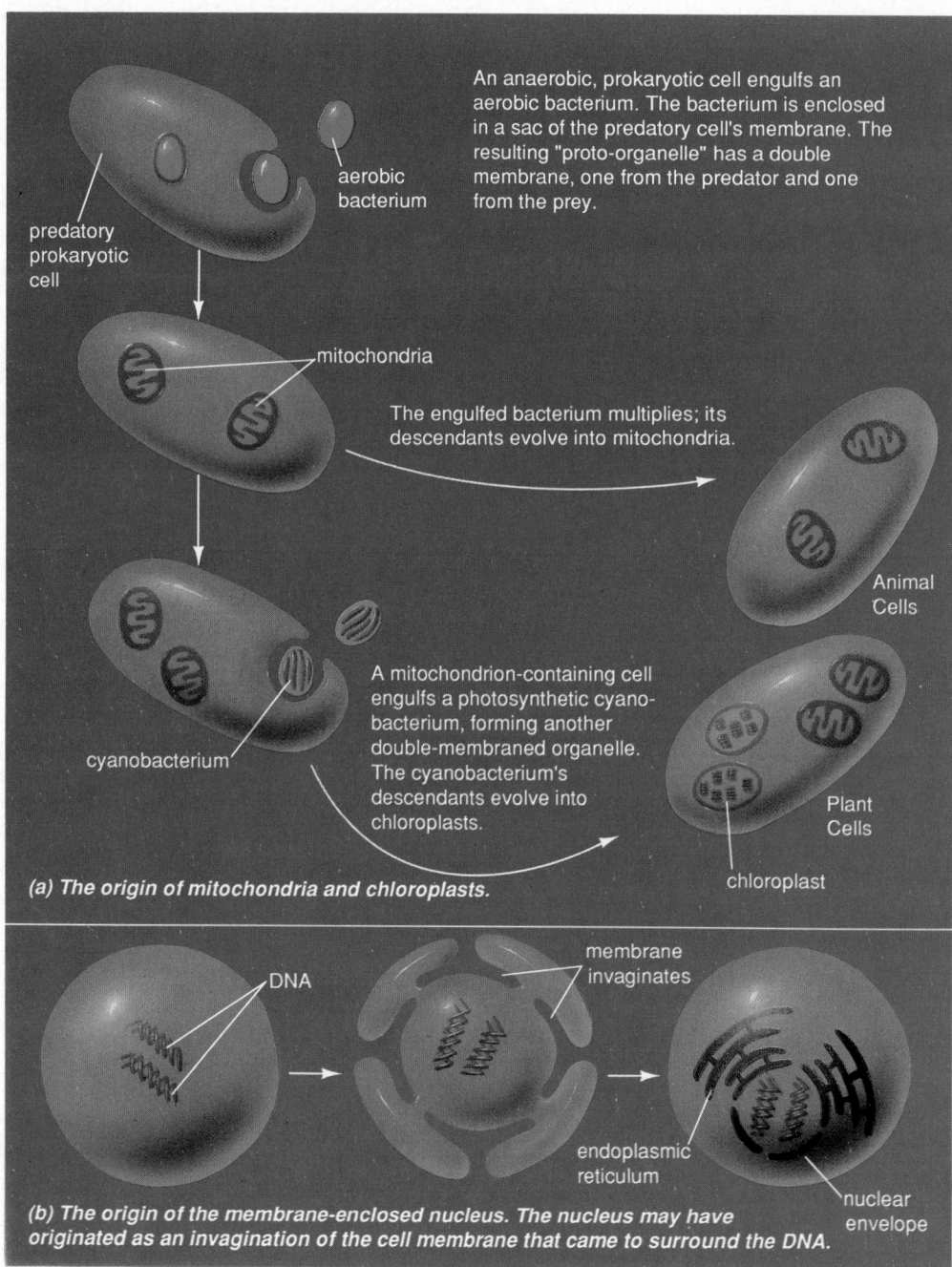

An anaerobic, prokaryotic cell engulfs an aerobic bacterium. The bacterium is enclosed in a sac of the predatory cell's membrane. The resulting "proto-organelle" has a double membrane, one from the predator and one from the prey.

aerobic bacterium

predatory prokaryotic cell

mitochondria

The engulfed bacterium multiplies; its descendants evolve into mitochondria.

Animal Cells

cyanobacterium

A mitochondrion-containing cell engulfs a photosynthetic cyanobacterium, forming another double-membraned organelle. The cyanobacterium's descendants evolve into chloroplasts.

Plant Cells

chloroplast

(a) The origin of mitochondria and chloroplasts.

DNA

membrane invaginates

endoplasmic reticulum

nuclear envelope

(b) The origin of the membrane-enclosed nucleus. The nucleus may have originated as an invagination of the cell membrane that came to surround the DNA.

Figure 19-5 Possible mechanisms for the appearance of organelles in eukaryotic cells.

unit volume of cytoplasm. There are only two ways that an organism larger than a millimeter or so in diameter can survive. First, it can be metabolically sluggish, so that it doesn't need much oxygen or produce much carbon dioxide. This seems to work for certain very large unicellular algae. Alternatively, an organism may be multicellular—that is, it may consist of many small cells packaged into a large unified body.

The fossil record reveals almost nothing about the origin of multicellularity, especially among animals.

The first unicellular eukaryotic fossils occur in rocks about 1.4 billion years old, and the first signs of multicellular animals occur as worm burrows and tracks in rocks 400 million years younger. The intervening animals almost certainly had no skeleton or other hard parts, and left few fossils. Consequently, we may never learn very much about them. Within another 500 million years, however, diverse types of multicellular animals appeared, leaving their fossilized remains in rocks in many parts of the world.

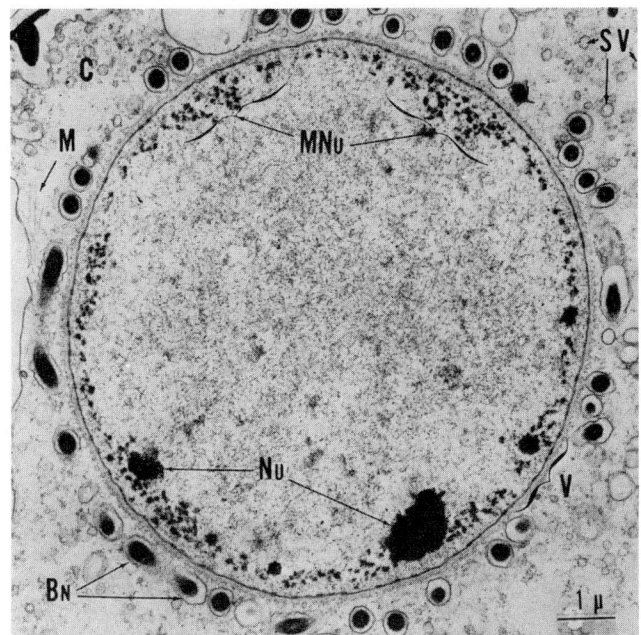

Figure 19-6 A Nucleus of *Pelomyxa palustris*. Unlike most eukaryotes, *Pelomyxa* lacks mitochondria. *Pelomyxa* compensates for this otherwise fatal defect with a resident population of aerobic bacteria surrounding its nucleus (bacteria are labelled BN in the lower left hand corner of the micrograph).

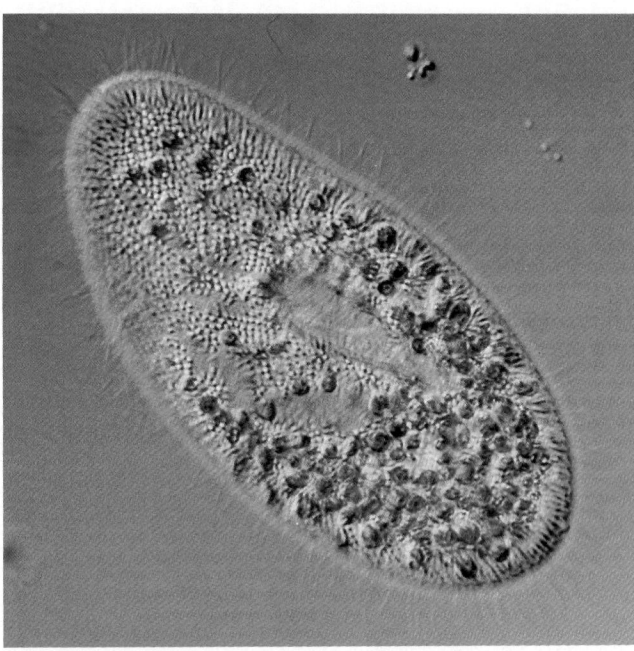

Figure 19-7 Green unicellular algae of the genus *Chlorella* live within the cytoplasm of a *Paramecium*. A similar symbiotic relationship may have given rise to the ancestors of modern chloroplasts.

Multicellular Life in the Sea

The first multicellular organisms almost certainly evolved in the sea.

Plants

Multicellular plants evolved from eukaryotic, chloroplast-containing unicellular organisms. Multicellularity would have provided at least two advantages for plants. First, large, multicellular plants would have been difficult for unicellular predators to swallow. Second, multicellularity and specialization of cells would have conferred the potential for staying in one place in the brightly lit waters of the shoreline, as root-like structures burrowed in sand or clutched onto rocks, while leaf-like structures floated above in the sunlight. The profusion of green, brown, and red algae lining our shores today, some over 200 feet long, are the descendants of these early multicellular algae.

Animals

In a great burst of evolution, a wide variety of invertebrate animals appeared in the sea near the beginning of the Cambrian period, about 600 million years

ago. For animals, one of the advantages of multicellularity is the potential for eating larger prey. This potential was probably first realized in a jellyfish-like animal, shaped vaguely like a vase (Fig. 19-8). A single opening served both as a mouth, to take in food, and as an anus, to expel indigestable remains. However, a half-digested prey filling its gut keeps such an animal from feeding again until it is finished with its first meal. Soon more efficient means of feeding evolved, employing a separate mouth and anus, found today in almost all animals (Fig. 19-9). With this design, the animal can feed more or less continuously, as earthworms and sea cucumbers do today.

Coevolution of predator and prey rapidly led to increased sophistication in many kinds of animals, ranging from the mud-skimming, armored trilobites, to their predators, the ammonites and the chambered nautilus, which still survives almost unchanged in deep Pacific waters (Fig. 19-10). A major trend at this time was toward greater mobility. Predators often need to travel over wide areas in search of suitable prey, while speedy escape is also an advantage for prey. Locomotion is usually accomplished by the contraction of muscles that move body parts through their attachments to some sort of skeleton. The dominant invertebrates at this time possessed either an internal hydrostatic skeleton, much like a water-filled

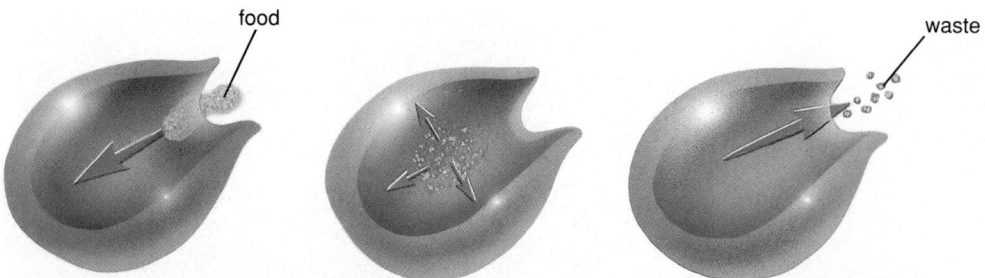

Figure 19-8 The earliest animals may have possessed a digestive system with only one opening to the outside world. Consequently, feeding would have had to be periodic. If one piece of prey was already in the digestive tract, a second piece would have to be stuffed on top. When the indigestible remnants of the first were expelled, they would force out the not-yet-digested second piece as well.

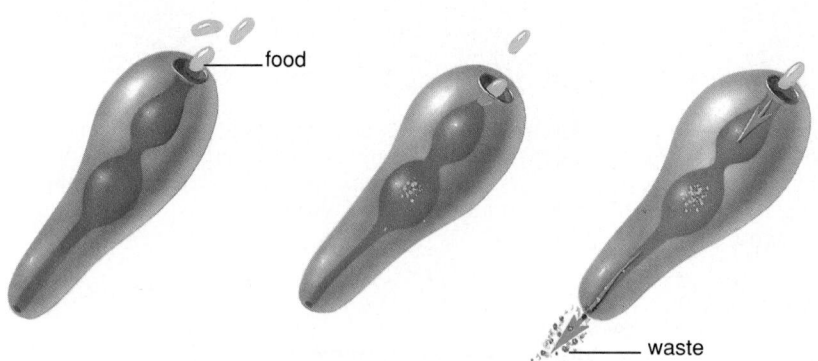

Figure 19-9 A major advance in feeding is to have separate openings for ingesting food and for ejecting the indigestible residue. In principle, this allows more or less continuous feeding.

tube (worms), or an external skeleton covering the body (arthropods such as trilobites). Rapid locomotion, of course, is of little use if an animal can't tell where it is going or what to do once it gets there. Greater sensory capabilities and more sophisticated nervous systems evolved along with locomotor abilities. Senses for detecting touch, chemicals, and light became highly developed. The senses were usually concentrated in the head end of the animal, along with a nervous system capable of handling the sensory information and directing appropriate behaviors.

About 500 million years ago, an entirely new type of animal appeared, possessing an internal skeleton. For a hundred million years these were inconspicuous members of the ocean community, but by 400 million years ago some had evolved into fishes. By and large, the fishes proved to be faster than their invertebrate compatriots, with more acute senses and larger brains. Eventually, they became the dominant predators of the open seas.

The Invasion of the Land

Between 600 and 400 million years ago, both plants and animals evolved greatly, but they remained cushioned by the equanimity of the sea. Life in the ocean provides buoyant support against gravity and ready access to life-sustaining water. Reproduction is also simple in the sea. At some point in their life cycles, both plants and animals produce sex cells that must fuse to form the beginnings of a new generation. Sea-dwelling organisms usually have mobile sperm and/or eggs that swim to each other through the water.

In contrast, on land an organism must bear its weight against the crushing force of gravity; it must

(a)

(b)

(c)

(d)

Figure 19-10 (a) Characteristic life of the oceans during the Silurian period. Among the most common fossils from that time are the trilobite **(b)** and its predators the ammonites **(c)** and the nautiloids. Although **(d)** illustrates a living *Nautilus*, the Silurian nautiloids were very similar in structure, showing that a successful body plan may exist virtually unchanged for hundreds of millions of years.

find adequate water; and to reproduce, it must ensure that its gametes, particularly the sperm, are protected from drying out. Nevertheless, the land offered great potential, particularly for plants. Water strongly absorbs light; even in the clearest water, photosynthesis is limited to the upper couple of hundred meters, and usually much less. Out of the water, the sun is dazzlingly bright, permitting rapid photosynthesis. Sea water also tends to be low in certain nutrients, particularly nitrogen and phosphorus. By comparison, terrestrial soils are rich storehouses of nutrients. Finally, by this time the sea swarmed with plant-eating animals. Since the land was devoid of animal life, plants that colonized the land would have had no predators.

Land Plants

In moist soils near the shore, a few small green algae began to grow, taking advantage of the sunlight and nutrients. They didn't have large bodies to support against the force of gravity, and living right in the film of water on the soil, they could easily obtain water. About 400 million years ago, some of these algae gave rise to the first multicellular land plants. Initially simple, low-growing forms, land plants rapidly evolved solutions to two of the main difficulties of plant life on land: obtaining and conserving water, and staying upright despite gravity and winds. Root-like structures delved into the soil, mining water and minerals, while waterproof coatings on the above-

Figure 19-11 A reconstruction of a swamp forest of the Carboniferous period. The tree-like plants are tree-ferns and giant club mosses, both now mostly extinct. Note the dragonfly at bottom center; some Carboniferous dragonflies had wing spans in excess of half a meter!

ground parts reduced water loss by evaporation. Specialized cells formed tubes called vascular tissues to conduct water from roots to leaves. Extra-thick walls surrounding certain cells enabled stems to stand erect.

Reproduction on Land

Reproduction out of water seems to have been harder to solve. We will examine plant reproduction in more detail in Chapters 24 and 28; the important point to note here is that, like animals, plants produce sperm and eggs. Primitive marine plants have swimming sperm and sometimes swimming eggs as well. The first land plants retained swimming sperm, which restricted them to swamps and marshes where the sperm and eggs could be released into the water, or to areas with abundant rainfall, where the ground would occasionally be covered with water.

The Rise of the Conifers

This strategy detailed above sufficed for millions of years. During the Carboniferous period, 360 to 286 million years ago, the climate was warm and moist, and great stretches of the land were covered with forests of giant tree ferns and club mosses that produced swimming sperm (see Fig. 19-11). The coal we mine

today is the fossilized remains of these forests. Meanwhile, some plants inhabiting drier regions evolved reproductive strategies that no longer depended on films of water. In these protoconifers, the eggs were retained in the parent plant, while the sperm were encased in drought-resistant pollen grains that blew on the wind from plant to plant. Landing on a female cone near the egg, the pollen released sperm cells directly into living tissue, eliminating the need for a surface film of water. About 250 million years ago, mountains rose, swamps drained, and the moist climate dried up. The swimming sperm of tree ferns and giant club mosses doomed most of them to extinction, while the conifers, which did not depend on water for reproduction, flourished and spread.

The Rise of Flowering Plants

About 130 million years ago, the flowering plants appeared, having evolved from a group of conifer-like plants. The initial advantage of the flowering plants seems to have been pollination by insects. The conifers are wind-pollinated, which demands an enormous pollen production, since the vast majority of pollen grains fail to reach their target. Flower pollination by insects wastes far less pollen. Flowering plants also evolved other advantages, including more rapid reproduction and, in some cases, much more rapid growth. Today, flowering plants dominate the land, except in cold boreal regions where conifers still prevail.

Land Animals

Soon after land plants evolved, providing potential food sources for animals, arthropods (probably early relatives of scorpions) emerged from the sea. Why arthropods? The answer seems to be that they were **preadapted** for land life: that is, they already possessed structures, evolved under totally different selective pressures, that were suited to life on land. Foremost among these was the exoskeleton, a hard covering surrounding the body such as the shell of a lobster or crab. Exoskeletons are both waterproof and strong enough to bear up a small animal under the stresses of gravity.

Land animals encounter another difficulty, also relatively easily solved by arthropods: breathing. Respiratory surfaces must be kept moist, and this is difficult in the dry air. Some arthropods, such as land crabs and spiders, evolved what amounts to an internal gill, kept moist within a waterproof sac (see Chapter 22). The insects developed tracheae, small branching tubes directly penetrating the body, with adjustable openings in the exoskeleton leading to the outside air.

For millions of years, arthropods had the land and its plants to themselves, and for tens of millions of years more, they were the dominant animals. Dragonflies with a wingspan of 70 centimeters flew among the Carboniferous tree ferns (See Fig. 19-11), while 2-meter-long millipedes munched their way across the swampy forest floor. Eventually, however, the arthropods' splendid isolation came to an end.

The Rise of the Amphibians

About 400 million years ago, a group of fishes called the lobefins appeared, probably in fresh water. Lobefins had two important preadaptations to land-life: stout, fleshy fins with which they crawled about on the bottoms of shallow, quiet waters, and an outpouching of the digestive tract that could be filled with air, like a primitive lung (Fig. 19-12). In one group of lobefins, the lung evolved into a swim bladder, with which they could regulate their buoyancy and remain suspended in the water without active exertion. Many of these migrated back to the sea, where they evolved into the modern bony fishes. Another group of lobefins colonized very shallow ponds and streams, which shrank during droughts and often became oxygen-poor. By taking air into

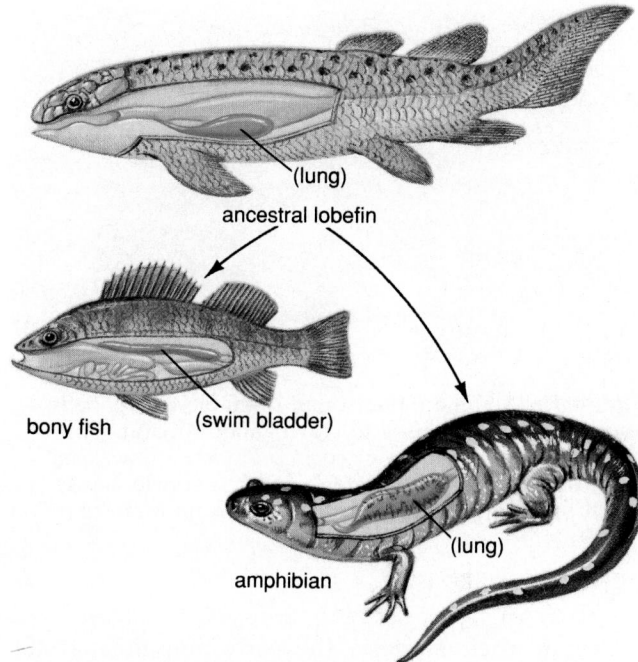

Figure 19-12 A group of primitive fish, called the *lobefins*, gave rise both to the bony fish and the amphibians. Lobefins had a pair of lungs, which arose as outpouchings of the digestive tract. Lobefins probably used their lungs for breathing air when the pools in which they lived became stagnant and foul. In one group of lobefins, one lung evolved into a swim bladder, used in bony fishes to regulate buoyancy. The other lung disappeared. Another group of lobefins became further adapted for air-breathing, evolving more complex, efficient lungs and sturdier fins for walking on land. These were the ancestors of the amphibians.

their lungs, these lobefins could obtain oxygen anyway. Some of their descendants began to use their fins to crawl from pond to pond in search of prey or water, as some modern fish can do today (Fig. 19-13).

As the arthropods discovered previously, the land is a rich source of food. Feeding on land and moving from pool to pool favored the evolution of fish that could stay out of water for longer periods and that could move about more effectively on land. With improvements in lungs and legs, the amphibians evolved from lobefins, first appearing in the fossil record about 350 million years ago. If an amphibian could have thought about such things, it would have thought that the Carboniferous swamp forests were heaven itself: no predators to speak of, abundant prey, and a warm, moist climate. As with the insects and millipedes, some amphibians evolved gigantic size, including salamanders over 3 meters long.

Figure 19-13 Some modern fish resemble their lobefin ancestors in their ability to crawl about on land. The walking catfish was imported into Florida a few years ago, escaped into local waters, and has spread across much of southern Florida by walking from pond to pond.

Despite their success, the early amphibians still lacked complete adaptation to life on land. Their lungs were simple sacs without very much surface area, so they had to obtain some of their oxygen through their skins. Therefore, the skin had to be kept moist, restricting amphibians to swampy habitats where they wouldn't dry out. Further, amphibians deposit their sperm and eggs in water. As with the tree ferns and club mosses, when the Carboniferous climate turned dry at the beginning of the Permian period, amphibians were in trouble.

The Rise of the Reptiles

Meanwhile, just as the conifers had been evolving on the fringes of the swamp forests, so too a group of amphibians were evolving adaptations to drier conditions. These became the reptiles, which achieved four great advances over the amphibians. First, they evolved internal fertilization: the reptilian female provides the watery environment for the sperm inside her reproductive tract. Thus, sperm transfer could occur on land, without venturing back to the dangerous swamps full of fish and amphibian predators. Second, they developed waterproof eggs enclosing their own supply of water for the developing embryo (see Chapter 22), again providing freedom from the swamps. Third, the proto-reptiles developed scaly, waterproof skin. Finally, accompanying the evolution of waterproof skin came improved lungs that could provide the entire oxygen supply for an active animal. As the climate dried in the Permian, reptiles became the dominant vertebrate land fauna, relegating the amphibians to swampy backwaters where most remain today.

A few tens of millions of years later, the climate returned to more moist and equable conditions, providing for lush plant growth. Once again, gigantism became the rage, as certain families of reptiles evolved into the dinosaurs (Fig. 19-14). These were among the most successful animals ever, if we consider length of dominance as a measure of success. They flourished for over a hundred million years, until about 65 million years ago when the last dinosaurs became extinct. No one is certain why they died, but a climate change, perhaps initiated by a gigantic meteorite impact, seems to have been the final blow (see Chapter 17). With less luxuriant plant growth, the herbivorous dinosaurs could not find enough food. Without large herbivores as prey, the carnivores were doomed too.

Even during the age of dinosaurs, many reptiles remained quite small. One major difficulty faced by small reptiles is keeping a high body temperature. Being active on land seems to require a rather warm body, maximizing the efficiency of the nervous system and muscles. However, a warm body loses heat to the environment unless the air is also warm. Small reptiles have a relatively large surface area through which heat is lost and a relatively small internal volume in which metabolic heat is generated. If the body is warmed metabolically when the air is cool, then an enormous amount of food must be consumed to provide sufficient energy. Apparently the food requirement is too high, for the naked-skinned small reptiles have a fairly low metabolism, not enough to keep their bodies warm in cool air, especially at night when there is no sun for radiant warmth. Two groups of small reptiles independently evolved insulation that minimizes heat loss: one group evolved feathers, while another group evolved hair.

The Rise of Birds and Mammals

In the protobirds, insulating feathers retained body heat. Consequently, these animals could be active in cooler habitats and during the night, when their scaly relatives became sluggish. Later, some protobirds developed longer, stronger feathers on their forelimbs, perhaps allowing them to glide from trees or to assist in jumping after insect prey. From this point, the evolution of flight became possible.

The hair evolved by the protomammals also provided insulation. Unlike the birds, which retained the reptilian habit of egg-laying, mammals evolved live birth and the ability to feed their young with secretions of the mammary glands. Since these structures do not fossilize, we may never know when the uterus, mammary glands, and hair first appeared, or what their intermediate forms looked like.

The earliest mammals were small creatures, proba-

Figure 19-14 A reconstruction of a Jurassic swamp. In the foreground we see, standing in the swamp, the gigantic *Apatosaurus*, 20 meters long and weighing over 30 tons. On the right, a carnivorous *Allosaurus* uses its meter-long jaws to tear flesh from its prey. In the background grazes the 10-ton *Stegosaurus*, an astoundingly dim-witted herbivore with a brain the size of a walnut. *Stegosaurus* had a swelling of its spinal cord in the hip region that formed a second "brain" of sorts, 20 times larger than the one inside its skull. The plates on its back probably functioned in temperature regulation. Depending on air temperature, the animal's orientation to the sun, and the amount of blood flow across the surface of the plates, they could be used as radiators to cool the blood or as solar panels to pick up warmth.

bly living in trees and being active mostly at night. When the dinosaurs became extinct, the mammals radiated out into the vast array of modern forms. While some stayed small and nocturnal, eating mostly seeds and insects, others evolved larger size and different habits, colonizing the habitats left empty by the extinction of the dinosaurs. One group remained in the trees, giving rise to the primates.

Human Evolution

Fossils of primates (lemurs, monkeys, apes, and humans) are relatively rare compared with those of many other animals. There are at least three reasons for this. First, most primates did not live in habitats that readily preserve fossils, such as swamps and shallow lagoons. Second, until recently primates were fairly small. Therefore predators and scavengers would be more likely to break up their bones into unrecognizable fragments than they would the bones of a bison or dinosaur. Third, ancestral primates may have had small populations, thus providing less material for fossilization in the first place.

Humans are intensely interested in their own origin and evolution, especially in trying to determine what conditions led to the evolution of the gigantic human brain. This obsession has led to sweeping speculations despite an often skimpy fossil record.

Therefore, while the outline of human evolution that we will present is a synthesis of current thought on the subject, it is by no means as well understood as, say, the genetic code. Paleontologists disagree about the interpretation of the fossil evidence, and many ideas may have to be revised as new fossils are found.

Primate Evolution

The first protoprimates were the insect-eating tree shrews, whose fossils are found in rocks about 80 million years old. Nimble, probably nocturnal animals, tree shrews were smaller than all but the tiniest modern primates. Over the next 50 million years, the descendants of the tree shrews evolved forms similar to the modern tarsiers, lemurs, and monkeys (Fig. 19-15). These primates stayed in the trees. Some remained nocturnal, but many become active during the day. Primates evolved several adaptations for life in the trees, feeding on fruits and leaves.

Grasping Hands

The tree shrews already possessed hand-like paws for holding on to small branches, and the primates further refined these appendages. Most primates are much larger than tree shrews, which means that only relatively large tree limbs could bear their weight. Long, grasping fingers that could wrap around and hold larger limbs made life in the trees a little safer.

The primate line that led to humans evolved hands with the ability to perform both delicate maneuvers using a *precision grip* (manipulating small objects, including writing and sewing) and powerful actions using a *power grip* (swinging a club, thrusting with a spear).

Binocular, Color Vision

One of the earliest primate adaptations seems to have been large, forward-facing eyes (see Fig. 19-15). Jumping from branch to branch is risky business unless an animal can accurately judge where the next branch is located. Accurate depth perception was made possible by binocular vision provided by forward-facing eyes with overlapping fields of view. Another adaptation was color vision. We cannot, of course, tell if a fossil animal had color vision, but modern primates have excellent color vision, and it seems reasonable to assume that earlier primates did too. Many primates feed on fruit, and color vision helps in detecting ripe fruit among a welter of green leaves.

Large Brain

Primates have brains that are larger, relative to their body size, than almost all other animals. No one really knows for certain what selective forces favored the evolution of large brains. However, it seems reasonable that controlling and coordinating binocular, color vision, rapid locomotion through trees, and dextrous movements of the hands would be facilitated by increased brain power. Most primates also have fairly complex social systems, which probably would require relatively high intelligence. If sociality promoted increased survival and reproduction, then this would provide selective pressures for the evolution of larger brains.

Hominid Evolution

Between 20 and 30 million years ago, in the moist tropical forests of Africa, a group of primates called the Dryopithecines diverged from the monkey line. The Dryopithecines appear to be ancestral to the hominids (humans and their fossil relatives) and pongids (the great apes). Around 18 million years ago, global climatic cooling began to shrink the vast expanses of forest, splitting up the woodlands into isolated islands dotted upon a sea of grassland. Diversification of habitat and isolation of small populations led to the diversification of Dryopithecines. One of these, perhaps *Pronconsul africanus*, gave rise to the later hominids and pongids.

(a)   **(b)**

(c)

Figure 19-15 Representative primates include the tarsier **(a)**, lemur **(b)**, and liontail macaque **(c)**. Note that all have relatively flat faces, with forward-looking eyes providing binocular vision. All also have color vision and grasping hands. These features served as preadaptations for tool and weapon use by early humans.

Essay: The Search for Eve

How do evolutionary biologists know that *Homo habilis* gave rise to *H. erectus*, which gave rise to *H. sapiens*, and that all of this happened in Africa? Well, they don't, and in fact not all of them agree with this sequence. Fossils, particularly hominid fossils, are rare. Further, there is no way of knowing whether any particular fossil hominid was a direct ancestor of modern humans. Is there another way of studying human evolution? There is—by studying the genes of modern humans. You may recall from Chapter 16 that there are various ways of analyzing DNA to estimate the degree of evolutionary relatedness among organisms. The principle behind these studies is that the DNA of closely related organisms will be more similar than the DNA of more distantly related organisms. DNA analysis is becoming widely used in evaluating relationships among species. In the mid-1980s, Allan Wilson and his colleagues at the University of California at Berkeley decided to apply DNA analysis to members of a single species: *H. sapiens*. Specifically, Wilson studied the DNA found in mitochondria.

Why mitochondria? Mitochondria are remarkable organelles. They contain their own store of DNA, they replicate, and they are passed from parent cell to daughter cells during cell division (see Chapter 5). All the mitochondria in every cell in your body are direct descendants of the mitochondria that were originally present in the fertilized egg from which you developed. But where did *those* mitochondria come from? It turns out that, in humans and many, but not all, other animals, *all of the mitochondria in the fertilized egg come from the mother, via the egg; none come from the father, via the sperm*. Sperm do have mitochondria, but they do not enter the egg cell during fertilization. This leads to a remarkable insight, with important implications for evolutionary biology: barring mutations, the DNA in your mitochondria is identical to the DNA in your mother's mitochondria, which is identical to the DNA in *her* mother's mitochondria, which is identical to the DNA in *her* mother's mitochondria, and so on back into prehistory. Theoretically, differences among mitochondrial DNA from modern humans could provide information about relationships among racial groups and about the likely amount of time that has passed since they shared a common (female) ancestor.

WIlson and his colleagues took samples of mitochondrial DNA from people native to Europe, Asia, Africa, New Guinea, and Australia. They then constructed a likely evolutionary tree based on DNA similarities. Although the methodology is complex, the conclusion is simple (and controversial): *all living humans are descended from a single woman who lived in Africa about 200,000 years ago.* Naturally, this hypothetical woman was soon dubbed Eve, "the mother of us all," in Wilson's words.

How can this be, and, if it is true, what does it mean? What happened to all the other people who must have lived in the same population as Eve, or for that matter, what happened to the father of her children? Remember, mitochondria are inherited strictly from the mother. Therefore, from the perspective of mitochondrial DNA, fathers don't count (more on that later). Further, a woman can pass mitochondrial DNA on to future generations only if she has daughters: her sons receive her mitochondria, but do not pass them on to the next generation. Thus any of Eve's female neighbors who bore only sons, or whose daughters, granddaughters, and so on, bore only sons, disappeared from mitochondrial history. Wilson's analysis indicates that, in fact, childlessness or son-only families occurred in the descendants of every other woman alive at Eve's time.

What does this mean genetically? Perhaps less than you might think. Don't forget that this analysis considers only mitochondrial DNA. Nothing whatever can be concluded about nuclear DNA, which is, after all, where the vast majority of our genes reside. For all anyone knows, Eve may have borne children by several different men. Her immediate female descendants may have borne children by men who were outside her social group. In principle, her remote descendants may have borne children fathered by men who evolved on entirely different continents (whether modern humans evolved solely in Africa or simultaneously on several continents is a hot controversy; see "Argument Over a Woman" in the Suggested Readings). Therefore, even if the "Eve hypothesis" is completely correct, it does *not* mean that all our genes are descended from Eve's. It is even possible that, with certain patterns of independent assortment of chromosomes during meiosis in the intervening millenia, virtually none of our genes are Eve's, except for those in our mitochondria.

Nevertheless, the notion of a single Eve, "the mother of us all," is emotionally appealing. Could there be a comparable Adam? Before you read on, think back to Chapter 11 a minute: what genetic material is passed on only from father to son? Right—the Y chromosome. The Y chromosome pairs with the X chromosome during meiosis, but exchanges little, if any, DNA with the X chromosome. Therefore, several researchers are using DNA analysis to try to find the "Y-chromosome Adam," a sort of genetic counterpart to the "mitochondrial Eve." Stay tuned.

The Australopithecines

A large gap ensues in the fossil record, until the first true hominids appear about 4 million years ago. These fossils are collectively called the Australopithecines ("southern ape," after their original discovery site in southern Africa; Fig. 19-16). Footprints almost 4 million years old, discovered in Tanzania by Mary Leakey, show that the Australopithecines could walk upright. However, the earliest Australopithecines had quite short legs—shorter, for their height, than any modern humans—so it is likely that they were not as good at walking upright as their later, longer-legged descendants became. The brains of Australopithecines were fairly large, although still much smaller than those of modern humans.

According to some interpretations, one early Australopithecine, called *Australopithecus afarensis* (the famous "Lucy" fossil was a member of *A. afarensis*) gave rise to all the other hominids. Later Australopithecines split into at least two distinct forms: the small, omnivorous *A. afarensis* and *A. africanus*, and the large, herbivorous *A. robustus* and *A. boisei*. Most authorities agree that one of the small "gracile" forms was the ancestor of modern hominids, including humans, while all of the large "robust" forms became extinct, without leaving descendants, about a million years ago.

Homo habilis

By about 2 million years ago, certain African populations of the smaller Australopithecines (probably *A. afarensis*) seem to have given rise to a new form, *Homo habilis* (see Fig. 19-16). *Homo habilis* had a considerably larger body and brain than the Australopithecines. *Homo habilis* was probably the first hominid that used stone and bone tools, mostly crude weapons of various sorts (Fig. 19-17). In a relatively short time, *H. habilis* spread throughout Europe, Asia, and Africa.

Homo erectus

About 1.8 million years ago, *H. erectus* appeared, probably a descendant of an African population of *H. habilis*. The brain of *H. erectus* was as large as the smallest modern adult human brains (around 1000 cubic centimeters). The face was notably different from that of modern humans (see Fig. 19-16), featuring large brow ridges, a slightly protruding face, and no chin (the protruding tip of the lower jaw).

Behaviorally, *H. erectus* seems to have been more advanced than *H. habilis*. *Homo erectus* fashioned sophisticated stone tools, ranging from hand axes used for cutting and chopping to points probably used on spears (see Fig. 19-17). These weapons suggest that *H. erectus* ate animal food, probably both hunting on

their own and scavenging for the remains of prey killed by lions and other predators. *Homo erectus*, like *H. habilis*, spread throughout most of Europe and Asia—Peking Man and Java Man were specimens of *H. erectus*. In China, evidence from 500,000 years ago suggest that *H. erectus* used fire, although for what purpose (to cook food, keep warm, or ward off predators) no one knows.

Homo sapiens

The first fossils of our own species, *H. sapiens*, date from about 200,000 years ago. Continuing the trend of our earlier ancestors, *H. sapiens* probably evolved in Africa (see the essay, "The Search for Eve"). The earliest fossils are fragmentary, but beginning about 100,000 years ago the Neanderthal variety left many complete fossils, found throughout the Old World. Contrary to the popular image of the hulking, stoop-shouldered "cave man," Neanderthals were actually quite similar to modern humans in many ways. It is true that they were very heavily muscled, but Neanderthals walked fully erect, were dextrous enough to manufacture finely crafted stone tools, and had brains, on the average, slightly larger than those of modern humans. Although many of the European fossils show heavy brow ridges and a broad, flat skull, others, particularly from the Near East, were remarkably like ourselves.

Neanderthal remains show evidence of modern behaviors, too, particularly ritualistic burial ceremonies (Fig. 19-18). Neanderthal skeletons have been discovered in clearly marked burial sites surrounded with stones, and often including offerings of flowers, bear skulls, and food. Altars found at other Neanderthal sites were probably used in "religious" rites associated with a bear cult.

Finally, about 90,000 years ago, fully modern humans appeared. Although the oldest specimens were discovered in Israel in 1987, these anatomically modern humans are called Cro-Magnon, after the district in France in which their remains were first discovered. Cro-Magnons had domed heads, smooth brows, and prominent chins. Their tools were precision instruments not very different from the stone tools used until recently in many parts of the world.

Figure 19-16 A possible evolutionary tree for human beings, with skills and facial reconstructions of representative specimens. Although many paleontologists consider this to be the most likely human family tree, there are several alternative interpretations of the known hominid fossils, each with its strong advocates.

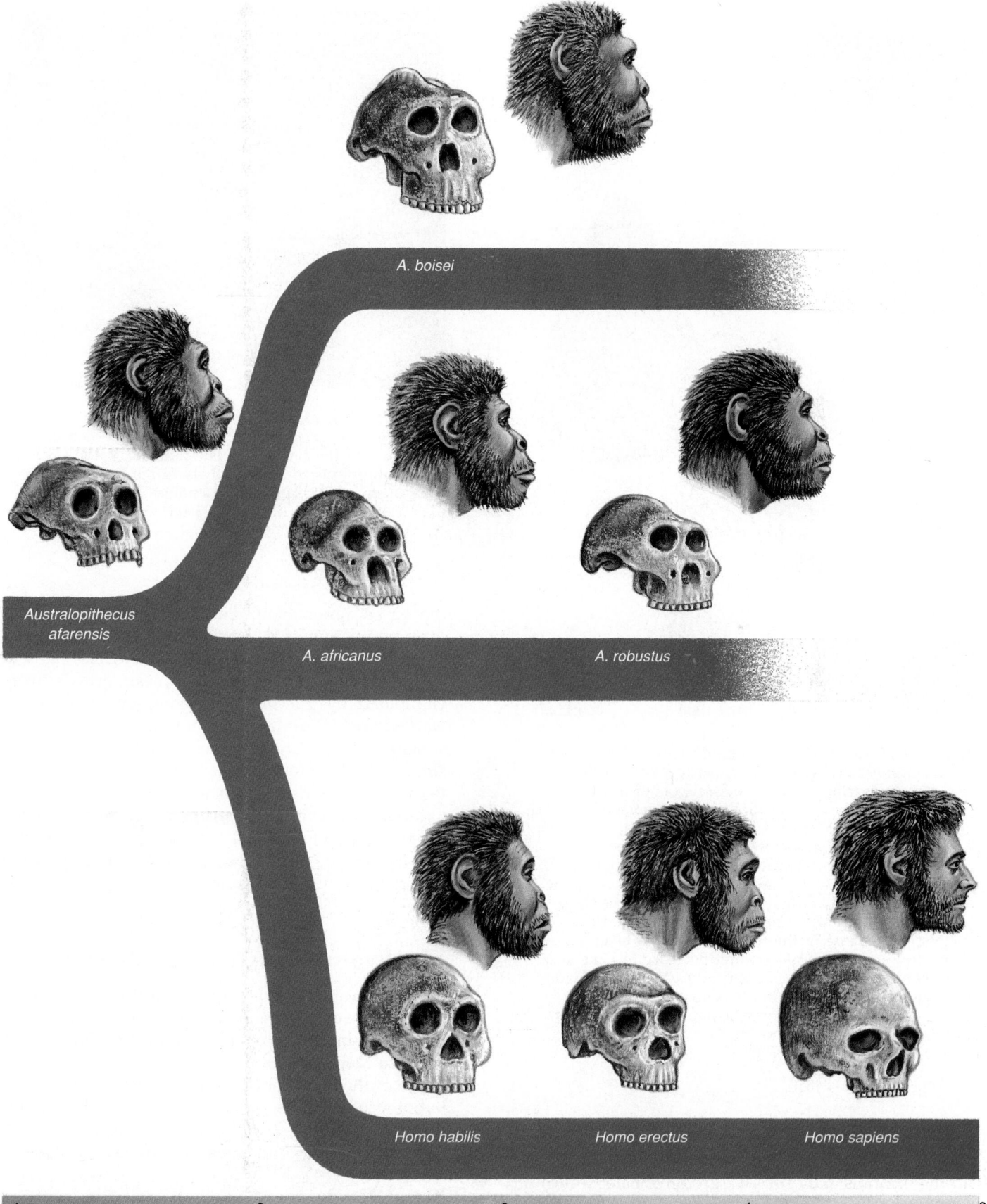

A. boisei

Australopithecus
afarensis

A. africanus

A. robustus

Homo habilis

Homo erectus

Homo sapiens

| 4 | 3 | 2 | 1 | 0 |

millions of years ago

Homo habilis

Homo erectus

Homo sapiens (Neanderthal)

Figure 19-17 Representative hominid tools. *Homo habilis* produced only fairly crude chopping tools called hand-axes, usually unchipped on one side to hold in the hand. *Homo erectus* manufactured much finer tools. Since the tools were often sharp all the way around the stone, at least some of these blades were probably tied to spears rather than held in the hand. Neanderthal tools were works of art, with extremely sharp edges made by flaking off tiny bits of stone. In comparing these *H. habilis*, *H. erectus*, and *H. sapiens* weapons, note the progressive increase in the number of flakes taken off the blades, and the corresponding decrease in flake size. Smaller, more numerous flakes produce a sharper blade, and suggest more insight into tool making (perhaps passed down culturally from experienced tool-makers to apprentices), more patience, finer control of hand movements, or perhaps all three.

Where did the Cro-Magnons come from and where did the Neanderthals go? One group of paleoanthropologists believes that Cro-Magnon evolved from the modern-looking Neanderthals of the Near East, and spread into Europe from there. The contrary view is

Figure 19-18 A reconstruction of a Neanderthal bear ceremony. Power and ferocity, and perhaps their ability to rise up on their hind legs like humans, made cave bears cult objects to Neanderthals. Some Neanderthals were buried with elaborate ceremonies, with bear skulls placed on the grave of the deceased.

that Cro-Magnon evolved independently, from a separate population of *H. erectus*. Both Neanderthal and Cro-Magnon probably coexisted for a while, particularly in Europe. Whether the Neanderthals were eliminated by interbreeding, competition, or open warfare with Cro-Magnon is unknown.

Behaviorally, Cro-Magnon seems to have been similar to, if more sophisticated than, Neanderthal. Perhaps the most remarkable accomplishment of Cro-Magnon is the magnificent art left in caves such as Altamira in Spain and Lascaux in France. Again, no one knows exactly why these paintings were made, but they attest to minds fully as human as our own.

The Evolution of Human Behavior

Perhaps the most contentious subject in human evolution is the development of human behavior. Except in rare instances, such as the Neanderthal burials and the Cro-Magnon paintings, there is no direct evidence of the behavior of prehistoric hominids. A few hypotheses have been offered by biologists and anthropologists, based on the fossil record and the behavior of modern humans and animals.

Brain Development

The fossil record shows that the development of truly huge brains occurred within the last couple of million years. What were the selective advantages of large brains? This subject continues to stir controversy among paleoanthropologists. However, the enlarging brain may have been selected both because it provided for improved hand–eye coordination and because it facilitated complex social interactions.

As the early hominids such as *Australopithecus* descended from the trees into the savanna, they began to walk upright. Bipedal locomotion allowed them to carry things in their hands as they walked. Hominid fossils show shoulder joints capable of powerful throwing motions and an opposable thumb that would enhance the ability to manipulate objects with great dexterity. These early hominids probably could see well and judge depth accurately with binocular vision. The brain expansion that occurred in the Australopithecines may have been at least partly related to integration of visual input and control of hand–arm movements.

Homo erectus, and perhaps *H. habilis* before it, was social. So, of course, are many monkeys and apes. The later hominids, however, seem to have engaged in a new type of social activity: cooperative scavenging and hunting. *H. erectus*, in particular, seems to have hunted extremely large game, which calls for cooperation in all phases of the hunt. Some paleoanthropologists believe that selection favored larger and more powerful brains as an adaptation for success in cooperative hunting. (Lest you become too impressed with such speculation, remember that lions and wolves also hunt cooperatively, yet are not noticeably more intelligent than their relatives, such as leopards and coyotes, who do not.)

Cultural Evolution

Isaac Newton once wrote, "If I have seen further, it is because I stand on the shoulders of giants." All humanity stands on the shoulders of giants, not only pre-Newtonian physicists, but Cro-Magnons, Neanderthals, the hairy reptiles of 100 million years ago, the first lobefinned fishes to crawl out of the swamps, and the first cell. We are the product of 3.5 billion years of biological evolution.

In recent millenia, however, biological evolution of humans has been outpaced by our **cultural evolution:** learned behaviors passed down from previous generations. Most human behaviors are strongly influenced by learning. The giants that loom tallest in our heritage are not necessarily our biological ancestors, but the people—living and dead—who shaped our behavior.

What is the relationship between biological and cultural evolution? How much of our behavior is cultural and how much is genetically determined? Some people believe that human nature is almost infinitely malleable; that biological evolution endowed us with a mind that can be molded into whatever form society wishes. Others maintain that millions of years of natural selection must have left us with some genetic tendencies to perform certain types of behaviors.

Which position, if either, is correct? It is likely that the truth lies somewhere in the middle ground, that most human behavior is the product both of genetic influences and extensive learning. First of all, no reasonable person doubts that human behavior is tremendously influenced by culture. We learn at least part of every behavior we perform. Second, humans do display some obviously instinctive behaviors, including suckling by infants, smiling, and flashing the eyebrows in greeting (see Chapter 42). Third, in recent years biologists have found that many forms of aberrant behavior are strongly influenced by heredity, including schizophrenia, manic depression, and Tourette's syndrome. If abnormal behavior can be influenced by defective genes, then it seems likely that normal behavior is influenced by the functioning of normal, nondefective genes. However, few, if any, behaviors, normal or abnormal, are completely controlled by heredity. Virtually all human behaviors can be enhanced, suppressed, or modified by culture. The roles of culture and heredity in human behavior will be explored further in Chapter 42.

SUMMARY OF KEY CONCEPTS

Origins

Before life arose, lightning, ultraviolet light, and heat formed organic molecules from water and the components of the primordial Earth's atmosphere. These molecules probably included nucleic acids, amino acids, short proteins, and lipids. By chance, some molecules of RNA may have had enzymatic properties, catalyzing the assembly of copies of themselves from nucleotides in the Earth's waters. These may have been the precursors of life. Protein–lipid microspheres enclosing these RNA molecules may have formed the first cell-like organisms.

The Age of Microbes

The first fossil cells are found in rocks about 3.5 billion years old. These cells were prokaryotes that fed by absorbing organic molecules that had been synthesized abiotically. Since there was no free oxygen in the atmosphere at this time, energy metabolism must have been anaerobic. As the cells multiplied, they depleted the organic molecules that had been formed by prebiotic synthesis. Some cells evolved the ability to synthesize their own food molecules using simple inorganic molecules and the energy of sunlight. These earliest photosynthetic cells were probably ancestors of today's cyanobacteria.

Photosynthesis releases oxygen as a by-product, and by about 2.2 billion years ago significant amounts of free oxygen had accumulated in the atmosphere. Aerobic metabolism, which generates more cellular energy than does anaerobic metabolism, probably arose about this time.

Eukaryotic cells evolved about 1.5 billion years ago. The first eukaryotic cells probably arose as symbiotic associations between predatory prokaryotic cells and bacteria. Mitochondria may have evolved from aerobic bacteria engulfed by predatory cells. Similarly, chloroplasts may have evolved from photosynthetic cyanobacteria.

Multicellularity

Multicellular organisms evolved from eukaryotic cells, first appearing about 1 billion years ago. Multicellularity offers several advantages, including increased speed of locomotion and increased size.

Multicellular Life in the Sea

The first multicellular organisms arose in the sea. In plants, increased size due to multicellularity offered some protection from predation. Specialization of cells allowed plants to anchor themselves in the nutrient-rich, well-lit waters of the shore. For animals, multicellularity allowed more efficient predation and more effective escape from predators. This in turn provided selection pressures for faster locomotion, improved senses, and greater intelligence.

The Invasion of the Land

The first land organisms were probably plants, appearing 500 to 600 million years ago. Although the land required special adaptations for support of the body, reproduction, and the acquisition, distribution, and retention of water, the land also offered abundant sunlight and protection from aquatic herbivores. Around 400 million years ago, arthropods invaded the land. Absence of predators and abundant land plants for food probably facilitated the invasion of the land by animals.

The earliest land vertebrates evolved from lobefinned fishes, which had leg-like fins and a primitive lung. A group of lobefins evolved into the amphibians about 350 million years ago. Reptiles evolved from amphibians, with several further adaptations for land life: internal fertilization, waterproof eggs that could be laid on land, waterproof skin, and better lungs. Around 150 million years ago, birds and mammals evolved independently from separate groups of reptiles. Major advances included a high, constant body temperature and insulation over the body surface.

Human Evolution

One group of mammals evolved into the tree-dwelling primates. Primates show several preadaptations for human evolution: forward-facing eyes for binocular vision, color vision, grasping hands and relatively large brains. Between 20 and 30 million years ago some primates descended from the trees; these were the ancestors of apes and humans. The Australopithecines arose in Africa about 4 million years ago. These hominids walked erect, had larger brains than their forebears, and made primitive tools. One group of Australopithecines evolved into true humans.

GLOSSARY

cultural evolution: changes in the behavior of a population of animals, especially humans, by learning behaviors acquired by members of previous generations.

endosymbiotic hypothesis: the hypothesis that certain organelles, especially chloroplasts and mitochondria, evolved from bacteria captured by ancient predatory prokaryotic cells.

microsphere: a small, hollow sphere formed from proteins or proteins complexed with other compounds.

preadaptation: a feature evolved under one set of environmental conditions that, purely by chance, helps an organism adapt to new environmental conditions.

prebiotic evolution: evolution before life existed; especially abiotic synthesis of organic molecules.

ribozyme: an RNA molecule that can catalyze certain chemical reactions, especially those involved in synthesis and processing of RNA itself.

STUDY QUESTIONS

1. What is the evidence that life might have originated from nonliving matter on the primordial Earth? What kind of evidence would you like to see before you would accept this hypothesis?

2. Explain the endosymbiotic hypothesis for the origin of chloroplasts and mitochondria.

3. Name two advantages of multicellularity in plants and animals.

4. What advantages and disadvantages would terrestrial existence have had for the first plants to invade the land? For the first land animals?

5. Outline the general trends in the evolution of vertebrates, from fish to amphibians to reptiles to birds and mammals. Explain how these adaptations increased the fitness of the various groups for life on land.

6. Outline the evolution of humans from early primates. Include in your discussion such features as binocular vision, grasping hands, bipedal locomotion, social living, tool making, and brain expansion.

DISCUSSION QUESTIONS

1. What is cultural evolution? Is cultural evolution more or less rapid than biological evolution? Why?

2. Do you think that studying our ancestors can shed light on the behavior of modern humans? Why or why not?

3. Many paleoanthropologists believe that *Australopithecus afarensis* was the ancestor to all later hominids, including, eventually, *Homo sapiens*. This position is based mostly upon two facts: (1) *A. afensis* fossils appear earlier than those of other Australopithecines or of fossils of the genus *Homo*, and (2) *A. afarensis* fossils show many characteristics that appear to be more primitive than, but possibly ancestral to, those of later hominids. Do these facts prove ancestry? Might other fossils, not yet found, displace *A. afarensis* in the descent of humanity? If so, what characteristics and age would you expect them to have?

4. Is the history of life on Earth a steady progression to more advanced life forms? Are organisms alive today "better" in some absolute sense than organisms alive 100 million years ago? Why or why not?

SUGGESTED READINGS

Diamond, J. "How to Speak Neanderthal." *Discover*, January 1990. Could the Neanderthal's speak? How could we tell? Jared Diamond lucidly explains the arguments and methodology behind a scientific dispute.

Hay, R. L., and Leakey, M. D. "The Fossil Footprints of Laetoli." *Scientific American*, February 1982. The actual footprints of a hominid family were discovered by Hay and Leakey in volcanic ash 3.5 million years old.

Horgan, J. "In the Beginning . . ." *Scientific American*, February 1991. An exploration of the controversies surrounding research into the origin of life.

Morell, V. "Announcing the Birth of a Heresy." *Discover*, March 1987. Paleontologists Robert Bakker and Jack Horner speculate that dinosaurs were not the plodding beasts of monster flicks, but warm-blooded, advanced animals that may have even cared for their young.

Shreeve, J. "Argument Over a Woman." *Discover*, August 1990. An account of the often acrimonious debate between evolutionary geneticists and paleontologists over human evolution, here specifically considering the controversy over "mitochondrial Eve."

Waters, T. "Almost Human." *Discover*, May 1990. Artist John Gurche reconstructs fossil hominids by adding clay "muscle" and "skin" to replicas of hominid skulls.

20

Taxonomy: Imposing Order on Diversity

In 1990, a new species of primate, the black-faced lion tamarin, was discovered in a small patch of dense rain forest on an island just off the east coast of Brazil. This squirrel-sized primate resembles the golden lion tamarin shown in this photograph, except for the black fur covering its head. The animal is so secretive that no clear photographs are available, and only 12 individuals of this new species have ever been spotted. Captive breeding may be its only hope for continued survival, for its rain-forest home is rapidly being destroyed.

"Systems of classification are not hat racks, objectively presented to us by nature. They are dynamic theories developed by us to express particular views about the history of organisms. Evolution has provided a set of unique species ordered by differing degrees of genealogical relationship. Taxonomy, the search for this natural order, is the fundamental science of history."

Stephen Jay Gould, Natural History *(1987)*

Taxonomic Categories

Taxonomy (from the Greek for "taxis," meaning "arrangement") is the science by which organisms are classified and placed into categories based on their structural similarities and evolutionary relationships. **Taxonomic categories form a hierarchy—that is, a series of levels each more inclusive than the last. There are seven major categories:** *kingdom, phylum* **or** *division, class, order, family, genus,* **and** *species.* **Each category from species to kingdom is increasingly more general and includes organisms whose common ancestor was increasingly remote in its evolutionary relationship.** Some examples of classifications of specific organisms are given in Table 20-1. The **scientific name** of an organism is actually formed from the two smallest of these categories: the genus and the species *sapiens.* The **genus** is a category that humans, *Homo sapiens,* places us in the genus *Homo* and the species *sapiens.* The genus is a category that includes very closely related organisms that do not usually interbreed. The **species** is a category limited to naturally interbreeding living things. Thus, the genus *Sialia* (bluebirds) includes the eastern bluebird (*Sialia sialis*), the western bluebird (*Sialia mexicana*), and the mountain bluebird (*Sialia currucoides*), very similar birds that do not normally interbreed (Fig.

20-1). Scientific names are always underlined or italicized, and the genus is always capitalized. These names are recognized by biologists worldwide, transcending language barriers and allowing very precise communication.

The Ancient Origins of Taxonomy

Aristotle (384–322 BC) was among the first to attempt to formulate a logical, standardized language for naming living things. Using characteristics such as structural complexity, behavior, and degree of development at birth, he classified about 500 different organisms into 11 categories. Aristotle placed organisms into a hierarchy of categories each more inclusive than the one before it, a concept that is still used today.

Building on this foundation over 2000 years later, the Swedish naturalist Carolus Linnaeus (1707–1778) laid the groundwork for the modern classification system. He placed each organism into a series of hierarchically arranged categories based on its resemblance to other life forms, and also introduced the

Table 20-1   Classification Reflects the Degree of Relatedness of Organisms[a]

	Human Being	Chimpanzee	Wolf	Fruit Fly	Sequoia Tree	Sunflower
Kingdom	**Animalia**	**Animalia**	**Animalia**	**Animalia**	**Plantae**	**Plantae**
Phylum	**Chordata**	**Chordata**	**Chordata**	Arthropoda	Coniferophyta	Anthophyta
Class	**Mammalia**	**Mammalia**	**Mammalia**	Insecta	Coniferosida	Dicotyledoneae
Order	**Primates**	**Primates**	Carnivora	Diptera	Coniferales	Asterales
Family	Hominidae	Pongidae	Canidae	Drosophilidae	Taxodiaceae	Asteraceae
Genus	*Homo*	*Pan*	*Canis*	*Drosophila*	*Sequoiadendron*	*Helianthus*
Species	*sapiens*	*troglodytes*	*lupus*	*melanogaster*	*giganteum*	*annuus*

[a]Boldface categories are those that are shared by more than one of the organisms classified. Genus and species are always italicized or underlined.

Figure 20-1 Male representatives of the genus *Sialia*, the bluebirds. The similarities among these species are obvious, but the species remain distinct because they do not interbreed.

Sialia sialis
(eastern bluebird)

Sialia mexicana
(western bluebird)

Sialia currucoides
(mountain bluebird)

scientific name based on genus and species. Nearly 100 years later, Charles Darwin (1809–1882) published *Origin of Species*, which added a new significance to these categories. Taxonomists then began to recognize that, ideally, taxonomic categories reflect the evolutionary relatedness of organisms. The more categories two organisms share, the closer their evolutionary relationship.

Modern Criteria for Classification

As mentioned earlier, organisms that reproduce sexually are classified as belonging to the same species if they normally interbreed. To classify organisms beyond the species level (and for those organisms that rarely engage in sexual reproduction—for example, many bacteria and some fungi), taxonomists use other criteria. Some of the most important criteria are anatomy, developmental stages, and biochemical similarities.

Historically, the most important and useful of these has been anatomy. In addition to obvious similarities in external body structure, such as those seen in members of the genus *Sialis* (see Fig. 20-1), taxonomists look carefully at details such as skeletons and

tooth structure. The presence of homologous structures (such as the finger bones of dolphins, bats, seals, and humans) provides evidence of an evolutionarily distant common ancestor (see Chapter 18). To distinguish between closely related species, taxonomists may use microscopes to discern finer details—the number and shape of the "teeth" on the tongue-like radula of a mollusk, the shape and configuration of spines on a marine worm, or the pattern of projections on a pollen grain of a flowering plant (Fig. 20-2).

Clues to common ancestry may also be provided by the developmental stages that animals undergo on their way to adulthood. For example, the tunicate, or "sea squirt," spends its adult life permanently attached to rocks on the ocean floor, resembling an undersea vase. But its active, free-swimming larva has a nerve cord, tail, and gills, placing it with the vertebrates in the phylum Chordata (see Fig. 22-29).

Modern taxonomists are also aided by sophisticated biochemical techniques. One such technique is **electrophoresis.** During this procedure, molecules are separated according to their electrical charge and molecular weight. Electrophoresis can be used to separate and identify molecules such as proteins, amino acids, and nucleic acids. The presence and relative abundance of specific molecules in various species provides a way to assess their relatedness. Another

(a)   *(b)*   *(c)*

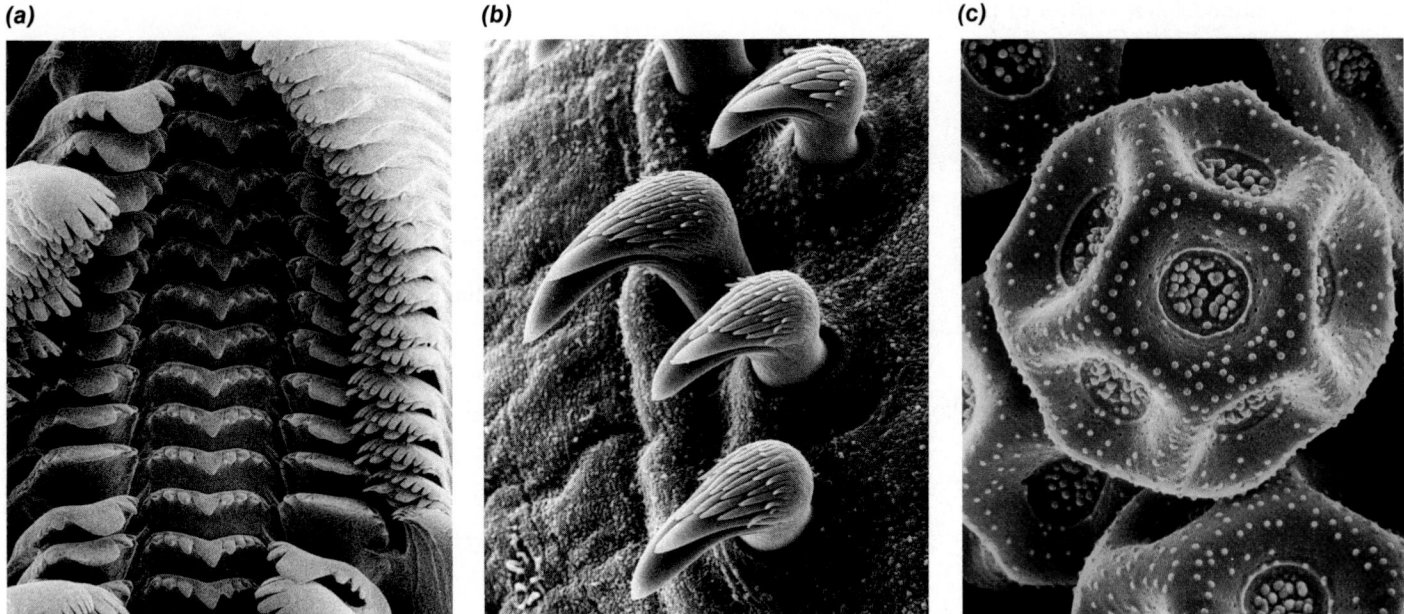

Figure 20-2 Microscopic structures help taxonomists classify organisms. **(a)** The "teeth" on the tonguelike radula of a snail or **(b)** the bristles on a marine worm are potential taxonomic criteria. **(c)** The shape and surface features of pollen grains help identify plants, particularly from fossil deposits and sediments.

technique is **DNA hybridization.** When heated, the double-stranded DNA separates, reforming when it is cooled. If DNA from two closely related species is combined and heated, hybrids of the two DNA types form upon cooling. Only complementary sequences can hybridize. Thus, the degree of hybridization is proportional to the degree of similarity between the DNA molecules from the two species. This, in turn, reflects how closely the two species are related. Further modern biochemical technology allows DNA, RNA, and amino acid sequencing, in which the exact sequence of nucleotides in DNA and RNA, and the amino acid sequences of proteins, can be determined. The greater the similarity in the sequence of the subunits in the genetic material or protein, the closer the evolutionary relationship. More sophisticated techniques, such as the polymerase chain reaction (which can amplify minuscule quantities of DNA, allowing it to be analyzed; see Chapter 14), permit scientists to sequence DNA from museum specimens of organisms that are now extinct and compare it with their living relatives.

The Five Kingdoms of Life

Prior to 1970, taxonomists classified all forms of life into two kingdoms: Animalia and Plantae. Bacteria, fungi, and photosynthetic protists were considered plants, while the protozoa were classified as animals. In 1969, Robert H. Whittaker proposed the five-kingdom classification scheme that is widely used today, and which we follow in this text.

Whittaker identified two kingdoms of primarily unicellular microorganisms based on whether they showed prokaryotic or eukaryotic cellular organization. The kingdom Monera consists of generally single prokaryotic cells, while the kingdom Protista consists of generally single eukaryotic cells. The remaining three kingdoms are all eukaryotic and multicellular (or at least multinucleate), and may be classified further based on their way of acquiring nutrients. Members of the kingdom Plantae photosynthesize, and members of the kingdom Fungi secrete enzymes that break down large organic molecules and then absorb the nutrients. In contrast, members of the kingdom Animalia ingest their food, then digest it within an internal cavity, or intracellularly. These and some other comparisons between the five kingdoms are presented in Table 20-2.

During evolution, the prokaryotic Monera gave rise to the eukaryotic Protista, which show all possible modes of nutrition. Different groups of protist then evolved into the multicellular plants, animals, and fungi, as illustrated in Fig. 20-3. Recently, a sixth

Table 20-2 Some Characteristics of the Five Kingdoms

Kingdom	Cell Type	Cell Number	Major Mode of Nutrition	Motility (Movement)	Cell Wall	Reproduction
Monera	Prokaryotic	Unicellular	Absorb or photosynthesize	Both motile and nonmotile	Present: peptidoglycan	Usually asexual, rarely sexual
Protista	Eukaryotic	Unicellular	Absorb, ingest, or photosynthesize	Both motile and nonmotile	Present in algal forms: varies	Both sexual and asexual
Plantae	Eukaryotic	Multicellular	Photosynthesize	Generally nonmotile	Present: cellulose	Both sexual and asexual
Fungi	Eukaryotic	Most multicellular	Absorb	Generally nonmotile	Present: chitin	Both sexual and asexual
Animalia	Eukaryotic	Multicellular	Ingest	All motile at some stage	Absent	Both sexual and asexual

kingdom of life has been proposed. Some microbiologists argue that the cellular structure of the Archaebacteria (see Chapter 21) is sufficiently different from either prokaryotic or eukaryotic cells to warrant placing them in their own kingdom. Since this is not yet widely accepted, we retain the five-kingdom classification scheme. The taxonomic categories described in this text are listed in Table 20-3.

Taxonomy: An Inexact Science

As the Archaebacteria debate suggests, taxonomic categories are not the unchanging authorities that we might wish them to be. For example, in 1973 ornithologists officially declared the Baltimore oriole and the Bullock's oriole to be a single species, which they named the northern oriole. The reason: where their ranges overlap, the two "species" of birds were interbreeding. Taxonomy is a challenging and frequently frustrating endeavor. When organisms reproduce asexually, the criterion of interbreeding cannot be applied to distinguish species. For example, some taxonomists recognize 200 species of the parthenogenetic British blackberry (a plant that can produce seeds without fertilization), while others recognize only 20 species. Most unicellular organisms reproduce asexually most of the time, making their classification difficult. For example, molecular biologists recently sequenced the nucleotides of the DNA of two different strains of the bacterium responsible for Legionnaire's disease, *Legionella pneumophila*. They found only a 50% correspondence between the strains. This degree of genetic dissimilarity within a

single species of bacterium is as great as that between mammals and fishes! In fact, because the taxonomy of bacteria is so controversial, we present only general descriptions of bacteria in Chapter 21.

Taxonomic categories are continuously debated and revised as taxonomists learn more about evolutionary relationships, particularly with the application of molecular biological techniques. Although the precise evolutionary relationships of many organisms continue to elude us, taxonomy is enormously helpful in ordering our thoughts and investigations into the diversity of life on Earth.

How Many Species Exist?

Scientists do not have the faintest idea how many species of organisms share our world! Each year, between 7000 and 10,000 new species are named, most of them insects, many from the tropical rain forests (see the chapter opener). The total number of named species is currently around 1.5 million. However, many scientists believe that 5 to 10 million species may exist, and estimates range as high as 30 million. Of the species that have been identified, about 5% are monerans and protists. An additional 22% are plants and fungi, and the rest are animals. This distribution may reflect the ease of identification and the relative number of scientists working on the different kingdoms as much as it does the relative abundance of species in these kingdoms.

Our ignorance of the full extent of life's diversity adds a new dimension to the tragedy of the destruction of the tropical rain forests, discussed in Chapter 46. Although these forests cover only about 6% of the

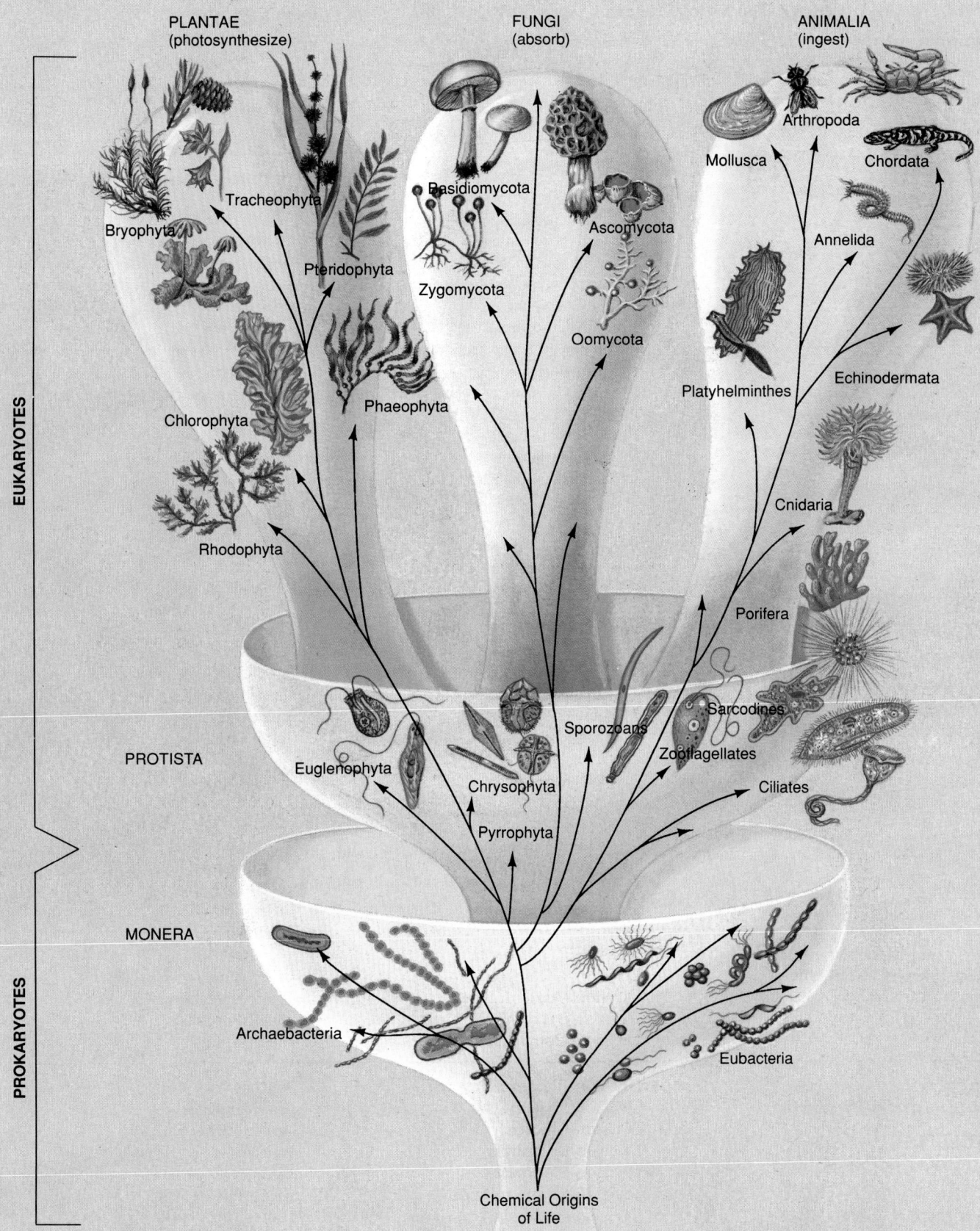

PLANTAE
(photosynthesize)

FUNGI
(absorb)

ANIMALIA
(ingest)

Tracheophyta

Bryophyta

Pteridophyta

Chlorophyta

Rhodophyta

Phaeophyta

Basidiomycota

Ascomycota

Zygomycota

Oomycota

Mollusca

Arthropoda

Chordata

Annelida

Platyhelminthes

Echinodermata

Cnidaria

Porifera

EUKARYOTES

PROTISTA

Euglenophyta

Chrysophyta

Pyrrophyta

Sporozoans

Sarcodines

Zooflagellates

Ciliates

MONERA

Archaebacteria

Eubacteria

PROKARYOTES

Chemical Origins
of Life

Figure 20-3 The five kingdoms of life.

Table 20-3   The Classification of the Major Groups of Organisms[a]

Kingdom	Division/Phylum	Common Name
Monera (unicellullar, prokaryotic)	Division Eubacteria	"true" bacteria
	Division Archaebacteria	"ancient" bacteria
Protista (unicellular, eukaryotic)	Division Pyrrophyta	dinoflagellates
	Division Chrysophyta	diatoms
	Division Euglenophyta	euglenoids
	Division Myxomycota	plasmodial slime molds
	Division Acrasiomycota	cellular slime molds
	Phylum Sarcomastigophora	zooflagellates, amoebae
	Phylum Apicomplexa	sporozoans
	Phylum Ciliophora	ciliates
Animalia (multicellular, heterotrophic, ingest nutrients)	Phylum Porifera	sponges
	Phylum Cnidaria	hydra, anemone jellyfish, corals
	Phylum Platyhelminthes	flatworms
	Phylum Nematoda	roundworms
	Phylum Annelida	segmented worms
	Class Oligochaeta	earthworms
	Class Polychaeta	tube worms
	Class Hirudinea	leeches
	Phylum Arthropoda	"jointed legs"
	Class Insecta	insects
	Class Arachnida	spiders, ticks
	Class Crustacea	crabs, lobsters
	Phylum Mollusca	"soft-bodied"
	Class Gastropoda	snails
	Class Pelecypoda	mussels, clams
	Class Cephalopoda	squid, octopus
	Phylum Echinodermata	sea stars, sea urchins
	Phylum Chordata	chordates
	Subphylum Vertebrata	vertebrates
	Class Agnatha	lampreys, hagfish
	Class Chondrichthyes	sharks, rays
	Class Osteichthyes	bony fish
	Class Amphibia	frogs, salamanders
	Class Reptilia	turtles, snakes, lizards
	Class Aves	birds
	Class Mammalia	mammals
Fungi (multicellular, heterotrophic, absorb nutrients)	Division Zygomycota	"zygote fungi"
	Division Ascomycota	"sac fungi"
	Division Deuteromycota	"imperfect fungi"
	Division Basidiomycota	"club fungi"
	Division Oomycota	"egg fungi"
Plantae (multicellular, photosynthetic)	Division Rhodophyta	red algae
	Division Phaeophyta	brown algae
	Division Chlorophyta	green algae
	Division Bryophyta	mosses
	Division Pteridophyta	ferns
	Division Coniferophyta	evergreens
	Division Anthophyta	flowering plants

[a]There is no agreement on taxonomic classification systems. This table lists only those taxonomic categories described in this text.

Earth's land area, they are believed to be home to two-thirds of the world's existing species. At current rates of destruction, most of these forests, with their undescribed wealth of wildlife, will be gone within the next century or two.

As we explore the bewildering variety of life in the following four chapters, keep in mind that science has only begun to recognize the full extent of the diversity of life on Earth.

SUMMARY OF KEY CONCEPTS

Taxonomic Categories
Taxonomy is the science by which organisms are classified and placed into hierarchical categories that reflect their evolutionary relationships. The seven major categories, in order of increasing inclusiveness, are species, genus, family, order, class, division (for plants, fungi, bacteria, and plantlike protists) or phylum (for animals and animallike protists), and kingdom. The scientific name of an organism is composed of its genus name and species name.

The Ancient Origins of Taxonomy
A hierarchical concept was first used by Aristotle, but Linnaeus in the mid-1700s laid the foundation for modern taxonomy. In the 1860s, evolutionary theory provided an explanation for the observed similarities and differences between organisms, and modern taxonomists attempt to classify organisms according to their evolutionary relationships.

Modern Criteria for Classification
Today, taxonomists use features such as anatomy, developmental stages, and biochemical similarities to categorize organisms. Molecular biological techniques are used to determine the types of proteins, amino acids, and nucleotides in organisms, as well as to determine the exact sequences of nucleotides in DNA and RNA and of amino acids in proteins. Biochemical similarities among organisms are a measure of evolutionary relatedness.

The Five Kingdoms of Life
Whittaker's 1969 proposal of a five-kingdom classification scheme is widely accepted today. The kingdom Monera consists of generally unicellular prokaryotic organisms. The kingdom Protista consists of generally unicellular eukaryotic organisms. The kingdom Animalia consists of multicellular eukaryotes that ingest their food. The kingdom Fungi consists of multicellular eukaryotes that absorb their food, and the kingdom Plantae consists of multicellular eukaryotes that photosynthesize.

Taxonomy: An Inexact Science
Taxonomic categories are controversial and subject to frequent revision, particularly in instances where sexual reproduction is absent and evolutionary relationships are uncertain. However, taxonomy is essential for precise communication and contributes immensely to our understanding of the origins and diversity of species.

How Many Species Exist?
Although only about 1.5 million species have been named, estimates of the total number of species range up to 30 million. New species are being identified at the rate of 7000 to 10,000 each year. Many will be driven to extinction by the destruction of the tropical rain forests before they are ever classified.

GLOSSARY

class: the taxonomic category composed of related genera. Closely related classes form a division or phylum.

division: the taxonomic category contained within a kingdom and consisting of related classes of plants, fungi, bacteria, or plantlike protists.

DNA hybridization: a technique by which DNA from two species is separated into single strands and then allowed to reform. Hybrid double-stranded DNA from the two species can occur where the sequence of nucleotides is complementary. The greater the degree of hybridization, the closer the evolutionary relatedness of the two species.

electrophoresis: a biochemical technique that separates molecules according to their electrical charge and molecular weight.

family: the taxonomic category contained within an order and consisting of related genera.

genus (jē-nis): the taxonomic category consisting of very closely related species.

kingdom: the broadest taxonomic category, consisting of phyla or divisions. We recognize five kingdoms in this text.

order: the taxonomic category contained within a class and consisting of related families.

phylum (fī-lum): the taxonomic category of animals and animallike protists contained within a kingdom and consisting of related classes.

species (spē-cēs): a group of organisms within a genus that interbreed under natural conditions, or, if asexually reproducing, are more closely related to one another than to other organisms within the genus.

taxonomy (tax-on-uh-mē): the science by which organisms are classified into hierarchically arranged categories that reflect their evolutionary relationships.

STUDY QUESTIONS

1. What contributions did Aristotle, Linnaeus, and Darwin each make to modern taxonomy?
2. What features would you study to determine whether a dolphin is more closely related to a fish or a bear?
3. What techniques might you use to determine whether the extinct cave bear is more closely related to a grizzly bear or to a black bear?
4. Only a small fraction of the total number of species on Earth has been scientifically described. Why?
5. On the basis of the information in the table, draw a tree diagram best describing the relationships between skunks (*Mephitis mephitis*), dogs (*Canis familiaris*), wolves (*Canis lupus*), brown bears (*Ursus arctos*), and polar bears (*Ursus maritimus*).

Order	Family	Genus	Species
Carnivora	Canidae	Canis	*Canis familiaris* (dog)
Carnivora	Ursidae	Ursus	*Ursus maritimus* (polar bear)
Carnivora	Mustilidae	Mephitis	*Mephitis mephitis* (skunk)
Carnivora	Ursidae	Ursus	*Ursus arctos* (brown bear)
Carnivora	Canidae	Canis	*Canis lupus* (wolf)

6. Consider the following classification information about houseflies, wolves, humans, and herring gulls:

Kingdom: all ANIMALIA
Phylum: houseflies are ARTHROPODA, all others are CHORDATA
Class: houseflies are INSECTA, gulls are AVES, wolves and humans are MAMMALIA
Order: houseflies are DIPTERA, gulls are CHARADRIIFORMES, wolves are CARNIVORA, humans are PRIMATA

Which two types of animals are most closely related? Which type is most closely related to mammals? Which type is least closely related to mammals?

DISCUSSION QUESTIONS

1. You are sailing with a companion after a major flood, and you see the topmost branch tips of a submerged tree. Your companion asks you to sketch the branches below the surface based solely on the location of the exposed tips. How is this situation similar to the problem facing taxonomists? What relative advantages do the taxonomists have?

2. What reasons can you think of for protecting species we now know nothing about?
3. Suppose Darwin's concept of evolutionary relatedness based on taxonomic similarity was proven to be false. Discuss several ways that this would change biology and society.

SUGGESTED READINGS

Avise, J. C. "Nature's Family Archives." *Natural History*, March 1989. Shows how evolutionary relationships can be determined by analyzing differences in DNA contained in mitochondrial organelles.

Lowenstein, J. M. "Molecular approaches to the identification of species." *American Scientist* 73: 541–547, 1985. A basic introduction to immunological methods of taxonomy.

Margulis, L., and K. Schwartz. *Five Kingdoms.* New York: Freeman, 1988. An illustrated paperback guide to the diversity of life.

Simpson, G. G. *Fossils and the History of Life.* New York: Scientific American Library, 1983. A good layman's introduction to fossils and how they can help us understand how life has evolved on earth.

Wilson, A. C. "The molecular basis of evolution." *Scientific American*, 1985. On using nucleotide and amino acid sequences to determine evolutionary relationships.

21

The Diversity of Life.
I. Microorganisms

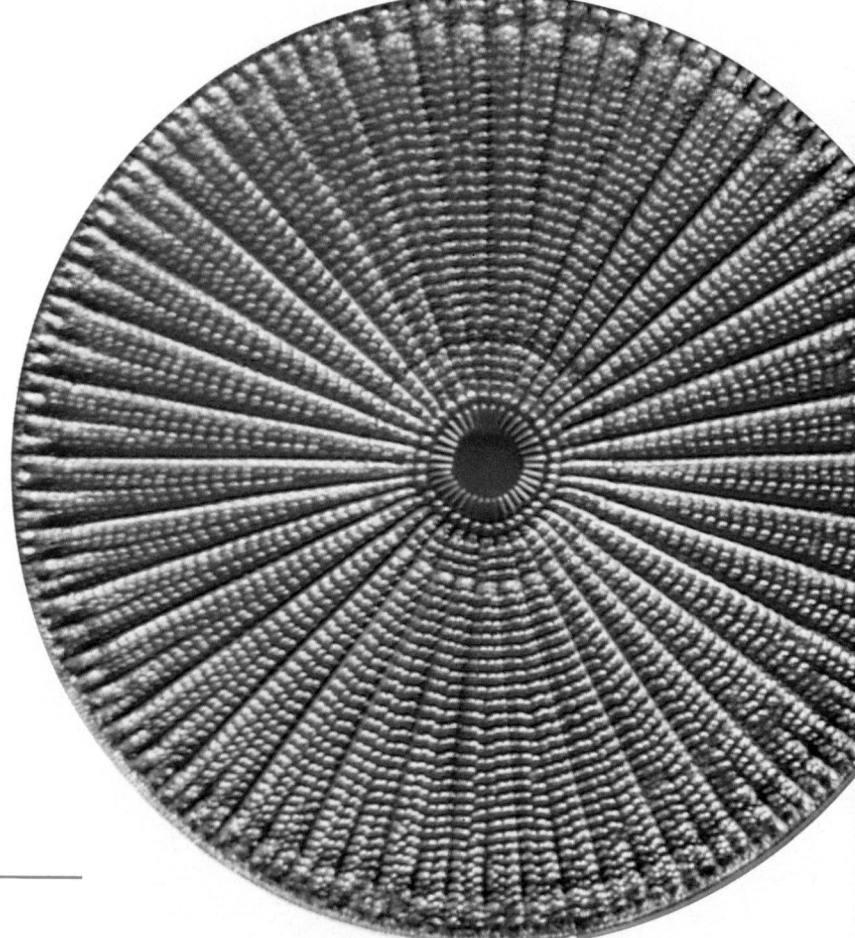

Plantlike protists, such as this diatom, are responsible for most of the photosynthetic activity on Earth.

If humans were microbes, the entire world population could thrive in a spadeful of garden soil. The water of a puddle, pond, or sea also teems with unicellular organisms invisible to the naked eye. We are surrounded, coated, and inhabited by life forms that we become aware of only if they make us ill. Although minuscule, microorganisms are of immense importance, and not just as agents of disease. Protists are responsible for the majority of photosynthetic activity on Earth, replenishing oxygen and capturing the sun's energy in food (chapter opener photo). Of all living things, only certain bacteria can capture atmospheric nitrogen and convert it into a nutrient used by plants. Were microorganisms suddenly to disappear, cows, sheep, and other ruminants might starve, unable to break down the energy-rich cellulose in their diet. Without bacteria to decompose them, bodies of animals and plants would accumulate, locking up their valuable nutrients and disrupting the recycling process on which life relies. Without microorganisms, life as we know it would cease.

This chapter introduces the unseen world of unicellular microorganisms. We start with viruses, puzzling parasitic particles that are not alive, and defy classification. We then move to the bacteria, the simplest form of life. The chapter concludes with the pro-

tists, among which are found the most complex cells on Earth. Figure 21-1 gives an idea of the relative size and complexity of these three groups.

Viruses

The existence of viruses reminds us of the imperfection of our taxonomic categories, since there is no kingdom into which they fit comfortably. **Viruses possess no membranes of their own, no ribosomes on which to make proteins, no cytoplasm, and no source of energy.** *They cannot move or grow, and they can reproduce only inside a host cell.* **The utter simplicity of viruses makes it impossible to call them cells, and indeed, seems to place them outside the realm of living things.** Of course, the viruses themselves are unaffected by our confusion regarding them, and continue their successful existence as the ultimate intracellular parasite.

Viral Structure and Reproduction

A virus particle is so small (0.05–0.2 micrometers [μM; 1/1000 of a centimeter] in diameter) that visual-

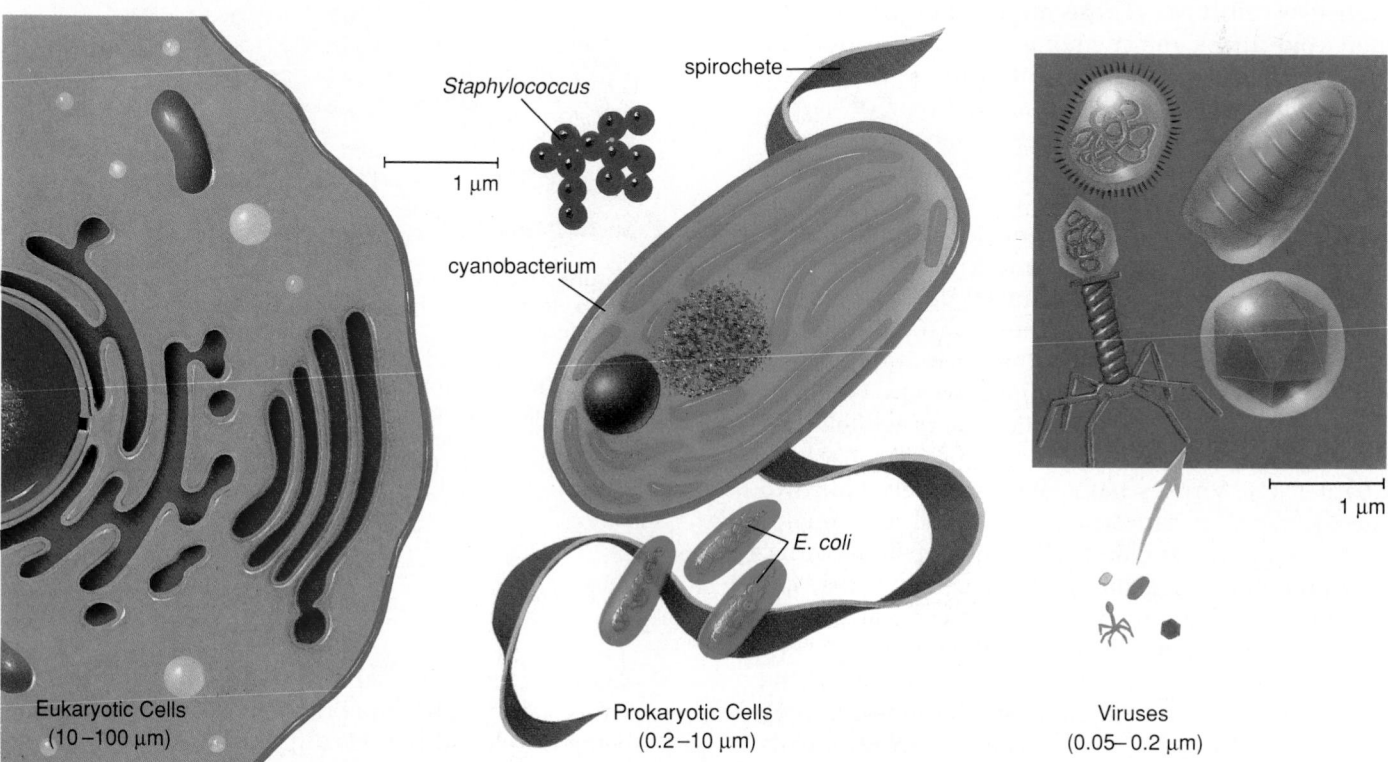

Staphylococcus

1 μm

spirochete

cyanobacterium

E. coli

Eukaryotic Cells
(10–100 μm)

Prokaryotic Cells
(0.2–10 μm)

Viruses
(0.05–0.2 μm)

1 μm

Figure 21-1 The relative sizes of eukaryotic cells, prokaryotic cells, and viruses.

izing it requires the enormous magnification of an electron microscope. Viruses consist of two major parts, a coat of protein surrounding a molecule of hereditary material, either DNA or RNA. An envelope formed from the membrane of the host cell may surround the protein coat (Fig. 21-2a). Even if placed in a rich broth of nutrients at optimal temperature, viruses remain inert, unable to grow or divide, since they lack the complex cellular organization that these activities require. The protein coat, however, is specialized to allow viruses to penetrate the cells of a specific host, where they assume a rather insidious semblance of life. After entering the host cells, the viral genetic material takes command. The host cells are forced to read the viral genes, and to use these instructions to produce the components of new viruses. The pieces are rapidly assembled (Fig. 21-2b), and an army of new viruses bursts forth to invade and conquer neighboring cells (this cycle is diagrammed for the bacterial virus in Fig. 10-5).

Viral Infections

Each type of virus is specialized to attack a specific host cell (Fig. 21-3), and probably no organism is immune to all viruses. Even bacteria fall victim to viral invaders called **bacteriophages** (Fig. 21-4). Within a particular organism, viruses specialize on particular cell types. Those responsible for the common cold attack the membranes of the respiratory tract, those causing measles infect the skin, and the rabies virus attacks nerve cells. One type of herpes virus specializes in the mucous membranes of the mouth and lips, causing cold sores, while a second type (transmitted through sexual contact) produces similar sores on or near the genitals. Herpes viruses, unfortunately, take up permanent residence in the body, erupting periodically (often during times of stress) as infectious sores. The devastating disease AIDS (acquired immune deficiency syndrome), which cripples the body's immune system, is caused by a virus that attacks a specific type of white blood cell that controls the body's immune response (see Chapter 34). Viruses have been definitely linked to specific types of cancer, such as T-cell leukemia, a cancer of the white blood cells. The papilloma virus, long known for its ability to cause genital warts, has recently been identified in 90% of cervical cancers sampled, suggesting a causal relationship.

Since viruses are intracellular parasites utilizing the cellular machinery of their host, the illnesses they cause are difficult to treat because antiviral agents may destroy host cells as well. The antibiotics so ef-

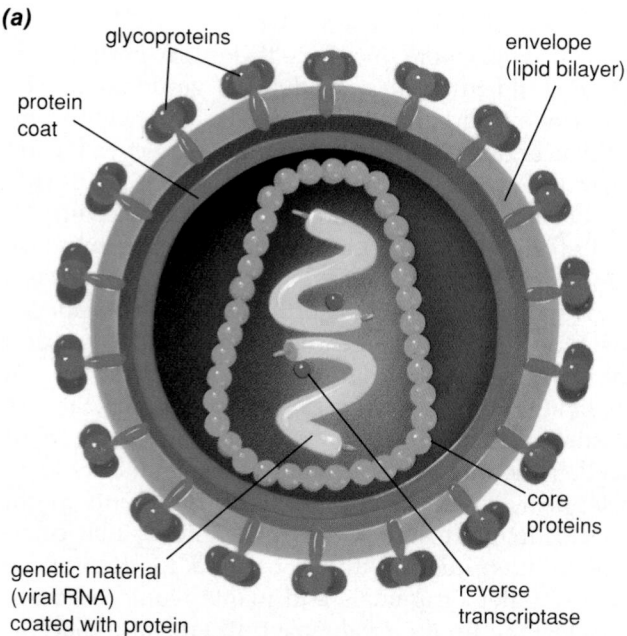

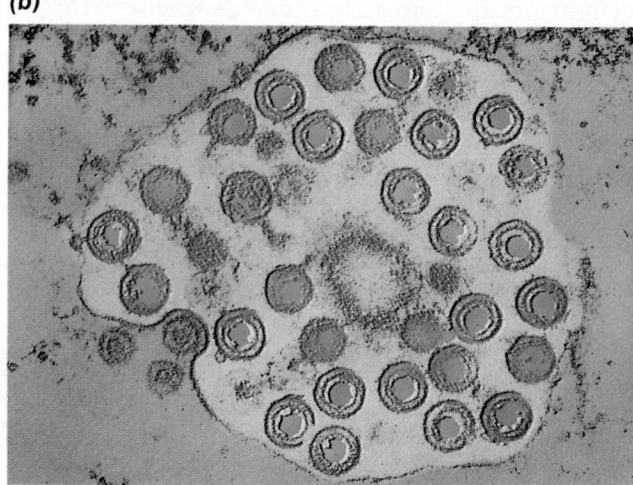

Figure 21-2 (a) Cross section of the virus that causes AIDS. Inside is genetic material, surrounded by a protein coat. Some viruses, including those causing herpes, rabies, and AIDS, have an outer envelope that may be formed from the membrane of the host cell. Spikes made of protein and carbohydrate may project from the envelope, and some viruses use these to attach to their host cell. **(b)** In this electron micrograph, herpes viruses are seen packed into an infected cell.

fective against bacterial infections are useless against viruses, although some promising antiviral drugs are being developed (see Health Watch: The War on Viruses).

HEALTH WATCH
The War on Viruses

The Best Defense: Prevention

As early as the eleventh century, Chinese doctors made a powder from the dried scabs of smallpox victims and blew it into the noses of their healthy patients to prevent infection. In 1796, Edward Jenner, an English country doctor, inoculated a child with pus from the sore of a milkmaid infected with cowpox, a disease closely related to smallpox but much less deadly (Fig. E21-1). His success heralded the most successful method ever developed for dealing with viral diseases—vaccination. In 1979, the World Health Organization declared that smallpox had been eradicated worldwide; the first and only infectious disease to be eliminated.

Vaccination (from the Latin "Vacca" for "cow"; referring to the original cowpox experiment) stimulates the immune system to produce long-lived cells that produce antibodies to viruses. Viral vaccines consist of weakened live virus or viruses that have been inactivated so that they are no longer infectious. Sometimes the immune system can be triggered to develop effective antibodies against a mere fragment of the virus—a section of its protein coat, for example. Such a vaccine could never infect the recipient, in contrast to weakened or "killed" viruses, which, in rare circumstances, cause disease.

Using recombinant DNA techniques, researchers have inserted the portion of the viral genome responsible for protein-coat production into bacteria, which then churn out large quantities of viral coat protein. Purified coat protein from the hepatitis B virus is now produced in this manner as a vaccine against this devastating disease. Work on genetically engineered vaccines against rabies, herpes, typhoid, diphtheria, and cholera is in progress.

Why, then, can we not eliminate the common cold, flu, and even AIDS, by similar techniques? The viruses that cause these diseases present a formidable challenge because of their incredibly rapid mutation rate. Influenza (type A) viruses mutate at a rate a million times greater than mammalian cells. These mutations alter the protein coat sufficiently that the antibodies made in response to a vaccine (or to last year's bout of flu) no longer recognize and attack the new strain. HIV (human immunodeficiency virus, the cause of AIDS) mutates even more rapidly than the influenza virus. Thus, a single AIDS victim may harbor several slightly different strains of the HIV virus, formed as the virus replicates and mutates within its human host. Such rapid mutation not only hinders efforts to produce a vaccine, it also increases the like-

Figure E21-1 This painting depicts the first vaccination. Edward Jenner has taken fluid from a cowpox lesion on the hand of a dairymaid (right) and is injecting it into the arm of a child. (Courtesy of Fisher Scientific Co.)

lihood that strains of virus will arise that resist the new antiviral drugs.

When Prevention Fails: Antiviral Drugs

Since virus particles are not living, and use the biochemical machinery of their host cells to replicate, the challenge is to develop compounds that selectively attack the virus without harming uninfected host cells. Two such drugs have been developed recently: acyclovir, a treatment for herpes, and AZT (azidothymidine), a treatment for AIDS. Both closely resemble a normal building block of DNA; they are incorporated into newly forming DNA strands, and disrupt further DNA synthesis. By a lucky chance, both of these compounds are much more readily incorporated into viral DNA (the DNA whose synthesis is directed by the virus) than into host DNA.

Unfortunately, these drugs have serious drawbacks. Both drugs are expensive and toxic. AZT is quite harmful to bone marrow cells, which produce blood cells. While it slows the progress of the disease and lessens its symptoms, AZT is *not* a cure for AIDS, nor can acyclovir cure herpes. Ominously, in some patients taking AZT for 6 months or longer, AZT-resistant strains of the HIV virus have appeared.

As you have probably concluded by now, the virus is a formidable enemy. Probably no medical advance would contribute more to human well-being than the development of truly effective antiviral drugs. Progress will depend on increased understanding of viral structure, the viral replication cycle, and the body's natural defenses against these nonliving invaders.

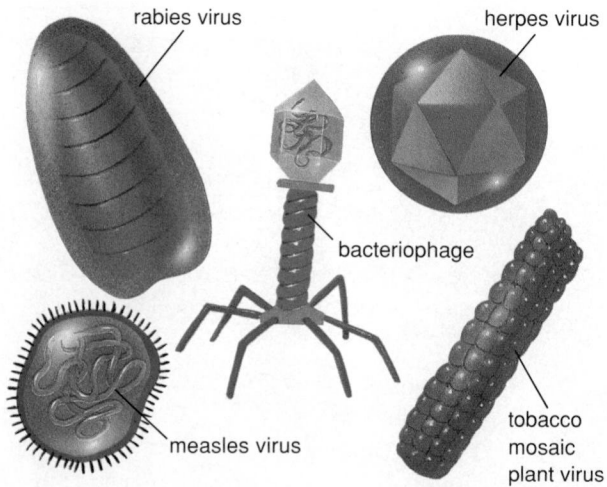

Figure 21-3 Viruses come in a variety of shapes, which are determined by their protein coats. The rabies and herpes viruses are surrounded by an extra envelope derived from membranes of the host cell.

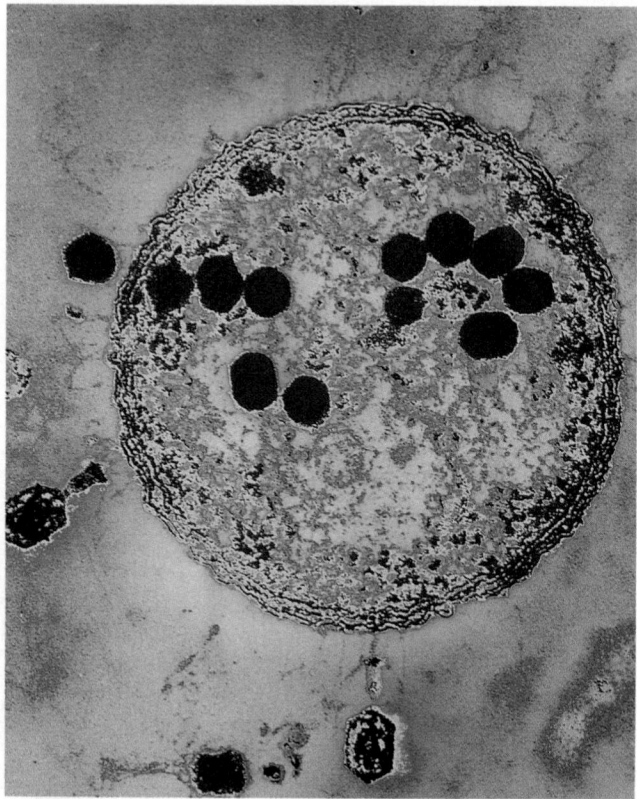

Figure 21-4 In this electron micrograph, bacteriophage viruses are seen attacking a bacterium. They have injected their genetic material inside, leaving their protein coats clinging to the bacterial cell wall. Black objects inside the bacterium are newly forming viruses.

Beyond Viruses: Viroids and Prions

Viroids

In the early 1970s, researchers discovered that some plant diseases were caused by particles only one tenth the size of normal plant viruses. Called **viroids,** these particles are merely short strands of RNA, lacking a protein coat. Like viruses, viroids apparently enter the nucleus of the infected cell, where they direct the synthesis of new viroids. About a dozen crop diseases have been attributed to viroids, including cucumber pale fruit disease, avocado sunblotch, and potato spindle tuber disease. Viroids may also be responsible for some diseases in animals, including humans, the causes of which have not yet been identified.

Prions

Many years ago, in a primitive, head-hunting tribe in New Guinea, doctors were puzzled by frequent cases of a fatal degenerative disease of the nervous system, which the people called **kuru.** The victims were children and adult women, who ritually consumed the brains of their tribespeople who had died from kuru. Adult men, who were prohibited by tribal law from eating human tissue, rarely contracted the disease. An infectious agent, transmitted in infected brain tissue, was clearly at work—but what was it? The symptoms of kuru showed a striking resemblance to those of scrapie, a disease of sheep. In 1982 a protein called a **prion** (proteinaceous infectious particle) was isolated from hamsters infected with scrapie (Fig. 21-5). Although the prion is infectious, there is no evidence of any genetic material in a prion, and there is no known method for proteins to replicate themselves. Perhaps the prion activates a gene that is normally turned off but that causes production of an enzyme that converts normal proteins to prions. Or perhaps it acts as an enzyme itself, catalyzing the formation of more prions. The discovery of prions has generated enormous interest among researchers. Not only are prions a totally unfamiliar means of infection, but prion-related diseases (which include scrapie, kuru, and Creutzfeldt–Jakob disease) are similar to other degenerative diseases of the nervous system, such as Alzheimer's disease, the leading cause of senility. Since the cause of Alzheimer's disease remains obscure, prions are a possible candidate.

Viral Origins

The origin of viruses, viroids, and prions is obscure. It is unlikely that these infectious particles are the

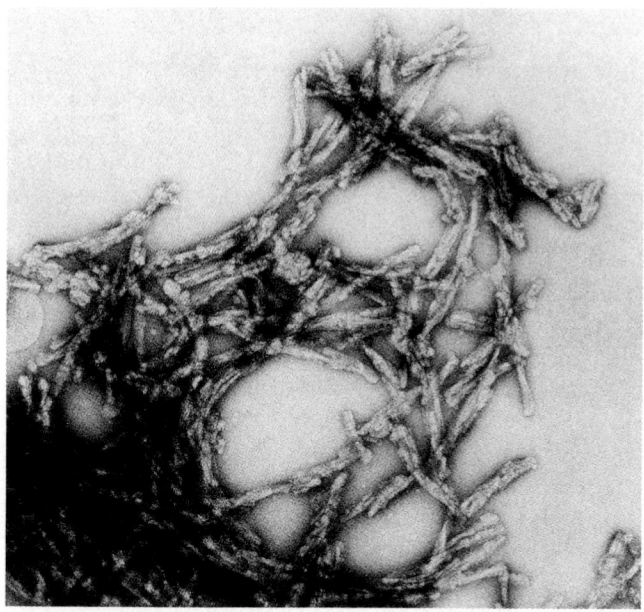

Figure 21-5 Prions (proteinaceous infectious particles) isolated from the brain of a hamster infected with scrapie, a degenerative disease of the nervous system. Courtesy of Dr. Stanley B. Prusiner.

forerunners of life, since they cannot reproduce without infecting more complex cells. Some scientists believe that viruses and viroids originated from simple parasitic cells that evolved such complete dependence on their hosts that they lost the ability to perform the basic processes of life. Or perhaps viruses and viroids originated as loose fragments of genetic material that took up an independent existence. Whatever their origin, the success of these parasitic particles poses a continuing challenge to living things.

Prokaryotic and Eukaryotic Cells

Members of the kingdom Monera, often called **bacteria,** are single, prokaryotic cells. They have changed very little in form from their fossil ancestors of roughly 3½ billion years ago, discovered in rocks. Prokaryotic cells lack organelles such as the nucleus, chloroplasts, and mitochondria (see Chapter 5 for a comparison of prokaryotic and eukaryotic cells). Bacterial cells are very small, ranging from about 0.2 to 10 micrometers in diameter, compared with eukaryotic cells, whose diameters range from about 10 to 100 micrometers. Two hundred and fifty thousand average-sized bacteria could congregate on the period at the end of this sentence.

The unicellular members of the kingdom Protista

and all multicellular organisms are composed of eukaryotic cells that evolved roughly 2 billion years after the first prokaryotic cells. Eukaryotic cells contain many organelles that are absent in prokaryotic cells. Organelles such as mitochondria and chloroplasts are similar in size to typical bacteria, and indeed probably evolved from bacteria (see Chapter 5). Within the kingdom Protista, we find some of the most complex eukaryotic cells in existence, with organelles taking on functions served by organs in multicellular organisms.

The Kingdom Monera

The Classification of Monerans

There are well over 1700 species of monerans (called bacteria from now on), which can be placed into two divisions: a large division called the **eubacteria** (Greek, "true bacteria") and a much smaller division, the **archaebacteria** (Greek for "ancient bacteria"). These two types of bacteria have striking structural and biochemical differences. Since bacterial reproduction is usually asexual, the classic definition of a species as a potentially interbreeding group cannot apply. In addition, the fossil record of bacteria is practically nonexistent. Consequently, taxonomists classify bacteria in a variety of ways, for example according to their shape, means of locomotion, pigments, staining properties, nutrient requirements, or the appearance of their colonies. Bacterial taxonomy is further confused by frequent revisions as more information is acquired. For simplicity, we have organized our discussion of the kingdom Monera around the general features and diverse lifestyles of bacteria, singling out two groups for special consideration: the cyanobacteria (members of the division Eubacteria) and the archaebacteria.

Bacterial Structure

Bacterial Shapes and the Cell Wall

Nearly all bacteria are encased in a porous but rigid cell wall that protects them from osmotic rupture in watery environments, and gives different types their characteristic shapes. The most common bacterial shapes are rodlike **bacilli,** spheres called **cocci,** and the corkscrew-shaped **spirilla** (Fig. 21-6). The cell wall contains a material called **peptidoglycan** that is unique to bacteria. Peptidoglycan is composed of chains of sugars cross-linked by peptides (short chains of amino acids).

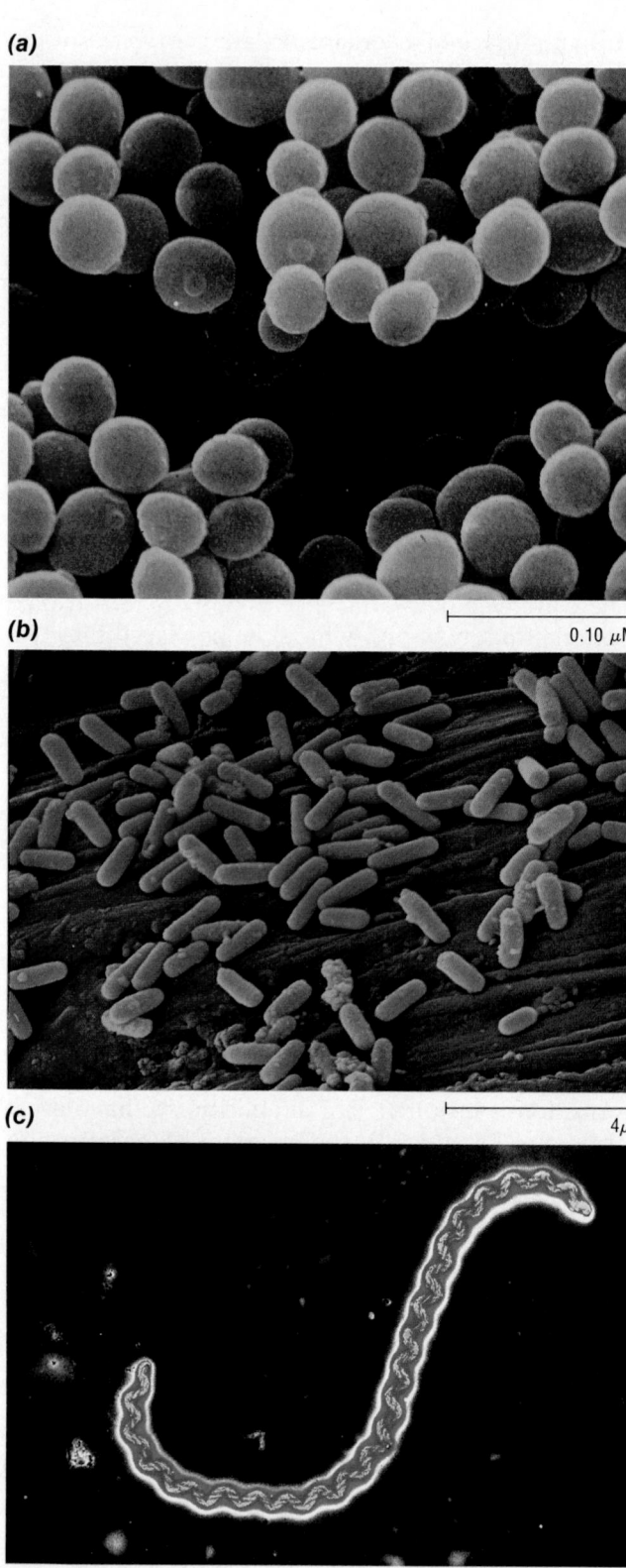

Figure 21-6 Three common bacterial forms as seen under the scanning electron microscope: **(a)** spherical bacteria, also called cocci, of the genus *Micrococcus;* **(b)** rods, also called bacilli, shown here growing on the point of a pin; and **(c)** corkscrew or spirilla-shaped bacteria; this species causes Leptospirosis in humans.

Capsules, Slime Layers, and Pili

Surrounding the cell walls of some bacteria are sticky **capsules** or **slime layers,** composed of polysaccharide or protein. Capsules help certain disease-causing bacteria escape detection by their victim's immune system. Slime layers allow the bacteria that cause tooth decay to adhere in masses to the smooth surface of a tooth. This slime forms the basis of dental plaque (Fig. 21-7).

Some bacteria cover themselves with a fuzz of hairlike projections called **pili.** Pili are made of protein, and generally serve to attach the bacterium to other cells. The pili of some infectious bacteria, such as those causing the venereal disease gonorrhea, attach to the cell membranes of their host, facilitating infection. Some bacteria produce special sex pili, described below (see Fig. 21-11).

Flagella and Bacterial Locomotion

Some bacteria are equipped with **flagella.** These are simpler in structure than the flagella seen in some eukaryotic cells. Bacterial flagella, which may either cover the cell or form a tuft at one end (Fig. 21-8a), can rotate rapidly, propelling the bacterium through its liquid environment. Recent research has revealed a unique wheellike structure embedded in the bacterial membrane and cell wall that allows the flagellum to rotate (Fig. 21-8b). Flagella allow bacteria to disperse into new habitats, to migrate toward nutrients, and to leave unfavorable environments.

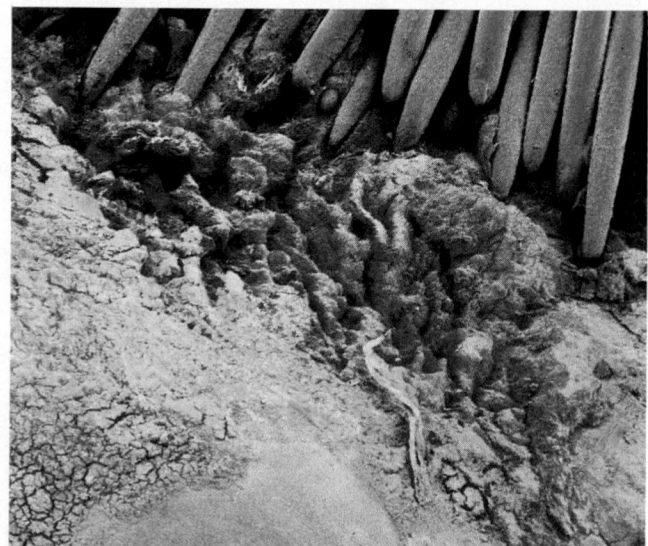

Figure 21-7 Slime layers allow decay-causing bacteria to adhere in masses to the enamel of teeth, as shown in this scanning electron micrograph. The green bristles of a toothbrush can be seen sweeping the bacteria away.

(a)

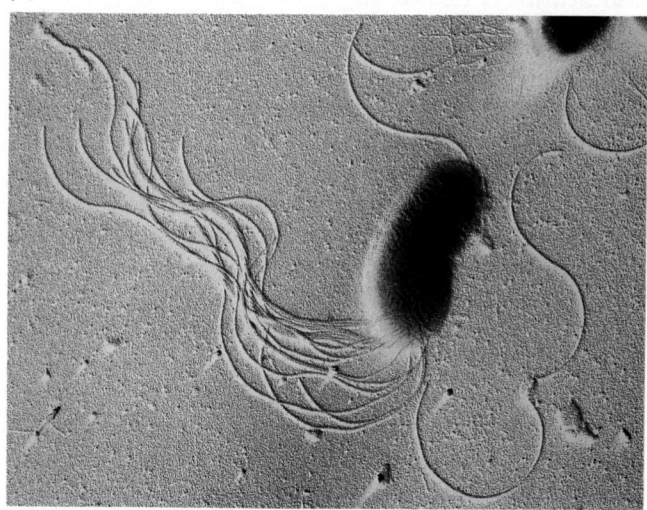

(b)

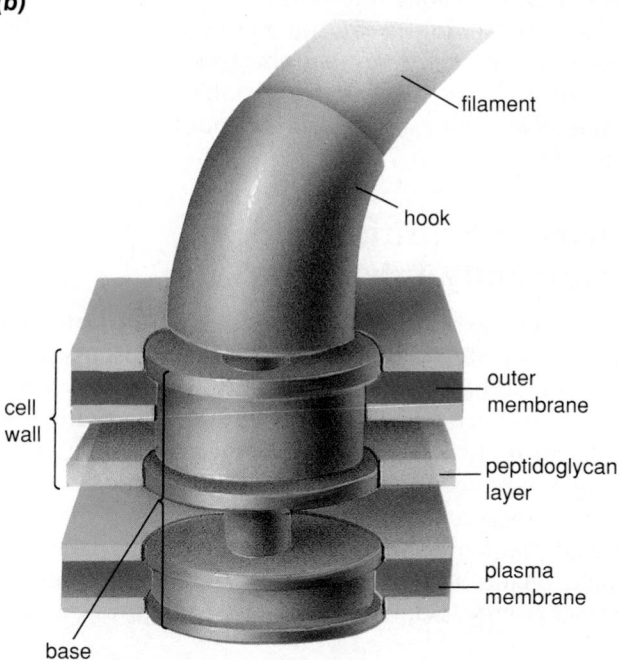

filament

hook

cell wall

outer membrane

peptidoglycan layer

plasma membrane

base

Figure 21-8 **(a)** A flagellated bacterium of the genus Pseudomonas, uses its flagellum to move toward favorable environments. **(b)** A unique "wheel and axle" arrangement anchors the bacterial flagellum within the cell wall and plasma membrane, allowing it to rotate rapidly.

In some flagellated bacteria, called **magnetotactic bacteria,** magnets formed within the cytoplasm from iron crystals allow the bacterium to orient according to the Earth's magnetic field. For organisms in the northern or southern hemispheres, the magnetic field of the Earth allows detection, not only of north and south, but of up and down as well. Magnetotactic bacteria often live in aquatic sediments where oxygen is minimal or absent. They use their unique sensory system to direct their beating flagella, moving downward into deep sediment layers where they find the low-oxygen conditions they require.

Spores

When environmental conditions become inhospitable, many bacteria form protective resting structures called **spores** (Fig. 21-9). The spore contains genetic material and a few enzymes within a thick protective coat of cell wall–substance and protein. Metabolic activity ceases. Spores can survive extremely unfavorable conditions. Some can withstand boiling for an hour or more while others, still alive, have been found in the intestines of mummies 2000 years old. Spores are important agents of bacterial dispersal, since they can be carried for long distances in air or water, germinating rapidly when they encounter favorable conditions.

Bacterial Reproduction

Bacteria reproduce asexually by a simple form of cell division called *fission* (see Chapter 9), which produces genetically identical copies of the original cell (Fig. 21-10). Under ideal conditions, a bacterium may divide about once every 20 minutes, potentially giving rise to sextillions (1×10^{21}) of offspring in a single day. This rapid reproduction allows bacteria to exploit

Figure 21-9 A resistant spore, also called an endospore, that has formed inside a bacterium of the genus *Clostridium,* responsible for the potentially fatal food poisoning called botulism.

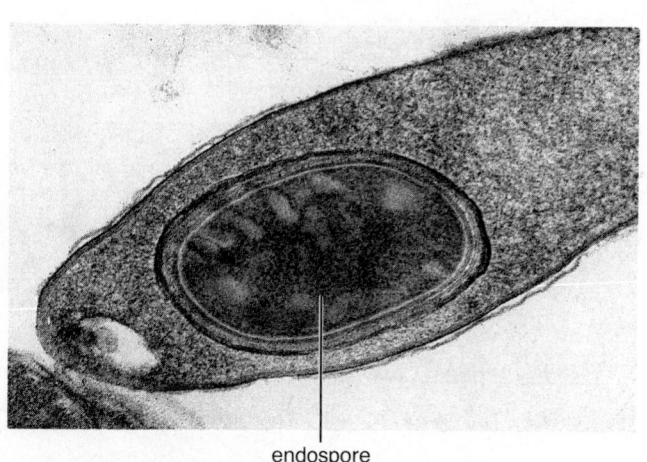

endospore

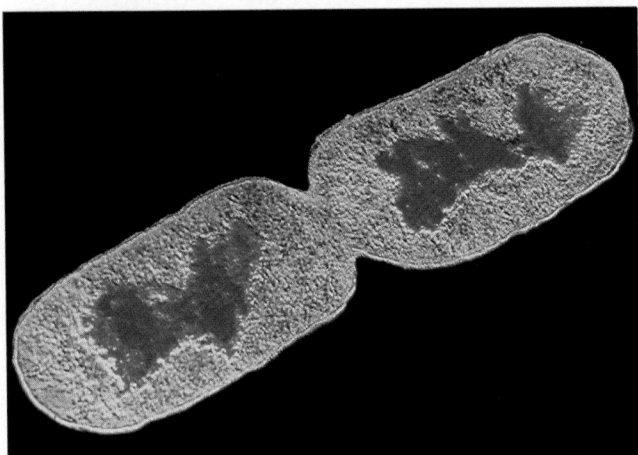

Figure 21-10 Prokaryotes reproduce by cell division, called fission or binary fission ("splitting into two"). This is illustrated by the electron micrograph of a dividing *Escherichia coli,* found abundantly in the human intestine. Red areas are genetic material.

temporary habitats such as a mud puddle or warm pudding. Recall that mutations, the source of genetic variability, occur as a result of mistakes in DNA replication during cell division (see Chapter 13). Thus, the rapid reproductive rate of bacteria provides ample opportunity for new forms to arise and also allows mutations that enhance survival to spread quickly (see Health Watch: "Unnatural" Selection—The Evolution of Drug-Resistant Pathogens).

In addition to simple cell division, some bacteria reproduce "sexually," transferring genetic material from a donor bacterium to a recipient during a process called **bacterial conjugation.** Conjugating bacteria use specialized hollow sex pili to transfer genetic material (Fig. 21-11). Conjugation produces new genetic combinations that may allow the offspring to survive under a greater variety of conditions.

Bacterial Habitats

Bacteria occupy a striking diversity of habitats. They have been isolated from snowy mountaintops and ocean depths of 4400 meters. Some thrive in hot springs such as those in Yellowstone National Park (Fig. 21-12). But the near-boiling temperatures of hot springs would be chilly for some of their relatives, who thrive near deep ocean vents where cracks in the Earth's crust spew superheated water. These bacteria can grow at temperatures of up to 110° C (230° F). Bacteria are also found in the Dead Sea, where the tremendous salt concentration (seven times that of the oceans) precludes all other life. They are found floating high in the atmosphere and in the fuel tanks of jetliners. Of course, rich bacterial communities are also found in and on the healthy human body. However, no single species of bacteria is as versatile as these examples may imply. Indeed, bacteria are specialists; those found in hot springs, for example, could thrive nowhere else. Bacteria found on the human body are also often specialized, inhabiting a single area such as the skin, the mouth, or the large intestine.

Bacterial Nutrition and Community Interactions

Bacterial invasion of diverse habitats is aided by their dietary versatility. Blue-green bacteria, discussed in more detail below, engage in plantlike photosynthesis. Other bacteria are **chemosynthetic,** deriving energy through reactions that combine oxygen with inorganic molecules such as sulfur, ammonia, or nitrite. In the process, they release sulfates or nitrates, crucial plant nutrients, into the soil. Many bacteria, called **anaerobes,** are not dependent on oxygen to extract energy. Some, such as the bacterium causing

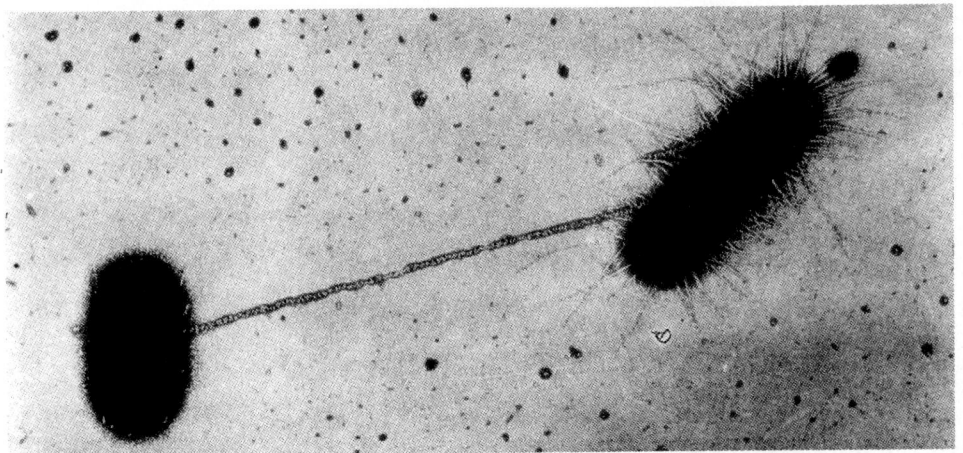

Figure 21-11 Conjugation—the transfer of genetic material—occurs through a special large, hollow sex pilus, which is shown here connecting a pair of *Escherichia coli.* One bacterium (seen here on the top right) acts as a donor, transferring DNA to the recipient. In this photo, the donor bacterium is bristling with nonsex pili that probably help it attach to surfaces.

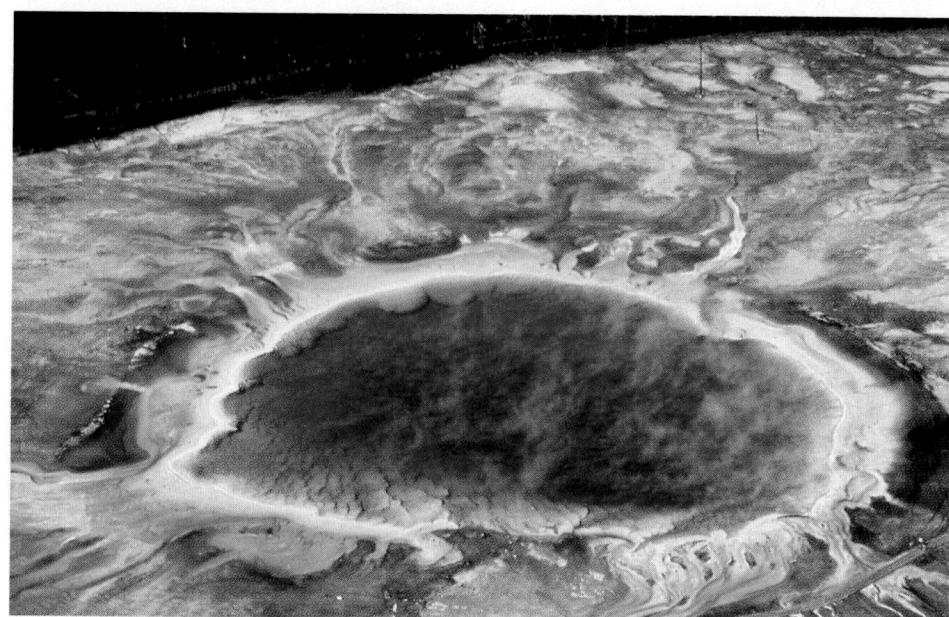

Figure 21-12 Hot springs harbor heat- and mineral-tolerant bacteria. Some cyanobacteria can tolerate temperatures up to 85° C (186° F). Several species of cyanobacteria paint these hot springs in Yellowstone National Park with vivid colors. The bacterial pigments aid in photosynthesis.

tetanus, are poisoned by oxygen. Others are opportunists, engaging in fermentation when oxygen is lacking and switching to cellular respiration (a more efficient process) when oxygen becomes available. Anaerobes such as the sulfur bacteria obtain energy from a unique type of bacterial photosynthesis. They use hydrogen sulfide (H_2S) instead of water (H_2O) in photosynthesis, releasing sulfur instead of oxygen.

Certain bacteria have the unusual ability to break down cellulose, the principal component of plant cell walls. Some of these have entered into a **symbiotic** (literally, "living together") relationship with cows, sheep, and goats, living in their digestive tracts and helping extract otherwise unavailable nutrients from the cell walls of plant fodder. You also host symbiotic bacteria that inhabit your intestines. These feed on undigested food and synthesize nutrients such as vitamin K and vitamin B_{12}, which your body absorbs. Another form of bacterial symbiosis of enormous ecological and economic importance is the growth of **nitrogen-fixing** bacteria in specialized nodules on the roots of certain plants (**legumes,** which include alfalfa, soybeans, lupines, and clover; Fig. 21-13). These bacteria capture nitrogen gas (N_2, which the plant cannot use directly) from air trapped in the soil and combine it with hydrogen to produce ammonium (NH_4^+), a form usable by the plant.

Most bacteria obtain energy by breaking down complex organic (carbon-containing) molecules, and the range of compounds attacked by bacteria is staggering. Nearly anything that human beings can synthesize, some bacteria can destroy. The term "biode-gradable" (meaning "broken down by living things") refers largely to the work of bacteria. Even oil is biodegradable. In the summer of 1989, the oil tanker Exxon-Valdez dumped 11 million gallons of oil into Prince William Sound, Alaska. Soon afterward, researchers from Exxon sprayed 70 miles of oil-soaked beaches around Prince William Sound with a fertilizer formulated to encourage the growth of naturally occurring oil-eating bacteria. Bacterial populations burgeoned, and within 15 days the oil deposits were noticeably reduced compared with unsprayed areas.

Bacteria have also become important in the production of human foods, including cheese, yogurt, and sauerkraut. The aging of meat tenderizes it through controlled bacterial digestion.

The appetite of some bacteria for nearly any organic compound is the key to their important role as decomposers in ecosystems. While feeding themselves, they break down the waste products and dead bodies of more complex life forms, freeing nutrients for reuse and allowing the recycling of nutrients (see Chapter 45) that provides the basis for continued life on Earth.

Bacteria and Human Health

The feeding habits of certain bacteria threaten our health and well-being. These bacteria, called **pathogens** (meaning "disease-producing"), synthesize toxic substances in the human body that cause disease symptoms. Some bacteria produce deadly toxins that enter the bloodstream and attack the nervous

(a)

(b)

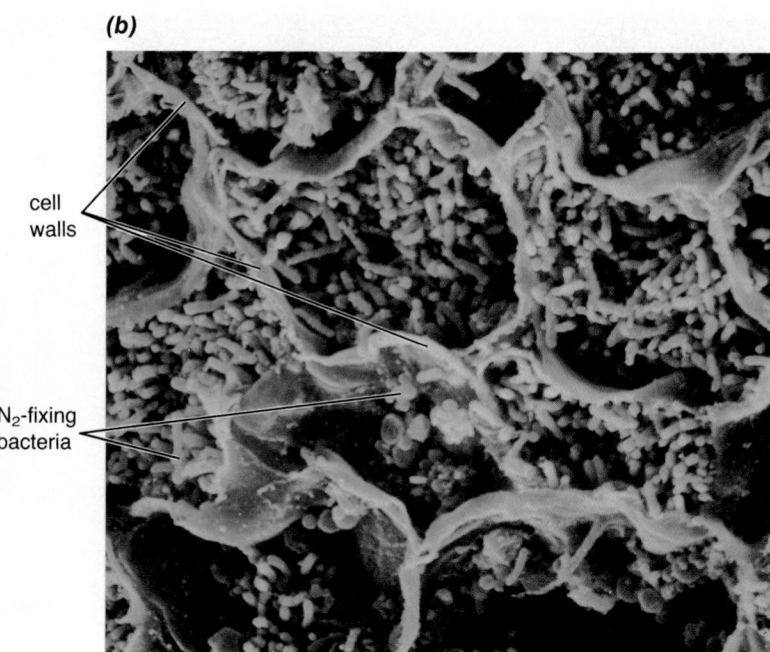

cell
walls

N₂-fixing
bacteria

Figure 21-13 (a) Special chambers called nodules on the roots of a legume (alfalfa) provide a protected and constant environment for nitrogen-fixing bacteria. **(b)** This scanning electron micrograph shows the nitrogen-fixing bacteria inside cells within the nodules.

system. One of these causes tetanus, and another botulism, a lethal form of food poisoning. These related bacteria are anaerobes that survive as spores until introduced into a favorable environment. A deep puncture wound protects tetanus bacteria from contact with oxygen, allowing them to multiply and release their paralyzing poison into the bloodstream. A sealed container of canned food that has been improperly sterilized provides a haven for botulism bacteria. These produce a toxin so potent that a single gram could kill 15 million people.

An allergic reaction to substances released by the bacterium *Streptococcus pneumoniae* results in the symptoms of pneumonia, in which the lungs become clogged with fluid. The plague, or "Black Death," which killed 100 million people during the fourteenth century, is caused by highly infectious bacteria spread by fleas carried by infected rats. Tuberculosis and leprosy are also bacterial diseases. Two bacterial diseases, **gonorrhea** and **syphilis,** transmitted through direct sexual contact, have reached epidemic proportions in modern society (see Chapter 39: "Health Watch").

The prevalence of **Lyme disease,** named after the town of Old Lyme, Connecticut, where it was first described in 1975, is rapidly increasing in the United States. This disease is caused by the spirochete bacte-

rium *Borrelia burgdorferi.* The bacterium is carried by the deer tick and transmitted to people who are bitten by the tick. Although the disease can be cured with antibiotics in its early stages, many cases have gone unidentified and untreated. Initial symptoms resemble flu symptoms, with chills, fever, and body aches. If untreated, weeks or months later the victim may experience rashes, bouts of arthritis, and sometimes abnormalities of the heart and nervous system. The disease has been reported in most states, but the vast majority of cases have originated in the Northeast and in Wisconsin and Minnesota. Fortunately, both physicians and lay people are becoming more familiar with the disease, so more victims are receiving treatment before serious symptoms develop.

These descriptions of bacterial assaults on the human body should not lead to an irrational hatred of all bacteria. There are far more bacterial cells living in and on a healthy human body than there are human cells composing it. Most of these bacteria are harmless, and some are beneficial. For example, the bacterial community in the female vagina creates an environment that is hostile to infections by parasites such as yeasts. As Lewis Thomas so aptly put it, "pathogenicity is, in a sense, a highly skilled trade, and only a tiny minority of all the numberless tons of microbes on the earth has ever been involved in it; most bacte-

HEALTH WATCH

"Unnatural" Selection—The Evolution of Drug-Resistant Pathogens

In the early 1950s, several antibacterial drugs, including penicillin, streptomycin, and tetracycline, became available to treat bacterial infections. (Penicillin and some other antibiotics, for example, interfere with the synthesis of the bacterial cell wall, causing the bacterium to swell and disintegrate [Fig. E21-2].) Penicillin was added to toothpaste, mouthwash, and chewing gum, and was also used indiscriminately to treat mild infections of all types.

It soon became apparent, however, that there were drawbacks to the indiscriminate use of antibiotics. Physicians found that certain bacteria, such as *Staphylococcus*, which can cause food poisoning, blood poisoning, and toxic shock syndrome, became increasingly difficult to kill with these drugs—the pathogens had developed antibiotic resistance.

Humans had unwittingly introduced a strong agent of natural selection into the microbial world. A variety of bacteria, such as *Staphylococcus*, which exists normally and harmlessly on human skin and in the nose and throat, were bathed frequently in weak antibiotic solutions. Those that, through mutation, developed resistance to the effects of the drugs (or a way to inactivate them) survived and flourished, while the less well-adapted bacteria perished. Excessive use of antibiotics unnecessarily increases the exposure of bacteria to these powerful selective agents and encourages the spread of resistant strains. When resistant bacteria then invade the body and cause disease, the antibiotics are useless. In the late 1950s, researchers discovered that the resistant bacteria could transfer their genes for drug resistance to other bacteria, even to members of other bacterial species. This dramatically shortens the time necessary for drug resistance to spread within and between bacterial populations.

In some countries, prescriptions for antibiotics are not required. One ominous outcome has been the emergence of a penicillin-resistant strain of gonorrhea as a result of the regular use of the drug as a preventive measure by prostitutes in southeast Asia. These resistant bacteria have now reached the United States and are spreading. The United States itself is far from blameless in hastening the selection of "super-germs." Both penicillin and tetracycline are still regularly added to animal feeds as growth promoters, despite years of protest by scientists and attempts by the FDA to ban this practice. A strain of *Salmonella* found in or on farm animals, which is responsible for a virulent form of food-poisoning, has now developed resistance to antibiotics. This is prob-

ably a result of continued exposure, as they reside on animals fed antibiotics. These bacteria contaminate raw meat, and are believed to have caused an outbreak of severe *Salmonella* poisoning in the midwest in 1983. Antibiotics continue to be prescribed inappropriately; for example, about two thirds of the prescriptions written for the treatment of colds are for antibiotics, even though most colds and sore throats are viral in origin and are not helped by these drugs. Amoxicillin, an antibiotic that has helped millions of babies overcome painful ear infections, is losing its effectiveness. Tetracycline can no longer be used to treat gonorrhea, and is losing its effectiveness against urinary tract infections, meningitis, and respiratory tract infections.

Researchers are working to develop new antibiotics, but drug development and testing take many years, while resistant bacterial strains spread rapidly. Moreover, most of the new drugs are more toxic and far more expensive than those they replace. Clearly we must restrict our use of antibiotics to situations in which they are urgently required, rather than relying on a steady flow of new ones. Awareness and restraint may yet enable our children to benefit from some of the same "miracle drugs" that protected our parents.

(a) **(b)**

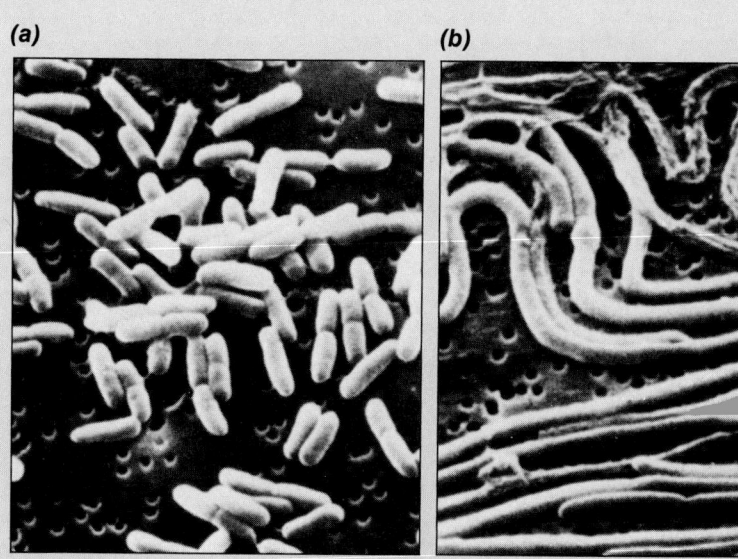

Figure E21-2 The effect of antibiotic on bacterial cell walls: **(a)** normal bacteria; **(b)** bacteria treated with antibiotic show dramatic elongation due to the antibiotic's interference with proper production of the cell wall.

ria are busy with their own business, browsing and recycling the rest of life."

Cyanobacteria

Like other bacteria, blue-green or cyanobacteria (*cyan* is Greek for "dark blue") are widespread, making their home on snowfields, in hot springs that may reach 85° C (186° F), in oceans, lakes, ponds, and in moist soils. Since most are aerobic and all are photosynthetic, they can exist only where light and oxygen are available. Like green plants, cyanobacteria possess chlorophyll and produce oxygen as a by-product of photosynthesis. Since the bacteria lack organelles, including chloroplasts, chlorophyll is located on special membranes inside the cell (Fig. 21-14a). In addition to trapping solar energy, most cyanobacteria can acquire nitrogen from the atmosphere, making them extremely self-sufficient nutritionally. Some cyanobacteria form chains of cells that are unique among prokaryotes because they have a rudimentary division of labor. In these filamentous colonies (Fig. 21-14b), a few cells capture atmospheric nitrogen while the rest photosynthesize. When new islands are formed from volcanic eruptions, cyanobacteria are among the first colonizers of the bare rock, their activities preparing the way for more complex and less nutritionally self-sufficient life forms.

Archaebacteria

Recently, a unique group of bacteria—the archaebacteria—have come under close scientific scrutiny. Experimental findings have revealed such dramatic differences between the archaebacteria and other Monera that some taxonomists believe they deserve the status of their own kingdom. Although this proposal is disputed, there is no question that the archaebacteria are different. The lipids of their cell membranes differ considerably from those of both eukaryotic and other prokaryotic cells, as do the composition of their cell walls and the sequence of subunits in their ribosomal RNA.

Archaebacteria include **methanogens,** anaerobic bacteria that convert carbon dioxide to methane (sometimes called "swamp gas"). Methanogens are found in such diverse habitats as swamps, sewage-treatment plants, hot springs, deep-sea vent communities (see Chapter 46), and the stomachs of cows. Other archaebacteria include **halophiles,** bacteria that thrive in concentrated salt solutions such as the Dead Sea, and **thermoacidophiles,** which, as their name implies, thrive in hot, acidic environments such as hot sulfur springs.

The extreme environments in which these organisms thrive, although rare now, were far more common when life first appeared on Earth—hence the name *archaebacteria* (meaning "ancient bacteria").

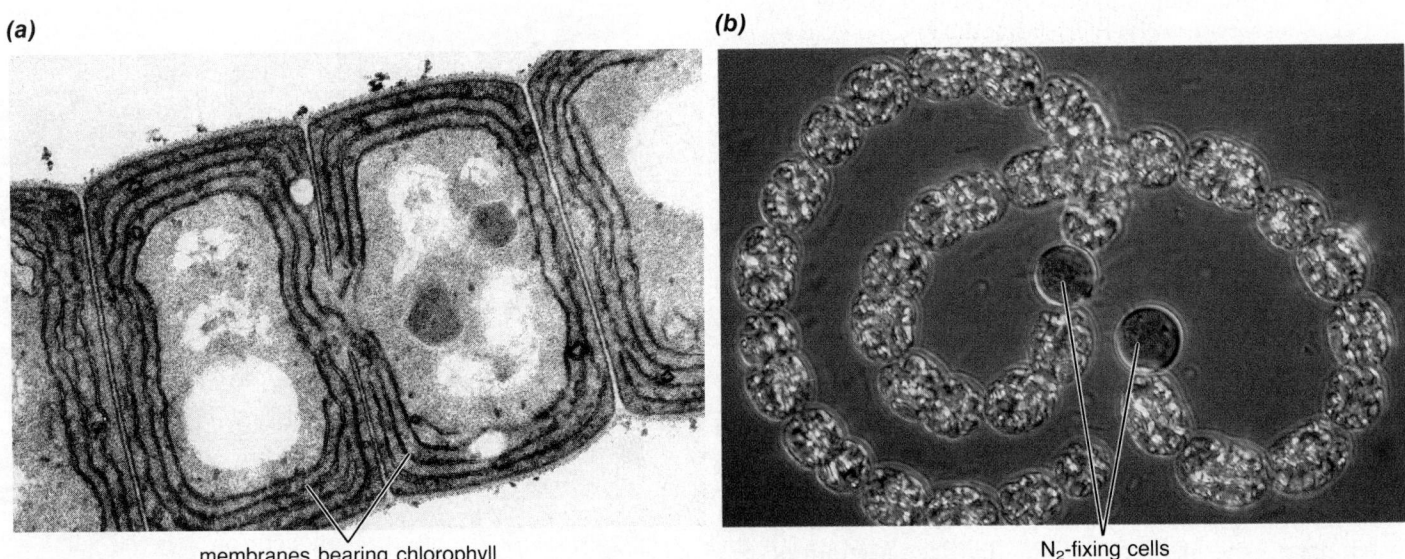

(a)   (b)

membranes bearing chlorophyll   N₂-fixing cells

Figure 21-14 (a) Electron micrograph of a section through a cyanobacterial filament (genus *Oscillatoria*). Chlorophyll is located on the membranes visible within the cells. **(b)** Simple division of labor, rare among prokaryotic cells, is seen in this filamentous cyanobacterium (genus *Nostoc*). The larger cells are specialized for nitrogen fixation.

However, recent analysis of archaebacterial RNA nucleotide sequences has revealed that they are actually more similar to eukaryotic cells than are the eubacteria, and probably evolved about half a billion years later.

The Kingdom Protista

Since Anton van Leeuwenhoek first observed protists through his simple homemade microscope in 1674, at least 50,000 different species have been described. Protists have one thing in common: *each consists of a single eukaryotic cell.*

Most protists can reproduce asexually by mitotic cell division, but many are capable of a form of sexual reproduction, called conjugation, as well (Fig. 21-15). All three major modes of nutrition are represented in this kingdom: the unicellular algae trap solar energy through photosynthesis; predatory protists ingest their food; and parasitic forms, some flagellates, and the versatile euglenoids can absorb nutrients from their surroundings.

The Classification of Protists

The ancestors of modern protists are believed to have given rise to the three kingdoms of multicellular organisms: the fungi, plants, and animals (see Fig. 20-

3). The kingdom Protista is an extremely diverse group, including animallike, plantlike, and fungus-like forms, which have sometimes been classified as animals, plants, or fungi. The presence of both phyla and divisions within this kingdom reflects earlier classification schemes.

To the confusion of taxonomists, many protists (such as the euglenoids) fit equally well into animal-like or plantlike categories. *Euglena* (see Fig. 21-19), for example, has a photoreceptor and can swim toward a stimulus, features commonly associated with animals, but it uses these abilities to seek light levels appropriate for photosynthesis, a process associated with plants.

Protistan taxonomy is still the subject of revision and controversy. Here we discuss members of the kingdom Protista in three categories: the plantlike unicellular algae, the funguslike slime molds, and the animallike protozoa (Table 21-1).

The Plantlike Protists: The Unicellular Algae

Often called **phytoplankton** (literally, "floating plants"), these photosynthetic protists are widely distributed in oceans and lakes. Although microscopic in size, their importance is immense. Marine

(a)

(b)

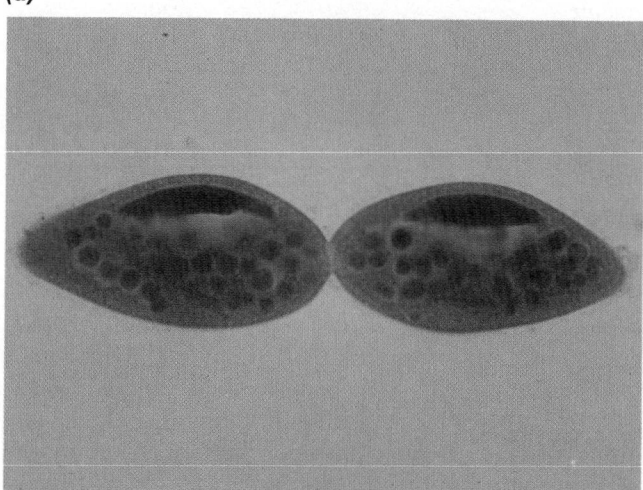
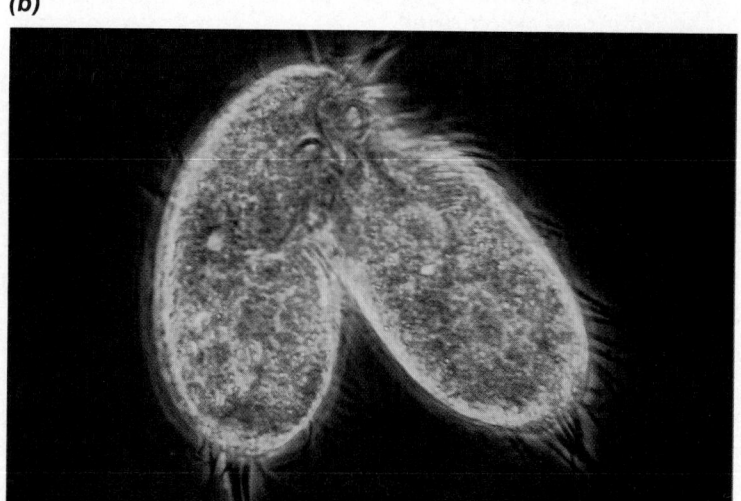

Figure 21-15 Two modes of reproduction in protists. **(a)** *Paramecium*, a ciliate, reproduces asexually by cell division that results in two daughters identical to the original parent. **(b)** Mating in a ciliate, *Euplotes*. Genetic material is exchanged across a cytoplasmic bridge. After the exchange occurs, new individuals formed by cell division will have gene combinations different from those of either parent cell.

Table 21-1 The Major Divisions of Protists

General Category	Division/Phylum	Locomotion	Nutrition	Other Features	Representative Genera
Plantlike protists: unicellular algae	Division Pyrrophyta—dinoflagellates	swim with two flagella	autotrophic; photosynthetic	many bioluminescent; often have cellulose wall; most marine	*Gonyaulax* (causes red tide)
	Division Chrysophyta—diatoms	glide along surfaces	autotrophic; photosynthetic	have silica shells; most marine	*Navicula* (glides toward light)
	Division Euglenophyta—euglenoids	swim with one flagellum	autotrophic; photosynthetic	have an eyespot; all freshwater	*Euglena* (common pond-dweller)
Funguslike protists: slime molds	Division Myxomycota—plasmodial slime molds	sluglike mass crawls over surfaces	heterotrophic	form multinucleate plasmodium	*Physarum* (forms a large bright orange mass)
	Division Acrasiomycota—cellular slime molds	amoeboid cells extend pseudopodia; sluglike mass crawls over surfaces	heterotrophic	form pseudoplasmodium with individual amoeboid cells	*Dictyostelium* (often used in laboratory studies)
Animallike protists: protozoa	Phylum Sarcomastigophora—zooflagellates	swim with flagella	heterotrophic	inhabit soil or water or may be parasitic	*Trypanosoma* (causes African sleeping sickness)
	and sarcodines	extend pseudopodia	heterotrophic	both naked and shelled forms exist	*Amoeba* (common pond-dweller)
	Phylum Apicomplexa—sporozoans	nonmotile	heterotrophic; all parasitic	form infectious spores	*Plasmodium* (causes malaria)
	Phylum Ciliophora—ciliates	swim with cilia	heterotrophic	most complex single cells	*Paramecium* (fast-moving pond-dweller)

phytoplankton account for nearly 70% of all the photosynthetic activity on Earth, thus supporting the complex web of aquatic life.

There are three major divisions of plantlike protists: the division Pyrrophyta, also called dinoflagellates; the division Chrysophyta, the diatoms; and the division Euglenophyta, the euglenoids.

Dinoflagellates

These photosynthetic protists are so named ("dino" is Greek for "two") because of their two whiplike **flagella.** One flagellum encircles the cell, located within a special groove, while the second projects behind it. Some dinoflagellates are bounded only by a cell membrane, while many are covered with cellulose walls that resemble armor plates (Fig. 21-16).

Although some live in fresh water, dinoflagellates are especially abundant in the ocean, where they are an important food source for larger organisms. Many dinoflagellates are bioluminescent, producing a brilliant blue-green light when disturbed. Clear waters inhabited by these protists take on a magical quality after sunset as the bodies of fish or swimmers are silhouetted in shimmering radiance. Specialized di-

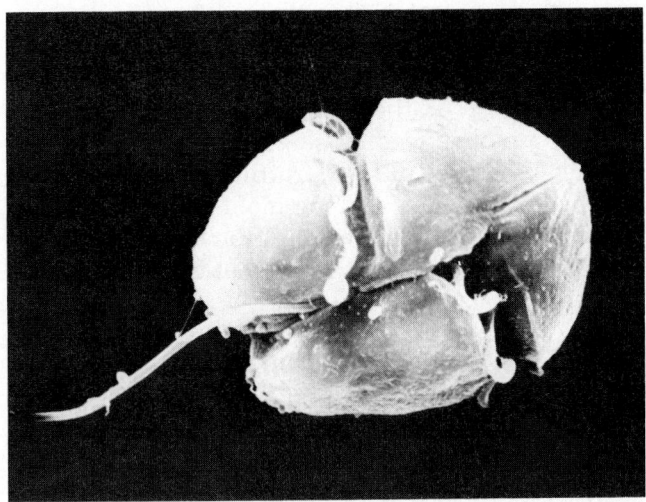

Figure 21-16 A dinoflagellate, covered with protective cellulose armor. Two flagella lie within the grooves encircling the body.

Figure 21-17 A red tide. The explosive reproductive rate of certain dinoflagellates under the right conditions can produce concentrations so great that their microscopic bodies dye the sea red, as in this bay in Mexico.

noflagellates, known as zooxanthellae, live within the tissues of corals and some clams, where they provide photosynthetic nutrients and remove carbon dioxide. Dinoflagellates limit the distribution of many corals to relatively shallow, well-lit waters.

The green chlorophyll in dinoflagellates is often masked by red pigments that help trap light energy. Under certain conditions, when the water is warm and rich in nutrients, a dinoflagellate population explosion occurs. These microorganisms can become so numerous that the waters are dyed red by the color of their bodies, causing a "red tide" (Fig. 21-17). Fish die by the thousands, suffocated by clogged gills or by oxygen depletion resulting from the decay of the bodies of billions of dinoflagellates. But oysters, mussels, and clams have a feast, filtering millions from the water for food. In the process, however, they concentrate a nerve poison produced by the dinoflagellates. During red tides, people or other animals feeding on these mollusks may be stricken with potentially lethal paralytic shellfish poisoning.

Diatoms

The photosynethic **diatoms,** found in both fresh and salt water, are so important to marine food webs that they have been called the "pastures of the sea." They produce glassy protective coverings, some of exceptional beauty (Fig. 21-18; see also chapter opener photo). These consist of top and bottom halves that fit together like a pillbox or Petri dish. Accumulations of the glassy walls of diatoms over thousands of years have produced fossil deposits of "diatomaceous

Figure 21-18 Some representative diatoms, illustrating the intricate, microscopic beauty of their glassy walls.

earth" that may be hundreds of meters thick. This slightly abrasive substance is widely used in products such as toothpaste and metal polish. Diatoms store reserve food as oil; their buoyancy in water helps their bodies to float near the surface, where light is abundant for photosynthesis. Prehistoric accumulations of diatoms and their stored oil may contribute to today's petroleum reserves.

Euglenoids

This group of protists is named after its best-known representative, *Euglena* (Fig. 21-19), a complex single cell that locomotes by whipping its flagellum through the water. Its simple light-sensing organelles consist of a photoreceptor, also called an *eyespot*, at the base of the flagellum, and an adjacent patch of pigment. The pigment shades the photoreceptor only when light impinges from certain directions, allowing *Euglena* to determine the direction of the light source. Using information from the photoreceptor, the flagellum propels the protist toward light levels appropriate for photosynthesis. All euglenoids live in fresh water, and in contrast to diatoms or dinoflagellates, they lack a rigid outer covering. This allows some to locomote by wriggling as well as by whipping the flagellum. If *Euglena* is maintained in darkness, it loses its chloroplasts, but can still absorb nutrients from its surroundings. In this state it closely resembles the animallike zooflagellates described below.

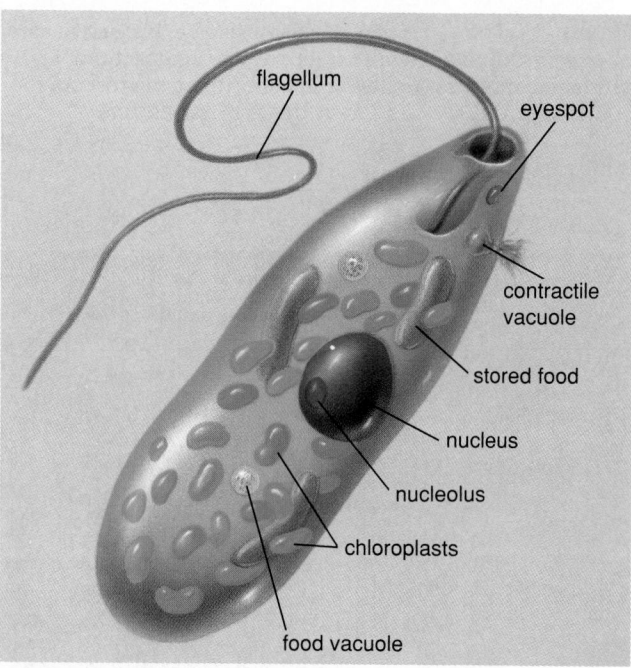

Figure 21-19 *Euglena*, a representative euglenoid, showing its elaborate, single-celled structure. The cell is packed with green chloroplasts, which will disappear if the protist is kept in darkness.

The Funguslike Protists: The Slime Molds

The life cycle of the slime mold consists of two phases: a mobile feeding stage, and a stationary reproductive stage called a **fruiting body.** There are two major divisions of slime molds: the acellular, or **plasmodial, slime molds** of the division Myxomycota, and the **cellular slime molds** of the division Acrasiomycota.

The Plasmodial Slime Molds

The acellular slime molds consist of a mass of cytoplasm that may spread thinly over an area of several square meters. Although the mass contains thousands of nuclei, the nuclei are not confined in discrete cells surrounded by cell membranes, as in most multicellular organisms. This structure, called a **plasmodium,** explains why these protists are described as "acellular" (without cells). The plasmodium oozes through decaying leaves and rotting logs, engulfing food such as bacteria and particles of organic material. The mass may be bright yellow or orange—a large plasmodium can be rather startling (Fig. 21-20a). Dry conditions or starvation stimulate the plasmodium to form a fruiting body, on which spores are produced (Fig. 21-20b). The spores are dispersed and germinate under favorable conditions, eventually giving rise to a new plasmodium.

The Cellular Slime Molds

The cellular slime molds live in soil as independent haploid amoeboid cells, extending **pseudopods** to engulf food such as bacteria. In the best-studied genus, *Dictyostelium*, individual cells release a chemical signal when food becomes scarce. This signal attracts nearby cells into a dense aggregation that forms a sluglike mass called a **pseudoplasmodium** ("false plasmodium"), since it consists of individual cells. The pseudoplasmodium then behaves like a multicellular organism. After crawling toward a source of light, the cells in the aggregation take on specific roles, forming a fruiting body. Haploid spores formed within the fruiting body are dispersed by the wind and germinate directly into new amoebalike individuals (Fig. 21-21).

The Animallike Protists: The Protozoa

The animallike protists, or **protozoa** (literally, "first animals") are placed into three major phyla: phylum Sarcomastigophora, which includes the zooflagellates and sarcodines (amoebas); phylum Apicomplexa, the

(a)

(b)

Figure 21-20 (a) The acellular slime mold *Physarum* oozes over a stone on the damp forest floor. **(b)** When food becomes scarce, the mass differentiates into black fruiting bodies in which spores are formed.

sporozoans; and phylum Ciliophora, the ciliates. All are unicellular, eukaryotic, and heterotrophic, but they differ in their methods of locomotion.

Zooflagellates

These protists all possess at least one flagellum. This versatile organelle may propel the organism, sense the environment, or ensnare food. The zooflagellates are a diverse group believed to be ancestral to the other protists. Many are free-living, inhabiting soil and water, while others are symbiotic, living inside other organisms in a relationship that may be either mutually beneficial or parasitic. One symbiotic form can digest cellulose and lives in the gut of termites, where it helps them extract energy from wood. A zooflagellate of the genus *Trypanosoma* is responsible for African sleeping sickness, a potentially fatal disease (Fig. 21-22). Like many parasites, this organism has a complex life cycle, part of which is spent in the tsetse fly, which transmits it to mammals while feeding on their blood. The parasite then develops in the host (which may be a person), entering the bloodstream. It may then be ingested by another tsetse fly that bites the host, thus beginning a new cycle of infection.

Another parasitic zooflagellate, *Giardia*, is an increasing problem in the United States, particularly to backpackers who drink from apparently pure mountain streams. Cysts of this flagellate are released in the feces of infected human beings or other animals (a single gram of feces may contain 300 million cysts) and enter freshwater streams and even community reservoirs. Cysts develop into the adult form (Fig. 21-23) in the small intestine of their mammalian host. In human beings, infections may cause severe diarrhea, dehydration, nausea, vomiting, and cramps. Fortunately, deaths are rare and infections may be cured with drugs.

Sarcodines

Also called *amoebae*, these protists possess flexible cell membranes that they can extend in any direction to form pseudopodia (literally, "false feet") used for locomotion and for engulfing food (Fig. 21-24). Amoebae lack many of the specialized organelles found in flagellates and ciliates, but their reputation as "blobs of jelly" is contradicted by their complex internal structure and their sophisticated ability to sense and capture prey. An important parasitic form, particularly common in warm climates, causes amoebic dysentery. Multiplying in the intestinal wall, this parasite causes severe diarrhea, and may perforate the intestine, occasionally causing fatal infections.

Heliozoans, a striking form of freshwater sarcodine, may be found floating in ponds or attached by stalks to an underwater plant or rock (Fig. 21-25). They have stiff, needlelike pseudopodia, each of which is supported internally by a bundle of microtubules. Some heliozoans cover themselves with intricate and delicate shells of silica.

The **foraminiferans** and **radiolarians** are primarily marine sarcodines that also produce beautiful and elaborate shells. Shells of foraminiferans are constructed mostly of calcium carbonate (chalk), while the radiolarians secrete shells of silica (glass) (Fig. 21-26). These elaborate shells are pierced by myriad openings through which pseudopods extend. The chalky shells of foraminiferans, accumulating over millions of years, have resulted in immense deposits of limestone such as form the famous white cliffs of Dover, England.

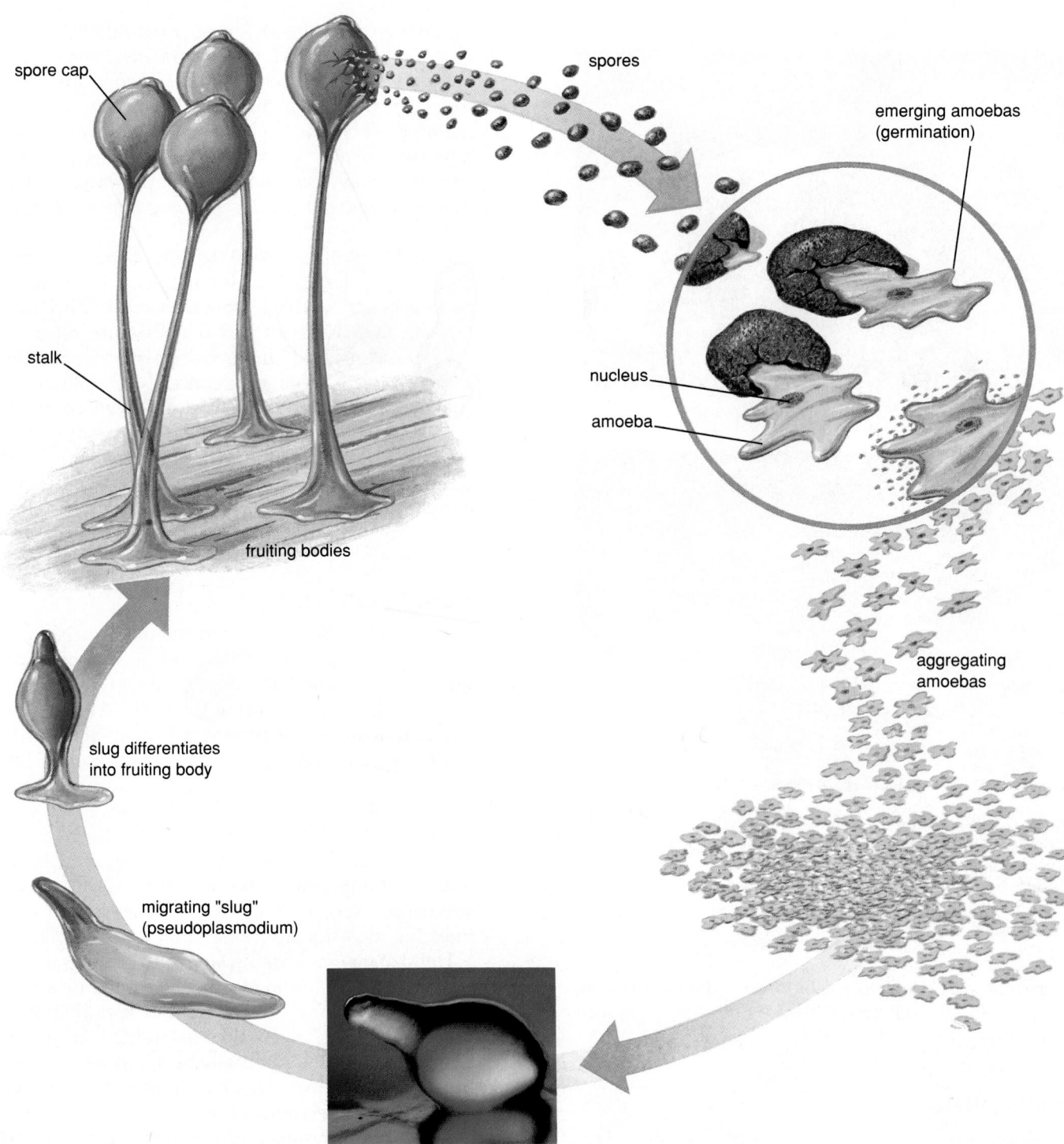

Figure 21-21 The life cycle of a cellular slime mold. Single amoebalike cells crawl and feed. When food becomes scarce, they aggregate into a sluglike mass called a pseudoplasmodium (inset). The pseudoplasmodium crawls toward the light, and then forms a fruiting body, in which haploid spores are produced.

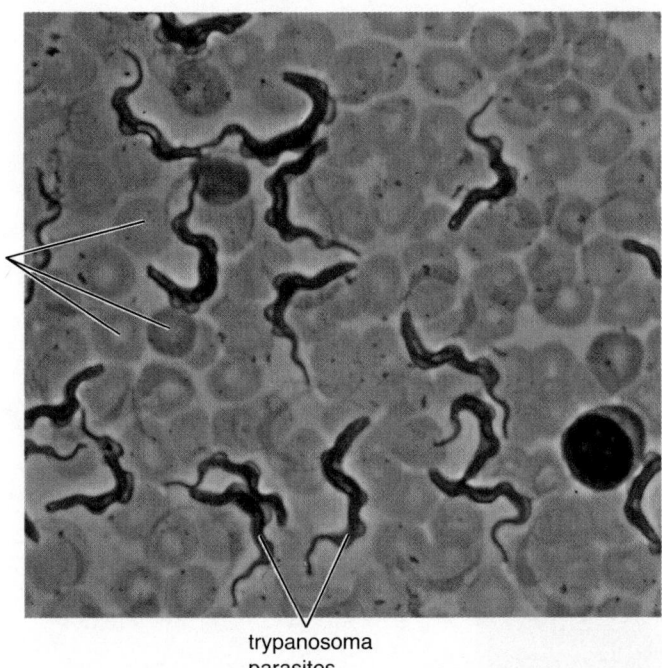

ood

trypanosoma
parasites

Figure 21-22 A photomicrograph showing human blood that is heavily infested with the parasitic zooflagellate. *Trypanosoma*, which causes African sleeping sickness.

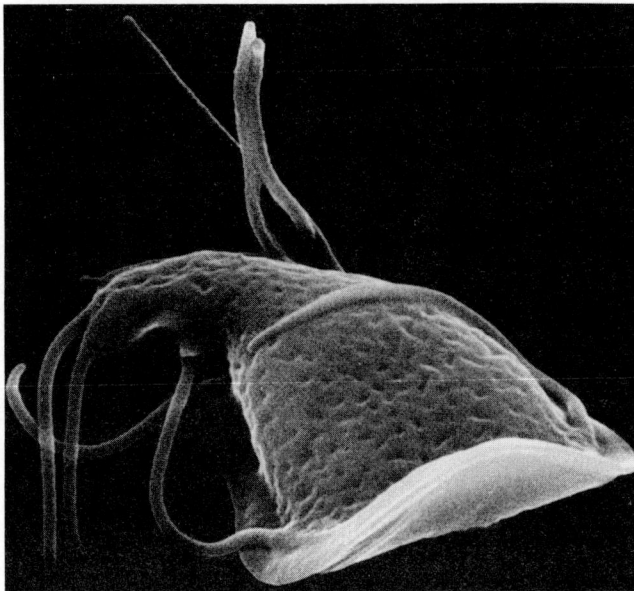

Figure 21-23 A zooflagellate (genus *Giardia*) that may infect drinking water, causing gastrointestinal disorders.

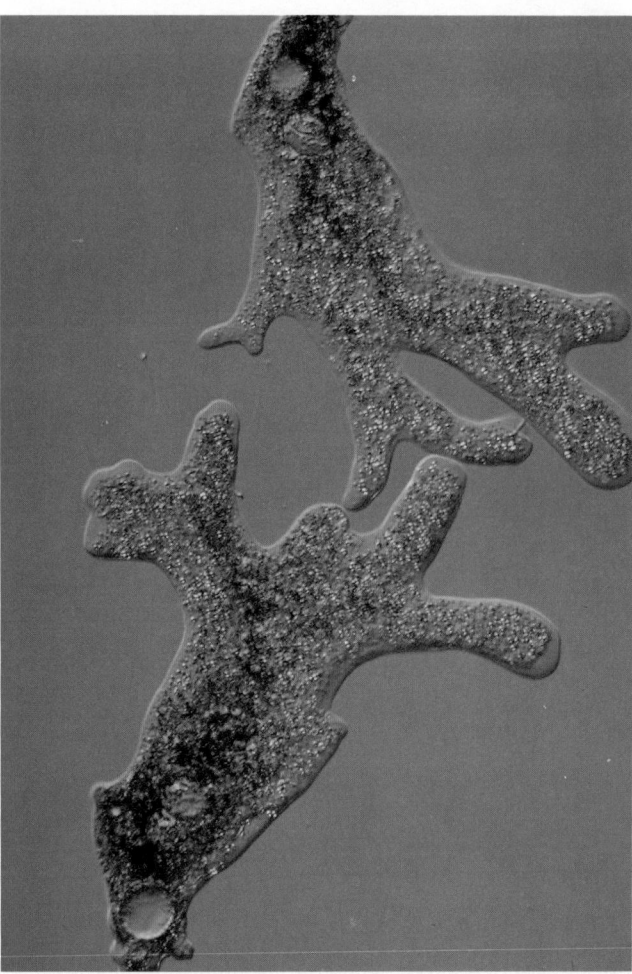

Figure 21-24 *Amoeba proteus* uses cytoplasmic projections called pseudopodia to move about and to capture prey.

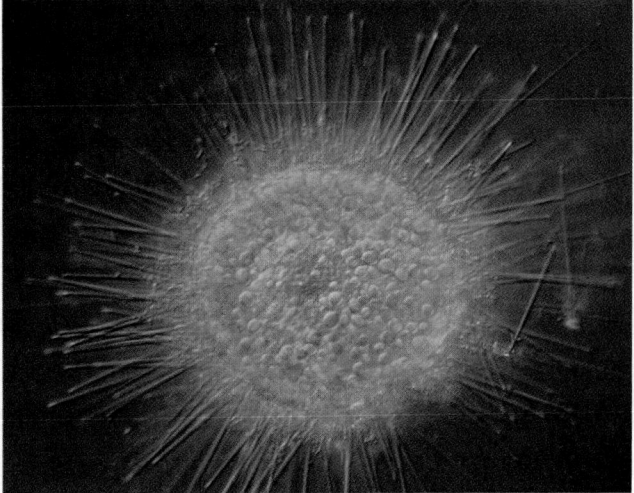

Figure 21-25 Heliozoans are beautiful freshwater sarcodines; the needlelike pseudopodia are clearly visible in this specimen (genus Acanthocystis).

(a)   **(b)**

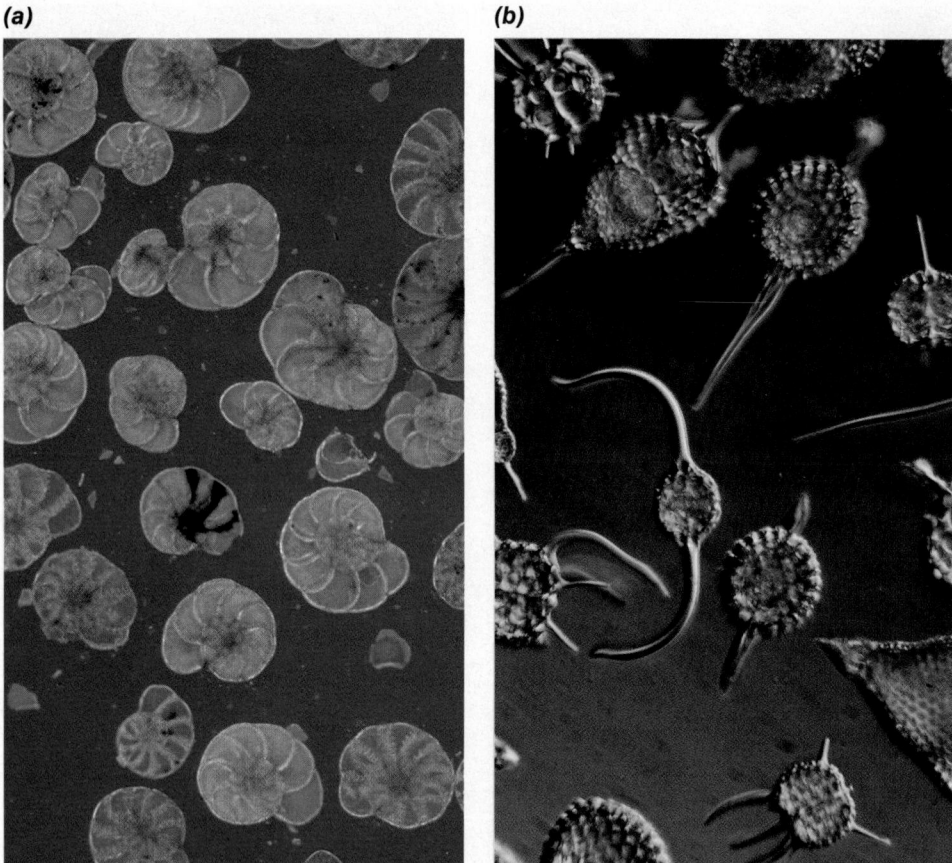

Figure 21-26 (a) The chalky shells of foraminiferans show numerous interior compartments. **(b)** The delicate, glassy shells of radiolarians. Pseudopodia extend out through the openings, sensing the environment and capturing food.

Sporozoans

These specialized protozoa are all parasites, living inside the bodies and sometimes inside the individual cells of their hosts. They are named after their ability to form infectious spores, resistant structures transmitted from one host to another through food, water, or the bite of an infected insect. As adults, sporozoans have no means of locomotion. Many have complex life cycles, a common feature of parasites. A well-known example is the malarial parasite *Plasmodium* (Fig. 21-27). Parts of its cycle are spent in the stomach, and later the salivary glands, of the *Anopheles* mosquito. When the mosquito bites a person, it passes the *Plasmodium* to the unfortunate victim. The sporozoan develops in the liver, then enters the blood, where it reproduces rapidly in human red blood cells. The synchronized release of spores through rupture of the blood cells causes the recurrent fever of malaria. Uninfected mosquitos may acquire the parasite by feeding on the blood of a malaria

victim, spreading the parasite when it bites another person, thus continuing the infectious cycle.

Although the drug chloroquine kills the malarial parasite, unfortunately, drug-resistant populations of *Plasmodium* are rapidly spreading throughout Africa, where the disease is prevalent. Programs to eradicate mosquitos have also failed, because the mosquitos rapidly evolve resistance to pesticides. Researchers hope to use genetic engineering to breed strains of mosquito that kill, rather than transmit, the sporozoan parasite.

Ciliates

These inhabitants of fresh or salt water represent the peak of unicellular complexity. They possess many specialized organelles, including the **cilia** after which they are named. Their cilia may cover the cell, or they may be localized, as in *Didinium* (see Fig. 21-29). In the well-known freshwater genus *Paramecium* (Fig. 21-28), rows of cilia cover the entire body sur-

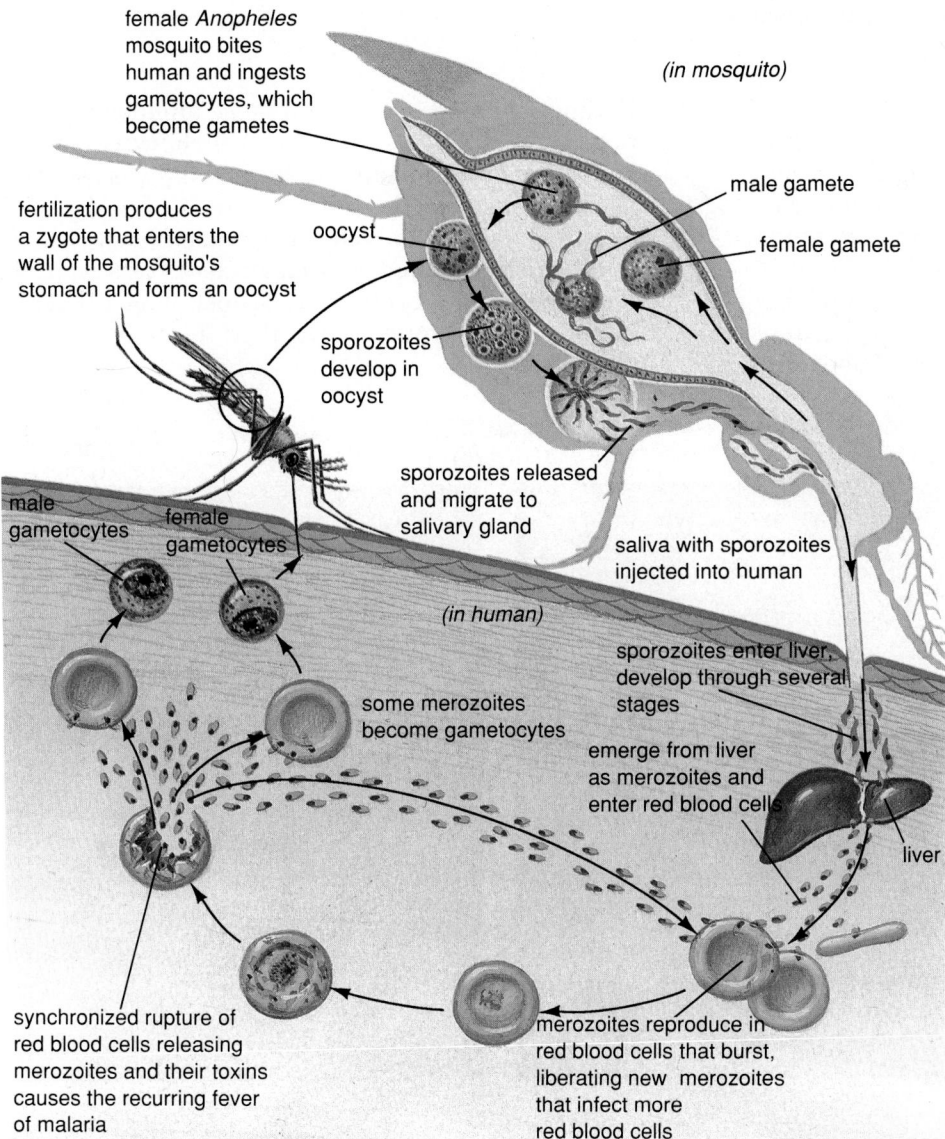

female *Anopheles* mosquito bites human and ingests gametocytes, which become gametes

(in mosquito)

fertilization produces a zygote that enters the wall of the mosquito's stomach and forms an oocyst

oocyst

male gamete

female gamete

sporozoites develop in oocyst

sporozoites released and migrate to salivary gland

male gametocytes

female gametocytes

(in human)

saliva with sporozoites injected into human

sporozoites enter liver, develop through several stages

some merozoites become gametocytes

emerge from liver as merozoites and enter red blood cells

liver

synchronized rupture of red blood cells releasing merozoites and their toxins causes the recurring fever of malaria

merozoites reproduce in red blood cells that burst, liberating new merozoites that infect more red blood cells

Figure 21-27 The life cycle of the malarial parasite (genus *Plasmodium*). In this complex life cycle, humans and mosquitos serve as alternate hosts.

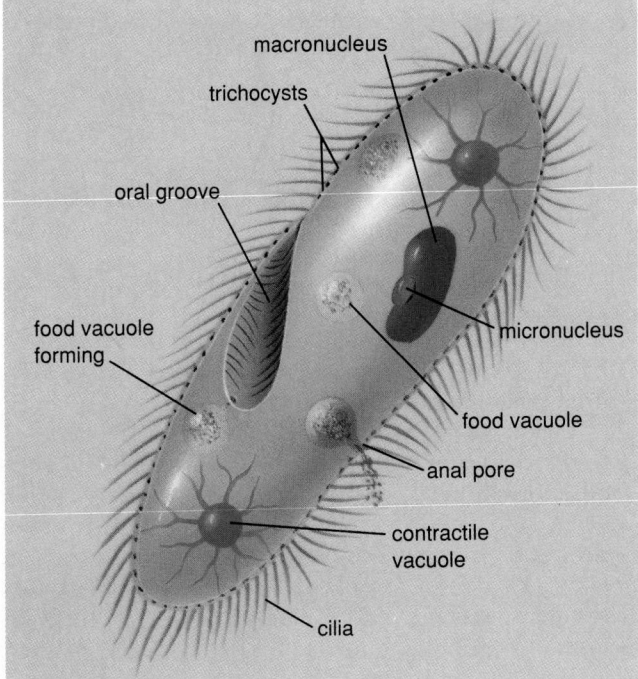

macronucleus

trichocysts

oral groove

micronucleus

food vacuole forming

food vacuole

anal pore

contractile vacuole

cilia

Figure 21-28 *Paramecium,* illustrating some important ciliate organelles. The oral groove acts as a mouth, and food vacuoles, miniature digestive systems, form at its apex. The contractile vacuoles regulate water balance. Trichocysts help this predator stun its prey, while cilia propel it rapidly through the water.

face. Their coordinated beating propels the cell through the water at a protistan speed record of a millimeter per second. Although only a single cell, *Paramecium* responds to its environment as if it had a well-developed nervous system. Confronted with some noxious chemical or with a physical barrier, the cell immediately backs up by reversing the beating of its cilia and then proceeds in a new direction. Ciliates are accomplished predators (Fig. 21-29). Some, in-

cluding *Paramecium* and *Didinium*, immobilize their prey with explosive darts called **trichocysts** embedded in the outer covering of the cell. Prey is escorted to a mouthlike opening, the oral groove. It is digested in a food vacuole, which forms a temporary "stomach," and is excreted by exocytosis. Excess water is accumulated in a contractile vacuole, which periodically contracts, emptying the fluid through a pore to the outside.

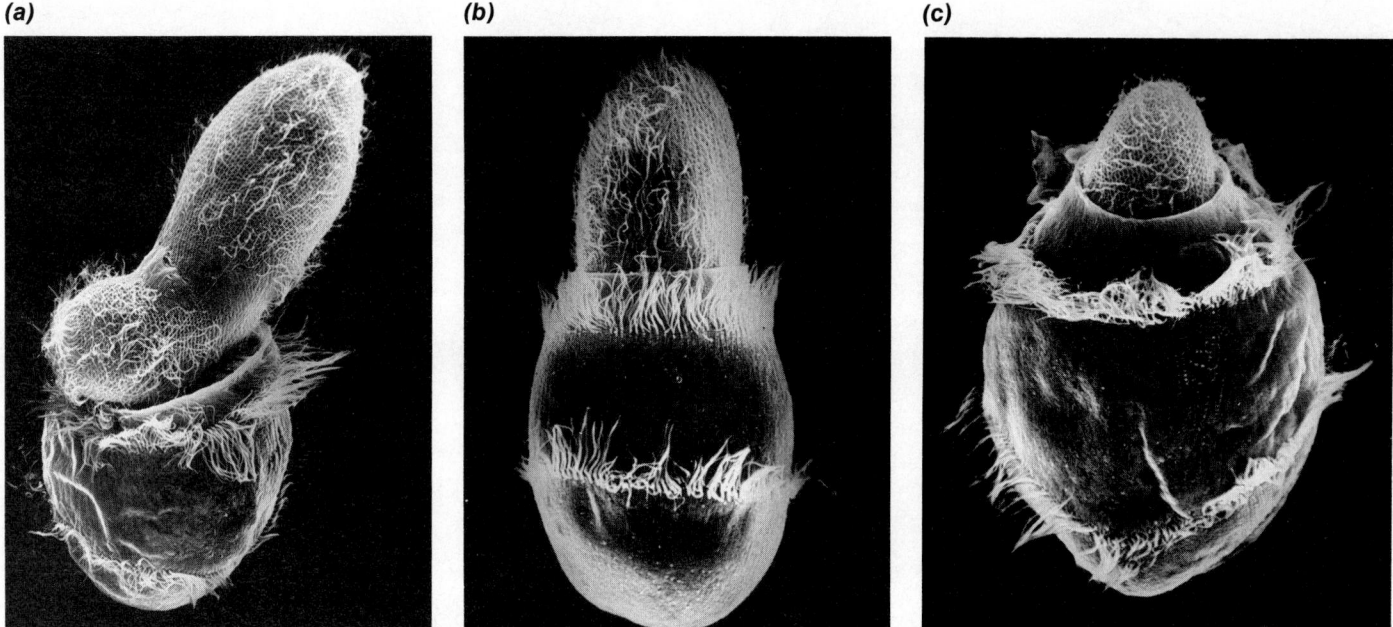

(a)   *(b)*   *(c)*

Figure 21-29 In this series of scanning electron micrographs. *Paramecium* is stung **(a)** and gradually engulfed **(b, c)** by another predatory ciliate, *Didinium*, whose cilia are confined to two bands encircling the egg-shaped body. This microscopic drama could occur on a pinpoint with room to spare.

Reflections on Our Unicellular Ancestors

In the kingdoms Monera and Protista, we find persisting today the types of cellular organization that gave rise to the complex multicellular organisms that we are most familiar with. Modern monerans have changed little from their ancestors, whose fossilized remains date back $3\frac{1}{2}$ billion years. Life might still consist of prokaryotic single cells if the protists, with their radical eukaryotic design, had not appeared on the scene nearly $1\frac{1}{2}$ billion years ago. As you learned in Chapters 5 and 19, eukaryotic cells may have originated when one moneran, perhaps a bacterium capable of cellular respiration, took up residence inside a partner, forming the first "mitochondrion." A separate but equally crucial merger may have

occurred when a photosynthetic bacterium (probably resembling a blue-green bacterium) took up residence within a nonphotosynthetic partner and became the first "chloroplast." The foundations of multicellularity were laid with the eukaryotic cell, whose intricacy allowed specialization of entire cells for specific functions within a multicellular aggregation. Thus primitive protists, some consuming their food in chunks, some photosynthesizing, and others absorbing nutrients from the environment, almost certainly followed divergent evolutionary paths that led to the three multicellular kingdoms predominating today: the animals, plants, and fungi, which are the subjects of the following three chapters.

SUMMARY OF KEY CONCEPTS

Microorganisms are placed into two kingdoms, Monera and Protista, which can be distinguished by their fundamentally different cell types. Monera consist of tiny prokaryotic cells lacking organelles, such as nuclei, mitochondria, and chloroplasts. Protists are eukaryotic cells that possess the full range of organelles and resemble the cells of multicellular organisms. Without the photosynthetic, nitrogen-trapping, and decomposing abilities of the bacteria, and the photosynthetic activities of protists, life as we know it would grind to a halt.

Viruses

Viruses are parasites consisting of a protein coat surrounding genetic material. They are noncellular and unable to move, grow, or reproduce outside a living cell. They invade cells of a specific host and use the host cell's energy, enzymes, and ribosomes to produce more virus particles, which are liberated when the cell ruptures. Many viruses are pathogenic to humans, including those causing colds and flu, herpes, AIDS, and certain forms of cancer.

Viroids are short strands of RNA that can invade the nucleus and direct synthesis of new viroids. To date, viroids are only known to cause certain diseases of plants.

Prions have been implicated in neurodegenerative diseases such as kuru and scrapie. Prions are unique in that they lack genetic material. They are composed solely of protein, which may act as an enzyme catalyzing the formation of more prions, or which may activate a gene to produce such an enzyme.

The Kingdom Monera

Members of the kingdom Monera—the bacteria—are unicellular and prokaryotic. A cell wall of peptidoglycan determines their characteristic shape: coccus (round), bacillus (rodlike), or spiral. Many form spores that disperse widely and withstand inhospitable environmental conditions. Bacteria obtain energy in a variety of ways. Some, including the cyanobacteria, rely on photosynthesis. Others are chemosynthetic, breaking down inorganic molecules to obtain energy. Heterotrophic forms are capable of consuming a wide variety of organic compounds. Many are anaerobic, able to obtain energy from fermentation when oxygen is not available.

Some bacteria are pathogenic, causing disorders including pneumonia, tetanus, botulism, and the venereal diseases gonorrhea and syphilis. Most, however, are harmless to humans and play important roles in natural ecosystems. Bacteria have colonized nearly every habitat on Earth. Some live in the digestive tracts of larger organisms such as cows and sheep, where they break down cellulose. Nitrogen-fixing bacteria enrich the soil and aid in plant growth, while many others live off the dead bodies and wastes of other organisms, liberating nutrients for reuse.

Cyanobacteria engage in plantlike photosynthesis and can also fix nitrogen, making them nutritionally very self-reliant.

The archaebacteria comprise a unique and diverse group that flourish under extreme conditions, including hot, acidic, very salty, and anaerobic environments. They differ in several ways from all other bacteria, including cell wall composition, ribosomal RNA sequence, and cell membrane lipid structure.

The Kingdom Protista

The kingdom Protista consists of organisms composed of single, highly complex eukaryotic cells. Protists may be placed in three major groups: the plantlike unicellular algae, the funguslike slime molds, and the animallike protozoa.

The unicellular algae are important photosynthetic organisms in marine and freshwater ecosystems. They include dinoflagellates, diatoms, and the exclusively freshwater euglenoids.

The acellular or plasmodial slime molds form a multinucleate plasmodium that crawls in amoeboid fashion, ingesting decaying organic matter. Drought or starvation stimulates the formation of a fruiting body on which spores are formed. Cellular slime molds exist as independent amoeboid cells. Under adverse conditions, they aggregate in response to a chemical signal and form a pseudoplasmodium that differentiates into a spore-forming fruiting body.

Protozoa are nonphotosynthetic protists that absorb or ingest their food. They are widely distributed in soil and water; some are parasitic. They include the zooflagellates, the amoeboid sarcodines, the parasitic sporozoans, and the predatory ciliates.

GLOSSARY

bacillus (buh-sil'-us; pl. bacilli): a rod-shaped bacterium.

bacterial conjugation: the exchange of genetic material between bacteria.

bacteriophage (bak-tir'-ē-ō-fāj): a virus specialized to parasitize bacteria.

bacterium (bak-tir'-ē-um; pl. bacteria): an organism consisting of a single prokaryotic cell surrounded by a complex polysaccharide coat.

capsule: a polysaccharide or protein coating that surrounds the cell walls of some bacteria.

cellular slime mold: a funguslike protist consisting of individual amoeboid cells that can aggregate to form a sluglike mass, which in turn forms a fruiting body.

chemosynthetic (kēm'-ō-sin-the-tic): capable of oxidizing inorganic molecules to obtain energy.

ciliate (sil'-ē-et): a category of protozoan characterized by cilia and a complex unicellular structure, including harpoonlike organelles called trichocysts. Members of the genus *Paramecium* are well-known ciliates.

Cilium (sil'-ē-um; pl. cilia): a short, hairlike organelle projecting through the cell membrane, usually numerous and engaged in coordinated beating, which moves a cell through a fluid environment or moves the fluid over the surface of the cell.

coccus (ka'-kus; pl. cocci): a spherical bacterium.

cyanobacteria: photosynthetic prokaryotic cells, utilizing chlorophyll and releasing oxygen as a photosynthetic byproduct, sometimes called "blue-green algae."

diatom (dī'-e-tom): a category of protist that includes photosynthetic forms with two-part glassy outer coverings that separate when the cell divides. Diatoms are important primary producers in fresh and salt water.

dinoflagellate (dī-nō-fla'-gel-et): a category of protist that includes photosynthetic forms in which two flagella project through armorlike plates. Abundant in oceans, these sometimes reproduce rapidly, causing "red tides."

euglenoid (yū'-gle-nōyd): a category of protist characterized by one or more whiplike flagella used for locomotion and a photoreceptor for detecting light. Euglenoids are photosynthetic, but some are capable of heterotrophic nutrition if deprived of chlorophyll.

flagellum (fla-gel'-um; pl. flagella): a motile hairlike organelle that propels a cell through a fluid.

fruiting body: a spore-forming reproductive structure found in certain protists, bacteria, and fungi.

gonorrhea (gon-a-rē'-uh): a sexually transmitted bacterial infection of the reproductive organs. Untreated gonorrhea may result in sterility.

halophile (hā'-lō-fīl): a salt-loving organism.

legume (leg'-oom): a family of dicotyledonous plants characterized by root swellings in which nitrogen-fixing bacteria are housed. Includes soybeans, lupines, alfalfa, and clover.

Methanogen (me-than'-ō-gen): a type of anaerobic archaebacterium capable of converting carbon dioxide to methane.

monera (mō'-ne-ra): a taxonomic kingdom consisting of unicellular prokaryotic organisms, including bacteria, archaebacteria, and cyanobacteria.

nitrogen-fixing: possessing the ability to remove nitrogen from the atmosphere and combine it with hydrogen to produce ammonia.

pathogen (path'-ō-gen): an organism capable of producing disease.

peptidoglycan (pep-tid-ō-glī'-can): material found in prokaryotic cell walls consisting of chains of sugars cross-linked by short chains of amino acids called peptides.

phytoplankton (fī'-tō-plank-ten): a general term describing photosynthetic protists that are abundant in marine and freshwater environments.

pili (pil'-ī): hair-like projections made of protein and found on the surface of certain bacteria, often used to attach the bacterium to another cell.

plasmodial slime mold: a funguslike protist consisting of a multinucleate mass or plasmodium that crawls and ingests decaying organic matter in amoeboid fashion and forms a fruiting body under adverse conditions.

plasmodium (plas-mō'-dē-um): an aggregation of individual amoeboid cells that form a sluglike mass.

prion (prē'-on): a proteinaceous infectious particle that may be responsible for certain neurodegenerative diseases.

protista (prō-tis'-tuh): a taxonomic kingdom including unicellular, eukaryotic organisms.

protozoan (prō-te-zō'-an; pl. protozoa): a nonphotosynthetic or animallike protist.

pseudoplasmodium (sū-dō-plas-mō'-dē-um): a sluglike mass of cytoplasm containing thousands of nuclei that are not confined within individual cells.

pseudopod (sūd'-ō-pod): extension of the cell membrane by which certain cells, such as amoebae, locomote and engulf prey.

sarcodine (sar-kō'-dīn): a category of nonphotosynthetic protist (protozoa) characterized by the ability to form pseudopodia. Some, such as amoebae, are naked, while others have elaborate shells.

spirilla (spī'-rilla): a spiral shaped bacterium.

spore (spōr): a resistant or resting structure that disperses readily and withstands unfavorable environmental conditions.

sporozoan (spōr-ō-zō'-en): a category of parasitic protist. Sporozoans have complex life cycles often involving more than one host, and are named for their ability to form infectious spores. A well-known member (genus *Plasmodium*) causes malaria.

syphilis (si'-ful-is): a sexually transmitted bacterial infection of the reproductive organs which, if untreated, can damage the nervous and circulatory systems.

thermoacidophile (ther-mō-a-sid'-eh-fīl): a form of archaebacterium that thrives in hot, acidic environments.

trichocyst (trik'-eh-sist): a stinging organelle of protists.

viroid (vī'-rōyd): a particle of RNA capable of infecting a cell and directing the production of more viroids; responsible for certain plant diseases.

virus (vī'-rus): a noncellular parasitic particle consisting of a protein coat surrounding a strand of genetic material. Viruses multiply only within cells of living organisms.

zooflagellate (zō-ō-fla'-gel-et): a category of nonphotosynthetic protist that move using flagella.

STUDY QUESTIONS

1. List the major differences between monerans and protistans.
2. Describe some of the ways in which bacteria obtain energy and nutrients.
3. What are nitrogen-fixing bacteria, and what role do they play in ecosystems?
4. What is a spore?
5. Why do bacteria readily become resistant to antibiotics, and what steps can people take to prevent this?
6. Describe some examples of bacterial symbiosis.
7. Describe the structure of a typical virus. How do viruses reproduce?
8. What is the importance of dinoflagellates in marine ecosystems? What happens when they reproduce rapidly?
9. What is the major ecological role played by unicellular algae?
10. What protozoan group consists entirely of parasitic forms?
11. Describe the life cycle and mode of transmission of the malarial parasite.
12. What protozoan group is responsible for forming the white limestone cliffs of Dover, England?
13. How do amoebae feed? What disease is caused by a parasitic amoeba?
14. The most complex single cells are found in which protozoan group? How do they feed? From what feature is their name derived?

DISCUSSION QUESTIONS

1. Argue for and against the statement: "Viruses are alive."
2. It is thought that the pre-biotic atmosphere of the earth did not contain free oxygen. Scientists also think that the sequence of metabolic evolution in primitive bacteria was first anaerobic respiration, then photosynthesis, and finally aerobic respiration. Why couldn't aerobic respiration have evolved before photosynthesis? Explain your answer to the following: Which stage of metabolic evolution had the most profound effect on the gaseous atmosphere of the earth and the possibility of life on land?
3. Why should the lives of multicellular animals be impossible if moneran and protistan organisms did not exist?

SUGGESTED READINGS

Adler, J. "The Sensing of Chemicals by Bacteria." *Scientific American*, April 1976.

Barghoorn, E. S. "The Oldest Fossils." *Scientific American*, May 1971.

Butler, P. J. G., and Klug, A. "The Assembly of a Virus." *Scientific American*, November 1978. (Offprint No. 1412).

Dixon, Bernard. "Overdosing on Wonder Drugs." *Science 86*, May 1986.

Gallo, R. C. "The AIDS Virus." *Scientific American*, January 1987. (Offprint No. 1577).

Grady, D. "The Mysterious Prion." *Discover*, April 1983.

Jannasch, H. W., and Wilson, C. O. "Microbial Life in the Deep Sea." *Scientific American*, June 1977.

Maranto, G. "Renaissance for Vaccines." *Discover*, September 1984.

Margulis, L. "Symbiosis and Evolution." *Scientific American*, August 1971.

McEvedy, C. "The Bubonic Plague." *Scientific American*, February 1988.

Simons K., Garoff, H., and Helenius, A. "How an Animal Virus Gets into and out of Its Host Cell." *Scientific American*, February 1982. (Offprint No. 1511).

Woese, C. "Archaebacteria." *Scientific American*, June 1981.

22

The Diversity of Life.
II. Animals

Brittle stars reveal their dazzling colors and graceful shapes. The arms of these echinoderms detach easily if grasped by a predator, hence their name.

Features of Animals

Animals have several features, which, taken together, distinguish them from all other kingdoms.

1. **Animals are all multicellular.**
2. **Animal cells lack a cell wall.**
3. **Animals can reproduce sexually.** Although animal species exhibit a tremendous diversity of reproductive styles, all are capable of sexual reproduction.
4. **Animals are heterotrophic**—that is, they obtain their energy by consuming the bodies of other organisms.
5. **Animals are motile during some stage in life.** Even the stationary sponges have a free-swimming larval stage.
6. **Animals are able to make rapid responses to external stimuli** as a result of the activity of neurons, contractile cells or muscles, or both.

Evolutionary Trends in Animal Body Plans

Trends in Overall Complexity

In our survey of the animal kingdom, we will present the major phyla in an order that approximates the sequence in which they evolved. As we progress from the sponges through the Cnidaria (jellyfish and their relatives) through the three phyla of worms, you will see a clear evolutionary trend toward increasing complexity. This complexity is reflected in increasing cellular organization and specialization, bilateral symmetry, **cephalization** (the concentration of sensory organs and a brain in a defined head region), the evolution of a body cavity, or **coelom, segmentation** (repeated, similar body parts), and a one-way digestive system in which material flows in a single direction, as discussed below. Some of these features are used in the simple evolutionary groupings of animals shown in Figure 22-1.

The evolutionary trend toward increasing complexity culminates in the arthropods, molluscs, echinoderms and chordates (see Table 22-1). The striking structural differences among these four phyla do not reflect further increases in complexity so much as adaptations to different environments and lifestyles. Within the chordates, however, another trend in complexity is clear: the size and sophistication of the brain.

Trends in Cellular Organization

As animals evolved, specialized cells became organized into **tissues,** for example, groups of muscle cells. Various tissues merged into **organs:** two or more tissues integrated to perform a specialized function, such as the kidney or the eye. Organs, in turn, formed **organ systems:** two or more organs that work together to perform a specific function, as illustrated by the digestive or excretory system. Although sponges have specialized cell types, the individual cells act rather independently of one another, and are not organized into tissues or organs. Cnidarians

Figure 22-1 A simple classification scheme for animals based on anatomical features. Animals show evolutionary trends from indeterminate shape to radial symmetry to bilateral symmetry. Bilaterally symmetrical animals form two groups: the simplest lack a body cavity called a coelom, while more complex phyla have a coelom.

show well-defined tissues, for example their nerve net, but lack organs. The flatworms possess not only organs, such as the gonads and eyespots, but also an organ system, the reproductive system. Organ systems are found in all the more complex animals.

A second trend in cellular organization is in the number of tissue layers, called **germ layers,** that arise during embryonic development. Flatworms and all the more complex animals have three germ layers: an inner layer of endoderm (forming most internal organs), a middle layer of mesoderm (forming muscle, and, when present, the circulatory and skeletal systems), and an outer layer of ectoderm (forming the epithelium and nervous tissue). The more primitive cnidarians, in contrast, have only two germ layers: endoderm and ectoderm. Although the cnidarian body wall has three layers, the middle consists of a jellylike substance, called **mesoglea,** with very few cells.

Trends in Symmetry

The simplest animals, the sponges, have many representatives whose growth pattern is free-form; their bodies are irregular and variable in shape, even within a given species. Other types of sponges show **radial symmetry,** in which any line through a central axis divides the animal into roughly equal halves (Fig. 22-2). This was the first type of symmetry to evolve in animals, and is also seen in cnidarians and in some adult echinoderms. **Bilateral symmetry** evolved later, and is characteristic of all more complex animals, including larval echinoderms. A bilaterally symmetrical animal can be divided into roughly mirror-image right and left halves, with an upper, or **dorsal,** surface and a lower, or **ventral,** surface.

Trends in Cephalization

Bilateral symmetry is accompanied by cephalization, which creates an **anterior** head end and a **posterior** tail end (see Fig. 22-2). During the evolution of cephalization, sensory cells, sensory organs, aggregations of neurons, and organs for ingesting food were increasingly concentrated at the anterior end of the animal. The flatworms are the simplest animals to show cephalization. Although they have a defined head with sensory organs, they still ingest food via a pharynx located near the middle of their bodies.

Trends in Body Cavities

As animals increased in complexity, they evolved a cavity located between the gut and body wall called a **coelom** (Fig. 22-3). Simple animals such as cnidarians

and flatworms lack any internal space between their body wall and their gut. Roundworms have a space, called a **pseudocoel,** between the body wall and gut. The prefix "pseudo" (meaning "false") refers to the fact that this opening has a different embryological origin than a true coelom. In earthworms and nearly all of the more complex animals, a true coelom is present, which can serve a variety of functions. In the earthworm, this fluid-filled cavity acts as a hydrostatic skeleton. Internal organs bulge into the coelom, which serves as a protective buffer between them and the outside world. Its presence has allowed the internal organs to evolve greater complexity and to move independently of the body wall. Thanks to your coelom, after a meal you may remain externally unmoved, even though your digestive tract is energetically churning.

Trends in Segmentation

Segmentation, or the presence in the body of similar repeated units, is first seen in the annelids (earthworms and their relatives) (see Fig. 22-13). Segmentation appears to be an evolutionary device for increasing body size with a minimum of new genetic information, since the genetic "blueprint" for each segment is similar. In more complex animals, the repeated segments have become specialized for specific functions, or are only visible in certain body parts, such as the series of similar vertebrae in the vertebrate spine.

Trends in Digestive Tracts

The simplest form of digestion is found in the sponges, which lack any specialized digestive tract. In sponges, digestion is entirely intracellular; individual cells trap, ingest, and digest smaller unicellular organisms. The cnidarians and flatworms have a digestive system consisting of a sac, called a **gastrovascular cavity,** with a single opening (see Figs. 22-7 and 22-9). This opening, politely called a mouth, is equally an anus through which undigested material is expelled. In the roundworms and all of the more complex animals, an efficient, tubular, one-way digestive system has evolved. A mouth, located at the anterior end near the sensory structures, allows food to be ingested as it is detected. The food is then processed in stages as it passes through a series of specialized regions. First it is physically broken down, then enzymatically digested, then absorbed. Wastes are voided through a separate anus, usually near the posterior end of the animal.

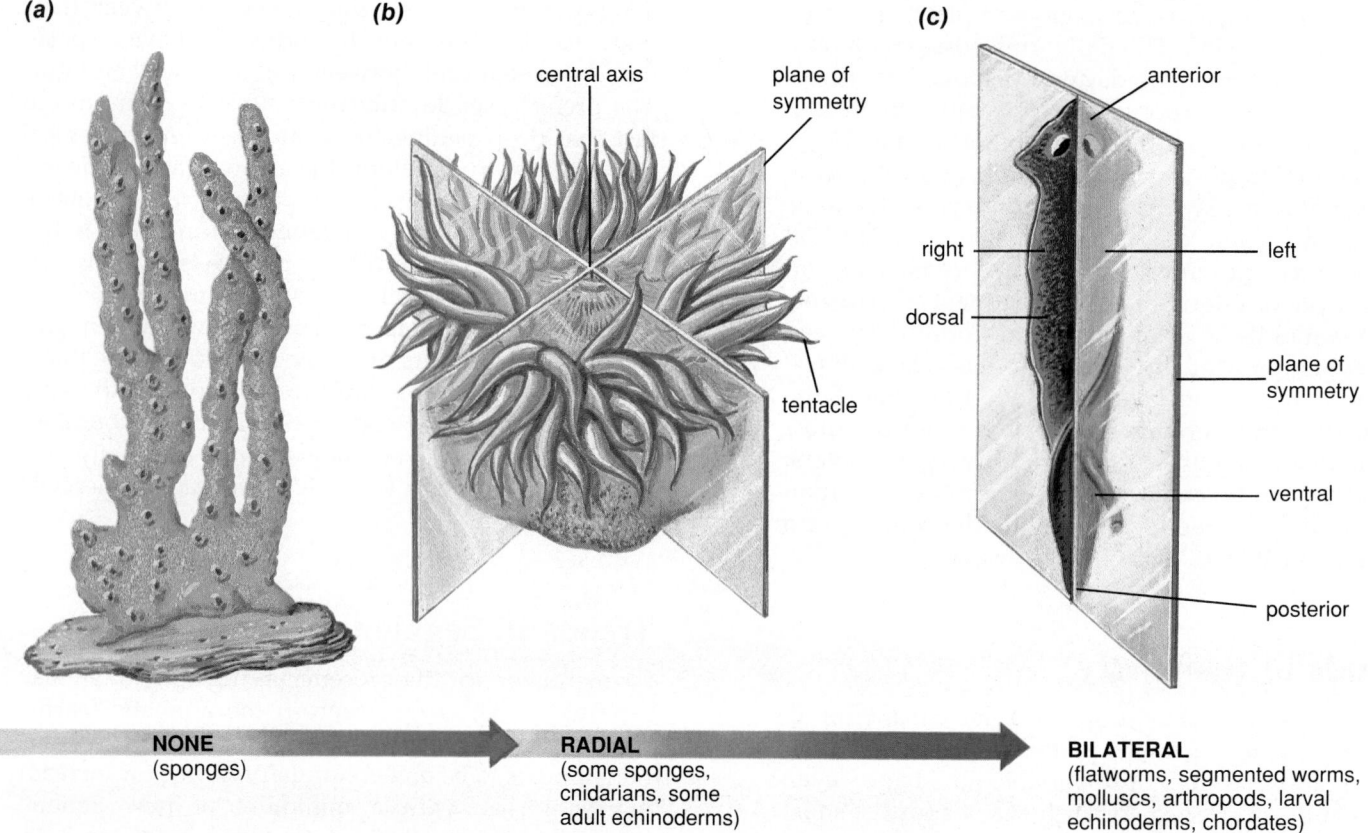

(a)

(b)

central axis   plane of symmetry

tentacle

(c)

anterior

right   left

dorsal

plane of symmetry

ventral

posterior

NONE
(sponges)

RADIAL
(some sponges,
cnidarians, some
adult echinoderms)

BILATERAL
(flatworms, segmented worms,
molluscs, arthropods, larval
echinoderms, chordates)

Figure 22-2 Trends in body symmetry and cephalization. **(a)** Sponges, the simplest animals, lack a head and most are asymmetrical. **(b)** Some sponges, cnidarians (anemones, hydra), and some adult echinoderms (sea urchins, sea stars) have bodies that are radially symmetrical. Here, any plane that passes through the central axis divides the body into mirror-image halves. Animals in these groups lack a well-defined head. **(c)** Nearly all the more complex animals, starting with flatworms (phylum Platyhelminthes) show bilateral symmetry. The body can be split into two mirror-image halves by a single plane running down the midline. Animals with bilateral symmetry also have an anterior head end, a posterior tail end, a dorsal upper surface, and a ventral underside.

The Major Animal Phyla

For convenience, biologists often place animals in one of two major categories: **vertebrates,** those with a backbone or vertebral column, and **invertebrates,** those lacking a backbone. You are probably most familiar with the vertebrates: fish, amphibians, reptiles, birds, and mammals. The invertebrates include everything else, from sponges to worms to snails to insects. Our human bias is clearly reflected in these

categories, since the invertebrates make up about 97.5% of all the animal species on Earth, including 27 different phyla. In contrast, vertebrates constitute only part of a single phylum, the phylum Chordata.

The first animals to evolve were invertebrates, probably originating from colonies of protozoa whose members had become specialized to perform distinct roles within the colonial body. In our survey of the kingdom Animalia, we will begin with the sponges, whose body plan most closely resembles the probable

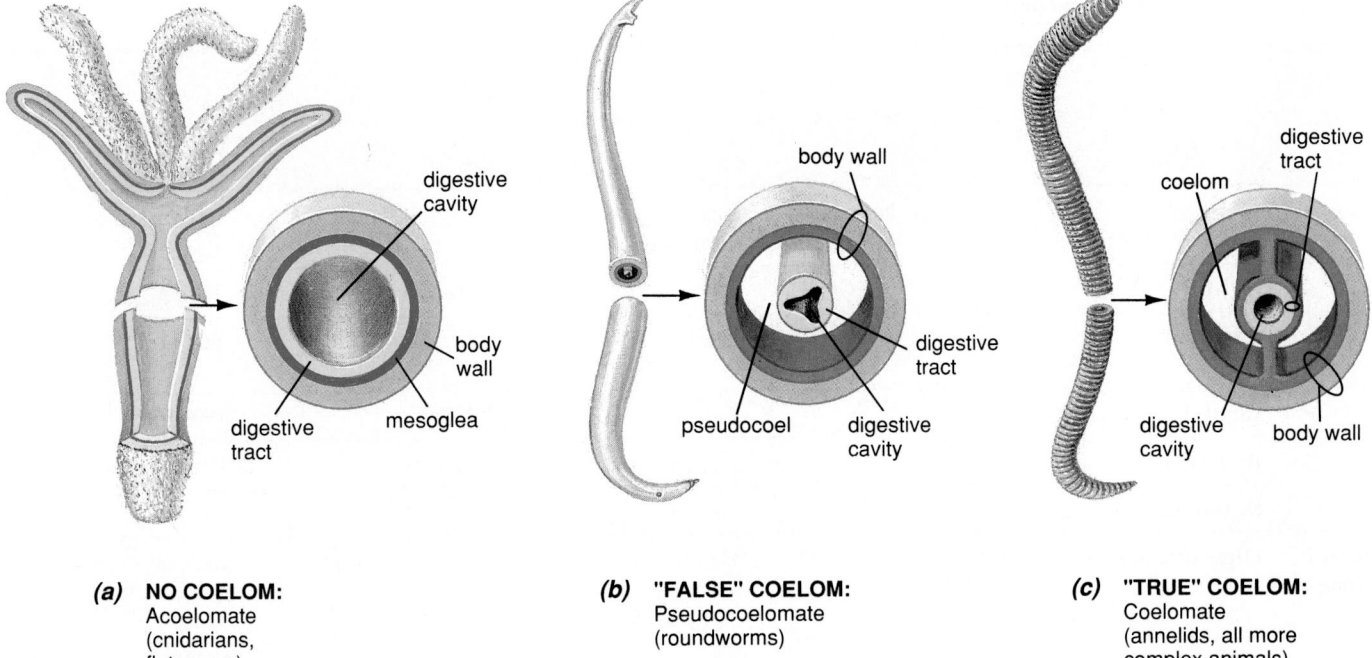

<table>
<tr><td>**(a) NO COELOM:**
Acoelomate
(cnidarians,
flatworms)</td><td>**(b) "FALSE" COELOM:**
Pseudocoelomate
(roundworms)</td><td>**(c) "TRUE" COELOM:**
Coelomate
(annelids, all more
complex animals)</td></tr>
</table>

Figure 22-3 Trends in body cavities. **(a)** Cnidarians and flatworms are acoelomate, lacking the space between the body wall and digestive tract called a coelom. **(b)** Roundworms have a space, called a pseudocoel, between the body wall and digestive tract. The pseudocoel resembles a coelom, but has a different embryological origin. **(c)** Annelids and most other complex animals have their digestive tracts and other internal organs suspended within a coelom.

ancestral protozoan colonies. Our discussion will roughly follow the order in which the various phyla of animals appeared on Earth, ending with the vertebrates in the phylum Chordata. The features of the major animal phyla are compared in Table 22-1.

Phylum Porifera: The Sponges

In 1907, the embryologist H. V. Wilson mashed a sponge through a piece of silk, dissociating it into single cells and cell clusters. After sitting in seawater for 3 weeks, the cells had reaggregated into a functional sponge. Because sponge cells are relatively independent, sponges resemble colonies in which single-celled organisms live together for mutual benefit, but lack specialized tissues. Sponges lack a nervous system and each cell engages in intracellular digestion. Sponges are, therefore, the simplest multicellular animals. Although some have a definite size and shape, others grow freely and indeterminately over rocks in their aquatic habitats (Fig. 22-4).

All sponges have a similar general body plan (Fig. 22-5). The body is perforated by numerous tiny pores through which water enters, and by fewer, large openings, called **oscula,** through which it is expelled. Within the hollow sponge, water travels through canals. During its passage, oxygen is extracted, microorganisms are filtered out and eaten, and wastes are released.

Sponges have three major cell types (see Fig. 22-5), each with a specialized role. Flattened epithelial cells cover their inner and outer surfaces. Some epithelial cells surround pores, controlling their size and regulating the flow of water. The pores are closed when harmful substances are present. **Collar cells** maintain a flow of water through the sponge, by beating a flagellum that extends into the inner canal. The collar that surrounds the flagellum acts as a fine sieve, filtering out microorganisms that are then ingested by the cell. Some of the food is passed to the third cell type, **amoeboid cells.** These roam freely between the epithelial and collar cells, digesting and distributing nutrients, producing reproductive cells, and secret-

Table 22-1   Comparison of the Major Animal Phyla

		Porifera (Sponges)	Cnidaria (Hydra, Anemones, Jellyfish)	Platyhelminthes (Flatworms)	Nematoda (Roundworms)
Body Plans	**Level of organization**	Cellular—lacks tissues and organs	Tissue—lacks organs	Organ system	Organ system
	Germ layers	Absent	Two	Three	Three
	Symmetry	Absent	Radial	Bilateral	Bilateral
	Cephalization	Absent	Absent	Present	Present
	Body cavity	Absent	Absent	Absent	Pseudocoel
	Segmentation	Absent	Absent	Absent	Absent
Internal Systems	**Digestive system**	Intracellular	Gastrovascular cavity; some intracellular	Gastrovascular cavity	Separate mouth and anus
	Circulatory system	Absent	Absent	Absent	Absent
	Respiratory system	Absent	Absent	Absent	Absent
	Excretory system (fluid regulation)	Absent	Absent	Canals with flame cells	Excretory gland cells
	Nervous system	Absent	Nerve net	Head ganglia with longitudinal nerve cords	Head ganglia with dorsal and ventral nerve cords
	Reproduction	Sexual; asexual (budding)	Sexual; asexual (budding)	Sexual (some hermaphroditic); asexual (body splits)	Sexual (some hermaphroditic)
	Support	Endoskeleton of spicules	Hydrostatic skeleton	Absent	Hydrostatic skeleton

Annelida (Segmented Worms)	Arthropoda (Insects, Arachnids, Crustacea)	Mollusca (Snails, Clams, Squid)	Echinodermata (Sea Stars, Sea Urchins)	Chordata (Tunicates, Vertebrates)
Organ system	Organ system	Organ system	Organ system	Organ system
Three	Three	Three	Three	Three
Bilateral	Bilateral	Bilateral	Bilateral larvae, radial adults	Bilateral
Present	Present	Present	Absent	Present
Coelom	Coelom	Coelom	Coelom	Coelom
Present	Present	Absent	Absent	Present (but reduced)
Separate mouth and anus	Separate mouth and anus	Separate mouth and anus	Separate mouth and anus (usually)	Separate mouth and anus
Closed	Open	Open	Absent	Closed
Absent	Tracheae, gills, or book lungs	Gills, lungs	Tube feet, skin gills, respiratory tree	Gills, lungs
Nephridia	Excretory glands resembling nephridia	Nephridia	Absent	Kidneys
Head ganglia with paired ventral cords; ganglia in each segment	Head ganglia with paired ventral nerve cords; ganglia in segments, some fused	Well-developed brain in some cephalopods; several paired ganglia, most in the head; nerve network in body wall	Head ganglia absent; nerve ring and radial nerves; nerve network in skin	Well-developed brain; dorsal nerve cord
Sexual (some hermaphroditic)	Usually sexual	Sexual (some hermaphroditic)	Sexual (some hermaphroditic); asexual by regeneration (rare)	Sexual
Hydrostatic skeleton	Exoskeleton	Hydrostatic skeleton	Endoskeleton of plates beneath outer skin	Endoskeleton of cartilage or bone

Figure 22-4 Sponges come in a wide variety of sizes, shapes, and colors, some over a meter tall, others growing in free-form pattern over undersea rocks. The large exit pores are visible in the yellow specimen.

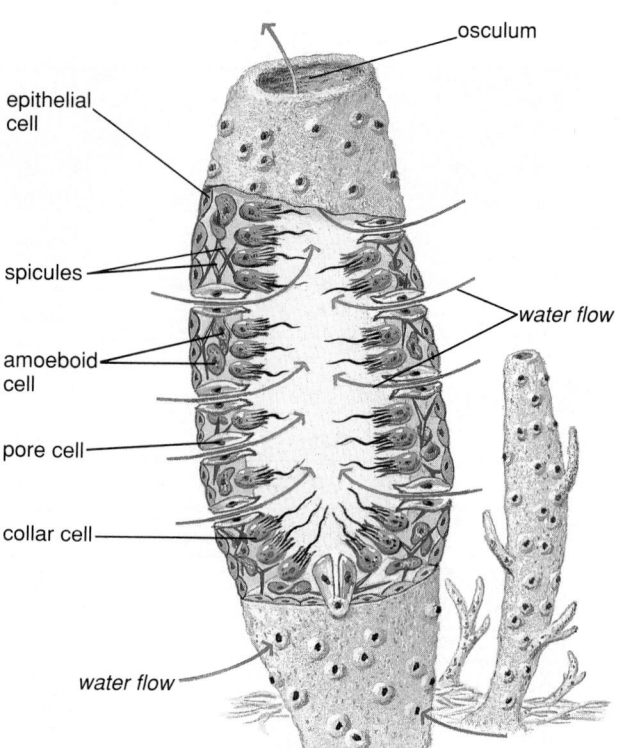

Figure 22-5 Sponges all have a similar body plan. Currents created by collar cells draw water in through numerous tiny pores. Microscopic food particles are filtered out by collar cells and shared among the various cell types. Water exits through larger pores, the oscula. Spicules form a supportive internal skeleton.

ing an internal skeleton. Sponges may grow to over a meter in height, and the skeleton, which is composed of **spicules,** provides support for the body (see Fig. 22-5). The spicules may be formed from calcium carbonate (chalk), silica (glass), or protein. The natural bath sponge, now rarely used, is a proteinaceous sponge skeleton.

About 5000 species of sponges have been identified; all are aquatic and most are marine. All adult sponges are **sessile,** attaching permanently to rocks

or other underwater surfaces. Sponges may reproduce asexually by **budding,** or sexually through the fusion of sperm and eggs. Fertilized eggs develop inside the adult into active larvae that escape through the oscula. Water currents disperse the larvae to new areas where they settle permanently and develop into adult sponges.

Phylum Cnidaria: The Hydra, Anemones, and Jellyfish

The Cnidaria are clearly a step above the Porifera in complexity. Their cells are organized into distinct **tissues,** including nerves, and contractile tissue that acts like muscle. Nerve cells are organized into a **nerve net** that branches through the body and controls the contractile tissue. Cnidarians lack true organs, however, and have no brain.

Members of the phylum Cnidaria come in a bewildering and beautiful variety of forms (Fig. 22-6), all of which are actually variations on two basic body plans: the **polyp** and the **medusa,** illustrated in Figure 22-7. The polyp, with its foot attached and its **tentacles** reaching upward, is adapted to a life spent quietly attached to rocks, awaiting prey like a predatory flower. The medusa ("jellyfish") swims weakly by contracting its bell-shaped body, but primarily is carried by ocean currents, trailing its tentacles like multiple fishing lines. Both forms are radially symmetrical, with body parts arranged in a circle around an axis drawn through the mouth and digestive cavity (see Fig. 22-7). This arrangement is particularly well suited to animals that are sessile, or carried every which way by water currents, since they are prepared to capture prey or defend themselves from any direction.

Although all cnidarians are predatory, none actively hunt. Instead, they rely on their victims blundering by chance into the grasp of their enveloping tentacles. Cnidarian tentacles are armed with **nematocysts,** cells containing poisonous or sticky

darts that are injected explosively into prey upon contact (Fig. 22-8). Stung and firmly grasped, the prey is forced through an expansible mouth into a digestive sac, the **gastrovascular cavity.** Digestive enzymes secreted into this cavity break down some of the food, and further digestion occurs within the cells lining the cavity. Digestion completed, the versatile mouth becomes an anus through which undigested material is expelled, since the gastrovascular cavity has only a single opening. Although this two-way traffic is inefficient because it prevents continuous feeding, it is adequate to support the low energy demands of these animals.

Cnidarians can reproduce both asexually and sexually. Some medusae and some polyps, such as hydra and sea anemones, bud off miniature replicas of themselves (see Fig. 39-1). An anemone, crawling slowly over a rock, may also leave behind pieces of

(a)

(b)

(d)

(c)

Figure 22-6 Cnidarian diversity. **(a)** A purple anemone spreads tentacles to snare prey in Monterey Bay, California. **(b)** A close-up of coral reveals bright yellow polyps in various stages of extension. At the left, areas where the coral has died expose the calcium carbonate skeleton that supports the polyps and forms the reef. A strikingly patterned crab (phylum Arthropoda; class Crustacea) sits atop the coral, holding tiny anemones in its claws. Their stinging tentacles help protect the crab. **(c)** A small medusa from the ocean off southern California. **(d)** The Portuguese Man-of-War, a colonial cnidarian dangerous to human beings. A stunned fish is trapped in its tentacles.

(a) POLYP

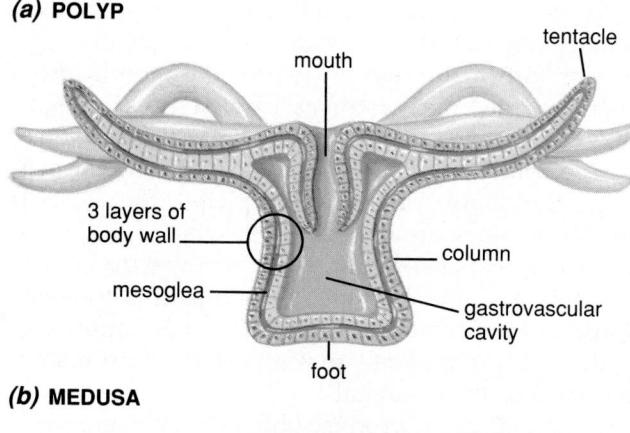

(b) MEDUSA

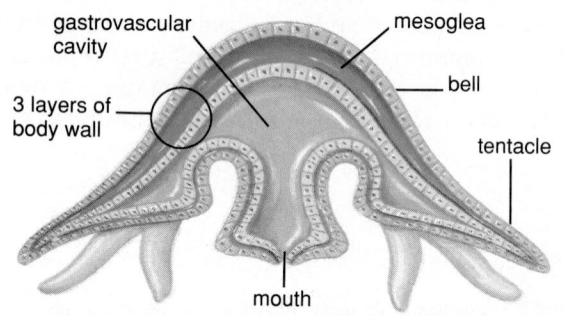

Figure 22-7 The two basic body forms of cnidarians are actually variations on a single, simple theme. **(a)** The polyp form is seen in hydra (see Fig. 22-8), sea anemones (Fig. 22-6a), and the individual polyps within a coral (Fig. 22-6b). **(b)** The medusa form, seen in the jellyfish (Fig. 22-6c), resembles an inverted polyp. Both forms exhibit radial symmetry, with body parts arranged in a circle around a central axis.

foot, which grow into new individuals. Sexual reproduction involves the fusion of sperm and eggs released into the water or retained within the parent. The fertilized egg often develops into a free-swimming ciliated larval stage that settles and becomes a tiny polyp.

Of the 9000 or more species of Cnidaria, all are aquatic and most are marine. One group, the corals, are of particular ecological importance (see Fig. 22-6b). These polyps secrete a hard protective "house" of limestone that persists long after their death, serving as a base for others. The cycle continues until, after thousands of years, massive coral reefs are formed. Corals are restricted to the warm, clear waters of the tropics, where their reefs form undersea habitats, the basis of an ecosystem of stunning diversity and unparalleled beauty (see Chapter 46).

Phylum Platyhelminthes: The Flatworms

Although flatworms do not look anything like cnidarians, the two have some features in common that have led biologists to speculate that they have evolved from a common ancestor. Both have a gastrovascular cavity with a single opening. Certain flatworms also show striking similarities to the larval stage of cnidarians.

Flatworms are clearly more complex than cnidarians. For one thing, they are bilaterally symmetrical (see Fig. 22-2). This body plan, found in all the more

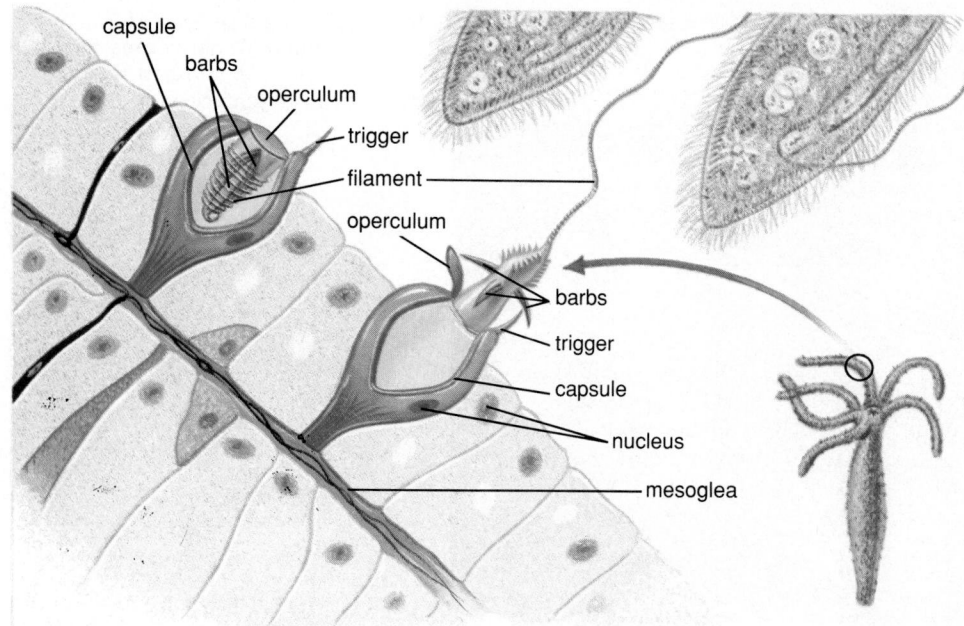

Figure 22-8 At the slightest touch to the trigger, the nematocyst of cnidarians, such as this hydra, violently expels the poisoned or sticky dart that lies coiled and inverted inside. During the process, the barbs and hollow filament actually turn inside out, impaling the prey and injecting a paralyzing venom. Nematocysts are microscopic, and only a few species inject venom in sufficient quantity to harm a person.

complex animals, is an adaptation to active movement. The anterior end first encounters the environment ahead. The sense organs are concentrated here, where they inform the organism to feed, forge onward, or retreat. In **free-living** (nonparasitic) flatworms such as the freshwater planarians (Fig. 22-9), sense organs consist of eyespots for detecting light and dark, and cells responsive to chemical and tactile stimuli. To process information, flatworms have aggregations of nerve cells called **ganglia** in the head that form a simple brain. A pair of nerve cords conducts nervous signals to and from the head ganglia.

Flatworms are the simplest organisms with true **organs,** in which tissues are grouped into functional units. When a free-living flatworm encounters food, usually smaller animals, it sucks up its prey using a muscular **pharynx** located in the middle of the ventral side of its body. The food is digested in an intricately branched gastrovascular cavity that distributes nutrients to all parts of the body (see Fig. 22-9). Free-living flatworms have a simple system for excreting and regulating body fluids. It consists of a network of canals ending in bulbs containing beating cilia. The flickering motion of the cilia has led to the descriptive name: **flame cells** (see Fig. 22-9). The beating cilia drive liquids through the system, emptying excess fluids to the outside through numerous tiny pores.

Flatworms lack both respiratory and circulatory systems. Nutrients are distributed by a branching digestive tract, from which they readily diffuse into nearby cells. Gas exchange between the cells and the environment by diffusion is aided by the flattened body, which ensures that all the cells are relatively close to the outside.

Although many flatworms, such as the planarians, are free-living, those of major importance to humans

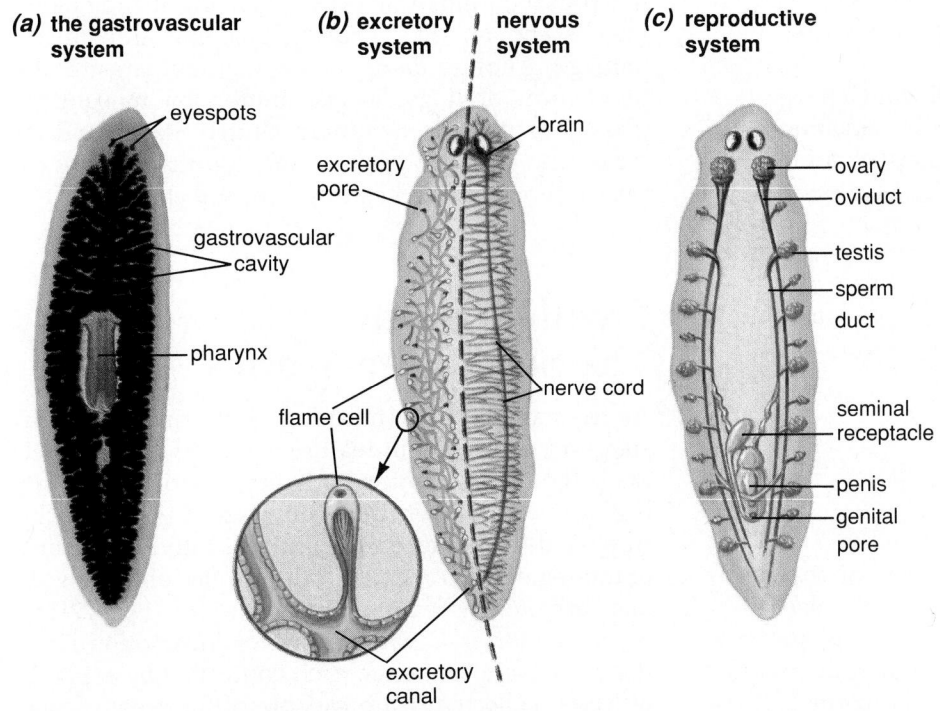

Figure 22-9 Flatworms such as planarians have well-developed organ systems. **(a)** The elaborately branched digestive system, the centrally located ventral pharynx, and eyespots in the head are clearly visible. **(b)** (left) The excretory system consists of branching tubes that conduct excess fluid to the outside via numerous pores. At intervals, flame cells, with their flickering tufts of cilia, keep the fluid moving. (right) The nervous system of flatworms shows clear cephalization, with eyes and a brain located within a well-defined head. Ladderlike nerve cords carry signals through the rest of the body. **(c)** Flatworm reproductive systems include both male and female reproductive organs.

are **parasites.** These include the **tapeworms,** several of which can infect humans. In most cases, people become infected by eating improperly cooked beef, pork, or fish infected by the worms, whose larvae form **cysts** in the muscles of these animals. The cysts hatch in the human digestive tract, where they attach to the intestine and mature. Here they may grow to a length of over 20 feet (7 meters), absorbing digested nutrients directly through their outer surface, and releasing packets of eggs that are shed in the host's feces. If pigs eat grass contaminated with infected human feces, the eggs hatch in the pig's digestive tract, releasing larvae that burrow into its muscles and form cysts, continuing the infective cycle (Fig. 22-10).

Another group of parasitic flatworms are the flukes. Of these, the most devastating are liver flukes (common in the Orient) and blood flukes, such as those of the genus *Schistosoma,* which cause schistosomiasis. Prevalent in Africa and parts of South America, this disease affects an estimated 200 million people; its symptoms include dysentery, anemia, and possible brain damage. Like most parasites, flukes have a complex life cycle that includes an intermediate host, in this case a snail. The irrigation ditches filled by the Aswan Dam in Egypt have contributed to the spread of schistosomiasis by creating an extensive new habitat for the snail host.

Flatworms can reproduce both sexually and asexually. Free-living forms may reproduce by cinching themselves around the middle until they separate into two halves, each of which regenerates its missing parts. All can reproduce sexually, and many are **hermaphroditic** (possessing both male and female sexual organs; see Fig. 22-10). This is a great advantage to parasitic forms because it allows an individual living alone in a host's intestine to self-fertilize.

Phylum Nematoda: The Roundworms

Few habitats on Earth lack representatives of this enormously successful phylum. Nematodes, also called roundworms, have a rather simple body plan. It consists of a tubular gut running from mouth to anus (a major evolutionary advance over flatworms), and a tough, flexible, nonliving cuticle that encloses and protects the thin, elongated body (Fig. 22-11). Sensory organs in the head transmit information to a simple ganglionic "brain" resembling that of flatworms. Reproduction is always sexual, and the sexes are separate, with the male (who is usually smaller) fertilizing the female internally.

Although only about 10,000 species of nematodes have been named, there may be as many as 500,000 species. Some parasitic forms reach a meter in length, but most are microscopic, as shown in Figure 22-11. Nematodes lack both circulatory and respiratory systems. Since most are extremely thin and all have low energy requirements, diffusion suffices for gas exchange and the distribution of nutrients. Free-living forms are important decomposers in terrestrial and aquatic environments. Although blissfully ignorant of their presence, you are surrounded by roundworms. A single rotting apple may contain 90,000 individuals. Nearly all plants and animals are host to several parasitic species. Chances are good that during your life you will be invaded by one of the 50 species that parasitize people, most without doing noticeable damage.

Although most roundworms are harmless, there are important exceptions. For example, hookworm larvae in soil may bore into human feet, enter the bloodstream, and travel to the intestine, where they cause continuous bleeding. The *Trichinella* worm (causing trichinosis) may be ingested by eating improperly cooked pork. Infected pork may contain up to 15,000 larval cysts per gram (Fig. 22-12a). The cysts hatch in the human digestive tract and invade blood vessels and muscles, causing bleeding and muscle damage. Another dangerous nematode parasite, the heartworm of dogs, is transmitted by mosquitoes (Fig. 22-12b). In the southern United States, and increasingly in other parts of the country, it poses a severe threat to the health of unprotected pets.

Phylum Annelida: The Segmented Worms

As the name *annelid* (meaning "little ring" in Latin) suggests, a prominent feature of the phylum Annelida is the division of the body into a series of repeating segments. Externally, these are demarcated by ringlike depressions on the surface. Internally, many of the segments contain identical copies of nerve ganglia, excretory structures, and muscles (Fig. 22-13). Segmentation is advantageous for locomotion since the body compartments, each controlled by separate muscles, collectively are capable of far greater complexity of movement than is seen in the nonsegmented worms.

A second evolutionary advance first seen in annelids is a fluid-filled coelom, between the body wall and the digestive tract (see Fig. 22-3). The incompressible fluid in the coelom in many annelids is confined by the partitions separating the segments and

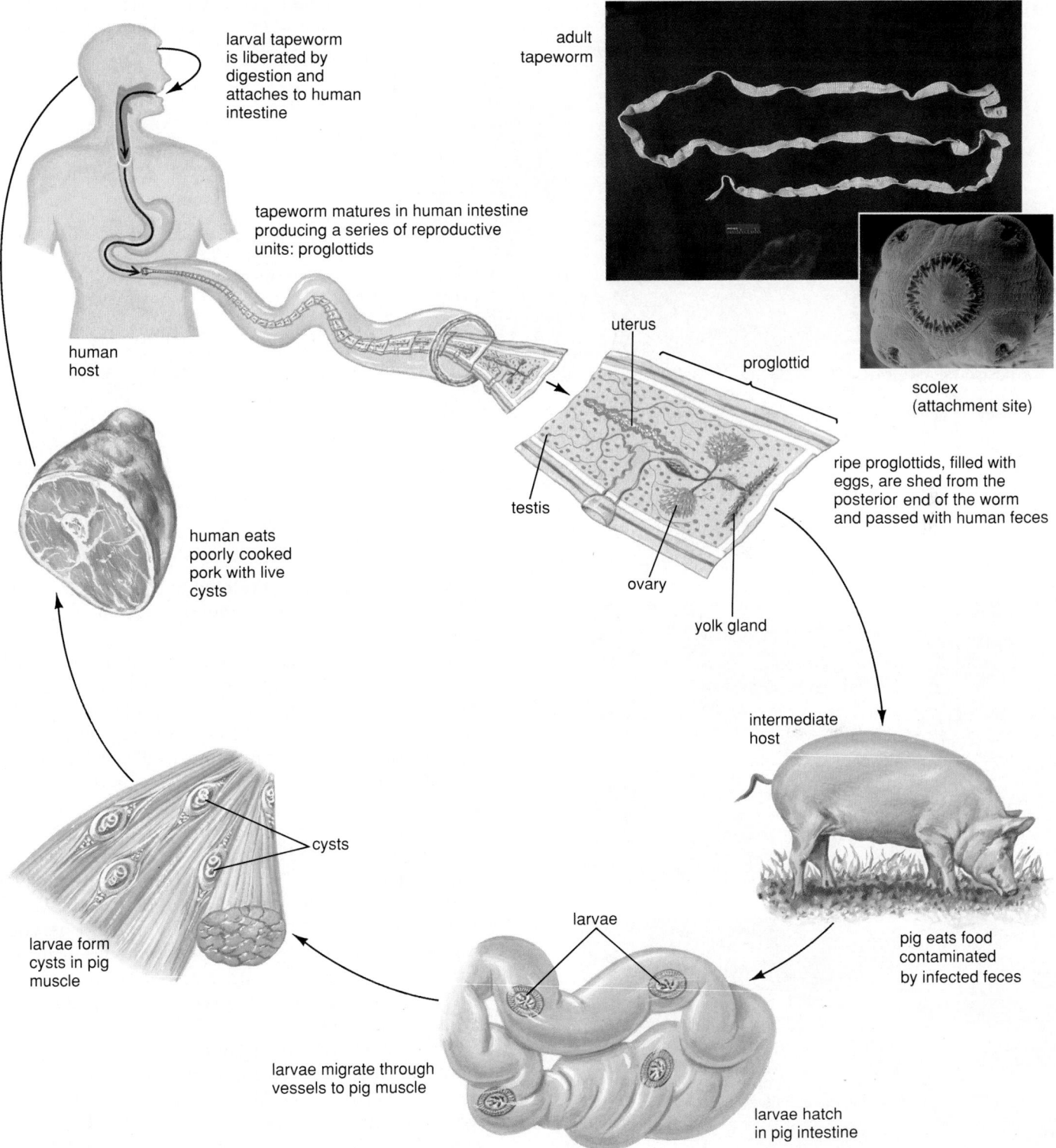

adult tapeworm

larval tapeworm is liberated by digestion and attaches to human intestine

tapeworm matures in human intestine producing a series of reproductive units: proglottids

human host

uterus

proglottid

scolex (attachment site)

testis

ovary

yolk gland

ripe proglottids, filled with eggs, are shed from the posterior end of the worm and passed with human feces

human eats poorly cooked pork with live cysts

intermediate host

cysts

pig eats food contaminated by infected feces

larvae form cysts in pig muscle

larvae

larvae migrate through vessels to pig muscle

larvae hatch in pig intestine

Figure 22-10 The life cycle of the human tapeworm. Each reproductive unit, or proglottid, is a self-contained reproductive factory including both male and female sex organs.

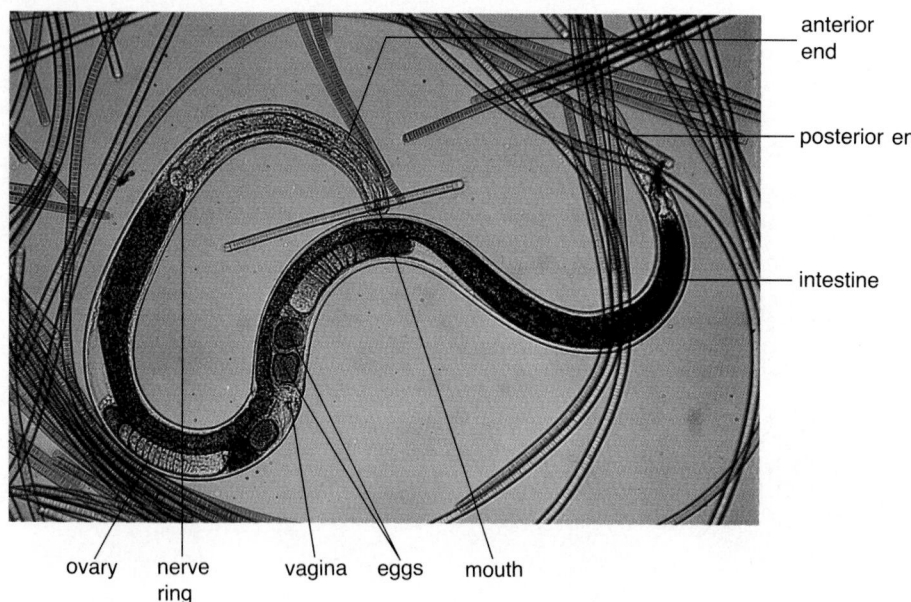

anterior end

posterior end

intestine

ovary   nerve ring   vagina   eggs   mouth

Figure 22-11 A freshwater nematode that feeds on algae. Eggs are visible in this female specimen.

(a)

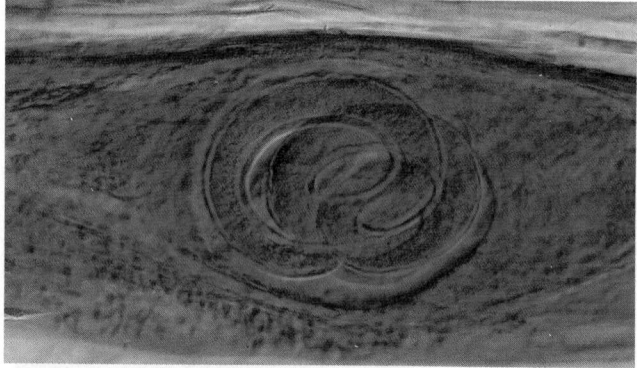

(b)

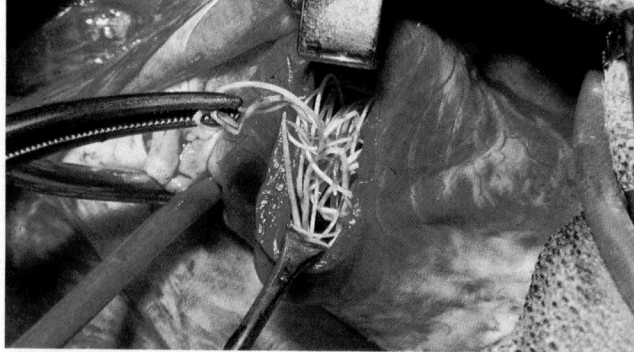

Figure 22-12 Some parasitic nematodes. **(a)** Encysted larva of the *Trichinella* worm in muscle tissue, where it may live for up to 20 years. **(b)** Adult heartworms in the heart of a dog. The juveniles are released into the bloodstream where they may be ingested by mosquitos and passed to another dog by the bite of the infected mosquito.

serves as a **hydrostatic skeleton,** a framework against which muscles can act, allowing feats such as burrowing through soil.

Annelids, in contrast to nematodes, have a well-developed **closed circulatory system** that distributes gases and nutrients throughout the body. In the earthworm, for example, blood with oxygen-carrying hemoglobin is pumped through well-developed vessels by five pairs of "hearts." These hearts are actually short, expanded segments of specialized blood vessel that contract rhythmically. The blood is filtered and wastes removed by excretory organs called *nephridia* that are found in many of the segments. Nephridia resemble the individual tubules of the vertebrate kidney (see Chapter 33). The annelid nervous system consists of a simple ganglionic brain in the head and a series of repeating paired segmental ganglia joined by a pair of nerve cords traveling the length of the body.

Digestion in annelids occurs in a series of compartments, each specialized for a different phase of food processing (see Fig. 22-13). For example, in the earthworm, a muscular pharynx draws in the food, consisting of bits of decaying plant and animal debris in soil. Food is conducted through the esophagus to a storage chamber, the crop, then released slowly into the muscular gizzard, where the food is ground into tiny particles assisted by muscular contractions of the gizzard and the sharp-edged sand grains it contains. Food then passes into the intestine, where it is digested and nutrients absorbed. Undigested food and soil exit through the anus.

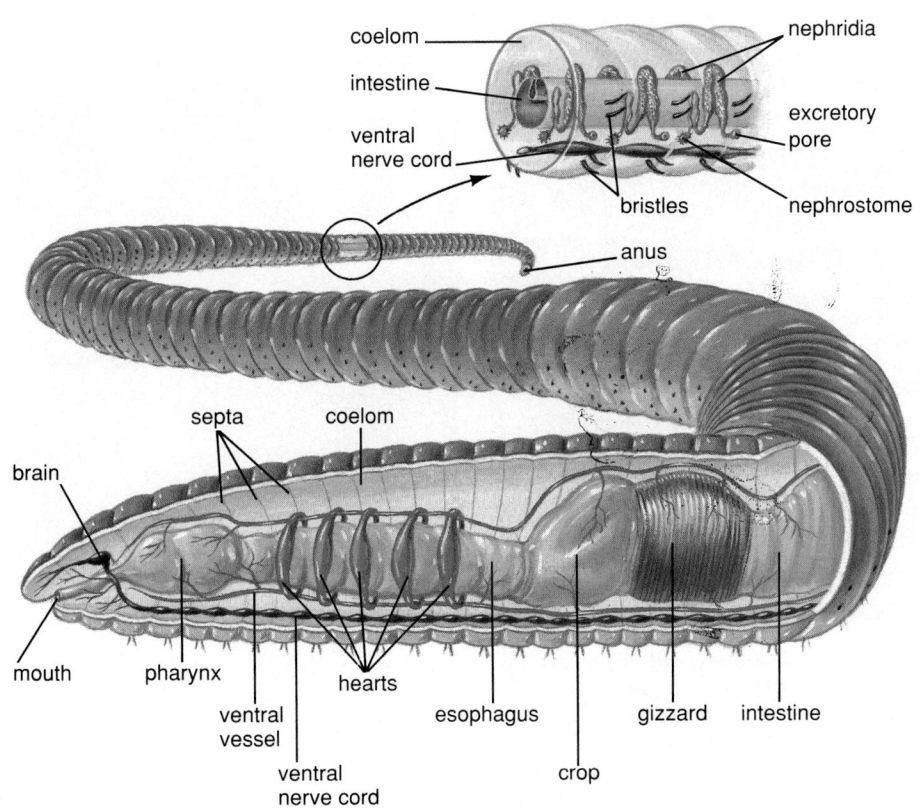

Figure 22-13 The earthworm, an annelid, showing an enlargement of segments, many of which are repeating similar units separated by partitions called septa. The digestive system, which has both a mouth and an anus, is divided into a series of compartments specialized to process food in an orderly sequence.

The phylum Annelida includes about 9000 species, including the familiar earthworm and its relatives (class Oligochaeta, meaning "few bristles"). In general, these exchange gas by diffusion through moist skin. The largest group of annelids (class Polychaeta, meaning "many bristles") is found primarily in the ocean. Some have numerous bristles and paired fleshy paddles on most of their segments, used in locomotion. Others live in tubes from which they project feathery gills that both exchange gases and sift the water for microscopic food (Fig. 22-14a, b). A third group of annelids (class Hirudinea) consists of the leeches (Fig. 22-14c). These worms, found in freshwater or moist terrestrial habitats, are either parasitic or **carnivorous,** some sucking the blood of larger animals, others preying on smaller invertebrates.

Phylum Arthropoda: The Insects, Arachnids, and Crustaceans

Spread your picnic tablecloth beneath a shading oak beside a stream-fed pond, and prepare to discover the diversity of arthropods. As you shoo flies from the potato salad, a yellowjacket may industriously attack your hamburger, flying off with a small piece of meat in its grasp. While a woolly caterpillar undulating up the tree trunk distracts you, ants will be discovering the cookie crumbs and a spider may lower itself into your midst, suspended by a gossamer thread. Watch carefully as you move a large sheltering stone in the stream for the sudden backward flipping of a crayfish. Dragonflies hover near the water's edge, their wings iridescent in the sunlight. Later, as dusk falls, you'll be glad you brought the mosquito repellent!

In numbers, both of individuals and species, arthropods are the dominant animals on Earth. About 1 million species have been discovered, and scientists estimate that up to 9 million remain undescribed. The phylum Arthropoda includes many classes, three of which are particularly large and important: class Insecta, class Arachnida (spiders and their relatives), and class Crustacea (crabs, shrimp, and their relatives). The success of this group can be attributed to several important adaptations that have allowed them to exploit nearly every possible habitat. These adaptations include an exoskeleton, segmentation, well-developed sensory and nervous systems, effi-

(a)   **(b)**   **(c)**

Figure 22-14 **(a)** A polychaete annelid projects brightly spiralling gills from a tube attached to rock. When the gills retract, the tube is covered by the trap door visible on the lower right. **(b)** The "fireworm" polychaete swims using paddles on each segment. The bristles on each paddle can deliver a fiery sting. **(c)** This leech, a freshwater annelid (class Hirudinea) found in a Georgia pond, shows numerous segments. The sucker encircles its mouth, allowing it to attach to its prey. Medicinal leeches were used by doctors up until the 1800s to suck the "tainted" blood from patients suffering from a variety of disorders.

cient gas-exchange mechanisms, and well-developed circulatory systems, described below.

The Exoskeleton

The **exoskeleton** (Greek, "outside skeleton") is an external skeleton that encloses the arthropod body like a suit of armor. In places, it is thin and flexible, to allow movement of the paired, jointed appendages from which the phylum Arthropoda (Greek, "jointed foot") derives its name. The exoskeleton is secreted by the epidermis and composed chiefly of protein and a polysaccharide called **chitin.** It provides an important defense against small predators and is responsible for the greatly increased agility of arthropods over their ancestors, the annelid worms. By providing rigid attachment sites for muscles together with stiff but flexible appendages, the exoskeleton makes possible the flight of the bumblebee and the intricate, delicate manipulations of the spider as it weaves its web (Fig. 22-15). The exoskeleton also contributed enormously to the arthropod invasion of dry terrestrial habitats by providing a watertight covering

Figure 22-15 A garden spider, having immobilized its prey with a paralyzing venom, rapidly encases it in web. Such dextrous manipulations are made possible by the exoskeleton and jointed appendages characteristic of arthropods.

for delicate, moist tissues, such as those used for gas exchange.

Although it offers some major advantages, the exoskeleton also poses some unique problems, sharing many of the undesirable traits of the suit of armor. First, because it cannot expand as the animal grows, the exoskeleton must be shed or **molted** periodically and replaced with a larger size (Fig. 22-16). This uses energy and leaves the animal temporarily vulnerable before the new skeleton hardens ("soft-shelled" crabs are eaten during this delicate period). The exoskeleton is also heavy; its weight increases exponentially as the animal grows. It is no coincidence that the largest arthropods are found among the crustaceans (crabs and lobsters), whose watery habitat supports much of their weight.

Segmentation

Segmentation in arthropods is evidence of annelid ancestry. Arthropod segments, however, tend to be

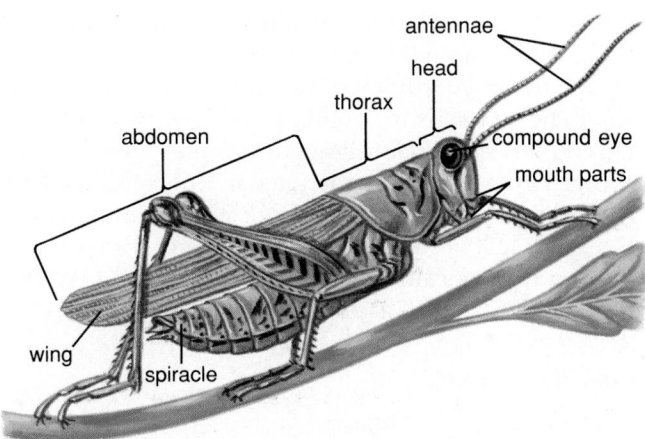

Figure 22-17 The grasshopper, an insect, shows fusion and specialization of body segments into a distinct head, thorax, and abdomen. Segments are visible beneath the wings on the abdomen.

reduced in number, fused, and specialized for distinct functions such as locomotion, feeding, and sensing the environment (Fig. 22-17).

Sensory and Nervous Systems

Most arthropods possess a well-developed sensory system, including complex **compound eyes** (Fig. 22-18) and acute chemical and tactile senses. Sensory

Figure 22-16 A newly emerged praying mantis (a predatory insect) hangs beside its outgrown exoskeleton (left).

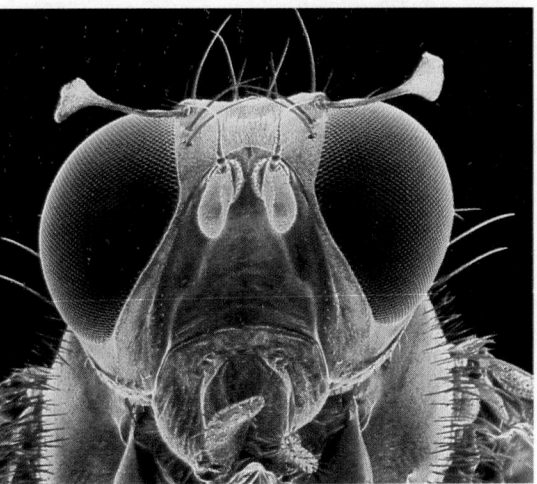

Figure 22-18 This scanning electron micrograph shows the compound eye of a fruitfly. Compound eyes consist of an array of similar light-gathering and sensing elements whose orientation gives the arthropod a panoramic view of the world. Insects have reasonable image-forming ability and good color discrimination.

information is processed by a nervous system similar to that of the annelids, but more complex. The capacity for finely coordinated movement combined with sophisticated sensory abilities and a well-developed nervous system has allowed complex behavior to evolve. In fact, the interactions among certain social insects such as the honeybee are more complex than those of most vertebrate societies. Here, communication and genetically programmed learning play important roles (see Chapter 42).

Gas Exchange

Efficient gas exchange is necessary for the rapid movement of arthropods. This is accomplished by **gills** in aquatic forms such as the crustacea, and either **tracheae** or **book lungs** in terrestrial forms (Fig. 22-19).

Circulatory Systems

Arthropods have well-developed circulatory systems with a feature not seen in annelids: the **hemocoel,** or blood cavity. Blood not only travels through vessels,

but also empties into the hemocoel, where it bathes internal organs directly. This arrangement, known as an **open circulatory system,** is also found in molluscs. Arthropod sexes are separate, and fertilization is internal.

Class Insecta

Insects are by far the most diverse and abundant arthropod class; the number of species is estimated at 800,000 (roughly the same as the total number of species in all other classes of animals combined, Fig. 22-20). Insects have three pairs of legs, usually supplemented by two pairs of wings. The capacity for flight distinguishes them from all other invertebrates and has contributed to their enormous success (Fig. 22-20c). As anyone who has unsuccessfully pursued a fly can testify, flight aids in escape from predators. It also allows the insect to find widely dispersed food. Locusts swarms have been traced from Saskatchewan, Canada, all the way into Texas on the trail of food. Flight requires rapid and efficient gas exchange. Insects use a network of narrow branching tubes

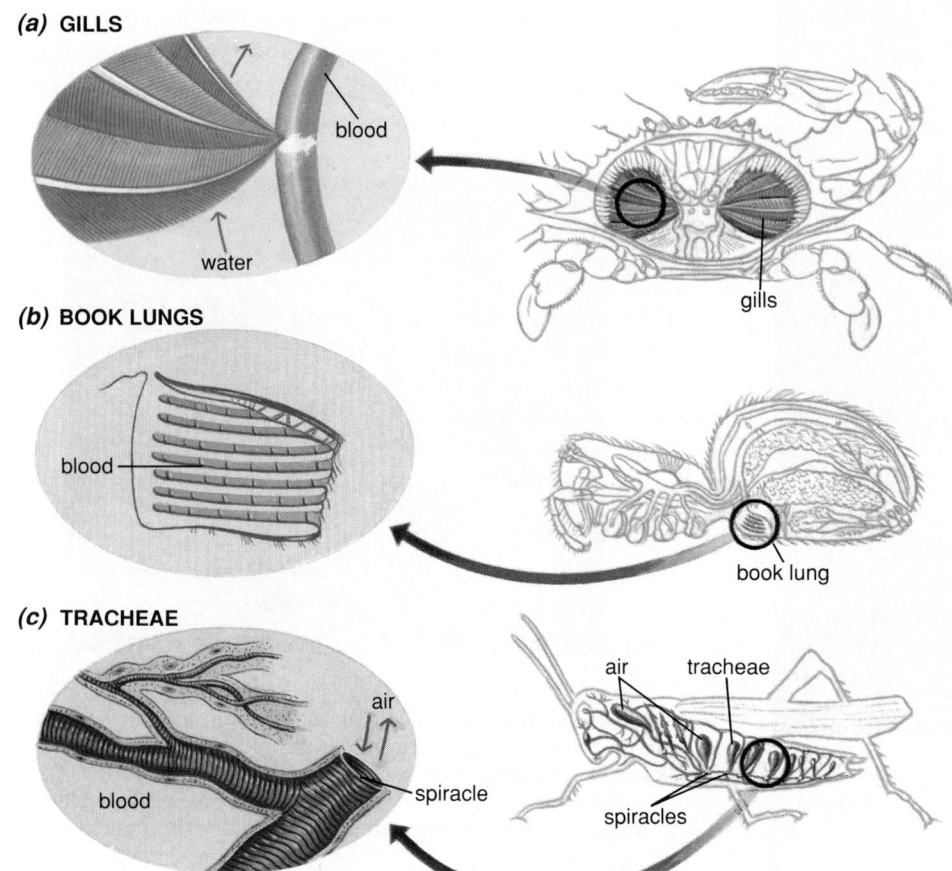

(a) GILLS

blood

water

(b) BOOK LUNGS

blood

(c) TRACHEAE

air

blood

spiracle

gills

book lung

air   tracheae

spiracles

Figure 22-19 Arthropod respiratory structures. **(a)** Gills, adapted for life in water, expose a large surface area of tissue rich in blood vessels to the water for gas exchange. Life on land demands protection of delicate, moist respiratory surfaces that are placed inside the body, with air entering through a small opening to minimize evaporation. **(b)** The book lungs of spiders resemble internal gills. **(c)** The internal tracheae of insects branch elaborately, carrying air close to each cell.

Figure 22-20 Insect diversity. **(a)** The rose aphid sucks sugar-rich juice from plants. **(b)** A mating pair of Hercules beetles. The large "horns" are found only on the male. **(c)** A may bug displays its two pairs of wings as it comes in for a landing. The outer wings protect the abdomen and inner wings, which are relatively thin and fragile. **(d)** The preying mantis waits motionless and camouflaged for its insect prey, which it will grasp swiftly in its forelegs. **(e)** Caterpillars are larval forms of moths or butterflies. This caterpillar larva of the Australian fruit-sucking moth displays large eye-spot patterns which may frighten potential predators, who mistake them for eyes of a large animal.

called **tracheae,** which conduct air to all parts of the body (see Fig. 20-19c).

During their development, insects undergo **metamorphosis,** which frequently involves a radical change in body form from juvenile to adult. The immature form is called a **larva,** which is wormlike in shape (e.g., the maggot of a housefly or the caterpillar of a moth or butterfly; see Fig. 22-20e). Metamorphosis may include a change in diet as well as shape, eliminating competition for food between adults and juveniles, and in some cases allowing the insect to exploit different foods when they are most available. For example, the caterpillar feeding on new green shoots in spring metamorphoses into the butterfly drinking nectar from summer flowers.

Class Arachnida

The arachnids comprise about 50,000 species of terrestrial arthropods, including spiders, mites, ticks, and scorpions (Fig. 22-21). All have eight walking legs, and most are carnivorous, many subsisting on a liquid diet consisting of blood or predigested prey. Spiders, the most numerous arachnids, first immobilize their prey with a paralyzing venom. They then inject digestive enzymes into the helpless victim (often an insect), and suck in the resulting soup. Arachnids breathe using tracheae or a specialized arachnid respiratory structure, **book lungs,** or both (see Fig. 22-19b). Arachnids have simple eyes, each with a single lens, in contrast to the compound eyes of insects and crustaceans. The eyes are particularly sensitive to movement, but in some species they probably can form images. Most spiders have eight eyes, whose placement gives them a panoramic view of predators and prey.

Class Crustacea

The roughly 30,000 species of crustaceans, including crabs, crayfish, lobster, shrimp, and barnacles, compose the only class of arthropods that is primarily aquatic (Fig. 22-22). Crustaceans range in size from the microscopic "water flea" (found in ponds) to the

(a)

(b)

(c)

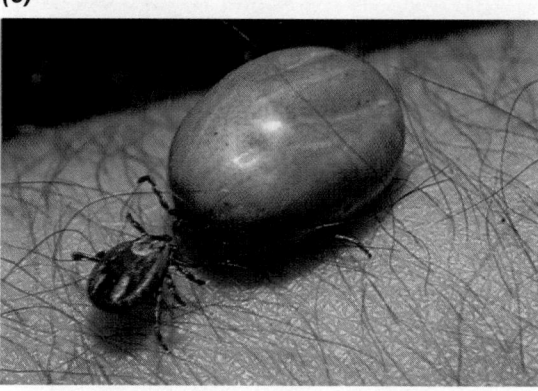

Figure 22-21 The diversity of arachnids. **(a)** The tarantula is among the largest spiders, but is relatively harmless. **(b)** Scorpions, found in warm climates including deserts of the American southwest, paralyze their prey with venom from a stinger at the tip of the abdomen. A few species can harm human beings. **(c)** Ticks before and after feeding on blood. The exoskeleton is flexible and folded, allowing the animal to become grotesquely bloated.

(a)

(b)

(c)

(d)

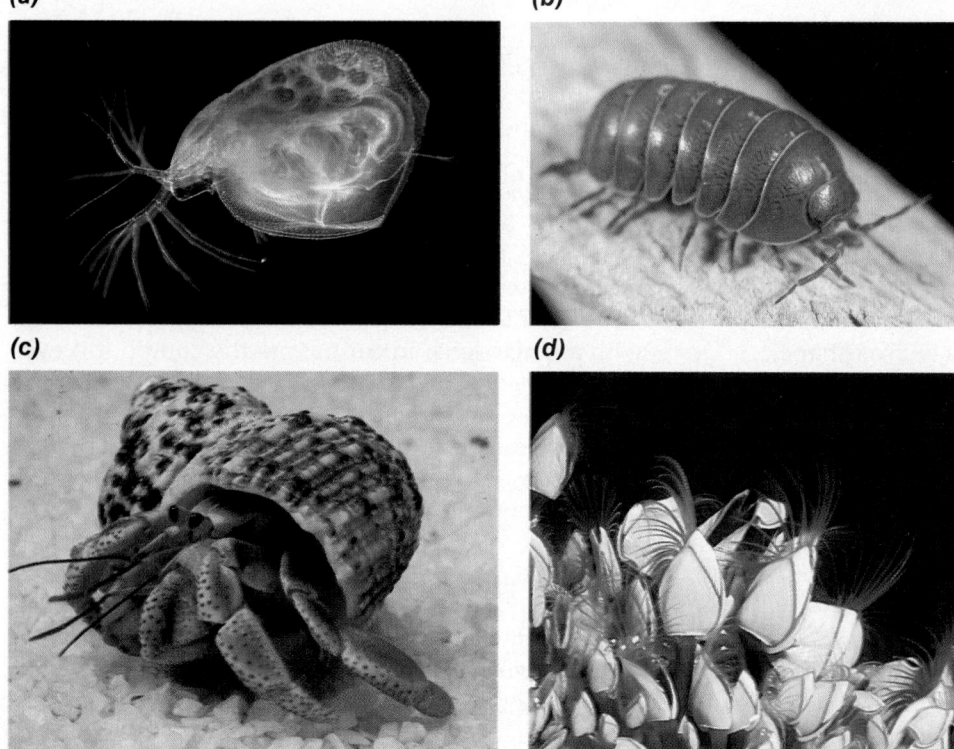

Figure 22-22 The diversity of crustacea (see also 22-6b). **(a)** The microscopic ''water flea'' *Daphnia* common in freshwater ponds. Notice the eggs developing in the dorsal brood chamber. **(b)** The sow bug, found in dark moist places such as under rocks, leaves, and decaying logs, is one of the few crustaceans to successfully invade the land. **(c)** The hermit crab protects its soft abdomen by inhabiting an abandoned snail shell. **(d)** The barnacle anchors itself to rocks, boats, or even animals such as whales. These goose barnacles attach using a tough, fleshy stalk. Other types of barnacles attach with shells resembling miniature volcanos (see Figure 22-24b). Early naturalists thought barnacles were molluscs until the jointed legs, seen extending into the water, were observed.

largest of all arthropods, the Japanese crab, with legs spanning up to 12 feet (3.7 meters). Crustaceans have two pairs of antennae, but the rest of their appendages are highly variable in form and number, depending on the habitat and life-style of the species. Most have compound eyes similar to those of insects, and nearly all respire using gills (see Fig. 22-19a).

Phylum Mollusca: The Snails, Clams, and Squid

Molluscs, like arthropods, may have arisen from an annelidlike ancestor. Their numbers and variety (about 100,000 species have been described) are second only to the arthropods. Molluscs (whose name comes from the Latin *mollis,* meaning "soft") have a moist, muscular body without a skeleton. Some protect their body with a shell of calcium carbonate, while others escape predation by tasting terrible or moving swiftly. The molluscan circulatory system is open, with blood directly bathing the organs in a hemocoel. The nervous system resembles that of annelids and arthropods, but many more of the ganglia are concentrated in the brain. Reproduction is always sexual; both separate sexes and hermaphrodites are represented.

Among the many classes of molluscs, three of outstanding importance will be discussed in more detail: the class Gastropoda (snails and their relatives), the class Pelecypoda (clams and their relatives), and the class Cephalopoda (octopuses and their relatives).

Class Gastropoda

This group of about 35,000 known species crawl on a muscular foot (gastropoda is from the Greek for "stomach foot"), and many have shells that vary widely in form and color. Some of the most beautiful gastropods, the "sea slugs," lack shells; their brilliant colors warn predators that they are poisonous or at least bad-tasting (Fig. 22-23). Gastropods feed with a **radula,** a flexible ribbon of tissue studded with spines that is used to scrape algae from rocks or grasp larger plants or prey (see Fig. 20-2c). Most gastropods respire using gills in addition to their moist skin, through which dissolved gases readily diffuse. The gills may be enclosed in a cavity beneath the shell, or exposed, as in the sea slugs. A few gastropods (including the destructive garden snails and slugs) live in moist terrestrial habitats. These terrestrial gastropods (and some freshwater forms that evolved from them) breathe using a simple lung.

Figure 22-23 The diversity of gastropods. **(a)** A Florida tree snail displays a brightly striped shell, and eyes at the tip of stalks that are retracted instantly if touched. **(b)** Spanish shawl sea slugs prepare to mate. The brilliant colors of many nudibranchs warn potential predators that they are distasteful.

Class Pelecypoda

Included in this class are the scallops, oysters, mussels, and clams. Not only do pelecypods lend exotic variety to the human diet, they are extremely important members of the marine intertidal community, where they attach to rocks that are alternately covered and exposed by the tides (Fig. 22-24). Pelecypods possess two shells connected by a flexible hinge. A strong muscle can clamp the shells closed in response to danger (this muscle is what you are served when you order scallops in a restaurant). Most pelecypods are sessile. Since the head is an adaptation to moving about in a directional manner, members of this group have lost their heads during evolution. Pelecypods are filter feeders, drawing water over gills covered with a thin layer of mucus that traps microscopic food particles. Food is conveyed to the mouth by beating cilia on the gills. A muscular foot is used by clams for burrowing in sand or mud. In mussels, the foot is reduced in size and used to help secrete a set of threads that anchor the animal to

(a)

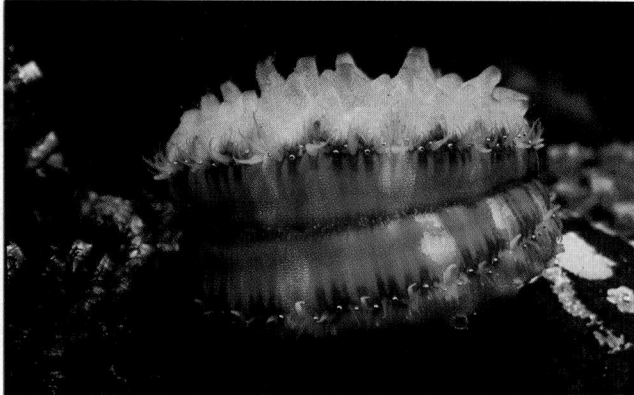

(b)

Figure 22-24 The diversity of pelecypods. **(a)** This swimming scallop from Vancouver parts its hinged shells, revealing an array of blue eyes. The upper shell is covered with an encrusting sponge. **(b)** Mussels attach to rocks in dense aggregations exposed at low tide. White barnacles are seen attached to the mussel shells and surrounding rock.

rocks. Scallops lack a foot, moving by a sort of whimsical jet propulsion achieved by flapping their shells together.

Class Cephalopoda

This fascinating group, including octopuses, squid, nautiluses, and cuttlefish (Fig. 22-25), includes the largest, swiftest, and smartest of all invertebrates. All cephalopods are predatory carnivores, and all are marine. The foot has evolved into tentacles with well-developed chemosensory abilities and suction disks for detecting and grasping prey. Prey grasped by tentacles may be immobilized by a paralyzing venom in the saliva before being torn apart by beaklike jaws. The cephalopod eye resembles our own in complexity and exceeds it in efficiency of design (see Chapter 37). Cephalopods move rapidly by jet propulsion, which

is accomplished by the forceful expulsion of water from the **mantle** cavity. The octopus may also travel along the seafloor using its tentacles like multiple, undulating legs. The cephalopod brain, especially that of the octopus, is exceptional, and in many ways resembles the brain of a vertebrate. It is enclosed in a skull-like case of cartilage, and endows the octopus with highly developed capabilities to learn and remember.

Phylum Echinodermata: The Sea Stars, Sea Urchins, and Sea Cucumbers

Although echinoderms have evolved a bewildering diversity of forms, they have never left their ancestral home on the ocean floor. Their descriptive common names reflect this: sand dollar, sea urchin, sea star (or starfish), sea cucumber, and sea lily (Fig. 22-26). Although their free-swimming embryos are bilaterally symmetrical, echinoderm adults have radial symmetry, an adaptation to a sluggish, or in some forms a sessile (immobile), existence. Most echinoderms lack a head and move very slowly and in any direction, feeding on algae or small particles sifted from sand or water. The sea star is a predator. It can slowly pursue prey (including pelecypod molluscs) from any direction. Echinoderms move on numerous tiny **tube feet,** delicate cylindrical projections that extend from the ventral surface of the body, terminating in a suction cup. Tube feet are part of a unique echinoderm feature, the **water-vascular system,** which functions in locomotion, respiration, and food capture (Fig. 22-27). Seawater enters through an opening (the sieve plate) on the animal's dorsal surface and is conducted through a ring canal that encircles the esophagus, from which branch a number of radial canals. These conduct water to the tube feet, each of which is controlled by a muscular squeeze bulb (ampulla). Contraction of the bulb forces water into the tube foot, causing it to extend. The suction cup may be pressed against the substrate or a food object, to which it adheres tightly until pressure is released.

Echinoderms have a relatively simple nervous system with no distinct brain. Movements are loosely coordinated using a nerve ring encircling the esophagus, radial nerves to the rest of the body, and a nerve network through the epidermis. In sea stars, simple receptors for light and chemicals are concentrated on the arm tips, and sensory cells are also scattered over the skin. The echinoderms lack a circulatory system, although movement of the fluid in their well-developed coelom serves this function. Gas exchange

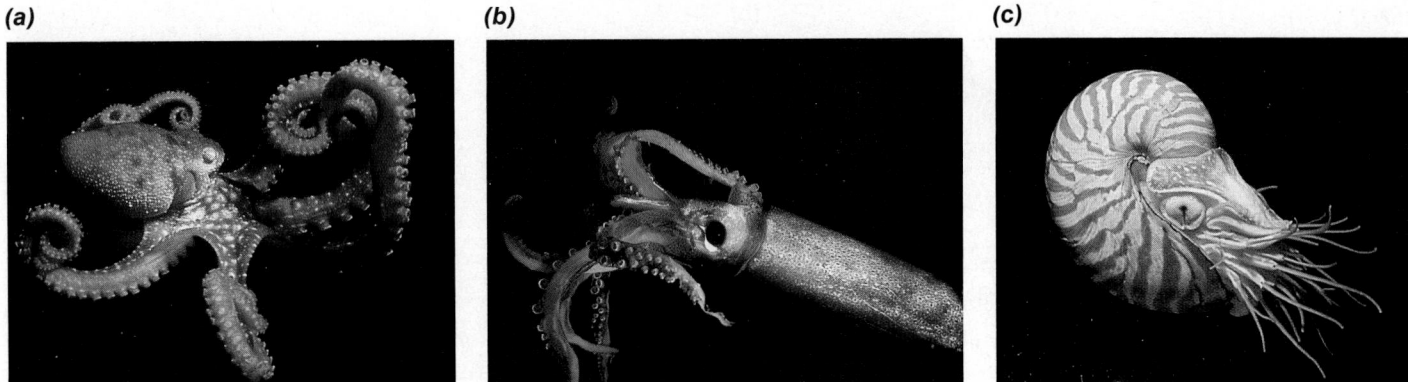

Figure 22-25 The diversity of cephalopods. **(a)** An octopus crawls using its eight suckered tentacles. It can alter its color and skin texture to blend with its surroundings. In emergencies this mollusk can jet backwards by vigorously contracting its mantle. Octopuses and squid can emit clouds of dark purple ink to confuse pursuing predators. **(b)** The squid moves entirely by jet propulsion by contracting its mantle, pushing the animal backwards through the water. The giant squid is the largest invertebrate, reaching a length of 15 meters, including tentacles. **(c)** The chambered nautilus secretes a shell with internal, gas-filled chambers providing buoyancy in the water. Note the well-developed eyes, and tentacles used to capture prey.

Figure 22-26 The diversity of echinoderms. **(a)** A sea cucumber off southern California feeds on debris in the sand. **(b)** The sea urchin's spines are actually projections of the internal skeleton. **(c)** The sea star has reduced spines, and often has five legs. This specimen is seen amidst colorful cnidarians called cup corals.

occurs through the tube feet, and in some forms, numerous tiny "skin gills" project through the epidermis. Sea cucumbers possess an internal system of canals called a *respiratory tree*. Most species reproduce by shedding sperm and eggs into the water, where fertilization occurs and a free-swimming larva develops. The sexes are usually separate. Sea stars have the ability to regenerate lost parts; new individuals may form from a single arm, provided that part of the central body is attached. When mussel fisherman

tried to rid their mussel beds of predatory sea stars by hacking them into pieces and throwing them back, needless to say, the strategy backfired!

Echinoderms possess an **endoskeleton** (Greek, "inside skeleton") composed of plates of calcium carbonate formed beneath the outer skin (Fig. 22-27a). The name *echinoderm* (meaning "hedgehog skin" in Greek) comes from projections of the endoskeleton that extend as bumps or spines through the epidermis. These are especially pronounced in the sea

(a)

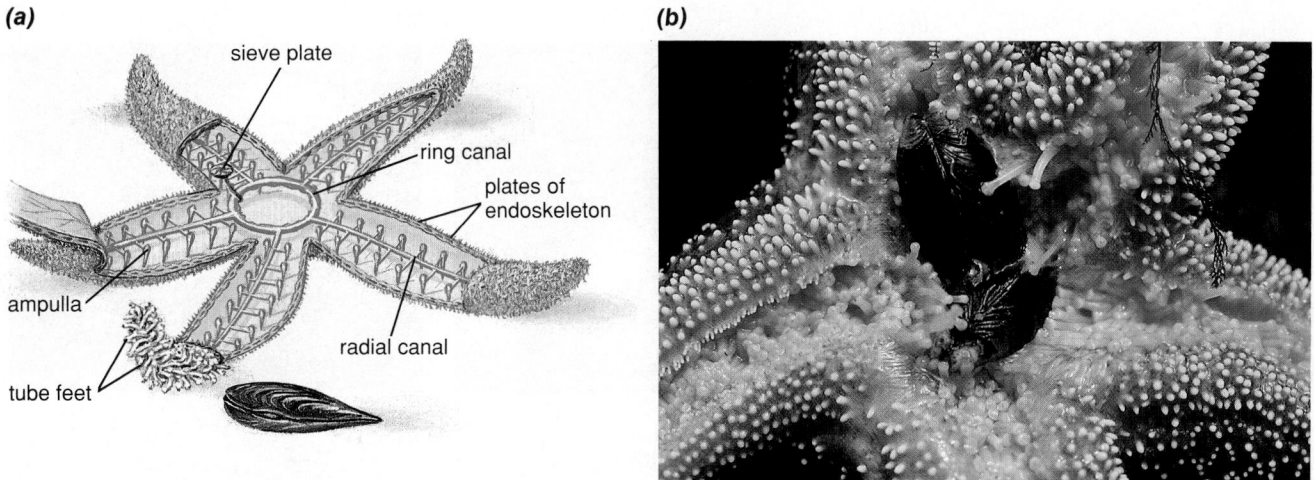

(b)

Figure 22-27 (a) The water-vascular system of echinoderms. Sea-water enters through the sieve plate and is transported into the ring canal, from which it is distributed to each of the arms through radial canals. The water inflates squeeze-bulb–like ampullae that expand and contract to extend or retract the tube feet. The plates of the endoskeleton can be seen embedded in the body wall. **(b)** The sea star often feeds on pelecypod molluscs such as this mussel. Numerous tube feet are attached to the shells, exerting a relentless pull. The sea star everts the delicate tissue of its stomach through the centrally located ventral mouth. A gape in the bivalve shells of less than 1 mm is sufficient for the stomach tissue to insinuate between the shells, secreting digestive enzymes which weaken the mollusc, causing it to gape further. Partially digested food is transported to the upper portion of the stomach where digestion is completed.

urchins, and much reduced in the sea stars and sea cucumbers.

Phylum Chordata: The Tunicates, Lancelets, and Vertebrates

The chordates are an extremely diverse group united by four features that all possess at some stage of their lives:

A notochord: A stiff but flexible rod that extends the length of the body and provides an attachment site for muscles.

A dorsal nerve cord: Lying dorsal to the digestive tract, this hollow, nervous structure develops a thickening at its anterior end that becomes a brain.

Pharyngeal gill grooves: Located in the pharynx (the cavity behind the mouth), these may form functional respiratory openings or may appear only as grooves during an early stage of development.

A tail: An extension of the body past the anus.

This list must seem particularly puzzling since humans are chordates, and at first glance we seem to lack every feature except the second. But evolutionary relationships are sometimes seen most clearly during early stages of development, and it is then that we develop, and lose, our **notochord,** our gill grooves, and finally, our tails (Fig. 22-28). We share these chordate features with other vertebrates and with two invertebrate chordate groups: the lancelets and the tunicates.

The Invertebrate Chordates

The invertebrate chordates lack a head, and of course, lack the backbone that distinguishes the vertebrates. The small (5 cm) fishlike lancelet (also called *Amphioxus*) is an invertebrate chordate that spends most of its time half-buried in the sandy sea bottom, filtering tiny food particles from the water. As seen in Figure 22-29a, all the typical chordate features are present in the adult organism.

The tunicates form a larger group of marine invertebrate chordates. It is difficult to imagine a less likely relative than this sessile, filter-feeding vase (Fig. 22-29b). Its ability to move is limited to a forceful contraction of the saclike body (sending a jet of seawater

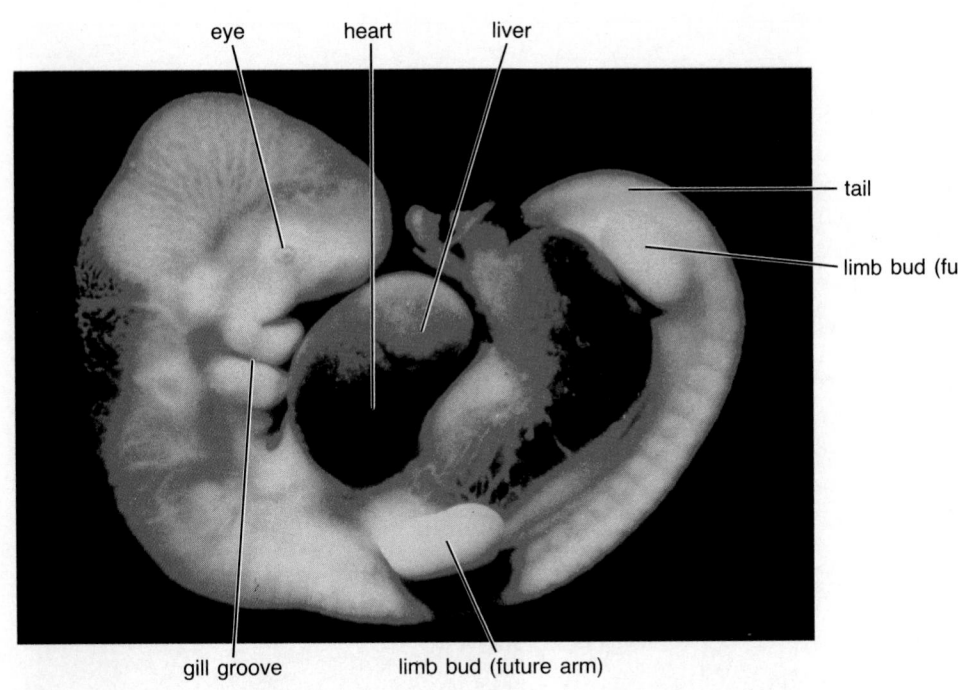

eye   heart   liver

tail

limb bud (future leg)

gill groove   limb bud (future arm)

Figure 22-28 The 5-week-old human embryo is about 1 cm in length and clearly shows external gill grooves and a tail. Although the tail will disappear completely, the gill grooves contribute to the formation of the lower jaw and the larynx.

into the face of anyone who plucks it from its undersea home, hence the common name "sea squirt"). However, tunicates produce actively swimming tadpolelike larvae that possess all the proper chordate features (see Fig. 22-29b). Since fossils of intermediate forms have never been discovered, the exact sequence of evolutionary events that led from invertebrate chordates to the first, fishlike vertebrates remains shrouded in mystery.

Subphylum Vertebrata

In the vertebrates, the embryonic notochord is replaced during development by a backbone, or **vertebral column,** composed of **cartilage** or bone. This structure provides support for the body, an attachment site for muscles, and protection for the delicate nerve cord and brain. The backbone is part of a living endoskeleton, capable of growth and self-repair. Because the internal skeleton provides support without the armorlike weight of the arthropod exoskeleton, it has allowed vertebrates to achieve great size and mobility and has contributed to their invasion of the land and the air. Today, vertebrates are represented by seven major classes: jawless fishes (class Agnatha), cartilaginous fishes (class Chondrichthyes), bony fishes (class Osteichthyes), amphibians (class Amphibia), reptiles (class Reptilia), birds (class Aves), and mammals (class Mammalia).

Class Agnatha

Vertebrates arose in the sea; the earliest vertebrate fossils are those of strange jawless fishes protected by bony armor plates. Today, two groups of jawless fishes survive: the hagfishes and the lampreys, which are similar in many features but are not closely related. Both have skeletons of cartilage and are eellike in shape. Both have unpaired fins, located along the midline of the body. Both lack scales, and their smooth, slimy skin is perforated by circular gill openings.

Hagfishes are exclusively marine (Fig. 22-30a). Purple to pink in color, they live in colonial burrows in the mud, feeding primarily on polychaete worms. They eagerly attack dying fish, however, and use pincerlike teeth that surround the tongue to burrow into the coelomic cavity and ingest the prey's soft internal organs. They are regarded with great disgust by fishermen for their ability to produce copious quantities of slime, which serves as a defense against predators.

Lampreys are found in both fresh and salt water, and include both parasitic and nonparasitic species. Even the marine forms return to freshwater to spawn. A parasitic lamprey has a suckerlike mouth lined with teeth that it uses to attach itself to larger fish. With other rasping teeth on its tongue, the lamprey excavates a hole in the host's body wall, through which it sucks blood and body fluids (Fig. 22-30b).

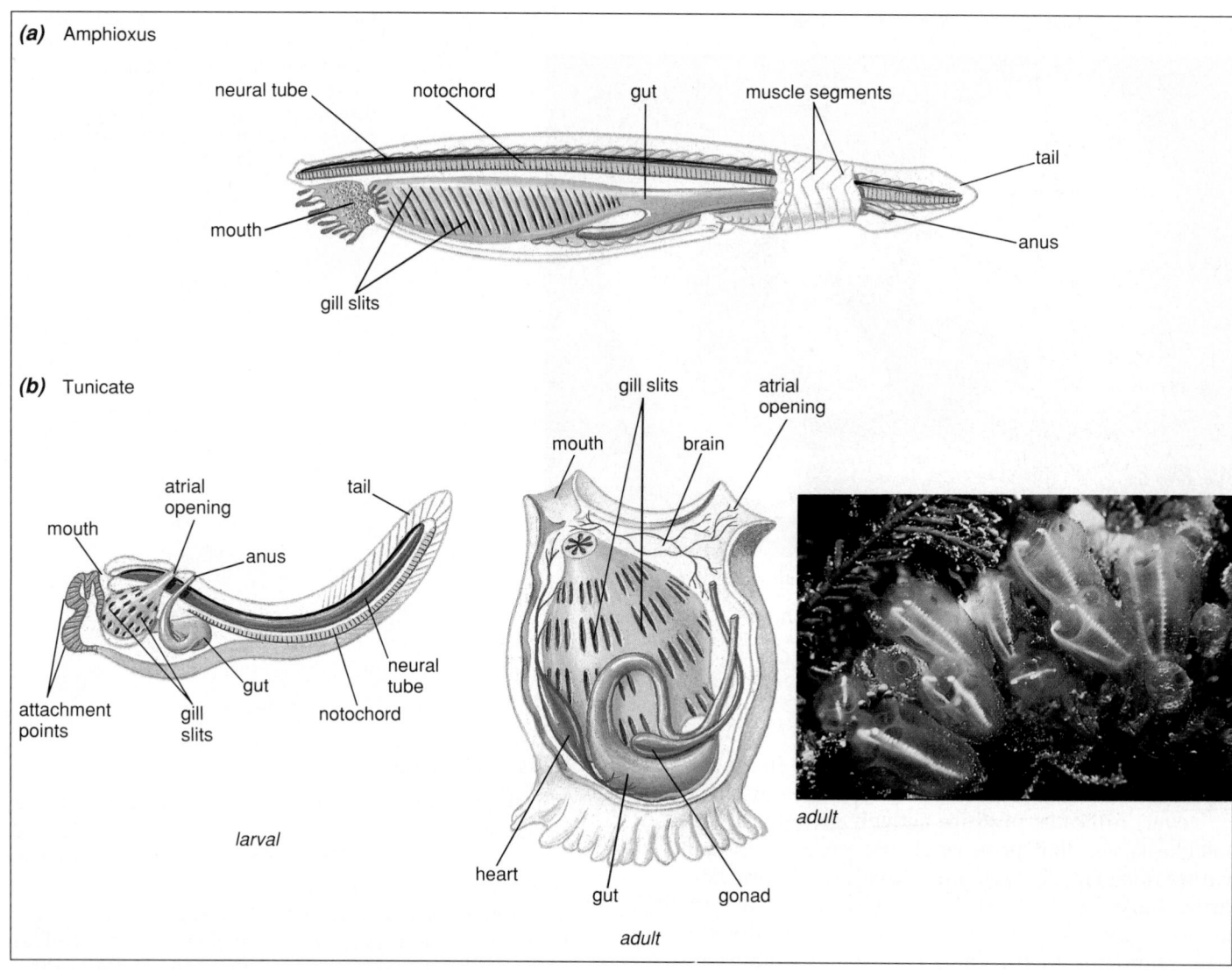

Figure 22-29 (a) *Amphioxus,* a fishlike invertebrate chordate. **(b)** The tunicate larva (left) has all the chordate features. The adult tunicate (middle) has lost the tail and notochord, and assumed a sedentary life as shown in the photo (right).

Beginning in the 1920s, lampreys spread into the Great Lakes where they have multiplied prodigiously and decimated commercial fish populations, including the lake trout. Only vigorous measures to control the lamprey population have allowed some recovery of the other fish populations of the Great Lakes.

About 425 million years ago, in the mid-Silurian period, primitive jawless fishes that were ancestral to the lampreys and hagfishes gave rise to a group of fish that possessed an important new structure found in all the more advanced vertebrates: jaws. Although these first jawed fishes have been extinct for 230 million years, they are the ancestors of the two major classes of jawed fishes that survive today: the class Chondrichthyes and the class Osteichthyes.

Class Chondrichthyes

This marine group of 625 species, whose name means "cartilage fishes," includes the sharks, skates, and rays (Fig. 22-31). These graceful predators lack any bone in their skeleton, which is formed entirely of flexible cartilage. The body is protected by a leathery skin roughened by tiny scales. Members of this group respire using gills. Although some must swim to circulate water through the gills, most can pump water

(a)

(b)

Figure 22-30 **(a)** The colorful but unattractive hagfishes live in communal burrows in the mud, feeding on polychaete worms. **(b)** Some lampreys are parasitic, attaching to fish such as this carp with suckerlike mouths lined with rasping teeth (inset).

(a)

(b)

Figure 22-31 Two members of the class Chondrichthyes. **(a)** The tropical blue-spotted sting ray swims by graceful undulations of lateral extensions of the body. **(b)** A sand tiger shark displaying several rows of teeth. As outer teeth are lost, they are replaced by new ones formed behind them. Both sharks and rays lack a swim bladder, and tend to sink toward the bottom when they stop swimming.

across their gills. These and all fish have a two-chambered heart (see Chapter 30). Sharks may have several rows of razor-sharp teeth, the back rows moving forward as the front teeth are lost. Although a few species consider us potential prey, most sharks are shy of humans. Sharks include the largest fishes; the gentle whale shark can grow to over 45 feet (15 meters). Skates and rays are also retiring creatures, although some can inflict dangerous wounds with a spine near their tail, and others produce a powerful electric shock that can stun their prey.

Class Osteichthyes

The name osteichthyes, literally ''bony fishes,'' refers to their skeleton, which is composed of bone rather than cartilage. From the snakelike moray eel to bi-

zarre, luminescent deep-sea forms to the streamlined tuna, this enormously successful group has spread to nearly every possible watery habitat, both freshwater and marine (Fig. 22-32). Although about 17,000 species have been identified, nearly twice this many may exist if the undescribed species from deep water and remote areas are considered. For example, a type of lobe-finned fish called a coelacanth, believed to have been extinct for 75 million years, was caught in deep water off the coast of South Africa in 1939 (Fig. 22-32d).

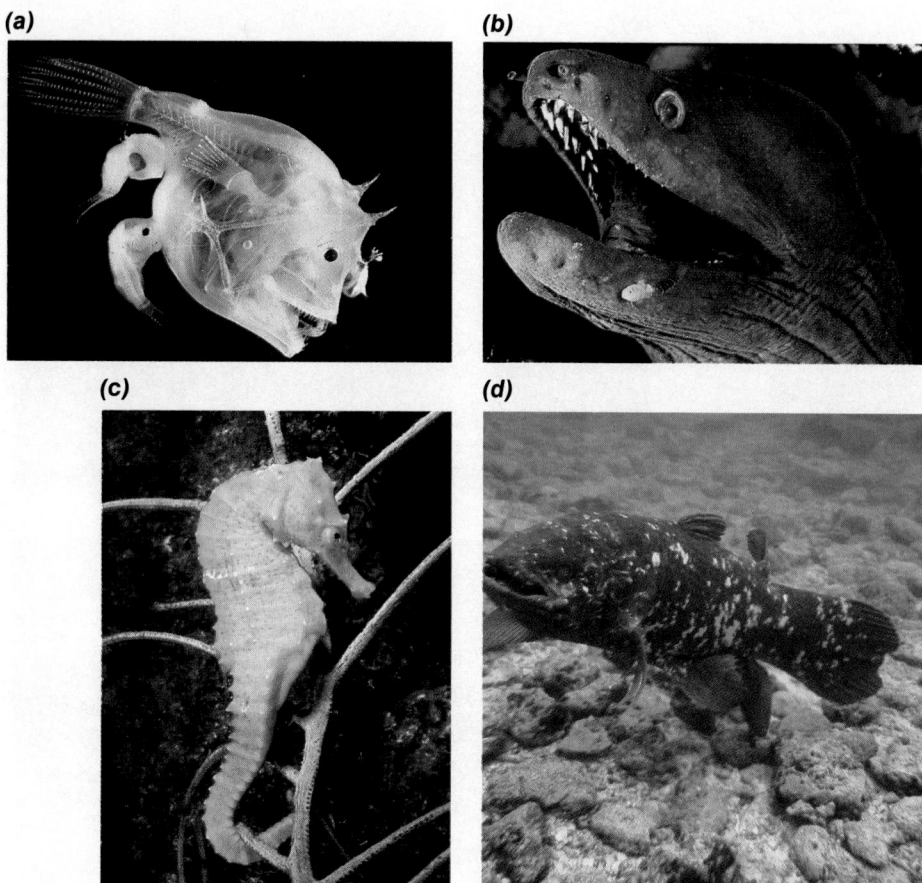

(a)

(b)

(c)

(d)

Figure 22-32 The diversity of the bony fishes attests to their successful invasion of nearly every aquatic habitat. **(a)** This female deep sea angler fish attracts prey with a living lure projecting just above her mouth. In the 2000-meter depth where anglers live, no light penetrates. Thus colors are superfluous, and the fish is ghostly white. Male deep sea anglers are extremely small, and attach to the female early in life. Here they remain as permanent parasites, always available to fertilize her eggs. Two parasitic males can be seen attached to this female. **(b)** The tropical green moray is rid of parasites by another fish, the banded cleaner goby. **(c)** The tropical sea horse may anchor itself with its prehensile tail while feeding on small crustaceans. **(d)** A rare photo of a coelacanth, a "living fossil," in its natural habitat.

A feature found in early representatives of this group and retained in a few modern species is the presence of lungs that supplement the gills. Lungs are adaptations that allow life in fresh water, which could become foul and stagnant or dry up entirely. The swimbladder, a sort of internal balloon that allows most osteichthyes to float effortlessly at any level, probably evolved from the lungs of freshwater ancestors. Some groups evolved another feature, modified fleshy fins that could be used (in an emergency) as legs, dragging the fish from a drying puddle to a deeper pool. From such ancestors arose a group that made the first tentative invasion of the land: the amphibians.

Emerging from the Sea

Land offered many advantages to those animals that first crawled from the water, including abundant food, shelter, and no predators. But the price was high. Deprived of water's support, the body was heavy and clumsy to drag along on modified fins or weak, poorly adapted legs. Unsupported by water,

gills collapse and become useless. The dry air and relentless sun suck vital water from unprotected skin and eggs, while temperature fluctuates dramatically compared with the sea. The successful colonization of the land depended on a series of adaptations that provided support for the body, waterproofing for the skin and eggs, protection of the respiratory membranes, control of body temperature, and efficient circulation.

Class Amphibia

This class of 2500 species, including frogs, toads, and salamanders, straddles the boundary between aquatic and terrestrial existence (Fig. 22-33). The limbs of amphibians show varying degrees of adaptation to movement on land, from the belly-dragging crawl of salamanders to the efficient leaping of frogs. Lungs replace gills in most adult forms, and a three-chambered heart (in contrast to the two-chambered heart of fishes) circulates blood more efficiently (see Chapter 30). However, the skin of frogs and salamanders must remain moist, since it serves as an additional respiratory organ that supplements poorly developed

Figure 22-33 The "double life" of amphibians is illustrated by the transition from tadpole **(a)** to bullfrog **(b)**. **(c)** The red salamander is restricted to moist habitats in the eastern United States.

lungs. This greatly restricts their habitats on land, and frogs are rarely found far from water. Fertilization of amphibian eggs is external and must occur in water so that the sperm can swim to the eggs. The eggs are particularly vulnerable to water loss, being surrounded with only a jellylike coating. Thus they are laid in water and develop into aquatic larvae—the tadpoles of frogs and toads, for example. The dramatic transition from completely aquatic larva to semiterrestrial adult gives the class Amphibia its name, meaning "double life." The double life and the thin, permeable skin of amphibians has made them particularly vulnerable to pollutants, as described in Planet Watch: Amphibians in Decline.

Class Reptilia

The approximately 7000 species of reptile include the lizards and snakes (by far the most successful of the modern groups), and the turtles, alligators, and crocodiles, which have survived virtually unchanged from prehistoric times (Fig. 22-34). Reptiles evolved from an amphibian ancestor about 250 million years ago. Their descendants—the dinosaurs—ruled the land for nearly 150 million years. Some reptiles, particularly desert dwellers such as tortoises and lizards, have achieved complete independence from their aquatic origins. This independence was achieved through a series of adaptations, of which three are outstanding.

Figure 22-34 The diversity of reptiles. **(a)** The mountain king snake has evolved a color pattern very similar to the poisonous coral snake, which potential predators avoid. This mimicry helps the harmless king snake avoid predation. **(b)** The American alligator, found in swampy areas of the south, has survived with little change for 150 million years. **(c)** The tortoises of the Galápagos Islands, Ecuador, may live to be over 100 years old.

PLANET WATCH
Amphibians in Decline

The earliest amphibians appeared on Earth 350 million years ago, and frogs and toads have been around for nearly 150 million years. But recently, herpetologists (biologists who study reptiles and amphibians) have been reporting an alarming trend: worldwide, thousands of species of frogs, toads, and salamanders are experiencing a dramatic decline in numbers. Although the causes for the losses are diverse and poorly understood, they can be traced to a single source: human modification of the biosphere—that portion of the Earth that sustains life.

Habitat destruction, particularly in the tropics, is one major cause of the decline. However, the unique biology of amphibians makes them vulnerable even where their habitat is not threatened. The "double life" of amphibians exposes them to toxins in a wider range of habitats, including water, air, and soil. Many amphibians' eggs develop in ponds and streams during the spring. Acid precipitation (see Chapter 45) has made spring a dangerous time for aquatic organisms. The melting of acid snow and ice causes a springtime "pulse" of intense acidity in freshwater ecosystems, just as many amphibian eggs are undergoing critical stages of development. Further, at all stages of life, amphibians are covered by a thin, permeable skin through which toxins carried in air or water can easily penetrate. Some also feed on insects that have accumulated insecticides in their bodies.

Many scientists believe that the decline in the amphibian population signals an overall deterioration of the Earth's ability to support life; and the decline is worldwide. Yosemite toads and yellow-legged frogs are disappearing from the mountains of California,

while tiger salamanders have been nearly wiped out of the Colorado Rockies. Leopard frogs, eagerly chased by rural children throughout the United States, are suddenly becoming rare. While logging destroys the habitats of amphibians from the Pacific Northwest to the tropics (Fig. E22-1), even those in preserves are dying. In the Monteverde Cloud Forest Preserve in Costa Rica, the golden toad (see Fig. 39-7) was common in the early 1980s, but has now almost disappeared. The gastric breeding frog fascinated biologists by swallowing its eggs, brooding them in its stomach, and later regurgitating fully formed offspring. The species was abundant and seemed safe within a national park in Australia. Then suddenly, in 1980, the gastric brooding frog disappeared and has never been seen since. Evidence is mounting that setting aside small islands of nature in preserves amidst surrounding environmental contamination and destruction is not enough to save species from extinction.

Amphibians are not just sensitive indicators of the health of the biosphere, they are themselves a crucial component of many ecosystems. They may keep insect populations in check, while in turn serving as food for larger carnivores. Their decline will further disrupt the balance of these delicate communities. Margaret Stewart, an ecologist at the State University of New York, Albany aptly summarized the problem: "There's a famous saying among ecologists and environmentalists: 'everything is related to everything else.' . . . You can't wipe out one large component of the system and not see dramatic changes in other parts of the system."

Figure E22-1 The corroboree toad, shown here with its eggs, is rapidly declining in its native Australia. Tadpoles can be seen developing within the eggs. The thin water- and gas-permeable skin of the adults, and the jellylike coating surrounding the eggs, provide inadequate protection against pollutants.

1. Reptiles evolved a tough, scaly skin that resists water loss and protects the body.
2. Reptiles evolved internal fertilization, in which the male deposits sperm within the female's body.
3. Reptiles evolved a shelled **amniotic egg** that can be buried in sand or dirt, far from water with its hungry predators. The shell prevents the egg from drying, while an internal membrane, the **amnion,** encloses the embryo in the watery environment that all developing animals require (Fig. 22-35).

Supplementing these features, reptiles evolved more efficient lungs, dispensing with the skin as a respiratory organ. The three-chambered heart improved to allow better separation of oxygenated and deoxygenated blood, and the limbs and skeleton were modified to provide better support and more efficient movement on land.

Class Aves

Having conquered the sea and the land, vertebrates took to the air, a source of abundant insect food and a haven from predators. The 8600 species of birds attest to the success of this strategy (Fig. 22-36). The first birdlike creatures, reptiles modified for flight, appeared roughly 150 million years ago (Fig. 22-37). Body scales were dramatically modified to form feathers, while those on the legs remained, testimony to their reptilian origin.

Many aspects of bird anatomy and physiology support the rigorous demands of flight. In contrast to reptiles, birds maintain an elevated body temperature. "Warm-bloodedness" allows both muscles and metabolic processes to operate at peak efficiency, regardless of the outside temperature, supplying the power and the energy necessary to fly. The high metabolic rate demands efficient oxygenation of tissues. To meet this need, birds have a four-chambered heart that completely separates oxygenated and deoxygenated blood. The respiratory system is supplemented by air sacs that supply oxygenated air to the lungs even as the bird exhales. Feathers protect and insulate the body; they also form lightweight extensions to the wings and tail for the lift and control demanded by flight. Hollow bones reduce the weight of the skeleton to a fraction of that in other vertebrates. Reproductive organs are considerably reduced in size during nonbreeding periods, and female birds possess only a single ovary, minimizing weight. The shelled egg that contributed to the reptiles' success on land frees the mother bird from carrying her developing offspring. The nervous system of birds accommodates the special demands of flight with extraordinary coordination and balance combined with acute eyesight.

Class Mammalia

As one line of reptiles was developing feathers, a different group, the mammals, was evolving hair, also a modification of scales. The mammals came into

(a)

(b)

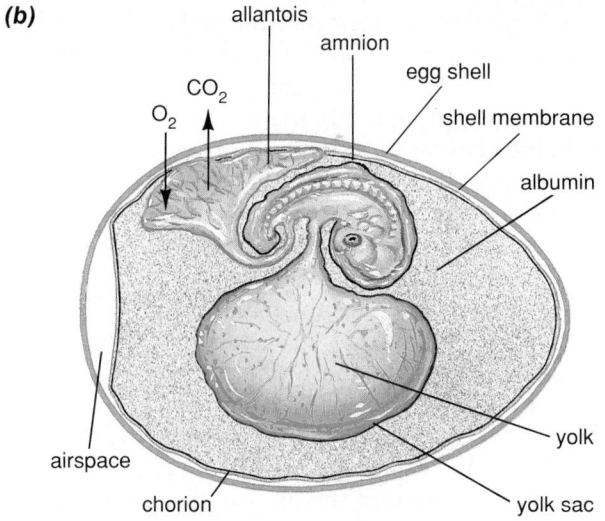

Figure 22-35 (a) An anole lizard from Carolina struggles free of its egg. **(b)** The amniotic egg of reptiles and birds is shown diagramatically. In addition to the shell, which helps prevent dehydration, the egg contains several membranes. Enclosing the embryo in a watery "pond" is the amnion. The allantois stores the urinary wastes of the embryo, and exchanges gases that can diffuse in and out through the shell. A yolk sac surrounds the fatty yolk, a high-energy food source. The chorion encloses the embryo with its membranes. The albumin is a protein food source.

(a)   **(b)**   **(c)**

Figure 22-36 The diversity of birds. **(a)** The delicate hummingbird beats its wings about 60 times per second and weighs about 4 grams. **(b)** This young frigate bird, a fish-eater from the Galápagos Islands, has nearly outgrown its nest. **(c)** The ostrich is the largest of all birds, weighing 317 pounds (144 kg) and producing eggs weighing over 3 pounds (1500 g).

prominence after the extinction of the dinosaurs roughly 70 million years ago, and today are represented by some 4500 species. Like birds, mammals are warm-blooded, with high metabolic rates. In most mammals, fur protects and insulates the warm body. Like birds, mammals have four-chambered hearts that increase the amount of oxygen delivered to the tissues. Legs designed for running rather than crawling make many mammals fast and agile. In contrast to birds, whose bodies are almost uniformly molded to the requirement of flight, mammals have evolved a remarkable diversity of form. The seal, bat, mole, impala, whale, monkey, and cheetah exemplify the radiation of mammals into nearly all habitats, with bodies finely adapted to their varied life-styles (Fig. 22-38).

This group is named for the **mammary glands** used by all members of this class to suckle their young (Fig. 22-38c). In addition to these unique milk-producing glands, the mammalian body is arrayed with sweat, scent, and sebaceous (oil-producing) glands, none of which are found in reptiles.

With the exception of the egg-laying **monotremes,** such as the platypus and spiny anteater (Fig. 22-39a), mammals give birth to live young that develop in the uterus. In one specialized group, the **marsupials** (including opossums, koalas, and kangaroos), the young develop briefly in the uterus, then crawl into a protective pouch (Fig. 22-39b). Here they firmly grasp a nipple and complete their development nourished by milk. Most mammals, called **placental** mammals, retain their young in the uterus for a much longer period.

The mammalian nervous system has contributed significantly to the success of this group by allowing behavioral adaptation to changing and varied environments. The cerebral cortex of the brain is more highly developed than in any other class, endowing mammals with unparalleled curiosity and learning ability. This allows them to alter their behavior based on experience, and helps them survive in a changing world. Recently, much of that change is due to the activities of a single mammalian species, *Homo sapiens,* whose intellectual development has led to domination of the environment.

Figure 22-37 *Archaeopteryx,* the "missing link" between reptiles and birds. This 150-million-year-old fossil shows a remarkable animal possessing a beak with sharp teeth, a long jointed tail, clawed wings, and feathers.

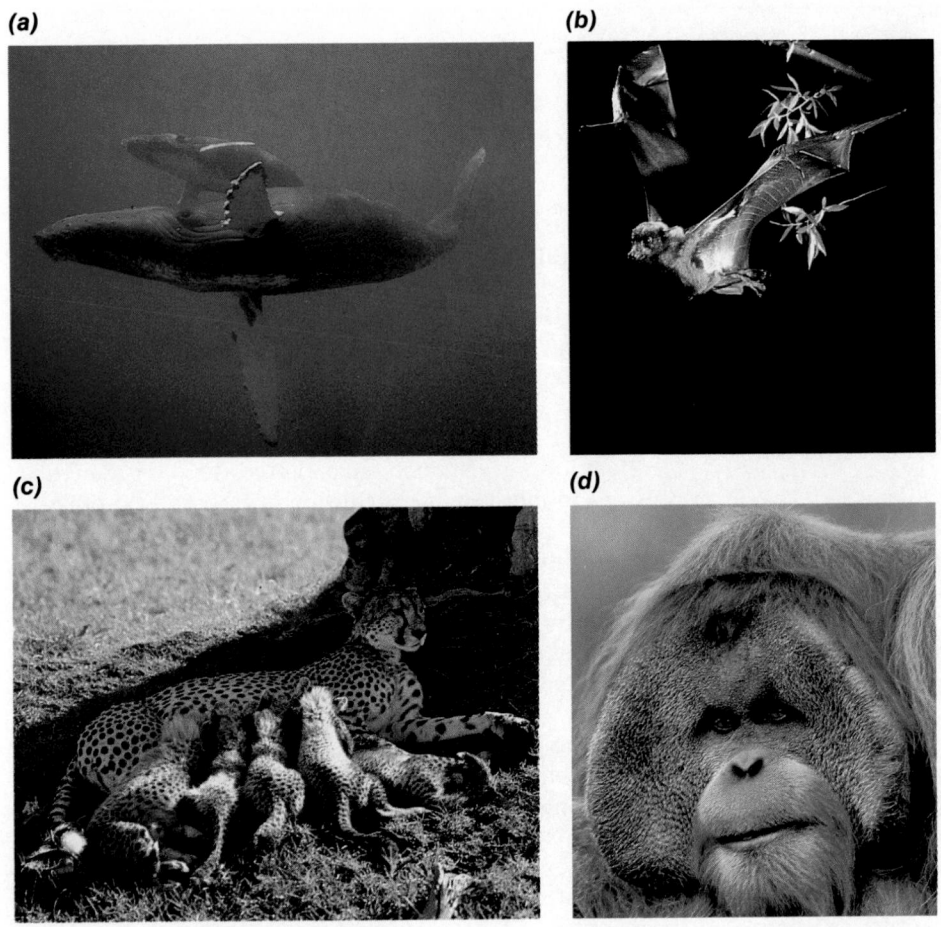

Figure 22-38 The diversity of mammals. **(a)** A humpback whale gives its offspring a boost. **(b)** A bat, the only mammal capable of true flight, navigates at night using a kind of sonar. Large ears aid in detecting echoes as its high-pitched cries bounce off nearby objects. **(c)** Mammals are named after the mammary glands with which females nurse their young, as illustrated by this mother cheetah. **(d)** The male orangutan can reach 165 pounds. These gentle, intelligent apes occupy swamp forests in limited areas of the tropics. They are endangered by hunting and habitat destruction.

Figure 22-39 Nonplacental mammals. **(a)** Monotremes, such as this platypus from Australia, lay leathery eggs resembling those of reptiles. The newly hatched young obtain milk from slitlike openings in the mother's abdomen. **(b)** Marsupials, such as the wallaby, give birth to extremely immature young who immediately grasp a nipple and develop within the mother's protective pouch (inset).

eflections: Are Humans a Biological Success?

". . . for now we have taken
The primal powers, creation and annihilation; we make
  new elements such as God never saw.
We can explode atoms and annul the fragments, noth-
  ing left but pure energy, we shall use it
In peace and in war . . .
We have minds like the tusks of those forgotten tigers
  hypertrophied and terrible . . ."
From Robinson Jeffers in *Passenger Pigeons* **(1963)**

Physically, human beings are fairly unimpressive biological specimens. We are not very strong for such large animals, nor very fast, and we lack the natural weapons of fang and claw. It is the human mind, with its tremendously developed cerebral cortex, that truly sets us apart from other animals. Our minds, in single bursts of brilliance, and in collective pursuit of common goals, have created wonders. No other animal could even appreciate the Parthenon, much less sculpt its graceful columns. We alone can eradicate smallpox and polio, domesticate other life forms, penetrate space with our rockets, and fly to the stars in our imaginations.

And yet, are we, as it appears at first glance, the most successful of all living things? The few-hundred-thousand-year duration of human existence is a mere instant in the $3\frac{1}{2}$-billion-year span of life on Earth. But during the last 300 years, our population has increased from 500 million to over 5 billion, and may double within the next 35 years. Such continued growth is unprecedented among natural populations. Is this a measure of our success? As we have expanded our range over the globe, we have driven at least 300 major species to extinction. Rapid destruction of tropical rain forests and other diverse habitats may wipe out millions of species of plants, invertebrates, and vertebrates within your lifetime, most of which we will never know.

Many of our activities have altered the environment in ways inimical to life, including our own. Acid pollutions from power plants and automobiles rains down on the land, threatening our forests and lakes—and eroding the Parthenon. Deserts spread as land is stripped of its cover by overgrazing and the demand for firewood. Parts of the land and the ocean have been rendered devoid of life by the poisonous wastes of modern civilization. Our aggressive tendencies, spurred by pressures of expanding wants and needs, their scope magnified by our technological prowess, have given us the capacity to destroy ourselves and most other life forms as well.

The human mind is the source of our power over the environment. Our brains, "hypertrophied and terrible," are simultaneously the source of our most pressing problems—and our greatest hope for solving them. Are we a phenomenal biological success, or are we a brilliant, but short-lived, "flash in the pan"? Perhaps the next few centuries will tell.

SUMMARY OF KEY CONCEPTS

Animals are multicellular, sexually reproducing, heterotrophic organisms, most of which can perceive and react rapidly to environmental stimuli. Several trends are apparent during animal evolution. There is a trend toward increasing overall complexity, culminating with the arthropods, molluscs, echinoderms, and chordates. Cellular organization increased from specialized cells through tissues to organs to organ systems. Three separate germ layers arose. Body forms evolved from asymmetry through radial symmetry to bilateral symmetry. Sense organs and aggregations of neurons became increasingly concentrated in the head, a process called cephalization. Body cavities were originally absent until pseudocoelomate and finally coelomate animals evolved. Segmentation, first seen in annelids, is present to some degree in most more complex animals. In the simple sponges, digestion is intracellular; later, a saclike gastrovascular cavity appeared, and finally, a one-way digestive tract.

Vertebrates, animals with backbones, comprise a single subphylum within the phylum Chordata. All other animals lack a backbone and are called invertebrates. Invertebrates comprise over 97% of the animal species on Earth.

Phylum Porifera: The Sponges

Sponge bodies are often free-form in shape and are always sessile. Sponges have relatively few types of cells, and although division of labor among the cell types is present, there is little coordination of activity. Sponges lack the muscles and nerves required for coordinated movement, and digestion occurs exclusively within the individual cells.

Phylum Cnidaria: The Hydra, Anemones, and Jellyfish

Far more coordination is evident among cells of cnidarians than occurs in sponges. A simple network of neurons is present, directing loosely coordinated activity of contractile cells, allowing coordinated movements. Digestion is extracellular, occurring in a central gastrovascular cavity with a single opening serving as mouth and anus. Cnidarians exhibit radial symmetry, an adaptation to the free-floating life-style of the medusa or the sedentary existence of the polyp.

Phylum Platyhelminthes, Phylum Nematoda, and Phylum Annelida: The Worms

Several advances in complexity are evident in the worms. Flatworms (phylum Platyhelminthes) are the simplest to show a distinct head with sensory organs and a simple brain. A system of canals forming a network through the body aids in excretion. Phylum Nematoda (the roundworms) are the first to possess a separate mouth and anus. The segmented worms (phylum Annelida) are the most complex, with a well-developed closed circulatory system and excretory organs resembling the basic unit of the vertebrate kidney. The segmented worms have a compartmentalized digestive system, like that of vertebrates, which processes food in a sequence. A final evolutionary advance found in annelids is the coelom, a fluid-filled space between the body wall and the internal organs, found in most complex animals, including vertebrates.

Phylum Arthropoda: The Insects, Arachnids, and Crustaceans

Arthropods are the most diverse and abundant organisms on Earth. They have invaded nearly every available terrestrial and aquatic habitat. Jointed appendages and well-developed nervous systems allow complex, finely coordinated behavior. The exoskeleton (which conserves water and provides support) and specialized respiratory structures (which remain moist and protected) allow the insects and arachnids to inhabit dry land. The diversification of insects has been enhanced by their ability to fly. Crustaceans are restricted to moist, usually aquatic habitats, respire using gills, and include the largest arthropods.

Phylum Mollusca: The Snails, Clams, and Squid

This highly successful group lacks a skeleton, sometimes protecting the soft, moist, muscular body with a single shell (as in many gastropods and a few cephalopods) or a pair of hinged shells (as in the pelecypods). The lack of a waterproof external covering limits this phylum to aquatic and moist terrestrial habitats. Although the body plan of gastropods and pelecypods limits the complexity of their behavior, the cephalopod's tentacles are capable of precisely controlled movements. The octopus has the most complex brain and the best-developed learning capacity of any invertebrate.

Phylum Echinodermata: The Sea Stars, Sea Urchins, and Sea Cucumbers

This is an exclusively marine group. Although like other complex invertebrates and chordates, echinoderm larvae are bilaterally symmetrical, the adults show radial symmetry. This, in addition to their primitive nervous system lacking any definite brain, adapts them to a relatively sedentary existence. Echinoderm bodies are supported by a nonliving internal skeleton that send projections through the skin. The water-vascular system, which functions in locomotion, feeding, and respiration, is a unique echinoderm feature.

Phylum Chordata: The Tunicates, Lancelets, and Vertebrates

The phylum Chordata includes two invertebrate groups, the lancelets and tunicates, as well as the familiar vertebrates. All possess a notochord, a dorsal nerve cord, gill grooves, and a tail at some stage in their development. Vertebrates are a subphylum of chordates that have a backbone which is part of a living endoskeleton. Vertebrate evolution is believed to have proceeded from the fishes to amphibians to reptiles, which gave rise to both birds and mammals. The heart increases in complexity from the two-chambered heart of fishes, to three in amphibians and most reptiles, to four in the warm-blooded birds and mammals. During the progression from fishes to amphibians to reptiles, a series of adaptations evolved that helped vertebrates colonize dry land. Amphibians have legs and most have simple lungs for air breathing, but most are confined to relatively damp terrestrial habitats by their moist skin, use of external fertilization, and the requirement that their eggs and larvae develop in water. Reptiles, with well-developed lungs, dry skin covered with relatively waterproof scales, internal fertilization, and the amniotic egg with its own water supply, are well adapted to the driest terrestrial habitats. Birds and mammals are also fully terrestrial and have additional adaptations, such as an elevated body temperature that allows the muscles to respond rapidly regardless of the temperature of the environment. The bird body is molded for flight, with feathers, hollow bones, efficient circulatory and respiratory systems, and well-developed eyes. Mammals have insulating hair and give birth to live young that are nourished with milk. The mammalian nervous system is the most complex in the animal kingdom, providing mammals with a unique learning ability that helps them adapt to changing environments.

GLOSSARY

amniotic egg (am-nē-ot'-ik): the egg of reptiles and birds. It contains an amnion that encloses the embryo in a watery environment; this allows the egg to be laid on dry land.

anterior (an-tēr'-ē-ur): the front, forward, or head end of an animal.

asexual reproduction: reproduction that does not involve the fusion of haploid sex cells. The parent body may divide and new parts regenerate, or a new, smaller individual may be formed attached to the parent, to drop off when complete.

bilateral symmetry: body plan in which only a single plane drawn through the central axis will divide the body into mirror-image halves.

book lungs: thin layers of tissue resembling pages in a book, enclosed in a chamber and used as a respiratory organ by certain types of arachnids.

budding: a form of asexual reproduction in which the adult produces miniature versions of itself that drop off and assume independent existence.

carnivorous (kar-niv'-e-rus): feeding on the bodies of other animals.

cartilage (kart'-lij): flexible, translucent tissue that serves as the forerunner of bone during embryonic development in most vertebrates. In the class Chondrichthyes, cartilage is retained and forms the entire skeleton.

cephalization (sef-al-ī-zā'-shun): the increasing concentration over evolutionary time of sensory structures and nerve ganglia at the anterior end of animals.

chitin (kī'-tin): a tough, flexible polysaccharide; an important constituent of the arthropod exoskeleton.

closed circulatory system: a type of circulatory system in which the blood is always enclosed in the heart and vessels.

coelom (sē'-lōm): a space or cavity within the body separating the body wall from the inner organs.

collar cells: specialized cells lining the inside channels of sponges. Flagella extend from a sievelike collar, creating a water current that draws microscopic organisms through the collar to be trapped.

compound eye: an image-forming eye consisting of numerous similar light-gathering and light-detecting elements.

cyst (sist): an encapsulated resting stage in the life cycle of certain invertebrates, such as parasitic flatworms and roundworms.

dorsal (dōr'-sul): the top, back, or uppermost surface of an animal oriented with its head forward.

endoskeleton: a supportive structure within the body; an internal skeleton. It may be nonliving, as in echinoderms and sponges, or living, as in vertebrates.

exoskeleton: an external, nonliving supporting structure; an external skeleton.

flame cells: cells in flatworms specialized for excretion and fluid regulation. They enclose a small chamber full of beating cilia, whose flickering appearance gives them their name.

free-living: not parasitic.

ganglion (gan'-glē-un): an aggregation of neurons.

gastrovascular cavity (gas'-trō-vas'-kū-lar): a saclike opening in the bodies of some invertebrates (such as cnidarians and flatworms) with a single opening serving as both mouth and anus.

germ layer: A tissue layer formed during early embryonic development.

hemocoel (hē'-mō-sēl): a blood cavity within the bodies of certain invertebrates in which blood bathes tissues directly. A hemocoel is part of an open circulatory system.

hermaphroditic (her-maf'-ruh-dit'-ik): possessing both male and female sexual organs. Some hermaphroditic animals can fertilize themselves; others must exchange sex cells with a mate.

hydrostatic skeleton (hī-drō-stat'-ik): the use of fluid contained in body compartments to provide support for the body and mass against which muscles can contract.

invertebrate (in-vert'-uh-bret): a category of animals that never possess a vertebral column.

larva (lar'-vuh): an immature form of an organism prior to metamorphosis into its adult form. The caterpillars of moths and butterflies, and the maggots of flies, are larvae.

mammary glands (mam'-uh-rē): milk-producing organs used by female mammals to nourish their young.

mantle (man'-tul): an extension of the body wall in certain invertebrates, such as molluscs. It may secrete a shell, protect the gills, and, as in cephalopods, aid in locomotion.

marsupial (mar-sū'-pē-ul): a type of mammal whose young are born at an extremely immature stage and undergo further development in a pouch while they remain attached to a mammary gland. Includes kangaroos, opossums, and koalas.

medusa (meh-dū'-suh): a bell-shaped, often free-swimming stage in the life cycle of many cnidarians. Jellyfish are one example.

mesoglea (mez-ō-glē'-uh): a middle, jellylike layer within the body wall of cnidarians.

metamorphosis (met-uh-mor'-fuh-ses): a dramatic change in body form during development, as seen in amphibians (tadpole to frog) and insects (caterpillar to butterfly).

molt: to shed an external body covering, such as an exoskeleton, skin, feathers, or fur.

monotreme: a type of mammal that lays eggs; for example, the platypus.

nematocyst (nēm-āt'-ō-sist): a specialized cell found in cnidarians which, when disturbed, ejects a sticky or poisoned thread. Used by cnidarians to trap and sting prey.

nerve cord: also called the spinal cord of vertebrates, a nervous structure lying along the dorsal side of the body of chordates.

nerve net: a loosely coordinated network of neurons.

notochord (nōt'-ō-kōrd): a stiff but somewhat flexible, supportive rod found in all members of the phylum Chordata at some stage of development.

open circulatory system: a type of circulatory system in arthropods and molluscs in which the blood is pumped through an open space (the hemocoel), where it bathes the internal organs directly.

organ: two or more tissues integrated to perform a specialized function, i.e., the kidney.

organ system: two or more organs that work together to perform a specific function, i.e., the digestive system.

osculum (os'-kū-lum): a relatively large opening in the sponge body through which water is expelled.

parasite (par'-uh-sīt): an organism that lives in or on the body of another organism, causing it harm as a result.

pharynx (fār'-inx): a portion of the digestive system between the mouth and the esophagus. In flatworms, it is developed as an extensible, muscular organ.

placenta (pluh-sen'-ta): a tissue rich in blood vessels that develops in the mammalian uterus during pregnancy. Here nutrients and oxygen from maternal blood are exchanged for wastes from the developing embryo.

polyp (pol'-ip): the sedentary, vase-shaped stage in the life cycle of many cnidarians. Hydra and sea anemones are examples.

posterior (post-tēr'-ē-ur): the tail, hindmost, or rear end of an animal.

pseudocoel (sū'-dō-sēl): "false coelom"; a body cavity with a different embryological origin than a coelom, but serving a similar function; found in roundworms.

radial symmetry: a body plan in which any plane drawn along a central axis will divide the body into approximately mirror-image halves. Cnidarians and many adult echinoderms show radial symmetry.

radula (ra'-dū-luh): a ribbon of tissue in the mouth of gastropod mollusks that bears numerous teeth on its outer surface and is used to scrape and drag food into the mouth.

segmentation (seg-men-tā'-shun): division of the body into repeated, often similar units.

sessile (ses'-ul): not free to move about, usually permanently attached to a surface.

spicule (spik'-ūle): subunits of the endoskeleton of sponges, made of protein, silica, or calcium carbonate.

tentacle (ten'-te-kul): an elongate, extensible projection of the body of cnidarians and cephalopod mollusks that may be used for grasping, stinging, and immobilizing prey, and locomotion.

tissue: a group of (usually similar) cells that together carry out a specific function, i.e., muscle.

tracheae: a system of air tubes that branch within the body of insects and some arachnids, carrying air close to every cell.

tube feet: cylindrical extensions of the water-vascular system of echinoderms, used for locomotion, grasping food, and respiration.

ventral (ven'-trul): the lower, or underside of an animal whose head is oriented forward.

vertebrate: an animal that possesses a vertebral column.

water-vascular system: a system in echinoderms consisting of a series of canals through which seawater is conducted and used to inflate tube feet for locomotion, grasping food, and respiration.

STUDY QUESTIONS

1. After each characteristic listed, state the phylum or phyla to which it applies.
   a. radial symmetry
   b. a gastrovascular cavity
   c. an open circulatory system
   d. some members are sessile
   e. lacks nerves and muscles
   f. a water-vascular system
   g. an internal skeleton
   h. an external skeleton
   i. some members have a radula
   j. some members have tentacles
   k. a notochord
   l. more species than all others combined
   m. segmentation
   n. includes both vertebrate and invertebrate members
   o. includes members that can fly

2. Of the phyla described in the text, list the first (simplest) to possess each of the following features:
   a. a gastrovascular cavity
   b. bilateral symmetry
   c. a coelom
   d. nerves
   e. a one-way digestive tract
   f. a closed circulatory system
   g. radial symmetry

3. Describe and compare respiratory systems in the three major arthropod classes.

4. Describe the advantages and disadvantages of the arthropod exoskeleton.

5. State in which of the three major mollusc classes each of the following characteristics is found.

   a. two hinged shells
   b. a radula
   c. tentacles
   d. some sessile members
   e. the best-developed brains
   f. numerous eyes

6. Give three functions of the water-vascular system of echinoderms.

7. Explain the origin of echinoderm spines.

8. To what life-style is radial symmetry an adaptation? bilateral symmetry?

9. List the vertebrate class (or classes) in which we find each of the following.
   a. a skeleton of cartilage
   b. a two-chambered heart
   c. a four-chambered heart
   d. lungs supplemented by air sacs
   e. the amniotic egg
   f. warm-bloodedness
   g. a placenta

10. Distinguish between vertebrates and invertebrates. List the major phyla found in each broad grouping.

11. List four distinguishing features of chordates.

12. Describe the ways in which amphibians are adapted to life on land, and in what ways they are still restricted to a watery or moist environment.

13. List the adaptations that distinguish reptiles from amphibians and help them adapt to life in dry terrestrial environments.

14. List the adaptations of birds that contribute to their ability to fly.

15. How do mammals differ from birds, and what adaptations do they share?

16. How has the mammalian nervous system contributed to the success of this group?

DISCUSSION QUESTIONS

1. In what ways has the mammalian nervous system and the placenta contributed to the success of this group?

2. Discuss at least three ways in which the ability to fly has contributed to the success and diversity of insects.

3. Discuss the many similarities and differences between birds and bats, two flying vertebrates.

4. Among reptiles and fish, some species are viviparous (they directly give birth to offspring) and some are oviparous (they lay eggs from which the young hatch). Discuss some advantages of each method?

SUGGESTED READINGS

Galdikas, B. M. F. "My Life with Orangutans." *International Wildlife*, March–April 1990.

Goreau, T. F., Goreau, N. I., and Goreau, T. J. "Corals and Coral Reefs." *Scientific American*, August 1979.

Hickman, C. P., Roberts, L. S., and Hickman, F. M. *Biology of Animals*. St. Louis: The C. V. Mosby Company, 1982.

Horridge, G. A., "The Compound Eye of Insects." *Scientific American*, July 1977.

McMenamin, M. A. S. "The Emergence of Animals." *Scientific American*, May 1987.

Phillips, K. "Frogs in Trouble." *International Wildlife*, November–December, 1990.

Rahn, H., Ar, A., and Paganelli, C. V. "How Bird Eggs Breathe." *Scientific American*, February 1979. (Offprint No. 1420).

Schwartz, D. M. "Hurray for Hedgehogs!" *International Wildlife*, March–April 1990.

23

The Diversity of Life.
III. Fungi

A ripe puffball touched by rain, a breeze, or a scurrying
animal releases clouds of spores that are dispersed far
and wide on air currents.

Although once classified as plants, fungi are as different from plants as they are from animals. Fungi are heterotrophic, feasting on other organisms. Parasitic fungi may attack food crops, with devastating economic consequences. A few cause disease in humans.

Fungi play diverse roles in natural ecosystems; one of the most important is decomposition. Frequently, the only noticeable portion of a fungus is its reproductive structure (see the chapter opening photograph). Most have filamentous bodies that penetrate deeply into the dead bodies and wastes of plants and animals. Here they secrete enzymes that break down complex molecules into simpler nutrients. Many of these nutrients are released into the air, water, and soil where they are available for reuse by other organisms.

The soft bodies of fungi do not fossilize well, and the evolutionary relationships of the fungi remain obscure. This chapter examines the "true fungi"; those that have cell walls, feed by absorption, and usually have filamentous bodies (yeasts are an exception). The cellular and acellular slime molds are covered in Chapter 21.

We begin with an overview of the general features and ecology of fungi, then move on to a brief description of the major fungal divisions.

Fungal Form and Function

The Fungal Body

The fungal body, with rare exceptions, consists of microscopically thin, threadlike filaments called **hyphae** (singular: hypha) that grow in an interwoven mass called a **mycelium** (Fig. 23-1). The hyphae of some fungi consist of single elongated cells with numerous nuclei, while in others the hyphae are subdivided by partitions called **septa** into many cells, each containing from one to many nuclei, depending on the species. Pores in the septa allow the cytoplasm to stream between cells, distributing nutrients.

In contrast to the nuclei found in animal cells, those of the fungal body are usually haploid (possessing only a single set of chromosomes). Like plants, fungal cells are surrounded by cell walls. Although some fungal cell walls contain cellulose, most are strengthened by chitin, the same substance found in the exoskeletons of arthropods (see Chapter 22).

The fungal mycelium insinuates itself into aging bread or cheese, beneath the bark of decaying logs, or into the soil. Periodically, the hyphae aggregate and differentiate into reproductive structures that project above the surface beneath which the mycelium grows. These structures, including mushrooms, puff-

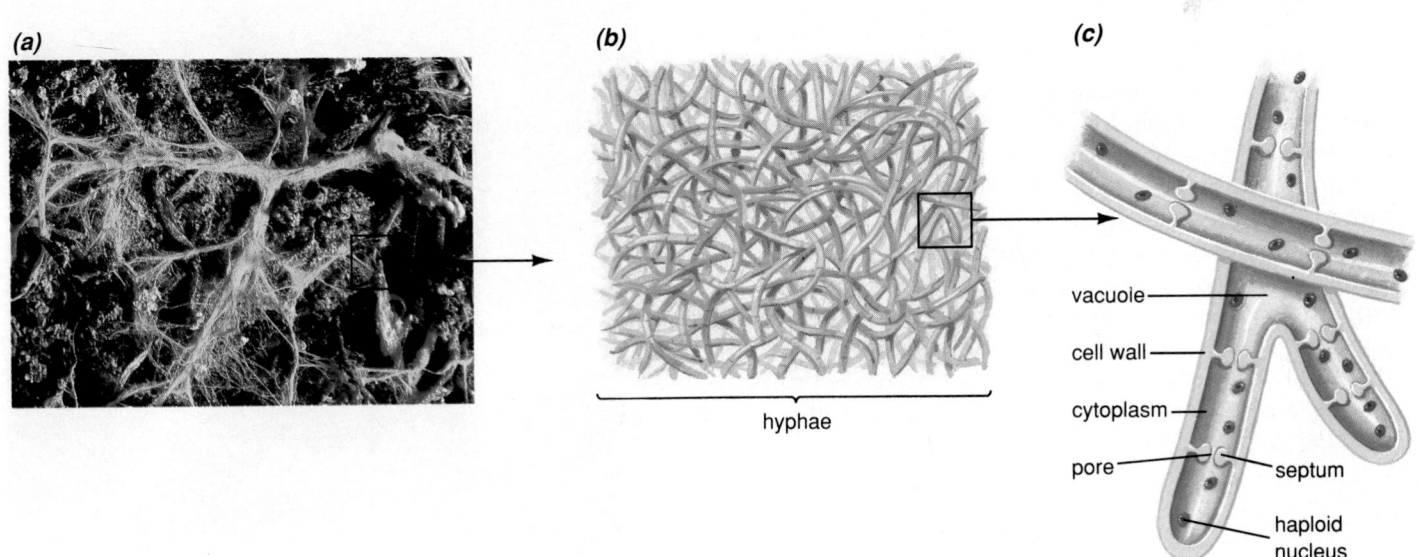

(a) **(b)** **(c)**

vacuole
cell wall
cytoplasm
pore
septum
haploid nucleus
hyphae

Figure 23-1 (a) A fungal mycelium, about the size of your hand, spreads over a decaying log. The mycelium is composed of a tangle of microscopic hyphae, drawn in **(b)** and enlarged in **(c)** to show their internal organization.

balls, or the powdery molds on food, are often the only part of the fungus that is easily seen.

Fungal Nutrition

Like animals, fungi are heterotrophic, living on nutrients stored in the bodies or wastes of other organisms. Some fungi are **saprobes;** these digest the bodies of dead organisms. Others, which are parasitic, feed on living organisms. A third group lives symbiotically with other organisms. Two well-known examples are **lichens** and **mycorrhizae,** discussed later.

Fungal nutrient absorption resembles that of bacteria; both have cell walls that prevent them from ingesting food. Like bacteria, fungi secrete enzymes outside their bodies, and digest complex macromolecules into smaller subunits that can be absorbed. The fungal body, composed of long filaments only one cell thick, presents an enormous surface area for nutrient absorption. Rapidly growing filaments penetrate deeply into a source of nutrients, hastening decomposition by digesting the material. The resulting small nutrient molecules may diffuse through the fungal cell walls and into the cells, but some remain outside, providing nourishment for other organisms.

Fungal Reproduction

Fungal reproduction is varied and often complex, involving both sexual and asexual processes. During simple asexual reproduction, a mycelium breaks into pieces, each of which grows into a new individual. Many fungi reproduce both asexually and sexually via different types of spores formed on or within special structures that project above the mycelium, allowing dispersal of the spores by wind or water (see Fig. 23-1). Haploid asexual spores are produced by mitotic divisions of haploid fungal cells. In favorable habitats, the asexual spores begin mitotic divisions that produce a new haploid mycelium. Sexual spore formation begins with the fusion of two haploid nuclei of compatible mating types, resulting in a diploid zygote. This zygote then undergoes meiosis to form haploid sexual spores. These are dispersed, germinate, and divide mitotically to form a new haploid mycelium. The reproductive capacity of fungi is prodigious: a single giant puffball may contain 5 trillion sexual spores (see Fig. 23-10c).

Economic and Ecological Impacts of Fungi

Although the sight of a brightly colored mushroom on the forest floor is a pleasant surprise, microscopic fungal filaments can be found throughout moist rich soils, penetrating decaying vegetation and the wastes and dead bodies of animals. As decomposers, fungi make an incalculable contribution to ecosystems. Their extracellular digestive activities liberate nutrients that can be used by plants, and fungal bodies provide food for small insects and worms. If fungi and bacteria were suddenly to disappear, nutrients would remain locked in the bodies of dead plants and animals, many nutrient cycles would grind to a halt, soil fertility would rapidly decline, and ecosystems would collapse.

Parasitic fungi, in contrast, are responsible for the majority of plant diseases. American chestnut and elm trees have been decimated by the fungi causing chestnut blight and Dutch elm disease. Fungal parasites such as those causing corn smut also cause billions of dollars in crop losses annually (see Fig. 23-10d). In contrast, farmers have discovered that fungal parasites of insects (Fig. 23-2) can be an important ally in pest control. These "fungal pesticides" are being used in Florida to control citrus mites, and it is hoped they will make further contributions as farmers be-

Figure 23-2 Although some parasitic fungi attack crops, others assist farmers by attacking insect crop pests.

come more aware of the need to reduce their reliance on chemical pesticides.

The human body can host several parasitic fungi. These cause a range of diseases that infect the skin (causing ringworm and athlete's foot), the lungs (causing valley fever and histoplasmosis, from soil fungi), and the vagina (causing candidiasis, from the yeast *Candida*).

Several types of fungi are edible. Some, such as the truffle, are rare delicacies. Fungi are also important in the production of many staple foods and beverages such as bread, cheese, wine, and beer.

Fungal Symbiotic Relationships

Lichens

Lichens are a unique **symbiotic** (literally "living together") association between fungi, usually ascomycetes (described below) and unicellular green algae or cyanobacteria (Fig. 23-3). Together, these organisms form a living unit so tough and undemanding of nutrients that it is among the first to colonize newly formed volcanic islands. The algal partner provides

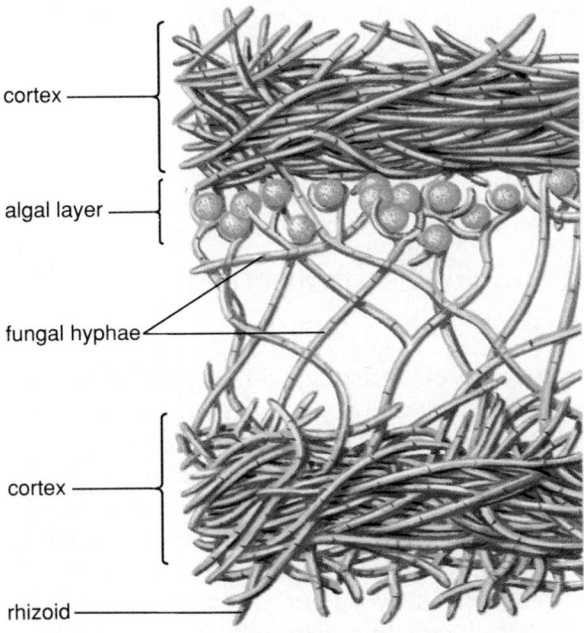

Figure 23-3 Most lichens have a layered structure bounded on the top and bottom by a cortex formed from fungal hyphae. Rhizoids, formed from fungal hyphae, emerge from the lower cortex and anchor the lichen to a surface, such as a rock or a tree. An algal layer in which the alga and fungus grow in close association is found beneath the upper cortex.

food formed by photosynthesis, while the fungus provides support and protection from dehydration. Lichens in a variety of bright colors can be found growing on bare rock, and have invaded habitats ranging from deserts to the Arctic. Based on their size and slow rate of growth, some lichens in the Arctic are believed to be 4000 years old. About 20,000 lichens have been identified, each with a different combination of fungus and algae or blue-green bacteria, and with a unique color and growth pattern (Fig. 23-4).

Mycorrhizae

Mycorrhizae are symbiotic associations between fungi and plant roots. Over 5000 species of mycorrhizal fungi (mostly basidiomycetes or ascomycetes, described below) are found growing in intimate association with the roots of about 80% of all vascular plants, including most trees. These associations benefit both the plant and its fungal partner. The hyphae of mycorrhizae surround the root and frequently invade the root cells (Fig. 23-5). Plants that participate in this unique relationship tend to grow larger and more vigorously than those deprived of the fungus, especially in poor soils. The fungus digests and absorbs organic nutrients from the soil, passing some of these nutrients directly into the plant root cells. The fungus also absorbs water and passes it to the plant, an advantage in dry, sandy soils. In return, food produced photosynthetically by the plant is passed from the root to the fungus.

Some scientists believe that mycorrhizal associations may have been important in the invasion of land by plants over 400 million years ago. Such a relationship between an aquatic fungus such as a water mold (division Oomycota) and a green alga (ancestral to terrestrial plants) could have helped the alga acquire the water and mineral nutrients it needed to survive out of water.

The Classification of Fungi

Although nearly 100,000 species of modern fungi have been described, biologists have only begun to comprehend fungal diversity—at least 1000 additional species are described yearly.

Fungal taxonomy, like that of microorganisms, is the subject of considerable controversy. For example, slime molds, which we classify as protists, are sometimes considered fungi. Like plants, fungi are grouped into divisions, comparable to animal phyla. The major divisions of fungi are the Zygomycota, Oomycota, Ascomycota, Basidiomycota, and Deuteromycota (Table 23-1).

(a)

(c)

(b)

Figure 23-4 Diverse lichens. **(a)** Goat's beard, a lichen found hanging from tree limbs. **(b)** A colorful encrusting lichen, growing on dry rock, illustrates the tough independence of this symbiotic combination of fungus and alga. **(c)** A leafy lichen grows from a dead tree branch.

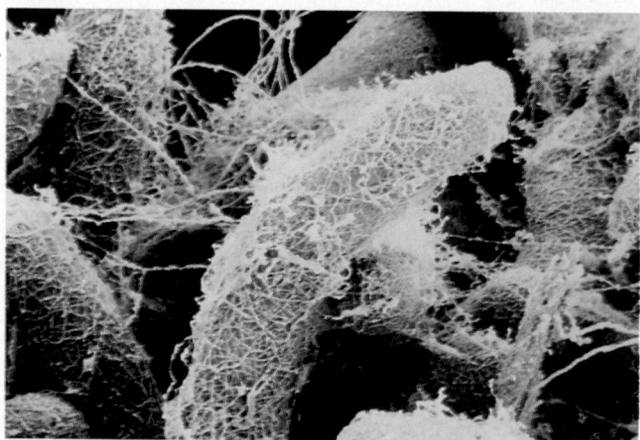

Figure 23-5 Here hyphae of mycorrhizae are seen entwining about the root of an aspen tree. Plants grow significantly better in a symbiotic association with these fungi, which help make nutrients and water available to the roots.

Division Zygomycota: The Zygote Fungi

The zygomycetes, also called the *zygote fungi*, include about 600 species. They are named for the ability of their haploid hyphae to "mate," fusing their nuclei to produce diploid **zygospores** (Fig. 23-6, inset). These resistant structures are dispersed through the air and can remain dormant until conditions are favorable for growth. They then undergo meiosis, producing haploid cells that grow into new hyphae. These hyphae may reproduce either asexually, by forming haploid spores in black spore cases called **sporangia** (Fig. 23-6, inset), or sexually, by fusing to produce more zygospores. The zygomycetes include the dung fungus *Pilobolus* (Greek, "cap-thrower;" see "A Closer Look at Evolutionary Ingenuity in Fungi," Fig. E23-2). More familiar and annoying zygomycetes are

Table 23-1   The Major Divisions of Fungi

Division	Reproductive Structures	Cellular Characteristics	Economic and Health Impacts	Representative Genera
Zygomycota: zygote fungi	Produce sexual diploid zygospores	Cell walls contain chitin; septa are absent	Causes soft fruit rot and black bread mold	*Rhizopus* (causes black bread mold); *Pilobolus* (dung fungus)
Oomycota: egg fungi	Sexual reproduction involves fertilization of a large egg; flagellated asexual zoospores require water to swim	Cell walls contain cellulose; septa may be present or absent	Causes downy mildew of grapes and late blight of potatoes	*Achyla* (water mold); *Phytophthora* (late blight of potatoes)
Ascomycota: sac fungi	Sexual spores formed in saclike ascus	Cell walls contain chitin; septa are present	Causes molds on fruit; can damage textiles; causes Dutch elm disease and chestnut blight; includes yeasts and morels	*Saccharomyces* (yeast); *Ceratocystis* (causes Dutch elm disease)
Basidomycota: club fungi	Sexual reproduction involves production of haploid basidiospores on club-shaped basidia	Cell walls contain chitin; septa are present	Causes smuts and rusts on crops; includes some edible mushrooms	*Amanita* (poisonous mushroom); *Polyporus* (shelf fungus)
Deuteromycota: imperfect fungi	Not observed engaging in sexual reproduction	Cell walls contain chitin; septa are present	Cause athlete's foot, ringworm, histoplasmosis; source of penicillin	*Penicillium* (produces penicillin); *Arthrobotrys* (nematode predator)

those of the genus *Rhizopus*, which cause soft fruit rot and black bread mold. The life cycle of a zygomycete, the black bread mold showing both sexual and asexual reproduction, is depicted in Figure 23-6.

Division Oomycota: The Egg Fungi

This group, whose name means "egg fungi," differs significantly from other true fungi. For example, oomycete cell walls often contain cellulose, the main ingredient in plant cell walls, instead of (or in addition to) chitin. Sexual reproduction involves the fertilization of a large egg cell, hence the name *egg fungi*. Oomycetes reproduce asexually using actively swimming flagellated cells called **zoospores** that are not found in other fungal divisions. The swimming spores require very moist conditions. A final distinc-

tion is that oomycetes are diploid (instead of haploid) throughout most of their life cycle.

This rather small division of 475 species includes inoffensive water molds that live in water and damp soil, as well as some species with profound economic importance. For example, an oomycete causes a disease called downy mildew of grapes (Fig. 23-7). Its inadvertent introduction into France from the United States in the late 1870s nearly destroyed the French wine industry. Another member of this division has destroyed millions of avocado trees in California, while still another is responsible for "late blight," a devastating disease of potatoes. When accidently introduced into Ireland, this fungus destroyed nearly the entire potato crop, on which most of the Irish lived. The resulting famine of 1845 to 1847 reduced by half the Irish population, from 8 to 4 million. Over 1 million starved outright, while the rest emigrated (many to the United States) or fell victim to diseases spreading through the weakened populace.

A CLOSER LOOK
At Evolutionary Ingenuity in Fungi

Natural selection, operating over millennia on the diverse forms of fungi, has produced some remarkable adaptations by which fungi solve the problems of dispersing their spores and obtaining nutrients. A few of these are highlighted in this essay.

The Rare, Sexy Truffle

Although many fungi are prized as food, none are as avidly sought as the truffle. A single specimen resembling a small, shriveled, blackened apple (Fig. E23-1) may sell for $100. Truffles are the reproductive structures of an ascomycete that forms a mycorrhizal association with the roots of oak trees. Formed underground, the truffle is faced with the problem of dispersing its spores. Its solution is to entice animals to dig it up. Although human beings cannot smell underground truffles, their odor serves as an irresistible attractant to certain mammals, especially wild pigs. Why? It was recently discovered that the truffle releases a chemical that closely resembles the pig's sex attractant. As aroused pigs dig up and devour the truffle, millions of spores are scattered to the winds.

Human truffle hunters use muzzled pigs to hunt their quarry; a good truffle-pig can smell an underground truffle 50 meters away! Although a truffle growth can be encouraged by sowing spores throughout oak groves, they have eluded commercial cultivation. Thus they remain so rare and costly that few of us will ever taste one.

The Shotgun Approach to Spore Dispersal

The delicate structures in Fig. E23-2 are actually fungal shotguns, the reproductive structures of the zygomycete *Pilobolus*. Only by closely scrutinizing piles of horse manure are you likely to observe this miniature beauty. Hyphae penetrating the dung send up clear bulbs capped with sticky black spore cases. As they mature, the sugar concentration in the bulbs increases, drawing in water by osmosis. Meanwhile, the bulb begins to weaken just below its cap. Suddenly, like an overinflated balloon, it bursts, blowing its spore-carrying top up to a meter away. *Pilobolus* bends toward the light. This may increase the probability that its spores will land on open pasture. Here they adhere to grass blades until consumed by a grazing herbivore, perhaps a horse. Later (some distance away) the spores are deposited unharmed in a pile of their favorite food: manure. Growing hyphae penetrate this rich source of nutrients, sending up new projectiles to continue this ingenious cycle.

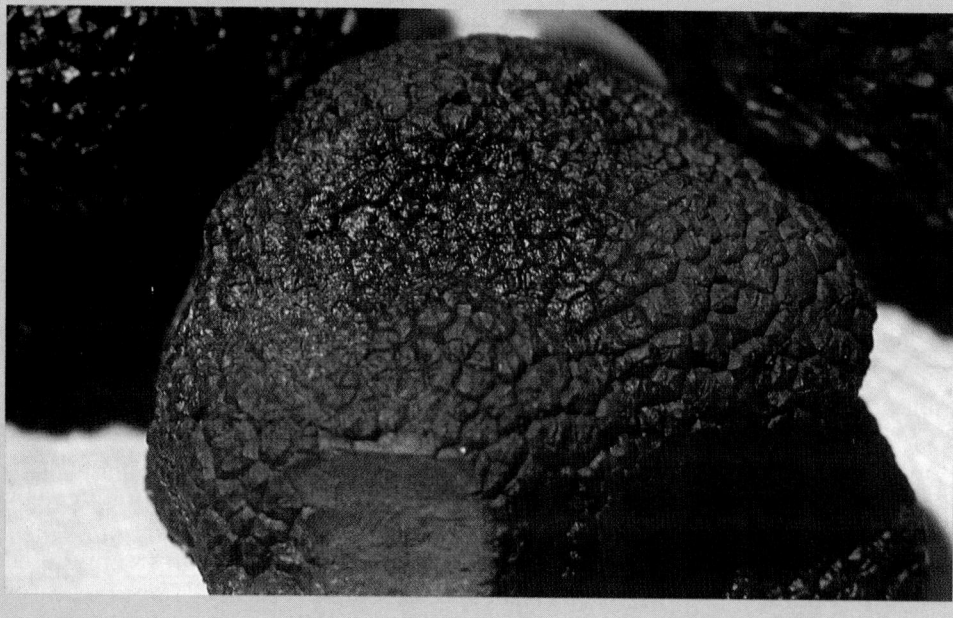

Figure E23-1 The truffle. This rare ascomycete is a gastronomic delicacy.

The Nematode Nemesis

You may recall from Chapter 22 that billions of microscopic nematode roundworms thrive in rich soil. Several types of fungi have evolved deadly snares that exploit this rich source of protein. Natural selection has resulted in a variety of ingenious modifications of nematode-nabbing hyphae, studied by George Barron of Guelph, Ontario. Some produce sticky pods that adhere to passing nematodes. These will germinate into hyphae that penetrate the nematode body, digesting it from within. Others ensnare the worms in sticky tangled hyphae. In 1978, Barron discovered a new species that shoots a harpoonlike projectile 1/10,000 of a centimeter in diameter into passing nematodes. The projectile develops into a new fungus inside the worm, using it as food. The fungal strangler, *Arthrobotrys*, a member of the fungi imperfecti, produces nooses formed from three hyphal cells. When a nematode blunders into the noose, its contact with the inner parts of the noose stimulates a sudden increase in permeability to water that causes the noose cells to swell (Fig. E23-3). The resulting constriction of the hole takes only one tenth of a second and anchors the worm firmly. Fungal hyphae then penetrate and feast on the hapless prey.

Figure E23-2 The delicate translucent reproductive structures of the zygomycete *Pilobolus* will literally blow their tops when ripe, dispersing the black caps with their payload of spores.

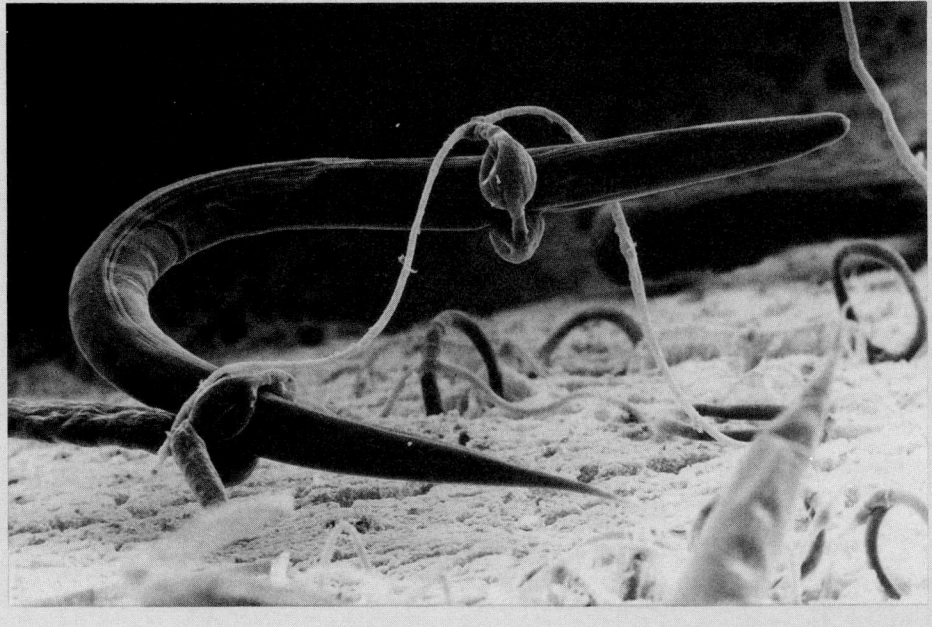

Figure E23-3 *Arthrobotrys,* the nematode strangler, traps its nematode (roundworm) prey in a modified hypha that swells when the inside of the loop is contacted.

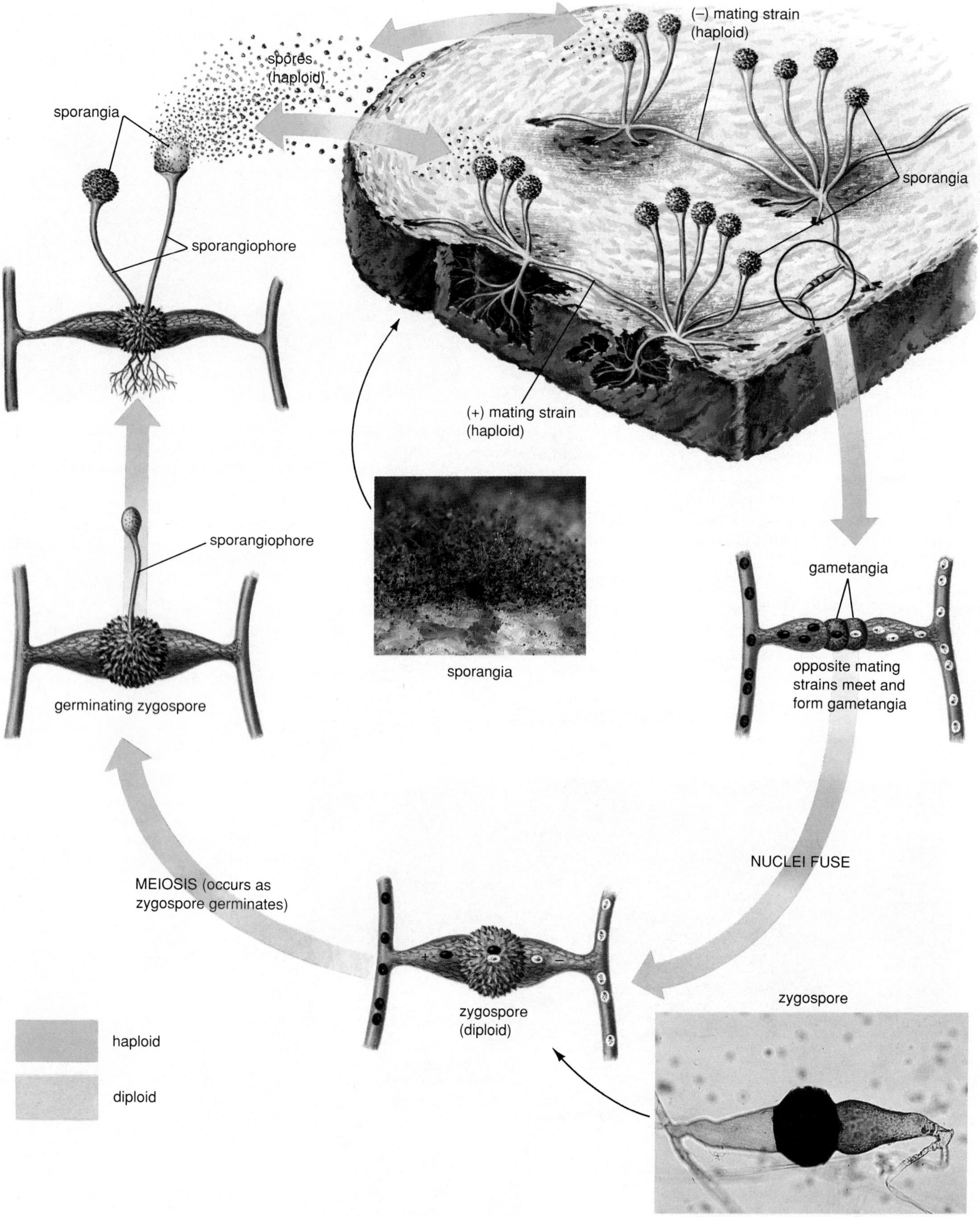

spores
(haploid)

sporangia

sporangiophore

(−) mating strain
(haploid)

sporangia

(+) mating strain
(haploid)

sporangia

sporangiophore

germinating zygospore

gametangia

opposite mating
strains meet and
form gametangia

NUCLEI FUSE

MEIOSIS (occurs as
zygospore germinates)

zygospore
(diploid)

zygospore

haploid

diploid

◀ **Figure 23-6** The life cycle of a zygomycete, the black bread mold (genus *Rhizopus*). During asexual reproduction (shown on top), haploid spores are produced within sporangia. These disperse and germinate on food such as bread, producing haploid hyphae. These in turn may produce additional spores within sporangia by mitosis.

During sexual reproduction (lower left and right) hyphae that appear identical but are actually of different mating types (here designated + and −) contact one another, and form specialized cells called gametangia at the point of contact. The gametangia, with their haploid nuclei, then fuse, producing a diploid zygospore (inset). The zygospore undergoes meiosis and germinates, producing a sporangiophore, on which spore cases, or sporangia, are formed (upper left). These produce and liberate haploid spores that begin the asexual reproductive cycle described above.

Figure 23-7 An oomycete that causes downey mildew on grapes nearly destroyed the French wine industry during the 1870s.

Division Ascomycota: The Sac Fungi

The 30,000 species of ascomycetes, also called *sac fungi*, are named after the saclike case, or **ascus**, in which spores form during sexual reproduction. This division includes many of the colorful molds that "decorate" decaying food or attack and destroy fruit crops. Ascomycetes secrete enzymes that include both cellulase and protease, which in warm, humid climates cause significant damage to textiles such as cotton and wool. The fungus causing Dutch elm disease, which has destroyed nearly all the elm trees in the United States, and the one causing chestnut blight, are both ascomycetes. Another ascomycete (*Claviceps purpurea*), which attacks rye plants, produces a variety of toxins, one of which is the active ingredient in LSD. If infected rye is made into flour and consumed, the toxins can produce convulsions, hallucinations, and death. This happened frequently in northern Europe in the Middle Ages, but modern agricultural techniques have essentially eliminated the disease. Another of its toxins, called ergot, is currently used in drugs that induce labor and control hemorrhaging after childbirth.

Some ascomycetes live in decaying forest vegetation and form beautiful cup-shaped reproductive structures or corrugated, mushroomlike fruiting bodies called *morels* (Fig. 23-8). In this group we also find the yeasts, one of the few unicellular fungi, without which we would lack not only our loaf of bread but our jug of wine as well. Yeast is also a common cause of vaginal infections (Fig. 23-9). Another gastronomic delicacy, the truffle, is also a member of this diverse division (see A Closer Look at Evolutionary Ingenuity in Fungi, Fig. E23-1).

(a)

(b)

Figure 23-8 Diverse ascomycetes. **(a)** The cup-shaped fruiting body of the scarlet cup fungus. **(b)** The morel, an edible delicacy. (But consult an expert before sampling any wild fungus—some are deadly!)

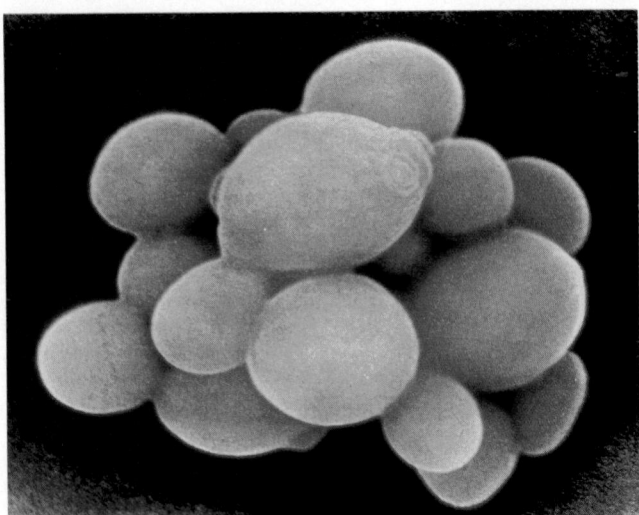

Figure 23-9 The yeast is a unique, nonfilamentous ascomycete, which reproduces most commonly by budding. The yeast shown here is *Candida*, a common cause of vaginal infections.

Division Basidiomycota: The Club Fungi

This division is called the *club fungi* because its members produce club-shaped reproductive structures. The division Basidiomycota consists of about 25,000 species, including the familiar mushrooms, puffballs, and shelf fungi ("monkey-stools"). It also includes some devastating plant pests descriptively called *rusts* and *smuts*, which cause billions of dollars worth of damage to grain crops yearly (Fig. 23-10d). Although several mushroom species are considered delicacies, mushrooms can be deadly. The good-tasting members of the genus *Amanita* (Fig. 23-10b) contain a number of potent and deadly toxins.

Basidiomycetes typically reproduce sexually; their life cycle is shown in Figure 23-11. Mushrooms and puffballs are actually reproductive structures: dense aggregations of hyphae that emerge under proper

(a)

(b)

(c)

(d)

Figure 23-10 Diverse basidiomycetes. **(a)** Shelf fungi, the size of dessert plates, are conspicuous on trees. **(b)** A poisonous mushroom of the genus *Amanita*. **(c)** The giant puffball *Lycopedon giganteum* may produce up to 5 trillion spores. **(d)** Corn smut causes major losses of this crop each year.

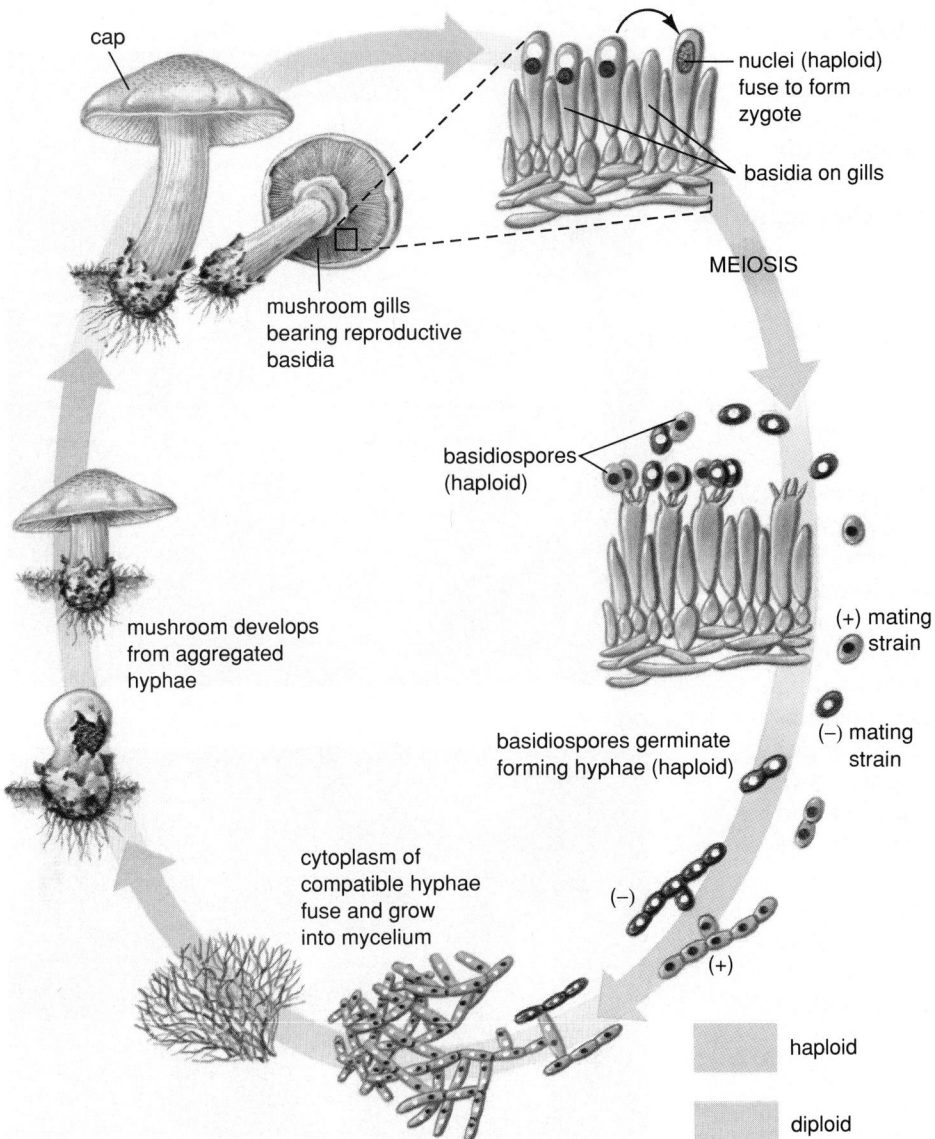

cap

nuclei (haploid) fuse to form zygote

basidia on gills

mushroom gills bearing reproductive basidia

MEIOSIS

basidiospores (haploid)

(+) mating strain

(−) mating strain

mushroom develops from aggregated hyphae

basidiospores germinate forming hyphae (haploid)

cytoplasm of compatible hyphae fuse and grow into mycelium

(−)

(+)

haploid

diploid

Figure 23-11 The life cycle of a "typical" basidiomycete. The mushroom (top) is a reproductive structure, formed from aggregated hyphae made up of cells that each contain two haploid nuclei. Within the cap, leaflike gills bear numerous special cells called *basidia*. Within each basidium, the two haploid nuclei fuse, producing a diploid zygote. The zygote then undergoes meiosis, forming haploid basidiospores that are released from the basidia (left). After being dispersed to new habitats by the wind and water, the basidiospores germinate, forming haploid hyphae of two different mating types [(+) and (−)]. When hyphae of the different mating types meet, some of the cells fuse. These cells, each containing two haploid nuclei, produce an extensive underground mycelium (bottom). In the presence of adequate water and nutrients, portions of the mycelium swell and differentiate, poking up through the soil as mushrooms, and completing the cycle.

conditions from a massive underground mycelium. On the undersides of mushrooms are leaflike gills that produce specialized club-shaped diploid cells called **basidia**. Basidia form haploid reproductive **basidiospores** by meiosis. These are released by the billions from the gills of mushrooms or the inner surface of puffballs to be dispersed by wind and water (see chapter opener photo).

Falling on fertile ground, a mushroom spore may germinate and produce an underground network of mycelia. These grow outward from the original spore in a roughly circular pattern as the older mycelia in the center die. The subterranean body periodically sends up numerous mushrooms, which emerge in a ringlike pattern called a *fairy ring*. The diameter of the fairy ring can reveal the approximate age of the fungus—some are estimated to be 700 years old (Fig. 23-12).

Division Deuteromycota: The Imperfect Fungi

Members of this division, also called the *imperfect fungi*, are so named because none have been observed engaging in sexual reproduction. In some species the sexual stage has been lost during evolution; in others it may exist but has not yet been observed. This large division includes about 25,000 described species of great diversity and considerable importance to humans. It was a member of this division that Alexander Fleming discovered contaminating and killing his bacterial cultures. His keen observations led to the isolation of penicillin from the fungus *Penicillium* (Fig. 23-13). To these fungi we also owe the indescribable flavor and aroma of Roquefort and Camembert cheeses. Other fungi imperfecti are human parasites, causing diseases such as ringworm, athlete's foot, and histoplasmosis. Some are not content to live on dead organisms or even to parasitize live ones—they act as predators, laying deadly traps for unsuspecting roundworms (see A Closer Look at Evolutionary Ingenuity in Fungi, Fig. E22-3).

(a)

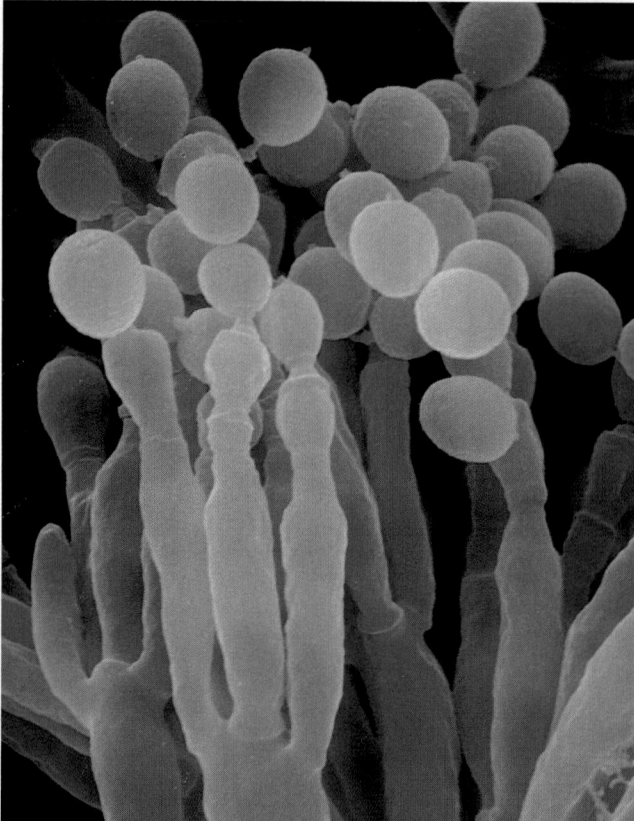

(b)

Figure 23-12 Mushrooms emerge in a "fairy ring" from an underground fungal mycelium growing outward from a central point where a spore germinated, perhaps centuries ago.

Figure 23-13 (a) *Penicillium* growing on an orange. Reproductive structures coat the fruit's surface, while hyphae draw nourishment from inside. **(b)** The spore-bearing structures of *Penicillium*, called conidiophores, are shown in this scanning electron micrograph. The spores, called conidia, are asexual.

SUMMARY OF KEY CONCEPTS

Fungal Form and Function

Fungal bodies generally consist of filamentous hyphae, which are either multicellular or multinucleate and form large, intertwined networks called *mycelia*. Fungal nuclei are generally haploid. A cell wall of chitin (or, in a few cases, cellulose) surrounds fungal cells.

All fungi are heterotrophic, either parasitic or saprobic. They secrete digestive enzymes outside their bodies and absorb the liberated nutrients.

Fungal reproduction is varied and complex, with the dominant generation usually haploid. Asexual reproduction can occur by fragmentation of the mycelium, or by asexual spore formation. Sexual spores form after fusion of compatible haploid nuclei to form a diploid zygote that undergoes meiosis to form haploid sexual spores. These produce haploid mycelia through mitosis.

Economic and Ecological Impacts of Fungi

Fungi are extremely important decomposers in ecosystems. Their filamentous bodies penetrate rich soil and decaying organic material, liberating nutrients through extracellular digestion.

The majority of plant diseases are caused by parasitic fungi. Some parasitic fungi can help control insect crop pests. Others can cause human diseases, including ringworm, athlete's foot, histoplasmosis, and candidiasis.

Fungal Symbiotic Relationships

A lichen is a symbiotic association between a fungus and a unicellullar alga or cyanobacterium. This self-sufficient combination can colonize bare rock. Mycorrhizae are associations between fungi and the roots of most vascular plants. The fungus derives photosynthetic nutrients from the plant roots, but carries water and nutrients into the root from the surrounding soil.

Division Zygomycota: The Zygote Fungi

The zygomycetes include the dung fungus *Pilobolus* and the black bread mold *Rhizopus.* They produce sexual zygospores.

Division Oomycota: The Egg Fungi

Oomycetes include members causing diseases of grape and potato crops, as well as water molds. Sexual reproduction involves fertilization of a large egg cell, and asexual reproduction is by flagellated zoospores. Oomycetes are diploid through most of their life cycle. Their cell walls contain cellulose.

Division Ascomycota: The Sac Fungi

The ascomycetes include yeasts, morels, truffles, molds, the fungus causing Dutch elm disease, and several crop parasites. They produce their spores in asci, which resemble small sacs, hence their common name.

Division Basidiomycota: The Club Fungi

The basidiomycetes include mushrooms, puffballs, and food crop parasites called *rusts* and *smuts*. Basidiomycetes produce diploid basidia, which form haploid basidiospores, released by the billions and dispersed by wind and water.

Division Deuteromycota: The Imperfect Fungi

The deuteromycetes are diverse and their sexual stages are either nonexistent or have never been observed. Included are fungi that produce penicillin, others causing ringworm and athlete's foot, and some predators of nematodes.

GLOSSARY

ascus (as′-kus): a saclike case in which sexual spores are formed by members of the fungal division Ascomycota.

basidium (bas-id′-ē-um): a diploid cell, often club-shaped, formed on the basidiocarp of members of the fungal division Basidiomycota, which produces basidiospores by meiosis.

basidiospore (ba-sid′-ē-ō-spōr): a sexual spore formed by members of the fungal division Basidiomycota.

hypha (hī′-pha; pl. hyphae): a threadlike structure, many of which make up the fungal body, that consists of elongated cells, often with many haploid nuclei.

lichen (lī′-ken): a symbiotic association between an alga or cyanobacterium and a fungus, resulting in a composite organism.

mycelium (mī-sēl′-ē-um): the body of a fungus, consisting of a mass of hyphae.

mycorrhiza (mī-kō-rī′-uh; pl. mycorrhizae): a symbiotic association between a fungus and a plant root. The fungus, often a basidiomycete or an ascomycete, grows around and often into the roots of most vascular plants in a mutually beneficial relationship.

saprobe (sap′-rōbe): an organism that derives its nutrients from the bodies of dead organisms.

septum (pl. septa): a partition that separates the fungal hypha into individual cells. Pores in the septa allow transfer of materials between cells.

sporangium (spōr-an′-gē-um; pl. sporangia): a structure in which spores are produced.

symbiosis (sim-bī-ō′-sis): a close relationship between two types of organisms. Either or both may benefit from the association, or, in the case of parasitism, one of the participants is harmed.

zoospore (zō′-ō-spōr): a flagellated motile asexual spore formed by members of the fungal division Oomycota.

zygospore (zī′-gō-spōr): produced by the division Zygomycota, a fungal spore surrounded by a thick, resistant wall, which forms from a diploid zygote.

STUDY QUESTIONS

1. Describe the structure of the fungal body. How do fungal cells differ from most plant and animal cells?
2. What portion of the fungal body is represented by mushrooms, puffballs, and similar structures? Why are these elevated above the ground?
3. What two plant diseases, caused by parasitic fungi, have had an enormous impact on U.S. forests? In what division are these found?
4. List some fungi that attack crops. In what division is each found?
5. Describe two different modes of asexual reproduction in fungi.
6. What is the major structural ingredient in the cell walls of fungi? In what other organisms is this substance most common?
7. List the major divisions of fungi, describe the feature that gives each its name, and give one example of each.
8. Describe how a "fairy ring" of mushrooms is produced. Why is the diameter related to its age?
9. Describe two different symbiotic associations between plants and fungi. In each case, explain how each partner in these associations is affected.

DISCUSSION QUESTIONS

1. Both asexual and sexual reproduction is possible in most types of fungi. What are the advantages and disadvantages of each type of reproduction?
2. Discuss the differences between fungi and animals in energy acquisition. Describe how the fungal body is specialized for its mode of nutrition.
3. If, by humans using a new and deadly fungicide, all fungi on earth were destroyed, what ecological consequences would occur?

SUGGESTED READINGS

Angier, N. "A stupid cell with all the answers." *Discover,* November 1986. Fascinating description of the uses of yeasts in molecular biology, including beautiful illustrations.

Brodie, H. J. *Fungi: Delight of Curiosity.* Toronto: University of Toronto Press, 1978. Highlights amazing fungal adaptations in a lively text written for the interested layperson.

Cooke, R. C. *Fungi, Man and His Environment.* New York: Longman, 1977. A discussion of fungi with emphasis on their impact on human activities.

Emerson, R. "Molds and Men." *Scientific American,* January 1952. Traces the impact of fungi on human civilization.

"The Society of Amoebas." *Science 84,* December, 1984. Beautiful photography and clear descriptions highlight this article for the interested layperson.

Vogel, S. "Taming the Wild Morel." *Discover,* May 1988. Describes the research that has allowed these rare delicacies to be cultivated in the laboratory.

24

The Diversity of Life. IV. Plants

The leaves of ferns emerge from gracefully coiled structures called "fiddleheads." This is the Ama'uma'u fern, photographed in Hawaii's Volcano National Park.

Members of the kingdom Plantae are eukaryotic, multicellular, photosynthetic organisms. Plants capture sunlight and synthesize organic molecules from water and carbon dioxide, as do the unicellular algae of the kingdom Protista and the blue-green bacteria of the kingdom Monera. Collectively, the energy these organisms harvest from sunlight powers nearly all life on Earth.

Early "Roots"—The Evolution of Multicellular Plants

Although eukaryotic cells resembling unicellular green algae have been found in fossil deposits dating back 1 billion years, clearly identifiable algae are first found in deposits from the Cambrian era, 500 to 600 million years ago. Terrestrial plants first appeared 430 to 500 million years ago. Several lines of evidence, discussed later in this chapter, point to green algae as the ancestors of all modern land plants.

Two groups of plants arose from the ancestral green algae. One group, the vascular plants (tracheophytes), evolved specialized vessels that transport water and nutrients and provide support for the plant body. These dominate the Earth today. Another group, the bryophytes, lack specialized vessels. They might be considered the "amphibians of the plant kingdom," since they straddle the boundary between aquatic and terrestrial existence. Bryophytes are represented by the mosses and liverworts.

To date, nearly 300,000 species of multicellular plants have been identified. The multicellular plants include aquatic algae of the divisions Rhodophyta, Phaeophyta, and Chlorophyta. Division Bryophyta consists of nonvascular terrestrial plants that require wet conditions for reproduction. All other major divisions fall under the general heading of tracheophytes: the vascular plants. This broad category consists of seedless plants such as the ferns (division Pteridophyta) and seed plants. Seed plants can be further subdivided into the nonflowering gymnosperms, primarily represented by the division Coniferophyta, and the flowering angiosperms of the division Anthophyta. The general features of the major divisions of plants are summarized in Table 24-1.

Evolutionary Trends in Plant Structure

Plant life arose in the sea, an environment that supports the plant body, provides a relatively constant temperature, and bathes the entire plant in nutrients.

Figure 24-1 Red coralline algae from the Pacific Ocean off California provide an anchoring site for bright yellow hydroids (phylum Cnidaria). Coralline algae, which deposit calcium carbonate within their bodies, contribute to coral reefs in tropical waters.

During their evolutionary history, plants underwent a series of changes that reflect their increasing adaptation to the terrestrial environment where plants are surrounded by dry, nonsupporting air.

1. **The evolution of roots or rootlike structures** that anchor the plant and absorb water and nutrients from the soil.
2. **The evolution of conducting vessels** that transport water and minerals upward from the roots and photosynthetic products from the leaves to the rest of the plant body.
3. **The evolution of the stiffening substance lignin** that impregnates the water and mineral conducting vessels and supports the plant body, allowing it to expose maximum surface to the sunlight.
4. **The evolution of a waxy cuticle** covering the surfaces of leaves and stems. This limits the evaporation of water.
5. **The evolution of pores called stomata** in the leaves and stems. These open to allow gas exchange but close when water is scarce, reducing the loss of water.

Table 24-1   Features of the Major Plant Groups

Major Divisions	Relationship of Sporophyte and Gametophyte	Transfer of Reproductive Cells	Early Embryonic Development	Dispersal	Water and Nutrient Transport Structures	Typical Habitat
Rhodophyta	Highly variable	Gametes or asexual zoospores released into water	Occurs independently after zygote or zoospore settles to substrate	By water currents	Absent	Mostly marine, some freshwater
Phaeophyta	Highly variable	Gametes or asexual zoospores released into water	Occurs independently after zygote or zoospore settles to substrate	By water currents	Absent (usually)	Almost entirely marine
Chlorophyta	Highly variable	Gametes or asexual zoospores released into water	Occurs independently after zygote or zoospore settles to substrate	By water currents	Absent	Mostly freshwater, some marine
Bryophyta (mosses)	Gametophyte dominant—sporophyte develops from zygote retained on gametophyte	Motile sperm swims to stationary egg retained on gametophyte	Occurs within archegonium of gametophyte	Haploid spores carried by wind	Absent	Moist terrestrial
Pteridophyta (ferns)	Sporophyte dominant—develops from zygote retained on gametophyte	Motile sperm swims to stationary egg retained on gametophyte	Occurs within archegonium of gametophyte	Haploid spores carried by wind	Present	Moist terrestrial
Coniferophyta (conifers)	Sporophyte dominant—microscopic gametophyte develops within sporophyte	Wind-dispersed pollen carries sperm to stationary egg in cone	Occurs within a protective seed containing a food supply	Seeds containing diploid sporophyte embryo dispersed by wind or animals	Present	Varied terrestrial habitats—dominate in dry, cold climates
Anthophyta (flowering plants)	Sporophyte dominant—microscopic gametophyte develops within sporophyte	Pollen, dispersed by wind or animals, carries sperm to stationary egg within flower	Occurs within a protective seed containing a food supply; seed encased in fruit	Fruit, carrying seeds, dispersed by animals, wind, or water	Present	Varied terrestrial habitats—dominant terrestrial plant

Evolutionary Trends in Plant Reproduction

For algae, the surrounding water provides an ideal medium for reproduction in addition to nourishing the developing plant. In some algae, gametes are carried passively by water currents, while in others, flagellated, actively swimming gametes and spores, called **zoospores,** are common. Not only do many algal gametes meet in the water, but zygotes and spores are dispersed by the water to suitable habitats.

The invasion of land necessitated new methods of transporting sperm to eggs. The developing embryo also required protection from drying and a means of dispersal independent of water. These requirements were met by the evolution of pollen, seeds, and later, the flower and the fruit.

The dry, microscopic pollen grains produced by nonflowering seed plants allowed wind, instead of water, to carry the male gametes. Later, attractive flowers evolved that enticed animal pollinators to deliver pollen more precisely than the wind. The seed provides waterproof protection and nourishment for the developing embryo, and modifications of the seed coat may enhance dispersal. Some fruits attract animal foragers who consume them and incidentally disperse the indigestible seeds.

Evolutionary Trends in the Plant Life Cycle

Plant life cycles generally exhibit **alternation of generations,** in which a multicellular haploid plant, the **gametophyte,** alternates with a multicellular diploid **sporophyte.** The gametophyte produces haploid sex cells by mitosis that unite to form a **zygote,** which then develops into the diploid sporophyte. The sporophyte in turn produces haploid spores by meiosis, and each spore can develop into a haploid gametophyte, completing the cycle (see Chapter 28).

From the evolutionarily ancient algae to the more recent seed plants, we see a general trend toward increased prominence of the sporophyte generation, and decreased size and duration of the gametophyte generation (see Table 24-1). Algal life cycles are quite diverse. However, in some types of green algae whose ancestors may have given rise to terrestrial plants (such as members of the genus *Coleochaete*), the sporophyte is absent and the only diploid stage is the zygote. In nonvascular plants such as mosses, the sporophyte is present. However, it is smaller than the gametophyte and remains attached to it (see Fig. 24-6). In the seedless vascular plants, such as ferns, the sporophyte is dominant, while the gametophyte is a much smaller, but still independent, plant. Finally, in seed plants, the male and female gametophytes are microscopic and barely recognizable as an alternate generation (see Figs. 24-11 and 24-14). These small gametophytes, however, still produce the eggs and sperm that unite and form the sporophyte.

Watery Origins—The Algae

The first plants—the algae—were relatively simple, partly because the sea provided so many of their needs. Having evolved in this equitable environment, algae remain simpler than terrestrial plants in several ways. For example, land plants show clear structural differences between their roots, stems, and leaves (described in more detail in Chapter 26). Although many algae have structures that appear rootlike or leaflike, they lack this internal differentiation. Algae also lack complex reproductive structures such as flowers and cones. Their gametes are usually shed directly into the water where they unite and develop.

Algal life cycles are complex, and vary considerably between, and even within, the algal divisions. In some genera, such as the green alga *Chara*, the haploid gametophyte is dominant. In *Fucus*, a brown alga, the diploid sporophyte is dominant; the only haploid cells are gametes. In other genera, the sporophyte and gametophyte may appear nearly identical, as in the green alga *Ulva* (see Fig. 24-4). Alternatively, they may be very different in size and appearance, as in the giant kelp, whose sporophyte can reach 250 feet in height (see Fig. 24-2), but whose gametophyte is filamentous and microscopic.

Algae may be classified into three divisions, each named for its characteristic color. Algae are colored by pigments that help them capture light energy for photosynthesis. These pigments are often red or brown, absorbing the green, violet, and blue light that most readily penetrates deep water. The combination of these pigments with green chlorophyll lend algae their distinctive colors, and their names: division Rhodophyta—the red algae; division Phaeophyta—the brown algae; and division Chlorophyta—the green algae.

Division Rhodophyta: The Red Algae

The 4000 species of red algae, ranging from bright red to nearly black, derive their color from red pigments that mask their green chlorophyll (Fig. 24-1). Red algae are mostly marine. They dominate in deep, clear tropical waters, where their red pigments absorb the deeply penetrating blue-green light and transfer this light energy to chlorophyll, where it is used in photosynthesis.

(a)

(b)

(c)

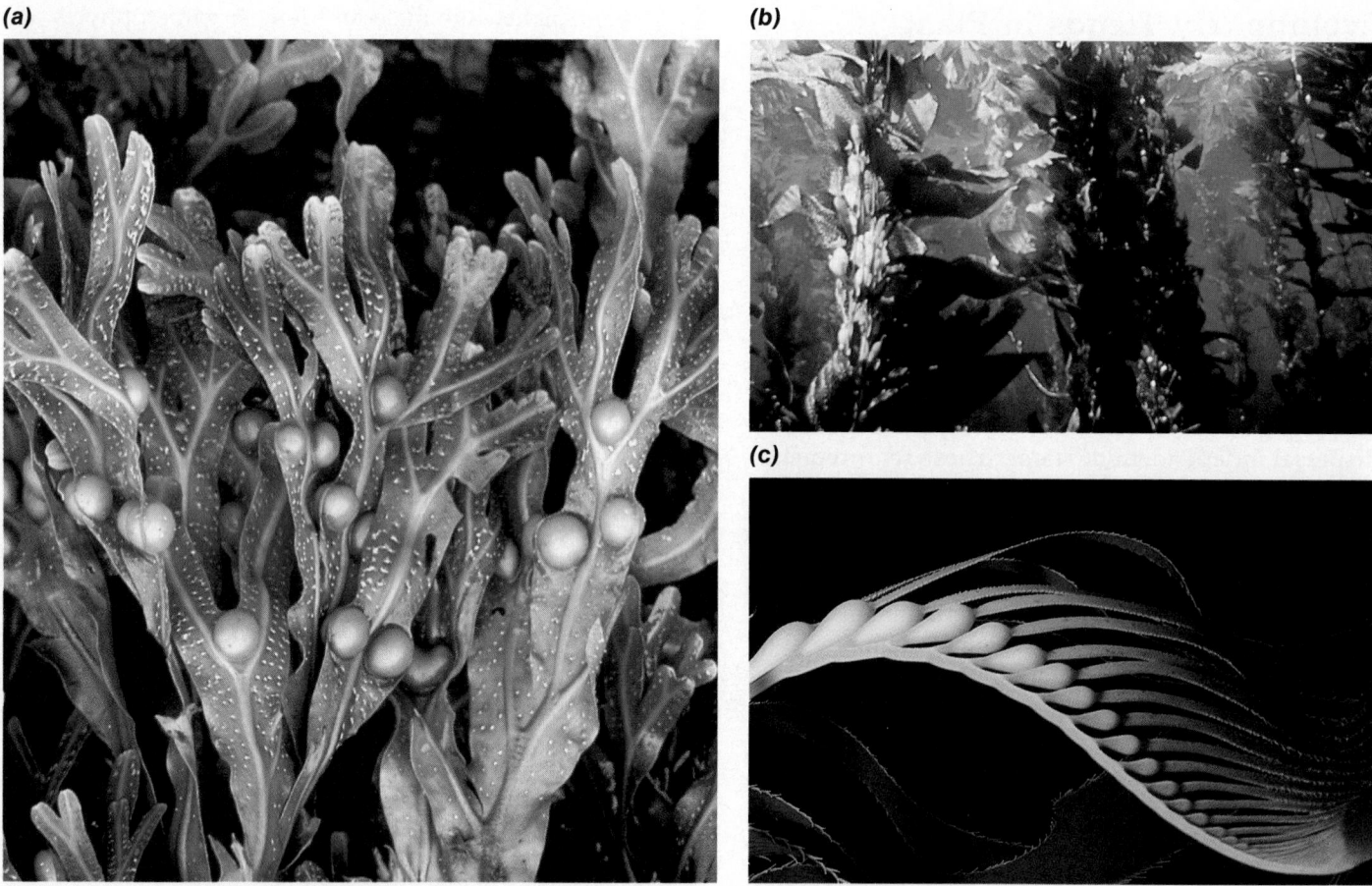

Figure 24-2 Diverse brown algae. **(a)** *Fucus*, an intertidal genus, is shown here exposed at low tide. Notice the gas-filled bladders, which confer buoyancy in water. **(b)** The sporophyte generation of the giant kelp *Macrocystis* forms underwater forests off southern California. **(c)** The growing tip of the giant kelp. Gas-filled bladders are clearly visible at the base of leaflike blades.

Some red algae deposit calcium carbonate (limestone) in their tissues, and may contribute to the formation of reefs. Others are harvested as food in the Orient, and from some the gelatinous substance carrageenan is extracted and used as an emulsifier in products such as paints, cosmetics, and ice cream. However, **the major importance of these and other algae is their photosynthetic ability: they contribute to the food-producing foundation of marine ecosystems.**

Division Phaeophyta: The Brown Algae

These plants increase their light-gathering ability with brownish-yellow pigments that (in combination with green chlorophyll) produce their brown to olive-green color. Like the red algae, the 1500 species of brown algae are almost entirely marine. This group forms the dominant "seaweed" found along rocky shores in the temperate (cooler) oceans of the world, including the east and west coasts of the United States. Brown algae are found from the intertidal zone, where they cling to rocks exposed at low tide, to far offshore. Several types use gas-filled floats to support their bodies (Fig. 24-2a, c). The giant kelp plants found along the Pacific coast occasionally reach heights of 100 meters, and may grow over 15 centimeters daily. With their dense growth and towering height, kelp forms undersea forests that provide food, shelter, and breeding areas for a variety of marine animals (Fig. 24-2b).

Division Chlorophyta: The Green Algae

This large (7000 species) and extremely diverse group is of special interest because green algal ancestors are believed to have given rise to the terrestrial plants. Three lines of evidence support the hypothesis that land plants evolved from green algal ancestors:

1. **Green algae use the same type of chlorophyll and accessory pigments as do land plants.**
2. **Green algae store food as starch, as do land plants, and have cell walls of similar composition.**
3. **Most green algae live in fresh water, where they form the dominant plant life.**

In contrast to the nearly constant conditions of the ocean, freshwater habitats (ponds, swamps, streams, and lakes) are highly variable. Dramatic fluctuations in temperature and rainfall over evolutionary time probably exerted intense selection pressure on ancient freshwater algae to withstand extremes of temperature and periods of dryness. The resulting adaptations served their descendants well in their invasion of the land.

Green algae are mainly multicellular, but include a few unicellular forms and **colonies**—clusters of cells that are somewhat interdependent. These colonies range from a few cells to a few thousand cells, as in members of the genus *Volvox*, where rudimentary division of labor is seen (Fig. 24-3a). Some green algae, such as members of the genus *Spirogyra*, form long chains of cells (Fig. 24-3b). Most are quite small, but some exceptions are found in the sea. The green alga *Ulva*, or "sea lettuce," is similar in size to the leaves of its namesake (Fig. 24-4).

(a)

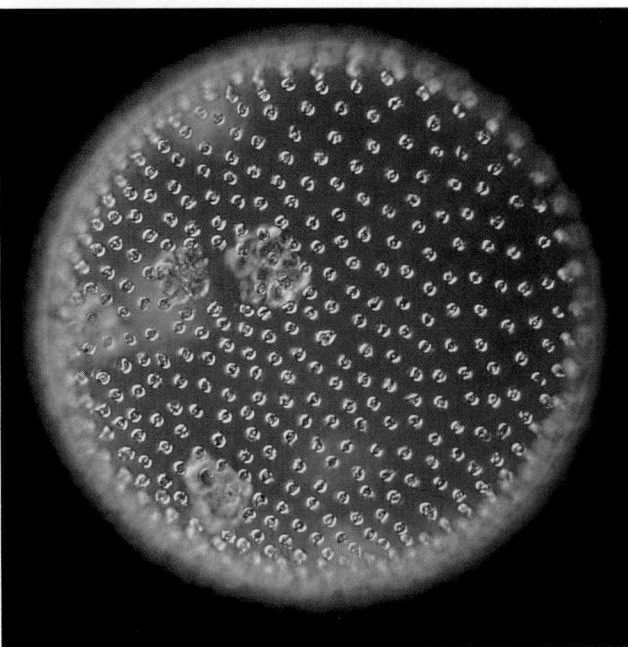

(b)

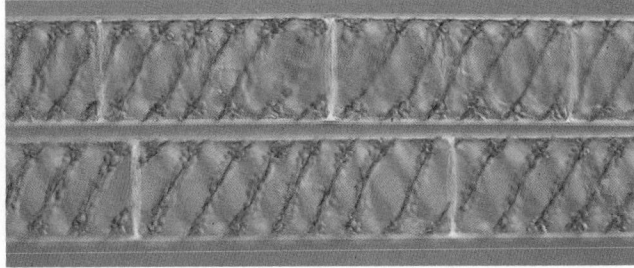

Figure 24-3 Two forms of microscopic green algae. **(a)** *Volvox*, a colonial green alga composed of a sphere of cells embedded in a gelatinous matrix. New daughter colonies can be seen developing inside the sphere. **(b)** *Spirogyra*, a filamentous green alga composed of strands only one cell thick. Inside these microscopic transparent cells, spiral chloroplasts are clearly visible.

Land—The New Frontier

Moving onto land provided significant advantages for the descendants of the ancient green algae. The raw materials for photosynthesis—carbon dioxide and sunlight—are present in far higher concentrations in the air than in water. The ponds and seas had grown crowded, teeming with hungry animals and with other plants competing for light and nutrients. The land, in contrast, offered abundant space and resources to the new colonists. But terrestrial environments are much less hospitable than aquatic environments. The supportive buoyancy of water is missing, and the air tends to dry things out. New structures to lend the body rigidity were needed, and vessels were required to transport nutrients. The terrestrial plant body also needed waterproofing. Special reproductive adaptations, including pollen, seeds, and flowers, further increased the success of certain terrestrial plants. These freed the plants from the need for water to carry their sex cells and disperse the fertilized eggs.

Two major groups of land plants arose from the ancient algal ancestors. One group, the **bryophytes,** straddles the boundary between aquatic and terrestrial life. The other group, the vascular plants, or **tracheophytes,** is completely adapted to life on land.

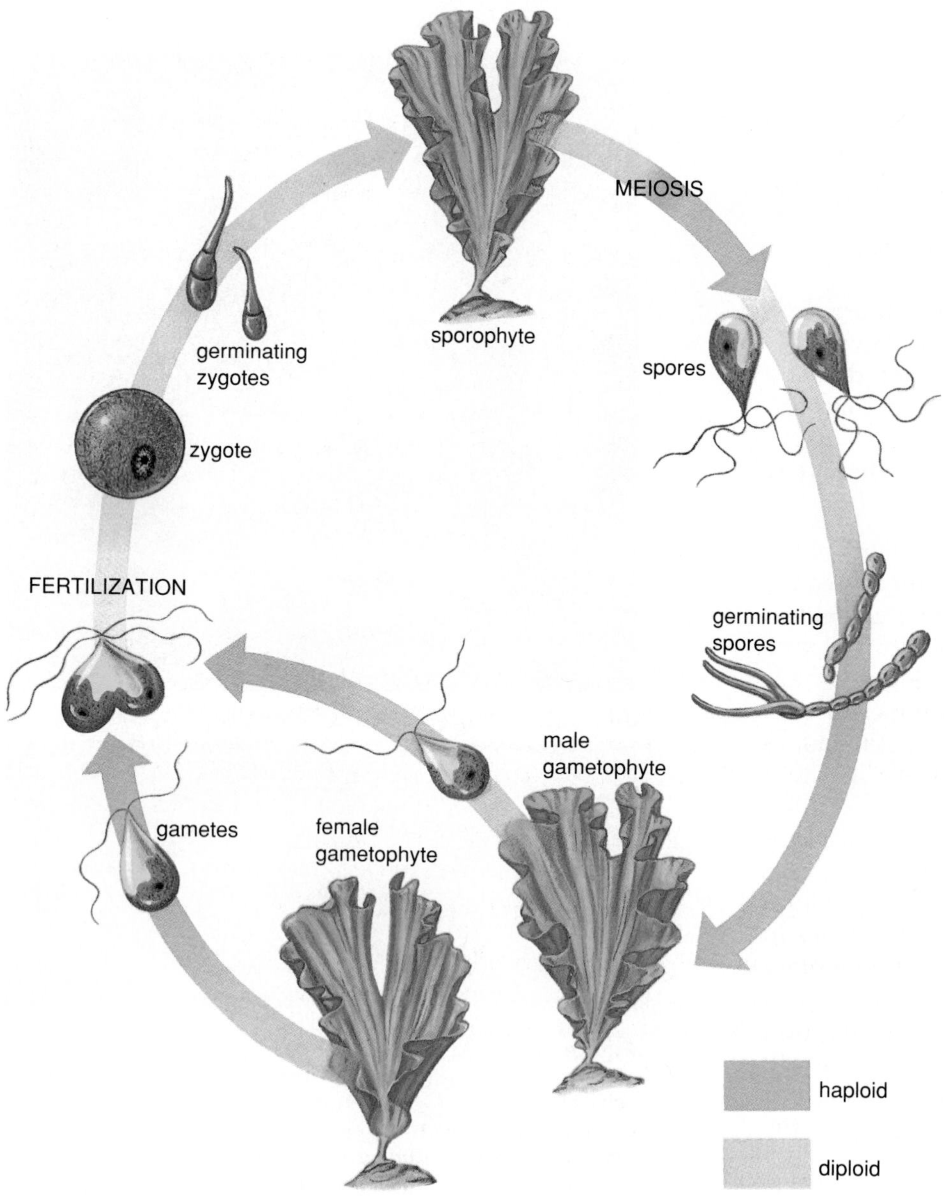

Figure 24-4 Life cycle of *Ulva*, a large marine alga. In this case the sporophyte and gametophyte plants are equal in size and nearly indistinguishable. However, the sporophyte is diploid and produces haploid spores through meiosis, while the male and female gametophytes are haploid and produce gametes by mitosis. The fusion of these gametes results in a diploid zygote that develops into the sporophyte plant.

Division Bryophyta: The Liverworts and Mosses

The major bryophyte representatives are the liverworts and mosses (Figs. 24-5 and 24-6). The 16,000 species of bryophytes show some, but not all, of the adaptations necessary for a completely terrestrial existence. The bryophytes, like the algae, lack true roots, leaves, and stems. Rootlike anchoring structures called **rhizoids** bring water and nutrients into the plant body. Bryophytes lack well-developed structures for conducting water and nutrients. Since they must rely on slow diffusion or poorly developed conducting tissues to distribute water and other nu-

trients, their body size is limited; most are less than 2 centimeters tall. The liverworts and most mosses are confined to moist areas. However, some mosses possess a waterproof **cuticle** that retains moisture, and stomata that can be closed, preventing water loss. Some mosses can survive in deserts, on bare rock, and in far northern and southern latitudes where water is scarce.

Bryophytes have adapted to terrestrial existence by evolving enclosed reproductive structures that protect the gametes from desiccation. These are the **archegonia,** in which eggs develop, and the **antheridia,** where sperm are formed. They may be located on the same plant, or the entire plant may be

Figure 24-5 Liverworts grow inconspicuously in moist, shaded areas. This is the female gametophyte plant, bearing umbrella-like archegonia, which hold the eggs. Sperm must swim up the stalks through a film of water to fertilize the eggs.

either male or female, depending on the species. In all bryophytes, the sperm must swim to the egg (which emits a chemical attractant) through a film of water. Bryophytes living in dry areas must time their reproduction to coincide with the infrequent rains.

The bryophyte life cycle is shown in Figure 24-6. As in many algae, the larger "leafy" plant body is the haploid gametophyte, which forms sperm and eggs by mitosis. Bryophytes retain the fertilized egg in the archegonium. Here the embryo grows and matures into a small diploid sporophyte attached to the parent gametophyte plant. At maturity, the sporophyte produces haploid spores by meiosis within a capsule. Spores are released by the opening of the capsule, and are dispersed by the wind. If a spore lands in a suitable environment, it may develop into a new, haploid plant.

Tracheophytes: The Vascular Plants

Although the bryophytes were quite successful, they left vast areas of the land unoccupied. When the moister regions were covered with the short green fuzz of bryophytes, any plant that could stand taller would benefit by basking in sunlight while shading its short competitors. To grow tall, two new features were needed—support for the body, and vessels to conduct water and nutrients absorbed by the roots into the upper portions of the plant. The **vascular** (Latin, "vessel-bearing") plants solved the two problems simultaneously by developing specialized groups of conducting cells (which we will call **vessels**) impregnated with a stiffening substance called **lignin**. These vessels not only conduct water but also provide support (see Chapter 26).

Seedless Vascular Plants: The Club Mosses, Horsetails, and Ferns

The seedless vascular plants reached treelike proportions and dominated the landscape during the Carboniferous era (from 265 to 355 million years ago). Their bodies—transformed by heat, pressure, and time—are burned today as coal. Their modern representatives, the club mosses, horsetails, and ferns, have diminished in size and importance and have largely been replaced by the more versatile seed plants.

The club mosses (division Lycophyta) are now limited to representatives a few centimeters in height (Fig. 24-7a). Their leaves are small and scalelike, resembling those of mosses. Club mosses of the genus *Lycopodium*, commonly known as *ground pine*, form a beautiful ground cover in some temperate coniferous and deciduous forests.

Modern horsetails (division Sphenophyta) form a single genus, *Equisetum*, with only 15 species, most less than 1 meter tall (Fig. 24-7b). The bushy appearance of some species has given them the common name *horsetails*. They may also be called *scouring rushes*, since they were used by early settlers to scour pots and floors. All species of *Equisetum* deposit large amounts of silica (glass) in their epidermal cells, giving them an abrasive texture.

The ferns (division Pterophyta), with 12,000 species, are far more successful (Fig. 24-7c and chapter opener photo). In the tropics, "fern trees" still reach heights reminiscent of their Carboniferous ancestors. Ferns are the only members of this group with broad leaves. Broad leaves can capture more sunlight, and this advantage over the small-leaved club mosses and horsetails may account for the relative success of modern ferns.

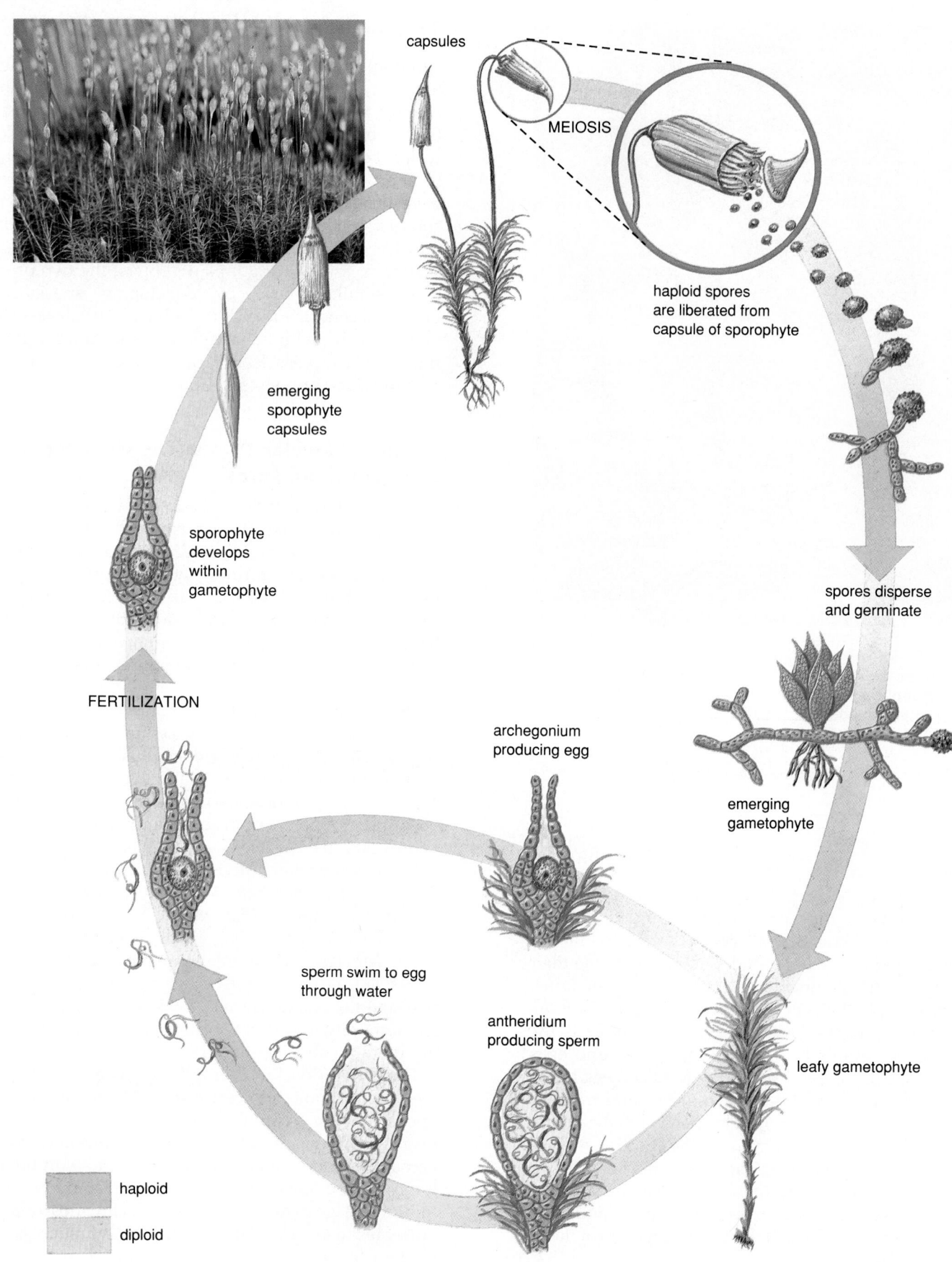

capsules

MEIOSIS

haploid spores
are liberated from
capsule of sporophyte

emerging
sporophyte
capsules

sporophyte
develops
within
gametophyte

spores disperse
and germinate

FERTILIZATION

archegonium
producing egg

emerging
gametophyte

sperm swim to egg
through water

antheridium
producing sperm

leafy gametophyte

haploid

diploid

Figure 24-6 Life cycle of a moss, showing alternation of diploid and haploid generations. The leafy green cushion (lower right) is actually the haploid gametophyte generation that produces sperm and eggs. The sperm develop in the antheridium and must swim through a film of water to the egg (which remains in the archegonium where it is formed). The zygote develops into a stalked, diploid sporophyte that emerges from the gametophyte plant. The sporophyte is topped by a brown capsule in which haploid spores are produced by meiosis. These are dispersed and germinate, producing another green gametophyte generation. (Inset) Moss plants showing both stages in the life cycle. The short, leafy green plants are the haploid gametophytes, while the reddish-brown stalks are the diploid sporophyte generation. The sporophytes are about 1 centimeter in height.

(a)

(b)

(c)

Figure 24-7 Some seedless vascular plants. All are found in moist woodland habitats. **(a)** The club mosses (sometimes called *ground pines*) grow in temperate forests. This specimen is liberating spores. **(b)** The giant horsetail (genus *Equisetum*) extends long, narrow branches in a series of rosettes. Its leaves are reduced to insignificant scales. On special stems, cone-shaped spore-forming structures called strobili are produced. **(c)** This deer fern, whose leaves emerge from coiled "fiddleheads," was photographed in Redwood National Park, California.

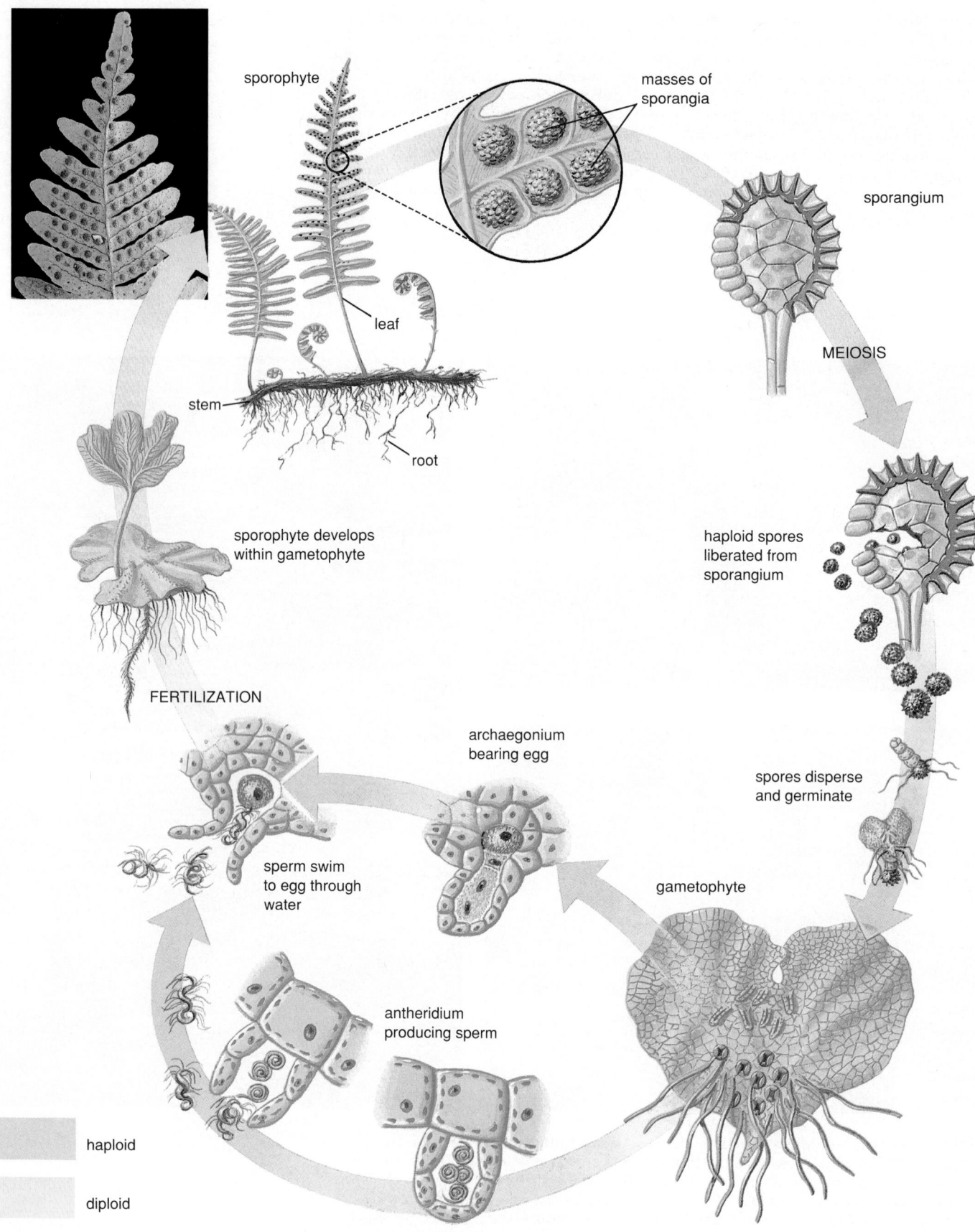

sporophyte

masses of
sporangia

sporangium

leaf

stem

root

MEIOSIS

sporophyte develops
within gametophyte

haploid spores
liberated from
sporangium

FERTILIZATION

archaegonium
bearing egg

spores disperse
and germinate

sperm swim
to egg through
water

gametophyte

antheridium
producing sperm

haploid

diploid

The life cycle of a fern is depicted in Figure 24-8. A major difference between vascular plants and bryophytes is that the diploid sporophyte is dominant. On special leaves of club mosses and ferns, and on conelike structures of horsetails, haploid spores are produced. These are dispersed by wind and give rise to tiny, haploid gametophyte plants, which produce sperm and eggs. Two traits in the seedless vascular plants are reminiscent of the bryophytes. First, the small gametophytes lack conducting vessels. Second, as in bryophytes, the sperm must swim through water to reach the egg, so these plants still depend on the presence of water for sexual reproduction.

The Seed Plants

Seed plants have dominated the land for the past 250 million years, and show no signs of yielding their supremacy. Their success can be attributed to their reproductive versatility. Freed from the requirement of water for reproduction, seed plants have invaded nearly all terrestrial habitats, from swamps to deserts. **Two major reproductive adaptations have given seed plants an edge over their seedless competitors: pollen and seeds.**

Pollen is all that remains of the male gametophyte of seed plants. In seed plants, both male and female gametophytes (which produce the sex cells) are greatly reduced in size, while the sporophyte is large. The female gametophyte is a small group of haploid cells that produces the egg, while the male gametophyte is the pollen grain. Sperm-producing cells are carried within the pollen grain, which is dispersed by wind or by animal pollinators such as bees. Thus, seed plants are not limited in their distribution by the need for water that allows sperm to swim to the egg; they are fully adapted to dry land.

The second reproductive adaptation is the seed itself (seed structure is covered in detail in Chapter 28). Seeds are somewhat analogous to the eggs of birds and reptiles, since seeds consist of an embryonic plant, a supply of food for the embryo, and a protective outer coat (Fig. 24-9). The seed coat maintains the embryo in a state of suspended animation or dormancy until conditions are proper for growth. The stored food helps sustain the emerging plant until it

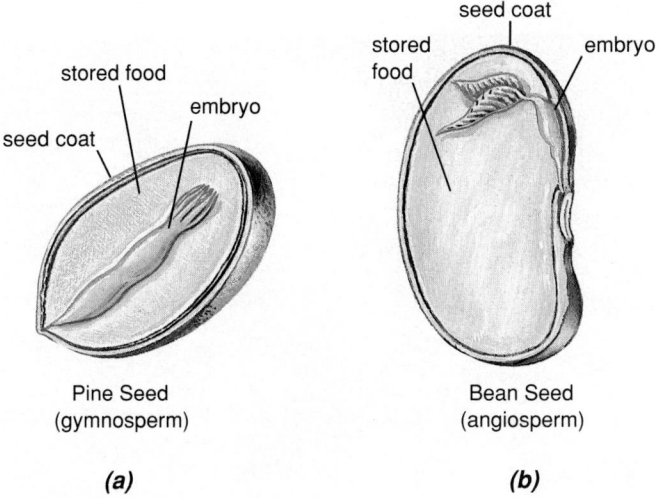

Figure 24-9 Seeds from **(a)** a gymnosperm (pine) and **(b)** an angiosperm (bean). Both consist of an embryonic plant and stored food confined within a seed coat.

develops roots and leaves and can make its own food by photosynthesis. Seeds possess elaborate adaptations that allow dispersal by wind, water, and animals. These adaptations have helped them invade nearly every nook and cranny of the world.

Seed plants may be considered in two general groups: **gymnosperms,** which lack flowers, and **angiosperms,** the flowering plants. Although these are not official taxonomic categories, they are useful in organizing our discussion of the seed plants.

The Gymnosperms: Nonflowering Seed Plants
Gymnosperms (whose name means "naked seed"), evolved earlier than the flowering plants. One group, the **conifers** (division Coniferophyta), with 500 species, still dominates large areas of the globe. Other gymnosperms, such as the cycads and ginkgos (divisions Cycadophyta and Ginkgophyta; Fig. 24-10) have declined to a small remnant of their former range and abundance.

The ginkgos were probably the first of the surviv-

◀ **Figure 24-8** The life cycle of a fern, showing alternation of generations. The dominant plant body (upper left) is the diploid sporophyte. Haploid spores, formed in sporangia located on the underside of certain leaves, are dispersed by the wind to germinate on the moist forest floor into inconspicuous haploid gametophyte plants. On the lower surface of these small, sheetlike gametophytes, male antheridia and female archegonia produce sperm and eggs. The sperm must swim to the egg, which remains in the archegonium. The zygote develops into the large sporophyte plant. (Inset) Underside of a fern leaf, showing clusters of sporangia.

(a)

(b)

Figure 24-10 Two uncommon gymnosperms. **(a)** A cycad. Common in the age of dinosaurs, these are now limited to about 100 species living in warm, moist climates. **(b)** The ginkgo, or "maidenhair tree," has been kept alive by cultivation in China and Japan. Relatively resistant to pollution, these have become popular in American cities. Both ginkgos and cycads have separate sexes. The ginkgo shown here is female and bears fleshy seeds the size of large cherries, which are noted for their foul smell when ripe.

ing seed plants to evolve, becoming widespread during the Jurassic period, which began 213 million years ago. Today they are represented by the single species *Ginkgo bilboa*, the "maidenhair tree." Ginkgo trees are either male or female; the female trees bear foul-smelling, fleshy fruits the size of cherries (see Fig. 24-10b). Ginkgos have been maintained by cultivation, particularly in the Orient; if not for this, they might be extinct today. Since they are more resistant to pollution than most other trees, gingkos (usually male trees) have been extensively planted in American cities.

Cycads resemble large ferns, from which they probably evolved (see Fig. 24-10a). Today there are about 160 species, found mostly in tropical or subtropical climates. Most are about 1 meter in height, although some species can reach 20 meters. Cycads are slow-growing and long-lived; one Australian specimen is estimated to be 5000 years old. The fleshy seeds of cycads were once a staple food in Guam, but they contain a toxin that may cause a neurological disorder resembling Parkinson's disease.

Conifers spread widely as the Earth became drier during the Permian period that followed the Carboniferous. Today they are most abundant in the cold latitudes of the far north and at high elevations where conditions are rather dry. Not only is rainfall limited in these areas, but soil water remains frozen and un-

available during the long winters. Conifers, including pines, firs, spruce, hemlocks, and cypresses, are adapted to withstand dry, cold conditions in several ways:

1. Conifer leaves are thin needles covered with a thick cuticle, whose small waterproofed surface minimizes evaporation.
2. Conifers (often called *evergreens*) retain their leaves throughout the year, enabling them to continue photosynthesizing and growing slowly during times when most other plants become dormant.
3. Conifers produce a resinous "antifreeze" in their sap that allows it to continue transporting nutrients in subfreezing temperatures. This gives them their fragrant "piney" scent.

Reproduction is similar in all conifers, with pines serving as a good illustration (Fig. 24-11). The tree itself is the diploid sporophyte. It develops male and female cones. The male cones are often found in clusters at the ends of lower branches, while the larger woody female cones may be higher in the tree (Fig. 24-12). The male cones are relatively small (usually 2 cm or less), delicate structures. They release clouds of pollen during the reproductive season, and then disintegrate. Each pollen grain is a male gametophyte, consisting of several specialized haploid cells, some

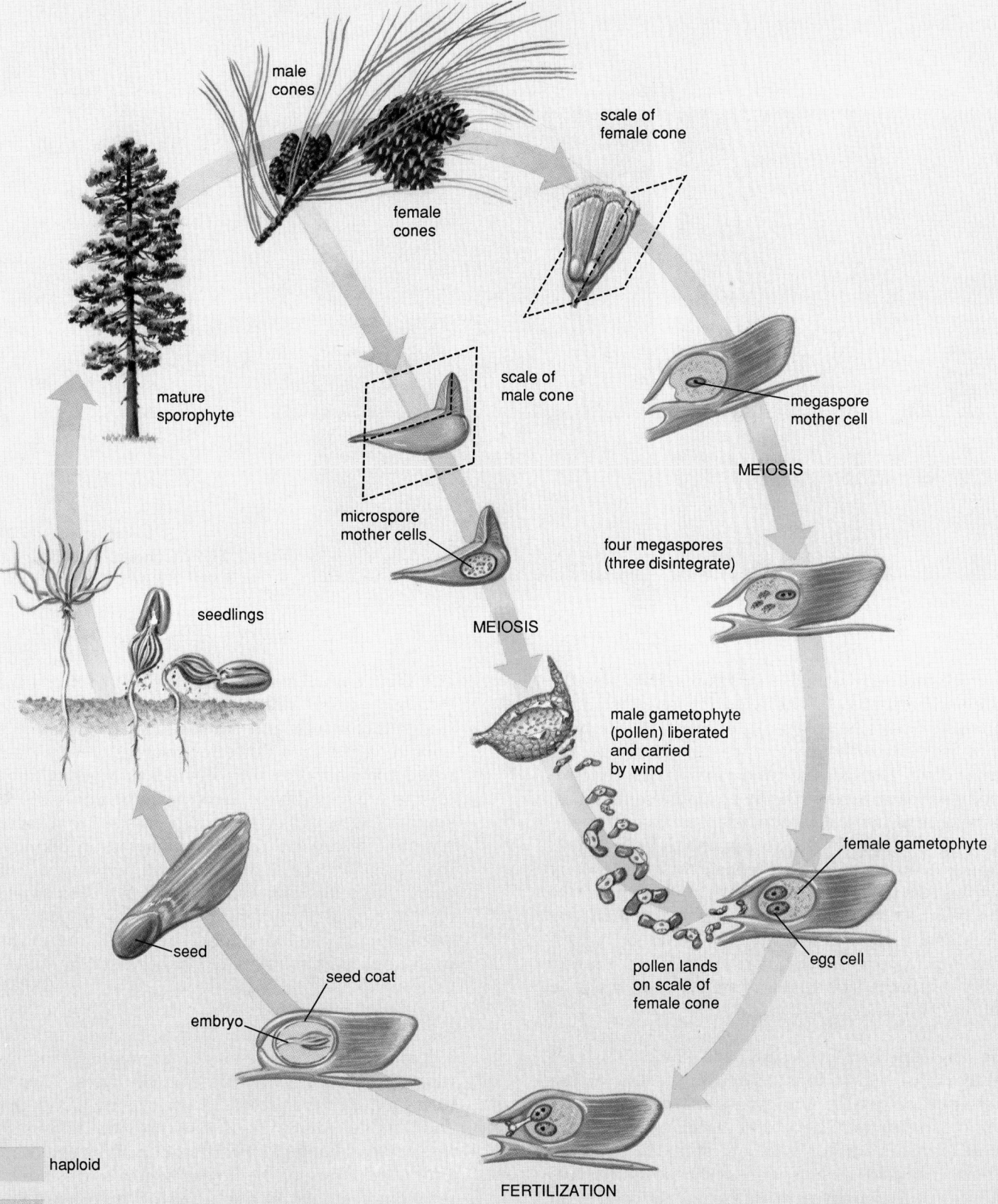

male
cones

scale of
female cone

female
cones

scale of
male cone

megaspore
mother cell

MEIOSIS

mature
sporophyte

four megaspores
(three disintegrate)

microspore
mother cells

seedlings

MEIOSIS

male gametophyte
(pollen) liberated
and carried
by wind

female gametophyte

seed

egg cell

seed coat

pollen lands
on scale of
female cone

embryo

FERTILIZATION

haploid

diploid

Figure 24-11 Life cycle of the pine, a seed plant of the division Coniferophyta. The tree is the sporophyte generation, bearing both male and female cones. Megaspore mother cells within the forming seeds on the scales of the female cones undergo meiosis to form megaspores, one of which develops into the female gametophyte. The female gametophyte in turn produces egg cells. Meanwhile, in the male cones, microspore mother cells undergo meiosis to produce the male gametophytes: the pollen. Pollen, carrying sperm nuclei, are dispersed by wind and land on the scales of the female cone. The pollen produces a pollen tube that penetrates the female gametophyte and conducts the sperm to the egg. The fertilized egg develops into an embryonic plant enclosed in a seed that is eventually released from the cone. The seed germinates and grows into the sporophyte tree.

(a)   *(b)*

Figure 24-12 Conifers bear both
(a) male and **(b)** female cones.

of which form tiny winglike structures that allow the pollen to be carried by the wind for long distances. Immense clouds of pollen are released by the male cones, so inevitably some lands by chance on the female cone. Each female cone consists of a series of woody scales arranged spirally around a central axis. At the base of the scale are two haploid female gametophytes, each producing an egg cell. A pollen grain landing nearby sends out a pollen tube that slowly burrows into the female gametophyte. After nearly 14 months, the tube finally reaches the egg cell and releases sperm that fertilize it. The fertilized egg becomes enclosed in a seed as it develops into a tiny embryonic plant. The seed is liberated when the cone matures and its scales separate.

The Angiosperms: Flowering Seed Plants

Three major adaptations have contributed to the enormous success of angiosperms: flowers, fruits, and broad leaves. The oldest fossils of flowering plants (division Anthophyta) are estimated to be 127 million years old. Angiosperms are believed to have evolved from gymnosperm ancestors that formed an association with animals (most likely insects), who carried their pollen from plant to plant. The insects benefitted by eating some of the protein-rich pollen, while the plant no longer had to produce prodigious quantities of pollen and send it flying on the fickle winds to ensure fertilization.

The relationship between ancient gymnosperms

and their animal pollinators was so beneficial that, through natural selection, plants evolved flowers that attract insects and other animals.** One hundred million years ago, flowering plants already dominated the Earth, as they do today. Modern angiosperms are incredibly diverse, including over 250,000 species (Fig. 24-13). They range in size from the diminutive duckweed (a few millimeters in diameter) that floats on ponds to the mighty eucalyptus tree, over 100 meters tall. From desert cactus to tropical orchids to grasses to parasitic mistletoe, angiosperms dominate the plant kingdom. The flower, one of their most important adaptations, is discussed in detail in Chapter 28.

The angiosperm life cycle is shown in Figure 24-14. Like their gymnosperm ancestors, angiosperms have a dominant sporophyte plant that produces and nurtures tiny male and female gametophytes. These in turn produce the sex cells. In angiosperms, both male and female gametophytes are formed within the flower (see Chapter 28). Fertilization of the egg occurs within the ovary of the flower, which surrounds the female gametophyte. The resulting zygote develops into an embryo enclosed in a seed. The ovary surrounding the seed matures into a **fruit.** The word *angiosperm* (Greek, "enclosed seed") refers to the enclosure of the seed within a fruit.

The fruit is a second adaptation that has contributed to the success of angiosperms. Just as flowers encourage animals to transport pollen, so too many

Figure 24-13 Diverse angiosperms. Both grasses **(a)**, and many trees such as this birch **(b)**, have inconspicuous flowers and rely on wind for pollination. **(c)** Flowers, such as those shown on this hedgehog cactus and eucalyptus tree (inset in e), entice insects to carry pollen between individual plants. **(d)** The smallest angiosperm is the duckweed, found floating on ponds. These specimens are about 1/8 inch in diameter. **(e)** The largest angiosperms are eucalyptus trees, which may reach 150 feet in height.

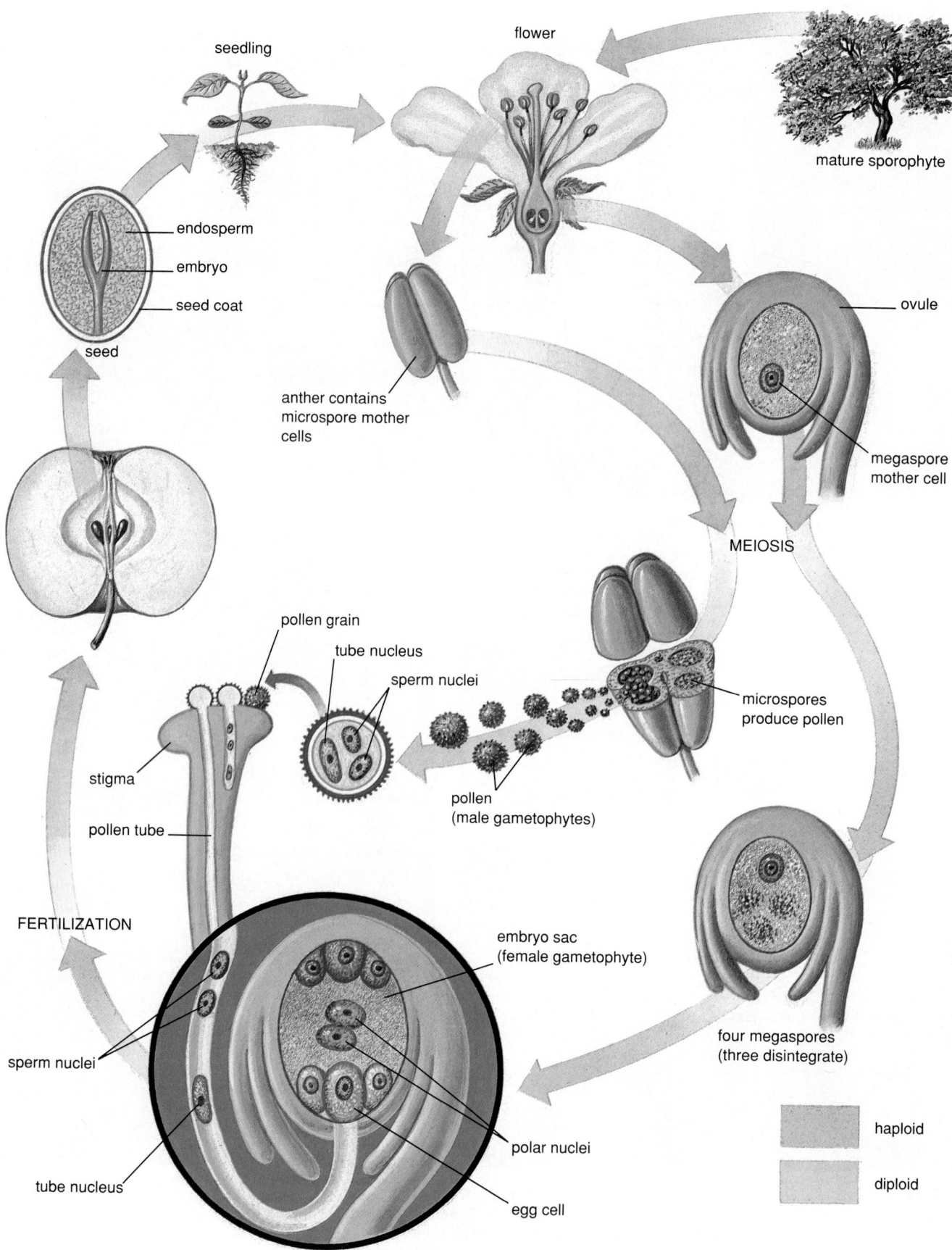

seedling

flower

mature sporophyte

endosperm

embryo

seed coat

seed

ovule

anther contains
microspore mother
cells

megaspore
mother cell

MEIOSIS

pollen grain

tube nucleus

sperm nuclei

microspores
produce pollen

stigma

pollen
(male gametophytes)

pollen tube

FERTILIZATION

embryo sac
(female gametophyte)

sperm nuclei

four megaspores
(three disintegrate)

tube nucleus

polar nuclei

egg cell

haploid

diploid

fruits entice them to disperse seeds. These seeds pass through animal digestive tracts unharmed; examples can easily be observed in bird droppings. As dog owners are well aware, some fruits (called burrs) disperse by clinging to animal fur. Others, like the fruits of maples, form wings that carry the seed through the air. The variety of dispersal mechanisms made possible by the fruit has helped the angiosperms invade nearly all possible habitats.

A third feature that gives angiosperms an adaptive advantage in warmer, wetter climates is broad leaves that are shed under adverse conditions, such as periods of cold and drought. When water is plentiful, as during the long, warm growing season of temperate and tropical climates, broad leaves give trees an advantage by collecting more sunlight for photosynthesis. The extra energy gained during the spring and summer allows the trees to drop their leaves and enter a dormant period during the fall and winter of temperate climates (or during the dry season in certain tropical climates). In the north, the period of warmth and moisture is considerably shorter than in the tropics. Here, angiosperm trees remain leafless and dormant for a much greater part of the year, avoiding water loss through their broad leaves. The water-conserving conifers continue photosynthesizing and growing slowly during the long winters, and for this reason they dominate in northern ecosystems (Fig. 24-15; see also Chapter 46).

The flowering plants are grouped into two broad classes based on their internal and external structure, discussed in more detail in Chapter 26. The **monocots** (class Monocotyledoneae) are a group of about 65,000 species including grasses, corn and other grains, irises, lilies, and palms. The dicots (class Dicotyledoneae) are a considerably larger group, with about 170,000 species, including most of the angiosperm trees, shrubs, and herbs.

Figure 24-15 Two ways of coping with the dryness of winter. The evergreen (a conifer) retains its needles throughout the year. The small surface area and heavy cuticle of the needles retard water loss. In contrast, the aspen (an angiosperm) sheds its leaves each fall. The dying leaves turn brilliant shades of gold as accessory pigments are exposed when the chlorophyll disintegrates.

◀ **Figure 24-14** Life cycle of a flowering plant. The dominant plant is the diploid sporophyte, whose flowers usually produce both male and female gametophytes. Male gametophytes (pollen grains) are produced within the anthers, where special cells (microspore mother cells) undergo meiosis to produce haploid microspores. These divide mitotically to produce pollen, in which a tube nucleus and two sperm nuclei are formed. The female gametophyte develops within the ovule of the ovary. Here special cells (megaspore mother cells) undergo meiosis to produce a haploid megaspore. The megaspore divides mitotically to produce the female gametophyte (also called the *embryo sac*) whose contents include one egg cell and two polar nuclei. After a pollen grain lands on the stigma of the carpel, it produces a tube that burrows through the style to the ovary and into the female gametophyte. There it releases its two sperm nuclei. One sperm nucleus fuses with the egg to form a zygote. The second fuses with the polar nuclei and forms the **endosperm,** a source of food for the developing embryo. The ovule gives rise to the seed containing endosperm and an embryo that develops from the zygote. The seed is dispersed, germinates, and develops into a mature sporophyte.

SUMMARY OF KEY CONCEPTS

The kingdom Plantae consists of photosynthetic, usually multicellular organisms. The ability of plants and other photosynthetic organisms to capture the energy of sunlight in high-energy molecules provides nearly all other forms of life on Earth with a source of usable energy.

Early "Roots"—The Evolution of Multicellular Plants

The first algae appeared between 500 and 600 million years ago. Green algal ancestors probably gave rise to two groups of terrestrial plants: vascular plants with specialized vessels that also provide support, and bryophytes that lack these vessels and are restricted to moist environments.

During evolution, plants became increasingly adapted to terrestrial existence. This required (1) rootlike structures for anchorage and for absorption of water and nutrients, (2) conducting vessels that transport water and nutrients throughout the plant, (3) a stiffening substance, called lignin, to impregnate the vessels and support the plant body, (4) a waxy cuticle to retard evaporative water loss, (5) stomata that can open, allowing gas exchange, and can also close, preventing water loss.

As plants invaded the land, new reproductive adaptations were required. Reduction of the male gametophyte to pollen allowed wind to replace water as a means of carrying sperm to eggs. Flowers attracted animals to carry pollen more precisely and efficiently than wind, while fruit enticed animals to disperse seeds. The seed nourishes, protects, and helps disperse the developing embryo.

Plants exhibit alternation of a haploid gametophyte generation with a diploid sporophyte generation. There has been a general evolutionary trend toward reduction of the haploid gametophyte, which is dominant in bryophytes but microscopic in seed plants.

Watery Origins—The Algae

The simplest plants are the aquatic algae. There are three major divisions of algae. The red algae (Rhodophyta) dominate in clear tropical waters, while the brown algae (Phaeophyta) populate temperate oceans. The green algae (Chlorophyta) are primarily small, freshwater forms, and are believed to be ancestral to modern land plants. Algal divisions are named for their predominant colors, resulting from a combination of green chlorophyll and accessory light-trapping pigments. Algae lack true roots, stems, and leaves, and rely on the water to carry sex cells.

Land—The New Frontier

Bryophytes, which include the liverworts and mosses, are small, simple land plants that lack conducting vessels. Although some have adapted to dry areas, most are found in moist habitats. Reproduction in bryophytes requires water through which the sperm must swim to the egg.

Vascular plants have evolved a system of vessels that also supports the body. Seedless vascular plants, including the club mosses (division Lycophyta), horsetails (division Sphenophyta), and ferns (division Pterophyta), can grow larger than bryophytes owing to this support system. Like bryophytes, the sperm must swim to the egg for sexual reproduction to occur, and the gametophyte lacks conducting vessels.

Vascular plants with seeds all have two major new adaptive features: pollen and seeds. They are often classified into two categories: gymnosperms and angiosperms. Gymnosperms include ginkgos, cycads, and the highly successful conifers. These were the first fully terrestrial plants to evolve. Their success on dry land is partially due to the evolution of the male gametophyte into the pollen grain. Pollen protects and transports the male gamete, eliminating the need for the sperm to swim to the egg. The seed, a protective resting structure containing an embryo and a supply of food, is a second important adaptation contributing to the success of seed plants.

Angiosperms, the flowering plants, dominate much of the land today. In addition to pollen and seeds, angiosperms also produce flowers and fruits. The flower allows angiosperms to utilize animals as pollinators. In contrast to wind dispersal, animals can carry pollen longer distances with greater accuracy and less waste. Fruits may attract animal consumers, who disperse the seeds in their feces.

GLOSSARY

accessory pigments: colored molecules other than chlorophyll that absorb light energy and pass it to chlorophyll.

algae (al'-gē; sing. alga): a general term for simple aquatic plants lacking vascular tissue.

alternation of generations: a life cycle typical of plants in which a diploid sporophyte (spore-producing) genera-tion alternates with a haploid gametophyte (gamete-producing) generation.

angiosperm (an'-gē-ō-sperm): a flowering vascular plant.

antheridium (an-ther-id'-ē-um): a structure in which male sex cells are produced, found in the bryophytes and certain seedless vascular plants.

archegonium (ar-ke-gō'-nē-um): a structure in which female sex cells are produced, found in the bryophytes and certain seedless vascular plants.

bryophyte (brī'-ō-fīt): a division of simple nonvascular plants, including mosses and liverworts.

conifer (kon'-eh-fer): a class of tracheophyte that reproduces using seeds formed inside cones and retains its leaves throughout the year.

cuticle (kū'-ti-kul): a waxy or fatty coating on the exposed epidermal cells of many land plants, which aids in the retention of water.

dicotyledon (dī'-kot-ul-ēd'-un): a class of angiosperm whose embryo has two cotyledons, or seed leaves.

endosperm (en'-dō-sperm): the stored food within a seed used to nourish the developing plant embryo.

flower: the reproductive structure of an angiosperm plant.

fruit: the mature ripened ovary of an angiosperm plant. This structure contains the seeds.

gametophyte (ga-mēt'-ō-fīt): a multicellular haploid plant that produces haploid sex cells by mitosis.

gymnosperms (jim'-nō-sperms): non-flowering seed plants such as conifers, cycads, and gingkos.

monocotyledon (mahn'-eh-kot-ul-ēd'-un): a class of angiosperm plant in which the embryo has one cotyledon, or seed leaf.

pollen: the male gametophyte of gymnosperms and angiosperms.

rhizoid: (rī'-zōyd): a rootlike structure found in bryophytes that anchors the plant and absorbs water and nutrients from the soil.

seed: the reproductive structure of a seed plant. The seed is protected by a seed coat and contains an embryonic plant and a supply of food for it.

sporophyte (spōr'-ō-fīt): the diploid form of a plant that produces haploid, asexual spores through meiosis.

tracheophyte (trā'-kē-ō-fīt): a category of plant that has conducting vessels; a vascular plant.

zoospore (zō'-ō-spōr): a nonsexual reproductive cell that swims using flagella.

zygote (zī'-gōt): a diploid cell resulting from the fusion of male and female sex cells.

STUDY QUESTIONS

1. What is meant by "alternation of generations"? What two generations are involved? How does each reproduce?

2. Explain the evolutionary trends in plant reproduction. How did these changes help plants adapt to increasingly dry environments?

3. Describe evolutionary trends in the life cycles of plants, with emphasis on the relative sizes of the gametophyte and sporophyte.

4. Assuming that green algae have fewer accessory pigments than red or brown algae, would you expect to find them in shallow or deep water? Where would you find the red algae, and why?

5. From which algal division did green plants probably arise? Explain the evidence supporting this hypothesis.

6. List the various adaptations necessary for the invasion by plants of dry habitats on land. Which of these are possessed by bryophytes? By ferns? By gymnosperms and angiosperms?

7. What single feature is probably most responsible for the enormous success of angiosperms? Explain why.

8. List the adaptations of gymnosperms that have helped them become the dominant tree in dry, cold climates.

9. What is a pollen grain? What role has it played in helping plants colonize dry land?

10. Compare the life cycle of a moss and a pine. Use information from Table 24-1 to help identify differences.

DISCUSSION QUESTIONS

1. Describe the major trends in plant evolution and discuss how they were adaptations to the dry environmental conditions encountered on land.

2. Describe evolutionary trends in the life cycles of plants with emphasis on the relative sizes and periods of duration of the gametophyte and sporophyte generations. Discuss why these trends were adaptive from the genetic point of view of haploid and diploid cells.

SUGGESTED READINGS

Bold, H. C., and LaClaire, J. W., II. *The Plant Kingdom*. 5th edition. New Jersey: Prentice-Hall, 1987. A concise summary of all forms of plant life.

Kaufman, P. B. *Plants—Their Biology and Importance*. New York: Harper and Row, 1989. Complete, readable coverage of all aspects of plant taxonomy, physiology, and evolution.

Milot, V. "Blueprint for Conserving Plant Diversity." *Bioscience*, June, 1989. Points out the importance of preserving genetic diversity in endangered plant species.

Niklas, K. J. "Computer-assisted Plant Evolution." *Scientific American*, March 1986. Shows how to use a desk-top computer to simulate trends in plant evolution.

Raven, P., Evart, R. F., and Eichhorn, S. E. *Biology of Plants*. 4th edition. New York: Worth, 1986. A beautifully illustrated and comprehensive botany text, stressing evolution and environmental adaptations.

25

Unifying Concepts in Physiology: Homeostasis and the Organization of Life

◀ Plants and animals face similar challenges, one of which is to obtain energy. Plants capture the energy of sunlight. All animals, either directly or indirectly, derive their energy from plants.

Nearly all physiological processes contribute to the perpetuation of life, through growth and reproduction or to the maintenance of a constant environment within the organism—homeostasis. In the first half of this chapter we explore the concept of homeostasis and the ways in which both plants and animals meet the challenges presented to all forms of life. In the second half, we discuss the specific types of animal tissues and organ systems that perform these life processes.

Homeostasis and the Challenges of Life

Homeostasis

In the mid-nineteenth century, the French physiologist Claude Bernard was the first to describe the concept now known as homeostasis, as the "constancy of the interior milieu." The term **homeostasis** was coined in 1932 by Walter Cannon, who suggested that nearly all physiological processes operated in a manner that tended to maintain the internal constancy required for the maintenance of life.

Although the term homeo*stasis* might imply a "static" or unchanging environment, in fact, the "internal milieu" seethes with activity as the body continuously responds and adjusts to internal and external changes. The term *dynamic equilibrium* better describes the internal conditon. Change does occur—but within the narrow range that cells require to function. Conditions are maintained within this range mainly by the continuous activation of negative feedback systems.

Negative Feedback Maintains Homeostasis

The most important principle governing the maintenance of homeostasis is **negative feedback,** in which a change initiates a series of events that counteract the change. A familiar example is your home thermostat (Fig. 25-1a). A drop in temperature from its setting is detected by a thermometer, which activates a switch that turns on a heating device. When the temperature is restored to the set point, the heater is turned off. Negative feedback requires a *set point* (the thermostat setting), a *sensor* (the thermometer), and an *effector*, which accomplishes the change (the heater). A biological example is the maintenance of your body at a temperature within 1° of 98.6° F (37° C). The set point is established by neurons in the hypothalamus. Other neurons in the hypothalamus,

abdomen, spinal cord, and large veins act as sensors for body temperature and transmit this information to the hypothalamus. When your body temperature drops, the hypothalamus activates various effector mechanisms including shivering (which generates heat through muscular activity), constriction of blood supply to the skin (which lessens heat loss), and elevation of metabolic rate (which generates heat). When body temperature is restored, the hypothalamus switches off these temperature control mechanisms (see Fig. 25-1b).

As you read the following chapters, you will encounter numerous examples of homeostatic control using negative feedback. For example, you will learn how the stomata of plants simultaneously control CO_2 entry and water loss by negative feedback, responding both to the demands of photosynthesis and to the water content of the plant tissues (Chapter 26). In Chapters 30, 31, and 35, you will encounter the mammalian hormones erythropoietin, ADH (antidiuretic hormone), and insulin, which regulate blood oxygen content, water balance, and blood sugar levels, respectively, all through negative feedback.

Positive Feedback and Homeostasis

Figure 25-1(c) illustrates a **positive feedback** system, in which a change initiates a series of events that amplify the original change. Positive feedback, as you can imagine, tends to create explosive events that must be carefully restricted. Nuclear fission is an example of positive feedback, in which multiple particles that are split from one atom each split another atom, and so on. Controlled, the reaction supplies nuclear power. When deliberately set out of control, it produces the atomic bomb. Population growth is a familiar biological example of positive feedback, as described in Chapter 43. The well-known ecologist Paul Ehrlich coined the apt expression *population bomb* to describe unchecked population growth.

In animal physiology, we find a few examples of self-limiting positive feedback events. Positive feedback occurs during childbirth (see Fig. 25-1d). The early contractions of labor begin to force the baby's head against the cervix at the base of the uterus, causing it to dilate. Stretch receptive neurons in the cervix signal the hypothalamus, which responds by triggering the release of a hormone that stimulates more and stronger uterine contractions. The positive feedback cycle is finally terminated by the expulsion of the baby and its placenta.

In the following section, we examine the challenges of life that all organisms must meet. We discuss how plants and animals have evolved either fundamen-

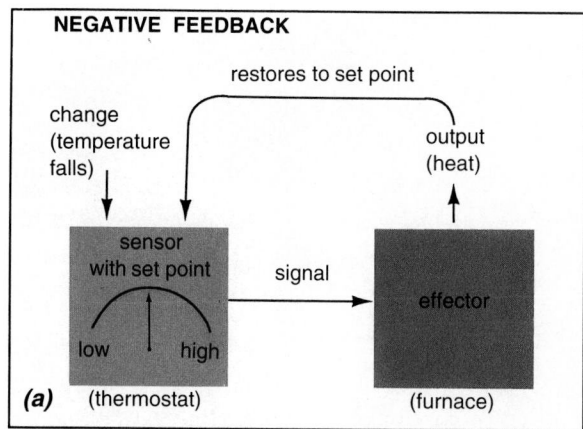

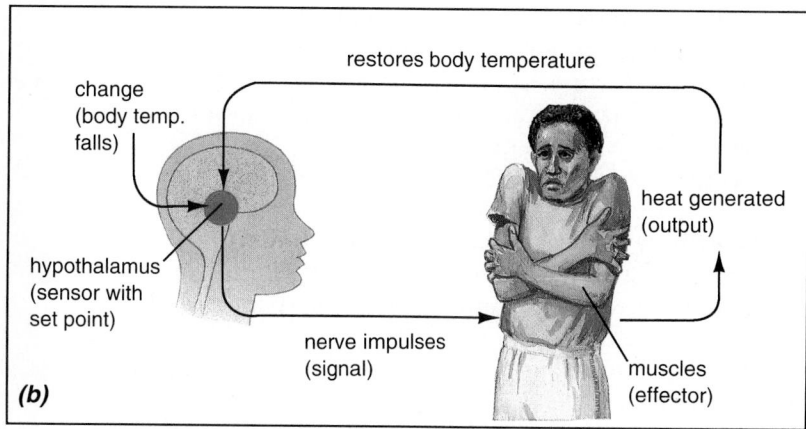

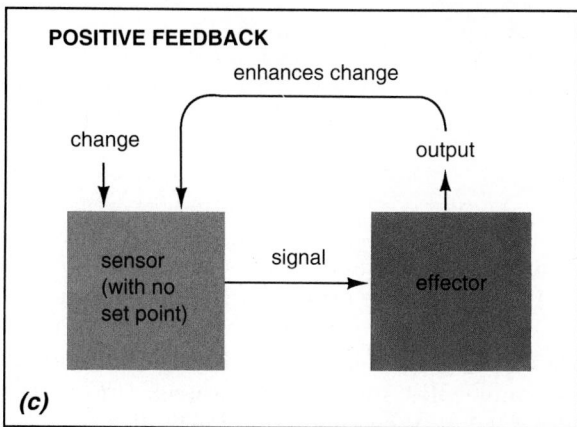

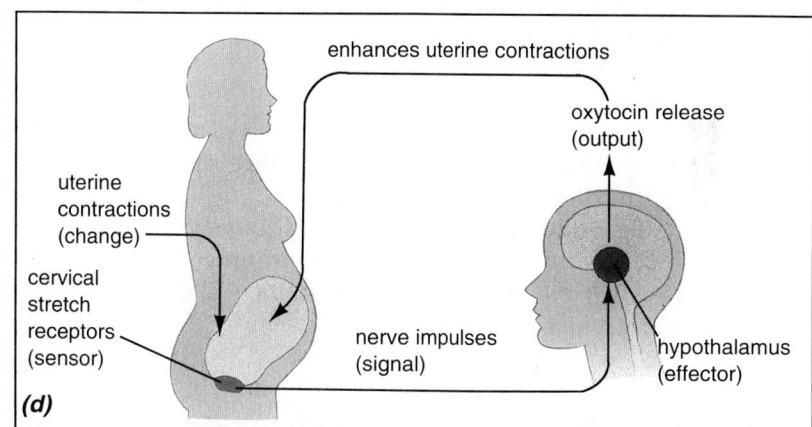

Figure 25-1 Negative and positive feedback.
(a) Negative feedback controls the temperature in most homes. When the system changes (the house cools below a specific temperature), the change is sensed by a sensor with a specific set point (the thermostat, set to 68° F). The sensor sends a signal to an effector (the furnace) whose output (heat) returns the system to its set point, turning off the signal, which turns off the effector.
(b) Body temperature is both set by and sensed by neurons in the hypothalamus (the sensor with a set point). A drop in body temperature causes the hypothalamic neurons to send a signal to effectors such as muscles, which begin shivering. Shivering generates heat, which helps restore body temperature to its set point, shutting off the signal, and causing shivering to cease.
(c) In positive feedback, a system is altered, and the change is sensed by a sensor that sends a signal to an effector. The output of the effector causes an even greater change in the system.
(d) At the onset of labor, uterine contractions are sensed by receptor neurons in the cervix, which send a signal to the hypothalamus that results in secretion of a hormone (oxytocin) by the posterior pituitary. The hormone, in turn, stimulates more and stronger uterine contractions.

tally similar or very different but equally effective ways of meeting these demands.

The Challenges of Life

As it goes through life, each organism requires certain things from its environment, and the environ- ment places stresses on the organism that must be overcome if the organism is to survive and repro- duce. **Whether unicellular or multicellular; bacte- rium, protist, plant, animal, or fungus; marine, freshwater, terrestrial, or parasitic, all organisms must meet a common set of challenges if they are to sustain life.** They must:

1. **Obtain materials** to construct the body.
2. **Obtain energy** for construction, maintenance, and reproduction.
3. **Exchange gases** (carbon dioxide and oxygen) with the environment.
4. **Regulate body composition** by eliminating excess materials and waste products.
5. **Distribute materials throughout the body.**
6. **Coordinate activities within the body.**
7. **Defend the body** against predators, parasites, and disease.
8. **Regulate growth and development.**
9. **Reproduce.**

Terrestrial organisms face additional challenges peculiar to life on land. They must also:

10. **Obtain water.**
11. **Reduce water loss.**
12. **Protect the body against the temperature extremes** encountered on land.
13. **Support the body against the force of gravity.**
14. **Reproduce without a watery environment** through which the sperm can swim to meet the egg, and in which the embryo can develop.

Although we will discuss each activity separately, many of these processes are interrelated. We will emphasize the adaptations of familiar plants and animals, the subjects of the next two units.

Challenges Faced by All Organisms

Obtaining Materials

Every organism must procure the materials it needs to construct its body. Animals typically obtain materials by ingesting food, usually plants, protists, or other animals (Fig. 25-2). The complex molecules of protein, fat, and carbohydrate that make up their prey are broken down in the digestive tract into simpler molecules such as amino acids, fatty acids, and sugars. The various parts of the body then use these simple molecules to synthesize new, larger molecules that are specific to the animal.

Plants obtain their materials as simple inorganic molecules of carbon dioxide, water, and minerals. Plants synthesize their amino acids, fatty acids, and sugars starting from scratch, so to speak, since they cannot acquire the ready-made "convenience foods" available to animals.

Obtaining Energy

Plants capture the energy of sunlight and, through photosynthesis, use it to drive the synthesis of or-

Figure 25-2 Animals, such as this tomato hornworm, obtain both nutrients and energy from other organisms, such as plants. Plants, in turn, obtain their nutrients from the air, soil, and water, and their energy from sunlight.

ganic molecules from minerals, carbon dioxide, and water. Animals, on the other hand, acquire both materials and energy from the food they eat. Animals capture the energy released by the breakdown of complex molecules such as fat, protein, and sugar and direct it into their own bodily activities.

Exchanging Gases

All animals and plants exchange oxygen and carbon dioxide with the air. Plants absorb carbon dioxide and release oxygen during photosynthesis. During cellular respiration, both plants and animals consume oxygen and produce carbon dioxide as a waste product (see Chapters 7 and 8).

For very small organisms, all parts of the body are close enough to the surface so that diffusion suffices to move gases to and from the cells. Plants absorb gases by diffusion through stomata. Large, active organisms, such as many animals, must have respiratory systems: elaborations of the body surface that provide a large area for gas exchange. Most also have some means of pumping air or water over the respiratory surface, ensuring maximum availability of oxygen. Plants lack respiratory systems. Their metabolic rate is much slower than that of animals, and the deep internal tissues of woody roots and stems are usually composed of nonliving substances that don't use oxygen.

Regulating Body Composition

Most organisms acquire too much of at least some materials, and most organisms also create toxic wastes as the result of normal metabolic processes. Since cells can function properly only within a narrow range of composition of both intracellular and extracellular fluids, the body must rid itself of wastes and surplus materials. Different organisms have evolved a variety of excretory organs, from the contractile vacuoles of some protists, for whom water is a surplus material, to the elaborate kidneys of terrestrial vertebrates, for whom water may be a scarce resource.

Plants sequester many surplus or toxic products in the central vacuoles of their cells. Some plants also excrete wastes in droplets of water that appear at the tips of leaves in the morning.

Distributing Materials

In a large body, most of the cells are located far from the sources of food and oxygen. Consequently, these materials must be distributed throughout the body. Even small and sluggish animals such as snails have circulatory systems in which a fluid is pumped around the body by contractions of the heart. Plants, too, have systems that transport sugars from leaves to roots and fruits, and move minerals and water from roots to leaves and fruits. The fluids in plants are not pumped, however. Instead, plants have

evolved mechanisms based on osmosis and the evaporation of water that transport fluids within rigid tubes.

Coordinating Activities Within the Body

Early in the history of life, cells evolved the ability to influence other cells by releasing chemicals that triggered responses in their recipients. Both plants and animals still use chemicals, called hormones, to coordinate the growth, development, and functioning of the body. Animals also possess nervous systems that, in essence, are rapid transmission lines that send information from one part of the body to another. At the receiving end, the nerve cells release a chemical called a neurotransmitter that evokes appropriate responses in target cells.

Defending the Body

"Big fleas have little fleas upon their backs to bite 'em," wrote Jonathan Swift, expressing the principle that every organism is food for another. Some plants and animals have evolved similar adaptations that discourage predation, including bad taste or poisons (Fig. 25-3a) and defensive armament (Fig. 25-3b, c). Many animals, being more motile than plants, have

(a)

(b)

(c)

Figure 25-3 Defensive adaptations in plants and animals.
(a) A monarch butterfly larva feeding on a milkweed leaf. The milkweed synthesizes toxic compounds that interfere with the functioning of the nervous systems and hearts of potential animal predators. Consequently, few animals eat milkweeds. One of the few that can is the caterpillar of the monarch butterfly. It incorporates the poisonous compounds in its tissues and advertises this fact with its distinctive coloration. As a result, birds that prey on other butterflies leave the monarch alone.
(b) The flesh of the cactus is often one of the few sources of moisture available in a desert. The leaves of cacti are reduced to thin spines, which both reduce evaporation and discourage predation.
(c) The porcupine displays an antipredator adaptation similar to that of the cactus.

also evolved more active techniques, such as running away, hiding, or fighting back. On the other hand, most plants can survive even if parts of them are eaten.

Equally serious is the threat of disease—that is, parasitism by microorganisms. Both plants and animals have relatively impervious outer coverings, and plants may respond to infections by walling off infected areas from the rest of the body. Animals also have roaming defensive cells that devour invaders, and circulating chemical warfare molecules, the antibodies, that destroy microbes or neutralize their toxins.

Regulating Growth and Development

In sexually reproducing organisms, all the structures of the adult develop from a fertilized egg. During embryonic development, the descendants of this single cell differentiate into specialized cell types and form organs that carry out specific bodily functions. Both plants and animals use chemical signals, either on the surfaces of cells or in the circulation, to govern cell differentiation and body development.

Reproduction

Most plants and animals reproduce sexually. Animals often seek out and court their mates, defend their nesting sites, perform elaborate courtship displays, and provide parental care for their offspring. Plants, being mostly stationary, use wind, water, and animals to carry their sperm or pollen (containing sperm) to the egg. Although plants have their own forms of "courtship" (pollen- and nectar-laden flowers that attract animal pollinators; see the Chapter Opener), territorial defense (for example, shading competitors from the sunlight needed for photosynthesis), and even passive parental care (packaging food together with the embryo in a seed), the adaptations involved are quite unlike those of animals.

Challenges Presented by Life on Land

The first organisms probably lived in the sea, and for over 2 billion years organisms adapted to continuous immersion in water. Besides ample water for cellular metabolism, the sea provides a relatively constant temperature, buoyant support for bodies against the pull of gravity, and a medium for the transport of sperm to eggs. When plants and animals invaded the land, they had to evolve adaptations to compensate for the loss of their watery habitat.

Obtaining Water

The first challenge for terrestrial organisms is to obtain sufficient water. Plants usually extract water from the soil with their roots, using the power of water evaporating from their leaves to pull water into the plant from the soil. Some water is also generated as a by-product of cellular respiration. Animals acquire water in three ways. Most drink liquid water. Some water is also obtained as part of the food that animals eat. Finally, as in plants, cellular respiration provides some water. This "metabolic water" actually supplies some desert animals with a significant percentage of their water needs.

Preventing Water Loss

Terrestrial organisms must minimize the amount of water their bodies lose. Virtually all land plants and animals are covered with a waterproof coating that reduces evaporation. Desert plants often have thick, fleshy bodies or roots that store water (Fig. 25-4a), and reduced leave size that minimizes evaporation (see Fig. 25-3b). Animals and plants also lose water during gas exchange, and desert dwellers usually have adaptations that minimize this loss. Many desert animals remain inactive during the heat of the day, when evaporation is greatest. Many desert plants open their stomata only at night, reducing water loss. Animals also lose significant amounts of water during excretion of urine. As you might expect, desert animals such as the kangaroo rat (Fig. 25-4b) have specialized kidneys that excrete wastes dissolved in the smallest possible volume of water.

Protecting Against Temperature Extremes

Both high and low temperatures can be fatal to cells. High temperatures denature proteins (as when egg whites are cooked), while freezing temperatures cause ice crystals to form within the cytoplasm. These sharp crystals may pierce through membranes and destroy the cell. Plants and animals have independently evolved such features as heat-resistant proteins and antifreeze molecules. Many plants and animals also produce heat- or cold-resistant reproductive structures (such as some eggs, pupae, spores, or seeds) that can survive the temperature extremes that kill the adult organisms. Many animals display other adaptations to cold. Mammals are insulated by fur and fat. Some, such as bears and ground squirrels, store fat and then **hibernate.** During hibernation, their metabolic processes slow and their body temperatures drop radically. Others mammals remain active and feeding (snowshoe hares, arctic foxes, wolves), maintaining high body temperatures even in

(a)

(b)

Figure 25-4 Since water is so scarce, desert organisms must reduce water loss to a minimum. **(a)** Succulents store water in fleshy leaves, covered with a thick waxy coating that reduces evaporation. **(b)** Kangaroo rats stay underground during the heat of the desert day, emerging out of their burrows to forage at night. The kidneys of kangaroo rats are remarkably efficient, producing extremely concentrated urine and thereby eliminating wastes without losing much water.

the coldest weather. Still others, such as most North American songbirds, migrate to warmer regions. High temperatures also evoke responses such as burrowing into the cool ground during the hottest part of the day, seeking out shade, and evaporating water from the skin or respiratory system to carry away excess heat (Fig. 25-5).

Supporting the Body Against Gravity

Lacking the buoyant support of water, plants and animals must have some type of skeleton. Plant cells are surrounded by strong cell walls made of cellulose.

Figure 25-5 High temperatures can be lethal to animals. If the temperature is very high, excess metabolic heat cannot be given off directly to the air. Surplus heat can be disposed of, however, by using its energy to evaporate water, as dogs do when they pant.

These cellulose "skeletons," together with the pressure of water pushing against the inside of the cell wall, support the bodies of many small plants such as herbs and grasses. Larger plants have tracts of specialized cells with extra-thick walls, forming the wood that makes up and supports the massive trunks of trees. Animal skeletons are diverse and include internal skeletons of bone (vertebrates), external skeletons of chitin (insects and other arthropods), and hydrostatic skeletons of water confined within a muscular tube (earthworms, cnidarians).

Reproducing on Land

Truly terrestrial animals require two reproductive adaptations: internal fertilization and a protected, watery environment in which the embryo can develop. During internal fertilization, the male deposits sperm directly into the reproductive tract of the female, which provides the watery environment the sperm needs to swim. The waterproof egg produced by insects, reptiles, and birds allows the embryo to develop in its own private pond until it can face the rigors of land existence. In mammals, the embryo develops within the mother's body, enclosed in the waters of the amniotic fluid.

The truly terrestrial plants, including grasses, wildflowers, and trees, have done away with swimming sperm entirely. Sperm cells are packaged in pollen grains that do not need water for survival. Most plants enclose the embryo in a seed. Both pollen and seeds can remain viable for weeks, months, or, rarely, even centuries. In some desert plants, seeds often remain dormant for years, through drought and frost, sprouting only when the infrequent rains provide adequate water (Fig. 25-6)

Figure 25-6 Flowers, such as this owl clover, may blanket the desert floor for a few brief weeks during early spring.

The Study of Plant and Animal Physiology

As you can see from our brief survey, plants and animals face similar challenges in similar environments. The solutions that they have evolved, however, are extremely different. Plant and animal physiology is unified by the similarity of function of the various systems, but separated by the enormously different mechanisms each has developed over millenia of independent evolution. For the sake of clarity, we discuss plant and animal physiology and the organization of plant and animal bodies separately. An overview of the tissues and organs of the animal body is provided in the following section, while the structure and organization of the plant body is covered in the following chapter.

As you progress through the chapters of Units IV and V, however, keep in mind that the different physiological solutions evolved by plants and animals are responses to virtually identical challenges posed by their environments.

The Organization of the Animal Body

Conceptually, the animal body is made up of a number of **organ systems** (for example, the digestive system and the excretory system), each made up of **organs** (the stomach and intestine of the digestive system; the kidney and urinary bladder of the excretory system). Each organ, in turn, consists of a number of **tissues** (such as epithelial, connective, nerve, or muscle tissue) composed of dozens to billions of

structurally similar cells that act in concert to perform a similar function (Fig. 25-7).

Animal Tissues

A tissue is composed of cells with similar structure and function, but it may also include noncellular components produced by these cells, as in the case of cartilage and bone. Here, we present a brief overview of four general categories of animal tissue: epithelial tissue, connective tissue, muscle tissue, and nerve tissue.

Epithelial Tissue

The cells of **epithelial tissues** form continuous sheets, or **membranes.** During development, glands are also formed from epithelial tissues. Epithelial membranes cover the body and line all the body cavities. Epithelial tissues form a barrier that either resists the movement of substances across it (such as the skin) or that allows only specific substances to cross it (such as the nutrients that pass through the lining of the small intestine). Epithelia serve as effective barriers because their cells are packed closely together and connected to one another by tight junctions (see Chapter 6). No blood vessels penetrate epithelial tissue; it is nourished by diffusion from capillaries penetrating the connective tissue that lies beneath it.

An important property of epithelial tissues is that they are continuously lost and replaced by cell division. Consider the abuse suffered, for example, by the epithelium lining your mouth. Scalded by coffee and scraped by corn chips, it would be destroyed within a few days if it were not continuously replacing itself. The stomach lining, abraded by food and attacked by acids and protein-digesting enzymes, is completely replaced every 2 to 3 days. Your skin epidermis is renewed about twice a month. Epithelial tissues are classified according to the shape of their cells and the number of cell layers present, as summarized in Table 25-1.

Connective Tissue

This category includes several diverse tissues, including the dermis, tendons, ligaments, cartilage, bone, fat, and blood. **Connective tissues** share a common feature: they all secrete large quantities of extracellular substances between the living cells. Some type of connective tissue underlies all epithelial tissue, containing capillaries and fluid-filled spaces that nourish the epithelium. Underlying the epidermis of the skin, for example, is connective tissue richly supplied with capillaries called the **dermis** (see Fig. 25-11). With the exception of blood, most connective tissue is inter-

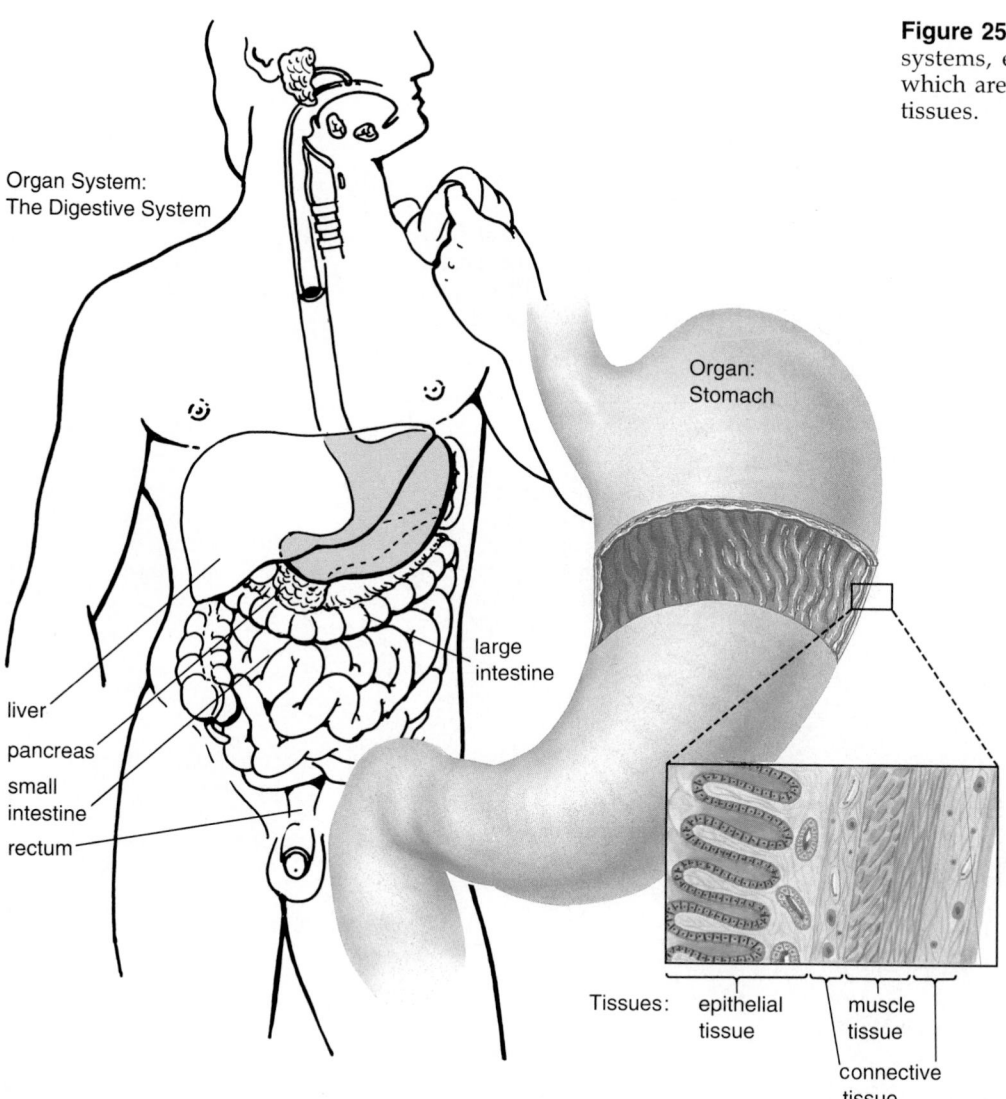

Organ System: The Digestive System

liver
pancreas
small intestine
rectum

large intestine

Organ: Stomach

Tissues: epithelial tissue   muscle tissue   connective tissue

woven with fibrous strands of an extracellular protein called **collagen** secreted by the cells.

Tendons and **ligaments** are forms of connective tissue that attach muscles to bones and bones to bones, respectively (see Fig. 38-11). They contain densely-packed collagen fibers in an orderly parallel arrangement. **Cartilage** consists of widely spaced cells called **chondrocytes** surrounded by a thick, nonliving matrix of collagen which they secrete (see Fig. 38-5). Cartilage covers the ends of bones at joints, provides the supporting framework for the respiratory passages, supports the ear and nose, and forms shock-absorbing pads between the vertebrae.

Bone resembles cartilage which has been hardened by deposits of calcium phosphate. During bone formation, bone cells or **osteocytes** become embedded in concentric layers of hardened collagen matrix. These layers surround a central canal containing a blood capillary that nourishes the bone cells. These concentric circles of bone each surrounding a canal are called

Haversian systems (Fig. 25-8). Both bone and cartilage are covered in detail in Chapter 38.

Fat cells, collectively called **adipose tissue,** are a form of connective tissue. Fat cells are specially modified to act as storage sacs for triglycerides (see Chapter 3), which can make up over 90% of their volume.

Blood is considered connective tissue because it is largely composed of extracellular substance, the plasma. Blood and its constituents are covered in detail in Chapter 30.

Muscle Tissue

Muscle is a specialized contractile tissue. Muscle cells derive their motility from the orderly arrangement of actin and myosin proteins, which can use energy to move relative to one another. There are three types of muscle: skeletal muscle, cardiac muscle, and smooth muscle. Muscle is covered in more detail in Chapter 38.

Table 25-1   Major Types of Epithelial Tissue

TISSUE TYPE	DIAGRAM	PROPERTIES AND FUNCTION	LOCATION IN BODY
Simple squamous epithelium.	Thin, flattened cells, one layer thick.	Allows movement of subtances. Found in areas where diffusion and filtration are occuring.	Lines lung alveoli; forms capillary walls; lines blood vessels.
Stratified squamous epithelium.	Thin, flattened cells, many layers thick.	Thick layer that is rapidly replaced by divisions of lower cells; has protective function; many secrete mucus.	Upper layer of skin; lines mouth anal canal, vagina.
Simple cuboidal epithelium.	Cube-shaped cells; one layer thick.	Functions in absorption and secretion.	Lines kidney tubules; has secretary role in salivary glands, thyroid, pancreas, and liver.
Simple columnar epithelium.	Elongated cells; one layer thick; nuclei lined up.	Forms a thick layer that functions in secretion and absorption. May secrete mucus.	Lines the esophagus, stomach, intestines, and uterus.
Pseudostratified columnar epithelium.	Elongated cells; one layer thick; nuclei at different levels.	Often possesses beating cilia; may secrete mucus. Functions in trapping and transporting particles out of respiratory surfaces; moving sex cells.	Lines the respiratory tract; lines the tubes of the reproductive system (oviducts, vas deferens).

Skeletal muscle is also called **striated muscle** because its orderly arrangement of thin (actin) and thick (myosin) protein filaments give it a striped appearance (Fig. 25-9a). Skeletal muscle is generally under voluntary or conscious control. As its name implies, its main function is to move the skeleton. **Cardiac muscle,** found only in the heart, also has a striated appearance (Fig. 25-9b). Unlike skeletal muscle, it is spontaneously active. Cardiac muscle cells are interconnected by gap junctions, which allow electrical

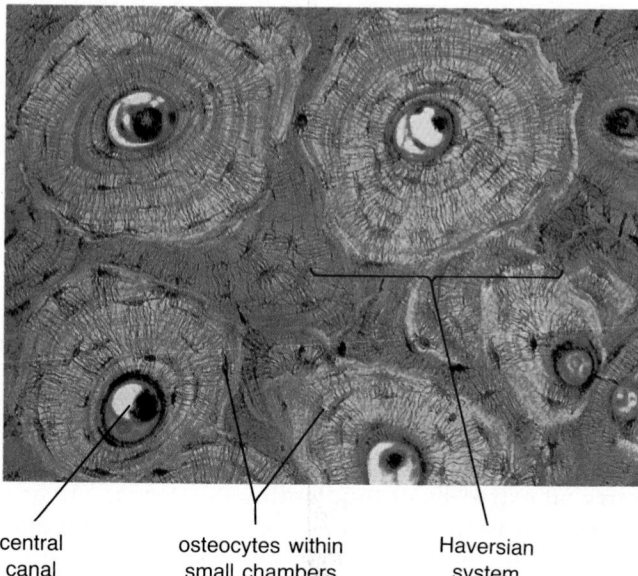

central canal | osteocytes within small chambers | Haversian system

Figure 25-8 Haversian systems, concentric circles of bone deposited around a central canal containing a blood vessel, are clearly visible in this micrograph. Bone cells, or osteocytes, are trapped within small chambers within the Haversian systems.

signals to spread rapidly through the heart. **Smooth muscle** lacks the orderly arrangement of thick and thin filaments, appearing more uniform under the microscope (Fig. 25-9c). Smooth muscle is embedded in the walls of the digestive tract, the uterus, the bladder, and the large blood vessels. It produces slow, sustained contractions, which are generally involuntary.

Nerve Tissue

Nerve tissue is composed of cells called **neurons.** Neurons are specialized to generate electrical signals and to conduct these signals along the cell to other neurons, muscles, or glands. A neuron has four major parts, each with a specialized function (Fig. 25-10). The **dendrites** receive signals from other neurons or from the external environment. The **cell body** directs maintenance and repair of the cell. The **axon** conducts the electrical signal to its target cell, and the **synaptic terminals** transmit the signal to the target cell at a specialized contact region called a **synapse.** Nerve tissue and the nervous system are covered in detail in Chapter 36.

Animal Organs

Organs are formed from at least two different tissue types that function together; some organs, such as the skin, described below, include all four of the tissues described above.

The Skin: A Representative Organ

The **epidermis** or outer layer of the skin is a specialized epithelial tissue (Fig. 25-11). It is covered by a

(a)   *(b)*   *(c)*

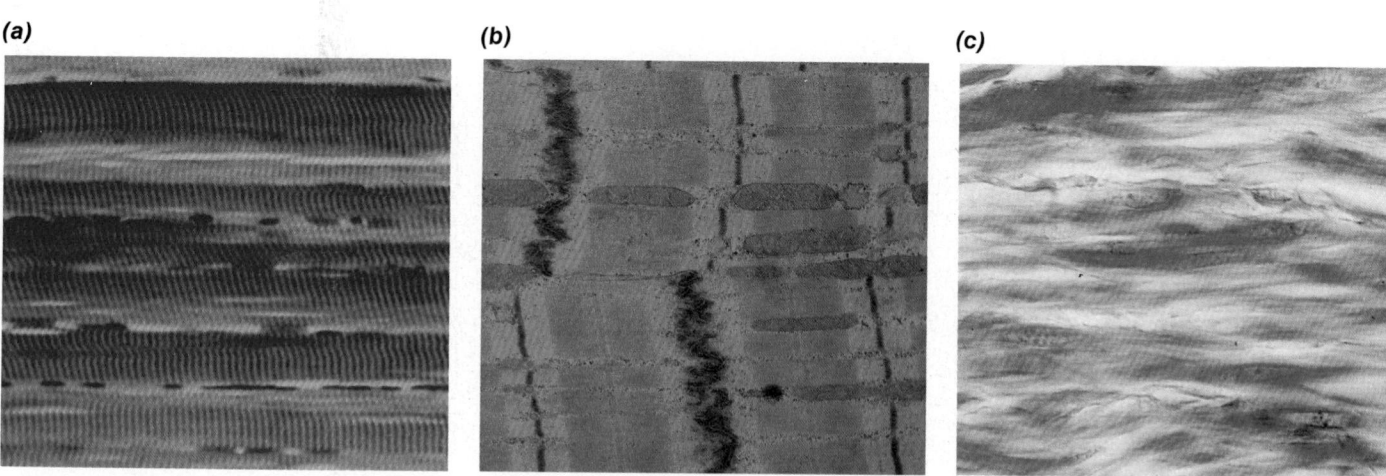

Figure 25-9 The three types of muscle tissue.
(a) Skeletal muscle has a striped or striated appearance in microscopic cross-section, due to the regular alignment of filaments containing actin and myosin.
(b) Cardiac muscle is also striated in appearance. Adjacent cardiac muscle cells are interconnected at areas where their membranes are interdigitated (dark, folded region in micrograph). Within these regions, called intercalated discs, the adjacent membranes are packed with connecting gap junctions.
(c) Smooth muscle lacks the orderly internal structure of skeletal and cardiac muscle.

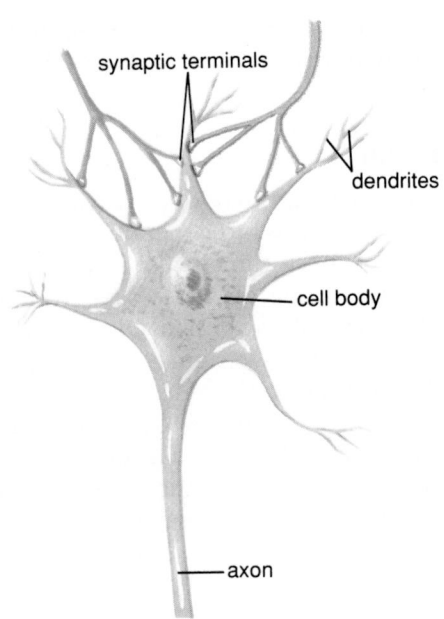

Figure 25-10 A nerve cell, or neuron, has four major parts, each specialized for a specific function.

protective layer of dead cells produced by underlying living epidermal cells. These are packed with the protein **keratin,** which helps keep the skin both airtight and relatively waterproof.

Immediately beneath the epidermis lies a layer connective tissue, the **dermis.** The loosely packed cells of the dermis are invaded by arterioles that feed blood into a dense meshwork of capillaries, which nourish both the dermal and epidermal tissue. Loss of heat through the skin is precisely regulated by neurons controlling the degree of dilation (expansion) of the arterioles. Dilation of the arterioles floods the capillary beds with blood, releasing excess heat, while heat is conserved by constriction of the arterioles supplying the skin capillaries. Lymph vessels collect and carry off extracellular fluid within the dermis. A variety of sensory nerve endings responsive to temperature, touch, pressure, vibration, and pain are scattered throughout the dermis and epidermis. The dermis is also packed with a variety of glands derived from epidermal tissue. **Hair follicles** produce hair from proteinaceous secretions. Sweat glands produce

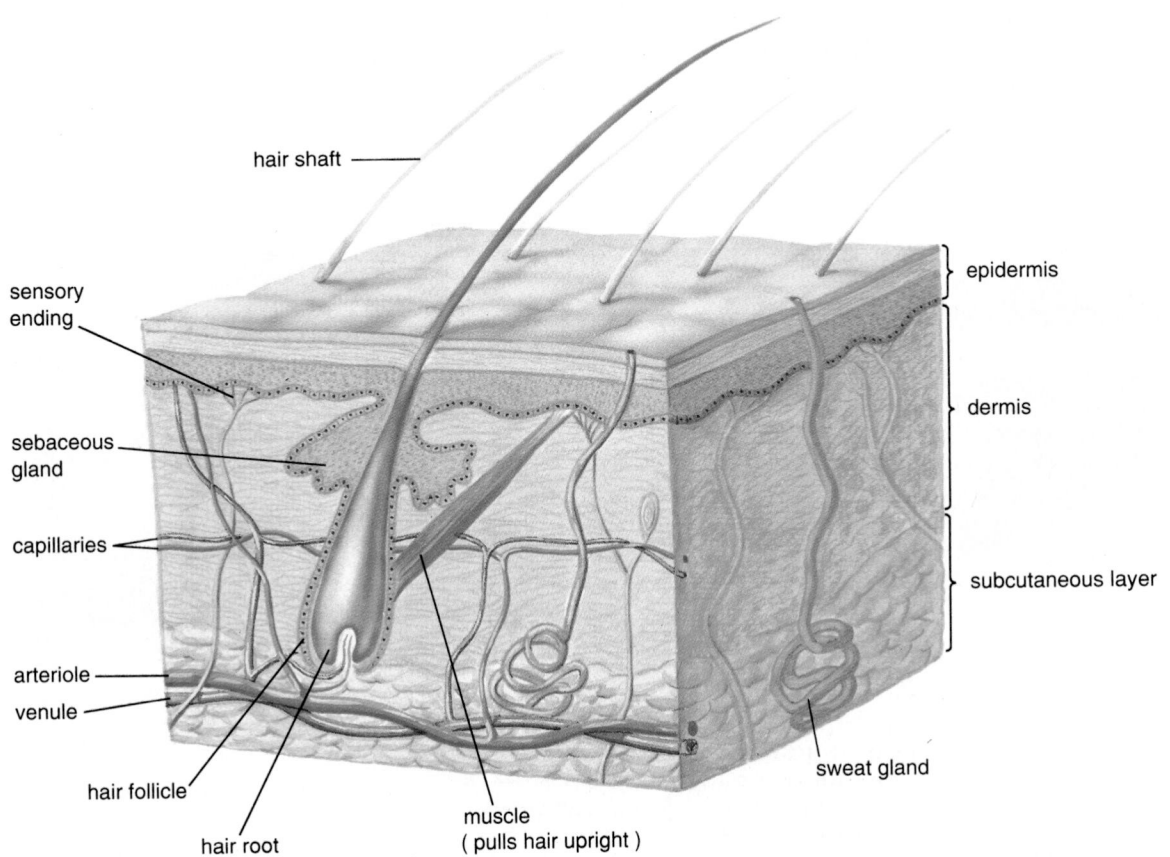

Figure 25-11 Mammalian skin, a representative organ, in cross-section.

watery secretions that cool the skin and excrete substances, including salts and urea. **Sebaceous glands** secrete an oily substance (sebum) that lubricates the epithelium.

In addition to the epithelial, connective, and nerve tissues already mentioned, the skin also contains muscle tissue. Tiny muscles attached to the hair follicles can cause the hairs of the skin to "stand on end" in response to signals from motor neurons. Although this is useless for heat retention in humans, most mammals are able to increase the thickness of their insulating fur in cold weather by erecting the individual hairs.

The structure of the skin is, in a general sense, representative of many organs. An outer epithelium is underlain by connective tissue containing a blood supply, a nerve supply, sometimes muscle, and glandular structures derived from the epithelium. If the organ is hollow, such as the bladder or blood vessels, its interior is also lined with epithelium underlain by connective tissue. Different organs have different types and proportions of glandular, muscular, and nervous tissue.

Animal Organ Systems

Organ systems consist of two or more individual organs (sometimes located in different regions of the body) that work together, performing a common function. The organ systems of the vertebrate body and their representative organs and functions are listed in Table 25-2. The structure and physiology of these organ systems are the subject of Unit V.

Table 25-2 Major Vertebrate Organ Systems

Organ System	Major Structures	Physiological Role
Digestive System	Mouth, esophagus, stomach, small and large intestines, glands producing digestive secretions.	Supplies the body with nutrients that provide energy and materials for growth and maintenance.
Excretory System	Kidneys, ureters, bladder, urethra.	Maintains homeostatic conditions within bloodstream. Filters out cellular wastes, certain toxins, and excess water and nutrients.
Respiratory System	Nose, trachea, lungs (mammals, birds, reptiles, amphibians), gills (fish and some amphibians).	Provides an area for gas exchange between the blood and the environment. Allows oxygen acquisition and carbon dioxide elimination.
Circulatory System	Heart, vessels, blood, lymph, lymphatic nodes and vessels.	Transport of nutrients, gases, hormones, metabolic wastes; also assists in temperature control.
Endocrine System	A variety of hormone-secreting glands including the hypothalamus, pituitary, thyroid, pancreas, and adrenals.	Controls physiological processes, often in conjunction with the nervous system.
Nervous System	Brain, spinal cord, peripheral nerves.	Controls physiological processes in conjunction with the endocrine system. Senses the environment, directs behavior.
Muscular System	Skeletal, smooth, and cardiac muscles.	Moves the skeleton; controls movement of substances through hollow organs (digestive tract, large blood vessels); initiates and implements the heart's contractions.
Skeletal System	Bones, cartilage.	Provides support for the body, attachment sites for muscles, and protection for internal organs.
Reproductive System	Male: testes, seminal vesicles, penis. Female (mammal): ovaries, oviducts, uterus, vagina, mammary glands.	Male: produces sperm, inseminates female Female (mammal): produces egg cells, nurtures developing offspring.

SUMMARY OF KEY CONCEPTS

Homeostasis and the Challenges of Life

Homeostasis describes the tendency of many physiological processes to maintain internal conditions within a narrow range that permits the continuation of life. In general, relatively constant internal conditions are maintained through negative feedback, in which a change initiates a series of events that counteracts the change and restores conditions to a set level. Temperature control as well as many hormone systems use negative feedback to maintain homeostasis. Positive feedback, in which a change initiates events that enhance the change, occurs relatively rarely and is self-limiting. For example, the uterine contractions that lead to childbirth are driven by positive feedback.

All organisms face many of the same challenges, including obtaining food, water, and energy; exchanging gases with the environment; regulating body composition; distributing materials throughout the body; coordinating bodily activities; defending themselves against predators, parasites, and disease; regulating the development of the body; and reproducing. Terrestrial organisms face further problems of water acquisition and loss; gravity stresses; temperature extremes; and protection of gametes and embryos from desiccation. In response to these challenges, plants and animals have evolved adaptations that, while they may be fundamentally similar in principle, are often strikingly different in detail.

The Organization of The Animal Body

The animal body is composed of organs systems that are made up of one or more organs. Organs, in turn, are composed of tissues, which are composed of cells. A tissue is a group of cells and extracellular material that form a structural and functional unit. Animal tissues include epithelial, connective, muscle, and nerve tissue. Epithelial tissue forms membranous coverings over internal and external body surfaces, and also gives rise to glands. Connective tissue usually contains considerable extracellular material, and includes dermal tissue, bone, cartilage, tendons and ligaments, fat, and blood. Muscle tissue is specialized for movement, using sliding filaments of actin and myosin protein. It includes skeletal, cardiac, and smooth muscle. Nerve tissue is specialized for the generation and conduction of electrical signals.

Organs include at least two tissue types that function together, such as mammalian skin. The epidermis, an epithelial tissue, covers and protects the dermis beneath it. The dermis contains blood and lymph vessels, a variety of glands, and tiny muscles that erect the hairs. Animal organ systems include the digestive, excretory, respiratory, circulatory, nervous, muscular, skeletal, endocrine, and reproductive systems, covered in detail in Unit V.

GLOSSARY

adipose tissue (a'-di-pōse): tissue composed of fat cells.

collagen (kol'-uh-gen): a fibrous protein found in connective tissue such as bone and cartilage.

connective tissue: a tissue type consisting of many diverse tissues, which generally include large amounts of extracellular material.

dermis (dur'-mis): the layer of skin lying beneath the epidermis, composed of connective tissue and containing blood vessels, muscles, nerve endings, and glands.

epidermis (ep-uh-der'-mis): specialized epithelial tissue that forms the outer layer of skin.

epithelial tissue (eh-puh-thē'-lē-ul): a tissue type that forms membranes that cover the body surface and line body cavities, and that also gives rise to glands.

fat: fat-storing connective tissue whose cells are packed with triglycerides (fats); also called adipose tissue.

hair follicle: a gland in the dermis of mammalian skin, formed from epithelial tissue, that produces a hair.

homeostasis (hō-mē-ō-stā'-sis): the maintenance of a relatively constant physiological state within an organism.

keratin (ker'-uh-tin): a fibrous protein found in hair, nails, and the epidermis of skin.

membrane: continuous sheets of epithelial cells that cover the body and line body cavities.

negative feedback: a situation in which a change initiates a series of events that tend to counteract the change and restore the original state. Negative feedback in physiological systems maintains homeostasis.

organ: a structure (such as the liver, kidney, or skin) composed of two or more distinct tissue types that function together.

organ system: two or more organs that work together to regulate a particular physiological process.

positive feedback: a situation in which a change initiates events that tend to amplify the original change.

sebaceous gland (se-bā'-shus): a gland in the dermis of skin, formed from epithelial tissue, that produces an oily substance called sebum that lubricates the epidermis.

tissue: a group of cells similar in structure and/or function. The tissue may include extracellular material produced by its cells.

STUDY QUESTIONS

1. Define homeostasis and explain how negative feedback helps maintain it. Explain one example of homeostasis in the human body.
2. Explain positive feedback and provide one physiological example. Explain why this type of feedback is relatively rare in physiological processes.
3. Choose four challenges of life and compare and contrast animal and plant mechanisms for meeting these challenges.

4. Describe the structure and functions of epithelial tissue.
5. What property distinguishes connective tissue from all other tissue types? List five types of connective tissue, and briefly describe the function of each type.
6. Describe the skin, a representative organ. Include the various tissues that comprise it and the role of each tissue.

DISCUSSION QUESTIONS

1. Why does life on land present more difficulties than life in water? What made it evolutionarily advantageous for organisms to colonize dry land?
2. The majority of regulatory mechanisms in animals are "autonomic," not requiring conscious control. Discuss several reasons why this is more advantageous to the animal than conscious regulation of homeostatic controls.

SUGGESTED READINGS

Readings in particular areas of plant and animal physiology are listed at the ends of the appropriate chapters. Here we offer a few interesting articles that explore adaptations to unusual body forms or rigorous environments.

Degabriele, R. "The Physiology of the Koala." *Scientific American*, June 1980 (Offprint No. 1476). Koalas eat poisonous eucalyptus leaves and almost never drink water. How can they survive on such a bizarre diet?

Roger, C. F. E., and Boss, K. J. "The Giant Squid." *Scientific American*, April 1982 (Offprint No. 1515). Yes, there really are giant squid in the ocean depths, doing battle with sperm whales. This article explores new findings on the anatomy, physiology, and ecology of this almost-mythical beast.

Schmidt-Nielson, K. "The Physiology of the Camel." *Scientific American*, December 1959. Do camels really store water in their humps? How long can one go without drinking?

Schmidt-Nielson, K., and Schmidt-Nielson, B. "The Desert Rat." *Scientific American*, July 1953 (Offprint No. 1050). The desert dweller *par excellence*, the kangaroo rat never needs to drink water.

Storey, K. B., and Storey, J. M. "Frozen and Alive." *Scientific American*, December 1990. Many animals have evolved the ability to withstand being frozen solid during the winter.

Went, F. W. "The Ecology of Desert Plants." *Scientific American*, April 1955. Like the kangaroo rat, desert plants must also acquire and conserve water.

UNIT IV

Plant Anatomy and Physiology

26

The Structure of Land Plants

A blue passion flower found in the rain forests of Central America.

efore the first animal crawled out of the sea onto the land, plants already stood there. When the glaciers advanced and retreated across the land, animals followed, not the glaciers themselves, but the plant communities that tracked suitable climates moving back and forth with the ice. The burst of human civilization that occurred about 10,000 years ago depended on the domestication of plants to provide a reliable food supply. As these examples suggest, terrestrial animals, including people, have always intimately depended on terrestrial plants.

Plants are so seemingly commonplace that many people take them for granted. We admire a field of wildflowers or a giant sequoia, but we seldom stop to think about the intricate structures and functions that allow plants to survive and prosper. Plants cannot move to escape enemies, to find food or water, to avoid the onslaught of winter, or to locate a mate. And yet spruces perch atop the permafrost in the Yukon, cypresses and mangroves stand immersed in swamps, and cactuses bloom in the searing heat of Death Valley. Redwoods and eucalyptus trees tower a hundred meters above the forest floor; grasses survive grazing year after year; and bristlecone pines in the White Mountains of California may live to be 4000 years old. Clearly, plants have evolved many extremely successful adaptations that enable them to thrive in a wide variety of habitats.

By far the most abundant land plants are the **flowering plants** (Division Anthophyta, also called **angiosperms**) and the conifers (Division Coniferophyta, which are usually classified as **gymnosperms;** see Chapter 24 for more information about taxonomy and evolutionary relationships among plants). In this Unit, we explore the adaptations that allow these plants to be so successful on land. Because of their overwhelming dominance in most terrestrial habitats and their importance to humans, we will focus on the flowering plants.

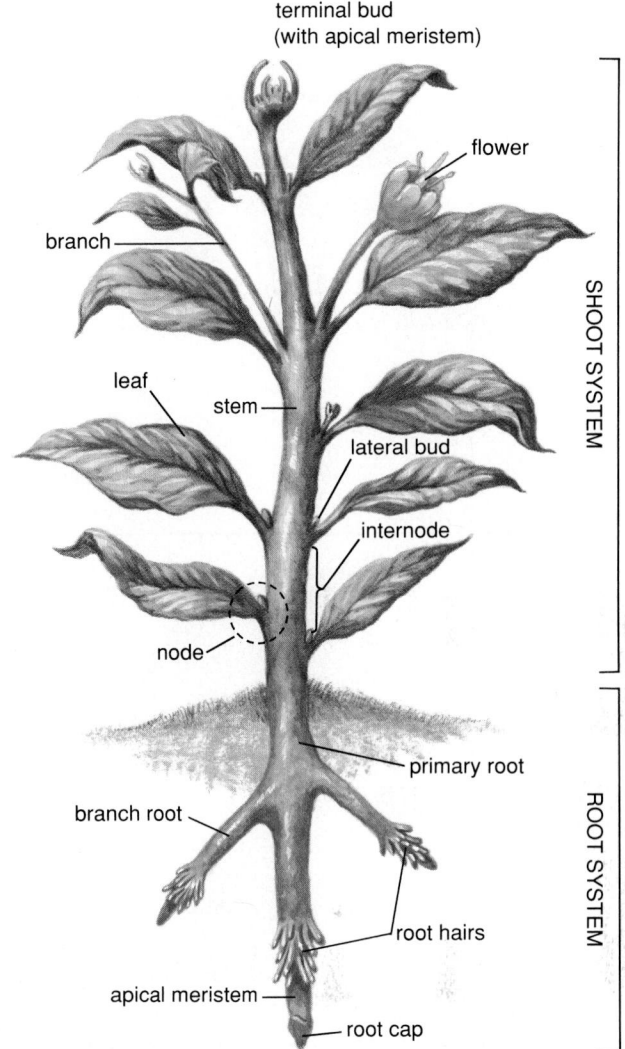

Figure 26-1 A flowering plant consists of shoot and root systems. The shoot system includes stems (often with branches), buds, and leaves. In the appropriate season, the shoot often bears flowers and fruit. The root system is usually highly branched, with many lateral roots growing from one or more main roots.

An Overview of Plant Structure

Flowering plants consist of two major regions, the root system and the shoot system (Fig. 26-1). The **root system** is usually below ground, and serves five functions. Roots (1) anchor the plant in the ground; (2) absorb water and minerals; (3) store surplus sugars manufactured during photosynthesis; (4) transport water, minerals, sugars, and hormones to and from the shoot; and (5) produce some hormones.

The rest of the plant is the **shoot system,** usually found above ground. The shoot system consists of stems (with branches), leaves, buds, and, in the appropriate season of the year, flowers and fruits. The functions of shoots include (1) photosynthesis, mainly in leaves and young green stems; (2) transport of materials between leaves, flowers, fruits, and roots; (3) reproduction; and (4) hormone synthesis.

Figure 26-2 illustrates the two types of flowering plants, **monocots** (Class Monocotyledonae, which includes grasses, lilies, and orchids) and **dicots** (Class Dicotyledonae, which includes deciduous trees and bushes, and most garden flowers). Formally, monocots and dicots are distinguished by the number of

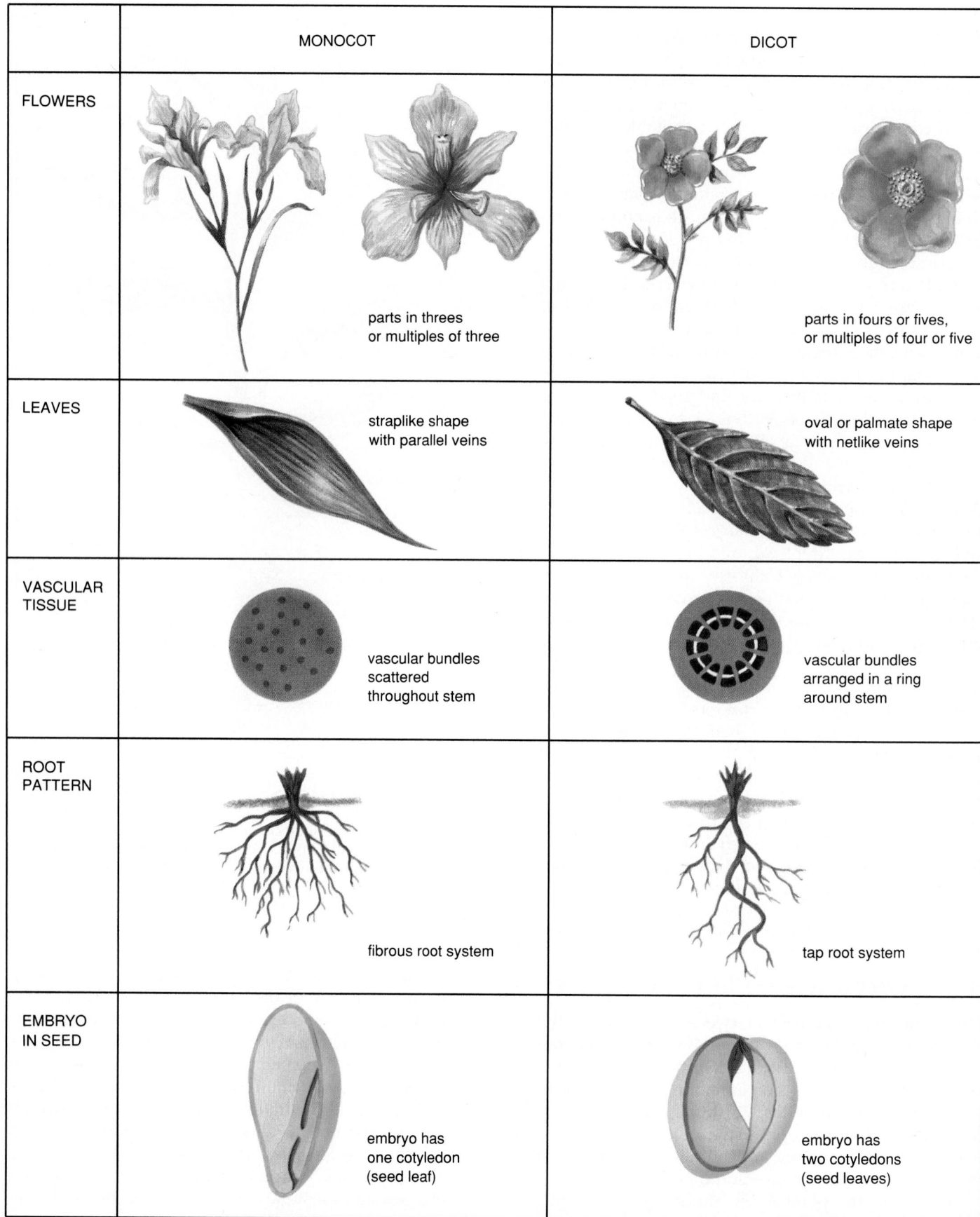

	MONOCOT	DICOT
FLOWERS	parts in threes or multiples of three	parts in fours or fives, or multiples of four or five
LEAVES	straplike shape with parallel veins	oval or palmate shape with netlike veins
VASCULAR TISSUE	vascular bundles scattered throughout stem	vascular bundles arranged in a ring around stem
ROOT PATTERN	fibrous root system	tap root system
EMBRYO IN SEED	embryo has one cotyledon (seed leaf)	embryo has two cotyledons (seed leaves)

Figure 26-2 Distinguishing traits of the two major classes of flowering plants, the monocots and the dicots.

"seed leaves," or cotyledons, found in their seeds (see Chapter 28), but they differ in a variety of other ways as well. Don't worry about terms that are not yet familiar to you; just look over the figure for now, and refer back to it later as we examine the parts of the flowering plants in more detail.

Plant Development

Animals and plants develop in dramatically different ways. One difference is very obvious: the timing and distribution of growth. As you grew from a baby to an adult, all parts of your body became larger. When you reached your adult height, you stopped growing (up, at least!). In contrast, flowering plants grow throughout their lives, never reaching a stable "adult" body form. What's more, most plants grow longer only at the tips of their branches and roots, while structures that developed earlier remain in exactly the same place: the tree branch that you duck your head under during your freshmen year will still be there, head-height, when you are a senior. Why do plants grow this way?

From the moment they sprout, plants are composites of two fundamentally different categories of cells: embryonic, undifferentiated **meristem cells** that are capable of cell division, and mature, **differentiated cells** that are specialized in structure and function, and that usually do not divide (e.g., cells that form the conducting systems of xylem and phloem). Nor-

mally, **each time a meristem cell divides, one daughter cell develops specialized structures and becomes a differentiated cell, while the other daughter remains meristematic** (Fig. 26-3). Continued divisions of meristem cells, then, keep a plant growing throughout its life, while their differentiated daughter cells form relatively permanent parts of the plant, such as mature leaves or the trunks of trees.

Plant Growth

Plants grow through the division and differentiation of meristem cells that are located in two regions of the plant: **apical meristems** at the tips of roots and shoots (including main stems and branches; see Figs. 26-10, 26-14, and 26-16) and **lateral meristems** or **cambia** (sing. **cambium**) forming cylinders that run parallel to the long axis of roots and stems (see Figs. 26-17 and 26-18).

Mitosis of apical meristem cells followed by differentiation of the resulting daughter cells is called **primary growth.** This occurs in young plants and in the growing tips of roots and shoots in older plants, as they simultaneously grow longer and develop the structures essential for life.

The stems and roots of most conifers and dicots become thicker and woody as they age. This is called **secondary growth,** and occurs through mitosis of lateral meristem cells and differentiation of their daughter cells. Although both stems and roots undergo secondary growth, we will discuss secondary growth only in stems.

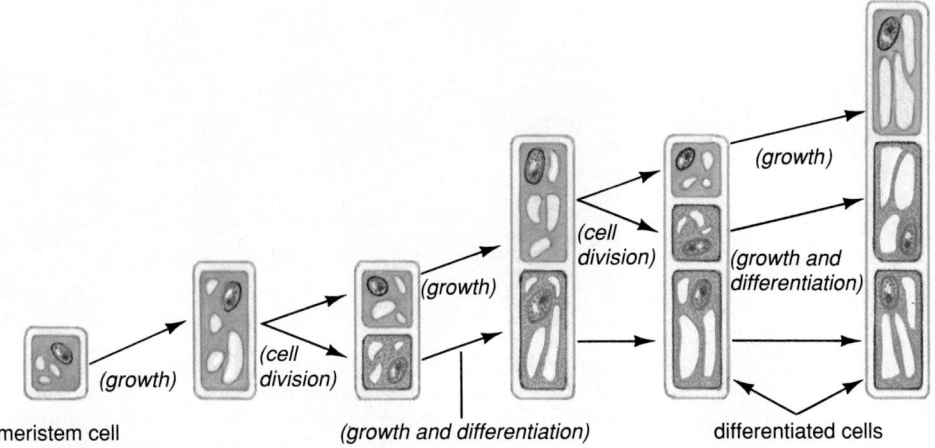

Figure 26-3 Plant development occurs through divisions of meristem cells and differentiation of one of the daughter cells. A meristem cell (green) takes in nutrients, grows, and divides. One daughter cell remains meristematic, while the other differentiates into a particular specialized cell type (tan). You may have noticed that many plants—trees, for example—grow longer only at the tips of their branches. This diagram shows why: at each division, the lower daughter cell becomes differentiated while the upper cell remains meristematic. Therefore, the meristem cells are carried along at the tip of the shoot as the plant grows.

Plant Tissues and Cell Types

The major structures of land plants, including roots, stems, and leaves, consist of three tissue systems: the dermal tissue system, the ground tissue system, and the vascular tissue system (Fig. 26-4). The **dermal tissue system** covers the outer surfaces of the plant body. The **ground tissue system,** which consists of all nondermal and nonvascular tissues, makes up most of the body of young plants; its functions include photosynthesis, support, and storage. The **vascular tissue system** transports water, minerals, sugars, and plant hormones throughout the plant. Each tissue system arises from divisions of meristem cells and differentiation of the daughter cells, and each consists of one or (usually) several types of differentiated cells.

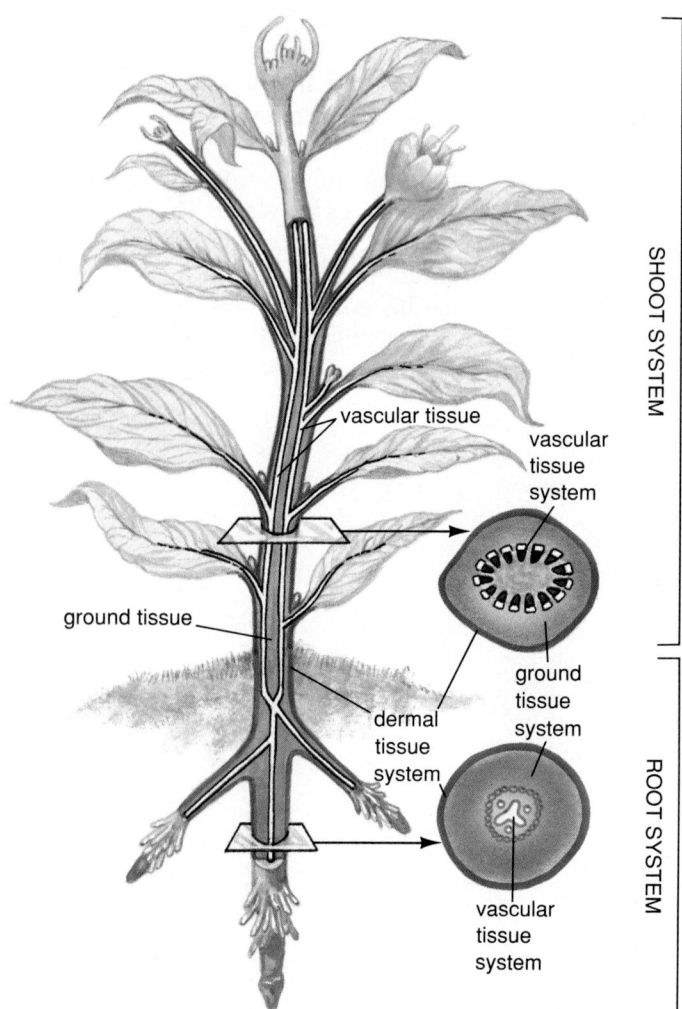

Figure 26-4 Both the root and shoot of a flowering plant consist of three tissue systems: the dermal tissue, ground tissue, and vascular tissue systems.

The Dermal Tissue System

The **dermal tissue system** is the outer covering of the plant body. Two types of tissues comprise this system: epidermal tissue and periderm (Fig. 26-5).

Epidermal tissue covers the leaves, stems, and roots of all young plants (Fig. 26-5a). It also covers flowers, seeds, and fruit. In herbaceous (soft-bodied) plants such as lettuce, grasses, and beans, the epidermis is retained as the outer covering of the entire plant body throughout its life. The epidermal tissue of the above-ground parts of a plant is generally composed of thin-walled cells packed tightly together and

(a)

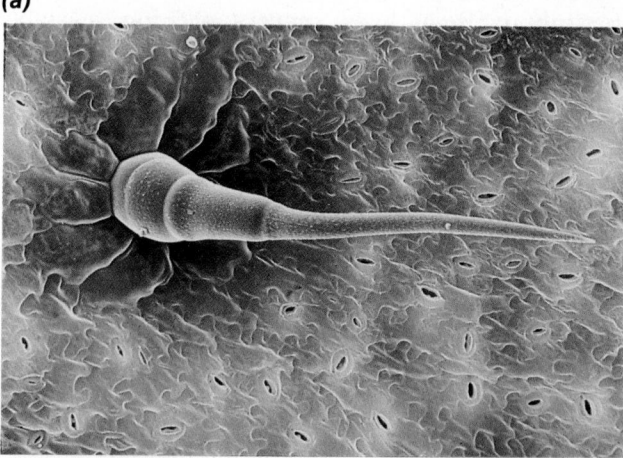

(b)

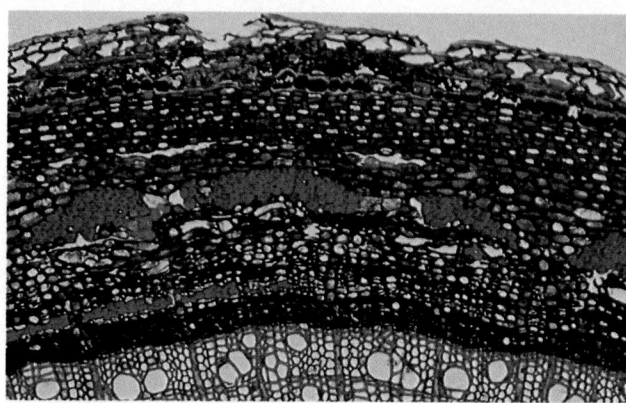

Figure 26-5 Dermal tissues cover the surfaces of plants. **(a)** The epidermis of a young root or shoot is a single layer of cells. In shoot epidermis, such as the epidermis of a zinnia leaf shown here, the outer surfaces of the cells are covered with a waxy waterproof coating, the cuticle, that reduces the evaporation of water. The "leaf hair" protruding out from the epidermis also reduces evaporation by impeding the moving of air across the surface of the leaf. **(b)** Woody stems and roots develop a thick, tough, waterproof periderm, which consists mostly of thick-walled cork cells.

covered with a waterproof, waxy **cuticle** that reduces evaporation of water from the plant. The epidermal cells of roots, in contrast, are not covered with cuticle; roots, of course, absorb water and minerals through the epidermis, and a waterproof cuticle would prevent this.

Some epidermal cells produce fine extensions, called hairs. Many root epidermal cells bear **root hairs,** which greatly increase the surface area of the root in contact with the soil. Epidermal hairs on the stems and leaves of desert plants reflect sunlight and produce an unstirred layer of air close to the epidermis that minimizes water loss.

Periderm replaces epidermal tissue on the roots and stems of woody plants as they age. It is composed primarily of **cork cells** that have thick walls and are dead at maturity (Fig. 26-5b). Cork cells form the outer layers of bark of trees and woody shrubs, and the woody covering of their roots.

The Ground Tissue System

The **ground tissue system,** which makes up the bulk of a young plant, consists of all nondermal and nonvascular tissues. There are three types of ground tissues: parenchyma, collenchyma, and sclerenchyma (Fig. 26-6).

Parenchyma tissue is the most abundant of the ground tissues. Parenchyma cells are thin-walled cells, alive at maturity, that typically carry out most of the metabolic activities of the plant (Fig. 26-6a). Depending on their location within the plant body, parenchyma cells have such diverse functions as photosynthesis, storage of sugars and starches, or secretion of hormones. Under the proper conditions, many parenchyma cells are capable of cell division.

Collenchyma tissue consists of elongated, polygonal (many-sided) cells with irregularly thickened cell walls (Fig. 26-6b). Collenchyma cells are alive at maturity, but usually cannot divide. Although strong, the cell walls of collenchyma are still somewhat flexible. In herbaceous plants, and in the leaf petioles and young growing stems of all plants, collenchyma tissue is an important source of support. The strings in celery stalks, for example, consist primarily of collenchyma cells in association with vascular tissue.

Sclerenchyma tissue consists of cells with thick, hardened secondary cell walls (Fig. 26-6c). Like collenchyma, sclerenchyma tissue acts to support and strengthen the plant body, but unlike collenchyma, sclerenchyma cells die as the last stage of differentiation. Sclerenchyma tissue can be found in many parts of the plant body, including xylem and phloem. Sclerenchyma cells comprise the fibrous portions of hemp and jute, which are used for making rope. Other types of sclerenchyma cells form nut shells, the outer covering of peach pits, and the gritty texture of pears.

The Vascular Tissue System

The **vascular tissue system** consists of two complex tissues: **xylem** and **phloem.** The major role of each tissue is the transport of materials (water and minerals in xylem, water and sugars in phloem) throughout the plant body.

Xylem

Xylem conducts water and minerals in tubes that are made from one of two types of cells: tracheids and vessel elements (Fig. 26-7). Most conifers have only tracheids, while flowering plants usually have both tracheids and vessel elements. As these cells differentiate, they develop thick cells walls that help to support the weight of the plant. In some trees (pines, for example), the bulk of the tree trunk consists of the thick cell walls of tracheids. The final step in the differentiation of both tracheids and vessel elements is death: the cytoplasm and plasma membrane disintegrate, leaving behind a hollow tube of cell wall.

Tracheids are thin cells with slanted ends like the tips of hypodermic needles. Tracheids are stacked atop one another with the slanted ends overlapping (see Fig. 26-7). The overlapping walls contain **pits** where secondary cell walls failed to form, so water and minerals can pass from one tracheid to the next by crossing only the thin and water-permeable primary cell wall.

Just as tracheids resemble double-ended needles, **vessel elements** resemble soup cans: larger in diameter, with blunt ends. Further, complete perforations form in the ends of abutting vessel elements, with both the primary and secondary cell walls disappearing. In some cases, the ends almost completely disintegrate, making an open pipe the same way that plumbers do, by welding together a series of open-ended cylinders (see Fig. 26-7). Vessel elements, then, form large-bore, relatively unobstructed pipelines from root to leaf.

Phloem

Phloem carries concentrated sugar solutions through tubes constructed of cells called **sieve-tube elements** (Fig. 26-8). As sieve-tube elements mature, most of their internal contents disintegrate, leaving behind only a thin rind of cytoplasm lining the plasma membrane. At the ends of sieve-tube elements, where adjacent cells meet, holes form in the cell walls, creating **sieve plates.** The plasma membranes of the two sieve-tube elements are fused around the lips of the

(a)

thin primary
cell wall

(b)

(c)

20 micrometers

20 micrometers

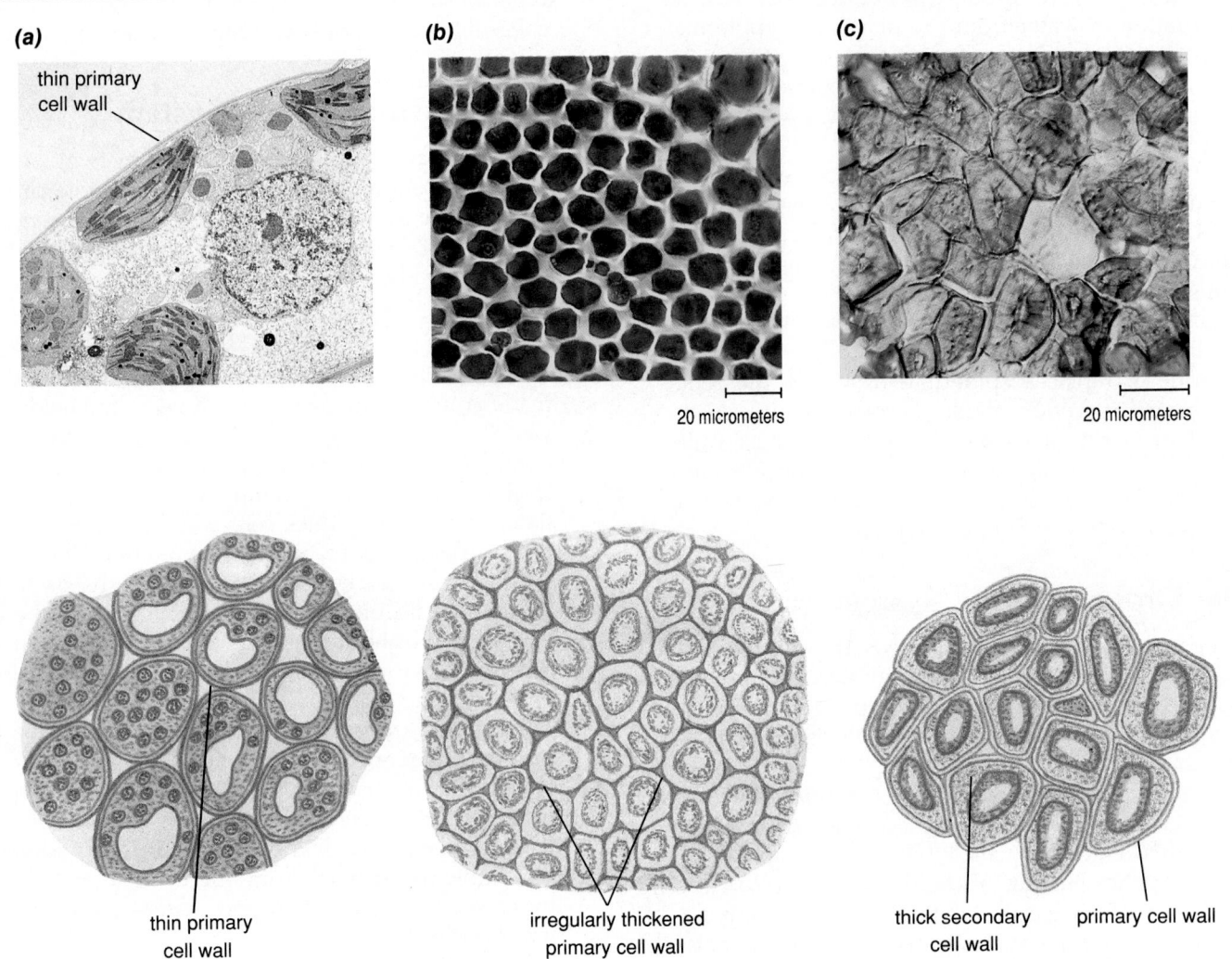

thin primary
cell wall

irregularly thickened
primary cell wall

thick secondary
cell wall

primary cell wall

Figure 26-6 Ground tissue consists of three cell types: parenchyma, collenchyma, and sclerenchyma.
(a) Parenchyma cells are living cells with thin, flexible cell walls. These parenchyma cells have abundant chloroplasts and carry out photosynthesis. Other parenchyma cells, particularly in stems and roots, store starches (see the potato cells in Fig. 3-4a).
(b) Collenchyma cells, such as these from a celery stalk, are living cells with irregularly thickened, but still somewhat flexible, walls. They help to support the plant body.
(c) Sclerenchyma cells, such as these "stone cells" in a pear, have thick, rigid secondary cell walls.

sieve plate pores, forming membrane-lined channels connecting the interiors of the two cells. A continuous conducting system is forged by many sieve-tube elements linking up end-to-end in this way.

A sieve-tube element has a plasma membrane, a few small mitochondria, and some endoplasmic reticulum, and is therefore considered to be alive, but it usually lacks ribosomes, Golgi apparatus, and a nucleus. How, then, can a sieve-tube element remain alive? *Each sieve-tube element is nourished by a smaller, adjacent* **companion cell.** Companion cells maintain the integrity of the sieve-tube elements by donating high-energy compounds and perhaps even by repairing the sieve-tube plasma membrane. As we will see in Chapter 27, *companion cells also regulate the movements of sugars into and out of the sieve tubes.*

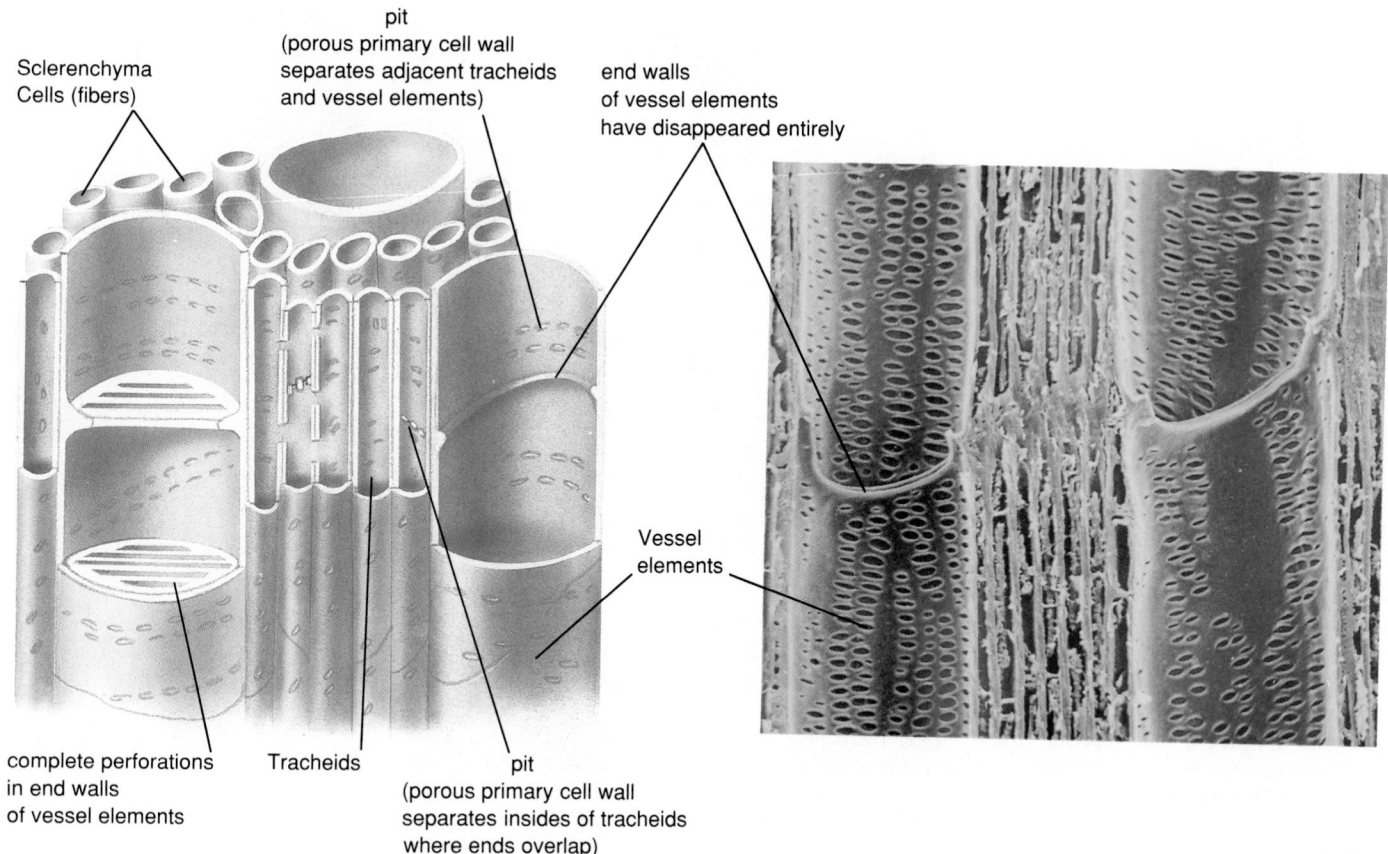

Sclerenchyma
Cells (fibers)

pit
(porous primary cell wall
separates adjacent tracheids
and vessel elements)

end walls
of vessel elements
have disappeared entirely

Vessel
elements

complete perforations
in end walls
of vessel elements

Tracheids

pit
(porous primary cell wall
separates insides of tracheids
where ends overlap)

Figure 26-7 Xylem is a mixture of cell types, including conducting cells, parenchyma and sclerenchyma. There are two types of conducting cells, tracheids and vessel elements. Tracheids are thin, with overlapping ends punctuated by pits. Notice that the pits are not holes all the way through from one cell to the next, but still have a primary cell wall separating the interiors of the two cells. The primary cell wall is very water permeable, and doesn't seriously impede water flow between tracheids. Vessels consist of vessel elements stacked atop one another. In many vessel elements, the end walls virtually disappear. Both tracheids and vessel elements have pits in their side walls, allowing water and dissolved minerals to move sideways between adjacent conducting tracts.

Roots: Anchorage, Absorption, and Storage

As a seed sprouts, the **primary root** grows down into the soil. Many dicots, such as carrots and dandelions, develop a **taproot system,** which consists of the primary root, which usually becomes longer and stouter with time, and many smaller lateral roots that grow out from the primary root (Fig. 26-9a). In monocots such as grasses, on the other hand, the primary root soon dies off, replaced by many new roots that emerge from the base of the stem. These secondary roots are nearly equal in size, forming a **fibrous root system** (Fig. 26-9b).

Primary Growth in Roots

In young roots of both tap and fibrous systems, divisions of the apical meristem give rise to four anatomically and functionally distinct regions (Fig. 26-10). At the very tip of the root, daughter cells produced on the "soil side" of the apical meristem differentiate into the root cap. The root cap protects the apical meristem from being scraped off as the root pushes down between the rocky particles of the soil. Root cap cells have thick cell walls and secrete a slimy lubricant that helps to ease the way between soil particles. Nevertheless, root cap cells wear away, and must be continuously replaced by new cells from the meristem.

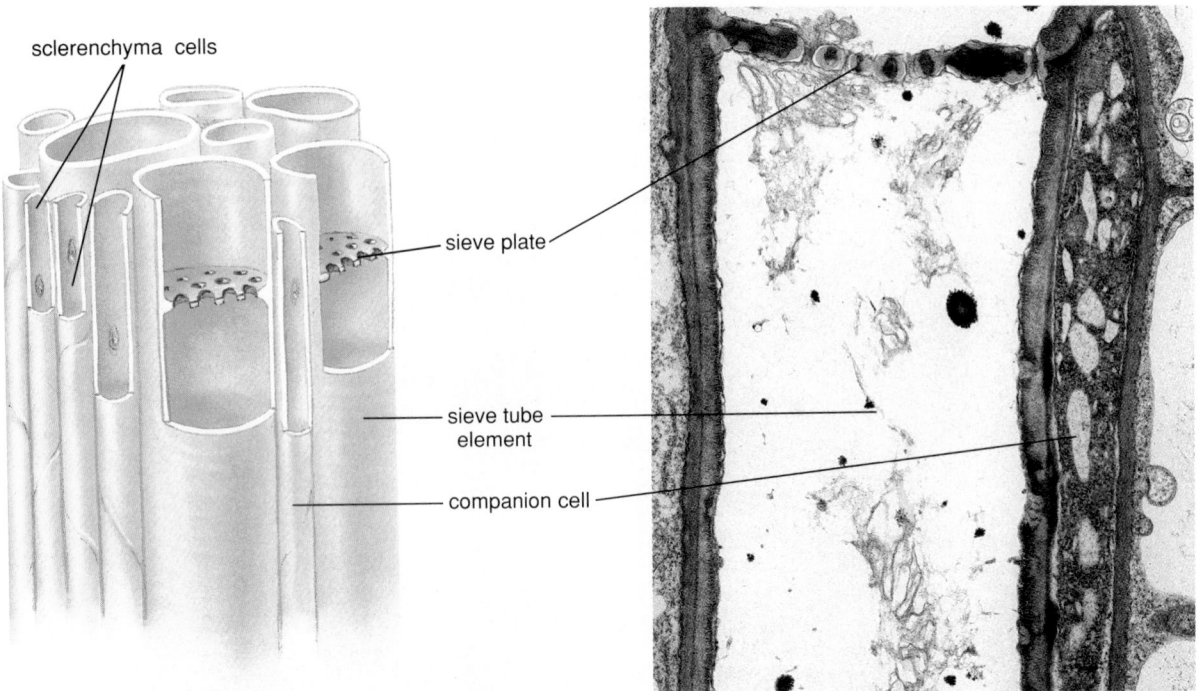

sclerenchyma cells

sieve plate

sieve tube element

companion cell

Figure 26-8 Phloem is a mixture of cell types, including parenchyma, sclerenchyma, sieve-tube elements, and companion cells. Sieve-tube elements contain mostly fluid, with only a rind of cytoplasm lining the cell membrane. Sieve-tube elements, stacked end to end, form the conducting system of phloem. Where they join, sieve-tube elements form sieve plates, where membrane-lined pores allow fluid to pass from cell to cell. Each sieve-tube element has a companion cell that nourishes it and regulates its function.

(a)   *(b)*

Figure 26-9 Typical root systems in dicots and monocots. **(a)** Dicots often have a taproot system, consisting of a long central root with many smaller, secondary roots branching from it. **(b)** Monocots usually have a fibrous system, with many roots of equal size.

Daughter cells produced on the "shoot side" of the apical meristem differentiate into an outer envelope of **epidermis,** a **vascular cylinder** at the core of the root, and, between the two, the **cortex** (Fig. 26-10).

Epidermis

The outermost covering of cells is the epidermis, which is in contact with the soil and any air or water trapped among the soil particles. The cell walls of the epidermal cells are highly water-permeable. Therefore, water can penetrate into the interior of the root either by passing through the membranes of the epidermal cells or by passing between cells of the epidermis, through the porous cell walls. Many epidermal cells grow long projections, called **root hairs,** into the surrounding soil (Fig. 26-11). By increasing the surface area, root hairs increase the root's ability to absorb water and minerals. Root hairs may add dozens of square meters of surface area to the roots of even small plants.

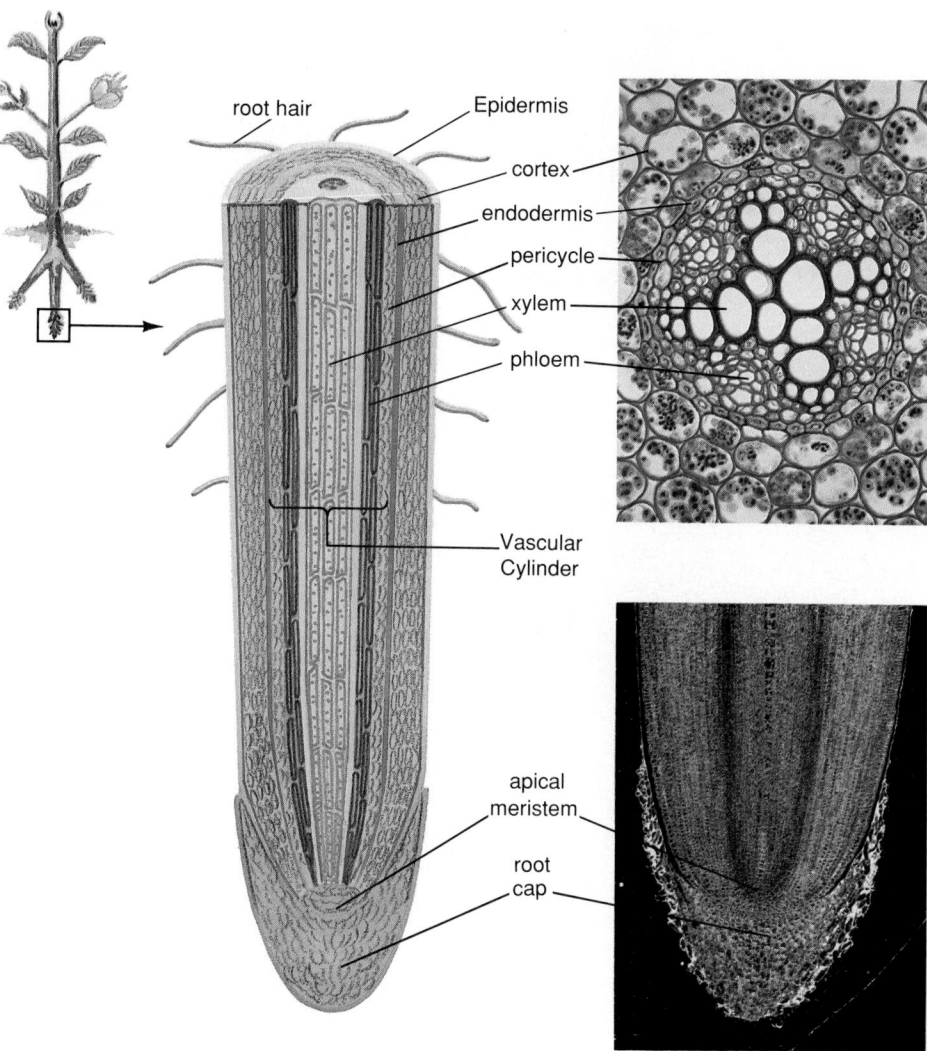

Figure 26-10 Primary growth in roots results from cell divisions in the apical meristem, located near the tip of the root, and subsequent elongation of daughter cells as they differentiate into epidermis, cortex, and vascular cylinder. The root cap forms at the very tip of the root, also from divisions of cells in the apical meristem.

Cortex

Cortex occupies most of the inside of a young root. The cortex consists of two very different regions: an outer mass of large, loosely packed parenchyma cells just beneath the epidermis, and an inner layer of smaller, close-fitting cells that form a ring around the vascular cylinder, called the **endodermis** (see Fig. 26-10). Sugars produced by photosynthesis in the shoot are transported down to the parenchyma cells of the cortex, where they are converted to starch and stored. These cells are particularly abundant in roots specialized for carbohydrate storage, such as the thick roots of carrots and dandelions.

The endodermis is a layer of cells with highly specialized cell walls (Fig. 26-12). Where endodermal

Figure 26-11 Root hairs, shown here in a germinating radish, greatly increase a root's surface area for the absorption of water and minerals from the soil.

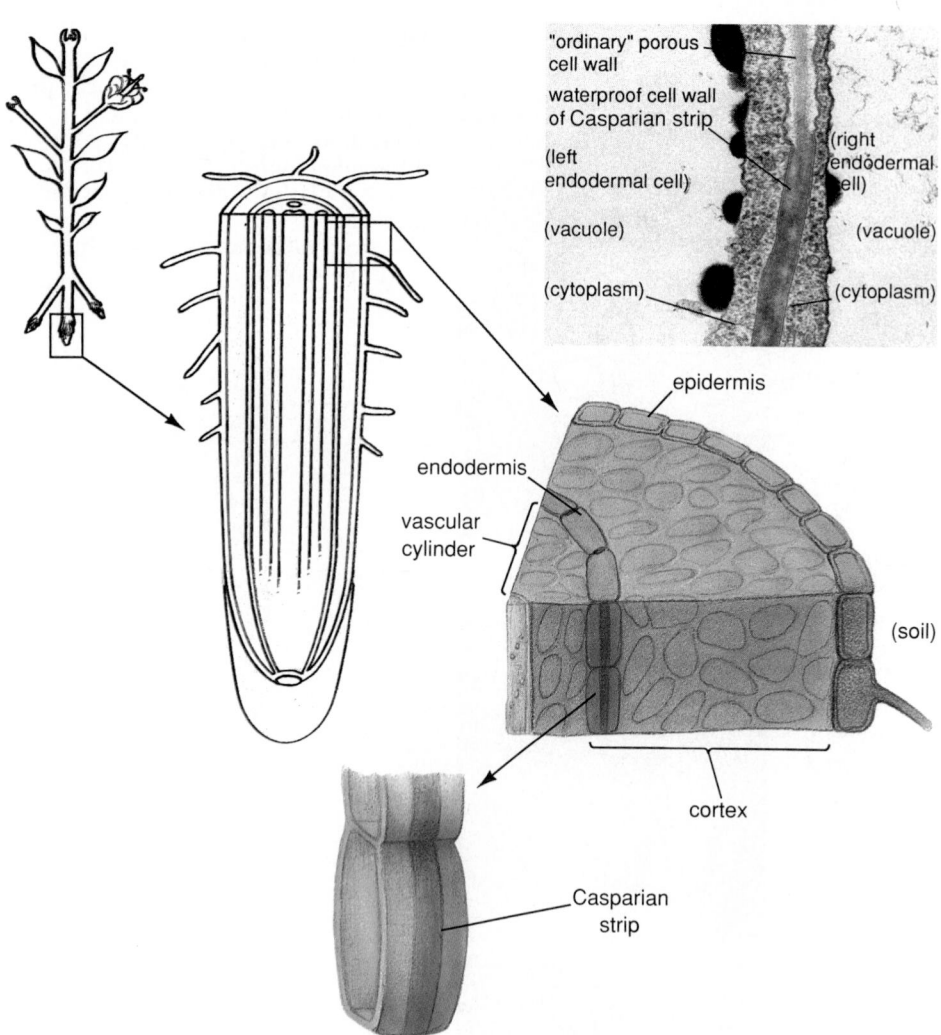

"ordinary" porous cell wall

waterproof cell wall of Casparian strip

(left endodermal cell)

(right endodermal cell)

(vacuole)

(vacuole)

(cytoplasm)

(cytoplasm)

Figure 26-12 The Casparian strip is a band of waterproof material in the walls between cells of the endodermis. In three dimensions, the Casparian strip is arranged like the mortar in a brick wall, sealing off the top, bottom, and sides of the endodermal cells, thereby preventing water movement between the cells.

epidermis

endodermis

vascular cylinder

(soil)

cortex

Casparian strip

cells contact each other, their cell walls are impregnated with a waxy material, forming the **Casparian strip.** In three dimensions, the Casparian strip resembles the mortar in a brick wall: the waxy waterproofing covers the top, bottom and sides of the endodermal cells, but not the outer surfaces (facing the rest of the cortex) or inner surfaces (facing the vascular cylinder). Although water and dissolved minerals can travel around both epidermal and cortex parenchyma cells by moving through their porous cell walls, the waxy Casparian strip effectively seals off the vascular cylinder from the outer parts of the root. As a result, water or minerals penetrating into the vascular cylinder must pass through the living parts of the endodermal cells, which regulate the types and amounts of materials that the roots can absorb. We will see how this works in Chapter 27.

Vascular Cylinder

The vascular cylinder contains the conducting tissues of xylem and phloem that transport water and dissolved materials within the plant. The outermost layer of the vascular cylinder, the pericycle, is a remnant of meristem that retains the capacity of cell division. Under the influence of plant hormones, pericycle cells divide and form the apical meristem of a **branch root** (Fig. 26-13). Branch root development is similar to primary root development except that the branch must first break out through the cortex and epidermis of the primary root. It does this partly by crushing the cells that lie in its path, and partly by secreting enzymes that digest them away. The vascular tissues of the branch root connect up with the vascular tissues of the primary root.

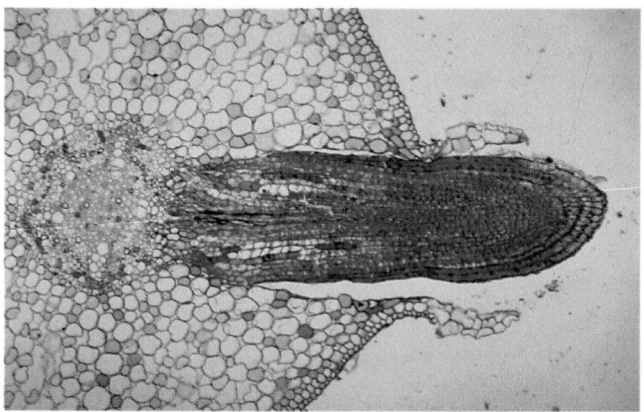

Figure 26-13 Branch roots emerge from the pericycle of a root. Notice that the central axis of the branch is already differentiating into vascular tissue.

Stems: Reaching for the Light

Primary Growth and the Structure of Stems

Like roots, stems develop from a small group of actively dividing cells, the **apical meristem,** that lies at the tip of the young shoot. The daughter cells of the apical meristem differentiate into the specialized cell types of stem, buds, leaves, and flowers.

Surface Structures of the Stem

As the shoot grows, small clusters of meristem cells are "left behind" at the surface of the stem. These meristem cells form the **leaf primordia** and **lateral buds** that appear at characteristic locations, called **nodes,** on the stem; regions of stem between these nodes are called **internodes** (Fig. 26-14). Leaf primordia develop into the mature leaves typical of the species of plant. Under appropriate conditions, lateral buds grow into branches. We will discuss the growth of branches shortly.

The Internal Organization of the Primary Stem

Most young stems are composed of four tissues: **epidermis** (dermal tissue), **vascular tissues, cortex,** and **pith** (the last two are both ground tissues). As Figure 26-15 illustrates, monocots and dicots differ somewhat in the arrangement of vascular tissues. We will discuss only dicot stems here.

Epidermis

You will recall that the epidermis of a young root is the primary pathway for water absorption. In the stem (and leaves), the epidermis is exposed to dry air, and is therefore a potential pathway for water loss, not gain. Epidermal cells of the stem, unlike those of the root, secrete a waxy covering, the **cuticle,** that reduces evaporation of water out of the stem. Unfortunately, the cuticle also reduces the diffusion of oxygen and carbon dioxide into and out of the plant. The epidermis, however, is often perforated with adjustable pores called **stomata** (sing. **stoma**) that regulate the diffusion of oxygen, carbon dioxide, and water vapor into and out of the stem. We will have more to say about stomata when we discuss the function of leaves in the next chapter.

Cortex and Pith

Cortex (located between the epidermis and vascular tissues) and pith (inside the vascular tissues at the center of the stem) are similar in most respects; in fact, in some stems it is difficult to tell where cortex ends and pith begins. Cortex and pith perform three major functions: support, storage, and, in some cases, photosynthesis.

1. *Support:* In very young stems, water filling the central vacuoles of cortex and pith cells causes turgor pressure (see Chapter 6). Turgor pressure pushes the cytoplasm up against the cell wall, stiffening the cells much like air inflates a tire. Just as an underinflated tire goes flat, lack of water causes the cells to go limp; if you forget to water your houseplants, their drooping tips show the importance of turgor pressure in keeping young stems erect. Somewhat older stems also have collenchyma or sclerenchyma cells with thickened cell walls. Because of their strong cell walls, these cells don't depend on turgor pressure for strength. That's why only the tips of a wilted plant droop: collenchyma and sclerenchyma cells support the older regions of the stem.
2. *Storage:* Parenchyma cells in both cortex and pith convert sugar into starch and store the starch as a food reserve.
3. *Photosynthesis:* In many stems, the outer layers of cortex cells contain chloroplasts and carry out photosynthesis. In some desert plants, such as cacti, the leaves are reduced or absent, and the cortex of the stem is the only green photosynthetic part of the plant.

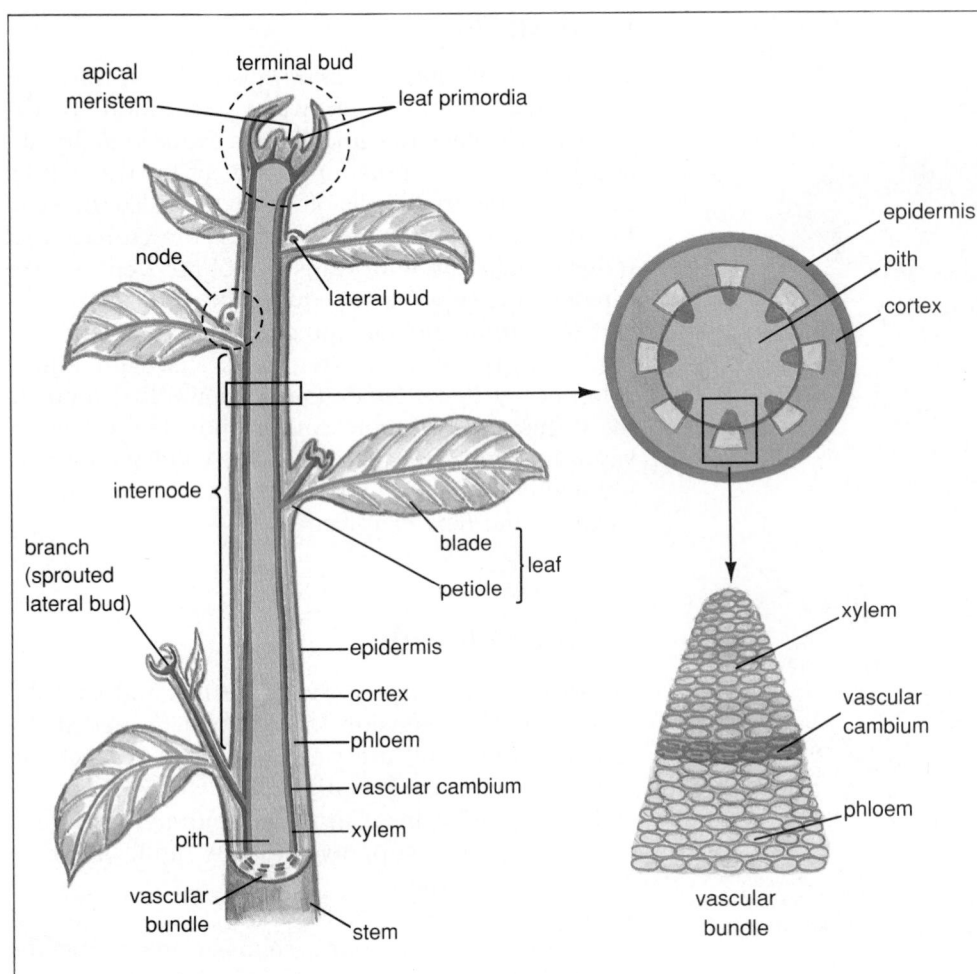

Figure 26-14 The structure of a young dicot shoot. At the tip of the stem, the terminal bud includes the apical meristem and several leaf primordia, produced by the meristem. Other daughter cells of the apical meristem differentiate into epidermis, cortex, pith, and vascular tissues. As the young stem grows, the leaf primordia develop into mature leaves. Meanwhile, epidermis, cortex, pith, and vascular cells elongate between the points of attachment of leaf to stem, effectively pushing the leaves apart. A remnant of meristem tissue, called a lateral bud, remains in the crotch between each leaf and the stem. Lateral buds may sprout into branches (see Fig. 26-16). Points on the shoot where leaves and lateral buds are located are called nodes; the naked stem between nodes is an internode.

Vascular Tissues

As in roots, the vascular tissues of stems transport water, minerals, sugars, and hormones. Vascular tissues are continuous in root, stem, and leaf, interconnecting all the parts of the plant. The **primary xylem** and **primary phloem** found in young stems arise from the apical meristem. In young dicot stems, the xylem and phloem are usually arranged as concentric cylinders (see Fig. 26-15a) or as a ring of bundles running up the stem, with each bundle containing both phloem and xylem (see Fig. 26-15b). Secondary growth in dicot stems, as we will discuss below, always results in concentric cylinders of xylem and phloem (see Fig. 26-17).

Stem Branching

A lateral bud is a cluster of dormant meristem cells left behind by the apical meristem as the stem grew. Lateral buds are located on the surface of the stem,

just above the attachment points of the leaves (see Fig. 26-14). When stimulated by the appropriate hormones (see Chapter 29), the meristem cells of a lateral bud break out of dormancy and the bud sprouts, growing into a branch (Fig. 26-16). As the meristem cells divide, they release hormones that change the developmental fate of the cells between the bud and the vascular tissues of the stem. Parenchyma cells of the cortex differentiate into xylem and phloem, ultimately connecting up with the main vascular systems in the stem. As the branch grows, it duplicates the development of the stem: it has an apical meristem at its tip and leaves behind its own nodes of leaf primordia and lateral buds as it grows.

Secondary Growth in Stems

In conifers and perennial dicots, stems last for several years, becoming thicker and stronger each year. This **secondary growth** in stem thickness results from cell

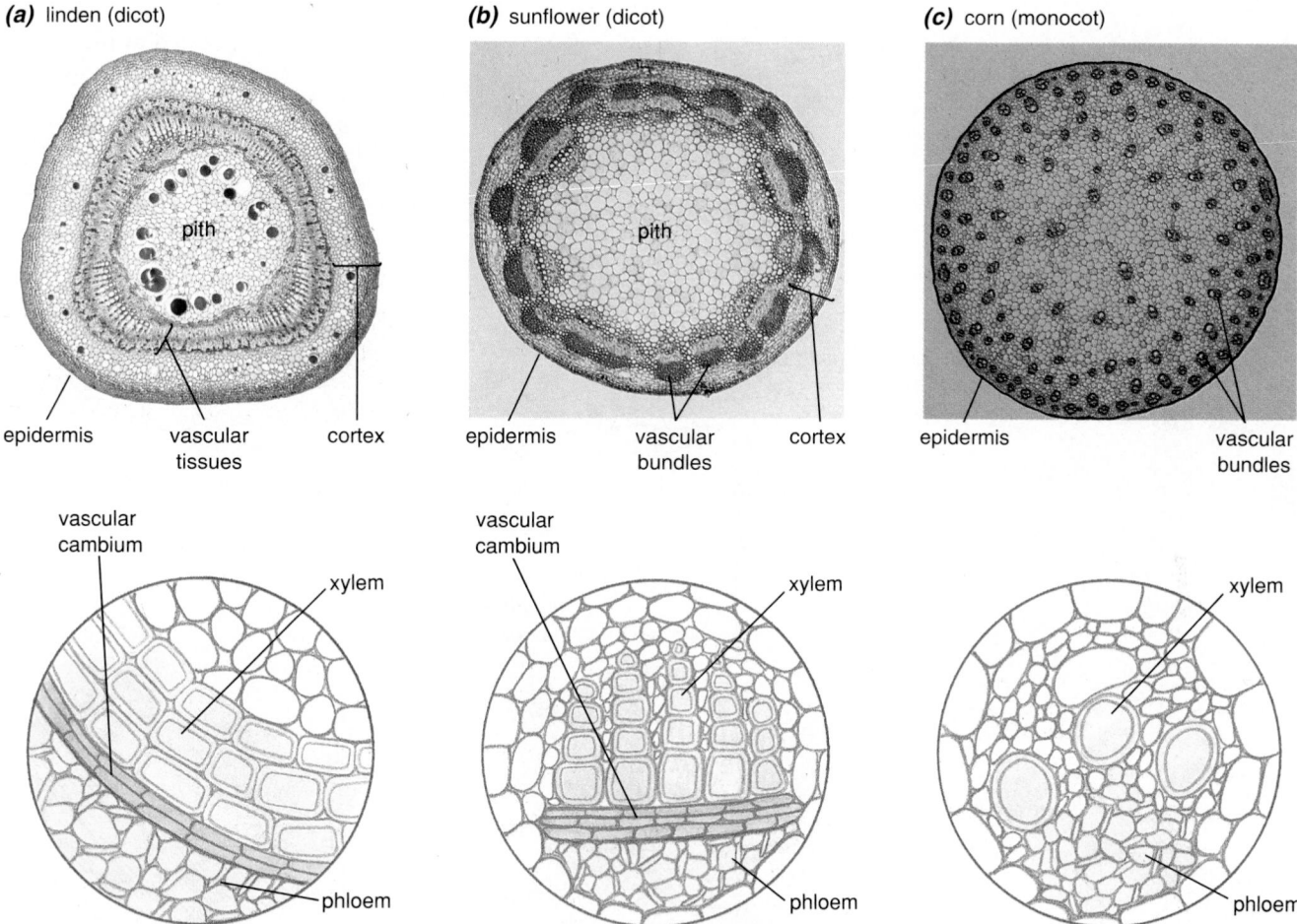

(a) linden (dicot)

(b) sunflower (dicot)

(c) corn (monocot)

Figure 26-15 Dicots **(a, b)** and monocots **(c)** have different arrangements of vascular tissues in their stems. Dicots have vascular tissues arranged in a cylinder surrounding a central "filler" of pith. The linden stem **(a)** has a continuous band of xylem on the inside lined with phloem on the outside, while the sunflower **(b)** has a beaded "bracelet" of bundles. Within each bundle of the sunflower bracelet, xylem lies to the inside and phloem to the outside. Monocots such as corn **(c)** have vascular bundles scattered throughout the stem.

division in lateral meristems known as the **vascular cambium** and **cork cambium** (Fig. 26-17).

Vascular Tissues

The **vascular cambium** is a cylinder of meristem cells located between the primary xylem and primary phloem. Daughter cells of the vascular cambium produced toward the inside of the stem differentiate into **secondary xylem,** while those produced toward the outside of the stem differentiate into **secondary phloem** (see Fig. 26-17). Since the center of the stem is already filled with pith and primary xylem, newly formed secondary xylem pushes the vascular cambium and all outer tissues further out, increasing the

diameter of the stem. The secondary xylem, with its thick cell walls, forms the wood that makes up most of the trunk of a tree. Young xylem (the sapwood just inside the vascular cambium) transports water and minerals; older xylem (the heartwood nearest the pith) only contributes to the strength of the trunk.

Phloem cells are much weaker than xylem. As they die with age, the sieve-tube elements and companion cells are crushed between the hard xylem on the inside of the trunk and the tough cork on the outside (see below). Only a thin strip of recently formed phloem remains alive and functioning.

In trees adapted to temperate latitudes, such as oaks and pines, cell division in the vascular cambium ceases during the cold of winter. In spring, the cam-

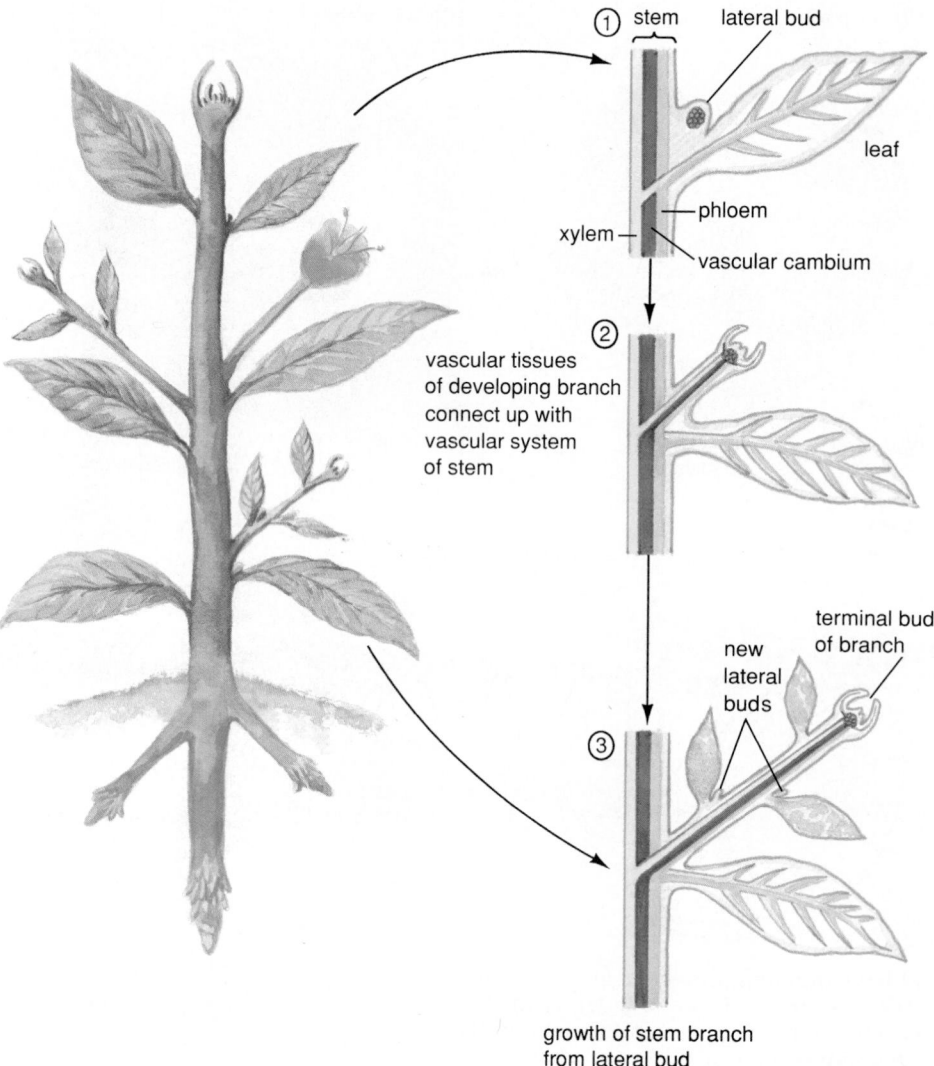

① stem   lateral bud

leaf

xylem ─   ─ phloem

vascular cambium

Figure 26-16 Stem branches grow from lateral buds located at the outer surface of a stem. The bud apical meristem generates an outward-growing branch, replicating the pattern of nodes and internodes characteristic of the type of plant. Meanwhile, cortex cells beneath the sprouting bud differentiate into vascular tissues and connect up with the vascular system of the stem.

② vascular tissues of developing branch connect up with vascular system of stem

terminal bud of branch

new lateral buds

③

growth of stem branch from lateral bud

bium cells divide to form new xylem and phloem. The young cells grow by absorbing water and swelling while the newly formed cell walls are still soft. As the cells mature, the cell walls thicken and harden, preventing further growth. Water is readily available in spring; therefore, young xylem cells take up a lot of water, swell considerably, and thus are large when mature. As the summer progresses and water becomes scarcer, new xylem cells cannot absorb as much water, and are smaller when they mature. As a result, early (spring) wood is pale (due to a thin layer of dark cell wall surrounding a large pale interior) while late (summer) wood is dark (with a thick cell wall and very little interior space; Fig. 26-18). This pattern of alternating light and dark xylem forms the familiar **annual rings** of growth in temperate trees.

Surface Tissues

You will recall that epidermal cells are mature, differentiated cells that no longer retain the capacity to divide. Therefore, as new xylem and phloem are added each year, enlarging the stem, the epidermis can't expand to keep up with the increasing circumference. The epidermis splits off and dies. Apparently prodded by hormones, some parenchyma cells in the cortex become rejuvenated and form a new lateral meristem, the **cork cambium** (see Fig. 26-17). These cells divide, forming daughter cells toward the outside of the stem. These daughter cells, called **cork,** develop tough, waterproof cell walls that protect the trunk from desiccation and abuse. Cork cells die as they mature, and may form a protective layer a foot thick

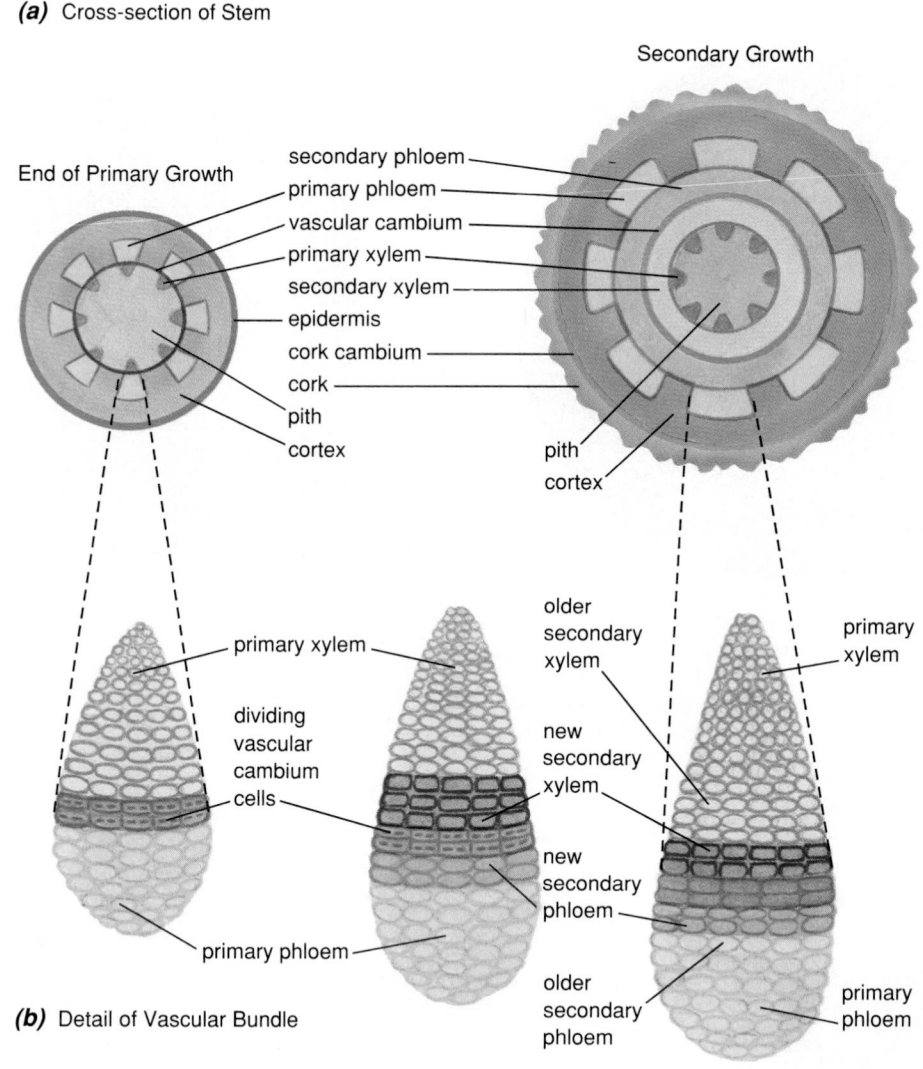

(a) Cross-section of Stem

End of Primary Growth

Secondary Growth

- secondary phloem
- primary phloem
- vascular cambium
- primary xylem
- secondary xylem
- epidermis
- cork cambium
- cork
- pith
- cortex

pith
cortex

Figure 26-17 Secondary growth in a dicot stem. **(a)** Cross-section of a dicot stem at the end of primary growth (left) and during early secondary growth (right). **(b)** Anatomical details of a vascular bundle during secondary growth. A vascular cambium forms between the primary xylem and primary phloem. When vascular cambium cells divide, daughter cells formed on the inside of the vascular cambium differentiate into xylem, while cells formed on the outside of the cambium differentiate into phloem. Note that, since xylem and pith already fill the inside of the stem, newly formed secondary xylem forces the cambium, phloem, and all outer tissues further out, increasing the diameter of the stem. The cork cambium produces cork cells that cover the outside of the stem.

- primary xylem
- dividing vascular cambium cells
- primary phloem

older secondary xylem
new secondary xylem
new secondary phloem
older secondary phloem

primary xylem
primary phloem

(b) Detail of Vascular Bundle

on some tree species, such as the fire-resistant sequoia (Fig. 26-19). As the trunk expands from year to year, the outermost layers of cork split apart or peel off, accommodating the growth. (The "cork" in the neck of a wine bottle is part of the outermost layer of cork from a certain type of oak, carefully peeled off by harvesters.)

The common term **bark** includes all the tissues outside the vascular cambium—namely, phloem, cork cambium, and cork. Complete removal of a strip of bark all the way around a tree, called girdling, is invariably fatal to a tree because it severs the phloem. With the phloem gone, sugars synthesized in the leaves cannot reach the roots. Hence, the roots die and no longer take up water and minerals, resulting in the death of the entire tree.

Leaves: Nature's Solar Collectors

Leaves are the major photosynthetic structures of most plants. As you may remember from Chapter 7, photosynthesis uses the energy of sunlight to convert water and carbon dioxide to sugars, releasing oxygen as a by-product:

$$6 H_2O + 6 CO_2 + energy \rightarrow$$
$$C_6H_{12}O_6 \text{ (glucose)} + 6 O_2$$

Therefore, the cells of a leaf must be provided with water and CO_2. Water is obtained from the soil and transported to the leaf through the xylem, but CO_2 must diffuse into the leaf from the air. Thus you might think that an ideal leaf should have a large surface area for gathering light and should be porous to

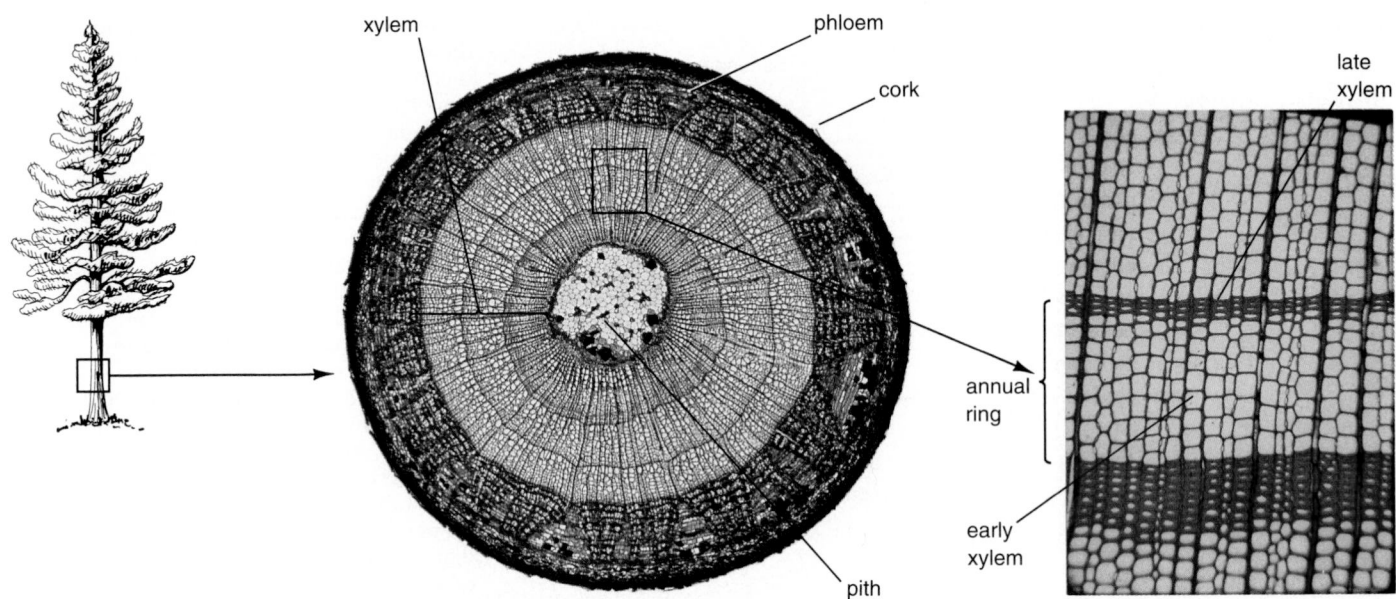

Figure 26-18 Many trees, such as this pine, form annual rings of xylem. As this micrograph shows, xylem cells formed during the wet spring are large, while xylem cells formed during the hotter, drier summer are small. The ratio of cell wall to "hole" (the now-empty interior of the cell) determines the color of the wood: early wood, formed during the spring, with lots of "hole," is pale; late wood, formed during the summer, with lots of wall, is dark.

Figure 26-19 An ancient sequoia in the Sierra Nevada of California. The cork cambium of a sequoia produces new layers of cork each year, eventually producing a protective, fire-resistant outer covering a foot or two thick. This massive cork layer contributes to a sequoia's great longevity; forest fires that kill lesser trees merely burn off a few inches of sequoia cork, leaving the living parts of the tree inside unharmed.

permit CO_2 to enter from the air for photosynthesis. However, a large, porous leaf would also lose large amounts of water through evaporation. Waterproofing the entire surface would reduce the diffusion of CO_2 into the leaf. The leaves of flowering plants represent an elegant compromise among these conflicting demands (Fig. 26-20).

Leaf Structure

A typical angiosperm leaf consists of a flat **blade** connected to the stem by a stalk called the **petiole.** The petiole positions the blade in space, usually orienting the leaf for maximum exposure to the sun. Inside the petiole are vascular tissues of xylem and phloem that are continuous with those in the stem, root, and blade. Within the blade, the vascular tissues branch into **vascular bundles** or **veins.**

The leaf epidermis consists of a layer of nonphotosynthetic, transparent cells that secrete a waxy waterproof cuticle on their outer surfaces. The epidermis and its cuticle are pierced by adjustable pores, the **stomata,** that regulate the diffusion of CO_2 and water into and out of the leaf. Each stoma is surrounded by

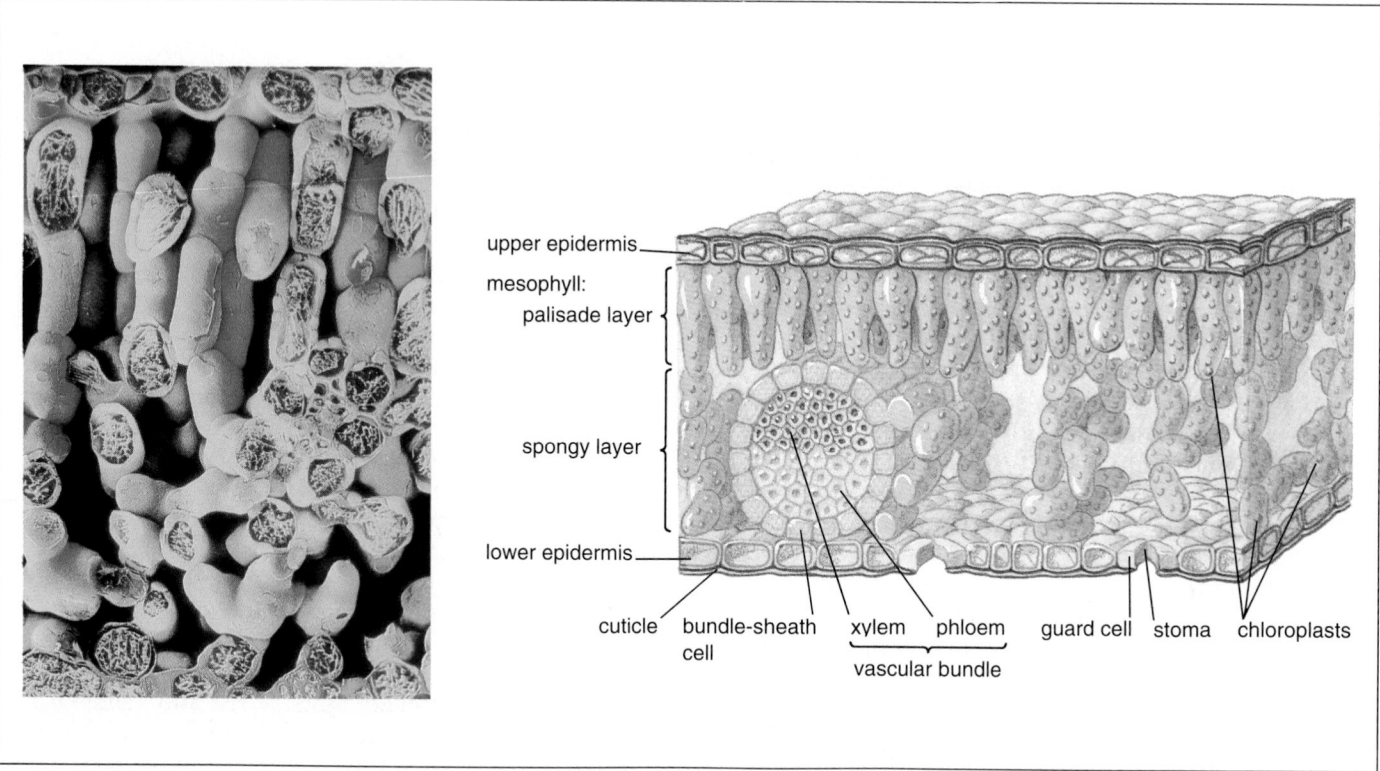

Figure 26-20 The structure of a typical dicot leaf. Note that the cells of the epidermis lack chloroplasts and are transparent, allowing sunlight to penetrate to the chloroplast-containing mesophyll cells beneath. The stomata that pierce the epidermis and the loose, open arrangement of the mesophyll cells ensure that CO_2 can diffuse into the leaf from the air and reach all the photosynthetic cells.

two sausage-shaped **guard cells** that regulate the size of the opening into the interior of the leaf (see Fig. 26-20). Unlike the surrounding epidermal cells, guard cells contain chloroplasts and can carry out photosynthesis. As we will see in Chapter 27, photosynthesis in the guard cells contributes to their ability to adjust the size of the pore.

Beneath the epidermis lies the loosely packed parenchyma cells of the **mesophyll** ("middle of the leaf"). In many leaves, mesophyll cells are of two types, a layer of columnar **palisade cells** just beneath the upper epidermis, and a layer of irregularly shaped **spongy cells** above the lower epidermis. Both palisade and spongy cells contain chloroplasts; these cells perform most of the photosynthesis of the leaf. The openness of the leaf interior (see Fig. 26-20) allows CO_2 to diffuse easily to all the mesophyll cells. Vascular bundles, each containing both xylem and phloem, are embedded within the mesophyll, with fine veins reaching very close to each photosynthetic cell. Thus, each mesophyll cell receives energy from sunlight transmitted through the clear epidermis; car-

bon dioxide from the air, diffusing through the stomata; and water from the xylem. The sugars it produces are carried away to the rest of the plant by the phloem.

Special Adaptations of Roots, Stems, and Leaves

Not all roots are sinuous fibers; not all stems are smooth and upright; and not all leaves are flat and fanlike. Just as evolution has changed the basic shape of the vertebrate forelimb to suit the demands of running, swimming, and flying, so too plant parts have become modified in response to environmental demands. You may be surprised to learn that many familiar structures are derived from unlikely parts of a plant.

Although we will highlight unusual adaptations, don't forget that *all plants are adapted to their environments.* The "typical" leaf of an oak or maple is just as

Essay: Autumn in Colorado

Those of us who live in temperate climates, including most of North America, are familiar with the yearly cycle of deciduous trees: the bare limbs of winter put forth new green leaves in spring, which turn flaming red and yellow in autumn, only to drop off, once again leaving the trees naked for winter. This cycle is an evolutionary response to temperature and water availability.

The leaves of deciduous trees present a large surface area to the sunlight, maximizing their photosynthetic abilities. These large leaves also transpire a lot of water, which must be replaced by water absorbed from the soil. What happens to water acquisition and loss when winter arrives? In temperate climates, the ground freezes for at least part of the winter, and roots cannot absorb water out of frozen soil. Simultaneously, *low temperatures reduce but do not eliminate evaporation of water from leaves.* By dropping their leaves, deciduous plants avoid the double bind of continued water loss without water replacement.

How are leaves shed? As a leaf grows, a layer of thin-walled cells develops at the base of the petiole, near its attachment to the stem (Fig. E26-1). In the fall, these cells, called the **abscission layer,** produce an enzyme that digests the cell walls holding them together. Only a few strands of xylem and phloem hold the leaf to the stem. These eventually break, allowing the leaf to fall.

Leaf fall is not, however, a direct response to freezing temperatures, nor the result of leaf death. Rather, leaf fall is an active event directed by the plant, evolutionarily programmed into its life cycle. This is shown by two observations. First, if a branch breaks from a tree in late summer, its leaves die and turn brown, but they do not drop off the branch. In fact, the branch may hold its dry, withered leaves all winter long and on into spring, until they disintegrate in the natural course of decay. Second, many deciduous trees have been transplanted to lawns and gardens far out of their native habitats. Maples in southern California are watered year round by their owners and never experience freezing weather, yet they drop their leaves in autumn anyway. Why? The trees seem to respond to daylength: abscission is triggered by the shorter days of autumn, which would be the sure sign of approaching winter in colder climes.

This leaves us with one unanswered question: Why the changing colors? In summer, leaves are green because of the chlorophyll in their chloroplasts, which reflects or transmits green light while absorbing red and blue (see Chapter 7). Chloroplasts also contain other pigments, notably red and yellow carotenoids, but these are normally almost completely masked by the chlorophyll. As the plant prepares for winter, it recycles many of the organic molecules in its leaves, including chlorophyll. Many of the atoms that formerly composed the chlorophyll molecules are reclaimed and stored in other, permanent, parts of the tree. As the chlorophyll disappears, the yellows and reds of carotenoids are unveiled. Meanwhile, blue and red anthocyanins are synthesized and stored in the central vacuoles of the leaf cells. Depending on the types and amount of carotenoids and anthocyanins, autumn displays the clear golds of birches, tuliptrees, and aspens (Fig. E26-2), the oranges and reds of maples, or the bronzes and purples of oaks. After the abscission layer breaks and the leaf falls off, it truly dies. The gaudy pigments are degraded, and the fallen leaves fade to the quiet brown of November.

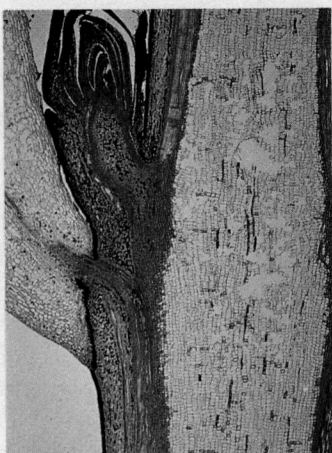

Figure E26-1 The abscission layer consists of a few rows of thin-walled cells in the leaf petiole close to its site of attachment to the stem.

Figure E26-2 Aspens, Sneffel Range, Colorado

much a "special adaptation" as a cactus spine or daffodil bulb.

Root Adaptations

Roots have probably undergone fewer unusual modifications of their basic structure than either stems or leaves. Some roots have extreme specializations for storage, such as the familiar beet, carrot, or sweet potato. Some of the most bizarre root adaptations occur in certain orchids that grow perched on trees. A few of these aerial orchids have green, photosynthetic roots; in fact, for some orchids, the green roots are the only photosynthetic part of the plant.

Stem Adaptations

Many plants have stems modified for functions very different from the original one of raising leaves up to the light. Strawberries, for example, grow horizontal **runners** that snake out over the soil, sprouting new strawberry plants where nodes touch the soil (Fig. 26-21). These new plants are connected with the "mother" plant, but once the plantlets form roots, they can live independently if the runner is severed.

Some plants, such as the saguaro cactus and the baobab tree (Fig. 26-22), store water in aboveground stems. Many other plants store carbohydrates in underground stems. The common white potato is actually a storage stem; each eye is a lateral bud, ready to send up a branch next year, using the energy stored

Figure 26-22 The baobab tree is usually not very tall, but develops an enormously fat, water-storing trunk. The baobab grows in arid regions, and when it rains, it is to the tree's advantage to store all the water it can get. Some of these trees have trunks so large that people have hollowed small houses out in them, in one case even a jail cell!

Figure 26-21 The beach strawberry can reproduce with horizontal stems called runners that course out over the surface of the sand. If a node of a runner touches the soil, it will sprout roots and develop into a complete plant.

as starch in the potato to power the growth of the branch. Irises have underground stems called **rhizomes** that store carbohydrates produced during the summer. Most rhizomes have ridges formed by the closely spaced nodes and internodes of the fat stem. Irises can be propagated by cutting up the rhizome; if it contains enough stored food, each piece with a node can generate a complete plant.

Many aboveground stems grow modified branches with special functions. One common branch adaptation is the **thorn,** usually growing from the normal branch location just above the site of attachment of a leaf (Fig. 26-23a). Thorns, of course, discourage animals from dining on the branches. Grapes and Boston ivy have some of their branches modified into grasping **tendrils** that hold the otherwise prostrate

(a)   **(b)**

Figure 26-23 The branches growing from a single stem may differ in structure or function.
(a) The hawthorn has both "regular" branches forming the crown of the tree, and many smaller branches modified into sharp thorns. That these thorns are actually branches in disguise is clearly shown by their position in the normal branch location, the crotch between a leaf and its stem, and by the fact that the thorns themselves often branch.
(b) Grape tendrils are long, soft, leafless branches. When a tendril contacts an object in the environment, it curls in the direction of the touching surface, wrapping itself firmly around the object.

plant up on supporting objects such as trees, trellises, or buildings, providing better access to the sunlight (Fig. 26-23b).

Leaf Adaptations

The most important environmental factors that affect the evolution of leaves are light, temperature, and water availability. For example, plants growing on the floor of a tropical rain forest have plenty of water year round, but very little light owing to the deep shade cast by several layers of trees above them. Consequently, their leaves tend to be extremely large, an adaptation demanded by the low light level and permitted by the abundant water (Fig. 26-24a).

At the other extreme, deserts receive bright sunlight virtually every day of the year, but have limited water and scorching temperatures. Desert plants have evolved two strikingly different adaptations to this situation. One group of plants, the succulents, have very thick leaves with large cells that store water from the infrequent rains against the inevitable long droughts (Fig. 26-24b). Succulent leaves are covered with a thick cuticle that greatly reduces the evaporation of water. The cacti display the opposite strategy, reducing the leaves into thin spines that protect the

plant from herbivores and reduce water loss (Fig. 26-24c). Photosynthesis in cacti occurs in cortex cells of the green, water-storing stems.

Modified leaves in other plants function in ways having nothing to do with photosynthesis or water conservation. The common edible pea, for example, climbs by grasping fences, mailbox posts, or other plants with clinging tendrils (Fig. 26-24d). Unlike the tendrils of grapes, which are derived from branches, pea tendrils are slender, supple leaflets. Some plants, such as onions, daffodils, and tulips, use thick, fleshy leaves as storage organs (Fig. 26-24e). A daffodil bulb consists of a short stem bearing thick, overlapping leaves that store nutrients over the winter.

Finally, a few plants have turned the table on the animals, and have become predators. Venus flytraps (Fig. 26-25a) and sundews (Fig. 26-25b) both have leaves modified into snares for unwary insects. These plants live in nitrogen-poor swamps, and derive most of their nitrogen supply from the bodies of their prey.

As varied and sometimes bizarre as these leaf specializations are, the most extreme and most important leaf modification is the flower. As we will see in the next chapter, these "reproductive leaves" enabled the flowering plants to become the dominant plants on land.

(a) *(b)* *(c)*

(d) *(e)*

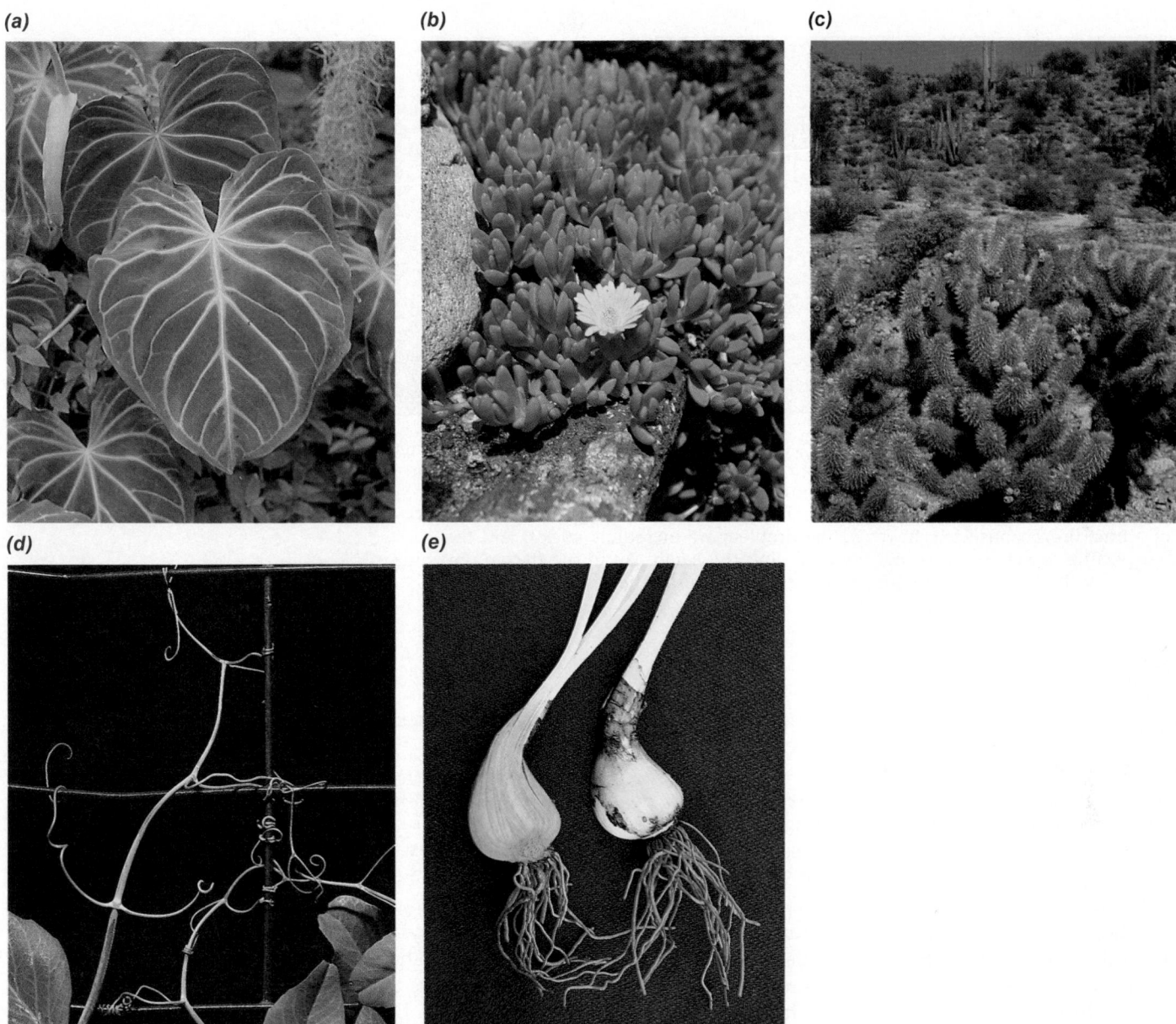

Figure 26-24 Differing environments have produced a myriad of leaf types. **(a)** In tropical rain forests, light levels near the ground are very low. Consequently, understory plants often bear huge, dark green leaves for maximal light absorption. Many jungle plants, such as this philodendron, have found new ecological niches in modern houses, where they are grown because of their deep green foliage and tolerance of dim light.

(b) Desert plants receive plenty of light but are chronically short on water. Succulents have evolved fleshy leaves that store water from the occasional rains, just as the baobab tree does in its trunk.

(c) The cactus takes the other extreme in desert adaptations: reduce the leaves to vanishingly thin lines, which cannot store water but don't have much surface area for evaporation either. These spines also have very sharp points that discourage animals who might be tempted to munch on the tasty stem beneath.

(d) Pea tendrils are modified leaflets that respond to contact with objects by coiling, much like stem tendrils do.

(e) Daffodil bulbs are enormous buds, with short central stems surrounded by thick, water- and food-storing leaves. Like other monocots, these bulbs form roots as outgrowths of the base of the stem.

(a)

(b)

Figure 26-25 Predatory leaves adorn the Venus flytrap **(a)** and the sundew **(b)**. When an insect blunders into the fringed, hinged leaves of the flytrap, the leaves close up rapidly, trapping their victim. Digestive enzymes reduce the insect to nourishing molecules that are absorbed by the plant. The sundew leaf bears glistening droplets that attract hungry insects. However, the droplets are incredibly sticky, and the unsuspecting insect becomes, not the eater, but the eaten. As in the flytrap, enzymes secreted by the leaf digest the insect and the leaf absorbs the resulting nutrients.

SUMMARY OF KEY CONCEPTS

An Overview of Plant Structure

The body of a land plant consists of root and shoot. Roots are usually underground, and have five functions. They (1) anchor the plant in the soil; (2) absorb water and minerals from the soil; (3) store surplus photosynthetic products; (4) transport water, minerals, photosynthetic products, and hormones; and (5) produce hormones. Shoots are usually aboveground, and consist of stem, leaves, and reproductive structures (including flowers and fruit in angiosperms). Shoot functions include: (1) photosynthesis; (2) transport of materials; (3) reproduction; and (4) hormone synthesis.

Plant Development

Plant bodies are composed of two main classes of cells. Meristem cells are embryonic cells that are capable of cell division throughout the life of the plant. Differentiated cells arise from divisions of meristem cells, become specialized for particular functions, and normally do not divide.

Most meristem cells are located in apical meristems at the tips of roots and shoots, and lateral meristems in the shafts of roots and shoots. Primary growth (growth in length and differentiation of parts) results from division and differentiation of cells from apical meristems, while secondary growth (growth in diam-

eter) results from division and differentiation of cells from lateral meristems.

Plant Tissues and Cell Types

Plant bodies consist of three tissue systems: the dermal, ground tissue, and vascular systems. The dermal tissue system forms the outer covering of the plant body. The dermal tissue system of leaves and of primary roots and stems is usually a single cell layer of epidermis. Dermal tissue after secondary growth is a multilayered covering of cork.

The ground tissue system consists of a variety of cell types, including parenchyma, collenchyma, and sclerenchyma, most of which are involved in photosynthesis, support, or storage. Ground tissue makes up most of a young plant during primary growth. During secondary growth of stems and roots, ground tissue becomes an increasingly small part of the plant body.

The vascular tissue system consists of xylem, which transports water and minerals, and phloem, which transports water and sugars.

Roots: Anchorage, Absorption, and Storage

Primary growth in roots results in a structure consisting of an outer epidermis, an inner vascular cylinder of conducting tissues, and cortex between the two.

The apical meristem near the tip of the root is protected by the root cap. Cells of the root epidermis absorb water and minerals from the soil. Root hairs are projections of epidermal cells that increase the surface area for absorption. Most cortex cells store surplus sugars produced through photosynthesis, usually as starch. The innermost layer of cortex cells is the endodermis, which forms a waterproof barrier preventing movement of water and minerals to and from the soil and the vascular cylinder via the extracellular space. The vascular cylinder contains the conducting tissues—xylem and phloem. The cells of the outermost layer of the vascular cylinder of roots, the pericycle, may divide and produce branch roots.

Stems: Reaching for the Light

Primary growth in dicot stems results in a structure consisting of an outer, waterproof epidermis; supporting and photosynthetic cells of cortex beneath the epidermis; vascular tissues of xylem and phloem; and supporting and storage cells of pith at the center of the stem. Leaves and lateral buds are found at locations, called nodes, along the surface of the stem, which are characteristic of the species of plant. Under the proper hormonal conditions, lateral buds may sprout into a branch.

Secondary growth in stems results from cell divisions in the vascular cambium and cork cambium. Vascular cambium produces secondary xylem and secondary phloem, increasing the diameter of the stem. Cork cambium produces waterproof cork cells that cover the outside of the stem.

Leaves: Nature's Solar Collectors

Most shoots produce leaves, the main photosynthetic organs of plants. The blade of a leaf consists of a waterproof outer epidermis surrounding mesophyll cells that have chloroplasts and carry out photosynthesis, and vascular bundles of xylem and phloem that carry water, minerals, and photosynthetic products to and from the leaf. The epidermis is punctuated by adjustable pores called stomata. A pair of guard cells forms the pore, and changes in their water content regulate the size of the opening.

Special Adaptations of Roots, Stems, and Leaves

Through evolution, the roots, stems, and leaves of many plants have been modified into diverse structures such as spines, tendrils, thorns, and bulbs. These unusual structures are often involved in water or energy storage, support, or protection.

GLOSSARY

abscission (ab-si'-shun): separation of leaves, flowers, or fruits from a stem, due to formation of a weakened layer of cells at the site of attachment to the stem.

abscission layer: a layer of thin-walled cells at the base of the petiole of a leaf, flower, or fruit, the usual site of separation from the stem.

angiosperm (an'-jē-ō-sperm): a flowering plant (Division Anthophyta); produces seeds enclosed within a ripened ovary.

annual ring: in the xylem of woody stems and roots, a pair of rings of light (early) and dark (late) wood, formed because of differential availability of water in different seasons of the year, usually spring and summer.

apical meristem (āp'-i-kul mer'-i-stem): the cluster of meristematic cells found at the tip of a shoot or root (or one of their branches).

bark: the outer layer of a woody stem, consisting of cork cells, cork cambium, and phloem.

blade: the flat part of a leaf.

branch root: a root that arises as a branch of a preexisting root, through divisions of pericycle cells and subsequent differentiation of the daughter cells.

bud: an embryonic shoot, usually very short and consisting of an apical meristem with several leaf primordia.

cambium (kam'-bē-um): a lateral meristem that causes secondary growth of woody plant stems and roots. See cork cambium; vascular cambium.

Casparian strip (kas-par'-ē-an): a waxy, waterproof band in the cell walls between endodermal cells in a root, which prevents the movement of water and minerals in and out of the vascular cylinder via extracellular space.

collenchyma (kōl-en'-ki-ma): a plant cell type that is alive at maturity, with irregularly thickened primary cell walls, and that provides support to the plant body.

companion cell: a cell adjacent to a sieve-tube element in phloem, involved in control and nutrition of the sieve-tube element.

cork cambium: a lateral meristem in woody roots and stems that gives rise to cork cells.

cork cell: a protective cell of the bark of woody stems and roots; at maturity, cork cells are dead, with thick, waterproofed cell walls.

cortex: the part of a primary root or stem located between the epidermis and the vascular cylinder.

cuticle: a waxy layer secreted by epidermal cells of shoots onto their outside surfaces.

dermal tissue system: a plant tissue system that makes up the outer covering of the plant body.

dicot: short for dicotyledon; a type of flowering plant characterized by embryos with two food-storage organs called cotyledons.

differentiated cell: a mature cell specialized for a specific function; in plants, differentiated cells usually do not divide.

endodermis (en-dō-der'-mis): the innermost layer of cells of the cortex of a root.

epidermis (ep-i-der'-mis): the outermost layer of cells of a leaf, young root, or young stem.

fibrous root system: a root system, commonly found in monocots, characterized by many roots of approximately the same diameter arising from the base of the stem.

ground tissue system: a plant tissue system consisting of parenchyma, collenchyma, and sclerenchyma cells that makes up the bulk of a leaf or young stem, excluding vascular or dermal tissues. Most ground tissue cells function in photosynthesis, support, or carbohydrate storage.

guard cell: one of a pair of specialized epidermal cells surrounding the central opening of a stoma of a leaf, which regulates the size of the opening.

gymnosperm (jim'-nō-sperm): vascular plant that produces seeds not enclosed in an ovary.

internode: the part of a stem between two nodes.

lateral bud: a bud located at a node of a stem, usually in the crotch between the stem and the petiole of the leaf found at the same node.

lateral meristem: also called cambium; a meristematic tissue in dicot stems and roots, usually found between the xylem and phloem (vascular cambium) and just outside the phloem (cork cambium).

leaf: an outgrowth of a stem, usually flattened and photosynthetic.

leaf primordium (pri-mōr'-dē-um): the outgrowth of a shoot that develops into a leaf.

meristem cell (mer'-i-stem): an undifferentiated cell that remains capable of cell division throughout the life of a plant.

mesophyll (mez'-ō-fil): cells located between the epidermal layers of a leaf.

monocot: short for monocotyledon; a type of flowering plant characterized by embryos with one food-storage organ called a cotyledon.

node: a region of a stem at which leaves and lateral buds are located.

palisade cells: mesophyll cells just beneath the upper epidermis of a leaf; usually elongated perpendicularly to the epidermis.

parenchyma (par-en'-ki-ma): a plant cell type that is alive at maturity, usually with thin primary cell walls, that carries out most of the metabolism of a plant. Most dividing meristem cells in a plant are parenchyma.

pericycle (per'-i-sī-kul): the outermost layer of cells of the vascular cylinder of a root.

periderm: the outer cell layers of a stem that has undergone secondary growth, consisting primarily of cork cambium and cork cells.

petiole (pet'-ē-ōl): the stalk that connects the blade of a leaf to the stem.

phloem (flō'-um): a conducting tissue of vascular plants that transports a concentrated sugar solution up and down the plant.

pit: an area in the cell walls between two plant cells in which secondary walls did not form, so that the two cells are separated only by a relatively thin and porous primary cell wall.

pith: cells at the center of a root or stem.

primary growth: growth in length and development of initial structures of plant roots and shoots, due to cell division of apical meristems and differentiation of the daughter cells.

primary phloem: phloem produced from an apical meristem.

primary root: the first root that develops from a seed.

primary xylem: xylem produced from an apical meristem.

rhizome (rī'-zōm): an undergound stem, usually horizontal and functioning in food storage.

root: the part of a plant, usually below ground, that anchors the plant in the soil, absorbs and transports water and minerals, stores food, and produces certain hormones.

root cap: a cluster of cells at the tip of a growing root, derived from the apical meristem. The root cap protects the growing tip from damage as it burrows through the soil.

root hair: a fine projection from an epidermal cell of a young root.

runner: a horizontally growing stem that may develop new plants at nodes that touch the soil.

sclerenchyma (skler-en'-ki-ma): a plant cell type with thick secondary cell walls, that usually dies as the last stage of differentiation, and provides both support and protection for the plant body.

secondary growth: growth in diameter of a stem or root due to cell division in lateral meristems.

secondary phloem: phloem produced from the vascular cambium.

secondary xylem: xylem produced from cells arising at the inside of the vascular cambium.

sieve plate: a part of the cell wall between two sieve-tube elements in phloem, with large pores interconnecting the cytoplasm of the elements.

sieve tube: in phloem, a single strand of sieve-tube elements, which transports sugar solutions.

sieve-tube element: one of the cells of a sieve tube.

shoot: all the parts of a vascular plant exclusive of the root. Usually aboveground, consisting of stem, leaves, and reproductive structures.

spongy cells: irregularly-shaped cells located just above the lower epidermis of a leaf.

stem: the normally vertical, aboveground part of a plant body that bears leaves.

stoma (stō'-ma; pl. stomata): an opening in the epidermis of a leaf, surrounded by a pair of guard cells.

taproot system: a root system commonly found in dicots, consisting of a long, thick main root that develops from the primary root, and numerous smaller lateral roots that grow from the primary root.

tendril: a slender outgrowth of a stem that coils about external objects and supports the stem; usually a modified leaf or branch.

terminal bud: the bud at the extreme end of a stem or branch.

thorn: a hard, pointed outgrowth of a stem; usually a modified branch.

tracheid (tra'-kē-id): an elongated xylem cell with tapering ends containing pits in the cell walls; forms tubes that transport water.

vascular bundle (vas'-kū-lar): a strand of xylem and phloem found in a leaf; commonly called a vein.

vascular cambium: a lateral meristem located between the xylem and phloem of a woody root or stem.

vascular cylinder: the centrally located conducting tissue of a young root, consisting of primary xylem and phloem.

vascular tissue system: a plant tissue system consisting of xylem (which transports water and minerals from root to shoot) and phloem (which transports water and sugars throughout the plant).

vessel: a tube of xylem composed of vertically stacked vessel elements, with perforated or missing end walls, leaving a continuous, uninterrupted hollow cylinder.

vessel element: one of the cells of a xylem vessel; elongated, dead at maturity, with thick lateral cell walls for support but with end walls either lacking entirely or heavily perforated.

xylem (zī'-lum): a conducting tissue of vascular plants that transports water and minerals from root to shoot.

STUDY QUESTIONS

1. List five functions of roots and four functions of shoots.
2. Distinguish between meristem cells and differentiated cells.
3. Describe the locations and functions of the three tissue systems in land plants.
4. Distinguish between primary growth and secondary growth, and describe the cell types involved in each. How do the tissue systems differ between a plant that has undergone only primary growth as compared with a plant that has undergone substantial secondary growth?
5. Diagram the internal structure of a root after primary growth, labeling and describing the function of epidermis, cortex, endodermis, pericycle, xylem, and phloem. What tissues are located in the vascular cylinder?
6. What types of cells form root hairs? What is the function of root hairs?
7. Diagram the internal structure of a shoot after primary growth. What is the function of each part of the shoot?
8. In what ways are the epidermal layers of roots and young shoots different? Relate these differences to the location and functions of roots and shoots.
9. What cell types make up the conducting tubes of xylem? Are these cells alive or dead at maturity? What cell types make up the conducting tubes of phloem? Are these cells alive or dead at maturity?
10. Diagram the internal structure of leaves. What structures regulate water loss and CO_2 absorption by a leaf?

DISCUSSION QUESTIONS

1. Discuss the structures and adaptations that might be found in the leaves of plants living in (1) dry, sunny habitats; (2) wet, sunny habitats; (3) dry, shady habitats; and (4) wet, shady habitats. Which of these habitats do you think would be most inhospitable (e.g., in which habitat would it be most difficult to design a functioning leaf)?
2. Discuss some advantages and disadvantages of having a "skeleton" made up of dead cells, such as the sclerenchyma cells of stems.
3. A ten-year-old boy carved his initials into the trunk of a five-year-old tree at about his eye level. Twenty years later, he returned home to find his initials still recognizable in the tree trunk. Did he have to look up or look down, or were the initials still at his eye level? Explain your answer in terms of how the man grew and how the tree grew.

SUGGESTED READINGS

Epstein, E. "Roots," *Scientific American*, May 1973. An in-depth look at root structure and function.

Frits, H. C. "Tree Rings and Climate." *Scientific American*, May 1972. The width of the annual rings of trees reflects the length of the growing season and the amount of rainfall, and can be used to determine prehistoric climate.

Raven, P., Evert, R. F., and Eichhorn, S. *Biology of Plants*. 4th ed. New York: Worth, 1986. A beautifully illustrated botany text, particularly strong on the evolution of flowering plants and the ecological adaptations of plants to their environments.

Shigo, A. L. "Compartmentalization of Decay in Trees." *Scientific American*, April 1985. Since so much of the body of a tree consists of dead cells, a tree cannot heal injuries the way most animals can. Instead, they wall off infected spots, isolating decay from the healthy parts of their bodies.

Stewart, D. "Green Giants." *Discover*, April 1990. Sequoias are the most massive living organisms on the planet, and among the oldest as well. Stewart describes how they can grow so large and live so long.

27

Nutrition and Transport in Land Plants

These giant redwoods extract water and minerals out of the soil and transport them hundreds of feet to their topmost leaves.

A Comparison of Plant and Animal Nutrition

In its broadest sense, **nutrition** is the sum of the activities by which living organisms acquire materials (called **nutrients**) from their environment and process those materials into the various molecules that make up their own bodies. We can group the activities of nutrition into four broad categories: (1) *acquisition* of nutrients; (2) *digestion* (if required); (3) *distribution* of nutrients from sites of acquisition and digestion to distant parts of the body; and (4) *synthesis* of the molecules that make up the organism's body. Plant and animal nutrition have both similarities and differences (Table 27-1).

Major similarities occur in the utilization of inorganic ions that are not normally incorporated into organic molecules. Sodium, potassium, chloride, calcium, and magnesium are usually acquired from the environment as ions dissolved in water. In most cases, they become important components of the cytoplasm or the extracellular fluid without further modification.

Plants and animals differ greatly, however, in their use of organic molecules and minerals that become incorporated into organic molecules, such as nitrogen, phosphorus, and sulfur. Aside from water and skeletal materials (such as bone), the vast bulk of both plant and animal bodies consists of relatively large molecules of carbohydrates, lipids, and proteins. These large molecules in turn are composed of smaller subunits: carbohydrates mostly of glucose, lipids of glycerol and fatty acids, and proteins of amino acids (see Chapter 3). Animals can synthesize many of these subunits from inorganic molecules; all animals, however, lack a few key enzymes that would be necessary to synthesize some of the subunits, such as certain amino acids or fatty acids. Consequently, these become essential nutrients that must be eaten as part of the animal's diet. In contrast, plants, with rare exceptions such as the predatory Venus flytrap and sundew (see Fig. 26-25), acquire all their nutrients as inorganic molecules from the soil, water, or air, and possess the enzymes needed to synthesize all their organic compounds from these inorganic precursors.

Energy production and use have both great similarities and fundamental differences. Adenosine triphosphate (ATP) is the energy currency of both plant and animal cells. However, animal cells can synthesize ATP only by metabolizing organic molecules that were acquired in the animal's diet. Plants, again with a few exceptions, obtain all their energy from sunlight: photosynthesis produces ATP directly in the light-dependent reactions, and also produces carbohydrates in the light-independent reactions (see Chapter 7). These carbohydrates may be used to make other molecules of the plant body, or they may be metabolized, as they would be in an animal cell, to produce ATP.

Plant Nutrition

Plants acquire nutrients as simple, inorganic compounds. Plants need a few elements in relatively large quantities, usually as constituents of common

Table 27-1 Plant and Animal Nutrition Compared

	Plant	Animal
Form in which nutrients are acquired	As individual inorganic molecules	Mostly in bulk as parts of prey
Digestion of nutrients	Usually none needed	Digest to component subunits and inorganic minerals
Transport of nutrients throughout body	Driven by evaporation and osmosis	Driven by hydrostatic pressure of pumping heart
Synthesis of molecules of own body	Synthesize all organic molecules from inorganic precursors	Synthesize some organic molecules from inorganic precursors; must obtain others in diet

organic molecules or as ions dissolved in the plant cytoplasm; these are called **macronutrients**. Most plants also need trace amounts of **micronutrients,** mostly elements that aid enzyme functioning. Table 27-2 briefly describes the sources and functions of macro- and micronutrients.

Carbon dioxide and oxygen normally enter a plant by diffusion from the air into leaves, stem, and roots. Roots extract water and all other nutrients (collectively called **minerals**) from the soil.

The Acquisition of Minerals

Soil consists of bits of pulverized rock, air, water, and organic matter (Fig. 27-1). Although both the rock particles and the organic matter contain many essential nutrients, only minerals dissolved in the soil water are accessible to the roots. The concentration of minerals in the soil water is very low, usually much lower than the concentration within plant cells and fluids. For example, potassium is the most abundant

Table 27-2 Essential Plant Nutrients

Element	Molecular Form Absorbed	Source	Major Functions
MACRONUTRIENTS (usually 0.1% or more of dry weight of plant)			
Carbon	CO_2	Air	Major element in organic molecules; photosynthesis
Oxygen	O_2	Air	Major element in organic molecules; cellular respiration
Hydrogen	H_2O	Soil	Major element in organic molecules
Nitrogen	NO_3^- or NH_3^+	Soil	Major element in proteins, nucleic acids, and chlorophyll
Potassium	K^+	Soil	Principal positive ion inside cells; control of stomatal opening and closing; enzyme activation
Phosphorus	$H_2PO_4^-$ or HPO_4^{2-}	Soil	Major element in nucleic acids, phospholipids, and electron carriers in chloroplasts and mitochondria
Calcium	Ca^{2+}	Soil	Component of adhesive compounds in cell walls; important in control of membrane permeability; enzyme activation
Magnesium	Mg^{2+}	Soil	Component of chlorophyll; enzyme activation; ribosome stability
Sulfur	SO_4^{2-}	Soil	Component of proteins and many coenzymes
MICRONUTRIENTS (usually less than 0.01% of dry weight of plant)			
Iron	Fe^{2+} or Fe^{3+}	Soil	Needed for synthesis of chlorophyll; component of many electron carriers
Chlorine	Cl^-	Soil	Required for photosynthesis
Manganese	Mn^{2+}	Soil	Required for photosynthesis; enzyme activation
Molybdenum	MoO_4^{2-}	Soil	Required for nitrogen metabolism
Copper	Cu^{2+}	Soil	Enzyme activation; component of electron carriers in chloroplasts
Boron	BO_3^{3-} or $B_4O_7^{2-}$	Soil	Involved in sugar transport
Zinc	Zn^{2+}	Soil	Enzyme activation; protein synthesis; hormone synthesis

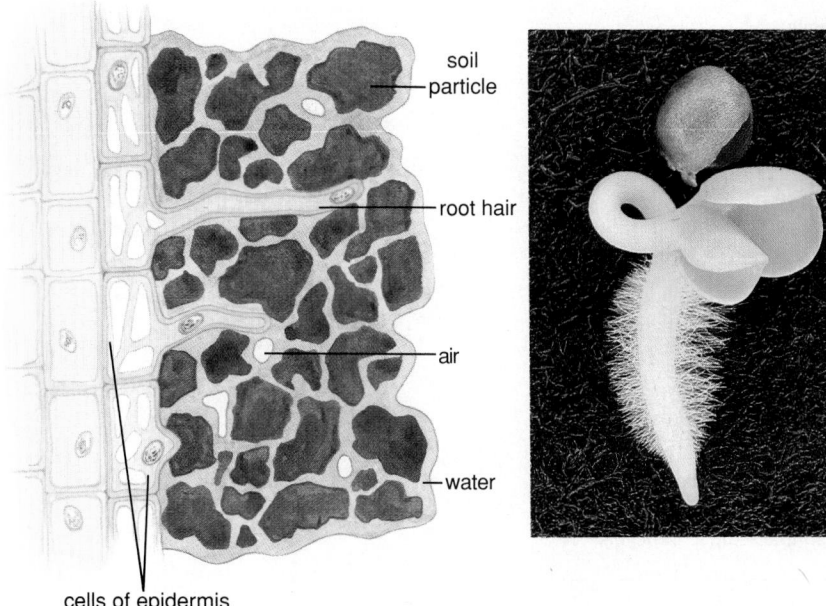

soil particle

root hair

air

water

cells of epidermis

Figure 27-1 Soil has a complex structure, consisting mostly of rock particles, water, and air, along with some organic matter and living organisms. Roots can take up only those substances that are dissolved in the soil water. Root hairs penetrate between rock particles and increase the surface area of the root that is in contact with the soil water. Inset: Root hairs cover the root of a sprouting radish seedling.

positive ion in the cytoplasm of most living cells. Its concentration within root cells is at least tenfold greater than in soil water. As a result, diffusion cannot move potassium into the root. As a general principle, **most minerals are moved into a root against their concentration gradients by active transport.** Since roots are below ground, they cannot photosynthesize. Sugar synthesized in the leaves is transported in the phloem to the roots, where mitochondria in the root cells produce ATP by cellular respiration. Some of this ATP is used to drive the active transport of minerals.

Mineral absorption by roots occurs in a four-step process, as diagrammed in Figure 27-2.

1. *Active transport into root hairs:* Root hairs projecting from the epidermal cells provide most of the surface area of the root and are in intimate contact with the soil water. The plasma membranes of the root hairs use the energy of ATP to transport minerals from the soil water, concentrating the minerals in the root hair cytoplasm (Step 1 in Fig. 27-2).

2. *Diffusion through cytoplasm to pericycle cells:* You may remember from Chapter 4 that the cytoplasm of adjacent living plant cells is interconnected by pores called **plasmodesmata.** Therefore, minerals can diffuse through plasmodesmata from the epidermal cells into the cortex, endodermis, and pericycle cells (Step 2).

3. *Active transport into the extracellular space of the vascular cylinder:* At the center of the vascular cylinder lies the xylem, into which the minerals must ultimately be transported. The tracheids and vessel elements of xylem are dead, without cytoplasm or plasma membrane—merely an outer skeleton of cell wall shot full of holes (see Fig. 26-7). Therefore, plasmodesmata do not connect pericycle cells with the insides of the xylem. On the other hand, any minerals that enter the extracellular space surrounding the xylem can easily diffuse into the xylem cells through the holes in their walls. Pericycle cells actively transport minerals out of their own cytoplasm into the extracellular space surrounding the xylem (Step 3).

4. *Diffusion into the xylem:* The active transport of minerals into the extracellular space of the vascular cylinder increases the concentration of minerals in the extracellular space. This high concentration creates a gradient that promotes the diffusion of minerals from the extracellular space into the tracheids and vessel elements of the xylem (Step 4).

The Role of the Casparian Strip

You can now appreciate one of the functions of the Casparian strip in the root endodermis. If water and minerals could pass between endodermal cells, then minerals would leak back out of the extracellular space of the vascular cylinder as fast as they were

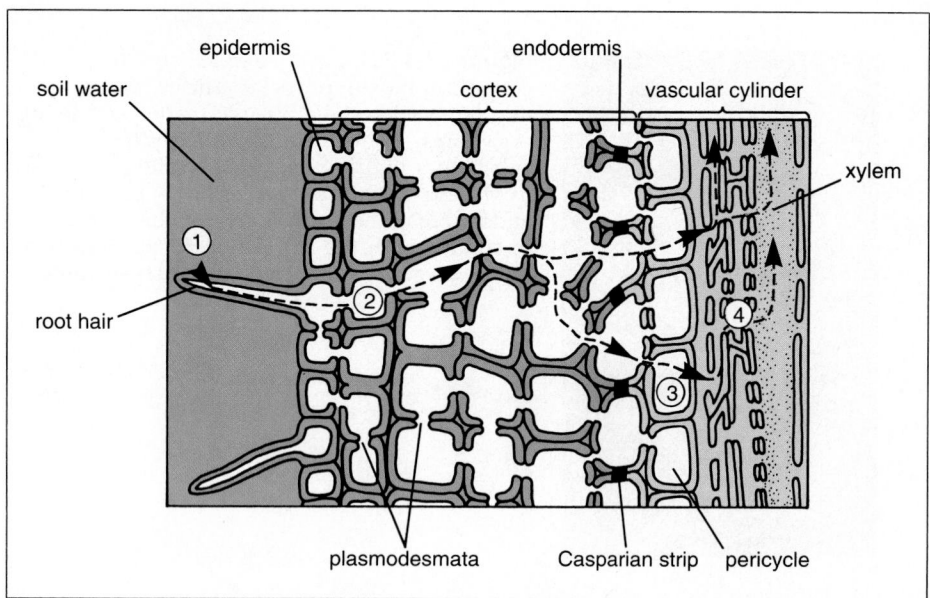

Figure 27-2 Mineral uptake by roots. The Casparian strip separates the extracellular space in the root into two compartments: an outer compartment (dark blue) that is continuous with the soil water, and an inner compartment (light blue) that is continuous with the inside of the conducting cells of the xylem. Mineral uptake occurs in a four-step process. (1) Active transport proteins in root hair membranes pump minerals into the root hair cytoplasm. (2) Minerals diffuse inward from cell to cell through plasmodesmata that interconnect the cytoplasm of all living cells in the root. (3) The last living cells at the root interior, the pericycle cells, actively transport minerals out of their cytoplasm into the extracellular space surrounding the xylem. (4) This raises the concentration of minerals in the extracellular space, so the minerals diffuse into the xylem cells through the pits in their walls.

pumped in. This would both waste the energy that was used to actively transport the minerals into the root in the first place, and would reduce the concentration gradient that allows the minerals to diffuse into the conducting cells of the xylem. The Casparian strip, however, effectively "leakproofs" the vascular cylinder: **The Casparian strip retains the concentrated mineral solution within the extracellular space of the vascular cylinder.**

Symbiotic Relationships in Plant Nutrition

Many minerals are too scarce in soil water to support plant growth, although there may be plenty bound up in the surrounding rock particles. Further, one nutrient—nitrogen—is almost always in short supply both in rock particles and soil water. Many plants enter into mutually beneficial relationships with other organisms that help them acquire these scarce nutrients.

Mineral Accessibility and Mycorrhizae

Under normal conditions, the spontaneous rate of release of water-soluble minerals from rock particles is very slow. Furthermore, the chemical forms of the minerals may not be suitable for uptake by the plasma membranes of plant root cells. Most plants form symbiotic relationships with fungi to form root–fungus complexes called **mycorrhizae** that facilitate mineral extraction and absorption. Fungal strands intertwine between the root cells and extend out into the soil (Fig. 27-3), sometimes even contacting the roots of other nearby plants. In some way that is not yet understood, the fungus renders nutrients accessible for uptake by the roots, perhaps by converting rock-bound minerals into simple soluble compounds that root plasma membranes can transport. The fungus, in return, receives sugars and amino acids from the plant. In this way, both the fungus and the plant can grow in places where neither could survive alone, including deserts and high-altitude, rocky soils that are low in nutrients.

(a) **(b)**

Figure 27-3 Mycorrhizae, a root-fungus symbiosis. **(a)** A tangled meshwork of fungal strands surrounds and penetrates into the root. **(b)** Seedlings growing with and without mycorrhizal fungi show the importance of mycorrhizae in plant nutrition.

Nitrogen Acquisition

Amino acids, nucleic acids, and chlorophyll all contain nitrogen, so plants need prodigious amounts of this element. Unfortunately, although nitrogen is abundant in the biosphere, most of it is not readily available to plants. About 80% of the atmosphere is molecular nitrogen, N_2, but plants can take up nitrogen only via their roots, in the form of ammonium ion (NH_4^+) or nitrate ion (NO_3^-). Although N_2 diffuses from the atmosphere into the air spaces in the soil, it cannot be used by plants because they don't have the enzymes needed to carry out **nitrogen fixation,** the conversion of N_2 into ammonium or nitrate.

A variety of **nitrogen-fixing bacteria** do have these enzymes. Some of these bacteria are free-living in the soil. However, nitrogen fixation is very costly, energetically speaking, using at least 12 ATPs per ammonium ion synthesized. Bacteria don't routinely manufacture a lot of extra ammonium and liberate it into the soil.

Some plants, particularly the **legumes** (peas, clover, and soybeans), enter into a mutually beneficial relationship with specific species of nitrogen-fixing bacteria. By secreting chemicals into the soil, legumes attract nitrogen-fixing bacteria to their roots (Fig. 27-4a). Once there, the bacteria enter the root hairs. The bacteria then digest channels through the cytoplasm of the epidermal cells and into adjacent cortex cells. Both bacteria and their host cortex cells multiply to form a **nodule** (Fig. 27-4b). A cooperative relationship ensues. The plant transports sugars from its leaves down to the cortex, just as it normally would for storage. The bacteria within the cortex cells take up the sugar and use its energy for all of their metabolic processes, including nitrogen fixation. The bacteria obtain so much energy that they produce more ammonium than they need. The surplus ammonium diffuses into the cytoplasm of their host cells, providing the plant with a steady supply of usable nitrogen.

The complex interplay among bacteria, plants, and animals, whereby nitrogen is fixed from the atmosphere, cycled through living matter, and returned to the atmosphere, is described in Chapter 45.

The Acquisition of Water

Once a high concentration of minerals builds up in the vascular cylinder, water absorption becomes very straightforward (Fig. 27-5). Water moves by osmosis from regions of high water concentration to regions of low water concentration. Dissolved minerals tie up water molecules, lowering the concentration of free water molecules (see Chapter 6). Therefore, the low-mineral solution in the soil water has a high free-water concentration, while the high-mineral solution in the vascular cylinder has a low free water concentration.

(a)

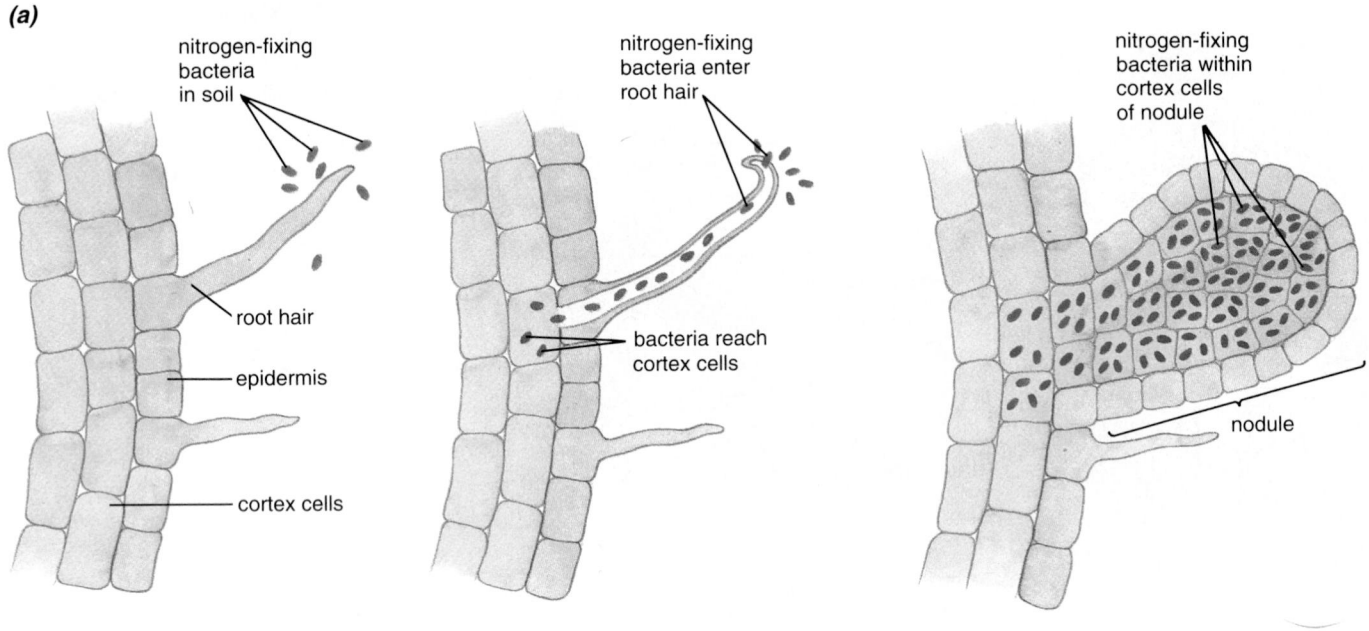

(b)

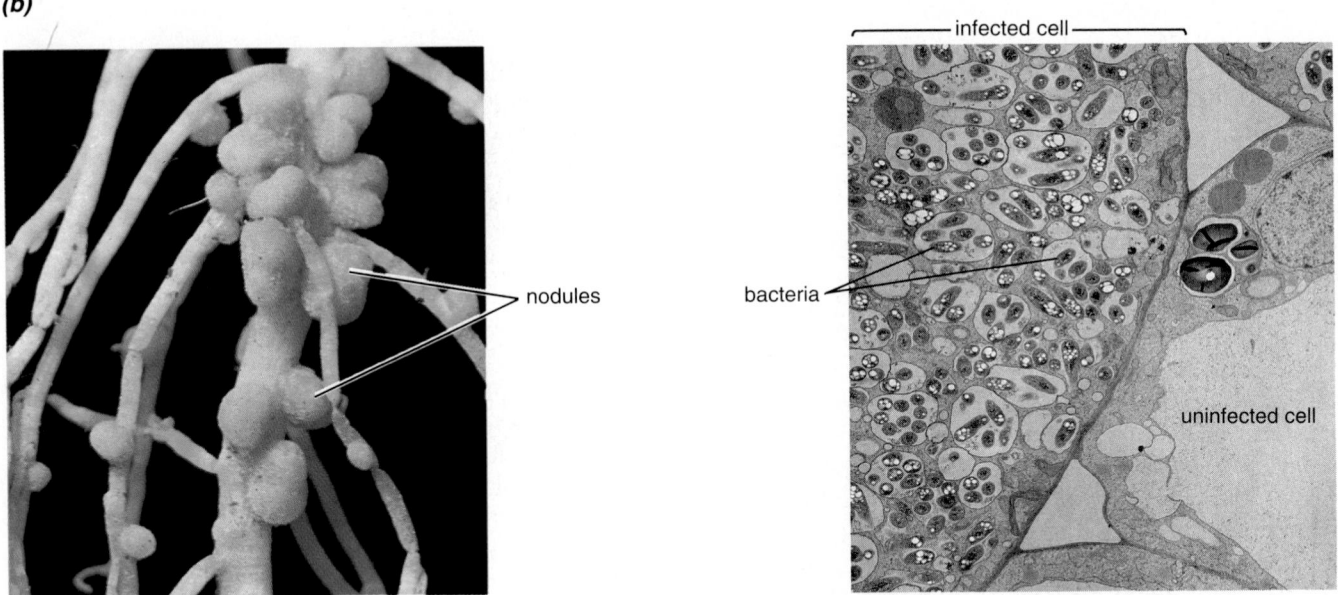

Figure 27-4 Nitrogen fixation in legumes.
(a) Nitrogen-fixing bacteria penetrate a legume root through the root hairs.
  (1) Bacteria in the soil are attracted to a root hair.
  (2) The bacteria digest their way into the root hair, through the cytoplasm and into underlying cortex cells.
  (3) The bacteria multiply and induce the cortex cells to multiply, forming a nodule.
(b) Nodules in legume roots consist of cortex cells filled with nitrogen-fixing bacteria.

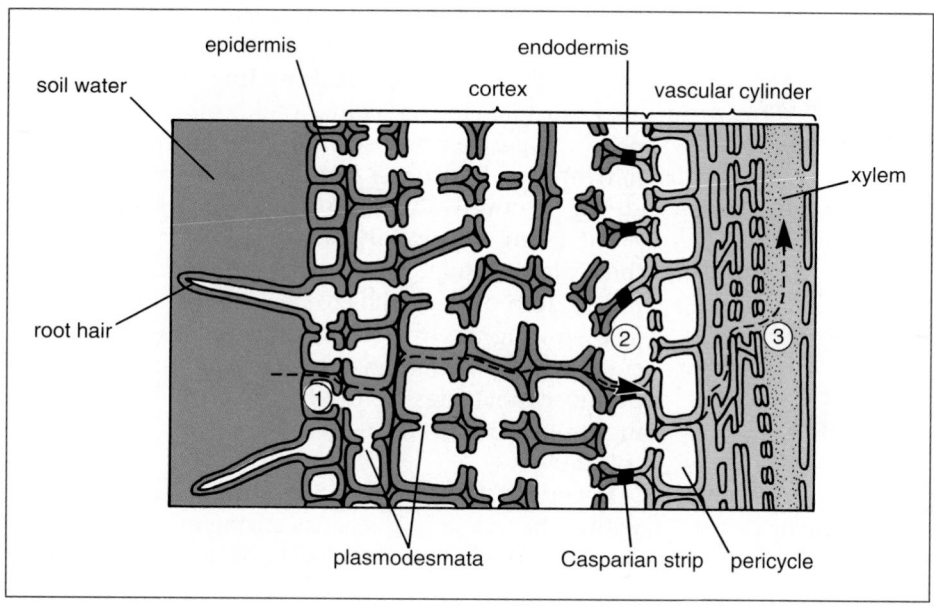

Figure 27-5 Water uptake by roots. (1) There is an uninterrupted pathway for water movement from the soil water, between cells of the epidermis and cortex, up to the waterproofing of the Casparian strip. Mineral uptake has created a higher concentration of minerals (and a lower concentration of water) in the extracellular space inside the Casparian strip (light blue) than in the extracellular space and soil water outside the Casparian strip (dark blue; see Fig. 27-2). (2) Therefore water moves by osmosis across the endodermal cells into the extracellular space of the vascular cylinder. (3) Water moves into the tracheids and vessel elements of root xylem through porous pits in their cell walls.

The epidermal and cortex cells of a young root are loosely packed and have highly porous cell walls, so the soil water has an uninterrupted pathway through the extracellular space of the outer layers of the root (Step 1 in Fig. 27-5). At the endodermis, the waterproofing of the Casparian strip blocks further movement of water between cells. The faces of the endodermal cells, however, are not waterproofed. **Therefore, the plasma membranes of the endodermis form a pair of semipermeable membranes separating an outer solution of high free-water concentration (in the cortex) from an inner solution of low free-water concentration (within the vascular cylinder). Water moves across the membranes by osmosis from high to low concentration—that is, from the extracellular space outside the Casparian strip into the extracellular space of the vascular cylinder inside the Casparian strip** (Step 2). Water can then freely move from the extracellular space within the vascular cylinder into the tracheids and vessel elements of the xylem, through porous pits in their cell walls (Step 3).

As you will see when we discuss transport in xylem, water is also pulled up the xylem, powered by the force of water evaporating from the leaves. This further lowers the water concentration within the vascular cylinder and promotes the entry of water across the endodermis.

The Transport of Water and Minerals

Once water and minerals enter the root xylem, the problem still remains of moving them to the uppermost reaches of the plant, which, in redwood trees, may be over a hundred meters away. The processes of active transport, diffusion, and osmosis that suffice for the few millimeters from soil to root xylem, however, would be hopelessly slow for getting water and minerals to the top of a tree. Land plants accordingly move fluids up the xylem from root to stem and leaf *en masse* by bulk flow.

Both minerals and water move up the xylem. However, the minerals are dissolved in the water, and are passively carried along as the water is transported. Therefore, we have to concern ourselves only with the mechanisms of water transport.

Water Movement in Xylem

According to the **cohesion–tension theory,** water is pulled up the xylem, powered by the evaporation of water from the leaves (Fig. 27-6). As its name suggests, this theory has two essential parts.

1. *Cohesion:* water within the xylem holds together like a solid rope.
2. *Tension:* this "water rope" is pulled up the xylem, with evaporation providing the necessary energy.

Let's briefly examine both of these propositions.

Cohesion Among Water Molecules

You will recall from Chapter 2 that water is a polar molecule, with the oxygen carrying a slight negative charge while the hydrogens carry a slight positive charge. As a result, nearby water molecules attract one another, forming weak **hydrogen bonds.** However, just as individually weak cotton threads together make a strong seam in your jeans, so too **the network of hydrogen bonds within water is very strong, giving water a high *cohesion*, or tendency to resist being separated.** This is essential, because if the water is pulled up the xylem from the top, then the column of water has to be strong enough to bear its own weight without breaking—all the water molecules within the xylem are "hanging on" to the uppermost molecules by the hydrogen bonds connecting this handful of molecules to the rest of the column.

Experiments have found that the column of water within the xylem is at least as strong as a steel wire of the same diameter, and therefore doesn't break (see

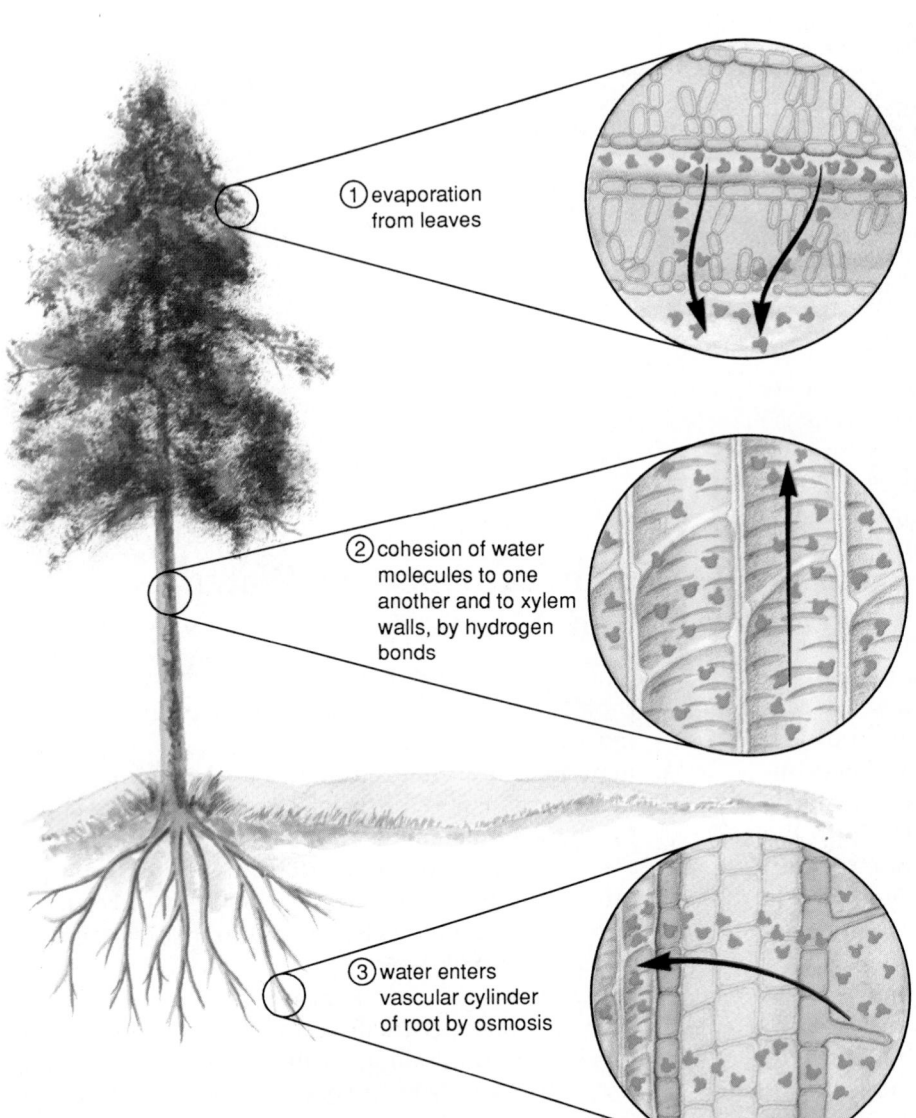

(1) evaporation from leaves

(2) cohesion of water molecules to one another and to xylem walls, by hydrogen bonds

(3) water enters vascular cylinder of root by osmosis

Figure 27-6 The cohesion–tension theory of water flow from root to leaf in xylem. (1) Water evaporates out of the leaves, and other water molecules replace them from the xylem of the leaf veins. (2) Within the xylem, hydrogen bonding holds nearby water molecules together so firmly that the column of water behaves very much like a rope. The top of the "water rope" is pulled up by evaporation, and the rest of the rope comes along as well, all the way down to the roots. (3) As the molecules of the water rope retreat up the xylem in the roots, the decreased water concentration within the root xylem and the surrounding extracellular space causes water to enter from the soil water by osmosis, thus steadily replenishing the bottom of the rope.

"Methods in Biology: How Do We Know What Goes on Inside Xylem?"). This is the "cohesion" part of the theory: hydrogen bonds among water molecules provide the cohesion that holds together a "rope" of water within the xylem.

Tension

Evaporation of water through the stomata of a leaf, a process called **transpiration,** provides the force for water movement—the "tension" part of the theory. As a leaf transpires, the concentration of water in the mesophyll cells falls. This lower water concentration causes osmosis of water from the xylem in the nearby veins into the dehydrating mesophyll cells. Water molecules leaving the xylem are attached to other water molecules in the same xylem tube by hydrogen bonds. Therefore, when one water molecule leaves, it pulls adjacent water molecules up the xylem. As these water molecules move upward, other water molecules further down move up to replace them. This process continues all the way to the roots, where water in the extracellular space around the xylem is pulled in through the holes in the walls of vessel elements and tracheids. This upward and inward movement of water finally causes water to move into the vascular cylinder by osmosis through the endodermal cells. The force generated by the evaporation of water from the leaves, transmitted down the xylem to the roots, is so strong that water can be absorbed from quite dry soils.

SUMMARY OF WATER TRANSPORT IN XYLEM:

Transpiration from the leaves (and, to a lesser extent, the stem) removes water from the top of a xylem tube. This water is replaced by water further down in the tube, so that water moves by bulk flow up the xylem. Finally, this upward flow removes water from the root xylem and the extracellular space surrounding it, which promotes osmosis of water from the soil water into the vascular cylinder of the root. **The flow of water in the xylem is unidirectional, from root to shoot, because only the shoot can transpire.**

The Control of Transpiration

Transpiration has both positive and negative effects on a plant. On the plus side, transpiration provides the force that transports water and minerals to the leaves at the top of the plant. On the minus side, transpiration is by far the largest source of water loss—a loss that may threaten the very survival of the plant, especially in hot, dry weather.

Most water transpires through the stomata of leaves and stem, so you might think that a plant could prevent water loss by simply closing its stomata. However, don't forget that photosynthesis requires carbon dioxide from the air, which diffuses into the leaf mainly through open stomata. **Therefore, a plant, by opening and closing its stomata, must achieve a balance between carbon dioxide uptake and water loss.**

Stoma Structure and Function

A **stoma** consists of a central opening surrounded by two kidney-shaped **guard cells** that regulate the size of the opening (Fig. 27-7). With some exceptions, stomata open during the day and close at night, but they will also close if the leaf begins to dehydrate. The selective value of this arrangement is obvious. First, the stomata open only during the day, when sunlight allows photosynthesis. Even then, the stomata close if water loss becomes too great.

Guard cells change the size of the opening between them by changing their own shape. How does this happen? There are two levels to this question: first, how does changing the shape of the guard cells open and close the opening; and second, what physiological processes cause the change in shape?

The Relationship Between Guard Cell Shape and Opening a Stoma

Stomata open when the guard cells take up water and swell, and close when guard cells lose water and shrink. This might seem paradoxical, since swollen guard cells must take up a larger volume than shrunken ones, and therefore you might expect that the potential central hole would be shut more tightly than ever. The key lies in the construction of the cell wall of guard cells (Fig. 27-8a). Cellulose fibers in the wall encircle the guard cells like a series of inelastic belts. Thus, when water enters the guard cells and their volume increases, they cannot become fatter but must become longer. Each pair of guard cells is attached at both ends, so the only way the cells can become longer is by bowing outward like a cooked sausage, opening a hole between them (Fig. 27-8b).

Osmosis of Water Into and Out of Guard Cells

According to the principles of osmosis, water will enter a guard cell if its cytoplasm has a lower water concentration than the cytoplasm of surrounding cells, and will leave the guard cell if it has a higher concentration of water. Large changes in potassium concentration within the guard cells cause correspondingly large changes in water concentration, driving the osmotic fluxes that open and close a

METHODS IN BIOLOGY
How Do We Know What Goes on Inside Xylem?

As you study biology, we hope you are not only absorbing a lot of facts about living organisms, but also learning how biologists discover those facts in the first place: what tools biologists use in their research, how they design experiments, how they analyze the resulting data, and how they draw conclusions. In this chapter, we presented the cohesion–tension theory for water transport in xylem from roots to leaves. You may have wondered how such a scheme could have been thought up at all, why it has been accepted by botanists, and whether other, simpler, explanations might not suffice.

The problem is how to get water from the soil to the topmost leaves of a plant, which might be as much as 100 meters away. Various hypotheses have been proposed, usually soon to be discarded. For example, consider capillarity. You may have noticed that if one end of a thin tube is immersed in water while the other end sticks out in the air, water rises a short way up the tube (Fig. E27-1). A little experimentation reveals that water ascends farther in thinner tubes. Perhaps in tubes as thin as those of xylem, water simply creeps up by capillarity. Xylem cells, however, with diameters of 30 to 50 microns (about 1/500th of an inch), would only allow capillarity to raise water up about a meter—not even close to the top of a tree.

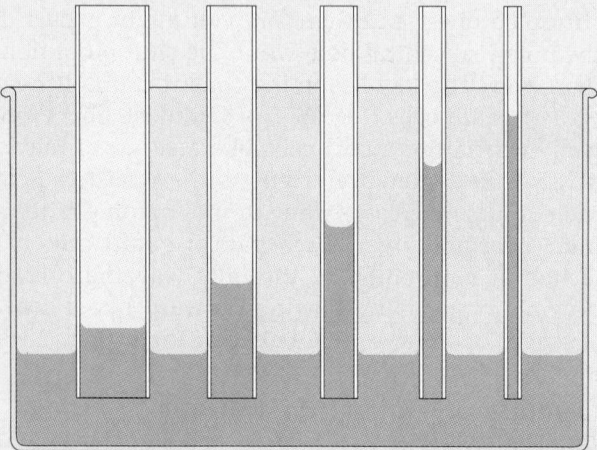

Figure E27-1 If a thin tube is touched to the surface of a dish of water, some water will spontaneously move up the tube. How high the water goes is a function of tube diameter: the thinner the tube, the farther the water rises.

A second, more serious, proposal is that the roots might pump water up the xylem, a phenomenon called **root pressure.** If you cut the top off a tomato plant and seal a pressure-monitoring device onto the remaining stump, you would find that the roots do indeed push water up: hard enough to reach 30 or 40 meters. Root pressure arises because of the accumulation of minerals in root xylem. As you recall, water moves by osmosis into the extracellular space and xylem tubes in a root vascular cylinder. This water has no place to go but up the plant, and generates root pressure. In some plants—strawberries, for example—root pressure occasionally forces water droplets out of the leaf xylem (Fig. E27-2). Under some conditions, and in some plants, root pressures contribute to xylem flow, but three crucial experiments rule it out as the major factor. First, it is too slow to account for measured xylem flow rates. Second, redwoods and other tall conifers usually have no measurable root pressures. Third, if you saw down a tree and stick the bottom end of the trunk in a bucket of water containing a suitable dye, the water and dye travel up to the leaves without any roots at all (this, of course, is why you put a Christmas tree in water).

We have already described the **cohesion–tension** theory for the movement of water up the xylem. Is there any positive evidence for this theory, or is it merely the only one left when the others have failed? Needless to say, no one has ever seen water molecules pulling one another up a xylem tube by their hydrogen bonds. However, as all good scientific theories must, the cohesion–tension theory offers predictions that can be tested. First, if water is to be pulled up to the top of a redwood, then a 100-meter-tall tube of water must be strong enough to hold up its own weight. Actual experiments show that the cohesion of water is more than strong enough to hold water molecules together on their way up a tree (Fig. E27-3).

Second, since water is being *pulled up*, the water inside the xylem must be under tension. As anyone who has ever cut down a tree knows, when xylem (wood) is cut with an axe, water does *not* flow out of the stump (as the root pressure hypothesis would predict). Water under tension should really *pull back into the cut* like a snapped rubber band. This is exactly what happens: if a drop of water is placed in a fresh axe cut, tension in the xylem will draw the water into the wood. Another prediction based on tension within the xylem is that the diameter of a tree trunk should decrease during periods

Figure E27-2 Under favorable conditions of high water availability in the soil and high humidity in the air, the root pressure of some plants will force water out the tips of the leaves. Some botanists speculate that the exuded water might carry wastes out of the plant.

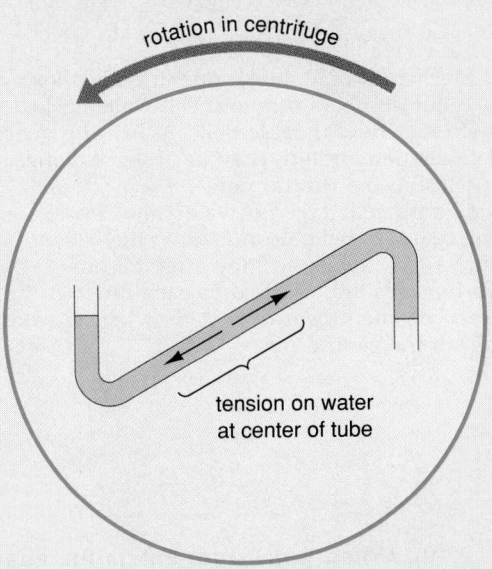

rotation in centrifuge

tension on water
at center of tube

Figure E27-3 An experiment to test whether water cohesion is strong enough to support the weight of a column of water in the xylem of a tall tree. A capillary tube is bent into a **Z** shape, filled with water, and spun at high speed in a centrifuge. Centrifugal forces push the water toward the ends of the **Z**. At high enough speeds, the column of water breaks apart in the middle. However, the forces required to break the water column in the Z tube are equivalent to the weight of water in a xylem tube more than 500 meters tall, far taller than any tree that ever lived.

of high flow rates, just as a soda straw collapses if you suck on it too hard. Although it is difficult to imagine a tree trunk shrinking, measurements do show that trunks are thinner during the day (when water is evaporating rapidly from the leaves and therefore is being rapidly pulled up the xylem) than at night (when evaporation and flow rates are low). Finally, botanists can measure the tension within a stem, and have found tensions strong enough to pull water up *200 meters!*

Clearly, the cohesion–tension theory has passed several tests with flying colors, while competing hypotheses have failed. Nevertheless, as with all scientific theories, future experiments may someday come up with data that the cohesion–tension theory cannot explain. Such is the nature of science: a theory stands only as long as experimental data allow it to. So far, the cohesion–tension theory stands as tall as the redwood trees whose water transport it explains.

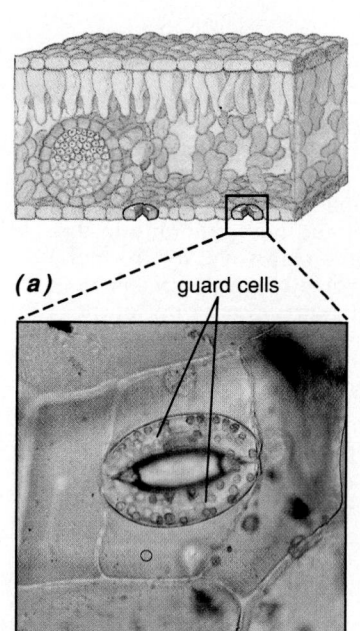

(b)

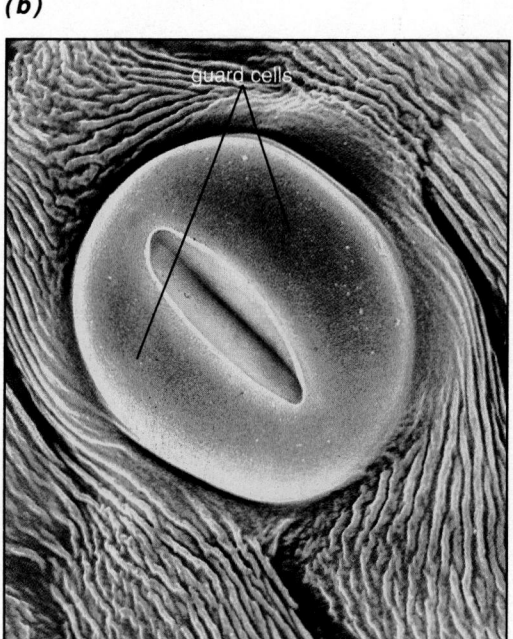

Figure 27-7 Stomata seen through the light microscope **(a)** and scanning electron microscope **(b).** In the light micrograph, note that the guard cells contain chloroplasts (the green ovals within the cells), but the other epidermal cells do not.

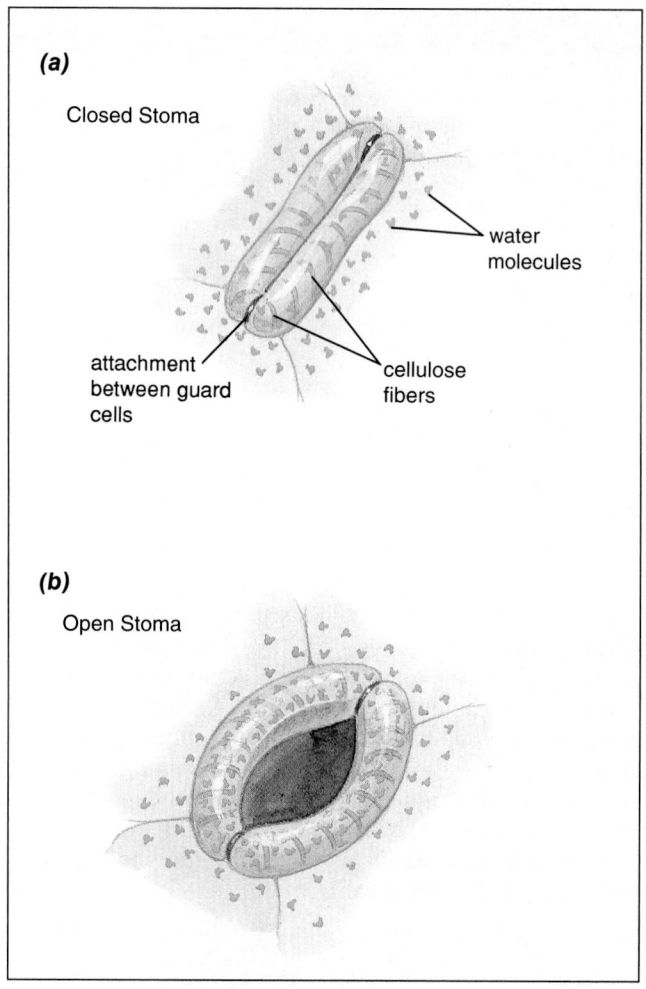

Figure 27-8 Mechanics of stomatal opening and closing. **(a)** Bands of cellulose fibers in guard cell walls enclose the cell like strong, inexpandable belts. When the guard cells are relatively dehydrated, they lie straight alongside one another, closing the central pore.
(b) When the guard cells take up water, they swell. However, the cellulose belts do not allow the cells to become fatter as they swell, so they must become longer. Since the cells are attached to each other at the ends, they can become longer only by bowing outward, thus opening up the central pore.

stoma (Fig. 27-9). **When potassium enters the guard cells, water follows by osmosis, opening the stoma; when potassium leaves the guard cells, water leaves again by osmosis, and the stoma closes.**

Several factors regulate the potassium concentration inside guard cells. The three most important are light, CO_2, and water levels within the leaf.

1. *Light reception:* Guard cells contain pigments that absorb light. When light strikes these pigments, they trigger a series of reactions that cause potassium to be actively transported from the extracellular fluid into the guard cells. At night, the pigments are no longer activated by light, so the potassium pumping stops, and the "extra" potassium within the guard cells diffuses back out.

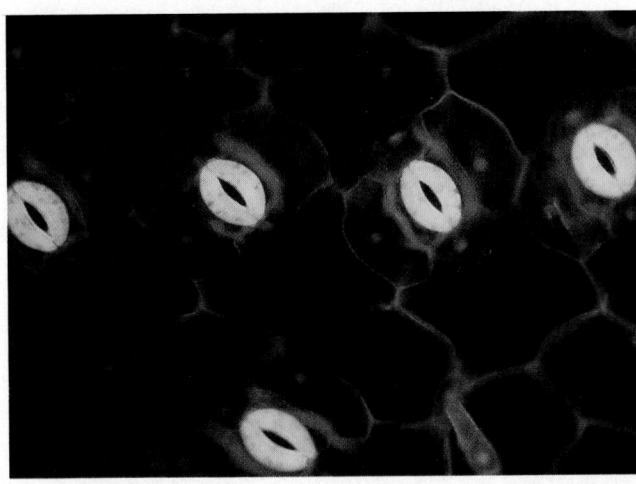

Figure 27-9 The concentration of potassium ions drives osmosis of water into and out of guard cells. In this micrograph, the potassium concentration is indicated by the brightness of the yellow dye, clearly showing that the guard cells surrounding open stomata have a higher potassium concentration than the other epidermal cells.

2. *Carbon dioxide concentration:* The concentration of CO_2 within the guard cells is regulated by the balance between photosynthesis and respiration. Remember that plant cells have mitochondria that generate ATP by aerobic respiration, using up O_2 and producing CO_2. Guard cells also have chloroplasts and carry out photosynthesis, which consumes CO_2 and produces O_2. During the day, photosynthesis uses up CO_2 faster than respiration produces it, so the CO_2 concentration in the guard cells drops. Low CO_2 concentrations stimulate the active transport of potassium into the

guard cells. At night, photosynthesis stops but respiration continues. As a result, the CO_2 level within the guard cells rises. Note that, under normal circumstances, light activates the potassium pumps both directly, via the light-receptor pigments, and indirectly, via photosynthesis, which reduces CO_2 concentrations.

3. *Water:* If a leaf loses water faster than it can be replaced from the xylem, and begins to wilt, the mesophyll cells release a hormone called **abscisic acid.** The hormone strongly inhibits the active transport of potassium into the guard cells. The inhibitory effects of abscisic acid override the stimulatory effects of light and low CO_2 levels, so potassium pumping stops. As potassium leaks out of the guard cells, water follows by osmosis, the guard cells shrink, and the stomata close.

The Transport of Sugars

Water and minerals transported into the leaves allow them to carry out photosynthesis, producing sugars from water and CO_2. These sugars must be moved to other parts of the plant, to nourish nonphotosynthetic structures such as roots or flowers, and to be stored in the cortex cells of the root and stem. Sugar transport is the function of phloem.

Botanists employ a most unlikely lab assistant in studying phloem function: the aphid. Aphids are insects that specialize in feeding on the fluid contained in phloem sieve tubes. An aphid inserts a pointed, hollow tube, the stylet, through the epidermis and cortex of a young stem into a sieve tube (Fig. 27-10). The aphid can then relax and let the plant do the

(a)

honeydew

(b)

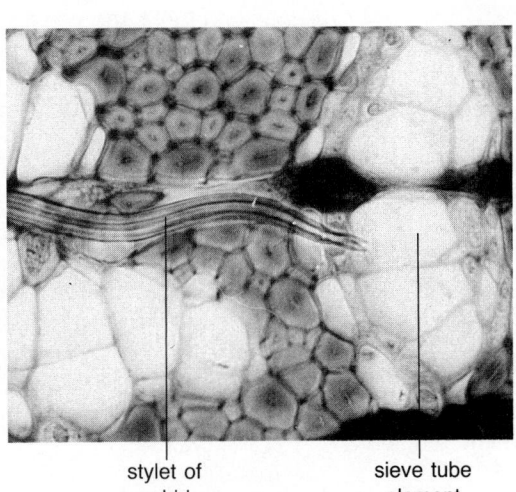

stylet of aphid

sieve tube element

Figure 27-10 Aphids feed on the sugary fluid in phloem sieve tubes. **(a)** When an aphid pierces a sieve tube, pressure in the tube forces the fluid out of the plant and into the digestive tract of the aphid. Sometimes the pressure is so great that fluid is forced completely through the aphid and out its anus, as "honeydew." This exuded fluid is collected by certain species of ants that act as "shepherds" to the aphids, defending them from predators in return for a diet of sweet honeydew. **(b)** The flexible stylet of an aphid, passing through many layers of cells to penetrate a sieve tube cell.

PLANET WATCH
The Relationship Between Plants and Water

A remarkable diversity of land plants exists on Earth. The distribution of plants over the planet is limited by environmental factors and the adaptations of the plants. Probably the most important environmental factor influencing plant distribution is water: cacti and succulents inhabit deserts because they can withstand drought, while orchids and mahogony trees need the frequent drenching rains of the rainforest. What people often overlook is the flip side of this plant–water relationship: plants, through transpiration, help to regulate the amount and distribution of rainfall, soil water, and even river flow.

Consider the Amazon rainforest (Fig. E27-4). An acre of soil supports hundreds of towering trees, each bearing millions of leaves. The surface area of the leaves dwarfs the surface area of the soil, so up to 75% of all the water evaporating from the acre of forest is transpiration from the leaves. This transpiration raises the humidity of the air and causes rain to fall. In fact, about half of the water transpired from the leaves falls again as rain, with the overall result that about a third of the total rainfall is water recycled by transpiration. Thus, in a very real sense, the high humidity and frequent showers that the rainforest needs to survive are partly *created by the forest itself!* If large tracts of rainforest are cut down, less water evaporates in that area, so less rain falls, and new rainforest tree seedlings cannot grow. An entirely different plant community would probably become established on the disturbed land, and might become a *permanent* new community.

As you can well imagine, by absorbing and transpiring so much water, plants can significantly deplete the water in the soil. The water pumped out of the soil is not available to run off into rivers, and so river flow is diminished. The Colorado River of the American southwest, for example, provides water both for millions of people and hundreds of thousands of acres of irrigated cropland. Years ago, some water management agencies actually proposed removing most of the plants along the river, to increase water supplies by eliminating transpiration. Although this was a terrible idea for other reasons, and was never seriously considered, it would have worked.

Plant transpiration might even have a moderating influence on some aspects of the "greenhouse effect" brought about by increasing atmospheric CO_2 levels, mostly due to burning fossil fuels and cutting down forests (see Chapter 45). The warmer planet predicted by greenhouse models would increase water evaporation, which in turn should lead to drier soils and the expansion of deserts. However, as you just learned, stomata open partly in response to low CO_2 levels within the guard cells. Elevated atmospheric CO_2 levels also raise CO_2 within the guard cells, and cause partial stomatal closing. Further, a recent study found that plants grown in an atmosphere with high CO_2 levels have fewer stomata per unit area of leaf than plants grown at preindustrial, low CO_2 levels. If there are fewer stomata, and they are chronically partially closed, then one might expect less transpiration from plants living in a "greenhouse future" than from "preindustrial" plants. Calculations by S. B. Idso and A. J. Brazel suggest that soil moisture and river flow in southwestern states might actually *increase* in response to increased atmospheric CO_2, despite the greenhouse warming effect.

These examples show that plants wield an enormous influence on what we often consider to be nonbiological aspects of the biosphere, such as humidity, rainfall, soil water, and stream flow. The responses of plants to human activities are not simple or easily predictable, and can have major impacts on ecosystems.

Figure E27-4 The Amazon rainforest, a plant community that helps to mold its own environment.

work. The fluid in the sieve tubes is under pressure, and actively pushes up through the stylet into the digestive tract of the aphid (sometimes with enough pressure to force its way out the other end!). By cutting off the aphid but leaving its stylet in place, botanists have collected sieve tube fluid and found that it consists mostly of sucrose and water, as much as 25% sucrose by weight. How are these high sugar concentrations moved about the plant, and in which directions?

Sugar Movement in Phloem

The most widely accepted mechanism for the transport of sugars in phloem is the **pressure flow theory** (Fig. 27-11). Let's illustrate this theory by following sucrose movements from a leaf to a developing fruit.

1. *Sucrose source—photosynthesis:* When a leaf is photosynthesizing rapidly, it manufactures lots of glucose, much of which is converted to sucrose.
2. *Phloem sieve tube loading:* Much of this sucrose is actively transported into companion cells of the phloem in the leaf veins. This raises the concentration of sucrose within the companion cells, so that sucrose then diffuses down its concentration gradient through plasmodesmata into an adjacent sieve tube element. This in turn raises the sucrose concentration in the leaf sieve tube.
3. *Osmosis into the leaf sieve tube:* The high sucrose concentration in the leaf sieve tube lowers the water concentration in the sieve tube. This causes water to enter the sieve tube by osmosis from nearby xylem.
4. *Sucrose sink—developing fruit:* Meanwhile, some

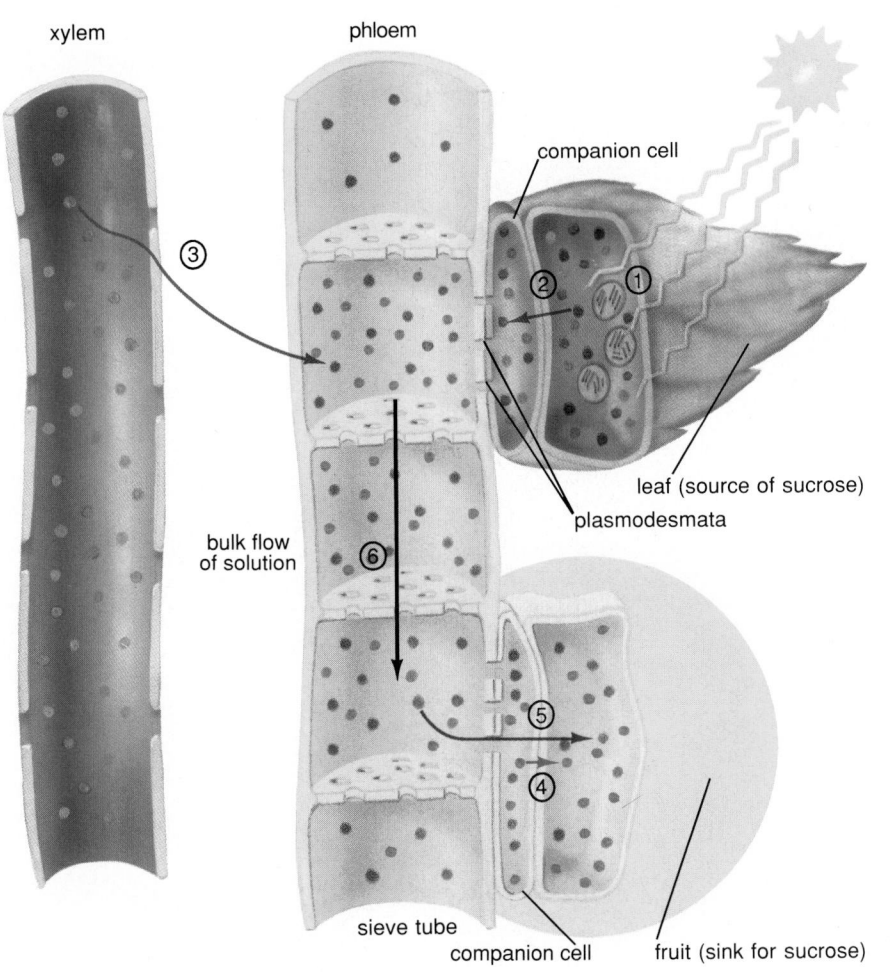

Figure 27-11 The pressure flow theory relies on differences in hydrostatic pressure to move fluid through phloem sieve tubes. (1) A photosynthesizing leaf manufactures sucrose (red dots) that (2) is actively transported into a nearby companion cell in phloem. The sucrose diffuses into the adjacent sieve tube element through plasmodesmata; this raises the concentration of sucrose in the sieve tube element. (3) Water (blue) leaves nearby xylem and moves into the "leaf end" of the sieve tube by osmosis, raising the hydrostatic pressure as increasing numbers of water molecules enter the fixed volume of the tube. (4) The same sieve tube connects to a developing fruit, where sugar is actively transported out of the companion cells and into the fruit cells. Sucrose follows out the "fruit end" of the sieve tube by diffusion through plasmodesmata into the companion cells that are being depleted of sucrose. (5) Water moves out of the tube by osmosis, lowering the hydrostatic pressure within the tube. (6) High pressure in the leaf end of the phloem and low pressure in the fruit end causes water, together with any dissolved solutes, to flow in bulk from leaf to fruit (thick black arrow).

distance away but connected by the same sieve tube, sucrose is actively transported out of the nearby sieve-tube elements and companion cells into the cells of a fruit. This raises the concentration of sugar in the fruit and lowers the concentration of sugar in the fruit end of the sieve tube.

5. *Osmosis out of the fruit sieve tube:* Water leaves the sieve tube by osmosis and follows the sugar into the fruit.

6. *Bulk flow, driven by the hydrostatic pressure gradient:* In a garden hose, water enters the faucet end of the hose and leaves the open end of the hose, driven by the difference in hydrostatic pressure between the faucet and the open end. The water moves by bulk flow between the two hose ends. The same process happens in phloem. Water enters the leaf end of a sieve tube and leaves the fruit end of the same tube, flowing in bulk from the leaf to the fruit, driven by the difference in hydrostatic pressure between the two ends of the sieve tube. The bulk flow of water carries the dissolved sugar along with it.

Sinks and Sources of Sucrose

As this example illustrates, **phloem flow is directed by sugar production and use.** Any structure that actively synthesizes sugar will be a **source** of phloem flow, and any structure that uses up sugar or converts sugar to starch will be a **sink** toward which phloem fluids will flow. A newly forming leaf will be a sink as it develops, with phloem flow up into it from more mature leaves located farther down the plant. When the leaf matures, it will photosynthesize and produce sugar, becoming a source for phloem flow to other newly developing leaves above it, to flowers or fruits, or to the roots below it. Therefore, fluid in phloem can move either up or down the plant, depending on the metabolic demands of the various parts of the plant at any given time.

SUMMARY OF KEY CONCEPTS

A Comparison of Plant and Animal Nutrition

An animal acquires both energy and materials from organic molecules eaten in its diet. A plant absorbs inorganic molecules and uses the energy of sunlight captured during photosynthesis to synthesize its organic molecules from these inorganic starting materials. Plants acquire carbon dioxide and oxygen from the air, and water and all other essential nutrients from the soil.

The Acquisition of Minerals

Most minerals are taken up from the soil water by active transport into the root hairs. These minerals diffuse from cell to cell into the root through plasmodesmata, to the pericycle just inside the vascular cylinder. The minerals are then actively transported out of the pericycle cells, into the extracellular space within the vascular cylinder. The minerals diffuse from the extracellular space into the tracheids and vessel elements of xylem.

Mineral accessibility is enhanced by fungi associated with the roots of many plants. Nitrogen, a necessary nutrient for plants, can be absorbed only in the form of ammonium or nitrate, and these ions are scarce in most soils. Some plants, the legumes, have evolved a cooperative relationship with nitrogen-fixing bacteria that invade legume roots. The plant provides the bacteria with sugars, and the bacteria use some of the energy to convert atmospheric nitrogen to ammonium that is then absorbed by the plant.

The Acquisition of Water

Because of the loose packing and porous walls of the cells of the root epidermis and cortex, water in the soil has a continuous, uninterrupted pathway through the outer layers of the root, up to the waterproofing layer of the Casparian strip between endodermal cells. Both mineral uptake and the upward movement of water in xylem contribute to an osmotic gradient across the endodermal cells, with a higher concentration of free water molecules in the extracellular space outside the endodermis than in the extracellular space inside the endodermis. Therefore, water moves by osmosis across the plasma membranes of the endodermal cells into the extracellular space of the vascular cylinder.

The Transport of Water and Minerals

The cohesion–tension theory explains xylem function: cohesion of water molecules to one another by hydrogen bonds holds together the water within xylem tubes almost as if it were a solid. As water molecules evaporate from the leaves, the hydrogen bonds pull other water molecules up the xylem to replace them. This movement is transmitted down the xylem to the root, where water loss from the vascular cylinder promotes water movement across the endodermis from the soil water by osmosis.

The Transport of Sugars

The pressure-flow theory explains sugar transport in phloem: parts of the plant that synthesize sugar (e.g., leaves) export sugar into the sieve tube. Increasing sugar concentrations attract water entry by osmosis, causing high hydrostatic pressure in that part of the phloem. Parts of the plant that consume sugar (e.g., fruits) remove sugar from the sieve tube. Loss of sugar causes loss of water by osmosis, resulting in low hydrostatic pressure. Water and dissolved sugar flow in the sieve tube from high to low pressure.

GLOSSARY

abscisic acid (ab-sis'-ik): a plant hormone that is generally inhibitory, enforcing dormancy in seeds and buds, and closing stomata.

cohesion–tension theory: a model for transport of water in xylem, which states that water is pulled up the xylem tubes, powered by the force of evaporation of water from the leaves (producing *tension*) and held together by hydrogen bonds between nearby water molecules (*cohesion*).

legume (leg'-yoom): a family of flowering plants, including peas, soybeans, and clover, most of which harbor nitrogen-fixing bacteria in nodules on their roots.

macronutrient: a nutrient needed in relatively large quantities (often defined as composing more than 0.1% of an organism's body).

micronutrient: a nutrient needed in relatively small quantities (often defined as composing less than 0.01% of an organism's body).

mineral: an inorganic substance, especially one found in rocks or soil.

mycorrhiza (mi-ko-ri'za; pl. mycorrhizae): a symbiotic relationship between a fungus and the roots of a land plant.

nitrogen fixation: the process of converting atmospheric nitrogen (N_2) to ammonium (NH_4^+).

nitrogen-fixing bacteria: bacteria that possess the ability to remove nitrogen (N_2) from the atmosphere and combine it with hydrogen to produce ammonium (NH_4^+).

nodule: a swelling on the root of a legume or other plant that consists of cortex cells inhabited by nitrogen-fixing bacteria.

nutrient: a substance acquired from the environment and needed for survival, growth, and development of an organism.

nutrition: the process of acquiring nutrients from the environment and, if necessary, processing them into a form that can be used by the body.

pressure flow theory: a model for transport of sugars in phloem, which states that movement of sugars into a phloem sieve tube causes water to enter the tube by osmosis, while movement of sugars out of another part of the same sieve tube causes water to leave by osmosis; the resulting pressure gradient causes bulk movement of water and dissolved sugars from the end of the tube into which sugar is transported toward the end of the tube from which sugar is removed.

root pressure: hydrostatic pressure generated by osmosis of water across the endodermis of a plant root, which can force water some distance up the xylem into the stem.

transpiration (trans'-per-ā-shun): evaporation of water from a leaf.

STUDY QUESTIONS

1. Compare and contrast animal and plant nutrition, especially with respect to their sources of energy and materials.

2. How are minerals and water taken up by roots? Diagram the structures involved, the pathways for water and minerals from soil water to xylem, and the transport processes at each step.

3. What is nitrogen fixation? Describe the formation of a root nodule in a legume, and the nature of the relationship between legume and bacteria in the nodule.

4. Describe the cohesion–tension theory of water movement in xylem.

5. Describe the pressure flow theory of movement in phloem.

6. Describe the daily cycle of guard cells, and explain how the movement of water into and out of guard cells causes them to open and close the central pore between them.

DISCUSSION QUESTIONS

1. A new mutant form of aphid, the klutzphid, has just appeared. Instead of feeding on phloem, it inserts its stylet into the vessel elements of xylem. What materials are found in the fluids of xylem? Would xylem fluid flow into the aphid? Justify your answer.

2. One of the foremost goals of plant molecular biologists is to insert the genes for nitrogen fixation, or the ability to enter into symbiotic relationships with nitrogen-fixing bacteria, into crop plants such as corn or wheat (see Chapter 14). Why would this be useful? What changes in farming practices would this allow?

3. We learned in Chapter 2 about the peculiar characteristics of water. Discuss several ways that the evolution of vascular plants has been greatly influenced by water's special characteristics.

4. When you buy cut flowers, the florist may advise you to do two things: Recut the stems while holding the stems under water, and add some table sugar to the water in which you place the arranged flowers so that they "stay fresh" longer. Discuss, in a technical manner, why both suggestions are good advice.

SUGGESTED READINGS

Brill, W. J. "Biological Nitrogen Fixation." *Scientific American*, March 1977. A description of the bacteria and algae that are the sole natural source of usable nitrogen for plant life.

Mansfield, T. A., and Davies, W. J. "Mechanisms for Leaf Control of Gas Exchange." *BioScience*, March 1985. How stomata control gas exchange through the surface of a leaf.

Raven, P., Evert, R. F., and Eichhorn, S. *Biology of Plants*. 4th ed. Worth; New York, 1986. A beautifully illustrated botany text, particularly strong on the evolution of flowering plants and the ecological adaptations of plants to their environments.

Zimmerman, M. H. "How Sap Moves in Trees." *Scientific American*, March 1963. A delightful description of the use of aphids as research tools in botany.

28

Plant Reproduction and Development

In early spring, wildflowers carpet the California hills. The colors, scents, and shapes attract insects that pollinate the flowers, ensuring a new display next year.

As you walk through a wildflower-strewn meadow, you may be tempted to think that the floral display was created just for your enjoyment. Unfortunately for the human ego, plants don't develop flowers for us, but for the birds and bees—and beetles, moths, and even bats. The flower, you see, is a sexual display that enhances the reproductive output of a plant. By enticing animals to transfer pollen from one plant to another, flowers enable stationary plants to "court" distant members of their own species. This critical selective advantage has allowed the flowering plants to become the dominant plants on land.

As you know, evolution commonly produces new structures by modifying old ones, and flowers are no exception. Flowers are not wholly new adaptations; rather, they are the most recent and most sophisticated elaboration of the reproductive strategies common to all plants. You will be able to understand the structure and function of flowers more easily if we begin, therefore, by briefly reviewing the essentials of the plant life cycle.

Reproduction and the Plant Life Cycle

Sexual versus Asexual Reproduction

Many plants can reproduce either sexually or asexually. Asexual reproduction in plants usually involves part of a single plant, say a stem, giving rise to a new plant. The cells that form an asexually produced offspring arise by mitosis from cells of the parent plant. Therefore, these offspring are genetically identical to the parent (see Chapter 9). In Chapter 26 we encountered several methods of asexual reproduction, including runners in strawberries and rhizome sprouting in irises. Asexual reproduction is often a highly effective reproductive strategy. For example, an offspring strawberry connected to its parent by a runner (see Fig. 26-21) draws nourishment from the parent until it grows large enough to fend for itself.

However, if an offspring is genetically identical to its parent, then the offspring is only as well adapted to the environment as its parent was. What if the environment changes? Sexually produced offspring usually combine genes from two different parents, and therefore they may be endowed with traits that differ from those of either parent. This new combination of traits may help the offspring adapt to the environment better than either parent could. As a result, most organisms reproduce sexually, at some time.

Plant Life Cycles

The sexual life cycle of plants is more complex than the familiar animal life cycle. In animals, cells in the gonads of a diploid adult undergo meiosis to produce haploid gametes, either sperm or eggs. Sperm and egg fuse to create a diploid fertilized egg, the **zygote**. Through repeated mitosis and differentiation of the daughter cells, the zygote develops into another diploid adult. **Plants have two distinct, multicellular "adult" forms, one diploid and one haploid, that give rise to each other. For this reason, the plant life cycle is named** *alternation of generations,* **as diploid adults alternate with haploid adults** (see Figs. 28-1 and 28-2).

Let's examine the life cycle of a fern (Fig. 28-1), starting with the diploid adult form. This stage of the life cycle, the **sporophyte** ("spore plant" in Greek), bears reproductive cells that undergo meiosis to produce haploid cells that are **spores,** not gametes. The difference between spores and gametes is that spores do not fuse together to reform a diploid cell. Instead, a fern spore is blown off the parent frond by the wind and lands on the soil. There the spore germinates, dividing repeatedly by mitosis to form a multicellular, haploid organism. This organism produces gametes, and hence is called the **gametophyte** (Greek for "gamete plant"). Since its cells are haploid already, the "gonads" of the gametophyte can produce sperm and eggs without further meiosis. Usually a single gametophyte produces both sperm and eggs, although often at different times, thereby preventing self-fertilization. Sperm and egg fuse to form a zygote that develops into a new diploid sporophyte plant.

Alternation of generations occurs in all plants. In primitive land plants, including mosses and ferns, the gametophyte is an independent, although usually small, plant. It liberates mobile sperm cells that reach an egg by swimming through thin films of water covering adjacent gametophytes, or by being splashed by raindrops from one plant to the next. Therefore, ferns and mosses can live only in moist habitats. Many terrestrial habitats, however, are not so liberally supplied with water. To reproduce in these drier places, a plant must surround its sperm in a desiccation-proof package, transport that package to another plant, and liberate the sperm directly into the egg-bearing structures of the second plant.

The seed plants (conifers and flowering plants) do just that. In the flowering plants, two types of spores

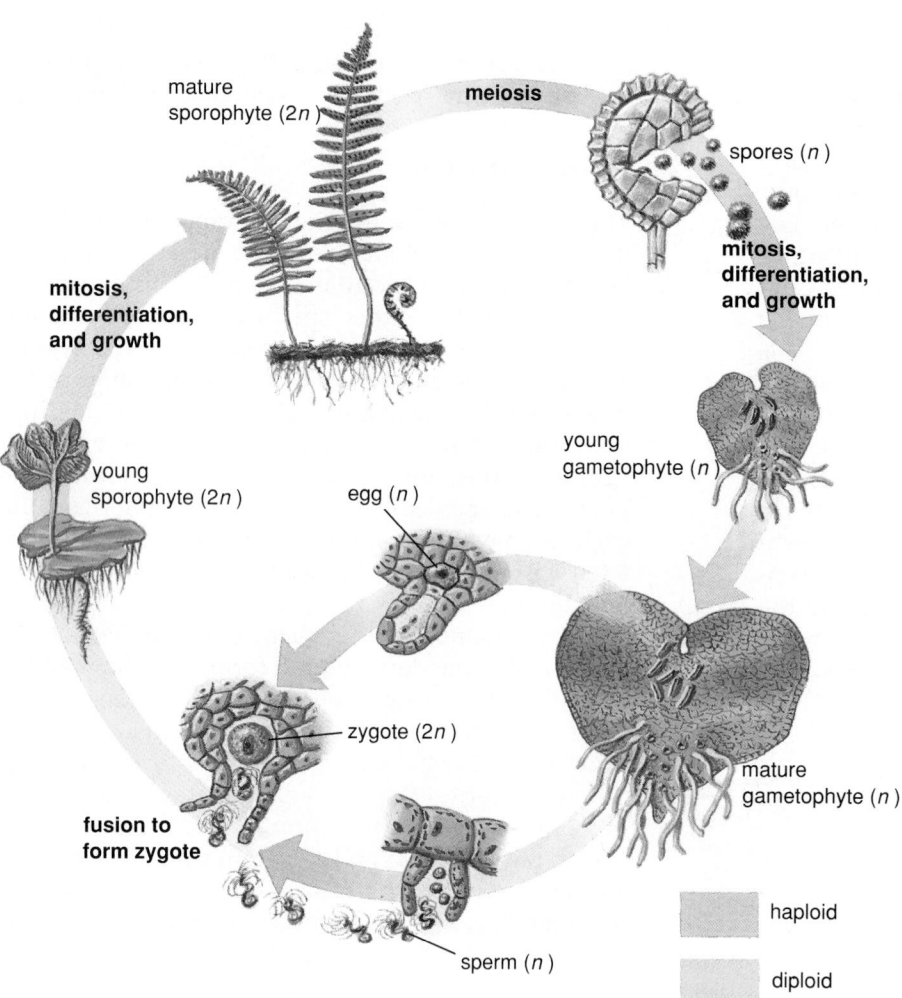

mature
sporophyte (2*n*)

meiosis

spores (*n*)

**mitosis,
differentiation,
and growth**

**mitosis,
differentiation,
and growth**

young
gametophyte (*n*)

young
sporophyte (2*n*)

egg (*n*)

zygote (2*n*)

mature
gametophyte (*n*)

**fusion to
form zygote**

sperm (*n*)

haploid

diploid

Figure 28-1 Ferns typify the alternation of generations life cycle found in all plants, in which separate multicellular haploid and multicellular diploid "adult" organisms occur at different parts of the life cycle. The various stages of the life cycle are described in the text.

are formed by meiosis within the flowers borne by the sporophyte generation (Fig. 28-2). These spores develop into gametophytes not in the soil, but within the flower. One type, the megaspore, undergoes a few mitotic cell divisions and develops into the female gametophyte, a small cluster of cells permanently retained within the flower. The other type of spore, the microspore, develops into the male gametophyte: a tough, watertight **pollen grain** containing two sperm. The pollen grain drifts on the wind or is carried by an animal from one flower to another. On the recipient flower, the pollen grain elongates, burrowing through the flower tissues to the female gametophyte within. This miniature male gametophyte liberates its sperm inside the female gametophyte, where fertilization occurs. The zygote becomes enclosed in a drought-resistant **seed** that may lie dormant for months, years, or even centuries waiting for favorable conditions for growth.

In this chapter, we examine sexual reproduction in flowering plants, from the evolution of the flower through the formation of the seed and the development of the new seedling.

The Evolution of Flowers

The earliest seed plants were the gymnosperms, represented today mainly by pines, firs, and other conifers. As we described in Chapter 24, conifers bear male and female gametophytes on separate cones. During early spring, the small male cones release hordes of pollen grains that waft about on the breezes (Fig. 28-3). Most blow uselessly away, but with so many grains floating around, some enter the pollen chambers located on the scales of the female cones, where they are captured by sticky coatings of sugars and resins. The pollen grains germinate and tunnel to the female gametophytes at the base of each scale. Sperm are liberated, fertilize the eggs within the female gametophyte, and a new generation begins.

Clearly, wind pollination can be an inefficient oper-

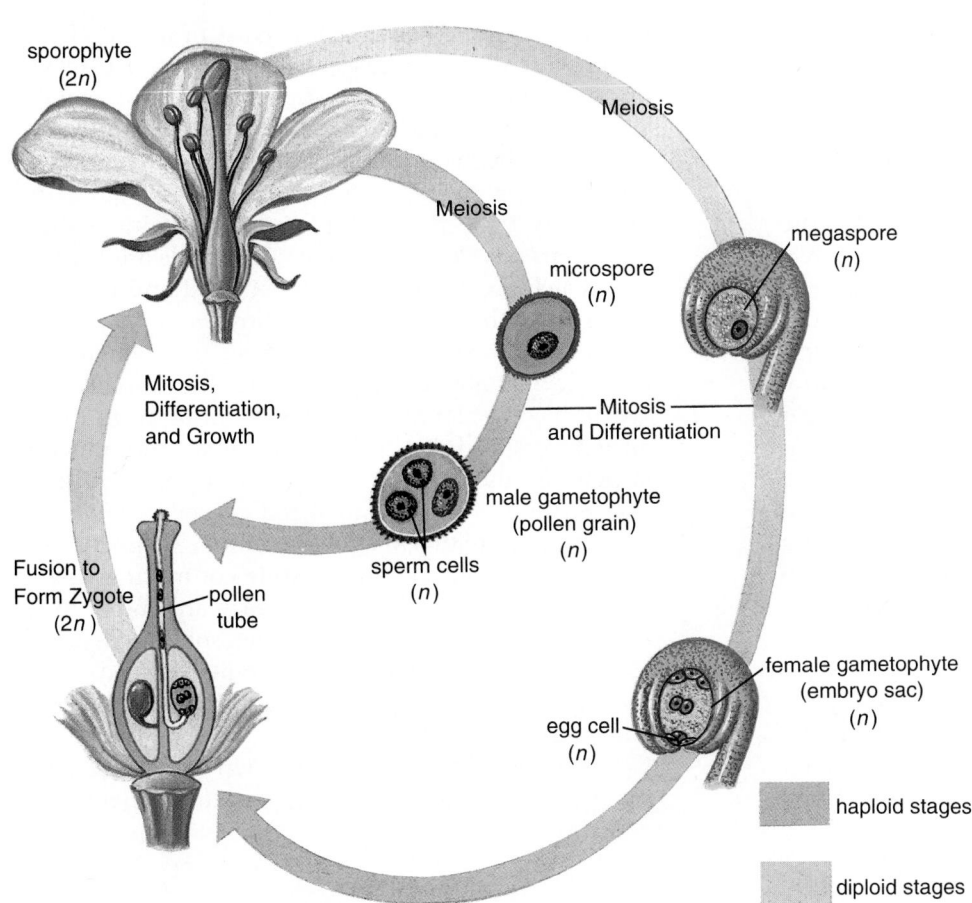

sporophyte
(2n)

Meiosis

Meiosis

Mitosis,
Differentiation,
and Growth

Mitosis —
and Differentiation

microspore
(n)

megaspore
(n)

male gametophyte
(pollen grain)
(n)

Fusion to
Form Zygote
(2n)

pollen
tube

sperm cells
(n)

female gametophyte
(embryo sac)
(n)

egg cell
(n)

haploid stages

diploid stages

Figure 28-2 The life cycle of a flowering plant shows the same stages as the life cycle of a fern (see Fig. 28-1). However, the haploid stages are considerably reduced in size and cannot live independently of the diploid plant.

ation, since the overwhelming majority of pollen grains are lost. In a world of stationary plants and mobile animals, if a gymnosperm could entice an animal to carry its pollen from male to female cone, it would greatly enhance its reproductive rate and hence its evolutionary success. As it happens, gymnosperms and insects were poised to establish just such a relationship about 150 million years ago.

Insects, especially beetles, are among the most abundant animals on Earth. They exploit nearly every possible food resource on land, including the reproductive parts of gymnosperms. About 150 million years ago, some beetles fed on both the protein-rich pollen of male cones and the sugar-rich secretions of female cones. Beetles can make quite a mess when they feed, and pollen-feeders often wind up with pollen dusted all over their bodies. If the same beetle were to visit one plant, eating pollen, and then wander over to another plant of the same species to dine on the sugary secretions of a female cone, some of the loose pollen would quite likely rub off on the female cone.

Figure 28-3 Conifers are wind-pollinated. Even slight breezes blow thick clouds of pollen from ripe male cones.

The stage was set for the evolution of flowering plants. Efficient pollination by insects requires that a given insect visit several plants of the same species, pollinating them on the way. For the plants, two key adaptations were necessary. First, enough pollen and/or sugary secretions (nectar) must be produced within the reproductive structures (the future flowers) so that insects would regularly visit them to feed. Second, the location and richness of these storehouses of pollen and nectar must be advertised to the insects, both to show them where to go and to entice them to specialize on that specific plant species. Any mutation that contributed to these adaptations would enhance the reproductive potential of the plant carrying the mutation, and would be favored by natural selection. By about 130 million years ago, flowers had evolved with exactly these adaptations. The advantages of flowers are so great that in today's temperate and tropical zones, flowering plants are overwhelmingly dominant, and a host of animals, including bees, moths, butterflies, hummingbirds, and even some mammals, feed almost exclusively at flowers.

Flower Structure

Flowers, like leaves, develop as outgrowths of stems, and in fact the various flower parts have evolved from leaves. **Complete flowers,** such as those of cro-

cuses, roses, and tomatoes, consist of a central axis upon which four successive sets of modified leaves are attached (Fig. 28-4). The first set, at the base of the flower, are the **sepals.** In dicots, the sepals are often green and leaflike, while in monocots they usually resemble the petals. In either case, sepals surround and protect the flower bud as the remaining three structures develop. Just above the sepals are the **petals,** which are usually brightly colored and fragrant, advertising the location of the flower.

The male reproductive structures, the **stamens,** are attached just above the petals. Stamens usually consist of a long slender **filament** bearing at its tip an **anther** that produces pollen. The female reproductive structures, the **carpels,** occupy the uppermost position in the flower. An idealized carpel is somewhat vase-shaped, with a sticky **stigma** for catching pollen mounted atop an elongated **style** connecting it with the bulbous **ovary.** Inside the ovary are one or more **ovules** in which the female gametophytes develop. When mature, the ovule will become the seed, while the ovary will develop into a protective, adhesive, or edible enclosure, the fruit.

As you may know from your own gardening experience, not all flowers are complete. **Incomplete flow-**

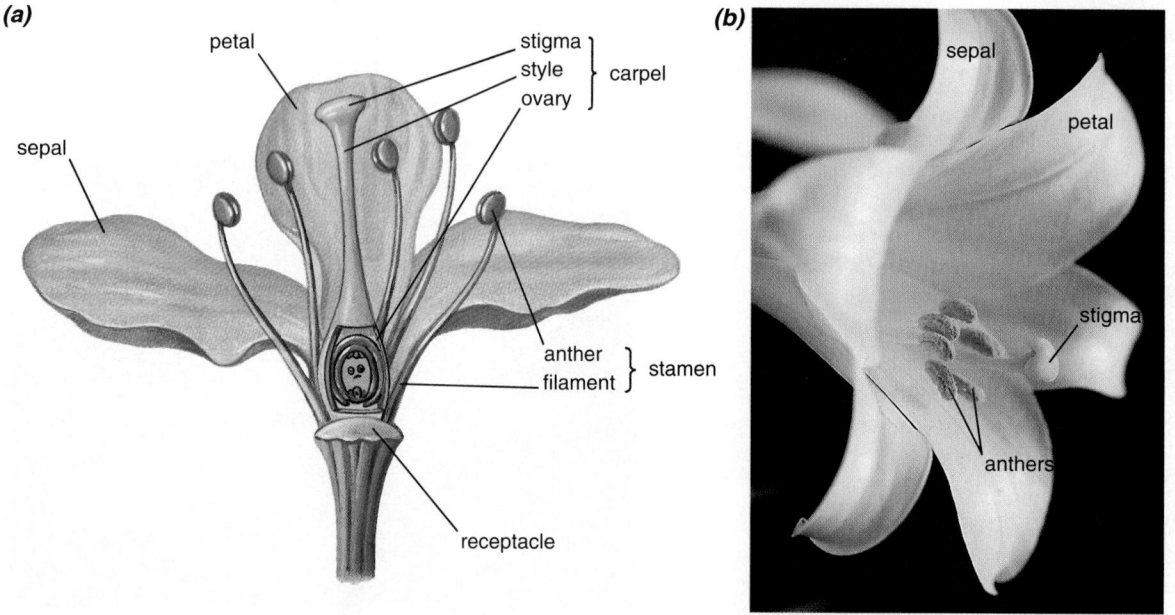

Figure 28-4 **(a)** A complete flower has four parts: sepals, petals, stamens (the male reproductive structures), and at least one carpel (the female reproductive structure). **(b)** The lily is a complete monocot flower, with three sepals (virtually identical to the petals), three petals, six stamens, and three fused carpels. Each stamen consists of a filament bearing an anther at its tip. The carpels consist of an ovary hidden in the base of the flower, with a long style protruding out, ending in a sticky stigma. Note that the anthers are considerably below the stigma. This is probably an adaptation preventing self-pollination: pollen cannot simply fall from the anther onto the stigma.

ers lack one or more of the four floral parts. For example, many plants have separate male and female flowers, which may be borne on the same plant, as in cucumbers and squashes (Fig. 28-5), or on different plants, as in the American holly. Male flowers lack carpels, while female flowers lack stamens. Incomplete flowers may also lack sepals or petals.

Coevolution of Flowers and Pollinators

Wind-pollinated flowers, such as those of grasses and oaks, are usually inconspicuous and unscented, often scarcely more than naked stamens that liberate pollen to the wind (Fig. 28-6). The beautiful flowers so much admired by humans, however, are pollinated by animals. The distinctive shapes, colors, and odors of animal-pollinated flowers and the sensory capabilities and life styles of their pollinators are an example of **coevolution**—evolution in two species that interact extensively with one another, so that each acts as a major force of natural selection on the other. Animal-pollinated flowers must attract useful pollinators and frustrate undesirable visitors who might eat nectar or pollen without fertilizing the flower in return. The animals, in their turn, have been under selective

Figure 28-6 The flowers of grasses and many deciduous trees are wind pollinated, with anthers (yellow structures hanging beneath flowers) exposed to the wind. Petals are usually reduced or absent.

Figure 28-5 Plants of the squash family, such as these zucchinis, bear separate female (left) and male (right) flowers. Obviously, individual flowers cannot be self-pollinated, but the plant could still be self-pollinated if an insect carried pollen from a male flower to a female flower of the same plant. However, each plant initially produces only male flowers, so some cross-pollination between plants that flower at slightly different times is virtually assured.

pressures to locate flowers quickly, identify the ones that can provide them with adequate nutrition, and extract the nectar or pollen with a minimum expenditure of energy.

Animal-pollinated flowers can be loosely grouped into three categories, depending on the benefits (real or imagined) that they offer to potential pollinators: food, sex, or a nursery.

Food

Many flowers provide food for foraging animals such as bees or hummingbirds. In return, the animals unwittingly distribute pollen from flower to flower.

Beetle-pollinated Flowers

Beetles dine on a wide variety of foods, including both living and dead plants and animals, and animal dung. Since animal tissue usually has more protein, fat, and calories than plant tissue, many beetles preferentially feed on animal material. Beetle-pollinated flowers often smell like rotting carrion or dung, which attracts scavenging beetles.

Most beetles are quite clumsy and unspecialized for feeding at flowers. Consequently, beetle-pollinated flowers are usually simple, open affairs, with everything except the ovaries in plain sight and within easy reach. A beetle may eat nectar, pollen, and even petals, strewing everything about in the process. Some pollen sticks to its body and may rub off on the next flower it visits. Flies have much the same taste in foods (or lack thereof!) as beetles, and most fly-pollinated flowers such as the carrion flower (Fig. 28-7) also emit a powerful stench.

Bee-pollinated Flowers

Bees are culinary specialists, often feeding only on nectar and pollen. Bees scout out prospective dinners

Figure 28-7 The carrion flower is pollinated by flies, which are attracted by its aroma of rotting meat.

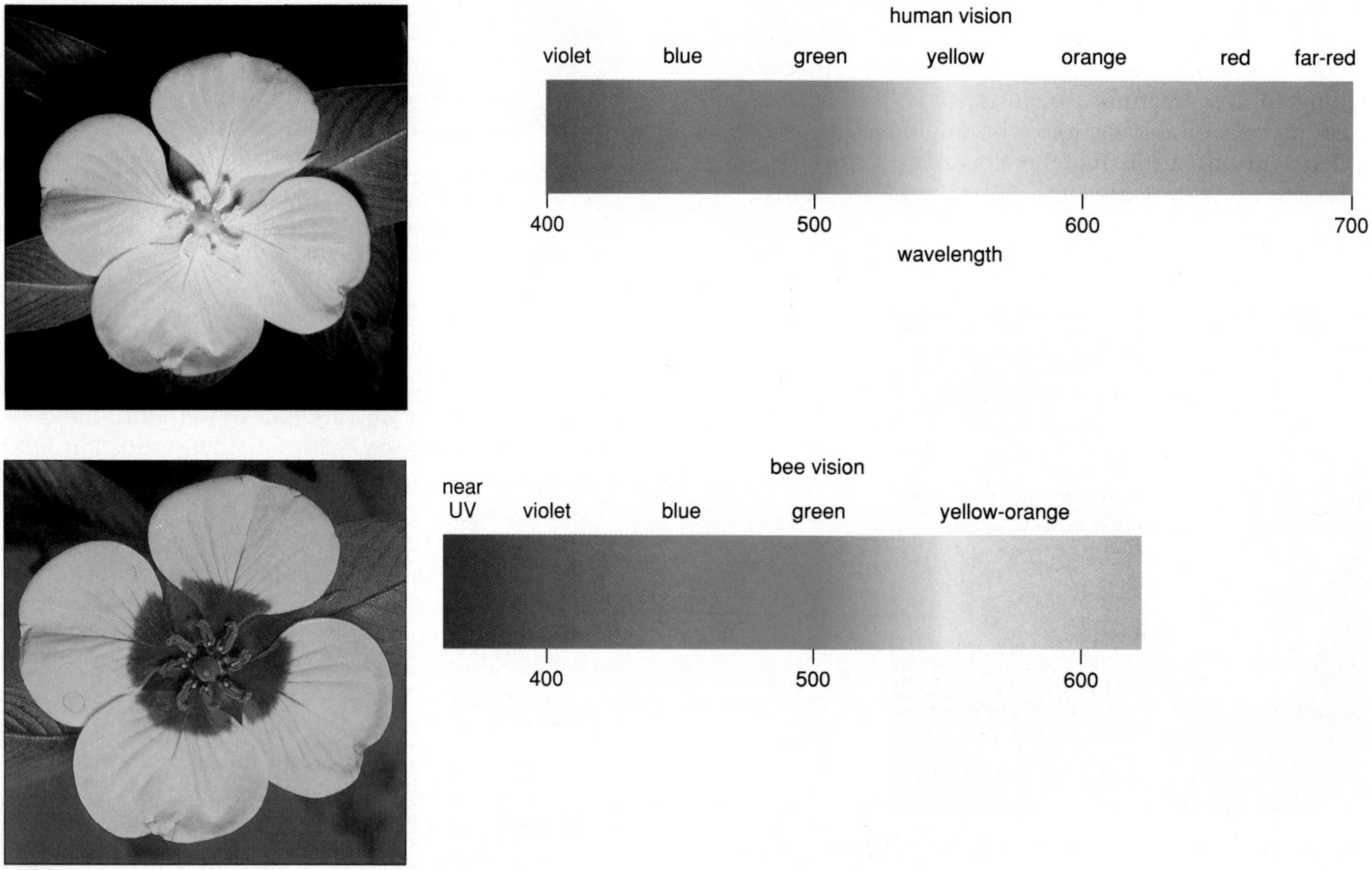

Figure 28-8 The spectra of color vision for humans and bees overlap considerably in the blue, green, and yellow ranges, but differ on the edges. Humans are sensitive to orange and red, which bees do not perceive, whereas bees can see ultraviolet, which is invisible to the human eye. Many flowers photographed under ordinary daylight (upper left) and under ultraviolet light (lower left) show striking differences in color patterns. Bees can see the ultraviolet patterns that presumably lead them to the nectar- and pollen-containing centers of the flowers.

from the air, using both scent and sight to locate and identify flowers. We can thank the bees for most of the sweet-smelling flowers, since sweet "flowery" odors attract these pollinators. Bees also have good color vision, but do not see exactly the same range of colors that humans do (Fig. 28-8). Although unable to distinguish red from gray or black, their color vision extends into the ultraviolet. To attract a bee from afar, bee-pollinated flowers must look brightly-colored *to a bee*. Typically, these flowers are white, yellow, or blue, and often have other markings, such as central spots or lines pointing toward the center, which reflect ultraviolet light (Fig. 28-8).

Bee-pollinated flowers have several structural adaptations that help to ensure pollen transfer. Many bee-pollinated flowers, such as nasturtiums and foxgloves, produce nectar at the bottom of a tube (Fig. 28-9). Either pollen-laden stamens (usually in newly opened flowers) or the sticky stigma of the carpel (in older flowers) protrude out of the top of the tube.

Figure 28-9 Many bee-pollinated flowers, such as these foxgloves, are tubular. Nectar is produced at the base of the tube, while the anthers and stigmas protrude out the open end. Often a lower "lip" on the tube serves as a landing platform for bees, since bees are not very good at hovering in mid-air.

When a bee visits a young flower, she lands on the lip of the flower and thrusts her head into the tube to reach the nectar. Simultaneously, the stamens brush pollen onto her back. She may then visit an older flower and repeat her foraging behavior. This time, she leaves pollen behind on the stigma.

Moth- and Butterfly-pollinated Flowers

Many moths and butterflies also feed on nectar. Flowers adapted for these pollinators are often superficially similar to bee-pollinated flowers, except that the nectar tubes are usually deep and narrow. Thus, while bees can neither crawl into the narrow tube nor reach the nectar with their shorter tongues, the long tongues of moths and butterflies can reach the nectar at the base of the tube. Day-flying butterflies are attracted by white, yellow, blue, and orange flowers with mild, sweet fragrances. Flowers pollinated by night-flying moths open only in the evening, are usually white, and exude strong, musky odors that help the moth locate the flower in the dark.

Hummingbird-pollinated Flowers

Hummingbirds are one of the few vertebrates that are important pollinators, although several mammals also visit flowers (Fig. 28-10). Birds have notoriously poor senses of smell, and hummingbird-pollinated flowers seldom synthesize fragrant chemicals. However, hummingbirds need lots of energy, and these flowers always produce large amounts of nectar. If they didn't, the hummers would go elsewhere and the flower would remain unpollinated. On the other hand, a large supply of nectar would also attract insects, and a bee might return again and again to the same flower, never transferring pollen to another flower. Not surprisingly, hummingbird-pollinated flowers have evolved several adaptations that keep insects from drinking their nectar. These flowers are always tubular, matching the long bills and tongues of hummers (Fig. 28-11). The tube is much too deep for bees to reach the nectar at its base. In addition, most are red or orange, which is brightly attractive to a bird but drab to a bee.

Sex

A few plants, most notably the orchids, take advantage of the insatiable libido and stereotyped behaviors of male wasps and flies to pollinate their flowers. Some orchid flowers mimic female wasps both in scent and shape (Fig. 28-12). The males land atop these "females" and attempt to copulate, but only get a packet of pollen for their efforts. Further rendezvous with other orchids result in pollination for the flowers.

Nursery Flowers

Perhaps the most elaborate relationships between plants and pollinators occur in a few cases in which insects fertilize a flower and then lay their eggs in the flower's ovary. This arrangement is found between milkweeds and milkweed bugs, figs and certain wasps, and yuccas and yucca moths (Fig. 28-13). The yucca moth performs a remarkable series of behaviors resulting in pollination of yuccas and a well-stocked pantry for its own offspring. A female moth visits a yucca flower, collects pollen, and rolls it into a compact ball. The moth flies off with the pollen ball to another yucca flower, drills a hole in the ovary wall, and lays its eggs inside the ovary. Then it takes its pollen ball and smears pollen all over the stigma of the flower! By pollinating the yucca, the moth ensures that the plant will provide a supply of developing seeds for its offspring caterpillars to eat. It is most unlikely that the moth "knows" that it must fertilize the yucca so that its young have enough to eat, but nonetheless it performs the genetically programmed behavior flawlessly. Since the caterpillars eat only a fraction of the seeds, the yucca also reproduces successfully. The mutual adaptation of yucca and moth is so complete that neither can reproduce without the other.

Figure 28-11 A hummingbird hovers before a flower. Note that the flower has a deep, tubular shape and lacks the landing platform commonly found on bee-pollinated flowers.

Figure 28-12 This male wasp is actually trying to copulate with an orchid flower. The result is successful reproduction, not for the wasp, but for the orchid.

(a)

(b)

◀ **Figure 28-10** Among the more unusual pollinators are bats and honeypossums. **(a)** A tropical bat feeds at a cluster of tubular flowers. Note the protruding stamens and stigma. As the bat hovers before the flower, the top of its head touches either the anthers or stigma or both, thus pollinating the flower. **(b)** As the honeypossum stuffs its face into this flower, pollen adheres to its muzzle and whiskers. A visit to another flower may result in transfer of pollen.

(a)

(b)

stamen   carpal

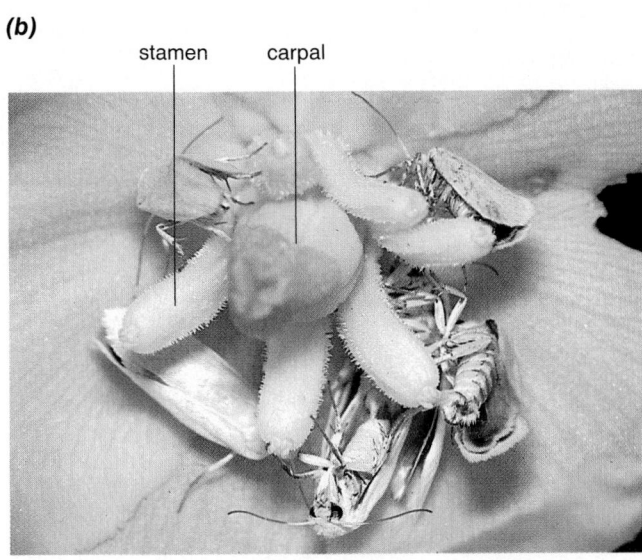

Figure 28-13 (a) Yuccas bloom on the dry plains of eastern Colorado in early summer. **(b)** Within many of the yucca flowers, yucca moths carry out their part in one of nature's most unusual and most effective cooperative relationships between plant and animal.

Gametophyte Development in Flowering Plants

As Figure 28-2 illustrates, in the life cycle of flowering plants, the familiar plant of meadow, garden, and farm is the diploid sporophyte. The **pollen grain** (male) and the **embryo sac** (female) are the haploid gametophytes that develop within the flowers borne by the sporophytes. Both are much smaller than the gametophyte stages of ferns and mosses, and cannot live independently of the sporophyte.

Pollen

An anther consists of four chambers called pollen sacs (Fig. 28-14). Within each sac, hundreds to thousands of diploid **microspore mother cells** develop. Each microspore mother cell undergoes meiosis (see Chapter 10) to produce four haploid **microspores.** Each microspore divides once, by mitosis, to produce a male gametophyte, or pollen grain, consisting of only two cells: a large **tube cell** and a smaller **generative cell** that resides *within the cytoplasm of the tube cell* (Fig. 28-14). A tough surface coat develops around the pollen grain, protecting the cells within during their journey to the carpel (Fig. 28-15).

When ripe, the pollen sacs split open. In wind-pollinated flowers such as those of grasses and oaks, the pollen spills out, a fortunate few to be carried by wind currents to other flowers of the same species. In animal-pollinated flowers, the pollen adheres weakly to the anther case until the pollinator comes along and brushes or picks it off.

Embryo Sac

Within an ovary, one or more dome-shaped masses of cells differentiate into **ovules.** Each ovule consists of outer layers of cells called **integuments** that surround a single, diploid **megaspore mother cell** (Fig. 28-16). The megaspore mother cell divides by meiosis to produce four large haploid **megaspores.** Three megaspores degenerate, and only one survives. This remaining megaspore undergoes an unusual set of mitotic divisions. Three nuclear divisions produce a total of eight haploid nuclei. Plasma membranes then divide up the cytoplasm into *seven*, not eight, cells: three small cells at each end, with one nucleus apiece, and one remaining large cell in the middle with two

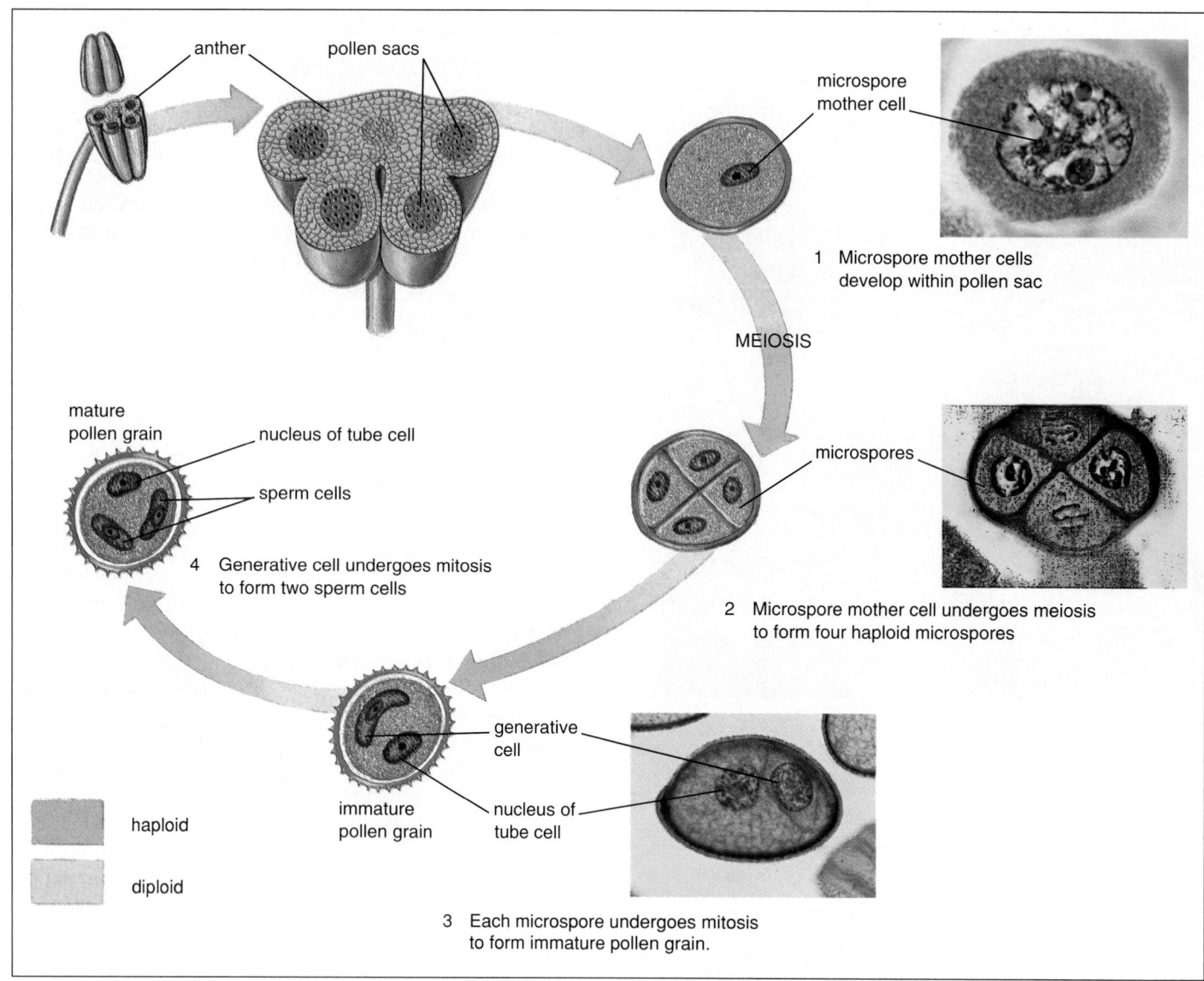

anther   pollen sacs

microspore
mother cell

1 Microspore mother cells
   develop within pollen sac

MEIOSIS

mature
pollen grain   nucleus of tube cell

sperm cells

4 Generative cell undergoes mitosis
   to form two sperm cells

microspores

2 Microspore mother cell undergoes meiosis
   to form four haploid microspores

generative
cell

haploid

diploid

immature   nucleus of
pollen grain   tube cell

3 Each microspore undergoes mitosis
   to form immature pollen grain.

Figure 28-14 Pollen development in angiosperms. (1) Within the anthers, diploid microspore mother cells form. (2) These cells undergo meiosis to produce four haploid microspores. (3) Through mitosis, each microspore divides into two (still haploid) cells, the tube cell and the generative cell. The entire generative cell resides within the cytoplasm of the tube cell. These two cells form the juvenile pollen grain (the male gametophyte stage of flowering plants). (4) Later (in some species, after pollination), the generative cell divides to form two sperm cells, which still remain within the tube cell cytoplasm. This three-celled "organism" is the mature male gametophyte.

nuclei. This seven-celled organism, called the **embryo sac,** is the haploid female gametophyte. The central, binucleate cell is the **primary endosperm cell.** The **egg** is one of the cells at the bottom of the embryo sac, near a pore in the integuments.

Pollination and Fertilization

When a pollen grain lands on the stigma of a flower of the same species of plant, a remarkable chain of events occurs (Fig. 28-17). The pollen grain absorbs

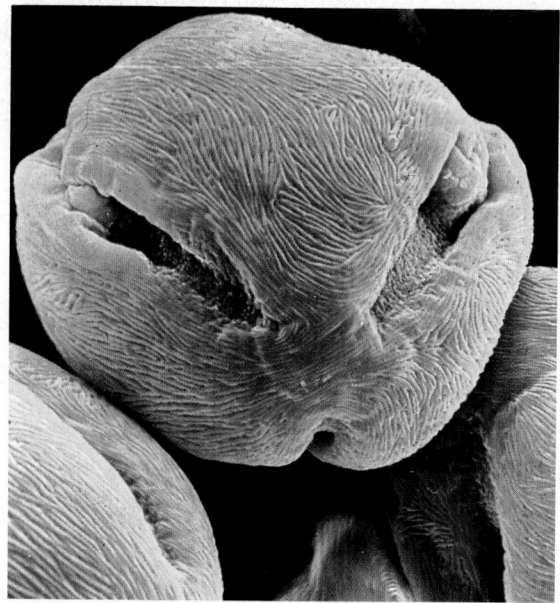

Figure 28-15 The tough outer coverings of pollen grains are often elaborately sculptured in species-specific shapes and patterns. This pollen grain from a pear tree shows three prominent furrows (the one on the "bottom" is hard to see). When the pollen grain germinates on the stigma of a recipient flower, the pollen tube will emerge through one of the furrows.

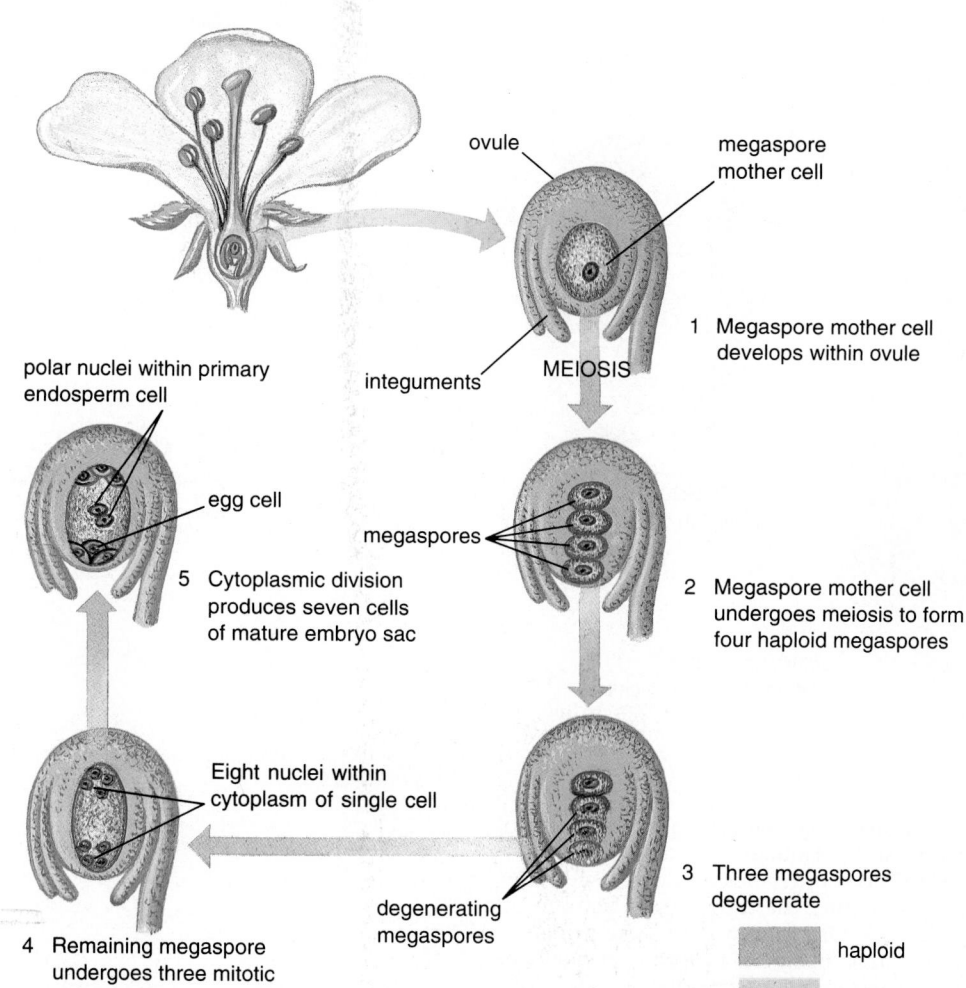

ovule

megaspore mother cell

integuments

MEIOSIS

1 Megaspore mother cell develops within ovule

polar nuclei within primary endosperm cell

egg cell

megaspores

5 Cytoplasmic division produces seven cells of mature embryo sac

2 Megaspore mother cell undergoes meiosis to form four haploid megaspores

Eight nuclei within cytoplasm of single cell

degenerating megaspores

4 Remaining megaspore undergoes three mitotic divisions to form eight nuclei

3 Three megaspores degenerate

haploid

diploid

Figure 28-16 Development of the female gametophyte. (1) A single diploid megaspore mother cell matures within the integuments of an ovule. (2) Through meiosis, it gives rise to four haploid megaspores. (3) Three of these degenerate. (4) The nucleus of the remaining megaspore then divides mitotically three times, producing eight haploid nuclei, four at each end of the as-yet-undivided cell. (5) Cytoplasmic division then forms seven cells, six with one nucleus apiece, and one with two nuclei. One of the uninucleate cells, near the pore in the integuments, is the egg cell. The other five uninucleate cells degenerate soon after fertilization. The large primary endosperm cell contains the remaining two nuclei.

water from the stigma. The tube cell elongates, growing down the style toward an ovule in the ovary. Meanwhile, the generative cell achieves puberty, so to speak, and divides mitotically to form two **sperm cells.** The resulting three-celled organism (one tube cell containing two sperm cells) is the mature male gametophyte.

If all goes well, the pollen tube reaches the pore in the integument of an ovule and breaks into the embryo sac. Its tip ruptures, releasing the two sperm. One sperm fertilizes the egg cell to form the diploid zygote that will develop into a new sporophyte. The second sperm enters the primary endosperm cell. Its nucleus fuses with *both endosperm nuclei,* forming a triploid nucleus. Through repeated mitotic divisions, the primary endosperm cell will develop into the triploid **endosperm,** a food storage organ within the seed. The fusion of the egg with one sperm and the primary endosperm cell with the second sperm is often called **double fertilization,** and is unique to flowering plants. The other five cells of the embryo sac degenerate soon after fertilization.

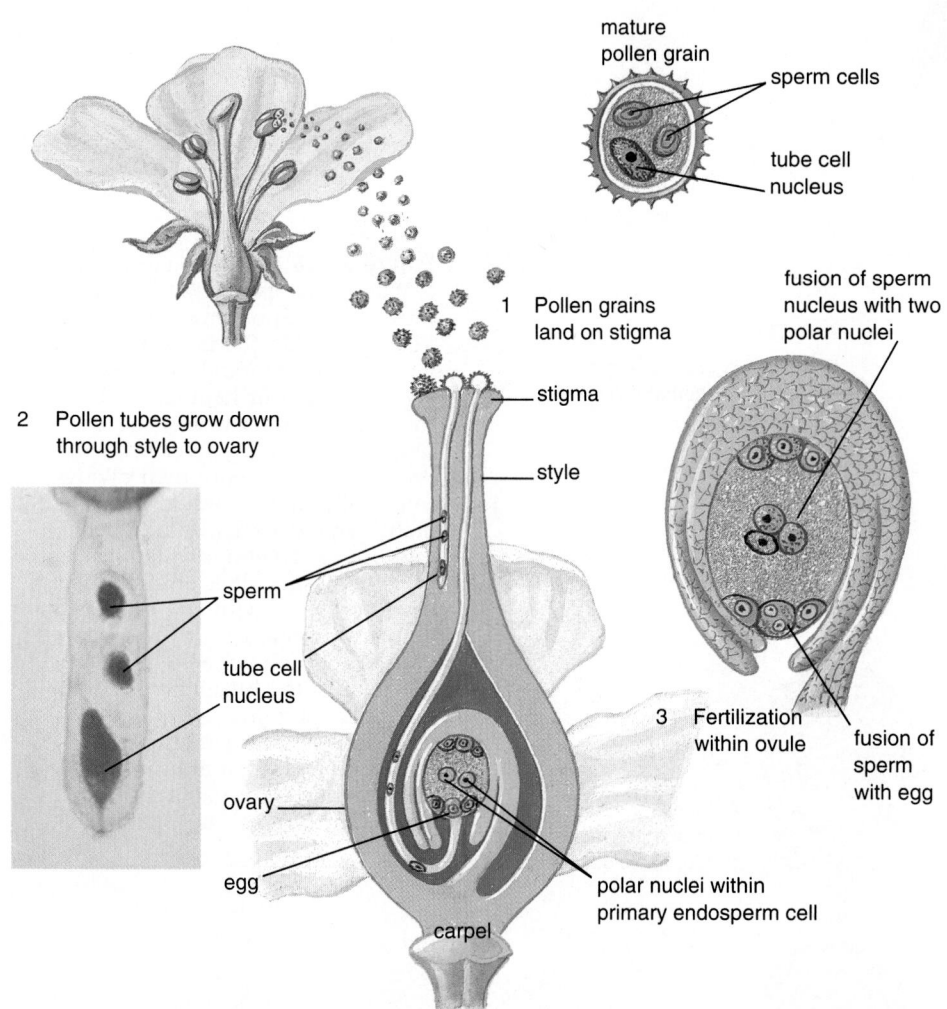

mature pollen grain

sperm cells

tube cell nucleus

1   Pollen grains land on stigma

2   Pollen tubes grow down through style to ovary

stigma

style

sperm

tube cell nucleus

ovary

egg

carpel

polar nuclei within primary endosperm cell

fusion of sperm nucleus with two polar nuclei

3   Fertilization within ovule

fusion of sperm with egg

Figure 28-17 Pollination and fertilization of a flower. (1) The pollen grain lands on the sticky, often textured surface of the stigma and germinates. (2) The tube cell elongates, burrowing down through the style. The sperm nuclei follow, inside the cytoplasm of the tube cell. (3) Upon reaching an ovule, the tube breaks through into the embryo sac, and the tube nucleus degenerates. One sperm enters the primary endosperm cell; its nucleus fuses with both endosperm nuclei to form a triploid cell. The other sperm fertilizes the egg cell to form the diploid zygote that will develop into the new embryonic plant. The other cells of the embryo sac degenerate.

Incidentally, you should note the distinction between pollination and fertilization. **Pollination** occurs when a pollen grain lands on a stigma, while **fertilization** is the fusion of sperm and egg. Although pollination is obviously a prerequisite for fertilization, they are two separate events. For example, pollination will not lead to fertilization if the tube cell fails to grow properly, if the embryo sac is sterile, or if a sperm from another pollen grain has already reached the egg.

The Development of Seeds and Fruits

Drawing upon the resources of the parent plant, the embryo sac develops into a seed, surrounded by the accessory tissues of a fruit (Fig. 28-18).

Seed Development

The integuments of the ovule develop into the **seed coat,** a thin, tough, waterproof outer covering. As we shall see, the characteristics of the seed coat play a role in regulating when the seed will sprout.

Meanwhile, within the integuments, two distinct developmental processes occur (Fig. 28-19a). First, the triploid endosperm cell divides rapidly. Its daughter cells absorb nutrients from the parent plant, forming a large, food-filled endosperm. Second, the zygote develops into the embryo. Both dicot and monocot embryos consist of three parts: the shoot, the root, and the **cotyledons,** or seed leaves. The cotyledons absorb food molecules from the endosperm and transfer them to the embryo. In dicots ("two cotyledons"), the cotyledons usually absorb most of the endosperm during seed development, so that the mature seed is virtually filled with embryo (Fig. 28-19b). In monocots ("one cotyledon"), the cotyledon absorbs some of the endosperm during seed development, but most of the endosperm remains in the mature seed (Fig. 28-19c).

The future primary root develops at one end of the main embryonic axis. The future shoot, at the other end, is usually divided into two regions at the site of

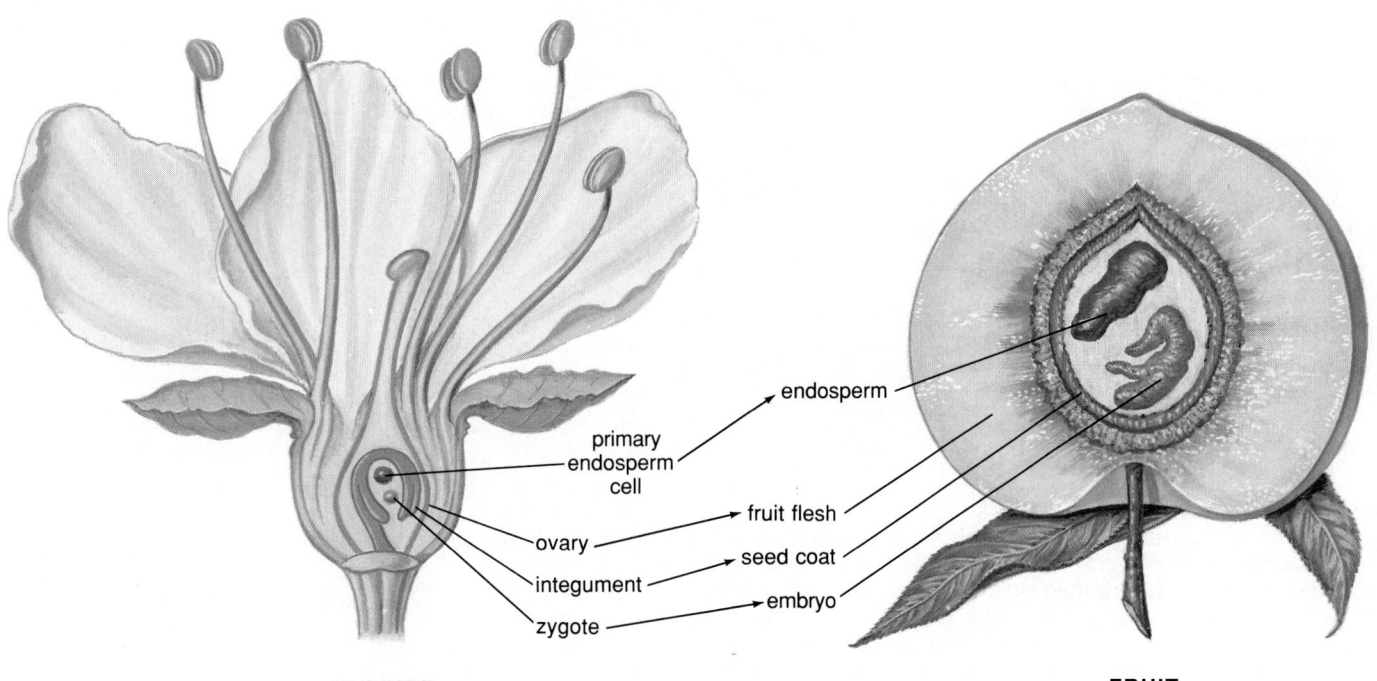

FLOWER **FRUIT**

Figure 28-18 The fruit and the seed within it develop from various parts of the flower. Starting at the outside, the ovary wall ripens into the fruit flesh, which may be soft and tasty like a peach, or variously hard, hooked, or tufted in other fruits. The integuments of the individual ovule, which surround the embryo sac, harden and become waterproof, forming the seed coat. These two fruit parts are derived from tissues of the parent sporophyte plant. Within the seed, two structures develop from cells of the fertilized female gametophyte. The triploid endosperm cell divides repeatedly, absorbs nutrients from the parent plant, and becomes the endosperm, a food storage structure within the seed. Finally, the zygote develops into the embryo.

(a)

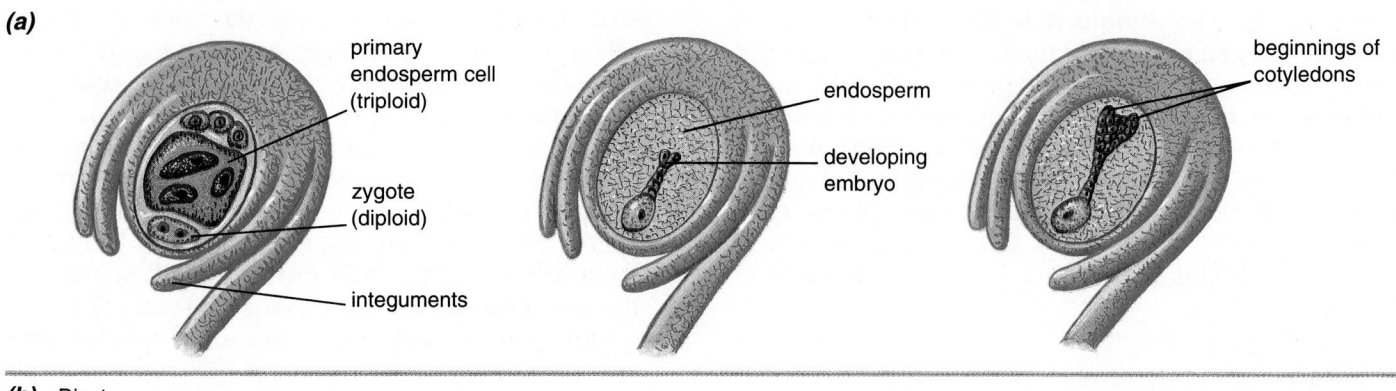

(b) Dicot

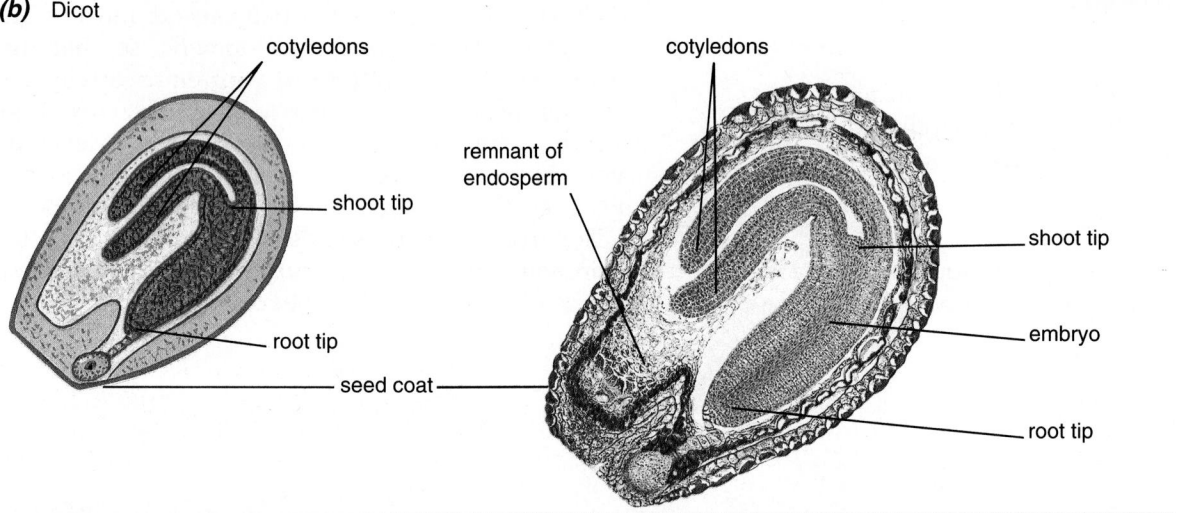

(c) Monocot

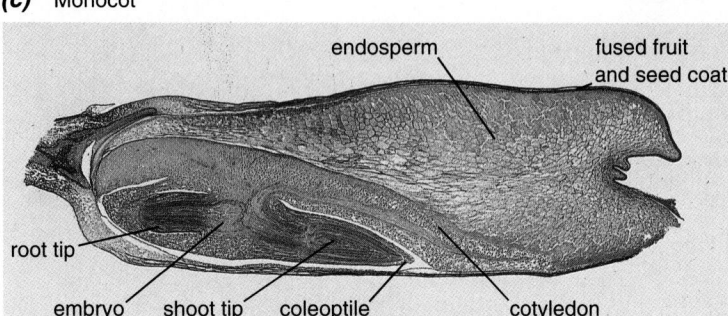

Figure 28-19 Seed development. **(a)** The endosperm develops first, absorbing nutrients from the parent plant. The embryo develops later, absorbing nutrients from the endosperm to fuel its growth. Monocot and dicot seeds are similar, but differ in three major respects: the number of cotyledons, the fate of the endosperm, and the means of protecting the tender shoot tip (see Fig. 28-20). **(b)** Dicot seeds such as the shepherd's purse have two cotyledons, and usually these absorb virtually the entire endosperm as the seed develops, so that the mature seed is mostly cotyledon (the seed shown here is not quite mature, and still has a remnant of endosperm left). **(c)** Monocot seeds, such as the corn kernel shown here, retain a large endosperm. (Corn meal is the ground-up endosperm of corn seeds.) The embryo produces a single cotyledon. As the seed germinates, the cotyledon absorbs the food reserves of the endosperm and transfers them to the growing embryo. The delicate shoot tip is protected within a tough sheath, the coleoptile, which pushes up through the soil when the seed sprouts.

attachment of the cotyledons. Below the cotyledons, but above the root, is the **hypocotyl** (*hypo* in Greek means "beneath" or "lower"), while above the cotyledons the shoot is called the **epicotyl** (*epi* means "above"). At the tip of the epicotyl lies the apical meristem of the shoot, often already with one or two developing leaves.

Fruit Development

The wall of the ovary develops into a **fruit** (see Fig. 28-18). There are a bewildering variety of fruits, with outer layers that are variously fleshy, hard, winged, or even spiked like a medieval mace. The selective forces favoring the evolution of all fruits, however, are similar: **fruits help to disperse the seeds to distant locations away from the parent plant** (see "A Closer Look At Adaptations for Seed Dispersal").

Seed Dormancy

All seeds need warmth and moisture to germinate. However, most newly matured seeds will not germinate immediately, even under ideal conditions. Instead, they enter a period of **dormancy** during which they will not sprout. Seed dormancy solves two problems, one intrinsic to the plant itself, and one related to environmental factors. First, many seeds develop inside juicy fruits, such as apples, grapes, or oranges. If a seed germinates while still enclosed in a fruit and hanging from the tree or vine, it may exhaust its food reserves before it ever touches the ground. When the fruit finally falls, the seedling would then lack the energy to burrow its roots into the soil, and would die. Second, environmental conditions of temperature and precipitation suitable for seedling growth may not coincide with seed maturation. Seeds that mature in late summer in temperate climates, for example, face the harsh winter to come. Spending the winter as a dormant seed is clearly preferable to death by freezing as a tender young sprout.

Enforcing and Breaking Dormancy

Plants have evolved many different mechanisms that produce seed dormancy, and the seeds of each species have their own set of requirements that must be met before germination can occur. Perhaps the three most common requirements are an initial desiccation, exposure to cold, and disruption of the seed coat.

1. **Desiccation:** Many seeds must dry out before they are able to germinate. This prevents the seed from germinating while still within the fruit. Such seeds are often dispersed by animals that eat fruit, but cannot digest the seeds, which are therefore excreted in their feces.
2. **Cold:** Seeds of many temperate and arctic plants will not germinate unless exposed to prolonged subfreezing temperatures, followed by warmth and moisture. This ensures that they stay dormant during balmy days in autumn, and sprout only after winter yields to spring.
3. **Disruption of the seed coat:** The seed coat itself is often a barrier to seed germination. Many seed coats are impermeable to water and oxygen, or bind the developing embryo so tightly that growth simply cannot occur, or contain chemicals that inhibit germination (see Chapter 29). In deserts, for example, rainfall is spotty and scarce. Years may go by without enough water for plants to germinate, grow, flower, and set more seed. Therefore, the seed must not sprout unless a given rainfall is heavy enough to allow the plant to complete its life cycle, since another rain may not fall in time. The seeds of most desert plants have water-soluble chemicals in the seed coat that inhibit germination. Only a hard rainfall can wash away enough of the inhibitors to allow sprouting.

Germination and Growth of Seedlings

During **germination**, the formerly dormant embryo resumes growth and emerges from the seed. The embryo absorbs water, which makes it swell and burst open its seed coat. The root is usually the first structure to emerge from the seed coat, growing rapidly and absorbing water and minerals from the soil. Much of the water is transported to cells in the shoot. As its cells elongate, the stem lengthens, pushing up through the soil.

Protecting the Shoot Tip

The growing shoot faces a serious difficulty: it must push through the soil without scraping away the apical meristem and tender leaflets at its tip. A root, of course, must always contend with tip abrasion, and its apical meristem is protected by a root cap (see Fig. 26-10). Shoots spend most of their time in the air and do not develop permanent protective caps. Instead,

A CLOSER LOOK

At Adaptations for Seed Dispersal

Successful plant reproduction requires more than fertilizing a flower and developing a seed. It also requires a suitable site for the seed to germinate and the young plant to grow. Gardeners usually think of "suitable sites" only in terms of fertile soil, equable weather, and moisture. From an evolutionary perspective, however, suitability is also a function of distance from the parent plant. The growth of a seedling that sprouts right next to its parent will be inhibited by the larger plant's shade. If the offspring should manage to survive, it will compete with its parent for water and nutrients, to the detriment of both plants. Finally, a plant species will become more widespread if its members send out some seeds to distant habitats. In flowering plants, seed dispersal is the function of fruits. A wide variety of fruits have evolved, each dispersing seeds in a different way.

Shotgun Dispersal

A few plants disperse their seeds in a very straightforward way: they develop explosive fruits that eject their seed meters away from the parent plant. Mistletoes, for example, are common parasites of trees. Mistletoe fruits shoot out sticky seeds. If one strikes a nearby tree, it adheres to the bark and germinates, sending rootlike fibers into the vascular tissues of its host, from which it draws its nourishment. Since the proper germination site for a mistletoe seed is not the ground, which it can reach simply by falling, but a tree limb many meters above, it is clearly useful to shoot the seeds up, up, and away.

Wind Dispersal

Dandelions, milkweeds, and maples (Fig. E28-1) produce lightweight fruits with surfaces that catch the wind. (Yes, each individual hairy tuft on a dandelion ball is a separate fruit!) Each fruit typically contains a single small seed, which reduces weight and lets the fruit remain aloft longer. These featherweight fruits aid the seed in traveling away from the parent plant, from a few meters for maples to miles for a milkweed or dandelion on a windy day.

Water Dispersal

Many fruits can float on water for a time, and may be dispersed by streams and rivers. The coconut fruit, however, is a floater *par excellence*. Round, buoyant, and watertight, the coconut drops off its parent palm, rolls to the sea, and floats for weeks or months until it washes ashore on some distant isle (Fig. E28-2). There it germinates, perhaps establishing a new coconut colony on a formerly barren island.

Animal Dispersal

Perhaps the majority of fruits use animals as agents of seed dispersal. Two quite distinct strategies have evolved for dispersal by animals: grab an animal as it passes by, or entice it to eat the fruit but not digest the seeds.

Anyone who takes a long-haired dog on a walk through an abandoned field knows about fruits that hitchhike on animal fur. Burdocks, burr clover, foxtails,

(a)

(b)

(c)

Figure E28-1 Wind-dispersed fruits usually contain only one or two lightweight seeds. Some, such as dandelions **(a)** and milkweeds **(b)** have filamentous tufts that catch the breezes. Others, such as maple fruits **(c)**, are actually miniature glider-helicopters, silently whirling away from the tree as they fall. To see how the wings aid in seed dispersal, take two maple fruits and pluck the wing off one. Hold both fruits over your head and drop them. The wingless fruit will fall at your feet, while the winged one will glide some distance away.

Figure E28-2 After a long journey at sea, this coconut was washed high onto a beach by a storm. The large size and massive food reserves of coconuts are probably adaptations required for successful germination and seedling growth on barren, sandy beaches.

away part of the seed coat. Besides transport away from its parent, a seed that is swallowed and excreted benefits in another way: it ends up with its own supply of fertilizer!

Figure E28-3 The burdock hitches a ride on animal fur. This mouse is trying to rid itself of its troublesome passengers.

and sticktights all develop fruits with prongs, hooks, spines, or adhesive hairs (Fig. E28-3). The parent plants hold these fruits very loosely, so that even slight contact with fur pulls the fruit free of the plant and leaves it stuck on the animal. Some of these fruits don't hold on to the fur very tightly either, and may fall off the next time the animal brushes against a tree or rock, or may come out when the animal grooms its fur. Others, such as burdocks, embed themselves very firmly, and it takes some determined effort to dislodge the fruit.

Unlike these hitchhiker fruits, edible fruits benefit both animal and plant. The plant stores sugars and tasty flavors in a fleshy fruit surrounding the seeds, luring hungry animals into eating the fruit (Fig. E28-4). Some fruits, such as peaches and plums, contain large, hard seeds that animals usually do not eat. After eating the flesh of the fruit, the animals discard the seeds. Other fruits, including blackberries, raspberries, strawberries, and tomatoes, have small seeds that are swallowed along with the fruit flesh. The seeds then pass through the animal's digestive tract without harm. In some cases, passing through an animal's gut may even be essential to seed germination, by abrading or digesting

Figure E28-4 These bright red fruits have attracted a cedar waxwing. Only ripe fruits with mature seeds inside have bright coloration, sweet tastes, and exotic flavors. Unripe fruits are usually green, hard, and bitter, which makes them unpalatable to birds and mammals. This too is an evolutionary adaptation. The immature seeds within unripe fruit may not survive passage through an animal's gut; if the fruits ripen before the seeds, the plant is doomed to reproductive failure.

germinating shoots have other mechanisms that cope with the abrasion of sprouting (Fig. 28-20). In monocots, a tough sheath, the **coleoptile,** encloses the shoot tip like a glove encloses a gardener's fingers (Fig. 28-20a). The coleoptile "glove" pushes aside the soil particles as it grows. Once out into the air, the coleoptile tip degenerates, allowing the tender "fin-

ger" of shoot to emerge. Dicots do not have coleoptiles. Instead, the dicot shoot forms a hook in either the hypocotyl or epicotyl (Fig. 28-20b). The bend of the hook, encased in epidermal cells with tough cell walls, leads the way through the soil, clearing the way for the downward-pointing apical meristem with its delicate new leaves.

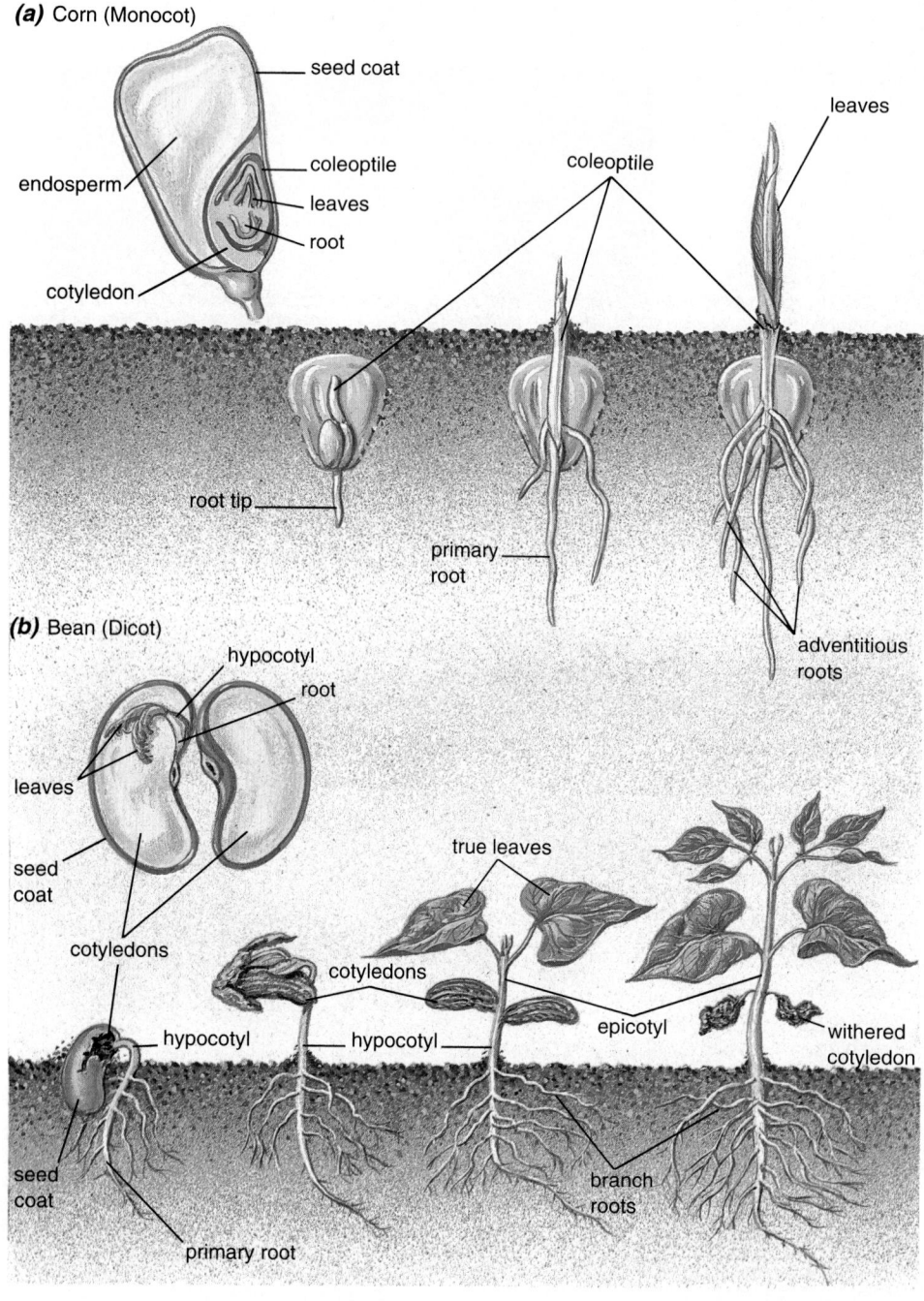

(a) Corn (Monocot)

seed coat

endosperm

coleoptile

leaves

root

cotyledon

coleoptile

leaves

root tip

primary root

adventitious roots

(b) Bean (Dicot)

hypocotyl

root

leaves

seed coat

cotyledons

hypocotyl

seed coat

primary root

cotyledons

hypocotyl

true leaves

cotyledons

epicotyl

withered cotyledon

branch roots

Figure 28-20 Seed germination in corn, a monocot **(a)**, and the common bean, a dicot **(b)**, differ in detail, but involve the same three principles: (1) use of stored food to provide energy for seedling growth until photosynthesis can take over; (2) rapid growth and branching of the root to absorb water and nutrients; and (3) protection of the delicate shoot tip as it moves upward through the soil.

The Fate of the Cotyledons

Food stored in the seed provides the energy for sprouting. You will recall that the cotyledons of dicots have already absorbed the endosperm while the seed was developing, and are now fat and full of food. In dicots with hypocotyl hooks, the elongating shoot carries the cotyledons out of the soil into the air. These above-ground cotyledons often become green and photosynthetic (Fig. 28-21), and transfer both previously stored food and newly synthesized sugars to the shoot. In dicots with epicotyl hooks, the cotyledons stay below ground, shrivelling up as the embryo absorbs their stored food. Monocots retain most of their food reserves in the endosperm until germination, when it is digested and absorbed by the cotyledon as the embryo grows. The cotyledon remains below ground in the remnants of the seed.

Controlling the Development of the Seedling

Once out in the air, the shoot rapidly spreads its leaves to the sun. Simultaneously, the root system delves into the soil. The apical meristem cells of shoot and root divide, giving rise to the mature structures discussed in Chapter 26. Eventually this plant, too, will mature, flower, and set seed, renewing the cycle of life. How this cycle is regulated—why shoots grow upward while roots grow downward, and how plants produce flowers at the proper time of year—is the subject of the next chapter.

Figure 28-21 In the squash family, a hypocotyl hook carries the cotyledons out of the soil. The cotyledons expand into photosynthetic leaves (the pair of smooth oval leaves). The first true leaf (crinkled single leaf) develops a little later. Eventually, the cotyledons shrivel up and die.

Reflections on the Coevolution of Plants and Animals

Flowering plants dominate terrestrial ecosystems largely because of the mutually beneficial relationships they forged with the animals that pollinate their flowers and disperse their seeds. Some plant and animal species have become totally dependent on one another. Such a relationship occurs between yuccas and yucca moths: the yucca is pollinated only by the moth, while the moth reproduces only in the fruit of the yucca. Absolute dependence is a risky proposition, however; if one partner becomes extinct, it may mean the extinction of the other as well.

A case in point is the story of the dodo bird and the Calvaria tree (Fig. E28-5). Dodos were flightless birds about the size of turkeys, found only on the island of Mauritius in the Indian Ocean. Dodos fed primarily on fruits and seeds. With their sizeable beaks, they could gulp down large fruits whole, crunching them up in their powerful gizzards. (Birds do not have teeth, but a specialized compartment of the digestive tract, the gizzard, serves the same function. The muscular gizzard is filled with stones and acts like a mill, churning food and stones together until the food is ground up.) Mariners of the sixteenth and seventeenth centuries found the large, slow dodos to be easy sources of fresh meat after a long journey at sea, and by 1681 they had hunted the dodo to extinction.

Calvaria trees produce large, edible fruit something like a peach, with a pulpy outside surrounding a stone-hard pit. Formerly one of the most common trees on Mauritius, only a few ancient specimens survive today. Judging by the ages of these patriarchs, no Calvaria tree has successfully reproduced in about 300 years, even though the remaining trees still set normal fruit each year. Ecologist Stanley Temple has found out why.

Temple believes that the Calvaria tree is endangered because it depends exclusively on the dodo for seed dispersal and germination. Eons ago, ancestral dodos fed on the fruits of ancestral Calvaria trees. Only fruits with strong pits had any chance of surviving the trial-by-gizzard within a dodo. As dodos became larger with ever-stronger gizzards, Calvarias were selected for

Figure E28-5 The dodo bird and Calvaria tree were formerly abundant denizens of the island of Mauritius in the Indian Ocean. Following the extinction of the dodo by European sailors, the Calvaria tree also declined, because its large, hard seeds must be processed through the digestive tract of the dodo before they can germinate.

harder and harder pits. Supremely strong pit walls, however, also inhibited germination. In fact, the Calvaria pit came to *require* processing through the dodo before it could germinate: the pit walls must be partially ground away before the seed can sprout. By eliminating the dodo, humans inadvertently doomed the Calvaria tree as well. Now that we know the problem, a few Calvaria trees can be propagated in nurseries, by filing away the pit wall, or, as Temple discovered, by force-feeding Calvaria pits to domestic turkeys. However, the forests of Calvarias that formerly clothed Mauritius, providing habitat for wildlife and lumber for people, are gone forever.

SUMMARY OF KEY CONCEPTS

Reproduction and the Plant Life Cycle

The sexual life cycle of plants, called alternation of generations, includes both a multicellular diploid form (the sporophyte generation) and a multicellular haploid form (the gametophyte generation). In the seed plants, the gametophyte stage is greatly reduced. The male gametophyte is the pollen grain, a drought-resistant structure that can be carried from plant to plant by wind or animals. The female gametophyte is also reduced, and is retained within the body of the sporophyte stage. In this way, seed plants can reproduce independently of liquid water.

The Evolution of Flowers

Flowering plants evolved from gymnosperms. In gymnosperms, pollen blows on the wind from male cones to female cones. In many habitats, flowering plants enjoy a selective advantage over gymnosperms because many types of flowers attract insects that carry pollen from plant to plant.

Flower Structure

A complete flower consists of four parts: sepals, petals, stamens (male reproductive structures), and carpels (female reproductive structures). The sepals form the outer covering of the flower bud. The petals (and sometimes the sepals) are usually brightly colored and attract pollinators to the flower. The stamen consists of a filament that bears at its tip an anther in which pollen (the male gametophyte) develops. The carpel consists of the ovary in which one or more embryo sacs (the female gametophytes) develop, and a style that bears at its end a sticky stigma to which pollen adheres during pollination. Incomplete flowers lack one or more of the four floral parts.

Coevolution of Flowers and Pollinators

Most flowers are pollinated by the wind or by animals, usually insects or birds. Flowers show specific adaptations to pollination by particular pollinators, due to the extensive coevolution between flowering plants and their pollinators.

Gametophyte Development in Flowering Plants

Pollen develops in the anthers. A diploid cell, the microspore mother cell, undergoes meiosis to produce four haploid microspores. Each of these divides mitotically to form pollen grains. An immature pollen grain consists of two cells: the tube cell and the generative cell. The generative cell divides once to produce two sperm cells.

The embryo sac develops within the ovules of the ovary. A diploid megaspore mother cell undergoes meiosis to form four haploid megaspores. Three of these degenerate; the fourth undergoes three sets of mitotic divisions to produce the eight nuclei of the embryo sac. These eight nuclei come to reside in only seven cells. One of these cells, with a single nucleus, is the egg cell; another, with two nuclei, is the primary endosperm cell. These two cells are involved in seed formation; the rest of the cells degenerate.

Pollination and Fertilization

Pollination is the transfer of pollen from anther to stigma. When a pollen grain lands on a stigma, its tube cell grows through the style down to the embryo sac. The generative cell divides to form two sperm cells that travel down the style within the tube cell, eventually entering the embryo sac. One sperm fuses with the egg to form a diploid zygote, which will give rise to the embryo. The other sperm fuses with the binucleate primary endosperm cell to produce a triploid cell. This cell will give rise to the endosperm, a food storage organ within the seed.

The Development of Seeds and Fruits

The embryo develops a root, shoot, and cotyledons. Cotyledons digest and absorb food from the endosperm, and transfer it to the growing embryo. Monocot embryos have one cotyledon, while dicot embryos have two. The seed is enclosed within a fruit that develops from the ovary wall. The function of the fruit is to disperse the seeds away from the parent plant.

Seeds often remain dormant for some time after fruit ripening. Environmental conditions involved in breaking dormancy may include an initial desiccation, exposure to cold, or disruption of the seed coat.

Germination and Growth of Seedlings

Seed germination requires warmth and moisture. Energy for germination comes from food stored in the endosperm, transferred to the embryo by the cotyledons.

GLOSSARY

anther (an'-ther): the uppermost part of the stamen in which pollen develops.

carpel (kar'pel): the female reproductive structure of a flower, composed of stigma, style, and ovary.

coevolution: the evolution of adaptations in two species due to their extensive interactions with one another, so that each species acts as a major force of natural selection upon the other.

coleoptile (kō-lē-op'tĭl): a protective sheath surrounding the shoot in monocot seeds.

complete flower: a flower that has all four floral parts (sepals, petals, stamens, and carpels).

cotyledon (kot-ul-ē'don): also called a seed leaf; a leaflike structure within a seed that absorbs food from the endosperm and transfers it to the growing embryo.

dormancy: a state in which an organism does not grow or develop; usually marked by lowered metabolic activity and resistance to adverse environmental conditions.

double fertilization: in flowering plants, a phenomenon in which two sperm nuclei fuse with the nuclei of two cells of the female gametophyte. One sperm fuses with the egg to form the zygote, while the second sperm nucleus fuses with the two haploid nuclei of the primary endosperm cell to form a triploid endosperm cell.

embryo sac: the haploid female gametophyte of flowering plants.

endosperm: a triploid food storage tissue found in the seeds of flowering plants.

epicotyl (ep'-ē-kot-ul): the part of the embryonic shoot located between the tip of the shoot and the attachment point of the cotyledons.

filament: in flowers, the stalk of a stamen, which bears an anther at its tip.

flower: the reproductive structure of a flowering plant.

fruit: in flowering plants, the ripened ovary (plus, in some cases, other parts of the flower).

gametophyte (ga-mēt'-ō-fīt): the multicellular haploid stage in the life cycle of plants.

generative cell: in flowering plants, one of the haploid cells of a pollen grain. The generative cell undergoes mitosis to form two sperm cells.

germination: the growth and development of a seed, spore, or pollen grain.

hypocotyl (hī'-pō-kot-ul): the part of the embryonic shoot located between the attachment point of the cotyledons and the root.

incomplete flower: a flower that is missing one of the four floral parts (sepals, petals, stamens, or carpels).

integument (in-teg'-ū-ment): the layers of the ovule surrounding the embryo sac; develops into the seed coat.

megaspore: a haploid cell formed by meiosis from a diploid megaspore mother cell. Through mitosis and differentia-

tion, the megaspore develops into the female gametophyte.

megaspore mother cell: a diploid cell contained within the ovule of a flowering plant, which undergoes meiosis to produce four haploid megaspores.

microspore: a haploid cell formed by meiosis from a microspore mother cell. Through mitosis and differentiation, the microspore develops into the male gametophyte.

microspore mother cell: a diploid cell contained within an anther of a flowering plant, which undergoes meiosis to produce four haploid microspores.

ovary: in flowering plants, a structure at the base of the carpel containing one or more ovules; develops into the fruit.

ovule: a structure within the ovary of a flower, inside which the female gametophyte develops. After fertilization, the ovule develops into the seed.

petal: part of a flower, often flat and brightly colored, serving to attract potential animal pollinators.

pollen: the male gametophyte of a gymnosperm or flowering plant.

pollination: in flowering plants, the deposition of pollen onto the stigma of a flower of the same species; in conifers, the deposition of pollen within the pollen chamber of a female cone of the same species.

primary endosperm cell: the central cell of the female gametophyte of a flowering plant, containing the polar nuclei (usually two). After fertilization, the primary endosperm cell undergoes repeated mitotic divisions to produce the endosperm of the seed.

seed: the reproductive stage of conifers and flowering plants, usually including an embryonic plant and a food reserve, enclosed within a resistant outer covering.

seed coat: the outermost covering of a seed, formed from the integuments of the ovule.

sepal (sē'-pul): one of the protective outer coverings of a flower bud, often opening into green, leaf-like structures when the flower blooms.

spore: in the alternation of generation life cycle of plants, a haploid cell produced through meiosis, that then undergoes repeated mitotic divisions and differentiation of daughter cells to produce a multicellular, haploid organism called the gametophyte.

sporophyte (spor'-ō-fīt): the multicellular diploid stage of the life cycle of plants.

stamen (stā'-men): the male reproductive structure of a flower, consisting of a filament and an anther in which pollen grains develop.

stigma (stig'-ma): the pollen-capturing tip of a carpel.

style: a stalk connecting the stigma of a carpel with the ovary at its base.

tube cell: the outermost cell of a pollen grain; the tube cell digests a tube through the tissues of the carpel, ultimately penetrating into the female gametophyte.

STUDY QUESTIONS

1. Diagram the plant life cycle, comparing ferns with flowering plants. Which stages are haploid and which are diploid? At which stage are gametes formed?
2. What are the advantages of the reduced gametophyte stages in flowering plants compared to the more substantial gametophytes of ferns?
3. Diagram a complete flower. Where are the male and female gametophytes formed? What are the male and female gametophytes called?
4. Diagram the development of the pollen grain. At what point does meiosis occur?
5. Diagram the development of the embryo sac. At what point does meiosis occur? How many cells and how many nuclei are in a mature embryo sac? Which cells and nuclei ultimately fuse with sperm?
6. Describe the characteristics you would expect to find in flowers that are pollinated by the wind, beetles, bees, and hummingbirds, respectively.
7. What is the endosperm? From which cell of the embryo sac is it derived? How much endosperm is usually present in a mature seed of a dicot? Of a monocot?
8. Describe the function of cotyledons.
9. Describe three mechanisms whereby seed dormancy is broken in different types of seeds. How do these mechanisms relate to the normal environment of the plant?
10. How do monocot and dicot seedlings protect the delicate shoot tip during seed germination?
11. Describe three types of fruits and the mechanisms whereby these fruit structures help to disperse their seeds.

DISCUSSION QUESTIONS

1. Grasses are flowering plants that are wind-pollinated rather than animal-pollinated. Wind pollination would seem to waste large amounts of pollen. What other properties of wind-pollinated flowers might at least partially compensate for this inefficiency? (Hint: Compare the grass flowers of Fig. 28-6 with the animal-pollinated flowers of Figs. 28-7 through 28-13.) How would the abundance (or scarcity) of a particular species of grass affect the likelihood of successful wind pollination?
2. Charles Darwin once described a flower that produced nectar at the bottom of a 25-centimeter-deep tube. He predicted that there must be a moth or other animal with a 25-centimeter-long tongue to match (he was right). Such specialization almost certainly means that this particular flower could be pollinated only by that specific moth. What are the advantages and disadvantages of such specialization?

SUGGESTED READINGS

Barth, F. G. *Insects and Flowers: The Biology of a Partnership.* Princeton, NJ: Princeton University Press, 1985. A fascinating, well-illustrated book that describes how flowers ensure their pollination through their interactions with insects.

Handel, S. N., and Beattie, A. J. "Seed Dispersal by Ants." *Scientific American,* August 1990. Many plants rely on ants for seed dispersal, even producing fat deposits on the outside of the seeds as a lure for the ants. Dispersal is not the only benefit for the seed: being "planted" in an anthill seems to provide an ideal environment for germination and growth, too.

Jordan, W. "The Bee Complex." *Science '84,* May 1984. Although mostly about bumblebees, this article vividly portrays the evolutionary forces behind insect–flower relationships.

Newman, C. "Pollen: Breath of Life and Sneezes." *National Geographic,* October 1984. An interesting look at pollen from the perspectives of plants and allergic humans.

Norstog, K. "Cycads and the Origin of Insect Pollination." *American Scientist,* May–June 1987. The author describes how cycads (a type of gymnosperm) help to unravel the evolution of insect pollination.

Raven, P. H., Evert, R. F., and Eichhorn. S. *Biology of Plants,* 4th ed. New York: Worth, 1986. Lavishly illustrated, with a particularly interesting section on the evolution of flowers and the coevolution between flowers and pollinating animals.

29

Control of the Plant Life Cycle

During autumn, gray squirrels bury many acorns, storing them for the long winter ahead. The acorns that the squirrels do not recover are the major source of new oak trees. How plants control their development, often by responding to specific cues in the environment, is the subject of this chapter.

The time is late fall, the place a woodlot in Ohio. Masses of acorns burden the oaks, and gray squirrels busily gather and bury the nuts, storing them against the coming winter. With so many acorns "squirrelled away," some are forgotten, or are not needed. They sprout next spring as the sun warms the forest soil.

The squirrels, of course, haven't planted the acorns so that new oaks will grow. Each acorn is oriented randomly, perhaps with the future stem end down and root end up. Buried beneath several inches of soil, how does the germinating seedling "know" which way is up? If it does sprout and send its shoot up into the air, how does it develop into a mature oak? How does it distinguish the seasons, flowering in spring and going dormant in fall?

Plants do not have specialized organs for perceiving the environment, comparable to our eyes or ears. Nevertheless, plants *do* perceive many features of their world: the direction of gravity; the direction, intensity, and duration of sunlight; the strength of the wind; and in some cases, even the touch of a fly upon a leaf. Plants also respond to these stimuli by regulating their growth and development in appropriate ways. Usually, plants control their bodily functions through the action of chemicals called hormones or growth regulators. In this chapter we explore how plants detect external stimuli, produce and distribute hormones in response, and how these hormones govern the development of the plant body.

The Discovery of Plant Hormones

Everyone who keeps houseplants on a windowsill knows that they bend toward the window as they grow, in response to the sunlight streaming in. Over a hundred years ago Charles Darwin and his son Francis studied this phenomenon of growth toward the light, or **phototropism.**

Hormonal Mechanisms in Phototropism

The Darwins illuminated grass coleoptiles from various angles (a coleoptile is the protective sheath surrounding a monocot seedling; see Fig. 28-20). They noted that a region of the coleoptile a few millimeters below the tip bent toward the light until the tip faced directly into the light source (Fig. 29-1). If they covered the very tip of the coleoptile with a light-proof cap, the coleoptile didn't bend. If they covered up the bending region, on the other hand, the coleoptile still bent toward the light. The Darwins concluded that (1) the tip of the coleoptile perceives the direction of light and (2) bending occurs further down the coleoptile; therefore (3) the tip must transmit information about the light direction down to the bending region.

How does the coleoptile bend? Although the Darwins didn't know this, **growth in the bending region is entirely due to elongation of preexisting cells. Therefore, the coleoptile must bend due to differential elongation of cells** (Fig. 29-2). The cells on the outside of the bend (away from the light) elongate more than the cells on the inside of the bend (toward the light). If one side of the coleoptile becomes longer than the opposite side, then the whole shaft must bend away from the longer side. Thus we can reinterpret the Darwins' results, and say that **the information transmitted from the tip to the bending region causes greater elongation of cells on the side of the coleoptile shaft away from the light.**

About 30 years later, Peter Boysen-Jensen showed that the information transferred down from the tip is chemical in nature (Fig. 29-3). He cut the tips off coleoptiles, and found that the remaining stump neither elongated nor bent toward the light. If he replaced the tip and put the patched-together coleoptile in the dark, it elongated straight up. In the light, it showed normal phototropism. Placing a thin layer of porous gelatin between the severed tip and the stump still allowed elongation and bending, but an impervious barrier of mica eliminated these responses. Boysen-Jensen concluded that a chemical is produced in the tip and moves down the shaft, causing cell elongation. In the dark, the chemical that causes the cells to elongate diffuses straight down from the tip and causes the coleoptile to elongate straight up. Presumably, light causes the chemical to become more concentrated on the "shady" side of the shaft, so that cells on the shady side elongate faster than cells on the "sunny" side, causing the shaft to bend toward the light.

The next step was to isolate and identify the chemical. In the 1920s, Frits Went devised a way to collect the elongation-promoting chemical. He cut off the tips of oat coleoptiles and placed them on a block of agar (a porous, gelatinous material) for a few hours (Fig. 29-4a). Went hoped that the chemical would migrate out of the coleoptiles into the agar. Removing the tips, he cut up the agar, now presumably loaded with the chemical, and placed small pieces on the tops of coleoptile stumps. If he put a piece of agar squarely atop a stump, the stump elongated straight up (Fig. 29-4b). If he placed a piece on one side of a cut stump, the stump would invariably bend away from the side with the agar (Fig. 29-4c). Went called

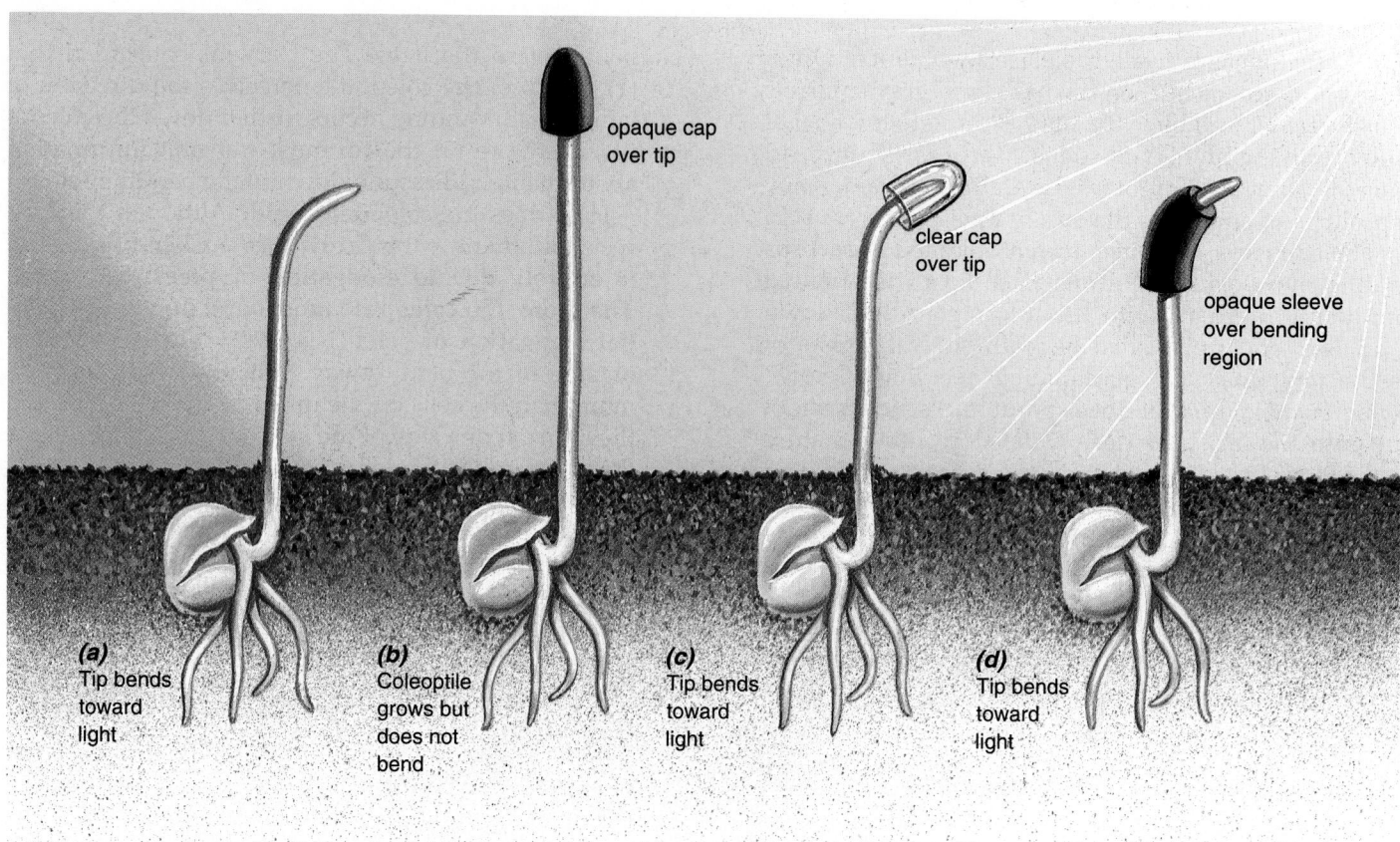

Figure 29-1 The phototropism experiments of Charles and Francis Darwin. **(a)** If light strikes a coleoptile from the side, a region a few millimeters down from the tip bends as it grows, until the tip points toward the light source. **(b)** If the tip is covered with a light-proof cap, the coleoptile does not bend. **(c)** If a clear cap is placed over the tip, bending still occurs, proving that the mere presence of a cap does not prevent bending. **(d)** If a flexible light-proof collar is placed around the bending region, the coleoptile still bends toward the light.

the chemical **auxin,** from a Greek word meaning "to increase." Kenneth Thimann later purified auxin and determined its molecular structure.

Plant Hormones and Their Actions

Animal physiologists have long recognized that chemicals called **hormones** are produced in one location and transported to other parts of the body, where they exert specific effects. By analogy, auxin and other plant-regulating chemicals are called **plant hormones.** So far, plant physiologists have identified five major classes of plant hormones: auxins, gibberellins, cytokinins, ethylene, and abscisic acid (Table 29-1). Several other types of hormones are thought to exist, but have not yet been isolated and identified.

Each hormone can elicit a variety of responses from plant cells, depending on the type of cell, its physiological state, and the presence of other hormones. Further, the roles of some plant hormones vary among plant species. (This should not be surprising; after all, in animals a single hormone can be involved in actions as diverse as a salmon's transition from fresh to salt water, a frog's metamorphosis from tadpole to adult, and a snake's shedding its skin [see the Reflection at the end of Chapter 35]).

Auxin, as we have seen, promotes elongation of cells in coleoptiles and other parts of the shoot. In roots, low concentrations of auxin stimulate elongation, while slightly higher concentrations inhibit elongation. Both light and gravity affect the distribution of auxin in roots and shoots, so that auxin plays a major role in both phototropism and gravitropism (directional growth with respect to gravity). Auxin

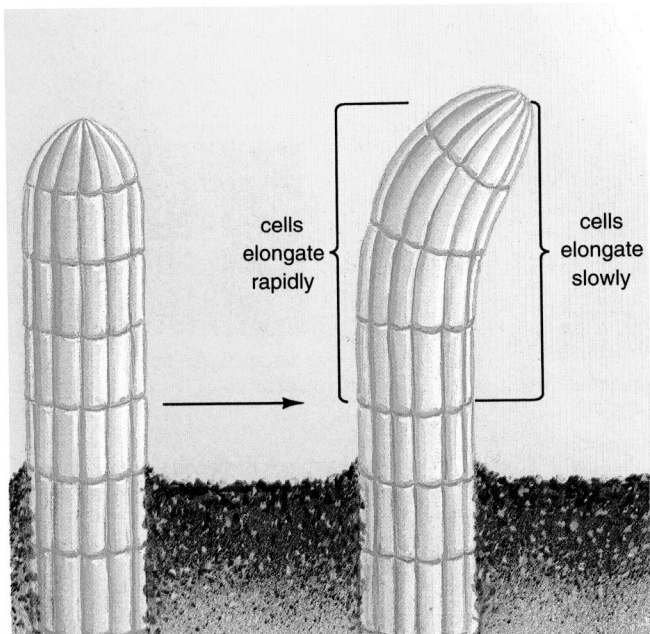

Figure 29-2 Coleoptiles bend by differential elongation of cells on opposite sides of the coleoptile. If cells on one side elongate more rapidly than cells on the other side, the coleoptile bends away from the longer side.

affects many other aspects of plant development, too. It stimulates root branching, the differentiation of vascular tissues, and the development of fruits. Auxin also prevents sprouting by lateral buds.

Gibberellins are a group of chemically similar molecules that, like auxin, promote elongation of cells in stems. In some plants, gibberellins stimulate flowering, fruit development, seed germination, and bud sprouting.

Cytokinins promote cell division in many plant tissues; consequently, they stimulate fruit, endosperm, and embryonic development, and the sprouting of buds. Cytokinins also stimulate plant metabolism, preventing or at least delaying the aging of plant parts, especially leaves.

Ethylene is the only known hormone in plants or animals that is a gas at normal environmental temperatures. Ethylene is best known, and most commercially valuable, for its ability to cause fruit to ripen. It also stimulates the separation of cell walls in abscission layers, allowing leaves, flowers, and fruit to drop off at the appropriate time.

Abscisic acid is an inhibitory hormone that helps plants to withstand unfavorable environmental conditions. It causes stomata to close when water avail-

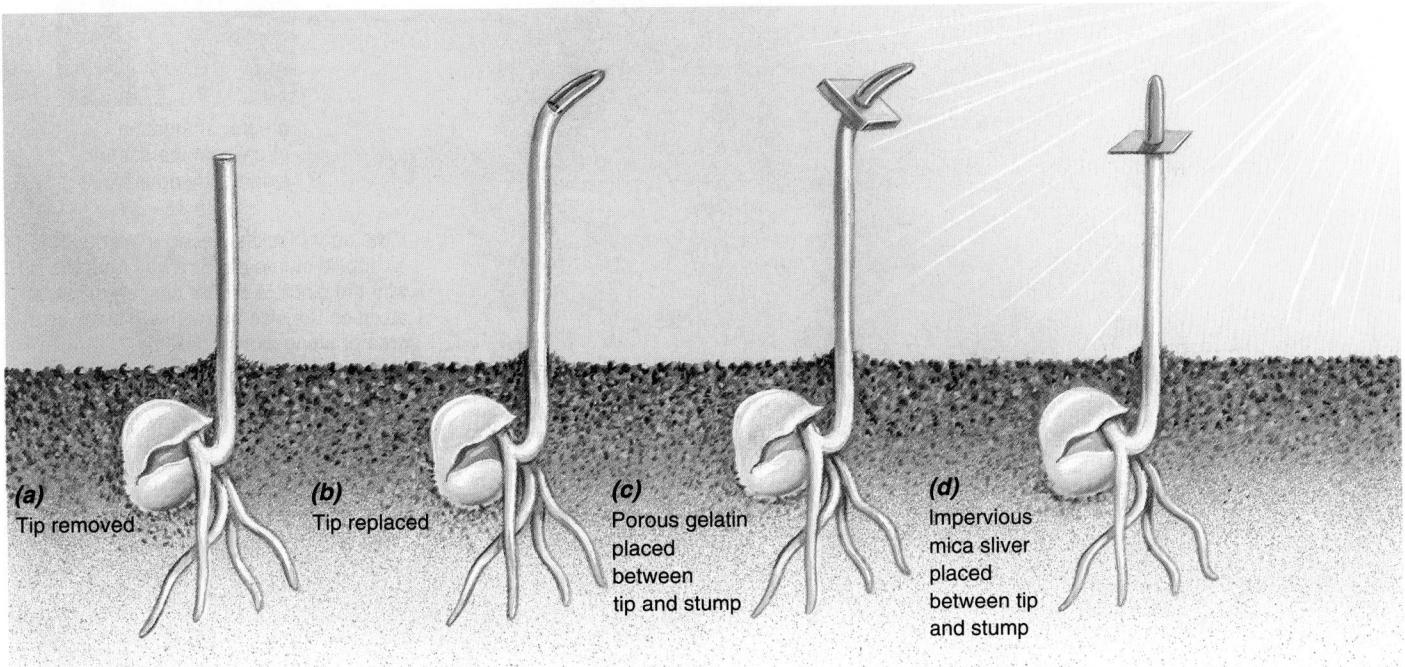

Figure 29-3 Boysen-Jensen's experiments demonstrating the chemical nature of the signal passing from the tip to the bending region of the coleoptile during phototropism. **(a)** If the tip is cut off a coleoptile, the remaining stump neither elongates nor bends toward the light. **(b)** Replacing the tip restores both elongation and phototropism. **(c)** A gelatin "filter" between the tip and stump does not prevent phototropism, but **(d)** an impenetrable slice of mica eliminates the response.

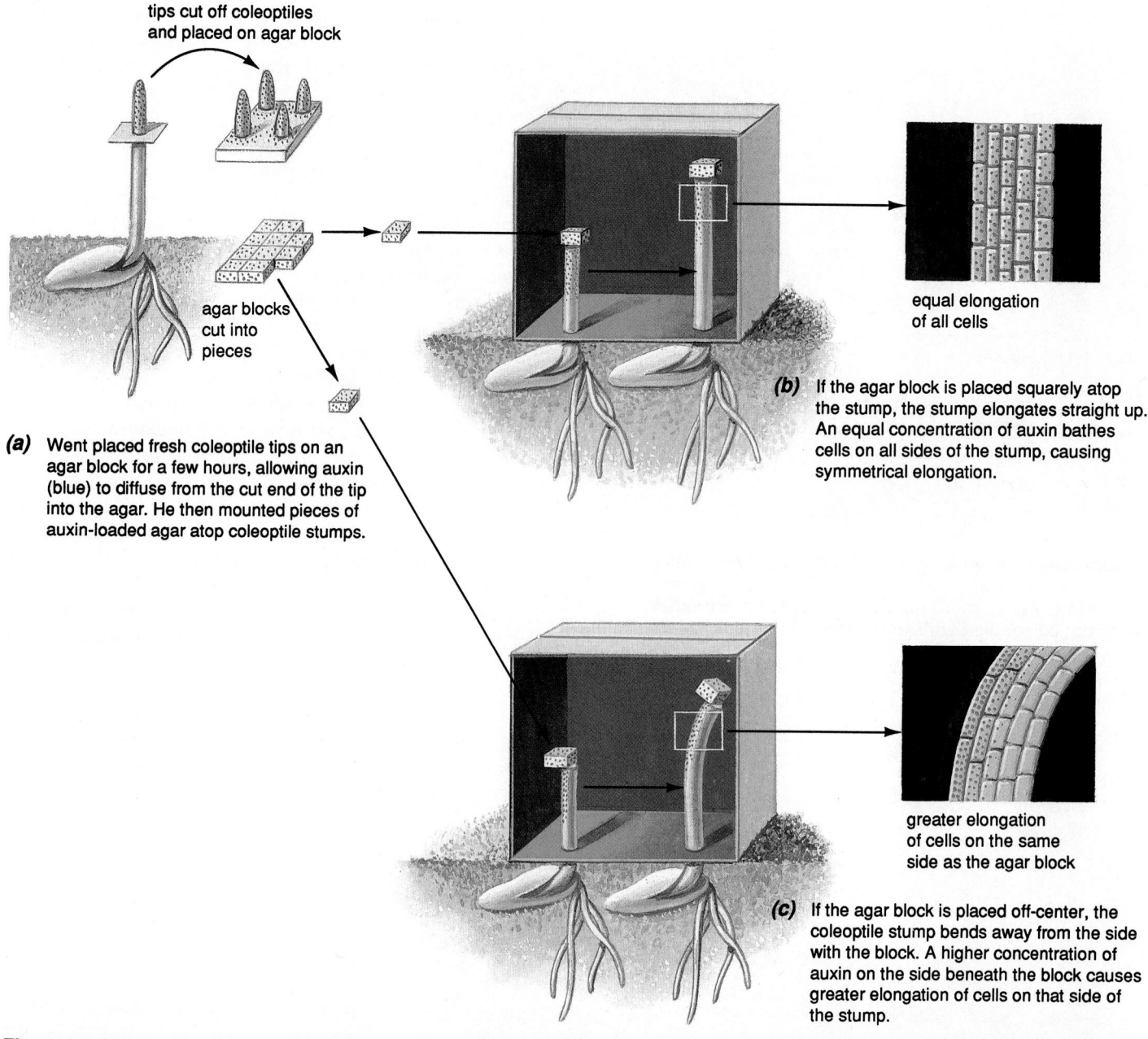

tips cut off coleoptiles
and placed on agar block

agar blocks
cut into
pieces

(a) Went placed fresh coleoptile tips on an
agar block for a few hours, allowing auxin
(blue) to diffuse from the cut end of the tip
into the agar. He then mounted pieces of
auxin-loaded agar atop coleoptile stumps.

equal elongation
of all cells

(b) If the agar block is placed squarely atop
the stump, the stump elongates straight up.
An equal concentration of auxin bathes
cells on all sides of the stump, causing
symmetrical elongation.

greater elongation
of cells on the same
side as the agar block

(c) If the agar block is placed off-center, the
coleoptile stump bends away from the side
with the block. A higher concentration of
auxin on the side beneath the block causes
greater elongation of cells on that side of
the stump.

Figure 29-4 Isolation of auxin by Frits Went.

ability is low. It inhibits the activity of gibberellin,
thus helping to maintain dormancy in buds and seeds
during times when germination would be dangerous.

In the rest of this chapter, we describe a year in the
life of a plant, illustrating how hormones regulate its
growth and development.

The Plant Life Cycle: Reception, Response, and Regulation

The life cycle of a plant results from a complex interplay between its genetic information and its environment. Hormones mediate many of the genetic deter-

Table 29-1 Hormone Actions in Plants

Hormone	Functions
Abscisic acid	Closing of stomata; seed dormancy; bud dormancy
Auxin	Elongation of cells in coleoptiles and shoots; phototropism; gravitropism in shoots and roots; root growth and branching; apical dominance; development of vascular tissue; fruit development; retarding senescence in leaves and fruit; ethylene production in fruit
Cytokinin	Promotion of sprouting of lateral buds; prevention of leaf senescence; promotion of cell division; stimulation of fruit, endosperm, and embryo development
Ethylene	Ripening of fruit; abscission of fruits, flowers, and leaves; inhibition of stem elongation; formation of hook in dicot seedlings
Gibberellin	Germination of seeds and sprouting of buds; elongation of stems; stimulation of flowering; development of fruit

minants of growth and development, as well as nearly all responses to environmental factors. At each stage in its life cycle, a plant produces a distinctive set of hormones that interact with one another in directing the growth of the plant body.

Seed Dormancy and Germination

As we pointed out in the previous chapter, a seed maturing within a juicy fruit on a warm autumn day has ideal conditions for germination, yet remains dormant until the following spring. In many seeds, *abscisic acid enforces dormancy.* Abscisic acid slows down the metabolism of the embryo within the seed, preventing its growth. The seeds of some desert plants contain high concentrations of abscisic acid; only a really hard rain can wash out the abscisic acid, freeing the embryo from its inhibitory effects and allowing the seed to germinate. Seeds of northern plants usually require a prolonged period of cold to break dormancy; in these seeds chilling induces the destruction of abscisic acid.

Germination is stimulated by other hormones, especially gibberellin. The same environmental conditions that cause the breakdown of abscisic acid also promote the synthesis of gibberellin. Gibberellin induces the transcription of genes that code for the enzymes that digest the food reserves of the endosperm and cotyledons, making sugars, lipids, and amino acids available to the growing embryo.

Growth of the Seedling

When the growing embryo breaks out of the seed coat, it immediately faces a crucial problem: which way is up? The roots must burrow downward, while the shoot must grow upward to emerge into the light. Auxin apparently controls the responses of both roots and shoots to light and gravity.

Shoot Growth

Let's begin by looking at the growth of a shoot as it first emerges from the seed, buried underground. As we described earlier, auxin is synthesized in shoot tips, moves down the shaft of the stem, and stimulates cell elongation. If the stem is not exactly vertical, organelles in the cells of the stem detect the direction of gravity and somehow cause auxin to accumulate on the stem's lower side (Fig. 29-5). Therefore the lower cells elongate rapidly, forcing the stem to bend upward. When the shoot tip is vertical, the auxin distribution becomes symmetrical. Now the stem grows straight up, emerging from the soil into the light.

Auxin also mediates phototropism. Ordinarily, the distribution of auxin caused by light is the same as the distribution caused by gravity, because the direction of brightest light (the sun) is directly opposite that of gravity. For example, if a young shoot still buried underground is close enough to the surface so that some light penetrates down to it, both light and gravity cause auxin to be transported to the lower side of the shoot, and promote upward bending.

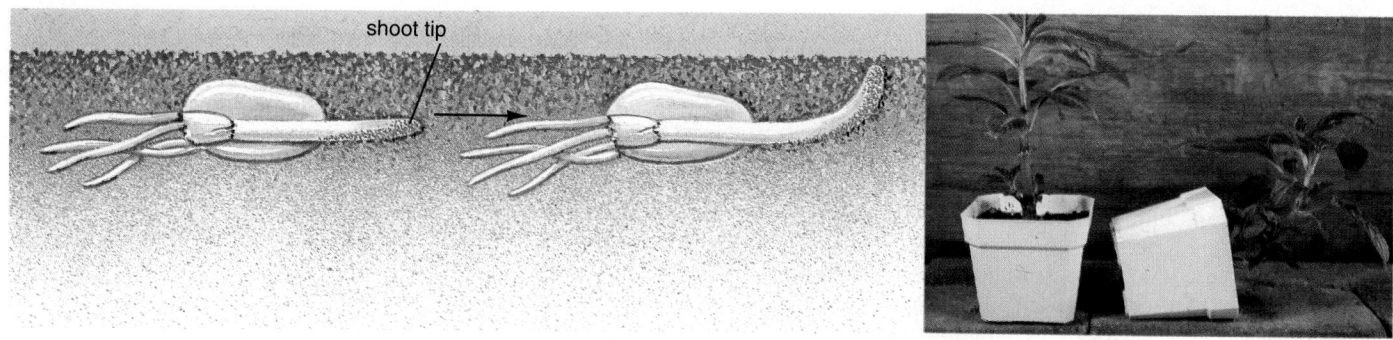

Figure 29-5 The mechanism of gravitropism in shoots. The photo at right shows two impatiens plants, one of which was placed on its side in the dark for 16 hours. In just this short time, faster elongation of cells on the lower side of the stem has brought the plant to a nearly vertical orientation. (Left) Auxin (blue) produced by the tip accumulates on the lower side of the shoot. Auxin promotes rapid elongation of cells on the lower side, causing the shoot to bend upward.

Thus, under normal conditions gravitropism and phototropism augment each other.

Root Growth

Gravitropism in roots is less well understood. According to one model, auxin controls the direction of root growth (Fig. 29-6). Auxin is transported from the shoot down to the root. *If the root is not vertical, the root cap senses the direction of gravity and causes the auxin to accumulate on the lower side of the root.* Unlike shoots, in which moderate concentrations of auxin *stimulate* cell elongation, in roots these same concentrations of auxin *inhibit* cell elongation. Therefore, cell elongation in the lower side of the root is inhibited, while cell elongation remains unaffected in the upper side

of the root. The result is that the root bends downward. When the root tip points directly downward, the auxin distribution becomes equal on all sides, and the root continues to grow straight down. Note that *auxin slows down but does not eliminate root cell elongation:* a vertical root continues to grow.

Development of the Mature Plant Form

As a plant grows, both its root and shoot develop branching patterns that are largely determined by its genetic heritage. For example, the stems of some plants, such as sunflowers, hardly branch at all; others, such as oaks and cottonwoods, branch profusely

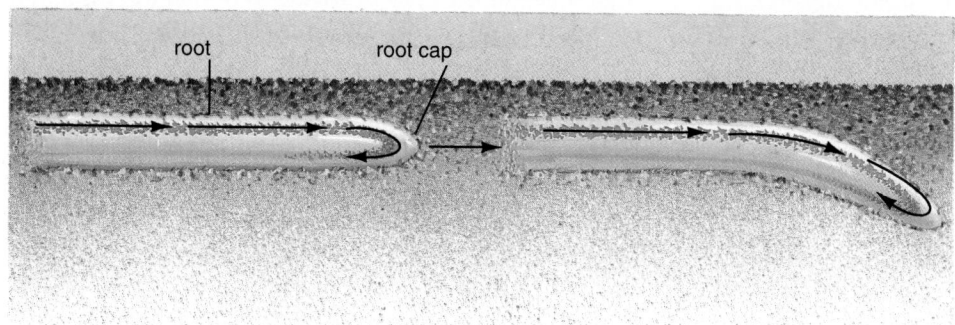

Figure 29-6 The mechanism of gravitropism in roots.
Auxin (blue) is transported to the root from the shoot. The root cap senses the direction of gravity and redirects the flow of auxin, so that auxin accumulates on the lower side of the root. Auxin inhibits cell elongation in roots. As cells on the top elongate faster than those on the bottom, the root bends downward.

in seeming confusion; still others branch in a very regular pattern, producing the conical shapes of firs and spruces.

The amount of growth in shoot and root systems must also be kept in balance. The shoot must be large enough to supply the roots with sugars, while the roots must be large enough to provide the shoot with water and minerals. Interactions between auxin and cytokinin regulate root and stem branching, thereby regulating the relative sizes of root and shoot systems.

Stem Branching

Gardeners know that pinching back the tip of a growing plant makes it become bushier. The botanical explanation for this practice is that the growing tip suppresses the sprouting of lateral buds, a phenomenon known as **apical dominance.** Although it is not known for sure how lateral bud sprouting is con-

trolled, there is some evidence that the proper levels of auxin and cytokinin must be present (Fig. 29-7). Auxin is produced by the shoot tip and transported down the stem. Cytokinin is produced by the roots, and is transported up the stem. Therefore, the relative concentrations of these two hormones will vary, depending on where a bud is located on the stem.

Auxin by itself appears to inhibit the sprouting of lateral buds, while auxin and cytokinin together stimulate bud sprouting. The lateral buds closest to the shoot tip receive a great deal of auxin, probably enough to inhibit their growth, but receive very little cytokinin because they are so far from the roots. Therefore, they remain dormant. Lower buds receive less auxin while receiving much more cytokinin. They are stimulated by optimal concentrations of both hormones, and so they sprout (Fig. 29-7). In many plants, this interaction between auxin and cytokinin produces an orderly progression of bud sprouting, from the bottom to the top of the shoot.

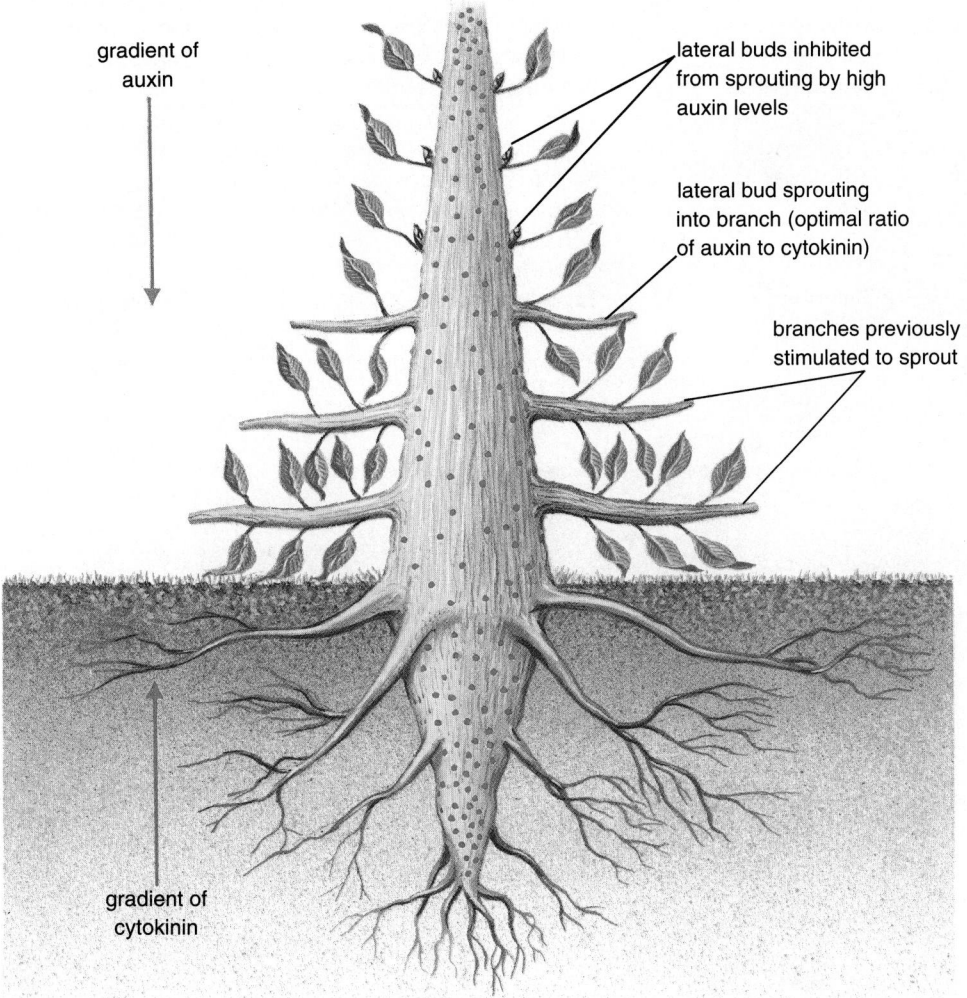

gradient of auxin

lateral buds inhibited from sprouting by high auxin levels

lateral bud sprouting into branch (optimal ratio of auxin to cytokinin)

branches previously stimulated to sprout

gradient of cytokinin

Figure 29-7 A simplified diagram of the interplay of auxin (blue dots) and cytokinin (red dots) in the control of sprouting of lateral buds. Auxin is produced by shoot tips and moves downward, while cytokinin is produced by root tips and moves upward. The auxin gradient therefore decreases with distance from the shoot tip, and the cytokinin gradient decreases with distance from the root tip. Sprouting of lateral buds depends on the ratio of cytokinin to auxin; the exact ratio needed for sprouting varies among species.

A CLOSER LOOK
At Rapid-Fire Plant Responses

All plants are alive, but some are definitely more lively than others. Poke a sycamore tree, for example, and nothing much seems to happen. But watch a fly brush against the sensory hairs in a Venus flytrap, and you will see a response that is almost animal-like in its purposefulness and speed of movement (Fig. E29-1). Why is it useful for a Venus flytrap to catch a fly, and how does it accomplish this task?

You may recall from Chapter 27 that many soils are nitrogen-poor: although the atmosphere contains vast amounts of molecular nitrogen, many soils contain little ammonium or nitrate, which are the only forms of nitrogen that can be taken up by plant roots. Acid bogs are especially nitrogen-deficient. As an evolutionary response to chronic nitrogen shortages, several bog plants, including the Venus flytrap, pitcher plant, and sundew, have resorted to carnivory. By snaring a beetle or fly now and then, the plant obtains nitrogen from the chitin, proteins, and nucleic acids of its prey. While the evolutionary advantage seems clear, the *mechanism* of movement is much less obvious. Let's examine the Venus flytrap more closely, and find the answers to two intriguing aspects of its rapid responses: (1) How does the plant perceive the touch of a fly? (2) How does it move its leaves to catch it?

Sensory Perception

Each lobe of the fringed trapping leaves of a Venus flytrap bears three sensory "hairs" on its inside surface (Fig. E29-1a). These hairs act as triggers, stimulating the leaf to close around its prey. However, a single touch to a single hair will not close the trap. This is clearly adaptive behavior. Dead leaves, twigs, or bits of dirt might fall into the open trap, but, being inanimate, would probably only brush against a hair once before settling to the bottom of the leaf. An insect, on the other hand, wanders around inside the trap, sipping the nectar secreted by the leaves. If it touches one hair twice in rapid succession, or touches two different hairs, the hairs initiate an electrical potential change analogous to the action potential of animal nerve cells. The electrical potential sets off a rapid chain of events that causes the trap to close (Fig. E29-1b).

Leaf Movement

In a beautiful set of experiments, botanists Stephen Williams and Alan Bennett found that the flytrap leaf closes because of *irreversible, differential growth*. The flytrap leaves can be pictured most simply as two layers of cells, outer and inner (Fig. E29-2). The electrical potential triggered by hair movement stimulates cells of the outer layer to pump hydrogen ions (H^+) extremely rapidly into their cell walls. Enzymes in the cell walls are activated by acid conditions, and loosen the cellulose fibers of the walls. As the walls weaken, the high osmotic pressure inside the cells causes them to absorb water from extracellular fluids, swiftly growing by about 25%. Since the outer layer expands while the inner layer does not, the leaf is pushed closed. Reopening the trap occurs much more slowly, taking several hours. However, the

(a)

(b)

Figure E29-1

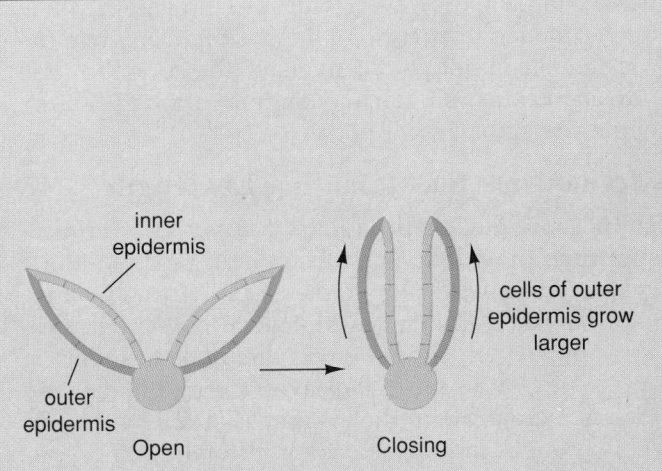

inner epidermis

outer epidermis

Open

cells of outer epidermis grow larger

Closing

Figure E29-2 How a Venus flytrap closes its leaf trap.

fundamental mechanism is similar: when the trap opens, the cells on the inside of the leaf expand, pushing apart the lobes of the trap.

So much energy is used up by the hydrogen ion pumps that *closing the trap consumes nearly a third of all the ATP within the entire leaf.* Since only the outer layer of cells is involved in closure, these cells must virtually empty themselves of ATP. It is therefore very important that something digestible actually be in the leaf before it closes the trap.

Puzzles to Solve

Although a lot has been learned about the mechanisms producing movement in lively plants such as the Venus flytrap, mysteries still remain. How is a touch stimulus transformed into an electrical stimulus by the sensory hairs? What is the nature of the electrical potential change? How does the electrical signal cause the cells to begin pumping hydrogen ions? As so often happens in biology, the answer to one question immediately poses several new, usually tougher, questions.

Root Branching

Auxin stimulates root branching, even in extremely low concentrations. As we described in Chapter 26, lateral roots arise from the pericycle layer of the vascular cylinder. Auxin, transported down from the stem, stimulates pericycle cells to divide and form a lateral root.

Balance Between Root and Shoot Systems

Through the interaction of auxin and cytokinin, the root and shoot systems regulate each other's growth. An enlarging root system synthesizes large amounts of cytokinin, which stimulates lateral buds to break dormancy and sprout. If the root system isn't keeping up, less cytokinin is produced. The lateral buds sprout later, slowing the growth of the shoot system. Simultaneously, as the stem grows and branches, it produces lots of auxin, which stimulates root branching and growth. Thus neither root nor shoot can get too far ahead or behind, and the plant is adequately supplied with all its needs.

Differentiation of Vascular Tissues

As a plant grows, its parts must be interconnected by the vascular tissues of xylem and phloem. Differentiation of cells into vascular tissues appears to be yet another function of auxin and perhaps gibberellin as well. Apical meristems release auxin and gibberellin into cells that are maturing just behind them. High hormone levels stimulate these cells to differentiate into xylem and phloem. As leaves grow and lateral buds sprout, they too release auxin and gibberellin. These hormones cause cortex cells just beneath the leaves and buds to differentiate into strands of xylem and phloem that connect with the main vascular systems of the stem.

Control of Flowering

Ultimately, the plant matures enough to reproduce. The timing of flowering and seed production are finely tuned to the physiology of the plant and the rigors of its environment. In temperate climates, plants must flower early enough so that their seeds can mature before the killing frosts of autumn. Depending on how quickly the seed and fruit develop, flowering may occur in spring, as it does in oaks, in summer, as in lettuce, or even in autumn, as in asters.

What environmental cues do plants use to determine the season? Most cues, such as temperature or water availability, are quite variable: October can be warm, a late snow may fall in May, or the summer might be unusually cool and wet. *The only reliable cue is daylength:* longer days always mean that spring and summer are coming, while shorter days foretell the onset of autumn and winter.

With respect to flowering, plants are classified as day-neutral, long-day or short-day plants (Fig. 29-8). A **day-neutral plant** is one that flowers as soon as it has grown and developed enough, regardless of the length of the day. Day-neutral plants include tomatoes, corn, and snapdragons. A **long-day plant** flowers when the daylength is *longer than some critical value*, while a **short-day plant** flowers when the daylength is *shorter than some critical value*. Thus spinach is classified as a long-day plant, because it flowers only if the day is *longer than* 13 hours, and cockleburs are short-day plants because they flower only if the day is *shorter than* 15.5 hours long. Note that both will flower with 14 hours of light and 10 hours of dark-

ness. Spinach, however, will also flower in much longer daylengths (e.g., 16 hours of light), while the cocklebur will not, and conversely the cocklebur will flower in much shorter daylengths (e.g., 12 hours), while the spinach will not.

Mechanisms for Measuring Daylength

To measure daylength, a plant needs a *clock* to measure time (how long has it been light or dark) and a *light-detecting system* to set the clock. Virtually all organisms have an internal **biological clock** that measures time even without environmental cues. For example, the sorrel raises its leaves during the day and lowers them again in the evening (Fig. 29-9). If a sorrel is brought indoors and kept in total darkness, its leaves still rise and fall on a daily cycle. Activities that recur approximately every 24 hours in the absence of external cues are called **circadian rhythms.**

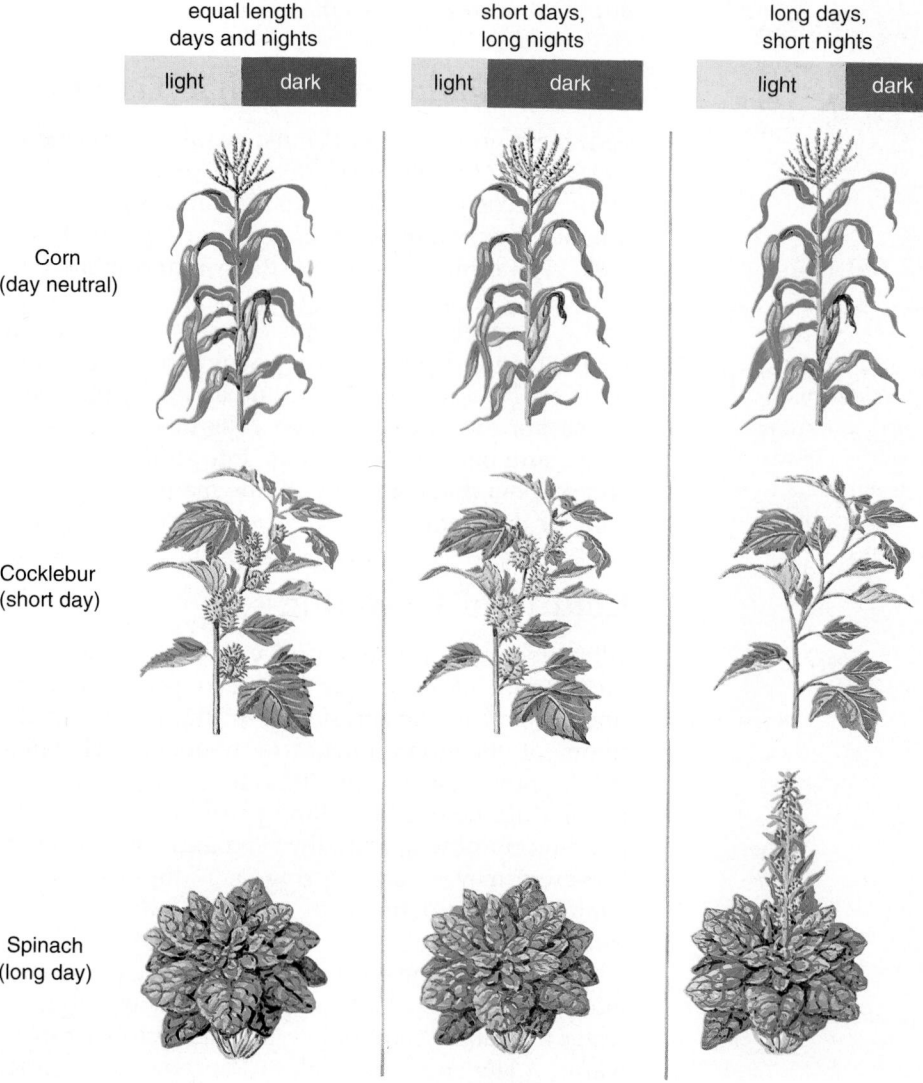

equal length days and nights	short days, long nights	long days, short nights
light / dark	light / dark	light / dark

Corn (day neutral)

Cocklebur (short day)

Spinach (long day)

Figure 29-8 The effects of daylength on flowering in plants. Yellow bars indicate day and black bars indicate darkness.

(a)

(b)

Figure 29-9 Under normal conditions, the sorrel *Oxalis* raises its leaves during the day (left) and lowers them at night (right). If the plant is placed in constant darkness it continues to raise its leaves during the "expected" daytime and lower them during the "expected" night, proving that the timing of leaf movement is intrinsic to the plant.

The light-detecting system of plants is a pigment in the leaves called **phytochrome** (meaning simply "plant color"). Phytochrome occurs in two interchangeable forms (Fig. 29-10). One form strongly absorbs red light, and is called P_r, while the other form absorbs far-red light (almost infrared), and is accordingly called P_{fr}. In most plants, P_{fr} is the active form of phytochrome; that is, a suitable concentration of P_{fr} stimulates or inhibits physiological processes, such as flowering or setting the biological clock. P_r has no effect on these same processes.

Phytochrome flips back and forth from one form to the other when it absorbs light of the appropriate color: when P_r absorbs red light, it is converted into P_{fr}, and when P_{fr} absorbs far-red light, it is transformed back into P_r. Daylight consists of all wavelengths of visible light, including both red and far-red. Therefore, during the day a leaf contains both forms of phytochrome. In the dark, P_{fr} rather rapidly breaks down or reverts to P_r.

Daylength and the Control of Flowering

Plants seem to use the phytochrome system and their internal biological clocks to control flowering. Cockleburs, for example, flower under a lighting regime of 8 hours of light and 16 hours of darkness. However, interrupting the middle of the dark period with just a minute or two of light prevents flowering. Thus, although cockleburs are usually classified as short-day plants, what really matters is not how long the day is, but how long the continuous darkness lasts. The color of the light used for the night flash is also important. A midnight flash of red light inhibits flowering, but a far-red flash allows flowering. This, of course, implicates phytochrome in the control of flowering. Unfortunately, no one knows how the response of phytochrome to light determines whether or not a plant will flower. It seems likely that the biological clock measures the length of the night, and that light reception by phytochrome tells the clock when sunrise and sunset have occurred, but this is not certain.

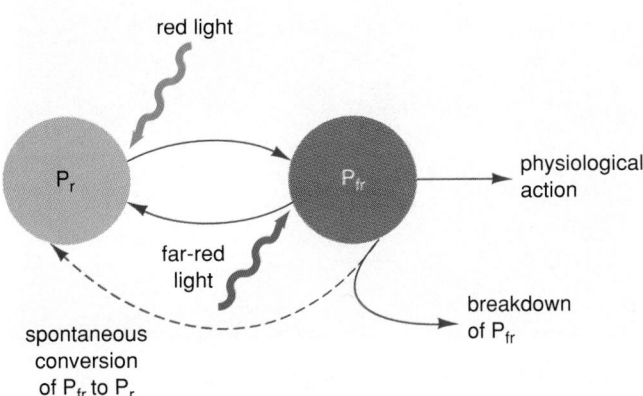

Figure 29-10 The light-sensitive pigment phytochrome exists in two forms, inactive (P_r) and active (P_{fr}). P_r is converted to P_{fr} by red light. P_{fr} may then participate in physiological responses, be converted to P_r by far-red light, revert spontaneously to P_r, or break down to other, inactive compounds.

Other Phytochrome-Mediated Processes

Phytochrome is involved in many plant responses. For example, P_{fr} inhibits elongation of seedlings, with profound and obviously adaptive results. Since P_{fr}

breaks down or reverts to P_r in the dark, seedlings germinating in the darkness of the soil contain no P_{fr}, and consequently elongate very rapidly, emerging out from the soil. Seedlings growing beneath other plants will be exposed largely to far-red light, because the green chlorophyll of the leaves above them will absorb most of the red light but transmit the far-red. Far-red light converts P_{fr} to P_r, so shaded seedlings grow rapidly, which may bring them out of the shade. Once out in the sunlight, P_{fr} forms. P_{fr} slows down elongation, which prevents the seedlings from becoming too spindly.

Other plant responses that are stimulated by P_{fr} include straightening the epicotyl or hypocotyl hook of dicot seedlings, leaf growth, and chlorophyll synthesis. As in the case of stem elongation, these responses are adaptations related to burial in the soil or shading by the leaves of other plants. For example, a newly germinating shoot needs to stay in its protective bend while still in the soil (i.e., in the dark), and only straighten out in the open air, where sunlight converts P_r to P_{fr}.

Development of Seeds and Fruit

When a flower is pollinated, auxin or gibberellin released by the pollen stimulates the ovary to begin developing into a fruit. If fertilization also occurs, the developing seeds release still more auxin and/or gibberellin into the surrounding ovary tissues. Cells of the ovary multiply and grow larger, often storing starches and other food materials, forming a mature fruit.

Seeds and fruits acquire nutrients for growth and development from their parent plant. If the seed is separated from the parent too soon, it may not complete its development. Not surprisingly, seed maturation and fruit ripening are closely coordinated. Unripe fruits are often inconspicuously colored (usually green like the rest of the plant), hard, bitter, and sometimes even poisonous. As a result, animals seldom eat unripe fruit. When the seeds mature, the fruit ripens: it becomes brightly colored, softer, and sweeter, and therefore more noticeable and attractive to animals (Fig. 29-11).

Ripening is stimulated by ethylene. Ethylene is synthesized by fruit cells in response to a surge of auxin that is released by the seeds. Since ethylene is a gas, a ripe fruit continually leaks ethylene into the air. In nature this probably doesn't make much difference. However, when people store fruit in closed containers, ethylene released from one fruit will hasten ripening in the rest, which is why "one rotten apple spoils the barrel." Although this may inconvenience the consumer whose entire carton of fruit

ripens in just a few days, the discovery of the role of ethylene in ripening revolutionized modern fruit and vegetable marketing. Bananas, for instance, are grown in Central America and shipped by boat to North American markets. By picking and shipping the bananas green and then exposing them to ethylene at their destination, grocers can market perfectly ripe fruit. Unfortunately, not all fruits seem to ripen properly when separated from the plant, resulting in such supermarket delicacies as the infamous pink cardboard tomato: ripening tomatoes probably continue to acquire nutrients from the parent plant, so gassing a green tomato with ethylene turns it red but doesn't really duplicate natural ripening.

Figure 29-11 Fruit ripening includes changes in color, texture, flavor, and sweetness. A strawberry fruit is green, hard, and bitter before it ripens, which discourages animals from eating it. After the seeds mature, the fruit becomes soft, red, and tasty, attracting animals such as this vole. The mature seeds are not harmed by the animal's digestive tract, and are dispersed in the animal's feces.

Senescence and Dormancy

The season is autumn. If animals haven't eaten the fruits yet, the time has come to let them drop to the ground. For perennial broadleafed plants, the leaves must be shed as well, because they will be a liability in winter, unable to photosynthesize but still evaporating water (see the Essay, "Autumn in Colorado," in Chapter 26). Both leaves and fruits undergo a rapid aging called **senescence**. The culmination of senescence is the formation of the **abscission layer** at the base of the petiole, allowing the leaf or fruit to drop off.

Senescence and abscission are complex processes controlled by several different hormones. In most plants, healthy leaves and developing seeds produce auxin, which in turn helps to maintain the health of the leaf or fruit. Simultaneously, the roots synthesize cytokinin, which is transported up the stem and out the branches. Cytokinin also prevents senescence—a leaf plucked from a tree and floated in a dilute solution of cytokinin stays green for weeks. As winter approaches, cytokinin production in the roots slows down, and fruits and leaves produce less auxin. Perhaps driven by these hormonal changes, much of the organic material in leaves is broken down to simple molecules that are transported to the roots for winter storage. Meanwhile, ethylene is released by both aging leaves and ripening fruit. Ethylene stimulates the production of enzymes that destroy the cell walls holding the abscission layer at the base of the petiole together. When the abscission layers weaken, leaves and fruits fall from the branches.

Other changes also occur that prepare the plant for winter. Buds, which developed into new leaves and branches during spring and summer, now become dormant, waiting out the winter tightly wrapped up. Dormancy in buds, as in seeds, is enforced by abscisic acid. Metabolism slows to a crawl, and the plant enters its long winter sleep, waiting for the signals of warmth and longer days in spring before awakening once again.

Reflections on Plant Physiology and Ecology

Plants have evolved the ability to regulate their growth and development in response to the demands of their environment. The mechanisms by which plants perceive environmental stimuli and respond to them are genetically determined, and vary considerably among plant species. These genetic differences in turn permit some species to thrive in a particular environment while other species cannot.

Consider, for example, that bane of hay-fever sufferers, ragweed (Fig. E29-3a). A ragweed seed germinates in spring, but the resulting plant doesn't mature enough to flower until summer. In addition, no matter how large the plant becomes, ragweed will not flower if the days are longer than about $14\frac{1}{2}$ hours. Now, summer days are longer the farther north you go (Fig. E29-3c). In nearly all of Canada, the longest day of the year (June 21) has more than 16 hours of daylight, and the daylength does not fall below $14\frac{1}{2}$ hours until mid-August. By that time, the killing frosts of autumn are not far away, and few ragweeds can complete seed development before dying. Therefore, most of Canada is free of ragweed. In New Jersey, on the other hand, summer days shorten to $14\frac{1}{2}$ hours by late July, and frosts occur later than they do in Canada. The result, as the authors can sneezily testify, is a prolific ragweed crop.

Other plants, such as the buffalo grass of the plains, can grow and reproduce successfully from Texas to Alberta. These plant species have geographically distinct varieties that differ in the critical daylength for flowering. Northern varieties of buffalo grass flower during the long days of early summer, which leaves plenty of time to complete seed development before frost. Southern varieties flower under much shorter days. A Canadian grass planted in Texas will never flower, while a transplanted Texan will continue blooming into late summer, spending energy on new flowers instead of maturing seeds.

The local environment of a plant also requires adaptations in flowering. Many wildflowers of the Eastern deciduous forests, such as the violet (Fig. E29-3b), have an extremely short growing season, sandwiched between the end of winter and the dense growth of leaves on the trees above them in spring. Once the trees have leafed out, not much light reaches the forest floor, and the violets cannot photosynthesize rapidly enough to set seeds. Thus, violets flower only when daylengths are short in early spring. By late spring, producing more flowers would waste energy better spent in maturing the seeds already set. Not surprisingly, the longer days of May inhibit flowering.

The daylength requirements for flowering found in various plants allow us to predict where they will be

Figure E29-3 Daylength requirements for flowering influences the distribution of plants. **(a)** Ragweed blooms in late summer when days are shorter than $14\frac{1}{2}$ hours. **(b)** Violets flower during short days in early spring, an adaptation to their habitat in deciduous forests, where the trees shade the forest floor by late spring. **(c)** The length of the longest day in June progressively increases with latitude, varying from just over 12 hours at the equator to continuous day at the North Pole.

found. For example, we can confidently predict that Alaska should be a haven for ragweed sufferers, because ragweed cannot set seeds there. Don't forget, however, that *daylength requirements have not evolved to prevent plants from colonizing new habitats.* Today, ragweed is restricted to places where summer days are not too long. On an evolutionary time scale, however, there

are probably other factors that would normally prevent completion of the ragweed life cycle in the far North. The genetically determined daylength requirements of ragweed and other plants have evolved, not to limit their distribution, but because they ensure that flowering is properly timed for the normal environment of the plant.

SUMMARY OF KEY CONCEPTS

The Discovery of Plant Hormones

Most responses of plants to their environment are produced through the actions of chemicals called plant hormones. The first plant hormone, auxin, was discovered as a result of many years of experiments on the mechanism of phototropism, the growth of plants toward the light.

Plant Hormones and Their Actions

Plant hormones are chemicals that are produced in one part of a plant body and exert actions on the same or distant parts of the plant. The five major classes of plant hormones are abscisic acid, auxin, cytokinin, ethylene, and gibberellin. The major functions of these hormones are summarized in Table 29-1.

The Plant Life Cycle: Reception, Response, and Regulation

Dormancy in seeds is enforced by abscisic acid. Falling levels of abscisic acid, and rising levels of gibberellin, trigger germination. As the seedling grows, it shows differential growth with respect to the direction of light (phototropism) and gravity (gravitropism). Auxin mediates phototropism and gravitropism in shoots, and gravitropism in roots. In shoots, auxin stimulates elongation of cells; by accumulating on the lower side of a shoot and the side away from light, auxin causes the shoot to bend away from gravity and toward the light. In roots, similar concentrations of auxin inhibit elongation. By accumulating on the lower side of a root, auxin causes the root to bend toward gravity.

Branching in stems results from the interplay of two hormones, auxin and cytokinin. High concentrations of auxin (produced in shoot tips and transported downward) inhibits the growth of lateral buds. An optimum concentration of both auxin and cytokinin (synthesized in roots and transported up the shoot) stimulates growth of lateral buds. Auxin also stimulates the growth of branch roots.

The timing of flowering is usually controlled by daylength. Plants appear to detect light and dark by changes in phytochrome, a pigment in the leaves. Plant processes influenced by phytochrome responses to light include flowering, straightening the epicotyl or hypocotyl hook, seedling elongation, leaf growth, and chlorophyll development.

Developing seeds produce auxin, which diffuses into the surrounding ovary tissues and causes production of a fruit. A surge of auxin as the seed matures stimulates fruit cells to release another hormone, ethylene, which causes the fruit to ripen. Ripening includes the conversion of starches to sugars, softening of the fruit, development of bright colors, and often the formation of an abscission layer at the base of the petiole.

Several changes prepare perennial plants of temperate zones for winter. Leaves and fruits undergo a rapid aging process called senescence, including formation of an abscission layer. Senescence occurs due to a fall in levels of auxin and cytokinin, and perhaps to a rise in ethylene concentrations. Other parts of the plant, including buds, become dormant. Dormancy in buds is enforced by high concentrations of abscisic acid.

GLOSSARY

abscisic acid (ab-sis'-ik): a plant hormone that generally inhibits the action of other hormones, enforcing dormancy in seeds and buds and causing closing of stomata.

apical dominance: the phenomenon whereby a growing shoot tip inhibits the sprouting of lateral buds.

auxin (awk'-sin): a plant hormone that influences many plant functions, including phototropism, apical dominance, and root branching. Auxin generally stimulates cell elongation and, in some cases, cell division and differentiation.

biological clock: a metabolic timekeeping mechanism found in most organisms, whereby the organism measures the approximate length of a (24 hour) day even without external environmental cues such as light and dark.

circadian rhythm (sir-kā'-dē-un): an event that recurs with a period of about 24 hours, even in the absence of environmental cues.

cytokinin (sī-tō-kī'-nin): a plant hormone that promotes cell division, fruit growth, sprouting of lateral buds, and prevents leaf aging and leaf drop.

day-neutral plant: a plant in which flowering occurs under a wide range of daylengths.

ethylene: a plant hormone that promotes ripening of fruits, and leaf and fruit drop.

gibberellin (jib-er-el'-in): a plant hormone that stimulates seed germination, fruit development, and cell division and elongation.

gravitropism: growth with respect to the direction of gravity.

hormone: in plants, a chemical produced by one group of cells that influences the growth or metabolic activity of other cells, often some distance away in the plant body.

long-day plant: a plant that will flower only if the length of daylight is greater than some species-specific duration.

phototropism: growth with respect to the direction of light.

phytochrome (fī'-tō-krōm): a light-sensitive plant pigment that mediates many plant responses to light, including flowering, stem elongation, and seed germination.

senescence: in plants, a specific aging process, often including deterioration and dropping of leaves and flowers.

short-day plant: a plant that will flower only if the length of daylight is shorter than some species-specific duration.

STUDY QUESTIONS

1. What two hormones are involved in seed dormancy and germination? What are their roles?
2. Describe the mechanisms of action of auxin in shoot phototropism and gravitropism and root gravitropism.
3. What is apical dominance? How do auxin and cytokinin interact in determining the growth of lateral buds?
4. Define day-neutral plant, long-day plant, and short-day plant. What pigment is thought to be involved in light perception in plants?
5. What is a biological clock?
6. Describe the role of phytochrome in stem elongation in seedlings growing in the shade of other plants. What is the likely adaptive significance of this response?
7. What hormone(s) causes fruit development? Where does this hormone come from? What hormone causes fruit ripening?
8. What hormone(s) are involved in leaf and fruit drop? In bud dormancy?

DISCUSSION QUESTIONS

1. Describe the experiments of the Darwins, Boysen-Jensen, and Went. Do these experiments truly prove that auxin is the hormone controlling phototropism? What other experiments would you like to see?
2. Many houseplants grow very tall, thin, weak stems. Relate this fact to the usual intensities of light outdoors versus indoors, the adaptive value of rapid elongation in dim light, and the likely role of phytochrome in this response. Assume that you had access to lamps that gave off light of any desired wavelength. Design a simple experiment to test your hypothesis about phytochrome effects on stem elongation.
3. Discuss, in detail, why plants get "bushier" when their stem tips are constantly pruned.
4. Explain why "one bad apple can spoil the whole barrel."
5. Suppose on July 4th, you discover that a short-day plant and a long-day plant, both growing in your garden, have bloomed. Discuss how this is possible.

SUGGESTED READINGS

Evans, M. L., Moore, R., and Hasenstein, K.-H. "How Roots Respond to Gravity." *Scientific American*, December 1986. Although botanists still dispute the mechanisms of root gravitropism, these authors convincingly argue that it is mediated by auxin.

Heslop-Harrison, Y. "Carnivorous Plants." *Scientific American*, February 1978. Some plants living in nitrogen-poor environments have turned carnivorous, evolving surprising adaptations for capturing insect food.

Raven, P. H., Evert, R. F., and Eichhorn, S. *Biology of Plants*, 4th ed. New York: Worth, 1986. One of the best botany texts available, with a lucid discussion of plant hormones.

UNIT V

Animal Anatomy and Physiology

30

Circulation

This plastic cast of blood vessels from a human head reveals its rich blood supply, in which details of the face can be discerned.

Billions of years ago, the first living cells were nurtured by the sea where they evolved. The water brought them nutrients that diffused into the cell, and washed away the wastes that diffused out. Today, microorganisms and some simple multicellular animals still rely almost exclusively on diffusion for exchange of wastes and nutrients with the environment. Sponges, for example, circulate seawater through pores in their bodies, bringing the environment within diffusing distance of each cell (see Fig. 32-2a). The threadlike bodies of nematode worms provide an enormous surface area for gas and nutrient exchange with their surroundings (see Fig. 32-4a).

But to satisfy the demands of a living cell, diffusion distances must be kept short. As larger, more complex animals evolved, individual cells became increasingly distant from the outside world. To avoid starving and stewing in their own wastes, a source of nutrients and a sink for wastes had to be brought within diffusing distance of each cell. With the evolution of the circulatory system, an internal sea was created, bringing each cell into close proximity with a source of food and oxygen. The circulatory system also provided a means to carry wastes away from cells. All circulatory systems have three major parts as listed below.

1. **A fluid called blood, which serves as a medium of transport.**
2. **A system of channels called vessels, which conduct the blood throughout the body.**
3. **A pump called a heart, which keeps the blood circulating.**

Two major types of circulatory systems are found in animals: open and closed.

Types of Circulatory Systems

Open circulatory systems include an open space within the body, the **hemocoel,** into which vessels empty and from which they pick up blood (Fig. 30-1a). Within this space, tissues are directly bathed in blood. Open circulatory systems are found in arthropods (insects, spiders, and crustaceans) and most molluscs (snails, clams).

In **closed circulatory systems,** blood is confined to the heart and a continuous series of vessels (Fig. 30-1b), allowing more rapid blood flow and more efficient transport than is possible in an open system. Closed systems are found in some invertebrates such as earthworms and cephalopod molluscs (squids, octopuses), and in all vertebrates.

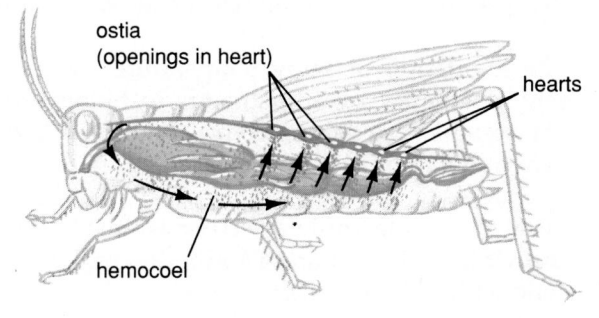

(a) Open circulatory system

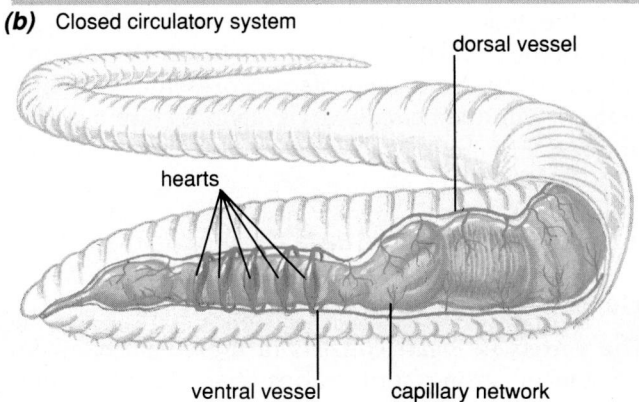

(b) Closed circulatory system

Figure 30-1 (a) In the open circulatory system of insects and other arthropods, a series of hearts pumps blood through vessels into the hemocoel, where blood directly bathes the organs. When the hearts relax, blood is sucked back into them through openings called ostia, guarded by one-way valves. When the hearts contract, the valves are pressed shut, forcing the blood to travel out through the vessels returning to the hemocoel. **(b)** In a closed circulatory system, blood remains confined to the heart and the blood vessels. In the earthworm, five contractile vessels serve as hearts.

The Vertebrate Circulatory System

The circulatory system has many diverse roles, and reaches its greatest development in the vertebrates. Some of the most important functions of the vertebrate circulatory system are listed below.

1. **The transport of oxygen from the lungs to the tissues and the transport of carbon dioxide from the tissues to the lungs.**
2. **The distribution of nutrients from the digestive system to all body cells.**
3. **The transport of waste products and toxic substances to the liver, where many are detoxified, and to the kidney for excretion.**

4. The distribution of hormones from the organs that produce them to the tissues on which they act.

5. The regulation of body temperature, which is achieved partly by adjustments in blood flow. For example, to cool the body, blood flow in the skin and extremities is increased.

6. The defense of the body against blood loss, through clotting, and protection against bacteria and viruses by circulating antibodies and phagocytic white blood cells.

In the following sections we examine the three parts of the circulatory system: the heart, vessels, and blood, using the human as a representative vertebrate. The human circulatory system is shown in Figure 30-2.

The Vertebrate Heart

Heart Structure

The vertebrate heart consists of muscular chambers capable of strong contractions that circulate blood through the body. During the course of vertebrate evolution, the heart has increased in complexity. Starting with the two-chambered heart of fish, it reaches its greatest complexity in the four-chambered hearts of birds and mammals (Fig. 30-3). These warm-blooded animals have high metabolic demands and require the complete separation of oxygenated and deoxygenated blood that the four-chambered heart provides. Mammalian and bird hearts consist of two separate pumps, each with two chambers. In each pump, an **atrium** receives and briefly stores the blood, passing it to a **ventricle** that propels it through the body (Figs. 30-4 and 30-5). One pump is for **pulmonary circulation** and consists of the right atrium and ventricle. Oxygen-depleted blood from the body is collected in the right atrium, transferred to the right ventricle, and pumped to the lungs where it picks up oxygen. The other pump, consisting of the left atrium and ventricle, powers **systemic circulation.** Newly oxygenated blood from the lungs is collected in the left atrium, then passed to the left ventricle, which sends it coursing through the rest of the body.

The Cardiac Cycle

The alternating contraction and relaxation of the heart chambers is called the **cardiac cycle.** The two atria contract in synchrony, emptying their contents into the ventricles. A fraction of a second later, the two ventricles contract simultaneously, forcing blood

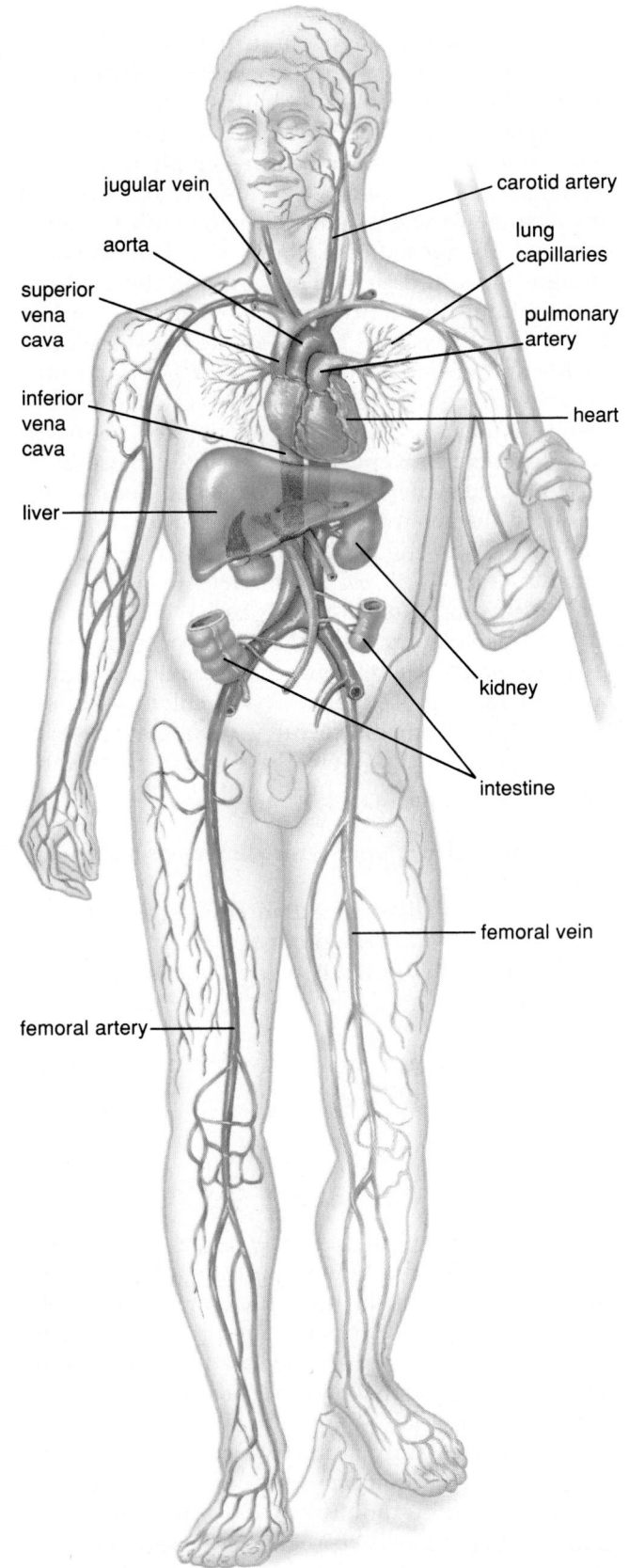

Figure 30-2 The human circulatory system. Oxygenated blood is shown in red, deoxygenated blood in blue.

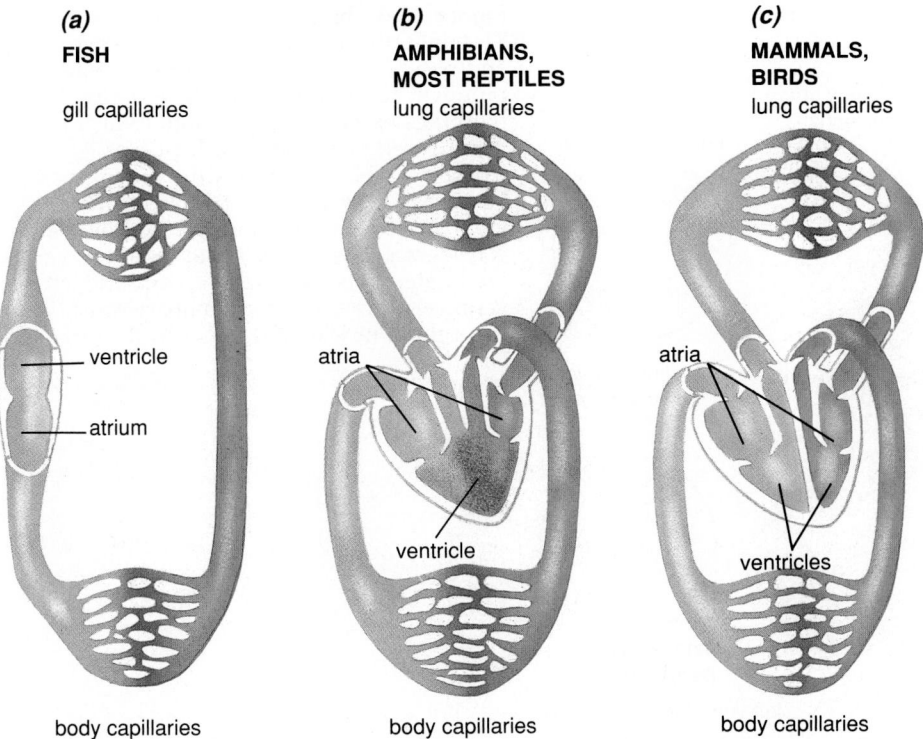

(a)
FISH

gill capillaries

ventricle

atrium

body capillaries

(b)
**AMPHIBIANS,
MOST REPTILES**

lung capillaries

atria

ventricle

body capillaries

(c)
**MAMMALS,
BIRDS**

lung capillaries

atria

ventricles

body capillaries

Figure 30-3 (a) The evolution of the vertebrate heart begins with the two-chambered heart of fishes. Blood from the body tissues is collected in the atrium and transferred to the single ventricle. Contraction of the ventricle sends blood through the gill capillaries, where it picks up oxygen and gives off carbon dioxide, and then to the body capillaries, where it delivers oxygen to the tissues and picks up carbon dioxide. **(b)** In amphibians and most reptiles, the heart has two atria, the left receiving oxygenated blood from the lungs, the right receiving deoxygenated blood from body tissues. Although both empty into a single ventricle, the deoxygenated blood tends to remain on the right, where it is directed to the lungs, while most of the oxygenated blood stays on the left and is sent to the body tissues. In reptiles, there is often a partial wall down the middle of the ventricle, enhancing this separation. **(c)** The hearts of birds and mammals are actually two separate pumps, with no possible mixing of oxygenated and deoxygenated blood.

into arteries leaving the heart. Both chambers then relax briefly before the cycle is repeated. The period of ventricular contraction is called **systole.** The rest of the cycle, including relaxation of all the chambers followed by contraction of the atria, is called **diastole.** At normal resting heart rate, systole lasts about 0.3 second and diastole about 0.5 second. Systole and the last phase of diastole are illustrated in Figure 30-5.

Coordination of Heart Activity

Coordinating the activity of the four heart chambers to maintain proper blood flow through the heart and its vessels poses some logistical challenges. First, when the ventricles contract, the blood must be di-

rected out through the arteries, and not back up into the atria. Then, once blood has entered the arteries, it must be prevented from flowing back as the heart relaxes. These problems are solved by four simple one-way valves (see Fig. 30-4). Pressure in one direction opens them readily, while reverse pressure forces them tightly closed. **Atrioventricular valves** separate the atria from the ventricles; a **tricuspid valve** (Latin, "three-pointed") separates the right ventricle and right atrium, and a **bicuspid** ("two-pointed") **valve** lies between the left atrium and left ventricle. Two **semilunar** (Latin, "half-moon") **valves** allow blood to enter the pulmonary artery and the aorta when the ventricles contract, but prevent it from returning as the ventricles relax.

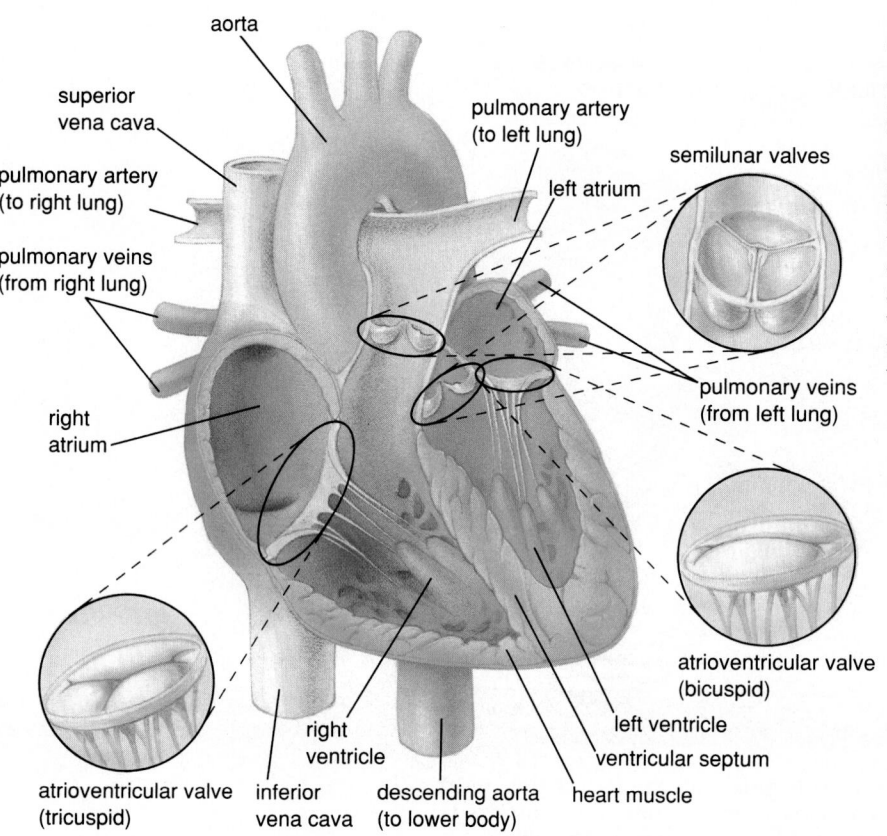

aorta

superior
vena cava

pulmonary artery
(to right lung)

pulmonary veins
(from right lung)

right
atrium

pulmonary artery
(to left lung)

left atrium

semilunar valves

pulmonary veins
(from left lung)

atrioventricular valve
(bicuspid)

left ventricle

ventricular septum

heart muscle

atrioventricular valve
(tricuspid)

inferior
vena cava

right
ventricle

descending aorta
(to lower body)

Figure 30-4 The human heart and its vessels. The right atrium receives deoxygenated blood and passes it to the right ventricle, which pumps it to the lungs. Blood returning from the lungs enters the left atrium, which passes it to the left ventricle, which pumps oxygenated blood through the rest of the body. Note the thickened walls of the left ventricle, which must pump blood over a considerably longer distance. One-way valves are located between the aorta and the left ventricle, between the pulmonary artery and the right ventricle, and between the atria and ventricles.

A second challenge is to create smooth, coordinated contractions of the muscle cells that make up each chamber. The individual muscle fibers of the heart contract spontaneously. This could lead to uncoordinated contractions, as happens when a heart goes into **fibrillation.** Fibrillation is soon fatal because blood is not pumped out of the heart but merely sloshed around. Coordinating the contractions of the heart muscle cells requires a **pacemaker,** an area of specialized muscle cells whose rapid contractions set the pace for the other muscle cells. The individual heart muscle cells communicate directly with one another through special pores in their adjacent membranes (Fig. 30-6). These pores allow electrical signals from the pacemaker to pass freely and rapidly between heart cells, initiating and coordinating their contractions. The heart's primary pacemaker is the **sinoatrial node** (SA node), a small mass of specialized muscle cells located in the wall of the right atrium (Fig. 30-7). The SA node generates electrical impulses at a higher rate than do the individual muscle fibers. Signals from the SA node spread rapidly through both the right and left atria, superceding the spontaneous contractions of individual fibers and causing the atria to contract in smooth synchrony.

A third challenge is to coordinate contractions of the four chambers. The atria must contract first, emptying their contents into the ventricles and then refilling while the ventricles contract. Thus, there must be a delay between the contractions of the atria and the ventricles. From the SA node, the wave of contraction sweeps through the atria until it reaches a barrier of inexcitable tissue separating the atria from the ventricles. Here, the excitation is channelled through a second small mass of specialized muscle cells, the **atrioventricular node** (AV node), located on the floor of the right atrium (see Fig. 30-7). The impulse is delayed at the AV node, postponing the ventricular contraction for about one tenth of a second after contraction of the atria. This delay gives the atria time to complete the transfer of blood into the ventricles before ventricular contraction begins. From the AV node, the signal to contract spreads to the base of the two ventricles along tracts of excitable fibers.

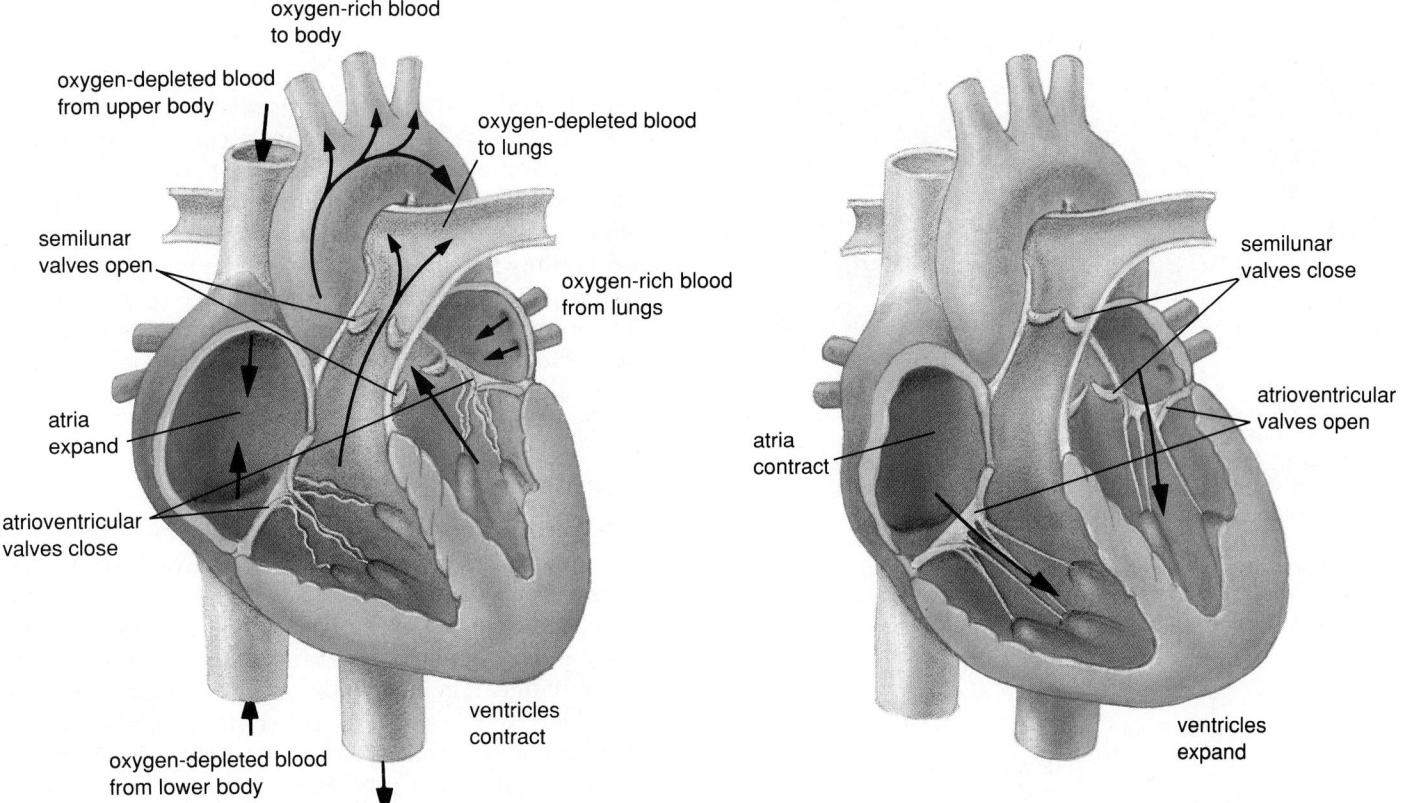

(a) SYSTOLE

oxygen-rich blood to body

oxygen-depleted blood from upper body

oxygen-depleted blood to lungs

semilunar valves open

oxygen-rich blood from lungs

atria expand

atrioventricular valves close

ventricles contract

oxygen-depleted blood from lower body

(b) DIASTOLE

semilunar valves close

atrioventricular valves open

atria contract

ventricles expand

Figure 30-5 The heart valves in action; arrows indicate the direction of blood flow. **(a)** During ventricular contraction (systole), the pressure within the ventricles forces the semilunar valves open, allowing blood to flow into the aorta and pulmonary arteries. The atrioventricular valves are simultaneously pressed shut, preventing blood flow back into the atria. **(b)** As the ventricles reexpand (diastole), they would tend to draw blood back from the arteries, but this backpressure forces the semilunar valves closed. Simultaneously, contraction of the atria forces open the atrioventricular valves, allowing blood to flow from the atria into the ventricles.

From these, the impulse travels rapidly through the communicating muscle fibers, causing the ventricles to contract in unison.

Outside Influences on Heart Rate

Left on its own, the SA node pacemaker would maintain a steady rhythm of about 100 beats per minute. However, the heart rate is significantly altered by the influence of nervous impulses and hormones. In the resting person, activity of the parasympathetic nervous system (which controls body functions during periods of rest; see Chapter 36) slows the heart rate to around 70 beats per minute. When exercise or stress creates a demand for greater blood flow to the mus-

cles, the parasympathetic influence is reduced and the sympathetic nervous system (which prepares the body for emergency action; see Chapter 36) accelerates the heart rate. Likewise, the hormone epinephrine increases heart rate as it mobilizes the entire body for response to threatening situations. When astronauts were landing on the moon, their heart rates were over 170 beats per minute, even though they were sitting still!

The Blood Vessels

Blood leaving the heart travels through a series of vessels in the following order: arteries to arterioles to

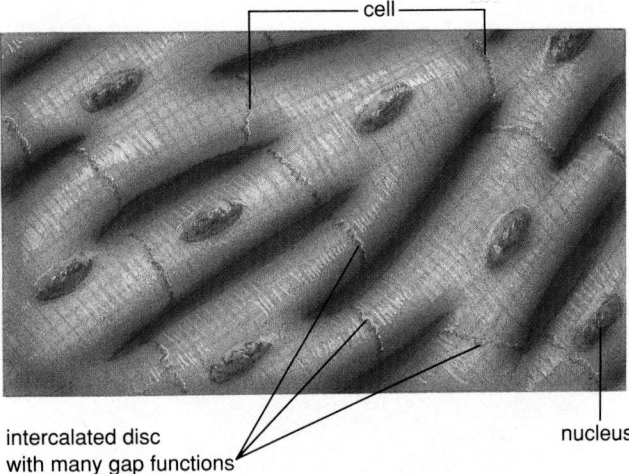

Figure 30-6 The structure of cardiac muscle. Cardiac muscle cells are branched. Adjacent cell membranes meet in folded areas called intercalated discs. These regions are densely packed with gap junctions (pores) that connect the interiors of adjacent cells. This allows direct transmission of electrical signals between the cells, coordinating their contractions.

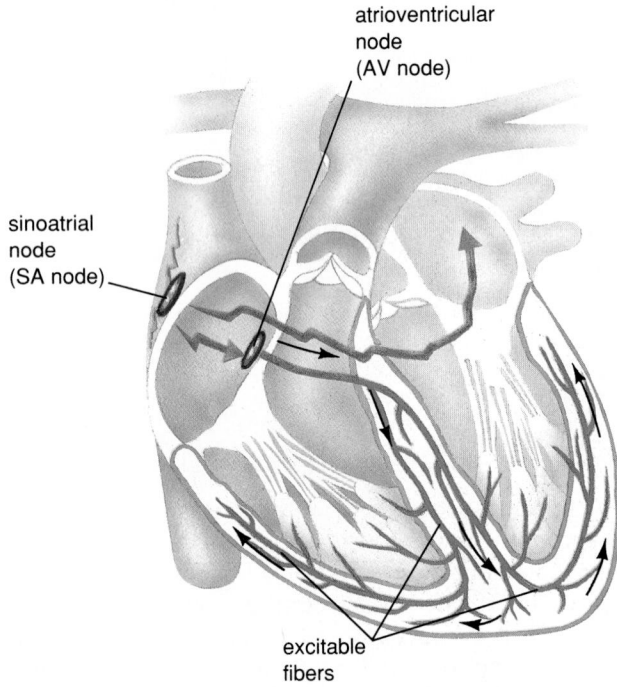

Figure 30-7 The pacemaker of the heart is a spontaneously active mass of modified muscle fibers in the right atrium called the sinoatrial (SA) node. The signal to contract spreads from the SA node through the muscle fibers of both atria, finally exciting the atrioventricular (AV) node in the right atrium. The AV node then transmits the signal to contract through bundles of excitable fibers which stimulate the ventricular muscle.

capillaries to venules to veins, which return it to the heart. Let's look at each in more detail.

Arteries and Arterioles

Blood leaving the heart enters large vessels called **arteries.** These have thick walls containing smooth muscle and elastic tissue (Fig. 30-8). With each surge of blood from the ventricles, the arteries expand slightly, like thick-walled balloons. Between heart beats, they recoil, helping pump the blood and maintain a steady flow through the smaller vessels. Arteries branch into vessels of smaller diameter called **arterioles,** which play a major role in determining how blood is distributed within the body, as described later.

Capillaries

One can envision the circulatory system as an elaborate device for getting blood into the **capillaries,** the tiniest of all vessels. Here, wastes, nutrients, gases, and hormones are exchanged between blood and the body cells. Capillaries are finely adapted to their role of exchange. Their walls are only a single cell thick (see Fig. 30-8). Dissolved substances readily diffuse through the capillary cell membranes or move through the spaces between adjacent capillary cells. Capillaries are so narrow that red blood cells must pass through them in single file (Fig. 30-9). This ensures that all the blood passes very close to the capillary walls where exchange occurs. In addition, capillaries are so numerous that no body cell is more than 100 micrometers from a capillary; this facilitates the exchange of materials by diffusion. It is estimated that the total length of capillaries in a human is over 50,000 miles, enough to encircle the globe twice if placed end-to-end! The speed of blood flow drops precipitously as blood is forced through this narrow, almost interminable network, and this is important to allow more time for diffusion to occur. The slower the blood flow, the greater the exchange of materials.

Venules and Veins

Blood from the capillaries drains into larger vessels called **venules** that empty into still larger **veins** (see Fig. 30-8). Veins provide a low-resistance pathway for blood to return to the heart. The walls of veins are much thinner and more distensible than those of arteries, although both contain a layer of smooth muscle. Blood pressure in the veins is low, and the return of blood to the heart is assisted by contractions of skeletal muscles during exercise and breathing. These muscular movements squeeze the veins, forcing blood through them. When veins are compressed, you might predict that blood would be forced away

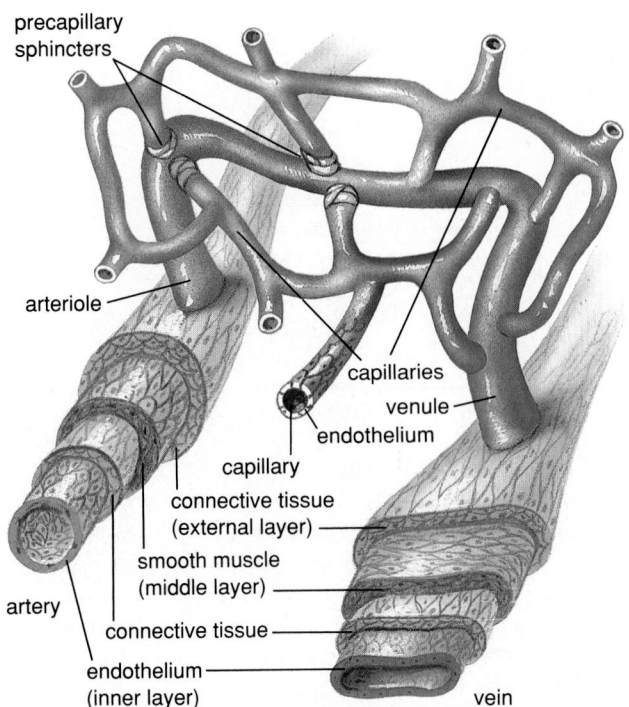

Figure 30-8 The cross-sectional structure and the interconnections of the types of blood vessels. Arteries and arterioles are more muscular and maintain a greater muscular tension than veins and venules, whose walls are thinner and more distensible. Capillaries have walls only a single cell thick, allowing movement of dissolved substances and white blood cells across the capillary wall. Oxygenated blood moves from arteries to arterioles to capillaries. Capillaries empty deoxygenated blood into venules, which empty into veins. The movement of blood from arterioles into capillaries is regulated by muscular rings called precapillary sphincters.

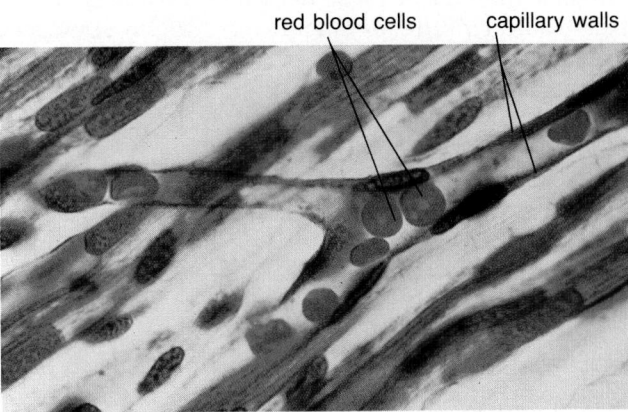

Figure 30-9 Capillaries are so narrow that red blood cells must pass through them single file. This facilitates exchange of gases by diffusion.

from the heart as well as towards it. To prevent this, veins are equipped with one-way valves that allow blood flow only toward the heart (Fig. 30-10). When you sit or stand for long periods, the lack of muscular activity allows blood to pool in the veins of the lower legs. This accounts for the swollen feet of airline passengers. It can also contribute to varicose veins, in which the valves become stretched and weakened.

If blood pressure should fall, for instance after extensive bleeding, veins can help restore it. The sympathetic nervous system stimulates contraction of the smooth muscles in the vein walls. This decreases their volume and raise blood pressure, speeding up the return of blood to the heart.

The Distribution of Blood Flow

The muscular walls of arterioles are under the influence of nerves, hormones, and chemicals produced by nearby tissues. They can therefore contract and relax in response to the changing needs of the tissues and organs they supply. For instance, as you read in your paperback thriller " . . . the blood drained from her face as she beheld the gruesome sight . . . ," keep in mind that the heroine is experiencing constriction of the arterioles that supply her skin with

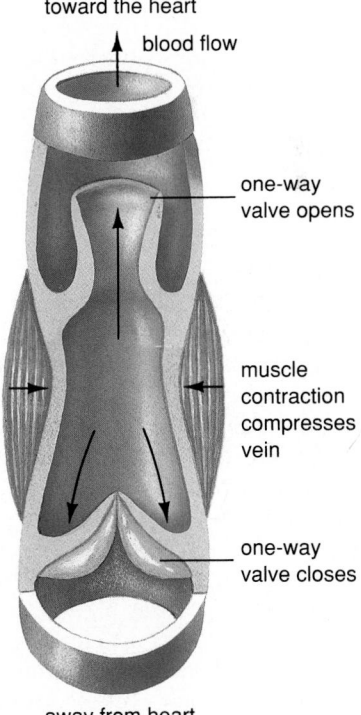

Figure 30-10 Veins and venules have one-way valves that maintain flow in the proper direction. When the vein is compressed by nearby muscles, the valves allow blood to flow toward the heart, but clamp shut to prevent backflow.

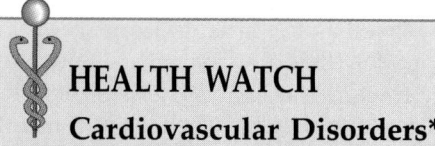

HEALTH WATCH
Cardiovascular Disorders*

Each year, nearly a million Americans die of heart attacks, strokes, and congestive heart failure. Consider that your heart muscle is expected to contract vigorously over 2.5 billion times during your lifetime without once stopping to rest, and that it is expected to force blood through a series of vessels whose total length would encircle the globe twice. Add to this the possibility that these vessels may become constricted, weakened, or clogged, and it is easy to see why the cardiovascular system is a prime target for malfunction.

Hidden Killers: High Blood Pressure and Atherosclerosis

Also called **hypertension,** high blood pressure is usually caused by constriction of the arterioles, increasing the resistance to blood flow. In the majority of the 45 million afflicted Americans, the cause of this constriction is unknown. Heredity seems to play a role. For people who are predisposed to hypertension, high salt intake in the diet may aggravate it, as may obesity. Although normal blood pressure tends to increase somewhat with age, an approximate borderline reading for high blood pressure is 140/90. The higher of the two readings (the systolic pressure) is a measure of the pressure exerted by the blood when the ventricles are contracting. The lower, or diastolic pressure, measures the pressure in the arteries between ventricular contractions. A diastolic pressure consistently over 90 is an indication of high blood pressure.

High blood pressure may give few warning signals, but it undermines the cardiovascular system in several ways. First, it causes strain on the heart by increasing resistance to blood flow. Although the heart may enlarge in response to this added demand, its own blood supply may not increase proportionately. The heart muscle is then inadequately supplied with blood, especially during exercise. Lack of sufficient oxygen to the heart can cause chest pain called **angina pectoris.** Second, high blood pressure contributes to "hardening of the arteries," or atherosclerosis, described below. Third, high blood pressure in conjunction with hardened arteries can lead to rupture of an artery and internal bleeding. Rupture of vessels supplying the brain causes **stroke,** in which brain function is lost in the area deprived of blood and the vital oxygen and nutrients it carries.

Hypertension can be treated in several ways. Mild hypertension may be alleviated by weight reduction, exercise, and reduction of dietary salt. Stress reduc-

tion therapy such as relaxation techniques, meditation, and biofeedback may also be helpful. For more severe cases, drugs may be prescribed. These include diuretics, which increase urination and reduce blood volume, and drugs that dilate the arteries and arterioles. Since 1971, the number of deaths due to hypertension has fallen by over one third, thanks to increased public awareness, early detection, and successful treatment.

Atherosclerosis contributes to the 660,000 deaths from heart attack and stroke each year in the U.S. Atherosclerosis causes a loss of elasticity in the large arteries and a thickening of the arterial walls. The thickening results from deposits called **plaques** composed of cholesterol and other fatty substances. These plaques are deposited within the wall of the artery between the smooth muscle and the inner lining (Fig. E30-1). Occasionally, the plaque ruptures through the lining into the interior of the vessel. This rupture stimulates the blood platelets to initiate blood clots. These clots further obstruct the vessels, and may completely block the artery. Arterial clots are responsible for the most serious consequence of atherosclerosis: heart attack.

Heart attacks occur when one of the coronary arteries (arteries that supply the heart muscle itself) (see Fig. E30-1a) is blocked. If a blood clot suddenly breaks loose, it may be carried to a narrower part of the artery, obstructing blood flow. Deprived of nutrients and oxygen, the heart muscle once served by the blocked artery rapidly and painfully dies. If the area is small, the heart may be able to continue without it and the victim may recover. Death of large areas of heart muscle is almost instantly fatal. Although heart attacks are the major cause of death from atherosclerosis, cholesterol deposits and clots form in arteries throughout the body. A clot or a plaque deposit that obstructs an artery supplying the brain can cause a stroke, with the same results as if the artery has burst (see Fig. E30-1c).

As with hypertension, the exact cause of atherosclerosis is unclear, but several factors are known to encourage it. These include hypertension, high blood cholesterol levels, cigarette smoking, genetic predisposition, obesity, diabetes, and a sedentary lifestyle. By exercising regularly, controlling weight, and not smoking, people can greatly reduce their chances of contracting atherosclerosis. Moderation in the consumption of animal and other saturated fats is also important, since recent evidence links dietary intake of these substances with blood cholesterol levels.

(a)

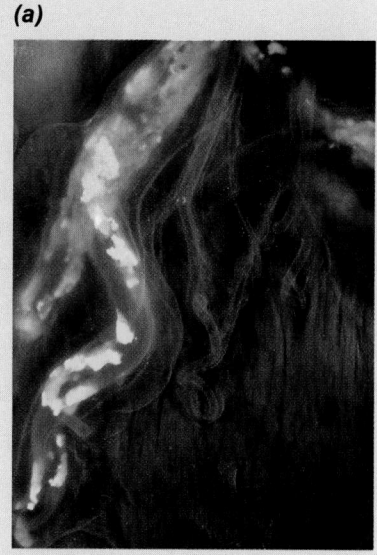

(b)

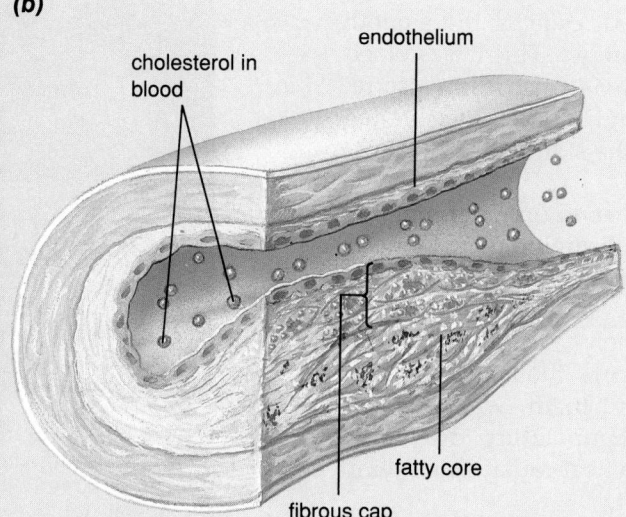

(c)

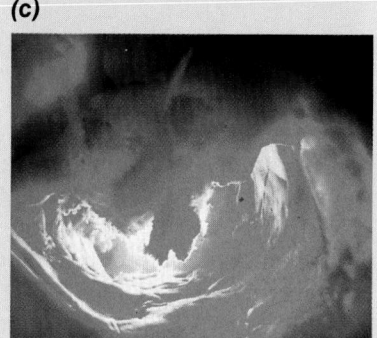

E30-1
(a) In this remarkable photo of coronary arteries, plaque deposits are seen in glowing yellow. If they block a coronary artery, a heart attack will occur.
(b) The anatomy of a plaque deposit is seen in this cross section of an artery. If the cap ruptures, a clot will form that may completely obstruct the artery, or the clot may break loose and clog a narrower artery "downstream."
(c) This remarkable view inside the aorta shows large plaque deposits on its wall. These deposits can become dislodged and be carried to a smaller artery which they may totally block, causing a heart attack or stroke.

There is also some evidence that dietary fiber may help reduce blood cholesterol.

Traditional treatment of atherosclerosis includes the use of drugs to lower blood pressure and blood cholesterol levels. In extreme cases, nitroglycerin is used to dilate blood vessels and ease the pain of angina caused by constriction of the coronary arteries. Coronary bypass surgery has been used increasingly in recent years, with over 330,000 operations per year in the U.S. This procedure consists of bypassing an obstructed coronary artery with a piece of vein, usually removed from the patient's leg. The new vessel may also clog eventually if the underlying problem— high blood cholesterol—is not corrected. Recent studies indicate that this radical operation is probably over-used; about 30% of patients might benefit equally well from other less drastic treatments.

Although prevention is by far the most cost-effective and successful strategy, a variety of other high-technology treatments are under development for this devastating condition. Blood clots are often dis-solved by injecting an enzyme, streptokinase, or another drug, TPA (tissue plasminogen activator), into the coronary artery. Both work by stimulating the production of an enzyme that breaks down fibrin, the protein that binds the clot together. When performed immediately after a heart attack, this treatment can significantly increase the victim's chances of survival. Another procedure involves squashing the clots flat against the artery walls by inserting a catheter with a tiny balloon into the obstructed artery. The balloon is inflated, crushing the deposits and restoring blood flow. Alternatively, a catheter bearing a tiny laser is manipulated up the artery to the clot and the laser is fired, destroying the obstruction. It is hoped that a combination of these new techniques in conjunction with important changes in our eating habits and lifestyle will significantly reduce early deaths from atherosclerosis in the future.

*Data are from the American Heart Association.

blood. In such threatening situations, the sympathetic nervous system (which prepares the body for emergency action) is activated, causing the smooth muscle of the arterioles to contract. This raises blood pressure generally, but selective constriction also redirects blood to the heart and muscles, facilitating rapid flight, and away from the skin, where it is less important.

In contrast, you become flushed on a hot summer day as skin arterioles expand, bringing more blood to the skin capillaries where heat is dissipated to the outside. In extreme cold, fingers and toes can become frostbitten because the arterioles supplying the extremities constrict. This shunts the blood to vital organs such as the heart and brain, which cannot function properly if their temperature drops. By minimizing blood flow to the heat-radiating extremities, the body conserves heat.

Capillary walls are only a single cell thick; lacking muscle, they act as passive tubes. The flow of blood in capillaries is regulated by tiny rings of smooth muscle called **precapillary sphincters** that surround the junctions between arterioles and capillaries (see Fig. 30-8). These sphincters open and close in response to local changes that signal the needs of nearby tissues. For example, accumulation of carbon dioxide, lactic acid, or other cellular wastes signals the need for increased blood flow to the tissues. This causes the precapillary sphincters as well as the muscles in nearby arterioles to relax, thus increasing blood flow through the capillaries.

The Blood

Blood is the medium in which dissolved nutrients, gases, hormones, and wastes are transported. It has two major components: specialized cells, including red blood cells, white blood cells, and platelets (Fig. 30-11; Table 30-1), and fluid, called **plasma,** in which they are suspended. On the average, the cellular components of blood account for 40% to 45% of its volume; the other 55% to 60% is plasma. The average person has 5 to 6 liters of blood, constituting about 8% of his or her total body weight.

Plasma

The straw-colored plasma is about 90% water, in which a number of substances are dissolved. Dissolved substances include proteins, hormones, nutrients (glucose, vitamins, amino acids, lipids), gases (carbon dioxide, oxygen), ions (sodium, calcium, potassium, magnesium) and wastes such as urea.

Plasma proteins are the most abundant of the dissolved substances. The three major plasma proteins

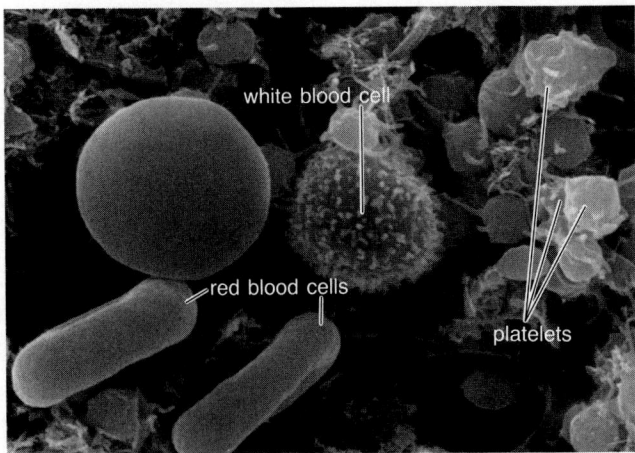

Figure 30-11 The three types of blood cells are all visible in this scanning electron micrograph. Red blood cells appear as large discs, the white blood cell resembles a fuzzy ball, while the platelets are small and irregular, with projecting filaments.

are (1) albumins, small proteins that help maintain the osmotic pressure of the blood; (2) globulins, which help transport nutrients and also function in immunity; and (3) fibrinogen, a major factor in blood clotting, discussed later in this chapter.

Red Blood Cells

Also called **erythrocytes,** the oxygen-carrying red blood cells make up about 99% of the cells in the blood. They comprise about 40% of the total blood volume in females, and 45% in males. Each milliliter of blood contains about 5 billion erythrocytes. Proteins on the surface of red blood cells differ among individuals, creating different blood types, as described in "Health Watch: Blood Types and Their Medical Implications." The red blood cell resembles a ball of clay squeezed between thumb and forefinger. Its shape provides a larger surface area than a spherical cell of the same volume (Fig. 30-12). The larger surface area maximizes the cell's ability to absorb and release oxygen through its membrane.

One of the most striking features of erythrocytes is their red color, caused by the pigment **hemoglobin** (Fig. 30-13). This large, iron-containing protein makes up about one third the weight of the blood cell. About 97% of the oxygen carried by the blood is bound to hemoglobin. The hemoglobin molecule picks up oxygen where the concentration is high, in the capillaries of the lungs, and releases it where the concentration is low, in other tissues of the body. After releasing its

Table 30-1 Blood Cells

Cell Type		Description	Average Number Present	Major Function
Red blood cell (erythrocyte)		Biconcave disk without nucleus, about one third hemoglobin Approximately 8 μm[b] in diameter	5,000,000 per mm^3 [a]	Transports oxygen and a small amount of carbon dioxide
White blood cells (leukocytes)			7,500 per mm^3	
1. Neutrophil		About twice the size of red cells, nucleus with two to five lobes	62% of white cells	Destroys relatively small particles by phagocytosis
2. Eosinophil		About twice the size of red cells, nucleus with two lobes	2% of white cells	Inactivates inflammation-producing substances; attacks parasites
3. Basophil		About twice the size of red cells, nucleus with two lobes	Less than 1% of white cells	Releases anticoagulant to prevent blood clots; and histamine, which causes inflammation
4. Monocyte		Two to three times larger than red cells, nuclear shape varies from round to lobed	3% of white cells	Gives rise to macrophage, which destroys relatively large particles by phagocytosis
5. Lymphocyte		Only slightly larger than red cell, nucleus nearly fills cell	32% of white cells	Functions in the immune response
Platelet		Cytoplasmic fragment of cells in bone marrow called megakaryocytes	250,000 per mm^3	Important in blood clotting

[a] mm^3 = cubic millimeter.
[b] μm = micrometer.

oxygen, some of the hemoglobin picks up carbon dioxide from the tissues for transport back to the lungs. The role of blood in gas exchange is discussed in more detail in Chapter 31.

The Life Cycle of Red Blood Cells

Red blood cells are formed in the marrow of bones, including those of the chest, upper arms and legs, and hips. During their development, mammalian red blood cells lose their nuclei and their ability to divide.

Each lives about 120 days. Every second, over 2 million red blood cells die and are replaced by new ones from the bone marrow. Dead or damaged red blood cells are removed from circulation, primarily in the liver and spleen, and are broken down to release their iron. The salvaged iron is carried in the blood to the bone marrow where it is used to make more hemoglobin and packaged into new red blood cells. Although the recycling process is efficient, small amounts of iron are excreted daily, and must be replenished by the diet. Bleeding from injury or menstruation also tends to deplete iron stores.

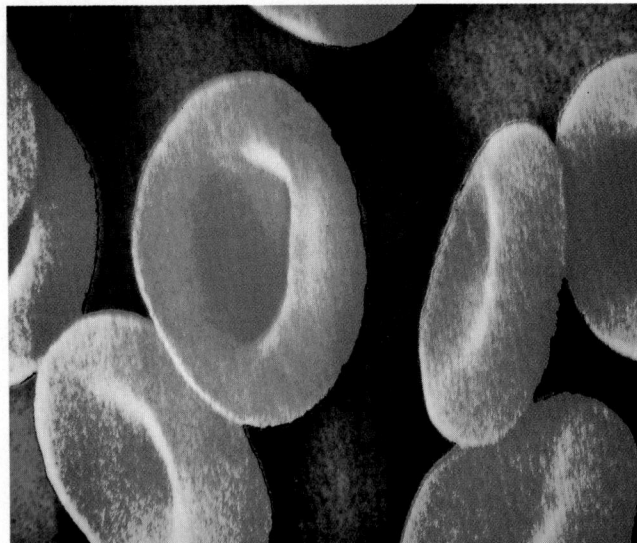

Figure 30-12 Under the scanning electron microscope, the biconcave disk shape of red blood cells is clearly visible. These cells have been artificially colored.

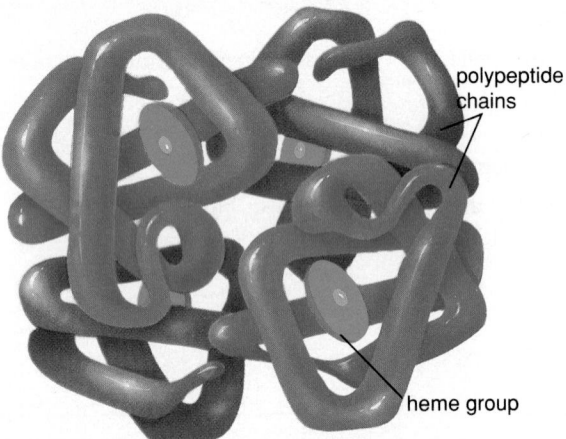

polypeptide chains

heme group

Figure 30-13 A molecule of hemoglobin is composed of four polypeptide chains, each surrounding a heme group. The heme group contains an iron atom and is the site of oxygen binding. When saturated, each hemoglobin molecule can carry four oxygen molecules (eight oxygen atoms).

Regulation of Red Blood Cell Numbers

The number of red blood cells is maintained at an adequate level through a negative feedback system that involves a hormone called **erythropoietin.** Erythropoietin is produced by the kidneys in response to

oxygen deficiency. A lack of oxygen may be caused by a loss of blood, insufficient production of hemoglobin, high altitude where the oxygen content of the air is reduced, or lung disease that interferes with gas exchange in the lungs. The hormone stimulates rapid production of new red blood cells by the bone marrow. When oxygen levels in the tissues become adequate, erythropoietin production ceases, and the rate of red blood cell production returns to normal. This negative feedback process is diagrammed in Figure 30-14.

White Blood Cells

There are five common types of white blood cells, or **leukocytes,** which together constitute less than 1% of the total cellular component of the blood. These cells, described and illustrated in Table 30-1, are distinguished from one another by their staining characteristics, size, and the shape of their nuclei. Most white blood cells function in some way to protect the body against foreign invaders, and use the circulatory system to travel to the site of invasion. For example, **monocytes** and **neutrophils** travel via capillaries to wounds where bacteria have gained entry. Here they ooze out through the capillary walls like tiny amoebas. In tissues, monocytes differentiate into **macrophages,** amoebalike cells that engulf foreign particles. Macrophages and neutrophils feed on the bacterial invaders (or other cells recognized as foreign, including cancer cells; Fig. 30-15). They often lose their lives in the process, and their dead bodies accumulate and contribute to the white substance we call pus, seen most abundantly at sites of infection. **Lymphocytes** are white blood cells responsible for the production of antibodies that help provide immunity

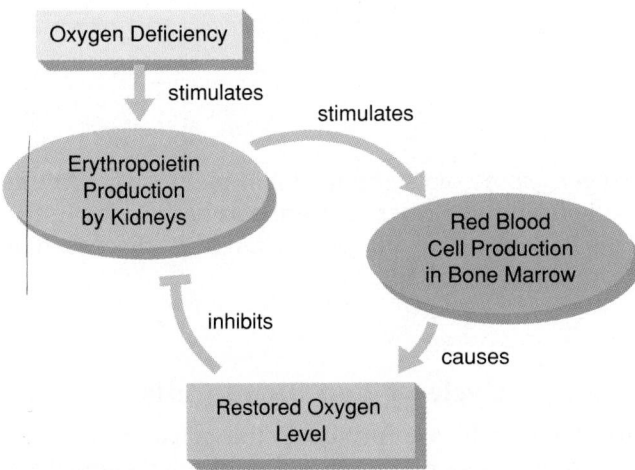

Figure 30-14 The production of red blood cells is regulated by a negative feedback system.

HEALTH WATCH
Blood Types and Their Medical Implications

The fact that all blood is not created equal was discovered painfully in the mid-1600s when blood from a lamb was transfused into a human—with fatal results. It was not until 1901 that the American biologist Karl Landsteiner identified different blood types among people, setting the stage for safe blood transfusions. Prior to the discovery of blood typing, transfusions between humans were either successful or disastrous, depending on chance for compatibility of the donor and recipient.

Blood is classified as type A, B, AB, or O depending on the presence or absence of specific glycoproteins (designated A and B) in the red blood cell membrane. Individuals with type A blood have the A glycoprotein on their red blood cells and have antibodies (see Chapter 34) to the B glycoprotein in their plasma. If they are transfused with type B blood, the B antibodies attack the transfused red blood cells, causing them to clump together and block small blood vessels. Type O blood cells lack both A and B glycoproteins, and antibodies to both A and B glycoproteins are present in the plasma. Type O blood can therefore be transfused into individuals with any blood type, but type O individuals can receive transfusions only of type O blood. Type AB blood cells have both glycoproteins and can be transfused only into another AB individual. But individuals with type AB blood can receive transfusions of other blood types, since their plasma lacks antibodies to either glycoprotein. Blood type is inherited; the genetics and properties of these major blood types are discussed in Chapter 11.

Nearly 40 years after his discovery of blood types, Landsteiner, working with Alexander Weiner, discovered the **Rh factor.** Rh-positive blood carries another specific type of protein in the red blood cell membrane, while Rh-negative blood lacks the protein. Although antibodies to A and B proteins develop spontaneously, antibodies to the Rh-positive antigen form only after massive exposure to the antigen. The first exposure—for example, transfusion of Rh-positive blood into an Rh-negative individual—generally causes no ill effects, but triggers the production of antibodies. Upon subsequent exposures, the antibodies attack the Rh-positive blood cells, causing them to clump and subsequently destroying them. This is a particular concern when an Rh-negative woman carries her second or third Rh-positive child. Rh-positive is a dominant genetic trait; about 85% of Caucasians and almost all blacks are Rh-positive. An Rh-negative woman has a 75% chance of bearing an Rh-positive child if the father is Rh-positive. Her first Rh-positive child suffers no ill effects, but may trigger her body to form antibodies against Rh-positive blood. During subsequent pregnancies in which the fetus is Rh-positive, there is an increasing prevalence of **erythroblastosis fetalis,** in which the mother's antibodies invade the fetus and attack its red blood cells. The baby is born severely anemic. Breakdown products from its own hemoglobin can cause jaundice, and in severe cases, permanent mental retardation. The newborn is treated immediately by transfusion with Rh-negative blood, which is not attacked by the antibodies. More transfusions are given during the infant's first few weeks of life. During this period, the mother's antibodies gradually break down. Gradually, the infant replaces the transfused Rh-negative blood with its own Rh-positive blood. In severe cases in which the fetus might not survive until delivery, transfusions are sometimes administered prior to birth. Under continuous ultrasound monitoring, a needle is inserted through the mother's abdominal wall, into the uterus, and through the abdomen of the fetus, until it rests in the peritoneal cavity (a space surrounding the abdominal organs). The needle is then replaced with a thin tube through which Rh-negative blood is transfused.

---

against disease (see Chapter 34). Least abundant are the **eosinophils** and **basophils.** Eosinophil production is stimulated by parasitic infections. Eosinophils converge on the parasitic invaders, releasing substances that kill the parasite. Basophils release both substances that inhibit blood clotting and chemicals such as histamine that participate in inflammatory and allergic reactions.

Platelets and Blood Clotting

The third group of blood "cells" are actually fragments of much larger cells. These large cells, called **megakaryocytes,** remain in the bone marrow, pinching off pieces of themselves which enter the blood. The fragments, called **platelets,** play a central role in blood clotting. Clot formation is a complex process. It

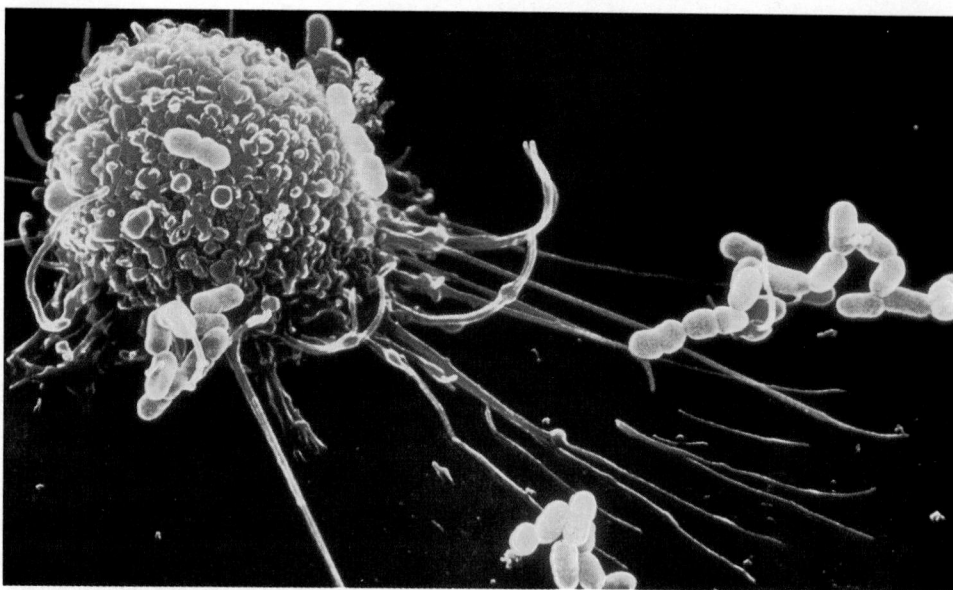

Figure 30-15 An ameoba-like white blood cell is seen capturing bacteria. These bacteria are *E. coli*, intestinal bacteria that can cause disease if they enter the bloodstream.

starts when platelets and other factors in the plasma contact an irregular surface, such as a damaged blood vessel. Platelets tend to stick to irregular surfaces, and they may build up and plug the damaged vessel if it is narrow enough. This mechanism is supplemented by blood coagulation, or clotting, which is the most important of the body's defenses against bleeding. The ruptured surface of an injured blood vessel not only causes platelets to adhere, it also initiates a complex sequence of events among circulating plasma proteins. These events culminate in production of the enzyme **thrombin.** Thrombin catalyzes the conversion of the plasma protein fibrinogen into stringlike molecules called **fibrin.** Fibrin molecules adhere to one another, end-to-end and side-to-side, forming a fibrous matrix. This protein web immobilizes the fluid portion of the blood, causing it to solidify much like cooling gelatin. The web traps red blood cells, increasing the density of the clot (Fig. 30-16). Platelets then adhere to the fibrous mass and send out sticky projections that attach to one another. Within half an hour, the platelets contract, pulling the mesh tighter and forcing liquid out. This creates a denser, stronger clot and also constricts the wound, pulling the damaged surfaces closer together to promote healing.

The Lymphatic System

The **lymphatic system** consists of a network of lymphatic capillaries that drain into larger lymphatic vessels, which in turn drain into the large blood veins just before they enter the heart (Fig. 30-17). The lymphatic system also includes numerous small lymph nodes, and two additional organs: the thymus and the spleen. The lymphatic system has several important functions:

1. **The removal of excess fluid and dissolved substances that leak from the capillaries.**
2. **The transport of fats from the intestine to the bloodstream.**
3. **Defense of the body by exposing bacteria and viruses to white blood cells.**

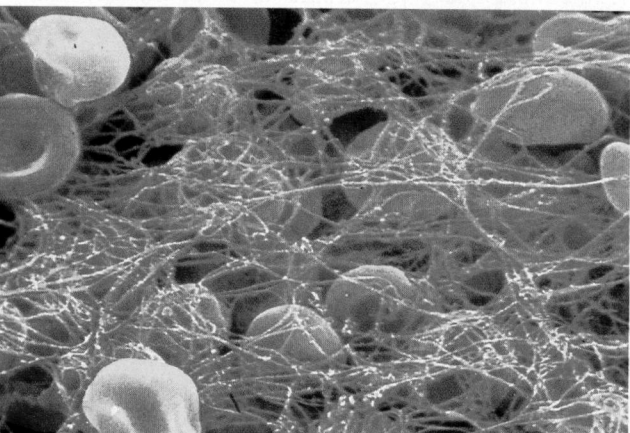

Figure 30-16 In response to damage to a blood vessel, threadlike proteins called fibrin form in the blood. These produce a tangled sticky mass that traps red blood cells and eventually forms a clot.

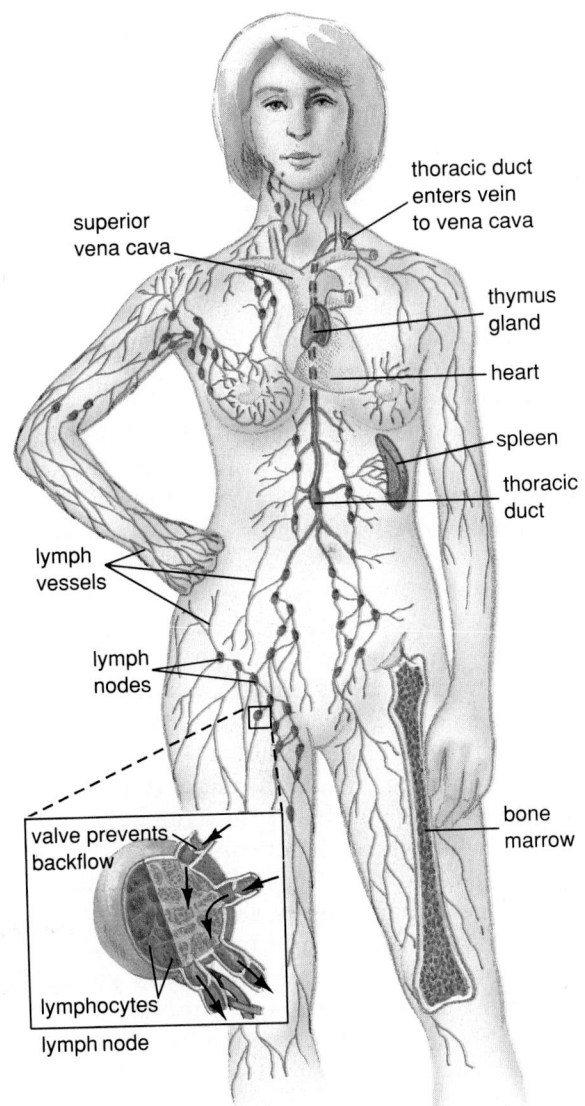

Figure 30-17 The human lymphatic system, illustrating lymph vessels, lymph nodes, and two auxiliary lymph organs, the thymus and spleen. Inset: A cross section of a lymph node. The node is filled with channels lined with white blood cells that attack foreign matter in the lymph.

Structure of the Lymphatic Vessels

Like blood capillaries, lymph capillaries form a complex network of thin-walled vessels into which substances can move readily. In contrast to blood capillaries, lymph capillary walls are composed of cells with openings between them that act as one-way valves, allowing relatively large particles to be carried into the capillary along with the fluid. Also unlike blood capillaries, lymph capillaries "dead end" in the

tiny spaces between cells (Fig. 30-18). Lymph is forced through lymph vessels as blood is through veins. Large lymph vessels have somewhat muscular walls, but most of the impetus for lymph flow comes from the contraction of nearby muscles, such as those used in breathing and walking. As in blood veins, the direction of flow is regulated by one-way valves (Fig. 30-19).

Return of Fluids to the Blood

The exchange of materials between capillary blood and nearby cells occurs through **interstitial fluid** that bathes nearly all the cells of the body. This fluid, derived from the blood plasma, leaks through the permeable walls of the capillaries. Interstitial fluid is primarily water in which are dissolved nutrients, hormones, gases, wastes, and small proteins from the blood. (The large plasma proteins, red blood cells, and platelets are unable to leave the capillaries because of their size, although white blood cells are able

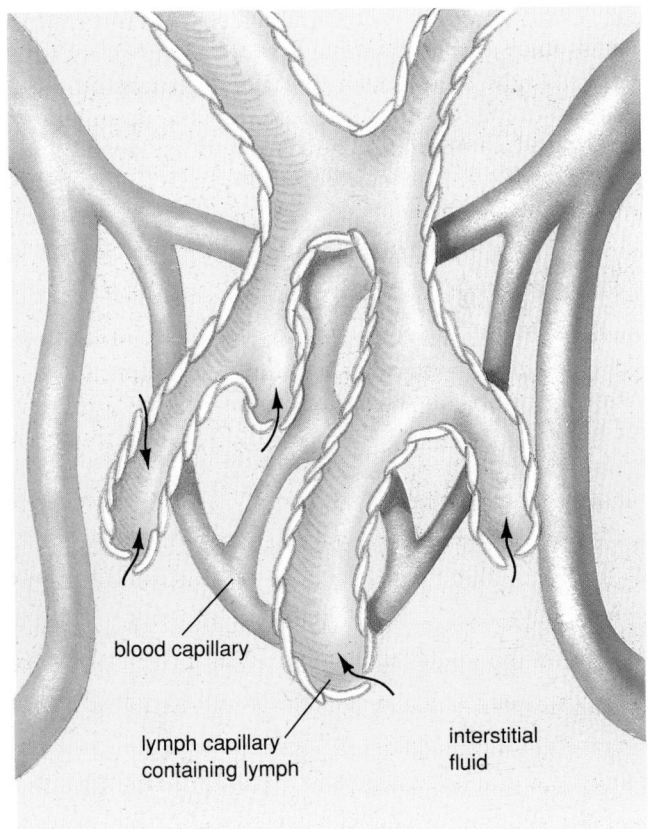

Figure 30-18 Lymph capillaries end blindly in the body tissues, where pressure from the accumulation of interstitial fluid forces the fluid into the lymph capillaries through valvelike openings between lymph capillary cells.

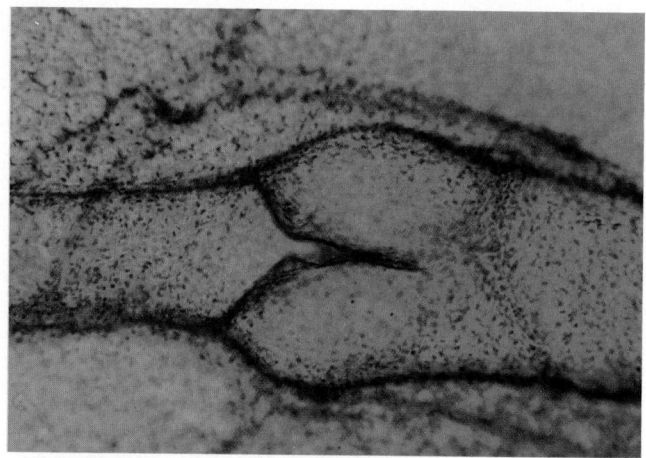

Figure 30-19 Like blood-carrying veins, lymph vessels have internal one-way valves that direct the flow of lymph toward the large veins into which they empty.

to ooze through tiny openings between capillary cells.) In an average person, about 3 liters more fluid leaves the blood capillaries than is reabsorbed each day. One role of the lymphatic system is to return this excess fluid and its dissolved proteins and other substances to the blood. As interstitial fluid accumulates, its pressure forces the fluid through the openings in the lymph capillaries. The lymphatic system transports this fluid, now called **lymph,** back to the circulatory system.

The Transport of Fats

As you will learn in Chapter 32, the small intestine is richly supplied with blindly ending lymph capillaries. After absorbing digested fats, intestinal cells release fat globules into the interstitial fluid. These globules are too large to diffuse into blood capillaries, but move easily through the spaces between lymph capillary cells. Once in the lymph, they are dumped into a large vein that enters the heart. After a fatty meal, these fat globules may make up 1% of the lymphatic fluid.

Defense of the Body

In addition to its other roles, the lymphatic system helps defend the body against foreign invaders such as bacteria and viruses. The large lymph vessels are interrupted periodically by kidney bean–shaped structures about 1 inch (2.5 cm) long called **lymph nodes** (see Fig. 30-17). Lymph is forced through channels within the nodes that are lined with masses of macrophages. Lymphocytes are also produced in the nodes. Both of these white blood cells recognize and destroy foreign particles such as bacteria and viruses, and are killed in the process. The painful swelling of lymph nodes in certain diseases such as mumps is largely a result of the accumulation of dead lymphocytes, dead macrophages, and the virus-infested cells they have engulfed.

Two additional organs, the **thymus** and the **spleen,** are often considered part of the lymphatic system, although they have no direct connection to the lymphatic vessels (see Fig. 30-17). The thymus is located beneath the sternum slightly above the heart, and is responsible for production of lymphocytes. The thymus is particularly active in infants and young children, but decreases in size and importance in early adulthood. The spleen is located in the left side of the abdominal cavity, between the stomach and diaphragm. Just as the lymph nodes filter lymph, the spleen filters blood, exposing it to macrophages and lymphocytes that destroy foreign particles and aged red blood cells.

SUMMARY OF KEY CONCEPTS

Types of Circulatory Systems

Unicellular and some simple multicellular organisms rely exclusively on diffusion to exchange wastes, nutrients, and gases with the environment. Larger, more active multicellular organisms evolved circulatory systems that keep diffusion distances small. These systems transport a fluid rich in dissolved nutrients and oxygen close to the cells, where nutrients can be released to the cells and wastes absorbed by diffusion. Invertebrates may have either open or closed circulatory system, while all vertebrates have closed systems. In open systems, blood is pumped into a hemocoel where it directly bathes internal organs. In closed systems, the blood is confined to the heart and blood vessels.

The Vertebrate Circulatory System

Vertebrate circulatory systems have many roles, including the transport of gases and wastes and the distribution of nutrients. In addition, circulatory systems transport hormones, help regulate body temperature, and defend the body against disease. The circulatory system has three parts: the heart, the vessels, and the blood. The lymphatic system is an important accessory system.

The Vertebrate Heart The vertebrate heart evolved from two chambers in fishes, to three in amphibians and most reptiles, to four in birds and mammals. In the four-chambered heart, blood is pumped separately to the lungs and through the body, maintaining complete separation of oxygenated and deoxygenated blood. Deoxygenated blood is collected from the body in the right atrium and passed to the right ventricle, which pumps it to the lungs. Oxygenated blood from the lungs enters the left atrium, is passed to the left ventricle and pumped to the rest of the body.

The heart cycle consists of two stages: systole, during which the ventricles contract, and diastole, when the ventricles relax. The direction of blood flow is maintained by atrioventricular valves located between the atria and ventricles, and semilunar valves located between the ventricles and the pulmonary artery and aorta.

Cardiac muscle cells are spontaneously active. Their contractions are coordinated by two specialized masses of muscle cells: the SA and AV nodes. The SA node of the right atrium serves as the heart's pacemaker. The wave of atrial contraction activates the AV node, also in the right atrium, from which it spreads through the ventricles along specialized fiber tracts. The heart rate can be modified by the nervous system and by hormones such as epinephrine.

The Blood Vessels Blood leaving the heart travels in sequence through arteries, arterioles, capillaries, venules, veins and then back to the heart. Each vessel is specialized for its specific role. Elastic, muscular arteries help pump the blood along. The thin-walled capillaries are the sites of exchange of materials between the body cells and the blood. Veins provide a low resistance path back to the heart, with one-way valves maintaining the direction of blood flow.

The distribution of blood is regulated by the constriction and dilation of arterioles under the influence of the sympathetic nervous system and local factors such as carbon dioxide in the tissues. Local factors also regulate precapillary sphincters that control blood flow in the capillaries.

The Blood The blood is composed of both fluid and cellular portions. The fluid plasma consists of water that contains proteins, hormones, nutrients, gases, and wastes. Red blood cells, or erythrocytes, are packed with a large iron-containing protein called hemoglobin, which carries oxygen. Erythrocytes are synthesized in bone marrow, lack a nucleus, and live about 120 days. Their numbers are regulated by the hormone erythropoietin. There are five types of white blood cells, or leukocytes, that fight infection (see Table 30-1). Platelets, which are fragments of megakaryocytes, are important for blood clotting.

The Lymphatic System

The human lymphatic system consists of vessels resembling veins and capillaries, lymph nodes, and the thymus and spleen. The lymphatic system removes excess interstitial fluid that leaks through blood capillary walls. It transports fats to the bloodstream from the intestine, and fights infection by filtering the lymph through lymph nodes where white blood cells ingest foreign invaders such as bacteria. The thymus, which is most active in young children, produces lymphocytes that function in immunity. The spleen filters blood past macrophages and lymphocytes, which remove bacteria and damaged blood cells.

GLOSSARY

angina pectoris (an-jī′-na pek-tōr′-is): chest pain associated with reduced blood flow to the heart muscle caused by obstruction of coronary arteries.

arteriole (ar-tēr′-ē-ōl): a small artery that empties into capillaries. Contraction of the arteriole regulates blood flow to various parts of the body.

artery (ar′-tur-ē): a vessel with muscular, elastic walls that conducts blood away from the heart.

atherosclerosis (ath′-er-ō-skler-ō′-sis): a disease characterized by obstruction of arteries by cholesterol deposits and thickening of the arterial walls.

atrioventricular node (ā′-trē-ō-ven-trik′-ū-lar nōd): a specialized mass of muscle at the base of the right atrium through which the electrical activity initiated in the sinoatrial node is transmitted to the ventricles.

atrioventricular valve: heart valves between the atria and the ventricles that prevent backflow of blood into the atria during ventricular contraction.

atrium (ā′-trē-um): a chamber of the heart that receives venous blood and passes it to a ventricle.

capillary: the smallest type of blood vessel, capillaries connect arterioles with venules. Capillary walls, through which exchange of nutrients and wastes occurs, are only one cell thick.

cardiac cycle: the alternation of contraction and relaxation of the heart chambers; systole and diastole.

closed circulatory system: the type of circulatory system found in certain worms and vertebrates in which the blood is always confined within the heart and vessels.

diastole (di-as′-tō-lē): the portion of the cardiac cycle during which the ventricles are relaxed; the lower of the two blood pressure readings.

erythroblastosis fetalis (ē-rith′-rō-blas-tō′-sis fē-tal′-is): a condition in which the red blood cells of a newborn Rh-positive baby are attacked by antibodies produced by its Rh-negative mother, causing jaundice and anemia. Retardation and death are possible consequences if treatment is inadequate.

erythrocytes (ē-rith′-rō-sītes): red blood cells active in oxygen transport, which contain the red pigment hemoglobin.

erythropoietin (ē-rith′-rō-pō-ē′-tin): a hormone produced by the kidneys in response to oxygen deficiency that stimulates production of red blood cells by the bone marrow.

fibrillation: rapid, uncoordinated, and ineffective contractions of heart muscle cells.

fibrin (fī′-brin): a clotting protein formed in the blood in response to a wound. Fibrin binds with other fibrin molecules and provides a matrix around which a blood clot forms.

hemocoel (hē′-mō-sēl): the blood cavity in an open circulatory system.

hemoglobin (hē′-mō-glō-bin): an iron-containing protein that gives red blood cells their color. Hemoglobin binds to oxygen in the lungs and releases it to the tissues.

hypertension: arterial blood pressure that is chronically elevated above the normal level.

interstitial fluid (in-tur-sti′-shul): fluid similar in composition to plasma (except lacking large proteins) which leaks from the capillaries and acts as a medium of exchange between the body cells and the capillaries.

leukocyte (loo′-kō-sīt): any of the white blood cells circulating in the blood.

lymph (limf): pale fluid within the lymphatic system composed primarily of interstitial fluid and lymphocytes.

lymphatic system: a system consisting of lymph vessels, lymph capillaries, lymph nodes, and the thymus and spleen. The system helps protect the body against infection, absorbs fats, and returns excess fluid and small proteins to the blood circulatory system.

lymph nodes: small structures that act as filters for lymph. These contain lymphocytes and macrophages, which inactivate foreign particles such as bacteria.

lymphocyte (lim′-fō-sīt): a white blood cell type important in the immune response.

macrophage (mak′-rō-faj): a cell derived from white blood cells called monocytes, whose function is to consume foreign particles including bacteria.

megakaryocyte (meg-a-kār′-ē-ō-sīt): a large cell type that remains in the bone marrow, pinching off pieces of itself. These fragments enter the circulation as platelets.

open circulatory system: a type of circulatory system found in some invertebrates, such as arthropods and molluscs, which includes an open space in which blood directly bathes body tissues.

plaque (plak): a deposit of cholesterol and other fatty substances within the wall of an artery.

plasma: the fluid, noncellular portion of the blood.

platelets (plāt′-lets): cell fragments formed from megakaryocytes in bone marrow. Platelets, which lack nuclei, circulate in the blood and play a role in blood clotting.

precapillary sphincter (sfink′-ter): a ring of smooth muscle between an arteriole and a capillary that regulates the flow of blood into the capillary bed.

Rh factor: a protein found on the red blood cells of some people (Rh-positive) but not others (Rh-negative). Exposure of Rh-negative individuals to Rh-positive blood triggers antibody production to Rh-positive blood cells.

semilunar valve: the type of valve between the ventricles of the heart and the pulmonary artery and aorta. Semilunar valves prevent backflow of blood into the ventricles when they relax.

sinoatrial node (sī′-nō-āt′-rē-ul nōd): also called the SA node, a small mass of specialized muscle in the wall of the right atrium. It generates electrical signals rhythmically and spontaneously and serves as the heart's pacemaker.

stroke: an interruption of blood flow to part of the brain, caused by the rupture of an artery or the blocking of an artery by a blood clot. Loss of blood supply leads to rapid death of the area of the brain affected.

systole (sis′-tō-lē): the portion of the heart cycle during which the ventricles contract; the higher of the two blood pressure readings.

thrombin: an enzyme produced in the blood as a result of injury to a blood vessel. Thrombin catalyzes the production of fibrin, a protein that assists in blood clot formation.

vein: a large-diameter, thin-walled vessel that carries blood from venules back to the heart.

ventricle (ven'-trē-kul): the lower muscular chamber on each side of the heart, which pumps blood out through the arteries. The right ventricle sends blood to the lungs, and the left to the rest of the body.

venule (ven'-yul): a narrow vessel with thin walls that carries blood from capillaries to veins.

STUDY QUESTIONS

1. Trace the flow of blood through the circulatory system, starting and ending with the right atrium.
2. List three types of blood cells and describe their principal functions.
3. What are five functions of the vertebrate circulatory system?
4. In what way do veins and lymph vessels resemble one another? Describe how fluid is transported in each of these vessels.
5. Describe three important functions of the lymphatic system.
6. Distinguish among plasma, interstitial fluid, and lymph.
7. Describe veins, capillaries, and arteries, noting their similarities and differences.
8. Trace the evolution of the vertebrate heart from two to four chambers.
9. What is the role of atherosclerosis in heart attack?
10. Describe the cardiac cycle, and relate systole and diastole to the two readings taken during the measurement of blood pressure.
11. Describe how red blood cell number is regulated by a negative feedback system. Include the hormone involved, its site of production, and its function.
12. Why is it important for a married couple to be aware of their Rh factors? Describe erythroblastosis fetalis and its mechanism. Predict what would happen if an Rh-positive female carried an Rh-negative baby.

DISCUSSION QUESTIONS

1. Discuss the steps you can take now and in the future to reduce your risk of developing heart disease.
2. The Rh factors of expectant parents become important due to the possibility of erythroblastosis fetalis in their offspring. Describe this disease, what causes it, and how it can be prevented. Predict what could happen if an Rh+ female carried an Rh− fetus, and if an Rh− female carried an Rh+ fetus.
3. Discuss why a four-chambered heart is much more efficient than a two-chambered heart in delivering oxygenated blood to the various body parts.

SUGGESTED READINGS

Brown, M. S. and J. L. Goldstein. "How LDL Receptors Influence Cholesterol and Atherosclerosis." *Scientific American*, November 1984. An easily understood article about an important health topic.

"Cholesterol: The Villian Revealed." *Discover*, March 1984. A beautifully illustrated three-part special report on cholesterol, its effects on human arteries, and methods of treatment.

Katzir, A. "Optical Fibers in Medicine." *Scientific American*, May 1989. Describes the use of fiber optics in medicine, with emphasis on their role in cardiovascular surgery.

Lillywhite, H. B. "Snakes, Blood Circulation, and Gravity." *Scientific American*, December 1988. Describes how the cardiovascular systems of snakes are adapted to their diverse lifestyles.

Perutz, M. "Hemoglobin Structure and Respiratory Transport." *Scientific American*, December 1978. How hemoglobin plays the dual role of transporting oxygen to the tissues and carbon dioxide back to the lungs.

"Replacing the Heart." *Discover*, February 1983. An excellent four-part report on the artificial heart.

Zucker, M. "The Functioning of Blood Platelets." *Scientific American*, June 1980. Describes the complex role of platelets in blood clotting.

31

Respiration

Respiratory structures require large, moist surfaces for gas exchange. Nudibranch (literally, "naked gill") molluscs extend feathery gills into the seawater. The flesh of nudibranchs is distasteful, so these delicate respiratory structures are safe from nibbling predators.

Each cell in the animal body is a tiny factory, demanding a continuous influx of energy to maintain itself. Most cells use a biochemical process called cellular respiration (see Chapter 8) to make this energy available from nutrients such as sugar. The production of energy by cellular respiration requires a steady supply of oxygen and generates carbon dioxide as a waste product. In most animals, these gases are transported in the blood and are exchanged between the blood and the cells by diffusion. Gases are also exchanged between the blood and the external environment (air or water) by diffusion. **Respiratory systems support cellular respiration by bringing a large, moist surface into intimate contact with both the blood and the external environment so that oxygen and carbon dioxide may be exchanged by diffusion between them.**

The Evolution of Respiratory Systems

Diffusion of oxygen occurs rapidly enough to support metabolically active cells only over very short distances. As natural selection produced increasingly large, active animals, elaborate respiratory systems became necessary to guarantee ample gas exchange throughout their bodies. Animal respiratory systems are amazingly diverse. However, respiratory systems all share two features that facilitate diffusion:

1. **The respiratory system must have a sufficiently large surface area in contact with the environment so that gas exchange by diffusion is adequate to support the demands of the body.**
2. **The respiratory exchange surface must remain moist, since gases must be dissolved in fluid when they enter or leave living cells.**

   In the following sections, we examine some respiratory systems of increasing complexity. As we touch on the diverse types of respiratory systems (illustrated schematically in Fig. 31-1), notice how each meets the demands for a sufficiently large, moist surface.

Respiration Without Specialized Respiratory Structures

A variety of animals that live in moist environments are able to exchange gases adequately without specialized respiratory systems. The bodies of these animals have one or more of a variety of features, listed below, that make respiratory systems unnecessary.

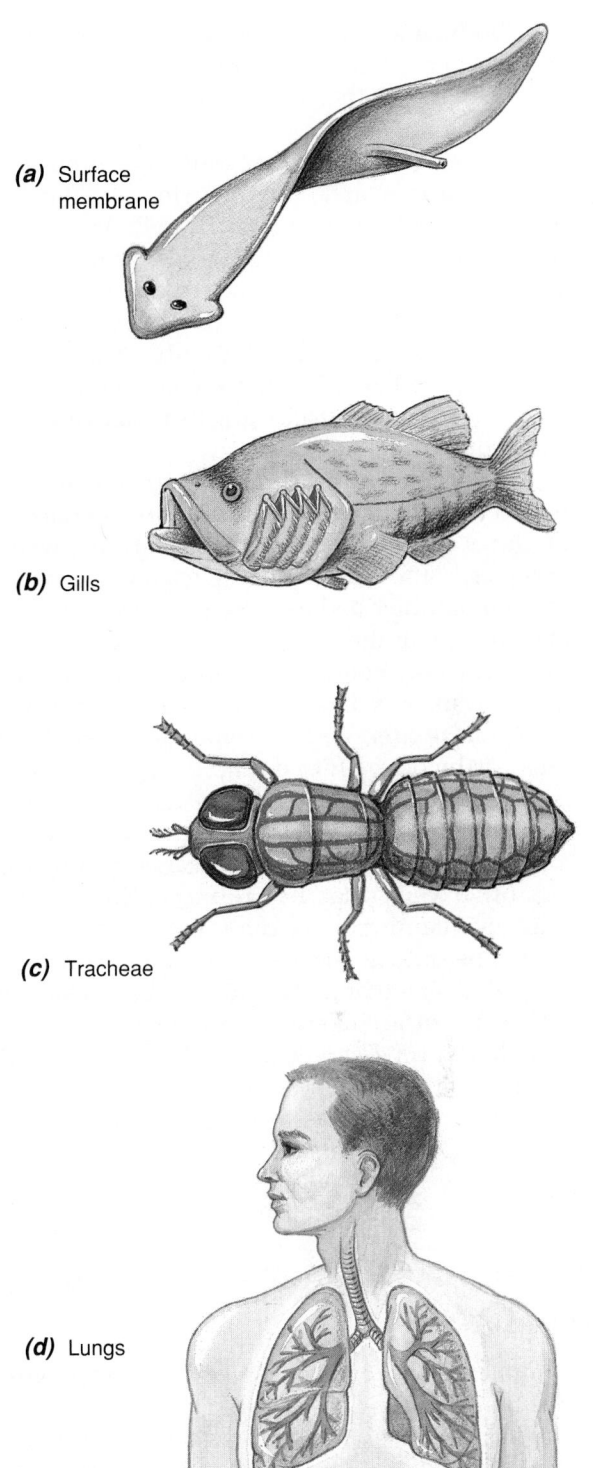

(a) Surface membrane

(b) Gills

(c) Tracheae

(d) Lungs

Figure 31-1 Schematic illustration of a variety of respiratory systems. **(a)** In the simplest case, gases are exchanged directly across the moist exterior of the body. **(b)** Gills evolved in larger animals that lived in water, bringing large surface areas richly supplied with blood capillaries into close contact with the surrounding water. As terrestrial animals evolved, the moist respiratory surfaces had to be protected within the body. **(c)** Insects evolved tracheae, and **(d)** vertebrates evolved lungs.

1. **The body may be extremely small.** In microscopic nematode worms, for example, gases have only a short distance to diffuse to reach all parts of the body.

2. **The body may be thin and flattened, producing a large surface area for diffusion.** In the flatworms, for example, most body cells are close to the moist skin through which gases can diffuse.

3. **The body may have cells with very low energy demands, so that relatively slow gas exchange is adequate.** For example, the body of a jellyfish may be quite large, but those cells that are relatively far from the surface are also relatively inert and require little oxygen.

4. **The body may bring the environment (usually water) close to its cells, allowing direct exchange of gases between body cells and the water.** Sponges, for example, circulate seawater throughout their bodies, where it comes reasonably close to all the cells, allowing gas exchange.

5. **The body may combine a well-developed circulatory system, a large skin surface area where diffusion occurs, and low energy demands.** In the earthworm, for example, gases diffuse through the moist skin and are distributed throughout the body by a well-developed circulatory system. Blood in skin capillaries rapidly carries off oxygen that has diffused through the skin, maintaining a concentration gradient that favors the diffusion of more oxygen inward. The elongated shape of the worm assures a relatively large skin surface relative to the internal volume, and the worm's sluggish metabolism demands relatively little oxygen. The skin must stay moist to remain effective as a gas-exchange organ; a dried-out earthworm will suffocate.

Respiratory Systems and Gas Exchange

Most animals have evolved specialized respiratory systems that work in close conjunction with their circulatory systems to exchange gases between the cells and the environment. The transfer of gases from the environment to the blood to the cells and back usually occurs in stages that alternate bulk flow with diffusion. During **bulk flow,** molecules of fluids or gases move in unison (in "bulk") through relatively large spaces, from areas of higher pressure to areas of lower pressure. Bulk flow contrasts with diffusion, in which molecules move individually from areas of high concentration to areas of lower concentration (see Chapter 6). In general, gas exchange in respiratory systems occurs in the following stages, which are illustrated for the human in Figure 31-9.

1. **Air or water containing oxygen is moved across a respiratory surface by bulk flow,** usually facilitated by muscular breathing movements.

2. **Oxygen and carbon dioxide are exchanged through the respiratory surface by diffusion.** Diffusion across the respiratory membranes carries oxygen into and carbon dioxide out of the capillaries of the circulatory system.

3. **Gases are transported between the respiratory system and the tissues by bulk flow of blood,** as it is pumped throughout the body by the heart.

4. **Gases are exchanged between the tissues and the circulatory system by diffusion.** At the tissues, oxygen diffuses out of the capillaries and carbon dioxide diffuses into the capillaries along their concentration gradients.

Respiration Using Gills

Gills are respiratory structures seen in a wide variety of aquatic animals. The simplest type of gill, found in certain molluscs and amphibians, consists of numerous projections of the body surface into the surrounding water (Fig. 31-2). Generally, gills are numerous or elaborately branched, maximizing surface area. In some animals, the size of the gill is determined by the availability of oxygen in the surrounding water. For example, salamanders living in stagnant water have larger gills than those that dwell in well-aerated water. Gills have a dense profusion of capillaries just beneath the delicate outer membrane. These bring blood close to the surface where gas exchange occurs.

Some gills are not extensions of the skin, but are protected inside the body. In the crab, gills are covered by a rigid exoskeleton (Fig. 31-3a). The gills of fish are covered by a protective flap, the **operculum.** This structure protects the delicate gill membranes from being nibbled off, and also streamlines the fish body, allowing faster swimming. Fish create a continuous current over the gills by pumping water into their mouths and ejecting it through the opercular openings (see Fig. E31-1). Fish can augment the flow of water by swimming with their mouths open; some fast swimmers such as the tuna and certain sharks may rely exclusively on swimming to ventilate their gills. The gills of fish contain an elaborate network of blood vessels, whose organization enhances the exchange of gases across the gill membrane. This arrangement, called **countercurrent flow,** is described in "A Closer Look at Gills and Gas Exchange: Countercurrent Flow" and shown in Figure E31-1.

Figure 31-2 A variety of aquatic animals respire using unprotected gills. **(a)** This type of mollusc is called a nudibranch (literally "naked gill"). Its white-tipped gills extend gracefully into the surrounding water. **(b)** The aquatic larval stage of the tiger salamander uses gills whose frills provide an enormous surface area for gas exchange.

As described in Chapter 22, the evolution of terrestrial animals required dramatic modifications of gas-exchange organs. Gills are useless out of water, collapsing and drying in the air. Terrestrial respiratory organs, therefore, need both support and protection from desiccation. Three solutions have evolved: book lungs and tracheae in arthropods, and lungs in vertebrates.

Respiration Using Book Lungs and Tracheae

Although the external skeleton of terrestrial arthropods helps retain body water, it eliminates the skin as a respiratory surface. Spiders (and some other arachnids such as scorpions) enclose a series of moist pagelike membranes within a chamber of the exoskel-

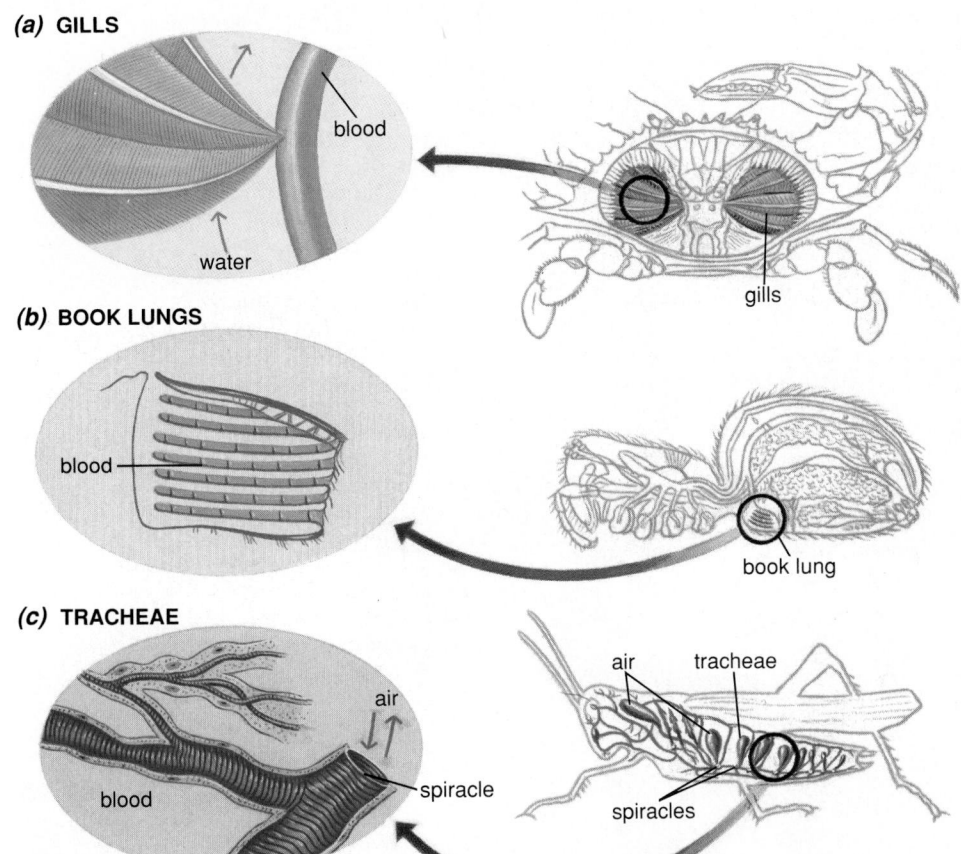

(a) GILLS

blood

water

(b) BOOK LUNGS

blood

book lung

(c) TRACHEAE

air

blood

spiracle

air

tracheae

spiracles

gills

Figure 31-3 Arthropod respiratory structures. **(a)** Crabs protect finely divided gills beneath a rigid exoskeleton. **(b)** The book lungs of spiders are sheetlike layers of thin tissue enclosed in a protective chamber. **(c)** The tracheae of insects open through spiracles in the abdominal wall and branch intricately throughout the body.

CLOSER LOOK
At Gills and Gas Exchange: Countercurrent Flow

The gills of fish (Fig. E31-1a) have evolved a very effective adaptation for removing the maximum amount of oxygen from the water flowing over them, called **countercurrent flow.** As the name implies, during countercurrent flow two types of fluids (in this case blood and water) with different concentrations of one or more dissolved substances flow in opposite directions past one another, separated by thin membranes. Countercurrent flow maintains a concentration gradient of a dissolved substance (such as oxygen) that promotes movement of the substance by diffusion between the fluids, down its concentration gradient.

Fish gills consist of a series of filaments supported by bony gill arches (Fig. E31-1b). Each filament is covered with thin folds of tissue called lamellae. Blood flows across each lamella within a dense network of capillaries. Within each lamella, countercurrent flow enhances diffusion by maintaining a concentration gradient of oxygen between the water (which is relatively high in oxygen) and the blood (lower in oxygen). As shown in Figure E31-1c, water is deflected over the lamellae in a direction opposite the flow of blood in the capillaries. Water highest in dissolved oxygen encounters the most highly oxygenated blood. As the water loses its oxygen to the blood, it flows past blood that is increasingly low in oxygen, so the gradient encouraging oxygen to move into the blood is maintained across each lamella. Countercurrent flow is so effective that some fish extract 85% of the oxygen from the water that flows over their gills.

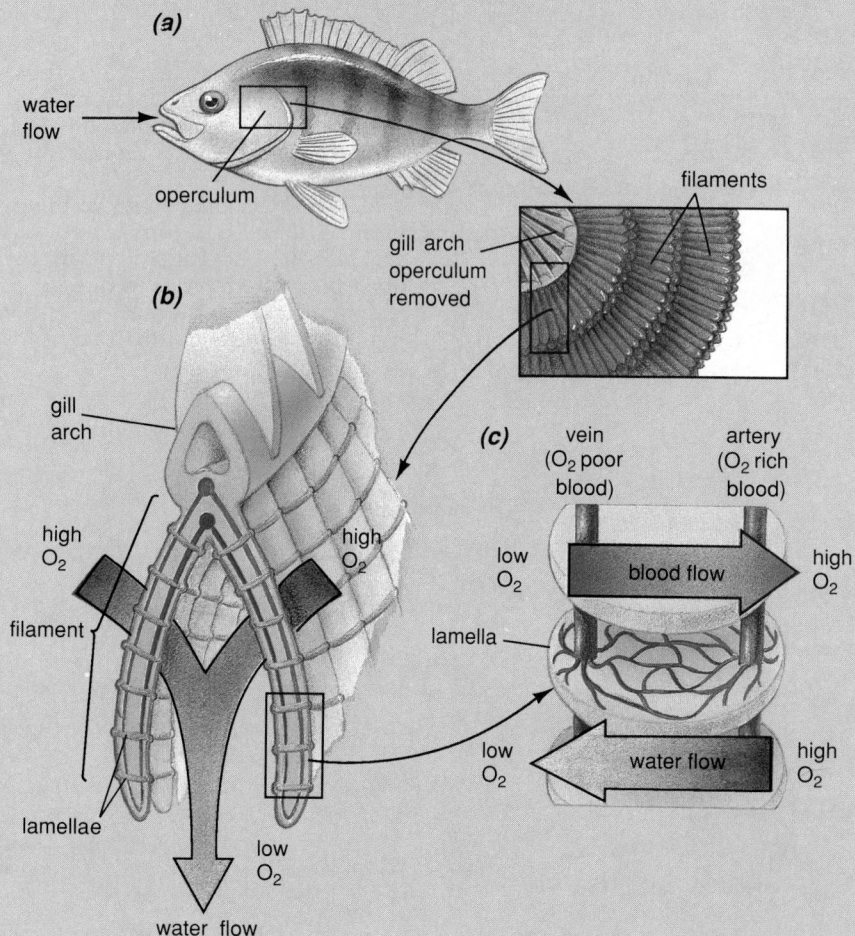

Figure E31-1 (a) The gill reaches its greatest complexity in the fish, where it is protected under a bony flap, the operculum. A one-way flow of water is maintained over the gill by pumping water through the mouth and out the opercular opening. **(b)** Each filament is composed of a series of platelike lamellae densely supplied with capillaries. Water is deflected across the gill filaments so that it flows directly across each lamella, as shown in (c). **(c)** Countercurrent flow occurs across each lamella. Water highest in dissolved oxygen encounters the most highly oxygenated blood. As the water loses its oxygen to the blood, it flows past blood that is increasingly low in oxygen, so the gradient encouraging oxygen to move into the blood is maintained.

eton, forming a structure called a book lung (Fig. 31-3b). Insects use a system of elaborately branching tubes called **tracheae** to convey air to each body cell (Fig. 31-3c). Tracheae subdivide into tiny channels and penetrate the entire body. Each body cell is close to a tracheal tube, minimizing diffusion distances. Tracheae usually communicate with the outside through openings called **spiracles** located along the side of the abdomen. Muscular pumping movements of the abdomen assist air movement through the tracheae in some large insects.

Respiration Using Lungs

Lungs protect moist, delicate respiratory surfaces deep within the body, where water loss is minimized and the body wall provides support. The first vertebrate lung probably appeared in a freshwater fish, and consisted of an outpocketing of the digestive tract. Gas exchange in this simple lung helped the fish survive in stagnant water where oxygen was scarce. Amphibians, straddling the boundary between aquatic and terrestrial life, may use lungs or gills at different stages of their lives. The aquatic tadpole exchanges its gills for lungs as it develops into a more terrestrial frog (Fig. 31-4). Frogs and salamanders use their moist skin as an additional respiratory surface.

The scales of reptiles (Fig. 31-5) limit loss of water through the skin and allow them to live in desert environments, but scales also prevent diffusion of gases through the skin. Thus, the lungs of reptiles are better developed than those of amphibia. Birds and mammals are exclusively lung-breathers. The bird lung is adapted for increased efficiency during strenuous flight (Fig. 31-6). Air sacs inflate during inhalation, storing fresh air. This air passes through the lung during exhalation, providing the bird with a continuous supply of fresh air during both phases of breathing.

The Human Respiratory System

The respiratory system in humans and other vertebrates can be divided into two parts: the **conducting portion** and the **gas-exchange portion.** The conducting portion consists of a series of passageways that carry air into the gas-exchange portion. Here gas is exchanged with the blood in tiny sacs called **alveoli.**

The Conducting Portion

Air enters through the nose or the mouth, passes through a common chamber, the **pharynx,** and then

(a)

(b)

Figure 31-4 (a) The bullfrog, an amphibian, begins life as a fully aquatic tadpole with feathery external gills **(b)**. During metamorphosis into an air-breathing adult frog, the gills are lost and replaced by simple saclike lungs. In both tadpole and adult, gas exchange also occurs by diffusion through the moist skin.

Figure 31-5 The fully terrestrial reptile, illustrated by this mangrove snake, is covered with dry scales that restrict gas exchange through the skin. Reptilian lungs are more efficient than those of amphibians.

(a)

(b)

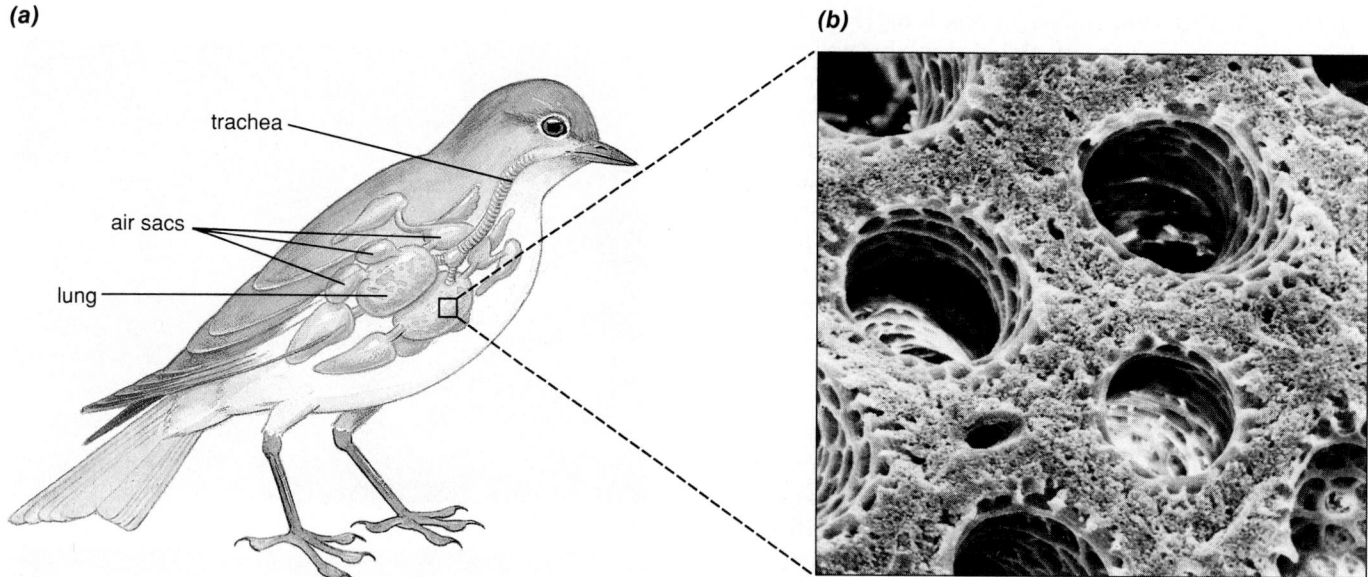

Figure 31-6 (a) The respiratory system of the bird allows extremely efficient gas exchange. When air is inhaled, some fills the lungs while the rest travels past the lungs to fill air sacs. As air is exhaled, the fresh air that has been temporarily stored in the air sacs fills the lungs on its way out. **(b)** The bird uses tubular gas-exchange organs called parabronchi rather than saclike alveoli. This allows the air to flow through the lung continuously.

travels through the **larynx** (Fig. 31-7). Within the larynx are the vocal cords, bands of elastic tissue controlled by muscles. Muscular contractions can cause them to partially obstruct the opening within the larynx, so that exhaled air causes them to vibrate, giving rise to the tones of speech or song. The tones are varied in pitch by stretch applied to the cords, and can be articulated into words by movements of the tongue and lips.

Inhaled air continues past the larynx into the **trachea,** a flexible tube whose walls are reinforced with semicircular bands of stiff cartilage (see Chapter 38). Within the chest, the trachea splits into two large branches called **bronchi** (sing. bronchus), one leading to each lung. Inside the lung, each bronchus branches repeatedly into ever-smaller tubes called **bronchioles.** These lead finally to the microscopic alveoli, tiny air pockets where gas exchange occurs (see Fig. 31-7).

During its passage through the conducting system, the air is warmed and moistened. Much of the dust and bacteria it carries is trapped in mucus secreted by cells lining the conducting passage. The mucus with its trapped debris is continuously swept upward toward the pharynx by cilia that line the bronchioles, bronchi, and trachea. Upon reaching the pharynx, the mucus is coughed up or swallowed. Smoking in-

terferes with this cleansing process by paralyzing the cilia (see "Health Watch: Smoking and Respiratory Disease").

The Gas Exchange Portion: The Alveoli

Each lung is packed with around 1.5 million alveoli. These tiny (0.2-mm diameter) chambers give magnified lung tissue the appearance of sponge cake (see Fig. E31-4). The thin-walled alveoli provide an enormous surface area for diffusion, about 800 square feet (75 m²) in the adult human. The alveoli, which cluster about the end of each bronchiole like a bunch of grapes, are entirely enmeshed in capillaries (see Fig. 31-7). Since the alveolar wall and the adjacent capillary walls are each only one cell thick, the air is extremely close to the blood in the capillaries. The lung cells remain moist, coated by a thin layer of water lining each alveolus. Gases dissolve in this water and diffuse through the alveolar and capillary membranes (Fig. 31-8).

Blood is pumped to the lungs by the heart after circulating through the body tissues. The incoming blood surrounding the alveoli is therefore low in oxygen (since the body cells have used it up) and high in carbon dioxide (released by the cells). Oxygen dif-

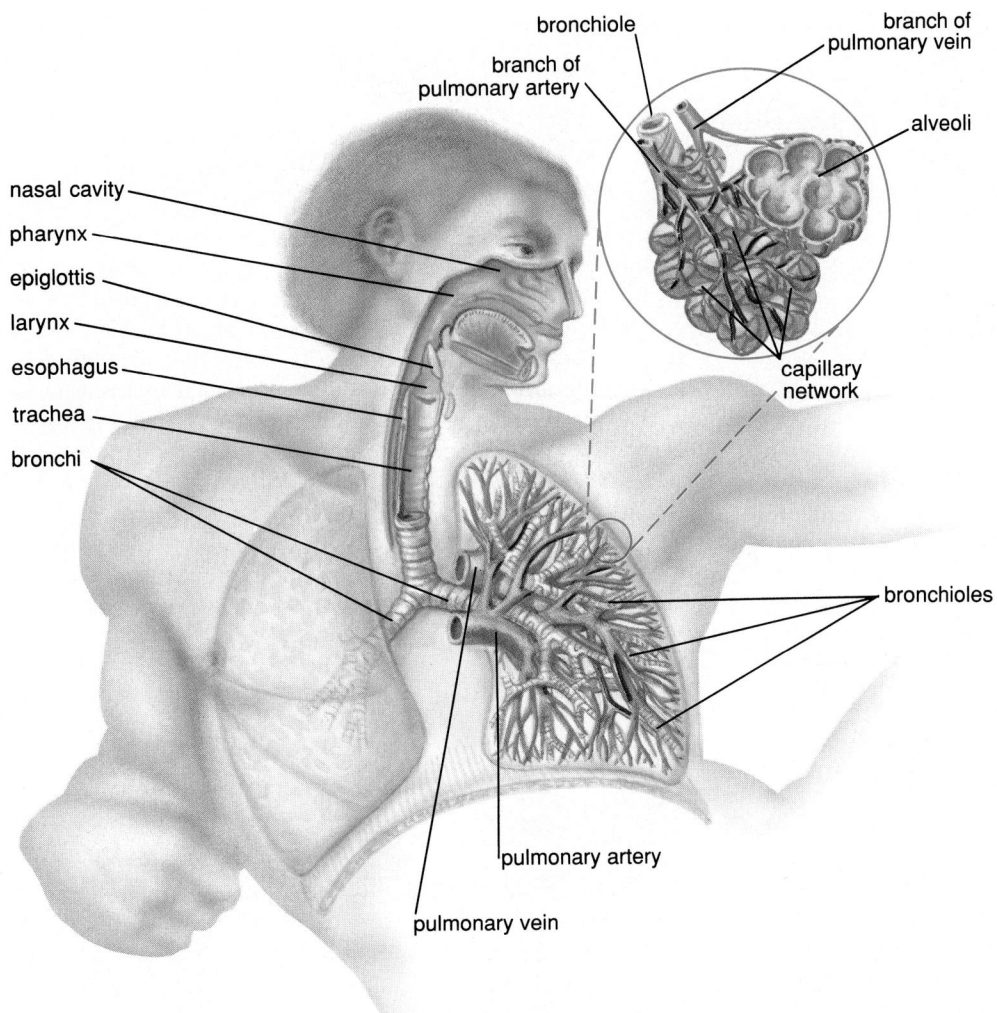

bronchiole

branch of
pulmonary artery

branch of
pulmonary vein

alveoli

nasal cavity

pharynx

epiglottis

larynx

esophagus

trachea

bronchi

capillary
network

bronchioles

pulmonary artery

pulmonary vein

Figure 31-7 The human respiratory system, showing a detail of the alveoli surrounded by capillaries. As blood flows through the capillaries, carbon dioxide is exchanged for oxygen.

fuses from the air in the alveoli, where its concentration is high, into the blood, where its concentration is low. Conversely, the carbon dioxide content of the blood as it enters the lung capillaries is higher than that of the air, so carbon dioxide diffuses out of the blood into the air in the alveoli (Fig. 31-9). Table 31-1 compares the concentrations of oxygen and carbon dioxide in air entering the lungs with air leaving the lungs.

Blood from the lungs, oxygenated and purged of carbon dioxide, returns to the heart, which pumps it to the body tissues. In the tissues, the concentration of oxygen is lower than in the blood, and oxygen diffuses into the cells. Carbon dioxide, which has built up in the cells, diffuses into the blood.

The Transport of Gases in the Blood

Inside the capillary, oxygen binds loosely and reversibly with a large, iron-containing protein called

hemoglobin in the red blood cells (see Fig. 30-13). Each hemoglobin molecule can bind up to four oxygen molecules (eight oxygen atoms). As hemoglobin binds oxygen, the protein undergoes a slight conformational change, which alters its color. Deoxygenated blood is dark maroon-red and appears blue

Table 31-1 Comparison of Inhaled and Exhaled Air

Gas	Inhaled Air	Exhaled Air
O_2	20.8%	15.7%
CO_2	0.04%	3.6%

CLOSER LOOK

At the Transport of Gases in the Blood

Carbon Dioxide Transport

Carbon dioxide (CO_2) is carried in the blood primarily as bicarbonate ion (HCO_3^-; 70%), or bound to hemoglobin (20%). Carbon dioxide binds most readily to hemoglobin that is not bound to oxygen (O_2), so as hemoglobin gives up its O_2 to the tissues, it becomes available to bind CO_2 picked up from the tissues.

Bicarbonate ions are formed when CO_2 diffuses into the blood from the tissues and reacts with water in the following reaction:

$$CO_2 + H_2O \rightarrow HCO_3^- + H^+$$

carbon dioxide   water   bicarbonate ion   hydrogen ion

The reaction producing HCO_3^+ occurs primarily within the red blood cells, where it is catalyzed by an enzyme called **carbonic anhydrase.** Most of the HCO_3^- then diffuses back into the plasma. Carbonic anhydrase is also embedded in the walls of lung capillaries and catalyzes the reverse reaction, helping CO_2 diffuse out of the capillaries and into the alveoli (Fig. E31-2).

Oxygen Transport Adaptations

The affinity of hemoglobin for O_2 is such that it readily binds O_2 at the high concentrations found in the alveoli, but readily gives up O_2 at the lower concentrations of the body tissues (Fig. E31-2). But the ability of hemoglobin to bind O_2 can be altered in response to changing

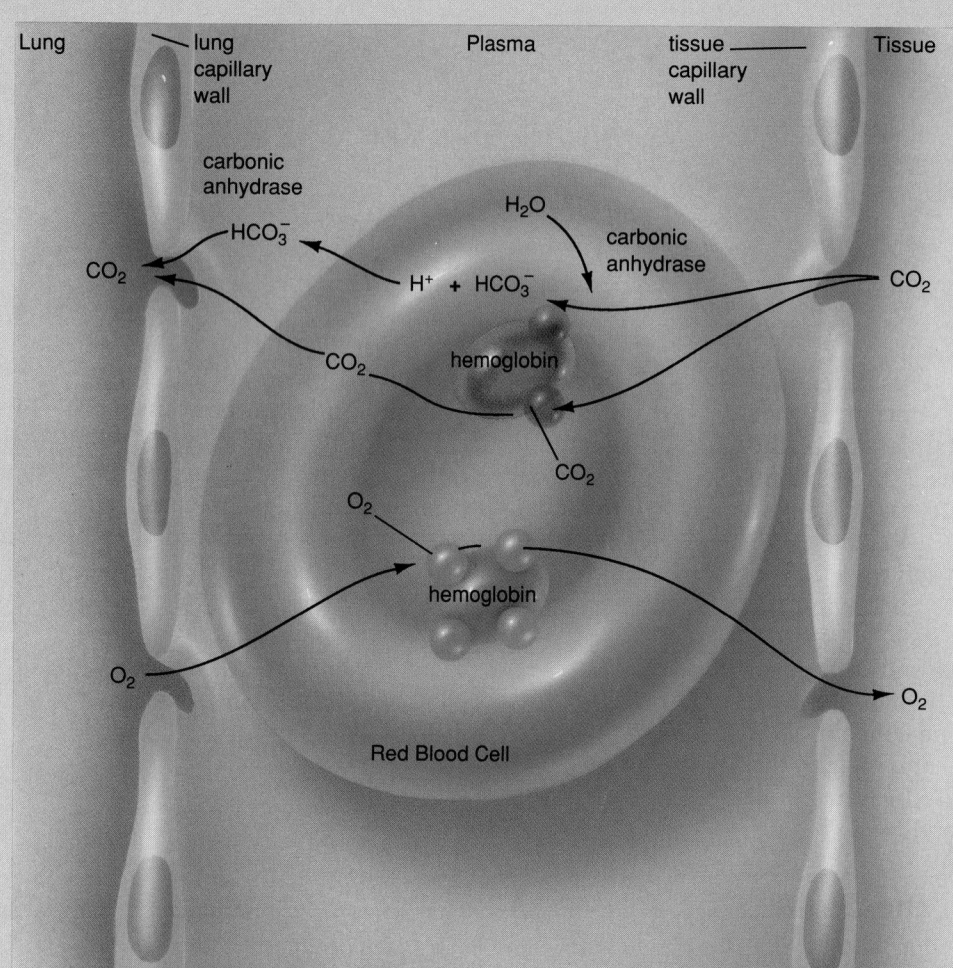

Figure E31-2 A diagrammatic illustration of the exchange of gases between capillary blood and the air in the alveoli (left) and capillary blood and body tissues (right).

conditions in the blood. For example, when the concentration of CO_2 increases, as occurs during exercise, the blood becomes more acidic due to the accumulation of H^+, as shown in the above equation. The slight acidity decreases the affinity of hemoglobin for O_2, so hemoglobin releases its O_2 more readily when the demand for oxygen is greatest. This phenomenon is called the **Bohr effect.**

Another adaptation is the presence of a compound called DPG (2,3-diphosphoglycerate) in the blood. DPG binds to hemoglobin and decreases its affinity for O_2. DPG levels rise when the blood becomes slightly acidic (such as occurs during exercise), or when hemoglobin levels drop (due to bleeding) or O_2 levels in the blood decline (such as at high elevation). For example, 24 hours after a person travels from sea level to an elevation of 10,000 feet, where O_2 is less available, his or her blood DPG levels will have increased by 10%. This causes the hemoglobin to release O_2 more readily to the tissues. However, as you might predict, the lower affinity of hemoglobin for O_2 caused by DPG also makes it more difficult for hemoglobin to pick up O_2 in the lungs. Compensating for this, the low O_2 will also stimulate the kidneys to produce a hormone called erythropoietin that increases the production of red blood cells (see Fig. 30-14). Working together, these two mechanisms allow more oxygen to reach the tissues when it is needed.

A third oxygen transport adaptation occurs in the fetus, whose oxygen demands are high. Fetal hemoglobin protein has a slightly different structure than adult hemoglobin, and a higher affinity for oxygen. This allows fetal hemoglobin to wrest oxygen from maternal hemoglobin, which does not cross the placenta (see Chapter 40). Fetal hemoglobin is gradually replaced by adult hemoglobin during the first 3 months after birth.

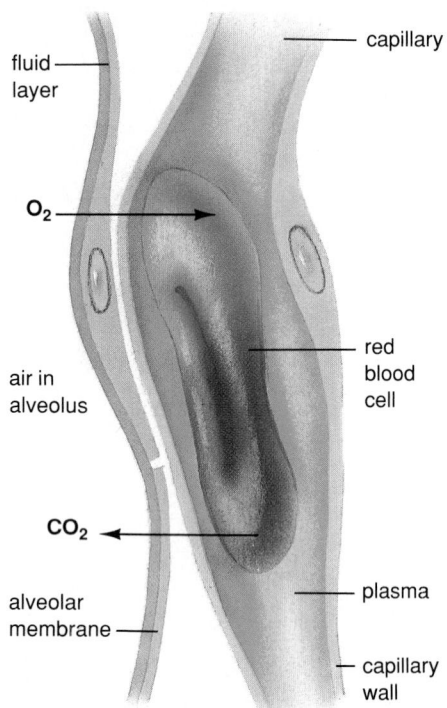

Figure 31-8 The relationship between the alveolar membranes and the capillary walls facilitates the exchange of oxygen (O_2) and carbon dioxide (CO_2) by diffusion.

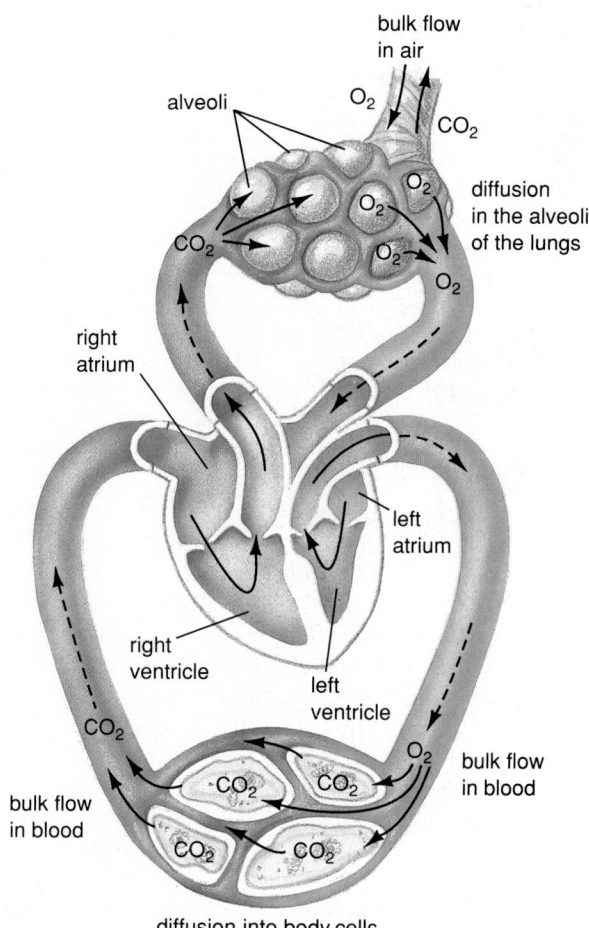

diffusion into body cells

Figure 31-9 A diagram of gas exchange between the air in the alveoli and the blood, and between the cells of the body and the blood. Oxygen is transported first by bulk flow as air is pumped in and out of the lungs. It moves by diffusion through the alveolar capillary walls, and then is transported again by bulk flow in the bloodstream. Finally, diffusion carries oxygen into the cells of the body. Carbon dioxide follows the opposite path, indicated by the arrows in the diagram. Red blood is oxygenated, blue is deoxygenated.

through the skin, while oxygenated blood is a bright cherry red. Unfortunately, carbon monoxide (CO) "fools" the hemoglobin, binding in place of oxygen and over 200 times as tenaciously as oxygen. The resulting hemoglobin is also bright red, but totally incapable of transporting oxygen. While most victims of asphyxiation have bluish lips and nail beds because their hemoglobin is deoxygenated, the lips and nail beds of victims of carbon monoxide poisoning are brighter red than normal.

Nearly all the oxygen carried by the blood is bound to hemoglobin. Thanks to hemoglobin, our blood can carry about 70 times as much oxygen as it could if the oxygen were simply dissolved in the plasma. Hemoglobin maintains a high concentration gradient of oxygen from the air to the blood by removing oxygen from solution in the plasma. This enhances the diffusion of oxygen from the alveoli into the blood.

Carbon dioxide transport is more complex. As hemoglobin releases its oxygen to the tissues, some of the hemoglobin binds carbon dioxide and carries it back to the lungs. But only about 20% of the carbon dioxide is returned to the lungs bound to hemoglobin. Most of the rest, about 70%, is carried as bicarbonate ion in the blood plasma, while a small amount remains dissolved in the plasma. This is described in

more detail in "A Closer Look at the Transport of Gases in the Blood."

The Mechanics of Breathing

Outside the lungs, the chest cavity is airtight, bounded by neck muscles and connective tissue on top and the dome-shaped muscular **diaphragm** on the bottom. Surrounding and protecting the lungs is the rib cage within the wall of the chest. Lining the chest and surrounding the lungs is a double layer of membranes called **pleural membranes.** These contribute to the airtight seal between the lungs and the chest wall.

Breathing occurs in two stages: **inspiration,** during which air is actively inhaled, and **expiration,** during which it is passively exhaled. Inspiration is accomplished by making the chest cavity larger. To do this, the diaphragm muscles are contracted, drawing it downward, and the rib muscles are contracted, lifting the ribs up and outward (Fig. 31-10). When the chest cavity is expanded, the lungs expand with it, since a vacuum holds them tightly against the inner wall of the chest. (If the chest is punctured and air leaks in, the lung will collapse.) As the lungs expand, air moves into them to fill the additional space. Expiration occurs automatically when the muscles causing inspiration are relaxed. The relaxed diaphragm domes upward, and the ribs fall down and inward, decreasing the size of the chest cavity and forcing air out of the lungs. Expiration can be assisted by contraction of abdominal muscles. After expiration, the lungs still contain air. This air prevents the thin alveoli from collapsing, and fills the space within the conducting portion of the respiratory system. A normal breath only moves about a pint (500 ml) of air into the respiratory system. Of this, only about 350 milliliters reaches the alveoli for gas exchange. Deeper breathing during exercise causes several times this volume to be exchanged.

The Control of Respiration

Breathing occurs rhythmically and automatically without conscious thought. But the muscles used in breathing are not self-activating; each contraction is stimulated by impulses from nerve cells. These impulses originate in the **respiratory center** located in the brain just above the spinal cord. Nerve cells in the respiratory center generate cyclic bursts of impulses causing alternating contraction and relaxation of the respiratory muscles.

The respiratory center receives input from a variety of sources; adjusting breathing rate and volume to meet the body's changing needs. For example, if the lungs become overexpanded, stretch receptors in the lungs inhibit further inspiration. In addition, the brainstem contains receptor neurons that monitor the concentration of carbon dioxide in the blood. Elevated carbon dioxide levels indicate an increase in cellular activity and a corresponding requirement for more oxygen. The receptors therefore stimulate an increase in the rate and depth of breathing. These receptors are extremely sensitive; an increase in carbon dioxide of only 0.3% can cause a doubling of the breathing rate. Thus, the respiratory rate is regulated to maintain a constant level of carbon dioxide in the blood. The respiratory rate is much less sensitive to changes in oxygen concentration, since normal

breathing supplies an overabundance of oxygen. However, should blood oxygen levels fall drastically, receptors in the aorta and carotid arteries stimulate the respiratory center. When a person begins strenuous activity, such as running, an increase in breathing rate *precedes* any changes in blood gas levels. Apparently, when higher brain centers activate muscles during heavy exercise, they simultaneously stimulate the respiratory center to increase breathing rate. Breathing activity is then "fine-tuned" by the receptors monitoring carbon dioxide concentrations.

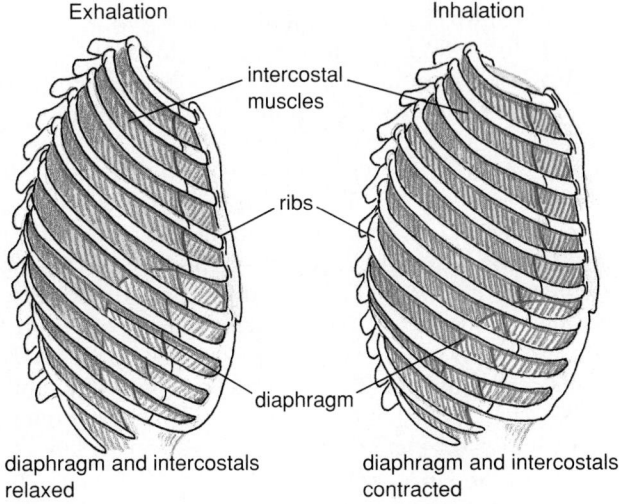

Figure 31-10 The mechanics of breathing. Rhythmic nerve impulses from the brain stimulate the diaphragm muscle to contract (pulling it downward) and the muscles surrounding the ribs to contract (moving them up and outward). The result is an increase in the size of the chest cavity, causing air to rush in. Relaxation of these muscles allows the diaphragm to dome upward, and the rib cage to collapse, forcing air out of the lungs.

HEALTH WATCH
Smoking and Respiratory Disease

Smoking causes about 340,000 deaths per year in the United States alone, nearly 1000 per day. Heavy smokers increase their chances of contracting lung cancer by a factor of 20, and some 94,000 of them will die from it this year (Fig. E31-3). The other 226,000 smoking deaths are from a combination of smoking-induced emphysema, chronic bronchitis, heart disease, and a variety of other cancers including cancer of the mouth, larynx, esophagus, pancreas, bladder, and kidney. Recently, researchers at the National Institute of Environmental Health Sciences reported that women who smoke more than a pack a day are 50% less fertile than nonsmokers. Women who smoke heavily during pregnancy have twice the rate of miscarriages. When they carry their babies to term, the babies weigh, on average, about half a pound less than those of nonsmoking women. Children of women who smoked heavily during pregnancy also show impairment in average achievement test scores during early childhood. A 1985 congressional study estimated that smoking cost American society $65 billion yearly, or $10 million per hour. This figure refers only to dollars spent on health-care services related to smoking and lost job productivity and wages. The human cost is incalculable.

Let us explore briefly the effects of tobacco smoke on the human respiratory tract. As smoke is inhaled through the nose, trachea, and bronchi, toxic substances such as nicotine and sulfur dioxide paralyze the cilia lining the respiratory tract; a single cigarette can inactivate them for a full hour. Since these ciliary sweepers remove inhaled particles, smoking poisons them just when they are most needed. The visible portion of cigarette smoke consists of billions of microscopic carbon particles. Adhering to them are a wide variety of toxic compounds, a dozen or more of which are carcinogenic (cancer-causing). With the cilia out of action, the particles stick to the walls of the respiratory tract or enter the lungs. Thus, smokers encounter a higher risk of cancer in all areas of the respiratory tract touched by smoke.

A second line of defense is the presence in the respiratory tract of large numbers of amoebalike white blood cells (macrophages), which engulf foreign particles and bacteria. Cigarette smoke also impairs these cells, allowing still more bacteria, dust, and smoke particles into the lungs. In response to the irritation of cigarette smoke, the respiratory tract increases production of mucus, a third method of trapping foreign particles. But without the cilia to sweep

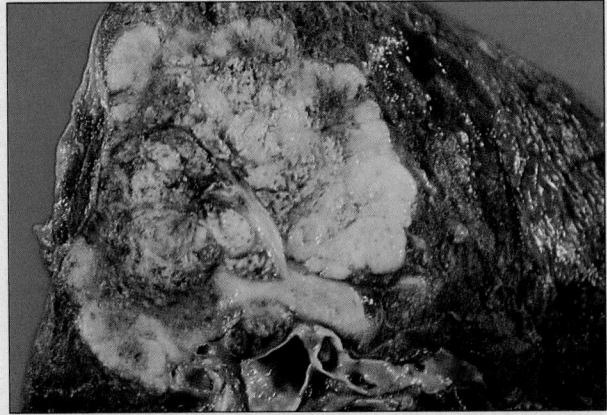

Figure E31-3 Lung cancer in a smoker. The cancer is seen as a whitish mass, while the tissue around it is blackened by trapped smoke particles.

it along, the mucus builds up and can obstruct the airways; the familiar "smoker's cough" is an attempt to clear the airways. Microscopic smoke particles find a secure lodging place in the tiny, moist alveoli deep within the lungs. There they accumulate over the years until the lungs of a heavy smoker are literally blackened (Fig. E31-4). The longer the delicate tissues of the lungs are exposed to the carcinogens on the trapped particles, the greater the chance of cancer developing.

Smoking can also cause chronic bronchitis. This persistent lung infection is characterized by cough, swelling of the lining of the respiratory tract, an increase in mucus production, and a decrease in the number and activity of cilia. The result: a decrease in air flow to the alveoli. Toxins in cigarette smoke, such as nitrogen oxides and sulfur dioxide, may cause the body to produce substances that reduce the elasticity and increase the brittleness of lung tissue. As the brittle alveoli rupture, the lung gradually loses its normal sponge-cake–like appearance, and more closely resembles blackened Swiss cheese (see Fig. E31-4). This condition is called **emphysema.** The loss of the alveoli, where gas exchange occurs, leads to oxygen deprivation of all body tissues. The emphysema victim's breathing is labored, and grows increasingly worse until death. Chronic bronchitis and emphysema kill about 50,000 persons yearly in the United States, and smoking is the leading contributing factor to each.

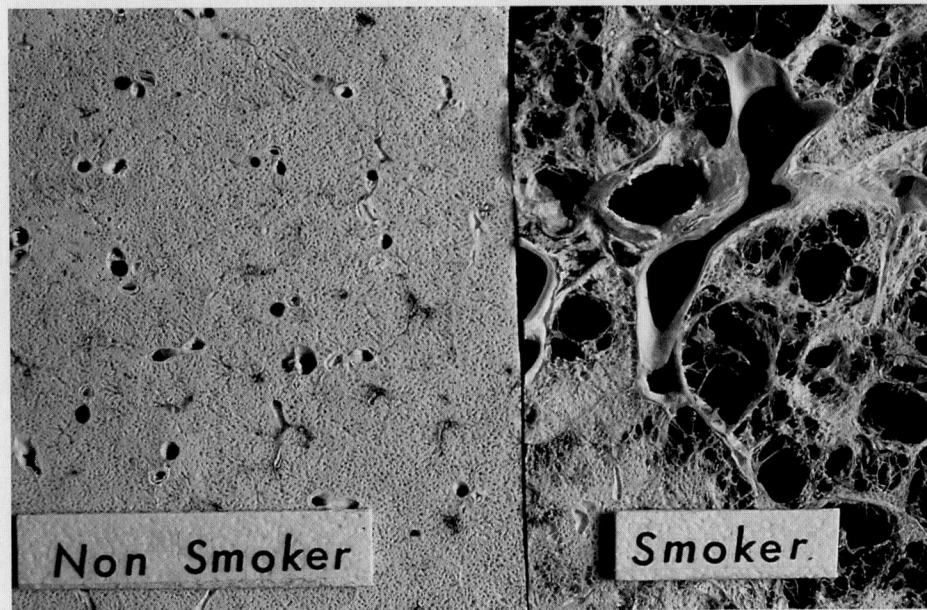

Figure E31-4 A section through a normal lung appears almost opaque, while the lung of a smoker suffering from emphysema is full of large holes, each caused by the rupture of hundreds of alveoli.

Meanwhile, carbon monoxide, present in high levels in cigarette smoke, is eagerly taken up by the red blood cells in place of oxygen. This inactivates the hemoglobin and reduces the blood's oxygen-carrying capacity, placing a strain on the heart. Chronic bronchitis and emphysema compound this difficulty by restricting the amount of oxygen that enters the lungs and by destroying the alveoli where oxygen enters the bloodstream. As a result, smokers are 70% more likely than nonsmokers to die of heart disease. Although the reasons are not fully understood, reduced oxygen in the blood supplying the heart muscle is undoubtedly a major contributor. The carbon monoxide in cigarette smoke may also contribute to the increased incidence of stillbirths in pregnant women who smoke, the lower birth weight of their babies, and the learning impairment in early childhood of children born to heavy smokers.

Evidence is accumulating rapidly that "passive smoking" (which occurs when nonsmokers are forced to breathe air polluted by cigarette smoke) causes real health hazards as well as emotional and physical discomfort. Several studies have concluded that infants whose mothers smoke are more likely to suffer from bronchitis and pneumonia. Another study revealed decreased lung capacity in children whose mothers smoked, and abnormally thickened and stiffened heart walls were found in 11- and 12-year-old boys whose parents smoked. People suffering from angina have been found to experience chest pain significantly more readily after sitting in a smoke-filled room. The *New England Journal of Medicine* reported a study of 2100 nonsmokers who were chronically exposed to cigarette smoke. These individuals showed impaired lung function similar to people smoking 1 to 10 cigarettes daily. A 1985 EPA study singled out "passive" tobacco smoke as the major cause of cancer caused by airborne carcinogens. The study estimates that passive smoking is responsible for between 500 and 5000 deaths of nonsmokers each year. A 1986 National Academy of Sciences study concluded that nonsmoking spouses of smokers face a 30% higher risk of lung cancer. Recently, after analyzing 11 different studies, researchers at the University of California at San Francisco concluded that nonsmokers who live with smokers also increase their risk of heart attack by 30%.

There is hope for smokers who stop before irreversible lung damage has occurred. Healing begins immediately after smoking stops, and the chances of a heart attack or lung cancer gradually drop. After about 10 smoke-free years, former smokers' risks of lung cancer and heart attacks approach those of people who never smoked.

SUMMARY OF KEY CONCEPTS

The Evolution of Respiratory Systems

Animals have evolved a diverse array of respiratory systems. Animals in moist environments whose bodies are small or flattened, and have very low metabolic demands and/or well-developed circulatory systems may lack specialized respiratory structures. Larger animals in aquatic environments have evolved gills, such as those of crustaceans, molluscs, fish, and many amphibians. On land, moist respiratory surfaces must be protected internally. This has led to the evolution of tracheae in insects, book lungs in arachnids, and lungs in terrestrial vertebrates.

Respiratory Systems and Gas Exchange

The transfer of gases between respiratory systems and tissues occurs in a series of stages that alternate bulk flow with diffusion. Bulk flow is used to move air or water past the respiratory surface, and to transport gases within the blood. Diffusion is used to move gases across membranes between the respiratory system and the capillaries and between the capillaries and the tissues.

The Human Respiratory System

The human respiratory system consists of a conducting portion and a gas-exchange portion. Air passes in sequence: first through the conducting portion of the nose and mouth, pharynx, larynx, trachea, bronchi and bronchioles, and then into the gas-exchange portion, composed of microscopic sacs called alveoli. Blood within a dense capillary network surrounding the alveoli releases carbon dioxide and absorbs oxygen from the air.

Most of the oxygen in the blood is bound to the protein hemoglobin within red blood cells. Hemoglobin binds oxygen at concentrations typical of those in alveolar capillaries and releases it at the lower oxygen concentrations found in the body tissues. Carbon dioxide diffuses into the blood from the tissues and is transported in three ways. Most is combined with water to form bicarbonate ion. Some is transported bound to hemoglobin, and a small amount is carried as dissolved carbon dioxide.

Breathing in humans involves actively drawing air into the lungs by contracting the diaphragm and the rib muscles, which expand the chest cavity. Relaxing these muscles causes the chest cavity to collapse, expelling the air passively.

Respiration is controlled by nerve impulses originating in the respiratory center of the brain, and its rate is modified by a variety of receptors, most importantly brain cells that monitor carbon dioxide levels in the blood.

GLOSSARY

alveolus (al-vē'-ō-lus; pl. alveoli): a tiny air sac within the lungs surrounded by capillaries where gas exchange with the blood occurs.

bohr effect (bor): the tendency of hemoglobin to release oxygen more readily when the blood is slightly more acidic than normal, as is caused by an increase in dissolved carbon dioxide.

bronchiole (bron'-kē-ōl): a narrow tube formed by repeated branching of the bronchi, which conducts air into the alveoli.

bronchus (bron'-kus): a tube that conducts air from the trachea to each lung.

bulk flow: the movement of many molecules of a gas or fluid in unison from an area of higher pressure to an area of lower pressre.

carbonic anhydrase (car-bon'-ik an-hī'-drās): an enzyme found in red blood cells that catalyzes the formation of bicarbonate ions from dissolved carbon dioxide and water.

countercurrent flow: a structural arrangement that enhances diffusion between two fluids that differ in their concentration of dissolved substances by moving the fluids past one another in opposite directions, separated by semipermeable membranes.

diaphragm (dī'uh-fram): a dome-shaped muscle forming the floor of the chest cavity. Contraction of this muscle pulls it downward, enlarging the cavity and causing air to be drawn into the lungs.

emphysema (em-fuh-sē'-muh): a condition in which the alveoli become brittle and rupture, causing decreased area for gas exchange.

expiration (ex-per-ā'-shun): the act of exhaling, which results from relaxation of the respiratory muscles.

gills: in aquatic animals, a branched tissue richly supplied with capillaries around which water is circulated for gas exchange.

hemoglobin (hē'mō-glō-bin): an iron-containing protein that gives red blood cells their color. Hemoglobin binds to oxygen in the lungs and releases it to the tissues.

inspiration: the act of inhaling air into the lungs by enlarging the chest cavity.

larynx (lār'-inx): that portion of the air passage between the pharynx and the trachea. The larynx contains the vocal cords.

operculum: an external flap, supported by bone, that covers and protects the gills of most fish.

pharynx (fār'-inx): a chamber at the back of the mouth shared by the digestive and respiratory systems.

respiratory center: a location in the brainstem that sends rhythmic bursts of nerve impulses to the respiratory muscles, resulting in breathing.

spiracles (spī'-re-kuls): openings in the abdominal segments of insects through which air enters the tracheae.

trachea (trā'-kē-uh): a rigid but flexible tube supported by rings of cartilage, which conducts air between the larynx and the bronchi.

tracheae (trā'-kē): elaborately branching tubes that ramify through the bodies of insects and carry air close to each body cell. Air enters the tracheae through openings called spiracles.

STUDY QUESTIONS

1. Describe three different arthropod respiratory systems and two different vertebrate respiratory systems.
2. Trace the route taken by air in the vertebrate respiratory system, listing the structures through which it flows and the point where gas exchange occurs.
3. What evolutionary changes in the "lifestyle" of animals led to the evolution of diverse respiratory structures? Explain how each of these adapts the animal to its environment.
4. How are human respiratory movements initiated? Modified?
5. What events occur during human inspiration? Expiration? Which is always an active process?
6. Trace the pathway of an oxygen molecule in the human body starting with the nose and ending with a body cell.
7. Describe the effects of smoking on the human respiratory system.
8. Explain how bulk flow and diffusion interact to promote gas exchange between air and blood, and between blood and tissues.
9. Compare carbon dioxide and oxygen transport in the blood. Include in your answer the source and destination of each.

DISCUSSION QUESTIONS

1. Discuss several advantages of obtaining oxygen from air rather than from water. Also discuss several disadvantages of this.
2. Discuss why a brief exposure to carbon monoxide is much more dangerous than a brief exposure to carbon dioxide.
3. Describe several adaptations that might evolve to help members of a species of mammal respire better if the population began living continuously for many generations at very high altitudes.

SUGGESTED READINGS

Comroe, J. "The Lung." *Scientific American*, February 1966. Describes the intricate relationship between air sacs and blood vessels in the human lung.

Feder, M. E., and W. W. Burggren. "Skin Breathing in Vertebrates," *Scientific American*, May 1985. Describes how some vertebrates supplement the action of their lungs or gills to better perform gas exchange.

Perutz, M. "Hemoglobin Structure and Respiratory Transport." *Scientific American*, December 1978. How hemoglobin plays the dual role of transporting oxygen to the tissues and carbon dioxide back to the lungs.

Randall, D. J., W. W. Burggren, A. P. Farrell, and M. S. Haswell. *The Evolution of Air Breathing in Vertebrates*. New York: Cambridge University Press, 1981. Very readable.

Schmidt-Nielsen, K. "How Birds Breathe." *Scientific American*, December 1971. Specializations of the bird respiratory system include additional air sacs and even hollow bones.

West, J. *Respiratory Physiology: The Essentials*. Baltimore: Williams and Wilkins, 3rd ed., paperback, 1985. A concise introduction to respiratory functions.

32

Nutrition and Digestion

For many animals, the first challenge of digestion is to capture elusive and unwilling prey.

In one sense, animals and plants have similar nutritional needs: materials to synthesize the components of their bodies and a source of energy to power that synthesis. However, plants and animals satisfy these needs in quite different ways. A plant takes water and minerals from the soil and carbon dioxide from the air, and uses the radiant energy of sunlight to form these materials into complex molecules such as starch and cellulose. Animals, on the other hand, eat plants or each other. **The food that animals eat provides both energy and the basic building blocks of complex molecules: amino acids, fatty acids, and simple sugars.** Each organism combines these building blocks in unique ways. The proteins of a mule deer, for example, are slightly different from those of the timber wolf that may consume it. Thus the deer proteins must first be broken down into amino acids before the wolf can use them to make its own proteins. **In general, the digestive systems of animals first grind up and then chemically break down the complex molecules of their food into simpler building blocks which are then absorbed.** These can then be reassembled into the unique molecules of the consumer.

Nutrition

Animal nutrients fall into five major categories: lipids, carbohydrates, proteins, minerals, and vitamins. These substances provide the body with its basic needs—energy, the building blocks such as amino acids to construct complex molecules unique to each animal, and vitamins and minerals that participate in a variety of metabolic reactions.

Sources of Energy

Each living cell in the animal body relies on a continuous expenditure of energy to maintain its incredible complexity and perform its specific functions within the body. Three nutrients provide dietary energy for animals: lipids, carbohydrates, and, if these are lacking, proteins. These molecules can be broken down during cellular respiration, providing chemical energy stored in adenosine triphosphate (ATP) (see Chapter 8). The energy in nutrients is measured in **calories.** A calorie is the amount of energy required to raise the temperature of one gram of water by one degree Celsius. The calorie content of foods is measured in units of 1000 calories (kilocalories), also known as **Calories** with a capital C. The human body at complete rest burns about 1550 Calories per day,

somewhat more for males and less for females. Exercise significantly boosts caloric requirements: well-trained athletes can temporarily raise their calorie consumption by a factor of 20 during vigorous exercise (Table 32-1).

Lipids

Lipids are a diverse group of molecules that are insoluble in water and generally contain long chains of carbon atoms. The principal types of lipids are fats or triglycerides, phospholipids, and cholesterol (see Chapter 3). Fats are used primarily as a source of energy. Phospholipids are important components of cell membranes, and also provide the insulating covering of neurons. Cholesterol is used in the synthesis of cell membranes, bile, and certain hormones. Some animals can synthesize all the specialized lipids they need. Others must acquire specific types of lipid building blocks, called **essential fatty acids,** from their food. For example, humans are unable to synthesize linoleic acid, required for the synthesis of certain phospholipids, so we need to obtain this essential fatty acid from our diet.

Lipids in the form of fats provide about 45% of the energy in the average American diet. Humans and most other animals store energy as fat. When the diet provides more energy than is expended, most of the excess carbohydrate, fat, or protein is converted to fat for storage; about 3600 Calories are contained in every pound. Fats have two major advantages as energy storage molecules. First, they are the most concentrated energy source, containing over twice the energy per unit weight of either carbohydrates or protein (about 9 Calories per gram for fats compared with about 4 Calories per gram for proteins and carbohydrates). Second, lipids are hydrophobic, meaning they do not mix with water. For this reason, fat deposits do not cause any extra accumulation of water in the body. For both these reasons, fats store more calories with less weight. Minimizing weight allows an animal to move faster (important for escaping predators and hunting prey) and to move using less energy (important when food supplies are limited). Mammals, who maintain an elevated body temperature, often make fat deposits do double duty, acting both as stored energy and as insulation. Fat is typically stored in a layer beneath the skin. Here it insulates the body, since fat conducts heat at only one third the rate of other body tissues. Mammals who live near the poles or in cold ocean waters are particularly dependent on this insulating layer, which reduces the energy they must expend to keep warm (Fig. 32-1).

Table 32-1   Energy Consumed (in Kilocalories per Minute) by Different Types of Activities

Activity	(weight in pounds)			
	105–115	127–137	160–170	182–192
Bicycling				
10 mph	5.41	6.16	7.33	7.91
Stationary, 10 mph	5.50	6.25	7.41	8.16
Calisthenics	3.91	4.50	7.33	7.91
Dancing				
Aerobic	5.83	6.58	7.83	8.58
Square	5.50	6.25	7.41	8.00
Gardening, weeding, and digging	5.08	5.75	6.83	7.50
Jogging				
5.5 mph	8.58	9.75	11.50	12.66
6.5 mph	8.90	10.20	12.00	13.20
8.0 mph	10.40	11.90	14.10	15.50
9.0 mph	12.00	13.80	16.20	17.80
Rowing, machine				
Easily	3.91	4.50	5.25	5.83
Vigorously	8.58	9.75	11.50	12.66
Skiing				
Downhill	7.75	8.83	10.41	11.50
Cross-country, 5 mph	9.16	10.41	12.25	13.33
Cross-country, 9 mph	13.08	14.83	17.58	19.33
Swimming, crawl				
20 yards per minute	3.91	4.50	5.25	5.83
40 yards per minute	7.83	8.91	10.50	11.58
55 yards per minute	11.00	12.50	14.75	16.25
Walking				
2 mph	2.40	2.80	3.30	3.60
3 mph	3.90	4.50	5.30	5.80
4 mph	4.50	5.20	6.10	6.80

Reprinted with permission from: Wm. C. Brown Publishing Co. *Human Physiology, 3rd ed.* Stuart Fox, ed., 1990.

Figure 32-1 This walrus withstands the icy waters of polar latitudes because it is insulated with a thick layer of fat beneath the skin.

Carbohydrates

Carbohydrates consist of sugars (monosaccharides and disaccharides) as well as longer chains of sugars called polysaccharides (see Chapter 3). Polysaccharides include starches, the principal energy storage material of plants; glycogen, a short-term energy storage molecule in animals; and cellulose, the major structural component of plant cell walls (see "A Closer Look at How Animals Cope with Cellulose"). During **digestion,** carbohydrates are broken down into sugars and absorbed. For practical purposes, body cells obtain their energy from a single sugar: glucose. Glucose can be derived from fats, amino acids, and the variety of carbohydrates consumed in the diet.

Carbohydrates, nearly all derived from plants, contribute about 45% of the energy in the average American diet, although it is possible to do without them entirely if sufficient calories and other nutrients are available. Eskimos, for example, traditionally thrived on a diet of seal, fish, and caribou, which is high in fat and protein, but supplies very little carbohydrate.

Animals, including humans, store a carbohydrate called **glycogen,** a large, highly branched polymer of glucose, in the liver and muscles. Although humans can store up to hundreds of pounds of fat, the glycogen stored in the body amounts to a few hundred grams, less than half a pound. During exercise such as running, the body draws on this store of glycogen as a source of quick energy. When the activity is prolonged, as in the case of a marathon runner, the stored glycogen can be totally depleted. "Hitting the wall" describes the extreme fatigue marathon runners may experience after exhausting their glycogen supply.

Protein as an Energy Source

Only about 10% of our energy requirement is normally obtained from protein. Each day, the body degrades 20 to 30 grams of its own protein, which is metabolized for energy. This is replaced by dietary intake, and any excess amino acids are broken down and used for energy or stored as fat. The breakdown of protein produces the waste product urea, which is filtered from the blood by the kidneys. Specialized diets in which protein is the major energy source place extra stress on the kidneys. The major role of dietary protein is as a source of amino acids to make new proteins, as described below and in Chapter 3.

Amino Acids

Amino acids are used to synthesize certain hormones and neurotransmitters, other amino acids, and new proteins. These proteins have diverse roles in the body, acting as enzymes, receptors on cell membranes, oxygen transport molecules (hemoglobin), structural components (hair and nails), hormones, antibodies, and muscle proteins.

In the digestive tract, dietary protein is broken down into its amino acid subunits. Then, in the body cells, the amino acids are linked in specific sequences to form new proteins. The human liver can synthesize (from other amino acids) half of the 20 different amino acids used in proteins. Those that cannot be synthesized, called **essential amino acids,** must be supplied by the diet in foods such as meat, milk, eggs, corn, beans, and soybeans. The 10 essential amino acids for human adults are listed in Table 32-2.

Table 32-2 The Essential Amino Acids

Threonine	Phenylalanine
Lysine	Tryptophan
Methionine	Leucine
Arginine	Isoleucine
Valine	Histidine

Minerals

Minerals are small inorganic molecules. A wide variety of minerals are required by animals (Table 32-3), and all must be obtained in the diet, either from food or dissolved in drinking water. Required minerals include calcium, magnesium, and phosphorus, which are major constituents of bones and teeth. Others, such as sodium and potassium, are essential for the conduction of nerve impulses and muscle contraction. Iron is used in the production of hemoglobin, and iodine is found in hormones produced by the thyroid gland. In addition, trace amounts of several other minerals, including zinc, copper, and selenium, are required, often as parts of enzymes.

Vitamins

Vitamins are a diverse group of organic compounds required in small amounts for normal bodily functioning. Generally, vitamins cannot be synthesized by the body and must be obtained from food. Since the ability of animals to manufacture substances varies considerably, the same compound might be a vitamin for one animal but not for another. For example, ascorbic acid is a vitamin (vitamin C) for guinea pigs, humans, and other primates who cannot manufacture it, but not to most other animals, whose bodies can synthesize it. Although our skin can manufacture some vitamin D when it is exposed to sunlight, because we spend so much time indoors most of us do not synthesize enough. Thus we must supplement it through our diet. Whether or not a substance is a vitamin, then, depends upon the animal species and the conditions under which it lives. The vitamins considered essential in human nutrition are listed in Table 32-4.

Human vitamins are often grouped into two categories, water soluble and fat soluble. Water-soluble vitamins include vitamin C and the 11 different compounds that make up the B vitamin complex. These substances dissolve in the water of the blood plasma and are excreted by the kidneys. They are not stored

CLOSER LOOK
At How Animals Cope with Cellulose

Among all the adaptations that evolution has produced, one that is conspicuously absent in most animals is a digestive enzyme to break down cellulose. Cellulose, like starch, consists of long chains of glucose molecules, but it differs from starch in the way these molecules are linked together (see Chapter 3). This linkage resists the attack of animal digestive enzymes. Cellulose surrounds each plant cell, and so is potentially one of the most abundant energy sources on Earth. But instead it passes untouched through most animal digestive systems.

In contrast to most animals, certain bacteria and protists have evolved the enzyme cellulase, which is able to break down cellulose into its component sugar molecules. Cows and other ruminants, cockroaches, and termites have all entered into symbiotic partnerships with microorganisms that digest cellulose. The animal provides ground-up cellulose, the microbe produces the necessary enzymes, and both share the abundant energy harvest.

The Remarkable Ruminants
Ruminant animals are the cud-chewers: cows, sheep, goats, camels, and hippos, to name a few. Rumination, or cud-chewing, is the process of regurgitating food and rechewing it, one of several adaptations of these animals for digesting plant material. The digestive systems of ruminants include three compartments preceding the

stomach (Fig. E32-1a). The first and largest of these, the rumen, is a massive fermentation vat. Here, microorganisms—including many species of bacteria and ciliates—thrive, their numbers reaching several hundred thousand per milliliter. These symbionts break down cellulose and other carbohydrates to sugar. Since conditions in the rumen are anaerobic (lacking oxygen), sugar is broken down further by microbial fermentation. Fermentation in the rumen produces several types of organic acids, methane, carbon dioxide, and water. The organic acids are absorbed into the bloodstream of the ruminant, and used as energy sources. To neutralize these acids and provide a suitable growth medium for its microscopic partners, the cow each day produces 100 to 200 liters of saliva consisting of a weak sodium bicarbonate solution.

After fermenting in the rumen, the plant material, now called *cud*, is regurgitated, chewed, and reswallowed (along with the saliva) to the rumen for further digestion. Gradually the processed and reprocessed cud is released into the rest of the digestive tract, passing into the reticulum, the omasum, and finally into the abomasum, or stomach.

In addition to digesting cellulose, the microscopic partners are able to synthesize proteins from urea, normally a waste product. These proteins, as well as those derived from digesting the microorganisms themselves,

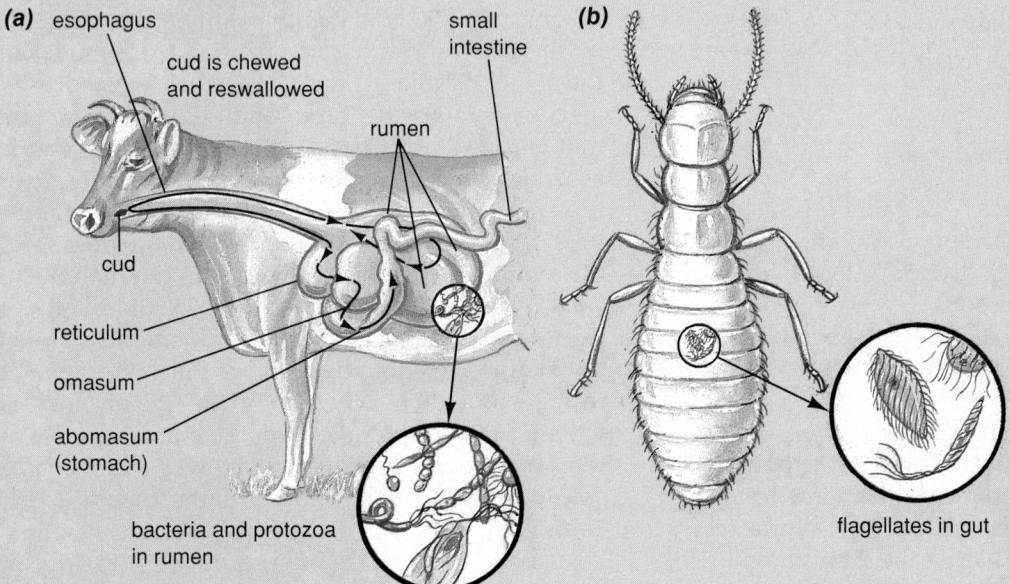

(a) esophagus
cud is chewed and reswallowed
small intestine
rumen
cud
reticulum
omasum
abomasum (stomach)
bacteria and protozoa in rumen
Cow

(b)
flagellates in gut
Termite

Figure E32-1 (a) The stomach of the cow, the abomasum, is preceded by three other chambers. The largest is the rumen, which houses a flourishing population of microorganisms that digest the cellulose in the cow's vegetarian diet. Arrows trace the path of food through the digestive tract. **(b)** The termite derives its dismaying ability to digest wood from the presence of several types of protists, called flagellates, in its intestine.

contribute significantly to the diet of the ruminant. The microorganisms in the digestive tract also produce the ruminant animal's entire supply of vitamin B_{12}.

The Terrible Termites

Termites are among the most remarkable of social insects, a fact easily overlooked as one surveys the wreckage they can make of a wood home. Some species derive their ability to digest wood from their partnership with several species of flagellates (protists) living packed inside their intestines (Fig. E32-1b). When the protists are removed and cultured separately, they can digest cellulose, but the termite cannot. Only after reinfection with its digestive partners is the termite able to survive on a diet of cellulose. As in ruminants, the microbial partners of termites are digested in large numbers, providing an important source of protein for the termite.

How Humans Benefit

People have neither the microorganisms nor the enzymes necessary to digest cellulose. Passing unscathed through the human intestine, cellulose provides important "roughage," or fiber. Dietary fiber increases the volume of feces, increases fecal water content (making them softer), and helps them pass rapidly through the intestine. This has several possible advantages. In addition to relieving constipation, fiber may help prevent diverticular disease, in which pressure in the large intestine causes saclike outpocketings of the intestinal wall. These pouches can become inflamed, infected, or they may even burst, with serious consequences. Adequate fiber may also help protect against intestinal cancer. Dietary fiber seems to decrease the production of cancer-causing chemicals by intestinal bacteria. Further, by speeding travel of wastes through the intestine, fiber may both limit the time available for carcinogens to form and reduce the time during which the intestinal cells are exposed to them.

in the body in any appreciable amount. Water-soluble vitamins generally work in the body in conjunction with enzymes to promote chemical reactions that supply energy or synthesize materials. Since each vitamin participates in several metabolic processes, a deficiency of a single vitamin can have wide-ranging effects (see Table 32-4).

The fat-soluble vitamins A, D, E, and K have even more varied roles, from the regulation of blood-clotting by vitamin K to the formation of visual pigment by vitamin A. Fat-soluble vitamins can be stored in body fat, and may accumulate in the body over time. Excessive intake of vitamins A and D can produce toxic effects (see Table 32-4).

The Challenge of Digestion

Animals eat the bodies of other organisms, bodies that may resist becoming food. The plant body, for example, armors each cell with a wall of indigestible cellulose (see "A Closer Look at How Animals Cope with Cellulose"). Animal bodies may be covered with equally indigestible fur, scales, or feathers. In addition, the complex lipids, proteins, and carbohydrates found in food are not in a form that can be used directly by the consumer. They must first be broken down before they can be absorbed and distributed to its own cells. As described in Chapter 22, different types of animals meet the challenge of acquiring nutrients with various types of digestive tracts, each finely tuned to a unique diet and life-style. Among this diversity, however, we find certain tasks that digestive systems must generally accomplish, listed below.

1. **Ingestion: The food must be brought into the digestive tract through an opening, usually called a *mouth*.**
2. **Mechanical breakdown: The food must be physically broken down into smaller pieces.** This is accomplished by gizzards or teeth as well as by the churning action of the digestive cavity itself. The particles produced by mechanical breakdown provide a large surface area for attack by digestive enzymes.
3. **Chemical breakdown: The particles must be exposed to a variety of digestive enzymes and other digestive fluids that cause large molecules to be broken down into smaller subunits.**
4. **Absorption: The small molecules must be transported out of the digestive cavity and into cells.**
5. **Elimination: Indigestible materials must be expelled from the body.**

Table 32-3 Human Mineral Requirements[a]

Mineral	RDA for Healthy Adult Male (milligrams)	Dietary Sources	Major Functions in Body	Deficiency Symptoms
Water	1.5 liters/day	Solid foods, liquids, drinking water	Transport of nutrients Temperature regulation Participates in metabolic reactions	Thirst, dehydration
Calcium	800	Milk, cheese, dark-green vegetables, dried legumes	Bone and tooth formation Blood clotting Nerve transmission	Stunted growth Rickets, osteoporosis Convulsions
Phosphorus	800	Milk, cheese, meat, poultry, grains	Bone and tooth formation Acid-base balance	Weakness, demineralization of bone Loss of calcium
Sulfur	(Provided by sulfur amino acids)	Sulfur amino acids (methionine and cystine) in dietary proteins	Constituent of most proteins	Related to intake and deficiency of sulfur amino acids
Potassium	2,500	Meats, milk, many fruits	Acid-base balance Body water balance Nerve function	Muscular weakness Paralysis
Chlorine	2,000	Table salt	Formation of gastric juice Acid-base balance	Muscle cramps Mental apathy Reduced appetite
Sodium	2,500	Table salt	Acid-base balance Body water balance Nerve function	Muscle cramps Mental apathy Reduced appetite
Magnesium	350	Whole grains, green leafy vegetables	Activates enzymes involved in protein synthesis	Growth failure Behavioral disturbances Weakness, spasms
Iron	10	Eggs, lean meats, legumes, whole grains, green leafy vegetables	Constituent of hemoglobin and enzymes involved in energy metabolism	Iron-deficiency anemia (weakness, reduced resistance to infection)
Fluorine	2	Drinking water, tea, seafood	May be important in maintenance of bone structure	Higher frequency of tooth decay
Zinc	15	Widely distributed in foods	Constituent of enzymes involved in digestion	Growth failure Small sex glands
Iodine	0.14	Seafish and shellfish, dairy products, many vegetables, iodized salt	Constituent of thyroid hormones	Goiter
Copper Silicon Vanadium Tin Nickel Selenium Manganese	Not established (trace amounts)	Widely distributed in foods	Some unknown; some work in conjunction with enzymes	Unknown

[a]Modified from Scrimshaw, N. S., and Vernon, V. R. "The Requirements of Human Nutrition." *Scientific American*, September 1976.

Table 32-4 Human Vitamin Requirements[a]

Vitamin	RDA for Healthy Adult Male (milligrams)	Dietary Sources	Major Functions in Body	Deficiency Symptoms	Symptoms of Excess
Water soluble					
Vitamin B_1 (thiamin)	1.5	Pork, organ meats, whole grains, legumes	Coenzyme in reactions involving the removal of carbon	Beriberi (peripheral nerve changes, edema, heart failure)	None reported
Vitamin B_2 (riboflavin)	1.8	Widely distributed in foods	Constituent of two coenzmes involved in energy metabolism	Reddened lips, cracks at corner of mouth, lesions of eye	None reported
Niacin	20	Liver, lean meats, grains, legumes (can be formed from tryptophan)	Constituent of two coenzymes involved in energy metabolism	Pellagra (skin and gastrointestinal lesions, nervous, mental disorders)	Flushing, burning, tingling around neck and hands
Vitamin B_6 (pyridoxine)	2	Meats, vegetables, whole-grain cereals	Coenzyme involved in amino acid metabolism	Irritability, convulsions, muscular twitching, dermatitis near eyes, kidney stones	None reported
Pantothenic acid	5–10	Widely distributed in foods	Constituent of coenzyme A, which plays a central role in energy metabolism	Fatigue, sleep disturbances, impaired coordination, nausea (rare in humans)	None reported
Folacin	0.4	Legumes, green vegetables, whole-wheat products	Coenzyme involved in nucleic acid and amino acid metabolism	Anemia, gastrointestinal disturbances, diarrhea, red tongue	None reported
Vitamin B_{12}	0.003	Muscle meats, eggs, dairy products (not present in plant foods)	Coenzyme involved in nucleic acid metabolism	Pernicious anemia, neurological disorders	None reported
Biotin	Not established Usual diet provides 0.15–0.3	Legumes, vegetables, meats	Coenzyme required for fat synthesis, amino acids, metabolism, and glycogen (animal-starch) formation	Fatigue, depression, nausea, dermatitis, muscular pains	None reported
Choline	Not established Usual diet provides 500–900	All foods containing phospholipids (egg yolk, liver, grains, legumes)	Constituent of phospholipids, precursor of the neutrotransmitter acetylcholine	None reported in humans	None reported

(continued on next page)

Table 32-4   Human Vitamin Requirements[a] *(cont'd.)*

Vitamin	RDA for Healthy Adult Male (milligrams)	Dietary Sources	Major Functions in Body	Deficiency Symptoms	Symptoms of Excess
Vitamin C (ascorbic acid)	45	Citrus fruits, tomatoes, green peppers	Maintains intercellular matrix of cartilage, bone and dentine, important in collagen synthesis	Scurvy (degeneration of skin, teeth, blood vessels, epithelial hemorrhages)	Relatively nontoxic, possibility of kidney stones
Fat soluble Vitamin A (retinol)	1	Provitamin A (beta-carotene) widely distributed in green vegetables. Retinol present in milk, butter, cheese, fortified margarine	Constituent of rhodopsin (visual pigment). Maintenance of epithelial tissues	Xerophthalmia (keratinization of ocular tissue), night blindness, permanent blindness	Headache, vomiting, peeling of skin, anorexia, swelling of long bones
Vitamin D	0.01	Cod-liver oil, eggs, dairy products, fortified milk and margarine	Promotes growth and mineralization of bones; increases absorption of calcium	Rickets (bone deformities) in children; osteomalacia in adults	Vomiting, diarrhea, loss of weight, kidney damage
Vitamin E (tocopherol)	15	Seeds, green leafy vegetables, margarines, shortenings	Functions as an antioxidant to prevent cell membrane damage	Possibly anemia	None reported
Vitamin K	0.03	Green leafy vegetables. Small amounts in cereals, fruits and meats	Important in blood clotting	Bleeding, internal hemorrhages	None reported

[a]Modified from Scrimshaw, N. S., and Young, V. R. "The Requirements of Human Nutrients." *Scientific American*, September 1976.

In the following sections, we will explore some of the diverse mechanisms by which digestive systems accomplish these functions.

Diverse Digestive Systems

Digestion Within Single Cells

Intracellular digestion occurs after microscopic food particles are engulfed by single cells. Once engulfed by a cell, the food is enclosed in a **food vacuole,** a space surrounded by membrane that serves as a temporary stomach. The vacuole is fused with small packets of digestive enzymes called *lysosomes,* and food is broken down within the vacuole into smaller molecules that can be absorbed into the cell cytoplasm. Undigested remnants remain in the vacuole, which eventually dumps its contents outside the cell. This system is found both in the single-celled protists and in the simplest animals. Sponges, for example, rely entirely on intracellular digestion (Fig. 32-2). This limits their menu to microscopic food particles, such as protists that they filter from the surrounding sea using the sievelike "collars" of special collar cells.

Digestion in a Simple Sac

Larger, more complex organisms evolved a chamber within the body where chunks of food could be broken down by enzymes acting outside the cells. This is called **extracellular digestion.** One of the simplest of these chambers is found in sea anemones, hydra, and jellyfish, members of the phylum Cnidaria. These animals possess a digestive sac called a **gastrovascular cavity,** with a single opening through which foods and wastes are ejected (Fig. 32-3). Although generally referred to as the mouth, this opening is equally an anus. Food captured by stinging tentacles is escorted into the gastrovascular cavity, where enzymes break it down. Cells lining the cavity absorb the nutrients and engulf small food particles. Further digestion occurs using the intracellular digestive processes described above. The undigested remains are eventually voided through the same opening by which they entered. While one meal is being digested, a second cannot be processed efficiently, since the same chamber is used. Thus, this type of digestive system is unsuited to active animals requir-

ing frequent meals, or to animals whose food supplies so little nutrition that they must feed continuously. The needs of these animals are met by a digestive system consisting of a one-way tube with an opening at each end.

Digestion in a Tube

Most animals, from nematode worms to earthworms, molluscs, arthropods, echinoderms, and vertebrates, have a digestive system that is basically a tube running through the body. In its simplest form, as seen in the threadlike nematode worm, the tube is relatively unspecialized along its length (Fig. 32-4a). In more complex organisms, such as the earthworm, the tube consists of a series of compartments, each with a specific role in the breakdown of food (Fig. 32-4b). The earthworm extracts nutrients from decaying organic material in the soil, and also feeds on particles of leaves and bits of animal remains or wastes that it gathers during nightly forays to the surface. A tubular digestive system is essential to the earthworm, which continuously ingests soil as it burrows through the earth, passing it out one end while taking it in the other. A muscular **pharynx** draws in soil and bits of vegetation, which are passed through the **esophagus** to a thin-walled storage organ, the **crop.** The crop collects the food and gradually passes it to the

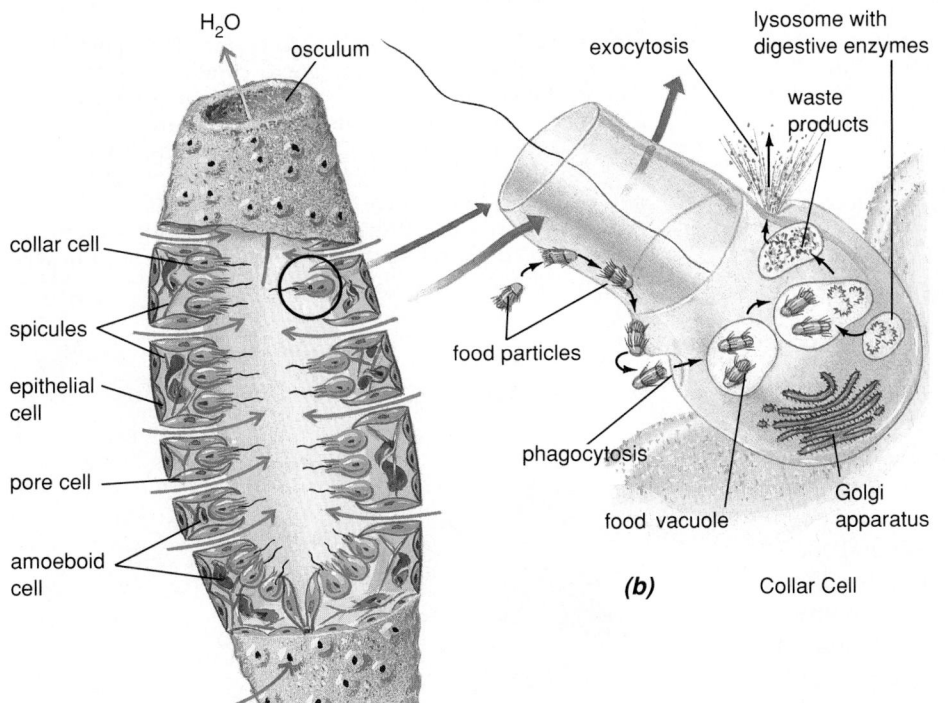

Figure 32-2 Intracellular digestion in a sponge. **(a)** Internal anatomy of a simple sponge showing the direction of water flow and the location of the collar cells. **(b)** Enlargement of a single collar cell. Water is filtered through the collar and food particles (single-celled organisms) are trapped. Within its cell body, food is engulfed and digested, and wastes are expelled.

(a)

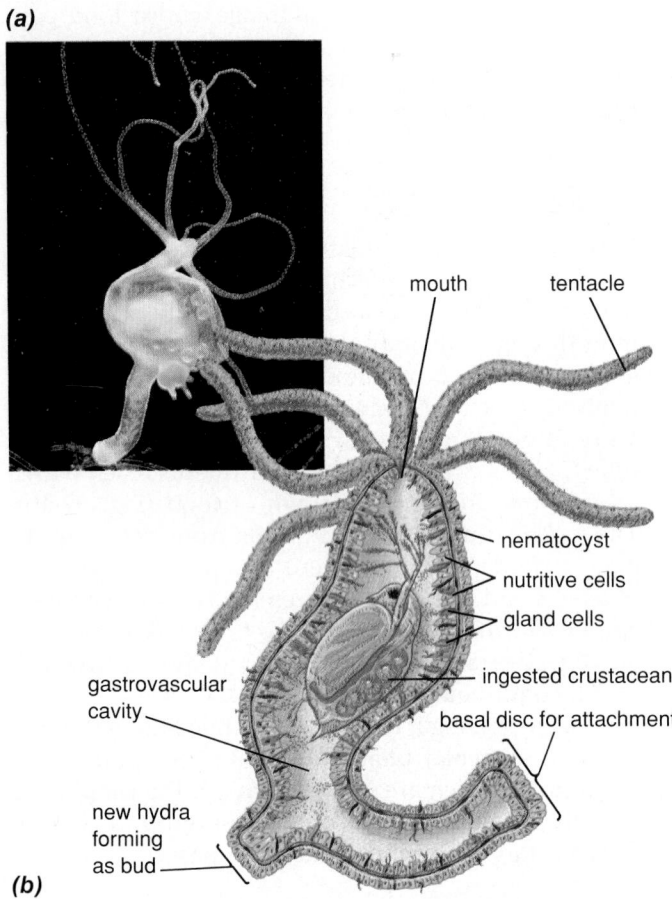

(b)

Figure 32-3 Digestion in a sac. **(a)** A *Hydra* has just ingested a water flea. (Note the small bud just beneath the flea, which will detach and form a new Hydra.) **(b)** Within the gastrovascular cavity, gland cells secrete enzymes that digest the prey into smaller particles and nutrients. Elongated cells lining the cavity ingest these particles (as described for the sponge see Fig. 32-2) and digestion is completed intracellularly. Undigested waste is then expelled through the single opening.

gizzard. Here, bits of sand and the contraction of muscles physically break the food down into smaller particles. Ground-up food from the gizzard then travels to the intestine, where enzymes break it down into simple molecules that can be absorbed by the cells lining the intestine. Animals with tubular digestive systems use extracellular digestion to dismantle their food outside the body cells.

As in the earthworm, humans and other vertebrates have tubular digestive tracts with several compartments in which food is first physically, then chemically broken down prior to absorption by individual cells. Vertebrate digestive tracts, as illustrated by those of the bird (Fig. 32-5), the cow (see Fig. E32-1), and the human (Fig. 32-6) are specialized for the particular diet of the animal.

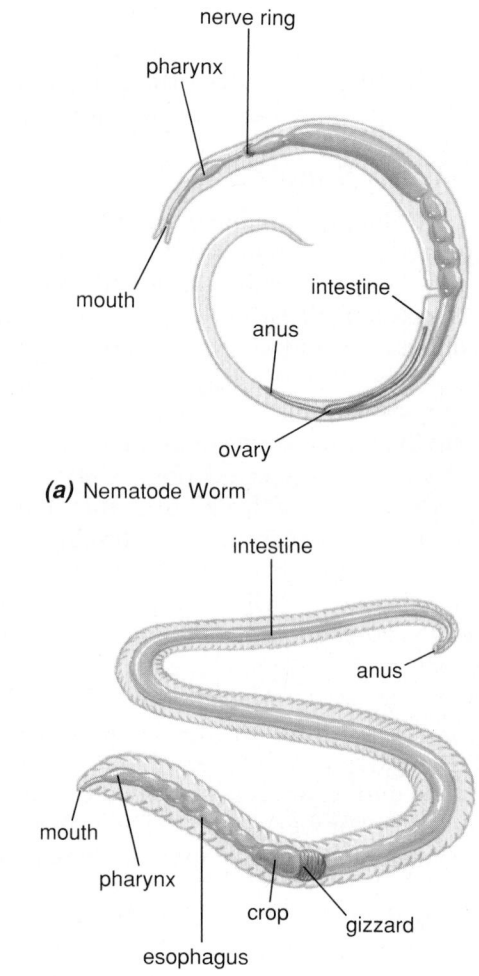

(a) Nematode Worm

(b) Earthworm

Figure 32-4 Tubular digestive tracts. **(a)** The nematode worm, an abundant, usually microscopic animal is among the simplest to have a one-way digestive system. Nematodes often live within their food source (several hundred thousand may be found in a rotting apple) and may eat almost continuously. **(b)** The earthworm has a one-way digestive system in which food is passed through a series of compartments. Each compartment is specialized to play a specific role in breaking food down and absorbing it.

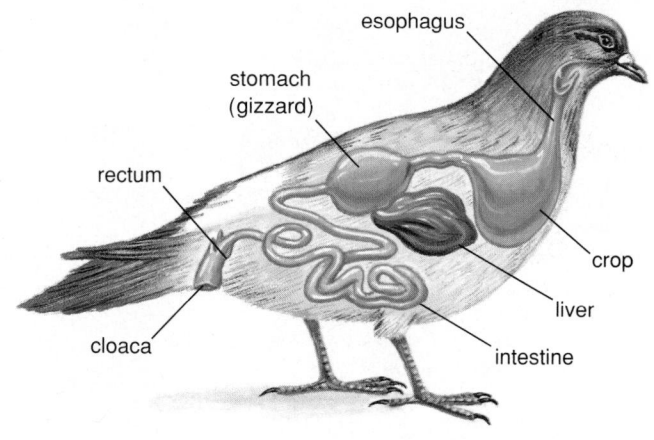

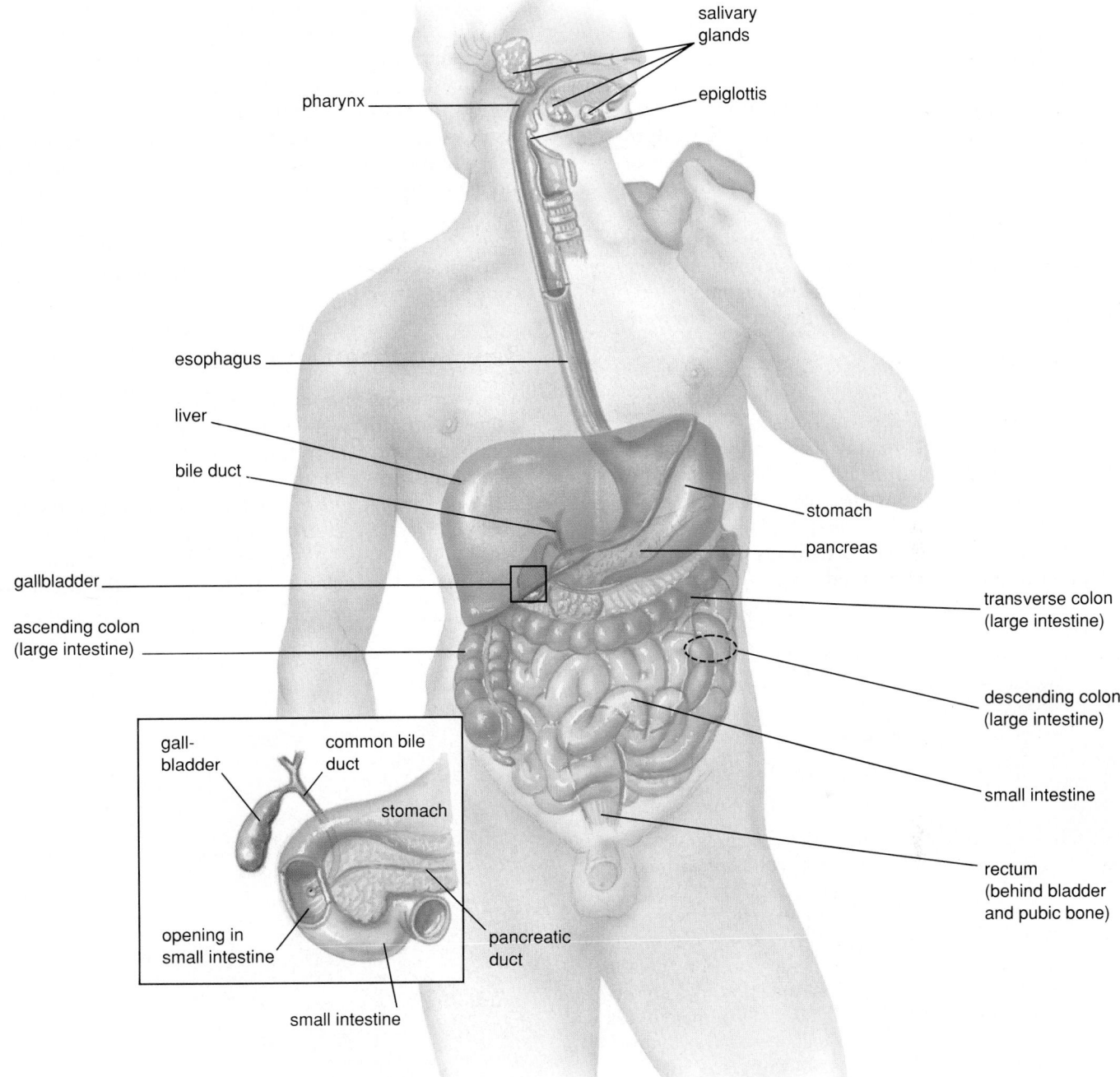

salivary glands

pharynx

epiglottis

esophagus

liver

bile duct

stomach

pancreas

gallbladder

transverse colon (large intestine)

ascending colon (large intestine)

descending colon (large intestine)

small intestine

rectum (behind bladder and pubic bone)

gall-bladder

common bile duct

stomach

opening in small intestine

pancreatic duct

small intestine

Figure 32-6 The human digestive tract. Some organs that produce and store digestive secretions, such as the salivary glands, liver, gallbladder, and pancreas, are included.

◀ **Figure 32-5** The digestive system of birds is adapted to the demands of flight. Teeth, present in primitive birds, have been lost through evolution. This shifts the weight from the head to a point closer to the bird's center of gravity. The expansible crop serves as a storage organ, allowing the bird to store food to meet the enormous caloric demands of flight. The gizzard replaces the teeth, using small stones that are stored in this organ, and muscular action to break down the hard seeds and insect exoskeletons prevalent in the diet of many birds.

Human Digestion

The Mouth

Both the mechanical and chemical breakdown of food begins in the mouth. In the adult human, 32 teeth of varying sizes and shapes cut and grind the food into small pieces. Human and other vertebrate teeth are specialized to the diet of the animal (Fig. 32-7). As the food is pulverized by the teeth, the first phase of chemical digestion occurs as three pairs of salivary glands pour out saliva in response to the smell, feel, taste, and (if you're hungry) even the thought of food (Fig. 32-8).

Saliva contains the digestive enzyme **amylase,** which begins the breakdown of starches into sugar (Table 32-5). Saliva has other functions as well. It contains a bacteria-killing enzyme and antibodies that help guard against infection, it lubricates the food to facilitate swallowing, and it dissolves some food molecules such as acids and sugars, carrying them to taste buds on the tongue. The taste buds help identify the type and quality of the food.

With the help of the muscular tongue, the food is manipulated into a mass and pressed backward into the **pharynx,** a cavity connecting the mouth with the esophagus (Fig. 32-9). The pharynx also connects the nose and mouth with the trachea, which conducts air to the lungs. As anyone who has ever choked on a piece of food can attest, this anatomical arrangement occasionally results in problems. Normally, however, the swallowing reflex elevates the larynx so that it meets the **epiglottis,** a flap of tissue that blocks off the respiratory passages. Food is thus directed into the esophagus (see Fig. 32-9).

The Esophagus and Stomach

The **esophagus** is a muscular tube that propels food from the mouth to the stomach. In cross section, the esophagus consists of four functional layers that are also found in the stomach and intestines. These are described in "A Closer Look at the Layers of the Digestive Tract." Circular muscles surrounding the esophagus contract in sequence above the swallowed food mass, squeezing it down toward the stomach

---

 **CLOSER LOOK**

At the Layers of the Digestive Tract

The four layers of the digestive tract are similar along its entire length, from the esophagus to the large intestine (Fig. E32-2). Each layer performs a specific function. The innermost layer is called the **mucosa.** It consists of a layer of epithelial cells that produce a variety of digestive secretions that differ in different portions of the tract (see Table 32-5). In the small intestine, the epithelial cells also absorb nutrients. Goblet cells in the mucosa secrete mucus. The epithelial layer is encased in connective tissue and a thin layer of smooth muscle that contracts to produce small folds in the lining of the digestive tract. Outside the mucosa lies the **submucosa,** a thick layer of connective tissue rich in capillaries. The capillaries nourish the mucosa and may take up nutrient molecules absorbed through the mucosa for transport throughout the body. Encircling the submucosa is a muscular layer, the **muscularis,** which contains layers of longitudinal and circular smooth muscle. Contractions of these muscles produce the peristaltic and churning movements that break down food and move it along. The outer wall, called the **serosa,** confines and protects the digestive tract with tough connective tissue coated with a layer of epithelium.

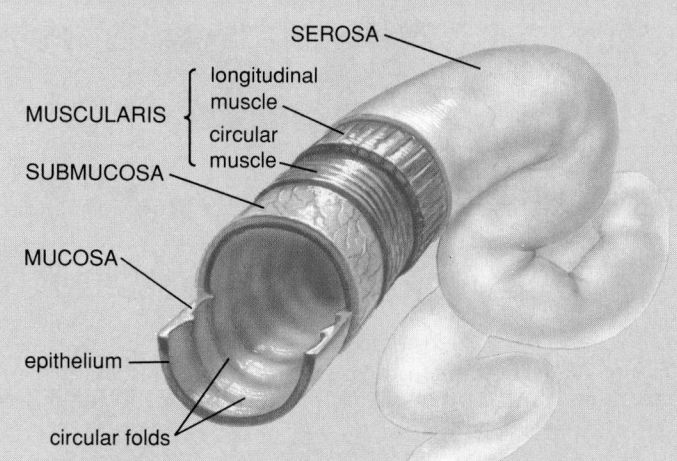

Figure E32-2 The four layers of the digestive tract.

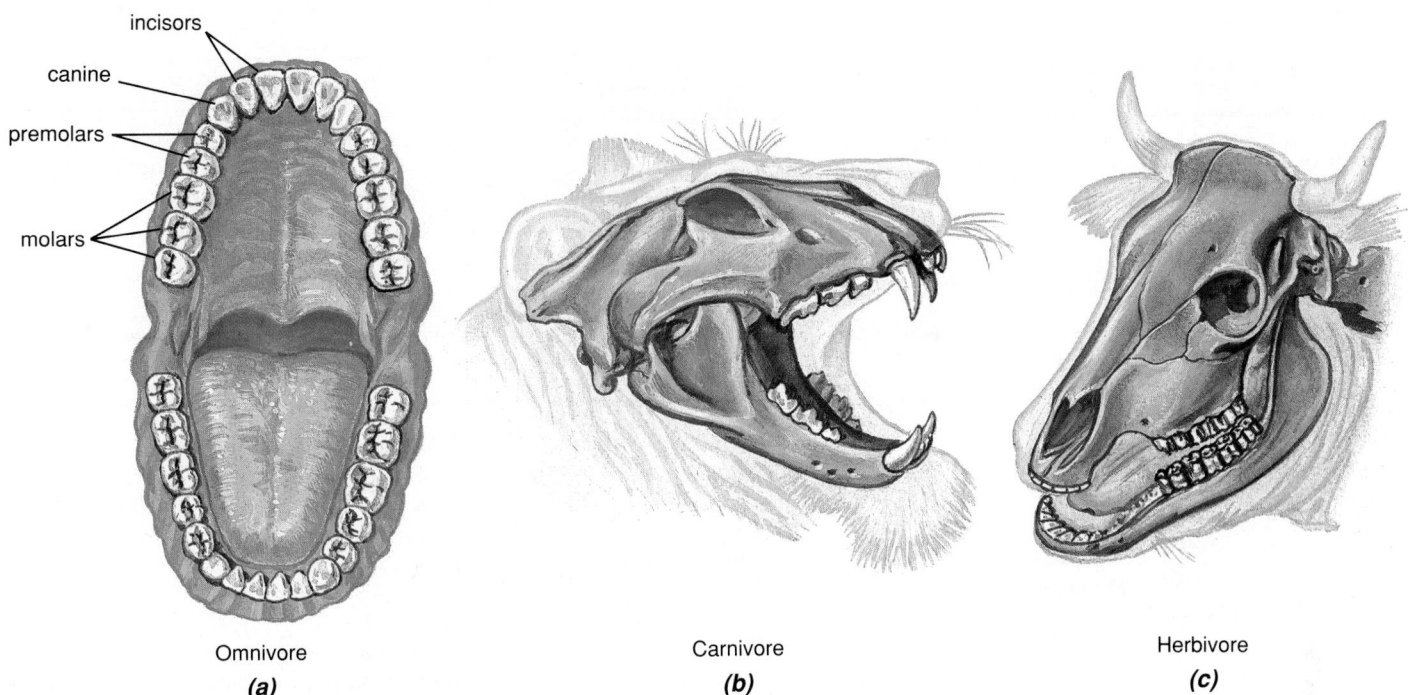

Omnivore
(a)

Carnivore
(b)

Herbivore
(c)

Figure 32-7 (a) The upper and lower jaws of a human adult showing teeth specialized for cutting and grasping (incisors and canines) and grinding (molars and premolars). The shape of human teeth reflects our varied omnivorous diet, which includes both plant and animal material. **(b)** The teeth of a lion, a carnivore (meat eater), are specialized for grasping and tearing flesh and for shearing bone. **(c)** The teeth of the herbivorous (plant-eating) cow have large flat surfaces for grinding tough plant material.

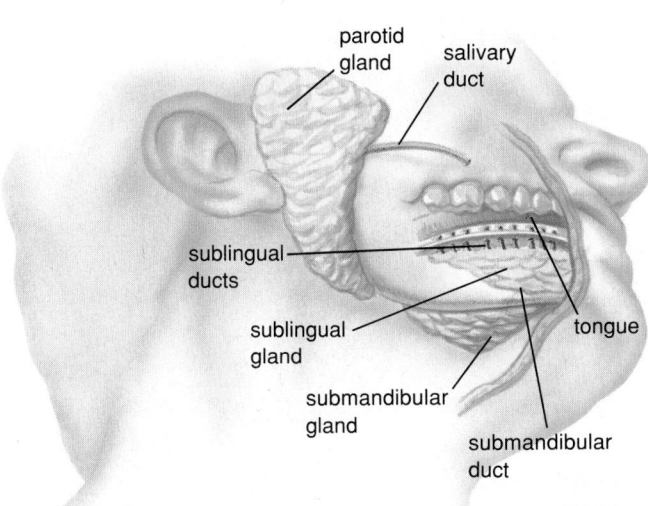

Figure 32-8 Digestion begins in the mouth as three large pairs of salivary glands and numerous smaller ones release saliva in response to the food stimuli.

(Fig. 32-10). This muscular action, called **peristalsis,** also occurs in the stomach and intestines, where it helps move food along the digestive tract. Peristalsis is so effective that a person can actually swallow when upside down. Mucus secreted by cells lining the esophagus helps protect it from abrasion and lubricates the food during its passage.

The **stomach** is an expansible muscular sac capable of holding from 2 to 4 liters (as much as a gallon) of food and liquids. Food is retained in the stomach by a ring of circular muscle separating the lower portion of the stomach from the upper small intestine. This muscle, called the **pyloric sphincter,** regulates the passage of food into the small intestine, as described later.

The stomach has three major functions:

1. **The stomach stores food and releases it gradually into the small intestine, at a rate suitable for proper digestion and absorption.** Thus, the stomach allows us to eat large, infrequent meals. Carnivores carry this to an extreme. A lion, for

Table 32-5   Digestive Secretions

Site of Digestion	Source of Secretion	Secretion	Role in Digestion
Mouth	Salivary glands	Amylase	Breaks down starch into disaccharides
	Salivary glands	Mucus, water	Lubricates, dissolves food
Stomach	Cells lining stomach	Hydrochloric acid	Allows pepsin to work, kills bacteria, solubilizes minerals
	Cells lining stomach	Pepsin	Breaks down proteins into large peptides
	Cells lining stomach	Mucus	Protects stomach
Small intestine	Pancreas	Sodium bicarbonate	Neutralizes acidic chyme from stomach
	Pancreas	Amylase	Breaks down starch into disaccharides
	Pancreas	Peptidases	Splits large peptides into small peptides
	Pancreas	Trypsin	Breaks down proteins into large peptides
	Pancreas	Chymotrypsin	Breaks down proteins into large peptides
	Pancreas	Lipase	Breaks down lipids into fatty acids and glycerol
	Liver	Bile	Emulsifies lipids
	Cells lining small intestine	Peptidases	Splits small peptides into amino acids
	Cells lining small intestine	Disaccharidases	Splits disaccharides into monosaccharides

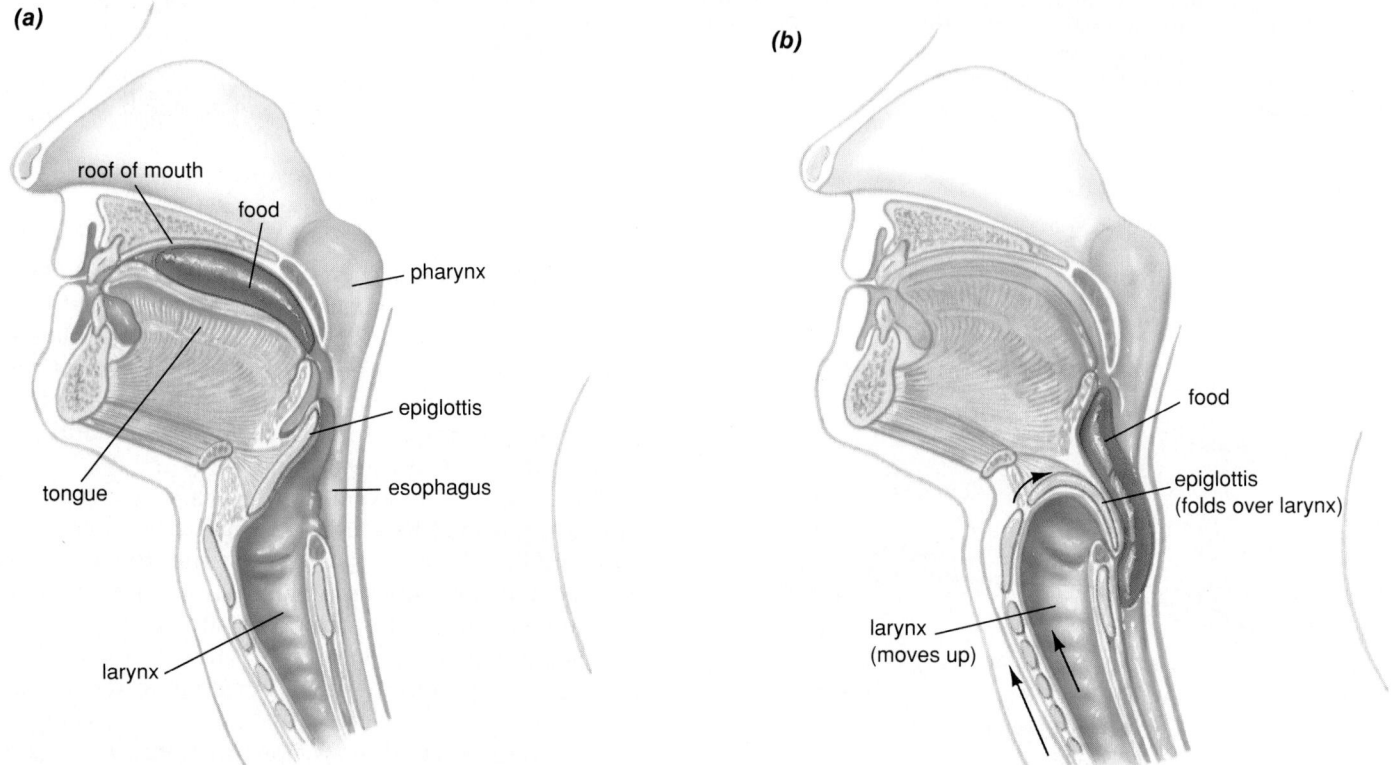

Figure 32-9 (a) Swallowing is complicated by the fact that both the esophagus and the larynx (part of the respiratory system) open into the pharynx. **(b)** During swallowing, the larynx moves upward beneath a small flap of cartilage, the epiglottis. The epiglottis folds down over the larynx, sealing off the opening to the respiratory system and directing food down the esophagus.

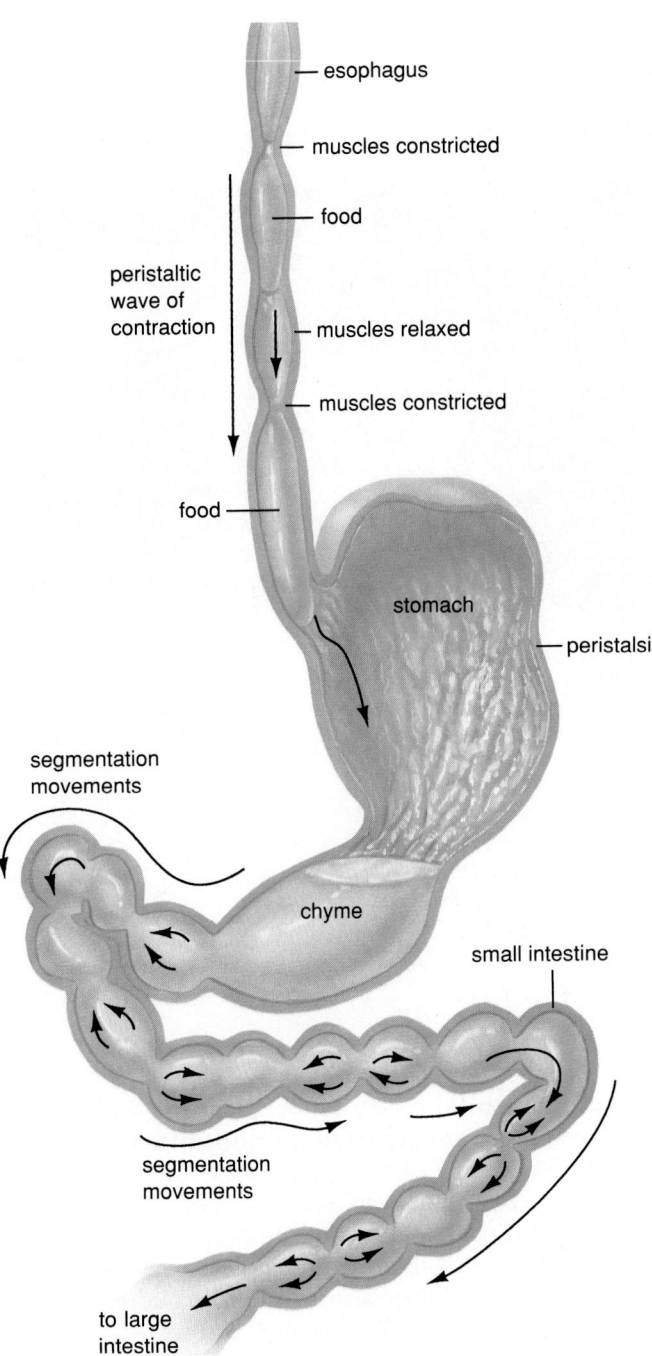

esophagus

muscles constricted

food

peristaltic
wave of
contraction

muscles relaxed

muscles constricted

food

stomach

peristalsis

segmentation
movements

chyme

small intestine

segmentation
movements

to large
intestine

Figure 32-10 Food is propelled through the digestive system by peristaltic contractions of circular muscles that proceed downward, forcing the food along in front of them. In the stomach and small intestine, a variety of contractions, such as segmentation movement of the small intestine, help break down food particles and speed absorption.

instance, may consume 40 pounds of meat at one meal, then spend the next few days quietly digesting it.

2. **The stomach contributes to the mechanical breakdown of food.** Its muscular walls undergo a variety of contracting, churning movements that help break apart large pieces of food.

3. **Glands in the lining of the stomach secrete enzymes and other substances that facilitate digestion.** These include gastrin, hydrochloric acid, pepsinogen, and mucus. Gastrin—a hormone—stimulates secretion of hydrochloric acid by specialized stomach cells. Others cells release pepsinogen, an inactive form of the protein-digesting enzyme pepsin. Pepsin, a **protease,** breaks proteins into shorter chains of amino acids called *peptides* (see Table 32-5). Pepsin is secreted in an inactive form to prevent it from digesting the cells that produce it. The highly acidic conditions in the stomach (pH 1) convert pepsinogen into pepsin, which functions best in an acidic environment.

As you may have noticed, the stomach produces all the ingredients necessary to digest itself, and indeed, this is what happens when a person develops ulcers (see "Health Watch: Ulcers—When the Digestive Tract Digests Itself"). However, cells lining the stomach normally produce copious quantities of thick mucus that coat the stomach lining, serving as a barrier to self-digestion. This process is not completely successful, however, and the cells lining the stomach are digested to some extent, needing replacement every few days.

Food in the stomach is gradually converted to a thick, acidic liquid called **chyme,** which consists of partially digested food and digestive secretions. Peristaltic waves, traversing the muscular stomach at a rate of about three per minute, propel the chyme toward the small intestine. The sphincter at the base of the stomach allows only about a teaspoon of chyme to be expelled with each contraction. It takes 2 to 6 hours, depending on the size of the meal, to empty the stomach completely. Then, the continued churning movements of the empty stomach are felt as hunger pangs.

Only a few substances can enter the bloodstream through the stomach wall, including water, some drugs, and alcohol. Alcohol consumed when the stomach is empty is immediately absorbed into the bloodstream, with strong and rapid effects. Since food in the stomach slows alcohol absorption, the advice "never drink on an empty stomach" is based on sound physiological principles.

HEALTH WATCH
Ulcers—When the Digestive Tract Digests Itself

Ulcers occur when the mucus barriers of the stomach and upper small intestine break down, and the inner lining and deeper layers of the digestive tract are eroded by digestive enzymes. People with ulcers tend to secrete excess stomach acid and protein-digesting pepsin. The upper small intestine is the most common site for ulcers (Fig. E32-3), because it is less protected by mucus than the stomach, but receives the highly acidic chyme. In some cases, decreased secretion of alkaline pancreatic juice, or decreased mucus secretion by the duodenum, contributes to the disease.

The tendency to develop ulcers is hereditary to a large extent. However, stress, which causes excess acid secretion, can also trigger them. Smoking, and the consumption of alcohol and aspirin all reduce the resistance of the digestive tract lining to the effects of pepsin and acids, and can aggravate ulcers. Reducing stress, eliminating smoking and drinking, and other dietary changes can help relieve ulcers. Ulcers are treated using antacids, which neutralize stomach acid, and drugs such as cimetidine or prostaglandins, which decrease stomach acid production.

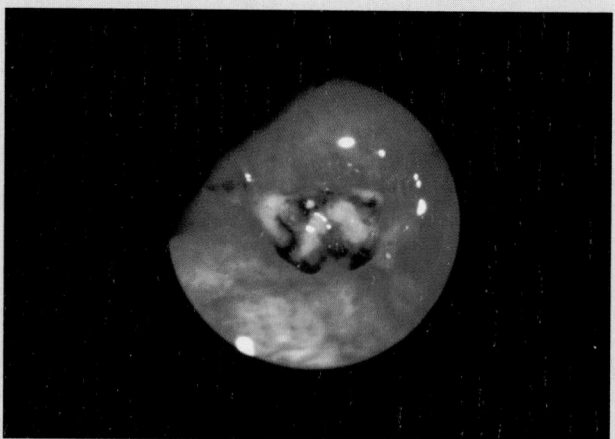

Figure E32-3 A bleeding ulcer in the stomach, photographed through a fiber-optic device. Red is fresh blood, black is clotted blood. The stomach muscosa has been digested away by acid and pepsin.

Digestion in the Small Intestine

The **small intestine** is a coiled, narrow (1 to 2 in. in diameter) tube that is about 9 feet long in a human adult. **The small intestine has two major functions: to digest food into small molecules, and to absorb these molecules, passing them to the bloodstream.**

The first role of the small intestine—digestion—is accomplished with the aid of digestive secretions from three sources: the liver, the pancreas, and the cells of the small intestine itself.

The Liver and Gallbladder

The **liver** is the largest and perhaps most versatile organ in the body. Its many functions include the storage of fats and carbohydrates for energy, the regulation of blood glucose levels, the synthesis of blood proteins, the storage of iron and certain vitamins, the conversion of toxic ammonia into urea, and the detoxification of other harmful substances such as nicotine and alcohol. **The role of the liver in digestion is to produce bile, which is stored in the gallbladder and released into the small intestine through the bile duct** (see Fig. 32-6).

Bile is a complex mixture composed of **bile salts,** water, other salts, and cholesterol. Bile salts are synthesized in the liver from cholesterol and amino acids. Although they assist in the breakdown of lipids, bile salts are not enzymes. Rather, they act as detergents or emulsifying agents to disperse fats. They suspend the lipids in the chyme as microscopic particles. These particles expose a large surface area for attack by **lipases,** lipid-digesting enzymes produced by the pancreas (see "A Closer Look at the Fate of Fats").

The Pancreas

The **pancreas** lies in the loop between the stomach and small intestine (see Fig. 32-6). It consists of two major cell types. One type produces hormones involved in blood sugar regulation (see Chapter 35), and the other type produces a digestive secretion called *pancreatic juice,* which is released into the small intestine. **The role of pancreatic juice is to neutralize the acidic chyme and to digest carbohydrates, lipids, and proteins.** About a quart of pancreatic juice is released into the small intestine each day. Pancreatic juice includes water, sodium bicarbonate, and a number of digestive enzymes (see Table 32-5). Sodium bicarbonate (the active ingredient in baking soda) neutralizes the acidic chyme in the small intestine, producing the slightly alkaline pH required for proper functioning of the pancreatic digestive enzymes. These enzymes are specialized to break down three major types of food: an amylase breaks down carbohydrates, lipases digest lipids, and several proteases disrupt proteins and peptides. The pancreatic proteases include trypsin, chymotrypsin, and carboxypeptidase. Both trypsin and chymotrypsin break proteins and peptides into shorter peptide chains. Carboxypeptidase completes protein digestion by liberating individual amino acids from the ends of the peptides. These proteases are secreted in an inactive form and become activated after they reach the small intestine.

The Intestinal Wall

The wall of the small intestine is studded with cells that are specialized to complete the digestive process and absorb the small molecules that result. These cells have various enzymes on their external membranes, which form the lining of the intestine. Some of these are proteases, which complete the breakdown of peptides into amino acids. Other enzymes include sucrase, lactase, and maltase, which break down disaccharides into monosaccharides (see Chapter 3), and small amounts of lipase. Since these enzymes are actually embedded in the membranes of the cells lining the small intestine, this final phase of digestion occurs as the nutrient is being absorbed into the cell. As is the stomach, the small intestine is protected from digesting itself by copious mucus secretions from specialized cells in its lining.

Absorption in the Small Intestine

The small intestine is not only the principal site of chemical digestion, it is also the major site of nutrient absorption into the blood. To facilitate absorption, the small intestine, in addition to its length, has an internal surface area that is increased 600-fold over that of a smooth tube by foldings and projections of its wall (Fig. 32-11). Not only is the wall itself folded, but covering its entire surface are minute, fingerlike projections called **villi** (literally, "shaggy hairs"). Villi range from 0.5 to 1.5 millimeter in length, and give

the intestinal lining a velvety appearance to the naked eye. They move gently back and forth amidst the chyme within the intestine. This movement increases their exposure to the molecules to be digested and absorbed. Further, the individual cells of the villi each bear a fringe of microscopic projections called **microvilli.** Altogether, these specializations of the small intestine wall give it a surface area of about 250 square meters, about the size of a tennis court. **Segmentation movements,** which are rhythmic contractions of the circular muscles of the intestine, also help bring the nutrients into contact with the absorptive surface of the small intestine (see Fig. 32-10). Segmentation movements are not synchronized, as is peristalsis, and they slosh the liquid back and forth. When absorption is complete, coordinated peristaltic waves conduct the remnants into the large intestine.

Nutrients absorbed by the small intestine include monosaccharides, amino acids and short peptides, fatty acids produced by lipid digestion, vitamins, and minerals. The mechanisms by which this absorption occurs are varied and complex, and are currently the subject of active investigation. In most cases, energy is expended to transport nutrients into the intestinal cells. The nutrients then diffuse out of the intestinal cells into the interstitial fluid, from whence they enter the bloodstream.

Each villus of the small intestine is provided with a rich supply of blood capillaries and a single lymph capillary called a **lacteal** to carry off the absorbed nutrients and distribute them throughout the body (see Fig. 32-11). Most of the nutrients enter the bloodstream via the capillaries, but fat subunits take a different route. After diffusing into the epithelial cells, they are resynthesized into fats, combined with other molecules, and then released as droplets into the interstitial fluid. Here, they enter the lymph capillary and are eventually delivered to the bloodstream when the lymph vessels empty into the veins (see "A Closer Look at the Fate of Fats").

The Large Intestine

The **large intestine** is not named for its length (about 5 ft in the adult human), but for its diameter (about 2.5 in.), which is considerably larger than the small intestine. It receives the leftovers of digestion: a mixture of water, undigested fats and proteins, and indigestible fibers such as the cell walls of vegetables and fruits. The large intestine contains a flourishing population of bacteria living on unabsorbed nutrients. These bacteria earn their keep by synthesizing vitamins B_{12}, thiamine, riboflavin, and most importantly, vitamin K, which is often deficient in a normal diet.

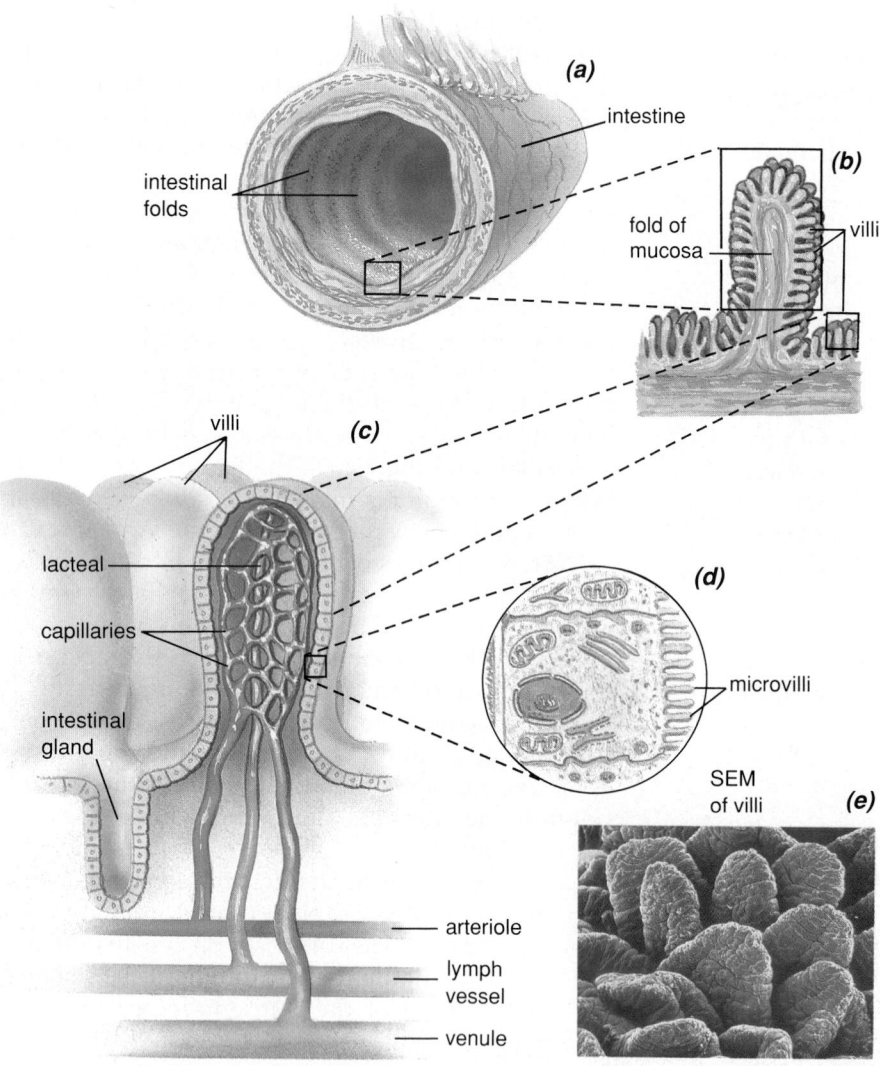

Figure 32-11 The surface area of the small intestine is greatly increased by **(a)** folds in the intestinal lining and **(b)** projections called *villi* (sing. villus) that extend from the folds, giving the lining a velvety appearance. The villi are enlarged in the micrograph in **(e)** and the diagram in **(c)**, showing the blood capillaries and lymph vessels that form a network within each villus where nutrients are absorbed. **(d)** Microvilli, which are microscopic projections of individual cell membranes, further increase the surface area for nutrient absorption.

Cells lining the large intestine absorb these vitamins, as well as leftover water and salts. After absorption is complete, the result is a semisolid feces, consisting of indigestible wastes and the dead bodies of bacteria, which account for about one third the dry weight of feces. The feces are transported by peristaltic movements until they reach the rectum. Distention of this chamber stimulates the desire to defecate. Although defecation is a reflex (new parents are constantly reminded of this) it is initiated voluntarily after around the age of two.

The Control of Digestion

Considerable coordination is required to break down a chef salad into amino acids and peptides, sugars, fatty acids, vitamins, minerals, and indigestible cellulose. As the mouth responds to the first bite, the stomach must be warned of the imminent influx of food. In addition, the enzymes of the stomach and small intestine require different environments for proper functioning (highly acidic in the stomach, slightly alkaline in the small intestine), and secretions into various parts of the digestive tract must be coordinated with the arrival of food. Not surprisingly, the secretions and activity of the digestive tract are coordinated by both nerves and hormones (Table 32-6). Here we will examine a few of these control mechanisms.

The initial phase of digestion is under the control of the nervous system and involves responses to signals originating in the head. These signals include the sight, smell, taste, and sometimes the thought of food, as well as the muscular activity of chewing. In response to these stimuli, saliva is secreted to the mouth, while nervous signals to the stomach walls initiate secretion of acid and a hormone called **gas-**

 **CLOSER LOOK**

At The Fate of Fats

Since digestion occurs in a watery environment using water-soluble enzymes, the digestion of water-insoluble lipids poses a particular challenge. Since fats and water don't mix, dietary fats in the chyme tend to aggregate into large globs that resist the attack of digestive enzymes. Bile salts, secreted by the liver and released by the gallbladder, provide the first level of attack. If you have ever poured detergent into a dishpan with oily water and watched the oil disperse, you've witnessed an effect very comparable to the action of bile salts on lipids. Like your dish detergent, bile salts are molecules with both hydrophilic (literally, "water-loving" or water-soluble) and hydrophobic ("water-fearing" or lipid-soluble) portions. The hydrophobic end dissolves in the lipid while the hydrophilic end dissolves in the surrounding watery fluid. This enables bile salts to disperse the fat globs into microscopic particles (Fig. E32-4). These particles expose a much greater surface area to

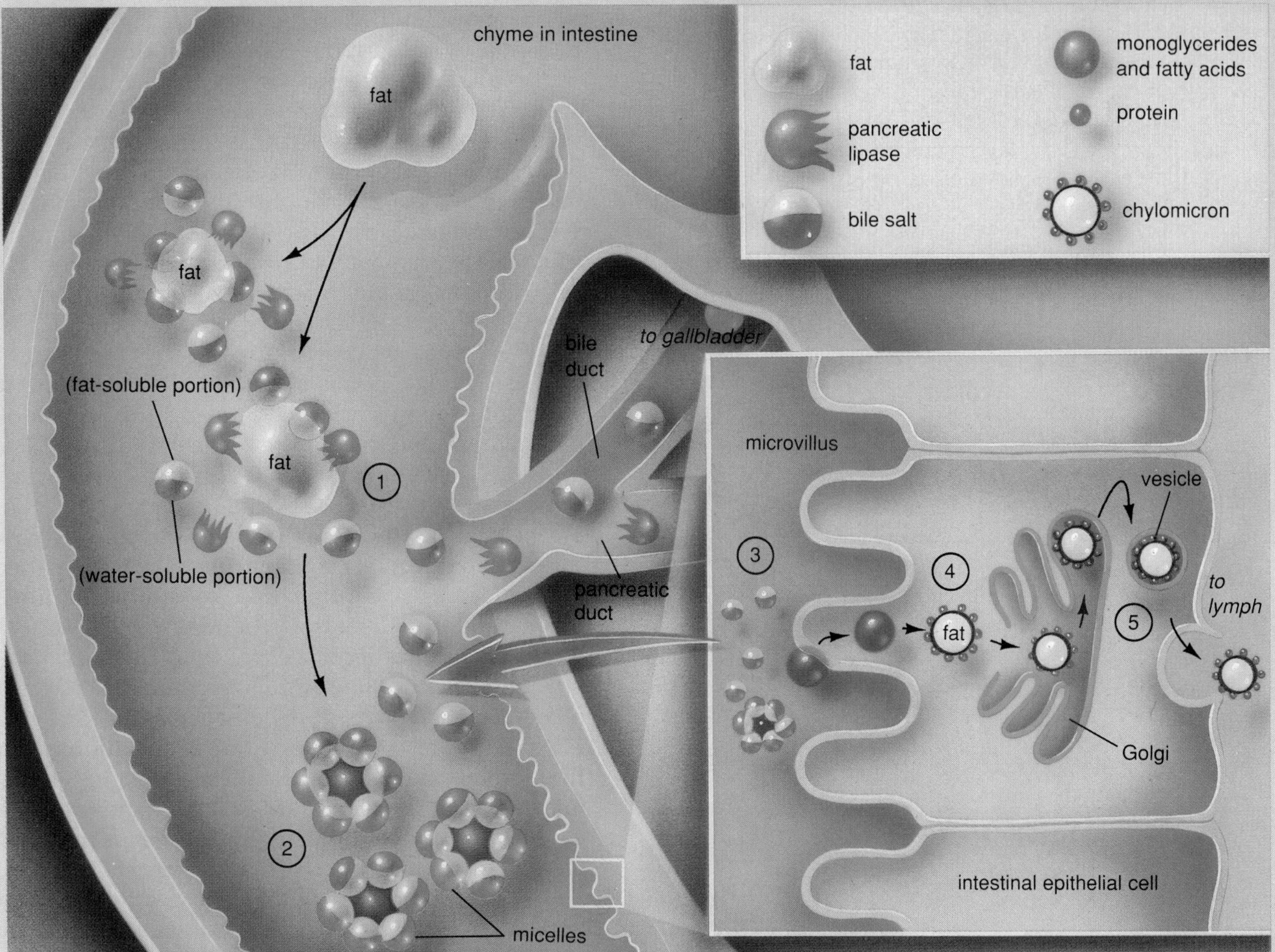

Figure E32-4 ① Bile salts disperse fats into small particles, which are broken into monoglycerides and fatty acids by lipase. ② Bile salts aggregate into clusters called *micelles*, which ferry the monoglycerides and fatty acids to the cells lining the intestine. ③ The fat subunits diffuse through the cell membranes of the intestinal cells, leaving the bile salts behind. ④ Within the intestinal cells, fats are resynthesized and coated with protein, forming droplets called *chylomicrons*. ⑤ Chylomicrons are packaged into vesicles by the Golgi, and released by exocytosis into the extracellular fluid, where they enter lymph vessels.

the surrounding fluid, which is rich in pancreatic lipases. As the lipases break triglycerides into fatty acids and monoglycerides, these subunits are absorbed into the lipid-soluble portion of tiny clusters of bile salts called **micelles**. The micelles escort the fatty acids and monoglycerides to the microvilli of the cells lining the intestine. There, the fatty acids and monoglycerides diffuse through the cell membranes into the intestinal cells, leaving the micelles behind where they continue emulsifying and transporting digested fats.

Within the intestinal cells, the fat subunits are reassembled into triglycerides, mixed with cholesterol and phospholipids, and collected into droplets coated with a thin layer of protein. These droplets, called **chylomicrons,** are packaged into vesicles by the Golgi apparatus and expelled by endocytosis into the interstitial fluid on the far side of the cell. Too large to enter capillaries, chylomicrons enter the lacteal, or lymph vessel, that projects into each intestinal villus. The fat is then transported in the lymph to the large veins of the neck where the lymphatic vessels empty their contents.

trin, which stimulates further acid secretion. The concentration of acid is regulated by a negative feedback mechanism (see Chapter 25). When acid levels reach a certain point, they inhibit gastrin secretion, thus inhibiting further acid production.

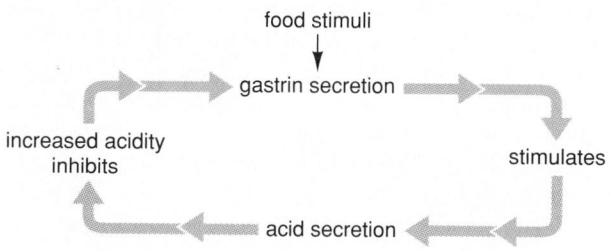

The arrival of food in the stomach triggers the second phase of digestion. Stimulation of the stomach wall causes copious production of mucus to protect against self-digestion. The acidity of the stomach converts pepsinogen to its active form, pepsin, which begins protein digestion. Protein in the food tends to buffer or reduce the concentration of stomach acid. The release of gastrin is no longer inhibited, and gastrin release stimulates further acid production. The cells secreting stomach acid are also activated by distension of the stomach and by the presence of peptides produced by protein digestion.

As the liquid chyme is gradually released into the small intestine, its acidity stimulates the release of a second hormone, **secretin,** by cells of the upper small intestine. Secretin causes the pancreas and liver to pour bicarbonate into the small intestine. Bicarbonate neutralizes the acidity of the incoming chyme, thus creating an environment in which the pancreatic enzymes can function. A third hormone, **cholecystokinin,** is also produced by cells of the upper small intestine in response to the presence of chyme. This hormone stimulates the release of various digestive enzymes by the pancreas into the small intestine. It also stimulates the gallbladder to contract, squeezing bile through the bile duct to the small intestine. Bile assists in fat breakdown, as described earlier. **Gastric inhibitory peptide**—a hormone secreted by the small intestine in response to fatty acids and sugars in chyme—inhibits acid production and peristalsis in the stomach. This slows down the rate at which chyme is dumped into the small intestine, providing additional time for digestion and absorption to occur.

Table 32-6 Some Important Digestive Hormones

Hormone	Site of Production	Stimulus for Production	Effect
Gastrin	Stomach	Food in mouth Distension of stomach Peptides in stomach	Stimulates acid secretion by cells in stomach
Secretin	Small intestine	Acid in small intestine	Stimulates bicarbonate production by pancreas and liver, increases bile output by liver
Cholecystokinin	Small intestine	Amino acids, fatty acids in small intestine	Stimulates secretion of pancreatic enzymes and release of bile by gallbladder
Gastric-inhibitory peptide	Small intestine	Fatty acids and sugars in small intestine	Inhibits stomach movements and release of stomach acid

SUMMARY OF KEY CONCEPTS

Nutrition
Each type of animal has specific nutritional requirements. These include molecules that can be broken down to liberate energy, amino acids that can be linked together to form proteins, minerals, and vitamins to facilitate the diverse chemical reactions of metabolism.

The Challenge of Digestion
Digestive systems must accomplish five tasks: ingestion, mechanical breakdown of food, chemical breakdown or digestion, absorption, and elimination of wastes.

Diverse Digestive Systems
Digestive systems are designed to convert the complex molecules of the bodies of other animals or plants into simpler molecules that can be utilized by the consumer. Animal digestion at its simplest is intracellular, as occurs within the individual cells of a sponge. Extracellular digestion, utilized by all more complex animals, occurs in a body cavity. The simplest form is a "dead end," saclike gastrovascular cavity in organisms such as flatworms and hydra. Still more complex animals utilize a tubular compartment with specialized chambers where food is processed in a well-defined sequence.

Human Digestion
In humans, digestion begins in the mouth, where food is physically broken down by chewing and chemical digestion is initiated by saliva. Food is then conducted to the stomach by peristaltic waves of the esophagus. In the acidic environment of the stomach, food is churned into smaller particles, and protein digestion begins. Gradually, the liquefied food, now called *chyme*, is released to the small intestine. Here it is neutralized by bicarbonate from the pancreas. Secretions from the pancreas, liver, and the cells of the intestine complete the breakdown of proteins, fats, and carbohydrates. The small intestine is also the site where the simple molecular products of digestion are absorbed into the bloodstream for distribution to the body cells. The large intestine absorbs the remaining water and converts indigestible material to feces.

Digestion is regulated by the nervous system and hormones. The smell and taste of food and the action of chewing causes salivary secretion and production of gastrin by the stomach. Gastrin stimulates stomach acid production. As chyme enters the small intestine, three additional hormones are produced by intestinal cells: secretin, which causes bicarbonate production to neutralize the acid chyme; cholecystokinin, which stimulates bile release and causes the pancreas to secrete digestive enzymes into the small intestine; and gastric inhibitory peptide, which inhibits acid production and peristalsis by the stomach. This slows the movement of food into the intestine.

GLOSSARY

absorption: the movement of nutrients into cells.

amylase (am'-ē-lās): an enzyme that catalyzes the breakdown of starch, found in saliva and pancreatic secretions.

bile (bīl): a liquid secretion of the liver stored in the gallbladder and released into the small intestine during digestion. Bile contains bile salts, whose role is to emulsify or disperse fats into small particles on which fat-digesting enzymes may act.

calorie (kal'-ōr-ē): a measure of the energy derived from food. When capitalized (i.e., Calorie) this unit is the amount of energy required to raise the temperature of 1 liter of water 1 degree Celsius. It represents 1000 calories (with a lowercase "c"). The energy content of foods is measured in Calories.

carbohydrate (kar-bō-hī'-drāt): a class of nutrient including monosaccharides, disaccharides, and polysaccharides (starches, glycogen, and cellulose). Sugars and starches are used by animal cells as a source of energy.

cholecystokinin (kō'-lē-sis-tō-ki'-nin): a digestive hormone produced by the small intestine that stimulates release of pancreatic enzymes.

chylomicron (kī-lō-mī'-kron): a droplet consisting of triglycerides, cholesterol, and phospholipids, and coated with a thin layer of protein. Chylomicrons are formed in the cells lining the small intestine, enter the lymphatic system, and eventually reach the circulatory system.

chyme (kīme): an acidic, souplike mixture of partially digested food, water, and digestive secretions that is released from the stomach into the small intestine.

crop: an organ found in both earthworms and birds in which ingested food is stored temporarily before passing to the gizzard, where it is pulverized.

digestion: the process by which food is physically and chemically broken down into molecules that can be absorbed by cells.

epiglottis (ep-eh-gla'-tis): a flap of cartilage in the lower pharynx that covers the opening to the larynx during swallowing. This directs the food down the esophagus.

esophagus (eh-sof'-eh-gus): a muscular passageway connecting the pharynx to the next chamber of the digestive tract, the stomach in humans and other mammals.

essential amino acids: amino acids which are required nutrients that the body is unable to manufacture and which must be supplied in the diet.

essential fatty acids: fatty acids which are required nutrients that the body is unable to manufacture and which must be supplied in the diet.

extracellular digestion: the physical and chemical breakdown of food that occurs outside of a cell, usually in a digestive cavity.

food vacuole: a membrane-bound space within a single cell in which food is enclosed. Digestive enzymes are released into the vacuole and intracellular digestion occurs here.

gallbladder: a small sac adjacent to the liver in which the bile secreted by the liver is stored and concentrated. Bile is released from the gallbladder via the bile duct to the small intestine.

gastrovascular cavity: a chamber that has both digestive and circulatory functions, found in simple invertebrates. A single opening serves as both mouth and anus, while the chamber provides direct access of nutrients to the cells.

gizzard: a muscular organ found in earthworms and birds in which food is mechanically broken down prior to chemical digestion.

glycogen (glī'-kō-jen): a long, branched polymer of glucose that is stored in the muscles and liver and metabolized as a source of energy.

intracellular digestion: the chemical breakdown of food occurring within single cells.

lacteal (lak-t'ēl): a single lymph vessel that penetrates each villus of the small intestine.

lipase (lī'-pāse): an enzyme that catalyzes the breakdown of lipids such as fats.

lymph (limpf): a fluid resembling blood plasma that collects in special lymph vessels and eventually returns to the bloodstream.

micelle (mī'-cēl): aggregations of bile salts with their lipophilic ends facing inward. Fatty acids and monoglycerides dissolve in the hydrophobic portions and are ferried to the cells of the lining of the small intestine.

microvilli (mī-krō-vi'-lī): a series of folded projections of the cell membrane that increase its surface area.

pancreas (pan'-krē-is): an organ lying adjacent to the stomach that secretes enzymes for fat, carbohydrate, and protein digestion into the small intestine. Other cells in the pancreas are responsible for production of the hormones glucagon and insulin.

pharynx (fār'-inx): a chamber located behind the mouth. In vertebrates, it is common to both the respiratory and digestive systems.

protease (prō'-tē-ās): an enzyme that digests proteins.

pyloric sphincter (pī-lor'-ik sfink'-ter): a circular muscle at the base of the stomach that regulates the passage of chyme into the small intestine.

segmentation movements: asynchronous contractions of the small intestine that result in mixing of the partially digested food and digestive enzymes. The movements also bring nutrients into contact with the absorptive intestinal wall.

villus (vi'-lus): fingerlike projections of the wall of the small intestine that increase its absorptive surface area.

vitamin: any one of a group of diverse chemicals that must be present in trace amounts in the diet to maintain health. Vitamins are used by the body in conjunction with enzymes in a variety of metabolic reactions; a few are involved in growth and differentiation.

STUDY QUESTIONS

1. List four general types of nutrients and describe the role of each in nutrition.
2. What general role do water-soluble vitamins play in human metabolism?
3. Vitamin C is a vitamin for humans but not for dogs. Explain.
4. Trace the pathway of an indigestible tomato seed through the human digestive tract.
5. List and describe the function of the three principal secretions of the stomach.
6. Explain how surface area is increased in the small intestine, and why this is important.
7. List the substances secreted into the small intestine and describe the origin and function of each.
8. Describe the function and origin of four major digestive hormones.
9. Name and describe the muscular movements that usher food through the human digestive tract.

DISCUSSION QUESTIONS

1. Some humans have difficulty in digesting the milk sugar lactose after the age of weaning, a condition known as lactose intolerance. Lactose-intolerant adults who drink milk experience a bloated feeling, cramps, and diarrhea due to lactose buildup in the intestines. Interestingly, human populations with a long history of dairy farming, and hence milk drinking past the age of weaning, have a high percentage of individuals with lactose tolerance, while those populations from non-dairying regions have a high percentage of lactose-intolerant individuals. Discuss a possible explanation for the frequencies of lactose intolerance in these populations. Which do you think is the more ancestral condition, lactose tolerance or intolerance? Explain.

2. Trace a ham and cheese with lettuce sandwich through the human digestive system, discussing what happens to each part of the sandwich as it passes through each region of the digestive tract.

3. One of the common remedies for human constipation (failure to eliminate feces for an abnormally long period of time) is a laxative solution containing magnesium salts. In the large intestine, magnesium salts are absorbed very slowly by the intestinal wall, leaving large amounts in the intestinal tract for long periods of time. This has an effect on water movement in the large intestine. How does the effect of the slowly-absorbed salt on water movement relate to the laxative action of magnesium salts?

SUGGESTED READINGS

Christian, J. L., and J. L. Gregor. *Nutrition for Living*, 2nd edition. Menlo Park, CA: Benjamin/Cummings, 1988. Clear, detailed presentation of nutritional requirements in relation to various lifestyles.

Davenport, H. "Why the Stomach Does Not Digest Itself." *Scientific American*, January 1972. How the stomach protects itself from its own strongly acidic secretions, and what happens when the defenses fail.

Eckert, R., Randall, D., and Augustine, G. *Animal Physiology: Mechanisms and Adaptations*. 3rd edition. New York: W. H. Freeman and Company, Publishers, 1988. An excellent and complete textbook of comparative animal physiology.

Hole, J. W., Jr. *Essentials of Human Anatomy and Physiology*. 2nd ed. Dubuque, Iowa: Wm. C. Brown Company, 1986. Descriptions of both human anatomy and physiology. Beautiful, clear illustrations.

Kretchmer, N. "Lactose and Lactase." *Scientific American*, April 1972. Discusses differences in tolerance to milk sugar among humans from different populations.

Moog, F. "The Lining of the Small Intestine." *Scientific American*, November 1981. A description of the structure and function of this intricate tissue, which is responsible for absorbing nutrients into the body.

Vander, A., and Sherman, J. *Human Physiology*. 4th edition. New York: McGraw-Hill Book Company, 1985. Complete coverage of human physiology.

33

Excretion

Animal excretory systems are finely adapted to the unique needs of each species. The beaver takes in excess water, and its kidneys produce dilute, watery urine.

Animal cells can function only under a relatively narrow range of conditions. Thus, while we load our digestive systems with pepperoni pizzas and hot fudge sundaes, our cells remain bathed in a precisely regulated solution of salts and nutrients, maintained by the body in spite of our dietary eccentricities. This precise internal regulation, called **homeostasis,** is discussed in Chapter 25. The cells of the digestive system, through which most substances enter the body, are relatively unselective; any molecule that can move into the body through the intestinal lining does so, including an excess of water, nutrients, salts, and minerals, as well as nonnutritive substances such as drugs. The burden of restoring and maintaining proper internal balance, then, falls on the organs of homeostasis, particularly those of the excretory system.

Excretory systems have two major functions: excretion of cellular waste products such as urea, and regulation and maintenance of the composition of body fluids. Both of these functions are performed simultaneously by filtering the blood. Excretory systems first collect the fluid portion of the blood. From this fluid, water and important nutrients are reabsorbed into the blood, while toxic substances, wastes, excess nutrients and hormones, and some water are left behind to be eliminated as urine. Urine is produced by the kidneys of vertebrates, and various simpler excretory organs of invertebrates, as described below. The type of urine produced is intimately related to the animals' need to conserve water, as discussed later in "A Closer Look at Excretion and Animal Lifestyle."

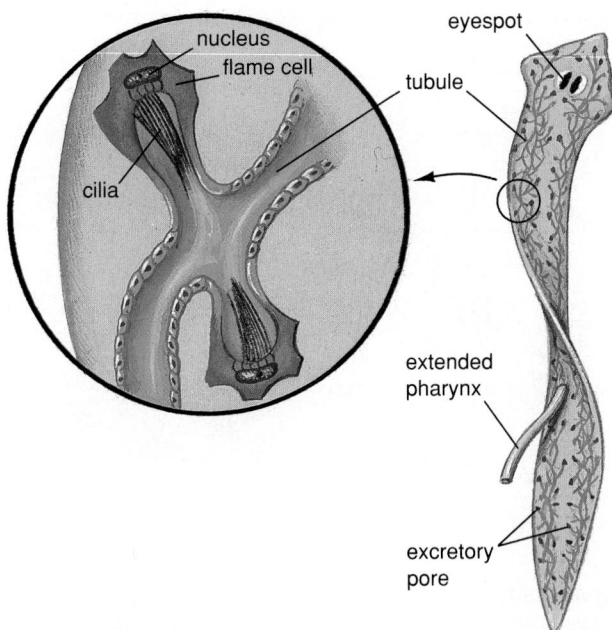

Figure 33-1 The simple excretory system of a flatworm. Hollow flame cells direct excess water and dissolved wastes into a network of tubes. The beating cilia of the flame cells help circulate the fluid to excretory pores.

Simple Excretory Systems

Flame Cells in Flatworms

One of the simplest animals with an excretory system is the flatworm, such as can be found under rocks in streams. The flatworm's excretory system consists of a network of tubes that branch throughout the body (Fig. 33-1). At intervals the tubes end blindly in single-celled bulbs called **flame cells,** named after the tuft of beating cilia extending into the hollow bulb. Under the microscope, the beating of the cilia resembles a flickering flame. Water and some dissolved wastes are filtered into the bulbs, where the beating cilia produce a current that conducts the fluid through the tubular network. Here, waste products are added to the filtrate, and nutrients withdrawn.

Eventually, the waste liquid reaches one of numerous pores that release it to the outside. Flatworms also have a large skin surface through which wastes leave by diffusion.

Nephridia in Earthworms

Earthworms, molluscs, and several other types of invertebrates have simple kidneys called **nephridia** (sing. nephridium). In the earthworm, fluid fills the body cavity, or coelom, that surrounds the internal organs. This coelomic fluid collects both wastes and nutrients from the blood and tissues. The fluid is conducted into the funnel-shaped opening called the **nephrostome** and swept by cilia along a narrow, twisted tube (Fig. 33-2). Here, salts and other dissolved nutrients are absorbed back into the blood, leaving water and wastes behind. The resulting urine is stored in an enlarged bladderlike portion of the nephridium, and then excreted through an opening, the **excretory pore,** in the body wall. The earthworm body is composed of repeating segments, nearly every one of which contains its own pair of nephridia.

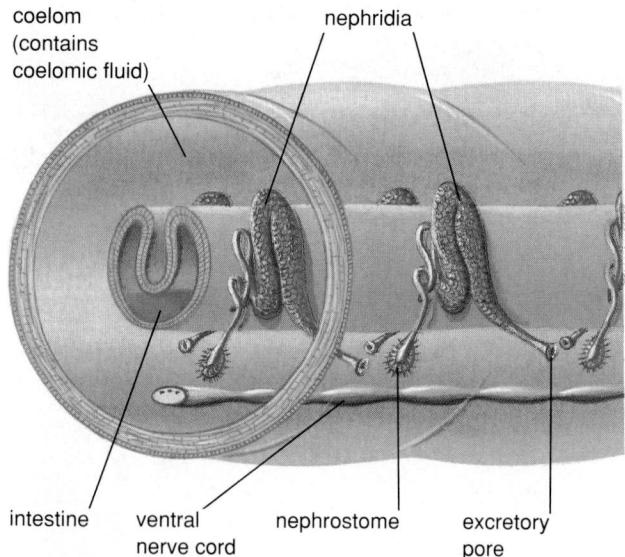

coelom
(contains
coelomic fluid)

nephridia

intestine

ventral
nerve cord

nephrostome

excretory
pore

Figure 33-2 The excretory system of the earthworm consists of structures called nephridia, one pair per segment. Coelomic fluid is drawn into the nephrostome and urine is released through the excretory pore. Each nephridium resembles a vertebrate nephron.

Human Excretion

The Human Excretory System

Humans and other vertebrates filter their blood through **kidneys,** complex organs that in some ways resemble dense collections of nephridia. The kidneys are part of a larger group of structures collectively called the excretory system (Fig. 33-3). While the kidneys actually produce the urine, other portions of the system transport, store, and eliminate it. First, we examine the major structures of the excretory system, tracing the pathway of waste products. Then we explain kidney function in greater detail, and finally we describe the role of the kidneys in homeostasis.

Human kidneys are paired, kidney bean-shaped organs located on either side of the spinal column and extending slightly above the waist. Each is approximately 5 inches (13 cm) long, 3 inches (8 cm) wide, and 1 inch (2.5 cm) thick. Blood carrying dissolved cellular wastes enters each kidney through a **renal artery.** After it has been filtered, the blood exits via the **renal vein** (Fig. 33-4). Urine (consisting of waste substances and water filtered from the blood) leaves each kidney through a narrow, muscular tube called the **ureter.** Using peristaltic contractions, the ureters transport urine to the **bladder.** This hollow,

muscular chamber collects and stores the urine. Urine completes its journey to the outside via the **urethra,** a single narrow tube about 1.5 inches (3.8 cm) long in the female and about 8 inches (20 cm) long in the male. The walls of the bladder, which contain smooth muscle, are capable of considerable expansion. Urine is retained in the bladder by two sphincter muscles located at its base just above the junction with the urethra. When the bladder becomes distended, receptors in the walls signal its condition and trigger reflexive contractions. The sphincter nearest the bladder, the internal sphincter, is opened during this reflex. The lower, or external, sphincter, however, is under voluntary control, so the reflex can be suppressed by the brain unless distension becomes acute. The average adult bladder will hold about a pint (approx. 500 ml) of urine, but the desire to urinate is triggered by accumulations of considerably less than this.

Human Kidney Structure and Function

Kidney Structure

The kidney contains a solid outer layer where urine is formed, and a hollow, inner chamber called the **renal pelvis.** The renal pelvis is a branched collecting chamber that funnels urine into the ureter (see Fig. 33-4). The outer layer of the kidney is divided into an inner **medulla** and an overlying **cortex.** Microscopic examination of these structures reveals an array of tiny individual filters, or **nephrons.** Over one million nephrons are packed into the cortex of each kidney, with many extending into the medulla.

The nephron has three major parts. In the direction of fluid flow, these are the **glomerulus,** which acts as a pressure filter for the blood; **Bowman's capsule,** which collects the filtrate, or fluid filtered from the blood; and a long, twisted **tubule** (Latin, "little tube"). The tubule is further subdivided first into the **proximal tubule,** then the **loop of Henle,** and finally the **distal tubule,** which leads to the collecting duct (Fig. 33-5; also see Fig. E33-1). In the tubule, nutrients are selectively reabsorbed from the filtrate back into the blood, while wastes and some of the water are left behind to form urine. Additional wastes are also secreted into the tubule from the blood. Different portions of the tubule selectively modify the filtrate as it travels through (Fig. 33-6). These processes are examined in the following sections, and in greater detail in "A Closer Look at the Nephron and Urine Formation."

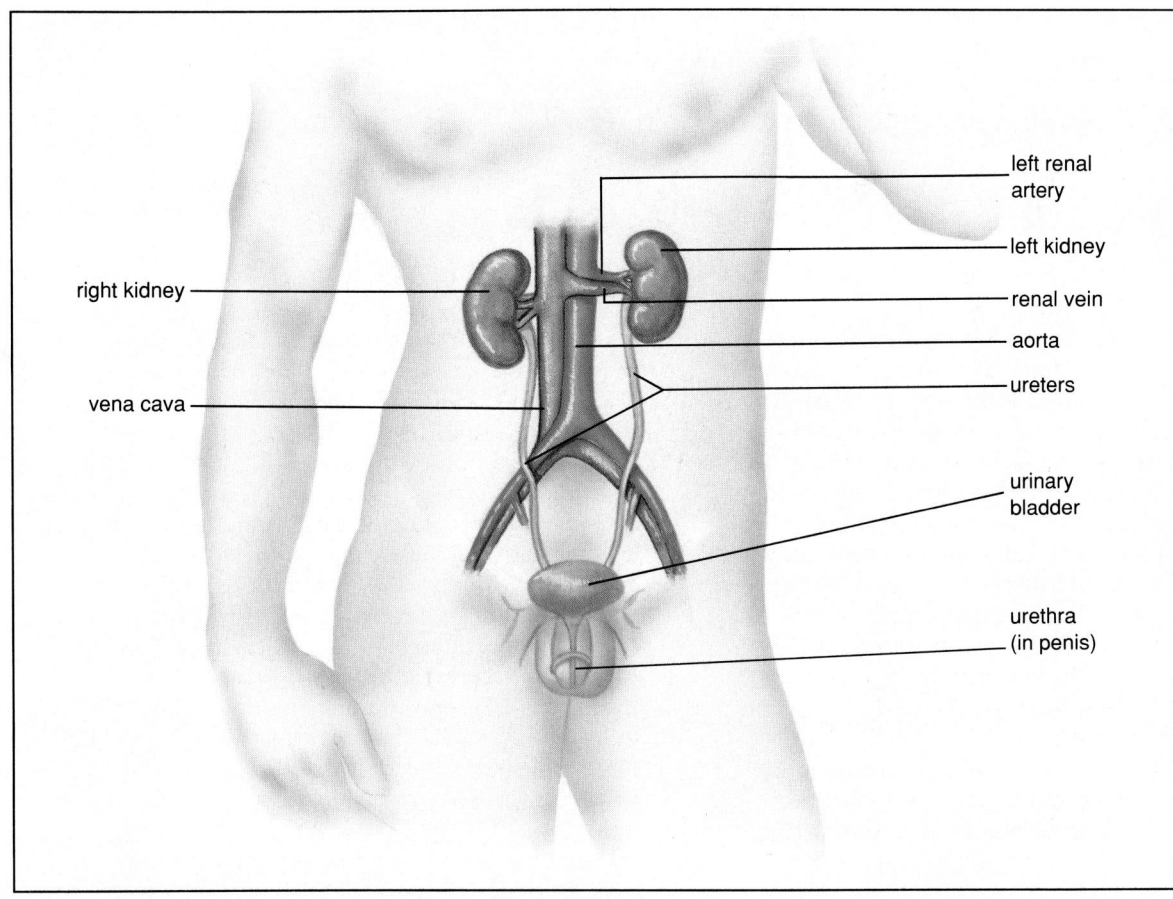

Figure 33-3 The human excretory system and its blood supply.

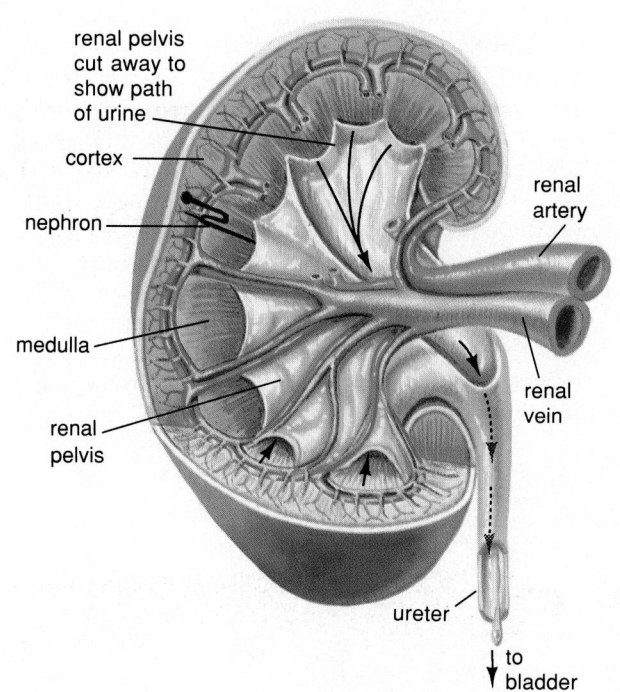

Figure 33-4 Cross section of a kidney, showing its blood supply and gross internal structure. The renal artery, which brings blood to the kidney, and the renal vein, which carries the filtered blood away, branch extensively within the kidney. The two are joined by a highly permeable capillary network through which substances are exchanged between the blood and the nephrons. A nephron, considerably enlarged, is drawn to show its orientation in the kidney.

A CLOSER LOOK
At The Nephron and Urine Formation

The complex structure of the nephron is finely adapted to its function. However, in spite of decades of research, the precise mechanism of urine formation is still controversial. In Figure E33-1, the nephron is presented diagramatically to illustrate the processing that occurs in each part. The numbers refer to the osmotic concentration of the filtrate (the higher the number, the greater the concentration of dissolved substances). The graph to the right shows the concentration of solutes in the surrounding fluid. Notice that the primary solutes are NaCl (salt) and urea, and that their concentrations increase toward the bottom of the loop of Henle. Circled numbers refer to the following descriptions.

(1) **Filtration.** Water and dissolved substances are forced out of the glomerular capillaries into Bowman's capsule, from which they are funneled into the tubule.

(2) **Tubular reabsorption.** In the proximal tubule, most of the important nutrients are actively pumped out through the walls of the tubule and are reabsorbed into the blood. These include about 75% of the salts and water, as well as amino acids, sugars, and vitamins. The proximal tubule is highly permeable to water, so water follows the nutrients, moving out by osmosis along its concentration gradient.

(3) The loop of Henle is unique to birds and mammals, and is essential for urine concentration. The loop of Henle maintains a salt concentration gradient in the extracellular fluid surrounding it, with the highest concentration at the bottom of the loop. The descending portion of the loop of Henle is very permeable to water, but not to salt or other dissolved substances. As the filtrate passes through the descending portion, water leaves by osmosis as the concentration of the surrounding fluid increases.

(4) The thin portion of the ascending loop of Henle is relatively impermeable to water and urea, but is permeable to salt, which moves out of the filtrate by diffusion. Why? Although the osmotic concentrations inside and outside the tubule are about equal, at this stage urea is higher outside, and salt is higher

inside. Thus, the diffusion gradient favors the movement of salt outward. Since water cannot follow it, the filtrate now becomes less concentrated than its surroundings.

(5) The thick portion of the ascending loop of Henle is also impermeable to water and urea. Here, salt is actively pumped out of the filtrate, leaving water and wastes behind.

(6) The watery filtrate, low in salt but retaining wastes such as urea, now arrives at the distal portion of the tubule, where more salt is pumped out. Since this portion is permeable to water, water follows by osmosis. **Tubular secretion** occurs throughout the tubule, but is especially active in the distal portion. Here, substances such as K^+, H^+, NH_3, and some drugs and toxins are actively pumped into the tubule.

(7) By the time the filtrate reaches the collecting duct, very little salt is left and about 99% of the water has been reabsorbed into the bloodstream. The collecting duct conducts the urine down through the increasing concentration gradient created by the loop of Henle. The collecting duct is very permeable to water when the hormone **antidiuretic hormone (ADH)** is present, so water moves out by osmosis as the concentration of the external fluid increases. If ADH is absent, the collecting duct remains impermeable to water, and the urine stays dilute and watery.

(8) The lower portion of the collecting duct is also permeable to urea. So as the filtrate moves further down the collecting duct, some urea diffuses out, contributing to the osmotic concentration of the surrounding fluid. As water (when ADH is present) and urea move out, the concentration in the collecting duct of dissolved wastes such as urea gradually approaches equilibrium with the high osmotic concentration of the external fluid.

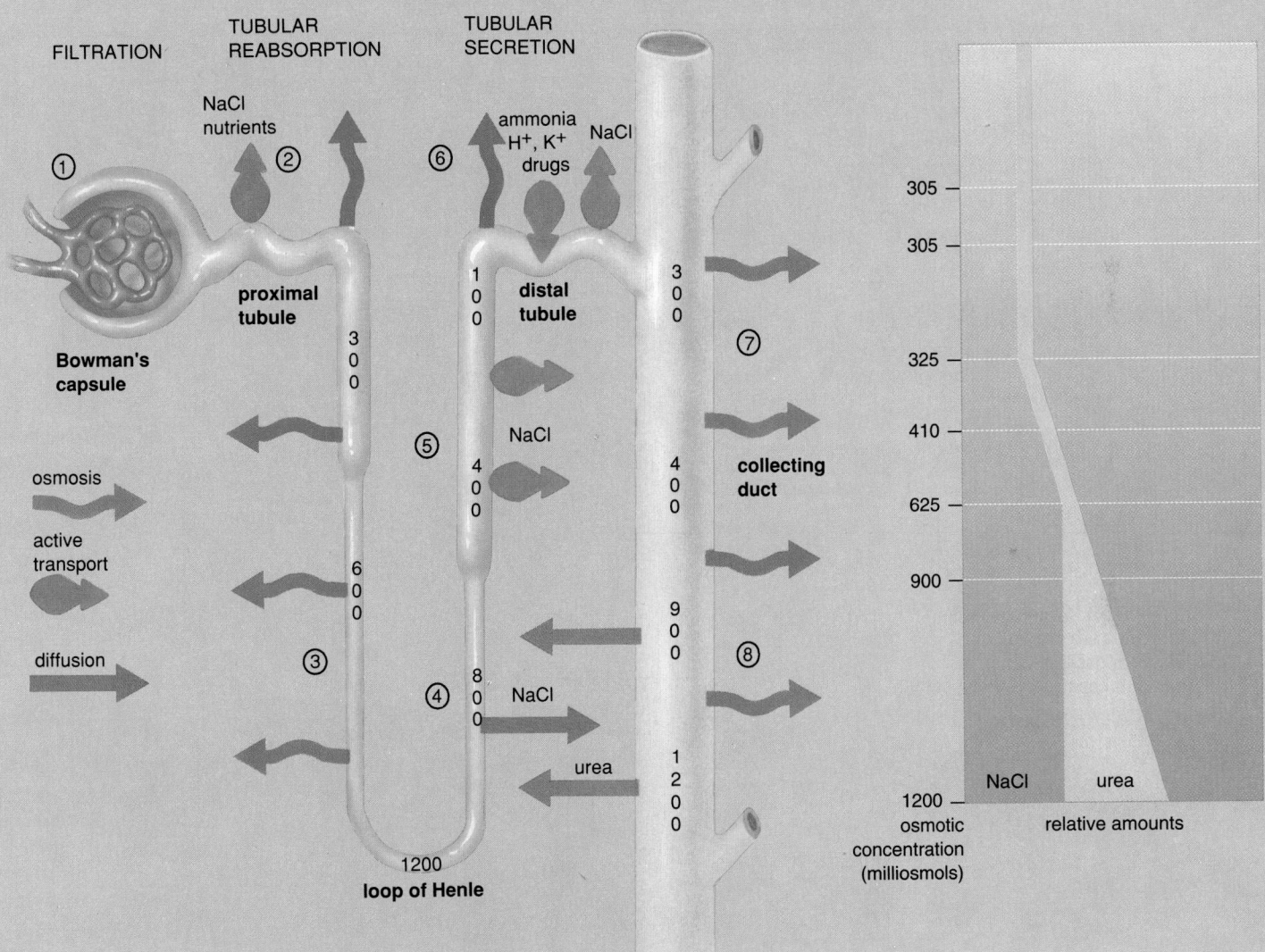

Figure E33-1 Diagram of a single nephron showing the movement of materials through different regions.

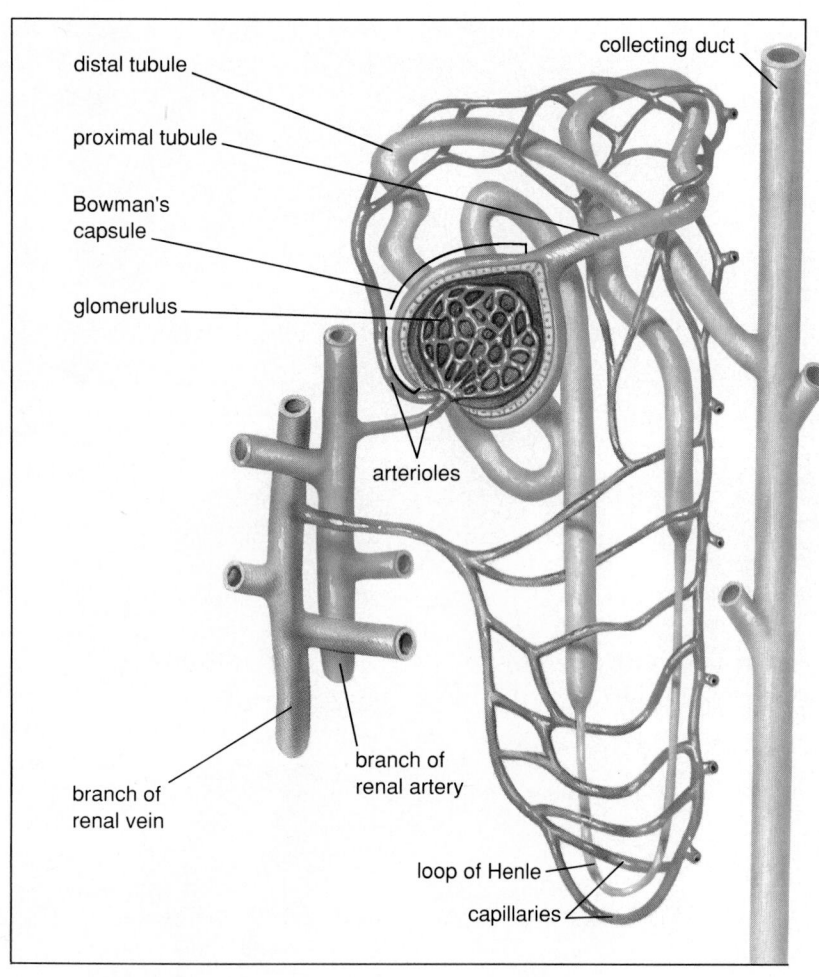

distal tubule

proximal tubule

Bowman's capsule

glomerulus

arterioles

branch of renal vein

branch of renal artery

collecting duct

loop of Henle

capillaries

Figure 33-5 An individual nephron and its blood supply.

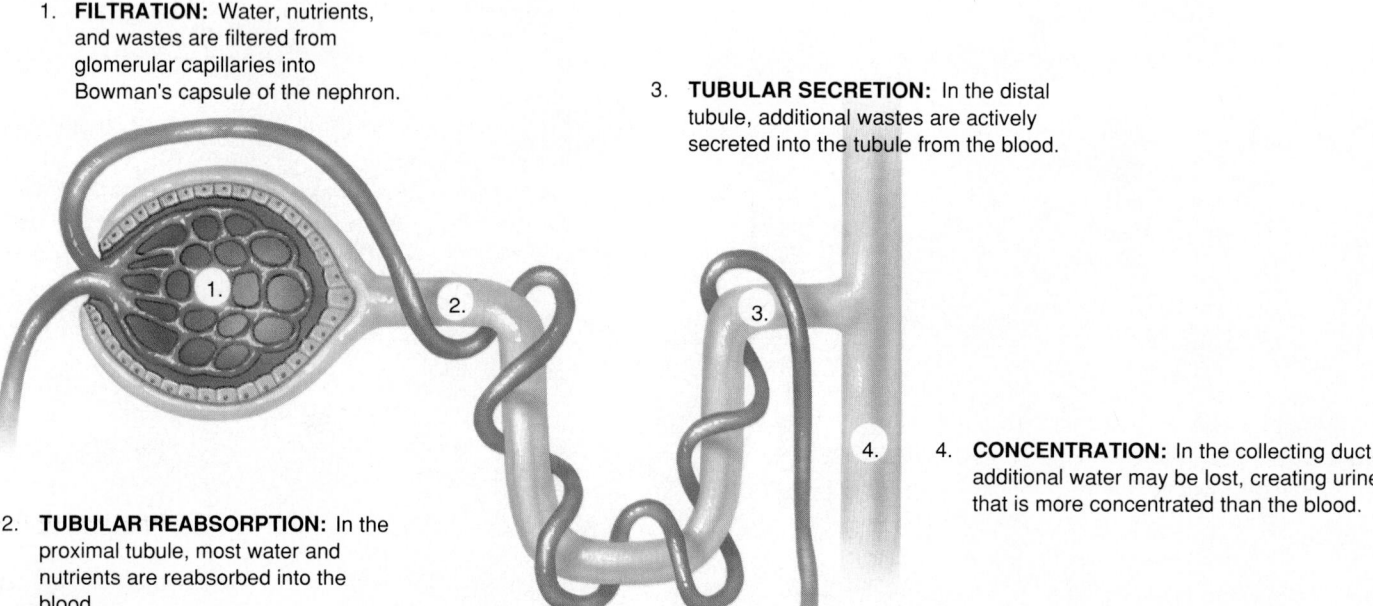

1. **FILTRATION:** Water, nutrients, and wastes are filtered from glomerular capillaries into Bowman's capsule of the nephron.

2. **TUBULAR REABSORPTION:** In the proximal tubule, most water and nutrients are reabsorbed into the blood.

3. **TUBULAR SECRETION:** In the distal tubule, additional wastes are actively secreted into the tubule from the blood.

4. **CONCENTRATION:** In the collecting duct, additional water may be lost, creating urine that is more concentrated than the blood.

Figure 33-6 A summary of events that occur during the formation of urine in the nephron.

Filtration by the Glomerulus

Blood is conducted to each nephron by an arteriole that branches from the renal artery. Within a cup-shaped portion of the nephron—Bowman's capsule—the arteriole subdivides into numerous microscopic capillaries that form an intertwined mass, the glomerulus (see Fig. 33-5). The walls of the glomerular capillaries are extremely permeable to water and dissolved substances. Past the glomerulus, the capillaries reunite to form an arteriole whose diameter is smaller than the incoming arteriole. The differences in diameter between the incoming and outgoing arterioles creates pressure within the glomerulus, driving water and many of the dissolved substances from the blood through the capillary walls. This process is called **filtration** (see Fig. 33-6). The watery filtrate, resembling blood plasma minus its proteins, is collected in Bowman's capsule for transport through the nephron. With the filtrate removed, the blood in the arteriole leaving the glomerulus is now very "concentrated." It retains substances too large to pass through the glomerular capillary walls, such as blood cells, large proteins, and fat droplets. Past the glomerulus, the arteriole branches into smaller, highly porous, capillaries. These capillaries surround the tubule, forming intimate contacts with it. Here water and nutrients are reabsorbed from the filtrate as it passes through the nephron, and returned to the blood.

Urine Formation in the Nephron

The blood filtrate collected in Bowman's capsule contains a mixture of both wastes and essential nutrients, including most of the blood's vital water. The nephron restores the nutrients and most of the water to the blood, while retaining wastes for elimination. This is accomplished by two processes: tubular reabsorption and tubular secretion (see Fig. 33-6).

Tubular reabsorption is the process by which cells of the tubule remove water and nutrients from the filtrate within the tubule and pass them back into the blood. Reabsorption of salts and other nutrients such as amino acids and glucose generally occurs by active transport (that is, cells of the tubule expend energy to transport these substances out of the tubule). These nutrients then enter adjacent capillaries by diffusion. Water is reabsorbed passively, following the nutrients out of the tubule in response to the osmotic gradient created in the fluid surrounding the tubule. Wastes such as urea (a product of protein digestion; see "A Closer Look at Excretion and Animal Lifestyle") remain in the tubule and become more concentrated as water leaves.

Tubular secretion is the process by which waste substances remaining in the blood are actively secreted *into* the tubule by tubule cells. These substances include hydrogen and potassium ions, as well as various foreign substances and drugs such as penicillin. During tubular secretion, excess substances that were not initially filtered out into Bowman's capsule are removed from the blood for excretion.

The Production and Concentration of Urine

The kidneys of mammals and birds are able to produce urine that is more concentrated (has a higher osmotic pressure) than the blood. The ability to concentrate urine is a result of the structure of both the nephron and the **collecting duct** into which several nephrons empty. Urine can become concentrated because there is an osmotic concentration gradient of salts and urea in the fluid surrounding the loop of Henle. This gradient is produced by the loop of Henle, and the longer the loop, the greater the concentration gradient. The most osmotically concentrated fluid, which is far more concentrated than blood, surrounds the bottom of the loop. The collecting duct passes through this osmotic gradient. As the filtrate passes through the portion of the collecting duct surrounded by the osmotically concentrated fluid, additional water leaves the filtrate by osmosis. So as the filtrate, now called urine, moves through the collecting duct, it can become as concentrated as the surrounding fluid. Since the rest of the excretory system does not allow water to enter or urea to escape, the urine remains concentrated.

It is important to produce concentrated urine when water is scarce (see "A Closer Look at Excretion and Animal Lifestyle"). However, when there is excess water in the blood, a dilute, watery urine is formed. The degree of concentration of the urine is regulated by a hormone called **antidiuretic hormone,** which regulates the permeability of the collecting duct to water. This determines how much water leaves the filtrate and how concentrated the urine becomes, as described below.

The Kidneys as Organs of Homeostasis

Each drop of blood in your body passes through a kidney about 350 times daily; thus, the kidney is able to "fine tune" the composition of the blood. The importance of this task is illustrated by the fact that kidney failure is rapidly fatal (see "Health Watch: When the Kidneys Collapse").

CLOSER LOOK
At Excretion and Animal Lifestyle

Most animals consume far more amino acids than they require for the synthesis of new proteins. Since these excess amino acids can't be stored, they are broken down to provide energy or converted to storable fats or carbohydrates. In each of these cases, the amino group ($-NH_2$) must be removed from the amino acid. This process occurs in the liver, and the initial waste product formed from the amino group is a highly poisonous substance: **ammonia** (NH_3). Because ammonia is so toxic, it must be excreted immediately or converted to a less toxic substance such as **urea** or **uric acid.** How this is done by different animals is intimately related to their environment. Aquatic animals—freshwater fish, certain amphibians, and many invertebrates—release ammonia into the water as it is formed. Land-dwellers, on the other hand, cannot afford to waste water by producing urine continuously, as is necessary for the excretion of ammonia. Instead, they convert the ammonia to uric acid or urea, substances that may be stored and concentrated.

All mammals produce urea, and since urea is highly soluble in water, some water must be excreted along with the urea. Mammalian kidneys are finely adapted to the availability of water. Mammals that must conserve water do so by producing hypertonic urine—that is, urine more concentrated than their blood. The degree of concentration that can be achieved is determined by the length of the loop of Henle. The longer the loop, the higher the salt concentration in the fluid surrounding it. The higher the salt concentration, the greater the urine concentration. As you might predict, animals living in very dry climates who have an urgent need to conserve their body water have the longest loops of Henle. In contrast, those in watery environments have relatively short loops. The beaver, for example, has all short-looped nephrons, and is unable to concentrate its urine to more than twice its plasma concentration. Human kidneys have a mixture of long- and short-looped nephrons and can concentrate urine to about four times the plasma concentration. The masters of urine concentration are desert rodents such as kangaroo rats, which can produce urine 14 times their plasma concentration (Fig. E33-2). Kangaroo rats (as you might predict) have only very long-looped nephrons. Because of their unique

Figure E33-2 The desert kangaroo rat of the southwestern United States can dispense with drinking partly because its long loops of Henle allow it to produce very concentrated urine.

Figure E33-3 This green sea turtle shows tearlike deposits caused by discharge of its salt glands.

ability to conserve water, they can completely dispense with drinking, relying entirely on water derived from their food.

Birds (who are not very efficient at concentrating their urine), reptiles, and insects produce uric acid. Uric acid is a white crystalline substance that is relatively insoluble in water and is excreted as a paste. This has three potential advantages. First, all of these animals lay eggs. The embryo, by depositing uric acid as insoluble crystals, avoids stewing in its own wastes. Second, the production of nearly dry urine allows flying insects and birds to avoid carrying the weight of extra water. Third, since much of the uric acid is not in solution, the urine can contain a great deal of it and still not be hypertonic to the blood. This is very important to reptiles, who lack a loop of Henle and are unable to produce hypertonic urine. By excreting uric acid in crystalline form, they are both producing a concentrated nitrogenous waste and conserving water. This ability helps reptiles thrive in dry desert environments.

Marine reptiles such as sea turtles, sea snakes, and marine iguanas, and birds such as seagulls have another problem: they take in excessive salt from their diet and their surroundings. Since they cannot produce concentrated urine, these groups have evolved other excretory organs that efficiently eliminate salt. These organs are located in the head, and contain cells that secrete salt by active transport. The highly concentrated salt solution may drain out through the nasal passages or from the corners of the eyes. Sea turtles may be seen "crying crocodile tears" as these glands discharge their contents near the eye (Fig. E33-3). Ironically, the salt glands of crocodiles discharge under the tongue, so the origin of this colorful expression is obscure.

Water Balance

One of the most important functions of the kidney is to regulate the water content of the blood (Fig. 33-7). Human kidneys filter about a half a cup (125 mL) of fluid from the blood each minute. This means that, without reabsorption of water, you would produce over 45 gallons (180 L) of urine daily! Water reabsorption occurs passively by osmosis as the filtrate travels through the tubule and the collecting duct.

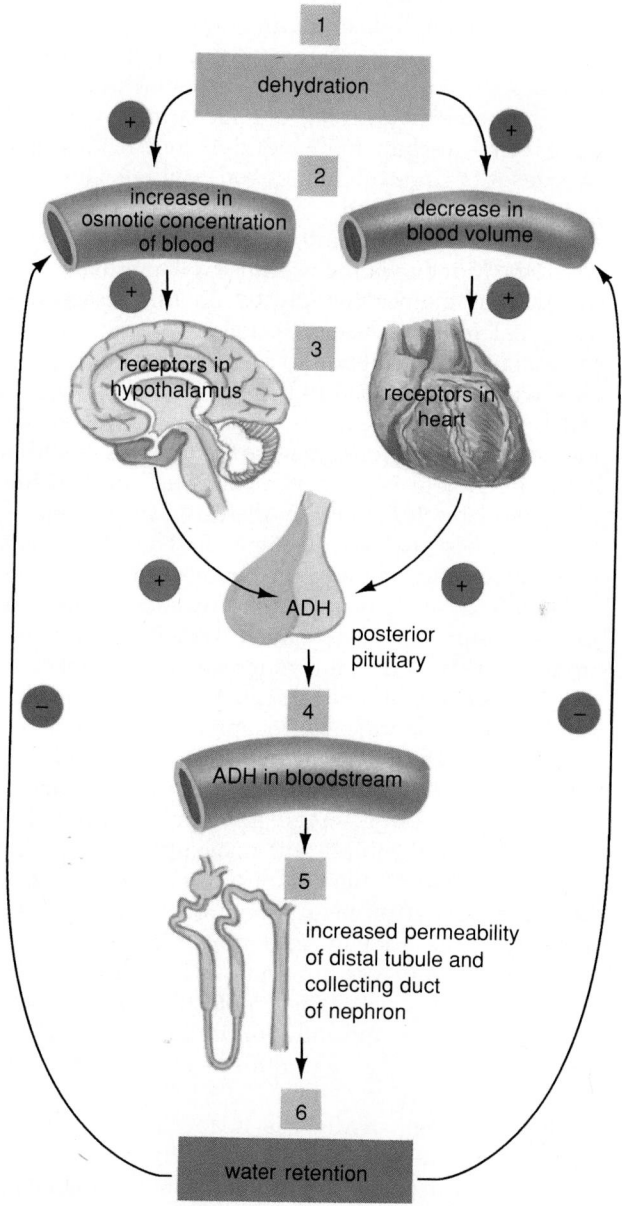

Figure 33-7 Regulation of the water content of the blood is hormonally controlled by ADH through a negative feedback process. Stimulation is indicated by (+), reduction by (−). Dehydration, in addition to stimulating ADH release, triggers the sensation of thirst, leading to increased water intake. This is necessary to restore blood volume and diminish ADH secretion.

HEALTH WATCH
When the Kidneys Collapse

The kidney is a relatively rugged organ. Normally it gives no indication of its industrious activity other than filling the bladder with urine. But if kidney function is eliminated, death occurs rapidly. The kidneys are vulnerable to attack from several sources. Toxins can destroy kidney tubules. These include heavy metals (mercury, arsenic), organic compounds such as solvents (carbon tetrachloride), insecticides, and overdoses of some antibiotics and pain relievers such as ibuprofen and aspirin. A second source of attack is bacteria, particularly intestinal bacteria that may reach the kidney via the urethra. Still another source of injury is glomerulonephritis. In this disease, an abnormal immune response to an infection causes the antigen and antibody to fuse into insoluble particles, which are trapped in the glomeruli, resulting in inflammation and death of the tubules. The damage may be temporary or it may be permanent. In either case, the patient is often treated using a remarkable piece of technology: kidney dialysis (Fig. E33-4).

The kidney dialysis machine, first used in 1945, operates on the simple principle of diffusion of substances from areas of high concentration to areas of low concentration. The patient's blood is diverted from the body, and run through small tubes made of a cellophane membrane, and suspended in dialyzing fluid. This semipermeable cellophane membrane has pores too small to permit the passage of blood cells and large proteins, but is permeable to small molecules such as water, sugar, salts, amino acids, and urea. To retain the important ions, sugars, and amino acids in the blood, the composition of the dialyzing fluid is adjusted to have normal quantities of these substances. As a result, only molecules whose concentration is in excess of normal will diffuse into the dialyzing fluid. Urea, a small molecule that is present in relatively high concentration in the blood and absent in the dialyzing fluid, diffuses rapidly across the membrane out of the blood.

One challenge of dialysis is to eliminate excess water. Since solutes are higher in the blood, water tends to diffuse *into* the blood from the dialyzing fluid. This process is counteracted by applying added pressure to the blood, which forces some of the small water molecules out of the blood against their concentration gradient.

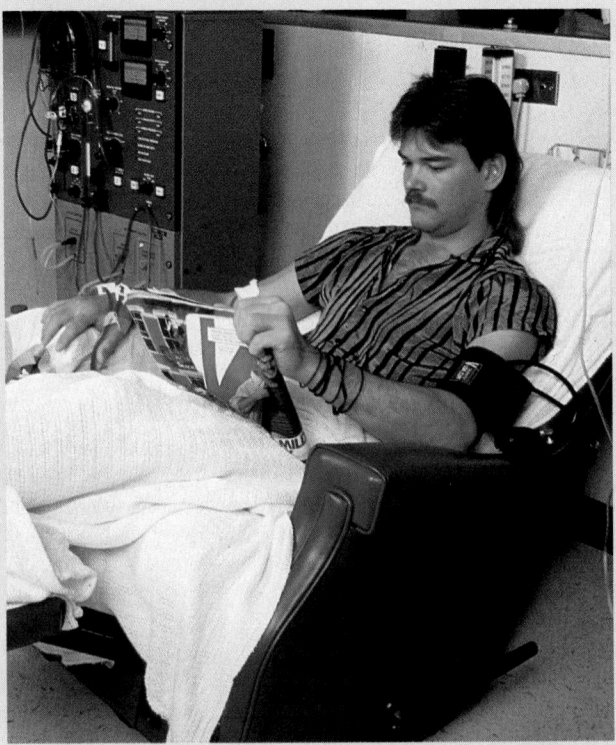

Figure E33-4 A patient undergoing kidney dialysis.

Roughly a pint (500 mL) of blood is in the machine at any one time, flowing at a rate of several hundred milliliters per minute. Still, the patient must remain attached to the dialysis machine for 4 to 6 hours three times a week. Although patients requiring dialysis have remained alive for up to 20 years, their lives are far from normal. Their diet and fluid intake must be carefully regulated. Between visits, their blood composition fluctuates and far more toxins accumulate than in the blood of a person with normal kidneys. In spite of its drawbacks, about 60,000 Americans are kept alive by dialysis. For those who are fortunate enough to find a compatible donor, kidney transplants are a far better solution; about 20,000 of these are performed yearly in the United States.

How much water is reabsorbed into the blood is controlled by the amount of **antidiuretic hormone** (ADH; also called vasopressin) circulating in the blood. This hormone increases the permeability of the distal tubule and the collecting duct to water, allowing more water to be reabsorbed from the urine. ADH is produced by cells in the hypothalamus, and is released by the posterior pituitary gland (see Chapter 35). ADH release is regulated by receptor cells in the hypothalamus that monitor the osmotic concentration of the blood, and by receptors in the heart that monitor blood volume. For example, as the lost traveller staggers through the searing desert sun, dehydration occurs. The osmotic concentration of his blood rises, and his blood volume falls, triggering the release of more ADH (Fig. 33-7). This increases water reabsorption, and produces urine more concentrated than the blood. In contrast, a partygoer overindulging in beer will experience a decrease in blood concentration and an increase in blood volume, and her receptors will cause a decrease in ADH output. Reduced ADH concentration will make the distal tubule and collecting duct less permeable to water. When ADH is very low, little water is reabsorbed after the urine leaves the loop of Henle, and the urine produced will be more dilute than the blood. In extreme cases, urine flow may exceed 1 quart (about 1 L) per hour. As the proper water level is restored, the increased osmotic concentration of the blood and decreased blood volume will stimulate increased ADH production. This is an example of a negative feedback system, which regulates the levels of most hormones (see Chapters 25 and 35).

Regulation of Dissolved Substances

As the kidney filters the blood, it monitors and regulates blood composition to maintain a constant internal environment. Substances regulated by the kidney, in addition to water, include nutrients such as glucose, amino acids, vitamins, urea, and a variety of ions including sodium, potassium, chloride, and sulfate. The kidney maintains a constant blood pH by regulating the content of hydrogen and bicarbonate ions. This remarkable organ also eliminates potentially harmful substances, including some drugs, food additives, pesticides, and some of the toxins produced by cigarette smoke.

SUMMARY OF KEY CONCEPTS

The digestive system plays a crucial role in homeostasis, the precise regulation of the composition of body fluids. The kidney is responsible for the excretion of cellular wastes such as urea, excess nutrients, hormones, water, and certain drugs and toxins.

Simple Excretory Systems

The simple excretory system of the flatworm consists of a network of tubules that branch through the body. Flame cells circulate body fluid through the tubules, where nutrients are reabsorbed. Wastes, including excess water, are excreted through numerous excretory pores.

Many of the more complex invertebrates, including earthworms and molluscs, utilize nephridia. In the earthworm, these are paired structures resembling vertebrate nephrons that are found in most of the earthworm's segments. Coelomic fluid is drawn into a ciliated opening, the nephrostome, and nutrients and water are reabsorbed. Wastes and excess water are released through the excretory pore.

Human Excretion

The human excretory system consists of elaborate blood filters called kidneys, and other structures that transport and store the urine. The ureters conduct urine from the kidneys to a distensible storage organ, the bladder. Distension of the muscular bladder wall triggers urination, a reflex that is also under voluntary control. From the bladder, urine traverses the urethra, a single narrow tube that opens to the outside.

Each kidney consists of over a million individual nephrons in an outer cortex, with many extending into an inner layer, the medulla. Urine formed in the nephrons enters collecting ducts that empty into the renal pelvis, a central chamber. From the renal pelvis, urine is funneled into the ureter. Each nephron is served by an arteriole that branches from the renal artery. The arteriole further branches into a mass of capillaries called the glomerulus. Here water and dissolved substances are filtered from the blood by pressure, which forces them through the porous capillary walls. The glomerulus is surrounded by a cuplike portion of the nephron called Bowman's capsule, which collects the filtrate From Bowman's capsule, the filtrate is conducted along the tubular portion of the nephron. During its transit, nutrients are actively pumped out of the filtrate through the walls of the tubule. Nutrients then enter capillaries that surround the tubule, and water follows by osmosis. Wastes and excess water remain in the filtrate. The tubule forms the loop of Henle, which creates a salt concentration gradient surrounding it. After completing its passage through the tubule, the filtrate enters the collecting duct, which passes through the concentration gradient. Final passage of the filtrate through this gradient via the collecting duct allows concentration of the urine.

The kidneys are important organs of homeostasis. The water content of the blood is regulated by antidiuretic hormone (ADH), produced in the hypothalamus and released by the posterior pituitary gland. Low blood volume and high osmotic concentration of the blood signal dehydration and stimulate release of ADH into the bloodstream. ADH increases the permeability to water of the distal tubule and the collecting duct, allowing more water to be reabsorbed into the blood. In addition to its role in water balance, the kidneys also control blood pH, remove toxins, and regulate ions such as sodium, chloride, potassium, and sulfate. Excess glucose, vitamins, and amino acids are also excreted by the kidneys.

GLOSSARY

ammonia: a highly toxic nitrogen-containing waste product of amino acid breakdown, which is converted to urea in the mammalian liver.

antidiuretic hormone (an-tē-dī-ūr-et'-ik): also called ADH; a hormone produced by the hypothalamus and released into the bloodstream by the posterior pituitary gland. It acts on the nephron of the kidney and causes more water to be reabsorbed into the bloodstream.

bladder: a muscular storage organ for urine.

Bowman's capsule: the portion of the nephron in which blood filtrate is collected from the glomerulus.

filtration: within Bowman's capsule in each nephron of a kidney, the process by which blood is pumped under pressure through permeable capillaries of the glomerulus, forcing out water, dissolved wastes, and nutrients.

flame cell: a specialized cell containing beating cilia that conducts water and wastes through the branching tubes that serve as an excretory system in flatworms.

glomerulus (glō-mer'-ū-lus): a dense network of thin-walled capillaries located within the Bowman's capsule of each nephron. Here blood pressure forces water and dissolved nutrients through capillary walls for filtration by the nephron.

homeostasis (hōm-ē-ō-stā'sis): the precise regulation of the composition of fluid bathing the body cells. The relatively constant environment required for optimal functioning of cells is maintained by the coordinated activity of numerous regulatory mechanisms, including the respiratory, endocrine, circulatory, and excretory systems.

kidney: one of a pair of organs of the excretory system located on either side of the vertebral column. Kidneys filter blood, removing wastes and regulating the ionic composition and water content of the blood.

loop of Henle (hen'-lē): a specialized portion of the tubule of the nephron in birds and mammals that creates an osmotic concentration gradient in the fluid immediately surrounding it. This in turn allows the production of urine more osmotically concentrated than blood plasma.

nephridium (nef-rid'-ē-um): a type of excretory organ found in earthworms, molluscs, and certain other invertebrates. A nephridium somewhat resembles a single vertebrate nephron.

nephron (nef'-ron): the functional unit of the kidney, where blood is filtered and urine formed.

tubular reabsorption: the process by which cells of the tubule of the nephron remove water and nutrients from the filtrate within the tubule and return them to the blood.

tubular secretion: the process by which cells of the tubule of the nephron remove additional wastes from the blood, actively secreting them into the tubule.

tubule (tūb'-ūle): the tubular portion of the nephron. It includes a proximal portion, the loop of Henle, and a distal portion. Urine is formed from the blood filtrate as it passes through the tubule.

urea (ū-rē'-uh): a water-soluble, nitrogen-containing waste product of amino acid breakdown that is one of the principal components of mammalian urine.

ureter (ū'-re-tur): a tube that conducts urine from each kidney to the bladder.

urethra (ū-rē'-thruh): a tube that conducts urine from the bladder to the outside of the body.

uric acid (ūr'-ik acid): a nitrogen-containing waste product of amino acid breakdown, which is a relatively insoluble white crystal. Uric acid is excreted by birds, reptiles, and insects.

STUDY QUESTIONS

1. Trace an amino acid from the renal artery to the renal vein.
2. Trace a urea molecule from the bloodstream to the external environment.
3. What is the function of the loop of Henle? The collecting duct? Antidiuretic hormone?
4. List two phyla whose members produce uric acid as a waste product and describe the advantages for each.
5. Why don't humans excrete ammonia?
6. Would you predict that the loop of Henle would be longer in a river otter or in a jackrabbit? Explain your answer.

DISCUSSION QUESTIONS

1. Discuss the differences in function of the two major capillary beds in the kidneys, namely the glomerulus capillaries and those around the tubules.
2. Desert animals need to conserve water. These animals have larger kidneys than animals that live in moist environments and thus need not conserve water. The larger kidneys allow for a greater distance between the glomerulus and the bottom of the loop of Henle. Discuss why this anatomical difference assists water conservation in the desert animals.
3. Some "quick weight loss" diets require ingestion of much protein-rich food and elimination of carbohydrates. Two side-effects of such diets are increased thirst and increased urination. Explain the connections between the diets and the side-effects.

SUGGESTED READINGS

Eckert, R., Randall, D., and Augustine, G. *Animal Physiology: Mechanisms and Adaptations.* Third edition. New York: W. H. Freeman and Company, Publishers, 1988. An excellent and complete textbook of comparative animal physiology.

Hole, Jr., J. W. *Essentials of Human Anatomy and Physiology.* Second edition. Dubuque, Iowa: Wm. C. Brown Company, 1986. Excellent illustrations.

34

Defenses Against Disease: The Immune Response

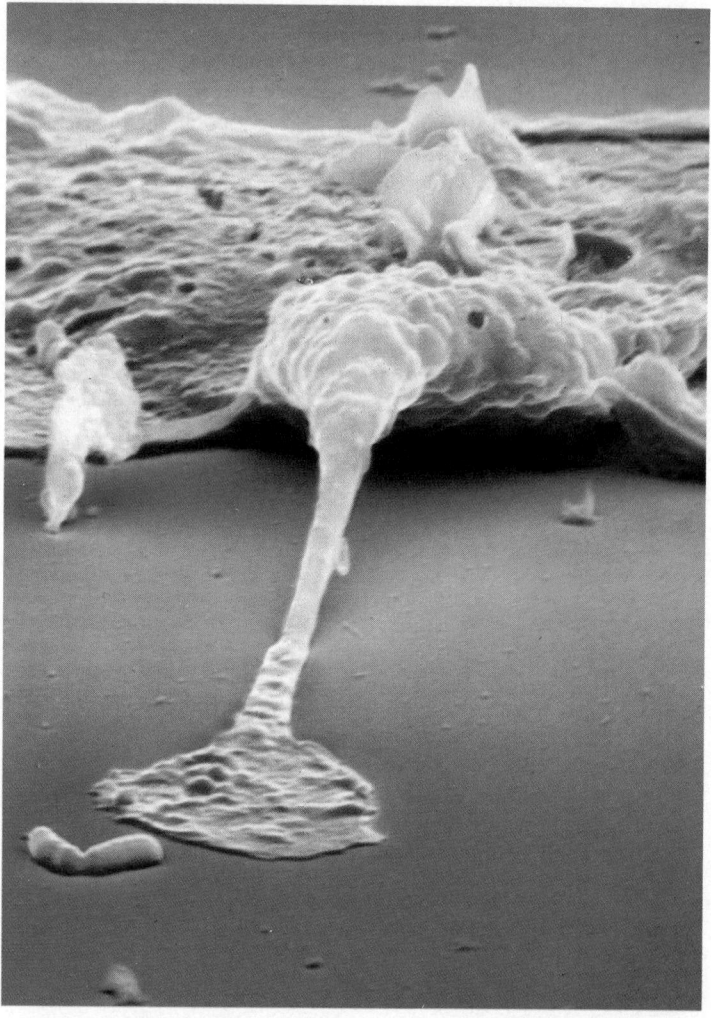

The drama reproduced in this culture dish is played out many times each second in your body. A macrophage (a type of phagocytic white blood cell, tinted gray in this micrograph) sends out a pseudopod to capture a bacterium (brown).

"If you wish to be well and keep well, take Braggs Vegetable Charcoal and Charcoal Biscuits. Absorbs all impurities in the stomach and bowels, effectually warding off cholera, smallpox, typhoid, and all malignant fevers. Eradicate worms in children. Sweeten the breath."

from an early 1900s newspaper ad

As incubators for microbial growth, human bodies are nearly ideal. They maintain a constant temperature of 37° C (98.6° F), and their cells contain abundant water and nutrients. Trillions of microbes lurk out there—in the air, in the water, in the foods we eat, and on the objects we touch—ready to colonize our bodies. Not even Braggs Vegetable Charcoal can protect us from all of them. Nevertheless, day after day we shrug off the assaults of these miniature parasites, usually completely unaware of them. Even when an infection does occur, we almost always fend it off after a few days. How does the human body keep from becoming a warm, moist incubator for microbes? If infected, how does it destroy its invaders?

Defenses Against Microbial Invasion

The human body has three basic mechanisms that defend it against microbial attack (Fig. 34-1): (1) external barriers to keep microbes out of the body; (2) nonspecific internal defenses that combat all invading microbes; and (3) the immune response that directs its assault against specific microbes.

Barriers to Entry

The first, and obviously best, defense is to keep microbes out of the body in the first place. The human body has two surfaces exposed to the environment: the **skin** and the **mucous membranes** of the digestive and respiratory tracts. These surfaces are barriers to microbial entry.

The Skin

The skin is both a physical barrier to microbial entry and an inhospitable environment for microbial growth. The outer surface of the skin consists of dry, dead cells filled with horny proteins similar to those in hair and nails. Consequently, most microbes that land on the skin cannot obtain the water and nutrients they need. Secretions from sweat glands and oil-producing sebaceous glands also cover the skin. These secretions contain acids and natural antibiotics, such as lactic acid, that inhibit the growth of bacteria and fungi. These multiple defenses make the unbro-

ken skin an extremely effective barrier against microbial invasion.

Mucous Membranes

The membranes of the digestive and respiratory tracts are also well defended. First, they secrete mucus that contains antibacterial enzymes such as lysozyme, which destroys bacterial cell walls. Second, the mucus physically traps microbes that enter the body through the nose or mouth (Fig. 34-2). Cilia on the membranes sweep up the mucus, microbes and all, until it is either coughed or sneezed out of the body, or swallowed. If microbes are swallowed, they enter the stomach, where they encounter a combination of extreme acidity (about pH 2) and protein-digesting enzymes, which can kill many types of microbes. Further along in the digestive tract, the intestine is inhabited by bacteria that are harmless to

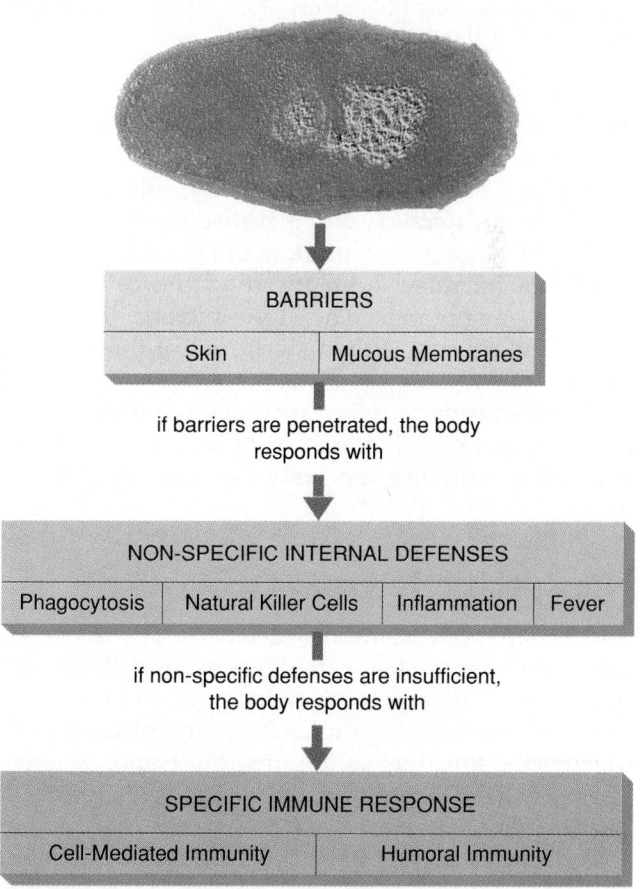

Figure 34-1 Levels of defense against infection.

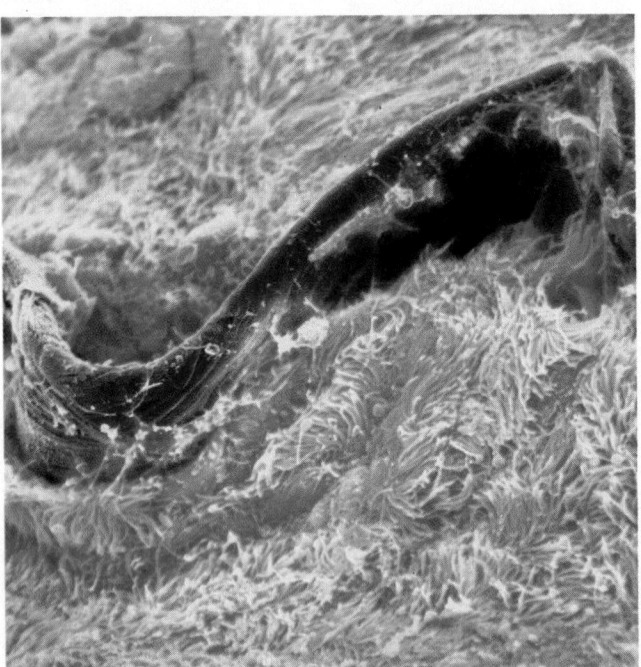

0.5 micrometers

Figure 34-2 Mucus traps microbes and debris in the respiratory tract (a brown strand of dirt is shown caught in mucus atop the red cilia). The cilia lining the walls of the respiratory tract then sweep both mucus and foreign matter out of the body.

the human body, but that secrete substances that destroy invading foreign bacteria or fungi.

Finally, a specific class of antibody, called IgA, resides in the respiratory and digestive tracts (see Table 34-2). IgA binds to the surfaces of microbes and prevents the microbes from invading the cells of the mucous membrane. The IgA–microbe complex, trapped in mucus, is then swept out of the body by ciliary action. Despite these defenses, the warm, moist mucous membranes are much more vulnerable than the dry, oily skin, and many disease organisms succeed in entering the body through these membranes.

Nonspecific Internal Defenses

Three nonspecific internal defenses are mustered against microbes that penetrate through these surfaces. These defenses are nonspecific because they attack a wide variety of microbes, rather than targeting specific invaders as the immune response does. The body has a standing army of **phagocytic cells** that destroy microbes and **natural killer cells** that destroy cells of the body that have been infected by viruses. The steady trickle of microbes that pass through the body's external barriers are mostly mopped up by

these cells. An injury, with its combination of tissue damage and relatively large-scale microbial influx, provokes an **inflammatory response.** The inflammatory response simultaneously recruits new members of the army of phagocytic cells and killer cells and walls off the injured area, isolating the infected tissue from the rest of the body. Finally, if a population of microbes succeeds in establishing a major infection, the body often produces a **fever,** which both slows down microbial reproduction and enhances the body's own fighting abilities.

Eaters and Killers

The body contains several types of amoeboid cells that can engulf and digest microbes. The most important of these are the **macrophages** (literally, "big eaters"), white blood cells that crawl around in the extracellular fluid. Macrophages ingest microbes by phagocytosis. As we will see shortly, besides the immediate effect of destroying an individual microbe, macrophages also play a crucial role in the immune response by "presenting" parts of the microbe to other cells of the immune system.

Natural killer cells are another class of white blood cells. Generally, natural killer cells do not directly attack invading microbes. Instead, natural killer cells strike at the body's own cells that have been invaded by viruses. Virus-infected cells bear some viral proteins on their surfaces. Natural killer cells recognize these proteins and kill the infected cell. In some way, natural killer cells also recognize and kill cancerous cells. Natural killer cells do not eat their victims; they strike from the outside (Fig. 34-3). Their weapons are proteins that they secrete onto the plasma membrane of the infected or cancerous cell. The proteins insert themselves into the target plasma membrane in a ring, much like making a barrel out of individual staves of wood, thereby opening up a large pore in the membrane. Shot full of holes, the target cell soon dies.

The Inflammatory Response

Large-scale breaches of the skin or mucous membranes, such as a cut or scrape, elicit an **inflammatory response.** Damaged cells release the chemical **histamine** into the wounded area. Histamine makes capillary walls leaky and relaxes the smooth muscle surrounding arterioles, leading to increased blood flow. With extra blood flowing through leaky capillaries, fluid seeps from the capillaries into the tissues around the wound. The wound becomes red, swollen, and warm ("inflammation" literally means "to set on fire"). Meanwhile, other chemicals released by injured cells initiate blood clotting (see Chapter 30), which "walls off" the wounded area from the rest of

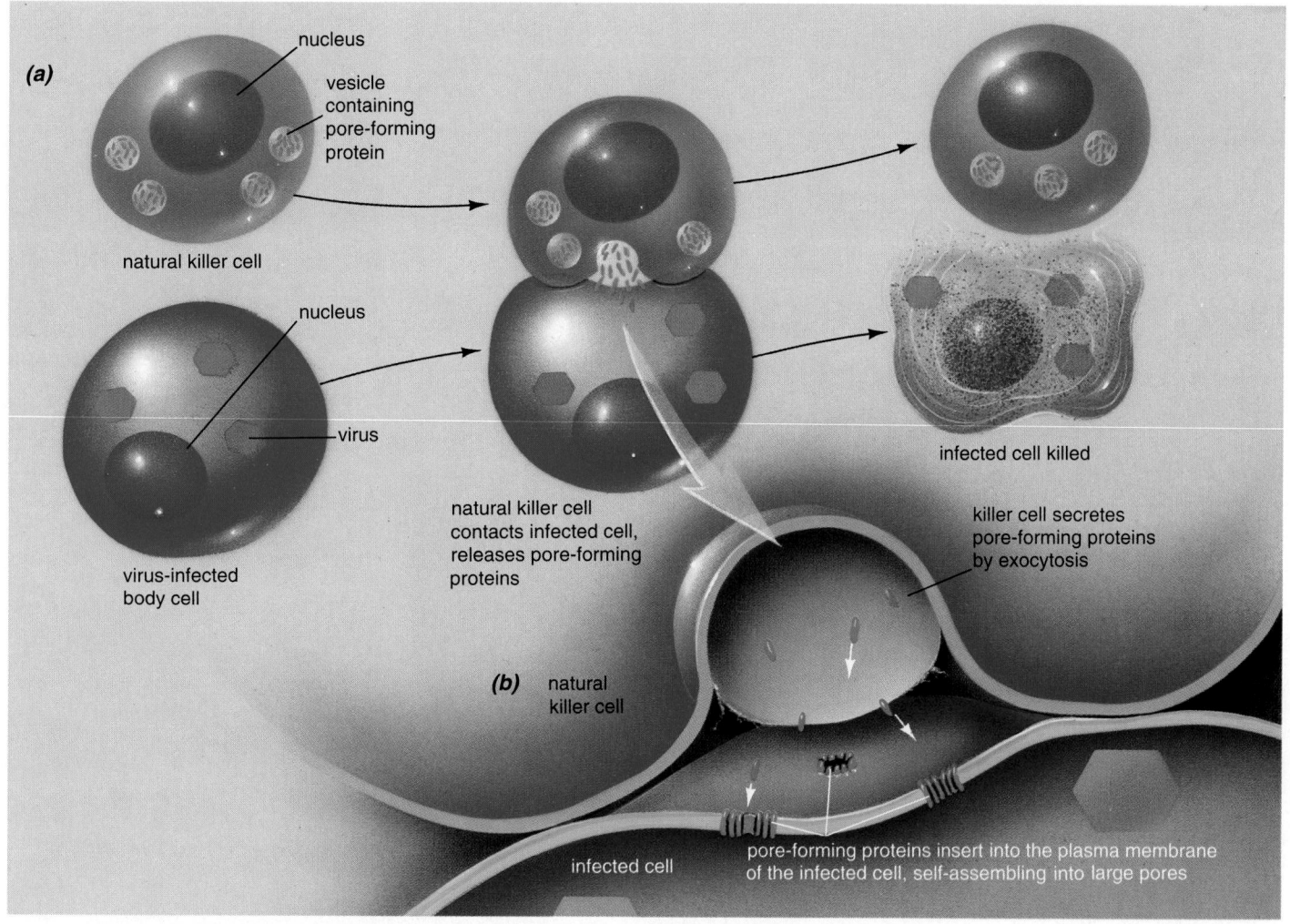

(a)

nucleus

vesicle containing pore-forming protein

natural killer cell

nucleus

virus

virus-infected body cell

natural killer cell contacts infected cell, releases pore-forming proteins

infected cell killed

killer cell secretes pore-forming proteins by exocytosis

(b) natural killer cell

infected cell

pore-forming proteins insert into the plasma membrane of the infected cell, self-assembling into large pores

Figure 34-3 How natural killer cells kill their targets.
(a) A natural killer cell contains vesicles full of pore-forming proteins. When it encounters a suitable target cell (here a virus-infected body cell), it briefly contacts the target and releases the pore-forming proteins by exocytosis. The pore-forming proteins self-assemble into large holes in the target cell's plasma membrane; its contents leak out, and the cell dies.
(b) A close-up view of the killing process. Pore-forming proteins leave the killer cell and insert themselves into the target cell's plasma membrane. The proteins line up next to one another like the staves in a barrel, forming large pores. The membrane of the killer cell is not affected; apparently it contains other proteins that bind to and inactivate any pore-forming proteins that try to insert themselves into the killer's membrane.

the body, thereby preventing microbes from escaping into the bloodstream.

Still other chemicals, some released by wounded cells and others produced by the microbes themselves, attract macrophages and other phagocytic cells to the wound. The phagocytic cells squeeze out through the capillary walls, enter the wound, and engulf bacteria, dirt, and tissue debris (Fig. 34-4).

Unfortunately, each phagocyte can eat just so many microbes, and then it dies. The pus that collects around a wound consists largely of bacteria, tissue debris, and living and dead white blood cells.

If the wound isn't too large and the microbes reproduce slowly enough, the inflammatory response will keep most of the invaders out of the bloodstream. The few that escape are eaten by other phagocytic

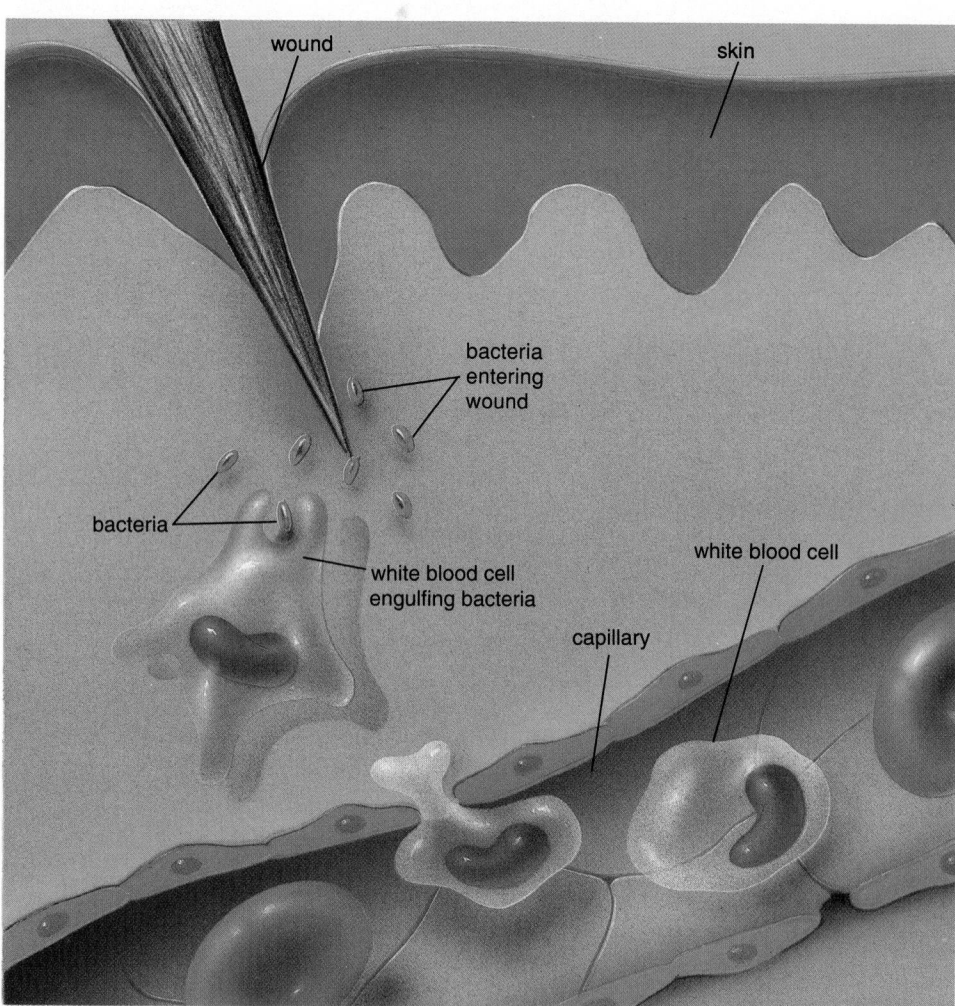

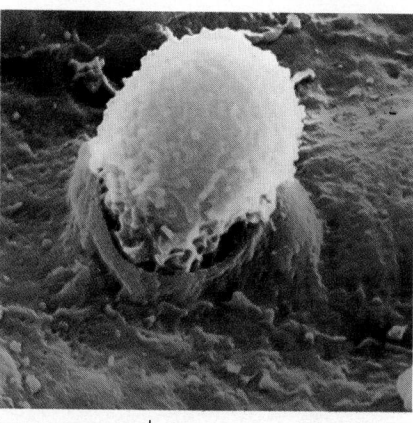

5 micrometers

Figure 34-4 During an inflammatory response, the cells forming capillary walls separate slightly. Here, a white blood cell (the fuzzy white ball) squeezes through a capillary to join the fray against bacteria that have entered a cut.

white blood cells in the blood vessels and lymph nodes, promptly halting the infection. However, inflammatory responses cannot handle all assaults. The wound may be large, or the microbes may multiply too quickly. Many viruses enter the body through the respiratory tract and do not provoke much, if any, immediate inflammatory response. In these cases, microbes may enter the bloodstream, reproducing rapidly and endangering the entire body. If this happens, the highly specific **immune response** comes into play, directing the destruction of the particular type of microbe that has passed through the initial defenses (see below).

Fever

Most Americans are so conditioned by advertisements and TV dramas that we automatically regard

fevers as dangerous and debilitating. We believe that fevers are part of a microbe's assault on our bodies. The facts are just the opposite: a fever is part of the body's assault on microbes. Severe fevers *are* dangerous, or even fatal, but "average" fevers of 38° or 39° C (100° to 102° F) are beneficial.

The hypothalamus, a part of the brain, contains temperature-sensing nerve cells that act as a thermostat for the body. Normally, the thermostat is set at 37° C (98.6° F). When disease organisms invade, however, the thermostat is turned up. Certain white blood cells, in responding to the infection, release hormones collectively called **endogenous pyrogens** ("self-produced fire-makers"). Pyrogens travel in the bloodstream to the hypothalamus and raise the thermostat's set point, triggering behaviors that increase body temperature: shivering, increased fat metabolism, or feeling cold so more clothing is put on. Pyrogens also cause other cells to reduce the concentration of iron and zinc in the blood.

Fever has both beneficial effects for the body's defenses and detrimental effects on the invading microbes. Many bacteria require more iron to reproduce at temperatures of 38° or 39° C than at 37° C, so fever and reduced iron in the blood combine to slow down their rate of reproduction. Simultaneously, fever increases the activity of phagocytic white blood cells that attack the bacteria, thereby producing a shorter and less serious infection.

Fever also helps to fight viral infections. When certain cells of the body are invaded by viruses, they synthesize and release a protein called **interferon.** Interferon travels to other cells and increases their resistance to viral attack. Fevers increase the production of interferon. In one study, patients with colds were treated with aspirin (to reduce fever) or a placebo (an inactive substance that looks like the drug in question, so the patients won't know whether they have been given the drug or not). Those given aspirin released far more viruses from their noses and throats than the placebo group. This means that the patients with lowered fevers weren't controlling their own infections very well and that they were significantly more infectious to other people.

As you can see, these defenses are all nonspecific; that is, their roles are to prevent and/or overcome any

and all microbial invasions of the body. However, these nonspecific defenses are not impregnable. When they fail to do the job, the body then mounts a highly specific **immune response** directed against the particular organism that has successfully invaded the body.

The Immune Response

Over 2000 years ago, the Greek historian Thucydides recognized the essential features of the immune response. He observed that occasionally someone contracts a disease, recovers, and thereafter is no longer susceptible to that particular disease—the person has become immune. With rare exceptions, however, immunity to one disease confers no protection against other diseases. Thus, the immune response attacks one specific type of microbe, overcomes it, and provides future protection against this microbe but no others.

The immune response results from the interactions of a number of different types of cells and molecules of the immune system. With such a large cast of characters, the theater of the immune response is difficult to follow without a program. Table 34-1 provides a brief summary of the major actors and their roles.

The immune response is extraordinarily complex in

Table 34-1 The Major Molecules and Cells of the Immune Response

Molecule	
Antigens	Large organic molecules, usually proteins, polysaccharides, or glycoproteins, that can trigger an immune response; often found on the surfaces of cells.
Antibodies	Proteins produced by the cells of the immune system that bind to antigens and either neutralize the antigenic molecule itself or mark cells that bear the antigens for destruction.
Major histocompatibility complex (MHC)	A set of proteins found on the surface of cells that "label" the cell as belonging to a unique individual organism.
Effector molecules	A diverse group of molecules, including histamine and the cell-destroying proteins of killer cells and complement (soluble proteins found in blood).
Regulatory molecules	Hormone-like molecules produced by cells of the immune system that regulate the immune response.
Cell	
Macrophages	Phagocytic white blood cells that both destroy invading microbes and help to alert other immune cells to the invasion.
B cells	Lymphocytes that produce antibodies; when stimulated, certain of their daughter cells (*plasma cells*) secrete large quantities of antibodies into the bloodstream.
T cells	A set of lymphocytes that regulate the immune response or kill certain types of cells. *Killer T cells* destroy specific targeted cells, usually either foreign eukaryotic cells, infected body cells, or cancerous body cells. *Helper T cells* stimulate immune responses by both B cells and killer T cells. *Suppressor T cells* inhibit immune responses by other lymphocytes.
Memory cells	A subset of the offspring of B and T cells that are long-lived and provide future immunity against a second invasion by the same antigen.

detail, but its essential features are conceptually simple. Therefore, we will present the fundamentals of the immune response in the text; additional explanations outlining how the immune system performs some of its more complicated tasks will be included in boxes entitled "A Closer Look . . ."

Fundamentals of the Immune Response

The immune response is provided chiefly by two types of lymphocytes, called **B cells** and **T cells** (Fig. 34-5). As with all white blood cells, B and T lymphocytes arise from precursor cells in the bone marrow (see Chapter 30). Early in embryonic development, the newly forming T cells migrate to the thymus (hence the name T cell), and differentiate into their mature forms. B cells mature in the bone marrow itself. As we shall see shortly, B and T cells play quite different roles in the immune response. However, immune responses produced by both B and T cells consist of the same three fundamental steps: (1) recognizing the invader, (2) launching a successful attack, and (3) retaining the memory of the invader to ward off future infections.

Recognition

To understand how the immune system responds to invading microbes, we must answer three related questions: (1) How do the immune cells recognize foreign molecules? (2) How can they produce specific responses to so many different molecules? (3) How do they determine that a substance is "foreign" and not "self," thereby avoiding destruction of the tissues of their own body?

Antibodies: Receptors for Foreign Molecules

The key to the immune system's ability to attack invading microbes lies in the structure and function of large proteins called **antibodies.** Antibodies are Y-shaped molecules composed of two pairs of peptide chains, one large (heavy) chain and one small (light) chain on each side of the Y (Fig. 34-6). Both heavy and light chains consist of a **constant region** that is similar in all antibodies, and a **variable region** that differs among antibodies. The combination of light and heavy chains results in an antibody with two functional parts: the "arms" and the "stem" of the Y.

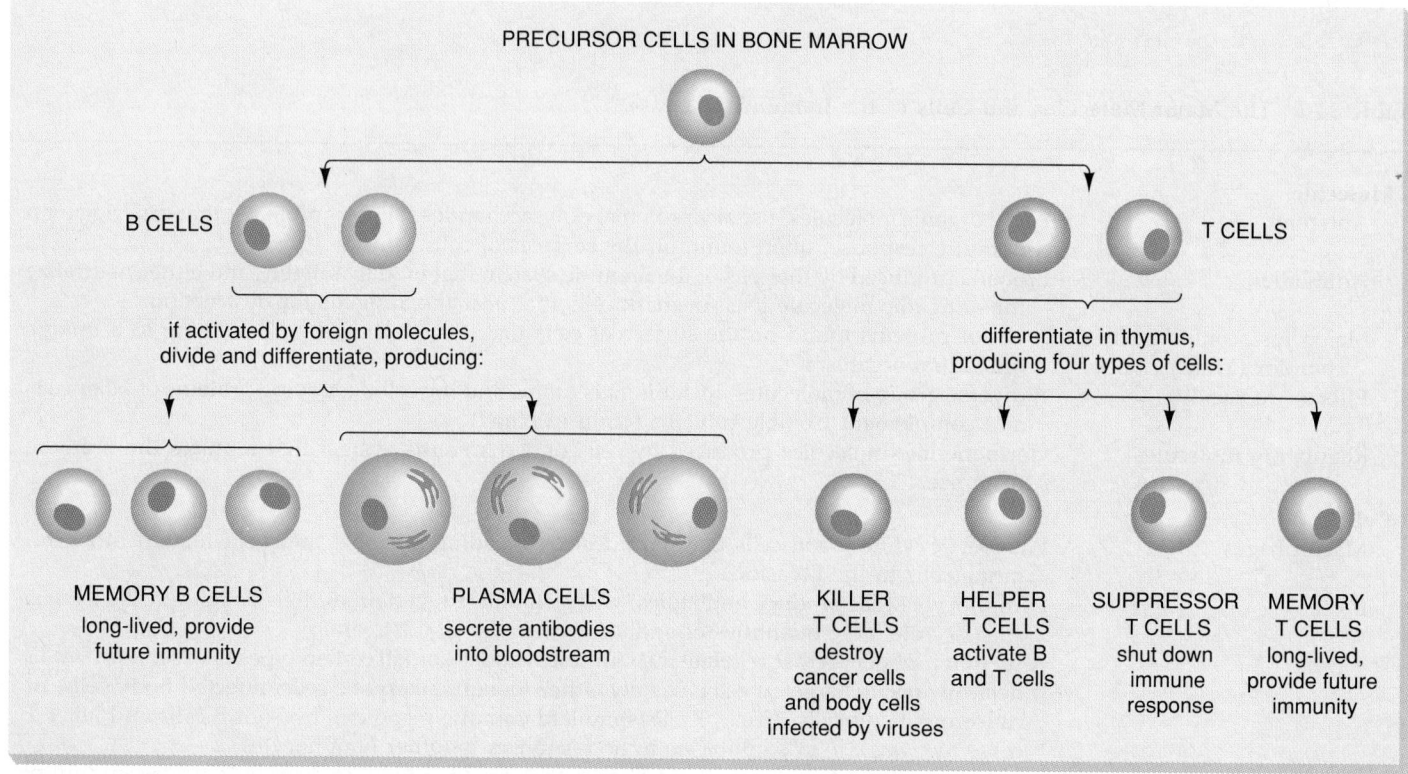

Figure 34-5 The major cells of the immune system and their roles in the immune response.

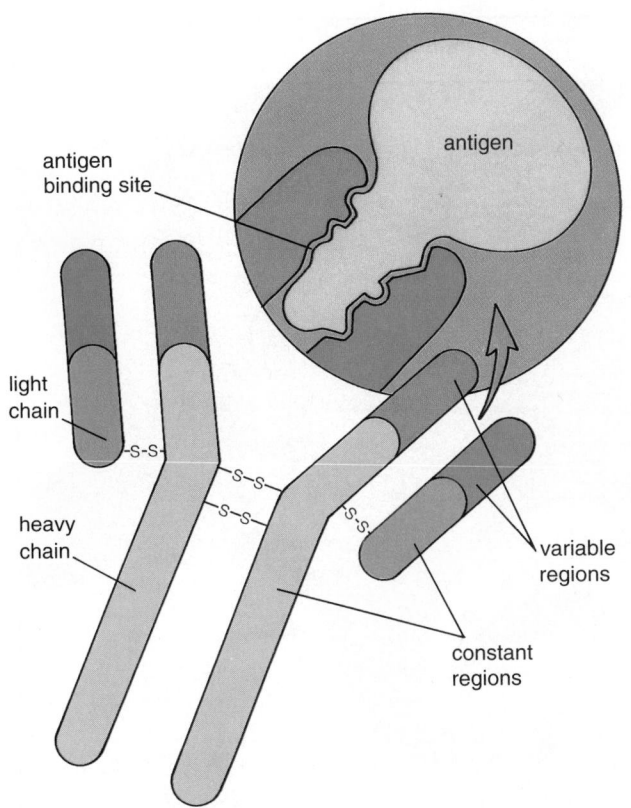

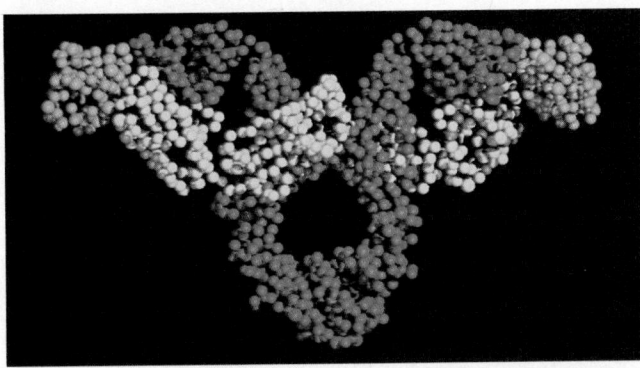

Figure 34-6 Antibodies are proteins composed of two pairs of peptide chains, called light and heavy chains, arranged something like the letter Y. The variable regions on the two chains form a specific binding site at the end of each arm of the Y. Different antibodies have different variable regions, forming unique binding sites. The human body synthesizes millions of distinct antibodies, each binding a different antigen.

The Arms: Specific Binding of Antigens

The variable regions that make up the arms form highly specific binding sites for large molecules.

These binding sites are a lot like the active sites of enzymes (Chapter 4): each binding site has its own peculiar shape and electrical charge, so only certain molecules can fit in and bind. **Antibody binding sites at the tips of the molecule's arms are so specific that each antibody can bind just a few different types of molecules, perhaps only one.**

Generally, only large, complex molecules, such as proteins, polysaccharides, and glycoproteins, can bind to antibodies. These molecules are called **antigens** (meaning "molecules that generate an antibody response"). Antigens may be attached to the surfaces of cells (e.g., the body's own cells or invading microbes) or may be dissolved in the blood or extracellular fluid (e.g., snake venom or a bacterial toxin). As we will discuss shortly, the binding of antigen to antibody triggers the immune response.

The Stem: Determining Antibody Function

The stem of the Y consists of part of the constant regions of the two heavy chains. **The stem determines the activity of an antibody.**

As we mentioned in our initial descrption, the constant regions are similar in different antibodies; they are not, however, the same. For example, the heavy chain constant regions that make up the stem of one antibody may attach the antibody to the plasma membrane of a cell; the constant regions of another antibody may bind to certain "killer proteins" in the blood **(complement)** to promote destruction of microbes.

Sources and Functions of Antibodies

Both B and T cells synthesize antibody-like molecules. Technically, only the molecules produced by B cells or their descendants are called antibod s. The related molecules produced by T cells are called T-cell receptors. However, since the structure and function of T-cell receptors and true antibodies are similar, we will simplify our terminology and refer to both the B cell and T cell molecules as antibodies.

Antibodies perform two fundamentally different functions in the immune response (Fig. 34-7). First, **antibodies are receptors.** Both B and T cells bear antibodies on their surfaces. In an antibody's role as a receptor, the stem attaches the antibody to the plasma membrane, while the two arms of the antibody protrude outward, sampling the blood and lymph for antigen. The binding of antigen to these "receptor antibodies" triggers responses in the lymphocytes bearing the antibodies.

Second, **antibodies are effectors.** As we will explain, certain descendants of B cells secrete antibodies into the bloodstream. There the antibodies play

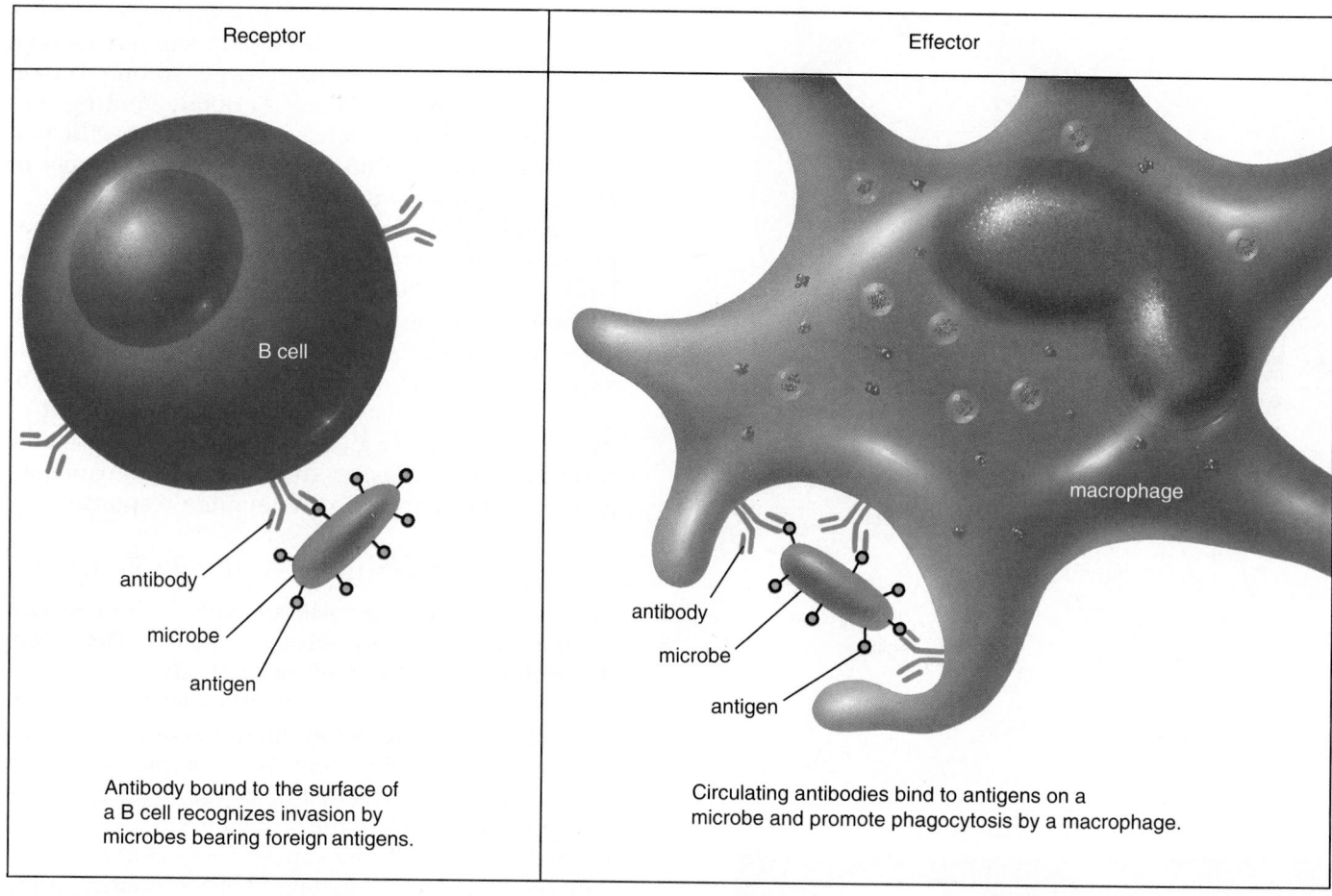

Receptor	Effector

Antibody bound to the surface of a B cell recognizes invasion by microbes bearing foreign antigens.

Circulating antibodies bind to antigens on a microbe and promote phagocytosis by a macrophage.

Figure 34-7 Examples of receptor and effector functions of antibodies. There are many other effector functions not illustrated here (see the text).

several roles in neutralizing poisonous antigens and or promoting the destruction of microbes bearing antigens. Small differences in the stem yield five types of true antibodies (not including T-cell receptors; Table 34-2). The stems of some antibodies form a coating on bound microbes that makes the microbes more easily ingested by macrophages (IgG); other stems promote secretion into the respiratory and digestive tracts (IgA); still other stems bind to "mast cells" lining the respiratory and digestive tracts, which release histamine during allergic responses (IgE).

Specific Recognition of a Multitude of Foreign Molecules

A human being may encounter millions of different types of invaders in a lifetime, from pollen grains and mold spores to flu viruses and botulinus toxin. The **immune system recognizes and responds to this multiplicity of antigens because its cells produce millions of different antibodies, each capable of binding a different antigenic molecule.**

This fact presented immunologists with a major problem: since antibodies are proteins, and proteins are encoded by the genes, it would seem that a human being must have millions of antibody genes. However, there are probably fewer than a hundred thousand genes in the entire human genome. How, then, can the genes code for millions of antibodies? Two distinct but complementary mechanisms join forces to produce an enormous diversity of antibodies from a relative handful of antibody genes.

1. **Genes encode parts of antibodies, not entire antibodies. These "antibody part" genes join together during immune cell development to form complete genes for the light and heavy chains of antibodies** (Fig. 34-8). The progenitors of B and T

Table 34-2   Antibodies and Their Roles

IgM		IgM is composed of five Y-shaped antibody "monomers" held together at their stems. IgM is usually the first antibody secreted during an immune response. In the bloodstream it agglutinates antigens (because of the large number of antigen binding sites), activates complement proteins, and stimulates phagocytosis of bound microbes by macrophages.
IgG		IgG, the most common antibody in the blood, is composed of a single antibody "monomer." It activates both complement and macrophages. Special transport processes carry IgG across the placenta, where it protects the developing fetus against disease.
IgA		IgA is a usually a dimer of two antibody monomers held together by their stems. An additional protein wound around the antibody stems helps IgA to be secreted from the bloodstream into saliva, tears, and mucus. Therefore, IgA is abundant on the surfaces of the respiratory and digestive tracts. It binds microbes on these surfaces, prevents them from entering the body, and allows them to be swept out of the body with mucus or other secretions. Thus, IgA provides a front line of defense for mucous membranes.
IgE		IgE, the "allergy antibody," is composed of a single antibody monomer. The IgE stem binds to mast cells in connective tissue, and to certain white blood cells, including basophils and eosinophils. Its normal function appears to be protection against parasites, which are expelled by the sneezing and coughing induced by mast cell activation, and weakened or killed by eosinophil activation. Allergic responses are produced when harmless substances bind to IgE. Allergy "shots" work by stimulating synthesis of IgG antibodies that bind the same antigen the IgE antibodies bind. If enough IgG antibodies are present, they bind up most of the allergy antigen, and keep the allergy antigens from contacting IgE.
IgD		IgD is a single antibody monomer, usually found bound to the plasma membranes of B cells. Its function is unknown; it may serve mainly as an antigen receptor.

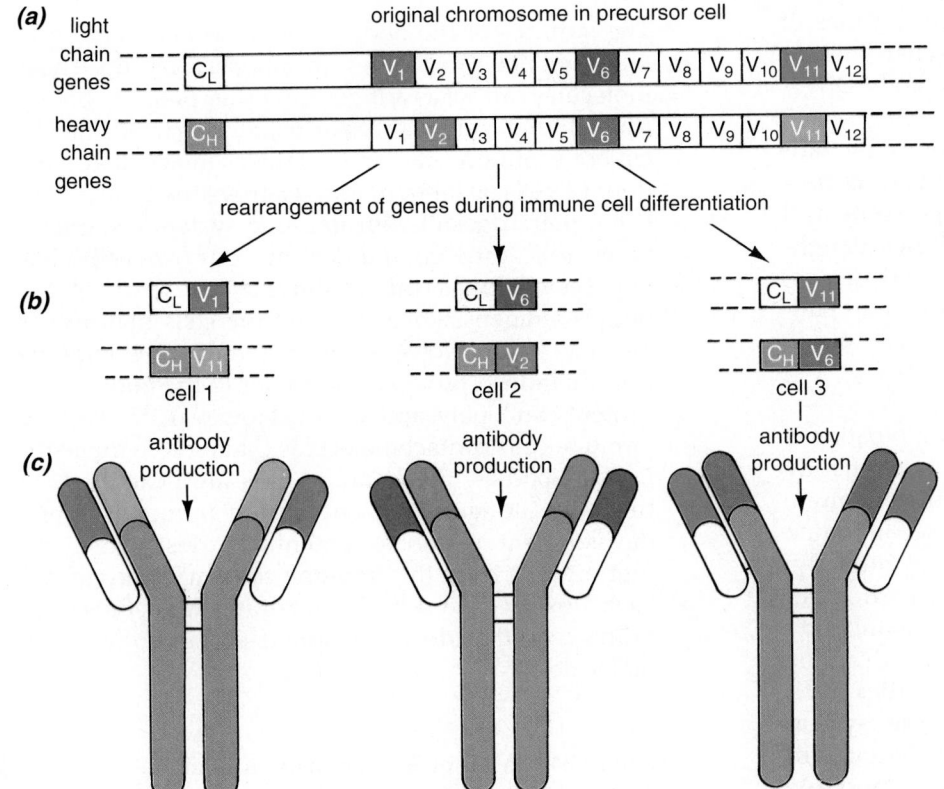

Figure 34-8 Recombination during the construction of antibody genes. (a) The precursor cells of the immune system each contain one or perhaps a few genes for the constant regions of the light and heavy chains of antibodies, and many genes for the variable regions.
(b) During the development of each immune cell, these genes are rearranged, moving one of the variable region genes next to a constant region gene. Each cell thus generates a "recombined antibody gene" for each chain, which differs from the recombined antibody genes generated by other immune cells.
(c) Representations of the different antibodies synthesized by each immune cell.

cells contain a few genes for the constant regions of antibodies and several hundred genes for the variable regions (see Fig. 34-8a). During the development of each individual B or T cell, massive chromosomal deletions bring one variable region gene next to one constant region gene. Each cell comes to possess one "constant + variable" light chain gene, and one "constant + variable" heavy chain gene (see Fig. 34-8b). Which variable region gene winds up alongside the constant region gene is probably random. Only the constant and variable genes that are next to one another are transcribed and ultimately provide the genetic information for antibody molecules; the rest remain unused. Therefore, each immune cell produces a single type of antibody, specified by the chance recombination of variable and constant region genes, that is different from the antibody produced by most or all other immune cells (see Fig. 34-8c).

It may help you to think of antibody gene formation in terms of card-playing. Each immune cell is dealt a "hand" of two variable region genes, one for the light chain and one for the heavy chain, randomly chosen from two large "decks" of genes. With each deck containing over a hundred "cards" (genes), virtually every cell will synthesize its own unique antibody.

Although this process may seem to be pretty fancy gene-shuffling, our description is really a simplification of a much more complex series of events. More detail on the actual mechanisms are presented in "A Closer Look at Antibody Diversity."

2. **The genes for certain antibody parts are incredibly prone to mutate, constantly generating new antibody genes.** This seems to happen only in B cells. As B cells reproduce, some of their daughter cells suffer mutations in their antibody genes. Thus, two sister cells may produce different antibodies.

K-Mart versus Custom Tailoring

The end result of mutation and gene recombination is that each immune cell has its own particular antibody genes, different from those of most other immune cells. Since the body contains billions of immune cells, each possibly synthesizing a different antibody, antigens almost always encounter antibodies that can bind them. A key fact about the antigen–antibody match, often misunderstood by students, is this: **The immune system does not "design" antibodies to fit invading antigens. Instead, the immune system randomly synthesizes billions of different antibodies. The fact that virtually every possible**

invading antigen binds to at least a few antibodies is due purely to the immense numbers of different antibodies present in the body.

A more familiar example may help to make this clear. When the Queen of England wants a dress, tailors go to Buckingham Palace, obtain the Queen's exact measurements, and custom-sew a dress to the Queen's specifications. Most of the rest of us go to a department store and look through the racks of ready-made clothes. If a store has a large enough selection, we will probably find something that fits reasonably well and is more or less the style we want.

The immune system is like a department store: the array of antibodies are simply there, waiting. They have *not* been designed to bind to any particular antigen. When an antigen enters the body, it randomly encounters many antibodies. The antigen bumps into the variable regions of each antibody. It cannot bind to most antibodies (in our shopping analogy, these antibodies don't "fit"), but, with millions of different antibodies around, the antigen inevitably encounters one that binds it pretty well. Antigen–antibody binding triggers changes in the immune cells, usually leading to the destruction of microbes bearing that antigen. We will examine these changes in more detail in a moment.

Distinguishing "Self" from "Non-Self"

The surfaces of the body's own cells bear large proteins and polysaccharides, just as microbes do. These molecules can act as antigens in other people's bodies (this is why transplants are usually rejected; the recipient's immune system recognizes molecules on the donor's cells as foreign, and destroys the transplant). Why, then, doesn't your immune system respond to your "self" antigens and destroy your own cells? The key seems to be the continuous presence of the body's antigens while the immune cells mature. As an embryo develops, some differentiating immune cells do indeed produce antibodies to the body's own proteins and polysaccharides. However, if *immature* immune cells contact molecules that bind to their surface antibodies, the immune cells are destroyed. In this way, antigens present during immune system development eliminate potentially destructive immune cells. Thus the immune system distinguishes "self" from "non-self" by retaining only those immune cells that do not respond to the body's own molecules.

Figure 34-9 A simplified summary of humoral and cell-mediated immune responses. See the text for details.

Attack

If the body is invaded by microbes, the immune system mounts two types of attack: B cells provide **humoral immunity,** which is provided by antibodies circulating in the bloodstream, while T cells produce **cell-mediated immunity,** as killer T cells destroy infected body cells (Fig. 34-9). The two types of re-

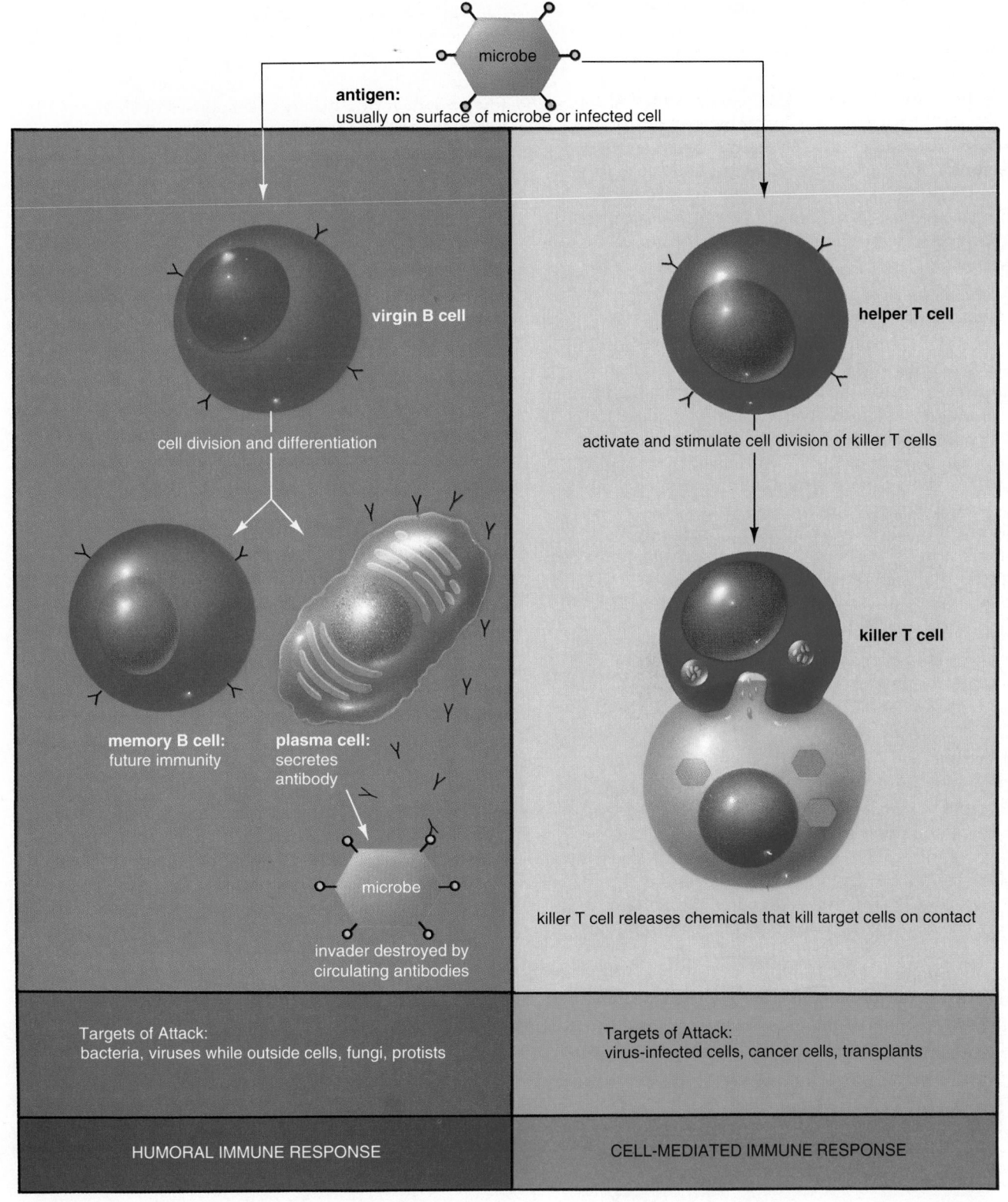

antigen:
usually on surface of microbe or infected cell

virgin B cell

cell division and differentiation

memory B cell:
future immunity

plasma cell:
secretes
antibody

microbe

invader destroyed by
circulating antibodies

helper T cell

activate and stimulate cell division of killer T cells

killer T cell

killer T cell releases chemicals that kill target cells on contact

Targets of Attack:
bacteria, viruses while outside cells, fungi, protists

Targets of Attack:
virus-infected cells, cancer cells, transplants

HUMORAL IMMUNE RESPONSE

CELL-MEDIATED IMMUNE RESPONSE

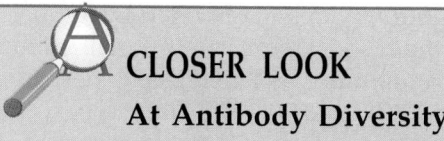

CLOSER LOOK
At Antibody Diversity

Our discussion of antibody diversity in the body of the text simplified a much more complex and remarkable process. Light and heavy chains are not composed of merely two regions, constant and variable. What we have lumped into the "variable" region actually consists of two sections of the light chain (usually called V for "variable" and J for "joining") and three sections of the heavy chain (V, J, and D, for "diversity," between the V and J regions). Still more diversity of antibody structure results from sloppy joining of the genes that encode each of these regions. In the remainder of this essay, we will focus on gene shuffling and recombination in the heavy chain only.

Heavy Chain Gene Structure

In humans, each chromosome 14 in each lymphocyte precursor cell in the bone marrow includes a long series of genes that code for the antibody heavy chain (Fig. E34-1a). Early in the differentiation process leading to B cells, either the maternal or paternal chromosome 14 is randomly "chosen" to be used to produce heavy chains (Fig. E34-1b). The antibody genes on the other chromosome 14 are apparently never used at all. The chosen chromosome 14 has a linear array of "antibody part" genes, consisting of perhaps 400 V genes, 20 D genes, 4 J genes, and 5 C genes.

Generating Antibody Diversity

Antibody diversity is produced in three ways: recombining V, D, and J genes; "splicing slop" at the V–D and D–J junctions; and mutation.

Recombination

Let us assume that the hypothetical cell shown in Figure E34-1 "decides" to make antibodies using V_{397}, D_{16}, and J_2 (how this "decision" is made is not known, but the overall process appears to be quite random). Enzymes in the nucleus cut out the DNA between V_{397} and D_{16}, and between D_{16} and J_2, and splice together the remaining genes (Fig. E34-1c, d). If this process is truly random, and using the numbers of V, D, and J genes given above, then different B cells could produce $400 \times 20 \times 4 = 32,000$ distinct combinations of heavy chain variable region genes.

Splicing Slop

When the V gene is joined to the D gene, and the D gene is joined to the J gene, a few nucleotides are *deleted* at the junction. This, of course, generates diversity of nucleotide sequence above and beyond the diversity already present in the different sequences of V, D, and J genes in the precursor cell. What's more, a few random nucleotides are then *inserted* at the junction, producing still more diversity. These deletions and insertions generate at least another 1000-fold increase in diversity. We now have over 30 million different types of heavy chains. The light chains have much less diversity than do the heavy chains, but 10,000 different types of light chain is a reasonable guess. If any light chain can combine with any heavy chain, then in principle the body could produce about 300 billion different antibodies.

Mutation

Antibody diversity doesn't even stop here. When activated by an antigen, B cells divide rapidly. Point mutations make small changes in the variable regions of the light and heavy chains. Although the mutations themselves are random, their consequences are not. It turns out that the rate of division of B cells is determined by how tightly their surface antibodies bind to antigens (see the "Humoral Immunity" section in the text). Mutations that cause looser binding result in slower division, so "worse binding" mutants don't proliferate much. Mutations that cause tighter binding result in faster division, so "better binding" mutants soon predominate. Mutations in the variable region genes, therefore, "fine tune" the immune response to a particular antigen.

The Benefits of Antibody Diversity

As we mentioned at the beginning of this chapter, the human body is potentially a warm, moist incubator for microbial growth. Further, each of us encounters trillions of potentially deadly microbes each day. However, with the vast assortment of antibodies produced by recombination, splicing slop, and mutation, our immune systems inevitably produce some antibodies that recognize any invading microbe. Except in the relatively rare instances in which invading microbes are especially numerous and/or potent, antibody diversity protects us from serious disease.

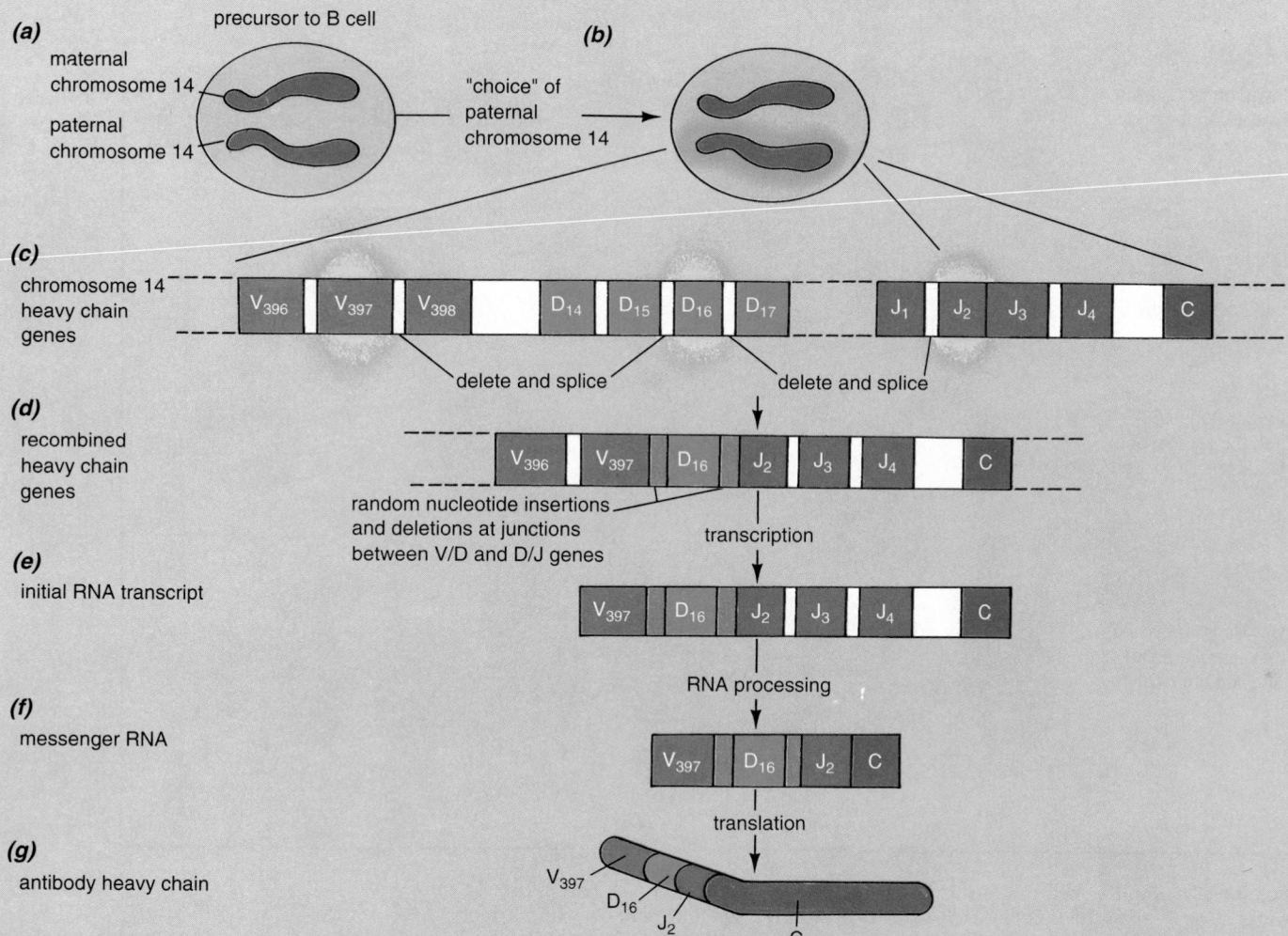

Figure E34-1 Generating diversity in heavy chain variable regions. In humans, the heavy chain genes are on chromosome 14.
(a) B cell precursors in the bone marrow have two homologs of chromosome 14, one inherited from the mother, one from the father.
(b) During B cell differentiation, one of these chromosomes is "chosen" to be used for all heavy chains made by a given cell.
(c) The arrangement of V, D, J, and C genes on chromosome 14. Dashed lines represent additional genes not shown in the diagram.
(d) V, D, and J genes are randomly selected (by an unknown process) to be used for antibody synthesis. The intervening DNA between the chosen V and D genes, and between the chosen D and J genes, is deleted, and the remaining chromosome parts are spliced together. Random nucleotides (hatched regions) are added and deleted at the junctions of V/D and D/J. Note that there are often "extra" J genes that have not been deleted.
(e) During transcription, an RNA copy is made beginning at the start of the chosen V gene and continuing through to the end of the C gene.
(f) The RNA transcript is processed to cut out the extra J nucleotides and splice the C nucleotides to the end of the chosen J region, forming the true messenger RNA.
(g) Translation of the mRNA results in the final antibody heavy chain.

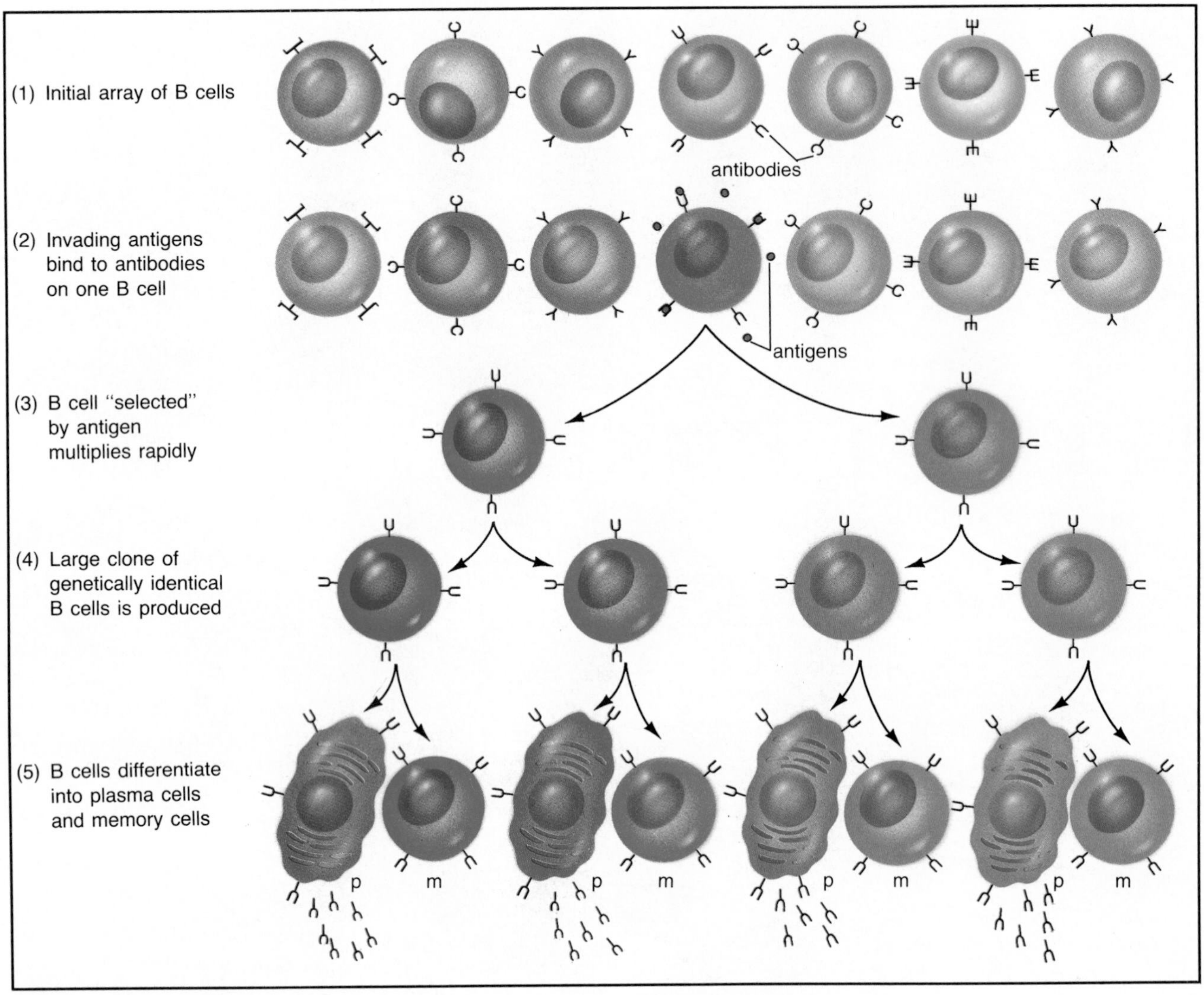

(1) Initial array of B cells

antibodies

(2) Invading antigens bind to antibodies on one B cell

antigens

(3) B cell "selected" by antigen multiplies rapidly

(4) Large clone of genetically identical B cells is produced

(5) B cells differentiate into plasma cells and memory cells

Figure 34-10 Clonal selection among B cells by invading antigen. (1) The body contains a large array of B cells, each of which synthesizes and bears on its surface a specific antibody, different from the antibodies borne by other B cells (see Fig. 34-8). (2) When an antigen enters the bloodstream, it binds to the antibodies on one (or a few) B cells. (3) Antigen–antibody binding causes the selected B cell to multiply rapidly. (4) A clone of B cells, all synthesizing the same antibody, is produced. (5) The daughter B cells differentiate into plasma cells (P) that secrete antibodies directed against the specific activating antigen or into long-lived memory cells (M).

sponses are not as distinct as this implies; however, the humoral and cellular responses are most easily understood when they are considered separately. The box, "A Closer Look at Cellular Communication During the Immune Response," describes some of the interactions among B and T cells that regulate the immune response.

Humoral Immunity

Each B cell bears a specific antibody on its surface (Fig. 34-10). When a microbial infection occurs, the

antibodies borne by a few B cells will bind to antigens on the microbe (in reality, the process is considerably more complex than this, but this is the end result). Antigen–antibody binding causes these B cells to divide rapidly (a type of T cell also helps to stimulate cell division in B cells; see below and the "Closer Look" box). The resulting population of cells differentiates into two cell types: **plasma cells** and **memory cells.** Plasma cells become enlarged and packed with endoplasmic reticulum, churning out huge quantities of that cell type's specific antibody (Fig. 34-11). These antibodies are released into the bloodstream (hence

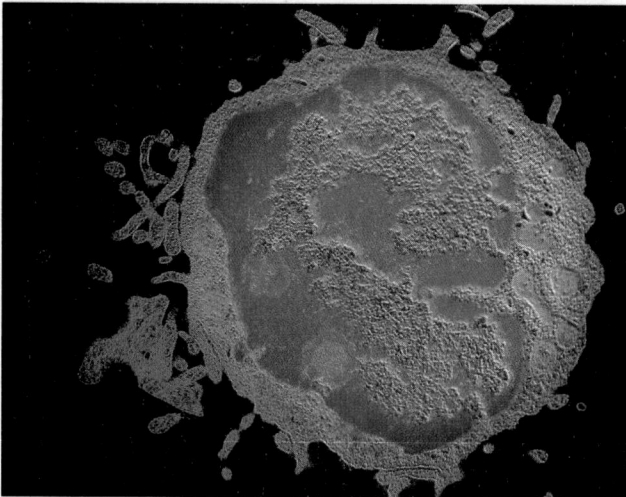

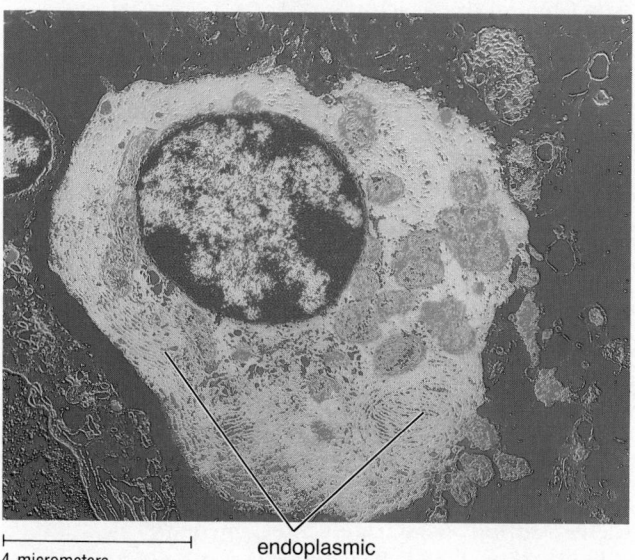

endoplasmic
reticulum

4 micrometers

Figure 34-11 False-color micrographs of B cells before (top) and after (bottom) conversion to plasma cells. The plasma cell is much larger than the B cell (note the difference in scale), and is virtually filled with rough endoplasmic reticulum that synthesizes antibodies.

the name "humoral" immunity; to the ancient Greeks, blood was one of the four body "humors"). Memory cells do not release antibodies. As we will see, memory cells play an important role in future immunity to this specific microbe.

Antibody Function

Antibodies in the bloodstream may affect antigenic molecules, microbes bearing antigens, and infected body cells in four ways:

1. **Neutralization:** the antibody may combine with or cover up the active site of a toxic antigen, thereby preventing the toxin from harming the body.
2. **Promotion of phagocytosis:** the antibody may coat the surface of a microbe. The protruding antibody stems seem to make it easier for phagocytic white blood cells to engulf the microbe.
3. **Agglutination:** each antibody has two binding sites for antigen, one on each arm (IgM antibodies, being pentamers of the unit antibody, have 10 binding sites). These binding sites may attach to antigens on two different microbes, holding them together. As more and more antibodies link up with antigens on different microbes, the microbes clump together, or agglutinate. Agglutination seems to enhance phagocytosis.
4. **Complement reactions:** the antibody–antigen complex on the surface of an invading cell may trigger a series of reactions with blood proteins called the **complement system.** When these proteins bind to the antibody stems, they attract phagocytic white blood cells to the site, promote phagocytosis of the foreign cells, or in some instances directly destroy the invaders by creating holes in their plasma membranes, much as natural killer cells do (see Fig. 34-3).

Cell-Mediated Immunity

The major function of cell-mediated immunity is to destroy the body's own cells when they have become cancerous or have been infected by viruses. The surfaces of T cells bear antibodies (T-cell receptors). When antigen binds to the antibodies on a T cell, the cell divides rapidly, producing two populations of offspring cells: effector cells and memory cells. There are several types of effector T cells: helper cells, killer cells, and suppressor cells. These all participate in the immune response, but they do not release antibodies into the bloodstream.

Functions of Effector T Cells

Helper T cells, when bound to antigen, release chemicals that assist other immune cells in their defense of the body. These hormone-like chemicals stimulate cell division and differentiation in both killer T cells and B cells that respond to the same microbial invasion. In fact, very little immune response, either cell-mediated or humoral, can occur without the boost provided by the helper T cells (see "A Closer Look at Cellular Communication"). This is why AIDS, which destroys helper T cells, is such a deadly disease. We will discuss AIDS more thoroughly shortly.

CLOSER LOOK
At Cellular Communication During the Immune Response

The immune system is a strange "system." Unlike the nervous system, for example, it is not composed of physically attached structures. Instead, as befits its mission of patrolling the entire body for microbial invaders, the immune system consists of an army of isolated cells. Nevertheless, the army of immune cells is highly coordinated. Figure E34-2 presents an overview of just some of the complex communications among antigen, antibody, hormones, receptors, and cells that coordinate the immune response.

When an antigen invades the body, say a protein on the surface of a virus (Step 1 in the figure), it sets off a cascade of events that can be loosely divided into three components.

Activation of Helper T Cells
One component of the immune response begins when macrophages ingest the virus (Step 2 in Fig. E34-2) and digest it. Antigens "chewed off" the virus become attached to certain "self-identification" proteins of the major histocompatibility complex (called MHC II) and are displayed on the surface of the macrophage. These antigen/MHC II complexes are recognized by virgin helper T cells (Step 3). When receptors ("antibodies") on helper T cells bind to the antigen/MHC II complex, the helper T cells release a hormone, called interleukin-2 (Step 4; "interleukin" means "between white blood cells," which is an appropriate name for a hormone that allows communication among immune cells). This hormone stimulates cell division (Step 5) in both the releasing cell and other T cells that have bound to antigen/MHC II complexes (the antigens could be the same one, different antigens from the same microbe, or even antigens from an entirely different invading organism). Some daughter helper T cells become memory cells, providing future immunity (Step 6); other daughter cells assist in stimulating the immune response of killer T cells and B cells (Step 7).

Activation of Killer T Cells: Cell-Mediated Immunity
Meanwhile, the virus is infecting ordinary body cells, such as those lining the respiratory tract (Step 8). Infected body cells display viral antigens on their surfaces, bound to a second set of MHC molecules (MHC I). Immature killer T cells bind to the antigen/MHC I complex (Step 9) and are simultaneously stimulated by interleukin-2 released by the activated helper T cells. This dual binding/stimulation causes the killer T cells to mul-

tiply and mature (Step 10). When mature killer T cells then encounter infected cells displaying the antigen/MHC I complex, they release toxic proteins that kill the infected cell (Step 11).

Activation of B Cells: Humoral Immunity
Some B cells bear antibodies on their surfaces that bind antigens on the surface of free viruses that have not yet invaded a body cell (Step 12). Simple antigen–antibody binding stimulates B cell division and maturation to some extent, but really turning B cells on requires a boost from helper T cells. B cells that have bound antigen ingest the antigen by receptor-mediated endocytosis (see Chapter 6), attach the antigen to MHC II molecules, and display the antigen/MHC II complex on their surfaces. The antigen/MHC II complex is recognized by activated helper T cells (Step 13), which then release several types of interleukin hormones that stimulate division and differentiation of antigen-binding B cells (Step 14). Some of their progeny become memory cells (Step 15), while others become plasma cells that secrete antibodies into the bloodstream (Step 16).

Summary of Immune Cell Communication
As you can see, this interlocking communication network is quite complex. We can summarize its essentials, however, in five generalizations:

1. **Macrophages bind to antigens, engulf them, and present them on their surfaces, along with "self-identification" MHC molecules, to helper T cells.**
2. **When helper T cells recognize the antigen/MHC II complex, they multiply rapidly.**
3. **Meanwhile, killer T cells and B cells recognize the same antigen.**
4. **Hormones released by helper T cells stimulate cell division and maturation of only those killer T cells and B cells that have also been stimulated by antigen binding.**
5. **The stimulated killer T cells and B cells then provide cell-mediated and humoral immunity, respectively.**

As you can see, helper T cells are essential in turning on both phases of the immune response. A loss of helper T cells, such as that caused by the AIDS virus, virtually eliminates the immune response to many diseases.

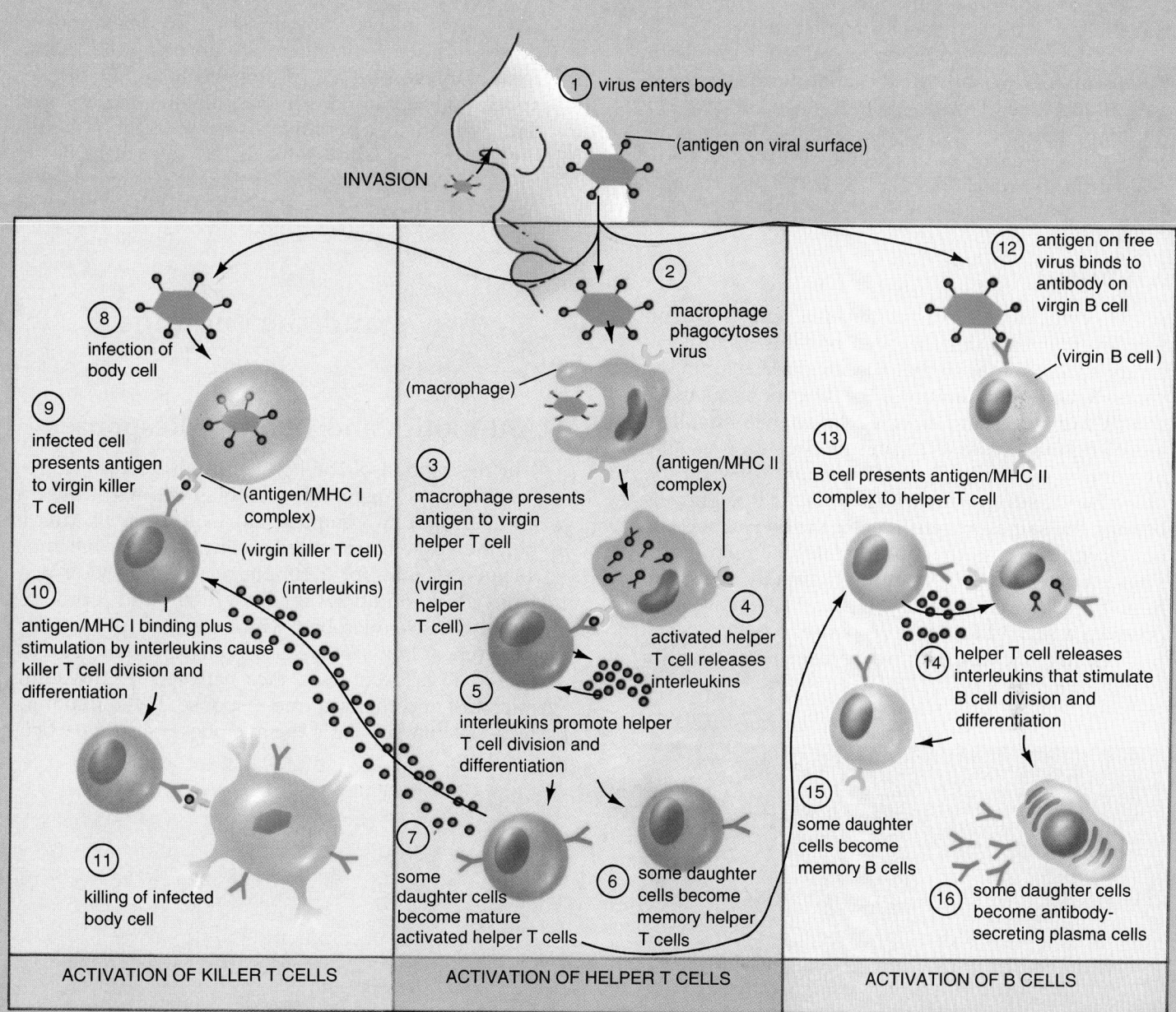

Figure E34-2 A simplified diagram of the interactions among immune cells during the immune response. The numbers are keyed to those used in the essay.

Antibodies on the surfaces of **killer T cells** bind to antigens on the surface of an infected cell. The killer T cell then releases proteins that disrupt the infected cell's plasma membrane (Fig. 34-12). In at least some instances, the killer T cell's proteins are similar or identical to those of the natural killer cell that create giant holes in the target cell's membrane.

After the infection has been conquered, **suppressor T cells** appear to shut off the immune response in both B and killer T cells.

Other offspring of activated T cells are memory cells. Like memory B cells, these protect the body against future infection.

Memory

As Thucydides observed 2 millenia ago, a person who overcomes a disease often remains immune to future encounters with that specific disease for many years. Retaining immunity is the function of memory cells. Plasma cells and killer T cells do the immediate job of fighting disease organisms, but they usually live only a few days. B and T memory cells, on the other hand, survive for many years. If foreign cells bearing the same antigens reenter the body, they will be recognized by the appropriate memory cells. These memory cells will multiply rapidly, generate huge populations of plasma cells and killer cells, and produce a second immune response.

In the first encounter with a disease microbe, only a

few B and T cells respond. Each of these, however, leave behind hundreds or thousands of memory cells. Further, memory cells respond to antigen much more rapidly than their progenitor B and T cells could. Therefore, the second immune response is very rapid (Fig. 34-13). In most instances, second or subsequent invasions by the same microbe are overcome so quickly that there are no noticeable symptoms of infection at all. Although cold and flu viruses appear to be exceptions to this rule, the immune system does in fact provide lasting protection against these viruses. The problem, as we describe in "Health Watch: Flu—The Unbeatable Bug," is that next year's flu isn't caused by the same virus that you fought off this year.

Medicine and the Immune Response

Antibiotics and Immune Responses

Our description of the body's responses to infection may seem to suggest that nothing can harm us: the immune system conquers all. Unfortunately this is not the case. If untreated, many diseases kill their victims, usually for a simple reason: the body provides ideal conditions for the growth and reproduction of disease microbes, which can multiply rapidly, sometimes dividing as fast as once an hour. The infection thus becomes a race between the invading microbes and the immune response. If the initial infection is massive, or if the microbes produce particu-

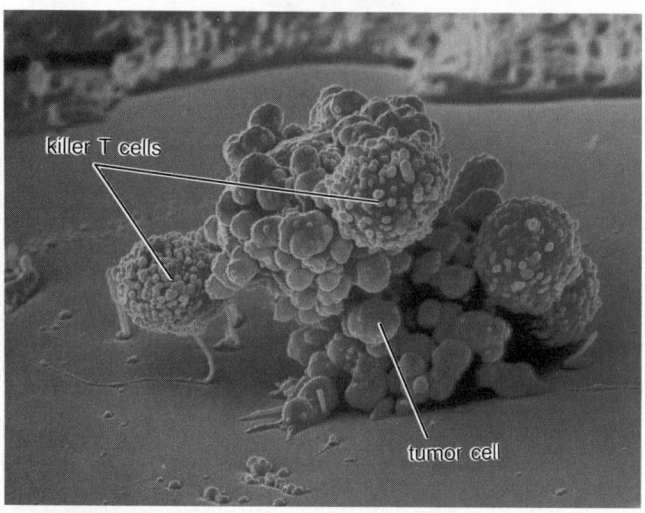

5 micrometers

Figure 34-12 Cell-mediated immunity at work. A killer T cell contacts a larger tumor cell and causes it to disintegrate. Note the membrane bubbles forming on the tumor cell as it is destroyed.

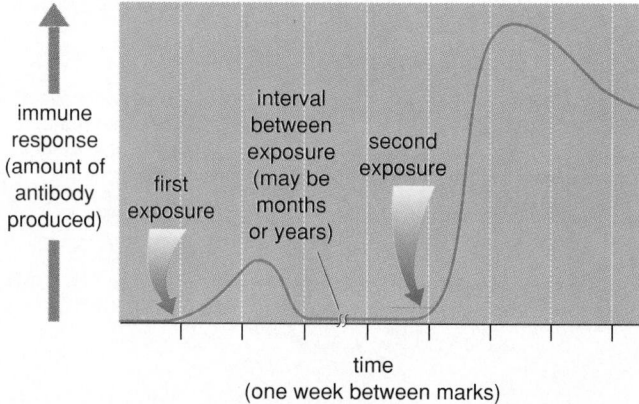

Figure 34-13 The immune response to the first exposure to a disease organism is fairly slow and not very large, as B and T cells are selected and multiply. A second exposure activates memory cells formed during the first response, and consequently the second response is both faster and larger.

larly toxic products, the full activation of the immune response may come too late.

Antibiotics help to combat infection by slowing down the growth and multiplication of many microbes, including bacteria, fungi, and protists (but not viruses). Although antibiotics don't destroy every single microbe, they give the immune system enough time to finish the job.

Vaccinations

As early as the year 1000, people in India, China, and Africa deliberately exposed themselves to mild cases of smallpox, to acquire immunity to the disease. In 1798, Edward Jenner discovered that being infected with cowpox conferred immunity to smallpox. This discovery initiated the modern practice of immunization. In the late 1800s, Louis Pasteur extended the use of immunization to several other diseases by injecting weakened or dead microbes into healthy people. The weakened microbes do not cause disease (or at least not a severe case) but bear antigens that elicit vigorous immune responses. These injections of weakened or killed microbes to confer immunity are called **vaccinations,** from the Latin word for cow, in honor of Jenner's pioneering efforts with cowpox. Today, many diseases, including polio, diphtheria, typhoid fever, and measles, can be controlled through vaccination. Smallpox, one of the most deadly diseases of all, has been completely eradicated from the Earth, owing to a program sponsored by the World Health Organization.

Through genetic engineering (see Chapter 14), we now enjoy the prospect of manufacturing tailor-made vaccines. One method is to synthesize the antigenic proteins from disease-causing microbes. These antigens can then be used as vaccines without having to raise, isolate, and weaken the disease microbes themselves, or to inject people with microbes at all. A vaccine against anthrax, a severe disease of livestock, has been manufactured with this procedure. A second technique may be to insert the genes for the antigens of, say, herpes into the genome of harmless microbes such as the cowpox virus. These designer microbes would produce herpes antigens without being able to cause the disease, and could be used for vaccination.

Allergies

Over 35 million Americans suffer from **allergies:** reactions to substances that are not harmful in themselves, and to which many other people do not respond. Common allergies include those to pollen, dust, mold spores, and bee stings. Allergies are actually a form of immune response (Fig. 34-14). A foreign substance, say a pollen grain, enters the bloodstream and is recognized as an antigen by a particular type of B cell. This B cell proliferates, producing plasma cells that pour out IgE antibodies against the pollen antigens. The stems of IgE antibodies attach to the plasma membranes of histamine-containing "mast cells" located in connective tissue throughout the body. When pollen grains encounter the attached IgE antibodies, they trigger the release of histamine, which causes increased mucus secretion, leaky capillaries, and other symptoms of inflammation. Since pollen grains most often enter the nose and throat,

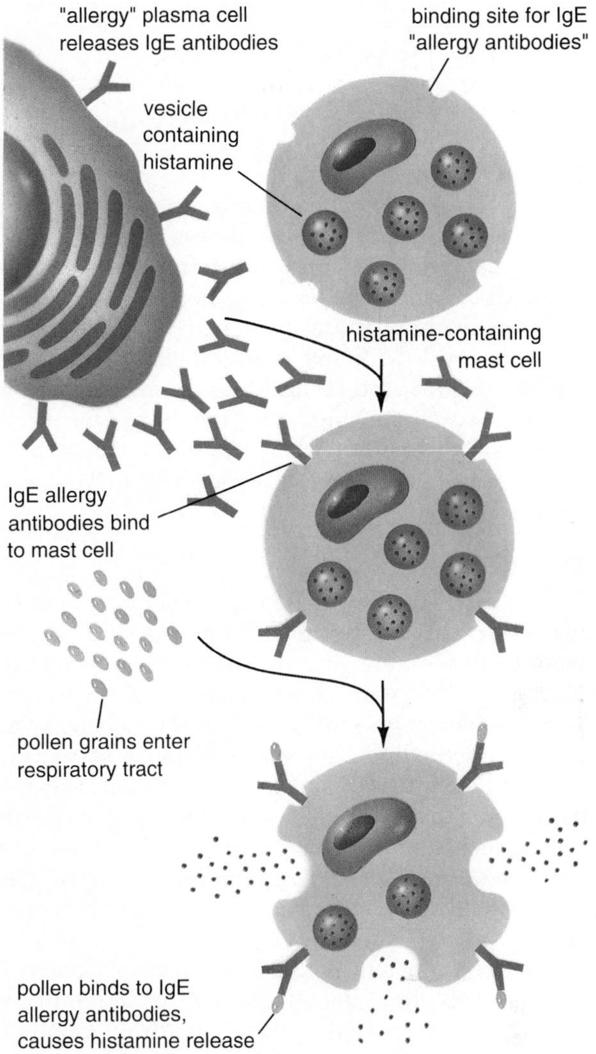

Figure 34-14 Allergic reactions. Upon exposure to certain antigens, for instance pollen grains, some plasma cells synthesize antibodies that bind both to histamine-containing cells (middle) and to the antigen (bottom). When this dual binding occurs, the cells release histamine into their surroundings, causing local inflammation and the symptoms of allergy.

HEALTH WATCH
Flu—The Unbeatable Bug

Every winter, a wave of influenza sweeps across the world. Thousands of the elderly, the newborn, and those already suffering from illness succumb, while hundreds of millions more suffer the fever and muscle aches of milder cases. Occasionally, devastating flu varieties appear. In the great flu pandemic of 1918, the worldwide toll was 20 million dead in one winter. In 1968, the Hong Kong flu infected 50 million Americans, causing 70,000 deaths in 6 weeks. As recently as the winter of 1984–85, the Centers for Disease Control estimate that about 57,000 Americans died from the flu.

The Flu Virus

Flu is caused by a virus that invades the cells of the respiratory tract, turning each one into a factory for manufacturing new viruses. The outer surface of the virus is studded with proteins, several of which serve as antigens recognized by the immune system. People survive the flu because their immune systems inactivate the viruses or kill off virus-infected body cells before the viruses finish reproducing. This is the same mechanism by which other viruses, such as mumps or measles, are conquered. So why don't people become immune to the flu, as they do to measles?

The answer lies in the flu virus' uncanny ability to change. The viral genes that code for the antigenic proteins mutate rapidly: there are 10 mutations in every million newly synthesized viruses. A single mutation usually doesn't change the properties of the antigen very much. Four or five, however, may change it enough so that the immune system doesn't fully recognize it as the same old flu that was beaten off last year. Some of the memory cells don't recognize it at all, and the immune response produced by the rest doesn't work as well as it should. The virus, although slowed down somewhat, gets a foothold in the body and multiplies until a new set of immune cells recognizes the mutated antigen and starts up a new immune response. And so you get the flu again this year.

Deadly New Strains

Far more serious are the dramatically new flu viruses that occasionally appear: the epidemic of 1918, the Asian flu of 1957, and the Hong Kong flu of 1968. In these, entirely new antigens seem to show up all at once. These are not just mutations of the old set, but completely novel antigens that the human immune system has never encountered before. Where do they come from? Believe it or not, they come from birds and pigs. Viruses strikingly similar to the human flu virus infect the intestinal tracts of birds, especially ducks, without causing any noticeable disease. The avian viruses don't infect people, and human flu viruses don't infect birds. However, both can infect pigs, with the result that both viruses may reproduce simultaneously in the same pig cell. Once in a great while (perhaps only three times this century), offspring viruses end up with a mixture of genes from human and bird viruses (Fig. E34-3). Some combine the worst genes of both types: from the human virus, the genes needed to subvert human cellular metabolism to produce new viruses; from the bird virus, genes for new surface antigens.

Ever wonder why the flu strains are called "Asian" or "Hong Kong"? Southeast Asia is usually the place where new strains crop up. The reason is that many farmers in Asia, especially in south China, have "integrated" farms. Crops are grown to feed pigs and ducks, and the feces from the pigs and ducks are used to fertilize fish ponds. This is a very efficient farming practice, producing large crops of carp and other fish without actually feeding the fish at all. Unfortunately, it also provides ideal mixing vessels for the flu virus (pigs) in close proximity to humans and ducks.

Effects of New Strains

If infected by a hybrid virus, the immune system must start from scratch, selecting out entirely new lines of B and T cells to attack the intruder. But the virus multiplies so rapidly in the meantime that many people die, or become so weakened that they contract some other disease and die of that. The rest of the people recover, with immune systems now primed to resist any further assault from the new virus. Next year a few point mutations allows a slightly altered strain to infect millions of people, but with a partial immune response ready, few fatalities occur. Once again, for most of us, the flu becomes a routine annoyance.

Until next time, somewhere, the improbable happens again. Maybe this year.

the major reactions occur in these locations, resulting in the runny nose, sneezing, and congestion typical of hay fever. Antihistamine drugs block some of the effects of histamine, relieving the symptoms of allergies. Food allergies cause equivalent symptoms in the digestive tract, including cramps and diarrhea.

The Function of IgE "Allergy Antibodies"

Why are some people allergic and not others? You can probably guess the answer. We are all exposed to the same antigens in pollens, molds, and foods. Therefore, people without allergies either must lack the genes for the allergy-causing antibodies, or do not produce as much antibody as allergic individuals do. From an evolutionary perspective, a more interesting question is: Why are *any* people allergic? What useful function do IgE antibodies serve, so that the obvious disadvantages have not caused their elimination through natural selection? It turns out that IgE antibodies confer protection against parasites.

Many parasites invade the body through natural body openings—usually the mouth, nose, or anus. The initial stage of infection occurs when a parasite attaches to the lining of the nasal passages, throat, or intestine. It then usually burrows through into the body tissues (see Chapter 22). The typical symptoms of allergies—increased mucus secretions, sneezing, coughing, intestinal convulsions, and diarrhea—are apparently properly directed against parasites, not pollen, and help to dislodge and expel the parasites. The stems of IgE antibodies also attach to white blood cells called eosinophils. When the variable region arms of the attached IgE antibodies bind to a parasite, the eosinophil secretes toxic proteins that damage or kill the parasite.

Autoimmune Diseases

A person's immune system does not normally respond to the antigens borne on the body's own cells. Occasionally, however, something goes awry, and "anti-self" antibodies are produced. The result is an **autoimmune disease,** in which the immune system attacks some component of one's own body. Some types of anemia, for example, are caused by antibodies that destroy a person's red blood cells. Many cases of insulin-dependent (juvenile-onset) diabetes occur because the insulin-secreting cells of the pancreas are the victims of a misdirected immune response. Unfortunately, at present there is no way to cure autoimmune diseases. For some types, replacement therapy can alleviate the symptoms, for instance by administering insulin to diabetics or blood transfusions to anemics. Alternatively, the autoimmune re-

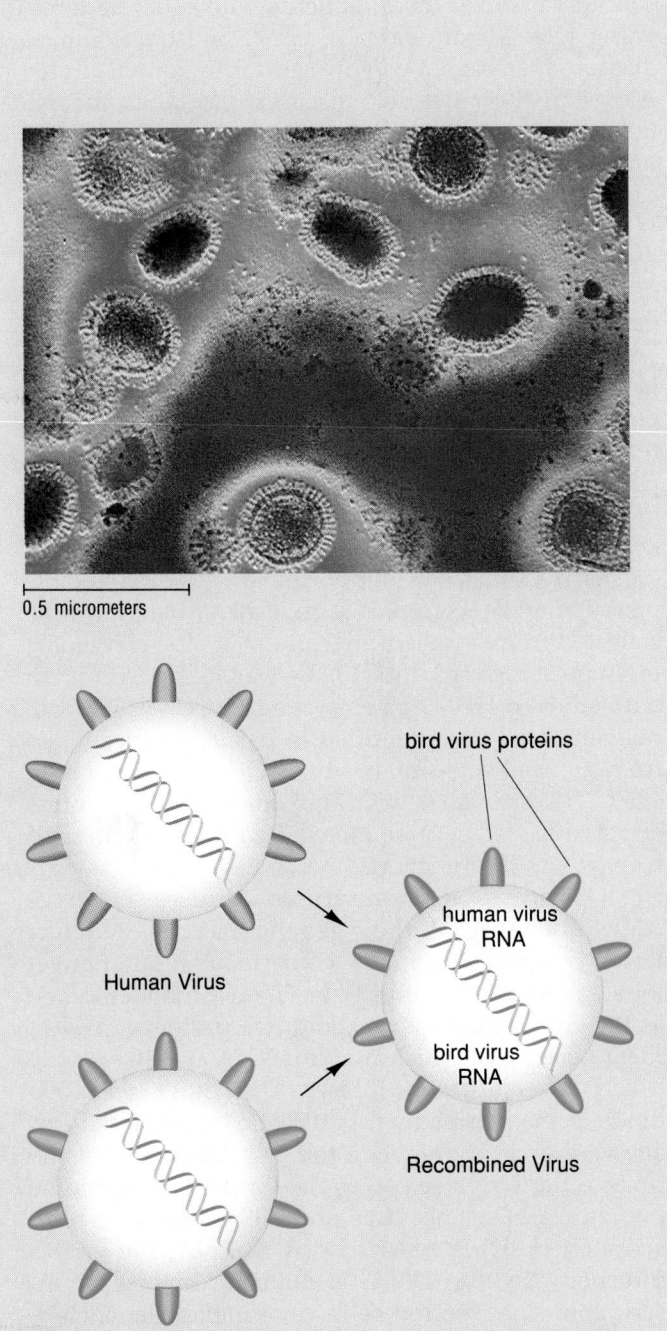

Figure E34-3 Recombination of genes from bird and human influenza viruses. The photograph is a false color electron micrograph of flu viruses. Note the protein "spikes" projecting from the virus coats. These attach to plasma membranes of cells in the human respiratory system, helping the virus gain entry into the cells.

sponse can be suppressed with drugs. Immune suppression, however, also reduces immune responses to the everyday assaults of disease microbes, so this therapy cannot be used except in the most life-threatening cases.

Immune Deficiency Diseases

Rarely, a child is born with a defect in which no immune cells, or very few, are formed. This condition is called severe combined immune deficiency, or SCID. Such a child may survive fetal life and even the first few months of postnatal life, protected by antibodies acquired from the mother during pregnancy or in her milk. Once these antibodies are lost, however, common bacterial infections may prove fatal. Some immune-deficient children have to live in a germ-proof "bubble," isolated from contact with every unsterilized object, including other people. One form of therapy is to transplant bone marrow (from which immune cells arise) from a normal donor into the child. In some children, marrow transplants have resulted in some antibody production, occasionally enough to confer normal immune responses. A group of researchers has recently started trials of injecting genetically engineered bone marrow cells into children with SCID (see Chapter 15).

AIDS

Probably the most common, and most devastating, immunodeficiency disease is acquired immunodeficiency syndrome, or AIDS. Two viruses, named human immunodeficiency viruses 1 and 2 (HIV-1 and HIV-2), cause AIDS by infecting and destroying helper T cells. If you refer to "A Closer Look at Cellular Communication During the Immune Response," you will note that helper T cells potently stimulate both the cell-mediated and humoral immune responses.

AIDS does not directly kill its victims, but as the helper T cell population declines, the AIDS patient becomes increasingly prone to other diseases. In fact, it was the incidence of unusual diseases that led to the recognition of AIDS in the first place, in 1981. In that year, a man entered the UCLA Medical Center with a fungal infection in his throat. A few weeks later, he developed a rare form of pneumonia (*Pneumocystis carinii*), one almost never seen except in patients with cancer and people whose immune system is being suppressed to prevent rejection of organ transplants. Following a series of infections, he died that December. Soon, doctors across the country en-

countered similar cases: patients who suffered debilitating effects from rare diseases, or from common diseases that are not usually serious in normal adults. Although the particular diseases varied, all the patients had one feature in common: a failure of the immune system to ward off invading microbes, because of a lack of helper T cells.

HIV: The AIDS Virus

Early in 1984, scientists at the Pasteur Institute in France and the National Institutes of Health in the United States isolated HIV-1 (Fig. 34-15). Based on gene sequences and usual rates of mutation, AIDS researchers believe that HIV-1 and HIV-2 arose independently in Africa 30 to 40 years ago, as mutations in separate viruses that normally infect African monkeys.

This hypothesis is supported by two remarkable pieces of evidence reported in 1990. A blood sample collected from a patient in Zaire in 1959—and amazingly preserved up until 1990—was found to contain antibodies to HIV. At about the same time, HIV genetic material was identified in tissue samples taken from an English seaman who died of pneumonia in 1959. The seaman's wife and one of his daughters apparently also became infected with HIV; both died later of similar infections.

AIDS viruses are **retroviruses**—viruses that have RNA, not DNA, as their genetic material. As illustrated in Figure 34-15, HIV consists of an outer envelope, taken from an infected cell's plasma membrane as the virus leaves the cell, and two protein capsules, the innermost of which contains RNA and an enzyme called **reverse transcriptase.** The outer envelope binds to the plasma membrane of a helper T cell and allows the virus to invade the cell. Once inside, the reverse transcriptase copies the virus's RNA genome over into DNA; since the "normal" direction of transcription in living cells is DNA to RNA, the virus's direction, RNA to DNA, is dubbed "reverse" transcription. The infected cell's own metabolic machinery then uses the viral "DNA copy" to make more viruses and more reverse transcriptase. Multiplication of HIV eventually kills the infected helper T cell.

In some respects, HIV is not a very powerful virus. It seems to be transmitted only by direct exchange of body fluids, including blood, saliva, and semen. Further, HIV can't survive very long outside the body. However, once a person is infected, AIDS seems to be invariably fatal, and AIDS is now an extremely widespread disease (Fig. 34-16). By the time that you read this, at least 10 million people, and probably many more, will have been infected with AIDS.

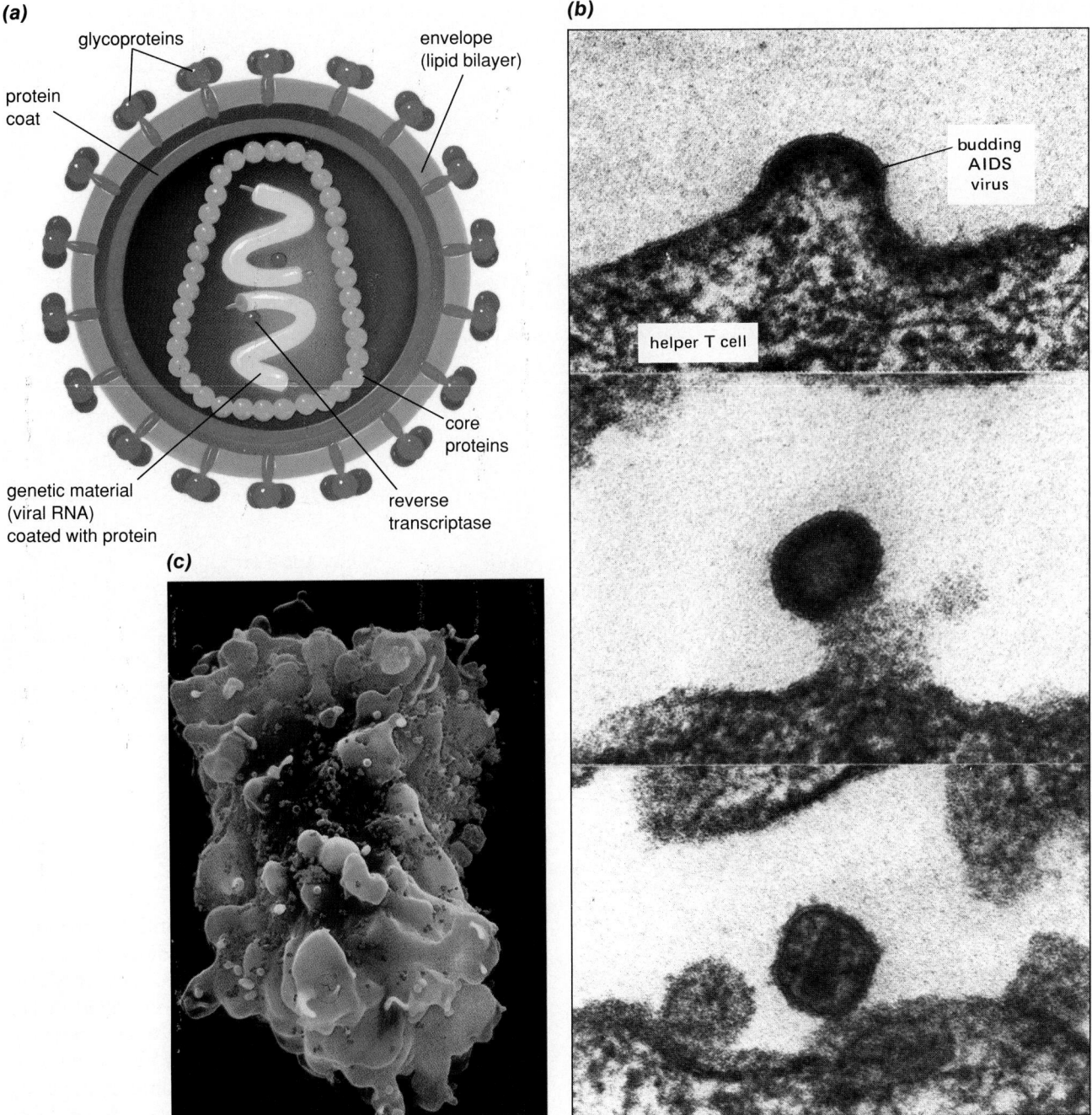

Figure 34-15 The human immunodeficiency virus (HIV), the cause of AIDS. **(a)** The structure of the AIDS virus. The virus consists of an outer envelope taken from the cells they infect, and an inner protein capsule that contains RNA (the genetic material of HIV) and the enzyme reverse transcriptase (that copies the RNA over into DNA when the virus infects a cell). The proteins protruding through the envelope attach to the plasma membranes of helper T cells. These proteins are potential targets for AIDS vaccines. **(b)** An HIV virus emerges from an infected helper T cell, coating itself with a bit of the T cell's plasma membrane as it leaves. **(c)** The blue specks in the false-color scanning electron micrograph are HIV viruses that have just emerged from the large helper T cell.

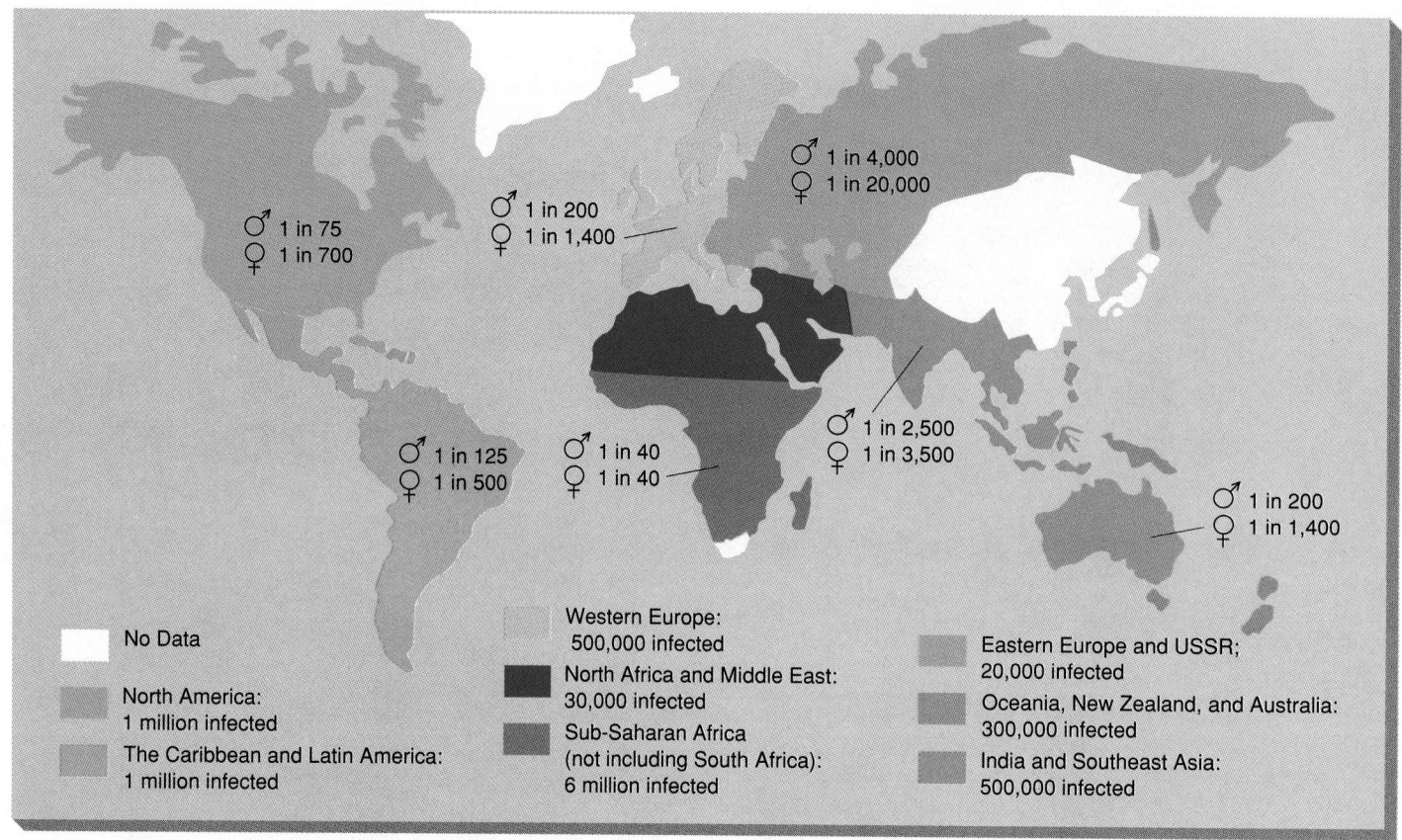

Figure 34-16 The worldwide incidence of infection with HIV, estimated by the World Health Organization in early 1991. The incidence figures are estimates for adults 15 to 49 years of age.

Victims of AIDS

In the United States, AIDS patients are not a typical cross-section of society. Initially, almost all AIDS victims in the United States were homosexual men or intravenous drug users. Some victims were hemophiliacs. The disease spreads through the homosexual population by sexual encounters. Drug users spread AIDS by sharing unsterilized hypodermic needles. Many hemophiliacs acquired AIDS through contaminated blood transfusions (before it became standard practice to screen all donated blood for anti-HIV antibodies).

Initially, many people were somewhat complacent about AIDS. They reasoned that they were not personally at risk unless they were homosexual, intravenous drug users, or needed frequent blood transfusions. Although these are still the main avenues for infection in the United States, women and heterosexual men who are not drug users have also contracted AIDS. Many health researchers therefore suspect that AIDS can be transmitted through heterosexual con-

tact, although probably not as efficiently as through homosexual contact. Further, "incidental" contact also accounts for a few cases of AIDS. As of 1990, at least 37 health-care workers have contracted AIDS from their patients, such as the nurse who picked up the virus through her chapped hands when she contacted blood from a patient undergoing emergency care. In 1990, the Centers for Disease Control concluded that a patient was infected by her dentist, himself an AIDS victim. Such cases are extremely rare, but because AIDS seems to be inevitably fatal, the public is understandably concerned.

The pattern of AIDS infection is very different in Africa. The World Health Organization believes that heterosexual intercourse is probably the most common means of infection among Africans. Further, AIDS is frighteningly common in parts of Africa (see Fig. 34-16). For example, in Abidjan, the capital of the Ivory Coast in West Africa, AIDS-related infections are now the leading cause of death among young men, and second only to complications of pregnancy and childbirth among young women.

Treatments for AIDS

For those already infected with AIDS, there are two categories of therapy. The opportunistic infections, such as Kaposi's sarcoma or *P. carinii* pneumonia, can be treated as they would in any patient. Within the past few years, more effective treatments for these diseases have prolonged the duration and improved the quality of life for AIDS patients. Second, the progress of AIDS can be slowed, but not stopped, by drugs such as ziduvidine (AZT) or dideoxyinosine (ddI). These drugs are "nucleotide mimics" that fool the viral reverse transcriptase. When incorporated into a growing DNA chain, they stop further DNA synthesis. Ideally, the viral RNA is thus never copied completely over into DNA, and therefore new virus cannot be synthesized. Unfortunately, these drugs are not completely successful in stopping reverse transcription; further, they also interfere to some extent with normal DNA replication, and in some patients they have very severe side effects.

What about normal immune responses to AIDS? As you probably know, AIDS patients produce anti-HIV antibodies (this is the basis for most blood testing). For some reason, these antibodies do little to prevent the progress of the infection. One hypothesis is that HIV spends so much of its life cycle inside cells that it is seldom exposed to the antibodies.

Enormous resources are now being devoted to preventing AIDS infections. Two show some promise. First, HIV gains entry into helpter T cells only after binding to a specific receptor on the T cell plasma membrane. Several researchers are exploring the possibility of flooding the body with other compounds that would bind to these receptors and prevent HIV from binding to them. Second, it may be possible to produce anti-HIV vaccines. This is a tricky business. As we just mentioned, the antibodies normally produced by the body do not prevent AIDS; therefore, vaccines would have to evoke a very different, and more effective, immune response than normal HIV infection does. Further, HIV has an incredible mutation rate, perhaps a thousand times faster than the flu virus. Nucleotide sequencing of virus isolated from AIDS patients shows that different people have remarkably different strains of HIV. Even more surprising, HIV can be very different even when isolated from the same patient, but at different times. Nevertheless, a few parts of HIV appear to be encoded by genes that mutate less rapidly, and clinical trials of AIDS vaccines are underway in several countries.

AIDS in Perspective

How should society respond to the AIDS epidemic? Perhaps of more immediate importance, how con-cerned should you be, personally? People in the high-risk categories should certainly take precautions against infection. Given the deadly nature of the disease, and the rising incidence of other sexually transmitted diseases, "safe sex" practices are advisable for everyone. Health care workers should exercise care when handling blood products and needles. Each year, thousands of health care workers suffer "accidental sticks" from needles. These can usually be avoided. Precautions against contamination by needle sticks not only protects against AIDS, but should reduce the incidence of hepatitis, which is a virtual epidemic among some categories of medical personnel. The odds of a patient acquiring AIDS from a physician or dentist are astronomically small, and no one should forego medical care for fear of AIDS.

On a global perspective, AIDS is a serious disease. The U.S. Public Health Service estimates that over a million Americans are already infected with HIV, and that there will have been over 400,000 full-blown cases by 1993, having caused a cumulative total of 300,000 deaths. In Africa, the figures could be much worse. On the other hand, many other diseases and disorders are much more common and cause many more deaths. For example, childbirth, malnourishment, and a host of infectious diseases cause far more deaths than AIDS in less-developed countries. In the United States, about one person in three will contract cancer, with over 500,000 deaths each year. Thus, although AIDS garners a lot of publicity, it is only one of a host of public health threats that require our attention and resources.

Cancer

Cancer, along with AIDS, is perhaps the most dreaded word in the English language, and with good reason. As we mentioned previously, nearly one out of three Americans will contract some form of cancer. For many, there will be no cure, only a slow wasting away to death. What *is* cancer? If we can prevent smallpox and polio, and cure dozens of other diseases, why can't we cure or prevent cancer?

Unlike most other diseases, cancer is not a straightforward invasion of the body by a foreign organism. Although some cancers may be triggered by viruses, in essence cancer is a malfunctioning of the growth controls of the body's own cells, a disease in which we destroy ourselves. Since it is a case of "self" fighting "self," most treatments designed to combat cancer also damage normal, healthy cells.

The usual development of any organ begins with rapid growth during embryonic life, slower growth as a juvenile, and finally maintenance of a constant size during adulthood. Individual cells may die and be replaced (as happens constantly in the stomach lining), but most organs remain about the same size throughout adult life. **A cancer is a population of cells that has escaped from normal regulatory processes and grows without control.** As a cancer grows, it uses increasing amounts of the body's energy and nutrient supplies, and literally squeezes out vital organs nearby.

Causes of Cancer

To learn the causes of cancer, we must answer two related but distinct questions: (1) What changes occur in a cancerous cell that allow it to escape normal growth controls? (2) What agents (genetic, viral, or environmental) initiate these cellular changes?

Cancer Genes

In the early 1980s, cancer researchers discovered **oncogenes:** genes that cause cancer. There seem to be two principal mechanisms by which oncogenes produce cancer.

First, **a potentially dangerous oncogene may be present in all cells, but only causes cancer when activated by some external trigger.** All cells have genes that can stimulate growth and cell division. These genes may be active during embryonic development, but usually they are turned off or transcribed more slowly in mature organisms. Other genes, such as those for the enzymes needed to metabolize glucose, are actively transcribed in many adult cells. Some oncogenes appear to be growth genes that have mistakenly been turned on full speed. For instance, chromosomes may become rearranged so that an embryonic growth gene is transferred to a part of a chromosome that is normally transcribed rapidly (Fig. 34-17). The protein synthesized under the direction of the growth gene then stimulates growth and cell division. The daughter cells inherit the same rearranged chromosome, resulting in explosive, cancerous growth.

Second, **a harmless gene may mutate into an oncogene.** Consider a gene that normally directs the synthesis of a protein that promotes cell reproduction at "maintenance levels," such as those needed to replace cells lost through normal body wear and tear. A mutation in this "pre-oncogene" may change the protein so that it greatly accelerates the rate of cell division. This would probably create a cancer.

What causes activation of oncogenes or mutations in "pre-oncogenes"? Some types of cancer are caused by viral infections (Fig. 34-18). As in HIV, these viruses have genes composed of RNA, which the virus forces the cell to "reverse transcribe" into DNA. This

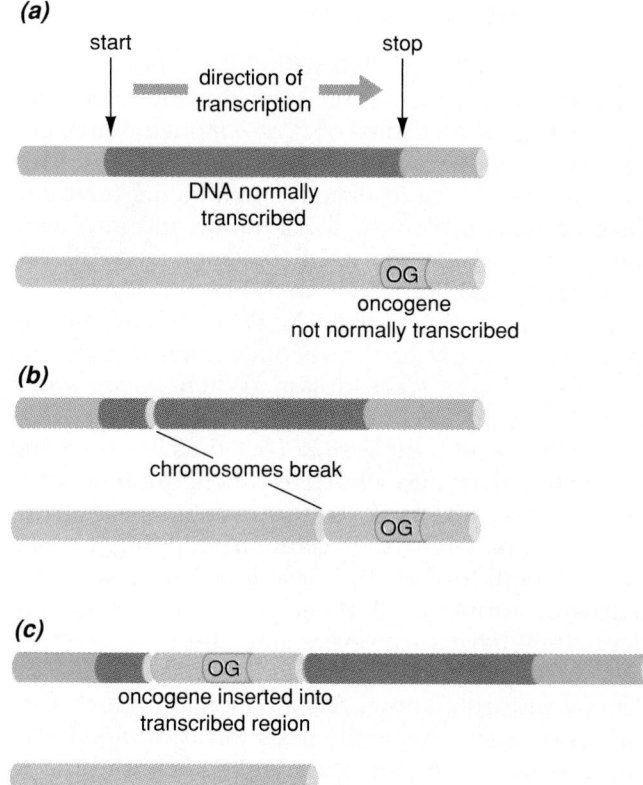

Figure 34-17 Activation of an oncogene (OG) by gene jumping. Only the hatched region of the blue chromosome is normally transcribed. The tan chromosome bearing the oncogene breaks and the oncogene is inserted within the transcribed region of the blue chromosome.

new DNA is then inserted into chromosomes of the host cell, where it is usually transcribed continuously to synthesize new viral RNA. Nearby DNA on the host chromosome may also be incidentally transcribed. If this host DNA happens to include a previously silent oncogene, the cell becomes cancerous.

Probably the most common causes of cancer are environmental insults, chiefly chemicals and radiation. We are besieged by cancer-causing chemicals (carcinogens), not only the eminently avoidable ones in cigarettes and various industrial processes but also those in the most innocent foods and even some synthesized in our own digestive tracts. Some chemicals and certain types of radiation induce point mutations in DNA. Others cause chromosomes to break in two and possibly rejoin in new and lethal combinations, by transferring oncogenes into actively transcribed regions of the chromosomes.

If these seemingly simple mechanisms cause cancer, why do cancers take so long to develop? No one

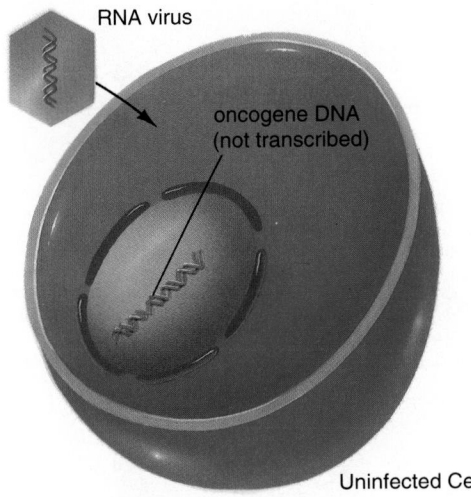

RNA virus

oncogene DNA
(not transcribed)

Uninfected Cell

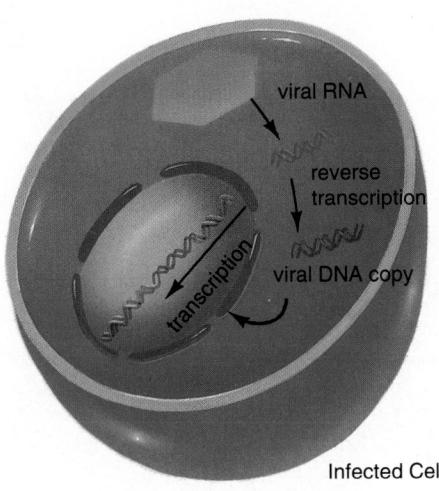

viral RNA

reverse
transcription

viral DNA copy

transcription

Infected Cell

Figure 34-18 Viral activation of an oncogene. Certain RNA viruses invade animal cells and cause "reverse transcription" of their RNA genes into DNA. This DNA then inserts itself into the DNA of the host cell. The viral DNA copy includes nucleotides that promote transcription by host RNA polymerase. Nearby host genes, including oncogenes in some cells, will be transcribed along with the viral DNA copy, and the cell may become cancerous.

knows the answer for certain, but most cancers seem to require two or more distinct steps (see "A Cancer Develops" below). Exposure to radiation early in life, perhaps, may mutate a "pre-oncogene" to a true oncogene. If this oncogene is located in a region of DNA that is not normally transcribed, no cancer will occur. Many years later, a viral infection or exposure to chemicals may move this oncogene to an active region of DNA, and cancer begins.

Tumor-Suppressor Genes

The flip side of the coin from oncogenes are **tumor-suppressor genes.** These genes suppress cellular growth and division. In some cases their protein products may turn off pre-oncogene transcription. In other cases, they may block hormonal signals that would otherwise stimulate cell division. Many cancers, including retinoblastoma and cancers of the bladder, bone, brain, breast, cervix, lung, and ovaries, appear to arise when such tumor-suppressor genes are damaged or lost. Freed from the restraint of the suppressor gene product, the affected cell becomes cancerous.

A Cancer Develops: Oncogenes and Tumor-Suppressor Genes Interact in Colon Cancer

The events that produce most cancers are not well understood. In the late 1980s, however, the sequence of genetic accidents that cause colon cancer began to be determined. Apparently, no fewer than one oncogene and three tumor-suppressor genes must mutate, in a single cell or its progeny, to produce malignant colon cancer (Fig. 34-19). The final step, damage to a tumor-suppressor gene on chromosome 17,

seems to be involved in dozens of different types of cancer.

Defenses Against Cancer

Cancer Prevention

Cancer cells form in our bodies every day, and not even the best of preventive measures can eliminate cancer completely. Gamma rays from the sun, radioactivity from the rocks beneath our feet, and naturally produced carcinogens in our food cannot be avoided. However, each of us can reduce his or her own chances of developing cancer. Some chemicals, including carotene and vitamins C and E, appear to offer protection against some forms of cancer. We can also avoid many well-known carcinogens. Cigarette smoking, for example, causes most of the lung cancers in the United States. Other chemicals, including those emitted from oil refineries and those used in certain industrial processes, can cause cancer in exposed workers. The ultraviolet rays from the sun that produce fashionable suntans are also a leading cause of skin cancer. Certain molds produce the most potent carcinogens known and can be avoided by storing food properly.

Fortunately, natural killer cells and killer T cells screen the body for cancer cells and destroy nearly all of them before they have a chance to proliferate and spread. Since cancer cells are "self" cells, and the immune system does not respond to "self," how are cancer cells weeded out? Probably the very processes that cause cancer also cause new and slightly different proteins to appear on the surfaces of cancer cells. Killer cells encounter these new proteins, recognize

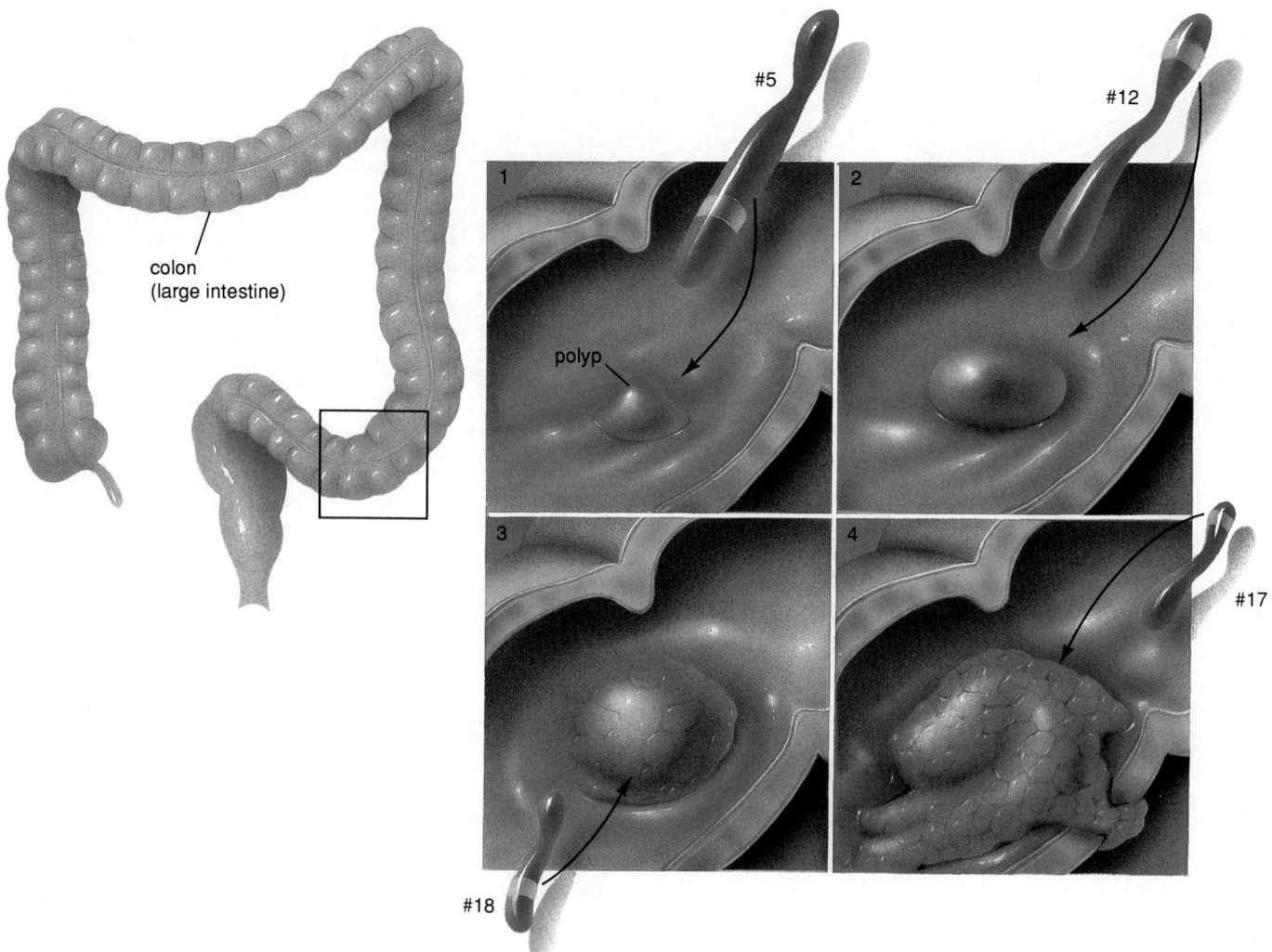

Figure 34-19 The development of cancer of the colon involves four genetic changes.
(1) A tumor-suppressor gene on chromosome 5 is mutated in one of the cells of the
colon. The cell multiplies, forming a polyp. (2) In one of the cells of the polyp, a
proto-oncogene on chromosome 12 mutates into a true oncogene. This cell multiplies
more rapidly, and the polyp becomes large. (3) A second tumor-suppressor gene, this
time on chromosome 18, mutates in one of the cells that already has the first two
mutations. This cell speeds up its multiplication rate, and the polyp becomes a
tumor. (4) Finally, a tumor-suppressor gene on chromosome 17 mutates in a cell with
all three previous mutations. Its daughter cells now multiply wildly, invading the
wall of the colon and spreading cancer to other parts of the body.

them as "non-self" antigens, and destroy the cancer
cells. Without constant surveillance by the immune
system, it is unlikely that any of us would survive
more than a few years.

Cancer Treatment

Sometimes, however, the immune system does *not*
recognize cancer cells as "non-self." Ignored by the

immune system, the cancer grows and spreads. What
can medical science do to cure cancer? The rate of
cure is increasing, but is still scarcely a third of all
cancers. The three main approaches taken are all
quite crude: burn the cancer out with radiation, cut it
out with surgery, or poison it with drugs.

If a cancer is discovered when it is small enough,
radiation or surgery may be able to eliminate it.
Breast cancer, for example, can almost always be

eliminated by surgery if detected early enough. However, while the body is rid of the cancer, surgery and radiation therapies may be traumatic, dangerous, and disfiguring.

In principle, chemotherapy might be able to destroy cancer cells without damaging normal cells, since cancer cells are, after all, different in many ways from normal cells. The most common chemotherapies involve drugs that are "nucleotide mimics" (different ones than are used to treat AIDS). These mimics are incorporated into DNA during chromosome replication. They then either prevent further replication or cannot be transcribed correctly. In either case, the cell dies or fails to reproduce. These drugs obviously have their main effect on dividing cells, and since cancer cells divide rapidly, they kill cancer cells. Unfortunately, other cells of the body divide too, such as those in the hair follicles and intestinal lining. Chemotherapy drugs damage those cells, producing the well-known side effects of nausea, vomiting, and hair loss.

Future drug therapies may use **monoclonal antibodies** directed against cancer cells (Fig. 34-20). A small patch of cancer cells may be snipped out of a patient and injected into a mouse. Recognizing the human cancer as "non-self," B cells of the mouse proliferate. Anticancer B cells are extracted from the mouse's spleen and fused with a particular strain of cancerous white blood cells, called myeloma cells. Some of the fused cells combine the desired properties of both original cell types: from the mouse B cells, specific antibody production against the patient's cancer, and from the myeloma cells, rapid cell division. All the descendant cells from one such cell (a clone from a single cell) produce the same anticancer antibody (hence the term *monoclonal antibody*). Massive amounts of antibodies specific for the cancer cells, and for no other cells of the patient, could be produced in the lab. Particularly lethal drugs may be attached to the antibodies, which are then injected into the patient. The antibodies bind only to the cancer cells, so the drug destroys them without harming normal body cells. Although this technique has not yet been perfected, it offers great promise.

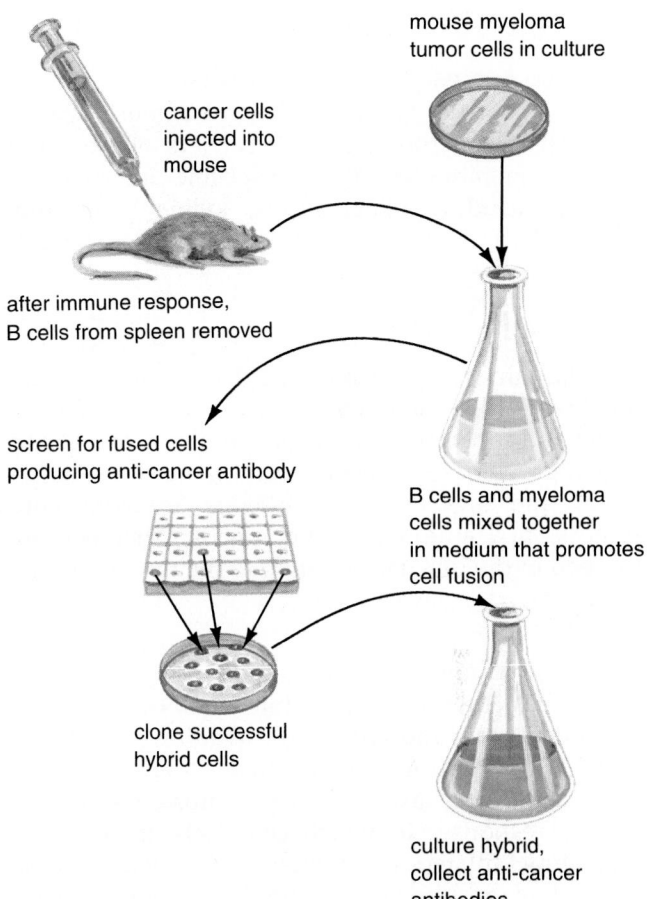

Figure 34-20 The production of monoclonal antibodies against a cancer.

In the early 1990s, clinical trials of an experimental therapy were begun. A special class of white blood cells, called tumor infiltrating lymphocytes, seeks out and invades tumors. Through genetic engineering, genes for a protein called tumor necrosis factor, which destroys cancerous cells, are inserted into a patient's own tumor infiltrating lymphocytes. These are then injected back into the patient, hopefully to seek and destroy the patient's cancer.

SUMMARY OF KEY CONCEPTS

Defenses Against Microbial Invasion

The human body has three lines of defense against invasion by microbes: (1) the barriers of skin and mucous membranes; (2) nonspecific internal defenses, including phagocytosis, killing by natural killer cells, inflammation, and fever; and (3) the immune response.

Barriers

The skin physically blocks the entry of microbes into the body. It is also covered with secretions from sweat and sebaceous glands that inhibit bacterial and fungal growth. The mucous membranes of the respiratory and digestive tracts secrete antibiotic substances, IgA antibodies, and mucus. Microbes are trapped in the mucus, swept up to the throat by cilia, and expelled or swallowed.

Nonspecific Internal Defenses

If microbes enter the body, white blood cells travel to the site of entry and engulf the invading cells. Natural killer cells secrete proteins that kill infected body cells or cancerous cells. Injuries stimulate the inflammatory response, in which chemicals are released that attract phagocytic white blood cells, increase blood flow, and make capillaries leaky. Later, blood clots wall off the injury site, preventing further spread of the microbes. Fever is caused by chemicals called pyrogens released by white blood cells in response to infection. High temperatures inhibit bacterial growth and accelerate the immune response.

The Immune Response

The immune response involves two types of lymphocytes, B cells and T cells. Plasma cells, which are descendants of B cells, secrete antibodies into the bloodstream, causing humoral immunity. Killer T cells destroy some microbes, cancer cells and virus-infected cells on contact, causing cell-mediated immunity. Helper T cells stimulate both the humoral and cell-mediated immune responses. Immune responses have three steps: recognition, attack, and memory.

Recognition

Each immune cell synthesizes only one type of antibody, unique to that particular cell and its progeny. The diversity of antibodies arises from gene shuffling and mutation of antibody genes during immune cell development. Each antibody has specific sites that bind one or a few types of antigen. Normally, only foreign antigens are recognized by the immune cells.

Attack

Antigens bind to and activate only those B and T cells with the complementary antibodies. In humoral immunity, B cells with the proper antibodies divide rapidly, producing plasma cells that synthesize massive quantities of the antibody. The circulating antibodies destroy antigens and antigen-bearing microbes by four mechanisms: direct neutralization, promotion of phagocytosis by white blood cells, agglutination, and complement reactions. In cellular immunity, T cells with the proper antibodies also divide rapidly. Their descendant killer T cells bind to antigens on microbes, infected cells, or cancer cells and kill the cells. Helper T cells stimulate, and suppressor T cells turn off, both the B and killer T cell responses.

Memory

Some progeny cells of both B and T cells are long-lived memory cells. If the same antigen reappears in the bloodstream, these memory cells are immediately activated, divide rapidly, and cause an immune response that is much faster and more effective than the original response.

Medicine and the Immune Response

Antibiotics kill microbes or slow down their reproduction, thus allowing the immune system more time to respond and exterminate the invaders. *Vaccinations* are injections of antigens from disease organisms, often the weakened or dead microbes themselves. An immune response is evoked by the antigens, providing memory and a rapid response should a real infection occur. *Allergies* are immune responses to normally harmless foreign substances, such as pollen or dust. Certain cells respond to the presence of these substances by releasing histamine, which causes a local inflammatory response. Some diseases are caused by defective immune responses. *Autoimmune diseases* arise when the immune system destroys some of the body's own cells. *Immune deficiency diseases* occur when the immune system cannot respond strongly enough to ward off normally minor diseases.

AIDS

AIDS (acquired immunodeficiency syndrome) is caused by one of two viruses, called human immunodeficiency viruses 1 and 2. These viruses invade helper T cells and destroy them. Without helper T cells to stimulate the immune responses of B cells and killer T cells, the AIDS victim is extremely susceptible to a wide assortment of diseases. These opportunistic infections eventually kill the patient.

Cancer

Cancer is a population of the body's cells that grows without control. Some cancers may be caused by activation of growth genes, called oncogenes, that cause cells to grow and multiply. These genes may be activated by viral infection or mutations caused by chemicals or radiation. Other cancers may be caused by loss or inactivation of tumor-suppressor genes, that turn off or suppress cell division in normal cells.

GLOSSARY

agglutination (a-glu-tin-ā'-shun): clumping of foreign substances or microbes, caused by binding with antibodies.

allergy: an inflammatory response produced by the body in response to invasion by foreign materials, such as pollen, which are themselves harmless.

antibody: a protein produced by cells of the immune system which combines with a specific antigen and usually facilitates its destruction.

antigen: a complex molecule, usually protein or polysaccharide, that stimulates the production of a specific antibody.

autoimmune disease: a disorder in which the immune system produces antibodies against the body's own cells.

B cell: a type of lymphocyte that gives rise to daughter cells (plasma cells) that secrete antibodies into the circulatory system.

cancer: a disease in which some of the body's cells grow without control.

cell-mediated immunity: an immune response in which foreign cells or substances are destroyed by contact with T cells.

complement: a group of blood-borne proteins that participate in the destruction of foreign cells to which antibodies have bound.

complement reactions: interactions among foreign cells, antibodies, and complement proteins, resulting in the destruction of the foreign cells.

constant region: part of an antibody molecule that is similar or identical in all antibodies.

endogenous pyrogen: a chemical produced by the body that stimulates the production of a fever (elevated body temperature).

fever: an elevation in body temperature caused by chemicals (pyrogens) released by white blood cells in response to infection.

helper T cell: a type of T cell that aids other immune cells to recognize and act against antigens.

histamine: a substance released by certain cells in response to tissue damage and invasion of the body by foreign substances. Histamine promotes dilation of arterioles and leakiness of capillaries, and triggers some of the events of the inflammatory response.

humoral immunity: an immune response in which foreign substances are inactivated or destroyed by antibodies circulating in the blood.

immune deficiency disease: a disorder in which the immune system is incapable of responding properly to invading disease organisms.

immune response: a specific response by the immune system to invasion of the body by a particular foreign substance or microorganism, characterized by recognition of the foreign material by immune cells and its subsequent destruction by antibodies or cellular attack.

inflammatory response: a nonspecific, local response to injury to the body, characterized by phagocytosis of foreign substances and tissue debris by white blood cells, and "walling off" of the injury site by clotting of fluids escaping from nearby blood vessels.

interferon: a protein released by certain virus-infected cells that increases the resistance of other, uninfected, cells to viral attack.

killer T cell: a type of T cell that directly destroys foreign cells upon contacting them.

macrophage: a type of white blood cell that engulfs microbes. Macrophages destroy microbes by phagocytosis and also present microbial antigens to T cells, helping to stimulate the immune response.

major histocompatibility complex (MHC): proteins, usually located on the surfaces of body cells, that identify the cell as "self"; MHC proteins are also important in stimulating and regulating the immune response.

memory cell: a long-lived descendant of a B or T cell that has been activated by contact with antigen. Memory cells are a reservoir of cells that rapidly respond to reexposure to the antigen.

monoclonal antibody: any of the antibodies produced by a clone of genetically identical cells; all antibodies produced by the clone of cells are identical.

mucous membrane: the lining of the inside of the respiratory and digestive tracts.

natural killer cell: a type of white blood cell that destroys some virus-infected cells and cancerous cells on contact. Part of the nonspecific internal defense against disease.

neutralization: the process of covering up or inactivating a toxic substance with antibody.

oncogene: a gene that, when transcribed, causes a cell to become cancerous.

plasma cell: an antibody-secreting descendant of a B cell.

retrovirus: a virus that uses RNA as its genetic material. When a retrovirus invades a eukaryotic cell, it "reverse transcribes" its RNA into DNA, which then directs the synthesis of more viruses, using the transcription and translation machinery of the cell.

reverse transcriptase: an enzyme found in retroviruses that catalyzes the synthesis of DNA from an RNA template.

suppressor T cell: a type of T cell that depresses the response of other immune cells to foreign antigens.

T cell: a type of lymphocyte that recognizes and destroys specific foreign cells or substances, or that regulates other cells of the immune system.

tumor-suppressor gene: a gene that encodes information for a protein that inhibits cancer formation, probably by regulating cell division in some way.

vaccine: a material injected into the body that contains antigens characteristic of a particular disease organism, and that stimulates an immune response.

variable region: part of an antibody molecule that differs among antibodies; the ends of the variable regions of the light and heavy chains form the specific binding site for antigen.

STUDY QUESTIONS

1. List the three lines of defense of the human body against invading microbes. Which are nonspecific (i.e., act against all types of invaders) and which are specific (i.e., act only against a particular type of invader)? Explain your answer.
2. Name three antiinfection properties of the skin and three of the mucous membranes.
3. How do natural killer cells and killer T cells destroy their targets?
4. Describe the inflammatory response.
5. Describe humoral immunity and cell-mediated immunity. Include in your answer the types of immune cells involved in each, the location of antibodies that attach to foreign antigens, and the mechanisms by which invading cells are destroyed.
6. How does the immune system construct so many different antibodies?
7. How does the body distinguish "self" from "non-self"?
8. Diagram the structure of an antibody. What parts bind

to antigens? Why does each antibody only bind to a specific antigen?
9. What are memory cells? How do they contribute to long-lasting immunity to specific diseases?
10. What is a vaccine? How does it confer immunity to a disease?
11. Describe the allergic reaction. What is the probable "normal function" of the IgE antibodies that cause allergies?
12. Distinguish between autoimmune diseases and immune deficiency diseases, and give one example of each.
13. Describe the causes, progression, and eventual outcome of AIDS. How do AIDS treatments work?
14. What is cancer? How do oncogenes cause cancer? How do tumor suppressor genes prevent cancer? How can environmental factors "turn on" oncogenes to cause cancer?

DISCUSSION QUESTIONS

1. Discuss why the following statement is true: Human cells do not have genomes of infinite size, but it is possible for the human immune system to make antibodies against a virtually infinite number of different pathogens.
2. Some types of cancer are partly hereditary; that is, members of certain families are more likely than the general population to develop certain cancers, but not all family members develop the cancer. Given the mul-

tistep mechanism for colon cancer outlined in Figure 34-19, could you propose a genetic model that would account for predisposition to cancer while simultaneously not requiring that all people carrying, for example, a defective tumor-suppressor gene necessarily develop cancer?
3. Why is it essential that antibodies bind only relatively large molecules (like proteins) and not relatively small molecules (like amino acids)?

SUGGESTED READINGS

Buisseret, P. "Allergy." *Scientific American*, August 1982. Allergy is now understood as a malfunctioning of the immune system.

Gallo, R. C. "The First Human Retrovirus," and "The AIDS Virus." *Scientific American*, December 1986 and January 1987. Readable descriptions of viruses of the immune system, by one of their chief discoverers.

Jaret, P. "The Wars Within." *National Geographic*, June 1986. Lucid diagrams of the complexity of the immune response, accompanied by incredible photographs by Lennart Nilsson.

Leder, P. "The Genetics of Antibody Diversity." *Scientific American*, May 1982. How only a few hundred genes can be used to make millions of antibodies.

Marrack, P., and Kappler, J. "The T Cell and Its Receptor." *Scientific American*, February 1986. The T cell, especially the helper T cell, is central to the entire immune response.

Milstein, C. "Monoclonal Antibodies." *Scientific American*, October 1980. One of the inventors of the monoclonal antibody technique explains the process and some of its uses.

Radetsky, P. "The Roots of Cancer." *Discover*, May 1991. A lucid discussion of the genetics of cancer.

Rennie, J. "The Body Against Itself." *Scientific American*, December 1990. Many diseases, including insulin-dependent diabetes, rheumatic fever, multiple sclerosis, and rheumatoid arthritis, are autoimmune diseases. This article describes what autoimmune diseases are, how they may develop, and potential strategies to fight them.

Tonegawa, S. "The Molecules of the Immune System." *Scientific American*, October 1985. A brief overview of the immune system.

Young, J. D–E., and Cohn, Z. A. "How Killer Cells Kill." *Scientific American*, January 1988. Killer T cells strike by secreting proteins that form large holes in the plasma membrane of their targets.

"What Science Knows About AIDS." *Scientific American*, October 1988. A single-topic issue devoted entirely to AIDS.

35

Chemical Control of the Animal Body: The Endocrine System

The changes that occur throughout the life cycles of all animals are under the control of hormones. One of the more dramatic events during the life cycle of certain insects is metamorphosis, illustrated here by a pair of red lacewing butterflies from Australia simultaneously emerging from their pupal cases.

During evolution, cells in multicellular organisms evolved many ways of communicating among themselves. Cell-to-cell communication is crucial to the control of movement, the maintenance of homeostasis, growth, and reproduction. One method of cellular communication is by direct contact. Molecules protruding from the surface membrane identify cells as belonging to an individual of a particular species, as parts of a unique individual organism, and as specific cell types, such as skin or liver. Surface contacts are important in the development of embryos, in which cells migrate around one another to arrive at their proper destination. Direct contact also plays a key role in defense against disease organisms, in which immune cells recognize invading foreign cells by their surface molecules (see Chapter 34).

In contrast to direct contact, cells can communicate with vast numbers of cells over large distances by releasing chemicals that influence other cells. Such chemical communication ranges from the "come hither" messages of slime molds (see "A Closer Look at How Chemical Communication Creates an Organism") to the sophisticated signals released by the mammalian endocrine and nervous systems.

Chemical Communication Within the Animal Body

Although it is convenient to discuss hormonal control separately from nervous control (Chapter 36), the two operate in strikingly similar ways. Both hormone-producing cells and nerve cells synthesize "messenger" chemicals that they release into extracellular spaces (Fig. 35-1). **There are four main differences between the hormonal and neural control of the animal body, using chemical messages:**

1. **The distance over which the chemical is transmitted.**
2. **The number of cells contacted by the chemical.**
3. **The speed at which the message is transmitted.**
4. **The duration of the message.**

First, nerve cells usually release their chemical messages (called **neurotransmitters**) very close to the cells they influence, often from less than a micrometer away. Hormone-producing cells, on the other hand, release their chemicals into the bloodstream, which carries them throughout the body, often to a specific target a considerable distance away. Second, a nerve cell very precisely bathes one or a few specific target cells with its chemicals, whereas hormones, carried in

the blood, bathe millions of cells indiscriminately. Third, a nerve cell speeds information from one part of the body to another via electrical signals traveling within the cell itself, and only then releases its neurotransmitter. Hormones, traveling in the bloodstream, move much more slowly. Finally, the effects of hormones generally far outlast those of neurotransmitters.

Even these differences blur as we learn more about the hormonal and nervous systems of animals. Some hormones are in fact produced and released into the bloodstream by nerve cells, and are appropriately called **neurohormones**. Other chemicals that biologists formerly believed were strictly hormones, such as insulin, have recently been discovered in the brain, where they are synthesized and released by neurons and act more like neurotransmitters.

In this chapter we emphasize the "classical" hormones and neurohormones. Although the functioning of the nervous system is described in the following chapter, remember that no system in your body works alone. The hormonal and nervous systems are closely coordinated in their control of bodily functions.

Hormone Function in Animals

A **hormone** is a chemical secreted by cells in one part of the body that is transported in the bloodstream to other parts of the body, where it affects particular target cells. As we shall see, the same hormone may have several different effects, depending on the nature of the target cells it contacts.

Hormone Actions on Target Cells

Since nearly all cells have a blood supply, once hormones enter the bloodstream, they reach nearly every cell of the body. But in order to exert their precise control, hormones must act only on certain target cells. **Hormone specificity is determined by receptors, specialized proteins located either inside the target cell or in its surface membrane. If a cell lacks a specific receptor for a hormone, the hormone will have no effect.** Receptors for hormones are found in three general locations on target cells: in the cell membrane, in the cytoplasm, and in the nucleus.

Hormones May Bind to Surface Receptors

Most peptide and protein hormones, as well as epinephrine and norepinephrine, are water-soluble but not lipid-soluble, and hence cannot cross cell membranes. These hormones react with protein receptors

(a) Hormone-Producing Cell

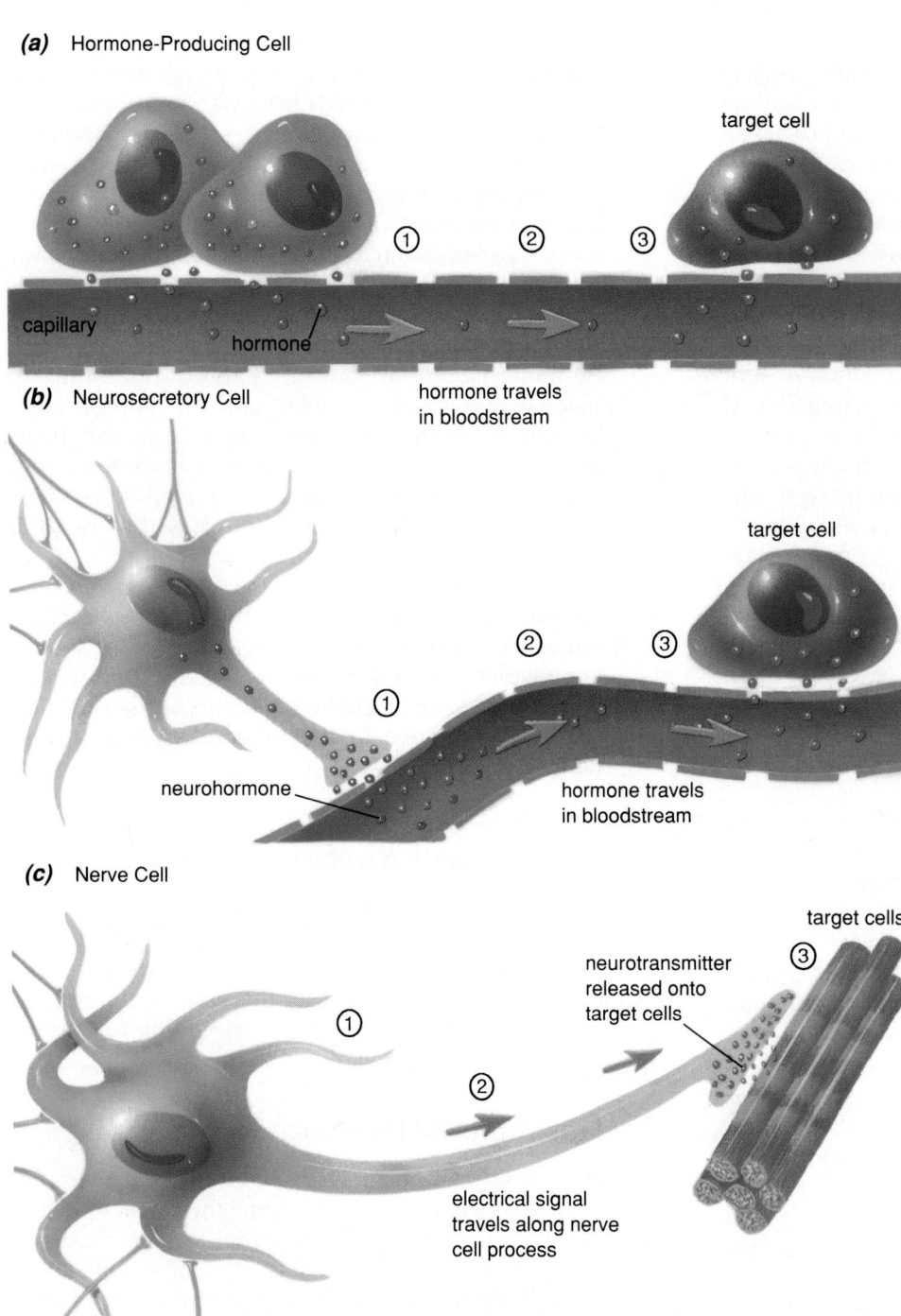

capillary

hormone

target cell

① ② ③

hormone travels
in bloodstream

(b) Neurosecretory Cell

neurohormone

② ③

①

target cell

hormone travels
in bloodstream

(c) Nerve Cell

①

②

electrical signal
travels along nerve
cell process

neurotransmitter
released onto
target cells

③

target cells

Figure 35-1 Three major types of "control cells" release chemicals that influence the activity of other cells of the body: **(a)** "classical" hormone-producing cells, **(b)** neurosecretory cells, and **(c)** ordinary nerve cells. Three common steps occur in each system: (1) The chemical-releasing cell is stimulated. (2) It sends a message to distant cells. (3) Selected target cells respond. Both classical hormone-producing cells and neurosecretory cells release their chemical messages into the bloodstream, which transports the chemicals to distant target cells. Regular nerve cells grow long processes to their target cells and release their chemical messages directly onto the target cell.

protruding from the outside surface of target cell membranes. **In general, hormones that bind to surface receptors trigger rapid, short-term responses.** Some receptors, such as those for epinephrine and norepinephrine, are directly linked to channels that are opened in response to the binding of the hormone. More frequently, a **second messenger** system

is used. When the hormone binds to the receptor, the shape of the receptor is altered, triggering a series of biochemical reactions that change the activity of the cell (Fig. 35-2a).

In many cases, the binding of the hormone to the receptor activates an enzyme. When activated, the enzyme catalyzes the conversion of ATP to **cyclic**

AMP (see Chapter 6), a nucleotide that regulates many cellular activities by activating enzymes. Cyclic AMP is often called a **second messenger,** since it transfers the signal from the first messenger, the hormone, to molecules within the cell. The formation of cyclic AMP initiates a series of reactions inside the cell. Each of these reactions involves an increasing number of molecules, amplifying the original signal. The end result varies with the target cell. For example, channels may be opened in the cell membrane, or substances may be synthesized or secreted.

Another second messenger that may be activated by hormones is **calmodulin.** The binding of certain hormones to their receptors triggers the opening of calcium channels, allowing an influx of calcium into the cell. Calmodulin is a protein in the cytoplasm that binds calcium ions, changing its configuration as a result. The altered shape of the calcium–calmodulin complex allows it to activate enzymes, acting in a manner similar to cyclic AMP.

Hormones May Bind to Intracellular Receptors

Steroid hormones and thyroid hormones are able to penetrate the cell membrane and bind to receptors inside the cell (Fig. 35-2b). **Both steroid and thyroid hormones alter the activity of the genes. It may take from minutes to days for these hormones to exert their full effects.** Steroid hormones bind to protein receptors in the cytoplasm. The receptor–hormone complex then travels into the nucleus where it binds to DNA and initiates the transcription of messenger RNA from specific genes. The messenger RNA then moves into the cytoplasm and directs the synthesis of specific new proteins. The thyroid hormones act in a similar manner, but enter the nucleus and bind to receptors that are already associated with the chromosomes. The receptor–hormone complex then initiates the transcription of specific genes, which will be translated into proteins on ribosomes in the cytoplasm. Many of these proteins are enzymes involved in cell growth and metabolic activity.

Hormones Maintain Homeostasis Through Negative Feedback

Animals usually regulate the release of hormones through *negative feedback.* **During negative feedback, the secretion of a hormone causes effects in target cells that inhibit further secretion of the hormone.** Negative feedback is one way of maintaining **homeostasis**—that is, keeping conditions within the body relatively constant over time (see Chapter 25). Most hormones exert such powerful effects on the body that it would be harmful to have too much hormone working for too long. For example, suppose you have jogged several miles on a hot, sunny day, and lost a liter of water through perspiration. In response to the loss of water from your bloodstream, your pituitary gland releases antidiuretic hormone (ADH), which causes your kidneys to reabsorb water and produce a very concentrated urine (see Chapter 33). However, if you arrive home and drink a half gallon of Gatorade, you will more than replace the water you lost in sweat. Continued retention of this water would overload the bloodstream, raising blood pressure and possibly damaging your heart. The negative feedback loop ensures that when your blood water level returns to normal, ADH secretion is turned off and your kidneys begin eliminating the excess water (see Fig. 33-7); negative feedback is also shown in the control of "milk letdown" (Fig. 35-6), and thyroxine secretion (Fig. 35-10). As this example illustrates, hormone secretion is regulated so that just the right amounts are released at the right times. Another example of precise regulation of hormone production is seen in the insect, as described in "A Closer Look at the Hormonal Control of the Insect Life Cycle."

Types of Animal Hormones

There are four classes of chemicals used as hormones in the animal kingdom as shown in Table 35-1.

Modified Amino Acids

A few hormones are chemically modified amino acids. The amino acid tyrosine forms the basis for the hormones epinephrine, norepinephrine, and the thyroid hormones.

Peptides and Proteins

Most types of hormones are peptides or proteins, chains of amino acids ranging from a few to over a hundred amino acids in length. These include all the hormones of the hypothalamus and the anterior and posterior pituitary, as well as insulin, antidiuretic hormone, and others.

Steroids

Steroid hormones all have a chemical structure resembling cholesterol, from which most of them are synthesized. Steroid hormones are secreted by the ovaries and placenta (estrogen and progesterone), the testes (testosterone), and the adrenal cortex (cortisol and aldosterone).

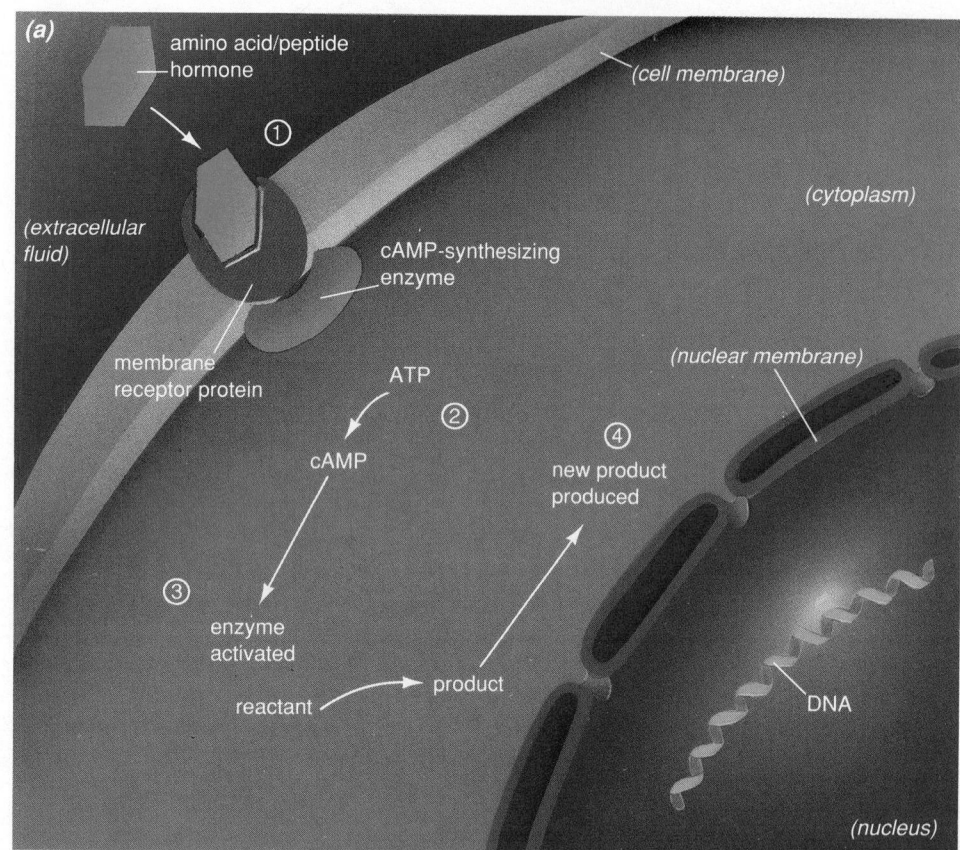

Figure 35-2 Modes of action of hormones.
(a) Amino acid and peptide hormones bind to a receptor on the outside of the target cell membrane ①. Hormone-receptor binding triggers synthesis of cyclic AMP (cAMP) ②. Cyclic AMP in turn activates specific enzymes ③ that promote specific cellular reactions that produce new products ④. This cyclic AMP "cascade" may generate a variety of responses. Examples include an increase in glucose synthesis induced by adrenalin and an increase in estrogen synthesis induced by luteinizing hormone.

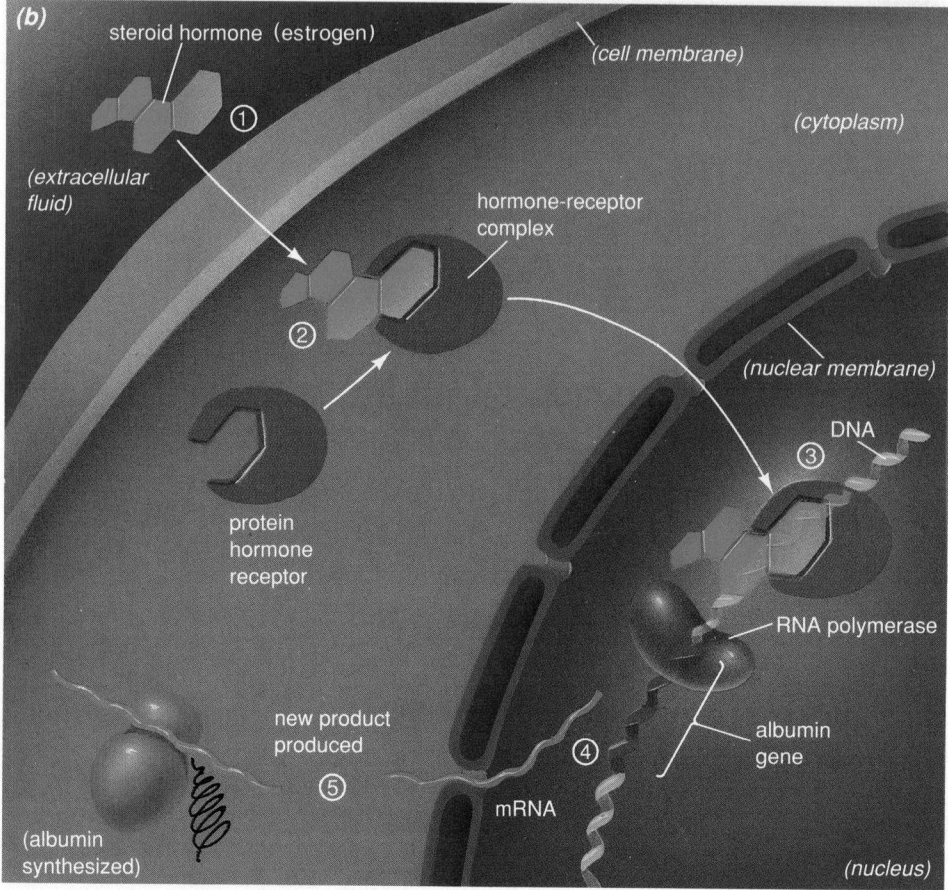

(b) Lipid-soluble steroid hormones diffuse readily through the cell membrane into the target cell ①, where they combine with a protein receptor molecule in the cytoplasm ② and travel to the nucleus. The steroid-receptor complex facilitates the binding of RNA polymerase to promoter sites on specific genes ③, accelerating transcription ④ of DNA into messenger RNA (mRNA). The mRNA then directs protein synthesis ⑤. In hens, for example, estrogen promotes transcription of the albumin gene, causing synthesis of albumin (egg white), which is packaged in the egg as a food supply for the developing chick.

Table 35-1 The Chemical Diversity of Vertebrate Hormones

Chemical Type	Examples
Modified amino acids (synthesized from **single amino acids**)	Noradrenalin
Peptides and proteins (synthesized from **multiple amino acids**)	Oxytocin
Steroids (synthesized from **cholesterol**)	Testosterone
	Estradiol
Prostaglandins (synthesized from **fatty acids**)	Prostaglandin E$_1$

Modified Fatty Acids

Nearly every type of cell in the body has been found to produce **prostaglandins**. These substances consist of two fatty acid carbon chains attached to a 5-carbon ring.

Mammalian Endocrine Systems

Endocrinologists (biologists who study the endocrine system) are still far from achieving a complete understanding of hormonal control in mammals. New hormones, or new roles for previously known hormones, are discovered virtually every year. What we might call the seven major endocrine systems, however, have been known for many years. These are the hypothalamus–pituitary complex, the thyroid, parathyroid, pancreas, adrenal cortex, adrenal medulla, and gonads (Fig. 35-3). Table 35-2 lists these and other glands, their major hormones, and their principal control functions.

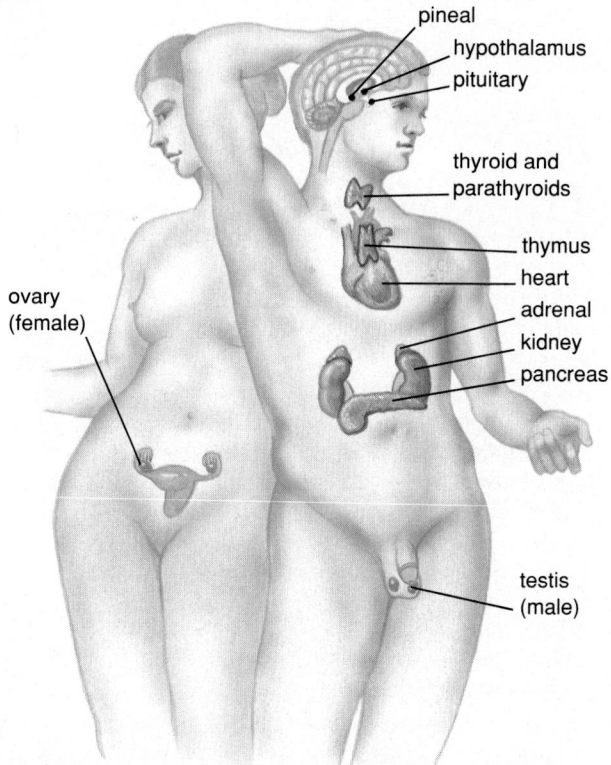

Figure 35-3 The major mammalian endocrine glands discussed in the text are the hypothalamus–pituitary complex, the thyroid and parathyroids, the adrenal glands, the pancreas, and the gonads (ovaries in females, testes in males). Other organs that secrete hormones include the pineal gland, thymus, heart, kidney, and digestive tract.

A CLOSER LOOK

At How Chemical Communication Creates an Organism

Most multicellular organisms, humans included, produce a variety of chemical secretions that are used for communication among cells. Biologists classify these secretions as hormones if they influence cells in the same body that produced them, and as pheromones if they influence cells in another animal's body (see Chapter 42). This dichotomy may be a product of our own biases, since each human is a self-contained organism. But consider the case of the cellular slime mold.

If you go out to your garden and pick up a pinch of soil, you will be holding millions of organisms, mostly bacteria, between your fingers. You will also be holding thousands of single-celled, amoebalike creatures called cellular slime molds. These amoeboid cells crawl along in the film of water surrounding each soil particle, engulfing bacteria for food. What happens to the slime molds if they eat bacteria faster than the bacteria can reproduce? Needless to say, the amoeboid cells cannot move very fast or very far in search of food. To survive, they must find some other way to travel the relatively vast distance to the nearest new food source. Cellular slime molds solve this problem by temporarily merging to create a multicellular organism.

If slime mold cells go long enough without food, they release a chemical into the soil water. This chemical is called *acrasin*, after Acrasia, a witch in Edmund Spenser's poem *Faerie Queene*. Acrasia attracted men and turned them into animals, which, as you will see, is very similar to what acrasin does to slime mold cells. The concentration of acrasin is highest right around the cell releasing it, and decreases with distance as it diffuses away (Fig. E35-1). Other amoeboid cells respond to acrasin by migrating up the gradient of increasing concentration until they encounter the releasing cell. The new cells remain with the original one, releasing their own acrasin, until thousands of cells have gathered. (Different species of slime molds release different acrasins. The acrasin of one common slime mold is cyclic AMP, the same chemical that mediates cellular responses to many animal hormones.) At this point, a remarkable transformation occurs. The mass of cells behaves as a coordinated organism, crawling around for a while (Fig. E35-2). Eventually the "slime slug" stops moving, and the cells begin to climb atop one another, creating first a lump, then a knob, and eventually a tall stalk (Fig. E35-3). The cells at the tip differentiate into weather-resistant spores that are dispersed to distant sites. A lucky spore may land on a patch of fertile soil or a rotting log, revert back to an amoeboid cell, and start the cycle over again. Viva la mold!

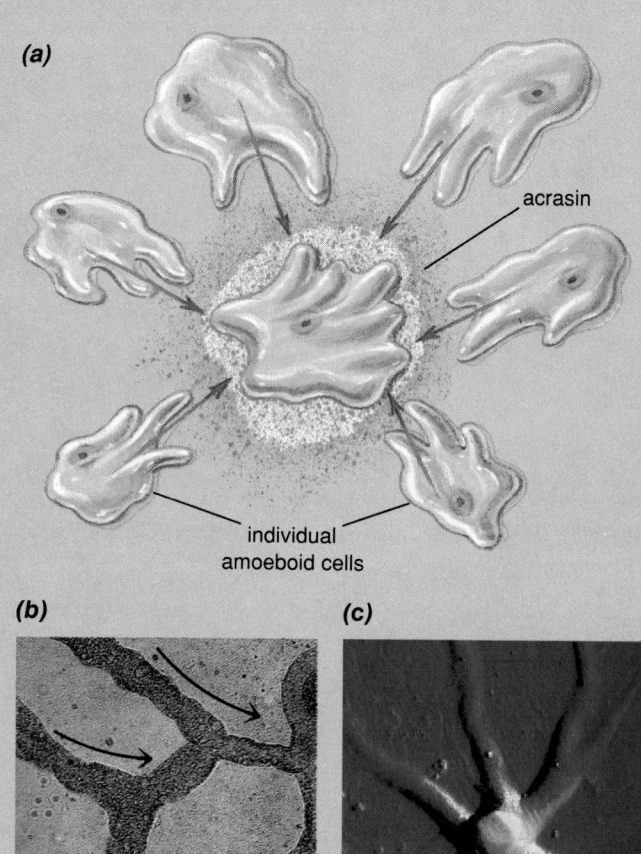

(a)

acrasin

individual amoeboid cells

(b)   *(c)*

Figure E35-1 (a and b) A slime mold cell releases acrasin into its surroundings, establishing a gradient that is followed by other cells. **(c)** Slime mold cells stream from all directions toward the acrasin-secreting center.

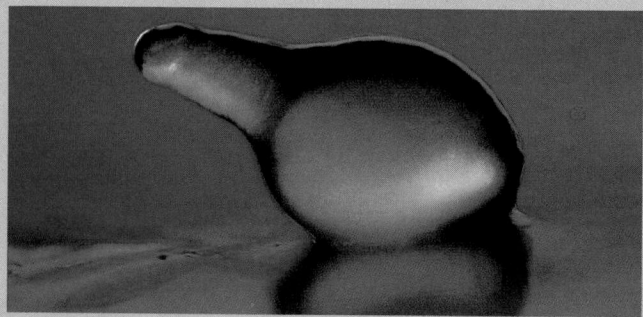

Figure E35-2 The aggregated mold cells form a crawling, sluglike mass.

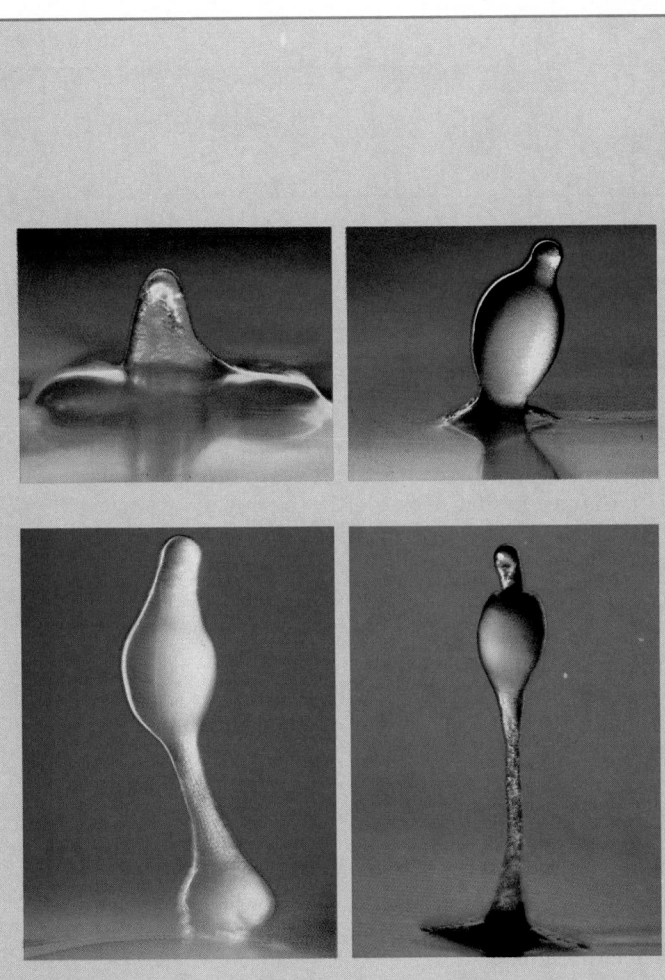

Figure E35-3 The dispersal stage of the cellular slime mold life cycle: out of the ooze rises a tall stalk of cells, upon which sits a capsule containing numerous spores.

Gland Structure and Function

Mammals have two types of glands: **exocrine glands** and **endocrine glands** (Fig. 35-4). Exocrine glands produce secretions that are released *outside the body* ("exo" means "out of" in Greek) or *into the digestive tract* (which is actually a hollow tube continuous with the outside world via mouth and anus). Exocrine gland secretions are released through tubes or openings called ducts. The exocrine glands include the sweat and sebaceous (oil-producing) glands of the skin, the lacrimal (tear-producing) glands of the eye, the mammary (milk-producing) glands, as well as glands producing digestive secretions. The endocrine glands, sometimes called ductless glands, release their hormones *within the body* ("endo" means "inside of"). An endocrine gland generally consists of clusters of hormone-producing cells embedded within a network of capillaries. The cells secrete their hormones into the extracellular fluid surrounding the capillaries. The hormones then enter the capillaries by diffusion and are distributed throughout the body via the bloodstream.

The Hypothalamus–Pituitary Complex

The **hypothalamus** is part of the brain (see Chapter 36) that contains clusters of specialized nerve cells, called **neurosecretory cells** (see Fig. 35-1b). Neurosecretory cells synthesize peptide hormones, store them, and release them when stimulated. The **pituitary** is a pea-sized gland that dangles from the hypothalamus by a stalk (Figs. 35-3 and 35-5). Anatomically, the pituitary consists of two distinct parts: the **posterior pituitary** and the **anterior pituitary**. The hypothalamus controls the release of hormones from both parts. The anterior pituitary is a true endocrine gland, composed of several types of hormone-secreting cells enmeshed in a network of capillaries. The posterior pituitary, on the other hand, is derived during fetal development from an outgrowth of the hypothalamus. Posterior pituitary hormones are secreted by the endings of neurosecretory cells whose cell bodies are in the hypothalamus. These nerve endings in the posterior pituitary lie in a capillary bed that carries their hormones into the bloodstream (Fig. 35-5).

The Posterior Pituitary

The posterior lobe of the pituitary contains the endings of two types of neurosecretory cells from the hypothalamus. These produce two peptide hormones, **antidiuretic hormone (ADH)** and **oxytocin.**

Table 35-2   Mammalian Endocrine Glands and Hormones

Source	Hormone	Type of Chemical	Principal Function
Major Endocrine Glands			
Hypothalamus (via posterior pituitary)	Antidiuretic hormone (ADH)	Protein	Promotes reabsorption of water in kidneys and sweat glands; constricts arterioles
	Oxytocin	Protein	In females, stimulates contraction of uterine muscles during childbirth, milk ejection, and maternal behaviors; in males, causes sperm ejection
Hypothalamus (to anterior pituitary)	Releasing and inhibiting hormones	Proteins	At least seven hormones; releasing hormones stimulate release of hormones from anterior pituitary; inhibiting hormones inhibit release of hormones from anterior pituitary
Anterior pituitary	Follicle-stimulating hormone (FSH)	Glycoprotein	In females, stimulates growth of follicle, secretion of estrogen, and perhaps ovulation; in males, stimulates spermatogenesis
	Luteinizing hormone (LH)	Glycoprotein	In females, stimulates ovulation, growth of corpus luteum, and secretion of estrogen and progesterone; in males, stimulates secretion of testosterone
	Thyroid-stimulating hormone (TSH)	Glycoprotein	Stimulates thyroid to release thyroxine
	Growth hormone (somatotropin)	Protein	Stimulates growth, protein synthesis, and fat metabolism; inhibits sugar metabolism
	Adrenocorticotropic hormone (ACTH)	Protein	Stimulates adrenal cortex to release hormones, especially glucocorticoids
	Prolactin	Protein	Stimulates milk synthesis in and secretion from mammary glands
Thyroid	Thyroxine	Modified amino acid	Increases metabolic rate of most body cells; increases body temperature; regulates growth and development
	Calcitonin	Protein	Inhibits release of calcium from bones
Parathyroid	Parathormone	Protein	Stimulates release of calcium from bone; promotes absorption of calcium by intestines; promotes reabsorption of calcium by kidneys
Adrenal medulla	Adrenalin and noradrenalin	Modified amino acids	Increase levels of sugar and fatty acids in blood; increase metabolic rate; increase rate and force of contractions of the heart; constrict some blood vessels
Adrenal cortex	Glucocorticoids	Steroid	Increase blood sugar; regulate sugar, lipid, and fat metabolism; anti-inflammatory effects
	Aldosterone	Steroid	Increases reabsorption of salt in kidney
	Testosterone	Steroid	Causes masculinization of body features, growth

Table 35-2—*Continued*

Source	Hormone	Type of Chemical	Principal Function
Major Endocrine Glands *Continued*			
Pancreas	Insulin	Protein	Decreases blood glucose levels by increasing uptake of glucose into cells and converting glucose to glycogen, especially in liver; regulates fat metabolism
	Glucagon	Protein	Converts glycogen to glucose, thereby raising blood glucose levels
Ovaries[a]	Estrogen	Steroid	Causes development of female secondary sexual characteristics and maturation of eggs; promotes growth of uterine lining; has general effects on metabolism
	Progesterone	Steroid	Stimulates development of uterine lining and formation of placenta
Testes[a]	Testosterone	Steroid	Stimulates development of genitalia and male secondary sexual characteristics; stimulates spermatogenesis and growth; has general effects on metabolism
Other Sources of Hormones			
Digestive tract[b]	Secretin, gastrin, cholecystokinin, and others	Proteins	Control secretion of mucus, enzymes and salts in digestive tract; regulate peristalsis
Thymus[c]	Thymosin	Protein	Stimulates maturation of cells of immune system
Pineal	Melatonin	Modified amino acid	Regulates biological clock; may regulate onset of puberty
Kidney	Renin	Protein	Acts on blood proteins to produce hormone (angiotensin) that regulates blood pressure
	Erythropoietin	Protein	Activates a blood protein to stimulate red blood cell synthesis in bone marrow
Heart	Atrial natriuretic hormone	Protein	Increases salt and water excretion by kidney; lowers blood pressure

[a] See Chapter 39.
[b] See Chapter 32.
[c] See Chapter 34.

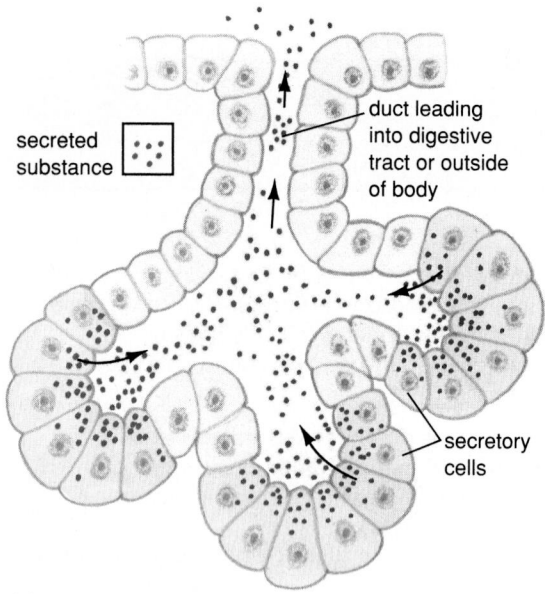

secreted substance

duct leading into digestive tract or outside of body

secretory cells

(a) Exocrine Gland

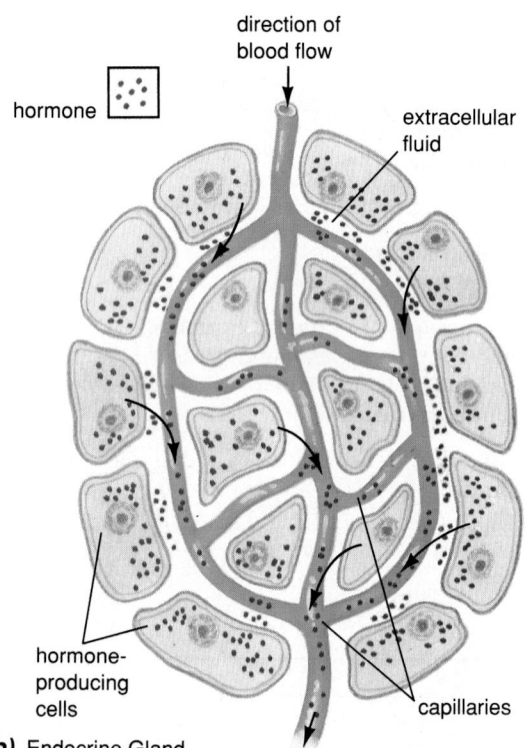

direction of blood flow

hormone

extracellular fluid

hormone-producing cells

capillaries

(b) Endocrine Gland

Figure 35-4 The structure of exocrine and endocrine glands. **(a)** The secretory cells of exocrine glands from the sources of ducts that usually open outside the body (sweat glands, mammary glands) or into the digestive tract (pancreas, salivary glands). **(b)** Endocrine glands consist of hormone-producing cells embedded with a network of capillaries. The cells secrete hormones into the extracellular space, from which they diffuse into the capillaries.

Antidiuretic hormone, which literally means "hormone that prevents urination," helps prevent dehydration. ADH, as you learned in Chapter 33, increases the permeability to water of the collecting ducts of nephrons in the kidney. This causes more water to be reabsorbed from the urine and retained in the body.

Other neurosecretory cells of the hypothalamus release oxytocin from their endings in the posterior pituitary. This hormone causes contraction of muscle cells within the breasts during lactation, triggering "milk letdown" by squeezing milk out of storage bulbs and into ducts leading to the nipples (Fig. 35-6). Oxytocin also causes contractions of the muscles of the uterus during childbirth, helping to expel the infant from the womb. Recent studies using laboratory animals indicate that oxytocin has behavioral effects too, inducing maternal behaviors in virgin females. In rats, for example, oxytocin injections cause virgin females to build a nest, lick pups, and retrieve pups that have strayed. Oxytocin may also have a role in male reproductive behavior. In several animals, oxytocin stimulates the contraction of muscles surrounding the tubes that conduct sperm from the testes to the penis, causing ejaculation.

The Anterior Pituitary

The anterior lobe of the pituitary produces six peptide hormones, four of which help to regulate hormone production in other endocrine glands. Two of these, **follicle-stimulating hormone (FSH)** and **luteinizing hormone (LH)**, stimulate production of sperm and testosterone in males, and eggs, estrogen, and progesterone in females. We will discuss the roles of FSH and LH in more detail in Chapter 39. **Thyroid-stimulating hormone (TSH)** stimulates the thyroid gland to release its hormones, while **ACTH**, or **adrenocorticotropic hormone** ("hormone that stimulates the adrenal cortex"), causes the release of hormones from the adrenal cortex. We discuss the effects of thyroid and adrenal cortical hormones later in this chapter.

The remaining two hormones of the anterior pituitary, prolactin and growth hormone, do not act on other endocrine glands. **Prolactin**, in conjunction with other hormones, stimulates the development of the mammary glands (which are exocrine glands) during pregnancy. Suckling by the newborn infant then stimulates further release of prolactin, which in turn stimulates milk secretion. When the infant no longer suckles, prolactin secretion is turned off, and the ability to produce milk is lost within a few days.

Growth hormone regulates the growth of the body. Growth hormone acts on all the body's cells, increas-

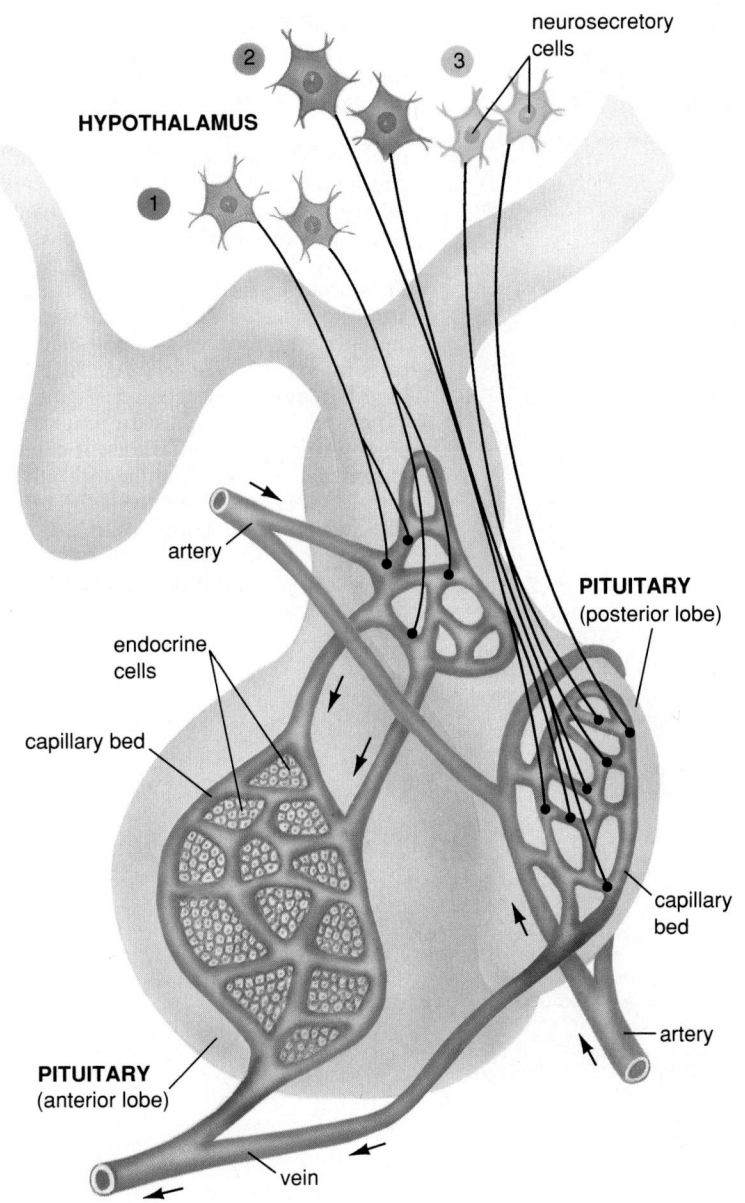

HYPOTHALAMUS

neurosecretory cells

artery

endocrine cells

capillary bed

PITUITARY
(posterior lobe)

capillary bed

artery

PITUITARY
(anterior lobe)

vein

Figure 35-5 Anatomical relationships between the hypothalamus and pituitary (top) and a table showing the hormones of the anterior and posterior pituitary and their target organs. The anterior lobe of the pituitary (left) consists of secretory cells enmeshed in a capillary bed. Release of hormones from these cells is controlled by releasing and inhibiting hormones produced by neurosecretory cells ① of the hypothalamus. These neurosecretory cells release their hormones into a capillary network directly "upstream" from the anterior pituitary (see Fig. 35-8). The posterior lobe of the pituitary (right) is an extension of the hypothalamus. Two types of neurosecretory cells [② and ③] send processes from the hypothalamus into the posterior lobe, where they end on a capillary bed into which they release their hormones.

ANTERIOR LOBE		POSTERIOR LOBE	
Hormones	Target Organs	Hormones	Target Organs
follicle-stimulating hormone (FSH)	gonads	oxytocin	uterus, mammary glands
luteinizing hormone (LH)		vasopressin (ADH)	kidney
thyroid-stimulating hormone (TSH)	thyroid		
adrenocorticotrpic hormone (ACTH)	adrenal cortex		
prolactin	mammary glands		
growth hormone (GH)	most cells		

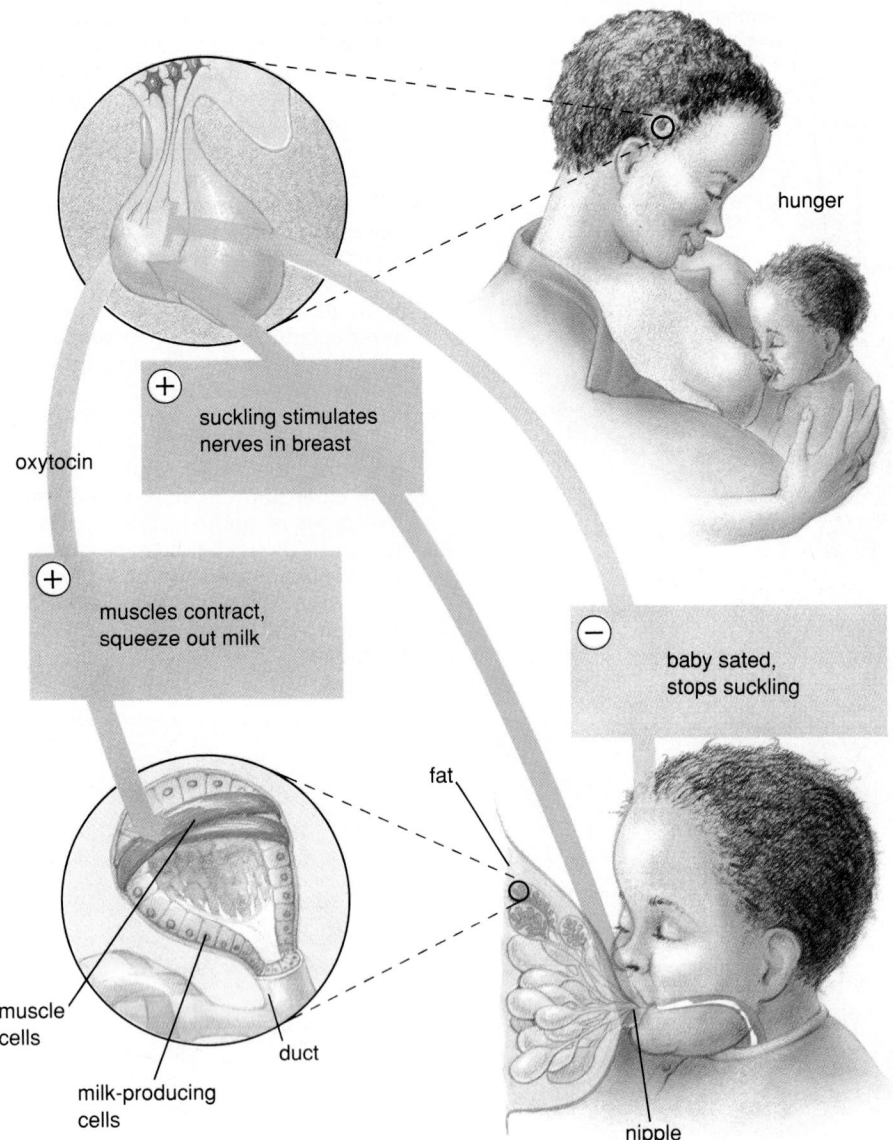

hunger

oxytocin

(+) suckling stimulates nerves in breast

(+) muscles contract, squeeze out milk

(−) baby sated, stops suckling

fat

muscle cells

milk-producing cells

duct

nipple

Figure 35-6 The control of milk letdown by oxytocin during breastfeeding is regulated by a complex negative feedback loop between the baby and its mother. The breast, or mammary gland, is an exocrine gland. Here, clusters of milk-producing cells surround hollow bulbs, where milk collects in lactating women. The bulbs are surrounded by muscle that can expel the milk through the nipple. This occurs when suckling by the baby stimulates nerve endings that send a signal to the mother's hypothalamus, causing secretion of oxytocin into the bloodstream by the posterior pituitary. When oxytocin reaches the muscles surrounding the milk ducts, it causes them to contract and expel milk through the nipple. This continues until the infant is full and stops suckling. With the nipple no longer being stimulated, oxytocin release stops, the muscles relax, and milk flow ceases.

ing protein synthesis, fat utilization, and the storage of carbohydrates. Its influence on the ultimate size of the adult organism is largely due to its stimulatory effect on the growth of bone during maturation. Much of the normal variation in human height is due to differences in secretion of growth hormone from the anterior pituitary. Too little growth hormone causes some cases of dwarfism, while too much can cause gigantism (Fig. 35-7). Although in adulthood the long bones lose their ability to grow, growth hormone continues to be secreted throughout life, helping to regulate protein, fat, and sugar metabolism. A major advance in the treatment of dwarfism occurred when molecular biologists successfully inserted the gene for human growth hormone into bacteria, which

churn out large quantities of the substance (see Chapter 14). Previously, tiny amounts were extracted from human cadavers at great cost. Now children with underactive pituitary glands who would previously have been dwarfs can achieve normal height.

Control of the Pituitary by Hypothalamic Hormones

Neurosecretory cells of the hypothalamus produce at least nine peptides that regulate the release of hormones from the anterior pituitary. These peptides are called **releasing hormones** or **inhibiting hormones** depending on whether they stimulate or prevent the release of pituitary hormone (Fig. 35-8). Releasing

Figure 35-7 An improperly functioning anterior pituitary can produce either too much or too little growth hormone. Too much can result in gigantism, while too little causes dwarfism.

and inhibiting hormones are synthesized in nerve cells in the hypothalamus, secreted into a capillary bed in the lower portion of the hypothalamus, and travel a short distance through blood vessels down the pituitary stalk to a second capillary bed surrounding the endocrine cells of the anterior pituitary (see Fig. 35-5). There, the releasers and inhibitors diffuse out of the capillaries and influence pituitary hormone secretion. Since the releasing and inhibiting hormones are secreted very close to the anterior pituitary, they are produced only in minute amounts. Not surprisingly, they were extremely difficult to isolate and study. Andrew Schally and Roger Guillemin, American endocrinologists who shared the Nobel Prize in 1977 for characterizing several of these hormones, used the brains of millions of sheep and pigs (obtained from slaughterhouses) to extract enough releasing hormone to analyze.

The Thyroid and Parathyroid Glands

In the front of the neck, nestled around the larynx, lies the **thyroid gland** (Fig. 35-9a). The four small disks of the **parathyroid glands** are embedded in the back of the thyroid.

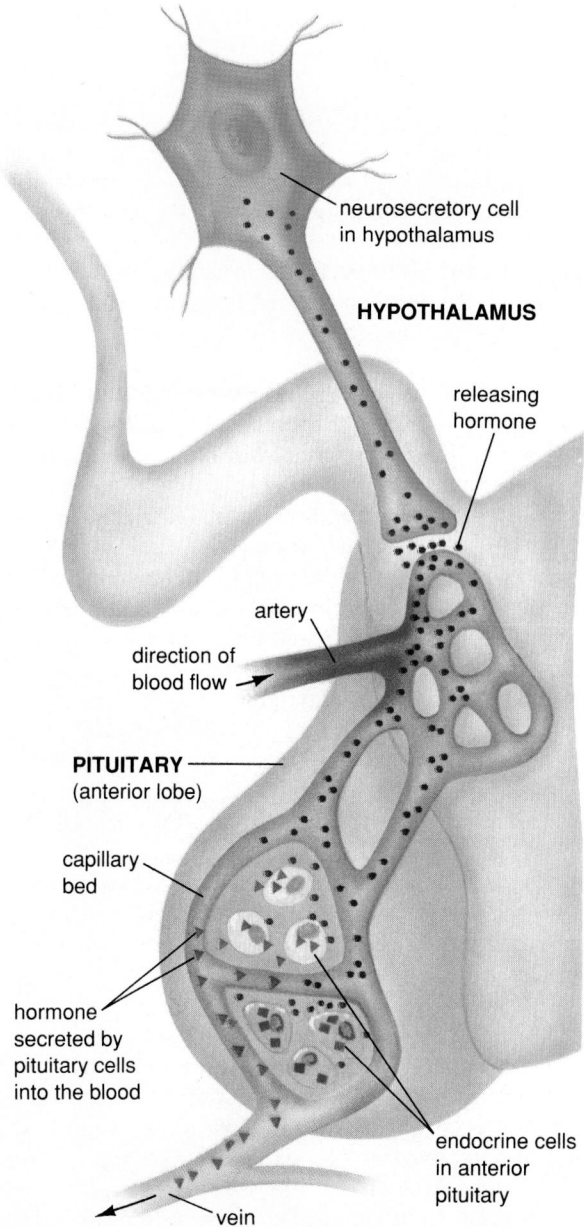

Figure 35-8 Hormone release from the anterior pituitary is under the control of releasing and inhibiting hormones from neurosecretory cells of the hypothalamus. Releasing hormones enter a capillary bed in the hypothalamus and travel downstream to capillaries in the anterior pituitary. There the releasing hormones contact the various endocrine cells of the pituitary. Only endocrine cells with matching cell membrane receptors respond to a given releasing hormone (see Fig. 35-2a), so each releasing hormone stimulates a particular type of endocrine cell to release its hormone, while leaving other types unaffected.

A CLOSER LOOK
At the Hormonal Control of the Insect Life Cycle

Just as hormones control many aspects of vertebrate life cycles, including reproduction, growth, and the development of adult sexual characteristics, they are also important in the life cycles of invertebrates. The most thoroughly studied example of hormonal regulation in invertebrates is the control of molting and maturation in insects.

Many insects, including flies, beetles, moths, and butterflies, have complex life cycles (Fig. E35-4). Females lay eggs that hatch out into wormlike larvae, variously called caterpillars, grubs, or maggots. The larvae grow and eventually form pupae, which, although they look dormant, seethe with activity as the larval body is rearranged and transformed into an adult. This complicated life cycle incurs three major challenges. First, all insects, even caterpillars, are enclosed in more-or-less rigid exoskeletons. Although the exoskeletons of larvae are somewhat expandable, eventually a larva must shed its old exoskeleton, or molt, and develop a new, larger one, thereby permitting continued growth. Second, there is a major rearrangement of body parts during the pupal stage. Finally, the adult must break through the pupal shell to emerge into the world. Each of these events is induced by a different hormone, as we can see by following the life cycle of a silk moth, beginning with a young caterpillar (Fig. E35-5).

Larval Molting

The caterpillar brain determines when molting is necessary, using several cues. One stimulus for larval molting is abdominal stretching: to a caterpillar, having its "clothing" get tight doesn't mean that it's time to quit eating, it means that it's time to molt. The brain secretes a protein hormone often called simply **brain hormone**. This hormone circulates in the blood to the prothoracic glands, which, in response, release **molting hormone** (ecdysone). Molting hormone is a steroid that triggers a series of behavioral and biochemical activities: the larva splits the old exoskeleton apart, wriggles out, and produces a new one. As with vertebrate steroid hormones, many of these effects are mediated by the actions of molting hormone on DNA, promoting transcription of selected genes.

Pupation

When it molts, a larval silk moth either becomes a new, larger caterpillar or develops into a pupa. Which of these events occurs depends on its level of **juvenile hormone**. A pair of glands, the corpora allata, attached to the brain and under its direct control, synthesize juvenile hormone. Early in the caterpillar stage, these glands

(a)

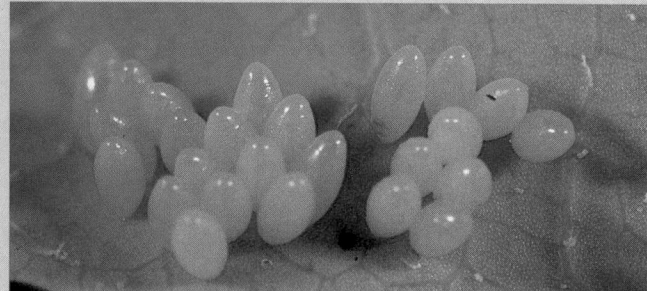

(b)

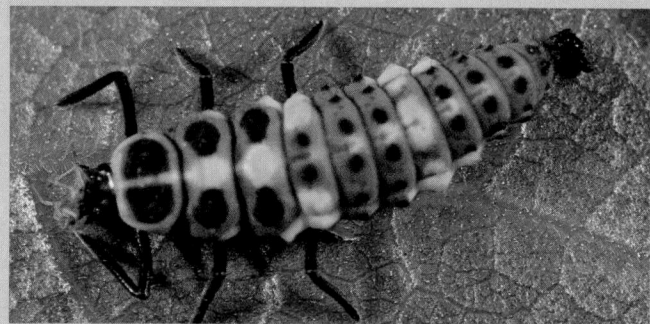

(c)

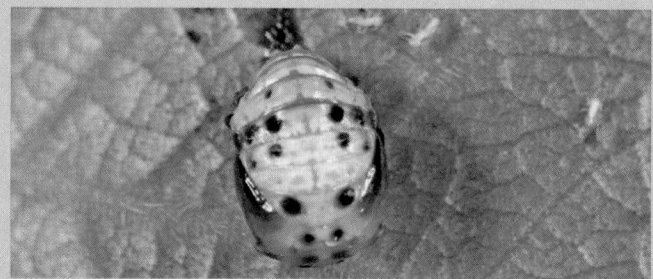

(d)

Figure E35-4 Many insects, such as this convergent ladybug, undergo complete metamorphosis. The egg **(a)** hatches into a larva **(b)**. The larva molts several times, growing at each molt, and finally forms a pupa **(c)**. During pupation, the larval body is reorganized into the adult ladybug **(d)**, seen here feeding on aphids.

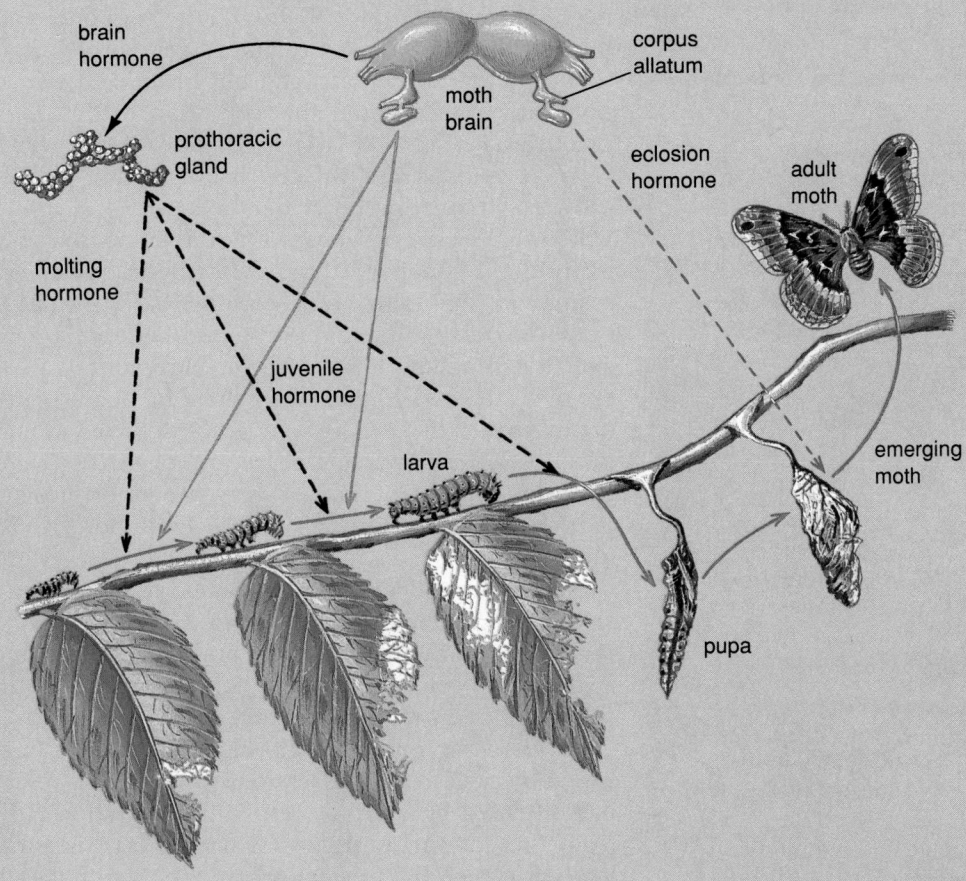

Figure E35-5 Hormone control of the silk moth life cycle. At each molt, the brain stimulates release of molting hormone (black arrows) by the prothoracic gland. During the caterpillar stage, the brain also stimulates secretion of juvenile hormone (blue arrows), which causes the caterpillar to molt into another caterpillar. Eventually, the brain inhibits further juvenile hormone release, and at the next molt the caterpillar forms a pupa. When the adult moth has completed development inside the pupa, its brain releases eclosion hormone (red arrow), causing emergence.

release lots of juvenile hormone, so at each molt a new, larger caterpillar is formed. When the caterpillar has matured, however, the brain inhibits further juvenile hormone release. Apparently, juvenile hormone prevents formation of the pupa; when it is no longer secreted, the caterpillar pupates at the next molt.

Emergence

During pupation, the insect undergoes incredible reorganization of its body, digesting away some parts while forming entirely new structures, such as wings. When the rebuilding program has been completed, the silk moth emerges from the pupa. The moth goes through a series of twists and turns, splitting open the pupal case and squirming out. The moth's nervous system is genet-

ically programmed with all the right moves, and this behavioral program is turned on by another peptide hormone from the brain, **eclosion hormone**. The brain secretes eclosion hormone as soon as development is complete, subject to one constraint: silk moths always emerge in the late afternoon.

As you can see, the principles underlying hormonal control in insects and humans are really very similar: most hormones are peptides or steroids; nerve cells in the brain produce a number of hormones, including some that control the release of other, nonnervous hormones; and major changes in life, such as maturation and reproduction, are under hormonal control. Although sometimes hidden beneath tremendous diversity, the unity of life on Earth is overwhelming.

(a)

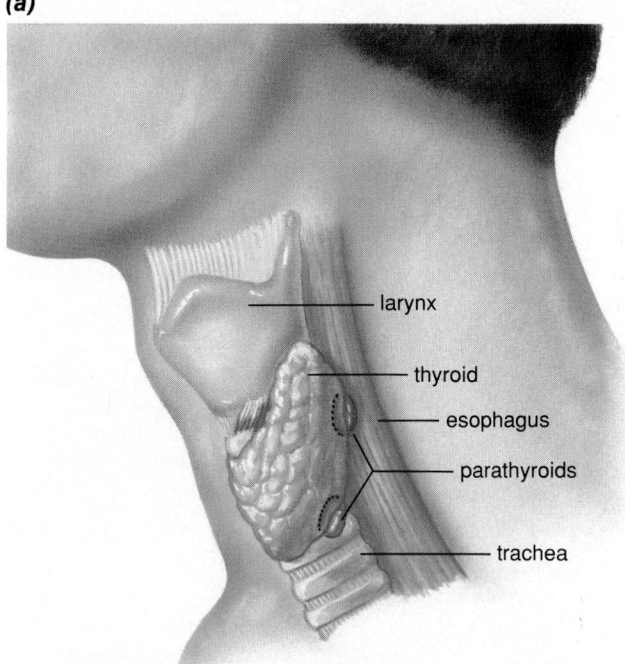

(b)

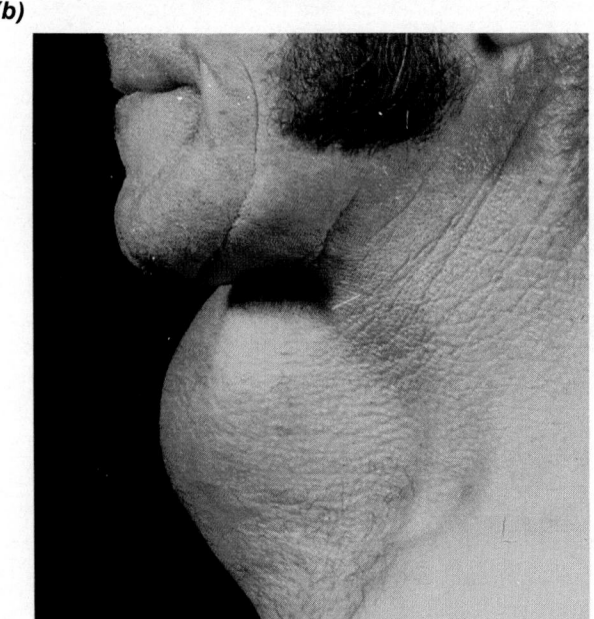

Figure 35-9 (a) The thyroid and parathyroid glands are located around the front of the larynx in the neck. Thyroxine contains four iodine atoms per molecule. **(b)** Individuals with iodine-deficient diets may suffer from goiter, a condition in which the thyroids become greatly enlarged.

The thyroid produces two hormones, **thyroxine** and **calcitonin**. Thyroxine is an iodine-containing, modified amino acid that raises the metabolic rate of most body cells. Iodine deficiency causes dramatic growth of the thyroid, resulting in a swelling of the neck called **goiter** (Fig. 35-9b). Goiter was once common in the Great Lakes region of the United States and Canada, but widespread use of iodized salt has now all but eliminated this condition in developed countries.

In juvenile animals, thyroxine helps regulate growth, stimulating both metabolic rate and the development of the nervous system. Cretinism, caused by undersecretion of thyroid hormone from birth, results in mentally retarded dwarfs. Fortunately, early diagnosis and thyroxine supplementation can avert this tragedy. Precocious development in a variety of vertebrate animals is triggered by oversecretion of thyroxine. In 1912, in one of the first demonstrations of hormone action, a physiologist discovered that thyroxine can induce early metamorphosis in tadpoles (see ''Reflections on the Evolution of Hormones'').

Thyroxine also elevates metabolic rate. In adults, an elevated metabolic rate seems to be involved in regulating body temperature and stress reactions. Exposure to cold, for example, greatly increases thyroid hormone production.

Levels of thyroxine in the bloodstream are finely tuned by negative feedback loops. Thyroxine release is stimulated by thyroid-stimulating hormone (TSH) from the anterior pituitary, which in turn is stimulated by a releasing hormone from the hypothalamus. The amount of TSH released from the pituitary is regulated by thyroxine levels in the blood (Fig. 35-10): high concentrations of thyroxine inhibit the secretion of both the releasing hormone and TSH, thus inhibiting further release of thyroxine from the thyroid.

Calcitonin, along with **parathormone**, the hormone secreted by the parathyroids, controls the concentration of calcium in the blood and other body fluids. Calcium is essential for many processes, including nerve and muscle function, so the calcium concentration in body fluids must be kept within narrow limits. Calcitonin and parathormone regulate calcium absorption and release by the bones, which serve both as a skeleton and as a bank into which calcium can be deposited or withdrawn as necessary. In response to low blood calcium, the parathyroids release parathormone, which causes release of calcium from bones. The parathyroids increase in size in pregnant and lactating women, enhancing parathormone output. This allows the mother's body to meet the extra demands for calcium by the developing fetus and, later, milk production. If blood calcium levels become too high,

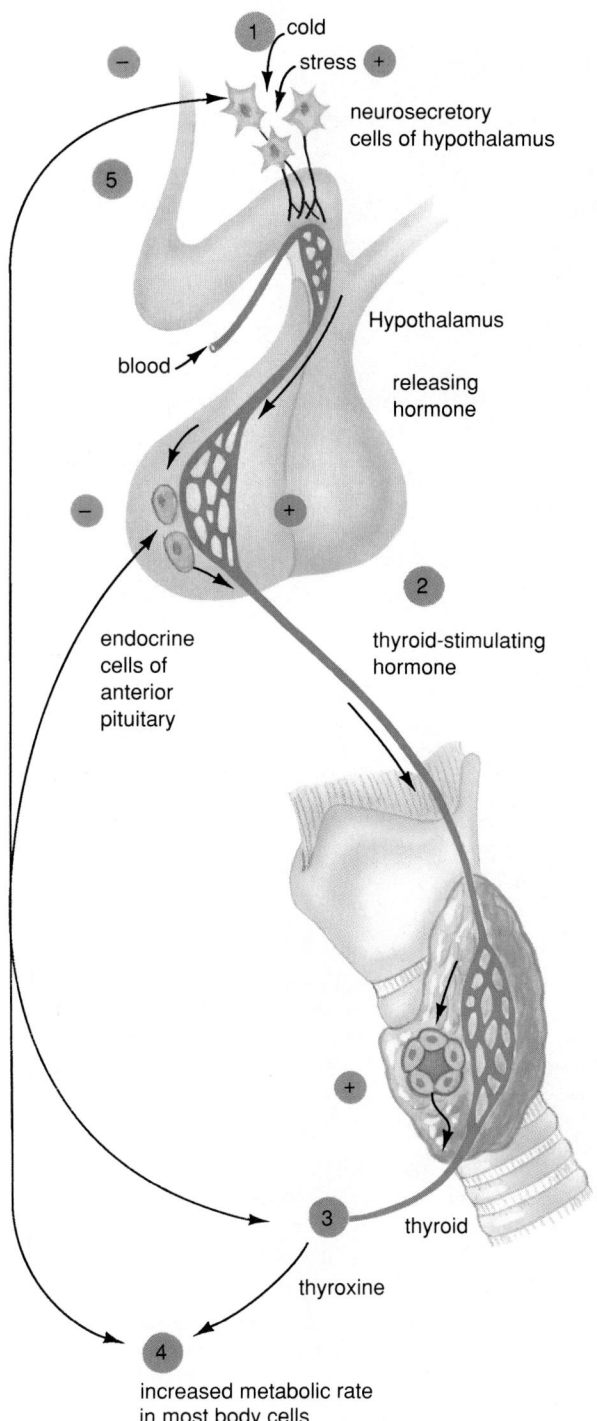

Figure 35-10 Negative feedback in thyroid function. The thyroid releases thyroxine in response to thyroid-stimulating hormone (TSH) from the anterior pituitary. TSH release is in turn controlled by releasing hormones of the hypothalamus. Low body temperature or stress stimulates neurosecretory cells of the hypothalamus ①, whose releasing hormones trigger TSH release in the pituitary ②. TSH then stimulates the thyroid to release thyroxine ③. Thyroxine causes increased metabolic activity in most cells of the body, generating ATP energy and heat ④. Both the raised body temperature and high thyroxine levels in the blood inhibit the releasing-hormone cells and the TSH-producing cells ⑤.

the thyroid releases calcitonin, which inhibits the release of calcium from bones.

The Pancreas

The **pancreas** is a double gland producing both exocrine and endocrine secretions (Fig. 35-11). The exocrine portion synthesizes digestive secretions that are released into the pancreatic duct and flow into the small intestine (see Chapter 32). The endocrine portion consists of clusters of cells, called **islet cells**, that produce peptide hormones. One type of islet cell produces the hormone **insulin**, while another type produces **glucagon**. Insulin and glucagon work in opposite ways to regulate carbohydrate and fat metabolism. When blood glucose rises (for example, after a meal), insulin is released. Insulin causes most of the cells of the body to take up glucose and either metabolize it for energy or convert it to fat or glycogen (a starchlike molecule) for storage. By far the most important storage organ for glycogen is the liver. When blood glucose levels drop (from, for example, skipping breakfast or running a 10-k race), glucagon is released. Glucagon activates a liver enzyme that breaks down glycogen, releasing glucose into the blood. It also promotes lipid breakdown, releasing fatty acids that are metabolized for energy. Insulin, then, reduces blood glucose levels, while glucagon increases them; together they help keep blood glucose levels nearly constant.

Defects in insulin production, release, or reception by target cells result in **diabetes mellitus**, a condition in which blood glucose levels are high and fluctuate wildly with sugar intake. The lack of functional insulin in diabetics causes the body to rely much more heavily on fats as an energy source, leading to high circulating levels of lipids, including cholesterol. Severe diabetes causes fat deposits in the blood vessels, resulting in high blood pressure and heart disease; diabetes is an important cause of heart attacks in the U.S. The fatty deposits in small vessels can also produce microscopic lesions that damage the retina, leading to blindness, and damage the kidneys, leading to kidney failure. Insulin supplements traditionally contained insulin extracted from the pancreases of cows and pigs, abundantly available from slaughterhouses. Recently, however, the gene for human insulin has been inserted into bacteria, allowing the production of large quantities of human insulin, which is now commercially available.

The Sex Organs

The male testes and female ovaries are important endocrine organs. The testes secrete several steroid

(a)

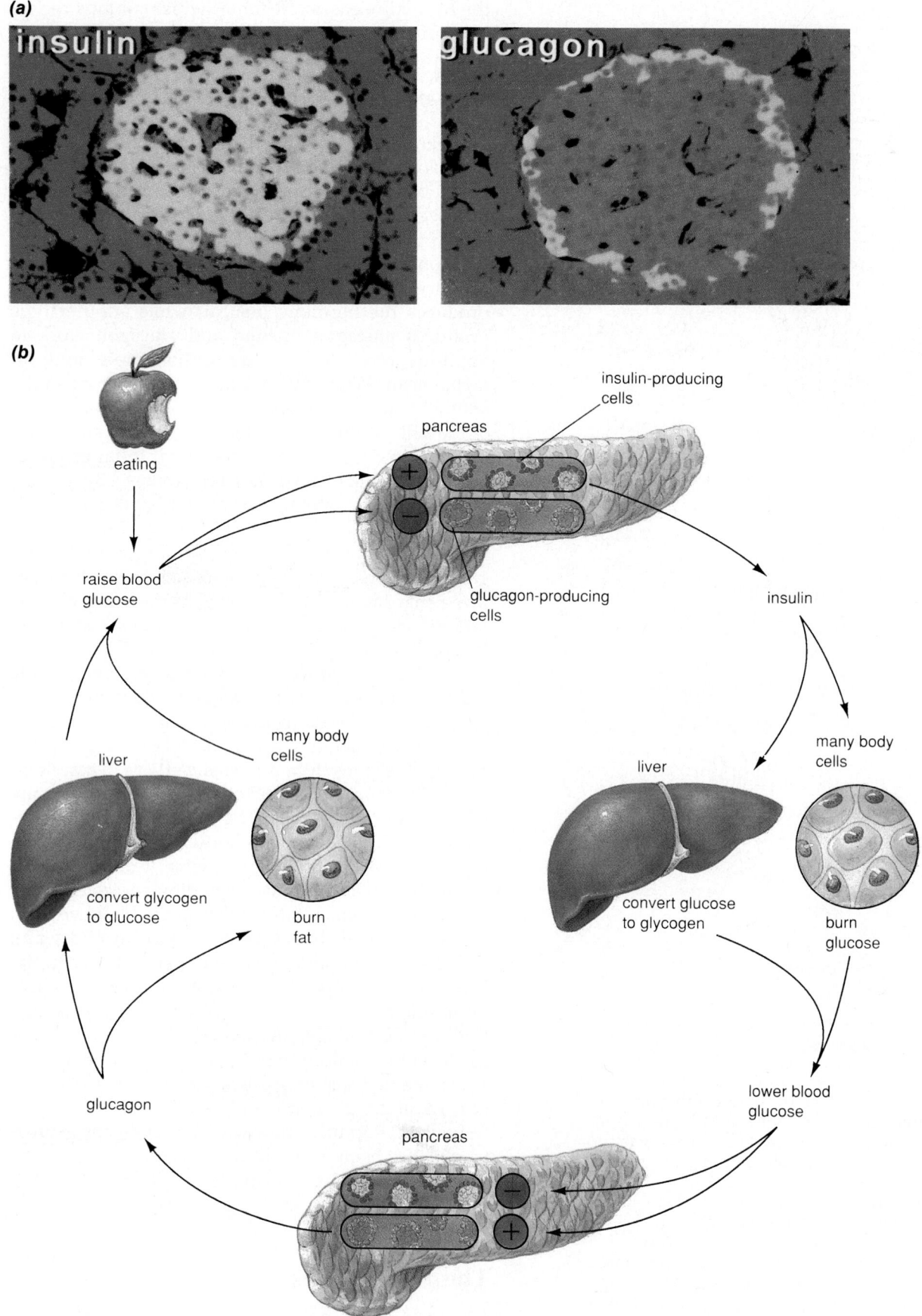

(b)

Figure 35-11 (a) The pancreatic islets contain two populations of hormone-producing cells, one producing insulin, the other glucagon. **(b)** These two hormones cooperate in a two-part negative feedback loop to control blood glucose concentrations. High blood glucose stimulates the insulin cells and inhibits the glucagon cells, while low blood glucose stimulates the glucagon cells and inhibits the insulin cells. This dual control quickly corrects either high or low blood glucose levels.

hormones, collectively called **androgens**. The most important of these is **testosterone**. The ovary secretes two types of hormones, **estrogen** and **progesterone**. The role of the sex hormones in development, the menstrual cycle, and pregnancy are discussed in Chapters 39 and 40.

The Adrenal Glands

Like the pituitary and pancreas, the adrenals (Latin for "on the kidney") are two glands in one: the **adrenal medulla** and the **adrenal cortex** (Fig. 35-12). The adrenal medulla is located in the center of each gland ("medulla" means "marrow" in Latin). It consists of secretory cells derived during development from nervous tissue, and its hormone secretion is controlled directly by the nervous system. The adrenal medulla produces two hormones, **adrenalin** and **noradrenalin** (also called *epinephrine* and *norepineph-rine*), in response to stress. These hormones, which are modified amino acids, prepare the body for emergency action. They increase the heart and respiratory rates, cause blood glucose levels to rise, and direct blood flow away from the digestive tract and toward the brain and muscles. The adrenal medulla is activated by the sympathetic nervous system, which prepares the body to respond to emergencies, as described in Chapter 36.

The outer layer of the adrenal gland forms the adrenal cortex ("cortex" is Latin for "bark"). The cortex secretes three types of steroid hormones synthesized from cholesterol, called **glucocorticoids**. These help to control glucose metabolism. Glucocorticoid release is stimulated by ACTH from the anterior pituitary; ACTH release in turn is stimulated by hypothalamic-releasing hormones that are produced in response to stressful stimuli, including trauma, infection, or exposure to extremes of temperature. In some respects, the glucocorticoids act similarly to glucagon, raising blood glucose levels by stimulating glucose production and promoting the use of fats instead of glucose for energy production. You may have noticed that many different hormones are involved in glucose metabolism: thyroxine, insulin, glucagon, adrenalin, and the glucocorticoids. Why? The reason is probably

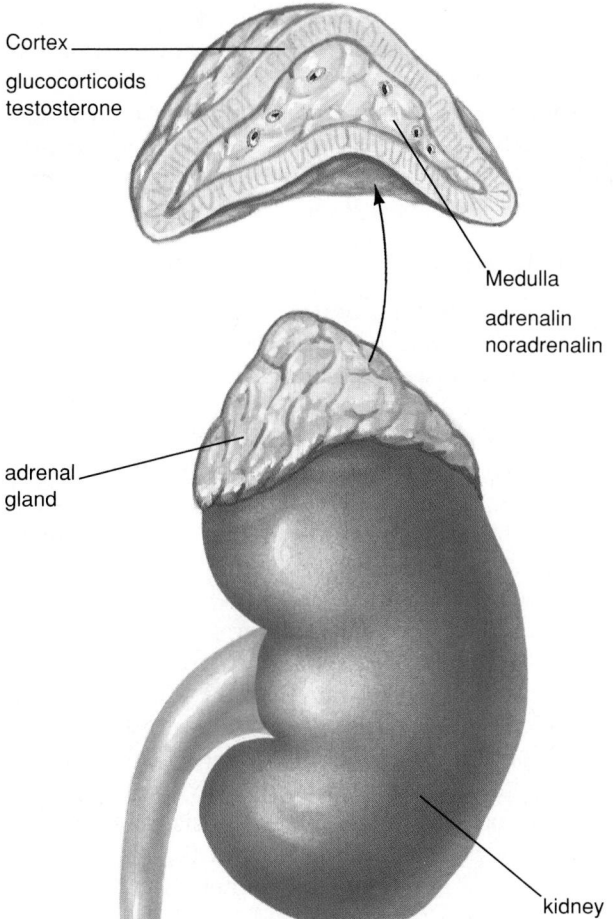

Figure 35-12 Atop each kidney sits an adrenal gland, which is a two-part gland composed of very dissimilar cells. The outer cortex consists of ordinary endocrine cells that secrete steroid hormones. The inner medulla is derived from nervous tissue during development and secretes adrenalin and noradrenalin.

a metabolic quirk of the brain. Although most body cells can produce energy from fats and proteins as well as carbohydrates, brain cells can only burn glucose. Thus, blood glucose levels cannot be allowed to fall too low, or brain cells rapidly starve, leading to unconsciousness and death.

The adrenal cortex also secretes **aldosterone**, which regulates the sodium content of the blood. If blood sodium falls, the cortex releases aldosterone, which causes the kidneys and sweat glands to retain sodium. Then salt and other sources of dietary sodium, combined with aldosterone-induced sodium conservation, raise blood sodium levels again, shutting off further aldosterone secretion.

Finally, the adrenal cortex produces the male sex

hormone testosterone, although normally in much smaller amounts than the testes produce. Tumors of the adrenal medulla sometimes lead to excessive testosterone release, causing masculinization of women. This malady used to be exploited by circus operators who displayed "bearded ladies" in their sideshows.

Prostaglandins

Unlike most other hormones, which are synthesized by a limited number of cells, **prostaglandins** are produced by many, perhaps all, cells of the body. They are modified fatty acids, synthesized by the cell from membrane phospholipids. Several prostaglandins are known, and a great many more probably await discovery. One prostaglandin causes arteries to constrict and stops bleeding from the umbilical cords of newborn infants. Another works in conjunction with oxytocin during labor, stimulating uterine contractions. Prostaglandin-soaked vaginal suppositories are currently used to induce labor. Menstrual cramps are caused by the overproduction of uterine prostaglandins. Some prostaglandins cause inflammation (such as occurs in arthritic joints) and also stimulate pain receptors. Drugs such as aspirin and ibuprofen, which inhibit prostaglandin synthesis, can provide relief. Some prostaglandins expand the air passages of the lungs, and may one day benefit asthma sufferers, while others stimulate production of the protective mucus that lines the stomach, a boon for ulcer patients. Research on this diverse and potent family of compounds is still in its infancy, and promises many health benefits in the future.

Other Endocrine Organs: The Pineal Gland, Kidneys, Heart, and Digestive Tract

The **pineal gland** is located between the cerebral hemispheres just above and behind the brainstem (see Fig. 35-3). Named for its resemblance to a pine cone, the pineal is smaller than a pea. Although its functions are still poorly understood, we have come a long way since 1646, when Rene Descartes described

it as "the seat of the rational soul." In some vertebrates, such as the frog, the pineal contains photoreceptive cells, and the skull above it is thin, allowing it to detect daylength. The pineal produces the hormone **melatonin**, a modified amino acid. This is secreted in a daily rhythm, which in mammals is regulated by the eyes. By responding to daylengths characteristic of different seasons, the pineal appears to regulate the seasonal reproductive cycles of many mammals. Despite years of research, the function of the human pineal and of melatonin is still unclear. It has been suggested to influence sleep–wake cycles and cause the depression that some people experience during the short days of winter.

The kidney, which plays a central role in maintaining body fluid homeostasis, has recently been recognized as an important endocrine organ as well. When the oxygen content of the blood drops the kidney produces **erythropoietin**, which increases red blood cell production, as described in Chapter 33. The kidney also produces a second hormone, **renin**, in response to low blood pressure, such as may be caused by bleeding. Renin is an enzyme that catalyzes the production of another hormone, **angiotensin**, from proteins in the blood. Angiotensin raises blood pressure by constricting arterioles. Constriction of the arterioles carrying blood to the kidneys also reduces the rate of blood filtration, causing the kidneys to retain water. This further raises blood pressure by increasing the volume of fluid in the bloodstream.

The heart seems an unlikely endocrine organ, but in 1981, a substance extracted from heart atrium was found to cause an increase in the output of salt and water by the kidneys, when injected into rats. Two years later, the active substance, **atrial natriuretic peptide (ANP)**, was described and its amino acid sequence determined. The pace of modern biomedical research is astonishing. Only five years after the factor was first detected in heart extract, clinical trials of synthetic ANP were started to test its value in treating high blood pressure and related problems.

The stomach and small intestine produce a variety of peptide hormones that help regulate digestion. These include **gastrin, secretin,** and **cholecystokinin,** discussed in Chapter 32.

℞eflections on the Evolution of Hormones

Not long ago, vertebrate endocrine systems were considered unique to our phylum, using chemicals that evolved expressly for their role in vertebrate physiology. In recent years, however, physiologists have discovered that hormones are evolutionarily ancient. Insulin, for example, is found not only in vertebrates but also in protists, fungi, and bacteria, although no one has an inkling as to the function of insulin in most of these organisms. Protists also manufacture ACTH, although of course they have no adrenal glands to stimulate. Yeasts have receptors for estrogen, but no ovaries. Thyroid hormones have been found in certain worms, insects, and molluscs. Even among the vertebrates, identical hormones, secreted by the same glands, may produce dramatically different effects. For example, let's look briefly at the diverse effects of thyroid hormone.

Some fish undergo radical physiological changes during their lifetimes. A salmon, for example, begins life in fresh water, migrates to the ocean, and finally returns to fresh water to spawn. In the stream where it hatched, fresh water tends to enter the fish's tissues by osmosis; in salt water, the fish tends to lose water, becoming dehydrated. The fish's migrations, therefore, require complete revamping of salt and water control. In salmon, one of the functions of thyroxine is to produce the metabolic changes necessary to go from life in streams to life in the ocean.

In amphibians, thyroxine has the dramatic effect of triggering metamorphosis. In 1912, in one of the first demonstrations of the action of any hormone, tadpoles were fed minced horse thyroid, and metamorphosed prematurely into miniature adult frogs (Fig. E35-6). In high mountain lakes in Mexico, where the water is deficient in the iodine needed to synthesize thyroxine, one species of salamander has evolved the ability to reproduce while still in its juvenile form.

In most vertebrates, thyroxine regulates seasonal molting. From snakes to birds to the family dog, surges of thyroxine stimulate the shedding of skin, feathers, or hair. In people (who neither metamorphose, molt, nor migrate), thyroxine has been relegated to the crucial but seemingly mundane role of regulating growth and metabolism.

The use of chemicals to regulate cellular activity is extremely ancient. The diversity of life of Earth rests upon a conservative foundation: a relative handful of chemicals coordinate activities within single cells and among groups of cells. Life's diversity originated in part by changing the systems used to deliver the chemicals, and by evolving new types of responses. Early in their evolution, animals developed a system for fast, precise delivery of chemical messages: the nervous system. As we explain in the next chapter, the nervous system permits rapid responses to environmental stimuli, flexibility in response options, and ultimately consciousness itself.

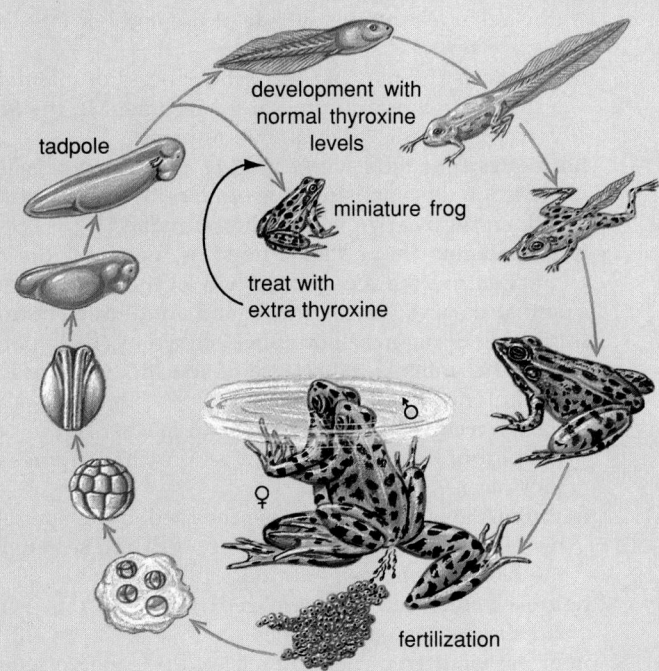

Figure E35-6 The life cycle of the frog includes fertilization of the eggs (bottom); development into an aquatic, fishlike tadpole; growth of the tadpole; and ultimately metamorphosis into an adult frog. Metamorphosis is triggered by a surge of thyroxine from the tadpole's thyroid gland. If a young tadpole is injected with extra thyroxine, it will metamorphose ahead of schedule into a miniature adult frog.

SUMMARY OF KEY CONCEPTS

Chemical Communication within the Animal Body

In multicellular organisms, cellular activity is often coordinated by chemicals that are released by one type of cell and cause effects in other types of cells. In animals, two main types of intercellular communication are found: hormonal and neural.

Hormone Function in Animals

A hormone is a chemical secreted by cells in one part of the body that is transported in the bloodstream to other parts of the body, where it affects the activity of specific target cells.

Four types of molecules are known to act as hormones: modified amino acids, peptides or proteins, steroids, and prostaglandins.

Most hormones act on their target cells in one of two ways. Peptide hormones and modified amino acids bind to receptors on the surfaces of target cells, and activate intracellular second messengers such as cyclic AMP and calmodulin. The second messengers then alter the metabolism of the cell. Steroid hormones diffuse through the cell membranes of the target cells, and bind with receptor proteins in the cytoplasm. The hormone–receptor complex travels to the nucleus and promotes transcription of specific genes.

Thyroid hormones also penetrate the cell membrane, but diffuse into the nucleus where they bind to receptors associated with the chromosomes and influence gene transcription.

Hormone action is often regulated through negative feedback, a process in which the hormone causes changes that inhibit further secretion of that hormone.

Mammalian Endocrine Systems

Hormones are produced by endocrine glands, which are clusters of cells embedded within a network of capillaries. Hormones are secreted into the extracellular fluid and diffuse into the capillaries.

The major endocrine glands of the human body are the hypothalamus–pituitary complex, thyroid, parathyroids, pancreas, adrenal cortex, adrenal medulla, and gonads. The hormones released by these glands and their actions are summarized in Table 35-2. Prostaglandins, unlike other hormones, are not secreted by discrete glands but are synthesized and released by many cells of the body.

Other endocrine organs include the pineal, kidneys, heart, stomach, and small intestine.

GLOSSARY

(*Note:* The major mammalian hormones, their sources, chemical nature, and functions are summarized in Table 35-2.)

adrenal gland: an endocrine gland consisting of an outer cortex and inner medulla. The cortex secretes steroid hormones that regulate metabolism and salt balance. The medulla secretes adrenalin and noradrenalin.

calmodulin (kal-mod'-ū-lin): a cytoplasmic protein that acts as a second messenger by binding calcium ions and changing configuration. The configuration change may activate enzymes, initiating a series of biochemical reactions.

cyclic AMP: a cyclic nucleotide formed within many target cells as a result of the reception of modified amino acid or protein hormones, and which causes metabolic changes in the cell; often called a second messenger.

diabetes mellitus (di-a-bē'-tes mel-ī'-tus): a disease characterized by defects in the production, release, or reception of insulin, characterized by high blood glucose levels that fluctuate with sugar intake, and glucose in the urine.

endocrine gland: a ductless, hormone-producing gland that releases its secretions into the extracellular fluid within the body, from which the secretions diffuse into nearby capillaries.

exocrine gland: a gland that releases its secretions into ducts that lead to the outside of the body or into the digestive tract.

homeostasis (hō-mē-ō-stā'-sis): the process of maintaining a relatively constant internal environment in the face of variations in the external environment.

hormone: a chemical synthesized by one group of cells and carried in the bloodstream to other cells, whose activity is influenced by reception of the hormone.

hypothalamus (hī-pō-thal'-a-mus): a region of the brain that controls the secretory activity of the pituitary gland, and also synthesizes oxytocin and antidiuretic hormone.

inhibiting hormone: a hormone secreted by the hypothalamus that inhibits the release of specific hormones from the anterior pituitary gland.

negative feedback: a type of control mechanism in which the output of a system causes actions that suppress further output.

neurohormone: a chemical synthesized by a specialized nerve cell (called a neurosecretory cell) and secreted into the bloodstream as a hormone.

neurosecretory cell: a specialized nerve cell that synthesizes and releases hormones.

neurotransmitter: a chemical released by a nerve cell close to a second nerve cell, a muscle, or a gland cell, and that influences the activity of the second cell.

pancreas (pan'-krē-as): a combined exocrine and endocrine gland located in the abdominal cavity. Its endocrine portion secretes the hormones insulin and glucagon, which regulate glucose concentrations in the blood.

parathyroids: a set of four small endocrine glands embedded in the surface of the thyroid gland that produce parathormone, which (with calcitonin from the thyroid) regulates calcium ion concentration in the blood.

pineal gland (pī-nē'-al): a small gland within the brain that secretes melatonin. The pineal controls the seasonal reproductive cycles of some mammals.

pituitary: an endocrine gland located at the base of the brain that produces several hormones, many of which influence the activity of other glands.

prostaglandin (pro-sta-glan'-din): a family of modified fatty acid hormones manufactured by many cells of the body.

releasing hormone: a hormone secreted by the hypothalamus that causes the release of specific hormones by the anterior pituitary gland.

second messenger: a term applied to intracellular chemicals, such as cyclic AMP, that are synthesized or released within a cell in response to the binding of a hormone or neurotransmitter (the first messenger) to receptors on the cell surface. Second messengers bring about specific changes in the metabolism of the cell.

target cell: a cell upon which a particular hormone exerts its effect.

thymus (thī'-mus): a gland located in the upper chest in front of the heart. The thymus functions in the immune system by secreting thymosin, which stimulates lymphocyte maturation. It begins to degenerate at puberty and has little function in the adult.

thyroid: an endocrine gland, located in front of the larynx in the neck, that secretes the hormones thyroxine (affecting metabolic rate) and calcitonin (regulating calcium ion concentration in the blood).

STUDY QUESTIONS

1. What are the four types of molecules used as hormones in vertebrates? Give an example of each.
2. What is the difference between an endocrine and an exocrine gland? Which ones release hormones?
3. Describe the mechanisms of action of peptide and steroid hormones. Since hormones bathe many or even all of the cells of the body, why do only specific target cells respond to the hormone?
4. Diagram the process of negative feedback, and give an example of negative feedback in the control of hormone action.
5. What are the major endocrine glands in the human body, and where are they located?
6. Describe the structure of the hypothalamus–pituitary complex. Which pituitary hormones are neurosecretory? What are their functions?
7. Describe how releasing hormones regulate the secretion of hormones by cells of the anterior pituitary. Name the hormones of the anterior pituitary and give one function of each.
8. Describe how the hormones of the pancreas act together to regulate the concentration of glucose in the blood.
9. What are the two parts of the adrenal gland? What hormones does each release?

DISCUSSION QUESTIONS

1. Describe the basic similarities and differences between the mechanisms of action of steroid and peptide hormones. Why is it that each type of hormone triggers a response only in very specific target cells?
2. Describe the very different roles of growth hormone in young children and in adults and discuss why these different effects occur.
3. Describe and discuss the negative feedback loops that exist among the thyroid and the anterior pituitary.

SUGGESTED READINGS

Berridge, M. J. "The Molecular Basis of Communication Within the Cell." *Scientific American*, October 1985. Both the "classical" cyclic AMP and more recently discovered second messengers convey information from cell surface receptors to DNA and cellular metabolism.

Guillemin, R., and Burgus, R. "The Hormones of the Hypothalamus." *Scientific American*, November 1972. The interaction between hypothalamus and pituitary is explored.

Pike, J. E., "Prostaglandins." *Scientific American*, November 1971 (Offprint No. 1235). Although somewhat out of date, this article provides a good introduction to this large and poorly understood class of chemicals.

Sapolsky, R. "Stress in the Wild," *Scientific American*, January 1990. An interesting study on baboons showing the effects of hormones on stress responses.

Snyder, S. H. "The Molecular Basis of Communication Between Cells." *Scientific American*, October 1985. Snyder describes the similarities and differences between neural and hormonal control systems in the body.

36

Information Processing: The Nervous System

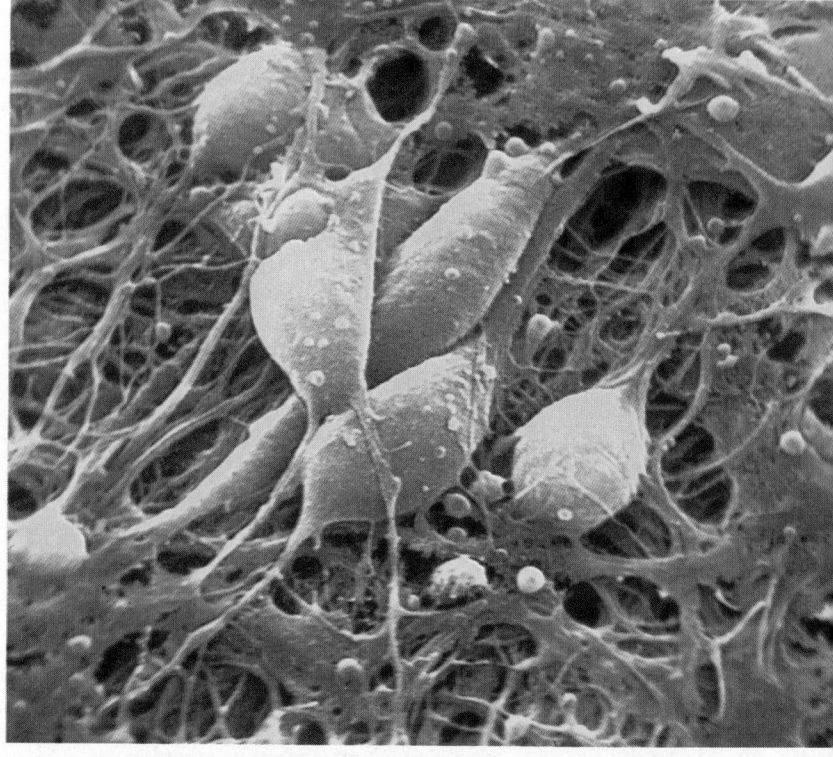

In this false-color scanning EM of a portion of the cerebral cortex, neuron cell bodies lie within a tangle of fibers that allow the neurons to communicate among themselves and with distant parts of the body.

s you read this, light reflected from this page bombards your eyes and sound waves from the stereo assault your ears. Your senses of smell and taste are stimulated as you drink coffee, touch-sensitive receptors all over your body are activated as your clothes, the chair, and the pages of this book all push at your skin. Meanwhile, your hands dextrously manipulate a pen as you take notes. You breathe, swallow, and shift positions in the chair. Untaxed by these ordinary but marvelous activities, your mind calls upon its reading skills, stores memories of what you read, and still has the capacity to appreciate the music from the stereo.

All this activity is controlled by your brain, an organ weighing about 3 pounds encased within your skull. Composed of billions of nerve cells, the brain is connected to the rest of your body by millions of slender cables, the **nerves.** In this chapter, we explore how nerve cells create electrical signals, how they use these stereotyped signals to convey information and communicate with one another, and how they activate muscles. In Chapters 37 and 38, we examine the senses through which the brain experiences the outside world, and the muscles that carry out its commands.

The Functions and Structure of Neurons

Our study of the nervous system begins with the individual nerve cell, or **neuron.**

The Functions of Neurons

As the fundamental unit of the nervous system, each neuron must perform five functions:

1. It must **receive information** from the internal or external environment or from other neurons.
2. It must **integrate the information** it receives and produce an appropriate output signal.
3. It must **conduct the signal** to its terminal ending.
4. It must **transmit** the signal to other nerve cells, glands, or muscles.
5. It must **coordinate the metabolic activities** that maintain the integrity of the cell.

The Structure of Neurons

Although neurons vary enormously in structure, **a "typical" vertebrate neuron has four distinct structural regions that carry out the functions listed**

previously. These are the dendrites, the cell body, the axon, and the synaptic terminals (Figure 36-1).

Dendrites

Information from the outside world or other neurons is received by the **dendrites,** a tangle of fibers that branch from the cell body. Whether the information is heat, light, pressure, or chemical stimuli (including chemical messages from other neurons), dendrites of neurons specialized to respond to these stimuli convert them into electrical signals. Dendrites that receive signals from other neurons have protein recep-

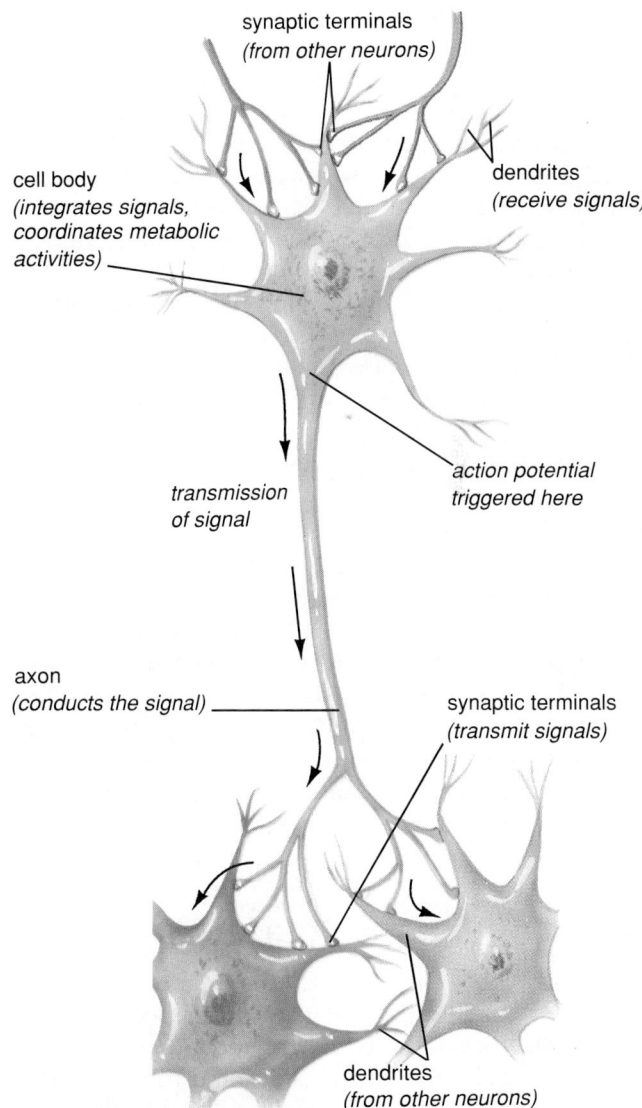

Figure 36-1 A nerve cell showing its specialized parts and their functions.

tors in their membranes that are specialized to receive a chemical released by another neuron.

Cell Body

Electrical signals from the dendrites converge upon the **cell body** of the neuron, which serves as an integration center. In its integrating role, the cell body adds up the various signals from the dendrites and "decides" whether to produce an **action potential,** the electrical output signal of the neuron. The cell body, containing the usual assortment of organelles, also synthesizes proteins, lipids, and carbohydrates, and coordinates the metabolic activities of the cell.

Axon

In a typical neuron, a long, thin fiber, called an **axon,** extends outward from the cell body. Axons make neurons the longest cells in the body. Single axons, for example, stretch from your spinal cord to your toes, a distance of about a yard (around 1 meter). Axons are distribution lines, carrying action potentials from the cell body to the synaptic terminals, located at the far end of each axon. Axons are usually bundled together into nerves, much like the strands of wire in an electric cable. In vertebrates, nerves emerge from the brain and spinal cord, extending out to all regions of the body. However, unlike electric power distribution cables (in which energy is lost along the way from power station to customer), the cell membranes of axons are specialized to conduct action potentials undiminished in size from the cell body to their synaptic terminals.

Synaptic Terminals

Signals are transmitted at **synaptic terminals,** which appear as swellings at the branched endings of axons (see Fig. 36-1). Most synaptic terminals contain a specific type of chemical, a **neurotransmitter,** that they release in response to a signal traveling down the axon. The synaptic terminals of one neuron may communicate with a gland, a muscle, or the dendrites or cell body of a second neuron, so that the output of the first cell becomes the input to the second.

Mechanisms of Neural Activity

About 40 years ago, using the giant axon of a mollusc—the squid—biologists developed ways to record electrical events inside individual neurons (Fig. 36-2). (The use of the squid and other invertebrates in neurobiology is explored in "Methods in Biology: Of Squids and Snails and Sea Hares.") They found that unstimulated, inactive neurons maintain a

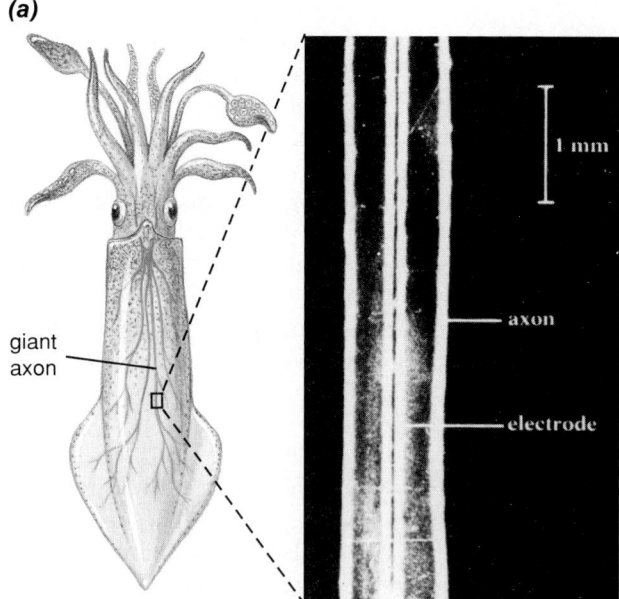

(a)

giant axon

1 mm

axon

electrode

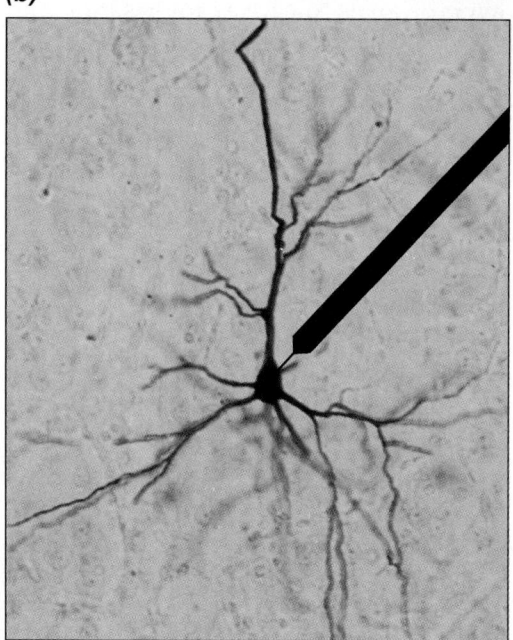

(b)

Figure 36-2 (a) Taking advantage of the giant axon of the squid, the British physiologists Bernard Katz, Alan Hodgkin, and Andrew Huxley pushed thin wires or narrow saline-filled tubes down the inside of the axon. These electrodes were connected to voltmeters to record the electrical potential difference between the inside and outside of the axon, about −70 millivolts. **(b)** Modern neurobiologists use hollow glass electrodes drawn to a needlelike tip less than 1 micrometer in diameter and filled with a salt solution. The sharp tip penetrates the neuron without damaging it.

METHODS IN BIOLOGY

Of Squids and Snails and Sea Hares (or, Why Do Biologists Study Such Bizarre Animals?)

If you wander through the biology laboratories at your college, you may find your professors studying some most unlikely organisms, such as insects, snails, bacteria, or algae. Sometimes, the reason is obvious: many insects, for example, are crop pests or carry diseases, and by learning about their life histories scientists may discover ways to control them. You might think, however, that neurobiologists would concentrate on humans, or at least mammals, since their primary interest is surely to find out how the human brain works. Nevertheless, many neurobiologists study snails, lobsters, or even leeches. Aside from the fact that if you work on lobsters, you can eat your experiment when you are through (no small consideration), why would anyone study such creatures?

Figure 36-2 provides a clue. While the largest mammalian axons are about 20 micrometers in diameter, the squid giant axon is about a *millimeter* in diameter! This is extremely convenient for neurobiologists. Tiny electrodes placed with painstaking precision are required to obtain good recordings from mammalian axons. In contrast, relatively crude electrodes can be inserted inside a squid axon without damaging it. Scientists have even taken little rubber rolling pins and squeezed the cytoplasm from a squid axon. This allowed them to analyze the chemical composition of the cytoplasm, and also to reinflate the axon with artificial cytoplasm to test theories about how action potentials are produced. Lobsters, crayfish, and some worms also have giant axons, and are studied for similar reasons.

Another advantage of certain animals is simplicity. Depending on whose guess you believe, the human brain has between 10 billion and 1 trillion tiny nerve cells. In contrast, some leeches and snails have only a few thousand neurons. Many of these neurons are very large and identifiable as individuals. If one wants to find out general principles about how circuits of neurons control behaviors, it is obviously easier to study a nervous system in which an entire behavior may be governed by only a few dozen neurons.

But, you may argue, once you have found out how a squid axon works, or what the circuit for snail feeding looks like, have you learned anything about human nervous systems? As unlikely as it may seem, the answer is unequivocally "yes!" The basis for action potential production in axons was worked using the giant axon of the squid in the early 1950s. Subsequent research has shown that with a few modifications, the same principles operate in mammalian axons. Without the guidance offered by the detailed analyses of squid axons, we still might not know how our own axons work. Similarly, synaptic function is understood largely through studies of the frog nerve–muscle synapse and a particularly large synapse in the squid.

Even complex behaviors such as learning may well have the same neuronal basis in humans and snails. In a simple form of learning called habituation, an animal ceases responding to a harmless, repeated stimulus. Habituation allows you to sleep through your roommate's snoring. Habituation has been studied thoroughly in the sea hare, *Aplysia* (see Fig. 36-12b). A few years later, psychologists examined habituation in the frog spinal cord and concluded that very similar mechanisms were at work.

Evolution is basically a conservative process. Organisms that are apparently very different share the same fundamental traits. Whether we consider humans or leeches, snails or crabs, the similarities are compelling. They include the structure of the gene, the mechanism of protein synthesis, the use of ATP for energy transfers, even hormone molecules. While the same mechanisms and molecules are used, they may be assembled in different ways to achieve very different results, such as the wings of the condor and the human hand. We should not be surprised if the human brain turns out to be another instance of an imposing new structure built with the standard issue of materials.

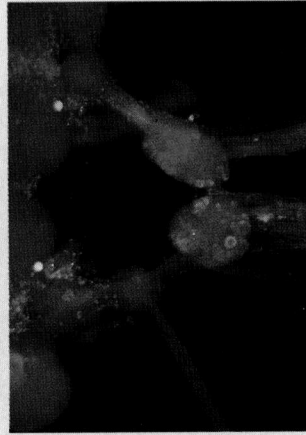

Figure E36-1 The brain of the pond snail, *Lymnaea stagnalis*, (left) is convenient for the study of neuronal function. Shown here are several ganglia from the brain (right). The circles within the ganglia are individual nerve cell bodies, often 100 to 200 micrometers in diameter. Some of these neurons can be identified as unique individuals and can be found in every snail brain.

constant electrical difference, or *potential*, across their cell membranes, similar to that found across the poles of a battery. This potential, called the **resting potential,** is always negative inside the cell, and ranges from −40 to −90 millivolts (thousandths of a volt). If the neuron is stimulated, either naturally or with an electric current, the negative potential inside the neuron can be altered. Depending on the nature of the stimulus, the inside potential can be made either more or less negative. If the potential is made sufficiently less negative, it reaches a level called **threshold** (roughly 15 millivolts less negative than the resting potential), at which an **action potential** is triggered. During the action potential, the neuron suddenly becomes 20 to 50 millivolts *positive inside.* Action potentials last a few milliseconds (thousandths of a second) before the cell restores its negative resting potential. Let's look more closely at these electrical signals, the language of the nervous system.

The Origin of the Resting Potential

How can a cell behave as a battery, separating electrical charge across its cell membrane? To understand this, recall two physical principles, *diffusion* and *electrical attraction,* and one property of cell membranes, *differential permeability,* all discussed in Chapter 6. These factors interact with concentration differences inside and outside the cell to produce the resting potential, as described below.

The cell membrane of a neuron encloses cytoplasm with various ions dissolved in it. The neuron itself is immersed in a salt solution, the extracellular fluid (Fig. 36-3). The ions of the cytoplasm consist mainly of positively charged potassium ions (K^+) and large, negatively charged organic molecules, such as proteins and the molecules of the citric acid cycle (Chapter 8). Outside the cell, the extracellular fluid contains mostly positively charged sodium ions (Na^+) and negatively charged chloride ions (Cl^-). We'll see how these concentration differences are maintained in a moment.

As you have learned, a cell membrane is a lipid ocean in which protein icebergs are embedded. Since charged particles cannot pass through the lipids, they must travel through tunnel-shaped proteins called channels extending through the membrane. In an unstimulated neuron, shown below, only potassium ions can cross the membrane. They travel through specific proteins called **potassium channels,** shown in yellow. Although **sodium channels** (shown in blue) are also present, in unstimulated neurons they remain closed. Since only potassium ions can cross the membrane, and potassium ions are most concen-

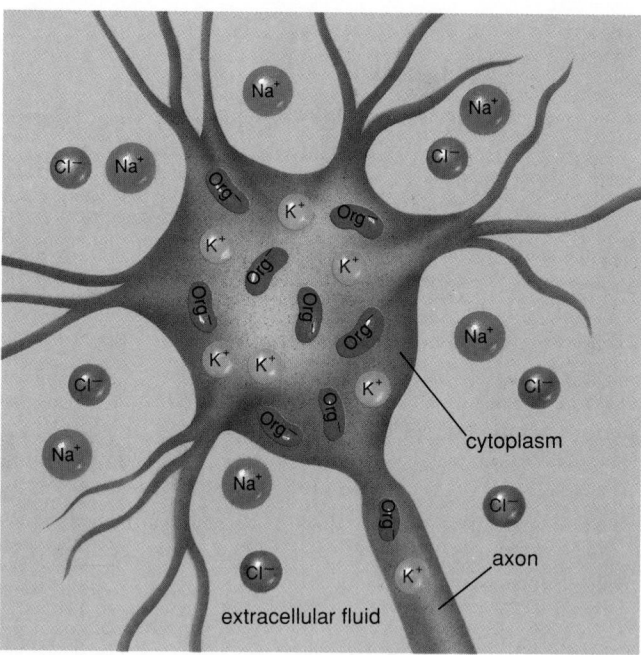

Figure 36-3 The ionic composition of the neuron cytoplasm is significantly different from the extracellular fluid. The neuron contains a high concentration of potassium ions (K^+) and large organic anions (Org^-), while the extracellular fluid is high in sodium chloride (Na^+ and Cl^-).

trated inside the cell, then potassium ions will diffuse out of the cell, leaving the large, negatively charged organic ions behind, as shown below.

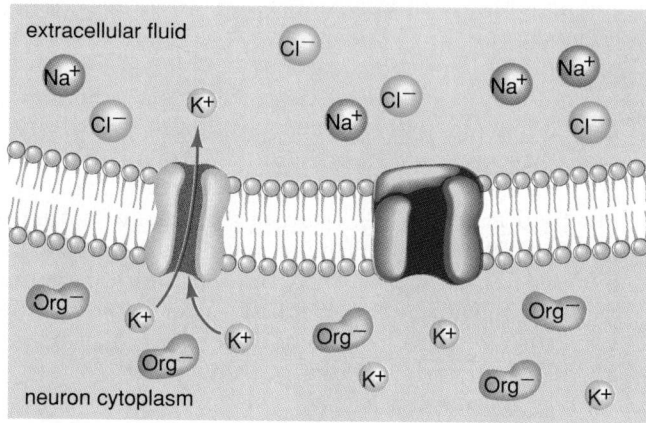

As more and more positively charged potassium ions leave, the inside of the cell becomes increasingly negative. But, since unlike charges attract one another, as potassium ions diffuse out, an electrical force develops that tends to pull them back inside. **At some point, the diffusion of potassium ions out of the neuron due to concentration differences will be balanced by the electrical attraction tending to pull**

them back inside. **At this point, there is no more net movement of potassium ions, and the cell reaches a stable resting potential, negative inside,** as illustrated below.

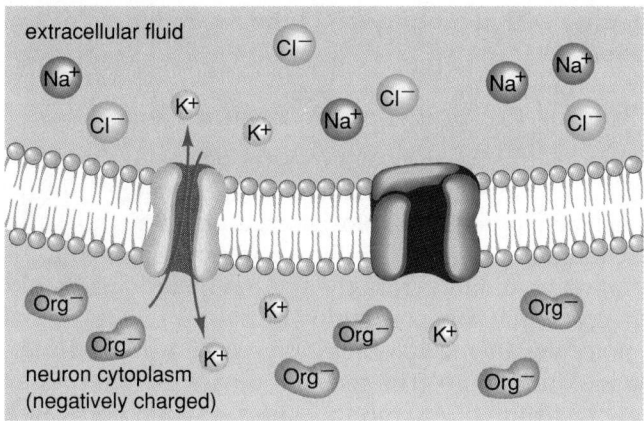

Establishing a resting potential in this way does not require significant changes in the potassium concentration inside and outside the cell. Only about 1/10,000 of the potassium ions initially inside our hypothetical cell must leave to set up a resting potential of −60 millivolts. Nearly all living cells maintain resting potentials. They are passed from cell to cell and generation to generation in an unbroken line that stretches back to the ancient ancestral cells from which they originated.

Action Potentials: Long-Distance Messages

An unchanging resting potential, like a single musical note, can't convey much information. **Nervous information is encoded in changes in potential in nerve cells. Two examples of changes in electrical potential in neurons are action potentials and postsynaptic potentials.**

If the potential inside the cell body of a neuron is made sufficiently less negative (such as by postsynaptic potentials, described later), the neuron may reach threshold, triggering an **action potential.** An action potential is a wave of positive charge that travels, undiminished in magnitude, along the axon to the synaptic terminal. The action potential is usually initiated at the point where the axon leaves the cell body, and travels like a wave along the axon to its synaptic terminal. Immediately after the action potential passes any point along the axon, the negative resting potential is restored within the axon. The events that occur during an action potential as re-

corded by an electrode inside the neuron are shown in Figure 36-4.

A neuron at rest, diagrammed below, is something like a loaded musket, charged and ready to fire if the trigger is pulled. In a neuron, the "explosive charge" is the concentration gradient of sodium ions, which are highest in concentration outside the cell. The "trigger" is a set of membrane proteins, the sodium channels (shown in blue). These proteins are selectively permeable to sodium, are closed in a resting neuron, and are specialized to open suddenly when threshold is reached. The energy to pull the trigger is provided by postsynaptic potentials (described below) that bring the neuron to threshold.

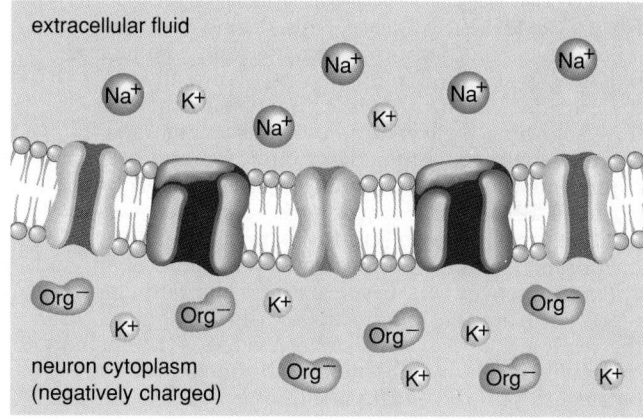

At threshold, sodium channels open, as shown below. Positively charged sodium ions flood into the cell, making the cell's interior momentarily positive.

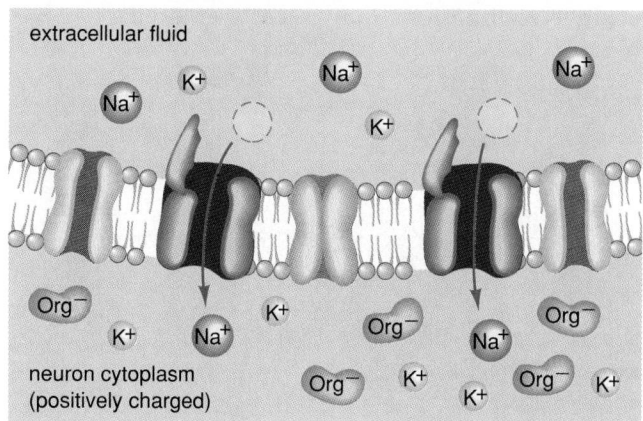

After a short time, the sodium channels spontaneously close, and a different set of potassium channels open, shown in orange below. Potassium ions now flow out of the cell through both types of potassium channels. The ions are driven out both by their diffu-

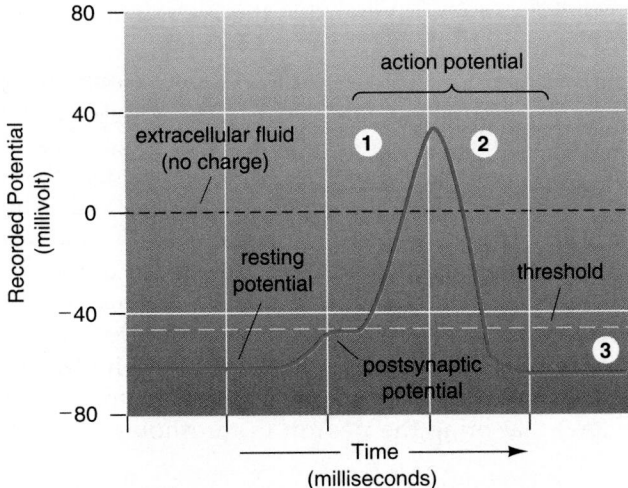

Figure 36-4 The electrical events during an action potential as recorded inside a nerve cell. The resting potential is about 60 millivolts negative with respect to the outside. When the cell is stimulated to reach threshold by a postsynaptic potential, membrane channels permeable to sodium open up, and sodium enters the cell, powered both by diffusion and by electrical attraction; the inside of the cell becomes positively charged (1). Shortly thereafter, other membrane channels permeable to potassium open, and potassium leaves (2), driven by diffusion and electrical repulsion from the now-positive inside of the cell, until the resting potential is reestablished (3). Active transport molecules in the membrane, called the sodium–potassium pump, continuously pump sodium out and potassium in maintaining the ionic gradient.

sion gradient and by electrical repulsion from the positive sodium ions that recently entered.

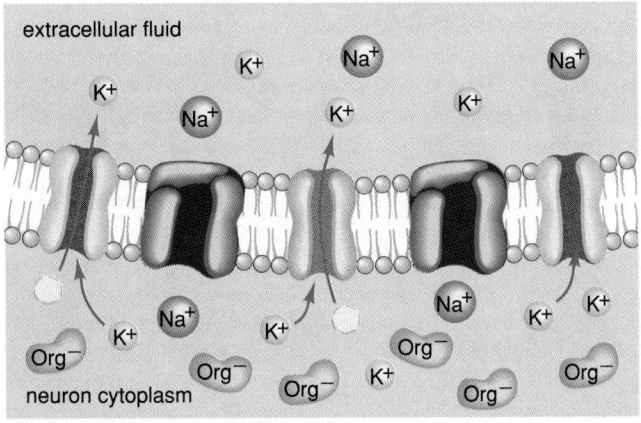

So many potassium ions leave that the inside once again becomes negative, reestablishing the resting potential. Thus, the action potential is a brief event; the neuron first becomes positive as sodium ions enter, and then becomes negative again as potassium ions flow out.

Action potentials are *all-or-none*—that is, they do not vary in magnitude. If the neuron does not reach threshold, there will be no action potential at all, but if threshold is reached, then a full-sized action potential will occur and travel the entire length of the axon.

Role of the Sodium–Potassium Pump

Only a tiny fraction of the total potassium and sodium in and around each neuron is exchanged during each action potential. However, after a few thousand action potentials, the sodium and potassium concentration gradients across the neuron membrane would be lost. This is prevented by a set of active transport molecules in the cell membrane called the **sodium–potassium pump.** The sodium–potassium pump uses energy from ATP to pump sodium out and potassium in, maintaining the concentration gradients of these ions across the cell membrane. Thus, the "explosive charge" we referred to earlier is created by the pump using the energy from ATP.

Conduction of the Action Potential

The action potential is a signal. To be effective, it must be transmitted along the axon to cells specialized to receive the message, including other neurons, muscle, or glandular cells. If the neuron is to conduct an action potential along its axon to its synaptic terminal, the action potential must not diminish in magnitude and die out along the way. The cell maintains the magnitude of the action potential by renewing it at each successive point along the axon, as illustrated in the following series of diagrams.

The action potential begins when threshold is reached, sodium channels open, and sodium ions enter the cell, making it positive inside the cell at that point.

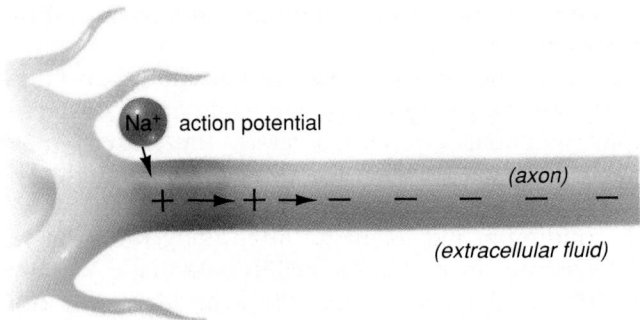

Although much of this positive charge leaks back out, some spreads passively and *almost instantaneously* along the inside of the axon, making the adjacent region less negative.

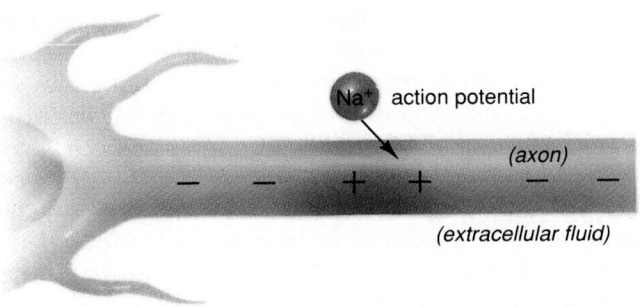

When the adjacent region of membrane reaches threshold, its sodium channels open, causing a further influx of sodium ions and an action potential here in the adjacent membrane.

The positive charge spreads still further, generating another action potential in the adjacent membrane. This process continues along the entire length of the axon. Meanwhile, the sodium channels at the site of the original action potential close, and the resting potential is reestablished there.

Saltatory Conduction

It is important that action potentials travel rapidly (a giraffe couldn't run from a lion if it took 10 seconds for a signal to travel from brain to hoof). However, the opening and closing of ion channels during action potentials is relatively slow. Therefore, for a signal to travel as rapidly as possible, as few channels as possible should be opened and closed. In vertebrates, axons that need to conduct rapidly (such as those that carry signals to muscles used in pursuit or escape) are wrapped with insulating layers of membrane called **myelin,** interrupted at intervals with naked areas called **nodes of Ranvier** (Fig. 36-5). These myelinated axons transmit signals extremely rapidly because ion channels are concentrated only at the nodes. When an action potential occurs in a myelinated axon, the positive charge that enters the axon near the cell body can't leak back out through the myelin, but instead travels almost instantaneously to the next node. At the next node, channels open and a new action potential is initiated. Positive charges enter at the node and flow immediately to the next node, and so on. By insulating the axon, myelin allows the rapid, passive spread of charge to continue as far as possible. But since the spread of charge beneath the myelin *does* diminish with distance, myelinated axons maintain the size of the signal by initiating new action potentials at each node. The transmission of an action potential along a myelinated axon is called **saltatory conduction** (literally, "jumping" conduction), since

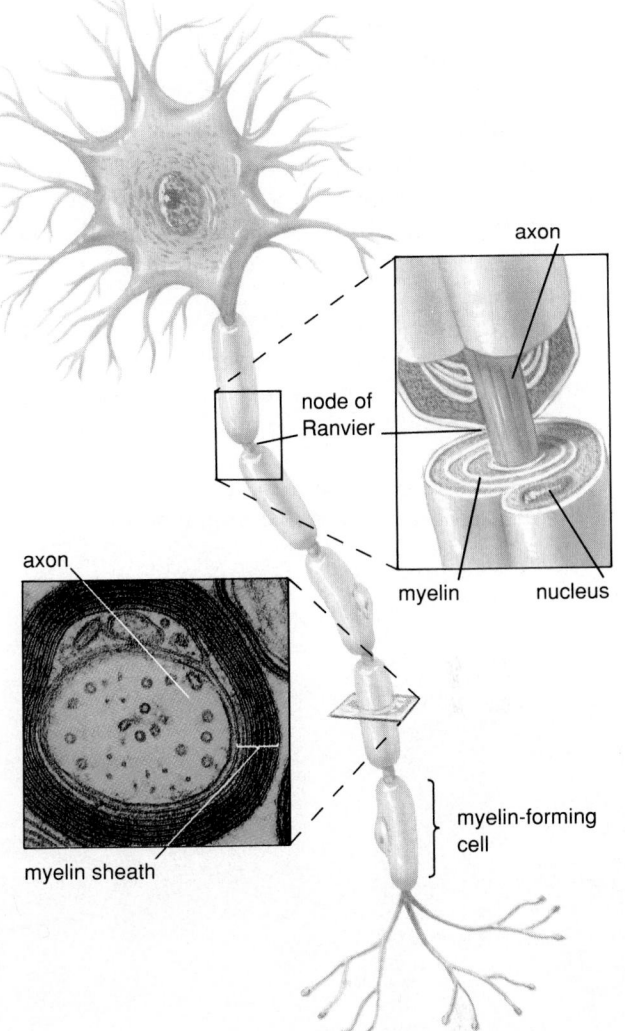

Figure 36-5 Some vertebrate axons are wrapped in a membrane "jellyroll" of myelin, formed from windings of membrane from specialized flattened cells. At intervals of around a millimeter, the wrapping is interrupted by bare places, called nodes of Ranvier, where action potentials occur.

the action potential appears to jump from node to node, speeding down the axon. Myelin is formed from specialized non-neural cells that flatten and wrap themselves around the axon (see Fig. 36-5).

Communication Between Neurons: Synapses and Postsynaptic Potentials

Once an action potential has been conducted to the synaptic terminal of the neuron, it must be transmitted to another cell, usually another neuron. This transmission occurs at specialized regions called synapses, and the signals transmitted at synapses are called postsynaptic potentials.

The Synapse

An action potential can travel undiminished down an axon several meters long (in giraffes and whales, for example). When this electrical signal reaches the synaptic terminal of the axon, it encounters a region, called a **synapse,** where two neurons are close together but do not touch one another. A minuscule gap, called a **synaptic cleft,** separates the synaptic terminal of the first, or **presynaptic,** neuron from the dendrite or cell body of the second, or **postsynaptic,** neuron (Fig. 36-6).

When an action potential reaches a synaptic terminal, the inside of the synaptic terminal becomes positively charged. This triggers the synaptic terminal to release a chemical neurotransmitter into the synaptic cleft. The neurotransmitter molecules rapidly diffuse across the gap and bind to receptors, specialized proteins in the membrane of the dendrites and cell body of the postsynaptic cell (see Fig. 36-6). The synapse can be defined as the region that includes the synaptic terminal of the presynaptic cell, the synaptic cleft, and the specialized membrane of the postsynaptic cell just across the cleft that contains receptors for neurotransmitters.

Postsynaptic Potentials

The receptors in the postsynaptic membrane have two roles. First, each type of receptor binds to a specific type of neurotransmitter. Second, after binding

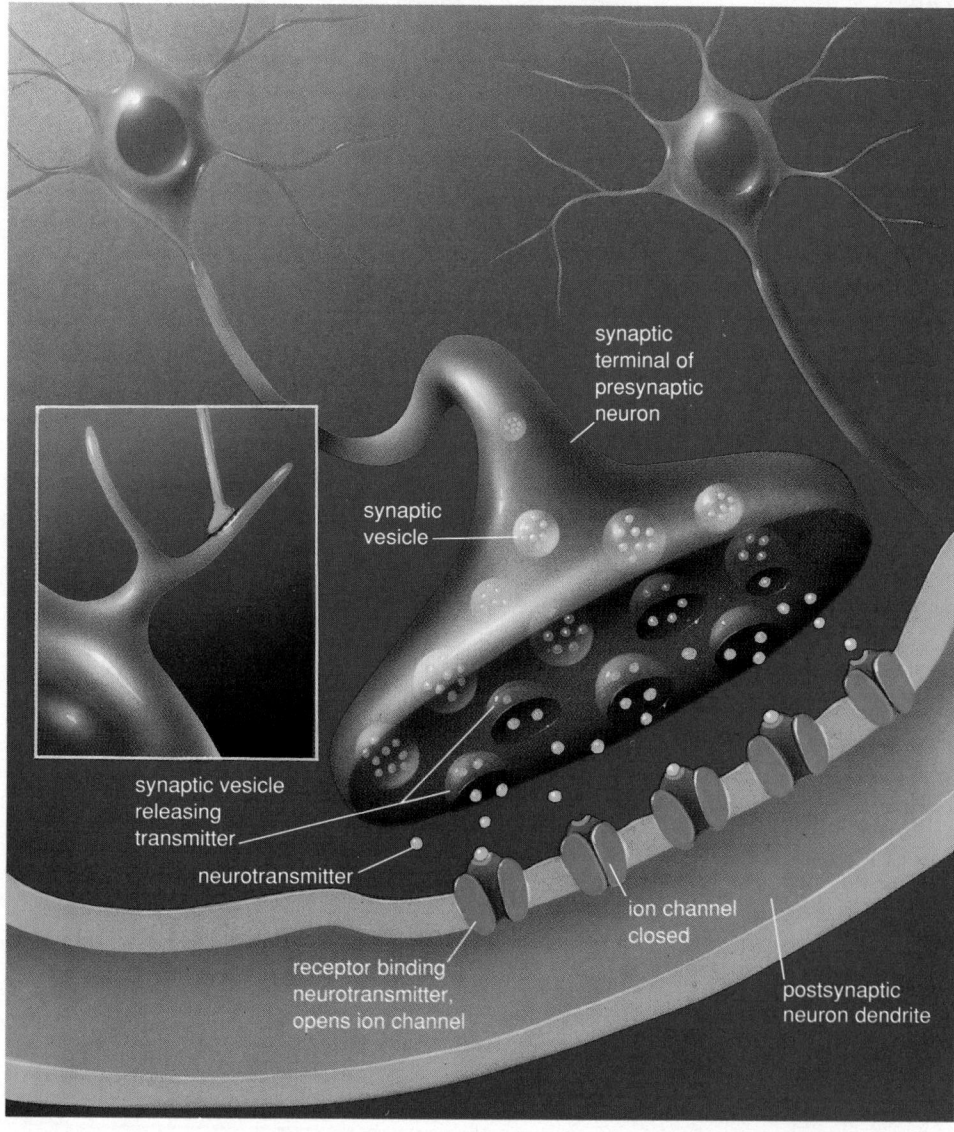

synaptic terminal of presynaptic neuron

synaptic vesicle

synaptic vesicle releasing transmitter

neurotransmitter

receptor binding neurotransmitter, opens ion channel

ion channel closed

postsynaptic neuron dendrite

Figure 36-6 The structure and function of the synapse. The ending of the presynaptic cell contains numerous membrane-bound spheres (called *synaptic vesicles*) that contain neurotransmitter. The postsynaptic cell has membrane receptors for the transmitter. When an action potential enters the synaptic terminal of the presynaptic cell, the vesicles dump their neurotransmitter into the space between the neurons. The neurotransmitter diffuses rapidly across the space, binds to the postsynaptic receptors, and causes ion channels to open. Ions flow through these open channels, causing a postsynaptic potential in the postsynaptic cell.

the neurotransmitter, the receptor causes specific types of ion channels in the membrane of the post-synaptic neuron to open. When ion channels are opened, ions flow across the cell membrane of the postsynaptic neuron along their concentration gradients. The flow of ions in the postsynaptic neuron causes a **postsynaptic potential** in the dendrites or cell body where the synapse occurs (Fig. 36-7). Depending on what type of channels are opened and what type of ions flow, **synaptic potentials** can either be excitatory, making the neuron less negative inside and more likely to fire an action potential, or inhibitory, making it more negative and less likely to fire. An excitatory postsynaptic potential is called an **EPSP,** while an inhibitory postsynaptic potential is called an **IPSP.** A synapse that produces EPSPs in the postsynaptic cell is called an **excitatory synapse,** while a synapse producing IPSPs is an **inhibitory**

synapse. Postsynaptic potentials cannot travel far in a neuron; after a few millimeters, at most, the ions leak back across the membrane and the signal is lost. However, postsynaptic potentials travel far enough to reach the cell body, where they determine whether or not an action potential will be produced, as described below.

The Integration of Synaptic Potentials

The dendrites and cell body of a single neuron can receive EPSPs and IPSPs from the synaptic terminals of thousands of presynaptic neurons. The postsynaptic potentials produced by different presynaptic neurons are then "added up," or **integrated,** in the cell body of the postsynaptic neuron. The postsynaptic cell will produce an action potential only if the excitatory and inhibitory potentials, when added together, raise the electrical potential inside the neuron above threshold (see Fig. 36-7).

The Fate of Neurotransmitters

Neurotransmitters act only briefly on the postsynaptic cell. A few transmitters are *destroyed by enzymes* released into the synaptic cleft; others are removed from the synaptic cleft by *active transport back into the presynaptic neuron* or simply *diffuse away* into the extracellular fluid. The various types of neurotransmitters are described later in this chapter.

Building and Operating a Nervous System

The individual neuron uses a language of action potentials. Yet somehow this basic language allows even simple animals to perform an impressive variety of complex behaviors. One key to the versatility of the nervous system is the presence of complex networks of neurons. These neural networks range from a few to billions of cells. As in computers, small, simple elements can perform amazing feats when connected properly.

Information Processing in the Nervous System

Before we delve into the basic anatomy of nervous systems, we should first examine the operating principles. **At a minimum, a nervous system must be able to perform four operations:**

1. **Signal the intensity of a stimulus,**
2. **Determine the type of stimulus,**
3. **Integrate information from many sources, and**
4. **Initiate and direct the response.**

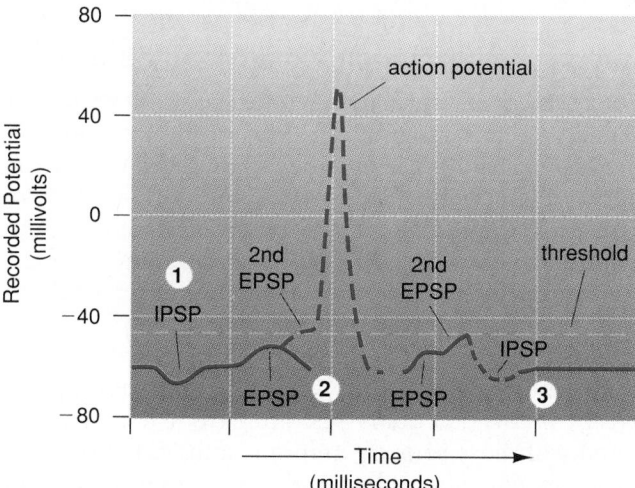

Figure 36-7 Electrical events recorded inside the cell body of a neuron that receives both excitatory and inhibitory postsynaptic potentials and integrates these signals. (1) A presynaptic neuron making an inhibitory synapse on the recorded cell causes an inhibitory postsynaptic potential (IPSP). This brings the potential inside the cell body farther away from threshold. (2) A different presynaptic neuron making an excitatory synapse on the recorded cell causes an excitatory postsynaptic potential (EPSP). This brings the potential closer to threshold. If a second EPSP from an excitatory synapse occurs at almost the same time, the two are added together, and (in this case) bring the potential in the cell body above threshold. This initiates an action potential in the axon. (3) If EPSPs from excitatory synapses and an IPSP from an inhibitory synapse occur at about the same time, they are all added together. The IPSP can prevent the EPSPs from bringing the potential inside the cell body to threshold, and so can prevent an action potential from occuring.

Let's examine each of these operations in more detail.

Signal the Intensity of a Stimulus

Since all action potentials are of the same magnitude and duration, no information about the **intensity** of a stimulus (e.g., the loudness of a sound) can be encoded in a single action potential. Instead, intensity is coded in two other ways. First, **intensity can be signaled by the frequency of action potentials in a single neuron.** The more intense the stimulus, the faster the neuron fires (Fig. 36-8). Second, a nervous system

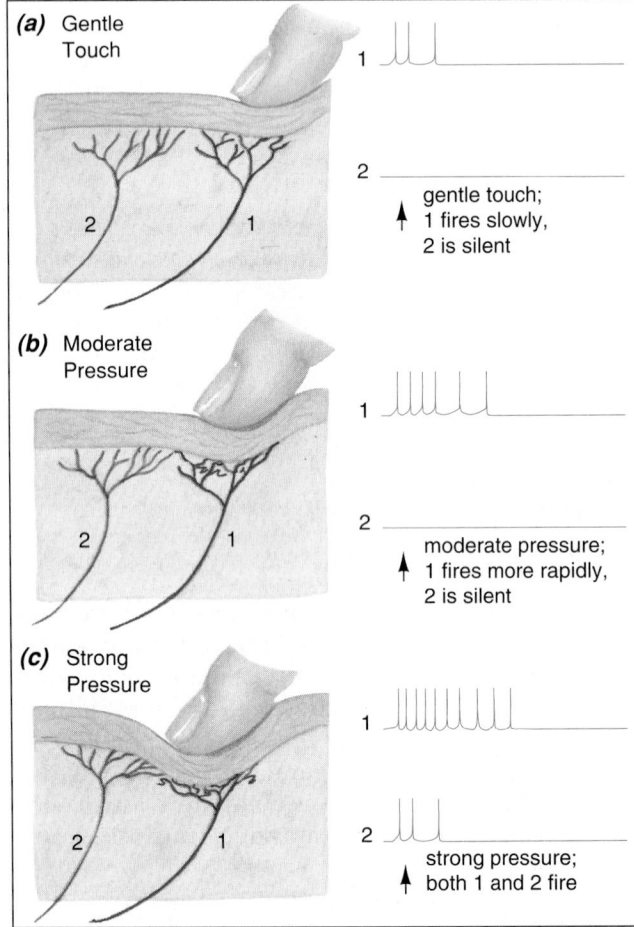

Figure 36-8 The intensity of a stimulus is signaled by the rate at which individual neurons produce action potentials, and by the number of neurons firing. For example, two touch receptors may have endings in adjacent patches of skin. **(a)** A gentle touch elicits only a few action potentials and from only one of the sensory neurons. **(b)** Moderate pressure still stimulates only one receptor, but this receptor now fires faster, informing the brain that the touch is more intense than before. **(c)** Strong pressure activates both receptors, firing one very fast, and the other more slowly, thus signaling to the brain that the pressure is very intense.

usually has many neurons that can respond to the same input. Stronger stimuli tend to excite more of these neurons, while weaker stimuli excite fewer. Thus, **intensity can also be signaled by the number of similar neurons firing at the same time** (see Fig. 36-8).

Determine the Type of Stimulus

Besides signaling intensity, the nervous system must also have a way of identifying the type of stimulus (e.g., light, touch, or sound). Here again, the type of stimulus cannot be coded by the properties of individual action potentials. Instead, the nervous system monitors which neurons are firing action potentials. Thus, your brain identifies action potentials occurring in the axons of your optic nerves (originating in the eye) as information about light, action potentials in olfactory nerves as odors, and so on. This genetic wiring may occasionally yield false information. Being poked in the eye may cause action potentials in the optic nerve. Even though the stimulus is mechanical, your brain nevertheless interprets all optic nerve activity as light, and you "see stars."

Integrate Information from Many Sources

Your brain is continuously bombarded by a variety of sensory stimuli originating both inside and outside the body. The brain must filter all these inputs, determine which ones are important, and decide what actions to take in response. Nervous systems integrate information much as do individual neurons, through **convergence.** In this process, many neurons funnel their signals to fewer neurons. For example, many sensory neurons may converge onto a smaller number of brain cells (Fig. 36-9). The brain cells add up the postsynaptic potentials resulting from the synaptic activity of these sensory neurons, and, depending on their relative strengths (and other internal factors such as hormones or metabolic activity), they produce appropriate outputs.

Initiate and Direct the Response

The output of the integrating cells is responsible for initiating activity. The actions directed by the brain may involve many parts of the body, and require **divergence,** the flow of electrical signals from a relatively small number of decision-making cells onto many different neurons controlling muscle or glandular activity (see Fig. 36-9).

Neural Networks

Most behaviors are controlled by nervous–muscular pathways composed of four elements:

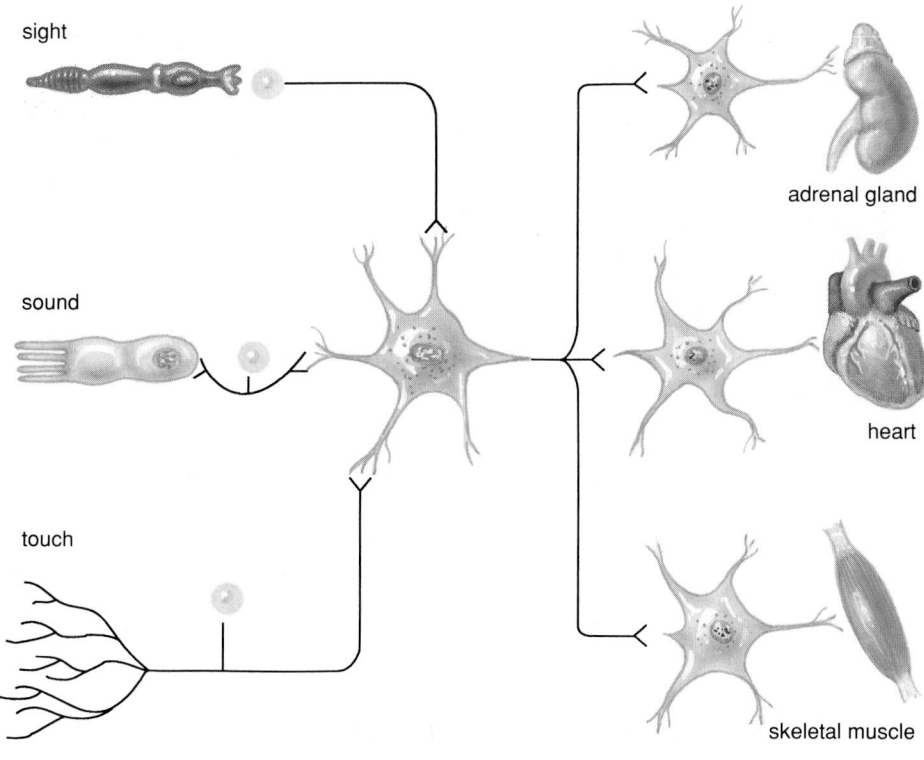

sight

sound

touch

adrenal gland

heart

skeletal muscle

CONVERGENCE

DECISION-MAKING ASSOCIATION NEURON IN BRAIN

DIVERGENCE

Figure 36-9 Integration of information and initiation of coordinated action involves convergence of inputs to and divergence of outputs from neurons. In this simplified example, inputs of sight, sound, and touch (say, being attacked by a swarm of bees) all converge on a "decision-making" neuron, which is strongly stimulated as a result. Its outputs go to the adrenal glands ("pump out adrenalin"), the heart ("beat faster and stronger"), and skeletal muscles ("move, legs!").

1. **Sensory neurons** that respond to a stimulus, either internal or external to the body;
2. **Association neurons** that "decide" what to do, based on input from many sensory neurons, stored memories, hormonal states, and other factors;
3. **Motor neurons** that receive instructions from the association neurons and activate the muscles; and
4. **Effectors,** usually muscles or glands, that perform the behavior.

From Neuron to Behavior

The simplest type of behavior is the **reflex,** a relatively involuntary movement of a body part in response to a stimulus. Examples of human reflexes include the familiar knee jerk and withdrawal reflex. The withdrawal reflex (which moves a body part away from a painful stimulus) is particularly instructive, since it uses only one neuron of each type (Fig. 36-10). Reflexes of this sort do not require the brain, although, as we know, other pathways do inform the

brain of pricked fingers and may in fact trigger other more complex behaviors (cursing, for example!).

Nearly all animals are capable of much more subtle and varied behavior than can be accounted for by simple reflexes. In principle, these more complex behaviors can be organized by *interconnected nervous pathways* in which several types of sensory input (along with memories, hormones, etc.) converge on a set of association neurons (see Fig. 36-9). By integrating the postsynaptic potentials from several sources, the association neurons can "decide" what to do, and stimulate the motor neurons to direct appropriate activity in muscles and glands.

Nervous System Design

In all the animal kingdom, there are really only two designs for nervous systems: *diffuse nervous systems,* found in the cnidarians (*Hydra,* jellyfish, and their relatives; Fig. 36-11), and *centralized nervous systems,* found to varying degrees in more complex organ-

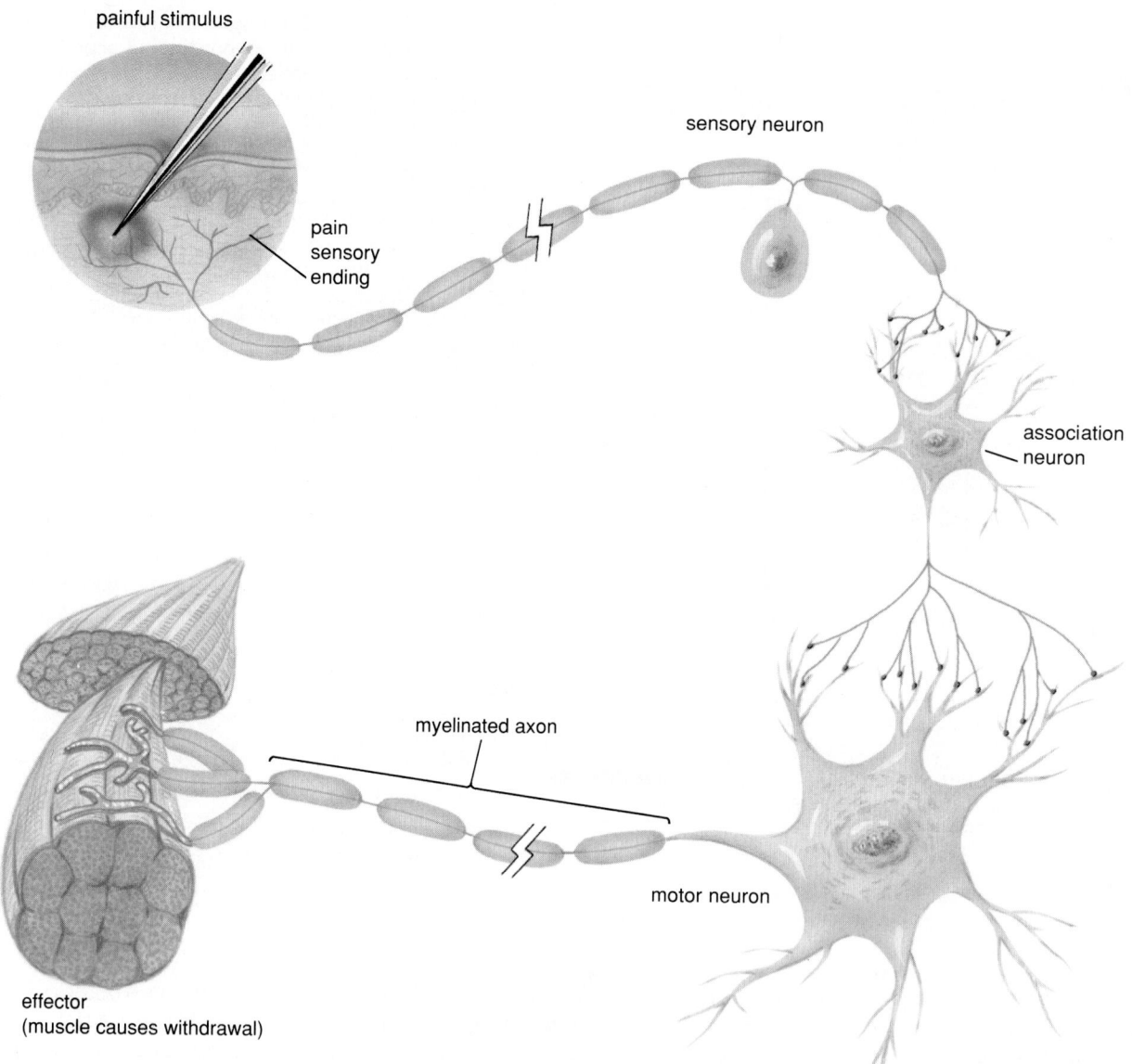

Figure 36-10 A diagram of a simple reflex, the pain–withdrawal reflex, illustrating the three types of neurons and an effector (muscle). Note that the neurons vary tremendously in size and shape, a reflection of their varied functions.

isms. Not surprisingly, nervous system design is highly correlated with the life-style of the animal. In the radially symmetrical cnidarians, there is no "front end," so there has been no evolutionary pressure to concentrate the senses in one place. A *Hydra* sits anchored to the substrate, and prey or danger are equally likely to come from any direction. Cnidarian nervous systems are composed of a network of neu-

rons, often called a **nerve net,** woven through the tissues of the animal. Here and there we can find a cluster of neurons, called a **ganglion** (plural *ganglia*), but nothing resembling a real brain.

Almost all other animals are bilaterally symmetrical, with definite head and tail ends. Since the head first encounters food, danger, and potential mates, it is advantageous to have sense organs concentrated

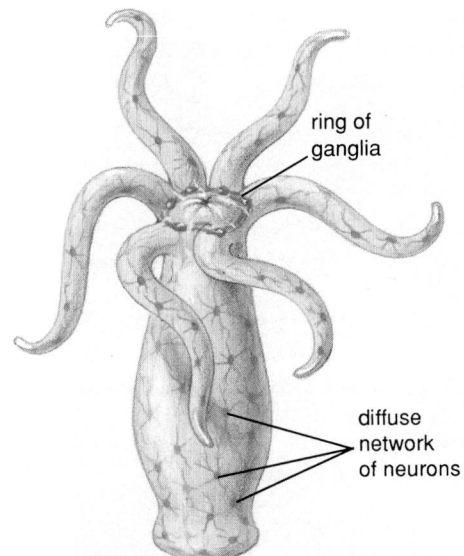

Figure 36-11 The diffuse nervous system of *Hydra* contains a few concentrations of neurons, particularly at the bases of the tentacles, but no brain. Conduction of neural signals may occur in virtually any direction throughout the body.

here. Sizable ganglia evolved that integrate the information gathered by the senses and initiate appropriate action. Over evolutionary time, the sense organs gathered in the head and the ganglia became centralized into a brain. This trend is clearly seen in the molluscs (Fig. 36-12). Centralization reaches its peak in the vertebrates, where nearly all the cell bodies of the nervous system are localized in the brain and spinal cord. The organization of the vertebrate nervous system is shown in Figure 36-13.

The Human Nervous System

The human nervous system may be divided into two parts. The **central nervous system** consists of a **brain** and a **spinal cord** that extends down the dorsal part of the torso. The **peripheral nervous system** consists of nerves connecting the central nervous system to the rest of the body (see Fig. 36-13).

The Peripheral Nervous System

The peripheral nervous system consists of nerves that link the brain and spinal cord to the rest of the body, including the muscles, the sensory organs, and the organs of the digestive, respiratory, excretory, and circulatory systems. Within the peripheral nerves are axons of sensory neurons that bring sen-

sory information *to* the central nervous system from all parts of the body. Peripheral nerves also contain the axons of motor neurons that carry signals *from* the central nervous system to the organs and muscles.

The motor portion of the peripheral nervous system may be subdivided into two parts: the **somatic nervous system** and the **autonomic nervous system.** Motor neurons of the somatic nervous system synapse on skeletal muscles, and control voluntary movement. Their cell bodies are located in the gray matter of the spinal cord (see Fig. 36-15), and their axons go directly to the muscles they control. (Muscles and their control are discussed in Chapter 38).

Motor neurons of the autonomic nervous system control involuntary responses. They synapse on the heart, smooth muscle, and glands. The autonomic nervous system is controlled both by the medulla and the hypothalamus of the brain, described below. It consists of two divisions, the **sympathetic nervous system** and the **parasympathetic nervous system** (Fig. 36-14). The two divisions of the autonomic nervous system generally make synaptic contacts with the same organs, and usually produce opposite effects.

The sympathetic nervous system acts on the organs in ways that prepare the body for stressful or highly energetic activity, such as fighting, escaping, or giving a speech. During such "fight or flight" activities, the sympathetic nervous system curtails activity of the digestive tract, redirecting some of its blood supply to be used by the muscles of arms and legs. Heart rate accelerates. The pupils of the eyes open wider, admitting more light, and the air passages in the lungs expand, accommodating more air.

The parasympathetic nervous system, on the other hand, dominates during maintenance activities that can be carried on at leisure, often called "rest and rumination." Under its control, the digestive tract becomes active, heart rate slows, and air passages in the lungs constrict.

Two differences in the organization of the sympathetic and parasympathetic nervous systems are evident in Figure 36-14. First, parasympathetic axons are found in nerves that originate from two separate locations, the brain (midbrain and medulla) and the base of the spinal cord. In contrast, sympathetic axons are found in nerves that originate from the middle and lower portions of the spinal cord. Second, in both the sympathetic and parasympathetic divisions, there are two neurons that carry messages in sequence from the central nervous system to each target organ, but they synapse at different locations. In the sympathetic nervous system, the synapse occurs in ganglia that are near the spinal cord. In the parasympathetic

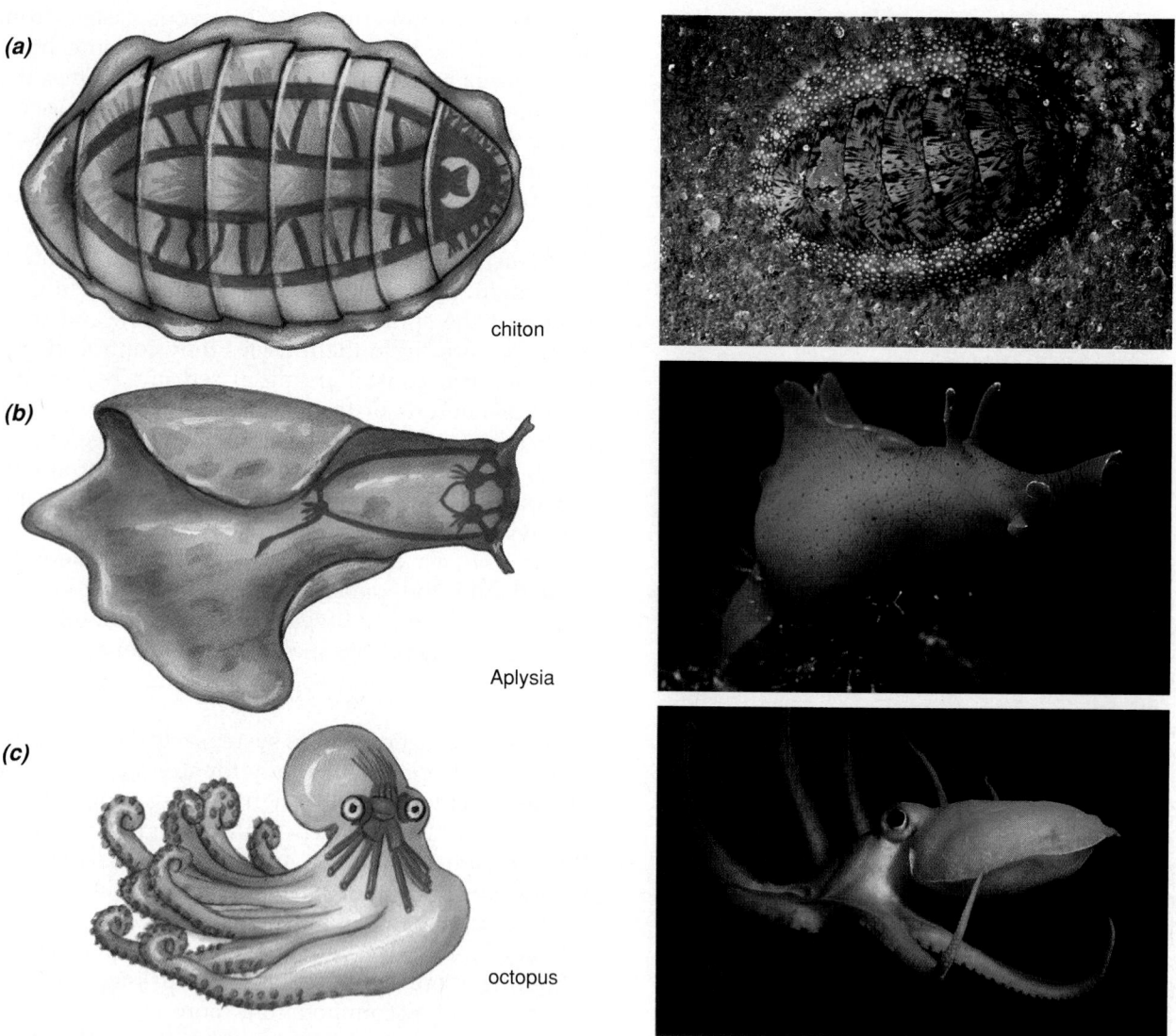

(a)

chiton

(b)

Aplysia

(c)

octopus

Figure 36-12 Bilaterally symmetrical animals usually have nervous systems concentrated in the head. The trend toward increasing concentration of the nervous system in the head is clearly illustrated by various molluscs (colored structures represent the nervous systems). **(a)** The chiton, although it does have a head end, seldom crawls in any direction and has had little selective pressure to concentrate sense organs and brain in the head. **(b)** Some marine snails, such as the shell-less *Aplysia*, can crawl quite rapidly, or even swim. Still more of their neurons are aggregated into a brain. **(c)** Mollusc mobility and intelligence culminates in *Octopus*, with its large, complex brain and behavioral capabilities rivaling those of some mammals.

nervous system, the synapse occurs in smaller ganglia located at or very near each target organ.

The Central Nervous System

The central nervous system consists of the brain and spinal cord. It is the integrating portion of the nervous system, where sensory information is received and processed, thoughts are generated, and responses are directed. The central nervous system consists primarily of association neurons—somewhere between 10 and 100 billion of them!

The brain and spinal cord are protected in three ways: The first line of defense is a bony armor—the skull surrounding the brain and the vertebral column surrounding the spinal cord. Beneath the bones lies a triple layer of connective tissue called **meninges**. Between the layers of the meninges, a clear lymphlike

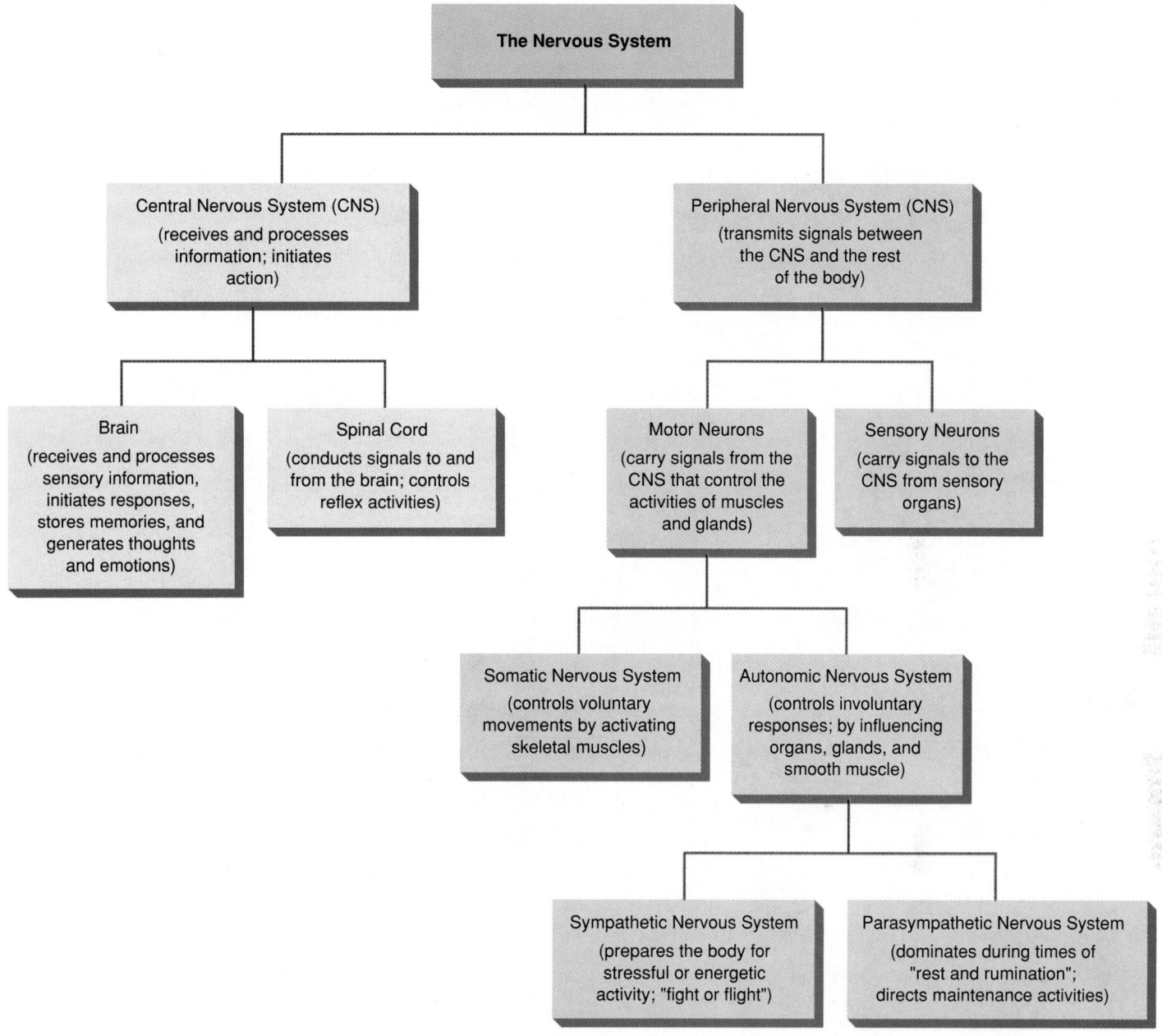

Figure 36-13 The organization of the vertebrate nervous system, showing its major parts and their subdivisions and functions.

liquid, the **cerebrospinal fluid,** cushions the brain and spinal cord. This fluid also fills spaces within the brain called **ventricles,** where it is produced (see Fig. 36-18).

The Spinal Cord

The spinal cord is a neural cable about as thick as your little finger that extends from the base of the brain to the hips, protected by the bones of the verte-

bral column (Fig. 36-15). Between the vertebrae, nerves called **dorsal roots** and **ventral roots** arise from the dorsal and ventral portions of the spinal cord, respectively; these merge to form the **spinal nerves.** In the center of the spinal cord are neuron cell bodies, which form a butterfly-shaped area of **gray matter.** These are surrounded by bundles of axons called **white matter** owing to their white insulating myelin coating (see Fig. 36-15). The spinal cord relays signals between the brain and the rest of the body, and con-

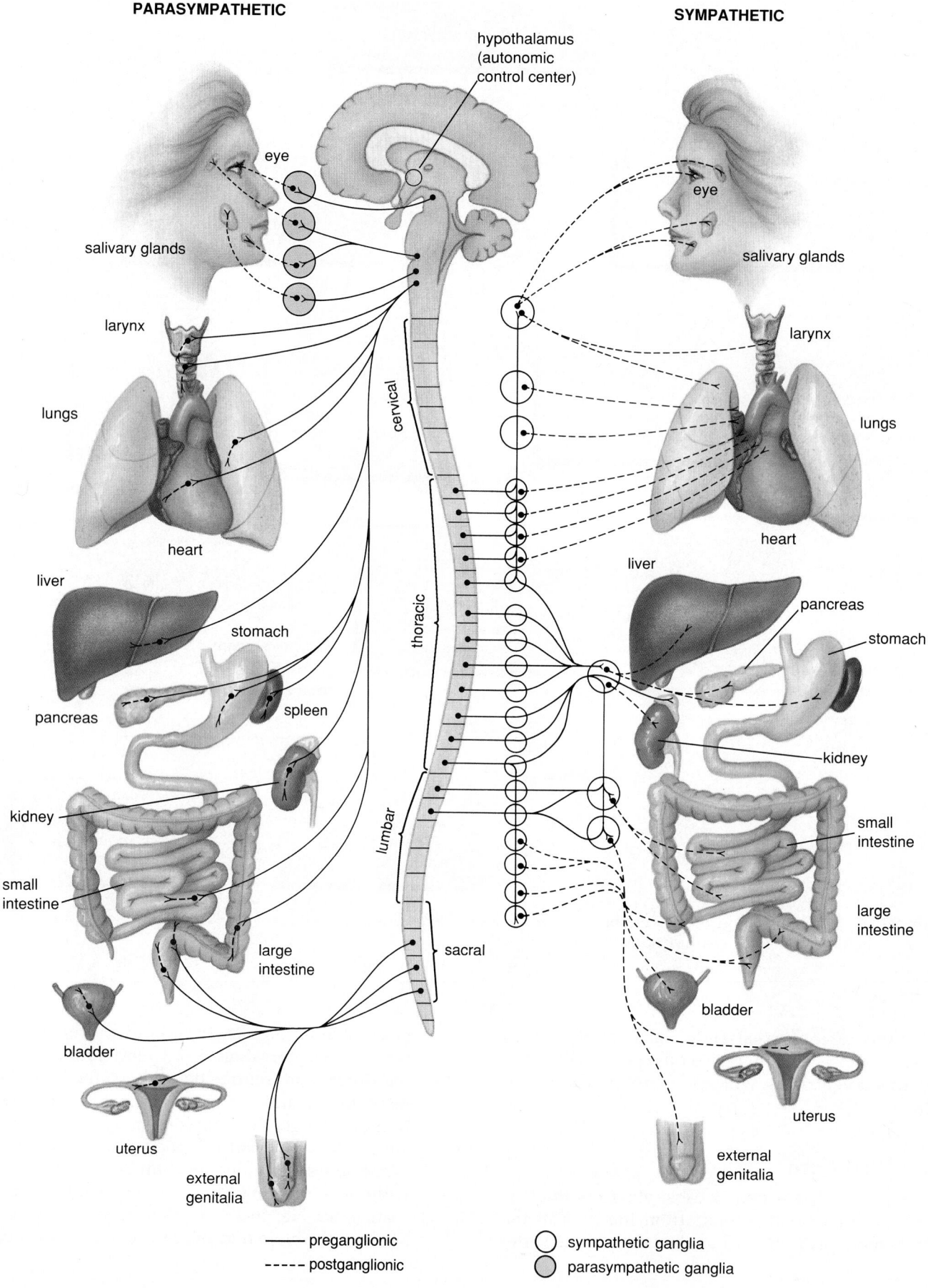

PARASYMPATHETIC

SYMPATHETIC

hypothalamus
(autonomic
control center)

eye

salivary glands

larynx

cervical

lungs

thoracic

heart

liver

stomach

pancreas

spleen

lumbar

kidney

small
intestine

large
intestine

sacral

bladder

uterus

external
genitalia

eye

salivary glands

larynx

lungs

heart

liver

pancreas

stomach

kidney

small
intestine

large
intestine

bladder

uterus

external
genitalia

—— preganglionic
---- postganglionic

◯ sympathetic ganglia
⬤ parasympathetic ganglia

◀ **Figure 36-14** The autonomic nervous system has two divisions, the sympathetic and parasympathetic. Both divisions supply nerves to many of the same organs, but produce opposite effects. Activation of the autonomic nervous system is mostly involuntary, produced by nervous outputs from the hypothalamus.

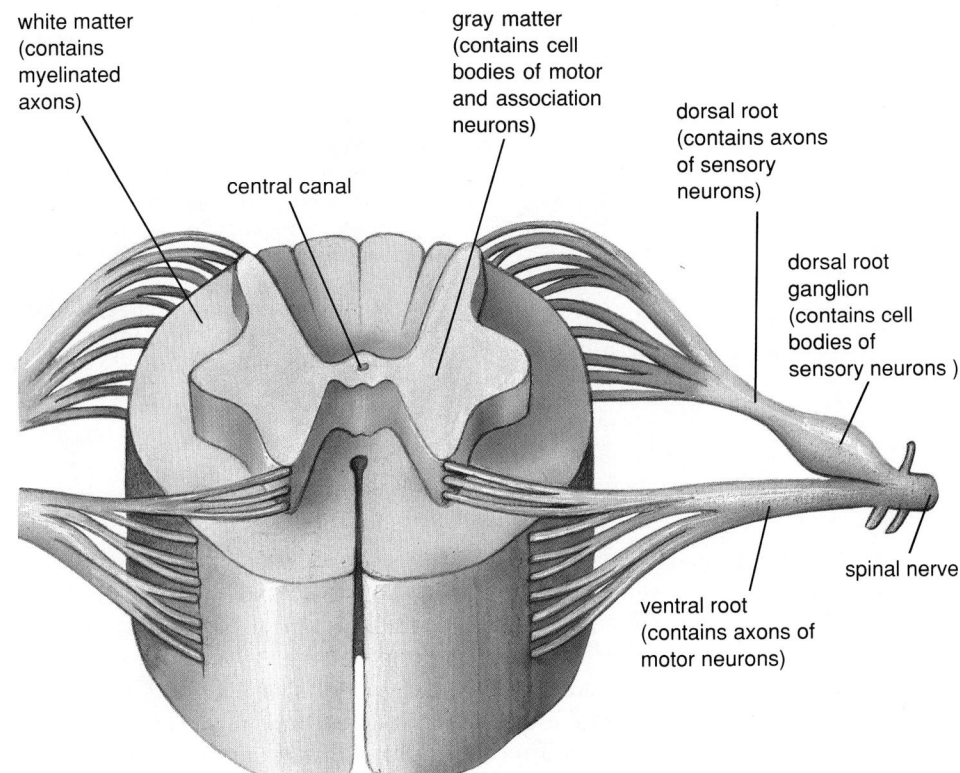

white matter (contains myelinated axons)

gray matter (contains cell bodies of motor and association neurons)

central canal

dorsal root (contains axons of sensory neurons)

dorsal root ganglion (contains cell bodies of sensory neurons)

spinal nerve

ventral root (contains axons of motor neurons)

Figure 36-15 The spinal cord runs from the base of the brain to the hips, protected by the vertebrae of the spine. Most of the body below the neck is supplied by paired spinal nerves that emerge from between the vertebrae. The nerves split before entering the spinal cord, with axons of sensory neurons running in the dorsal root, while motor axons comprise the ventral root. A cross section of the spinal cord reveals an outer region of myelinated axons (white matter) traveling to and from the brain, surrounding an inner, butterfly-shaped region of dendrites and the cell bodies of association and motor neurons (gray matter). The cell bodies of the sensory neurons are located outside the cord in the dorsal root ganglion.

tains the neural circuitry for certain behaviors, including reflexes.

To illustrate some of the functions of the parts of the spinal cord, let's examine a simple spinal reflex, the pain–withdrawal reflex, which involves neurons of both the central nervous system and the peripheral nervous system (Fig. 36-16; see also Fig. 36-10). The cell bodies of the sensory neurons from the skin (in this case signaling pain) are found just outside the spinal cord in a row of ganglia. Each of these **dorsal root ganglia** is located on a spinal nerve and nestled close to the vertebral column. Both association and motor neuron cell bodies are found in the gray matter in the center of the spinal cord. The axons in the surrounding white matter communicate with the brain. Association neurons for the pain reflex, for example, not only synapse on motor neurons, but have axons extending up to the brain. Signals carried along these axons alert the brain to the painful event. The brain, in turn, sends impulses down axons in the white

matter to cells in the gray matter. These signals can modify spinal reflexes. With sufficient motivation, you can suppress the pain–withdrawal reflex; to rescue a child from a burning building, for example, you could reach into the flames.

In addition to simple reflexes, the entire program for operating some fairly complex activities also resides within the spinal cord. All the neurons and interconnections needed to walk and run, for example, are found within the cord. In these cases, the role of the brain is to initiate and guide the activity of spinal neurons. The advantage of this semi-independent arrangement is probably an increase in speed and coordination, since messages do not have to travel all the way up the cord to the brain and back down again (in the case of walking) merely to swing forward one of your legs.

The motor neurons of the spinal cord also control the muscles involved in conscious, voluntary activities such as eating, writing, or playing tennis. Axons

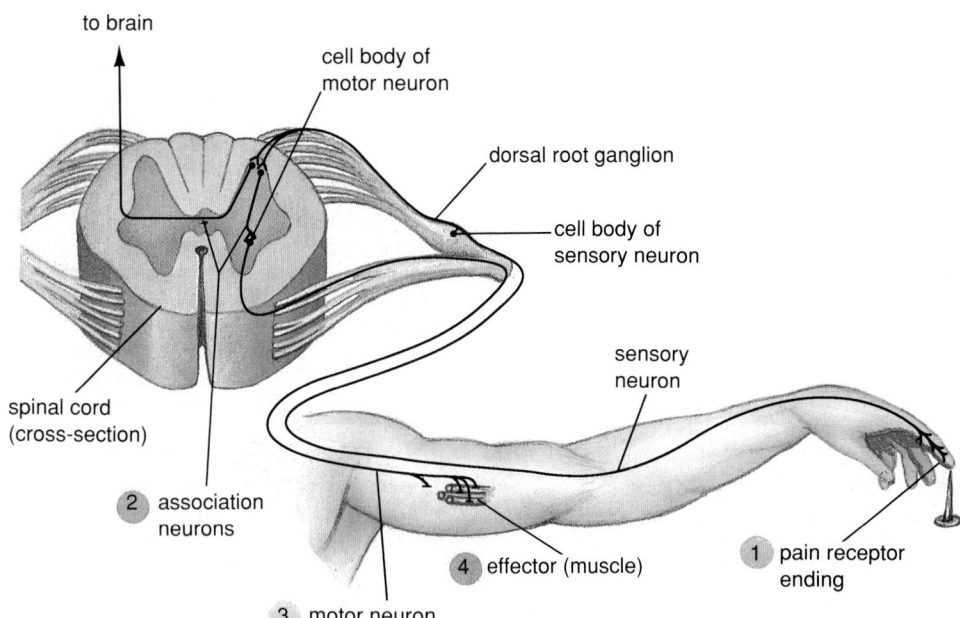

to brain

cell body of
motor neuron

dorsal root ganglion

cell body of
sensory neuron

sensory
neuron

spinal cord
(cross-section)

2 association
neurons

4 effector (muscle)

1 pain receptor
ending

3 motor neuron

Figure 36-16 The vertebrate pain–withdrawal reflex circuit includes one each of the four elements of a nervous pathway. The sensory neuron has pain-sensitive endings in the skin (1), and a long fiber leading to the spinal cord. The sensory neuron stimulates an association neuron in the spinal cord (2), which in turn stimulates a motor neuron, also in the cord. The axon of the motor neuron (3) carries action potentials to muscles (4), causing them to contract and withdraw the body part from the damaging stimulus. Note that the sensory neuron also makes a synapse on other association neurons not directly involved in the reflex, which carry signals to the brain, informing it of the danger below.

of the brain cells directing these activities carry signals down the cord and stimulate the appropriate motor cells.

The Brain

All vertebrate brains have the same general structure, with major modifications corresponding to life-style and intelligence. Embryologically, the vertebrate brain begins as a simple tube, which soon develops into three parts: the hindbrain, midbrain, and forebrain (Fig. 36-17a). It is believed that in the earliest vertebrates, these three anatomical divisions were also functional divisions: the **hindbrain** governed automatic behaviors such as breathing and heart rate, the **midbrain** controlled vision, and the **forebrain** largely dealt with the sense of smell. In nonmammalian vertebrates, these three divisions remain prominent. However, in mammals, and particularly in humans, the brain regions are significantly modified. Some have been reduced in size, and others, especially the forebrain, greatly enlarged (Fig. 36-17b).

The hindbrain In humans, the hindbrain is represented by the medulla, the pons, and the cerebellum (Fig. 36-18). In both structure and function, the **medulla** is very much like an enlarged extension of the spinal cord. Like the spinal cord, the medulla has neuron cell bodies at its center, surrounded by a layer of myelin-covered axons. The medulla controls several automatic functions, such as breathing, heart

rate, blood pressure, and swallowing. Certain neurons in the **pons,** located above the medulla, appear to influence transitions between sleep and wakefulness, and between stages of sleep. Others influence the rate and pattern of breathing. The **cerebellum** is crucially important in coordinating movements of the body. It receives information from command centers in the higher, conscious areas of the brain that control movement and from position sensors in muscles and joints. By comparing what the command centers ordered with information from the position sensors, the cerebellum guides smooth, accurate motions and body position. Not surprisingly, the cerebellum is largest in animals whose activities require fine coordination. It is best developed in birds (see Fig. 36-17b), who engage in the complex activity of flight.

The midbrain The midbrain is extremely reduced in humans, but an important relay center, the **reticular formation,** passes through it (Fig. 36-19). The neurons of the reticular formation extend all the way from the central core of the medulla, through the pons, the midbrain, and on into lower regions of the forebrain. It receives input from virtually every sense and every part of the body, and from many areas of the brain as well. The reticular activating formation plays a role in sleep and arousal, emotion, muscle tone, and certain movements and reflexes. It filters sensory inputs before they reach the conscious regions of the brain, although the selectivity of the filtering seems to be set by higher brain centers.

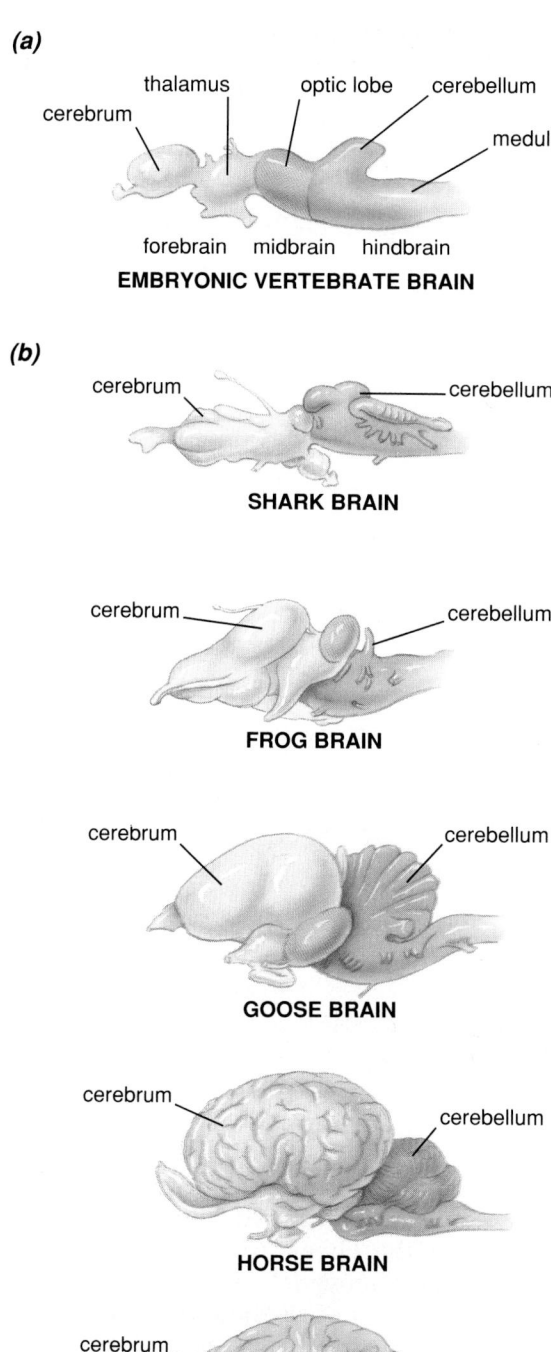

(a)

cerebrum · thalamus · optic lobe · cerebellum · medulla

forebrain · midbrain · hindbrain

EMBRYONIC VERTEBRATE BRAIN

(b)

cerebrum · cerebellum

SHARK BRAIN

cerebrum · cerebellum

FROG BRAIN

cerebrum · cerebellum

GOOSE BRAIN

cerebrum · cerebellum

HORSE BRAIN

cerebrum · cerebellum

HUMAN BRAIN

Figure 36-17 (a) The embryonic vertebrate brain shows three distinct regions: the forebrain, midbrain, and hindbrain. **(b)** This basic structure persists in all adult brains, but the relative sizes and importance of the parts varies enormously.

Through a combination of genetically determined wiring and learning, the reticular formation "decides" which stimuli require attention. Important stimuli are forwarded to the conscious centers for processing, while unimportant stimuli are suppressed. The fact that a mother wakens upon hearing the faint cry of her infant, but sleeps through loud traffic noise outside her window, testifies to the effectiveness of the reticular formation in screening inputs to the brain and to the role of learning in determining the importance of sensory stimulation.

The forebrain The forebrain can be roughly divided into three functional parts.

1. A relay center, the **thalamus,** shuttles sensory information to the limbic system and the cerebrum.
2. The **limbic system** consists of groups of neurons that control instincts and emotions.
3. The **cerebrum** is a brain region that receives sensory information, processes it, and initiates actions.

In fish, amphibians, and reptiles, these three areas are of roughly equal size and importance. In mammals, however, the cerebrum is enlarged, culminating in the enormous cerebral hemispheres of humans, which almost completely envelop the thalamus and limbic system (see Fig. 36-17).

1. **The thalamus.** Most of the neural information that reaches the cerebrum is channeled through the **thalamus** (see Fig. 36-18). This includes sensory input from auditory and visual pathways, as well as sensations from the skin and from within the body. Inputs from the cerebellum and limbic system are also channeled through this busy thoroughfare. Very little information processing goes on in the thalamus.
2. **The limbic system.** Anatomically, the **limbic system** is a diverse group of structures located in an arc between the thalamus and the cerebrum (Fig. 36-20). These structures work together to produce our most basic and primitive emotions, drives, and behaviors, including fear, rage, tranquility, hunger, thirst, pleasure, and sexual responses. The limbic system includes the **hypothalamus,** portions of the thalamus, the **amygdala,** and the **hippocampus.**

The **hypothalamus** (literally "under the thalamus") contains many different clusters of neurons. Some of these are neurosecretory cells that release hormones (see Chapter 35). Through its hormone production and neural connections, the

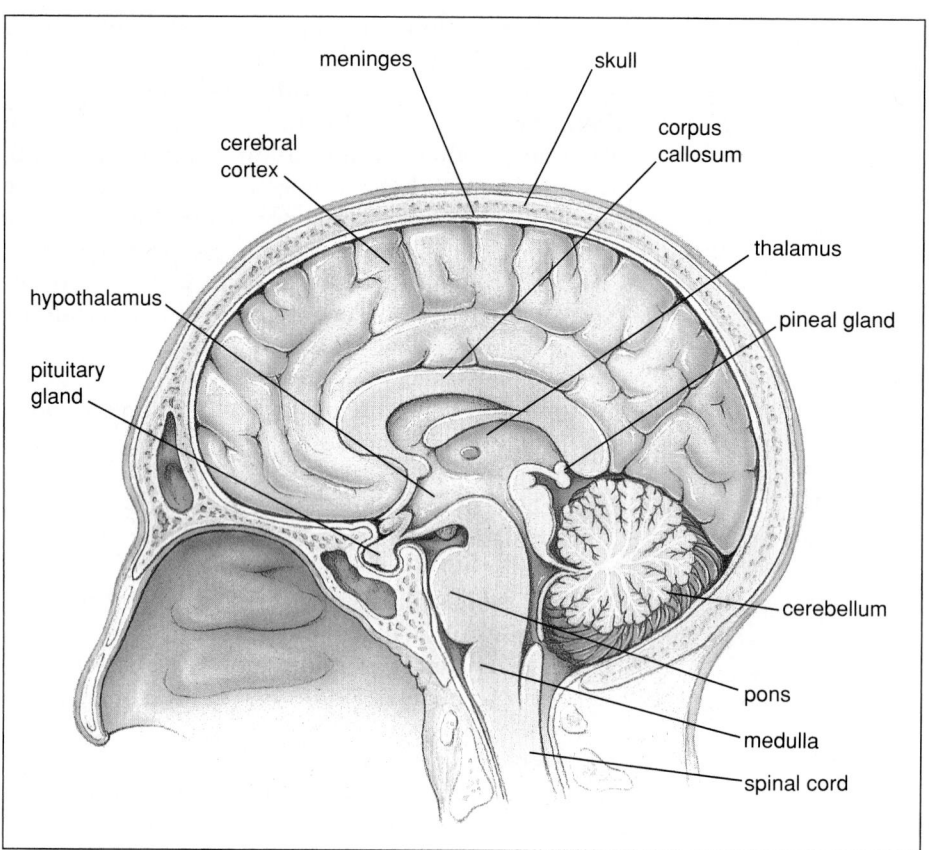

hypothalamus acts as a major coordinating center, controlling body temperature, hunger, the menstrual cycle, water balance, and the autonomic nervous system. In addition, stimulation of specific areas of the hypothalamus elicits emotions such as rage, fear, pleasure, and sexual arousal.

The **amygdala** is believed to be responsible for the production of appropriate behavioral responses to environmental stimuli. It receives input from many sources, including the auditory and visual areas of the cerebral cortex. Different clusters of neurons in the amygdala produce sensations of pleasure, punishment, or sexual arousal when stimulated. By stimulating different portions of the amygdala, researchers can either reduce or enhance aggressive behavior. Conscious humans whose amygdalas are electrically stimulated have reported feelings of rage or fear.

The shape of the **hippocampus** as it curves around the thalamus inspired its name, which is derived from the Greek word meaning "seahorse." Like the amygdala and hypothalamus, stimulation of portions of the hippocampus can elicit behaviors that reflect a variety of emotions, including rage and sexual arousal. The hippocampus also plays an important role in the formation of long-term memory, and thus is required for learning, discussed in more detail below.

3. **The cerebrum.** In humans, by far the largest part of the brain is the **cerebrum,** which contains about half the nerve cells in the brain. Ironically, it is the area of the brain that scientists know least about. Anatomically, the cerebrum is split into two halves, called **cerebral hemispheres,** which communicate with each other via a large band of axons, the **corpus callosum.** The cerebral hemispheres are the most sophisticated information

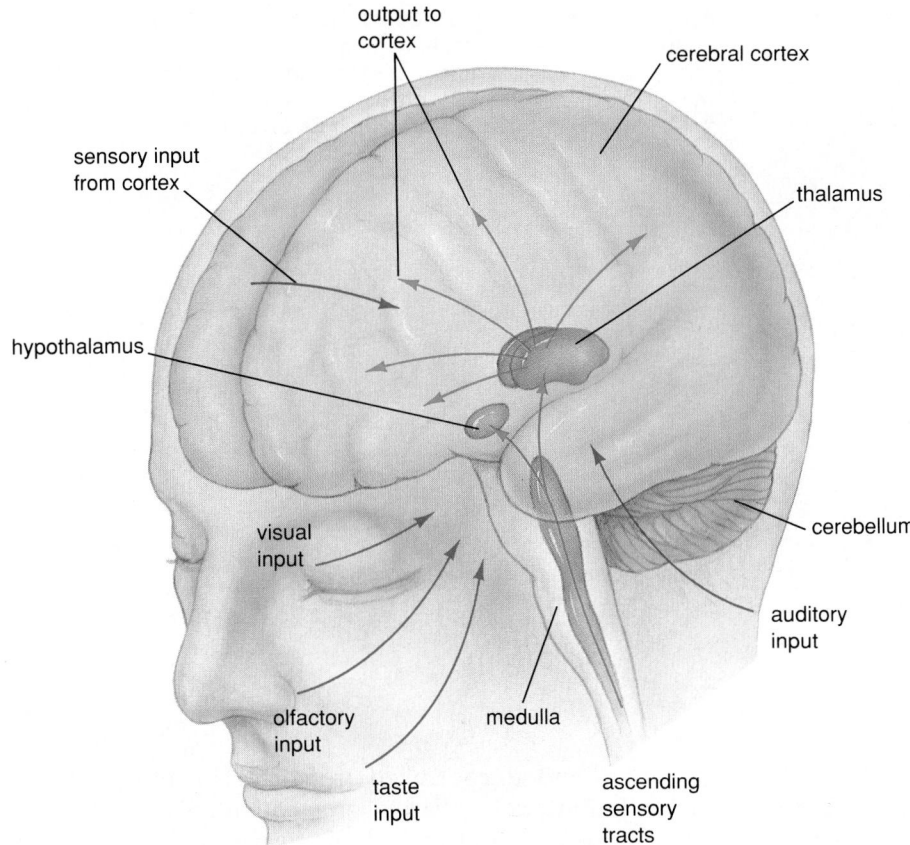

output to cortex

cerebral cortex

sensory input from cortex

thalamus

hypothalamus

cerebellum

visual input

auditory input

olfactory input

medulla

taste input

ascending sensory tracts

Figure 36-19 The reticular activating formation is a diffuse network of neurons running through the lower regions of the brain from the hindbrain, through the midbrain, and up into the thalamus and hypothalamus of the forebrain. It receives input from most of the senses and sends outputs to many higher brain centers, filtering the sensory information that reaches the conscious brain.

processing centers known. Roughly *50 to 100 billion* neurons are packed into a thin layer at the surface, called the **cerebral cortex.** To accommodate this profusion of cells, the cortex is thrown into folds, called **convolutions,** that greatly increase its area. In the cortex, cell bodies of neurons predominate, giving this outer layer of the brain a gray appearance, similar to that seen inside the spinal cord. These neurons receive sensory information, process it, store some in memory for future use, and direct voluntary motor output.

The functions of the cerebral cortex are localized in discrete regions (Fig. 36-21). Damage to the cortex due to trauma, stroke, or a tumor results in specific deficits, such as problems with speech, difficulty reading, or the inability to sense or move specific parts of the body. Since brain cells cannot reproduce, once a brain region is destroyed it cannot be repaired or replaced, so these deficits are often permanent. Fortunately,

however, in some cases diligent training can cause undamaged regions of the cortex to take control over and restore some of the lost functions.

Neurotransmitters and Neuromodulators

Twenty-five years ago, neurobiologists thought that nervous systems operated with just a few excitatory and inhibitory neurotransmitters. Since that time, investigators have become increasingly aware that the brain is a teeming cauldron; its neurons synthesize and respond to a vast array of chemicals, including many of the hormones we once thought were unique to the endocrine system. For example, hormones that control digestive tract secretions are now known to be synthesized in the brain, where they influence appetite. At least 50 neurotransmitters and neuromodulators have been identified, and more are added to the list each year. In the following sections,

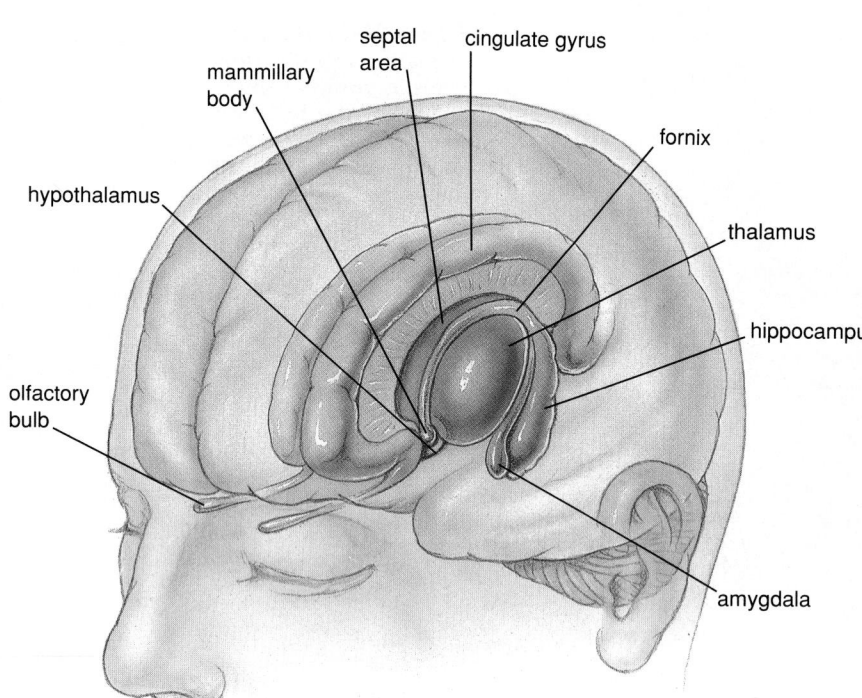

septal area
mammillary body
cingulate gyrus
hypothalamus
fornix
thalamus
hippocampus
olfactory bulb
amygdala

Figure 36-20 The limbic system extends through several brain regions. It seems to be the center of most unconscious, emotional behaviors, such as love, hate, hunger, sex, and fear.

we first discuss a few "classic neurotransmitters" that have been recognized for many years and whose roles are partially understood. Then we describe a few recently recognized **neuromodulators,** substances released by neurons that alter the activity of groups of neurons over longer periods of time.

Acetylcholine

The neurotransmitter **acetylcholine** is found in many areas of the brain, and is the only transmitter found at the synapses between motor neurons and skeletal muscles, where it is always excitatory. The drug curare, isolated from the South American poison arrow frog, blocks acetylcholine receptors on the postsynaptic membrane. This prevents muscle contraction, causing paralysis, and, sometimes, death. In contrast, many insecticides poison insects by inhibiting an enzyme (found in the synaptic cleft) that breaks down acetylcholine. This causes the muscles to contract uncontrollably, producing seizures and death. In the human central nervous system, degeneration of specific groups of acetylcholine-producing neurons is found in patients with Alzheimer's disease, also called *senile dementia.*

Dopamine

Dopamine is an important neurotransmitter in the brain, where its effects are largely inhibitory. The degeneration of dopamine-producing neurons causes Parkinson's disease, characterized by muscular rigidity (due to the continuous contraction of some mus-

cles) and uncontrolled tremors. The drug L-dopa, administered to Parkinson's patients, is used by unharmed neurons in the brain to synthesize dopamine, partially replacing that which has been lost. Recent evidence indicates that schizophrenia may be due to an overabundance of dopamine receptors. Drugs that block these receptors relieve some schizophrenic symptoms, but, unfortunately, may induce symptoms of Parkinson's disease.

Serotonin

The neurotransmitter **serotonin** acts in the brain and spinal cord. It can inhibit pain sensory neurons in the spinal cord, and electrical devices that stimulate these neurons are sometimes implanted in patients suffering from chronic pain. Serotonin is also believed to affect sleep and mood. Animals in which serotonin production is blocked are unable to sleep normally. Too little serotonin (and noradrenaline, see below) may cause depression. Some drugs used to treat depression block either the enzymatic destruction of both serotonin and noradrenaline or their transport back into the presynaptic cell, thus prolonging the effects of these transmitters. There is evidence that the drug LSD, among its many other effects, may block serotonin receptors.

Noradrenaline

Noradrenaline (also called *norepinephrine*) is chemically very similar to the hormone adrenalin secreted by the adrenal glands (see Chapter 35). Noradrena-

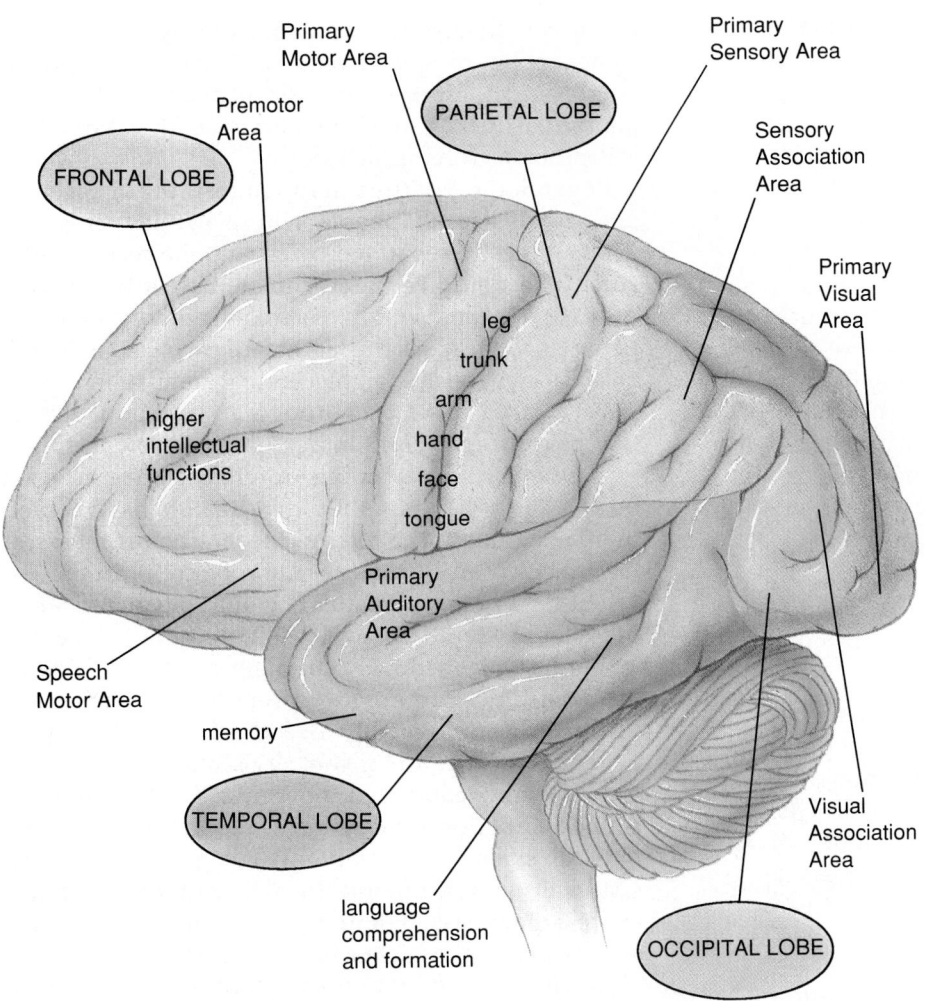

Primary Motor Area

Premotor Area

FRONTAL LOBE

PARIETAL LOBE

Primary Sensory Area

Sensory Association Area

Primary Visual Area

leg

trunk

arm

hand

face

tongue

higher intellectual functions

Primary Auditory Area

Speech Motor Area

memory

TEMPORAL LOBE

language comprehension and formation

OCCIPITAL LOBE

Visual Association Area

Figure 36-21 Structural (colored) and functional regions of the human left cerebral cortex. A map of the right cerebral cortex would be similar, except that speech and language are less well developed on the right side (see Fig. 36-22).

line is released by neurons of the sympathetic nervous system onto many organs (such as the heart, digestive system, and lungs; see Fig. 36-14). The effects of noradrenaline, which may be either inhibitory or excitatory, prepare the body to respond to stressful situations. Cocaine and amphetamines act, in part, by stimulating release of noradrenaline and dopamine. Drugs that block the secretion of noradrenaline often cause depression.

Neuromodulators

In addition to these and several other classic neurotransmitters, dozens of peptides are synthesized and released by neurons. Many may be released along with neurotransmitters, acting as neuromodulators. Neuromodulators modify the properties of synapses,

making them more or less effective, and can cause long-term changes in the excitability of neurons. Neuromodulators function over a longer time period than classic neurotransmitters, and may influence many neurons at once.

Examples of peptide neuromodulators are the **opioids** (literally "opiate-like substances") including **endorphin** and **enkephalin.** For centuries the pain-relieving effects of opiates (morphine, opium, codeine, heroin, and others) have been recognized. In the early 1970s, neurobiologists discovered that these opiates bind to specific receptors on neurons of the central nervous system. Since opiates are plant products, it seemed unlikely that the human brain would have evolved receptors specifically for these chemicals. Perhaps, they reasoned, the opiates happen to resemble unknown substances produced by the brain that diminish pain perception. This hypothesis led to a search for these substances, which was rewarded in 1975 with the discovery of the opioids.

Some opioids act with serotonin in the spinal cord to block perception of pain. They are partially responsible for the suppression of pain in times of extreme stress, such as on a battlefield (or a football field!). Opioids are released during strenuous exercise, and may account for the well-known "runner's high." The analgesic effects of acupuncture are apparently caused by its ability to stimulate the release of opioids. Opioids and other peptide neuromodulators participate in the maintenance of body temperature and blood pressure. They have also been implicated in the regulation of behavioral states such as hunger, sexual excitement, anger, or depression, and they may even be involved in learning.

Brain and Mind

Historically, people have always had difficulty reconciling the physical presence of a few pounds of grayish material in the skull with the range of thoughts, emotions, and memories of the human mind. This "mind–brain problem" has occupied generations of philosophers and, more recently, neurobiologists. Beginning with observations of patients with head injuries and progressing to sophisticated surgical, physiological, and biochemical experiments, the outlines of how the brain creates the mind are beginning to emerge. Here, we will only be able to touch upon a few of the more fascinating features.

Left Brain–Right Brain

The human brain appears bilaterally symmetrical, particularly the cerebrum, which consists of two extremely similar-looking hemispheres. However, it has been known since the early 1900s that this symmetry does not extend to brain function (Fig. 36-22). Much of what is known of the differences in hemisphere function comes from two sources: studies of accident victims with localized damage to one hemisphere, and studies of patients who have had the corpus callosum (which connects the two hemispheres) severed. This surgical procedure was performed in rare cases of uncontrollable epilepsy to prevent the spread of seizures through the brain.

Studies based on selective damage to the left cerebral hemisphere had led to the belief that the right hemisphere was relatively retarded, lacking the ability to speak, write, recognize words, or reason. For example, people suffering damage to localized areas of the left hemisphere, but not the right, often became unable to speak, read, or understand spoken language. In addition, the left hemisphere for most people is superior in mathematical ability, and in logical problem-solving tasks.

Roger Sperry, of the California Institute of Technology, worked with people whose hemispheres had been surgically separated by cutting the corpus callosum. In his studies, Sperry made use of the knowledge that axons within each optic nerve follow a pathway that causes the left half of each visual field to be projected on the right cerebral hemisphere, and vice versa (Fig. 36-22). Through an ingenious device that projected different images onto the left and right visual fields (thus sending different signals to each hemisphere), he and other investigators have gained more insight into the roles of the two hemispheres. If he projected an image of a nude figure onto the left visual field only, the patients would blush and smile but would claim to have seen nothing, since the image had reached only the nonverbal right side of the brain! The same figure projected onto the right visual field was readily described verbally. These experiments, begun in the 1960s and refined since then, have revealed that the right side of the brain is actually superior to the left in several areas, including musical skills, artistic ability, recognition of faces, spatial visualization, and the ability to recognize and express emotions. For his pioneering work, Sperry was awarded the Nobel Prize in 1981.

Recent experiments indicate that the left–right dichotomy is not as rigid as was once believed. Patients who have suffered a stroke that disrupted blood supply to the left hemisphere typically show symptoms such as loss of speaking ability. Frequently, however, training can partially overcome these speech or reading deficits, even though the left hemisphere itself has not recovered. This suggests that the right hemisphere has some latent language capabilities. Interestingly, female stroke victims recover lost abilities more often than males, and females also have a larger corpus callosum. These findings suggest a sex difference in the degree of specialization of the two hemispheres and the extent of their interconnections.

Learning and Memory

Although theories abound as to the cellular mechanisms of learning and memory, we are a long way from understanding these phenomena. In mammals, and particularly in humans, however, we do know a fair amount about two other aspects of learning and memory: the time course of learning, and some of the brain sites involved in learning, memory storage, and recall.

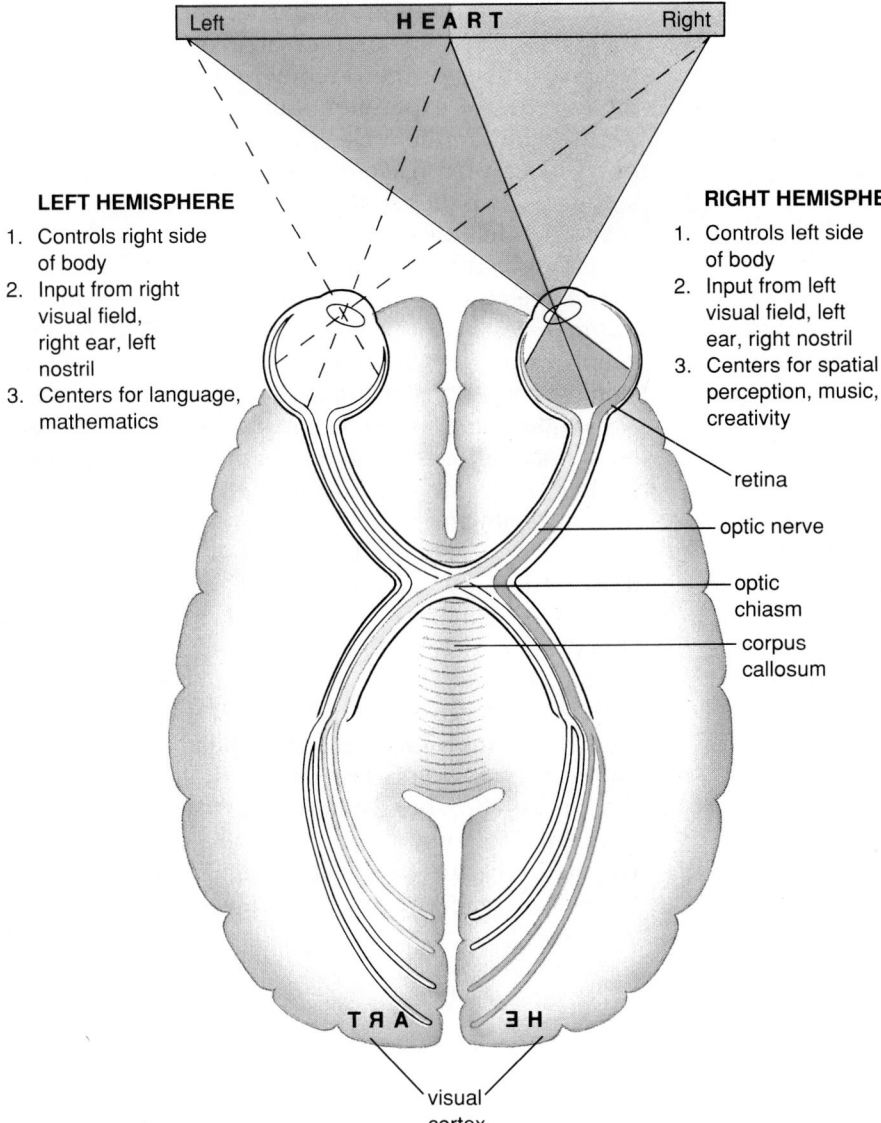

LEFT HEMISPHERE

1. Controls right side of body
2. Input from right visual field, right ear, left nostril
3. Centers for language, mathematics

RIGHT HEMISPHERE

1. Controls left side of body
2. Input from left visual field, left ear, right nostril
3. Centers for spatial perception, music, creativity

retina

optic nerve

optic chiasm

corpus callosum

visual cortex

Figure 36-22 Specializations of the two cerebral hemispheres. Generally, each hemisphere controls sensory and motor functions of the opposite side of the body. Further, the left side seems to predominate in rational and computational activities, while the right side governs creative and spatial abilities.

Time Course of Learning

Experiments show that learning occurs in two phases: an initial **short-term memory** followed by **long-term memory.** For example, if you look up a number in the phone book, you will probably remember the number long enough to dial but forget it promptly thereafter. This is short-term memory. But if you call the number frequently, eventually you will remember the number more or less permanently. This is long-term memory.

Short-term memory seems to be *electrical* in nature, involving the repeated activity of a particular neural circuit in the brain. As long as the circuit is active, the memory stays. If the brain is distracted by other thoughts, or if electrical activity is interrupted, such as by electroconvulsive shock or by a concussion, the memory disappears and cannot be retrieved no matter how hard you try.

Long-term memory, on the other hand, seems to be *structural,* involving, perhaps, the formation of new, permanent synaptic connections between specific neurons, or the strengthening of existing but weak synaptic connections. These new or strengthened synapses last indefinitely, and the long-term memory persists unless certain brain structures are destroyed. Short-term memory can be converted to long-term memory, but how this actually happens is not well understood.

Learning, Memory, and Retrieval Sites in the Brain

Learning, memory, and the retrieval of memory seem

to be separate phenomena, controlled by separate areas of the brain, as shown below.

LEARNING AND MEMORY

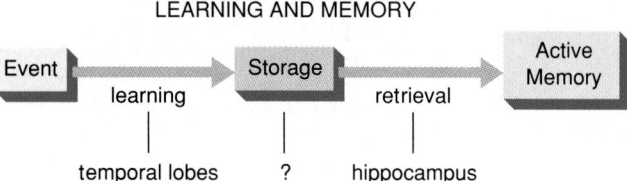

Ample evidence shows that the hippocampus (part of the limbic system) is involved in learning. For example, intense electrical activity occurs in the hippocampus during learning. Even more striking are the results of hippocampal damage. People whose hippocampi are destroyed retain most of their past memories, but are unable to learn anything that occurs after the loss. One victim was still unable to recall his address or find his way home after 6 years at the same residence. He could be entertained indefinitely by reading the same magazine over and over, and people whom he saw regularly required reintroduction at each encounter. People with extensive hippocampal damage can recall events momentarily, but the memory rapidly fades, as does the memory of a dream upon awakening. This has led to the hypothesis that the hippocampus is responsible for transferring information from short-term into long-term memory.

Retrieval or recall of established long-term memories is localized in another area of the brain, the **temporal lobes** of the cerebral hemispheres. In a famous series of experiments in the 1940s, neurosurgeon Wilder Penfield electrically stimulated the temporal lobes of conscious patients undergoing brain surgery. The patients did not merely *recall* past memories, but felt that they were *experiencing* the past events right there in the operating room!

The site of *storage* of complex long-term memories is much less clear. The psychologist Karl Lashley spent many years training rats and subsequently lesioning parts of their brains in an effort to learn the site of the memory trace, but failed. None of Lashley's small lesions could erase a memory completely. In 1950, a frustrated Lashley wrote: "I sometimes feel, in reviewing the evidence on the localization of the memory trace, that the necessary conclusion is that learning just is not possible."

Some researchers suggest that each memory is stored in numerous distinct places in the brain. Or perhaps memories are stored like a hologram image, both everywhere and nowhere at the same time: the memory is more precise if the whole brain is intact, but each "bit" (probably several thousands of neurons) of cerebral hemisphere can store an essentially complete memory. While further research may provide definitive answers, the storage site of memories, for now, remains an unsolved mystery.

Reflections on Brain and Mind

Humans have always been intensely interested in the workings of their own minds. But until about 100 years ago, the mind was more appropriately a subject for philosophers than scientists, because the tools to study the brain did not yet exist. Through the first half of the twentieth century, the mind was treated by psychologists as a "black box," whose internal workings could be deduced only through the investigation of how past and present experiences were interpreted and influenced behavior. New discoveries, however, are rapidly changing our views of the workings of the brain.

During recent decades, we have begun to understand the neural bases of at least some psychological phenomena. Many forms of mental illness, such as schizophrenia, manic depression, and autism, once thought to be due to childhood trauma or inept parenting, are now recognized as resulting from biochemical imbalances in the brain. Studies are revealing a strong heritability factor (and hence, a biological basis) for traits that were once considered entirely learned, such as shyness and alcoholism.

A striking illustration of how the physical structure of the brain is related to personality was unwittingly provided by Phineas Gage in 1848. An explosion propelled a large metal rod through his skull, removing his left temporal lobe. He miraculously survived, but his personality changed radically. Prior to the accident, Phineas was conscientious, industrious, and well-liked. Following his recovery, he became impetuous, profane, and incapable of planning or working toward a goal. Subsequent research has implicated the frontal lobe in emotional expression and the ability to work for delayed rewards. Other sites of damage have revealed additional anatomical specializations. One patient with very localized left frontal lobe damage found himself unable to name fruits and vegetables (although he could name everything else). One science writer quipped: "Does the brain have a produce section?" Similarly, damage to certain areas on the underside of the brain results in a selective inability to recognize faces.

To date, much of our understanding of the human mind–brain connection comes from serendipity—a stroke, trauma, tumor, or surgical procedure damages part of the brain. If the victim is cooperative, and if the case comes to the attention of an interested researcher, a battery of tests may be administered to define the change or loss of ability. Often, the exact extent of the

damage remains unknown until revealed by autopsy.

New techniques, however, are freeing researchers from the chains of chance and permitting insight into the functioning of normal, as well as diseased, brains. The most spectacular is the PET scan (positron emission tomography, described in Chapter 2), which allows a glimpse into the working brain. Researchers can watch as the brain responds to a specific odor, or a visual or auditory stimulus (Fig. E36-2a). The contrast between normal and disturbed brain functioning can also be observed, as can the cyclic changes of brain function in bipolar (manic-depressive) disorder (Fig. E36-2b).

These and increasingly sophisticated techniques of the future are creating ever-larger windows into the "black box" that is the human brain. The next decades will see a gradual merging of the fields of psychology and neurophysiology, and, it is hoped, a clearer understanding of how the human brain generates the human mind.

(a)
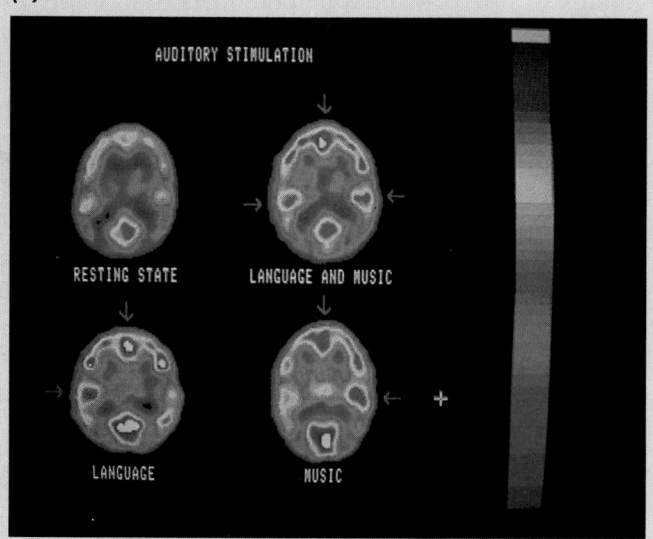

(b)
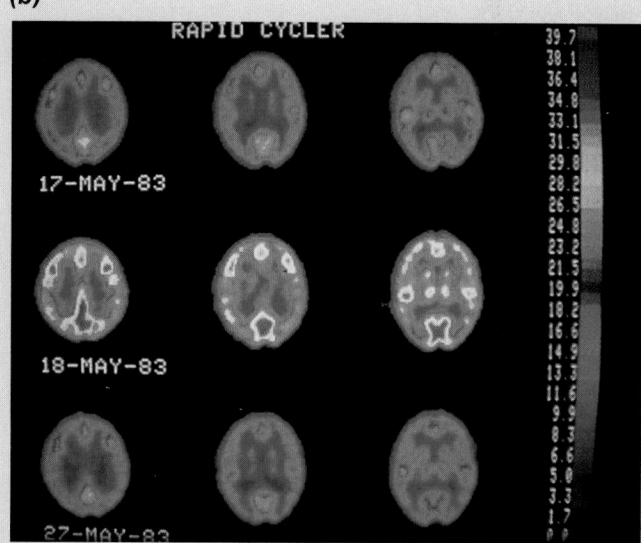

Figure E36-2 PET scan images reveal changes in regional brain activity in response to **(a)** stimuli and **(b)** mood. Brain activity is measured as a function of glucose utilization, which is color-coded from red (highest) to purple (lowest).
(a) The listener shows specializations of the cerebral hemispheres for language (note the red area on the left, but not the right half of the brain) and music (red area on right only). In combination, these auditory stimuli activate both hemispheres roughly equally.
(b) A series of three scans of the brain of a patient whose mental state cycles between depression and near normalcy. Scans reveal extremely low levels of brain activity in the depressed state (May 17th and May 27th) alternating with approximately normal levels (May 18th). (Numbers are the rate of glucose utilization in micromoles of glucose consumed per minute per 100 grams of brain tissue.)

SUMMARY OF KEY CONCEPTS

The Functions and Structure of Neurons

Nervous systems are composed of billions of individual cells called neurons. A neuron has five major functions: (1) to receive information; (2) to integrate the information and produce a signal; (3) to conduct the signal; (4) to transmit the signal to other cells; and (5) to coordinate metabolic activities.

Each neuron's structure reflects these functions. The cell body coordinates the cell's metabolic activities. Dendrites receive information from the environment or other neurons. The cell body integrates this information and "decides" whether to produce an output signal—the action potential. The axon conducts the signal to its output terminal—the synapse. Synaptic terminals transmit the signal to other nerve cells, glands, or muscles.

Mechanisms of Neural Activity

The electrical activity in a neuron can be summarized as follows.

1. An electrical potential exists between the interiors of neurons and the extracellular fluid surrounding them. In an unstimulated neuron, the inside of the cell is negative with respect to the extracellular fluid; this is called the *resting potential.*

2. Signals received by the dendrites from other neurons take the form of small, rapidly fading changes in potential called *postsynaptic potentials.* Some postsynaptic potentials are inhibitory, making the neuron more negative inside. Others are excitatory, making the neuron less negative inside and bringing it closer to threshold.

3. If postsynaptic potentials, added together within the cell body, make the neuron substantially less negative, the neuron may reach threshold. This opens sodium channels and triggers an action potential. The action potential is a wave of positive charge that travels, undiminished in magnitude, to the synaptic terminals at the end of the axon. The speed at which the signal travels is in-

creased in neurons whose axons are insulated with myelin.

4. The resting potential is restored when the sodium channels close spontaneously and additional potassium channels open, allowing potassium to diffuse out, restoring the negative resting potential.

A synapse is a region at which two neurons communicate, consisting of the synaptic terminal of the presynaptic neuron and a region of dendrite or cell body of the postsynaptic neuron, separated by the synaptic cleft. Neurotransmitter from the presynaptic neuron, released in response to an action potential, binds to receptors on the postsynaptic cell membrane. This causes ion channels to open, allowing ions to flow, and producing either an excitatory or inhibitory postsynaptic potential, depending on the type of channels opened. The postsynaptic neuron adds up the postsynaptic potentials from many presynaptic cells and produces an action potential if the postsynaptic potentials bring it above threshold.

Building and Operating a Nervous System

Information processing in the nervous system requires four operations. The nervous system must: (1) signal the intensity of the stimulus, (2) determine the type of stimulus, (3) integrate information, and (4) initiate and direct the response. Nervous systems signal intensity by the frequency of action potentials in single neurons and by the number of similar neurons firing at the same time. The type of stimulus is determined by which neurons produce action potentials. The nervous system collects and processes sensory information from many sources. These may converge on fewer neurons whose activity in response to these inputs determines action. The "decision" to act may then be transmitted to many more neurons (divergence), which direct the activity.

Neural pathways normally have four elements: sensory neurons, association neurons, motor neurons, and effectors. Overall, nervous systems consist of numerous interconnected nervous pathways, and may be diffuse or centralized.

The Human Nervous System

The nervous system of humans and other vertebrates consists of the central nervous system (brain and spinal cord) and the peripheral nervous system (nerves leading from the central nervous system to the rest of the body).

The peripheral nervous system is further subdivided into sensory and motor portions. The motor portions consist of the somatic nervous system (which controls voluntary movement) and the autonomic nervous system (directing involuntary responses). The autonomic nervous system has both a sympathetic and parasympathetic component. The sympathetic nervous system prepares the body to "fight or take flight" while the parasympathetic nervous system promotes maintenance activities such as digestion.

Within the central nervous system, the spinal cord contains (1) neural pathways for reflexes and certain simple behaviors; (2) motor neurons controlling voluntary muscles; and (3) axons leading to and from the brain. The brain consists of three parts—the hindbrain, midbrain, and forebrain, each further subdivided into distinct structures.

The hindbrain in humans consists of the medulla and pons, which control involuntary functions (such as breathing), and the cerebellum, which coordinates muscular activities (such as walking). In humans, the small midbrain contains the reticular activating formation, a filter and relay for sensory stimuli. The forebrain includes the thalamus, a sensory relay station that shuttles information to and from higher conscious centers in the forebrain. The limbic system of the forebrain is a collection of diverse structures involved in emotion and the control of instinctive behaviors such as sex, feeding, and aggression. The last part of the forebrain, the cerebrum, is the center for information processing, memory, and initiation of voluntary actions. It includes primary sensory and motor areas, and association areas that analyze sensory information and plan movements.

The human brain teems with chemical neurotransmitters and neuromodulators. Classical neurotransmitters include acetylcholine, dopamine, norepinephrine, and serotonin. Neuromodulators modify synaptic transmission, and may influence many neurons over a longer time span than classic transmitters. Examples are peptides called *opioids* that bind to the same receptors as do opiates. These influence pain perception, appetite, emotional states, and perhaps learning.

Brain and Mind

The cerebral hemispheres are each specialized. Generally, the left hemisphere is dominant in speech, reading, writing, language comprehension, mathematical ability, and logical problem-solving. The right hemisphere specializes in recognizing faces and spatial relationships, artistic and musical abilities, and recognition and expression of emotions.

Memory takes two forms. Short-term memory is electrical, apparently linked to continuous activation of a neural circuit. Long-term memory, in contrast, probably involves structural changes that increase the effectiveness of synapses. The hippocampus is an important site for learning, and for the transfer of information into long-term memory. The temporal lobes are important for memory retrieval. The site of memory storage is unknown.

GLOSSARY

acetylcholine (ah-sēt'-il-kō'-lēn): a neurotransmitter used at vertebrate skeletal neuromuscular junctions, and found in the brain.

action potential: a rapid change from a negative to a positive electrical potential in a nerve cell. This signal travels along an axon without change in size.

amygdala (am-ig'-da-la): part of the forebrain of vertebrates, involved in control of emotions and instinctive behaviors.

association neuron: in nervous circuits, a nerve cell that is postsynaptic to a sensory neuron and presynaptic to a motor neuron. In actual circuits, there may be many association neurons between individual sensory and motor neurons.

autonomic nervous system: part of the peripheral nervous system of vertebrates that innervates mostly glands and internal organs and produces largely involuntary responses.

axon: a long process of a nerve cell, extending from the cell body to synaptic endings on other nerve cells or on muscles.

brain: the part of the central nervous system of vertebrates enclosed within the skull.

cell body: part of a nerve cell in which most of the common cellular organelles are located. Also often a site of integration of inputs to the nerve cell.

central nervous system: in vertebrates, the brain and spinal cord.

cerebellum (ser-uh-bel'-um): part of the hindbrain of vertebrates, concerned with coordination of motor activities.

cerebral cortex (ser-ē'-brel kōr'-tex): a thin layer of neurons on the surface of the vertebrate cerebrum, in which most neural processing and coordination of activity occurs.

cerebral hemisphere: one of two nearly symmetrical halves of the cerebrum, connected by a broad band of axons, the corpus callosum.

cerebrospinal fluid: A clear fluid produced within the ventricles of the brain that fills the ventricles and surrounds the brain and spinal cord. A major function of cerebrospinal fluid is to cushion the brain and spinal cord.

cerebrum (ser-ē'-brum): part of the forebrain of vertebrates concerned with sensory processing, direction of motor output, and coordination of most bodily activities. The cerebrum consists of two nearly symmetrical halves (the hemispheres) connected by a broad band of axons, the corpus callosum.

convergence: a condition in which a large number of nerve cells provide input to a smaller number of cells.

convolutions: foldings of the cerebral cortex of the brain.

corpus callosum (kōr'pus kal-ō'-sum): the tract of axons that connect the two cerebral hemispheres of vertebrates.

dendrite (den'-drīt): the site of signal input to a nerve cell, usually takes the form of branched fibers located close to the cell body.

divergence: a condition in which a small number of nerve cells provide input to a larger number of cells.

dopamine (dōp'-uh-mēn): a transmitter in the brain whose actions are largely inhibitory. Degeneration of dopamine-containing neurons leads to Parkinson's disease.

dorsal root ganglion: a ganglion located on the dorsal (sensory) branch of each spinal nerve, containing the cell bodies of sensory neurons.

effector (ē-fek'-tōr): a part of the body (usually a muscle or gland) that carries out responses as directed by the nervous system.

endorphin (en-dōr'-fin): one of a group of peptides in the vertebrate brain that mimics some of the actions of opiates. Endorphins reduce the sensation of pain.

excitatory synapse: a synapse between two nerve cells in which the resting potential of the postsynaptic cell becomes less negative due to the activity of the presynaptic cell.

forebrain: during development, the anterior portion of the brain. In mammals, the forebrain differentiates into the thalamus, the limbic system, and the cerebrum. In humans, the cerebrum contains about half of all the neurons in the brain.

ganglion (gang'-lē-un): a collection of nerve cells.

graded potential: in a nerve cell, an electrical response due to sensory input or synaptic input from another nerve cell. Graded potentials may be positive or negative and vary in amplitude with the strength of stimulation.

hindbrain: the posterior portion of the brain, containing the medulla, pons, and cerebellum

hippocampus (hip-ō-cam'-pus): part of the forebrain of vertebrates, important in motivation, emotion, and especially learning.

hypothalamus (hī-pō-thal'-uh-mus): part of the forebrain of vertebrates, located just below the thalamus, involved in regulation of hormonal activities (especially of the pituitary gland), and many behaviors such as feeding, drinking, sex, aggression, and fear responses, largely through activation of the autonomic nervous system.

inhibitory synapse: a synapse between two nerve cells in which the resting potential of the postsynaptic cell becomes more negative as a result of the activity of the presynaptic cell.

integration: in nerve cells, the process of adding up electrical signals from sensory inputs or other nerve cells, to determine the overall electrical activity of the nerve cell.

intensity: the strength of stimulation or response.

limbic system: a diverse group of brain structures, mostly in the lower forebrain, including the thalamus, hypothalamus, amygdala, hippocampus, and parts of the cerebrum, involved in emotion, motivation, and learning.

medulla (med-oo'-la): part of the hindbrain of vertebrates that controls automatic activities such as breathing, swallowing, and heartbeat.

meninges (men-in'-gēs): three layers of connective tissue that surround the brain and spinal cord.

midbrain: during development, the central portion of the brain, which contains the reticular formation.

motor neuron: a neuron that carries information from the central nervous system and stimulates effector organs such as muscles or glands.

myelin (mī'-eh-lin): a wrapping of lipid-rich membranes of specialized nonneural cells around the axon of a vertebrate nerve cell. Myelin increases the speed of conduction of action potentials.

nerve: a bundle of axons of nerve cells, bound together in a sheath.

nerve net: a simple form of nervous systems consisting of a network of neurons that extend throughout the body of organisms such as cnidarians.

neuromodulator: a chemical, usually a peptide, released by neurons at synapses along with neurotransmitters. Neuromodulators alter the properties of synapses.

neuron (nur'-on): a single nerve cell.

neurotransmitter: a chemical released by a presynaptic cell at a synapse, which binds to receptors on the postsynaptic cell, causing changes in the electrical potential of the second cell.

node of Ranvier (nōd of ron'-vē-ā): an interruption of the myelin on a vertebrate myelinated axon, at which action potentials are generated.

noradrenaline (nor-ad-ren'-a-lin): also called norepinephrine, a neurotransmitter released onto organs by neurons of the parasympathetic nervous system, which prepares the body to respond to stressful situations.

opioid (ōp'-ē-ōyd): a group of peptides found in the vertebrate brain that mimic some of the actions of opiates (such as opium, heroin, morphine). Besides analgesia (pain relief), opioids seem to be involved in many behaviors, including emotion, learning, and the control of appetite.

parasympathetic nervous system: the division of the autonomic nervous system that produces largely involuntary responses related to maintenance of normal body functions, such as digestion.

peripheral nervous system: in vertebrates, that part of the nervous system located outside the brain and spinal cord, consisting of the nerves leading to and from the brain and spinal cord, and the ganglia of the autonomic nervous system.

pons: a portion of the hindbrain just above the medulla containing neurons that influence sleep, and control the rate and pattern of breathing.

postsynaptic: referring to the nerve cell at a synapse which changes its electrical potential in response to a chemical (the neurotransmitter) released by another (presynaptic) cell.

postsynaptic potential: an electrical signal produced in a postsynaptic cell by transmission across the synapse. It may be excitatory (EPSP), making the cell more likely to produce an action potential, or inhibitory (IPSP), tending to inhibit an action potential.

potassium channel: a hollow-cored protein that spans a nerve cell membrane, forming a pore, and which only allows potassium ions to flow through the pore.

presynaptic: referring to a nerve cell that releases a chemical (the neurotransmitter) at a synapse, which causes changes in the electrical activity of another (postsynaptic) cell.

receptor: (1) a cell that responds to an environmental stimulus (chemicals, sound, light, pH, etc.) by changing its electrical potential; (2) a protein molecule in or on a cell membrane that reacts with another molecule (hormone, neurotransmitter, odorous compound, etc.) to trigger metabolic or electrical changes in the cell.

reflex: a simple, automatic, usually unconscious behavior performed by part of the body in response to a stimulus.

resting potential: a negative electrical potential found in unstimulated nerve cells.

reticular activating formation (reh-tik'-ū-lar): a diffuse network of neurons extending from the hindbrain, through the midbrain, and into the lower reaches of the forebrain, involved in filtering sensory input and regulating what information is relayed to higher centers in the cerebrum for further attention.

saltatory conduction (sal'-ta-tōr-ē): literally "jumping conduction", the transmission of an action potential along a myelinated axon, during which the action potential seems to jump from one node to the next.

sensory neuron: a nerve cell that carries information about internal or external environmental conditions to the central nervous system.

serotonin (ser-uh-tō'-nin): a neurotransmitter in the central nervous system, which is involved in mood, sleep, and the inhibition of pain.

sodium channel: a protein that spans a nerve cell membrane, forming a pore, and which allows only sodium ions to flow through the pore.

sodium–potassium pump: an enzyme-like protein in nerve cell membranes which actively transports potassium ions into the cell and sodium ions out of the cell.

spinal cord: part of the central nervous system of vertebrates, extending from the base of the brain to the hips, protected by the vertebrae of the spine; contains the cell bodies of motor neurons innervating skeletal muscles, the circuitry for some simple reflex behaviors, and axons communicating with the brain.

sympathetic nervous system: the division of the autonomic nervous system that produces largely involuntary responses that prepare the body for stressful situations.

synapse (sin'-apz): the site of communication between nerve cells. One cell (presynaptic) usually releases a chemical (the neurotransmitter) that changes the electrical potential of the second (postsynaptic) cell.

synaptic cleft: a narrow space in a chemical synapse that separates the pre- and postsynaptic cells.

synaptic potential: at a synapse between two nerve cells, the electrical change occurring in the postsynaptic cell as a result of reception of neurotransmitter released by the presynaptic cell.

synaptic terminal: the terminal branches of an axon, where the axon forms a synapse, usually with the dendrites of the second nerve cell.

temporal lobe: part of a cerebral hemisphere of the human brain, involved in recall of learned events.

thalamus: part of the forebrain, the thalamus serves as a relay network between other parts of the nervous system and the cerebrum.

threshold: the electrical potential (less negative than the resting potential) at which an action potential is initiated.

STUDY QUESTIONS

1. What are five functions of individual neurons? Which structural parts of neurons are specialized for each of these functions?
2. Describe the events during an action potential. Be sure to include the terms "threshold" and "postsynaptic potential."
3. What is the difference between postsynaptic potential and an action potential?
4. Diagram the structure of a synapse. How are signals transmitted from one neuron to another at a synapse?
5. How does the brain perceive the intensity of a stimulus? The type of stimulus?
6. Name three ways in which neurotransmitters are eliminated from a synapse.
7. What are the four elements to a simple nervous pathway? Describe how these elements function in the human pain reflex.
8. Describe and distinguish between convergence and divergence in nervous systems.
9. What is the difference between a diffuse and a concentrated nervous system? What types of animals possess each type?
10. Describe the autonomic nervous system. What are its two subdivisions, and what role does each play in daily life?
11. Draw a cross section of the spinal cord. What types of neurons are located in the spinal cord? What types of behaviors may have their neural circuitry located entirely within the cord?
12. What are the three embryological divisions of the vertebrate brain? In the earliest vertebrates, what were the functions of each part? What structures in the human brain comprise each part?
13. Describe the functions of the following parts of the human brain: medulla, cerebellum, reticular activating formation, thalamus, limbic system, cerebrum.
14. What structure connects the two cerebral hemispheres? Describe the evidence that the two hemispheres are specialized for distinct intellectual functions.
15. Distinguish between long-term and short-term memory.
16. What is an opioid? What types of functions may opioids serve in the human brain?

DISCUSSION QUESTION

1. Discuss in as much detail as possible how the resting potential is produced. Include both chemical molecules and physical forces in your dicussion.

SUGGESTED READINGS

Angier, N. "Storming the Wall." Discover, May 1990. Describes the blood–brain barrier and new methods of penetrating it with drugs.

Bloom, F. E., and Lazerson, A. *Brain, Mind, and Behavior.* 2nd edition. San Francisco: W. H. Freeman and Co., 1988. A very readable introduction to neurophysiology and neuropsychology oriented toward the beginning student. Superb illustrations of the brain and its parts.

The Brain. (1979). San Francisco: W. H. Freeman and Co., 1979. The complete September 1979 issue of *Scientific American*, devoted exclusively to brain structure and function.

Hall, S. S. "Aplysia and Hermissenda." Science 85, May 1985. Describes the use of simple molluscan nervous systems to unlock the secrets of learning and memory.

Holloway, M. "Rx for Addiction." Scientific American, March 1991. By studying drug addiction, neurobiologists are learning more about the brain, and new ways to counteract addiction.

Lester, H. "The Response of Acetylcholine." Scientific American, February 1977. (Offprint #1352.) Lester clearly explains the interactions of neurotransmitters with receptors, and how ion permeabilities lead to changes in the activity of postsynaptic neurons.

Montgomery, G. "The Mind in Motion." Discover, March 1989. Describes the use of the PET scan to view the workings of the living brain.

Iversen, L. "Chemicals to Think By." New Scientist, May 30, 1985. Reviews our expanding knowledge of the diversity of chemicals used by the brain.

Routtenberg, A. "The Reward System of the Brain." Scientific American, November 1978. (Offprint #584.) This article vividly portrays the crucial roles of limbic system structures in pleasure, learning, and memory.

Souchurek, H. "Medicine's New Vision." National Geographic, January 1987. Striking photography and interesting case histories highlight this readable overview of new noninvasive techniques of viewing inside the body.

37

Perception: The Senses

Molluscs such as this octopus have evolved eyes that are very similar in structure to vertebrate eyes, with one improvement: the receptors are at the front of the retina, facing the light, and the optic nerve leaves via the back.

"Nothing is understood by the intellect which is not first perceived by the senses."

Aristotle (384–322 BC)

In the previous chapter we examined the machinery of the mind, starting with neurons and their unique electrical signal-making properties and ending with the structure of the brain. Despite continuous advances in the neurosciences, it is possible that the human mind may never fully comprehend itself. However, the raw material for the intellect, the basis of its thoughts and dreams, is provided by the senses, as Aristotle observed over 2000 years ago. In this chapter we discuss the mechanisms underlying sensory perception, and introduce the major types of receptors found in animals.

Receptor Mechanisms

The term *receptor* is used in several contexts in biology. **In the most general sense, a *receptor* is a structure that changes when it is acted upon by a stimulus from its surroundings, causing a signal to be produced. All receptors are *transducers:* structures that convert signals from one form to another.** The receptor may be a membrane protein that changes configuration when it binds a specific hormone or neurotransmitter, as discussed in previous chapters. Alternatively, as described in this chapter, a receptor may be an entire cell (often a neuron) that is specialized to produce an electrical response to particular stimuli in its environment. Receptor cells are named after the stimulus to which they respond, as summarized in Table 37-1. All receptors produce electrical signals, but each receptor type is specialized so that it produces its signal only in response to a particular type of environmental stimulus, as discussed below.

Receptor Potentials

Stimulation of a **sensory receptor** causes a **receptor potential,** an electrical signal whose size is proportional to the strength of the stimulus (Fig. 37-1). Receptor potentials are translated into action potentials by one of two mechanisms. The first mechanism is seen in sensory neurons with axons that carry the signal for a relatively long distance, such as pain receptors in the skin that transmit messages all the way to the spinal cord. If the stimulus is sufficiently strong, the receptor potential in these neurons will exceed threshold, initiating action potentials. The frequency of these action potentials is related to the size of the receptor potential, and communicates the intensity of the stimulus, as shown in Figure 37-1. The second mechanism is used by receptor cells such as those specialized to detect tastes, sounds, and light. These receptors do not have axons or produce action potentials. Instead, their receptor potentials cause the release of transmitter onto a postsynaptic neuron,

Table 37-1 Vertebrate Receptor Types

Type of Receptor	Specific Sensory Cell Type	Stimulus	Location
Mechanoreceptor	Hair cell	Vibration, motion, gravity	Inner ear
	Variety of specialized endings in skin (Pacinian corpuscle; Merkel's disk)	Vibration, pressure, touch	Skin
	Free nerve endings in muscles or joints (muscle spindle, Golgi tendon organ)	Stretch	Muscles, tendons
Pain receptor	Free nerve ending	Chemicals released by tissue injury caused by excessive pressure, laceration, heat, or cold	Widespread in body
Thermoreceptor	Free nerve ending	Heat, cold	Skin
Chemoreceptor	Olfactory receptor	Odor (airborne molecules)	Nasal cavity
	Gustatory receptor	Taste (waterborne molecules)	Tongue
Photoreceptor	Rod, cone	Light	Retina of eye

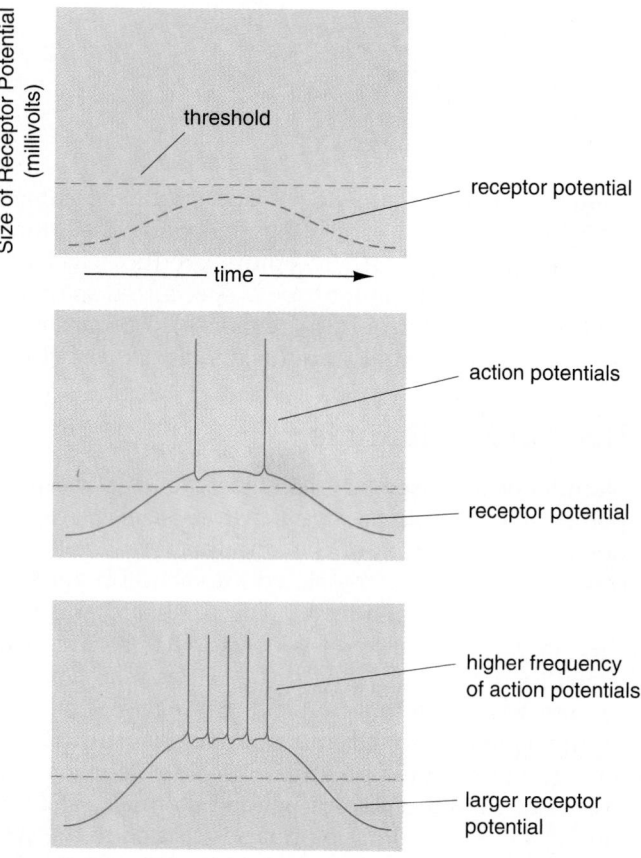

Figure 37-1 The magnitude of a receptor potential is proportional to the intensity of the stimulus. If the receptor potential exceeds threshold, action potentials are produced whose frequency is proportional to the size of the receptor potential.

producing a postsynaptic potential as described in Chapter 36. The postsynaptic neuron, in turn, signals the brain via action potentials. Most receptors produce receptor potentials in response either to mechanical deformation of their membranes or to stimuli that alter specific receptor molecules embedded in their cell membranes, as described below.

Mechanical Deformation

Cells called **mechanoreceptors** respond to mechanical deformation, which causes receptor potentials in one of two ways: by stretching the membrane of the receptor cell (as in receptors for touch or pressure) or by bending "hairs" that project from the receptor cell membrane. Receptors for sound, motion, and gravity bear hairlike structures and are called **hair cells.** Currents of fluid, motion, or the weight of dense objects bend the hairs, initiating a receptor potential.

Membrane Receptor Molecules

Receptors for light, chemicals (tastes or odors), and pain possess extensive areas of membrane studded with specialized receptor molecules. When light energy or a chemical stimulus hits a receptor molecule specialized to respond to that stimulus, the receptor molecule changes shape. This alters the permeability of the cell membrane, generating a receptor potential.

Other Receptor Mechanisms

Other mechanisms are used by the receptors responsible for perceiving temperature **(thermoreceptors),** electrical fields, and magnetic fields. To take one intriguing example, many animals, including insects, birds, and dolphins, have miniature magnets in or near their brains. Although no one knows how these animals "read" their internal magnetic compasses, behavioral experiments have shown that many animals can navigate using the Earth's magnetic fields (see "Road Maps of the Mind").

Thermoreception

Thermoreceptors sense temperature by responding to infrared radiation. Vertebrates have both "warm" receptors that respond to an increase in skin temperature, and "cold" receptors that respond to decreasing temperature. Scientists hypothesize that changes in the metabolic rate of the receptor—which is directly affected by temperature—cause the receptor potential. In human skin, warm and cold receptors consist of branched free nerve endings located just beneath the epidermis (see "A Closer Look at the Receptors of the Skin and Joints"; Fig. E37-3). Thermoreceptors are most responsive to temperature *changes*, which trigger a burst of action potentials that gradually decline in frequency if the temperature remains constant. This is why a swimming pool may feel uncomfortably cold or a hot tub uncomfortably hot when you first step in, but feels comfortable a few minutes later.

Thermoreceptors reach their highest degree of sensitivity in the **pit organs** of vipers (rattlesnakes, cottonmouths, and copperheads; Fig. 37-2). Within the pit organs are aggregations of free nerve endings so sensitive that they respond to the body heat of a mouse 16 inches away. Since there is a pit organ on each side of the head, they can sense the direction of the prey.

Figure 37-2 A pit viper showing the pit organs that are exquisitely sensitive thermoreceptors. These are used to detect warm-blooded prey.

Mechanoreception

The outer surface of humans and most other animals is exquisitely sensitive to touch. Embedded in and directly beneath human skin are several distinct types of mechanoreceptor neurons, each of which contains a sensory ending that produces a receptor potential when its membrane is deformed and stretched (see "A Closer Look at the Receptors of the Skin and Joints"; Fig. E37-3). Some of these endings are enclosed in layers of connective tissue called capsules—for example, the **Pacinian corpuscle,** which responds to rapid changes in pressure such as may be produced by a poke or vibrations. Some receptor neurons end in many fine branches, called **free nerve endings,** some of which respond to touch and pressure, others to heat and cold, and others to pain (described below). Free nerve endings in the skin also produce sensations of itching and tickling. Although free nerve endings are abundant in the skin, they are also found in other body tissues sensitive to touch and pressure, such as the cornea of the eye. The density of receptors in the skin varies tremendously over the surface of the body: each square centimeter of fingertip has dozens of touch receptors, while on the back there may be fewer than one receptor per square centimeter.

Mechanoreceptors in the walls of the stomach, rectum, and bladder signal fullness by responding to stretch. Other mechanoreceptive endings in the joints and muscles sense the orientation and direction of movement of various body parts (see Fig. E37-4). These position sensors, collectively called **proprioceptors,** allow you to walk without watching your feet or eat without watching the fork on its way to your mouth. Proprioceptors are concentrated in the ligaments and tissue surrounding the joints, in tendons, and in skeletal muscle.

The Auditory System

Virtually all animals have receptors sensitive to vibration: earthworms and insects detect ground vibrations as humans stomp by, while fish detect vibrations of water produced by approaching predators or fleeing prey. Sound, of course, is merely high-frequency vibration of air or water, and animals hear by sensing these vibrations. Both vertebrate and invertebrate sound receptors, although evolutionarily unrelated, use the same basic structure: a flexible membrane with receptor cells connected to the membrane. Sound waves cause the membrane to vibrate, stimulating the receptor cells. This basic mechanism is found in a variety of sense organs that detect movement of water, air, or the body, described below.

Lateral Line Organs

In vertebrates, the organs that detect sound, gravity, and movement almost certainly evolved from the **lateral line organ,** which is found in all fish and in the aquatic forms of amphibians. The lateral line organ, which detects water movement, consists of a series of clusters of hair cells located in pits or tubes that form a strip beginning in the head and extending along either side of the body. As shown in Figure 37-3, the "hairs" of the hair cells are embedded in a gelatinous cap, the cupula, which is deflected by water currents, causing the hairs to bend. The receptor potentials caused by the bending produce action potentials that travel to the brain. Lateral line organs are used to detect water currents, water movements caused by prey or predators, and possibly low-frequency sounds.

In the following sections, we describe the cochlea and vestibular apparatus of terrestrial vertebrates, using the human as an example. As you read about the mechanisms of sound, gravity, and movement detection, note the similarities to the lateral line organ from which they evolved. In each case notice that the "hairs" of hair cells are embedded in some sort of gelatinous structure that is moved by the stimulus.

Sound Detection by the Human Ear

Human Ear Structure

The human ear is a remarkable elaboration on the theme described above. The ear consists of three

Road Maps of the Mind: The Role of the Senses in Animal Orientation

Humans have devised elaborate means of orientation: street names, house numbers, odometers to measure distance, and road maps. Without a map, many of us could not drive across town without getting lost. Yet many birds and fish routinely travel hundreds of miles during migrations, often at night, in cloudy weather, or (perhaps most remarkably of all) through or above the seemingly endless sameness of the ocean. How do they find their way?

Orientation by Vision

To a human, the most obvious way to navigate is by sight. Close to home, at least, many animals, including some insects, learn the appearance of their surroundings. For example, the female digger wasp lays eggs in a burrow in the soil. She then flies off to capture insect prey, which she stores in the burrow to feed her offspring when they hatch. The famous ethologist Niko Tinbergen (see Chapter 41) studied the cues used by the female to relocate her burrow. While the female was inside, he surrounded the nest with pine cones. When the wasp emerged, she flew around the nest before departing. While she was gone, Tinbergen moved the cones about a foot away. The returning wasp still sought her nest within the ring of cones. Although the nest was in plain sight nearby, she was unable to locate it because the visual landmarks on which she relied had been shifted.

Birds may also use landmarks, such as rivers and seashores, to find their way, but several species migrate at night or over large expanses of ocean, using the position of the sun or stars to tell direction. Many species seem to have genetically programmed information about the direction of the sun at various times of day, and also possess a biological clock that measures off a roughly 24-hour day. Other birds have the remarkable ability to "read" the night sky. Indigo buntings, for example, seem to have a built-in star map that enables them to find north during spring migration by looking at the stars.

Orientation by Scent

It is difficult to see far under water, and aquatic animals tend to rely on cues other than vision to travel long distances. Pacific salmon, for example, migrate hundreds of miles using scent to locate their final destination. Salmon hatch in freshwater streams, migrate downstream to the ocean to feed and mature, then return to their native stream to spawn (Fig. E37-1). Experiments by Arthur Hasler have conclusively shown that salmon find their home stream by scent. Hasler raised young salmon in a hatchery with a trace

Figure E37-1 Salmon migrating home upstream to spawn. They will arrive ravaged by the rigors of migration, and die shortly after spawning, their mission complete.

of an odorous chemical, morpholine, added to the water. After a month, the salmon were released into the ocean. When they were scheduled to return, Hasler added morpholine to a stream past which the fish would swim on their migration. The salmon stopped at the treated stream, even though they had never encountered that stream before. Other salmon not raised in morpholine ignored the odor.

Orientation by Sound

Animals who hunt in darkness or murky water often rely heavily on their auditory abilities to recognize nearby objects. Some mammals have even evolved a type of sonar, which produces an auditory image of their nearby surroundings. This ability, called **echolocation**, is highly developed in bats and porpoises.

Bats can navigate and hunt insect prey in total darkness using echolocation. An echolocating bat may emit pulses of noise at ultrasonic frequencies (20,000–80,000 cycles per second) that bounce back from nearby objects. The intensity of this sound would make it unpleasantly loud if we could detect it. The patterns of returning sound convey accurate information as to the size, shape, surface texture, proximity, and location of objects in the environment. This system is remarkably sensitive. Little brown bats can detect wires only 1 millimeter thick from a distance of 2 meters. Several specializations of the bat's auditory system contribute to its effectiveness. The enormous, elaborately folded outer ears of the bat collect the returning echos and help the bat locate their source (Fig. E37-2a). As the bat emits its cry, muscles attached to the bones of the middle ear briefly

(a)

(b)

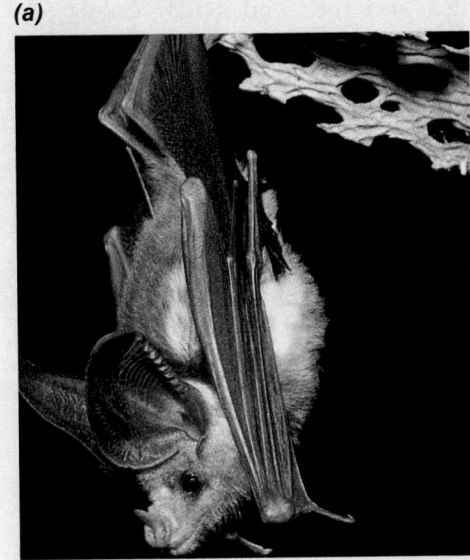

Figure E37-2 Echolocation by bats and porpoises is based on very different anatomical structures. **(a)** A yellow-winged bat, showing the elaborate folds of the external pinnae that help it localize returning echos. **(b)** The bottlenose porpoise focuses ultrasonic clicks using the oil-filled bulge in the front of its head.

contract, temporarily reducing the bones' vibrations and preventing the bat from being deafened by its own calls. The tympanic membrane and bones of the middle ear are exceptionally light and easily vibrated by the faint returning echos.

Porpoises produce ultrasonic clicks within their nasal passages, and emit them through the front of their heads (Fig. E37-2b). Here, a large, flexible, oil-filled sac acts like an acoustic lens, directing the sound forward in a broad beam (for navigation) or a narrow beam (to locate prey). Echolocating porpoises are able to pick up a 5-millimeter object from the floor of their tanks, and distinguish different species of fish. The narrowly focused beam may also be used to stun fish with a blast of sound, making them easier to capture.

Orientation by Magnetic Fields

Radar operators have observed flocks of birds migrating at night under heavy cloud cover, when landmarks, sun, and stars were all unavailable, and at altitudes where scent could not possibly be used. Eels of eastern North America and western Europe swim out of streams and rivers into the Atlantic Ocean, and migrate to the Sargasso Sea to spawn. In all that expanse of ocean, it seems unlikely that there could be any consistent chemical cues, and vision certainly cannot be used. How do these birds and eels find their way? The answer seems to lie in responses to the Earth's magnetic field.

Homing pigeons are famous for their abilities to fly home after being released some distance away. They can accurately locate their home roost even under cloudy skies in terrain with few landmarks. If a small

magnet is strapped to a pigeon's back, it still homes successfully in sunny weather, but loses its way under overcast skies. Apparently, pigeons can navigate either by the sun or by magnetic fields. In sunny weather, the pigeon can orient by the sun, ignoring the magnetic field. In cloudy weather, the magnets throw off the pigeon's magnetic compass, leaving them no way to find home. How do the pigeons detect magnetic fields? No one knows for sure, but in 1979 it was discovered that pigeons have deposits of magnetite (a magnetic iron compound) located just beneath the skull. These deposits may act as a built-in magnet that the pigeons use to tell direction.

Eels probably also use magnetic fields for navigation, but in a different way. You may recall from high school physics that when an electrical conductor is moved through a magnetic field, an electric current is induced in the conductor (this is how we generate electricity commercially). Seawater, with its high salt concentration, is a fairly good conductor, and the currents of the Gulf Stream provide movement through the Earth's magnetic field. The Gulf Stream generates extremely weak electric fields, roughly equivalent to a potential created by a one-volt battery with its poles 20 kilometers apart. At first, this was thought to be far too weak to be detected by eels or any other animals. Remarkable behavioral experiments have shown, however, that eels can do much better. Eels were trained to slow their heart rates in response to electrical fields, and the researchers found that eels can detect electric fields as weak as a one-volt battery with poles *over 3000 miles* (5000 kilometers) apart! For an eel, finding the Gulf Stream must be a piece of cake!

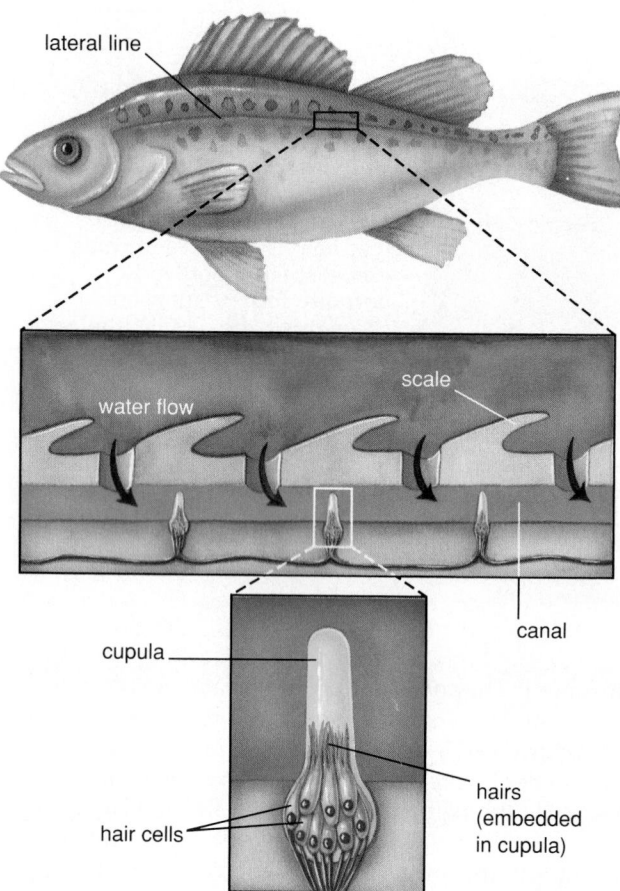

Figure 37-3 The lateral line organ of fish and aquatic amphibians consists of clusters of specialized hair cells whose hairs are embedded in a gelatinous structure, the cupula. These clusters are embedded at intervals within a canal along the length of the body. Openings along the canal allow water currents to enter, which deflect the cupula. This bends the hairs of the hair cells, causing a receptor potential.

parts: the outer, the middle, and the inner ear (Fig. 37-4). The **outer ear** includes all the structures outside the **tympanic membrane** (eardrum). Sound waves in the air are funneled by the **external ear** into the **auditory canal.** The **middle ear** is also air-filled. The **eustachian tube** connects the middle ear to the pharynx, allowing equalization of air pressure between the middle ear and the atmosphere. Within the middle ear, vibrations are transmitted through membranes and bones. Sound first vibrates the tympanic membrane, which in turn vibrates a series of three bones: the **hammer,** the **anvil,** and the **stirrup.** These bones transmit vibrations from the tympanic membrane to a much smaller membrane, the oval window. The oval window covers the opening of the **inner ear,** in which sound vibrations travel through

fluid. The fluid-filled hollow bones of the inner ear form the cochlea and the semicircular canals (described below) where vibrations are translated into neural signals.

The Mechanism of Sound Detection

The receptor cells for hearing are located within the inner ear, in the spiral-shaped cochlea (Fig. 37-4a). If we mentally straighten out the cochlea (Fig. 37-4b), in longitudinal-section it consists of two fluid-filled tubes, an outer U-shaped canal and a straight **central canal.** The central canal contains the **basilar membrane,** on top of which are located the receptors, or hair cells. Protruding into the central canal is another membrane, the **tectorial membrane,** a gelatinous structure in which the "hairs" of the hair cells are embedded.

How do these structures allow the perception of sound? When sound waves enter the ear, they vibrate first the tympanic membrane, then the bones of the middle ear, the membrane of the oval window, and finally the fluid in the cochlea. The vibrating fluid in the cochlea vibrates the basilar membrane, causing it to move relative to the tectorial membrane. This bends the hairs spanning the gap between the membranes (Fig. 37-4c), producing receptor potentials in the hair cells. The hair cells release transmitter onto neurons of the **auditory nerve.** Action potentials are triggered in the auditory nerve axons and travel to the brain.

The inner ear also allows us to perceive loudness and pitch. A weak sound causes small vibrations, which bend the hairs only slightly. This produces small receptor potentials in the hair cells and a low frequency of action potentials in the auditory nerve axons. A loud sound causes large vibrations, which cause greater bending of the hairs and a larger receptor potential. This leads to a high frequency of action potentials in axons of the auditory nerve. Loud sounds sustained for a long time can actually damage the hairs (Fig. 37-5), resulting in hearing loss, a fate suffered by many prominent rock musicians and their fans.

The perception of pitch is a little more complex. Humans can detect vibration frequencies from about 30 cycles per second (very low pitched) up to around 20,000 cycles per second (very high pitched). The basilar membrane is stiff and narrow at the end near the oval window and more flexible and wider near the tip of the cochlea. This progressive change in structure causes different parts of the membrane to resonate best to particular frequencies of sound. High-frequency sound waves cause the greatest vibration toward the end of the basilar membrane near the oval

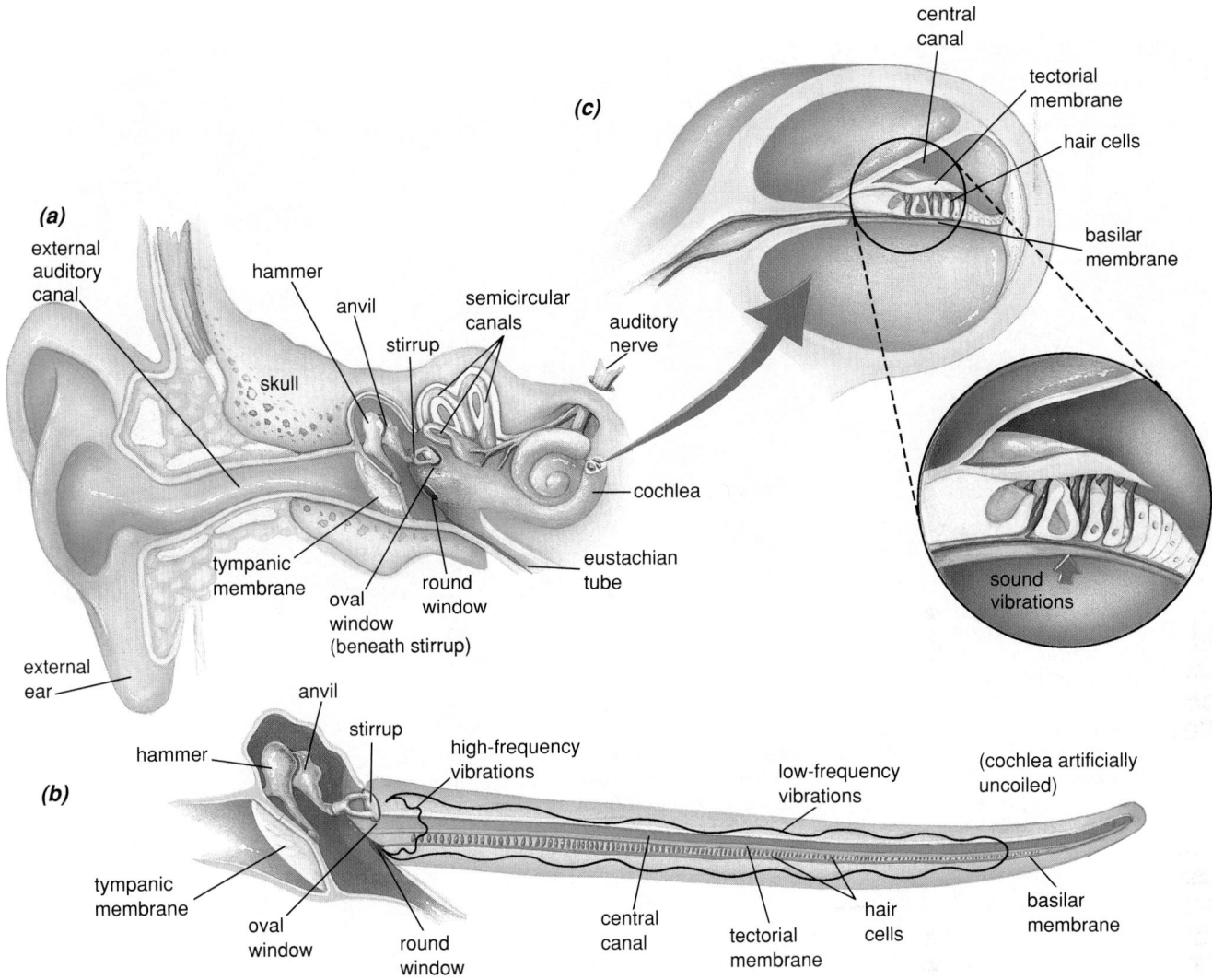

Figure 37-4 The human ear contains structures that detect sound, gravity, and movement.
(a) Sound waves enter the auditory canal and vibrate the tympanic membrane. These vibrations are transmitted through the bones of the middle ear to the membrane of the oval window connected to the fluid-filled cochlea, surrounded by thin bone.
(b) Uncoiled, the cochlea consists of an outer tube surrounding the central canal. Vibrations of the fluid filling the cochlea cause the membranes of the central canal to vibrate, stimulating the hair cells, shown in (c). The basilar membrane varies in width and tension along its length, causing lower frequency vibrations (low-pitched sounds) to activate receptors toward the tip of the (uncoiled) cochlea, and higher frequency vibrations (higher-pitched sounds) to activate receptors closer to the oval window.
(c) The hairs of hair cells span the gap between the basilar and tectorial membranes in the central canal. Sound vibrations move the membranes relative to one another, bending the hairs and producing a receptor potential in the hair cells. The hair cells then cause action potentials in the auditory nerve.

window, while increasingly lower frequencies produce vibration progressively further toward the opposite end (see Fig. 37-4b). Thus, where the basilar membrane vibrates most and consequently which receptors are stimulated most varies with the fre-

quency of sound. The brain interprets signals from receptors near the oval window as high-pitched sound. Signals from receptors further along toward the ends of the uncoiled cochlea are interpreted as lower in pitch.

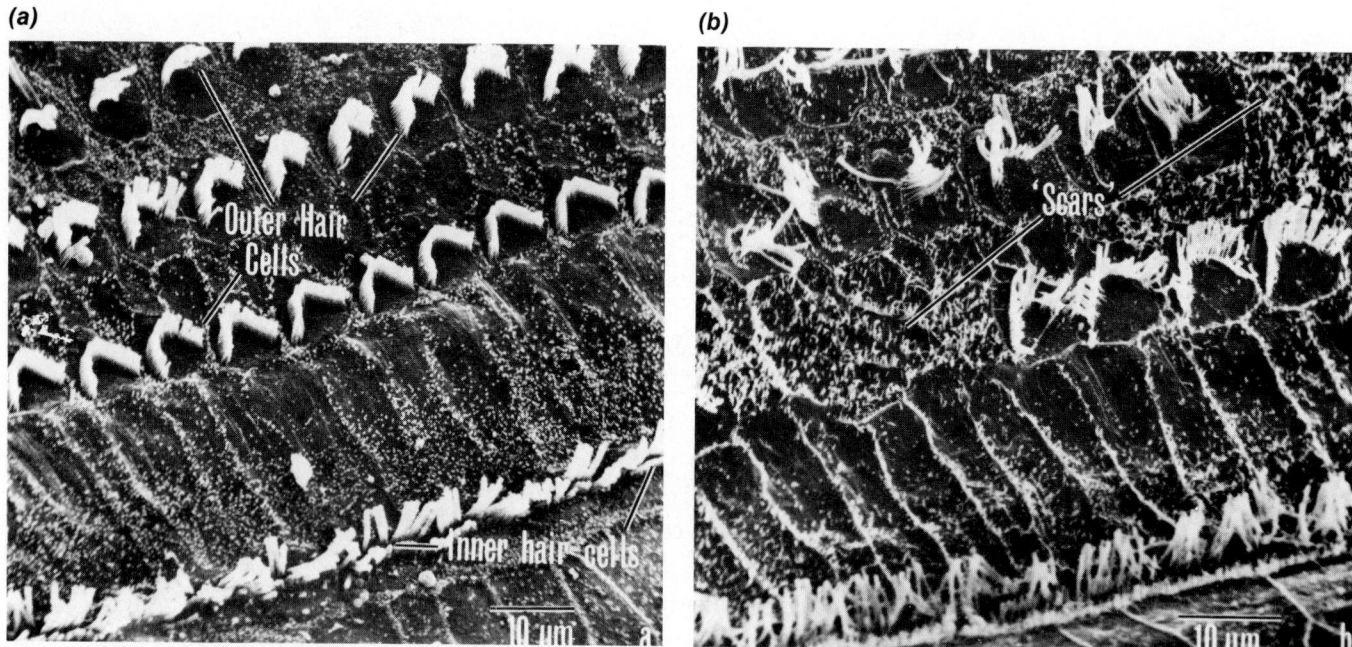

(a)

(b)

Figure 37-5 Scanning electron micrographs show the effect of intense sound on the hair cells of the inner ear. **(a)** Organ of Corti of a normal guinea pig, showing three rows of outer hair cells with the hairs of each receptor arranged in a V-shaped pattern. **(b)** Organ of Corti of a guinea pig after 24-hour exposure to a sound level approached by loud rock music (2000 hertz at 120 decibels). Note that many of the hairs are damaged or missing entirely. Since hair cells do not normally regenerate, hearing loss is permanent. (Scanning electron micrographs by Robert S. Preston, courtesy of Professor J. E. Hawkins, Kresge Hearing Research Institute, University of Michigan Medical School.)

The Vestibular Apparatus: Gravity and Movement Perception

Besides hearing, the mammalian inner ear is the site of two other modified mechanical senses, one for detecting motion and the other for detecting gravity. These are both found within the **vestibular apparatus** housed in a set of bony canals adjacent to the cochlea (Fig. 37-6a). Motion is detected by the **semicircular canals,** a set of three curving, fluid-filled, bone-covered tubes. In swellings at the base of each canal are clusters of hair cells similar to those of the inner ear, their hairs embedded in a gelatinous cupula very similar to that of the lateral line organ of fish (Fig. 37-6b). Sudden acceleration of the head causes the cupula and surrounding fluid to lag behind the movement of the canals, bending the hairs in one direction. Cessation of motion causes the fluid to slosh ahead, bending the hairs in the opposite direction. Since each of the canals is oriented at right angles to the others, they allow us to sense movements in all directions (see Fig 37-6a).

Gravity perception is the function of two chambers below the semicircular canals, the **utricle** and the **saccule,** each containing clusters of hair cells. The hairs of these receptors are embedded in a gelatinous mass, the **otolith membrane,** which contains crystals of calcium carbonate (otolith is from the Greek, meaning "ear stone"). The heavy crystals are pulled downward by gravity, bending the hairs according to the orientation of the head (see Fig. 37-6b). So sensitive are the gravity-detecting organs that we are aware of as little as half a degree of deviation from an upright posture.

Vision

Animal vision varies in acuity, and several types of eyes have evolved independently. All forms of vision, however, use **photoreceptors.** These sensory cells contain receptor molecules called **photopigments** (because they are colored) that absorb light and chemically change in the process. This chemical

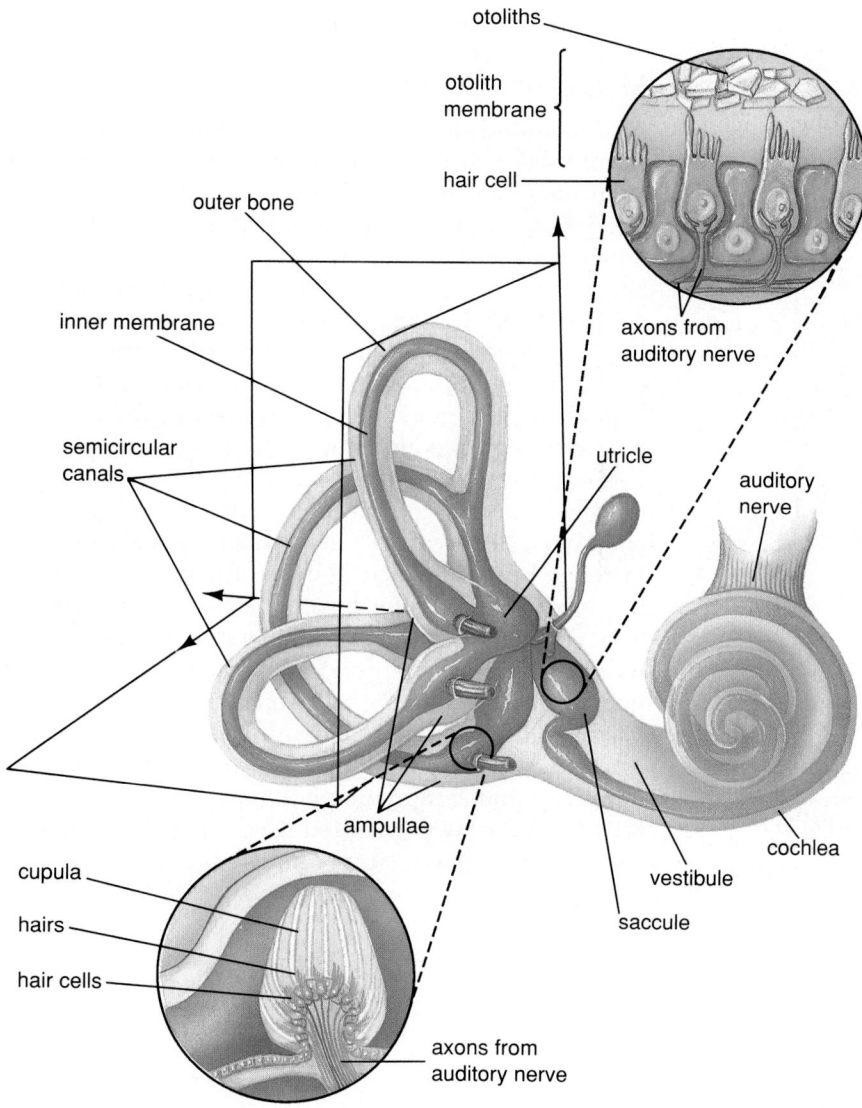

otoliths

otolith
membrane

hair cell

outer bone

inner membrane

semicircular
canals

utricle

auditory
nerve

axons from
auditory nerve

ampullae

vestibule

saccule

cochlea

cupula

hairs

hair cells

axons from
auditory nerve

Figure 37-6 The vestibular system showing its relation to the cochlea. Both the semicircular canals and the utricle and saccule have clusters of hair cells whose hairs are embedded in gelatinous membranes. The cupula in semicircular canals detects movement of the fluid, while the otolith membrane of the utricle and saccule (weighted with stones of calcium carbonate) exerts pressure on the hairs in response to the pull of gravity.

change alters ion channels in the receptor cell membrane, producing a receptor potential.

Types of Eyes

The simplest animal light detector is the **eyespot,** found in flatworms (Fig. 37-7). The eyespot has no lens and cannot focus light or form an image. Flatworms can distinguish light from dark, and may also perceive the direction and intensity of the light. This information can be important to the animal: a passing shadow (which could be a predator) triggers withdrawal or escape responses.

The arthropods (insects, spiders, and crustaceans)

evolved **compound eyes** that consist of a mosaic of many individual light-sensitive subunits called **ommatidia** (Fig. 37-8). Although much more complex than an eyespot, each ommatidium functions similarly as an on/off, bright/dim detector. Using a large number of individual units (up to 36,000 per eye in a dragonfly), most arthropods probably obtain a reasonably faithful, although grainy, image of the world.

The cephalopod molluscs (see Chapter 22) and the vertebrates independently evolved a third type of eye, often called the **camera eye** (Fig. 37-9; see also the opener photo). The camera eye consists of three basic parts: a light sensitive layer (the retina), a lens for focusing light, and a set of muscles for adjusting focus by moving or changing the shape of the lens.

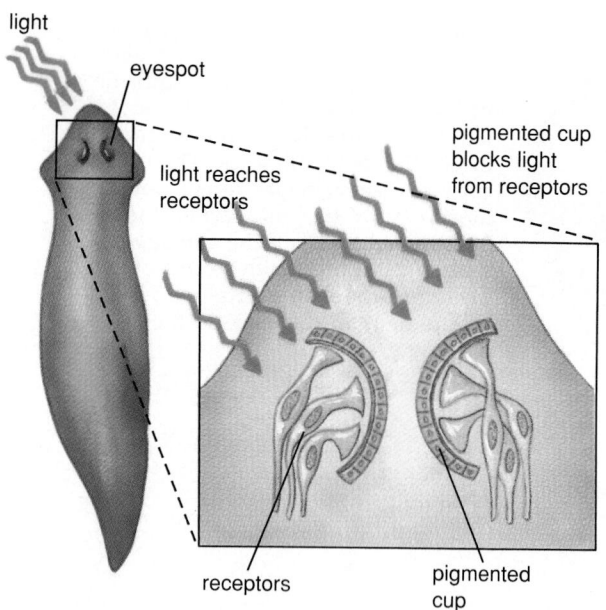

Figure 37-7 The most primitive eyes in the animal kingdom are eyespots, such as those found in flatworms. Eyespots consist of photoreceptors in a pigmented cup. Light entering the open end of the cup stimulates the photoreceptors, but the pigmented back side of the cup prevents light from other directions from reaching them. This allows detection of the direction of light, guiding simple behaviors such as finding a dark place to hide.

The Vertebrate Visual System

Structure of the Eye

As illustrated in Fig. 37-9, incoming light first encounters the **cornea,** a transparent covering over the front of the eyeball. Behind the cornea is a chamber filled with a watery fluid called **aqueous humor,** which provides nourishment for the lens. The amount of light entering the eye is adjusted by a muscular tissue, the **iris,** whose circular opening, the **pupil,** can be expanded or contracted. Light passing through the pupil encounters the **lens,** a structure resembling a flattened sphere and composed of transparent proteinaceous fibers. The lens is suspended behind the pupil by ligaments and muscles that regulate its shape. Behind the lens is another, much larger chamber filled with a clear jellylike substance, the **vitreous humor,** which helps maintain the shape of the eye. After passing through the vitreous humor, light reaches the **retina,** a multilayered nervous tissue where the light energy is converted into electrical nerve impulses that are transmitted to the brain (Fig. 37-9b). The retina is richly supplied with blood vessels and contains a layer of pigment that absorbs stray light rays that escape the photoreceptors. Behind the retina is a darkly pigmented tissue, the choroid. The choroid's rich blood supply helps nourish the cells of

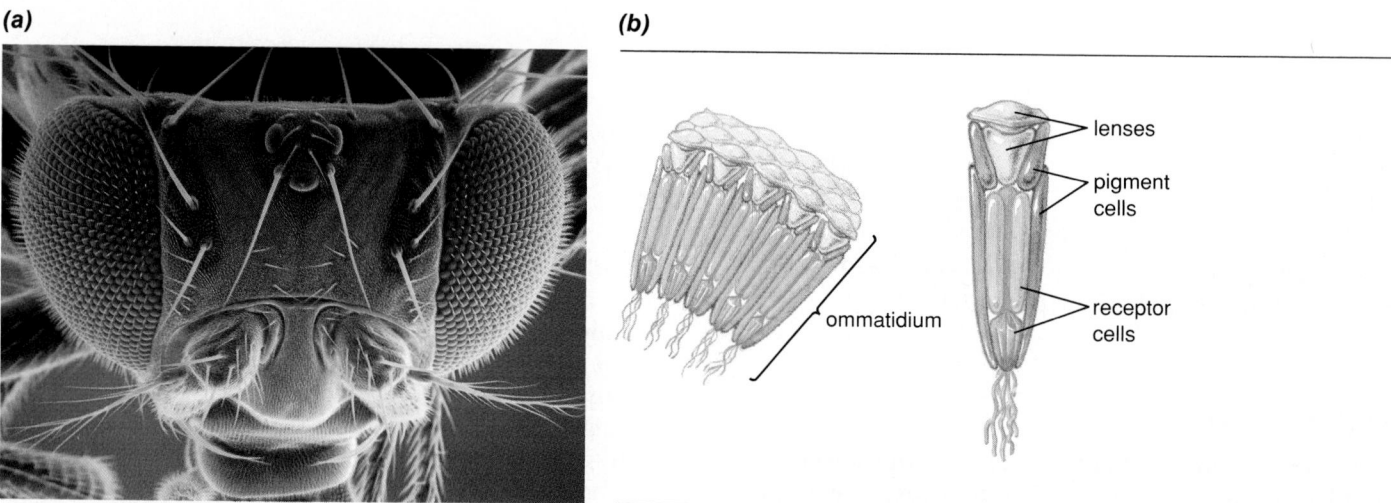

Figure 37-8 (a) Scanning electron micrograph of the compound eye of a fruit fly. **(b)** Each eye is made up of numerous individual light-receptive ommatidia. Within each ommatidium are several receptor cells, capped by a lens. Pigmented cells surrounding each ommatidium prevent light from passing through to adjacent receptors.

(a)

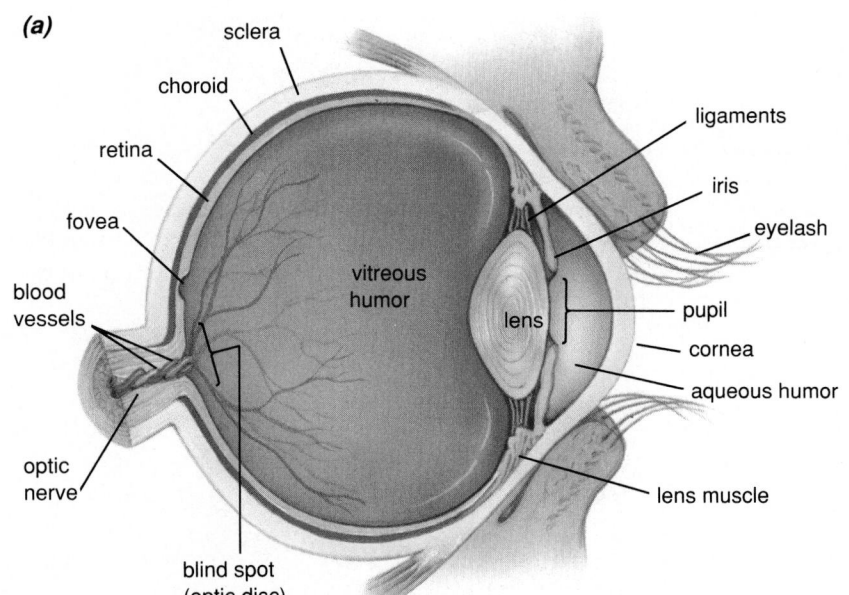

Figure 37-9 (a) The anatomy of the human eye. **(b)** The human retina has rods and cones (photoreceptors), integrating cells, and ganglion cells. Each rod and cone bears a long extension packed with membranes in which the light-sensitive molecules are embedded.

(b)

LAYERS OF THE RETINA

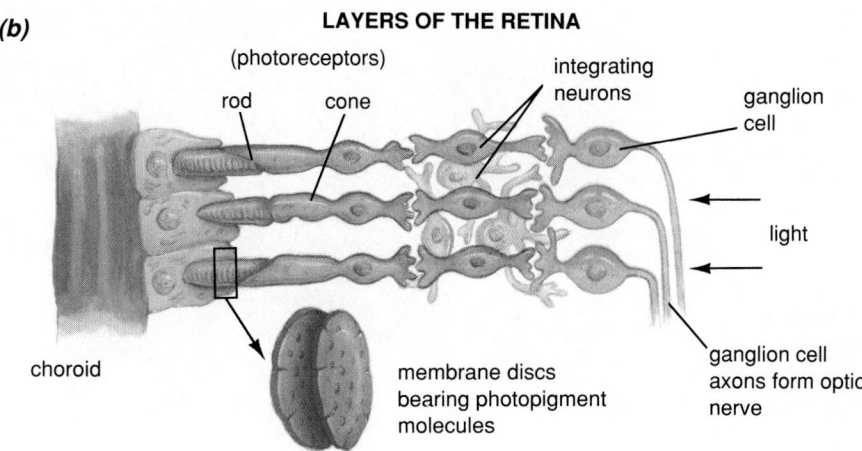

the retina. Its dark pigment also absorbs stray light whose reflection inside the eyeball would interfere with clear vision. Surrounding the outer portion of the eyeball is a tough connective tissue layer, the sclera, visible as the white of the eye.

In vertebrates (such as deer) that are most active at dusk, when little light is available, the choroid may be modified to reflect light, rather than absorb it. By reflecting light that escaped the photoreceptors during its initial passage, the choroid gives the receptors a second chance to detect it, maximizing the animal's ability to see in dim light. Reflective choroids give the eyes of these animals an eerie glow when bright light

(such as produced by car headlights) is reflected back through the wide-open pupil.

Focusing in the Human Eye

The visual image is focused most sharply on a small area of the retina called the **fovea.** Focusing is aided by the cornea, which contributes significantly to the bending of incoming light rays, producing an image of approximately the right size in the general vicinity of the retina. However, the shape of the cornea cannot be adjusted, and the lens is responsible for final, sharp focusing.

The shape of the lens can be adjusted so that it is

CLOSER LOOK

At the Receptors of the Skin and Joints

Some of the diverse receptors of human skin are illustrated in Figure E37-3. The simplest are branched free nerve endings that are specialized to respond either to touch, pain, or temperature. The endings of some receptor neurons are wrapped around the base of hairs. Called **hair end-organs,** they detect the slightest movement of the hair, such as would be produced by a small crawling insect. **Ruffini's end-organs,** located in the

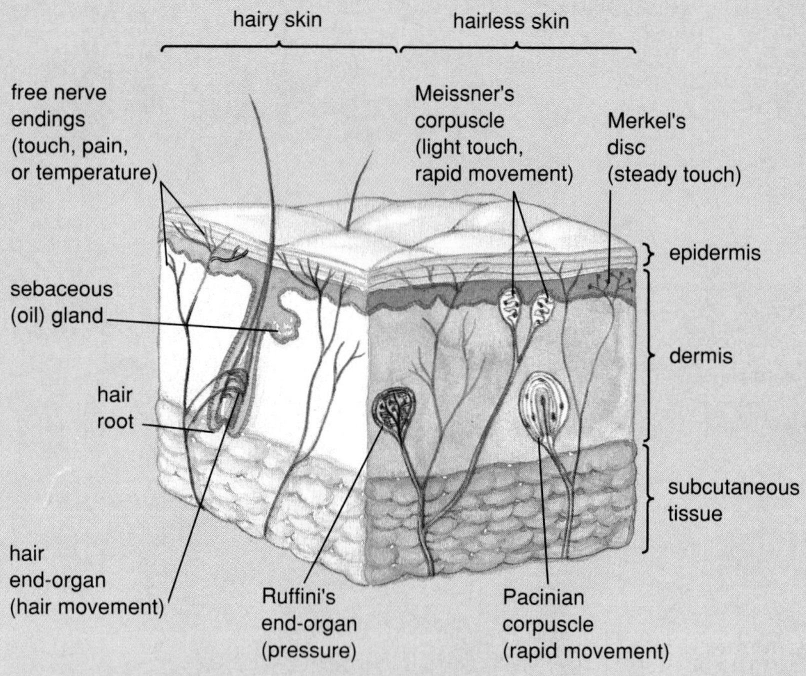

Figure E37-3 A sampling of receptors within the skin (consisting of the epidermis and dermis; see Chapter 25) and immediately beneath the skin (subcutaneous tissue) that respond to mechanical stimuli, temperature, and pain.

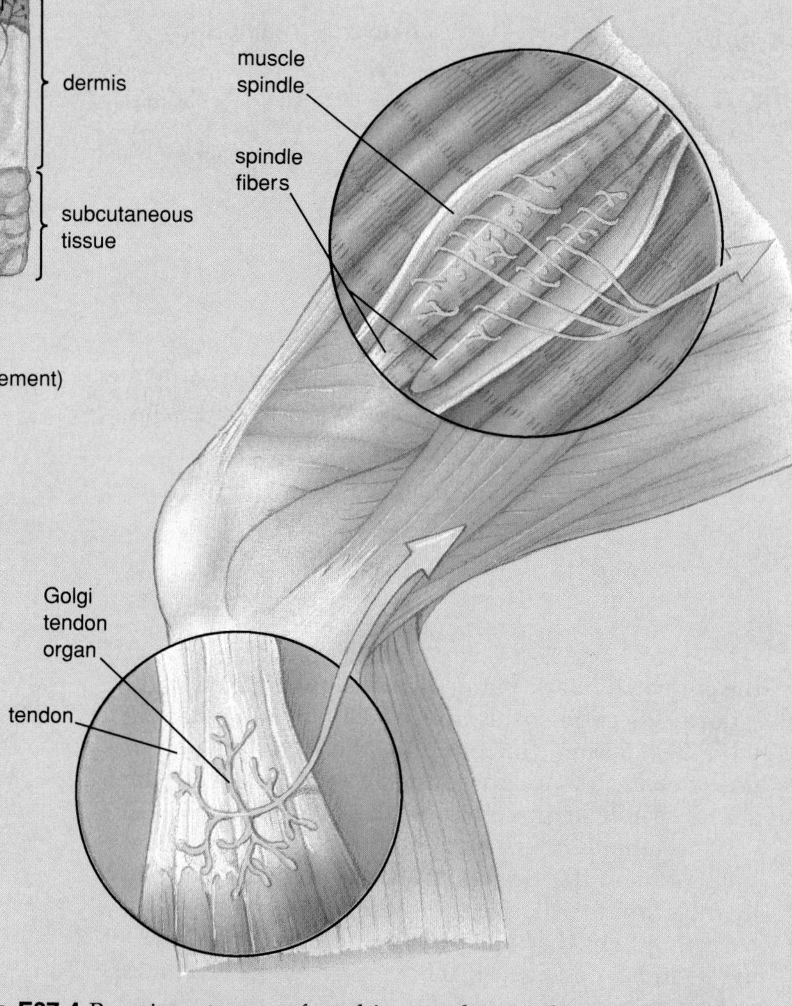

Figure E37-4 Proprioceptors are found in muscles, tendons, and tissues surrounding joints. They signal pressure and stretch and inform the brain of the position of the limbs.

dermis, respond to heavy, continuous pressure, as well as to changes in pressure. **Merkel's disks,** particularly abundant in the fingertips, have flattened receptive endings that are specialized to respond to steady touch. **Pacinian corpuscles** are located in subcutaneous tissue (as well as in connective tissue deep within the body), and **Meissner's corpuscles** are found just below the epidermis. These both respond to rapid movements, while Meissner's corpuscles respond to light touch as well.

Proprioceptors also come in a variety of forms. Ruffini's endings (also found in the skin) are abundant in ligaments and tissues around the joint. **Golgi tendon organs** found in tendons (which connect muscles to bone; see Chapter 38), and **muscle spindles,** found in skeletal muscle, respond to stretch (Fig. E37-4). They are activated by movement of the joints, which causes a sudden change in the degree of stretching of muscles and tendons, and also by the steady-state pull that occurs in any position of the limb.

more rounded or more flattened when viewed from the side. This adjustment is accomplished by a circular muscle surrounding the lens. The size of the opening encircled by the muscle is largest when the muscle is relaxed. Therefore, in its relaxed state, the muscle applies maximum pull on the lens, stretching and flattening it. In this configuration, the lens focuses distant objects, over 20 feet away. When focusing nearby, the muscle contracts, relaxing tension on the lens, and allowing the lens to resume its more rounded shape. The curved lens bends light rays more, allowing the eye to focus on closer objects (Fig. 37-10a).

Nearsighted people cannot focus on distant objects, while farsighted people cannot focus on nearby objects. These conditions, which can be caused by abnormally long or short eyeballs, are corrected by external lenses of the appropriate shape (Fig. 37-10b and c). As humans age, the lens stiffens. By their mid-forties, most people have lost much of their ability to change the shape of the lens, and become farsighted, requiring glasses for close work such as reading.

Neural Processing in the Retina

The vertebrate eye provides the sharpest vision in the animal kingdom, even though the retina, a complex, multilayered structure, is "built backward" (see Fig. 37-9b). The photoreceptors, called **rods** and **cones** after their shapes, have their light-gathering elements farthest away from the light, at the rear of the retina. Between the receptors and incoming light are several layers of neurons that process the signals from the photoreceptors. The outermost layer consists of ganglion cells, whose axons make up the **optic nerve.** The receptor potential from the photoreceptors is processed in complex ways by the retinal neurons. This processing results in an enhancement of our ability to detect edges, movement, dim light, and changes in light intensity.

The much-modified signal from the photoreceptors is finally converted to action potentials transmitted along the ganglion cell axons in the optic nerve to the brain. Here, further processing ultimately results in the sensation of vision. Ganglion cell axons emerge from the front of the retina, then must pass back through the retina to reach the brain. The point where axons pass through the retina is called the **optic disk** or **blind spot** (Fig. 37-11; see also Fig. 37-9). This area lacks receptors, and objects focused here seem to disappear. You can locate your blind spot by closing your left eye, and focusing on the star on page 825 with your right eye. Start with the book about a

(a) NORMAL EYE

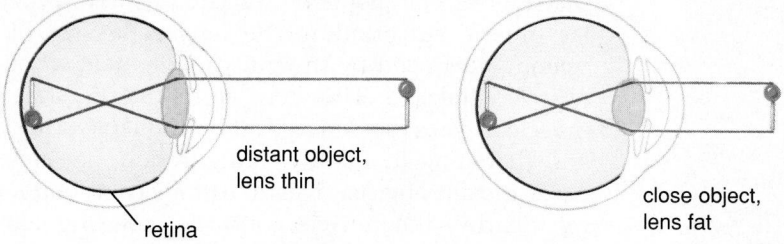

distant object,
lens thin

retina

close object,
lens fat

(b) NEARSIGHTED EYE

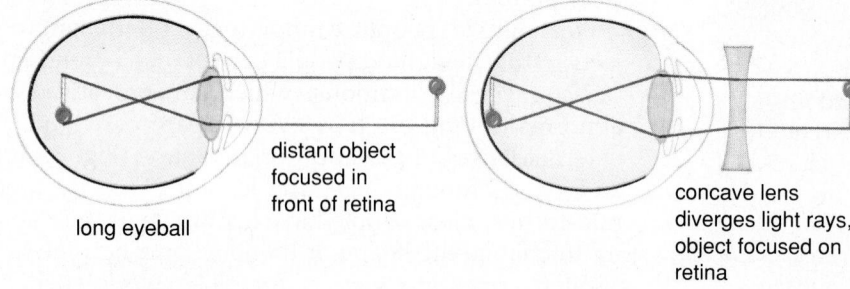

distant object
focused in
front of retina

long eyeball

concave lens
diverges light rays,
object focused on
retina

(c) FARSIGHTED EYE

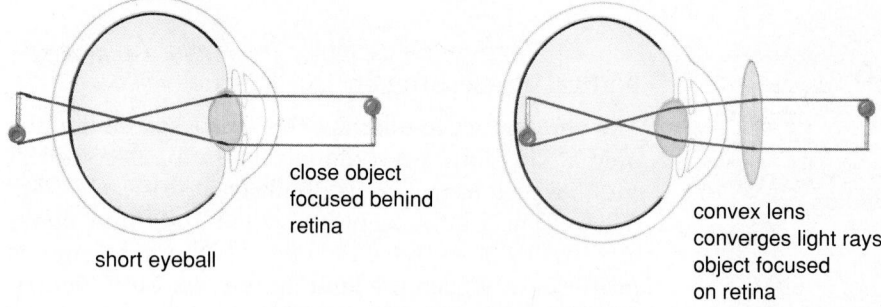

close object
focused behind
retina

short eyeball

convex lens
converges light rays,
object focused
on retina

Figure 37-10 **(a)** Focusing in the human eye. Left: to focus on a distant object, the lens is made thinner, causing relatively little bending of the light rays. Right: to focus on a nearby object, the lens assumes a more nearly spherical shape, bending the light rays more sharply. **(b)** Left: nearsighted people usually have eyeballs that are too long. Light rays focus in front of the retina and are out of focus again by the time they strike the retina. Right: eyeglasses with a concave lens cause the light rays to diverge slightly so the focal point falls on the retina. **(c)** Left: farsighted people have eyeballs that are too short. The lens cannot bend incoming light rays enough to focus on the retina, so the focal point falls behind the retina. Right: eyeglasses with a convex lens converge light rays, causing the focal point to fall on the retina.

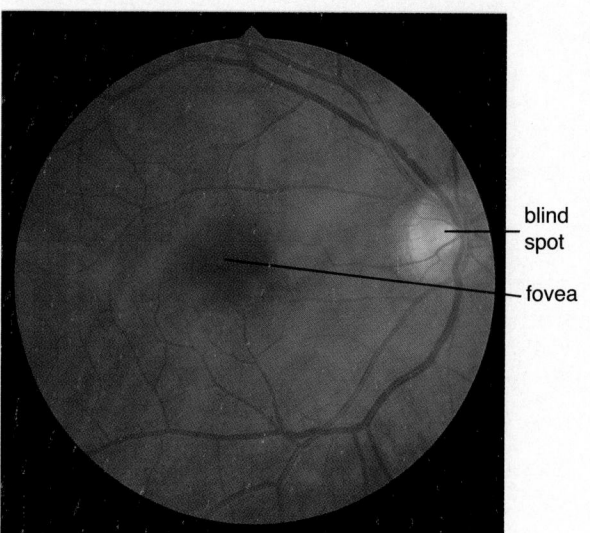

blind
spot

fovea

Figure 37-11 A photograph of the human retina, taken through the cornea and lens of a living person. Blood vessels supply oxygen and nutrients. The blind spot (optic disk) and fovea are visible.

foot away, and gradually move it closer. At one point, the spot will disappear—this occurs when its image falls on the optic disc.

Interestingly, the camera eye of molluscs such as the octopus, which evolved independently from ours, has its photoreceptors in the outermost layer of the retina, so that incoming light impinges immediately on the photoreceptors. This eliminates the blind spot, and increases the availability of light to the receptors.

The Photoreceptors: Rods and Cones

Photoreception in both rods and cones begins with absorption of light by photopigment molecules embedded in the membranes of the photoreceptors. These membranes form flattened, hollow disks in rods, and are deeply folded in cones, giving them a large surface area on which to bear the photopigments (see Fig. 37-9).

Light hitting the photopigment molecule causes the molecule to change shape. The altered shape of the molecule initiates a series of biochemical reactions inside the photoreceptor. These result in a change in the permeability of the receptor membrane to ions, producing a receptor potential in the photoreceptor cell. The receptor potential is then transmitted as described above.

Rods are far more numerous than cones (125 million rods versus 5 million cones), and dominate in the peripheral portions of the retina. Compared with cones, rods have much deeper stacks of pigment-bearing membrane, and are about 100 times more sensitive to light (see Fig. 37-9). Studies have shown that a rod is capable of responding to a single photon, the smallest possible unit of light. Thus, our vision in dim light is almost entirely due to rods. Unlike cones, rods do not distinguish colors. So in moonlight, which is too dim to activate the cones, the world appears black and white.

Although cones are found throughout the retina, they are concentrated in the fovea, where the lens focuses images most sharply (see Fig. 37-9). The fovea, which consists entirely of densely packed cones, appears as a depression near the center of the retina because the layers of neurons normally covering the cones are pushed aside here, while still retaining their synaptic connections. Thus, light reaches the cones of the fovea with less interference.

Unlike rods, cones respond to different wavelengths (colors) of light. Human cones come in three varieties, each containing a slightly different photo-

pigment. Each type of photopigment is most strongly stimulated by a particular wavelength of light, corresponding roughly to red, green, or blue. The brain distinguishes color according to the relative intensity of stimulation of different cones. For example, the sensation of yellow is caused by fairly equal stimulation of red and green cones. About 8% of all males lack normal color vision. Although described as "color-blind," they are actually only color-deficient. The most common abnormality is red–green color deficiency due to a recessive allele on the X chromosome (see Chapter 15) that codes for a defective photopigment in the red cones. The altered red photopigment has about the same light-absorbing properties as the green photopigment, so the affected individual has trouble distinguishing red from green.

Not all animals have both rods and cones. Animals active almost entirely during the day (certain lizards, for example) may have all-cone retinas, while night-active animals (such as the ferret) or those dwelling in dimly lit habitats (such as deep-sea fishes) often have mostly rods.

Binocular Vision

Most animals are bilaterally symmetrical, and possess a pair of eyes. Having two eyes is useful in several ways—for example, two eyes are essential to allow a flatworm to determine the direction of a light source (see Fig. 37-7). Among vertebrates, we find two basically different eye placements—herbivores usually have one eye on each side of the head, while predators have both eyes facing forward (Fig. 37-12). The forward-facing eyes of predators and omnivores such as humans each have slightly different but extensively overlapping visual fields. This **binocular vision** allows depth perception and more accurate judgment of the size and distance of an object from the eyes. This is important to a cat about to pounce on a mouse, or a monkey leaping from branch to branch.

In contrast, the widely spaced eyes of herbivores have little overlap in their visual fields; accurate depth perception is sacrificed in favor of nearly a 360 degree field of view. This allows these animals, who are frequently preyed upon, to spot a predator approaching from any direction.

The Chemical Senses: Olfaction and Taste

Through chemical senses, animals find food, avoid poisonous materials, and may locate homes or find mates (Fig. 37-13). Virtually all animals have **chemoreceptors** that sample the chemical composi-

(a)

(b)

Figure 37-12 There are two usual placements for the eyes—one on each side of the head or both in front. Among the mammals, the location of the eyes is related to the life-style of the animal.
(a) Herbivorous prey animals, such as rabbits, mice, horses, and deer, tend to have eyes placed at the sides, the better to scan all around for possible predators. **(b)** Predators and primates tend to have eyes in front, and both can be brought to bear on a target. Each eye gets a slightly different view of the target, which allows size and distance to be judged fairly accurately.

(a)

(b)

Figure 37-13 Male moths **(a)** find females not by sight, but by following airborne scents (pheromones) released by the females. These odors are sensed by receptors on the male's huge antennae, whose enormous surface area maximizes the chances of detecting the female scent. **(b)** When dogs meet they usually sniff each other about the base of the tail. Scent glands there seem to broadcast information about sex (both type and interest in) and status, and influence what behaviors follow.

tion of the environment. Terrestrial vertebrates have two separate chemical senses: one for air-borne molecules, called smell or **olfaction,** and one for chemicals dissolved in an aqueous medium (water or saliva), the sense of **taste.**

Olfaction

In most vertebrates, receptors for olfaction are nerve cells located in tissues lining the back of the nasal cavity (Fig. 37-14). These sensory neurons have hair-

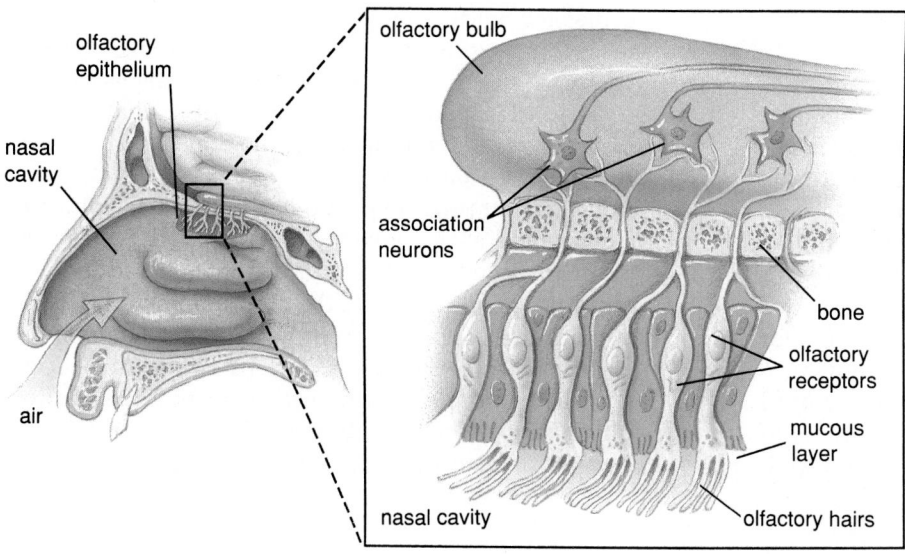

Figure 37-14 The receptors for olfaction in humans are neurons that bear hairlike projections protruding into the nasal cavity. The projections are embedded in a mucus layer, in which odor molecules dissolve before contacting the receptors.

like dendrites that protrude into the cavity and sample incoming air. The olfactory neurons are specialized, bearing different receptor molecules on their hairs. Each responds only to one or a few types of chemicals. Individual odors are perceived by the brain as patterns of activity in particular olfactory neurons.

Taste

The taste buds, located on the tongue, consist of small clusters of taste receptors and supporting cells (Fig. 37-15). Taste receptor cells protrude microvilli through a small pore. Dissolved chemicals enter the pore and bind to special receptor molecules on the microvilli. Although we probably perceive hundreds of distinct tastes, there are only four types of taste receptors: sweet, sour, salty, and bitter. The great variety of tastes is produced in two ways. First, a particular substance may stimulate two or more receptor types to different degrees and taste, for example, "salty–bitter." More important, material being tasted usually also gives off molecules into the air inside the mouth. These odor molecules diffuse to the olfactory receptors (remember, the mouth and nasal passages are connected). The fact that what we call "taste" is mostly due to our sense of smell is clearly illustrated by the blandness of normally tasty foods when a cold plugs up our nasal passages.

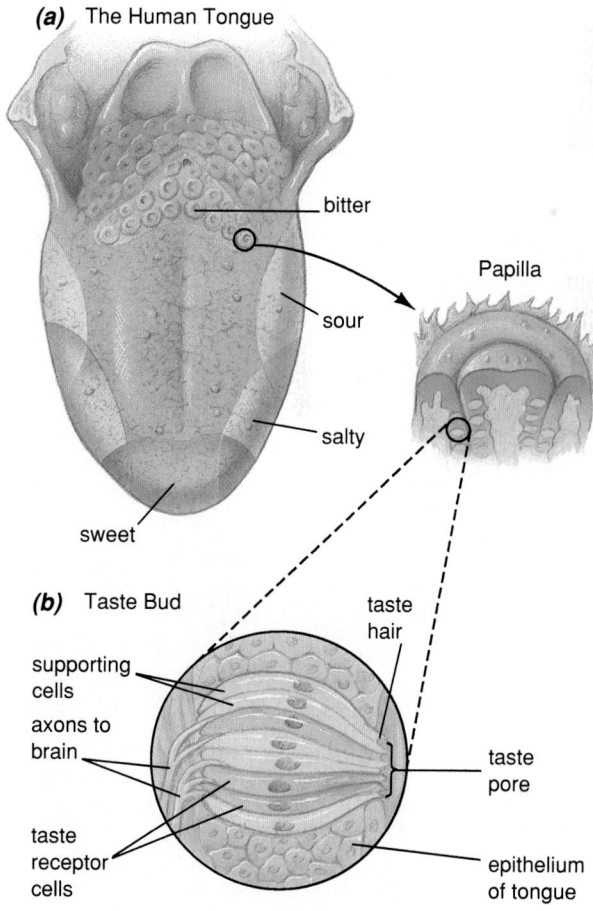

Figure 37-15 (a) The human tongue bears numerous bumps in which are clustered masses of taste buds. **(b)** Each taste bud consists of supporting cells and several taste receptors, whose microvilli bind the tasty molecules. Axons from neurons in the brain receive synapses from the receptor cells and transmit messages to the brain.

Pain

Pain is an unusual sense in that it is somewhat non-specific. Whether you burn, cut, or crush a fingertip, you will feel pain. This gives us a clue to the nature of pain perception: most pain is produced by tissue damage, regardless of the cause. Over the past few years, researchers have found that pain perception is actually a special kind of chemical sense (Fig. 37-16). When cells are broken open by a cut or a burn, for example, their contents flow into the extracellular fluid and blood. The cell contents include enzymes that convert certain blood proteins into a chemical called **bradykinin.** Pain receptors, which are dendrites of specific sensory neurons, have receptor molecules for bradykinin. Binding of bradykinin to these receptors results in action potentials that are interpreted as pain by the brain. Since each part of the body has a separate set of pain neurons that provide input to particular brain cells, the brain knows where the pain is occurring.

Drugs that provide pain relief, such as morphine or Demerol, block synapses in the pain pathways of the brain or spinal cord. In ways that we are just beginning to understand, the brain can modulate its perception of pain through its own narcotic-like **endorphins** (see Chapter 36). In critical situations, such as combat or during escape from a fire, endorphins may allow us to function by blocking our perception of pain until the emergency is over.

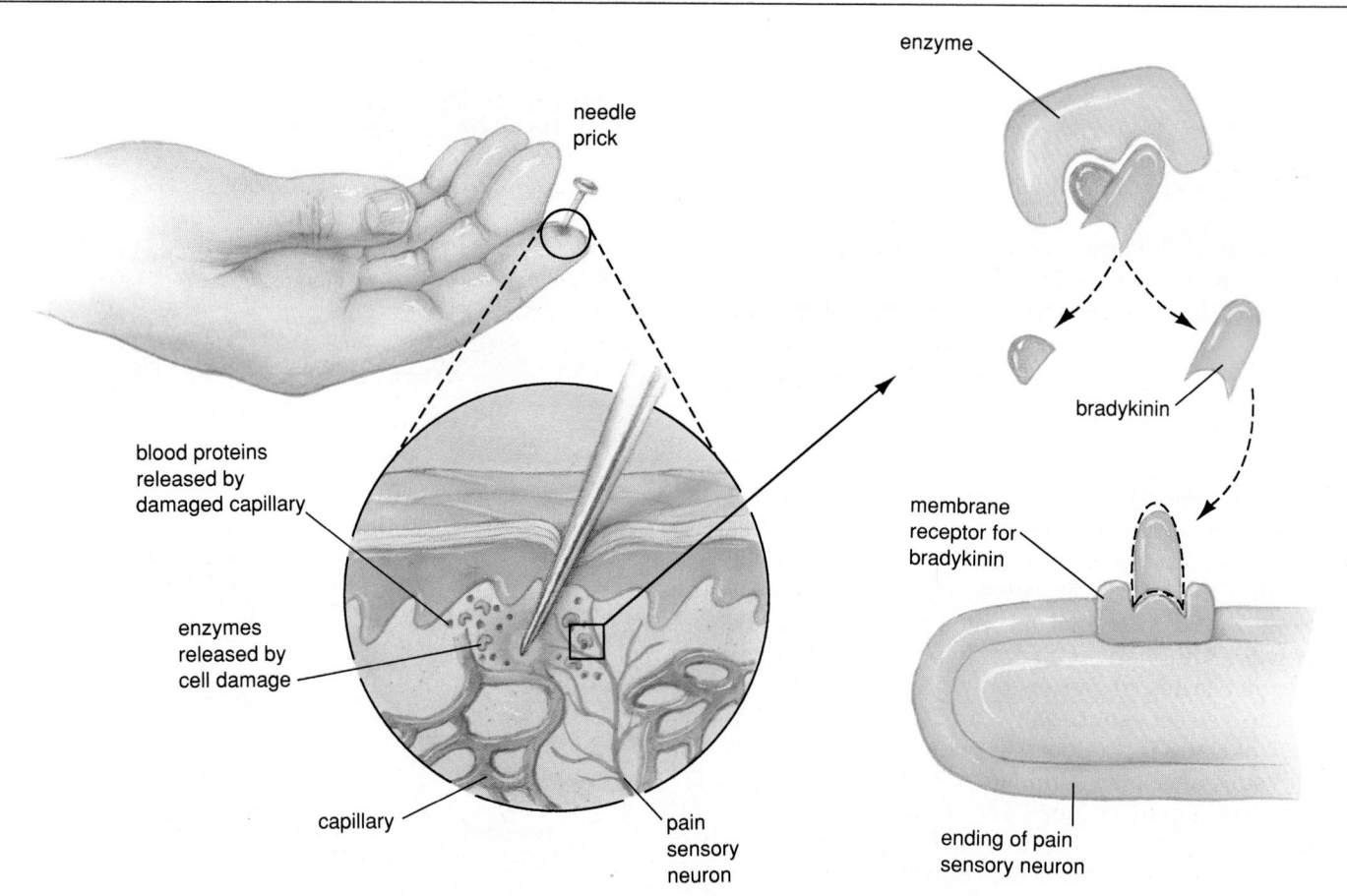

Figure 37-16 Pain perception is a specialized chemical sense. An injury, such as stabbing the finger with a needle, damages both cells and blood vessels. The cells release enzymes that convert certain blood proteins into bradykinin. Bradykinin stimulates pain-sensitive neurons by binding with membrane receptor molecules.

℞ Reflections on Perception

Since birth, you have experienced your surroundings in ways specified by your nervous system and coded by your genes. The word "tree" may evoke a visual image of branching form, green and brown color, perhaps the sound of rustling leaves, the rough feel of bark, or the scent of pine needles. No doubt you feel secure in the belief that your senses provide accurate information about "the way things really are."

But think for a moment about the nature of the signals that the sense organs send to the brain. Even though each sense organ is specialized to respond to a different form of energy from the environment, sense organs don't transmit light, sound, or pressure. Instead, all sense organs transmit a single form of energy: electrical action potentials, virtually indistinguishable from one another. We must conclude, then, that the sensations we perceive as a result of these action potentials (color, light, sound, pain) are *purely a creation of the brain.* The brain translates the action potentials into sensations that it invents to allow us to distinguish properties of objects in our environment.

The subjective sensations that the brain uses to code for light, sound, and so on, arise from different areas of the brain, which are stimulated by the action potentials carried by the different sensory nerves. For example, signals carried by the optic nerve eventually reach the visual cortex of the brain, which creates the sensations of vision. If action potentials from the optic nerve could be directed into the auditory cortex of the brain, we would presumably hear complex chords when we directed our eyes toward a tree or a sunset, but we would see nothing. Our senses do provide us with accurate and useful information about our environment, but the quality of our perceptions arises strictly from the neural connections within the brain.

Because of the incredible complexity of the brain, scientists are a long way from understanding how these subjective perceptions are formed. If life has evolved independently on other planets, it is reasonable to suspect that these alien life forms have evolved entirely different ways of interpreting environmental stimuli. But what of life on Earth? Do all animals, or do all humans, for that matter, perceive the world in the same way? The basic anatomical similarity among human brains, and

our ability to communicate easily with one another about our sensory perceptions, leads us to assume that human brains interpret the world in much the same way. Although scientists are less certain of what other vertebrates perceive, all vertebrates share a common ancestor. This evolutionary relationship has resulted in similarities in the structure of vertebrate sense organs and brains, which suggests that the quality of all vertebrate perceptions may be somewhat similar. However, keep in mind that differing selective pressures on different animals have modified the sensitivity of their receptors and the range of stimuli to which they respond. For example, dogs can hear sounds as high as 40,000 cycles per second, and bees can see light into the near ultraviolet range.

We need not travel to other universes, however, to find modes of perception totally alien to our own. Place yourself in the sensory world of an echolocating bat as it hunts insects on a moonless night. You "see" using sound waves! In fact, the auditory image may give rise to a sensation in the bat's brain that you cannot even imagine, much less interpret. Now imagine yourself as a migrating eel, orienting via the Earth's magnetic field. How do you perceive that field? As a wavering note, or a bluish tinge in the water? Again, more likely the sensation is impossible to imagine.

Invertebrate perception is another mystery. Animals such as insects, snails, and worms possess sense organs that operate on principles similar to our own. However, their brains, where perceptions are created, are totally unlike ours and enormously simpler. Do they create subjective impressions of sight and sound at all? No one knows. Perhaps their sensory neurons respond only to relevant stimuli. Then the receptors could be connected directly to other neurons that produce behavior. This kind of reflexive nervous system would allow appropriate behavioral responses with far fewer neurons, since none would be required to create subjective perceptions.

To use a sensory metaphor, science "opens our eyes to the world." Consider that we share the world with millions of other life forms whose perceptions of it (if we could enter their brains) would make us feel as though we had entered an alien landscape!

SUMMARY OF KEY CONCEPTS

Receptor Mechanisms

Receptors are transducers, converting signals from one form to another. In response to external stimuli, sensory receptor cells produce a graded receptor potential. In sensory neurons, this may cause action potentials whose frequency is proportional to the size of the receptor potential. Alternatively, some receptor cells release transmitter onto a postsynaptic neuron in amounts proportional to the size of the receptor potential. This results in action potentials in the postsynaptic cell whose frequency reflects the intensity of the stimulus. The most common receptor types respond either to mechanical deformation (touch, stretch, hearing, gravity) or to stimuli that influence receptor molecules in the receptor cell membrane (taste, odor, light, and damaging stimuli causing pain).

Thermoreception

Thermoreceptors are free nerve endings located in the skin that respond to infrared radiation. Warm and cold receptors respond to increases and decreases in temperature, respectively. Pit vipers have thermoreceptors concentrated in pit organs near the eyes that allow detection of warm-blooded prey.

Mechanical Senses

The skin contains dendrites of many different types of sensory neurons, which respond to mechanical stimuli such as pressure, touch, or vibration. Joints and muscles have similar receptors, collectively called proprioceptors, sensitive to stretch, that inform the brain about the position of the body. Many internal organs, including the stomach, rectum, and bladder, have stretch receptors that signal fullness.

The Auditory System

Hearing is a modified mechanical sense specialized for reception of vibrations of the air or water. Vertebrate auditory and vestibular systems arose from the lateral line system of fish, used to detect water movement. In the mammalian ear, air vibrates the tympanic membrane, which transmits vibrations to the bones of the middle ear, and then to the oval window of the fluid-filled cochlea. Within the cochlea, vibrations bend the hairs of hair cells, which are receptors located between the basilar and tectorial membranes.

This produces receptor potentials in the hair cells that cause action potentials in the axons of the auditory nerve to the brain.

The inner ear also contains the vestibular apparatus, consisting of bony, fluid-filled chambers with hair cell receptors. Hair cells located in clusters at the base of each of three semicircular canals detect changes in movement in any direction. Below the semicircular canals are chambers, the utricle and saccule, containing hair cells that can detect the pull of gravity.

Vision

All eyes have photoreceptor cells that contain a pigment which, upon absorption of light, causes a receptor potential. Simple eyespots in invertebrates detect the direction and intensity of light, but do not form images. The eye of octopus, in contrast, is nearly as complex as our own, and forms sharp images. Arthropods have evolved compound eyes, with numerous ommatidia, which probably form grainy images.

In the vertebrate eye, light enters the cornea and passes through the pupil to the lens, which focuses an image on the fovea of the retina. Two types of photoreceptor, rods and cones, are found deep in the retina. They produce receptor potentials in response to light. These signals are processed through several layers of neurons in the retina, and finally translated into action potentials in ganglion cells. Ganglion cell axons form the optic nerve, which carries action potentials to the brain. Rods are more abundant and more light-sensitive than cones, and provide vision in dim light. Cones, which are concentrated in the fovea, provide color vision.

The Chemical Senses

Terrestrial vertebrates detect chemicals in the external environment either by olfaction (for airborne sources of chemicals) or taste (chemicals dissolved in saliva). Each olfactory or taste receptor cell type responds to only one or a few specific types of molecules, allowing discrimination among tastes and odors. Olfactory neurons of vertebrates are found in a tissue lining the nasal cavity. Taste receptors are found in clusters called taste buds on the tongue. Pain is a special type of chemical sense, in which sensory neurons respond to chemicals released by damaged cells.

GLOSSARY

aqueous humor (ā'-kwē-us): clear, watery fluid between the cornea and lens of the eye.

auditory canal (aw'-dih-tōry): a canal within the outer ear between the external ear and the eardrum.

auditory nerve: the nerve leading from the mammalian cochlea to the brain, carrying information about sound.

basilar membrane (bas'-eh-lar): a membrane in the cochlea that bears hair cells that respond to the vibrations produced by sound.

bradykinin (brā'-dē-kī'-nin): a chemical formed during tissue damage that binds to receptor molecules on pain nerve endings, giving rise to the sensation of pain.

camera eye: the type of eye found in vertebrates and molluscs, in which a lens focuses an image on the retina.

chemoreceptor: a sensory receptor that responds to chemicals from the environment; chemoreceptors mediate taste and smell.

choroid (kōr'-ōyd): a layer of tissue behind the retina that contains blood vessels and pigment that absorbs stray light.

cochlea (kōk'-lē-uh): a coiled, bony, fluid-filled tube found in the mammalian inner ear, which contains receptors (hair cells) that respond to the vibration of sound.

compound eye: a type of eye found in arthropods, composed of numerous independent subunits, called ommatidia. Each ommatidium apparently contributes a single piece of a mosaic-like image perceived by the animal.

cone: a cone-shaped photoreceptor cell in the vertebrate retina, not as sensitive to light as the rods. The three types of cones are most sensitive to different colors of light, and provide color vision. See also rod.

cornea (kōr'-nē-uh): the clear outer covering of the eye in front of the pupil and iris.

cupula (kūp'-ū-luh): a gelatinous structure found in the lateral line system of fish, and the semicircular canals of the vestibular system of other vertebrates. Hairs are embedded in the cupula, which is deflected by movement of the fluid filling the canals.

eustachian tube (ū-stā'-shin): a tube connecting the middle ear with the pharynx; allows pressure to equilibrate between the middle ear and the outside.

external ear: the fleshy portion of the ear that extends outside the skull: also called the pinna.

eyespot: a simple, lensless eye found in various invertebrates, including flatworms and jellyfish. Eyespots provide information about light vs. dark, and sometimes the direction of light, but cannot form an image.

fovea (fō'-vē-uh): the central region of the vertebrate retina, upon which images are focused. The fovea contains closely packed cones (about 150,000 per mm^2).

ganglion cell (gang'-lē-un): a cell type comprising the innermost layer of the vertebrate retina whose axons form the optic nerve.

hair cell: The receptor cell type found in the inner ear. Hair cells bear hairlike projections. Bending of the hairs between two membranes causes the receptor potential.

inner ear: the innermost part of the mammalian ear, composed of the bony, fluid-filled tubes of the cochlea and the vestibular system.

iris: the pigmented part of the vertebrate eye, surrounding an opening, the pupil.

lateral line organ: a receptor organ in fish and aquatic amphibians that detects water movement. It utilizes clusters of hair cells whose hairs are bent by the deflection of a gelatinous cupula.

lens: a clear object that bends light rays; in eyes, a flexible or movable structure used to focus light upon a layer of photoreceptor cells.

mechanoreceptor: a receptor that responds to mechanical deformation such as is caused by pressure, touch, or vibration.

middle ear: part of the mammalian ear composed of the tympanic membrane and three bones (malleus, incus, stapes) that transmit vibrations from the auditory canal to the oval window.

olfaction (ōl-fak'-shun): a chemical sense, the sense of smell; in terrestrial vertebrates the result of detection of airborne molecules.

ommatidium (ōm-ma-tid'-ē-um): an individual light-sensitive subunit of a compound eye. Each ommatidium consists of a lens and several (usually eight) receptor cells.

optic disk (op'-tik): the area of the retina at which the axons of the ganglion cell merge to form the optic nerve; the blind spot of the retina.

optic nerve (op'-tik): the nerve leading from the eye to the brain, carrying visual information.

otolith membrane (ō'-tō-lith): a gelatinous membrane in which are embedded the hairs of hair cells of the utricle and saccule of the inner ear. The otolith membrane is heavier than the surrounding fluid due to calcium carbonate crystals embedded in it, and thus responds to the pull of gravity.

outer ear: the outermost part of the mammalian ear, including the external ear and auditory canal leading to the tympanic membrane.

oval window: the membrane-covered entrance to the inner ear.

Pacinian corpuscle (pas-in'-ē-an): a receptor of the skin and joints whose ending is wrapped in a layered capsule of connective tissue, and which responds to rapid movement.

photopigment (fō'-tō-pig-ment): a chemical substance in photoreceptor cells (rhodopsin in rods) that changes molecular conformation when struck by light.

photoreceptors: receptor cells that respond to light; in vertebrates, rods and cones.

pit organ: a thermoreceptive organ found in certain poisonous snakes (vipers) consisting of a pair of pits located between each eye and the nose; used to detect warm-blooded prey.

proprioceptor (prō'-prē-ō-cep-tōr): a receptor that monitors the position of the parts of the body and their direction of movement by responding to stretch and pressure.

pupil: the adjustable opening in the center of the iris through which light enters the eye.

receptor: (1) a cell that responds to an environmental stimulus (chemicals, sound, light, pH, etc.) by changing its electrical potential. (2) a protein molecule in a cell membrane that binds to another molecule (hormone, neurotransmitter, odorous compound, etc.) triggering metabolic or electrical changes in a cell.

receptor potential: a graded electrical potential change in a receptor cell produced in response to reception of an environmental stimulus (chemicals, sound, light, pH, heat, cold, etc.). The size of the receptor potential is proportional to the intensity of the stimulus.

retina (ret'-in-a): a layer of nerve tissue at the rear of camera-type eyes, composed of photoreceptor cells plus associated nerve cells that refine the photoreceptor information and transmit it to the optic nerve.

rod: a rod-shaped photoreceptor cell in the vertebrate retina, sensitive to dim light, but not involved in color vision. See also cone.

saccule (sak'-ūle): a portion of the vestibular system responsible for detecting gravity.

sclera (sklāra): a tough white connective tissue layer that covers the outside of the eyeball and forms the white of the eye.

semicircular canal: one of three fluid-filled, semicircular tubes of the inner ear which functions in the detection of rotational movements of the head.

sensory receptor: a cell specialized to respond to particular internal or external environmental stimuli by producing an electrical potential.

taste: a chemical sense; in mammals, perceptions of sweet,

sour, bitter, or salt produced by stimulation of receptors on the tongue.

tectorial membrane (tek-tōr'-ē-ul): one of the membranes of the cochlea, in which the hairs of the hair cells are embedded. During sound reception, movement of the basilar membrane relative to the tectorial membrane bends the cilia.

thermoreceptor: a sensory receptor that responds to changes in temperature.

transducer: a device that converts signals (forms of energy, for example) from one form to another. Sensory receptors are transducers that convert environmental stimuli, such as heat, light, vibration, etc. into electrical signals (such as action potentials) recognized by the nervous system.

tympanic membrane (tim-pan'-ik): the eardrum; a membrane stretched across the opening of the ear, which transmits vibration of sound waves to bones of the middle ear.

utricle (ū'-tri-cul): a portion of the vestibular system responsible for detecting gravity.

vestibular apparatus (ves-tib'-ū-lar): the portion of the inner ear of mammals that responds to gravity and changes in the direction and speed of movement.

vitreous humor (vit'-rē-us): a clear jellylike substance that fills the large chamber of the eye between the lens and retina.

STUDY QUESTIONS

1. Describe the series of events leading from an environmental stimulus to an action potential in a sensory neuron.
2. Describe the anatomy of two types of receptor cell responsive to mechanical deformation; one found in the skin, the other in the inner ear.
3. How are chemicals detected by receptor cells? How could a body distinguish among many different chemicals?
4. For any sense, describe how the brain receives information about the type of stimulus and about stimulus intensity.
5. What are proprioceptors? For what types of behavior are they important?
6. Describe the structure and function of the various parts of the human ear. Do this by tracing a sound wave from

the air outside the ear to the cells causing action potentials in the auditory nerve.
7. How does the structure of the inner ear allow for the perception of pitch?
8. Diagram the overall structure of the human eye. Label the cornea, iris, lens, sclera, retina, and choroid. Describe the function of each structure.
9. How does the lens change shape to allow focusing of faraway objects? What defect makes focusing on faraway objects impossible, and what is this condition called? What type of lens can be used to correct it, and why?
10. List the similarities and differences between rods and cones.
11. Distinguish between taste and olfaction.
12. Describe how pain is signalled by tissue damage.

DISCUSSION QUESTIONS

1. Describe an animal sense that humans lack altogether and explain why it is needed by the animal but not by humans.
2. Discuss what happens when another person speaks and you hear the speech. Fully describe what happens from

the time the air leaves the larynx of the speaker until your brain becomes aware of the sounds.
3. Discuss (I.) why we do not see objects "upside down" and (II.) why we lose the sense of taste when we have a head cold.

SUGGESTED READINGS

Fenton, M. B., and Fullard, J. H, "Moth Hearing and the Feeding Strategies of Bats." *American Scientist*, May/June 1981. Explores coevolutionary interactions between bats and moths.

Horridge, G. A. "The Compound Eye of Insects." *Scientific American*, July 1977. Compound eyes provide fairly grainy vision, yet insects navigate, find food and mates, and even migrate long distances.

Koretz, J. F., and Handelman, G. H. "How the Human Eye Focuses." *Scientific American*, July 1988. Describes focusing in the human eye with an emphasis on the loss of focusing ability that occurs with age.

McKean, K. "Pain." *Discover*, October 1986. Describes the discovery of bradykinin and its role in pain perception.

Montgomery, G. "Color Perception: Seeing with the Brain." *Discover*, December 1988. Describes, in layman's terms, how the higher centers of the brain interpret signals from the receptors to arrive at a perception of color. Beautifully illustrated.

Nathans, J. "The Genes for Color Vision." *Scientific American*, February 1989. Good discussion of the physiology and genetics of color blindness.

Parker, D. E. "The Vestibular Apparatus." Scientific American, *November* 1980. Perception of gravity and motion requires not only the utricle, saccule, and semicircular canals, but also input from other senses.

Ratliff, F. "Contour and Contrast." *Scientific American*, June 1972. Interactions among retinal cells provides us with more visual contrast than actually exists in the light entering our eyes.

Regan, D., Beverley, K., and Cynader, M. "The Visual Perception of Motion in Depth." *Scientific American*, July 1979 (Offprint No. 586). Binocular vision provides us with two views of the world, which the brain fuses into a coherent three-dimensional representation.

Roeder, K. D. *Nerve Cells and Insect Behavior*. 2nd ed. Cambridge, MA: Harvard University Press, 1967. An exceptionally readable record of Roeder's early explorations into the nervous systems of a variety of insects.

38

Action and Support: The Muscles and Skeleton

The muscular and skeletal systems work in close harmony to provide the balance, strength, and flexibility shown by these dancers.

Gleaming muscles bulging, the weightlifter crouches, adjusting his grip on the barbell. Taking three deep breaths, he makes his move. In an explosive but fluid and continuous motion, he raises the weights above his head on outstretched arms. Then slowly, muscles quaking under the strain, he stands upright. The ability of a 200-pound human to lift well over twice his own weight testifies to the functional design and precise interaction of the muscular and skeletal systems. This chapter begins by describing muscle tissue and the complex mechanisms underlying muscle contraction, and then describes the tissues and mechanics of the skeleton.

Muscle

Many animal cells are capable of some type of movement (see Chapter 5). At the cellular level, motion is often based on the relative movements of two types of protein strands, **actin** and **myosin.** In animals, the evolution of muscles has been a remarkable elaboration of this pre-existing system. Animals have a variety of muscle types, each specialized to perform a particular function. Mammals have evolved three distinct types of muscle: **skeletal, cardiac,** and **smooth.** All work on the same basic principles, but differ in function, appearance, and control (Table 38-1).

Skeletal muscle, also called **striated muscle** because of its striped appearance under the microscope, is used to move the skeleton. Skeletal muscles are under the direct control of the nervous system, and can produce contractions ranging from quick twitches (as in blinking) to powerful, sustained tension (as in carrying an armload of textbooks). **Cardiac muscle** is found only in the heart, is spontaneously active, and also has a striped appearance. **Smooth muscle** is found in the walls of the digestive tract and large blood vessels, and produces slow, sustained contractions. Our discussion of muscles begins with and emphasizes skeletal muscle, the most abundant form of muscle in the body, then provides an overview of cardiac and smooth muscle.

The Anatomy of Skeletal Muscle

Individual muscle cells, called **muscle fibers,** are among the largest cells in the human body. Ranging from 10 to 100 micrometers in diameter, each muscle fiber runs the entire length of the muscle, which may be as much as 35 centimeters (about 14 in.) in a human thigh (Fig. 38-1a, b). Each muscle fiber, in

turn, contains many individual contractile subunits, the **myofibrils,** extending from one end of the fiber to the other. Each cylindrical myofibril is surrounded by **sarcoplasmic reticulum.** Like the endoplasmic reticulum from which it is derived, the sarcoplasmic reticulum is a series of double sheets of membrane forming interconnected hollow tubes (Fig. 38-1c). The fluid within the sarcoplasmic reticulum stores high concentrations of calcium ions. Deep indentations of the muscle cell membrane, called **transverse tubules** or **T tubules,** extend down into the muscle fiber, passing very close to portions of the sarcoplasmic reticulum. This arrangement of T tubules and sarcoplasmic reticulum is crucial to the control of muscle contraction, as described later.

Within each myofibril is a beautifully precise arrangement of filaments of actin and myosin, organized into subunits called **sarcomeres** (Fig. 38-1d, e, f). Sarcomeres are attached end-to-end throughout the length of the myofibril; their junction points are called **Z lines.** Attached to the Z lines are strands composed of actin plus two accessory proteins. These three proteins form the **thin filaments.** Suspended between the thin filaments are **thick filaments** composed of myosin protein. The thick and thin filaments are lined up in all the myofibrils, giving the cell its striped appearance (Fig. 38-1f). The strands of myosin extend small arms, called **cross-bridges,** contacting the thin filaments. The complex structure of the thin filament is crucial to the regulation of muscle contraction. The actin protein is formed from a double chain of subunits resembling a twisted double strand of pearls. Each subunit has a binding site for a myosin cross-bridge. In a relaxed muscle, however, these sites are covered by thin strands formed by one type of accessory protein, held in place by a second globular accessory protein (Fig. 38-1e). These accessory proteins prevent the myosin cross-bridges from attaching.

The Sliding Filament Model of Muscle Contraction

When a muscle contracts, the accessory proteins of the thin filament are moved aside, exposing the binding sites on the actin. As soon as the sites are exposed, myosin cross-bridges attach. Using energy from splitting adenosine triphosphate (ATP), the cross-bridges repeatedly bend, release, and reattach farther along, much like a sailor pulling in an anchor line hand over hand (Fig. 38-2a). The thin filaments are pulled past the thick filaments, shortening the sarcomere and contracting the muscle (Fig. 38-2b).

Table 38-1   Location, Characteristics, and Functions of the Three Muscle Types

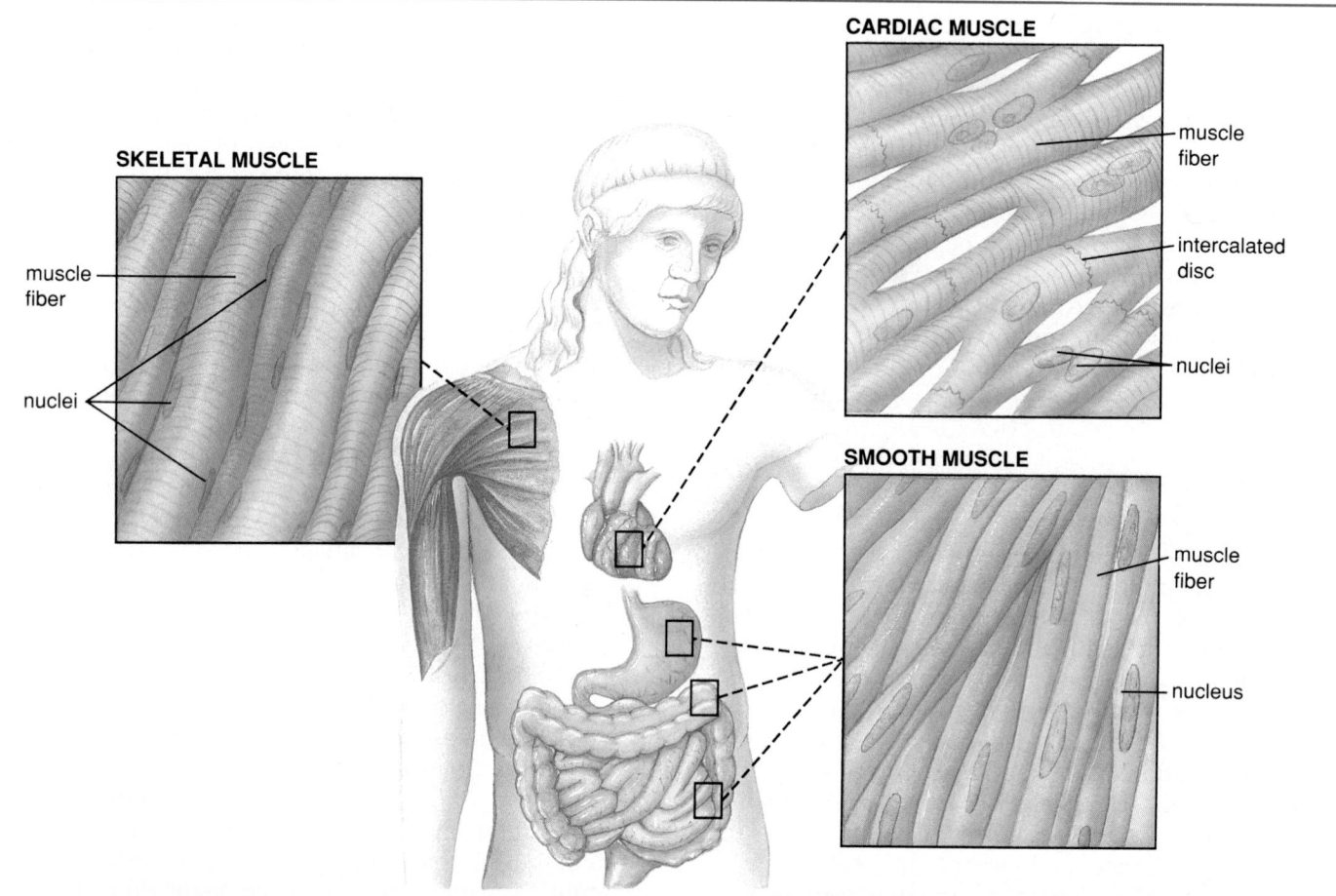

| Property | Type of Muscle | | |
	Smooth	Cardiac	Skeletal
Muscle appearance	Unstriped	Irregular stripes	Regular stripes
Cell shape	Spindle	Branched	Spindle or cylindrical
Number of nuclei	One per cell	Many per cell	Many per cell
Speed of contraction	Slow	Intermediate	Slow to rapid
Contraction caused by	Spontaneous, stretch, nervous system, hormones	Spontaneous	Nervous system
Function	Controls movement of substances through hollow organs	Pumps blood	Moves the skeleton
Voluntary control	Usually no[a]	Usually no[a]	Yes

[a]Smooth and cardiac muscle normally contract without conscious control. However, in some cases their contractions may be initiated or modified voluntarily (for example, heart rate can be voluntarily slowed after biofeedback training, and bladder contractions are initiated consciously).

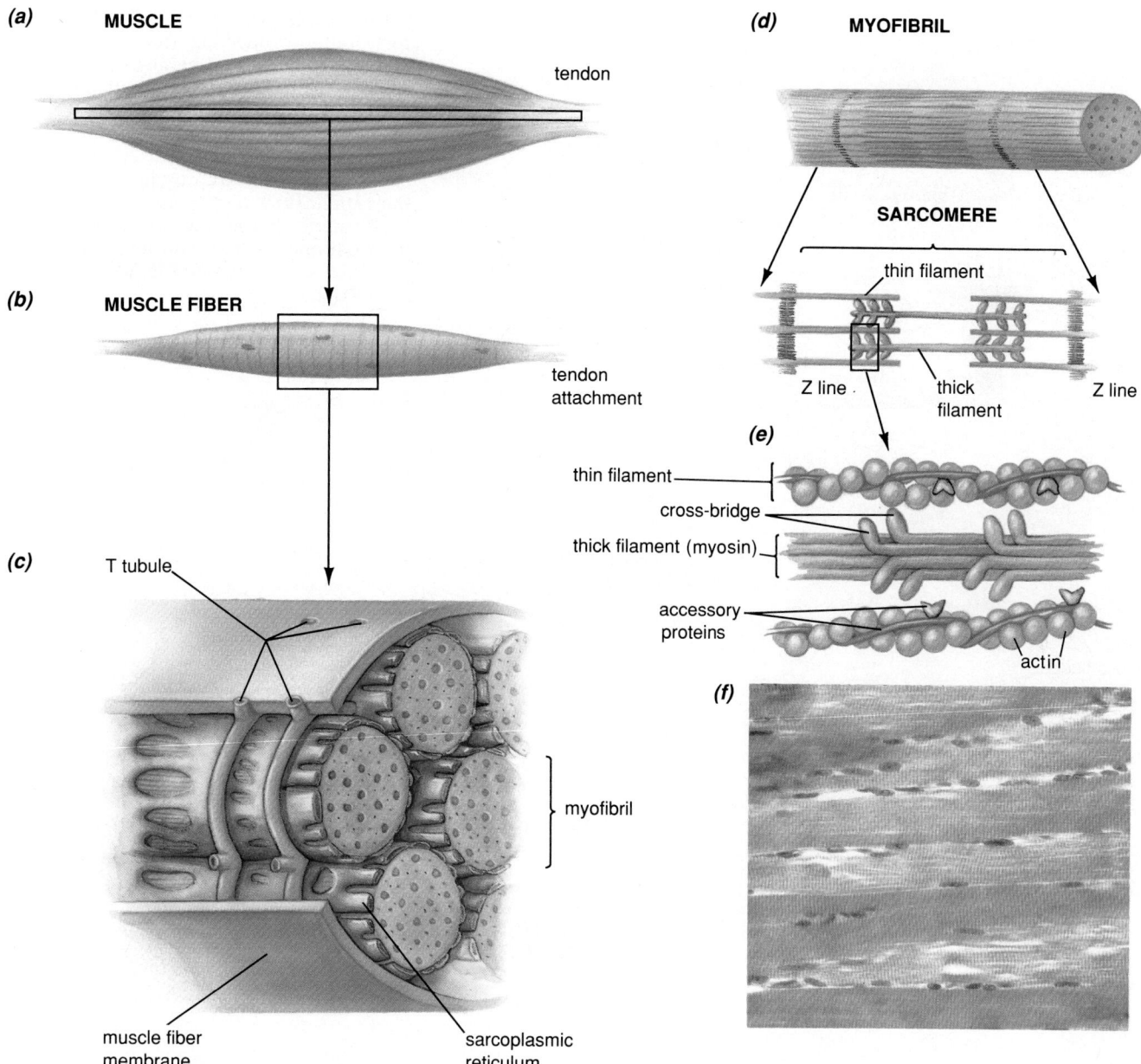

(a) MUSCLE

tendon

(b) MUSCLE FIBER

tendon
attachment

(c)

T tubule

myofibril

muscle fiber
membrane

sarcoplasmic
reticulum

(d) MYOFIBRIL

SARCOMERE

thin filament

Z line

thick
filament

Z line

(e)

thin filament

cross-bridge

thick filament (myosin)

accessory
proteins

actin

(f)

Figure 38-1 A muscle **(a)** is made up of individual muscle cells, called fibers, shown in **(b)**. The sarcoplasmic reticulum subdivides each muscle fiber into smaller cylinders called myofibrils **(c)**. The myofibril consists of a series of subunits called sarcomeres, attached end to end, shown in **(d)**. Within each sarcomere are alternating thick and thin filaments **(e)** that can be connected by cross-bridges (projections of the myosin molecules comprising the thick filaments). In vertebrate skeletal muscles, the actin and myosin filaments are in register across the muscle fiber, giving it a striped appearance, as seen in **(f)**.

The Control of Muscle Contraction

Muscle contracts when the binding sites on actin are exposed, and relaxes when the binding sites are covered by accessory thin filament proteins. But what

regulates the position of these accessory proteins? The answer is the concentration of calcium ions around the filaments, which in turn is under the control of the nervous system.

In many respects, skeletal muscle cells are much

(a)

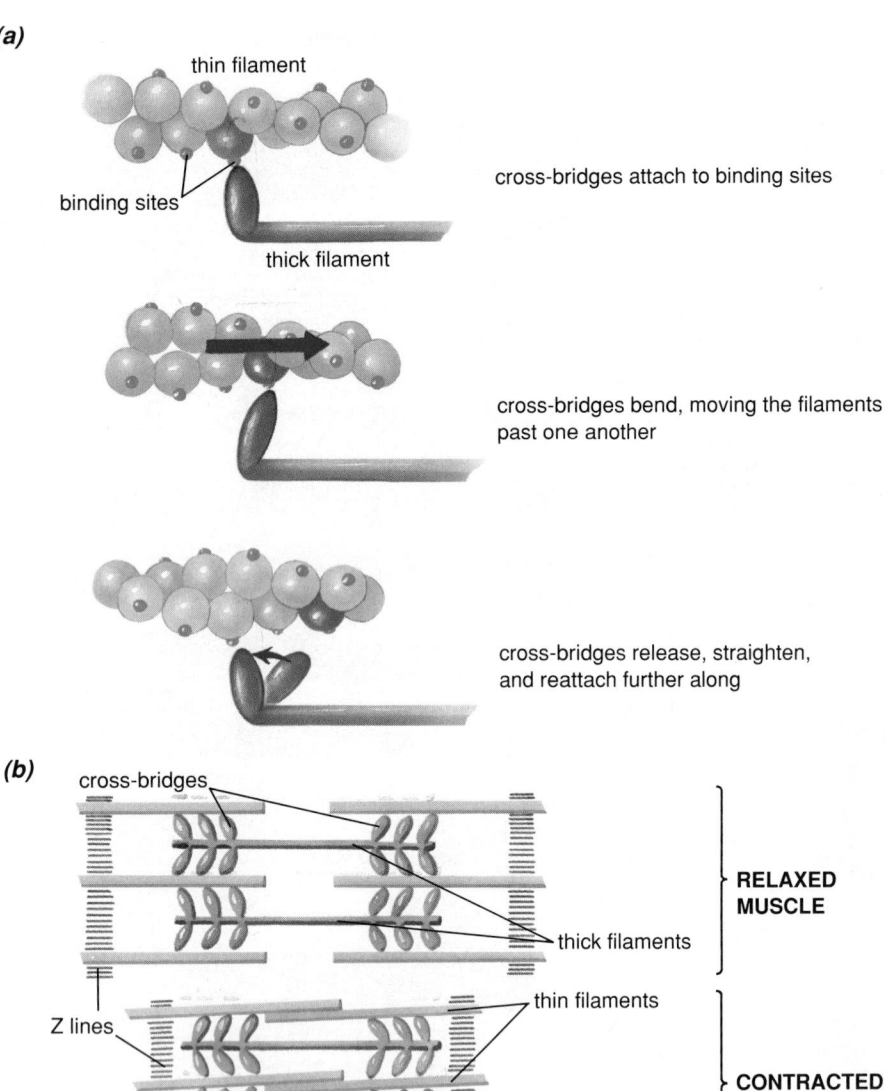

cross-bridges attach to binding sites

cross-bridges bend, moving the filaments past one another

cross-bridges release, straighten, and reattach further along

(b)

Figure 38-2 Muscle movement. **(a)** The cross-bridges connecting actin to myosin swivel as if on hinges, pulling the actin filaments (which are attached to the ends of the sarcomere) toward the middle. Repeated cycles of attachment, swivelling, release, and reattachment result in muscle contraction. **(b)** Muscle contraction causes the thick and thin filaments to slide past one another, shortening the individual sarcomeres and hence the muscle cell.

like neurons. They have resting potentials, they are stimulated by synaptic contact with neurons, and they produce action potentials when stimulated above threshold. Skeletal muscles contract as a result of their own action potentials, which are caused by synaptic stimulation from motor neurons. The cell bodies of the motor neurons (most of which reside in the spinal cord) send axons out the spinal nerves to the muscles, where they form synapses on the muscle fibers (Fig. 38-3). These **neuromuscular junctions** work much like any other synapse: when an action potential reaches the synapse, the motor neuron releases a neurotransmitter (acetylcholine) that diffuses across the synaptic cleft to receptors on the muscle fiber membrane, producing an excitatory postsynaptic potential. Neuromuscular junctions differ from most other synapses in that every action potential in a motor neuron causes a large enough excitatory synaptic potential to evoke a muscle action potential (Fig. 38-4).

The muscle action potential invades the interior of the muscle cell by passing down the T tubules (see Fig. 38-1c). The change in potential causes the sarcoplasmic reticulum to release calcium ions into the myofibrils. Calcium ions bind to the globular accessory proteins on the thin filament, altering their shape and causing them to pull the thin-stranded accessory proteins off the binding sites on the actin. This allows the myosin cross-bridges to attach to the actin, initiating contraction. As soon as the action potential fades away, active transport proteins in the sarcoplasmic reticulum membrane pump the calcium

(a)

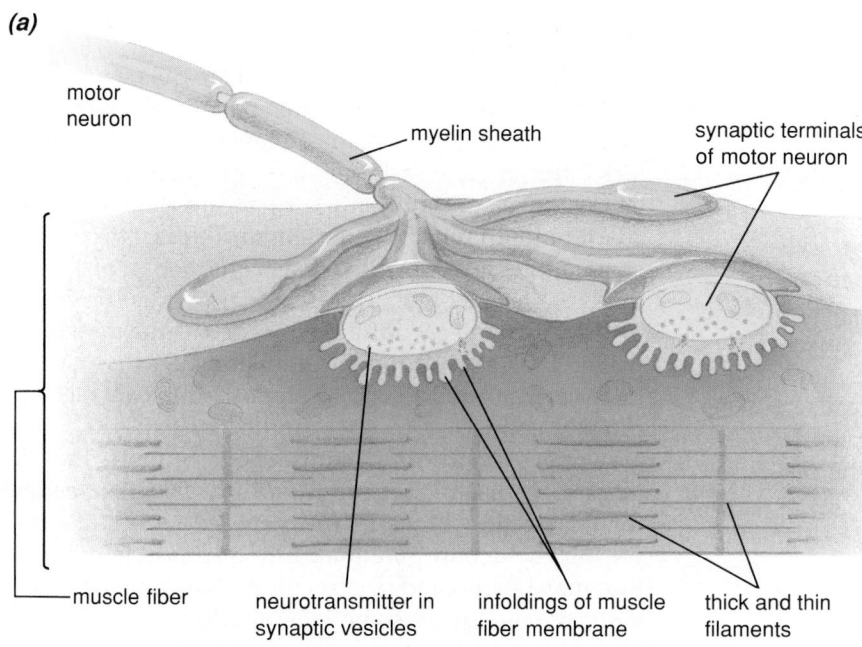

motor
neuron

myelin sheath

synaptic terminals
of motor neuron

muscle fiber

neurotransmitter in
synaptic vesicles

infoldings of muscle
fiber membrane

thick and thin
filaments

Figure 38-3 The neuromuscular junction. **(a)** Diagram of a neuromuscular junction in cross section. Action potentials in the motor neuron cause transmitter release from the synaptic terminals. Transmitter binds to receptors on the muscle cell membrane, which is folded beneath the terminal to allow more surface area for receptors. **(b)** A scanning electron micrograph shows motor neuron terminals synapsing on muscle fibers.

(b)

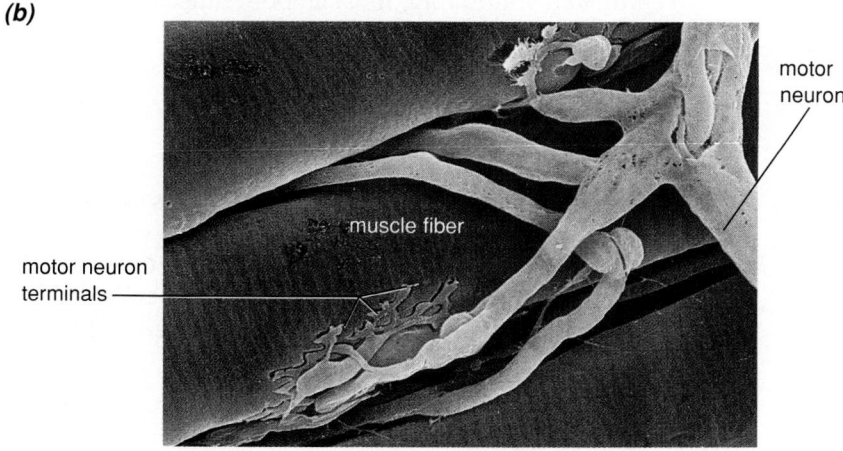

muscle fiber

motor
neuron

motor neuron
terminals

back inside the reticulum. In the absence of calcium, the accessory proteins move back into place on the thin filaments, preventing further cross-bridge binding and ending the contraction.

If each action potential in a motor neuron elicits an action potential in a muscle fiber, causing all its sarco- meres to contract, how does the nervous system control the strength and degree of muscle contraction? **The strength and degree of muscle contraction depends on the number of muscle fibers stimulated and the frequency of action potentials in each fiber.**

Most motor neurons innervate more than one mus-

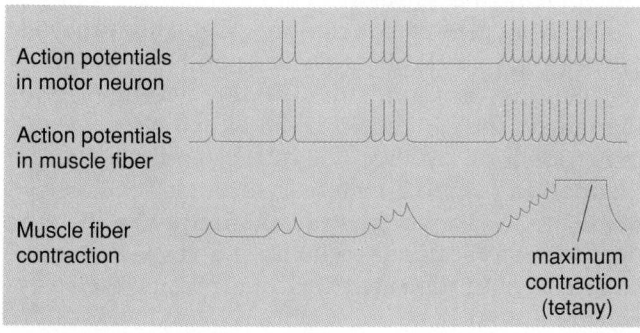

Action potentials
in motor neuron

Action potentials
in muscle fiber

Muscle fiber
contraction

maximum
contraction
(tetany)

◀ **Figure 38-4** The strength of contraction of a single muscle fiber depends on how fast it is stimulated by its motor neuron. Each individual muscle action potential causes only a partial contraction of the muscle fiber. Up to a point, more rapid action potentials cause the individual contractions to summate, producing larger overall contractions of the muscle. Continued rapid firing of motor neurons gives rise to tetany: sustained, maximal contraction of the muscle.

cle fiber. The group of fibers on which a single motor neuron synapses is called a **motor unit.** The number of muscle fibers in a motor unit varies from muscle to muscle. Large muscles used for gross movement, such as those of the thigh or buttocks, may have hundreds of muscle fibers in each motor unit. In muscles used for fine control of small body parts, such as those of the lips, eyes, and tongue, only a few muscle cells may be innervated by each motor neuron. Therefore, when a single motor neuron fires an action potential, it may cause contraction of a few muscle cells or of many, depending on the size of the motor unit. In any case, one action potential doesn't cause a muscle cell to contract fully. After an action potential occurs, the concentration of calcium around the filaments is high for just a short time, and only a few cross-bridge movement cycles can occur before the calcium is pumped away.

How much the cells of a given motor unit contract also depends on the frequency at which the motor neurons are firing: higher frequency firing produces stronger contractions, since the contractions produced by each action potential add to one another. This build-up of contraction caused by rapid firing of the motor neuron is called **summation.** If rapid firing is prolonged, the muscle produces a sustained maximal contraction called **tetany** (see Fig. 38-4). The weightlifter's muscles in the chapter opener photo are in tetany, as are yours with an armful of books like this one.

Cardiac Muscle

Cardiac muscle is found only in the heart. Cardiac muscle, like skeletal muscle, is striated due to the regular arrangement of sarcomeres with their alternating thick and thin filaments. Action potentials spreading into the cell via the T tubules cause release of calcium from the sarcoplasmic reticulum, but, in contrast to skeletal muscle, calcium also enters from the extracellular fluid. Unlike skeletal muscle fibers, cardiac muscle fibers can initiate their own contractions. This quality is particularly well-developed in the specialized cardiac muscle fibers of the sinoatrial node, which serves as the heart's pacemaker (see Chapter 30). Action potentials originating in the pacemaker spread rapidly throughout the heart. This is accomplished by **intercalated discs,** specialized areas in which membranes of adjacent muscle fibers are connected with numerous gap junctions. Intercalated discs allow electrical potentials to travel from one cell to the next, synchronizing their contractions (see Table 38-1). This coupling is absent in skeletal muscle, where each cell is individually innervated by a branch of a motor neuron.

Smooth Muscle

Smooth muscle surrounds blood vessels and most hollow organs, including the uterus, bladder, and digestive tract. As its name suggests, smooth muscle lacks the regular arrangement of sarcomeres that characterizes skeletal and cardiac muscle (see Table 38-1). Smooth muscle generally produces either slow, sustained contractions (such as constriction of the arteries to elevate blood pressure during times of stress) or slow, wave-like contractions (such as the peristaltic waves that move food through the digestive tract). Like cardiac muscle, most smooth muscle cells are directly connected to one another by gap junctions, allowing synchronized contraction. Smooth muscle lacks sarcoplasmic reticulum; all the calcium needed for contraction flows in from the extracellular fluid during the action potential. Smooth muscle contraction may be initiated by stretch, by hormones, by nervous signals, or by some combination of these stimuli. Although contractions of the bladder can be initiated voluntarily, most smooth muscle contraction is under involuntary control.

Cartilage and Bone

The living framework of the body, the skeleton, is composed primarily of two types of tissue: cartilage and bone. Both cartilage and bone are types of rigid connective tissue. Both consist of living cells embedded in a matrix of a protein called **collagen** (see Chapter 25).

Cartilage

Cartilage plays many roles in the human skeleton. It covers the ends of bones at joints, supports the flexible portion of the nose and external ears, connects the ribs to the sternum (breastbone), and provides the framework for the larynx, trachea, and bronchi of the respiratory system. It forms tough pads that act as shock absorbers. These are found in the knee joints and also form the **intervertebral discs** between the vertebrae of the backbone.

The living cells of cartilage are called **chondrocytes.** These secrete a flexible, elastic, nonliving matrix of collagen that surrounds the chondrocytes and forms the bulk of the cartilage (Fig. 38-5). No blood vessels penetrate into cartilage; chondrocytes must rely on gradual diffusion of materials through the collagen matrix to exchange wastes and nutrients. As you might predict, cartilage cells have a very slow metabolic rate, and damaged cartilage repairs itself very slowly, if at all.

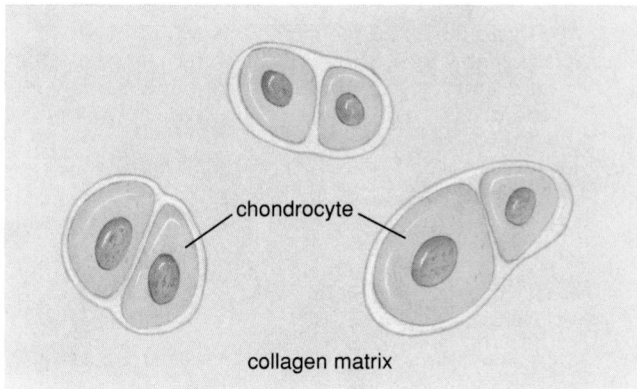

Figure 38-5 In cartilage, the cartilage cells, or chondrocytes, are embedded within an extracellular matrix of the protein collagen, which they secrete.

During the development of the embryo, the skeleton is first formed from cartilage that is later replaced by bone (Fig. 38-6).

Bone

Bones, such as those supporting your arms and legs, consist of an outer shell of **compact bone,** with **spongy bone** in the interior (Fig. 38-7). Compact bone is dense and strong and provides an attachment site for muscle. Spongy bone is lightweight, rich in blood vessels, and highly porous. Bone marrow, where blood cells are formed, is found in cavities of spongy bone.

Bone is the most rigid form of connective tissue. While bone resembles cartilage, the collagen fibers of bone are hardened by deposits of calcium phosphate. In contrast to cartilage, bone is well-supplied with blood capillaries. There are three types of cells associated with bone: **osteoblasts** (bone-forming cells), **osteocytes** (mature bone cells), and **osteoclasts** (bone-dissolving cells). Early in development, when bone is replacing cartilage, the cartilage is invaded by osteoclasts that dissolve it, then osteoblasts replace the cartilage with bone.

Bone-producing osteoblasts form a thin layer covering the outside of the bone. The osteoblasts secrete a hardened matrix of bone and gradually become entrapped within it. They then stop secreting matrix, and become **osteocytes.** Osteocytes are nourished by nearby capillaries, and are connected to other osteocytes by processes that extend through narrow channels in the bone. Although unable to produce more bone, osteocytes may secrete substances that control the continuous remodeling of bone.

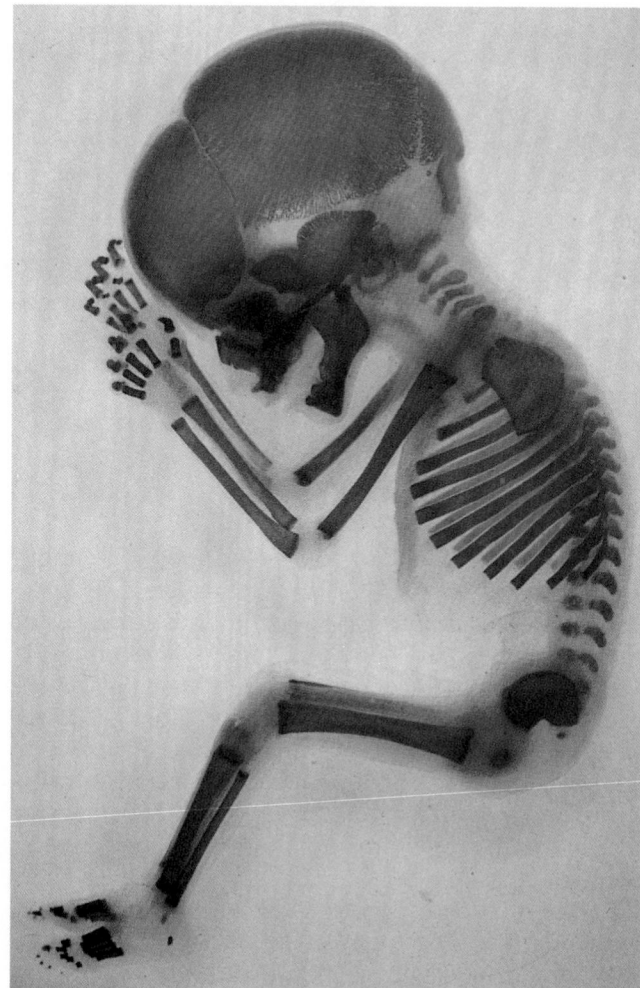

Figure 38-6 Bone is stained magenta, while cartilage appears clear in this 16-week old human embryo.

Remodeling of Bone

Each year 5% to 10% of all the bone in your body is dissolved away and replaced. This allows your skeleton to subtly alter its shape in response to the demands placed upon it, for example by increasing the thickness of bones that carry heavy loads or are subjected to extra stress. For example, archaeologists excavating skeletons from Pompeii were able to identify archers because the bones of their right and left arms differed considerably in thickness. Normal stresses are a major factor in maintaining bone strength. If an arm or leg is immobilized in a cast and protected from these stresses, its bones rapidly loose significant amounts of calcium.

Bone remodeling causes bone to be replaced as it ages and becomes brittle. As people age, the remod-

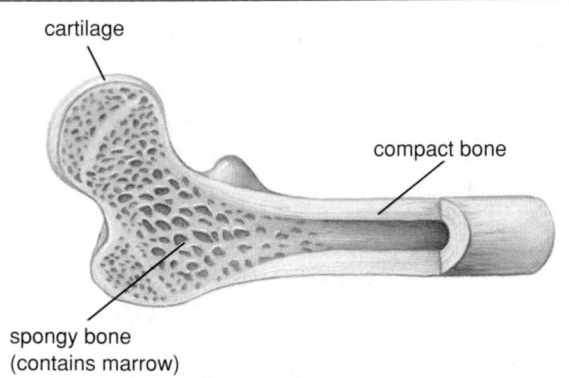

cartilage

compact bone

spongy bone
(contains marrow)

Figure 38-7 A bone such as is found in the arms and legs, showing the outer compact bone and inner spongy bone, where bone marrow is found.

eling process slows, and bones tend to become more fragile as a result (see "Health Watch: Osteoporosis—The Hidden Crippler"). The continuous turnover of bone also allows the body to maintain constant levels of calcium in the blood; calcium from bones is retained in the blood if blood calcium drops, but returned to bone if blood calcium levels are adequate or high. This process is regulated by hormones; calcitonin and parathormone cause bones to sequester and to release calcium into the blood, respectively (see Chapter 35).

Bone remodeling is the result of the coordinated activity of two types of cells: **osteoclasts** that dissolve bone, and osteoblasts that rebuild it. Osteoclasts cling to the bone surface, secreting acids and enzymes that dissolve the hard matrix. Working in small groups, osteoclasts tunnel into the bone, creating channels. These channels are invaded by capillaries and by osteoblasts. The osteoblasts fill the channel with concentric deposits of new bone matrix, leaving only a small opening for the capillary. As a result of this process, in cross-section, hard bone is made up of tightly packed units, called **Haversian systems,** each consisting of concentric layers of bone with embedded osteocytes. The concentric deposits surround a central canal through which traverses a capillary (Fig. 38-8). Osteoclasts and osteocytes also play a crucial role in the repair of bone fractures, as described in "Health Watch: Sticks and Stones—The Repair of Broken Bones."

The Skeleton

A skeleton can be broadly defined as a supporting framework. Within the animal kingdom, skeletons come in three radically different forms: **internal endoskeletons, hydrostatic skeletons** made of fluid,

and **exoskeletons** on the outside of the animal. **Endoskeletons,** the internal skeletons of humans and other vertebrates, are found only in echinoderms and chordates (see Chapter 22) and are actually the least common type of skeleton.

The **hydrostatic skeletons** of worms, molluscs, and cnidarians are the simplest, consisting of a fluid-filled sac (Fig. 38-9a). Fluid, which is incompressible, provides excellent support, but since it is formless, these animals rely on surrounding muscles in the body wall to determine their shape. The sinuous movements of a burrowing earthworm, alternately extending to stringlike thinness then fattening as it contracts, provide an excellent illustration of the flexibility of hydrostatic skeletons.

Exoskeletons (literally, "outside skeletons"), encase the bodies of arthropods (such as spiders, crustaceans, and insects). Exoskeletons vary tremendously in thickness and rigidity, from the thin flexible covering of many insects and spiders to the armorlike covering of many crustacea (Fig. 38-9b). All exoskeletons are thin and flexible at the joints, allowing complex and dexterous movements such as those of a web-spinning spider. Exoskeletons and hydrostatic skeletons are discussed further in Chapter 22.

The Vertebrate Skeleton

The bony endoskeleton of most vertebrates serves a wide variety of functions.

1. **The skeleton provides a rigid framework that supports the body and protects the internal organs.** The central nervous system, for example, is almost completely enclosed within the skull and vertebral column, while the rib cage protects the lungs and the heart with its major blood vessels.

2. **Bones produce red blood corpuscles, white**

HEALTH WATCH
Osteoporosis—The Hidden Crippler

As many as 20 million Americans, most of them women past the age of menopause, are victims of an insidious and preventable form of bone loss called **osteoporosis** (literally, "porous bones"; Fig. E38-1a). Bone density increases steadily from birth through middle adulthood, reaching a peak at around age 35. From this point on, the activity of osteoclasts exceeds that of osteoblasts, and bone density begins a slow, natural decline.

Women are eight times as likely to suffer from osteoporosis as men. Why? One reason is that the bones of women are about 30% less massive than those of men to start with, so they can afford to lose less bone. Another factor is dietary calcium, which is lower in women's diets than in men's. Two thirds of women between the ages of 18 and 30 get less than the RDA (Recommended Daily Allowance) of calcium, and women tend to consume even less calcium as they become older. As a result, when women reach the age when bone loss begins naturally, their bones may already be more fragile than they should be. Another factor unique to women is the role of estrogen. In women, estrogens stimulate osteoblasts and help maintain bone density. After menopause, when estrogen production drops dramatically, bone loss is accelerated. Half of all women over age 65 are estimated to have some degree of osteoporosis.

Another important factor contributing to osteoporosis is lack of weight-bearing exercise such as walking, dancing, or running. Bones thrive on moderate stress. Being bedridden (or being weightless, as the Skylab astronauts discovered) results in rapid loss of bone calcium. In contrast, exercise, even in elderly people, can reverse bone loss and even increase bone mass.

Although some bone loss is normal, in osteoporosis the loss is sufficient to weaken the bones, making them vulnerable to fractures and deformities. Frequently, the vertebrae of osteoporosis victims compress, causing a hunch-backed appearance (Fig. E38-1b). In extreme cases, simple activities such as lifting a shopping bag, opening a window, or sneezing can break a bone. Of women living to age 85, nearly one third will fracture a hip weakened by osteoporosis. As a result of hip fractures, the elderly often become much less self-sufficient and active, and frequently die of complications such as pneumonia.

To slow the progress of osteoporosis, supplemental hormones (estrogen or a combination of estrogen and progesterone) may be prescribed for women past menopause. Sodium fluoride, in combination with

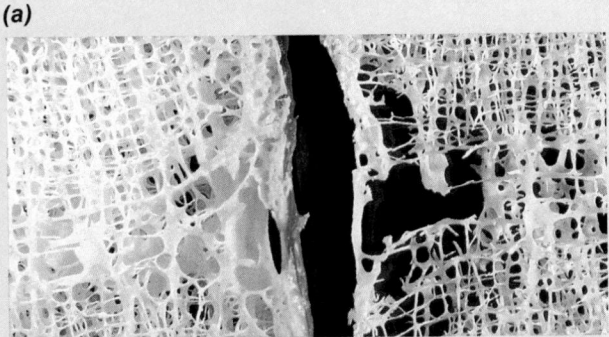

Figure E38-1 (a) Cross-section of a normal bone (left) compared with a bone from a woman suffering from osteoporosis (right). **(b)** The devastating effects of osteoporosis extend beyond the obvious deformities. Its victims are also at high risk for bone fractures.

calcium and vitamin D supplements, is sometimes used, under a doctor's care, to increase bone deposition. However, such treatments may have undesirable side effects. Fortunately, much of the pain, incapacitation, and expense (estimated at $3.8 billion per year) caused by fractures due to osteoporosis can be prevented. A combination of regular exercise and adequate dietary calcium will assure that bone mass is as high as possible before natural, age-related losses begin, and will also minimize such losses in old age. Adequate amounts of vitamin D, which is crucial to the proper metabolism of calcium, are also necessary. Alcoholism and smoking also contribute to osteoporosis. In fact, easy measures taken to prevent osteoporosis can enhance the quality of life at any age.

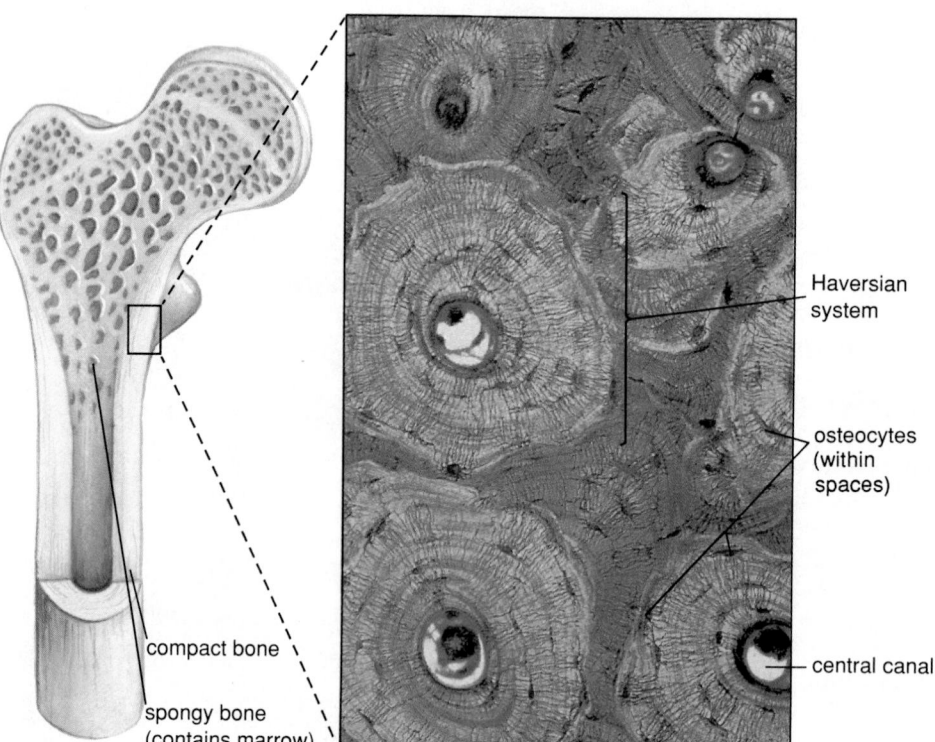

Figure 38-8 In this micrograph, Haversian systems are clearly visible. Each includes a central canal containing a capillary. The capillary nourishes the osteocytes, seen embedded in the concentric rings of bone material.

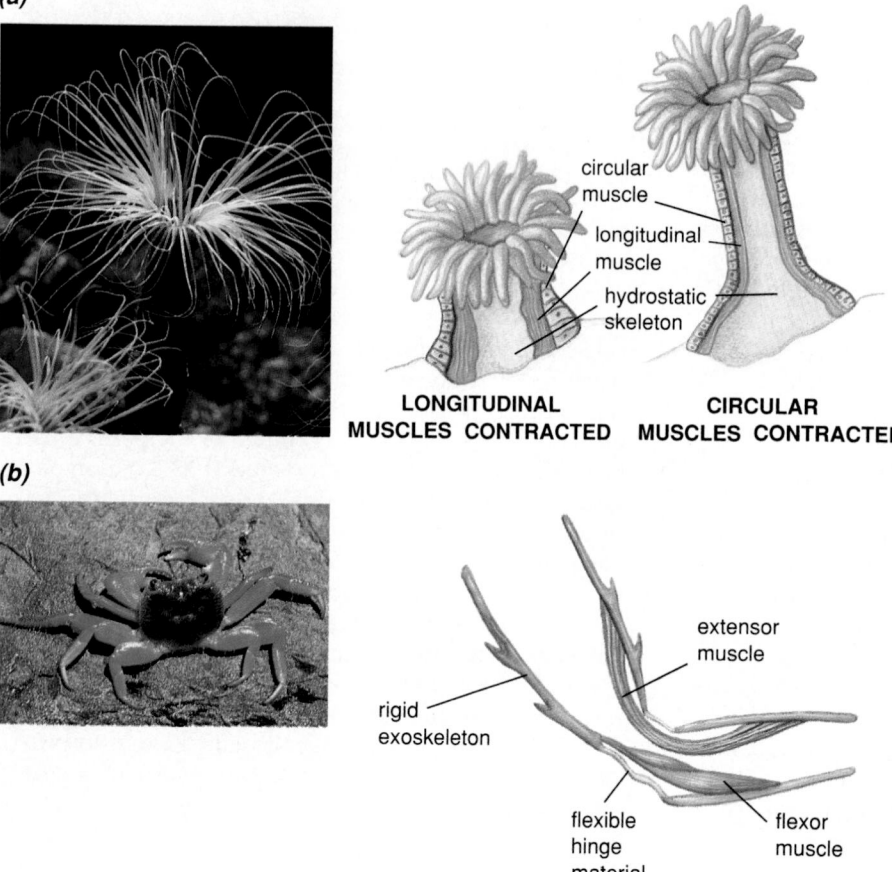

Figure 38-9 (a) *Hydrostatic skeletons:* The skeleton of worms, many molluscs and cnidarians such as this anemone is essentially a fluid-filled tube with soft walls. Two layers of muscles are arranged perpendicularly to one another: a layer of circular muscles forms a band around the circumference of the tube and a layer of longitudinal muscles runs lengthwise. Since fluids are incompressible, if the circular muscles contract, the animal will become long and thin; if the longitudinal muscles contract, it will become short and fat. **(b)** *Exoskeletons:* Arthropods have armorlike skeletons on the outsides of their bodies. Joints allow movement, produced by pairs of antagonistic muscles spanning the joint.

HEALTH WATCH
Sticks and Stones—The Repair of Broken Bones

With a thud accompanied by a sickening *crack,* a boy falls from high in a tree amidst a crackling of small branches and a shower of leaves. The rapidly swelling, awkwardly bent forearm confirms his parents' worst fears: a bone is broken. Let's examine the sequence of events that occur during the next 6 weeks or so, as the bone heals.

First, blood from ruptured vessels forms a large clot surrounding the break. Phagocytic cells and osteoclasts in the blood ingest and dissolve the cellular debris and bone fragments (Fig. E38-2).

Figure E38-2

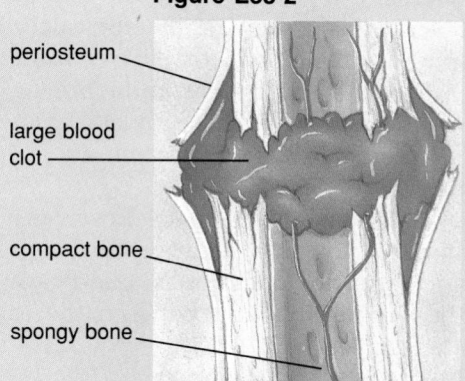

periosteum

large blood clot

compact bone

spongy bone

Second, bones are normally covered with a thin layer of connective tissue (the **periosteum**), rich in capillaries, osteoblasts, and osteoblast-forming cells. A fracture ruptures the periosteum and stimulates the production and release of numerous osteoblasts. These, in conjunction with cartilage-forming cells, secrete a porous mass of bone and cartilage called a **callus** surrounding the break. The callus replaces the original blood clot, and holds the ends of the bones together while remodeling processes re-form the original shape of the bone (Fig. E38-3).

Figure E38-3

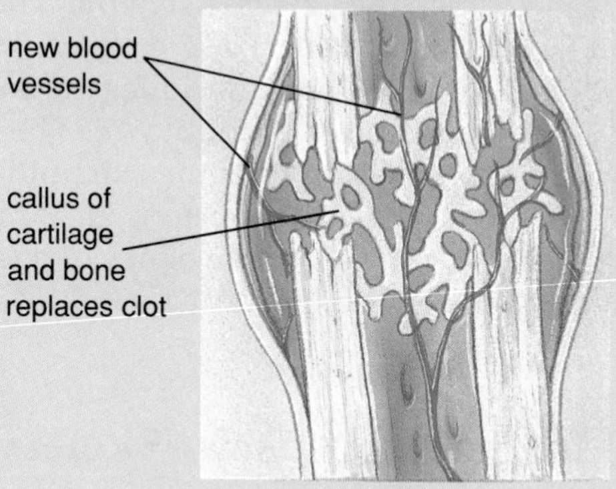

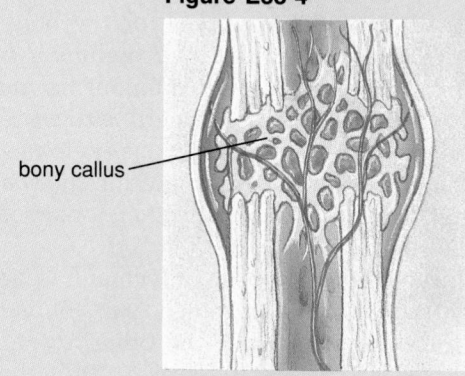

new blood vessels

callus of cartilage and bone replaces clot

Third, osteoclasts, osteoblasts, and capillaries invade the callus. Nourished by the capillaries, osteoclasts break down the cartilage, while osteoblasts replace it with bone (Fig. E38-4).

Figure E38-4

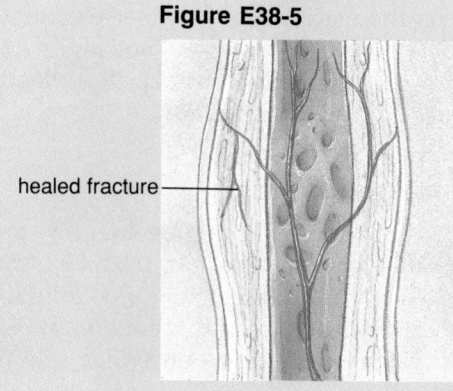

bony callus

Finally, osteoclasts remove excess bone, restoring the original shape (E38-5).

Figure E38-5

healed fracture

blood cells, and platelets (see Chapter 30). In adults, these cells of the circulatory system are produced by red bone marrow, found in porous areas of bone in the sternum, ribs, upper arms and legs, and hips.

3. **Bone serves as a storage site for calcium and phosphorus.** Bone contains 99% of the calcium and 90% of the phosphorus in the human body. It absorbs and releases these minerals as needed, maintaining a constant concentration in the blood.

4. **The skeleton even participates in sensory transduction.** As you may recall from the previous chapter, three tiny bones of the middle ear transmit the sound vibrations between the eardrum and the cochlea.

The 206 bones of the human skeleton can be placed in two categories: the **axial skeleton** and the **appendicular skeleton.** The axial skeleton, whose bones form the axis of the body, includes the bones of the head, vertebral column, and rib cage. The bones of the appendicular skeleton form the extremities and their attachments to the axial skeleton, including the pectoral and pelvic girdles and the bones of the arms, legs, hands, and feet (Fig. 38-10).

Body Movement: Muscle–Skeletal Interactions

In addition to providing support for the body, the skeleton facilitates movement by providing a framework that muscles can move. Antagonistic muscles alter the configuration of the skeleton, either by causing movement around joints (in exoskeletons and vertebrate endoskeletons), or by altering the shape of the internal fluid (in hydrostatic skeletons; see Fig. 38-9).

In all three skeletal types, movement is accomplished by the action of antagonistic pairs of muscles, one of which contracts while the other is extended (Fig. 38-11; see also Fig. 38-9). Earlier in this chapter, we described the mechanism for muscle contraction, but not for extension. This was not an oversight: **Muscles can only actively contract. To extend, the muscle must be pulled out again.**

Movement Around Joints

The vertebrate skeleton allows movement by providing attachment points for skeletal muscles. Muscles move the skeleton around regions called **joints,** flexible attachment sites between adjacent bones. Skeletal muscles are attached to bones on either side of the

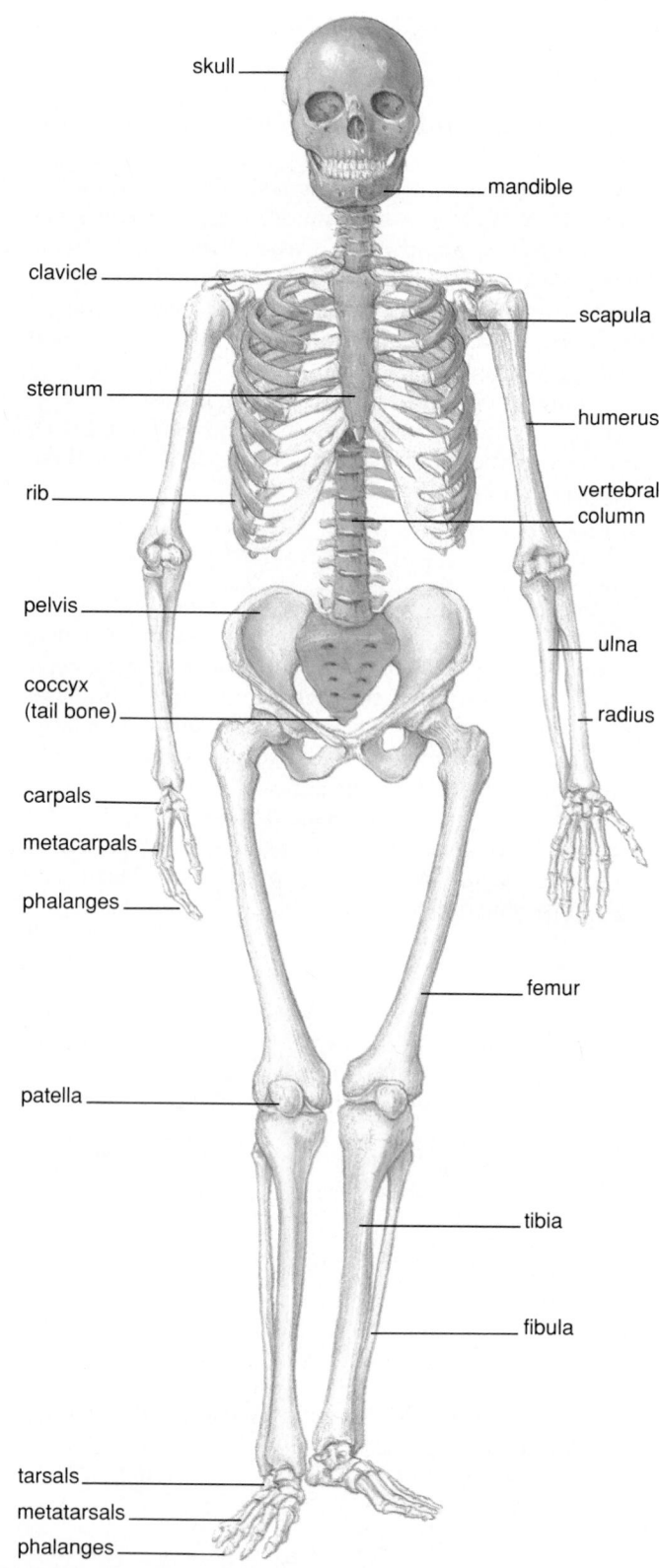

Figure 38-10 The human skeleton, showing the axial skeleton (tinged in blue) and the appendicular skeleton (bone color).

joint by bands of tough, fibrous connective tissue called **tendons.** The portion of each bone that forms the joint is typically coated with a layer of cartilage, whose smooth, resilient surface allows the bone surfaces to slide past one another. Bones are joined to one another at joints by bands of fibrous connective tissue called **ligaments** (see Fig. 38-11).

Most skeletal muscles are arranged in antagonistic pairs, one called a **flexor** and the other an **extensor,** on opposite sides of a joint (see Fig. 38-9b and 38-11). When one muscle contracts, it moves one bone with respect to the other, and simultaneously stretches out the opposing muscle. In the most common types of joints, skeletal muscles span the joint; their contraction moves one bone while the other remains fixed. Many joints, such as those in the elbow, knee, or fingers, called **hinge joints,** are moveable in only two

dimensions. In hinge joints, pairs of muscles lie in roughly the same plane as the joint (see Fig. 38-11). One end of each muscle, called the **origin,** is fixed to a relatively immovable bone on one side of the joint, while the other end, the **insertion,** is attached to a mobile bone on the far side of the joint. When the flexor muscle contracts, it bends the joint; when the extensor muscle contracts, it straightens the joint. Thus, alternate contractions of flexor and extensor muscles cause the moveable bone to pivot back and forth at the joint. Other joints, such as that of the hip, are **ball-and-socket joints,** in which the rounded end of one bone fits into a hollow depression in another. Ball-and-socket joints allow movement in several directions. Such joints have at least two pairs of muscles, oriented perpendicularly to each other, that provide flexibility of movement.

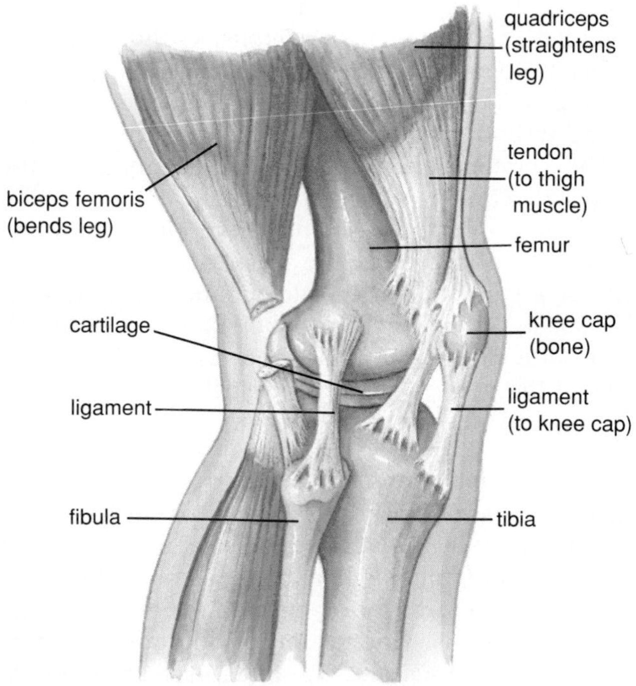

Figure 38-11 The human knee, a hinge joint, showing tendons, ligaments, and antagonistic muscles. The complexity of this joint, coupled with the extreme stresses placed on it during activities such as jumping, running, or skiing, make it very susceptible to injury.

SUMMARY OF KEY CONCEPTS

The Anatomy of Skeletal Muscle
Skeletal muscle fibers consist of numerous subunits—myofibrils—surrounded by calcium-storing sarcoplasmic reticulum. Sarcomeres within each myofibril are formed by alternating thick filaments of myosin and thin filaments of actin and two accessory proteins.

The Control of Muscle Contraction
During skeletal muscle contraction, the thick filaments form cross-bridges to exposed sites on the thin filaments. Using ATP, the bridges bend, release, and reattach in a way that slides the filaments past each other, shortening the fiber. Muscles produce action potentials when stimulated by motor neurons at nueromuscular junctions. Calcium, stored in the sarcoplasmic reticulum, is released when an action potential invades the muscle fiber. Calcium binds to one accessory protein, which pulls the other off the binding sites on the thin filaments, allowing contraction to occur. The strength and degree of muscle contraction is determined by the number of muscle fibers stimulated and the frequency of action potentials in each fiber. Rapid firing causes a build-up of contraction called summation, which can result in maximum contraction, or tetany, if sustained.

Cardiac and Smooth Muscle
Cardiac, or heart, muscle also consists of sarcomeres containing alternating thick and thin filaments. Its cells tend to contract rhythmically and spontaneously, but these contractions are synchronized by electrical signals produced by specialized muscle fibers in the sinoatrial node. Cardiac muscle fibers are interconnected electrically by gap junctions located in membrane regions called intercalated discs, allowing coordinated contraction.

Smooth muscle lacks organized sarcomeres, but like cardiac muscles, its cells are electrically coupled by gap junctions. Smooth muscle surrounds hollow organs (uterus, digestive tract, bladder) and blood vessels, producing slow sustained or rhythmic contractions, which are usually involuntary.

Cartilage
Cartilage is formed by chondrocytes, which surround themselves with a matrix of fibrous collagen. Cartilage is found at the ends of bones, and supports the nose, ears, and respiratory passages. Cartilage also forms pads in the knees and the intervertebral discs. During embryological development, cartilage is the precursor of bone.

Bone
Bone is formed by osteoblasts that secrete a collagen matrix that becomes hardened by calcium phosphate. A typical bone consists of an outer shell of compact, hard bone, to which muscles are attached, and inner spongy bone, which provides a space for bone marrow. Remodeling of bone occurs continuously. Osteoclasts tunnel through the bone using acids and enzymes. Nourishing capillaries invade the tunnels, and osteoblasts fill the space with concentric layers of new bone, leaving a small central space for the capillaries. This process produces Haversian systems. Osteoblasts trapped within the bone are called osteocytes.

The Skeleton
Three types of skeletons are found in animals. Hydrostatic skeletons (found in cnidarians, molluscs, and worms) use fluid confined in a chamber and surrounded by muscle. Exoskeletons, found in arthropods, are outer coverings with flexible joints. Endoskeletons, including the bony vertebrate skeleton, are found in both echinoderms and chordates.

The vertebrate skeleton provides support for the body, attachment sites for muscles, and protection for internal organs. Red and white blood cells and platelets are formed in the marrow of bones. Bone acts as a storage site for calcium and phosphorous. The axial skeleton includes the skull, vertebral column, and rib cage. The appendicular skeleton consists of the pectoral and pelvic girdles and the bones of the arms, legs, hands, and feet.

Muscle–Skeletal Interactions
Skeletal muscles form antagonistic pairs that move the skeleton. In the vertebrate skeleton, movement occurs around joints, where bones are joined by ligaments. Muscles attach to bones on either side of the joint by tendons. Contraction of one muscle bends the joint and straightens its antagonistic muscle. At hinge joints, muscles are attached to the immovable bone at their origins. Their insertions attach to the mobile bone. Contraction of the flexor muscle bends the joint, while contraction of its antagonistic extensor straightens it.

GLOSSARY

actin (ak'-tin): one of the major proteins of muscle, whose interactions with myosin produce contractions; found in the thin filaments of the muscle fiber. See also myosin.

appendicular skeleton (ap-pen-dik'-ū-lur): that portion of the skeleton consisting of the bones of the appendages and their attachments; the pectoral and pelvic girdles, the arms, legs, hands, and feet.

axial skeleton: the skeleton forming the body axis, including the skull, vertebral column, and rib cage.

cardiac muscle (kar'-dē-ak): specialized muscle of the heart, able to initiate its own contraction independent of the nervous system.

cartilage (kar'-teh-lij): a form of connective tissue forming portions of the skeleton, consisting of chondrocytes and their extracellular secretion of collagen. Cartilage resembles flexible bone.

chrondrocyte (kon'-drō-sīt): a cell type which, with its extracellular secretions of collagen, forms cartilage.

compact bone: the hard outer bone; composed of Haversian systems.

cross-bridge: in muscles, an extension of myosin that binds to and pulls upon actin to produce contraction of the muscles.

endoskeleton (en'-dō-skel'-uh-tun): a rigid *internal* skeleton with flexible joints to allow for movement.

exoskeleton (ex'-ō-skel'-uh-tun): a rigid *external* skeleton with flexible joints to allow for movement.

extensor muscle: a muscle that straightens a joint.

flexor muscle: a muscle that flexes (decreases the angle of) a joint.

Haversian system (ha-ver'-sē-un): unit of hard bone consisting of concentric layers of bone matrix with embedded osteocytes surrounding a small central canal containing a capillary.

hydrostatic skeleton (hī-drō-sta'-tic): a skeleton composed of fluid contained within a flexible, usually tubular, covering.

insertion: the site of attachment of a muscle to the relatively moveable bone on one side of a joint.

intercalated discs (in-tur'-cal-ā-ted): specialized interdigitating membranes containing large numbers of gap junctions that connect adjacent cardiac muscle fibers.

intervertebral discs (in-tur-ver-tē'-brul): pads of cartilage found between the vertebrae that act as shock absorbers.

joint: a flexible region between two rigid units of an exoskeleton or endoskeleton, to allow for movement between the units.

ligament: a tough connective tissue band connecting two bones.

motor unit: a single motor neuron and all the muscle fibers on which it synapses.

muscle fiber: an individual muscle cell.

myofibril (mī-ō-fī'-bril): a cylindrical subunit of each muscle cell, consisting of a series of sarcomeres. Myofibrils are surrounded by sarcoplasmic reticulum.

myosin (mī'-ō-sin): one of the major proteins of muscle that interacts with actin to produce contraction; found in the thick filaments of the muscle fiber. See also actin.

neuromuscular junction: the synapse formed between a motor neuron and muscle fiber.

origin: the site of attachment of a muscle to the relatively stationary bone on one side of a joint.

osteoblast (os'-tē-ō-blast): a cell type that produces bone.

osteoclast: a cell type that dissolves bone.

osteoporosis (os'-tē-ō-pōr-ō'-sis): a condition in which bones become porous, weak, and easily fractured; most common in elderly women.

sarcomere (sark'-ō-mēr): the unit of contraction of a muscle fiber; a subunit of the myofibril, consisting of actin and myosin filaments and bounded by Z-lines.

sarcoplasmic reticulum (sark'-ō-plas'-mik re-tik'-ū-lum): specialized endoplasmic reticulum found in muscle cells. The sarcoplasmic reticulum stores calcium ions and releases them into the interior of the muscle cell to initiate contraction.

skeletal muscle: also called striated or voluntary muscle; the type of muscle that is attached to and moves the skeleton.

skeleton: a supporting structure for the body, upon which muscles act to change the body configuration.

smooth muscle: type of muscle found around hollow organs, such as the digestive tract, bladder, and blood vessels, normally not under voluntary control.

striated muscle (strī'-ā-ted): muscle with a striped appearance due to the orderly arrangement of thick and thin filaments; both skeletal and cardiac muscle are striated.

summation: a steady increase in the degree of contraction of a muscle in response to rapid firing by its motor neuron.

T tubules: deep infoldings of the muscle cell membrane that conduct the action potential inside the cell.

tendon: a tough connective tissue band connecting a muscle to a bone.

tetany: smooth, sustained contraction of a muscle in response to rapid firing by its motor neuron.

thick filaments: bundles of myosin protein within the sarcomere which interact with thin filaments to produce muscle contraction.

thin filaments: proteinaceous strands within the sarcomere which interact with thick filaments to produce muscle contraction. Composed primarily of actin, with the accessory proteins tropomyosin and troponin.

tropomyosin (trō-pō-mī'-ō-sin): a double-stranded accessory protein found in thin filaments which, in the relaxed muscle, covers the binding sites on the actin molecules and prevents contraction.

troponin (trō-pō'-nin): a globular accessory protein in thin filaments that anchors the tropomyosin. During muscle contraction, troponin temporarily binds calcium, changes its conformation, and pulls tropomyosin off the binding sites of the actin molecules.

Z lines: fibrous protein structures to which the thin filaments of skeletal muscle are attached, forming the boundaries of sarcomeres.

STUDY QUESTIONS

1. Diagram the anatomy of a muscle and its various subunits, starting with the whole muscle and working down to the level of actin and myosin filaments. Label your drawings, and briefly define each part.
2. Describe the process of skeletal muscle contraction, beginning with an action potential in a motor neuron and ending with the relaxation of the muscle. Your answer should include the following words: neuromuscular junction, T tubule, sarcoplasmic reticulum, calcium, thin filaments, binding sites, thick filaments, sarcomere, and active transport.
3. What are the three types of skeletons found in animals? For one of these, describe how the muscles are arranged around the skeleton and how contractions of the muscles result in movement of the skeleton.
4. Describe the structure of a long bone. What roles are served by spongy bone? By compact bone?
5. Explain the functions of osteoblasts, osteoclasts, and osteocytes.
6. Explain the similarities and the differences between the structure of collagen and bone.
7. Where is collagen found in the body, and what functions does it serve?
8. Describe a hinge joint and how it is moved by antagonistic muscles.

DISCUSSION QUESTION

1. Discuss some of the problems that would result if the human heart were made of skeletal muscle instead of cardiac muscle.

SUGGESTED READINGS

Cohen, C. "The Protein Switch of Muscle Contraction." *Scientific American*, November 1975 (Offprint No. 1329). Cohen describes the interaction of calcium ions and the accessory proteins of the thin filament, which together control the contraction of muscles.

Huyghe, P. "No Bone Unturned." *Discover*, December 1988. The story of a remarkable forensic anthropologist who uncovers the secrets hidden in unidentified skeletons.

Merton, P. A. "How We Control the Contraction of Our Muscles." *Scientific American*, May 1972. Useful muscular activity requires precise simultaneous adjustments of the contractions of many muscles.

Moore, K. "Little Is Beyond His Scope." *Discover*, March 1985. A description of intricate arthroscopic surgery on joints.

Smith, K. K., and Kier, W. M. "Trunks, Tongues, and Tentacles: Moving with Skeletons of Muscle." *American Scientist*, January-February 1989. In certain organs of many animals, including vertebrates, muscles provide support as well as movement.

39

Animal Reproduction

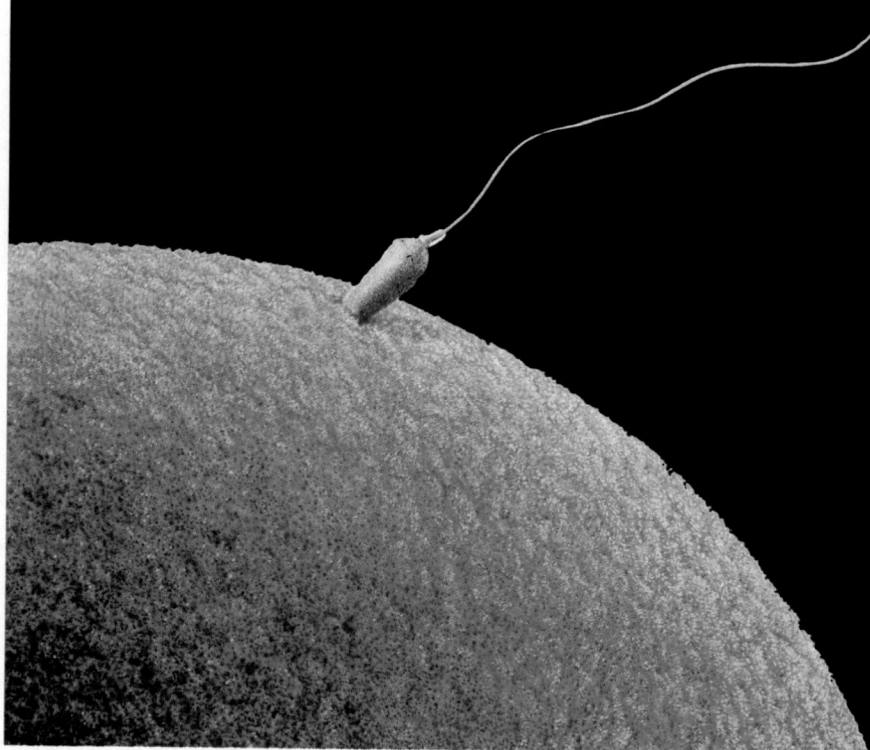

A human sperm, lured by chemical attractants, dives into the layer of cells surrounding a human egg.

The word *reproduction* may bring to mind images of courtship, cute babies, and cuddly kittens. However, from an evolutionary perspective, romance and the universal appeal of babies are frills that have evolved only because they further the real goal: to pass on one's genes to another generation, in a sense cheating death and achieving immortality through one's offspring. From this viewpoint, an animal's life can be divided into three stages. First, it is born or hatched from an egg, and grows to sexual maturity. Second, it gathers the resources needed to reproduce, which may include stores of food, impressive strength or weaponry, or a territory. Finally, it finds a mate (if necessary) and reproduces, which may include caring for its offspring until they can fend for themselves. The marvelous adaptations that we have discussed in the previous chapters, such as sophisticated sensory equipment or complex digestive systems, have evolved through millenia of mutation and natural selection because they have allowed successful reproduction by the animals that possess them. Reproduction is the key to the continued existence of the species.

Reproductive Strategies

Animals reproduce either sexually or asexually. As you learned in Chapter 10, in **sexual reproduction** an animal produces haploid gametes through meiosis. Two gametes, usually from separate parents, fuse to form a diploid offspring. Since an offspring receives genes from two parents, it is genetically different from both of them. In most forms of **asexual reproduction,** on the other hand, a single animal produces offspring through repeated mitosis of cells in some part of its body. Therefore, the offspring are genetically identical to the parent. We humans reproduce sexually, and we tend to regard sexual reproduction as the normal, best way to do it. From a biological standpoint, by bringing together genes from two different parental organisms, sexual reproduction allows for new gene combinations that may enhance the survival and reproduction of the offspring. Nevertheless, asexual reproduction is more efficient, since there is no need to find a mate, court, and fend off rivals, and no waste of sperm and eggs that never unite to form an offspring. Not surprisingly, a number of animals reproduce asexually, at least some of the time. Let's begin, then, with a brief survey of asexual reproduction among animals.

Asexual Reproduction

Budding

Many sponges and coelenterates, such as *Hydra* and some anemones, reproduce by **budding** (Fig. 39-1). A miniature version of the animal (a **bud**) grows directly on the body of the adult, drawing nourishment from its parent. When it has grown large enough, the bud breaks off and becomes independent.

Regeneration

Regeneration from body fragments is a potential form of reproduction in some animals such as sea stars (Fig. 39-2a). If sea stars are cut up, fragments that contain part of the central disc can regrow the rest of the star. (In an effort to reduce predation by sea stars on their oyster beds, oyster "ranchers" used to catch sea stars, hack them to pieces, and throw the parts back into the sea. Much to their dismay, this merely resulted in more sea stars than ever, as the fragments regenerated entire animals.) A few brittle-stars routinely reproduce in a similar fashion, by splitting apart and each half regenerating a complete animal. Despite these asexual capabilities, sea stars usually reproduce sexually, casting huge numbers of sperm and eggs into the sea.

Fission

Some animals reproduce by **fission.** A few corals can divide longitudinally to produce two smaller but

Figure 39-1 The offspring of some cnidarians, such as the anemone shown here, grow as buds upon the body of the parent. When sufficiently developed, the buds break off and assume independent existence.

(a) Regeneration

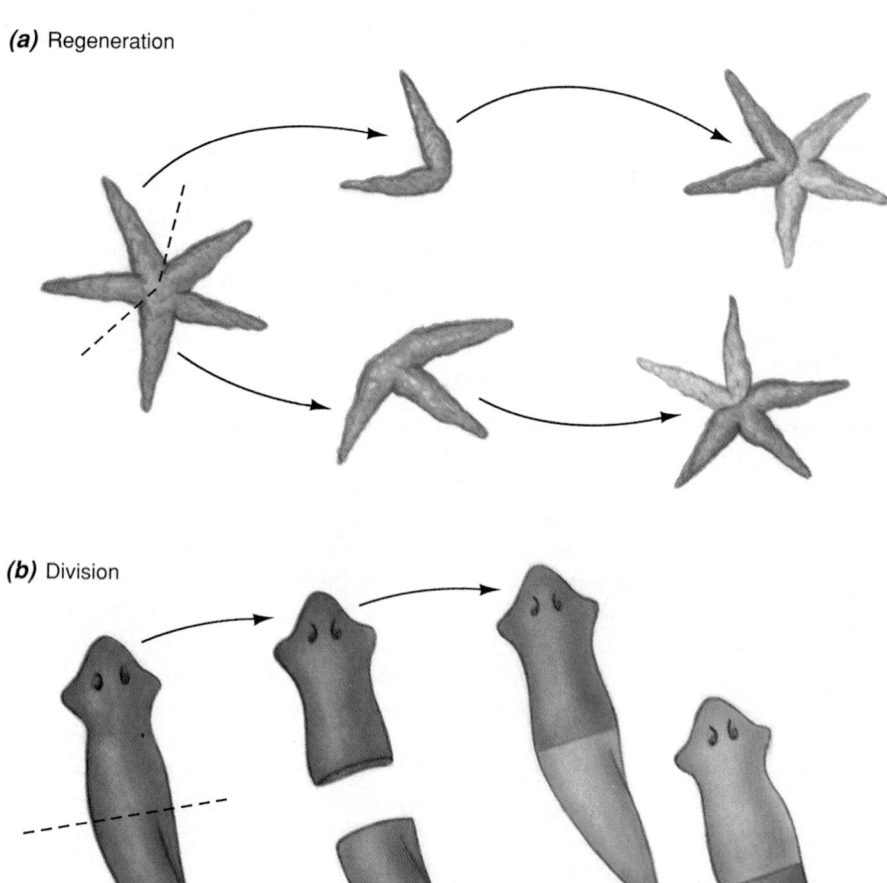

Figure 39-2 Methods of asexual reproduction. **(a)** Many sea stars can regenerate new individuals from fragments if the fragment includes part of the center of the body. **(b)** Certain flatworms divide transversely. At first, each offspring is missing half the adult body, but these are regrown from cells near the broken edge.

(b) Division

complete individuals. Some flatworms and annelids divide transversely and regenerate the missing parts (Fig. 39-2b). This, of course, means that the "tail half" of the animal must regenerate the head, including the brain!

Parthenogenesis

The females of some animal species can reproduce by a process known as **parthenogenesis,** in which haploid egg cells develop into adults without being fertilized. In some animals, parthenogenetically produced offspring remain haploid. Male honeybees, for example, are haploid, developing from unfertilized eggs; their diploid sisters develop from fertilized eggs (Fig. 39-3). On the other hand, some fish, amphibians, and reptiles regain the diploid number of chromosomes in parthenogenetically produced offspring. This is ac-

complished by duplicating all the chromosomes either before or after meiosis. The resulting offspring are all females. Some species of fish, including relatives of the mollies and platies found in tropical fish stores, and some lizards, such as the whiptail, have done away with males completely. Their populations consist entirely of parthenogenetically reproducing females. Still other animals, such as the aphid, can reproduce either sexually or parthenogenetically, depending on environmental factors such as the season of the year or the availability of food (Fig. 39-4).

Sexual Reproduction

In animals, sexual reproduction occurs when a haploid sperm fertilizes a haploid egg, generating a diploid offspring. In most animal species, an individual is either male or female. These species are termed

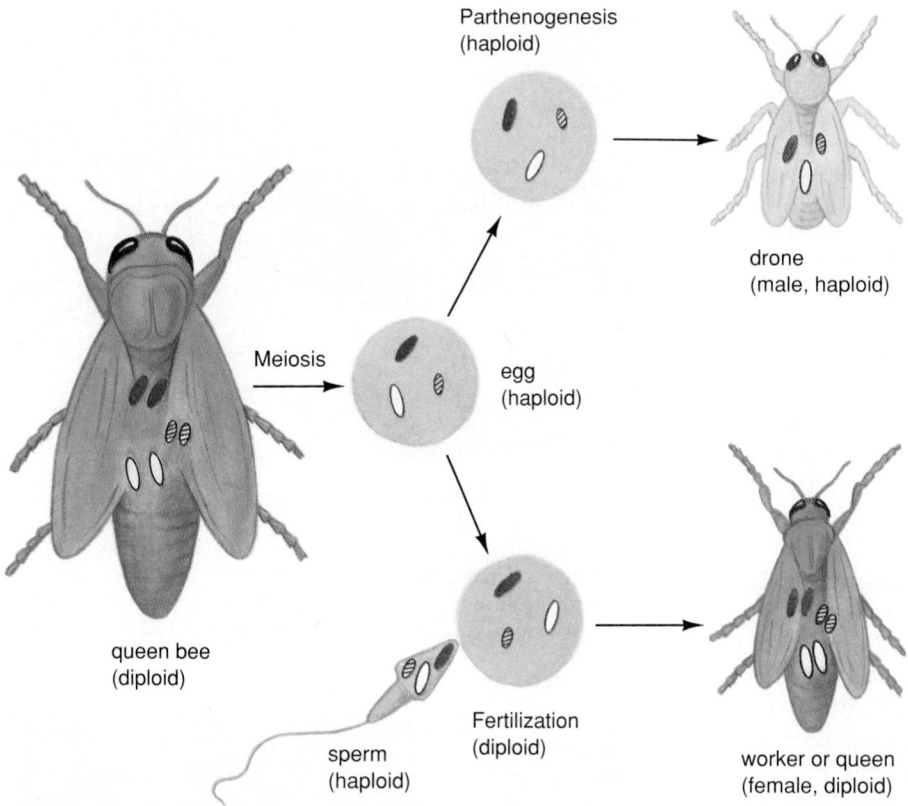

Parthenogenesis
(haploid)

Meiosis

egg
(haploid)

queen bee
(diploid)

drone
(male, haploid)

sperm
(haploid)

Fertilization
(diploid)

worker or queen
(female, diploid)

Figure 39-3 Queen honeybees produce haploid eggs by meiosis. The eggs then take one of two developmental paths. If unfertilized, the eggs develop parthenogenetically into haploid drones (males). Fertilized eggs develop into females, either workers or (rarely) new queens. Only three chromosome pairs are shown for clarity. (Haploid = yellow; diploid = orange.)

dioecious (Greek for "two houses"). The sexes are defined by the type of gamete that each produces. Females produce **eggs,** which are large, nonmotile cells containing substantial food reserves. Males produce small, motile **sperm** that have almost no cytoplasm and hence no food reserves. In **monoecious** ("one house") species, such as earthworms and many snails, single individuals produce both sperm and eggs. Such individuals are commonly called **hermaphrodites** (after Hermaphroditos, a male Greek god whose body was merged with that of a female water nymph). Although most hermaphrodites exchange sperm with other individuals if they have the opportunity (retaining the advantages of genetic exchange), some hermaphrodites can fertilize their eggs with their own sperm if necessary. These animals, including tapeworms and many pond snails, are relatively immobile and may find themselves isolated from other members of their species. Obviously, the ability to fertilize oneself is advantageous under these circumstances.

For dioecious species and for hermaphrodites that cannot self-fertilize, successful reproduction requires that sperm and eggs from different animals be

emerging offspring

Figure 39-4 A female aphid gives live birth. In spring and early summer, when food is abundant, aphid females reproduce parthenogenetically. In fact, the development of the ovaries proceeds so rapidly that females are born pregnant! In fall, reproduction becomes sexual, as the females mate with males. Aphids have thus evolved the ability to exploit the advantages of asexual reproduction (rapid population growth during times of abundant food, no energy spent in seeking a mate, no wasted gametes) and sexual reproduction (genetic recombination).

brought together for fertilization. This is accomplished in a variety of ways, depending on the mobility of the animals and on whether they breed in water or on land.

External Fertilization

In **external fertilization,** the parents release sperm and eggs into water, through which the sperm swim to reach an egg. This procedure, called **spawning,** is obviously restricted to animals that breed in water. Since sperm and egg are relatively short-lived, spawning animals must synchronize their reproductive behaviors, both *temporally* (male and female spawn at the same time) and *spatially* (male and female spawn in the same place). Animals employ a combination of environmental cues, pheromones, and behaviors to synchronize spawning.

Most spawning animals rely on environmental cues to some extent. Breeding usually occurs only during certain seasons of the year, but more precise synchrony is required to coordinate the actual release of sperm and egg. Grunion, fish that inhabit southern California coastal waters, time their unusual reproductive rituals by the season, time of day, and phase of the moon (Fig. 39-5a). On nights of the highest tides in fall (which occur during a full moon), they swim up onto the beach. Writhing masses of males and females release their gametes into the wet sand, and then swim back out to sea on the next wave. Many corals of Australia's Great Barrier Reef also synchronize spawning by the phase of the moon. On the fourth or fifth night after the full moons of November and December, all the corals of a particular species on an entire reef release a blizzard of sperm and eggs into the water (Fig. 39-5b).

Other animals communicate their sexual readiness to one another by releasing pheromones into the water. A **pheromone** is a chemical released from the body of one animal that affects the behavior of a second animal. Pheromones synchronize spawning in many immobile or sluggish invertebrates, such as mussels and sea stars. Usually, when a female is ready to spawn, she releases eggs and a pheromone into the water. Nearby males, detecting the mating pheromone, quickly release millions of sperm.

Temporal synchrony alone does not guarantee efficient reproduction. Corals, sea stars, and mussels all waste enormous quantities of sperm and eggs because they are released too far apart. In mobile animals, both temporal and spatial synchrony can be ensured by mating behaviors. Most fish, for example, have some sort of courtship ritual in which the male and female come very close together and release

(a) At the highest tides of fall, grunion swarm ashore on the few undeveloped beaches left in southern California. The fish burrow slightly into the sand and release sperm and eggs. The eggs hatch in the warm sand and develop over the following 2 weeks. When the next highest tide comes, the juveniles wash out of the sand back into the ocean.

(b) Along the Great Barrier Reef of Australia, thousands of corals spawn simultaneously, creating this "blizzard" effect. The inset photo shows a package of sperm and eggs erupting from a spawning hermaphroditic coral. Spawning in these corals is linked to the phase of the moon.

Figure 39-5 Some animals use environmental cues to synchronize spawning.

their gametes in the same place and at the same time (Fig. 39-6). Frogs carry this one step further, by assuming a characteristic mating pose called **amplexus** (Fig. 39-7). At the edges of ponds and lakes, the male frog mounts upon the back of the female and prods her in the side. This stimulates her to release eggs, which he immediately fertilizes.

Figure 39-7 Amplexus in golden toads, as the smaller male rides atop the female and stimulates her to release eggs. The large eggs surrounded by a transparent jelly coat make the eggs of frogs and toads ideal for studies of embryonic development.

Figure 39-6 Violent courtship rituals among Siamese fighting fish *(Betta splendens)* ensure fertilization of the female's eggs, as male and female curl about one another, releasing sperm and eggs together. The male retrieves the eggs as they fall, spits them into his bubble nest, and cares for the offspring during their first few weeks of life.

Internal Fertilization

In **internal fertilization,** sperm are taken into the body of the female, where fertilization occurs. This method has two advantages. First, sperm are provided with a direct fluid path to reach the eggs. In terrestrial environments, this path can be guaranteed only inside the body of the female. Second, even in aquatic environments, internal fertilization increases the likelihood that most eggs will be fertilized, since the sperm are not left to thrash about in a large volume of water to find the eggs. Internal fertilization usually occurs by **copulation,** in which the penis of the male is inserted into the body of the female, where it releases sperm (Fig. 39-8). In a variation of internal fertilization, males of some animals package their sperm in a container called a *spermatophore* (Greek for "sperm carrier"). Males of some species of mites and scorpions simply drop the spermatophore on the ground. If found by a female, she then fertilizes herself by inserting the spermatophore into her

reproductive cavity. The male squid is somewhat more careful with his spermatophore, which he picks up with a tentacle and inserts into the female. In either case, the sperm are then liberated inside the female's reproductive tract.

Just because sperm are deposited in the body of the female does not guarantee fertilization. Fertilization can occur only if an egg is mature and released into the female reproductive tract during the limited time when sperm are present. In most mammals, copulation occurs only at certain seasons of the year, or when the female signals readiness to mate, which often coincides with ovulation. Copulation itself triggers ovulation in a few animals, such as rabbits. An alternative strategy, employed by many snails and insects, is to store sperm for days, weeks, or even months, thus assuring a supply of sperm whenever eggs are ready.

Mammalian Reproduction

In mammals, male and female reproductive systems are found in separate individuals. Many mammals reproduce only during certain seasons of the year, and consequently produce sperm and eggs only at that time. Human reproduction is similar to that of other mammals, except for a loss of seasonality. Men

(a)

(b)

(c)

Figure 39-8 Internal fertilization is essential for reproduction on land. (a) Ladybugs mate on a dandelion flower; (b) South American tortoises must cope with confining shells; (c) King penguins mate comfortably in the snow.

produce sperm more-or-less continuously, while women ovulate about once a month. Our discussion of mammalian reproduction will emphasize humans.

The Human Male Reproductive Tract

The male reproductive tract consists of the paired **gonads,** where sperm are produced, and accessory structures that store the sperm, produce secretions that activate and nourish them, and finally conduct them to the inside of the female reproductive tract (Fig. 39-9 and Table 39-1).

The Testes

The male gonads, the **testes** (sing. testis), produce both sperm and male sex hormones. The testes are located in the **scrotum,** a pouch that hangs outside the main body cavity. This location keeps the testes about 4° C cooler than the core of the body and provides the optimal temperature for sperm development. (Tight jeans may look sexy, but they push the scrotum up against the body, raising the temperature of the testes. Some researchers think that this may reduce sperm counts and hence reduce fertility. This is not, however, a reliable means of birth control!) Coiled, hollow **seminiferous tubules,** in which sperm are produced, nearly fill each testis (Fig. 39-10a). In the spaces between the tubules are the **interstitial cells,** which synthesize the male hormone testosterone.

Just inside the wall of each seminiferous tubule lie the diploid germ cells, or **spermatogonia,** from which all the sperm will eventually arise, and the much larger **Sertoli cells** (Fig. 39-10b, c). Each time a spermatogonium divides, it can take one of two developmental paths. First, it may undergo mitosis, thereby providing a steady supply of new spermatogonia throughout life. Second, it may undergo **spermatogenesis**—that is, meiosis followed by differentiation into sperm (Fig. 39-10d). Spermatogenesis begins with growth and differentiation of spermatogonia into **primary spermatocytes.** These are large diploid cells that will develop into sperm. The primary spermatocytes then undergo meiosis (see Chapter 10). At the end of meiosis I, each primary spermatocyte gives rise to two haploid **secondary spermatocytes.** Each secondary spermatocyte divides again during meiosis II to produce two **spermatids,** for a total of four spermatids per primary spermatocyte. Spermatids undergo radical rearrangements of their cellular components as they differentiate into sperm.

Sertoli cells regulate the process of spermatogenesis and nourish the developing sperm. The spermatogonia, spermatocytes, and spermatids are embedded

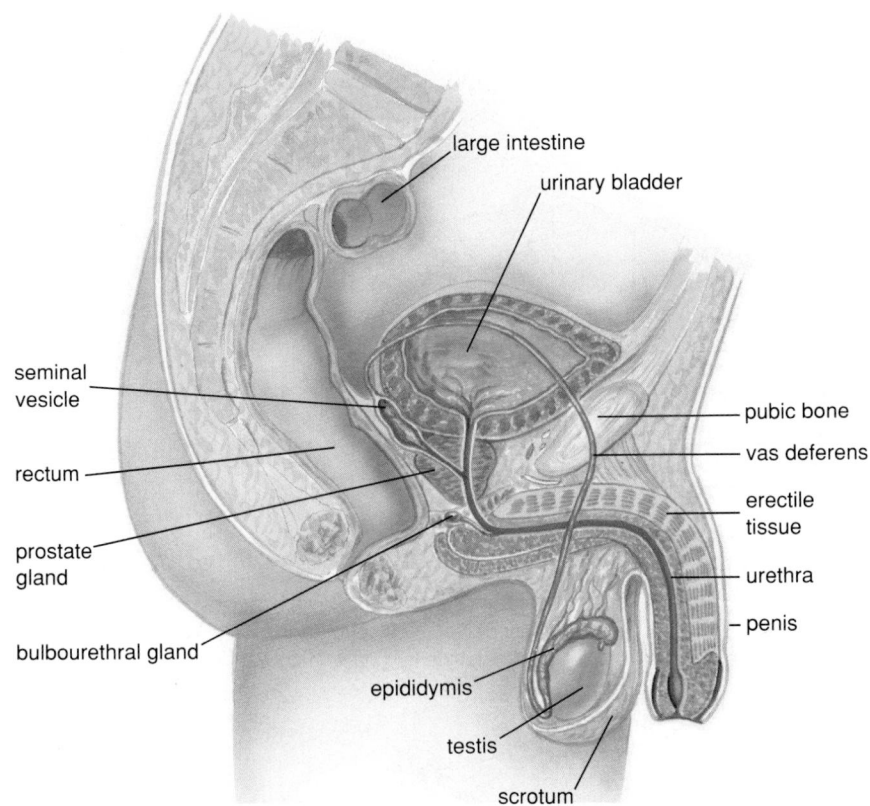

large intestine

urinary bladder

seminal
vesicle

rectum

prostate
gland

bulbourethral gland

pubic bone

vas deferens

erectile
tissue

urethra

penis

epididymis

testis

scrotum

Figure 39-9 The human male reproductive tract. The male gonads, the testes, hang beneath the abdominal cavity in the scrotum. Sperm pass from the seminiferous tubules of a testis to the epididymis, and thence through the vas deferens and urethra to the tip of the penis. Along the way, fluids are added from three sets of glands, the paired seminal vesicles and bulbourethral glands and the unpaired prostate gland.

Table 39-1   Structures and Functions of the Human Male Reproductive Tract

Structure	Type of Organ	Function
Testis	Gonad	Produces sperm and testosterone
Epididymis and vas deferens	Ducts	Store sperm; conduct sperm from testes to penis
Urethra	Duct	Conducts semen from vas deferens and urine from urinary bladder to the tip of the penis
Penis	External "appendage"	Deposits sperm in female reproductive tract
Seminal vesicles	Gland	Secrete fluids that contain fructose (energy source) and prostaglandins (possibly cause "upward" contractions of vagina, uterus, and oviducts, assisting sperm transport to oviducts); fluids may wash sperm out of ducts of male reproductive tract into vagina
Prostate	Gland	Secretes fluids that are basic (neutralize acidity of vagina) and contain factors that enhance sperm motility
Bulbourethral glands	Gland	Secrete mucus (may lubricate penis in vagina)

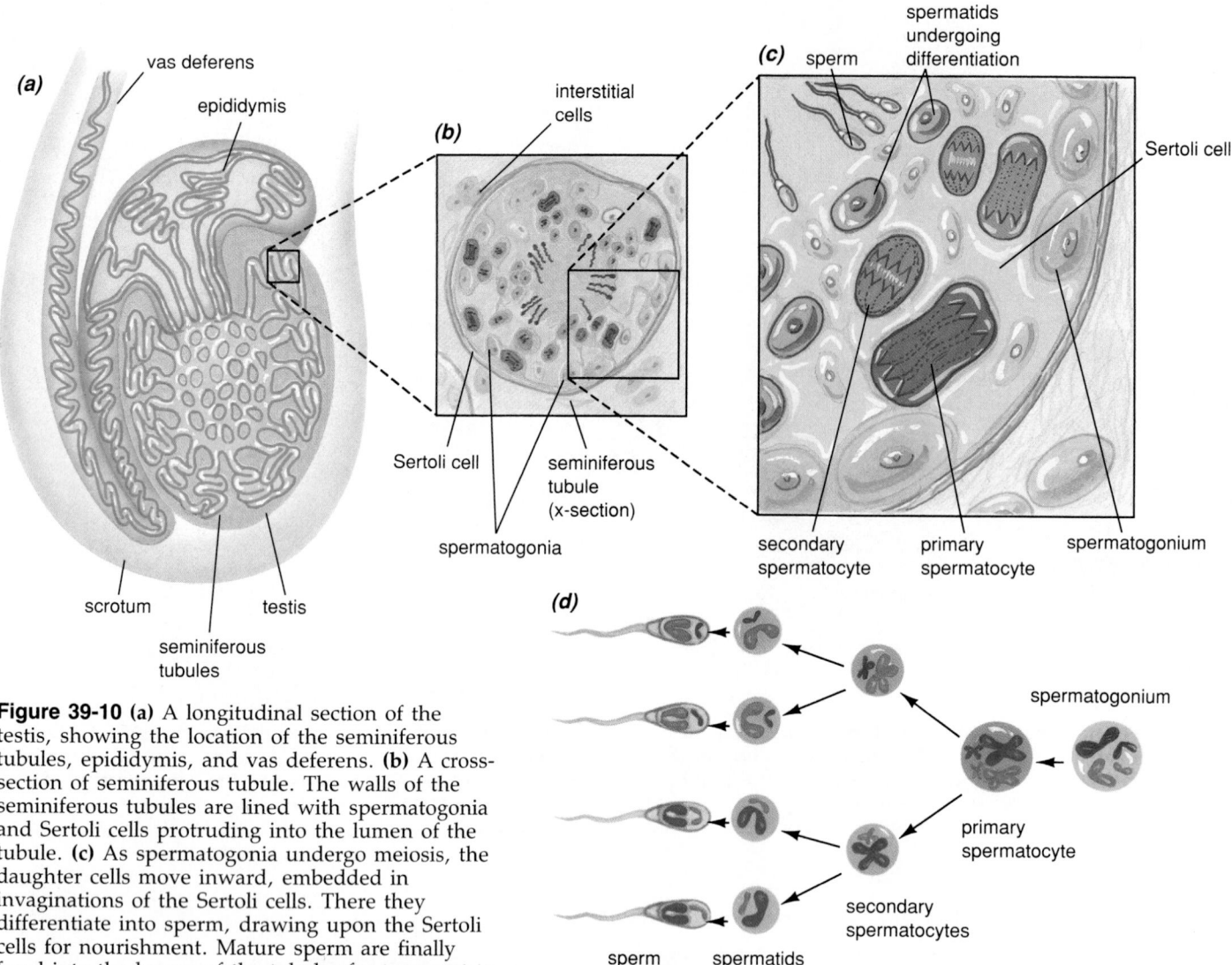

Figure 39-10 (a) A longitudinal section of the testis, showing the location of the seminiferous tubules, epididymis, and vas deferens. **(b)** A cross-section of seminiferous tubule. The walls of the seminiferous tubules are lined with spermatogonia and Sertoli cells protruding into the lumen of the tubule. **(c)** As spermatogonia undergo meiosis, the daughter cells move inward, embedded in invaginations of the Sertoli cells. There they differentiate into sperm, drawing upon the Sertoli cells for nourishment. Mature sperm are finally freed into the lumen of the tubules for transport to the penis. Testosterone is produced by the interstitial cells found in the spaces between tubules. **(d)** Spermatogenesis is accomplished by meiotic divisions that produce haploid sperm (compare with the actual locations shown in (c)). Although four chromosomes are shown for clarity, in humans the diploid number is 46 and the haploid number is 23.

in invaginations of the Sertoli cells, and migrate up from the outermost edge of the seminiferous tubule to the lumen (tubular cavity) at the center as spermatogenesis proceeds (Fig. 39-10c). The mature sperm, several hundred million a day, are finally liberated into the lumen.

A human sperm (Fig. 39-11) is unlike any other cell of the body. Most of the cytoplasm disappears, leaving a haploid nucleus nearly filling the head. Atop the nucleus lies a specialized lysosome, called the **acrosome.** This contains enzymes that will be needed

to dissolve protective layers around the egg, enabling the sperm to enter and fertilize it. Behind the head is the midpiece, which is packed with mitochondria that provide the energy needed to move the tail that protrudes out the back. Sculling movements of the tail, which is really a long flagellum, propel the sperm along inside the female reproductive tract.

Spermatogenesis begins in puberty as a result of the interplay of **luteinizing hormone** (LH) and **follicle-stimulating hormone** (FSH) from the anterior pituitary, and **testosterone** from the testes themselves

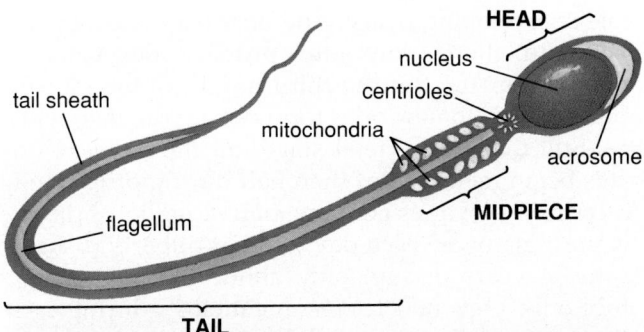

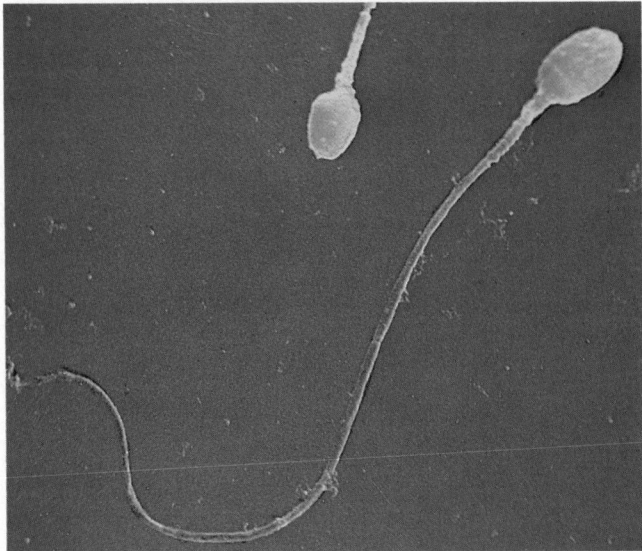

Figure 39-11 A mature sperm is a stripped-down cell equipped with only the essentials: a haploid nucleus containing the male genetic contribution to the future zygote, a lysosome (called the acrosome) containing enzymes that will digest the barriers surrounding the egg, mitochondria for energy production, and a tail (actually a long flagellum) for locomotion. The photo is a false-color electron micrograph of human sperm.

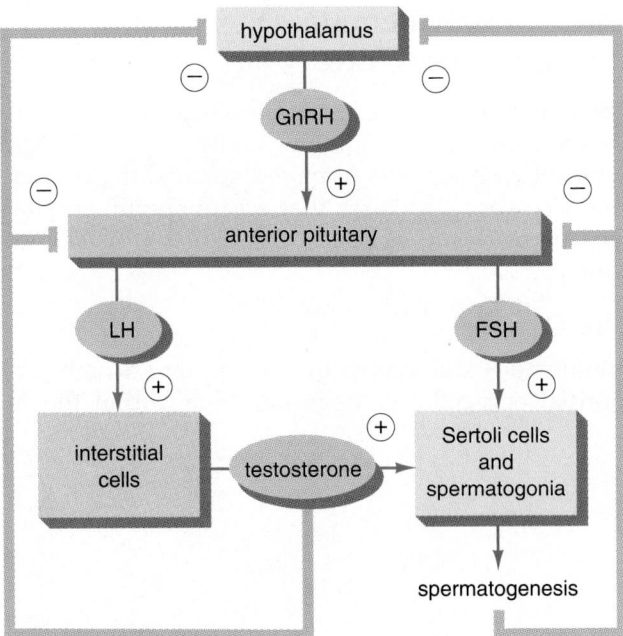

Figure 39-12 Gonadotropin releasing hormone (GnRH) from the hypothalamus stimulates the pituitary to release LH and FSH. LH stimulates the interstitial cells to produce testosterone. Testosterone and FSH stimulate the Sertoli cells and the spermatogonia, causing spermatogenesis. Testosterone and chemicals produced during spermatogenesis inhibit further release of FSH and LH, forming a negative feedback loop that keeps the rate of spermatogenesis and the concentration of testosterone in the blood nearly constant. $\oplus$ = stimulates; $\ominus$ = inhibits.

(Fig. 39-12). Luteinizing hormone stimulates the interstitial cells to produce testosterone. The combination of testosterone and FSH stimulates the Sertoli cells and spermatogonia, causing spermatogenesis. Testosterone also stimulates the development of secondary sexual characteristics, maintains sexual drive, and is required for successful intercourse. Sperm, however, are not involved in these functions. Therefore, if one could suppress FSH release (blocking spermatogenesis) but not LH release (thereby allowing continued testosterone production), a man would be infertile but not impotent. Efforts are underway to develop a drug to do just that, as a form of male birth control.

Accessory Structures

The seminiferous tubules merge to form a single convoluted tube, the **epididymis** (see Fig. 39-9). The epididymis becomes the **vas deferens,** which leaves the scrotum and enters the abdominal cavity. Most of the hundreds of millions of sperm produced each day are stored in the vas deferens and epididymis. The vas deferens joins the **urethra,** leading from the bladder to the tip of the penis. This final common path is time-shared by sperm (during ejaculation) and urine (during urination).

The fluid ejaculated from the penis, called **semen,** consists of sperm mixed with secretions from three glands that empty into the vas deferens or urethra: the **seminal vesicles,** the **prostate gland,** and the **bulbourethral glands.** The secretions activate swimming by the sperm, provide energy for swimming, and neutralize the acidic fluids of the vagina (see Table 39-1).

The Human Female Reproductive Tract

The female reproductive tract is almost entirely contained within the abdominal cavity (Fig. 39-13 and Table 39-2). It consists of paired gonads, the **ovaries,** and accessory structures that accept sperm, conduct the sperm to the egg, and nourish the developing embryo.

The Ovaries

While she is still a fetus in her mother's womb, primordial egg cells, or **oogonia,** form within the female's developing ovary. The oogonia divide by mitosis and then grow into **primary oocytes.** No oogonia remain after the third month of fetal development, and no new ones form during the rest of her life. Still during the fetal stage, all the primary oocytes begin meiosis, but then halt during prophase I. At birth, the ovaries contain about 2 million primary oocytes; many die each day, until at puberty (usually 11 to 14 years of age) only about 400,000 remain. Since only a few oocytes resume meiosis during each month of a woman's reproductive span (from puberty to menopause at age 45 to 55), there is no shortage of oocytes (Fig. 39-14).

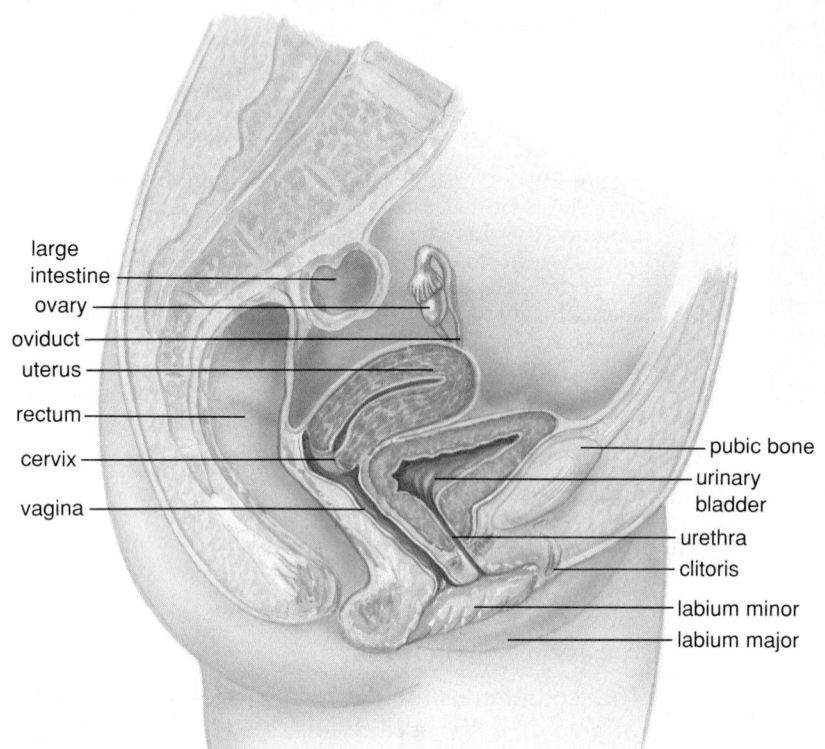

large
intestine
ovary
oviduct
uterus
rectum
cervix
vagina

pubic bone
urinary bladder
urethra
clitoris
labium minor
labium major

Figure 39-13 The human female reproductive tract, in side view. Eggs are produced in the ovaries and swept into the oviduct. A male deposits sperm in the vagina, from which they move up through the cervix and uterus into the oviduct. Sperm and egg usually meet in the oviduct, where fertilization occurs. The fertilized egg attaches to the lining of the uterus, where the embryo develops.

Table 39-2   Structures and Functions of the Human Female Reproductive Tract

Structure	Type of Organ	Function
Ovary	Gonad	Produces eggs, estrogen, and progesterone
Fimbria	Mouth of duct	Cilia sweep egg into oviduct
Oviduct	Duct	Conducts egg to uterus; site of fertilization
Uterus	Muscular chamber	Site of development of fetus
Cervix	Connective tissue ring	Closes off lower end of uterus, supports fetus, and prevents foreign material from entering uterus
Vagina	Large "duct"	Receptacle for semen; birth canal

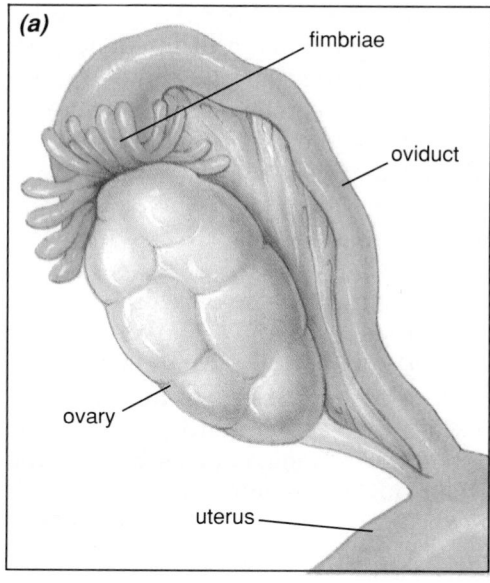

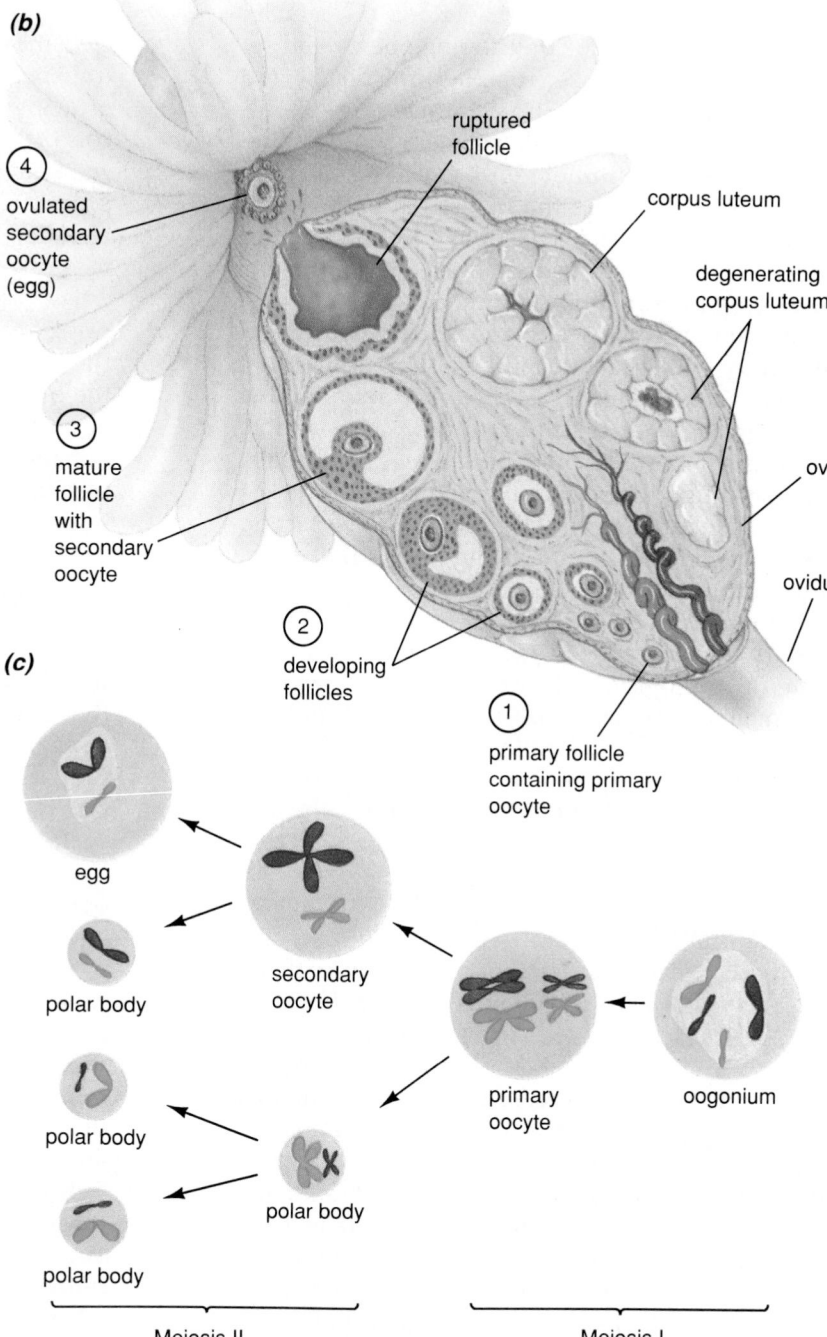

Figure 39-14 Oogenesis in human females. **(a)** External view of the ovary and oviduct. **(b)** The development of follicles in an ovary, portrayed in a time sequence going counterclockwise from the lower left. (1) A primary oocyte begins development within a follicle. (2), (3) The follicle grows, providing both hormones and nourishment for the enlarging oocyte. (4) At ovulation, the secondary oocyte, or egg, bursts through the ovary wall, surrounded by some follicle cells (now called the corona radiata). The remaining follicle cells develop into a secretory organ, the corpus luteum. If fertilization does not occur, the corpus luteum degenerates after a few days. **(c)** The cellular stages of oogenesis. The oogonium enlarges to form the primary oocyte. At meiosis I, almost all the cytoplasm is included in one daughter cell, the secondary oocyte. The other daughter cell is a small polar body that contains chromosomes but little cytoplasm. At meiosis II, almost all the cytoplasm of the secondary oocyte is included in the egg, and a second small polar body discards the remaining "extra" chromosomes. The first polar body sometimes also undergoes the second meiotic division. In humans, meiosis II does not occur unless the egg is fertilized.

Surrounding each oocyte is a layer of much smaller cells that both nourish the developing oocyte and secrete female sex hormones. Together, the oocyte and these accessory cells make up a **follicle** (Fig. 39-14). Approximately once a month during a woman's reproductive years, she undergoes a menstrual cycle, described below. During the menstrual cycle, pituitary hormones stimulate development of a dozen or more follicles, although usually only one completely matures. The primary oocyte completes the first meiotic division to become a single **secondary oocyte** and a **polar body,** which is little more than a discarded set of chromosomes (Fig. 39-14b, c). Meanwhile, the small cells of the follicle multiply and secrete **estrogen.** As it matures, the follicle grows, eventually erupting through the surface of the ovary and releasing the secondary oocyte (Fig. 39-15). The second meiotic division occurs, not in the ovary, but in the oviduct, and then only if the secondary oocyte is fertilized. For convenience, we will refer to the ovulated secondary oocyte as the "egg."

Some of the follicle cells accompany the egg, but most remain behind in the ovary. These cells enlarge and become glandular, forming the **corpus luteum,** which secretes estrogen and a second hormone, **progesterone.** If fertilization does not occur, the corpus luteum degenerates a few days later.

Accessory Structures

Each ovary is adjacent to, but not continuous with, an **oviduct** (often called the Fallopian tube in humans; see Fig. 39-13). The open end of the oviduct is fringed with ciliated "fingers" called **fimbriae** that nearly surround the ovary. The cilia create a current that sweeps the egg into the mouth of the oviduct. Fertilization usually occurs in the oviduct, and the new zygote gently tumbles down the oviduct into the pear-shaped **uterus,** or womb, in which it will develop for the next 9 months. The wall of the uterus has two layers that correspond to its dual functions of nourishment and childbirth. The inner lining, or **endometrium,** is richly supplied with blood vessels. (This lining will form the mother's contribution to the **placenta,** the structure that transfers oxygen, carbon dioxide, nutrients and wastes between fetus and mother; see Chapter 40). The outer muscular wall of the uterus, the **myometrium,** contracts strongly during delivery, expelling the infant out into the world.

Developing follicles secrete estrogen, which stimulates the uterine lining to grow an extensive network of blood vessels and nutrient-producing glands. After ovulation, estrogen and progesterone released by the corpus luteum promote continued growth of the endometrium. Thus, if an egg is fertilized, it encounters a rich environment for growth. If the egg is not fertilized, however, the corpus luteum disintegrates, estrogen and progesterone levels fall, and the overgrown endometrium disintegrates as well. The uterus contracts, squeezing out the excess endometrial tissue (and sometimes causing menstrual cramps in the process). The resulting flow of tissue and blood is called **menstruation** (from the Greek "mensis," meaning month).

The outer end of the uterus is nearly closed off by a ring of connective tissue, the **cervix.** The cervix holds the developing baby in the uterus, expanding only at the onset of labor to permit passage of the child. Beyond the cervix lies the **vagina,** which opens to the outside of the body. The vagina serves both as the receptacle for the penis during intercourse and as the birth canal.

The Menstrual Cycle

The human male produces sperm continuously, thereby increasing the number of potential offspring that a man can father. In contrast, it is fruitless for a woman to ovulate unless her reproductive tract is prepared for pregnancy. The human female reproductive system goes through a complex **menstrual cycle,** in which hormonal interactions among the hypothalamus, pituitary gland, and ovary coordinate ovulation and the preparation of the uterus to receive and nourish the fertilized egg.

You may recall from Chapter 35 that hormone release by the anterior pituitary gland is controlled by neurosecretory cells in the hypothalamus. Some of these neurosecretory cells produce **gonadotropin-releasing hormone** (GnRH), which stimulates endocrine cells in the anterior pituitary to release FSH and LH.

A key to understanding the menstrual cycle is that these neurosecretory cells spontaneously release GnRH all the time, unless actively prevented from doing so by other hormones, notably progesterone. We will begin our discussion of the menstrual cycle with the spontaneous release of GnRH.

GnRH stimulates the anterior pituitary to release FSH and LH (Step 1 in Fig. 39-16). FSH and LH circulate in the bloodstream and initiate the development of several follicles within the ovaries. The follicle cells surrounding the developing oocyte are stimulated by FSH and LH to secrete estrogen. Under the combined influences of FSH, LH, and estrogen, the follicles grow during the next two weeks (Step 2). Simultaneously, the primary oocyte within each follicle enlarges, storing both food and regulatory substances (mostly proteins and messenger RNA) that will be needed by the fertilized egg during early develop-

(a)

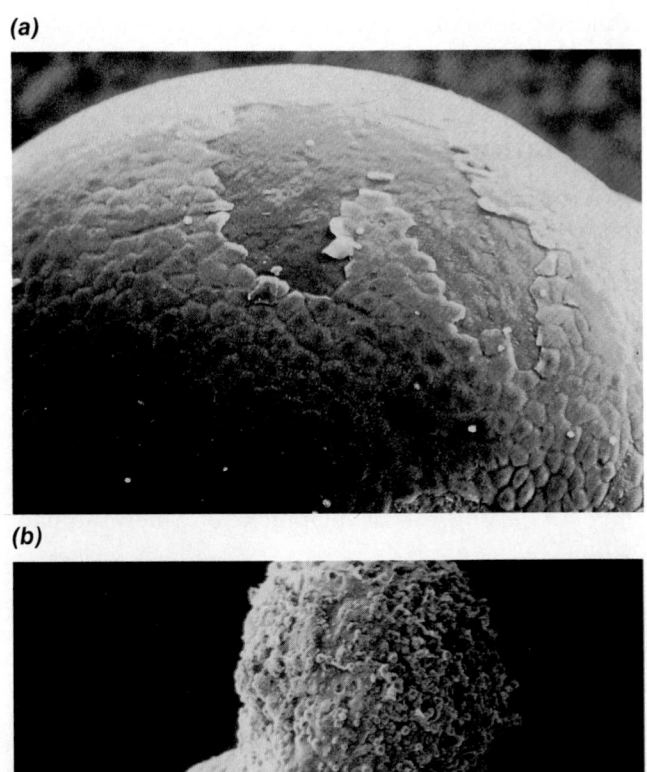

(b)

Figure 39-15 The mature follicle grows so large, and is filled with so much fluid, that it moves to the surface of the ovary **(a)** and literally bursts through the ovary wall like a miniature volcano **(b)**.

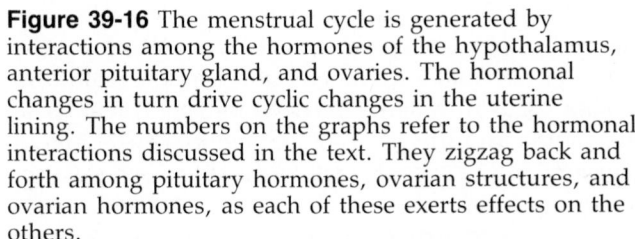

Figure 39-16 The menstrual cycle is generated by interactions among the hormones of the hypothalamus, anterior pituitary gland, and ovaries. The hormonal changes in turn drive cyclic changes in the uterine lining. The numbers on the graphs refer to the hormonal interactions discussed in the text. They zigzag back and forth among pituitary hormones, ovarian structures, and ovarian hormones, as each of these exerts effects on the others.

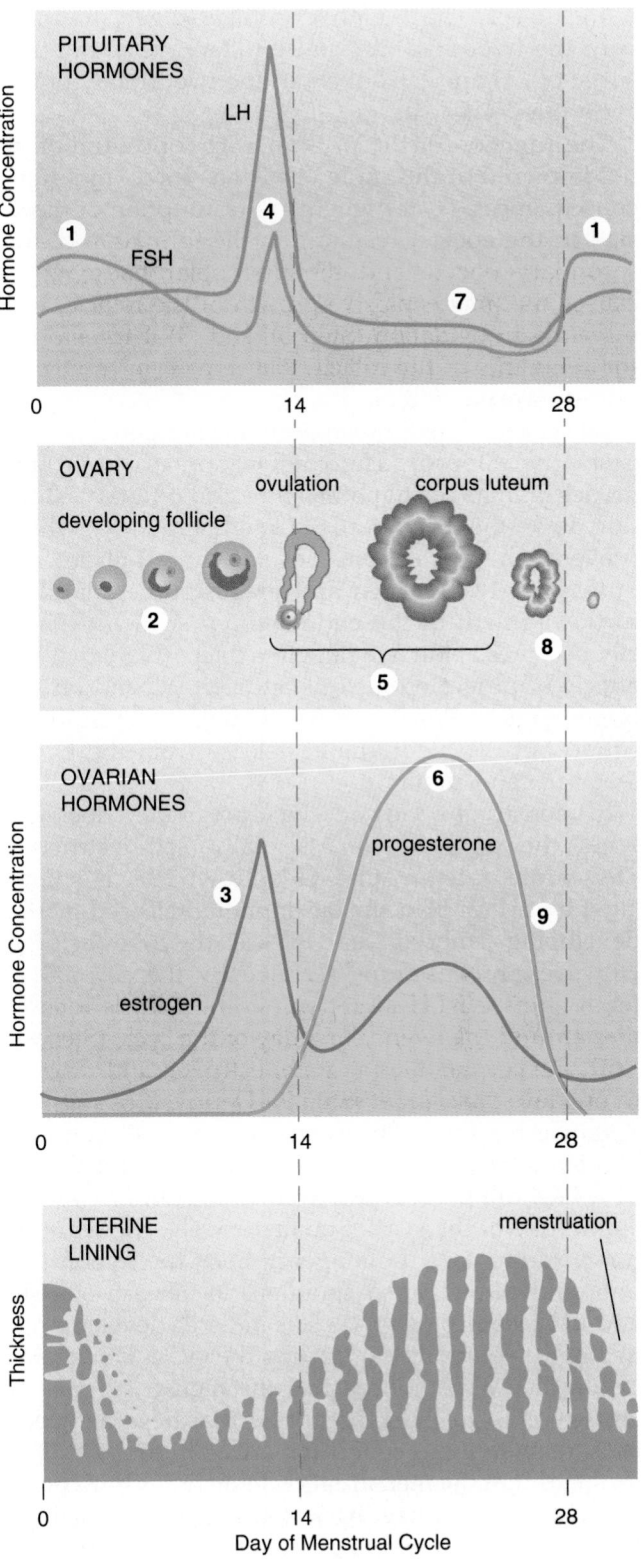

ment (see Chapter 40). For reasons that are not completely understood, only one, or rarely two, follicles complete development each month. As the maturing follicle enlarges, it secretes ever greater amounts of estrogen (Step 3). This estrogen has three effects. First, it promotes the continued development of the follicle itself and the primary oocyte contained within it. Second, it stimulates growth of the endometrium of the uterus. Third, high levels of estrogen stimulate both the hypothalamus and pituitary, resulting in a surge of LH and FSH at about the twelfth day of the cycle (Step 4).

The function of the peak in FSH concentration is not known, but the surge of LH has three important consequences: (1) it triggers the resumption of meiosis I in the oocyte, resulting in the formation of the secondary oocyte and the first polar body; (2) it causes the final explosive growth of the follicle, culminating in ovulation (Step 5); and (3) it transforms the remnants of the follicle that remain in the ovary into the corpus luteum.

The corpus luteum secretes both estrogen and progesterone (Step 6). The combination of these hormones inhibits the hypothalamus and pituitary, shutting down the release of FSH and LH (Step 7), which prevents the development of any more follicles. Simultaneously, estrogen and progesterone stimulate further growth of the endometrium, which eventually becomes about 5 millimeters thick. (Note that the effects of progesterone and estrogen depend on the target organ. Progesterone *stimulates* the endometrium, but *inhibits* hormone release from the hypothalamus and pituitary.)

In menstrual cycles in which pregnancy does not occur, the corpus luteum essentially "self-destructs." The corpus luteum survives only while it is stimulated by LH (or by a similar hormone released by the developing embryo, as we will describe below). However, progesterone secreted by the corpus luteum shuts off LH secretion, so the corpus luteum dies around the twenty-first day of the cycle (Step 8). With the corpus luteum gone, estrogen and progesterone levels plummet (Step 9). Deprived of stimulation by estrogen and progesterone, the endometrium of the uterus also dies, sloughing off as the menstrual flow beginning about the twenty-seventh or twenty-eighth day of the cycle. Simultaneously, the reduced progesterone level no longer inhibits the hypothalamus and pituitary, so spontaneous release of FSH and LH resumes (Step 1). This initiates development of a new set of follicles, starting the cycle over again.

You may have noticed that the menstrual cycle is an exception to the general rule that negative feedback regulates the concentration of hormones. The reason is that **the menstrual cycle includes both positive and negative feedback.** During the first half of the cycle, FSH and LH stimulate estrogen production. High levels of estrogen *stimulate* the midcycle surge of FSH and LH release (positive feedback). During the second half of the cycle, estrogen and progesterone together *inhibit* the release of FSH and LH (negative feedback). The early positive feedback causes hormone concentrations to reach high levels, while the later negative feedback shuts the system down again unless pregnancy intervenes.

Copulation and Fertilization

As terrestrial mammals, humans use internal fertilization to deposit the sperm in the moist environment of the female reproductive tract. To do this, the penis is inserted into the vagina, where sperm are released during ejaculation. The sperm swim upward in the female reproductive tract, from the vagina through the opening of the cervix into the uterus, and on up into the oviducts. If the female has ovulated within the past day or so, the sperm will meet an egg in one of the oviducts. Only one sperm can succeed in fertilizing it, starting the development of a new human being.

Copulation

The male role in copulation begins with erection of the penis. Before erection, the penis is flaccid, because the arterioles supplying it are constricted, allowing little blood flow (Fig. 39-17a). Under the dual influences of psychological and physical stimulation, the arterioles dilate and blood flows into vascular spaces within the penis. As these swell, they squeeze off the veins that drain the penis (Fig. 39-17b). Pressure builds up, causing an erection. After the penis is inserted into the vagina, movements further stimulate touch receptors on the penis, triggering ejaculation. Ejaculation occurs when muscles encircling the epididymis, vas deferens, and urethra contract, forcing semen out of the penis and into the vagina. The average ejaculate consists of 3 or 4 milliliters of semen, containing 300 to 400 million sperm. Male orgasm causes both ejaculation and a feeling of intense pleasure and release.

Similar changes occur in the female. Sexual excitement causes increased blood flow to the vagina and external genitalia, including the labia and clitoris (see Fig. 39-13). The clitoris, which is embryologically similar to the penis, becomes erect. Stimulation by the penis of the male often, but not always, results in female orgasm, a series of rhythmic contractions of the vagina and uterus accompanied by sensations of pleasure and release. Female orgasm is not necessary for fertilization.

The intimate contact involved in copulation creates a situation where disease organisms can be readily transmitted. Since the "sexual revolution," which began in the 1960s, more people have had multiple sexual partners. This has greatly increased the incidence of sexually transmitted diseases (see "Health Watch: Sexually Transmitted Diseases").

Fertilization

Neither sperm nor egg lives very long. An egg may remain viable for a day, while sperm, under ideal conditions, may live for two. Therefore, fertilization can succeed only if copulation occurs within a couple of days before or after ovulation. In some animals, such as rabbits, synchrony is assured by reflex ovulation, in which stimulation of the vagina by the penis triggers ovulation. In humans, things seem to be left more to chance, but in light of our rapidly burgeoning population, in most cases this is not a handicap to reproduction.

You will recall that the egg leaves the ovary surrounded by follicle cells (Fig. 39-18a). These cells, now called the **corona radiata,** form a barrier between the sperm and the egg. A second barrier, the jellylike

(a)

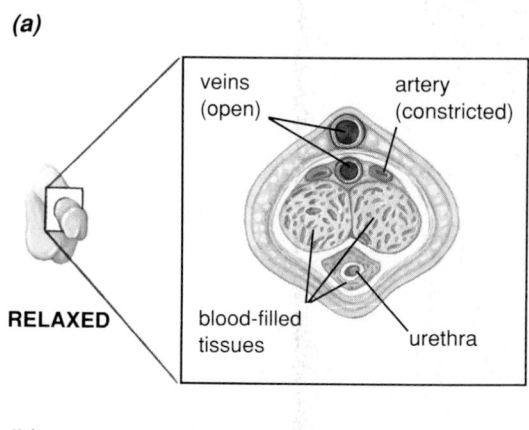

(b)

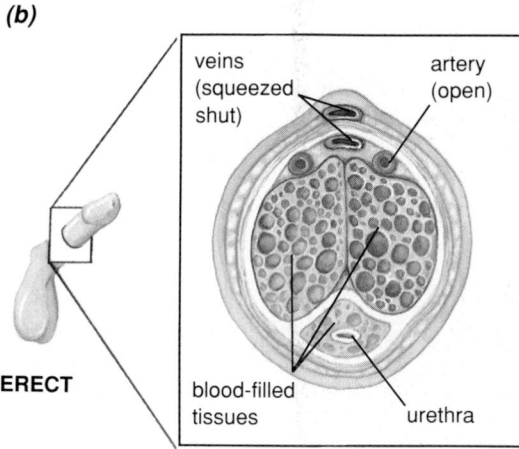

Figure 39-17 Changes in blood flow within the penis cause erection. **(a)** Normally, smooth muscles encircling the arteries leading into the penis are contracted, limiting blood flow. **(b)** During sexual excitement, these muscles relax, and blood flows into spaces within the penis. The swelling penis squeezes off the veins leaving the penis, raising the hydrostatic pressure and causing the penis to become elongated and rigid.

(a)

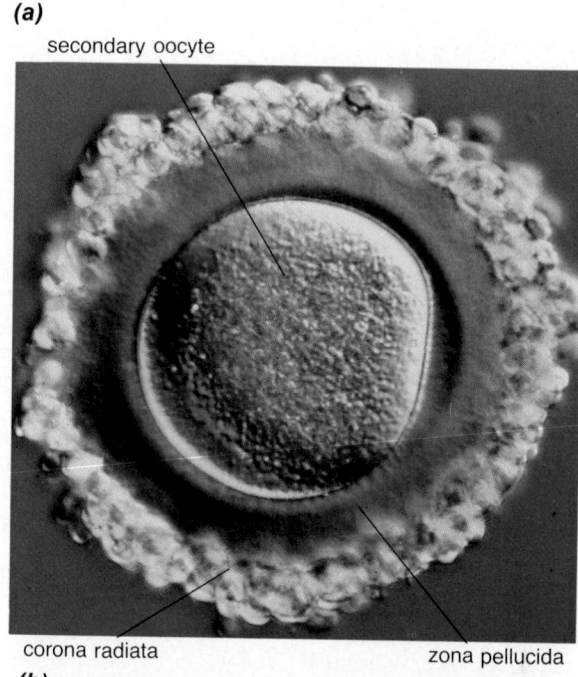

(b)

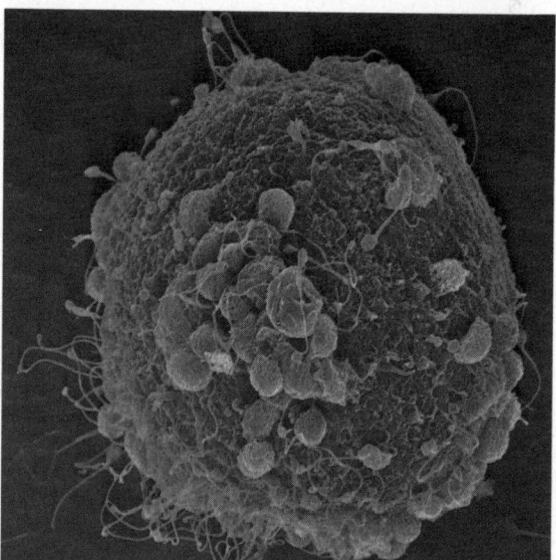

Figure 39-18 (a) A human secondary oocyte shortly after ovulation. Sperm must digest their way through the small follicular cells of the corona radiata and the clear zona pellucida to reach the oocyte itself. **(b)** Sperm surround the oocyte, attacking its defensive barriers.

HEALTH WATCH
Sexually Transmitted Diseases

Sexually transmitted diseases, caused by viruses, bacteria, protists, or arthropods that infect the sexual organs and reproductive tract, are a serious and growing health problem worldwide. As their name implies, these diseases are transmitted either exclusively or primarily through sexual intercourse. Here we discuss some of the more common of these diseases.

Bacterial Infections

Gonorrhea

Gonorrhea, an infection of the genital and urinary tract, is one of the most common of *all* infectious diseases in the United States, estimated to infect at least 2 million people per year. The bacterium, which cannot survive outside the body, is transmitted almost exclusively by intimate contact. It penetrates the membranes lining the urethra, anus, cervix, uterus, oviducts, tubes, and throat. In males, inflammation of the urethra results in a discharge of pus from the penis and painful urination. About 10% of infected males show no obvious symptoms and do not seek treatment. They become carriers who can readily spread the disease. Females typically have milder symptoms, and only about 50% seek medical help, providing an even larger number of carriers. Gonorrhea can lead to infertility by blocking the fallopian tubes with scar tissue. Infants born to infected mothers may acquire the bacterium during delivery. The bacterium attacks the eyes of newborns, and was once a major cause of blindness. Today, most newborns are immediately given antibiotic eyedrops to kill the bacterium. Treatment by penicillin was formerly highly successful, but penicillin-resistant strains now require use of other antibiotics.

Syphilis

Syphilis is a far more dangerous, though less prevalent, disease than gonorrhea. It is caused by a spiral-shaped bacterium that enters the mucus membranes of the genitals, lips, anus, or breasts. Like gonorrhea, it is readily killed by exposure to air, and is spread only by intimate contact. Syphilis begins with a sore at the site of infection. Syphilis may be cured with antibiotics. If untreated, syphilis bacteria spread through the body, multiplying and damaging many organs, including the skin, kidneys, heart, and brain, sometimes with fatal results. About 0.4% of newborns in the United States have been infected with syphilis before birth. The skin, teeth, bones, liver, and central nervous system of such infants may be damaged.

Chlamydia

Chlamydia causes inflammation of the urethra in males and the cervix in females. Like gonorrhea, *Chlamydia* can infect and sometimes block the oviducts, resulting in sterility. It can cause eye inflammations in infants born to infected mothers.

Viral Infections

AIDS

Acquired immunodeficiency syndrome, or AIDS, is caused by the HIV virus. Since the virus does not survive exposure to air, it is spread primarily by sexual activity, and by contaminated blood and needles. The HIV virus attacks the immune system, leaving the victim vulnerable to a variety of infections, which almost invariably prove fatal. Children born to mothers with AIDS sometimes become infected before or during birth. There is no cure, although certain drugs, such as AZT, can prolong life (see Chapter 21). AIDS is discussed in detail in Chapter 34.

Genital Herpes

This viral infection reached epidemic proportions during the 1970s, and continues to spread to over a million new victims yearly. It causes painful blisters on the genitals and surrounding skin. The herpes virus never leaves the body, but resides in certain nerve cells, emerging unpredictably, but possibly in response to stress. The virus is transmitted only when blisters are present. The first outbreak is the most serious; subsequent outbreaks produce fewer blisters and may be quite infrequent. The drug acyclovir, discussed in Chapter 21, may reduce the severity of the initial infection. Pregnant women with an active case of genital herpes may transmit the virus to the developing child, causing severe mental or physical disability or stillbirth. Herpes may also be transmitted from mother to infant if the infant contacts blisters as it is born, so Caesarean sections are often performed on women who have herpes blisters at the time of delivery.

Protists and Arthropod Infections

Trichomonas

This flagellated protist colonizes the mucus membranes lining the urinary tract and genitals of both males and females. The symptoms are a discharge caused by inflammation in response to the parasite. The protist is spread by intercourse, but can also be acquired through contaminated clothing and toilet articles. Lengthy untreated infections can result in sterility.

Crab Lice

Also called pubic lice, these microscopic arachnids lay their eggs in pubic hair. Their mouthparts are adapted for penetrating the skin and sucking blood and body fluids, a process that causes severe itching. "Crabs" are not only irritating, they may also spread infectious diseases. They can be controlled through careful hygiene and chemical treatments.

zona pellucida ("clear area"), lies between the corona radiata and the egg. Recent research suggests that the human egg releases a chemical attractant that guides the sperm toward it. In the oviduct, scores of sperm reach the egg and encircle the corona radiata, each sperm releasing enzymes from its acrosome (Fig. 39-18b). These enzymes weaken both the corona radiata and the zona pellucida, allowing the sperm to wriggle through to the egg. If there aren't enough sperm, not enough enzymes are released, and none of the sperm will reach the egg. This may be the selective pressure for the ejaculation of so many sperm. Perhaps 1 in 100,000 reach the oviduct, and 1 in 20 of those find the egg, so only a couple of hundred sperm join the attack on the surrounding barriers.

In humans, males who have fewer than 20 million sperm per milliliter of semen (about one fifth the normal amount) usually cannot fertilize a woman during intercourse because too few sperm reach the egg. If the sperm are otherwise normal, such men can father children by artificial insemination, in which a large quantity of their semen is injected directly into the oviduct. In other cases, a blocked oviduct may prevent sperm from reaching the egg. Today, some couples seek high technology help in the form of *in vitro* fertilization (see "Methods in Biology: *In Vitro* Fertilization").

In the normal course of events, one sperm finally contacts the surface of the egg. The cell membranes of egg and sperm fuse, and the sperm head passes into the cytoplasm of the egg. As the sperm enters the egg, it triggers two vital changes in the egg. First, vesicles near the surface of the egg release chemicals into the zona pellucida, reinforcing it and preventing further sperm from entering the egg. Second, the egg undergoes its second meiotic division, producing a haploid gamete at last. Fertilization occurs as the haploid nuclei of sperm and egg fuse, forming a diploid nucleus that contains all the genes of a new human being.

Pregnancy

The zygote begins to divide, all the while drifting down the oviduct to the uterus (Fig. 39-19). By about 1 week after fertilization, the zygote has developed into a hollow ball of cells, the **blastocyst** (Fig. 39-19b). The **inner cell mass** will become the embryo itself, while the sticky outer ball will adhere to the uterus and burrow into the endometrium, a process termed **implantation.** Blood from ruptured uterine vessels plus glycogen secreted by glands in the endometrium nourish the growing embryo. The details of embryonic development are presented in Chapter 40.

In a menstrual cycle not culminating in pregnancy, the corpus luteum disintegrates about 1 week after ovulation (see Fig. 39-16). This removes the source of estrogen and progesterone needed to sustain the enlarged endometrium, and it too disintegrates, resulting in menstruation. During pregnancy, the embryo itself prevents these changes from occurring. Shortly after the outer cells of the blastocyst begin to implant in the endometrium, they start secreting an LH-like hormone called **chorionic gonadotropin** (CG). This hormone travels in the bloodstream to the ovary, where it prevents degeneration of the corpus luteum. The corpus luteum continues to secrete estrogen and progesterone, and the uterine lining continues to grow, nourishing the embryo. So much CG is released by the embryo that it is excreted by the mother in her urine; most tests for pregnancy are assays for CG in maternal urine.

Obtaining nutrients directly from the nearby cells of the endometrium will suffice only for the first week or two of embryonic growth. During this time, the placenta begins to form, composed of interlocking tissues of the embryo and the endometrium. Through the placenta, the embryo will receive nutrients and oxygen and dispose of wastes into the maternal circulation (see Chapter 40).

Lactation

As the fetus grows, nourished by nutrients diffusing through the placenta, changes are occurring in the mother's breasts that prepare her to continue nourishing her child after it is born. The breast contains milk glands arranged radially around the nipple, each with a duct leading to the nipple. Although the breasts begin enlarging at puberty under the influence of estrogen, their glandular structure does not fully develop until pregnancy occurs (Fig. 39-20). Then, large quantities of estrogen and progesterone (acting together with several other hormones) stimulate the milk glands to grow, branch, and develop the capacity to secrete milk. The actual secretion of milk is promoted by the pituitary hormone prolactin (see Chapter 35), which rises steadily from about the fifth week of pregnancy until birth.

Immediately after birth, estrogen and progesterone levels plummet, and prolactin, which stimulates milk secretion, takes over. Suckling stimulates nerve endings in the nipples. These signal the hypothalamus, which in turn triggers an extra surge of prolactin and oxytocin from the pituitary. Oxytocin causes muscles surrounding the milk glands to contract, ejecting the milk into the ducts leading to the nipples (see Chapter 35; Fig. 35-6).

During the first few days after birth, the milk

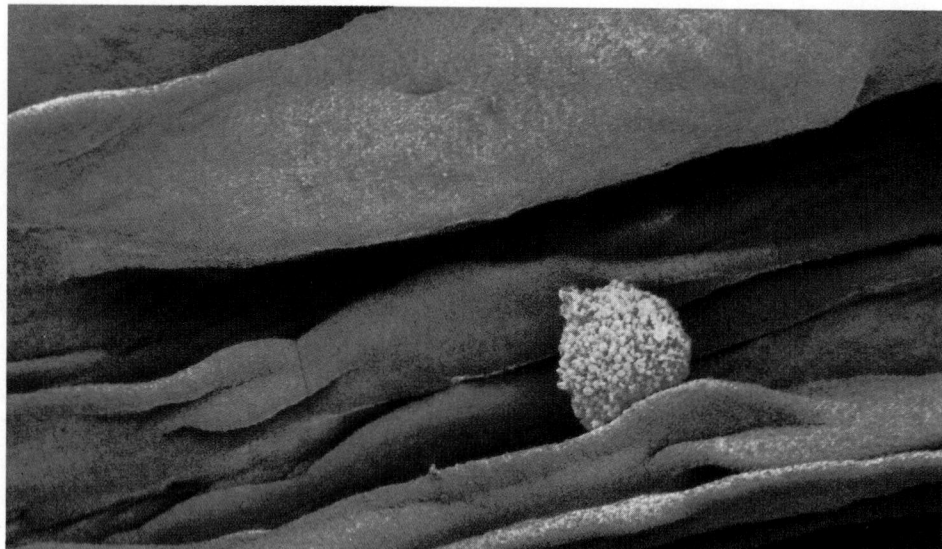

Figure 39-19 (a) The egg, surrounded by the corona radiata, travels down the oviduct toward the uterus. It emits chemicals that attract sperm, increasing its chances of being fertilized. **(b)** The egg is fertilized in the oviduct and slowly travels down to the uterus. Along the way, a few cell divisions occur, until a hollow blastocyst is formed. The inner cell mass will form the embryo proper, while the surrounding cells will adhere to the uterine endometrium, burrow in, and begin forming the placenta.

(a)

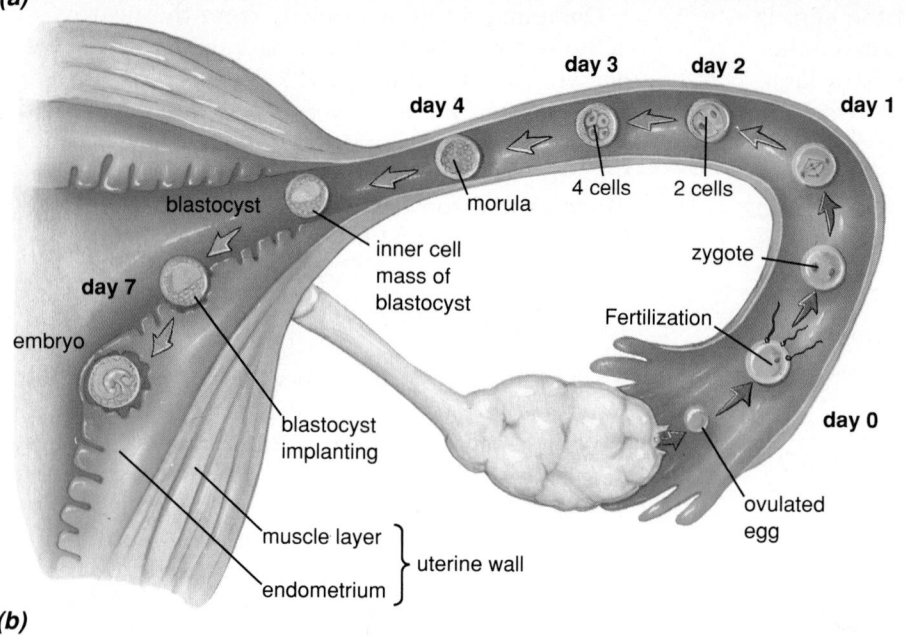

(b)

glands secrete a thin, yellowish fluid called **colostrum.** Colostrum is high in protein and contains antibodies that are absorbed directly through the infant's intestine, and help protect the newborn against some diseases. Colostrum is gradually replaced by mature milk, which is higher in fat and milk sugar (lactose) and lower in protein.

Delivery

Near the end of the ninth month, give or take a few weeks, the process of birth normally begins (Fig. 39-21). It is thought that the infant itself somehow signals "readiness for birth," perhaps via a hormone, but no one knows for certain what the signal is. However they are triggered, the immediate causes of delivery seem to be a complex interplay among prostaglandins released by the uterus and the fetal membranes, the stretching of the uterus and cervix as the baby grows and is delivered, and oxytocin released by the posterior pituitary gland.

Unlike skeletal muscles, uterine muscles can contract spontaneously, and these contractions are enhanced by stretching. As the baby grows, it weighs

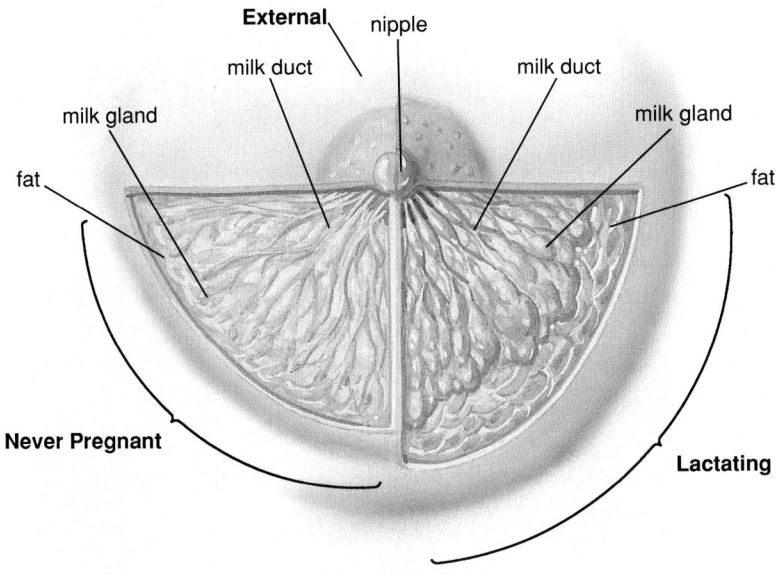

External nipple
milk duct
milk duct
milk gland
milk gland
fat
fat

Never Pregnant

Lactating

Figure 39-20 The structure of the mammary glands. During pregnancy, both fatty tissue and the milk-secreting glands and ducts increase in size.

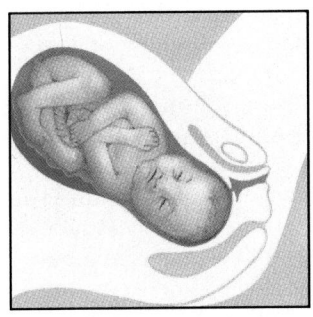 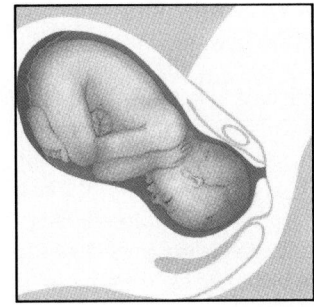 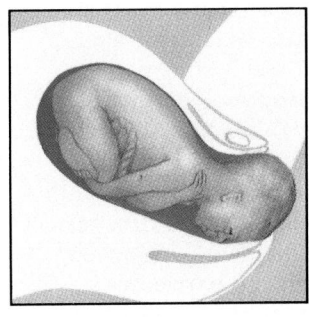 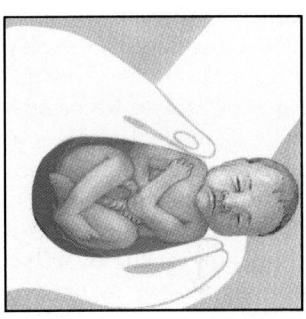

(a) The baby is oriented head downward, facing the mother's side. The cervix is beginning to thin (efface) and expand in diameter (dilate).

(b) The cervix is completely dilated to 10 cm and the baby's head has entered the vagina, or birth canal. The baby has rotated to face the mother's back.

(c) The baby's head is emerging (crowning).

(d) The baby has rotated to the side once again as the shoulders emerge.

Figure 39-21 Delivery.

more and fills up the uterus, stretching the uterine muscles. These occasionally contract weeks before delivery. In apparent response to the fetal "signal," prostaglandins are released both by the uterus and the membranes surrounding the fetus. These hormones cause intense contractions to begin, signaling the onset of labor. As the contractions proceed, the baby's head pushes against the cervix, making it dilate. Stretch receptors in the walls of the cervix send signals to the hypothalamus, triggering oxytocin release. Under the dual stimulation of prostaglandin and oxytocin, the uterus continues to contract, finally expelling the baby. After a brief rest, uterine contractions resume, so that the uterus shrinks remarkably.

METHODS IN BIOLOGY

In Vitro Fertilization

Louise Brown, the first "test tube baby," was born in England in 1978. Since that well-publicized event, over 200 centers for *in vitro* fertilization (IVF) have been established in the United States; these have produced well over 5000 healthy babies. The demand for IVF is fueled by an epidemic of infertility. Surveys indicate that one out of every six American couples has been unable to conceive after trying for one year or more; a rate that has tripled in the past 20 years. One reason for the increase in infertility is that modern couples more often delay childbearing, and fertility declines with age. A second reason is a higher incidence of sexually transmitted diseases such as *Chlamydia* and gonorrhea that can scar and block the oviducts or sperm ducts. IVF can overcome both blocked ducts and low sperm counts, since oocytes are removed directly from the ovaries, and their meeting with sperm is guaranteed within the confines of a small glass dish (*in vitro* means "in glass" in Latin). The popularity of IVF, which costs around $5000 per attempt, is a testimony to the strong biological drive to have children. With 25% considered an excellent success rate, the average child conceived by this method will cost its parents $20,000 even before he or she is born.

Although the technique is simple in concept, the procedure is complex and delicate. First, the woman is given daily injections of drugs and/or hormones to stimulate multiple ovulation. Using blood tests and ultrasound imaging of the ovaries, doctors determine when the time is ripe for ovulation. Then the women is injected with human chorionic gonadotropin, which begins the process of expelling the oocytes from the follicles. Just prior to the ejection of the oocytes, surgeons insert a thin fiberoptic viewing device through a small abdominal incision to locate the mature follicles. Next, they insert a long, hollow needle into each ripe follicle and suck out the follicular cells and fluid, which are examined under a microscope. With luck, an oocyte is present (Fig. E39-1). Usually, at least four oocytes are harvested and incubated in a glass dish to which freshly collected sperm are then added. In 48 hours, about two thirds of the oocytes will have been fertilized and have reached the 8-cell stage. These early embryos are sucked into a tube and expelled very gently into the uterus. Transplanting multiple embryos increases the success rate for implantation, but at the same time increases the probability (and the risks) of multiple births.

Recently, IVF has become a weapon in the fight to save endangered species. The National Zoo in Washington D.C. has been working since 1984 to adapt IVF technology to help endangered species reproduce. They have developed a mobile IVF laboratory that can travel

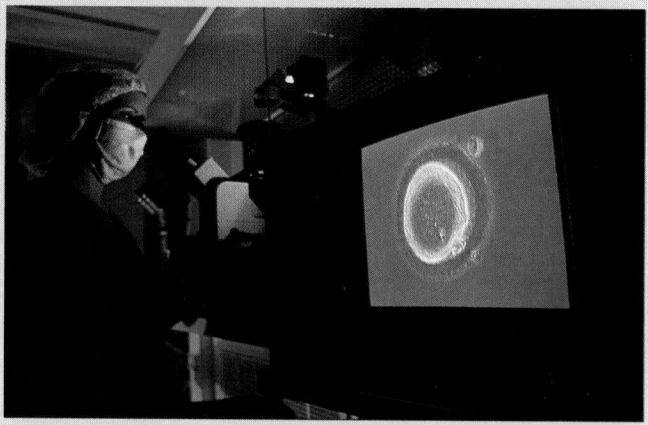

Figure E39-1 The microscopic egg cell is found in the follicular fluid and its image is projected onto a screen for examination.

to zoos where endangered species are housed. A tremendous advantage of IVF is that it will allow sperm from an endangered male to be transported between continents, if necessary, to fertilize an appropriate female. This eliminates the danger and trauma of transporting the animals themselves. IVF also overcomes the very real probability that, once together, the animals will refuse to mate. In April 1990, the first "test-tube tiger" was born (Fig. E39-2). Only 200 of this rare Siberian subspecies remain in the wild. IVF increases the chances that we may save this and other precious forms of life on Earth.

Figure E39-2 The world's first test-tube tiger, born in 1990, is the first successful use of IVF in the fight to save endangered species.

During these contractions, the placenta is sheared off from the uterus and is expelled through the vagina as the "afterbirth."

Further release of prostaglandins in the umbilical cord causes the muscles surrounding fetal blood vessels in the cord to contract, shutting off blood flow. (Tying off the cord is standard practice but is not usually necessary; if it were, other mammals would not survive birth!) Although he or she is still intimately dependent on the parents for survival, a new human being has been born.

On Limiting Fertility

Successful reproduction is essential if any species, including our own, is to endure. During most of human evolution, child mortality was high. Therefore, natural selection favored people who tended to produce many children. With a few tragic exceptions, people today enjoy low infant mortality and a life span triple that of ancient times. Parents need not have a dozen children to ensure that a few survive to adulthood. Today's humans have sexual drives appropriate to prehistoric times, but live under vastly different circumstances. As discussed in Chapter 43, every 4 days a million new people are added to our increasingly overcrowded planet, and controlling human births has become crucial. On the individual level, limiting reproduction allows people to complete their educations, in some cases pursue careers, and generally provide the best opportunities for their children.

Historically, it has not been easy to separate the acts of copulation and conception. Primitive cultures tried such inventive, if bizarre, techniques as swallowing froth from the mouth of a camel or using vaginal suppositories of crocodile dung. Even 50 years ago there were no reliable methods of birth control. Over the past 20 years, however, several effective techniques have been developed for preventing pregnancy (Table 39-3).

Permanent Contraception

The most effective, and, in the long run, most effortless method of contraception is **sterilization,** in which the pathways through which sperm or egg must travel are interrupted. In men, the vas deferens leading from each testis may be severed in an operation called a **vasectomy** (Fig. 39-22a). Sperm are still produced, but they cannot reach the penis during ejacu-

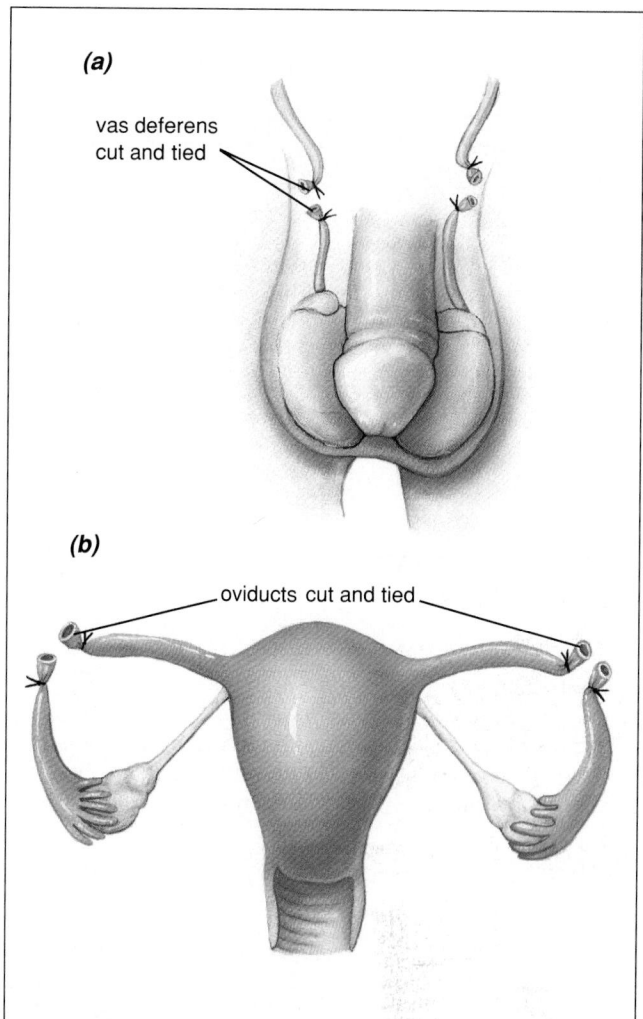

Figure 39-22 Sterilization is the most effective form of birth control. It is not dependably reversible. **(a)** In the male, the vas deferens is reached through a small slit in the scrotum. It is cut and each end is tied off. **(b)** In the female, the oviduct is reached through an abdominal incision. It is cut, and each end is tied off.

lation. The surgery is performed under a local anesthetic, and vasectomy has no known physical side effects on health or sexual performance. The slightly more complex operation of **tubal ligation** renders a woman infertile by cutting her oviducts (Fig. 39-22b). Over 23% of United States women of childbearing age have opted for this form of birth control. Ovulation still occurs, but sperm cannot travel to the egg, nor can the egg reach the uterus. Occasionally, tubal ligation leads to irregular menstrual periods and increased menstrual flow. Sterilization is generally permanent. However, sometimes, in a delicate and ex-

Table 39-3   Birth Control Techniques

Method	Mechanism of Action	Possible Side Effects	Failure Rate (pregnancies per 100 women per year)
Sterilization			
Vasectomy	Sever vas deferens, so sperm cannot reach penis	None known	0
Tubal ligation	Sever oviduct, so egg cannot reach uterus	Occasional menstrual irregularity	0
Prevent Ovulation			
Birth control pill	Prevents ovulation	Blood clotting, breast soreness, water retention, nausea, vitamin deficiencies	0–5
Barrier Methods			
Diaphragm (with spermicide)	Block opening of cervix, kill sperm	Spermicide may irritate vaginal lining	3–20
Contraceptive sponge	Block opening of cervix, kill sperm	Spermicide may irritate vaginal lining	? (prob. low)
Condom	Cover penis, prevent sperm from entering vagina	Reduced sensation in penis	3–20
Spermicide alone	Kill sperm	Spermicide may irritate vaginal lining	3–30
Withdrawal	Remove penis from vagina before ejaculation	None	10–30
Douching	Wash sperm out of vagina before they can pass through cervical opening	Douching solution may irritate vaginal lining	20–40
Rhythm	Avoid intercourse during probable time of ovulation	None	5(?)–30
Prevent Implantation			
Intrauterine device	Prevent implantation of fertilized egg	May irritate or perforate uterine lining	1–5
"Morning after" pill	Prevent implantation of fertilized egg	Breast soreness, water retention, nausea, blood clotting, cancer(?)	? (prob. low)

pensive operation, a surgeon can reconnect the vas deferens or oviducts.

Temporary Contraception

Temporary contraception techniques fall into three general categories: preventing ovulation, preventing sperm and egg from meeting when ovulation does occur, and preventing implantation of a fertilized egg in the uterus.

Preventing Ovulation

As you learned earlier in this chapter, during a normal menstrual cycle ovulation is triggered by a mid-cycle surge of LH. An obvious way to prevent ovulation is to suppress LH release by providing a continuing supply of estrogen and progesterone. Estrogen and progesterone (usually in synthetic form) are the components of **birth control pills.** "The Pill" is extremely effective, but must be taken daily, usually for 21 days each menstrual period. During the first few months, the hormones in the Pill can induce side effects such as water retention, soreness of the breasts, nausea, and especially in women over 40, a slightly increased risk of blood clots. These side effects are extremely rare in women taking pills that have very low doses of estrogen.

A new form of long-term contraception, called Norplant, already available in 16 countries, was approved for use in the United States in December 1990. It consists of six $1\frac{1}{3}$ inch silicone rubber rods inserted under the skin of the upper arm. These contain a synthetic hormone (levonorgestrel) that prevents ovulation. The rods provide gradual, steady diffusion of hormone into the bloodstream for 5 years. In extensive tests, Norplant has proven slightly more effective than the birth control pill. Women using Norplant become fertile within months after the capsule is removed.

Barrier Methods

There are several effective barrier methods that prevent the encounter of sperm and egg. One is the **diaphragm,** a rubber cap that fits snugly over the cervix, preventing sperm from entering the uterus. In conjunction with a spermicide, diaphragms are very effective and have no known side effects. The **contraceptive sponge** is another type of barrier. A soft, spermicide-impregnated plug is inserted in the vagina up against the cervical opening, physically blocking the opening and killing any sperm that approach. Alternatively, a **condom** may be worn over the penis, preventing sperm from being deposited in the vagina. Diaphragms, sponges, and especially condoms must be applied shortly before intercourse, often at a time when the participants would rather be thinking about something else. Furthermore, if the diaphragm or condom happens to have even a small hole, sperm may still enter the uterus and fertilize the egg. (It seems a cruel irony that a man ejaculating 50 million sperm will often be infertile, but if just a few drops of semen escape past a diaphragm or condom, pregnancy occasionally results.)

Other, less effective, procedures include **spermicides alone, withdrawal** (removal of the penis from the vagina just before ejaculation), and **douching** (washing sperm out of the vagina, hopefully before they have had a chance to enter the uterus). Spermicides have some contraceptive effect, but withdrawal and douching are essentially useless. A final method of preventing fertilization is the **rhythm method:** abstinence from intercourse during the ovulatory period of the menstrual cycle. In practice, rhythm usually has a high failure rate, due to lack of

discipline on the part of the users or inaccuracies in determining the menstrual cycle, which varies somewhat from month to month. Since a slight rise in body temperature usually coincides with ovulation, the rhythm method can be made more effective if the woman's body temperature is recorded daily and sexual activity is regulated accordingly.

Preventing Implantation

Even if an egg is fertilized, pregnancy will not occur unless the blastocyst implants in the uterus. The **intrauterine device** (IUD) is a small copper or plastic loop, squiggle, or shield that is inserted into the uterus and remains there until removed by a physician. Although highly effective (if it stays in place), IUDs seem to work by irritating the uterine lining so that it cannot receive the embryo. IUDs occasionally cause discomfort and uterine infection, and in rare instances, IUDs have punctured the uterine wall. A second method of preventing implantation is the "morning after" pill, which contains a massive dose of estrogen. For some women, preventing implantation has the major drawback that it is, in effect, an extremely early abortion.

Abortion

When contraception fails, the pregnancy may be terminated by an **abortion.** Abortion commonly involves dilating the cervix and removing the embryo and placenta by suction. The recently discovered compound RU-486, taken as a pill, is now routinely used in France to terminate pregnancy during the first 2 months. This substance binds to progesterone receptors and blocks the actions of progesterone, which are essential to the maintenance of pregnancy. Abortions are expensive medical procedures that are much more dangerous to a woman's health than the contraceptive techniques described above. In addition, abortions kill a human fetus. Science can describe fetal development during pregnancy, but cannot provide judgments about when a fetus becomes a "person," legally and ethically, nor about the relative merits of fetal versus maternal rights. Therefore, abortion is likely to remain a controversial procedure.

Future Contraceptive Methods

Further advances in contraception are on the way. One possibility is a removable plug for the oviduct made of silicone rubber. Such a plug might be removable if the woman later wishes to become pregnant. There are also once-a-month contraceptive pills under development, most of which prevent the uterus from becoming fully prepared for implantation. Another possibility is a contraceptive vaccine that would remain effective for one to several years. Two different types of vaccine are under development. One, currently being tested in India, induces antibodies against human chorionic gonadotropin, which is essential for the implantation of the embryo in the uterus. The second, under development at the University of Virginia, induces the formation of antibodies to a protein called SP-10 that is unique to sperm. The vaccine is currently being tested on baboons, whose sperm also carries SP-10.

You may have noticed that most of the contraceptive techniques are directed at the woman, not the man. It is much easier to interfere with ovulation than with sperm formation. A woman ovulates only once a month, and ovulation itself does not apparently influence a woman's sexual drives. In contrast, testosterone is essential both for sperm formation and sexual performance; early "male pills" caused not only infertility but also impotence. Nevertheless, there is a major research effort underway to develop male contraceptives equivalent to the Pill. A promising contraceptive is a daily dose of testosterone and a modified form of gonadotropin-releasing hormone, which together seem to block sperm production without affecting sexual performance. Clinical trials are now underway.

SUMMARY OF KEY CONCEPTS

Reproductive Strategies

Animals reproduce either sexually or asexually. Sexual reproduction involves the union of haploid gametes, usually from two separate parents, which produces an offspring that is genetically different from either parent. In asexual reproduction, offspring are usually genetically identical to the parent. Asexual reproduction may occur by regeneration, fission, budding, or parthenogenesis.

Among animals that engage in sexual reproduction, the female produces large, nonmotile eggs, while the male produces small, motile sperm. Animals may be either monoecious, in which a single animal produces both sperm and eggs, or dioecious, in which a single animal produces one type of gamete. The union of sperm and egg, called fertilization, may occur outside the bodies of the animals (external fertilization) or inside the body of the female (internal fertilization). External fertilization must occur in water, so that the sperm can swim to meet the egg. Most internal fertilization is through copulation, in which the male deposits sperm directly into the female reproductive tract.

Mammalian Reproduction

The human male reproductive tract consists of paired testes that produce sperm and testosterone, and accessory structures that conduct the sperm to the female's reproductive tract and contribute fluids, nutrients, and activating factors required for sperm motility. In human males, spermatogenesis and testosterone production are stimulated by FSH and LH, secreted the anterior pituitary. Spermatogenesis and testosterone production are nearly continuous, beginning at puberty and lasting until death.

The human female reproductive tract consists of paired ovaries that produce eggs and the hormones estrogen and progesterone, and accessory structures that conduct sperm to the egg and receive and nourish the embryo during prenatal development. In human females, oogenesis, hormone production, and development of the lining of the uterus vary in a monthly menstrual cycle. The cycle is controlled by hormones from the hypothalamus (gonadotropin releasing factor), anterior pituitary (FSH and LH), and ovaries (estrogen and progesterone).

During copulation, the male inserts his penis into the female's vagina and ejaculates semen. The sperm move through the vagina and uterus into the oviduct, where fertilization usually takes place. The unfertilized egg is surrounded by two barriers, the corona radiata and the zona pellucida. Enzymes released from the acrosomes at the tips of sperm digest these layers, permitting sperm to reach the egg. Only one sperm enters the egg and fertilizes it.

The fertilized egg undergoes a few cell divisions in the oviduct and then implants in the uterine lining. Implantation and subsequent release of chorionic gonadotropin by the embryo maintains the integrity of the corpus luteum and the endometrium during early pregnancy, preventing further menstrual cycles.

During pregnancy, milk glands in the mother's breasts are enlarging under the influence of estrogen, progesterone, and other hormones. After birth, milk secretion is triggered by prolactin and oxytocin, whose release is triggered by suckling.

After about 9 months, uterine contractions are triggered by a complex interplay of uterine stretch and prostaglandin and oxytocin release. These expel first the baby, and then the placenta.

On Limiting Fertility

Permanent contraception can be achieved by sterilization: severing the vas deferens in males (vasectomy) or the oviducts in females (tubal ligation). Temporary contraception techniques include those that prevent ovulation: birth control pills, and Norplant, a new 5-year subdermal contraceptive. Barrier methods prevent sperm and egg from meeting. These include the diaphragm, the contraceptive sponge, and the condom, accompanied by spermicide. Spermicide alone is less effective, while withdrawal and douching are poor techniques. The rhythm method involves sexual abstinence around the time of ovulation. Intrauterine devices prevent implantation of the blastocyst. Abortion, which causes the expulsion of the developing embryo, is the most hazardous and controversial means of preventing birth.

GLOSSARY

acrosome (ak'-rō-sōm): an enzyme-containing vesicle located at the tip of an animal sperm.

amplexus (am-pleck'-sus): a form of external fertilization found in amphibians, in which the male holds the female during spawning and releases his sperm directly onto her eggs.

asexual reproduction: reproduction not involving the union of genetic material from two different organisms. Usually, asexual reproduction produces genetically identical copies of the parent organism.

blastocyst (blas'-tō-sist): an early stage of human embryonic development, consisting of a fluid-filled ball with walls one cell layer thick, enclosing a mass of cells attached to its inner surface.

bud: in animals, a small copy of an adult that develops on the body of the parent. It eventually breaks off and becomes independent.

budding: asexual reproduction by growth of a miniature copy, or bud, of the adult animal on the body of the parent. The bud breaks off to begin independent existence.

bulbourethral gland (bul-bō-ū-rē'-thrul): in male mammals, a gland that secretes a basic, mucus-containing fluid that forms part of the semen.

cervix (ser'-vicks): a ring of connective tissue at the outer end of the uterus, leading into the vagina.

chorionic gonadotropin: a hormone secreted by the chorion (one of the fetal membranes), which maintains the integrity of the corpus luteum during early pregnancy.

colostrum (kō-los'-trum): a yellowish fluid high in protein and containing antibodies, that is produced by the female breasts before milk secretion begins.

copulation: reproductive behavior in which the penis of the male is inserted into the body of the female, where it releases sperm.

corona radiata (ka-rō'-na rā-dē-a'-ta): the layer of cells surrounding an egg after ovulation.

corpus luteum (kor'-pus lū'-tē-um): in the mammalian ovary, a structure derived from the follicle after ovulation, which secretes the hormones estrogen and progesterone.

dioecious (dī-ē'-shus): pertaining to organisms in which male and female gametes are produced by separate individuals.

egg: the haploid female gamete, usually large and nonmotile, containing food reserves for the developing embryo and regionally localized gene-regulating substances that direct early development.

endometrium (en-dō-mē'-trē-um): the nutritive inner lining of the uterus.

epididymis (e-pi-di'-dē-mus): tubes that connect with and receive sperm from the seminiferous tubules of the testis.

estrogen: in vertebrates, a female sex hormone produced by follicle cells of the ovary, which stimulates follicle development, oogenesis, development of secondary sex characteristics, and growth of the uterine lining.

external fertilization: union of sperm and egg outside the body of either parental organism.

fimbria (fim'-brē-a; pl. fimbriae): in female mammals, the ciliated, fingerlike projections of the oviduct that sweep the ovulated egg from the ovary into the oviduct.

fission: asexual reproduction by dividing the body into two smaller, complete organisms.

follicle: in the ovary of female mammals, the oocyte and its surrounding accessory cells.

follicle-stimulating hormone (FSH): a hormone produced by the anterior pituitary gland that stimulates spermatogenesis in males and development of the follicle in females.

hermaphrodite (her-ma'-frō-dīt): an organism that produces both male and female gametes.

implantation: the process whereby the early embryo embeds itself within the lining of the uterus.

internal fertilization: union of sperm and egg inside the body of the female.

interstitial cells (in-ter-sti'-shul): in the vertebrate testis, testosterone-producing cells located between the seminiferous tubules.

luteinizing hormone (LH): a hormone produced by the anterior pituitary gland that stimulates testosterone production in males and development of the follicle, ovulation, and production of the corpus luteum in females.

menstrual cycle: in humans, a complex 28-day cycle during which hormonal interactions among the hypothalamus, pituitary gland, and ovary coordinate ovulation and the preparation of the uterus to receive and nourish the fertilized egg. If pregnancy does not occur, the uterine lining is shed during menstruation.

menstruation: in females of some primate species, the monthly discharge of uterine tissue and blood from the vagina.

monoecious (mon-ē'-shus): pertaining to organisms in which male and female gametes are produced in the same individual.

myometrium (mī-ō-mē'-trē-um): the muscular outer layer of the uterus.

oogonium (ō-ō-gō'-nē-um): a diploid cell in female animals that gives rise to a primary oocyte.

ovary: the gonad of female animals.

oviduct: in mammals, the tube leading from the ovary to the uterus.

parthenogenesis (par-the-nō-gen'-i-sis): a specialization of sexual reproduction, in which an egg undergoes development without fertilization.

placenta: in mammals, a structure formed of both embryonic and maternal tissues, which serves to exchange nutrients and wastes between embryo and mother.

polar body: in oogenesis, a small cell containing a nucleus but virtually no cytoplasm.

primary oocyte (ō'-ō-sīt): a diploid cell, derived from the oogonium by growth and differentiation, which undergoes meiosis to produce the egg.

primary spermatocyte (sper-ma'-tō-sīt): a diploid cell, derived from the spermatogonium by growth and differentiation, which undergoes meiosis to produce four sperm.

progesterone (prō-ge'-ster-ōn): a hormone produced by the corpus luteum that promotes development of the uterine lining.

prostate gland (prō'-stāt): a gland that produces part of the fluid component of semen. The prostate fluid is basic and contains a chemical that activates sperm movement.

regeneration: (1) regrowth of a body part after loss or damage; (2) asexual reproduction by regrowth of an entire body from a fragment.

scrotum (skrō'-tum): the pouch of skin containing the testes of male mammals.

secondary oocyte (ō'-ō-sīt): a large haploid cell derived by meiosis I from the diploid primary spermatocyte.

semen: the sperm-containing fluid produced by the male reproductive tract.

seminal vesicle: in male mammals, a gland that produces a basic, fructose-containing fluid that forms part of the semen.

seminiferous tubules (semi-i-ni'-fer-us): a series of tubes in the vertebrate testis in which sperm are produced.

sertoli cell: a large cell in the seminiferous tubule that regulates spermatogenesis and nourishes the developing sperm.

sexual reproduction: a form of reproduction in which genetic material from two parental organisms is combined in the offspring. Usually, two haploid gametes fuse to form a diploid zygote.

spawning: a method of external fertilization in which male and female parents shed gametes into the water, and sperm must swim through the water to reach the eggs.

sperm: the haploid male gamete, usually small, motile, and containing little cytoplasm.

spermatid: a haploid cell derived from the secondary spermatocyte by meiosis II. The mature sperm is derived from the spermatid by differentiation.

spermatogenesis: the formation of sperm.

spermatogonium (pl. spermatogonia): a diploid cell lining the walls of the seminiferous tubules that gives rise to a primary spermatocyte.

testis (pl. testes): the gonad of male mammals.

testosterone: in vertebrates, a hormone produced by the interstitial cells of the testis; stimulates spermatogenesis and the development of male secondary sex characteristics.

urethra (ū-rē'-thra): the tube leading from the urinary bladder to the outside of the body; in males, the urethra also receives sperm from the vas deferens and conducts both sperm and urine (at different times) to the tip of the penis.

uterus: in female mammals, the part of the reproductive tract that houses the embryo during pregnancy.

vagina: the passageway leading from the outside of the body to the cervix of the uterus.

vas deferens (vas de'-fer-ens): the tube connecting the epididymis of the testis with the urethra.

zona pellicida (pel-ū'-si-da): a clear, noncellular layer between the corona radiata and the egg.

zygote: the fertilized egg.

STUDY QUESTIONS

1. Distinguish between sexual and asexual reproduction, and list at least one advantage of each.
2. Describe three types of asexual reproduction and give an example of an animal showing each type.
3. Describe external fertilization, spawning, and amplexus.
4. How do animals using external fertilization synchronize the release of eggs and sperm?
5. Why is internal fertilization essential for terrestrial animals?
6. List the structures, in order, through which a sperm passes on its way from the seminiferous tubules of the testis to the oviduct of the female.
7. Name the three accessory glands of the male reproductive tract. What are the functions of the secretions they produce?

8. Draw the structure of a mature human sperm, and list the functions of each part.
9. What is the corpus luteum? From what structure in the ovary is it derived? What determines its survival after ovulation?
10. Diagram the menstrual cycle and describe the interactions among hormones produced by the pituitary gland and ovaries that produce the cycle.
11. Diagram the structure of the ovulated egg. How do the sperm penetrate the barriers surrounding the egg?
12. What event early in pregnancy prevents the degeneration of the corpus luteum?
13. Describe the changes in the breast that prepare a mother to nurse her child. How do hormones influence these changes and stimulate milk production?

DISCUSSION QUESTIONS

1. Identify and discuss some of the ethical issues involved in *in vitro* fertilization and surrogate motherhood.
2. Discuss the most appropriate method of birth control for each of the following couples: Couple A has intercourse three times a week but does not ever want to have children; Couple B has intercourse only once a month, and may want to have children someday; and couple C has intercourse three times a week and wants to have children someday.

SUGGESTED READINGS

Eberhard, W. G. "Runaway Sexual Selection." *Natural History,* December 1987. Interesting article on the evolution of male genitalia.

Gold, M. "The Baby Makers." *Science 85,* April 1985. In vitro fertilization is the only way that some otherwise infertile couples can conceive. Gold describes the various techniques fertility clinics use to promote pregnancy in their patients.

Grady, D. "Pregnancies That Can Kill." *Discover,* June 1983. Describes the dangers of ectopic pregnancies and the reasons why they are on the increase.

Grobstein, C. "External Human Fertilization." *Scientific American* June 1979. The technique of in vitro fertilization raises both ethical and legal questions.

Nilsson, L., et al. *A Child is Born.* New York: Delacorte Press/Seymour Lawrence, 1986. Embryonic development depicted in a series of fantastic photographs.

Ulmann, A., Teutsch, G., and Philibert, D. "RU 486." *Scientific American,* June 1990. Describes the controversial abortion pill now widely used in France, and other applications for this drug.

Wassarman, P. W. "Fertilization in Mammals." *Scientific American,* December 1988. Describes the events leading to the penetration of the egg by the sperm with emphasis on the role and composition of the zona pellucida.

40

Animal Development

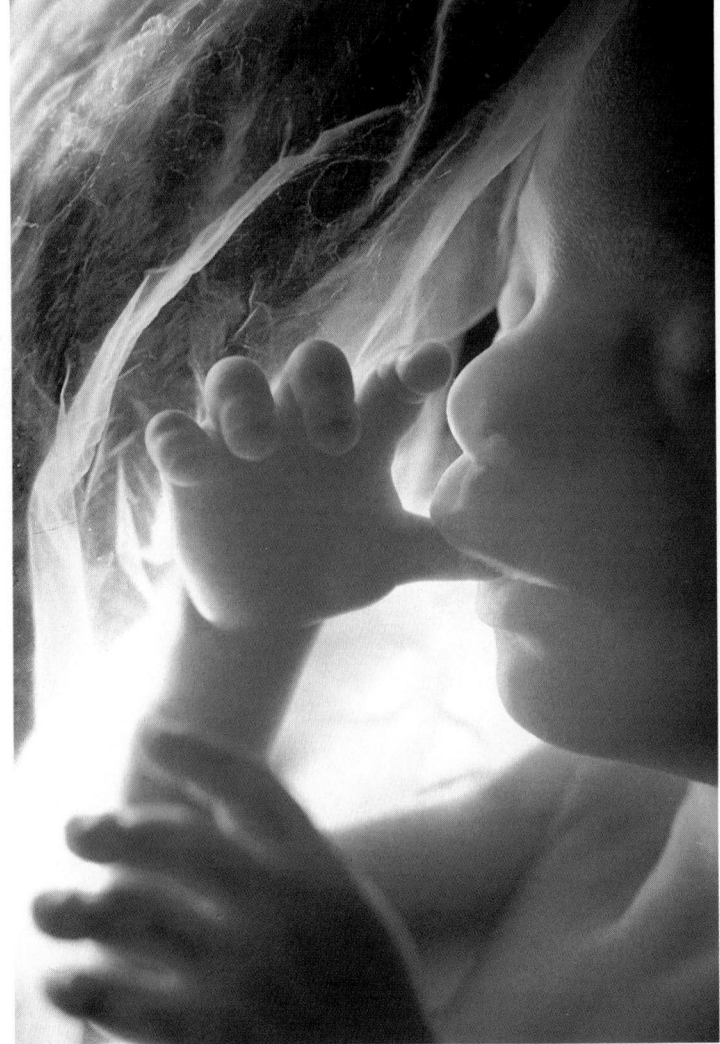

Behaviors as well as bodies are molded during the complex process of development, as demonstrated by this 18-week human fetus.

Out of millions of contestants, a single sperm fuses with the egg. The two haploid nuclei—one from the egg, contributed by the mother, and one from the sperm, contributed by the father—fuse to create a diploid cell, the beginning of a new generation. How does this single cell give rise to the trillions of cells of the adult body? It cannot be a simple matter of cell division after cell division, for that would merely give rise to a massive lump of identical cells. Rather, as the cells divide, they also must **differentiate**—that is, they must specialize to become particular cell types, such as liver, brain, or muscle.

How do the cells of the body become different from one another, when all are descended from the same fertilized egg? How do all the organs of the adult body assume their correct locations, and connect with one another in the proper way? What governs the onset of puberty and reproductive behavior? Why do animals age? Is death inevitable? These questions inspire the study of **development,** the process by which an organism proceeds from fertilized egg through adulthood to eventual death. In this chapter, we examine animal development, especially the early embryonic stages, and discuss some of the mechanisms that control it. Finally, we describe the development of the human embryo.

Differentiation

How do cells become differentiated from one another during development? Since the characteristics of each cell are ultimately determined by its genes, one possibility might be that differentiation results from a progressive loss of genes. By this scenario, the zygote would contain all the genes needed to direct the construction of the whole organism, and each differentiated cell would lose those genes that aren't needed for its particular function in the body. In an ingenious experiment, the British molecular biologist J. B. Gurdon showed that gene loss cannot be the mechanism of differentiation. Gurdon implanted nuclei from intestinal cells of tadpoles of the African clawed frog, *Xenopus*, into unfertilized eggs whose own nuclei had been destroyed (Fig. 40-1). Although the operation was very difficult and most of the "patients" died, some of the eggs with intestinal nuclei developed into normal adult frogs. Since the nucleus from an intestinal cell provided all the genetic information necessary to form a normal tadpole, these experiments led to the following conclusion: **Differentiated cells each contain all the genetic information needed for the development of the entire organism.** Cell types differ, then, not in what genes they contain, but in what

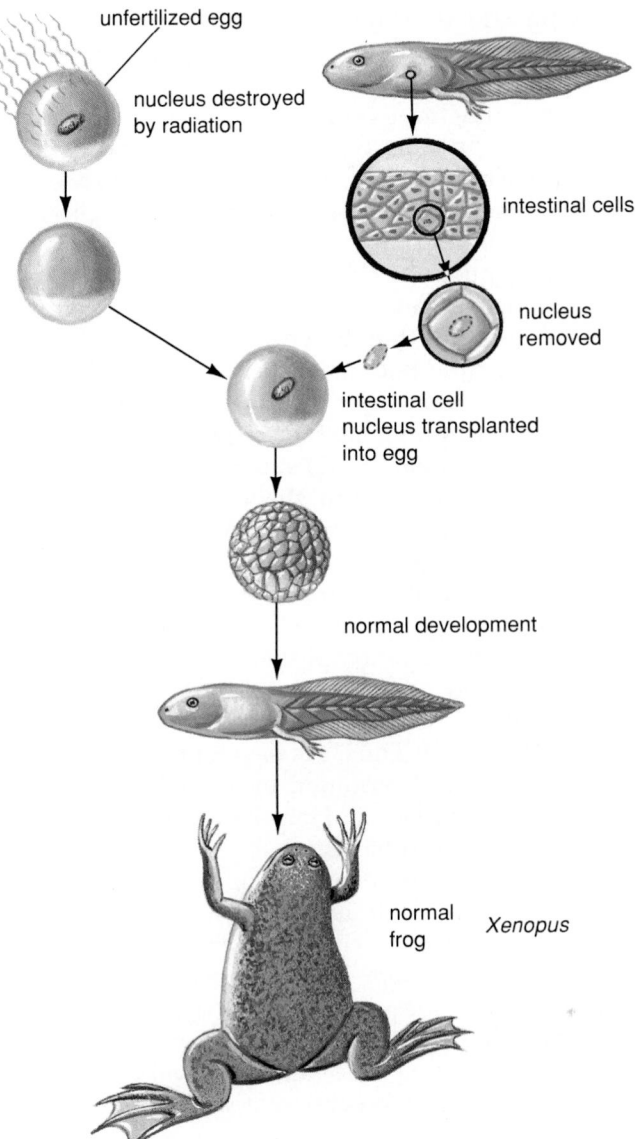

Figure 40-1 J. B. Gurdon's experiment proving that cells do not lose genes as they differentiate. Gurdon destroyed the DNA of unfertilized frog eggs, and then transplanted nuclei of intestinal cells from a tadpole into the egg. The resulting egg cells developed into normal tadpoles and eventually adult frogs, demonstrating that the intestinal cells retained all the genes necessary for normal development.

genes they make use of. In other words, cells types differ because different genes are transcribed to messenger RNA and translated into proteins.

Gene Regulation During Development

In Chapter 13, we discussed some of the mechanisms that control gene transcription, the production of messenger RNA using the gene as a blueprint. Although geneticists are just beginning to unravel the details of how transcription is controlled, the principles are straightforward. Cellular materials, often proteins or proteins combined with activating substances such as steroid hormones, travel to the nucleus and bind to the chromosomes. These proteins then block transcription of certain genes or promote the transcription of other genes (see "A Closer Look at Homeoboxes and the Control of Body Form"). Which genes are transcribed largely determines the shape, structure, and activity of the cell.

There are two major sources of chemical substances that control gene transcription and hence direct cell differentiation: (1) substances positioned in specific locations within the egg cytoplasm, and (2) chemical messages received from other cells.

During oogenesis, various gene-regulating substances become concentrated in different specific places in the egg cytoplasm (scientists do not yet know how this is accomplished). The fertilized egg then divides in particular orientations; its daughter cells, therefore, receive different gene-regulating substances (Fig. 40-2). **Thus, the developmental fate of a daughter cell is determined by the part of the egg cytoplasm it receives, and hence by the gene-regulating substances it inherits.**

During later embryonic development and continuing throughout adult life, cells constantly receive chemical messages, including nutrients, hormones, and neurotransmitters, from other cells of the body. These chemical messages can alter the transcription of genes and the activity of enzymes within a cell. As we proceed through this chapter, we will describe the effects of these two processes at work during development.

Indirect and Direct Development

All animal eggs contain food reserves of **yolk**, rich in lipids and protein. This is crucial to the early development of the **embryo.** Since a zygote has no mouth or digestive tract, its ability to acquire food from its environment is limited. The yolk provides nourishment for the early embryo until it develops into a form that can obtain food from outside sources. The amount of yolk in an egg corresponds closely with the life history of the animal. Animal development usually proceeds down one of two paths: **indirect development** or **direct development.**

Indirect Development

In indirect development, the juvenile animal that hatches from the egg differs significantly from the adult, as a caterpillar differs from a butterfly. Indirect development occurs in most of the invertebrates, including insects and echinoderms, and in a few vertebrates, notably the amphibians. **Animals with indirect development typically produce huge numbers of eggs, and each egg has only a small amount of yolk.** The yolk nourishes the developing embryo during a rapid transformation into a small, sexually immature feeding stage called a **larva** (Fig. 40-3a). Some larval animals occupy entirely different habitats and most larvae feed on different organisms than they will as adults. This is illustrated by the dragonfly, whose aquatic larva feeds on aquatic organisms such as tadpoles, but whose adult form is terrestrial and feeds on insects (Fig. 40-3b). Eventually the larvae undergo a revolution in body form, or **metamorphosis,** and become sexually mature adults. Although people tend to regard the adult form as the "real animal" and larvae as "preparatory stages," in some animals, especially insects, most of the life cycle is spent as a larva. The adult may live for only a few days, reproducing frantically and in some cases not even eating. The mayfly, for example, metamorphoses from aquatic larvae that may have spent a year or more feeding and growing. Emerging in huge swarms from freshwater streams, ponds, and lakes, adult mayflies may live a few hours or at most a few days. The sole occupation of the adult is to mate and lay eggs; their frag-

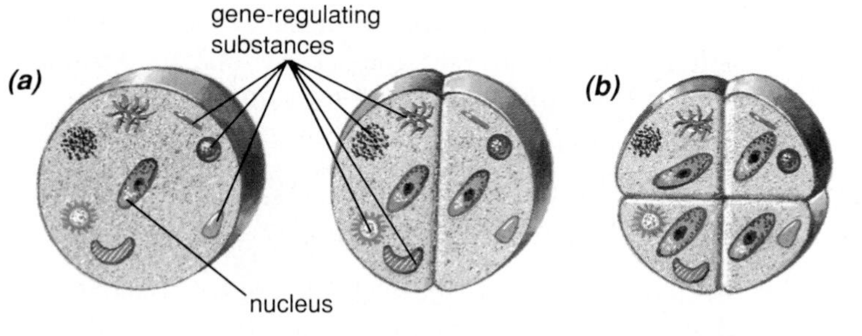

(a)   gene-regulating substances   *(b)*

nucleus

Figure 40-2 Inheritance of gene-regulating substances during division of a fertilized egg. **(a)** Different gene-regulating substances (represented by the various symbols) are positioned in particular places in the egg cytoplasm during oogenesis. **(b)** As the egg divides, these materials remain in about the same positions, so that the daughter cells inherit different substances.

CLOSER LOOK

At Homeoboxes and the Control of Body Form

In 1948, Edward Lewis of the California Institute of Technology began a systematic analysis of mutations in the fruit fly *Drosophila* that cause one body part to be replaced by another; for example, legs grow where antennae should be (Fig. E40-1). Lewis discovered that these organ substitutions could be caused by mutations in single genes, which are apparently "master" genes controlling the activity of the many other genes necessary to produce the misplaced organ.

In 1983, researchers found that the different master genes in *Drosophila* (and in other invertebrates as well) shared a nearly identical DNA sequence, now called a **homeobox.** Homeobox gene segments have a common evolutionary origin. Recently, researchers have found homeoboxes in humans and mice that are nearly identical to those of the fruitfly. By altering the expression of homeobox genes, researchers have produced developmental defects in both embryonic frogs and mice.

Each homeobox gene segment codes for a 60 amino acid protein. The proteins encoded by each homeobox are similar, but have small crucial differences that allow them to specify which body regions will develop into which organs. These proteins are found in the cell nucleus, where they bind to specific genes, turning them on or off. **Homeoboxes, by coding for proteins that activate or inactivate other genes, seem to directly specify the course of differentiation of embryonic cells.** Homeoboxes determine the head-to-tail axis of the embryo and establish the appropriate sequence of structures along its length, thereby determining the overall shape of the body and the location and shape of its parts. This is why mutations in the homeobox can cause legs to replace antennae in the fruitfly.

Researchers studying a variety of developmental questions from the formation of color patterns on butterfly wings to the development of the forelimb in chickens and frogs have found that **chemical gradients often specify the fate of cells. In other words, during development, cells may take different forms in response to different concentrations of a regulatory substance.** As a result, a gradient (such as occurs by diffusion from a concentrated source) of a single substance can produce an entire sequence of morphologic characteristics (such as the sequence of structures from the shoulder to the fingers of a forelimb). One source of such chemical gra-

Figure E40-1 A mutant in a homeobox segment of a gene in this *Drosophila* has caused legs to grow where antennae should be.

dients is the homeobox. For example, during the development of the clawed frog *Xenopus*, a forelimb bud forms from mesoderm that expresses an identified homeobox gene. The homeobox codes for a protein that forms a concentration gradient: high on the "thumb side" of the bud, and low on the "pinkie side." As the limb elongates, this protein forms a second gradient: high at the shoulder, and lower toward the hand. Gradients of identical proteins have been found in the forelimbs of developing chick and mouse embryos as well.

The development of a single cell, the fertilized egg, into the fantastic complexity of a tadpole or a human infant is simultaneously one of life's great mysteries and one of biology's most exciting fields of research. The discovery of the homeobox has brought us a step closer to understanding this incredible journey.

(a)

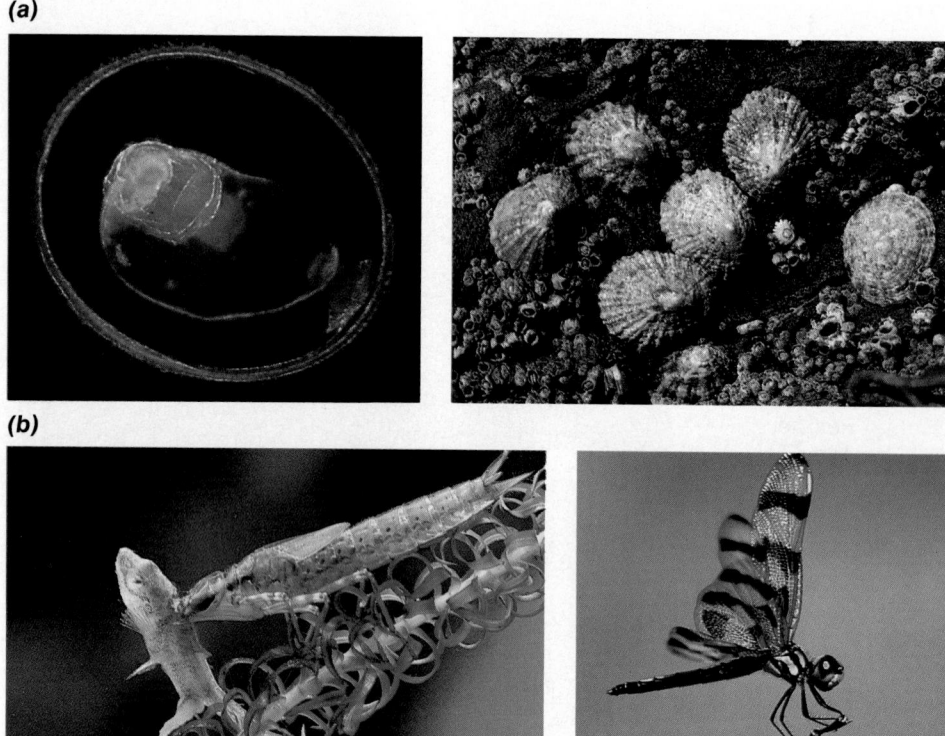

(b)

Figure 40-3 Many animals, including limpets **(a)** and dragonflies **(b)**, undergo indirect development. The larval stage is often very different from the adult in size, appearance, diet, and life-style.

ile dead bodies then accumulate in piles to be swept away by the wind.

Direct Development

Other animals, including such diverse groups as reptiles, birds, mammals, and land snails, show **direct development,** in which the newborn animal, or juvenile, is a sexually immature, miniature version of the adult (Fig. 40-4). These juveniles are typically much larger than larvae, and consequently need much more nourishment before emerging into the world. Two different strategies have evolved that meet the embryo's food requirement. Reptiles, birds, and land snails produce large eggs containing large amounts of yolk: for example, an ostrich egg weighs several pounds. Mammals, some snakes, and a few fish have relatively little yolk in their eggs, but instead nourish the developing embryo within the body of the mother (Fig. 40-4d). Either way, providing food for directly developing embryos places great demands on the mother, and relatively few offspring are produced.

Reptiles, Birds, Mammals, and Membranes

Both birds and mammals evolved from reptiles. As an adaptation to direct development in a terrestrial envi-

ronment, reptile and bird embryos produce four membranes, called **extraembryonic membranes.** The **chorion** lines the shell and exchanges oxygen and carbon dioxide through the shell. The **amnion** encloses the embryo in a watery environment, the **allantois** surrounds wastes, and the **yolk sac** contains the stored food. Although mammalian eggs contain almost no yolk, much of the reptilian genetic program for development still persists, including the four extraembryonic membranes. Table 40-1 compares the structures and functions of these extraembryonic membranes in reptiles and mammals.

Stages of Animal Development

Animal development (particularly in the vertebrates) may be loosely divided into several stages. The initial stages of cleavage, gastrulation, organogenesis, and growth occur during embryonic life (Fig. 40-5), during which nearly all the organs that will be present in the adult are formed. After birth, the animal undergoes further growth, achieves sexual maturity, reproduces, ages, and finally dies. Let's look at each of these events.

(a)

(b)

(c)

(d)

Figure 40-4 The offspring of animals with direct development closely resemble their parents from the moment of birth, except of course in size. Lizards **(a)**, land snails **(b)**, and birds **(c)** hatch from large, yolk-filled eggs, while mammalian mothers **(d)** nourish their young within their bodies for weeks or months. The availability of food supplies during early development permits the offspring to be born as miniature versions of the adult, bypassing a larval feeding stage.

Cleavage: Distributing Gene-Regulating Substances

Development begins with *cleavage,* **which refers to the division of the fertilized egg without an increase in size. Cleavage reduces the cell size and distributes gene-regulating substances to the newly formed cells.** As you know, an egg is a very large cell. Unlike most cell divisions that proceed through the cycle—

divide, grow, duplicate genetic material, then divide again—embryonic cells skip the growth phase during cleavage. Consequently, as cleavage progresses, the available cytoplasm is split up into ever-smaller cells whose sizes approach those of the adult organism. Finally, a solid ball of small cells, the **morula,** is formed. The morula is still about the same size as the zygote. Then a cavity opens within the morula, so that the cells become the outer covering of a hollow

Table 40-1   Vertebrate Embryonic Membranes

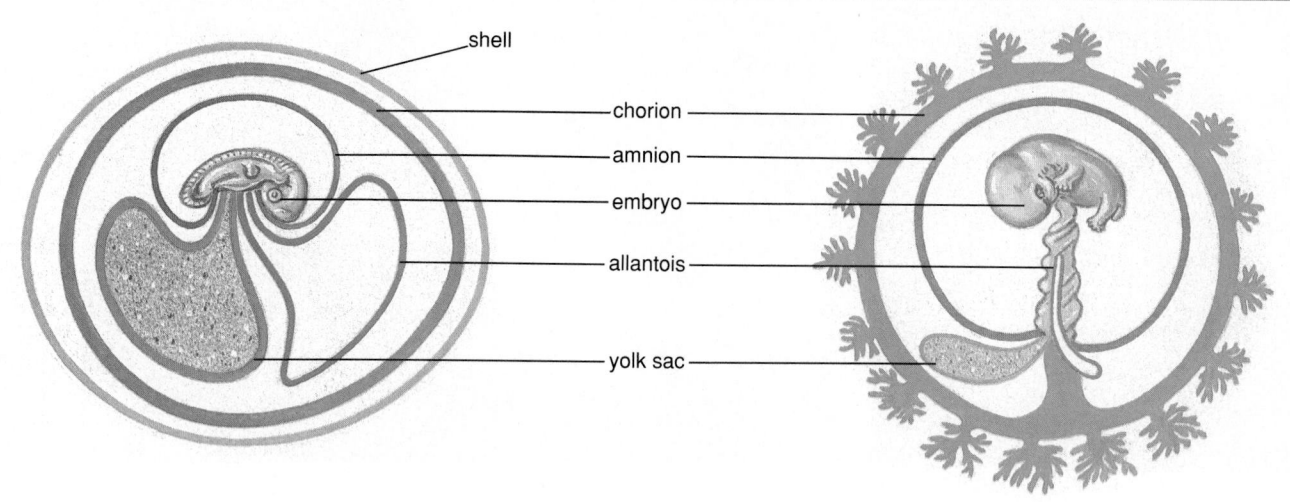

REPTILE   MAMMAL

| Membrane | Reptilian Embryo | | Mammalian Embryo | |
	Structure	Function	Structure	Function
Chorion	Membrane lining inside of shell	Acts as respiratory surface; regulates exchange of gases and water between embryo and air	Fetal contribution to placenta	Provides surface for exchange of gases, nutrients, and wastes between embryo and mother
Amnion	Sac surrounding embryo	Encloses embryo in fluid	Sac surrounding embryo	Encloses embryo in fluid
Allantois	Sac connected to embryonic urinary tract; capillary-rich membrane lining inside of chorion, with blood vessels connecting to embryonic circulation	Stores wastes (esp. urine); acts as respiratory surface	Provides blood vessels of umbilical cord	Carries blood between embryo and placenta
Yolk sac	Membrane surrounding yolk	Contains yolk as food; digests yolk and transfers nutrients to embryo; forms part of digestive tract	"Empty" membranous sac	Forms part of digestive tract

(often spherical) structure, the **blastula.** To a considerable extent, the pattern of cleavage is controlled by the amount of yolk, because yolk hinders the division of the cytoplasm (cytokinesis). The almost yolkless eggs of sea urchins divide symmetrically, but eggs with extremely large yolks, such as a hen's egg, don't even divide all the way through. Nevertheless, a hollow blastula is always produced, though in reptiles and birds the blastula is flattened rather than spherical.

| Vertebrate Class | Fertilized Egg | Cleavage | | Gastrulation | Organogenesis |
		Morula	Blastula or Blastocyst	Gastrula	Late Embryo
Amphibian					
Reptile					
Mammal					

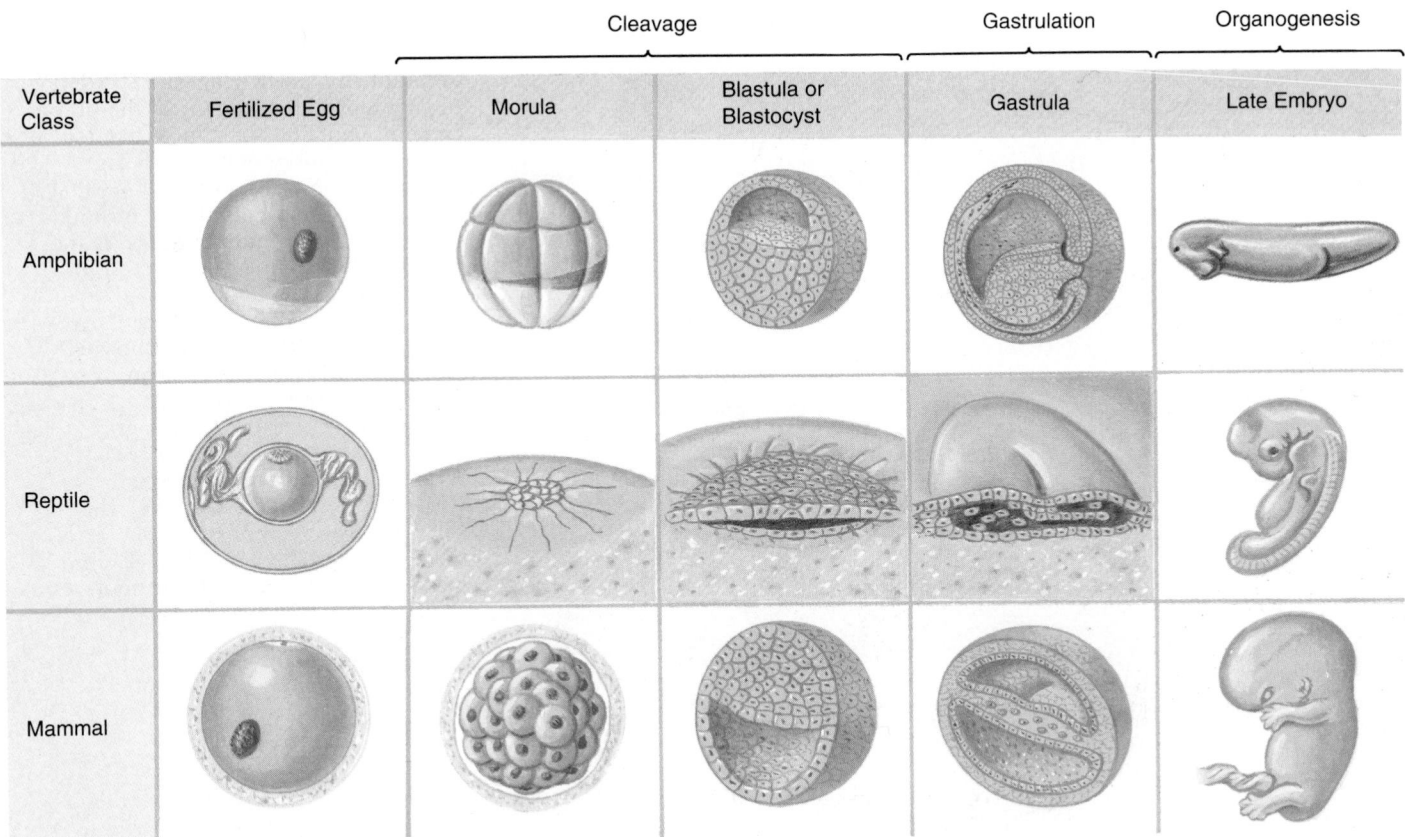

Figure 40-5 Stages in the development of amphibians, reptiles, and mammals. Use these illustrations as a guide to accompany the descriptions to follow in the text.

During cleavage, different gene-regulating materials become incorporated into different daughter cells (see Fig. 40-2). This was discovered decades ago in experiments on frog eggs (Fig. 40-6). The unfertilized frog egg has pale yolk on the "bottom" or vegetal pole, and pigmented cytoplasm on the "top" or animal pole. At fertilization, some of the pigment shifts toward the animal pole, leaving behind a **gray crescent** of intermediate pigmentation. The first cleavage division of frog eggs normally passes through the center of the gray crescent, so that each daughter cell receives roughly half the crescent (Fig. 40-6a). If the two cells are gently separated, each will develop into a normal tadpole. But if embryologists force the first division to occur in a plane rotated 90 degrees, one daughter cell receives the entire gray crescent (Fig. 40-6b). If the cells are then separated, the cell with the gray crescent develops into a tadpole, while the one without any crescent material merely forms a lump of cells that soon dies. Clearly, gene-regulating substances in the gray crescent region are required for the normal development of the tadpole.

Gastrulation

In the next step of development, an indentation called the **blastopore** forms on one side of the blastula. Blastula cells migrate in a continuous sheet in through the blastopore, much as if you punched in an underinflated basketball (Fig. 40-7). The enlarging dimple is destined to become the digestive tract; the cells that line its cavity are now called **endoderm** (Greek for "inner skin"). The cells remaining on the outside will form the epidermis of the skin and the nervous system, and are called **ectoderm** ("outer skin"). Meanwhile, some cells migrate between the endoderm and ectoderm, forming a third layer, the **mesoderm** ("middle skin"). Mesoderm gives rise to muscles, skeleton, and the circulatory system (Table 40-2). This process of cell movement is called **gastrulation,** and the three-layered embryo that results is the **gastrula.**

Influences from the Cellular Environment

During gastrulation, the developmental fate of most of the embryo's cells is determined by chemical mes-

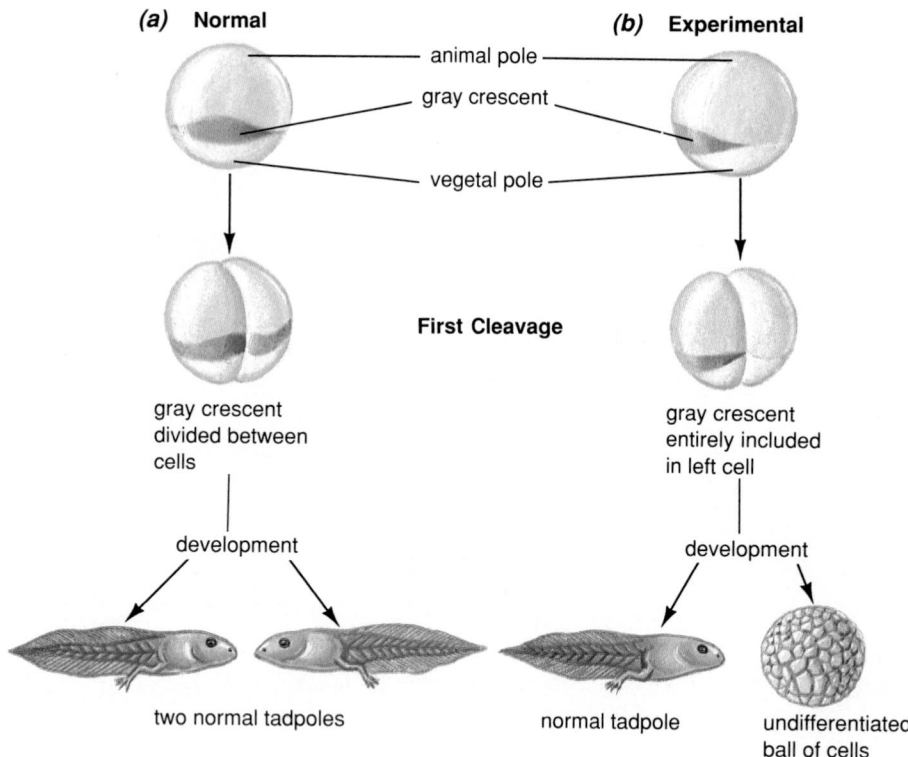

(a) Normal

animal pole
gray crescent
vegetal pole

First Cleavage

gray crescent
divided between
cells

development

two normal tadpoles

(b) Experimental

gray crescent
entirely included
in left cell

development

normal tadpole

undifferentiated
ball of cells

Figure 40-6 Cleavage distributes gene-regulating substances differentially. In frog eggs, a pigmented region known as the gray crescent contains substances needed for normal embryonic growth and differentiation. **(a)** Normally, the first cleavage division cuts neatly through the center of the gray crescent. If the two daughter cells are separated in the lab, each cell contains gray crescent material, and each can subsequently develop into a normal tadpole. **(b)** Embryologists can experimentally force the first cleavage division to miss the gray crescent, so that one daughter cell receives the entire gray crescent, while the other receives none. The cell with the crescent material develops normally, but the cell lacking crescent substance cannot.

sages received from other cells, a process called **induction.** In amphibian embryos, cells derived from the gray crescent region of the zygote form the site of dimpling as the blastula is transformed into the gastrula. This area (called the dorsal lip of the blastopore) controls the developmental fate of the cells around it, as Hilde Mangold and Hans Spemann showed in the

1920s (Fig. 40-8a). They transplanted the dorsal lip of the blastopore from one embryo to another. The transplanted dorsal lip then induced the nearby cells of the host to form a second embryo, showing that **dorsal lip tissue controls differentiation in the surrounding cells.** One can also perform a control experiment, transplanting cells from regions of the gastrula other than the dorsal lip of the blastopore (Fig. 40-8b). These cells give rise to tissues appropriate for the region into which they were transplanted rather than tissues appropriate for the region from which they were taken. These experiments clearly show that gene use within differentiating cells (nonblastopore cells) can be regulated by substances originating in other cells (the dorsal lip of the blastopore).

Organogenesis: Developing Adult Structures

Gradually, ectoderm, mesoderm, and endoderm rearrange themselves into the organs characteristic of the animal species (see Table 40-2). This process, called **organogenesis,** also usually occurs by induction. This was demonstrated for eye formation in frog embryos by Warren Lewis. During normal development, the brain develops a pair of swellings, the optic cups, that grow out toward the epidermis of the head. Where an optic cup touches the epidermis,

Table 40-2   Derivation of Adult Tissues from Embryonic Cell Layers

Embryonic Layer	Adult Tissue
Ectoderm	Epidermis of skin; lining of mouth and nose; hair; glands of skin (sweat, sebaceous, and mammary glands); nervous system; lens of eye; inner ear
Mesoderm	Dermis of skin; muscle, skeleton; circulatory system; gonads; kidneys; outer layers of digestive and respiratory tracts
Endoderm	Lining of digestive and respiratory tracts; liver; pancreas

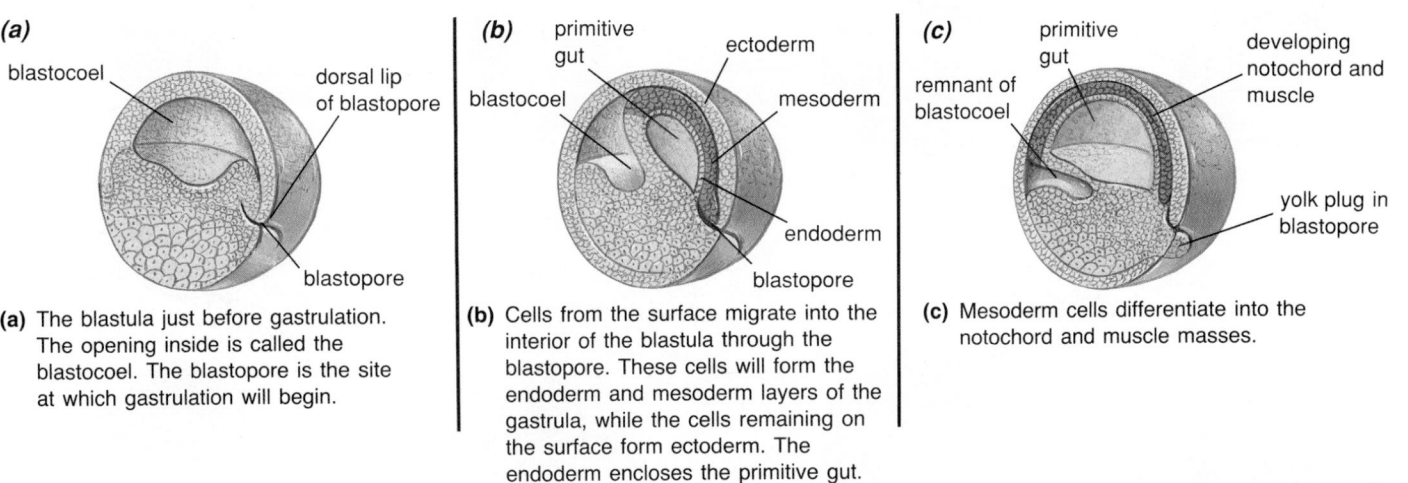

(a)

blastocoel

dorsal lip of blastopore

blastopore

(a) The blastula just before gastrulation. The opening inside is called the blastocoel. The blastopore is the site at which gastrulation will begin.

(b)

primitive gut

ectoderm

blastocoel

mesoderm

endoderm

blastopore

(b) Cells from the surface migrate into the interior of the blastula through the blastopore. These cells will form the endoderm and mesoderm layers of the gastrula, while the cells remaining on the surface form ectoderm. The endoderm encloses the primitive gut.

(c)

primitive gut

developing notochord and muscle

remnant of blastocoel

yolk plug in blastopore

(c) Mesoderm cells differentiate into the notochord and muscle masses.

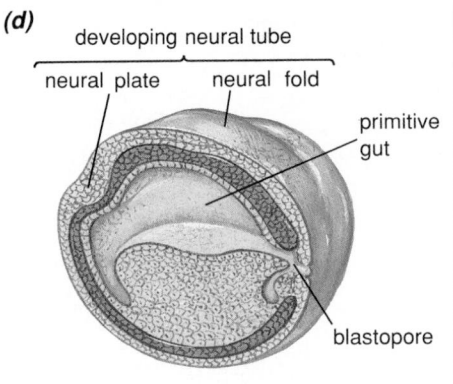

(d)

developing neural tube

neural plate   neural fold

primitive gut

blastopore

(d) The notochord induces ectoderm cells lying directly above it to form the neural tube.

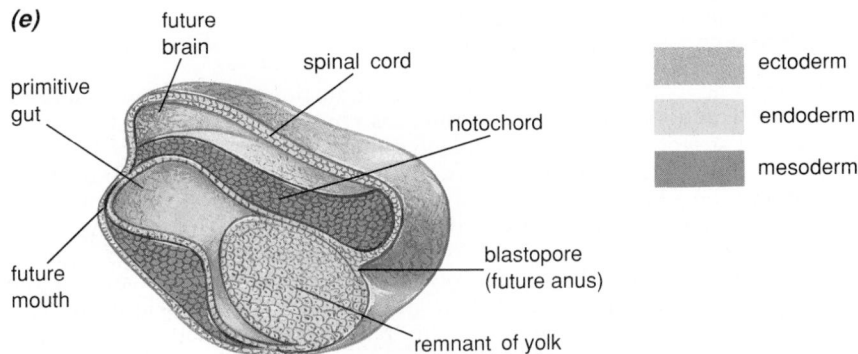

(e)

future brain

spinal cord

primitive gut

notochord

future mouth

blastopore (future anus)

remnant of yolk

ectoderm

endoderm

mesoderm

(e) The neural tube enlarges and differentiates into brain and spinal cord. A future mouth is produced when the opening formed by the primitive gut breaks through at the end of the embryo opposite the blastopore. The blastopore is the future anus.

Figure 40-7 Gastrulation in the frog.

cells of the epidermis divide, invaginate, and form the future lens. Lewis found that if he transplanted an optic cup into the tail of the embryo, a lens would form from the overlying epidermis of the tail, while no lens would form on the side of the head from which the optic cup had been removed. Therefore, the optic cup induces formation of the lens from epidermal tissue.

In some cases, adult structures are, in effect, "sculpted" from excess cells produced during embryonic development. For some cells, death is pro-

grammed to occur at a precise time during development. At least two mechanisms seem to be at work in different tissues. First, **some cells die during development unless they receive a "survival signal."** Embryonic vertebrates, for example, have far more motor neurons in their spinal cords than adult animals do. Motor neurons are programmed to die unless they successfully innervate a skeletal muscle, which releases a chemical that prevents the death of its own motor neuron. For other cells, the situation is just the reverse: **some cells live unless they receive a**

(a)

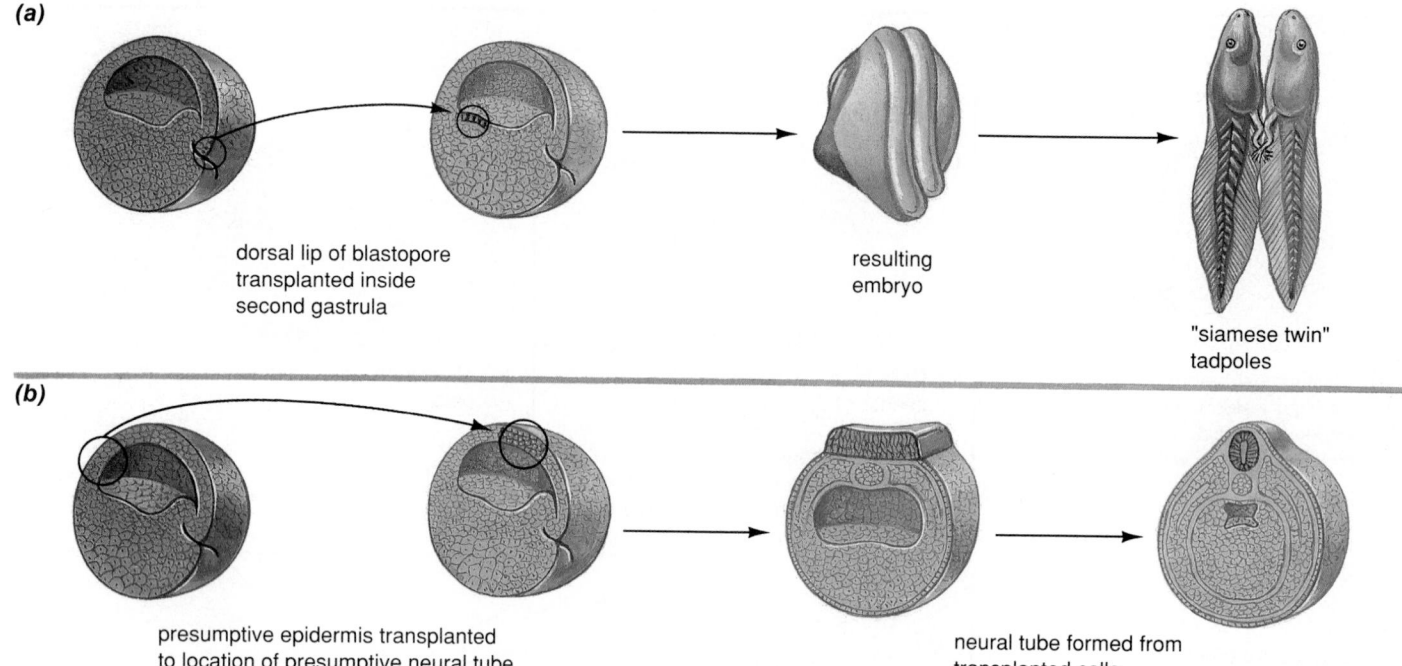

dorsal lip of blastopore
transplanted inside
second gastrula

resulting
embryo

"siamese twin"
tadpoles

(b)

presumptive epidermis transplanted
to location of presumptive neural tube

neural tube formed from
transplanted cells

Figure 40-8 Experiments demonstrating the role of induction in the differentiation of frog embryos. **(a)** The dorsal lip of the blastopore is removed from the gastrula of a heavily pigmented strain of frog and transplanted into a different region of the gastrula from a lightly pigmented strain of frog. The transplanted dorsal lip of the blastopore induces the adjacent regions of the "host" gastrula to form a separate, nearly complete tadpole with the dark pigmentation of the transplanted tissue. **(b)** As the first experiment suggests, many cells of a gastrula are "followers," guided along certain developmental paths by their proximity to inducing tissues such as the dorsal lip of the blastopore. Here, cells that would normally become skin from the gastrula of a darkly pigmented strain of frog are transplanted to another site on a "host" gastrula of a lighter strain. Their developmental fate is determined by the region into which they are transplanted, and not upon the region from which they were removed. Here, the host gastrula has induced cells that would normally have formed skin to produce a neural tube. Such donor cells can be induced to form tissues as diverse as brain or muscle.

"death signal" from other cells of the developing animal. Many embryonic structures disappear during development. Vertebrates all pass through developmental stages with tails and webbed hands and feet. In humans, this can be clearly seen in the 5-week embryo (Fig. 40-13). Two weeks later, the webbing cells have died, revealing separate fingers, and the tail is regressing (Fig. 40-9). In frogs, the tail is lost during metamorphosis from its tadpole larva. Thyroid hormone, which triggers metamorphosis, stimulates cells in the tail to synthesize enzymes that digest the tail away. If the thyroid gland is surgically removed, the frog retains its tail.

Sexual Maturation

Development does not stop at birth; animals continue to change throughout their lives. Animals become sexually mature at an age that is determined by both genes and environment. Many animals undergo months to years of growth and development before they can become sexually mature. This developmental process is largely regulated by their genes. Once the animal has reached the appropriate age, the precise onset of sexual maturity is often triggered by environmental stimuli. Songbirds, for example, almost always become sexually mature in the spring,

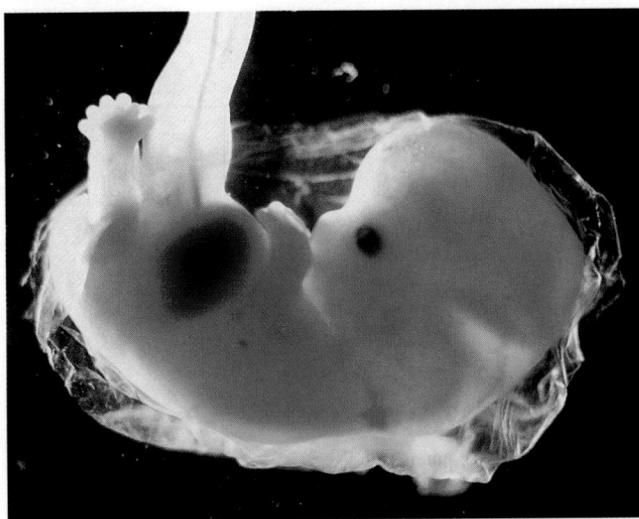

Figure 40-9 By the seventh week, the human form has been more clearly defined by selective death of cells that connect the fingers and toes, and form the tail. The tail has nearly disappeared, and the fingers and toes have separated from one another.

stimulated by the increasingly long days. Internal, and perhaps social, factors also influence maturation in many species. The age of puberty among women, for instance, has dropped substantially during the past few centuries. This is due in part to improved nutrition, but social stimulation may also be involved.

Aging

No animal is static. The cells in your stomach, exposed to strong acids and protein-digesting enzymes, are replaced every few weeks. Even in the carefully regulated environment of your central nervous system, some nerve cells die each day, and are never replaced. Many other cells will function less efficiently, or divide more slowly, as you age. Why do these cells deteriorate and die? Is death, for cells and for entire organisms, a programmed part of life?

Even if a cell survives the drastic changes of embryonic life, it still has a finite life span. If cells are removed from normal tissues and grown in culture in the laboratory, they divide a few times, then stop, and eventually die. This seems to imply that a cell lineage (a parent cell and all its offspring for all generations) has a built-in maximum life span. Recent research has suggested that this maximum life span varies from species to species, and that longevity depends on the ability of the cell to repair damage to

its DNA. Longer-lived cells and longer-lived animals are better at repairing damaged DNA. Nevertheless, all normal cells seem to die eventually.

Not all cells, however, are "normal." Cancer cells survive and reproduce indefinitely in cell culture or in the body (which is what makes cancer so catastrophic). Some lines of cancer cells have been reproducing in culture for decades. Although scientists do not understand the mechanism that regulates the life span of a cell, cancer cells provide evidence that this mechanism can be bypassed. But cultured cancer cells frequently mutate and change their characteristics. Can we discover a method of defusing the self-destruct mechanism while retaining proper controls over the cells in our bodies? No one knows, but we can be certain it will not happen soon.

Human Development

Human development is controlled by the same mechanisms that control the development of other animals. In fact, our development strongly reflects our evolutionary heritage, a fact that we shall emphasize in the brief discussion to follow. Figure 40-10 summarizes the stages of human embryonic development. You may want to refer back to this figure as we go along, and, perhaps years from now, if you have a child of your own.

The First Two Months

A human egg is normally fertilized in a woman's oviduct and undergoes a few cleavage divisions on its way to the uterus. Just before being implanted in the uterus, the embryo, now called a **blastocyst** (the mammalian version of a blastula), consists of a thin-walled, hollow ball with a thicker **inner cell mass** on one side (Fig. 40-11). The thin outer wall becomes the **chorion,** and will form the embryonic contribution to the placenta, while the inner cell mass develops into the embryo and the three other extraembryonic membranes.

After implantation, the inner cell mass grows and splits, forming two fluid-filled sacs that are separated by a double layer of cells called the **embryonic disk** (Fig. 40-11). One sac, bounded by the **amnion,** forms the amniotic cavity. The amnion eventually grows around the embryo, enclosing it in what one author has called its "private aquarium," providing the watery environment needed by all animal embryos. The **yolk sac,** homologous to the yolk sac of reptiles and birds, forms the second cavity, although in humans it

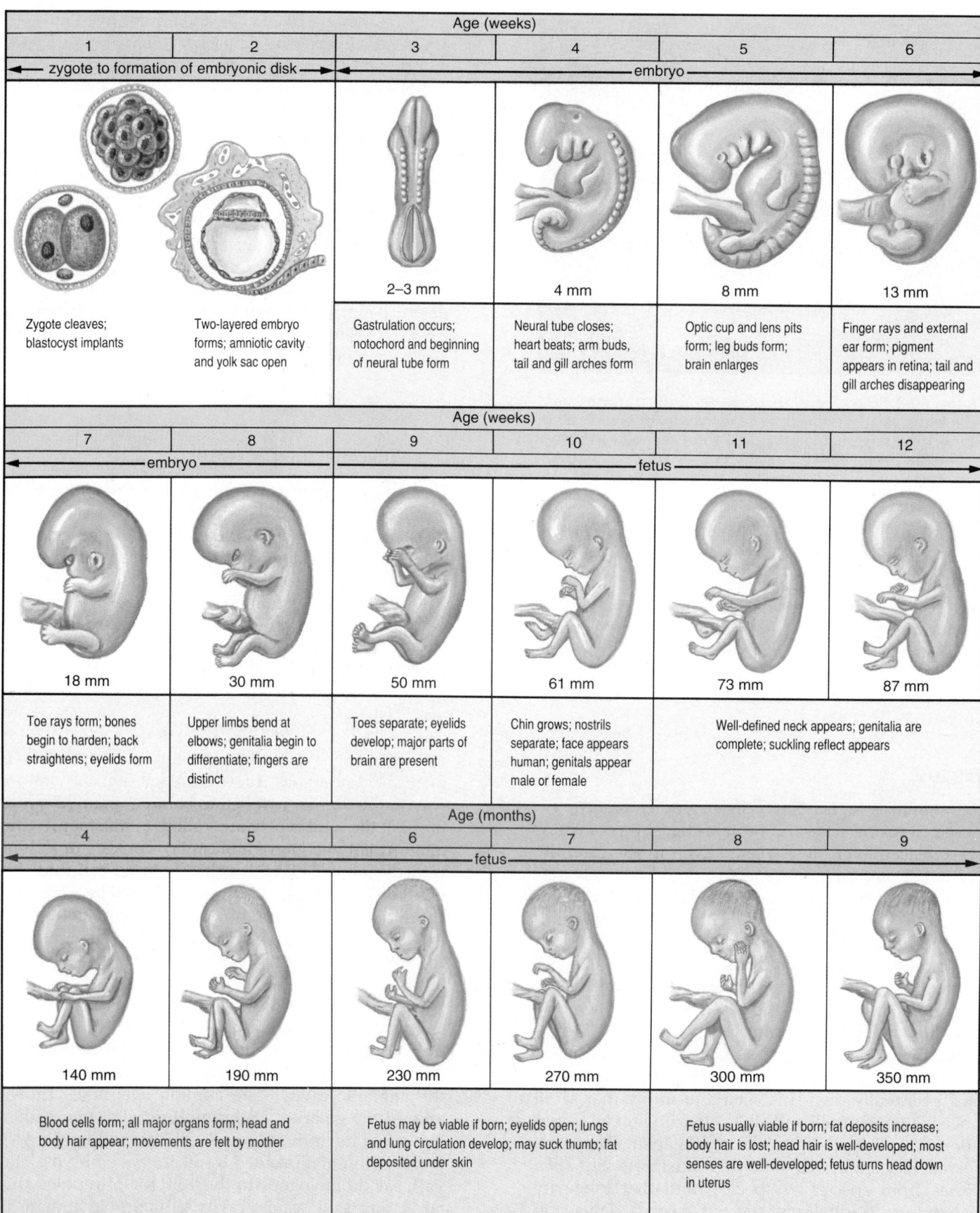

Age (weeks)					
1	2	3	4	5	6
◄── zygote to formation of embryonic disk ──►		◄──────────────── embryo ────────────────►			
		2–3 mm	4 mm	8 mm	13 mm
Zygote cleaves; blastocyst implants	Two-layered embryo forms; amniotic cavity and yolk sac open	Gastrulation occurs; notochord and beginning of neural tube form	Neural tube closes; heart beats; arm buds, tail and gill arches form	Optic cup and lens pits form; leg buds form; brain enlarges	Finger rays and external ear form; pigment appears in retina; tail and gill arches disappearing

Age (weeks)					
7	8	9	10	11	12
◄──── embryo ────►		◄──────────────── fetus ────────────────►			
18 mm	30 mm	50 mm	61 mm	73 mm	87 mm
Toe rays form; bones begin to harden; back straightens; eyelids form	Upper limbs bend at elbows; genitalia begin to differentiate; fingers are distinct	Toes separate; eyelids develop; major parts of brain are present	Chin grows; nostrils separate; face appears human; genitals appear male or female	Well-defined neck appears; genitalia are complete; suckling reflect appears	

Age (months)					
4	5	6	7	8	9
◄──────────────── fetus ────────────────►					
140 mm	190 mm	230 mm	270 mm	300 mm	350 mm
Blood cells form; all major organs form; head and body hair appear; movements are felt by mother		Fetus may be viable if born; eyelids open; lungs and lung circulation develop; may suck thumb; fat deposited under skin		Fetus usually viable if born; fat deposits increase; body hair is lost; head hair is well-developed; most senses are well-developed; fetus turns head down in uterus	

Figure 40-10 A calendar of human embryonic development, from blastula to birth.

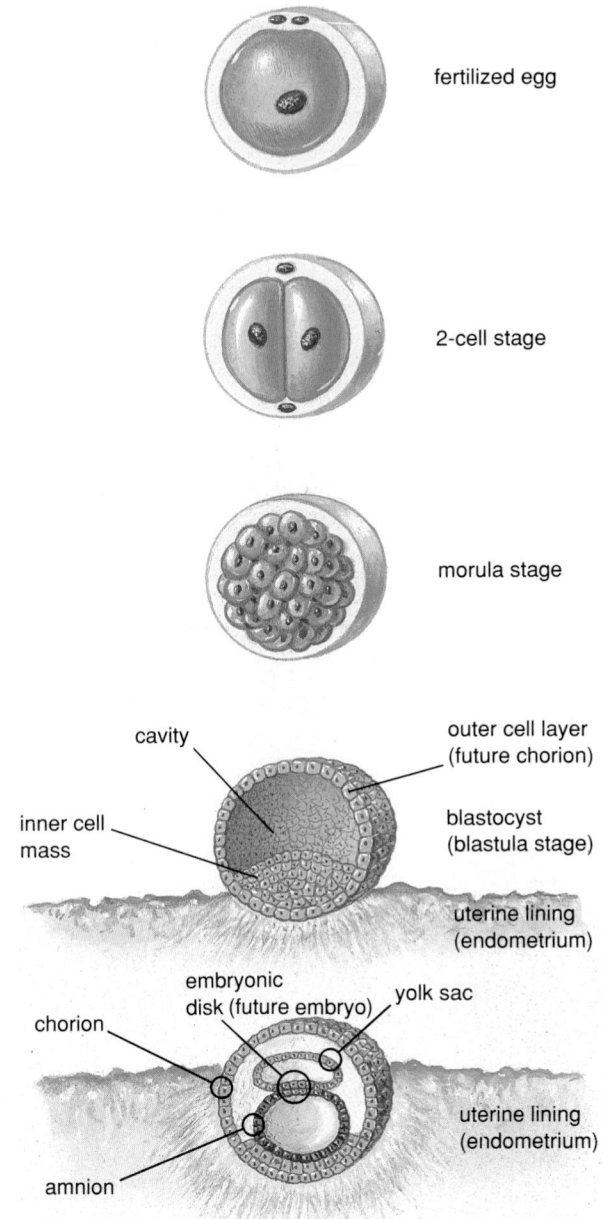

fertilized egg

2-cell stage

morula stage

cavity

inner cell mass

outer cell layer (future chorion)

blastocyst (blastula stage)

uterine lining (endometrium)

embryonic disk (future embryo)

yolk sac

chorion

uterine lining (endometrium)

amnion

Figure 40-11 As it travels through the oviduct, the fertilized egg undergoes cleavage, forming a morula. In the uterus, the morula becomes a blastocyst (blastula) and implants in the uterine lining. At this stage it consists of an outer layer of cells surrounding an inner cell mass. As it burrows into the uterine lining, the outer cell layer forms the chorion, the embryonic contribution to the placenta. The inner cell mass forms the amnion, yolk sac, and the embryonic disk, which will become the embryo.

contains no yolk. At this stage, the embryonic disk consists of a layer of ectoderm cells (on the side facing the amniotic cavity) and a layer of endoderm cells (on the side facing the yolk sac).

Gastrulation begins about the fifteenth day after fertilization (Fig. 40-12a). The endoderm and ectoderm split apart slightly, and a slit, the **primitive streak** (analogous to the blastopore), appears in the center of the ectoderm. Ectoderm cells migrate through the primitive streak into the interior of the embryo, forming mesoderm. One of the earliest mesoderm structures to develop is the notochord, a supporting rod found at some stage in all chordates (Fig. 40-12b).

During the third week of development, the embryo and its amniotic sac begin to curl ventrally (Fig. 40-12c). As the embryo grows, the endoderm curves around, forming a tube that will become the gut (Fig. 40-12d). Simultaneously, the notochord induces formation of a groove in the overlying ectoderm, which invaginates and then closes over to become the **neural tube,** the forerunner of the brain and spinal cord. By the end of the fourth week, the amnion completely surrounds the embryo, punctured only by the umbilical cord connecting the embryo to the placenta.

By the end of the sixth week, the embryo clearly displays its chordate ancestry (see Chapter 22), having developed a notochord, primitive gill arches, and a prominent tail (Fig. 40-13). These, of course, disappear as gestation continues. The embryo already has the rudimentary beginnings of the eyes, a beating heart, and separating fingers on its tiny hands. Especially notable at this stage is the rapid growth of the brain, which is nearly as large as the rest of the body. In fact, many of the structures of the adult brain are already recognizable.

As the second month draws to an end, nearly all the major organs have formed, and the embryo begins to look quite human (Fig. 40-14). The gonads appear and develop into testes or ovaries, depending on the presence or absence of the Y chromosome. Sex hormones are secreted, testosterone from the testes or estrogen from the ovaries, and these hormones affect the future development of the embryonic organs, including not only the genitalia, but also certain regions of the brain. After the second month of development, the embryo is called a **fetus,** denoting that it has taken on a generally human appearance.

These first 2 months of pregnancy are times of extremely rapid differentiation and growth for the embryo, and also times of considerable danger. Although the developing child is vulnerable throughout gestation, rapidly developing organs are most sensitive to environmental insults, such as drugs or certain medications taken by the mother.

The Placenta

During the first few weeks of pregnancy, embryonic cells burrow into the thickened lining of the uterus.

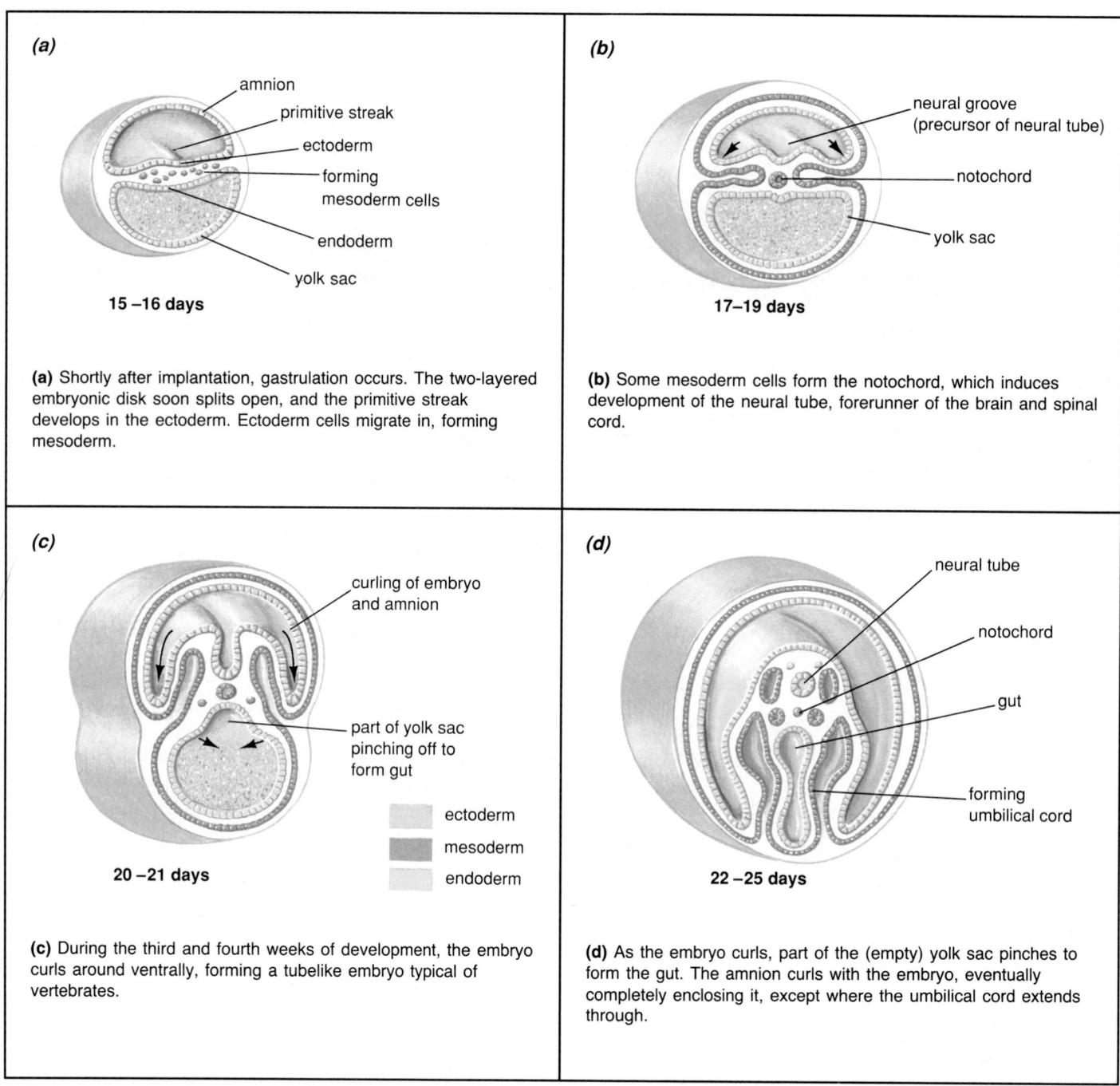

(a)

amnion
primitive streak
ectoderm
forming
mesoderm cells
endoderm
yolk sac

15–16 days

(a) Shortly after implantation, gastrulation occurs. The two-layered embryonic disk soon splits open, and the primitive streak develops in the ectoderm. Ectoderm cells migrate in, forming mesoderm.

(b)

neural groove
(precursor of neural tube)
notochord
yolk sac

17–19 days

(b) Some mesoderm cells form the notochord, which induces development of the neural tube, forerunner of the brain and spinal cord.

(c)

curling of embryo
and amnion
part of yolk sac
pinching off to
form gut

ectoderm
mesoderm
endoderm

20–21 days

(c) During the third and fourth weeks of development, the embryo curls around ventrally, forming a tubelike embryo typical of vertebrates.

(d)

neural tube
notochord
gut
forming
umbilical cord

22–25 days

(d) As the embryo curls, part of the (empty) yolk sac pinches to form the gut. The amnion curls with the embryo, eventually completely enclosing it, except where the umbilical cord extends through.

Figure 40-12 Early human development.

The outer cells of the embryo form the chorion, which penetrates the uterine lining with fingerlike projections called **chorionic villi.** From this complex interweaving of tissues arises the **placenta,** a marvelously intricate organ. **The placenta has two major functions: it secretes hormones, and it allows the selec-** **tive exchange of materials between the mother and the fetus.**

As the placenta develops during the first 2 months of pregnancy, it begins secreting estrogen and progesterone. Estrogen stimulates growth of the uterus and mammary glands, while progesterone also stimulates

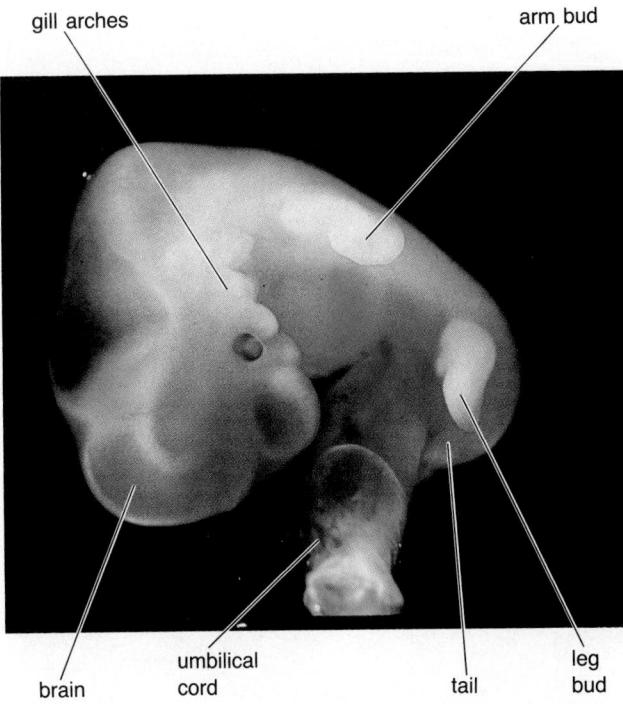

gill arches

arm bud

brain

umbilical cord

tail

leg bud

Figure 40-13 At the end of the fifth week, the human embryo is about half head. The feet and hands have begun to develop fingers, and the tail is clearly visible.

the mammary glands and inhibits premature contractions of the uterus.

The placenta also regulates the exchange of materials between the blood of the mother and the blood of the fetus without allowing the two to mix. The chorionic villi contain a dense network of fetal capillaries, and are bathed in pools of maternal blood (Fig. 40-15). This arrangement permits diffusion of many small molecules between the fetal and maternal blood. Oxygen diffuses from maternal blood to fetal blood, and carbon dioxide from fetal blood to maternal blood. Nutrients travel from mother to fetus, some aided by active transport. Fetal urea diffuses into the mother's blood, to be filtered out by the mother's kidneys.

While allowing exchange by diffusion, the membranes of the capillaries and chorionic villi act as barriers to the passage of some disease organisms, some large proteins, and most cells. In spite of this, some fetal red blood cells cross the placenta (see "Health Watch: Blood Types and Their Medical Implications"; Chapter 30). In addition, many injurious substances can penetrate the placental barrier as described in "Health Watch: The Placenta Provides Only Partial Protection."

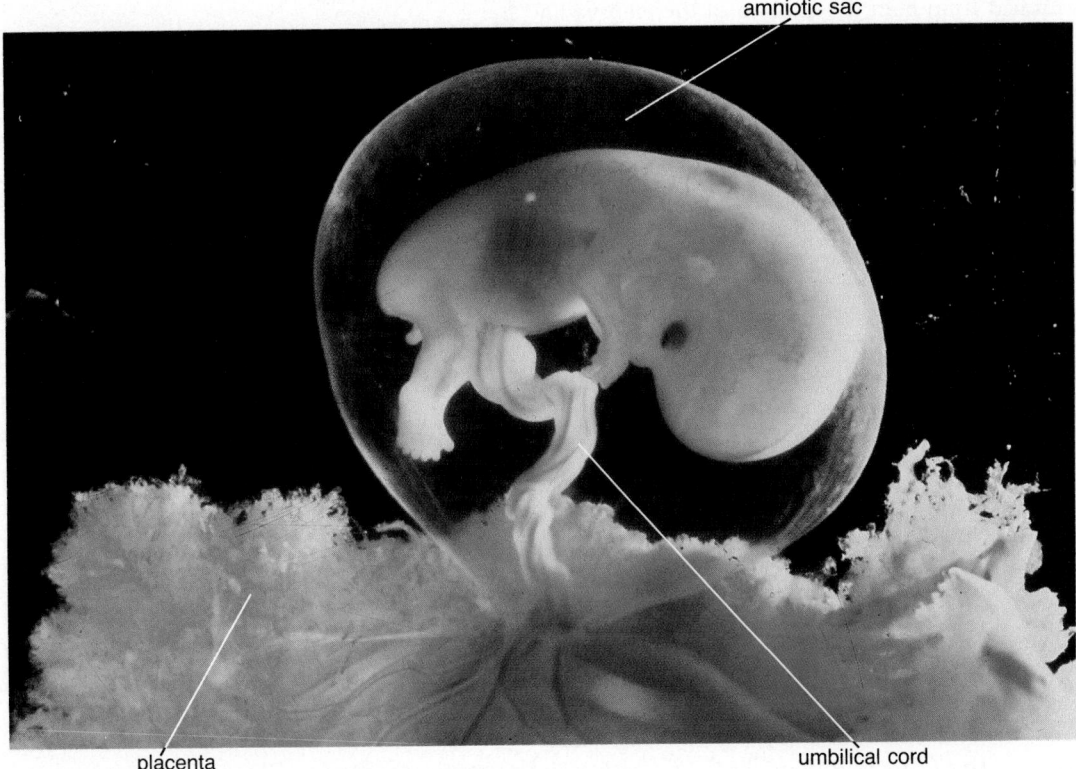

amniotic sac

placenta

umbilical cord

Figure 40-14 At the end of the eighth week, the embryo is clearly human in appearance and is now termed a fetus. Most of the major organs of the adult body have begun to develop.

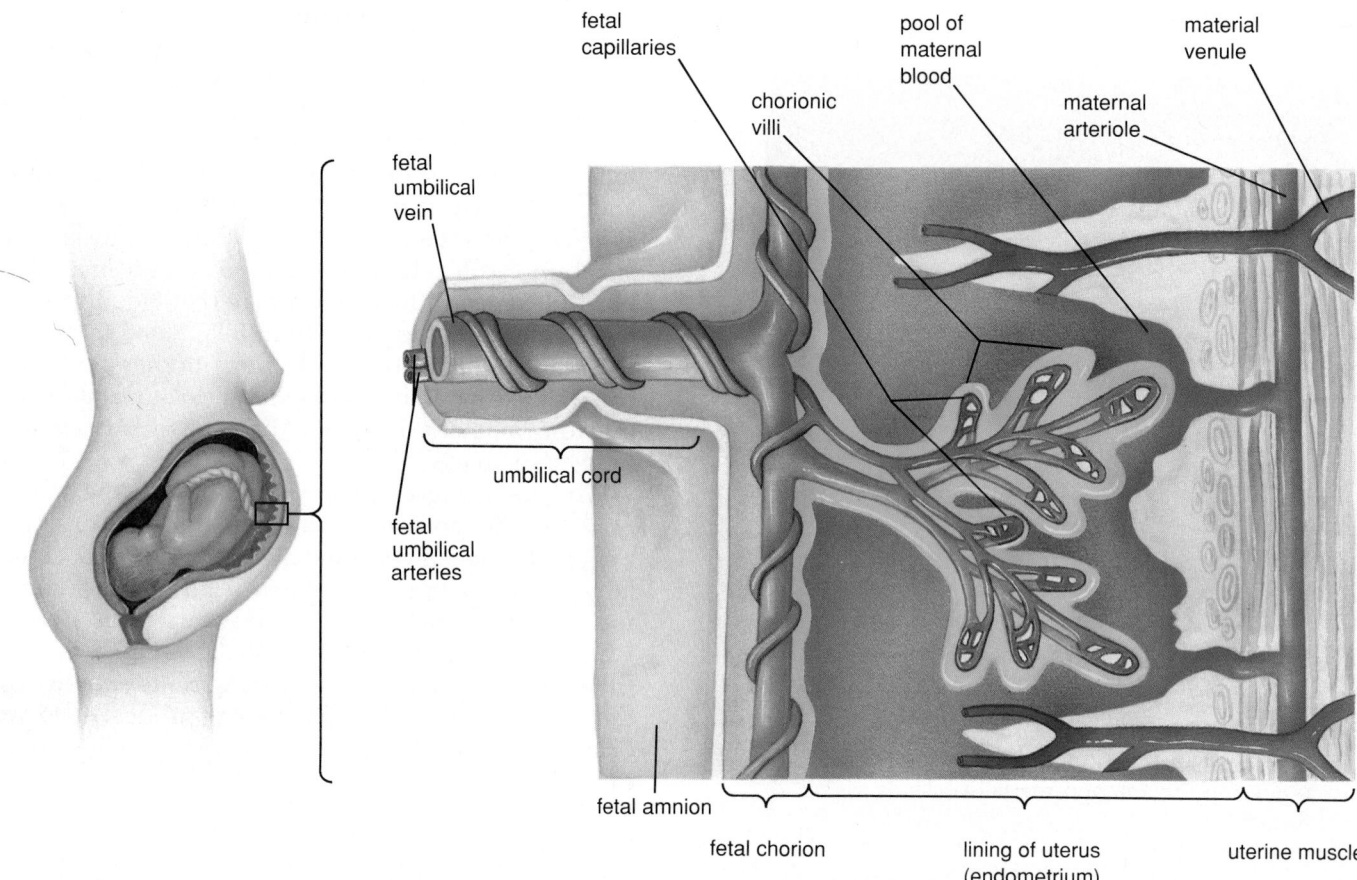

Figure 40-15 The placenta is formed from both the chorion of the embryo and the endometrium of the mother. Capillaries of the endometrium break down, releasing blood to form pools within the placenta. Meanwhile, the chorion develops projections (the chorionic villi) extending into these pools of maternal blood. Blood vessels from the umbilical cord branch extensively within the villi. The resulting structure separates the maternal and fetal blood supplies, while generating a large surface area for diffusion of oxygen, carbon dioxide, nutrients, and wastes between the fetal capillaries and the maternal blood pools.

The Last Seven Months

The fetus continues to grow and develop for another 7 months. Although the rest of the body is "catching up" with the head in size, the brain continues to develop rapidly, and the head remains disproportionately large. Nearly every nerve cell ever formed during the entire human life span develops during embryonic life. As the brain and spinal cord grow, they begin to generate noticeable behaviors. As early as the third month of pregnancy, the fetus begins to move about and respond to stimuli. Some instinctive behaviors appear, such as sucking, which will have

obvious importance soon after birth (see the Chapter Opener). Structures that the fetus will need when it emerges from the womb, such as the lungs, stomach, intestine, and kidneys, enlarge and become functional, though they will not be used until after birth.

Fetuses 7 months or older can usually survive outside the womb, but larger and more mature fetuses have a much greater chance of survival. Evolutionarily speaking, it appears that human fetuses stay in the womb as long as possible, but after about 9 months of development, the head is so large that it can barely fit through the mother's pelvis. The skull is compressed into a slightly conical shape as it passes through the birth canal.

HEALTH WATCH
The Placenta Provides Only Partial Protection

When your grandparents were bearing children, doctors assumed that the placenta protected the developing fetus from most of the substances in maternal blood that could harm it. We know now that this is far from true. In fact, most medications and drugs and even some disease organisms readily penetrate the placental barrier and affect the fetus.

Infections May Cross the Placenta

The German measles virus can cross the placenta and attack the fetus, causing potentially severe retardation and a number of other defects, a strong case for vaccination against this disease. As mentioned in Chapter 39, the virus causing genital herpes (during active outbreaks) and the bacterium causing syphilis can cause mental or physical defects in the developing fetus. The virus causing AIDS may also cross the placenta, so that infants may be born with this incurable disease.

Drugs Readily Cross the Placenta

A tragic example of drugs crossing the placenta is the tranquilizer thalidomide, commonly prescribed in Europe in the early 1960s (Fig. E40-2). Thalidomide's devastating effects on embryos were discovered only when many babies were born with missing or extremely abnormal limbs. More recently, in the late 1980s, the anti-acne drug Accutane was found to cause gross deformities in babies born to women using it (Accutane contains retinoic acid, which has been shown to activate homeobox genes; see "A Closer Look at Homeoboxes and the Control of Body Form"). Although these are extreme examples, **any drug, including aspirin, has the potential to harm the fetus, and any woman who thinks she may be pregnant should seek medical advice about any drugs she takes.** The use of so-called "recreational" drugs, such as heroin and cocaine, has devastating effects on the developing fetus. Babies of heroin addicts are often born addicted. In the United States, hundreds of thousands of infants have been born in recent years to users of "crack" cocaine. These children may be impaired, both behaviorally and emotionally.

The Effects of Smoking

Probably the most common toxins to which fetuses are exposed are those in cigarette smoke. Since many women who smoke are so addicted that they are unwilling to stop, even during pregnancy, nearly a million human embryos are exposed to the poisons (including nicotine, carbon monoxide, and a host of carcinogens) each year in the United States alone. Women who smoke have a higher incidence of miscarriages, and give birth to smaller infants who are more likely to die shortly after birth. There is some evidence that children born to heavy smokers may be somewhat retarded both socially and intellectually.

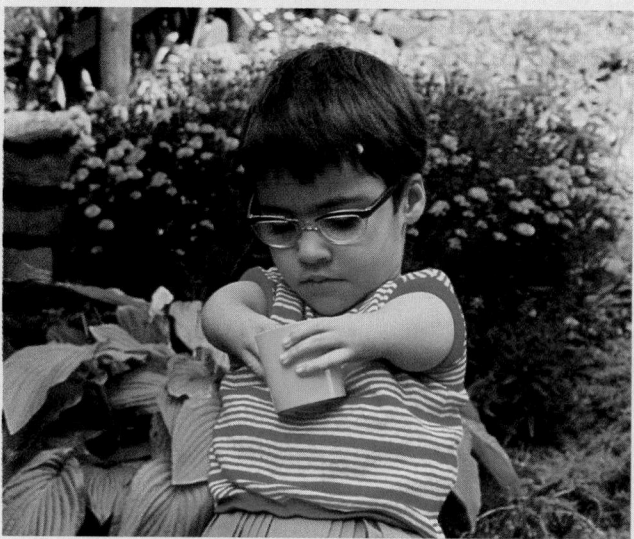

Figure E40-2 Children born to mothers taking the tranquilizer Thalidomide cope heroically with missing and deformed limbs. Drugs with no obvious ill effects on the mother may have devastating impacts on her developing fetus. © UPI/Bettmann

Fetal Alcohol Syndrome

The effects of alcohol on the developing fetus can be devastating. When a pregnant woman drinks, alcohol in the blood of her unborn child reaches levels as high as her own. Children born to alcoholic mothers often exhibit **fetal alcohol syndrome (FAS).** Such children are retarded, and may be hyperactive and irritable. FAS children have small heads, characteristic facial features, and a higher incidence of defects of the heart and other organs. Their growth is also inhibited. If a pregnant woman goes on an alcoholic binge during a critical period of her pregnancy, her infant may be born retarded. Even moderate drinking may produce infants with some of the symptoms of FAS. A woman who takes 1 or 2 alcoholic drinks per day during the first 3 months of pregnancy significantly increases her chances of having a miscarriage. An estimated 1800 to 3600 babies are born each year with FAS, and 36,000 newborns each year are born with milder alcohol-induced problems. **Researchers have not established any safe level of alcohol consumption during pregnancy, nor have they identified any time period during pregnancy when drinking is safe. The U.S. Surgeon General has advised pregnant women and those who are likely to become pregnant to avoid all alcohol consumption.**

In summary, a pregnant woman should assume that whatever chemicals she ingests will find their way into the bloodstream of her developing infant. The mother's choices during this critical 9-month period can strongly influence her child's future well-being.

Birth

Normally, during the last months of pregnancy the fetus becomes positioned head downward in the uterus, with the crown of the skull resting against (and being held up by) the cervix (see Chapter 39). Usually delivered head first, the infant is in for a rude awakening. Formerly, all the world was soft, fluid-cushioned, and warm. Suddenly, the baby must obtain oxygen and eliminate carbon dioxide by breathing. It must regulate its own body temperature, and it must suckle to obtain food. Considering that they have just experienced what by all odds is a change for the worse, babies are often remarkably cheerful (Fig. 40-16)!

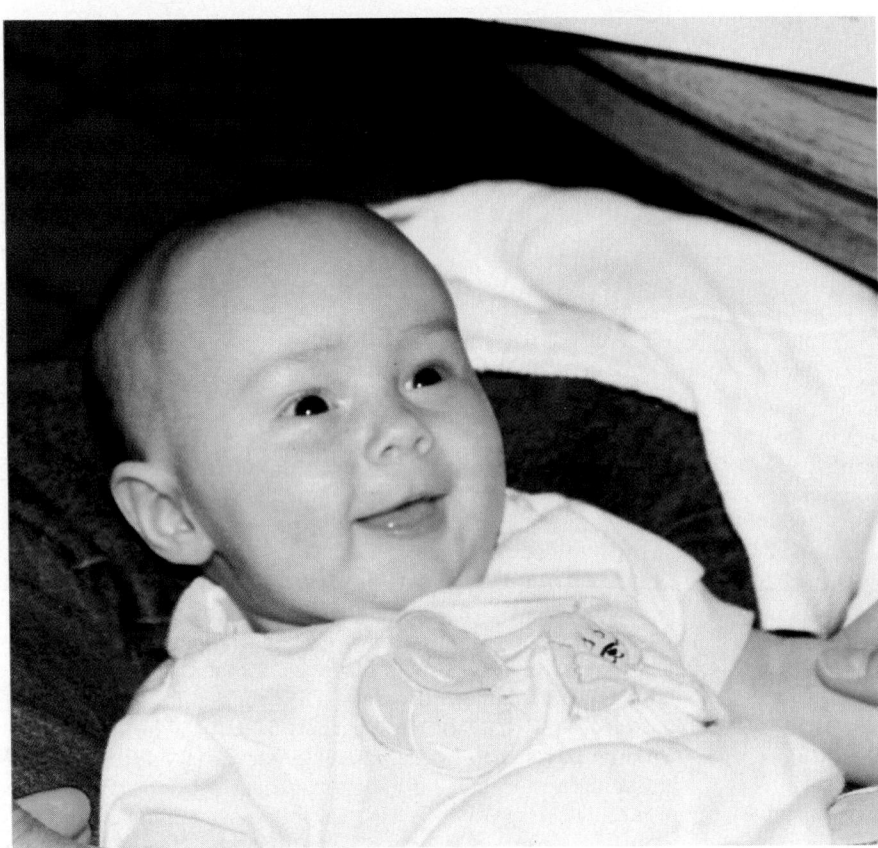

Figure 40-16 On her own.

SUMMARY OF KEY CONCEPTS

Differentiation

All the cells of an animal body contain a full set of genetic information, yet each cell is specialized for a particular function. During development, cells differentiate by stimulating and repressing transcription of specific genes. Gene transcription is regulated in two ways: (1) The egg cytoplasm contains gene-regulating substances that are incorporated into specific daughter cells during the first few cleavage divisions. The developmental fate of each daughter cell is determined by which substances it receives. (2) Later in development, certain cells produce chemical messages that induce other cells to differentiate into particular cell types.

Indirect and Direct Development

Animals undergo either direct or indirect development. In indirect development, eggs (usually with relatively little yolk) hatch into larval feeding stages that later metamorphose into adults with body forms notably different from that of the larva. In direct development, the newborn animal is sexually immature, but otherwise resembles a small adult. Animals with direct development usually either have large, yolk-filled eggs, or nourish the developing embryo within the mother's body.

Stages of Animal Development

Animal development occurs in several stages. *Cleavage:* The zygote undergoes cell divisions with little intervening growth, so that the egg cytoplasm is partitioned into smaller cells. Cleavage divisions result in the formation of a solid ball of cells, the morula. A cavity then opens up within the morula, forming a hollow ball of cells, the blastula. *Gastrulation:* A dimple forms in the blastula, and cells migrate from the surface into the interior of the ball, eventually forming a three-layered gastrula. The three cell layers of ectoderm, mesoderm, and endoderm give rise to all the adult tissues (see Table 40-2). *Organogenesis:* The cell layers of the gastrula form organs characteristic of the animal species. *Maturation:* The juvenile animal increases in size and achieves sexual maturity. *Aging:* Cells begin to function less efficiently, at least partly as a result of failure to repair their DNA, and eventually the animal dies.

Human Development

The human developmental program clearly reflects our reptilian ancestry. Reptilian embryos develop four extraembryonic membranes, the chorion, amnion, allantois, and yolk sac, that function in gas exchange, provision of the watery environment needed for development, waste storage, and storage of yolk, respectively. These membranes are retained by the human embryo, but are modified to suit development inside the mother's womb (see Table 40-1). The chorion invades the endometrium and together these tissues form the placenta. Here the exchange of materials occurs by diffusion between maternal and fetal blood, while the blood supplies remain separate. By the end of the second month, the major organs have formed, the embryo appears human, and is called a fetus. Human embryonic development follows the same principles as the development of other mammals. The stages of human development are summarized in Fig. 40-10.

GLOSSARY

allantois (al-an-tō'-is): one of the embryonic membranes of reptiles, birds, and mammals. In reptiles and birds, the allantois serves as a waste-storage organ and a respiratory surface. In mammals, the allantois forms most of the umbilical cord.

amnion (am'-nē-on): one of the embryonic membranes of reptiles, birds, and mammals, enclosing a fluid-filled cavity that envelops the embryo.

blastocyst: in mammalian embryonic development, a hollow ball of cells formed at the end of cleavage.

blastopore: the site at which a blastula invaginates to form a gastrula.

blastula: in animals, the embryonic stage attained at the end of cleavage, in which the embryo usually consists of a hollow ball with a wall one or several cell layers thick.

chorion (kor'-ē-on): the outermost embryonic membrane in reptiles, birds, and mammals. In birds and reptiles, the chorion functions mostly in gas exchange. In mammals, the chorion forms most of the embryonic part of the placenta.

chorionic villi (kor-e-on-ik): in mammals, fingerlike projections of the chorion of the embryo that penetrate the uterine lining and form the embryonic portion of the placenta.

cleavage: the early cell divisions of embryos, in which little or no growth occurs between divisions.

differentiation: the process whereby relatively unspecialized cells, especially of embryos, become specialized into particular tissue types.

direct development: a developmental pathway in which the offspring is born as a miniature version of the adult and does not radically change its body form as it grows and matures.

ectoderm (ek′-tō-derm): the outermost embryonic tissue layer, which gives rise to structures such as hair, the epidermis of the skin, and the nervous system.

embryo: the stages of development of animals that begin with fertilization of the egg cell and end with hatching or birth. In mammals, embryo usually refers to the early stages in which the developing animal does not yet resemble the adult of the species.

embryonic disk: in human embryonic development, the flat, two-layered group of cells derived from the inner cell mass of the blastocyst, which will develop into the embryo proper.

endoderm (en′-dō-derm): the innermost embryonic tissue layer, which gives rise to structures such as the lining of the digestive and respiratory tracts.

fetal alcohol syndrome: a cluster of symptoms including retardation and physical abnormalities that occur in infants born to mothers who consumed large amounts of alcoholic beverages during pregnancy.

fetus: the later stages of mammalian embryonic development, when the developing animal has come to resemble the adult of the species.

gastrula (gas′-trū-la): in animal development, a three-layered embryo with ectoderm, mesoderm, and endoderm cell layers. The endoderm layer usually encloses the primitive gut.

gastrulation (gas-trū-lā′-shun): the process whereby a blastula develops into a gastrula.

homeobox (hō′-mē-ō-box): a sequence of DNA coding for special, 60 amino acid proteins. These proteins activate or inactivate genes that control development. Homeoboxes thus specify embryonic cell differentiation.

indirect development: a developmental pathway in which a free-living offspring goes through radical changes in body form as it matures.

induction: the process by which a group of cells causes other cells to differentiate into a specific tissue type.

inner cell mass: in human embryonic development, the cluster of cells on one side of the blastocyst, which will develop into the embryo.

larva: an immature form of an animal with indirect development, often much different in body form from the adult.

mesoderm (mes′-ō-derm): the middle embryonic tissue layer, lying between the endoderm and ectoderm, and usually the last to develop. Mesoderm gives rise to structures such as muscle and skeleton.

metamorphosis (met-a-mōr′-fō-sis): in animals with indirect development, a radical change in body form from one larval stage to another or from larva to adult.

morula (mor′-ū-la): in animals, an embryonic stage during cleavage, when the embryo consists of a solid ball of cells.

neural tube: a structure derived from ectoderm during early embryonic development that later becomes the brain and spinal cord.

organogenesis (or-gan-ō-jen′-i-sis): the process by which the germ layers of the gastrula develop into organs.

placenta: in mammals, a structure formed partly from the uterine lining and partly from the embryonic membranes, especially the chorion; functions in gas, nutrient, and waste exchange between embryonic and maternal circulatory systems.

primitive streak: in reptiles, birds, and mammals, the region of the ectoderm of the two-layered embryonic disk through which cells migrate to form mesoderm.

yolk: protein- or lipid-rich substances contained in eggs as food for the developing embryo.

yolk sac: one of the embryonic membranes of reptile, bird, and mammalian embryos. In birds and reptiles, the yolk sac is a membrane surrounding the yolk in the egg. In mammals, the yolk sac is empty, but forms part of the umbilical cord and the gut.

STUDY QUESTIONS

1. Define differentiation. How do cells differentiate; that is, how is it that adult cells express some but not all the genes of the fertilized egg?
2. Describe the two processes of differentiation during development: the influence of the egg cytoplasm and induction by other cells.
3. Distinguish between indirect and direct development, and give examples of each.
4. What is yolk? How does it influence cleavage?
5. Name two structures derived from each of the three embryonic germ layers—endoderm, ectoderm, and mesoderm.
6. What is gastrulation? Describe gastrulation in frogs and humans.
7. Describe the process of induction and give two examples.
8. How does cell death contribute to development?
9. List the four extraembryonic membranes and the function of each in a reptile and a human embryo.
10. Explain how the structure of the placenta prevents mix-

ing of fetal and maternal blood while allowing the exchange of substances between the mother and the fetus.

11. List and explain two very different functions of the placenta.

12. Is the placenta an effective barrier against substances that can harm the fetus? Describe two different types of harmful agents that can cross the placenta and their effects on the fetus.

DISCUSSION QUESTION

1. When fetal ectoderm tissue from the mouth of one embryo is transplanted into a region of mesoderm of an embryo of another species, the mesoderm induces the ectoderm to form mouth structures of the type of organism from which the ectoderm tissue was taken. When this type of experiment is done using chick ectoderm implanted into a mouse embryo, the chicken tissue produces teeth. Discuss what this experiment tells us about why the adage "scarce as hen's teeth" is true. Also, what does this experiment tell us about the evolutionary origin of chickens?

SUGGESTED READINGS

Anderson, R., and Anderson K. The effects of alcohol consumption during pregnancy. J Am Assoc Occup Health Nurs 1986; 34: 88–91. Clearly describes the range of alcohol's devastating effects on the human embryo.

Beaconsfield, P., Birdwood, G., and Beaconsfield R. "The Placenta." *Scientific American,* August 1980 (Offprint No. 1478). The placenta is one of the most remarkable of mammalian structures, allowing internal development of offspring within the mother.

Cooke, J. "The Early Embryo and the Formation of Body Pattern." *American Scientist,* January–February 1988. Although the developmental patterns of worms, fruit flies, and humans are superficially very different, they are probably based on common mechanisms.

DeRobertis, E. M., and Gurdon, J. B. "Gene Transplantation and the Analysis of Development." *Scientific American,* December 1979. Genes can be injected into amphibian eggs, allowing direct observation of the effects of specific gene products on development.

DeRobertis, E. M., Oliver, G., and Wright C. V. E. "Homeobox Genes and the Vertebrate Body Plan." *Scientific American,* July 1990. Excellent description of the research that led to the current understanding of homeobox genes.

Fjose, A. "Spatial Expression of Homeotic Genes in *Drosophila.*" *Bioscience,* September 1986. A lucid discussion of the genetic control of insect body pattern formation.

Hall, S. S. "The Fate of the Egg." *Science 85,* November 1985. A look at some of the attempts to understand development in everything from fruit flies to people.

Ingelman-Sundberg, A. *A Child Is Born.* New York: Dell Publishing Company, 1966. Simple, clear explanations of the stages of human development, with magnificent photographs by Lennart Nilsson.

Nijhout, H. F. "The Color Patterns of Butterflies and Moths." *Scientific American,* November 1981. The structure and development of wing patterns is a fascinating study for its own sake, as well as for what it may tell us about developmental principles.

Saunders, J. W., Jr. *Developmental Biology.* New York: Macmillan Publishing Company, 1982. Excellent descriptions and illustrations of developmental processes on both the morphological and molecular levels.

41

Animal Behavior. I. Principles of Individual Behavior

Animals show a fascinating variety of feeding behaviors. Frozen by a high-speed camera, this chameleon captures a butterfly with a flick of its enormously extensible tongue.

"As I watched the geese, it appeared to me as little short of a miracle that a hard, matter-of-fact scientist should have been able to establish a real friendship with wild, free-living animals, and the realization of this fact made me strangely happy. It made me feel as though man's expulsion from the Garden of Eden had thereby lost some of its bitterness."

Konrad Lorenz, in King Solomon's Ring

As you hike through a forest, roll a damp, decaying log to reveal the hidden pillbug, slug, and salamander. Listen for the scolding of squirrel and bird as you pass through their territories. Consider that the existence of these life forms today is the result of eons of successful reproduction. Reproduction requires that the animal live to maturity, find a mate, and gather the resources needed to produce offspring. These activities demand a lifetime of behavior finely tuned to the animal's environment, both physical and social. Anatomy, physiology, genetic programming, and learning all must mesh properly to produce the correct action at the appropriate time.

Historically, diverse disciplines have contributed to the science of animal behavior. **Ethologists,** for example, study behavior as it occurs under natural conditions in a wide variety of organisms. An important principle of **ethology** is that behavior has evolved as an adaptive trait through natural selection. Thus, ethologists view behavior as a genetically programmed response to specific stimuli in the natural environment. Beginning with the observational and intuitive approach of Konrad Lorenz, whose investigations spanned the first half of this century, ethology was molded into an experimental science by Niko Tinbergen and Karl von Frisch, whose works are mentioned later in this and the next chapter.

Comparative psychologists, in contrast, treat animal behavior as a laboratory science. The evolutionary aspects and adaptive value of behavior are minimized. These investigators generally use a few types of laboratory animals (such as rats and pigeons) under highly controlled conditions. Learning and its mechanisms are the prime focus of their studies. Their ultimate goal is to decipher underlying principles that may be applicable to humans.

Behavioral neurobiologists attempt to decipher the mechanisms by which the brain directs behavior. Advances in technology that make the inner workings of the brain more accessible have contributed to the rapid growth and success of this endeavor. This discipline incorporates elements of psychology, neuroanatomy, brain biochemistry, and the properties of individual neurons. Some investigators use invertebrates to provide insight into how relatively simple nervous systems direct behavior (see "Of Squids and Snails and Sea Hares" in Chapter 36).

Recently, some prominent investigators of animal behavior have adopted a highly theoretical, evolutionary approach. Their studies focus on the adaptive value of broad classes of behavior that do not appear to enhance the survival of the individual. For example, an animal showing **altruism** may endanger itself to protect others in its social group. Since these individuals are more likely to be killed, why aren't the genes directing this behavior eliminated from the gene pool of the population? This question is addressed in Chapter 42 in the essay on "Altruism and the Selfish Gene" and in the essay on kin selection in Chapter 17.

In this chapter, we introduce basic concepts of animal behavior, emphasizing individual behavior. Its examples are drawn largely (but not exclusively) from the work of ethologists. These early studies provide insight into the fascinating diversity of animal behavior, its adaptiveness, and its causes. In Chapter 42, we explore animal communication systems and their use in social behavior.

The Genetic Basis of Behavior

All behavior has some genetic basis. For example, much behavior consists of responses to environmental stimuli. These stimuli are received and filtered by the sense organs. The design of these organs, coded by the genes, determines what stimuli the animal perceives. Because sensory capabilities differ dramatically, each species perceives a different world (see "Roadmaps of the Mind: The Role of the Senses in Animal Orientation" in Chapter 37). Bats, for example, produce and respond to sounds far higher in pitch than the human ear can detect. A bee observing a flower sees patterns of reflected ultraviolet light, invisible to the human eye, that guide it toward nectar (Fig. 41-1). The pit organs of rattlesnakes can detect the body heat of a nearby mouse (see Fig. 37-2). The tick is exquisitely sensitive to the odor of butyric acid, produced by the skin of mammals. The scent will cause it to drop off its perch, usually landing on the passing animal. Clearly, behavior is governed by perception.

(a)
(b)

Figure 41-1 Sensory abilities control perception. Bees see ultraviolet wavelengths as color, while humans do not, so humans and bees perceive many flowers differently. **(a)** The ultraviolet-reflecting center of this flower guides bees toward the nectar and encourages them to pollinate the flower. **(b)** Humans are unable to see this pattern.

Sense organs are extensions of the genetically coded nervous system. Besides sending the impulses that directly drive behavior, the nervous system, by its complexity and its specific neural connections, determines how much and what type of learning is possible. Another influence of the genes on behavior is seen in the radically differing body forms of different species. Imagine the streamlined body of the dolphin slicing through the water, and the dextrous manipulations of the long-legged spider. Now try to imagine either of them performing the behavior of the other! It is only within these genetically determined limits that behavior can be influenced by the environment through learning.

As we will see in the descriptions that follow, heredity sets very narrow boundaries for behaviors that are largely **innate** or **instinctive,** and wider boundaries for learned behaviors.

Innate Behavior

A fruit fly, placed in a tunnel with light at one end and darkness at the other, will fly toward the light; a hungry newborn human infant, touched on the side of her mouth, will turn her head and attempt to suckle. These are examples of **innate** or instinctive behaviors. **Innate behavior is performed in reasonably complete form the first time an animal of the right age and motivational state encounters a particular stimulus. It is programmed by the genes and passed from one generation to the next. Since innate behavior is performed without learning or prior experience, it tends to be highly stereotyped (performed similarly each time).** Innate behaviors fall

into four categories: (1) kineses, (2) taxes, (3) reflexes, and (4) fixed action patterns.

Kineses

A **kinesis** is a behavior in which an organism changes its speed of random movement in response to an environmental stimulus. As a result, it ends up in a favorable environment by stopping when it blunders there by chance, and escapes hostile conditions by speeding up. For example, the pillbug (a land-dwelling crustacean) uses a kinesis to reach the moist areas it needs to survive. As the air surrounding it becomes drier, the pillbug moves faster (but in no particular direction) until it encounters a damper area, where it slows and eventually stops. This kinesis results in congregations of pillbugs under damp leaves and rotting logs.

Taxes

In contrast to a kinesis, a **taxis** is a *directed* movement toward or away from a stimulus. A moth flying toward a light is one example (Fig. 41-2). Much of the behavior of very simple organisms, including single-celled protists, are taxes. For example, the protist *Euglena* (see Chapter 21) is photosynthetic, but its chlorophyll can be damaged by intense light. As you might predict, *Euglena* shows a positive taxis toward dim light, but a negative taxis toward intense light. Mosquito behavior involves several taxes. Males orient toward the high-pitched whine of the female. Female mosquitos (only females suck blood) show taxes to the warmth, humidity, and carbon dioxide exuded by their prey. Mosquito repellents probably act by blocking the receptors for these stimuli, rendering the insect incapable of sensing her victim. Taxes can be influenced by the internal state of the organism. The grayling butterfly shows a taxis toward bright light, but only when it is pursued. A grayling fleeing a predator flies directly toward the sun, temporarily blinding its pursuer.

Vertebrates may also show taxes, although vertebrate taxes are more difficult to detect because they interact with other behaviors. For example, fish swim upright by orienting their dorsal surface away from the force of gravity, and toward light (normally the water surface). If the gravity-detecting organ of a fish is removed, or the light enters the side of a fishtank, the animal becomes disoriented, as shown in Figure 41-3.

Figure 41-2 Moths and other night-flying insects show a positive taxis toward light, resulting in congregations such as this.

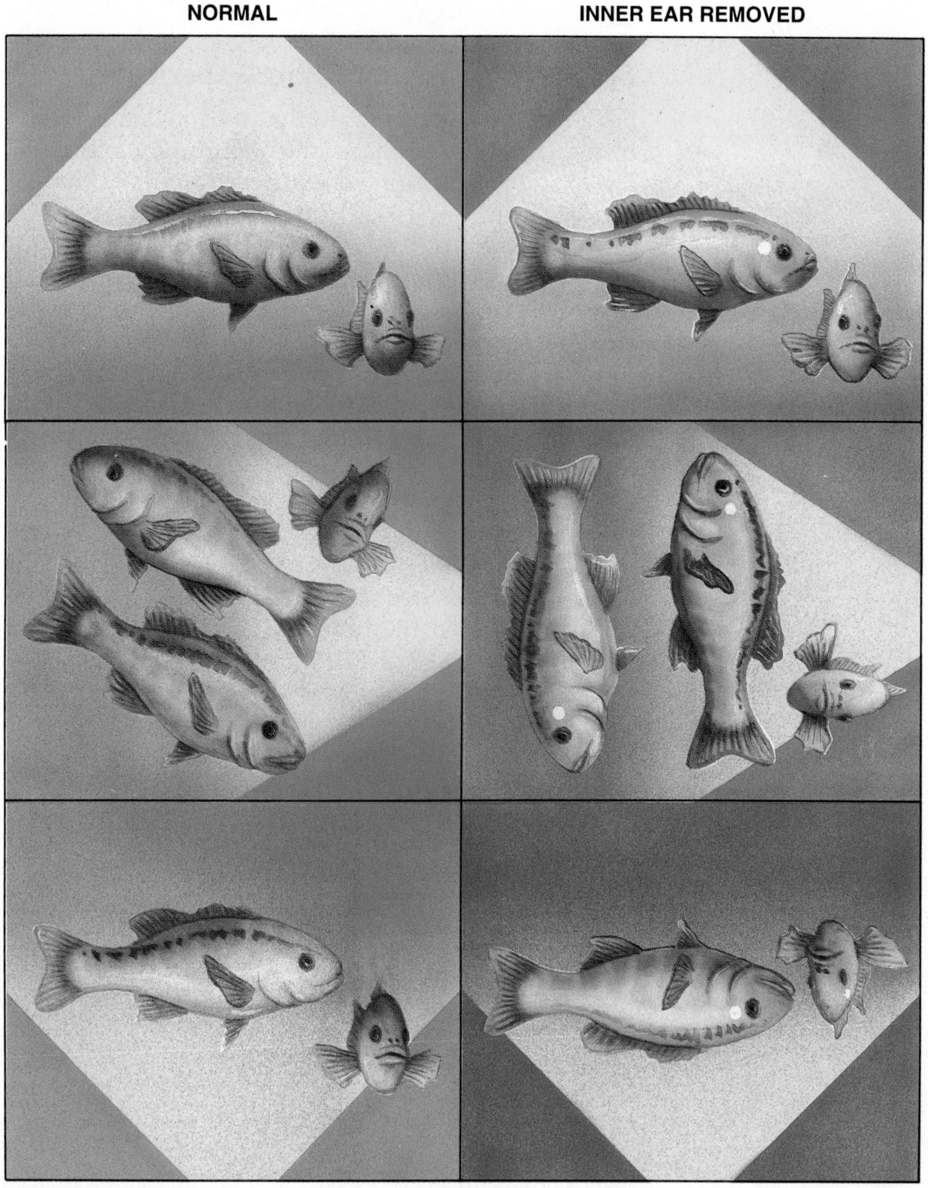

NORMAL   **INNER EAR REMOVED**

Figure 41-3 Taxes to light and gravity in the fish *Crenilabrus*. Fish on the left are normal, while those on the right have had their inner ears removed, eliminating their response to gravity. The fish unable to detect gravity orient to the light alone.

Reflexes

In contrast to taxes or kineses, which involve orientation of the entire body, a **reflex** is a movement of a body part, such as blinking an eye or withdrawing the hand from a hot stove. Reflexes are usually stereotyped and rapid, since the conscious brain is not involved. Take the knee-jerk reflex, for example. Tapping the patellar (kneecap) ligament stretches the thigh muscle to which the kneecap is attached, just as would occur if your knee suddenly buckled. A stretch receptor in the muscle carries an emergency message directly to a motor neuron in the spinal cord. This commands the thigh muscle to contract, suddenly straightening the leg (Fig. 41-4). Had the knee actually buckled, this reflex could prevent a dangerous fall. Your brain, which is informed of this activity via the eyes and spinal cord, does not participate in the reflex. Instead, it observes the reflex response to the doctor's hammer tap with detached interest and curiosity.

Fixed Action Patterns

A **fixed action pattern** is a stereotyped and often complex series of movements. If the animal is at the right developmental stage and properly motivated, a fixed action pattern will be performed correctly the first time the appropriate stimulus, also called a **releaser,** is presented. Fixed action patterns may be so complex and so appropriate that they appear learned. How can we differentiate these complex instinctive acts from learned behaviors? Three approaches are suggested below.

Performance Without Prior Experience

Scientists can demonstrate that a behavior is innate by depriving the animal of the opportunity to learn it. For example, red squirrels bury extra nuts in the fall for retrieval during the winter. Red squirrels can be raised from birth in a bare cage on a liquid diet, providing them with no experience with nuts, digging, or burying. Presented with nuts for the first time, such a squirrel (after eating several) will carry one to the corner of its cage, then make covering and patting motions with its forefeet. Nut-burying is therefore an innate fixed action pattern, and the nut serves as a releaser.

Fixed action patterns can also be recognized by their occurrence immediately after birth, before there is any opportunity for learning. The cuckoo bird, for example, lays its egg in the nest of another bird species, to be raised by the unwitting adoptive parent. Immediately after hatching, the cuckoo chick shoves the nest-owner's eggs (or baby birds) out of the nest, eliminating its competitors for food (Fig. 41-5).

Manipulating the Releaser

Since a fixed action pattern is a rigid response to its releaser, scientists can elicit inappropriate behavior by providing the releaser in the wrong context. An example was inadvertently provided by bird-banding. Parent birds clean their nests, throwing out light-colored objects such as broken egg shells or excrement. Biologists who had placed shiny metal identification bands on the legs of baby birds were dismayed to find the parents attempting to throw the band out of the nest, baby bird and all! The band was a releaser for the fixed action pattern of nest-cleaning, regardless of the consequences.

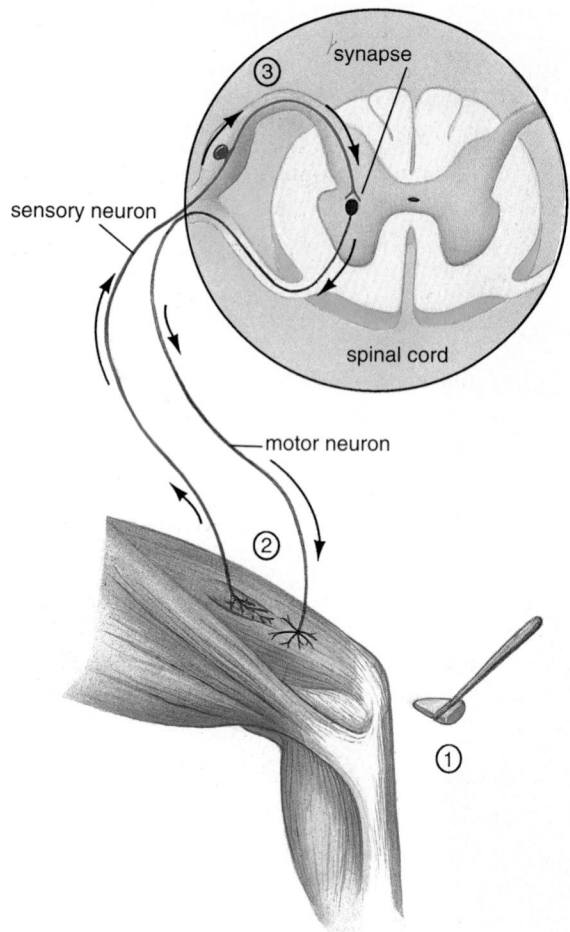

Figure 41-4 The knee-jerk reflex. (1) Tapping the patellar ligament causes a sudden stretching of the thigh muscle. (2) This activates stretch receptors in the muscle that communicate to the spinal cord. (3) In the cord, the impulse is transmitted directly to a motor neuron, which causes the thigh muscle to contract, jerking the lower leg upward. The brain is not involved in this simple behavior.

(a)

(b)

Figure 41-5 (a) The cuckoo chick, just hours after it hatches, and before its eyes have opened, evicts the eggs of its foster parents from the nest. **(b)** The parents feed the single, oversized cuckoo chick, unaware that it is not their own.

Fixed action patterns are often released only by particular characteristics of the animal toward which they are directed. The female redwinged blackbird signals her readiness to mate by the angle of her tail, which serves as a releaser for mating by the male. He will even attempt to copulate with the stuffed tail of the female, provided the feathers are angled appropriately (Fig. 41-6a). Konrad Lorenz described another interesting example. He was attacked by tame and normally friendly jackdaw birds (European relatives of crows) as he walked toward his swimming hole. The limp black bathing trunks he carried released the same behavior as would a dead jackdaw in the clutches of a predator (Fig. 41-6b).

Scientists can use models to exaggerate releasers. Such exaggerated releasers, called **supernormal stimuli,** are far more effective than their natural counterparts. By selectively exaggerating specific features of a stimulus, ethologists can determine which features serve as releasers. For example, the herring gull chick, studied by Niko Tinbergen, instinctively pecks at a red spot on the parent gull's beak (Fig. 41-7). This fixed action pattern, which chicks perform immediately after hatching, causes the parent to regurgitate food. Tinbergen found that the releasing features of the bill are its long, thin shape, the red color, and the presence of color contrasts. When he offered chicks a thin red rod with white stripes painted on it, they pecked at it more often than at a real beak (Fig. 41-8). Herring gulls, which nest on the ground, show a fixed action pattern of egg retrieval released by the sight of an egg that has rolled out of the nest. Scientists placed various model eggs outside the nest and found that the most effective models are speckled, rounded, and very large. As shown in Figure 41-9, the gulls also preferred to brood these monster egg models. This behavior is apparently common in ground-nesting birds. Tinbergen offered nesting geese a choice between their own egg and a volleyball, both placed just outside the nest. Invariably, the geese would attempt to retrieve the volleyball, whose size presented a truly supernormal stimulus.

Breeding Experiments: Behavioral Genetics

Sometimes scientists can produce hybrids by mating closely related species that differ in a particular behavior. If the behavior is genetic, the hybrid offspring may try to perform a combination of the two behaviors. One excellent example was provided by W. Dilger, who mated two different species of African lovebirds. One species carries nesting material to its nest site in its beak, one piece at a time. The second species tucks several pieces of nesting material into its tail feathers before flying to the nest (Fig. 41-10). Their hybrid offspring tried to place nesting material in their tailfeathers, but positioned it improperly, or failed to release it, apparently trying to carry it in both their tails and their bills.

Another well-known example was provided by breeding two strains of honeybees with different behaviors. These elegant experiments (described in "Methods in Biology: Tracing the Genetic Basis of Behavior") provide evidence that a behavior may be controlled by a single recessive allele.

The Importance of Innate Behavior

Innate behaviors are adaptive for a variety of reasons. First, survival may depend on the proper performance of behavior the first time the stimulus occurs. A good example is a camouflaged animal "freezing" at the sight of a predator. In such cases, even rapid

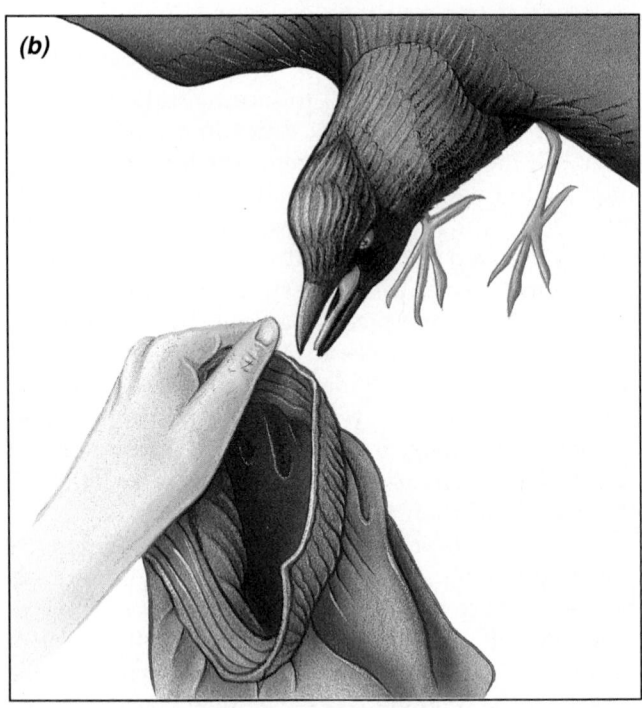

Figure 41-6 (a) A male redwinged blackbird is stimulated to mate by the angle of the female's tail, whether or not she is attached to it. **(b)** Konrad Lorenz was attacked by his tame jackdaws as he carried his black bathing trunks. The trunks, resembling a dead jackdaw in the clutches of a predator, acted as a releaser for this defensive behavior.

Figure 41-7 A herring gull chick pecks at the red spot on its mother's bill, causing her to regurgitate food for it. Niko Tinbergen and others have shown that the thin shape of the bill and the contrasting spot serve as releasers for pecking, which occurs immediately after hatching. The pecking then serves as a releaser for food regurgitation by the parent.

learning is too slow; the animal may not have a second chance. Second, animals with simple nervous systems may not be capable of learning behaviors important to survival. Thus, the simpler the organism the more it relies on innate behavior. Third, social interactions important for survival may depend on the animal rigidly performing specific roles. Complex insect societies such as those of termites and bees are possible only because insect behavior is almost entirely innate; any show of individuality could cause the society to collapse. Mating rituals provide another example. These innate rituals (described in Chapter 42) allow animals of the same species to recognize and respond to one another automatically. Since the sexual behavior of different species is usually incompatible, they avoid wasteful mating between different species.

Learning

The eviction of banded baby birds by their nest-cleaning parents shows that instinctive behavior has its drawbacks, especially when the environment is not entirely predictable. Survival is enhanced if an animal can modify its behavior based on experience. Thus, nervous systems have evolved with varying degrees of flexibility. In general, the more complex an animal's nervous system, the more its behavior can be modified by experience. This capacity is called

Model   Response
       (no. of pecks)

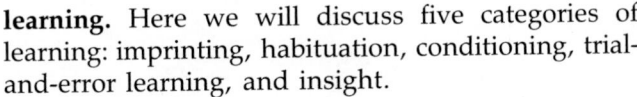

Figure 41-8 Models are used to determine which features of the bill are releasers for pecking behavior. The length of the bar is proportional to the number of pecks delivered to the corresponding model. A brightly contrasting spot improves performance; the color red further increases pecking; but the red knitting needle with contrasting stripes was most effective. Here the long, thin shape, the red color, and the contrast are all exaggerated to produce a supernormal stimulus.

Figure 41-9 A herring gull will ignore her own eggs if offered a supernormal brooding stimulus.

Figure 41-10 Lovebirds tearing off strips of nesting material and tucking them among their tailfeathers for transport to the nest site.

learning. Here we will discuss five categories of learning: imprinting, habituation, conditioning, trial-and-error learning, and insight.

Imprinting

As discussed earlier, learning always occurs within boundaries specified by the animal's genes. This is strikingly illustrated by the unique form of learning called **imprinting**. Imprinting refers to a strong association learned during a particular stage, called a sensitive period, in an animal's life. The process is best known in birds such as geese, ducks, and chickens. These birds learn to follow the animal or object that they most frequently encounter during a sensitive period (for mallard ducks this occurs about 13 to 16 hours after hatching). In nature, the mother is the object of imprinting. In the laboratory, however, these birds can be forced to imprint on a toy train or other moving object, although, if given a choice, they select a duck. Konrad Lorenz coined the word "imprinting" in the 1930s, and demonstrated that the object of imprinting did not have to be appropriate (Fig. 41-11). During sexual imprinting, the young bird learns the characteristics of its future mate. Sexual imprinting occurs slightly later than the imprinting that causes the bird to follow a moving object. During imprinting, the animal is actually learning a releaser

METHODS IN BIOLOGY

Deciphering the Genetic Basis of Behavior

A famous example of behavior genetics comes from N. Rothenbuhler's elegant studies of the honeybee, performed in the 1960s. Honeybee eggs are laid in special chambers within the hive called brood cells. The brood cells are capped to protect the developing larvae. Honeybees perform many genetically programmed behaviors. One of these is uncapping the brood cells that contain diseased larvae and removing the larvae. This prevents disease from spreading, and is called "hygienic" behavior. Certain strains of honeybee (called "unhygienic") neither remove the caps nor remove the diseased larvae. When Rothenbuhler crossed homozygous (true-breeding; see Chapter 11) hygienic with homozygous unhygienic strains, the resulting offspring were *all unhygienic*, indicating that hygienic behavior must be a recessive trait. However, the complete behav-

ior involves two steps: first, removing the wax cap from the comb cell (uncapping), and second, removing the diseased larvae. Rothenbuhler made an interesting discovery by crossing a member of the heterozygous F_1 generation produced above with the hygienic strain (homozygous recessive). Four behaviors appeared in different offspring. Of 29 offspring, nine uncapped diseased larval cells but failed to remove the larvae, six did not uncap but would remove diseased larvae if the combs were already uncapped, eight would neither uncap nor remove diseased larvae (unhygienic), and the remaining six both uncapped and removed diseased larvae (hygienic). The simplest interpretation of the data is that uncapping and removal are each under the control of a single recessive allele, as described in Figure E41-1.

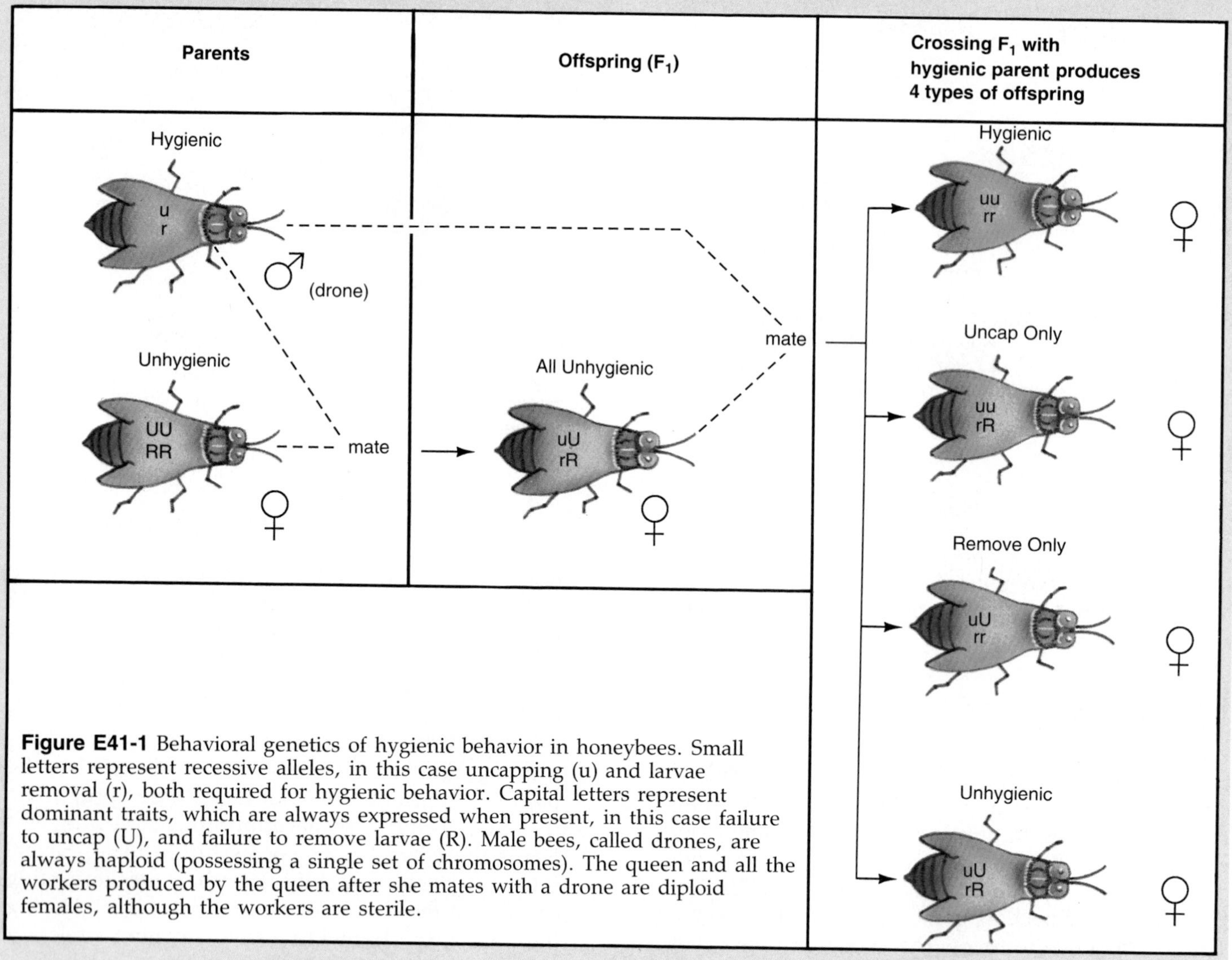

Parents	**Offspring (F₁)**	**Crossing F₁ with hygienic parent produces 4 types of offspring**
Hygienic — u r (drone) ♂ Unhygienic — UU RR ♀ — mate	All Unhygienic — uU rR ♀ — mate	Hygienic — uu rr ♀ Uncap Only — uu rR ♀ Remove Only — uU rr ♀ Unhygienic — uU rR ♀

Figure E41-1 Behavioral genetics of hygienic behavior in honeybees. Small letters represent recessive alleles, in this case uncapping (u) and larvae removal (r), both required for hygienic behavior. Capital letters represent dominant traits, which are always expressed when present, in this case failure to uncap (U), and failure to remove larvae (R). Male bees, called drones, are always haploid (possessing a single set of chromosomes). The queen and all the workers produced by the queen after she mates with a drone are diploid females, although the workers are sterile.

Figure 41-11 Konrad Lorenz, known as the "father of ethology," is followed by goslings that imprinted on him shortly after they hatched. They now treat him as they would their mother.

(the object followed and later courted) for a fixed action pattern (following, and later, courtship) during a limited period in its development, the sensitive period. Parents may also imprint on their offspring. Herring gulls imprint on their young during the first 2 days after hatching. During this 2-day period, they will accept foster young placed in their nest. Thereafter, the parents will attack and kill baby chicks who are not their own.

Habituation

A more common form of simple learning is **habituation,** defined as a decline in response to a harmless, repeated stimulus. The ability to habituate prevents an animal from wasting its energy and attention on irrelevant stimuli. This form of learning is shown by the simplest animals, and has even been demonstrated in the single-celled members of the kingdom Protista. For example, the protist *Stentor* contracts to touch (Fig. 41-12a), but gradually stops retracting if touching is continued. The sea anemone, which also lacks a brain, shows a similar response to touch (Fig. 41-12b). The ability to habituate is clearly adaptive. If an anemone contracted every time it was brushed by a strand of waving seaweed, it would waste a great deal of energy, and its retracted posture

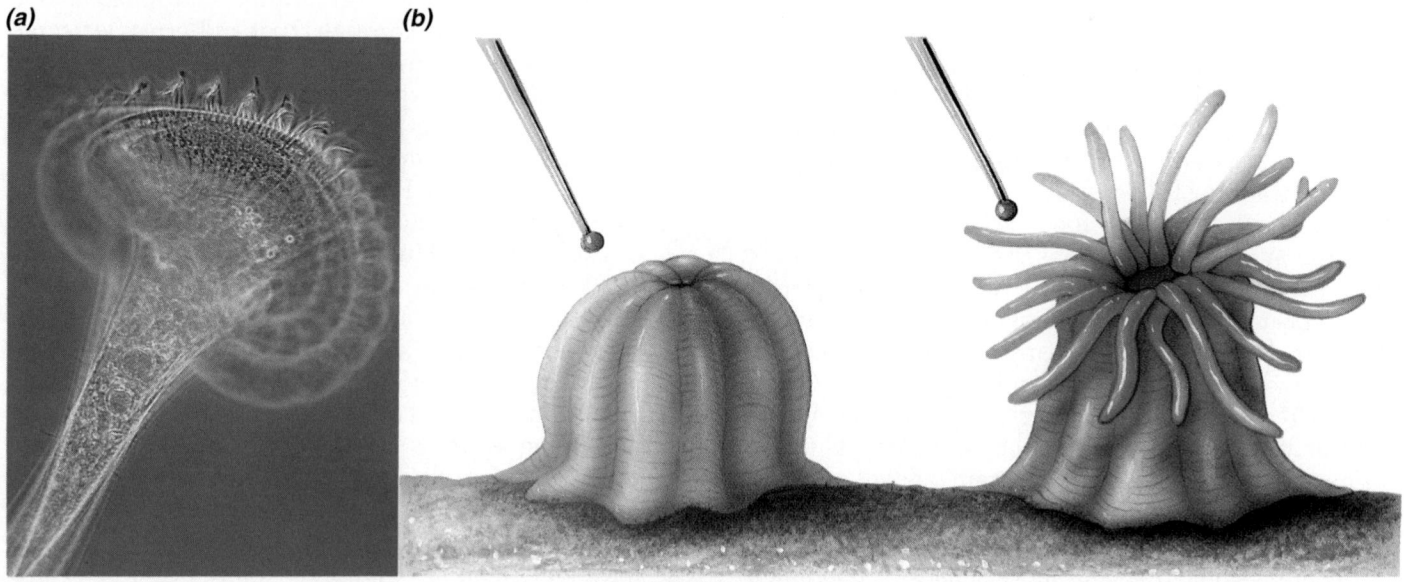

(a)   *(b)*

Stentor   First touch   After many touches

Figure 41-12 (a) The protist *Stentor* is a complex single cell that can show habituation, a simple form of learning. **(b)** Habituation in the sea anemone. (Left) The animal is touched for the first time, causing withdrawal. (Right) After many touches, the anemone habituates to this harmless stimulus. Learning occurs even though the anemone possesses only a simple network of neurons and lacks a brain.

would prevent it from snaring food. Humans habituate to a variety of stimuli: city dwellers to nighttime traffic sounds, and country dwellers to choruses of crickets and tree frogs. Each may initially find the other's habitat unbearably noisy at night, but habituate after a time.

Classical and Operant Conditioning

A more complex form of learning, most commonly seen in the laboratory, is called conditioning. During **classical conditioning** an animal learns to perform a response (normally caused by one stimulus) to a new stimulus. Ivan Pavlov, a Russian physiologist, is sometimes called the "father of conditioning." In the early 1900s, he performed his famous experiment on dogs. Pavlov first placed dried meat powder into a dog's mouth, causing reflex salivation. Then he rang a bell just before the meat powder was presented. After several repetitions, the dog would salivate to the bell alone. The dog who leaps up at the sight of her leash or drools on the floor at the sound of a can opener has been classically conditioned, however unintentionally. This form of learning can also be demonstrated in simple organisms. For example, some investigators have reported that flatworms can learn to associate a flash of light with an electric shock, which causes them to contract. After several exposures to light paired with shock, they contract to the light alone.

During **operant conditioning** an animal learns to perform a behavior (such as pushing a lever or pecking a button) to receive a reward or avoid punishment. This technique is most closely associated with the American comparative psychologist B. F. Skinner. Skinner designed the "Skinner Box" in which an animal (often a white rat or pigeon) is isolated and allowed to train itself (Fig. 41-13). The box might contain a lever that ejects a food pellet when pressed. The animal in its explorations inevitably bumps the lever and is rewarded. Soon it will be pressing the lever repeatedly for food. Skinner boxes can be quite complex. A pigeon might learn that one lever will provide food only when a green light is shining, but when a red light is on, a specific spot on the wall must be pecked to avoid a shock. The Coast Guard has used operant conditioning to train pigeons for sea search-and-rescue missions. The birds are trained to peck a button when they see orange, the color used for lifejackets. Pigeons spot more lifejackets bobbing in the sea and make fewer mistakes than the best human observers.

Trial-and-Error Learning

In the natural environment, animals are faced with naturally occurring rewards and punishments and

Figure 41-13 The "Skinner Box," used for operant conditioning.

learn by **trial and error.** (Operant conditioning is actually a special form of trial-and-error learning that occurs under artificial and carefully controlled laboratory conditions.) Through trial-and-error learning, animals acquire new and appropriate responses to stimuli through experience. Much learning of this type occurs during play and exploratory behavior in animals with complex nervous systems (see "A Closer Look at the Purpose of Play"). Trial-and-error learning makes a major contribution to the behavior of young children (not to mention adults). It is how a child learns which foods taste good or bad, that a stove may be hot, and not to pull the cat's tail.

In some instances, trial-and-error learning can modify the releaser for innate behavior, thus making it more adaptive. For example, to the hungry toad any flying insect is a releaser for feeding. Although this usually results in a meal, occasionally the toad captures a mouthful of trouble: a bee. Having its tongue stung results in trial-and-error learning that takes only a single experience (Fig. 41-14). Learning modifies the releaser to exclude bees and even other insects that resemble them.

Insight

The most complex form of learning is called **insight** or reasoning. Insight involves manipulating concepts in

A CLOSER LOOK

At the Purpose of Play

One of the most appealing aspects of animal behavior, and one of the most puzzling to the investigator, is play. Play has been observed in many birds and in most mammals, from cows to lions to mice. Pygmy hippopotami push one another, shake and toss their heads, splash in the water, and pirouette on their hind legs. Otters delight in elaborate acrobatics. Bottlenose dolphins balance fish on their snouts, throw objects, and carry them in their mouths while swimming. Even baby vampire bats have been observed chasing, wrestling, and slapping each other with their wings.

Play varies enormously. Solitary play often involves a single animal manipulating an object, like a cat with a ball of yarn, or the dolphin with its fish, or a monkey hopping on and off a feather. Play may also be social. Often young of the same species play together (Fig. E41-2), but parents may join them, as may members of other species (cats and dogs from the same household often play together). Social play often includes chasing, fleeing, wrestling, kicking, and gentle biting.

What are the general characteristics of play? (1) Play seems to lack any clear goal. (2) Animals play when there are no other immediate demands on their time and attention. Play is abandoned in favor of escaping from danger, feeding, and courtship. (3) Play seems to involve feelings of pleasure. (4) Young animals play more frequently than adults. (5) Play often involves movements borrowed from other behaviors (attacking, fleeing, stalking, etc.). (6) Play uses considerable energy. (7) Play is potentially dangerous. Young humans and other animals are frequently injured, at least slightly, during play. In addition, play may distract the animal from the presence of danger while making it conspicuous to predators. So why do animals play?

Probably the most intuitively reasonable theory of play is the "practice theory," first suggested by K. Groos in 1898. He suggested that play allows young animals to gain experience in a variety of behaviors that they will use as adults. By performing these acts repeatedly in a nonserious context, the animal gains skill that will later be important in hunting, fleeing, or social interactions.

A somewhat different interpretation is that play is a mechanism of building cardiovascular fitness, coordination, and endurance. Studies using rats, rabbits, and humans have shown that, in response to equal amounts of exercise, young organisms show a greater increase in cardiovascular fitness than do adults. The stimulation of pleasure centers by play is adaptive because it encourages young animals (including humans) to engage in hard work, "just for the fun of it".

The exercise theory does not account for children playing quietly with dolls or blocks. It also fails to ex-

Figure E41-2 Young red foxes at play.

plain the strong social component of many forms of play. Manipulating objects and interacting with other individuals may serve as "exercise" for the developing nervous system. Animals raised in environments where their experiences were limited were found to have more poorly developed neural connections than those raised in a diverse environment. By manipulating objects, an animal may learn new skills. For example, W. Kohler, who studied chimp behavior, offered his chimp "Sultan" a pair of sticks that could be linked together to form a longer stick. Only the linked sticks would allow Sultan to reach a suspended banana. Sultan repeatedly tried to reach the banana with the longer of the two sticks, then finally gave up. Only later, playing with the sticks, did he happen to link them together. He then immediately used his new tool to obtain the food. Considerable trial-and-error learning occurs during play, both social and solitary.

Play must be a highly adaptive behavior, probably for a combination of the reasons suggested above. Regardless of its adaptive significance, humans and other animals (as any child will tell you) play because it's fun!

(a) A naive toad is presented with a bee.

(b) While trying to eat the bee, the toad is stung painfully on the tongue.

(c) Presented with a harmless robber fly, which resembles a bee, the toad cringes.

(d) The toad is presented with a dragonfly.

(e) The toad immediately eats the dragonfly, demonstrating that the learned aversion is specific to bees and insects resembling them.

Figure 41-14 Trial-and-error learning in the toad.

the mind to arrive at adaptive behavior (Fig. 41-15). It can be considered a kind of mental trial-and-error learning. Insight arises from the ability to remember a variety of past experiences and to apply these lessons in creative ways to new situations.

In 1917, W. Kohler showed that a hungry chimp, without any training, would stack boxes to reach a banana suspended from the ceiling. (Fig. 41-15b). This type of mental problem-solving was once believed to be limited to very intelligent animals such as primates. In 1984, however, similar insight (moving a box beneath a suspended banana, then standing on the box to reach the banana) was described in pigeons by R. Epstein and associates at Harvard. As investigators learn more about how to design appropriate experiments, we may find that insight is much more common than was originally suspected.

The Instinct to Learn and the Learning of Instincts

You have probably concluded by now that neither "innate" nor "learned" adequately describes the behavior of any given organism. Especially in an adult animal, every behavior is an intimate mixture of the two. In some cases, the nature and timing of learning is so rigidly programmed by the genes that we might use the term "innate learning" to describe it.

An example of innate learning is imprinting: learning rigidly programmed to occur at a certain critical period of development. Adult bees also use a type of innate learning to allow them to locate their hive after a foraging trip. Ethologists discovered that bees memorize the location of their hive with respect to certain landmarks *only* on their first flight of the day. If the hive is moved even a short distance later in the day, the bees will have trouble finding it. If a bee-keeper wishes to move the hive, he or she must do it during the night. The bees memorize the new location on their first flight, and return to it later.

We have seen how genes influence learning, but learning may also modify innate behavior. Recall the example of the herring gull chicks pecking their parent's beak for food. Immediately after hatching, the chicks prefer a striped knitting needle to the real beak. Within a few days, however, the chicks learn enough about the appearance of their parents that they begin pecking more frequently at models more closely resembling the parents. After 1 week, young gulls have learned enough about how their parents

(a)   *(b)*

Figure 41-15 **(a)** A dog, lacking insight in this situation, fails a detour problem on the first try. It will eventually learn by trial and error. **(b)** Insight in a chimpanzee. Unable to reach the bananas, the chimp stacks boxes beneath them to extend its reach.

look to prefer models of their own species to models of a closely related species.

Habituation is another example of learning that can fine-tune an organism's innate responses to environmental stimuli. Early investigators were impressed to see young birds crouching when a hawk flew over, while ignoring harmless birds such as geese. They hypothesized that crouching was only released by the very specific shape of predatory birds. Niko Tinbergen and Konrad Lorenz tested this using an ingenious model. As shown in Figure 41-16, when moved in one direction the model resembled a goose, which the chicks ignored. When its movement was reversed, it resembled a hawk, and elicited crouching. Further research revealed that naive chicks instinctively crouch when *any* object moves over their heads. Over time, their response habituates to things that soar by harmlessly and frequently, such as leaves, songbirds, and geese. Predators are much less common, and the novel shape of a hawk still elicits instinctive crouching. Thus, learning modifies the innate response, making it more adaptive.

Trial-and-error learning can also result in more appropriate responses to releasers for fixed action patterns, as in the case of our bee-catching toad (see Fig. 41-14). The naive toad instinctively snaps at all flying insects of appropriate size, but from painful experience learns to make certain exceptions!

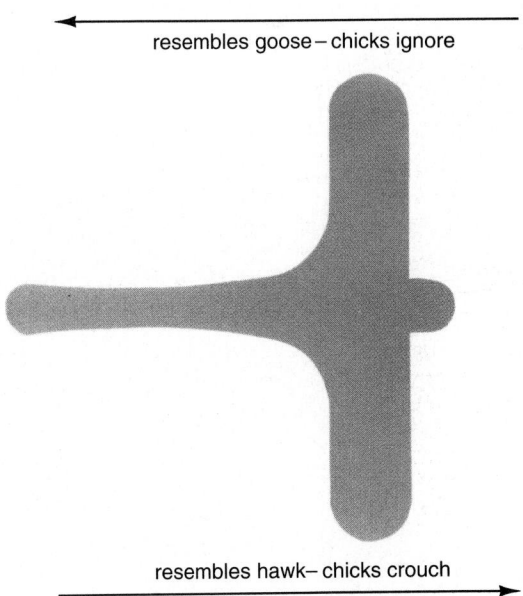

resembles goose – chicks ignore

resembles hawk– chicks crouch

Figure 41-16 The model used by Konrad Lorenz and his student Niko Tinbergen to investigate the response of chicks to the shape of objects flying overhead. The chicks' response depended on which direction the model moved. Moving toward the right, the model resembles a predatory hawk, but moving left it resembles a harmless goose.

Reflections on Ethology

Why study animal behavior? The more carefully one observes animals, the more fascinating they become. Whether it's the dog next door, a salmon thrashing upstream to spawn, or a honeybee sipping nectar from a flower, a knowledge of ethological principles can enhance our appreciation of other animals and thereby enrich our own lives.

Ethology also has its practical side, for example, in agriculture. Understanding the reproductive and other social behavior of domestic animals helps us raise them efficiently in large numbers. Knowledge of the communication system of certain insect pests is allowing the development of natural controls to increase agricultural yields while reducing our use of chemical pesticides.

Humans are extending their influence over the face of the Earth, altering it radically in the process. We are destroying the habitat and interfering with the social interactions necessary for the reproduction of thousands of different species. Many of these will become extinct unless a special effort is made to maintain conditions in which they can thrive and reproduce. For example, we have placed dams in many of the rivers in which salmon swim upstream to lay their eggs. To allow the fish to bypass the dams, people constructed "fish ladders" around the dams, only to find that the fish don't use them. Ethological studies showed that salmon tend to swim into the strongest current, leading them into the outflow from the dam rather than into the ladders. This has led to the design of better ladders and in some cases to netting and trucking the salmon around the dams. In attempting to preserve endangered species from the rapidly disappearing tropical rainforests, knowledge of behavior is crucial. For example, we must learn how much territory each species requires to forage and reproduce successfully. This will allow us to establish a minimum size for preserves.

Another practical reason to study animal behavior is for the insight it can provide into the workings of the nervous system. The study of reflexes in vertebrates and fixed action patterns such as feeding and swimming in invertebrates is used to probe the mechanisms by which nerve cells interact to produce behavior.

Finally, the techniques of detached observation necessary to understand other animals can yield valuable insights into human behavior. One of the greatest contributions of ethology has been the realization that all animals, including humans, are constrained in their behavior. Our options are extremely broad, but they are still circumscribed by the structure of our minds and bodies. The study of animal behavior can help us to decide what questions to ask about human behavior, and how to ask them, to learn more about ourselves. Human behavior is explored further in Chapter 42.

SUMMARY OF KEY CONCEPTS

The Genetic Basis of Behavior

All behavior has some genetic basis. Both the complexity of the nervous system that directs the behavior and the physical structure of the body that performs the behavior are genetically determined. Some types of learning, such as imprinting, are rigidly constrained by the genes. The relative contributions of heredity and learning to behavior vary between animal species, and between behaviors within the individual.

Innate Behavior

Innate, or instinctive, behaviors have several possible advantages over learned behaviors. First, they allow the animal to respond properly the first time it encounters a stimulus. Second, they allow animals with simple nervous systems to perform behaviors too complex for them to learn. Third, innate behavior causes members of the same species to perform and respond correctly during courtship and mating.

Innate behaviors can be placed into four categories: kineses, taxes, reflexes, and fixed action patterns. Animals showing kineses orient by varying the speed of essentially random movements, stopping when they encounter favorable conditions. In contrast, taxes are directed movements toward or away from specific stimuli. Reflexes are rapid movements of part of the body. Some, such as the knee-jerk reflex, are directed by the spinal cord and do not involve higher brain centers.

A fixed action pattern is a complex innate behavior elicited by a specific stimulus called a releaser. Learning can sometimes modify the releasers for fixed action patterns. Fixed action patterns can be distinguished from learned behaviors (1) if they are performed without prior experience, (2) if manipulating the releaser can result in inappropriate behavior, and (3) if a genetic basis can be determined through breeding experiments.

Learning

Learning is the ability to make changes in behavior as a result of experience. Learning is especially adaptive in changing and unpredictable environments and may modify innate behavior to make it more appropriate. Forms of learning include imprinting, habituation, conditioning, trial and error, and insight.

Imprinting occurs during a genetically programmed sensitive period. It often involves attachment between parent and offspring, or learning the features of a future mate. Habituation is the decline in response to a harmless stimulus that is repeated frequently. It often modifies innate escape or defensive responses. During classical conditioning, an animal learns to make a reflexive response, such as withdrawal or salivation, to a stimulus that did not originally elicit that response. During operant conditioning, an animal learns to make a new response, such as pressing a button, to obtain a reward or avoid punishment. Trial-and-error learning may modify innate behavior or may produce new behavior as a result of rewards and punishments provided by the environment. Insight can be considered a form of mental trial-and-error learning. An animal showing insight makes a new and adaptive response to an unfamiliar situation.

The Instinct to Learn and the Learning of Instincts

In animals with complex nervous systems, learning and instinct interact to produce adaptive behavior. Certain types of learning occur instinctively, during a rigidly defined time span. Imprinting, song acquisition in some birds, and location of the hive by bee foragers are examples.

Instinctive responses are often relatively unselective. Examples include baby gulls pecking at long thin objects, toads feeding on all flying insects, and chicks crouching whenever large birds fly overhead. Learning allows animals to modify these innate responses so that they occur only to appropriate stimuli.

GLOSSARY

altruism (al'-trū-izm): a behavior that benefits another organism, usually at some risk to the altruistic organism.

classical conditioning: a training procedure in which an animal learns to make a reflexive response (such as salivation) to a new stimulus that did not elicit that response originally (such as a sound). This is accomplished by pairing a stimulus that elicits the response automatically (in this case, food) with the new stimulus.

ethology (ē-thol'-ō-gē): the study of animal behavior under natural or near-natural conditions.

fixed action pattern: stereotyped, rather complex behavior that is genetically programmed (innate); often triggered by a stimulus called a releaser.

habituation (heh-bich-ū-ā'-shun): simple learning characterized by a decline in response to a harmless, repeated stimulus.

imprinting: the process by which an animal forms an association with another animal or object in the environment during a sensitive period.

innate (in-nāt'): inborn; instinctive; determined by the genetic makeup of the individual.

insight: a complex form of learning in which the solution to a problem is reached through reasoning.

instinctive: innate; inborn; determined by the genetic makeup of the individual.

kinesis (kin-nē'-sis): an innate process by which an organism achieves an orientation to a stimulus by altering its speed of movement in response to the stimulus.

learning: an adaptive change in behavior as a result of experience.

operant conditioning: a laboratory training procedure in which an animal learns to make a response (such as pressing a lever) through reward or punishment.

reflex: a simple, stereotyped movement of part of the body that occurs automatically in response to a stimulus.

releaser: a stimulus that triggers a fixed action pattern.

supernormal stimulus: a stimulus that exaggerates crucial elements of the releaser, making it more effective than the normal releaser.

taxis: (tax'-is): innate movement of an organism toward or away from a stimulus such as heat, light, or gravity.

trial-and-error learning: a process by which adaptive responses are learned through rewards or punishments provided by the environment.

STUDY QUESTIONS

1. List four types of innate behavior and provide an example of each.
2. Distinguish between taxes and kineses.
3. What is the name for a stimulus that elicits a fixed action pattern? Give three examples of this type of stimulus.
4. Compare classic conditioning, operant conditioning, and trial-and-error learning. Give an example of each.
5. Compare insight with trial-and-error learning.
6. Give examples in which habituation and trial-and-error learning make an innate response more appropriate to a particular situation.
7. Explain why imprinting could be described as genetically controlled learning.

DISCUSSION QUESTIONS

1. Contrast the approach to animal behavior of B. F. Skinner and N. Tinbergen. Name the behavioral "school" of each and describe an important contribution of each scientist.
2. Assume you observe two male fish of the same species fighting whenever they approach one another. Design a series of experiments to determine whether this behavior is innate or learned.
3. Assume your experimental results in Question 2 support the hypothesis that fighting in these fish is a fixed action pattern. Design a series of experiments to identify the releaser for this behavior.

SUGGESTED READINGS

Benzer, S. "Genetic Dissection of Behavior." *Scientific American*, December 1973. Mutant fruit flies are used to study the genetic bases of behavior.

Boycott, B. B. "Learning in the Octopus." *Scientific American*, March 1965. Long- and short-term memory are investigated using the learning ability of the octopus.

Hailman, J. P. "How an Instinct Is Learned." *Scientific American*, December 1969. How instinct and learning interact in the development of feeding in gull chicks.

Gould, J. L. and Gould, C. G. "The Instinct to Learn." *Science 81*, May 1981. Birds, bees, and perhaps even humans are genetically programmed to learn specific things are particular times.

Jolly, A. "A New Science that Sees Animals as Conscious Beings." *Smithsonian*, March 1985. The new thinking regarding animal thought is clearly explained in this exciting article.

Lorenz, K. *King Solomon's Ring: New Light on Animal Ways*. New York: Thomas Y. Cromwell, 1952. Beautifully written and full of interesting anecdotes; provides important insights into early ethology.

Tinbergen, N. (1951) *The Study of Instinct*. Introduction to the work of an experimental ethologist.

42

Animal Behavior. II. Principles of Social Behavior

Ants discovering food on foraging expeditions leave a chemical trail that attracts their nest-mates to harvest the bounty.

(a)

(b)

Figure 42-1 Drawings by Darwin, a perceptive student of animal behavior, showing aggressive and submissive displays of dogs. **(a)** Upright stance, erect hair, ears, and tail, and a direct stare combine to make the aggressor appear formidable indeed. **(b)** Notice how these displays are reversed in the submissive pose.

Your picnic is complete—ants have discovered the bread crumbs and are advancing in single file, then returning by the same path, booty held high. The activity seems efficient and purposeful. Its apparent intelligence is surprising in an animal with a relatively simple nervous system. The ant "bread line" passes over a fallen leaf. When you lift the leaf, immediately the scene changes from one of purposeful intelligence to one of blind chaos! Ants reaching the point where the leaf used to lie abruptly stop their march, searching back and forth. Ants coming and going mill in apparent confusion until the leaf-sized gap is crossed by chance and the march resumed. You have witnessed complex social behavior in a simple organism. Removing the leaf disrupted the communication system—a chemical trail laid by successful foragers—on which this behavior is rigidly based.

Communication

Social behavior is exhibited to some degree by all but the simplest organisms. The basis of all social behavior is communication, and the ultimate goal is survival and reproduction. **Communication** may be defined as the production of a signal by one organism that causes another to change its behavior in a way beneficial to one or both. Although communication can cross species lines, most occurs between members of the same species. Because members of the same species have the same needs, it is these individuals who compete most directly with one another for food, space, and mates. Communication must resolve the conflicts that result from these competitive interactions while minimizing harmful encounters.

The mechanisms by which animals communicate are astonishingly diverse and use all the senses. In this chapter we discuss basic types of communication, consider some of the mechanisms for resolving competitive interactions, and describe some complex, cooperative societies.

Visual Communication

Animals with well-developed eyes, from insects to mammals, use vision to communicate. Visual signals may be **active,** in which a specific movement or posture conveys a message (Fig. 42-1). Alternatively, they may be **passive.** In passive visual signals, the size, shape, or color of the animal conveys important information, often concerning its reproductive state. Active and passive signals may be combined, as illustrated by the lizard in Figure 42-2. Like all forms of communication, visual signals have both advantages

and disadvantages. On the plus side, they are instantaneous. The rapidity with which they can be sent and received allows them to impart a great deal of information in a short time. Visual signals may be

Figure 42-2 Aggressive display of the South American Anolis lizard. Raising his head high in the air, he extends a brilliantly colored throat pouch, a warning to others to keep their distance.

graded, allowing the animal to convey the intensity of its motivational state (Fig. 42-3). Visual communication is quiet and unlikely to alert distant predators. A negative feature is that the signaller makes itself conspicuous to nearby predators. Visual signals are limited to close-range communication, and aren't effective in the dark (fireflies excepted), through murky water, or in dense vegetation.

Communication by Sound

The use of sound overcomes many of the deficiencies of visual displays. Like visual displays, sounds are almost instantaneous. But unlike visual signals, sound can be transmitted through darkness, dense forests, and even water, as the intricate songs of humpback whales testify. If the animal is sufficiently energetic, its call may carry much farther than the eye can see. For example, the howls of a wolf pack carry for miles on a still night. The low, resonant song of the humpback whale could theoretically be detected by other whales hundreds of miles away.

Auditory signals are similar to visual displays in that they can convey rapidly changing information (think of words and of the emotional nuances of human vocal dynamics during a conversation). In other animals, changes in motivation may be signaled by a change in the loudness or pitch of the sound. Different messages may also be conveyed by variations in the pattern, volume, and pitch of the sound made by the same individual. Thomas Struhsaker studied vervet monkeys in Kenya in the 1960s and found that they produced different calls in response to threats from each of their major predators: snakes, leopards, and eagles. In 1980, other researchers reported that the response of other vervet monkeys to each of these calls is appropriate to the particular predator. The "bark" that warns of a leopard or other four-legged carnivore causes monkeys on the ground to take to trees and those in trees to climb higher. The "rraup" call signaling an eagle or other hunting bird causes monkeys on the ground to look upward and take cover, while monkeys already in trees drop to the shelter of lower, denser branches. The "chutter" call indicates a snake, and causes the monkeys to stand up and search the ground for this slower predator.

The use of sound is by no means limited to birds and mammals. Male crickets produce species-specific songs that attract female crickets of the same species. The whine of the female mosquito as she prepares to bite alerts nearby males that she will soon have the blood meal necessary for laying eggs. From these rather simple signals to the virtuoso performance of human language, sound is one of the most important forms of communication.

Communication by Chemicals

Chemical substances produced by an individual that influence the behavior of others of its species are called **pheromones.** Chemicals may carry messages over long distances, and, unlike sound, take very little energy to produce. Pheromones may not even be

(a)   **(b)**

Figure 42-3 (a) A relaxed, nonaggressive wolf. **(b)** The wolf signals aggression by lowering the head, ruffling the fur on its neck and along its back, facing the opponent with a direct stare, and exposing its fangs. These signals may vary in intensity, communicating different levels of aggression.

detected by other species, including predators who might be attracted to visual or auditory displays. Like a signpost, a pheromone persists over time, and can convey a message after the animal has departed. As anyone who has walked a dog can attest, dog urine carries such a chemical message. Wolf packs, hunting over areas up to 1000 square kilometers, mark the boundaries of their travels with pheromones in urine, thus warning other packs of their presence.

Generally, fewer different messages are communicated with chemicals than with sight or sound, since the animal must have the ability to synthesize and respond to a different chemical for each message. Thus, pheromone signals are unlikely to possess the variety and gradation of auditory or visual signals.

Pheromones may act in one of two ways. **Releaser pheromones** cause an immediate, overt behavior in the animal detecting them. They convey messages such as "this area is mine" or "I am ready to mate now." The ants mentioned in the opening paragraph were responding to a releaser pheromone laid down in a trail by successful foragers. Its message: "food is this way." **Primer pheromones** cause a physiological change in the animal detecting them. They may enter through the nose or they may actually be eaten by the receiver of the message. Most primer pheromones affect the reproductive state of the receiver. They are crucial to the maintenance of complex insect societies such as those of termites, ants, and bees. In the honeybee, the queen produces a primer pheromone called **queen substance,** that is eaten by her hive-mates and prevents other females in the hive from becoming sexually mature. Primer pheromones are produced by mammals such as mice. In certain mice, the urine of mature males contains a primer pheromone that influences female reproductive hormones. This pheromone initiates the first estrus of a newly-mature female. It will also cause a female mouse newly pregnant by another male to abort her litter and become sexually receptive to the new male. There is indirect evidence (discussed later) that primer pheromones may even influence human reproductive cycles.

Some pheromones have been chemically analyzed and synthesized in the laboratory. The sex attractant pheromones of agricultural pests such as the Japanese beetle and gypsy moth have been synthesized. They can be used to lure these insects into traps, or sprayed in an infested area to confuse and disrupt mating. The advantages of pest control using pheromones are enormous. Unlike pesticides, which kill beneficial as well as harmful organisms, pheromones are specific to one species and harmless to others. In contrast to pesticides, insects will never become resistant to the attraction of a pheromone.

Communication by Touch

Physical contact between individuals is used in several ways, one of the most important being to establish and maintain social bonds among group members. Primates, including humans, are "contact species" in which a variety of gestures including kissing, nuzzling, patting, petting, and grooming play an important social function (Fig. 42-4a,b). In wolves and dogs, a greeting ceremony involves mutual licking, sniffing, and gentle nipping around the mouth. The bond between parent and offspring is often cemented by close physical contact, and sexual activity is frequently preceded by ritualized contact (Fig. 42-4c).

Dramatic effects of touch on the health of premature human infants have recently been reported. In a controlled study, stroking and moving the limbs of premature infants for 45 minutes daily resulted in more rapid weight gain, a higher activity level, more alertness and responsiveness, and greater emotional stability. Their progress was so striking that they left the hospital earlier than infants treated normally.

The Many Roles of Communication

Much of communication cannot be categorized simply as sight, sound, scent, or touch. Many social interactions use a combination of these as illustrated by the waggle dance of the honeybee (see Fig. E42.1a). Nearly all social interactions, however, can be classified according to their functions in the lives of animals. The following sections will discuss communication used as animals compete for limited resources, reproduce, and cooperate in complex societies.

Competition for Resources

Aggression

One of the most obvious manifestations of competition for resources such as food, space, or mates is aggressive behavior between members of the same species. Although the expression "survival of the fittest" evokes images of the strongest animal emerging triumphantly from amongst the dead bodies of its competitors, in reality most aggressive encounters between members of the same species are rather harmless. **Natural selection has favored the evolution of symbolic displays or rituals for resolving conflicts.** One reason may be that in a serious fight, even the victorious animal may be injured. Serious fighters are therefore less likely to survive and propagate their genes. Aggressive displays, in contrast,

(a)   **(b)**   **(c)**

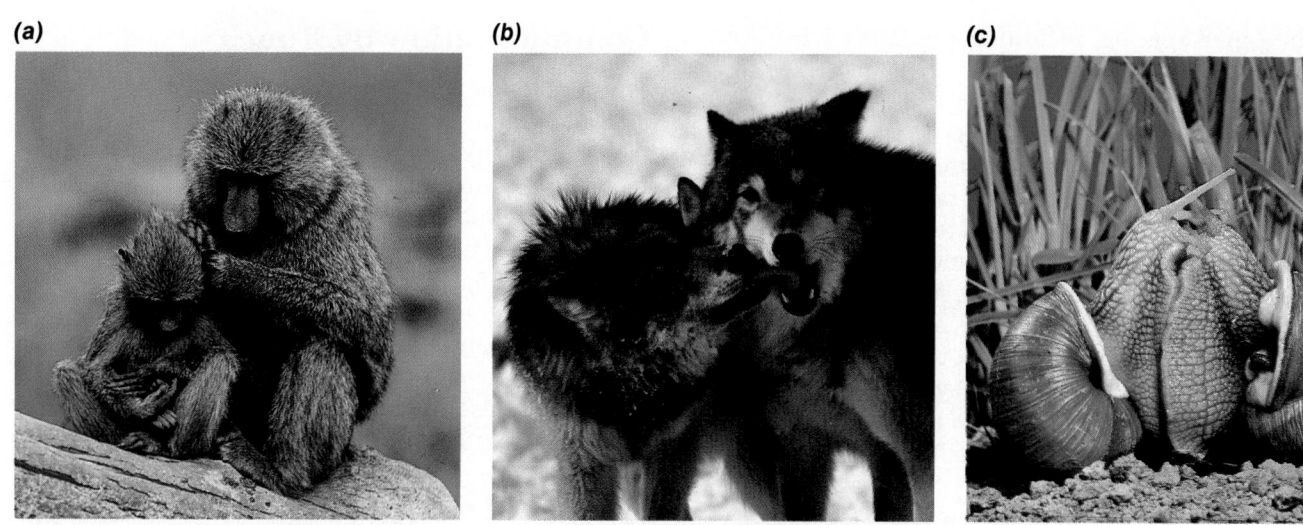

Figure 42-4 (a) An adult olive baboon grooms a juvenile. Grooming not only reinforces social relationships but also removes debris and parasites from the fur. **(b)** Wolves from the same pack engage in a greeting ceremony that involves nuzzling and licking. This behavior may indicate submission, as when a subordinate animal licks a dominant one. Wolf pups use muzzle-licking to beg for food. **(c)** Touch is also important in sexual communication. These land snails *(Helix)* engage in courtship behavior that will culminate in mating.

allow the competitors to assess each other and acknowledge a winner on the basis of its size, strength, and motivation rather than on the wounds it can inflict. Visual aggressive displays exhibit weapons, such as fangs and claws, and often include behaviors that make the animal appear larger (Fig. 42-5). These include standing upright, and erecting the fur, feathers, ears, or fins (see Figs. 42-1a, 42-2, and 42-3). The displays may be accompanied by intimidating sounds (growls, croaks, roars, chirps) whose loudness can help decide the winner. Fighting is usually a last resort when displays fail to resolve the dispute.

In addition to visual and vocal displays of strength, many animal species engage in ritualized combat. Deadly weapons may clash harmlessly (Fig. 42-6), or may not be used at all. Frequently these encounters involve shoving rather than slashing. Again, the strength and motivation of the combatants is determined and the loser slinks away in a submissive posture that minimizes the size of its body (see Fig. 42-1b).

Dominance Hierarchies

Aggressive interactions use a great deal of energy, may cause injury, and can disrupt other important tasks such as finding food, watching for predators,

courting a mate, or raising young. So there are adaptive advantages to resolving conflicts with minimal aggression. One example is the **dominance hierarchy,** in which each animal establishes a rank that determines its social status. Domestic chickens, after a period of squabbling, sort themselves into a reasonably stable "pecking order." Thereafter, when

Figure 42-5 The aggressive display of the male fighting fish includes elevating the fins and flaring the gill covers, thus making the body appear larger.

Figure 42-6 Ritualized combat of male impalas, a type of African antelope. The deadly horns, which could gore a predator, clash harmlessly. Eventually one impala, sensing greater vigor in his opponent, will often retreat unharmed.

competition for food occurs, all hens defer to the dominant bird, all but the dominant bird give way to the second, and so on. Conflict is minimized because each bird knows its place. Dominance in male bighorn sheep is reflected in the size of their horns (Fig. 42-7). Wolf packs are organized so that each sex has a dominant or "alpha" individual to whom all others are subordinate. Although aggressive encounters occur frequently during the establishment of a hierarchy, they are subsequently reduced by the submission of subordinate individuals. The dominant individuals obtain most access to the resources needed for reproduction, including food, space, and mates.

Figure 42-7 The dominance hierarchy of the male bighorn sheep is signaled by the size of the horns; these rams increase in status from right to left. These backward-curving horns are clearly not designed to inflict injury and are used in ritualized combat.

Territoriality

Territoriality is the defense of an area where important resources are located. The defended resources may include places to mate, raise young, feed, or store food. Territorial animals generally restrict some or all of their activities to the defended area, and advertise their presence there. Territories may be defended by males, females, a mated pair, or by entire social groups (as in the case of the defense of their nest by social insects). However, territorial behavior is most often seen in adult males, and territories are usually defended against members of the same species, those who compete most directly for the resources being protected. Territories are as diverse as the animals defending them. They include small depressions in the sand used as nesting sites by *Oreochromis* fish (Fig. 42-8); a hole in the sand used as a home by a crab; a tree where a woodpecker stores acorns (Fig. 42-9); or an area of forest providing food for a squirrel.

Acquiring and defending a territory requires considerable time and energy, yet territoriality is seen in diverse animals including worms, arthropods, fish, birds, and mammals. This striking example of similar behaviors evolving independently suggests that territoriality provides some important adaptive advan-

territory

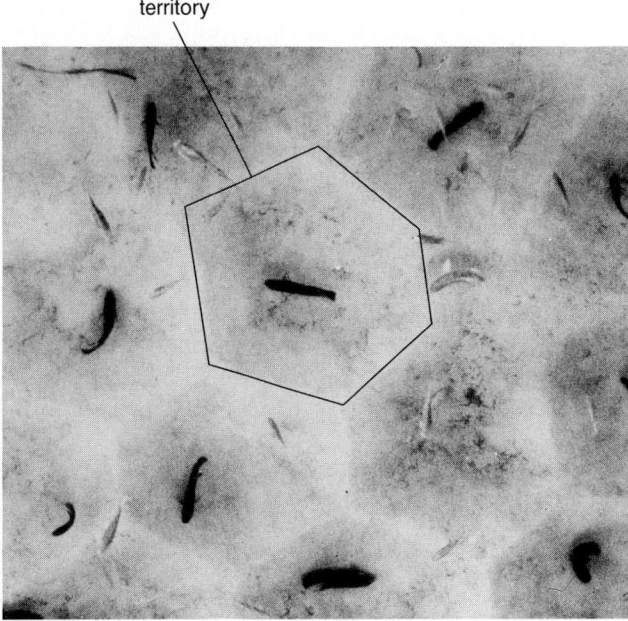

Figure 42-8 Nesting males of the mouthbrooding cichlid fish (*Oreochromis mossambicus*) guard territories. When space is abundant, these territories are circular. When the population density rises, the territories approach hexagonal shapes because they are so closely packed. A female visits the territories and selects a male, who fertilizes the eggs she releases. She then gathers the eggs in her mouth (where they will develop) and departs.

Figure 42-9 The acorn woodpecker excavates acorn-sized holes in dead trees, stuffing them with green acorns for dining during the lean winter months. He defends the trees vigorously against other acorn woodpeckers and against acorn-eating birds of other species, such as jays.

tages. Although the benefits depend on the species and the type of territory it defends, some broad generalizations are possible. First (as with dominance hierarchies), once a territory is established through aggressive interactions, relative peace prevails as boundaries are recognized and respected. The saying "good fences make good neighbors" also applies to nonhuman territories. One reason for this respect is that an animal is highly motivated to defend its territory, and will often defeat larger, stronger animals if they attempt to invade it. Conversely, an animal outside its territory is much less secure and more easily defeated. This principle was demonstrated by Niko Tinbergen's experiment using the stickleback fish, described in Figure 42-10.

Defending reproductive territories is advantageous for certain species. The reproductive success of these animals is enhanced by a high-quality breeding territory, which might have features such as large size, abundant food, and secure nesting areas. Males who defend more desirable territories have a greater chance of mating and passing on their genes. For example, experiments have shown that male sticklebacks who defend large territories are more successful in attracting mates. Females who select males with the best territories increase their reproductive success and pass their genetic traits (often including their mate-selection preferences) to their offspring.

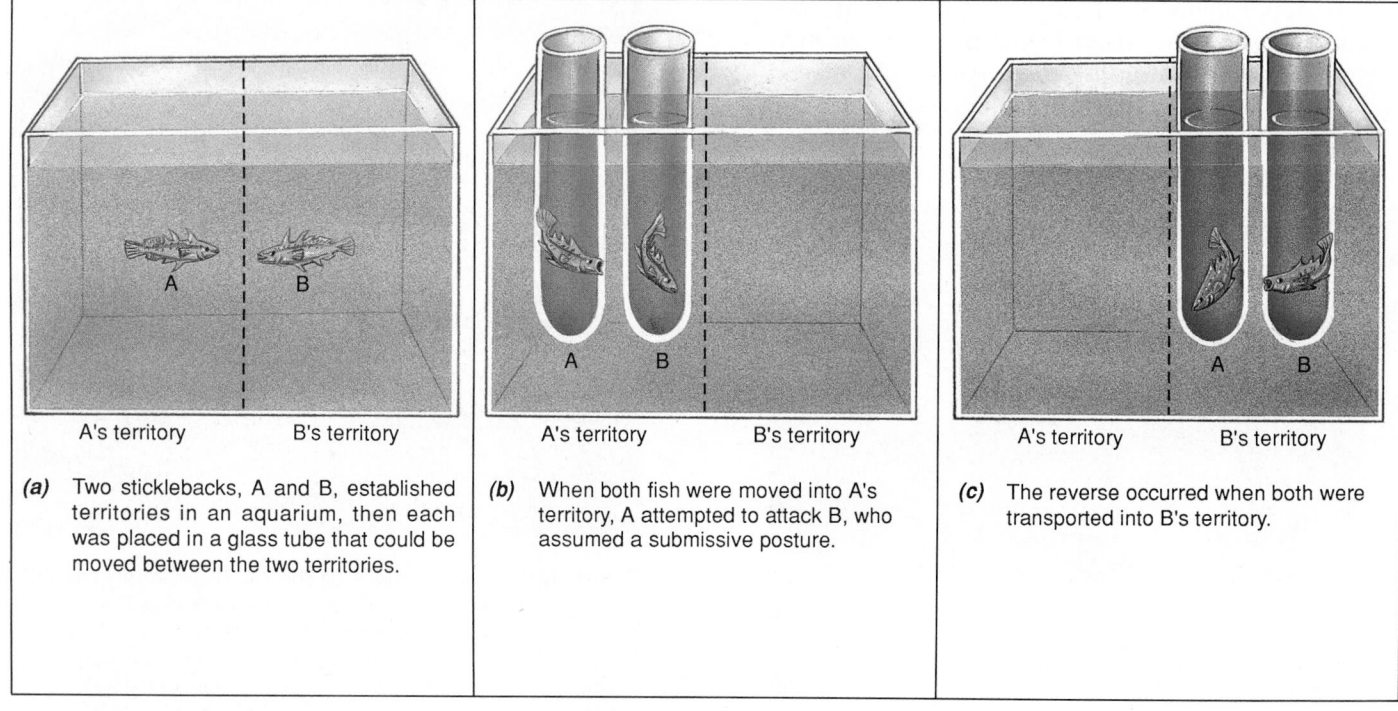

(a) Two sticklebacks, A and B, established territories in an aquarium, then each was placed in a glass tube that could be moved between the two territories.

(b) When both fish were moved into A's territory, A attempted to attack B, who assumed a submissive posture.

(c) The reverse occurred when both were transported into B's territory.

Figure 42-10 Niko Tinbergen's experiment demonstrating the effect of territory ownership on aggressive motivation.

In some species, territoriality limits the population size, helping keep it within the limits set by the available resources. For example, the Scottish red grouse defends feeding and nesting territories. Scientists found variations in territory size corresponding with the availability of food. In years when food was abundant, territories were smaller, and the population density was correspondingly greater. In lean years, territories were larger, allowing fewer pairs to breed in the same area, thus reducing the population.

Territories are advertised through sight, sound, and smell. If the territory is small enough, the owner's mere presence, reinforced by aggressive displays at intruders, may be sufficient defense. In mammals, when the owner cannot always be present, it may scent-mark the boundaries using pheromones. Male rabbits use pheromones secreted by chin and anal glands to mark their territories. Hamsters rub the areas around their dens with secretions from a special gland on their flanks.

Vocal displays are a common form of territorial advertisement. Male sea lions defend a strip of beach by swimming up and down in front of it, calling continuously. Male crickets produce a specific pattern of chirps to warn other males away from their burrows. Birdsong is a striking example of territorial defense. The cheerful melody of the male meadowlark is part of an aggressive display, warning other males to steer clear of his territory (Fig. 42-11). In songbirds, the loudest singer generally defends the largest territory, and is most successful in attracting a mate and at driving away intruders. In ingenious experiments using the woodthrush, W. Dilger of Cornell University "invaded" the territories of male woodthrushes with a stuffed woodthrush accompanied by a loudspeaker through which he played recordings of the thrush's territorial song. When the volume was turned low, the resident thrushes attacked the stuffed bird, but they retreated rapidly when the volume was raised.

Reproduction

Successful reproduction requires that several criteria be satisfied. Animals must identify one another as members of the same species, as members of the opposite sex, and as being sexually receptive. Many animals resist the close approach of another individual; this must be overcome before mating can occur. Some animals, such as frogs and many fish, must release eggs and sperm at precisely the same moment for fertilization to occur. The need to fulfill all these requirements has resulted in exceedingly complex, diverse, and fascinating courtship behavior.

Figure 42-11 A male meadowlark announces ownership of his territory to all listeners.

Individuals who mate with members of other species, or members of the same sex, waste considerable energy and do not pass on their genes. Thus, animals have evolved elaborate ways to communicate their species and sex, often using vocalizations. The raucous, nighttime chirping that can keep campers awake is probably a chorus of male tree frogs, each singing a species-specific song. Male grasshoppers and crickets advertize species and sex by their calls, as does the female mosquito with her high-pitched whine. Male birds use song to attract a mate as well as defend a territory. For example, the male bellbird uses its deafening call to defend large territories and attract females from great distances. The females fly from one territory to another, alighting near the male in his tree. The male, beak gaping, leans directly over the flinching female and utters an earshattering note. It is believed that the female endures this to compare volumes of the various males, choosing the loudest (who would also be the best defender of a territory) as a mate. In other species, male and female birds join in elaborate duets that help to synchronize reproductive readiness and reinforce the bond between them.

Many species court using visual displays. The firefly, for example, flashes a message identifying its sex and species. Male fence lizards bob their heads in a species-specific rhythm, and females distinguish and prefer the rhythm of their own species. The tail of the male peacock and the scarlet throat of the male frigate bird serve as flashy advertisements of sex and species (Fig. 42-12). In contrast, the females are often quite drab. Since females are often in close association with their young, eye-catching (and predator-attracting) displays would be maladaptive.

Species and sex recognition and the synchronization of reproductive behavior often require a complex series of signals, both active and passive, by both sexes. This is beautifully illustrated by the complex underwater "ballet" executed by the male and female stickleback fish (Fig. 42-13).

Pheromones can play an important role in reproductive behavior. The sexually receptive female silkmoth, for example, sits quietly and releases a chemical message so powerful that it may be detected by males 4 or 5 kilometers (2.5 to 3 miles) away. The exquisitely sensitive and selective receptors on the antennae of the male respond to just a few molecules of the substance, allowing him to travel upwind along a concentration gradient to find the female (Fig. 42-14). Water is an excellent medium for dispersing chemical signals, and fish often use a combination of pheromones and elaborate courtship movements to ensure synchronous release of gametes. Mammals, with their highly developed sense of smell, often rely on pheromones released by the female during her fertile periods to attract males. The irresistible attraction of a female dog in heat to nearby males is one example. The primer pheromone in male mouse urine is another.

In solitary animals, most encounters between individuals are competitive and aggressive. During the brief mating season, their reluctance to allow others to approach closely must be overcome for their mate, but retained toward others. These conflicting needs introduce an element of tension into sexual encounters that may be overcome by submissive signals. The female Siamese fighting fish appeases the aggressive male with a submissive, head-down posture. In several species of birds, either the male or female defuses aggressive impulses by mimicking juvenile behavior such as begging. Courting male hamsters emit high-pitched cries like baby hamsters, eliciting a maternal response from the female.

Presenting "gifts" seems to inhibit aggression in some species. Finches present their mates with nesting material, terns give fish, and flightless cormorants offer one another gifts of seaweed (Fig. 42-15). In one species of fly, the female sometimes devours the male when he makes sexual advances. The males of a closely related species present the female with a gift: a dead insect. While she eats, the male mates and runs. In another species, the male gains extra time to mate by "giftwrapping" the insect in a silken web that she must remove before she can eat the gift.

Cooperation

Social groupings of animals are conspicuous, but by no means universal. Group living has both advantages and disadvantages, and the relative weight of the positive and negative factors varies considerably among different animal species. **On the negative side, social animals may encounter:**

(a)

(b)

Figure 42-12 (a) The extravagant tail of the male peacock is displayed during courtship. This oversized tail hampers flight and increases the male's vulnerability to predators. It probably evolved as females consistently selected the most flashy birds as mates, preferring this exaggerated releaser to a more practical tail. **(b)** The male frigate bird of the Galápagos Islands inflates a scarlet throat pouch to attract passing females.

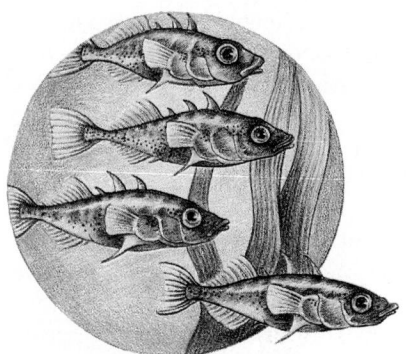

(a) A male, inconspicuously colored, leaves the school of males and females to establish a breeding territory.

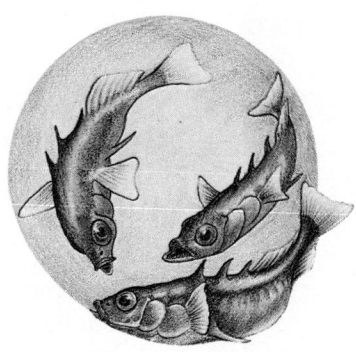

(b) As his belly takes on the red color of the breeding male, he displays aggressively at other red-bellied males, exposing his red underside.

(c) Having established a territory, the male begins nest construction by digging a shallow pit which he will fill with bits of algae cemented together by a sticky secretion from his kidneys.

(d) After tunnelling through the nest to make a hole, his back begins to take on the blue courting color which makes him attractive to females.

(e) An egg-carrying female displays her enlarged belly to him by assuming a head-up posture. Her swollen belly and his courting colors are passive visual displays.

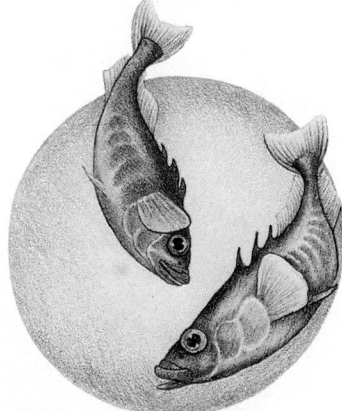

(f) He leads her to the nest using a zig-zag dance.

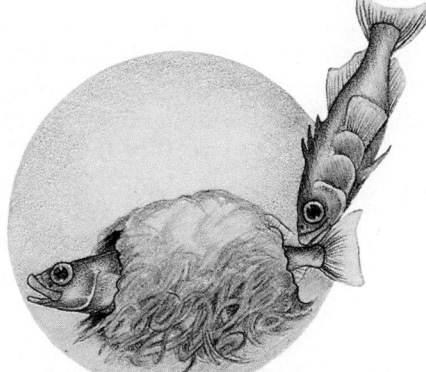

(g) After she enters, he stimulates her to release eggs by prodding at the base of her tail.

(h) He enters the nest as she leaves and deposits sperm to fertilize the eggs.

Figure 42-13 Courtship of the three-spined stickleback.

Figure 42-14 The antennae of the male silkmoth are plumelike structures specialized to detect the female's sex pheromone.

Figure 42-15 A flightless cormorant from the Galápagos Islands returns to its nest bearing a "gift" of seaweed for its mate, who will aggressively snatch it away. A bird who fails to bring a gift will be driven away by its mate. The gift appears to allow the nesting bird to take out its aggressive impulses harmlessly.

1. **Increased competition** within the group for limited resources.
2. **Increased risk of infection** from contagious diseases.
3. **Increased risk that offspring will be killed** by other members of the group.
4. **Increased risk of being spotted by predators.**

For a species to have evolved social behavior, the benefits must outweigh the costs. **Benefits to social animals include:**

1. **Increased ability to detect, repel, and confuse predators.**
2. **Increased hunting efficiency** or increased ability to spot localized food resources.
3. **Advantages resulting from the potential for division of labor** within the group.
4. **Conservation of energy.**
5. **Increased likelihood of finding mates.**

The degree to which animals cooperate varies significantly from one species to the next. Some, such as the mountain lion, are basically solitary; interactions between adults consist of brief aggressive encounters and mating. Some animals cooperate based on changing needs. For example, the coyote is solitary when food is abundant, but hunts in packs in the winter when food becomes scarce. Loose social groupings such as herds of musk oxen (Fig. 42-16), pods of dolphins, schools of fish, and flocks of birds provide a variety of benefits. For example, the characteristic spacing of fish in schools creates a hydrodynamic advantage for the individuals, reducing the energy required to swim. The characteristic V pattern of geese in flight provides a similar aerodynamic advantage, and also conserves energy. Scientists hypothesize that schooling fish confuse predators—their myriad flashing bodies prevent the predator from focusing on any single individual.

At the far end of the social spectrum are a few highly integrated cooperative societies, found primarily among the insects and mammals. As you read this section, you may notice that some cooperative societies are based on behavior that sometimes seems to sacrifice the individual for the good of the group. How could such behavior evolve? In the essay "A Closer Look at Altruism, Kin Selection, and the Selfish Gene" we explore the evolutionary basis for self-sacrificing behaviors that contribute to the success of cooperative societies. In the following section, we focus on examples of these complex societies that represent the epitome of cooperation.

Insect Societies

The most rigidly organized, most complex societies (humans excepted) are found among the social insects. In these communities, the individual is a mere cog in an intricate, smoothly running machine; it could not function by itself. Social insects are born into one of several castes within the society. These castes are groups of similar individuals genetically programmed to perform a specific function.

Honeybees

Honeybees emerge from their larval stage into one of three major preordained roles. One role is that of

Figure 42-16 Cooperation in loosely organized social groups. A herd of musk oxen functions as a unit when threatened by predators such as wolves. Males form a circle, horns pointed outward, around the females and young.

queen. Only one queen is tolerated in a hive at any time. Her functions are to produce eggs (up to 1000 per day for a lifetime of 5 to 10 years) and to regulate the lives of the workers. Male bees, called drones, serve merely as mates for the queen. Lured by her sex pheromones, drones mate with the queen during her first week of life, perhaps as many as fifteen times. This relatively brief "orgy" supplies her with sperm that will last a lifetime, enough to fertilize over three million eggs. Their sexual chore accomplished, the drones become superfluous and are eventually driven out of the hive or killed. The hive is run by the third class of bees, sterile female workers. The tasks of the worker are determined by her age and conditions in the colony (Fig. 42-17). The newly emerged worker starts as a waitress, carrying food such as honey and pollen to the queen, other workers, and developing larvae. As she matures, special glands begin wax production, and she becomes a builder, constructing perfectly hexagonal cells of wax where the queen will deposit her eggs and the larvae will develop. She will take a shift as maid, cleaning the hive and removing the dead, and as a guard, protecting the hive against intruders. Her final role in life is that of a forager gathering pollen and nectar, food for the hive. She will spend nearly half of her 2-month life in this role. Acting as a forager scout, she will seek new and rich sources of nectar, and having found one will return to the hive and communicate its location to other foragers using the **waggle dance,** an elegant form of symbolic communication (see "Methods in Biology: Deciphering Symbolic Communication in the Honeybee").

Pheromones play a major role in regulating the lives of social insects. In the honeybee, drones are drawn irresistibly to the queen's sex pheromone

(queen substance), which she releases during her mating flights. Back at the hive, she maintains her position as the only fertile female using the same substance (now acting as a primer pheromone). The queen substance is licked off her body and passed among all the workers, rendering them sterile. The queen's presence and health are signaled by her continuing production of queen substance; a decrease in production (which occurs normally in the spring) alters the behavior of the workers. Almost immediately they begin building extra large "royal cells" and feeding the larvae that develop in them a special glandular secretion known as "royal jelly." This unique food alters the development of the growing larvae so that, instead of a worker, a new queen emerges from the royal cell. The old queen will then leave the hive, taking a swarm of workers with her to establish residence elsewhere. If more than one new queen emerges, a battle to the death ensues, with the victorious queen taking over the hive.

Vertebrate Societies

Vertebrates possess far more complex nervous systems than insects, and one might therefore expect vertebrate societies to be proportionately more complex. With the exception of human society, however, this is not the case. Perhaps because the vertebrate brain *is* more complex, vertebrate societies tend to be simpler than those of the social insects such as honeybees, army ants, and termites. Each individual is unique, and this uniqueness is enhanced because vertebrates exhibit more flexible learned behavior. Although much social behavior has an innate component, vertebrates show a great deal more flexibility (and thus unpredictability) and less of the robotic

(a)

(b)

(c)

Figure 42-17 Some stages in the life of a worker bee. **(a)** Workers crowd around the queen (center), feeding her and licking the pheromone called queen substance from her body. **(b)** Workers construct hexagonal cells made of wax to enlarge the honeycomb. The wax is secreted by a gland in the abdomen, passed to the mouth and chewed to a workable consistency. **(c)** A forager collects pollen and nectar from a flower. Note the yellow pollen baskets on her legs.

METHODS IN BIOLOGY
Deciphering Symbolic Communication in the Honeybee

One of the most fascinating examples of symbolic language among nonhuman animals is seen in the honeybee as it communicates a source of nectar to its hivemates. The Austrian ethologist Karl von Frisch spent 35 years investigating this unique language, beginning in 1915.

A forager bee, on discovering a distant source of nectar, returns to the hive and communicates the nectar's location to other workers using the waggle dance. The informative part of the dance is called the "straight run" during which the bee moves in a straight line while shaking her abdomen back and forth ("waggling") and buzzing her wings. The bee repeats the run over and over, circling back in alternating directions (Fig. E42-1b). The duration of the straight run communicates the distance of the nectar source from the hive. The longer the run, the greater the distance. The symbolic conversion of run duration into distance is innate and varies among subspecies. Egyptian, Italian, and German bees are different subspecies, and each speaks a slightly different language.

The straight run with its waggling and buzzing is symbolic flight, sensed by the foragers as they crowd around the dancer (identified in Fig. E42-1a by her pollen baskets) in the dark hive. The smell of the flowers on her body tells them what scent to search for. The forager also communicates in which direction they should fly. On a vertical wall of the hive (Fig. E42-1b), the angle that the straight run deviates from the vertical represents the angle between the sun and the flowers. In the dark hive, straight up symbolizes the sun, regardless of its actual location in the sky. If the dance is performed on a horizontal surface outside, the straight run is aimed directly at the flowers (Fig. E42-1c). Both the performance of the waggle dance and the bees' ability to interpret it correctly are innate and can be performed by bees raised in isolation.

The experiments of Karl von Frisch were both simple and informative. To demonstrate that the waggle dance communicates *direction*, he trained foragers to go to a particular food source, and allowed them to communicate its location to their hivemates. He then set out plates with the same food scent in a fanlike arrangement at equal distances from the hive. He showed that most of the foragers flew to the plate at the exact location visited by the original forager, even with identical plates nearby. Then he asked: How do bees determine the *distance* to the nectar source? He hypothesized that they might be translating the effort they had to expend to get to the nectar into the waggling portion of their dance. To test the effort hypothesis, von Frisch artificially increased the effort expended by foragers by attaching tiny lead weights to their bodies. As predicted, these overburdened foragers overstated the distance to nectar.

Karl von Frisch made careful field observations followed by simple, controlled experiments that tested competing hypotheses. For his work in deciphering the symbolic communication of the honeybee, as well as other major contributions to ethology, von Frisch shared the Nobel Prize in 1973.

(a)

(c)

(b)

Figure E42-1 (a) A forager, with full pollen "baskets" on her legs, performs the waggle dance as her hivemates crowd around her. **(b)** On the wall of the hive, the angle that the straight run deviates from the vertical represents the angle between the sun and the flowers. **(c)** On a horizontal surface outside, the straight run is aimed directly toward the flowers.

precision that makes complex insect societies possible.

Bullhead Catfish

The social interactions of bullhead catfish, described by John Todd of the Woods Hole Oceanographic Institution in Massachusetts, provide a fascinating illustration of a relatively simple vertebrate in which complex social interactions are based almost entirely on pheromones. Todd observed these nocturnal fish in large aquariums in the laboratory. He discovered that when a group of them were housed together, territories were staked out, and a dominance hierarchy was established with the dominant fish defending the largest and best-protected area of the tank. Contests between tank-mates consisted of aggressive displays with open mouths, alternately approaching and retreating. Once a fish became dominant, its aggressive displays caused the subordinate fish to flee. Actual violence occurred only when a stranger was introduced into a tank with an established hierarchy. In this case the established group also exhibited cooperative behavior. The dominant fish allowed others to take refuge in his protected territory, then ventured forth to engage the intruder in combat (Fig. 42-18). When the newcomer was defeated and the danger past, the dominant fish chased the others back out of his territory.

Todd discovered that blinding the bullheads did not cause any appreciable change in their social interactions. When their sense of smell was temporarily destroyed, however, the fish acted like permanent strangers. Their aggressive behavior continued for weeks until their sense of smell returned. The status of an individual is also communicated by scent, as is a change in status. If a dominant fish is removed from his tank and later returned, usually both his territory and his status are remembered and respected by his tankmates. However, if he is removed and subjected to defeat in the tank of a more aggressive fish, his pheromones are somehow altered. Upon return to his home tank, he will be attacked by his former subordinates, who smell the change caused by his defeat.

Under certain circumstances in the wild, or in the laboratory when a large number of newly caught fish are placed in the same tank, bullheads may form a dense and peaceful community lacking territories or dominance hierarchies. Todd established such a community in one tank, and placed a pair of aggressive rival fish in an adjacent tank. When "community water" was pumped continuously from the community tank into the adjacent tank containing the aggressive bullheads, they too became peaceful, only resuming their fighting when the flow was stopped. Under the crowded conditions of the community tank, an "antiaggression pheromone" is apparently produced, minimizing conflict.

Prairie Dogs

The prairie dog "town" of the American western plains is a relatively complex mammalian society. Black-tailed prairie dogs (which are actually large rodents) live in towns of up to 1000 individuals, but the unit of social behavior is a small group known as the coterie. A coterie usually consists of about 10 individuals, including a few adult males and females and some juveniles and younger pups. These share the same burrows and recognize one another as members of the same coterie. Together they defend a small territory around their burrow, advertising their ownership by vigorous barks. The members of a coterie

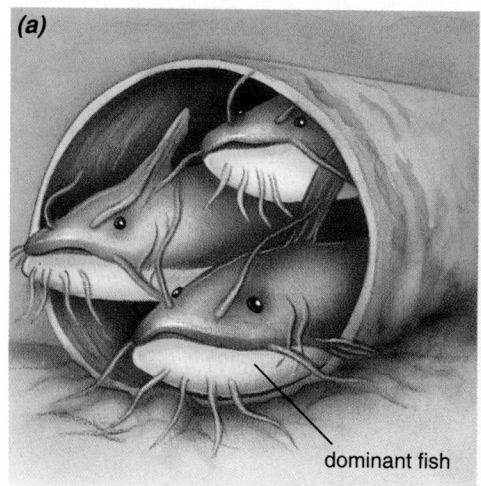

(a)

dominant fish

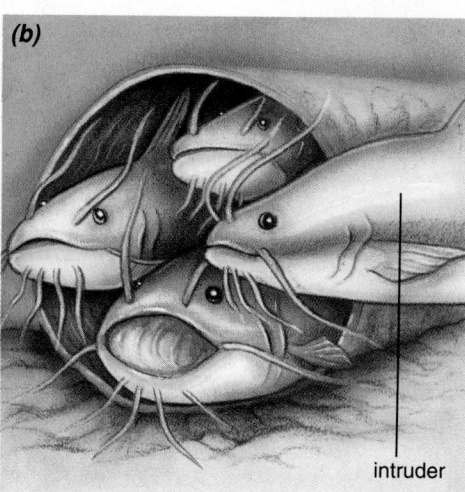

(b)

intruder

Figure 42-18 Three bullhead catfish occupy a section of pipe in a laboratory tank. The largest of the three fish is recognized by the other two as dominant. He is usually the exclusive occupant of the pipe, which is part of his territory. In **(a)**, however, he has allowed two subordinate fish to seek refuge in the pipe after detecting the odor of an unfamiliar bullhead. The approach of the strange bullhead elicits an aggressive response from the dominant fish, as is shown by its gaping mouth display **(b)**. The tankmates are content to allow the dominant fish to attack the intruder, while they remain safely in the shelter.

change from month to month as individuals are born, die, or emigrate, but the boundaries of the territory remain the same, passed on by learning from one generation to the next. Social ties are strengthened by a great deal of contact. When these social rodents encounter one another, they greet with an opened-mouth "kiss" (Fig. 42-19), and if they belong to the same coterie, this may be followed by mutual grooming of the fur.

Living in exposed areas of the plains, prairie dogs benefit from community life, where many eyes can watch for danger. When a predator such as a golden eagle is spotted by an alert "watch dog," it warns nearby animals with a distinctive yipping bark that is quickly passed through the town.

Hamadryas Baboons

Among nonhuman mammals, the primates, our closest relatives, are among the most social. African Hamadryas baboons, for example, exhibit social organization on three levels. The first level, the smallest unit of baboon society, is known as the *one-male unit*. This is an extended family consisting of a male and his "harem" of two to five females, and their young. The male, who is twice the females' size, dominates them using slaps, bites, or aggressive stares. He leads the group from place to place and, like a protective father or a jealous husband, intervenes if any of his followers attempt to associate with strangers, fight among themselves, or stray too far from him. A young male baboon begins forming his unit by adopting or kidnapping juvenile females from their families. They remain with him permanently, becoming his mates when they reach sexual maturity. This harem system inevitably results in an excess of males who gather together in all-male *bachelor units.*

At the end of each day, several one-male units congregate to form a *band* in which they travel to waterholes and sleeping cliffs. At the large sleeping cliffs, several bands may gather to form the largest social grouping, the *troop,* which may include hundreds of individuals. When aggressive interactions occur—for example, when strange bands meet, the males defend the band. Scientists who filmed these encounters found that they consist largely of elaborate bluffs. The males display their formidable fangs (Fig. 42-20) and use their hands in a slapping motion, but rarely contact one another. A defeated male can quickly end a dispute by exposing his neck to his opponent (a submissive gesture).

Human Ethology

When scientists who study animal behavior observe a specific activity, they often seek its basis in the animal's genetic makeup and past experience, and attempt to determine how the behavior helps the animal survive and reproduce. Sociobiologists and human behavioral geneticists believe that many human tendencies also have a genetic basis, and are attempting to study human ethology. Because it is new as a rigorous discipline, and because it deals with broad tendencies in large groups of diverse individuals, human ethology is still relatively speculative

Figure 42-19 A greeting kiss is exchanged when two prairie dogs meet.

Figure 42-20 Threat display of a male baboon shows formidable weapons indeed. Despite potentially lethal fangs (seen only in males), aggressive encounters between baboons rarely result in injury.

A CLOSER LOOK

At Altruism, Kin Selection, and the Selfish Gene

Darwin's concept of the survival of individuals best adapted to their environment and able to leave the largest number of offspring remains the foundation of evolutionary theory. Social animals, however, often behave in ways that appear to endanger the individual, decreasing its chance to survive and reproduce. There are many examples. Worker honeybees do not reproduce, but care for the offspring of the queen. Worker ants die in defense of their nest. In the Florida scrub jay, young mature males may remain at their parents' nest and help them raise subsequent broods, instead of breeding themselves. Young adult jackals may also help their parents raise a new litter. In colonies of Belding ground squirrels, individuals may sacrifice their lives to warn the group of an approaching predator. These behaviors fulfill the biologist's definition of **altruism**: behavior that may decrease the reproductive success of one individual to the benefit of another.

How can such behavior be reconciled with the survival of the fittest? Why aren't the individuals performing such self-sacrificing deeds rapidly eliminated from the population, taking the genes contributing to this behavior with them? In fact, the laws of natural selection operate in such a way that "selfish" behavior (that is, behavior that increases the chance of perpetuating itself) will always be most successful. So, in some way, altruism must be "selfish," but how? **A major insight into the selfishness of altruistic behavior is supplied by the theory that natural selection does not operate exclusively on the individual, but operates at the level of the gene.** From this theoretical viewpoint, individuals are short-lived carriers of genes, while genes may persist for millions of years, passing from one generation to the next. **One important way in which genes may be preserved is by individuals performing innate behavior, even self-sacrificing behavior, that enhances the survival of others carrying the same genes.** Some researchers like to say that genes promote the survival of copies of themselves in other individuals, and have coined the term *selfish gene* to describe this.

Statistically, closely related individuals are most likely to carry the same genes. Thus, an individual promotes survival of the types of genes it carries through inherited behaviors that maximize survival not only of that individual, but of its close relatives as well. This concept, called **kin selection**, is also discussed in Chapter 17. Kin selection helps to explain a variety of altruistic behaviors. In honeybees, for example, the males are haploid while females are diploid. Female workers share, on the average, 75 percent of their sister's genes. If they were to mate and reproduce, they would only share 50 per-

Figure E42-2 A Belding ground squirrel sounds an alarm as danger approaches.

cent of the genes of their offspring. To maximize survival of their own genes, then, rather than reproducing, they are better off helping their mother (the queen) raise more of their sisters, who share a greater percentage of their genes than would their own offspring. In fact, this is what happens in the honeybee societies described in this chapter.

Alarm calls of Belding ground squirrels (Fig. E42-2) may also be explained by kin selection. Paul Sherman studied colonies of these rodents for several years, marking individuals so they could be recognized. He observed that the squirrel calling an alarm often drew the attention of the predator and was sometimes killed, but allowed other members of the colony to seek shelter. In ground squirrel colonies, males leave the burrow in which they were born and establish new burrows some distance away. Females, in contrast, stay close to home. So females in a given area are usually closely related, while males are not. Kin selection theory predicts that females should give more alarm calls than males, since their calls benefit relatives who share more of their genes, justifying the danger. It would also predict that females surrounded by close relatives would call more often than those without close relatives nearby. Sherman's data confirmed both of these predictions, supporting the theory of kin selection.

Kin selection provides important insights into how self-sacrificing, seemingly nonadaptive behaviors can evolve and persist within a species.

and inexact. To demonstrate that human behavior, like that of all other animals, does have some innate basis, scientists have attempted to isolate the genetic components of human behavior. The following sections review some of these studies.

Studies of Young Children

Much of the behavior of very young infants is likely to have a large innate component, since there has been little time for learning to occur. The rhythmic movements of a baby's head in search of the mother's breast is a human fixed action pattern that may be observed during the first days after birth. Suckling, which can be observed even in the human fetus, is equally instinctive (Fig. 42-21). Other fixed action patterns seen in newborns or even premature infants include walking movements when the body is supported and grasping with the hands and feet. Another example is smiling, which may occur soon after birth. Initially, smiling can be released by almost any object looming over the newborn. Before an infant is 2 months old, an exaggerated releaser (see Chapter 41) may be constructed consisting of two dark, eye-sized spots on a light background. This elicits smiling even more successfully than an accurate representa-

tion of a human face. As the child matures, learning and further development of the nervous system interact to limit the response to increasingly accurate representations of a face.

Another way to minimize the effects of learning is to observe children who are blind, deaf, or both, and have been unable to learn through sight and sound. Without ever having seen or heard them, these children produce normal smiles and laughter, and expressions of frustration and anger.

As discussed in the previous chapter, some animals have a strong innate tendency to learn specific things during certain periods of development, a behavior called imprinting. The human fetus begins responding to sounds during the third trimester of pregnancy, and by 6 weeks after birth is able to distinguish a variety of consonant sounds, strong evidence that the human brain is programmed to interpret language. Babies are notoriously difficult experimental subjects, and ingenious methods have been devised to test them. One of the most successful uses the infant's ability to make sucking movements. The baby responds to the presentation of various consonant sounds by sucking on a pacifier containing a force transducer that records the sucking rate. After hearing one sound (such as "ba") repeatedly, the infant becomes bored with it and decreases her sucking rate. A new sound (such as "pa") causes an increase in sucking, revealing that the infant perceives these as different.

The sucking technique has recently been used to demonstrate that newborns in their first 3 days of life can be conditioned to produce certain rhythms of sucking using their mother's voice as reinforcement. The mother's voice was clearly preferred to other female voices, as indicated by the sucking rhythm. The ability of the infant to learn his or her mother's voice and respond positively to it within days of birth has strong parallels to imprinting, and may help initiate bonding with the mother.

The ability of young children to acquire language rapidly and almost effortlessly is almost certainly an example of developmentally programmed learning. Between age 1 and 8, children typically acquire a vocabulary of 28,000 words whose meanings they recognize. If you are currently struggling with a foreign language class, you are probably painfully aware that your critical period for language acquisition is long past.

Exaggerating Human Releasers

A behavior probably has an instinctive component if the stimulus that causes it (the releaser) can be exaggerated beyond the bounds of reality and elicit an

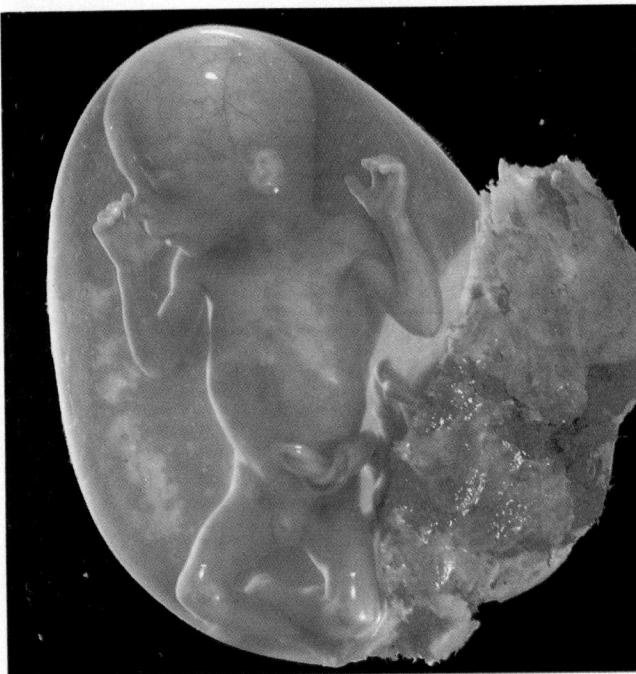

Figure 42-21 Thumb sucking is a difficult habit to discourage in young children, since suckling on appropriately sized objects is an instinctive, food-seeking behavior. This fetus (which is about 8 inches in length) sucks its thumb at the gestational age of 4½ months.

even stronger response. The eyespots that cause young infants to smile are one example. In turn, the smile of an infant, along with certain characteristic baby features, may release protective feelings in adults. These features include a relatively large head with a domed forehead, chubby cheeks, small nose, short arms and legs, and a small, rounded body. Even 3-year-old children respond to these features with "mothering" behavior. The marketplace has exploited the releasing aspects of these features by exaggerating them in representations of baby animals and people and using their innate appeal to sell dolls, posters, calendars, and cards (Fig. 42-22).

One human signal with an innate, physiological basis is the involuntary enlargement of the pupil of the eye when viewing something pleasant, be it a loved one or a hot fudge sundae. We also react to this signal in others. To test this, male subjects were shown identical photographs of smiling women which had been retouched to enlarge or contract the pupils. They overwhelmingly preferred those with large pupils, although none were consciously aware of the pupil size or able to pinpoint the reason for their reaction. The positive response of men to enlarged pupils has long been recognized. In the middle ages, women artificially enlarged their pupils by using the drug belladonna (meaning "beautiful woman" in Italian). We react positively to someone who gazes at us with dilated pupils (although our recognition of this feature is subconscious), since it implies interest and attraction.

Comparative Cultural Studies

Another way to study the instinctive bases of adult human behavior is to compare simple acts performed by people from isolated and diverse cultures. This comparative approach, pioneered by the ethologist Eibl-Eibesfeldt, has revealed several gestures that seem to form a universal, and therefore probably innate, human language. Such gestures include a variety of facial expressions for pleasure, rage, and disdain, and movements such as the "eye flash" and a hand upraised in greeting (Fig. 42-23).

Human Pheromones?

Humans may have unconscious responses to **pheromones.** Our sense of smell is poorly developed

Figure 42-22 We instinctively respond to certain features associated with infants and very young children, such as big eyes, small noses, rounded faces, and large heads. These are sometimes exaggerated to produce supernormal stimuli, as in this doll.

Figure 42-23 Gestures that have similar meanings in diverse and isolated cultures may be evidence of a common biological heritage. Here motion pictures freeze the "eye-flash greeting" (in which the eyes are widely opened and the eyebrows rapidly elevated) in a person from New Guinea (left) and Bali (right). Watch for this possibly innate response in yourself when you encounter a friend. These photos were taken by I. Eibl-Eibesfeldt, who conducted this research.

as compared with many other mammals, and the role of odor as a means of human communication is largely unknown. However, an interesting study by Martha McClintock of Harvard provided indirect evidence that primer pheromones may influence female reproductive physiology. She studied the menstrual cycles of 135 women living in a college dormitory and found that the menstrual cycles of roommates and close friends became significantly more synchronous over a 6-month period. The cycles of women randomly chosen from the dormitory did not. In a further study, she divided the women into groups based on how often during an average week they associated with men. She found that the women reported spending time with men less than three times weekly had significantly longer cycles than those who associated with men more frequently. While far from conclusive, these findings are tantalizing and call for further investigation of the role of both primer and releaser pheromones in human behavior.

Studies of Twins

By studying identical and fraternal twins, investigators can come as close as possible to controlled breeding experiments in humans. Fraternal twins arise from two individual eggs and are no more similar genetically than other siblings. However, they are exactly the same age, and share a very similar environment. Identical twins, arising from a single fertilized egg, have identical genes. The most fascinating twin findings are based on anecdotal observations of identical twins separated soon after birth, reared in different environments, and reunited for the first time as adults. They have been found to share nearly identical taste in jewelry, clothing, humor, food, and names for children and pets. Personal idiosyncrasies such as giggling, nail-biting, drinking patterns, hypochondria, and mild phobias may be shared by these unacquainted twins. More rigorous studies are also supporting the heritability of many human behavioral traits. These have documented a significant genetic component for traits such as activity level, alcoholism, sociability, anxiety, intelligence, dominance, and even political attitudes. Based on tests designed to measure many aspects of personality, identical twins are about twice as similar in personality as fraternal twins. Further, identical twins reared apart were found to be as similar in personality as those reared together, indicating that the differences in their environments had little influence on their personality development.

The field of human behavioral genetics is controversial, since it challenges the long-held belief that environment is the most important determinant of human behavior. Investigators have progressed past the nature–nurture debate in their investigations of nonhuman animals. As discussed in Chapter 41, we now recognize that much behavior has some genetic basis, and that complex behavior often combines elements of both learned and innate behavior. In the case of our own behavior, the debate continues. Human ethology is not yet recognized as a rigorous science, and it will always be hampered because humans can neither view themselves with detached objectivity nor treat each other as laboratory animals. In spite of this, there is much to be learned about the interaction of learning and innate tendencies in people. Such information could be usefully applied to areas as diverse as childrearing, education, understanding interpersonal relationships, and possibly one day even using synthetic pheromones to regulate reproductive physiology.

SUMMARY OF KEY CONCEPTS

Communication

Communication, an action by one animal that alters the behavior of another, is the basis of all social behavior. It allows animals of the same species to interact effectively in their quest for mates, food, shelter, and other resources. Animals communicate through visual signals, sound, chemicals (pheromones), and touch.

Visual communication is quiet, and can convey subtle, rapidly changing information. Visual signals may be active (body movements) or passive (body shape and color).

Sound communication can also carry a wide range of rapidly changing information, and is effective where vision is impossible. Although sound may attract predators, the animal may remain hidden while communicating.

Chemical signals take the form of primer or releaser pheromones. Primer pheromones alter the physiological state of the recipient, while releaser pheromones influence the recipient's behavior. Pheromones are detected only by others of the same species. Releaser pheromones can be detected after the sender has departed, conveying a message over time. Physical contact reinforces social bonds and helps synchronize mating in a variety of animals, from mammals to molluscs.

Competition for Resources

Although competitive interactions are often resolved through aggression, serious injuries are rare. Most aggressive encounters are settled using displays that communicate the motivation, size, and strength of the combatants.

Some species establish dominance hierarchies that minimize aggression and regulate access to resources. Based on initial aggressive encounters, each animal acquires a status in which it defers to more dominant individuals, and dominates subordinates. When resources are limited, dominant animals obtain the largest share, and are more likely to reproduce.

Territoriality, a behavior in which animals defend areas where important resources are located, also allocates resources and minimizes aggressive encounters. In general, territory boundaries are respected, and the best-adapted individuals defend the richest territories and produce the most offspring.

Reproduction

Successful reproduction requires that animals recognize the species, sex, and sexual receptivity of potential mates. In some cases, they must also overcome a resistance to close approach by another individual. These requirements have resulted in the evolution of a wide variety of sexual displays that use all possible forms of communication.

Cooperation

Social living has both advantages and disadvantages, and species show a wide variation in the degree to which their members cooperate. Some form cooperative societies. The most rigid and highly organized are those of the social insects such as the honeybee, where the members follow rigidly prescribed roles throughout life. These roles are maintained both through genetic programming and the influence of primer pheromones. Nonhuman vertebrates also form complex, but less rigid, societies, such as are found among bullhead catfish, prairie dogs, and baboons.

Human Ethology

The degree to which human behavior is genetically influenced is highly controversial. Because humans cannot be treated as laboratory animals, and because learning plays a major role in nearly all human behavior, investigators must rely on studies of newborn infants, observation of responses to exaggerated stimuli, comparative cultural studies, correlations between certain behaviors and physiology (which suggest a role for pheromones), and studies of identical and fraternal twins. Evidence is accruing that our genetic heritage plays a role in personality, intelligence, simple universal gestures, our responses to certain stimuli, and our tendency to learn specific things such as language at particular stages of development.

GLOSSARY

active visual signal: a movement or posture that communicates information.

aggression: antagonistic behavior, usually between members of the same species, often resulting from competition for resources.

altruism: a type of behavior that may harm the individual performing it but benefits other individuals.

communication: the act of producing a signal that causes another animal, usually of the same species, to modify its behavior in a way beneficial to one or both participants.

dominance hierarchy: a social arrangement in which animals, usually through aggressive interactions, establish a rank for some or all of the members of the social unit.

kin selection: the concept that natural selection selects for altruistic behaviors that benefit close relatives of the individual performing the behavior.

passive visual signal: a body shape, color, or size that communicates information, even in the absence of specific behaviors.

pheromone (fār'-uh-mōn): a chemical produced by an organism that alters the behavior or physiological state of another of the same species.

primer pheromone: a chemical produced by an organism that alters the physiological state of another of the same species.

queen substance: a chemical produced by a queen bee that can act as both a primer and a releaser pheromone.

releaser pheromone: a chemical produced by one organism that alters the behavior of another of the same species.

territoriality: the defense of an area in which important resources are located.

waggle dance: symbolic communication used by honeybee foragers to communicate the location of a food source to their hivemates.

STUDY QUESTIONS

1. List four senses through which animals communicate. After each, present both advantages and disadvantages of that form of communication. Also give one example of each.
2. Distinguish between passive and active visual signals, providing an example of each. Which can convey the most rapidly changing information?
3. What are graded visual signals, and what is their purpose?
4. Define and distinguish between primer and releaser pheromones. Give an example of each.
5. A songbird will ignore a squirrel in its territory, but act aggressively toward a member of its own species. Explain why.
6. Describe some ways in which animals advertise species and sex during reproductive behavior. Include examples using sight, sound, and chemicals. From an evolutionary standpoint, why is such communication important?
7. What type of animal tends to form the most complex societies, and why?
8. A forager honeybee has just discovered flowers directly below the projection of the sun on the horizon (0 degrees from the sun). Compare the orientation of her waggle dance if performed on a horizontal platform outside the hive to that performed on a vertical wall inside the hive.
9. Describe one of the experiments that reveal the importance of chemical communication in bullhead catfish. Suggest why visual communication is not highly developed in this species.
10. How do male Hamadryas baboons acquire females for their harems? Can you suggest why such a system might evolve?

DISCUSSION QUESTIONS

1. Many aggressive encounters are relatively harmless. In what way does this promote the "survival of the fittest"?
2. Describe and give an example of a dominance hierarchy. What role does it play in social behavior? Give a human parallel and describe its role in human society. Are the two roles similar? Why or why not? Now repeat this exercise for territorial behavior.
3. Humans perform many altruistic behaviors, both for their relatives and for unrelated individuals. Discuss possible reasons why people might make sacrifices for unrelated individuals, and whether or not this behavior might be adaptive.

SUGGESTED READINGS

Gould, C. G. "Out of the Mouths of Beasts." *Science 83,* April 1983. Animal communication discussed in an interesting and informative way.

Macdonald, D., Brown, R. "The Smell of Success." *New Scientist,* May 1985. Describes the amazing diversity of mammalian pheromones.

Pennisi, E. "Not Just Another Pretty Face." *Discover,* March 1986. Describes the newly investigated society of the naked mole rat, a vertebrate whose social behavior resembles that of some social insects.

Seeley, T. D. "The Honey Bee Colony as a Superorganism." *American Scientist,* November–December 1989. The honeybee society exemplifies the concept of kin selection in which the society, rather than the individual, serves as a "vehicle for the survival of genes."

Wilson, E. O. "Empire of the Ants." *Discover,* March 1990. This entertaining article discusses the diversity of ant behavior that has contributed to their enormous success.

UNIT VI

Ecology

43

Population Growth and Regulation

This dense population of King penguins on an antarctic shore attests to the richness of the food supply in the surrounding waters.

"There is no exception to the rule that every organic being naturally increases at so high a rate that, if not destroyed, the earth would soon be covered by the progeny of a single pair."

Charles Darwin in On the Origin of Species

Introduction to Ecology

In eastern Colorado lies a remnant of shortgrass prairie, dominated by buffalo grass and blue grama grass. In spring and summer it is ablaze with wildflowers: paintbrush, vetch, sunflower, and bladderpod (Fig. 43-1). Prairie dogs stand upright by their burrows, alerting the "town" with vigorous high-pitched cries as a single hawk soars lazily overhead. The prairie is an **ecosystem,** a complex, interrelated network of living organisms and their nonliving surroundings. An ecosystem can be as small as a puddle or as large as

Figure 43-1 The prairie ecosystem, showing representatives of interacting populations, which together constitute the prairie community.

an ocean. Within our prairie ecosystem, the wild-flowers, prairie dogs, the grass on which they feed, the hawk that preys on them, and the myriad microscopic organisms that keep the soil fertile, constitute a **community** of interacting organisms. The community, in turn, is composed of **populations,** each consisting of all the members of a particular species, be they hawks, grasshoppers, buffalo grass, or bacteria.

Just to the north, a farm was abandoned nearly 2 years ago. The community here is quite different, with Russian thistle, pigweed, amaranth, and cheatgrass invading the new habitat. A "For Sale" sign in the field foreshadows the high-density housing that will soon displace both farmland and prairie. The first human residents will be delighted to see a hawk, a rare and magnificent bird of prey, practically in their backyards. But as the prairie dog town is bulldozed, this large predator will also disappear.

How has the prairie (in contrast to the failed farm) sustained itself for centuries without artificial fertilizer or irrigation? Why are predators such as the hawk rare relative to their prey? What keeps prairie dogs from overpopulating their habitat and starving? What happens when two organisms compete for the same resources? Why is the abandoned farm community different from that of the untouched prairie? What will the new community look like in 40 years, left to itself? Why has the human population continued to increase, while other populations remain stable or decline in the face of human expansion? These are the questions of **ecology,** the science that deals with the interrelationships among living things and their environment. The environment includes the nonliving components of soil, water, and weather, called the **abiotic** portion, and a **biotic** component, including all forms of life within the ecosystem. Ecology, whose name is derived from the Greek word *oikos,* meaning "a place to live," is a tremendously diverse, complex, and relatively young scientific discipline.

In preceding chapters we have studied the anatomy, physiology, and behavior of individual organisms. The science of ecology begins at the next level: the population. From this starting point we will proceed to increasing levels of complexity, first to communities and the interactions within them, and finally to entire ecosystems.

Population Growth

Studies of ecosystems undisturbed by human beings show that many populations tend to remain relatively stable over time. Yet we are vividly aware from the human example that populations can readily increase. Let's first examine how and why populations grow, then look at the factors that normally control this growth.

Three factors determine whether and how much the size of a population changes: births, deaths, and migration. Organisms join a population through birth or **immigration** (migration in) and leave it through death or **emigration** (migration out). A population remains stable if, on the average, as many individuals leave as join. Population growth occurs when the number of births plus immigrants exceeds the number of deaths plus emigrants. Populations decline when the reverse occurs. A simple equation for the change in population size is:

(births − deaths) + (immigrants − emigrants) = population change

In many natural populations, organisms moving in and out contribute relatively little to population change, leaving birth and death rates as the primary factors influencing population growth.

The ultimate size of any population (discounting migration) is the result of a balance between two major opposing factors. The first is **biotic potential,** or the maximum rate at which the population could increase, assuming ideal conditions allowing a maximum birth rate and minimum death rate. Opposing this potential for growth are limits set by the living and nonliving environment. These limits include the availability of food and space, competition with other organisms, and interactions among species such as predation and parasitism. Collectively, these limits are called **environmental resistance.** Environmental resistance can both decrease the birth rate and increase the death rate. **The interaction between biotic potential and environmental resistance usually results in a balance between population size and available resources.** To understand how populations grow and how their size is regulated, we must examine each of these forces in more detail.

Biotic Potential: Exponential Growth

Changes in population size (ignoring migration) are functions of the birth rate, the death rate, and the number of individuals in the original population. Rates of change in populations may be expressed as changes per individual per unit time. For example, the birth rate may be expressed as the number of births per individual per unit time.

The **rate of growth** (r) of a population is determined by subtracting the death rate (d) from the birth rate (b):

$$r \quad = \quad b \quad - \quad d$$

growth rate = birth rate − death rate

To determine the number of individuals added to a population of size N in a given time period, the growth rate (r) is multiplied by the original population size (N):

population growth = rN

For example, the annual growth rate of a population of 10,000 in which 1500 births and 500 deaths occur yearly can be calculated as follows:

$$r = \text{birth rate} - \text{death rate}$$
$$r = 1500/10{,}000 - 500/10{,}000 = 0.10 \quad \text{or}$$
$$r = 0.15 - 0.05 = 0.10$$

Population growth (rN) equals 0.10 × 10,000 = 1000. If this rate of increase persists, then the following year, r must be multiplied by a larger population size (N + rN = 11,000), resulting in an increase of 1,100 individuals, which in turn is added to N, and so on. This is **exponential growth.** During exponential growth, the population grows (during a given time period) by a fixed percentage of its size at the beginning of that time period. Thus an increasing number of individuals is added to the population during each succeeding time period, causing population size to grow at an ever-accelerating pace. Births will exceed deaths if, on the average, each individual produces more than one surviving offspring during its lifetime. This causes an accelerating increase in population size. The curve formed by exponential population growth is often called a J-shaped growth curve (or a **J-curve**) after its shape, clearly illustrated in Figures 43-2 and 43-3.

Although the number of offspring produced by an individual each year varies from millions for an oyster to one or fewer for a human, each organism, whether working alone or as part of a sexually reproducing pair, has the potential to replace itself manyfold during its lifetime. This capacity, called **biotic potential,** has evolved because it helps assure that at least one offspring survives to bear its own young. Several factors influence biotic potential. These include

1. The age at which the organism first reproduces.
2. The frequency with which reproduction occurs.
3. The average number of offspring produced each time.
4. The length of the reproductive life span of the organism.
5. The death rate of individuals under ideal conditions.

Examples in which these factors differ will be used to illustrate the concept of exponential growth (Fig. 43-2).

The bacterium *Staphylococcus* is a normally harmless resident in and on the human body. But in an ideal culture medium such as warm custard, each bacterial cell can divide every 20 minutes, doubling the population three times each hour. (The by-products of the bacteria's metabolism can result in serious food poisoning under these conditions.) The biotic potential of bacteria is so great that, were nutrients unlimited, the offspring of a single bacterium could cover the Earth over 7 feet deep within 48 hours! In contrast, the golden eagle is a relatively long-lived, rather slowly reproducing species. Let's assume that the golden eagle can live 30 years, reaches sexual maturity at 4 years, and that each pair of eagles produces two offspring per year for the remaining 26 years. Figure 43-2 compares the potential population growth of eagles to that of bacteria, assuming no deaths occur in either population during the time graphed. Notice that the shapes of the curves are virtually identical. Although the time scale differs, population sizes eventually become astronomically large. Figure 43-2 also shows what happens if eagle reproduction begins at 6 years instead of 4. Exponential growth still occurs, but the time required to reach a particular size is increased considerably. **This has important implications for the human population: delayed childbearing significantly slows population growth. If each woman has only three children, but has them in her early teens, the population will grow much faster than if women each have *five* children, but begin having them at age 30!**

So far we have looked only at birthrates. Even under ideal conditions, however, some mortality occurs. To illustrate the effect of differing death rates, three bacterial populations are compared in Figure 43-3. Again, the shapes of the curves are the same. In each case the population eventually approaches infinite size; only the time required to reach any given population size differs.

Boom and Bust Cycles and Exponential Growth

In nature, exponential growth curves are observed only under special circumstances, and only for limited periods. Exponential growth is seen temporarily in populations that undergo regular cycles, where rapid population growth is followed by a massive die-off. These **boom and bust cycles** are observed in a variety of organisms for complex and varied reasons. Many short-lived, rapidly reproducing species, from algae to insects, have seasonal population cycles that are linked to predictable changes in rainfall, temperature, or nutrient availability (Fig. 43-4). In temperate climates, insect populations grow rapidly during the spring and summer, then crash with the killing hard

(a)

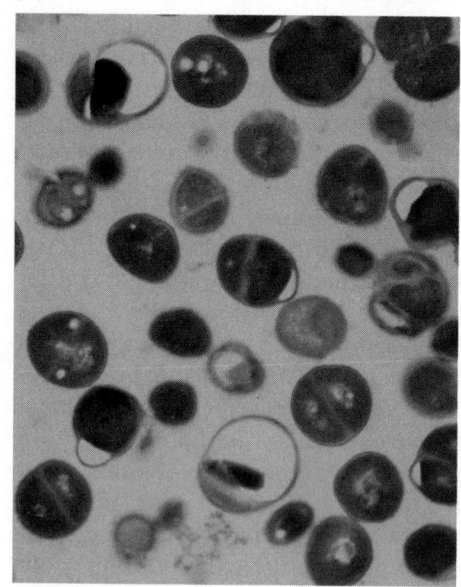

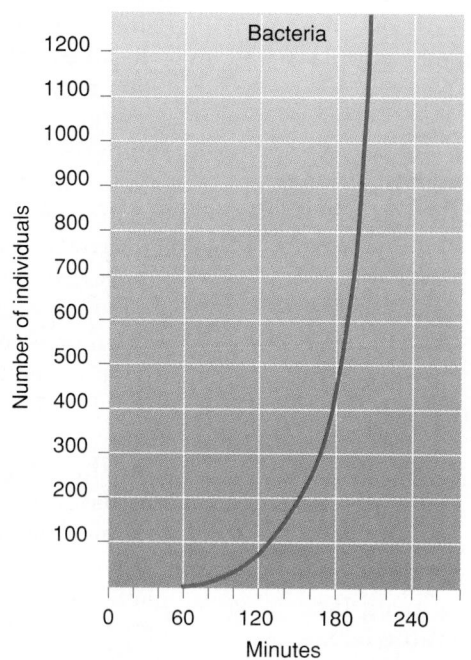

Time (minutes)	Number of Bacteria
0	1
20	2
40	4
60	8
80	16
100	32
120	64
140	128
160	256
180	512
200	1024
220	2048

(b)

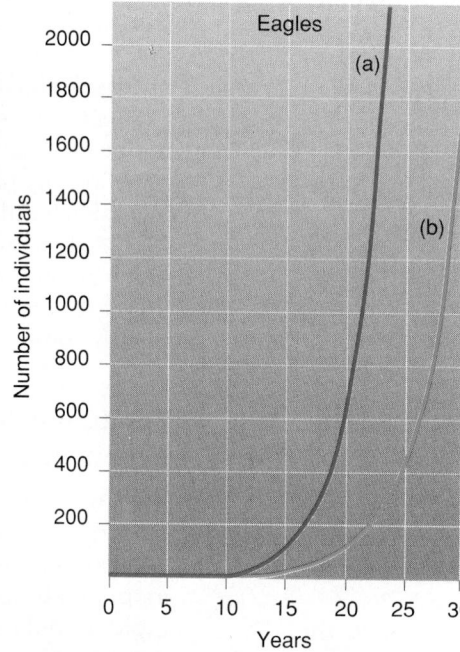

Time (years)	Number of Eagles (a)	Number of Eagles (b)
0	2	2
2	2	2
4	4	2
6	8	4
8	14	8
10	28	12
12	52	18
14	100	32
16	190	54
18	362	86
20	630	142
22	1314	238
24	2504	392
26	4770	644
28	9088	1066
30	17314	1764

Figure 43-2 Exponential growth curves all share a similar J shape; the major difference is the time scale. **(a)** The growth of a population of bacteria, assuming one individual to start with and a doubling time of 20 minutes. **(b)** Growth of an eagle population, starting with a single pair of hatchlings, and assuming that the age at first reproduction is 4 years (solid line) and 6 years (dashed line). Notice that after 26 years, those that began reproducing at 4 years have seven times the population of those that began reproducing at 6 years.

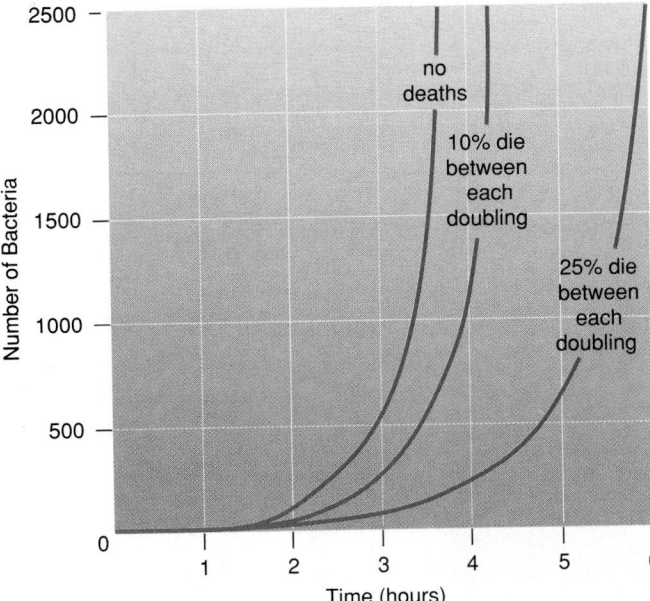

Figure 43-3 The effect of differing death rates on bacterial population growth, assuming the population doubles every 20 minutes. Notice that the population in which a quarter of the bacteria die every 20 minutes reaches 2500 only 2 hours and 20 minutes later than that in which no deaths occur.

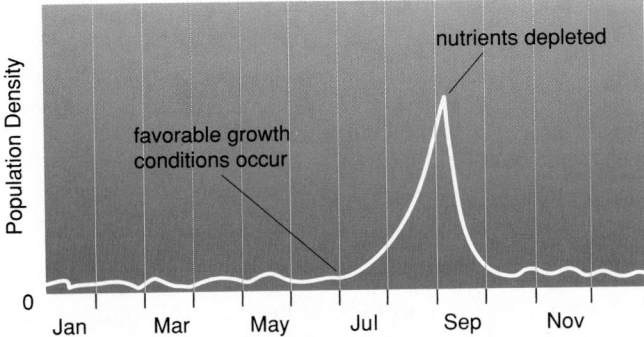

Figure 43-4 The population of blue-green algae in a boom-and-bust cycle. Algae survive at a low level through the fall, winter, and spring. Early in July, conditions become favorable for growth, and exponential growth occurs through August. After this time nutrients become depleted and the population "goes bust."

frosts of winter. More complex factors produce roughly 4-year cycles of voles and lemmings, and much longer population cycles in hares, muskrats, grouse, and ptarmigan.

Lemming populations, for instance, may grow until they overgraze their fragile arctic tundra ecosystem. Lack of food, increasing populations of predators, and social stress caused by overpopulation may all contribute to a sudden high mortality. A subsequent decrease in predator numbers (see "A Closer Look at Cycles in Predator and Prey Populations") and recovery of the plant community set the stage for another round of exponential growth in the lemming population (Fig. 43-5)

In noncycling populations, exponential growth may occur temporarily under special circumstances, for example, if population-controlling factors such as predators or parasites are eliminated or the food supply is increased. Exponential growth can also occur when individuals invade a new habitat where conditions are favorable and competition is scarce. This has happened repeatedly when people have introduced foreign or **exotic** species into ecosystems, often with tragic results (see "Planet Watch: Growth Without Resistance—The Problem of Introduced Species"). As you will learn in the next section, all exponential growth must eventually either level off or crash.

Environmental Resistance: Limits to Growth

Exponential growth carries with it the seeds of its own destruction. As individuals join the population, competition for resources intensifies. Predators may increase in number or they may make this abundant prey a larger part of their diet. Parasites and diseases spread more readily due to crowding and weakness caused by lack of food or stress caused by adverse social interactions. Consequently, after a period of exponential growth, populations tend to stabilize at or below the maximum size that the environment can sustain, called **carrying capacity** (see next section). The rate of growth drops precipitously, fluctuating around zero. This type of population growth, which is typical of long-lived organisms colonizing a new area, is represented graphically by a "sigmoid" or **"S-curve"** (Fig. 43-6).

Carrying Capacity

Populations may stabilize at a level called the **carrying capacity** of the ecosystem. The carrying capacity is the maximum number of organisms that an

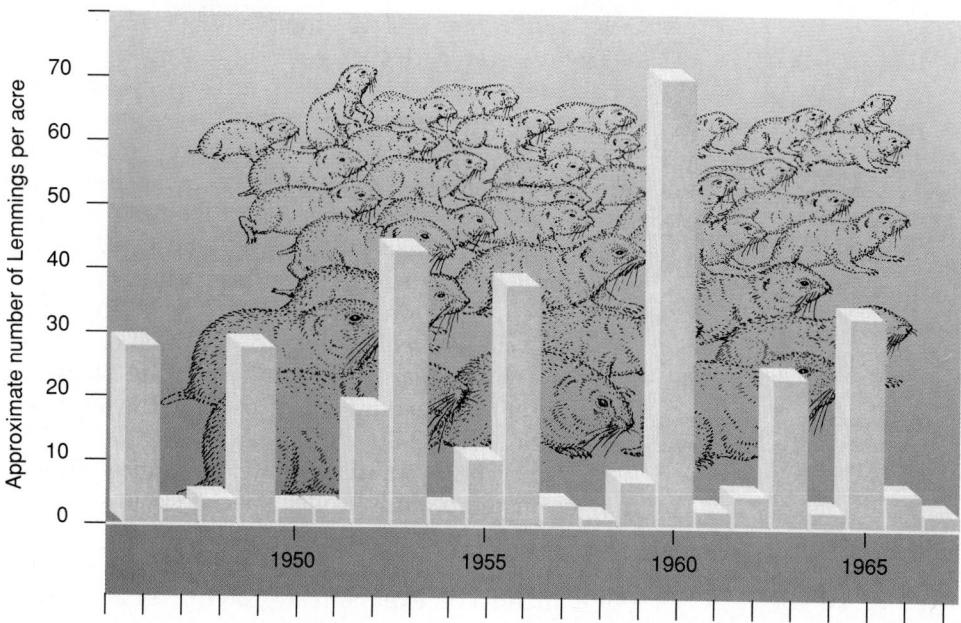

Figure 43-5 Lemming population density follows roughly a 4-year cycle, as illustrated by this data from Point Barrow, Alaska.

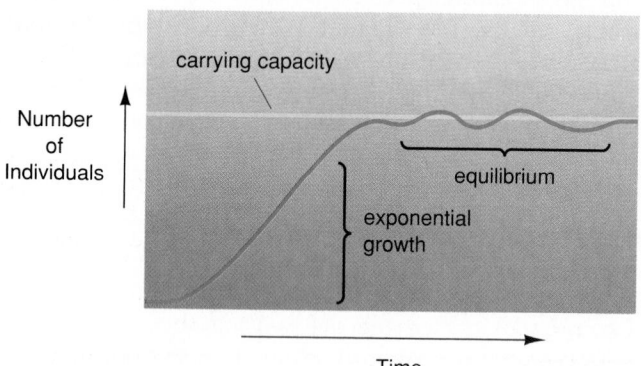

Figure 43-6 The S-curve, showing a population first growing exponentially, then fluctuating around carrying capacity.

area can support on a sustained basis. It is determined primarily by the availability of two types of resources: those that are replenished by natural processes such as nutrients, water, and light; and a nonrenewable resource: space. If space requirements are exceeded, animals may emigrate, but often to less suitable areas where their death rate will be higher. Reproduction will decline, since animals may not find adequate breeding sites or the seeds of plants may not reach a suitable place to germinate. If demands on renewable resources such as food, water, and light (the energy source for plants) are too high, organisms will starve. Excess demands may damage ecosystems, reducing their carrying capacity. The result is a population decline either until the ecosystem recovers or a permanently reduced population. For example, overgrazing by cattle on dry western grasslands has given sagebrush (which cattle will not eat)

a competitive advantage. Once established, sagebrush thrives, replacing edible grasses and reducing the carrying capacity of the land for cattle. Other dramatic cases of overgrazing have occurred when herbivores such as reindeer have been introduced onto islands without large predators, as shown in Figure 43-7. In nature, populations are maintained at or *below* the carrying capacity of their environment by environmental resistance. Factors of environmental resistance may be classified into two broad categories: **density independent** and **density dependent** limits to growth.

Density-Independent Limits to Growth

Density-independent factors limit population size and growth regardless of the density of the population. Perhaps the most important density-independent factor is weather. For example, many insects

A CLOSER LOOK

At Cycles in Predator and Prey Populations

If we assume that prey such as rabbits are eaten exclusively by a particular predator, such as the lynx, it seems logical that both populations might show cyclic changes, with increases in the predator population lagging behind increases in prey population size. For example, a large rabbit population would provide abundant food for lynx and their offspring, which will survive in large numbers, increasing the lynx population and increasing predation on the rabbits. This will cause a decline in the rabbit population, which will allow fewer lynx to survive and reproduce, causing the lynx population to decline shortly thereafter.

Does this out-of-phase population cycle of predators and prey actually occur in nature? A classic example of such a cycle was observed using the ingenious method of counting all the pelts from northern Canada lynx and snowshoe hares purchased by the Hudson Bay Company between 1845 and 1935. The availability of pelts (which presumably reflects population size) showed dramatic, closely linked cycles between these predators and their prey (Fig. E43-1). Unfortunately, as with any field investigation, many variables could influence the relationship between hare and lynx. One problem is that hare populations have been shown to fluctuate even without lynx present, possibly owing to overshooting carrying capacity and reducing their food supply. Further, lynx do not feed exclusively on hares, but can eat a variety of small mammals. Environmental variables such as exceptionally severe winters could adversely affect both populations and produce similar cycles.

To test this hypothesis more scientifically, investigators turned to controlled laboratory studies on pop-

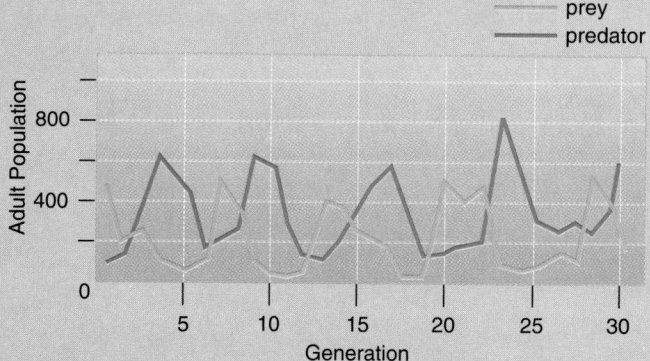

Figure E43-2 Out-of-phase fluctuations in laboratory populations of the azuli bean beetle and its braconid wasp predator.

ulations of small predators and their prey. The study illustrated here involved only braconid wasp predators and their bean weevil prey. As predicted, the two populations showed regular cycles, with the predator population rising and falling slightly later than the prey population (Fig. E43-2). The wasps lay their eggs on weevil larvae, which provide food for the newly hatched wasps. A large weevil population assures a high survival rate for wasp offspring, increasing the predator population. Then, under intense predation pressure, the weevil population plummets, reducing the population size of the next generation of wasps. Reduced predation then allows the weevil population to increase rapidly, and so on.

It is highly unlikely that such a straightforward example will ever be found in nature, but this type of predator–prey interaction clearly can contribute to the fluctuations observed in many natural populations.

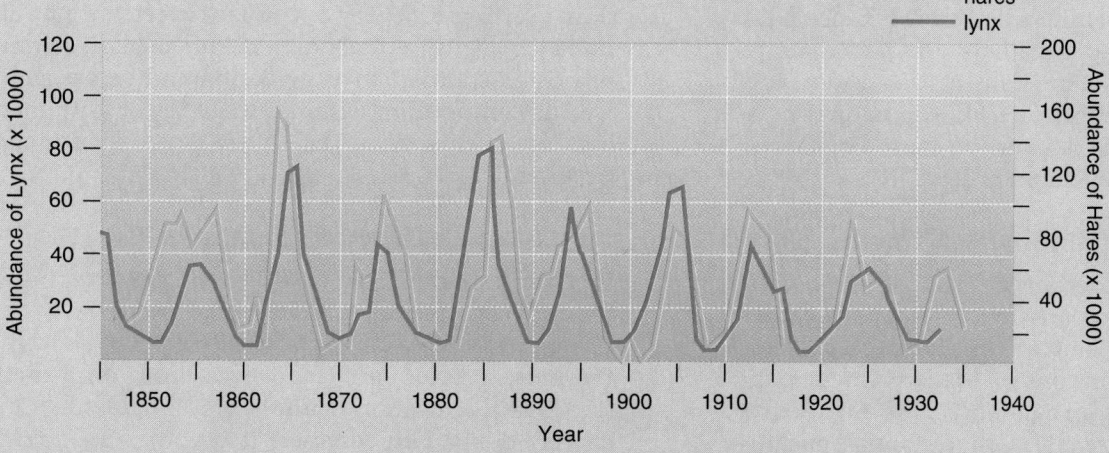

Figure E43-1 Population cycles of snowshoe hares and their lynx predators based on the number of pelts received by the Hudson Bay Company.

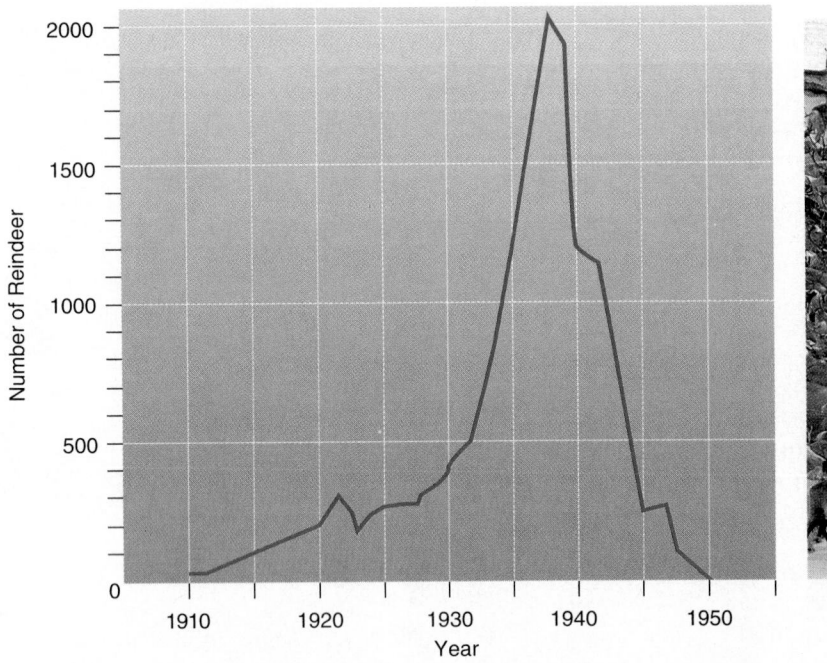

Figure 43-7 Exceeding the carrying capacity can damage the ecosystem, reducing its ability to support the population. In 1911, 25 reindeer were introduced onto one of the Pribilof Islands (St. Paul) in the Bering Sea off Alaska. No predators of reindeer were found on the island. The herd grew exponentially (note the initial J shape) until it reached 2000 reindeer in 1938. At this point, the small island was seriously overgrazed, and the population declined precipitously. By 1950, only eight reindeer survived.

and annual plant populations are limited in size by the number of individuals that can be produced before the first hard freeze. Weather is largely responsible for the boom-and-bust population curves described above. Such populations typically do not reach carrying capacity, since density-independent factors intervene first. Human activities can also limit the growth of natural populations, in ways that are independent of population density. Pesticides and pollutants may cause drastic declines in natural populations, as does habitat destruction caused by farms, roads, and housing developments.

Density-Dependent Limits to Growth

Organisms that live several years have evolved various mechanisms to compensate for seasonal weather changes, thus circumventing this form of density-independent population control. Many mammals, for example, develop thick coats and store fat for the winter; some also hibernate. Other animals, including many birds, migrate long distances to find food and a hospitable climate. Plants may survive the rigors of winter by entering a period of dormancy, drop-

ping their leaves and drastically slowing their metabolic activities.

By far the most important elements of environmental resistance for these long-lived species are density-dependent factors: those that become increasingly effective as population density increases, thus exerting a negative feedback effect on population size. Density-dependent factors include community interactions such as **predation** and **parasitism** as well as **competition** within the species or with members of other species. These are covered further in Chapter 44.

Predation and Prey Populations

Predators are organisms that kill and eat other organisms. This broad definition includes the swift and violent capture of elk by wolves (Fig. 43-8) as well as the more prosaic munching by a moth on a cactus (Fig. 43-9). Predation controls prey populations in a density-dependent manner. It becomes an increasingly important factor in population control as prey populations increase, simply because the more dense the prey, the more often they are encountered by

PLANET WATCH
Growth Without Resistance—The Problem of Introduced Species

Exotics, or species introduced into ecosystems where they did not evolve, may show rapid population growth in the absence of normal environmental resistance. Exotics may find no predators or parasites in their new environment, and their new prey may have little defense against them. The unchecked population growth of these invaders may seriously damage the ecosystem as they displace, outcompete, and prey on native species. Both starlings and English sparrows have spread dramatically since their deliberate introduction to the eastern United States in the 1890s (Fig. E43-3). Their success threatens the bluebird and other native American birds. A Japanese vine called kudzu was introduced to the South as an ornamental in 1876, and planted extensively in the 1940s to control erosion. Today kudzu is a major pest, overgrowing and killing trees and underbrush, and engulfing small houses while their owners are vacationing (Fig. E43-4). The water hyacinth, introduced from South America, now clogs about 2 million acres of lakes and waterways, impeding boat traffic and displacing natural vegetation (Fig. E43-5).

Figure E43-3 Massive flocks of introduced European starlings are a nuisance to homeowners, and are outcompeting native species.

Figure E43-4 The Japanese vine kudzu will rapidly cover entire trees and houses.

Figure E43-5 The beauty that became a beast—water hyacinths, originally from South America, today infest waterways in our southern states.

Figure 43-8 Grey wolves have brought down an elk, who may have been weakened by old age or parasites. Predators will often switch to the most abundant prey, limiting prey populations.

(a)

(b)

Figure 43-9 (a) A pasture in Queensland, Australia, is blanketed by the imported prickly pear cactus, whose spread was unchecked by predators. **(b)** The same site 3 years after the introduction of a predatory cactus moth appropriately named *Cactoblastis cactorum*. (Courtesy of the Department of Lands, Queensland, Australia.)

predators. Many predators will eat a variety of prey depending on what is most abundant and easiest to find. Coyotes may eat more mice when the mouse population is high, but switch to eating more ground squirrels as the mouse population declines, incidentally allowing the mouse population to recover.

Predators may also exert density-dependent effects by increasing in number as their prey population grows. For instance, predators that feed heavily on lemmings, such as the arctic fox and snowy owl, regulate the number of offspring they produce according to the abundance of lemmings. The snowy owl may produce up to 13 chicks when lemmings are abundant, but not reproduce at all in years when lemmings are scarce. In other cases, the increase in predators may cause a crash of the prey population, which is then followed by a decline in the predator population. This can result in out-of-phase population cycles of both predators and prey (see "A Closer Look at Cycles in Predator and Prey Populations").

The influence of predators on prey populations varies considerably. Some predators feed primarily on prey made vulnerable because their populations have exceeded carrying capacity. Such prey may be weakened by lack of food, or may be exposed owing to lack of appropriate shelter. In such cases, predation may maintain prey populations near their optimal density.

In other cases, predators may maintain their prey at well below carrying capacity. A dramatic example of this is the prickly pear cactus, which was introduced into Australia. Lacking natural predators, it spread uncontrollably, destroying millions of acres of valuable pasture and range land. Finally, in the 1920s, a cactus moth (a predator of the prickly pear) was imported from Argentina and released to feed on the cacti. Within a few years, the cacti were virtually eliminated. Today the moth continues to maintain its prey at very low population densities (see Fig. 43-9).

Parasitism and Host Populations

In contrast to predators, **parasites** live on or in their prey, also called the host. Although they weaken their hosts, most parasites do not kill them outright. However, parasites may dramatically increase the death rate in a population where individuals are weakened by the effects of overcrowding. Further,

infestation by parasites makes organisms more vulnerable to predators. Parasitism is density-dependent. Most parasites have limited motility and spread more readily between hosts at high population densities, as illustrated by the spread of the pine bark beetle within densely growing pine forests (Fig. 43-10). Another example is the bubonic plague, a fatal bacterial disease that devastated the population of Europe during the fourteenth century (see Fig. 43-15). The spread of this bacterial parasite by infected rat fleas, and by direct contact with infected humans, was greatly facilitated by the dense human and rat populations of European cities.

Although enemies to their victims, both predators and parasites can have beneficial effects on the prey population as a whole. As you will learn in the next chapter, predators, parasites, and their prey coevolve. Parasites and predators destroy the least fit of the prey, leaving the better-adapted prey to reproduce. The result is usually a balance in which the prey population is regulated but not eliminated.

However, when a predator or parasite is introduced into an area where it did not evolve, the local prey species are often not adapted for it, and the consequences can be disastrous. The conquest of much of the globe by Europeans can be traced partly to the

(a)   **(b)**

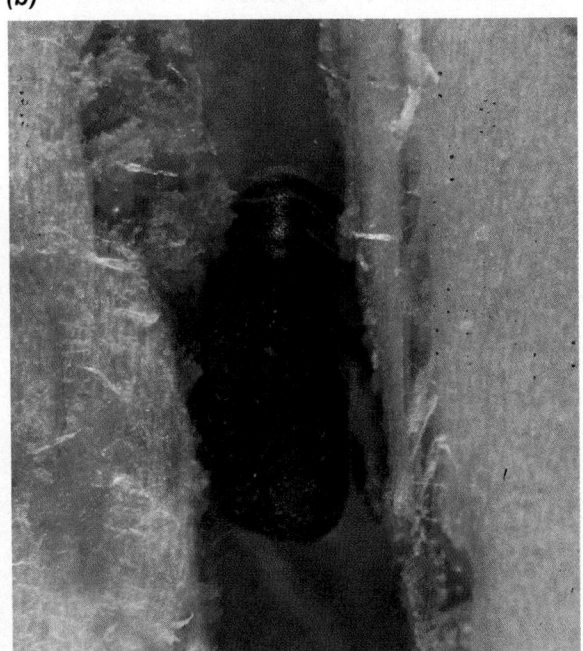

Figure 43-10 (a) A forest of lodgepole pines devastated by the parasitic pine bark beetle. After clearcutting, forests tend to regrow in dense stands of same-age trees. Trees in the uniform, crowded forests are more likely to die of parasitic infestations, since they are weakened by competition for water and nutrients. Thus, parasites help regulate populations in a density-dependent manner. **(b)** Pine bark beetles are seen tunnelling destructively beneath the bark of a lodgepole pine. Eggs are visible along the edges of the tunnel.

disease organisms they carried with them as they invaded continents whose populations had no previous exposure or resistance to these pathogens. Smallpox, for example, imported by Europeans, ravaged the native population of Hawaii, the Amerindians of Argentina, and the Aborigines of Australia. Equally destructive to natural ecosystems is the case of species introduced into new areas where they have no natural predators or parasites (see Planet Watch: Growth Without Resistance—The Problem of Introduced Species).

Competition and Population Size

Because the resources that determine carrying capacity (space, nutrients, light) are limited, use by one individual limits their availability to another. For this reason, **competition** limits population size in a density-dependent manner. There are two major forms of competition: **interspecific competition** (between members of different species; see Chapter 44), and **intraspecific competition** (between members of the same species). Because the needs of members of the same species for water and nutrients, shelter, breeding sites, light, and other resources are almost identical, intraspecific competition is more intense. Organisms have evolved several ways to deal with this. Some, including most plants and many insects, engage in **scramble competition,** a kind of free-for-all with resources as the prize. For example, when a plant disperses its seeds in a small area, hundreds

may germinate. However, as they grow, the larger ones begin to shade the smaller, those with the most extensive roots absorb most of the water, and the weaker individuals eventually wither and die.

Many animals (and even a few plants) have evolved **contest competition,** which helps regulate population size and reduce direct competition. Contest competition consists of social or chemical interactions used to limit access to important resources, including mates (Fig. 43-11; for further discussion, see Chapter 42). Territorial species—such as wolves, many fish, rabbits, and songbirds—defend an area containing important resources such as food or nesting sites. When the population begins to exceed the available resources, only the best-adapted individuals are able to defend adequate territories. Those without territories often do not reproduce and are also easy prey. The creosote bush, which secretes a chemical into the ground that prevents germination of seeds nearby, could be considered a territorial plant (see Fig. 43-13b).

Dominance hierarchies, or "pecking orders," are another form of contest competition observed in many social animals. High-ranking individuals have first access to food, breeding sites, and mates. When resources are limited, only dominant individuals obtain what they need to reproduce successfully. Thus, dominance hierarchies can limit population size.

As population densities increase and competition becomes more intense, some animals react by emi-

Figure 43-11 Contest competition is illustrated by this elephant seal, whose gurgling bellow warns other males away from his "harem" of females.

Figure 43-12 Emigration in response to overcrowding reduces local populations. Locust swarms provide a dramatic example.

grating. Large numbers leave their homes to colonize new areas, and many, sometimes most, die in the quest. The massive movements of lemmings, which sometimes end in suicidal marches into the sea, may be attempts to migrate in response to overcrowding. Migrating swarms of locusts plague the African continent, stripping all vegetation in their path (Fig. 43-12).

In laboratory studies, overcrowding of small mammals—such as mice, rats, and voles—causes **social stress.** This results in a decrease in the size of reproductive organs, reduced reproductive rate, slower growth, reduced resistance to disease, and cannibalism of young, all of which would reduce population size. In the wild, however, population densities rarely reach the levels achieved in the laboratory, so these findings may have limited applicability to natural populations.

The size of any population is a result of complex interactions between density-independent and density-dependent forms of environmental resistance. For example, a stand of pines weakened by drought (a density-independent factor) may more readily fall victim to the pine bark beetle (density-dependent). Likewise, a caribou weakened by parasites (density-dependent) is more likely to be killed by an exceptionally cold winter (density-independent).

Distribution of Populations

Organisms may live in flocks, herds, pairs, or as solitary individuals, or they may cluster around re-

sources such as water holes. Distribution may vary with time, changing with the breeding season, for example. Ecologists recognize three major types of spatial distribution: **aggregated, uniform,** and **random** (Fig. 43-13).

Aggregated Distribution

The most common pattern of distribution is aggregated, in which members of the population tend to live in groups. Many species form family or social groupings, such as elephant herds, wolf packs, prides of lions, flocks of birds, or schools of fish (Fig. 43-13a). What are the advantages of aggregation? Flocks provide many eyes on the lookout for localized food, such as a tree full of fruit. Schooling fish avoid predation by confusing the predator with myriad flashing bodies darting in all directions. Some species form temporary aggregations for mating. Other plant or animal populations cluster, not for social reasons, but because resources such as nutrients, shelter, or water are localized. Cottonwood trees, for example, grow along streams and rivers in grasslands. Animals also aggregate around water holes in the dry savanna of Africa.

Uniform Distribution

Uniformly distributed organisms maintain a relatively constant distance between individuals. Spacing is often the result of defending scarce resources, such as breeding sites, nutrients, or water. This distribution occurs most frequently among animals that defend territories. Male Galápagos iguanas establish evenly spaced breeding territories. Shorebirds are also often found in evenly spaced nests, just out of reach of one another. Other territorial species, such as the tawny owl, mate for life and continuously occupy well-defined, relatively uniformly spaced territories (for breeding animals, the spacing refers to pairs, not individuals). Desert plants growing in poor soil with limited water, such as the creosote bush (Fig. 43-13b), have chemical spacing mechanisms that assure adequate resources for each individual.

Random Distribution

Random distribution is the least common. Here individuals do not form social groups, the resources they need are more or less equally available throughout the area they inhabit, and resources are not scarce enough to require territorial spacing. Trees and other plants in rain forests come close to being randomly distributed (Fig. 43-13c). There are probably no vertebrate species that maintain random distribution throughout the year because they must breed, a behavior that makes social interaction inevitable.

(a)

aggregated

(b)

uniform

(c)

random

Figure 43-13 Population patterns in space.
(a) Aggregated, as illustrated by this gathering of caterpillars. **(b)** Uniform, seen in the spacing of these desert bushes. **(c)** The random distribution of trees in a rain forest.

Survivorship in Populations

Population patterns can be considered from the perspective of time as well as space. Over time, populations show characteristic patterns of deaths or (more optimistically) survivorship. These patterns, called **survivorship curves,** are revealed when the number of individuals of each age is graphed against time. Three different types of survivorship curve, **convex, constant,** and **concave,** are shown in Figure 43-14.

Populations with convex survivorship curves have relatively low infant mortality, and most individuals survive to old age. This curve is characteristic of humans and many other large animals, such as Dall mountain sheep. These species produce relatively few offspring, which are protected by the parents. Species with constant survivorship curves have an equal chance of dying at any time during their life span. This phenomenon is seen in the American robin, the gull, and laboratory populations of organisms that reproduce asexually, such as hydra and bacteria. The concave curve is characteristic of organisms producing large numbers of offspring that are left to compete on their own. Mortality is very high among the offspring, but those that reach adulthood have a good chance of surviving to old age. Most invertebrates, most plants, and many fish exhibit concave survivorship curves. In some populations of black-tailed deer, 75% of the population dies within the first 10% of its life span, giving this mammalian population a concave curve as well.

The Human Population

Your first child received free schooling and free medical care, and you were offered longer maternity leave and a larger pension if you signed a promise to have no more children. Pregnant with your second child, you face fines equivalent to all the benefits provided to you for the first child, up to a full year's pay. Each night, delegates come to your home to tell you of the harm you are doing to society. Your neighbors shun you, and the collective pressure finally drives you to the hospital and a state-sponsored abortion. A distant future scenario from a pessimistic sci-fi novel? Hardly. Similar policies were instituted in China. With a population of over 1 billion and an increase of over 15 million people in 1983 alone, the government took drastic measures. China's coercive policies have been eased somewhat, and birth rates are rising again. What is China's demographic future? What of the rest of the world? Let's examine human population in light of what we know of exponential growth and carrying capacity.

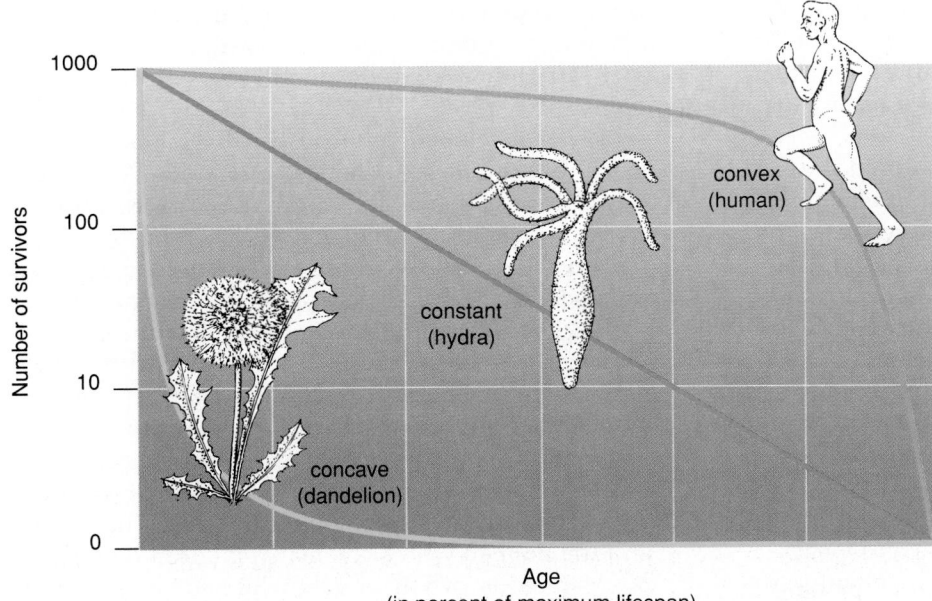

Figure 43-14 Population patterns in time are illustrated by these three types of survivorship curve.

Human Population Growth

Look now at the growth of the human population graphed in Fig. 43-15 and compare it with Fig. 43-2. The time span is different, but our growth curve is exponential. It took over 1 million years for the

human population to reach 1 billion, the second billion was added in 100 years, the third in 30, the fourth in 15, and in 1986, the 5-billionth human joined our planet, only 11 years after the 4 billion mark was reached. The sixth will probably come even faster. **World population currently grows by over 93**

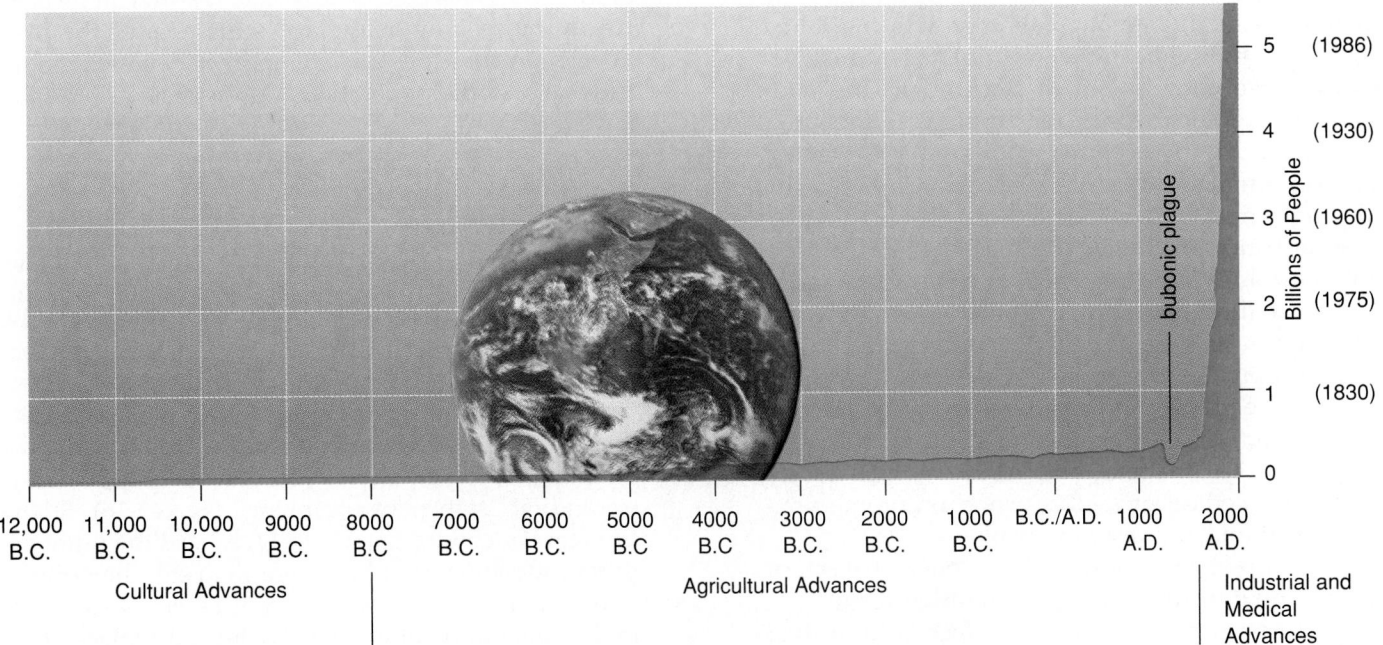

Figure 43-15 The human population from the Stone Age to the present has shown continued exponential growth. Note the dip in the fourteenth century caused by the black plague. Note also the steadily decreasing intervals at which additional billions are added. Inset: The Earth is an island of life in a sea of emptiness; its space and resources are limited.

million yearly, *over a million people added every 4 days.* The number will be higher as you read this. Why hasn't environmental resistance put an end to exponential growth? What is the carrying capacity of the world for humans?

Like all populations, ours has encountered environmental resistance, but unlike other populations, we have responded to resistance by overcoming it rather than reaching a balance with it. As a result, the human population has grown exponentially for an unprecedented time span; to accommodate our growing numbers, we have altered the face of the globe. Human population growth has been spurred by a series of "revolutions" which conquered various aspects of environmental resistance and increased the Earth's carrying capacity for people.

Primitive people produced a **cultural revolution** when they discovered fire, invented tools and weapons, built shelters, and designed protective clothing. Weapons meant an increased food supply; clothing and shelter increased the habitable areas of the globe, and population grew as they increased the Earth's carrying capacity for humans. Starting about 8000 BC, the **agricultural revolution,** in which the raising of domesticated crops and animals supplanted hunting and gathering, provided people with a much more dependable food supply. This further increased the carrying capacity. Increased food resulted in increased longevity and a longer reproductive span, but a high death rate from disease still restricted the rate of growth. Human population growth continued slowly for thousands of years until the **industrial-medical revolution** began in England in the mid-eighteenth century, spreading through Europe and North America in the nineteenth century. Medical advances dramatically decreased the death rate by reducing environmental resistance from disease. These advances included the discovery of bacteria and their role in infection, leading to control of bacterial disease through improved sanitation and the use of antibiotics; and the discovery of viruses, leading to the development of vaccines for diseases such as smallpox. The revolution continues today as research proceeds on vaccines against such major killers as schistosomiasis, malaria, and AIDS, as well as sophisticated medical procedures such as coronary bypass operations and organ transplants.

In developed countries, such as those of western Europe, the industrial-medical revolution resulted in an initial rise in population due to decreased deaths, but a decline in birth rate followed. This decline can be attributed to many factors, including better education, increased availability of contraceptives, a shift to a primarily urban life-style, and more career options for women. In developed countries such as Switzer-land, Sweden, Austria, Germany, and England, populations have more or less stabilized, but these countries are home to only about 4% of the world's population.

In less-developed countries, such as most of those in Central and South America, Africa, and Asia, medical advances have decreased death rates and increased life span, but a major decline in birth rate has not occurred. These countries have not experienced the increase in wealth that was partly responsible for the decline in birth rate of developed countries. Children serve as a form of social security in Third World nations since they may be the only support for parents in their old age. In agricultural societies, children are an important source of labor. In extreme poverty, children may be the parents' major source of pride. Social traditions offer prestige to the man who fathers and the woman who bears many children. For example, in Nigeria, when women who had already given birth to *nine* or more children were asked if they wanted to stop having children, only 16% said yes. Lack of education and lack of contraceptives further impede progress in curbing population growth. Thus, birth rates remain high, while death rates have been lowered dramatically by medical advances. Of the 6.3 billion people projected for the year 2000, 5 billion will reside in less-developed countries. The prospects for population stabilization in the near future are nil, barring major catastrophes that might dramatically increase deaths. The reason can be seen clearly by looking at the age structures of the less-developed countries and comparing them with countries with stable populations, as described in the following section.

Age Structure and Population Growth

Age-structure diagrams graphically illustrate the number of males and females of various ages comprising the population. All age-structure diagrams come to a peak at the top, since relatively few people live into their nineties. The shape of the rest of the diagram, however, shows whether the population is expanding, stable, or shrinking (Fig. 43-16). If the numbers of children (age 0 to 14) exceed the numbers of reproducing individuals (age 15 to 45), the population is growing and the diagram resembles a pyramid. Stable populations have achieved replacement-level fertility, and the number of children is about equal to the number of reproducing adults. In shrinking populations, there are fewer children than reproducing adults and the figure is constricted at the base. Figure 43-17 shows the average age structures of the populations of less-developed and developed countries. The outermost boundaries represent the projected population structure for the year 2025; the inner ones are for 1990. Each graph has been divided into three parts to show individuals who are prereproductive (0 to 15 years old), reproductive (16 to 44 years), and postreproductive (45 and older). In 1990, the less-developed countries (Asia, Africa, India, and South and Central America) had an average annual growth rate of 2.1%, and the developed countries (United States, Europe, USSR) showed an average annual growth rate of 0.5%. In the developed countries, projections for the year 2025 are only slightly higher than present levels (see Fig. 43-17). In contrast, each year in the less-developed countries, increasing numbers of people enter their reproduc-

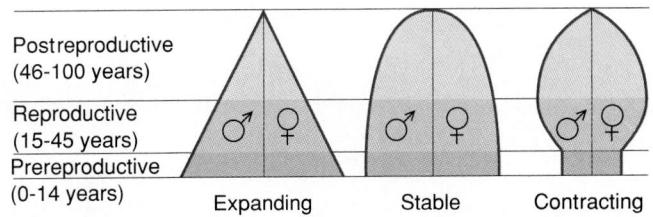

Figure 43-16 Idealized age-structure diagrams for expanding, stable, and contracting populations. The width of each figure is proportional to the number of individuals, with the male population on the left and the female population on the right. The vertical axis shows increasing age. The number of children (prereproductive, age 0 to 15 years) produced by the reproductive-age population determines whether the population is expanding (more children than reproductive-age adults), stable (children equal to reproductive-age adults), or contracting (fewer children than reproductive-age adults).

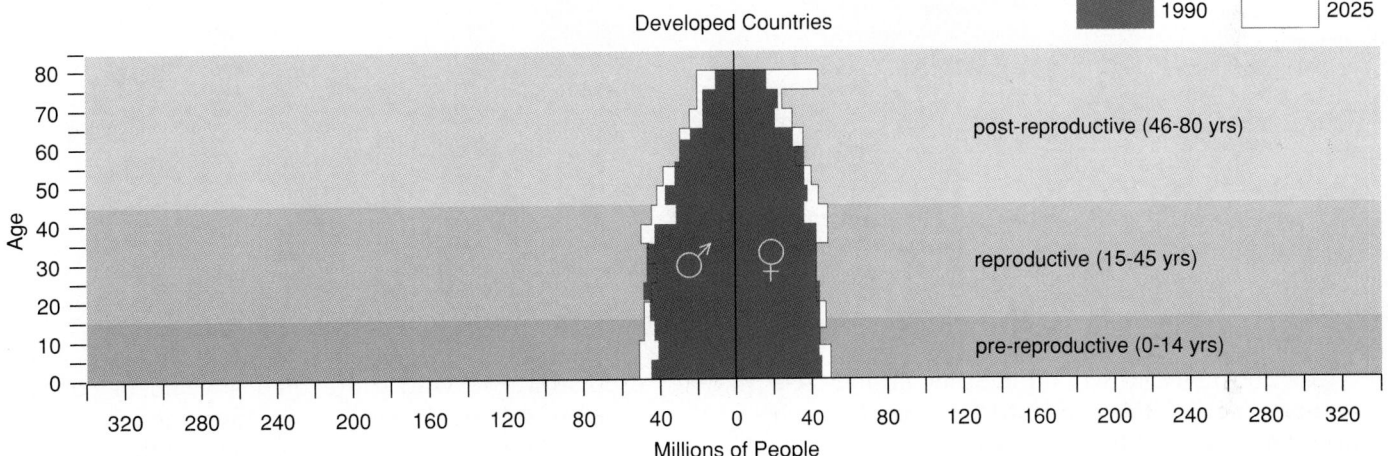

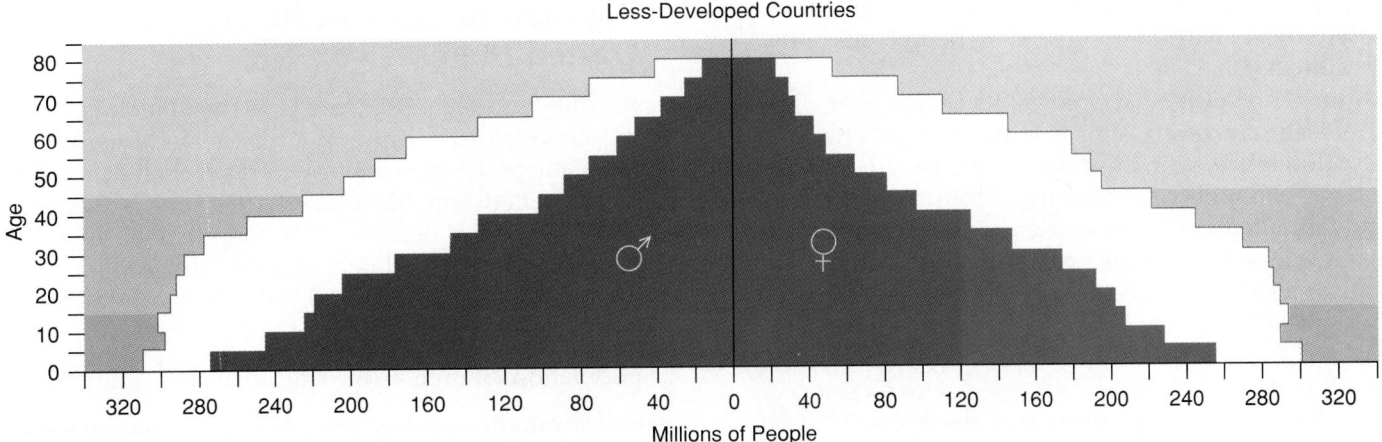

Figure 43-17 Actual age-structure diagrams for developed countries (top) and less-developed countries (bottom). Compare these with the idealized diagrams shown in Figure 43-16. The outer lines are projections to the year 2025. Notice how similar the current and projected age structures are. This suggests that rapid population growth will continue in the less-developed countries for the foreseeable future.

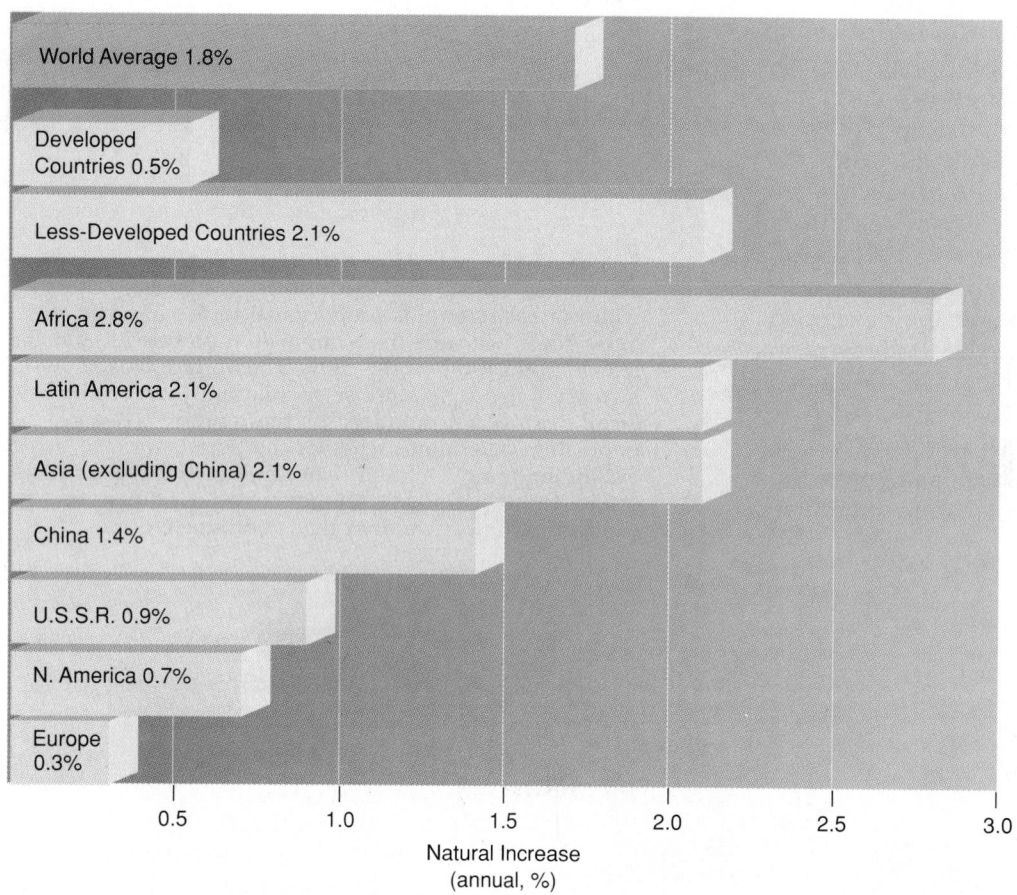

Figure 43-18 The percent population growth caused by natural increase (the excess of births over deaths) per year for various regions of the world. These figures do not include immigration or emigration.

tive years and give birth to an ever-increasing base of infants. Even if these countries were to reach **replacement-level fertility** immediately (i.e., if people of reproductive age had only sufficient children to replace themselves), population growth would continue for decades because of built-in momentum caused by increasing numbers of people reaching reproductive age.

Notice, however, that predictions for the year 2025 show a structure very similar to the present one, the only difference being far more people in all age classes. Although several of these countries have recently initiated family-planning programs, parents in less-developed countries will probably continue to have larger than replacement-level families. The percent growth per year caused by natural increase (births − deaths) for various parts of the world is shown in Figure 43-18.

Ironically, population growth in these countries is helping to perpetuate the poverty and ignorance that in turn tends to sustain high birth rates. The relationship of income and education to birth rate has been documented in the United States. Here, women who do not complete high school have twice as many chil-

dren as those with more than 4 years of college. Women whose family income averages less than $10,000 have twice as many children as those whose income is $35,000 and higher.

Population Growth in the United States

As shown in Fig. 43-19, the U.S. population is growing exponentially. In fact, at over 1% annually, we have one of the highest growth rates of all developed nations. Between November, 1989 and November, 1990, for example, the U.S. population grew by 2,679,000 (nearly 300 people per hour), bringing the total U.S. population to 250,969,000. Recall the equation:

population change = (births − deaths) + migration

Let's examine each component of the equation to determine why the United States is growing so rapidly.

If each American woman had 2.1 children, we would have replacement-level fertility (RLF). RLF is slightly higher than 2 since the parents must replace both themselves and the children who die before

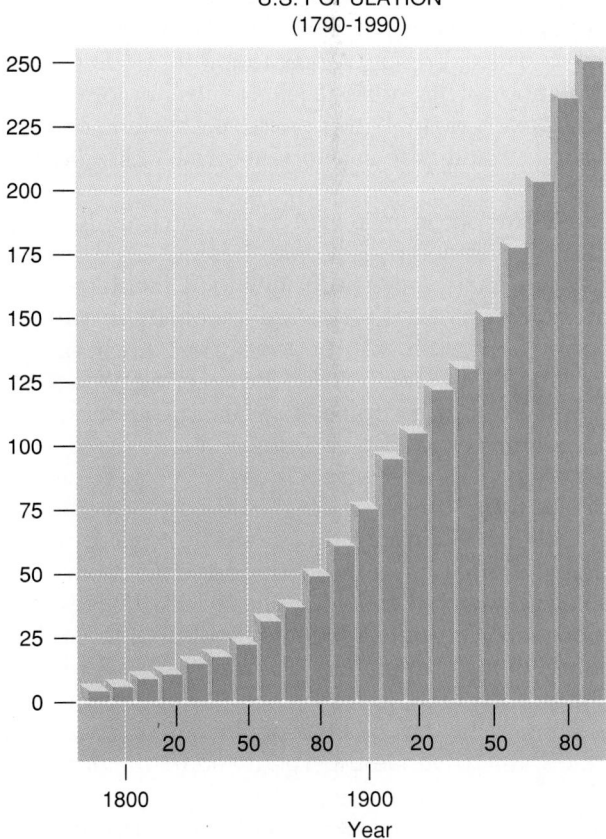

Figure 43-19 Population growth in the United States since 1790 shows the J-shaped curve characteristic of an exponentially growing population.

measure. Even using conservative estimates, legal and illegal immigration to the United States is responsible for 30% to 40% of our annual population growth.

According to Census Bureau projections, the U.S. population will not stabilize for decades. Figure 43-20 shows three scenarios, each based on the assumption of continued legal immigration at a level of about 450,000 per year, but varying fertility rates. Even if fertility drops immediately to an average of 1.6 births per woman, the population will continue to grow for the next 50 years to 275 million before beginning a gradual decline. At 1.9 births per woman (a birth rate that is still below RLF), the population would stabilize in 50 years at about 315 million, sustained by the steady influx of immigrants. But what about the world? When and how will human numbers ever stabilize? How many people can the Earth support?

World Population and Carrying Capacity

A glance at the age structure of developing countries where most of the world's population resides shows a tremendous momentum for continued growth. World population in the year 2000 is predicted to be 6.3 billion, and the population will then be growing at a rate of 100 million annually. A medium-level United Nations (UN) projection is that the human population may stabilize in the year 2110 at 10.5 billion. Can the Earth support over twice its current population?

reaching maturity. The U.S. birth rate is actually about 2.0 children per woman, slightly *below* RLF. Why then do we continue to grow? Two factors are contributing to the rapid growth of the U.S. population: immigration and the past "baby boom."

Part of our current growth rate is a legacy of our recent past. Parents of the late 1940s through the 1960s had larger-than-replacement-level families, resulting in a "baby boom," and a momentum in population growth that has not yet subsided. For example, in 1980 over a million more women entered their reproductive years than did in 1970. Even though these women are averaging 2.0 children each, because there are more women having children, our population swells.

A second crucial component of the U.S. population equation is immigration, which contributes more to population growth here than in any other nation in the world. Legal immigration averages about 500,000 per year, causing about 20% of our total growth. Illegal immigration by its very nature is impossible to

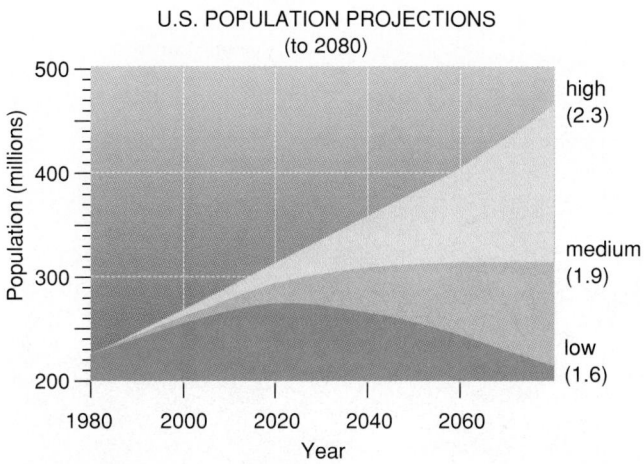

Figure 43-20 Three population scenarios for the United States based on differing fertility rates and immigration of 450,000 persons per year. (Redrawn from Westoff, C. F. "Fertility in the United States." *Science* 234:554–559 (1986).)

Earlier we defined carrying capacity as the maximum population that could be indefinitely sustained. This requires that the ecosystem not be damaged in ways that lower its ability to provide necessary resources. By this definition we may have already exceeded the Earth's carrying capacity for people. Each year it is estimated that an area of once-productive land the size of Maine is being turned into desert through overgrazing and deforestation, especially in less-developed countries. The Earth's desert area is projected to increase by 20% by the year 2000 as a result of human activities. In a world where between 500 million and 1 billion people are chronically undernourished, a UN survey estimated that 20% of the world's farmland is being degraded through erosion and mismanagement (Fig. 43-21). The UN's Food and Agricultural Organization reports that Africa's food production per capita has dropped by 20% since 1960, and predicts another 30% drop in the next 25 years. Each year the United States is losing 1 million acres of farmland to urban sprawl. World wood production per capita has declined by 9% or more since peaking in the mid-1960s. The demand for wood in underdeveloped countries is far outstripping production, and forested areas are being reduced by an area the size of Cuba each year. This in turn causes erosion of precious topsoil, runoff of much-needed fresh water, and the spread of deserts. The fish harvest per capita has fallen by 13% or more since 1970, even though expenditures on fishing fleets and fish farming have steadily increased. The destruction of tropical rain forests may exterminate 1 million species of animals and plants by the year 2000, due to the loss of their habitat. These are clear indications that our present population, at its present level of technology, is already "overgrazing" the world ecosystem and decreasing its ability to support all forms of life, including people.

Mass starvation in recent years, such as the 1985 and 1987 famines in Ethiopia, has occurred not because the Earth lacks sufficient food for its population, but because resources are unequally distributed among, and even within, countries. Each natural population must live within the carrying capacity of the ecosystem it occupies. In contrast, human populations often expand well beyond the ability of their local ecosystem to support them, relying on wealth derived from other sources (such as minerals) to allow them to import the food they need. When human populations exceed the carrying capacity of the soil on which they live, and also lack the wealth to import and distribute adequate food, disaster strikes. A burgeoning population, coerced by poverty and ignorance into using destructive farming techniques that damage the land, is trapped in a downward economic spiral driven and perpetuated by the demands of population growth.

In estimating how many people the Earth can support, we must keep in mind that humans desire more out of existence than a minimum caloric intake each day. It has been estimated that for everyone on Earth to live as Americans do, world population would

(a)

(b)

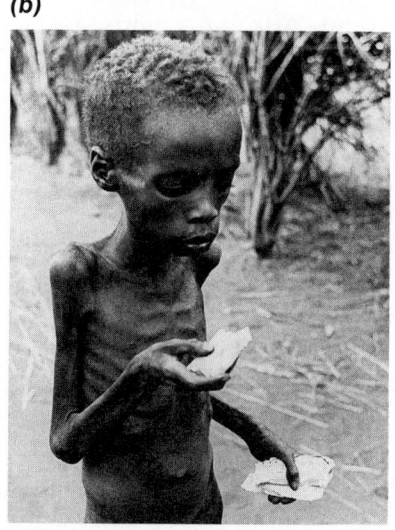

Figure 43-21 (a) Desertification. Human activities, including overgrazing livestock, deforestation, and poor agricultural practices, convert once-productive land into barren desert. **(b)** The loss of productive land, when combined with an expanding human population, can lead to tragedy.

have to be *reduced* to 500 million, *about one tenth the present population.* For all to achieve a standard of living similar to that in Europe, where the life-style is less extravagant and untouched wilderness is almost nonexistent, *the present population would have to be reduced by half.* Merely "redistributing the wealth" is obviously not an adequate answer.

There are some who predict that technological advances will continue to increase the Earth's carrying capacity for humans into the indefinite future. In evaluating this forecast, keep in mind that technology takes time, wealth, and education to develop and implement. Rapid population growth in less-developed countries increases poverty, overloads the educational system, and hampers technological development, while simultaneously making its need more urgent. Ironically, the countries that so desperately need these predicted technological advances are often unable even to take advantage of today's level of technology. The high-yield crops developed in the 1960s (the so-called "Green Revolution") have helped considerably, but their impact has been lessened be-

cause they require expensive equipment, fertilizer, pesticides, and irrigation water to cultivate. Less-developed countries often cannot afford these, and donated farm equipment has been left to rust in the fields for lack of fuel as well as the parts and know-how to maintain it.

Hope for the future lies in using the intelligence that has allowed us to overcome environmental resistance to see signs of "overgrazing" and act before we have irrevocably damaged our world ecosystem. Human population *will* stop its exponential growth. Either we will voluntarily reduce our birth rate, or various forces of environmental resistance will increase our death rate; the choice is ours. Facing the problem of how to limit births is politically and emotionally difficult, but continued failure to do so will be disastrous. **Our dignity and intelligence, and our role as self-appointed stewards of life on Earth, demand that we make the decision to halt population growth ourselves before we have irrevocably decreased the Earth's ability to support all life, including our own.**

SUMMARY OF KEY CONCEPTS

Introduction to Ecology
Ecology is the study of the interrelationships between organisms and their environment. Ecology may be approached at the level of the population, the community with its complex interactions, or an entire ecosystem, including communities and their abiotic environment.

Population Growth
Individuals join populations through births or immigration and leave through death or emigration. Thus

population change = (births − deaths) + (immigrants − emigrants).

The ultimate size of a stable population is the result of interactions between biotic potential (the maximum possible growth rate) and environmental resistance, which limits population growth.

Biotic Potential
All organisms have the biotic potential to more than replace themselves over their lifetime, resulting in population growth. Populations tend to grow exponentially, since in a growing population, the number of reproducing individuals is constantly increasing. Ignoring migration, population growth rate (r) can be expressed by the equation:

$$r = b - d$$

where b is the birth rate and d is the death rate. The actual number of individuals added to the population is $r \times N$, where N is the population size. During exponential growth, increasing numbers of individuals are added during each successive period. Factors influencing the maximum rate of growth of a given population include the age at first reproduction, frequency of reproduction, number of offspring produced each time, and the death rate under ideal conditions. Populations cannot continue to grow exponentially indefinitely; they either stabilize or undergo periodic oscillations as a result of environmental resistance.

Environmental Resistance
Environmental resistance restrains population growth by increasing the death rate or reducing the reproductive rate. The maximum size at which a population may be sustained indefinitely by an ecosystem is called the carrying capacity, determined by limited resources such as space, nutrients, and light.

Populations are maintained at or below carrying capacity by density-independent forms or environmental resistance such as weather, and density-dependent forms of resistance that include predation, parasitism, and competition.

Predators often eat the most abundant prey, and may switch prey based on the size of prey population. When predators concentrate on a single prey, both predator and prey populations may cycle. Parasites spread most rapidly in dense populations, and can increase death rates when animals are stressed due to overcrowding. Interspecific competition limits both the size and distribution of populations. Intraspecific competition may be resolved directly through scramble competition, or indirectly through chemical interactions or social behaviors, collectively called contest competition. Emigration is another possible response.

Population Distribution and Survivorship

Populations are spatially distributed and can be classified as aggregated, uniformly distributed, or randomly distributed. Aggregations may occur for social reasons or around limited resources. Uniform distribution is usually the result of territorial spacing. Random distribution is rare, occurring only when individuals do not interact socially and when resources are abundant and evenly distributed.

Populations show specific survivorship curves that express the likelihood of survival at a given age. Convex curves are characteristic of long-lived species with few offspring that receive parental care. Species with constant curves have an equal chance of dying at any age. Concave curves are typical of organisms that produce numerous offspring, most of which die.

The Human Population

The human population has exhibited exponential growth for an unprecedented time by overcoming certain aspects of environmental resistance and increasing the Earth's carrying capacity for people. This has been accomplished by the use of tools, agriculture, industry, and medical advances.

Age-structure diagrams depict numbers of males and females in various age groups comprising a population. Expanding populations have pyramidal age structures, stable populations show rather straight-sided figures, while decreasing populations are illustrated by figures that are constricted at the base.

Today, most of the world's people live in less-developed countries with rapidly expanding populations, where a variety of social and cultural conditions encourage large families. The U.S. is the fastest growing of the developed countries, due to high immigration rates and the postwar "baby boom."

The carrying capacity of the Earth for humans is unknown, but with a population of over 5 billion, resources are already too limited for all to be supported at a high living standard. A steady decline in productive land, wood, and fish harvests suggests that we are already damaging our world ecosystem and decreasing its ability to sustain us.

GLOSSARY

abiotic (ā-bī-ah'-tik): nonliving.

age structure: the distribution of males and females in a population according to age categories; often represented graphically.

aggregated distribution: characteristic distribution of populations in which individuals are clustered into groups. These may be social or based on the need for a localized resource.

biotic (bī-ah'-tik): living.

biotic potential: the most rapid potential growth rate of a population, assuming a maximum birth rate and minimum death rate.

boom-and-bust cycle: a population cycle characterized by rapid exponential growth followed by a sudden major decline in population size, seen in seasonal species and some populations of small rodents, such as lemmings.

carrying capacity: the maximum population size that an ecosystem can maintain on a sustained basis. Determined primarily by the availability of space, nutrients, water, and light.

community: all the interacting populations within an ecosystem.

competition: interaction that occurs between individuals when both attempt to utilize a resource (e.g., food or space) that is limited relative to the demand for it.

contest competition: a mechanism for resolving intraspecific competition using social or chemical interactions.

density dependent: description of any factor that limits population size more effectively as the population density increases.

density independent: description of any factor such as freezing weather that limits a sensitive population without regard to its size.

dominance hierarchy: a system of social ranks within a population that determines which individuals obtain first access to limited resources.

ecology (ē-kol'-uh-gē): the study of the interrelationships of organisms with each other and with their nonliving environment.

ecosystem (ē'kō-sis-tem): all the organisms and their nonliving environment within a defined area.

emigration (em-uh-grā'shun): movement of individuals out of an area.

environmental resistance: any factor that tends to counteract biotic potential, limiting population size.

exponential growth: a continuously accelerating increase in population size.

immigration (im-uh-grā'-shun): movement of individuals into an area.

interspecific competition: competition between individuals of different species.

intraspecific competition: competition between individuals of the same species.

J-curve: the shape of the growth curve of an exponentially growing population, in which increasing numbers of individuals join the population with each succeeding time increment.

parasite (par'-a-sīt): an organism that feeds on the body of a larger organism, called a host, causing it harm.

parasitism (par'-a-sit-ism): the process of feeding on a larger organism without killing it immediately or directly.

population: a group of interbreeding organisms (organisms of the same species) within an ecosystem.

predation (pre-dā'-shun): the act of killing and eating another living organism.

predator: an organism that kills and eats other organisms.

random distribution: spacing in which the probability of finding an individual is equal in all parts of an area.

replacement-level fertility: the average birth rate at which a reproducing population exactly replaces itself during its lifetime.

scramble competition: direct interactions between individuals attempting to acquire the same limited resource.

S-curve: the growth curve that describes a population introduced into a new area. It consists of an initial period when numbers remain relatively low, followed by a period of exponential growth, followed by decreasing growth rate, and finally, relative stability.

survivorship curve: a curve (convex, constant, or concave) resulting when the number of individuals in a population is graphed against their age, usually expressed as a percentage of their maximum life span.

territoriality: a behavior in which individuals defend an area in which important resources are located from others of their species.

uniform distribution: a relatively regular spacing of individuals within a population, often as a result of territorial behavior.

STUDY QUESTIONS

1. Define *biotic potential*. Explain natural selection in terms of biotic potential and environmental resistance.
2. Draw the growth curve of a population before it encounters significant environmental resistance. What is the name of this type of growth, and what is its distinguishing characteristic?
3. Distinguish between density-independent and density-dependent forms of environmental resistance.
4. Describe two major forms of intraspecific competition, and give an example of each.
5. Describe (or draw a graph illustrating) what is likely to happen to a population that far exceeds the carrying capacity of its ecosystem. Explain your answer.
6. List three density-dependent forms of environmental resistance and explain why each is density-dependent.
7. Distinguish between populations showing concave and convex survivorship curves. Which is characteristic of Americans, and why?
8. Given that the U.S. birth rate is currently below replacement-level fertility, why is our population growing?

DISCUSSION QUESTIONS

1. The United States has a long history of accepting large numbers of immigrants. Discuss the implications of immigration for population stabilization.
2. What factors encourage rapid population growth in less-developed countries? What will it take to change this?
3. Contrast age structure in rapidly growing versus stable human populations. Why is there a momentum in population growth built into a rapidly growing population?
4. Discuss the potential consequences of continued exponential growth of the human population.

SUGGESTED READINGS

Brown, L. R., and others. *State of the World.* New York: W. W. Norton & Company, Inc., 1991. Annually updated collection of articles concerning global resources, pollution, and population.

Ehrlich, P. R., and Erlich, A. *Extinction.* New York: Random House, Inc. 1981. The causes and consequences of the disappearance of species.

Hillfry, E. *Ecology 2000. The Changing Face of Earth.* New York: Beaufort Books, Inc., 1980. Engaging reading on the major environmental problems facing the world.

Keyfitz, N. "The Growing Human Population" *Scientific American* 261:118–127 (1989). Discusses recent trends in population growth and the economic and environmental problems associated with continued growth.

Myers, J. H., and Krebs, C. J. "Population Cycles in Rodents." *Scientific American*, June 1974. Natural populations of small rodents show periodic 3- to 4-year cycles in population size.

44

Community Interactions

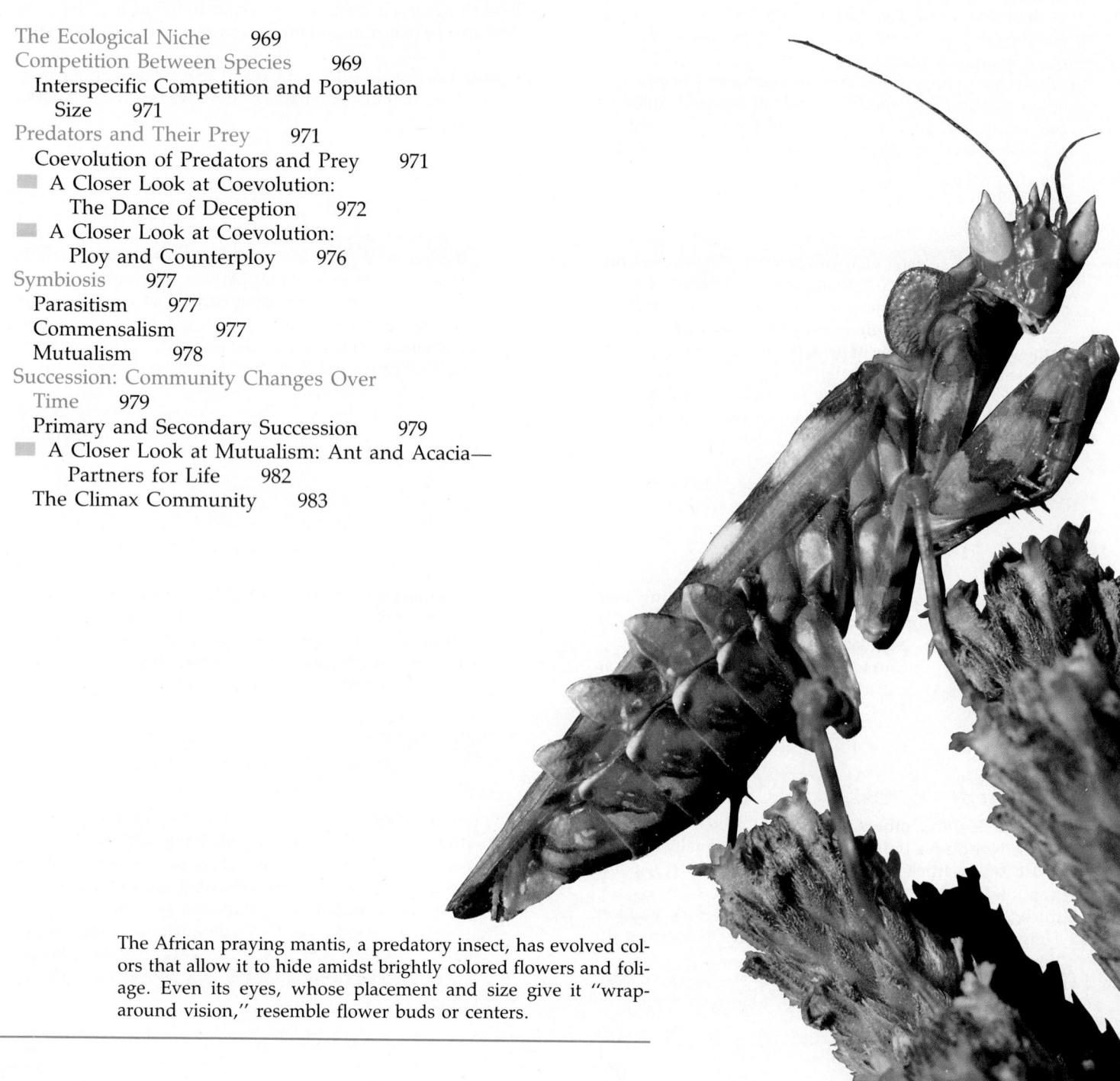

The African praying mantis, a predatory insect, has evolved colors that allow it to hide amidst brightly colored flowers and foliage. Even its eyes, whose placement and size give it "wraparound vision," resemble flower buds or centers.

An ecological **community** consists of all the interacting populations within an ecosystem. Typically, populations within communities have coevolved. During **coevolution,** different species act as agents of natural selection on one another. For example, predators and parasites limit their prey populations without eliminating them. Animals that are preyed upon have evolved elaborate defenses that help them survive. Herbivores have digestive specializations that allow them to eat local plants. The plants in turn grow rapidly or defend themselves by chemical or physical means, keeping one step ahead of their predators.

Community interactions fall into three major categories: **competition,** in which populations compete for limited resources; **predation,** in which one organism kills and eats another; and **symbiosis,** in which two species live together in close association over an extended time. Symbiotic interactions may be further classified according to how they affect each of the species involved (Table 44-1).

Long-established communities are generally complex and self-sustaining, with interacting populations that remain relatively stable. Self-sustaining communities originate, however, over long periods of gradual change. During these periods, one type of community gives way to another in a process called **succession**, until the self-sustaining community is established.

In this chapter we focus on the variety of community interactions in stable ecosystems and the community changes that occur during succession.

The Ecological Niche

Although the word *niche* may call to mind a small cubbyhole, in ecology it means much more. Each species occupies a unique **ecological niche** that encompasses all aspects of its way of life. The concept of the ecological niche includes the organism's physical home or habitat. The habitat of a white-tailed deer, for example, is the eastern deciduous forest. However, the ecological niche goes far beyond habitat; it specifies the organism's food, behavior, and predators, what might be called the organism's role or "occupation" within its ecosystem. In addition, the niche includes all the physical environmental factors necessary for survival, such as the range of temperatures under which the organism can survive, the amount of moisture it requires, the pH of the water or soil it may inhabit, the type of soil nutrients required, and the degree of shade it can tolerate. **Although different types of organisms share many aspects of their niche**

Table 44-1   Interactions Among Organisms

Type of Interaction	Effect on Organism A	Effect on Organism B
Competition between A and B	Harms	Harms
Predation by A on B	Benefits	Harms
Symbiosis		
Parasitism by A on B	Benefits	Harms
Commensalism of A with B	Benefits	No effect
Mutualism between A and B	Benefits	Benefits

with others, no two species ever occupy exactly the same ecological niche.

Competition Between Species

Competition between different species, or **interspecific competition,** is an important form of community interaction in which two or more species attempt to use the same limited resources, particularly food and/or space. The intensity of interspecific competition depends on how similar the requirements of the two species are. In other words, **the degree of competition is proportional to the amount of niche overlap between the competing species.**

Just as no two organisms can occupy exactly the same physical space at the same time, no two species can inhabit the same ecological niche. This important concept, often called the **competitive exclusion principle** was formulated in 1934 by the Russian microbiologist G. F. Gause. If two species with the same niches are placed together and forced to compete for limited resources, inevitably one will outcompete the other, and the less well-adapted of the two will die out. Gause demonstrated this using two different species of the protist *Paramecium*. In laboratory flasks, *Paramecium aurelia* and *P. caudatum* both thrived on bacteria, and fed in the same parts of the flask. Grown separately, each population flourished. But when Gause placed the two species together, one always eliminated or "competitively excluded" the other (Fig. 44-1). Gause then repeated the experiment, replacing *P. caudatum* with another species, *P. bursaria*, which tended to feed at the bottom of the flask. In this case, the two species of *Paramecium* were

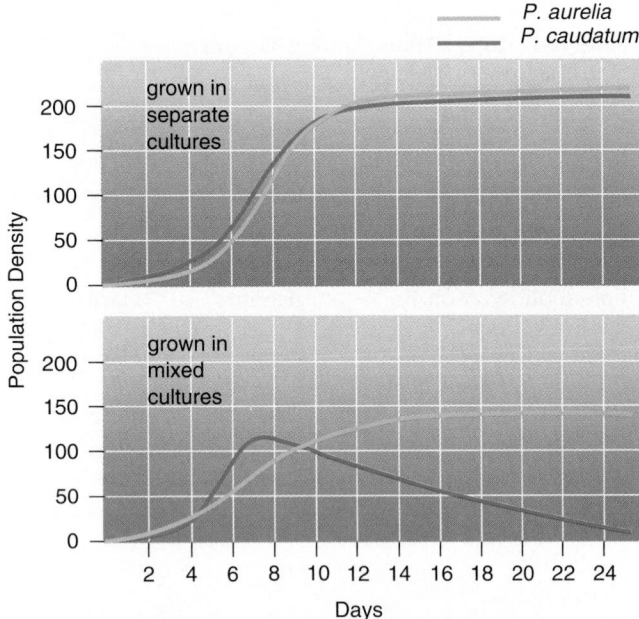

Figure 44-1 Top: Raised separately with a constant food supply, both *Paramecium aurelia* and *P. caudatum* show the S-curve typical of a population that grows, then stabilizes. Bottom: Raised together and forced to occupy the same niche, *P. aurelia* consistently outcompetes *P. caudatum*. (Modified from Gause, G. F. *The Struggle for Existence*. Baltimore: Williams & Wilkins, 1934.)

able to coexist indefinitely, because they occupied slightly different niches.

The ecologist R. MacArthur tested Gause's laboratory findings under natural conditions by investigating five species of North American warbler. These birds all hunt for insects and nest in the same type of spruce tree. Although the niches of these birds appear to overlap considerably, MacArthur found that each species concentrates its search in specific areas of the tree, employs different hunting tactics, and nests at slightly different times. By partitioning the resource, the warblers minimize niche overlap and reduce competition (Fig. 44-2).

As MacArthur discovered, when species with similar requirements coexist, they often occupy a smaller niche than either would if it were by itself. This is called **resource partitioning.** Resource partitioning is the outcome of the coevolution of species with extensive (but not total) niche overlap. **Coevolution** (introduced in Chapter 16) occurs when two or more species act as agents of natural selection on one another. Since natural selection favors individuals with fewer competitors, over evolutionary time the competing species develop physical and behavioral adaptations that minimize their competitive interactions. A dra-

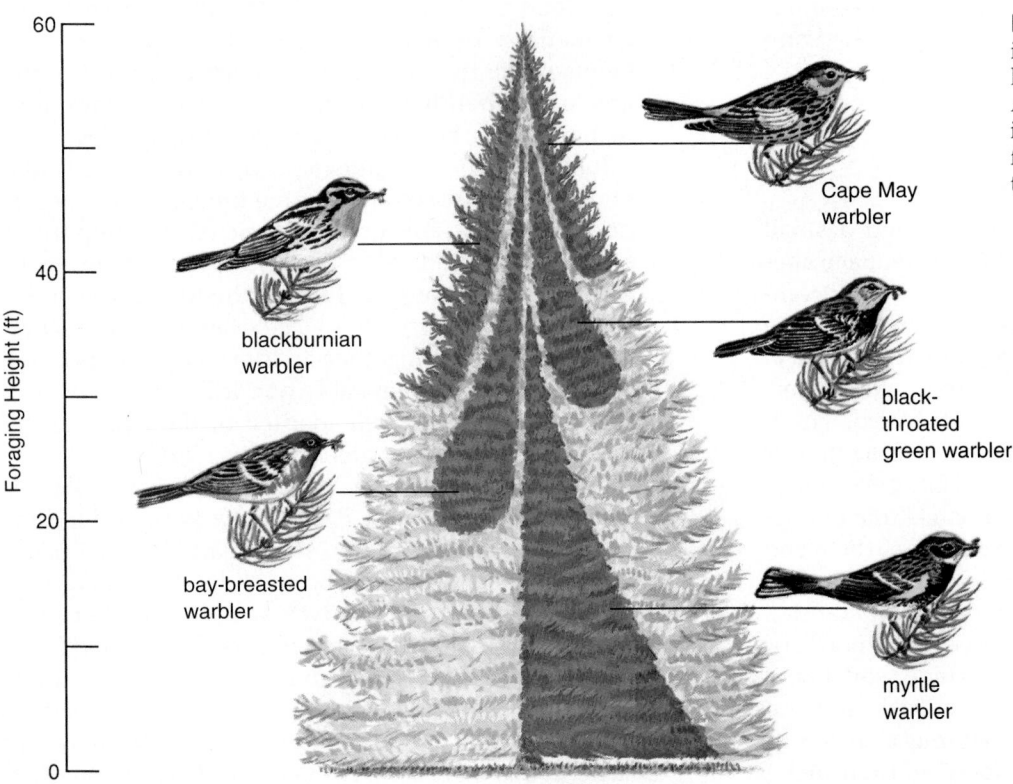

Figure 44-2 Resource partitioning is demonstrated by the feeding habits of five species of North American warblers. Each of these insect-eating species searches for food in different regions of spruce trees.

matic example of this was discovered by Darwin among finches of the Galápagos Islands (see Chapter 16).

Interspecific Competition and Population Size

Although coevolution tends to minimize niche overlap, closely related species still compete directly for limited resources. This may restrict the size and distribution of the competing populations. For example, barnacles of the genus *Chthamalus* share the rocky shores of Scotland with another genus of barnacle, *Balanus*, and their niches overlap considerably. The ecologist J. Connell found that *Chthamalus* dominates the upper shore and *Balanus* dominates the lower. When he scraped off *Balanus*, the *Chthamalus* population increased, spreading downward into the area its competitor had once inhabited. Likewise, removal of *Chthamalus* allowed *Balanus* to spread upward. Where the habitat is appropriate for both species, *Balanus* conquers, since it is larger and faster growing. But *Chthalamus* tolerates drier conditions, so on the upper shore, where only high tides submerge the barnacles, it has a competitive advantage. As this example illustrates, **interspecific competition limits both the size and the distribution of the competing populations.**

Predators and Their Prey

Although predators are most commonly considered to be organisms that kill and eat other organisms, herbivorous animals are often included in this category even though they may kill only part of the plants they consume. For our purposes, then, predators include the antelope munching on sagebrush, the exotic sundew plant entangling and digesting its insect prey, as well as the bat swooping down on the frog and the bear catching a salmon (Fig. 44-3). Predators are usually either larger than their prey or hunt collectively, as wolves do when bringing down a moose. Predators are also less abundant than their prey, as discussed in the next chapter.

Coevolution of Predators and Prey

To survive, predators must feed and prey must avoid becoming food. Therefore, these populations exert intense selective pressure on one another, resulting in coevolution. As prey become more difficult to catch, predators must become more adept at hunting. Coevolution has endowed the cheetah with speed and camouflage spots and its zebra prey with speed

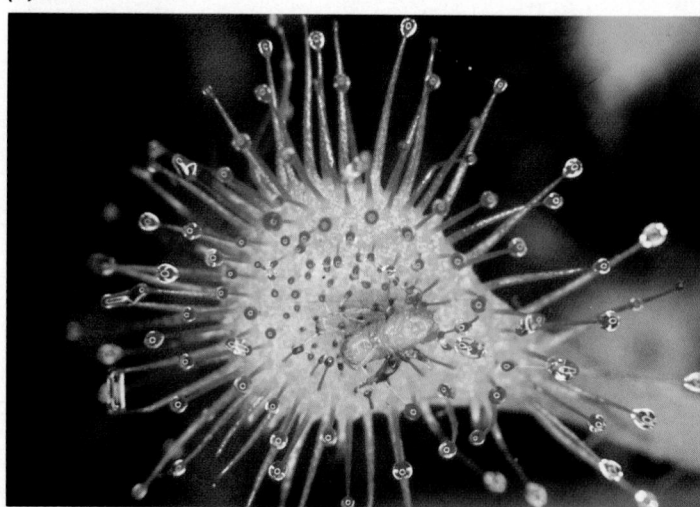

(a)

(b)

Figure 44-3 Forms of predation. **(a)** A predatory plant, the sundew, attracts and binds insects with its glistening sticky knobs. The insects provide supplemental nitrogen for the plant. **(b)** A few tropical bats feed on frogs or fish. Predatory bats home in on the mating calls of this tree frog.

and camouflage stripes. It has produced the keen eyesight of the hawk and the warning call of the prairie dog, the stealth of the jumping spider and the remarkable spider mimicry of the fly it stalks. This complex coevolutionary adaptation is described in "A Closer Look at Coevolution: The Dance of Deception." In the following sections, we examine some of the results of predator–prey coevolution.

Herbivores and Plants

Plants have evolved a variety of adaptations that deter predators. Many, such as the milkweed, syn-

A CLOSER LOOK

At Coevolution:
The Dance of Deception

A jumping spider stalks a snowberry fly, approaching with catlike stealth. Suddenly spotting the spider, the fly spreads its wings, brings them forward slightly, and moves back and forth in a jerky dance. The spider hesitates, then flees, leaving the fly untouched. Careful examination of the wing markings of the fly show an uncanny resemblance to the legs of a spider (Fig. E44-1). Further, the jerky movements of the threatened fly resemble those a spider makes when driving another spider from its territory. Natural selection has finely tuned both the behavior and the appearance of the fly to avoid predation by jumping spiders.

Figure E44-1 In response to the approach of a jumping spider (top), the snowberry fly spreads its wings, revealing a pattern resembling spider legs (bottom). The fly enhances the effect by performing a jerky, side-to-side dance that resembles the leg-waving display of another jumping spider defending its territory.

thesize toxic and distasteful chemicals. Animals rapidly learn not to eat foods that make them sick, so milkweeds and other toxic plants suffer little browsing. Consequently, such plants are often very abundant, and any animal immune to the plant poisons enjoys a bountiful food supply. As plants evolved toxic chemicals for defense, certain insects evolved increasingly efficient ways to detoxify or even make use of the chemicals (see "A Closer Look at Coevolution: Ploy and Counterploy"). The result is that nearly every toxic plant is eaten by at least one type of insect. For example, monarch butterfly caterpillars consume the toxic milkweed. The caterpillars not only tolerate the milkweed poison, they store it in their tissues as a defense against their own predators. The stored toxin is even retained in the metamorphosed monarch butterfly.

Grasses have evolved tough silicon (glassy) substances in their blades, deterring all predators except those with strong, grinding teeth and powerful jaws. Thus, grazing animals have come under selective pressure for longer, harder teeth. An example is the coevolution of horses and the grass they eat. As grasses evolved tougher blades that reduce predation, horses evolved longer teeth with thicker enamel coatings that resist wear.

Bats and Moths

Most bats are nocturnal hunters that navigate and locate prey by echolocation. They emit extremely high-frequency and high-intensity pulses of sound, localizing nearby objects by analyzing the returning echos. In response to this unique prey-location system, certain moths (a favored prey of bats) have evolved simple ears that are particularly sensitive to the frequencies used by echolocating bats. When they hear a bat, these moths take evasive action, flying erratically or dropping to the ground. The bats may counter this defense by switching the frequency of their sonar away from the moth's sensitive range. Further, some moths have evolved a way to jam the bats' sonar by producing their own ultrasonic clicks. In response, when hunting a clicking moth, the bat may counter by turning off its own sonar temporarily and homing in on the moth by following the moth's clicks. These interactions dramatically illustrate the complexity of coevolutionary adaptations.

Camouflage

An old maxim of detective novels is that the best hiding place may be right out in plain sight! Both predators and prey have evolved colors and patterns called **camouflage** that render them inconspicuous even out in the open. Often, camouflaged animals resemble their general surroundings (Fig. 44-4). Some, however, closely resemble specific (but uninteresting) objects such as leaves, twigs, or even bird droppings (Fig. 44-5). For camouflaged animals, behavior is as important as coloration. Camouflaged animals tend to remain motionless rather than fleeing their predators, since, for example, a moving bird dropping would appear quite suspicious. While camouflaged animals often resemble plants (see Fig. 44-5), a few types of plants have evolved to resemble rocks, which are ignored by their herbivorous predators (Fig. 44-6).

Predators who ambush their prey rather than chasing it are also aided by camouflage. For example, the angler fish closely resembles the algae-covered rocks and algae on which it sits motionless, dangling a small lure from its upper lip (Fig. 44-7). While the predator's body is camouflaged, its lure might be considered a type of aggressive mimicry, described below.

Warning Coloration

Some animals have evolved very differently, exhibiting bright **warning coloration** (Fig. 44-8; see Fig. E44-2). These animals are usually distasteful and often extremely poisonous. Since poisoning your predator is small consolation if you have already been eaten, the bright colors declare: "eat me at your own risk." After a single unpleasant experience, predators avoid these conspicuous prey, as illustrated by the toad and bee in Figure 41-14. In some instances, different species of dangerous animals have evolved similar warning coloration, for example, the conspicuous stripes on bees, hornets, and yellow jackets and the bright wing patterns of monarch and viceroy butterflies. A common color pattern results in faster learning by predators, and less predation on all similarly colored species.

Mimicry

Once warning coloration evolved, there arose a selective advantage for tasty animals to resemble poisonous ones. For example, predators rapidly learn to avoid the poisonous orange and black monarch butterfly and which have evolved to closely resemble each other. The deadly coral snake has brilliant warning coloration, and the harmless mountain king snake avoids predation by resembling it (Fig. 44-9). These are examples of **mimicry,** in which two species resemble one another.

Some predators have evolved **aggressive mimicry**. Resembling a harmless animal, predators such as

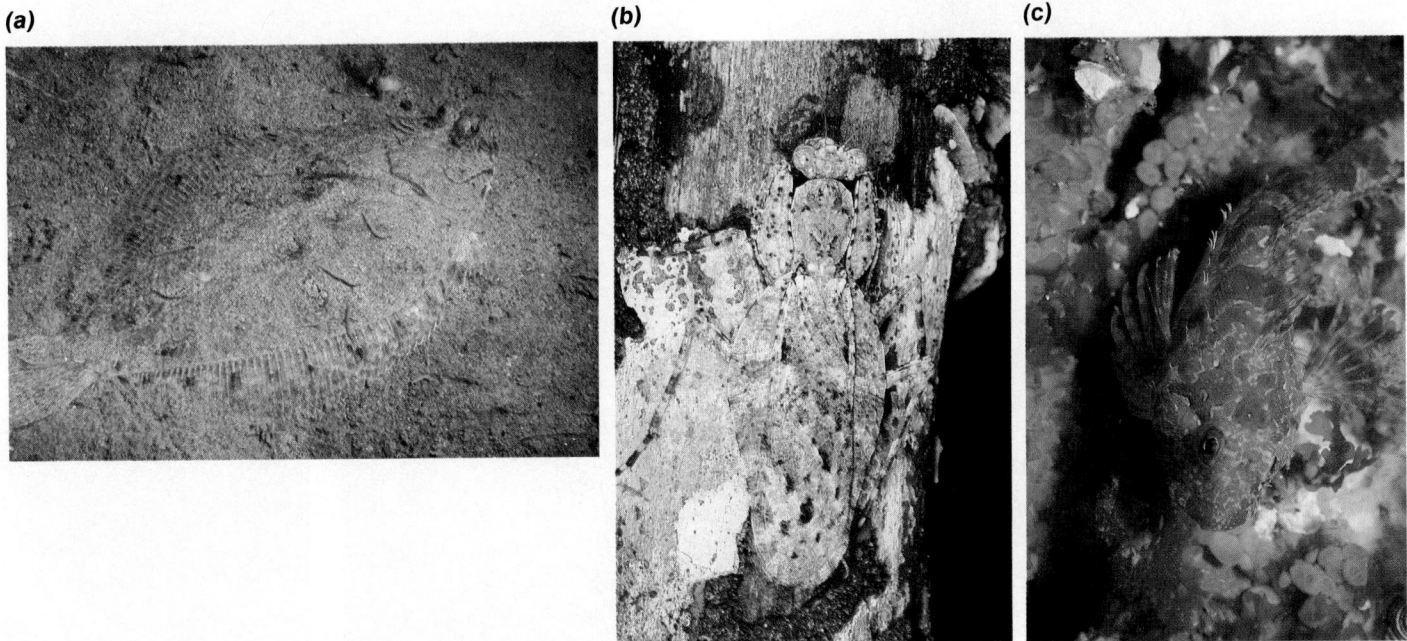

(a)   *(b)*   *(c)*

Figure 44-4 Camouflage renders potential prey inconspicuous. **(a)** The flounder's flat shape and its behavior (flipping sand over its back, then lying perfectly still) help the fish resemble a sandy bottom. The flounder can also alter its color and pattern somewhat to resemble its surroundings. The leeches attached to it are not deceived, since they locate their host by scent. **(b)** The tree bark mantis from Malaysia blends perfectly with the bark of this tree. **(c)** The coralline sculpin, found along the Pacific coast, hides among blotches of pink coralline algae, after which it is named.

(a) *(b)* *(c)*

Figure 44-5 Resembling uninteresting parts of the environment allows some animals to avoid predation. **(a)** A leaf katydid from Costa Rica perfectly resembles a leaf, including an irregularity where the "leaf" has apparently been gnawed. **(b)** The leafy sea dragon (an Australian "seahorse" fish) has evolved extension of its body that precisely duplicates the algae in which it hides. **(c)** A moth on a Missouri mulberry closely resembles a bird dropping.

Figure 44-6 This cactus of the American southwest is appropriately called the "living rock cactus."

Figure 44-7 The angler fish waits in ambush, its body camouflaged to match the algae-encrusted rock on which it normally rests. Above its mouth dangles a lure, closely resembling a small fish. This will attract other would-be predators, who will find themselves prey.

Figure 44-8 Warning coloration. **(a)** The tropical lionfish, with its venomous spines, and **(b)** the south American poison arrow frog, with its poisonous skin, each advertise their unpalatability with bright and contrasting color patterns.

(a) *(b)*

Figure 44-9 Mimicry of the warning coloration of the distasteful monarch butterfly **(a)** by the equally distasteful viceroy. **(b)** After attempting to eat either of these species, a bird will avoid them both. The warning coloration of the poisonous coral snake **(c)** is mimicked by the harmless king snake **(d)**.

the saber-toothed blenny fish that imitates a harmless cleaner wrasse provide real-life examples of the classic "wolf-in-sheep's-clothing" (Fig. 44-10; see also Fig. 44-7).

Certain prey species use still another form of mimicry called **startle coloration**. Several insects, including certain moths and caterpillars, have independently evolved patterns of color that closely resemble the eyes of a much larger, and possibly dangerous, animal (Fig. 44-11). If a predator gets too close, the prey suddenly flashes its eye-spots, startling the predator and making it escape. A sophisticated variation on the theme of prey who mimic predators was recently discovered among several related fly species

who mimic the territorial displays of their spider predators (see "A Closer Look at Coevolution: The Dance of Deception").

Chemical Warfare

Both predators and prey have evolved a variety of toxic chemicals for attack and defense. As mentioned above, many plants produce defensive toxins. Lupines produce alkaloids, which deter attack by the blue butterfly, whose larvae feed on the lupine's buds. In fact, different individuals of the same species of lupine produce different forms of this toxin, thus deterring the butterflies from evolving resistance to it. The venom of spiders and poisonous snakes

A CLOSER LOOK

At Coevolution: Ploy and Counterploy

The intricate evolutionary adaptations of predators and prey are beautifully illustrated by the relationship of the assassin bug and the southwestern desert camphor weed, studied by Thomas Eisner of Cornell and his collaborators. The camphor weed exudes a sticky, noxious resin from its leaves that discourages herbivorous predators. The assassin bugs (Fig. E44-2) is undeterred by the camphor "glue." The female assassin bug collects the substance and smears it on her abdomen, where it coats her eggs as they emerge, making them undesirable to predators. But the story is not yet over. Soon after hatching, young assassin bugs laboriously scrape the "glue" from their discarded eggshells and transfer it to their forelegs, where it will aid in the capture of their own prey. In this complex association, the defenses evolved by a plant are utilized by the assassin bug both defensively against its own predators and offensively against its prey.

Figure E44-2 A female assassin bug scrapes noxious resin from a camphor plant. Her warning coloration alerts potential predators of her painful sting.

such as the coral snake (see Fig. 44-9) serves both to paralyze prey and to deter its predators.

Certain molluscs, including squid, octopus, and some sea slugs, emit clouds of ink when attacked. These colorful chemical "smoke-screens" confuse their predators and mask their escape.

A dramatic example of chemical defense is seen in the bombardier beetle. In response to the bite of an ant, the beetle releases secretions from special glands into an abdominal chamber. Here, enzymes catalyze an explosive chemical reaction that shoots a toxic, boiling-hot spray onto the attacking ant (Fig. 44-12).

(a)

(b)

Figure 44-10 Aggressive mimicry. **(a)** Some fish, such as this marine wrasse, obtain food by eating parasites off the bodies of larger fish, a mutualistic relationship called a *cleaning symbiosis*. Fish welcome the approach of this predator of parasites. **(b)** The saber-toothed blenny has evolved to closely resemble the cleaner wrasse. Fish that allow the blenny to approach usually get a chunk bitten out of them—hence the term *aggressive mimicry*.

(a)

(b)

(c)

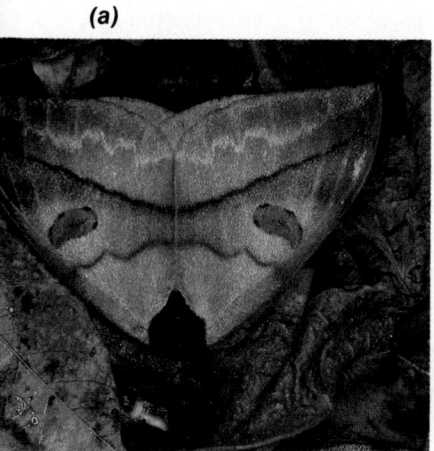

Figure 44-11(a) The peacock moth from Trinidad is well camouflaged, but should a predator approach too closely, it suddenly opens its wings to reveal spots resembling large eyes **(b)**. This startles the predator, giving the moth a chance to flee. **(c)** Would-be predators of this caterpillar larva of the swallowtail butterfly are deterred by its close resemblance to a snake. Note that the caterpillar's head is the "snake's nose."

Figure 44-12 The bombardier beetle sprays a hot toxic brew in response to a leg pinch.

Symbiosis

Symbiosis, which literally means "living together," is defined as a close interaction between organisms of different species for an extended time. Considered in its broadest sense, symbiosis includes parasitism, commensalism, and mutualism. Although one species always benefits in symbiotic relationships, the second species may be harmed, not affected, or benefited (see Table 44-1).

Parasitism

Although it is sometimes difficult to distinguish clearly between a predator and a **parasite**, parasites are generally much smaller and more numerous than their prey (also called hosts). Parasites live in or on their hosts, often without killing them. Familiar parasites include tapeworms, fleas, and numerous disease-causing protozoa, bacteria, and viruses. Many parasites, particularly worms and protozoa, have complex life cycles involving two or more hosts, as described in Chapter 22. There are few parasitic vertebrates; the lamprey eel (see Fig. 22-30) is a rare example.

The variety of infectious bacteria and viruses and the impressive precision of the immune system that counters their attacks are evidence of the powerful forces of coevolution between parasitic microorganisms and their hosts. A good example is **Trypanosoma**, a parasitic protozoan that causes both human sleeping sickness and a disease in cattle called nagana. African antelope, which coevolved with this parasite, are relatively unaffected by it. Cattle that have been bred in an infested area for many generations suffer but usually survive infection, while newly imported cattle generally die if untreated.

Commensalism

Commensalism occurs when the relationship between two species benefits one without affecting the other. For example, as herds of bison graze, they disturb insects dwelling in the grass. Birds follow the bison, eating the insects. The birds, which benefit, are considered commensal with the bison, who are unaffected. Birds and the trees in which they nest are involved in a commensal relationship. The birds obtain shelter and protection without affecting the trees. Many orchids attach to trees without harming them. The orchid obtains support and access to sunlight high above the dim tropical forest floor. Barnacles are harmless hitchhikers on marine animals, including whales and manatees, but benefit from a free ride through food-rich waters (Fig. 44-13).

Mutualism

When two organisms interact in a way that benefits both, the relationship is called **mutualism**. The mutualistic interactions between flowering plants and their pollinators are discussed in Chapter 28. Mutualistic associations occur in the digestive tracts of cows and termites, where protists and bacteria find food and shelter while helping their hosts extract nutrients, and in our own intestines, where bacteria synthesize certain vitamins (see Chapter 32). The nitrogen-fixing bacteria inhabiting special chambers on the roots of legume plants are another important example (see Fig. 21-13). The bacteria obtain food and shelter from the plant, and in return trap nitrogen in a form the plant can utilize. Some mutualistic partners have coevolved to the extent that neither can survive alone. An interesting example is the ant–acacia mutualism described in "A Closer Look at Mutualism: Acacia and Ant—Partners for Life."

Mutualistic relationships involving vertebrates are rare and typically are less intimate and extended, as in the relationship of the cleaner wrasse (see Fig. 44-10) and the fish it cleans. Another is seen in the clown fish, which takes shelter among the venomous tentacles of the anemone. The fish derive shelter and protection, and at least occasionally bring bits of food to their anemone host (Fig. 44-14).

Figure 44-14 The clown fish snuggles unharmed among the stinging tentacles of the anemone. The fish obtains protection, and sometimes brings food to the anemone in this mutualistic relationship.

(a)

(b)

Figure 44-13 Commensal relationships. **(a)** This tropical orchid clings to a rain forest tree. Its perch allows it access to sunlight without harming the tree. **(b)** Clusters of small barnacles hitchhike harmlessly on the hides of these manatees.

Succession: Community Changes Over Time

In a mature terrestrial ecosystem, the populations comprising the community interact with one another and with their nonliving environment in intricate ways. But this tangled web of life did not spring fully formed from bare rock or naked soil; rather, it emerged in stages over a long period, a process called succession. **Succession** is a change in a community and its nonliving environment over time. It is a kind of "community relay" in which assemblages of plants and animals replace one another in a sequence that is at least somewhat predictable.

Succession occurs under a variety of circumstances but is most easily observed in terrestrial and freshwater ecosystems. Freshwater ponds and lakes tend to undergo a series of changes that transform them first into marshes and eventually to dry land (see Fig. 44-18). Shifting sand dunes are stabilized by creeping plants and may eventually support a forest. Volcanic eruptions may, as in the case of Mt. St. Helens, wipe out previously existing ecosystems, or they may create new islands which are soon colonized. Forest fires create a nutrient-rich environment that encourages rapid invasion of new life (Fig. 44-15).

The precise changes occurring during succession are as diverse as the environments in which succession occurs, but certain general stages can be recognized. In each case, succession is begun by a few hardy invaders called **pioneers** and ends with a diverse and relatively stable **climax community**. As the community progresses from the pioneers to the climax, the organisms gradually alter the nonliving environment. Ironically, these changes favor competitors, which displace the existing populations. The climax community differs from earlier successional stages because it no longer alters the environment significantly. The climax community will persist unless external forces (such as a gradual change in climate or human activities) alter it.

During terrestrial succession certain general trends occur in ecosystem structure:

1. The soil increases in depth and in its content of organic material.

2. The overall productivity (the amount of organic material produced in a given area over a given time) increases.

3. The number of different species increases, as does the number of interactions within the community.

4. Longer-lived species come to dominate the ecosystem, and the rate at which populations replace one another slows. The presence of longer-lived species combined with more complex community interactions results in a community that is more stable and resistant to change.

5. In the climax community, the total weight of living organisms reaches a maximum, and the species present no longer alter the ecosystem in ways that encourage the growth of their competitors.

Primary and Secondary Succession

Succession takes two major forms: primary and secondary. During **primary succession,** an ecosystem is forged from bare rock, sand, or a clear glacial pool where there is no trace of a previous community. The formation of an ecosystem "from scratch" is a process often requiring thousands or even tens of thousands of years. During **secondary succession,** a new ecosystem develops after an existing ecosystem is disturbed, as in the case of a forest fire or an abandoned farm field. It happens much more rapidly because the previous community has left its mark in the form of soil and seeds. Succession in an abandoned farm field in the southeastern United States can reach its climax after two centuries. In the examples below, we examine these processes in more detail.

Primary Succession On Bare Rock

Figure 44-16 illustrates primary succession as studied on Isle Royale, Michigan, an island in Lake Superior. Bare rock, such as that exposed by a retreating glacier, begins to liberate nutrients by weathering. Cracks form as the rock alternately freezes and thaws, contracting and expanding. For lichens (symbiotic associations of fungi and algae), the weathered rock provides a place to attach where there are no competitors and plenty of sunlight. Lichens can photosynthesize, and they obtain minerals by dissolving some of the rock with an acid they secrete. As the lichens spread over the rock, drought-resistant, sun-loving mosses begin growing in the cracks. Fortified by nutrients liberated by the lichens, the moss forms a dense mat that traps dust, tiny rock particles, and bits of organic debris. The death of some of the moss adds to a growing nutrient base, while the moss mat itself acts as a sponge, trapping moisture. Within the moss, seeds of larger plants germinate. Eventually, their bodies contribute to a growing layer of soil. As woody shrubs such as blueberry and juniper take advantage of the newly formed soil, the moss and lichens may be shaded out and buried by decaying leaves and vegetation. Eventually, trees such as jack pine and blue spruce take root in the deeper crevices, and the sun-loving shrubs are shaded out. Within the

(a)

(b)

(c)

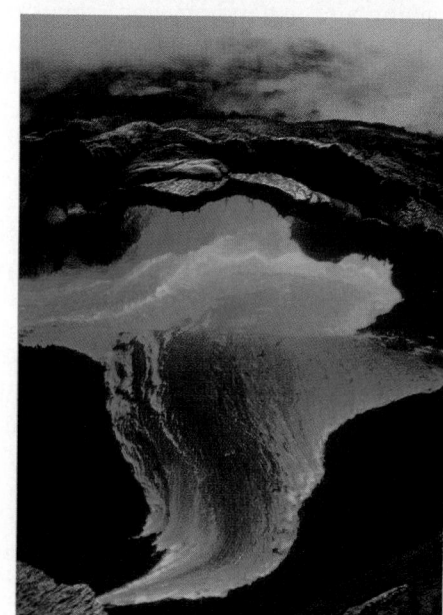

Figure 44-15 Succession in progress.
(a) Left: On May 18, 1980, the explosion of Mt. St. Helens in Washington devastated the coniferous forest ecosystem on its flanks. Right: Nine years later, pioneering evergreens are 4 to 6 inches tall, and shrubs carpet the nutrient-rich ash.
(b) Left: In the summer of 1988, extensive fires swept through the forests of Yellowstone National Park in Wyoming. Right: Two years later, flowering plants have sprung up in the sunlight and wildlife populations are rebounding.
(c) Left: The Hawaiian volcano Mt. Kilauea has erupted repeatedly during the past 8 years, sending rivers of lava over the surrounding countryside. Right: Sword ferns, lichens, and mosses are the first pioneers to colonize the barren landscape.

Figure 44-16 Primary succession on bare rock in upper Michigan.

lichen on bare rock   bluebell, yarrow   blueberry, juniper   jack pine, black spruce aspen   balsam fir, paper birch, white spruce climax forest

Time (years)

0 ———————————————————————————— 1000

depths of the forest, shade-tolerant seedlings of taller or faster growing trees, such as balsam fir, paper birch, and white spruce, thrive. Eventually these overtower and replace the original trees, which are intolerant of shade. After a thousand years or more, a tall climax forest thrives on what was once bare rock.

Secondary Succession on an Abandoned Field

Figure 44-17 illustrates succession on an abandoned southeastern farm. The pioneers are fast-growing annual weeds such as crabgrass, ragweed, and sorrel, which root in the rich soil already present and thrive in direct sunlight. A few years later, perennial new-comers such as asters and goldenrod, broomsedge grass, and woody shrubs such as blackberry invade. These proliferate and dominate for the next few decades. Eventually, they are replaced by pines and fast-growing deciduous trees such as tulip poplar and sweetgum, which sprout from windblown seeds. These become prominent after about 25 years, and a pine forest dominates the field for the rest of the first century. Meanwhile, shade-resistant, slow-growing hardwoods such as oak and hickory take root beneath the pines. After the first century, these begin to overtower and shade the pines, which eventually die from lack of sun. A relatively stable climax forest dominated by oak and hickory is present by the end of the second century.

Figure 44-17 Secondary succession on a plowed, abandoned southeastern farm field.

plowed field   crabgrass and other grasses   asters, ragweed, goldenrod   blackberry   Virginia pine   oak-hickory climax forest

Time (years)

0 ———————————————————————————— 200

A CLOSER LOOK

At Mutualism: Acacia and Ant—Partners for Life

Daniel Janzen of the University of Pennsylvania, then a doctoral student, was walking down a road in Veracruz, Mexico, when he saw a flying beetle alight on a thorny tree, only to be driven off by an ant. Further observation revealed that the tree, a bull's-horn acacia, was covered with ants. A large ant colony of the genus *Pseudomyrmex* made its home inside the enlarged thorns of the plant, whose soft pulpy interiors are easily excavated to provide secure shelter (Fig. E44-3).

To determine how important the ants are to the tree, Janzen began stripping the thorns by hand until he found and removed the thorn housing the ant queen, thus destroying the colony. He later turned to more efficient but dangerous methods, eliminating all the ants on a large stand of acacias with the insecticide parathion. The acacias were unharmed by the poison, Janzen became ill from it, and the ants were all killed. Within a year of the spraying, the recovered Janzen found nearly all the acacia trees dead, consumed by insects and other herbivores, and shaded out by competing plants. The ground surrounding the trees, which the ants normally kept pruned, was overgrown. The trees were apparently dependent on their resident ants for survival.

Wondering if the ants could survive off the tree, Janzen painstakingly peeled the ant-inhabited thorns off 100 acacia trees, suffering multiple stings in the process. He housed each ant colony in a jar provided with local nonacacia vegetation and insects for food. The colonies all starved. Close inspection of the acacia revealed swollen structures filled with sweet syrup at the base of the leaves and protein-rich capsules on the leaf tips (Fig. E44-3, inset). Together, these provide a balanced diet for the ants.

Janzen's experiments strongly suggest that this species of ant and acacia have an obligatory mutualistic relationship—that is, that neither can survive without the other. Of course, further observations were required to confirm this. The fact that the ants starved in Janzen's jars did not rule out that they might survive successfully elsewhere, but, in fact, this species of ant is never found living independently. Similarly, the bull's-horn acacia is never found without its resident ant colony. Thus, a chance observation followed by careful research led to the discovery of an important mutualistic association.

Figure E44-3 Yellow, protein-rich capsules are produced at the tips of certain acacia leaves. These provide food for the resident ants. Inset: A hole in the enlarged thorn of the bull's horn acacia provides shelter for members of the ant colony. The ant entering the thorn is carrying a food capsule produced by the acacia. As the ant colony grows, more thorns are invaded.

Succession in Freshwater Ponds

In freshwater ponds or lakes, succession may result primarily from an influx of nutrients from outside the ecosystem. This is particularly apparent in small freshwater lakes, ponds, and bogs, which gradually undergo succession to dry land. Sediment and nutrients are carried in by runoff from adjacent land (Fig. 44-18).

The Climax Community

Succession ends with a relatively stable climax community. In your travels, you have undoubtedly noticed that the type of climax community varies dramatically from one area to the next. For example, Colorado has shortgrass prairie as a climax on its eastern plains, spruce-pine forests in its mountains, tundra on their uppermost reaches, and sagebrush-dominated climax in its western valleys.

The exact nature of the climax community is determined by numerous geological and climatic variables, including temperature, rainfall, elevation, latitude, type of rock (which influences the type of nutrients available), exposure to sun and wind, and many more. Human activities may also dramatically alter the climax vegetation. Large tracts of grasslands in the west, for example, are now dominated by sagebrush due to overgrazing. The grass that normally outcompetes sagebrush is selectively eaten by cattle, allowing the sagebrush to prosper.

Some ecosystems are not allowed to reach the climax stage but are maintained in an earlier stage called a **subclimax**. The tallgrass prairie that once covered northern Missouri and Illinois is actually a subclimax of an ecosystem whose climax community is deciduous forest. The prairie was maintained by periodic fires, some set by lightning and others deliberately set by Indians to increase grazing land for buffalo. Forest now encroaches and limited prairie preserves are maintained by carefully managed burning.

Agriculture also depends on the artificial maintenance of carefully selected subclimax communities. Grains are specialized grasses characteristic of early

(a)

(c)

(b)

Figure 44-18 Succession in a small freshwater pond is the result of an influx of materials from the surroundings. **(a)** In this small pond, dissolved minerals carried by runoff from the surroundings support aquatic plants, whose seeds or spores were carried here by the winds or by birds and other animals. **(b)** Over time, the decaying bodies of aquatic plants build up soil that provides anchorage for more terrestrial plants. **(c)** Finally, the pond is entirely converted to dry land.

successional stages, and much energy goes into preventing competitors (weeds and shrubs) from taking over. The suburban lawn is also a painstakingly maintained subclimax ecosystem. Mowing destroys woody invaders, and selective herbicides kill pioneers such as crabgrass and dandelions.

To study succession is to study the variations in communities over *time.* The climax communities that form during succession are strongly influenced by climate and geology: the distribution of ecosystems in *space.* Over broad geographical regions, climax communities such as deserts, grasslands, and deciduous forests occur; these are called **biomes.** In Chapter 46 we explore some of the great biomes of the world. Although the communities comprising the various biomes differ radically in the types of populations they support, communities worldwide are structured according to general rules. These principles of ecosystem structure are described in Chapter 45.

SUMMARY OF KEY CONCEPTS

The Ecological Niche
The ecological niche defines all aspects of a species' habitat and interactions with its living and nonliving environment. Each species occupies a unique ecological niche.

Competition Between Species
Interspecific competition occurs when the niches of two populations within a community overlap; the amount of competition is proportional to the amount of niche overlap. When two species are forced under laboratory conditions to occupy the same niche, one species always outcompetes the other. Species within natural communities have evolved in ways that avoid excessive niche overlap, with behavioral and physiological adaptations that allow resource partitioning.

Predators and Their Prey
Predators eat other organisms, and are usually both larger and less abundant than their prey.

Predators and prey act as strong agents of selection on one another. Plants that are preyed upon have evolved elaborate defenses ranging from poisons to thorns to overall toughness. These defenses, in turn, have selected for predators that can detoxify poisons, ignore thorns, and grind down tough tissues. Prey animals have evolved a variety of protective colorations that render them either inconspicuous or startling to their predators. Some prey have become poisonous, and often exhibit warning coloration by which they are readily recognized and avoided. Animals that have evolved to resemble one another are called mimics.

Symbiosis
Symbiotic relationships involve two different species that interact closely over an extended period. Symbiosis includes parasitism, in which the parasite feeds on a larger, less abundant host, usually harming it but not killing it immediately. In commensal symbiotic relationships, one species benefits, often by finding food more easily in the presence of another species, which is not affected by the association. Mutualism benefits both symbiotic species.

Succession
A directional change in the types of populations inhabiting an ecosystem over time is called succession. Primary succession, which may take thousands of years, occurs where no remnant of a previous community exists. Primary succession could occur on bare rock scoured by a glacier or cooled from molten lava, a sand dune, or in a newly formed glacial lake. Secondary succession occurs much more rapidly, since it builds on the remains of a disrupted community, such as the aftermath of a fire, or a plowed and abandoned field. During succession, pioneer organisms invade and alter the environment by their presence. Community changes occur as one group of organisms, particularly plants, alters the environment in ways that encourage growth of their competitors. These replace the first colonizers and are subsequently replaced by others. Some ecosystems, including tallgrass prairie and farm fields, are maintained in relatively early stages of succession by periodic disruptions. Uninterrupted succession ends with a relatively stable group of organisms called the climax community.

GLOSSARY

aggressive mimicry (mim'ik-rē): the evolution of a predatory organism to resemble a harmless one, thus gaining access to its prey.

biome (bī'-ōm): a general type of ecosystem occupying extensive geographical areas, characterized by similar plant communities: for example, deserts. (See Chapter 46.)

camouflaged (cam'-a-flaged): a term used to describe organisms that resemble their environment.

climax community: a relatively stable community of plants and animals that does not change appreciably over long periods of time unless influenced by external forces.

coevolution: the process by which two interacting species act as agents of natural selection on one another over evolutionary time.

commensalism (kum-en'-sal-ism): a symbiotic relationship between two species in which one benefits while the other is neither harmed nor benefited.

community: all the interacting populations within an ecosystem.

competitive exclusion principle: the concept that no two species can simultaneously and continuously occupy the same ecological niche.

ecological niche (nitch): the role of a particular species within an ecosystem, including all aspects of its interaction with the biotic and abiotic environment.

mimic (mim'-ik): an organism that has evolved to resemble another.

mutualism (mū'-chū-al-ism): a symbiotic relationship in which both participating species benefit.

parasitism: a symbiotic relationship in which one organism (usually smaller and more numerous than its host) benefits by feeding on the other, usually without killing it immediately.

pioneer: an organism that is among the first to colonize an unoccupied habitat in the first stages of succession.

primary succession: succession that occurs in an environment, such as bare rock, in which no remnants of a previous community are present.

secondary succession: succession that occurs after an existing community is destroyed, in an environment modified by that community: for example, after a forest fire.

startle coloration: a color pattern (often resembling large eyes) that can be displayed suddenly by a prey organism when approached by a predator.

subclimax: a community where succession is arrested before the climax community is reached, usually maintained by regular disturbance: for example, tallgrass prairie maintained by regular fires.

succession (suk-se'-shun): the directional change in the community structure of an ecosystem over time. Community changes alter the ecosystem in ways that favor competitors, and species replace one another in a somewhat predictable manner until a stable climax community is reached.

symbiosis (sim'-bī-ō'sis): a close association between different species usually over an extended period. *Symbiosis* as used in this book includes mutualism, commensalism, and parasitism.

STUDY QUESTIONS

1. Define an ecological community, and list three important types of community interactions.
2. An ecologist visiting an island finds two very closely related species of birds, one of which has a slightly larger bill than the other. Interpret this finding with respect to the competitive exclusion principle and the ecological niche, defining both terms.
3. Describe three very different examples of coevolution. In each case, describe the selection pressure that has resulted in the evolutionary adaptations present.
4. List three different types of symbiosis; define and provide an example of each.
5. What type of succession would occur on a clear cut in a national forest, and why?
6. List two subclimax and two climax communities. How do they differ?
7. Define *succession* and explain why it occurs.

DISCUSSION QUESTIONS

1. Herbivorous animals that eat seeds are considered by some ecologists to be predators of plants, while herbivorous animals that eat leaves are considered to be parasites of plants. Discuss the validity of this classification scheme.
2. An interesting interspecific relationship exists between the tarantula spider and the tarantula hawk wasp. Such wasps attack tarantulas, paralyzing them with venom from their stingers. The wasps then lay eggs on the paralyzed spider. The eggs hatch and the young eat the living, immobilized tissues of the spider. Discuss whether this relationship between spider and wasp exemplifies parasitism or predation.

SUGGESTED READINGS

Amos, W. H. "Hawaii's Volcanic Cradle of Life." *National Geographic* July, 1990. A naturalist explores succession on lava flows.

Brower, L. P. "Ecological Chemistry." *Scientific American,* 1969, pp. 22–29. Insects feeding on toxic plants become unpalatable to bird predators.

Daniels, P. "How Flowers Seduce the Bugs and the Bees." *International Wildlife,* November/December 1984, pp. 4–11. Beautifully illustrated description of plants and their mutualistic pollinators.

Horn, H. S. "Forest Succession." *Scientific American,* May 1975 (Offprint No. 1321). Describes succession in eastern deciduous forests.

45

The Structure and Function of Ecosystems

"A thing is right, when it tends to preserve the integrity, stability, and beauty of the biotic community. It is wrong when it tends otherwise."

Aldo Leopold in A Sand County Almanac *(1949)*

In Chapters 7 and 8, you learned how energy is trapped by photosynthesis, released by cellular respiration, and used to construct the complex molecules of life. In Chapter 44, you learned some of the complex ways in which organisms interact within ecological communities. In this chapter, we relate some of these basic principles to the functioning of ecosystems.

Nearly all the activity of life is powered by the energy of sunlight, from the leaping of the jackrabbit to the active transport of molecules through a cell membrane. Each time this energy is used, some of it is lost as heat. But while solar energy continuously bombards the Earth and is continuously lost as heat, nutrients remain. They may change in form and distribution but they do not leave the world ecosystem, and are continuously recycled. Thus, **the basic laws of ecosystem function are that energy flows through ecosystems, while nutrients cycle and recycle (Fig. 45-1). These laws shape the complex interactions within living communities.**

The Flow of Energy

Primary Productivity

Ninety-three million miles away, the sun fuses hydrogen into helium, releasing tremendous quantities of energy. A tiny fraction of this energy reaches the Earth in the form of electromagnetic waves. Of the energy that reaches Earth, much is reflected by the atmosphere, clouds, and the Earth's surface. Still more is absorbed as heat by the Earth and its atmosphere, leaving only about 1% to power all life on Earth. Of this 1%, green plants capture 3% or less. The teeming life on Earth is thus supported by less than 0.03% of the energy reaching it from the sun. But how does this energy enter the biological community? **The energy that powers ecosystems enters through the process of photosynthesis.**

In photosynthesis (performed by plants, plantlike protists, and cyanobacteria), pigments such as chlorophyll absorb certain wavelengths of sunlight. This solar energy is used to combine carbon dioxide and water into sugar, a compound that stores energy in chemical bonds. Some of this energy is used to power other chemical reactions, converting sugars into starches, cellulose, fats, vitamins, pigments, and proteins (Fig. 45-2).

Photosynthetic organisms are called **autotrophs** (Greek, "self-feeders"), or **producers,** since they produce food not only for themselves but for nearly all other life as well. The organisms that rely on the high-energy molecules made by autotrophs are called **heterotrophs** (Greek, "other feeders"), or **consumers.**

The amount of life an ecosystem can support is determined by the energy captured by the producers. The energy that photosynthetic organisms make available to other members of the community is called **net primary productivity.** Net primary productivity can be measured in units of energy (calories), or as the dry weight of organic material per unit area per year. The productivity of an ecosystem is influenced by many environmental variables, including the amount of nutrients available to the autotrophs, the amount of sunlight reaching them, the availability of water, and the temperature. In the desert, for example, lack of water limits productivity, while in the open ocean, light is limited in deep water and nutrients are limited in surface water. When resources are abundant, as in estuaries or tropical rain forests, productivity is high. Some average productivities for a variety of ecosystems are shown in Figure 45-3.

Trophic Levels

Living things may be categorized according to their role in the flow of energy through communities. Energy flows through communities from photosynthetic producers through several levels of consumers. Each category of organism is called a **trophic level** (Greek, "feeding level"). The producers, from redwood trees to cyanobacteria, form the first trophic level, obtaining their energy directly from sunlight (see Fig. 45-1).

◀ A microcosm is reflected in a dewdrop on the spore case of a moss. The sparkling energy of sunlight, captured by photosynthetic organisms, powers nearly all life on Earth.

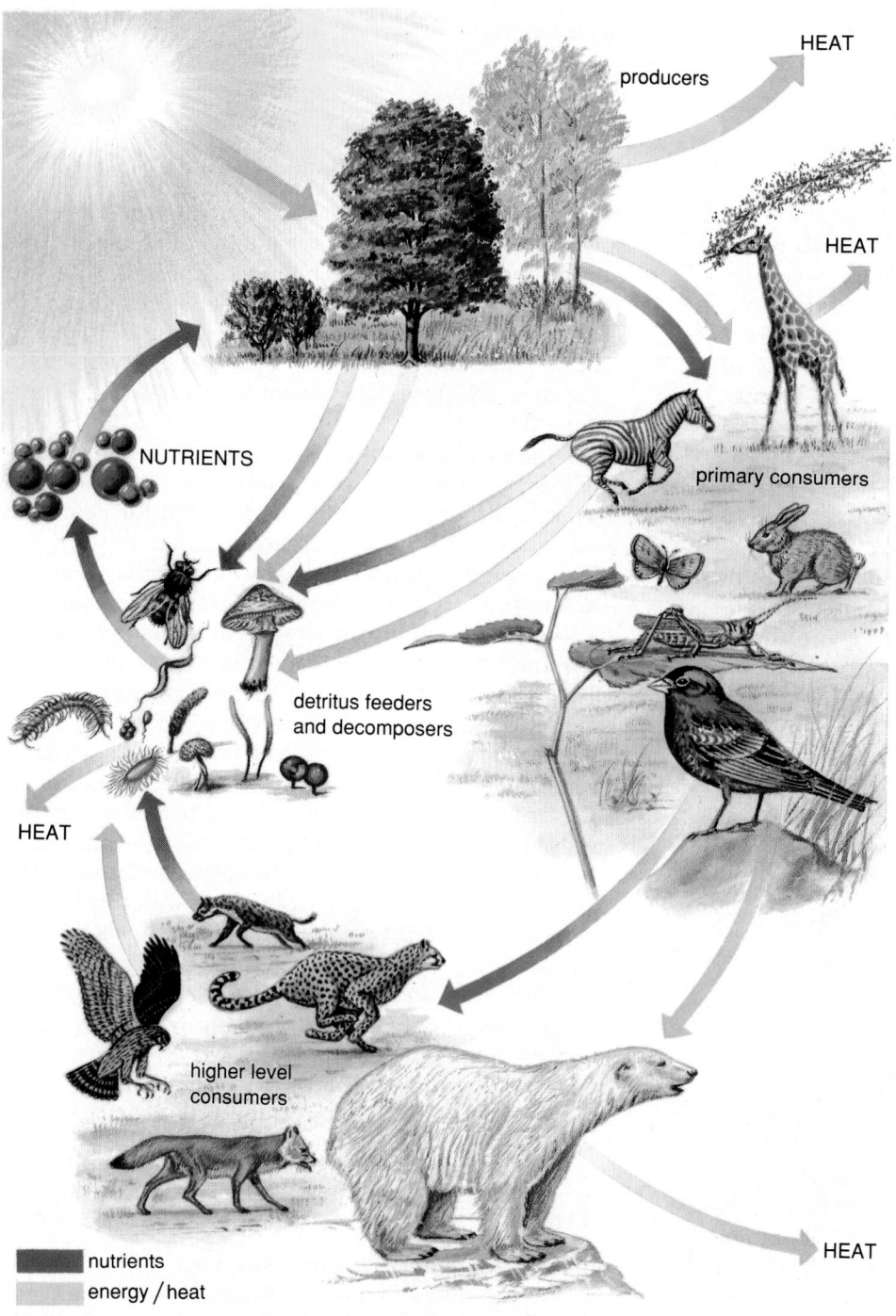

HEAT

producers

HEAT

NUTRIENTS

primary consumers

detritus feeders
and decomposers

HEAT

higher level
consumers

HEAT

nutrients
energy / heat

Figure 45-1 Energy flow, nutrient cycling, and feeding relationships in ecosystems. Note that nutrients (purple) neither enter nor leave the cycle, while energy (yellow) is lost at each level as heat and must be constantly replenished by sunlight.

Consumers form several trophic levels, and certain consumers even switch trophic levels by eating organisms from different levels. Some consumers have evolved to feed directly and exclusively on producers, the most abundant living energy source in the ecosystem. These range from grasshoppers to giraffes. They are the **herbivores,** or **primary consumers,** and form the second trophic level. **Carnivores,** such as the spider, hawk, and wolf, are flesh-eaters, feeding primarily on herbivores; they are called **secondary consumers** and form the third trophic level. Some carnivores occasionally eat other carnivores, and when doing so they form the fourth trophic level: **tertiary consumers.**

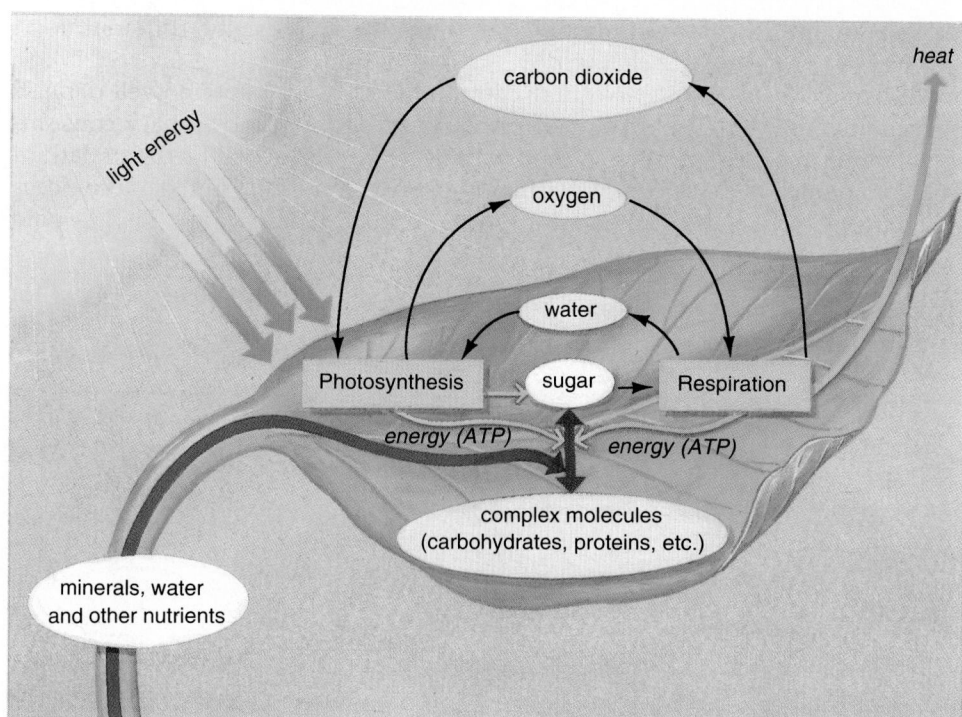

Figure 45-2 During photosynthesis, plants capture the energy of sunlight and store it in ATP, sugar, and other high-energy carbohydrates synthesized from carbon dioxide and water. Oxygen is released as a by-product. Some of the energy stored in sugar is liberated during respiration and used in the construction of more complex molecules. In the process, oxygen is used, water is released, and some energy is lost as heat.

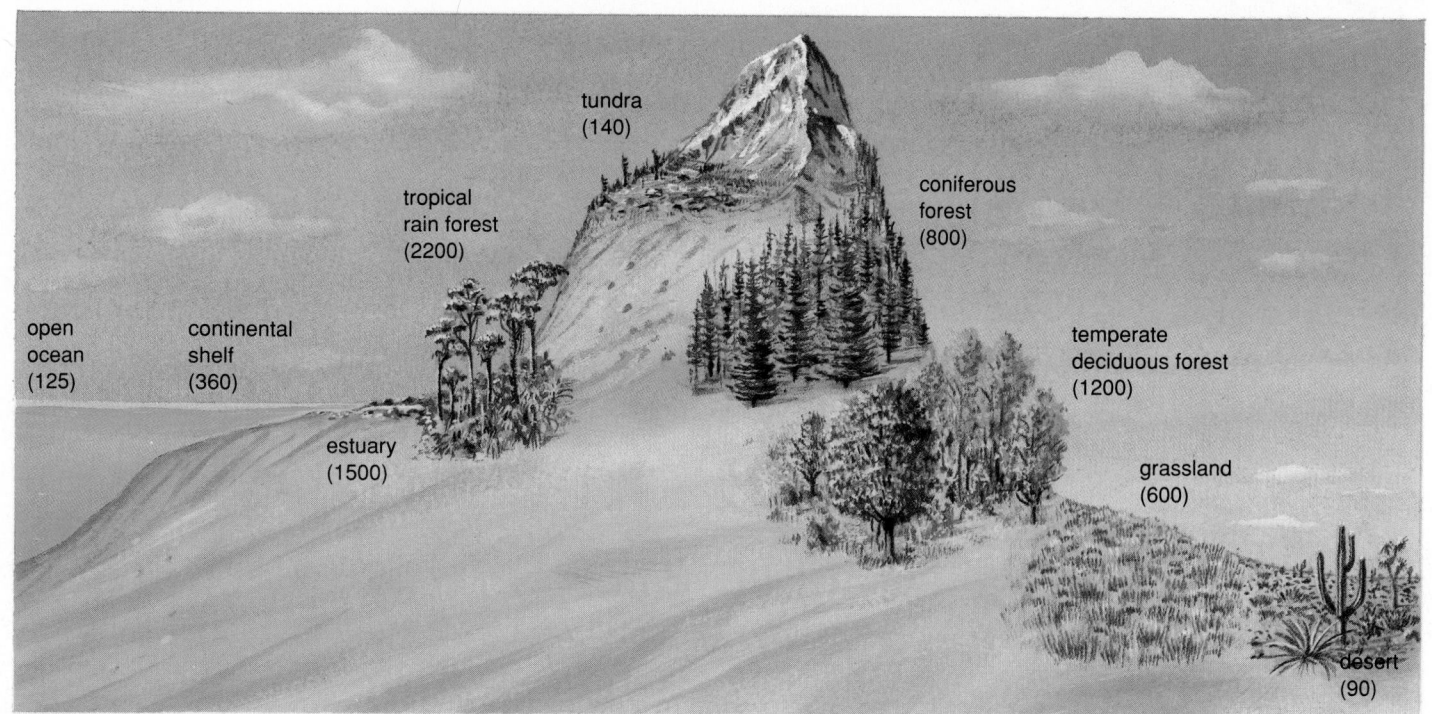

Figure 45-3 Average net primary productivity in grams of organic material per square meter per year of some terrestrial and aquatic ecosystems.

Food Chains and Food Webs

To illustrate the feeding relationships in an eco-system, it is common to identify a representative of each trophic level that eats the representative of the level below it. This linear feeding relationship is called a **food chain.** As illustrated in Figure 45-4,

different ecosystems have radically different food chains.

Natural communities rarely contain well-defined groups of primary, secondary, and tertiary consum-ers. A **food web** describes the actual feeding relation-ships within a given community much more accu-rately than does a food chain (Fig. 45-5). Some

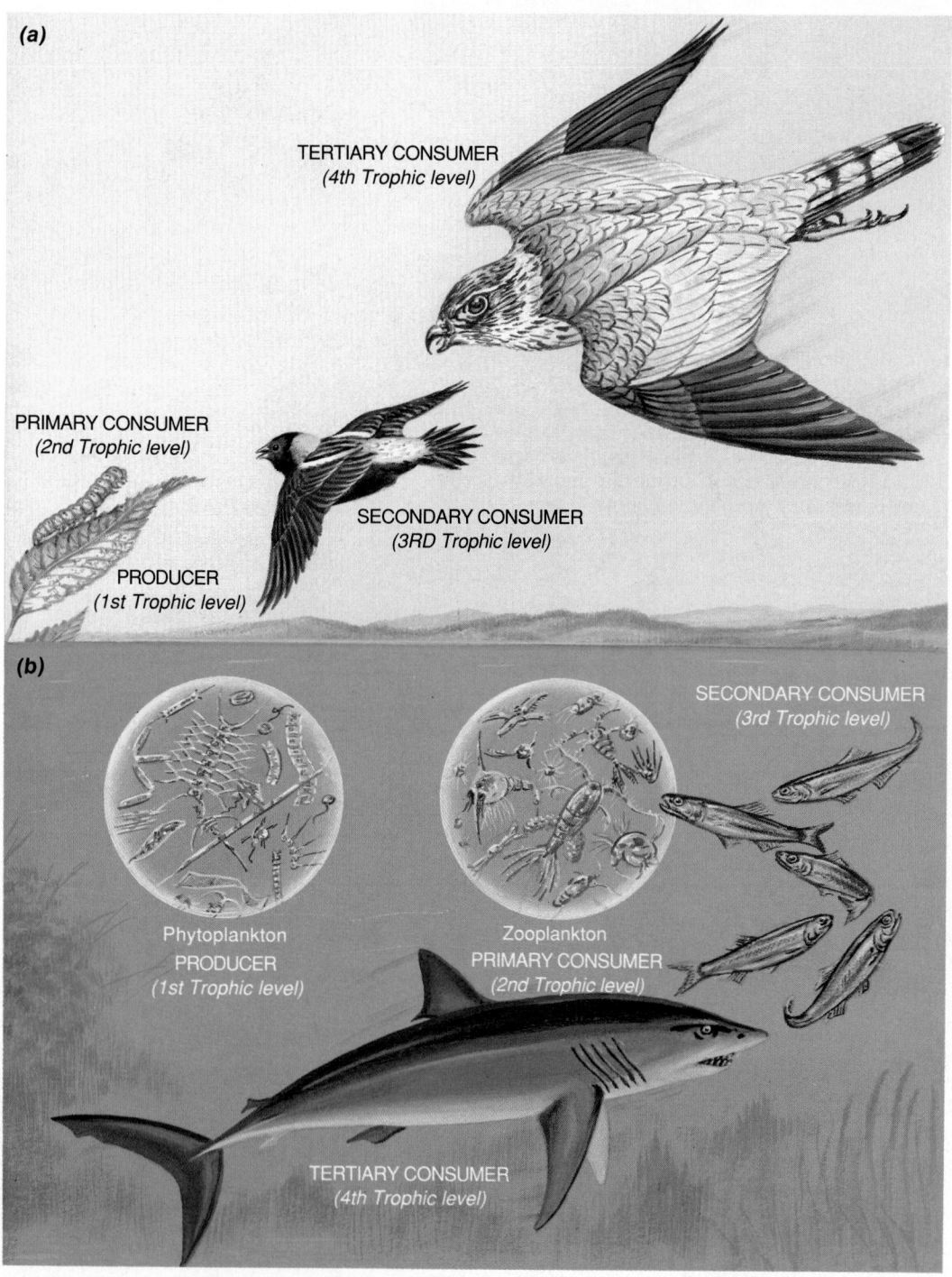

Figure 45-4 (a) A simple terrestrial food chain. **(b)** A simple marine food chain.

animals, such as raccoons, bears, rats, and humans, are omnivores (Latin, "eating all"), acting at different times as primary, secondary, and occasionally tertiary consumers. Many carnivores will eat either herbivores or other carnivores, thus acting either as secondary or tertiary consumers. An owl, for instance, is a secondary consumer when it eats a mouse, but a tertiary consumer when it eats a shrew, which feeds on insects. If the shrew ate a carnivorous insect, it would be a tertiary consumer, and the owl that fed on it would then be a quaternary (fourth-level) consumer. A carnivorous plant such as the sundew, when digesting a spider, can tangle the web hopelessly by serving simultaneously as a producer and a secondary consumer!

Detritus Feeders and Decomposers

Among the most important strands in the food web are the decomposers and detritus feeders. By liberating nutrients for reuse, they form a vital link in the nutrient cycles of ecosystems. The **decomposers** are primarily fungi and bacteria, which digest food outside their bodies, absorb the nutrients they need, and free the remaining nutrients into the soil or water. The **detritus feeders** are an army of small and often unnoticed animals and protists that live on the refuse of life: molted exoskeletons, fallen leaves, wastes, and dead bodies. The network of detritus feeders is extremely complex, including earthworms, mites, protists, centipedes, some insects and crustaceans, nematode worms, and even a few large vertebrates such as vultures. These organisms consume dead organic matter, extract some of the energy stored within it, and excrete it in a still further decomposed state. Their excretory products serve as food for other detritus feeders and decomposers, until most of the stored energy has been utilized. The once-living substance is reduced to simple molecules such as carbon dioxide and water that return to the atmosphere, and minerals and organic acids that return to the soil. In some ecosystems, such as deciduous forests, more energy passes through the detritus feeders and decomposers than through the primary, secondary, and tertiary consumers. This inconspicuous portion of the food web is absolutely essential to life on Earth. If the detritus feeders and decomposers were to disappear suddenly, communities would gradually be smothered by accumulated wastes and dead bodies. The nutrients stored in these bodies would be unavailable and the soil would become poorer and poorer until plant life could no longer be sustained. With plants eliminated, energy would cease to enter the community and the higher trophic levels would disappear as well.

Energy Flow Through Trophic Levels

A basic law of thermodynamics is that energy use is never completely efficient. For example, as your car converts the energy stored in gasoline to the energy of movement, about 75% is immediately lost as heat. So too in living systems; as anyone who has exercised vigorously is well aware, the energy of muscular contraction produces heat as a by-product. So does the germination of a seed and the thrashing of the tail of a sperm.

The transfer of energy from one trophic level to the next is also quite inefficient. When a caterpillar (a primary consumer) feeds on a shrub (a producer), only some of the solar energy originally trapped by the plant is available to the insect. Some was used by the plant to grow and maintain life. Some was converted into the chemical bonds of molecules such as cellulose, which the caterpillar cannot break down. Some energy remains in the shrub. Therefore, only a fraction of the energy captured by the first trophic level is available to organisms in the second trophic level. The energy consumed by the caterpillar in turn is partially used to power crawling and the gnashing of mouthparts. Some is used to construct the indigestible chitinous exoskeleton, and much is liberated as heat. All this energy is unavailable to the songbird in the third trophic level when it eats the caterpillar. The bird loses energy as body heat, uses more in flight, and converts a considerable amount into indigestible feathers, beak, and bone. All this energy will be unavailable to the hawk that catches it. A simplified model of energy flow through the trophic levels in a deciduous forest ecosystem is illustrated in Figure 45-6.

Energy Pyramids

Studies of a variety of ecosystems indicate that net energy transfer between trophic levels is roughly 10% efficient, although different ecosystems vary significantly. This is called the "10% Law." This means that the energy stored in primary consumers (herbivores) is only about 10% of the energy stored in the bodies of producers. The bodies of secondary consumers, in turn, possess roughly 10% of the energy stored in primary consumers. In other words, for every 100 calories of solar energy captured by grass, only about 10 calories are converted into herbivores, and only 1 calorie into carnivores. An **energy pyramid** illustrates these relationships graphically (Fig. 45-7).

What does this mean for ecosystems? If you wander through an undisturbed ecosystem, you will notice that the predominant organisms are plants; these have the most energy available to them because they

Figure 45-5 A simple terrestrial food web.

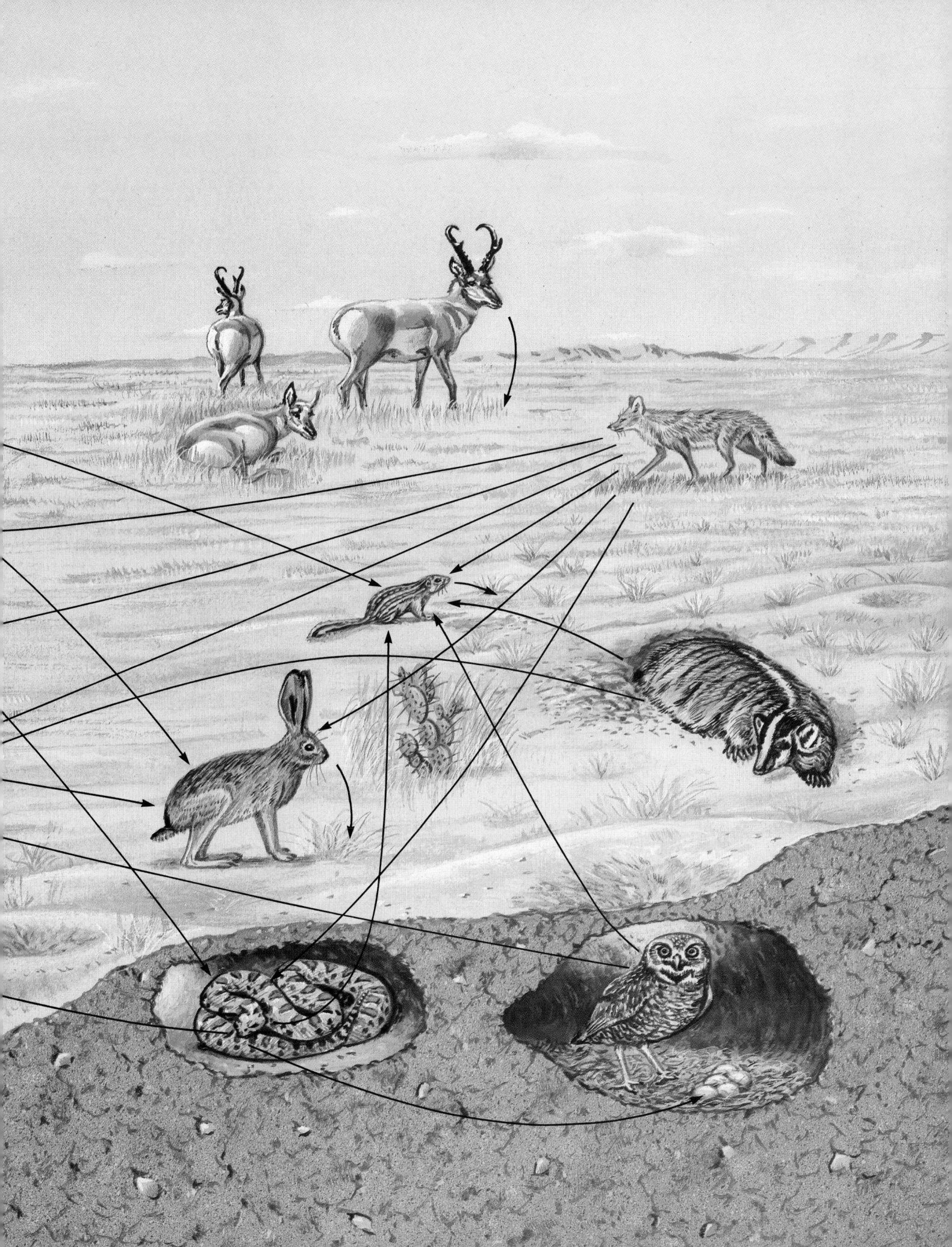

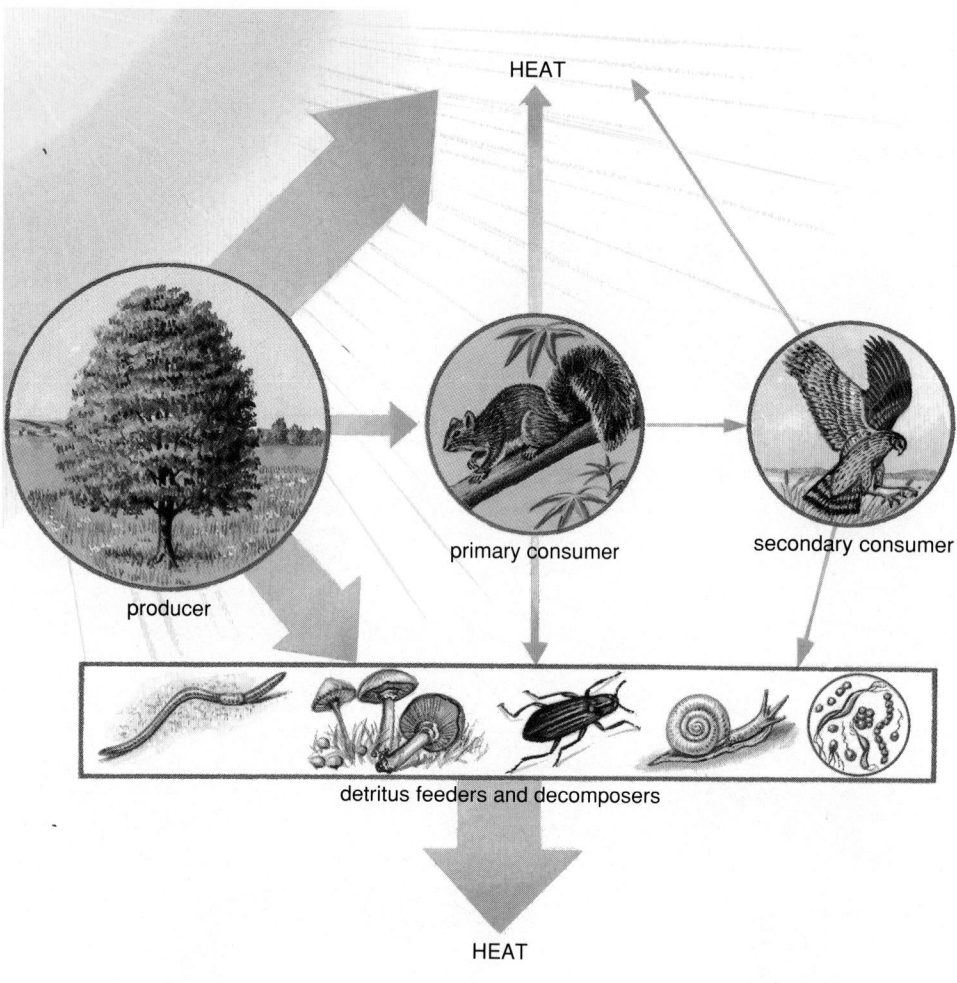

HEAT

producer

primary consumer

secondary consumer

detritus feeders and decomposers

HEAT

Figure 45-6 A diagram showing the loss of energy during its transfer between trophic levels in a forest community. The width of the arrows is roughly proportional to the quantity of energy transferred or lost.

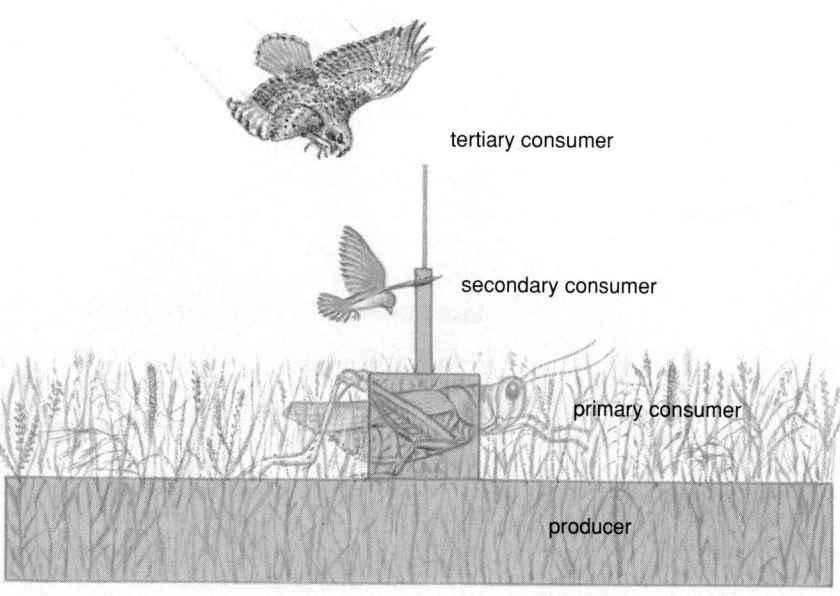

tertiary consumer

secondary consumer

primary consumer

producer

Figure 45-7 An energy pyramid for a prairie ecosystem. Each trophic level from producer to tertiary consumer has less energy stored in it. The width of the rectangles represents energy found in the organisms at each trophic level.

trap it directly from sunlight. The most abundant animals will be those feeding on plants, while carnivores will always be relatively rare. The inefficiency of energy transfer also has important implications for human food production. The lower the trophic level we utilize, the more food energy will be available to us; far more people can be fed on grain than on meat.

An unfortunate side effect of the inefficiency of energy transfer, coupled with human production and release of toxic chemicals, is the phenomenon of **biological magnification.** This is the concentration of certain persistent toxic chemicals in the bodies of carnivores, including ourselves (see "Planet Watch: DDT, Biological Magnification, and the Complex Web of Life").

The Cycling of Nutrients

In contrast to the energy of sunlight, nutrients do not flow down onto the Earth in a steady stream from above. For practical purposes, the same pool of nutrients has been supporting life for over 3 billion years. Nutrients are all the chemical building blocks of life. Some, called **macronutrients,** are required by organisms in large quantities. These include water, carbon, hydrogen, oxygen, nitrogen, phosphorus, and calcium. **Micronutrients,** including zinc, molybdenum, iron, selenium, and iodine, are required only in trace quantities. **Nutrient cycles** describe the pathways these substances follow as they move from the living to the nonliving portions of ecosystems and back again to living tissues.

The major source, or **reservoir,** of important nutrients is generally in the nonliving environment. For example, the major reservoir of carbon and nitrogen is the atmosphere, so these are called **atmospheric cycles.** The reservoir of phosphorus is rock, making this a **sedimentary cycle.** In the following section we briefly describe the cycles of water, carbon, nitrogen, and phosphorus.

The Carbon Cycle— An Atmospheric Cycle

Chains of carbon atoms form the framework of all organic molecules. Carbon enters the living community through capture of carbon dioxide (CO_2) during photosynthesis. The major reservoir of this gaseous compound is the atmosphere, where it comprises 0.033% of the total gases.

Carbon enters food webs through producers, which trap CO_2 during photosynthesis. Some CO_2 is returned to the atmosphere through cellular respira-

tion, while some that is incorporated into the plant body is later passed to herbivores. They, in turn, respire some of it and incorporate some into their tissues. All living things are eventually consumed by predators, detritus feeders, and decomposers, and ultimately most carbon is returned to the atmosphere as CO_2 (Fig. 45-8).

Some carbon is cycled much more slowly. For example, mollusks extract carbon dioxide dissolved in water and combine it with calcium to form calcium carbonate ($CaCO_3$), from which they construct their shells. Shells of dead mollusks collect in undersea deposits and may eventually be converted to limestone. Limestone may dissolve gradually as it is exposed to water, making the carbon available to living organisms once more. Another long-term cycle is the production of fossil fuels. Fossil fuels are formed from the remains of ancient plants and animals. The carbon found in the organic molecules of these ancient organisms remains in these deposits, transmuted by high temperature and pressure over geological time to coal, oil, or natural gas. The energy of prehistoric sunlight is also trapped in fossil fuels and is released by combustion. Human activities, including burning fossil fuels and cutting and burning the Earth's great forests where much carbon is stored, are increasing the amount of carbon dioxide in the atmosphere, as described later in this chapter.

The Nitrogen Cycle— An Atmospheric Cycle

The atmosphere is about 79% nitrogen gas (N_2), but neither plants nor animals can use this gas directly. Instead, plants must be supplied with nitrates (NO_3^-) or ammonia (NH_3). But how is atmospheric nitrogen converted to these molecules? Ammonia is synthesized by certain bacteria and cyanobacteria that engage in **nitrogen fixation,** a process that combines nitrogen with hydrogen. Some of these bacteria are found living in water and soil, while others have entered a symbiotic association with plants called **legumes** (a group including soybeans, clover, and peas) where they live in special swellings on the roots (see Chapter 21). Decomposer bacteria can also produce ammonia from amino acids and urea found in dead bodies and wastes. Still other bacteria convert ammonia to nitrates. Nitrates are also produced by electrical storms and by other forms of combustion that cause nitrogen to react with atmospheric oxygen. In human-dominated ecosystems such as farm fields, gardens, and suburban lawns, ammonia and nitrates are supplied by chemical fertilizers; which are pro-

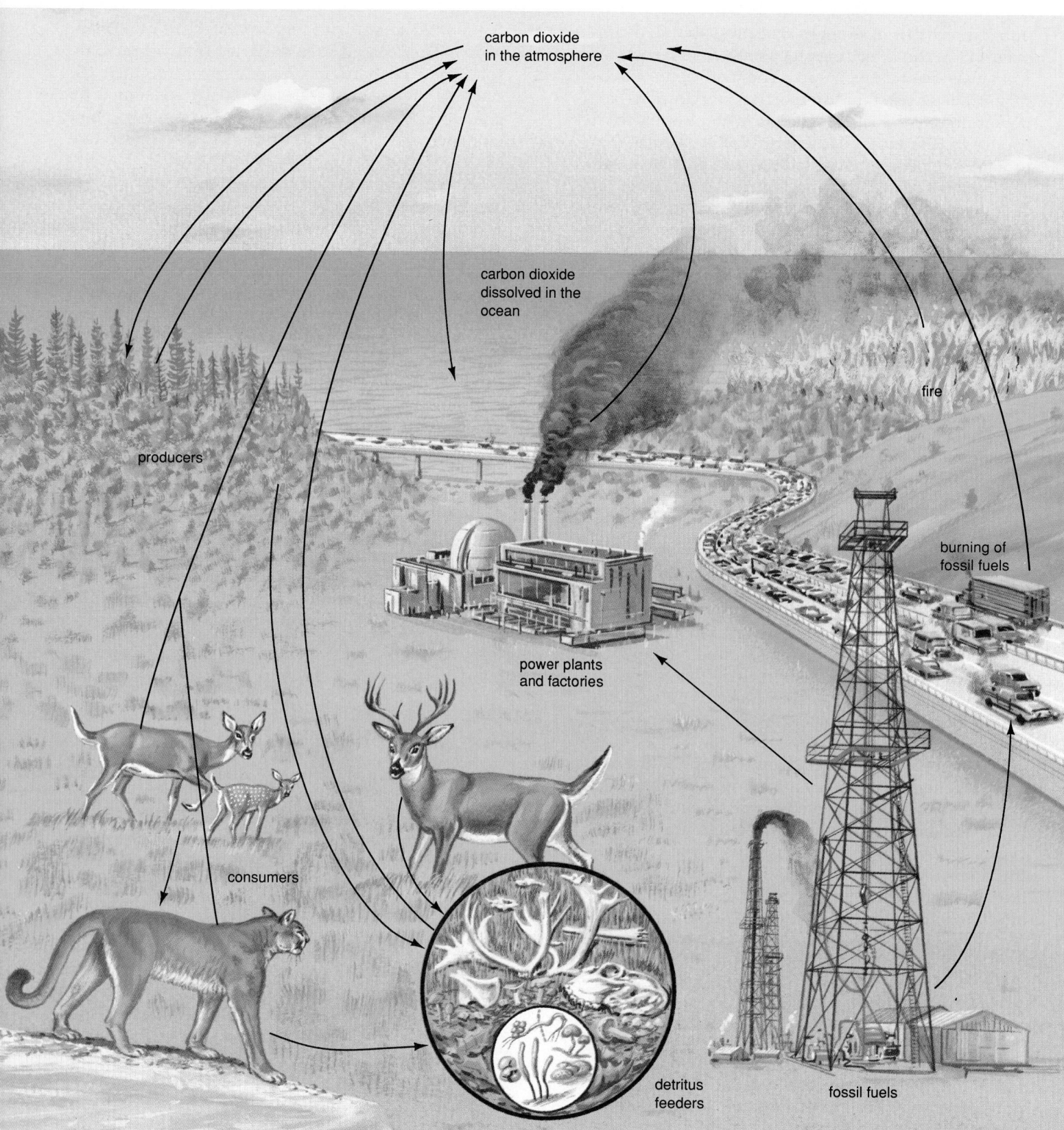

Figure 45-8 A simplified carbon cycle. Carbon is captured from the atmosphere during photosynthesis and passed up through the trophic levels. It is released during respiration from all trophic levels, and by the burning of forests and fossil fuels.

duced by using the energy in fossil fuels to artificially "fix" atmospheric nitrogen.

Plants incorporate the nitrogen from ammonia and nitrates into amino acids, proteins, nucleic acids, and vitamins. These nitrogen-containing molecules from the plant are eventually consumed, either by primary consumers, detritus feeders, or decomposers. As it is passed through the food web, some of the nitrogen is liberated in wastes and dead bodies, which decomposer bacteria convert back to nitrates and ammonia. The cycle is balanced by a continuous return of nitrogen to the atmosphere by **denitrifying bacteria.** These residents of mud, bogs, and estuaries break down nitrates, releasing nitrogen gas back to the atmosphere (Fig. 45-9).

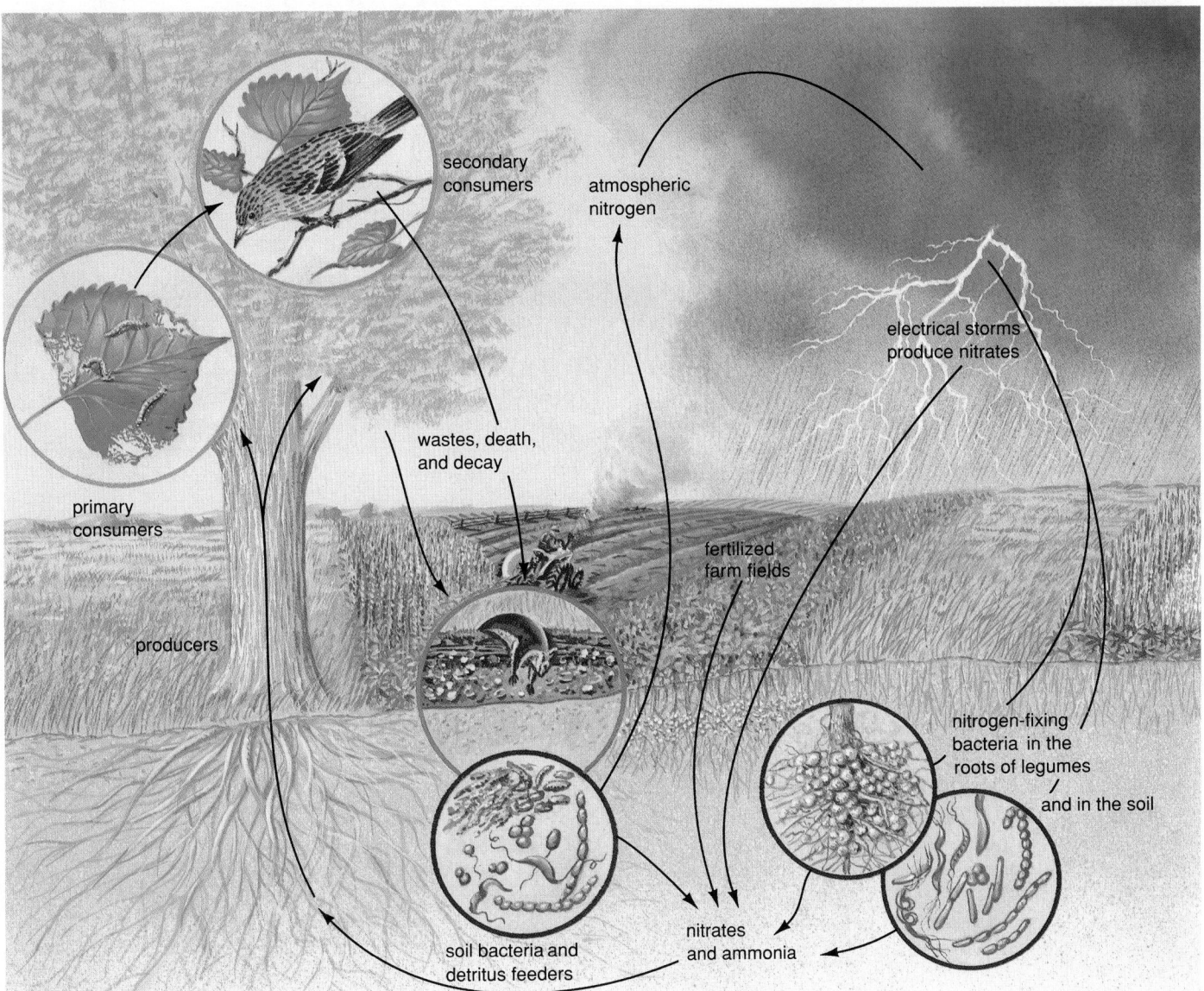

Figure 45-9 The nitrogen cycle. From its reservoir in the atmosphere, nitrogen gas is combined with oxygen to form nitrates by lightning or by the burning of forests or fossil fuels and carried to Earth dissolved in rain. Nitrogen-fixing bacteria produce ammonia. Nitrates and ammonia are also synthesized by humans for use in fertilizers. These are absorbed by plants and other producers and incorporated into biological molecules that are passed up through the trophic levels. Nitrates and ammonia are released by excretion or by decomposer bacteria. Other bacteria convert these molecules back to atmospheric nitrogen, completing the cycle.

PLANET WATCH
DDT, Biological Magnification, and the Complex Web of Life

In the 1940s, the properties of the new insecticide DDT seemed close to miraculous. In less-developed countries, especially those in the tropics, DDT saved millions of lives by killing the mosquitos that spread malaria. Increased crop yields resulting from DDT's destruction of insect pests saved millions more from starvation. DDT is long-lasting, so a single application keeps on killing. The inventor was awarded the Nobel Prize, and people looked forward to a new age of freedom from insect pests. Little did they realize that the indiscriminant use of this pesticide was unravelling the complex web of life.

In the mid 1950s, the World Health Organization sprayed DDT on the island of Borneo to control malaria. Soon, unexpected side effects appeared. Many insects inhabiting thatched-roofed houses dropped dead, to the delight of the islanders. Unfortunately, however, a caterpillar that fed on the thatch was relatively unaffected, while the wasp that preyed on it was destroyed. Delight turned to dismay as thatched roofs collapsed, eaten by the uncontrolled caterpillars. Since DDT is persistent, gecko lizards that ate the poisoned insects accumulated high concentrations of DDT in their bodies. Both they, and the village cats that ate the geckos, died of DDT poisoning. With the cats eliminated, the rat population exploded and the village was threatened with an outbreak of plague. This was avoided only by airlifting new cats to the villages.

Meanwhile, in the United States, wildlife biologists during the 1950s and 1960s witnessed an alarming decline in populations of several predatory birds, especially fish-eaters such as bald eagles, cormorants, ospreys, and brown pelicans. These top predators are never abundant, and the decline pushed the brown pelican close to extinction. Suspicion fell on DDT. The aquatic ecosystems supporting these birds had been sprayed to control insects, but the amounts of DDT used did not seem high enough to account for the toxic effects. Scientists who analyzed the bodies of predatory birds were amazed to find concentrations of DDT in their bodies up to *one million* times the amount sprayed over the water to kill insects. This led to the discovery of **biological magnification,** the process by which toxins accumulate in increasingly high concentrations at higher trophic levels. In 1973, DDT was banned in the United States, although it is still used in some less-developed countries.

The explanation for biological magnification re-

quires an understanding both of the pyramid of energy and of certain properties of this insecticide. DDT and other substances that undergo biological magnification have two insidious properties: (1) they do not readily break down into harmless substances, and (2) they are fat-soluble, but not water-soluble. Thus, they are not excreted in the watery urine, but rather accumulate in the fat of animals. Since the transfer of energy from lower to higher trophic levels is extremely inefficient, herbivores must eat large quantities of plant material (which may have been sprayed with DDT), carnivores must eat many herbivores, and so on. DDT is not excreted, so the predator gets the accumulated dosage from all its prey over a long period of time. Thus DDT reaches its highest levels in top predators, as illustrated by the following example.

For many years, DDT was sprayed on the marshes of Long Island to control mosquitos. Instead of being washed out to sea and diluted, the toxic residues accumulated on detritus and built up in detritus feeders such as shrimp. Since fish such as minnows must eat many shrimp each day, they accumulated all the DDT from thousands of shrimp during their lifetimes. Predatory fish such as needlefish and pickerel consume many minnows daily, acquiring all the DDT that each minnow had accumulated and storing it in fat. Cormorants and merganser ducks, feeding on a variety of fish, received massive doses of DDT with each meal. Table E45-1 shows the increasing levels of DDT in animals in increasingly high trophic levels.

Table E45-1   Biological Magnification of DDT on Long Island

Found In	DDT (parts per million)
Water	0.00005
Plankton	0.04
Sheepshead minnow	0.94
Pickerel	1.33
Needlefish	2.07
Merganser duck	22.8
Cormorant	26.4

DDT in sufficiently high doses will kill adult birds directly; at sublethal concentrations its effects are less direct but equally disastrous. For example, DDT interferes with the deposition of calcium in eggshells, resulting in extremely fragile eggs that are frequently crushed by the parents in the nest (Fig. E45-1). Unable to reproduce, predatory bird populations rapidly declined. Today, the eggshells of peregrine falcons are still, on the average, 14% to 18% thinner than they were in 1947, before the introduction of DDT. This serves as dramatic testimony to the persistence of this pesticide in the food web.

Although DDT is the best known example, a number of substances are subject to biological magnification, including other pesticides related to DDT (called **chlorinated hydrocarbons**), mercury, and some radioactive compounds. When the Atomic Energy Commission's Hanford plant in eastern Washington released trace quantities of radioactive phosphorus into the Columbia River in the 1950s, geese nesting downstream later laid eggs in which radioactive phosphorus was concentrated two million-fold over the amount in the river. Mercury from ocean pollution reached such high levels in swordfish (top predators) in the 1960s that the sale of swordfish meat was temporarily banned.

Understanding the properties of pollutants and the workings of food webs has become increasingly important to prevent both human health hazards and widespread loss of wildlife. Humans have considerable cause for concern, because we feed high on the food chain. When we eat a tuna sandwich, for example, we are tertiary or even quartenary consumers. In addition, our long life span provides more time for substances stored in the body to accumulate to toxic levels. For example, prior to its ban, DDT in nursing mothers' milk sometimes exceeded the safe limit for human consumption.

Figure E45-1 DDT interferes with calcium deposition in the eggs of predatory birds, who receive high doses due to their position in the food chain. The resulting fragile eggs, such as that seen on the left, are often accidentally crushed in the nest by the brooding parent.

The Phosphorus Cycle— A Sedimentary Cycle

Phosphorus is a crucial component of biological molecules, including the energy transfer molecules (ATP) and (NADP), nucleic acids, and the phospholipids of cell membranes. It is a major component of vertebrate teeth and bones. The reservoir of phosphorus in ecosystems is crystalline rock, where it is found as phosphate (PO_4^{3-}). Since phosphorus does not enter the atmosphere, the phosphorus cycle is called a **sedimentary cycle.** As phosphate-rich rocks are exposed and eroded, rainwater dissolves the phosphate ion. Dissolved phosphate is readily absorbed through the roots of plants, and by other autotrophs such as photosynthetic protists and cyanobacteria. From these producers, phosphorus is passed through food webs (Fig. 45-10). At each level, excess phosphate is excreted. Ultimately, decomposers return the phosphorus remaining in dead bodies back to the soil and water in the form of phosphate. Here it may be reabsorbed by autotrophs, or it may become bound to sediment and eventually reincorporated into rock. Some of the phosphate dissolved in fresh water is carried to the oceans. Although much of this phosphate ends up in marine sediments, some is absorbed by marine producers and eventually incorporated into the bodies of invertebrates and fish. Some of these, in turn, are consumed by seabirds, who excrete large quantities of phosphorus back onto the land. The guano deposited by seabirds along the western coast of South America was mined and provided a major source of the world's phosphorus. Phosphate-rich rock is also mined, and the phosphate is incorporated into fertilizer. Soil that erodes from fertilized fields carries large quantities of phosphates into lakes, streams, and the ocean, where it stimulates the growth of producers.

The Water Cycle

The water cycle, or **hydrologic cycle** (Fig. 45-11), differs from most other nutrient cycles since most water remains chemically unchanged throughout the cycle. The major reservoir of water is the ocean, which covers about three quarters of the Earth's surface and contains over 97% of the available water. The hydrologic cycle is driven by solar energy, which evaporates water, and by gravity, which draws the water back to earth in the form of precipitation (rain, snow, sleet, dew). Most evaporation occurs from the oceans, and much water returns directly to them by precipitation. Water falling on land takes more varied paths. Some is evaporated from the soil, lakes, and

streams. A portion runs off the land back to the oceans, and a small amount enters underground reservoirs. Since the bodies of living things are roughly 70% water, some of the water in the hydrologic cycle enters the biotic portion of ecosystems. It is absorbed by the roots of plants, and much of this is evaporated back to the atmosphere from their leaves. A small amount is combined with carbon dioxide during photosynthesis to produce high-energy molecules. Eventually these are broken down during cellular respiration, releasing water back to the environment. Heterotrophs get water from their food or by drinking.

Although the hydrologic cycle would continue in the absence of living things, the distribution of life and the composition of biological communities depends on and, to a great extent is determined by, patterns of precipitation and evaporation. As is discussed in more detail in Chapter 46, the operation of the hydrologic cycle is considerably different over a desert and a tropical rain forest, and this difference is reflected in the composition of the communities in these ecosystems.

Human Intervention in Energy Flow and Nutrient Cycling: Acid Rain and the Greenhouse Effect

Many of the environmental problems that beset modern society are the result of human interference in the natural processes of ecosystem functioning. Primitive people were sustained solely by the energy flowing from the sun, and produced wastes that were readily assimilated back into the nutrient cycles. But as the population grew and technology increased, humans began to act more and more independently of these natural processes. Mining the Earth, we procured substances such as lead, arsenic, cadmium, mercury, oil, and uranium that are foreign to natural ecosystems and toxic to many of the organisms in them (Fig. 45-12). In our factories, we synthesize substances never before found on Earth: pesticides, solvents, and a wide array of other industrial chemicals harmful to many forms of life. The Industrial Revolution, which began in the mid-nineteenth century, resulted in a tremendous increase in our reliance on energy from fossil fuels (rather than sunlight) for heat, light, transportation, industry, and even agriculture.

In this section we describe two environmental problems of global proportions that are a direct result of human reliance on fossil fuels: acid rain and the greenhouse effect. **By intervening to augment the**

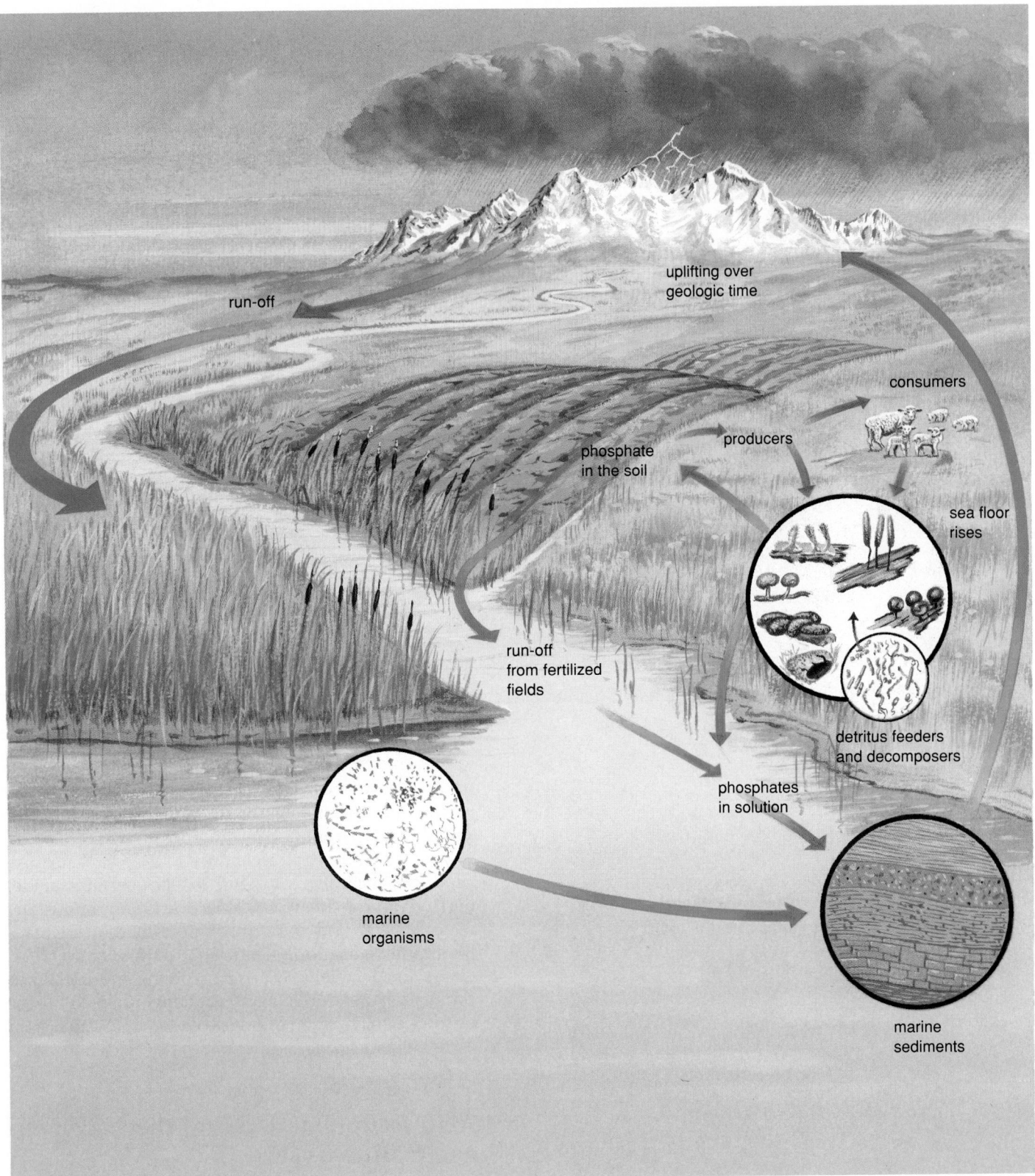

Figure 45-10 The phosphorus cycle. Phosphate leaches from phosphate-rich rocks or from fertilizers and enters plants and other producers where it is incorporated into biological molecules. These are passed through the trophic levels. Phosphate is excreted or returned to the soil and water by decomposer bacteria. It may then be reused by producers or eventually incorporated into rock.

precipitation
over land

evaporation from land
and transpiration
from plants

evaporation
from the
ocean

groundwater

precipitation
over the ocean

surface
run-off

ocean

groundwater
seepage

Figure 45-11 The hydrologic cycle is the simplest of the nutrient cycles.

Figure 45-12 An oil-soaked sea bird is cleaned by rescuers following an oil spill off the coast of Alaska.

natural flow of energy, we have also disrupted the cycling of certain nutrients, with unexpected and possibly catastrophic consequences.

When coal and oil are burned, they produce by-products; including nitrogen oxides, sulfur dioxide, and carbon dioxide (CO_2). These are found naturally in ecosystems and, in normal amounts, they are readily assimilated into the nitrogen, sulfur, and carbon cycles, respectively. But our growing industrialized population produces such vast quantities of these nutrients that we have far exceeded the capacity of ecosystems to process them.

Acid Rain: Overloading the Nitrogen and Sulfur Cycles

Each year, the United States alone discharges about 30 million tons of sulfur dioxide into the atmosphere, two thirds of it from power plants burning coal or oil (Fig. 45-13). The rest is largely a by-product of industrial boilers, smelters, and refineries. Although volca-

Figure 45-13 Power plants burning high-sulfur coal with inadequate emission controls are a major source of atmospheric sulfur dioxide, the prime contributor to acid rain.

nos and hot springs also release sulfur dioxide, human industrial activities account for 90% of the sulfur dioxide in the atmosphere. Twenty-five million tons of nitrogen oxides are also released by the United States each year, 40% from transportation vehicles such as automobiles, trucks, and planes, 30% by power plants, and most of the rest by industrial sources.

Both sulfur and nitrogen are crucial to life; for example, both are found in amino acids. However, despite their importance as nutrients, excess production of these substances was identified in the late 1960s as the cause of an insidious environmental threat: acid rain. **Combined with water vapor in the atmosphere, nitrogen oxides are converted to nitric acid, and sulfur dioxide to sulfuric acid. Days later, and often hundreds or thousands of miles from the source, the corrosive acids fall, either dissolved in rain or as microscopic particles. The acid can damage trees and crops, render lakes lifeless, and damage statues and buildings.**

The term *acid rain* was coined in 1872 by British chemist Robert Angus Smith, who observed that the fallout from industrial English cities was actually corroding away buildings. The industrial acids were carried to Scandinavia, where Norwegian officials noted a decline in fish production in increasingly acidic lakes and rivers. It wasn't until 1967, however, that the Swedish soil scientist Svante Oden (now called the "father of acid rain") developed a comprehensive theory that encompassed the origin, transport, and effects of acid rain.

While both sulfuric and nitric acids form solutions in water vapor, sulfuric acid may also form particles under dry conditions. These particles, which may visibly cloud the air, are deposited even in the absence of rain, snow, or fog. **Acid deposition** is the term used to describe both the wet and dry acid assaults on the environment.

Although sources of sulfur dioxide are widely distributed, in the United States they tend to be concentrated in the upper Ohio Valley, Indiana, and Illinois, where high-sulfur coal is burned in old power plants with few emission controls. As complex chemical reactions in the atmosphere produce sulfuric and nitric acids, winds carry them toward New England, where lakes and forests are particularly vulnerable. Some areas of the country have rock rich in calcium carbonate, a natural buffer for acidity (and a major ingredient in some remedies sold for "acid indigestion"). In New England, however, hard rock such as granite and basalt predominates, and cannot neutralize the acid rain. Although the Northeast has been the most seriously damaged it is not the only sensitive area; about 75% of the United States is at least moderately vulnerable to acid rain owing to a lack of buffering capacity in the rocks and soil (Fig. 45-14).

Damage to Aquatic Ecosystems

In the Adirondack Mountains of New York, acid rain has rendered about 25% of all the lakes and ponds too acidic to support fish. But before the fish die, much of the food web that sustains them is destroyed. Clams, snails, crayfish, and insect larvae die first, then amphibians, and finally fish. The result is a crystal clear lake, beautiful—but dead. The impact is not limited to aquatic organisms. The loss of insect larvae and crustaceans has probably contributed to a dramatic decline in the population of black ducks, which rely on these freshwater invertebrates as a critical source of protein for growth and reproduction. In a desperate attempt to counteract the acidity, researchers recently dumped 1,000 tons of limestone on the land

Figure 45-14 The most acidic rain is concentrated in the northeast, but thousands of lakes in the west are also vulnerable. (Source: National Acid Precipitation Assessment Program.)

pH Levels

- under 4.3
- 4.3 - 4.4
- 4.5 - 4.6
- 4.7 - 5.0
- over 5.0

surrounding Wood's Lake in the Adirondack Mountains. The results are still uncertain.

Damage to Forests

Acid rain is not only a threat to lakes. Recent studies under laboratory conditions have shown that it also interferes with the growth and yield of many farm crops. Acid water percolating through the soil leaches out essential nutrients such as calcium and potassium. It may also kill decomposer microorganisms, thus preventing the return of nutrients to the soil. Plants, poisoned and deprived of nutrients, become weak and more susceptible to infection and insect attack. High in the Green Mountains of Vermont, scientists have witnessed an alarming 50% mortality among red spruces, 47% among beeches, and 32% among sugar maples since 1965. The snow, rain, and heavy fog that commonly cloak these eastern mountaintops is highly acidic. At a monitoring station atop Mt. Mitchell in North Carolina, the pH of fog even on good days is 2.9, more acidic than vinegar. In this fragile and stressful environment, acid precipitation may well be tipping the balance against these forests (Fig. 45-15). In the famed Black Forest of Germany, the devastation is even greater, with over half the trees affected. Many of the firs have turned yellow or brown, trunks have become gnarled and roots have shrunk, and plagues of insects and fungal diseases are attacking the weakened trees. It is almost impossible to trace the source of the problem, but German

officials are convinced that acid rain from the burning of high-sulfur coal, in conjunction with other pollutants, is the cause. Vigorous pollution-control programs have begun, but many fear that the forests have been irrevocably damaged. In 1984, scientists confirmed that the growth rate of eastern U.S. forests had slowed dramatically over the past several decades. Germany's problems may foreshadow similar disasters for the forests of the United States.

Increased Exposure to Toxins

Acid deposition not only harms animals and plants directly, it also increases their exposure to toxic metals, which are relatively inert in natural ecosystems. Metals, including aluminum, lead, nickel, mercury, cadmium, and copper, are far more soluble in acidified water. Metals such as aluminum may inhibit plant growth. Other metals, absorbed by plants along with acidified groundwater, may threaten animals higher on the food chain. In Sweden, the kidneys and livers of moose grazing on plants grown in acidified water sometimes contain enough cadmium to be lethal to a person eating them. In acid lakes and streams, aluminum leached from the surrounding rock and soil causes mucus to accumulate in the gills of fish, suffocating them. Drinking water in some households has been found dangerously contaminated with lead dissolved by acid water from lead pipes. Boston's water supply must be chemically treated to counteract the acidity. Fish in acidified

Figure 45-15 Researcher examines the devastation caused by acid rain atop Mt. Mitchell in North Carolina.

water have been found to have dangerous levels of mercury in their bodies. The combination of acid and mercury results in the formation of methylmercury, an extremely toxic compound that is subject to biological magnification as it is passed through trophic levels.

Damage to Structures

Acid deposition is corroding our bridges, eating away at our buildings and monuments, and destroying ancient and priceless works of art at a cost estimated at $5 billion annually in the United States alone (Fig. 45-16). The east wall of the Capitol building in Washington, D.C., is pitted with small craters and looks as if it were hit by shrapnel. In Baltimore, the famous statue "The Thinker" by Rodin has been moved indoors to protect it from the corrosive rain. The stone and statuary of ancient Rome cannot be protected and are steadily being dissolved. The damage to both natural ecosystems and property caused by this insidious form of pollution is incalculable.

The Greenhouse Effect: Short-Circuiting the Carbon Cycle

Sources of Additional CO$_2$

Between 345 and 280 million years ago, under the unique conditions of the Carboniferous period, huge quantities of carbon were diverted from the carbon cycle when the bodies of plants and animals were buried in sediments, escaping decomposition. There,

heat and pressure over time converted their bodies into fossil fuels such as coal, oil, and natural gas. Since the Industrial Revolution, modern cultures have increasingly relied on the energy stored in these

Figure 45-16 This marble statue shows the corrosive effects of acid rain.

fuels. As we burn them in our power plants, factories, and cars, we release CO_2 into the atmosphere. Without human intervention, this carbon would have remained essentially untouched for millions of years, but human activities are releasing this carbon as CO_2 at a furious pace.

A second important source of additional CO_2 is global deforestation, the cutting of tens of millions of acres of forests each year. Deforestation is occurring principally in the tropics, where rain forests are rapidly being eliminated to increase the availability of agricultural land. The carbon stored in the massive trees in these forests returns to the atmosphere after they are cut, either through burning or decomposition. During the past 200 years, human activities have released as much excess CO_2 as is currently stored in all life on Earth.

The CO_2 content of the atmosphere has increased

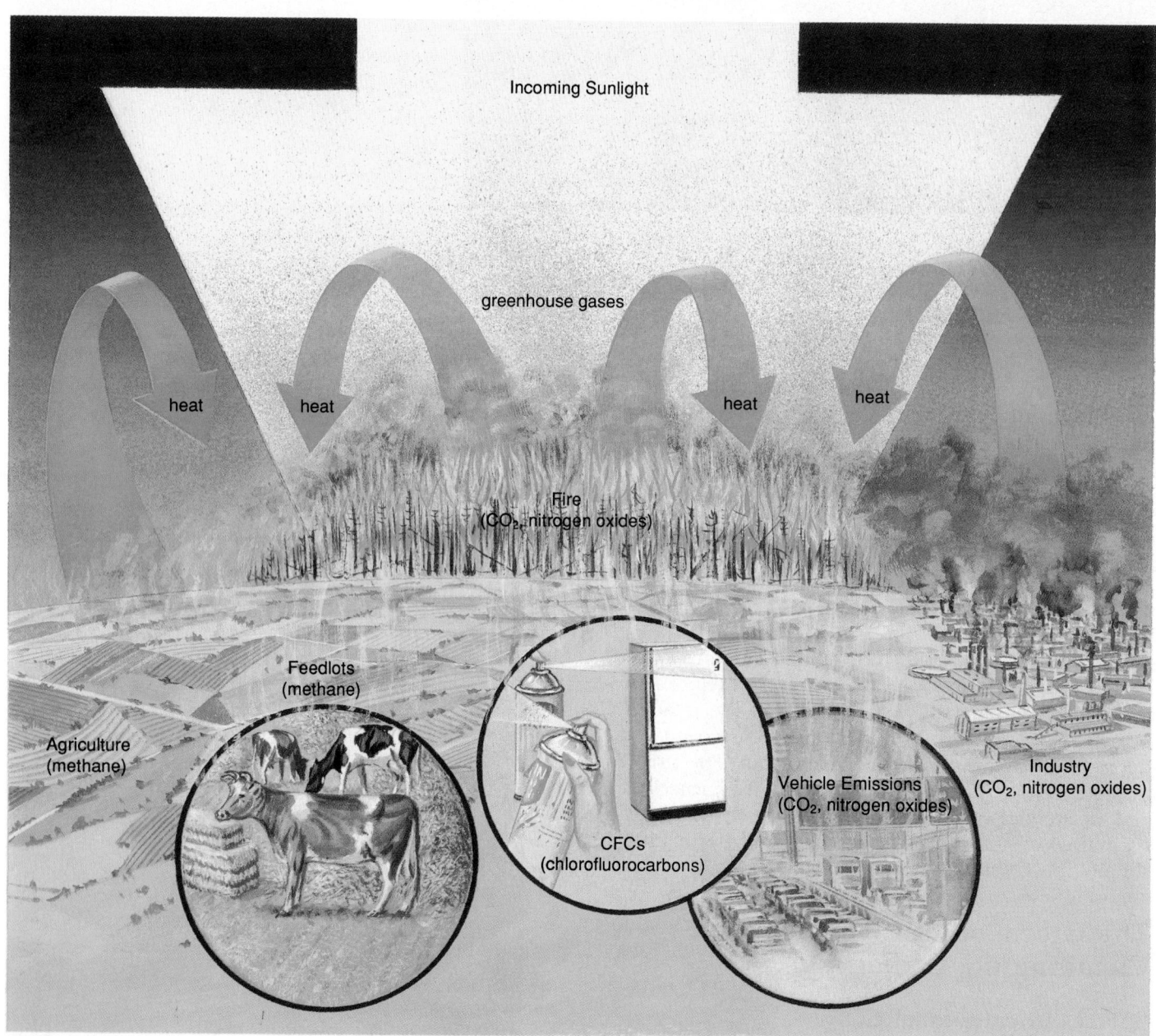

Figure 45-17 Incoming sunlight warms the surface of the Earth and is radiated back to the atmosphere. Greenhouse gases (such as CO_2 and nitrous oxides from combustion, chlorofluorocarbons from industrial activities, and methane from farms and feedlots) absorb some of this heat, trapping it in the atmosphere. The result is a gradual rise in average global temperatures.

25% since the start of the Industrial Revolution in 1850. The burning of fossil fuels is the single greatest cause of this increase. The present CO_2 content of the atmosphere is projected to double within the next century, and possibly within the next 50 years.

How the Greenhouse Effect Works

Carbon dioxide still comprises only a tiny fraction of the Earth's atmosphere, about .035%. However, CO_2 has an important property that makes its build-up a cause for concern: it traps heat. Atmospheric CO_2 acts something like the glass in a greenhouse, allowing energy in the form of sunlight to enter, but absorbing and holding that energy once it has been converted to heat (Fig. 45-17). Several other "greenhouse gases" share this property, including methane, chlorofluorocarbons, and nitrous oxide. This **greenhouse effect** could cause a rise in average global temperature of about 4° C by the middle of the next century. Although this may not seem like much, the temperature change since the peak of the last ice age was about 5° C. This change moved the Atlantic Ocean inland by about 100 miles, radically changed the species composition of forests throughout North America, and created the Great Lakes. Within the next century, the Earth may be hotter than it has been any time in the past million years, with the change occurring more rapidly than any change in Earth's history.

Consequences of Global Warming

As the ice cap and glaciers melt in response to atmospheric warming, sea level will rise, threatening coastal cities and flooding coastal wetlands. These threatened ecosystems are the breeding grounds for many species of birds, fish, shrimp, and crabs, whose populations may be severely diminished.

Another serious consequence of the warming trend is a shift in the global distribution of temperature and rainfall. Even small temperature changes can dramatically alter the paths of major air and water currents. This would change precipitation patterns in unpredictable ways. Some land that is currently cultivated only with the help of irrigation might become too dry for agriculture. Other areas might receive more rain. Agricultural disruption could be disastrous for human populations already on the brink of starvation.

The impact of warming on forests could be catastrophic. The distribution of tree species is exquisitely sensitive to average annual temperature, and small changes could dramatically alter the extent and species composition of our forests. A recent study of eastern hemlock, yellow birch, beech, and sugar maple found that the temperature increases caused by a doubling of CO_2 concentration would result in a 500 to 1500 kilometer shift northward in the range of these species. Sugar maple could disappear from the United States, except for a remnant population in Maine. Beech, currently ranging from Georgia to southern Canada, might also be nearly eliminated from the United States by the warming climate. Although temperatures further north would become hospitable for these trees, the ability of forests to move northward is limited by slow seed dispersal. In addition, the patterns of rainfall and the composition of the soil might prevent northern movement, causing some species to die out entirely. Meanwhile, the great forests of Mississippi and Georgia might be replaced by grassland.

Is the greenhouse effect merely a speculative future scenario? Recent NASA studies have revealed a global temperature increase of 0.5 to 0.7° C since 1860, paralleling the rise in atmospheric CO_2 and methane levels (Fig. 45-18). Much of this increase has occurred in the past decade; the 1980s included the 6 warmest years ever recorded, and 1990 set still another record for high temperatures world wide.

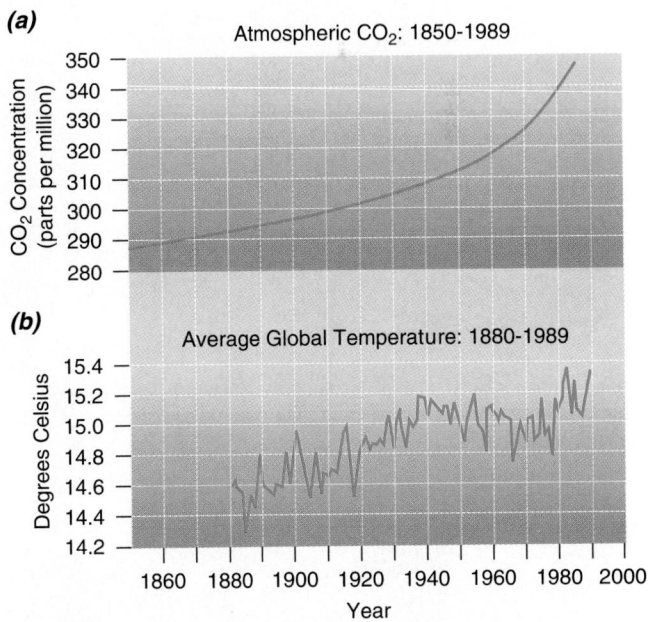

Figure 45-18 (a) The CO_2 concentration of the atmosphere has shown a steady increase since the Industrial Revolution. **(b)** Average global temperatures have also shown a gradual increase during the past century, paralleling the increasing atmospheric CO_2. (Data from NASA's Goddard Institute for Space Studies.)

ℝℝeflections: What Can We Do?

Unfortunately, neither acid rain nor the greenhouse effect are problems about which we can afford to "wait and see." We are already beginning to experience their effects. In full force, they will be both catastrophic and essentially irreversible. A major Environmental Protection Agency study released in 1989 recommends shifting to energy-efficient technologies, reversing forest destruction, starting a major tree-planting campaign, and increasing fuel prices to encourage efficiency. Americans must take the lead in this effort. In our extravagant use of energy, we produce 18 tons of CO_2 per person each year, nearly five times the world average. All pro-

grams that conserve energy will help reduce acid rain and postpone the greenhouse effect (Fig. E45-2). A car that gets 20 miles per gallon emits 1 pound of CO_2 per mile, so we must increase the fuel efficiency of our cars, reduce our driving, and develop and use efficient mass transit. Electricity generated in fossil-fuel–fired power plants is a tremendous contributor of CO_2, sulfur dioxide, and nitrogen oxides. In addition to developing other sources of electricity, such as photovoltaic cells, we must purchase efficient appliances, turn off lights, and reduce our use of electricity for heat and cooling. Insulating and weatherproofing our homes and offices

(a)

(b)

will significantly reduce fuel consumption, as will incorporating solar energy features into all new building designs. Recycling is also a tremendous energy saver—for example, 95% of the energy used to produce an aluminum can is conserved when it is recycled. Finally, all countries must work to stabilize their populations, since all environmental advances are undermined by the steady growth of the human population.

Unfortunately, governments (and many citizens) tend to be unwilling to even begin to take the steps necessary to reverse acid rain and the greenhouse effect. This is ironic, since conserving energy will both save money and increase our quality of life. Imagine a city with clean air, shaded by abundant trees no longer threatened by acid rain and pollution. Imagine reading newspapers as we travel to work on fast monorails instead of sitting in traffic jams, breathing exhaust fumes. Imagine a countryside no longer scarred by mines, forest clearcuts, and landfills because people (like natural ecosystems) now recycle materials. Old habits die hard, but we have everything to gain by making the effort, while we risk losing everything by inaction.

(c)

Figure E45-2 Some solutions. **(a)** Trees not only absorb and store CO_2, they also provide shade, reducing the desire for air conditioning. **(b)** Reducing our reliance on fossil fuels will require alternative methods of generating power and heating our homes. **(c)** Efficient mass transit not only conserves energy and reduces pollution, it can save time and alleviate frustration.

SUMMARY OF KEY CONCEPTS

Ecosystems are sustained by a continuous flow of energy from sunlight and a constant recycling of nutrients.

The Flow of Energy

Primary Producitivity Energy enters the biotic portion of ecosystems when it is harnessed by autotrophs during photosynthesis. Primary productivity is the amount of energy that autotrophs store in organic material over a given period of time. It is extremely high in ecosystems such as estuaries and tropical rain forests, where light, nutrients, appropriate temperature, and adequate water are available, and low in deserts and the open ocean, where the availability of some of these factors is restricted.

Trophic Levels, Food Chains, and Food Webs Trophic levels describe feeding relationships in ecosystems. Autotrophs are the producers, the lowest trophic level. Herbivores occupy the second level as primary consumers. Carnivores act as secondary consumers when they prey on herbivores, or tertiary or higher-level consumers when they eat other carnivores.

Feeding relationships in which each trophic level is represented by one organism are called food chains. In natural ecosystems, feeding relationships are far more complex and are described as food webs.

Detritus feeders and decomposers, which digest dead bodies and wastes, use and release the energy stored in these substances and free nutrients for recycling.

Energy Flow Through Trophic Levels and Energy Pyramids In general, only about 10% of the energy captured by organisms at one trophic level is converted to the bodies of organisms in the next highest level. The higher the trophic level, the less energy is available to sustain it. As a result, plants are more abundant than herbivores, and herbivores more common than carnivores. The storage of energy at each trophic level is illustrated graphically as an energy pyramid.

The Cycling of Nutrients

A nutrient cycle depicts the movement of a particular nutrient from its reservoir (usually in the abiotic portion of the ecosystem) through the biotic portion of the ecosystem and back to its reservoir, where it is again available to the producers.

The Carbon Cycle—An Atmospheric Cycle The reservoir for carbon is CO_2 gas, found in the atmosphere. Carbon enters the producers through photosynthesis. From these autotrophs it is passed through the food web and released to the atmosphere as CO_2 during cellular respiration.

The Nitrogen Cycle—An Atmospheric Cycle Although the atmosphere is mainly nitrogen gas, producers are unable to use nitrogen in this form. First, it must be converted to ammonia or nitrate. Nitrogen gas is converted to ammonia by nitrogen-fixing bacteria in soil or in root nodules of legumes. Other soil bacteria convert ammonia to nitrates. Humans synthesize ammonia and nitrates from nitrogen gas for fertilizer. Lightning and other forms of combustion produce nitrates. Nitrogen passes from producers to consumers and is returned to the environment through excretion and the activities of detritus feeders and decomposers. Some release nitrates, some ammonia, and others convert nitrates back to nitrogen gas.

The Phosphorus Cycle—A Sedimentary Cycle Phosphorus is found in biological molecules, teeth, and bones. Its reservoir is in rocks in the form of phosphate, which dissolves in rainwater. Dissolved phosphate is absorbed by photosynthetic organisms, then passed through food webs. Some is excreted and the rest is returned to the soil and water by decomposers. Some reenters the cycle, while some is carried to the ocean to be deposited in marine sediments. Humans mine phosphate-rich rock to produce fertilizer.

The Water Cycle The major reservoir of water is the oceans. Water is evaporated by solar energy, and returned to Earth as precipitation. The water flows into lakes, underground reservoirs, and in rivers to the ocean. Water is absorbed directly by plants and animals, and is also passed through food webs. A relatively minuscule amount is combined with CO_2 during photosynthesis to form sugars.

Human Intervention:
Acid Rain and the Greenhouse Effect

Environmental disruption occurs when humans interfere with the natural functioning of ecosystems. Pollution occurs when human activities produce more nutrients than nutrient cycles can efficiently process. It also occurs when human beings produce and release chemicals that are foreign and toxic to the natural community. By augmenting the flow of energy through massive consumption of fossil fuels, we have disrupted the natural cycles of carbon, sulfur, and nitrogen, with serious environmental consequences.

Acid Rain: Overloading the Nitrogen and Sulfur Cycles Acid deposition occurs when oxides of nitrogen and sulfur, produced by burning fossil fuels, combine with water in the atmosphere to form nitric and sulfuric acids. These acids are deposited, sometimes at distant sites. Rock containing calcium carbonate can help neutralize the acidity, but large parts of the country, and particularly New England, lack buffering rocks and soils and are vulnerable to damage. Acid deposition harms aquatic ecosystems by destroying aquatic communities. Forests, both here and in Germany, have been seriously harmed as well. Crop production is reduced. Acidified water dissolves heavy metals and increases the exposure of plants and animals (including humans) to these toxins. Buildings, statues, and bridges are also being corroded, at tremendous cost.

The Greenhouse Effect: Short-Circuiting the Carbon Cycle The tremendous increase in the burning of fossil fuels, which began in the mid-1800s, has caused a 25% increase in the CO_2 content of the atmosphere since that time. Global deforestation has also contributed, since trees absorb and store CO_2. Atmospheric CO_2 and other greenhouse gases such as methane trap solar heat, and are believed to be responsible for a slight but significant increase in global world temperature since 1860. Based on CO_2 production rates, atmospheric CO_2 levels could double within the next 50 to 100 years, causing a rise in temperature that would severely alter the Earth's climate. This would cause a rise in sea level, massive agricultural disruption, and a shift in the range of plant and animal communities that could result in a major loss of species.

Solutions to both acid rain and the greenhouse effect lie in reducing our use of fossil fuels and undertaking a major reforestation effort.

GLOSSARY

acid deposition: the deposition of nitric or sulfuric acid, either in precipitation or in particulate form, as a result of the production of nitrogen oxides or sulfur dioxide through burning, primarily of fossil fuels.

atmospheric cycle: a nutrient cycle for a substance whose primary reservoir is the atmosphere.

autotroph (aut'-ō-trof): a "self-feeder," usually meaning a photosynthetic organism.

biological magnification: the increasing accumulation of a toxic substance in increasingly high trophic levels.

carnivore (kar'-neh-vōr): literally "meat eater," a predatory organism feeding on other heterotrophs.

chlorinated hydrocarbon: an organic compound that includes a chain of carbon atoms, some of which have chlorine atoms bonded to them, including DDT and several other pesticides.

consumer: an organisms that eats other organisms; a heterotroph.

decomposers: a group of decay organisms, mainly fungi and bacteria. These digest organic material by secreting digestive enzymes into the environment. In the process they liberate nutrients into the environment.

denitrifying bacteria (dē-nī'-treh-fī-ing): bacteria that break down nitrates, releasing nitrogen gas to the atmosphere.

detritus feeders (de-trī'-tus): a diverse assemblage of organisms ranging from worms to vultures that live off the wastes and dead remains of other organisms.

energy pyramid: a graphical representation of the energy contained in succeeding trophic levels, with maximum energy at the base (primary producers) and steadily diminishing amounts at higher levels.

food chain: an illustration of feeding relationships in an ecosystem using a single representative from each of the trophic levels.

food web: a relatively accurate representation of the complex feeding relationships within an ecosystem, including many organisms at various trophic levels, with many of the consumers occupying more than one level simultaneously.

greenhouse effect: the ability of certain gases such as carbon dioxide and methane to trap sunlight energy in the atmosphere as heat. The glass in a greenhouse does the same, hence the name. This warming of the atmosphere is being enhanced by human production of these gases.

herbivore (erb'-i-vōr): literally "plant-eater"; an organism that eats plants.

heterotroph (het'-er-ō-trōf'): an organism that eats other organisms.

legumes (leg'-yūm): a group of plants (including alfalfa, clover, and soybeans) that harbor colonies of nitrogen-fixing bacteria in swellings or nodules on their roots.

macronutrients: molecules used in relatively large quantities in the metabolic activities of organisms.

micronutrients: molecules required by organisms in trace quantities.

net primary productivity: the energy stored in the primary producers of an ecosystem over a given period.

nitrogen fixation: the process of converting atmospheric nitrogen into ammonia.

nutrient cycle: a description of the movement of a specific inorganic nutrient (carbon, nitrogen, water, etc.) through the living and non-living portions of an ecosystem.

primary consumer: an organism that feeds on producers; an herbivore.

producer: a photosynthetic organism, synonymous with autotroph.

reservoir: the major source of any particular nutrient in an ecosystem, usually in the abiotic portion.

secondary consumer: an organism that feeds on primary consumers; a carnivore.

sedimentary cycle: a nutrient cycle for a substance whose primary reservoir is sediment, such as soil or rock.

tertiary consumer (ter'-shē-ār-ē): A carnivore that feeds on other carnivores.

STUDY QUESTIONS

1. What are the two basic laws of ecosystem function concerning energy and nutrients?
2. What is an autotroph? What trophic level does it occupy, and what is its importance in ecosystems?
3. Define primary productivity. Based on information presented here and in Chapter 44, would you predict higher productivity in a farm pond or an alpine lake? Defend your answer.
4. List the first three trophic levels. Among the consumers, which are most abundant? Relate your answer to the "10% Law."
5. How do food chains and food webs differ? Which is the most accurate representation of actual feeding relationships in ecosystems?

6. Define detritus feeders and decomposers and explain their importance in ecosystems.
7. Trace the movement of carbon from its reservoir, through the biotic community and back to the reservoir. How have human activities altered the carbon cycle, and what are the implications for future climate?
8. How does nitrogen get from the air to a plant?
9. Trace the movement of a water molecule from the moment it leaves the ocean until it eventually reaches a plant root, then a plant stomata, and then makes its way back to the ocean.
10. Trace a phosphorus molecule from a phosphate-rich rock into the DNA of a carnivore.

DISCUSSION QUESTIONS

1. What could your college or university do to reduce its contribution to acid rain and the greenhouse effect? Be specific and, if possible, offer practical alternatives to current practices.
2. Make a list of manufactured products and agricultural products that you use but that contribute to the greenhouse effect. Based on this list, discuss actions you could take to reduce your contribution to this global environmental problem.
3. Relate fossil fuel consumption to (a) the decline of forests in the Northeast and (b) the loss of aquatic life in lakes in the Northeast and Canada. Trace each step from the burning of gasoline in a car or power plant to the death of organisms.
4. Define and give an example of biological magnification. What qualities are present in materials that undergo biological magnification. In what trophic level are the problems worst, and why?
5. Discuss the contribution of population growth to (a) acid rain, (b) the greenhouse effect.

SUGGESTED READINGS

Cloud, P., and Gibor, A. "The Oxygen Cycle." *Scientific American*, September 1970. Traces the movement of oxygen through the living and nonliving parts of ecosystems.

The Earthworks Group. "Fifty Simple Things You Can Do to Save the Earth." Earthworks Press, Berkeley, 1989. Provides practical, simple advice on how individuals can make a difference.

Ehrlich, P. R. *The Machinery of Nature*. New York: Simon and Schuster, 1986. An eloquent description of the workings of the natural world, written for the educated layperson.

"Energy for Planet Earth." *Scientific American*, September 1990.

Gosz, J. R., Holmes, R. T., Likens, G. E., and Bormann, F. H. "The Flow of Energy in a Forest Ecosystem." *Scientific American*, March 1978. Gives experimental data about a specific ecosystem, relating it to the general principles of energy flow through ecosystems.

Graedel, T. E., and Crutzen, P. J. "The Changing Atmosphere." *Scientific American*, September 1989. Describes the deleterious effects of human activities on the chemistry of the atmosphere.

Houghton, R. A., and Woodwell, G. M. "Global Climate Change." *Scientific American*, April 1989. An authoritative overview of the evidence that the greenhouse effect is altering the global climate.

Hutchinson, G. E. "The Biosphere." *Scientific American*, September 1970 (Offprint No 11). Overview of the Earth's thin film of life and how it is supported by the flow of energy and cycling of nutrients.

Matthews, S. W. "Under the Sun." *National Geographic*, October 1990. A comprehensive, readable exploration of the greenhouse effect, its causes, and its consequences.

May, R. M. "The Evolution of Ecological Systems." *Scientific American*, September 1978 (Offprint No 1404). Co-evolution of species is examined in light of their interrelationships within complex food webs in specific ecosystems.

Peakall, D. B. "Pesticides and the Reproduction of Birds." *Scientific American*, April 1970. How chlorinated hydrocarbon insecticides interfere with the reproduction of predatory birds.

Penman, H. L. "The Water Cycle." *Scientific American*, September 1970. Traces the movement of water through the biosphere.

Schneider, S. H. "The Changing Climate." *Scientific American*, September 1989. Discussion of how rising CO_2 levels threaten to dramatically change the Earth's climate.

46

The Earth's Diverse Ecosystems

"Living creatures press up against all barriers, they fill every possible niche all the world over. . . . We see life persistent and intrusive—spreading everywhere, insinuating itself, adapting itself, resisting everything, defying everything, surviving everything!"

Sir John Arthur Thompson, a Scottish biologist (1920)

Patterns Within Diversity

In preceding chapters, we discussed the dynamics of population growth, the flow of energy, and the cycling of nutrients, along with the basic structure of ecosystems within which these processes occur. Ecosystems are extraordinarily diverse, yet clear patterns exist within this diversity. If you travel around the world, you may observe that, although the plant *species* may differ, very similar groups of plants are found wherever a particular climate exists. Deciduous forests, deserts, grasslands, tropical rain forests, coniferous forests, coral reefs, and estuaries all have distinguishing features that identify them no matter where they are found or what particular species inhabit them.

The reason we are able to place these communities into distinct categories is that each is dominated by organisms that are specifically adapted for a particular environment. The desert community, for example, is dominated by plants with unique attributes that are adaptations to heat and drought. The cacti of the American Mohave desert are strikingly similar to the euphorbia of South Africa; their spinelike leaves and thick, green, water-storing stems are adaptations for water conservation (Fig. 46-1), although these plants are in separate families and are only distantly genetically related. Likewise, the plants of the arctic tundra and those of the alpine tundra of the Rocky Mountains show growth patterns clearly recognizable as adaptations to a cold, dry, windy climate. Thus, in regions of the Earth where the environmental conditions are similar, we find similar types of organisms organized into similar types of communities.

Environments vary in the relative abundance of

(a)

(b)

Figure 46-1 Convergent evolution in response to similar environments has molded the bodies of **(a)** American cacti and **(b)** South African euphorbs into nearly identical shapes, although they are in completely different families.

◀ Coral reefs, such as that seen here surrounding the island of Bora Bora in the South Pacific, are shallow ocean ecosystems which are extremely beautiful, diverse, and fragile.

four basic resources that provide the requirements of life: nutrients, energy, water, and temperatures appropriate to metabolic reactions. The availability of sunlight, water, and appropriate temperatures determine the climate of a given region.

Climate

Life on Earth, particularly life on land, is dramatically affected by both weather and climate. **Weather** refers to short-term fluctuations in temperature, humidity, cloud cover, wind, and precipitation that vary over periods of hours or days. **Climate,** in contrast, refers

to overall patterns of weather that prevail from year to year and even century to century in a particular region. While weather affects individual organisms, climate influences and limits the overall distribution of entire species.

Solar Radiation and Climate

Both climate and weather are driven by a great thermonuclear engine: the sun. The solar energy that reaches Earth drives the wind, the ocean currents, and the global water cycle.

Before it reaches the Earth's surface, sunlight is modified by the atmosphere. An atmospheric layer

Figure 46-2 Solar energy that reaches the Earth is partially reflected, partially absorbed as heat, and then reradiated; a tiny fraction is captured by plants during photosynthesis.

high in the stratosphere that is relatively rich in ozone absorbs much of the high-energy ultraviolet radiation that can damage biological molecules (see Planet Watch: The Ozone Layer—Puncturing the Protective Shield). Dust, water vapor, and clouds scatter light, reflecting some of the energy back into space. Carbon dioxide, water vapor, ozone, methane, and other trace gases selectively absorb infrared wavelengths (heat), trapping warmth in the atmosphere. This natural greenhouse effect has been augmented by human activities, as described in Chapter 45. Only about half the solar radiation reaching the atmosphere actually strikes the Earth's surface. Of this, small fractions are reflected directly back into space and used in photosynthesis, while the rest is absorbed as heat (Fig. 46-2). Eventually, all the incoming solar energy is either reflected or reradiated as heat. The solar energy absorbed as heat by the atmosphere and the Earth's surface (about 66% of the total) is what maintains the Earth's relative warmth.

Factors That Influence Climate

Many factors influence climate. Among the most important are latitude, air currents, ocean currents, and the presence of irregularly shaped continents.

Latitude

The amount of solar radiation that strikes a given area of the Earth's surface has a major effect on average yearly temperatures. At the equator, sunlight hits the Earth's surface nearly at a right angle, making the weather there constantly warm, with little seasonal variation. Further north or south, the Earth's surface is oblique to the sun's rays, so that a given amount of sunlight is spread over a larger area, resulting in lower overall temperatures (Fig. 46-3). Because the Earth tilts on its axis as it revolves around the sun, higher latitudes also experience considerable variation in the directness of sunlight as the year progresses, resulting in pronounced seasons.

Air Currents

As the sun's direct rays fall on the tropics, heated air rises, laden with water evaporated by solar heat. As the water-saturated air rises, it cools somewhat. Since cool air cannot retain as much moisture, water condenses from the rising air and falls as rain. This creates a band around the equator, called the tropics, which is both the warmest and wettest region on Earth. The cooler but still relatively warm dry air then flows north and south from the equator, cooling further as it travels. At around 30° north and south lati-

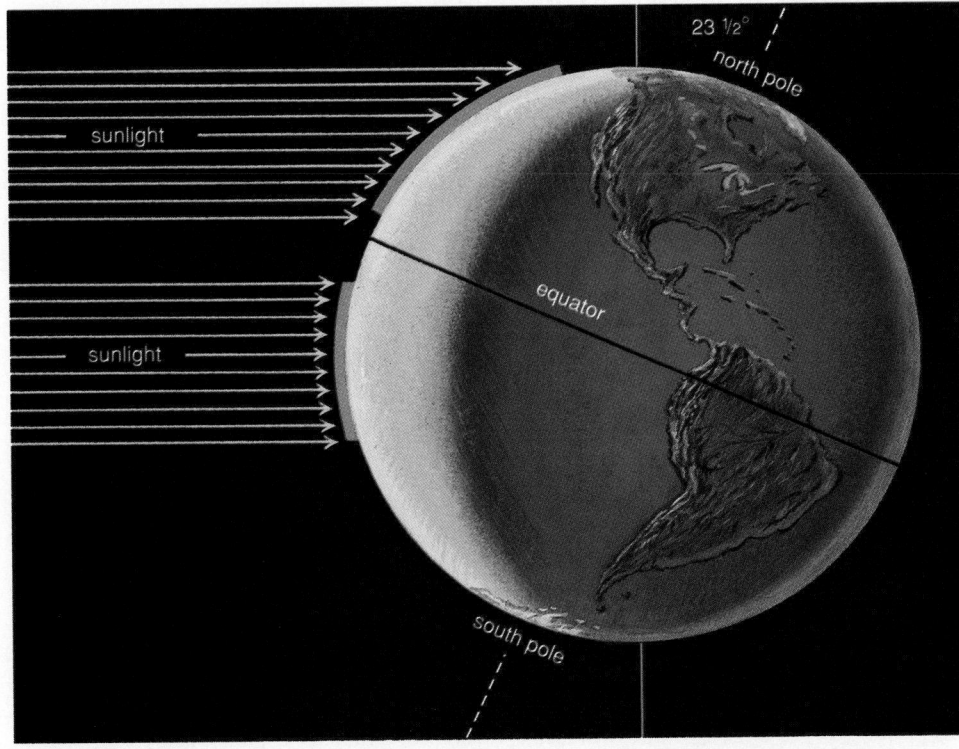

Figure 46-3 Near the equator, sunlight falls perpendicularly on the Earth's surface, so its warmth is concentrated onto a relatively small area. Toward the poles, the same amount of sunlight falls over a much larger surface area. This results in a temperature gradient, with maximum temperature at the equator and minimum temperature at the poles.

tudes, the cooled air begins to sink. As it sinks, the air is warmed both by compression and by heat radiated from the Earth. By the time it reaches the surface, it is both warm and very dry. Not surprisingly, the major deserts of the world are found at these latitudes (Fig. 46-4b). This air then flows back toward the equator. Further north and south, this general circulation pattern is repeated, dropping moisture at around 60° north and south latitudes and creating extremely dry conditions at the poles (Fig. 46-4a).

Ocean Currents

Ocean currents are driven by the Earth's rotation, the wind, and direct heating of the water by the sun. Continents interrupt the currents, breaking them into roughly circular patterns called **gyres,** which rotate clockwise in the northern hemisphere and counterclockwise in the southern hemisphere (Fig. 46-5). Since water both heats and cools more slowly than land (see Chapter 2), ocean currents tend to reduce

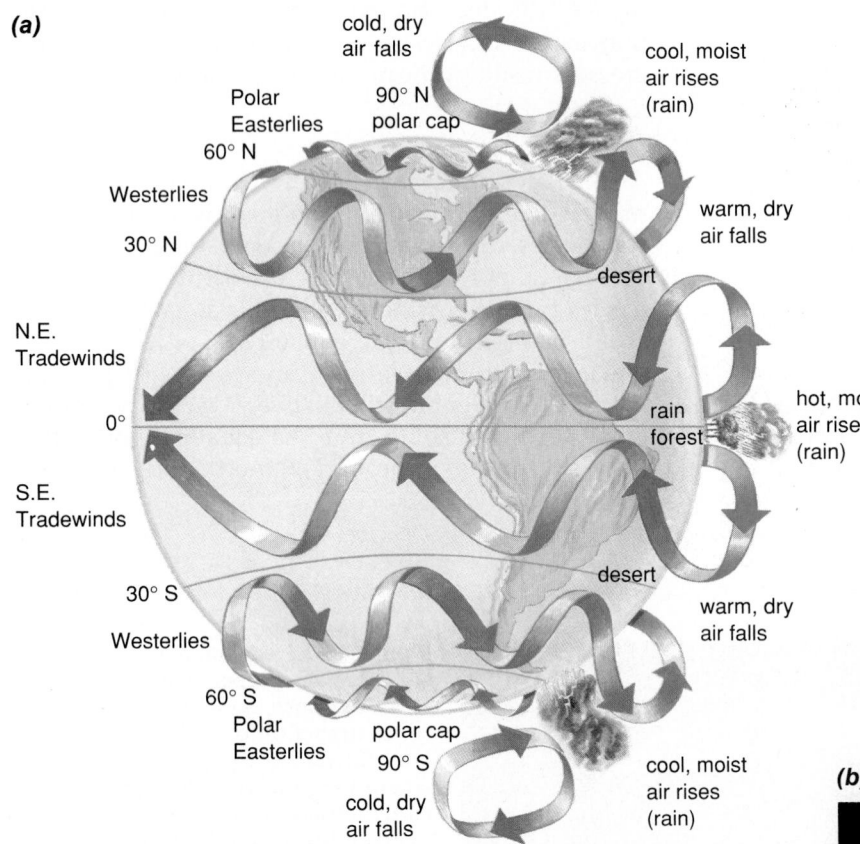

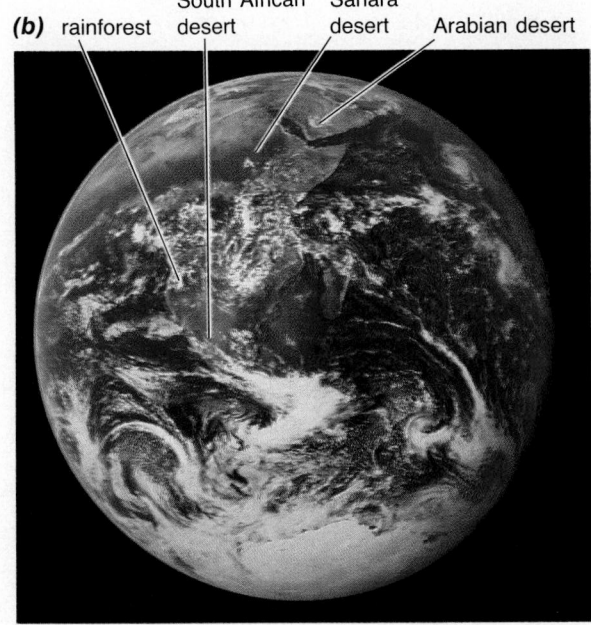

Figure 46-4 (a) The distribution of rainfall is primarily determined by the temperature gradient described in Figure 46-3, combined with the Earth's rotation. These create air currents that rise and fall predictably with latitude, producing broad general climatic regions. **(b)** Some of these regions are visible in this photograph of the African continent taken by Apolo 11 astronauts. Along the equator, there are heavy clouds dropping moisture on the rainforests of the Congo. Note the lack of clouds over the Sahara and Arabian deserts near 30° north latitude, and the South African desert near 30° south.

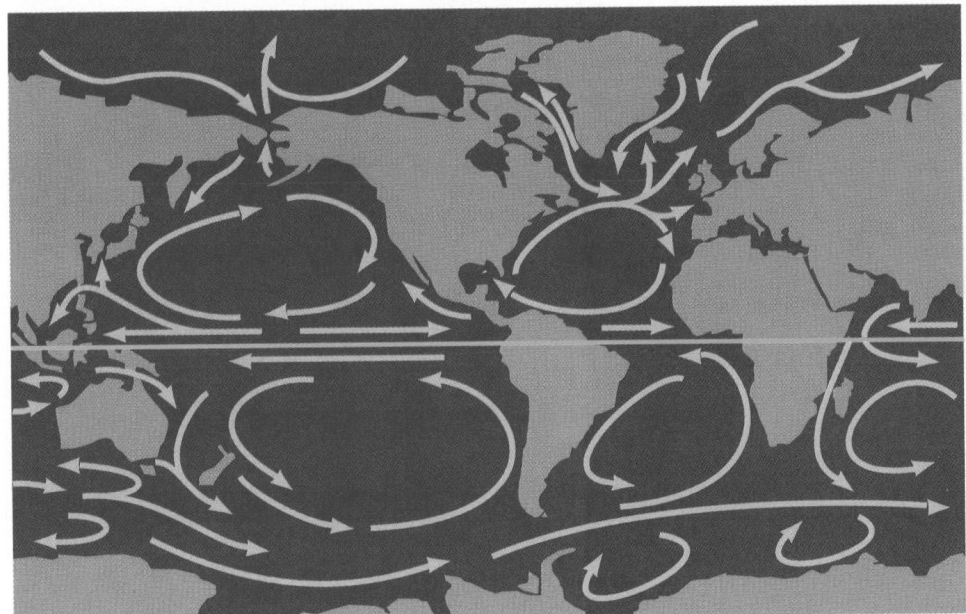

Figure 46-5 Ocean circulation patterns, called gyres, travel clockwise in the northern hemisphere and counterclockwise in the southern hemisphere. These tend to distribute warmth from the equator to northern and southern coastal areas.

temperature extremes. Coastal areas, then, generally have less variable climates than areas near the center of continents. For example, a gyre in the Atlantic Ocean brings warm water from equatorial regions north along the eastern coast of North America, creating a warmer, moister climate than is found further inland. It then carries the still-warm water further north and east, warming the western coast of Europe before returning south (see Fig. 46-5).

Continents and Mountains

If the Earth's surface were uniform, we would find climate zones occurring in bands corresponding to latitude. These zones would result from the interaction of temperature and rainfall, which are determined by the rise and fall of air masses (see Fig. 46-4). The presence of irregularly shaped continents (which heat and cool relatively quickly) and oceans (which heat and cool more slowly) alters the flow of wind and water and contributes to the irregular distribution of ecosystems (see Fig. 46-8).

Variations in elevation further complicate the picture. With increasing elevation, the atmosphere becomes thinner and retains less heat. There is an approximately 0.6° C decrease in temperature for every 100 meters of elevation (3.5° F per 1000 feet), so that snow-capped mountains may be found even in the tropics (Fig. 46-6).

Mountains also modify rainfall patterns. When water-laden air is forced to rise as it meets a mountain, it cools. Cooling reduces the air's ability to retain water, which condenses as rain or snow on the windward (near) side of the mountain. The cool, dry air is warmed again as it travels down the far side of the mountain, and absorbs water from the land, creating a local dry area called a **rain shadow.** For example, mountain ranges such as the Sierra Nevada of the western United States wring the moisture from the westerly winds coming off the Pacific Ocean, leaving deserts in the rain shadow on their eastern sides (Fig. 46-7).

The Requirements of Life

From the lichens on bare rock to the thermophilic (Greek for "heat-loving") algae in the hot springs of Yellowstone to the bacteria thriving under the pressure-cooker conditions of a deep-sea vent, the Earth teems with life.

Four fundamental resources are required for life on Earth:

1. **Nutrients from which to construct living tissue.**
2. **Energy to power that construction.**
3. **Liquid water to serve as a solvent for metabolic reactions.**
4. **Appropriate temperatures in which to carry out these processes.**

As we shall see in the following sections, these resources are very unevenly distributed. Their availability limits and defines the types of organisms that can exist within the various ecosystems on Earth.

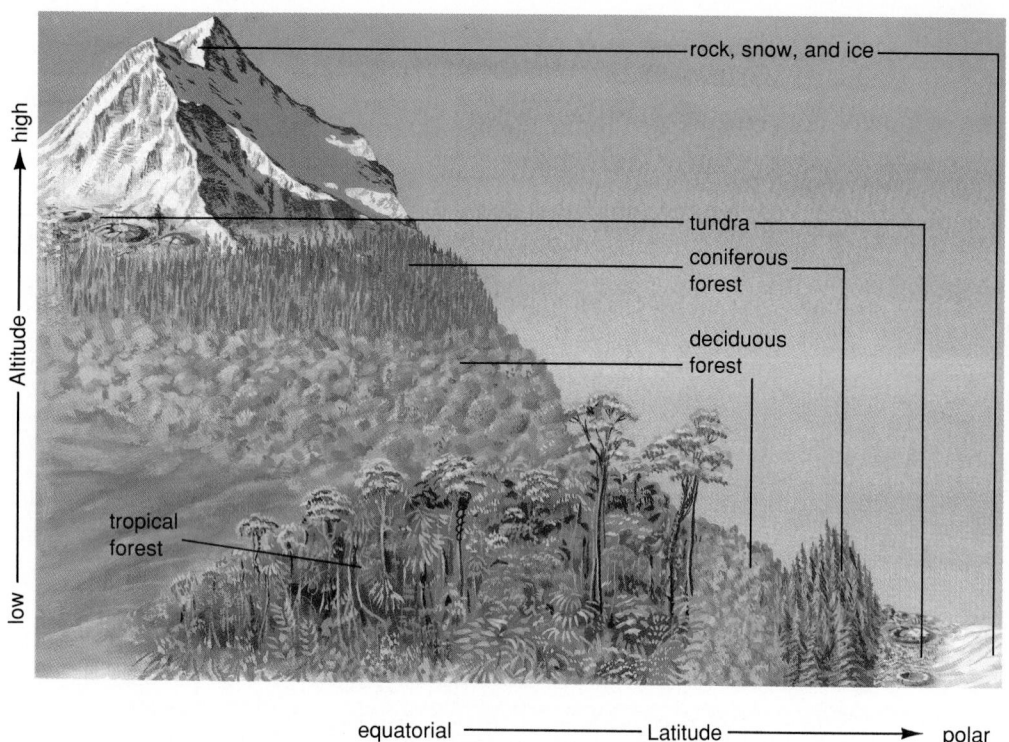

Figure 46-6 Climbing a mountain in some ways is like going north; in both cases, increasingly cool temperatures produce a similar series of biomes.

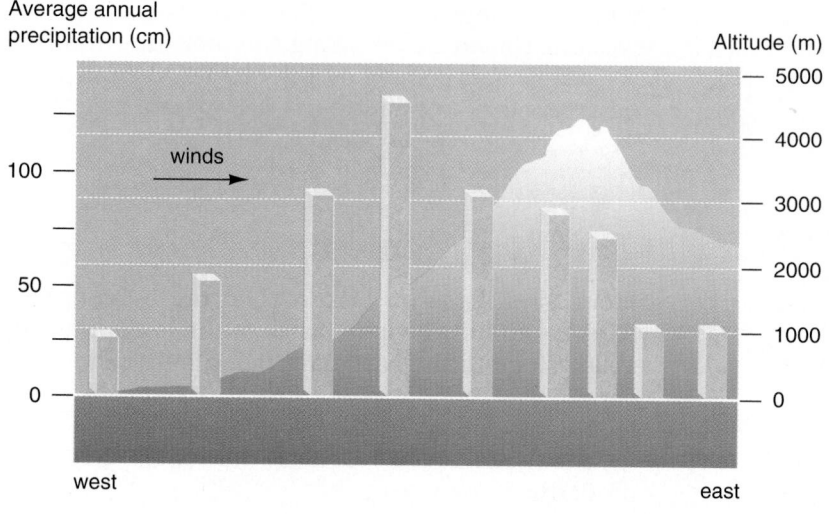

Figure 46-7 The rain shadow of the Sierra Nevada is illustrated graphically by charting rainfall against altitude. Westerly (eastward moving) winds deposit their moisture on the western slope, leaving a desert to the east.

Life on Land

Terrestrial organisms are restricted in their distribution largely by the availability of water and appropriate temperatures. Terrestrial ecosystems receive plenty of light, even on an overcast day, and the soil provides abundant nutrients. Water, however, is limited and very unevenly distributed, both in place and in time. Terrestrial organisms must be adapted to obtain water when it is available and conserve it when it is scarce.

As with water, favorable temperatures are also very unevenly distributed in place and time. At the South Pole, even in summer, the average temperature is well below freezing, and (not surprisingly) life is scarce. Other places, such as central Alaska, have favorable temperatures only during certain seasons of the year, while the tropics have a uniformly warm, moist climate.

Terrestrial Biomes

Terrestrial communities are dominated and defined by their plant life. Since plants can't escape from drought, sun, or winter weather, they tend to be precisely adapted to the climate of a particular region. Large land areas with similar environmental conditions and characteristic plant communities are called **biomes** (Fig. 46-8), and are usually named after the major type of vegetation found there. The predominant vegetation of each biome is determined by the complex interplay of rainfall and temperature (Fig.

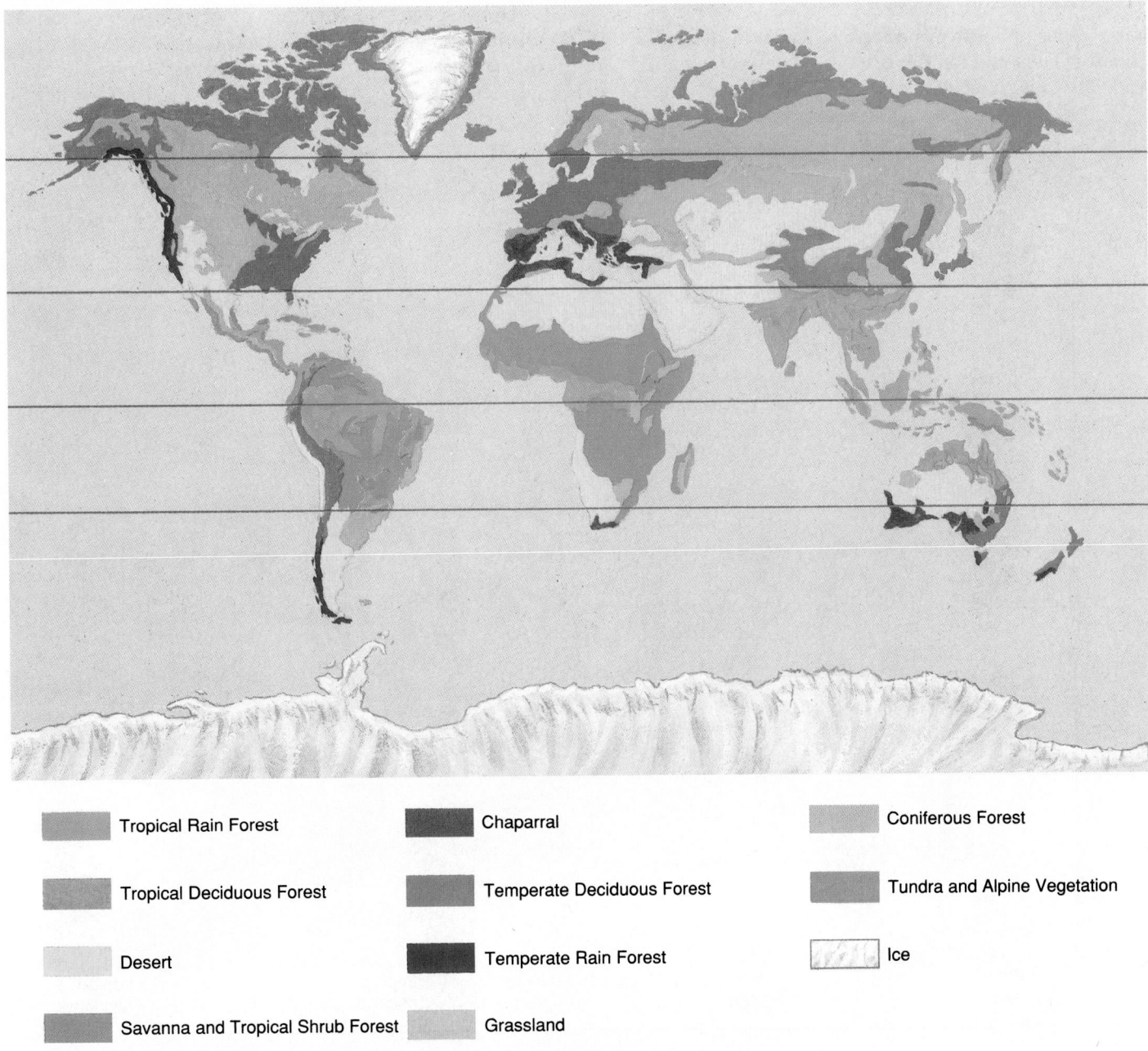

Tropical Rain Forest	Chaparral	Coniferous Forest
Tropical Deciduous Forest	Temperate Deciduous Forest	Tundra and Alpine Vegetation
Desert	Temperate Rain Forest	Ice
Savanna and Tropical Shrub Forest	Grassland	

Figure 46-8 The distribution of biomes. Although mountain ranges and the sheer size of the continents complicate their pattern, note the overall consistencies. Tundra and coniferous forest always occur in the northernmost parts of the northern hemisphere, while the deserts of Mexico, the Sahara, Saudi Arabia, South Africa, and Australia are located around 20° to 30° north and south latitude.

46-9). In addition to the total amount of rainfall and the overall yearly average temperature, the *variability* of rain and temperature over the year also determines which plants can grow in an area. Arctic tundra plants, for example, must be adapted to survive marshy conditions in the early summer but cold desert conditions for much of the rest of the year, when water is solidly frozen and unavailable.

Tropical Forests

Large areas of South America and Africa lie along the equator. Here the temperature averages between 25° and 30° C with little variation, and rainfall ranges

from 250 to 400 cm (100–160 in) yearly. These evenly warm, evenly moist conditions combine to create the most diverse biome on Earth, the **tropical rain forest,** dominated by huge broadleaf evergreen trees (Fig. 46-10).

Although rain forests cover only 6% of the Earth's total land area, they are believed to be home to at least five million species, representing half to two thirds of the world's total. A recent survey of a 2.5-acre site in the upper Amazon basin revealed 283 different species of trees, most of which were represented by a single individual.

Tropical rain forests typically are stratified, with several layers of vegetation. The tallest trees reach 50

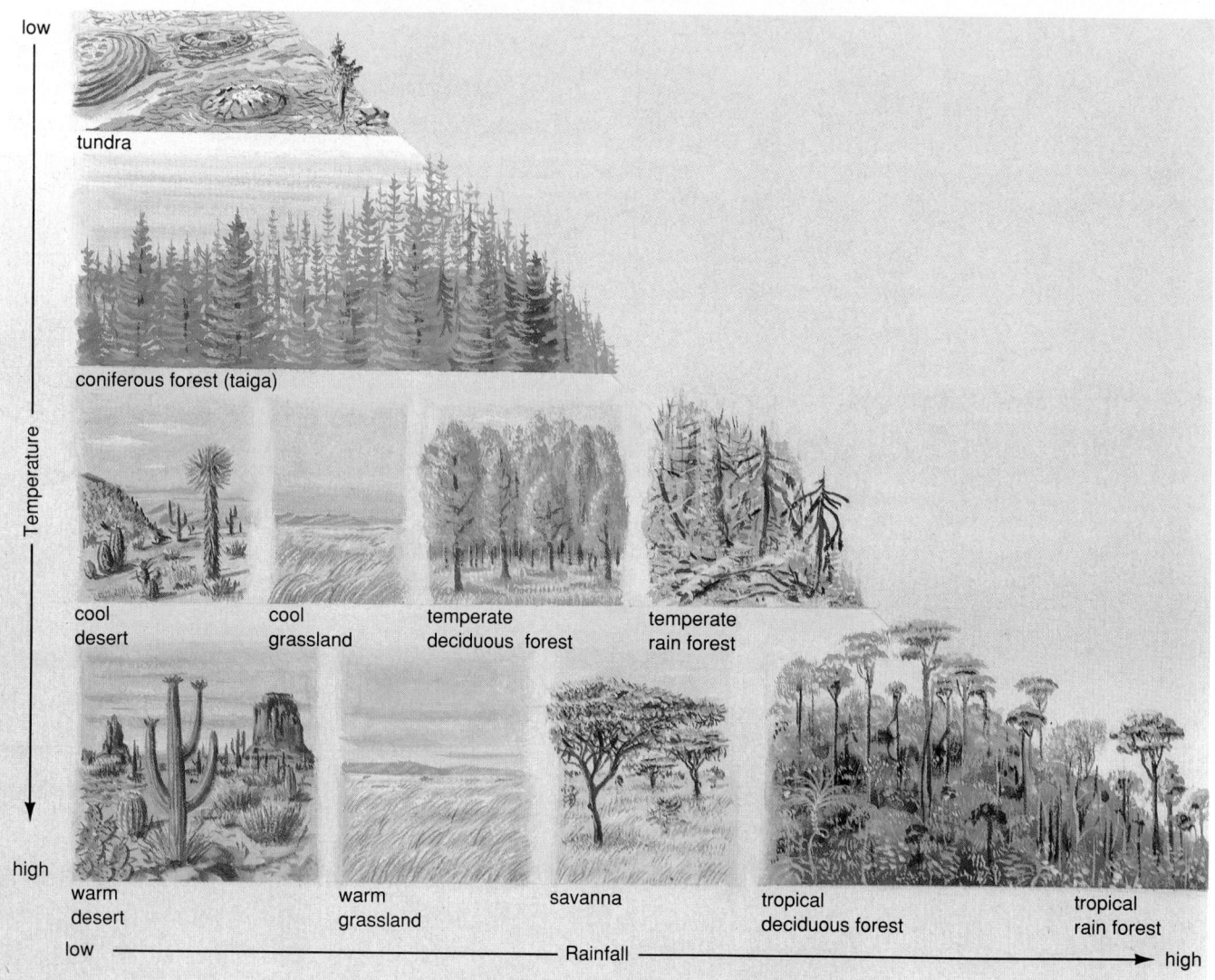

Figure 46-9 The distribution of biomes on land depends largely on the interplay of temperature and rainfall, which together determine the available soil moisture needed for plant growth.

Figure 46-10 Towering trees draped with vines reach for the light in the dense tropical rain forest. Since food and light energy are found high in trees, amidst their branches dwells the most diverse assortment of animals on Earth, including a fruit-eating toucan, an iridescent butterfly, an epiphytic orchid, and a howler monkey (insets).

meters (150 ft), towering above the rest of the forest. Below these is a fairly continuous canopy of treetops at about 30 to 40 meters. Still closer to the forest floor is often another layer of shorter trees. Huge woody vines, often 100 meters or more in length, grow up the trees, reaching the sunlight far above. These layers of vegetation block out most of the sunlight. Plants that live in the dim green light that filters through to the forest floor often have enormous leaves to trap the little available energy.

Due to the relative scarcity of edible plant material close to the ground, much of the animal life in tropical rain forests is arboreal (living in the trees), including numerous birds, monkeys, and insects. Competition for the nutrients that do reach the ground is intense, both among animals and plants. Even such

PLANET WATCH
The Ozone Layer—Puncturing the Protective Shield

A small fraction of the radiant energy produced by the sun, called ultraviolet or UV radiation, is so highly energetic that it can damage biological molecules. In small quantities, UV radiation allows human skin to produce vitamin D, and causes tanning in fair-skinned people. But in larger doses, UV causes sunburn and premature wrinkling of the skin. By damaging DNA, UV radiation can produce skin cancer. UV radiation also contributes to cataracts, in which the lens of the eye becomes cloudy, and evidence suggests that it may also damage the immune system.

Fortunately for the evolution of life as we know it, most of the UV radiation is filtered out in a layer of atmosphere extending from 10 to 50 kilometers (6 to 30 miles) above the Earth, called the **ozone layer.** In pure form, ozone (O_3) is a bluish, explosive, and highly poisonous gas. Ozone is found in relatively high concentrations in the ozone layer (0.1 parts per million, compared with 0.02 parts per million in the lower atmosphere). UV light striking ozone and oxy-

gen causes reactions that both break down and regenerate ozone. In the process, the UV radiation is converted to heat, and the overall level of ozone remains constant—normally.

In 1985, British atmospheric scientists published a startling discovery: the springtime levels of stratospheric ozone over Antarctica had declined by over 40% since 1977. In 1987, scientists visiting Antarctica measured a 50% decline covering an area the size of the United States. A hole had been punctured in the protective shield (Fig. E46-1). Global trends were quickly charted, revealing that since 1969 ozone had decreased by 2% over the entire northern hemisphere and 3% over the most heavily populated areas. Although the percentages seem small, the impact is significant. It is estimated that for every 1% decrease in ozone, the UV radiation reaching Earth increases by 2%. This 2% increase in UV radiation, in turn, could cause a 4% to 6% rise in some types of skin cancer. Human health effects are only one cause for concern. Plankton, the producers for marine ecosystems, could be damaged by further ozone depletion, ultimately reducing fish harvests. Some types of trees and farm crops are also harmed by increased UV radiation.

Researchers immediately suspected chlorofluorocarbons (CFCs), which contain chlorine, fluorine, and carbon. Developed some 60 years ago, these gases found widespread use as coolants in refrigerators and air conditioners, aerosol spray propellants, agents for producing rigid foams such as styrofoam, and cleansers for electronic parts. Because they are nonreactive and very stable, these chemicals were considered extremely safe. Their stability, however, proved to be a major problem, since they remain chemically unchanged as they gradually work their way high into the stratosphere. Here, under intense bombardment by UV light, the CFCs degrade, releasing chlorine atoms. Chlorine acts as a catalyst, causing the breakdown of ozone, ultimately forming oxygen gas. Because catalysts are unchanged during chemical reactions, each chlorine atom could cause the breakdown of 100,000 ozone molecules. Over the arctic and antarctic regions, clouds of ice particles in the stratosphere hasten the reaction.

Although CFC production has not been halted, in

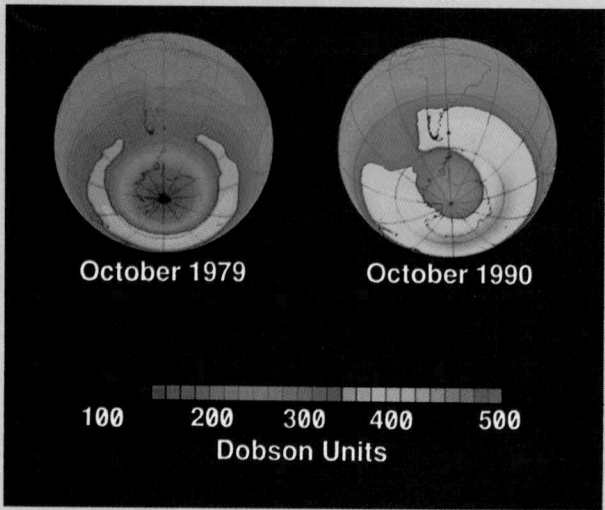

Figure E46-1 The decline in antarctic ozone levels between 1979 and 1990 is charted using data from a NASA satellite. A Dobson unit is a hundredth of a millimeter, and is used to measure the thickness of the layer of ozone in each vertical column of atmosphere if the ozone were purified and held at standard temperature and pressure. The ozone levels over Antarctica were twice as high just 10 years ago.

Montreal in 1987, 23 nations signed a treaty that would reduce their production of CFCs by 50% by the year 1999. While this was an important step, many of the less-developed countries have not agreed to participate. Further, as shown in Fig. E46-2, the treaty will allow a steady rise in the levels of chlorine in the atmosphere well beyond the next century. Since these highly stable compounds can persist 75 to 100 years, the millions of tons already released will continue their destruction even if CFC production were to cease immediately. To stabilize the current level of chlorine would require an immediate reduction of 85% of the CFCs currently produced.

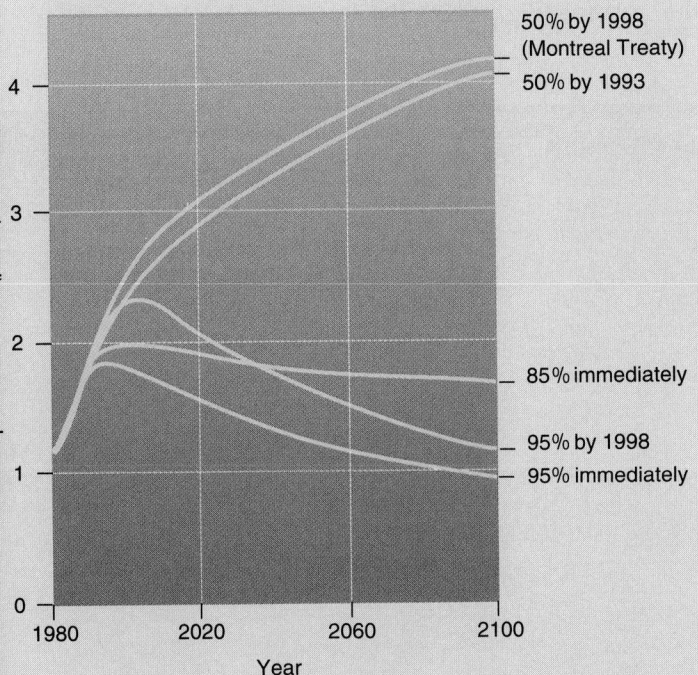

Figure E46-2 Future atmospheric chlorine levels under various scenarios of CFC reduction. The uppermost line reflects the present treaty, which will allow a steady increase in the concentration of this ozone-destroying chemical in the atmosphere.

unlikely sources of food as the droppings of monkeys are in great demand, in this case by dung beetles that feed and lay their eggs on droppings. When ecologists attempted to collect droppings of the South American howler monkeys to find out what the monkeys had been eating, they found themselves in a race with the beetles, hundreds of which would arrive within minutes after a dropping hit the ground!

Almost as soon as bacteria or fungi release any nutrients from dead plants or animals into the soil, rain forest trees and vines absorb the nutrients. This is one of the reasons why, despite the teeming vegetation, agriculture is very risky and usually destructive in rain forests. Virtually all the nutrients in a rain forest are tied up in the vegetation, so the soil is very infertile. If the land is cleared, and especially if the trees are removed, few nutrients remain to support crops. Further, even if the nutrients are released by burning the vegetation, the heavy year-round rainfall quickly dissolves and erodes them away, leaving the soil infertile after a few seasons of cultivation. The exposed soil, which is rich in iron and aluminum, then takes on an impenetrable, bricklike quality as it bakes in the tropical sun. As a result, secondary succession on cleared rain forest land is slowed significantly, and the lush tropical growth is, for practical human purposes, irrevocably lost.

Human Impact In spite of their unsuitability for agriculture, rain forests are being felled for lumber and agriculture at a rate of roughly 50 million acres per year, or about 100 acres per minute (Fig. 46-11).

Figure 46-11 Fire engulfs a tropical rain forest, which will be converted to ranching or agriculture, both doomed to failure.

About 40% of the original rain forests are now gone, and at current rates of cutting, all the forests not specifically protected in parks could disappear within the next 30 years, along with nearly half the world's species. Currently protected regions cover less area than is cut in 1 year. The removal of these forests is contributing significantly to the buildup of CO_2 in the atmosphere, making the greenhouse effect more severe. In West Africa, burning the forests is causing acid rain, which is damaging the remaining trees.

Although disastrous losses have occurred and are continuing, recognition of the problem is growing. Some preserves have been protected, and a few reforestation efforts are underway, although currently only 1 acre is replanted for every 10 denuded. Local residents are becoming more involved in conservation efforts. In Brazil, where 20 million acres of forest are destroyed each year, a strong rubber-tappers union (people who collect natural rubber from tree sap) is fighting to preserve large tracts of land open only to rubber production and the harvesting of fruits and nuts. This suggests the ultimate solution, which is tragically slow in coming: sustainable use—

harvesting products without permanently damaging the trees or the ecosystem they support.

Tropical Deciduous Forests

Slightly farther away from the equator, the rainfall is not nearly as constant, and there are pronounced wet and dry seasons. In these areas (for example, India, much of southeast Asia and parts of South and Central America), **tropical deciduous forests** grow. During the dry season, the trees cannot get enough water from the soil to compensate for evaporation from their leaves. As a result, the plants have adapted to the dry season by shedding their leaves, thereby minimizing water loss. If the rains fail to return on schedule, the trees merely wait until times are better.

Savanna

Along the edges of the tropical deciduous forest, the trees gradually become more widely spaced, with grasses growing between them. Eventually, grasses become the dominant vegetation, with only scattered trees and thorny scrub forests here and there; this is the **savanna** (Fig. 46-12). Savanna grasslands typically

Figure 46-12 On the African savanna, elephants roam while a cheetah feasts on its prey. Large herds of grazing animals, such as these (inset), can still be seen on African preserves. The herds of herbivores provide food for the greatest assortment of large carnivores on the planet, including lions (inset), leopards, cheetahs, hyenas, and wild dogs.

have a rainy season during which virtually all of the year's precipitation falls (30 cm or less; 12 in or less). When the dry season arrives, it comes with a vengeance. No rain may fall for months, and the soil becomes hard, dry, and dusty. Grasses are well adapted to this type of climate, growing very rapidly during the rainy season and dying back to drought-resistant roots during the dry times. Only a few specialized trees, such as the thorny acacia or the water storing baobab, can survive the devastating savanna dry seasons. In areas in which the dry season becomes even more pronounced, virtually no trees at all can grow, and the savanna imperceptibly grades into tropical grassland.

The African savanna has probably the most diverse and impressive array of large mammals on Earth, including numerous herbivores such as antelope, wildebeest, buffalo, elephants, and giraffes, and carnivores such as the lion, leopard, hyena, and wild dog.

Human Impact Africa's rapidly expanding human population threatens the wildlife of the savanna. Poaching has driven the black rhinoceros to the brink of extinction, and endangers the African elephant (Fig. 46-13). The abundant grasses that make the savanna a suitable habitat for so much wildlife also make it suitable for grazing domestic cattle. As the human population of East Africa increases, so does the pressure of cattle grazing upon the savanna. Fences increasingly disrupt the migration of the great herds of herbivores in search of food and water. Ecologists have discovered that the native herbivores are much more efficient at converting grass into meat than are cattle. Perhaps the future African savanna may support herds of domesticated antelope and other large native grazers in place of cattle.

Deserts

Even drought-resistant grasses need at least 25 to 50 cm (10 to 20 in) of rain a year, depending on its seasonal distribution and the average temperature. When less rain than this falls, many grasses fail and **desert** biomes occur. These biomes are found on every continent, often around 20° to 30° north and south latitude, and also in the rain shadows of major mountain ranges. As with all biomes, the designation ''desert'' includes a variety of environments. At one extreme are certain areas of the Sahara or Chile, where it virtually never rains and there is no vegetation at all (Fig. 46-14a). The more common deserts, however, are characterized by intermittent vegetation and large areas of bare ground. The plants are often spaced evenly, as if planted by hand (Fig. 46-14b). Frequently, the perennial plants are bushes or cacti

Figure 46-13 Rhinoceros horns, believed by some to have aphrodisiac properties, fetch staggering prices and encourage poaching. The black rhino is now nearly extinct.

with large, shallow root systems. The shallow roots quickly soak up the soil moisture after the infrequent desert storms, while the rest of the plant is typically covered with a waterproof, waxy coating to prevent evaporation of precious water. Water is stored in the thick stems of cacti and other succulents (Fig. 46-14c). The spines of cacti are leaves modified for protection and water conservation, since they present almost no surface area for evaporation. In many deserts, all the rain falls in just a few storms, and specialized annual wildflowers take advantage of the brief period of moisture to race through germination, growth, flowering, and seed production in a month or less (Fig. 46-15).

The animals of the deserts, like the plants, are specially adapted to survive on little water. Most deserts appear to be completely devoid of animal life during the day, because the animals seek relief from the sun and heat in cool underground burrows. After dark, when the deserts cool down considerably, reptiles such as horned lizards and snakes emerge to feed, as do mammals such as the kangaroo rat (see Fig. 46-14d) and birds such as the burrowing owl. Most of the smaller animals survive without ever drinking at all, deriving all the water they need from their food and that produced during cellular respiration in their tissues. Larger animals, such as desert bighorn sheep, are dependent on permanent water holes during the driest times of the year.

Human Impact While human activities are reducing the extent of many biomes, they are causing the

(a)

(b)

(c)

(d)

Figure 46-14 Desert biomes. **(a)** Under the most extreme conditions of heat and drought, deserts can be almost devoid of life, such as these sand dunes of the Sahara desert in Africa. **(b)** Throughout much of Utah and Nevada, the Great Basin Desert presents a monotonous landscape of widely spaced shrubs, such as sagebrush and greasewood. These shrubs often secrete a growth inhibitor from their roots, preventing germination of nearby plants and thus reducing competition for water. **(c)** The Sonoran Desert in southern Arizona and northern Mexico is typified by the saguaro cactus, prickly pear, and the spindly ocotillo, whose red flowers are pollinated by hummingbirds. The body of the saguaro is pleated and can expand after a rain to store water. **(d)** The kangaroo rat is an elusive denizen of the deserts of North America.

spread of deserts, a process call **desertification.** A dramatic example is occurring in the Sahel, which borders the southern edge of the Sahara desert in Africa. Twenty five years of below-average rainfall, coupled with rapid growth of the human population, have caused a steady southward spread of the desert.

Western technology allowed the Sahelians to drill deep wells, abandon their nomadic way of life, and increase their livestock herds. The natural vegetation was destroyed as livestock competed for forage during the drought (Fig. 46-16). Native shrubs were uprooted by goats, preventing regeneration. The grow-

Figure 46-15 The Mojave desert in southern California in spring is carpeted with wildflowers. Through much of the year, annual wildflower seeds lie dormant waiting for the spring rains to fall.

Figure 46-16 Overgrazing in the Sahel region of Africa is degrading the already marginal productivity of the land and contributing to the southward spread of the Sahara desert.

ing human population began cultivating increasingly fragile land without allowing time for the soil to recover between crops. As the desert spread southward, so did the people, with their livestock and crops, causing further deforestation, soil destruction, and desertification of once productive land.

The Sahel is an example of a human population exceeding carrying capacity. The loss of productivity of the ecosystem is nearly irreversible, and massive famines, such as occurred in Ethiopia in the mid-1980's, are the tragic result.

Chaparral

In many coastal regions that border on deserts, such as southern California and much of the Mediterranean, a unique type of vegetation grows, called **chaparral**. The annual rainfall in these regions is similar to that of a desert, but the proximity of the sea provides a slightly longer rainy season in the winter and frequent fogs during the spring and fall, which reduce evaporation. Chaparral consists of small trees or large bushes with thick waxy or fuzzy evergreen leaves that conserve water. These shrubs are also able to withstand the frequent summer fires started by light-ning (Fig. 46-17).

Human Impact   As is often the case with deserts, grasslands, and forests, the extent of chaparral vege-tation is influenced by human activities. About 400 BC, the Greek philosopher Plato wrote that "there are mountains in Attica which can now keep nothing but bees, but which were clothed, not so very long ago, with timber suitable for roofing very large build-ings." The original Greek forests were cut down not only for ceiling beams but also for the great Athenian naval fleets. Subsequently, heavy grazing by goats prevented regrowth of the forests. Only the chaparral plants, which could survive in the hot sun and were unpalatable or poisonous to goats, regrew. The for-ests have never returned.

Grasslands

In the temperate regions of North America, deserts occur in the rain shadows east of the mountain ranges, such as the Sierra Nevada and Rockies. Far-ther east, as the rainfall gradually increases, the land begins to support more and more grasses, giving rise to the prairies of the Midwest. These **grassland** or prairie biomes usually have a continuous cover of grass and virtually no trees at all except along the rivers. From the tallgrass prairies of Iowa and Illinois (Fig. 46-18) to the shortgrass prairies of eastern Colo-rado, Wyoming, and Montana (Fig. 46-19), the North American grassland once stretched across almost half the continent.

Water and fire are the crucial factors in the competi-tion between grasses and trees. The hot dry summers and frequent droughts of the shortgrass prairies can be tolerated by grass but are fatal to trees. Trees can grow in the more eastern prairies, but historically the trees were destroyed by frequent fires, often set by the Indians to maintain grazing land for the bison. Although the tops of the grasses are killed by fire, their root systems usually survive; trees, on the other hand, are killed outright. The resulting grasslands of North America once supported huge herds of bison, as many as 60 million in the early nineteenth century. Grasses growing and decomposing for thousands of years produced perhaps the most fertile soil in the world.

Human Impact   With the elimination of the bison and the development of plows that could break the dense turf, the former prairie has become the bread-basket of North America. The tallgrass prairie has been converted to agricultural land, except for tiny protected remnants.

Figure 46-17 The chaparral biome is limited primarily to coastal mountains in dry regions, such as the San Gabriel Mountains in southern California. Chaparral is maintained by frequent fires set by summer lightning. Although the tops of the plants may be burned off, the roots send up new sprouts the next spring.

Figure 46-18 Tallgrass prairie in Missouri. In the central United States, moisture-bearing winds out of the Gulf of Mexico produce summer rains, allowing a lush growth of tall grasses and wildflowers such as these coneflowers.

Figure 46-19 The lands east of the Rocky Mountains receive relatively little rainfall, and shortgrass prairie results, characterized by low-growing bunch grasses such as buffalo grass and grama grass, and a wealth of wildflowers such as this coneflower (inset). Pronghorn antelope, prairie dogs, and protected bison herds occupy this biome (insets).

On the western shortgrass prairie, cattle and sheep have replaced the bison and pronghorn antelope. As a result of their overgrazing the grasses, the boundary between the cool deserts and the grassland has often been altered in favor of desert plants. Much of the sagebrush desert of the American West is actually the result of overgrazing shortgrass prairie (Fig. 46-20). Cattle prefer grass to sagebrush, so heavy grazing destroys the grass. This leaves moisture in the soil that the grass would have absorbed, encouraging the growth of the woody sagebrush. Since the cattle will not eat sagebrush, it soon becomes the dominant vegetation; thus the prairie grasses are replaced by plants characteristic of the cool desert.

Temperate Deciduous Forests

At their eastern edge the North American grasslands merge into the **temperate deciduous forest** biome (Fig. 46-21). Higher precipitation occurs there than in the grasslands (75 to 150 cm, 30 to 60 in), and in particular more summer rainfall occurs. The soil therefore retains enough moisture for trees to grow, and the resulting forest shades out grasses. In contrast to the tropical forests, the temperate deciduous forest

Figure 46-20 Sagebrush desert or shortgrass prairie? Biomes are not only influenced by temperature, rainfall, and soil, but also by human activities. The shortgrass prairie field on the right has been overgrazed by cattle, causing the grasses to be replaced by sagebrush.

Figure 46-21 The temperate deciduous forest of the eastern United States, in which the whitetail deer (inset) is the largest herbivore and birds such as the blue jay (inset) are abundant. In spring, a profusion of woodland wildflowers (such as these hepaticas, inset) blooms briefly before the trees leaf out.

biome has cold winters, usually with at least several hard frosts and often long periods of subfreezing weather. Winter in this biome has an effect on the trees similar to that of the dry season in the tropical deciduous forests: during periods of subfreezing temperatures, liquid water is not available to the trees. To reduce evaporation when water is in short supply, the trees drop their leaves in the fall. They leaf out again in the spring, when liquid water becomes available. During the brief time in spring when the ground has thawed but the trees have not yet blocked off all the sunlight, numerous wildflowers grace the forest floor.

Insects and other arthropods are numerous and conspicuous in deciduous forests. The decaying leaf litter on the forest floor also provides food and habitat for bacteria, earthworms, fungi, and small plants.

Many arthropods feed on these or on each other. A variety of vertebrates, including mice, shrews, squirrels, raccoons, and many species of birds dwell in the deciduous forests.

Human Impact Large mammals such as black bear, deer, wolves, bobcats, and mountain lions were formerly abundant, but the predators have been largely exterminated by humans. Clearing for lumber and agriculture has dramatically reduced America's deciduous forests, and virgin forests are now almost nonexistent.

Temperate Rain Forests

On the Pacific coast, from the lowlands of the Olympic peninsula in Washington to southeast Alaska, is the **temperate rain forest** biome (Fig. 46-22). As in the

Figure 46-22 The Hoh River temperate rain forest in Olympic National Park. The coniferous trees do not block off the light as effectively as do broadleaf trees, and so ferns, mosses, and wildflowers grow in the pale green light of the forest floor. The dead feeds the living, as new trees grow from the decay of this fallen giant, called a "nurse log" (inset), and fungi and flowering foxglove find ideal conditions amidst the moist, decaying vegetation.

tropical rain forest, there is no shortage of liquid water year round. This is due to two factors. First, there is a tremendous amount of rain. The Hoh River rain forest in Olympic National Park receives over 400 cm (160 in) of rain each year, over 60 cm (24 in) in the month of December alone. Second, the moderating influence of the Pacific Ocean prevents severe frost from occurring along the coast, so the ground seldom freezes.

The abundance of water means that the trees have no need to shed their leaves in the fall, and almost all the trees are evergreens. In contrast to the broadleaf evergreens of the tropics, the temperate rain forests are dominated by conifers. The ground and often the trunks of the trees are covered with mosses and ferns. As in the tropical rain forests, so little light reaches the forest floor that tree seedlings usually cannot become established. Whenever one of the forest giants falls, however, it opens up a patch of light, and new seedlings quickly sprout, often right atop the fallen log. This produces a "nurse log," as shown in Figure 46-22.

Taiga

North of the grasslands and temperate forests, the **northern coniferous forest,** also called the **taiga** (Fig. 46-23), stretches horizontally across the entire continent of North America, mainly in southern Canada. Conditions in the taiga are harsher than those in the temperate deciduous forest. In the taiga, the winters are longer and colder and the growing season is shorter. The few months of warm weather are too short to allow trees the luxury of regrowing leaves in the spring. As a result, the taiga is populated almost entirely by evergreen coniferous trees with small, waxy needles. The waxy coating and small surface area of the needles reduce water loss by evaporation during the cold months, and the leaves remain on the trees year round. Thus the tree can grow slowly for much of the year and is instantly ready to take advantage of good growing conditions when spring arrives.

Because of its harsh climate, the diversity of life in the taiga is much lower than in many other biomes. Vast stretches of central Alaska, for example, are cov-

Figure 46-23 The taiga. The small needles and pyramidal shape of conifers allows them to shed heavy snows. Winter is not only a challenge for the trees, but for animals such as this snowshoe hare and the bobcat that preys on it (inset). The hare is also prey for the snowy owl (inset). Taiga animals face diminished food supply but increased energy requirements during subfreezing weather.

ered by a somber forest composed almost exclusively of black spruce and an occasional birch. Large mammals such as the wood bison, grizzly bear, moose, and wolf, mostly exterminated from the southern regions of their original range, still roam the taiga, along with smaller animals such as the wolverine, marten, foxes, snowshoe hare, and deer.

Human Impact Owing to the remoteness of the northernmost taiga and the severity of its climate, a greater percentage of the taiga remains in primeval condition than any other North American biome except the tundra. Nonetheless, the taiga is a major source of lumber for construction, and huge expanses of forest have been clearcut (Fig. 46-24). In the Pacific Northwest, the cutting of virgin forests threatens the spotted owl, an endangered species.

Tundra

The last biome encountered before reaching the polar ice caps is the arctic **tundra**, a vast treeless region bordering the Arctic Ocean (Fig. 46-25). Conditions in the tundra are severe. Winter temperatures in the

Figure 46-24 Clearcutting as seen in this Oregon forest, is relatively simple and cheap, but its environmental costs are high. Erosion will diminish the fertility of the soil, slowing new growth. As described in Chapter 43, the dense stands of trees that often regrow are more susceptible to attack by parasites.

Figure 46-25 Life on the tundra is adapted to cold. Plants such as dwarf willows and perennial wildflowers (such as this dwarf clover; inset) grow low to the ground, escaping the chilling tundra wind. Tundra animals such as caribou and arctic foxes (insets) can regulate blood flow in their legs, keeping them just warm enough to prevent frostbite, while preserving precious body heat for the brain and vital organs.

arctic tundra often reach −40° C or below, winds howl at 50 to 100 kilometers per hour, and precipitation averages 25 cm (10 in) or less per year, making this a freezing desert. Even during the summer, the temperatures can drop to freezing, and the growing season may last only a few weeks before a freeze-up recurs. Somewhat less cold but similar conditions produce alpine tundra on mountaintops above the treeline.

The cold climate of the arctic tundra results in **permafrost,** a permanently frozen layer of soil often no more than half a meter below the surface. As a result, when summer thaws come, the melted snow and ice cannot soak into the ground, and the tundra becomes a huge marsh. Trees cannot survive in the tundra for several reasons, one of which is that the permafrost limits root growth to the topmost foot or so of soil, which turns to a cold swamp in summer.

Nevertheless, the tundra supports a surprising abundance and variety of life. The ground is carpeted with small perennial flowers and dwarf willows no more than few inches tall, and often with a large lichen called reindeer moss, a favorite food of caribou. The standing pools provide superb mosquito habitat, and tundra summers produce a veritable blizzard of mosquitos. The mosquitos and other insects provide food for numerous birds, most of which migrate long distances to nest and raise their young during the brief summer feast. The tundra vegetation supports lemmings (a type of rodent), which are eaten by wolves, snowy owls, arctic foxes, and even grizzly bears.

Human Impact The tundra is perhaps the most fragile of all the biomes because of its short growing season. A 4-inch-high willow may have a 3-inch-diameter trunk and be 50 years old. Human activities in the tundra leave scars that persist for centuries. Fortunately for the tundra inhabitants, the impact of civilization is localized around oil drilling sites, pipelines, mines, and military bases.

Rainfall, Temperature, and Vegetation

Terrestrial biomes are greatly influenced by both temperature and rainfall, whose effects interact. Temperature strongly influences the effectiveness of rainfall in providing soil moisture for plants and standing water for animals to drink. The hotter it is, the more rapidly water evaporates, both from the ground and from plants. As a result of this interaction of temperature with rainfall (and to a lesser extent, the distribution of rain throughout the year), areas that receive

almost exactly the same rainfall can have startlingly different vegetation, all the way from desert to taiga. Take a trip with us from southern Arizona to northern Alaska as we visit ecosystems that each receive around 30 centimeters (about 12.5 in) of rain annually.

The Sonoran Desert near Tucson, Arizona (see Fig. 46-14c) receives 28 centimeters of rain yearly, with an average annual temperature of 20° C. The landscape is dominated by giant saguaro cactus and low-growing, drought-resistant bushes. Fifteen hundred kilometers north, in eastern Montana, rainfall is about the same, but we have passed into the shortgrass prairie biome (see Fig. 46-19), largely because the average temperature is much lower, about 7° C.

Central Alaska receives the same annual rainfall (about 28 cm), yet is covered with taiga forest (see Fig. 46-23). Due to the low average annual temperature (about −4° C), permafrost underlies much of the ground. During the summer thaw, the taiga earns its Russian name "swamp forest," although its rainfall is the same as the Sonoran desert.

Life In Water

Although this chapter has emphasized the terrestrial biomes, aquatic ecosystems cover nearly 72% of the Earth's surface. The saltwater oceans and seas are the largest ecosystems on Earth, covering about 71% of its surface, while freshwater ecosystems cover less than 1%.

The unique properties of water lend some common features to aquatic ecosystems. First, since water is slower to heat and cool than air, temperatures in aquatic ecosystems are more moderate than in terrestrial ecosystems. Second, although water may appear quite transparent, it absorbs a considerable amount of the light energy that sustains life. Even in the clearest water, the intensity of light decreases rapidly with depth, so that at depths of 200 meters (600 ft) or more, little light is left to power photosynthesis. If the water is at all cloudy—for example, due to suspended silt or microorganisms—the depth to which light can penetrate is greatly reduced. Third, nutrients in aquatic ecosystems tend to be concentrated near the bottom, where light levels are often too low to support photosynthesis. This separation of energy and nutrients limits aquatic life. **Of the four requirements for life, aquatic ecosystems provide abundant water and appropriate temperatures. Thus, the major factors that determine the quantity and type of life in aquatic ecosystems are energy and nutrients.**

Although they share some common features, aquatic ecosystems are extremely diverse. Freshwater ecosystems include rivers, streams, ponds, lakes, and marshes, while marine (saltwater) ecosystems include estuaries, tidepools, the open ocean, and coral reefs. In the following sections, we cover a few of these important aquatic ecosystems.

Freshwater Lakes

Freshwater lakes vary tremendously in size, depth, and nutrient content. Although each lake is unique, moderate to large lakes in temperate climates share some common features, including distinct life zones and temperature stratification.

Life Zones

The distribution of life in lakes depends largely on access to light, to nutrients, and to a substrate for attachment (the bottom). The life zones of lakes, then, correspond to specific locations within the lake.

Near the shore is the **littoral zone.** Here the water is shallow, and plants find abundant light, anchorage, and adequate nutrients from the bottom sediments. Not surprisingly, littoral zone communities are the most diverse. Here we find cattails and bulrushes nearest the shore, and water lilies and entirely submerged vascular plants and algae at the deepest reaches of the littoral zone (Fig. 46-26). The plant

bodies of the littoral zone trap sediments carried in by streams and by runoff from the surrounding land, increasing the nutrient content in this region. Living among the anchored plants are microscopic **phytoplankton** (Greek, "drifting plants") such as photosynthetic protists, bacteria, and algae; **zooplankton** (Greek, "drifting animals") such as protozoa; and the greatest diversity of animals in the lake. Littoral invertebrate animals include small crustacea, insect larvae, snails, flatworms, and hydra, while vertebrates include frogs, minnows, and aquatic snakes and turtles.

As the water increases in depth further from shore, plants are unable to anchor to the bottom and still collect enough light for photosynthesis. This open-water area is divided into two regions: the upper limnetic zone and the lower profundal zone (see Fig. 46-26).

In the **limnetic zone,** enough light penetrates to support photosynthesis. Here phytoplankton and cyanobacteria (also called blue-green algae) serve as producers. These are eaten by protozoa and small crustacea, which in turn are consumed by fish.

Below the limnetic zone lies the **profundal zone,** where light is insufficient to support photosynthesis. This area is nourished mainly by detritus that falls from the littoral and limnetic zones, and by incoming

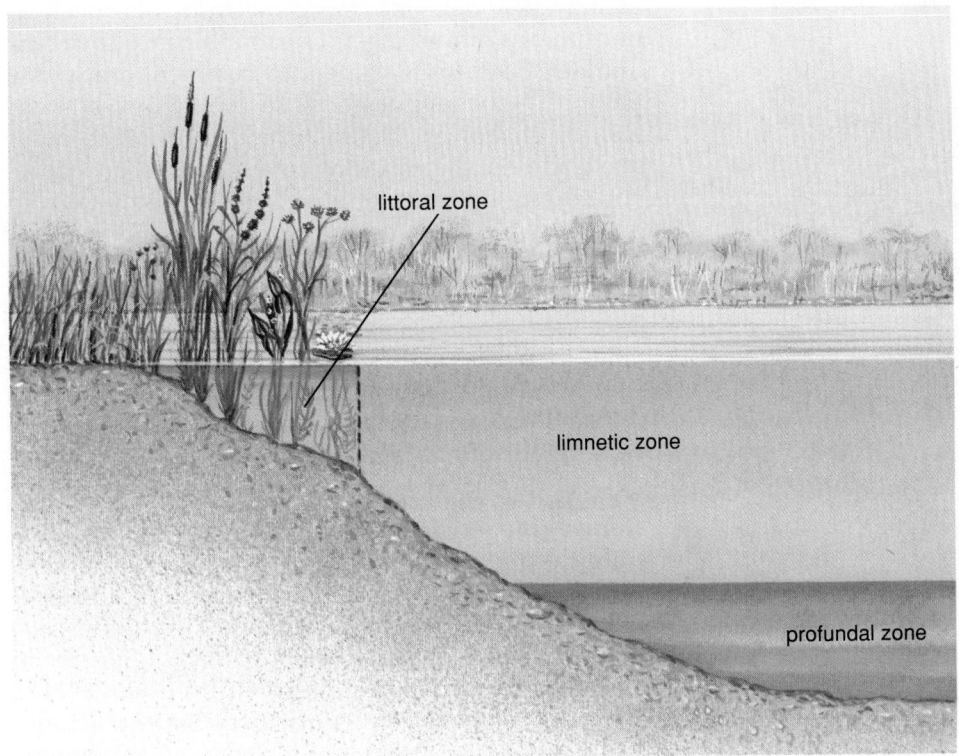

Figure 46-26 There are three life zones in a "typical" lake: a near-shore littoral zone with rooted plants, an open-water limnetic zone, and a deep, dark profundal zone.

littoral zone

limnetic zone

profundal zone

sediment. It is inhabited primarily by decomposer and detritus feeders such as snails and certain insect larvae and fish.

Temperature Stratification

Lakes in regions that experience seasonal temperature changes become stratified during the summer (Fig. 46-27), with a layer of relatively warm, less-dense water "floating" atop a layer of denser cold water (around 4° C; recall from Chapter 2 that water reaches its highest density at 4° C) (Fig. 46-28). The layers persist through the summer, and restrict the mixing of surface and bottom waters. Since nutrients are concentrated in the bottom sediments, the upper region, where photosynthetic organisms flourish, becomes depleted of nutrients as the summer progresses. Simultaneously, the heterotrophic organisms near the bottom use up available oxygen, which is not replenished from the surface. So by late summer, the surface populations are limited by lack of nutrients, while those in the profundal zone are limited by lack of oxygen. (The surface growth of blue-

green algae in Lake Erie followed by a die-off due to nutrient depletion is shown in Figure 43-4.)

In the fall, surface temperatures drop and the stratification breaks down, allowing mixing of the nutrient-rich lower layer and the oxygen-rich upper layer. A similar situation occurs in the spring as the floating ice (at 0° C) melts, and surface waters warm to 4° C and mix with the deep water. Fall and spring turnovers maintain the lake's capacity to support its natural communities, and cause spurts of growth among phytoplankton and zooplankton, particularly in the spring when temperature and hours of sunlight are increasing.

Types of Lakes

Although each lake is unique, freshwater lakes may be classified as either oligotrophic (Greek, "poorly fed") or eutrophic (Greek, "well fed"), based on their nutrient content.

Oligotrophic lakes are very low in nutrients. They are often formed by glaciers that scour depressions in bare rock, and they are fed by mountain streams carrying little sediment. Because there is little sediment or microscopic life to cloud the water, oligotrophic lakes are clear, and light penetrates deeply. This allows photosynthesis (and oxygenation) in deeper water, so the limnetic zone may extend all the way to the bottom. Oxygen-loving fish, such as trout, thrive in oligotrophic lakes.

Eutrophic lakes receive larger inputs of sediments, organic material, and inorganic nutrients (such as phosphorus) from their surroundings. They are murkier, both from suspended sediment and dense phytoplankton populations, so the lighted limnetic zone is shallower. Dense "blooms" of algae occur seasonally in the limnetic zone. Their dead bodies fall into the profundal zone, which becomes very oxygen-depleted as decomposer organisms metabolize the detritus.

Lakes are transient ecosystems, although very large lakes may persist for millions of years. Over time, they gradually fill with sediment, undergoing succession to dry land (see Fig. 44-18). As nutrient-rich sediment accumulates, oligotrophic lakes tend to become eutrophic, a process called eutrophication.

Human Impact Human activities may greatly accelerate the process of eutrophication, since nutrients are carried into lakes from farms, feedlots, and sewage. Even if solid wastes are removed, water discharged from sewage treatment plants is often rich in phosphates and nitrates dissolved from wastes and detergents. Excessively enriched water also washes off fertilized farm fields and feedlots, where the manure of thousands of cattle accumulates. These added

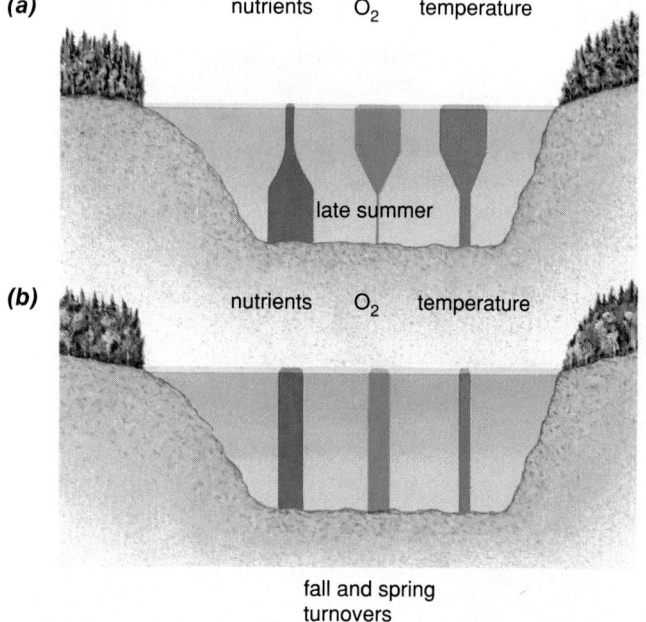

(a)   nutrients O_2 temperature

late summer

(b)   nutrients O_2 temperature

fall and spring
turnovers

Figure 46-27 (a) Temperature stratification during the summer results in high surface temperatures and oxygen levels, while surface nutrients become depleted due to lack of mixing. **(b)** During the spring and fall turnovers, temperature stratification breaks down and water from all levels mixes, resulting in uniform temperature, oxygen, and nutrient levels.

nutrients support excessive growth of phytoplankton. These producers, particularly blue-green algae, form a scum on the lake surface, depriving the submerged plants of sunlight. The plant bodies are decomposed by bacteria, depleting the dissolved oxygen. Deprived of oxygen, fish, snails, and insect larvae die, and their bodies fuel more bacterial growth, further depleting oxygen. Even without oxygen, certain bacteria thrive, fouling the water by producing noxious-smelling gases. Although it is full of life, the nutrient-polluted lake smells dead. Most of the trophic levels, including the fish, have been eliminated, and the community is dominated by bacteria and microscopic algae.

Acid rain, caused by the burning of fossil fuels, poses a very different threat to freshwater ecosystems. Acidified lakes appear clear and oligotrophic, since few organisms can withstand the low pH. Acid rain is discussed in detail in Chapter 45.

Marine Ecosystems

In the oceans, the upper layer of water, in which the light is strong enough to support photosynthesis, is called the **photic zone.** Below the photic zone lies the **benthic zone,** where the only energy comes from the excrement and bodies of organisms that sink or swim down.

As in lakes, most of the nutrients in the oceans are at or near the bottom where there is not enough light for photosynthesis. Nutrients dissolved in the water of the photic zone are constantly being incorporated into the bodies of living organisms. When these organisms die, some sink into the benthic zone, providing its organisms with nutrients. If no new nutrients entered the photic zone, life (both in the photic and the benthic zones) would eventually cease.

Fortunately, there are two sources of new nutrients: the land, from which rivers constantly remove nutrients and carry them to the oceans, and **upwelling,** where cold, nutrient-laden water from the ocean depths is brought to the surface. This often occurs along coastlines, as in California, Peru, and West Africa, where prevailing winds displace surface water, causing it to be replaced by water from below. Upwellings also occur around Antarctica. Not surprisingly, the major concentrations of life in the oceans are found where abundant light is combined with a source of nutrients. This happens most often in regions of upwelling and in shallow coastal waters, including coral reefs.

Coastal Waters

The most abundant life in the oceans is found in a narrow strip surrounding the Earth's land masses, where the water is shallow and a steady flow of nutrients is washed off the land. Coastal waters include the **intertidal zone,** the area that is alternately covered and uncovered by water with the rising and falling of the tides, and the **nearshore zone,** bays, salt marshes, estuaries, and shallow subtidal areas (Fig. 46-28). Here is the only part of the ocean where large plants can grow, anchored to the bottom. In addition, the abundance of nutrients and sunlight promotes the growth of a veritable soup of photosynthetic phytoplankton. Associated with these plants and protists are animals from nearly every phylum: annelid worms, sea anemones, jellyfish, sea urchins, starfish, mussels, snails, fish, and sea otters, to name just a few. A large number and variety of organisms live permanently in coastal waters, but many that spend most of their lives in the open ocean come into the coastal waters to reproduce. Bays, salt marshes, and estuaries in particular are the breeding grounds for a wide variety of organisms such as crabs, shrimp, and a host of fish, including most of our commercially important species.

Human Impact Coastal regions are of great importance not only to the organisms that live or breed there, but also to humans who wish to use them for food sources, recreation, mineral and oil extraction, or living places. Like rain forests, wetlands of all types once covered 6% of the Earth's surface, but nearly half of them have been obliterated during the past century by human activities. In the United States, 100 million of our 215 million acres of wetlands have been destroyed by agricultural and industrial activities, and by urban sprawl. As populations increase in our coastal states and resources such as oil become increasingly scarce, the conflict between preservation of our coastal wetlands as wildlife and animal habitat and development of these areas for housing, marinas, and energy extraction will become increasingly intense. Since much of the life of the entire ocean is inextricably dependent upon the well-being of the coastal waters, it is imperative that we protect these fragile, vital areas.

Coral Reefs

Coral reefs are actually created by animals and plants. In warm tropical waters, with just the right combination of bottom depth, wave action, and nutrients, specialized algae and corals (members of the phylum Cnidaria) build reefs from their own calcium carbonate skeletons. Coral reefs are most abundant in tropical waters of the Pacific and Indian Oceans, the Caribbean, and the Gulf of Mexico as far north as southern Florida, where the maximum water temperatures range between 22° C and 28° C (72° F to 82° F).

(a)

(b)

(c)

(d)

Figure 46-28 The variety of environments at the meeting place of land and sea. **(a)** A salt marsh in the eastern United States. Expanses of shallow water fringed by marsh grass *(Spartina)* provide excellent habitat and breeding grounds for many marine organisms and shorebirds. **(b)** Although the shifting sands present a challenge to life, grasses stabilize them, and (inset) animals such as this *Emerita* crab burrow in the sandy intertidal zone. **(c)** A rocky intertidal shore in Oregon, where animals and algae grip the rock against the pounding waves and resist drying during low tide. Inset: Colorful sea stars cling to the rocks surrounded by a seaweed, fucus. **(d)** Towering kelp sway through the clear water off southern California, providing the basis for a diverse community of invertebrates, fishes, and an occasional sea otter (inset).

Reef-building corals are involved in a mutualistic relationship with unicellular algae called dinoflagellates, who live embedded in the coral tissue. They grow best at depths of less than 40 meters (130 ft), where light can penetrate and allow the algal symbionts to photosynthesize. The algae benefit from the high CO_2 concentration in the coral tissues, which they use during photosynthesis. They assist the coral by removing this waste product and aiding in the deposition of the calcium carbonate skeleton. These reefs provide an anchoring place for many other algae, a home for bottom-dwelling animals, and shelter and food for the most diverse assemblage of invertebrates and fish to be found in the oceans (Fig. 46-29). The Great Barrier Reef in Australia is home to more than 200 species of coral alone, and a single reef may harbor 3000 different species of fish, invertebrates, and algae.

Human Impact Coral reefs are extremely sensitive to certain types of disturbance, especially silt eroding from nearby land. As silt clouds the water, light is diminished and photosynthesis reduced, hampering growth of the corals. The reef may eventually become buried in mud, the corals smothered, and the entire marvelous community of diverse organisms destroyed. Several reefs near Honolulu have been lost to siltation owing to erosion from construction, roadways, and poor land management. The reef inhabitants have been replaced by large numbers of sediment-feeding invertebrates such as sea cucumbers. In the Philippines, logging has dramatically increased

(a)

(b)

(c)

(d)

Figure 46-29 (a) Coral reefs, composed of the skeletons of corals and algae, provide habitat for an extremely diverse community of extravagantly colored animals. **(b)** Many fish, including this blue tang, feed on coral (note the bright yellow corals in background). A host of invertebrates such as this sponge **(c)** and blue-ringed octopus **(d)** live among the corals of Australia's Great Barrier Reef. This tiny (6 inches, fully extended) octopus is one of the world's most venomous creatures, possessing enough venom to kill 10 adult men.

erosion, so reefs (as well as rain forests) are being destroyed.

A second threat to the reef communities is over-fishing. In at least 80 countries, a variety of species including mollusks, turtles, fish, crustaceans, and corals themselves are being harvested from reefs faster than they can replace themselves. Many of these are sold to shell collectors and aquarists in the developed countries. In 40 different tropical countries, dynamite is used to kill coral reef fish, destroying the entire community. Many tropical fish collectors use cyanide to stun the fish before collecting them, leaving nine out of ten dead. Removal of predators from reefs may disrupt the ecological balance of the community, allowing an explosion in populations of coral-eating sea urchins, which are destroying reefs along the Kenyan coast in Africa.

As with rain forests, both protection and sustainable use are crucial to the survival of these diverse underwater ecosystems. Carefully regulated harvesting and tourism produce far more economic benefits than activities that destroy the reefs. Some countries have recognized this; there are about 300 protected areas in 65 different countries, and many more are under consideration for protection.

The Open Ocean

Beyond the coastal regions lie vast areas of the ocean in which the bottom is too deep to allow plants to anchor and still receive enough light to grow. Most life in the open ocean is limited to the upper photic zone, in which the life forms are **pelagic,** that is, free-swimming or floating for their entire lives. The base upon which life exists in the open sea is the **plankton,** a collection of often microscopic protists (phytoplankton) and animals (zooplankton) (Fig. 46-30). The photosynthetic phytoplankton (mainly diatoms and di-

Figure 46-30 In the open ocean, rare humpback whales breach, porpoise skim the surface, and fish such as these blue jacks swim (insets). The photosynthetic phytoplankton **(a)** are the producers on which most other life ultimately depends. Phytoplankton are eaten by zooplankton **(b)**, represented by this microscopic crustacean, a copepod. The spiny projections on these planktonic creatures help keep them from sinking below the photic zone.

noflagellates) are the ultimate food source for the other inhabitants of the open ocean. Most of the planktonic animals are minute crustaceans, relatives of crabs and lobsters that feed on the phytoplankton or on each other. In turn, they serve as food for larger invertebrates, small fish, and marine mammals such as the humpback whale (Fig. 46-30).

One challenge faced by denizens of the open ocean is how to stay afloat in the photic zone where sunlight and food are abundant. Many members of the planktonic community have elaborate flotation devices such as oil droplets in their cells or long projections to slow down their rate of sinking (see Fig. 46-31). Most fish have swim bladders that can be filled with gas to regulate their buoyancy. Some animals, and even some of the phytoplankton such as the dinoflagellates (see Chapter 21), actively swim to maintain their position in the photic zone. Some small crustaceans migrate to the surface at night to feed, sinking into the dark depths during daylight, thus avoiding visual predators such as fish.

The amount of pelagic life varies tremendously from place to place. The blue clarity of tropical waters is a result of a lack of nutrients, which limits the concentration of plankton in the water. Nutrient-rich waters that support a large plankton community are greenish and relatively murky, as sunlight is scattered by the microscopic bodies of myriad protists and animals.

Below the photic zone, the only available energy in most cases comes from the excrement and dead bodies raining down from above. Nevertheless, a surprising quantity and variety of life exists in the benthic zone, including fishes of bizarre shape, worms, sea cucumbers, sea stars, and mollusks.

Vent Communities

In 1977, a new and unusual source of nutrients, forming the basis of a spectacular undersea community, was discovered in the benthos. Geologists exploring the Galápagos Rift (an area of the Pacific floor where plates forming the Earth's crust are separating) found vents spewing superheated water, black with sulfur and minerals. Surrounding these vents was a rich community of pink fish, blind white crabs, enormous mussels, white clams, sea anemones, and tube worms (Fig. 46-31). Sixteen previously undescribed families of invertebrates have been found in these novel communities.

In this unique ecosystem, sulfur bacteria serve as the primary producers, harvesting the energy stored in hydrogen sulfide spewed from beneath the Earth's crust. This process, called chemosynthesis, replaces photosynthesis at these depths, 2500 meters (over 7500 ft) below the surface. The bacteria proliferate in the warm water surrounding the vents, covering nearby rocks with thick, matlike colonies. These provide the food on which the animals of the vent community thrive. Many vent animals consume the bacteria directly. Others, such as the tube worm **Riftia** (see Fig. 46-31), which lacks a digestive tract, harbor the bacteria within cells in their bodies. In this mutualistic association, the bacteria provide high-energy carbon compounds, while the tube worm provides sulfide, which it transports to the bacteria via its bloodstream using a unique form of hemoglobin. **The world still holds wonders and mysteries for those who seek them, and we have only begun to explore the versatility and diversity of life on Earth.**

Figure 46-31 Vent communities in the ocean depths include giant 12-foot-tall worms, red with oxygen-trapping hemoglobin. These worms have no digestive tract, but are host to sulfur bacteria, which provide the worms with energy by oxidizing hydrogen sulfide.

Reflections: Humans and Ecosystems

The expanding human population has left relatively few ecosystems undisturbed. Our impacts on natural ecosystems are so diverse and wide ranging that they far exceed the scope of this book. However, we can identify some general characteristics of ecosystems dominated by humans and contrast them to the characteristics of undisturbed ecosystems. Below are listed six characteristic differences and a few ideas for minimizing them.

First, ecosystems dominated by people tend to be simpler—that is, to have fewer species and fewer community interactions—than undisturbed ecosystems. Although a city street bustles with life and apparent complexity, count the number of different species you encounter in an average block and compare it with the number you encounter while hiking in a wilderness for a similar distance. As humans enter an ecosystem, animals in the highest trophic levels are the first to go. Carnivores are always relatively rare, and their specialized needs are most easily disrupted. Many big carnivores, such as the wolf and mountain lion, require large undisturbed hunting territories. Human beings often selectively destroy large predators, believing them a threat to people or their livestock. As a result, even in relatively undisturbed areas they have often been eliminated. Grizzlies and wolves no longer roam the Rocky Mountains. On the western shortgrass prairie, prairie dog towns fall as human towns and ranches rise (Fig. E46-3). The black-footed ferret, a predator of prairie dogs, faces possible extinction. Farm fields have been deliberately simplified from the original prairie biome to eliminate competition and predation and allow the maximum productivity of a single crop species. Nowhere is the contrast between human and natural ecosystems greater than in the tropical rain forests, whose unparalleled diversity is replaced by failed attempts at farming.

There is probably no practical way to restore to human ecosystems the great diversity found in undisturbed areas, nor is this always desirable from a human standpoint. Agriculture, for example, demands a simplified ecosystem (although not all biomes are suited to it). Cities concentrate human activities and culture and may lessen the human impact on the surrounding countryside. However, as we recognize the benefits of artificially simplified ecosystems, we must be aware of the need to preserve intact as many natural communities as possible. The undisturbed forest traps and purifies water and can reduce air pollution. Swamps and estuaries, when not dredged or filled, contain a wealth of detritus feeders and decomposers that purify water. Coastal wetlands are breeding places for millions of birds and spawning sites for the majority of our commercially important fish and crustacean species. In undisturbed, diverse ecosystems we find aesthetic pleasure, as well as a storehouse of species whose commercial or medicinal values are not yet recognized. For example, nearly half the medicines in use today were originally discovered in plants, and human beings have examined only a small fraction of existing plants for possible medical uses.

Second, whereas natural ecosystems run on sunlight, human ecosystems have become dependent on nonrenewable energy from fossil fuels. From the suburbanite pouring gas into a lawn mower to the farmer driving a combine, managing a simplified ecosystem is an energy-intensive proposition. Energy must be expended to oppose the tendency of the natural system to restore complexity. Fertilizers also require large amounts of energy to produce, and farms must be heavily fertilized because nutrient cycles have been disrupted.

Figure E46-3 Loss of habitat due to human activities is a major threat to most of the Earth's wildlife.

To counteract this trend, some farmers are returning to organic farming. Plant and animal waste and natural nutrient cycles are used to maintain soil fertility. Alternating legume crops such as soybeans and alfalfa helps maintain nitrogen in the soil. Mulching can help retain fertility, reduce water loss, and control weeds. The use of natural insect predators and limited pesticide spraying at critical times during a pest's life cycle can dramatically reduce the need for poisons. Organic farms can be as productive as more conventional farms while using 15% to 50% less energy per quantity of food produced.

In our homes and commercial structures, better insulation and increased use of solar heat can result in dramatic energy savings. By increasing our use of renewable energy sources such as wind and especially sunlight, we can conserve fossil fuels, dramatically reduce pollution, and move human ecosystems a step closer to those that occur naturally.

Third, natural ecosystems recycle nutrients, while human ecosystems tend to lose nutrients. Walk around a suburban neighborhood on trash-pickup day. Grass clippings and leaves are packed in plastic bags to be hauled away. To compensate for this loss, suburban lawns and gardens are often heavily fertilized. A similar trend has occurred in modern farming. The exposed soil is eroded away by wind and water, removing crucial nutrients and requiring large inputs of fertilizer to replace them. Runoff from the field, carrying fertile topsoil and artificial fertilizers, may pollute nearby rivers, streams, and lakes. Pesticides may kill detritus feeders and decomposers, further disrupting natural nutrient cycles. Thus, while fertile soil accumulates in many natural ecosystems, it tends to be lost in those dominated by human beings. Some three billion tons of topsoil are eroded from farms each year in the United States. The Mississippi River alone carries off 40 tons per hour.

Again, organic farming can help reverse this trend, and we can apply the same principles in our own lawns and gardens. Farmers can counteract erosion by using contour planting, in which row crops are oriented so they retard the flow of water instead of funneling it down a slope (Fig. E46-4). Row crops, which tend to allow erosion, can be alternated in strips with dense, soil-catching crops such as wheat. Planting rows of trees as windbreaks helps prevent soil loss from blowing wind, as well as creating a more diverse ecosystem and a nesting place for insect-eating birds. Farm fields need not be plowed in the fall and left unplanted to erode during the winter; in fact, an increasing number of farmers are planting crops in the stubble from the previous season, reducing erosion.

Fourth, natural ecosystems tend to store water and purify it through biological processes, whereas human ecosystems tend to pollute water and shed it rapidly. A thundershower strikes a forest, an adjacent city, and a farm. The rich soil of the forest sponges up the water, which gradually percolates into the ground, filtered by the soil and purified by decomposers that break down organic contaminants. In the nearby city, water pours from sidewalks, rooftops, and streets, picking up soot, silt, oil, lead, and refuse. It races down gutters into storm sewers, and a weakly toxic soup gushes into the nearest stream or river. Farm runoff carries priceless topsoil, expensive fertilizer, and animal manure into rivers and lakes, where these resources become pollutants.

Preventing erosion will simultaneously conserve water and reduce water pollution from farm runoff. Manure from cattle feedlots, which is a significant source of both groundwater and surface water pollution, could be placed on fields, where it would restore needed nutrients.

Although water will continue to run off our cities, the pollutants it carries can be reduced by minimizing our reliance on fossil fuels. We must eliminate leaded gasoline and tighten standards for emissions from diesel and gasoline engines and smokestacks. Efficient rapid-transit systems will reduce pollution and the massive frustration caused by congested traffic. Insulation will reduce power consumption in our homes and offices, as will increased reliance on solar heat.

Fifth, simple human ecosystems such as farms tend to be extremely unstable. Natural ecosystems have many species and tend to remain stable over time. Herbivorous insects are controlled by natural predators, including other insects, birds, and shrews. Insect popula-

Figure E46-4 Contour planting on sloping fields minimizes erosion.

tions are also limited since their preferred plants are interspersed among many others rather than growing in a pure stand, as on a farm. Although both farm pests and their natural predators are exposed to pesticides, the pests often develop resistance to the poison, while their predators are killed. A resistant pest, its predators destroyed and its favorite food surrounding it, will reproduce rapidly and devastate a crop. In these simplified communities, the introduction of an exotic species (see Chapter 43) is often disastrous, since most of its potential predators and competitors have been removed. Simplified human ecosystems are also more vulnerable to unfavorable environmental conditions—a hailstorm, too much or too little rain, or an early or late freeze.

Farmers can counteract this trend by planting smaller fields with a wider variety of crops. Alternating crops not only helps maintain soil fertility but also helps prevent the proliferation of disease and insect pests that are specialized for a particular crop. Populations of corn borers, for example, will die off during years when the field is planted in alfalfa. Use of biological controls such as natural insect predators and insect diseases can reduce reliance on pesticides.

Finally, human ecosystems are characterized by continuously growing populations, whereas nonhuman populations in natural ecosystems are relatively stable. As our population expands, the spread of human-dominated ecosystems presents a growing threat to the diversity of species and to the delicate balance that has evolved over the three-billion-year history of life on Earth.

In summary, natural ecosystems tend to be complex, stable, and self-sustaining, powered by solar energy and nourished by recycling nutrients. They provide diverse habitats for wildlife, tend to purify contaminants through the action of decomposers, and build up nutrient-rich soil. Modern human ecosystems are relatively simple and are sustained by large inputs of energy from fossil fuels. They tend to minimize wildlife habitat, contaminate soil and water, and lose nutrients and fertile soil. These problems are compounded by continued population growth, which is causing expansion of human-dominated ecosystems at the expense of undisturbed ones.

As we have pointed out, human ecosystems do not have to be as disruptive and alien to the operation of natural ecosystems as we have allowed them to become. Through understanding, education, and commitment; appropriate use of technology; and stabilization of our population, we can reverse many of these destructive trends.

SUMMARY OF KEY CONCEPTS

Patterns Within Diversity
Throughout the world, areas with similar climates have similar assemblages of organisms. Although species differ, adaptations to climate create recognizable distinguishing features, particularly among plants.

Climate
Sunlight that reaches the Earth may either be reflected or absorbed by the atmosphere, land, and water which reradiates it as heat, with a small amount captured by photosynthetic organisms. This heat maintains the temperature of the Earth, but is eventually radiated back into space.

Latitude determines the angle at which sunlight strikes the earth. Equal amounts of solar energy are spread over a smaller surface at the equator than farther north and south, making the equator relatively warm. The Earth's tilt on its axis causes dramatic seasonal variation at northern and southern latitudes.

Warm air currents at the equator and at 60° north and south latitudes rise and release moisture as they cool, increasing precipitation at these latitudes. Cooler, dry air sinks at around 30° north and south latitudes, and at the poles, creating areas of very low moisture. These patterns are modified by the continents. Ocean gyres carry warm equatorial waters into northern and southern latitudes, moderating the climate of coastal areas.

Climate is modified by the rapid heating and cooling of continents as compared with the oceans, and by elevation. Air cools as it rises, so high elevations are cooler. Cool air holds less moisture, so rain falls on the windward side of mountain ranges, and rain shadows occur on the far sides.

The Requirements of Life
The requirements for life on Earth include nutrients, energy, liquid water, and a reasonable temperature. The differences in the form and abundance of living things in various locations on Earth are largely attributable to differences in the interplay of these four factors.

Life on Land

On land, the crucial limiting factors are temperature and liquid water. Large regions of the continents which have similar climates will have similar vegetation, determined by the interaction of temperature and rainfall. These regions are called biomes. Biomes are influenced by the temperature and rainfall patterns, which vary with latitude. The equatorial tropical rain forests have a consistently warm, moist climate, while from about 20° to 30° north and south latitude, the Earth's great deserts are found. Farther north, grasslands and temperate deciduous forests predominate, giving way to the taiga or northern coniferous forest. The last stronghold of life before the polar icecap is the tundra. Continental topography also influences terrestrial biomes. In North America, deserts are often found in the eastern rain shadows of mountain ranges. Since temperature decreases both with increasing latitude and increasing elevation, the vegetation changes observed going up a mountain are similar to those seen while traveling from the equator toward the poles.

Tropical forests are warm and wet, dominated by huge broadleaf evergreen trees. Most nutrients are tied up in vegetation, and most animal life is arboreal. These areas, which are home to the most diverse assemblage of plants and animals on Earth, are rapidly being cut for agriculture, although the soil is extremely poor.

The African savanna is an extensive grassland with pronounced wet and dry seasons. It is home to the world's most diverse and extensive herds of large mammals.

Deserts, hot and dry, are found primarily between 20° and 30° of latitude and in the rain shadows of mountain ranges. Plants are intermittent and have adaptations to conserve water. Animals tend to be small and nocturnal, also adapted to drought.

Chaparral exists in desertlike conditions, which are moderated by their proximity to a coastline. This allows small trees and bushes to thrive.

Grasslands, concentrated in the centers of continents, have a continuous grass cover and few trees. They produce the world's richest soils, and have largely been converted to agriculture.

Temperate deciduous forests, whose broadleaf trees drop their leaves in winter to conserve moisture, dominate the eastern half of the United States. The wet temperate rainforests, dominated by evergreens, are found on the North Pacific coast of the United States.

The taiga, or northern coniferous forest, covering much of the northern United States and southern Canada, is dominated by conifers whose small waxy needles are adapted for water conservation and year-round photosynthesis.

The tundra is a frozen desert where permafrost prevents the growth of trees, and bushes remain stunted. Nonetheless, a diverse array of animal life and perennial plants flourishes in this fragile biome.

Life in Water

Energy and nutrients are the most common limiting factors in the distribution and abundance of life in aquatic ecosystems. Water absorbs sunlight, and murkiness limits light penetration. Nutrients are found in bottom sediments and are washed in from surrounding land, concentrating them near shore and in deep water.

Freshwater lakes have three life zones. The littoral zone, near shore, is rich in energy and nutrients and supports the most diverse community. The limnetic zone is the lighted region of open water where photosynthesis can occur. The profundal zone is the deep water, where light is inadequate for photosynthesis and the community is dominated by heterotrophic organisms.

Lakes in temperate climates become stratified during the summer and winter, with a layer of dense water (at 4° C) at the bottom and less dense, warmer water (in summer) or colder water (in winter) in an upper layer. During the spring and fall turnovers when temperatures equalize, the layers mix, distributing oxygen and nutrients.

Oligotrophic lakes are clear, low in nutrients, and support sparse communities. Eutrophic lakes are rich in nutrients and support dense communities. During succession to dry land, lakes tend to go from an oligotrophic to a eutrophic condition.

Most life in the oceans is found in shallow water, where sunlight can penetrate, and is concentrated near the continents and in areas of upwelling, where nutrients are most plentiful.

Coastal waters, consisting of the intertidal zone and the nearshore zone, contain the most abundant life. Producers consist of aquatic plants anchored to the bottom and photosynthetic protists called phytoplankton.

Coral reefs are confined to warm, shallow seas. The calcium carbonate reefs form a complex habitat supporting the most diverse undersea ecosystem.

In the open ocean, most life is found in the photic zone, where light supports phytoplankton. In the lower benthic zone, life is supported by nutrients that rain down from the photic zone. Specialized vent communities, supported by bacteria, thrive at great depths in the superheated waters where the Earth's crustal plates are separating.

GLOSSARY

benthic zone (ben'-thik): the ocean depths, into which insufficient sunlight penetrates to support photosynthesis.

biome (bī'-ōm): a terrestrial region with a characteristic climate and vegetation. Regions with similar climates will have similar vegetation and belong to the same biome (grassland, desert) even though widely separated and inhabited by different species.

chaparral (shap-eh-rel'): a temperate coastal biome with hot dry summers and cool, somewhat rainy winters with frequent fogs. The typical vegetation consists of small trees or large bushes which are drought and fire resistant.

climate: prevailing weather patterns over a particular region.

coral reef: the most diverse marine ecosystem, formed by the bodies of corals in warm, relatively shallow water.

desert: a biome in which potential evaporation greatly exceeds rainfall. Perennial vegetation is widely spaced and has drought-resistant adaptations such as waxy or spiny leaves.

desertification (deh-ser'-ti-fi-kā-shun): the conversion of productive land in warm, dry areas into unproductive land through loss of soil and vegetation, often as a result of overgrazing or other improper farming methods.

eutrophic lake (ū'-trōf-ik): a lake that is rich in nutrients and supports dense communities of organisms.

grassland: a biome characterized by a relatively continuous ground cover of grasses, with few or no trees except along watercourses. Typically, grassland biomes have a prolonged dry season and frequent fires, both of which favor grasses over trees.

gyre (jīre): a massive, roughly circular ocean current. Gyres flow clockwise in the northern hemisphere and counterclockwise in the southern hemisphere.

intertidal zone: the area along the edges of land masses that is alternately flooded and left exposed by the rising and falling of the tides.

limnetic zone (lim-net'-ik): the upper, lighted, open-water region of a lake, where phytoplankton can carry out photosynthesis.

littoral zone (lit'-ōr-ul): the near-shore, shallow-water region of a lake where anchored plants grow.

nearshore zone: shallow-water areas of the ocean just below the low-tide line. Due to the abundance of nutrients and strong sunlight, the nearshore zone often has the greatest concentration of life in the ocean.

northern coniferous forest: a northern biome dominated by conifers; taiga.

oligotrophic lake (ol-i-gō-trōf'-ik): a lake that is low in nutrients and supports sparse communities of organisms.

ozone layer: a layer of atmosphere between 10 and 50 kilometers above the Earth, which contains relatively high concentrations of ozone (O_3). Ozone absorbs harmful ultraviolet radiation.

pelagic (pel-a'-jik): referring to organisms that spend their lives in open water, floating or swimming.

permafrost: soil layer a short distance below the surface of the tundra that stays frozen year round.

photic zone (fō'-tik): the surface waters of the ocean, into which sunlight penetrates well enough for plants to grow.

phytoplankton (fī'-tō-plank-ton): microscopic photosynthetic protists that form the basis of ocean food webs.

plankton (plank'-ton): microscopic pelagic protists and animals found in the photic zone of the oceans, whose movement is determined primarily by movement of the water. Phytoplankton are the main producers in the open ocean.

profundal zone (prō-fun'-dul): the depths of a lake where light is inadequate for photosynthesis.

rain shadow: an area of low rainfall on the side of a mountain facing away from the prevailing winds.

savanna: a transition biome between the tropical deciduous forest and tropical grassland, characterized by a nearly continuous cover of grasses with occasional drought-resistant trees. The savanna climate includes such a severe dry season that most tree species cannot grow.

taiga (tī'-ga): the cold coniferous forest biome, characterized by needleleaf evergreen trees. The waxy needles of the trees reduce evaporative water loss during winter.

temperate deciduous forest: a biome having cold winters but a fairly long, moist growing season. The dominant vegetation is broadleaf deciduous trees, which shed their leaves in winter as a protection against water loss.

temperate rain forest: a biome of very limited distribution, along certain coasts where the climate is fairly mild year round and rainfall is very high, characterized by coniferous evergreen trees, often of enormous size.

tropical deciduous forest: a broadleaf deciduous forest biome found in areas with a warm climate year round but a pronounced wet and dry season. The trees drop their leaves in the dry season.

tropical rain forest: a broadleaf evergreen forest biome found in the tropics, where rainfall is adequate and temperatures are warm year round. The tropical rain forest has the greatest diversity of both plants and animals to be found anywhere on land.

tundra: a treeless biome characterized by permafrost and an extremely short growing season. Most of the plants are low-growing perennial shrubs, grasses, and wildflowers.

upwelling: movement of deep, nutrient-rich water to the surface.

weather: fluctuations in temperature, humidity, precipitation, and cloud cover occurring over a period of hours or days.

zooplankton (zō'-plank-ton): a diverse assemblage of small pelagic protists and animals dominated by crustaceans.

STUDY QUESTIONS

1. Trace the fate of the solar energy that reaches the atmosphere. Why doesn't the Earth continuously get hotter and hotter (ignoring the greenhouse effect)?
2. Explain how air currents contribute to the formation of the tropics and the large deserts.
3. What are large, roughly circular ocean currents called? What effect do they have on climate, and where is that effect most strong?
4. What are the four major requirements for life? Which two are most often limiting in terrestrial ecosystems? In ocean ecosystems?
5. Explain why traveling up a mountain takes you through biomes similar to those you would encounter traveling north for a long distance.
6. Where are the nutrients of the tropical forest biome concentrated?
7. Explain two different undesirable effects of agriculture in the tropical rain forest biome.
8. Why is life in the tropical rain forest concentrated high above the ground?
9. List some adaptations of (a) desert plants and (b) desert animals to heat and drought.
10. How and where does desertification occur?
11. On what biome does much of the world's agriculture occur?
12. How are trees of the taiga adapted to a lack of water and a short growing season?
13. Where is life in the oceans most abundant, and why?
14. Why is the diversity of life so high in coral reefs? What human impacts threaten them?
15. Distinguish between the limnetic, littoral, and profundal zones of lakes in terms of their location and the communities they support.
16. Distinguish between oligotrophic and eutrophic lakes. Describe (a) a natural scenario and (b) a human-created scenario under which an oligotrophic lake might be converted to a eutrophic lake.
17. What is the reason and importance of the spring and fall overturn in temperate lakes?
18. Distinguish between the photic and benthic zones. How do organisms in the photic zone obtain nutrients? How are nutrients obtained in the benthic zone?
19. What unusual primary producer forms the basis for rift-vent communities?
20. List the terrestrial biomes and some dominant features of each.
21. How do deciduous and coniferous biomes differ?
22. What single environmental factor best explains why there is shortgrass prairie in Colorado, tallgrass prairie in Illinois, and deciduous forest in Ohio?
23. Where are the world's largest populations of large herbivores and carnivores found?

DISCUSSION QUESTIONS

1. List at least six differences between human-dominated and undisturbed ecosystems, and discuss in some detail how these differences can be minimized.
2. Discuss some of the advantages you personally enjoy as a member of an affluent industrialized society. Now discuss some of the drawbacks of life in such a society. Comment on whether the advantages outweigh the drawbacks.
3. In which terrestrial biome is your college or university located? Discuss similarities and differences between your location and the general description of that biome in the text. If you are living in a city, how has the urban environment modified your interaction with the biome?

SUGGESTED READINGS

Bell, R. H. V. "A Grazing Ecosystem in the Serengeti." *Scientific American*, July 1971. Describes the great migrations of the herbivores of the African savanna.

Brownlee, S. "Bizarre Beasts of the Abyss." *Discover*, July 1984. Beautiful photography and informative narrative about the rift-vent communities.

Goreau, T. F., Goreau, N. I., and Goreau, T. J. "Corals and Coral Reefs." *Scientific American*, August 1979. Discusses the ecology of the great reefs built by tiny polyps harboring symbiotic protists.

Isaacs, J. D. "The Nature of Oceanic Life." *Scientific American*, September 1969 (Offprint No. 884). Describes the oceanic food chain with its microscopic primary producers.

Richards, P. W. "The Tropical Rain Forest." *Scientific American*, December 1973 (Offprint No. 1286). Describes this reservoir of genetic diversity and its rapid destruction.

Stolarski, R. "The Antarctic Ozone Hole." *Scientific American*, January 1988. Describes the loss of this protective atmospheric layer, its causes, and the consequences.

Ward, F. "Coral Reefs are Imperiled." *National Geographic*, July 1990. A disturbing look at Florida's "undersea jewels," now threatened by overuse by sightseers and near-shore development.

Wilson, E. O. "Threats to Biodiversity." *Scientific American*, September 1989. The elimination of habitats, particularly in the rain forest, is driving plants and animals to extinction in record numbers.

Glossary

Abiotic (ā-bī-ah′-tik): nonliving.

Abscisic acid (ab-sis′-ik): a plant hormone that generally inhibits the action of other hormones, enforcing dormancy in seeds and buds and causing closing of stomata.

Abscission (ab-si′-shun): separation of leaves, flowers, or fruits from a stem, due to formation of a weakened layer of cells at the site of attachment to the stem.

Abscission layer: a layer of thin-walled cells at the base of the petiole of a leaf, flower, or fruit, the usual site of separation from the stem.

Absorption: the movement of nutrients into cells.

Accessory pigments: colored molecules other than chlorophyll that absorb light energy and pass it to chlorophyll.

Acetylcholine (ah-sēt′-il-kō′-lēn): a neurotransmitter used at vertebrate skeletal neuromuscular junctions, and found in the brain.

Acid: a substance that releases hydrogen ions (H^+) into solution; a solution with a pH of less than 7.

Acid deposition: the deposition of nitric or sulfuric acid, either in precipitation or in particulate form, as a result of the production of nitrogen oxides or sulfur dioxide through burning, primarily of fossil fuels.

Acrosome (ak′-rō-sōm): an enzyme-containing vesicle located at the tip of an animal sperm.

Actin (ak′-tin): one of the major proteins of muscle, whose interactions with myosin produce contractions; found in the thin filaments of the muscle fiber. See also myosin.

Action potential: a rapid change from a negative to a positive electrical potential in a nerve cell. This signal travels along an axon without change in size.

Activation energy: in a chemical reaction, the energy needed to force the electron clouds of reactants together, prior to the formation of products.

Active site: the region of an enzyme molecule that binds substrates and performs the catalytic function of the enzyme.

Active transport: the movement of materials across a membrane through the use of cellular energy, usually against a concentration gradient.

Active visual signal: a movement or posture that communicates information.

Adaptation: a specific structure, physiological mechanism, or behavior that promotes the survival and/or reproduction of an organism; also the process of acquiring such characteristics.

Adaptive radiation: extensive speciation occurring among related organisms as a result of adaptation to a wide variety of habitats.

Adenosine triphosphate (a-den′-ō-sen trīfos′-fāt; ATP): a nucleotide composed of ribose sugar, adenine, and three phosphate groups. The last two phosphate groups are attached by energy-carrier "high-energy bonds." ATP is the major energy carrier in cells.

Adipose tissue (a′-di-pōse): tissue composed of fat cells.

Adrenal gland: an endocrine gland consisting of an outer cortex and inner medulla. The cortex secretes steroid hormones that regulate metabolism and salt balance. The medulla secretes adrenalin and noradrenalin.

Aerobic: using oxygen.

Age structure: the distribution of males and females in a population according to age categories; often represented graphically.

Agglutination (a-glu-tin-ā′-shun): clumping of foreign substances or microbes, caused by binding with antibodies.

Aggregated distribution: characteristic distribution of populations in which individuals are clustered into groups. These may be social or based on the need for a localized resource.

Aggression: antagonistic behavior, usually between members of the same species, often resulting from competition for resources.

Aggressive mimicry (mim′ik-rē): the evolution of a predatory organism to resemble a harmless one, thus gaining access to its prey.

Algae (al′-gē; sing. alga): a general term for simple aquatic plants lacking vascular tissue.

Allantois (al-an-tō′-is): one of the embryonic membranes of reptiles, birds, and mammals. In reptiles and birds, the allantois serves as a waste-storage organ and a respiratory surface. In mammals, the allantois forms most of the umbilical cord.

Allele (al-ēl'): one of several alternative forms of a particular gene.

Allele frequency: for any given gene, the relative proportion of each allele of that gene found in a population.

Allergy: an inflammatory response produced by the body in response to invasion by foreign materials, such as pollen, which are themselves harmless.

Allopatric speciation: (al-ō-pat'-rik): speciation that occurs when two populations are separated by a physical barrier that prevents gene flow between them (geographical isolation).

Allosteric inhibition (al-ō-ster'-ik): enzyme regulation in which an inhibitor molecule binds to an enzyme at a site away from the active site, changing the shape or charge of the active site so that it can no longer bind substrate molecules.

Alternation of generations: a life cycle typical of plants in which a diploid sporophyte (spore-producing) generation alternates with a haploid gametophyte (gamete-producing) generation.

Altruism: a behavior that benefits another organism, usually at some risk to the altruistic organism.

Alveolus (al-vē'-ō-lus; pl. alveoli): a tiny air sac within the lungs surrounded by capillaries where gas exchange with the blood occurs.

Amino acid: the individual subunit of which proteins are made, composed of a central carbon atom to which is bonded an amino group ($-NH_2$), a carboxylic acid group ($-COOH$), a hydrogen atom, and a variable group of atoms denoted by the letter "R."

Ammonia: a highly toxic nitrogen-containing waste product of amino acid breakdown, which is converted to urea in the mammalian liver.

Amniocentesis (am-nē-ō-sen-tē'-sis): a procedure for sampling the amniotic fluid surrounding a fetus. A sterile needle is inserted through the abdominal wall, uterus, and amniotic sac of a pregnant woman, into the amniotic fluid. Ten to twenty milliliters of amniotic fluid is withdrawn into a syringe. Various tests may be performed on the fluid and the fetal cells suspended in it to provide information on the developmental and genetic state of the fetus.

Amnion (am'-nē-on): one of the embryonic membranes of reptiles, birds, and mammals, enclosing a fluid-filled cavity that envelops the embryo.

Amniotic egg (am-nē-ot'-ik): the egg of reptiles and birds. It contains an amnion that encloses the embryo in a watery environment; this allows the egg to be laid on dry land.

Amplexus (am-pleck'-sus): a form of external fertilization found in amphibians, in which the male holds the female during spawning and releases his sperm directly onto her eggs.

Amygdala (am-ig'-da-la): part of the forebrain of vertebrates, involved in control of emotions and instinctive behaviors.

Amylase (am'-ē-lās): an enzyme that catalyzes the breakdown of starch, found in saliva and pancreatic secretions.

Anaerobic: not using oxygen.

Analogous structures: structures that have similar functions and superficially similar appearance but very different anatomy, such as the wings of insects and birds. The similarities are due to similar selective pressures.

Anaphase (an'-a-fāz): the stage of mitosis and meiosis II in which the sister chromatids of each chromosome separate from one another and are moved to opposite poles of the cell. In meiosis I, the stage in which homologous chromosomes are separated.

Angina pectoris (an-jī'-na pek-tōr'-is): chest pain associated with reduced blood flow to the heart muscle caused by obstruction of coronary arteries.

Angiosperm (an'-jē-ō-sperm): a flowering plant (Division Anthophyta); produces seeds enclosed within a ripened ovary.

Annual ring: in the xylem of woody stems and roots, a pair of rings of light (early) and dark (late) wood, formed because of differential availability of water in different seasons of the year, usually spring and summer.

Anterior (an-tēr'-ē-ur): the front, forward, or head end of an animal.

Anther (an'-ther): the uppermost part of the stamen in which pollen develops.

Antheridium (an-ther-id'-ē-um): a structure in which male sex cells are produced, found in the bryophytes and certain seedless vascular plants.

Antibody: a protein produced by cells of the immune system which combines with a specific antigen and usually facilitates its destruction.

Anticodon: a sequence of three nucleotides in transfer RNA that is complementary to the three nucleotides of a codon of messenger RNA.

Antidiuretic hormone (an-tē-dī-ūr-et'-ik): also called ADH; a hormone produced by the hypothalamus and released into the bloodstream by the posterior pituitary gland. It acts on the nephron of the kidney and causes more water to be reabsorbed into the bloodstream.

Antigen: a complex molecule, usually protein or polysaccharide, that stimulates the production of a specific antibody.

Apical dominance: the phenomenon whereby a growing shoot tip inhibits the sprouting of lateral buds.

Apical meristem (āp'-i-kul mer'-i-stem): the cluster of meristematic cells found at the tip of a shoot or root (or one of their branches).

Appendicular skeleton (ap-pen-dik'-ū-lur): that portion of the skeleton consisting of the bones of the appendages and their attachments; the pectoral and pelvic girdles, the arms, legs, hands, and feet.

Aqueous humor (ā'-kwē-us): clear, watery fluid between the cornea and lens of the eye.

Archegonium (ar-ke-gō'-nē-um): a structure in which female sex cells are produced, found in the bryophytes and certain seedless vascular plants.

Arteriole (ar-tēr'-ē-ōl): a small artery that empties into capillaries. Contraction of the arteriole regulates blood flow to various parts of the body.

Artery (ar'-tur-ē): a vessel with muscular, elastic walls that conducts blood away from the heart.

Ascus (as'-kus): a saclike case in which sexual spores are formed by members of the fungal division Ascomycota.

Asexual reproduction: reproduction that does not involve the fusion of haploid sex cells. The parent body may divide and new parts regenerate, or a new, smaller individual may be formed attached to the parent, to drop off when complete. Usually, asexual reproduction produces genetically identical copies of the parent organism.

Association neuron: in nervous circuits, a nerve cell that is postsynaptic to a sensory neuron and presynaptic to a motor neuron. In actual circuits, there may be many association neurons between individual sensory and motor neurons.

Aster: during cell division in animals and some protists, a star-shaped array of microtubules extending in all directions outward from the centrioles.

Atherosclerosis (ath'-er-ō-skler-ō'-sis): a disease characterized by obstruction of arteries by cholesterol deposits and thickening of the arterial walls.

Atmospheric cycle: a nutrient cycle for a substance whose primary reservoir is the atmosphere.

Atom: the smallest particle of an element that retains the properties of the element.

Atomic number: the number of protons in the nuclei of all atoms of a particular element.

Atrioventricular node (ā'-trē-ō-ven-trik'-ū-lar nōd): a specialized mass of muscle at the base of the right atrium through which the electrical activity initiated in the sinoatrial node is transmitted to the ventricles.

Atrioventricular valve: heart valves between the atria and the ventricles that prevent backflow of blood into the atria during ventricular contraction.

Atrium (ā'-trē-um): a chamber of the heart that receives venous blood and passes it to a ventricle.

Auditory canal (aw'-dih-tōry): a canal within the outer ear between the external ear and the eardrum.

Auditory nerve: the nerve leading from the mammalian cochlea to the brain, carrying information about sound.

Autoimmune disease: a disorder in which the immune system produces antibodies against the body's own cells.

Autonomic nervous system: part of the peripheral nervous system of vertebrates that innervates mostly glands and internal organs and produces largely involuntary responses.

Autosome (aw'-tō-sōm): a chromosome found in homologous pairs in both males and females, and which does not bear the genes determining sex.

Autotroph (aw'-tō-trōf): an organism that can manufacture all its high-energy organic molecules (e.g., sugars, proteins) from simple inorganic molecules (such as carbon dioxide, water, and minerals), using a nonliving energy source (usually sunlight); an organism that does not have to consume organic molecules as food.

Auxin (awk'-sin): a plant hormone that influences many plant functions, including phototropism, apical dominance, and root branching. Auxin generally stimulates cell elongation and, in some cases, cell division and differentiation.

Axial skeleton: the skeleton forming the body axis, including the skull, vertebral column, and rib cage.

Axon: a long process of a nerve cell, extending from the cell body to synaptic endings on other nerve cells or on muscles.

B cell: a type of lymphocyte that gives rise to daughter cells (plasma cells) that secrete antibodies into the circulatory system.

Bacillus (buh-sil'-us; pl. bacilli): a rod-shaped bacterium.

Bacterial conjugation: the exchange of genetic material between bacteria.

Bacteriophage (bak-tēr'-ē-ō-fāj): a virus that infects bacteria.

Bacterium (bak-tir'-ē-um; pl. bacteria): an organism consisting of a single prokaryotic cell surrounded by a complex polysaccharide coat.

Balanced polymorphism: the prolonged maintenance of two or more alleles in a population, usually because each allele is favored by a separate selective force.

Bark: the outer layer of a woody stem, consisting of cork cells, cork cambium, and phloem.

Barr body: an inactive X chromosome found in somatic cells of mammals that have at least two X chromosomes (usually females). The Barr body usually appears as a dark spot in the nucleus.

Basal body: the organelle, structurally identical to a centriole, that gives rise to the microtubules of cilia and flagella.

Base: a substance that is capable of combining with and neutralizing H^+ ions, producing a solution with a pH greater than 7. In molecular genetics, one of the nitrogen-containing, single- or double-ringed structures that distinguish one nucleotide from another. In DNA, the bases are adenine, guanine, cytosine, and thymine.

Basidiospore (ba-sid'-ē-ō-spōr): a sexual spore formed by members of the fungal division Basidiomycota.

Basidium (bas-id'-ē-um): a diploid cell, often club-shaped, formed on the basidiocarp of members of the fungal division Basidiomycota, which produces basidiospores by meiosis.

Basilar membrane (bas'-eh-lar): a membrane in the cochlea that bears hair cells that respond to the vibrations produced by sound.

Behavioral isolation: lack of mating between species of animals that differ substantially in courtship and mating rituals.

Benthic zone (ben'-thik): the ocean depths, into which insufficient sunlight penetrates to support photosynthesis.

Bilateral symmetry: body plan in which only a single plane drawn through the central axis will divide the body into mirror-image halves.

Bile (bīl): a liquid secretion of the liver stored in the gallbladder and released into the small intestine during digestion. Bile contains bile salts, whose role is to emulsify or disperse fats into small particles on which fat-digesting enzymes may act.

Biological clock: a metabolic timekeeping mechanism found in most organisms, whereby the organism measures the approximate length of a (24 hour) day even without external environmental cues such as light and dark.

Biological magnification: the increasing accumulation of a toxic substance in increasingly high trophic levels.

Biome (bī'-ōm): a terrestrial region with a characteristic climate and vegetation. Regions with similar climates will have similar vegetation and belong to the same biome (grassland, desert) even though widely separated and inhabited by different species.

Biosphere (bī'-ō-sfēr): that part of the Earth inhabited by living organisms; includes both the living and nonliving components.

Biotic (bī-ah'-tik): living.

Biotic potential: the most rapid potential growth rate of a population, assuming a maximum birth rate and minimum death rate.

Bladder: a muscular storage organ for urine.

Blade: the flat part of a leaf.

Blastocyst (blas'-tō-sist): an early stage of human embryonic development, consisting of a fluid-filled ball with walls one cell layer thick, enclosing a mass of cells attached to its inner surface.

Blastopore: the site at which a blastula invaginates to form a gastrula.

Blastula: in animals, the embryonic stage attained at the end of cleavage, in which the embryo usually consists of a hollow ball with a wall one or several cell layers thick.

Bohr effect (bor): the tendency of hemoglobin to release oxygen more readily when the blood is slightly more acidic than normal, as is caused by an increase in dissolved carbon dioxide.

Book lungs: thin layers of tissue resembling pages in a book, enclosed in a chamber and used as a respiratory organ by certain types of arachnids.

Boom-and-bust cycle: a population cycle characterized by rapid exponential growth followed by a sudden major decline in population size, seen in seasonal species and some populations of small rodents, such as lemmings.

Bowman's capsule: the portion of the nephron in which blood filtrate is collected from the glomerulus.

Bradykinin (brā'-dē-kī'-nin): a chemical formed during tissue damage that binds to receptor molecules on pain nerve endings, giving rise to the sensation of pain.

Brain: the part of the central nervous system of vertebrates enclosed within the skull.

Branch root: a root that arises as a branch of a preexisting root, through divisions of pericycle cells and subsequent differentiation of the daughter cells.

Bronchiole (bron'-kē-ōl): a narrow tube formed by repeated branching of the bronchi, which conducts air into the alveoli.

Bronchus (bron'-kus): a tube that conducts air from the trachea to each lung.

Bryophyte (brī′-ō-fīt): a division of simple nonvascular plants, including mosses and liverworts.

Bud: in plants, an embryonic shoot, usually very short and consisting of an apical meristem with several leaf primordia. In animals, a small copy of an adult that develops on the body of the parent. It eventually breaks off and becomes independent.

Budding: a form of asexual reproduction in which the adult produces miniature versions of itself that drop off and assume independent existence.

Buffer: a compound that minimizes changes in pH by reversibly taking up or releasing H^+ ions.

Bulbourethral gland (bul-bō-ū-rē′-thrul): in male mammals, a gland that secretes a basic, mucus-containing fluid that forms part of the semen.

Bulk flow: the movement of many molecules of a gas or fluid in unison from an area of higher pressure to an area of lower pressre.

C_3 cycle: the cyclic series of reactions whereby carbon dioxide is fixed into carbohydrates during the light-independent reactions of photosynthesis. Also called Calvin-Benson cycle.

C_4 pathway: the series of reactions in certain plants that fixes carbon dioxide into organic acids for later use in the C_3 cycle of photosynthesis.

Calmodulin (kal-mod′-ū-lin): a cytoplasmic protein that acts as a second messenger by binding calcium ions and changing configuration. The configuration change may activate enzymes, initiating a series of biochemical reactions.

Calorie (kal′-ōr-ē): When capitalized (i.e., Calorie) this unit is the amount of energy required to raise the temperature of 1 liter of water 1 degree Celsius. It represents 1000 calories (with a lowercase "c"). The energy content of foods is measured in Calories.

Calvin-Benson cycle: see C_3 cycle.

Cambium (kam′-bē-um): a lateral meristem that causes secondary growth of woody plant stems and roots. See cork cambium; vascular cambium.

Camera eye: the type of eye found in vertebrates and molluscs, in which a lens focuses an image on the retina.

Camouflaged (cam′-a-flaged): a term used to describe organisms that resemble their environment.

Cancer: a disease in which some of the body's cells grow without control.

Capillary: the smallest type of blood vessel, capillaries connect arterioles with venules. Capillary walls, through which exchange of nutrients and wastes occurs, are only one cell thick.

Capsule: a polysaccharide or protein coating that surrounds the cell walls of some bacteria.

Carbohydrate (kar-bō-hī′-drāt): a class of nutrient including monosaccharides, disaccharides, and polysaccharides (starches, glycogen, and cellulose). Sugars and starches are used by animal cells as a source of energy.

Carbon fixation: the initial steps in the C_3 cycle, in which carbon dioxide reacts with ribulose bisphosphate to form a stable organic molecule.

Carbonic anhydrase (car-bon′-ik an-hī′-drās): an enzyme found in red blood cells that catalyzes the formation of bicarbonate ions from dissolved carbon dioxide and water.

Cardiac cycle: the alternation of contraction and relaxation of the heart chambers; systole and diastole.

Cardiac muscle (kar′-dē-ak): specialized muscle of the heart, able to initiate its own contraction independent of the nervous system.

Carnivore (kar′-neh-vōr): literally "meat eater," a predatory organism feeding on other heterotrophs.

Carnivorous (kar-niv′-e-rus): feeding on the bodies of other animals.

Carotenoid (ka-rot′-en-oyd): red, orange, or yellow pigments found in chloroplasts that serve as accessory light-gathering molecules in thylakoid photosystems.

Carpel (kar′pel): the female reproductive structure of a flower, composed of stigma, style, and ovary.

Carrier: an individual who is heterozygous for a recessive condition. Carriers display the dominant phenotype but can pass on their recessive allele to their offspring.

Carrier protein: a membrane protein that facilitates diffusion of specific substances across the membrane. The molecule to be transported binds to the outer surface of the carrier protein; the protein then changes shape, allowing the molecule to move across the membrane through the protein.

Carrying capacity: the maximum population size that an ecosystem can maintain on a sustained basis. Determined primarily by the availability of space, nutrients, water, and light.

Cartilage (kar′-teh-lij): a form of connective tissue forming portions of the skeleton, consisting of chondrocytes and their extracellular secretion of collagen. Cartilage resembles flexible bone.

Casparian strip (kas-par′-ē-an): a waxy, waterproof band in the cell walls between endodermal cells in a root, which prevents the movement of water and minerals in and out of the vascular cylinder via extracellular space.

Catalyst (cat′-a-list): a substance that speeds up a chemical reaction without itself being permanently changed in the process. Catalysts lower the activation energy of a reaction.

Catastrophism: the hypothesis that the Earth has experienced a series of geological catastrophes, much like Noah's Flood, probably imposed by a supernatural being.

Causality: the scientific principle that natural events occur as a result of preceding natural causes.

Cell: the smallest unit of life, consisting, at a minimum, of an outer membrane enclosing a watery medium containing organic molecules, including genetic material composed of DNA.

Cell body: part of a nerve cell in which most of the common cellular organelles are located. Also often a site of integration of inputs to the nerve cell.

Cell cycle: the sequence of events in the life of a cell, from one division to the next.

Cell division: in eukaryotes, the process of reproduction of single cells, usually into two identical daughter cells, by mitosis accompanied by cytokinesis.

Cell plate: in plant cell division, a series of vesicles that fuse to form the new plasma membranes and cell wall separating the daughter cells.

Cell wall: a layer of material, usually made up of cellulose or celluloselike materials, found outside the plasma membranes of plants, fungi, bacteria, and plantlike protists.

Cell-mediated immunity: an immune response in which foreign cells or substances are destroyed by contact with T cells.

Cellular respiration: the oxygen-requiring reactions occurring in mitochondria that break down the end products of glycolysis into carbon dioxide and water, while capturing large amounts of energy as ATP.

Cellular slime mold: a funguslike protist consisting of individual amoeboid cells that can aggregate to form a sluglike mass, which in turn forms a fruiting body.

Cellulose: an insoluble carbohydrate composed of glucose subunits; forms the cell wall of plants.

Central nervous system: in vertebrates, the brain and spinal cord.

Central vacuole: a large, fluid-filled vacuole that occupies most of the volume of many plant cells. See also *turgor pressure.*

Centriole (sen′-trē-ōl): in animal cells, a microtubule-containing structure found at the microtubule organizing center and the base of each cilium and flagellum. Gives rise to the microtubules of cilia and flagella, and may be involved in spindle formation during cell division.

Centromere (sen'-trō-mēr): the region of a replicated chromosome at which the sister chromatids are held together.

Cephalization (sef-al-ī-zā'-shun): the increasing concentration over evolutionary time of sensory structures and nerve ganglia at the anterior end of animals.

Cerebellum (ser-uh-bel'-um): part of the hindbrain of vertebrates, concerned with coordination of motor activities.

Cerebral cortex (ser-ē'-brel kōr'-tex): a thin layer of neurons on the surface of the vertebrate cerebrum, in which most neural processing and coordination of activity occurs.

Cerebral hemisphere: one of two nearly symmetrical halves of the cerebrum, connected by a broad band of axons, the corpus callosum.

Cerebrospinal fluid: A clear fluid produced within the ventricles of the brain that fills the ventricles and surrounds the brain and spinal cord. A major function of cerebrospinal fluid is to cushion the brain and spinal cord.

Cerebrum (ser-ē'-brum): part of the forebrain of vertebrates concerned with sensory processing, direction of motor output, and coordination of most bodily activities. The cerebrum consists of two nearly symmetrical halves (the hemispheres) connected by a broad band of axons, the corpus callosum.

Cervix (ser'-vicks): a ring of connective tissue at the outer end of the uterus, leading into the vagina.

Channel protein: a membrane protein that forms a channel or pore completely through the membrane, and that is usually permeable to one or a few a water-soluble molecules, especially ions.

Chaparral (shap-eh-rel'): a temperate coastal biome with hot dry summers and cool, somewhat rainy winters with frequent fogs. The typical vegetation consists of small trees or large bushes which are drought and fire resistant.

Chemical bond: the force of attraction between neighboring atoms that holds them together in a molecule.

Chemical equilibrium: the condition in which the "forward" reaction of reactants to products proceeds at the same rate as the "backward" reaction from products to reactants, so that no net change in chemical composition occurs.

Chemiosmosis (kē-mē-os-mō'-sis): a process of ATP generation in chloroplasts and mitochondria. The movement of electrons down an electron transport system is used to pump hydrogen ions across a membrane, thereby building up a concentration gradient of hydrogen ions across the membrane. The hydrogen ions diffuse back across the membrane through the pores of ATP-synthesizing enzymes. The energy of their movement down their concentration gradient drives ATP synthesis.

Chemoreceptor: a sensory receptor that responds to chemicals from the environment; chemoreceptors mediate taste and smell.

Chemosynthetic (kēm'-ō-sin-the-tic): capable of oxidizing inorganic molecules to obtain energy.

Chiasma (kē-as'-ma; pl. chiasmata): during prophase I of meiosis, a point at which a chromatid of one chromosome crosses with a chromatid of the homologous chromosome. Exchange of chromosomal material between chromosomes takes place at a chiasma.

Chitin (kī'-tin): a tough, flexible polysaccharide found in the cell walls of fungi and the exoskeletons of insects and some other arthropods, composed of chains of nitrogen-containing, modified glucose molecules.

Chlorinated hydrocarbon: an organic compound that includes a chain of carbon atoms, some of which have chlorine atoms bonded to them, including DDT and several other pesticides.

Chlorophyll (klōr'-ō-fil): a pigment found in chloroplasts that captures light energy during photosynthesis.

Chloroplast (klōr'-ō-plast): the organelle of plants and plantlike protists that is the site of photosynthesis; surrounded by a double membrane and containing an extensive internal membrane system bearing chlorophyll.

Cholecystokinin (kō'-lē-sis-tō-ki'-nin): a digestive hormone produced by the small intestine that stimulates release of pancreatic enzymes.

Chorion (kor'-ē-on): the outermost embryonic membrane in reptiles, birds, and mammals. In birds and reptiles, the chorion functions mostly in gas exchange. In mammals, the chorion forms most of the embryonic part of the placenta.

Chorionic gonadotropin: a hormone secreted by the chorion (one of the fetal membranes), which maintains the integrity of the corpus luteum during early pregnancy.

Chorionic villi (kor-e-on-ik): in mammals, fingerlike projections of the chorion of the embryo that penetrate the uterine lining and form the embryonic portion of the placenta.

Chorionic villus sampling: a procedure for sampling cells from the chorionic villi produced by a fetus. A tube is inserted into the uterus of a pregnant woman, and a small sample of villi are suctioned off for genetic and biochemical analysis.

Choroid (kōr'-ōyd): a layer of tissue behind the retina that contains blood vessels and pigment that absorbs stray light.

Chromatid (krō'-ma-tid): one of the two identical strands of DNA and protein forming a replicated chromosome. The two sister chromatids are joined at the centromere.

Chromatin (krō'-ma-tin): the complex of DNA and proteins that makes up the chromosomes of eukaryotic cells.

Chromosome (krō'-mō-sōme): in eukaryotes, a linear strand composed of DNA and protein, found in the nucleus of a cell, that contains the genes; in prokaryotes, a circular strand composed solely of DNA.

Chrondrocyte (kon'-drō-sīt): a cell type which, with its extracellular secretions of collagen, forms cartilage.

Chylomicron (kī-lō-mī'-kron): a droplet consisting of triglycerides, cholesterol, and phospholipids, and coated with a thin layer of protein. Chylomicrons are formed in the cells lining the small intestine, enter the lymphatic system, and eventually reach the circulatory system.

Chyme (kīme): an acidic, souplike mixture of partially digested food, water, and digestive secretions that is released from the stomach into the small intestine.

Ciliate (sil'-ē-et): a category of protozoan characterized by cilia and a complex unicellular structure, including harpoonlike organelles called trichocysts. Members of the genus *Paramecium* are well-known ciliates.

Cilium (sil'-ē-um; pl. cilia): a short, hairlike projection from the surface of certain eukaryotic cells, containing microtubules in a 9 + 2 arrangement. Movement of cilia may propel cells through a fluid medium or move fluids over a stationary surface layer of cells.

Circadian rhythm (sir-kā'-dē-un): an event that recurs with a period of about 24 hours, even in the absence of environmental cues.

Citric acid cycle: a cyclic series of reactions in which the acetyl groups from the pyruvic acids produced by glycolysis are broken down to CO_2, accompanied by the formation of ATP and electron carriers. Occurs in the matrix of mitochondria.

Class: the taxonomic category composed of related genera. Closely related classes form a division or phylum.

Classical conditioning: a training procedure in which an animal learns to make a reflexive response (such as salivation) to a new stimulus that did not elicit that response originally (such as a sound). This is accomplished by pairing a stimulus that elicits the response automatically (in this case, food) with the new stimulus.

Cleavage: the early cell divisions of embryos, in which little or no growth occurs between divisions.

Climate: prevailing weather patterns over a particular region.

Climax community: a relatively stable community of plants and animals that does not change appreciably over long periods of time unless influenced by external forces.

Closed circulatory system: the type of circulatory system found in certain worms and vertebrates in which the blood is always confined within the heart and vessels.

Coccus (ka'-kus; pl. cocci): a spherical bacterium.

Cochlea (kōk'-lē-uh): a coiled, bony, fluid-filled tube found in the mammalian inner ear, which contains receptors (hair cells) that respond to the vibration of sound.

Codominance: the relation between two alleles of a gene, such that both alleles are phenotypically expressed in heterozygous individuals.

Codon: a sequence of three nucleotides of messenger RNA that specifies a particular amino acid to be incorporated into a protein. Certain codons also signal the beginning and end of protein synthesis.

Coelom (sē'-lōm): a space or cavity within the body separating the body wall from the inner organs.

Coenzyme (kō-en'-zīm): an organic molecule that assists enzymes in their actions.

Coevolution: the evolution of adaptations in two species due to their extensive interactions with one another, so that each species acts as a major force of natural selection upon the other.

Cohesion: the tendency of the molecules of a substance to hold together.

Cohesion–tension theory: a model for transport of water in xylem, which states that water is pulled up the xylem tubes, powered by the force of evaporation of water from the leaves (producing *tension*) and held together by hydrogen bonds between nearby water molecules (*cohesion*).

Coleoptile (kō-lē-op'tīl): a protective sheath surrounding the shoot in monocot seeds.

Collagen (kol'-uh-gen): a fibrous protein found in connective tissue such as bone and cartilage.

Collar cells: specialized cells lining the inside channels of sponges. Flagella extend from a sievelike collar, creating a water current that draws microscopic organisms through the collar to be trapped.

Collenchyma (kōl-en'-ki-ma): a plant cell type that is alive at maturity, with irregularly thickened primary cell walls, and that provides support to the plant body.

Colostrum (kō-los'-trum): a yellowish fluid high in protein and containing antibodies, that is produced by the female breasts before milk secretion begins.

Commensalism (kum-en'-sal-ism): a symbiotic relationship between two species in which one benefits while the other is neither harmed nor benefited.

Communication: the act of producing a signal that causes another animal, usually of the same species, to modify its behavior in a way beneficial to one or both of the participants.

Community: all the interacting populations within an ecosystem.

Compact bone: the hard outer bone; composed of Haversian systems.

Companion cell: a cell adjacent to a sieve-tube element in phloem, involved in control and nutrition of the sieve-tube element.

Competition: a relationship between individuals or species in which both require the same resource, (e.g., food or space) that is limited relative to the demand for it.

Competitive exclusion principle: the concept that no two species can simultaneously and continuously occupy the same ecological niche.

Competitive inhibition: in enzyme-catalyzed reactions, a condition in which two molecules (at least one a substrate for the enzyme) compete for entry into the active site of the enzyme, thus slowing the rate of reaction.

Complement: a group of blood-borne proteins that participate in the destruction of foreign cells to which antibodies have bound.

Complement reactions: interactions among foreign cells, antibodies, and complement proteins, resulting in the destruction of the foreign cells.

Complementary: referring to a nucleotide that can pair with another nucleotide via hydrogen bonding; in DNA, adenine is complementary to thymine, and guanine is complementary to cytosine.

Complete flower: a flower that has all four floral parts (sepals, petals, stamens, and carpels).

Compound: a substance composed of two or more elements that can be broken into its constituent elements by chemical means.

Compound eye: a type of eye found in arthropods, composed of numerous independent subunits, called ommatidia. Each ommatidium apparently contributes a single piece of a mosaic-like image perceived by the animal.

Concentration: the number of particles of a dissolved substance per unit volume of fluid.

Concentration gradient: the difference in concentration of a substance between two parts of a fluid or across a barrier such as a membrane.

Condensation reaction: a chemical reaction in which two molecules are joined by a covalent bond with the simultaneous removal of a hydrogen from one molecule and a hydroxyl group from the other, forming water.

Cone: a cone-shaped photoreceptor cell in the vertebrate retina, not as sensitive to light as the rods. The three types of cones are most sensitive to different colors of light, and provide color vision. See also rod.

Conifer (kon'-eh-fer): a class of tracheophyte that reproduces using seeds formed inside cones and retains its leaves throughout the year.

Connective tissue: a tissue type consisting of many diverse tissues, which generally include large amounts of extracellular material.

Constant region: part of an antibody molecule that is similar or identical in all antibodies.

Consumer: an organisms that eats other organisms; a heterotroph.

Contest competition: a mechanism for resolving intraspecific competition using social or chemical interactions.

Contractile vacuole: a fluid-filled vacuole found in certain protists that takes up water from the cell cytoplasm, contracts, and expels the water outside the cell via a pore in the plasma membrane.

Control: that portion of an experiment in which all possible variables are held constant; in contrast to the "experimental" portion, in which a particular variable is altered.

Convergence: a condition in which a large number of nerve cells provide input to a smaller number of cells.

Convergent evolution: the independent evolution of similar structures among unrelated organisms, due to similar selective pressures. See *analogous structures*.

Convolutions: foldings of the cerebral cortex of the brain.

Copulation: reproductive behavior in which the penis of the male is inserted into the body of the female, where it releases sperm.

Coral reef: the most diverse marine ecosystem, formed by the bodies of corals in warm, relatively shallow water.

Cork cambium: a lateral meristem in woody roots and stems that gives rise to cork cells.

Cork cell: a protective cell of the bark of woody stems and roots; at maturity, cork cells are dead, with thick, waterproofed cell walls.

Cornea (kōr′-nē-uh): the clear outer covering of the eye in front of the pupil and iris.

Corona radiata (ka-rō′-na rā-dē-a′-ta): the layer of cells surrounding an egg after ovulation.

Corpus callosum (kōr′pus kal-ō′-sum): the tract of axons that connect the two cerebral hemispheres of vertebrates.

Corpus luteum (kor′-pus lū′-tē-um): in the mammalian ovary, a structure derived from the follicle after ovulation, which secretes the hormones estrogen and progesterone.

Cortex: the part of a primary root or stem located between the epidermis and the vascular cylinder.

Cotyledon (kot-ul-ē′don): also called a seed leaf; a leaflike structure within a seed that absorbs food from the endosperm and transfers it to the growing embryo.

Countercurrent flow: a structural arrangement that enhances diffusion between two fluids that differ in their concentration of dissolved substances by moving the fluids past one another in opposite directions, separated by semipermeable membranes.

Coupled reaction: a pair of reactions, one exergonic and one endergonic, that are linked together so that the energy produced by the exergonic reaction provides the energy needed to drive the endergonic reaction.

Covalent bond (ko-vā′-lent): a chemical bond between atoms in which electrons are shared.

Creationism: the hypothesis that all species of organisms on Earth were created in essentially their present form by a supernatural Being, and that significant modification of those species, specifically their transformation into new species, cannot occur through natural processes.

Crista (kris′-ta; pl. cristae): a fold in the inner membrane of a mitochondrion.

Crop: an organ found in both earthworms and birds in which ingested food is stored temporarily before passing to the gizzard, where it is pulverized.

Cross-bridge: in muscles, an extension of myosin that binds to and pulls upon actin to produce contraction of the muscles.

Cross-fertilization: union of sperm and egg from two different individuals of the same species.

Crossing over: the exchange of corresponding segments of the chromatids of two homologous chromosomes during meiosis.

Cultural evolution: changes in the behavior of a population of animals, especially humans, by learning behaviors acquired by members of previous generations.

Cupula (kūp′-ū-luh): a gelatinous structure found in the lateral line system of fish, and the semicircular canals of the vestibular system of other vertebrates. Hairs are embedded in the cupula, which is deflected by movement of the fluid filling the canals.

Cuticle (kū′-ti-kul): a waxy or fatty coating on the exposed epidermal cells of many land plants, which aids in the retention of water.

Cyanobacteria: photosynthetic prokaryotic cells, utilizing chlorophyll and releasing oxygen as a photosynthetic by-product, sometimes called ''blue-green algae.''

Cyclic AMP: a cyclic nucleotide formed within many target cells as a result of the reception of modified amino acid or protein hormones, and which causes metabolic changes in the cell; often called a second messenger.

Cyclic nucleotide (sik′-lik nū′-klē-ō-tīd): a nucleotide in which the phosphate group is bonded to the sugar at two points, forming a ring. Cyclic nucleotides serve as intracellular messengers.

Cyst (sist): an encapsulated resting stage in the life cycle of certain invertebrates, such as parasitic flatworms and roundworms.

Cytokinesis (sī-tō-ki-nē′-sis): division of the cytoplasm and organelles into two daughter cells during cell division. Usually cytokinesis occurs during telophase of mitosis.

Cytokinin (sī-tō-kī′-nin): a plant hormone that promotes cell division, fruit growth, sprouting of lateral buds, and prevents leaf aging and leaf drop.

Cytoplasm (sī′-tō-plazm): the material contained within the plasma membrane of a cell, exclusive of the nucleus.

Cytoskeleton: a network of protein fibers in the cytoplasm that gives shape to a cell, holds and moves organelles, and is often involved in cell movement.

Cytosol (sī′-tō-sol): the fluid part of the cytoplasm.

Day-neutral plant: a plant in which flowering occurs under a wide range of daylengths.

Decomposers: a group of decay organisms, mainly fungi and bacteria. These digest organic material by secreting digestive enzymes into the environment. In the process they liberate nutrients into the environment.

Degeneracy: the property of the genetic code whereby several codons may specify the same amino acid.

Deletion: a mutation in which one or more nucleotides are removed from a gene.

Dendrite (den′-drīt): the site of signal input to a nerve cell, usually takes the form of branched fibers located close to the cell body.

Denitrifying bacteria (dē-nī′-treh-fī-ing): bacteria that break down nitrates, releasing nitrogen gas to the atmosphere.

Density dependent: description of any factor that limits population size more effectively as the population density increases.

Density independent: description of any factor such as freezing weather that limits a sensitive population without regard to its size.

Deoxyribonucleic acid (dē-ox-ē-rī-bō-nū-klā′-ik; DNA): a molecule composed of deoxyribose nucleotides that encodes the genetic information of all living cells.

Dermal tissue system: a plant tissue system that makes up the outer covering of the plant body.

Dermis (dur′-mis): the layer of skin lying beneath the epidermis, composed of connective tissue and containing blood vessels, muscles, nerve endings, and glands.

Desert: a biome in which potential evaporation greatly exceeds rainfall. Perennial vegetation is widely spaced and has drought-resistant adaptations such as waxy or spiny leaves.

Desertification (deh-ser′-ti-fi-kā-shun): the conversion of productive land in warm, dry areas into unproductive land through loss of soil and vegetation, often as a result of overgrazing or other improper farming methods.

Desmosome (dez′-mō-sōm): a strong cell-to-cell junction that functions in attaching cells to one another.

Detritus feeders (de-trī′-tus): a diverse assemblage of organisms ranging from worms to vultures that live off the wastes and dead remains of other organisms.

Diabetes mellitus (di-a-bē′-tes mel-ī′-tus): a disease characterized by defects in the production, release, or reception of insulin, characterized by high blood glucose levels that fluctuate with sugar intake, and glucose in the urine.

Diaphragm (dī′uh-fram): a dome-shaped muscle forming the floor of the chest cavity. Contraction of this muscle pulls it downward, enlarging the cavity and causing air to be drawn into the lungs.

Diastole (di-as′-tō-lē): the portion of the cardiac cycle during which the ventricles are relaxed; the lower of the two blood pressure readings.

Diatom (dī′-e-tom): a category of protist that includes photosynthetic forms with two-part glassy outer coverings that separate when the cell divides. Diatoms are important primary producers in fresh and salt water.

Dicot: short for dicotyledon; a type of flowering plant characterized by embryos with two food-storage organs called cotyledons.

Dicotyledon (dī'-kot-ul-ēd'-un): a class of angiosperm whose embryo has two cotyledons, or seed leaves.

Differential permeability: the property of a membrane by which some substances can permeate more readily than other substances.

Differential reproduction: differences in reproductive output among individuals of a population, usually as a result of genetic differences.

Differentiated cell: a mature cell specialized for a specific function; in plants, differentiated cells usually do not divide.

Differentiation: the process whereby relatively unspecialized cells, especially of embryos, become specialized into particular tissue types.

Diffusion: the net movement of particles from a region of high concentration of that particle to a region of low concentration, driven by the concentration gradient. May occur entirely within a fluid or across a barrier such as a membrane.

Digestion: the process by which food is physically and chemically broken down into molecules that can be absorbed by cells.

Dihybrid cross: a breeding experiment involving parents that differ in two distinct, genetically determined traits.

Dinoflagellate (dī-nō-fla'-gel-et): a category of protist that includes photosynthetic forms in which two flagella project through armorlike plates. Abundant in oceans, these sometimes reproduce rapidly, causing "red tides."

Dioecious (dī-ē'-shus): pertaining to organisms in which male and female gametes are produced by separate individuals.

Diploid (dip'-loyd): referring to a cell with pairs of homologous chromosomes.

Direct development: a developmental pathway in which the offspring is born as a miniature version of the adult and does not radically change its body form as it grows and matures.

Directional selection: a type of natural selection in which one extreme phenotype is favored over all others.

Disaccharide (dī-sak'-a-rīd): a carbohydrate formed by the covalent bonding of two monosaccharides.

Disruptive selection: a type of natural selection in which both extreme phenotypes are favored over the average phenotype.

Disulfide bridge: the covalent bond formed between the sulfur atoms of two cysteines in a protein; often causes the protein to fold, by bringing otherwise distant parts of the protein close together.

Divergence: a condition in which a small number of nerve cells provide input to a larger number of cells.

Divergent speciation: speciation in which subpopulations of a species split to form two or more new species that exist simultaneously. See phyletic speciation.

Division: the taxonomic category contained within a kingdom and consisting of related classes of plants, fungi, bacteria, or plantlike protists.

DNA hybridization: a technique by which DNA from two species is separated into single strands and then allowed to reform. Hybrid double-stranded DNA from the two species can occur where the sequence of nucleotides is complementary. The greater the degree of hybridization, the closer the evolutionary relatedness of the two species.

DNA library: a complete set of all the DNA of a particular organism, usually cloned into bacterial plasmids.

DNA polymerase: an enzyme that covalently bonds DNA nucleotides together into a continuous strand, using a preexisting DNA strand as a template. DNA polymerase catalyzes the replication of the DNA of chromosomes during interphase prior to mitosis and meiosis.

Dominance hierarchy: a social arrangement in which animals, usually through aggressive interactions, establish a rank for some or all of the members of the social unit. The rank that determines which individuals obtain first access to limited resources.

Dominant: an allele that can determine the phenotype of heterozygotes completely, so that they are indistinguishable from individuals homozygous for the allele. In the heterozygotes, the expression of the other (recessive) allele is completely masked.

Dopamine (dōp'-uh-mēn): a transmitter in the brain whose actions are largely inhibitory. Degeneration of dopamine-containing neurons leads to Parkinson's disease.

Dormancy: a state in which an organism does not grow or develop; usually marked by lowered metabolic activity and resistance to adverse environmental conditions.

Dorsal (dōr'-sul): the top, back, or uppermost surface of an animal oriented with its head forward.

Dorsal root ganglion: a ganglion located on the dorsal (sensory) branch of each spinal nerve, containing the cell bodies of sensory neurons.

Double covalent bond: a covalent bond that occurs when two atoms share two pairs of electrons.

Double fertilization: in flowering plants, a phenomenon in which two sperm nuclei fuse with the nuclei of two cells of the female gametophyte. One sperm fuses with the egg to form the zygote, while the second sperm nucleus fuses with the two haploid nuclei of the primary endosperm cell to form a triploid endosperm cell.

Down syndrome: a genetic disorder caused by the presence of three copies of chromosome 21. Common characteristics include mental retardation, abnormally shaped eyelids, a small mouth with protruding tongue, short fingers, heart defects, and unusual susceptibility to infectious diseases.

Ecological isolation: lack of mating between organisms belonging to different populations that occupy distinct habitats within the same general area.

Ecology (ē-kol'-uh-gē): the study of the interrelationships of organisms with each other and with their nonliving environment.

Ecosystem (ē'kō-sis-tem): all the organisms and their nonliving environment within a defined area.

Ectoderm (ek'-tō-derm): the outermost embryonic tissue layer, which gives rise to structures such as hair, the epidermis of the skin, and the nervous system.

Effector (ē-fek'-tōr): a part of the body (usually a muscle or gland) that carries out responses as directed by the nervous system.

Egg: the haploid female gamete, usually large and nonmotile, containing food reserves for the developing embryo and regionally localized gene-regulating substances that direct early development.

Electron: a subatomic particle, found in the orbitals outside the nucleus of an atom, bearing a unit of negative charge and very little mass.

Electron carrier: a molecule that can reversibly gain and lose electrons. Electron carriers generally accept high-energy electrons produced during an exergonic reaction and donate the electrons to acceptor molecules that use the energy to drive endergonic reactions.

Electron shell: all the electron orbitals at a given distance from the nucleus of an atom.

Electron transport system: a series of molecules found in the inner membrane of mitochondria and the thylakoid membranes of chloroplasts that extract energy from electrons and generate ATP or other energetic molecules.

Electrophoresis: a biochemical technique that separates molecules according to their electrical charge and molecular weight.

Element: a substance that cannot be broken down to a simpler substance by ordinary chemical means.

Embryo: the stages of development of animals that begin with fertilization of the egg cell and end with hatching or birth. In mammals, embryo usually refers to the early stages in which the developing animal does not yet resemble the adult of the species.

Embryo sac: the haploid female gametophyte of flowering plants.

Embryonic disk: in human embryonic development, the flat, two-layered group of cells derived from the inner cell mass of the blastocyst, which will develop into the embryo proper.

Emigration (em-uh-grā′shun): movement of individuals out of an area.

Emphysema (em-fuh-sē′-muh): a condition in which the alveoli become brittle and rupture, causing decreased area for gas exchange.

Endergonic (en-der-gon′-ik): pertaining to a chemical reaction that requires an input of energy to proceed; an "uphill" reaction.

Endocrine gland: a ductless, hormone-producing gland that releases its secretions into the extracellular fluid within the body, from which the secretions diffuse into nearby capillaries.

Endocytosis (en-dō-sī-tō′-sis): the movement of material into a cell by a process in which the plasma membrane engulfs extracellular material, forming membrane-bound sacs that enter the cytoplasm.

Endoderm (en′-dō-derm): the innermost embryonic tissue layer, which gives rise to structures such as the lining of the digestive and respiratory tracts.

Endodermis (en-dō-der′-mis): the innermost layer of cells of the cortex of a root.

Endogenous pyrogen: a chemical produced by the body that stimulates the production of a fever (elevated body temperature).

Endometrium (en-dō-mē′-trē-um): the nutritive inner lining of the uterus.

Endoplasmic reticulum (en-dō-plaz′-mik re-tik′-ū-lum; ER): a system of membranous channels within eukaryotic cells; the site of most protein and lipid biosynthesis.

Endorphin (en-dōr′-fin): one of a group of peptides in the vertebrate brain that mimics some of the actions of opiates. Endorphins reduce the sensation of pain.

Endoskeleton: a supportive structure within the body; an internal skeleton. It may be nonliving, as in echinoderms and sponges, or living, as in vertebrates.

Endosperm: a triploid food storage tissue found in the seeds of flowering plants.

Endosymbiotic hypothesis: the hypothesis that chloroplasts and mitochondria (and perhaps some other organelles, such as centrioles) arose as mutually beneficial associations between the ancestors of eukaryotic cells and captured bacteria living within the cytoplasm of the pre-eukaryotic cell.

End-product inhibition: in enzyme-mediated chemical reactions, the condition in which the product of a reaction inhibits one or more of the enzymes involved in synthesizing the product.

Energy: the capacity to do work.

Energy carrier: a molecule that stores energy in "high-energy" chemical bonds and releases the energy again to drive coupled endothermic reactions. ATP is the most common energy carrier in cells.

Energy level: the specific amount of energy characteristic of a given electron shell in an atom.

Energy pyramid: a graphical representation of the energy contained in succeeding trophic levels, with maximum energy at the base (primary producers) and steadily diminishing amounts at higher levels.

Enhancer: in eukaryotes, a stretch of DNA that influences the access of RNA polymerase to the promoter region of a structural gene.

Entropy (en′-trō-pē): a measure of the amount of randomness and disorder in a system.

Environmental resistance: any factor that tends to counteract biotic potential, limiting population size.

Enzyme (en′zīm): a protein catalyst that speeds up the rate of specific biological reactions.

Epicotyl (ep′-ē-kot-ul): the part of the embryonic shoot located between the tip of the shoot and the attachment point of the cotyledons.

Epidermis (ep-uh-der′-mis): specialized epithelial tissue that forms the outer layer of skin. In plants, the outermost layer of cells of a leaf, young root, or young stem.

Epididymis (e-pi-di′-dē-mus): tubes that connect with and receive sperm from the seminiferous tubules of the testis.

Epiglottis (ep-eh-gla′-tis): a flap of cartilage in the lower pharynx that covers the opening to the larynx during swallowing. This directs the food down the esophagus.

Epistasis (ep-i-stā′-sis): a pattern of inheritance in which the actions of one gene are required for another gene to be expressed.

Epithelial tissue (eh-puh-thē′-lē-ul): a tissue type that forms membranes that cover the body surface and line body cavities, and that also gives rise to glands.

Equilibrium population: a population in which allele frequencies do not change from generation to generation.

Erythroblastosis fetalis (ē-rith′-rō-blas-tō′-sis fē-tal′-is): a condition in which the red blood cells of a newborn Rh-positive baby are attacked by antibodies produced by its Rh-negative mother, causing jaundice and anemia. Retardation and death are possible consequences if treatment is inadequate.

Erythrocytes (ē-rith′-rō-sītes): red blood cells active in oxygen transport, which contain the red pigment hemoglobin.

Erythropoietin (ē-rith′-rō-pō-ē′-tin): a hormone produced by the kidneys in response to oxygen deficiency that stimulates production of red blood cells by the bone marrow.

Esophagus (eh-sof′-eh-gus): a muscular passageway connecting the pharynx to the next chamber of the digestive tract, the stomach in humans and other mammals.

Essential amino acids: amino acids which are required nutrients that the body is unable to manufacture and which must be supplied in the diet.

Essential fatty acids: fatty acids which are required nutrients that the body is unable to manufacture and which must be supplied in the diet.

Estrogen: in vertebrates, a female sex hormone produced by follicle cells of the ovary, which stimulates follicle development, oogenesis, development of secondary sex characteristics, and growth of the uterine lining.

Ethology (ē-thol′-ō-gē): the study of animal behavior under natural or near-natural conditions.

Ethylene: a plant hormone that promotes ripening of fruits, and leaf and fruit drop.

Euglenoid (yū′-gle-nōyd): a category of protist characterized by one or more whiplike flagella used for locomotion and a photoreceptor for detecting light. Euglenoids are photosynthetic, but some are capable of heterotrophic nutrition if deprived of chlorophyll.

Eukaryotic (ū-kar-ē-ot′-ik): referring to cells of organisms of the kingdoms Protista, Fungi, Plantae, and Animalia. Eukaryotic cells have their genetic material enclosed within a membrane-bound nucleus and contain other membrane-bound organelles and are usually larger than prokaryotic cells.

Eustachian tube (ū-stā'-shin): a tube connecting the middle ear with the pharynx; allows pressure to equilibrate between the middle ear and the outside.

Eutrophic lake (ū'-trōf-ik): a lake that is rich in nutrients and supports dense communities of organisms.

Evolution: the descent of modern organisms from preexisting life forms; strictly speaking, any change in the proportions of different genotypes in a population from one generation to the next.

Excitatory synapse: a synapse between two nerve cells in which the resting potential of the postsynaptic cell becomes less negative due to the activity of the presynaptic cell.

Exergonic (ex-er-gon'-ik): pertaining to a chemical reaction that liberates energy (either heat energy or in the form of increased entropy); a "downhill" reaction.

Exocrine gland: a gland that releases its secretions into ducts that lead to the outside of the body or into the digestive tract.

Exocytosis (ex-ō-sī-tō'-sis): the movement of material out of a cell by a process in which intracellular material is enclosed within a membrane-bound sac that moves to the plasma membrane and fuses with it, releasing the material outside the cell.

Exon: a segment of DNA in a eukaryotic gene that codes for amino acids in a protein (see also *intron*).

Exoskeleton: an external, nonliving supporting structure; an external skeleton with flexible joints to allow for movement.

Expiration (ex-per-ā'-shun): the act of exhaling, which results from relaxation of the respiratory muscles.

Exponential growth: a continuously accelerating increase in population size.

Extensor muscle: a muscle that straightens a joint.

External ear: the fleshy portion of the ear that extends outside the skull; also called the pinna.

External fertilization: union of sperm and egg outside the body of either parental organism.

Extinction: the death of all members of a species.

Extracellular digestion: the physical and chemical breakdown of food that occurs outside of a cell, usually in a digestive cavity.

Eyespot: a simple, lensless eye found in various invertebrates, including flatworms and jellyfish. Eyespots provide information about light vs. dark, and sometimes the direction of light, but cannot form an image.

Facilitated diffusion: diffusion of molecules across a membrane, assisted by protein pores or carriers embedded in the membrane.

Family: the taxonomic category contained within an order and consisting of related genera.

Fat: a lipid composed of three saturated fatty acids covalently bonded to glycerol; fats are solid at room temperature. Also, fat-storing connective tissue whose cells are packed with triglycerides (fats); also called adipose tissue.

Fatty acid: an organic molecule composed of a long chain of carbon atoms, with a carboxylic acid (—COOH) group at one end. Fatty acids may be saturated (all single bonds between the carbon atoms) or unsaturated (one or more double bonds between the carbon atoms).

Fermentation: anaerobic reactions that convert the pyruvic acid produced by glycolysis into lactic acid or alcohol and CO_2.

Fertilization: the fusion of male and female haploid gametes to form a zygote.

Fetal alcohol syndrome: a cluster of symptoms including retardation and physical abnormalities that occur in infants born to mothers who consumed large amounts of alcoholic beverages during pregnancy.

Fetus: the later stages of mammalian embryonic development, when the developing animal has come to resemble the adult of the species.

Fever: an elevation in body temperature caused by chemicals (pyrogens) released by white blood cells in response to infection.

Fibrillation: rapid, uncoordinated, and ineffective contractions of heart muscle cells.

Fibrin (fī'-brin): a clotting protein formed in the blood in response to a wound. Fibrin binds with other fibrin molecules and provides a matrix around which a blood clot forms.

Fibrous root system: a root system, commonly found in monocots, characterized by many roots of approximately the same diameter arising from the base of the stem.

Filament: in flowers, the stalk of a stamen, which bears an anther at its tip.

Filtration: within Bowman's capsule in each nephron of a kidney, the process by which blood is pumped under pressure through permeable capillaries of the glomerulus, forcing out water, dissolved wastes, and nutrients.

Fimbria (fim'-brē-a; pl. fimbriae): in female mammals, the ciliated, fingerlike projections of the oviduct that sweep the ovulated egg from the ovary into the oviduct.

First Law of Thermodynamics: a principle of physics that within any isolated system, energy can be neither created nor destroyed, but can be converted from one form to another.

Fission: asexual reproduction by dividing the body into two smaller, complete organisms.

Fitness: the reproductive success of an organism, usually expressed in relation to the average reproductive success of all individuals in the same population.

Fixed action pattern: stereotyped, rather complex behavior that is genetically programmed (innate); often triggered by a stimulus called a releaser.

Flagellum (fla-jel'-um; pl. flagella): a long, hairlike extension of the cell membrane. In eukaryotic cells, contains microtubules arranged in a 9 + 2 pattern. Movement of flagella propel some cells through fluid media.

Flame cell: a specialized cell containing beating cilia that conducts water and wastes through the branching tubes that serve as an excretory system in flatworms.

Flexor muscle: a muscle that flexes (decreases the angle of) a joint.

Flower: the reproductive structure of an angiosperm plant.

Fluid: a liquid or gas.

Fluid mosaic model: a model of membrane structure. According to this model, membranes are composed of a double layer of phospholipids in which is embedded a variety of proteins. The phospholipid bilayer is a somewhat fluid matrix that allows movement of proteins within it.

Fluid-phase endocytosis: nonselective movement of extracellular fluid into a cell, enclosed within a vesicle formed from the plasma membrane.

Follicle: in the ovary of female mammals, the oocyte and its surrounding accessory cells.

Follicle-stimulating hormone (FSH): a hormone produced by the anterior pituitary gland that stimulates spermatogenesis in males and development of the follicle in females.

Food chain: an illustration of feeding relationships in an ecosystem using a single representative from each of the trophic levels.

Food vacuole: a membrane-bound space within a single cell in which food is enclosed. Digestive enzymes are released into the vacuole and intracellular digestion occurs here.

Food web: a relatively accurate representation of the complex feeding relationships within an ecosystem, including many organisms at various trophic levels, with many of the consumers occupying more than one level simultaneously.

Forebrain: during development, the anterior portion of the brain. In mammals, the forebrain differentiates into the thalamus, the limbic system, and the cerebrum. In humans, the cerebrum contains about half of all the neurons in the brain.

Fossil: the remains of an organism, usually preserved in rock. Fossils include petrified bones or wood; shells; impressions of body forms such as feathers, skin, or leaves; and markings made by organisms such as footprints.

Founder effect: a type of genetic drift in which an isolated population founded by a small number of individuals may develop allele frequencies that are very different from those of the parent population, because of chance inclusion of disproportionate numbers of certain alleles in the founders.

Fovea (fō'-vē-uh): the central region of the vertebrate retina, upon which images are focused. The fovea contains closely packed cones (about 150,000 per mm²).

Free-living: not parasitic.

Fruit: in flowering plants, the ripened ovary (plus, in some cases, other parts of the flower). This structure contains the seeds.

Fruiting body: a spore-forming reproductive structure found in certain protists, bacteria, and fungi.

Functional group: one of several groups of atoms commonly found in organic molecules, including hydrogen, hydroxyl, amino, carboxyl, and phosphate groups.

Gallbladder: a small sac adjacent to the liver in which the bile secreted by the liver is stored and concentrated. Bile is released from the gallbladder via the bile duct to the small intestine.

Gamete (gam'-ēt): a haploid sex cell formed in sexually reproducing organisms.

Gametic incompatibility: the inability of sperm from one species to fertilize eggs of another species.

Gametophyte (ga-mēt'-ō-fīt): a multicellular haploid plant that produces haploid sex cells by mitosis.

Ganglion (gan'-glē-un): an aggregation of neurons.

Ganglion cell (gang'-lē-un): a cell type comprising the innermost layer of the vertebrate retina whose axons form the optic nerve.

Gap junction: a type of cell-to-cell junction in animals in which channels connect the cytoplasm of adjacent cells.

Gastrovascular cavity (gas'-trō-vas'-kū-lar): a chamber in the bodies of some invertebrates (such as cnidarians and flatworms) that has both digestive and circulatory functions. A single opening serves as both mouth and anus, while the chamber provides direct access of nutrients to the cells.

Gastrula (gas'-trū-la): in animal development, a three-layered embryo with ectoderm, mesoderm, and endoderm cell layers. The endoderm layer usually encloses the primitive gut.

Gastrulation (gas-trū-lā'-shun): the process whereby a blastula develops into a gastrula.

Gene: a unit of heredity containing the information for a particular characteristic. A gene is a segment of DNA located at a particular place on a chromosome.

Gene flow: the movement of alleles from one population to another owing to migration of individual organisms.

Gene pool: for a single gene, the total of all the alleles of that gene that occur in a population; the total gene pool is the total of all alleles of all genes in the population.

Generative cell: in flowering plants, one of the haploid cells of a pollen grain. The generative cell undergoes mitosis to form two sperm cells.

Genetic code: the collection of codons of mRNA, each of which directs the incorporation of a particular amino acid into a protein during protein synthesis.

Genetic drift: a change in the allele frequencies of a small population purely by chance.

Genetic recombination: the recombining of alleles on homologous chromosomes, due to exchange of DNA during crossing over.

Genotype (jēn'-ō-tīp): the genetic composition of an organism; the actual alleles of each gene carried by the organism.

Genus (jē-nis): the taxonomic category consisting of very closely related species.

Geographical isolation: the separation of two populations by a physical barrier.

Germ layer: A tissue layer formed during early embryonic development.

Germination: the growth and development of a seed, spore, or pollen grain.

Gibberellin (jib-er-el'-in): a plant hormone that stimulates seed germination, fruit development, and cell division and elongation.

Gills: in aquatic animals, a branched tissue richly supplied with capillaries around which water is circulated for gas exchange.

Gizzard: a muscular organ found in earthworms and birds in which food is mechanically broken down prior to chemical digestion.

Glomerulus (glō-mer'-ū-lus): a dense network of thin-walled capillaries located within the Bowman's capsule of each nephron. Here blood pressure forces water and dissolved nutrients through capillary walls for filtration by the nephron.

Glucose: the most common monosaccharide, with the molecular formula $C_6H_{12}O_6$. Most polysaccharides, including cellulose, starch, and glycogen, are made of glucose subunits covalently bonded together.

Glycerol (glis'-er-ol): a three-carbon alcohol to which fatty acids are covalently bonded to make fats and oils.

Glycogen (glī'-kō-jen): a long, branched polymer of glucose that is stored in the muscles and liver and metabolized as a source of energy in animals.

Glycolysis (glī-kol'-i-sis): reactions carried out in the cytosol that break down glucose into two molecules of pyruvic acid, producing two ATP molecules. Glycolysis does not require oxygen, but can proceed when oxygen is present.

Glycoprotein: a protein to which a carbohydrate is attached.

Golgi complex (gōl'-jē): a stack of membranous sacs found in most eukaryotic cells, which is the site of processing and separation of membrane components and secretory materials.

Gonorrhea (gon-a-rē'-uh): a sexually transmitted bacterial infection of the reproductive organs. Untreated gonorrhea may result in sterility.

Graded potential: in a nerve cell, an electrical response due to sensory input or synaptic input from another nerve cell. Graded potentials may be positive or negative and vary in amplitude with the strength of stimulation.

Gradient: a difference in concentration, pressure, or electrical charge between two regions of space.

Gradualism: a model of evolution, stating that morphological change and speciation are slow, gradual processes that are not necessarily simultaneous or linked.

Granum (gra'-num; pl. grana): in chloroplasts, a stack of thylakoids.

Grassland: a biome characterized by a relatively continuous ground cover of grasses, with few or no trees except along watercourses. Typically, grassland biomes have a prolonged dry season and frequent fires, both of which favor grasses over trees.

Gravitropism: growth with respect to the direction of gravity.

Greenhouse effect: the ability of certain gases such as carbon dioxide and methane to trap sunlight energy in the atmosphere as heat. The glass in a greenhouse does the same, hence the name. This warming of the atmosphere is being enhanced by human production of these gases.

Ground tissue system: a plant tissue system consisting of parenchyma, collenchyma, and sclerenchyma cells that makes up the bulk of a leaf or young stem, excluding vascular or dermal tissues. Most ground tissue cells function in photosynthesis, support, or carbohydrate storage.

Guard cell: one of a pair of specialized epidermal cells surrounding the central opening of a stoma of a leaf, which regulates the size of the opening.

Gymnosperms (jim'-nō-sperms): non-flowering seed plants such as conifers, cycads, and gingkos.

Gyre (jīre): a massive, roughly circular ocean current. Gyres flow clockwise in the northern hemisphere and counterclockwise in the southern hemisphere.

Habituation (heh-bich-ū-ā'-shun): simple learning characterized by a decline in response to a harmless, repeated stimulus.

Hair cell: The receptor cell type found in the inner ear. Hair cells bear hairlike projections. Bending of the hairs between two membranes causes the receptor potential.

Hair follicle: a gland in the dermis of mammalian skin, formed from epithelial tissue, that produces a hair.

Halophile (hā'-lō-fīl): a salt-loving organism.

Haploid (hap'-loyd): referring to a cell that has only one member of each pair of homologous chromosomes.

Haversian system (ha-ver'-sē-un): unit of hard bone consisting of concentric layers of bone matrix with embedded osteocytes surrounding a small central canal containing a capillary.

Heat of fusion: the energy that must be removed from a compound to transform it from a liquid into a solid at its freezing temperature.

Heat of vaporization: the energy that must be supplied to a compound to transform it from a liquid into a gas at its boiling temperature.

Helix (hē'-licks): a spiral structure similar to a corkscrew or a spiral staircase; a type of secondary structure of a protein.

Helper T cell: a type of T cell that aids other immune cells to recognize and act against antigens.

Hemocoel (hē'-mō-sēl): a blood cavity within the bodies of certain invertebrates in which blood bathes tissues directly. A hemocoel is part of an open circulatory system.

Hemoglobin (hē'-mō-glō-bin): an iron-containing protein that gives red blood cells their color. Hemoglobin binds to oxygen in the lungs and releases it to the tissues.

Hemophilia: a recessive, sex-linked disease in which the blood fails to clot normally.

Herbivore (erb'-i-vōr): literally "plant-eater"; an organism that eats plants.

Hermaphrodite (her-ma'-frō-dīt): an organism possessing both male and female sexual organs. Some hermaphroditic animals can fertilize themselves; others must exchange sex cells with a mate.

Heterotroph (het'-er-ō-trōf): an organism that cannot use inanimate energy sources (such as sunlight) to synthesize all its energy-rich organic molecules; hence heterotrophs must acquire energy-rich organic molecules manufactured by other living organisms as food.

Heterozygote (het-er-ō-zī'-gōt): an organism carrying two different alleles of the gene in question; sometimes called a hybrid.

Hindbrain: the posterior portion of the brain, containing the medulla, pons, and cerebellum

Hippocampus (hip-ō-cam'-pus): part of the forebrain of vertebrates, important in motivation, emotion, and especially learning.

Histamine: a substance released by certain cells in response to tissue damage and invasion of the body by foreign substances. Histamine promotes dilation of arterioles and leakiness of capillaries, and triggers some of the events of the inflammatory response.

Homeobox (hō'-mē-ō-box): a sequence of DNA coding for special, 60 amino acid proteins. These proteins activate or inactivate genes that control development. Homeoboxes thus specify embryonic cell differentiation.

Homeostasis (hō-mē-ō-stā'-sis): the process of maintaining a relatively constant internal environment in the face of variations in the external environment. The relatively constant environment required for optimal functioning of cells is maintained by the coordinated activity of numerous regulatory mechanisms, including the respiratory, endocrine, circulatory, and excretory systems.

Homologous structures: structures that may differ in function but that have similar anatomy, presumably because of descent from common ancestors.

Homologue (hō'-mō-log): a chromosome that is similar in appearance and genetic information to another chromosome with which it pairs during meiosis. Also called homologous chromosome.

Homozygote (hō-mō-zī'-gōt): an organism carrying two copies of the same allele of the gene in question; also called a true-breeding organism.

Hormone: a chemical synthesized by one group of cells and carried in the bloodstream to other cells, whose growth or metabolic activity is influenced by reception of the hormone.

Humoral immunity: an immune response in which foreign substances are inactivated or destroyed by antibodies circulating in the blood.

Hybrid: an organism that is the offspring of parents differing in at least one genetically determined characteristic; also used to refer to the offspring of parents of different species.

Hybrid infertility: reduced fertility (often complete sterility) in hybrid offspring of two different species.

Hybrid inviability: the failure of a hybrid offspring of two different species to survive to maturity.

Hydrogen bond: the weak attraction between a hydrogen atom bearing a partial positive charge (due to polar covalent bonding with another atom) and another atom, usually oxygen or nitrogen, bearing a partial negative charge. Hydrogen bonds may form between atoms of a single molecule or of different molecules.

Hydrolysis (hī-drol'-i-sis): the chemical reaction that breaks a covalent bond through the addition of hydrogen to the atom forming one side of the original bond, and a hydroxyl group to the atom on the other side.

Hydrophilic (hī-drō-fil'-ik): pertaining to a substance that dissolves readily in water, or to parts of a large molecule that form hydrogen bonds with water.

Hydrophobic (hī-drō-fō'-bik): pertaining to a substance that does not dissolve in water.

Hydrophobic interaction: the tendency for hydrophobic molecules to cluster together when immersed in water.

Hydrostatic skeleton (hī-drō-stat'-ik): the use of fluid contained in body compartments to provide support for the body and mass against which muscles can contract.

Hypertension: arterial blood pressure that is chronically elevated above the normal level.

Hypertonic (hī-per-ton'-ik): referring to a solution that has a higher concentration of dissolved particles (and therefore a lower free water concentration) than the cytoplasm of a cell.

Hypha (hī'-pha; pl. hyphae): a threadlike structure, many of which make up the fungal body, that consists of elongated cells, often with many haploid nuclei.

Hypocotyl (hī'-pō-kot-ul): the part of the embryonic shoot located between the attachment point of the cotyledons and the root.

Hypothalamus (hī-pō-thal'-uh-mus): part of the forebrain of vertebrates, located just below the thalamus, involved in regulation of hormonal activities (especially of the pituitary gland), and many behaviors such as feeding, drinking, sex, aggression, and fear responses, largely through activation of the autonomic nervous system.

Hypothesis (hī-poth'-e-sis): in science, a supposition based on previous observations, which is offered as an explanation for an event, and used as the basis for further observations or experiments.

Hypotonic (hī-pō-ton'-ik): referring to a solution that has a lower concentration of dissolved particles (and therefore a higher free water concentration) than the cytoplasm of a cell.

Immigration (im-uh-grā'-shun): movement of individuals into an area.

Immune deficiency disease: a disorder in which the immune system is incapable of responding properly to invading disease organisms.

Immune response: a specific response by the immune system to invasion of the body by a particular foreign substance or microorganism, characterized by recognition of the foreign material by immune cells and its subsequent destruction by antibodies or cellular attack.

Implantation: the process whereby the early embryo embeds itself within the lining of the uterus.

Imprinting: the process by which an animal forms an association with another animal or object in the environment during a sensitive period.

Inactivation: a process whereby certain chromosomes or parts of chromosomes are converted into a dense mass, preventing transcription.

Inclusive fitness: the reproductive success of all organisms bearing a given allele, usually expressed in relation to the average reproductive success of all individuals in the same population. Compare with fitness.

Incomplete dominance: a pattern of inheritance in which heterozygotes have a phenotype intermediate between those of the two homozygotes.

Incomplete flower: a flower that is missing one of the four floral parts (sepals, petals, stamens, or carpels).

Independent assortment: a pattern of inheritance of multiple traits, in which the distribution of alleles for one trait into the gametes does not affect the distribution of alleles for other traits. Occurs with genes that are located on different chromosomes.

Indirect development: a developmental pathway in which a free-living offspring goes through radical changes in body form as it matures.

Induced fit: a model of enzyme activity proposing that binding of substrates to an enzyme active site changes the shape or charge of both the substrates and the active site.

Induction: the process by which a group of cells causes other cells to differentiate into a specific tissue type.

Inflammatory response: a nonspecific, local response to injury to the body, characterized by phagocytosis of foreign substances and tissue debris by white blood cells, and "walling off" of the injury site by clotting of fluids escaping from nearby blood vessels.

Inheritance: the transmission of inborn characteristics from parent to offspring.

Inheritance of acquired characteristics: the hypothesis that organisms' bodies change during their lifetimes by use and disuse, and that these changes are inherited by their offspring.

Inhibiting hormone: a hormone secreted by the hypothalamus that inhibits the release of specific hormones from the anterior pituitary gland.

Inhibitory synapse: a synapse between two nerve cells in which the resting potential of the postsynaptic cell becomes more negative as a result of the activity of the presynaptic cell.

Innate (in-nāt'): inborn; instinctive; determined by the genetic makeup of the individual.

Inner cell mass: in human embryonic development, the cluster of cells on one side of the blastocyst, which will develop into the embryo.

Inner ear: the innermost part of the mammalian ear, composed of the bony, fluid-filled tubes of the cochlea and the vestibular system.

Inorganic molecule: any molecule that does not contain both carbon and hydrogen.

Insertion: a mutation in which one or more nucleotides are inserted within a gene. Also, the site of attachment of a muscle to the relatively moveable bone on one side of a joint.

Insight: a complex form of learning in which the solution to a problem is reached through reasoning.

Inspiration: the act of inhaling air into the lungs by enlarging the chest cavity.

Instinctive: innate; inborn; determined by the genetic makeup of the individual.

Integration: in nerve cells, the process of adding up electrical signals from sensory inputs or other nerve cells, to determine the overall electrical activity of the nerve cell.

Integument (in-teg'-ū-ment): the layers of the ovule surrounding the embryo sac; develops into the seed coat.

Intensity: the strength of stimulation or response.

Intercalated discs (in-tur'-cal-ā-ted): specialized interdigitating membranes containing large numbers of gap junctions that connect adjacent cardiac muscle fibers.

Interferon: a protein released by certain virus-infected cells that increases the resistance of other, uninfected, cells to viral attack.

Intermediate filament: part of the cytoskeleton of eukaryotic cells, probably functioning mainly for support.

Intermembrane compartment: the fluid-filled space between the inner and outer membranes of a mitochondrion.

Internal fertilization: union of sperm and egg inside the body of the female.

Internode: the part of a stem between two nodes.

Interphase: the stage of the cell cycle between cell divisions. During interphase, chromosomes are replicated, and other cell functions occur, such as growth, movement, and acquisition of nutrients.

Interspecific competition: competition between individuals of different species.

Interstitial cells (in-ter-sti'-shul): in the vertebrate testis, testosterone-producing cells located between the seminiferous tubules.

Interstitial fluid (in-tur-sti'-shul): fluid similar in composition to plasma (except lacking large proteins) which leaks from the capillaries and acts as a medium of exchange between the body cells and the capillaries.

Intertidal zone: the area along the edges of land masses that is alternately flooded and left exposed by the rising and falling of the tides.

Intervertebral discs (in-tur-ver-tē'-brul): pads of cartilage found between the vertebrae that act as shock absorbers.

Intracellular digestion: the chemical breakdown of food occurring within single cells.

Intraspecific competition: competition between individuals of the same species.

Intron: a segment of DNA in a eukaryotic gene that does not code for amino acids in a protein.

Invertebrate (in-vert'-uh-bret): a category of animals that never possess a vertebral column.

Ion (ī'-on): an atom or molecule that has either an excess of electrons (and hence is negatively charged) or has lost electrons (and is positively charged).

Ionic bond: a chemical bond formed by the electric attraction between positively and negatively charged ions.

Iris: the pigmented part of the vertebrate eye, surrounding an opening, the pupil.

Isotonic (ī-sō-ton′-ik): referring to a solution that has the same concentration of dissolved particles (and therefore the same free water concentration) as the cytoplasm of a cell.

Isotope: one of several forms of a single element, the nuclei of which contain the same number of protons but different numbers of neutrons.

J-curve: the shape of the growth curve of an exponentially growing population, in which increasing numbers of individuals join the population with each succeeding time increment.

Joint: a flexible region between two rigid units of an exoskeleton or endoskeleton, to allow for movement between the units.

Keratin (ker′-uh-tin): a fibrous protein found in hair, nails, and the epidermis of skin.

Kidney: one of a pair of organs of the excretory system located on either side of the vertebral column. Kidneys filter blood, removing wastes and regulating the ionic composition and water content of the blood.

Killer T cell: a type of T cell that directly destroys foreign cells upon contacting them.

Kin selection: the concept that natural selection selects for altruistic behaviors that benefit close relatives of the individual performing the behavior.

Kinesis (kin-nē′-sis): an innate process by which an organism achieves an orientation to a stimulus by altering its speed of movement in response to the stimulus.

Kinetic energy: the energy of movement; includes light, heat, mechanical movement, and electricity.

Kinetochore (kī-nēt′-ō-kōr): a protein structure that forms at the centromere regions of chromosomes; attaches the chromosomes to the kinetochore microtubules of the spindle.

Kinetochore microtubule: a microtubule of the spindle apparatus that connects the kinetochore of a chromosome with a spindle pole; involved in movement of the chromosome to the spindle pole during anaphase.

Kingdom: the most inclusive category in the classification of organisms. The five kingdoms are Monera, Protista, Fungi, Plantae, and Animalia.

Klinefelter's syndrome: a set of characteristics typically found in individuals who have two X chromosomes and one Y chromosome. These individuals are phenotypically males, but sterile, and have several femalelike traits, including narrow shoulders, broad hips, and partial breast development.

Krebs cycle: the citric acid cycle (in honor of Hans Krebs, who discovered many of its biochemical details).

Lacteal (lak-t′ēl): a single lymph vessel that penetrates each villus of the small intestine.

Lactose (lak′-tōs): a disaccharide composed of glucose and galactose; found in mammalian milk.

Larva (lar′-vuh): The caterpillars of moths and butterflies, and the maggots of flies, are larvae. An immature form of an animal with indirect development, often much different in body form from the adult.

Larynx (lār′-inx): that portion of the air passage between the pharynx and the trachea. The larynx contains the vocal cords.

Lateral bud: a bud located at a node of a stem, usually in the crotch between the stem and the petiole of the leaf found at the same node.

Lateral line organ: a receptor organ in fish and aquatic amphibians that detects water movement. It utilizes clusters of hair cells whose hairs are bent by the deflection of a gelatinous cupula.

Lateral meristem: also called cambium; a meristematic tissue in dicot stems and roots, usually found between the xylem and phloem (vascular cambium) and just outside the phloem (cork cambium).

Leaf: an outgrowth of a stem, usually flattened and photosynthetic.

Leaf primordium (pri-mōr′-dē-um): the outgrowth of a shoot that develops into a leaf.

Learning: an adaptive change in behavior as a result of experience.

Legumes (leg′-yūm): a group of plants (including alfalfa, lupines, peas, clover, and soybeans) that harbor colonies of nitrogen-fixing bacteria in swellings or nodules on their roots.

Lens: a clear object that bends light rays; in eyes, a flexible or movable structure used to focus light upon a layer of photoreceptor cells.

Leukocyte (loo′-kō-sīt): any of the white blood cells circulating in the blood.

Lichen (lī′-ken): a symbiotic association between an alga or cyanobacterium and a fungus, resulting in a composite organism.

Life cycle: the events in the life of an organism from one generation to the next.

Ligament: a tough connective tissue band connecting two bones.

Light-dependent reactions: the first stage of photosynthesis, in which the energy of light is captured as ATP and NADPH; occurs in thylakoids of chloroplasts.

Light-harvesting complex: in photosystems, the assembly of pigment molecules (chlorophyll and often carotenoids or phycocyanins) that absorb light energy and transfer the energy to electrons.

Light-independent reactions: the second stage of photosynthesis, in which the energy obtained by the light-dependent reactions is used to fix carbon dioxide into carbohydrates; occurs in the stroma of chloroplasts.

Limbic system: a diverse group of brain structures, mostly in the lower forebrain, including the thalamus, hypothalamus, amygdala, hippocampus, and parts of the cerebrum, involved in emotion, motivation, and learning.

Limnetic zone (lim-net′-ik): the upper, lighted, open-water region of a lake, where phytoplankton can carry out photosynthesis.

Linkage: the inheritance of certain genes as a group because they are parts of the same chromosome. Linked genes do not show independent assortment.

Lipase (lī′-pāse): an enzyme that catalyzes the breakdown of lipids such as fats.

Lipid (li′-pid): one of a number of water-insoluble organic molecules, containing large regions composed solely of carbon and hydrogen. Lipids include oils, fats, waxes, phospholipids, and steroids.

Littoral zone (lit′-ōr-ul): the near-shore, shallow-water region of a lake where anchored plants grow.

Lock-and-key: a model of enzyme activity proposing that substrates fit precisely into the active site of an enzyme.

Locus: the physical location of a gene on a chromosome.

Long-day plant: a plant that will flower only if the length of daylight is greater than some species-specific duration.

Loop of Henle (hen′-lē): a specialized portion of the tubule of the nephron in birds and mammals that creates an osmotic concentration gradient in the fluid immediately surrounding it. This in turn allows the production of urine more osmotically concentrated than blood plasma.

Luteinizing hormone (LH): a hormone produced by the anterior pituitary gland that stimulates testosterone production in males and development of the follicle, ovulation, and production of the corpus luteum in females.

Lymph (limf): pale fluid within the lymphatic system composed primarily of interstitial fluid and lymphocytes.

Lymph nodes: small structures that act as filters for lymph.

These contain lymphocytes and macrophages, which inactivate foreign particles such as bacteria.

Lymphatic system: a system consisting of lymph vessels, lymph capillaries, lymph nodes, and the thymus and spleen. The system helps protect the body against infection, absorbs fats, and returns excess fluid and small proteins to the blood circulatory system.

Lymphocyte (lim'-fō-sīt): a white blood cell type important in the immune response.

Lysosome (lī'-sō-sōm): a membrane-bound organelle containing intracellular digestive enzymes.

Macronutrient: a molecule used in relatively large quantities in the metabolic activities of organisms.

Macrophage (mak'-rō-faj): a type of white blood cell that engulfs microbes. Macrophages destroy microbes by phagocytosis and also present microbial antigens to T cells, helping to stimulate the immune response.

Major histocompatibility complex (MHC): proteins, usually located on the surfaces of body cells, that identify the cell as "self"; MHC proteins are also important in stimulating and regulating the immune response.

Maltose (mal'-tōs): a disaccharide composed of two glucose molecules.

Mammary glands (mam'-uh-rē): milk-producing organs used by female mammals to nourish their young.

Mantle (man'-tul): an extension of the body wall in certain invertebrates, such as mollusks. It may secrete a shell, protect the gills, and, as in cephalopods, aid in locomotion.

Marsupial (mar-sū'-pē-ul): a type of mammal whose young are born at an extremely immature stage and undergo further development in a pouch while they remain attached to a mammary gland. Includes kangaroos, opossums, and koalas.

Matrix: the fluid contained within the inner membrane of a mitochondrion.

Matter: the material of which the universe is made.

Mechanical incompatibility: the inability of male and female organisms to exchange gametes, usually because of incompatibility of the reproductive structures.

Mechanoreceptor: a receptor that responds to mechanical deformation such as is caused by pressure, touch, or vibration.

Medulla (med-oo'-la): part of the hindbrain of vertebrates that controls automatic activities such as breathing, swallowing, and heartbeat.

Medusa (meh-dū'-suh): a bell-shaped, often free-swimming stage in the life cycle of many cnidarians. Jellyfish are one example.

Megakaryocyte (meg-a-kār'-ē-ō-sīt): a large cell type that remains in the bone marrow, pinching off pieces of itself. These fragments enter the circulation as platelets.

Megaspore: a haploid cell formed by meiosis from a diploid megaspore mother cell. Through mitosis and differentiation, the megaspore develops into the female gametophyte.

Megaspore mother cell: a diploid cell contained within the ovule of a flowering plant, which undergoes meiosis to produce four haploid megaspores.

Meiosis (mī-ō'-sis): a type of cell division found in eukaryotic organisms, in which a diploid cell divides twice to produce four haploid cells.

Membrane: in cells, a thin sheet of lipids and proteins that surrounds the cell or its organelles, separating them from their surroundings. Also, continuous sheets of epithelial cells that cover the body and line body cavities.

Memory cell: a long-lived descendant of a B or T cell that has been activated by contact with antigen. Memory cells are a reservoir of cells that rapidly respond to reexposure to the antigen.

Meninges (men-in'-gēs): three layers of connective tissue that surround the brain and spinal cord.

Menstrual cycle: in humans, a complex 28-day cycle during which hormonal interactions among the hypothalamus, pituitary gland, and ovary coordinate ovulation and the preparation of the uterus to receive and nourish the fertilized egg. If pregnancy does not occur, the uterine lining is shed during menstruation.

Menstruation: in females of some primate species, the monthly discharge of uterine tissue and blood from the vagina.

Meristem cell (mer'-i-stem): an undifferentiated cell that remains capable of cell division throughout the life of a plant.

Mesoderm (mes'-ō-derm): the middle embryonic tissue layer, lying between the endoderm and ectoderm, and usually the last to develop. Mesoderm gives rise to structures such as muscle and skeleton.

Mesoglea (mez-ō-glē'-uh): a middle, jellylike layer within the body wall of cnidarians.

Mesophyll (mez'-ō-fil): cells located between the epidermal layers of a leaf.

Messenger RNA (mRNA): a strand of RNA, complementary to the DNA of a gene, that conveys the genetic information in DNA to the ribosomes to be used during protein synthesis. Sequences of three nucleotides (codons) in mRNA specify particular amino acids to be incorporated into a protein.

Metabolic pathway: a sequence of chemical reactions within a cell, in which the products of one reaction are the reactants for the next.

Metabolism: the sum of all chemical reactions occurring within a single cell or within all the cells of a multicellular organism.

Metamorphosis (met-a-mōr'-fuh-sis): in animals with indirect development, a radical change in body form from one larval stage to another or from larva to adult, seen in amphibians (tadpole to frog) and insects (caterpillar to butterfly).

Metaphase (met'-a-fāz): the stage of mitosis or meiosis in which the chromosomes, attached to kinetochore microtubules, line up along the equator of the cell.

Methanogen (me-than'-ō-gen): a type of anaerobic archaebacterium capable of converting carbon dioxide to methane.

Micelle (mī'-cēl): aggregations of bile salts with their lipophilic ends facing inward. Fatty acids and monoglycerides dissolve in the hydrophobic portions and are ferried to the cells of the lining of the small intestine.

Microfilament: part of the cytoskeleton of eukaryotic cells, composed of the proteins actin and (sometimes) myosin; functions in the movement of cell organelles and in locomotion by pseudopodia.

Micronutrient: a molecule required by organisms in relatively small or trace quantities.

Microsphere: a small, hollow sphere formed from proteins or proteins complexed with other compounds.

Microspore: a haploid cell formed by meiosis from a microspore mother cell. Through mitosis and differentiation, the microspore develops into the male gametophyte.

Microspore mother cell: a diploid cell contained within an anther of a flowering plant, which undergoes meiosis to produce four haploid microspores.

Microtubule: a hollow, cylindrical strand found in eukaryotic cells, composed of the protein tubulin; part of the cytoskeleton used in movement of cell organelles, cell growth, and construction of cilia and flagella.

Microtubule organizing center: a region of a eukaryotic cell at which tubulin is assembled into microtubules.

Microvilli (mī-krō-vi'-lī): a series of folded projections of the cell membrane that increase its surface area.

Midbrain: during development, the central portion of the brain, which contains the reticular formation.

Middle ear: part of the mammalian ear composed of the tympanic membrane and three bones (malleus, incus, stapes) that transmit vibrations from the auditory canal to the oval window.

Middle lamella: a thin layer of pectin and other carbohydrates that separates and sticks together the primary cell walls of adjacent plant cells.

Migration: in population genetics, the flow of genes between populations.

Mimic (mim'-ik): an organism that has evolved to resemble another.

Mineral: an inorganic substance, especially one found in rocks or soil.

Mitochondrion (mī-tō-kon'-drē-un): an organelle, bounded by two membranes, that is the site of the reactions of aerobic metabolism.

Mitosis (mī-tō'-sis): a type of nuclear division found in eukaryotic cells. Chromosomes are duplicated during interphase before mitosis. During mitosis, one copy of each chromosome moves into each of two daughter nuclei. The daughter nuclei are therefore genetically identical to each other.

Mixture: a substance composed of two or more elements in variable proportions.

Molecule (mol'-e-kūl): a particle composed of one or more atoms held together by chemical bonds. A molecule is the smallest particle of a compound that displays all the properties of that compound.

Molt: to shed an external body covering, such as an exoskeleton, skin, feathers, or fur.

Monera (mō'-ne-ra): a taxonomic kingdom consisting of unicellular prokaryotic organisms, including bacteria, archaebacteria, and cyanobacteria.

Monoclonal antibody: any of the antibodies produced by a clone of genetically identical cells; all antibodies produced by the clone of cells are identical.

Monocot: short for monocotyledon; a type of flowering plant characterized by embryos with one food-storage organ called a cotyledon.

Monocotyledon (mahn'-eh-kot-ul-ēd'-un): a class of angiosperm plant in which the embryo has one cotyledon, or seed leaf.

Monoecious (mon-ē'-shus): pertaining to organisms in which male and female gametes are produced in the same individual.

Monohybrid cross: a breeding experiment in which the parents differ in only one genetically determined trait.

Monomer (mo'-nō-mer): a small organic molecule, several of which may be bonded together to form a chain called a polymer.

Monosaccharide (mo-nō-sak'-a-rīd): the basic molecular unit of all carbohydrates, usually composed of a chain of carbon atoms to which are bonded hydrogen and hydroxyl groups.

Monotreme: a type of mammal that lays eggs; for example, the platypus.

Morula (mor'-ū-la): in animals, an embryonic stage during cleavage, when the embryo consists of a solid ball of cells.

Motor neuron: a neuron that carries information from the central nervous system and stimulates effector organs such as muscles or glands.

Motor unit: a single motor neuron and all the muscle fibers on which it synapses.

Mucous membrane: the lining of the inside of the respiratory and digestive tracts.

Multicellular: a term describing an organism whose body consists of many cells.

Muscle fiber: an individual muscle cell.

Mutation: a change in a gene; usually refers to a genetic change that is significant enough to change the appearance or function of the organism.

Mutualism (mū'-chū-al-ism): a symbiotic relationship in which both participating species benefit.

Mycelium (mī-sēl'-ē-um): the body of a fungus, consisting of a mass of hyphae.

Mycorrhiza (mī-kō-rī'-uh; pl. mycorrhizae): a symbiotic association between a fungus and a plant root. The fungus, often a basidiomycete or an ascomycete, grows around and often into the roots of most vascular plants in a mutually beneficial relationship.

Myelin (mī'-eh-lin): a wrapping of lipid-rich membranes of specialized nonneural cells around the axon of a vertebrate nerve cell. Myelin increases the speed of conduction of action potentials.

Myofibril (mī-ō-fī'-bril): a cylindrical subunit of each muscle cell, consisting of a series of sarcomeres. Myofibrils are surrounded by sarcoplasmic reticulum.

Myometrium (mī-ō-mē'-trē-um): the muscular outer layer of the uterus.

Myosin (mī'-ō-sin): one of the major proteins of muscle that interacts with actin to produce contraction; found in the thick filaments of the muscle fiber. See also actin.

Natural killer cell: a type of white blood cell that destroys some virus-infected cells and cancerous cells on contact. Part of the nonspecific internal defense against disease.

Natural selection: the unequal survival and reproduction of organisms due to environmental forces (e.g., physical factors such as climate and living organisms such as predators or prey) that act differently upon genetically different members of a population.

Nearshore zone: shallow-water areas of the ocean just below the low-tide line. Due to the abundance of nutrients and strong sunlight, the nearshore zone often has the greatest concentration of life in the ocean.

Negative feedback: a situation in which a change initiates a series of events that tend to counteract the change and restore the original state. Negative feedback in physiological systems maintains homeostasis.

Nematocyst (nēm-āt'-ō-sist): a specialized cell found in cnidarians which, when disturbed, ejects a sticky or poisoned thread. Used by cnidarians to trap and sting prey.

Nephridium (nef-rid'-ē-um): a type of excretory organ found in earthworms, molluscs, and certain other invertebrates. A nephridium somewhat resembles a single vertebrate nephron.

Nephron (nef'-ron): the functional unit of the kidney, where blood is filtered and urine formed.

Nerve: a bundle of axons of nerve cells, bound together in a sheath.

Nerve cord: also called the spinal cord of vertebrates, a nervous structure lying along the dorsal side of the body of chordates.

Nerve net: a simple form of nervous systems consisting of a network of neurons that extend throughout the body of organisms such as cnidarians.

Net primary productivity: the energy stored in the primary producers of an ecosystem over a given period.

Neural tube: a structure derived from ectoderm during early embryonic development that later becomes the brain and spinal cord.

Neurohormone: a chemical synthesized by a specialized nerve cell (called a neurosecretory cell) and secreted into the bloodstream as a hormone.

Neuromodulator: a chemical, usually a peptide, released by neurons at synapses along with neurotransmitters. Neuromodulators alter the properties of synapses.

Neuromuscular junction: the synapse formed between a motor neuron and muscle fiber.

Neuron (nur'-on): a single nerve cell.

Neurosecretory cell: a specialized nerve cell that synthesizes and releases hormones.

Neurotransmitter: a chemical released by a nerve cell close to a second nerve cell, a muscle, or a gland cell, that influences the activity of the second cell.

Neutral mutation: a mutation (change in DNA sequence) that has little or no phenotypic effect.

Neutralization: the process of covering up or inactivating a toxic substance with antibody.

Neutron: a subatomic particle found in the nuclei of atoms, bearing no charge and having mass approximately equal to that of a proton.

Niche (nitch): the role of a particular species within an ecosystem, including all aspects of its interaction with the biotic and abiotic environment.

Nitrogen fixation: the process of removing nitrogen from the atmosphere and combining it with hydrogen to produce ammonia.

Nitrogen-fixing bacteria: bacteria that possess the ability to remove nitrogen (N_2) from the atmosphere and combine it with hydrogen to produce ammonium (NH_4^+).

Node: a region of a stem at which leaves and lateral buds are located.

Node of Ranvier (nōd of ron'-vē-ā): an interruption of the myelin on a vertebrate myelinated axon, at which action potentials are generated.

Nodule: a swelling on the root of a legume or other plant that consists of cortex cells inhabited by nitrogen-fixing bacteria.

Nondisjunction: an error in meiosis in which chromosomes fail to segregate properly into the daughter cells.

Nonpolar covalent bond: a covalent bond with equal sharing of electrons.

Noradrenaline (nor-ad-ren'-a-lin): also called norepinephrine, a neurotransmitter released onto organs by neurons of the parasympathetic nervous system, which prepares the body to respond to stressful situations.

Northern coniferous forest: a northern biome dominated by conifers; taiga.

Notochord (nōt'-ō-kōrd): a stiff but somewhat flexible, supportive rod found in all members of the phylum Chordata at some stage of development.

Nuclear envelope: the double membrane system surrounding the nucleus of eukaryotic cells. The outer membrane is often continuous with the endoplasmic reticulum.

Nucleic acid (nū-klā'-ik): an organic molecule composed of nucleotide subunits. The two common types of nucleic acids are ribonucleic acids (RNA) and deoxyribonucleic acids (DNA).

Nucleoid (nū-klē-oyd): the location of the genetic material in prokaryotic cells; not membrane-enclosed.

Nucleolus (nū-klē'-ō-lus): the region of the eukaryotic nucleus engaged in ribosome synthesis, consisting of the genes encoding ribosomal RNA, newly synthesized ribosomal RNA, and ribosomal proteins.

Nucleotide: an individual subunit of which nucleic acids are composed. A single nucleotide consists of a phosphate group covalently bonded to a sugar (deoxyribose in DNA), which is in turn covalently bonded to a nitrogenous base (adenine, guanine, cytosine, or thymine). Nucleotides are linked together by covalent bonds between the phosphate of one and the sugar of the next, to form a strand of nucleic acid.

Nucleus (nū-klē-ús): the membrane-bound organelle of eukaryotic cells that contains the cell's genetic material. Also, the central region of an atom, composed of protons and neutrons.

Nutrient: a substance acquired from the environment and needed for survival, growth, and development of an organism.

Nutrient cycle: a description of the movement of a specific inorganic nutrient (carbon, nitrogen, water, etc.) through the living and non-living portions of an ecosystem.

Nutrition: the process of acquiring nutrients from the environment and, if necessary, processing them into a form that can be used by the body.

Oil: a lipid composed of three fatty acids, some of which are unsaturated, covalently bonded to a molecule of glycerol. Oils are liquid at room temperature.

Olfaction (ōl-fak'-shun): a chemical sense, the sense of smell; in terrestrial vertebrates the result of detection of airborne molecules.

Oligotrophic lake (ol-i-gō-trōf'-ik): a lake that is low in nutrients and supports sparse communities of organisms.

Ommatidium (ōm-ma-tid'-ē-um): an individual light-sensitive subunit of a compound eye. Each ommatidium consists of a lens and several (usually eight) receptor cells.

Oncogene: a gene that, when transcribed, causes a cell to become cancerous.

One-gene, one-protein hypothesis: the proposition that each gene encodes the information for the synthesis of a specific protein.

Oogonium (ō-ō-gō'-nē-um): a diploid cell in female animals that gives rise to a primary oocyte.

Open circulatory system: a type of circulatory system in arthropods and mollusks in which the blood is pumped through an open space (the hemocoel), where it bathes the internal organs directly.

Operant conditioning: a laboratory training procedure in which an animal learns to make a response (such as pressing a lever) through reward or punishment.

Operator: in prokaryotes, a segment of DNA that controls access of RNA polymerase to the promoter.

Operculum: an external flap, supported by bone, that covers and protects the gills of most fish.

Operon: a unit of organization of prokaryotic chromosomes, in which several genes that specify related functions (e.g., enzymes in the same biosynthetic pathway) are grouped together on the chromosome, are transcribed at the same time, and are regulated together.

Opioid (ōp'-ē-ōyd): a group of peptides found in the vertebrate brain that mimic some of the actions of opiates (such as opium, heroin, morphine). Besides analgesia (pain relief), opioids seem to be involved in many behaviors, including emotion, learning, and the control of appetite.

Optic disk (op'-tik): the area of the retina at which the axons of the ganglion cell merge to form the optic nerve; the blind spot of the retina.

Optic nerve (op'-tik): the nerve leading from the eye to the brain, carrying visual information.

Orbital: the region of an atom, outside the nucleus, in which an electron is likely to be found.

Order: the taxonomic category contained within a class and consisting of related families.

Organ: two or more tissues integrated to perform a specialized function, i.e., the kidney, liver, or skin.

Organ system: two or more organs working together in the execution of a specific bodily function (e.g., digestive tract).

Organelle (or-ga-nel'): a structure found in the cytoplasm of a cell that performs a specific function; sometimes used to refer specifically to membrane-bound structures such as the nucleus or endoplasmic reticulum of eukaryotic cells.

Organic molecule: a molecule that contains both carbon and hydrogen.

Organism (ōr'-ga-niz-em): an individual living thing.

Organogenesis (or-gan-ō-jen'-i-sis): the process by which the germ layers of the gastrula develop into organs.

Origin: the site of attachment of a muscle to the relatively stationary bone on one side of a joint.

Osculum (os'-kū-lum): a relatively large opening in the sponge body through which water is expelled.

Osmosis (oz-mō'-sis): the diffusion of water across a differentially permeable membrane, usually down a concentration gradient of free water molecules. Water moves into the solution that has a lower free water concentration from the solution with the higher free water concentration.

Osmotic pressure: a measure of the tendency of water to move from a solution with a lower concentration of free water molecules into one with a higher concentration of free water molecules; the physical pressure which must be applied to a solution to prevent water movement into it from pure water.

Osteoblast (os'-tē-ō-blast): a cell type that produces bone.

Osteoclast: a cell type that dissolves bone.

Osteoporosis (os'-tē-ō-pōr-ō'-sis): a condition in which bones become porous, weak, and easily fractured; most common in elderly women.

Otolith membrane (ō'-tō-lith): a gelatinous membrane in which are embedded the hairs of hair cells of the utricle and saccule of the inner ear. The otolith membrane is heavier than the surrounding fluid due to calcium carbonate crystals embedded in it, and thus responds to the pull of gravity.

Outer ear: the outermost part of the mammalian ear, including the external ear and auditory canal leading to the tympanic membrane.

Oval window: the membrane-covered entrance to the inner ear.

Ovary: the gonad of female animals. In flowering plants, a structure at the base of the carpel containing one or more ovules; develops into the fruit.

Oviduct: in mammals, the tube leading from the ovary to the uterus.

Ovule: a structure within the ovary of a flower, inside which the female gametophyte develops. After fertilization, the ovule develops into the seed.

Ozone layer: a layer of atmosphere between 10 and 50 kilometers above the Earth, which contains relatively high concentrations of ozone (O_3). Ozone absorbs harmful ultraviolet radiation.

Pacinian corpuscle (pas-in'-ē-an): a receptor of the skin and joints whose ending is wrapped in a layered capsule of connective tissue, and which responds to rapid movement.

Palisade cells: mesophyll cells just beneath the upper epidermis of a leaf; usually elongated perpendicularly to the epidermis.

Pancreas (pan'-krē-is): an organ lying adjacent to the stomach that secretes enzymes for fat, carbohydrate, and protein digestion into the small intestine. Other cells in the pancreas are responsible for production of the hormones glucagon and insulin.

Parasite (par'-a-sīt): an organism that feeds on the body of a larger organism, called a host, causing it harm.

Parasitism (par'-a-sit-ism): a relationship in which an organism, usually smaller and more numerous than its host, feeds on another, usually without killing it immediately.

Parasympathetic nervous system: the division of the autonomic nervous system that produces largely involuntary responses related to maintenance of normal body functions, such as digestion.

Parathyroids: a set of four small endocrine glands embedded in the surface of the thyroid gland that produce parathormone, which (with calcitonin from the thyroid) regulates calcium ion concentration in the blood.

Parenchyma (par-en'-ki-ma): a plant cell type that is alive at maturity, usually with thin primary cell walls, that carries out most of the metabolism of a plant. Most dividing meristem cells in a plant are parenchyma.

Parthenogenesis (par-the-nō-gen'-i-sis): a specialization of sexual reproduction, in which an egg undergoes development without fertilization.

Passive transport: movement of materials across a membrane down a gradient of concentration, pressure, or electrical charge without using cellular energy.

Passive visual signal: a body shape, color, or size that communicates information, even in the absence of specific behaviors.

Pathogen (path'-ō-gen): an organism capable of producing disease.

Pedigree: a diagram showing genetic relationships among a set of individuals, usually with respect to a specific genetic trait.

Pelagic (pel-a'-jik): referring to organisms that spend their lives in open water, floating or swimming.

Peptide (pep'-tīd): a chain composed of two or more amino acids linked together by peptide bonds.

Peptide bond: the covalent bond between the amino group nitrogen of one amino acid and the carboxyl group carbon of a second amino acid, which joins the two amino acids together in a peptide or protein.

Peptidoglycan (pep-tid-ō-glī'-can): material found in prokaryotic cell walls consisting of chains of sugars cross-linked by short chains of amino acids called peptides.

Pericycle (per'-i-sī-kul): the outermost layer of cells of the vascular cylinder of a root.

Periderm: the outer cell layers of a stem that has undergone secondary growth, consisting primarily of cork cambium and cork cells.

Peripheral nervous system: in vertebrates, that part of the nervous system located outside the brain and spinal cord, consisting of the nerves leading to and from the brain and spinal cord, and the ganglia of the autonomic nervous system.

Permafrost: soil layer a short distance below the surface of the tundra that stays frozen year round.

Permeate (per'-mē-āt): to pass through, as through pores in a membrane.

Petal: part of a flower, often flat and brightly colored, serving to attract potential animal pollinators.

Petiole (pet'-ē-ōl): the stalk that connects the blade of a leaf to the stem.

pH scale: a scale with values from 0 to 14, used for measuring the relative acidity of a solution. At pH 7 a solution is neutral, pH 0 to 7 is acidic, and pH 7 to 14 is basic. Each unit on the pH scale represents a tenfold change in the concentration of hydrogen ions.

Phagocytosis (fa-gō-sī-tō'-sis): a type of endocytosis in which extensions of a plasma membrane engulf extracellular particles and transport them into the interior of the cell.

Pharynx (fār'-inx): a chamber at the back of the mouth shared by the digestive and respiratory systems.

Phenotype (fēn'-ō-tīp): the physical properties of an organism. Phenotype can be defined as outward appearance (e.g., flower color), as behavior, or in molecular terms (e.g., ABO glycoproteins on red blood cells).

Phenylketonuria (fen-ul-kē-tō-nū'-rē-a): a recessive disease in which the enzyme that catalyzes the conversion of the amino acid phenylalanine to tyrosine is faulty.

Pheromone (fār'-uh-mōn): a chemical produced by an organism that alters the behavior or physiological state of another of the same species.

Phloem (flō'-um): a conducting tissue of vascular plants that transports a concentrated sugar solution up and down the plant.

Phospholipid (fos-fō-li'-pid): a lipid consisting of glycerol to which two fatty acids and one phosphate group are bonded. The phosphate group bears another group of atoms, often containing nitrogen, and often bearing an electric charge.

Phospholipid bilayer: a double layer of phospholipids that forms the basis of all cellular membranes; the phospholipid heads face the water of extracellular fluid or the cytoplasm, and the tails are buried in the middle of the bilayer.

Photic zone (fō'-tik): the surface waters of the ocean, into which sunlight penetrates well enough for plants to grow.

Photon (fō'-ton): the smallest unit of light.

Photophosphorylation (fō-tō-fos-for-i-lā'-shun): ATP synthesis in the light-dependent reactions of photosynthesis; light energy ("photo") is used to add a phosphate group to ("phosphorylate") ADP to yield ATP.

Photopigment (fō'-tō-pig-ment): a chemical substance in photoreceptor cells (rhodopsin in rods) that changes molecular conformation when struck by light.

Photoreceptors: receptor cells that respond to light; in vertebrates, rods and cones.

Photosynthesis: the complete series of chemical reactions in which the energy of light is used to synthesize high-energy organic molecules, usually carbohydrates, from low-energy inorganic molecules, usually carbon dioxide and water.

Photosystem: in thylakoid membranes, a light-harvesting complex and its associated electron transport system.

Phototropism: growth with respect to the direction of light.

Phycocyanin (fī-kō-sī'-a-nin): a bluish pigment found in the membranes of chloroplasts and used as an accessory light-gathering molecule in thylakoid photosystems.

Phyletic speciation: a form of evolution in which the gene pool of a species changes so greatly over time that a new species is formed. Old and new species do not exist simultaneously. See divergent speciation.

Phylum (fī-lum): the taxonomic category of animals and animallike protists contained within a kingdom and consisting of related classes.

Phytochrome (fī'-tō-krōm): a light-sensitive plant pigment that mediates many plant responses to light, including flowering, stem elongation, and seed germination.

Phytoplankton (fī'-tō-plank-ten): a general term describing photosynthetic protists that are abundant in marine and freshwater environments.

Pili (pil'-ī): hair-like projections made of protein and found on the surface of certain bacteria, often used to attach the bacterium to another cell.

Pineal gland (pī-nē'-al): a small gland within the brain that secretes melatonin. The pineal controls the seasonal reproductive cycles of some mammals.

Pinocytosis (pī-nō-sī-tō'-sis): nonselective movement of extracellular fluid into a cell, enclosed within a vesicle formed from the plasma membrane.

Pioneer: an organism that is among the first to colonize an unoccupied habitat in the first stages of succession.

Pit: an area in the cell walls between two plant cells in which secondary walls did not form, so that the two cells are separated only by a relatively thin and porous primary cell wall.

Pit organ: a thermoreceptive organ found in certain poisonous snakes (vipers) consisting of a pair of pits located between each eye and the nose; used to detect warm-blooded prey.

Pith: cells at the center of a root or stem.

Pituitary: an endocrine gland located at the base of the brain that produces several hormones, many of which influence the activity of other glands.

Placenta (pluh-sen'-ta): in mammals, a structure formed partly from the uterine lining and partly from the embryonic membranes, especially the chorion; functions in gas, nutrient, and waste exchange between embryonic and maternal circulatory systems.

Plankton (plank'-ton): microscopic pelagic protists and animals found in the photic zone of the oceans, whose movement is determined primarily by movement of the water. Phytoplankton are the main producers in the open ocean.

Plaque (plak): a deposit of cholesterol and other fatty substances within the wall of an artery.

Plasma: the fluid, noncellular portion of the blood.

Plasma cell: an antibody-secreting descendant of a B cell.

Plasma membrane: the outer membrane of a cell, composed of a bilayer of phospholipids in which proteins are embedded.

Plasmid (plaz'-mid): a small, circular piece of DNA found in the cytoplasm of many bacteria; usually does not carry genes required for the normal functioning of the bacterium, but may carry genes that assist bacterial survival in certain environments, e.g., antibiotic resistance.

Plasmodesma (plaz-mō-dez'-ma; pl. plasmodesmata): a cell-to-cell junction in plants that connects the cytoplasm of adjacent cells.

Plasmodial slime mold: a funguslike protist consisting of a multinucleate mass or plasmodium that crawls and ingests decaying organic matter in amoeboid fashion and forms a fruiting body under adverse conditions.

Plasmodium (plas-mō'-dē-um): an aggregation of individual amoeboid cells that form a sluglike mass.

Plastid (plas'-tid): in plant cells, an organelle bounded by two membranes that may be involved in photosynthesis (chloroplasts), pigment storage, or food storage.

Platelets (plāt'-lets): cell fragments formed from megakaryocytes in bone marrow. Platelets, which lack nuclei, circulate in the blood and play a role in blood clotting.

Pleated sheet: a type of secondary structure of a protein. In a pleated sheet, protein chains lie side-by-side, bonded to one another by hydrogen bonds.

Pleiotropy (plē'-ō-trō-pē): a situation in which a single gene influences more than one phenotypic characteristic.

Point mutation: a mutation in which a single base pair in DNA has been changed.

Polar body: in oogenesis, a small cell containing a nucleus but virtually no cytoplasm.

Polar covalent bond: a covalent bond with unequal sharing of electrons, so that one atom is relatively negative while the other is relatively positive.

Polar microtubule: a microtubule of the spindle apparatus that extends from one spindle pole to the equator of the cell; involved in moving the spindle poles farther apart during anaphase.

Pollen: the male gametophyte of gymnosperms and angiosperms.

Pollination: in flowering plants, the deposition of pollen onto the stigma of a flower of the same species; in conifers, the deposition of pollen within the pollen chamber of a female cone of the same species.

Polygenic inheritance: a pattern of inheritance in which the interactions of two or more genes determine phenotype.

Polymer (pa'-li-mer): a molecule composed of three or more (may be thousands) smaller subunits called monomers. The monomers that make up a single polymer may be identical (e.g., the glucose monomers of starch) or different (e.g., the amino acids of a protein).

Polymerase chain reaction: a method of producing virtually unlimited numbers of copies of a specific piece of DNA, starting with as little as one copy of the desired DNA.

Polyp (pol'-ip): the sedentary, vase-shaped stage in the life cycle of many cnidarians. Hydra and sea anemones are examples.

Polyploid (pol'-ē-ployd): having more than two homologous chromosomes of each type.

Polysaccharide (pol-ē-sak'-a-rīd): a large carbohydrate molecule composed of branched or unbranched chains of repeating monosaccharide subunits, usually glucose or modified glucose molecules. Polysaccharides include starches, cellulose, and glycogen.

Pons: a portion of the hindbrain just above the medulla containing neurons that influence sleep, and control the rate and pattern of breathing.

Population: a group of individuals of the same species, found in the same time and place, and actually or potentially interbreeding.

Population bottleneck: a form of genetic drift in which a population becomes extremely small, which may lead to differences in allele frequencies as compared with other populations of the species, and to a loss in genetic variability.

Population genetics: the study of the frequency, distribution, and inheritance of alleles in a population of organisms.

Positive feedback: a situation in which a change initiates events that tend to amplify the original change.

Posterior (post-tēr'-ē-ur): the tail, hindmost, or rear end of an animal.

Postmating isolating mechanism: any structure, physiological function, or developmental abnormality that prevents organisms of two different populations, once mating has occurred, from producing vigorous, fertile offspring.

Postsynaptic: referring to the nerve cell at a synapse which changes its electrical potential in response to a chemical (the neurotransmitter) released by another (presynaptic) cell.

Postsynaptic potential: an electrical signal produced in a postsynaptic cell by transmission across the synapse. It may be excitatory (EPSP), making the cell more likely to produce an action potential, or inhibitory (IPSP), tending to inhibit an action potential.

Potassium channel: a hollow-cored protein that spans a nerve cell membrane, forming a pore, and which only allows potassium ions to flow through the pore.

Potential energy: stored energy; includes chemical energy and the energy of position within a gravitational field.

Preadaptation: a feature evolved under one set of environmental conditions that, purely by chance, helps an organism adapt to new environmental conditions.

Prebiotic evolution: evolution before life existed; especially abiotic synthesis of organic molecules.

Precapillary sphincter (sfink'-ter): a ring of smooth muscle between an arteriole and a capillary that regulates the flow of blood into the capillary bed.

Predation (pre-dā'-shun): the act of killing and eating another living organism.

Predator: an organism that kills and eats other organisms.

Premating isolating mechanism: any structure, physiological function, or behavior that prevents organisms of two different populations from exchanging gametes.

Pressure flow theory: a model for transport of sugars in phloem, which states that movement of sugars into a phloem sieve tube causes water to enter the tube by osmosis, while movement of sugars out of another part of the same sieve tube causes water to leave by osmosis; the resulting pressure gradient causes bulk movement of water and dissolved sugars from the end of the tube into which sugar is transported toward the end of the tube from which sugar is removed.

Presynaptic: referring to a nerve cell that releases a chemical (the neurotransmitter) at a synapse, which causes changes in the electrical activity of another (postsynaptic) cell.

Primary cell wall: cellulose and other carbohydrates secreted by a young plant cell between the middle lamella and the plasma membrane.

Primary consumer: an organism that feeds on producers; an herbivore.

Primary endosperm cell: the central cell of the female gametophyte of a flowering plant, containing the polar nuclei (usually two). After fertilization, the primary endosperm cell undergoes repeated mitotic divisions to produce the endosperm of the seed.

Primary growth: growth in length and development of initial structures of plant roots and shoots, due to cell division of apical meristems and differentiation of the daughter cells.

Primary oocyte (ō'-ō-sīt): a diploid cell, derived from the oogonium by growth and differentiation, which undergoes meiosis to produce the egg.

Primary phloem: phloem produced from an apical meristem.

Primary root: the first root that develops from a seed.

Primary spermatocyte (sper-ma'-tō-sīt): a diploid cell, derived from the spermatogonium by growth and differentiation, which undergoes meiosis to produce four sperm.

Primary structure: the amino acid sequence of a protein.

Primary succession: succession that occurs in an environment, such as bare rock, in which no remnants of a previous community are present.

Primary xylem: xylem produced from an apical meristem.

Primer pheromone: a chemical produced by an organism that alters the physiological state of another of the same species.

Primitive streak: in reptiles, birds, and mammals, the region of the ectoderm of the two-layered embryonic disk through which cells migrate to form mesoderm.

Prion (prē'-on): a proteinaceous infectious particle that may be responsible for certain neurodegenerative diseases.

Producer: a photosynthetic organism, synonymous with autotroph.

Product: an atom or molecule resulting from a chemical reaction.

Profundal zone (prō-fun'-dul): the depths of a lake where light is inadequate for photosynthesis.

Progesterone (prō-ge'-ster-ōn): a hormone produced by the corpus luteum that promotes development of the uterine lining.

Prokaryotic (prō-kar-ē-ot'-ik): referring to cells of the kingdom Monera. Prokaryotic cells do not have their genetic material enclosed within a membrane-bound nucleus, lack other membrane-bound organelles, and are usually smaller and simpler in structure than eukaryotic cells.

Promoter: a specific sequence of DNA to which RNA polymerase binds, initiating gene transcription.

Prophase (prō-'fāz): the first stage of mitosis or meiosis, in which the chromosomes first become visible with a light microscope as thickened, condensed threads and the spindle begins to form. In meiosis I, the homologous chromosomes pair up and exchange parts at chiasmata.

Proprioceptor (prō'-prē-ō-cep-tōr): a receptor that monitors the position of the parts of the body and their direction of movement by responding to stretch and pressure.

Prostaglandin (pro-sta-glan'-din): a family of modified fatty acid hormones manufactured by many cells of the body.

Prostate gland (prō'-stāt): a gland that produces part of the fluid component of semen. The prostate fluid is basic and contains a chemical that activates sperm movement.

Protease (prō'-tē-ās): an enzyme that digests proteins.

Protein: an organic molecule composed of one or more chains of amino acids.

Protista (prō-tis'-tuh): a taxonomic kingdom including unicellular, eukaryotic organisms.

Proton: a subatomic particle found in the nuclei of atoms, bearing a unit of positive charge and a relatively large mass roughly equal to the mass of the neutron.

Protozoan (prō-te-zō'-an; pl. protozoa): a nonphotosynthetic or animallike protist.

Pseudocoel (sū'-dō-sēl): "false coelom"; a body cavity with a different embryological origin than a coelom, but serving a similar function; found in roundworms.

Pseudoplasmodium (sū'-dō-plas-mō'-dē-um): a sluglike mass of cytoplasm containing thousands of nuclei that are not confined within individual cells.

Pseudopod (sū'-dō-pod): a temporary extension of the plasma membrane used for locomotion or phagocytosis in certain cells such as the protist *Amoeba* or white blood cells of vertebrates.

Punctuated equilibrium: a model of evolution, stating that morphological change and speciation are rapid, simultaneous events (the "punctuation"), separated by long periods during which a species remains unchanged (the "equilibrium").

Pupil: the adjustable opening in the center of the iris through which light enters the eye.

Pyloric sphincter (pī-lor'-ik sfink'-ter): a circular muscle at the base of the stomach that regulates the passage of chyme into the small intestine.

Quaternary structure (kwat'-er-na-rē): the complex three-dimensional structure of a protein that is composed of more than one peptide chain.

Queen substance: a chemical produced by a queen bee that can act as both a primer and a releaser pheromone.

Radial symmetry: a body plan in which any plane drawn along a central axis will divide the body into approximately mirror-image halves. Cnidarians and many adult echinoderms show radial symmetry.

Radioactive: pertaining to an atom with an unstable nucleus that spontaneously disintegrates with the emission of radiation.

Radula (ra'-dū-luh): a ribbon of tissue in the mouth of gastropod mollusks that bears numerous teeth on its outer surface and is used to scrape and drag food into the mouth.

Rain shadow: an area of low rainfall on the side of a mountain facing away from the prevailing winds.

Random distribution: spacing in which the probability of finding an individual is equal in all parts of an area.

Reactant: an atom or molecule that is used up in a chemical reaction to form a product.

Reaction center: in the light-harvesting complex of a photosystem, the chlorophyll molecule to which light energy is transferred by the antenna pigments. The captured energy ejects an electron from the reaction center chlorophyll, and the electron is transferred to the electron transport system.

Receptor: (1) a cell that responds to an environmental stimulus (chemicals, sound, light, pH, etc.) by changing its electrical potential; (2) a protein molecule in or on a cell membrane that reacts with another molecule (hormone, neurotransmitter, odorous compound, etc.) to trigger metabolic or electrical changes in the cell.

Receptor potential: a graded electrical potential change in a receptor cell produced in response to reception of an environmental stimulus (chemicals, sound, light, pH, heat, cold, etc.). The size of the receptor potential is proportional to the intensity of the stimulus.

Receptor protein: a protein or glycoprotein, located on a membrane (or in the cytoplasm), that recognizes and binds to specific molecules. Binding by receptors often triggers a response by a cell, such as endocytosis, protein synthesis, or exocytosis.

Receptor-mediated endocytosis: selective uptake of molecules from the extracellular fluid, by binding to a receptor located at a coated pit on the plasma membrane, and pinching off the coated pit into a vesicle that moves into the cytosol.

Recessive: an allele expressed only in homozygotes, and which is completely masked in heterozygotes.

Recognition protein: a protein or glycoprotein protruding from the outside surface of a plasma membrane that identifies a cell as belonging to a particular species, to a specific

individual of that species, and often to a specific organ within the individual.

Reflex: a simple, automatic, usually unconscious behavior performed by part of the body in response to a stimulus.

Regeneration: (1) regrowth of a body part after loss or damage; (2) asexual reproduction by regrowth of an entire body from a fragment.

Regulatory gene: a gene that controls the timing or rate of transcription of other genes.

Releaser: a stimulus that triggers a fixed action pattern.

Releaser pheromone: a chemical produced by one organism that alters the behavior of another of the same species.

Releasing hormone: a hormone secreted by the hypothalamus that causes the release of specific hormones by the anterior pituitary gland.

Replacement-level fertility: the average birth rate at which a reproducing population exactly replaces itself during its lifetime.

Repression: in the lactose operon, the condition that the genes of the operon are not transcribed unless specifically activated by the presence of lactose.

Repressor protein: in the lactose operon, a protein that binds to the operator site, thereby preventing access of RNA polymerase to the promoter.

Reproductive isolation: the failure of organisms of one population to breed successfully with members of another population; may be due to premating or postmating isolating mechanisms.

Reservoir: the major source of any particular nutrient in an ecosystem, usually in the abiotic portion.

Respiratory center: a location in the brainstem that sends rhythmic bursts of nerve impulses to the respiratory muscles, resulting in breathing.

Resting potential: a negative electrical potential found in unstimulated nerve cells.

Restriction enzyme: an enzyme, usually isolated from bacteria, that cleaves double-stranded DNA at a specific nucleotide sequence; the cleavage sequence differs for different restriction enzymes.

Restriction fragment: a piece of DNA that has been isolated by cleaving a larger piece of DNA with restriction enzymes.

Restriction fragment length polymorphisms (RFLPs): differences in the lengths of restriction fragments, produced by cutting samples of DNA from different individuals of the same species with the same set of restriction enzymes; occurs because of differences in nucleotide sequences among different individuals of the same species.

Reticular activating formation (reh-tik'-ū-lar): a diffuse network of neurons extending from the hindbrain, through the midbrain, and into the lower reaches of the forebrain, involved in filtering sensory input and regulating what information is relayed to higher centers in the cerebrum for further attention.

Retina (ret'-in-a): a layer of nerve tissue at the rear of camera-type eyes, composed of photoreceptor cells plus associated nerve cells that refine the photoreceptor information and transmit it to the optic nerve.

Retrovirus: a virus that uses RNA as its genetic material. When a retrovirus invades a eukaryotic cell, it "reverse transcribes" its RNA into DNA, which then directs the synthesis of more viruses, using the transcription and translation machinery of the cell.

Reverse transcriptase: an enzyme found in retroviruses that catalyzes the synthesis of DNA from an RNA template.

Rh factor: a protein found on the red blood cells of some people (Rh-positive) but not others (Rh-negative). Exposure of Rh-negative individuals to Rh-positive blood triggers antibody production to Rh-positive blood cells.

Rhizoid: (rī'-zōyd): a rootlike structure found in bryophytes that anchors the plant and absorbs water and nutrients from the soil.

Rhizome (rī'-zōm): an undergound stem, usually horizontal and functioning in food storage.

Ribonucleic acid (rī-bō-nū-klā-ik; RNA): A single-stranded nucleic acid molecule composed of nucleotides. Includes messenger RNA, transfer RNA and ribosomal RNA. RNA is also the genetic material of certain viruses.

Ribosomal RNA (rRNA): a type of RNA that combines with proteins to form ribosomes.

Ribosome: an organelle consisting of two subunits, each composed of ribosomal RNA and protein. Ribosomes are the site of protein synthesis, in which the sequence of nucleotides of messenger RNA is translated into the sequence of amino acids in a protein.

Ribozyme: an RNA molecule that can catalyze certain chemical reactions, especially those involved in synthesis and processing of RNA itself.

RNA polymerase: an enzyme that catalyzes the covalent bonding of free RNA nucleotides into a continuous strand, using RNA nucleotides that are complementary to those of a strand of DNA.

Rod: a rod-shaped photoreceptor cell in the vertebrate retina, sensitive to dim light, but not involved in color vision. See also cone.

Root: the part of a plant, usually below ground, that anchors the plant in the soil, absorbs and transports water and minerals, stores food, and produces certain hormones.

Root cap: a cluster of cells at the tip of a growing root, derived from the apical meristem. The root cap protects the growing tip from damage as it burrows through the soil.

Root hair: a fine projection from an epidermal cell of a young root.

Root pressure: hydrostatic pressure generated by osmosis of water across the endodermis of a plant root, which can force water some distance up the xylem into the stem.

Rough endoplasmic reticulum: endoplasmic reticulum lined on the outside with ribosomes.

Runner: a horizontally growing stem that may develop new plants at nodes that touch the soil.

Saccule (sak'-ūle): a portion of the vestibular system responsible for detecting gravity.

Saltatory conduction (sal'-ta-tōr-ē): literally "jumping conduction", the transmission of an action potential along a myelinated axon, during which the action potential seems to jump from one node to the next.

Saprobe (sap'-rōbe): an organism that derives its nutrients from the bodies of dead organisms.

Sarcodine (sar-kō'-dīn): a category of nonphotosynthetic protist (protozoa) characterized by the ability to form pseudopodia. Some, such as amoebae, are naked, while others have elaborate shells.

Sarcomere (sark'-ō-mēr): the unit of contraction of a muscle fiber; a subunit of the myofibril, consisting of actin and myosin filaments and bounded by Z-lines.

Sarcoplasmic reticulum (sark'-ō-plas'-mik re-tik'-ū-lum): specialized endoplasmic reticulum found in muscle cells. The sarcoplasmic reticulum stores calcium ions and releases them into the interior of the muscle cell to initiate contraction.

Saturated: referring to a fatty acid with as many hydrogen atoms as possible bonded to the carbon backbone; a fatty acid with no double bonds in its carbon backbone.

Savanna: a transition biome between the tropical deciduous forest and tropical grassland, characterized by a nearly continuous cover of grasses with occasional drought-resistant trees. The savanna climate includes such a severe dry season that most tree species cannot grow.

Sclera (sklāra): a tough white connective tissue layer that covers the outside of the eyeball and forms the white of the eye.

Sclerenchyma (skler-en'-ki-ma): a plant cell type with thick secondary cell walls, that usually dies as the last stage of differentiation, and provides both support and protection for the plant body.

Scramble competition: direct interactions between individuals attempting to acquire the same limited resource.

Scrotum (skrō'-tum): the pouch of skin containing the testes of male mammals.

S-curve: the growth curve that describes a population introduced into a new area. It consists of an initial period when numbers remain relatively low, followed by a period of exponential growth, followed by decreasing growth rate, and finally, relative stability.

Sebaceous gland (se-bā'-shus): a gland in the dermis of skin, formed from epithelial tissue, that produces an oily substance called sebum that lubricates the epidermis.

Second Law of Thermodynamics: a principle of physics that any change in an isolated system causes the quantity of concentrated, useful energy to decrease, and the amount of randomness and disorder (entropy) to increase.

Second messenger: a term applied to intracellular chemicals, such as cyclic AMP, that are synthesized or released within a cell in response to the binding of a hormone or neurotransmitter (the first messenger) to receptors on the cell surface. Second messengers bring about specific changes in the metabolism of the cell.

Secondary cell wall: a thick layer of cellulose and other polysaccharides secreted by certain plant cells between the primary cell wall and the plasma membrane.

Secondary consumer: an organism that feeds on primary consumers; a carnivore.

Secondary growth: growth in diameter of a stem or root due to cell division in lateral meristems.

Secondary oocyte (ō'-ō-sīt): a large haploid cell derived by meiosis I from the diploid primary spermatocyte.

Secondary phloem: phloem produced from the vascular cambium.

Secondary structure: a repeated, regular structure assumed by protein chains, held together by hydrogen bonds; usually either a helix or a pleated sheet.

Secondary succession: succession that occurs after an existing community is destroyed, in an environment modified by that community: for example, after a forest fire.

Secondary xylem: xylem produced from cells arising at the inside of the vascular cambium.

Sedimentary cycle: a nutrient cycle for a substance whose primary reservoir is sediment, such as soil or rock.

Seed: the reproductive stage of conifers and flowering plants, usually including an embryonic plant and a food reserve, enclosed within a resistant outer covering.

Seed coat: the outermost covering of a seed, formed from the integuments of the ovule.

Segmentation (seg-men-tā'-shun): division of the body into repeated, often similar units.

Segmentation movements: asynchronous contractions of the small intestine that result in mixing of the partially digested food and digestive enzymes. The movements also bring nutrients into contact with the absorptive intestinal wall.

Segregation: a principle of inheritance, that the two alleles of a given gene separate from each other during gamete formation, with the result that each gamete receives one allele; occurs because of the separation of homologous chromosomes during meiosis.

Self-fertilization: union of sperm and egg from the same individual.

Semen: the sperm-containing fluid produced by the male reproductive tract.

Semicircular canal: one of three fluid-filled, semicircular tubes of the inner ear which functions in the detection of rotational movements of the head.

Semiconservative replication: the process of replication of the DNA double helix; the two DNA strands separate, and each is used as a template for the synthesis of a complementary DNA strand. Each daughter double helix therefore consists of one parental strand and one new strand.

Semilunar valve: the type of valve between the ventricles of the heart and the pulmonary artery and aorta. Semilunar valves prevent backflow of blood into the ventricles when they relax.

Seminal vesicle: in male mammals, a gland that produces a basic, fructose-containing fluid that forms part of the semen.

Seminiferous tubules (semi-i-ni′-fer-us): a series of tubes in the vertebrate testis in which sperm are produced.

Senescence: in plants, a specific aging process, often including deterioration and dropping of leaves and flowers.

Sensory neuron: a nerve cell that carries information about internal or external environmental conditions to the central nervous system.

Sensory receptor: a cell specialized to respond to particular internal or external environmental stimuli by producing an electrical potential.

Sepal (sē′-pul): one of the protective outer coverings of a flower bud, often opening into green, leaf-like structures when the flower blooms.

Septum (pl. septa): a partition that separates the fungal hypha into individual cells. Pores in the septa allow transfer of materials between cells.

Serotonin (ser-uh-tō′-nin): a neurotransmitter in the central nervous system, which is involved in mood, sleep, and the inhibition of pain.

Sertoli cell: a large cell in the seminiferous tubule that regulates spermatogenesis and nourishes the developing sperm.

Sessile (ses′-ul): not free to move about, usually permanently attached to a surface.

Sex chromosome: one of the pair of chromosomes that differ between the sexes and usually determine the sex of an individual; e.g., human females have similar sex chromosomes (XX) while males have dissimilar ones (XY).

Sex linkage: a pattern of inheritance characteristic of genes located on one type of sex chromosome (e.g., X) and not found on the other type (e.g., Y) (see also *X linkage*).

Sex-influenced inheritance: a mode of inheritance in which traits of a nonsexual nature are more common in one sex than in the other, often due to differing levels of sex hormones.

Sex-linked inheritance: inheritance of traits controlled by genes carried on the X chromosome. Females show the dominant trait unless they are homozygous recessive, whereas males will express whatever allele is found on their single X chromosome.

Sexual recombination: the formation of new combinations of alleles owing to inheritance of chromosomes from two different parental organisms during sexual reproduction.

Sexual reproduction: a form of reproduction in which genetic material from two parental organisms is combined. In eukaryotes, two haploid gametes fuse to form a diploid zygote.

Sexual selection: a type of natural selection in which the choice of mates by one sex is the selective agent.

Shoot: all the parts of a vascular plant exclusive of the root. Usually aboveground, consisting of stem, leaves, and reproductive structures.

Short-day plant: a plant that will flower only if the length of daylight is shorter than some species-specific duration.

Sickle-cell anemia: a recessive disease caused by a single amino acid substitution in the hemoglobin molecule. Sickle-cell hemoglobin molecules tend to cluster together in long chains, distorting the red blood cell shape and causing them to break and clog the capillaries.

Sieve plate: a part of the cell wall between two sieve-tube elements in phloem, with large pores interconnecting the cytoplasm of the elements.

Sieve tube: in phloem, a single strand of sieve-tube elements, which transports sugar solutions.

Sieve-tube element: one of the cells of a sieve tube.

Simple diffusion: diffusion of water, dissolved gases, or lipid-soluble molecules through the phospholipid bilayer of a membrane.

Single covalent bond: a covalent bond that occurs when two atoms share a single pair of electrons.

Sinoatril node (sī′-nō-āt′-rē-ul nōd): also called the SA node, a small mass of specialized muscle in the wall of the right atrium. It generates electrical signals rhythmically and spontaneously and serves as the heart's pacemaker.

Skeletal muscle: also called striated or voluntary muscle; the type of muscle that is attached to and moves the skeleton.

Skeleton: a supporting structure for the body, upon which muscles act to change the body configuration.

Smooth endoplasmic reticulum: endoplasmic reticulum without ribosomes.

Smooth muscle: type of muscle found around hollow organs, such as the digestive tract, bladder, and blood vessels, normally not under voluntary control.

Sodium channel: a protein that spans a nerve cell membrane, forming a pore, and which allows only sodium ions to flow through the pore.

Sodium–potassium pump: an enzyme-like protein in nerve cell membranes which actively transports potassium ions into the cell and sodium ions out of the cell.

Solvent: a liquid that is capable of dissolving (uniformly dispersing) other substances in itself.

Spawning: a method of external fertilization in which male and female parents shed gametes into the water, and sperm must swim through the water to reach the eggs.

Speciation: the process whereby two populations achieve reproductive isolation.

Species (spē-cēs): a group of organisms within a genus that are potentially capable of interbreeding under natural conditions, or, if asexually reproducing, are more closely related to one another than to other organisms within the genus.

Specific heat: the amount of energy required to raise the temperature of 1 gram of a substance 1° C.

Sperm: the haploid male gamete, usually small, motile, and containing little cytoplasm.

Spermatid: a haploid cell derived from the secondary spermatocyte by meiosis II. The mature sperm is derived from the spermatid by differentiation.

Spermatogenesis: the formation of sperm.

Spermatogonium (pl. spermatogonia): a diploid cell lining the walls of the seminiferous tubules that gives rise to a primary spermatocyte.

Spicule (spik′-ūle): subunits of the endoskeleton of sponges, made of protein, silica, or calcium carbonate.

Spinal cord: part of the central nervous system of vertebrates, extending from the base of the brain to the hips, protected by the vertebrae of the spine; contains the cell bodies of motor neurons innervating skeletal muscles, the circuitry for some simple reflex behaviors, and axons communicating with the brain.

Spindle apparatus: a structure found in dividing eukaryotic cells, composed of microtubules, which appears to guide the movement of chromosomes to opposite poles of a cell during anaphase of meiosis and mitosis.

Spiracles (spī′-re-kul): openings in the abdominal segments of insects through which air enters the tracheae.

Spirilla (spī′-rilla): a spiral shaped bacterium.

Spongy cells: irregularly-shaped cells located just above the lower epidermis of a leaf.

Spontaneous generation: the proposal that living organisms can arise from nonliving matter.

Sporangium (spōr-an'-gē-um; pl. sporangia): a structure in which spores are produced.

Spore (spōr): in the alternation of generation life cycle of plants, a haploid cell produced through meiosis, that then undergoes repeated mitotic divisions and differentiation of daughter cells to produce a multicellular, haploid organism called the gametophyte. Also, a resistant or resting structure that disperses readily and withstands unfavorable environmental conditions.

Sporophyte (spor'-ō-fīt): the multicellular diploid stage of the life cycle of plants that produces haploid, asexual spores through meiosis..

Sporozoan (spōr-ō-zō'-en): a category of parasitic protist. Sporozoans have complex life cycles often involving more than one host, and are named for their ability to form infectious spores. A well-known member (genus *Plasmodium*) causes malaria.

Stabilizing selection: a type of natural selection in which those organisms displaying extreme phenotypes are selected against.

Stamen (stā'-men): the male reproductive structure of a flower, consisting of a filament and an anther in which pollen grains develop.

Starch: a polysaccharide composed of branched or un-branched chains or glucose molecules, used by plants as a carbohydrate-storage molecule.

Start codon: a codon in messenger RNA that signals the beginning of protein synthesis on a ribosome.

Startle coloration: a color pattern (often resembling large eyes) that can be displayed suddenly by a prey organism when approached by a predator.

Stem: the normally vertical, aboveground part of a plant body that bears leaves.

Steroid: a lipid composed of four fused rings of carbon atoms to which functional groups are attached.

Stigma (stig'-ma): the pollen-capturing tip of a carpel.

Stoma (stō'-ma; pl. stomata): an adjustable opening in the epidermis of a leaf, surrounded by a pair of guard cells. Most gas exchange between leaves and the air occurs through the stomata.

Stop codon: a codon in messenger RNA that stops protein synthesis and causes the completed protein chain to be released from the ribosome.

Striated muscle (strī'-ā-ted): muscle with a striped appearance due to the orderly arrangement of thick and thin filaments; both skeletal and cardiac muscle are striated.

Stroke: an interruption of blood flow to part of the brain, caused by the rupture of an artery or the blocking of an artery by a blood clot. Loss of blood supply leads to rapid death of the area of the brain affected.

Stroma (strō'-ma): the semi-fluid material of chloroplasts in which the grana are embedded.

Structural gene: a gene that codes for a protein used by the cell for purposes other than gene regulation. Structural genes code for enzymes or for proteins that are structural parts of a cell.

Style: a stalk connecting the stigma of a carpel with the ovary at its base.

Subatomic particle: the particles of which atoms are made: electrons, protons, and neutrons.

Subclimax: a community where succession is arrested before the climax community is reached, usually maintained by regular disturbance: for example, tallgrass prairie maintained by regular fires.

Substrate: the atoms or molecules that are the reactants for an enzyme-catalyzed chemical reaction.

Subunit: a small organic molecule, several of which may be bonded together to form a larger molecule. See also monomer.

Succession (suk-se'-shun): the directional change in the community structure of an ecosystem over time. Community changes alter the ecosystem in ways that favor competitors, and species replace one another in a somewhat predictable manner until a stable climax community is reached.

Sucrose: a disaccharide composed of glucose and fructose.

Sugar: a simple carbohydrate molecule, either a monosaccharide or a disaccharide.

Summation: a steady increase in the degree of contraction of a muscle in response to rapid firing by its motor neuron.

Supernormal stimulus: a stimulus that exaggerates crucial elements of the releaser, making it more effective than the normal releaser.

Suppressor T cell: a type of T cell that depresses the response of other immune cells to foreign antigens.

Surface tension: the property of a liquid to resist penetration by objects at its interface with the air, due to cohesion between molecules of the liquid.

Survivorship curve: a curve (convex, constant, or concave) resulting when the number of individuals in a population is graphed against their age, usually expressed as a percentage of their maximum life span.

Symbiosis (sim'-bī-ō'sis): a close association between different species usually over an extended period. *Symbiosis* as used in this book includes mutualism, commensalism, and parasitism.

Sympathetic nervous system: the division of the autonomic nervous system that produces largely involuntary responses that prepare the body for stressful situations.

Sympatric speciation (sim-pat'-rik): speciation of populations that are not physically divided; usually due to ecological isolation or chromosomal abnormalities (e.g., polyploidy).

Synapse (sin'-apz): the site of communication between nerve cells. One cell (presynaptic) usually releases a chemical (the neurotransmitter) that changes the electrical potential of the second (postsynaptic) cell.

Synaptic cleft: a narrow space in a chemical synapse that separates the pre- and postsynaptic cells.

Synaptic potential: at a synapse between two nerve cells, the electrical change occurring in the postsynaptic cell as a result of reception of neurotransmitter released by the presynaptic cell.

Synaptic terminal: the terminal branches of an axon, where the axon forms a synapse, usually with the dendrites of the second nerve cell.

Syphilis (si'-ful-is): a sexually transmitted bacterial infection of the reproductive organs which, if untreated, can damage the nervous and circulatory systems.

Systole (sis'-tō-lē): the portion of the heart cycle during which the ventricles contract; the higher of the two blood pressure readings.

T cell: a type of lymphocyte that recognizes and destroys specific foreign cells or substances, or that regulates other cells of the immune system.

T tubules: deep infoldings of the muscle cell membrane that conduct the action potential inside the cell.

Taiga (tī'-ga): the cold coniferous forest biome, characterized by needleleaf evergreen trees. The waxy needles of the trees reduce evaporative water loss during winter.

Taproot system: a root system commonly found in dicots, consisting of a long, thick main root that develops from the primary root, and numerous smaller lateral roots that grow from the primary root.

Target cell: a cell upon which a particular hormone exerts its effect.

Taste: a chemical sense; in mammals, perceptions of sweet, sour, bitter, or salt produced by stimulation of receptors on the tongue.

Taxis: (tax′-is): innate movement of an organism toward or away from a stimulus such as heat, light, or gravity.

Taxonomy (tax-on-uh-mē): the science by which organisms are classified into hierarchically arranged categories that reflect their evolutionary relationships.

Tay–Sachs disease: a recessive disease caused by a deficiency in enzymes regulating lipid metabolism in the brain.

Tectorial membrane (tek-tōr′-ē-ul): one of the membranes of the cochlea, in which the hairs of the hair cells are embedded. During sound reception, movement of the basilar membrane relative to the tectorial membrane bends the cilia.

Telophase (tēl′-ō-fāz): the last stage of meiosis and mitosis, in which a nuclear envelope reforms around each new daughter nucleus, the spindle disappears, and the chromosomes relax from their condensed form.

Temperate deciduous forest: a biome having cold winters but a fairly long, moist growing season. The dominant vegetation is broadleaf deciduous trees, which shed their leaves in winter as a protection against water loss.

Temperate rain forest: a biome of very limited distribution, along certain coasts where the climate is fairly mild year round and rainfall is very high, characterized by coniferous evergreen trees, often of enormous size.

Temporal isolation: the inability of organisms to mate if they have significantly different breeding seasons.

Temporal lobe: part of a cerebral hemisphere of the human brain, involved in recall of learned events.

Tendon: a tough connective tissue band connecting a muscle to a bone.

Tendril: a slender outgrowth of a stem that coils about external objects and supports the stem; usually a modified leaf or branch.

Tentacle (ten′-te-kul): an elongate, extensible projection of the body of cnidarians and cephalopod mollusks that may be used for grasping, stinging, and immobilizing prey, and locomotion.

Terminal bud: the bud at the extreme end of a stem or branch.

Territoriality: the defense of an area in which important resources are located.

Tertiary consumer (ter′-shē-ār-ē): A carnivore that feeds on other carnivores.

Tertiary structure (ter′-shē-ār-ē): the complex three-dimensional structure of a single peptide chain. The tertiary structure is held in place by disulfide bonds between cysteine amino acids, by attraction and repulsion among amino acid side groups, and by interaction between the cellular environment (water or lipids) and the amino acid side groups.

Test cross: a breeding experiment in which an individual showing the dominant phenotype is mated with an individual that is homozygous recessive for the same gene. The ratio of offspring with dominant versus recessive phenotypes can be used to determine the genotype of the phenotypically dominant individual.

Testis (pl. testes): the gonad of male mammals.

Testosterone: in vertebrates, a hormone produced by the interstitial cells of the testis; stimulates spermatogenesis and the development of male secondary sex characteristics.

Tetany: smooth, sustained contraction of a muscle in response to rapid firing by its motor neuron.

Thalamus: part of the forebrain, the thalamus serves as a relay network between other parts of the nervous system and the cerebrum.

Theory: in science, an explanation for natural events that is based on a large number of observations and is in accord with scientific principles, especially causality.

Thermoacidophile (ther-mō-a-sid′-eh-fīl): a form of archaebacterium that thrives in hot, acidic environments.

Thermoreceptor: a sensory receptor that responds to changes in temperature.

Thick filaments: bundles of myosin protein within the sarcomere which interact with thin filaments to produce muscle contraction.

Thin filaments: proteinaceous strands within the sarcomere which interact with thick filaments to produce muscle contraction. Composed primarily of actin, with the accessory proteins tropomyosin and troponin.

Thorn: a hard, pointed outgrowth of a stem; usually a modified branch.

Threshold: the electrical potential (less negative than the resting potential) at which an action potential is initiated.

Thrombin: an enzyme produced in the blood as a result of injury to a blood vessel. Thrombin catalyzes the production of fibrin, a protein that assists in blood clot formation.

Thylakoid (thī′-la-koyd): a disk-shaped, membranous sac found in chloroplasts, the membranes of which contain chlorophyll and the photosystems and ATP-synthesizing enzymes used in the light-dependent reactions of photosynthesis.

Thymus (thī′-mus): a gland located in the upper chest in front of the heart. The thymus functions in the immune system by secreting thymosin, which stimulates lymphocyte maturation. It begins to degenerate at puberty and has little function in the adult.

Thyroid: an endocrine gland, located in front of the larynx in the neck, that secretes the hormones thyroxine (affecting metabolic rate) and calcitonin (regulating calcium ion concentration in the blood).

Tight junction: a type of cell-to-cell junction in animals that prevents the movement of materials through the spaces between cells.

Tissue: a group of (usually similar) cells that together carry out a specific function, i.e., muscle. The tissue may include extracellular material produced by its cells.

Trachea (trā′-kē-uh): a rigid but flexible tube supported by rings of cartilage, which conducts air between the larynx and the bronchi.

Tracheae (trā′-kē): elaborately branching tubes that ramify through the bodies of insects and some arachnids and carry air close to each body cell. Air enters the tracheae through openings called spiracles.

Tracheid (tra′-kē-id): an elongated xylem cell with tapering ends containing pits in the cell walls; forms tubes that transport water.

Tracheophyte (trā′-kē-ō-fīt): a category of plant that has conducting vessels; a vascular plant.

Transcription: the synthesis of an RNA molecule from a DNA template.

Transducer: a device that converts signals (forms of energy, for example) from one form to another. Sensory receptors are transducers that convert environmental stimuli, such as heat, light, vibration, etc. into electrical signals (such as action potentials) recognized by the nervous system.

Transfer RNA (tRNA): a type of RNA that (1) binds to a specific amino acid and (2) bears a set of three nucleotides (the anticodon) complementary to the mRNA codon for that amino acid. Transfer RNA carries its amino acid to a ribosome during protein synthesis, recognizes a codon of mRNA, and positions its amino acid for incorporation into the growing protein chain.

Transformation: a method of genetic recombination whereby DNA from one bacterium (usually released after the death of the bacterium) becomes incorporated into the DNA of another, living, bacterium.

Translation: the process whereby the sequence of nucleotides of messenger RNA is converted into the sequence of amino acids of a protein.

Transpiration (trans'-per-ā-shun): evaporation of water from a leaf.

Transport protein: a protein in a plasma membrane that binds specific molecules and facilitates their transport across the membrane. See also *channel protein, facilitated diffusion,* and *active transport.*

Trial-and-error learning: a process by which adaptive responses are learned through rewards or punishments provided by the environment.

Trichocyst (trik'-eh-sist): a stinging organelle of protists.

Triglyceride: a lipid composed of three fatty acid molecules bonded to a single glycerol molecule.

Triple covalent bond: a covalent bond that occurs when two atoms share three pairs of electrons.

Trisomy 21: see *Down syndrome.*

Trisomy X: a condition of females who have three X chromosomes instead of the normal two. Most of these women are phenotypically normal, and are fertile.

Tropical deciduous forest: a broadleaf deciduous forest biome found in areas with a warm climate year round but a pronounced wet and dry season. The trees drop their leaves in the dry season.

Tropical rain forest: a broadleaf evergreen forest biome found in the tropics, where rainfall is adequate and temperatures are warm year round. The tropical rain forest has the greatest diversity of both plants and animals to be found anywhere on land.

Tropomyosin (trō-pō-mī'-ō-sin): a double-stranded accessory protein found in thin filaments which, in the relaxed muscle, covers the binding sites on the actin molecules and prevents contraction.

Troponin (trō-pō'-nin): a globular accessory protein in thin filaments that anchors the tropomyosin. During muscle contraction, troponin temporarily binds calcium, changes its conformation, and pulls tropomyosin off the binding sites of the actin molecules.

True-breeding: pertaining to an individual all of whose offspring produced through self-fertilization are identical to the parental type. True-breeding individuals are homozygous for the trait in question.

Tube cell: the outermost cell of a pollen grain; the tube cell digests a tube through the tissues of the carpel, ultimately penetrating into the female gametophyte.

Tube feet: cylindrical extensions of the water-vascular system of echinoderms, used for locomotion, grasping food, and respiration.

Tubular reabsorption: the process by which cells of the tubule of the nephron remove water and nutrients from the filtrate within the tubule and return them to the blood.

Tubular secretion: the process by which cells of the tubule of the nephron remove additional wastes from the blood, actively secreting them into the tubule.

Tubule (tūb'-ūle): the tubular portion of the nephron. It includes a proximal portion, the loop of Henle, and a distal portion. Urine is formed from the blood filtrate as it passes through the tubule.

Tumor-suppressor gene: a gene that encodes information for a protein that inhibits cancer formation, probably by regulating cell division in some way.

Tundra: a treeless biome characterized by permafrost and an extremely short growing season. Most of the plants are low-growing perennial shrubs, grasses, and wildflowers.

Turgor pressure: pressure developed within a cell (especially the central vacuole of plant cells) as a result of osmotic water entry.

Turner's syndrome: a set of characteristics typical of a woman with only one X chromosome. These women are sterile, failing to develop normal ovaries. They also tend to be very short and to lack normal female secondary sexual characteristics.

Tympanic membrane (tim-pan'-ik): the eardrum; a membrane stretched across the opening of the ear, which transmits vibration of sound waves to bones of the middle ear.

Unicellular: a term describing an organism whose body consists of a single cell.

Uniform distribution: a relatively regular spacing of individuals within a population, often as a result of territorial behavior.

Uniformitarianism: the hypothesis that the Earth developed gradually through natural forces similar to those at work today.

Unsaturated: referring to a fatty acid with fewer than the maximum number of hydrogen atoms bonded to its carbon backbone; a fatty acid with one or more double bonds in its carbon backbone.

Upwelling: movement of deep, nutrient-rich water to the surface.

Urea (ū-rē'-uh): a water-soluble, nitrogen-containing waste product of amino acid breakdown that is one of the principal components of mammalian urine.

Ureter (ū'-re-tur): a tube that conducts urine from each kidney to the bladder.

Urethra (ū-rē'-thra): the tube leading from the urinary bladder to the outside of the body; in males, the urethra also receives sperm from the vas deferens and conducts both sperm and urine (at different times) to the tip of the penis.

Uric acid (ūr'-ik acid): a nitrogen-containing waste product of amino acid breakdown, which is a relatively insoluble white crystal. Uric acid is excreted by birds, reptiles, and insects.

Uterus: in female mammals, the part of the reproductive tract that houses the embryo during pregnancy.

Utricle (ū'-tri-cul): a portion of the vestibular system responsible for detecting gravity.

Vaccine: a material injected into the body that contains antigens characteristic of a particular disease organism, and that stimulates an immune response.

Vacuole (vak'-ū-ōl): a vesicle, often large, consisting of a single membrane enclosing a fluid-filled space.

Vagina: the passageway leading from the outside of the body to the cervix of the uterus.

Variable: a condition, particularly in a scientific experiment, that is subject to change.

Variable region: part of an antibody molecule that differs among antibodies; the ends of the variable regions of the light and heavy chains form the specific binding site for antigen.

Vas deferens (vas de'-fer-ens): the tube connecting the epididymis of the testis with the urethra.

Vascular bundle (vas'-kū-lar): a strand of xylem and phloem found in a leaf; commonly called a vein.

Vascular cambium: a lateral meristem located between the xylem and phloem of a woody root or stem.

Vascular cylinder: the centrally located conducting tissue of a young root, consisting of primary xylem and phloem.

Vascular tissue system: a plant tissue system consisting of xylem (which transports water and minerals from root to shoot) and phloem (which transports water and sugars throughout the plant).

Vein: a large-diameter, thin-walled vessel that carries blood from venules back to the heart.

Ventral (ven'-trul): the lower, or underside of an animal whose head is oriented forward.

Ventricle (ven'-trē-kul): the lower muscular chamber on each side of the heart, which pumps blood out through the arteries. The right ventricle sends blood to the lungs, and the left to the rest of the body.

Venule (ven'-yul): a narrow vessel with thin walls that carries blood from capillaries to veins.

Vertebrate: an animal that possesses a vertebral column.

Vesicle (ves'-i-kul): a small membrane-bound sac within the cytoplasm.

Vessel: a tube of xylem composed of vertically stacked vessel elements, with perforated or missing end walls, leaving a continuous, uninterrupted hollow cylinder.

Vessel element: one of the cells of a xylem vessel; elongated, dead at maturity, with thick lateral cell walls for support but with end walls either lacking entirely or heavily perforated.

Vestibular apparatus (ves-tib'-ū-lar): the portion of the inner ear of mammals that responds to gravity and changes in the direction and speed of movement.

Vestigial structures (ves-tij'-ē-ul): structures with no known function, but which are homologous to functional structures in related organisms.

Villus (vi'-lus): fingerlike projections of the wall of the small intestine that increase its absorptive surface area.

Viroid (vī'-röyd): a particle of RNA capable of infecting a cell and directing the production of more viroids; responsible for certain plant diseases.

Virus (vī'-rus): a noncellular parasitic particle consisting of a protein coat surrounding a strand of genetic material. Viruses can multiply only within the cells of living organisms.

Vitamin: any one of a group of diverse chemicals that must be present in trace amounts in the diet to maintain health. Vitamins are used by the body in conjunction with enzymes in a variety of metabolic reactions; a few are involved in growth and differentiation.

Vitreous humor (vit'-rē-us): a clear jellylike substance that fills the large chamber of the eye between the lens and retina.

Waggle dance: symbolic communication used by honeybee foragers to communicate the location of a food source to their hivemates.

Water-vascular system: a system in echinoderms consisting of a series of canals through which seawater is conducted and used to inflate tube feet for locomotion, grasping food, and respiration.

Wax: a lipid composed of fatty acids covalently bonded to long-chain alcohols.

Weather: fluctuations in temperature, humidity, precipitation, and cloud cover occurring over a period of hours or days.

X-linkage: referring to a gene carried on the X chromosome, in organisms in which the X chromosome carries many non–sex-related genes, while the Y chromosome carries few if any homologous genes (see also *sex linkage*).

X-linked inheritance: see *sex-linked inheritance*.

Xylem (zī'-lum): a conducting tissue of vascular plants that transports water and minerals from root to shoot.

Yolk: protein- or lipid-rich substances contained in eggs as food for the developing embryo.

Yolk sac: one of the embryonic membranes of reptile, bird, and mammalian embryos. In birds and reptiles, the yolk sac is a membrane surrounding the yolk in the egg. In mammals, the yolk sac is empty, but forms part of the umbilical cord and the gut.

Z lines: fibrous protein structures to which the thin filaments of skeletal muscle are attached, forming the boundaries of sarcomeres.

Zona pellicida (pel-ū'-si-da): a clear, noncellular layer between the corona radiata and the egg.

Zooflagellate (zō-ō-fla'-gel-et): a category of nonphotosynthetic protist that move using flagella.

Zooplankton (zō'-plank-ton): a diverse assemblage of small pelagic protists and animals dominated by crustaceans.

Zoospore (zō-ō-spōr): a nonsexual reproductive cell that swims using flagella.

Zygospore (zī'-gō-spōr): produced by the division Zygomycota, a fungal spore surrounded by a thick, resistant wall, which forms from a diploid zygote.

Zygote (zī'-gōt): in sexual reproduction, a diploid cell formed by the fusion of two haploid cells: the fertilized egg.

Photo Acknowledgments

5.12b © Dr. Barry King, Department of Human Anatomy, School of Medicine, University of California at Davis/BPS

5.13 © D. W. Fawcett/Photo Researchers, Inc.

5.14 Zdenek Hruban, University of Chicago

E5.6 © Dr. Don Fawcett/T. Kanaseki/Photo Researchers, Inc.

5.16 © Micrograph by W. P. Wergin, Courtesy of E. H. Newcomb, University of Wisconsin/BPS

5.17 © Keith R. Porter/Laboratory of Cell Biology, University of Maryland

5.18 © Biophoto Associates/Photo Researchers, Inc.

5.19a © Dr. Gopal Murti/SPL/Science Source/Photo Researchers, Inc.

5.19b © Dr. Peter Dawson/SPL/Science Source/Photo Researchers, Inc.

5.19c © Dr. Gopal Murti/SPL/Science Source/Photo Researchers, Inc.

5.20 Keith R. Porter, Laboratory of Cell Biology, University of Maryland

5.21a © W. L. Dentler/University of KS/BPS

5.21b © Don W. Fawcett/Photo Researchers, Inc.

Chapter 6

p. 116 © Eric Grave/Science Source/Photo Researchers, Inc.

6.1 © Larry Ulrich/DRK Photo

6.2 © Dr. G. F. Leedale, Biophoto Associates/Photo Researchers, Inc.

6.8a © Dr. Joseph Kurantsin-Mills

6.8b © Dr. Joseph Kurantsin-Mills

6.8c © Dr. Joseph Kurantsin-Mills

6.9a © Dr. Thomas Eisner

6.9b © Dr. Thomas Eisner

6.13a–d © From Perry, M. M., & Gilbert, A. B., 1979, *Journal of Cell Science*, 39:257–272.

6.14a–c © Eric V. Grave/Photo Researchers, Inc.

6.15 From, Hufnagel, L. A., 1991, "Ultrastructural Aspects of Chemoreception in Ciliated Protists (Ciliaphora)," *Journal of Electrical Microscience and Technology*, in press. Photomicrograph by Jurgen Bohmer and Linda Hufnagel, University of Rhode Island

6.16 © G. Raviola/Don Fawcett/P. Eliase/D. Friend/Photo Researchers, Inc.

6.17 © L. Andrew Staehelin

6.18 © Dr. Don Fawcett/K. Hama/D. Albertini Photo Researchers, Inc.

6.19 © W. P. Wergin/University of Wisconsin/BPS

E6.1 © S. J. Kraseman/DRK Photo

Chapter 7

p. 140 © John Netherton/The Image Bank

7.2a © Ken W. Davis/Tom Stack & Associates

7.2b Micrograph by W.P. Wergin, Courtesy of E. H. Newcomb, University of Wisconsin/BPS

7.2c Micrograph by W. P. Wergin, Courtesy of E. H. Newcomb, University of Wisconsin/BPS

Chapter 8

p. 156 © Marcia W. Griffen/Animals/Animals

8.3a © Douglas C. Pizac/AP Wide World Photos

8.3b © Jonathan Watts/SPL/Science Source/Photo Researchers, Inc.

Unit II Opener, p. 172–173 © **Lennart Nilsson**

Chapter 9

p. 174 © CNRI/SPL/Photo Researchers, Inc.

9.3 © Dr. Gopal Murti/SPL/Photo Researchers, Inc.

9.4 © CNRI/SPL/Photo Researchers, Inc.

9.6 © Biophoto Associates/Photo Researchers, Inc.

9.7 © Audesirk

9.8a–h © Dr. Andrew S. Bayer/University of Oregon

9.10a © Barry King/University of California, Davis/BPS

9.10b © John D. Cunningham/Visuals Unlimited

9.13 © T. E. Schroeder, University of Washington/BPS

9.14a © Biophoto Associates/Photo Researchers, Inc.

9.15a © David M. Phillips/Visuals Unlimited

9.15b © Carolina Biological Supply Co.

9.16 © Audesirk

Chapter 10

p. 192 © Frans Lanting/Minden Pictures

10.4 © James Kezer/University of Oregon

Chapter 11

p. 208 © Ron Kimball

11.1 © Archiv/Photo Researchers, Inc.

11.9 © Darwin Dale/Photo Researchers, Inc.

11.11a © Carolina Biological Supply Co.

11.11b © Carolina Biological Supply Co.

11.13a © James Keger/University of Oregon

11.15 © Audesirk

11.19 © Jane Burton/Bruce Coleman, Inc.

Chapter 12

p. 238 © Loren Williams/MIT

12.3 © Thomas Broker/CNRI/Phototake, Inc.

E12.1 © Photo Researchers, Inc.

12.6 © Rosalind Franklin/Science Source/Photo Researchers, Inc.

Chapter 13

p. 260 © E. V. Kiseleva, 1989, FEBS LETTS, 257:251–153.

13.5 From Miller, O. L., 1969, "Visualization of Nuclear Genes," 164:955. Copyright © 1969 by AAAS.

13.14 © Science Source/Photo Researchers, Inc.

13.15a © Dr. M. L. Barr

13.15b © Dr. M. L. Barr

13.16 © Audesirk

Chapter 14

p. 288 © Granada Bioscience, Inc.

14.1 © Prof. S. Cohen/SPL/Photo Researches, Inc.

14.4 © Matt Meadows/Peter Arnold, Inc.

14.6 © Hank Morgan/Photo Researchers, Inc.

E14.2a © Gary Braasch Photography

E14.2b © Joseph R. Newhouse, Ph.D.

14.8 © Applied Biosystems/Peter Arnold, Inc.

14.9 © Steve Grove/The Bettman Archive

Chapter 15

p. 308 © Nicholas Descoise/Photo Researchers, Inc.

E15.1 © Audesirk

15.3a © Marlene & Bob Lippman

15.4a © Omikron/Science Source/Photo Researchers, Inc.

15.4b © Omikron/Science Source/Photo Researchers, Inc.

15.5a © The Bettman Archive

15.5b From Folstein, S. E., 1989, *Huntington's Disease: A Disorder of Families*, Baltimore, Md.: Johns Hopkins University Press, Fig. 5.6.

15.7 © Mary Evans Picture Library/Photo Researchers, Inc.

15.10 © Lawrence Migdale Photography

21.23 © Dr. S. L. Erlandsen & D. E. Feely

21.24 © M. I. Walker/Photo Researchers, Inc.

21.25 © M. I. Walker/Photo Researchers, Inc.

21.26a © M. I. Walker/Photo Researchers, Inc.

21.26b © Eric Grave/Photo Researchers, Inc.

21.29a © Biophoto Associates/Science Source/Photo Researchers, Inc.

21.29b © Biophoto Associates/Science Source/Photo Researchers, Inc.

21.19c © Biophoto Associates/Science Source/Photo Researchers, Inc.

Chapter 22

p. 460–461 © Kjell B. Sandved

22.4a © Audesirk

22.4b © Brian Parker/Tom Stack & Associates

22.4c © Charles Seaborn/Odyssey/Chicago

22.6a © Stephen J. Krasemann/DRK Photo

22.6b © Charles Seaborn/Odyssey/Chicago

22.6c © Audesirk

22.6d © John D. Cunningham/Visuals Unlimited

22.9 © Carolina Biological Supply Co.

22.10 © Martin M. Rotker/Phototake

22.10 © S. Flegler/Visuals Unlimited

22.11 © Tom E. Adams/Peter Arnold, Inc.

22.12a © Carolina Biological Supply Co.

22.12b Courtesy of Howard Shiang, D.V.M./American Veterinarian Med. Assoc.

22.14a © Kjell B. Sandved

22.14b © David Bull

22.14c © J. H. Robinson/Photo Researchers, Inc.

22.15 © Audesirk

22.16 © Dwight Kuhn

22.18 © Peter Scharf/Peter Arnold, Inc.

22.20a © Carolina Biological Supply Co.

22.20b P. J. Bryant, University of California, Irvine/BPS

22.20c © Stephen Dalton/Photo Researchers, Inc.

22.20d © Robert & Linda Mitchell

22.20e © Stanley Breeden/DRK Photo

22.21a © Dwight R. Kuhn

22.21b © Robert & Linda Mitchell

22.21c © Audesirk

22.22a © Tom Branch/Photo Researchers, Inc.

22.22b P. J. Bryant, University of California, Irvine/BPS

22.22c © Carolina Biological Supply Co.

22.22d © Alex Kerstitch

22.23a © Ray Coleman/Photo Researchers, Inc.

22.23b © Alex Kerstitch

22.24.a © Fred Bavendam/Peter Arnold, Inc.

22.24B © Ed Reschke/Peter Arnold, Inc.

22.25a © Jeff Rotman

22.25b © Fred Bavendam/Peter Arnold, Inc.

22.25c © Alex Kerstitch

22.26a © Audesirk

22.26b © Jeff Foott

22.26c © Chris Newbert/Bruce Coleman, Inc.

22.27b © Michael Male/Photo Researchers, Inc.

22.28 © John Giannicchi/Science Source/Photo Researchers, Inc.

22.29b © Audesirk

22.30a © Tom McHugh/Photo Researchers, Inc.

22.30b © Tom Stack/Tom Stack & Associates

22.30b © Breck P. Kent

22.31a © David Hall/Photo Researchers, Inc.

22.31b © Jeff Rotman

22.32a © Peter David/Planet Earth Pictures

22.32b © Mike Neumann/Photo Researchers, Inc.

22.32.c © Doug Perrine/DRK Photo

22.32d © Peter Scoones/Planet Earth Pictures

22.33a © Breck P. Kent/Animals Animals

22.33b © Joe Mc Donald/Tom Stack & Associates

22.33c © Cosmos Black/Photo Researchers, Inc.

22.34a © D. G. Barker/Tom Stack & Associates

22.34b © BPS

22.34c © Frans Lanting/Minden Pictures

E22.1 © Stanley Breeden/DRK Photo

22.35a © Carolina Biological Supply Co.

22.36a © Walter E. Harvey/Photo Researchers, Inc.

22.36b © Carolina Biological Supply Co.

22.36c © Ray Ellis/Photo Researchers, Inc.

22.37 © Carolina Biological Supply Co.

22.38a © Flip Nicklin/Minden Pictures

22.38b © Jonathan Watts/SNL/Photo Researchers, Inc.

22.38c © Arthus-Bertrand/Peter Arnold, Inc.

22.38d © Brian Parker/Tom Stack & Associates

22.39a © Jacana/Photo Researchers, Inc.

22.39b © Mark Newman/Auscape International

22.39b Inset © D. Parer & E. Parer-Cook/Auscape International

Chapter 23

p. 500 © Michael Fogden/DRK Photo

23.1a © Dwight R. Kuhn

23.2 © Dr. Charles Brewer-Carias

23.4a © Gregory G. Dimijian/Photo Researchers, Inc.

23.4b © Jeff Foott

23.4c © Robert & Linda Mitchell

23.5 © Stanley L. Flegler/Visuals Unlimited

E23.1 © Doisneau-Rapho/Photo Researchers, Inc.

E23.2 © G. L. Barron/University of Guelph/BPS

E23.3 © N. Allen & G. Barron/University of Guelph/BPS

23.6 Inset #1 © Carolina Biological Supply Co.

23.6 Inset #2 © Breck P. Kent

23.7 Prof. Wm. Merrill, Department of Plant Pathology/Pennsylvania State University

23.8a © W. K. Fletcher/Photo Researchers, Inc.

23.8b © D. Dvorak Jr.

23.9 © John Durham/SPL Photo Researchers, Inc.

23.10a © David M. Dennis/Tom Stack & Associates

23.10b © Scott Blackman/Tom Stack & Associates

23.10c © Scott Camazine/Photo Researchers, Inc.

23.10d © David M. Dennis/Tom Stack & Associates

23.12 © Robert & Linda Mitchell

23.13a © Audesirk

23.13b © Manfred Kage/Peter Arnold, Inc.

Chapter 24

p. 516 © Larry Ulrich/DRK Photo

24.1 © Jeff Rotman

24.2a © D. P. Wilson/Eric & David Hosking/Photo Researchers, Inc.

24.2b Courtesy Catalina Marine Science Center

24.2c © Bob Evans/Peter Arnold, Inc.

24.3a © Michael Abbey/Science Source/Photo Researchers, Inc

24.3b © Eric V. Grave/Photo Researchers, Inc.

24.5 © John Gerlach/Tom Stack & Associates

24.6 © Carolina Biological Supply Co.

24.7a © Dwight R. Kuhn

24.7b © Milton Rand/Tom Stack & Associates

24.7c © Larry Ulrich/DRK Photo

24.8 Inset © David M. Dennis/Tom Stack & Associates

24.10a © Audesirk

24.10b © A-Z Botanical Collection
24.12a © Audesirk
24.12b © Audesirk
24.13a © Dwight R. Kuhn
24.13b © Audesirk
24.13c © Audesirk
24.13d © Dwight R. Kuhn
24.13e © David Dare Parker/ Auscape International
24.13e *Inset* © Michael Jensen/Auscape International
24.15 © T. K. Kitchin/Tom Stack & Associates

Chapter 25

p. 536–537 © Gaul Shumway
25.2 © Audesirk
25.3a © J. H. Robinson/ Photo Researchers, Inc.
25.3b © Audesirk
25.3c © Jeff Foott
25.4a © Robert & Linda Mitchell
25.4b © Jeff Foott/Valan Photos
25.5 © Audesirk
25.6 © Stephen J. Krasemann/Valan Photos
25.8 © Manfred Kage/ Peter Arnold, Inc.
25.9a © Triarch/Visuals Unlimited
25.9b © Don Fawcett/ Visuals Unlimited
25.9c © David M. Phillips/ Visuals Unlimited

Unit IV Opener, p. 552–553
© Margarette Mead/Image Bank

Chapter 26

p. 554 © Michael Fogden/ DRK Photo
26.5a © Dr. Jeremy Burgess/SPL/Photo Researchers, Inc.
26.5b © John D. Cunningham/Visuals Unlimited
26.6a © Biology Media/ Photo Researchers, Inc.
26.6b © George Wilder/ Visuals Unlimited
26.6c © George Wilder/ Visuals Unlimited
26.7 © G. Shih-R. Kessel/ Visuals Unlimited

26.8 © Dr. G. F. Leedale/ Biophoto Assoc./ Photo Researchers, Inc.
26.9a © Lynwood M. Chase/Photo Researchers, Inc.
26.9b © Dwight R. Kuhn
26.10a © Ed Reschke/Peter Arnold, Inc.
26.10b © E. R. Degginger/ Earth Scenes
26.11 © Audesirk
26.12 © Biophoto Associates/Photo Researchers, Inc.
26.13 © J. R. Waeland, University of Washington/BPS
26.15a © Bruce Iverson, BSc
26.15b © Ed Reschke/Peter Arnold, Inc.
26.15c © Carolina Biological Supply Co.
26.18 *left* © Carolina Biological Supply Co.
26.18 *right* © John D. Cunningham/Visuals Unlimited
26.19 © John D. Cunningham/Visuals Unlimited
26.20 © Dr. Jeremy Burgess/SPL/Photo Researchers, Inc.
E26.1 © Biophoto Associates/Sci. Source/Photo Researchers, Inc.
E26.2 © Pat O'Hara/DRK Photo
26.21 © Audesirk
26.22 © Belinda Wright/ DRK Photo
26.23a © Audesirk
26.23b © Carolina Biological Supply Co.
26.24a © John D Cunningham/Visuals Unlimited
26.24b © Audesirk
26.24c © Audesirk
26.24d © Dwight R. Kuhn
26.24e © Audesirk
26.25a © Jeff Lepore/Photo Researchers, Inc.
26.25b © Ken W. Davis/ Tom Stack & Associates

Chapter 27

p. 580 © Michel Viard/Peter Arnold, Inc.
27.1 © Robert & Linda Mitchell
27.3a © Biophoto Associates/Science

Source/Photo Researchers, Inc.
27.3b © R. Ronacordi/ Visuals Unlimited
27.4b.1 © Hugh Spencer/ Photo Researchers, Inc.
27.4b.2 © E. H. Newcomb/ S. R. Tandon, University of Wisconsin/BPS
E27.2 © James L. Castner
27.7a © Ray Simon/Photo Researchers, Inc.
27.7b © Dr. Jeremy Burgess/SPL/Photo Researchers, Inc.
27.9 © P. Dayanandan/ Photo Researchers, Inc.
27.10a–b © Dr. Martin H. Zimmerman, Courtesy of Harvard Forest, Harvard University
E27.4 © Randall Hyman

Chapter 28

p. 600 © Larry Ullrich/DRK Photo
28.3 © Robert & Linda Mitchell
28.4b © James L. Castner
28.5 © Audesirk
28.6 © Audesirk
28.7 © James L. Castner
28.8a © Thomas Eisner
28.8b © Thomas Eisner
28.9 © John Gerlach/Tom Stack & Associates
28.10a © Merlin D. Tuttle/ Photo Researchers, Inc.
28.10b © Oxford Scientific Films/Animals Animals
28.11 © Luiz Claudio Marigo/Peter Arnold, Inc.
28.12 © Dr. J. A. Cooke/ Animals Animals
28.13a © Tom Bean/DRK Photo
28.13b © Robert & Linda Mitchell
28.14a © Carolina Biological Supply Co.
28.14b © Carolina Biological Supply Co.
28.14c © Carolina Biological Supply Co.
28.15 © CNRI/SPL/Photo Researchers, Inc.
28.17 © Carolina Biological Supply Co.
28.19b © Carolina Biological Supply Co.

28.19c © Carolina Biological Supply Co.
E28.1a © Audesirk
E28.1b © D. Cavagnaro/ DRK Photo
E28.1c © Carolina Biological Supply Co.
E28.2 © Audesirk
E28.3 © Dwight R. Kuhn
E28.4 © Jeff Lepore/Photo Researchers, Inc.
28.21 © Audesirk
E28.5 © George Holton/ Photo Researchers, Inc.

Chapter 29

p. 624 © Zig Leszczynski/ Animals Animals
E29.3a © Dwight R. Kuhn
29.5 © Runk Schoenberger/Grant Heilman
E29.1a © Carolina Biological Supply Co.
E29.1b © Carolina Biological Supply Co.
29.9a © Audesirk
29.9b © Audesirk
29.11 © Dwight R. Kuhn
E29.3a © Virginia P. Weinland/Photo Researchers, Inc.
E29.3b © John Shaw/Tom Stack & Associates

Unit V Opener, p. 642–643
© Frans Lanting/Minden Pictures

Chapter 30

p. 644 © Manfred Kage/ Peter Arnold, Inc.
30.9 © Lennart Nilsson
E30.1a © Lou Lainey/Time Magazine
E30.1c © Lennart Nilsson
30.11 © BPS
30.12 © Bill Longcore/ Science Source/Photo Researchers, Inc.
30.15 © Lennart Nilsson/ Boehringer Ingelheim Int'l GmbH
30.16 © CNRI/Science Source/Photo Researchers, Inc.
30.19 © John D. Cunningham/Visuals Unlimited

Chapter 31

p. 664 © Alex Kerstitch
31.2a © Charles Seaborn/ Odyssey, Chicago
31.2b © T. A. Weiwandt/ DRK Photo

31.4a © Breck P. Kent
31.4b © Edward D. Degginger/Bruce Coleman, Inc.
31.5 © David D. Dennis/ Tom Stack & Associates
31.6b H. R. Duncker/ Department of Anatomy/University of Giessen, Germany
E31.3 Bill Travis, M.D./ National Cancer Institute
E31.4a © O. Averbach/ Visuals Unlimited

Chapter 32

p. 680 © Weldon Lee
32.1 © Johnny Johnson/ Allstock
32.3a © Dwight R. Kuhn
E32.3 © Dr. Beer-Gabel/ CNRI/Science Source/ Photo Researchers, Inc.
32.11e © Lennart Nilsson/ The Incredible Machine

Chapter 33

p. 704 © Diana Stratton/ Tom Stack & Associates
E33.2 © Tom McHugh/ Allstock
E33.3 © M. P. Kahl/Photo Researchers, Inc.
E33.4 © SIU/Visuals Unlimited

Chapter 34

p. 718 © Lennart Nilsson/ The Incredible Machine © Boehringer Ingelheim International GmbH
34.1 © CNRI/SPL/Photo Researchers, Inc.
34.2 © Lennart Nilsson © Boehringer Ingelheim International GmbH
34.4 © NIBSC/SPL/Photo Researchers, Inc.
34.6 © Leonard Lessin/ Peter Arnold, Inc.
34.11a © Secchi/Lecaque/ Roussel/CNRI Science Source/Photo Researchers, Inc.
34.11b © Dr. Brian Eyden/ SPL/Photo Researchers, Inc.
34.12 © Lennart Nilsson/ Boeringer Ingelheim International GmbH

E34.3 © CNRI/SPL/Photo Researchers, Inc.
34.15b © Dr. Matthew A. Gonda Ph.D./Head, Laboratory of Cellular and Molecular Structure/ National Cancer Institute
34.15c © Lennart Nilsson/ Boeringer Ingelheim International GmbH

Chapter 35

p. 752 © Belinda Wright/ DRK Photo
E35.1b © Robert Knauft/ Photo Researchers, Inc.
E35.1c © Carolina Biological Supply Co.
E35.2 © Carolina Biological Supply Co.
E35.3a © Carolina Biological Supply Co.
E35.3b © Carolina Biological Supply Co.
E35.3c © Carolina Biological Supply Co.
E35.3d © Carolina Biological Supply Co.
35.7 © The Bettman Archive
E35.4a © Bryant/BPS
E35.4b © BPS
E35.4c © Bryant/BPS
E35.4d © BPS
35.9b © Biophoto Associates/Photo Researchers, Inc.
35.11a From Orci, L., 1982, *Diabetes* 31: 538–565.

Chapter 36

p. 776 © CNRI/SPL/Science Source/Photo Researchers, Inc.
36.2a © *Journal of Physiology*
36.2b © Carolina Biological Supply Co.
E36.1a © Audesirk
E36.1b © Audesirk
36.5 © C. Raines/Visuals Unlimited
36.12a © Jeff Foott/Tom Stack & Associates
36.12b © Norbert Wu/Peter Arnold, Inc.
36.12c © Corp. Lou Barr/ Allstock
E36.2a Courtesy of M. E. Phelps & John C. Mazziota, UCLA School of Medicine
E36.2b © Dr. Lewis R. Baxter, UCLA/1985 American Medical Association

Chapter 37

p. 810 © Charles Seaborn/ Odyssey/Chicago
37.2 © M.P.L. Fogden/ Bruce Coleman, Inc.
E37.1 © John Shaw/Tom Stack & Associates
E37.2a © Gilbert Grant/ Photo Researchers, Inc.
E37.2b © Brian Parker/Tom Stack & Associates
37.5a–b © Robert S. Preston/ Courtesy Prof. J. E. Hawkins, Krege Hearing Research Institute, University of Michigan Medical School
37.8 © Manfred Kage/ Peter Arnold, Inc.
37.11 © Susan Eichorst
37.12a © L. West/Photo Researchers, Inc.
37.12b © Audesirk
37.13a © Dwight R. Kuhn
37.13b © Audesirk

Chapter 38

p. 835 © Jack Mitchell
38.1f © Dr. Brian Eyden/ SPL/Photo Researchers, Inc.
38.3b © Don Fawcett/ Science Source/Photo Researchers, Inc.
38.6 © SPL/Photo Researchers, Inc.
E38.1a © Michael Klein/ Peter Arnold, Inc.
E38.1b © Erika Stone/Peter Arnold, Inc.
38.8 © Manfred Kage/ Peter Arnold, Inc.
38.9a © Steinhart Aquarium/Tom McHugh/Photo Researchers, Inc.
38.9b © Stanley Breeden/ DRK Photo

Chapter 39

p. 852 © Francis Leroy, Biocosmos/SPL/Photo Researchers, Inc.
39.1 © Fred Bavendam/ Allstock
39.4 © P. J. Bryant, University of California at Irvine/ BPS
39.5a © Tom McHugh/ Photo Researchers, Inc.
39.5b © Audesirk
39.5b *Inset* © Audesirk
39.6 © Anne et Jacques Six
39.7 © Michael Fogden/ Animals Animals

39.8a © Stephen J. Krasemann/DRK Photo
39.8b © Tom McHugh/ Photo Researchers, Inc.
39.8c © Frans Lanting/ Minden Pictures
39.11 © CNRI/SPL/Photo Researchers, Inc.
39.15a © G. Martin/Visuals Unlimited
39.15b © G. Martin/Visuals Unlimited
39.18a © Lennart Nilsson
39.18b © Lennart Nilsson
39.19a © Lennart Nilsson
E.39.1 © Hank Morgan/ Science Source/Photo Researchers, Inc.
E39.2 © Wide World Photos, Inc.

Chapter 40

p. 880 © Lennart Nilsson
E40.1 © David Scharf/Peter Arnold, Inc.
40.3a *left* © Oxford Scientific Films/Animals Animals
40.3a *right* © Oxford Scientific Films/Animals Animals
40.3b *left* © Stephen Dalton/ Photo Researchers, Inc.
40.3b *right* © L. West/Photo Researchers, Inc.
40.4a © K. H. Switak/ Photo Researchers, Inc.
40.4b © J.A.L. Cooke/ Oxford Scientific Films/Animals Animals
40.4c © Breck P. Kent
40.4d © Frans Lanting/ Minden Pictures
40.9 © Petit Format/ Nestle/Science Source/Photo Researchers, Inc.
40.13 © Lennart Nilsson
40.14 © Lennart Nilsson
E40.2 Leonard McCombe/ Life Magazine © Time Warner Inc.
40.16 © Audesirk

Chapter 41

p. 902 © Stephen Dalton/ Photo Resarchers, Inc.
41.1a © Terry Domico/ Earth Images
41.1b © Terry Domico/ Earth Images
41.2 © G. I. Bernard/ Oxford Scientific

Films/Animals Animals

41.5a, b © Eric & David Hosking

41.7 © Frans Lanting/ Minden Pictures

41.9 © Time Warner Inc., Thomas McAvoy

41.10 © William Vandivert

41.11 © Time Warner Inc., Thomas McAvoy

41.12a © Manfred Kage/ Peter Arnold, Inc.

41.13 Nina Leen/Life Magazine © Time Warner, Inc.

E41.2a © Stephen J. Krasemann/DRK Photo

41.14a–e © Lee Boltin Picture Library

Chapter 42

p. 920 © Gregory G. Dimijian/Photo Researchers, Inc.

42.1 N.Y.P.L.

42.2 © Richard K. Laval/ Animals Animals

42.3a © Tom McHugh/ Photo Researchers, Inc.

42.3b © Tom McHugh/ Photo Researchers, Inc.

42.4a © Stephen J. Krasemann/Photo Researchers, Inc.

42.4b © Jim Brandenburg/ Minden Pictures

42.4c © Hans Pfletschinger/ Peter Arnold, Inc.

42.5 © Anne et Jacques Six

42.6 © Jeff Foott

42.7 © William Ervin

42.8 Courtesy George W. Barlow, University of California, Berkeley, 1974, from "Hexagonal Territories," *Animal Behavior*, 22:876–878.

42.9 © John D. Cunningham/Visuals Unlimited

42.11 © Harold Hoffman/ Photo Researchers, Inc.

42.12a © Audesirk

42.12b © Frans Lanting/ Minden Pictures

42.14 © J. H. Robinson/ Photo Researchers, Inc.

42.15 © Des & Jen Bartlett/ Bruce Coleman, Inc.

42.16 © Denver Museum of Natural History

42.17a © Scott Camazine/ Photo Researchers, Inc.

42.17b © Carolina Biological Supply Co.

42.17c © Hans Pfletschinger/ Peter Arnold, Inc.

E42.1a © Carolina Biological Supply Co.

42.19 © W. Perry Conway

42.20 © M. P. Kahl/DRK Photo

E42.2 © Richard R. Hansen/Photo Researchers, Inc.

42.21 © Lennart Nilsson

42.22 © Todd Merritt Haiman/The Stock Market

42.23 © Dr. I. Eibl-Eibesfeldt

Unit VI Opener, p. 942–943 © Bruno Barby/Magnum Photos

Chapter 43

p. 944 © Art Wolfe/AllStock

43.1 © W. P. Perry Conway

43.1 Inset #1 © Jeff Foott/ Tom Stack & Associates

43.1 Inset #2 © Gary Milburn/Tom Stack & Associates

43.1 Inset #3 © Rod Planck/Tom Stack & Associates

43.1 Inset #4 © Gary R. Zahm/DRK Photo

43.2a © G. Musil/Visuals Unlimited

43.2b © Thomas Kitchin/ Tom Stack & Associates

E43.3 © R. J. Erwin/DRK Photo

E43.4 © Chuck Pratt/Bruce Coleman, Inc.

E43.5 © Robert & Linda Mitchell

43.7 © B&C Alexander

43.8 © Tom McHugh/ Photo Researchers, Inc.

43.9a Courtesy of the Department of Lands, Queensland, Australia

43.9b Courtesy of the Department of Lands, Queensland, Australia

43.10a © Breck P. Kent

43.10b © USDA-Forest Service, Ogden, Utah

43.11 © Frans Lanting/ Minden Pictures

43.12 © Gianni Tortoli/ Photo Researchers, Inc.

43.13a © Robert & Linda Mitchell

43.13b © Tom Bean/DRK Photo

43.13c © Gary Braasch

43.21a © Mark Boulton/ Photo Researchers, Inc.

43.21b © A. Vollan/Visuals Unlimited

Chapter 44

p. 968 © K. G. Preston-Mafham/ Premaphotos Wildlife

44.3a © Carolina Biological Supply Co

44.3b © Merlin D. Tuttle/ Bat Conservation, Int'l Photo Researchers, Inc.

E44.1 Dr. Monica Mather/ Simon Fraser University

44.4a © Ronald I. Shimek

44.4b © Robert & Linda Mitchell

44.4c © Charles Seaborn/ Odyssey

44.5a © Michael Fogden/ DRK Photo

44.5b © Paul A. Zahl/ Photo Researchers, Inc.

44.5c © Audesirk

44.6 © Robert & Linda Mitchell

44.7 T. W. Pietsch & D. B. Grobecker/ School of Fisheries, Seattle, Washington

44.8a © A. Kerstitch

44.8b © James L. Castner

44.9a © R. Humbert/BPS

44.9b © R. Humbert/BPS

44.9c © Z. Leszcynski/ Animals Animals

44.9d © Breck P. Kent

E44.2 © Thomas Eisner

44.10a © Zig Leszczynski/ Animals Animals

44.10b © Zig Leszczynski/ Animals Animals

44.11a © James L. Castner

44.11b © James L. Castner

44.11c © James L. Castner

44.12 © Thomas Eisener & Daniel Aneshansley, Cornell University

44.13a © Frans Lanting/ Minden Pictures

44.13b © Jeff Foott

44.14 © Charles Seaborn Odyssey/Frerk/ Chicago

44.15a left © Gary Braasch

44.15a right © Gary Braasch

44.15b © Wendy Shattil/Bob Rozinski © 1989/Tom Stack & Associates

44.15b right © Jeff Foott

44.15c left © John I. Kjargaard

44.15c right © William H. Amos/ National Geographic Magazine, 7/90

E44.3 © Gilbert Grant/ Photo Researchers, Inc.

E44.3 © Carol Hughes/ Bruce Coleman, Inc.

44.18a–c © Robert & Linda Mitchell

Chapter 45

p. 986 © Dwight R. Kuhn

E45.1 Michigan Department of Natural Resources

45.12 © Bob Hallinen/ Gamma-Liaison

45.13 © V.Leloup/Figaro/ Gamma-Liaison

45.15 © Will McIntyre/ Photo Researchers, Inc.

45.16 © Doug Plummer/ Photo Researchers, Inc.

E45.2a Tree People, Beverley Hills, California

E45.2b © Paul Mozell/Stock Boston

E45.2c © Wes Thompson/ The Stock Market

Chapter 46

p. 1014 © Nicholas Devore/ Photographers/Aspen

46.1a © Audesirk

46.1b © Audesirk

46.4b Courtesy of NASA72-HC-928

46.10 © Loren McIntyre

46.10 Inset #1 © Kevin Schafer/Peter Arnold, Inc.

46.10 Inset #2 © Bruce Lyon/Valan Photos

46.10 Inset #3 © Michael Fogden/DRK Photo

46.10 Inset #4 © Frans Lanting/Minden Pictures

46.11 © Jacques Janqoux/ Peter Arnold, Inc.

46.12 © Perry Conway